T0324090

The Periodic Table: Nature's Building Blocks

An Introduction to the Naturally Occurring Elements, Their Origins, and Their Uses

The Periodic Table: Nature's Building Blocks

An Introduction to the Naturally Occurring Elements, Their Origins, and Their Uses

J. Theo Kloprogge
School of Earth and Environmental Sciences, University of Queensland, St Lucia, QLD, Australia
Department of Chemistry, College of Arts and Sciences, University of the Philippines Visayas, Miagao, Philippines

Concepcion P. Ponce
Department of Chemistry, College of Arts and Sciences, University of the Philippines Visayas, Miagao, Philippines

Tom A. Loomis
Dakota Matrix, Rapid City, SD, United States

Elsevier
Radarweg 29, PO Box 211, 1000 AE Amsterdam, Netherlands
The Boulevard, Langford Lane, Kidlington, Oxford OX5 1GB, United Kingdom
50 Hampshire Street, 5th Floor, Cambridge, MA 02139, United States

British Library Cataloguing-in-Publication Data
A catalogue record for this book is available from the British Library

Library of Congress Cataloging-in-Publication Data
A catalog record for this book is available from the Library of Congress

ISBN: 978-0-12-821279-0

For Information on all Elsevier publications
visit our website at https://www.elsevier.com/books-and-journals

Publisher: Susan Dennis
Acquisitions Editor: Emily M. McCloskey
Editorial Project Manager: Liz Heijkoop
Production Project Manager: Swapna Srinivasan
Cover Designer: Miles Hitchen

Typeset by MPS Limited, Chennai, India

Dedication

To my late wife, Vicki Klar-Loomis (1950–2018). She supported my out-of-control hobby, which turned into a business. She was a trooper through the decades of rugged field collecting always with a smile and a twinkle in her eye. Her encouragement and inspiration will never be forgotten. Thank you my dear.

Contents

Preface

The year 2019 marked the 150th anniversary of the periodic table developed by the Russian chemist Dimitri Mendeleev in 1869. Mendeleev classified the elements by increasing atomic weight and discovered a *periodicity* of the properties of known elements when arranged as such, which became known as the periodic law. This discovery enabled Mendeleev to redefine the properties of certain elements and also lead to the prediction of undiscovered elements. Over the 150 years since this discovery, the periodic table has become the international standard of the classification of nature's elements and is used in all textbooks concerning, but not limited to, chemistry, mineralogy, physics, astronomy, ecology, and biology. The value of the periodic table to these sciences cannot be underestimated when realizing that the so-called "rare earth elements," which were discovered soon after Mendeleev's discovery in 1869, would over 100 years later, become an essential "ingredient" in the manufacture of computers, batteries, magnets, smoke detectors, and cell phones to name just a few.

Out of necessity, humans have used rock since Antiquity and metal (copper, lead, gold, silver, iron) perhaps dating back to 9000 BCE. The use of naturally occurring chemical compounds for medicinal purposes dates back to the Greco-Roman period (332 BCE to CE 395). Whether it be a rock or a native metal used through the millenniums, an element such as silicon, oxygen, carbon, iron, or copper was utilized in the manufacture of pottery, tools, weapons, and other implements and each of these elements was later incorporated into Mendeleev's periodic table. Fast forward to 2020, and we now know that there are 118 elements of which the first 92 elements are found on Earth (the remaining elements 93–118 are only known from laboratory experiments) and certain combinations of these elements produce over 5400 naturally occurring compounds called "minerals." This book introduces the student and reader to how these elements were discovered, where and how they occur geologically, how they are mined, and most importantly how they are used in our modern world.

The Periodic Table—Nature's Building Blocks was conceived by the principal author Dr. Theo Kloprogge as a way to introduce the reader to the fact that the periodic table is not just a stand-alone table of elements to be referenced only now and then, but is in fact an incredible source of information, which can be used to study all of the natural sciences. Unfortunately, in today's busy scholastic environment there is generally no time to discuss the historical and scientific context of the periodic table and its elements. Nor is there time for the chemistry student, for example, to become familiar with the natural occurrences of the chemical compounds and minerals of which they study. Likewise, a geology student today relies upon the modern analytical equipment to identify minerals and thus fails to become versed in the simple chemistry of the elements which compose the minerals upon which they study. As such, this book will perhaps bridge that knowledge gap between what we learn and what we need to learn.

June 17, 2020

J. Theo Kloprogge
Concepcion P. Ponce
Tom A. Loomis

Chapter 1

History of the periodic table and the structure of the atom

1.1 History of the period table

Several physical elements (such as platinum, mercury, tin, and zinc) have been known since antiquity, as they could be found in their native form and were relatively simple to mine with primitive tools. Around 330 BCE, the Greek philosopher Aristotle proposed that everything is made up of a mixture of one or more roots, an idea that had first been proposed by the Sicilian philosopher Empedocles. The four roots, which were later renamed as elements by Plato, were earth, water, air, and fire. Similar ideas about these four elements also existed in other ancient traditions, such as Indian philosophy.

1.1.1 Hennig Brand (c.1630–1692 or c.1710)

The history of the periodic table is at the same time a history of the discovery of the elements. The first person in history to discover a new element was Hennig Brand, a bankrupt German merchant. He tried to discover the Philosopher's Stone—a mythical object that was supposed to turn inexpensive base metals into gold. In 1669 (or later) his experiments with distilled human urine lead to the formation of a glowing white substance, which he called "cold fire" (kaltes feuer). He kept his finding secret until 1680, when Robert Boyle (January 25, 1627 to December 31, 1691) rediscovered phosphorus and published his results. Boyle is largely regarded today as the first modern chemist, and therefore one of the founders of modern chemistry, and one of the pioneers of modern experimental scientific method. The discovery of phosphorus helped to raise the question of what it meant for a substance to be an element. In 1661 Boyle defined an element as "those primitive and simple Bodies of which the mixed ones are said to be composed, and into which they are ultimately resolved."

1.1.2 Antoine-Laurent de Lavoisier (August 26, 1743 to May 8, 1794)

Lavoisier's Traité Élémentaire de Chimie (*Elementary Treatise of Chemistry*) written in 1789, and first translated into English by the Scottish surgeon and scientific writer Robert Kerr (October 20, 1757 to October 11, 1813), is thought of as the first modern textbook on chemistry (Lavoisier, 1790). Lavoisier (Fig. 1.1) was a French nobleman and chemist who was central to the 18th-century chemical revolution and who had a large influence on both the history of chemistry and the history of biology. He is widely considered in popular literature as the "father of modern chemistry." He defined an element as a substance that cannot be broken down into a simpler substance by a chemical reaction. This simple definition served for a century and lasted until the discovery of subatomic particles. His book listed several "simple substances" that he believed could not be broken down further, which included oxygen, nitrogen, hydrogen, phosphorus, mercury, zinc, and sulfur. These formed the basis for the modern list of elements. His list also included "light" and "caloric," which at the time were believed to be material substances. He classified these substances into metals and nonmetals. While many leading chemists rejected his new revelations, the *Elementary Treatise* was written well enough to convince the younger generation. Nevertheless, his descriptions of his elements were deficient in completeness, as he only classified them as metals and nonmetals.

The Periodic Table: Nature's Building Blocks. DOI: https://doi.org/10.1016/B978-0-12-821279-0.00001-7

FIGURE 1.1 Antoine-Laurent de Lavoisier. *Line engraving by Louis Jean Desire Delaistre, after a design by Julien Leopold Boilly.*

FIGURE 1.2 Painting of William Prout, the chemist and physician. *From a miniature by Henry Wyndham Phillips.*

1.1.3 William Prout (January 15, 1785 to April 9, 1850)

William Prout (Fig. 1.2) was an English chemist, physician, and natural theologian. He is remembered today chiefly for what is known as Prout's hypothesis. In 1815 based on the tables of atomic weights existing at the time, he anonymously hypothesized that the atomic weight of every element is an integer multiple of that of hydrogen, suggesting that the hydrogen atom is the only truly fundamental particle (which he called protyle), and that the atoms of the other elements are made of groupings of various numbers of hydrogen atoms (Prout 1815,1816). While Prout's hypothesis was not borne out by later more-accurate measurements of atomic weights, it was a sufficiently fundamental insight into the structure of the atom that in 1920 Ernest Rutherford (August 30, 1871 to October 19, 1937) chose the name of the newly discovered proton to, among other reasons, to give credit to Prout.

1.1.4 Johann Wolfgang Döbereiner (December 13, 1780 to March 24, 1849)

In 1817 Johann Wolfgang Döbereiner (Fig. 1.3), a German chemist who invented the first lighter known as the Döbereiner's lamp, began to formulate one of the earliest attempts to classify the elements. In 1829 he found that he could form some of the elements into groups of three, with the members of each group having related properties (Döbereiner, 1829). He termed these groups triads. Chemically analogous elements arranged in increasing order of their atomic weights formed well-marked groups of three called Triads in which the atomic weight of the middle element was found to be generally the arithmetic mean of the atomic weight of the other two elements in the triad. Additionally, the densities for some of these triads followed a similar pattern.

- chlorine, bromine, and iodine,
- calcium, strontium, and barium,
- sulfur, selenium, and tellurium, and
- lithium, sodium, and potassium.

1.1.5 Alexandre-Emile Béguyer de Chancourtois (January 20, 1820 to November 14, 1886)

Alexandre-Emile Béguyer de Chancourtois, a French geologist and mineralogist (He was also a professor of mine surveying at the École Nationale Supérieure des Mines de Paris. Likewise, he was the Inspector of Mines in Paris and was widely responsible for realizing many mine safety regulations and laws in his time.), was the first scientist to notice the periodicity of the elements—similar elements occurring at regular intervals when they are ordered by their atomic weights. In 1862, he created an early form of the periodic table, which he called Vis tellurique (the "telluric helix"), after the element tellurium, which fell near the center of his diagram. With the elements organized in a spiral on a cylinder in order of increasing atomic weight, he observed that elements with similar properties lined up vertically (Béguyer de Chancourtois, 1862). His 1863 publication included a chart (which contained ions and compounds, in addition to elements), but his original paper in the Comptes Rendus de l'Académie des Sciences used geological rather than chemical

FIGURE 1.3 Engraving of a painting of Johann Wolfgang Döbereiner. *Carl August Schwerdgeburth [1785–1878 (engraver)] and Fritz Ries [1826–57 (painter)].*

	No.		No.		No.		No.		No.		No.		No.		No.
H	1	F	8	Cl	15	Co & Ni	22	Br	29	Pd	36	I	42	Pt & Ir	50
Li	2	Na	9	K	16	Cu	23	Rb	30	Ag	37	Cs	44	Os	51
G	3	Mg	10	Ca	17	Zn	24	Sr	31	Cd	38	Ba & V	45	Hg	52
Bo	4	Al	11	Cr	19	Y	25	Ce & La	33	U	40	Ta	46	Tl	53
C	5	Si	12	Ti	18	In	26	Zr	32	Sn	39	W	47	Pb	54
N	6	P	13	Mn	20	As	27	Di & Mo	34	Sb	41	Nb	48	Bi	55
O	7	S	14	Fe	21	Se	28	Ro & Ru	35	Te	43	Au	49	Th	56

FIGURE 1.4 Newlands' law of octaves.

terms and did not include a diagram (Béguyer de Chancourtois, 1863). As a result, his ideas received little attention until after the work of Dmitri Mendeleev had been published (see below).

1.1.6 John Newlands (November 26, 1837 to July 29, 1898)

In 1864, the English chemist John Newlands classified the 62 known elements into eight groups, based on their physical properties (Newlands, 1864a,b). He observed that numerous pairs of similar elements existed, which differed by some multiple of eight in mass number and was the first to assign them an atomic number. When his "law of octaves" was published in Chemistry News, comparing this periodicity of eights to the musical scale, it was derided by some of his colleagues (Newlands, 1865) (Fig. 1.4). His lecture to the Chemistry Society on March 1, 1866 was not published, the Society defending their decision by stating that such "theoretical" topics might be controversial.

The importance of his analysis was in the end recognized by the Chemistry Society with a Gold Medal 5 years after they recognized Mendeleev's work. It was not until the next century, with Gilbert N. Lewis's valence bond theory (1916) and Irving Langmuir's octet theory of chemical bonding (1919), that the importance of the periodicity of eight would be recognized. The Royal Chemistry Society acknowledged his contribution to science in 2008, when they put a Blue Plaque on the house where he was born, which described him as the "discoverer of the Periodic Law for the chemical elements." He contributed the word "periodic" in chemistry.

1.1.7 Julius Lothar Meyer (August 19, 1830 to April 11, 1895)

Meyer (Fig. 1.5), a German chemist, observed that (as Newlands did in England) when the elements were arranged in the order of their atomic weights, they fell into groups of comparable chemical and physical properties recurring at periodic intervals. According to him, if the atomic weights were plotted as ordinates and the atomic volumes as abscissas a plot would be obtained with a series of maxima and minima with the most electro-positive elements appearing at the peaks of the curve in the order of their atomic wieghts. His book, Die modernen Theorien der Chemie (*The Modern Theories of Chemistry*), which he started writing in Breslau in 1862 and was published in 1864, contained an early version of the periodic table containing 28 elements. It classified elements into six families based on their valence for the first time. Earlier works on organizing the elements by atomic weight until then had been stymied by inaccurate measurements of the atomic weights. He published articles about a classification table of the elements in horizontal form (1862, 1864) and vertical form (1870), in which the series of periods are properly ended by an element of the alkaline earth metal group. In 1869 a few months later than Mendeleev, he published a revised and expanded version of his 1864 table independently, which was similar to that published by Mendeleev (he had been sent a copy of Mendeleev's table earlier; Mendeleev had sent it to many well-known chemists of his day) and a paper showing graphically the periodicity of the elements as a function of atomic weight. In 1882 both Meyer and Mendeleev received the Davy Medal from the Royal Society in recognition of their work on the Periodic Law.

1.1.8 Dmitri Ivanovich Mendeleev [February 8, 1834 to February 2, 1907 (OS January 27, 1834 to January 20, 1907)]

The Russian chemist Dmitri Mendeleev (Fig. 1.6) arranged the elements by atomic mass, corresponding to relative molar mass. It is occasionally said that he played "chemical solitaire" on long train trips, using cards with various facts about the known elements. On March 1 (OS February 17) 1869 he put a date on his first table and sent it for publication. On March 18 (OS March 6) 1869 he gave a formal lecture titled *The Dependence Between the Properties of the*

FIGURE 1.5 Julius Lothar Meyer (January 11, 1883). *Photo by Wilhelm Hornung (1834–84).*

FIGURE 1.6 Photo of Dmitri Ivanovich Mendeleev in 1897.

Atomic Weights of the Elements to the Russian Chemical Society. In 1869 the table was published in an obscure Russian journal and then republished in a German journal, Zeitschrift für Chemie (Mendeleev 1869a,b). In it, he specified that (translated from German):

1. The elements, if arranged according to their atomic mass, exhibit an apparent periodicity of properties.
2. Elements which are similar as regards to their chemical properties have atomic weights that are either of nearly the same value (e.g., Pt, Ir, Os) or which increase regularly (e.g., K, Rb, Cs).
3. The arrangement of the elements, or of groups of elements in the order of their atomic masses, corresponds to their so-called valencies, as well as, to some extent, to their distinctive chemical properties, as is apparent among other series in that of Li, Be, B, C, N, O, and F.

4. The elements which are the most widely diffused have small atomic weights.
5. The magnitude of the atomic weight determines the character of the element, just as the magnitude of the molecule determines the character of a compound body.
6. We must expect the discovery of many yet unknown elements—for example, elements analogous to aluminum and silicon—whose atomic weight would be between 65 and 75.
7. The atomic weight of an element may sometimes be amended by knowledge of those of its contiguous elements. Thus the atomic weight of tellurium must lie between 123 and 126 and cannot be 128.
8. Certain characteristic properties of elements can be foretold from their atomic masses.

The periodic table enabled Mendeleev to predict the discovery of new elements and left spaces for them, namely eka-silicon (germanium, discovered in 1885), eka-aluminum (gallium, 1875), and eka-boron (scandium, 1879). Thus there was no problem to fit these elements in the periodic table. It was also used by him to point out that some of the atomic weights being used at the time were incorrect. It provided for variance from atomic weight order (Fig. 1.7).

1.1.9 William Odling (September 5, 1829 to February 17, 1921)

In 1864, the English chemist William Odling (Fig. 1.8) also devised a periodic table that was remarkably like the table published by Mendeleev (Odling, 1864). Odling solved the tellurium-iodine problem and even succeeded in getting thallium, lead, mercury, and platinum into the right groups, which is something that Mendeleev failed to do in his first attempt. Odling failed to achieve recognition, however, since it was suspected that he, as Secretary of the Chemical Society of London, was influential in discrediting Newlands' earlier work on the periodic table. One such unrecognized aspect was for the suggestion he made in a lecture he gave at the Royal Institution in 1855 entitled The

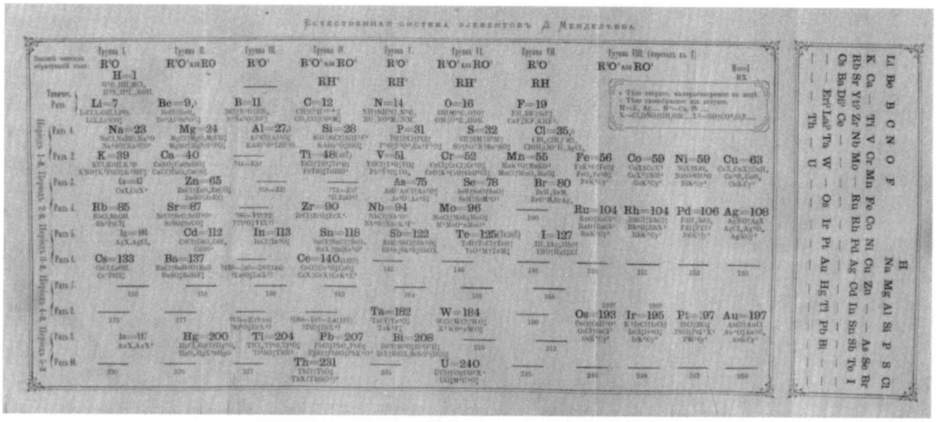

FIGURE 1.7 The periodic table as published in Osnovy khimīi by Mendeleev. Dashes: unknown elements; group I–VII: modern groups 1 and 2 and 3–7 with transition metals added; some of these extend into a group VIII [noble gases unknown (and unpredicted)].

FIGURE 1.8 William Odling.

Constitution of Hydrocarbons in which he proposed a methane type for carbon (Proceedings of the Royal Institution, 1855, Vol. 2, pp. 63–66). Perhaps influenced by Odling's paper, August Kekulé (September 7, 1829 to July 13, 1896, German organic chemist) made a similar suggestion in 1857, and then in a subsequent paper later that same year proposed that carbon is a tetravalent element.

1.1.10 Shortcomings of early versions of the periodic table so far

The periodic table so far was not able to forecast the existence of the noble gases but did leave spaces for yet-to-be discovered elements. Time proved this method correct. When the entire group of noble gases was discovered, primarily by William Ramsay (October 2, 1852 to July 23, 1916), he added them to the table as Group 0, without disturbing the basic concept of the periodic table. He received the Nobel Prize in Chemistry in 1904 "in recognition of his services in the discovery of the inert gaseous elements in air" (along with his collaborator, John William Strutt, 3rd Baron Rayleigh, who received the Nobel Prize in Physics that same year for their discovery of argon). A single position could not be assigned to hydrogen, which could be placed either in the alkali metals group, the halogens group, or separately above the table between boron and carbon. The lanthanides were difficult to fit into the table.

1.1.11 Frederick Soddy (September 2, 1877 to September 22, 1956)

Frederick Soddy was an English radiochemist who elucidated, with Ernest Rutherford, that radioactivity is due to the transmutation of elements, now known to comprise nuclear reactions. By 1912 almost 50 different radioactive elements had been found, too many for the periodic table. Frederick Soddy in 1913 found that although they emitted different radiation, many elements were alike in their chemical characteristics, so they had to share the same place on the periodic table (Soddy, 1913a,b). They became known as isotopes, from the Greek eisos topos ("same place"). He received the Nobel Prize in Chemistry in 1921 and the same year he was elected member of the International Atomic Weights Committee. A small crater on the far side of the moon as well as the radioactive uranium mineral soddyite is named after him.

1.1.12 Henry Moseley (November 23, 1887 to August 10, 1915)

Henry Gwyn Jeffreys Moseley (Fig. 1.9) was an English physicist, whose involvement in the science of physics provided the justification from physical laws of previous empirical and chemical concepts of the atomic number. This stemmed from his development of Moseley's law in X-ray spectra, which advanced atomic physics, nuclear physics, and quantum physics by providing the first experimental proof in favor of Niels Bohr's theory, aside from the hydrogen atom spectrum which the Bohr theory was designed to reproduce. That theory refined Ernest Rutherford's and Antonius Johannes van den Broek's (May 4, 1870, Zoetermeer – October 25, 1926, Bilthoven, Dutch amateur physicist) model, which proposed that the atom contains in its nucleus a number of positive nuclear charges that is equal to its (atomic) number in the periodic table. This remains the accepted model until today. In 1914, a year before Moseley was killed in action at Gallipoli, he found a relationship between the X-ray wavelength of an element and its atomic number. He was then able to reorder the periodic table by nuclear charge, rather than by atomic weight. Before this discovery, atomic

FIGURE 1.9 Henry Moseley (photo: 1914).

numbers were sequential numbers based on an element's atomic weight. His discovery proved that atomic numbers were in fact based upon experimental measurements. Using information about their X-ray wavelengths, he placed argon (with an atomic number Z = 18) before potassium (Z = 19), even though argon's atomic weight of 39.9 is greater than the atomic weight of potassium (39.1) (Moseley, 1913, 1914). The new sequence agreed with the chemical properties of these elements, since argon is a noble gas and potassium is an alkali metal. Likewise, he placed cobalt before nickel and was able to explain that tellurium occurs before iodine, without revising the experimental atomic weight of tellurium, as had been proposed by Mendeleev. Moseley's research showed that there were gaps in the periodic table for two elements at atomic numbers 43 and 61, which are now identified as technetium and promethium, respectively.

1.1.13 Glenn T. Seaborg (April 19, 1912 to February 25, 1999)

Glenn Theodore Seaborg was an American chemist whose involvement in the synthesis, discovery, and investigation of 10 transuranium elements earned him a share of the 1951 Nobel Prize in Chemistry. During his Manhattan Project research in 1943, he experienced unexpected difficulties in isolating the elements americium and curium. These elements, in addition to the elements from actinium to plutonium, were believed to form a fourth series of transition metals. He wondered if these elements fitted in a different series, which would elucidate why their chemical properties, specifically the instability of the higher oxidation states, were different from predictions. In 1945, against the advice of colleagues, he proposed an important change to Mendeleev's periodic table: the actinide series. His actinide concept of heavy element electronic structure, predicting that the actinides formed a transition series equivalent to the rare earth series of lanthanide elements, is now generally accepted and included in the periodic table. The actinide series is the second row of the f-block (5f series). In both the actinide and lanthanide series, an inner electron shell is being filled. The actinide series comprises the elements from actinium to lawrencium. Seaborg's subsequent elaborations of the actinide concept theorized a series of superheavy elements in a transactinide series comprising elements from 104 to 121 and a superactinide series of elements from 122 to 153. Seaborg was the principal or codiscoverer of 10 elements: plutonium, americium, curium, berkelium, californium, einsteinium, fermium, mendelevium, nobelium, and element 106, which, while he was still living, was named seaborgium in his honor. He also discovered more than 100 atomic isotopes and was credited with important contributions to the chemistry of plutonium, originally as part of the Manhattan Project where he developed the extraction process used to isolate the plutonium fuel for the second atomic bomb. Seaborg anticipated an extended periodic table with an additional period of 50 elements (thus reaching element 168); this eighth period was derived from an extrapolation of the Aufbau principle and placed elements 121–138 in a g-block, in which a new g subshell would be filled. His model, though, did not consider relativistic effects ensuing from high atomic number and electron orbital speed. Burkhard Fricke in 1971 and Pekka Pyykkö in 2010 used computer modeling to calculate the positions of elements up to Z = 172 and observed that several element positions were different from those predicted by Seaborg. While models from Pyykkö, Fricke, and Nefedov and coworkers generally place element 172 as the next noble gas, there is no clear agreement on the electron configurations of elements beyond 120 and consequently their assignment in an extended periodic table. It is now believed that as a result of relativistic effects, such an extension will feature elements that break the periodicity in known elements, hence resulting in another possible obstacle to future periodic table constructs. The discovery of tennessine in 2010 filled the last outstanding place in the seventh period. Any newly discovered elements will therefore be positioned in an eighth period. Notwithstanding the completion of the seventh period, experimental chemistry of some transactinides has been observed to be inconsistent with the periodic law. In the 1990s, Ken Czerwinski at the University of California, Berkeley found parallels between rutherfordium and plutonium and dubnium and protactinium, rather than a clear continuation of periodicity in groups 4 and 5. More recent researches on copernicium and flerovium have produced inconsistent results, some of which seem to indicate that these elements behave more like the noble gas radon rather than mercury and lead, their respective congeners. As such, the chemistry of many superheavy elements has yet to be well characterized, and it is still not clear whether the periodic law can still be used to extrapolate the properties of undiscovered elements. Elements 95–118 have only been synthesized in laboratories or nuclear reactors. The synthesis of elements having higher atomic numbers is currently being pursued: these elements would begin an eighth row, and theoretical work has been done to suggest possible candidates for this extension. Numerous synthetic radionuclides of naturally occurring elements have also been produced in laboratories.

1.2 Short history and basic properties of the atom

It is impossible to see the history of the periodic table and the discovery of the elements separate from the history of the atom and its particles and subparticles. Therefore this section will try to provide a short introduction into its long history as well as a description of its basic properties.

1.2.1 History

The concept that matter consists of discrete units is a very old idea, appearing in many ancient cultures such as Greece and India. The word atomos, meaning "uncuttable," was "invented" by the ancient Greek philosophers Leucippus and his student Democritus (c.460–370 BCE) (Fig. 1.10). Democritus taught that atoms were infinite in number, uncreated, and eternal, and that the qualities of an object result from the kind of atoms that compose it. Democritus's atomism was refined and expanded by the later philosopher Epicurus (341–270 BCE). In the period of the Early Middle Ages, atomism was mostly forgotten in western Europe, but survived among some groups of Islamic philosophers. During the 12th century, atomism became known again in western Europe because of references found in the newly rediscovered writings of Aristotle. In the 14th century, the rediscovery of major works describing atomist teachings, including the Roman poet and philosopher Titus Lucretius Carus' De rerum natura (c. October 15, 99 BCE to c.55 BCE) and biographer of Greek philosophers Diogenes Laërtius's Lives and Opinions of Eminent Philosophers, caused an increased scholarly attention on the subject. However, since atomism was linked to the philosophy of Epicureanism, which contradicted orthodox Christian teachings, belief in atoms was not considered acceptable. The French Catholic priest, philosopher, astronomer, and mathematician Pierre Gassendi (January 22, 1592 to October 24, 1655) revived Epicurean atomism with amendments, stating that atoms were created by God and, though extremely numerous, are not infinite. His modified theory of atoms was spread in France by the physician and traveler François Bernier (September 25, 1620 to September 22, 1688) and in England by the natural philosopher and writer Walter Charleton (February 2, 1619 to April 24, 1707). The Anglo-Irish natural philosopher, chemist, physicist, and inventor Robert Boyle (January 25, 1627 to December 31, 1691) and the English mathematician, physicist, astronomer, theologian, and author Isaac Newton (December 25, 1642 to March 20, 1727) both defended atomism and, by the end of the 17th century, it was accepted by parts of the scientific community.

In the early 1800s English chemist, physicist, and meteorologist John Dalton (September 6, 1766 to July 27, 1844) (Fig. 1.11) used the idea of atoms to describe why elements always react in ratios of small whole numbers (the law of multiple proportions). For instance, there are two types of tin oxide: one is 88.1% tin and 11.9% oxygen and the other is 78.7% tin and 21.3% oxygen [tin(II) oxide and tin dioxide, respectively]. This means that 100 g of tin can either

FIGURE 1.10 Democritus.

FIGURE 1.11 John Dalton by Charles Turner (1773–1857) after James Lonsdale (1777–1839) (Mezzotint).

combine with 13.5 or 27 g of oxygen. 13.5 and 27 form a ratio of 1:2, a ratio of small whole numbers. This general pattern in chemistry indicated to Dalton that elements react in multiples of discrete units—in other words, atoms. In the case of tin oxides, one tin atom will combine with either one or two oxygen atoms. Dalton also held the belief that the atomic theory could explain why water absorbs different gases in different proportions. For example, he found that water absorbs carbon dioxide far better than it absorbs nitrogen. Dalton hypothesized that this was due to the differences between the masses and configurations of the gases' respective particles, and carbon dioxide molecules (CO_2) are heavier and larger than nitrogen molecules (N_2).

In 1827, Scottish botanist and paleobotanist Robert Brown (December 21, 1773 to June 10, 1858) used a microscope to observe dust grains floating in water and discovered that they moved about erratically, a phenomenon that is now known as "Brownian motion." This was thought to be caused by water molecules knocking the grains about. In 1905 German-born theoretical physicist Albert Einstein (March 14, 1879 to April 18, 1955) demonstrated the reality of these molecules and their motions by producing the first statistical physics analysis of Brownian motion. French physicist Jean Perrin (September 30, 1870 to April 17, 1942) applied Einstein's work to experimentally measure the mass and dimensions of atoms, thereby definitively confirming Dalton's atomic theory.

The physicist Sir Joseph John Thomson (December 18, 1856 to August 30, 1940) determined the mass of cathode rays, showing they consisted of particles, but were about 1800 times lighter than the lightest atom, hydrogen (Thomson, 1901) (Fig. 1.12). Hence, they could not be atoms, but a new particle, the first subatomic particle to be discovered, which he initially called "corpuscle" but was later renamed electron, after particles postulated by Irish physicist George Johnstone Stoney (February 15, 1826 to July 5, 1911) in 1874. Thomson also proved that they were identical to particles given off by photoelectric and radioactive materials. It was soon recognized that they are the particles that carry electric currents in metal wires and carry the negative electric charge within atoms. He received the 1906 Nobel Prize in Physics for this work. With this work he overturned the belief that atoms are the indivisible, ultimate particles of matter. In addition, Thomson erroneously postulated that the low mass, negatively charged electrons were distributed throughout the atom in a uniform sea of positive charge. This became known as the plum pudding model.

In 1909 German physicist Johannes Wilhelm "Hans" Geiger (September 30, 1882 to September 24, 1945) and English-New Zealand physicist Ernest Marsden (February 19, 1889 to December 15, 1970), under the direction of New Zealand physicist Ernest Rutherford (August 30, 1871 to October 19, 1937), bombarded a metal foil with alpha

FIGURE 1.12 J.J. Thomson (pre-1915).

particles (which we now know consists of two protons and two neutrons bound together into a particle identical to a helium-4 nucleus) to determine how they scattered. They anticipated all the alpha particles to pass straight through with little deflection, since Thomson's model predicted that the charges in the atom are so diffuse that their electric fields should not affect the alpha particles much. Nevertheless, Geiger and Marsden observed alpha particles being deflected by angles greater than 90 degrees, something that was thought to be impossible based on Thomson's model. To explain this, Rutherford proposed that the positive charge of the atom is concentrated in a tiny nucleus at the center of the atom.

While investigating the products of radioactive decay, in 1913 English radiochemist Frederick Soddy (September 2, 1877 to September 22, 1956) found that there seemed to be more than one type of atom at each position on the periodic table (Soddy, 1913a-c) (Fig. 1.13). He also explained, with Ernest Rutherford, that radioactivity is due to the transmutation of elements, now known to involve nuclear reactions. The term isotope, Greek for at the same place, was suggested by Scottish doctor and writer Margaret Todd (April 23, 1859 to September 3, 1918) as a suitable name for different atoms that belong to the same element. Todd was a family friend of chemist Frederick Soddy, and a lecturer at the University of Glasgow. In 1913 Soddy explained to her the research on radioactivity for which he later won the Nobel Prize in Chemistry in 1921. The term isotope was accepted and used by Soddy and has become standard scientific nomenclature. J.J. Thomson created a technique for isotope separation through his work on ionized gases, which subsequently led to the discovery of stable isotopes [The term stable isotope has a meaning similar to stable nuclide but is preferably used when speaking of nuclides of a specific element. Therefore the plural form "stable isotopes" usually refers to isotopes of the same element. The relative abundance of such stable isotopes can be determined experimentally (isotope analysis), resulting in an isotope ratio that can be used as a research tool. Theoretically, such stable isotopes can include the radiogenic daughter products of radioactive decay, used in radiometric dating. Nevertheless, the expression stable-isotope ratio is preferably used to refer to isotopes whose relative abundances are affected by isotope fractionation in nature. This field is known as stable-isotope geochemistry.].

In 1913 the Danish physicist Niels Bohr (October 7, 1885 to November 18, 1962) (Fig. 1.14) developed a model in which the electrons of an atom were assumed to orbit the nucleus but could only do so in a finite set of orbits (similar to planets around the sun) and could jump between these orbits only with discrete changes of energy corresponding to absorption or radiation of a photon (Fig. 1.15). This quantization was used to explain why the electron orbits are

FIGURE 1.13 Frederick Soddy [Nobel Prize in Chemistry (1921)].

FIGURE 1.14 Niels Bohr (c.1922).

stable (given that normally, charges in acceleration, including circular motion, lose kinetic energy that is emitted as electromagnetic radiation) and why elements absorb and emit electromagnetic radiation in discrete spectra (Bohr, 1922a,b). Later in that same year English physicist Henry Gwyn Jeffreys Moseley (November 23, 1887 to August 10, 1915) showed additional experimental proof supporting Niels Bohr's theory. These results refined Ernest Rutherford's and Dutch lawyer and amateur physicist Antonius Van den Broek's model (May 4, 1870 to October 25, 1926), which suggested that the atom contains in its nucleus a number of positive nuclear charges that is equal to its (atomic) number

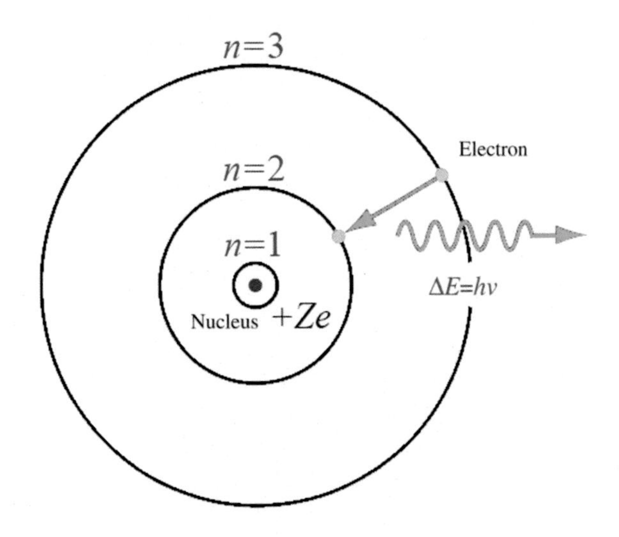

FIGURE 1.15 Bohr's model of an atom.

in the periodic table. Until these experiments, atomic number was not known to be a physical and experimental quantity. That it is equal to the atomic nuclear charge is still the accepted atomic model today.

Chemical bonds between atoms could now be explained, by American physical chemist and a former Dean of the College of Chemistry at University of California, Berkeley Gilbert Newton Lewis [October 25 (or 23), 1875 to March 23, 1946] in 1916, as the interactions between their constituent electrons (Lewis, 1916). Since it was known that the chemical properties of the elements largely repeat themselves consistent with the periodic law, in 1919 the American chemist and physicist Irving Langmuir (January 31, 1881 to August 16, 1957) proposed that this could be explained if the electrons in an atom were connected or clustered in some manner (Langmuir, 1919a,b). Groups of electrons were believed to occupy a set of electron shells around the nucleus [In chemistry and atomic physics, an electron shell, or a principal energy level, may be thought of as an orbit followed by electrons around an atom's nucleus. The closest shell to the nucleus is called the "1 shell" (also called "K shell"), followed by the "2 shell" (or "L shell"), then the "3 shell" (or "M shell"), and so on farther and farther from the nucleus. The shells correspond with the principal quantum numbers ($n = 1, 2, 3, 4\ldots$) or are labeled alphabetically with letters used in the X-ray notation (K, L, M,. . .). Each shell can contain only a fixed number of electrons: the first shell can hold up to two electrons; the second shell can hold up to eight ($2 + 6$) electrons; the third shell can hold up to 18 ($2 + 6 + 10$); and so on. The general formula is that the nth shell can in principle hold up to $2(n^2)$ electrons. Since electrons are electrically attracted to the nucleus, an atom's electrons will generally occupy outer shells only if the more inner shells have already been completely filled by other electrons. However, this is not a strict requirement: atoms may have two or even three incomplete outer shells. The electrons in the outermost occupied shell (or shells) determine the chemical properties of the atom; it is called the valence shell. Each shell consists of one or more subshells, and each subshell consists of one or more atomic orbitals.].

The Stern–Gerlach experiment of 1922 conducted by the German-American physicist Otto Stern (February 17, 1888 to August 17, 1969) and German physicist Walter Gerlach (August 1, 1889 to August 10, 1979) provided further proof of the quantum nature of atomic properties. When a beam of silver atoms was passed through a specially shaped magnetic field, the beam was split in a way correlated with the direction of an atom's angular momentum, or spin. As this spin direction is initially random, the beam would be expected to deflect in a random direction. Instead, the beam was split into two directional components, corresponding to the atomic spin being oriented up or down with respect to the magnetic field. Historically, this experiment was decisive in convincing physicists of the reality of angular-momentum quantization in all atomic-scale systems. In 1925 German theoretical physicist Werner Heisenberg (December 5, 1901 to February 1, 1976) (Fig. 1.16) reported the first consistent mathematical formulation of quantum mechanics (Matrix Mechanics) (Heisenberg, 1925). One year earlier, in 1924, French physicist Louis Victor Pierre Raymond de Broglie (August 15, 1892 to March 19, 1987) had suggested that all particles behave to an extent like waves and, in 1926, Austrian physicist Erwin Rudolf Josef Alexander Schrödinger (August 12, 1887 to January 4, 1961) (Fig. 1.17) used this idea to come up with a mathematical model of the atom (Wave Mechanics) that described the electrons as three-dimensional waveforms rather than point particles (Schrödinger, 1926). A consequence of

FIGURE 1.16 Werner Heisenberg. *Bundesarchiv, Bild 183-R57262/Unknown/CC-BY-SA 3.0.*

FIGURE 1.17 Erwin Schrodinger (1933).

utilizing waveforms to define particles is that it is mathematically impossible to obtain precise values for both the position and momentum of a particle at a given point in time; this became known as the uncertainty principle, formulated by Werner Heisenberg in 1927. In this concept, for a given accuracy in measuring a position one could only obtain a range of probable values for momentum, and vice versa. This model could explain observations of atomic behavior that previous models were not able to, such as certain structural and spectral patterns of atoms larger than hydrogen. Thus the planetary model of the atom (Bohr's model) was discarded in favor of one that described atomic orbital zones around the nucleus where a given electron is most likely to be observed. The advances in mass spectrometry allowed the mass of atoms to be determined with increased accuracy. A mass spectrometer uses a magnet to bend the trajectory

of a beam of ions, and the amount of deflection is determined by the ratio of an atom's mass to its charge. The English chemist and physicist Francis William Aston (September 1, 1877 to November 20, 1945) proved with this instrument that isotopes had different masses. The atomic mass of these isotopes varied by integer amounts, called the whole number rule. The explanation for these different isotopes had to wait for the discovery of the neutron, an uncharged particle with a mass similar to the proton, by the British physicist James Chadwick (October 20, 1891 to July 24, 1974) in 1932 (Chadwick, 1935) (Fig. 1.18). Isotopes were then explained as elements with the same number of protons, but different numbers of neutrons within the nucleus. In 1938, the German chemist Otto Hahn (March 8, 1879 to July 28, 1968), a student of Rutherford, directed neutrons onto uranium atoms expecting to get transuranium elements. In its place, his chemical experiments showed barium as a product. A year later, Austrian-Swedish physicist Lise Meitner (November 7, 1878 to October 27, 1968) and her nephew Austrian physicist Otto Frisch (October 1, 1904 to September 22, 1979) confirmed that Hahn's results were the first experimental nuclear fission (Meitner and Frisch, 1939). In 1944, Hahn received the Nobel Prize in Chemistry. Despite Hahn's efforts, the contributions of Meitner and Frisch were not recognized. Based on their original correspondence, many historians have documented their view of the discovery of nuclear fission and belief that Meitner should have been awarded the Nobel Prize with Hahn, a classic example of the Matilda effect [a bias against acknowledging the achievements of those women scientists whose work is attributed to their male colleagues. This effect was first described by suffragist and abolitionist Matilda Joslyn Gage (1826–98) in her essay, "Woman as Inventor." The term "Matilda effect" was coined in 1993 by science historian Margaret W. Rossiter.]. At the end of World War II in 1945, Hahn was suspected of working on the German nuclear weapon project to develop an atomic reactor or an atomic bomb, but his only connection was the discovery of fission; he did not work on the program. In April 1945, Hahn and nine leading German physicists (including Max von Laue, Werner Heisenberg, and Carl Friedrich von Weizsäcker) were taken into custody by the Alsos Mission and interned at Farm Hall, Godmanchester, near Cambridge, England, from 3 July 1945 to 3 January 1946. Hahn was still being detained at Farm Hall when the announcement was made; thus his whereabouts were a secret, and it was impossible for the Nobel committee to send him a congratulatory telegram. Instead, he learned about his award through the Daily Telegraph newspaper. His fellow interned German scientists celebrated his award on November 18 by giving speeches, making jokes, and composing songs. On December 4, Hahn was persuaded by two of his captors to write a letter to the Nobel committee accepting the prize but also stating that he would not be able to attend the award ceremony. He could not participate in the Nobel festivities on December 10 since his captors would not allow him to leave Farm Hall. In the 1990s, the records of the Nobel Prize committee that decided on that prize were opened. Based on this information, several scientists and journalists have called her exclusion "unjust," and Meitner has received many posthumous honors, including naming chemical element 109 meitnerium in 1992. Despite not having been awarded the Nobel Prize, Lise Meitner was invited to attend

FIGURE 1.18 James Chadwick (c.1945). © *Copyright Triad National Security, LLC. All Rights Reserved.*

the Lindau Nobel Laureate Meeting in 1962 (annual, scientific conferences held in Lindau, Bavaria, Germany since 1951. Their aim is to bring together Nobel laureates and young scientists to foster scientific exchange between different generations and cultures.). In the 1950s, the development of improved particle accelerators and particle detectors allowed scientists to study the impacts of atoms moving at high energies. Neutrons and protons were found to be hadrons, or composites of smaller particles called quarks. The standard model of particle physics developed so far has successfully explained the properties of the nucleus in terms of these subatomic particles and the forces that govern their interactions.

1.2.2 Structure and properties

Though the word atom originally represented a particle that cannot be divided into smaller particles, in modern scientific usage the atom is composed of a variety of subatomic particles. These fundamental particles of an atom are the electron (e or e^-), the proton (p or p^+) and the neutron (n); all three are fermions (a particle that follows Fermi–Dirac statistics. These particles obey the Pauli exclusion principle. Fermions include all quarks and leptons, as well as all composite particles made of an odd number of these, such as all baryons and many atoms and nuclei. Fermions differ from bosons, which obey Bose–Einstein statistics.). Nevertheless, the hydrogen-1 atom has no neutrons and the hydron ion has no electrons [general name for a cationic form of atomic hydrogen, represented with the symbol H^+. However, this term is avoided and instead "proton" is used, which strictly speaking refers to the cation of protium, the most common isotope of hydrogen. The term "hydron" includes cations of hydrogen regardless of their isotopic composition; thus it refers collectively to protons ($^1H^+$) for the protium isotope, deuterons ($^2H^+$ or D^+) for the deuterium isotope, and tritons ($^3H^+$ or T^+) for the tritium isotope. Unlike most other ions, the hydron consists only of a bare atomic nucleus.]. The electron is the least massive of these particles at 9.11×10^{-31} kg, with a negative electrical charge and a size that is too small to be measured using currently available techniques. It was the lightest particle with a positive rest mass determined, until the discovery of the neutrino mass. Under normal conditions, electrons are bound to the positively charged nucleus due to the attraction created by opposite electric charges. If an atom has more or fewer electrons than its atomic number, then it becomes, respectively, negatively or positively charged as a whole; a charged atom is called an ion (positive charge – cation, negative charge – anion). Electrons have been known since the late 19th century, mostly because of J.J. Thomson's research. Protons have a positive charge and a mass 1836 times that of the electron, at 1.6726×10^{-27} kg. The number of protons in an atom is known as its atomic number. Ernest Rutherford detected that nitrogen under alpha-particle bombardment ejects what seemed to be hydrogen nuclei. By 1920 he had recognized that the hydrogen nucleus forms a distinct particle within the atom and called it a proton. Neutrons have no electrical charge and have a free mass of 1839 times the mass of the electron, or 1.6749×10^{-27} kg. Neutrons are the heaviest of the three fundamental particles, but their mass can be reduced by the nuclear binding energy (the minimum energy that would be required to disassemble the nucleus of an atom into its component parts. The binding is always a positive number, as it is necessary to spend energy in moving these nucleons, attracted to each other by the strong nuclear force, away from each other. The mass of an atomic nucleus is less than the sum of the individual masses of the free constituent protons and neutrons, according to Einstein's equation $E = mc^2$, where m is the mass loss and c is the speed of light (in vacuum). This "missing mass" is known as the mass defect and characterizes the energy that was released when the nucleus was formed.). Neutrons and protons (collectively known as nucleons) have comparable dimensions—on the order of 2.5×10^{-15} m—although the "surface" of these particles is not sharply defined.

In the standard model of physics, electrons are truly elementary particles with no internal structure. In contrast, both protons and neutrons are composite particles consisting of elementary particles called quarks. Quarks combine to form composite particles called hadrons, the most stable of which are protons and neutrons, the components of atomic nuclei. Due to a phenomenon known as color confinement, quarks are never directly observed or found in isolation; they can be found only within hadrons, which include baryons (such as protons and neutrons) and mesons [In particle physics, mesons are hadronic subatomic particles composed of one quark and one antiquark, bound together by strong interactions. Because mesons are composed of quark subparticles, they have physical size, notably a diameter of roughly 1 fm, which is about 1.2 times the size of a proton or neutron. All mesons are unstable, with the longest-lived lasting for only a few hundredths of a microsecond. Charged mesons decay (sometimes through mediating particles) to form electrons and neutrinos. Uncharged mesons may decay to photons. Both of these decays indicate that color is no longer a property of the byproducts.]. Therefore, much of what is known about quarks has been based upon observations of hadrons. Quarks have several intrinsic properties, including electric charge, mass, color charge, and spin. They are the only elementary particles in the standard model of particle physics to experience all four fundamental interactions, also known as fundamental forces (electromagnetism, gravitation, strong interaction, and weak interaction), in addition to being the only known particles whose electric charges are not integer multiples of the elementary charge. There are six types, known as flavors, of quarks: up, down, strange, charm, bottom, and top. Up and down quarks have the lowest masses of all quarks. The heavier quarks rapidly change into up and down quarks through a process of particle decay: the transformation from a higher mass state to a lower mass state. Because of this, up and down quarks are generally stable and the most common in the universe, whereas strange, charm, bottom, and top quarks can only be produced in high-energy collisions (such as those involving cosmic rays and in particle accelerators). For every quark flavor there is a corresponding type of antiparticle, known as an antiquark, that differs from the quark only in that some of its properties (such as the electric charge) have equal magnitude but opposite sign. There are two types of quarks in atoms, each having a fractional electric charge. Protons are composed of two up quarks (each with charge $+$) and one down quark (with a charge of $-$). Neutrons consist of one up quark and two down quarks. This difference accounts for the difference in mass and charge between the two particles. The quarks are held together by the strong interaction (or strong force), which is mediated by gluons. A gluon is an elementary particle that acts as the exchange particle (or gauge boson) for the strong force between quarks. It is analogous to the exchange of photons in the electromagnetic force between two charged particles. In simple terms, they "glue" quarks together, forming hadrons such as protons and neutrons. In technical terms, gluons are vector gauge bosons that mediate strong interactions of quarks in quantum chromodynamics (QCD). Gluons themselves carry the color charge of the strong interaction. This is unlike the photon, which mediates the electromagnetic interaction but lacks an electric charge. Gluons therefore participate in the strong interaction in addition to mediating it, making QCD significantly harder to analyze than quantum electrodynamics (QED). The protons and neutrons, in turn, are held to each other in the nucleus by the nuclear force, which is a residuum of the strong force that has slightly different range properties. The gluon is a member of the family of gauge bosons, which are elementary particles that mediate physical forces. All the bound protons and neutrons in an atom make up a tiny atomic nucleus and are collectively called nucleons. The radius of a nucleus is approximately equal to 1.07 fm. This is much smaller than the radius of the atom, which is on the order of 10^5 fm. The nucleons are bound together by a short-ranged attractive potential known as the residual strong force. At distances less than 2.5 fm this force is much stronger than the electrostatic force that causes positively charged protons to repel each other. Atoms of the same element have the same number of protons, called the atomic number. Within a single element, the number of neutrons may vary, determining the isotope of that element. The total number of protons and neutrons determines the nuclide. The number of neutrons relative to the protons determines the stability of the nucleus, with certain isotopes undergoing radioactive decay. The proton, the electron, and the neutron are classified as fermions. Fermions follow the Pauli exclusion principle, which forbids identical fermions, such as multiple protons, from occupying the same quantum state at the same time. Thus every proton in the nucleus must occupy a quantum state different from all other protons, and the same applies to all neutrons of the nucleus and to all electrons of the electron cloud. A nucleus that has a different number of protons than neutrons can potentially drop to a lower energy state through a radioactive decay that causes the number of protons and neutrons to more closely match. As a result, atoms with matching numbers of protons and neutrons are more stable against decay. However, with increasing atomic number, the mutual repulsion of the protons requires an increasing proportion of neutrons to preserve the stability of the nucleus, which slightly modifies this trend of equal numbers of protons to neutrons. The number of protons and neutrons in the atomic nucleus can be altered, although this can require very high energies because of the strong force. Nuclear fusion occurs when multiple atomic particles join to form a heavier nucleus, such as through the energetic collision of two nuclei. For example, at the core of the Sun, protons require energies of 3–10 keV to overcome their mutual repulsion—the Coulomb barrier—and fuse together into a single nucleus. Nuclear fission is the opposite process, causing a nucleus to split into two smaller nuclei—typically through radioactive decay. The nucleus can also be modified through bombardment with high-energy subatomic particles or photons. If this changes the number of protons in a nucleus, the atom changes to a different chemical element. If the mass of the nucleus following a fusion reaction is less than the sum of the masses of the separate particles, then the difference between these two values can be emitted as a type of usable energy (such as a gamma ray, or the kinetic energy of a beta particle), as described by Albert Einstein's mass–energy equivalence formula, $E = mc^2$. This deficit is part of the binding energy of the new nucleus, and it is the nonrecoverable loss of the energy that causes the fused particles to remain together in a state that requires this energy to separate. The fusion of two nuclei that create larger nuclei with lower atomic numbers than iron and nickel—a total nucleon number of about 60—is usually an exothermic process that releases more energy than is required to bring them together. It is this energy-releasing process that makes nuclear fusion in stars a self-sustaining reaction. For heavier nuclei, the binding energy per nucleon in the nucleus begins to decrease. That means fusion processes producing nuclei that have atomic numbers higher than about 26, and atomic masses higher than about 60, are endothermic processes. These more massive nuclei cannot undergo an energy-producing fusion reaction that can sustain the hydrostatic equilibrium of a star.

The electrons in an atom are attracted to the protons in the nucleus by the electromagnetic force. This force binds the electrons inside an electrostatic potential well surrounding the smaller nucleus, which means that an external source of energy is necessary for the electron to escape (Fig. 1.19). The closer an electron is to the nucleus, the larger the attractive force. Therefore electrons bound near the center of the potential well need more energy to escape than those at greater separations. Electrons, like other particles, have properties of both a particle and a wave. The electron cloud is a region inside the potential well where each electron forms a type of three-dimensional standing wave—a waveform that does not move relative to the nucleus. This behavior is defined by an atomic orbital, a mathematical function that characterizes the probability that an electron appears to be at a certain location when its position is measured. Only a discrete (or quantized) set of these orbitals exist around the nucleus, as other possible wave patterns rapidly decay into a more stable form. Orbitals can have one or more ring or node structures and differ from each other in size, shape, and orientation. Each atomic orbital corresponds to a particular energy level of the electron. The electron can change its state to a higher energy level by absorbing a photon with enough energy to boost it into the new quantum state. Similarly, through spontaneous emission, an electron in a higher energy state can drop to a lower energy state while radiating the excess energy as a photon. These characteristic energy values, defined by the differences in the energies of the quantum states, are responsible for atomic spectral lines. The amount of energy needed to remove or add an electron—the electron binding energy—is far less than the binding energy of nucleons. For example, it requires only 13.6 eV to strip a ground-state electron from a hydrogen atom, compared to 2.23 MeV for splitting a deuterium nucleus. Atoms are electrically neutral if they have the same number of protons and electrons. Atoms that have either a deficit or a surplus of electrons are called ions. Electrons that are farthest from the nucleus may be transferred to other nearby atoms or shared between atoms. By this mechanism, atoms can form bonds in molecules and other types of chemical compounds such as ionic and covalent network crystals.

By definition, any two atoms with the same number of protons in their nuclei belong to the same chemical element. Atoms with equal numbers of protons but a different number of neutrons are known as different isotopes of the same element. For example, all hydrogen atoms admit exactly one proton, but isotopes exist with no neutrons (hydrogen-1, by far the most common form, also called protium), one neutron (deuterium), two neutrons (tritium), and more than two neutrons. The known elements form a set of atomic numbers, from the single proton element hydrogen up to the 118-proton element oganesson. All known isotopes of elements with atomic numbers greater than 82 are radioactive, although the radioactivity of element 83 (bismuth) is so slight as to be practically negligible. Around 339 nuclides occur naturally on Earth, of which 254 (about 75%) have not been observed to decay and are known as "stable isotopes." Yet, only 90 of these nuclides are stable to all decay, even in theory. Another 164 (bringing the total to 254) have not been observed to decay, although in theory it is energetically possible. These are also formally classified as "stable." A further 34 radioactive nuclides have half-lives longer than 80 million years and are long-lived enough to have been present since the birth of the solar system. This collection of 288 nuclides is known as primordial nuclides. Finally, extra 51 short-lived nuclides are known to occur naturally, as daughter products of primordial nuclide decay (such as radium from uranium), or else as products of natural energetic processes on Earth, such as cosmic ray bombardment (e.g., carbon-14). For 80 of the chemical elements, at least one stable isotope exists. As a rule, there are only a limited number of stable isotopes for each of these elements, the average being 3.2 stable isotopes per element. Twenty-six elements have only a single stable isotope, while the largest number of stable isotopes observed for any element is 10, for the

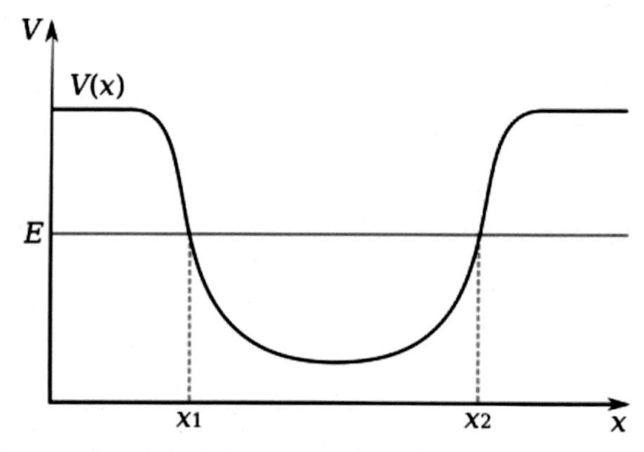

FIGURE 1.19 A potential well, showing, according to classical mechanics, the minimum energy $V(x)$ needed to reach each position x. Classically, a particle with energy E is constrained to a range of positions between x_1 and x_2.

element tin. Elements 43, 61, and all elements numbered 83 or higher have no stable isotopes. Stability of isotopes is affected by the ratio of protons to neutrons, in addition to the presence of certain "magic numbers" of neutrons or protons that represent closed and filled quantum shells. These quantum shells correspond to a set of energy levels within the shell model of the nucleus; filled shells, such as the filled shell of 50 protons for tin, confers unusual stability on the nuclide. Of the 254 known stable nuclides, only four have both an odd number of protons and odd number of neutrons: hydrogen-2 (deuterium), lithium-6, boron-10, and nitrogen-14. Similarly, only four naturally occurring, radioactive odd–odd nuclides have a half-life over a billion years: potassium-40, vanadium-50, lanthanum-138, and tantalum-180m. Most odd–odd nuclei are highly unstable with respect to beta decay, because the decay products are even–even and are therefore more strongly bound, due to nuclear pairing effects [In nuclear physics, beta decay (β-decay) is a type of radioactive decay in which a beta particle (fast energetic electron or positron) is emitted from an atomic nucleus. For example, beta decay of a neutron transforms it into a proton by the emission of an electron accompanied by an antineutrino, or conversely a proton is converted into a neutron by the emission of a positron (positron emission) with a neutrino, thus changing the nuclide type. Neither the beta particle nor its associated (anti-)neutrino exist within the nucleus prior to beta decay but are created in the decay process. Through this process, unstable atoms obtain a more stable ratio of protons to neutrons. The probability of a nuclide decaying due to beta and other forms of decay is determined by its nuclear binding energy. The binding energies of all existing nuclides form what is called the nuclear band or valley of stability. For either electron or positron emission to be energetically possible, the energy release or Q value must be positive (Fig. 1.20). Beta decay is a consequence of the weak force, which is characterized by relatively lengthy decay times. Nucleons are composed of up quarks and down quarks, and the weak force allows a quark to change type by the exchange of a W boson and the creation of an electron/antineutrino or positron/neutrino pair. For example, a neutron, composed of two down quarks and an up quark, decays to a proton composed of a down quark and two up quarks. Decay times for many nuclides that are subject to beta decay can be thousands of years.].

Most of the atom's mass comes from the protons and neutrons. The total number of these particles (called "nucleons") in any given atom is called the mass number. It is a positive integer and dimensionless (instead of having dimension of mass), because it expresses a count. An example of use of a mass number is "carbon-12," which has 12 nucleons (six protons and six neutrons). The actual mass of an atom at rest is frequently expressed using the unified atomic mass unit (u), also known as Dalton (Da). This unit is defined as a twelfth of the mass of a free neutral atom of carbon-12, which is approximately 1.66×10^{-27} kg. Hydrogen-1 (the lightest isotope of hydrogen, which is also the nuclide with the lowest mass) has an atomic weight of 1.007825 u. The value of this number is called the atomic mass. A given atom has an atomic mass approximately equal (within 1%) to its mass number times the atomic mass unit. Yet, this number will not be exactly an integer except in the case of carbon-12. As even the most massive atoms are far too light to work with directly, chemists in its place use the unit of moles. One mole of atoms of any element always has the same number of atoms (about 6.022×10^{23}). This number was chosen so that if an element has an atomic mass of 1 u, a mole of atoms of that element has a mass close to 1 g. Because of the definition of the unified atomic mass unit, each carbon-12 atom has an atomic mass of exactly 12 u, and so a mole of carbon-12 atoms weighs exactly 0.012 kg.

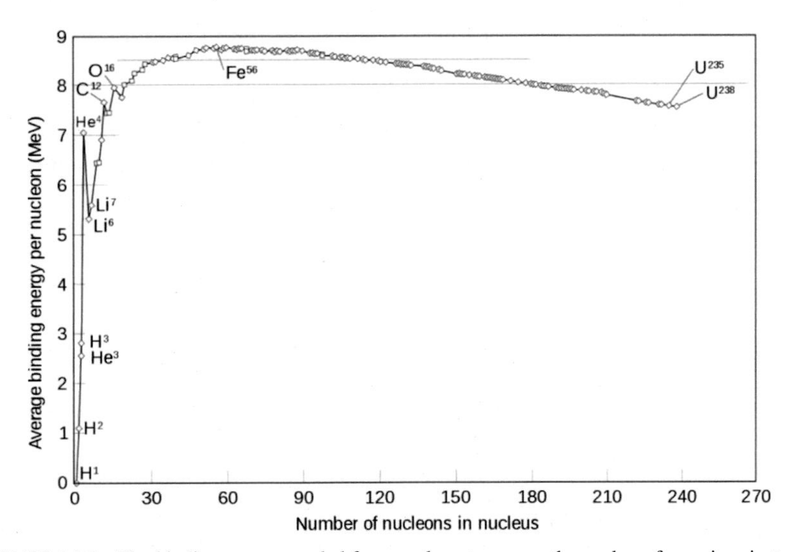

FIGURE 1.20 The binding energy needed for a nucleon to escape the nucleus, for various isotopes.

Atoms have no well-defined outer boundary, so their dimensions are generally described in terms of an atomic radius. This is a measure of the distance out to which the electron cloud extends from the nucleus. Nevertheless, this is under the assumption that the atom exhibits a spherical shape, which is only true for atoms in vacuum or free space. Atomic radii may be derived from the distances between two nuclei when the two atoms are joined in a chemical bond. The radius varies with the location of an atom on the atomic chart, the type of chemical bond, the number of neighboring atoms (coordination number), and a quantum mechanical property known as spin. On the periodic table of the elements, atom size tends to increase when moving down columns, but decrease when moving across rows (left to right). Therefore the smallest atom is helium with a radius of 32 pm, while one of the largest is cesium at 225 pm. When subjected to external forces, such as electrical fields, the shape of an atom may deviate from spherical symmetry. The deformation depends on the field magnitude and the orbital type of outer shell electrons, as shown by group-theoretical considerations. Aspherical deviations might be elicited for instance in crystals, where large crystal-electrical fields may occur at low-symmetry lattice sites. Significant ellipsoidal deformations have been shown to occur for sulfur ions and chalcogen ions in pyrite-type compounds (pyrite itself is a mineral with composition FeS_2). Atomic dimensions are thousands of times smaller than the wavelengths of light (400–700 nm), so they cannot be viewed using an optical microscope. Nevertheless, these days individual atoms can be observed using a scanning tunneling microscope. To visualize the minuteness of the atom, consider that a typical human hair is about 1 million carbon atoms in width. A single drop of water contains about 2 sextillion (2×10^{21}) atoms of oxygen, and twice the number of hydrogen atoms. A single carat diamond with a mass of 2×10^{-4} kg contains about 10 sextillion (10^{22}) atoms of carbon.

Every element has one or more isotopes that have unstable nuclei that are subject to radioactive decay, causing the nucleus to emit particles or electromagnetic radiation. Radioactivity can occur when the radius of a nucleus is large compared with the radius of the strong force, which only acts over distances on the order of 1 fm. The three most common forms of radioactive decay are:

1. Alpha (α) decay: this process is caused when the nucleus emits an alpha particle, which is a helium nucleus consisting of two protons and two neutrons. The result of the emission is a new element with a lower atomic number.
2. Beta (β) decay (and electron capture): these processes are regulated by the weak force and result from a transformation of a neutron into a proton, or a proton into a neutron. The neutron to proton transition goes together with the emission of an electron and an antineutrino, while proton to neutron transition (except in electron capture) causes the emission of a positron and a neutrino. The electron or positron emissions are called beta particles. Beta decay either increases or decreases the atomic number of the nucleus by one. Electron capture is more common than positron emission, since it requires less energy. In this type of decay, an electron is absorbed by the nucleus, rather than a positron emitted from the nucleus. A neutrino is still emitted in this process, and a proton changes to a neutron.
3. Gamma (γ) decay: this process is caused by a change in the energy level of the nucleus to a lower state, resulting in the emission of electromagnetic radiation. The excited state of a nucleus which results in gamma emission usually occurs following the emission of an alpha or a beta particle. Thus, gamma decay usually follows alpha or beta decay.

Other rarer types of radioactive decay include ejection of neutrons or protons or clusters of nucleons from a nucleus, or more than one beta particle. An analog of gamma emission, which allows excited nuclei to lose energy in a different way, is internal conversion—a process that produces high-speed electrons that are not beta rays, followed by production of high-energy photons that are not gamma rays. A few large nuclei explode into two or more charged fragments of varying masses plus several neutrons, in a decay called spontaneous nuclear fission. Each radioactive isotope has a characteristic decay time period—the half-life—that is determined by the amount of time needed for half of a sample to decay (Fig. 1.21). This is an exponential decay process that steadily decreases the proportion of the remaining isotope by 50% every half-life. Therefore, after one half-life has passed 50% of the isotope is left, after two half-lives have passed only 25%, after three half-lives 12.5%, and so forth. Since these are statistical probabilities, it is impossible to predict exactly which atom will decay at what time.

Elementary particles have an intrinsic quantum mechanical property known as spin. This is comparable to the angular momentum of an object that is spinning around its center of mass, although strictly speaking these particles are thought to be point-like and cannot be said to be rotating. Spin is measured in units of the reduced Planck constant ($\hbar$), with electrons, protons and neutrons all having spin ½ $\hbar$, or "spin -½." In an atom, electrons in motion around the nucleus have orbital angular momentum in addition to their spin, while the nucleus itself possesses angular momentum due to its nuclear spin. The magnetic field produced by an atom—its magnetic moment—is determined by these different forms of angular momentum, just as a rotating charged object classically produces a magnetic field. Nevertheless, the most important contribution comes from electron spin. Due to the nature of electrons to obey the Pauli exclusion

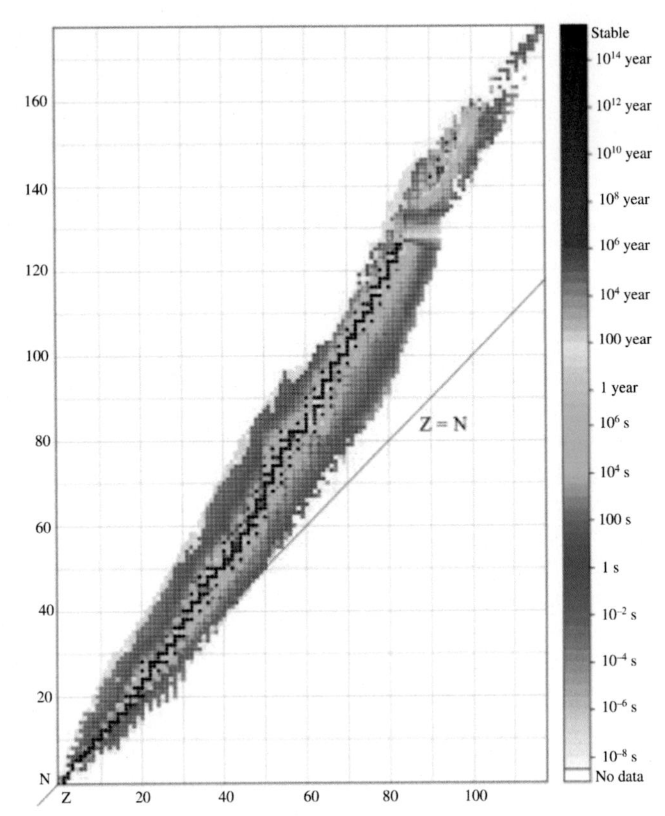

FIGURE 1.21 Diagram showing the half-life (T½) of various isotopes with Z protons and N neutrons.

principle, in which no two electrons may be found in the same quantum state, bound electrons pair up with each other, with one member of each pair in a spin up state and the other in the opposite, spin down state. Consequently, these spins cancel each other out, reducing the total magnetic dipole moment to zero in some atoms with even number of electrons. In ferromagnetic elements such as iron, cobalt, and nickel, an odd number of electrons lead to an unpaired electron and a net overall magnetic moment. The orbitals of neighboring atoms overlap, and a lower energy state is achieved when the spins of unpaired electrons are aligned with each other, a spontaneous process known as an exchange interaction. When the magnetic moments of ferromagnetic atoms are lined up, the material can produce a measurable macroscopic field. Paramagnetic materials have atoms with magnetic moments that line up in random directions when no magnetic field is present, but the magnetic moments of the individual atoms line up in the presence of a field. The nucleus of an atom will have no spin when it has even numbers of both neutrons and protons, but for other cases of odd numbers, the nucleus may have a spin. Normally nuclei with spin are aligned in random directions because of thermal equilibrium. However, for certain elements (such as xenon-129) it is possible to polarize a significant proportion of the nuclear spin states so that they are aligned in the same direction—a condition called hyperpolarization.

The potential energy of an electron in an atom is negative; its dependence of its position reaches the minimum (the most absolute value) inside the nucleus and vanishes when the distance from the nucleus goes to infinity, roughly in an inverse proportion to the distance. In the quantum mechanical model, a bound electron can only occupy a set of states centered on the nucleus, and each state corresponds to a specific energy level. An energy level can be measured by the amount of energy needed to free the electron from the atom and is usually given in units of electronvolts (eV). The lowest energy state of a bound electron is called the ground state, that is, stationary state, while an electron transition to a higher level results in an excited state. The electron's energy raises when n (principal quantum number) increases because the (average) distance to the nucleus increases. Dependence of the energy on ℓ (angular momentum quantum number) is caused not by electrostatic potential of the nucleus, but by interaction between electrons. For an electron to transition between two different states, for example, ground state to first excited state, it must absorb or emit a photon at an energy equal to the difference in the potential energy of those levels, consistent with the Niels Bohr model, what can be precisely calculated using the Schrödinger equation. Electrons jump between orbitals in a particle-like fashion.

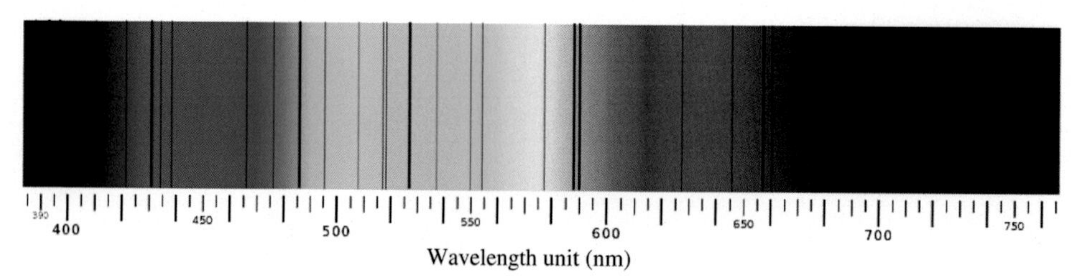

FIGURE 1.22 Example of absorption lines in a spectrum.

For example, if a single photon strikes the electrons, only a single electron changes states in response to the photon. The energy of an emitted photon is proportional to its frequency, so these specific energy levels appear as distinct bands in the electromagnetic spectrum. Each element has a characteristic spectrum that can depend on the nuclear charge, subshells filled by electrons, the electromagnetic interactions between the electrons, and other factors. When a continuous spectrum of energy is passed through a gas or plasma, some of the photons are absorbed by atoms, causing electrons to change their energy level. Those excited electrons that remain bound to their atom spontaneously emit this energy as a photon, traveling in a random direction, and so drop back to lower energy levels. Hence, the atoms behave like a filter that forms a series of dark absorption bands in the energy output. (An observer viewing the atoms from a view that does not include the continuous spectrum in the background, in contrast, sees a series of emission lines from the photons emitted by the atoms.) Spectroscopic measurements of the strength and width of atomic spectral lines allow the composition and physical properties of a substance to be determined.

Close examination of the spectral lines shows that some display a fine structure splitting. This happens due to spin–orbit coupling, which is an interaction between the spin and motion of the outermost electron. When an atom is in an external magnetic field, spectral lines become split into three or more components; a phenomenon called the Zeeman effect. This is caused by the interaction of the magnetic field with the magnetic moment of the atom and its electrons. Some atoms can have multiple electron configurations with the same energy level, which thus appear as a single spectral line. The interaction of the magnetic field with the atom shifts these electron configurations to somewhat different energy levels, resulting in multiple spectral lines (Fig. 1.22). The existence of an external electric field can result in a comparable splitting and shifting of spectral lines by modifying the electron energy levels, a phenomenon called the Stark effect. If a bound electron is in an excited state, an interacting photon with the proper energy can cause stimulated emission of a photon with a matching energy level. For this to occur, the electron must drop to a lower energy state that has an energy difference matching the energy of the interacting photon. The emitted photon and the interacting photon then move off in parallel and with matching phases. That is, the wave patterns of the two photons are synchronized. This physical property is used to make lasers, which can emit a coherent beam of light energy in a narrow frequency band.

Valency is the combining power of an element. The outermost electron shell of an atom in its uncombined state is known as the valence shell, and the electrons in that shell are called valence electrons. The number of valence electrons determines the bonding behavior with other atoms. Atoms tend to chemically react with each other in a manner that fills (or empties) their outer valence shells. For example, a transfer of a single electron between atoms is a useful approximation for bonds that form between atoms with one-electron more than a filled shell, and others that are one-electron short of a full shell, such as occurs in the compound sodium chloride (NaCl) and other chemical ionic salts. Nevertheless, many elements exhibit multiple valences or tendencies to share differing numbers of electrons in different compounds. Thus chemical bonding between these elements takes many forms of electron-sharing that are more than simple electron transfers. Examples of these include the element carbon and the organic compounds. The chemical elements are often displayed in a periodic table that is laid out to display recurring chemical properties, and elements with the same number of valence electrons form a group that is aligned in the same column of the table. (The horizontal rows correspond to the filling of a quantum shell of electrons.) The elements at the far right of the table have their outer shell completely filled with electrons, which results in chemically inert elements known as the noble gases (He, Ne, Ar, Kr, Xe, and Rn).

Quantities of atoms are found in different states of matter that depend on the physical conditions, such as temperature and pressure. By varying the conditions, materials can transition between solids, liquids, gases, and plasmas. Within a state, a material can also exist in different allotropes. An example of this is solid carbon, which can exist as graphite or diamond. Gaseous allotropes exist as well, such as dioxygen and ozone. At temperatures close to absolute zero (−273.15°C), atoms can form a Bose–Einstein condensate, at which point quantum mechanical effects, which are normally only observed at the atomic scale, become apparent on a macroscopic scale. This supercooled collection of atoms then behaves as a single super atom, which may allow fundamental checks of quantum mechanical behavior.

References

Béguyer de Chancourtois, A.-É. (1862). Tableau du classement naturel des corps simples, dit vis tellurique. Comptes Rendus de l'Académie des Sciences, *55*, 600–601.

Béguyer de Chancourtois, A.-É. (1863). Vis tellurique. Classement des corps simples ou radicaux, obtenu au moyen d'un système de classification hélicoïdal et numérique. Paris: Mallet-Bachelier.

Bohr, N. (1922a). Nobel lecture. The structure of the atom. <NobelPrize.org>. Nobel Media AB. <https://www.nobelprize.org/prizes/physics/1922/bohr/lecture/> Accessed 05.07.19.

Bohr, N. (1922b). *The theory of spectra and atomic constitution; three essays*. London: Cambridge University Press.

Chadwick, J. (1935). Nobel lecture. The neutron and its properties. <NobelPrize.org>. Nobel Media AB. <https://www.nobelprize.org/prizes/physics/1935/chadwick/lecture/>.

Döbereiner, J. W. (1829). Versuch zu einer gruppirung der elementaren stoffe nach ihrer analogie. *Annalen der Physik und Chemie*, *15*, 301–307, 2nd series.

Heisenberg, W. (1925). *Über quantentheoretishe Umdeutung kinematisher und mechanischer Beziehungen, Zeitschrift für Physik, 33*, 879–893.

Langmuir, I. (1919a). The arrangement of electrons in atoms and molecules. *Journal of the American Chemical Society*, *41*, 868–934.

Langmuir, I. (1919b). The structure of atoms and the octet theory of valence. *Proceedings of the National Academy of Sciences of the United States of America*, *5*, 252–259.

Менделеев Д.М. [Mendeleev, D.I.] (1869a). Соотношение свойств с атомным весом элементов [Relationship of elements' properties to their atomic weights]. Журнал Русского Химического Общества *(Journal of the Russian Chemical Society)*, *1*, 60–77.

Mendeleev, D. (1869b). Ueber die beziehungen der eigenschaften zu den atomgewichten der elemente. *Zeitschrift für Chemie*, *12*, 405–406.

Менделеев Д.М. [Mendeleev, D.I.]. Периодический закон. ЛитРес, (2018 [1869]), Moscow.

Meyer, L., Die modernen Theorien der Chemie und ihre Bedeutung für die chemische Statistik. 1864, Breslau: Maruschke & Berendt.

Moseley, H. G. J. (1913). The high-frequency spectra of the elements. *Philosophical Magazine*, *26*, 1012–1034, 6th series.

Moseley, H. (1914). The high-frequency spectra of the elements, Part II. *Philosophical Magazine Series*, *7*, 703–713.

Newlands, J. A. R. (1864a). On relations among the equivalents. *The Chemical News*, *10*, 94–95.

Newlands, J. A. R. (1864b). Relations between equivalents. *The Chemical News*, *10*, 59–60.

Newlands, J. A. R. (1865). On the law of octaves. *The Chemical News*, *12*, 83.

Odling, W. (1864). On the proportional numbers of the elements. *Quarterly Journal of Science*, *1*, 642–648.

Prout, W. (1815). On the relation between the specific gravities of bodies in their gaseous state and the weights of their atoms. *Annals of Philosophy*, *6*, 321–330.

Prout, W. (1816). Correction of a mistake in the essay on the relation between the specific gravities of bodies in their gaseous state and the weights of their atoms. *Annals of Philosophy*, *7*, 111–113.

Schrödinger, E., (1926). Quantisierung als Eigenwertproblem, *Annalen der Physik, 384*, 273–376. Sciolist Online Etymology Dictionary. <https://www.etymonline.com/> Accessed 21.06.19.

Soddy, F. (1913a). Intra-atomic charge. *Nature*, *92*, 399–400.

Soddy, F. (1913b). The radio-elements and the periodic law. *The Chemical News*, *107*, 97–99.

Soddy, F. (1913c). Radioactivity. *Annual Reports on the Progress of Chemistry*, *10*, 262–288.

Thomson, J. J. (1901). On bodies smaller than atoms. *The Popular Science Monthly*, 323–335.

Lavoisier, A. L., Elements of chemistry, in a new systematic order, containing all the modern discoveries: Illustrated with thirteen copperplates (trans: Kerr R). Traité élémentaire de chimie (English).1790, Printed for William Creech; By G.G. and J.J. Robinsons [sic], Edinburgh; and sold in London.

Lewis, G. N. (1916). The atom and the molecule. *Journal of the American Chemical Society*, *38*, 762–786.

Meitner, L., & Frisch, O. (1939). Disintegration of uranium by neutrons: A new type of nuclear reaction. *Nature*, *143*, 239–240.

Further reading

Asimov, I. (1966). *The noble gases*. New York: Basic Books.

Aston, F. W. (1920). The constitution of atmospheric neon. *Philosophical Magazine*, *39*, 449–455, 4th series.

Audi, G., Wapstra, A. H., & Thibault, C. (2003). The AME2003 atomic mass evaluation: (II). Tables, graphs and references. *Nuclear Physics A*, *729*, 337–676.

Aycan, S. (2005). Chemistry education and mythology. *Journal of Social Sciences*, *1*, 238–239.

Barbalace, K. (1995–2019). Periodic table of elements. <https://environmentalchemistry.com/yogi/periodic/> Accessed 27.06.19.

Bell, R. E., & Elliott, L. G. (1950). Gamma-rays from the reaction $H^1(n,\gamma)D^2$ and the binding energy of the deuteron. *Physical Review*, *79*, 282–285.

Bethe, H. (1929). Termaufspaltung in Kristallen. *Annalen der Physik*, *3*, 133–208.

Beyer, H. F., & Shevelko, V. P. (2002). *Introduction to the physics of highly charged ions*. Boca Raton, FL: CRC Press.

Birkholz, M. (1995). Crystal-field induced dipoles in heteropolar crystals I: Concept. *Zeitschrift für Physik B Condensed Matter, 96*, 325–332.

Birkholz, M. (2014). Modeling the shape of ions in pyrite-type crystals. *Crystals, 4*, 390–403.

Birkholz, M., & Rudert, R. (2008). Interatomic distances in pyrite-structure disulfides – A case for ellipsoidal modeling of sulfur ions. *Physica Status Solidi B, 245*, 1858–1864.

Bowden, M. E. (1997). *Chemical achievers: The human face of the chemical sciences*. Philadelphia, PA: Chemical Heritage Foundation.

Boyle, R. (1661). The sceptical chymist or, chymico-physical doubts & paradoxes: Touching the spagyrist's principles commonly call'd hypostatical, as they are wont to be propos'd and defended by the generality of alchymists. Whereunto is praemis'd part of another discourse relating to the same subject. London: J. Crooke.

Bury, C. R. (1921). Langmuir's theory of the arrangement of electrons in atoms and molecules. *Journal of the American Chemical Society, 43*, 1602–1609.

Chaptal, J. A. (1800). In (3rd ed.). W. Nicholson (Ed.), *Elements of chemistry* (Vol. 1). London: G.G. and J. Robinson.

Clark, D. L., & Hobart, D. E. (2000). Reflections on the legacy of a legend: Glenn T. Seaborg, 1912–1999. *Los Alamos Science, 26*, 56–61.

Considine, G. D. (2005). *Van Nostrand's Encyclopedia of Chemistry*. Wiley-Interscience.

Coplen Tyler, B., & Peiser, H. S. (1998). History of the recommended atomic-weight values from 1882 to 1997: A comparison of differences from current values to the estimated uncertainties of earlier values (Technical Report). *Pure and Applied Chemistry, 70*, 237–257.

Crawford, E., Sime, R. L., & Walker, M. (1997). A nobel tale of postwar injustice. *Physics Today, 50*, 26–32.

Daintith, J., Mitchell, S., Tootill, E., & Gjertsem, D. (1994). *Biographical encyclopedia of scientists* (2nd ed.). Taylor & Francis, 2 Volume Set.

De Gregorio, A. (2003). A historical note about how the property was discovered that hydrogenated substances increase the radioactivity induced by neutrons. *Nuovo Saggiatore, 19*, 41–47.

DeKosky, R. K. (1973). Spectroscopy and the elements in the late nineteenth century: The work of Sir William Crookes. *The British Journal for the History of Science, 6*, 400–423.

de Laeter John, R., Böhlke John, K., De Bièvre, P., Hidaka, H., Peiser, H. S., Rosman, K. J. R., & Taylor, P. D. P. (2003). Atomic weights of the elements. Review 2000 (IUPAC Technical Report). *Pure and Applied Chemistry, 75*, 683.

Demtröder, W. (2006). *Atoms, molecules and photons: an introduction to atomic molecular and quantum physics*. Berlin: Springer.

Ede, A., & Cormack, L. B. (2016). (3rd ed.). *A history of science in society, Volume I: From the ancient Greeks to the scientific revolution*, (Vol. 1). Toronto: University of Toronto Press.

Thomas Jefferson National Accelerator Facility-Office of Science Education. It's elemental – The periodic table of elements. <https://education.jlab.org/itselemental/index.html> Accessed 27.06.19.

Einstein, A. (1905). Über die von der molekularkinetischen Theorie der Wärme geforderte Bewegung von in ruhenden Flüssigkeiten suspendierten Teilchen. *Annalen der Physik, 322*, 549–560.

Emsley, J. (2001). *Nature's building blocks*. Oxford: Oxford University Press.

Enghag, P. (2004). *Encyclopedia of the elements: Technical data - History - Processing - Applications*. Weinheim: Wiley-VCH.

Fermi, E. (1938). Artificial radioactivity produced by neutron bombardment: Nobel Lecture. *Royal Swedish Academy of Sciences*. Available from https://www.nobelprize.org/uploads/2018/06/fermi-lecture.pdf, Accessed 26.06.19.

Fernando D. (1998). Alchemy: An illustrated A to Z. Blandford.

Fewell, M. P. (1995). The atomic nuclide with the highest mean binding energy. *American Journal of Physics, 63*, 653–658.

Fontani, M., Costa, M., & Orna, M. V. (2014). *The lost elements: The periodic table's shadow side*. Oxford: Oxford University Press.

Fowles, G. R. (1989). *Introduction to modern optics*. New York: Dover Publications.

Fricke, B., Greiner, W., & Waber, J. T. (1971). The continuation of the periodic table up to Z = 172. The chemistry of superheavy elements. *Theoretica Chimica Acta, 21*, 235–260.

Ghosh, D. C., & Biswas, R. (2002). Theoretical calculation of absolute radii of atoms and ions. Part 1. The atomic radii. *International Journal of Molecular Sciences, 3*, 87–113.

Goodstein, D. L. (1975). *States of matter*. Englewood Cliffs, NJ: Prentice-Hall.

Greenwood, N. N., & Earnshaw, A. (1996). *Chemistry of the elements*. Oxford: Elsevier Science and Technology Books.

Hampel, C. A. (1968). *The encyclopedia of the chemical elements*. New York: Reinhold Book Corp.

Heilbron, J. L. (1966). The work of H. G. J. Moseley. *Isis, 57*, 336–364. Heisenberg, W. *Über quantentheoretishe Umdeutung kinematisher und mechanischer Beziehungen, Zeitschrift für Physik*, **33**, 1925, 879–893.

Hoffman, D. C., Ghiorso, A., & Seaborg, G. T. (2000). *The transuranium people: The inside story*. London: Imperial College Press.

Ihde, A. J. (1984). *The development of modern chemistry*. New York: Dover Publications.

Jevremovic, T. (2005). *Nuclear principles in engineering*. New York: Springer.

Kaji, M. (2002). D. I. Mendeleev's concept of chemical elements and the principles of chemistry. *Bulletin for the History of Chemistry, 27*, 4–16.

L'Annunziata, M. F. (2012). *Handbook of radioactivity analysis*. Amsterdam: Academic Press.

MacGregor, M. H. (1992). *The enigmatic electron*. Oxford: Oxford University Press.

Mazo, R. M. (2002). *Brownian motion: Fluctuations, dynamics, and applications*. Oxford: Clarendon Press.

McNeil, I. (2002). *An encyclopedia of the history of technology*. New York: Taylor & Francis.

Meija, J., Coplen Tyler, B., Berglund, M., Brand Willi, A., De Bièvre, P., Gröning, M., ... Prohaska, T. (2016). Atomic weights of the elements 2013 (IUPAC Technical Report). *Pure and Applied Chemistry, 88*, 265–291.

Mel'nikov, V. P. (1982). Some details in the prehistory of the discovery of element 72. *Centaurus, 26*, 317–322.

Mills, I., Cvitaš, T., Homann, K., Kallay, N., & Kuchitsu, K. (1993). *Quantities, units, and symbols in physical chemistry*. Oxford: Blackwell Scientific Publications.

Morris, R. (2003). *The last sorcerers: The path from alchemy to the periodic table*. Washington, DC: National Academies Press.

Mulliken, R. S. (1967). Spectroscopy, molecular orbitals, and chemical bonding. *Science*, *157*, 13–24.

Nefedov, V. I., Trzhaskovskaya, M. B., & Yarzhemskii, V. G. (2006). Electronic configurations and the periodic table for superheavy elements. *Doklady Physical Chemistry*, *408*, 149–151.

Newlands, J. A. R. (1863). On relations among the equivalents. *The Chemical News*, *7*, 70–72.

Newlands, J. A. R. (1884). *On the discovery of the periodic law and on relations among the atomic weights*. London: E. & F.N. Spon.

Nobel Media A.B. (2019). Frederick Soddy, Biographical. Nobel Media AB. <https://www.nobelprize.org/prizes/chemistry/1921/soddy/biographical/> Accessed 05.06.19.

Odling, W. (1857a). On the natural groupings of the elements. Part 1. *Philosophical Magazine*, *13*, 423–440, 4th series.

Odling, W. (1857b). On the natural groupings of the elements. Part 2. *Philosophical Magazine*, *13*, 480–497, 4th series.

Odling, W. (1864). On the hexatomicity of ferricum and aluminium. *Philosophical Magazine*, *27*, 115–119, 4th series.

Öhrström, L., & Reedijk, J. (2016). Names and symbols of the elements with atomic numbers 113, 115, 117 and 118 (IUPAC Recommendations 2016). *Pure and Applied Chemistry*, *88*, 1225–1229.

Pais, A. (1986). *Inward bound: Of matter and forces in the physical world*. Oxford: Clarendon Press.

Paneth, F. (1922). *Das periodische system der chemischen elemente, . Ergebnisse der exakten naturwissenschaften* (Vol. 1). Berlin: Springer-Verlag.

Patterson, G. (2007). Jean Perrin and the triumph of the atomic doctrine. *Endeavour*, *31*, 50–53.

Pullman, B., & Reisinger, A. R. (2001). *The atom in the history of human thought*. Oxford: Oxford University Press.

Pyykkö, P. (2011). A suggested periodic table up to $Z \leq 172$, based on Dirac–Fock calculations on atoms and ions. *Physical Chemistry Chemical Physics*, *13*, 161–168.

Rocke, A. J. (1984). *Chemical atomism in the nineteenth century: From Dalton to Cannizzaro*. Columbus, OH: Ohio State University Press.

Rutherford, E. (1911). The scattering of α and β particles by matter and the structure of the atom. *Philosophical Magazine*, *21*, 669–688, 4th series.

Rutherford, E., & Owens, R. B. (1899). Thorium and uranium radiation. *Transactions of the Royal Society of Canada*, *2*, 9–12.

Scerri, E. (2013). Cracks in the periodic table. *Scientific American*, *308*, 68–73.

Scerri, E. R. (2007). *The periodic table: Its story and its significance*. Oxford: Oxford University Press.

Scully, M. O., Lamb, W. E., & Barut, A. (1987). On the theory of the Stern-Gerlach apparatus. *Foundations of Physics*, *17*, 575–583.

Shannon, R. D. (1976). Revised effective ionic radii and systematic studies of interatomic distances in halides and chalcogenides. *Acta Crystallographica A*, *32*, 751–767.

Shaviv, G. (2012). *The synthesis of the elements* (p. 38) Berlin: Springer-Verlag.

Shultis, J. K., & Faw, R. E. (2002). *Fundamentals of nuclear science and engineering*. New York: Marcel Dekker.

Smirnov, B. M. (2005). *Physics of atoms and ions*. New York: Springer.

Stillman, J. M. (2008). *The story of alchemy and early chemistry*. Kila: Kessinger Pub. Co.

Stwertka, A. (1999). *A guide to the elements*. Oxford: Oxford University Press.

Thoennessen, M. (2016). *The discovery of isotopes: A complete compilation*. Switzerland: Springer.

Thomson, J. J. (1913). Rays of positive electricity. *Proceedings of the Royal Society A*, *89*, 1–20.

Trigg, G. L. (1995). *Landmark experiments in twentieth century physics*. New York: Dover Publications.

Van Melsen, A. G. (2004). *From atomos to atom: The history of the concept atom*. New York: Dover Publications.

Weeks, M. E., & Leichester, H. M. (1968). *Discovery of the elements. Journal of Chemical Education*. Easton, PA.

Wu, C. S. (1957). Experimental test of parity conservation in beta decay. *Physical Review*, *105*, 1413–1415.

Wurzer, F. (1817). Auszug eines Briefes vom Hofrath Wurzer, Prof. der Chemie zu Marburg. Annalen der Physik, **56**, 1817, 331–334.

Zumdahl, S. S. (2003). *Introductory chemistry: A foundation*. Boston, MA: Houghton Mifflin.

Chapter 2

Minerals, ores, and mining

2.1 Short introduction to mineralogy

Mineralogy is a subject of geology (or earth sciences) specializing in the scientific study of the chemistry, crystal structure, and physical (including optical) properties of minerals and mineralized artifacts. Specific studies within mineralogy include the processes of mineral origin and formation, classification of minerals, their geographical distribution, as well as their utilization. A simple definition of a mineral can be given as a naturally occurring, homogeneous inorganic solid substance having a definite chemical composition and characteristic crystalline structure, color, and hardness.

A few minerals are chemical elements, including sulfur, copper, silver, and gold, but the vast majority are compounds. Systematic mineralogy is the identification and classification of minerals by their properties. Historically, mineralogy was heavily concerned with taxonomy of the rock-forming minerals. In 1959, the International Mineralogical Association (IMA) formed the Commission of New Minerals and Mineral Names (CNMMN) to rationalize the nomenclature and regulate the introduction of new names. In July 2006, it was merged with the Commission on Classification of Minerals to form the Commission on New Minerals, Nomenclature, and Classification (CNMNC). There are currently over 5000 accepted minerals, and about 100 new minerals are discovered each year. Based on the work by the American geologist, mineralogist, volcanologist, and zoologist James Dwight Dana (February 12, 1813 to April 14, 1895) (Fig. 2.1) the minerals are now commonly placed in the following classes: native elements, sulfides, sulfosalts, oxides and hydroxides, halides, carbonates, nitrates and borates, sulfates, chromates, molybdates and tungstates, phosphates, arsenates and vanadates, and silicates (Table 2.1). Since 1960, most chemical analyses are done using instruments. One of these, atomic absorption spectroscopy (AAS), is similar to wet chemistry in that the sample must still be dissolved, but it is much faster and cheaper. The solution is vaporized, and its absorption spectrum is measured in the visible and ultraviolet range. Other techniques are X-ray fluorescence (XRF), electron microprobe (EMP) analysis, and optical emission spectrometry.

The crystal structure is the arrangement of atoms in a crystal (Fig. 2.2). It is represented by a lattice of points, which repeats a basic pattern, called a unit cell, in three dimensions. The lattice can be characterized by its symmetries and by the dimensions of the unit cell. The lattice remains unchanged by certain symmetry operations about any given point in the lattice: reflection, rotation, inversion, and rotary inversion (a combination of rotation and reflection). Together, they make up a mathematical object called a crystallographic point group or crystal class. There are 32 possible crystal classes. In addition, there are operations that displace all the points: translation, screw axis, and glide plane. In combination with the point symmetries, they form 230 possible space groups. Most geology departments have X-ray powder diffraction equipment to analyze the crystal structures of minerals. X-rays have wavelengths that are the same order of magnitude as the distances between atoms.

Diffraction, the constructive and destructive interference between waves scattered at different atoms, leads to distinctive patterns of high and low intensity that depend on the geometry of the crystal. In a sample that is ground to a powder, the X-rays sample a random distribution of all crystal orientations. Powder diffraction can distinguish between minerals that may appear the same in a hand sample, for example, quartz and its polymorphs tridymite and cristobalite.

An initial step in identifying a mineral is to examine its physical properties, many of which can be measured on a hand sample. These can be classified into density (often given as specific gravity); measures of mechanical cohesion (hardness, tenacity, cleavage, fracture, parting); macroscopic visual properties (luster, color, streak, luminescence, diaphaneity); magnetic and electric properties; radioactivity; and solubility in hydrogen chloride (HCl).

Hardness is determined by comparison with other minerals (preferably of known hardness). In the Mohs scale, a standard set of minerals are numbered in order of increasing hardness from 1 to 10: talc 1, gypsum 2, calcite 3, fluorite 4, apatite 5, orthoclase 6, quartz 7, topaz 8, corundum 9, and diamond 10. A harder mineral will scratch a softer, so an unknown mineral can be placed in this scale by which minerals it scratches, and which scratch it. A few minerals such as kyanite have a hardness that depends significantly on direction. Hardness can also be measured on an absolute scale using a sclerometer; compared to the absolute scale (e.g., Vickers or Knoop), the Mohs scale is nonlinear (Figs. 2.3).

The Periodic Table: Nature's Building Blocks. DOI: https://doi.org/10.1016/B978-0-12-821279-0.00002-9

FIGURE 2.1 James Dwight Dana. Photo in 1865 by George Kendall Warren, Cambridgeport, Massachusetts.

TABLE 2.1 Dana classification—primary groups.

1	NATIVE ELEMENTS AND ALLOYS
2	SULFIDES
3	SULFOSALTS
4	SIMPLE OXIDES
5	OXIDES CONTAINING URANIUM OR THORIUM
6	HYDROXIDES AND OXIDES CONTAINING HYDROXYL
7	MULTIPLE OXIDES
8	MULTIPLE OXIDES CONTAINING NIOBIUM, TANTALUM, OR TITANIUM
9	NORMAL HALIDES
10	OXYHALIDES AND HYDROXYHALIDES
11	HALIDE COMPLEXES
12	COMPOUND HALIDES
13	ACID CARBONATES
14	ANHYDROUS NORMAL CARBONATES
15	HYDRATED NORMAL CARBONATES
16a	ANHYDROUS CARBONATES CONTAINING HYDROXYL OR HALOGEN
16b	HYDRATED CARBONATES CONTAINING HYDROXYL OR HALOGEN
17	COMPOUND CARBONATES
18	NORMAL NITRATES
19	NITRATES CONTAINING HYDROXYL OR HALOGEN
20	COMPOUND NITRATES
21	NORMAL IODATES
22	IODATES CONTAINING HYDROXYL OR HALOGEN
23	COMPOUND IODATES
24	ANHYDROUS BORATES
25	ANHYDROUS BORATES CONTAINING HYDROXYL OR HALOGEN
26	HYDRATED BORATES CONTAINING HYDROXYL OR HALOGEN
27	COMPOUND BORATES
28	ANHYDROUS ACID AND NORMAL SULFATES
29	HYDRATED ACID AND NORMAL SULFATES
30	ANHYDROUS SULFATES CONTAINING HYDROXYL OR HALOGEN
31	HYDRATED SULFATES CONTAINING HYDROXYL OR HALOGEN

(Continued)

TABLE 2.1 (Continued)

32	COMPOUND SULFATES
33	SELENATES AND TELLURATES
34	SELENITES, TELLURITES, AND SULFITES
35	ANHYDROUS CHROMATES
36	COMPOUND CHROMATES
37	ANHYDROUS ACID PHOSPHATES, ARSENATES, AND VANADATES
38	ANHYDROUS NORMAL PHOSPHATES, ARSENATES, AND VANADATES
39	HYDRATED ACID PHOSPHATES, ARSENATES, AND VANADATES
40	HYDRATED NORMAL PHOSPHATES, ARSENATES, AND VANADATES
41	ANHYDROUS PHOSPHATES, ETC. CONTAINING HYDROXYL OR HALOGEN
42	HYDRATED PHOSPHATES, ETC. CONTAINING HYDROXYL OR HALOGEN
43	COMPOUND PHOSPHATES, ETC.
44	ANTIMONATES
45	ACID AND NORMAL ANTIMONITES AND ARSENITES
46	ANTIMONITES AND ARSENITES CONTAINING HYDROXYL OR HALOGEN
47	VANADIUM OXYSALTS
48	ANHYDROUS MOLYBDATES AND TUNGSTATES
49	HYDRATED MOLYBDATES AND TUNGSTATES
50	ORGANIC COMPOUNDS
51	NESOSILICATES Insular SiO_4 Groups Only
52	NESOSILICATES Insular SiO_4 Groups and O, OH, F, and H_2O
53	NESOSILICATES Insular SiO_4 Groups and Other Anions or Complex Cations
54	NESOSILICATES Borosilicates and Some Beryllosilicates
55	SOROSILICATES Si_2O_7 Groups, Generally with no Additional Anions
56	SOROSILICATES Si_2O_7 Groups, with Additional O, OH, F, and H_2O
57	SOROSILICATES Si_3O_{10} Groups and Larger Noncyclic Groups
58	SOROSILICATES Insular, Mixed, Single, and Larger Tetrahedral Groups
59	CYCLOSILICATES Three-Membered Rings
60	CYCLOSILICATES Four-Membered Rings
61	CYCLOSILICATES Six-Membered Rings
62	CYCLOSILICATES Eight-Membered Rings
63	CYCLOSILICATES Condensed Rings
64	CYCLOSILICATES Rings with Other Anions and Insular Silicate Groups
65	INOSILICATES Single-Width, Unbranched Chains ($W = 1$)
66	INOSILICATES Double-Width, Unbranched Chains ($W = 2$)
67	INOSILICATES Unbranched Chains with $W > 2$
68	INOSILICATES Structures with Chains of More Than One Width
69	INOSILICATES Chains with Side Branches or Loops
70	INOSILICATES Column or Tube Structures
71	PHYLLOSILICATES Sheets of Six-Membered Rings
72	PHYLLOSILICATES Two-Dimensional Infinite Sheets with Other Than Six-Membered Rings
73	PHYLLOSILICATES Condensed Tetrahedral Sheets
74	PHYLLOSILICATES Modulated Layers
75	TECTOSILICATES Si Tetrahedral Frameworks
76	TECTOSILICATES Al-Si Framework
77	TECTOSILICATES Zeolites
78	Unclassified Silicates

Tenacity is used to describe the way a mineral behaves when it is broken, crushed, bent, or torn. A mineral can be brittle, malleable, sectile, ductile, flexible, or elastic. An important influence on tenacity is the type of chemical bond between atoms in the crystal structure (e.g., ionic or metallic). Of the other measures of mechanical cohesion, cleavage is the tendency to break along certain crystallographic planes. It is described by the quality (e.g., perfect, distinct or indistinct) and the orientation of the plane in crystallographic nomenclature. Parting is the tendency to break along planes of weakness due to pressure, twinning, or exsolution. Where these two kinds of break do not occur, fracture is a less orderly form that may be conchoidal (having smooth curves resembling the interior of a shell), fibrous, splintery, hackly (jagged with sharp edges), or uneven.

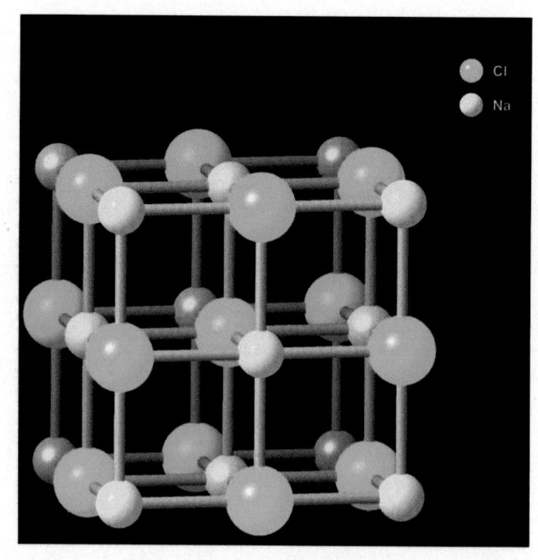

FIGURE 2.2 Example of a crystal structure, in this case the mineral halite (also known as rock salt) NaCl, with the yellow balls representing Na atoms and the larger green balls Cl atoms.

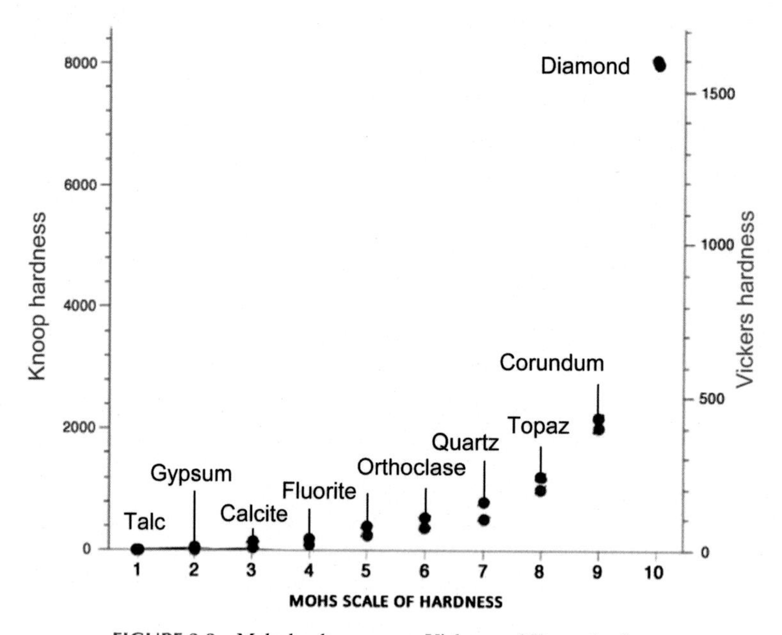

FIGURE 2.3 Mohs hardness versus Vickers and Knoop hardness.

If the mineral is well crystallized, it will also have a distinctive crystal habit (e.g., hexagonal, columnar, botryoidal) that reflects the crystal structure or internal arrangement of atoms (Fig. 2.4). It is also affected by crystal defects and twinning (Fig. 2.5). Many crystals are polymorphic, having more than one possible crystal structure depending on factors such as pressure and temperature.

In addition to macroscopic properties such as those described above, minerals have properties that require a polarizing microscope to observe. When light passes from air or a vacuum into a transparent crystal, some of it is reflected at the surface and some refracted. The latter is a bending of the light path that occurs because the speed of light changes as it goes into the crystal. Snell's law relates the bending angle to the refractive index (RI), the ratio of speed of light in

FIGURE 2.4 Quartz crystals.

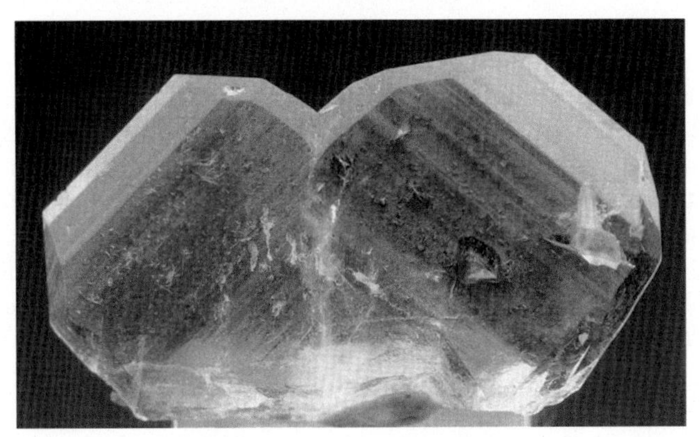

FIGURE 2.5 Quartz, Japanese twin.

a vacuum to the speed of light in the crystal. Crystals in the cubic system are isotropic: the RI does not depend on direction. All other crystals are anisotropic: light passing through them is broken up into two plane-polarized rays that travel at different speeds and refract at different angles. A polarizing microscope is like an ordinary microscope, but it has two plane-polarized filters, a polarizer below the sample and an analyzer above it, polarized perpendicular to each other. Light passes successively through the polarizer, the sample, and the analyzer. If there is no sample, the analyzer blocks all the light from the polarizer. However, an anisotropic sample will generally change the polarization, so some of the light can pass through. Thin sections and powders can be used as samples.

The environments of mineral formation and growth are highly varied, ranging from slow crystallization at the high temperatures and pressures of igneous melts deep within the Earth's crust to the low-temperature precipitation from a saline brine at the Earth's surface. Various possible types of formation include:

- sublimation from volcanic gases
- deposition from aqueous solutions and hydrothermal brines
- crystallization from an igneous magma or lava
- recrystallization due to metamorphic processes and metasomatism
- crystallization during diagenesis of sediments
- formation by oxidation and weathering of rocks exposed to the atmosphere or within the soil environment

2.2 Advanced chemical analytical techniques

In present day research on the chemical composition of minerals (both major elements and trace elements) a number of different analytical techniques are available. These include techniques such as electron microscopy (both scanning and transmission as well as microprobe) with energy-dispersive or wavelength-dispersive X-ray spectrometry, XRF spectrometry, X-ray photoelectron spectroscopy (XPS), inductively coupled plasma atomic emission spectroscopy (ICP-AES), also known as inductively coupled plasma-optical emission spectroscopy (ICP-OES) and inductively coupled plasma-mass spectrometry (ICP-MS).

2.2.1 Electron microscopy

An electron microscope is a microscope that utilizes a beam of accelerated electrons as a source of illumination. Since the wavelength of an electron can be up to 100,000 times shorter than that of visible light photons, electron microscopes have a higher resolving power than light microscopes and can reveal the structure of much smaller objects. For example, a scanning transmission electron microscope (TEM) has attained better than 50 pm resolution in annular dark-field imaging mode and magnifications of up to about 10,000,000 ×, while most light microscopes are limited by diffraction to about 200 nm resolution and useful magnifications below 2000 ×. Electron microscopes can be applied to study the ultrastructure of a wide range of biological and inorganic specimens including microorganisms, cells, large molecules, biopsy samples, metals, and crystals. Industrially, electron microscopes are often used for quality control and failure analysis. Modern electron microscopes produce electron micrographs using specialized digital cameras and frame grabbers to capture the images.

2.2.1.1 Transmission electron microscope

The original form of the electron microscope, the TEM, applies a high-voltage electron beam to illuminate the specimen and create an image. The electron beam is produced by an electron gun, normally containing a tungsten filament cathode as the electron source. The electron beam is accelerated by an anode typically at +100 keV (40−400 keV) with respect to the cathode, focused by electrostatic and electromagnetic lenses, and transmitted through the specimen which is in part transparent to electrons and in part scatters them out of the beam. After it appears from the specimen, the electron beam carries information about the structure of the specimen that is magnified by the objective lens system of the microscope. The spatial variation in this information (the "image") can be viewed directly by projecting the magnified electron image onto a fluorescent viewing screen coated with a phosphor or scintillator material such as zinc sulfide. On the other hand, the image can be photographically recorded by exposing a photographic film or plate directly to the electron beam, or a high-resolution phosphor may be coupled by means of a lens optical system or a fiber optic light guide to the sensor of a digital camera. The image detected by the digital camera may be displayed on a monitor or computer.

The resolution of TEMs is limited principally by spherical aberration, but a new generation of hardware correctors can decrease spherical aberration to increase the resolution in high-resolution transmission electron microscopy (HRTEM) to below 0.5 Å (50 pm), permitting magnifications above 50 million times. The ability of HRTEM to determine the positions of atoms within materials is useful for nanotechnologies research and development. TEMs are frequently used in electron diffraction mode. The advantages of electron diffraction over X-ray diffraction are that the specimen does not have to be a single crystal or even a polycrystalline powder, and also that the Fourier transform reconstruction of the object's magnified structure occurs physically and therefore circumvents the need for solving the phase problem faced by the X-ray crystallographers after obtaining their X-ray diffraction patterns.

The most important disadvantage of the TEM is the requirement of extremely thin sections of the specimens, typically about 100 nm. Producing these thin sections for biological and materials specimens can technically be very challenging. Solid material, such as those of rocks and minerals, thin sections can be produced by applying a focused ion beam.

2.2.1.2 Scanning electron microscope

The scanning electron microscope (SEM) produces images by probing the surface of a specimen with a focused electron beam that is scanned across a rectangular area of the specimen (raster scanning). When the electron beam interacts with the specimen, it loses energy by several different mechanisms. The lost energy is transformed into alternative forms such as heat, emission of low-energy secondary electrons and high-energy backscattered electrons, light emission (cathodoluminescence), or X-ray emission, all of which create signals providing information about the properties of the specimen surface, such as its topography and composition. The SEM image displays in a sense a map of the varying intensity of any of these signals into the image in a position corresponding to the position of the beam on the specimen when the signal was generated.

Generally, the image resolution of an SEM is lower than that of a TEM. Nevertheless, because the SEM images the surface of a sample rather than its interior, the electrons do not have to travel through the sample. This reduces the need for extensive sample preparation to thin the specimen to electron transparency. The SEM can image bulk samples limited in size only by the size of its stage and still be moved, including a height less than the working distance being used, often 4 mm for high-resolution images. The SEM also has a great depth of field and so can produce images that are good representations of the three-dimensional surface shape of the sample, for example, crystals of minerals. Another advantage of SEMs comes with environmental SEMs that can produce images of good quality and resolution with hydrated samples or in low, rather than high, vacuum or under chamber gases.

2.2.1.3 Sample preparation

Embedding materials (for electron microprobe analysis)—after embedding in resin, the specimen is usually ground and polished to a mirror-like finish using ultra-fine abrasives. The polishing process must be performed carefully to minimize scratches and other polishing artifacts that reduce image quality.

Ion beam milling (TEM)—thins samples until they are transparent to electrons by firing ions (typically argon) at the sample's surface from an angle and sputtering material from the surface. A subclass of this is focused ion beam milling, where gallium ions are used to produce an electron transparent membrane in a specific region of the sample. Ion beam milling may also be used for cross section polishing prior to SEM analysis of materials that are difficult to prepare using mechanical polishing.

Conductive coating (SEM and electron microprobe)—an ultrathin coating of electrically conducting material, deposited either by high vacuum evaporation or by low vacuum sputter coating of the sample. This is done to prevent the accumulation of static electric fields at the specimen due to the electron irradiation required during imaging. The coating materials include gold, gold/palladium, platinum, tungsten, graphite, etc.

2.2.1.4 Electron microprobe

An EMP, also known as an electron probe microanalyzer (EPMA) or electron microprobe analyzer (EMPA), is an analytical technique based on the principles of the SEM used to nondestructively determine the chemical composition of small volumes of solid materials. It works similarly to an SEM: the sample is bombarded with an electron beam, emitting X-rays at wavelengths characteristic to the elements being analyzed. This allows the concentrations of elements present within small sample volumes (typically 10–30 cubic micrometers or less) to be determined. The concentrations of elements from beryllium to plutonium can be measured at levels as low as 100 parts per million (ppm). Recent models of EPMAs can accurately measure elemental concentrations of approximately 10 ppm.

Low-energy electrons are produced from a tungsten filament, a lanthanum hexaboride crystal cathode or a field emission electron source and accelerated by a positively biased anode plate to 3–30,000 electron volts (keV). The anode plate has central aperture and electrons that pass through it are collimated and focused by a series of magnetic lenses and apertures. The resulting electron beam (approximately 5 nm to 10 μm diameter) may be scanned across the sample or used in spot mode to produce excitation of various effects in the sample. Among these effects are: phonon excitation (heat), cathodoluminescence (visible light fluorescence), continuum X-ray radiation (Bremsstrahlung), characteristic X-ray radiation, secondary electrons (plasmon production), backscattered electron production, and Auger electron production.

When the beam electrons (and scattered electrons from the sample) interact with bound electrons in the innermost electron shells of the atoms of the various elements in the sample, they can scatter the bound electrons from the electron shell producing a vacancy in that shell (ionization of the atom). This vacancy is unstable and must be filled by an electron from either a higher energy bound shell in the atom (producing another vacancy which is in turn filled by electrons from yet higher energy bound shells) or by unbound electrons of low energy. The difference in binding energy (BE) between the electron shell in which the vacancy was produced and the shell from which the electron comes to fill the vacancy is emitted as a photon. The energy of the photon is in the X-ray region of the electromagnetic spectrum. As the electron structure of each element is unique, the series of X-ray line energies produced by vacancies in the innermost shells is characteristic of that element, although lines from different elements may overlap (interference). As the innermost shells are involved, the X-ray line energies are generally not affected by chemical effects produced by bonding between elements in compounds except in low atomic number (Z) elements (B, C, N, O, and F for Kα and Al to Cl for Kβ) where line energies may be shifted as a result of the involvement of the electron shell from which vacancies are filled in chemical bonding.

The characteristic X-rays are used for chemical analysis. Specific X-ray wavelengths or energies are selected and counted, either by wavelength-dispersive X-ray spectroscopy (WDS) or energy-dispersive X-ray spectroscopy (EDS). WDS uses Bragg diffraction from crystals to select X-ray wavelengths of interest and direct them to gas-flow or sealed proportional detectors. In contrast, EDS uses a solid-state semiconductor detector to accumulate X-rays of all wavelengths produced from the sample. While EDS yields more information and typically requires a much shorter counting time, WDS is normally a more exact technique with lower detection limits due to its superior X-ray peak resolution and better peak to background ratio.

The chemical composition is determined by comparing the intensities of characteristic X-rays from the sample material with intensities from known composition (standards). Counts from the sample must be corrected for matrix effects (such as depth of production of the X-rays, absorption, and secondary fluorescence) to yield quantitative chemical compositions. The resulting chemical information is gathered in textural context. Variations in chemical composition within a material (zoning), such as a mineral grain or metal, can be readily determined.

2.2.1.5 Limitations

WDS is useful for higher atomic numbers; therefore WDS cannot determine elements below number 5 (boron). This limitation sets restrictions to WDS when analyzing geologically important elements such as H, Li, and Be. Notwithstanding the improved spectral resolution of elemental peaks, some peaks exhibit significant overlaps that result in analytical challenges (e.g., VKα and TiKβ). WDS analyses are not able to distinguish among the valence states of elements (e.g., Fe^{2+} vs. Fe^{3+}) such that this information must be obtained by other techniques (e.g., Mössbauer spectroscopy or electron energy loss spectroscopy). The multiple masses of an element (i.e., isotopes) cannot be determined by WDS, but rather are most commonly obtained with a mass spectrometer.

This technique is most commonly used by mineralogists and petrologists. Most rocks are aggregates of small mineral grains. These grains may retain chemical information related to their formation and subsequent alteration. This information may provide information regarding geologic processes such as crystallization, lithification, volcanism, metamorphism, orogenic events (mountain building), and plate tectonics. The change in elemental composition from the center (also known as core) to the edge (or rim) of a mineral grain or crystal can provide information about the history of the crystal's formation, including the temperature, pressure, and chemistry of the surrounding medium. Quartz crystals, for example, incorporate a small, but measurable amount of titanium into their structure as a function of temperature, pressure, and the amount of titanium available in their environment. Changes in these parameters are recorded by titanium as the crystal grows. This technique is also used for the study of extra-terrestrial rocks (i.e., meteorites) and provides chemical data that are vital to understanding the evolution of the planets, asteroids, and comets.

2.2.2 X-ray fluorescence

XRF is the emission of characteristic "secondary" (or fluorescent) X-rays from a material that has been excited by being bombarded with high-energy X-rays or gamma rays. The phenomenon is widely used for elemental analysis and chemical analysis, particularly in the investigation of metals, glass, ceramics, and building materials, and for research in geochemistry, forensic science, archeology, and art objects such as paintings and murals. When materials are bombarded with short-wavelength X-rays or gamma rays, ionization of their component atoms can take place. Ionization consists of the ejection of one or more electrons from the atom and may occur if the atom is exposed to radiation with an energy greater than its ionization energy. X-rays and gamma rays can be energetic enough to expel tightly held electrons from the inner orbitals of the atom. The ejection of such an electron makes the electronic structure of the atom unstable, and electrons in higher orbitals "fall" into the lower orbital to fill the hole left behind. During this process, energy is released in the form of a photon, the energy of which is equal to the energy difference of the two-electron orbitals involved. Thus the material emits radiation, which has energy characteristic of the atoms present. The term fluorescence is applied to phenomena in which the absorption of radiation of a specific energy results in the reemission of radiation of different energy (usually lower).

In order to excite the atoms, a source of radiation is required, with enough energy to expel tightly held inner electrons. Conventional X-ray generators are most frequently used, as their output can readily be "tuned" for this technique, and since higher power can be deployed relative to other techniques. However, gamma-ray sources can be utilized without the need for an elaborate power supply, permitting an easier use in small portable instruments (e.g., useful during geological field work). When the energy source is a synchrotron or the X-rays are focused by an optic like a polycapillary, the X-ray beam can be very small and very intense. As a result, atomic information on the submicrometer scale can be obtained. X-ray generators in the range between 20 and 60 kV are used, which allow excitation of a broad range of atoms. The continuous spectrum consists of "Bremsstrahlung" radiation: radiation produced when high-energy electrons passing through the tube are progressively decelerated by the material of the tube anode (the "target"). In energy dispersive analysis, the fluorescent X-rays emitted by the material sample are directed into a solid-state detector which produces a "continuous" distribution of pulses, the voltages of which are proportional to the incoming photon energies. This signal is processed by a multichannel analyzer which produces an accumulating digital spectrum that can be processed to obtain analytical data. In wavelength dispersive analysis, the fluorescent X-rays emitted by the material sample are directed into a diffraction grating monochromator. The diffraction grating used is usually a single crystal. By varying the angle of incidence and take-off (θ) on the crystal, a single X-ray wavelength (λ) can be selected. The wavelength obtained is given by Bragg's law:

$$n\lambda = 2d\sin(\theta)$$

where d is the spacing of atomic layers parallel to the crystal surface and n is a positive integer. In energy dispersive analysis, dispersion and detection are a single operation, as already mentioned above. Proportional counters or various types of solid-state detectors [e.g., PIN diode, Si(Li), Ge(Li), silicon drift detector (SDD)] are used. They are all based on the same detection principle: an incoming X-ray photon ionizes many detector atoms with the amount of charge produced being proportional to the energy of the incoming photon. The charge is then collected, and the process repeats itself for the next photon. Detector speed is clearly critical, as all charge carriers measured must come from the same photon to measure the photon energy correctly (peak length discrimination is used to eliminate events that seem to have been produced by two X-ray photons arriving almost simultaneously). The spectrum is then built up by dividing the energy spectrum into discrete bins and counting the number of pulses registered within each energy bin. Energy dispersive X-ray fluorescence detector types vary in resolution, speed, and the means of cooling (a low number of free charge carriers is critical in the solid-state detectors): proportional counters with resolutions of several hundred eV cover the low end of the performance spectrum, followed by PIN diode detectors, while the Si(Li), Ge(Li) and SDD occupy the high end of the performance scale. In wavelength dispersive analysis, the single-wavelength radiation produced by the monochromator is passed into a photomultiplier, a detector comparable to a Geiger counter, which counts individual photons as they pass through. The counter is a chamber containing a gas that is ionized by X-ray photons. A central electrode is charged at (typically) +1700 V with respect to the conducting chamber walls, and each photon triggers a pulse-like cascade of current across this field. The signal is amplified and transformed into an accumulating digital count. These counts are then processed to obtain analytical data.

The fluorescence process is inefficient, and the secondary radiation is significantly weaker than the primary radiation. Additionally, the secondary radiation from the lighter elements is of relatively low energy (long wavelength), has low penetrating power, and is severely attenuated if the beam passes through air for any distance. As a result of this, for high-performance analysis, the path from tube to sample to detector is maintained under vacuum (around 10 Pa residual pressure). This means in practice that most of the working parts of the instrument must be located inside a large vacuum chamber. The difficulties of maintaining moving parts in vacuum, and of rapidly introducing and withdrawing the sample without losing vacuum, form major challenges for the design of the instrument. For less demanding applications, or when the sample is damaged by a vacuum (e.g., a volatile sample), a helium-swept X-ray chamber can be substituted, with some loss of low-Z (Z = atomic number) intensities.

The use of a primary X-ray beam to excite fluorescent radiation from the sample was first proposed by Glocker and Schreiber in 1928. These days, the method is generally applied as a nondestructive analytical tool, and as a process control technique in many extractive and processing industries. In principle, the lightest element that can be analyzed is beryllium ($Z = 4$), but due to instrumental limitations and low X-ray yields for the light elements, it is often difficult to quantify elements lighter than sodium ($Z = 11$), unless background corrections and very comprehensive interelement corrections are made.

2.2.2.1 *Energy dispersive spectrometry*

In energy dispersive spectrometers (EDX or EDS), the detector allows the determination of the energy of the photon when it is detected. Detectors historically have been based on silicon semiconductors, in the form of lithium-drifted silicon crystals, or high-purity silicon wafers. Si(Li) detectors contain essentially a 3–5 mm thick silicon junction type p-i-n diode (same as PIN diode) with a bias of −1000 V across it. The lithium-drifted center part forms the nonconducting i-layer, where Li compensates the residual acceptors which would otherwise make the layer p-type. When an X-ray photon passes through, it causes a swarm of electron–hole pairs to form, and this results in a voltage pulse. To obtain sufficiently low conductivity, the detector must be maintained at low temperature, and liquid-nitrogen cooling has to be used for the best resolution. With some loss of resolution, the much more convenient Peltier cooling can be employed. More recently, high-purity silicon wafers with low conductivity have become routinely available. Cooled by the Peltier effect, this provides a cheap and convenient detector, although the liquid nitrogen–cooled Si(Li) detector still has the best resolution (i.e., ability to distinguish different photon energies). The pulses generated by the detector are processed by pulse-shaping amplifiers. It takes time for the amplifier to shape the pulse for optimum resolution, and there is therefore a trade-off between resolution and count rate: long processing time for good resolution results in "pulse pile-up" in which the pulses from successive photons overlap. Multiphoton events are, however, typically more drawn out in time (photons do not arrive exactly at the same time) than single-photon events and pulse-length discrimination can therefore be applied to filter most of these events out. Nevertheless, a small number of pile-up peaks will persist, and pile-up correction should be built into the software in applications that require trace analysis. To make the most efficient use of the detector, the tube current should be reduced to keep multiphoton events (before discrimination) at a reasonable level, for example, 5%–20%. Considerable computer power is necessary for correcting for pulse pile-up and for extraction of data from poorly resolved spectra. These elaborate correction processes tend to be based on empirical relationships that may change with time, so that continuous vigilance is required in order to obtain chemical data of adequate precision.

EDX spectrometers differ from WDX spectrometers in that they are generally smaller, simpler in design, and have fewer engineered parts, but the accuracy and resolution of EDX spectrometers are lower than for WDX. EDX spectrometers can also use miniature X-ray tubes or gamma sources, which make them cheaper and allow miniaturization and portability. This type of instrument is commonly used for portable quality control screening applications, such as during geological field work. On the other hand, the low resolution and problems with low count rate and long dead-time make them inferior for high-precision analysis. They are, however, very effective for high-speed, multielemental analysis. Field portable XRF analyzers currently on the market weigh less than 2 kg and have detection limits in the order of 2 ppm (parts per million) of lead (Pb) in pure sand.

2.2.2.2 *Wavelength dispersive spectrometry*

In wavelength dispersive spectrometers (WDX or WDS), the photons are separated by diffraction on a single crystal before they are detected. While wavelength dispersive spectrometers are sometimes used to scan a wide range of wavelengths, producing a spectrum plot as in EDS, they are usually set up to obtain measurements only at the wavelength of the emission lines of the elements of interest. This is achieved in two different ways—"simultaneous" or "sequential." "Simultaneous" spectrometers have several "channels" dedicated to analysis of a single element, each consisting of a fixed-geometry crystal monochromator, a detector, and processing electronics. This allows a number of elements to be measured simultaneously, and in the case of high-powered instruments, complete high-precision analyses can be achieved in less than 30 seconds. Additional advantage of this arrangement is that the fixed-geometry monochromators have no continuously moving parts and so are very reliable. Reliability is imperative in production environments where instruments are expected to function without interruption for months at a time. Disadvantages of simultaneous spectrometers include relatively high cost for complex analyses, as each channel used is expensive. The number of elements that can be measured is limited to between 15 and 20, due to space limitations on the number of monochromators that can be crowded around the fluorescing sample. The requirement to house multiple monochromators means that a rather open arrangement around the sample is necessary, leading to relatively long tube-sample-crystal distances, which results in lower detected intensities and more scattering. The instrument is inflexible, because if a new element is to be analyzed, a new measurement channel must be bought and installed. "Sequential" spectrometers have a single variable-geometry monochromator (but usually with an arrangement for selecting from a choice of crystals), a single detector assembly (but usually with more than one detector arranged in tandem), and a single electronic pack. The instrument is programmed to move through a sequence of wavelengths, in each case selecting the appropriate X-ray tube power, the appropriate crystal, and the appropriate detector arrangement. The length of the measurement program is basically unlimited, so this arrangement is very flexible. Since there is only a single monochromator, the tube-sample-crystal distances can be kept very short, resulting in minimal loss of detected intensity. The clear disadvantage is relatively long analysis time, especially when a large number of elements are being analyzed, not only because the elements are measured in sequence, but also because a certain amount of time is taken in readjusting the monochromator geometry between measurements. Additionally, the activity of the monochromator during an analysis program is a challenge for its mechanical reliability. Nevertheless, modern sequential instruments can achieve reliability nearly as good as that of simultaneous instruments, even in continuous-usage applications.

2.2.2.3 Sample preparation

In order to keep the geometry of the tube-sample-detector assembly constant, the sample is normally prepared as a flat disk, typically of diameter 20–50 mm. This flat disk is then located at a standardized, small distance from the tube window. Because the X-ray intensity follows an inverse-square law, the tolerances for this placement and for the flatness of the surface must be very tight in order to maintain a repeatable X-ray flux. Ways of preparing the sample disks vary: metals may be machined to shape, minerals may be finely ground and pressed into a tablet, and glasses may be cast to the required shape. An additional reason for obtaining a flat and representative sample surface is that the secondary X-rays from lighter elements often only emit from the top few micrometers of the sample. In order to further reduce the effect of surface irregularities, the sample is usually spun at 5–20 rpm. It is essential to ensure that the sample is sufficiently thick to absorb the entire primary beam. For higher-Z materials, a few millimeters thickness is adequate, but for a light-element matrix such as coal, a thickness of 30–40 mm is needed.

2.2.2.4 Extracting analytical results

Superficially, the translation of X-ray photon count-rates into elemental concentrations would seem to be straightforward: WDX separates the X-ray lines efficiently, and the rate of generation of secondary photons is proportional to the element concentration. However, the number of photons leaving the sample is also affected by the physical properties of the sample: so-called "matrix effects." These fall broadly into three categories:

- X-ray absorption
- X-ray enhancement
- sample macroscopic effects

All elements absorb X-rays to some extent. Each element has a characteristic absorption spectrum which consists of a "saw-tooth" succession of fringes, of which each step-change has a wavelength close to an emission line of the element. Absorption attenuates the secondary X-rays leaving the sample. For example, the mass absorption coefficient of silicon at the wavelength of the aluminum Kα line is 50 m^2/kg, whereas that of iron is 377 m^2/kg. This means that a given concentration of aluminum in a matrix of iron gives only one-seventh of the count rate compared with the same concentration of aluminum in a silicon matrix. Fortunately, mass absorption coefficients are well known and can be calculated. Yet, to calculate the absorption for a multielement sample, the composition must be known. For the analysis of an unknown sample, an iterative procedure is therefore used. Hence, to derive the mass absorption accurately, data for the concentration of elements not measured by XRF may be needed, and various strategies are employed to estimate these. Enhancement can occur when the secondary X-rays emitted by a heavier element are sufficiently energetic to stimulate additional secondary emission from a lighter element. This phenomenon can also be modeled, and corrections can be made on condition that the full matrix composition can be deduced. Sample macroscopic effects consist of effects due to inhomogeneities of the sample, and unrepresentative conditions at its surface. Samples are ideally homogeneous and isotropic, but they often deviate from this ideal. Mixtures of multiple crystalline components in mineral powders can result in absorption effects that deviate from those calculable from theory. When a powder is pressed into a tablet, the finer minerals tend to concentrate at the surface. Spherical grains tend to migrate to the surface more than do angular grains. In machined metals, the softer components of an alloy tend to smear across the surface. Considerable care and ingenuity are necessary to minimize these effects. Since they are artifacts as a result of the method of sample preparation, these effects cannot be compensated by theoretical corrections, and must be "calibrated in." This means that the calibration materials and the unknowns must be compositionally and mechanically similar, and a given calibration is appropriate only to a limited range of materials. Glasses most closely resemble the ideal of homogeneity and isotropy, and for accurate work, minerals are typically prepared by dissolving them in a borate glass and casting them into a flat disk or "bead." Prepared in this form, a virtually universal calibration is applicable. Other corrections that are often used include background correction and line overlap correction. The background signal in an XRF spectrum originates mainly from scattering of primary beam photons by the sample surface. Scattering varies with the sample mass absorption, being greatest when the mean atomic number is low. When measuring trace amounts of an element, or when measuring on a variable light matrix, background correction becomes necessary. This is only possible on a sequential spectrometer. Line overlap is a common problem, bearing in mind that the spectrum of a complex mineral can contain several hundred measurable lines. Occasionally it can be avoided by measuring a less-intense, but overlap-free line, but in certain instances a correction is inevitable. For instance, the Kα is the only usable line for measuring sodium, and it overlaps the zinc Lβ (L_2-M_4) line. Thus zinc, if present, must be analyzed in order to properly correct the sodium value.

2.2.3 X-ray photoelectron spectroscopy

XPS is a surface-sensitive quantitative spectroscopic technique that measures the elemental composition at the parts per thousand range, empirical formula, chemical state, and electronic state of the elements that exist within a material. Put more simply, XPS is a useful measurement technique because it not only shows what elements are within a sample but also what other elements they are bonded to. But, at most the instrument will only probe 20 nm into a sample. XPS spectra are obtained by irradiating a material with a beam of X-rays while at the same time measuring the kinetic energy and the number of electrons that escape from the top 0 to 10 nm of the material being analyzed. XPS requires high vacuum ($P \sim 10^{-8}$ millibar) or ultra-high vacuum (UHV; $P < 10^{-9}$ millibar) conditions, although a current area of development is ambient-pressure XPS, in which samples are analyzed at pressures of a few tens of millibar. XPS can be used to analyze the surface chemistry of a material in its as-received state, or after some treatment, for example, fracturing, cutting, grinding, or scraping in air or UHV to expose the bulk chemistry, ion beam etching to clean off some or all of the surface contamination (with mild ion etching), or to intentionally expose deeper layers of the sample (with more extensive ion etching) in depth-profiling XPS, exposure to heat to study the changes due to heating, exposure to reactive gases or solutions, exposure to ion beam implant, and exposure to ultraviolet light.

XPS is also known as ESCA (electron spectroscopy for chemical analysis), an abbreviation introduced by Kai Siegbahn's research group to accentuate the chemical (rather than merely elemental) information that the technique provides. In principle XPS detects all elements. In practice, using typical laboratory-scale X-ray sources, XPS detects all elements with an atomic number (Z) of 3 (lithium) and above. It cannot easily detect hydrogen ($Z = 1$) or helium ($Z = 2$). Detection limits for most of the elements (on a modern instrument) are in the parts per thousand range. Detection limits of parts per million (ppm) are possible but require special conditions: concentration at top surface or very long collection time (overnight). XPS is routinely used to analyze inorganic compounds (including minerals, see e.g. Kloprogge and Wood, 2020), metal alloys, semiconductors, polymers, elements, catalysts, glasses, ceramics, paints, papers, inks, woods, plant parts, make-up, teeth, bones, medical implants, biomaterials, viscous oils, glues, ion-modified materials, and many others.

For such samples, XPS is used to gain information on:

- elemental composition of the surface (top 0–10 nm usually), or bulk elemental composition if fine powder is analyzed,
- empirical formula of pure materials,
- elements that contaminate a surface,
- chemical or electronic state of each element in the surface,
- uniformity of elemental composition across the top surface (or line profiling or mapping), and
- uniformity of elemental composition as a function of ion beam etching (or depth profiling).

XPS can be performed using a commercially built XPS system, a privately built XPS system, or a synchrotron-based light source combined with a custom-designed electron energy analyzer. Commercial XPS instruments generally use either a focused 20- to 500-micrometer-diameter beam of monochromatic Al Kα X-rays, or a broad 10- to 30-mm-diameter beam of nonmonochromatic (polychromatic) Al Kα X-rays or Mg Kα X-rays. A few specially designed XPS instruments can analyze volatile liquids or gases, or materials at pressures of roughly 1 torr (1.00 torr = 1.33 millibar), but there are relatively few of these types of XPS systems. The ability to heat or cool the sample during or prior to analysis is relatively common. Since the energy of an X-ray with particular wavelength is known (for Al Kα X-rays, $E_{\text{photon}} = 1486.7$ eV), and since the emitted electrons' kinetic energies are measured, the electron BE of each of the emitted electrons can be calculated by using the equation:

$$E_{\text{binding}} = E_{\text{photon}} - (E_{\text{kinetic}} + \varphi)$$

where E_{binding} is the BE of the electron, E_{photon} is the energy of the X-ray photons being used, E_{kinetic} is the kinetic energy of the electron as measured by the instrument, and φ is the work function dependent on both the spectrometer and the material. This equation is essentially a conservation of energy equation. The work function term φ is an adjustable instrumental correction factor that accounts for the few eV of kinetic energy given up by the photoelectron as it becomes absorbed by the instrument's detector. It is a constant that hardly ever needs to be adjusted in practice.

A typical XPS spectrum is a plot of the number of electrons detected (sometimes per unit time) (*Y*-axis, ordinate) versus the BE of the electrons detected (*X*-axis, abscissa). Each element gives a characteristic set of XPS peaks at characteristic BE values that directly identify each element that is present in or on the surface of the material being analyzed. These characteristic spectral peaks correspond to the electron configuration of the electrons within the atoms, for example, 1s, 2s, 2p, 3s, etc. The number of detected electrons in each of the characteristic peaks is directly related to the amount of the element within the XPS sampling volume in the sample. To calculate atomic percentage values, each raw XPS signal must be corrected by dividing its signal intensity (number of electrons detected) by a "relative sensitivity factor" (RSF) and normalized over all of the elements detected. As hydrogen is not detected, these atomic percentages exclude hydrogen. To count the number of electrons during the acquisition of a spectrum with a minimum of error, XPS detectors must be operated under UHV conditions because electron-counting detectors in XPS instruments are typically 1 m away from the material irradiated with X-rays. This long path length for detection necessitates such extremely low pressures. XPS detects only those electrons that have escaped from the sample into the vacuum of the instrument and reach the detector. In order to escape from the sample into vacuum, a photoelectron must travel through the sample. Photo-emitted electrons can undergo inelastic collisions, recombination, excitation of the sample, recapture or trapping in various excited states within the material, all of which can reduce the number of escaping photoelectrons. These effects appear as an exponential attenuation function as the depth increases, making the signals detected from analytes at the surface much stronger compared to the signals detected from analytes deeper below the sample surface. Therefore the signal measured by XPS is an exponentially surface-weighted signal, and this fact can be used to estimate analyte depths in layered materials.

The principal components of a commercially made XPS system include a source of X-rays, an UHV stainless steel chamber with UHV pumps, an electron collection lens, an electron energy analyzer, Mu-metal magnetic field shielding (a nickel–iron soft ferromagnetic alloy with very high permeability, which is used for shielding sensitive electronic equipment against static or low-frequency magnetic fields), an electron detector system, a moderate vacuum sample introduction chamber, sample mounts, a sample stage, and a set of stage manipulators. Monochromatic aluminum Kα X-rays are typically produced by diffracting and focusing a beam of nonmonochromatic X-rays off of a thin disk of natural, crystalline quartz with a <1010> orientation. The resulting wavelength is 8.3386 Å (0.83386 nm) which corresponds to a photon energy of 1486.7 eV. Aluminum Kα X-rays have an intrinsic full width at half maximum (FWHM) of 0.43 eV, centered on 1486.7 eV ($E/\Delta E = 3457$). For a properly optimized monochromator, the energy width of the monochromic aluminum Kα X-rays is 0.16 eV. However, energy broadening in common electron energy analyzers (spectrometers) produces an ultimate energy resolution on the order of FWHM = 0.25 eV, which, in effect, is the ultimate energy resolution of most commercial systems. When working under practical, everyday conditions, high-energy resolution settings will produce peak widths (FWHM) between 0.4 and 0.6 eV for various pure elements and some compounds. Nonmonochromatic magnesium X-rays have a wavelength of 9.89 Å (0.989 nm), which corresponds to a photon energy of 1253 eV. The energy width of the nonmonochromic X-ray is roughly 0.70 eV, which, in effect is the ultimate energy resolution of a system using nonmonochromatic X-rays. Nonmonochromatic X-ray sources do not use any crystals to diffract the X-rays, which allows all primary X-rays lines and the full range of high-energy Bremsstrahlung X-rays (1–12 keV) to reach the surface. The ultimate energy resolution (FWHM) when using a nonmonochromatic Mg Kα source is 0.9–1.0 eV, which includes some contribution from spectrometer-induced broadening.

2.2.3.1 Chemical states and chemical shift

The capacity to produce chemical state information (as distinguished from merely elemental information) from the topmost few nm of any surface makes XPS a unique and valuable tool for understanding the chemistry of any surface, either as received, or after physical or chemical treatment(s). In this context, "chemical state" denotes the local bonding environment of a species in question. The local bonding environment of a species in question is affected by its formal oxidation state, the identity of its nearest-neighbor atom(s), its bonding hybridization to that nearest-neighbor atom(s), and in some cases even the bonding hybridization between the atom in question and the next-nearest-neighbor atom. Thus while the nominal BE of the C 1s electron is 284.6 eV (some also use 285.0 eV as the nominal value for the BE of carbon), subtle but reproducible shifts in the actual BE, the so-called chemical shift, provide the chemical state information referred to here.

Chemical state analysis of, for example, the surface of a silicon wafer readily reveals chemical shifts due to the presence or absence of the chemical states of silicon in its different formal oxidation states, such as n-doped silicon and p-doped silicon, silicon suboxide (Si_2O), silicon monoxide (SiO), Si_2O_3, and silicon dioxide (SiO_2).

2.2.3.2 Quantitative accuracy and precision

XPS is extensively used to obtain an empirical formula as it readily yields excellent quantitative accuracy from homogeneous solid-state materials. Quantification can be divided into two categories: absolute quantification and relative quantification. The former generally requires the use of certified (or independently verified) standard samples, is usually more challenging, and is generally less common. Relative quantification is more common and comprises comparisons between several samples in a set for which one or more analytes are varied, while all other components (the sample matrix) are held constant. Quantitative accuracy is contingent on several parameters such as signal-to-noise ratio, peak intensity, accuracy of RSFs, correction for electron transmission function, surface volume homogeneity, correction for energy dependence of electron mean free path, and degree of sample degradation due to analysis. Under optimum conditions, the quantitative accuracy of the atomic percent (at.%) values calculated from the major XPS peaks is between 90% and 95% for each major peak. If a high-level quality control protocol is used, the accuracy can be further improved. Under routine work conditions, where the surface is a mixture of contamination and expected material, the accuracy ranges from 80% to 90% of the value reported in atomic percent values. The quantitative accuracy for the weaker XPS signals, which have peak intensities of 10%–20% of the strongest signal, is between 60% and 80% of the true value and depends upon the amount of effort used to improve the signal-to-noise ratio (e.g., by signal averaging). Quantitative precision (the ability to repeat a measurement and obtain the same result) is an essential consideration for proper reporting of quantitative results. Standard statistical tests, such as the Student's t-test for comparison of means, should be used to determine confidence levels in the average value from a set of replicate measurements, and when comparing the average values of two or more different sets of results. In general, a P value (an output of the Student's t-test) of .05 or less indicates a level of confidence (95%) that is accepted in the field as significant.

2.2.3.3 Analysis time

The analysis time typically ranges from 1 to 20 minutes for a broad survey scan that measures the amount of all detectable elements, typically 1–15 minutes for each element's high-resolution scan that reveals chemical state differences (for a high signal-to-noise ratio for count area result often requires multiple sweeps of the region of interest), 1–4 hours for a depth profile that measures 4–5 elements as a function of etched depth (this process time can vary the most as many factors will play a role).

2.2.3.4 Detection limits

Detection limits can vary greatly with the cross section of the photoelectron line of interest and the background signal level, which is a function of the matrix material. In general photoelectron cross sections increase with atomic number, while the background is a function of the composition of the matrix material and the BE. Background signals generally increase with atomic number of the matrix material and decrease with increasing kinetic energy. For example, in the case of gold on silicon where the high cross section Au 4f peak is at a higher kinetic energy than the major silicon peaks, it sits on a very low background and detection limits of 1 ppm or better may be achieved with reasonable acquisition times. Conversely for silicon on gold, where the modest cross section Si 2p line sits on the large background below the Au 4f lines, detection limits would be much worse for the same acquisition time. Detection limits are often quoted as 0.1%–1.0% atomic percent (0.1% = 1 part per thousand = 1000 ppm) for practical analyses, but lower limits may be achieved in many circumstances.

2.2.3.5 Degradation during analysis

This depends on the sensitivity of the material to the wavelength of X-rays used, the total dose of the X-rays, the temperature of the surface and the level of the vacuum. Most minerals, metals, alloys, ceramics, and most glasses are not measurably degraded by either nonmonochromatic or monochromatic X-rays. Some, but not all, polymers, catalysts, certain highly oxygenated compounds, various inorganic compounds, and fine organics are degraded by either monochromatic or nonmonochromatic X-ray sources. Nonmonochromatic X-ray sources create a significant amount of high-energy Bremsstrahlung X-rays (1–15 keV of energy), which can directly degrade the surface chemistry of various materials. In addition, nonmonochromatic X-ray sources produce a significant amount of heat (100°C–200°C) on the surface of the sample because the anode that produces the X-rays is typically only 1–5 cm away from the sample surface. This level of heat, when combined with the Bremsstrahlung X-rays, acts synergistically to increase the amount and rate of degradation for certain materials. Monochromatic X-ray sources, because they are far away (50–100 cm) from the sample, do not produce any heat effects. Because the vacuum removes various gases (e.g., O_2, CO) and liquids (e.g., water, alcohol, solvents, etc.) that were initially trapped within or on the surface of the sample, the chemistry and morphology of the surface will continue to change until the surface achieves a steady state. This type of degradation is sometimes difficult to detect.

2.2.3.6 Peak identification

The number of peaks produced in a survey scan by a single element varies from 1 to more than 20. Tables of BEs that identify the shell and spin-orbit of each peak produced by a given element are included with modern XPS instruments and can be found in various handbooks and websites. Because these experimentally determined BEs are characteristic of specific elements, they can be directly used to identify experimentally measured peaks of a material with unknown elemental composition. Before beginning the process of peak identification, the analyst must determine if the BEs of the unprocessed survey spectrum (0–1200 eV) have or have not been shifted due to a positive or negative surface charge. This is most often done by looking for two peaks that are due to the presence of carbon and oxygen.

2.2.3.7 Charge referencing insulators

Charge referencing is necessary when a sample suffers either a positive (+) or negative (−) charge induced shift of the experimental BEs. Charge referencing is required to obtain meaningful BEs from both wide-scan, high-sensitivity (low-energy resolution) survey spectra (0–1100 eV), and also narrow-scan, chemical state (high-energy resolution) spectra. Charge referencing is performed by adding or subtracting a "Charge Correction Factor" to each of the experimentally measured BEs. In general, the BE of the hydrocarbon peak of the C 1s signal is used to charge reference (charge correct) all BEs obtained from nonconductive (insulating) samples or conductors that have been deliberately insulated from the sample mount. Charge induced shifting is usually due to a modest excess of low-voltage (−1 to −20 eV) electrons attached to the surface, or a modest shortage of electrons (+1 to +15 eV) within the top 1–12 nm of the sample caused by the loss of photo-emitted electrons. The degree of charging depends on several factors. If, by chance, the charging of the surface is excessively positive, then the spectrum might appear as a series of rolling hills, not sharp peaks. The C 1s BE of the hydrocarbon species (moieties) of the "adventitious" carbon that appears on all, air-exposed, conductive and semiconductive materials' surfaces is normally found between 284.5 and 285.5 eV. For convenience, the C 1s BE of hydrocarbon moieties is defined to appear between 284.6 and 285.0 eV. A value of 284.8 eV has become popular in recent years. However, some recent reports indicate that 284.9 or 285.0 eV represents hydrocarbons attached on metals, not the natural native oxide. The 284.8 eV BE is routinely used as the "Reference BE" for charge referencing insulators. When the C 1s BE is used for charge referencing, then the charge correction factor is the difference between 284.8 eV and the experimentally measured C 1s BE of the hydrocarbon moieties. Conductive materials and most native oxides of conductors should never need charge referencing. Conductive materials should never be charge referenced unless the topmost layer of the sample has a thick nonconductive film.

2.2.3.8 Peak-fitting

The process of peak-fitting high-energy resolution XPS spectra is still a mixture of art, science, knowledge, and experience. The peak-fit process is affected by instrument design, instrument components, experimental settings (in other words the analysis conditions), and sample variables. Most instrument parameters are constant while others depend on the choice of experimental settings. Before starting any peak-fit effort, the analyst performing the peak-fit needs to know if the topmost 15 nm of the sample is expected to be a homogeneous material or is expected to be a mixture of materials. If the top 15 nm is a homogeneous material with only very minor amounts of adventitious carbon and adsorbed gases, then the analyst can use theoretical peak area ratios to enhance the peak-fitting process. Variables that affect or define peak-fit results include:

- FWHM
- chemical shifts
- peak shapes
- instrument design factors
- experimental settings
- sample factors

The FWHM values are useful indicators of chemical state changes and physical influences. That is, broadening of a peak may indicate a change in the number of chemical bonds contributing to a peak shape, a change in the sample condition (X-ray damage) and/or differential charging of the surface (localized differences in the charge state of the surface). Though it should be noted that the FWHM also depends on the detector and can also increase due to the sample getting charged. When using high-energy resolution experiment settings on an XPS equipped with a monochromatic Al Kα X-ray source, the FWHM of the major XPS peaks ranges from 0.3 to 1.7 eV. Chemical shift values depend on the degree of electron bond polarization between nearest-neighbor atoms. A specific chemical shift is the difference in BE values of one specific chemical state versus the BE of one form of the pure element, or of a particular agreed-upon chemical state of that element. Component peaks derived from peak-fitting a raw chemical state spectrum can be assigned to the presence of different chemical states within the sampling volume of the sample. The peak shape depends on instrument parameters, experimental parameters, and sample characteristics

2.2.4 Inductively coupled plasma atomic emission spectroscopy

ICP-AES, also known as ICP-OES, is an analytical technique used for the detection of chemical elements. It is a category of emission spectroscopy that uses an inductively coupled plasma to produce excited atoms and ions that emit electromagnetic radiation at wavelengths characteristic of a particular element. It is a flame technique with a flame temperature in a range from 6000 to 10,000K. The intensity of this emission is indicative of the concentration of the element within the sample.

The ICP-AES is composed of two parts: the ICP and the optical spectrometer. The ICP torch consists of three concentric quartz glass tubes. The output or "work" coil of the radio-frequency (RF) generator surrounds part of this quartz torch. Argon gas is generally used to create the plasma. When the torch is turned on, an intense electromagnetic field is created within the coil by the high-power RF signal flowing in the coil. This RF signal is created by the RF generator, which is, effectively, a high-power radio transmitter driving the "work coil" the same way a typical radio transmitter drives a transmitting antenna. Typical instruments run at either 27 or 40 MHz. The argon gas flowing through the torch is ignited with a Tesla unit that creates a brief discharge arc through the argon flow to initiate the ionization process. Once the plasma is "ignited," the Tesla unit is turned off. The argon gas is ionized in the intense electromagnetic field and flows in a particular rotationally symmetrical pattern toward the magnetic field of the RF coil. A stable, high-temperature plasma of about 7000K is then formed due to the inelastic collisions created between the neutral argon atoms and the charged particles. A peristaltic pump delivers an aqueous or organic sample into an analytical nebulizer where it is changed into a mist of extremely fine droplets and introduced directly inside the plasma flame. The sample immediately collides with the electrons and charged ions in the plasma and is itself broken down into charged ions. The various molecules break up into their respective atoms, which then lose electrons and recombine repeatedly in the plasma, giving off radiation at the characteristic wavelengths of the elements involved.

In some designs, a shear gas, typically nitrogen or dry compressed air is used to "cut" the plasma at a specific spot. One or two transfer lenses are then used to focus the emitted light on a diffraction grating where it is separated into its component wavelengths in the optical spectrometer. In other designs, the plasma impinges directly upon an optical interface, which consists of an orifice from which a constant flow of argon emerges, deflecting the plasma and providing cooling while at the same time allowing the emitted light from the plasma to enter the optical chamber. Still other designs use optical fibers to convey some of the light to separate optical chambers. Within the optical chamber(s), after the light is separated into its different wavelengths (colors), the light intensity is measured with a photomultiplier tube or tubes physically positioned to "view" the specific wavelength(s) for each element line involved, or, in more modern units, the separated colors fall upon an array of semiconductor photodetectors such as charge-coupled devices. In units using these detector arrays, the intensities of all wavelengths (within the system's range) can be measured simultaneously, allowing the instrument to analyze for every element to which the unit is sensitive all at once. Consequently, samples can be analyzed very rapidly. The intensity of each line is then compared to previously measured intensities of known concentrations of the elements, and their concentrations are then computed by interpolation along the calibration lines. In addition, special software generally corrects for interferences caused by the presence of different elements within a given sample matrix.

ICP-AES is commonly used in minerals processing to provide the data on grades of various streams, for the construction of mass balances. ICP-AES is often used for analysis of trace elements in soil, and it is for that reason that it is often used in forensics to ascertain the origin of soil samples found at crime scenes or on victims. It is also fast becoming the analytical method of choice for the determination of nutrient levels in agricultural soils. This information is then used to calculate the amount of fertilizer required to maximize crop yield and quality.

2.2.5 Inductively coupled plasma-mass spectrometry

ICP-MS is a type of mass spectrometry, which is capable of detecting metals and several nonmetals at concentrations as low as one part in 10^{15} (part per quadrillion, ppq) on noninterfered low-background isotopes. This is achieved by ionizing the sample with an inductively coupled plasma and then using a mass spectrometer to separate and quantify those ions. In comparison to AAS and atomic emission spectroscopy (ICP-AES), ICP-MS has greater speed, precision, and sensitivity. However, compared with other types of mass spectrometry, such as thermal ionization mass spectrometry (TIMS) and glow discharge mass spectrometry, ICP-MS introduces many interfering species: argon from the plasma, component gases of air that leak through the cone orifices, and contamination from glassware and the cones. The variety of applications exceeds that of ICP-AES and includes isotopic speciation.

For coupling to mass spectrometry, the ions from the plasma are extracted through a series of cones into a mass spectrometer, usually a quadrupole. The ions are separated based on their mass-to-charge ratio and a detector receives an ion signal proportional to the concentration. The concentration of a sample can be obtained through calibration with certified reference material such as single or multielement reference standards. ICP-MS also lends itself to quantitative determinations through isotope dilution, a single point method based on an isotopically enriched standard. Other mass analyzers coupled to ICP systems include double-focusing magnetic-electrostatic sector systems with both single and multiple collector, as well as time-of-flight systems (both axial and orthogonal accelerators have been used).

ICP-MS is widely used in the geochemistry field for radiometric dating, in which it is used to analyze relative abundance of different isotopes, in particular uranium and lead, as well as for stable isotope studies (e.g., in paleoclimate research). ICP-MS is more suitable for this application than the previously used TIMS, as species with high ionization energy such as osmium and tungsten can be easily ionized. For high-precision ratio work, multiple collector instruments are generally used to reduce the noise effect on the calculated ratios. A growing trend in the world of elemental analysis has revolved around the speciation, or determination of oxidation state of certain metals such as chromium and arsenic. One of the primary techniques to achieve this is to separate the chemical species with high-performance liquid chromatography or field-flow fractionation and then measure the concentrations with ICP-MS. The ICP-MS allows determination of elements with atomic mass ranges 7 to 250 (Li to U), and sometimes higher. Some masses are prohibited such as 40 due to the abundance of argon in the sample. Other blocked regions may include mass 80 (due to the argon dimer) and mass 56 (due to ArO), the latter of which greatly hinders Fe analysis unless the instrumentation is fitted with a reaction chamber. Such interferences can be reduced by using a high-resolution ICP-MS, which uses two or more slits to constrict the beam and distinguish between nearby peaks. This comes at the cost of sensitivity. For example, distinguishing iron from argon requires a resolving power of about 10,000, which may reduce the iron sensitivity by around 99%. A single collector ICP-MS may use a multiplier in pulse counting mode to amplify very low signals, an attenuation grid or a multiplier in analog mode to detect medium signals, and a Faraday cup/bucket to detect larger signals. A multicollector ICP-MS can have more than one of any of these, normally Faraday buckets that are much less expensive. With this combination, a dynamic range of 12 orders of magnitude, from 1 ppq to 100 ppm is possible.

Unlike AAS, which can only determine a single element at a time, ICP-MS has the capability to scan for all elements simultaneously. This allows rapid sample processing. An ICP-MS may use multiple scan modes, each one striking a different balance between speed and precision. Using the magnet alone to scan is slow, due to hysteresis, but is precise. Electrostatic plates can be used in addition to the magnet to increase the speed, and this, combined with multiple collectors, can allow a scan of every element from Lithium 6 to Uranium Oxide 256 in less than a quarter of a second. For low detection limits, interfering species and high precision, the counting time can increase considerably. The rapid scanning, large dynamic range, and large mass range are ideally suited to measuring multiple unknown concentrations and isotope ratios in samples that have had minimal preparation (an advantage over TIMS), for example, seawater and digested whole rock or mineral samples. It also lends well to laser-ablated rock samples, where the scanning rate is so quick that a real-time plot of any number of isotopes is possible. This also allows easy spatial mapping of mineral grains.

2.2.5.1 Sample introduction

The first step in analysis is the introduction of the sample. This has been realized in ICP-MS through a variety of means. The most common method is the use of analytical nebulizers. Nebulizer converts liquids into an aerosol, and that aerosol can then be swept into the plasma to create the ions. Nebulizers work best with simple liquid samples (i.e., solutions). However, there have been examples of their use with more complex materials such as slurry. Many varieties of nebulizers have been coupled to ICP-MS, including pneumatic, cross-flow, Babington, ultrasonic, and desolvating types. The aerosol generated is often treated to limit it to only the smallest droplets, commonly by means of a Peltier-cooled double pass or cyclonic spray chamber. The application of autosamplers makes this easier and faster, especially for routine work and large numbers of samples. A desolvating nebulizer (DSN) may also be used; this uses a long, heated capillary, coated with a fluoropolymer membrane, to remove most of the solvent and reduce the load on the plasma. Matrix removal introduction systems are sometimes used for samples, such as seawater, where the species of interest are at trace levels, and are surrounded by much more abundant contaminants. Laser ablation is another method. While being less common in the past, it is rapidly becoming popular as a means of sample introduction, thanks to increased ICP-MS scanning speeds. In this method, a pulsed UV laser is focused on the sample and creates a plume of ablated material, which can be swept into the plasma. This allows geochemists to spatially map the isotope composition in cross sections of rock samples, a tool which is lost if the rock is digested and introduced as a liquid sample. Lasers for this task are built to have highly controllable power outputs and uniform radial power distributions, to produce craters that are flat-bottomed and of a chosen diameter and depth. For both laser ablation and DSNs, a small flow of nitrogen may also be introduced into the argon flow. Nitrogen exists as a dimer, so it has more vibrational modes and is more efficient at receiving energy from the RF coil around the torch. Other methods of sample introduction are also used. Electrothermal vaporization and in-torch vaporization (ITV) use hot surfaces (graphite or metal, generally) to vaporize samples for introduction. These can use very small amounts of liquids, solids, or slurries. Other methods like vapor generation are also known.

2.2.5.2 Transfer of ions into vacuum

The carrier gas is sent through the central channel and into the very hot plasma. The sample is then exposed to RF, which converts the gas into a plasma. The high temperature of the plasma is enough to cause a very large portion of the sample to form ions. This fraction of ionization can approach 100% for some elements (e.g., sodium), but this is dependent on the ionization potential. A fraction of the formed ions passes through a ~1 mm hole (sampler cone) and then a ~0.4 mm hole (skimmer cone). The purpose of this is to allow a vacuum that is required by the mass spectrometer. The vacuum is created and maintained by a series of pumps. The first stage is typically based on a roughing pump, most commonly a standard rotary vane pump. This removes most of the gas and typically reaches a pressure of around 133 Pa. Later stages have their vacuum generated by more powerful vacuum systems, most often turbomolecular pumps. Older instruments may use oil diffusion pumps for high-vacuum regions.

2.2.5.3 Ion optics

Before mass separation, a beam of positive ions must be extracted from the plasma and focused into the mass analyzer. It is critical to separate the ions from UV photons, energetic neutrals and from any solid particles that may have been carried into the instrument from the ICP. Traditionally, ICP-MS instruments have used transmitting ion lens arrangements for this purpose. Examples include the Einzel lens, the Barrel lens, Agilent's Omega Lens, and Perkin-Elmer's Shadow Stop. Another method is to use ion guides (quadrupoles, hexapoles, or octopoles) to guide the ions into mass analyzer along a path away from the trajectory of photons or neutral particles. Yet another method is the Varian patented and used by Analytik Jena ICP-MS 90-degrees reflecting parabolic "Ion Mirror" optics, which are claimed to provide more efficient ion transport into the mass analyzer, resulting in better sensitivity and reduced background. A sector ICP-MS will usually have four sections: an extraction acceleration region, steering lenses, an electrostatic sector, and a magnetic sector. The first region takes ions from the plasma and accelerates them using a high voltage. The second region may use a combination of parallel plates, rings, quadrupoles, hexapoles, and octopoles to steer, shape, and focus the beam so that the resulting peaks are symmetrical, flat-topped, and have high transmission. The electrostatic sector may be before or after the magnetic sector depending on the instrument and reduces the spread in kinetic energy caused by the plasma. This spread is particularly large for ICP-MS, being larger than glow discharge and much larger than TIMS. The geometry of the instrument is chosen so that the combined focal points of the electrostatic and magnetic sectors are at the collector, known as double focusing. If the mass of interest has a low sensitivity and is just below a much larger peak, the low mass tail from this larger peak can intrude onto the mass of interest. A retardation filter might be used to reduce this tail. This sits near the collector and applies a voltage equal but opposite to the accelerating voltage; any ions that have lost energy while flying around the instrument will be decelerated to rest on the filter.

2.2.5.4 Collision reaction cell and integrated collisional reaction cell

The collision/reaction cell is used to remove interfering ions through ion/neutral reactions. Collision/reaction cells are known under several names. The dynamic reaction cell is located before the quadrupole in the ICP-MS device. The chamber has a quadrupole and can be filled with reaction (or collision) gases (ammonia, methane, oxygen, or hydrogen), with one gas type at a time or a mixture of two of them, which reacts with the introduced sample, eliminating some of the interference. The integrated collisional reaction cell (iCRC) used by Analytik Jena ICP-MS is a mini-collision cell installed in front of the parabolic ion mirror optics that removes interfering ions by injecting a collisional gas (He), or a reactive gas (H_2), or a mixture of the two, directly into the plasma as it flows through the skimmer cone and/or the sampler cone. The iCRC removed interfering ions using a collisional kinetic energy discrimination phenomenon and chemical reactions with interfering ions similarly to traditionally used larger collision cells.

2.3 Natural abundances of the elements in the Earth's crust

The abundance of the chemical elements is a quantification of the occurrence of the chemical elements relative to all other elements in a given environment, for example, the Earth's crust. Abundance can be reported in one of three different ways: by the mass fraction (the same as weight fraction); by the mole fraction (fraction of atoms by numerical count, or sometimes fraction of molecules in gases); or by the volume fraction. Volume fraction is a common abundance measure in mixed gases such as planetary atmospheres and is similar in value to molecular mole fraction for gas mixtures at relatively low densities and pressures and ideal gas mixtures.

The abundance of chemical elements in the universe is dominated by the large amounts of hydrogen and helium, which were formed in the Big Bang. Remaining elements, making up only about 2% of the universe, were mostly formed in supernovae and certain red giant stars. Lithium, beryllium, and boron are rare since, even though they are formed through nuclear fusion, they are subsequently destroyed by other reactions in the stars. The elements from carbon to iron are relatively more abundant in the universe due to the ease of forming them in supernova nucleosynthesis. Elements of higher atomic number than iron (element 26) become increasingly rarer in the universe, since they increasingly absorb stellar energy in their production. As well, elements with even atomic numbers are usually more common than their neighbors in the periodic table, due to favorable energetics of formation.

The abundance of elements in the Sun and outer planets is similar to that in the universe as a whole. As a result of solar heating, the elements of Earth and the inner rocky planets of the Solar System have experienced an additional depletion of volatile hydrogen, helium, neon, nitrogen, and carbon (which volatilizes as methane). The crust, mantle, and core of the Earth exhibit evidence of chemical segregation plus some sequestration by density. Lighter silicates of aluminum are found in the crust, with more magnesium silicate in the mantle, while metallic iron and nickel compose the core. The abundance of elements in specialized environments, such as the atmosphere, oceans, or even the human body, is principally a product of chemical interactions with the medium in which they reside.

2.3.1 Earth

The Earth formed from the same cloud of matter that formed the Sun, but the planets developed different compositions during the formation and evolution of the solar system. Subsequently, the geological history of the Earth caused parts of the planet to have varying concentrations of the elements. The mass of the Earth is approximately 5.98×10^{24} kg. In bulk, by mass, it is composed mostly of iron (32.1%), oxygen (30.1%), silicon (15.1%), magnesium (13.9%), sulfur (2.9%), nickel (1.8%), calcium (1.5%), and aluminum (1.4%), with the remaining 1.2% consisting of trace amounts of other elements. The overall composition of the Earth by elemental mass is roughly the same as the gross composition of the solar system, with the most important differences being that Earth is missing a large part of the volatile elements hydrogen, helium, neon, and nitrogen, as well as carbon, which has been lost as volatile hydrocarbons. The remaining elemental composition is roughly characteristic of the "rocky" inner planets that formed in the thermal zone where solar heat drove volatile compounds into space. The Earth retains oxygen as the second-largest component of its mass (and largest atomic fraction), mainly due to this element being retained in silicate minerals that have a very high melting point and low vapor pressure.

2.3.2 Crust

The upper part of the crust, entailing the materials near the Earth's surface, consists of a relatively large percentage of sedimentary rocks and unconsolidated material. Nevertheless, this sedimentary cover forms only a thin layer on an underlying basement of igneous and metamorphic rocks that are exposed in mountain belts and on the seafloor. It is estimated that the upper 16 km (10 miles) of the crust contains 95% igneous rocks (or their metamorphic equivalents), 4% shale, 0.75% sandstone, and 0.25% limestone. Consequently, the average composition of igneous rocks would closely resemble the average crustal composition. The mass abundance of the nine most abundant elements in the Earth's crust is about 46% oxygen, 28% silicon, 8.2% aluminum, 5.6% iron, 4.2% calcium, 2.5% sodium, 2.4% magnesium, 2.0% potassium, and 0.61% titanium. Other elements occur at less than 0.15%. For a complete list, see Table 2.2. Many of the elements shown in Fig. 2.6 are classified into (partially overlapping) categories:

- rock-forming elements (major elements in green field, and minor elements in light green field);
- rare earth elements (lanthanides, La-Lu, and Y; labeled in blue);
- major industrial metals (global production $> \sim 3 \times 10^7$ kg/year; labeled in red);
- precious metals (labeled in purple);

TABLE 2.2 Abundance of elements in Earth's crust.

Rank	Z	Element	Abundance in crust (ppm) by source					
			1	2	3	4	5	Average
1	8	Oxygen (O)	466,000	474,000	460,000	467,100	461,000	46,5620
2	14	Silicon (Si)	277,200	277,100	270,000	276,900	282,000	27,6640
3	13	Aluminum (Al)	81,300	82,000	82,000	80,700	82,300	81,660
4	26	Iron (Fe)	50,000	41,000	63,000	50,500	56,300	52,160
5	20	Calcium (Ca)	36,300	41,000	50,000	36,500	41,500	41,060
6	11	Sodium (Na)	28,300	23,000	23,000	27,500	23,600	25,080
7	12	Magnesium (Mg)	20,900	23,000	29,000	20,800	23,300	23,400
8	19	Potassium (K)	25,900	21,000	15,000	25,800	20,900	21,720
9	22	Titanium (Ti)	4400	5600	6600	6200	5600	5680
10	1	Hydrogen (H)	1400		1500	1400	1400	1425
11	15	Phosphorus (P)	1200	1000	1000	1300	1050	1110
12	25	Manganese (Mn)	1000	950	1100	900	950	980
13	9	Fluorine (F)	800	950	540	290	585	633
14	56	Barium (Ba)	500	340	340	500	425	421
15	6	Carbon (C)	300	480	1800	940	200	744
16	38	Strontium (Sr)		370	360		370	367
17	16	Sulfur (S)	500	260	420	520	350	410
18	40	Zirconium (Zr)		190	130	250	165	184
19	23	Vanadium (V)	100	160	190		120	143
20	17	Chlorine (Cl)	500	130	170	450	145	279
21	24	Chromium (Cr)	100	100	140	350	102	158
22	37	Rubidium (Rb)	300	90	60		90	135
23	28	Nickel (Ni)		80	90	190	84	111
24	30	Zinc (Zn)		75	79		70	75
25	29	Copper (Cu)	100	50	68		60	70
26	58	Cerium (Ce)		68	60		66.5	65
27	60	Neodymium (Nd)		38	33		41.5	38
28	57	Lanthanum (La)		32	34		39	35
29	39	Yttrium (Y)		30	29		33	31
30	7	Nitrogen (N)	50	25	20		19	29
31	27	Cobalt (Co)		20	30		25	25
32	3	Lithium (Li)		20	17		20	19
33	41	Niobium (Nb)		20	17		20	19
34	31	Gallium (Ga)		18	19		19	19
35	21	Scandium (Sc)		16	26		22	21
36	82	Lead (Pb)		14	10		14	13
37	62	Samarium (Sm)		7.9	6		7.05	6.98
38	90	Thorium (Th)		12	6		9.6	9.20
39	59	Praseodymium (Pr)		9.5	8.7		9.2	9.13
40	5	Boron (B)		950[a]	8.7		10	9.35
41	64	Gadolinium (Gd)		7.7	5.2		6.2	6.37
42	66	Dysprosium (Dy)		6	6.2		5.2	5.80

43	72	Hafnium (Hf)		5.3	3.3	3	3.87
44	68	Erbium (Er)		3.8	3	3.5	3.43
45	70	Ytterbium (Yb)		3.3	2.8	3.2	3.10
46	55	Cesium (Cs)		3	1.9	3	2.63
47	4	Beryllium (Be)		2.6	1.9	2.8	2.43
48	50	Tin (Sn)	0	2.2	2.2	2.3	1.68
49	63	Europium (Eu)		2.1	1.8	2	1.97
50	92	Uranium (U)		0	1.8	2.7	1.50
51	73	Tantalum (Ta)		2	1.7	2	1.90
52	32	Germanium (Ge)		1.8	1.4	1.5	1.57
53	74	Tungsten (W)		160.6[a]	1.1	1.25	1.18
54	42	Molybdenum (Mo)		1.5	1.1	1.2	1.27
55	33	Arsenic (As)		1.5	2.1	1.8	1.80
56	67	Holmium (Ho)		1.4	1.2	1.3	1.30
57	65	Terbium (Tb)		1.1	0.94	1.2	1.08
58	69	Thulium (Tm)		0.48	0.45	0.52	0.48
59	35	Bromine (Br)		0.37	3	2.4	1.92
60	81	Thallium (Tl)		0.6	0.53	0.85	0.66
61	71	Lutetium (Lu)				0.5	0.50
62	51	Antimony (Sb)		0.2	0.2	0.2	0.20
63	53	Iodine (I)		0.14	0.49	0.45	0.36
64	48	Cadmium (Cd)		0.11	0.15	0.15	0.14
65	47	Silver (Ag)		0.07	0.08	0.075	0.075
66	80	Mercury (Hg)		0.05	0.067	0.085	0.067
67	34	Selenium (Se)		0.05	0.05	0.05	0.050
68	49	Indium (In)		0.049	0.16	0.25	0.153
69	83	Bismuth (Bi)		0.048	0.025	0.0085	0.027
70	52	Tellurium (Te)		0.005	0.001	0.001	0.002
71	78	Platinum (Pt)		0.003	0.0037	0.005	0.004
72	79	Gold (Au)		0.0011	0.0031	0.004	0.003
73	44	Ruthenium (Ru)		0.001	0.001	0.001	0.001
74	46	Palladium (Pd)		0.0006	0.0063	0.015	0.0073
75	75	Rhenium (Re)		0.0004	0.0026	0.0007	0.0012
76	77	Iridium (Ir)		0.0003	0.0004	0.001	0.0006
77	45	Rhodium (Rh)		0.0002	0.0007	0.001	0.0006
78	76	Osmium (Os)		0.0001	0.0018	0.0015	0.0011

1, Darling (2016); 2, Barbalace (1995–2019); 3, WebElements (1993–2019); 4, Israel Science and Technology Directory (1999–2018); 5, Thomas Jefferson National Accelerator Facility—Office of Science Education.

[a]*Dubious.*

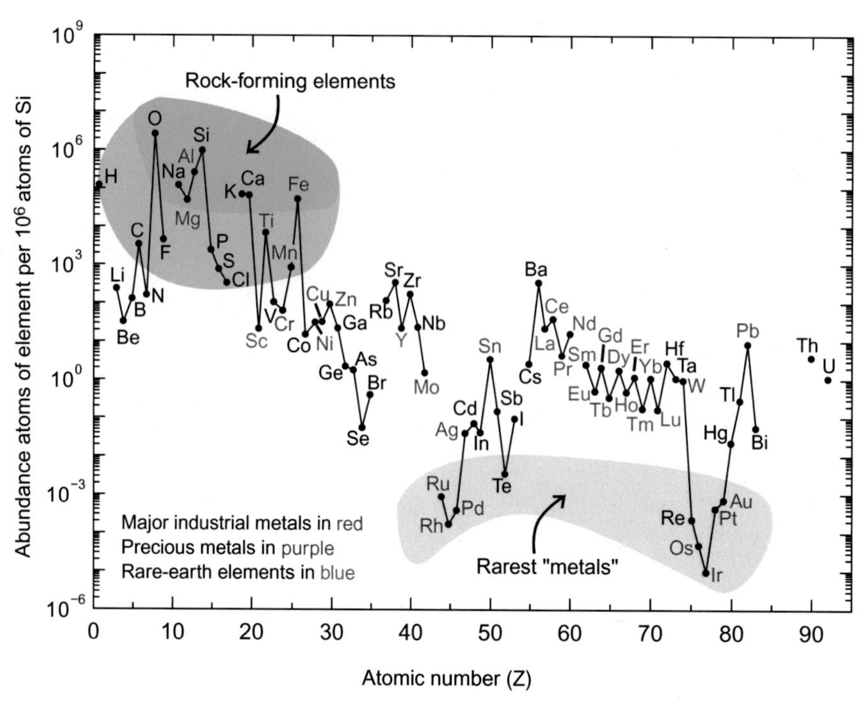

FIGURE 2.6 Abundance (atom fraction) of the chemical elements in Earth's upper continental crust as a function of atomic number. The rarest elements in the crust (shown in yellow) are not the heaviest but rather the siderophile (iron-loving) elements. These have been depleted by being relocated deeper into the Earth's core. Te and Se are concentrated as sulfides in the core and have also been depleted by preaccretional sorting in the nebula that caused them to form volatile hydrogen selenide and hydrogen telluride.

- the nine rarest "metals"—the six platinum group elements (PGE) plus Au, Re, and Te (a metalloid)—in the yellow field. These are rare in the crust due to being soluble in iron and are therefore strongly concentrated in the Earth's core. Tellurium is the single most depleted element in the silicate Earth relative to its cosmic abundance, since in addition to being concentrated in dense chalcogenides in the Earth's core it was severely depleted by preaccretional sorting in the nebula as volatile hydrogen telluride.

There are two breaks where the unstable (radioactive) elements technetium (atomic number 43) and promethium (atomic number 61) would be. These elements both have relatively short half-lives ($\sim$4 million years and $\sim$18 years, respectively) and are therefore extremely rare, as any primordial initial fractions of these in presolar system materials have long since decayed. These two elements are now only formed naturally through the spontaneous fission of very heavy radioactive elements (e.g., uranium, thorium, or the trace amounts of plutonium that exist in uranium ores), or by the interaction of certain other elements with cosmic rays. There are also breaks in the abundance graph where the noble gases should be, as they are not chemically bound in the Earth's crust, and they are only formed through decay chains from radioactive elements in the crust and are hence extremely rare there. The eight naturally occurring extremely rare, highly radioactive elements (polonium, astatine, francium, radium, actinium, protactinium, neptunium, and plutonium) are not included in the graph, because any of these elements that were present at the formation of the Earth have decayed away billions of years ago, and their total amount today is negligible and is only produced from the radioactive decay of uranium and thorium.

Oxygen and silicon are the most common elements in the crust. On Earth and in rocky planets in general, silicon and oxygen are far more common than their cosmic abundance. This is due to the fact that they combine with each other to form silicate minerals. As such they are the lightest of all of the 2% "astronomical metals" (i.e., nonhydrogen and helium elements) to form a solid that is refractory to the Sun's heat, and thus cannot boil away into space. All elements lighter than oxygen have been removed from the crust in this way, as have the heavier chalcogens sulfur, selenium, and tellurium. The Earth's crust as a result consists almost entirely of silicate, carbonate, oxide, hydroxide, phosphate, and sulfate minerals. One could imagine the Earth's crust on an atomic scale to consist in essence of a close packing of oxygen anions with interstitial metal ions, primarily silicon. Hence, the minerals referred to as the rock-forming minerals in the crust are, with a few exceptions, members of the silicate, oxide, and carbonate groups. Quantitative chemical analyses of minerals allow for the grouping of elements on the basis of their abundance. When chemical elements occur in large amounts ($>$1 wt.%), they are considered to be major elements. Minor elements occur in lower concentrations (0.1–1.0 wt.%), while other elements that occur in minor amounts ($<$0.1 wt.%) are known as trace elements.

2.3.3 Rare earth elements

"Rare" earth elements can be considered to be a historical misnomer. The persistence of the term shows unfamiliarity rather than true rarity. The more abundant rare earth elements have similar abundances in the crust as commonplace industrial metals such as chromium, nickel, copper, zinc, molybdenum, tin, tungsten, or lead. The two least abundant rare earth elements (thulium and lutetium) are approximately 200 times more common than gold. Still, in contrast to the ordinary base and precious metals, rare earth elements show a very low tendency to become concentrated in exploitable ore deposits. As a result, most of the world's supply of rare earth elements comes from only a handful of sources. Furthermore, the rare earth metals are all quite chemically similar to each other, and they are thus quite difficult to separate into quantities of the pure elements. Differences in abundances of individual rare earth elements in the upper continental crust of the Earth are due to the superposition of two effects, one nuclear and one geochemical. First, the rare earth elements with even atomic numbers ($_{58}Ce$, $_{60}Nd$,...) have greater cosmic and terrestrial abundances than the neighboring rare earth elements with odd atomic numbers ($_{57}La$, $_{59}Pr$,...). Second, the lighter rare earth elements are more incompatible (as they have larger ionic radii) and hence more strongly concentrated in the continental crust than the heavier rare earth elements. In most rare earth elements ore deposits, the first four rare earth elements—lanthanum, cerium, praseodymium, and neodymium—form 80%–99% of the total amount of rare earth metal that can be found in the ore.

2.3.4 Mantle

The mass abundance of the eight most abundant elements in the Earth's mantle is approximately 45% oxygen, 23% magnesium, 22% silicon, 5.8% iron, 2.3% calcium, 2.2% aluminum, 0.3% sodium, and 0.3% potassium. The mantle differs in elemental composition from the crust in having a higher concentration of magnesium and significantly more iron, while having much less aluminum and sodium. Therefore the upper mantle is dominated by the mineral olivine with lesser amounts of pyroxene and only trace amounts of aluminous minerals, such as feldspar, spinel, and garnet. At greater depth and higher pressure, the mantle transition zone is marked by discontinuities associated with changes in material properties without major changes in overall chemical composition. At about 400 km, olivine, Mg_2SiO_4, will isochemically transform to a denser structure with closer packing, a spinel structure, named ringwoodite. The intermediate phase in this reaction is another spinel structure, wadsleyite. The lower mantle starts at about 660 km at still higher pressures and silicates undergo another radical mineralogical and structural rearrangement. Silica goes from tetrahedral coordination [4] to octahedral coordination [6]. This depth of about 660 km is also where spinel structures, and other Ca- and Mg-silicates, transform to the perovskite structure together with other oxide structures. In the perovskite structure, six oxygen atoms are grouped around the Mg. Such a structure is in sharp contrast to those of silicate minerals in the crust, where four oxygen atoms surround each Si in tetrahedral coordination.

2.3.5 Core

The transition from the lower mantle to the core is a definite chemical discontinuity. The core is extremely dense and represents 30% of the Earth's mass but only 17% of its volume. Due to mass segregation, the core of the Earth is believed to be primarily composed of iron (88.8%), with smaller amounts of nickel (5.8%), sulfur (4.5%), and less than 1% trace elements. The liquid outer core, from 2900 to 5100 km, consists primarily of iron plus about 2% nickel. Its density of 9.9 g/cm^3 is slightly too low to represent the density of pure iron and incorporation of 9%–12% silica, or other light elements, produces a better fit with the known density. The solid inner core from 5100 to 6371 km also consists of a Fe-Ni alloy, containing about 20% nickel.

2.4 Ore genesis

Most of the mineral deposits are in the crust, which covers the mantle and contains the greatest variety of Earth's rocks. It is divided into two main parts: the oceanic crust, which is 5–10 km thick and consists of mafic igneous rocks, largely basalt, and the continental crust, which is 20–70 km thick and consists of felsic igneous and metamorphic rocks overlain by sedimentary rocks. The oceanic crust is difficult to reach from an engineering standpoint, and mineral extraction only takes place from where it has been moved onto the continents by faulting. It is the continental crust that hosts most of the mineral resources. Based on the previous section on natural abundances of the elements in the crust it will be clear that these concentrations are far too low for any element to be mineable. Hence a significant natural enrichment must take place. This geological process is commonly known as ore genesis. Various theories of ore genesis explain how the various types of mineral deposits form within the Earth's crust. Ore-genesis theories vary depending on the mineral or commodity examined. Ore-genesis theories generally involve three components: source, transport or conduit, and trap.

1. Source is essential because metal must originate somewhere and be liberated by some process.
2. Transport is needed first to move the metal-bearing fluids or solid minerals into their current position and refers to the act of physically moving the metal, as well as to chemical or physical phenomenon, which encourages movement.
3. Trapping is necessary to concentrate the metal via some physical, chemical, or geological mechanism into a concentration, which forms mineable ore.

The biggest deposits form when the source is large, the transport mechanism is efficient, and the trap is active and ready at the right time.

2.4.1 Ore-forming processes

2.4.1.1 Endogenous

2.4.1.1.1 Magmatic processes

There are numerous different types of ore deposits associated with igneous rocks. There is also a wide variety of igneous rock compositions with which ore deposits are linked. Magmas intend to inherit their metal content from the source area from which they are partially melted. Productive source areas, such as metasomatized mantle or sedimentary rock, are frequently themselves a product of some sort of metal concentration process. Felsic magmas crystallize to form granites, or their extrusive equivalents and are associated with concentrations of elements such as Sn, W, U, Th, Li, Be, and Cs, as well as Cu, Mo, Pb, Zn, and Au. Incompatible elements in felsic magmas are concentrated into the products of very small degrees of partial melting or into residual magma at an advanced state of crystallization. Such processes do not, however, very often result in economically viable ore deposits.

Fractional crystallization: Crystal fractionation in mafic magmas results in important concentrations of elements such as Cr, Ti, Fe, and V. It separates ore and non-ore minerals according to their crystallization temperature. As early crystallizing minerals form from magma, they incorporate certain elements, some of which are metals. These crystals may settle onto the bottom of the intrusion, concentrating ore minerals there. Chromite and magnetite are ore minerals that form in this way.

Liquid immiscibility: Sulfide ores containing Cu, Ni, or PGE may form from this process. As a magma changes, parts of it may separate from the main body of magma. Two liquids that will not mix are called immiscible; oil and water are an example. In magmas, sulfides may separate and sink below the silicate-rich part of the intrusion or be injected into the rock surrounding it. These deposits are found in mafic and ultramafic rocks.

Layered mafic intrusions are very significant exploration targets for this suite of metals worldwide. Primary diamond deposits represent a very uncommon situation where deep-seated mafic magma vents to the surface as explosive diatreme-maar type volcanoes, bringing with them older, xenocrystic diamond from the mantle. In both mafic and felsic magmas, the later stages of crystallization are accompanied by the exsolution of a dominantly aqueous and carbonic fluid phase that ultimately plays a very important role in ore formation.

All magmas contain the constituents that, on crystallization, combine to exsolve discrete fluid and vapor phases. Most magmas will exsolve substantial amounts (up to several wt.%) of water, as well as an order of magnitude or so less of carbon dioxide (CO_2), and these two are the dominant magmatic-hydrothermal fluids. Water has the ability to dissolve significant amounts of anionic substances, especially Cl^-, which in turn promotes the dissolution of other alkali and transition metal cations. The magmatic aqueous phase can exist as liquid, vapor, or homogenous supercritical fluid. The process of water saturation can be achieved in two ways, either by decreasing the pressure of the system or by progressive crystallization of the magma. Water saturation is especially relevant to ore-forming processes during the emplacement and crystallization of granitic magmas at moderate to shallow crustal levels. This environment results in the formation of a wide variety of important ore deposit types including porphyry Cu and Mo deposits, polymetallic skarn ores (interaction between magmatic fluids and calcareous sediments), granite-related Sn-W deposits, and the family of volcanic-related epithermal Au-Ag-(Cu) deposit types. Many metals will strongly partition into the liquid or vapor that forms on water saturation and, in such cases, mineralization accompanies the alteration of the host rocks, both within and external to the intrusion.

2.4.1.1.2 Hydrothermal processes

These processes are the physicochemical phenomena and reactions caused by movement of hydrothermal water within the crust, often because of magmatic intrusion or tectonic upheavals. The foundations of hydrothermal processes are the source-transport-trap mechanism. Sources of hydrothermal solutions also include seawater and meteoric water circulating through fractured rock, formational brines (water trapped within sediments at deposition), and metamorphic fluids created by dehydration of hydrous minerals during metamorphism.

Metamorphic fluids, resulting from volatiles liberated during prograde mineral reactions, are typically aqueo-carbonic in composition and are associated worldwide with orogenic Au deposits that are particularly well developed in Archean (4–2.5 billion years ago) and Phanerozoic (541–0 million years ago) rocks. Connate fluids formed during diagenesis interact with either reduced (forming Pb-Zn dominant ores) or oxidized (forming Cu-dominant ores) sedimentary environments. These fluids are implicated in the formation of Mississippi Valley-type (Pb-Zn) and stratiform sediment-hosted Cu ores. Circulation of near-surface meteoric waters can dissolve labile constituents such as the uranyl ion, giving rise to a variety of different sediment-hosted uranium deposits. Finally, seawater circulating through the oceanic crust in the vicinity of (mid-oceanic ridge-related) fracturing and volcanic activity is vented onto the seafloor as black smokers, providing a modern analog for the formation of Cu-Zn-dominated volcanic massive sulfide (VMS) deposits. Similar exhalative processes also occur in different tectonic settings and with different metal assemblages, giving rise to sediment-hosted Zn-Pb dominated sedimentary exhalative deposits (SEDEX)-type deposits.

Metal sources may include a large variety of rocks. However, most metals of economic importance are carried as trace elements within rock-forming minerals, and so may be liberated by hydrothermal processes. This occurs because of:

1. incompatibility of the metal with its host mineral, for example, Zn in calcite ($CaCO_3$), which favors aqueous fluids in contact with the host mineral during diagenesis;
2. solubility of the host mineral within nascent hydrothermal solutions in the source rocks, for example, mineral salts (halite, NaCl), carbonates (cerussite, $PbCO_3$), phosphates [monazite, (Ce,La,Nd,Th)PO_4, and thorianite (ThO_2)], and sulfates (baryte, $BaSO_4$);
3. elevated temperatures causing decomposition reactions of minerals.

Transport by hydrothermal solutions usually requires a salt or other soluble species that can form a metal-bearing complex. These metal-bearing complexes facilitate transport of metals within aqueous solutions, generally as hydroxides, but also by processes comparable to chelation (type of bonding of ions and molecules to metal ions. It involves the formation or presence of two or more separate coordinate bonds between a polydentate (multiple bonded) ligand and a single central atom. These ligands are called chelants, chelators, chelating agents, or sequestering agents.). This process is particularly well understood in gold metallogeny where various thiosulfate, chloride, and other gold-carrying chemical complexes (particularly tellurium-chloride/sulfate or antimony-chloride/sulfate) are used. Most metal deposits formed by hydrothermal processes include sulfide minerals, indicating sulfur is an important metal-carrying complex.

2.4.1.2 Sulfide deposition

Sulfide deposition within the trap zone occurs when metal-carrying sulfate, sulfide, or other complexes become chemically unstable due to one or more of the following processes:

1. falling temperature, which renders the complex unstable or metal insoluble;
2. loss of pressure, which has a similar effect;
3. reaction with chemically reactive wall rocks, usually of reduced oxidation state, such as iron-bearing rocks, mafic or ultramafic rocks, or carbonate rocks;
4. degassing of the hydrothermal fluid into a gas and water system, or boiling, which alters the metal-carrying capacity of the solution and even destroys metal-carrying chemical complexes.

Metal can also precipitate when temperature and pressure or oxidation state favor different ionic complexes in the water, for instance, the change from sulfide to sulfate, oxygen fugacity, exchange of metals between sulfide and chloride complexes, etc.

2.4.1.3 Classification of ore deposits

Classification of hydrothermal ore deposits is also achieved by classifying according to the temperature of formation, which roughly also correlates with particular mineralizing fluids, mineral associations, and structural styles. This scheme, proposed by Waldemar Lindgren (1933) classified hydrothermal deposits as hypothermal, mesothermal, epithermal, and telethermal.

1. Hypothermal hydrothermal rocks and minerals ore deposits are formed at great depth under conditions of high temperature.
2. Mesothermal mineral deposits are formed at moderate temperature and pressure, in and along fissures or other openings in rocks, by deposition at intermediate depths, from hydrothermal fluids.
3. Epithermal mineral ore deposits are formed at low temperatures (50°C–200°C) near the Earth's surface (<1500 m), which fill veins, breccias, and stockworks.
4. Telethermal mineral ore deposits are formed at shallow depth and relatively low temperatures, with little or no wall-rock alteration, presumably far from the source of hydrothermal solutions.

2.4.2 Metamorphic processes

Lateral secretion: Ore deposits formed by lateral secretion are formed by metamorphic reactions during shearing, which liberate mineral constituents such as quartz, sulfides, gold, carbonates, and oxides from deforming rocks, and focus these constituents into zones of reduced pressure or dilation such as faults. This may occur without much hydrothermal fluid flow, and this is typical of podiform chromite deposits. Metamorphic processes also control many physical processes that form the source of hydrothermal fluids, outlined above.

2.4.3 Sedimentary or surficial processes (exogenous)

Surficial processes are the physical and chemical phenomena which cause concentration of ore material within the regolith, generally by the action of the environment. This includes placer deposits, laterite deposits, and residual or eluvial deposits. The physical processes of ore deposit formation in the surficial realm include:

1. erosion;
2. deposition by sedimentary processes, including winnowing and density separation (e.g., gold placers);
3. weathering via oxidation or chemical attack of a rock, either liberating rock fragments or creating chemically deposited clays, laterites, or supergene enrichment;
4. deposition in low-energy environments in beach environments.

The chemical processes that contribute to weathering include hydration and dissolution, hydrolysis and acid hydrolysis, oxidation, and cation exchange. Surficial and supergene ore-forming processes are related essentially to pedogenesis (soil formation), which can be simplified into an upper zone of eluviation (leaching of labile constituents and residual concentration of immobile elements) and an underlying zone of illuviation (precipitation of labile constituents from above). Laterites are the result of intense weathering in humid, warm intertropical regions and are important hosts to bauxite ores, in addition to concentrations of metals such as Ni, Au, and the PGE. Residual concentrations of alumina, and formation of bauxitic ores, occur in high rainfall areas where Eh (oxidation-reduction potential) and pH are such that both Si and Fe in laterites are more soluble than Al. Ni enrichments are found above ultramafic intrusions in the illuviated laterite zone where the metal is concentrated in phyllosilicate (layer silicate) minerals by cation exchange. Au and Pt enrichment are also found in laterites above earlier mineralized terranes, forming in the presence of highly oxidized, acidic, and saline groundwaters. Fixation of the precious metals occurs through reduction or adsorption, in the presence of carbonaceous matter or Fe and Mn oxyhydroxides. Oxidized, acid groundwaters are able to leach metals, not only during laterite formation, but in any environment where such fluids are present. Supergene enrichment of Cu can be very important in the surficial environment above hypogene porphyry styles of mineralization. Enrichment of Cu again occurs in the illuviated zone, although in this case both Cu-sulfide minerals (in the relatively reduced zone beneath the water table) and Cu-oxide minerals (above the water table) form. The formation of viable clay deposits such as kaolinite, is a product of progressive acid hydrolysis, essentially of plagioclase feldspar. Calcrete is a pedogenic product of high evaporation environments but can also form in paleo-drainage channels in now arid climatic regions. In uranium-fertile drainage systems, calcrete channels represent zones where concentrations of secondary uranium minerals precipitate together with calcite by evaporation-induced processes.

Sedimentation is a fundamental geological process that occurs over most of the Earth's surface, in response to the pattern of global tectonic cycles. Several very important mineral commodities (as well as fossil fuels) are concentrated during the formation of sedimentary rocks. Placer processes are important during clastic sedimentation and form by the

TABLE 2.3 Summary of geological processes that form mineral deposits.

Type of process	Type of deposits	Metals/minerals concentrated
Subsurface		
Magmatic	Crystal fractionation	Cr, Ti, Fe, V
	Sulfide immiscibility	Ni, Cu, Co, PGE
Water	Magmatic water	Porphyry Cu, Cu-Mo, skarn
	Metamorphic water	Au, Cu
	Groundwater and related deposits	U, S
	Basinal brines	Mississippi Valley-type, SEDEX
	Seawater	VMS, SEDEX
Surface		
Weathering	Laterite deposits	Bauxite, Ni, Au, clay
Physical sedimentation: flowing water (stream or beach)	Placer deposits	Au, PGE, diamond, ilmenite, rutile, zircon, sand, gravel
Physical sedimentation: wind	Dune deposits	Sand
Chemical sedimentation: precipitation in or from water	Evaporite deposits	Halite, sylvite, borates, nitrates
	Chemical deposits	Iron, VMS, SEDEX
Organic sedimentation	Other deposits	S, P

sorting of light from heavy particles. Various hydrodynamic mechanisms, such as settling, entrainment, shear sorting, and transport sorting, are responsible for the concentration of different commodities in a variety of sedimentary micro- and meso-environments. At a larger scale, placer deposits form mainly in fluvial and beach-related environments and include concentrations of Au, Sn, Zr, Ti, and diamonds. Chemical sedimentation, where dissolved components precipitate out of solution from brines, also results in a wide range of important natural resources. Ferric iron precipitation from seawater is one of the main mechanisms for the formation of iron ore deposits. Banded iron-formations formed in a variety of oceanic settings, mainly in the Paleoproterozoic (2500–1600 million years ago) and prior to the onset of significant levels of oxygen in the atmosphere. Ironstone deposits tend to form mainly in the Phanerozoic times (541 million years ago until present). Other types of metal concentration linked to syn-sedimentary and early diagenetic processes occur in carbonaceous black shales (Ni, Co, Cr, Cu, Mn, Zn, Ag, etc.) in Mn nodules on the ocean floor. Accumulation of P and the development of phosphorites in ocean settings is a process that is related both to direct chemical precipitation from seawater and to biological mediation. The upwelling of cold currents onto the continental shelf and biological productivity are processes implicated in the formation of most of the world's important phosphate deposits. Chemical sediments formed by high rates of evaporation of seawater and lacustrine brines host most of the world's Na, K, borate, nitrate, and sulfate resources.

Ore deposits rarely fit neatly into the categories in which geologists wish to place them. Many may be formed by one or more of the basic genesis processes above, creating ambiguous classifications and much argument and conjecture. Often ore deposits are classified after examples of their type, for instance, Broken Hill–type lead-zinc-silver deposits or Carlin-type gold deposits. Table 2.3 gives a simplified overview of the different process and types of deposits.

2.5 Mining

2.5.1 History

2.5.1.1 Prehistoric mining

Since the start of civilization, people have utilized stone, ceramics and, later, metals found on or near the Earth's surface. These were used to make early tools and weapons; for example, high-quality flint (a hard, sedimentary cryptocrystalline form of the mineral quartz, categorized as the variety of chert that occurs in chalk or marly limestone) found in northern France, southern England, southern Netherlands, northeast Belgium and Poland was used to create flint tools. Flint mines have been found in chalk areas where seams of the stone were followed underground by shafts and galleries. The mines at Grime's Graves and Krzemionki are especially famous, and like most other flint mines, are Neolithic in origin (c.4000–3000 BCE) [Notes: Grime's Graves is a large Neolithic flint mining complex in Norfolk England. It lies 8 km northeast from Brandon, Suffolk in the East of England. It was worked between c.2600 and c.2300 BCE, although production may have continued well into the Bronze and Iron Ages (and later) owing to the low cost of flint compared with metals. Krzemionki (also Krzemionki Opatowskie), on the other hand, is a Neolithic and early Bronze Age complex of flint mines for the extraction of Upper Jurassic (Oxfordian) banded flints located about 8 km northeast of the Polish city of Ostrowiec Świętokrzyski. It is one of the largest known complexes of prehistoric flint mines in Europe]. Other hard rocks mined or collected for axes included the greenstone of the Langdale axe industry based in the English Lake District. The Langdale axe industry is the name given by archeologists to specialized stone tool manufacturing centered at Great Langdale in England's Lake District during the Neolithic period (beginning about 4000 BCE in Britain). The existence of a production site was originally suggested by chance discoveries in the 1930s, which were followed by more systematic searching in the 1940s and 1950s by Clare Fell and others. The finds were mainly reject axes, rough-outs, and blades created by knapping large lumps of the rock found in the scree or perhaps by simple quarrying or opencast mining. Hammerstones have also been found in the scree and other lithic debitage from the industry such as blades and flakes. The area has outcrops of fine-grained greenstone or hornstone suitable for making polished stone axes. Such axes have been found distributed across Great Britain. The rock is an epidotized greenstone quarried or perhaps just collected from the scree slopes in the Langdale Valley on Harrison Stickle and Pike of Stickle. The oldest known mine on archeological record is the Ngwenya Mine in Swaziland, which is about 43,000 years old based on radiocarbon dating. At this site Paleolithic humans mined hematite (Fe_2O_3) to make the red pigment ochre [a natural clay earth pigment which is a mixture of ferric oxide and varying amounts of clay and sand. It ranges in color from yellow to deep orange or brown. It is also the name of the colors produced by this pigment, especially a light brownish-yellow. A variant of ochre containing a large amount of hematite, or dehydrated iron oxide, has a reddish tint known as "red ochre" (or, in some dialects, ruddle)]. Mines of a similar age in Hungary are believed to be sites where Neanderthals may have mined flint for weapons and tools.

2.5.1.2 Ancient Egypt

Ancient Egyptians mined malachite ($Cu_2CO_3(OH)_2$) at Maadi. At first, Egyptians applied the bright green malachite in ornamentations and pottery. Later, between 2613 and 2494 BCE, large building projects required expeditions abroad to the area of Wadi Maghareh in order to secure minerals and other resources not available in Egypt itself. Quarries for turquoise ($CuAl_6(PO_4)_4(OH)_8 \cdot 4H_2O$) and copper were also found at Wadi Hammamat (a dry riverbed in Egypt's Eastern Desert, about halfway between Al-Qusayr and Qena. It was a major mining region and trade route east from the Nile Valley in ancient times, while these days 3000 years of rock carvings and graffiti make it a major scientific and tourist site), Tura (the primary quarry for limestone in ancient Egypt. The site, which was known by the ancient Egyptians as Troyu or Royu, is located about halfway between modern-day Cairo and Helwan. Its ancient Egyptian name was misinterpreted by the Ancient Greek geographer Strabo who thought that it meant it was inhabited by Trojans, thus the Hellenistic city was named Troia. The site is located by the modern town of Tora in the Cairo Governorate.), Aswan and various other Nubian sites on the Sinai Peninsula and at Timna [The Timna Valley (תִּמְנָע) is located in southern Israel in the southwestern Arava/Arabah, approximately 30 km north of the Gulf of Aqaba and the city of Eilat. The area is rich in copper ore and has been mined since the 5th millennium BCE. There is controversy about whether the mines were active during the famous biblical United Kingdom of Israel and its second ruler, King Solomon]. Mining in Egypt occurred in the earliest dynasties. The gold mines of Nubia were among the largest and most extensive of any in Ancient Egypt. These mines are described by the Greek historian Diodorus Siculus (fl. 1st century BCE), who mentions fire-setting (a method of traditional mining used most commonly from prehistoric times up to the Middle Ages. Fires were set against a rock face to heat the stone, which was then doused with liquid, causing the stone to fracture by thermal shock. Some experiments have suggested that the water (or any other liquid) did not have a noticeable effect on the rock, but rather helped the miners' progress by quickly cooling down the area after the fire.) as one method used to break down the hard rock holding the gold. The miners crushed the ore and ground it to a fine powder before washing the powder for the gold dust.

2.5.1.3 Ancient Greek and Roman mining

Mining in Europe has a very long history. Examples include the silver mines of Laurium [or Lavrio or Lavrion (Greek: Λαύριο; Ancient Greek: Λαύριον; before early 11th century BCE: Θορικός Thorikos; from Middle Ages until 1908: Εργαστήρια Ergastiria) is a town in southeastern part of Attica, Greece.], which helped support the Greek city-state of Athens. Although they had over 20,000 slaves working the mines, their technology was essentially identical to their Bronze Age predecessors. At other mines, such as on the island of Thassos [(Greek: Θάσος, Thásos) is a Greek island, geographically part of the North Aegean Sea, but administratively part of the Kavala regional unit], marble was quarried by the Parians [Paros (Greek: Πάρος; Venetian: Paro) is a Greek island in the central Aegean Sea. One of the Cyclades island group, it lies to the west of Naxos, from which it is separated by a channel about 8 km wide. It lies approximately 150 km southeast of Piraeus] after they arrived in the 7th century BCE. The marble was shipped away and was later found by archeologists to have been used in buildings including the tomb of Amphipolis. Philip II of Macedon [(Greek: Φίλιππος Β' ὁ Μακεδών; 382–336 BCE) the king (basileus) of the kingdom of Macedon from 359 BCE until his assassination in 336 BCE.], the father of Alexander the Great, captured the gold mines of Mount Pangeo [The Pangaion Hills (Greek, Παγγαίο, ancient forms: Pangaeon, Pangaeum, Homeric name: Nysa) are a mountain range in Greece, approximately 40 km from Kavala. The highest elevation is 1956 m. Pangaion is very often referred by ancient Greek and Latin sources. It was famous for silver and gold mines, as well as for shipyard wood and the oracle of Dionysus.] in 357 BCE to fund his military campaigns. He also captured gold mines in Thrace for minting coinage, eventually producing 26 tons per year.

Yet, it was the Romans who developed large-scale mining methods, particularly the use of large volumes of water brought to the minehead by numerous aqueducts (a watercourse constructed to carry water from a source to a distribution point far away). The water was used for a variety of purposes, including removing overburden and rock debris, called hydraulic mining (a form of mining that uses high-pressure jets of water to dislodge rock material or move sediment), as well as washing comminuted (comminution is the reduction of solid materials from one average particle size to a smaller average particle size, by crushing, grinding, cutting, vibrating, or other processes), or crushed, ores and driving simple machinery. The Romans applied hydraulic mining methods on a large scale to prospect for the veins of ore, especially a now-obsolete form of mining known as hushing, an ancient and historic mining method using a flood or torrent of water to reveal mineral veins. The method was applied in several ways, both in prospecting for ores, and for their exploitation. Mineral veins are often hidden below soil and subsoil, which must be stripped away to discover the ore veins. A flood of water is very effective in moving soil as well as working the ore deposits when combined with other methods such as fire-setting. They built numerous aqueducts to supply water to the minehead. There the water was stored in large reservoirs and tanks. When a full tank was opened, the flood of water sluiced away the overburden to expose the bedrock underneath and any gold veins. The rock was then worked upon by fire-setting to heat the rock, which would be quenched with a stream of water. The resulting thermal shock cracked the rock, enabling it to be removed by further streams of water from the overhead tanks. The Roman miners used similar methods to work cassiterite (SnO_2) deposits in Cornwall and lead ore in the Pennines. The methods had been established by the Romans in Spain in CE 25 to exploit large alluvial gold deposits, the largest site being at Las Médulas, where seven long aqueducts tapped local rivers and sluiced the deposits. Spain was one of the most important mining regions, but all regions of the Roman Empire were exploited. In Great Britain the local population had mined minerals for thousands of years, but after the Roman conquest, the scale of the operations increased significantly, as the Romans needed Britannia's resources, especially gold, silver, tin, and lead.

Roman mining methods were not limited to surface mining. They followed the ore veins underground once opencast mining was no longer feasible. At Dolaucothi (The Dolaucothi Gold Mines, also known as the Ogofau Gold Mine, are ancient Roman surface and underground mines located in the valley of the River Cothi, near Pumsaint, Carmarthenshire, Wales. The gold mines are located within the Dolaucothi Estate which is now owned by the National Trust) they stoped out (the process of extracting the desired ore or other minerals from an underground mine, leaving behind an open space known as a stope. Stoping is used when the country rock is sufficiently strong not to collapse into the stope, although in most cases artificial support is also provided) the veins and drove adits (an entrance to an underground mine which is horizontal or nearly horizontal, by which the mine can be entered, drained of water, ventilated, and minerals extracted at the lowest convenient level. Adits are also used to explore for mineral veins) through bare rock to drain the stopes. The same adits were also used to ventilate the workings, especially important when fire-setting was used. At other parts of the site, they penetrated the water table and dewatered the mines using several kinds of machines, especially reverse overshot water wheels. These were used extensively in the copper mines at Rio Tinto in Spain, where one sequence included 16 such wheels arranged in pairs and lifting water about 24 m. They were worked as treadmills with miners standing on the top slats. Many examples of such devices have been found in old Roman mines.

2.5.1.4 Medieval Europe

Mining as an industry experienced dramatic changes in medieval Europe. The mining industry in the early Middle Ages was mostly focused on the extraction of copper and iron. Other precious metals were also used, mainly for gilding or coinage. Initially, many metals were obtained through open-pit mining, and ore was mostly extracted from shallow depths, rather than through deep mine shafts. Around the 14th century, the growing use of weapons, armor, stirrups, and horseshoes greatly increased the need for iron. Medieval knights, for example, were often laden with up to 100 pounds (45 kg) of plate or chain link armor in addition to swords, lances, and other weapons. The overwhelming reliance on iron for military purposes prompted the increase in iron production and extraction processes. The silver crisis of 1465 happened when all mines had reached depths at which the shafts could no longer be pumped dry with the at that time available methods. While an increased use of banknotes, credit and copper coins during this period did decrease the value of, and dependence on, precious metals, gold and silver continued to be vital to the story of medieval mining. Due to differences in the social structure of society, the increasing extraction of mineral deposits spread from central Europe to England in the mid-16th century. On the continent, mineral deposits were owned by the crown, and this regalian right was stoutly preserved. However, in England, royal mining rights were limited to only gold and silver (of which England had virtually no deposits) by a judicial decision of 1568 and a law in 1688. England, however, had iron, zinc, copper, lead, and tin ores. Landlords who owned the base metals and coal under their estates therefore had a strong incentive to extract these metals or to lease the deposits and collect royalties from mine operators. English, German, and Dutch capital combined to finance extraction and refining. Hundreds of German technicians and skilled workers were brought over; in 1642 a colony of 4000 foreigners was mining and smelting copper at Keswick in the northwestern mountains.

Use of waterpower in the form of water mills was widespread. The water mills were employed in crushing ore, raising ore from shafts, and ventilating galleries by powering giant bellows. Black powder was the earliest known chemical explosive consisting of a mixture of sulfur (S), charcoal (C), and potassium nitrate (saltpeter, KNO_3) (The sulfur and charcoal act as fuels while the saltpeter is an oxidizer) and was first used in mining in Selmecbánya, Kingdom of Hungary (now Banská Štiavnica, Slovakia) in 1627. Black powder allowed blasting of rock and earth to loosen and reveal ore veins. Blasting was much faster than fire-setting and permitted the mining of previously impenetrable metals and ores. The first mining school in the Kingdom of Hungary was founded there in 1735 by Samuel Mikovíny. Beginning in 1763, the Hofkammer in Vienna, with support from Queen Maria Theresa [Maria Theresa Walburga Amalia Christina (May 13, 1717 to November 29, 1780) was the only female ruler of the Habsburg dominions and the last of the House of Habsburg. She was the sovereign of Austria, Hungary, Croatia, Bohemia, Transylvania, Mantua, Milan, Lodomeria and Galicia, the Austrian Netherlands, and Parma. By marriage, she was Duchess of Lorraine, Grand Duchess of Tuscany, and Holy Roman Empress.], transformed the school into the Academy of Mining, the world's first mining academy. In 1807, a Forestry Institute was "established under the decision of Emperor Franz I" (August 18, 1830 to November 21, 1916, Emperor of Austria, King of Hungary, King of Bohemia, and monarch of many other states of the Austro-Hungarian Empire, from 2 December 1848 to his death); in 1848 the school was renamed the Academy of Mining and Forestry, "the first technical university in the world." In 1919, after the creation of Czechoslovakia, the Academy was moved to Sopron in Hungary. The widespread adoption of agricultural innovations such as the iron plowshare, as well as the increasing use of metal as a building material, was also a driving force in the enormous growth of the iron industry during this period. Inventions such as the arrastra were often used by the Spanish to pulverize ore after being mined. [The arrastra is a primitive mill for grinding and pulverizing (typically) gold or silver ore. The simplest form of the arrastra is two or more flat-bottomed drag stones placed in a circular pit paved with flat stones and connected to a center post by a long arm. With a horse, mule, or human providing power at the other end of the arm, the stones were dragged slowly around in a circle, crushing the ore. Some arrastras were powered by a water wheel; a few were later on powered by steam or gasoline engines, and even electricity.] Much of the knowledge of medieval mining techniques comes from books such as Biringuccio's De la pirotechnia and probably most importantly from Georg Agricola's De re metallica (1556). These books detail many different mining methods used in German and Saxon mines. A prime issue in medieval mines, which Agricola explains in detail, was the removal of water from mining shafts. As miners dug deeper to access new veins, flooding became a very real obstacle. The mining industry became dramatically more efficient and prosperous with the invention of mechanical and animal-driven pumps.

2.5.1.5 The Americas

Throughout prehistoric times, large quantities of copper were mined along Lake Superior's Keweenaw Peninsula and in nearby Isle Royale (Michigan); metallic copper was still existing near the surface in colonial times. The indigenous people used Lake Superior copper as early as 5000 years ago; copper tools, arrowheads, and other artifacts that were part of a widespread native trade network have been found. Furthermore, obsidian, flint, and other minerals were mined, worked, and traded. Early French explorers who came across the sites made no use of the metals due to the difficulties of transporting them, but the copper was in due course traded throughout the continent along major river routes. In the early colonial history of the Americas, "native gold and silver was quickly confiscated and sent back to Spain in fleets of gold- and silver-laden galleons." The gold and silver came mostly from mines in Central and South America. Turquoise ($CuAl_6(PO_4)_4(OH)_8{\cdot}4H_2O$) dated at CE 700 was mined in pre-Columbian America, in the Cerillos Mining District in New Mexico. Estimates are that "about 15,000 tons of rock had been removed from Mt. Chalchihuitl using stone tools before 1700." In 1727, Louis Denys (Denis) de La Ronde (1675–1741)—brother of Simon-Pierre Denys de Bonaventure (June 22, 1659 to February 7, 1711), an officer in the colonial troupes de la marine of New France, and the son-in-law of French-Canadian poet René-Louis Chartier de Lotbinière (1641–1709)—took command of Fort La Pointe at Chequamegon Bay; where natives informed him of an island of copper. La Ronde got permission from the French crown to operate mines in 1733, becoming "the first practical miner on Lake Superior." Seven years later, mining was ceased due to an outbreak of unrest between Sioux and Chippewa tribes. Mining in the United States became widespread in the 19th century, and the General Mining Act of 1872 was passed to boost mining of federal lands. As with the California Gold Rush in the mid-19th century, mining for minerals and precious metals, along with ranching, was a driving factor in the Westward Expansion to the Pacific coast. With the exploration of the West, mining camps were established and "expressed a distinctive spirit, an enduring legacy to the new nation;" Gold Rushers would experience similar difficulties as the Land Rushers of the transient West that went before them. Supported by railroads, many traveled West for work prospects in mining. Western cities such as Denver and Sacramento originated as mining towns. When new areas were explored, it was usually the gold (placer and then lode) and then silver that were taken into possession and mined first. Other metals would often wait for railroads or canals, as coarse gold dust and nuggets do not require smelting and are easy to identify and transport.

2.5.1.6 Asia

The ancient Chinese began prospecting for useful minerals in antiquity. Coal carvings have been found in the ruins of the Fushun coalfield, Liaoning Province, with a radiometric age of over 6000 years (Neolithic). The oldest known copperware in northern China is also Neolithic in age and bronze was widely utilized during the Shang Dynasty (16th–11th century BCE). In 1973 a 3000-year-old copper mine with smelting facilities was discovered in Tonglushan on Mount Verdigris in Daye County. The Daye copper mines in Hubei Province started some 2800 years ago and they are still in production. All these finds indicate that mining activities in China began at least 5000 years ago. Largely, Chinese mining history can be divided into two major periods. The time from about 6000 years ago to the first Opium War (1839–42) is considered the primary stage of Chinese mining development, characterized by simple tools and ancient manual techniques. From the Opium War to the present, Chinese mining can be characterized by a combination of both Western mining techniques and traditional Chinese methods. In the period of the Xia Dynasty by 2000 BCE, copper weapons and tools were already widely used. In the Yin ruins at Anyang, Henan Province, a major archeological site of the Xia and Shang Dynasties (21st–11th century BCE), gold, copper, tin, and lead tools and containers have been discovered. For example, the Simuwu Ding (a sacrificial vessel) discovered in the Yin ruins consists of 84.77% copper, 11.64% tin, and 2.79% lead, a composition very close to that of the bronze alloy with the highest hardness known to modern metallurgy. Nephrite and serpentine jewelry have also been unearthed in sites dating to the same period. Significant advancement had taken place in mineral identification and mineral usage during the Zhou Dynasty (11th century BCE to 221 BCE), as can be read in the book Mountains and Seas (Shan Hai Jing, third to first century BCE). In this book, 89 different kinds of minerals and rocks from 309 localities are described. The author describes physical properties such as hardness, color, luster, transparency, and texture, and observes that magnetism and medical properties can be used to identify mineral species and rock types. Indicator types (those minerals that frequently can be found near gold or iron ore deposits) are also discussed. The book also references the use of well-developed mining techniques, while it does not suggest an inherited culture of mining as such. Botanical indicators of mineral deposits are also recorded, for example, "in a place, one to one and a half kilometers from a hill named Huang Shan, where many Huitong plants grow, gold ore can be found below." Around this time, the Tonglushan mine at Daye in Hubei Province, one of the world's oldest copper mines, was opened. It mined the oxidized zone of a high-grade copper deposit: the total length of the ancient trenches and shafts is estimated at 8 km. Next to the old workings, a large open-pit mine for copper is still being worked to this day. The astronomer and scientist Shen Kuo (1031–95) of the Song Dynasty substantially researched minerals and rocks. In a famous work called Chat by Dream Creek, mineral prospecting, mining activities, and mineral usage in his time are discussed. For example, Shen indicated that bluestone (chalcanthite, $CuSO_4{\cdot}5H_2O$) contains water and can be refined into copper, and he predicted that oil (petroleum) obtained from rocks would be extensively used in the future. The Chinese, with their long history of geological observation and mining, could have played a leading role in world history: unfortunately, between the time of the Qingshi Emperor (1722–35) and the mid-19th century, China closed its borders to Western nations and rejected significant advanced theories and technology relating to mining, geology, and mineralogy developed in the West during the Industrial Revolution. Since the end of the Opium Wars in 1860, Western science and technology have strongly influenced the development of Chinese mineralogy and the mining industry. Modern geological surveys for the mining industry in China started with exploration for iron ore, as iron was needed for its national defense purposes. Between the revolution in 1911 and World War II, foreigners ran various mines in China, including the world's largest antimony and tungsten mines. Systematic geological surveys were started by Westerners, and the most important large-scale mines, such as the Xikuangshan antimony mine in Hunan and the Wanshan mercury mine in Guizhou, were run by Western experts. After the war against the Japanese (1930s–1940s), most foreign-owned mines were taken over by the Chinese government, but by then most of them were either depleted or completely exhausted. Since the 1950s, important advances have taken place in three phases. From 1950 to 1967, Chinese mining leaped forward, driven by the large-scale reconstruction and redevelopment of all industrial sectors after the civil war. All the mines were owned either by the central government or by local governments. The period from 1968 to 1977 was a period of economic depression caused by the Cultural Revolution and other political movements. Most mines decreased their production, and some mines suspended their production altogether. After 1978, in China a new period started characterized by political liberalization and economic reforms—changes that permitted the rapid growth of the mining industry. By the late 1990s, the commercial mineral output of China included 168 different products mined from more than 20,000 deposits. The abundant raw materials to produce iron, aluminum, and cement, as well as many other industrial products, came to be important in world markets. These days, lead, zinc, nickel, tungsten, and antimony are crucial for Chinese industry, and China has also become the world's leading supplier of politically and economically crucial rare earth elements. More than 7 million workers are currently employed in the mining industry in China. The mining techniques in use in most mines remain backward compared to modern standards, some being appallingly like those used in Europe in the 16th and 17th centuries. Vertical and inclined shafts predominate and modern adits (horizontal tunnels) are rare in underground mining, as the wood and steel they require for support are expensive. In many small mines, oil lamps, hand hammers, human transport, and rope hoists are the rule. However, a large number of these small mines were forced to close down at the start of the 21st century due to the huge problems with safety, the adverse effects of mining on miners' health, and environmental degradation. Tremendously low recovery rates have wasted large quantities of valuable resources, and sprawling mining activities have resulted in serious environmental problems.

India has been highly endowed with ore deposits. There is much evidence that exploitation of minerals such as iron ore, copper, and lead-zinc has been going on in the country from time immemorial. Flint was known and exploited by the inhabitants of the Indus Valley Civilization by the 3rd millennium BCE. Several Harappan quarries were discovered in archeological excavations. The quarries were described as: "From the surface the quarries consisted of almost circular empty areas, representing the quarry pits, filled with eolian sand, blown from the Thar Desert dunes, and heaps of limestone block, deriving from the prehistoric mining activity. All around these structures flint workshops were noticed, represented by scatters of flint flakes and blades among which typical Harappan-elongated blade cores and characteristic bullet cores with very narrow bladelet detachments." Radiocarbon dating of Zyzyphus cf. nummularia charcoal found in the quarries has yielded evidence that the activity continued into 1870–1800 BCE. Old Sanskrit texts note the use of bitumen, rock salt, yellow orpiment, chalk, alum, bismuth, calamine, realgar, stibnite, saltpeter, cinnabar, arsenic, sulfur, yellow and red ochre, black sand, and red clay in prescriptions. Among the metals used were gold, silver, copper, mercury, iron, iron ores, pyrite, tin, and brass. Mercury seemed to have been the most often used and is called by several names in the texts. No source for mercury or its ores has been found. This suggests that it may have been imported. However, the first recorded recent history of mining in India dates to 1774 when an English Company was granted permission by the East India Company for mining coal in Raniganj. M/s John Taylor & Sons Ltd. started gold mining in Kolar Gold Fields in 1880. Mining activities in the country though continued to be primitive in nature and modest in scale up till the beginning of this century. Subsequently, with progressive industrialization the demand for and hence the production of various minerals gradually increased. After India became independent, the growth of mining under the impact of successive five-year plans has been very fast. India has significant sources of bauxite, titanium ore, chromite, natural gas, diamonds, petroleum, and limestone. According to the 2008 Ministry of Mines estimates, "India has stepped up its production to reach the second rank among the chromite producers of the world. Besides, India ranks second in barytes, fourth in iron ore, fifth in bauxite and crude steel, seventh in manganese ore, and eighth in aluminum." India accounts for 12% of the world's known and economically available thorium. It is the world's largest producer and exporter of mica, accounting for almost 60% of the net mica production in the world, which it exports to the United Kingdom, Japan, United States, etc.

Mining in Japan is minimal because Japan does not have many onshore mineral deposits. Production of copper in 1917 was 108,000 tons, in 1921 54,000 tons, and in 1926 63,400 tons, but this production was augmented to 70,000 tons in 1931–37. Gold production in Korea was 6.2 tons in 1930 rising to 26.1 tons/year at its peak. In rivers and mines, other deposits were in Saganoseki (Ōita) Honshū, Kyushu, and North Formosa. Important iron deposits were Muroran (Hokkaidō) and Kenji (Korea). Total reserves were 90 M tons of their own, 10 M or 50 M in Korea (Kenjiho) and Formosa. The main silver mines were found at Kosaki, Kawaga, and Hitachi, and others in Karafuto with pyrite (FeS_2). The production of gold was curbed in 1943 by Order for Gold Mine Consolidation to concentrate on the minerals more critical for the munitions production. Cobalt, copper, gold, iron, lead, manganese, silver, tin, tungsten, and zinc are common and were extensively mined in Japan. Barium, beryllium, bismuth, cadmium, chromium, indium, lithium, mercury, molybdenum, nickel, titanium, uranium, and vanadium are uncommon but still were mined in Japan. Antimony, arsenic, boron, germanium, graphite, and sulfur were all mined in Japan. The Japanese mining industry began to rapidly decline in the 1980s.

Sri Lanka's gem industry has a very long and colorful history. Sri Lanka was lovingly known as Ratna-Dweepa meaning Gem Island, reflecting its natural wealth. Marco Polo wrote that the island had the best sapphires, topazes, amethysts, and other gems in the world. Ptolemy, the 2nd-century astronomer, noted that beryl and sapphire were the backbone of Sri Lanka's gem industry. Records from sailors that visited the island indicate that they brought back "jewels of Serendib." Serendib was the ancient name given to the island by middle—eastern and Persian traders that crossed the Indian Ocean to trade gems from Sri Lanka to the East during the 4th and 5th centuries. Sri Lanka is geologically an extremely old country. Ninety percent of the rocks of the island are of Precambrian age, 560 million to 2400 million years ago. The gems are found in sedimentary residual gem deposits, eluvial deposits, metamorphic deposits, skarn and calcium-rich rocks. Nearly all the gem formations in Sri Lanka are in the central high-grade metamorphic terrain of the Highland Complex. The mineralogy of the gem deposits varies widely with, among others, corundum (sapphire, ruby), chrysoberyl, beryl, spinel, topaz, zircon, tourmaline, and garnet being common. Residual deposits are mainly found in flood plains of rivers and streams. The metamorphic types of gems constitute 90% of the gem deposits in Sri Lanka. It has been estimated that approximately 25% of the total land area of Sri Lanka is possibly gem-bearing, making Sri Lanka one of the countries with the highest density of gem deposits compared to its landmass. Ratnapura, a major city in Sri Lanka, contains the most gem deposits and derived its name from the gem industry. Ratnapura means "city of gems."

Indonesia is a significant player in the global mining industry, with significant production of copper, gold, tin, bauxite, and nickel. A good example is the island of Sumatra. Mineral exploration and mining activities in Sumatra, going as far back as the prehistory, have been dominated by gold, involving both the local population and mostly foreign companies. The first documented mining activity was the reopening of the ancient silver-rich Salido gold mine in West Sumatra in 1669 by the Vereenigde Oost-Indische Compagnie, the Dutch trading company that for some 200 years monopolized trade between Europe and Asia. The government of the Netherlands East Indies started geological investigations combined with mineral exploration in 1850 and private industry followed about 30 years later. Between 1899 and 1940, 14 gold mines were opened, including two alluvial dredging operations, most of which were short-lived and uneconomic. Total production between 1899 and 1940 amounted to 101 tons of Au and 1.2 million tonnes of Ag. Only Lebong Donok and Lebong Tandai in Bengkulu, which up to 1940 produced together 79.2 tons of Au and 651 tons of Ag, were highly profitable. Furthermore, alluvial tin and Fe/base-metal ± Au skarn deposits were mined on a small scale. During the Japanese occupation, its aftermath, and the first 20 years of Indonesia's independence, there was very limited mining activity. Introduction of new foreign investment and mining laws by the New Order Government in 1967 heralded a new era of exploration and mining, which continues to the present day. It witnessed several peaks in exploration activity, namely, 1969–73 (porphyry copper), 1985–90 (gold), 1995–99 (gold), and 2006–10 (multicommodity). Several types of mineralization were found that were previously unknown, such as porphyry Cu, high-sulfidation gold, sediment-hosted gold, and SEDEX and Mississippi Valley-type Pb-Zn. Mining in recent decades has been limited to the reopening of Lebong Tandai (1985–95), and development of three small low/intermediate-sulfidation epithermal gold-silver vein deposits (Bukit Tembang, 1997–2000; Way Linggo, 2010–13; Talang Santo, starting in 2014), and a medium-sized high-sulfidation epithermal gold deposit (Martabe, starting 2012). Moreover, some Indonesian companies have been mining small Fe skarn deposits in recent years. Notwithstanding its long exploration and mining history Sumatra is underexplored by world standards.

Mining in the Philippines began around 1000 BCE. The early Filipinos worked various mines of gold, silver, copper, and iron. Jewels, gold ingots, chains, calombigas (armlets of wrought gold), and earrings were handed down from antiquity and inherited from their ancestors. Gold dagger handles, gold dishes, tooth plating, and huge gold ornaments were also used. In Laszlo Legeza's "Tantric elements in pre-Hispanic Philippines Gold Art," he mentioned that gold jewelry of Philippine origin was found in Ancient Egypt. According to Antonio Pigafetta, the people of Mindoro possessed great skill in mixing gold with other metals and gave it a natural and perfect appearance that could deceive even the best of silversmiths. The natives were also known for the jewelries made of other precious stones such as carnelian, agate, and pearl. Some outstanding examples of Philippine jewelry included necklaces, belts, armlets, and rings placed around the waist. Gold was also bartered, through the Arab world, with merchants all over Asia and Europe in the pre-Islamic and Islamic periods. Many merchants from Luzon (Northern Philippines), Brunei and Jolo frequently traveled all throughout Mindanao in search of slaves and gold. The first commercial mine in the Philippines was in Benguet, in Central Luzon, opened by the Benguet Mining Corporation. The Spanish colonizers took advantage of whatever mineral resources they could get. In fact, gold was the main reason why Spain colonized the Philippines, mainly for their so-called Royal Service. They even instituted a special law, called Inspeccion de Minas, to inspect the existence of minerals in the archipelago. It was the Americans, however, who made strategic steps to exploit the minerals of the Philippines. They performed an extensive geological survey, which confirmed the Philippines as a mineral-rich country, and issued Act 468, a law that essentially gave the government the right to reserve mineral lands for its own purposes. They claimed several areas as "reserved areas" for future mining, hence the commercialization of the Benguet gold mining. In the year 1914 in the south, Surigao and other parts of the Caraga region were declared as an "iron reserved" area for future mining. At that time, the mining industry in the Philippines was on its way to boom and the Commonwealth US government took a much stronger hold of it, forming a Mining Bureau to control all potential operations in the future. Up until 1921, there was no large-scale mining, though many were making a living from small-scale gold mining. Between 1933 and 1941, gold was the dominant and most important mineral in the mining industry. Under the tyranny of the Japanese, Filipinos in many regions of the country were forced into mining for metals to be used for weapons in the Japanese imperial army. This paved the way for further commercialization, exploitation, and degeneration of the Philippines. Large-scale copper mining reached its maximum in the 1960s and 1970s. By the late 1980s, world demand for copper decreased in favor of gold. However, several gold mining companies closed in that period due to law violations and as a result gold mining went into decline. With the support of the World Trade Organization, the International Monetary Fund, and the World Bank, the neo-colonized Philippines was again forced to change its economic policies to follow neo-liberal policies. By 1994, pro-development politicians, such as Gloria Macapagal Arroyo, lobbied for a Mining Bill, which would later become the Republic Act 7942 or the Philippine Mining Act of 1995. This law basically gives power over land, resources, and life to corporations; as a result, many areas became mining hot spots. By 1996, the Philippine mining industry got back on track, permitting offshore companies to operate fully in the "reserved areas"—a disaster for a number of places in the Philippines. In March 1996, the Marcopper tunnel in Marinduque collapsed. It has been estimated that roughly 1.6 million cubic meters of mine tailings flowed from the mine pit to the Makulapnit and Boac river, trapping 4400 people in 20 villages. As a result, the government declared the Boac river officially dead. The disaster caused massive siltation in downstream communities and coastal areas. In 2004, another disaster took place in Surigao Del Norte, Mindanao. This time it was caused by one of the largest and longest-standing mining corporations in the Philippines, the Manila Mining Corporation. Three disastrous incidents happened, where around 5 million cubic meters of waste materials containing high levels of mercury were released, damaging local people's agricultural lands and temporarily poisoning the adjacent Placer Bay. Currently, 20 large-scale mining operations, 10 medium-scale, and more than 2000 nonmetallic small-scale mining operations exist in the Philippines. Yet, hundreds of mining applications are pending to prey on what is left of the country's resources. However, the current Duterte government has been reluctant to allow further mining due to its environmental impact.

2.5.1.7 Modern period

In the early 20th century, the gold and silver rush to the western United States also stimulated mining for coal as well as base metals such as copper, lead, and iron. Areas in modern Montana, Utah, Arizona, and later Alaska turned into leading suppliers of copper to the world, which was increasingly needing copper for electrical and household goods. Canada's mining industry grew more slowly than did the United States' due to limitations in transportation, capital, and US competition. Ontario was the major producer of the early 20th century with nickel, copper, and gold.

Around the same time, Australia experienced the Australian gold rushes and by the 1850s was producing 40% of the world's gold, followed by the establishment of large mines such as the Mount Morgan Mine, which ran for nearly 100 years (a copper, gold, and silver mine in Queensland, Australia. Mining began at Mount Morgan in 1882 and continued until 1981. Over its lifespan, the mine yielded approximately 262 metric tons of gold, 37 metric tons of silver and 387,000 metric tons of copper. The mine was once the largest gold mine in the world.), Broken Hill ore deposit (one of the largest zinc-lead ore deposits) in western New South Wales, and the iron ore mines at Iron Knob on the Eyre Peninsula in South Australia. Charles Sturt made a pencil sketch of the Broken Hill area in 1884 and noted iron ore along an isolated hill. In 1866 the Mount Gipps sheep station named their paddock, which embraced the lode outcrop, Broken Hill. However, the hill was supposed to be mullock. On September 5, 1883, a Mount Gipps boundary rider named Charles Rasp staked a claim on the outcrop because it resembled tin oxide as described in his prospecting guidebook. With six others he staked the entire outcrop. In 1884 the syndicate reorganized as the Broken Hill Mining Company. Horn silver (chlorargyrite, AgCl) was first found in 1885 and the Broken Hill Proprietary Company Limited was organized. Charles Rasp discovered the gossan or weathered sulfide outcrop of massive lead-zinc sulfides on a feature known as Broken Hill. Rasp described finding massive galena (PbS), sphalerite (ZnS), cerussite ($PbCO_3$), and other oxide minerals, but was most interested in the galena, a primary source of lead. His reports, thought to be exaggerated at the time, of masses of lead in the desert, soon was shown to be true and started a "lead rush" similar to the gold rushes. Broken Hill was at first exploited by a small group of prospectors working the gossan for easily won galena, and soon dozens of shafts were sunk. Ore was transported to South Australia by camel trains, wagons, and pack mules. A major secondary source of income became apparent, with extremely high silver grades recovered, including native silver, and other rare silver minerals present in large quantities. Open-pit mining of silver ores was the norm from 1885 to 1898, with local smelting. From 1898 until 1915, the lead-zinc-silver sulfide minerals were mined and crushed with the concentrates treated overseas. From 1915 onwards, the concentrates were treated entirely in Australia. The central part of the lode was exhausted by 1940 and production was concentrated in the north and south ends. Properties in the 1950s included North Broken Hill Limited, Broken Hill South Limited, The Zinc Corporation Limited, and New Broken Hill Consolidated Limited. Through 1946, 63,795,241 tons of ore have produced 8,580,226 tons of lead, 5,194,125 tons of zinc, 538,040,847 ounces of silver, and 164,658 ounces of gold. Mining slowly moved away from the initial small group of prospectors, consistent with the experience in all other major mineral fields, toward gradual consolidation of claims and tenure, an increase in tenure and mine size and efficiencies in operations resulting in smaller workforces. This has accelerated in the last part of the 20th century via the formation of the Broken Hill Proprietary Company—now BHP Billiton—and its exit from Broken Hill, toward only two operators at present, using highly efficient bulk underground mechanized mining. Broken Hill South Limited ceased mining in 1972 and its leases were acquired by Minerals Mining and Metallurgy Limited (MMM). Underground mining stopped in 1976 and open-pit mining is concentrated in the Blackwood Open Cut, which began in 1973, and the Kintore Open Cut. Stope fill, plus remnant supergene silver, and sulfide ore characterize the ore reserve. Pasminco Limited took over the leases of The Zinc Corporation (ZC) Limited and New Broken Hill Consolidated Limited. Their main shafts included ZC, New Broken Hill Consolidated, and Southern Cross. These ZC Mines Leases lie south of the MMM Mining Leases along the Broken Hill "Line of Lode."

The name Iron Knob first appeared on pastoral lease maps of 1854, and the first mineral claim in the area was pegged by the Broken Hill Proprietary Company in 1897. Mining started in 1900 and iron ore was transported by bullock wagon to Port Augusta then by rail to Port Pirie where it was used as a flux in the lead smelter there. In 1901 a tramway from Iron Knob to Hummock Hill (later renamed Whyalla) was completed, followed by wharves in 1903. These permitted the direct loading of ships that could transport the ore across Spencer Gulf to Port Pirie. Iron Knob's iron ore proved to be of such high quality (upwards of 60% purity) that it resulted in the development of the Australian steel industry. It supplied iron to the steelworks established at Newcastle and Port Kembla in the 1910s and 1920s and Whyalla in the 1930s. The iron ore was transported by railway to Whyalla where it was either smelted or dispatched by sea. Twenty-one percent of the steel necessary for the construction of the Sydney Harbour Bridge was quarried at Iron Knob and smelted at Port Kembla, New South Wales. The remaining 79% was imported from England. In the 1920s, iron ore from Iron Knob was exported to the Netherlands and the United States of America. In the 1930s, customers included Germany and the United Kingdom. Prior to World War II, iron ore from Iron Knob was also exported to Japan. In the financial year 1935–36, 291,961 tons of ore from Iron Knob was shipped there via the seaport of Whyalla. This became a controversial matter in the late 1930s due in part to Australia's known reserves at the time being limited to Iron Knob and Yampi Sound in Western Australia. Japan was also thought of as an "aggressor" nation following acts of war against China in 1937. Waterfront workers and seamen protested against the export of iron ore to Japan, leading to strikes and arrests. In 1937, output from the Middleback Range, mostly from Iron Monarch was estimated at 2 mtpa. In 1939, it was referred to in England as the highest-grade deposit of iron ore known in the world. In 1943, the Iron Knob deposit was still producing an average ore grade of 64% metallic content. In 1949, 99% of Australian demand for iron ore was met by supply from Iron Knob and associated mines in South Australia, having risen from 95% in 1943. Additional deposits of iron ore were developed by the Broken Hill Proprietary Company further south along the Middleback Range. These included Iron Baron, Iron Prince, and Iron Queen (discovered in 1920) and Iron Knight, Iron Duchess, and Iron Duke (discovered in 1934). Quarrying for iron at Iron Knob and Iron Monarch ceased in 1998. After the quarrying stopped, the town population decreased to 200 and Iron Knob was in danger of becoming a ghost town. In 2010, OneSteel (now Arrium) declared that it would return to Iron Knob to reopen the Iron Monarch mine. The Iron Monarch mine was prepared for reopening by Arrium in 2013. As of 2015, both Iron Monarch and Iron Duke continue to produce iron ore for export and for smelting at the Whyalla steelworks. Now, in the early 21st century, Australia remains a major world mineral producer.

At the start of the 21st century, a globalized mining industry of large multinational corporations has arisen. Peak minerals and environmental impacts have also become a concern. Different elements, particularly rare earth minerals, have begun to increase in demand as a result of new technologies.

2.5.2 Mine development and life cycle

The process of mining from discovery of an ore body through extraction of minerals and finally to returning the land to its natural state contains several distinct steps. The first is discovery of the ore body, which is carried out through prospecting or exploration to find and then define the size, location, and value of the ore body. This results in a mathematical resource estimation with respect to the size and grade of the deposit. This estimation is used to perform a prefeasibility study to determine the theoretical economics of the ore deposit. This identifies, early on, whether further investment in estimation and engineering studies is reasonable and identifies key risks and areas for further work. The next step is to conduct a feasibility study to evaluate the financial viability, the technical and financial risks, and the robustness of the project. This is when the mining company decides whether to develop the mine or to walk away from the project. This includes mine planning to evaluate the economically recoverable portion of the deposit, the metallurgy and ore recoverability, marketability and profitability of the ore concentrates, engineering concerns, milling and infrastructure costs, finance and equity necessities, and an analysis of the proposed mine from the initial excavation all the way through to reclamation at the end. The proportion of a deposit that is economically recoverable hinges on the enrichment factor of the ore in the area. To access the mineral deposit within an area it is frequently necessary to mine through or remove waste material that is not of immediate interest to the miner. The total movement of ore and waste constitutes the mining process. Generally, more waste than ore is mined during the life of a mine, depending on the nature and location of the ore body. Waste removal and placement are a major cost to the mining operator, so a detailed characterization of the waste material is a vital part of the geological exploration program for a mining operation. When the analysis shows that a given ore body is worth recovering, development begins to create access to the ore body. The mine buildings and processing plants are built, and any required equipment is obtained. The operation of the mine to recover the ore begins and continues as long as the company operating the mine finds it economical to do so. Once all the ore that the mine can produce profitably is recovered, reclamation begins to make the land used by the mine suitable for future use.

2.5.3 Mining techniques

Mining methods can be divided into two common excavation types: surface mining and subsurface (underground) mining. Nowadays, surface mining is much more common and produces, for example, 85% of minerals (excluding petroleum and natural gas) in the United States, including 98% of metallic ores. Targets are divided into two general groups of materials: placer deposits, consisting of valuable minerals contained within river gravels, beach sands, and other unconsolidated materials; and lode deposits, where valuable minerals are found in veins, in layers, or in mineral grains generally distributed throughout a mass of actual rock (see previous section on ore-forming processes). Both types of ore deposit, placer or lode, are mined by both surface and underground methods. Some mining, including much of the rare earth elements and uranium mining, is performed by less-common methods, such as in situ leaching: this method involves digging neither at the surface nor underground. The extraction of the target minerals by this method requires that they are soluble, for example, potash, potassium chloride (KCl), sodium chloride (NaCl), and sodium sulfate (Na_2SO_4), which dissolve in water. Some minerals, such as copper minerals and uranium oxide, necessitate the use of acid or carbonate solutions to dissolve.

2.5.3.1 Surface mining

Surface mining involves removing (stripping) the surface vegetation, dirt, and, if necessary, layers of bedrock in order to reach the buried ore deposits. Surface mining techniques include: open-pit mining, which is the recovery of ore from an open pit in the ground, quarrying, identical to open-pit mining except that it refers to sand, stone, and clay; strip mining, which involves stripping surface layers off to reveal ore/seams underneath; and mountaintop removal, commonly associated with coal mining, which involves taking the top of a mountain off to reach ore deposits at depth. Most (but not all) placer deposits, due to their shallowly buried nature, are mined by surface methods. Finally, landfill mining involves sites where landfills are excavated and processed. Landfill mining has been thought of as a solution to dealing with long-term methane emissions and local pollution.

2.5.3.2 Underground mining

Subsurface mining consists of digging tunnels or shafts into the earth to reach buried ore deposits. Ore, for processing, and waste rock, for disposal, are brought to the surface through the tunnels and shafts. Subsurface mining can be classified by the type of access shafts used, the extraction method, or the technique used to reach the mineral deposit. Drift mining utilizes horizontal access tunnels, slope mining uses diagonally sloping access shafts, and shaft mining utilizes vertical access shafts. Mining in hard and soft rock formations require different techniques. Other techniques include shrinkage stope mining, which is mining upward, creating a sloping underground room, longwall mining, which is grinding a long ore surface underground, and room and pillar mining, which is removing ore from rooms while leaving pillars in place to support the roof of the room. Room and pillar mining often leads to retreat mining, in which supporting pillars are removed as miners retreat, allowing the room to cave in, thereby loosening more ore. Additional subsurface mining methods include hard rock mining, which is mining of hard rock (igneous, metamorphic or sedimentary) materials, borehole mining, drift and fill mining, long hole slope mining, sublevel caving, and block caving.

2.5.3.3 Processing

Once the mineral is extracted, it often has to be processed. The science of extractive metallurgy is a specialized area in the science of metallurgy that studies the extraction of valuable metals from their ores, especially through chemical or mechanical means. Mineral processing (or mineral dressing) is a specialized area in the science of metallurgy that studies the mechanical means of crushing, grinding, and washing that enable the separation (extractive metallurgy) of valuable metals or minerals from their gangue (waste material). Processing of placer ore material generally consists of gravity-dependent techniques of separation, such as sluice boxes. Only minor shaking or washing may be necessary to disaggregate the sands or gravels before processing. Processing of ore from a lode mine, whether it is a surface or subsurface mine, entails that the rock ore has to be crushed and pulverized before extraction of the valuable minerals starts. After the lode ore is crushed, recovery of the valuable minerals is performed through one, or a combination of several, mechanical, and chemical techniques. As most metals are present in ores as oxides or sulfides, the metal has to be reduced to its metallic form. This can be achieved using (thermo)chemical means such as smelting or through electrolytic reduction, as in the case of aluminum.

2.6 Environmental impact

Environmental impacts of mining can be found at local, regional, and global scales through direct and indirect mining practices. Impacts can cause erosion, sinkholes, loss of biodiversity, or the contamination of soil, groundwater, and surface water by the chemicals emitted from mining processes. In addition, these processes can have an effect on the atmosphere from the emissions of carbon, which can impact the quality of human health and biodiversity. Some mining techniques can have such major environmental and public health effects that mining companies in some countries are required to adhere to strict environmental and rehabilitation codes to ensure that the mined areas have returned to their original state.

2.6.1 Erosion

Erosion of exposed hillsides, mine dumps, tailings dams and resultant siltation of drainages, creeks, and rivers can impact strongly on the surrounding areas, a major example being the giant Ok Tedi Mine in Papua New Guinea. In wilderness areas mining may result in destruction and disturbance of ecosystems and habitats, and in farming areas it may disturb or destroy productive grazing and croplands. In urbanized environments mining may result in noise pollution, dust pollution, and visual pollution.

2.6.2 Sinkholes

A sinkhole at or near a mine site is generally the result of the collapse of a mine roof from the extraction of resources, weak overburden or geological discontinuities. The overburden at the mine site can develop cavities in the subsoil or rock, which can be filled up with sand and soil from the overlying strata. These cavities in the overburden have the potential in due course to cave in, resulting in a sinkhole at the surface. This type of sudden failure can result in a large depression at the surface without any warning, which can be a serious hazard to both life and property. Sinkholes at a mine site can be mitigated with the correct design of infrastructure such as mining supports and better construction of walls to act as a barrier around an area prone to sinkholes. Back-filling and grouting can be done to stabilize abandoned underground workings.

2.6.3 Water pollution

Mining can have a harmful influence on both surrounding surface and groundwater. If appropriate precautions are not taken, unnaturally high concentrations of chemicals, for example, arsenic, sulfuric acid, and mercury, can affect a significant area of surface or subsurface water. With large quantities of water used for mine drainage, mine cooling, aqueous extraction, and other mining processes, the chance for these chemicals to contaminate groundwater and surface water increases. Since mining produces large amounts of wastewater, disposal methods are limited due to contaminants within the wastewater. Runoff containing these chemicals can result in severe damage to the surrounding vegetation. The dumping of the runoff in surface waters or in a lot of forests is the worst option. Land storage and refilling of the mine after it has been depleted is better option, provided that no forests need to be cleared for the storage of debris. The contamination of watersheds resulting from the leakage of chemicals also influences the health of the local population. In well-regulated mines, hydrologists and geologists carefully analyze the water in order to be able to take the necessary precautions to prevent any type of water contamination that could be the result of the mine's operations. The minimization of environmental degradation is enforced in America mining practices by federal and state law, by restricting operators to meet standards for the protection of surface and groundwater from contamination. This is best done using nontoxic extraction processes such as bioleaching.

2.6.4 Acid rock drainage

Subsurface mining often progresses below the water table, consequently water must be continually pumped out of the mine in order to stop flooding. After a mine is abandoned and the pumping ceases water will flood the mine. This introduction of water is the first step in most acid rock drainage (ARD) situations. ARD can be found naturally within some environments as part of the rock weathering process, but it is exacerbated by large-scale earth disturbances characteristic of mining and other large construction activities, typically within rocks comprising an abundance of sulfide minerals. In many localities, the liquid that drains from coal stocks, coal handling facilities, coal washeries, and coal waste tips can be highly acidic, and in such cases, it is treated as acid mine drainage (AMD).

After being exposed to air and water, oxidation of metal sulfides (often pyrite, FeS_2) within the surrounding rock and overburden creates acidity. Colonies of bacteria and archaea greatly accelerate the dissolution of metal ions, though the reactions also occur in an abiotic environment. These microbes, known as extremophiles for their ability to survive in harsh conditions, occur naturally, but limited water and oxygen supplies usually restrict their numbers. Special extremophiles known as acidophiles specifically favor the low pH levels of abandoned mines. Especially, *Acidithiobacillus ferrooxidans* is a major contributor to pyrite oxidation. Mines can produce highly acidic discharges when the ore is a sulfide mineral or is associated with pyrite. In these situations, the predominant metal ion may not be iron but rather zinc, copper, or nickel. The most commonly mined ore of copper, chalcopyrite ($CuFeS_2$), occurs with a range of other sulfide minerals. Consequently, copper mines are often major producers of AMD. At some of these mines, acidic drainage is noticed within 2–5 years after mining started, while at other mines, it is not observed for several decades. Furthermore, acidic drainage may be generated for decades or even centuries after it is first noticed. Therefore AMD is thought of as a serious long-term environmental problem associated with mining. The oxidation chemistry of pyrite, the production of ferrous ions and subsequently ferric ions, is very complex, and this complexity has significantly hindered the design of effective treatment methods. Although a host of chemical processes contributes to AMD, pyrite oxidation is undoubtedly the largest contributor. A general equation for this process is:

$$2FeS_2(s) + 7O_2(g) + 2H_2O(l) \rightarrow 2Fe^{2+}(aq) + 4SO_4^{2-}(aq) + 4H^+(aq)$$

The oxidation of the sulfide to sulfate solubilizes the ferrous iron [iron(II)], which is then oxidized to ferric iron [iron(III)]:

$$4Fe^{2+}(aq) + O_2(g) + 4H^+(aq) \rightarrow 4Fe^{3+}(aq) + 2H_2O(l)$$

Either of these reactions can occur spontaneously or can be catalyzed by microorganisms that derive energy from the oxidation reaction. The ferric cations produced can also oxidize additional pyrite and reduce into ferrous ions:

$$FeS_2(s) + 14Fe^{3+}(aq) + 8H_2O(l) \rightarrow 15Fe^{2+}(aq) + 2SO_4^{2-}(aq) + 16H^+(aq)$$

The net effect of these reactions is to release H^+, which lowers the pH and preserves the solubility of the ferric ion. Water temperatures as high as 47°C have been measured underground at the Iron Mountain Mine (also known as the Richmond Mine at Iron Mountain, is a mine near Redding in Northern California, United States. Geologically classified as a "massive sulfide ore deposit," the site was mined for iron, silver, gold, copper, zinc, and pyrite intermittently from the 1860s until 1963), and the pH can be as low as −3.6. Organisms that cause AMD can thrive in waters with pH very close to zero. Negative pH occurs when water evaporates from already acidic pools, thereby increasing the concentration of hydrogen ions even further. When the pH of AMD is increased past 3, either through contact with freshwater or neutralizing minerals, previously soluble iron(III) ions precipitate as iron(III) hydroxide, a yellow-orange solid colloquially known as "yellow boy." Other types of iron precipitates are also possible, including iron oxides and oxyhydroxides. All these precipitates can discolor water and smother plant and animal life on the streambed, disrupting stream ecosystems. The process also produces additional hydrogen ions, which can further decrease pH. In some instances, the concentrations of iron hydroxides in "yellow boy" are so high; the precipitate can be recovered for commercial use in pigments. Many acid rock discharges also carry elevated levels of potentially toxic metals, especially nickel and copper together with lower levels of a range of trace and semimetal ions such as lead, arsenic, aluminum, and manganese. The elevated levels of dissolved heavy metals can only occur in waters that have a low pH, as is found in the acidic waters produced by pyrite oxidation. In the coal belt around the south Wales valleys in the United Kingdom highly acidic nickel-rich discharges from coal stocking sites have proved to be particularly troublesome. The same type of chemical reactions and processes can occur due to the disturbance of acid sulfate soils formed under coastal or estuarine conditions after the last major sea-level rise and constitutes a similar environmental hazard. The five main methods used to monitor and control water flow at mine sites are: (1) diversion systems, (2) containment ponds, (3) groundwater pumping systems, (4) subsurface drainage systems, and (5) subsurface barriers. In the case of AMD, contaminated water is generally pumped to a treatment facility that neutralizes the contaminants.

2.6.5 Treatment

1. Lime neutralization

   Without a doubt, the most frequently used commercial method for treating AMD is lime (CaO) precipitation in a high-density sludge (HDS) process. In this application, a slurry of lime is dispersed into a tank containing AMD and recycled sludge to increase water pH to about 9. At this pH, most toxic metals become insoluble and precipitate, aided by the presence of recycled sludge. Optionally, air may be introduced in this tank to oxidize iron and manganese and assist in their precipitation. The resulting slurry is moved to a sludge-settling vessel, such as a clarifier. In that vessel, clean water will overflow for release, whereas settled metal precipitates (sludge) will be recycled to the AMD treatment tank, with a sludge-wasting side stream. Several variations of this process exist, as dictated by the chemistry of ARD, its volume, and other factors. Generally, the products of the HDS process also contain gypsum ($CaSO_4 \cdot 2H_2O$) and unreacted lime, which enhance both its settleability and resistance to reacidification and metal mobilization. A general equation for this process is:

$$H_2SO_4 + CaO \rightarrow CaSO_4 + H_2O$$

or more precisely in aqueous solution:

$$SO_4^{2-} + 2H^+ + Ca^{2+} + O^{2-}(aq) \rightarrow Ca^{2+} + SO_4^{2-}(aq) + 2H^+ + O^{2-}(aq)$$

   Less complex variants of this technique, such as simple lime neutralization, may involve no more than a lime silo, mixing tank and settling pond. These systems are cheaper to build, but are also less efficient (i.e., longer reaction times are needed, and they produce a discharge with higher trace metal concentrations, if present). They would be suitable for relatively small flows or less complex AMD.

2. Calcium silicate neutralization

   A calcium silicate feedstock, produced from processed steel slag, can also be used to neutralize active acidity in AMD systems by removing free hydrogen ions from the bulk solution, thus increasing the pH. As the silicate anion captures H^+ ions (raising the pH), it forms monosilicic acid (H_4SiO_4), a neutral solute. Monosilicic acid remains in the bulk solution to play many roles in correcting the negative effects of acidic conditions. While its mode-of-action is quite different from limestone, the ability of calcium silicate to neutralize acid solutions is equivalent to limestone as evidenced by its calcium carbonate equivalent value of 90%–100% and its relative neutralizing value of 98%. In the presence of heavy metals, calcium silicate reacts in a different way than limestone. As limestone increases the pH of the bulk solution, and if heavy metals are present, precipitation of the metal hydroxides (with extremely low solubilities) is generally accelerated and the potential of armoring of limestone particles increases significantly. In the calcium silicate aggregate, as silicic acid species are absorbed onto the metal surface, the formation of silica layers (mono- and bi-layers) resulting in colloidal complexes with neutral or negative surface charges. These negatively charged colloids create an electrostatic repulsion with each other (as well as with the negatively charged calcium silicate granules) and the sequestered metal colloids are stabilized and remain in a dispersed state—effectively interrupting metal precipitation and reducing vulnerability of the material to armoring.
3. Carbonate neutralization

   Commonly, limestone or other calcareous strata that could neutralize acid are lacking or deficient at sites that produce acidic rock drainage. Limestone chips may be transported to mine sites to use for their neutralizing effect. Where limestone has been used, such as at Cwm Rheidol in mid-Wales, the positive impact has been much less than anticipated because of the creation of an insoluble calcium sulfate layer on the surface of the limestone chips, binding the material and preventing further neutralization.
4. Ion exchange

   Cation exchange methods have previously been investigated as a potential treatment for AMD. The principle is that an ion exchange resin can remove potentially toxic metals (cationic resins), or chlorides, sulfates, and uranyl sulfate complexes (anionic resins) from mine water. Once the contaminants are adsorbed, the exchange sites on resins must be regenerated, which typically requires acidic and basic reagents and generates a brine that contains the pollutants in a concentrated form.
5. Constructed wetlands

   Constructed wetlands systems have been suggested during the 1980s to treat AMD generated by the abandoned coal mines in Eastern Appalachia in the United States. Usually, the wetlands receive near-neutral water, after it has been neutralized by (typically) a limestone-based treatment process. Metal precipitation occurs from their oxidation at near-neutral pH, complexation with organic matter, and precipitation as carbonates or sulfides. The latter is caused by sediment-borne anaerobic bacteria capable of reverting sulfate ions into sulfide ions. These sulfide ions can in turn bind with heavy metal ions, precipitating heavy metals out of solution and efficiently reversing the entire process. The attractiveness of a constructed wetlands solution is due to its relative low cost. However, they are limited by the metal loads they can handle (either from high flows or metal concentrations); nevertheless, current practitioners have been successful in developing constructed wetlands that treat high volumes and/or highly acidic water (with suitable pretreatment). Typically, the effluent from constructed wetland receiving near-neutral water will be well buffered at between 6.5 and 7.0 and can readily be discharged. Nevertheless, some of metal precipitates retained in sediments can be unstable when exposed to oxygen (e.g., copper sulfide or elemental selenium), and it is therefore extremely important that the wetland sediments remain largely or permanently submerged. An example of an effective constructed wetland is on the Afon Pelena in the River Afan valley above Port Talbot (Wales, United Kingdom) where highly ferruginous discharges from the Whitworth mine have been successfully treated.
6. Precipitation of metal sulfides

   Most base metals in acidic solution form a precipitate in contact with free sulfide, for example, from H_2S or NaHS. Solid–liquid separation after reaction would produce a base metal–free effluent that can be discharged or further treated to reduce sulfate, and a metal sulfide concentrate with possible economic value. As an alternative, the precipitation of metals using biogenic sulfide has been studied. In this method, sulfate-reducing bacteria oxidize organic matter using sulfate, instead of oxygen. Their metabolic products include bicarbonate, which can neutralize water acidity, and hydrogen sulfide, which forms highly insoluble precipitates with many toxic metals. Although showing promise, this process has been slow in being adopted for a variety of technical reasons.

2.6.6 Effect on biodiversity

The opening of a mine is a major habitat modification, and smaller perturbations occur on a larger scale than the exploitation site itself, for example, mine-waste residuals contamination of the environment. Negative effects can be found long after the mining activities stopped. Destruction or drastic modification of the original mine site and anthropogenic substances release can have a major impact on biodiversity in the area around the mine site. Destruction of the local habitat is the principal component of biodiversity losses, but direct poisoning caused by mine-extracted material, and indirect poisoning through food and water, can also have a negative influence on animals, vegetation, and microorganisms. Habitat modifications such as pH and temperature modification disturb communities in the surrounding area. Endemic species are particularly sensitive, since they need very specific environmental conditions. Destruction or even just slight modification of their habitat can put them at the risk of extinction. Concentrations of heavy metals are known to decrease with distance from the mine, and the influence on biodiversity has a tendency to follow the same pattern. Impacts can vary greatly depending on mobility and bioavailability of the contaminant: less-mobile molecules will stay inert in the environment while highly mobile molecules will easily move into another compartment or be taken up by organisms. For example, speciation of metals in sediments could modify their bioavailability, and thus their toxicity for aquatic organisms. Biomagnification plays an important role in polluted habitats: mining impacts on biodiversity, if concentration levels are not high enough to directly kill exposed organisms, should be greater to the species on top of the food chain because of this phenomenon. Negative mining effects on biodiversity is to a large extent subject to the nature of the contaminant, the concentration level at which it can be found in the environment, and the nature of the ecosystem itself. Some species are rather resistant to anthropogenic disturbances, whereas some others will completely disappear from the contaminated zone. Time on its own does not seem to allow the habitat to recover completely from the contamination. Remediation practices take time and in most situations will not result in the recovery of the original biodiversity as found before the mining activity.

2.6.6.1 Aquatic organisms

The mining industry can affect aquatic biodiversity in various ways. One way can be direct poisoning; a higher risk for this is present when contaminants are mobile in the sediment or bioavailable in the water. Mine drainage can modify water pH, making it difficult to distinguish direct impact on organisms from impacts caused by pH changes. Effects can however be observed and proven to be due to pH modifications. Contaminants can also affect aquatic organisms through physical effects: streams with high concentrations of suspended sediment can limit light penetration, consequently decreasing algae biomass. Metal oxide deposition can limit biomass by coating algae or their substrate, thus preventing colonization. Issues that affect communities in AMD sites can vary temporarily and seasonally: temperature, rainfall, pH, salinization, and metal quantity all show variations on the long term and can heavily affect communities. Changes in pH or temperature can affect metal solubility, and thereby the bioavailable quantity that directly impacts organisms. Moreover, contamination persists over time: 90 years after a pyrite mine closure, water pH was still very low and microorganism populations consisted mainly of acidophil bacteria. Algae communities show less diversity in acidic water containing high zinc concentrations, and mine drainage stress reduces their primary production. Diatom communities are strongly modified by any chemical change, pH, phytoplankton assemblage, while high-metal concentration decreases the abundance of planktonic species. Some diatom species may grow in high metal–concentration sediments. In sediments close to the surface, cysts suffer from corrosion and heavy coating. Under very polluted conditions of the water, total algae biomass is quite low, and the planktonic diatom community is absent. Water insect and crustacean communities change around a mine, resulting in a low trophic completeness and their community being dominated by predators. Though biodiversity of macroinvertebrates can remain high, provided that sensitive species are replaced with tolerant ones. When diversity within the area is diminished, while there is sometimes no effect of stream contamination on abundance or biomass, it indicates that tolerant species are fulfilling the same function replacing sensitive species in polluted sites. pH decrease in addition to elevated metal concentration can also have adverse effects on macroinvertebrates' behavior, showing that direct toxicity is not the only issue. Fish can also be affected by pH, temperature variations, and chemical concentrations.

2.6.6.2 Terrestrial organisms

Soil texture and water content can be strongly modified in disturbed sites, resulting in plant community changes in the surrounding area. Most of the plants have a low concentration tolerance for metals in the soil, though sensitivity differs between species. Grass diversity and total coverage are less affected by high contaminant concentration than forbs [an herbaceous flowering plant that is not a graminoid (grasses, sedges, and rushes). The term is used in biology and in vegetation ecology, especially in relation to grasslands and understory] and shrubs. Mine-waste materials or trace elements due to mining activity can be found near the mine and sometimes further away from the source. Established plants cannot move away from disturbed areas and will sooner or later die if their habitat is contaminated by heavy metals or metalloids at a concentration that is too high for their physiology. Some species are more resistant and will survive these contaminant levels, and some nonnative species that can tolerate these concentrations in the soil will migrate in the surrounding areas of the mine to occupy the ecological niche. Plants can be affected through direct poisoning, for example, arsenic soil content decreases bryophyte [an informal group consisting of three divisions of nonvascular land plants (embryophytes): the liverworts, hornworts, and mosses] diversity. Soil acidification through pH diminution by chemical contamination can also result in a lower species number. Contaminants can modify or disturb microorganisms, thus changing nutrient availability and resulting in a loss of vegetation in the area. Some tree roots divert away from deeper soil layers in order to avoid the contaminated zone, consequently lacking anchorage within the deep soil layers, resulting in the potential uprooting by the wind when their height and shoot weight increase. Overall, root exploration is diminished in contaminated areas compared to nonpolluted ones. Plant species diversity will remain lower in reclaimed habitats than in undisturbed areas. Cultivated crops might be a problem near mines. Most crops can grow on weakly contaminated sites, but their yield is usually lower than it would have been under regular growing conditions. Plants also have a tendency to accumulate heavy metals in their aerian organs, possibly entering the human food supply in fruits and vegetables. Regular consumption of contaminated crops might result in health problems caused by long-term metal exposure.

Habitat destruction is one of the major problems of mining activity. Large areas of natural habitat are destroyed during mine construction and exploitation, forcing animals to leave the site. Animals can be poisoned directly by mine products and residuals. Bioaccumulation in the plants or the smaller organisms they eat can also result in metal poisoning, for example, horses, goats, and sheep are exposed in certain areas to potentially toxic concentration of copper and lead in grass. There are fewer ant species in soil containing high copper concentrations, in the vicinity of a copper mine. If fewer ants are found, chances are higher that other organisms living in the surrounding landscape are strongly affected by the high copper levels as well. Microorganisms are extremely sensitive to environmental changes, such as modified pH, temperature changes, or chemical concentrations because of their size. For example, the presence of arsenic and antimony in soils has led to a reduction in total soil bacteria. Like waters' sensitivity, a small change in the soil pH can result in the remobilization of contaminants, in addition to the direct effect on pH-sensitive organisms. Microorganisms have a wide variety of genes among their total population, so there is a better chance of survival of the species due to the resistance or tolerance genes that some colonies have, provided that the modifications are not too extreme. Yet, survival in these conditions will lead to a big loss of gene diversity, resulting in a decreased potential for adaptations to subsequent changes. Undeveloped soil in heavy metal contaminated areas could be a sign of diminished activity by microfauna and microflora in the soil, pointing to a decreased number of individuals or reduced activity. It has been shown for certain mining areas that 20 years after disturbance, even in a rehabilitation area, microbial biomass is still significantly diminished in comparison to the original undisturbed habitat. Arbuscular mycorrhizal fungi [a type of mycorrhiza (a symbiotic association between a fungus and a plant) in which the symbiont fungus penetrates the cortical cells of the roots of a vascular plant forming arbuscules] are especially sensitive to the presence of chemicals, and the soil is occasionally so disturbed that they are no longer able to associate with root plants. However, some fungi possess contaminant accumulation capacity and soil cleaning ability by changing the bioavailability of pollutants, and this can protect plants from possible damages that could be the result of these chemicals. Their occurrence in contaminated sites could avert loss of biodiversity due to mine-waste contamination or allow for bioremediation, the removal of undesired chemicals from contaminated soils. In contrast, some microbes can worsen the environment, which can result in elevated sulfate levels in the water, and can also increase microbial production of hydrogen sulfide, a toxin for many aquatic plants and organisms.

2.6.7 Waste

Mining processes produce an excess of waste materials known as tailings. These materials are left over after separating the valuable fraction from the uneconomic fraction of the ore. These large amounts of waste are a mixture of water, sand, clay, crushed gangue minerals, or residual bitumen. Tailings are usually stored in tailings ponds created in naturally existing valleys or in large engineered dams and dike systems. Tailings ponds can remain part of an active mine operation for 30–40 years. This lets tailings deposits settle or allows for storage and water recycling. Tailings have enormous potential to damage the environment by releasing toxic metals by AMD or dam/dike failure, thereby damaging aquatic wildlife. They necessitate constant monitoring and treatment of water passing through the dam. Tailings ponds are typically formed by locally derived fills (soil, coarse waste, or overburden from mining operations and tailings), and the dam walls are often built up on to withstand greater amounts of tailings. The lack of regulation for design criteria of the tailing ponds is what put the environment at risk for flooding from the tailings ponds. A spoil tip is a pile of accumulated overburden that was removed from a mine site during the extraction of coal or ore. These waste materials are composed of ordinary soil and rocks, with the potential to be contaminated with chemical waste. Spoil is much different from tailings, as it is processed material that remains after the valuable components have been extracted from ore.

2.6.8 Effects of mine pollution on humans

Humans are also affected by mining. There are many diseases that can come from the pollutants that are released into the air and water during the mining process. For example, during smelting operations large quantities of air pollutants, such as the suspended particulate matter, SO_x, arsenic particles, and cadmium, are emitted. Metals are usually emitted into the air as particulates as well. There are also many occupational health hazards that miners face. Most miners suffer from various respiratory and skin diseases such as asbestosis, silicosis, or black lung disease.

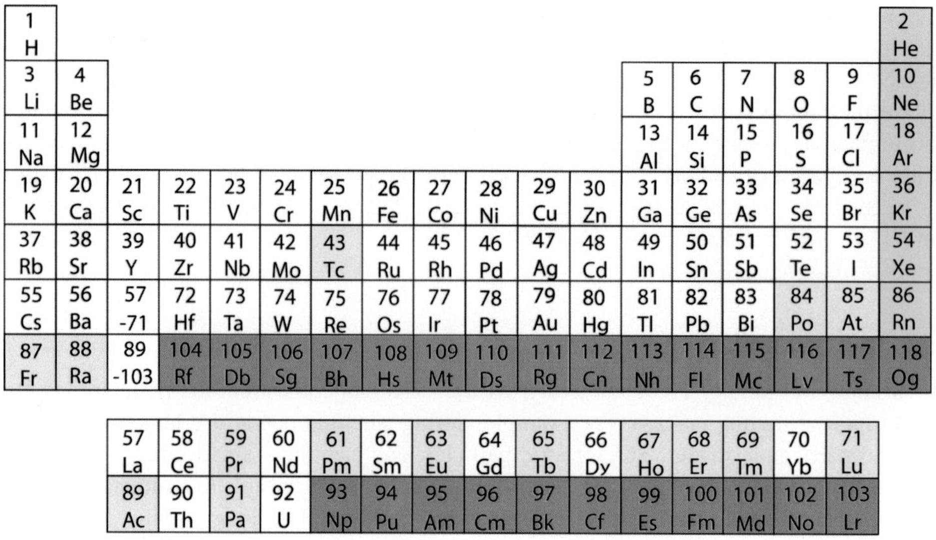

1 H																	2 He
3 Li	4 Be											5 B	6 C	7 N	8 O	9 F	10 Ne
11 Na	12 Mg											13 Al	14 Si	15 P	16 S	17 Cl	18 Ar
19 K	20 Ca	21 Sc	22 Ti	23 V	24 Cr	25 Mn	26 Fe	27 Co	28 Ni	29 Cu	30 Zn	31 Ga	32 Ge	33 As	34 Se	35 Br	36 Kr
37 Rb	38 Sr	39 Y	40 Zr	41 Nb	42 Mo	43 Tc	44 Ru	45 Rh	46 Pd	47 Ag	48 Cd	49 In	50 Sn	51 Sb	52 Te	53 I	54 Xe
55 Cs	56 Ba	57 -71	72 Hf	73 Ta	74 W	75 Re	76 Os	77 Ir	78 Pt	79 Au	80 Hg	81 Tl	82 Pb	83 Bi	84 Po	85 At	86 Rn
87 Fr	88 Ra	89 -103	104 Rf	105 Db	106 Sg	107 Bh	108 Hs	109 Mt	110 Ds	111 Rg	112 Cn	113 Nh	114 Fl	115 Mc	116 Lv	117 Ts	118 Og

57 La	58 Ce	59 Pr	60 Nd	61 Pm	62 Sm	63 Eu	64 Gd	65 Tb	66 Dy	67 Ho	68 Er	69 Tm	70 Yb	71 Lu
89 Ac	90 Th	91 Pa	92 U	93 Np	94 Pu	95 Am	96 Cm	97 Bk	98 Cf	99 Es	100 Fm	101 Md	102 No	103 Lr

FIGURE A.1 Period table of the elements. White—elements that form minerals; green—noble gases; yellow—only present in very low concentration on Earth; orange-red—only known from laboratories and nuclear reactors.

A.1 Elements not included (not occurring as minerals or only existing as synthetic elements)

For several reasons some of the 118 known elements are not included in this book as they do not occur as building blocks in minerals. These include noble gases, which do not react with any other element, that is, are inert (green in the periodic table), are so rare that there is not enough naturally occurring that it can form a mineral (yellow), or are only known from laboratories and nuclear reactors (orange-red) (Fig. A.1).

A.2 What is described for each element?

A.2.1 Discovery

This section gives a short history of the discovery of each element. When, where, and how was it discovered and by whom (as far as that information is available)?

A.2.2 Mining, production, and major minerals

This section describes the mining processes, production of the element, and major minerals that contain the element. Some elements form only a few minerals, while other elements are so common that they appear in many minerals. In the latter case a selection will be presented that can be considered as characteristic.

A.2.3 Chemistry

This section describes first the isotopes of the element followed by general properties of the element. This is followed by a general description of the chemistry of each element in terms of reactions with e.g. oxygen, sulfur, halides, etc. The section finishes with the organic chemistry of the element.

A.2.4 Major uses

This section describes how the element is used in our current economy.

References

Barbalace, K. (1995–2019). Periodic table of elements. <https://environmentalchemistry.com/yogi/periodic/> Accessed 27.06.19.

Darling, D. (2016). Terrestrial abundance of elements. <http://www.daviddarling.info/encyclopedia/E/elterr.html> Accessed 27.06.19.

Israel Science and Technology Directory. (1999–2018). Chemistry: List of periodic table elements sorted by: Abundance in earth's crust. <https://www.science.co.il/elements/?s = Earth> Accessed 27.06.19.

Kloprogge, J.T. & Wood, B.J. (2020). *Handbook of Mineral Spectroscopy. Volume 1 X-ray Photoelectron Spectra*. Amsterdam: Elsevier.

Lindgren, W. (1933). *Mineral deposits*. New York: McGraw-Hill.

Thomas Jefferson National Accelerator Facility – Office of Science Education. It's elemental – The periodic table of elements. <https://education.jlab.org/itselemental/index.html> Accessed 27.06.19.

WebElements. (1993–2019). Abundance in earth's crust (by weight). <https://www.webelements.com/periodicity/abund_crust/> Accessed 27.06.19.

Further reading

Adams, F., Grijbels, R., & Grieken, R. V. (1988). *Inorganic mass spectrometry*. New York: Wiley.

Agricola, G. (1556). De re metallica; tr. from the 1st Latin ed. of 1556, with biographical introduction, annotations and appendices upon the development of mining methods, metallurgical processes, geology, mineralogy & mining law, from the earliest times to the 16th century (trans: Hoover, H., Hoover, L.H.), 1912, London: Mining Magazine.

Agricola, G. (1546). De Natura Fossilium (Textbook of mineralogy): Translated from the First Latin Ed. of 1546 by Mark Chance Bandy and Jean A. Bandy for the Mineralogical Society of America (trans: Bandy, M.C., Bandy, J.A.), 1955 (1546), Boulder: Geological Society of America.

Anderson, D. L. (1989). *Theory of the earth*. Boston, MA: Blackwell Scientific Publications.

Bachmann, K., Frenzel, M., Krause, J., & Gutzmer, J. (2017). Advanced identification and quantification of in-bearing minerals by scanning electron microscope-based image analysis. *Microscopy and Microanalysis*, *23*, 527–537.

Barshick, C. M., Duckworth, D. C., & Smith, D. H. (2000). *Inorganic mass spectrometry: Fundamentals and applications*. New York: Dekker.

Becker, S. (2008). *Inorganic mass spectrometry: Principles and applications*. Chichester: John Wiley & Sons.

Beckhoff, B., Kanngießer, B., Langhoff, N., Wedell, R., & Wolff, H. (2007). *Handbook of practical X-ray fluorescence analysis*. Heidelberg: Springer.

Bless, D., Park, B., Nordwick, S., Zaluski, M., Joyce, H., Hiebert, R., & Clavelot, C. (2008). Operational lessons learned during bioreactor demonstrations for acid rock drainage treatment. *Mine Water and the Environment*, *27*, 241–250.

Blodau, C. (2006). A review of acidity generation and consumption in acidic coal mine lakes and their watersheds. *Science of the Total Environment*, *369*, 307–332.

Boumans, P. W. J. M. (1987). *Inductively coupled plasma emission spectroscopy, . Methodology, instrumentation, and performance, part 1* (Vol. 1). New York: Wiley.

Broekaert, J. A. C. (2006). *Analytical atomic spectrometry with flames and plasmas*. Weinheim: Wiley-VCH.

Cabri, L. J., & Vaughan, D. J. (1998). *Modern approaches to ore and environmental mineralogy, . Mineralogical Society of Canada Short Course Series* (Vol. 27). Ottawa: Mineralogical Society of Canada.

Chang, L. L. Y. (2002). *Industrial mineralogy. Materials, processes, and uses*. Upper Saddle River: Prentice Hall.

Craddock, P. T. (1996). The use of firesetting in the granite quarries of South India. *The bulletin of the Peak District Mines Historical Society*, *13*.

Craig, J. R., Vaughan, D. J., & Skinner, B. J. (2001). *Resources of the earth: Origin, use, and environmental impact* (3rd ed.). Upper Saddle River: Prentice Hall.

De Viguerie, L., Sole, V. A., & Walter, P. (2009). Multilayers quantitative X-ray fluorescence analysis applied to easel paintings. *Analytical and Bioanalytical Chemistry*, *395*, 2015–2020.

DeKosky, R. K. (1973). Spectroscopy and the elements in the late nineteenth century: The work of Sir William Crookes. *The British Journal for the History of Science*, *6*, 400–423.

del Pilar Ortega-Larrocea, M., Xoconostle-Cazares, B., Maldonado-Mendoza, I. E., Carrillo-Gonzalez, R., Hernandez-Hernandez, J., Dıaz Garduno, M., & Chavez, M. (2010). Plant and fungal biodiversity from metal mine wastes under remediation at Zimapan, Hidalgo, Mexico. *Environmental Pollution*, *158*, 1922–1931.

Dill, H. G. (2010). The "chessboard" classification scheme of mineral deposits: Mineralogy and geology from aluminum to zirconium. *Earth-Science Reviews*, *100*, 1–420.

Drexel, J. F. *Mining in South Australia: A pictorial history*, 1982, S. Aust. Department of Mines and Energy, Adelaide.

Ek, A. S., & Renberg, I. (2001). Heavy metal pollution and lake acidity changes caused by one thousand years of copper mining at Falun, central Sweden. *Journal of Paleolimnology*, *26*, 89–107.

Endlich, F. M. (1888). On some interesting derivations of mineral names. *The American Naturalist*, *22*, 21–32.

Erickson, R. L. Crustal abundance of elements, and mineral reserves and resources. In *US Geological Survey Professional Paper 820, 1973*, (pp. 21–25).

Evans, A. M. (2009). *Ore geology and industrial minerals: An introduction* (3rd ed.). Malden: Blackwell Publishing.

Fortey, R. A. (2004). *The earth: An intimate history*. London: Harper Collins.

Fowles, G. R. (1989). *Introduction to modern optics*. New York: Dover Publications.

Franks, D. M., Boger, D. V., Côte, C. M., & Mulligan, D. R. (2011). Sustainable development principles for the disposal of mining and mineral processing wastes. *Resources Policy*, *36*, 114–122.

Gerhardt, A., Janssens de Bisthoven, L., & Soares, A. M. V. M. (2004). Macroinvertebrate response to acid mine drainage: Community metrics and on-line behavioural toxicity bioassay. *Environmental Pollution*, *130*, 263–274.

Groat, L. A. (2007). *Geology of gem deposits, . Mineralogical Society of Canada Short Course Series* (Vol. 37). Quebec: Mineralogical Society of Canada.

Groves, D. I., & Bierlein, F. P. (2007). Geodynamic settings of mineral deposit systems. *Journal of the Geological Society, 164*, 19–30.

Guilbert, J. M., & Park, C. F. (2007). *The Geology of ore deposits*. Long Grove: Waveland Press.

Gusek, J. J., Wildeman, T. R., Conroy, K. W. Conceptual methods for recovering metal resources from passive treatment systems. In *Paper presented at the proceedings of the 7th international conference on acid rock drainage (ICARD)*, March 26–30, 2006, St. Louis, MO.

Gustafson, J. K., Burrell, H. C., Garretty, M. D. *Geology of the Broken Hill ore deposit, Broken Hill, NSW* (Vol. Bulletin No. 20). 1952, Bureau of Mineral Resources, Geology and Geophysics, Dept. of National Dev., Commonwealth of Australia.

Hammarstrom, J. M., Sibrell, P. L., & Belkin, H. E. (2003). Characterization of limestone reacted with acid-mine drainage. *Applied Geochemistry, 18*, 1710–1714.

Hartman, H. L. (1992). *SME mining engineering handbook* (Vol. 1, pp. 3–42). Society for Mining, Metallurgy, and Exploration, Littleton.

Heiss, A. G., & Oeggl, K. (2008). Analysis of the fuel wood used in Late Bronze Age and Early Iron Age copper mining sites of the Schwaz and Brixlegg area (Tyrol, Austria). *Vegetation History and Archaeobotany, 17*, 211–221.

Hill, S. J. (2006). *Inductively coupled plasma spectrometry and its applications*. Oxford: Blackwell.

Hoostal, M. J., Bidart-Bouzat, M. G., & Bouzat, J. L. (2008). Local adaptation of microbial communities to heavy metal stress in polluted sediments of Lake Erie. *FEMS Microbiology Ecology, 65*, 156–168.

Hüfner, S. (1995). *Photoelectron spectroscopy: Principles and applications*. Berlin: Springer Verlag.

Jambor, J. L., Blowes, D. W., & Ritchie, A. I. M. (2003). *Environmental aspects of mine wastes, . Mineralogical Society of Canada Short Course Series* (Vol. 31). Ottawa: Mineralogical Society of Canada.

Jenkins, R., Gould, R. W., & Gedcke, D. (1995). *Quantitative X-ray spectrometry*. New York: Marcel Dekker.

Jung, M. C., & Thornton, I. (1996). Heavy metals contamination of soils and plants in the vicinity of a lead-zinc mine, Korea. *Applied Geochemistry, 11*, 53–59.

Kanazawa, Y., & Kamitani, M. (2006). Rare earth minerals and resources in the world. *Journal of Alloys and Compounds, 408–412*, 1339–1343.

Kesler, S. E. (1994). *Mineral resources, economics and the environment*. New York: Macmillan College Publishing Company, Inc.

Kimura, S., Bryan, C. G., Hallberg, K. B., & Johnson, D. B. (2011). Biodiversity and geochemistry of an extremely acidic, low-temperature subterranean environment sustained by chemolithotrophy. *Environmental Microbiology, 13*, 2092–2104.

Klein, C., Dutrow, B. (2008). *The 23rd edition of the manual of mineral science* (after James D. Dana), Hoboken: John Wiley & Sons, Inc.

Kogel, J. E., Trivedi, N. C., Barker, J. M., & Krukowski, S. T. (2006). *Industrial minerals & rocks: Commodities, markets, and uses*. Littleton: Society for Mining, Metallurgy, and Exploration.

Krebs, R. E. (2006). *The history and use of our earth's chemical elements: A reference guide*. Westport: Greenwood Press.

Krook, J., Svensson, N., & Eklund, M. (2012). Landfill mining: A critical review of two decades of research. *Waste Management, 32*, 513–520.

Langmuir, C. H., & Broecker, W. (2012). *How to build a habitable planet: The story of earth from the big bang to humankind*. Princeton, NJ: Princeton University Press.

Lankton, L. D. (1993). *Cradle to grave: Life, work, and death at the Lake Superior copper mines*. New York: Oxford University Press.

Letterman, R., & Mitsch, W. (1978). Impact of mine drainage on a mountain stream in Pennsylvania. *Environmental Pollution, 17*, 53–73.

Lynch, M. (2003). *Mining in world history*. London: Reaktion Books.

Mackenzie, D. H., & Davies, R. H. (1990). Broken Hill lead-silver-zinc deposit at Z.C. mines. In F. E. Hughes (Ed.), *Geology of the mineral deposits of Australia and Papua New Guinea* (pp. 1079–1084). Melbourne: The Australasian Institute of Mining and Metallurgy.

Malmqvist, B., & Hoffsten, P.-O. (1999). Influence of drainage from old mine deposits on benthic macroinvertebrate communities in central Swedish streams. *Water Research, 33*, 2415–2423.

Michelutti, N., Laing, T. E., & Smol, J. P. (2001). Diatom assessment of past environmental changes in lakes located near the Noril'sk (Siberia) Smelters. *Water, Air, & Soil Pollution, 125*, 231–241.

Mielke, R. E., Pace, D. L., Porter, T., & Southam, G. (2003). A critical stage in the formation of acid mine drainage: Colonization of pyrite by *Acidithiobacillus ferrooxidans* under pH-neutral conditions. *Geobiology, 1*, 81–90.

Morgan, J. W., & Anders, E. (1980). Chemical composition of Earth, Venus, and Mercury. *Proceedings of the National Academy of Sciences of the United States of America, 77*, 6973–6977.

Mummey, D. L., Stahl, P. D., & Buyer, J. S. (2002). Soil microbiological properties 20 years after surface mine reclamation: Spatial analysis of reclaimed and undisturbed sites. *Soil Biology and Chemistry, 34*, 1717–1725.

Nekrasov, I. Y. *Geochemistry, mineralogy and genesis of gold deposits* (trans: Rao, P.M.), 1996, Rotterdam: A.A. Balkema.

Nesse, W. D. (2000). *Introduction to mineralogy*. New York: Oxford University Press.

Niyogi, D. K., William, M. L. J., & McKnight, D. M. (2002). Effects of stress from mine drainage on diversity, biomass, and function of primary producers in mountain streams. *Ecosystems, 6*, 554–567.

Nordstrom, D. K., Alpers, C. N., Ptacek, C. J., & Blowes, D. W. (2000). Negative pH and extremely acidic mine waters from Iron Mountain, California. *Environmental Science & Technology, 34*, 254–258.

Perkins, D. (2002). *Mineralogy* (2nd ed.). Upper Saddle River: Prentice Hall.

Piestrzynski, A. (2001). *Mineral deposits at the beginning of the 21st century*. Lisse: A.A. Balkema Publishers.

Pohl, W. L. (2011). *Economic geology of metals. Economic geology principles and practice* (pp. 149–284). Chichester: John Wiley & Sons.

Putnis, A. (2003). *Introduction to mineral sciences*. Cambridge: Cambridge University Press.

Pyatt, F. B., Gilmore, G., Grattan, J. P., Hunt, C. O., & McLaren, S. (2000). An imperial legacy? An exploration of the environmental impact of ancient metal mining and smelting in Southern Jordan. *Journal of Archaeological Science*, *27*, 771–778.

Rasmussen, K., & Lindegaard, C. (1988). Effects of iron compounds on macroinvertebrate communities in a Danish Lowland River System. *Water Research*, *22*, 1101–1108.

Reed, S. J. B. (1993). *Electron microprobe analysis* (2nd ed.). Cambridge: Cambridge University Press.

Reilly, M. (2007). The last place on earth to preserve a piece of Earth's original crust. *New Scientist*, *194*, 38–39.

Rickard, T. A. (1932). *A history of American mining*. New York: McGraw-Hill.

Rico, M., Benito, G., & Díez-Herrero, A. (2008). Floods from tailings dam failures. *Journal of Hazardous Materials*, *154*, 79–87.

Robb, L. (2005). *Introduction to ore-forming processes*. Malden: Blackwell Publishing.

Rösner, T., & van Schalkwyk, A. (2000). The environmental impact gold mine tailings footprints in the Johannesburg region, South Africa. *Bulletin of Engineering Geology and the Environment*, *59*, 137–148.

Rudenberg, H. G., & Rudenberg, P. G. (2010). Origin and background of the invention of the electron microscope. *Advances in Imaging and Electron Physics*, *160*, 207–286.

Salonen, V.-P. S., Tuovinen, N., & Valpola, S. (2006). History of mine drainage impact on Lake Orijärvi algal communities, SW Finland. *Journal of Paleolimnology*, *35*, 289–303.

Shaw, I. (2003). *The Oxford history of ancient Egypt*. Oxford: Oxford University Press.

Siegbahn, K. (1981). Electron spectroscopy for atoms, molecules and condensed matter: Nobel lecture. November 8, 1981.

Siegbahn, K., & Edvarson, K. (1956). β-Ray spectroscopy in the precision range of 1:10^5. *Nuclear Physics*, *1*, 137–159.

Singh, K. B. (1997). Sinkhole subsidence due to mining. *Geotechnical & Geological Engineering*, *15*, 327–341.

Sonter, L. J., Ali, A. H., & Watson, J. E. M. (2018). Mining and biodiversity: Key issues and research needs in conservation science. *Proceedings of the Royal Society B: Biological Sciences*, *285*.

Steinhauser, G., Adlassnig, W., Lendl, T., Peroutka, M., Weidinger, M., Lichtscheidl, I. K., & Bichler, M. (2009). Metalloid contaminated microhabitats and their biodiversity at a former antimony mining site in Schlaining, Austria. *Open Environmental Sciences*, *3*, 26–41.

Suess, H., & Urey, H. (1956). Abundances of the elements. *Reviews of Modern Physics*, *28*, 53–74.

Tarras-Wahlberga, N. H., Flachier, A., Lanec, S. N., & Sangforsd, O. (2001). Environmental impacts and metal exposure of aquatic ecosystems in rivers contaminated by small scale gold mining: The Puyango River basin, southern Ecuador. *Science of the Total Environment*, *278*, 239–261.

Taylor, S. R., & McLennan, S. M. (1985). *The continental crust: Its composition and evolution*. Oxford: Blackwell Scientific Publications.

Turekian, K. K., & Wedepohl, K. H. (1961). Distribution of the elements in some major units of the Earth's crust. *Geological Society of America Bulletin*, *72*, 175–192.

Ure, A., Karmarsch, K. (1843). *Technisches wörterbuch oder Handbuch der Gewerbskunde ...: Bearb. nach Dr. Andrew Ure's Dictionary of arts, manufactures and mines* (Vol. 1), Prague: G. Haase.

Van der Heide, P. (2011). *X-ray photoelectron spectroscopy: An introduction to principles and practices*. Hoboken: Wiley-Blackwell.

van der Heyden, A., & Edgecombe, D. R. (1990). Silver-lead-zinc deposit at south mine, Broken Hill. In F. E. Hughes (Ed.), *Geology of the mineral deposits of Australia and Papua New Guinea* (pp. 1073–1077). Melbourne: The Australasian Institute of Mining and Metallurgy.

Van Grieken, R. E., & Markowicz, A. A. (2001). *Handbook of X-ray spectrometry*. New York: Marcel Dekker.

von Knorring, O., & Condliffe, E. (1987). Mineralized pegmatites in Africa. *Geological Journal*, *22*, 253–270.

Wedepohl, H. K. (1995). The composition of the continental crust. *Geochimica et Cosmochimica Acta*, *59*, 1217–1232.

Wenk, H.-R., & Bulakh, A. (2012). *Minerals: Their constitution and origin* (6th ed.). Cambridge: Cambridge University Press.

West, G. A. (1980). *Copper: Its mining and use by the aborigines of the Lake Superior Region – Report of the McDonald-Massee Isle Royale Expedition 1928*. Westport: Greenwood Press.

Wise, M. A. (1995). Trace element chemistry of lithium-rich micas from rare-element granitic pegmatites. *Mineralogy and Petrology*, *55*, 203–215.

Wong, H. K. T., Gauthier, A., & Nriagu, J. O. (1999). Dispersion and toxicity of metals from abandoned gold mine tailings at Goldenville, Nova Scotia, Canada. *Science of the Total Environment*, *228*, 35–47.

Young, O. E. (1965). The Spanish tradition in gold and silver mining. *Arizona and the West*, *7*, 299–314.

Zheng, Z., & Lin, C. (1996). The behaviour of rare-earth elements (REE) during weathering of granites in southern Guangxi, China. *Chinese Journal of Geochemistry*, *15*, 344–352.

Zuckerman, B., & Malkan, M. A. (1996). *The origin and evolution of the universe*. Boston, MA: Jones and Bartlett Publishers.

Chapter 3

Periods 1 and 2

3.1 1 H — hydrogen

3.1.1 Discovery

In 1671, Robert Boyle (January 25, 1627–December 31, 1691, Anglo-Irish natural philosopher, chemist, physicist, and inventor) discovered and described the reaction between iron filings and dilute acids, that produced hydrogen gas (Boyle, 1672) (Fig. 3.1). In 1766, Henry Cavendish (October 10, 1731–February 24, 1810, English natural philosopher, scientist, and an important experimental and theoretical chemist and physicist) was the first to identify hydrogen gas as a separate substance, by naming the gas from a metal-acid reaction "inflammable air." (Cavendish, 1766) (Fig. 3.2) He speculated that "inflammable air" was the same as the hypothetical substance termed "phlogiston" and additionally finding in 1781 that the gas formed water when burned. He is generally given credit for the discovery of hydrogen as an element. In 1783, Antoine Lavoisier named the element hydrogen (from the Greek ὑδρο- hydro meaning "water" and -γενής genes meaning "creator") when he and Laplace replicated Cavendish's observation that water is produced when hydrogen is burned.

3.1.2 Mining, production, and major minerals

Under normal conditions on Earth, elemental hydrogen exists as the diatomic gas, H_2. Yet, hydrogen gas is very rare in the Earth's atmosphere (1 ppm by volume) because of its light weight, which enables it to escape from Earth's gravity more easily than heavier gases. Still, hydrogen is the third most abundant element on the Earth's surface, mostly in the form of chemical compounds such as hydrocarbons, minerals, and water. Hydrogen gas is produced by some bacteria and algae and is a natural component of flatus, as is methane, itself a hydrogen source of increasing importance.

FIGURE 3.1 The Shannon Portrait of the Hon. Robert Boyle by Johann Kerseboom, 1689.

The Periodic Table: Nature's Building Blocks. DOI: https://doi.org/10.1016/B978-0-12-821279-0.00003-0

FIGURE 3.2 Frontispiece of the Life of the Hon. Henry Cavendish by George Wilson, 1851.

The electrolysis of H_2O is a simple process of producing H_2. A low voltage current is applied through the water, and gaseous O_2 is produced at the anode while at the same time gaseous H_2 is produced at the cathode. Typically, the cathode consists of Pt or another inert metal when producing H_2 for storage. If, though, the gas is to be burnt on site, O_2 is desirable to assist the combustion, and hence both electrodes would consist of inert metals. The theoretical maximum efficiency (electricity used vs. energetic value of H_2 formed) is between 88% and 94%.

$$2H_2O(l) \rightarrow 2H_2(g) + O_2(g)$$

H_2 is frequently produced from natural gas, which comprises the removal of hydrogen from hydrocarbons at very high temperatures. Commercial bulk H_2 is usually formed through steam reforming of natural gas (mainly CH_4). At high temperatures (700°C–1100°C), steam (water vapor) reacts with methane to yield carbon monoxide and H_2.

$$CH_4 + H_2O \rightarrow CO + 3H_2$$

This reaction is favored at low pressures but is however performed at high pressures (2.0 MPa, or 20 atm), since high-pressure H_2 is the most marketable product and pressure swing adsorption (PSA) purification systems perform better at increased pressures. The product mixture is commonly known as "synthesis gas" as it is regularly used straight for the production of methanol and related compounds. Hydrocarbons other than methane can also be utilized to make synthesis gas with variable product ratios. One of the many difficulties to this extremely optimized process is coke or carbon formation:

$$CH_4 \rightarrow C + 2H_2$$

Therefore, steam reforming typically uses an excess of H_2O. Additional H_2 can be recovered from the steam through the use of carbon monoxide (CO) through the water gas shift reaction, particularly with an Fe oxide catalyst. This reaction is also a common industrial source of carbon dioxide (CO_2):

$$CO + H_2O \rightarrow CO_2 + H_2$$

Other important processes for H_2 production comprise partial oxidation of hydrocarbons:

$$2CH_4 + O_2 \rightarrow 2CO + 4H_2$$

and the coal reaction, which can serve as a precursor to the water gas shift reaction mentioned above:

$$C + H_2O \rightarrow CO + H_2$$

In the laboratory, H_2 is usually prepared by the reaction of dilute nonoxidizing acids on some reactive metals such as zinc with Kipp's apparatus (also called Kipp generator, an apparatus designed for preparation of small volumes of

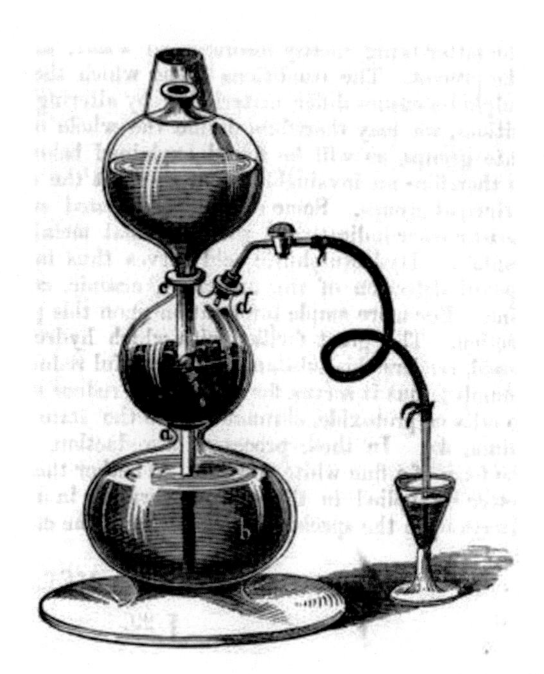

FIGURE 3.3 Kipp apparatus, 1859. Elementary instruction in chemical analysis, Page 33, Carl Remigius Fresenius.

gases, Fig. 3.3). It was invented around 1844 by the Dutch pharmacist, chemist and instrument maker Petrus Jacobus Kipp (March 5, 1808–February 3, 1864) and widely used in chemical laboratories and for demonstrations in schools into the second half of the 20th century. The apparatus is typically made of glass, or occasionally of polyethylene, and comprises three vertically stacked chambers. The upper chamber extends downward as a tube that passes through the middle chamber into the lower chamber. There is no direct path between the middle and upper chambers, but the middle chamber is separated from the lower chamber by a retention plate, such as a conical piece of glass with small holes, which allows the passage of liquid and gas. The solid material (e.g., zinc or iron sulfide) is placed into the middle chamber in lumps large enough to prevent falling through the retention plate. The liquid, such as an acid, is poured into the top chamber. Although the acid is free to flow down through the tube into the bottom chamber, it is prevented from rising there due to the pressure of the gas contained above it, which can leave the apparatus only by a stopcock near the top of the middle chamber. This stopcock may be opened, initially to allow the air to leave the apparatus, allowing the liquid in the bottom chamber to rise through the retention plate into the middle chamber and react with the solid material. Gas is evolved from this reaction, which may be drawn off through the stopcock as desired. When the stopcock is closed, the pressure of the evolved gas in the middle chamber rises and pushes the acid back down into the bottom chamber, until it is not in contact with the solid material anymore. At that point the chemical reaction comes to a stop, until the stopcock is opened again, and more gas is drawn off.

$$Zn + 2H^+ \rightarrow Zn^{2+} + H_2$$

Aluminum can also produce H_2 upon reaction with bases:

$$2Al + 6H_2O + 2OH^- \rightarrow 2Al(OH)_4^- + 3H_2$$

An alloy of Al and Ga in pellet form added to water can be utilized to produce H_2. The reaction also forms alumina, but the expensive gallium, which prevents the formation of an oxide skin on the pellets, can be re-used. This has important potential implications for an H_2 economy, as H_2 can be produced on-site and there is no need for transport.

Hydrogen occurs in several forms in minerals. Most common are hydroxyl (OH) groups and water (H_2O). Chemically speaking the simplest minerals are found as hydroxides and oxyhydroxides, for example, brucite ($Mg(OH)_2$) (Fig. 3.4), gibbsite and bayerite ($Al(OH)_3$), diaspore (AlOOH), goethite (FeOOH) (Fig. 3.5), manganite (MnOOH) (Fig. 3.6), and söhngeite ($Ga(OH)_3$) (Fig. 3.7). In the carbonate class one finds minerals such as azurite ($Cu_3(CO_3)_2(OH)_2$), malachite ($Cu_2CO_3(OH)_2$), hydrotalcite ($Mg_6Al_2(CO_3)(OH)_{16}\cdot 4H_2O$), etc. In the borate class minerals such as borax ($Na_2(B_4O_5)(OH)_4\cdot 8H_2O$), colemanite ($Ca[B_3O_4(OH)_3]\cdot H_2O$), inderite ($MgB_3O_3(OH)_5\cdot 5H_2O$), kernite ($Na_2[B_4O_6(OH)_2]\cdot 3H_2O$), and ulexite ($NaCa[B_5O_6(OH)_6]\cdot 5H_2O$), etc., contain not

FIGURE 3.4 Brucite, $Mg(OH)_2$, lemon-yellow botryoidal crystal group about 2 cm across. Killa Saifullah dist., Baluchistan, Pakistan.

FIGURE 3.5 Goethite, FeOOH, botryoidal dark brown with internal banded golden-brown cross-sections. 3 cm. Mather B shaft, Mather mine, Negaunee, Marquette iron range, Marquette Co., Michigan, United States.

only hydroxyl groups but also water molecules. Major minerals in the sulfate class are: alunite ($KAl_3(SO_4)_2(OH)_6$), copiapite ($Fe^{2+}Fe_4^{3+}$ $(SO_4)_6(OH)_2{\cdot}20H_2O$), cyanotrichite ($Cu_4Al_2(SO_4)(OH)_{12}{\cdot}2H_2O$), epsomite ($MgSO_4{\cdot}7H_2O$), ettringite ($Ca_6Al_2(SO_4)3(OH)_{12}{\cdot}26H_2O$), gypsum ($CaSO_4{\cdot}2H_2O$), and jarosite ($KFe_3^{3+}(SO_4)_2(OH)_6$). In the phosphate class one finds many minerals containing OH or H_2O: such as adamite ($Zn_2(AsO_4)(OH)$), annabergite ($Ni_3(AsO_4)_2{\cdot}8H_2O$), autunite ($Ca(UO_2)_2(PO_4)_2{\cdot}11H_2O$), bayldonite ($PbCu_3(AsO_4)_2(OH)_2$), brazilianite ($NaAl_3(PO_4)_2(OH)_4$), conichalcite ($CaCu(AsO_4)(OH)$), descloizite ($PbZn(VO_4)(OH)$), hydroxylapatite ($Ca_5(PO_4)_3(OH)$), erythrite ($Co_3(AsO_4)_2{\cdot}8H_2O$), lazulite ($MgAl_2(PO_4)_2(OH)_2$), legrandite ($Zn_2(AsO_4)(OH){\cdot}H_2O$), libethenite ($Cu_2(PO_4)(OH)$), olivenite ($Cu_2(AsO_4)(OH)$), torbernite ($Cu(UO_2)_2(PO_4)_2{\cdot}12H_2O$), turquoise ($Cu(Al,Fe^{3+})_6(PO_4)_4(OH)_8{\cdot}4H_2O$), variscite ($AlPO_4{\cdot}2H_2O$), vivianite ($Fe_3^{2+}$ $(PO_4)_2{\cdot}8H_2O$), and wavellite ($Al_3(PO_4)_2(OH,F)_3{\cdot}5H_2O$). Within the silicates the majority of the phyllosilicates (such as the micas, for example, biotite series ($K(Mg,Fe)_3AlSi_3O_{10}(OH)_2$), muscovite ($KAl_2AlSi_3O_{10}(OH)_2$), phlogopite ($KMg_3AlSi_3O_{10}(OH)_2$), etc., and clay minerals, for example, kaolinite ($Al_2Si_2O_5(OH)_4$), montmorillonite ($(Na,Ca)_{0.3}(Al,Mg)_2Si_4O_{10}(OH)_2{\cdot}nH_2O$), vermiculite ($(Mg,Fe,Al)_3(Si,Al)_4O_{10}(OH)_2{\cdot}4H_2O$, etc.) and all the amphiboles contain OH groups (e.g., actinolite ($Ca_2(Mg,Fe)_5Si_8O_{22}(OH)_2$), glaucophane ($Na_2Mg_3Al_2Si_8O_{22}(OH)_2$), pargasite ($NaCa_2Mg_4AlSi_6Al_2O_{22}(OH)_2$) and tremolite ($Ca_2Mg_5Si_8O_{22}(OH)_2$)). In addition there are silicate minerals such as analcime ($Na(AlSi_2O_6){\cdot}H_2O$), axinite-(Fe) ($Ca_2FeAl_2BSi_4O_{15}OH$), chabazite-(Ca) ($(Ca,K_2,Na_2)_2[Al_2Si_4O_{12}]_2{\cdot}12H_2O$), chrysocolla ($Cu_{2-x}Al_x(H_{2-x}Si_2O_5)(OH)_4{\cdot}nH_2O$), dioptase ($CuSiO_3{\cdot}H_2O$), epidote ($\{Ca_2\}\{Al_2Fe^{3+}\}(Si_2O_7)(SiO_4)O(OH)$), fluorapophyllite-(K) ($KCa_4(Si_8O_{20})(F,OH){\cdot}8H_2O$),

FIGURE 3.6 Manganite, MnOOH, highly reflective and lustrous prismatic crystals to 4 cm. Ilfeld, Harz Mountains, Thuringia, German.

FIGURE 3.7 Söhngeite, $Ga(OH)_3$, 2 mm crystals, extremely rare. Tsumeb mine, Tsumeb, Namibia.

natrolite ($Na_2Al_2Si_3O_{10}{\cdot}2H_2O$), prehnite ($Ca_2Al_2Si_3O_{10}(OH)_2$), stilbite-(Ca) ($NaCa_4[Al_9Si_{27}O_{72}]{\cdot}nH_2O$), topaz ($Al_2SiO_4(F,OH)_2$), tourmaline group (e.g., elbaite, $Na(Li_{1.5}Al_{1.5})Al_6(Si_6O_{18})(BO_3)_3(OH)_3(OH)$), vesuvianite ($(Ca,Na,\square)_{19}(Al,Mg,Fe^{3+})_{13}(\square,B,Al,Fe^{3+})_5(Si_2O_7)_4(SiO_4)_{10}(OH,F,O)_{10}$), and zoisite ($Ca_2Al_3[Si_2O_7][SiO_4]O(OH)$). A small group of minerals contain hydrogen in combination with nitrogen (mostly as NH_4, but also NH_3 or NH_2). In most of these minerals the NH_4 can be seen as a substitution for K. Minerals include for example: ammoniojarosite ($(NH_4)Fe^{3+}_3SO_4)_2(OH)_6$), buddingtonite ($(NH_4)(AlSi_3)O_8$), koktaite ($(NH_4)_2Ca(SO_4)_2{\cdot}H_2O$), lecontite ($(NH_4,K)NaSO_4{\cdot}2H_2O$), struvite ($(NH_4)Mg(PO_4){\cdot}6H_2O$), tobellite ($(NH_4,K)Al_2(AlSi_3O_{10})(OH)_2$), and urea ($CO(NH_2)_2$). Finally, there is a small group of organic minerals that either contain water (e.g., the oxalate group minerals such as whewellite, $Ca(C_2O_4){\cdot}H_2O$) or C-H bonds (hydrocarbons group minerals such as Kratochvílite, $C_{13}H_{10}$).

3.1.3 Chemistry

Hydrogen has three naturally occurring isotopes, designated ^1H, ^2H, and ^3H. Other, highly unstable nuclei (^4H to ^7H) have been synthesized in the laboratory but have not been observed in nature. ^1H is the most common hydrogen isotope with

TABLE 3.1 Hydrogen properties.

Appearance	Colorless gas
Standard atomic weight $A_{r,std}$	1.008
Block	s-Block
Element category	Reactive nonmetal
Electron configuration	$1s^1$
Phase at STP	Gas
Melting point	−259.16°C
Boiling point	−252.879°C
Density at STP	0.08988 g/L
Heat of fusion	(H_2) 0.117 kJ/mol
Heat of vaporization	(H_2) 0.904 kJ/mol
Molar heat capacity	(H_2) 28.836 J/(mol · K)
Oxidation states	−1, +1
Ionization energies	1st: 1312.0 kJ/mol
Atomic radius	31 ± 5 pm
Covalent radius	120 pm

STP, Standard temperature and pressure.
Bold font indicates main oxidation state.

an abundance of >99.98%. Since the nucleus of this isotope consists of only a single proton, it is given the descriptive but seldom used formal name protium. ^2H, the other stable hydrogen isotope, is known as deuterium and has one proton plus one neutron in the nucleus. All deuterium in the universe is believed to have been formed at the time of the Big Bang and has endured since that time. Deuterium is not radioactive and does not form a noteworthy toxicity hazard. Water enriched in molecules that include deuterium instead of normal hydrogen is known as heavy water. Deuterium and its compounds are used as a nonradioactive label in chemical experiments and in solvents for ^1H-NMR spectroscopy. Heavy water is utilized as a neutron moderator and coolant in nuclear reactors. In addition, deuterium is a potential fuel for commercial nuclear fusion. ^3H is also called tritium and has one proton plus two neutrons in the nucleus. It is radioactive, decaying into helium-3 through beta decay with a half-life of 12.32 years. Small amounts of tritium are formed naturally through the interaction of cosmic rays with atmospheric gases; tritium has also been released into the atmosphere during nuclear weapons tests. It is used in nuclear fusion reactions, as a tracer in isotope geochemistry, and in specialized self-powered lighting devices. In addition, tritium has been utilized in chemical and biological labeling experiments as a radiolabel. Hydrogen is the only element that currently has different names for its isotopes in general use. During early research on radioactivity, various heavy radioactive isotopes were given their own names, but such names are no longer used, except for deuterium and tritium. The symbols D and T (instead of ^2H and ^3H) are occasionally used for deuterium and tritium, but the corresponding symbol for protium, P, is already in use for phosphorus and thus is not available for protium. In its nomenclatural guidelines, the International Union of Pure and Applied Chemistry (IUPAC) permits any of D, T, ^2H, and ^3H to be used, even though ^2H and ^3H are preferred (Table 3.1).

Notwithstanding its very simple electronic configuration ($1s^1$) hydrogen can, unexpectedly, exist in more than 50 different forms most of which have been well characterized. This array of forms is the result firstly from the existence of atomic, molecular and ionized species in the gas phase: H, H_2, H^+, H^-, H_2^+, H_3^+, ..., H_{11}^+; secondly, from the existence of three isotopes, ${}_1^1H$, ${}_2^1H$ (D), and ${}_3^1H$ (T), and correspondingly of D, D_2, HD, DT, etc.; and, finally, from the existence of nuclear spin isomers for the homonuclear diatomic species, that is, *ortho*- and *para*-dihydrogen, -dideuterium, and -ditritium. Hydrogen is a colorless, tasteless, odorless gas which has only low solubility in liquid solvents. It is comparatively unreactive at room temperature though it combines with fluorine even in the dark and readily reduces aqueous solutions of palladium(II) chloride:

$$PdCl_2(aq) + H_2 \rightarrow Pd(s) + 2HCl(aq)$$

H_2 gas (dihydrogen or molecular hydrogen, also known as diprotium when comprising specifically a pair of protium atoms) is highly flammable and will burn in air at an extremely wide range of concentrations between 4% and 75% by volume. The enthalpy of combustion is −286 kJ/mol:

$$2H_2(g) + O_2(g) \rightarrow 2H_2O(l) + 572\ kJ$$

H_2 gas forms explosive mixtures with air in concentrations from 4% to 74% and with chlorine at 5%−95%. The explosive reactions may be caused by a spark, heat, or even sunlight. The H_2 autoignition temperature, that is, the temperature of spontaneous ignition in air, is 500°C. Pure hydrogen-oxygen flames emit ultraviolet light and with high oxygen mix are almost invisible to the naked eye. Hydrogen flames in other conditions are blue, similar to blue natural gas flames. H_2 will react with all oxidizing elements. It can react at room temperature spontaneously and violently with chlorine and fluorine to form the corresponding hydrogen halides, HCl and HF, which are potentially dangerous acids.

There are two different spin isomers of hydrogen diatomic molecules that are different based on the relative spin of their nuclei. In the *orthohydrogen* form, the spins of the two protons are parallel forming a triplet state with a molecular spin quantum number of 1 (1/2 + 1/2); in the *parahydrogen* form, the spins are antiparallel forming a singlet with a molecular spin quantum number of 0 (1/2−1/2). At standard temperature and pressure, H_2 gas consists of approximately 25% of the *para* form and 75% of the *ortho* form, also called the "normal form." The equilibrium ratio of *orthohydrogen* to *parahydrogen* is dependent on temperature, but since the *ortho* form is an excited state and has a higher energy than the *para* form, it is not stable and can therefore not be purified. At extremely low temperatures, the equilibrium state consists almost completely of the *para* form. The liquid- and gas-phase thermal properties of pure *parahydrogen* are significantly different from those of the normal form due to the differences in rotational heat capacities. The *ortho/para* dissimilarity also exists in other H-containing molecules or functional groups, such as water and methylene, but is of minimal consequence with respect to their thermal properties. The uncatalyzed interconversion between *para* and *ortho* H_2 increases with increasing temperature; consequently, quickly condensed H_2 has large amounts of the high-energy *ortho* form which converts very slowly to the *para* form. The *ortho/para* ratio in condensed H_2 is a significant consideration in the preparation and storage of liquid H_2: the conversion from *ortho* to *para* is exothermic and releases enough heat to evaporate some of the H_2 liquid, resulting in loss of liquefied material. Catalysts for the *ortho-para* interconversion, such as ferric oxide, activated carbon, platinized asbestos, rare earth metals, uranium compounds, chromic oxide, or some nickel compounds, are utilized during H_2 cooling.

3.1.3.1 Covalent and organic compounds

Though H_2 is not extremely reactive under standard conditions, it does produce compounds with most elements. Hydrogen can form compounds with elements that are more electronegative, for example, halogens (F, Cl, Br, I, etc.), or oxygen; in all these compounds the hydrogen takes on a partial positive charge. Once bonded to fluorine, oxygen, or nitrogen, it can participate in a form of medium-strength noncovalent bonding with the hydrogen of other comparable molecules, a phenomenon known as hydrogen bonding which is critical to the stability of a large number of biological molecules. In addition, hydrogen can form compounds with less electronegative elements, such as metals and metalloids, taking on a partial negative charge. These compounds are often known as hydrides. Hydrogen forms an extremely large number of compounds with carbon called the hydrocarbons, and an even larger variety with heteroatoms that, due to their general association with living things, are known as organic compounds. Millions of hydrocarbons have currently been identified, and they are usually formed by complex synthetic pathways that rarely involve elemental hydrogen.

3.1.3.2 Hydrides

Compounds of hydrogen are frequently called hydrides, a term that is used loosely. The term "hydride" suggests that the H atom has developed a negative or anionic character, symbolized H^-, and is applied when hydrogen forms a compound with a more electropositive element. The presence of the hydride anion, suggested by Gilbert N. Lewis in 1916 for group 1 and 2 salt-like hydrides, was shown by Moers in 1920 by the electrolysis of molten lithium hydride (LiH), producing a stoichiometric quantity of hydrogen at the anode. For hydrides other than group 1 and 2 metals, the term is rather misleading, taking into account the low electronegativity of hydrogen. An exception in group 2 hydrides is BeH_2, which is polymeric. In lithium aluminum hydride, the AlH_4^- anion carries hydridic centers strongly attached to the Al(III). While hydrides can be formed with nearly all main-group elements, the number and combination of potential compounds varies extensively; for example, over 100 binary borane hydrides have been identified, but only one binary aluminum hydride. Binary indium hydride has not yet been found, though larger complexes exist. In inorganic chemistry, hydrides can also act as bridging ligands linking two metal centers in a coordination complex. This function is especially common in group 13 elements, particularly in boranes (boron hydrides) and aluminum complexes, in addition to clustered carboranes.

3.1.3.3 Protons and acids

Oxidation of hydrogen removes its electron and gives H^+, which contains no electrons and a nucleus which is usually composed of one proton. That is the reason why H^+ is commonly called a proton. The hydrogen atom has a high ionization energy (1312 kJ mol^{-1}) and as such it is similar to the halogens rather than the alkali metals. Removal of the 1s electron leaves a bare proton which, with a radius of only approximately 1.5×10^{-3} pm, is not a stable chemical entity in the condensed phase. But, when bonded to other species it is well known in solution and in solids, such as H_3O^+,

NH_4^+, etc. The proton affinity of water and the enthalpy of solution of H^+ in water have been estimated with typical values that are currently accepted:

$$H^+(g) + H_2O(g) \rightarrow H_3O^+(g) - \Delta H \sim 710\ kJ \cdot mol^{-1}$$
$$H^+(g) \rightarrow H_3O^+(aq) - \Delta H \sim 1090\ kJ \cdot mol^{-1}$$

It follows that the heat of solution of the oxonium ion in water is $-380\ kJ \cdot mol^{-1}$, intermediate between the values calculated for Na^+ ($-405\ kJ \cdot mol^{-1}$) and K^+ ($-325\ kJ \cdot mol^{-1}$).

This H^+ species is fundamental in the discussion of acids. Under the Brønsted–Lowry acid-base theory, acids are proton donors, while bases are proton acceptors. A bare proton, H^+, cannot exist in solution or in ionic crystals due to its persistent attraction to other atoms or molecules with electrons. Apart from the high temperatures related to the existence of plasmas, such protons cannot be removed from the electron clouds of atoms and molecules and will continue to be attached to them. However, the term "proton" is occasionally used loosely and figuratively to denote positively charged or cationic hydrogen attached to other species in this fashion, and as such is denoted "H^+" without any suggestion that any single protons exist freely as a species. To avoid the suggestion of the naked "solvated proton" in solution, acidic aqueous solutions are occasionally considered to consist of a less improbable fictitious species, called the "hydronium ion" (H_3O^+). Still, even in this situation, such solvated hydrogen cations are more truthfully considered as being organized into clusters forming species closer to $H_9O_4^+$. Other oxonium ions are detected when water is in acidic solution in the presence of other solvents. While exotic on Earth, one of the most common ions in the universe is the H_3^+ ion, known as protonated molecular hydrogen or the trihydrogen cation.

The species H_2^+ and H_3^+ are significant as model systems for chemical bonding theory. The hydrogen molecule ion H_2^+ comprises two protons and one electron and is extremely unstable even in a low-pressure gas discharge system; the energy of dissociation and the internuclear distance (with the corresponding values for H_2 in parentheses) are:

$$\Delta H(\text{dissoc})\ 255\ (436)\ kJ \cdot mol^{-1};$$
$$r(H - H)\ 106\ (74.2)\ pm$$

The "observed" equilateral triangular 3-center, 2-electron structure has a higher stability than the hypothetical linear H_3^+ structure, and the relative stability of the species is revealed by the following gas-phase enthalpies:

$$H + H + H^+ \rightarrow H_3^+ - \Delta H\ 855.9\ kJ \cdot mol^{-1}$$
$$H_2 + H^+ \rightarrow H_3^+ - \Delta H\ 423.8\ kJ \cdot mol^{-1}$$
$$H + H_2^+ \rightarrow H_3^+ - \Delta H\ 600.2\ kJ \cdot mol^{-1}$$

The H_3^+ ion is the simplest possible illustration of a three-center two-electron bond. A series of ions H_n^+ with n-odd up to 15 and n-even up to 10 have been detected by mass-spectrometry. The odd-numbered species are far more stable compared to the even-numbered members. The structures of H_5^+, H_7^+, and H_9^+ are associated with that of H_3^+ with H_2 molecules added perpendicularly at the corners, while those of H_4^+, H_6^+, and H_8^+ include an added H atom at the first corner. The structures of higher members of the series, with $n \geq 10$ are currently unknown but may comprise further loosely bonded H_2 molecules above and below the H_3^+ plane. Enthalpies of dissociation are ΔH^0_{300}, ($H_5^+ \leftrightarrow H_3^+ + H_2$) 28 $kJ \cdot mol^{-1}$, and ΔH^0_{300} ($H_7^+ \leftrightarrow H_5^+ + H_2$) 13 $kJ \cdot mol^{-1}$.

3.1.3.4 pH

In chemistry, pH is a scale used to specify how acidic or basic a water-based solution is. At room temperature (25°C), pure water is neither acidic nor basic and has a pH of 7. The pH scale is logarithmic and approximates the negative of the base 10 logarithm of the molar concentration (measured in units of moles per liter) of H^+ ions in a solution. More precisely it is the negative of the base 10 logarithm of the activity (a) of the hydrogen ion. At 25°C, solutions with a pH less than 7 are acidic and solutions with a pH greater than 7 are basic. The neutral value of the pH depends on the temperature, being lower than 7 if the temperature increases. The pH value can be less than 0 for very strong acids, or higher than 14 for very strong bases. The idea of pH was first established by the Danish chemist Søren Peder Lauritz Sørensen (January 9, 1868–February 12, 1939) at the Carlsberg Laboratory in 1909 and amended to the modern pH in 1924 to accommodate definitions and measurements in terms of electrochemical cells. In the first papers, the notation had the "H" as a subscript to the lowercase "p," as so: p_H. The exact meaning of the "p" in "pH" is unclear, as Sørensen did not clarify why he used it. He describes a way of measuring it utilizing potential differences. All the words for these start with p in French, German, and Danish, all languages Sørensen published in; Carlsberg Laboratory was French-speaking, German was the leading language of scientific publishing, and Sørensen was Danish. He also utilized "q" in a similar fashion elsewhere in the paper. So, the "p" could stand for the French puissance, German Potenz, or Danish potens, meaning "power," or it could imply "potential." He could also simply have labeled the test solution "p" and the reference solution "q" arbitrarily; these letters are often combined. There is not much to back the suggestion that "pH" comes from the Latin terms pondus hydrogenii (quantity of hydrogen) or potentia hydrogenii (power of hydrogen). These days in chemistry, the p stands for "decimal cologarithm of," and is also used in the term pKa, used for acid dissociation constants and pOH, the equivalent for hydroxide (OH) ions. pH is defined as the decimal logarithm of the reciprocal of the hydrogen-ion activity, a_{H^+}, in a solution.

$$pH = -\log_{10}(a_{H^+}) = \log_{10}\left(\frac{1}{a_{H^+}}\right)$$

Note that pH depends on temperature. For example, at 0°C the pH of pure water is 7.47. At 25°C it is 7.00, and at 100°C it is 6.14. This definition was adopted since ion-selective electrodes, which are employed to measure pH, respond to activity. Ideally, electrode potential, E, follows the Nernst equation, which, for the H^+ ion is written as

$$E = E_0 + \frac{RT}{F}\ln(a_{H^+}) = E_0 - \frac{2.303RT}{F}\text{pH}$$

where E is a measured potential, E_0 is the standard electrode potential, R is the gas constant, T is the temperature in kelvins, F is the Faraday constant. For H^+ the number of electrons transferred is one. It follows that electrode potential is proportional to pH when pH is defined in terms of activity. Precise measurement of pH is given in International Standard ISO 31-8 as: A galvanic cell is set up to measure the electromotive force (e.m.f.) between a reference electrode and an electrode sensitive to the hydrogen-ion activity when they are both immersed in the same aqueous solution. The reference electrode may be a silver chloride electrode or a calomel electrode. The hydrogen-ion selective electrode is a standard hydrogen electrode.

Reference electrode | concentrated solution of KCl || test solution | H_2 | Pt

First, the cell is filled with a solution of known hydrogen-ion activity and the emf, E_S, is measured. Then the emf, E_X, of the same cell containing the solution of unknown pH is measured.

$$pH(X) = pH(S) + \frac{E_S - E_X}{z}$$

The difference between the two measured emf values is proportional to pH. This method of calibration avoids the need to know the standard electrode potential. The proportionality constant, 1/z is ideally equal to $\frac{1}{2.303RT/F}$, that is, the "Nernstian slope." To use this process in practice, a glass electrode is employed instead of the awkward hydrogen electrode. A combined glass electrode has an in-built reference electrode. It is calibrated against buffer solutions of known hydrogen-ion activity. IUPAC has proposed the use of a set of buffer solutions of known H^+ activity. Two or more buffer solutions are employed to accommodate the fact that the "slope" may differ slightly from ideal. To apply this methodology to calibration, the electrode is first immersed in a standard buffer solution and the reading on a pH meter is adjusted to be equal to the standard buffer's value. The reading from a second standard buffer solution is then adjusted, using the "slope" control, to be equal to the pH for that solution. When more than two buffer solutions are employed the electrode is calibrated by fitting observed pH values to a straight line with respect to standard buffer values. Commercial standard buffer solutions usually come with information on the value at 25°C and a correction factor to be applied for other temperatures. The pH scale is logarithmic and consequently pH is a dimensionless quantity.

3.1.3.5 Hydrogen preparation

Hydrogen can be formed through the reaction of water or dilute acids on electropositive metals such as the alkali metals, alkaline earth metals, the metals of Groups 3, 4 and the lanthanoids. The reaction can be explosively violent. Suitable laboratory methods use sodium amalgam or calcium with water, or zinc with hydrochloric acid. The reaction of aluminum or ferrosilicon with aqueous sodium hydroxide has also been applied. For small-scale preparations the hydrolysis of metal hydrides is appropriate, and this produces twice the amount of hydrogen as contained in the hydride, for example:

$$CaH_2 + 2H_2O \rightarrow Ca(OH)_2 + 2H_2$$

Electrolysis of acidified water using Pt electrodes is a suitable source of hydrogen (and oxygen) and, on a larger scale, very pure hydrogen ($>99.95\%$) can be obtained from the electrolysis of warm aqueous solutions of barium hydroxide between Ni electrodes. The process is costly but becomes economical on an industrial scale, for example, when integrated with the chloralkali industry. Other bulk methods comprise the (endothermic) reaction of steam on hydrocarbons or coke:

$$CH_4 + H_2O \xrightarrow{1000^\circ C} CO + 3H_2$$
$$C + H_2O \xrightarrow{1000^\circ C} CO + H_2 \text{ (water gas)}$$

In both reactions the CO can be converted to CO_2 by passing the gases and steam over an iron oxide or cobalt oxide catalyst at 400°C, in so doing generating even more hydrogen:

$$CO + H_2O \xrightarrow{400^\circ C,\text{catalyst}} CO_2 + H_2$$

This is the so-called water gas shift reaction ($-\Delta G^0_{rt}$ 19.9 kJ mol^{-1}) and it can also be affected by low-temperature homogeneous catalysts in aqueous acid solutions. The extent of following purification of the hydrogen depends on the use to which it will be put.

3.1.4 Major uses

Large volumes of H_2 are required in the petroleum and chemical industries. The largest use of H_2 is for the processing of fossil fuels, and in the production of ammonia. The most important reactions with H_2 in the petrochemical plant include hydrodealkylation, hydrodesulfurization, and hydrocracking. H_2 has several additional important applications. H_2 is used as a hydrogenating agent, chiefly in increasing the level of saturation of unsaturated fats and oils (present in substances such as margarine), and in the production of methanol. It is also the source of hydrogen in the production of hydrochloric acid. H_2 is also used as a reducing agent of metallic ores. Apart from its use as a reactant, H_2 has wide applications in physics and engineering. It is used as a shielding gas in welding methods such as atomic hydrogen welding. In the tungsten arc welding process, the hydrogen gas is present as shielding atmosphere between two tungsten electrodes. The electric arc breaks the H_2 molecules, which later recombine while producing a large amount of heat, with temperatures reaching to 4000°C. This heat is enough to weld tungsten, considered to be one of the metals most resistant to heat and wear. H_2 is used as the rotor coolant in electrical generators at power stations, because it has the highest thermal conductivity of any gas at 0.168 W/(m K), in addition to low density, and high specific heat. This makes possible for power plants to produce a lot more megawatts from a significantly smaller and less expensive generator. Liquid H_2 is used in cryogenic research, including superconductivity studies.

Since H_2 is lighter than air, having a little over 1/14 of the density of air, it was once widely used as a lifting gas in balloons and airships. However, it reacts explosively with oxygen even in air and its future in filling airships ended when the German passenger airship Hindenburg caught fire on May 6, 1937 in Manchester Township, New Jersey, United States, during its attempt to dock with its mooring mast at Naval Air Station Lakehurst. On board were 97 people (36 passengers and 61 crewmen); there were 36 fatalities (13 passengers and 22 crewmen, 1 worker on the ground). It is still used as a lifting gas for weather balloons especially in areas where a safe on-site hydrogen generator is available. In more recent applications, H_2 is used pure or mixed with nitrogen as a tracer gas for minute leak detection. It is present only in very minute quantities in air (low background); is an environmentally friendly and renewable natural resource, nontoxic, noncorrosive; and is inexpensive. To avoid its flammability limitations, it is usually supplied as a standard industrial gas mixture of 5% hydrogen in nitrogen (often called forming gas). Applications can be found in the automotive, chemical, power generation, aerospace, and telecommunications industries. H_2 is an authorized food additive (E 949) that allows food package leak testing among other antioxidizing properties. H_2 is used in gaseous form, stored in pressure vessels, alongside nickel to form a rechargeable electrochemical power source that has characteristic long life and is able to handle more than 20,000 charge cycles with 85% energy efficiency and 100% faradaic efficiency. This property makes the nickel-hydrogen battery ideal for the energy storage of electrical energy in satellites and space probes. Notable examples of these satellites and space probes using NiH_2 batteries include the Hubble Space Telescope, International Space Station (ISS), Mercury Messenger, Mars Odyssey, and Mars Global Surveyor. Hydrogen is used as a protective atmosphere for making flat glass sheets. Here, the hydrogen is used to blanket a tin bath and prevents it from oxidizing. Molten glass flows onto the molten tin to produce glass sheets of uniform thicknesses and very flat surfaces. Without the hydrogen blanket, oxygen reacts with the metal surface producing imperfections on the surface of the tin bath. In the electronics industry, hydrogen is used in the manufacturing of semiconductor, display, LED and photovoltaic components. Since hydrogen has excellent heat transfer capabilities and efficient reducing and etching capacities, it is regularly used in the following processes: (1) annealing—where Si wafers are heated to temperatures over 1000°C to anneal the crystal structure, hydrogen transfers the heat uniformly and reacts to remove any oxide that is present; (2) epitaxy—as an excellent reducing agent, hydrogen is used to deposit a new crystalline overlayer on a crystalline substrate; (3) deposition—hydrogen can be incorporated into thin films to reduce crystallinity; and (4) stabilizing—adding hydrogen increases the shelf life of some chemicals (e.g., diborane and digermane) used in the electronics industry.

The high flammability and the associated technical difficulties of producing, storing and transporting H_2 limited its use as combustion fuel. Nevertheless, proponents of the hydrogen economy now see H_2 gas as the clean fuel of the future as it forms only water as a waste product and a substantial amount of energy when it is burned.

$$2H_2(g) + O_2(g) \rightarrow 2H_2O(l) + 572\ \mathrm{kJ}\left(286\ \mathrm{kJ/mol}\right)$$

Hydrogen-powered fuel cells are increasingly considered as "pollution-free" sources of energy and are now being used in some buses and cars. In its liquid form and in combination with a strong oxidizer such as liquid oxygen, hydrogen burns with extreme intensity and provides the highest specific impulse or the highest efficiency relative to the amount of propellant used. The technical difficulties in harnessing the potential of liquid hydrogen as propulsion fuel and as the fuel of choice for space exploration engines were enormous but not insurmountable. To appreciate these difficulties, one has to understand that hydrogen can only be liquefied at very low temperatures and must be stored at very low temperatures as well. In addition, being the smallest, lightest and most flammable element, it has to be handled with utmost care—it must be insulated from all sources of heat during flight; protected from the radiant heat of the sun when in space; vented to prevent the tank from exploding due to rapid expansion of hydrogen when heated but at the same time this process of venting must not expose metals from the extreme cold of hydrogen to prevent them from getting brittle. The American government invested a vast amount of technical expertise to solve these problems and tame a very dangerous fuel, but it helped them propel men to the moon. Nowadays, liquid hydrogen is the signature fuel of the American space program. Occasionally, it is used by other countries to propel satellites into space.

Deuterium is used in nuclear fission reactors, usually in the form of D_2O (heavy water) to slow neutrons without worrying much about high neutron absorption problems as is the case with ordinary hydrogen. In research reactors, liquid deuterium is used to moderate neutrons to very low energies and wavelengths appropriate for scattering experiments. When combined with tritium, it is the most common nuclide in prototype nuclear fusion reactor designs because of the large reaction rate and energy yield. Tritium (^3H) is present in very small traces in nature but is now made in large amounts in the laboratory. It is a radioactive isotope, emitting β^- particles with 18.6 keV maximal energy. Its half-life is 12.32 years. It is widely used as a tracer in both industry and research. It is popularity as a tracer stems from the fact that hydrogen is present in a wide variety of compounds. One of the best examples of its use as tracer is in the study of the extent of groundwater pollution. In such studies, a scientist can make up a sample of water made with tritium instead of hydrogen. As that water moves through the soil, its path can be followed by means of the radioactivity the tritium gives off. The attributes that make tritium an ideal tracer of groundwater pollution include: (1) no interaction with aquifer materials during movement through the porous media; (2) geochemically conservative; (3) can be measured at low level via electrolytic enrichment prior to quantification; (4) allows identification of groundwater contamination at less than 1%; and (5) allows detection of pollution migration to groundwater even at greater distances.

Though, thankfully, it has never been used in a war, the most terrible weapon of all is the hydrogen bomb (i.e., a thermonuclear explosion). At its core is an atomic bomb that, when detonated, releases sufficient neutrons and energy to generate a high enough temperature for hydrogen fusion to take place among the enclosing lithium deuteride (LiD) mass, which also comprises some tritium (T). When a Li atom absorbs a neutron, more tritium is formed (i.e., $Li + n \rightarrow He + T$); this, plus the original tritium, at that point fuses with the deuterium to release a massive amount of explosive energy, equivalent to millions of tons of TNT (2,4,6-trinitrotoluene, $C_6H_2(NO_2)_3CH_3$), large enough to annihilate a major city of millions of people.

3.2 3 Li — Lithium

3.2.1 Discovery

Petalite ($LiAlSi_4O_{10}$), a phyllosilicate (layer silicate), was first found in 1800 by the Brazilian chemist, poet, and statesman José Bonifácio de Andrada e Silva (June 13, 1763–April 6, 1838) in a mine on the island of Utö, Sweden (Fig. 3.8). But it was not until 1817 that the Swedish chemist Johan August Arfwedson (alternative spelling Arfvedson) (January 12, 1792–October 28, 1841) (Fig. 3.9), at that time working in the laboratory of the Swedish chemist Baron Jöns Jakob Berzelius (August 20, 1779–August 7, 1848), observed the presence of a new element while analyzing petalite ore (Arfwedson, 1818a,b). This element formed compounds like those of sodium and potassium, but its carbonate and hydroxide

FIGURE 3.8 José Bonifácio de Andrada e Silva.

FIGURE 3.9 Pre-1849 lithograph from Fehr and Miller of Stockholm of Johan August Arfwedson.

were less soluble in water and less alkaline. Berzelius named the alkaline material "lithion/lithina," from the Greek word λιθος (transliterated as lithos, meaning "stone"), to reflect its discovery in a mineral, contrary to potassium, which had been discovered in plant ashes, and sodium, which was known partly for its high abundance in animal blood. He called the metal inside the mineral "lithium." Arfwedson later showed that this same element was present in the minerals spodumene and lepidolite. In 1818, the German chemist Christian Gottlob Gmelin (October 12, 1792–May 13, 1860) was the first to detect

FIGURE 3.10 William Thomas Brande.

that lithium salts give a bright red color to a flame (Gmelin, 1818). Though both Arfwedson and Gmelin tried, they were unsuccessful in isolating the pure element from its salts. This did not happen until 1821, when the English chemist William Thomas Brande (January 11, 1788–February 11, 1866) (Fig. 3.10) obtained it by electrolysis of lithium oxide, a process that had earlier been used by the chemist Sir Humphry Davy to isolate the alkali metals potassium and sodium (Brande, 1821).

3.2.2 Mining, production, and major minerals

Estimations for the Earth's crustal lithium content range from 20 to 70 ppm by weight. Consistent with its name, lithium forms a minor component in igneous rocks, with the highest concentrations found in granites. Granitic pegmatites have the greatest abundance of lithium-containing minerals, with spodumene and petalite as the most commercially viable sources. Another important mineral of lithium is lepidolite, which is now an obsolete name for a series formed between the endmembers polylithionite, $KLi_2AlSi_4O_{10}F_2$, and trilithionite, $KLi_{1.5}Al_{1.5}Si_3O_{10}F_2$. A newer source for lithium is the smectite group clay mineral hectorite, $Na_{0.3}(Mg,Li)_3Si_4O_{10}(F,OH)_2$. Lithium salts are extracted from water in mineral springs, brine pools, and brine deposits. Brine extraction is probably the only technology widely used today, as actual mining of lithium ores is much more expensive and has been priced out of the market. Another possible source of lithium is the leachates of geothermal wells, which are carried to the surface. Recovery of lithium has been demonstrated in the field by simple filtration. The process and environmental costs are primarily those of the already-operating well; net environmental impacts may thus be positive.

Li occurs in a limited number of halides such as griceite (LiF) and simmonsite (Na_2LiAlF_6) and in the oxides class with minerals such as lithiophorite ($(Al,Li)MnO_2(OH)_2$) or lithiowodginite ($LiTa_3O_8$). Only one carbonate contains Li: zabuyelite (Li_2CO_3). Similarly, the only borate containing lithium is walkerite ($Ca_{16}(Mg,Li,\square)_2[B_{13}O_{17}(OH)_{12}]_4Cl_6\cdot28H_2O$). A larger group is found in the phosphate class including minerals such as: amblygonite ($LiAl(PO_4)F$) (Figs. 3.11 and 3.12), lithiophilite ($LiMn^{2+}PO_4$), tiptopite ($K_2(Na,Ca)_2Li_3Be_6(PO_4)_6(OH)_2\cdot H_2O$), and triphylite ($LiFe^{2+}PO_4$). The largest group of minerals, however, is found in the silicate class, for example, cookeite ($(Al_2Li)Al_2(AlSi_3O_{10})(OH)_8$), elbaite ($Na(Li_{1.5}Al_{1.5})Al_6(Si_6O_{18})(BO_3)_3(OH)_3(OH)$), eucryptite ($LiAlSiO_4$), holmquistite ($\square\{Li_2\}\{Mg_3Al_2\}(Si_8O_{22})(OH)_2$), neptunite ($Na_2KLiFe^{2+}_2Ti_2Si_8O_{24}$), petalite ($LiAl(Si_4O_{10})$) (Fig. 3.13), and spodumene ($LiAlSi_2O_6$) (Box 3.1).

3.2.3 Chemistry

Naturally occurring lithium consists of two stable isotopes, ^6Li and ^7Li, the latter being the more abundant at 92.5%. Both natural isotopes have anomalously low nuclear binding energy per nucleon (compared to the neighboring elements

FIGURE 3.11 Solid white mass of amblygonite, $LiAl(PO_4)F$,-montebrasite, $LiAl(PO_4)(OH)$, series. Tin Mountain Mine, Custer Co., South Dakota, United States.

FIGURE 3.12 Montebrasite, $LiAl(PO_4)(OH)$, crystal about 1 cm. Telirio mine, Linopolis, Minas Gerais, Brazil.

on the periodic table, helium and beryllium); Li is the only low numbered element that can produce net energy via nuclear fission. The two Li nuclei have lower binding energies per nucleon than any other stable nuclides other than deuterium and helium-3. Due to this, though very light in atomic weight, lithium is less common in the Solar System than 25 of the first 32 chemical elements. Seven radioisotopes have been characterized, the most stable being ^{8}Li with a half-life of 838 ms and ^{9}Li with a half-life of 178 ms. All the other radioactive isotopes possess half-lives that are shorter than 8.6 ms. The shortest-lived isotope of lithium is ^{4}Li with a half-life of 7.6×10^{-23} s, which decays through proton emission. Li isotopes fractionate considerably during a wide variety of natural processes, including mineral formation (chemical precipitation), metabolism, and ion exchange. Li^{+} ions substitute for magnesium and iron in octahedral sites in clay minerals, where ^{6}Li is preferred to ^{7}Li, resulting in enrichment of the light isotope in processes of hyperfiltration and rock alteration.

Lithium is a malleable, soft silvery metal and the first element belonging to the alkali metals. It is small and the lightest among all metals. Its chemistry and that of the rest of the alkali metals is attributed mostly to the ease of involving the ns^1 electron in bonding, coupled with the difficulty of removing the second electron from the inner orbitals. As such, lithium as with other alkali metals, react vigorously with water and has an oxidation state that never exceeds +1 (Table 3.2). Lithium's interaction with water is not as explosive and violent as the other alkali metals. This reaction forms hydroxides and hydrogen gas:

$$Li(s) + H_2O(l) \rightarrow LiOH(aq) + H_2(g)$$

FIGURE 3.13 Petalite, $LiAl(Si_4O_{10})$, 2.5 cm colorless partial crystal naturally etched but also with some well-developed faces. Urubu mine, Monte Belo, Itinga, Minas Gerais, Brazil.

BOX 3.1 Spodumene ($LiAlSi_2O_6$) and petalite ($LiAlSi_4O_{10}$)

Spodumene's name comes from the Greek *spodumenos*, meaning "ash-colored." It is characterized by two perfect prismatic cleavages {110}, well-developed parting on {100}, hardness of 6½-7, luster vitreous, transparency transparent to translucent, color highly variable (colorless, white, gray, pink, green, yellow, purple, or tan white), streak white. Spodumene is monoclinic *2/m*, space group *C2/c*. It is optically biaxial (+). Crystals with well-developed {100} faces showing vertical striations are common. Crystal are often polysynthetically twinned. A small amount of Na usually substitutes for Li. It is mainly used as a gemstone. The green variety hiddenite is named after W.F. Hidden (Fig. 3.14), while the pink variety kunzite is named after G.F. Kunz (Fig. 3.15).

FIGURE 3.14 Spodumene, variety hiddenite, $LiAlSi_2O_6$, 3 cm. Etta mine, Pennington Co., South Dakota, United States.

(*Continued*)

BOX 3.1 (Continued)

FIGURE 3.15 6.5 cm lilac to pink color zoned crystal of gemmy kunzite, the pink gem variety of spodumene, $LiAlSi_2O_6$. The crystal is naturally etched. Mt. Kharitha, Kala, Darrah Pech, Kunar prov., Afghanistan.

Petalite's name comes from the Greek word *petalos,* meaning leaf, alluding to its cleavage. The cleavage on {001} is perfect, {201} is good. Petalite is brittle, hardness between 6 and 6½, luster vitreous, pearly on {001}, color varies from colorless to white and gray. It is transparent to translucent. Petalite is monoclinic 2/*m*. Optically it is biaxial(+). It occurs usually massive or in foliated cleavable masses. Colorless petalite is sometimes cut as a gem.

TABLE 3.2 Lithium properties.

Appearance	Silver-white
Standard atomic weight $A_{r,std}$	6.941
Block	s-Block
Element category	Alkali metal
Electron configuration	[He] $2s^1$
Phase at STP	Solid
Melting point	180.50°C
Boiling point	1330°C
Density (near rt)	0.534 g/cm^3
Heat of fusion	3.00 kJ/mol
Heat of vaporization	136 kJ/mol
Molar heat capacity	24.860 J/(mol · K)
Oxidation states	+1
Ionization energies	1st: 520.2 kJ/mol 2nd: 7298.1 kJ/mol 3rd: 11815.0 kJ/mol
Atomic radius	Empirical: 152 pm
Covalent radius	128 ± 7 pm
Van der Waals radius	182 pm

STP, Standard temperature and pressure.
Bold font indicates main oxidation state.

Its small size (both for the atom and for the ion) and high polarizing power give its compounds and alloys special, sometimes anomalous properties. For example, lithium is harder and has higher melting and boiling point than other metals in the group. It is miscible with sodium (Na) only at temperatures above 380°C and immiscible with the other members of the alkali metals. In contrast, all the other alkali metal pairs are miscible with each other at all proportions.

Lithium shows many similarities with magnesium (Mg) owing to their similar atomic and ionic radius. The ionic radius of Li^+ is 76 pm while that of Mg^{2+} is 72 pm, compared with the 102 pm ionic radius of Na^+. Chemical resemblances between these two include direct reaction with N_2 to form the nitrides; ignition with O_2 to form oxides and a small percentage of peroxides; and formation of salts with similar solubilities. This similarity in chemistry of Li and Mg is termed a diagonal relationship since their position in the periodic table is diagonal to each other.

3.2.3.1 Reaction in air, oxygen, nitrogen, and hydrogen

The reactivity of lithium makes it impossible to find it in its pure form in nature. Even in moist air, it tarnishes rapidly to form a black coating of lithium hydroxide (LiOH) and hydrated lithium hydroxide ($LiOH{\cdot}H_2O$). The LiOH initially formed also reacts with the carbon dioxide in moist air to form $Li_2CO_3{\cdot}LiOH$ is soluble in water and slightly soluble in ethanol, and is available commercially in anhydrous form and as the monohydrate ($LiOH{\cdot}H_2O$). LiOH is formed in a metathesis reaction between lithium carbonate and calcium hydroxide:

$$Li_2CO_3 + Ca(OH)_2 \rightarrow 2LiOH + CaCO_3$$

The initially formed hydrate is dehydrated by heating under vacuum up to 180°C. In the laboratory, lithium hydroxide can be produced through the action of water on lithium or lithium oxide.

$$2\ Li + 2\ H_2O \rightarrow 2\ LiOH + H_2$$
$$Li_2O + H_2O \rightarrow 2\ LiOH$$

Typically, these reactions are avoided. Even though Li_2CO_3 is more commonly used, the hydroxide is an effective precursor to produce lithium salts, for example,

$$LiOH + HF \rightarrow LiF + H_2O$$

It is one of the few metals that can react directly with nitrogen (N_2) even under room conditions to form lithium nitride (Li_3N).

$$2LiOH(s) + CO_2(g) \rightarrow Li_2CO_3(s) + H_2O(l)$$
$$6Li(s) + N_2(g) \rightarrow 2Li_3N(s)$$

As an alternative to burning lithium metal in an atmosphere of nitrogen, a solution of lithium in liquid sodium metal can be reacted with N_2. Lithium nitride reacts violently with water forming ammonia:

$$Li_3N + 3H_2O \rightarrow 3LiOH + NH_3$$

Lithium is flammable and burns with a striking crimson flame. When ignited sufficiently well, it gives a brilliant silver flame. The resulting products are mainly lithium oxide (Li_2O) and some lithium peroxide (Li_2O_2), both white powders. Lithium reacts directly with hydrogen especially at temperatures above 600°C to form lithium hydride (LiH).

$$4Li(s) + O_2(g) \rightarrow 2Li_2O(s)$$
$$2Li(s) + O_2(g) \rightarrow 2Li_2O_2(s)$$
$$2Li(s) + H_2(g) \rightarrow 2LiH(s)$$

3.2.3.2 Reaction with halogens

Lithium metal reacts with all the halogens to form lithium halides. So, it reacts with fluorine, F_2, chlorine, Cl_2, bromine, I_2, and iodine, I_2, to form respectively lithium fluoride, LiF, lithium chloride, LiCl, lithium bromide, LiBr, and lithium iodide, LiI.

$$2Li(s) + X_2(g) \rightarrow 2LiX(s) \text{ where } X = F,\ Cl,\ Br,\ I$$

LiF is a colorless solid, that transitions to white with decreasing crystal size. Although odorless, lithium fluoride has a bitter-saline taste. Its structure is similar to that of NaCl, but has a much lower solubility in water. It is primarily used as a component of molten salts. LiF formation from the elements releases one of the highest energy per mass of reactants, second only to that of BeO. LiF is produced from LiOH or Li_2CO_3 with HF. It can be also formed through the reaction of sulfur hexafluoride with metallic lithium, but this method is not used industrially because of the high cost of reagents. Naturally occurring LiF is known as the extremely rare mineral griceite. LiCl is a typical ionic compound, although the small size of the Li^+ ion results in properties not found with other alkali metal chlorides, such as extraordinary solubility in polar solvents (83.05 g/100 mL of water at 20°C) and its hygroscopic properties. The salt forms crystalline hydrates, unlike the other alkali metal chlorides, such as mono-, tri-, and pentahydrates. The anhydrous salt can be regenerated by heating the hydrates. LiCl also absorbs up to four equivalents of ammonia/mol. Similar to any other ionic chloride, solutions of lithium chloride can serve as a source of chloride ion, for example, forming a precipitate upon treatment with silver nitrate:

$$LiCl + AgNO_3 \rightarrow AgCl + LiNO_3.$$

Lithium chloride is prepared by the treatment of lithium carbonate with hydrochloric acid. It can in principle also be formed via the highly exothermic reaction of lithium metal with either chlorine or anhydrous hydrogen chloride gas. Anhydrous LiCl is produced from the hydrate by heating with a stream of hydrogen chloride. Its extreme hygroscopic character makes LiBr useful as a desiccant in certain air conditioning systems. LiBr is produced in the reaction of lithium carbonate with hydrobromic acid. The salt forms several crystalline hydrates, in contrast to the other alkali metal bromides. The anhydrous salt forms cubic crystals similar to common salt (NaCl). The reaction of lithium hydroxide and hydrobromic acid (aqueous solution of hydrogen bromide) will result in the precipitation of lithium bromide in the presence of water.

$$LiOH + HBr \rightarrow LiBr + H_2O$$

When exposed to air, LiI becomes yellow in color, due to the oxidation of iodide to iodine. It crystallizes in the NaCl motif. It can form various hydrates. Lithium iodide is employed as an electrolyte for high-temperature batteries. It is also utilized for long-life batteries as required e.g. by artificial pacemakers. It is used as a phosphor for neutron detection. In addition, it is utilized, in a complex with iodine, in the electrolyte of dye-sensitized solar cells. In organic synthesis, LiI is useful for cleaving C—O bonds. For instance, it can be employed to convert methyl esters to carboxylic acids:

$$RCO_2CH_3 + LiI \rightarrow RCO_2Li + CH_3I$$

Comparable reactions apply to epoxides and aziridines. Lithium iodide was used as a radio-contrast agent for CT scans. Its use was discontinued due to renal toxicity. Current iodinated contrast agents consist of organoiodine compounds.

3.2.3.3 Reaction with acids

Lithium reacts with acids in a complex manner since the acid solutions contain water, which was shown to vigorously react with lithium forming lithium hydroxide (LiOH). LiOH, in turn, reacts with the acid to form salt and water. Thus, when lithium is added to dilute sulfuric acid, the reaction forms lithium sulfate and hydrogen. But when a small amount of lithium is added to concentrated sulfuric acid, the very exothermic reaction ignites the hydrogen gas initially formed and gives a bright carmine flame. The final products are lithium sulfate, sulfur dioxide and water. Similarly, lithium added to concentrated nitric acid forms lithium nitrate, nitrogen dioxide and water.

$$2Li(s) + H_2SO_4(aq) \rightarrow 2Li_2SO_4(aq) + H_2(g)$$
$$2Li(s) + 2H_2SO_4(conc.) \rightarrow Li_2SO_4(aq) + SO_2(g) + 2H_2O(l)$$
$$Li(s) + 2HNO_3(conc.) \rightarrow LiNO_3(aq) + NO_2(g) + H_2O(l)$$

3.2.3.4 Organolithium compounds

Lithium has a rich organometallic chemistry. It forms a direct bond with carbon to form an important class of reagents called organolithium compounds. They can be prepared using a variety of ways. The easiest is by direct reaction of metallic lithium with alkyl halides. As both the lithium and resulting organolithium are very reactive, air and moisture must be rigorously excluded from the reaction atmosphere. The reaction is thus carried out in solvents such as light petroleum, cyclohexane, benzene or ether.

$$2Li + RX \xrightarrow{solvent} LiR + LiX \text{ where } X = \text{halide and } R = \text{alkyl}$$

The direct reaction with lithium metal is used for the formation of simple alkyllithium compounds. For more complex organolithium compounds, other methods of preparations are used. Alkynyllithium reagents are best prepared by metalation or lithium hydrogen exchange. In this method, the acidity of the hydrogen in the terminal alkyne allows for easy deprotonation and exchange with lithium.

$$R{-}C{-}C{-}H + R':Li \longrightarrow R{-}C{-}C:Li + R'{-}H$$

terminal alkyne, stronger acid — alkyllithium, stronger base — alkynyllithium, weaker base — alkane, weaker acid

For aromatic compounds, the position of lithium in the aryllithium compound is determined by the substituent groups. Alkoxy, amido, sulfoxide, and sulfonyl groups direct the lithium compound to the ortho position. Another method of preparing organolithium reagents is through lithium halogen exchange. This method is mostly used to convert aryl and alkenyl iodides and bromides to the corresponding organolithium compounds. Examples of the two reactions are shown below.

For the preparation of allylic, benzylic, and propargylic lithium reagents, which are difficult to produce by other routes, lithium-metal exchange (transmetalation) is used.

Organolithium reagents are important in organic synthesis for the preparation of elastomers, which are polymers with viscoelasticity, weak intermolecular forces, generally low Young's modulus, and high failure strain compared with other materials. Organolithium compounds are also used in the asymmetric synthesis of pharmaceutical compounds.

3.2.4 Major uses

The most important application of Li is in rechargeable batteries for mobile phones, laptops, digital cameras, and electric vehicles. Lithium is also used in some nonrechargeable batteries in applications such as heart pacemakers, toys, watches, and clocks. Because of its low atomic mass, it has a high charge- and power-to-weight ratio. A typical lithium-ion battery can generate approximately 3 volts per cell, compared with 2.1 volts for lead-acid and 1.5 volts for zinc-carbon. Lithium-ion batteries, which are rechargeable and have a high-energy density, differ from lithium batteries, which are disposable (primary) batteries with lithium or its compounds as the anode. Other rechargeable batteries that use lithium include the lithium-ion polymer battery, lithium iron phosphate battery, and the nanowire battery.

Lithium is utilized in alloys with aluminum and magnesium, improving their strength and making them lighter. A magnesium-lithium alloy is used for armor plating. Aluminum-lithium alloys are used in aircraft, bicycle frames and high-speed trains. Lithium (e.g., as lithium carbonate) is applied as an additive to continuous casting mold flux slags where it increases fluidity. Lithium compounds are also used as additives (fluxes) to foundry sand for iron casting to reduce veining. Lithium (as lithium fluoride) is used as an additive to aluminum smelters (Hall–Héroult process, see Aluminum), lowering melting temperature and increasing electrical resistance. When applied as a flux for welding or soldering, metallic lithium promotes the fusing of metals and prohibits the formation of oxides by absorbing any impurities. Lithium oxide is widely used as a flux for processing silica, reducing the melting point and viscosity of the material and leading to glazes with improved physical properties including low coefficients of thermal expansion. Worldwide, besides batteries this is one of the largest uses for lithium compounds. Glazes containing lithium oxides are used for ovenware. Lithium carbonate (Li_2CO_3) is generally used in this application because it converts to the oxide upon heating. Lithium produces glasses which are resistant to sudden heating or cooling because of its low thermal expansion coefficient. The lithium amide ($LiNH_2$) and lithium hydride (LiH) systems are used as a means of storing hydrogen for use as a fuel. The mixture is shown to have a reversible hydrogen storage capacity of 6.5% with working temperatures of around 285°C. The mechanism proposed for the release of hydrogen is given as:

$$2LiNH_2 \rightarrow Li_2NH + NH_3$$

and

$$LiH + NH_3 \rightarrow LiNH_2 + H_2$$

The third most common use of lithium is in greases. Lithium hydroxide (LiOH) is a strong base and, when heated with a fat, results in the formation of a soap made of lithium stearate. Lithium soap can thicken oils, and it is used in the production of all-purpose, high-temperature lubricating greases. Lithium imparts a bright crimson color to fireworks. It is also used as the oxidizer in red fireworks and flares.

Lithium chloride (LiCl) and lithium bromide (LiBr) are hygroscopic and are used as desiccants for gas streams. Lithium hydroxide (LiOH) and lithium peroxide (Li_2O_2) are the compounds most used in confined spaces, such as found in spacecraft and submarines, for carbon dioxide (CO_2) removal and air purification. Lithium hydroxide absorbs carbon dioxide from the air by forming lithium carbonate and is favored over other alkaline hydroxides for its low weight. Lithium peroxide (Li_2O_2) in presence of moisture not only reacts with carbon dioxide to form lithium carbonate, but in addition forms oxygen:

$$2Li_2O_2 + 2CO_2 \rightarrow 2Li_2CO_3 + O_2$$

Some of these compounds, as well as lithium perchlorate ($LiClO_4$), are used in oxygen candles that supply submarines with oxygen. These can also include small amounts of B, Mg, Al, Si, Ti, Mn, and Fe. Since lithium chloride is one of the most hygroscopic materials known it is used in air conditioning and industrial drying systems (as is lithium bromide).

Synthetic lithium fluoride (LiF), grown as crystal, is clear and transparent and is regularly used in specialist optics for IR, UV, and VUV (vacuum UV) applications. It possesses one of the lowest refractive indexes and the furthest transmission range in the deep UV of most common materials. Finely divided LiF powder has been used for thermoluminescent radiation dosimetry (TLD): a sample exposed to radiation accumulates crystal defects which, when heated, resolve through the release of bluish light the intensity of which is proportional to the absorbed dose, consequently allowing quantification. LiF is occasionally used in focal lenses of telescopes. The high nonlinearity of lithium niobate ($LiNbO_3$) makes it a valuable material for nonlinear optics applications. It is used widely in telecommunication products such as mobile phones and optical modulators, for such components as resonant crystals.

Organolithium molecules are extensively used in the production of polymer and fine chemicals. In the polymer industry, which is the leading consumer of these reagents, alkyl lithium compounds are catalysts/initiators in, for example, anionic polymerization of unfunctionalized olefins. To produce fine chemicals, organolithium compounds have a role as strong bases and as reagents for the formation of C-C bonds. Organolithium compounds are synthesized from lithium metal and alkyl halides. Many other lithium compounds are used as reagents to prepare organic compounds, for example, lithium aluminum hydride ($LiAlH_4$), lithium triethylborohydride ($Li(C_2H_5)_3BH$), n-butyllithium (IUPAC name butyllithium, tetra-μ3-butyl-tetralithium, C_4H_9Li) and tert-butyllithium (preferred IUPAC name tert-butyllithium, LiC_4H_9) are commonly used as extremely strong bases called superbases. Lithium is useful in the treatment of bipolar disorder. Lithium salts may also be helpful for related diagnoses, such as schizoaffective disorder and cyclic major depression. Accordingly, lithium affects the flow of sodium through nerve and muscle cells in the body. Since sodium is known to affect excitation or mania, lithium carbonate, thus, can affect (prevent or lessen) manic symptoms such as hyperactivity, rushed speech, poor judgment, reduced sleep, aggression, and anger. It may increase the risk of developing Ebstein's cardiac anomaly in infants born to women who take lithium during the first trimester of pregnancy. In addition, lithium is being studied as a possible treatment for cluster headaches.

3.3 4 Be — Beryllium

3.3.1 Discovery

The mineral beryl, which contains beryllium, has been known at least since the Ptolemaic dynasty of Egypt (305 to 30 BCE). In the 1st century CE, Roman naturalist Pliny the Elder (born Gaius Plinius Secundus, 23–79 CE) stated in his encyclopedia Natural History that beryl and emerald ("smaragdus") were similar. The Papyrus Graecus Holmiensis, written in the 3rd or 4th century CE, has notes on how to prepare artificial emerald and beryl. Early analyses of emeralds and beryls by Martin Heinrich Klaproth (German chemist, December 1, 1743–January 1, 1817), Torbern Olof Bergman (Swedish chemist and mineralogist, March 20, 1735–July 8, 1784), Franz Karl Achard [German (Prussian) chemist, physicist, and biologist, April 28, 1753–April 20, 1821], and Johann Jakob Bindheim (German pharmacist and chemist, March 5, 1740–January 17, 1825) always yielded similar elements, leading to the erroneous conclusion that both materials were aluminum silicates. French priest and mineralogist René Just Haüy (February 28, 1743–June 3, 1822) discovered that both crystals are geometrically indistinguishable, and he asked chemist Louis-Nicolas Vauquelin (French pharmacist and chemist, May 16, 1763–November 14, 1829) for a chemical analysis (Fig. 3.16). In a 1798 paper read before the Institut de France, Vauquelin described how he found a new "earth" by dissolving aluminum hydroxide from emerald and beryl in an additional alkali (Vauquelin, 1798). The editors of the journal Annales de Chimie et de Physique called the new earth "glucine" for the sweet taste of some of its

FIGURE 3.16 Louis-Nicolas Vauquelin, published in 1824.

compounds. Klaproth favored the name "beryllina" since yttria also formed sweet salts. The name "beryllium" was first used by the German chemist Friedrich Wöhler (July 31, 1800–September 23, 1882) in 1828 (Fig. 3.17). Friedrich Wöhler and Antoine Bussy independently isolated beryllium in 1828 by the chemical reaction of metallic potassium with beryllium chloride (Wöhler, 1828; Bussy, 1828):

$$BeCl_2 + 2K \rightarrow 2KCl + Be$$

Using an alcohol lamp, Wöhler heated alternating layers of beryllium chloride and potassium in a wired-shut platinum crucible. The above reaction happened instantly and caused the crucible to become white hot. Upon cooling and washing the resulting gray-black powder he observed consisted of fine particles with a dark metallic luster. The highly reactive potassium had been produced by the electrolysis of its compounds, a process discovered 21 years before. The chemical method using potassium yielded only small grains of beryllium from which no ingot of metal could be cast or hammered. Electrolysis of a combination of beryllium fluoride and sodium fluoride was utilized to obtain beryllium during the 19th century. The metal's high melting point makes this process more energy-consuming than equivalent processes used for the alkali metals. At the start of the 20th century, the production of beryllium by the thermal decomposition of beryllium iodide was studied following the success of a comparable method for the production of zirconium, but this method was shown to be uneconomical for volume production. The direct electrolysis of a molten mixture of beryllium fluoride and sodium fluoride by the French chemist Paul Lebeau (December 19, 1868–November 18, 1959) in 1898 resulted in the first pure (99.5%–99.8%) samples of beryllium.

3.3.2 Mining, production, and major minerals

Beryllium is generally extracted from the mineral beryl (and bertrandite) (Box 3.2), which is either sintered using an extraction agent or melted into a soluble mixture. The sintering process comprises mixing beryl with sodium fluorosilicate, $Na_2(SiF_6)$ and soda at 770°C to form sodium tetrafluoroberyllate, Na_2BeF_4, aluminum oxide, Al_2O_3, and silicon dioxide, SiO_2. Beryllium hydroxide, $Be(OH)_2$, is precipitated from a solution of sodium tetrafluoroberyllate and sodium

FIGURE 3.17 Friedrich Wöhler, Lithography by Rudolf Hoffmann, 1856, after a photograph by Petri (Göttingen).

BOX 3.2 Beryl ($Be_3Al_2(Si_6O_{18})$) and its gemstone varieties

The name beryl is of ancient origin, derived from the Greek word *beryllos* that was applied to green gemstones. Beryl is found in granitic rocks, notably in pegmatites. It may also be found in schists and in rare ore deposits. The cleavage {0001} is imperfect, hardness 7½–8, luster vitreous, transparent to translucent, color high variable (commonly bluish-green or light yellow, may be deep emerald-green, blue, gold-yellow, pink, white, colorless, and red) (Figs. 3.20–3.23). Beryl is hexagonal with point group *6/m2/m2/m* (space group *P6/m2/c2/c*) with a strong prismatic habit, terminated faces are pinacoids. Pyramidal forms are rare. Optically it is uniaxial(–). The color serves as the basis for several variety names of gem beryl. Variety names are not accepted mineral names but are commonly used in the jewelry industry. Aquamarine is the pale bluish transparent variety. Emerald, the most desired and expensive variety, is transparent emerald-green (can be per carat (0.2 g) more expensive than diamond). Morganite is pale pink to deep rose, while pale, almost colorless beryl is known as goshenite. Heliodor is a transparent-golden-yellow variety. Red beryl (or bixbyte) is the red-colored variety only occurring in a single locality at Thomas Range, Utah, the United States. The different colors are commonly ascribed to impurities known as chromophores. The most common chromophores are Cr and V for emerald, Fe for aquamarine, Mn and Fe for morganite, and Mn, Fe, and Ti for heliodor. Maxixe-type beryl is characterized by a bright blue color which fades in the sunlight. Vorobyevite is a caesium-bearing variety of beryl. Riesling beryl is a strongly dichroic (showing different colors in different crystallographic directions: pale green and golden-yellow in this case) beryl. Trapiche beryl is a variety showing six-spoked growth features. "Trapiche" is the Spanish name for a machine with large wooden spokes, used for crushing juice out of sugar cane.

hydroxide, NaOH, in water. Extraction of Be using the melt method comprises grinding beryl into a powder and heating it to 1650°C. The melt is quickly cooled with water and then reheated to 250°C–300°C in concentrated sulfuric acid, mostly yielding beryllium sulfate, $BeSO_4 \cdot 4H_2O$ and aluminum sulfate, $Al_2(SO_4)_3$. Aqueous ammonia is then used to eliminate the Al and S, leaving $Be(OH)_2$. $Be(OH)_2$ formed using either the sinter or melt method is then converted into BeF_2 or $BeCl_2$. To synthesize the fluoride, aqueous ammonium hydrogen fluoride is added to $Be(OH)_2$ to produce a precipitate of ammonium tetrafluoroberyllate, $(NH_4)_2BeF_4$, which is heated to 1000°C to form BeF_2. Next, heating the BeF_2 to 900°C in the presence of magnesium results in finely divided Be, and additional heating to 1300°C produces the compact metal. Heating $Be(OH)_2$ forms the oxide, which becomes beryllium chloride when combined with carbon and chlorine. Electrolysis of molten beryllium chloride is then employed to obtain the metal.

Beryllium is known to occur in ten different oxide minerals such as behoite ($Be(OH)_2$) and the well-known chrysoberyl (under different light sources color-changing gemstone variety known as alexandrite) ($BeAl_2O_4$) (Fig. 3.18). Niveolanite ($NaBe(CO_3)(OH)\cdot 2H_2O$) is the only known carbonate. Four borates are known to include Be: berborite ($Be_2(BO_3)(OH,F)\cdot H_2O$), hambergite ($Be_2(BO_3)(OH)$), londonite ($(Cs,K,Rb)Al_4Be_4(B,Be)_{12}O_{28}$), and rhodizite ($(K,Cs)Al_4Be_4(B,Be)_{12}O_{28}$). A larger group of minerals is found in the phosphate class, such as beryllonite ($NaBePO_4$), herderite ($CaBePO_4(F,OH)$) (Fig. 3.19), tiptopite ($K_2(Na,Ca)_2Li_3Be_6(PO_4)_6(OH)_2\cdot H_2O$), and zanazziite ($Ca_2Mg_5Be_4(PO_4)_6(OH)_4\cdot 6H_2O$). However, the biggest group

FIGURE 3.18 Chrysoberyl, $BeAl_2O_4$, yellowish green, translucent twinned crystal. 2 × 2 cm. Espirito Santo, Brazil.

FIGURE 3.19 Hydroxylherderite, $CaBe(PO_4)(OH,F)$, pale green 2 cm crystal on microcline. Sabsar, Skardu dist., Baltistan, Gilgit-Baltistan, Pakistan.

forms the Be silicates with minerals such as: bavenite ($Ca_4Be_2Al_2Si_9O_{26}(OH)_2$), beryl (which includes gemstone varieties emerald (green), aquamarine (blue) and heliodore (yellow)) ($Be_3Al_2(Si_6O_{18})$) (Figs. 3.20–3.23), euclase ($BeAl(SiO_4)(OH)$) (Fig. 3.24), genthelvite ($Be_3Zn_4(SiO_4)_3S$), helvite ($Be_3Mn_4^{2+}(SiO_4)_3S$) (Fig. 3.25), milarite ($K_2Ca_4Al_2Be_4Si_{24}O_{60}\cdot H_2O$), pezzottaite ($Cs(Be_2Li)Al_2(Si_6O_{18})$), phenakite ($Be_2SiO_4$) (Fig. 3.26), and tugtupite ($Na_4BeAlSi_4O_{12}Cl$).

3.3.3 Chemistry

Both stable and unstable isotopes of Be are created in stars, but the radioisotopes do not live long. It is thought that most of the stable Be in the universe was initially formed in the interstellar medium when cosmic rays induced fission in heavier elements found in interstellar gas and dust. Primordial Be has only one stable isotope, ^{9}Be, and hence Be is a monoisotopic element. Radioactive cosmogenic ^{10}Be is formed in the Earth's atmosphere through the cosmic ray spallation of oxygen (also known as the x-process, is a set of naturally occurring nuclear reactions causing nucleosynthesis; it refers to the formation of chemical elements from the impact of cosmic rays on an object. Cosmic rays are highly energetic charged particles from beyond Earth, ranging from protons, alpha particles, and nuclei of many heavier elements. About 1% of cosmic rays also consist of free electrons.). ^{10}Be accumulates at the soil surface, where its relatively long half-life (1.36 million years) allows for a long residence time before decaying to boron-10. Therefore, ^{10}Be and its daughter products are used to study phenomena such as natural soil erosion, soil formation and the development of lateritic soils, and as a proxy for measurement of the variations in solar activity and the age of ice cores. The

FIGURE 3.20 Beryl, $Be_3Al_2(Si_6O_{18})$, variety aquamarine, about 2 cm. Near Camp Lake, Elmore Co., Idaho, United States.

FIGURE 3.21 Beryl, $Be_3Al_2(Si_6O_{18})$, gemmy aquamarine crystal to 5 cm with a complex termination and subtly color zoned from yellow to yellow green. Magli, Nigeria.

formation of ^{10}Be is inversely proportional to solar activity, since increased solar wind during periods of high solar activity decreases the flux of galactic cosmic rays that reach the Earth. Nuclear explosions also form ^{10}Be by the reaction of fast neutrons with ^{13}C in carbon dioxide (CO_2) in the atmosphere. The isotope ^{7}Be (half-life 53 days) is also cosmogenic, and shows an atmospheric abundance linked to sunspots, much like ^{10}Be. ^{8}Be has a very short half-life of about 7×10^{-17} s that contributes to its significant cosmological role, as elements heavier than Be could not have been formed through nuclear fusion in the Big Bang, because of the lack of enough time during the Big Bang's nucleosynthesis phase to produce carbon by the fusion of ^{4}He nuclei and the very low concentrations of available ^{8}Be. The English astronomer Sir Fred Hoyle (June 24, 1915–August 20, 2001) first proved that the energy levels of ^{8}Be and ^{12}C allow carbon formation via the so-called triple-alpha process in helium-fueled stars where more nucleosynthesis time is available. This process allows carbon to be produced in stars, but not in the Big Bang. Star-created carbon (the basis of carbon-based life) is thus a component in the elements in the gas and dust ejected by AGB stars (the asymptotic giant branch (AGB) is a region of the Hertzsprung–Russell diagram populated by evolved cool luminous stars. This is a period of stellar evolution undertaken by all low- to intermediate-mass stars (0.6–10 solar masses) late in their lives) and supernovae, as well as the formation of all other elements with atomic numbers larger than that of carbon. The 2s electrons of beryllium may contribute to chemical bonding. Therefore, when ^{7}Be decays by L-electron capture, it does

FIGURE 3.22 Beryl, $Be_3Al_2(Si_6O_{18})$, a small 3.3 cm, light colored, crystal nested at the center of an albite, $NaAlSi_3O_8$, crystal with an attached quartz, SiO_2, crystal. Dassu, Braldu Valley, Balistan, Northern Areas, Pakistan.

FIGURE 3.23 Beryl, $Be_3Al_2(Si_6O_{18})$, variety red beryl (bixbite), Rare hexagonal, prismatic crystal 4 mm long × 2.5 wide. Wah Wah Mts, Beaver Co., Utah, United States.

so by taking electrons from its atomic orbitals that may be participating in bonding. This makes its decay rate dependent to a measurable degree upon its chemical environments—a rare occurrence in nuclear decay. The shortest-lived known isotope is ^{13}Be with a half-life of 2.7×10^{-21} s which decays through neutron emission. ^{6}Be also has a very short half-life of 5.0×10^{-21} s. The exotic isotopes ^{11}Be and ^{14}Be are known to exhibit a nuclear halo.

Beryllium is the lightest member of the alkaline earth metals family, which have electronic configuration of ns^2. The predominant oxidation state of beryllium is +2 where the beryllium atom has lost both of its valence electrons. However, lower oxidation states have been found in, for example, bis(carbene) compounds (Table 3.3). The chemistry

FIGURE 3.24 Euclase, $BeAl(SiO_4)(OH)$, euhedral, striated crystal to 8 mm with blue inclusions. Alto Santo, Equador, Rio Grande do Norte, Brazil.

FIGURE 3.25 Helvite, $Be_3Mn_4^{2+}(SiO_4)_3S$, well-formed crystals to 7 mm on albite, $NaAlSi_3O_8$, variety cleavelandite. Navegador mine, Conselheiro Pena, Minas Gerais, Brazil.

of beryllium is due largely to its small atomic and ionic radii and the resulting very high ionization potentials and strong polarization while bonded to other atoms. This is the reason why all beryllium compounds are covalent. Similar to the diagonal relationship discussed for Li and Mg, Be and aluminum (Al) also exhibit such a relationship. Points of similarity include: they have the same electronegativity and polarizing power; they have the tendency to form covalent compounds; they both dissolve in strong alkali to form complexes; their chlorides can both act as Lewis acids and are both used as Friedel-Crafts catalysts; their resistance to acid due to formation of protective oxide coating; their oxides and hydroxides are amphoteric and so on. Beryllium metal is relatively unreactive at room temperature, especially in its bulk form. It does not react with water or steam even if the metal is heated to red heat.

3.3.3.1 Reaction of beryllium in air, oxygen, nitrogen, and hydrogen

Beryllium is a hard, brittle metal with a silvery white surface. The bulk metal reacts briefly with oxygen to form a thin layer of oxide. Once this oxide forms, Be no longer oxidizes in air below 600°C. When ground to a powder, it does ignite in air into a brilliant flame giving beryllium oxide (BeO) and beryllium nitride (Be_3N_2).

$$2Be(s) + O_2(g) \rightarrow 2BeO(s)$$
$$3Be(s) + N_2(g) \rightarrow Be_3N_2(s)$$

FIGURE 3.26 Phenakite, Be_2SiO_4, a small 7 mm colorless to white crystal on matrix. Mt. Antero, Chaffe Co., Colorado, United States.

TABLE 3.3 Beryllium properties.

Appearance	White-gray metallic
Standard atomic weight $A_{r,std}$	9.012
Block	s-Block
Element category	Alkaline earth metal
Electron configuration	[He] $2s^2$
Phase at STP	Solid
Melting point	1287°C
Boiling point	2469°C
Density (near rt)	1.85 g/cm^3
Heat of fusion	12.2 kJ/mol
Heat of vaporization	292 kJ/mol
Molar heat capacity	16.443 J/(mol · K)
Oxidation states	+1, **+2** (amphoteric oxide)
Ionization energies	1st: 899.5 kJ/mol
	2nd: 1757.1 kJ/mol
	3rd: 14,848.7 kJ/mol
Atomic radius	Empirical: 112 pm
Covalent radius	96 ± 3 pm
Van der Waals radius	153 pm

STP, Standard temperature and pressure.
Bold font indicates main oxidation state.

Beryllium does not react with hydrogen. However, beryllium hydride can still be prepared from beryllium(II) compounds. For example, it was prepared by treating dimethylberyllium, $Be(CH_3)_2$ with lithium aluminum hydride, $LiAlH_4$. Highly pure samples of BeH_2 are prepared by reacting beryllium borohydride, $Be(BH_4)_2$ and triphenylphosphine, PPh_3.

$$Be(BH_4)_2 + 2PPh_3 \rightarrow 2Ph_3PBH_3 + BeH_2$$

3.3.3.2 Reaction with the halogens

Beryllium metal reacts with chlorine, Cl_2, or bromine, Br_2, at high temperatures (above 600°C) to form beryllium(II) chloride and beryllium(II) bromide. The beryllium halides formed have a linear monomeric molecular structure in the gas phase.

$$Be(s) + Cl_2(g) \rightarrow BeCl_2(s)$$
$$Be(s) + Br_2(g) \rightarrow BeBr_2(s)$$

The fluoride is synthesized by reaction of $Be(OH)_2$ with HF or via heating of $(NH_4)_2BeF_4$ above 500°C (see Reaction with bases).

$$Be(OH)_2 + 2HF \rightarrow BeF_2 + H_2O$$

Beryllium halides have polymeric structure analogous to SiO_2 for BeF_2 and SiS_2 for the other halides. This structure, however, can be broken with coordinating solvents or ligands except for BeF_2 which reacts only in water or liquid ammonia. Beryllium readily forms tetrahedral beryllate anions $[BeX_4]^{2-}$ with fluoride and chloride, which are less reactive toward hydrolysis. Nevertheless, the high affinity of beryllium for oxygen makes all its halides react with water.

3.3.3.3 Aqueous chemistry

Beryllium salts are amphoteric, dissolving in both acidic and basic conditions. Beryllium is so oxophilic that few anions can compete with H_2O as ligand. In water, the Be^{2+} ion is present as the tetraaquoberyllium(II) cation, $[Be(H_2O)_4]^{2+}$. When the pH is increased to 4–6, polynuclear aggregates formed such as the six-membered cyclic hexaaquo-trihydroxo-triberyllate trication. Upon further increase in pH to 6–9, beryllium hydroxide, $Be(OH)_2$, precipitates. Making the solutions even more basic results in redissolution of $Be(OH)_2$ to form several polynuclear, hydroxo-bridged anions.

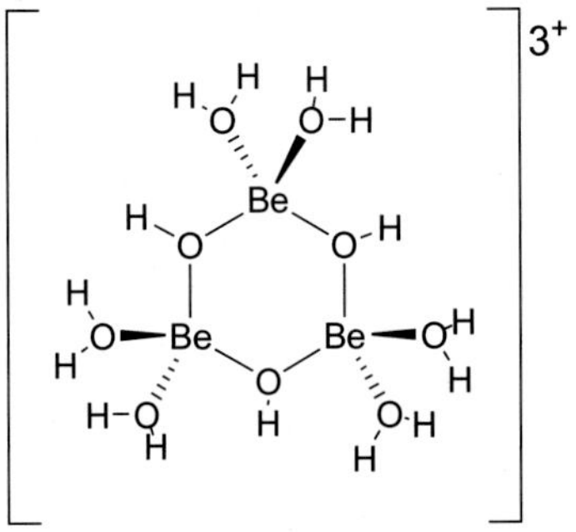

The only nonchelating ligand that can compete with water and hydroxide anions as ligand for beryllium is fluoride. BeF_2 dissolves and form $[BeF_2(OH_2)_2]$ in equilibrium with $[BeF(OH_2)_3]^+$ and $[BeF_3(OH_2)]^-$. Species that are predominant at low fluoride concentrations are $[BeF(OH_2)_3]^+$ and $[Be(OH_2)_4]^{2+}$ while the species that are dominant at high fluoride concentrations are $[BeF_3(OH_2)]^-$ and $[BeF_4]^{2-}$. Even at high fluoride concentrations, water molecules are still able to compete successfully with F^- as ligands.

3.3.3.4 Reaction with acids

Beryllium surfaces are covered by a thin layer of oxide that protects it from further oxidation such as through attack by acids. Cold, concentrated nitric acid, passivates Be but the metal dissolves readily in dilute aqueous acids (H_2SO_4, HCl, HNO_3) forming hydrogen gas and aqueous Be^{2+}.

$$Be(s) + H_2SO_4(aq) \rightarrow Be^{2+}(aq) + SO_4^{2-}(aq) + H_2(g)$$

3.3.3.5 Reaction with bases

Beryllium differs from the other alkaline earth metals in reacting with aqueous alkalis (NaOH, KOH) with evolution of hydrogen.

$$Be + 2NaOH + 2H_2O \rightarrow Na_2\left[Be(OH)_4\right] + H_2$$

The metal, as does beryllium hydroxide $Be(OH)_2$, reacts with ammonium bifluoride (NH_4HF_2) to generate ammonium tetrafluoroberyllate, $NH_4)_2BeF_4$· This reaction is technologically important in the preparation of BeF_2 and purified Be.

$$Be(OH)_2 + 2(NH_4)HF_2 \rightarrow (NH_4)_2BeF_4 + 2H_2O$$
$$(NH_4)_2BeF_4 \rightarrow 2NH_3 + 2HF + BeF_2$$

3.3.3.6 *Organometallic chemistry*

Beryllium halides dissolve readily in ethers to form BeX_2 etherates. These adducts are less reactive and the ether can only be removed under extreme conditions. With crown ethers, Be^{2+} can be forced into a fivefold coordination. With alcohols, beryllium forms beryllium alcoholates. The coordination number of beryllium in these complexes is generally four unless strained by steric factors, in which the coordination number is forced to be less than four. With dimethylsulfoxide, it forms structures similar to the tetrahydroxo beryllium di-cation. The coordination with carboxylates, dicarboxylates and hydroxyl-carboxylates are well examined. The latter is especially well-studied since beryllium can form stable complexes with hydroxyl-carboxylates as six-membered ring systems with Be. These complexes are stable at pH close to the biologic pH making them especially interesting in chelate therapy for acute beryllium poisoning. Dimethylthioetherates of $BeCl_2$, $BeBr_2$ and $BePh_2$ are easily prepared by dissolving the corresponding beryllium (II) compound in dimethylsulfide. The structures formed are analogous to etherates, with Be tetrahedrally coordinated to two thioethers and two halides or phenyl groups. These thioetherates, are however, less stable than their etherate counterparts due to the softer S-donor. This is shown by the dimerization of $(SMe_2)_2BeCl_2$ in noncoordinating solvents to form chloro-bridged $[S(Me_2)BeCl_2]_2$ with the loss of dimethylsulfide. The halides, except fluoride, in beryllium halides are completely displaced by ammonia to form tetraamine beryllium di-cation. The structure is similar to $[Be(OH)_4]^{2+}$, but the strong ligand water, even in trace amounts, can lead to the formation of hydroxyl-bridged species and six-membered triberyllacycle compounds. Ammonia is not as strong in affinity to Be as water, thus, it is not able to replace fluoride in BeF_2.

Owing to the cost and toxicity of beryllium, its organic chemistry is limited to scholarly work. Organoberyllium compounds are known to be highly reactive. For example, beryllium dialkyls (BeR_2, R = Me, Et, Pr'', Pr', Bu', etc.), which are colorless solids or viscous liquids, spontaneously burn in air and are explosively hydrolyzed by water. Beryllium dialkyls can be prepared by reacting lithium alkyls or Grignard reagents with $BeCl_2$ in ether. However, clean-up of the products from the solvent ether is difficult. Thus, when pure compounds are required, Be metal is heated with the appropriate mercury dialkyl:

$$BeCl_2 + 2LiMe \xrightarrow{Et_2O} BeMe_2{\cdot}nEt_2O + 2LiCl$$
$$BeCl_2 + MeMgCl \xrightarrow{Et_2O} BeMe_2{\cdot}nEt_2O + 2MgCl_2$$
$$Be + HgMe_2 \xrightarrow{110^{\circ}C} BeMe_2 + Hg$$

From BeR_2, alkylberyllium alkoxides (RBeOR') can be prepared by a variety of routes such as alcoholysis with R'OH, addition to carbonyls, cleavage of peroxides $R'OOR'$ or redistribution with the appropriate dialkoxide Be $(OR')_2$· As with other organoberyllium compounds, the alkylberyllium alkoxides $(RBeOR')_4$ are reactive, low-melting solids (mp for R' = Me 25°C, Et 30°C, Pr'' 40°C, Pr' 136°C, Bu' 93°C). Bulky substituents may reduce the degree of oligomerization and reaction with coordinating solvents or strong ligands can also lead to depolymerization.

3.3.4 Major uses

Due to its low atomic number and very low absorption for X-rays, the oldest and even now one of the most significant uses of beryllium is in radiation windows for X-ray tubes. Extreme purity and cleanliness of Be are necessary to avoid artifacts in the X-ray images. Thin Be foils are utilized as radiation windows for X-ray detectors, and the very low absorption minimizes the heating effects due to the high intensity, low energy X-rays typical of synchrotron radiation. Vacuum-tight windows and beam-tubes for radiation experiments on synchrotrons are made solely from Be. In experimental setups for a variety of X-ray emission studies (e.g., energy-dispersive X-ray spectroscopy), the sample holder typically consists of beryllium as its emitted X-rays have much lower energies ($\approx$100 eV) than the X-rays produced by most studied materials. Ultra-thin beryllium foil is also used in X-ray lithography. Moreover, beryllium is used in nuclear reactors as a reflector or moderator of neutrons.

Due to its stiffness, lightweight, and dimensional stability over a large temperature range, Be metal is used for lightweight structural components in the defense and aerospace industries in, for example, high-speed aircraft, guided missiles, spacecraft, and satellites. Some liquid-fuel rockets had rocket nozzles made out of pure Be. Beryllium is used in alloys with Cu or Ni to make gyroscopes, springs, electrical contacts, spot-welding electrodes and nonsparking tools. Mixing Be with these metals significantly increases the electrical and thermal conductivity of these alloys. Applications comprise nonsparking tools that are used near flammable gases (beryllium nickel), in springs and membranes (beryllium nickel and beryllium iron) used in surgical instruments and high-temperature devices. As little as 50 ppm of Be alloyed with liquid Mg results in a substantial increase in oxidation resistance and decrease in flammability.

Beryllium mirrors are of specific interest. For example, large-area mirrors, often with a honeycomb support structure are used in meteorological satellites where low weight and long-term dimensional stability are of utmost importance. Smaller Be mirrors are utilized in optical guidance systems and in fire-control systems, for example, in battle tanks. In these systems, very fast movement of the mirror is essential which once more necessitates low mass and high rigidity. Typically, the Be mirror is coated with hard electroless Ni plating which can be more easily polished to a finer optical finish than Be. In some applications, though, the beryllium blank is polished without any coating. This is predominantly relevant for cryogenic usage where thermal expansion mismatch can cause the coating to buckle. The James Webb Space Telescope (JWST), which will be launched in 2021, will have 18 hexagonal Be sections for its mirrors. Since JWST will face a temperature of $-240.15°C$ (33K), the mirror is made of gold-plated beryllium, capable of dealing with the extreme cold much better than glass. Beryllium contracts and deforms less than glass—and remains more uniform—at such low temperatures. For the same reason, the optics of the Spitzer Space Telescope completely consist of Be metal. Owing to its low weight and high rigidity, Be is useful as material for high-frequency speaker drivers. It is also used in phonograph cantilevers for improved tracking while reducing mass of the device. However, its cost, toxicity when mishandled and rigidity limit its applications to professional audio, public address applications and sophisticated homes.

Beryllium is nonmagnetic and shown to have excellent use as structural materials. As such, they are often used in structures and devices near magnetic equipment and facilities such as near magnetic resonance imaging and near naval mines. Beryllium is also used in tools to tune magnetic components of devices. Beryllium is a p-type dopant in III-V compound semiconductors. It is extensively utilized in materials such as GaAs, AlGaAs, InGaAs, and InAlAs grown by molecular beam epitaxy (MBE). Cross-rolled Be sheet is an exceptional structural support for printed circuit boards in surface-mount technology. In critical electronic applications, Be is both a structural support and heat sink. It also involves a coefficient of thermal expansion that is well matched to the alumina and polyimide-glass substrates. The beryllium-beryllium oxide composite "E-Materials" have been specifically designed for these electronic applications and have the added benefit that the thermal expansion coefficient can be matched to diverse substrate materials.

Beryllium oxide is valuable in numerous applications that necessitate the combined properties of an electrical insulator and an exceptional heat conductor, with high strength and hardness, and a very high melting point. Beryllium oxide is regularly utilized as an insulator base plate in high-power transistors in radio frequency transmitters for telecommunications. Beryllium oxide is also being researched for application in increasing the thermal conductivity of uranium dioxide nuclear fuel pellets. Beryllium compounds were used in fluorescent lighting tubes, but this application was discontinued due to the disease known as berylliosis (or chronic beryllium disease (CBD), is a chronic allergic-type lung response and chronic lung disease caused by exposure to beryllium and its compounds, a form of beryllium poisoning. It is distinct from acute beryllium poisoning, which became rare following occupational exposure limits established around 1950) which developed in the workers who were making the tubes.

Beryl and chrysoberyl are beryllium silicate: when these are of gem quality, they are known as emerald (green), aquamarine (blue), and heliodore (yellow) for the beryl varieties, and alexandrite for chrysoberyl. The alexandrite variety shows a color change (alexandrite effect) dependent upon the nature of ambient lighting (metamerism). It is the phenomenon of an observed color change from greenish to reddish with a change in source illumination. It is due to a small-scale replacement of Al by Cr ions in the crystal structure, which causes intense absorption of light over a narrow range of wavelengths in the yellow region (580 nm) of the visible light spectrum. Because human vision is most sensitive to green light and least sensitive to red light, alexandrite appears greenish in daylight where the full spectrum of visible light is present, and reddish in incandescent light which emits less green and blue spectrum. This color change is independent of any change of hue with viewing direction through the crystal that would arise from pleochroism.

3.4 5 B — Boron

3.4.1 Discovery

Borax, its mineral form then known as tincal, glazes were used in China from as early as 300 CE, and some crude borax reached the West, where the Perso-Arab alchemist Jābir ibn Hayyān (c.721–c.815) apparently mentioned it in 700 CE (Fig. 3.27). Marco Polo (Venetian merchant, explorer, and writer, 1254–January 8–9, 1324) brought some glazes back to Italy in the 13th century from his travels through Asia (including China). Georgius Agricola (German humanist scholar, mineralogist, and metallurgist, March 24, 1494–November 21, 1555), around 1600, described the use of borax as a flux in metallurgy. In 1777, boric acid was documented in the hot springs (soffioni) near Florence, Italy, and became known as sal sedativum, with principally medical applications. This rare mineral is called sassolite, H_3BO_3, which is found at Sasso, Italy. Sasso was the foremost source of European borax from 1827 to 1872, after which American sources replaced it. Boron compounds were hardly ever used until the late 1800s when Francis Marion Smith's Pacific Coast Borax Company first popularized and produced them in volume at low cost. Francis Marion Smith (February 2, 1846–August 27, 1931) (once known nationally and internationally as "Borax Smith" and "The Borax King") was an American miner, business magnate, and civic builder in the Mojave Desert, the San Francisco Bay Area, and Oakland, California. The word boron was coined from borax, $Na_2B_4O_5(OH)_4{\cdot}8H_2O$, the mineral from which it was isolated, by analogy with carbon, which boron bears a resemblance to chemically. Boron was not known as an element until it was isolated by Sir Humphry Davy (Cornish chemist and inventor, December 17, 1778–May 29, 1829) (Fig. 3.28) and by Joseph Louis Gay-Lussac (French chemist and physicist, December 6, 1778–May 9, 1850) and Louis Jacques Thénard (French chemist, May 4, 1777–June 21, 1857) (Davy, 1809; Gay-Lussac and Thénard, 1808, 1811a,b). In 1808 Davy noticed that electric current sent through a solution of borates resulted in a brown precipitate on one of the electrodes (Davy, 1809). In his follow-up experiments, he used potassium to reduce boric acid rather than electrolysis. He formed enough boron to confirm a new element and called the element boracium. Gay-Lussac and Thénard employed iron to reduce boric acid at high temperatures. By oxidizing boron with air, they proved that boric acid is an oxidation product of boron. Swedish chemist Jöns Jakob Berzelius (August 20, 1779–August 7, 1848)

FIGURE 3.27 Jabir ibn Hayyan, from a 15th-century European portrait of Geber, Codici Ashburnhamiani 1166.

FIGURE 3.28 Sir Humphrey Davy, oil on canvas by Thomas Phillips (died 1845).

recognized boron as an element in 1824 (Berzelius, 1824a,b). Pure boron was arguably first synthesized by the American chemist Ezekiel Weintraub (Belarusian chemist, July 4, 1874–??) in 1909 (Weintraub, 1910).

3.4.2 Mining, production, and major minerals

Economically significant sources of boron are the minerals colemanite, kernite, ulexite and borax. Together these constitute 90% of mined boron-containing ore. The largest global borax deposits known, many still unexploited, are in Central and Western Turkey, including the provinces of Eskişehir, Kütahya, and Balıkesir. Globally proven boron mineral mining reserves surpass one billion metric tons, compared to a yearly production of about four million tons. Turkey and the United States are the principal producers of boron products. The earliest routes to obtain elemental boron involved the reduction of boric oxide with metals such as magnesium or aluminum. However, the product is nearly always contaminated with borides of those metals. Pure boron can be obtained through reduction of volatile boron halides with hydrogen at high temperatures (Box 3.3). Ultrapure boron for use in the semiconductor industry is produced by the decomposition of diborane (B_2H_6) at high temperatures and then further purified by the zone melting or Czochralski processes (a method of crystal growth mainly used to obtain single crystals of semiconductors (e.g., silicon, germanium, and gallium arsenide), metals (e.g., palladium, platinum, silver, gold), salts and synthetic gemstones. The process is named after Polish scientist Jan Czochralski, who invented the method in 1915 while investigating the crystallization rates of metals). The production of boron compounds does not involve the formation of elemental boron but exploits the convenient availability of borates.

The majority of the minerals containing boron are found in the borates class. However, a handful of rather rare minerals are found in the halide class, for example, avogadrite ($(K,Cs)[BF_4]$), sulfates, for example, sturmanite ($Ca_6(Fe^{3+},Al,Mn^{3+})_2(SO_4)_2[B(OH)_4](OH)_{12}\cdot 25H_2O$).

Borates can, similar to C in organic chemistry and the silicates, form different polymeric groups, resulting in a rather large group of minerals. The best known of these include minerals such as: boracite ($Mg_3(B_7O_{13})Cl$), borax ($Na_2(B_4O_5)(OH)_4\cdot 8H_2O$), colemanite ($Ca[B_3O_4(OH)_3]\cdot H_2O$) (Fig. 3.29), hambergite ($Be_2(BO_3)(OH)$), inderite ($MgB_3O_3(OH)_5\cdot 5H_2O$), kernite ($Na_2[B_4O_6(OH)_2]\cdot 3H_2O$), londonite ($(Cs,K,Rb)Al_4Be_4(B,Be)_{12}O_{28}$)

BOX 3.3 Borax production

Boron comes almost completely from lacustrine evaporites. One of the largest deposits in southern California consists of hydrous sodium borate minerals. The boron in this deposit is believed to have originated from hot springs that flowed into the lakes before and during evaporation. Boron is also enriched in the Searles Lake brines in California. Boron production was made famous by the 20-mule team wagons that were used to haul borates out of Death Valley during the early part of the 1900s. Current operations in the area are much more modern, and include underground mining in order to limit surface disturbance at the edge of Death Valley National Monument. Elsewhere in southern California, mining is by conventional surface methods. Mined ore is crushed and milled (ground) as a preliminary step, followed by calcination (to remove water), screening, and air separation (to remove clay) in order to produce a powder that is subsequently dissolved in weak and hot borax liquor. Large insoluble particles are separated when the dissolved ore solution is passed over vibrating screens. The liquor and the fines are pumped into a series of large counter-current thickeners. Here the clarified liquor is filtered and pumped to vacuum crystallizers. Crystals of borax from these crystallizers are removed from the spent liquor in automatic centrifuges and then dried as refined borax decahydrate or borax pentahydrate. Anhydrous boric acid and anhydrous borax are produced by heat treatment in gas-fired furnaces. Boron production from brines simply avoids the crushing, grinding, and dissolution, but must deal with a more complex, dilute solution. This can be done by one of two processes: evaporation or carbonation. In the evaporation process, the brine is treated by rapid controlled cooling which selectively removes the crystallized NaCl, Na_2CO_3, and KCl. The remaining borax-containing liquor is fed into large crystallizers, where it is mixed with a thick bed of borax seed crystals. The resultant borax crystals are separated from the liquor in cyclones followed by drying to produce the final products. In the carbonation process, carbon dioxide from lime kiln gases or boiler flue gases is bubbled through the brine to convert Na_2CO_3 to $NaHCO_3$, which, being slight soluble in the brine, crystallizes and is filtered out. The filtrate is cooled in vacuum crystallizers in the presence of borax seed. The crystallized borax is dewatered in centrifuges and dried to produce borax decahydrate and borax pentahydrate.

FIGURE 3.29 Water clear colemanite, $Ca[B_3O_4(OH)_3]\cdot H_2O$, crystal to 1 cm showing complex terminations. Billie mine, Death Valley, California, United States.

(Fig. 3.30), tincalconite ($Na_2(B_4O_7)\cdot 5H_2O$), and ulexite ($NaCa[B_5O_6(OH)_6]\cdot 5H_2O$) (Fig. 3.31) (Box 3.4). In addition, there is a significant group of boron-containing silicates, which includes minerals such as: axinite-(Fe) ($Ca_2Fe^{2+}Al_2BSi_4O_{15}OH$), danburite ($CaB_2Si_2O_8$) (Fig. 3.32), datolite ($CaB(SiO_4)(OH)$) (Fig. 3.33), dumortierite ($(Al,Fe^{3+})_7(SiO_4)_3(BO_3)O_3$), kornerupine ($Mg_3Al_6(Si,Al,B)_5O_{21}(OH)$), tourmaline group (e.g., dravite $Na(Mg_3)Al_6(Si_6O_{18})(BO_3)_3(OH)_3(OH)$, elbaite $Na(Li_{1.5}Al_{1.5})Al_6(Si_6O_{18})(BO_3)_3(OH)_3(OH)$ (Fig. 3.34), schorl $Na(Fe_3^{2+})Al_6(Si_6O_{18})(BO_3)_3(OH)_3(OH)$), vesuvianite ($(Ca,Na,\square)_{19}(Al,Mg,Fe^{3+})_{13}(\square B,Al,Fe^{3+})_5(Si_2O_7)_4(SiO_4)_{10}(OH,F,O)_{10}$), and wiluite ($Ca_{19}(Al,Mg)_{13}(B,\square,Al)_5(Si_2O_7)_4(SiO_4)_{10}(O,OH)_{10}$) (Fig. 3.35).

FIGURE 3.30 Londonite, $(Cs,K,Rb)Al_4Be_4(B,Be)_{12}O_{28}$, unusual bluish-green crystal to 1.4 cm, twinned and shows six good isometric crystal faces. Antsongomvato mine, Betroka dist., Antananarivo, Madagascar.

FIGURE 3.31 Solid piece of satiny-white ulexite, $NaCa[B_5O_6(OH)_6]\cdot 5H_2O$, composed of tightly compact optical fiber-like crystals, 6 × 4 × 3.5 cm. The Boron mine in Kern Co., California is a large open-pit mine, the largest in California and likely the largest Borate mine in the World, that has been mined since the deposit's discovery in the 1920s.

3.4.3 Chemistry

Boron exist as two natural stable isotopes, ^{11}B (80.1%) and ^{10}B (19.9%). The mass difference produces a wide range of $\delta^{11}B$ values (defined as a fractional difference between ^{11}B and ^{10}B and usually reported in parts per thousand), in natural waters ranging from -16 to $+59$. Thirteen isotopes of boron are known to exist. The shortest-lived isotope is ^{7}B with a half-life of 3.5×10^{-22} s which decays through proton emission and alpha decay. Isotopic fractionation of B

BOX 3.4 Kernite, $Na_2[B_4O_6(OH)_2]{\cdot}3H_2O$, and borax, $Na_2(B_4O_5)(OH)_4{\cdot}8H_2O$

Kernite is named after its only major occurrence, in Kern County, California. Kernite is typically in massive or coarse aggregates that cleave into long splintery fragments, may appear fibrous due to intersecting perfect cleavages. Crystals are nearly equant. Kernite is monoclinic *2/m* (space group *P2/a*). It has a hardness of 2½ to 3, perfect prismatic cleavage (100) and (001), poor (010) and uneven fracture, the luster is vitrous to pearly, alters to earthy tincalconite, transparent. The color is colorless to white, while the streak is white. Optically it is biaxial(-).

The name borax is derived from the Persian word *burah*, meaning white. Euhedral crystals are stubby prisms with 8-sided cross-sections and with complex combinations of terminating faces. It is common in massive or granular aggregates. It is soluble in water; it has a slightly alkaline taste. It easily alters by dehydration to crumbly tincalconite on exposure to air. Borax is monoclinic *2/m* (space group *C2/c*). The hardness is 2–2½. It has a perfect {100} and good {110} cleavage and conchoidal fracture. The luster is vitreous to resinous, while the transparency is translucent. The color varies from white or gray to rarely light blue or green. The streak is white. Optically it is biaxial(–).

FIGURE 3.32 Danburite, $CaB_2Si_2O_8$, single 6 cm terminated pink crystal. San Luis Potosi, Mexico.

is controlled through the exchange reactions of the B species $B(OH)_3$ and $[B(OH)_4]^-$. Boron isotopes are also fractionated in the process of mineral crystallization, during H_2O phase changes in hydrothermal systems, and during hydrothermal alteration of rocks. The latter alteration produces a preferential removal of the $[^{10}B(OH)_4]^-$ ion by clays and solutions enriched in $^{11}B(OH)_3$ and thus may be the reason for the large ^{11}B enrichment in seawater relative to both oceanic and continental crusts. The exotic ^{17}B has a nuclear halo, that is, its radius is significantly larger than that

FIGURE 3.33 Datolite, $CaB(SiO_4)(OH)$, lime green, glassy crystals to 2.3 cm. Dal'negorsk, Primorsky Kray, Russia.

FIGURE 3.34 A 2 cm red to green zoned elbaite tourmaline, $Na(Li_{1.5}Al_{1.5})Al_6(Si_6O_{18})(BO_3)_3(OH)_3(OH)$, terminated on one end. Paprok, Kunar prov., Afghanistan.

predicted by the liquid drop model. The ^{10}B isotope is suitable for capturing thermal neutrons. The nuclear industry enriches natural occurring boron to almost pure ^{10}B. The less-valuable by-product, depleted boron, consists of almost pure ^{11}B. Due to its high neutron cross-section, ^{10}B is frequently used to control fission reactions in nuclear reactors as

FIGURE 3.35 Wiluite, $Ca_{19}(Al,Mg)_{13}(B,\square,Al)_5(Si_2O_7)_4(SiO_4)_{10}(O,OH)_{10}$, partial crystal to 1.2 cm - rare member of the vesuvianite group. Wilui River Basin, Yamutia, Siberia, Russia.

TABLE 3.4 Boron.

Appearance	Black-brown
Allotropes	α-, β-rhombohedral, β-tetragonal (and more)
Standard atomic weight $A_{r,std}$	10.81
Block	p-Block
Element category	Metalloid
Electron configuration	[He] $2s^2$ $2p^1$
Phase at STP	Solid
Melting point	2076°C
Boiling point	3927°C
Density (at m.p.)	2.08 g/cm^3
Heat of fusion	50.2 kJ/mol
Heat of vaporization	508 kJ/mol
Molar heat capacity	11.087 J/(mol · K)
Oxidation states	−5, −1, +1, +2, **+3**
Ionization energies	1st: 800.6 kJ/mol
	2nd: 2427.1 kJ/mol
	3rd: 3659.7 kJ/mol
Atomic radius	Empirical: 90 pm
Covalent radius	84 ± 3 pm
Van der Waals radius	192 pm

STP, Standard temperature and pressure.
Bold font indicates main oxidation state.

a neutron-capturing substance. Several industrial-scale enrichment methods have been established; but only the fractionated vacuum distillation of the dimethyl ether (methoxymethane, CH_3OCH_3) adduct of boron trifluoride (DME-BF_3) and column chromatography of borates are being used.

Boron is the lightest element with an electron in the p-orbital (electron configuration is $1s^22s^22p^1$) in its ground state (Table 3.4). It is the first element in Group 13 (aluminum family) of the periodic table but differs markedly from the other members of the group, which consists of aluminum, gallium, indium, and thallium. Boron is unique as it is the only nonmetal of the group and that it exhibits properties that are similar to its neighbor, carbon and its diagonal relative, silicon. Thus, like C and Si, boron forms covalent, molecular compounds. However, unlike those two, boron is more electron deficient and rarely obeys the octet rule. Given that it has three valence electrons, it typically forms trivalent neutral compounds such as

BF_3 wherein boron is surrounded by six electrons, assumes sp^2 hybridization and, adopts a trigonal planar geometry. This confers electron-pair accepting (Lewis acidity) and multicenter bonding properties to boron.

3.4.3.1 Reaction with acids

Although the number of boron compounds is extensive, elemental boron itself is rare and poorly studied because of the extreme difficulty of the preparation. Most of the existing samples of elemental boron are contaminated with small amounts of carbon. Crystalline boron is chemically inert and resistant to attack by boiling hydrofluoric or hydrochloric acid. It does react with hot concentrated nitric acid, hot sulfuric acid or hot mixture of sulfuric and chromic acids when finely divided.

$$2B + 2H_2SO_4(hot) \rightarrow B_2O_3 + 3SO_2 + 3H_2O$$

3.4.3.2 Reaction in air, oxygen, hydrogen, and nitrogen

The oxidation of boron in air is dependent on crystallinity, particle size, purity, and temperature. It does not react at room temperature but burns at higher temperature to form boron trioxide (B_2O_3).

$$4B + 3\,O_2 \rightarrow 2B_2O_3$$

B_2O_3 is produced by treating borax with sulfuric acid in a fusion furnace. At temperatures above 750°C, the molten boron oxide layer separates out from sodium sulfate. It is then decanted, cooled and obtained in 96%–97% purity. Another method is heating boric acid above ~300°C. Boric acid will initially decompose into steam ($H_2O(g)$) and metaboric acid (HBO_2) at around 170°C, and further heating above 300°C will produce more steam and diboron trioxide. The reactions are:

$$H_3BO_3 \rightarrow HBO_2 + H_2O \text{ and}$$
$$2HBO_2 \rightarrow B_2O_3 + H_2O$$

Boric acid goes to anhydrous microcrystalline B_2O_3 in a heated fluidized bed. Carefully controlled heating rate avoids gumming as water evolves. Molten boron oxide attacks silicates. Internally graphitized tubes via acetylene thermal decomposition are passivated. Crystallization of molten α-B_2O_3 at ambient pressure is strongly kinetically disfavored (compare liquid and crystal densities). Threshold conditions for crystallization of the amorphous solid are 10 kbar and ~200°C. Its proposed crystal structure in enantiomorphic space groups $P3_1$; $P3_2$ (e.g., γ-glycine) has been revised to enantiomorphic space groups $P3_121$; $P3_221$ (e.g., α-quartz). Boron oxide will also form when diborane (B_2H_6) reacts with oxygen in the air or trace amounts of moisture:

$$2B_2H_6(g) + 3O_2(g) \rightarrow 2B_2O_3(s) + 6H_2(g) \text{ and}$$
$$B_2H_6(g) + 3H_2O(g) \rightarrow B_2O_3(s) + 6H_2(g)$$

Boron monoxide (B_2O) is a chemical compound of boron and oxygen. Two experimental studies have proposed existence of diamond-like and graphite-like B_2O, as for boron nitride and carbon solids. However, a later, systematic, experimental study of boron oxide phase diagram suggests that B_2O is unstable. The instability of the graphite-like B_2O phase was also predicted theoretically. Boron reacts with nitrogen at 900°C–1000°C to form boron nitride (BN).

$$2B + N_2 \rightarrow 2BN$$

BN exhibit structures analogous to the various allotropes of carbon, including graphite, diamond, and nanotubes. In the diamond-like structure, boron atoms exist in the tetrahedral structure but one in every four B-N bonds can be viewed as a coordinate covalent bond wherein the nitrogen acts as Lewis base donating its pair of electrons to a Lewis acidic boron(III) center. In the graphitic analog, the positively charged boron and negatively charged nitrogen atoms in the hexagonal plane lie adjacent to the oppositely charged atom in the next plane. As a result of this arrangement, planes slip past each other easily and thus, graphite-like BN is a relatively poor electrical and thermal conductor in the planar directions in contrast to graphite.

Boron reacts with hydrogen gas to produce a colorless gas called borane, BH_3.

$$2B + 3H_2 \rightarrow 2BH_3$$

Boranes do not occur in nature as they readily oxidize on contact with air, some violently. BH_3 is found only in the gaseous state, and dimerizes to form diborane, B_2H_6. The larger boranes all consist of boron clusters that are polyhedral, some of which exist as isomers. The number of compounds containing boron and hydrogen are vast and their classifications are presented in a separate section.

3.4.3.3 Reaction with halogens

Boron undergoes halogenation to give trihalides;

$$2B + 3F_2 \rightarrow 2BF_3$$
$$2B + 3Cl_2 \rightarrow 2BCl_3$$
$$2B + 3Br_2 \rightarrow 2BBr_3$$

BF_3 is produced by the reaction of boron oxides with HF:

$$B_2O_3 + 6HF \rightarrow 2BF_3 + 3H_2O$$

Typically the HF is formed *in situ* from sulfuric acid and fluorite (CaF_2). For laboratory scale reactions, BF_3 is usually prepared *in situ* using boron trifluoride etherate, which is a commercially available liquid. There are many laboratory methods to obtain the solvent-free materials. One well-known method is the thermal decomposition of diazonium salts of BF_4^-:

$$PhN_2BF_4 \rightarrow PhF + BF_3 + N_2$$

Another method involves the reaction of sodium tetrafluoroborate, boron trioxide, with sulfuric acid:

$$6NaBF_4 + B_2O_3 + 6H_2SO_4 \rightarrow 8BF_3 + 6NaHSO_4 + 3H_2O$$

The geometry of BF_3 molecule is trigonal planar. Its D_{3h} symmetry conforms with the prediction of VSEPR (Valence Shell Electron Pair Repulsion) theory. The molecule has no dipole moment due to its high symmetry. The molecule is isoelectronic with the carbonate anion, CO_3^{2-}. BF_3 is generally referred to as "electron deficient," a description that is reinforced by its exothermic reactivity toward Lewis bases. In the boron trihalides, BX_3, the length of the B–X bonds (1.30 Å) is shorter than would be expected for single bonds, and this shortness may point to stronger B–X π-bonding in the fluoride. A simple explanation relies on the symmetry-allowed overlap of a p orbital on the B atom with the in-phase combination of the three similarly oriented p orbitals on F atoms. Others point to the ionic nature of the bonds in BF_3. Boron monofluoride or fluoroborylene, BF, was discovered as an unstable gas and only in 2009 found to be a stable ligand combining with transition metals, similar to carbon monoxide. It is a subhalide, containing fewer than the normal number of fluorine atoms, compared with boron trifluoride. It can also be called a borylene, as it contains boron with two unshared electrons. BF is isoelectronic with carbon monoxide and dinitrogen; each molecule has 14 electrons. Boron monofluoride can be formed by passing boron trifluoride gas at 2000°C over a boron rod. It can be condensed at liquid nitrogen temperatures (−196°C). The experimental B–F bond length is 1.26267 Å. Despite being isoelectronic to the triple-bonded species CO and N_2, computational studies in general concur that the actual bond order is much lower than 3. The lowest computed bond order is 1.4, compared with 2.6 for CO and 3.0 for N_2. BF can react with itself to form polymers of boron-containing fluorine with between 10 and 14 boron atoms. BF reacts with BF_3 to form B_2F_4. BF and B_2F_4 further combine to form B_3F_5. B_3F_5 is unstable above −50°C and forms B_8F_{12}. This substance is a yellow oil. BF reacts with acetylenes to form the 1,4-diboracyclohexadiene ring system. BF can condense with 2-butyne forming 1,4-difluoro-2,3,5,6-tetramethyl-1,4-diboracyclohexadiene. In addition, it reacts with acetylene to form 1,4-difluoro-1,4-diboracyclohexadiene. Propene reacts producing a mix of cyclic and non-cyclic molecules which may contain BF or BF_2. BF hardly reacts with C_2F_4 or SiF_4. BF does react with arsine, carbon monoxide, phosphorus trifluoride, phosphine, and phosphorus trichloride to make adducts like $(BF_2)_3B{\cdot}AsH_3$, $(BF_2)_3B{\cdot}CO$, $(BF_2)_3B{\cdot}PF_3$, $(BF_2)_3B{\cdot}PH_3$, and $(BF_2)_3B{\cdot}PCl_3$. BF reacts with oxygen:

$$BF + O_2 \rightarrow OBF + O;$$

with chlorine:

$$BF + Cl_2 \rightarrow ClBF + Cl;$$

and with nitrogen dioxide

$$BF + NO_2 \rightarrow OBF + NO.$$

Boron trichloride, BCl_3, is a colorless gas used as a reagent in organic synthesis. It is highly reactive toward water. Boron reacts with halogens to form the corresponding trihalides. BCl_3 is, however, manufactured commercially by direct chlorination of boron oxide and carbon at 500°C.

$$B_2O_3 + 3C + 3Cl_2 \rightarrow 2BCl_3 + 3CO$$

The carbothermic reaction is similar to the Kroll process for the conversion of TiO_2 to $TiCl_4$. In the laboratory BF_3 reacted with $AlCl_3$ produces BCl_3 via halogen exchange. BCl_3 is a trigonal planar molecule like the other boron trihalides, and has a bond length of 175pm. A degree of π-bonding has been suggested to explain the short B−Cl distance, but there is some debate as to its extent. It does not dimerize. However, NMR studies of mixtures of boron trihalides indicate the existence of mixed halides. The absence of dimerization is in contrast to the tendencies of $AlCl_3$ and $GaCl_3$, which form dimers or polymers with 4 or 6 coordinate metal centers. BCl_3 hydrolyzes readily to give hydrochloric acid and boric acid:

$$BCl_3 + 3H_2O \rightarrow B(OH)_3 + 3HCl$$

Alcohols behave in a similar manner forming the borate esters, e.g. trimethyl borate. Reduction of BCl_3 to elemental boron is performed commercially. In the laboratory, boron trichloride can be converted to diboron tetrachloride by heating with copper metal:

$$2BCl_3 + 2Cu \rightarrow B_2Cl_4 + CuCl$$

Likewise, B_4Cl_4 can be produced in a similar way. Colorless diboron tetrachloride (m.p. −93°C) is a planar molecule as a solid, similar to dinitrogen tetroxide, but in the gas phase the structure is staggered. It decomposes at room temperature to produce a series of monochlorides with the general formula $(BCl)_n$, in which n may be 8, 9, 10, or 11. The compounds with formulas B_8Cl_8 and B_9Cl_9 are known to have closed cages of boron atoms.

Boron tribromide, BBr_3, is a colorless, fuming liquid. Commercial samples usually are amber to red/brown, due to weak bromine contamination. It is decomposed by water and alcohols. The first synthesis was performed by Corsican chemist Antoine-Baudoin Poggiale (February 9, 1808 to August 26, 1879) in 1846 by reacting boron trioxide with carbon and bromine at high temperatures:

$$B_2O_3 + 3C + 3Br_2 \rightarrow 2BBr_3 + 3CO$$

An improvement of this method was developed by German chemist Friedrich Wöhler (July 31, 1800 to September 23, 1882) and French chemist Henri Étienne Sainte-Claire Deville (March 11, 1818 to July 1, 1881) in 1857. By starting from amorphous boron the reaction temperatures are lower and no carbon monoxide is produced:

$$2B + 3Br_2 \rightarrow 2BBr_3$$

The reaction of boron carbide with bromine at temperatures over 300°C results in the formation of boron tribromide. The product can be purified by vacuum distillation. It is an outstanding demethylating or dealkylating agent for the cleavage of ethers, also with subsequent cyclization, often in the manufacturing of pharmaceuticals. The mechanism of dealkylation of tertiary alkyl ethers proceeds via the formation of a complex between the boron center and the ether oxygen followed by the elimination of an alkyl bromide to produce a dibromo(organo)borane.

$$ROR + BBr_3 \rightarrow RO^+(^-BBr_3)R \rightarrow ROBBr_2 + RBr$$

Aryl methyl ethers (as well as activated primary alkyl ethers), on the other hand are dealkylated through a bimolecular mechanism involving two BBr_3-ether adducts.

$$RO^+(^-BBr_3)CH_3 + RO^+(^-BBr_3)CH_3 \rightarrow RO(^-BBr_3) + CH_3Br + RO^+(BBr_2)CH_3$$

The dibromo(organo)borane can then undergo hydrolysis to form a hydroxyl group, boric acid, and hydrogen bromide as products.

$$ROBBr_2 + 3H_2O \rightarrow ROH + B(OH)_3 + 2HBr$$

In addition, it finds usage in olefin polymerization and in Friedel-Crafts chemistry as a Lewis acid catalyst. The electronics industry uses boron tribromide as a boron source in predeposition processes for doping in the production of semiconductors. Boron tribromide also mediates the dealkylation of aryl alkyl ethers, for example, demethylation of 3,4-dimethoxystyrene into 3,4-dihydroxystyrene.

Boron triiodide, BI_3, has a trigonal planar molecular geometry. It is a crystalline solid, which reacts vigorously with water to form hydroiodic acid and boric acid. At extremely high pressures, BI_3 becomes metallic at ~23 GPa and is a superconductor above ~27 GPa. Boron triiodide can be produced via the reaction of boron with iodine at 209.5°C. It can also be prepared by another method:

$$3HI + BCl_3 \rightarrow BI_3 + 3HCl \text{ (though this reaction requires high temperature)}$$

The trihalides are electron deficient and adopts a planar trigonal structure. Boron trihalides are volatile, highly reactive compounds. All three lighter boron trihalides, BX_3 (X = F, Cl, Br) form stable adducts with common Lewis bases. Their relative Lewis acidities can be evaluated in terms of the relative exothermicities of the adduct-forming reaction. Such measurements have shown the following sequence for the Lewis acidity:

$$BF_3 < BCl_3 < BBr_3 \text{ (strongest Lewis acid)}$$

This trend is generally attributed to the degree of π-bonding in the planar boron trihalide that would be lost upon pyramidalization of the BX_3 molecule, which follows this trend:

$$BF_3 > BCl_3 > BBr_3 \text{ (most easily pyramidalized)}$$

The criteria for evaluating the relative strength of π-bonding are not clear, however. One suggestion is that the F atom is small compared to the larger Cl and Br atoms, and the lone pair electron in p_z of F is readily and easily donated and overlapped to empty p_z orbital of boron. As a result, the π donation of F is greater than that of Cl or Br. In an alternative explanation, the low Lewis acidity for BF_3 is attributed to the relative weakness of the bond in the adducts $F_3B{-}L$. For example, the BF_3 can react with ammonia to form a Lewis acid-base adduct.

$$BF_3 + :NH_3 \rightarrow F_3B:NH_3$$

3.4.3.4 Boron compounds

The chemistry of boron is due primarily to its small size and high ionization energy coupled with the similarity in electronegativity of B, C, and H. As such, boron has an extensive and unusual type of covalent (molecular) chemistry. It has affinity for oxygen leading to an extensive chemistry of its borate and oxo-complexes. Boron's small size allows the preparation of interstitial alloy-type metal borides. It has a propensity to form branched and unbranched chains, planar networks and intrinsically stable three-dimensional arrays.

Boron compounds can be classed into five types based on the type of bonding involved which results into distinguishable structures and reactions:

1. Metal borides
2. Boron hydrides and their derivatives
3. Boron trihalides (see above)
4. Oxo compounds
5. Organoboron and B-N compounds

3.4.3.5 Metal borides

The borides comprise a very large group (over 200) of binary compounds with a diverse stoichiometry and structural types, for example, metal borides in the form of M_5B, M_4B, M_3B, M_5B_2, M_7B_3, M_2B, M_5B_3, M_3B_2, $M_{11}B_8$, MB, $M_{10}B_{11}$, M_3B_4, M_2B_3, M_3B_5, MB_2, M_2B_5, MB_3, MB_4, MB_6, M_2B_{13}, MB_{10}, MB_{12}, MB_{15}, MB_{16}, and MB_{66}. There are also many ternary and more complex phases in which more than one metal combines with boron (e.g., $Nd_2Fe_{14}B$). Metal-rich borides in which atoms are either isolated, form dumbbells or chains can be distinguished from the boron-rich borides that display exotic framework structures. There are also compounds with more balanced metal to boron ratio where the structures are layered structures with two-dimensional arrays of boron atoms. The degree of dimensionality of boron atom arrangement correlates closely with the metallic/nonmetallic character of the borides leading to extreme variability in electrical properties between metallic and ceramic. The interests in borides stem from the inherent academic challenge of understanding such unusual compounds as well as from the extensive industrial interest generated by their unique combination of desirable physical and chemical properties. Borides can be prepared in a variety of ways including direct combination of the elements. For example:

$$Cr + nB \xrightarrow{1150^\circ C} CrB_n$$

Transition elements tend to combine with boron to form metal-rich borides. Boron-rich bromides are mostly formed by more electropositive elements in Groups 1–3, the lanthanides and actinides. Diborides are common to both group of borides.

$$4FeSO_4 + 8NaBH_4 + 18H_2O \rightarrow 2Fe_2B + 6\,B(OH)_3 + 25H_2 + 4Na_2SO_4$$

$$Eu_2O_3 + 3B_4C \xrightarrow{1600^\circ C} 2EuB_6 + 3CO$$

Boron in metal borides is assigned a negative oxidation state. For example, in magnesium boride (MgB_2), each boron atom has a formal -1 charge and magnesium is assigned a formal charge of $+2$. MgB_2 has the boron centers in trigonal planar geometry with an extra double bond for each boron, forming sheets akin to graphite. The delocalized electrons in magnesium diboride allow it to conduct electricity unlike the previously discussed BN.

3.4.3.6 Boron hydrides (boranes)

Boranes are composed of boron and hydrogen, with the generic formula of B_xH_y. Over 50 neutral boranes, B_xH_y, and an even larger number of borane anions $B_xH_y^{n-}$ have been characterized. In these compounds, the formal oxidation number of boron is positive. This is based on the assumption that hydrogen is counted as -1 as in active metal hydrides. Thus, the mean oxidation number for the borons is then simply the ratio of hydrogen to boron in the molecule. For example, in diborane B_2H_6, the boron oxidation state is $+3$, but in decaborane B_5H_9, it is $^9/_5$ or $+1.8$. In these compounds the oxidation state of boron is often not a whole number.

The vast number of boranes can be classified into five series according to structure and stoichiometry—*closo*-boranes, *nido*-boranes, *arachno*-boranes, *hypho*-boranes, and *conjuncto*-boranes. Only a few examples will be enumerated here. *Closo*-boranes have complete, closed polyhedral clusters of n boron atoms with the general formula $B_xH_x^{2-}$ ($x = 6-12$). Neutral boranes of the formula B_xH_{x+2} are not known. *Nido*-boranes have nonclosed structures in which B_x cluster occupies x corners of an $(x + 1)$-cornered polyhedron. The general formula for neutral *nido*-boranes is B_xH_{x+4} such as B_2H_6, B_5H_9, and $B_{10}H_{14}$ whereas the anionic members of this series have general formulas of $B_xH_{x+3}^-$ such as $B_4H_7^-$, $B_5H_8^-$ $B_9H_{12}^-$, etc. *Arachno*-boranes have even more open clusters with the B atoms occupying x contiguous corners of an $(x + 2)$-cornered polyhedrons. The general formula for the neutral boranes is B_xH_{x+6}, and for the anionic boranes are $B_xH_{x+5}^-$ and $B_xH_{x+4}^{2-}$. Examples are B_4H_{10}, B_5H_{11}, B_6H_{12}, $B_2H_7^-$, $B_9H_{14}^-$, and $B_{10}H_{14}^{2-}$. *Hypho*-boranes have the most open clusters in which the B atoms are in x corners of $(x + 3)$-cornered polyhedrons. The general formula is B_xH_{x+8}. The known compounds B_8H_{16} and $B_{10}H_{18}$ follow this general formula although they are not clearly established to follow the defined structure for *hypho*-boranes. *Conjuncto*-boranes have structures formed by linking two (or more) of the preceding types of cluster together. They have the general formula B_xH_y and are classified into five structure types: (1) fused clusters sharing a single B atom such as $B_{15}H_{23}$; (2) two clusters sharing a 2-center B-B sigma (σ) bond such as $(B_4H_9)_2$; (3) two clusters fused via 2 B atoms at a common edge such as $B_{13}H_{19}$; (4) two clusters fused via 3 B atoms at a common face such as $(MeCN)_2B_{20}H_{16}\cdot MeCN$; and (5) more extensive fusion involving 4 B atoms in various configurations such as $B_{20}H_{16}$ and $B_{20}H_{18}^{2-}$.

3.4.3.7 B-O compounds

Boron invariably occurs in nature as oxo compounds and is never found as the element or even directly bonded to any other element than oxygen. As with borides and boranes, the structure and reactivity of B-O compounds are extraordinarily complex and diverse. As already presented earlier in this section, boron reacts with oxygen to form B_2O_3. This oxide is difficult to crystalize and is only found in the vitreous state. The structure is believed to be a network of partially ordered trigonal BO_3 units in which the 6-membered $(BO)_3$ ring predominates. Above its melting point (450°C), it forms polar $-B=O$ groups. Fused B_2O_3 dissolves many metal oxides which impart color to borate glasses.

Orthoboric acid $B(OH)_3$ is a weak, monobasic Lewis acid and acts as hydroxyl ion acceptor.

$$B(OH)_3 + 2H_2O \rightarrow H_3O^+ + B(OH)_4^-$$

It dissolves in anhydrous sulfuric acid according to the reaction:

$$B(OH)_3 + 6H_2SO_4 \rightarrow 3H_3O^+ + 2HSO_4^- + B(HSO_4)_4^-$$

It can be reacted with alcohols in the presence of dehydrating agent such as concentrated sulfuric acid to form borate esters ($B(OR)_3$, where R is an alkyl or aryl group):

$$B(OH)_3 + 3ROH \rightarrow B(OR)_3 + 3H_2O$$

Other reaction of orthoboric acid include coordination with NaH in tetrahydrofuran (THF) to produce a powerful reducing agent $Na[BH(OR)_3]$ and reaction with H_2O_2 to form peroxoboric acid solutions which may contain the monoperoxoborate anion $[B(OH)_3OOH]^-$.

Borate is the name for a large number of boron-oxygen compounds usually containing oxyanions. The structural unit of borates varies from mononuclear (1 B atom), bi-, tri-, tetra-, pentanuclear or polydimensional networks including glasses. A widely known borate is sodium tetraborate or borax ($Na_2B_4O_7$), which has many applications as will be discussed later. Borax can easily be converted to boric acid or even other borates.

$$Na_2B_4O_7{\cdot}10H_2O + 2HCl \rightarrow 4B(OH)_3 + 2NaCl + 5H_2O$$

The guiding principles in the borate bonding are enumerated below.

1. Boron can form triangle structures or a tetrahedron by linking to either three or four atoms, respectively.
2. Boron-oxygen triangles and tetrahedra can share corners in a manner that form compact insular groups.
3. Protonable oxygen atoms in hydrated borates will protonate in a sequence: available protons are first assigned to free O^{2-} ions to convert these to free OH^-, succeeding protons are then assigned to tetrahedral oxygens then to triangular oxygens in the borate ion; and lastly, any remaining proton are assigned to free OH^- to form H_2O molecules.
4. Polymerization can occur in various ways by splitting water, which may be accompanied by breaking of B-O bonds within the framework.
5. Modification of the complex borate polyanions can be done by attaching an individual side group.
6. Isolated or polymeric forms of $B(OH)_3$ may exist in the presence of other anions.

3.4.3.8 *Organoboron chemistry*

Organoboron reagents continue to be pivotal in the development and understanding of new chemistries involving boron and are widely studied and applied throughout organic synthesis and catalysis. Organoborons mostly contain trigonal planar B although several 4-coordinate complexes also exist. Owing the immensity of both the compounds and its reactions, only a few examples will be shown here. The reader is referred to the references herein to read more on the recent innovations in organic synthesis involving boron compounds. Suzuki cross-coupling is a seminal reaction that have stood the test of time and is included in the examples below.

1. Migration of organic groups to electrophilic sites on adjacent atoms.

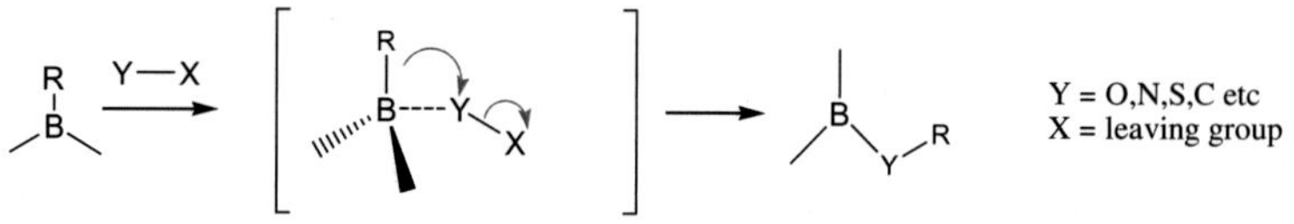

R B Y—X B----Y X B Y R

Y = O,N,S,C etc
X = leaving group

2. Suzuki coupling for C-C bond formation.

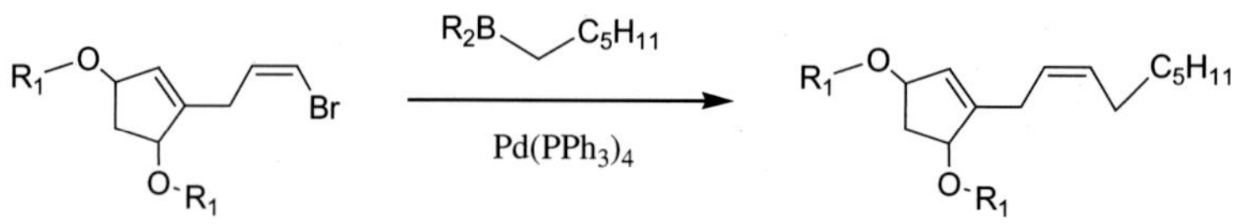

R_1 O Br O R_1 R_2B C_5H_{11} $Pd(PPh_3)_4$ R_1 O C_5H_{11} O R_1

3. Enol, allyl, and propargyl boranes will transfer the group on boron to suitable electrophiles.

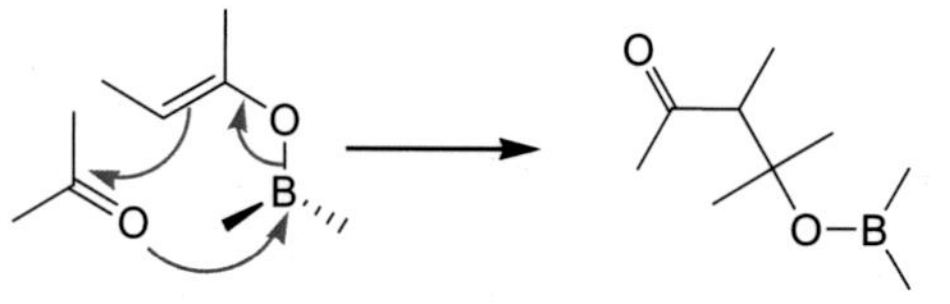

O B O → O O—B

3.4.4 Major uses

Boron compounds are important in many industries, such as in making glass and detergents, and in agriculture with thousands of tons added to fertilizer each year as boron is essential to plant growth. The most important compounds are borax (sodium borate), boric oxide, and boric acid. The chief global industrial-scale use of boron compounds (nearly half of end-use) is in manufacturing of glass fiber for B-containing insulating and structural fiberglass, especially in Asia. Boron is added to the glass as borax pentahydrate or boron oxide, to affect the strength or fluxing qualities of the glass fibers. Boron fibers are high-strength, lightweight materials that are used primarily for advanced aerospace structures as a component of composite materials, as well as limited production consumer and sporting goods, for example, golf clubs and fishing rods. The fibers can be made through chemical vapor deposition (CVD) of boron on a tungsten filament. Boron fibers and submillimeter sized crystalline boron springs are formed through laser-assisted CVD. Translation of the focused laser beam allows the manufacture of even more complex helical structures. Such structures exhibit good mechanical properties (elastic modulus 450 GPa, fracture strain 3.7%, fracture stress 17 GPa) and can be used as a reinforcement in ceramics or in micromechanical systems. Fiberglass is a fiber-reinforced polymer consisting of plastic reinforced by glass fibers, usually woven into a mat. The glass fibers used in the material consist of a variety of types of glass dependent on the fiberglass application. These glasses all have silica or silicate, with variable amounts of oxides of Ca, Mg, and sometimes B. The B is present as borosilicate, borax, or boron oxide, and is added to the glass to increase its strength, or as a fluxing agent to decrease the melting temperature of silica, which is too high to be easily worked in its pure form to make glass fibers. Borosilicate glass, which consists typically of 12%–15% B_2O_3, 80% SiO_2, and 2% Al_2O_3, has a low thermal expansion coefficient resulting in a good resistance to thermal shock. Schott AG's Duran and Owens-Corning's trademarked Pyrex are two major brand names for this glass, used both in laboratory glassware and in consumer cookware and bakeware, primarily for this resistance.

Several boron compounds have been recognized for their extreme hardness and toughness. Boron carbide is a ceramic material which is formed via decomposing B_2O_3 with carbon in an electric furnace:

$$2B_2O_3 + 7C \rightarrow B_4C + 6CO$$

Its structure is only approximately B_4C, and it exhibits a distinct depletion of carbon from this stoichiometric ratio. This is because of its very complex structure. The repeating polymer plus semicrystalline structure of boron carbide results in great structural strength per weight. It is used in tank armor, bulletproof vests, and numerous other structural applications. Boron carbide's capacity to absorb neutrons without forming long-lived radionuclides (particularly when doped with additional boron-10) makes this ceramic of interest as an absorbent material for neutron radiation arising in nuclear power plants, including shielding, control rods, and shut-down pellets. Within control rods, boron carbide is frequently powdered, to increase its surface area. In pressurized water reactors a variable concentration of boronic acid in the cooling water is used to compensate the variable reactivity of the fuel: when new rods are inserted the concentration of boronic acid is maximal, and then reduced during the lifetime. Boron carbide and cubic boron nitride powders are extensively used as abrasives. Boron nitride is a material isoelectronic to carbon. Like carbon, it has both hexagonal (soft graphite-like h-BN) and cubic (hard, diamond-like c-BN) forms. h-BN is utilized as a high-temperature component and lubricant. c-BN, also known under commercial name borazon, is a superior abrasive. Its hardness is only slightly smaller than, but its chemical stability is superior, to that of diamond. Heterodiamond (also called BCN) is another diamond-like boron compound. It is a superhard material containing B, C, and N (BCN). It is formed at high temperatures and high pressures, for example, by application of an explosive shock wave to a mixture of diamond and cubic boron nitride (c-BN). The heterodiamond is a polycrystalline material coagulated with nano-crystallites. The heterodiamond has both the high hardness of diamond and the excellent heat resistance of cubic BN. These characteristic properties are due to the diamond structure combined with the sp^3 σ-bonds among carbon and the heteroatoms.

Borax is used in various household laundry and cleaning products. It is also present in some tooth bleaching formulas. Sodium perborate serves as a source of active oxygen in many detergents, laundry detergents, cleaning products, and laundry bleaches. Boric acid is used as an insecticide, particularly against ants, fleas, and cockroaches. Borates are used as environmentally benign wood preservatives. The most important compounds of boron—boric (or boracic) acid, borax (sodium borate), and boric oxide —can be found in eye drops as mild antiseptics. Borax is also used as food preservative.

Boron is a useful dopant for such semiconductors as Si, Ge, and silicon carbide. Having one fewer valence electron than the host atom, it donates a hole resulting in p-type conductivity. The traditional process of introducing boron into semiconductors was through atomic diffusion at high temperatures. This process utilized either solid (B_2O_3), liquid (BBr_3), or gaseous boron sources (B_2H_6 or BF_3). However, after the 1970s, it was generally replaced by ion implantation, which relies mostly on BF_3 as a B source. In addition, boron trichloride gas is an important chemical in semiconductor industry, though not for doping but instead for plasma etching of metals and their oxides. Triethylborane is also injected into vapor deposition reactors as a source of B, for example, in the plasma deposition of B-containing hard carbon films, silicon nitride-boron nitride films, and for doping of diamond film with B.

Borates are used as flame retardants. Upon heating sodium borates release water from their crystalline structure, which help as a fire retardant. In addition, boric acid releases water acting as fire retardant while at the same time acting as smolder suppressant. When added to more acidic fire-retardant formulations, borates act as buffer. It also provides synergistic effects with other fire retardants compounds (for example, aluminum trihydrate, aluminum sulfate, ammonium phosphate, ammonium sulfate, borax/boric acid combined in a species of di-sodium octaborate tetrahydrate, calcium sulfate (gypsum), guanyl-urea phosphate, urea, zinc chloride, and zinc phosphate, etc.) by increasing the flame retardancy of the borate-enhanced fire retardants.

3.5 6 C — Carbon

3.5.1 Discovery

The English name carbon comes from the Latin carbo for coal and charcoal, similarly in French charbon, meaning charcoal. In German, Dutch, and Danish, the names for carbon are Kohlenstoff, koolstof, and kulstof respectively, all literally meaning coal-substance. Carbon was discovered in prehistory and was known in the forms of soot and charcoal to the earliest humans. Diamonds were known possibly as early as 2500 BCE in China, while carbon in the form of charcoal was made around Roman times by the same chemical process, which is still used today, by heating wood in a pyramid covered with clay to exclude air. In 1722, René Antoine Ferchault de Réaumur (French entomologist and writer who contributed to many different fields, February 28, 1683–October 17, 1757) proved that iron was changed into steel through the absorption of some substance, now known to be carbon. In 1772, Antoine Lavoisier (French nobleman and chemist, August 26, 1743–May 8, 1794) showed that diamonds are a form of carbon; when he burned samples of charcoal and diamond and found that neither formed any water and that both released the same amount of carbon dioxide, CO_2, per gram. In 1779, Carl Wilhelm Scheele (Swedish Pomeranian and German pharmaceutical chemist, December 9, 1742–May 21, 1786) (Fig. 3.36) proved that graphite, which was believed to be a form of lead, was instead the same

FIGURE 3.36 Carl Wilhelm Scheele from Svenska Familj-Journalen 1874.

FIGURE 3.37 Claude Louis Berthollet.

as charcoal but with a small amount of iron, and that it gave "aerial acid" (his name for carbon dioxide) when oxidized with nitric acid. In 1786, three French scientists Claude Louis Berthollet (chemist, December 9, 1748–November 6, 1822) (Fig. 3.37), Gaspard Monge (mathematician, May 9, 1746–July 28, 1818) and Alexandre-Théophile Vandermonde (mathematician, musician, and chemist, February 28, 1735–January 1, 1796) confirmed that graphite was mostly carbon by oxidizing it in oxygen similar to what Lavoisier had done with diamond. Some iron was once more left behind, which the French scientists believed to be a necessary component in the graphite crystal structure. In their paper they proposed the name carbone (Latin carbonum) for the element found in graphite which was given off as a gas upon burning graphite. Antoine Lavoisier subsequently listed carbon as an element in his 1789 textbook (Lavoisier, 1790).

3.5.2 Mining, production, and major minerals

Commercially viable deposits of graphite can be found in many parts of the world, but the economically most significant deposits are in China, India, Brazil and North Korea. Graphite deposits are of metamorphic origin, found in association with quartz, mica and feldspars in schists, gneisses, and metamorphosed sandstones and limestone as lenses or veins, sometimes of a meter or more in thickness. There are three major types of natural graphite—amorphous, flake or crystalline flake, and vein or lump graphite. Amorphous graphite is the lowest quality and most abundant. Different to science, in industry "amorphous" here refers to very small crystal size rather than complete absence of crystal structure. Amorphous graphite is used for lower value graphite products and is the lowest priced. Large amorphous graphite deposits can be found in China, Europe, Mexico, and the United States. Flake graphite is less common and of higher quality than amorphous graphite; it occurs as separate plates that crystallized in metamorphic rocks (Fig. 3.38). Flake graphite can be four times the price of amorphous. Good quality flakes can be processed into expandable graphite for many applications, for example, flame retardants. The principal deposits are found in Austria, Brazil, Canada, China, Germany, and Madagascar. Vein or lump graphite is the rarest, highest quality and therefore most valuable, type of natural graphite. It occurs in veins along intrusive contacts in solid lumps, and it is only commercially mined in Sri Lanka.

Historically diamonds were known only from alluvial deposits in southern India. India led the world in diamond production from the time of their discovery in approximately the 9th century BCE to the mid-18th century CE, but the alluvial deposits were basically exhausted by the late 18th century and at that time Brazil became the world's most important producer, after the first non-Indian diamonds were found in 1725. Diamond production of primary deposits (kimberlites and lamproites) only began in the 1870s after the discovery of the diamond fields in South Africa. Production has grown over time and now an accumulated total of 4.5 billion carats (1 carat = 0.2 grams) have been

FIGURE 3.38 Silvery, lamellar graphite, C, crystals to 1.5 cm. Amity, Orange Co., New York, United States.

FIGURE 3.39 A yellow 24pt (1/4 carat) uncut, natural diamond, C, crystal showing many complex forms with a smooth, silky luster. Crater of Diamond, Pike Co., Arkansas, United States.

mined since that date. Today, most commercially viable diamond deposits are being mined in Russia, Botswana, Australia, and the Democratic Republic of Congo. There are also commercial deposits being actively mined in the Northwest Territories of Canada and Brazil. The diamond supply chain is controlled by a small number of powerful businesses (e.g., De Beers) and is also highly concentrated in a limited number of locations around the world. Only a very small fraction of the diamond ore consists of actual diamonds. The ore is crushed, during which care must be taken to prevent larger diamonds from being destroyed in this process after which the particles are sorted by density. Today, diamonds are located in the diamond-rich density fraction with the help of X-ray fluorescence, after which the final sorting steps are done by hand. Before the use of X-rays became routine, the separation was performed using grease belts; diamonds have a stronger tendency to stick to the grease than the other minerals in the ore.

Carbon is the first element in the periodic table that occurs in nature as a native element in graphite (C) (Fig. 3.38) and diamond (C) (Fig. 3.39) (Box 3.5). Due to the difference in which the C-C bonds are formed, graphite is rather soft as they form layers with strong bonds within the layer but rather weak bonds between the layers, whereas in diamond a

BOX 3.5 Synthetic diamond

The first claim of the synthesis of diamond was in 1878 by the Scottish chemist James B. Hannay. He placed in a sealed steel tube a mixture of bone oil and paraffin to which metallic lithium had been added. The mixture was heated in a furnace for about 14 hours. He conducted a total of 80 experiments and in 3 of them he found small transparent particles he thought to be diamond. A dozen of these were sent to the Keeper of Minerals in the British Museum, M.H. Story-Maskelyne, who confirmed they were diamonds. The particles were preserved and tested in 1943 by X-ray diffraction. While 11 of the 12 particles were shown to be diamond, they have since been proven to be natural and not synthetic. The next was the French mineralogical chemist Henri Moissan, who in 1904 announced that he had made diamonds by placing pure carbon and iron in a crucible, melting the mixture in an electric furnace, and subsequently quenching the molten mass. The crystals he was able to recover had a specific gravity of 3–3.5 and produced CO_2 when burned in oxygen. Hence Moissan concluded he had made diamond. However, it later turned out that his assistant had added diamond chips to the mixture. Based on current knowledge of the graphite-diamond phase diagram it is impossible for either to have produced diamond. It was not until 1955 that scientists of General Electric Company were successful in synthesizing diamonds using an apparatus that could create and maintain temperatures of more than 2500°C and pressures up to 1.5×10^6 psi (1 atmosphere = 14.7 psi). By 1958 they were producing synthetic diamond abrasives at competitive prices compared to natural diamonds. On May 28, 1970 the same company announced that they had successfully synthesized transparent, gem-quality diamonds, some of them weighing as much as 1 carat. The method used for synthesizing industrial-grade synthetic diamonds was adapted so that the required temperature and pressure could be maintained for as long as several days. This allowed the crystals to grow slowly, essential for nearly flawless crystals of large size. The pressure chamber consisted of a ring-shaped tungsten-carbide piece supported by a series of tight belts. The pressure was generated by forcing together conical tungsten-carbide pistons that wedged into a die, with a gasket of pyrophyllite ($Al_2Si_4O_{10}(OH)_2$). The reaction zone container was also made of pyrophyllite. The pressure applied to the cell was 60,000 atmospheres. the reaction zone was inside a carbon tube heater, which was heated through passage of an electric current from one piston, through the heater, and out through the opposite piston. Inside the reaction cell is a source of carbon in the form of industrial diamond. On each side of this carbon source is a catalyst, such as iron or nickel, that becomes molten under operating conditions. At the ends are diamond seed crystals. A temperature difference of about 10°C–15°C was maintained between the center and the ends of the cell. The carbon dissolved in the hotter central part of the cell is deposited on the diamond seeds in the cooler regions. A typical experiment applied a temperature of 1455°C in the center, while the temperature around the growing crystals was around 1425°C. These temperatures were maintained for several days at 60,000 atm. In 1985 Sumitomo Electric Industries reported that they were producing gem-quality diamonds up to 1.2 carats. According to De Beers officials they have been growing diamonds on strictly an experimental basis with none being marketed. The largest gem-quality diamond grown by De Beers Diamond Research Laboratory weighed 11.14 carats. The reports by Sumitomo and De Beers seem to indicate that, at present, it is easier to grow yellow synthetic diamonds than colorless or blue ones.

three-dimensional network is present accounting for its extreme hardness (10 on Mohs scale). A third mineral is lonsdalite (C), In addition, a couple of carbides can be formed under extreme conditions, such as cohenite (Fe_3C), moissanite (SiC), qusongite (WC) or tongbaite (Cr_3C_2). Four halides are known to contain the carbonate group (CO_3), for example, barstowite ($Pb_4Cl_6(CO_3)·H_2O$). Similarly, a handful of oxides contain the carbonate group, for example, hydrocalumite ($Ca_4Al_2(OH)_{12}(Cl,CO_3,OH)_2·4H_2O$) or mroseite ($CaTe^{4+}(CO_3)O_2$). A number of borates are also known with this group, such as borcarite ($Ca_4Mg(B_4O_6(OH)_6)(CO_3)_2$) and canavesite ($Mg_2(HBO_3)(CO_3)·5H_2O$). Similarly for the sulfate class, one find for example, hanksite ($Na_{22}K(SO_4)_9(CO_3)_2Cl$) and thaumasite ($Ca_3(SO_4)[Si(OH)_6](CO_3)·12H_2O$). The best-known phosphate-containing carbonate is tyrolite ($Ca_2Cu_9(AsO_4)_4(CO_3)(OH)_8·11H_2O$), while carbonate-rich apatite is also known to exist where the carbonate substitutes for the phosphate group. The most important and largest group of minerals is formed by the class containing the carbonates. The best-known minerals include: ankerite ($Ca(Fe^{2+},Mg)(CO_3)_2$), aragonite ($CaCO_3$) (Fig. 3.40), azurite ($Cu_3(CO_3)_2(OH)_2$), calcite ($CaCO_3$) (Fig. 3.41), cerussite ($PbCO_3$) (Fig. 3.42), dawsonite ($NaAlCO_3(OH)_2$), dolomite ($CaMg(CO_3)_2$) (Fig. 3.43), hydrotalcite ($Mg_6Al_2(CO_3)(OH)_{16}·4H_2O$), kutnohorite ($CaMn^{2+}(CO_3)_2$), magnesite ($MgCO_3$), malachite ($Cu_2(CO_3)(OH)_2$), parisite-(Ce) ($CaCe_2(CO_3)_3F_2$), phosgenite ($Pb_2CO_3Cl_2$) (Fig. 3.44), rhodochrosite ($MnCO_3$) (Fig. 3.45), rosasite ($(Cu,Zn)_2(CO_3)(OH)_2$), siderite ($FeCO_3$) (Fig. 3.46), sphaerocobaltite ($CoCO_3$), strontianite ($SrCO_3$) (Fig. 3.47) welagonite ($Na_2Sr_3Zr(CO_3)_6·3H_2O$), and witherite ($BaCO_3$) (Fig. 3.48). Second in terms of number of minerals containing carbonate are the silicates. Minerals in this class include, for example, cancrinite ($(Na,Ca,\square)_8(Al_6Si_6O_{24})(CO_3,SO_4)_2·2H_2O$), carletonite ($KNa_4Ca_4Si_8O_{18}(CO_3)_4(OH,F)·H_2O$), and meionite ($Ca_4Al_6Si_6O_{24}CO_3$). The final group containing carbon are the organic compounds. These can be separated in 4 groups: (1) oxalates, for example, whewellite ($CaC_2O_4·H_2O$), weddellite ($CaC_2O_4·2H_2O$), and humboltine ($FeC_2O_4·2H_2O$), (2) mellitates, citrates, cyanates, acetates and formates, for example, mellite ($Al_2C_6(COO)_6·16H_2O$),

FIGURE 3.40 Aragonite, $CaCO_3$, colorless hexagonal, twinned crystals to 1.5 cm. Molina de Aragón, Guadalajara, Castile-La Mancha, Spain.

FIGURE 3.41 A 9 cm amber-orange twinned calcite, $CaCO_3$, crystal with sharp terminations. Elmwood mine, Smith Co., Tennessee, United States.

earlandite ($Ca_3(C_6H_5O_7)_2 \cdot 4H_2O$), and kafehydrocyanite ($K_4Fe(CN)_6 \cdot 3H_2O$), (3) hydrocarbons, for example, fichtelite ($C_{19}H_{34}$), evenkite ($C_{24}H_{50}$), and carpathite ($C_{24}H_{12}$), and (4) miscellaneous, for example, uricite ($C_5H_4N_4O_3$), guanine ($C_5H_3(NH_2)N_4O$), and urea ($CO(NH_2)_2$).

3.5.3 Chemistry

Carbon isotopes are atomic nuclei that have 6 protons plus several neutrons (between 2 and 16). Carbon has two stable, naturally occurring isotopes. The isotope ^{12}C forms 98.93% of the carbon on Earth, while ^{13}C forms the remaining 1.07%. The ^{12}C concentration is further selectively increased in biological materials due to biochemical reactions discriminating against ^{13}C. In 1961, the IUPAC accepted the isotope ^{12}C as the basis for atomic weights. ^{14}C is a naturally formed radioisotope, produced in the upper atmosphere (lower stratosphere and upper troposphere) by interaction of nitrogen with cosmic rays. It is observed in trace amounts on Earth of 1 part per trillion (0.0000000001%) or more, generally limited to the atmosphere and near-surface deposits, mostly of peat and other organic materials. This isotope decays by 0.158 MeV β^-

FIGURE 3.42 White 1–5 mm "jack-straw" cerussite, $PbCO_3$, crystals. Flux mine, Patagonia, Santa Cruz Co., Arizona, United States.

FIGURE 3.43 Dolomite, $CaMg(CO_3)_2$, pink saddle-shaped 9 mm crystals. Black Rock, Lawrence Co., Arkansas, United States.

emission. Due to its relatively short half-life of 5730 years, ^{14}C is effectively absent in ancient rocks. The concentration of ^{14}C in the atmosphere and in living organisms is virtually constant but decreases predictably in their bodies after death. This principle is applied in radiocarbon dating, developed in 1949, which is applied extensively to determine the age of carbonaceous materials with ages up to about 40,000 years. There are 15 known isotopes of carbon and the shortest-lived of these is ^{8}C with a half-life of 1.98739×10^{-21} s and decays through proton emission and alpha decay. The exotic ^{19}C has a nuclear halo, meaning its radius is noticeably larger than would be predicted assuming the nucleus was a sphere of constant density.

Carbon is the sixth element in the periodic table and its ground-state electronic configuration is $[He]2s^22p^2$ (Table 3.5). The carbon atom is usually tetravalent with the s and the three degenerate p orbitals hybridizing to form sp* (* = 1, 2, or 3) hybrid states. However, divalent species in which carbon uses only two of its valence electrons while the remaining two retain as a lone electron pair (e.g., carbenes) can also form. The chemistry of carbon is dominated by three factors: (1) it forms unusually strong bond with itself in varying bond orders—1 (single bonds), 2 (double bonds) or 3 (triple bonds); (2) it has an electronegativity (EN = 2.55) that is too small to allow formation of C^{4-} ions and too large to form C^{4+} and so, carbon forms covalent bonds with many other elements; and (3) it forms strong double and triple bonds

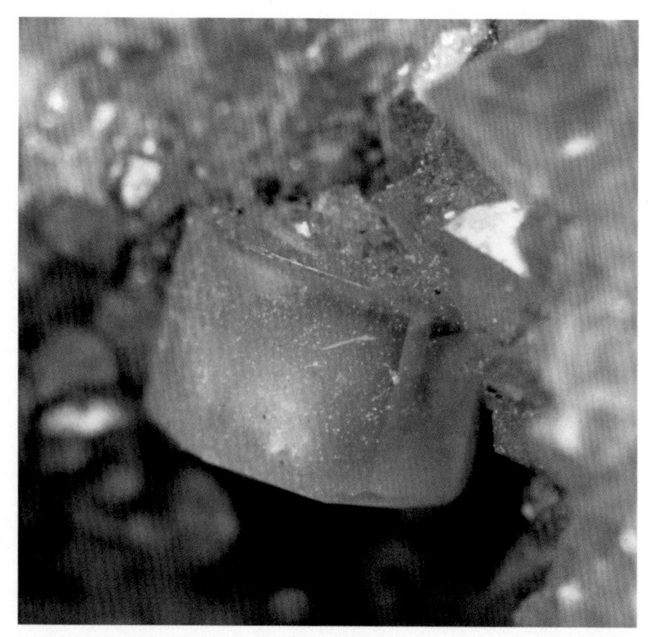

FIGURE 3.44 Phosgenite, $Pb_2CO_3Cl_2$, a 3 mm phosgenite crystal in an open vein filled with clear quartz, SiO_2, crystals. Newporth, Falmouth, Cornwall, England.

FIGURE 3.45 Rhodochrosite, $MnCO_3$, partial cherry red rhombic crystal to just about 1.5 cm. Associations include colorless quartz, SiO_2, and tetrahedrite, $Cu_6Cu_4(Fe^{2+},Zn)_2Sb_4S_{13}$. Sweet Home mine, near Alma, Park Co., Colorado, United States.

with a number of other nonmetals, including N, O, P, and S. The unique ability to form strong bonds with itself and with many other elements leads to a vast number of its compounds — there are probably more than one million known today. As such it is impossible to do justice to the chemistry of carbon in just a few paragraphs. This section will only deal with the general chemistry of carbon and will omit topics recognized as traditionally organic chemistry. Its organometallic chemistry will also be skipped since examples are presented in the organic chemistry of some metals.

3.5.3.1 Reaction in air, oxygen, hydrogen, and nitrogen

Given the large number of carbon compounds that exist, it might be expected to be very reactive. However, at room temperature solid carbon is relatively unreactive. At standard temperature and pressure, it resists oxidation and is

FIGURE 3.46 Siderite, $FeCO_3$, brown discoidal crystals the largest to over 2.5 cm. There are several smaller crystals. Eagle Mine, Gilman, Eagle Co., Colorado, United States.

FIGURE 3.47 Strontianite, $SrCO_3$, pseudohexagonal crystals up to 1.7 cm with orange brown terminations. Oberdorf, Austria.

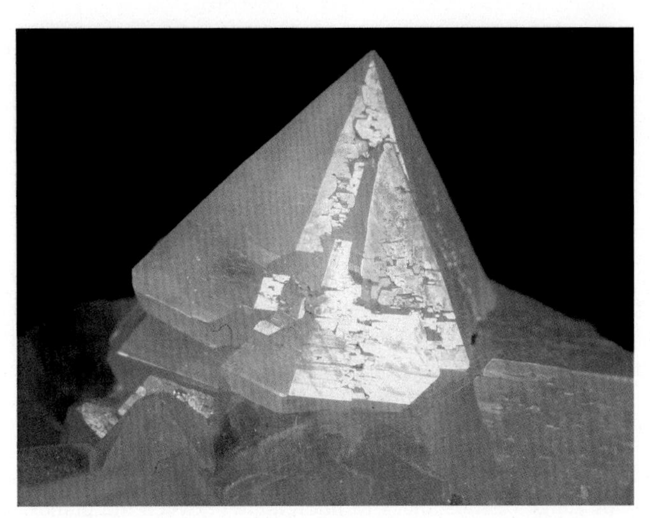

FIGURE 3.48 White to clear and translucent witherite, $BaCO_3$, crystal to 1.3 cm. Alston Moor, Cumbria, England.

TABLE 3.5 Carbon properties.

Appearance	Graphite: black
	Diamond: clear
Allotropes	Graphite, diamond, others
Standard atomic weight $A_{r,\ std}$	12.011
Block	p-Block
Element category	Reactive nonmetal, sometimes considered a metalloid
Electron configuration	[He] $2s^2\ 2p^2$
Phase at STP	Solid
Sublimation point	3642°C
Density (near r.t.)	Amorphous: 1.8–2.1 g/cm^3
	Graphite: 2.267 g/cm^3
	Diamond: 3.515 g/cm^3
Heat of fusion	Graphite: 117 kJ/mol
Molar heat capacity	Graphite: 8.517 J/(mol · K)
	Diamond: 6.155 J/(mol · K)
Oxidation states	**−4**, −3, −2, −1, 0, +1, +2, +3, **+4**
Ionization energies	1st: 1086.5 kJ/mol
	2nd: 2352.6 kJ/mol
	3rd: 4620.5 kJ/mol
Covalent radius	sp^3: 77 pm
	sp^2: 73 pm
	sp: 69 pm
Van der Waals radius	170 pm

STP, Standard temperature and pressure.
Bold font indicates main oxidation state.

unreactive to hydrochloric acid, sulfuric acid, chlorine, and the alkali metals. It becomes reactive only at high temperatures, and especially in the vapor phase.

Carbon forms carbon monoxide (CO) when heated in a limited supply of air and forms carbon dioxide when heated in a sufficient supply of air. The equilibrium between CO and CO_2 is also known.

$$2C\ (s) + O_2(g) \rightarrow 2CO\ (g)$$
$$C\ (s) + O_2\ (g) \rightarrow CO_2\ (g)$$

Carbon is unreactive to hydrogen under ordinary conditions at any appreciable rate. However, hydrogen can react with powdered carbon in the presence of a finely divided nickel catalyst to form methane.

$$C(s) + 2H_2(g) \rightarrow CH_4(g)$$

The reaction of carbon with hydrogen and oxygen proceeds rapidly below 1000°C, but its reaction with nitrogen does not occur appreciably under comparable conditions. Carbon and nitrogen form the highly toxic cyanogen gas at temperatures above 2200°C.

$$2C(s) + N_2(g) \rightarrow C_2N_2(g)$$

3.5.3.2 Reaction with halogens

At high temperatures, graphite reacts readily with fluorine, F_2, but not with other halogens. With an atmosphere of F_2 at temperatures between 400°C and 500°C it gives CF_x, where $x = 0.68–0.99$. This species is black when x is low, silvery at $x = 0.9$, and colorless when x is about 1. Above 600°C, the reaction proceeds explosively to form a mixture of carbon tetrafluoride, CF_4, together with some C_2F_6 and C_5F_{12}.

$$C(s) + \text{excess}\ F_2(g) \rightarrow CF_4(g) + C_2F_6 + C_5F_{12}$$

Graphite has large interlayer distances between parallel planes of C atoms. This means that the interlayer bonding is relatively weak and enables a wide range of substances to intercalate between the planes to give lamellar compounds of variable composition.

The halogens show a curious alternation of intercalation behavior toward graphite. F_2 gives the compounds CF, C_2F, and C_4F whereas liquid Cl_2 reacts slowly to give C_8Cl. Br_2 readily intercalates in several stages to give compounds of formula C_8Br, $C_{12}Br$, $C_{16}Br$, and $C_{20}Br$, while I_2 appears not to intercalate at all.

3.5.3.3 Reaction with oxidizing acids

Carbon is oxidized by hot concentrated HNO_3 to form mellitic acid, $C_6(CO_2H)_6$, in which planar-hexagonal C_{12} units are preserved. It reacts with a suspension of $KClO_4$ in a 1:2 mixture (by volume) of conc. HNO_3/H_2SO_4 to give "graphite oxide", an unstable, pale lemon-colored product of variable stoichiometry and structure.

3.5.3.4 Carbides

Carbides are compounds composed of carbon and a less electronegative element. They can be formed by direct combination of the elements (C and element of lower electronegativity) at temperatures above ~2000°C. For example, the pale-yellow aluminum carbide, Al_4C_3 is prepared by direct reaction of the elements in an electric furnace. This compound is sometimes called a methanide because it forms methane upon hydrolysis.

$$4Al\,(s) + 3C(s) \rightarrow Al_4C_3(s)$$
$$Al_4C_3(s) + H_2O(l) \rightarrow Al(OH)_3(aq) + CH_4(g)$$

A metal oxide reacted with carbon at high temperature also forms carbides. For example, calcium carbide, CaC_2, a compound with many applications including production of ethyne gas is produced from the reaction between calcium oxide, CaO and C at around 2200°C–2500°C. Hydrolysis of CaC_2 is highly exothermic and must be carefully controlled.

$$CaO\,(s) + 3C \rightarrow CaC_2 + CO$$
$$CaC_2(s) + 2H_2O(l) \rightarrow C_2H_2(g) + Ca(OH)_2(aq); \quad \Delta H = -120\ kJ\ mol^{-1}$$

Carbides containing a C_2 unit can also form from the reaction of acetylene with electropositive metals in liquid ammonia. Carbides of the alkali metals, alkaline earth metals, MnC_2, lanthanoids (LnC_2, Ln_2C_3, $Ln_4(C_2)_3$, group 11 (Cu, Ag and Au) and group 12 (Zn, Cd, Hg) are examples. These carbides are colorless crystalline compounds which react violently with water and oxidize to the carbonate on being heated in air.

$$Ca\,(liq\ NH_3) + 2C_2H_2 \xrightarrow{-80^\circ C} H_2 + CaC_2{\cdot}C_2H_2 \xrightarrow{325^\circ C} CaC_2 + C_2H_2$$

3.5.3.5 Carbon-oxygen compounds

The simplest but most important oxides are carbon monoxide, (CO) and carbon dioxide (CO_2). Other oxides are known such as C_3O_2, C_5O_2, and $C_{12}O_9$ and a host of poorly characterized oxides.

Carbon monoxide is a toxic, flammable, colorless, and odorless gas. It is formed during incomplete combustion of carbon. Commercially, it is produced in the form of producer gas and water gas. Producer gas is a mixture of 25 % CO, 70% N_2, 4% CO_2, and small amounts of H_2, CH_4, and O_2. It is obtained by blowing air through incandescent coke. The reactions involved are:

$$2C + O_2 \rightarrow 2CO; \quad \Delta H = -221\ kJ \cdot mol^{-1}$$
$$C + O_2 \rightarrow CO_2; \quad \Delta H = -393.5\ kJ \cdot mol^{-1}$$

Water gas (a mixture of 50% H_2, 40% CO, 5% CO_2, 5% $N_2 + CH_4$) is made by blowing steam through a bed of incandescent coke; the reaction is endothermic:

$$C + H_2O \rightarrow CO + H_2; \quad \Delta H = +131.3\ kJ \cdot mol^{-1}$$

CO reacts with alkali hydroxides only at elevated temperatures, combining to give formates. It reacts similarly with alkali methoxides forming acetates—salts where (unlike carbon monoxide) the carbon is formally quadrivalent.

$$CO + NaOH \rightarrow NaCO_2H$$
$$CO + NaOCH_3 \rightarrow NaCO_2CH_3$$

CO reacts with alkali metals in liquid NH_3 leading to reductive coupling to give colorless crystals of the salt $Na_2C_2O_2$ which contains linear groups NaOC=CONa packed in chains. CO reacts with Cl_2 and Br_2 to give COX_2. It also reacts with liquid sulfur to form COS.

Many reactions of CO have many applications in industry especially in bulk chemicals manufacturing. Examples include production of methanol by the catalytic reduction of CO at temperatures of 230°C–400°C and at pressures of 50–100 atm; amination with ammonia over zeolite catalysts at 350°C–400°C to give methylamine (and some dimethylamine); carbonylation of methanol to produce acetic acid; and many more.

$$CO + 2H_2 \rightarrow CH_3OH$$

$$CO + 3NH_3 \xrightarrow{HZSM-5} CH_3NH_2 + H_2O + N_2 + H_2 \text{ (HZSM} - 5 = \text{zeolite catalyst)}$$

$$H_3COH + CO \xrightarrow{catalyst} H_3COOH$$

The other important oxide of carbon, CO_2, is a colorless gas with a sufficiently pungent odor at high concentrations. Its industrial importance, however, is anchored mostly on its physical properties. Its more important chemistry is related to the CO_2-water system. CO_2 reacts slowly with water at pH < 8 to form carbonic acid:

$$CO_2 + H_2O \rightarrow H_2CO_3 \quad \text{(slow)}$$

$$H_2CO_3 + OH^- \rightarrow HCO_3^- + H_2O \quad \text{(fast)}$$

When the pH is increased to >10, the reactions are

$$CO_2 + OH^- \rightarrow HCO_3^- \quad \text{(slow)}$$

$$HCO_3^- + OH^- \rightarrow CO_3^{2-} + H_2O \quad \text{(fast)}$$

CO_2 reacts with more electropositive metals to form, depending on the conditions, C and the metal carbonate, oxalate or oxide. Less electropositive metals yield carbides and oxygen. CO_2 is a weak electrophile. Thus, it reacts with nucleophiles, such as carbanions in organolithium compounds, to form carboxylates;

$$CO_2 + \text{Li-}R \rightarrow RCOOLi \text{ where } R = \text{alkyl or aryl}$$

One important reaction of CO_2 is the manufacture of urea via ammonium carbamate:

$$CO_2 + 2\,NH_3 \xrightarrow{185^\circ C,\ 200\ atm} NH_2CONH_2 \xrightarrow{-H_2O} CO(NH_2)_2$$

As mentioned, there are several known oxides of carbon. Only one, the tricarbon dioxide, C_3O_2, usually known as "carbon suboxide," will be discussed here. C_3O_2 has been prepared in relatively pure form by the thermal decomposition of malonic acid, $C_3H_4O_4$. It was also shown to form from CO in an ozoniser. C_3O_2 reacts with ammonia to form malonamide, $CH_2(CONH_2)_2$, whose derivatives are now finding industrial importance. It shows considerable importance in organic synthesis as it can form pyrones, which form the central core of several natural chemical compounds, with acetylacetone, benzoylacetone, and acetone.

3.5.3.6 Carbon-nitrogen compounds

The reaction of carbon with nitrogen to form the highly toxic cyanogen was shown to occur above 2200°C. In the laboratory, cyanogen is produced by the reaction of copper (II) sulfate with potassium cyanide in aqueous medium at 60°C.

$$2CuSO_4 + 4KCN \rightarrow Cu_2(CN)_2 + 2K_2SO_4 + (CN)_2$$

Cyanogen is stable up to 850°C when pure but polymerizes at 350°C–500°C to a black mass called paracyanogen when impure. When it is dissolved in solvents such as water, alcohol, and ether, it decomposes to give a mixture of hydrogen cyanide (HCN), cyanic acid and urea.

Another C-N compound is hydrogen cyanide, HCN. This is an extremely poisonous gas, having a maximum tolerable vapor composition of 10 ppm. HCN can only be prepared from its elements, H_2, N_2, and carbon at temperatures greater than 1800°C. It is commonly prepared by catalytic oxidation of a mixture of methane and ammonia.

$$CH_4 + NH_3 \xrightarrow{1200^\circ C - 1300^\circ C,\ Pt} HCN + 3H_2$$

Despite the toxicity of this compound, world production exceeds 1 million tons per annum due to its industrial importance.

3.5.4 Major uses

Carbon is essential to all known living systems. It is present in a vast number of molecules in living bodies. It is found in the backbone of biomolecules vital to life such as carbohydrates, lipids, proteins, and nucleic acids. Most of our food—carbohydrates, fats, proteins, and fiber—consist of compounds of carbon. During digestion these compounds are broken down into simpler molecules that can be adsorbed through the wall of the stomach and intestines into the blood stream. Most of the ingested carbon compounds are oxidized to release the energy they contain, and we breathe out the carbon as carbon dioxide. Released into the atmosphere it can again be extracted by plants and become part of the natural carbon cycle. Cellulose is a naturally occurring, C-containing polymer produced by plants in the form of wood, cotton, linen, and hemp. Cellulose is used primarily for maintaining structure in plants. Economically valuable C-polymers of animal origin include wool, cashmere and silk.

Carbon is unique among the elements in its ability to form strongly bonded chains, sealed off by hydrogen atoms. These hydrocarbons, extracted naturally as fossil fuels (coal, oil, and natural gas), are mostly used as fuels. Crude oil is distilled in refineries by the petrochemical industry to produce gasoline, diesel, kerosene, and other products. A small but important fraction is used as a feedstock for the petrochemical industries producing polymers, fibers, paints, solvents, plastics, etc. Impure carbon in the form of charcoal (from wood) and coke (from coal) is used in metal smelting. It is particularly important in the iron and steel industries. The uses of carbon and its compounds are extremely varied. It can form alloys with iron, of which the most common is carbon steel. Graphite is combined with clays to produce the "lead" in pencils used for writing and drawing. In addition, it is used as a lubricant and a pigment, as a molding material in glass production, in electrodes for dry batteries and in electroplating and electroforming, in brushes for electric motors and as a neutron moderator in nuclear reactors. Charcoal is used as a drawing material in artwork, barbecue grilling, iron smelting, and in many other applications. Historically, charcoal was used together with sulfur and saltpeter (KNO_3) to make gunpowder and was used for many centuries as an explosive in guns and cannons, and in early mining. Today it is no longer used in weapons, though it is still the basis of many fireworks. Activated charcoal is used for purification and filtration. It can be found in respirators and kitchen extractor hoods, and in medicine to absorb toxins, poisons, or gases from the digestive system.

Carbon fiber is produced via pyrolysis of extruded and stretched filaments of polyacrylonitrile (PAN) and other organic compounds. The crystallographic structure and mechanical properties of the fiber depend on the type of starting material, and on the following processing. Carbon fibers made from PAN have a structure similar to narrow filaments of graphite, but thermal processing may re-order its structure into a continuous rolled sheet. The resulting fibers have higher specific tensile strength than steel. Carbon black is used as the black pigment in printing ink, artist's oil paint and water colors, carbon paper, automotive finishes, India ink and laser printer toner. In addition, it is added as a filler in rubber products such as tires and in plastic compounds. Carbon is used in chemical reduction at high temperatures. Coke is used to reduce iron ore into iron (smelting). Case hardening of steel is realized by heating finished steel components in carbon powder. Carbides of Si, W, B, and Ti, are some of the hardest known materials, and are applied as abrasives in cutting and grinding tools. Carbon compounds make up most of the materials used in clothing, such as natural and synthetic textiles and leather, and nearly all the interior surfaces in the built environment other than glass, stone, and metal. The more recent discovery of carbon nanotubes, other fullerenes and atom-thin sheets of graphene has revolutionized hardware developments in the electronics industry and in nanotechnology generally.

The diamond industry can be divided into two types: one dealing with gem-grade diamonds and the other, with industrial-grade diamonds. Although a large trade in both types of diamonds exists, the two markets function in very different ways. Different to precious metals such as gold or platinum, gem diamonds are not traded as a commodity: there is a considerable mark-up in the sale of diamonds, and there is not a very active market for resale of diamonds. Industrial diamonds are valued mostly for their hardness and heat conductivity, with the gemmological qualities of clarity and color being mostly irrelevant (Box 3.6). About 80% of mined diamonds (equal to about 100 million carats or 20 tons per year) cannot be cut as gemstones and are therefore relegated to industrial use (commonly called bort). Synthetic diamonds (See Box 3.5), invented in the 1950s, found virtually straight away industrial applications; these days some 3 billion carats (600 tons) of synthetic diamond is produced per year. The foremost industrial application of industrial diamonds is in cutting, drilling, grinding, and polishing. The majority of these applications do not need large diamonds; in fact, most diamonds of gem quality that are too small can be used industrially. Diamonds are embedded in drill tips or saw blades or ground into a powder for use in grinding and polishing applications. Specialized applications include use in laboratories as containment for high-pressure experiments (e.g., in the so-called diamond anvil cell), high-performance bearings, and limited use in specialized windows. With the ongoing developments in the manufacture of synthetic diamonds, new applications are becoming viable. Potential applications may be for diamonds

BOX 3.6 The four Cs of gemstones

The four Cs stand for carat, clarity, color, and cut. Together these determine the value of any gemstone including diamonds. Carat is a measure of weight and is equal to 0.2 grams. Since the specific gravity of different minerals (i.e., different gemstones) is unique, this means that 1 carat of diamond has a different size than 1 carat emerald (beryl). Generally, a hand loupe (10x) is used to study the surface and internal features of a gemstone, such as mineral and fluid inclusions, fractures, etc. This is important to determine the gemstone's clarity (the visible absence of such irregularities at 10x magnification). The color is an important factor to determine the price of a gemstone. For example, a deep green is desirable for emerald, while for most diamonds colorless is most desirable (except for what is known as rare fancy colors such as blue, pink, and yellow). Finally, the cut, in other words the final form of the cut and polished gemstone must be taken into account for the determination of the price. The cut of a diamond, for example, determines the internal reflection of the visible light and hence the "sparkle" of colors visible when the light leaves through the top surface.

as a semiconductor suitable for microchips, and due to its exceptional heat conductance property, as a heat sink in electronics.

3.6 7 N — Nitrogen

3.6.1 Discovery

Nitrogen compounds have an extensive history, ammonium chloride was already known to the Greek historian Herodotus (c. 484 BCE–circa 425 BCE). They were well known by the Middle Ages. Alchemists knew nitric acid as aqua fortis (strong water), in addition to other nitrogen compounds such as ammonium salts and nitrate salts. The mixture of nitric and hydrochloric acids is known as aqua regia (royal water), renowned for its ability to dissolve gold, the king of metals. The discovery of nitrogen is ascribed to the Scottish physician, chemist, and botanist Daniel Rutherford (November 3, 1749–December 15, 1819) in 1772, who named it noxious air (Rutherford, 1772) (Fig. 3.49). Though he did not recognize it as a completely different chemical compound, he evidently distinguished it from Joseph Black's "fixed air," or carbon dioxide (Joseph Black was a Scottish physicist and chemist, April 16, 1728–December 6, 1799). That there was a component in air that did not support combustion was obvious to Rutherford, though he did not recognize it as an element. Nitrogen was also a matter of research around the same time by Carl Wilhelm Scheele (Swedish Pomeranian and German pharmaceutical chemist, December 9, 1742–May 21, 1786), Henry Cavendish (English natural philosopher, scientist, and an important experimental and theoretical chemist and physicist, October 10, 1731–February 24, 1810), and Joseph Priestley (English Separatist theologian, natural philosopher, chemist, innovative grammarian, multisubject educator, and liberal political theorist, March 24, 1733–February 6, 1804), who referred to it as burnt air or phlogisticated air (Scheele, 1777; Cavendish, 1766; Priestley, 1772, 1775). Nitrogen gas was inert enough that Antoine Lavoisier (French nobleman and chemist, August 26, 1743–May 8, 1794) mentioned it as "mephitic air" or azote, from the Greek word ἀζωτικός (azotikos), "no life." In an atmosphere of pure nitrogen, animals died, and flames were extinguished. Though Lavoisier's name was not accepted in English, as it was shown that nearly all gases (with the only exception of oxygen) are mephitic, it is used in many languages (French, Italian, Portuguese, Polish, Russian, Albanian, Turkish, etc.; the German Stickstoff and Dutch stikstof, likewise refer to the same characteristic, viz. sticken or stikken "to choke or suffocate") and still can be found in English in the common names of many nitrogen compounds, such as hydrazine and compounds of the azide ion. Finally, it resulted in the name "pnictogens" for the group headed by nitrogen, from the Greek πνίγειν "to choke".

The English word nitrogen (1794) entered the language from the French nitrogène, coined in 1790 by French chemist, physician, agronomist, industrialist, statesman, educator, and philanthropist Jean-Antoine Chaptal, comte de Chanteloup (June 5, 1756–July 30, 1832) (Fig. 3.50), from the French niter (potassium nitrate, also called saltpeter) and the French suffix -gène, "producing," from the Greek -γενής (-genes, "begotten") (Chaptal, 1800). Chaptal's connotation was that nitrogen is the vital part of nitric acid, which in turn was produced from niter, KNO_3. In earlier times, niter had been confused with Egyptian "natron" (sodium carbonate, Na_2CO_3)—called νίτρον (nitron) in Greek—which, despite the name, contained no nitrate.

3.6.2 Mining, production, and major minerals

Nitrogen is the most common pure element in the earth, making up 78.1% of the entire volume of the atmosphere. Regardless of this, it is not very abundant in Earth's crust, making up only 19 ppm of this, similar to niobium, gallium,

FIGURE 3.49 Daniel Rutherford, 1896 from "The Gases of the Atmosphere" by William Ramsay.

FIGURE 3.50 Jean-Antoine Chaptal, comte de Chanteloup, oil on canvas painting by Anicet Charles Gabriel Lemonnier (1743–1824).

and lithium. The only important nitrogen minerals are niter (potassium nitrate, saltpeter KNO_3) and nitratine (sodium nitrate, Chilean saltpeter, $NaNO_3$). However, these have not been an important source of nitrates since the 1920s, when the industrial synthesis of ammonia and nitric acid became common. Nitrogen gas is an industrial gas produced by the fractional distillation of liquid air, or by mechanical means using gaseous air (pressurized reverse osmosis membrane or PSA). Nitrogen gas generators using membranes or PSA are generally more cost and energy efficient than bulk delivered nitrogen. Commercial nitrogen is frequently a by-product of air-processing for industrial concentration of oxygen for steelmaking and other applications. When supplied compressed in cylinders it is often known as oxygen-free nitrogen (OFN). Commercial-grade nitrogen has at most 20 ppm oxygen, and specially purified grades containing at most 2 ppm oxygen and 10 ppm argon are also available. Almost all of the world ammonia production comes from the Haber-Bosch process, developed by German chemists Fritz Haber (December 9, 1868–January 29, 1934) (Fig. 3.51) and Carl Bosch (August 27, 1874–April 26, 1940) (Fig. 3.52), in which air reacts with hydrogen from natural gas with

FIGURE 3.51 Fritz Haber, circa 1919.

FIGURE 3.52 Carl Bosch, circa 1929.

the aid of a catalyst to make ammonia. The most popular catalysts are based on iron promoted with K_2O, CaO, SiO_2, and Al_2O_3. The reaction mechanism, involving the heterogeneous catalyst, is believed to involve the following steps:

1. $N_2(g) \rightarrow N_2(adsorbed)$
2. $N_2(adsorbed) \rightarrow 2N(adsorbed)$
3. $H_2(g) \rightarrow H_2(adsorbed)$
4. $H_2(adsorbed) \rightarrow 2H(adsorbed)$
5. $N(adsorbed) + 3H(adsorbed) \rightarrow NH_3(adsorbed)$
6. $NH_3(adsorbed) \rightarrow NH_3(g)$

Reaction 5 occurs in three steps, forming NH, NH_2, and then NH_3. Experimental evidence points to reaction 2 as being the slow, rate-determining step. This is not unexpected since the bond broken, the nitrogen triple bond, is the strongest of the bonds that must be broken. Before this process became commercially viable in the 1920s, nitrogen was obtained from natural deposits, including guano, the bird and bat excrement that was a major source of phosphate, and nonmarine evaporite deposits. Nitrate evaporite deposits are found only in the Atacama Desert of northern Chile, one of the world's driest parts, with an average rainfall of less than 1 mm. Extreme dryness is essential for the formation of the highly soluble nitrate minerals (Box 3.7).

Nitrogen occurs in several forms in minerals. Most minerals contain the nitrate (NO_3) group, while some minerals contain NH_x groups (mostly NH_4). Finally, some of the organic minerals contain N as part of their structure. The NH_4^+ group has a similar size and charge as K^+ and nearly all the ammonium-containing minerals have an equivalent K-containing mineral. A good example of this is the ammonium-feldspar buddingtonite, which has the same structure as the K-feldspar orthoclase.

In the class of native elements there are a handful of nitrides, such as nierite (Si_3N_4), osbornite (TiN) or siderazot (Fe_5N_2). There is only a single sulfide mineral containing the ammonium group, ambrinoite, with the formula $[K,(NH_4)]_2(As,Sb)_6(Sb,As)_2S_{13}{\cdot}H_2O$. Within the class of the halides there are several minerals, the best known of which are: buttgenbachite ($Cu_{19}(NO_3)_2(OH)_{32}Cl_4{\cdot}2H_2O$) (Fig. 3.53), cryptohalite ($(NH_4)_2[SiF_6]$), kleinite ($(Hg_2N)(Cl,SO_4){\cdot}nH_2O$) (Fig. 3.54) and salammoniac (NH_4Cl) (Fig. 3.55). Only a single oxide mineral is known to contain N: melanophlogite with the formula $46SiO_2{\cdot}6(N_2,CO_2){\cdot}2(CH_4,N_2)$. The group of the nitrates is a subclass under the carbonates and contains minerals such as: gerhardtite ($Cu_2(NO_3)(OH)_3$), niter (KNO_3), nitratine ($NaNO_3$), nitrobarite ($Ba(NO_3)_2$), and nitrocalcite ($Ca(NO_3)_2{\cdot}4H_2O$). Two borate minerals contain the ammonium group: ammonioborite ($(NH_4)_2[B_5O_6(OH)_4]_2{\cdot}H_2O$) and larderellite ($(NH_4)B_5O_7(OH)_2{\cdot}H_2O$). Within the sulfate class there is a relatively large group of minerals containing the ammonium group, for example, ammonioalunite ($(NH_4)Al_3(SO_4)_2(OH)_6$), ammoniojarosite ($(NH_4)Fe_3^{3+}(SO_4)_2(OH)_6$), koktaite ($(NH_4)_2Ca(SO_4)_2{\cdot}H_2O$), lecontite ($(NH_4,K)NaSO_4{\cdot}2H_2O$), and tschermigite ($(NH_4)Al(SO_4)_2{\cdot}12H_2O$) (Fig. 3.56), while three minerals contain the nitrate group, for example, darapskite ($Na_3(SO_4)(NO_3){\cdot}H_2O$), humberstonite ($Na_7K_3Mg_2(SO_4)_6(NO_3)_2{\cdot}6H_2O$), and ungemachite ($K_3Na_8Fe(SO_4)_6(NO_3)_2{\cdot}6H_2O$). Within the phosphate class a total of 14 ammonium-containing minerals are found, the best known of them being phosphammite ($(NH_4)_2(HPO_4)$) and struvite ($(NH_4)Mg(PO_4){\cdot}6H_2O$) (Fig. 3.57). Only 5 minerals are known in the silicate class, 4 of them containing the ammonium group and 1 containing the $N(CH_3)_4$ unit: ammonioleucite ($(NH_4,K)(AlSi_2O_6)$), buddingtonite ($(NH_4)(AlSi_3)O_8$), suhailite ($(NH_4)Fe_3^{2+}(AlSi_3O_{10})(OH)_2$), tobellite ($(NH_4,K)Al_2(AlSi_3O_{10})(OH)_2$), and tsaregorodtsevite ($(N(CH_3)_4)(AlSi_5O_{12})$). Finally, 8 minerals in the organic class contain N in various forms, for example, guanine ($C_5H_5N_5O$), julienite ($Na_2[Co(SCN)_4]{\cdot}8H_2O$), oxammite ($(NH_4)_2(C_2O_4){\cdot}H_2O$), and urea ($CO(NH_2)_2$).

BOX 3.7 Beneficiation of caliche ore

In Chile two major types of nitrate ore, together known as caliche can be identified. The alluvial caliche in which the saline minerals are present as a cement, usually rest on poorly cemented debris, and the bedrock caliche, in which the saline minerals form impregnations, veins, and masses in fractured bedrock. The typical alluvial-type nitrate deposit consists of several layers—*chuca* (overburden of the nitrate deposit), *costra* (hard and brittle material similar to the underlying caliche, or poorly cemented friable material), *caliche*, *conjelo*, and *coba* (loose uncemented regolith, which is in direct contact with caliche as the conjelo is absent in the Chilean nitrate deposits)—each having characteristic chemical and physical features. Caliche is generally 1–3 m thick. The bed is as thin as 50 cm and as thick as 5 m exist locally. Veins and layers of high-purity, white, nitrate-rich saline material, known as *caliche blanco*, are widespread in some caliche, with thickness ranging from 10 to 50 cm. The caliche often contains a wide variety of complex salts such as blödite ($Na_2Mg(SO_4)_2{\cdot}4H_2O$), polyhalite ($K_2Ca_2Mg(SO_4)_4{\cdot}2H_2O$), darapskite ($Na_3(SO_4)(NO_3){\cdot}H_2O$), and syngenite ($K_2Ca(SO_4)_2{\cdot}H_2O$). In addition, borates, iodates, and perchlorates may be present in small amounts. Leaching $NaNO_3$ from caliche is a very complex process. The ore contains only 7–10 wt.% $NaNO_3$ and needs multiple stage crushing and screening before leaching using either the Shanks or Guggenheim process, to reach a satisfactory yield. Both processes use the fact that the solubility of $NaNO_3$ increases with increasing temperature, while that of NaCl remains basically constant. The Shanks process leaches by means of a boiling solution (135°C–140°C), which is subsequently cooled to 10°C. The crude ore must contain at least 13% $NaNO_3$ for profitable operation using this method. The Guggenheim process instead uses a warm solution (40°C) for leaching, followed by refrigeration to about 5°C, using a heat exchanger on the cooling system and power plant for much of the heat input.

FIGURE 3.53 Buttchenbachite, $Cu_{19}(NO_3)_2(OH)_{32}Cl_4 \cdot 2H_2O$, acicular 3 mm crystals. Los Azules mine, Quebrada San Miguel, Copiapó Prov., Atacama, Chile.

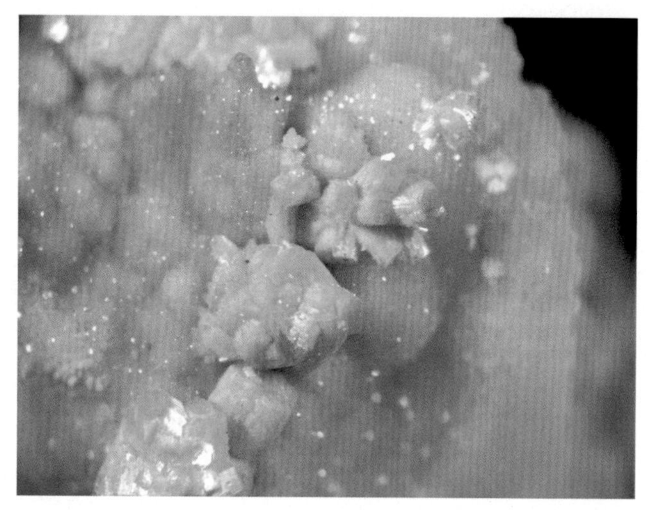

FIGURE 3.54 Kleinite, $(Hg_2N)(Cl,SO_4) \cdot nH_2O$, prismatic bright yellow crystals to 2 mm. McDermitt Mine, Humboldt Co., Nevada, United States.

FIGURE 3.55 Salammoniac, NH_4Cl, a 6.5 cm group of white mostly formless crystals of salammoniac on yellow rosickyite, S. Volcano Island, Lipari, Eolie Islands, Messina Prov., Sicily, Italy.

FIGURE 3.56 Colorless tschermigite, $(NH_4)Al(SO_4)_2 \cdot 12H_2O$, columnar crystals in 1.5 cm thick parallel seams. Cermniky, Bohemia, Czech Republic.

FIGURE 3.57 Struvite, $(NH_4)Mg(PO_4) \cdot 6H_2O$, well-formed single crystal to 0.9 mm. St Nikolai church, Hamburg, Germany.

3.6.3 Chemistry

Nitrogen has two natural stable isotopes: ^{14}N making up 99.634% and ^{15}N (which is slightly heavier) makes up the remaining 0.366%. Both stable isotopes are produced in the CNO cycle in stars [The CNO cycle (for carbon–nitrogen–oxygen) is one of the two known sets of fusion reactions by which stars convert hydrogen to helium, the other being the proton–proton chain reaction (pp-chain reaction). Unlike the latter, the CNO cycle is a catalytic cycle. It is dominant in stars that are more than 1.3 times as massive as the Sun.], though ^{14}N is more common since its neutron capture forms the rate-limiting step. ^{14}N is one of the five stable odd–odd nuclides (a nuclide having an odd number of protons and neutrons); the other four are ^{2}H, ^{6}Li, ^{10}B, and ^{180m}Ta (m denotes a metastable state). The relative abundance of ^{14}N and ^{15}N is nearly constant in the atmosphere but can differ elsewhere, because of natural isotopic fractionation due to biological redox reactions and the evaporation of natural ammonia or nitric acid. Biologically controlled reactions (e.g., assimilation, nitrification, and denitrification) strongly affect nitrogen dynamics in the soil. These reactions characteristically produce a ^{15}N enrichment of the substrate and corresponding depletion of the product. The ^{15}N:^{14}N ratio is frequently utilized in stable isotope analysis in the fields of geochemistry, hydrology, paleoclimatology, and paleoceanography. Of the 10 other isotopes synthetized, ranging from ^{12}N to ^{23}N, ^{13}N has a half-life of ten minutes and the remaining isotopes have half-lives on the order of seconds (^{16}N and ^{17}N) or even milliseconds. Given the half-life difference, ^{13}N is the most significant nitrogen radioisotope, being relatively long-lived enough to be applied in positron emission tomography (PET), even though its half-life is still short and therefore it must be produced at the PET venue, for example, in a cyclotron through proton bombardment of ^{16}O producing ^{13}N and an alpha particle. The radioisotope ^{16}N is the main radionuclide in the coolant of pressurized water reactors or boiling water reactors during normal operation, and therefore it is a sensitive and instant indicator of leaks from the primary coolant system to the secondary steam cycle and is the principal method of detection of such leaks. It is produced from ^{16}O (in water) through an (n,p) reaction in which the ^{16}O atom captures a neutron and expels a proton. It has a short half-life of approximately 7.1 s, but through its decay back to ^{16}O produces high-energy gamma radiation (5–7 MeV). Due to this, access to the primary coolant piping in a pressurized water reactor has to be restricted during reactor power operation.

Nitrogen is the first element in Group 15 and is found in the first row of p-block elements. The 2p elements especially nitrogen, oxygen, and fluorine exhibit anomalous properties (Table 3.6). This is because the 2p subshell is very small and similar in size to the 2s shell. This results in easy orbital hybridization. This also causes very large electrostatic attraction between nucleus and the valence electrons in the 2s and 2p subshells. As such, nitrogen, oxygen, and fluorine have high electronegativities. Nitrogen has an electronegativity of approximately 3.0, surpassed only by oxygen, fluorine, and the bigger chlorine atom in the third row. The ground-state electronic configuration of N atom is $1s^2 2s^2 2p_x^{\ 1} 2p_y^{\ 1} 2p_z^{\ 1}$ with three unpaired electrons (^4S). A nitrogen atom can be surrounded by an octet of electrons by sharing three pairs of electrons with another nitrogen atom. This N≡N triple bond is so strong that the N_2 molecule is very unreactive. Nitrogen combines with very few elements at room temperature. It does not react with air, water, halogens, acids, bases, and metals (except lithium) under normal conditions. Nevertheless, nitrogen is interesting since its compounds have tremendous importance in living organisms and in industrial applications. It is found combined in almost every element in the periodic table except the lighter noble gases. This seemingly contradictory observation is because N_2 becomes significantly more reactive as the temperature increases. Moreover, several catalysts found in nature can overcome the inertness of N_2 at low temperatures.

3.6.3.1 Reaction in air, oxygen, and hydrogen

An active form of nitrogen, believed to contain free nitrogen atoms, can form in electric sparks such as lightning. This reactive nitrogen combines with most other elements to form binary species. In air or in oxygen, nitrogen reacts to form NO and subsequently, NO_2.

$$N_2(g) + O_2(g) \rightarrow 2NO\ (g)$$
$$2\ NO(g) + O_2(g) \rightarrow 2NO_2(g)$$

TABLE 3.6 Nitrogen properties.

Appearance	Colorless gas, liquid or solid
Standard atomic weight $A_{r,std}$	14.007
Block	p-block
Element category	Reactive nonmetal
Electron configuration	[He] $2s^2\ 2p^3$
Phase at STP	Gas
Melting point	−210.00°C
Boiling point	−195.795°C
Density (at STP)	1.2506 g/L at 0°C, 1013 mbar
when liquid (at b.p.)	0.808 g/cm^3
Heat of fusion	(N_2) 0.72 kJ/mol
Heat of vaporization	(N_2) 5.56 kJ/mol
Molar heat capacity	(N_2) 29.124 J/(mol · K)
Oxidation states	**−3**, −2, −1, +1, +2, **+3**, +4, **+5**
Ionization energies	1st: 1402.3 kJ/mol 2nd: 2856 kJ/mol 3rd: 4578.1 kJ/mol
Covalent radius	71 ± 1 pm
Van der Waals radius	155 pm

STP, Standard temperature and pressure.
Bold font indicates main oxidation state.

It was shown to form cyanogen, $(CN)_2$, when coke is heated to incandescence and to ammonia, NH_3, in the presence of catalysts and at elevated temperature.

$$N_2(g) + 3H_2(g) \rightarrow 2NH_3(g)$$

3.6.3.2 Reaction with halogens

Nitrogen does not react with the halogens at room temperature but all four nitrogen trihalides, NX_3, are known. Of the four, only nitrogen trifluoride, NF_3, can be prepared directly from the elements, provided uncommon conditions are met such as the presence of an electric discharge. Commercially, NF_3 is often prepared by the electrolysis of molten NH_4F/HF. The other three common NX_3 compounds are also prepared from compounds of nitrogen. NCl_3 is prepared by passing chlorine gas in an ammonium salt such as ammonium nitrate. NBr_3 is prepared by reacting bis(trimethylsilyl)amine bromide with bromine at low temperatures (−87°C). NI_3 is prepared by reacting boron nitride with iodine monofluoride in trichlorofluoromethane at −30°C in low yield. All the heavier nitrogen trihalides are explosives with NI_3 being the most reactive, detonating even when touched only lightly. NF_3, on the other hand, is far less reactive, being unaffected by water of dilute aqueous acid or alkali. However, when heated or exposed to fire, the containers may explode violently. Other nitrogen halides are also known, and their chemistry will be discussed later.

$$NH_4F + 2HF \xrightarrow{electrolysis} NF_3 + 3H_2$$
$$NH_4Cl + Cl_2 \rightarrow NCl_3 + HCl$$
$$(Me_3Si)_2NBr + 2BrCl \rightarrow NBr_3 + 2Me_3SiCl$$
$$BN + 3IF \rightarrow NI_3 + BF_3$$

3.6.3.3 Nitrides

Nitrides are compounds where nitrogen is combined with an element of similar or lower electronegativity. Lithium reacts with nitrogen at room temperature to form the only stable metal nitride. At higher temperatures, nitrogen is combined with a host of other metals such as Be, the alkaline earths, B, Al, Si, Ge, Se, Y, lanthanoids, Zr, Hf, V, Cr, Mo, W, Th, U, and Pu. In the presence of electric discharge, nitrogen molecules can form very reactive nitrogen atoms. The N atom can react with several elements to form nitrides or can associate with another N atom to form an N_2^* molecule in the excited state. The excited N_2^* molecule is also very reactive and can cause the dissociation of atoms normally unreactive to ground-state N_2 or even N atoms.

$$N_2^* + H_2O \rightarrow N_2 + OH + H$$

The nitrides formed by the reaction of nitrogen with other elements can be classified into ionic nitrides, interstitial nitrides, and covalent nitrides. Ionic nitrides, which can be considered to consist of metal cations and nitride anions, undergo hydrolysis to produce ammonia and metal hydroxide. Interstitial nitrides are formed when nitrogen atoms occupy the interstices in the lattice of metal ions, typically those of transition metals. These are chemically inert, undergoing hydrolysis very slowly (may even require acid) to produce ammonia and nitrogen gas. Covalent nitrides, for example, boron nitride, cyanogen, phosphorus nitride, and nitrides of sulfur are discussed under the element that is paired with nitrogen and will not be presented here.

Other elements combine with three nitrogen atoms to form azides, $-N_3$. Examples are sodium azide, NaN_3, and organic derivatives such as alkyl and aryl azides, RN_3. Most azides are shock-sensitive and are even used as initiating explosives in detonators and percussion caps. Azide salts decompose with sodium nitrite when acidified releasing nitrogen.

$$2NaN_3 + 2HNO_2 \rightarrow 3N_2 + 2NO + 2NaOH$$

3.6.3.4 Hydrides

The hydrides of nitrogen include azane commonly known as ammonia (NH_3), diazane or hydrazine (N_2H_4), and hydrogen trinitride or hydrogen azide (HN_3). These three are the most industrially important. Ammonia is a colorless, pungent gas. It is industrially prepared by the Haber process, where nitrogen reacts with hydrogen at 400°C−500°C and 10^2−10^3 atm in the presence of a catalyst. The high electronegativity of N makes NH_3 capable of hydrogen bonding. This confers high melting and high boiling points to ammonia. Moreover, this makes the liquid phase of the compound a good solvent with high dielectric constant. Its most notable property is the ability to dissolve alkali metals to form highly colored, electrically conducting solutions. It has a lower dielectric constant than water and is thus, a poorer solvent for ionic compounds. However, when complexation with ammonia is more favorable than water, compounds can dissolve better in ammonia as in the case of silver halides. Ammonia burns in air forming nitrogen and water, but with difficulty. This is due to the low heat of combustion, high auto ignition temperature, and low flammability range (15%–25%) of the compound. In the presence of a platinum catalyst at 800°C, it burns to produce the thermodynamically less-favored NO and NO_2.

$$4NH_3 + 3O_2 \xrightarrow{burn} 2N_2 + 6H_2O$$
$$4NH_3 + 5O_2 \xrightarrow{Pt/800^\circ C} 4NO + 6H_2O$$
$$2NO + O_2 \xrightarrow{Pt/800^\circ C} 2NO_2$$

Hydrazine is a colorless and flammable liquid with an ammonia-like odor. It is prepared by the reaction between ammonia and hydrogen peroxide.

$$2NH_3 + H_2O_2 \rightarrow H_2NNH_2 + H_2O$$

Hydrazine is kinetically stable against decomposition, but it burns in air with considerable evolution of heat. It is a good reductant as the by-products are normally nitrogen gas and water. It also undergoes a variety of redox reactions undergoing a 1-, 2-, or 4-electron oxidation and whether this is in acid or alkaline solution.

Hydrogen azide is a potentially explosive, highly toxic, foul-smelling, very hygroscopic covalent azide. It is formed by the acidification of an azide salt like sodium azide.

$$NaN_3 + HCl \rightarrow HN_3 + NaCl$$

Aqueous solutions of hydrogen azide (hydrazoic acid) are as acidic as acetic acid with a K_a of around 1.8×10^{-5}. They may react with carbonyl derivatives (aldehydes, ketones, carboxylic acids) to give an amine or amide with release of nitrogen. They dissolve in the strongest acids to produce explosives containing the $H_2N{=}N{=}N^+$ ion. It decomposes to N_2 and H_2 when triggered by shock, friction, and sparks.

3.6.3.5 Oxides

Nitrogen forms no fewer than eight molecular oxides—nitric oxide (NO), nitrous oxide (N_2O), nitrogen dioxide (NO_2), dinitrogen trioxide (N_2O_3), dinitrogen tetroxide (N_2O_4), dinitrogen pentoxide (N_2O_5), nitrosylazide (N_4O), trinitramide $N(NO_2)_3$ and possibly oxatetrazole (N_4O). All are thermodynamically unstable with respect to decomposition into N_2 and O_2. The oxides—N_2O, NO, and NO_2—were among the first gaseous compounds to be isolated and identified.

Nitric oxide is the simplest thermally stable odd-electron molecule known. It is synthesized in the laboratory by the mild reduction of acidified nitrites with iodide or ferrocyanide or by the disproportionation of nitrous acid in the presence of dilute sulfuric acid.

$$2KNO_2 + 2KI + 2H_2SO_4 \rightarrow 2NO + 2K_2SO_4 + 2H_2O + I_2$$

NO reacts rapidly with molecular O_2 to give brown NO_2. Nitric oxide reacts with the halogens to give XNO. Reactions with sulfides, polysulfides, sulfur oxides and the oxoacids of sulfur are complex, and the products depend markedly on reaction conditions. NO readily reacts with many transition metal compounds to give nitrosyl complexes and these are also frequently formed in reactions involving other oxo-nitrogen species.

Nitrous oxide is a moderately unreactive gas comprised of linear unsymmetrical molecules. It can be made by the careful thermal decomposition of molten NH_4NO_3 at about 250°C:

$$NH_4NO_3 \rightarrow N_2O + 2H_2O$$

N_2O (like N_2 itself) can act as a ligand by displacing H_2O from the aquo complex $[Ru(NH_3)_5(H_2O)]^{2+}$.

3.6.4 Major uses

Nitrogen occurs in all organisms, primarily in amino acids (and thus proteins), in the nucleic acids (DNA and RNA) and in the energy transfer molecule adenosine triphosphate. The human body contains about 3% nitrogen by mass, the 4th most abundant element in the human body after oxygen, carbon, and hydrogen. Numerous industrially important compounds, such as ammonia, nitric acid, organic nitrates (propellants and explosives), and cyanides, all have nitrogen in their structures. The very strong triple bond in elemental nitrogen ($N{\equiv}N$), the second strongest bond in any diatomic molecule after carbon monoxide (CO), largely controls nitrogen chemistry. This extremely strong bond results in difficulties for both organisms and industry alike in converting N_2 into useful compounds, but then again at the same time it means that burning, exploding, or decomposing nitrogen compounds to form nitrogen gas releases large quantities of frequently useful energy. Synthetically manufactured ammonia and nitrates are important industrial fertilizers, while fertilizer nitrates form important pollutants in the eutrophication of water systems. Besides its use in fertilizers and energy-stores, nitrogen forms part of organic compounds as varied as Kevlar used in high-strength fabric and cyanoacrylate used in superglue. Other industrial products that are produced directly from ammonia and nitric acid are plastics such as polyurethane, fibers such as nylon, and the brilliantly colored azo-dyes. Other nitrogen compounds produced on an industrial scale include hydrazine (N_2H_4), which is used as a rocket propellant and an antioxidant, and hydroxylamine (NH_2OH), a mildly reducing agent which is used in photography and dyeing.

Nitrogen forms a constituent of all major pharmacological drug classes, including antibiotics. Many pharmacological drugs are mimics or prodrugs of natural nitrogen-containing signal molecules in the human body: for example, the organic nitrates nitroglycerin and nitroprusside control blood pressure by metabolizing into nitric oxide. Many prominent nitrogen-containing drugs, such as the natural caffeine and morphine or the synthetic amphetamines, act on receptors of animal and human neurotransmitters.

About two-thirds of nitrogen manufactured by industry is sold as the gas and the remaining one-third as the liquid. The gas is generally used as an inert atmosphere when the oxygen in the air could form a fire, explosion, or oxidizing hazard. It is utilized, for example, as a modified atmosphere, pure or mixed with carbon dioxide (CO_2), to nitrogenate and preserve the freshness of packaged or bulk foods. Pure nitrogen as food additive is labeled in the European Union with the E number E941. It is also used, for example, in incandescent light bulbs as a low-cost alternative to Ar, in fire suppression systems for Information technology (IT) equipment, in the production of stainless steel, in the case hardening of steel by nitriding, in some aircraft fuel systems to reduce fire hazard, and to inflate race car and aircraft tires, minimizing the difficulties caused by moisture and oxygen in normal air, etc.

Liquid nitrogen is a cryogenic liquid. When insulated in proper containers such as Dewar flasks (vacuum flask which acts as an insulating storage vessel that greatly lengthens the time over which its contents remain hotter or cooler than the flask's surroundings. Invented by Scottish chemist and physicist Sir James Dewar (September 20, 1842–March 27, 1923) in 1892, the vacuum flask consists of two flasks, placed one within the other and joined at the neck. The gap between the two flasks is partially evacuated of air, creating a near-vacuum which significantly reduces heat transfer by conduction or convection.), it can be transported without much evaporative loss. Like dry ice (frozen, solid CO_2), the principal application of liquid nitrogen is as a refrigerant. Besides other things, it is used for the cryopreservation of blood, reproductive cells (sperm and egg), and other biological samples and materials. It is used in the clinical setting in cryotherapy to remove cysts and warts on the skin by freezing them. In addition, it is used in cold traps for certain laboratory equipment and to cool infrared or X-ray detectors. Other uses are, for example, freeze-grinding, and machining materials that are soft or rubbery at room temperature, shrink-fitting and assembling engineering components, and more generally to reach very low temperatures whenever essential (around $-196°C$). Due to its low cost, liquid nitrogen is also regularly used even when such low temperatures are not absolutely required, for example, refrigeration of food, freeze-branding livestock, freezing pipes to halt flow when valves are not present, and consolidating unstable soil by freezing whenever excavation is going on underneath.

As already mentioned in the section on uses of carbon, nitrogen in the form of saltpeter (KNO_3) was historically used as part of gunpowder. This was followed in 1846 by the discovery of the explosive compound nitroglycerin, which was produced through the reaction of glycerine (now called glycerol) with nitric and sulfuric acids. It was ideal for blasting in quarries and for tunneling but was notoriously unpredictable. Subsequently, in 1867 the Swedish businessman, chemist, engineer, inventor, and philanthropist Alfred Nobel (October 21, 1833–December 10, 1896) discovered that by absorbing it on kieselguhr (diatomaceous earth, a naturally occurring, soft, siliceous sedimentary rock that is easily crumbled into a fine white to off-white powder. It has a particle size ranging from less than 3 μm to more than 1 mm, but typically 10–200 μm. Diatomaceous earth consists of fossilized remains of diatoms, a type of hard-shelled protist) it became safe to handle and would only explode when it was exposed to a smaller explosive from a detonator. Hence, dynamite was invented. The third nitrogen compound to be used as an explosive was TNT (2,4,6-trinitrotoluene, $C_6H_2(NO_2)_3CH_3$), but today there are much more sophisticated ones, in that they are safer to handle, such as HMX, also called octogen, (cyclotetramethylenetetranitramine, IUPAC preferred name 1,3,5,7-tetranitro-1,3,5,7-tetrazoctane, $C_4H_8N_8O_8$) and RDX (cyclotrimethylenetrinitramine, IUPAC preferred name 1,3,5-trinitro-1,3,5-triazinane, $C_3H_6N_6O_6$), which are widely used by armed forces around the world for munitions.

3.7 8 O — Oxygen

3.7.1 Discovery

One of the first known experiments on the relationship between combustion and air was conducted by the 2nd century BCE Greek engineer physicist and writer on mechanics, Philo of Byzantium (ca.280 BCE–ca.220 BCE). In his work Pneumatica, Philo detected that inverting a vessel over a burning candle and surrounding the vessel's neck with water resulted in some water rising into the neck. Philo wrongly deduced that parts of the air in the vessel were changed into the classical element fire and hence were able to escape through pores in the glass. Many centuries later the Italian polymath Leonardo da Vinci (April 15, 1452–May 2, 1519) built on Philo's work by detecting that a portion of air was used up during combustion and respiration (Fig. 3.58). In the late 17th century, Anglo-Irish natural philosopher, chemist, physicist, and inventor Robert Boyle (January 25, 1627–December 31, 1691) demonstrated that air is essential for combustion. English chemist, physician, and physiologist John Mayow (1641–79) refined this work by showing that fire needs only a part of air that he called spiritus nitroaereus (Fig. 3.59). In one experiment, he found that placing either a mouse or a lit candle in a closed container over water caused the water to rise and substitute one-fourteenth of the air's volume before extinguishing the subjects. From this he deduced that nitroaereus was used up in both respiration and combustion. Mayow detected that antimony increased in weight when heated and suggested that the nitroaereus must have combined with it. He also believed that the lungs separate nitroaereus from air and pass it into the blood and that animal heat and muscle movement were the result of the reaction of nitroaereus with certain substances in the body. Descriptions of these and other experiments and ideas were published in 1668 in his work *Tractatus duo* in the tract "De respiration." Robert Hooke (English natural philosopher, architect and polymath, July 28, 1635–March 3, 1703), Ole Borch (Danish scientist, physician, grammarian, and poet, 1626–1690), Mikhail Lomonosov (Russian polymath, scientist, and writer, November 19, 1711–April 15, 1765), and Pierre Bayen (French chemist, 1725–1798) all formed oxygen in experiments in the 17th and the 18th century but they did not recognize it as a chemical element. This may have been to some extent due to the prevalence of the philosophy of combustion and corrosion called the phlogiston theory, which was at that time the favored description of those processes. Developed in 1667 by the German

FIGURE 3.58 Leonardo da Vinci, after 1510. Red chalk on paper by Francesco Melzi (1491–1568).

FIGURE 3.59 John Mayow, engraving first published 1674.

physician, alchemist, scholar, and adventurer Johann Joachim Becher (May 6, 1635–October 1682), and revised by the German chemist, physician, and philosopher Georg Ernst Stahl (October 22, 1659–May 24, 1734) by 1731, phlogiston theory specified that all combustible materials were made of two parts. One part, called phlogiston, was given off when the substance containing it was burned, while the dephlogisticated part was supposed to be its true form, or calx. Highly combustible materials that left little residue, such as wood or coal, were thought to be made mostly of phlogiston; noncombustible substances that corrode, such as iron, contained very little. Air was no factor in phlogiston theory, nor were any preliminary quantitative experiments conducted to test the idea. In its place, it was based on observations of what happens when something burns, that most common substances appear to become lighter and appear to lose something in the process. Polish alchemist, philosopher, and physician Michael Sendivogius (1566–1636) in his work De Lapide Philosophorum Tractatus duodecim e naturae fonte et manuali experientia depromti (1604) defined a substance contained in air, naming it as "cibus vitae" (food of life), and this substance is equal to oxygen (Fig. 3.60). Sendivogius, during his experiments performed between 1598 and 1604, correctly recognized that the substance is equal to the gaseous by-product released by the thermal decomposition of potassium nitrate. The isolation of oxygen and the proper association of the substance to that part of air, which is required for life, offers adequate weight to the discovery of oxygen by Sendivogius. This discovery of Sendivogius though was regularly denied by generations of scientists and chemists who came after him. It is also frequently claimed that oxygen was first discovered by Swedish Pomeranian and German pharmaceutical chemist Carl Wilhelm Scheele (December 9, 1742–May 21, 1786). He had formed oxygen gas by heating mercuric oxide, HgO, and various nitrates in 1771–72. Scheele named the gas "fire air" because it was at that time the only known agent to sustain combustion. He wrote an account of his discovery in a manuscript titled "Treatise on Air and Fire," which he sent to his publisher in 1775 and was published in 1777 (Scheele, 1777). Meanwhile, on August 1, 1774, an experiment performed by the British clergyman, separatist theologian, natural philosopher, chemist, innovative grammarian, multisubject educator, and liberal political theorist Joseph Priestley (March 24, 1733–February 6, 1804) focused sunlight on mercuric oxide (HgO) contained in a glass tube, which liberated a gas he named "dephlogisticated air." (Priestley, 1775) (Fig. 3.61). He observed that candles burned brighter in the gas and that a mouse was more active and lived longer while breathing it. After breathing the gas himself, Priestley wrote: "The feeling of it to my lungs was not sensibly different from that of common air, but I fancied that my breast felt peculiarly

FIGURE 3.60 Michael Sendivogius, Tygodnik Ilustrowany, 1862.

FIGURE 3.61 Joseph Priestley, 1794 (possibly 1797), pastel by Ellen Sharples (1769–1849).

light and easy for some time afterwards." Priestley published his discoveries in 1775 in a paper titled "An Account of Further Discoveries in Air," which was contained within in the second volume of his book titled "Experiments and Observations on Different Kinds of Air." Since he published his findings first, Priestley is generally given priority in the discovery. The French chemist Antoine Laurent Lavoisier (August 26, 1743–May 8, 1794) later also claimed to have discovered the new substance independently. However, Priestley visited Lavoisier in October 1774 and told him about his experiment and how he liberated the new gas. Scheele also sent a letter to Lavoisier on September 30, 1774, that described his discovery of the previously unknown substance, but Lavoisier never acknowledged receiving it (a copy of the letter was found in Scheele's belongings after his death). Lavoisier did, however, perform the first adequate quantitative experiments on oxidation and provided the first accurate explanation of how combustion works. He used these and comparable experiments, all started in 1774, to discredit the phlogiston theory and to demonstrate that the substance discovered by Priestley and Scheele was a chemical element. In one experiment, Lavoisier detected that there was no overall increase in weight when tin and air were heated in a closed container. He observed that air rushed in when he opened the container, which showed that part of the trapped air had been used up. He also observed that the tin had increased in weight and that the increase was equal to the weight of the air that rushed back in. This and other experiments on combustion were documented in his book "Sur la combustion en general," which was published in 1777. In that work, he showed that air was a mixture of two gases; "vital air," which is critical to combustion and respiration, and azote (Gk. ἄζωτον "lifeless"), which did not support either. Azote later became nitrogen in English, although it has kept the earlier name in French and several other European languages. Lavoisier renamed "vital air" to oxygène in 1777 from the Greek roots ὀξύς (oxys) (acid, literally "sharp," from the taste of acids) and -γενής (-genēs) (producer, literally begetter), because he wrongly thought that oxygen was a constituent of all acids. Chemists (e.g., Sir Humphry Davy in 1812) in the end determined that Lavoisier was erroneous in this regard (H forms the basis for acid chemistry), but by then the name was too well established. Oxygen became part of English despite opposition from English scientists and the fact that the Englishman Priestley had first isolated the gas and written about it. This is partly due to a poem praising the gas titled "Oxygen" in the popular book "The Botanic Garden" (1791) by Erasmus Darwin, grandfather of Charles Darwin.

3.7.2 Mining, production, and major minerals

Oxygen is the most abundant chemical element by mass in the Earth's biosphere, air, sea, and land (Box 3.8). Oxygen constitutes 49.2% of the Earth's crust by mass as part of oxide compounds such as silicon dioxide, SiO_2, and is the most abundant element by mass in the Earth's crust. More than one hundred million tons of O_2 are extracted each year from air for industrial uses by two principal methods. The most common method is fractional distillation of liquefied air, with N_2 distilling as a vapor while O_2 is left as a liquid. The other primary method of producing O_2 is passing a stream of clean, dry air through one bed of a pair of identical zeolite molecular sieves, which absorbs the nitrogen and delivers a gas stream that contains 90%–93% O_2. At the same time, N_2 gas is released from the other nitrogen-saturated zeolite bed, by reducing the chamber operating pressure and diverting part of the oxygen gas from the producer bed through it, in the reverse direction of flow. After a set cycle time the operation of the two beds is interchanged, in so doing permitting a continuous supply of gaseous oxygen to be pumped through a pipeline. This is called PSA. Oxygen gas is more and more

BOX 3.8 Oxygen build up in the Earth's atmosphere

Free O_2 was basically absent in Earth's atmosphere before photosynthetic archaea and bacteria evolved, probably some 3.5 billion years ago. Free O_2 first appeared in substantial amounts during the Paleoproterozoic eon (3.0 to 2.3 billion years ago). For the first billion years, any free oxygen produced by these organisms reacted with dissolved Fe in the oceans to form banded iron formations. After these oxygen sinks became saturated, free O_2 began to outgas from the oceans 3 to 2.7 billion years ago, reaching ~10% of its current level about 1.7 billion years ago. The existence of large amounts of dissolved and free O_2 in the oceans and atmosphere probably drove most of the existing anaerobic organisms to extinction during the Great Oxygenation Event (~2.4 billion years ago). Cellular O_2 respiration allows aerobic organisms to make much more ATP than anaerobic organisms. Cellular O_2 respiration takes place in all eukaryotes, including all complex multicellular organisms. Since the start of the Cambrian period 540 million years ago, atmospheric O_2 levels have varied between 15 and 30 vol.%. Near the end of the Carboniferous period (~300 million years ago) atmospheric O_2 levels reached a maximum of 35 vol.%, which may have partially contributed to the large size of insects and amphibians at this time. Atmospheric O_2 level changes have affected past climates. Falling O_2 levels caused atmospheric density decreases and surface evaporation increases, resulting in precipitation increases and higher temperatures.

produced by these noncryogenic technologies. Oxygen gas can also be produced through electrolysis of water into molecular oxygen and hydrogen. DC electricity must be used: if AC is used, the gases in each limb consist of hydrogen and oxygen in the explosive ratio 2:1. A comparable method is the electrocatalytic O_2 evolution from oxides and oxoacids. Chemical catalysts can be applied as well, such as in chemical oxygen generators or oxygen candles that are used as part of the life-support equipment on submarines and are still part of standard equipment on commercial airliners in case of depressurization emergencies. An alternative air separation method forces air to dissolve through ceramic membranes based on zirconium dioxide by either high pressure or an electric current, to produce nearly pure O_2 gas.

Oxygen (besides H) is the most common element in minerals and forms the basis of the anionic groups (carbonates, nitrates, borates, sulfates, chromates, phosphates, arsenates, vanadates, molybdates, tungstates, silicates, etc.). The chemically simplest minerals are found in the oxide/hydroxide class, for example, anatase (TiO_2), bixbyite (Mn_2O_3), böhmite (AlOOH), brookite (TiO_2) (Fig. 3.62), brucite ($Mg(OH)_2$) (Fig. 3.63), cassiterite (SnO_2) (Figs. 3.64), chromite ($Fe^{2+}Cr_2^{3+}O_4$), chrysoberyl ($BeAl_2O_4$), corundum (Al_2O_3) (Figs. 3.65), cuprite (Cu_2O), diaspore (AlOOH), franklinite ($ZnFe_2^{3+}O_4$), gibbsite ($Al(OH)_3$), hematite (Fe_2O_3) (Fig. 3.66), ilmenite ($FeTiO_3$), magnetite ($Fe^{2+}Fe_2^{3+}O_4$), manganite (MnOOH), opal ($SiO_2{\cdot}nH_2O$), quartz (SiO_2) (Fig. 3.67 and Fig. 3.68), rutile (TiO_2) (Fig. 3.69), spinel ($MgAl_2O_4$), uraninite (UO_2), and zincite (ZnO). For other classes see the element forming the anion with oxygen: for example, C (carbonates), N (nitrates), S (sulfates), etc.

3.7.3 Chemistry

Naturally occurring oxygen consists of 3 stable isotopes, ^{16}O, ^{17}O, and ^{18}O, with ^{16}O being the most abundant at 99.762%. Most ^{16}O is formed at the end of the helium fusion process (The triple-alpha process is a set of nuclear fusion reactions by which 3 ^{4}He nuclei (alpha particles) are transformed into carbon. As a side effect of the process, some carbon nuclei fuse with additional helium to form a stable isotope of oxygen and energy: $^{12}C + {}^{4}He \rightarrow {}^{16}O + \gamma$) in massive stars but some is produced in the neon-burning process (The neon-burning process (nuclear decay) is a set of nuclear fusion reactions that take place in massive stars. Neon-burning requires high temperatures and densities. At such high

FIGURE 3.62 Brookite, TiO_2, a brownish orange 4 cm terminated crystal. Mohammad Ghawos mine, Kharan dist., Baluchistan, Pakistan.

FIGURE 3.63 Brucite, $Mg(OH)_2$, yellow botryoidal crystal groups to 2 cm across. Killa Saifullah dist., Baluchistan, Pakistan.

FIGURE 3.64 Gray, adamantine, tetragonal cassiterite, SnO_2, crystals from 1 cm. Peerless Mine, Pennington Co., South Dakota, United States.

temperatures photodisintegration becomes a significant effect, so some neon nuclei decompose, releasing alpha particles: $^{20}Ne + \gamma \rightarrow {}^{16}O + {}^{4}He$). ^{17}O is mainly formed through the burning of hydrogen into helium during the CNO cycle [The CNO cycle (carbon–nitrogen–oxygen) is one of the 2 known sets of fusion reactions by which stars convert H to He; the CNO cycle is a catalytic cycle. It is dominant in stars that are more than 1.3 times as massive as the Sun.]; as a result, it is a common isotope in the hydrogen burning zones of stars. Most ^{18}O is produced when ^{14}N (formed abundantly from CNO burning) captures a ^{4}He nucleus, hence ^{18}O is common in the helium-rich zones of evolved, massive stars. Fourteen radioisotopes are known. The most stable are ^{15}O with a half-life of 122.24 seconds and ^{14}O with a half-life of 70.606 seconds. All of the other radioactive isotopes have half-lives that are less than 27 seconds and most have half-lives that are less than 83 ms. The most common mode of decay of the isotopes lighter than ^{16}O is via β^+ decay to form N, and for the isotopes heavier than ^{18}O is β decay to produce F.

Oxygen is a member of the chalcogen group on the periodic table. The electronic configuration of the free O atom is $[He]2s^2 2p_x^2 2p_y^1 2p_z^1$. Oxygen is normally divalent, though, other oxidation states such as 1/2, 0, $-1/3$, $-1/2$, -1 and -2 in isolable compounds are also known (Table 3.7). It has a very high electronegativity of 3.5, which is exceeded

FIGURE 3.65 Corundum, Al_2O_3, variety sapphire, bluish purple prismatic crystal to 3 cm on biotite schist. Zazafotsy quarry, Ihosy dist., Horombe, Fianarantsoa, Madagascar.

FIGURE 3.66 Sharp, reflective near black thick tabular crystals of hematite, Fe_2O_3, with the largest crystal to 3.5 cm. Bahia, Brazil.

only by fluorine. It is a colorless, odorless, tasteless highly reactive gas. Much of the chemistry of oxygen can be attributed to its electronic structure, high electronegativity and small size. Oxygen shows many similarities to nitrogen in its covalent chemistry, and its propensity to form H bonds and p_π double bonds. Similarities to fluorine and fluorides are also notable.

Several allotropes of oxygen exist, each with varying degrees of reactivity. Atomic oxygen, O, is an extremely reactive, fugitive species which cannot be isolated free from other substances. This allotrope is a strong oxidizing agent and it is an important reactant in the chemistry of the upper atmosphere. Some reactions include:

$$N_2 + O \rightarrow N_2O$$
$$NO + O \rightarrow NO_2$$
$$OH + O \rightarrow O_2 + H$$
$$Cl_2 + O \rightarrow Cl_2O,\ ClO_2$$
$$CO + O \rightarrow CO_2$$
$$HCN + O \rightarrow CO,\ CO_2,\ NO,$$
$$H_2S + O \rightarrow H_2O,\ SO_2,\ SO_3,\ H_2SO_4$$

FIGURE 3.67 Quartz, SiO_2, variety agate. Specimen of rich red and grayish blue cut and polished Roselle agate, 6.5 × 4.5 × 5 cm. Agate Creek, Queensland, Australia.

FIGURE 3.68 Quartz, SiO_2, variety amethyst, a 4.5 cm prismatic, terminated crystal. Rio Grande do Sul, Brazil.

Dioxygen molecule, O_2, is a reactive substance and must be stored away from combustible materials. Its ground-state configuration has two unpaired electrons occupying two degenerate molecular orbitals. The two electrons in the degenerate molecular orbitals can have parallel spins leading to a triplet ground state. This prevents ground-state oxygen from reacting directly at room temperature with many molecules, which are often in the singlet state, unless the molecule is a very reactive one like white phosphorus. At higher temperatures or in the presence of catalysts, the reaction with most molecules proceed readily. When the two electrons described above pair their spins even if they remain in two degenerate orbitals, the total quantum spin becomes 0 and the molecule is referred to be in a singlet state (only one possible arrangement of the spins). Singlet oxygen lies 94.72 kJ/mol above in energy than the triplet ground state.

FIGURE 3.69 Golden straw-like crystals of rutile, TiO_2, to 9 mm in sprays around hematite, Fe_2O_3. Novo Horizonte, Bahia, Brazil.

TABLE 3.7 Oxygen properties.

Appearance	Colorless gas
Allotropes	O_2, O_3 (ozone)
Standard atomic weight $A_{r,std}$	15.999
Block	p-Block
Element category	Reactive nonmetal
Electron configuration	[He] $2s^2$ $2p^4$
Phase at STP	Gas
Melting point	−218.79°C
Boiling point	−182.962°C
Density (at STP)	1.429 g/L
Heat of fusion	(O_2) 0.444 kJ/mol
Heat of vaporization	(O_2) 6.82 kJ/mol
Molar heat capacity	(O_2) 29.378 J/(mol · K)
Oxidation states	**−2**, −1, +1, +2
Ionization energies	1st: 1313.9 kJ/mol
	2nd: 3388.3 kJ/mol
	3rd: 5300.5 kJ/mol
Covalent radius	66 ± 2 pm
Van der Waals radius	152 pm

STP, Standard temperature and pressure.
Bold font indicates main oxidation state.

It is more reactive than ground-state oxygen and even participating in Diels-Alder [4 + 2]- and [2 + 2]-cycloaddition reactions. It is particularly important in the photochemical reactions with unsaturated or aromatic organic compounds.

One of the most important reactions of dioxygen is that with the protein hemoglobin which forms the basis of oxygen transport in blood. In cells of aerobic organisms, oxygen plays a crucial role as electron acceptor in the mitochondria to generate chemical energy in the form of adenosine triphosphate (ATP) during oxidative phosphorylation. It can be said that aerobic respiration is the opposite of photosynthesis except that photons are not generated but rather, 2880 kJ of chemical energy is released per mole of glucose. The simplified version of this process is:

$$C_6H_{12}O_6 + 6O_2 \rightarrow 6CO_2 + 6H_2O + 2880\ kJ/mol$$

Ozone, O_3, is an unstable, blue diamagnetic gas with a characteristic pungent odor. It condenses to a deep blue liquid (bp −111.9°C) and to a violet almost black solid (mp −192.5°). Both the liquid and the solid are explosive. Two most characteristic chemical properties of ozone are its strongly oxidizing nature and its tendency to transfer an O atom with coproduction of O_2. The highly reactive nature of O_3 is further typified by the following reactions:

$$CN^- + O_3 \rightarrow OCN^- + O_2$$
$$2NO_2 + O_3 \rightarrow N_2O_5 + O_2$$
$$PbS + 4O_3 \rightarrow PbSO_4 + 4O_2$$
$$3I^- + O_3 + 2H^+ \rightarrow I_3^- + O_2 + H_2O$$

Ozone adds readily to unsaturated organic compounds and can cause unwanted crosslinking in rubbers and other polymers with residual unsaturation. It also adds to alkenes to form "ozonides" which can be converted to aldehydes and ketones.

From the reactive nature of all the forms of oxygen, important compounds including water are formed. These compounds will be treated in the section for individual elements where oxygen is bonded.

3.7.3.1 Organic compounds

Oxygen is present in some of the most important classes of organic compounds such as alcohols (ROH), ether (ROR). ketones (RCOR), aldehydes (RCHO), carboxylic acids (RCOOH), esters (RCOOR), acid anhydrides (RCOOCOR) and amides (RC(O)NR). These organic compounds containing oxygen include industrially important organic solvents such as acetone, methanol, ethanol, isopropanol, furan, THF, diethyl ether, dioxane, ethyl acetate, DMF, DMSO, acetic acid, and formic acid. Most important, oxygen is found in major classes of biomolecules—carbohydrates, lipids, proteins, and nucleic acids that are important to life.

3.7.4 Major uses

Oxygen is used in cellular respiration and many important groups of organic molecules in living organisms including humans contain oxygen, for example, proteins, nucleic acids, carbohydrates, and fats, as do the most important constituent inorganic compounds of animal shells, teeth, and bone (calcite, aragonite, apatite, etc.). Most of the mass of living organisms contains oxygen as a part of water, H_2O. Oxygen is continuously replaced in the Earth's atmosphere by photosynthesis in plants, using the energy of sunlight to produce oxygen from H_2O and CO_2. Uptake of O_2 from the air is the vital purpose of respiration for humans, so oxygen supplementation is utilized in medicine. Treatment not only increases oxygen levels in the patient's blood, but in addition has the secondary effect of lowering resistance to blood flow in various types of diseased lungs, easing the workload of the heart. Oxygen therapy is used to treat emphysema, pneumonia, some heart disorders (congestive heart failure), some disorders causing increased pulmonary artery pressure, and any disease impairing the body's capacity to take up and use gaseous oxygen. Special oxygen chambers are used in hyperbaric (high-pressure) medicine in order to increase the partial O_2 pressure around the patient and, if necessary, the treating medical staff. Carbon monoxide (CO) poisoning, gas gangrene, and decompression sickness (the "bends") are sometimes treated using hyperbaric chambers. Increased O_2 concentration in the lungs helps to displace carbon monoxide (CO) from the heme group of hemoglobin. O_2 gas is poisonous to the anaerobic bacteria that cause gas gangrene, hence increasing its partial pressure helps kill these bacteria. Decompression sickness occurs in divers who decompress too quickly when coming up to the surface after a dive, resulting in bubbles of inert gas, mostly nitrogen and helium, forming in the blood. Increasing the O_2 pressure as quickly as possible assists in redissolving these gas bubbles back into the blood so that these excess gases can be exhaled naturally via the lungs. A special application of O_2 as a low-pressure breathing gas can be found in modern space suits, which surround the astronaut's body with the breathing gas. These devices use almost pure O_2 at about one-third normal pressure, resulting in a normal blood partial pressure of O_2. This trade-off of higher O_2 concentration for lower pressure is required to preserve suit flexibility. Scuba and surface-supplied underwater divers and submariners also rely on artificially delivered O_2. Submarines, submersibles, and atmospheric diving suits generally work at normal atmospheric pressure. Breathing air is scrubbed of carbon dioxide (CO_2) through chemical extraction and O_2 is replaced to sustain a constant partial pressure. Ambient pressure divers breathe air or gas mixtures with an oxygen fraction appropriate for the operating depth. Pure or nearly pure O_2 use in diving at pressures higher than atmospheric is typically restricted to rebreathers, or decompression at relatively shallow depths ($\sim$6 m depth, or less), or medical treatment in recompression chambers at pressures up to 2.8 bar, where acute oxygen toxicity can be managed without the risk of drowning. Deeper diving involves substantial dilution of O_2 with other gases, such as nitrogen or helium, to prevent oxygen toxicity.

Smelting of iron ore into steel accounts for about 55% of total commercially produced O_2. In this process, O_2 is injected through a high-pressure lance into molten iron, which removes sulfur impurities and excess carbon as the corresponding oxides, SO_2 and CO_2. These reactions are exothermic, so the temperature increases to 1700°C. About 25% of commercially produced O_2 is used by the chemical industry. For example, ethylene is reacted with O_2 to form ethylene oxide, which subsequently is transformed into ethylene glycol; the primary starting compound used to produce a range of different products, such as antifreeze and polyester polymers (the precursors of many plastics and fabrics), etc. Most of the remaining 20% of commercially produced O_2 is utilized in medical applications (see above), metal cutting and welding, as an oxidizer in rocket fuel, and in water treatment. Oxygen is used in oxyacetylene welding burning acetylene with O_2 to produce an extremely hot flame. In this process, metal up to 60 cm thick is first heated with a small oxy-acetylene flame and then quickly cut by a large stream of O_2.

Ozone, or trioxygen, is an inorganic molecule with the chemical formula O_3. It is a pale blue gas with a distinctively pungent smell. The largest application of ozone is found in the manufacturing of pharmaceuticals, synthetic lubricants, and various other commercially useful organic compounds, where it is used to break carbon-carbon bonds. In addition, it can be used for bleaching substances and for killing microorganisms in air and water sources. Many municipal drinking water systems kill bacteria with ozone as an alternative to the more common chlorine. Ozone possesses a very high oxidation potential. Ozone does not form organochlorine compounds, and it does not remain in the water after treatment. Where electrical power is readily available, ozone is a cost-effective treatment method of water, as it can be manufactured on demand and does not necessitate the transportation and storage of hazardous chemicals. Once it has decayed, it leaves no taste or odor in drinking water. Even though low levels of ozone have been advertised to be of some disinfectant use in residential homes, the ozone concentration in dry air needed to have a rapid, significant effect on airborne pathogens exceeds safe levels recommended by, for example, the US Occupational Safety and Health Administration and Environmental Protection Agency. Humidity control can significantly improve both the ozone killing power and the decay rate back to oxygen (more humidity allows more effectiveness). Industrially, ozone is used in a variety of applications, for example, to disinfect laundry in hospitals, food factories, care homes, etc., to disinfect water in place of chlorine, to deodorize air and objects, such as after a fire (this process is extensively used in fabric restoration), to kill bacteria on food or on contact surfaces, to disinfect cooling towers and control legionella with reduced chemical consumption, water bleed-off and increased performance, to sanitize swimming pools and spas, to kill insects in stored grain, to scrub yeast and mold spores from the air in food processing plants, to wash fresh fruits and vegetables to kill yeast, mold and bacteria, to chemically attack contaminants in water (Fe, As, hydrogen sulfide, nitrites, and complex organics, etc.), to provide an aid to flocculation (agglomeration of molecules, which aids in filtration, where the Fe and As are removed), for the manufacturing of chemical compounds via chemical synthesis, to clean and bleach fabrics, to act as an antichlor in chlorine-based bleaching, to assist in processing plastics to allow adhesion of inks, to age rubber samples to determine the useful life of a batch of rubber, to eradicate water borne parasites such as *Giardia lamblia* and *Cryptosporidium* in surface water treatment plants. Water intense industries such as breweries and dairy plants can effectively use dissolved ozone as an alternative to chemical sanitizers such as peracetic acid, hypochlorite or heat. Ozone is a reagent in many organic reactions in both the laboratory and industry. For example, ozonolysis is the cleavage of an alkene to carbonyl compounds. Many hospitals around the world use large ozone generators to decontaminate operating rooms between surgeries. To achieve this the rooms are first cleaned and then sealed airtight before being filled with ozone which effectively kills or neutralizes all remaining bacteria. Ozone is utilized as a substitute for chlorine or chlorine dioxide in the bleaching of wood pulp. It is frequently used in combination with oxygen and hydrogen peroxide in order to remove the need for chlorine-containing compounds in the production of high-quality, white paper. Ozone can be used to detoxify cyanide wastes (e.g., from gold and silver mining) by oxidizing cyanide to cyanate and eventually to carbon dioxide.

Devices generating high levels of ozone, some of which use ionization, are utilized to sanitize and deodorize uninhabited buildings, rooms, ductwork, woodsheds, boats and other vehicles. Ozonated water is used to launder clothes and to sanitize food, drinking water, and surfaces in the home. As stated by the US Food and Drug Administration (FDA), it is "amending the food additive regulations to provide for the safe use of ozone in gaseous and aqueous phases as an antimicrobial agent on food, including meat and poultry." Studies have shown that levels as low as 0.3 μmol/mol of ozone dissolved in filtered tap water can result in a reduction of more than 99.99% in such food-borne microorganisms such as *salmonella*, *E. coli* 0157:H7 and *Campylobacter*. Ozone can be used to eliminate pesticide residues from fruits and vegetables. Ozone is used in homes and hot tubs to kill bacteria in the water and to reduce the amount of chlorine or bromine necessary by reactivating them to their free state. Since ozone does not remain in the water long enough, ozone by itself is ineffective at avoiding cross-contamination among bathers and has to be used in combination with halogens. Gaseous ozone produced by ultraviolet light or by corona discharge is injected into the water. Ozone is also extensively used in treatment of water in aquariums and fishponds. Its use can minimize bacterial growth, control parasites, eliminate transmission of some diseases, and reduce or eliminate "yellowing" of the water. Ozone must not come in contact with fishes' gill structures. Natural saltwater (with life forms) provides enough "instantaneous demand" that controlled amounts of ozone activate bromide ions to hypobromous acid (HOBr), and the ozone completely decays in a few seconds to minutes. If oxygen-fed ozone is used, the water will be higher in dissolved oxygen and fishes' gill structures will atrophy, making them dependent on oxygen-enriched water.

Ozone use on freshly cut pineapple and banana resulted in an increase in flavonoids and total phenol contents with exposure up to 20 minutes. At the same time a decrease in ascorbic acid (one form of vitamin C) content was found but the positive effect on total phenol content and flavonoids can negate the negative effect. Tomatoes upon ozone treatment exhibited an increase in β-carotene, lutein, and lycopene. Yet, ozone application on strawberries in preharvest period showed a decrease in ascorbic acid concentration. Ozone enables the extraction of some heavy metals from soil using EDTA (ethylenediaminetetraacetic acid, IUPAC name 2,2′,2″,2‴-(ethane-1,2-diyldinitrilo)tetraacetic acid, $C_{10}H_{16}N_2O_8$). EDTA can form strong, water-soluble coordination complexes with some heavy metals (Pb, Zn) thus making it possible to dissolve them out from contaminated soil. Some studies have shown that when contaminated soil is pretreated with ozone, the extraction efficacy of Pb, Am, and Pu increased by as much as 11.0%–28.9%, 43.5% and 50.7%, respectively.

3.8 9 F — Fluorine

3.8.1 Discovery

In 1529, Georgius Agricola (March 24, 1494–November 21, 1555) wrote how fluorite was used as an additive to lower the melting point of metals during smelting (Fig. 3.70). He used the Latin word fluorés (fluor, flow) for fluorite rocks. The name later changed into fluorspar (still regularly used) and then fluorite. The chemical composition of fluorite was later determined to be calcium difluoride, CaF_2. Hydrofluoric acid was used in glass etching after the 1720s. The German chemist Andreas Sigismund Marggraf (March 3, 1709–August 7, 1782) first characterized it in 1764 when he heated fluorite with sulfuric acid, and the subsequent solution corroded its glass container (Marggraf, 1770). Swedish Pomeranian and German pharmaceutical chemist Carl Wilhelm Scheele (December 9, 1742–May 21, 1786) reproduced the experiment in 1771 and called the acidic product fluss-spats-syran (fluorspar acid) (Scheele, 1771). In 1810, the French physicist and mathematician André-Marie Ampère (January 20, 1775–June 10, 1836) proposed that hydrogen and an element similar to chlorine formed hydrofluoric acid. The Cornish chemist and inventor Sir Humphry Davy (December 17, 1778–May 29, 1829) suggested that this so far unknown substance be called fluorine from fluoric acid and the -ine suffix of other halogens (Davy, 1813). This word, with amendments, is used in most European languages; Greek, Russian, and some

FIGURE 3.70 Georgius Agricola.

FIGURE 3.71 Nobel Prize photo of Henri Moissan, 1906, published in 1907 in Les Prix Nobel.

others (following Ampère's suggestion) use the name ftor or derivatives, from the Greek φθόριος (phthorios, destructive). The New Latin name fluorum gave the element its current symbol F; Fl was used in early papers. Isolation of elemental fluorine was extremely difficult due to the extreme corrosiveness of both elemental fluorine itself and hydrogen fluoride, along with the absence of a simple and suitable electrolyte. The French chemist Edmond Frémy (February 28, 1814–February 3, 1894) hypothesized that electrolysis of pure hydrogen fluoride to produce fluorine was achievable and developed a method to produce anhydrous samples from acidified potassium bifluoride, KF_2; instead, he found that the subsequent (dry) hydrogen fluoride did not conduct electricity. Frémy's former student Henri Moissan (French chemist, September 28, 1852–February 20, 1907) persisted, and after ample trial and error observed that a mixture of potassium bifluoride and dry hydrogen fluoride was a conductor, enabling electrolysis (Moissan, 1886). He won the Nobel Prize in Chemistry in 1906 for his work on isolating fluorine from its compounds (Fig. 3.71).

3.8.2 Mining, production, and major minerals

Fluorine is the13th most common element in Earth's crust at 600–700 ppm by mass. It is found only in minerals, of which fluorite, fluorapatite, and cryolite are the most industrially important. Fluorite or fluorspar (CaF_2), colorful and abundant worldwide, is fluorine's main source; China and Mexico are the main providers. The United States led winning in the early 20th century but stopped mining in 1995. While fluorapatite ($Ca_5(PO_4)_3F$) comprises most of the world's fluorine, its low mass fraction of 3.5% means that most of it is used as a phosphate. In the United States small amounts of fluorine compounds are obtained via fluorosilicic acid, a phosphate industry by-product. Cryolite (Na_3AlF_6), once used directly in aluminum manufacture, is the rarest and most concentrated of these three minerals. The chief commercial mine on Greenland's west coast stopped production in 1987, and most cryolite is now synthesized. Fluorite mining, which supplies most global fluorine, was at its maximum production in 1989 when 5.6 million metric tons of ore were extracted. Chlorofluorocarbon restrictions lowered this to 3.6 million tons in 1994; production has since recovered. Froth flotation splits mined fluorite into two main metallurgical grades of equal proportion: 60%–85% pure metalspar (metspar) is almost all used in iron smelting whereas 97% + pure acidspar is mainly converted to the key industrial intermediate hydrogen fluoride (HF) (Fig. 3.72).

The method developed by French chemist Ferdinand Frédéric Henri Moissan (September 28, 1852–February 20, 1907) is used to produce industrial quantities of fluorine, via the electrolysis of a potassium fluoride/hydrogen fluoride mixture. In this process hydrogen and fluoride ions are reduced and oxidized at a steel container cathode and a carbon block anode, under 8–12 volts, to produce hydrogen and fluorine gas respectively. Temperatures are elevated, $KF \cdot 2HF$ melting at 70°C and being electrolyzed at 70°C–130°C. KF, which acts as catalyst, is critical as pure HF cannot be

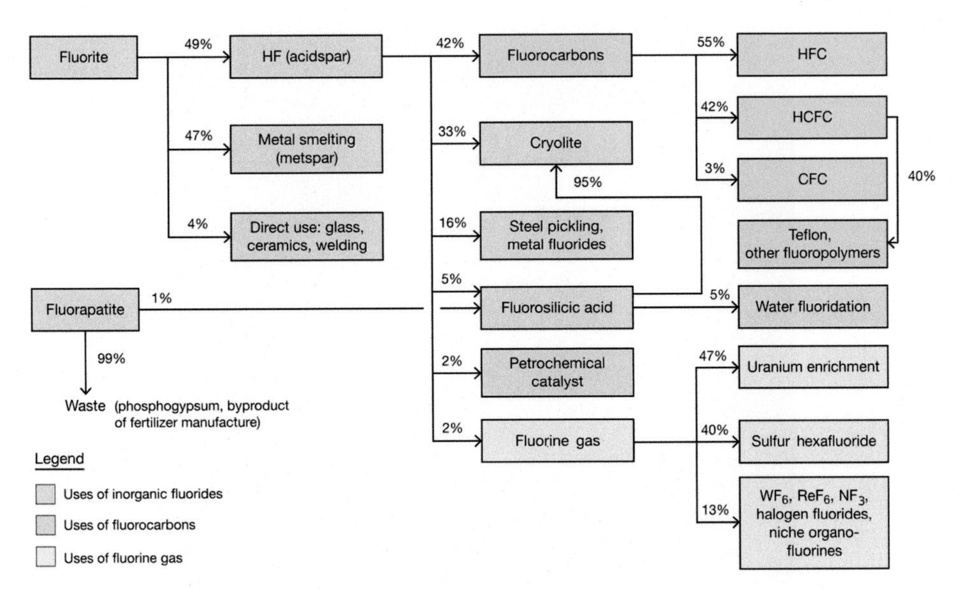

FIGURE 3.72 Major uses and sources of fluorine: Applications involving fluorine gas are shown in yellow. Percentage indicates mass volumes, 2003 data. https://en.wikipedia.org/wiki/Fluorine#Industrial_applications CC S-A 3.0.

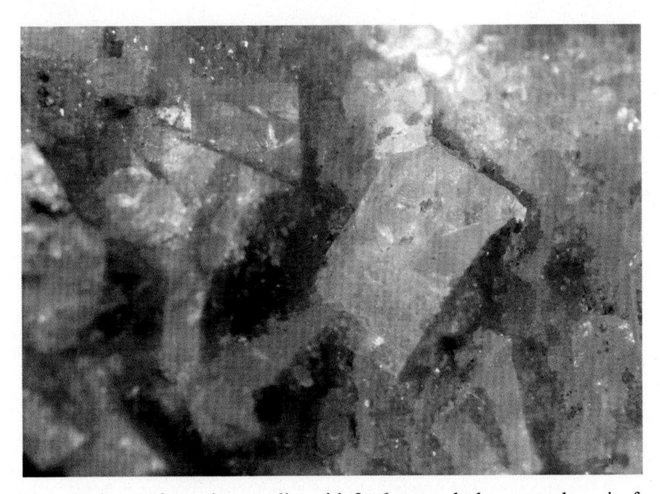

FIGURE 3.73 Cryolite, Na_2NaAlF_6, a specimen of massive cryolite with 3–6 mm colorless crystals on its face. Ivigtut Cryolite deposit, Ivigtut, Arsuk Fjord, Sermersooq, Greenland.

electrolyzed. Fluorine can be stored in steel cylinders that have passivated interiors, at temperatures below 200°C; otherwise nickel can be used. Regulator valves and pipework are made of nickel, the latter possibly using Monel (a group of nickel alloys, principally composed of Ni (from 52% to 67%) and Cu, with minor amounts of Fe, Mn, C, and Si) instead. Regular passivation, along with the strict exclusion of water and greases, must be undertaken.

Fluorine forms a group of halide class minerals. It is observed in a number of minerals where F forms a unique part of the structure, but in many OH groups—containing minerals it is found to substitute for the OH-group. Within the halide group the best-known minerals are: creedite ($Ca_3SO_4Al_2F_8(OH)_2{\cdot}2H_2O$), cryolite ($Na_2NaAlF_6$) (Fig. 3.73), fluorite (CaF_2) (Fig. 3.74), sellaite (MgF_2), and villiaumite (NaF) (Fig. 3.75) (Box 3.9). In the oxide class one finds a handful of minerals such as fersmite ($(Ca,Ce,Na)(Nb,Ta,Ti)_2(O,OH,F)_6$), fluornatromicrolite ($(Na_{1.5}Bi_{0.5})Ta_2O_6F$), and landauite ($NaMnZn_2(Ti,Fe)_6Ti_{12}O_{38}$). The carbonates contain minerals such as the bastnäsite series ($(Ce,La,Nd,Y)(CO_3)F$), parisite-(Ce) ($CaCe_2(CO_3)_3F_2$), and the synchysite series ($Ca(Ce,Nd,Y)(CO_3)_2F$). There are only five borates that contain F,

FIGURE 3.74 Fluorite, CaF_2, green, cubic, sharp crystals to 2.2 cm on edge. Okorusu mine, Otjiwarongo dist., Otjozondjupa, Namibia.

FIGURE 3.75 The villiaumite, NaF, in this specimen has crystallized in a skeletal or arborescent habit, about 3.5 cm. It is of graphitic nature in which the villiaumite was probably crystallized simultaneously with the microcline, $K(AlSi_3O_8)$. This type of intergrowth has also been termed "elatolith." Palitra pegmatite, Mt. Kedykverpakhk, Lovozero Massif, Kola Peninsula, Russia.

the best known are fluoborite ($Mg_3(BO_3)(F,OH)_3$) and jeremejevite ($Al_6(BO_3)_5(F,OH)_3$). The sulfate class contains 12 minerals, none of which are particularly well known. A few examples are: grandreefite ($Pb_2(SO_4)F_2$), sulfohalite ($Na_6(SO_4)_2FCl$), and wilcoxite ($MgAl(SO_4)_2F\cdot17H_2O$). A larger group of minerals is found in the phosphate class, for example, amblygonite ($LiAl(PO_4)F$), fluellite ($Al_2(PO_4)F_2(OH)\cdot7H_2O$), fluorapatite ($Ca_5(PO_4)_3F$) (Fig. 3.76), herderite ($CaBePO_4(F,OH)$), minyulite ($KAl_2(PO_4)_2(OH,F)\cdot4H_2O$), and wavellite ($Al_3(PO_4)_2(OH,F)_3\cdot5H_2O$). A large group of minerals containing F can be found in the silicate class either as direct part of the structure or as substitution for OH. Well-known minerals include: astrophyllite ($K_2NaFe^{2+}_7Ti_2Si_8O_{28}(OH)_4F$), chondrodite ($(Mg,Fe^{2+})_5(SiO_4)_2(F,OH)_2$), clinohumite ($Mg_9(SiO_4)_4F_2$), fluorapophyllite ($(K,Na)Ca_4(Si_8O_{20})(F,OH)\cdot8H_2O$), humite ($(Mg,Fe^{2+})_7(SiO_4)_3(F,OH)_2$), narsarsukite ($Na_4(Ti,Fe)_2[Si_8O_{20}](O,OH,F)_2$), topaz ($Al_2(SiO_4)(F,OH)_2$), and vesuvianite ($(Ca,Na,\square)_{19}(Al,Mg,Fe^{3+})_{13}(\square,B,Al,Fe^{3+})_5(Si_2O_7)_4(SiO_4)_{10}(OH,F,O)_{10}$). In addition, F for OH substitution is regularly observed in the tourmaline, mica and amphibole series.

BOX 3.9 Fluorite, CaF_2, and cryolite, Na_3AlF_6

Fluorite is a common mineral in hydrothermal mineral deposits associated with sulfides (pyrite, galena, sphalerite, etc.), carbonates, and baryte. Fluorite can also be found as a minor mineral in granite, pegmatites, syenite, and greisen. It may occur as veins in carbonate sediments, as detrital grains or (rarely) as a cement in sandstone. The name fluorite comes from the Latin word *fluere*, mining "to flow" referring to the ease with which it melts. Fluorite is isometric (cubic) *4/m2/m* (space group *F4/m2/m*). Cubic crystals are common (sometimes modified with octahedron or other forms), octahedral cleavage fragments may appear to be crystals. Other forms alone are rare. Penetration twins are common. It has a hardness of 4. Fluorite shows a perfect {111} cleavage, while its fracture is conchoidal and splintery. The luster is vitreous while the transparency varies from transparent to translucent. The color can vary from colorless, blue, purple, or green to almost any other color. The streak is white. Often it shows fluorescence in both long- and short-wave UV light and some varieties are phosphorescent.

Cryolite is a rare mineral. The most significant samples are from Ivigtut on the west coast of Greenland, where it is in ore deposits hosted by granitic rocks. Its name comes from the Greek words *kryos* meaning frost and *lithos* meaning stone, referring to its icy appearance. Cryolite is monoclinic *2/m* (space group $P2_1/n$) Individual pseudocubic crystals are rare. Aggregates are massive, lamellar or columnar, often exhibiting pseudocubic parting. Its hardness is 2½. The luster vitreous to greasy, transparent to translucent. It has no cleavage and uneven fracture, cubic parting. The color varies from colorless to snow white, while the streak is white. Optically it is biaxial(+).

FIGURE 3.76 Fluorapatite, $Ca_5(PO_4)_3F$, purple prismatic crystal to 1.3 cm. King Lithia mine, Pennington Co., South Dakota, United States.

3.8.3 Chemistry

Just one isotope of F can be found naturally in abundance, the stable isotope ^{19}F. It has a high magnetogyric ratio (The ratio of the angular momentum to magnetic moment is known as the gyromagnetic ratio. Certain nuclei can for many purposes be imagined as spinning round an axis like the Earth or like a top. In general, the spin provides them with angular momentum and with a magnetic moment; the first because of their mass, the second because all or part of their electric charge may be rotating with the mass.) and extraordinary sensitivity to magnetic fields. Seventeen radioisotopes with mass numbers from 14 to 31 have been synthesized, of which ^{18}F is the most stable with a half-life of 109.77 minutes. Other radioisotopes have much shorter half-lives of less than 70 seconds; most of them decay in less than 0.5 seconds. The isotopes ^{17}F and ^{18}F undergo β^+ decay and electron capture, while lighter isotopes decay via proton emission, and those heavier than ^{19}F through β^- decay (the heaviest ones with delayed neutron emission). Two metastable isomers of fluorine are known, ^{18m}F, with a half-life of 162(7) nanoseconds, and ^{26m}F, with a half-life of 2.2(1) milliseconds.

Fluorine is the first member of the halogens. Its electronic configuration is $1s^2 2s^2 2p_x^2 2p_y^2 2p_z^1$. It is the most electronegative element and is exclusively univalent (Table 3.8). Its compounds are formed either by gain of one electron to give F^- ion and become isoelectronic with neon ($[He]2s^2 2p_x^2 2p_y^2 2p_z^2$) or by sharing one electron in a covalent single bond. The lone pairs of electrons surrounding fluorine also permit both the fluoride ion and certain molecular fluorides to act as Lewis bases in which the coordination number of F is greater than 1.

Fluorine is the most reactive of all elements. It reacts, under appropriate conditions, with every other element in the periodic table except He, Ar and Ne; most of these reactions are very vigorous that they become explosive. The ability of F_2 to react with Xe under mild conditions to produce crystalline xenon fluorides is a testament to the reactivity of fluorine. This reactivity made the isolation of the element by chemical reactions or electrolysis very difficult. The first successful attempt was made in June 26, 1886 by H. Moissan, who applied electrolysis on a cooled solution of KHF_2 in an anhydrous liquid HF, using Pt/Ir electrodes sealed into a platinum U-tube with fluorspar caps. For achieving this elusive feat along with his development of the electric furnace, Moissan was awarded the Nobel Prize in Chemistry in 1906. Since then, fluorine technology and the applications of fluorine-containing compounds have developed especially during the 20th century.

The reactivity of fluorine can be attributed to a number of factors: (1) small dissociation energy leading to low activation energies of reaction; (2) very strong bonds formed by fluorine with other elements; and (3) strong oxidizing property of fluorine, engendering unusually high oxidation states in the elements it reacts, for example, IF_7, PuF_6, BiF_5, etc. Like all elements in the 2p row, fluorine is atypical of the elements in its group and for the same reasons: (1) small size of the atoms; (2) tightly bound electrons that makes ionization and distortion difficult; and (3) absence of low-lying d orbitals for bonding.

Elemental fluorine reacts with water to form oxygen and hydrofluoric acid. It reacts with oxygen to form OF_2. It is very reactive that it reacts with noble gases such as xenon to form fluorides.

$$2F_2 + 2H_2O \rightarrow O_2 + 4HF$$
$$2F_2 + O_2 \rightarrow 2OF_2$$
$$F_2 + Xe \rightarrow XeF_2$$

It also reacts with acids and bases according to the reactions below.

$$F_2 + 2NaOH \rightarrow O_2 + 2NaF + H_2$$
$$4F_2 + HCl + H_2O \rightarrow 3HF + OF_2 + ClF_3$$
$$F_2 + 2HNO_3 \rightarrow 2NO_3F + H_2$$

TABLE 3.8 Fluorine properties.

Appearance	Gas: very pale-yellow liquid: bright yellow solid: alpha is opaque, beta is transparent
Allotropes	Alpha, beta
Standard atomic weight $A_{r,std}$	18.998
Block	p-Block
Element category	Reactive nonmetal
Electron configuration	[He] $2s^2$ $2p^5$
Phase at STP	Gas
Melting point	−219.67°C
Boiling point	−188.11°C
Density (at STP)	1.696 g/L
When liquid (at b.p.)	1.505 g/cm^3
Heat of vaporization	6.51 kJ/mol
Molar heat capacity	Cp: 31 J/(mol · K) (at 21.1°C) Cv: 23 J/(mol · K) (at 21.1°C)
Oxidation states	**−1**
Ionization energies	1st: 1681 kJ/mol 2nd: 3374 kJ/mol 3rd: 6147 kJ/mol
Covalent radius	64 pm
Van der Waals radius	135 pm

STP, Standard temperature and pressure.
Bold font indicates main oxidation state.

3.8.3.1 Compounds of fluorine

Hydrogen fluoride (HF) is a colorless volatile liquid and an oligomeric H-bonded gas. It is an acute poison that may immediately and permanently damage the respiratory tract and the eyes. Anhydrous HF is prepared from the reaction between concentrated sulfuric acid, H_2SO_4, and fluorspar, CaF_2.

$$H_2SO_4 + CaF_2 \rightarrow CaSO_4 + 2HF$$

Anhydrous HF fluorinates metals, nonmetals, hydrides, oxides, and many other classes of compound to form fluorides. Alkali metals, except beryllium, form soluble, ionic fluorides. Alkaline earth metals form insoluble, ionic difluorides. The rare earth elements and many other metals form mostly ionic trifluorides. Zirconium, hafnium, and some actinides form ionic tetrafluorides, whereas, those of titanium, vanadium and niobium are covalent. Compound with up to seven fluorine atoms attached to the metal center are known and are observed to be particularly reactive. The noble gases are mostly unreactive but were found to form compounds with fluorine. Xenon hexafluroplatinate, xenon difluoride, tetrafluoride, hexafluoride and multiple oxyfluorides are known to exist. Krypton is also found to form a difluoride and radon is suspected to react with fluorine to form solid radon difluoride. The lighter noble gases also form difluorides, but they are very unstable.

Organofluorines are stable because of the exceptional strength of the carbon-fluorine bond. When the hydrogen atoms in alkanes are substituted by fluorine atoms, several properties change. These substitutions result in melting and boiling points decreases, solubility in hydrocarbons decreases, density increases, and overall stability increases.

3.8.4 Major uses

There was no commercial production of fluorine until the Second World War, when the development of the atom bomb, and other nuclear energy projects, made it necessary to produce large quantities. Before this, fluorine salts, known as fluorides, were for a long time used in welding and for frosting glass. The largest application of fluorine gas is in the preparation of UF_6 for the nuclear fuel cycle. Fluorine is used to fluorinate uranium tetrafluoride (UF_4), itself formed from the reaction between uranium dioxide (UO_2) and hydrofluoric acid (HF). Fluorine is monoisotopic, therefore any mass differences between UF_6 molecules are due to the presence of ^{235}U or ^{238}U, allowing uranium enrichment via gaseous diffusion or gas centrifuge. A significant amount is used to produce the inert dielectric SF_6 for high-voltage transformers and circuit breakers, removing the necessity for dangerous polychlorinated biphenyls associated with oil-filled devices. Some fluorine compounds are used in electronics: for example, rhenium and tungsten hexafluoride in chemical vapor deposition, tetrafluoromethane in plasma etching and nitrogen trifluoride in cleaning equipment. In addition, fluorine is used in the synthesis of organic fluorides, but its reactivity often requires first transformation to the gentler ClF_3, BrF_3, or IF_5, which together allow calibrated fluorination. Fluorinated pharmaceuticals use sulfur tetrafluoride instead.

Similar to other Fe alloys, around 3 kg metspar (fluorite, CaF_2) is added to each metric ton of steel; the fluoride ions decrease its melting point and viscosity. Together with its role as an additive in materials like enamels and welding rod coats, most acidspar (CaF_2) is reacted with sulfuric acid to form hydrofluoric acid, which is utilized in steel pickling, glass etching, and alkane cracking. One-third of HF is used for the production of cryolite (Na_3AlF_6) and aluminum trifluoride (AlF_3), both used as fluxes in the Hall–Héroult process for aluminum extraction; replacement is required due to their infrequent reactions with the smelting apparatus. Each metric ton of Al requires about 23 kg of flux. Fluorosilicates consume the second largest amount, with sodium fluorosilicate used in water fluoridation and laundry effluent treatment, and as an intermediate in the production of cryolite and silicon tetrafluoride. Other important inorganic fluorides include those of cobalt, nickel, and ammonium.

Organofluorides consume over 20% of mined fluorite and over 40% of hydrofluoric acid. Surfactants form a minor group. Due to the danger from direct hydrocarbon–fluorine reactions above −150°C, industrial fluorocarbon manufacture is indirect, typically via halogen exchange reactions such as Swarts fluorination, in which chlorocarbon chlorines are substituted for fluorines by hydrogen fluoride using a catalyst. Electrochemical fluorination electrolyzes hydrocarbons in hydrogen fluoride, and the Fowler process treats them with solid fluorine carriers like cobalt trifluoride. Halogenated refrigerants, commonly known as Freons, are identified by R-numbers that signify the amount of fluorine, chlorine, carbon, and hydrogen present. Chlorofluorocarbons (CFCs) like R-11, R-12, and R-114 once dominated organofluorines, with maximum production in the 1980s. Used for air conditioning systems, propellants and solvents, their manufacture was less than 10% of this maximum by the early 2000s, after extensive international prohibition. Hydrochlorofluorocarbons (HCFCs) and hydrofluorocarbons (HFCs) were designed as their replacements; the production uses more than 90% of the fluorine in the organic industry. Important HCFCs include R-22, chlorodifluoromethane, and R-141b. The main HFC is R-134a with a new type of molecule HFO-1234yf, a hydrofluoroolefin (HFO) is becoming more and more important due to its global warming potential of less than 1% that of HFC-134a.

Fluoropolymers can only be formed by polymerizing free radicals. Polytetrafluoroethylene (PTFE), sometimes called by its DuPont trade name Teflon, forms 60%–80% by mass of global fluoropolymer manufacture. The main use is in electrical insulation as PTFE is an excellent dielectric. It is also used in the chemical industry where corrosion resistance is required, in coating pipes, tubing, and gaskets. One more major application is in the use of PFTE-coated fiberglass cloth for stadium roofs. The major consumer application is for nonstick cookware. Jerked PTFE film becomes expanded PTFE (ePTFE), a fine-pored membrane occasionally referred to by the brand name Gore-Tex and used in rainwear, protective apparel, and filters. In addition, ePTFE fibers may be made into seals and dust filters. Other fluoropolymers, including fluorinated ethylene propylene, mimic PTFE's properties and can substitute for it; they are more moldable, but also more expansive and exhibit a lower thermal stability. The chemically resistant (though costly) fluorinated ionomers are used as electrochemical cell membranes, of which the first and most important example is Nafion. Developed in the 1960s, it was originally used as fuel cell material in spacecraft and subsequently replaced Hg-based chloralkali process cells. Lately, the fuel cell application has re-emerged with attempts to install proton exchange membrane fuel cells into automobiles. Fluoroelastomers such as Viton are crosslinked fluoropolymer mixtures mainly used in O-rings; perfluorobutane (C_4F_{10}) is used as a fire-extinguishing agent. Fluorosurfactants are small organofluorine molecules used for repelling water and stains. They form a minority in the overall surfactant market, most of which is taken up by much cheaper hydrocarbon-based products.

Around 30% of agrichemicals have fluorine, most of them herbicides and fungicides together with a few crop regulators. Fluorine substitution, usually of a single atom or at most a trifluoromethyl group, is a robust change with effects similar to fluorinated pharmaceuticals: better biological stay time, membrane crossing, and altering of molecular recognition. Sodium monofluoroacetate (1080) is a mammalian poison in which two acetic acid hydrogens are substituted with fluorine and sodium; it disrupts cell metabolism by replacing acetate in the citric acid cycle. First synthesized in the late 19th century, it was recognized as an insecticide in the early 20th century. New Zealand, the largest user of 1080, applies it to protect kiwis from the invasive Australian common brushtail possum. Europe and the United States have banned the use of 1080 though.

Population studies from the mid-20th century onwards showed topical fluoride reduce dental caries. This was initially attributed to the conversion of tooth enamel hydroxyapatite into the more durable fluorapatite, but studies on prefluoridated teeth refuted this hypothesis, and current theories involve fluoride aiding enamel growth in small caries. After studies of children in regions where fluoride was naturally present in drinking water, controlled public water supply fluoridation to fight tooth decay began in the 1940s and is now applied to water supplying 6% of the world population. Sodium monofluorophosphate and occasionally sodium or tin(II) fluoride are frequently found in fluoride toothpastes, first introduced in the United States in 1955 and now abundant in developed countries, together with fluoridated mouthwashes, gels, foams, and varnishes.

About 20% of modern pharmaceuticals contain fluorine. One of these, the cholesterol reducer atorvastatin (Lipitor), made more revenue than any other drug until it became generic in 2011. The combination asthma prescription Seretide contains 2 active ingredients, of which fluticasone is fluorinated. Many drugs are fluorinated to delay inactivation and lengthen dosage periods as the C-F bond is very stable. Fluorination also increases lipophilicity since the bond is more hydrophobic than the C-F bond, and this frequently helps in cell membrane penetration and therefore bioavailability. Tricyclics and other pre-1980s antidepressants had numerous side effects due to their nonselective interference with neurotransmitters besides the serotonin target; the fluorinated fluoxetine was selective and one of the first to avoid this problem. Many recent antidepressants were adjusted in a similar manner, including the selective serotonin reuptake inhibitors: citalopram, its isomer escitalopram, and fluvoxamine and paroxetine. Quinolones are synthetic broad-spectrum antibiotics that are frequently fluorinated to boost their effects, for example, ciprofloxacin and levofloxacin. Fluorine is also used in steroids: fludrocortisone is a blood pressure-raising mineralocorticoid, and triamcinolone and dexamethasone are strong glucocorticoids. Most inhaled anesthetics are strongly fluorinated; the prototype halothane is much more inert and potent than its contemporaries. Newer compounds, for example, the fluorinated ethers sevoflurane and desflurane, are better than halothane and are nearly insoluble in blood, allowing faster waking times. Fluorine-18 is frequently present in radioactive tracers for PET, as its half-life of almost two hours is long enough to allow for its transport from production facilities to imaging centers. The most common tracer is fluorodeoxyglucose which, after intravenous injection, is adsorbed by glucose-requiring tissues such as the brain and most malignant tumors; computer-assisted tomography (CAT) can then be used for detailed imaging.

References

Arfwedson, A. (1818a). Undersökning af några vid Utö Jernmalmsbrott förekommende Fossilier, och af ett deri funnet eget Eldfast Alkali. *Afhandlingar i Fysik, Kemi och Mineralogi*, *6*, 145–172.

Arfwedson, A. (1818b). Untersuchung einiger bei der Eisen-Gru be von Utö vorkommenden Fossilien und von einem darin gefundenen neuen feuerfesten Alkali. *Journal für Chemie und Physik*, *22*, 93–117.

Berzelius, J. J. (1824a). II. Untersuchungen über Flussspathsäure und deren merkwürdigsten Verbindungen. *Annalen der Physik*, *77*, 169–230.

Berzelius, J. J. (1824b). Undersökning af flusspatssyran och dess märkvärdigaste föreningar. *Kongliga Vetenskaps-Academiens Handlingar*, *12*, 46–98.

Boyle, R. (1672). Tracts written by the Honourable Robert Boyle containing New experiments, touching the relation betwixt flame and air, and about explosions, an hydrostatical discourse occasion"d by some objections of Dr. Henry More against some explications of new experiments made by the author of these tracts: to which is annex"t, An hydrostatical letter, dilucidating an experiment about a way of weighing water in water, new experiments, of the positive or relative levity of bodies under water, of the air"s spring on bodies under water, about the differing pressure of heavy solids and fluids, Printed for Richard Davis, book-seller in Oxon, London.

Brande, W. T. (1821). (2 edn, pp. 57–58). *A manual of chemistry*, (Vol 2, pp. 57–58). London: John Murray.

Bussy, A. (1828). D'une travail qu'il a entrepris sur le glucinium. *Journal de Chimie Medicale*, *4*, 456–457.

Cavendish, H. (1766). Three papers, containing experiments on factitious air, by the Hon. Henry Cavendish, F. R. S. *Philosophical Transactions*, *56*, 141–184.

Chaptal, J. A. (1800). *Elements of chemistry* (3rd ed.), (W. Nicholson, Trans.), (Vol 1). London: G.G. and J. Robinson.

Davy, H. (1809). An account of some new analytical researches on the nature of certain bodies, particularly the alkalies, phosphorus, sulphur, carbonaceous matter, and the acids hitherto undecomposed: With some general observations on chemical theory. *Philosophical Transactions of the Royal Society of London*, *99*, 39–104.

Davy, H. (1813). Some experiments and observations on the substances produced in different chemical processes on fluor spar. *Philosophical Transactions of the Royal Society*, *103*, 263–279.

Gay-Lussac, J. L., & Thénard, L. J. (1808). Sur la décomposition et la recomposition de l'acide boracique. *Annales de chimie*, *68*, 169–174.

Gay-Lussac, J. L., & Thénard, L. J. (1811a). *Recherches physico-chimiques: faites sur la pile, sur la préparation chimique et les propriétés du potassium et du sodium, sur la décomposition de l'acide boracique, sur les acides fluorique, muriatique et muriatique oxigéné, sur l'action chimique de la lumière, sur l'analyse végétale et animale, etc* (Vol 2). Paris: Deterville.

Gay-Lussac, J. L., & Thénard, L. J. (1811b). *Recherches physico-chimiques: faites sur la pile, sur la préparation chimique et les propriétés du potassium et du sodium, sur la décomposition de l'acide boracique, sur les acides fluorique, muriatique et muriatique oxigéné, sur l'action chimique de la lumière, sur l'analyse végétale et animale, etc* (Vol 1). Paris: Deterville.

Gmelin, C. G. (1818). Von dem Lithon. *Annalen der Physik*, *59*, 238–241.

Lavoisier, A. L. (1790). *Elements of chemistry, in a new systematic order, containing all the modern discoveries: illustrated with thirteen copperplates [English] [Traité élémentaire de chimie]. (R. Kerr, Trans.). Printed for William Creech; By G. G. and J. J. Robinsons [sic]*. Edinburgh: And sold in London.

Marggraf, A. S. (1770). Observation concernant une volatilisation remarquable d'une partie de l'espece de pierre, à laquelle on donne les noms de flosse, flüsse, flus-spaht, et aussi celui d'hesperos; laquelle volatilisation a été effectuée au moyen des acides. *Mémoires de l'Académie royale des sciences et belles-lettres*, *XXIV*, 3–11.

Moissan, H. (1886). Action d'un courant électrique sur l'acide fluorhydrique anhydre. *Comptes rendus hebdomadaires des séances de l'Académie des sciences*, *102*, 1543–1544.

Priestley, J. (1772). Observations on different kinds of air. *Philosophical Transactions of the Royal Society of London*, *62*, 147–256.

Priestley, J. (1775). An account of further discoveries in air. *Philosophical Transactions*, *65*, 384–394.

Rutherford, D. (1772). Dissertatio Inauguralis de aere fixo, aut mephitico (Inaugural dissertation on the air [called] fixed or mephitic). University of Edinburgh, Edinburgh, Scotland.

Scheele, C. W. (1771). Undersökning om fluss-spat och dess syra. *Kungliga Svenska Vetenskapsademiens Handlingar*, *32*, 129–138.

Scheele, C. W. (1777). *Chemische Abhandlung von der Luft und dem Feuer*. Upsala, Sweden and Leipzig, Germany: Magnus Swederus, Siegfried Lebrecht Crusius.

Vauquelin, L.-N. (1798). De l'Aiguemarine, ou Béril; et découverie d'une terre nouvelle dans cette pierre. *Annales de Chimie*, *26*, 155–169.

Weintraub, E. (1910). Preparation and properties of pure boron. *Transactions of the American Electrochemical Society*, *16*, 165–184.

Wöhler, F. (1828). Ueber das Beryllium und Yttrium. *Annalen der Physik und Chemie*, *89*, 577–582.

Further reading

Acott, C. (1999). Oxygen toxicity: A brief history of oxygen in diving. *South Pacific Underwater Medicine Society Journal*, *29*, 150–155.

Agricola, G. (1546/1955). *De Natura Fossilium (Textbook of Mineralogy): Translated from the First Latin Ed. of 1546 by Mark Chance Bandy and Jean A. Bandy for the Mineralogical Society of America (M. C. Bandy, J. A. Bandy, Trans.)*. Boulder: Geological Society of America.

Aida, M., Fujii, Y., & Okamoto, M. (1986). Chromatographic enrichment of 10B by using weak-base anion-exchange resin. *Separation Science and Technology*, *21*, 643–654.

Aigueperse, J., Mollard, P., Devilliers, D., Chemla, M., Faron, R., Romano, R., & Cuer, J. P. (2000). *Fluorine compounds, inorganic. Ullmann's encyclopedia of industrial chemistry*. Weinheim: Wiley-VCH.

Alothman, M., Kaur, B., Fazilah, A., Bhat, R., & Karim, A. A. (2010). Ozone-induced changes of antioxidant capacity of fresh-cut tropical fruits. *Innovative Food Science and Emerging Technologies, 11*, 666–671.

Ampère, A.-M. (1816). Suite d'une classification naturelle pour les corps simples. *Annales de chimie et de physique, 2*, 1–5.

Asimov, I. (1966). *The noble gases*. New York, NY: Basic Books.

Babu, R. S., & Gupta, C. K. (1988). Beryllium extraction – A review. *Mineral Processing and Extractive Metallurgy Review, 4*, 39–94.

Banks, R. E. (1986). Isolation of fluorine by Moissan: Setting the scene. *Journal of Fluorine Chemistry, 33*, 3–26.

Barth, S. (1997). Boron isotopic analysis of natural fresh and saline waters by negative thermal ionization mass spectrometry. *Chemical Geology, 143*, 255–261.

Berger, L. I. (1996). *Semiconductor materials* (pp. 37–43). Boca Raton: CRC Press.

Berner, R. A. (1999). Atmospheric oxygen over phanerozoic time. *Proceedings of the National Academy of Sciences of the United States of America, 96*, 10955–10957.

Berzelius, J. J. (1817). Ein neues mineralisches Alkali und ein neues Metall. *Journal für Chemie und Physik, 21*, 44–48.

Best, N. W. (2015). Lavoisier's 'reflections on phlogiston' I: Against phlogiston theory. *Foundations of Chemistry, 17*, 137–151.

Bethe, H. A. (1939). Energy production in stars. *Physical Review, 55*, 434–456.

Bowman, S. (1995). *Radiocarbon dating*. London: British Museum Press.

Bunsen, R. (1855). Darstellung des Lithiums. *Annalen der Chemie und Pharmacie, 94*, 107–111.

Butterfield, N. J. (2009). Oxygen, animals and oceanic ventilation: An alternative view. *Geobiology, 7*, 1–7.

Carlson, D. P., & Schmiegel, W. (2000). *Fluoropolymers, organic. Ullmann's encyclopedia of industrial chemistry*. Weinheim: Wiley-VCH.

Cook, G. A., & Lauer, C. M. (1968). Oxygen. In C. A. Hampel (Ed.), *The encyclopedia of the chemical elements* (pp. 499–512). New York: Reinhold Book Corporation.

Cooke, T. F. (1991). Inorganic fibers—A literature review. *Journal of the American Ceramic Society, 74*, 2959–2978.

Crowe, S. A., Døssing, L. N., Beukes, N. J., Bau, M., Kruger, S. J., Frei, R., & Canfield, D. E. (2013). Atmospheric oxygenation three billion years ago. *Nature, 501*, 535–538.

Daintith, J., Mitchell, S., Tootill, E., & Gjertsem, D. (1994). *Biographical encyclopedia of scientists* (2nd ed.). Taylor & Francis, 2 Volume Set.

D'Andrada, J. B. (1800). Des caractères et des propriétés de plusieurs nouveaux minérauxde Suède et de Norwège, avec quelques observations chimiques faites sur ces substances. *Journal de chimie et de physique, 51*, 239–246.

de Boer, H. E. L., van Elzelingen-Dekker, C. M., van Rheenen-Verberg, C. M. F., & Spanjaard, L. (2006). Use of gaseous ozone for eradication of methicillin-resistant *Staphylococcus aureus* from the home environment of a colonized hospital employee. *Infection Control and Hospital Epidemiology, 27*, 1120–1122.

Emsley, J. (2001). *Nature's building blocks*. Oxford: Oxford University Press, Oxford.

Enghag, P. (2004). *Encyclopedia of the elements: Technical data - history - processing - applications*. Weinheim: Wiley-VCH.

Erisman, J. W., Sutton, M. A., Galloway, J., Klimont, Z., & Winiwarter, W. (2008). How a century of ammonia synthesis changed the world. *Nature Geoscience, 1*, 636–639.

Ferchault de Réaumur, R.-A. (1722). *L'art de convertir le fer forgé en acier, et l'art d'adoucir le fer fondu, ou de faire des ouvrages de fer fondu aussi finis que le fer forgé*. Paris: Michel Brunet.

Filler, R., & Saha, R. (2009). Fluorine in medicinal chemistry: A century of progress and a 60-year retrospective of selected highlights. *Future Medicinal Chemistry, 1*, 777–791.

Flanagan, L. B., Ehleringer, J. R., Pataki, D. E. (2005). *Stable isotopes and biosphere-atmosphere interactions*, Processes and biological controls.

Gaines, R. V., Skinner, H. C., Foord, E. F., Mason, B., & Rosenzweig, A. (1997). *Dana's new mineralogy* (8th edn). New York, NY: John Wiley & Sons, Inc.

Gannes, L. Z., Del Rio, C. M., & Koch, P. (1998). Natural abundance variations in stable isotopes and their potential uses in animal physiological ecology. *Comparative Biochemistry and Physiology – Part A: Molecular & Integrative Physiology, 119*, 725–737.

Garrett, D. E. (1998). *Borates: handbook of deposits, processing, properties, and use*. San Diego, CA: Academic Press.

Greenwood, N. N., & Earnshaw, A. (1996). *Chemistry of the Elements*. Oxford: Elsevier Science & Technology Books.

Grot, W. (2011). *Fluorinated ionomers* (2 edn). Oxford: Elsevier.

Gurov, Y. B., Aleshkin, D. V., Behr, M. N., Lapushkin, S. V., Morokhov, P. V., Pechkurov, V. A., ... Tschurenkova, T. D. (2004). Spectroscopy of superheavy hydrogen isotopes in stopped-pion absorption by nuclei. *Physics of Atomic Nuclei, 68*, 491–497.

Hagmann, W. K. (2008). The many roles for fluorine in medicinal chemistry. *Journal of Medicinal Chemistry, 51*, 4359–4369.

Hansen, P. G., Jensen, A. S., & Jonson, B. (1995). Nuclear halos. *Annual Review of Nuclear and Particle Science, 45*, 591–634.

Harlow, G. E. (1998). *The nature of diamonds*. Cambridge: Cambridge University Press.

Hildebrand, G. H. (1982). *Borax pioneer: Francis Marion Smith*. Berkeley: Howell-North Books.

Ihde, A. J. (1984). *The development of modern chemistry*. New York: Dover Publications.

Jaccaud, M., Faron, R., Devilliers, D., & Romano, R. (2000). Fluorine. In F. Ullmann (Ed.), *Ullmann's encyclopedia of industrial chemistry* (Vol 15, pp. 381–395). Weinheim: Wiley-VCH.

Janse, A. J. A. (2007). Global rough diamond production since 1870. *Gems and Gemology, XLIII*, 98–119.

Jastrow, J. (1936). *The story of human error*. New York: Books for Libraries Press.

Johansson, S., Schweitz, J.-Å., Westberg, H., & Boman, M. (1992). Microfabrication of three-dimensional boron structures by laser chemical processing. *Journal of Applied Physics, 72*, 5956–5963.

Kamienski, C. W., McDonald, D. P., Stark, M. W., & Papcun, J. R. (2004). *Lithium and lithium compounds. Kirk-Othmer encyclopedia of chemical technology*. John Wiley & Sons.

Kar, Y., Şen, N., & Demİrbaş, A. (2006). Boron minerals in Turkey, their application areas and importance for the country's economy. *Minerals & Energy – Raw Materials Report, 20*, 2–10.

Katzenberg, M. A. (2010). Chapter 13: Stable isotope analysis: A tool for studying past diet, demography, and life history. In M. A. Katzenberg, & S. R. Saunders (Eds.), *Biological anthropology of the human skeleton* (pp. 411–441). Hoboken, NJ: John Wiley & Sons.

Kirsch, P. (2004). *Modern fluoroorganic chemistry: Synthesis, reactivity, applications*. Weinheim: Wiley-VCH.

Klaproth, M. H. (1802). *Beiträge zur chemischen Kenntniss der Mineralkörper* (Vol. 3). Posen: Decker.

Klein, C., & Dutrow, B. (2008). *The 23rd edition of the manual of mineral science (after James D. Dana)*. Hoboken, NJ: John Wiley & Sons, Inc.

Kogel, J. E., Trivedi, N. C., Barker, J. M., & Krukowski, S. T. (2006). *Industrial minerals & rocks: Commodities, markets, and uses*. Littleton: Society for Mining, Metallurgy, and Exploration.

Koivula, J. I., & Fryer, C. W. (1984). Identifying gem-quality synthetic diamonds: An update. *Gems and Gemology, 20*, 146–158.

Komatsu, T., Samedima, M., Awano, T., Kakadate, Y., & Fujiwara, S. (1999). Creation of superhard B–C–N heterodiamond using an advanced shock wave compression technology. *Journal of Materials Processing Technology, 85*, 69–73.

Korsheninnikov, A. A., Nikolskii, E. Y., Kuzmin, E. A., Ozawa, A., Morimoto, K., Tokanai, F., . . . Zhukov, M. V. (2003). Experimental evidence for the existence of ^{7}H and for a specific structure of ^{8}He. *Physical Review Letters, 90*, 082501.

Krebs, R. E. (2006). *The history and use of our Earth's chemical elements: A reference guide*. Westport: Greenwood Press.

Lavoisier, A. (1799). *Elements of chemistry: In a new systematic order, containing all the modern discoveries* (4th ed.), (R. Kerr, Trans.). Edinburgh, Scotland: William Creech.

Lestan, D., Hanc, A., & Finzgar, N. (2005). Influence of ozonation on extractability of Pb and Zn from contaminated soils. *Chemosphere, 61*, 1012–1019.

Lorenz, V. (2007). Argyle in Western Australia: The world's richest diamantiferous pipe; its past and future. *Gemmologie, Zeitschrift der Deutschen Gemmologischen Gesellschaft, 56*, 35–40.

Marples, F. J. A. Michael Sendivogius, Rosicrucian, and Father of Studies of Oxygen. Societas Rosicruciana in Civitatibus Foederatis, Nebraska College. <http://www.masonic.benemerito.net/msricf/papers/marples/marples-michael.sendivogius.pdf> Accessed 15.06.19.

Meyer, B. S. (2005) Nucleosynthesis and galactic chemical evolution of the isotopes of oxygen. In: *Proceedings of the NASA Cosmochemistry Program and the Lunar and Planetary Institute*. Paper presented at the Workgroup on Oxygen in the Earliest Solar System. Gatlinburg, Tennessee. (September 19–21). Gatlinburg, Tennessee.

Morgenthaler, G. W., Fester, D. A., & Cooley, C. G. (1994). An assessment of habitat pressure, oxygen fraction, and EVA suit design for space operations. *Acta Astronautica, 32*, 39–49.

Morris, R. (2003). *The last sorcerers: The path from alchemy to the periodic table*. Washington, DC: National Academies Press.

Nassau, K. (1980). *Gems made by man*. Radnor, PA: Chilton.

Nesse, W. D. (2000). *Introduction to mineralogy*. New York, NY: Oxford University Press.

Norton, J. J. (1973). Lithium, cesium, and rubidium—The rare alkali metals. In D. A. Brobst, & W. P. Pratt (Eds.), *United States mineral resources* (pp. 365–378). Washington, DC: U.S. Geological Survey Professional, Paper 820.

Ogden, J. M. (1999). Prospects for building a hydrogen energy infrastructure. *Annual Review of Energy and the Environment, 24*, 227–279.

Okazoe, T. (2009). Overview on the history of organofluorine chemistry from the viewpoint of material industry. *Proceedings of the Japan Academy, Series B, 85*, 276–289.

Papanelopoulou, F. (2013). Louis Paul Cailletet: The liquefaction of oxygen and the emergence of low-temperature research. *Notes and Records of the Royal Society of London, 67*, 355–373.

Perkins, D. (2002). *Mineralogy* (2nd ed.). Upper Saddle River, NJ: Prentice Hall.

Piantadosi, C. A. (2004). Carbon monoxide poisoning. *Undersea and Hyperbaric Medicine Journal, 31*, 167–177.

Pizzo, G., Piscopo, M. R., Pizzo, I., & Giuliana, G. (2007). Community water fluoridation and caries prevention: A critical review. *Clinical Oral Investigations, 11*, 189–193.

Plaue, J. W., & Czerwinski, K. R. (2003). The influence of ozone on ligand-assisted extraction of ^{239}Pu and ^{241}Am from rocky flats soil. *Radiochimica Acta, 91*, 309–313.

Poulsen, C. J., Tabor, C., & White, J. D. (2015). Long-term climate forcing by atmospheric oxygen concentrations. *Science, 348*, 1238–1241.

Proudfoot, A. T., Bradberry, S. M., & Vale, J. A. (2006). Sodium fluoroacetate poisoning. *Toxicological Reviews, 25*, 213–219.

Ramkumar, J. (2012). Nafion perfluorosulphonate membrane: Unique properties and various applications. In S. Banerjee, & A. K. Tyagi (Eds.), *Functional materials: Preparation, processing and applications* (pp. 549–578). London: Elsevier.

Reich, M., & Kapenekas, H. (1957). Nitrogen purification. Pilot plant removal of oxygen. *Industrial & Engineering Chemistry, 49*, 869–873.

Renfrew, C. (2011). *Before civilization: The radiocarbon revolution and prehistoric Europe*. New York, NY: Random House.

Rossotti, H. (1998). *Diverse atoms. Profiles of the chemical elements*. Oxford: Oxford University Press.

Scerri, E. R. (2007). *The periodic table: Its story and its significance*. Oxford: Oxford University Press.

Schmitz, A., Kälicke, T., Willkomm, P., Grünwald, F., Kandyba, J., & Schmitt, O. (2000). Use of fluorine-18 fluoro-2-deoxy-D-glucose positron emission tomography in assessing the process of tuberculous spondylitis. *Journal of Spinal Disorders, 13*, 541–544.

Sciolist Online Etymology Dictionary. <https://www.etymonline.com/> Accessed 21.06.19.

Seitz, H. M., Brey, G. P., Lahaye, Y., Durali, S., & Weyer, S. (2004). Lithium isotopic signatures of peridotite xenoliths and isotopic fractionation at high temperature between olivine and pyroxenes. *Chemical Geology, 212*, 163–177.

Shigley, J. E., Fritsch, E., Stockton, C. M., Koivula, J. I., Fryer, C. W., & Kane, R. E. (1986). The gemological properties of the Sumitomo gem-quality synthetic yellow diamonds. *Gems and Gemology, 22*, 192–208.

Shigley, J. E., Fritsch, E., Stockton, C. M., Koivula, J. I., Fryer, C. W., Kane, R. E., ... Welch, C. W. (1987). The gemological properties of the De Beers gem-quality synthetic diamonds. *Gems and Gemology, 23*, 187–207.

Siegemund, G., Schwertfeger, W., Feiring, A., Smart, B., Behr, F., Vogel, H., ... Kirsch, P. (2016). *Fluorine compounds, organic. Ullmann's encyclopedia of industrial chemistry*. Weinheim: Wiley-VCH.

Sinkankas, J. (1981). *Emerald and other beryls*. Radnor, PA: Chilton Book Company.

Stephenson, R. N., Mackenzie, I., Watt, S. J., & Ross, J. A. (1996). Measurement of oxygen concentration in delivery systems used for hyperbaric oxygen therapy. *Undersea and Hyperbaric Medicine Journal, 23*, 185–188.

Strutt, R. J. (1911). Bakerian lecture. A chemically active modification of nitrogen, produced by the electric discharge. *Proceedings of the Royal Society A, 85*, 219–229.

Stwertka, A. (1999). *A guide to the elements*. Oxford: Oxford University Press.

Taylor, S. R., & McLennan, S. M. (1985). *The continental crust: Its composition and evolution*. Oxford: Blackwell Scientific Publications.

Theodoridis, G. (2006). Fluorine-containing agrochemicals: An overview of recent developments. In A. Tressaud (Ed.), *Fluorine and the environment: Agrochemicals, archaeology, green chemistry & water* (pp. 121–176). Amsterdam: Elsevier.

Toon, R. (2011). The discovery of fluorine. *Education in Chemistry, 48*, 148–151.

Urey, H. C., Brickwedde, F. G., & Murphy, G. M. (1933). Names for the hydrogen isotopes. *Science, 78*, 602–603.

Vangioni-Flam, E., & Cassé, M. (2000). Cosmic lithium-beryllium-boron story. In M. Spite (Ed.), *Galaxy evolution: Connecting the distant universe with the local fossil record: Proceedings of a colloquium held at the Observatoire de Paris-Meudon from 21-25 September, 1998* (pp. 77–86). Netherlands, Dordrecht: Springer.

Vigoureux, P. (1961). The gyromagnetic ratio of the proton. *Contemporary Physics, 2*, 360–366.

Villalba, G., Ayres, R. U., & Schroder, H. (2007). Accounting for fluorine: Production, use, and loss. *Journal of Industrial Ecology, 11*, 85–101.

Walsh, K. A., & Vidal, E. E. (2009). Sources of beryllium. In K. A. Walsh (Ed.), *Beryllium chemistry and processing* (p. 575). Materials Park, OH: ASM International.

Watson, A. (1999). Beaming into the dark corners of the nuclear kitchen. *Science, 286*, 28–31.

Weeks, M. E., & Leichester, H. M. (1968). *Discovery of the elements*. Easton, PA: Journal of Chemical Education.

Weimer, A. W. (1997). *Carbide, nitride and boride materials synthesis and processing*. London: Chapman & Hall.

Whitehead, N., Endo, S., Tanaka, K., Takatsuji, T., Hoshi, M., Fukutani, S., ... Zondervan, A. (2008). A preliminary study on the use of ^{10}Be in forensic radioecology of nuclear explosion sites. *Journal of Environmental Radioactivity, 99*, 260–270.

Winter, M. (2007). *Hydrogen: Historical information*. WebElements Ltd. <http://education.jlab.org/itselemental/ele001.html> Accessed 14 June 2019.

Wise, M. A. (1995). Trace element chemistry of lithium-rich micas from rare-element granitic pegmatites. *Mineralogy and Petrology, 55*, 203–215.

Zamboni, W. A., Riseman, J. A., & Kucan, J. O. (1990). Management of Fournier's gangrene and the role of hyperbaric oxygen. *Journal of Hyperbaric Medicine, 5*, 177–186.

Zbayolu, G., & Poslu, K. (1992). Mining and processing of borates in Turkey. *Mineral Processing and Extractive Metallurgy Review, 9*, 245–254.

Chapter 4

Period 3

4.1 11 Na — Sodium

4.1.1 Discovery

For the reason of its importance in human metabolism, salt has long been a vital product, as shown by the English word salary, which stems from the Latin word salarium, the wafers of salt occasionally given to Roman soldiers together with their other wages. In medieval Europe, a substance of sodium with the Latin name of sodanum was used as a headache medicine. The term sodium is thought to come from the Arabic word suda, meaning headache, as the headache-alleviating properties of sodium carbonate or soda were well known in early times. Although sodium, occasionally called soda, had long been documented in compounds, the metal itself was not isolated until 1807 by the Cornish chemist and inventor Sir Humphry Davy (December 17, 1778 to May 29, 1829) using the electrolysis of sodium hydroxide, NaOH (Davy, 1808a,b). In 1809, the German physicist and chemist Ludwig Wilhelm Gilbert (August 12, 1769 to March 7, 1824) suggested the names Natronium for Davy's "sodium" and Kalium for Davy's "potassium." (Davy, 1809a,b, Footnote on page 157) (Fig. 4.1). The chemical abbreviation for sodium was first published in 1814 by Swedish chemist Jöns Jakob Berzelius (August 20, 1779 to August 7, 1848) in his system of atomic symbols and is an abbreviation of the element's New Latin name natrium, in reference to the Egyptian natron, a natural mineral salt mainly consisting of hydrated sodium carbonate, $Na_2CO_3.10H_2O$ (Berzelius, 1814, page 87). Historically natron had some significant industrial and household applications, later overtaken by other sodium compounds.

FIGURE 4.1 Ludwig Wilhelm Gilbert, circa 1810.

The Periodic Table: Nature's Building Blocks. DOI: https://doi.org/10.1016/B978-0-12-821279-0.00004-2

4.1.2 Mining, production, and major minerals

The earth's crust contains 2.27% Na, making it the seventh most abundant element on earth and the fifth most abundant metal, behind Al, Fe, Ca, and Mg and ahead of K. Sodium's estimated oceanic abundance is 1.08×10^4 mg/L. Because of its high reactivity, it is never found as a pure element. It is found in many different minerals, some very soluble, such as halite, NaCl, and natron, $Na_2CO_3 \cdot 10H_2O$ (Box 4.1), others much less soluble, such as amphiboles and zeolites. Used only in relatively specialized applications, only around 100,000 tons of metallic sodium are produced per year. Metallic sodium was first produced commercially in the late 19th century by the carbothermal reduction of sodium carbonate at 1100°C, as the first step of the Deville process to produce aluminum:

$$Na_2CO_3 + 2C \rightarrow 2Na + 3CO$$

The high demand for aluminum created the need to produce sodium. The introduction of the Hall–Héroult process to produce aluminum using electrolysis in a molten salt bath ended the need for large amounts of sodium. A related process based on the reduction in sodium hydroxide, NaOH, was established in 1886. Sodium is now produced by commercially applying the electrolysis of molten sodium chloride, NaCl, applying a patented process (1924). This is achieved in a Downs cell where the NaCl is mixed with calcium chloride, $CaCl_2$, in order to lower the melting point below

BOX 4.1 Brine purification

Salt can be mined in three different methods: conventional underground mining, solar mining (evaporation of brine from seawater, salt lakes, and springs), and solution mining. Solution mining is the most modern and economically viable process of extracting salt from underground deposits (e.g., salt domes). For this method to be applied, the salt deposit must have sufficient thickness and lateral extent, must be highly pure, and must not contain KCl. The design and installation of the brine well comprise an outer casing that is cemented to the wall of the borehole and two inner concentric pipes that are introduced into the casing. During operation, fresh water is pumped into the borehole through the annular space between the two inner pipes and dissolves the salt from the walls of the cavern below the borehole. As the salt concentration increases, the fresh water changes to brine with higher specific gravity. It sinks to the bottom and then rises to the surface in the central pipe because of the applied pressure. The most common impurities in crude brines are Ca^{2+}, Mg^{2+}, and SO_4^{2-} ions. They are usually removed by chemical treatment; evaporation, or recrystallization as well as other methods. In chemical treatment, Mg^{2+} ions are precipitated from the crude brine as $Mg(OH)_2$ by adding $Ca(OH)_2$ or NaOH:

$$MgCl_2 + 2NaOH \text{ or } Ca(OH)_2 \rightarrow Mg(OH)_2 + 2NaCl \text{ or } CaCl_2$$

Ferric iron can also be removed by this method, if it is present in the crude brine. Ca^{2+} ions are precipitated as $CaCO_3$ by adding Na_2CO_3:

$$CaSO_4 + Na_2CO_3 \rightarrow CaCO_3 + Na_2SO_4$$

and

$$CaCl_2 + Na_2CO_3 \rightarrow CaCO_3 + 2NaCl$$

or by pumping CO_2 through the brine. Salt is removed from the purified brine by evaporation, which is stopped before the start of Na_2SO_4 crystallization. The solubility of NaCl increases only slightly with increasing temperature, and it can be crystallized out in an evaporative crystallizer at 50°C–150°C. The most common method currently utilized is the multistage evaporation process. Here the vapors from the boiling brine in the first stage are used to boil the brine in the second stage, which is at a lower pressure than the first. The vapors from each subsequent stage are circulated through the following stage, which is in turn at a lower pressure and enables it to boil at a correspondingly lower temperature. Three different types of crystallizers are generally used. Type 1 is an evaporator with forced circulation and external heating. Type 2 consists of an evaporator with internal heaters and a circulating pump in the central pipe that produces forced circulation. Type 3 is known as the Oslo crystallizer in which the recirculating brine and fresh brine are heated in a heat exchanger and evaporated in a crystallizer, and from there passes upward through a bed of crystals. Type 1 crystallizers normally produce cubic crystals with 50% larger than 400 μm, while type 2 crystallizers produce cubic crystals with rounded corners with 50% larger than 650 μm. Finally, type 3 crystallizers produce a granular product. The recrystallization process is based on the fact that the solubility of $CaSO_4$ in brine decreases with increase in temperature while that of NaCl increases slightly. The whole process contains three different stages: (1) steam is introduced into the dissolving vessel, where fine salt and mother liquor are heated to ~105°C. Sodium chloride dissolves, but $CaSO_4$ remains in the mother liquor as a solid, (2) $CaSO_4$ is removed from the brine by filtration, and (3) NaCl is precipitated during cooling when the brine is pumped to an expansion evaporator.

FIGURE 4.2 A partial, cleaved crystal of blue halite, NaCl, 9 × 7 × 5 cm. Intrepid Potash East mine, Carlsbad, Eddy Co., New Mexico, United States.

FIGURE 4.3 Dawsonite, $NaAlCO_3(OH)_2$, thin white needles in clusters to 8 mm. Terlano, Bolzano, Italy.

700°C. Since calcium is less electropositive than sodium, no calcium will be deposited at the cathode. This method is less expensive than the previous Castner process (the electrolysis of sodium hydroxide).

Na is a relatively common element in minerals. The sulfide class contains eight minerals containing Na, for example, erdite ($NaFeS_2{\cdot}2H_2O$) or ottensite ($(Na,K)_3Sb_6^{3+}(Sb^{3+}S_3)O_9{\cdot}3H_2O$). A larger group can be found in the halide class: for example, cryolite (Na_2NaAlF_6), halite (NaCl) (Fig. 4.2), pachnolite ($NaCa[AlF_6]{\cdot}H_2O$), and villiaumite (NaF). Within the oxide class, more than 40 minerals contain Na: for example, birnessite ($(Na,Ca)_{0.5}(Mn^{4+},Mn^{3+})_2O_4{\cdot}1.5H_2O$), franconite ($Na(Nb_2O_5)(OH){\cdot}3H_2O$), lueshite ($NaNbO_3$), natroniobite ($NaNbO_3$), natrotantite ($Na_2Ta_4O_{11}$), and todorokite ($(Na,Ca,K,Ba,Sr)_{1-x}(Mn,Mg,Al)_6O_{12}{\cdot}3\text{-}4H_2O$). Na forms an essential part of a large number of carbonates, such as dawsonite ($NaAlCO_3(OH)_2$) (Fig. 4.3), gaylussite ($Na_2Ca(CO_3)_2{\cdot}5H_2O$), natron ($Na_2CO_3{\cdot}10H_2O$), nitratine ($NaNO_3$), trona ($Na_3H(CO_3)_2{\cdot}2H_2O$), and welagonite ($Na_2Sr_3Zr(CO_3)_6{\cdot}3H_2O$). The borate class contains 21 minerals with sodium as part of their structure. These include well-known minerals such as borax ($Na_2(B_4O_5)(OH)_4{\cdot}8H_2O$), kernite ($Na_2[B_4O_6(OH)_2]{\cdot}3H_2O$), tincalconite ($Na_2(B_4O_7){\cdot}5H_2O$), and ulexite ($NaCa[B_5O_6(OH)_6]{\cdot}5H_2O$). The sulfate class contains minerals such as alum-(Na) ($NaAl(SO_4)_2{\cdot}12H_2O$), blödite ($Na_2Mg(SO_4)_2{\cdot}4H_2O$), glauberite ($Na_2Ca(SO_4)_2$), hanksite ($Na_{22}K(SO_4)_9(CO_3)_2Cl$), kröhnkite ($Na_2Cu(SO_4)_2{\cdot}2H_2O$), lecontite ($(NH_4,K)NaSO_4{\cdot}2H_2O$), mirabilite ($Na_2SO_4{\cdot}10H_2O$), and natrojarosite ($NaFe_3(SO_4)_2(OH)_6$). The second largest group of minerals containing Na is found in the phosphate class, for example,

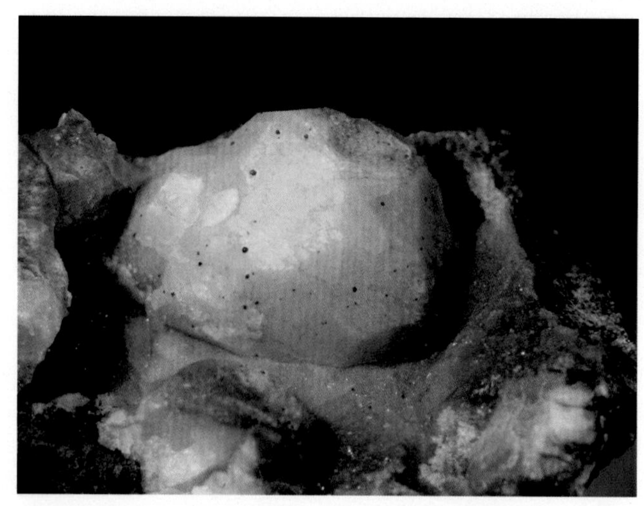

FIGURE 4.4 Analcime, $Na(AlSi_2O_6){\cdot}H_2O$, a 2 cm crystal with red staining. Tyrol, Austria.

FIGURE 4.5 White to colorless prismatic natrolite, $Na_2Al_2Si_3O_{10}{\cdot}2H_2O$, crystals to 8 mm on albite matrix with micaceous polylithionite. Mont Saint-Hilaire, Rouville Co., Quebec, Canada.

beryllonite ($NaBePO_4$), brazilianite ($NaAl_3(PO_4)_2(OH)_4$), lavendulan ($NaCaCu_5(AsO_4)_4Cl{\cdot}5H_2O$), natrodufrénite $\left(NaFe^{2+}Fe_5^{3+}(PO_4)_4(OH)_6{\cdot}2H_2O\right)$, sampleite ($NaCaCu_5(PO_4)_4Cl{\cdot}5H_2O$), and wardite ($NaAl_3(PO_4)_2(OH)_4{\cdot}2H_2O$). By far the largest group, however, is found in the silicate class with more than 500 minerals containing Na, for example, aegirine ($NaFe^{3+}Si_2O_6$), albite ($Na(AlSi_3O_8)$), analcime ($Na(AlSi_2O_6){\cdot}H_2O$) (Fig. 4.4), cancrinite ($(Na,Ca,\Box)_8(Al_6Si_6O_{24})(CO_3,SO_4)_2{\cdot}2H_2O$), elpidite ($Na_2ZrSi_6O_{15}{\cdot}3H_2O$), eudialyte ($Na_{15}Ca_6(Fe^{2+},Mn^{2+})_3Zr_3[Si_{25}O_{73}](O,OH,H_2O)_3(OH,Cl)_2$), glaucophane ($\Box[Na_2][Mg_3Al_2]Si_8O_{22}(OH)_2$), haüyne ($(Na,K)_3(Ca,Na)(Al_3Si_3O_{12})(SO_4,S,Cl)$), jadeite ($Na(Al,Fe^{3+})Si_2O_6$), mesolite ($Na_2Ca_2Si_9Al_6O_{30}{\cdot}8H_2O$), natrolite ($Na_2Al_2Si_3O_{10}{\cdot}2H_2O$) (Fig. 4.5), nepheline ($Na_3K(Al_4Si_4O_{16})$), pectolite ($NaCa_2Si_3O_8(OH)$), schorl $\left(Na\left(Fe_3^{2+}\right)Al_6(Si_6O_{18})(BO_3)_3(OH)_3(OH)\right)$, and sodalite ($Na_8(Al_6Si_6O_{24})Cl_2$). Finally, there are six minerals in the organic class that contain Na such as julienite ($Na_2[Co(SCN)_4]{\cdot}8H_2O$) or natroxalate ($Na_2C_2O_4$).

4.1.3 Chemistry

Twenty isotopes of sodium have been recognized, but only ^{23}Na is stable. ^{23}Na is formed in the carbon burning process in stars by fusing two carbon atoms together ($^{12}C + ^{12}C \rightarrow ^{23}Na + ^{1}H$); this necessitates temperatures above 600 megakelvins and a star of no less than three solar masses. Two radioactive, cosmogenic isotopes are the by-product of cosmic

TABLE 4.1 Sodium properties.

Appearance	Silvery-white metallic
Standard atomic weight $A_{r,std}$	22.989
Block	S-block
Element category	Alkali metal
Electron configuration	[Ne] $3s^1$
Phase at STP	Solid
Melting point	97.794°C
Boiling point	882.940°C
Density (near r.t.)	0.968 g/cm^3
When liquid (at m.p.)	0.927 g/cm^3
Heat of fusion	2.60 kJ/mol
Heat of vaporization	97.42 kJ/mol
Molar heat capacity	28.230 J/(mol·K)
Oxidation states	−1, **+1**
Ionization energies	1st: 495.8 kJ/mol
	2nd: 4562 kJ/mol
	3rd: 6910.3 kJ/mol
Atomic radius	empirical: 186 pm
Covalent radius	166 ± 9 pm
Van der Waals radius	227 pm

STP, Standard temperature and pressure.
Bold font indicates main oxidation state.

ray spallation (also known as the x-process, is a set of naturally occurring nuclear reactions causing nucleosynthesis; it refers to the formation of chemical elements from the impact of cosmic rays on an object. Cosmic rays are highly energetic charged particles from beyond earth, ranging from protons, alpha particles, and nuclei of many heavier elements.): ^{22}Na has a half-life of 2.602 years and decays via β^+ to ^{22}Ne while ^{24}Na with a half-life of 14.96 hours decays via β^- to ^{24}Mg; all other isotopes have half-lives below one minute. Two nuclear isomers have been found, the longer-lived metastable one being ^{24m}Na with a half-life of approximately 20.2 milliseconds. Acute neutron radiation exposure, as from a nuclear criticality accident, transforms some of the stable ^{23}Na in human blood to ^{24}Na; the neutron radiation dosage of a victim can be calculated by determining the ratio of ^{24}Na relative to ^{23}Na.

Sodium is the second member of the alkali metals and has an electronic configuration [Ne]$3s^1$. It is a soft, silvery-white, low-melting metal which crystallizes with *bcc* lattices (Table 4.1). As a member of the alkali group, sodium is very reactive which is largely due to the ease of removing the ns^1 electron. The second ionization of sodium is very high since the electron would have to be removed from an inner orbital. Thus, sodium forms ionic compounds involving the Na^+ cation. Sodium is more reactive than lithium but less reactive than the heavier potassium. Its strong tendency to lose the ns^1 electron makes it a very strong reducing agent.

Sodium metal used to be produced commercially by the electrolysis of a fused eutectic mixture of 40% NaCl, 60% $CaCl_2$. The process is carried out in a Downs cell, which is an electrolytic cell containing a carbon anode and iron cathode, introduced by du Pont in 1921 to lower the melting point to around 600°C. In this process, oxidation of the chloride ions to chlorine gas occurs in the carbon anode while reduction in the sodium ions occurs at the iron cathode. The calcium ion is not reduced in this process since its reduction potential (−2.87 volts) is lower than that of the sodium ion (−2.71 volts).

$$\text{Anode: } 2Cl^- \rightarrow Cl_2(g) + 2e^-$$

$$\text{Cathode: } 2Na^+ + 2e^- \rightarrow 2Na(l)$$

4.1.3.1 Reaction of sodium with water

Sodium metal reacts violently with water to form a colorless solution of sodium hydroxide (NaOH) and hydrogen gas (H_2). The reaction is exothermic, and the heat released may ignite the hydrogen produced. The reaction is less violent than that of potassium (immediately below sodium in the periodic table) but more vigorous than that of lithium (immediately above sodium in the periodic table).

$$2Na(s) + 2H_2O(l) \rightarrow 2NaOH(aq) + H_2(g)$$

4.1.3.2 *Reaction of sodium with air*

Sodium has a shiny surface when freshly cut but easily dulls because of the action of air and moisture. If sodium is burned in air, the result is white sodium peroxide, Na_2O_2, together with some sodium oxide, Na_2O, which is also white.

$$2Na(s) + O_2(g) \rightarrow 2Na_2O_2(s)$$

$$4Na(s) + O_2(g) \rightarrow 2Na_2O(s)$$

The alkali metal oxides M_2O (M = Li, Na, K, Rb) crystallize in the antifluorite structure (fluorite CaF_2). In this structure the positions of the anions and cations are reversed relative to their positions in CaF_2, with Na ions tetrahedrally coordinated to 4 O ions and O cubically coordinated to 8 Na ions. Sodium oxide is formed through the reaction of sodium with sodium hydroxide, sodium peroxide, or sodium nitrite:

$$2NaOH + 2Na \rightarrow 2Na_2O + H_2$$
$$Na_2O_2 + 2Na \rightarrow 2Na_2O$$
$$2NaNO_2 + 6Na \rightarrow 4Na_2O + N_2$$

Most of these reactions rely on the reduction of a molecule by sodium, whether it is hydroxide, peroxide, or nitrite. Burning sodium in air will produce Na_2O and about 20% sodium peroxide Na_2O_2.

$$6Na + 2O_2 \rightarrow 2Na_2O + Na_2O_2$$

Sodium peroxide is a yellowish solid. It is a strong base. This metal peroxide occurs in several hydrates and peroxyhydrates, such as $Na_2O_2{\cdot}2H_2O_2{\cdot}4H_2O$, $Na_2O_2{\cdot}2H_2O$, $Na_2O_2{\cdot}2H_2O_2$, and $Na_2O_2{\cdot}8H_2O$. The octahydrate, which is easy to prepare, is white, in contrast to the anhydrous material. The octahydrate is formed by reacting sodium hydroxide with hydrogen peroxide. Sodium peroxide can be manufactured on a large scale through the reaction of metallic sodium with oxygen at 130°C–200°C, a process that generates sodium oxide, which in a separate stage absorbs oxygen

$$2Na_2O + O_2 \rightarrow 2Na_2O_2$$

It may also be prepared by passing ozone gas over solid sodium iodide inside a platinum or palladium tube. The ozone oxidizes the sodium to form sodium peroxide. The iodine can be sublimed by mild heating. The platinum or palladium catalyzes the reaction and is not attacked by the sodium peroxide. Sodium peroxide crystallizes with hexagonal symmetry. Upon heating, the hexagonal form undergoes a transition into a phase of unknown symmetry at 512°C. With further heating above the 657°C boiling point, the compound decomposes to Na_2O, releasing O_2.

4.1.3.3 *Reaction of sodium with the halogens*

Sodium metal reacts vigorously with all the halogens to form sodium halides. It reacts with fluorine, F_2; chlorine, Cl_2; bromine, Br_2; and iodine, I_2, to form sodium fluoride (NaF), sodium chloride (NaCl), sodium bromide (NaBr), and sodium iodide (NaI), respectively.

4.1.3.4 *Reaction of sodium with acids*

Sodium metal dissolves readily in dilute sulfuric acid to form solutions containing the Na^+ ion together with hydrogen gas, H_2.

$$2Na(s) + H_2SO_4(aq) \rightarrow 2Na^+(aq) + SO_4^{2-}(aq) + H_2(g)$$

The decahydrate of sodium sulfate is known as Glauber's salt after the Dutch/German alchemist and apothecary Johann Rudolf Glauber (March 10, 1604 to March 16, 1670), who discovered it in 1625 in Austrian spring water. He called it *sal mirabilis* (miraculous salt), due to its medicinal properties: the crystals were used as a general purpose laxative, until more sophisticated alternatives came about in the 1900s. In the 18th century, Glauber's salt began to be utilized as a starting material for the industrial production of soda ash (sodium carbonate), by reaction with potash (potassium carbonate). Demand for soda ash increased and the supply of sodium sulfate had to increase as well. Hence, in the 19th century, the large scale Leblanc process, producing synthetic sodium sulfate as a primary intermediate, became the principal method of soda ash production. Sodium sulfate is a typical electrostatically bonded ionic sulfate. The existence of free sulfate ions in solution is indicated by the easy formation of insoluble sulfates when these solutions are treated with, for example, Ba^{2+} or Pb^{2+} salts:

$$Na_2SO_4 + BaCl_2 \rightarrow 2\ NaCl + BaSO_4$$

Sodium sulfate is unreactive toward most oxidizing or reducing agents. At high temperatures, it can be converted to sodium sulfide by carbothermal reduction (high- temperature heating with charcoal, etc.):

$$Na_2SO_4 + 2C \rightarrow Na_2S + 2CO_2$$

This reaction was employed in the Leblanc process, a defunct industrial route to sodium carbonate. Sodium sulfate reacts with sulfuric acid to give the acid salt sodium bisulfate:

$$Na_2SO_4 + H_2SO_4 \rightleftharpoons 2NaHSO_4$$

Sodium sulfate exhibits a moderate tendency to form double salts. The only alums formed with common trivalent metals are $NaAl(SO_4)_2$ (unstable above 39°C) and $NaCr(SO_4)_2$, in contrast to potassium sulfate and ammonium sulfate which form many stable alums. Double salts with some other alkali metal sulfates are known, including $Na_2SO_4{\cdot}3K_2SO_4$ which occurs naturally as the mineral aphthitalite. Formation of glaserite by the reaction of sodium sulfate with potassium chloride has been employed as the basis of a method for producing potassium sulfate, a fertilizer. Other double salts are, for example, $3Na_2SO_4{\cdot}CaSO_4$, $3Na_2SO_4{\cdot}MgSO_4$ (vanthoffite) and $NaF{\cdot}Na_2SO_4$.

4.1.3.5 Reaction with bases

Sodium reacts with water to form basic solutions of sodium hydroxide. In the presence of anhydrous liquid ammonia, sodium metal dissolves to give bright blue, metastable solutions that have interesting properties. These properties include blue color, high electrical conductivity, and magnetic susceptibility similar to species having one free electron per metal atom. The metastable solution is unstable with respect to amide formation:

$$2Na + 2NH_3 \rightarrow 2NaNH_2 + H_2$$

NaH is formed by the direct reaction of hydrogen and liquid sodium. Pure NaH is colorless, although samples typically appear grey. NaH is ca. 40% denser than Na (0.968 g/cm^3). NaH, like LiH, KH, RbH, and CsH, has the NaCl crystal structure. In this structure, each Na^+ ion is surrounded by six H^- centers in an octahedral geometry. The ionic radii of H^- (146 pm in NaH) and F^- (133 pm) are comparable, as judged by the Na − H and Na − F distances. A very unusual situation is found in a compound unofficially named "inverse sodium hydride", which contains Na^- and H^+ ions. Na^- is an alkalide, and this compound differs from ordinary sodium hydride in having a much higher energy content because of the net displacement of two electrons from hydrogen to sodium. A derivative of this "inverse sodium hydride" arises in the presence of the base adamanzane. [Adamanzanes (abbreviated *Adz*) are compounds containing four nitrogen atoms linked by carbons (analogous to adamantane with nitrogen at the branched position). Often coordinated to a central ligand, the nitrogens occupy the vertices of a tetrahedron, with potentially four faces and six edges, with the carbon chains running approximately along the edges. They can have a "bowl" or "cage" structure, with varying lengths or omission of the carbon chains.] This molecule irreversibly encapsulates the H^+ and shields it from interaction with the alkalide Na^-. Theoretical studies have indicated that even an unprotected protonated tertiary amine complexed with the sodium alkalide might be metastable under certain solvent conditions, though the barrier to reaction would be small and finding a suitable solvent might be difficult. NaH is a base of wide scope and utility in organic chemistry. As a superbase, it can deprotonate a range of even weak Brønsted acids to produce the corresponding sodium derivatives. Typical "easy" substrates contain O−H, N−H, S−H bonds, including alcohols, phenols, pyrazoles, and thiols. NaH especially deprotonates carbon acids (i.e., C − H bonds) such as 1,3-dicarbonyls such as malonic esters. The resulting sodium derivatives can be alkylated. NaH is widely used to promote condensation reactions of carbonyl compounds via the Dieckmann condensation, Stobbe condensation, Darzens condensation, and Claisen condensation. Other carbon acids susceptible to deprotonation by NaH are, for example, sulfonium salts and

DMSO. NaH is used to make sulfur ylides, which are then used to convert ketones into epoxides, as in the Johnson–Corey–Chaykovsky reaction. NaH reduces certain main group compounds, but similar reactivity is very rare in organic chemistry. In particular, boron trifluoride reacts to give diborane and sodium fluoride:

$$6NaH + 2BF_3 \rightarrow B_2H_6 + 6NaF$$

Si-Si and S-S bonds in disilanes and disulfides are also reduced. A series of reduction reactions, including the hydrodecyanation of tertiary nitriles, reduction of imines to amines, and amides to aldehydes, can be effected by a composite reagent consisting of sodium hydride and an alkali metal iodide (NaH:MI, M = Li, Na).

4.1.3.6 Compounds of sodium

Sodium ions bond with anions to form a complete range of compounds with various industrial importance. These compounds include table salt (NaCl), soda ash (Na_2CO_3), baking soda ($NaHCO_3$), caustic soda (NaOH), sodium nitrate ($NaNO_3$), sodium phosphates (Na_3PO_4, Na_2HPO_4), sodium thiosulfate ($Na_2S_2O_3$), and borax ($Na_2B_4O_7 \cdot 10H_2O$).

4.1.4 Major uses

In humans, Na is an essential element that regulates blood volume, blood pressure, osmotic equilibrium, and pH. The minimum physiological amount needed daily for Na is 500 milligrams. Sodium chloride (NaCl) is the primary source of Na in the diet and is used as seasoning and preservative. In the western world, most sodium chloride comes from processed foods. Other sources of Na are its natural occurrence in food and such food additives as monosodium glutamate (MSG, IUPAC name sodium 2-aminopentanedioate, $C_5H_8NO_4Na$), sodium nitrite ($NaNO_2$), sodium saccharin (IUPAC name sodium;1,1-dioxo-1,2-benzothiazol-2-id-3-one, $C_7H_4NNaO_3S$), baking soda (sodium bicarbonate, $NaHCO_3$), and sodium benzoate ($C_7H_5NaO_2$).

Even though metallic Na has some essential applications, the main ones for Na utilize compounds; millions of tons of sodium chloride, hydroxide, and carbonate are manufactured per year. NaCl is widely utilized for antiicing and deicing and as a preservative; $NaHCO_3$ is used in, for example, baking, as a raising agent, and soda-blasting. Together with K, numerous vital medicines have Na added to increase their bioavailability; despite K being the better ion in most instances, Na is used due to its lower price and atomic weight. Sodium hydride (NaH) is utilized as a base for a variety of reactions [e.g., the aldol reaction, a means of forming C-C bonds in organic chemistry. Discovered independently by the Russian chemist Alexander Borodin (November 12, 1833 to February 27, 1887) in 1869 and by the Alsation French chemist Charles-Adolphe Wurtz (November 26, 1817 to May 10, 1884) in 1872, the reaction combines two carbonyl compounds (the original experiments used aldehydes) to form a new β-hydroxy carbonyl compound. These products are known as aldols, from the aldehyde + alcohol, a structural motif seen in many of the products. Aldol structural units are found in many important molecules, whether naturally occurring or synthetic.] in organic chemistry, and as a reducing agent in inorganic chemistry.

Metallic Na is used primarily for the manufacture of sodium borohydride ($NaBH_4$), sodium azide (NaN_3), indigo ($C_{16}H_{10}N_2O_2$), and triphenylphosphine ($C_{18}H_{15}P$). A once-common application was the production of tetraethyllead (TEL, $C_8H_{20}Pb$) and tetramethyllead as petro-fuel additive (antiknock agent) and Ti metal; because of the move away from TEL on environmental grounds and new titanium production methods, the manufacture of Na dropped after 1970. In addition, Na is used as an alloying metal, an antiscaling agent, and as a reducing agent for metals where other materials are unsuccessful. The free element is not used as a scaling agent, instead ions in the water are exchanged for Na^+ ions. Na plasma ("vapor") lamps are frequently used for street lighting in cities, producing light that ranges in color from yellow-orange to peach as the pressure increases. Alone or in combination with K, Na is a desiccant; it produces an intense blue color with benzophenone (diphenylmethanone, $C_{13}H_{10}O$) when the dessicant is dry. In organic synthesis, Na is used in various reactions such as the Birch reduction [the reaction was reported in 1944 by the Australian organic chemist Arthur Birch (August 3, 1915 to December 8, 1995) working in the Dyson Perrins Laboratory at the University of Oxford, building on earlier work by Wooster and Godfrey published in 1937 (Birch, 1944). It converts aromatic compounds having a benzenoid ring into a product, 1,4-cyclohexadienes, in which two hydrogen atoms have been attached to opposite ends of the molecule. It is the organic reduction in aromatic rings in liquid ammonia with sodium, lithium, or potassium and an alcohol, such as ethanol and tert-butanol. This reaction is quite unlike catalytic hydrogenation, which usually reduces the aromatic ring all the way to a cyclohexane.], and the sodium fusion test is conducted to qualitatively analyze compounds [or Lassaigne's test is used in elemental analysis for the qualitative determination of the presence of foreign elements, namely, halogens, nitrogen, and sulfur, in an organic compound. It was developed by French chemist J.

L. Lassaigne (September 22, 1800 to March 18, 1859)]. Sodium reacts with alcohol giving alkoxides, and when sodium is dissolved in ammonia solution, it can be utilized to reduce alkynes to trans-alkenes. Solid sodium carbonate is required to manufacture glass. A solution of sodium carbonate is the principal acid-neutralizer in water treatment. It is also added to fizzy drinks to boost the solubility of carbon dioxide (CO_2). Fire extinguishers that produce a jet of water or foam do so through generating carbon dioxide gas from the reaction of sulfuric acid on sodium bicarbonate.

Liquid Na is applied as a heat transfer fluid in some nuclear reactor types as it has the high thermal conductivity and low neutron absorption cross section necessary to achieve a high neutron flux in the reactor. The high boiling point of Na permits the reactor to run at ambient (normal) pressure, but the disadvantages comprise its opacity hindering visual maintenance, and its explosive properties. Radioactive ^{24}Na may be formed by neutron bombardment during operation, causing a minor radiation hazard; the radioactivity stops within a few days after removal from the reactor. If a reactor needs to be shut down regularly, NaK is used; since NaK is a liquid at room temperature and does not solidify in the pipes. In this situation, the pyrophoricity of potassium [A pyrophoric substance, from Greek πυροφόρος, pyrophoros, "fire-bearing," is a substance that ignites spontaneously in air at or below 54°C (for gases) or within 5 minutes after coming into contact with air (for liquids and solids)] necessitates additional safeguards to prevent and detect leaks. Another heat transfer function can be found in poppet valves in high-performance internal combustion engines; the valve stems are partially filled with Na and function as a heat pipe to cool the valves.

4.2 12 Mg — Magnesium

4.2.1 Discovery

The name magnesium comes from the Greek word for a district in Thessaly called Magnesia. It is related to magnetite and manganese, which also originated from this area and mandated a distinction as separate substances. In 1618, a farmer at Epsom in England tried to give his cows water from a well there. The cows did not want to drink because the water tasted bitter; nevertheless, the farmer saw that the water appeared to heal scratches and rashes. The substance became known as Epsom salts and its reputation spread. It was ultimately determined to be a hydrated magnesium sulfate, $MgSO_4 \cdot 7H_2O$. The metal itself was first isolated by the Cornish chemist and inventor Sir Humphry Davy (December 17, 1778 to May 29, 1829) in England in 1808 (Davy, 1808a). He used electrolysis on a mixture of magnesia and mercuric oxide. The French chemist Antoine Alexandre Brutus Bussy (May 29, 1794, Marseille to February 1, 1882, Paris) synthesized it in coherent form in 1831 (Fig. 4.6). Davy's first proposal for a name was magnium, but the name magnesium is now used.

FIGURE 4.6 Antoine Alexandre Brutus Bussy, from photograph by Pierson.

4.2.2 Mining, production, and major minerals

Magnesium is the eight-most-abundant element in the earth's crust by mass and tied in seventh place with Fe in molarity. It is found in large deposits of magnesite, $MgCO_3$, dolomite, $CaMg(CO_3)_2$, and other minerals, and in mineral waters, where magnesium ion is soluble. Although magnesium is found in more than 60 minerals, only dolomite, $CaMg(CO_3)_2$, magnesite, $MgCO_3$, brucite, $Mg(OH)_2$, carnallite, $KMgCl_3{\cdot}6H_2O$, talc, $Mg_3Si_4O_{10}(OH)_2$, and olivine (forsterite), Mg_2SiO_4, are of commercial importance. The Mg^{2+} cation is the second-most-abundant cation in seawater, which makes seawater (and brines) and sea salt attractive viable sources for Mg. To extract the magnesium, calcium hydroxide, $Ca(OH)_2$, is added to seawater to form magnesium hydroxide precipitate.

$$MgCl_2 + Ca(OH)_2 \rightarrow Mg(OH)_2 + CaCl_2$$

Magnesium hydroxide (naturally occurring as the mineral brucite) is insoluble in water and can be filtered out and reacted with hydrochloric acid to produced concentrated magnesium chloride.

$$Mg(OH)_2 + 2HCl \rightarrow MgCl_2 + 2H_2O$$

The salt is then electrolyzed in the molten state. At the cathode, the Mg^{2+} ion is reduced by two electrons to magnesium metal:

$$Mg^{2+} + 2e^- \rightarrow Mg$$

At the anode, each pair of Cl^- ions is oxidized to chlorine gas, releasing two electrons to complete the circuit:

$$2Cl^- \rightarrow Cl_2(g) + 2e^-$$

A new method, solid oxide membrane technology, comprises the electrolytic reduction of MgO. At the cathode, Mg^{2+} ions are reduced by two electrons to magnesium metal. The electrolyte is yttria-stabilized zirconia (YSZ). The anode is a liquid metal. At the YSZ/liquid metal anode O^{2-} is oxidized. A layer of graphite borders the liquid metal anode, and at this interface carbon and oxygen react to form carbon monoxide. When silver is used as the liquid metal anode, there is no reductant carbon or hydrogen needed, and only oxygen gas is evolved at the anode. It has been reported that this method provides a 40% reduction in cost per kilo over the electrolytic reduction method. This method is also more environmentally friendly than other methods as much less carbon dioxide is emitted. China is almost completely dependent on a different method, the silicothermic Pidgeon process (the reduction of the oxide at high temperatures with silicon, frequently provided by a ferrosilicon alloy in which the iron is but a spectator in the reactions), to produce the metal. The process can also be carried out with carbon at approximately 2300°C:

$$2MgO(s) + Si(s) + 2CaO(s) \rightarrow 2Mg(g) + Ca_2SiO_4(s)$$

$$MgO(s) + C(s) \rightarrow Mg(g) + CO(g)$$

Magnesium is a common element in many minerals with more than 800 minerals containing this element, either on its own or in combination with similar metals in terms of charge and size such as Fe and Mn substituting for each other

FIGURE 4.7 Thin platy, tannish brucite, $Mg(OH)_2$, on matrix, 6.5 × 4 × 4 cm. Basic Refractories mine, Gabbs, Nye Co., Nevada, United States.

FIGURE 4.8 Spinel, $MgAl_2O_4$, 6 cm large crystal. Mandalay region, Myanmar.

BOX 4.2 Dolomite, $CaMg(CO_3)_2$

The term "dolomite" is used for both the mineral and the rock composed predominantly of the mineral. Though the term "dolostone" has been suggested for the rock to avoid confusion with the mineral, it is not widely used. Dolomite rocks are common sedimentary rocks and most of them are monomineralic consisting of >90% of the mineral dolomite. Rocks intermediate between dolomite and limestone can be classified as: dolomite (<10% calcite, >90% dolomite), calcitic dolomite (10%–50% calcite, 50%–90% dolomite), dolomitic limestone (50%–90% calcite, 10%–50% dolomite), magnesian limestone (90%–95% calcite, 5%–10% dolomite), and limestone (>95% calcite, <5% dolomite). Normally the composition of dolomite is fairly close to $CaMg(CO_3)_2$ but many dolomites contain some Fe^{2+} in place of the Mg^{2+}. The substitution produces a complete series to ankerite, $Ca(Mg,Fe)(CO_3)_2$. Trace amounts of Zn, Co, and Mn can also be present (Fig. 4.11). The crystal structure of dolomite may be described as retaining the calcite structure but substituting every other Ca layer with a Mg layer, producing an ordered arrangement. A random distribution of 2+ cations in both layers will produce a disordered dolomite. Dolomite with excess Ca is known. Dolomite was named in 1791 by Nicolas Théodore de Saussure in honor of the French mineralogist and geologist, Déodat (Dieudonné) Guy Silvain Tancrède Gratet de Dolomieu (June 24, 1750, Dolomieu, near Tour-du-Pin, Isère, France - November 26, 1801, Château-Neuf, Sâone-et-Loire, France). de Dolomieu wrote numerous books on observations on geology, notably about the Alps and Pyrenees, in addition to theoretical books about the internal structure of the Earth. He discovered a specimen of what would eventually be called dolomite during his participation in Napoleon Bonaparte's expedition into Egypt in 1798. The color varies from colorless, white, gray, reddish-white, brownish-white, to pink, while its streak is white. It is transparent to translucent and has a luster that varies from vitreous, sub-vitreous, resinous, waxy, to pearly. The hardness is 3.5 to 4. It has a perfect cleavage on $\{10\bar{1}1\}$, while parting has been noted lamellar twins on $\{02\bar{2}1\}$. Twin gliding on $\{02\bar{2}1\}$. The fracture is sub-conchoidal, while the tenacity is brittle. Optically it is uniaxial(-). Dolomite has a trigonal crystal structure with point group $\bar{3}$ - rhombohedral (space group $R\bar{3}$).

on the same position in the crystal structure. Within the sulfide group, only a handful minerals contain Mg, for example, niningerite ($(Mg,Fe^{2+},Mn^{2+})S$), oldhamite ($(Ca,Mg)S$), and tochilinite ($Fe^{2+}_{5-6}(Mg,Fe^{2+})_5S_6(OH)_{10}$). The halides contain 18 minerals with Mg, such as carnallite ($KMgCl_3{\cdot}6H_2O$), chloromagnesite ($MgCl_2$), and sellaite (MgF_2). More than 50 oxide minerals contain Mg in their structure. Here one finds minerals such as brucite ($Mg(OH)_2$) (Fig. 4.7), magnesiochromite ($MgCr_2O_4$), periclase (MgO), and spinel ($MgAl_2O_4$) (Fig. 4.8). In the carbonate class, 41 minerals can be found that contain Mg, for example, artinite ($Mg_2(CO_3)(OH)_2{\cdot}3H_2O$), dolomite ($CaMg(CO_3)_2$) (Box 4.2), hydrotalcite ($Mg_6Al_2(CO_3)(OH)_{16}{\cdot}4H_2O$), magnesite ($MgCO_3$) (Fig. 4.9), pyroaurite $\left(Mg_6Fe_2^{3+}(OH)_{16}[CO_3]{\cdot}4H_2O\right)$, and stichtite $\left(Mg_6Cr_2^{3+}(OH)_{16}[CO_3]{\cdot}4H_2O\right)$. Over 50 borate minerals contain Mg, for example, boracite ($Mg_3(B_7O_{13})Cl$), harkerite ($Ca_{12}Mg_4Al(BO_3)_3(SiO_4)_4(CO_3)_5{\cdot}H_2O$), inderite ($MgB_3O_3(OH)_5{\cdot}5H_2O$), kurnakovite ($MgB_3O_3(OH)_5{\cdot}5H_2O$), ludwigite

FIGURE 4.9 Large rhombs of lustrous, colorless magnesite, $MgCO_3$, to 3.5 cm across associated with water-clear quartz, SiO_2. Brumado, Bahia, Brazil.

FIGURE 4.10 Well-formed green diopside, $CaMgSi_2O_6$, crystal 7 cm long. Betroka dist., Anosy, Tuléar prov., Madagascar.

($Mg_2Fe^{3+}(BO_3)O_2$), and sinhalite ($MgAl(BO_4)$). About the same number of sulfate minerals exist that contain Mg, for example, botryogen ($MgFe^{3+}(SO_4)_2(OH)\cdot 7H_2O$), epsomite ($MgSO_4\cdot 7H_2O$), hexahydrite ($MgSO_4\cdot 6H_2O$), magnesiocopiapite $\left(MgFe_4^{3+}(SO_4)_6(OH)_2\cdot 20H_2O\right)$, pickeringite ($MgAl_2(SO_4)_4\cdot 22H_2O$), and polyhalite ($K_2Ca_2Mg(SO_4)_4\cdot 2H_2O$). The phosphate class has more than 100 minerals that have Mg in their structure. Here one finds minerals such as collinsite ($Ca_2Mg(PO_4)_2\cdot 2H_2O$), gordonite ($MgAl_2(PO_4)_2(OH)_2\cdot 8H_2O$), lazulite ($MgAl_2(PO_4)_2(OH)_2$), montgomeryite ($Ca_4MgAl_4(PO_4)_6(OH)_4\cdot 12H_2O$), saléeite ($Mg(UO_2)_2(PO_4)_2\cdot 10H_2O$), and struvite ($(NH_4)Mg(PO_4)\cdot 6H_2O$). The largest group of Mg-containing minerals can be found in the silicate class with more than 250 minerals, for example, actinolite ($\square\{Ca_2\}\{Mg_{4.5\text{-}2.5}Fe_{0.5-2.5}\}(Si_8O_{22})(OH)_2$), antigorite/chrysotile/lizardite ($Mg_3(Si_2O_5)(OH)_4$), augite ($(Ca_xMg_yFe_z)(Mg_{y1}Fe_{z1})Si_2O_6$), cordierite ($(Mg,Fe)_2Al_3(AlSi_5O_{18})$), diopside ($CaMgSi_2O_6$) (Fig. 4.10), dravite ($Na(Mg_3)Al_6(Si_6O_{18})(BO_3)_3(OH)_3(OH)$), enstatite ($MgSiO_3$), forsterite ($Mg_2SiO_4$), phlogopite ($KMg_3(AlSi_3O_{10})(OH)_2$), pyrope ($Mg_3Al_2(SiO_4)_3$), talc ($Mg_3Si_4O_{10}(OH)_2$), tremolite ($\square\{Ca_2\}\{Mg_5\}(Si_8O_{22})(OH)_2$), vermiculite ($Mg_{0.7}(Mg,Fe,Al)_6(Si,Al)_8O_{20}(OH)_4\cdot 8H_2O$), and vesuvianite ($(Ca,Na,\square)_{19}(Al,Mg,Fe^{3+})_{13}(\square,B,Al,Fe^{3+})_5(Si_2O_7)_4(SiO_4)_{10}(OH,F,O)_{10}$).

4.2.3 Chemistry

Magnesium has three stable isotopes: ^{24}Mg, ^{25}Mg, and ^{26}Mg. All are present in significant amounts. About 79% of Mg is ^{24}Mg, while ^{25}Mg covers about 10% and ^{26}Mg the remaining 11%. The isotope ^{28}Mg is radioactive and in the 1950s to 1970s was manufactured in several nuclear power plants for application in scientific research. This isotope has a relatively short half-life (21 hours), and its application was restricted by shipping times. The nuclide ^{26}Mg has found

FIGURE 4.11 A purplish-pink or magenta colored cobaltoan dolomite, $CaMg(CO_3)_2$. The rhombs are to 6 mm. Bou Azzer dist., Anti-Atlas, Morocco.

application in isotopic geology, like that of aluminum. ^{26}Mg is a radiogenic daughter product of ^{26}Al, which has a half-life of 717,000 years. Excessive amounts of ^{26}Mg have been found in the Ca-Al-rich inclusions of some carbonaceous chondrite meteorites. This anomalous abundance has been ascribed to the decay of its parent ^{26}Al in the inclusions, and scientists think that such meteorites were formed in the solar nebula before the ^{26}Al had decayed. These are among the oldest objects in the solar system and comprise preserved information about its early history. It is conventional to plot $^{26}Mg/^{24}Mg$ versus an Al/Mg ratio. In an isochron dating plot, the Al/Mg ratio plotted is $^{27}Al/^{24}Mg$. The slope of the isochron has no age implication but points to the initial $^{26}Al/^{27}Al$ ratio in the sample at the time when the systems were separated from a common reservoir. Isochron dating is a common technique of radiometric dating and is used to date certain events, such as crystallization, metamorphism, shock events, and differentiation of precursor melts, in the history of rocks. Isochron dating can be divided into mineral and whole-rock isochron dating; both techniques are applied regularly to date terrestrial and extra-terrestrial rocks (meteorites). The advantage of isochron dating as compared to simple radiometric dating techniques is that no assumptions are required about the original concentration of the daughter nuclide in the radioactive decay sequence. The original concentration of the daughter product can be calculated using isochron dating. This technique can be used provided that the daughter element has at least one stable isotope other than the daughter isotope into which the parent nuclide decays.

Mg is a light but tough silvery-white metal and the second member of the alkaline earth metals with the electronic configuration of $[Ne]3s^2$. It crystallizes in an *hcp* arrangement, which confers a marked anisotropy in its properties. The predominant +2 oxidation state of magnesium and other group 2 metals can be attributed to their electronic configuration, ionization energies, and size (Table 4.2).

Magnesium reacts with water at room temperature forming bubbles of hydrogen gas and magnesium hydroxide. Increasing the temperature speeds up this reaction.

$$Mg(s) + 2H_2O(l) \rightarrow Mg(OH)_2 + H_2(g)$$

4.2.3.1 *Reaction with air, oxygen, nitrogen, and hydrogen*

Magnesium tarnishes in air, forming an oxide layer around itself to prevent it from further oxidation. However, when ignited in air and especially when it is in strips or powder and in pure oxygen, it burns with a brilliant white light. The flame from burning magnesium can reach up to 3100°C and is not easily extinguished once started. This is because the combustion continues with nitrogen at temperatures around 800°C and form magnesium nitride; with carbon dioxide to form magnesium oxide (MgO) and carbon; and with water to form magnesium hydroxide and hydrogen. The flammability of magnesium is reduced by a small amount of calcium in magnesium alloys.

$$2Mg(s) + O_2(g) \rightarrow 2MgO(s)$$
$$3Mg(s) + N_2(g) \rightarrow Mg_3N_2(s)$$
$$2Mg(s) + CO_2(g) \rightarrow 2MgO(s) + C(s)$$
$$Mg(s) + 2H_2O(g) \rightarrow Mg(OH)_2 + H_2(g)$$

TABLE 4.2 Magnesium properties.

Appearance	Shiny gray solid
Standard atomic weight	24.305
Block	S-block
Element category	Alkaline earth metal
Electron configuration	[Ne] $3s^2$
Phase at STP	Solid
Melting point	650°C
Boiling point	1091°C
Density (near r.t.)	1.738 g/cm^3
When liquid (at m.p.)	1.584 g/cm^3
Heat of fusion	8.48 kJ/mol
Heat of vaporization	128 kJ/mol
Molar heat capacity	24.869 J/(mol·K)
Oxidation states	+1, **+2**
Ionization energies	1st: 737.7 kJ/mol 2nd: 1450.7 kJ/mol 3rd: 7732.7 kJ/mol
Atomic radius	empirical: 160 pm
Covalent radius	141 ± 7 pm
Van der Waals radius	173 pm

STP, Standard temperature and pressure.
Bold font indicates main oxidation state.

When Mg is exposed to hydrogen at high temperature and pressure (500°C, 200 atm.), magnesium turns into magnesium hydride (MgH_2). MgH_2 can be prepared at moderate temperatures and pressures when nanoscale magnesium in the presence of catalysts is used, but challenges still remain before the system can be of practical use.

$$Mg(s) + H_2(g) \rightarrow MgH_2(s)$$

4.2.3.2 *Reaction with halogens*

Magnesium is more electropositive than the amphoteric Be above it and reacts more readily with most of the nonmetals. It ignites with the halogens, particularly when they are moist, to give MgX_2. For example, when Mg is heated in the presence of chlorine gas, it ignites to form magnesium chloride ($MgCl_2$).

$$Mg(s) + Cl_2(g) \rightarrow MgCl_2(s)$$

Magnesium fluoride, MgF_2, is produced from magnesium oxide with sources of hydrogen fluoride such as ammonium bifluoride:

$$MgO + (NH_4)HF_2 \rightarrow MgF_2 + NH_3 + H_2O$$

Related metathesis reactions are also possible. The compound crystallizes as tetragonal birefringent crystals. The structure of the compound is similar to that in rutile (TiO_2), with octahedral Mg^{2+} centers and 3-coordinate fluoride centers. Magnesium chloride, $MgCl_2$ and its various hydrates $MgCl_2(H_2O)_x$, are typical ionic halide salts, being highly soluble in water. The hydrated magnesium chloride can be extracted from brine or sea water. In North America, magnesium chloride is produced mainly from Great Salt Lake brine. It is extracted in a similar process from the Dead Sea in the Jordan Valley. Magnesium chloride, as the natural mineral bischofite, is also extracted (by solution mining) out of ancient seabeds, for example, the Zechstein seabed in northwest Europe. Some magnesium chloride is produced by solar evaporation of seawater. Anhydrous magnesium chloride is the principal precursor to magnesium metal, which is produced on a large scale. Hydrated magnesium chloride is the form most readily available. $MgCl_2$ crystallizes in the cadmium chloride structure, which has octahedral Mg centers. Several hydrates are known with the formula $MgCl_2(H_2O)_x$, and each loses water at higher temperatures: x = 12 (−16.4°C), 8 (−3.4°C), 6 (116.7°C), 4 (181°C), 2 (about 300°C). In the hexahydrate, the Mg^{2+} is also octahedral, but is coordinated to six water ligands. The thermal dehydration of the hydrates $MgCl_2(H_2O)_x$ (x = 6, 12) does not occur straightforwardly. Anhydrous $MgCl_2$ is manufactured industrially by heating the chloride salt of the hexammine complex $[Mg(NH_3)_6]^{2+}$. As indicated by the existence of some hydrates, anhydrous $MgCl_2$ is a Lewis acid, although a weak one. In the Dow process, magnesium chloride is regenerated from magnesium hydroxide using hydrochloric acid:

$$Mg(OH)_2(s) + 2HCl(aq) \rightarrow MgCl_2(aq) + 2H_2O(l).$$

It can also be produced from magnesium carbonate by a comparable reaction. Derivatives with tetrahedral Mg^{2+} are less common. Examples include salts of (tetraethylammonium)$_2MgCl_4$ and adducts such as $MgCl_2$(TMEDA). $MgCl_2$ is the principal precursor to metallic magnesium. The conversion is effected by electrolysis:

$$MgCl_2 \rightarrow Mg + Cl_2$$

This process is used on a substantial scale. Magnesium iodide is the name for the chemical compounds with the formulas MgI_2 and its various hydrates $MgI_2(H_2O)_x$. It has few commercial uses, but can be used to prepare compounds for organic synthesis. Magnesium iodide can be produced from magnesium oxide, magnesium hydroxide, and magnesium carbonate by reaction with hydroiodic acid:

$$MgO + 2HI \rightarrow MgI_2 + H_2O$$
$$Mg(OH)_2 + 2HI \rightarrow MgI_2 + 2H_2O$$
$$MgCO_3 + 2HI \rightarrow MgI_2 + CO_2 + H_2O.$$

4.2.3.3 Reaction with acids

Magnesium readily dissolves in acids and forms solutions that have both the Mg^{2+} ions and hydrogen gas. When reacted with bases, magnesium form basic hydroxide solutions.

$$Mg(s) + H_2SO_4(aq) \rightarrow Mg^{2+}(aq) + SO_4^{2-}(aq) + H_2(g)$$
$$Mg(s) + 2HCl(aq) \rightarrow Mg^{2+}(aq) + 2Cl^-(aq) + H_2(g)$$
$$Mg(s) + H_2CO_3(aq) \rightarrow Mg^{2+}(aq) + CO_3^{2-}(aq) + H_2(g)$$

4.2.3.4 Organomagnesium compounds

Organic compounds of magnesium are commonly in the form of Grignard reagents. Grignard reagents are formed by the reaction of magnesium metal with alkyl of alkenyl halides following a single electron transfer mechanism as shown below.

$$R - X + Mg \rightarrow R - X^{\bullet -} + Mg^{\bullet +}$$
$$R - X^{\bullet -} \rightarrow R^{\bullet} + X^-$$
$$R^{\bullet} + Mg^{\bullet +} \rightarrow RMg^+$$
$$RMg^+ + X^- \rightarrow RMgX$$

Other organic compounds of magnesium such as magnesium dialkyls (MgR_2), magnesium diphenyls ($MgPh_2$), and organosilylmagnesium compound $[Mg(SiMe_3)_2]\cdot(\text{-}CH_2OMe)_2$ are known. Grignard reagents, however, are the most important organometallic compounds of Mg because of their easy preparation and synthetic versatility.

4.2.4 Major uses

When burning in air, magnesium produces a brilliant white light that includes strong ultraviolet wavelengths. Historically, Mg powder (flash powder) was used for subject illumination at the start of photography. Later, Mg filament was used in electrically ignited single-use photography flashbulbs. Mg powder is still used in fireworks and marine flares where a brilliant white light is essential. It was also used for various theatrical effects, for example, lightning, pistol flashes, and supernatural appearances.

Mg is the third most commonly used structural metal, following Fe and Al. The key applications of Mg are in sequence: Al alloys, die-casting (alloyed with Zn), removing sulfur in the manufacture of iron and steel, and the manufacturing of Ti metal in the Kroll process. Mg is utilized in super strong, lightweight materials and alloys, for example, when infused with silicon carbide (SiC) nanoparticles, it exhibits extremely high specific strength.

Traditionally, Mg was one of the principal aerospace construction metals and was used in Germany for military aircraft as early as World War I as well as extensively in World War II. They created the name "Elektron" for Mg alloy, a term which is still currently in use. In the commercial aerospace industry, Mg was largely limited to engine-related components, because of fire and corrosion hazards. Research and development of new Mg alloys continues, particularly Elektron 21, which (in test) has shown to be suitable for aerospace engine, internal, and airframe components. In the form of thin ribbons, Mg is used for the purification of solvents, for example, for the preparation of superdry ethanol. Due to its low weight and good mechanical and electrical properties, Mg is extensively used for making mobile phones, laptop, and tablet computers, cameras, and other electronic components.

Magnesium, being readily available and comparatively nontoxic, has a wide range of applications: Mg is flammable, burning at a temperature of around 3100°C, and the autoignition temperature of Mg ribbon is approximately 473°C, it creates intense, bright, white light when burning, its high combustion temperature makes it a useful tool for starting emergency fires; Mg is also frequently used to ignite thermite [a pyrotechnic composition of metal powder, which serves as fuel, and metal oxide. When ignited by heat, thermite undergoes an exothermic reduction-oxidation (redox) reaction. Most varieties are not explosive but can create brief bursts of heat and high temperature in a small area.] or other materials that need a high ignition temperature; in the form of turnings or ribbons, to prepare Grignard reagents, which are useful in organic synthesis [an organometallic chemical reaction in which alkyl, vinyl, or aryl-magnesium halides (Grignard reagent) add to a carbonyl group in an aldehyde or ketone. This reaction is important for the formation of C-C bonds.]; as an additive agent in conventional propellants and the manufacture of nodular graphite in cast iron; as a reducing agent to separate uranium and other metals from their salts; as a sacrificial (galvanic) anode to protect metallic structures, for example, boats, underground tanks, pipelines, buried structures, and water heaters, etc.; alloyed with Zn to produce the Zn sheet used in photoengraving plates in the printing industry, dry-cell battery walls, and roofing; as a metal, its principal application is as an alloying additive to Al with these Al-Mg alloys primarily being used for making beverage cans, sports equipment such as golf clubs, fishing reels, and archery bows and arrows. Specialty, high-grade car wheels of magnesium alloy are known as "mag wheels," though the term is frequently misapplied to Al wheels. Many car and aircraft manufacturers have produced engine and body parts from Mg. Mg batteries have been commercialized as primary batteries and are an active topic of research for rechargeable secondary batteries.

Magnesium compounds, mainly magnesium oxide (MgO), are used as a refractory material in furnace linings for producing iron, steel, nonferrous metals, glass, and cement. In addition, MgO and other Mg compounds are utilized in agricultural, chemical, and construction industries. MgO from calcination is used as an electrical insulator in fire-resistant cables. Mg salts are included in a variety of foods, fertilizers (Mg is a component of chlorophyll), and microbe culture media. Mg sulfite is used in the manufacture of paper (sulfite process). Mg phosphate is used to fireproof wood used in construction. Mg hexafluorosilicate is used for moth-proofing textiles. Mg hydroxide (milk of magnesia), sulfate (Epsom salts), chloride, and citrate are all used in medicine.

4.3 13 Al — Aluminum

4.3.1 Discovery

The early history of aluminum has been dictated by the use of alum ($XAl(SO_4)_2 \cdot 12H_2O$, where X is a monovalent cation such as potassium or ammonium). The first written record of alum, made by Greek historian Herodotus (c.484 BCE to c. 425 BCE), dates to the 5th century BCE. It is known from the records that the ancients used alum as a dyeing fixative and for city defense. After the Crusades, alum, an important compound in the European fabric industry, was an internationally commercial product. It was imported to Europe from the eastern Mediterranean until the mid-15th century. The nature of alum was still not identified. Around 1530, Swiss physician, alchemist, and astrologer Paracelsus (1493/4 to September 24, 1541, born Philippus Aureolus Theophrastus Bombastus von Hohenheim) proposed alum was a salt of an earth of alum (Fig. 4.12). In 1595 German doctor and chemist Andreas Libavius (c.1555 to July 25, 1616) experimentally established this. In 1722 German physician and chemist Friedrich Hoffmann (February 19, 1660 to November 12, 1742) proclaimed his belief that the base of alum was a distinct earth. In 1754 German chemist Andreas Sigismund Marggraf (March 3, 1709 to August 7, 1782) synthesized alumina by boiling clay in sulfuric acid and then adding potash (Potash is the term for various mined and manufactured salts that contain potassium in water-soluble form. The name derives from pot ash, which refers to plant ashes soaked in water in a pot, the primary means of manufacturing the product before the industrial era.). The earliest efforts to produce aluminum metal started around 1760. The first successful attempt, though, was accomplished in 1824 by Danish physicist

FIGURE 4.12 Paracelsus, Engraving by Wenceslaus Hollar (1607–1677).

FIGURE 4.13 Hans Christian Ørsted, before 1851.

and chemist Hans Christian Ørsted (August 14, 1777 to March 9, 1851) (Fig. 4.13). He used a reaction of anhydrous aluminum chloride, $AlCl_3$, with potassium amalgam, resulting in a lump of metal that resembled tin. He presented his results and showed a sample of the new metal in 1825 (Ørsted, 1825). In 1827 the German chemist Friedrich Wöhler (July 31, 1800 to September 23, 1882) repeated Ørsted's experiments but did not detect any aluminum. (The cause for this discrepancy was only discovered in 1921.) He performed a comparable reaction in 1827 by mixing anhydrous aluminum chloride with potassium and formed a powder of aluminum (Wöhler, 1827). In 1845 he produced some small pieces of the metal and described some physical properties of this metal. For many years afterward, Wöhler was attributed the discovery of aluminum. As Wöhler's method could not produce large amounts of aluminum, the metal remained rare; its cost surpassed that of gold. French chemist Henri Étienne Sainte-Claire Deville (March 11, 1818 to July 1, 1881) announced an industrial method of aluminum production in 1854 at the

FIGURE 4.14 Henri Étienne Sainte-Claire Deville.

Paris Academy of Sciences (Sainte-Claire Deville, 1859) (Fig. 4.14). Aluminum trichloride could be reduced by sodium, which was more convenient and less expensive than potassium, which Wöhler had used. In 1856 Deville and colleagues set up the world's first industrial production of aluminum. From 1855 to 1859, the price of aluminum dropped from US$500 to $40 per kilogram. Still, aluminum was not of high purity and each batch of aluminum produced differed in properties. The first industrial large-scale production method was independently developed in 1886 by French engineer Paul Louis-Toussaint Héroult (April 10, 1863 to May 9, 1914) and American engineer, inventor, businessman, and chemist Charles Martin Hall (December 6, 1863 to December 27, 1914); these days it is known as the Hall–Héroult process. The Hall–Héroult process converts alumina into the metal. In 1889 the Austrian chemist Carl Joseph Bayer (March 4, 1847 to October 4, 1904) found a method to purify bauxite to yield alumina, now known as the Bayer process. Bauxite is a sedimentary rock with a relatively high aluminum content. It is the world's main source of aluminum. Bauxite consists mostly of the aluminum minerals gibbsite ($Al(OH)_3$), boehmite (γ-AlO(OH)), and diaspore (α-AlO(OH)), mixed with the two iron oxides goethite (FeOOH) and hematite (Fe_2O_3), the aluminum clay mineral kaolinite ($Al_2Si_2O_5(OH)_4$), and small amounts of anatase (TiO_2) and ilmenite ($FeTiO_3$). In 1821, the French geologist and mining engineer Pierre Berthier (July 3, 1782 to August 24, 1861) found bauxite near the village of Les Baux in Provence, southern France (Berthier, 1821). Current manufacture of aluminum metal is based on both the Bayer and Hall–Héroult processes.

4.3.2 Mining, production, and major minerals

Overall, the earth contains about 1.59% aluminum by mass (seventh in abundance by mass). Aluminum occurs in greater proportion on Earth than in the universe because aluminum easily forms the oxide and becomes bound into rocks and aluminum stays in the earth's crust while less reactive metals sink to the core. In the earth's crust, aluminum is the most abundant (8.3% by mass) metallic element and the third most abundant of all elements (after oxygen and silicon). Many silicates in the earth's crust contain aluminum. In contrast, the earth's mantle contains only 2.38% aluminum by mass. Although aluminum is a common and widespread element, not all aluminum minerals are economically viable sources of the metal. Almost all metallic aluminum is produced from the ore bauxite (gibbsite ($Al(OH)_3$), boehmite (γ-AlO(OH)), and diaspore (α-AlO(OH)). Bauxite occurs as a weathering product of low iron and silica bedrock in tropical climatic conditions (Box 4.3). Most bauxite is mined in Australia, China, Guinea, and India. Aluminum production consumes an enormous amount of energy, and so the producers tend to locate smelters in places where electric power is both abundant and inexpensive. The world's largest smelters of aluminum can be found in China, Russia, Bahrain, United Arab Emirates, and South Africa.

BOX 4.3 Bauxite

Bauxite is an earthy aggregate of Al-hydroxide minerals, mainly a combination of gibbsite ($Al(OH)_3$), boehmite, and diaspore (both AlOOH), showing a range of colors from light gray, cream, pink, or yellow, to dark brown and dark red. The color is mainly dependent on the type and particle size of the prevalent iron mineral. Highly dispersed goethite (FeOOH) tends to produce yellow or brown colors, whereas dark red tones typically are associated with coarser hematite (Fe_2O_3). Associated minerals can include quartz (SiO_2), Ti minerals (rutile (TiO_2), anatase (TiO_2), leucoxene (alteration product and mixture of Fe-Ti oxides)), Fe sulfides (pyrite, marcasite (both FeS_2)), and carbonates (calcite ($CaCO_3$, siderite ($FeCO_3$), dolomite ($CaMg[CO_3]_2$)). Kaolinite ($Al_2Si_2O_5(OH)_4$) and halloysite ($Al_2Si_2O_5(OH)_4{\cdot}nH_2O$) are by far the most common associated clay minerals. The solubility of amorphous Al-hydroxides exceeds the surrounding gibbsite values by a factor of 10^4. Hence, clay minerals will normally form from the reaction of silica with amorphous Al-hydroxides. Typically, mixtures of gibbsite and boehmite are common, boehmite and diaspore less common, and gibbsite and diaspore rare. Bauxite shows mostly a cellular, porous, or fine-grained compact texture and conchoidal or uneven fractures. Oölitic and pisolitic textures are common, with rounded concretionary grains embedded in a clayey mass and may extend through a deposit or series of deposits. Geologically old diaspore bauxites are very hard and can reach a specific gravity of 3.6, while young bauxites with mostly gibbsite are soft and have a much lower specific gravity of around 2.0–2.5. The chemical composition of bauxite depends upon the type of Al-hydroxide minerals it consists of. There exist different theories to explain the origin of bauxite formation, but it is generally thought that most bauxite deposits are residual products which are the result of high intensified weathering of high alumina rocks by the solution and removal of constituents other than alumina such as Na, K, Ca, Mg, and Si. The favorable geological conditions necessary for bauxite formation are as follows: (1) high rock permeability to permit desilication, (2) tropical climate with high rainfall and alternating wet and dry seasons to promote leaching, (3) low to moderate topographical relief to allow drainage and fluctuation in groundwater levels, and (4) low rate of erosion and lengthy period of stability to allow the accumulation of weathered products. Based upon the mode of occurrence and parent-rock, bauxite deposits are divided into two groups: lateritic bauxite for deposits derived from the weathering of igneous or other rocks rich in aluminum silicates, and karst or limestone bauxites for deposits occurring as weathering residue of limestone and dolomite. Bauxite occurs in all geological ages from Precambrian to Recent. The formation of bauxite was particularly abundant in the Tertiary period with gibbsite as the principal mineral. Mesozoic bauxites, which are the second most common, are always derived from limestone, with boehmite as the dominant mineral. Palaeozoic deposits are restricted with mainly diaspore and boehmite.

4.3.2.1 Bayer process

Bauxite is converted to aluminum oxide by what is known as the Bayer process developed by Austrian chemist Carl Josef Bayer (March 4, 1847 to October 4, 1904). Bauxite is blended for uniform composition and subsequently ground. The resulting slurry is mixed with a hot solution of sodium hydroxide; the mixture is then treated in a digester vessel at an increased pressure well above one atmosphere, dissolving the aluminum hydroxide in the bauxite while at the same time converting any impurities into relatively insoluble compounds:

$$Al(OH)_3 + Na^+ + OH^- \rightarrow Na^+ + [Al(OH)_4]^-$$

After this reaction, the slurry is at a temperature above its atmospheric boiling point. It is cooled by removing steam as pressure is reduced. The bauxite residue is separated from the solution and discarded. The solution, now free of any solid impurities, is seeded with small crystals of aluminum hydroxide; this causes decomposition of the $[Al(OH)_4]^-$ ions to aluminum hydroxide ($Al(OH)_3$). After about half of aluminum has precipitated, the mixture is sent to classifiers. Small crystals of aluminum hydroxide are collected to serve as seeding agents in future batches; coarse particles are converted to aluminum oxide by heating; excess solution is removed by evaporation (if needed), purified, and recycled.

4.3.2.2 Hall–Héroult process

Prior to the Hall–Héroult process, aluminum metal was obtained by heating ore along with elemental sodium or potassium in a vacuum. The process was complicated and consumed materials that were in themselves expensive at that time. This meant the cost to obtain a small amount of aluminum in the early 19th century was very high, higher than for gold or platinum. Bars of aluminum were exhibited alongside the French crown jewels at the Exposition Universelle of 1855, and Emperor Napoleon III (April 20, 1808 to January 9, 1873) of France was said to have reserved his few sets of aluminum dinner plates and eating utensils for his most honored guests. Production costs using older methods

FIGURE 4.15 Charles Martin Hall, circa 1880s.

FIGURE 4.16 Paul (Louis-Toussaint) Héroult.

did come down, but when aluminum was selected as the material for the cap/lightning rod to sit atop the Washington Monument in Washington, D.C., it was still more expensive than silver. The Hall–Héroult process was invented independently and almost simultaneously in 1886 by the American chemist, inventor and businessman Charles Martin Hall (December 6, 1863 to December 27, 1914) (Fig. 4.15) and by the French scientist Paul (Louis-Toussaint) Héroult (April 10, 1863 to May 9, 1914) (Fig. 4.16)—both only 22 years old. Some authors claim Hall was assisted by his sister Julia Brainerd Hall (November 11, 1859 to September 4, 1926); however, the extent to which she was involved has been disputed. In 1888 Hall opened the first large-scale aluminum production plant in Pittsburgh. It later became the Aluminum Company of America (AlCOA) corporation. The conversion of alumina to aluminum metal is realized through the Hall–Héroult process. In this energy-intensive process, a solution of alumina in a molten (950°C and 980°C) mixture

of cryolite (Na_3AlF_6) with calcium fluoride is electrolyzed to yield metallic aluminum. In the Hall–Héroult process, the following simplified reactions take place at the carbon electrodes:
Cathode:

$$Al^{3+} + 3e^- \rightarrow Al$$

Anode:

$$O^{2-} + C \rightarrow CO + 2e^-$$

Overall:

$$Al_2O_3 + 3C \rightarrow 2Al + 3CO$$

In reality, much more CO_2 is formed at the anode than CO:

$$2Al_2O_3 + 3C \rightarrow 4Al + 3CO_2$$

The heavier liquid aluminum metal sinks to the bottom of the solution and is tapped off, and usually cast into large blocks called aluminum billets for further processing. Anodes of the electrolysis cell are made of carbon—the most resistant material against fluoride corrosion—and either baked at the process or are prebaked. The former, also called Söderberg anodes, are less power efficient, and fumes released during baking are costly to collect, which is why they are being replaced by prebaked anodes even though they save the power, energy, and labor to prebake the cathodes. Carbon for anodes should be, if possible, of high purity so that neither aluminum nor the electrolyte is contaminated with ash. Despite carbon's resistivity against corrosion, it is still consumed at a rate of 0.4–0.5 kg per each kilogram of produced aluminum. Cathodes are made of anthracite; high purity for them is not required because impurities leach only very slowly. Cathode consumption is at a rate of 0.02–0.04 kg per each kilogram of produced aluminum. A cell is typically terminated after 2–6 years following a failure of the cathode. The Hall–Héroult method produces aluminum with a purity better than 99%. Further purification, if necessary, can be performed using the Hoopes process [the process was patented by William Hoopes, a chemist of the Aluminum Company of America (ALCOA) in 1925]. This method involves the electrolysis of molten aluminum with a sodium, barium, and aluminum fluoride electrolyte. The resulting aluminum has a purity of 99.99%. Electric power represents about 20%–40% of the total cost of producing aluminum, depending on the location of the smelter. Aluminum production consumes about 5% of total electricity generated in the United States.

Aluminum is a common element in many minerals. Overall, more than 1150 minerals are known to contain Al as structural element or as a substituting element for similar metals such as Fe, Mg, and Ti, while in the silicates it

FIGURE 4.17 Corundum, Al_2O_3, variety ruby with a single crystal to 1.5 cm. Kola Peninsula, Russia.

FIGURE 4.18 Spinel, $MgAl_2O_4$, 6 cm large crystal. Mandalay region, Myanmar.

FIGURE 4.19 Dawsonite, $NaAlCO_3(OH)_2$, thin white needles in clusters to 8 mm. Terlano, Bolzano, Italy.

can also substitute for Si. Though rare, it occurs as a native element (Al) and in three alloys: cupalite ((Cu,Zn)Al), decagonite ($Al_{71}Ni_{24}Fe_5$), and khatyrkite ($(Cu,Zn)Al_2$). Three sulfide class minerals are known to contain Al: valleriite ($(Fe^{2+},Cu)_4(Mg,Al)_3S_4(OH,O)_6$), vyalsovite ($FeCaAlS(OH)_5$), and yushkinite ($(Mg,Al)V^{4+}S_2(OH)_2$). In total 42 halides have Al as part of their chemical composition, for example, creedite ($Ca_3SO_4Al_2F_8(OH)_2{\cdot}2H_2O$), cryolite ($Na_2NaAlF_6$), pachnolite ($NaCa[AlF_6]{\cdot}H_2O$), and thomsenolite ($NaCa[AlF_6]{\cdot}H_2O$). The most important minerals in terms of ore minerals for Al are found in the oxide class and include minerals such as bayerite ($Al(OH)_3$), boehmite (AlOOH), chrysoberyl ($BeAl_2O_4$), corundum (including gemstone varieties sapphire and ruby) (Al_2O_3) (Fig. 4.17), diaspore (AlOOH), gibbsite ($Al(OH)_3$), and spinel ($MgAl_2O_4$) (Fig. 4.18). Among the carbonate minerals, 25 are known to have Al in their structure, for example, dawsonite ($NaAlCO_3(OH)_2$) (Fig. 4.19), dundasite ($PbAl_2(CO_3)_2(OH)_4{\cdot}H_2O$), hydrotalcite ($Mg_6Al_2(CO_3)(OH)_{16}{\cdot}4H_2O$), and takovite ($Ni_6Al_2(OH)_{16}[CO_3]{\cdot}4H_2O$). There are only 15 borates with aluminum, such as harkerite ($Ca_{12}Mg_4Al(BO_3)_3(SiO_4)_4(CO_3)_5{\cdot}H_2O$), jeremejevite ($Al_6(BO_3)_5(F,OH)_3$), londonite ($(Cs,K,Rb)Al_4Be_4(B,Be)_{12}O_{28}$), painite ($CaZrAl_9(BO_3)O_{15}$), and rhodizite ($(K,Cs)Al_4Be_4(B,Be)_{12}O_{28}$). More than 50 sulfate class minerals are known with Al in their structure, such as alunite ($KAl_3(SO_4)_2(OH)_6$), ettringite ($Ca_6Al_2(SO_4)_3(OH)_{12}{\cdot}26H_2O$), halotrichite

FIGURE 4.20 Rhombic crystals of green augelite, $Al_2(PO_4)(OH)_3$, the largest to 7 mm, perched on quartz matrix. Mundo Nuevo, La Libertad dept., Peru.

FIGURE 4.21 Brazilianite, $NaAl_3(PO_4)_2(OH)_4$, yellow-green 2 cm crystal with complex crystal faces. Telirio mine, Linopolis, Minas Gerais, Brazil.

($FeAl_2(SO_4)_4{\cdot}22H_2O$), and sturmanite ($Ca_6(Fe^{3+},Al,Mn^{3+})_2(SO_4)_2[B(OH)_4](OH)_{12}{\cdot}25H_2O$). Even more minerals, over 150, are found in the phosphate class to contain Al, for example, amblygonite ($LiAl(PO_4)F$), augelite ($Al_2(PO_4)(OH)_3$) (Fig. 4.20), brazilianite ($NaAl_3(PO_4)_2(OH)_4$) (Fig. 4.21), eosphorite ($Mn^{2+}Al(PO_4)(OH)_2{\cdot}H_2O$), lazulite ($MgAl_2(PO_4)_2(OH)_2$), turquoise ($Cu(Al,Fe^{3+})_6(PO_4)_4(OH)_8{\cdot}4H_2O$), variscite ($AlPO_4{\cdot}2H_2O$) (Fig. 4.22),

FIGURE 4.22 Dark pink spherical clusters of variscite, $AlPO_4 \cdot 2H_2O$, to 2 mm. Boa Vista, Galileia, Minas Gerais, Brazil.

FIGURE 4.23 Spheroidal wavellite, $Al_3(PO_4)_2(OH,F)_3 \cdot 5H_2O$, to 12mm with a large cross-sectional crystal 13 mm across. Dug Hill, Avant, Garland Co., Arkansas, United States.

vauxite ($Fe^{2+}Al_2(PO_4)_2(OH)_2 \cdot 6H_2O$), and wavellite ($Al_3(PO_4)_2(OH,F)_3 \cdot 5H_2O$) (Fig. 4.23). A handful of organic compounds contain Al, for example, mellite ($Al_2[C_6(COO)_6] \cdot 16H_2O$) and zhemchuzhnikovite ($NaMgAl(C_2O_4)_3 \cdot 8H_2O$). The vast majority of minerals containing Al are found under the silicates with more than 500 minerals. Some well-known examples include albite ($Na(AlSi_3O_8)$), anorthite ($Ca(Al_2Si_2O_8)$), beryl ($Be_3Al_2(Si_6O_{18})$), cordierite ($(Mg,Fe)_2Al_3(AlSi_5O_{18})$), epidote ($\{Ca_2\}\{Al_2Fe^{3+}\}(Si_2O_7)(SiO_4)O(OH)$), grossular ($Ca_3Al_2(SiO_4)_3$), jadeite ($Na(Al,Fe^{3+})Si_2O_6$), kaolinite ($Al_2(Si_2O_5)(OH)_4$), kyanite ($Al_2(SiO_4)O$) (Fig. 4.24), microcline (Fig. 4.25) and orthoclase ($K(AlSi_3O_8)$), phlogopite ($KMg_3(AlSi_3O_{10})(OH)_2$), prehnite ($Ca_2Al_2Si_3O_{10}(OH)_2$), schorl ($Na\left(Fe_3^{2+}\right)Al_6(Si_6O_{18})(BO_3)_3(OH)_3(OH)$), topaz ($Al_2(SiO_4)(F,OH)_2$), vesuvianite ($(Ca,Na,\square)_{19}(Al,Mg,Fe^{3+})_{13}(\square,B,Al,Fe^{3+})_5(Si_2O_7)_4(SiO_4)_{10}(OH,F,O)_{10}$), and zoisite ($Ca_2Al_3[Si_2O_7][SiO_4]O(OH)$).

FIGURE 4.24 Deep blue blades of kyanite, $Al_2(SiO_4)O$, to 7 cm. Note the darkest blue color is a band through the center, typical for kyanite. Barra do Salinas, Coronel Murta, Minas Gerais, Brazil.

FIGURE 4.25 Bluish-green 3 cm microcline, $K(AlSi_3O_8)$, var. amazonite crystal on albite. Florissant, Teller Co., Colorado, United States.

4.3.3 Chemistry

^{27}Al is the only stable aluminum isotope, consistent with Al having an odd atomic number. It is the only Al isotope that has existed on earth in its current form since the earth's creation. Almost all Al on earth exists as this isotope, making aluminum a mononuclidic element and this means that its standard atomic weight is virtually equal to that of the isotope. All other isotopes of aluminum are radioactive. The most stable of these is ^{26}Al with a half-life of 720,000 years and thus could not have survived since earth's creation. Nevertheless, ^{26}Al is produced from argon in the atmosphere by spallation (in planetary physics, spallation describes meteoritic impacts on a planetary surface and the effects of stellar winds and cosmic rays on planetary atmospheres and surfaces) caused by cosmic ray protons. The ratio of ^{26}Al to ^{10}Be has been utilized for radiodating of geological processes over 10^5 to 10^6-year time scales, especially transport, deposition, sediment storage, burial times, and erosion. Most meteorite researchers think that the energy released via the decay of ^{26}Al (β^+ to ^{26}Mg, ε to ^{26}Mg, and γ) was responsible for the melting and differentiation of some asteroids after their formation 4.55 billion years ago. The other isotopes of aluminum, with mass numbers ranging from 21 to 43, all have half-lives well below one hour. Three metastable states have been recognized, all with half-lives under a minute.

Aluminum is an element just below boron. Unlike boron, which is a nonmetal, aluminum and the heavier members of the group are metals. Aluminum is a silvery-white, soft, nonmagnetic, and ductile metal with an electron configuration of $[Ne]3s^23p^1$. It has three electrons beyond a stable noble gas configuration. As such, the first three ionization energies of aluminum are far lower than the fourth ionization energy, which requires removal of an electron from a

TABLE 4.3 Aluminum properties.

Appearance	Silvery gray metallic
Standard atomic weight $A_{r,std}$	26.981
Block	P-block
Element category	Posttransition metal, sometimes considered a metalloid
Electron configuration	[Ne] $3s^2$ $3p^1$
Phase at STP	Solid
Melting point	660.32°C
Boiling point	2470°C
Density (near r.t.)	2.70 g/cm^3
When liquid (at m.p.)	2.375 g/cm^3
Heat of fusion	10.71 kJ/mol
Heat of vaporization	284 kJ/mol
Molar heat capacity	24.20 J/(mol·K)
Oxidation states	−2, −1, +1, +2, **+3**
Ionization energies	1st: 577.5 kJ/mol
	2nd: 1816.7 kJ/mol
	3rd: 2744.8 kJ/mol
Atomic radius	empirical: 143 pm
Covalent radius	121 ± 4 pm
Van der Waals radius	184 pm

STP, Standard temperature and pressure.
Bold font indicates main oxidation state.

stable noble gas configuration. The size of the atom is 143 pm but decreases markedly to 39 pm or 53.5 pm in a four-coordinated and 6-coordinated aluminum compound, respectively (Table 4.3).

Aluminum combines with most nonmetallic elements when heated to give compounds such as AlN, Al_2S_3, and AlX_3 (X = halogen). It also forms intermetallic compounds with elements from all groups of the periodic table that contain metals.

4.3.3.1 Reaction with air, oxygen, nitrogen, and hydrogen

Finely powdered Al metal explodes on contact with liquid O_2. However, for normal samples of the metal, initial reaction with oxygen in air forms a coherent protective oxide film that protects the metal from further attack by air. Thus, as long as this protective layer is intact, aluminum does not react with oxygen in air. The moment this oxide film is damaged, Al will burn in air and/or oxygen with a brilliant white light to form aluminum oxide (Al_2O_3). Aluminum forms one stable oxide, Al_2O_3, commonly called alumina. It occurs naturally as the mineral corundum, α-alumina; there exists also a γ-alumina phase. The two main oxide-hydroxides, AlO(OH), are the minerals boehmite and diaspore. There are three main trihydroxide minerals: bayerite, gibbsite, and nordstrandite, which differ in their crystalline structure (polymorphs). Many other intermediate and related structures have also been observed. Most are produced from ores by a variety of wet processes using acid and base. Heating the hydroxides leads to formation of corundum. These materials are of key importance to the production of aluminum and are themselves extremely useful. In addition, some mixed oxide phases are very useful, such as spinel ($MgAl_2O_4$), Na-β-alumina ($NaAl_{11}O_{17}$), and tricalcium aluminate ($Ca_3Al_2O_6$, an important mineral phase in Portland cement). Very simple Al(II) compounds have been invoked or detected in the reactions of Al metal with oxidants. For instance, aluminum monoxide, AlO, has been detected in the gas phase after explosion and in stellar absorption spectra. More thoroughly studied are compounds of the general formula R_4Al_2 which contain an Al–Al bond and where R is a large organic ligand. In aqueous solution, Al^{3+} exists as the hexaaqua cation $[Al(H_2O)_6]^{3+}$, which has a pK_a of $\sim 10^{-5}$. Such solutions are acidic as this cation can act as a proton donor, progressively hydrolysing to $[Al(H_2O)_5(OH)]^{2+}$, $[Al(H_2O)_4(OH)_2]^+$, etc. As the pH increases these mononuclear species begin to aggregate together through the formation of hydroxide bridges,resulting in many different oligomeric ions, such as the Keggin ion $[Al_{13}O_4(OH)_{24}(H_2O)_{12}]^{7+}$. The process ends with precipitation of aluminum hydroxide, $Al(OH)_3$. Increasing the pH even further leads to the hydroxide dissolving again as the aluminate anion, $[Al(H_2O)_2(OH)_4]^-$, is formed. Aluminum hydroxide forms both salts and aluminates and dissolves in acid and alkali, as well as on fusion with acidic and basic oxides. Aluminum reacts with pure nitrogen to form aluminum nitride at high temperatures, normally from 800°C to 1200°C. Aluminum does not react with hydrogen under normal conditions. It does react at high pressures (10 GPa) and at high temperatures (600°C) to form α-AlH_3. Aluminum hydride in ether solutions is known to degrade in three days and thus must be used immediately upon preparation.

$$4Al(s) + 3O_2(g) \rightarrow 2Al_2O_3(s)$$
$$2Al(s) + N_2(g) \rightarrow 2AlN(s)$$
$$2Al(s) + 3H_2(l) \rightarrow 2AlH_3(s)$$

4.3.3.2 Reaction with halogens

Aluminum metal reacts vigorously with all the heavier halogens to form aluminum halides. Therefore, it reacts with chlorine (Cl_2), bromine (Br_2), and iodine (I_2) to form the respective halides as shown below. Aluminum fluoride (AlF_3) has a very high heat of formation and is often produced by routes other than direct reaction of the elements.

$$2Al(s) + 3Cl_2(g) \rightarrow 2AlCl_3(s)$$
$$2Al(s) + 3Br_2(l) \rightarrow 2AlBr_3(s)$$
$$2Al(s) + 3I_2(s) \rightarrow 2AlI_3(s)$$

All four trihalides are well known. Unlike the structures of the three heavier trihalides, aluminium fluoride (AlF_3) features six-coordinate aluminium, which explains its involatility and insolubility as well as high heat of formation. Each aluminium atom is surrounded by six fluorine atoms in a distorted octahedral arrangement, with each fluorine atom being shared between the corners of two octahedra in a structure related to but distorted from that of ReO_3. Such $\{AlF_6\}$ units also exist in complex fluorides such as cryolite, Na_3AlF_6, but should not be considered as $[AlF_6]^{3-}$ complex anions as the Al–F bonds are not significantly different in type from the other M–F bonds. Such differences in coordination between the fluorides and heavier halides are not unusual, occurring in Sn(IV) and Bi(III) as well for example; even bigger differences occur between CO_2 and SiO_2. With heavier halides, the coordination numbers are lower. The other trihalides are dimeric or polymeric with tetrahedral four-coordinate aluminium centers. Aluminium trichloride ($AlCl_3$) has a layered polymeric structure below its melting point of 192.4°C (378°F), but transforms on melting to Al_2Cl_6 dimers with a concomitant increase in volume by 85% and a near-total loss of electrical conductivity. These still predominate in the gas phase at low temperatures (150–200°C), but at higher temperatures increasingly dissociate into trigonal planar $AlCl_3$ monomers similar to the structure of BCl_3. Aluminium tribromide and aluminium triiodide form Al_2X_6 dimers in all three phases and hence do not show such significant changes of properties upon phase change. These materials are prepared by treating aluminium metal with the halogen. The aluminium trihalides form many addition compounds or complexes; their Lewis acidic nature makes them useful as catalysts for the Friedel–Crafts reactions. Aluminium trichloride has major industrial uses involving this reaction, such as in the manufacture of anthraquinones and styrene; it is also often used as the precursor for many other aluminium compounds and as a reagent for converting nonmetal fluorides into the corresponding chlorides (a transhalogenation reaction). AlF, AlCl, AlBr, and AlI exist in the gas phase when the respective trihalide is heated with aluminum, and at cryogenic temperatures. Their instability in the condensed phase is because of their ready disproportionation to aluminum and the respective trihalide: the reverse reaction is favored at high temperature (but even then they are still short-lived), explaining why AlF_3 is more volatile when heated in the presence of aluminium metal, as is aluminium metal when heated in the presence of $AlCl_3$. A stable derivative of aluminum monoiodide is the cyclic adduct formed with triethylamine, $Al_4I_4(NEt_3)_4$.

4.3.3.3 Reaction with acids

The oxide protecting surfaces of aluminum prevents it from reacting with water or dilute acids. However, amalgamation with Hg or contact with solutions of salts of certain electropositive metals destroys the film and permits further reaction. Further, aluminum readily dissolves in hot concentrated acids. Concentrated nitric acid, however, passivates aluminum metal and prevents it from further reacting.

$$2Al(s) + 3H_2SO_4(aq) \rightarrow 2Al^{3+}(aq) + 2SO_4^{2-}(aq) + 3H_2(g)$$
$$2Al(s) + 6HCl(aq) \rightarrow 2Al^{3+}(aq) + 6Cl^-(aq) + 3H_2(g)$$

4.3.3.4 Reaction with bases

Aluminum also reacts with aqueous NaOH at room temperature. The reaction is highly exothermic once started and is accompanied by the rapid evolution of hydrogen gas. The reaction can be written to show the formation of sodium aluminate or expressed in the ionic form of the aluminate.

$$2Al(s) + 2NaOH + 2H_2O(l) \rightarrow 2NaAlO_2(s) + 3H_2(g)$$
$$2Al(s) + 2NaOH(aq) + 6H_2O \rightarrow 2Na^+(aq) + 2[Al(OH)_4]^- + 3H_2(g)$$

4.3.3.5 Organoaluminum compounds

Organoaluminum compounds are those containing a direct bond between aluminum and carbon. The trialkyls and triaryls of aluminum are very reactive, volatile, or low-melting solids. They ignite spontaneously in air and react violently in water. As such, utmost care is needed when handling these compounds.

The simple aluminum alkyls of the type Al_2R_6 are prepared in two steps. For example, in the preparation of the triethylaluminum compound, the first step involves the alkylation of the aluminum powder. The second step involves reduction in the ethylaluminum sesquichloride, $(CH_3CH_2)_3Al_2Cl_3$, to the triorganoaluminum compound

$$(1) \quad 2Al + 3CH_3CH_2Cl \rightarrow (CH_3CH_2)_3Al_2Cl_3$$
$$(2) \quad 2(CH_3CH_2)_3Al_2Cl_3 + 6Na \rightarrow (CH_3CH_2)_6Al_2 + 2Al + 6NaCl$$

The higher trialkyls are readily prepared by the hydroalumination of alkene route:

$$2Al(s) + 3H_2(g) + 2Al_2Et_6 \rightarrow \{6Et_2AlH\} \xrightarrow{6CH_2CH_2, 70^\circ C} 3Al_2Et_6$$

Organoaluminum compounds are important in reactions involving its Lewis acidity, reactivity toward electrophiles, and its ability to catalyze alkene polymerizations to olefins.

4.3.4 Major uses

Aluminum is the most widely used nonferrous metal. Al is used in a huge variety of products including cans, foils, kitchen utensils, beer kegs, building, and construction (window frames, doors, siding, building wire, sheathing, roofing, etc. Since steel is cheaper, aluminum is used when lightness, corrosion resistance, or engineering features are important), and transportation (automobiles, aircraft, trucks, railway cars, marine vessels, bicycles, spacecraft, etc.). It is also used in electricity-related applications (conductor alloys, motors and generators, transformers, capacitors, etc.). Aluminum is used because it is relatively cheap, highly conductive, has adequate mechanical strength and low density, and resists corrosion. Besides, it exhibits high thermal conductivity, excellent corrosion resistance, and can easily be cast, machined, and formed. It is also nonmagnetic and nonsparking. It is the second most malleable and the sixth most ductile metal.

Most of aluminum oxide (around 90%) is converted to metallic Al. As it is a very hard material (Mohs hardness 9, mineral corundum, Al_2O_3), alumina is widely used as an abrasive. Since it is also extremely chemically inert, it is useful in highly reactive environments such as high-pressure sodium lamps. Al_2O_3 is frequently used as a catalyst or catalyst carrier (meaning that the expensive catalyst material is dispersed over a surface of the inert alumina) for industrial processes; for example, the Claus process to convert hydrogen sulfide to sulfur in refineries and to alkylate amines. Another major use is as a drying agent or absorbent. Several aluminum sulfates have industrial and commercial application. About 65% is used in water treatment. The following main application is in the production of paper. In addition, it is used as a mordant (or dye fixative, a substance used to set (i.e., bind) dyes on fabrics by forming a coordination complex with the dye, which then attaches to the fabric or tissue) in dyeing, in pickling seeds, deodorizing of mineral oils, in leather tanning, and in manufacture of other aluminum compounds. Two kinds of alum, ammonium alum and potassium alum, were previously used as mordants and in leather tanning, but their use has substantially decreased following the availability of high-purity aluminum sulfate. Anhydrous aluminum chloride is used as a catalyst in the chemical and petrochemical industry, the dyeing industry, and in the synthesis of various inorganic and organic compounds. Aluminum hydroxychlorides (including the Keggin structure Al_{13}, $(^{IV}AlO_4{}^{VI}Al_{12}(OH)_{24}(H_2O)_{12})^{7+}$) are used in purifying water, in the paper industry, and as antiperspirants. Sodium aluminate is used for treating water and as an accelerator of solidification of cement.

Numerous Al compounds have niche applications, for example, Al acetate in solution is used as an astringent; Al phosphate is used in the manufacture of glass, ceramic, pulp, and paper products, in cosmetics, paints, varnishes, and in dental cement; Al hydroxide is used as an antacid, and mordant, it is also used in water purification, the manufacture of glass and ceramics, and in the waterproofing of fabrics; Li-Al-hydride is a powerful reducing agent used in organic chemistry; organoaluminums are used as Lewis acids and cocatalysts; methylaluminoxane ($(Al(CH_3)O)_n$) is a cocatalyst for Ziegler–Natta olefin polymerization to produce vinyl polymers such as polyethene ($(C_2H_4)_n$); aqueous Al ions (such as aqueous Al sulfate) are used to treat against fish parasites such as *Gyrodactylus salaris*; and in many vaccines, some Al salts act as an immune adjuvant (immune response booster) to allow the protein in the vaccine to achieve sufficient potency as an immune stimulant. When evaporated in a vacuum, aluminum forms a highly reflective coating for both light and heat. It does not deteriorate, like a silver coating would. These aluminum coatings have many uses, including telescope mirrors, decorative paper, packages, and toys.

4.4 14 Si — Silicon

4.4.1 Discovery

In 1787 French nobleman and chemist Antoine Lavoisier (August 26, 1743 to May 8, 1794) thought that silica could be an oxide of a fundamental chemical element, but the chemical affinity of silicon for oxygen is high enough that he could not reduce the oxide and isolate the element. After an attempt to isolate silicon in 1808, Cornish chemist and inventor Sir Humphry Davy (December 17, 1778 to May 29, 1829) suggested the name "silicium" for silicon, from the Latin silex, silicis for flint, and adding the "-ium" ending because he thought it to be a metal (Davy, 1808a). Most other languages use transliterated forms of Davy's name, sometimes adapted to local phonology (e.g., German Silizium, Turkish silisyum). A few others use instead a calque of the Latin root (e.g., Russian кремний, from кремень "flint"; Greek πυριτιο from πυρ "fire"; Finnish pii from piikivi "flint"). The French chemists Joseph Louis Gay-Lussac (December 6, 1778 to May 9, 1850) (Fig. 4.26) and Louis-Jacques Thénard (May 4, 1777 to June 21, 1857 (Fig. 4.27)) are believed to have synthesized impure amorphous silicon in 1811, through the heating of freshly isolated potassium metal with silicon tetrafluoride, but they did not purify and characterize the product nor recognize it as a new element (Gay-Lussac and Thénard, 1811a,b). Silicon was given its current name in 1817 by Scottish chemist and mineralogist Thomas Thomson (April 12, 1773 to August 2, 1852). He kept part of Davy's name but added "-on" because he believed that silicon was a nonmetal comparable to boron and carbon (Thomson, 1817). In 1823 Swedish chemist Jöns Jacob Berzelius (August 20, 1779 to August 7, 1848) prepared amorphous silicon using nearly the same method as Gay-Lussac (reducing potassium fluorosilicate with molten potassium metal) but he purified the resulting product to a brown powder by repeatedly washing it (Berzelius, 1824a,b). Therefore, he is generally given credit for the discovery

FIGURE 4.26 Joseph Louis Gay-Lussac, by François Séraphin Delpech (1778–1825).

FIGURE 4.27 Louis-Jacques Thénard.

of silicon. That same year, Berzelius was the first to successfully prepare silicon tetrachloride; silicon tetrafluoride had previously been prepared in 1771 by Swedish Pomeranian and German pharmaceutical chemist Carl Wilhelm Scheele (December 9, 1742 to May 21, 1786) by dissolving silica in hydrofluoric acid. Silicon in its more common crystalline form was not prepared until 31 years later, by the French chemist Henri Étienne Sainte-Claire Deville (March 11, 1818 to July 1, 1881) (Sainte-Claire Deville, 1855). He electrolyzed a mixture of sodium chloride and aluminum chloride containing approximately 10% silicon and obtained a slightly impure allotrope (property of some chemical elements to exist in two or more different forms, in the same physical state, known as allotropes of the elements) of silicon in 1854. More recently, more cost-effective procedures have been established to isolate several allotrope forms, the most recent being silicene in 2010 (Silicene is a two-dimensional allotrope of silicon, with a hexagonal honeycomb structure like that of graphene. Contrary to graphene, silicene is not flat but has a periodically buckled topology; the coupling between layers in silicene is much stronger than in multilayered graphene; and the oxidized form of silicene, 2D silica, has a very different chemical structure from graphene oxide.). In the meantime, research on the chemistry of silicon continued; the German chemist Friedrich Wöhler (July 31, 1800 to September 23, 1882) found the first volatile hydrides of silicon, synthesizing trichlorosilane, $HSiCl_3$, in 1857 and silane, SiH_4, itself in 1858, but a full study of the silanes was only performed in the early 20th century by the German inorganic chemist Alfred Stock (July 16, 1876 to August 12, 1946), notwithstanding early conjecture on the matter dating as far back as the early stages of synthetic organic chemistry in the 1830s. Similarly, the first organosilicon compound, tetraethylsilane, $Si(C_2H_5)_4$, was synthesized by French chemist and mineralogist Charles Friedel (March 12, 1832 to April 20, 1899) and American chemist James Crafts (March 8, 1839 to June 20, 1917) in 1863, but thorough characterization of organosilicon chemistry was only performed in the early 20th century by English chemist Frederic Kipping (August 16, 1863 to May 1, 1949).

4.4.2 Mining, production, and major minerals

In the universe, Si is the seventh most abundant element, after H, He, C, N, O, and Ne. These abundances are not replicated well on earth due to considerable separation of the elements during the formation of the Solar System. Silicon makes up 27.2% of the earth's crust by weight, second only to oxygen at 45.5%, with which it always is bonded in nature. Further fractionation took place during the formation of the earth by planetary differentiation: the earth's core, which makes up 31.5% of the mass of the earth, has an estimated composition of $Fe_{25}Ni_2Co_{0.1}S_3$; the mantle makes up 68.1% of the earth's mass and is composed mostly of denser oxides and silicates, an example being olivine $(Mg,Fe)_2SiO_4$; while the lighter siliceous minerals such as aluminosilicates rise to the surface and form the crust, making up only 0.4% of the earth's mass. Silicon of 96%–99% purity is produced by reducing quartzite (a metamorphic rock consisting mainly of the mineral quartz) or sand (mainly quartz grains) with highly pure coke. The reduction is performed in an electric arc furnace, with an excess of SiO_2 used to prevent silicon carbide (SiC) from accumulating:

$$SiO_2 + 2C \rightarrow Si + 2CO$$
$$2SiC + SiO_2 \rightarrow 3Si + 2CO$$

This reaction, known as carbothermal reduction of silicon dioxide, generally is performed together with scrap iron with low concentrations of phosphorus and sulfur, producing ferrosilicon. Ferrosilicon, an iron-silicon alloy that encompasses varying ratios of elemental silicon and iron, accounts for about 80% of the world's production of elemental silicon, with China, the leading supplier of elemental silicon, followed by Russia, Norway, Brazil, and the United States. Ferrosilicon is mainly used by the iron and steel industry with principal use as alloying addition in iron or steel and for deoxidation of steel in integrated steel plants. Another reaction, sometimes used, is aluminothermal reduction of silicon dioxide:

$$3SiO_2 + 4Al \rightarrow 3Si + 2Al_2O_3$$

Leaching crushed 96%–97% pure silicon with water produces about 98.5% pure silicon, which is used in the chemical industry. Yet, even higher purity is necessary for semiconductor applications, and this is achieved through the reduction in tetrachlorosilane or trichlorosilane. The former is made by chlorinating scrap silicon, and the latter is a by-product of silicone production. These compounds are volatile and therefore can be purified by repetitive fractional distillation, followed by reduction to elemental silicon with very pure zinc metal as the reducing agent. The spongy particles of silicon thus formed are melted and then grown to form cylindrical single crystals, before being purified by zone refining (method of purifying crystals, in which a narrow region of a crystal is melted, and this molten zone is moved along the crystal. The molten region melts impure solid at its forward edge and leaves a wake of purer material solidified behind it as it moves through the crystal. The impurities concentrate in the melt and are moved to one end of the crystal. Zone refining was invented by John Desmond Bernal (May 10, 1901 to September 15, 1971) and further developed by William Gardner Pfann (October 27, 1917 to October 22, 1982) in Bell Labs as a method to prepare high-purity materials). Other methods use the thermal decomposition of silane or tetraiodosilane. Another process used is the reduction in sodium hexafluorosilicate, a common waste product of the phosphate fertilizer industry, by metallic sodium: this reaction is highly exothermic and therefore needs no external fuel source. Hyperfine silicon is made at a higher purity than almost any other material: transistor production requires impurity levels in silicon crystals less than 1 part per 10^{10}, and in special applications impurity levels below 1 part per 10^{12} can be desirable and be achieved.

Silicate and aluminosilicate minerals can have many different crystal structures and varying stoichiometry, but they are all classified following some general principles based on the degree of polymerization. Tetrahedral $\{SiO_4\}$ units are found in nearly every mineral, either as discrete tetrahedra or combined into larger units by the sharing of corner oxygen atoms. These may be divided into neso-silicates (discrete $\{SiO_4\}$ units) sharing no oxygen atoms, soro-silicates (discrete $\{Si_2O_7\}$ units) sharing one oxygen atom, cyclo-silicates (closed ring structures) and ino-silicates (continuous chain or ribbon structures) both sharing two oxygen atoms, phyllo-silicates (continuous sheets) sharing three oxygen atoms, and tecto-silicates (continuous three-dimensional frameworks) sharing four oxygen atoms. The lattice of oxygen atoms that results is usually close packed, or close to it, with the charge being balanced by other cations in a variety of different polyhedral sites according to size and coordination.

Minerals containing silicon are the largest group of all minerals with more than 1600 known minerals of the roughly 5000 accepted minerals containing Si, the majority in the form of silicates. In the class of elements besides Si as a native element, several alloys exist, all rather rare, for example, hapkeite (Fe_2Si), moissanite (SiC), nierite (Si_3N_4), and zangboite ($TiFeSi_2$). In the halides class, eight minerals are known, for example, asisite ($Pb_7SiO_9Cl_2$) and cryptohalite ($(NH_4)_2[SiF_6]$). Under the oxides, one finds the polymorphs of quartz (SiO_2) (Figs. 4.28–4.30) (Box 4.4), such as coesite, cristobalite, seifertite, stishovite, and trydimite as well as opal ($SiO_2 \cdot nH_2O$). There are only two carbonates containing Si: defernite ($Ca_6(CO_3)_{2-x}(SiO_4)_x(OH)_7(Cl,OH)_{1-2x}$ ($x < 0.5$)) and lepersonnite-(Gd) ($Ca(Gd,Dy)_2(UO_2)_{24}(SiO_4)_4(CO_3)_8(OH)_{24} \cdot 48H_2O$). Three borates are known with Si in their structure: harkerite ($Ca_{12}Mg_4Al(BO_3)_3(SiO_4)_4(CO_3)_5 \cdot H_2O$), howlite ($Ca_2B_5SiO_9(OH)_5$), and wiserite ($(Mn,Mg)_{14}[Cl|(OH)_8|(Si,Mg)(O,OH)_4|(B_2O_5)_4]$). In the sulfate class, eight minerals are known, such as hemihedrite ($Pb_{10}Zn(CrO_4)_6(SiO_4)_2(OH)_2$), thaumasite ($Ca_3(SO_4)[Si(OH)_6](CO_3) \cdot 12H_2O$), and wherryite ($Pb_7Cu_2(SO_4)_4(SiO_4)_2(OH)_2$). The group of phosphate minerals is slightly bigger with 16 containing Si, for example, dixenite ($CuMn^{2+}_{14}Fe^{2+}(SiO_4)_2(As^{5+}O_4)(As^{3+}O_3)_5(OH)_6$), krasnoite ($Ca_3Al_{7.7}Si_3P_4O_{23.5}(OH)_{12.1}F_2 \cdot 8H_2O$), and perhamite ($Ca_3Al_{7.7}Si_3P_4O_{23.5}(OH)_{14.1} \cdot 8H_2O$). As already indicated above, the largest group of minerals is the silicates with more than

FIGURE 4.28 Quartz, SiO_2, variety amethyst, on matrix. The largest amethyst crystal is 3 cm. Piedras Parado, Tatatlia, Veracruz, Mexico.

FIGURE 4.29 A 2 cm dark smoky quartz, SiO_2, crystal on white microcline, $K(AlSi_3O_8)$. Bighorn Crags, Lemhi Co., Idaho, United States.

FIGURE 4.30 Quartz, SiO_2, variety agate, overall 5 cm banded polished agate. Teepee Canyon, Black Hills, South Dakota, United States.

BOX 4.4 SiO_2 polymorphs

In most silicates, the silicon atom shows tetrahedral coordination, with four oxygen atoms surrounding a central Si atom. The most common example is seen in the quartz polymorphs. It is a three-dimensional network solid in which each Si atom is covalently bonded in a tetrahedral manner to 4O atoms. For example, in the unit cell of α-quartz, the central tetrahedron shares all four of its corner O atoms, the two face-centered tetrahedra share two of their corner O atoms, and the four edge-centered tetrahedra share only one of their O atoms with other SiO_4 tetrahedra. SiO_2 exists as several distinct crystalline forms (polymorphs) as well as amorphous forms. Except for stishovite, all the crystalline forms involve tetrahedral SiO_4 units linked together by shared vertices in different arrangements. Si–O bond lengths vary between the different crystal forms; for example, in α-quartz (hexagonal) the bond length is 161 pm (picometer), whereas in α-tridymite (orthorhombic) it is in the range 154–171 pm. The Si-O-Si angle also varies from as low as 140° in α-tridymite to 180° in β-tridymite (hexagonal). In α-quartz, the Si-O-Si angle is 144°. Stishovite (tetragonal), the higher-pressure form, in contrast, has a rutile-like structure where Si is in 6-coordination (octahedral). The density of stishovite is 4.287 g/cm^3, which compares to α-quartz, the densest of the low-pressure forms, which has a density of 2.648 g/cm^3. The difference in density can be attributed to the increase in coordination as the six shortest Si-O bond lengths in stishovite (four Si–O bond lengths of 176 pm and two others of 181 pm) are larger than the Si–O bond length (161 pm) in α-quartz. The change in the coordination results in an increase in the the ionicity of the Si–O bond. More significantly, any deviations from these standard parameters constitute microstructural differences or variations, which signify an approach to an amorphous, vitreous, or glassy solid. The only stable form under normal conditions is α-quartz, in which crystalline SiO_2 is usually encountered. In nature, impurities in crystalline α-quartz can give rise to colors (e.g., amethyst, smoky quartz, citrine). The high-temperature minerals, cristobalite (tetragonal) and tridymite, have both lower densities and refractive indices than quartz. Since their composition is the same, these discrepancies must be caused by the increased spacing in the high-temperature minerals. As is common with most substances, the higher the temperature, the larger the distance between atoms, due to the increased vibration energy. The transformation from α-quartz to β-quartz takes place abruptly at 573°C at 1 atm. and is reversed upon cooling below this temperature. Since the transformation is accompanied by a significant change in volume, it can easily induce fracturing of ceramics or rocks passing through this temperature limit. The high-pressure minerals, seifertite (orthorhombic), stishovite, and coesite, though, have higher densities and refractive indices than quartz. This is possibly due to the intense compression of the atoms occurring during their formation, resulting in more condensed structure. Metastable occurrences of the high-pressure forms coesite and stishovite have been found around meteorite impact structures and associated with eclogites formed during ultra-high-pressure metamorphism. The high-temperature forms of tridymite and cristobalite are known from silica-rich volcanic rocks. Opal, $SiO_2{\cdot}nH_2O$, although it is essentially amorphous, has been shown to have a regular structure consisting of a regular three-dimensional array of hexagonal or cubic closed packing of silica spheres. Air or water occupies the space between the spheres. In precious opal, large domains are made up of regular packed spheres of the same size. Their diameters vary from one opal to the other and range from 1500 to 3000 Å. When white light passes through the essentially colorless opal and strikes planes of voids between the spheres, certain wavelengths are diffracted and flash out of the stone as nearly pure spectral colors.

1250 different minerals. A few of the best-known silicate minerals include actinolite ($\square\{Ca_2\}\{Mg_{4.5\text{-}2.5}Fe_{0.5-2.5}\}(Si_8O_{22})(OH)_2$), albite ($Na(AlSi_3O_8)$), almandine ($Fe_3^{2+}Al_2(SiO_4)_3$), andalusite ($Al_2(SiO_4)O$), apophyllite ($(Na,K)Ca_4(Si_8O_{20})(F,OH).8H_2O$) (Fig. 4.31) beryl ($Be_3Al_2(Si_6O_{18})$) (Fig. 4.32), cavansite ($Ca(VO)Si_4O_{10}{\cdot}4H_2O$) (Fig. 4.33, chrysocolla ($Cu_{2\text{-}x}Al_x(H_{2\text{-}x}Si_2O_5)(OH)_4{\cdot}nH_2O$), danburite ($CaB_2Si_2O_8$), diopside ($CaMgSi_2O_6$) (Fig. 4.34), dioptase ($CuSiO_3{\cdot}H_2O$) (Fig. 4.35), elbaite (tourmaline) ($Na(Li_{1.5}Al_{1.5})Al_6(Si_6O_{18})(BO_3)_3(OH)_3(OH)$), epidote ($\{Ca_2\}\{Al_2Fe^{3+}\}(Si_2O_7)(SiO_4)O(OH)$) (Fig. 4.36), hemimorphite ($Zn_4Si_2O_7(OH)_2{\cdot}H_2O$) (Fig. 4.37), kyanite ($Al_2(SiO_4)O$), leucite ($K(AlSi_2O_6)$), microcline ($K(AlSi_3O_8)$) (Fig. 4.38), muscovite ($KAl_2(AlSi_3O_{10})(OH)_2$) (Fig. 4.39), natrolite ($Na_2Al_2Si_3O_{10}{\cdot}2H_2O$), "olivine" (fayalite-forsterite series) ($(Fe,Mg)_2SiO_4$), pentagonite ($Ca(VO)Si_4O_{10}{\cdot}4H_2O$) (Fig. 4.40), phenakite (Be_2SiO_4), prehnite ($Ca_2Al_2Si_3O_{10}(OH)_2$) (Fig. 4.41), sillimanite ($Al_2(SiO_4)O$), spodumene ($LiAlSi_2O_6$), stellerite ($Ca_4(Si_{28}Al_8)O_{72}{\cdot}28H_2O$) (Fig. 4.42), talc ($Mg_3Si_4O_{10}(OH)_2$), topaz ($Al_2(SiO_4)(F,OH)_2$) (Fig. 4.43), vesuvianite ($(Ca,Na,\square)_{19}(Al,Mg.Fe^{3+})_{13}(\square,B,Al,Fe^{3+})5(Si_2O_7)_4(SiO_4)_{10}(OH,F,O)_{10}$) (Fig. 4.44), willemite (Zn_2SiO_4), wollastonite ($CaSiO_3$) (Fig. 4.45), and zircon ($Zr(SiO_4)$) (Fig. 4.46).

4.4.3 Chemistry

Naturally occurring Si consists of three stable isotopes, ^{28}Si (92.23%), ^{29}Si (4.67%), and ^{30}Si (3.10%). Among these, only ^{29}Si is useful in NMR and EPR spectroscopy, since it is the only isotope with a nuclear spin (I = 1/2). All three are formed in stars through the oxygen-burning process (a set of nuclear fusion reactions that take place in massive stars

FIGURE 4.31 Colorless to white, striated apophyllite, $(Na,K)Ca_4(Si_8O_{20})(F,OH)\cdot 8H_2O$, crystals to 2 cm. Mumbai district, Maharashtra, India.

FIGURE 4.32 Prismatic and hexagonal crystals of emerald green beryl, $Be_3Al_2(Si_6O_{18})$, to 1.5 cm. Kamar Safed, Bazarak dist., Panjsher prov., Afghanistan.

FIGURE 4.33 Blue clusters of cavansite, $Ca(VO)Si_4O_{10}\cdot 4H_2O$, to 8 mm set on pearly white stilbite crystals. Poonah dist., Maharashtra, India.

FIGURE 4.34 Deep, rich green prismatic chromian diopside, $CaMgSi_2O_6$, crystals to 2 cm. Outokumpu, Finland.

FIGURE 4.35 Dioptase, $CuSiO_3{\cdot}H_2O$, deep emerald green crystal to 1.2 cm with two other crystals to 5 mm. Tsumeb mine, Tsumeb, Namibia.

that have used up the lighter elements in their cores. Oxygen-burning, $^{16}O + ^{16}O \rightarrow ^{28}Si + ^{4}He$, is preceded by the neon-burning process and succeeded by the silicon-burning process), with ^{28}Si being made as part of the alpha process (one of two classes of nuclear fusion reactions by which stars convert helium into heavier elements, $^{24}Mg + ^{4}He \rightarrow ^{28}Si + \gamma$) and hence the most abundant. The fusion of ^{28}Si with alpha particles by photodisintegration rearrangement in stars is known as the silicon-burning process (a very brief sequence of nuclear fusion reactions that occur in massive stars with a minimum of about 8–11 solar masses. Silicon burning is the final stage of fusion for massive stars that have run out of the fuels that power them for their long lives in the main sequence on the Hertzsprung-Russell diagram, series of reactions that start with $^{28}Si + ^{4}He \rightarrow ^{32}S$); it is the last stage of stellar nucleosynthesis before the rapid collapse and violent explosion of the star in question is a type II supernova. Twenty radioisotopes have been observed, the two most stable radioisotopes being ^{32}Si with a half-life of about 153 years, which decays by β^- to ^{32}P and then stable ^{32}S, while ^{31}Si with a half-life of 2.62 hours decays by β^- to ^{31}P. All the other radioisotopes have half-lives less than seven seconds, and most have half-lives that are less than 0.1 seconds. Silicon does not have any known nuclear isomers. ^{31}Si

FIGURE 4.36 Prismatic dark green epidote, $\{Ca_2\}\{Al_2Fe^{3+}\}(Si_2O_7)(SiO_4)O(OH)$, crystals to 4.5 cm. Huancavelica, Peru.

FIGURE 4.37 Botryoidal hemimorphite, $Zn_4Si_2O_7(OH)_2 \cdot H_2O$, 1.5 cm across. 79 mine, Gila Co., Arizona, United States.

may be formed via the neutron activation of natural Si and is therefore useful for quantitative analysis; it can simply be detected by its characteristic β^- decay to stable ^{31}P, in which the emitted electron carries up to 1.48 MeV of energy. The known Si isotopes range in mass number from 22 to 44. The most common decay mode of the isotopes with mass numbers lower than the three stable isotopes is inverse β decay, mainly forming aluminum isotopes (13 protons) as decay products. The most common decay mode for the heavier unstable isotopes is β decay, mainly forming phosphorus isotopes (15 protons) as decay products.

Silicon, a hard and brittle crystalline solid with a blue-gray metallic luster, is a member of group 14 in the periodic table, with carbon just above it and germanium just below it (Table 4.4). Its ground-state electron configuration is [Ne] $3s^23p^2$. It has four valence electrons, occupying the 3s orbital and two of the 3p orbitals. It can complete its octet and obtain the stable noble gas configuration of argon by forming sp^3 hybrid orbitals, forming tetrahedral SiX_4 derivatives.

FIGURE 4.38 Microcline, $K(AlSi_3O_8)$, variety amazonite, opaque green crystal to 3 cm on light smoky quartz, SiO_2. Crystal Peak area, Teller Co., Colorado, United States.

FIGURE 4.39 Plate of crystallized ruby-colored muscovite, $KAl_2(AlSi_3O_{10})(OH)_2$. The largest crystal is 7.5 cm across. Jaguaraçu pegmatite, Jaguaraçu, Minas Gerais, Brazil.

FIGURE 4.40 Neon blue radial cluster of pentagonite, $Ca(VO)Si_4O_{10}{\cdot}4H_2O$, to 9 mm sitting protected in zeolite vug. Wagholi, near Pune, Maharashtra, India.

FIGURE 4.41 Prehnite, $Ca_2Al_2Si_3O_{10}(OH)_2$, a 5.5 cm light mint green botryoidal crystal group. O & G Woodbury Quarry, Litchfield Co., Connecticut, United States.

FIGURE 4.42 Stellerite, $Ca_4(Si_{28}Al_8)O_{72}{\cdot}28H_2O$, orange crystal groups to 1 cm. Sarbaiskoe mine, Rudny, Kostanay, Kazakhstan.

FIGURE 4.43 Gem quality, light sherry-colored topaz, $Al_2(SiO_4)(F,OH)_2$, crystal to 2.5 cm set on matrix of white albite, $Na(AlSi_3O_8)$, and colorless quartz, SiO_2. Yuno, Shigar Valley, Baltistan, Northern Areas, Pakistan.

FIGURE 4.44 Vesuvianite, $(Ca,Na,\square)_{19}(Al,Mg,Fe^{3+})_{13}(\square,B,Al,Fe^{3+})_5(Si_2O_7)_4(SiO_4)_{10}(OH,F,O)_{10}$, pink to yellow crystal group to 2 cm. Jeffrey mine, Asbestos, Quebec, Canada.

FIGURE 4.45 Wollastonite, $CaSiO_3$, white compact acicular crystals in divergent habit, overall 8 cm. British Rail Quarry, Meldon, Okehampton, Devon, England.

FIGURE 4.46 Zircon, $Zr(SiO_4)$, well-formed creamy brown crystal to 9 mm. Tigerville Prospect, Greenville Co., South Carolina, United States.

TABLE 4.4 Silicon properties.

Appearance	Crystalline, reflective with bluish-tinged faces
Standard atomic weight $A_{r,std}$	28.085
Block	P-block
Element category	Metalloid
Electron configuration	[Ne] $3s^2$ $3p^2$
Phase at STP	Solid
Melting point	1414°C
Boiling point	3265°C
Density (near r.t.)	2.3290 g/cm^3
When liquid (at m.p.)	2.57 g/cm^3
Heat of fusion	50.21 kJ/mol
Heat of vaporization	383 kJ/mol
Molar heat capacity	19.789 J/(mol·K)
Oxidation states	**−4**, −3, −2, −1, +1, +2, +3, **+4**
Ionization energies	1st: 786.5 kJ/mol
	2nd: 1577.1 kJ/mol
	3rd: 3231.6 kJ/mol
Atomic radius	Empirical: 111 pm
Covalent radius	111 pm
Van der Waals radius	210 pm

STP, Standard temperature and pressure.
Bold font indicates main oxidation state.

Like those of carbon, the SiX_4 compounds have the central silicon atom sharing an electron pair with each of the four atoms it is bonded to. Unlike the 2p subshell of carbon, however, the 3p subshell of silicon is more diffuse and does not hybridize so well with the 3s subshell. This results in differences in the chemistry of silicon and heavier members of group 14 compared to that of carbon. For example, the poor overlap of the 3p orbitals results in a weakening of the Si-Si bond. Thus, silicon exhibits a much lower tendency for catenation than carbon. Si-Si bond energy is only 226 kJ/mol. In contrast, C-C bond energy is 356 kJ/mol. This makes multiple-bonded silicon compounds generally less stable than their carbon counterparts. On the other hand, silicon has 3d orbitals in the valence shell, which allows for hypervalency. Thus, silicon has five- and six-coordinated compounds.

Silicon crystallizes in a giant covalent structure at standard conditions, specifically in a diamond cubic lattice. As a consequence, it has a high melting point of 1414°C, as a large amount of energy is required to break the strong covalent bonds and melt the solid. Once the melting point of silicon is reached, it becomes extremely reactive, alloying with most metals to form silicides, and reducing most metal oxides. Thus, liquid silicon must be stored in refractory, unreactive containers such as zirconium dioxide or group 4, 5, and 6 borides.

Bulk silicon is relatively inert except at high temperatures. Oxygen, water, and steam all have little effect probably because of the formation of a very thin, continuous, protective surface layer of SiO_2 a few atoms thick that protects the silicon from oxidation.

4.4.3.1 Reaction with air, oxygen, nitrogen, and hydrogen

Silicon does not measurably react with the air below 900°C, but formation of the vitreous dioxide rapidly increases between 950°C and 1160°C and when 1400°C is reached, atmospheric nitrogen also reacts to give the nitrides SiN and Si_3N_4.

$$Si(s) + O_2(g) \rightarrow SiO_2(s)$$
$$2Si(s) + N_2(g) \rightarrow 2SiN(s)$$
$$3Si(s) + 2N_2(g) \rightarrow Si_3N_4(s)$$

The silanes consist of a homologous series of silicon hydrides with a general formula of Si_nH_{2n+2}. They are all strong reducing agents. Unbranched and branched chains are known up to n = 8, and the cycles Si_5H_{10} and Si_6H_{12} have also been observed. The first two, silane and disilane, are colorless gases; the heavier members of the series are volatile liquids. All silanes are very reactive and catch fire or explode spontaneously in air. They become less thermally stable with room temperature, so that only silane is indefinitely stable at room temperature, although disilane does not decompose very quickly (only 2.5% of a sample decomposes after the passage of 8 months). They decompose to produce polymeric polysilicon hydride and hydrogen gas. As expected from the difference in atomic weight, the silanes are less volatile than the corresponding alkanes and boranes, but more so than the corresponding germanes. They are much more reactive than the corresponding alkanes, due to the larger radius of silicon compared to carbon facilitating nucleophilic attack at the silicon, the greater polarity of the Si–H bond compared to the C–H bond, and the ability of silicon to expand its octet and thus form adducts and lower the reaction's activation energy. Silane pyrolysis produces polymeric species and finally elemental silicon and hydrogen. Although the thermal decomposition of alkanes begins with the breaking of a C–H or C–C bond and the formation of radical intermediates, polysilanes decompose by eliminating silylenes: SiH_2 or : SiHR, as the activation energy of this process (~210 kJ/mol) is much less than the Si–Si and Si–H bond energies. Although pure silanes do not react with pure water or dilute acids, traces of alkali catalyze immediate hydrolysis to hydrated silicon dioxide. If the reaction is performed in methanol, controlled solvolysis results in the products $SiH_2(OMe)_2$, $SiH(OMe)_3$, and $Si(OMe)_4$. The Si–H bond also adds to alkenes, a reaction which proceeds slowly and speeds up with increasing substitution of the silane involved. At 450°C, silane participates in an addition reaction with acetone, as well as a ring-opening reaction with ethylene oxide. Direct reaction of the silanes with chlorine or bromine results in explosions at room temperature, but the reaction of silane with bromine at −80°C is controlled and produces bromosilane and dibromosilane. The monohalosilanes may be prepared by reacting silane with the appropriate hydrogen halide with an Al_2X_6 catalyst, or by reacting silane with a solid silver halide in a heated flow reactor:

$$SiH_4 + 2AgCl \xrightarrow{260^\circ C} SiH_3Cl + HCl + 2Ag.$$

Among the derivatives of silane, iodosilane (SiH_3I) and potassium silanide ($KSiH_3$) are very useful synthetic intermediates in the preparation of more complicated silicon-containing compounds: the latter is a colorless crystalline ionic solid containing K^+ cations and SiH_3^- anions in the NaCl structure and is produced by the reduction of silane by potassium metal. Likewise, the reactive hypervalent species SiH_5^- is also known. With suitable organic substituents it is possible to produce stable polysilanes: they have surprisingly high electric conductivities, due to σ delocalisation of the electrons in the chain. Silicon nitride, Si_3N_4, may be produced through directly reacting silicon with nitrogen above 1300°C, but a more economical route is by heating silica and coke in a stream of nitrogen and hydrogen gas at 1500°C. It would make a promising ceramic if not for the difficulty of working with and sintering it: chemically, it is near-totally inert, and even above 1000°C it keeps its strength, shape, and continues to be resistant to wear and corrosion. It is very hard (9 on the Mohs hardness scale), dissociates only at 1900°C at 1 atm, and is quite dense (density 3.185 g/cm^3), because of its compact structure similar to that of the mineral phenacite (Be_2SiO_4). A similar refractory material is Si_2N_2O, produced by heating silicon and silica at 1450°C in an argon stream containing 5% nitrogen gas, involving 4-coordinate silicon and 3-coordinate nitrogen alternating in puckered hexagonal tilings interlinked by non-linear Si–O–Si linkages to each other.

4.4.3.2 Reaction with halogens

The protective oxide layer of silicon, nevertheless, does not prevent reaction with the halogens. Silicon reacts vigorously with all the halogens to form silicon tetrahalides. Fluorine attacks silicon vigorously at room temperature, chlorine does so at about 300°C, and bromine and iodine at about 500°C.

$$Si(s) + 2F_2(g) \rightarrow SiF_4(g)$$
$$Si(s) + 2Cl_2(g) \rightarrow SiCl_4(g)$$
$$Si(s) + 2Br_2(g) \rightarrow SiBr_4(l)$$
$$Si(s) + 2I_2(g) \rightarrow SiI_4(s)$$

Silicon and silicon carbide easily react with all four stable halogens, producing the colorless, reactive, and volatile silicon tetrahalides. Silicon tetrafluoride may also be produced via fluorinating the other silicon halides and is formed by the attack of hydrofluoric acid on glass. In addition, heating two different tetrahalides together results in a random mixture of mixed halides, which may also be formed by halogen exchange reactions. The melting and boiling points of these species generally increase with increasing atomic weight, though there are many exceptions, for example, the melting and boiling points drop as one passes from $SiFBr_3$ through $SiFClBr_2$ to $SiFCl_2Br$. The shift from the hypoelectronic elements in group 13 and earlier to the group 14 elements is illustrated by the change from an infinite ionic structure in aluminum fluoride to a lattice of simple covalent silicon tetrafluoride molecules, as dictated by the lower electronegativity of aluminum than silicon, the stoichiometry (the +4 oxidation state being too high for true ionicity), and the smaller size of the silicon atom in comparison to the aluminum atom. Silicon tetrachloride is produced on a huge scale as a precursor to the production of pure silicon, silicon dioxide, and some silicon esters. The silicon tetrahalides hydrolyse easily in water, in contrast to the carbon tetrahalides, again due to the larger size of the silicon atom rendering it more open to nucleophilic attack and the ability of the silicon atom to expand its octet which carbon lacks. The reaction of silicon fluoride with excess hydrofluoric acid produces the octahedral hexafluorosilicate anion SiF_6^{2-}. Similar to the silanes, halopolysilanes Si_nX_{2n+2} also are known. While catenation in carbon compounds is maximized in the hydrogen compounds rather than the halides, the opposite is true for silicon, so that the halopolysilanes are known up to at least $Si_{14}F_{30}$, Si_6Cl_{14}, and Si_4Br_{10}. A possible explanation for this phenomenon is the compensation for the electron loss of silicon to the more electronegative halogen atoms by π backbonding from the filled p_π orbitals on the halogen atoms to the empty d_π orbitals on silicon: this is comparable to the situation of carbon monoxide in metal carbonyl complexes and explains their stability. These halopolysilanes may be formed by comproportionation of silicon tetrahalides with elemental silicon, or by condensation of lighter halopolysilanes (trimethylammonium is a useful catalyst for this reaction).

4.4.3.3 Reaction with acids and bases

Silicon does not react with most aqueous acids but is oxidized and fluorinated by a mixture of concentrated nitric acid and hydrofluoric acid. It is also attacked by bases such as aqueous sodium hydroxide to give silicates, highly complex species containing the anion $[SiO_4]^{4-}$.

$$Si(s) + 6HF(aq) \rightarrow [SiF_6]^{2-}(aq) + 2H^+(aq) + 2H_2(g)$$
$$Si(s) + 4NaOH(aq) \rightarrow [SiO_4]^{4-}(aq) + 4Na^+(aq) + 2H_2(g)$$

4.4.3.4 Organosilicon compounds

Organosilicon compounds are organometallic compounds containing carbon–silicon bonds. Most organosilicon compounds derive from organosilicon chlorides $(CH_3)_4$-$xSiCl_x$. These chlorides are produced by the reaction of methyl chloride with a silicon-copper alloy. The main product is dimethyldichlorosilane:

$$2CH_3Cl + Si \rightarrow (CH_3)_2SiCl_2$$

Another method for the formation of Si-C bonds is hydrosilylation (also called hydrosilation). In this process, compounds with Si-H bonds (hydrosilanes) add to unsaturated substrates such as alkenes, alkynes, imines, ketones, and aldehydes in the presence of catalysts. More than 100,000 organosilicon compounds have been synthesized. Many of these are industrially important for their considerable thermal stability and chemical inertness.

4.4.4 Major uses

Most silicon is used industrially without purification and frequently with relatively little processing. More than 90% of the earth's crust is composed of silicate minerals. Numerous silicate minerals have direct commercial applications, for example, clays, silica sand, and most kinds of building stone. Hence, most uses for silicon are as structural compounds, either as the silicate minerals or silica (crude silicon dioxide). Silicates are utilized in producing Portland cement (made mostly of calcium silicates) which is used in building mortar and modern stucco, but more importantly, in combination with silica sand, and gravel (usually containing silicate minerals such as granite), to produce concrete which currently forms the basis of most of the very large industrial building projects. In addition, silica is used to manufacture fire brick, a kind of ceramic. Silicate minerals are also in whiteware ceramics, an important class of products usually containing various types of fired clay minerals (natural aluminum phyllosilicates), for example, porcelain, which is made from the clay mineral kaolinite. Traditional glass (silica-based soda-lime glass) is produced in a similar fashion and is used for windows and containers. In addition, specialty silica-based glass fibers are used for optical fiber as well as to produce fiberglass for structural support and glass wool for thermal insulation. Silicones are silicon-oxygen polymers with methyl groups attached. Silicones are regularly used in waterproofing treatments, molding compounds, mold-release agents, mechanical seals, high-temperature greases and waxes, and caulking compounds. Silicone oil is a lubricant and is added to some cosmetics and hair conditioners. Silicone rubber is used as a waterproof sealant in bathrooms and around windows, pipes, and roofs. Silicone is occasionally used in breast implants, contact lenses, explosives, and pyrotechnics. Silly Putty was initially produced by combining boric acid with silicone oil. Additional silicon compounds serve as high-technology abrasives (carborundum, nearly as hard as diamond) and new high-strength ceramics based on silicon carbide (SiC). Silicon is a component of some superalloys.

Elemental Si is added to molten cast iron as ferrosilicon or silicocalcium alloys to enhance performance in casting thin sections and to inhibit the formation of cementite (or iron carbide, a compound of iron and carbon, more precisely an intermediate transition metal carbide with the formula Fe_3C) where exposed to outside air. The elemental Si present in the molten Fe functions as a sink for oxygen, so that the steel carbon content, which has to be maintained within narrow limits for each steel type, can be carefully controlled. Ferrosilicon manufacture and use, although this type of elemental Si is rather impure, accounts for around 80% of the global free Si usage. Silicon forms an important element of electrical steel, modifying its resistivity and ferromagnetic properties. The properties of Si can be utilized to modify alloys with metals other than Fe. "Metallurgical-grade" silicon has between 95% and 99% purity. Approximately 55% of the global use of metallurgical purity Si is used in the manufacture of Al-Si alloys (silumin alloys) for Al part casts, primarily for use in the automotive industry. Silicon's importance in Al casting is that a considerably high amount (12%) of Si in Al forms a eutectic mixture that solidifies with very little thermal contraction. This significantly reduces tearing and cracks formed from stress as the casting alloys solidify. Silicon also substantially enhances the hardness and hence the wear resistance of Al. There are several forms of sodium silicate produced by heating soda (sodium carbonate) and sand (quartz, SiO_2) to 1500°C. The product, a glass-like solid, is subsequently dissolved in water under pressure, and sodium silicates can be crystallized from the resultant solution. They are used in detergents, adhesive, and textile bleaching and to enhance oil recovery. By pumping a solution of sodium silicate down an exhausted oil well, it is possible for more oil to be pumped out of a nearby production well. The alkalinity of the solution neutralizes acidic compounds in the oil-bearing strata and this releases more oil.

Most elemental Si produced is used in ferrosilicon alloys, and only about 20% is refined to metallurgical-grade purity. Approximately 15% of the global production of metallurgical-grade Si is further refined to semiconductor purity. This typically is the "nine-9" or 99.9999999% purity, nearly defect-free single crystalline material. Monocrystalline silicon of such purity is usually produced by the Czochralski process (Box 4.5) and is used to produce silicon wafers used in the semiconductor industry, in electronics, and in some high-cost and high-efficiency photovoltaic applications.

Pure Si is an intrinsic semiconductor, which means that different to metals, it conducts electron holes and electrons released from atoms by heat; its electrical conductivity increases with higher temperatures. Pure Si has too low a conductivity (i.e., too high a resistivity) to be applied as a circuit element in electronics. In fact, pure Si is doped with small concentrations of particular other elements, which significantly enhance its conductivity and adjust its electrical response by adjusting the number and charge (positive or negative) of activated carriers. Such control is required for transistors, solar cells, semiconductor detectors, and other semiconductor devices used in the computer industry and other technical uses. In Si photonics, Si may be utilized as a continuous wave Raman laser medium to generate coherent light. In common integrated circuits, a wafer of monocrystalline silicon acts as a mechanical support for the circuits, which are made by doping and insulated from each other by thin layers of silicon oxide, an insulator that is easily created on Si surfaces by processes of thermal oxidation or local oxidation (LOCOS), which comprise exposing the element to oxygen under the appropriate conditions that can be predicted by the Deal–Grove model [which mathematically describes the growth of an oxide layer on the surface of a material. In particular, it is used to predict and interpret thermal oxidation of silicon in semiconductor device fabrication. The model was first published in 1965

BOX 4.5 Czochralski process

The Czochralski process is a method of crystal growth used to obtain single crystals of semiconductors (e.g., silicon, germanium, and gallium arsenide), metals (e.g., palladium, platinum, silver, and gold), salts, and synthetic gemstones. The process is named after Polish chemist Jan Czochralski (October 23, 1885 to April 22, 1953), who developed the process in 1915 while researching the crystallization rates of metals. He discovered it by accident: instead of dipping his pen into his inkwell, he dipped it in molten tin and drew a tin filament, which later turned out to consist of a single crystal. High-purity, semiconductor-grade Si (only a few parts per million of impurities) is melted in a crucible at 1425°C, usually made of quartz (SiO_2). Dopant impurity atoms such as B or P can be added to the molten Si in precise amounts to dope the Si, thus changing it into p-type or n-type Si, with different electronic properties. A precisely oriented rod-mounted seed crystal is dipped into the molten silicon and then slowly pulled upwards and rotated simultaneously. By accurately controlling the temperature gradients, rate of pulling, and speed of rotation, it is possible to extract a large, single-crystal, cylindrical ingot from the melt. The existence of unwanted instabilities in the melt can be prevented by investigating and visualizing the temperature and velocity fields during the crystal growth process. This process is typically performed in an inert atmosphere, such as Ar, in an inert chamber, such as quartz (SiO_2).

by Bruce Deal and Andrew Grove, of Fairchild Semiconductor, and served as a step in the development of complementary metal-oxide-semiconductor (CMOS) devices and the fabrication of integrated circuits.]. Silicon has become the most prevalent material for both high power semiconductors and integrated circuits since it can resist the highest temperatures and greatest electrical activity without suffering avalanche breakdown (an electron avalanche is created when heat results in the production of free electrons and holes, which then pass more current, which produces more heat). Furthermore, the insulating oxide of Si is not water-soluble, giving it an advantage over Ge (which has similar properties that can also be used in semiconductor devices) in certain manufacturing techniques. Monocrystalline silicon is costly to manufacture and is typically warranted only in producing integrated circuits, where tiny crystal imperfections can interfere with tiny circuit paths. For other applications, other types of pure silicon may be used, for example, hydrogenated amorphous silicon and upgraded metallurgical-grade silicon (UMG-Si) used in the making of low-cost, large-area electronics in applications such as liquid crystal displays and of large-area, low-cost, thin-film solar cells. Such semiconductor grades of silicon are either of somewhat lower purity or are polycrystalline instead of monocrystalline and are produced in comparable quantities as the monocrystalline silicon. The market for the lower grade is growing more rapidly than for monocrystalline Si.

Several SiO_2 varieties are better known by their variety names as gemstones. Precious opal is basically silica, consisting of a structure of ordered amorphous silica spheres with interstitial water, which exhibits an attractive iridescence. There are two broad classes of opal: precious and common. Precious opal displays play-of-color (iridescence), and common opal does not. Play-of-color is defined as "a pseudochromatic optical effect resulting in flashes of colored light from certain minerals, as they are turned in white light." The internal structure of precious opal causes it to diffract light, resulting in play-of-color. Depending on the conditions in which it formed, opal may be transparent, translucent, or opaque and the background color may be white, black, or nearly any color of the visual spectrum. Black opal is considered to be the rarest, whereas white, gray, and green are the most common. Among the silica varieties, one finds agate (banded cryptocrystalline silica, chiefly chalcedony) and the best-known purple-variety amethyst. The color of amethyst has been demonstrated to result from substitution by irradiation of trivalent iron (Fe^{3+}) for silicon in the structure (and not by traces of Mn as is originally thought and still, incorrectly, cited in some literature), in the presence of trace elements of large ionic radius, and, to a certain extent, the amethyst color can naturally result from displacement of transition elements even if the iron concentration is low. Natural amethyst is dichroic in reddish violet and bluish violet, but when heated, turns yellow-orange, yellow-brown, or dark-brownish and may resemble citrine, but loses its dichroism, unlike genuine citrine. When partially heated, amethyst can result in ametrine. Amethyst can fade in tone if overexposed to light sources and can be artificially darkened with adequate irradiation.

4.5 15 P — Phosphorus

4.5.1 Discovery

The name phosphorus in Ancient Greece was the name for the planet Venus and comes from the Greek words (φ ς = light, φέρω = carry), which roughly translates as light-bringer or light carrier. The discovery of phosphorus, the first element to be found that was not known since ancient times, is attributed to the German merchant, pharmacist, and alchemist Hennig Brand (c.1630–1692 or c.1710) (Fig. 4.47) in 1669, even though other alchemists may have discovered phosphorus in the same period. Brand investigated reactions with urine, which contains significant amounts of dissolved phosphates from normal metabolism (Weeks, 1932). Working in Hamburg, Brand tried to create the legendary philosopher's stone through the distillation of some salts by evaporating urine, and in the process formed a white material that glowed in the dark and burned brilliantly. It was called phosphorus mirabilis ("miraculous bearer of light"). Brand's process initially involved letting urine stand for days until it produced an awful smell. At that time, he boiled it down to a paste, heated this paste to a high temperature, and bubbled the resulting vapors through water, where he anticipated they would condense to gold. In its place, he formed a white, waxy substance that glowed in the dark. Brand had discovered phosphorus. It is now known that Brand actually produced ammonium sodium hydrogen phosphate $(NH_4)NaHPO_4$. While the amounts were in essence accurate (it took about 1,100 L of urine to make about 60 g of phosphorus), it was superfluous to let the urine rot first. Later scientists observed that fresh urine produced the same amount of phosphorus. Brand initially tried to keep the process secret, but later sold it for 200 thalers to Johann Daniel Krafft (1624–1697) from Dresden, who could now produce it as well, and toured much of Europe with it, including England, where he met with Anglo-Irish natural philosopher, chemist, physicist, and inventor Robert Boyle (January 25, 1627 to December 31, 1691). The secret that it was produced from urine leaked out and first chemist and apothecary Johann von Löwenstern-Kunckel (1630—prob. March 20, 1703) in Sweden (1678) and later Boyle in London (1680)

FIGURE 4.47 Henning Brand discovering phosphorus. https://wellcomecollection.org/works/fff6s7vk, CC-BY-4.0.

also were successful to make phosphorus, the latter possibly with the aid of his assistant, Ambrose Godfrey-Hanckwitz (1660–January 15, 1741), who later made a business of the production of phosphorus. Boyle said that Krafft provided no information as to the synthesis of phosphorus except that it was obtained from "somewhat that belonged to the body of man." This gave Boyle an important hint, so that he, too, achieved to produce phosphorus and published the method of its production. Later he enhanced Brand's process by using sand (mainly quartz, SiO_2, grain) in the reaction (still using urine as base material),

$$4NaPO_3 + 2SiO_2 + 10C \rightarrow 2Na_2SiO_3 + 10CO + P_4$$

Robert Boyle was the first to use phosphorus to ignite sulfur-tipped wooden splints, forerunners of the modern matches, in 1680. Phosphorus was the 13th element to be discovered. For this reason, and because of its use in explosives, poisons, and nerve agents, it is occasionally mentioned as "the Devil's element." In 1769 Johan Swedish chemist and metallurgist Gottlieb Gahn (August 19, 1745 to December 8, 1818) and Swedish Pomeranian and German pharmaceutical chemist Carl Wilhelm Scheele (December 9, 1742 to May 21, 1786) proved that calcium phosphate ($Ca_3(PO_4)_2$) is found in bones, and they produced elemental phosphorus from bone ash. French nobleman and chemist Antoine Lavoisier (August 26, 1743 to May 8, 1794) identified phosphorus as an element in 1777. Bone ash was the main source of phosphorus until the 1840s. The process began by roasting bones and then used clay retorts enclosed in a very hot brick furnace to distill out the highly toxic elemental phosphorus product. Alternately, precipitated phosphates could be produced from ground-up bones that had been degreased and treated with strong acids. White phosphorus could at that point be produced by heating the precipitated phosphates, mixed with ground coal or charcoal in an iron pot, and distilling off phosphorus vapor in a retort (Thomson, 1870, page 416). Carbon monoxide and other flammable gases formed as part of the reduction process were burned off in a flare stack. In the 1840s, world phosphate production became based on the mining of tropical island deposits formed from bird and bat guano (Box 4.6). These developed into a significant resource of phosphates for fertilizer in the latter half of the 19th century. Phosphate rock, which usually contains calcium phosphate, was first used in 1850 to produce phosphorus, and subsequent to the introduction of the electric arc furnace by Scottish chemist James Burgess Readman (c.1850–1927) in 1888 (patented 1889), elemental phosphorus production transferred from the bone ash heating, to electric arc production from

BOX 4.6 Phosphate minerals in guano

Guano is the accumulation of feces of certain animals in time and space. In caves, it mainly comes from birds, crickets, and bats. Guano from bats is the more common and most diversified type occurring in caves, particularly in the Neotropical regions. Bird guano regularly has an undulating layered appearance and a spongy microstructure consisting of dark yellowish cryptocrystalline apatite, $Ca_5(PO_4)_3(Cl,F,OH)$. Microcrystalline struvite, $(NH_4)Mg(PO_4).6H_2O$, can also be a major constituent of fresh guano. In contrast, fresh bat guano does not contain apatite, but it is one of its first alteration products and can occur as groundmass material, light yellow nodules, or incomplete infillings of planar voids. Like bird guano deposits, the apatite is microcrystalline. Ardealite, $Ca_2(HPO_4)(SO_4)\cdot 4H_2O$, brushite, $Ca(HPO_4)\cdot 2H_2O$ (Fig. 4.48), struvite, and newberyite, $Mg(HPO_4)\cdot 3H_2O$, are other minerals related to guano, but they are rarely preserved. Newberyite has an acicular crystal habit with rosette-like aggregates; it has been found in altered cave sediments from archeological sites. Phosphate deposits derived from guano represent only a small amount of the world reserves of phosphate rock. Insular deposits are common in warm- or semiarid regions with large bird populations either currently or in the recent past. The most important deposits, now essentially depleted, are on larger islands over 50 m above sea-level, such as Nauru and Christmas Island, and are believed to be older than $\sim$1 million years. Here solutions derived from overlying bird droppings have percolated into the bedrock, where minerals such as apatite, whitlockite, $Ca_9Mg(PO_4)_6(HPO_4)$, crandallite, $CaAl_3(PO_4)(PO_3OH)(OH)_6$, and millisite, $(Na,K)CaAl_6(PO_4)_4(OH)_9\cdot 3H_2O$ have crystallized. Cave deposits resulting from bat droppings are more of interest for the unusual mineralogy than for their commercial value. Such deposits are mostly in limestone caves, with some in lava-tube caves. The reactions involved in forming these minerals are complex but typically begin with leaching of very soluble nitrogen from the guano. This leaves phosphorus to react with whatever cations are available from the surrounding rocks. The ensuing sequence of minerals can be stratified within the guano. Typical cave phosphates, for example, brushite, carbonate-rich hydroxylapatite, $Ca_5(PO_4,CO_3)_3(OH,O)$ and fluorapatite, $Ca_5(PO_4,CO_3)_3(F,OH)$, taranakite, $(K,NH_4)Al_3(PO_4)_3(OH)\cdot 9H_2O$, and variscite, $AlPO_4\cdot 2H_2O$, generally occur as powdery nodules within the guano or as coatings on bedrock or cave walls. Distinct crystals of minerals, such as newberyite and struvite, are rare, with a notable occurrence in lava caves at Skipton, Victoria.

FIGURE 4.48 Scanning electron microscope image of synthetic brushite, $Ca(HPO_4)\cdot 2H_2O$, crystals.

phosphate rock. After the exhaustion of world guano sources, around the same time, mineral phosphates turn into the foremost source of phosphate fertilizer production. Phosphate rock production significantly increased after World War II and is still the principal worldwide source of phosphorus and phosphorus chemicals today.

4.5.2 Mining, production, and major minerals

Phosphorus has a concentration in the earth's crust of about 0.1% by weight. It is not found free in nature but is widely found in many minerals, usually as phosphates. Inorganic phosphate rock, which is partially made of apatite (a group of minerals being, generally, pentacalcium triorthophosphate fluoride (hydroxide), $Ca_5(PO_4)_3(F,OH)$), is today the principal commercial source of this element. About 50% of the worldwide phosphorus reserves are in the Arab nations. Large deposits of apatite are also in China, Russia, Morocco, Florida, Idaho, Tennessee, Utah, and elsewhere. Most of the production of phosphorus-containing material is for use as agriculture fertilizers. For this purpose, phosphate minerals are converted to phosphoric acid. There are two distinct chemical processes, the major one being the reaction of phosphate minerals with sulfuric acid. The other process uses white phosphorus, which may be produced by reaction and distillation from very low-grade phosphate sources. The white phosphorus is subsequently oxidized to phosphoric acid and then neutralized with a base to give the phosphate salts. Phosphoric acid produced from white phosphorus is relatively pure and is the main route to produce phosphates for all purposes, including detergent production. In the early 1990s, Albright and Wilson's purified wet phosphoric acid business (Albright and Wilson was founded in 1856 as a United Kingdom manufacturer of potassium chlorate and white phosphorus for the match industry. For much of its first 100 years of existence, phosphorus-derived chemicals formed most of its products.) was being badly affected by phosphate rock sales by China and the entry of their long-standing Moroccan phosphate suppliers into the purified wet phosphoric acid business. Calcium phosphate (phosphate rock), mostly mined in Florida and North Africa, can be heated to 1,200–1,500 with sand, which is mostly SiO_2, and coke (refined coal) to produce vaporized P_4. The product is then condensed into a white powder under water to prevent oxidation by air. Even under water, white phosphorus is slowly converted to the more stable red phosphorus allotrope. The chemical equation for this process when starting with fluorapatite, a common phosphate mineral, is:

$$4Ca_5(PO_4)_3F + 18SiO_2 + 30C \rightarrow 3P_4 + 30CO + 18CaSiO_3 + 2CaF_2$$

Side products from this process include ferrophosphorus, a crude form of Fe_2P, resulting from iron impurities in the mineral precursors. The silicate slag is a suitable construction material. The fluoride is occasionally recovered for use in water fluoridation. More challenging is a "mud" containing substantial amounts of white phosphorus. Production of white phosphorus takes place in large facilities partly due to the fact that it is energy-intensive. The white phosphorus is transported in molten form. An alternative process through which elemental phosphorus is produced includes applying at high temperatures (1500°C):

$$2Ca_3(PO_4)_2 + 6SiO_2 + 10C \rightarrow 6CaSiO_3 + 10CO + P_4$$

Historically, prior to the advancement of mineral-based extractions, white phosphorus was extracted on an industrial scale from bone ash. In this process, the tricalcium phosphate in bone ash is converted to monocalcium phosphate with sulfuric acid:

$$Ca_3(PO_4)_2 + 2H_2SO_4 \rightarrow Ca(H_2PO_4)_2 + 2CaSO_4$$

The monocalcium phosphate is subsequently dehydrated to the corresponding metaphosphate:

$$Ca(H_2PO_4)_2 \rightarrow Ca(PO_3)_2 + 2H_2O$$

After igniting to a white heat with charcoal, calcium metaphosphate produces two-thirds of its weight of white phosphorus while one-third of the phosphorus is left in the residue as calcium orthophosphate:

$$3Ca(PO_3)_2 + 10C \rightarrow Ca_3(PO_4)_2 + 10CO + P_4$$

In total there are just under 700 minerals containing P, the majority of them in the form of phosphates. Within the class of the elements, a number of phosphides can be found, such as barringerite ($(Fe,Ni)_2P$) and nickelphosphide ($(Ni,Fe)_3P$). The only halide containing P is bøggildite ($Na_2Sr_2Al_2PO_4F_9$). Similarly, the only oxide is fritzscheite ($Mn(UO_2)_2(PO_4,VO_4)_2\cdot10H_2O$). The carbonate class contains five minerals with phosphate groups, for example, bradleyite ($Na_3Mg(CO_3)(PO_4)$) and crawfordite ($Na_3Sr(CO_3)(PO_4)$). Two borates are known with phosphate groups in their structure, lüneburgite ($Mg_3[B_2(OH)_6](PO_4)_2\cdot6H_2O$) and seamanite $\left(Mn_3^{2+}[B(OH)_4](PO_4)(OH)_2\right)$. Within the sulfate class, five minerals contain P, for example, embreyite ($Pb_5(CrO_4)_2(PO_4)_2\cdot H_2O$) and vauquelinite ($Pb_2Cu(CrO_4)(PO_4)(OH)$). The silicate class has 18 minerals containing P, for example, bornemanite ($Na_6BaTi_2Nb(Si_2O_7)_2(PO_4)O_2(OH)F$), lomonosovite ($Na_5Ti_2(Si_2O_7)(PO_4)O_2$), and steenstrupine-(Ce) $\left(Na_{14}Mn_2^{2+}(Fe^{3+},Mn^{3+})_2Ce_6(Zr,Th)(Si_6O_{18})_2(PO_4)_6(HPO_4)(OH)_2\cdot2H_2O\right)$. The largest group of minerals is found under the phosphates with more than 500 different minerals. Some of the best-known of these include minerals such as amblygonite ($LiAl(PO_4)F$), augelite ($Al_2(PO_4)(OH)_3$) (Fig. 4.49), autunite ($Ca(UO_2)_2(PO_4)_2\cdot11H_2O$) (Fig. 4.50), brazilianite ($NaAl_3(PO_4)_2(OH)_4$) (Fig. 4.51), cacoxenite ($Fe_{24}^{3+}AlO_6(PO_4)_{17}(OH)_{12}\cdot75H_2O$), fluorapatite ($Ca_5(PO_4)_3F$) (Fig. 4.52), herderite ($CaBePO_4(F,OH)$), lazulite ($MgAl_2(PO_4)_2(OH)_2$), monazite series ($(Ce,La,Nd,Sm)(PO_4)$), montebrasite ($Li,Al(PO_4)(OH)$) (Fig. 4.53), pyromorphite ($Pb_5(PO_4)_3Cl$) (Fig. 4.54), scholzite

FIGURE 4.49 Augelite, $Al_2(PO_4)(OH)_3$, apple green rhombic crystal to 4 mm on quartz, SiO_2. Mundo Nuevo, La Libertad dept., Peru.

FIGURE 4.50 Autunite, $Ca(UO_2)_2(PO_4)_2 \cdot 11H_2O$, bright yellow-green crystals to 5 mm. Saint-Symphorien, near Autun, Saône-et-Loire, Burgundy, France.

($CaZn_2(PO_4)_2 \cdot 2H_2O$) (Fig. 4.55), sincosite ($Ca(VO)_2(PO_4)_2 \cdot 5H_2O$) (Fig. 4.56), strengite ($FePO_4 \cdot 2H_2O$), torbernite ($Cu(UO_2)_2(PO_4)_2 \cdot 12H_2O$), turquoise ($Cu(Al,Fe^{3+})_6(PO_4)_4(OH)_8 \cdot 4H_2O$) (Fig. 4.57), variscite ($AlPO_4 \cdot 2H_2O$) (Fig. 4.58), vauxite ($Fe^{2+}Al_2(PO_4)_2(OH)_2 \cdot 6H_2O$), vivianite ($Fe^{2+}{}_3(PO_4)_2 \cdot 8H_2O$), and wavellite ($Al_3(PO_4)_2(OH,F)_3 \cdot 5H_2O$) (Fig. 4.59).

4.5.3 Chemistry

Twenty-three isotopes of P have been recognized, ranging from ^{24}P to ^{46}P. Of these only ^{31}P is stable and is hence found at 100% abundance. The half-integer nuclear spin and high abundance of ^{31}P make ^{31}P NMR spectroscopy a

FIGURE 4.51 Brazilianite, $NaAl_3(PO_4)_2(OH)_4$, yellow-green 2 cm crystal. Telirio mine, Linopolis, Minas Gerais, Brazil.

FIGURE 4.52 Fluorapatite, $Ca_5(PO_4)_3F$, purple, well-formed crystals to 2.5 mm. King Lithia mine, Pennington Co., South Dakota, United States.

very valuable analytical tool in research on P-containing samples. Two radioactive isotopes of P have half-lives appropriate for biological research applications. ^{32}P, a β^--emitter forming ^{32}S with a half-life of 14.28 days, routinely used in life-science laboratories, mostly to produce radiolabelled DNA and RNA probes, for example, for use in Northern blots or Southern blots. ^{33}P, a β^--emitter forming ^{33}S with a half-life of 25.3 days. The latter is used in life-science laboratories in applications where lower energy β emissions are beneficial such as DNA sequencing. The high energy β particles from ^{32}P can penetrate the skin and corneas and any ^{32}P ingested, inhaled, or absorbed is quickly incorporated into bone and nucleic acids. Therefore, the Occupational Safety and Health Administration in the United States, and similar institutions in other developed countries require employees working with ^{32}P to wear lab coats, disposable gloves, and safety glasses or goggles to protect the eyes, and avoid working directly over open containers. In addition, monitoring personal, clothing, and surface contamination is essential. Shielding necessitates special consideration. The high energy of the β particles gives rise to secondary emission of X-rays via Bremsstrahlung in dense shielding

FIGURE 4.53 Faint yellow montebrasite, $LiAl(PO_4)(OH)$, to 3.5 cm, typically considered a primary phosphate. Telirio mine, Linopolis, Minas Gerais, Brazil.

FIGURE 4.54 Pyromorphite, $Pb_5(PO_4)_3Cl$, dark grass green, lustrous, stubby 1 mm crystals. Chaillac, Indre, Centre, France.

materials (e.g., lead). Therefore, the radiation has to be shielded with low-density materials, for example, acrylic or other plastics, water, or (when transparency is not essential), even wood.

Phosphorus is a solid nonmetal at room temperature and a poor conductor of heat and electricity. It belongs to group 15 and has the electronic configuration $[Ne]3s^2 3p_x^1 3p_y^1 3p_z^1$. It has three unpaired electrons just like nitrogen above it. However, P has available low-lying vacant 3d orbitals and this accounts for the predominant oxidation states III and V in phosphorus chemistry (Table 4.5). This allows for valency expansion where P is bonded to 5 or 6 atoms and is surrounded by 10 or 12 pairs of electrons, in exception to the octet rule. Thus, PF_5 or PF_6 are known but the analogous NF_5 or NF_6 have not been prepared. The chemistry of phosphorus is extensive and varied transcending the traditional boundaries of inorganic chemistry due to its propensity to form organophosphorus compounds and due to the number of crucial roles it plays in the biochemistry of living things. Only a select few examples of this extensive chemistry is featured in this section.

Phosphorus exists in several allotropic forms; the most common ones are called white, red, and black phosphorus based on how these forms appear. The reactivity of phosphorus depends markedly upon which allotrope is being

FIGURE 4.55 Divergent group of colorless scholzite, $CaZn_2(PO_4)_2{\cdot}2H_2O$, crystals to 1.5 cm. Reaphook Hill, Flinders Ranges, South Australia, Australia.

FIGURE 4.56 Green thin platy sincosite, $Ca(VO)_2(PO_4)_2{\cdot}5H_2O$, crystals to 0.5 mm. South Rasmussen Ridge mine, Soda Springs, Caribou Co., Idaho, United States.

studied. Moreover, an increase in catenation in the structure notably diminishes both reactivity and solubility as in the case of the polymeric red and black forms. The most common form is the waxy, cubic, white form, α-P_4 (d 1.8232 $g{\cdot}cm^{-3}$ at 20°C). This is usually formed by condensation from the gaseous or liquid states. This is also the

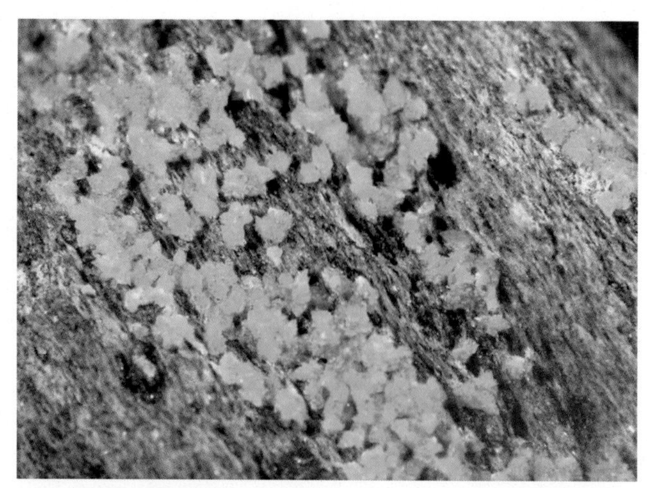

FIGURE 4.57 Turquoise, $Cu(Al,Fe^{3+})_6(PO_4)_4(OH)_8{\cdot}4H_2O$, less than 1 mm crystals. Lynch Station, Campbell Co., Virginia, United States.

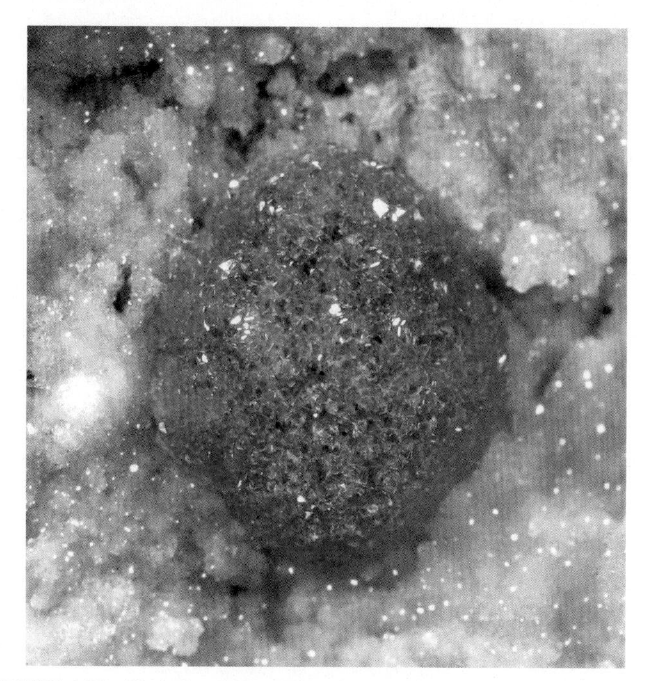

FIGURE 4.58 Variscite, $AlPO_4{\cdot}2H_2O$, a crystal "ball" to 9 mm. Goiás, Brazil.

most volatile and reactive solid form and thermodynamically the least stable. The alpha form of P_4 is converted to the white, hexagonal, β-form (d 1.88 g cm^{-3}). The solubility of white phosphorus varies in different solvents. It is not soluble in water and most organic solvents but exceedingly soluble in carbon disulfide (CS_2). It also dissolves in PCl_3, $POCl_3$, liquid SO_2, liquid NH_3, and benzene. Heating white P_4 for several days in the absence of air converts it to amorphous red phosphorus. It is denser than white P_4 (d 2.16 g cm^{-3}), has a much higher melting point, and is much less reactive. Suitable heat treatment converts amorphous red phosphorus to various crystalline forms. Black phosphorus is the thermodynamically most stable form of the element. The three crystalline forms of this allotrope have higher density than red phosphorus (orthorhombic 2.69, rhombohedral 3.56, and cubic 3.88 g cm^{-3}). The properties of this allotrope are analogous to graphite having puckered sheets of linked atoms, appearing flaky black, and showing good electrical conductivity.

FIGURE 4.59 Blue wavellite, $Al_3(PO_4)_2(OH,F)_3{\cdot}5H_2O$, with green zones. The largest ball at the center measures 1 cm across. Dug Hill, Avant, Garland Co., Arkansas, United States.

TABLE 4.5 Phosphorus properties.

Appearance	Colorless, waxy white, yellow, scarlet, red, violet, black
Allotropes	White, red, several others
Standard atomic weight $A_{r,std}$	30.973
Block	P-block
Element category	Reactive nonmetal
Electron configuration	[Ne] $3s^2$ $3p^3$
Phase at STP	Solid
Melting point	White 44.15°C, red ~590°C
Boiling point	White 280.5°C
Density (near r.t.)	White: 1.823 g/cm^3 Red: ≈2.2–2.34 g/cm^3 Violet: 2.36 g/cm^3 Black: 2.69 g/cm^3
Heat of fusion	White: 0.66 kJ/mol
Heat of vaporization	White: 51.9 kJ/mol
Molar heat capacity	White: 23.824 J/(mol·K)
Oxidation states	**−3**, −2, −1, +1, +2, **+3**, +4, **+5**
Ionization energies	1st: 1011.8 kJ/mol 2nd: 1907 kJ/mol 3rd: 2914.1 kJ/mol
Covalent radius	107 ± 3 pm
Van der Waals radius	180 pm

STP, Standard temperature and pressure.
Bold font indicates main oxidation state.

4.5.3.1 Reaction in air, oxygen, nitrogen, and hydrogen

White phosphorus reacts with moist air to show chemiluminescence (glowing in the dark). It ignites spontaneously in air at around 35°C to form tetraphosphorus decaoxide. When the oxidation atmosphere is controlled to contain less oxygen such as a mixture of 75% O_2 and 25% N_2 at 90 mm Hg and 50°C, tetraphosphorus hexaoxide is formed.

$$P_4(s) + 5O_2(g) \rightarrow P_4O_{10}(s)$$
$$P_4(s) + 3O_2(g) \rightarrow P_4O_6(s)$$

Red and yellow phosphorus can be made to react with hydrogen using a plasma stream of electronically excited hydrogen, H_2^* to form phosphine, PH_3.

$$P_4(s) + 6H_2^*(g) \rightarrow 4PH_3(g)$$

The PN molecule is considered unstable, but is a product of crystalline phosphorus nitride decomposition at 827°C (1100 K). Likewise, H_2PN is considered unstable, and phosphorus nitride halogens like F_2PN, Cl_2PN, Br_2PN, and I_2PN oligomerise into cyclic polyphosphazenes. For instance, compounds with the general formula $(PNCl_2)_n$ exist primarily as rings such as the trimer hexachlorophosphazene. The phosphazenes are formed by reacting phosphorus pentachloride with ammonium chloride:

$$PCl_5 + NH_4Cl \rightarrow 1/n(NPCl_2)_n + 4HCl$$

When the chloride groups are replaced by alkoxide (RO^-), a family of polymers is formed with potentially useful properties. Binary nitrides or compounds containing only P and N have been reported but their preparations follow routes other than the direct combination of the elements. The high stability of the N_2 molecule makes preparation of binary phosphorus-nitrogen compounds from the elements difficult, requiring high temperatures and a catalyst. For example, triphosphorus pentanitride, P_3N_5, has been prepared from the elements in electric discharges. More conveniently, it is prepared by the reaction between phosphorus(V) and nitrogen compounds such as sodium azide. Phosphorus mononitride, PN, is a product of the decomposition of P_3N_5.

$$3P_4 + 10N_2 \rightarrow 4P_3N_5$$
$$3PCl_5 + 15NaN_3 \rightarrow P_3N_5 + 15NaCl + 5N_2$$
$$2P_3N_5 \rightarrow 6PN + 2N_2$$

4.5.3.2 Reaction with halogens

All four symmetrical trihalides are well known: gaseous PF_3, the yellowish liquids PCl_3 and PBr_3, and the solid PI_3. These compounds are moisture sensitive, hydrolysing to form phosphorous acid. White phosphorus, P_4, reacts vigorously with all the halogens at room temperature to form phosphorus trihalides.

$$P_4(s) + 6F_2(g) \rightarrow 4PF_3(g)$$
$$P_4(s) + 6Cl_2(g) \rightarrow 4PCl_3(l)$$
$$P_4(s) + 6Br_2(g) \rightarrow 4PBr_3(l)$$
$$P_4(s) + 6I_2(g) \rightarrow 4PI_3(g)$$

Phosphorus trifluoride (PF_3), a colorless and odorless gas, is highly toxic, comparable to phosgene (IUPAC name carbonyl dichloride, $COCl_2$), and reacts slowly with water. Its key application is as a ligand in metal complexes. As a ligand, it parallels carbon monoxide in metal carbonyls. Its toxicity is due to its binding with the iron in blood hemoglobin in a comparable manner to carbon monoxide. Phosphorus trifluoride is typically produced from phosphorus trichloride through halogen exchange utilizing different fluorides such as hydrogen fluoride, calcium fluoride, arsenic trifluoride, antimony trifluoride, or zinc fluoride:

$$2PCl_3 + 3ZnF_2 \rightarrow 2PF_3 + 3ZnCl_2$$

PF_3 hydrolyzes in particular at high pH, but it is less hydrolytically sensitive than PCl_3. It does not attack glass except at high temperatures, and anhydrous potassium hydroxide may be employed to dry it with little loss. With hot metals, phosphides and fluorides are formed. With Lewis bases such as ammonia addition products (adducts) are created, while PF_3 is oxidized by oxidizing agents such as bromine or potassium permanganate. As a ligand for transition metals, PF_3 is a strong π-acceptor. It can form numerous metal complexes with metals in low oxidation states. PF_3 forms several complexes for which the corresponding CO derivatives are unstable or nonexistent. Thus, $Pd(PF_3)_4$ can be formed, but $Pd(CO)_4$ not. Such complexes are typically produced directly from the related metal carbonyl compound, with loss of CO. However, nickel metal reacts directly with PF_3 at 100°C under 35 MPa pressure to form $Ni(PF_3)_4$, which is similar to $Ni(CO)_4$. $Cr(PF_3)_6$, the analogue of $Cr(CO)_6$, can be formed from dibenzenechromium:

$$Cr(C_6H_6)_2 + 6PF_3 \rightarrow Cr(PF_3)_6 + 2C_6H_6$$

PF_3 has an F–P–F bond angle of ~96.3°. PF_3 is mostly used as a ligand in transition metal complexes. Phosphorus trichloride, PCl_3, is a toxic and volatile liquid which reacts violently with water while releasing HCl gas. It has a trigonal pyramidal shape, owing to the lone pairs on the phosphorus. PCl_3 is a precursor to other phosphorus compounds, undergoing oxidation to phosphorus pentachloride (PCl_5), thiophosphoryl chloride ($PSCl_3$), or phosphorus oxychloride ($POCl_3$). PCl_3 is the precursor to organophosphorus compounds that contain one or more P(III) atoms, most notably phosphites and phosphonates. These compounds do not usually have the chlorine atoms found in PCl_3. PCl_3 reacts vigorously with water producing phosphorous acid, H_3PO_3 and HCl:

$$PCl_3 + 3H_2O \rightarrow H_3PO_3 + 3HCl$$

A large number of comparable substitution reactions have been shown to exist, the most important of which is the production of phosphites by reaction with alcohols or phenols. For instance, with phenol, triphenyl phosphite is formed:

$$3PhOH + PCl_3 \rightarrow P(OPh)_3 + 3HCl$$

where "Ph" stands for phenyl group, $-C_6H_5$. Alcohols such as ethanol react likewise in the presence of a base such as a tertiary amine:

$$PCl_3 + 3EtOH + 3R_3N \rightarrow P(OEt)_3 + 3R_3NH^+Cl^-$$

In the absence of a base, however, the reaction proceeds with the following stoichiometry to form diethylphosphite:

$$PCl_3 + 3EtOH \rightarrow (EtO)_2P(O)H + 2HCl + EtCl$$

PCl_3 undergoes many substitution reactions and is the main source of organophosphorus compounds. Compounds derived from PCl_3 notably include PR_3, PR_nCl_{3-n}, $PR_n(OR)_{3-n}$, $(PhO)_3PO$, and $(RO)_3PS$, which are produced in the ton scale per annum.

Phosphorus tribromide, PBr_3, is a colorless liquid that fumes in moist air due to hydrolysis and has a penetrating odor. It is utilized for the conversion of alcohols to alkyl bromides. PBr_3 is produced by reacting red phosphorus with bromine. An excess of phosphorus is used in order to stop the formation of PBr_5:

$$2P + 3Br_2 \rightarrow 2PBr_3$$

Since the reaction is extremely exothermic, it is frequently conducted in the presence of a diluent such as PBr_3. PBr_3, similar to PCl_3 and PF_3, has both properties of a Lewis base and a Lewis acid. For instance, with a Lewis acid such as boron tribromide it produces stable 1:1 adducts for example, $Br_3B{\cdot}PBr_3$. PBr_3 can also react as an electrophile or Lewis acid in many of its reactions, for instance with amines. The most significant reaction of PBr_3 is with alcohol, where it replaces an OH group with a bromine atom to form an alkyl bromide. All three bromides can be transferred.

$$PBr_3 + 3ROH \rightarrow 3RBr + HP(O)(OH)_2$$

The mechanism involves formation of a phosphorus ester (to form a good leaving group), followed by an S_N2 substitution. [The S_N2 reaction is a type of reaction mechanism that is common in organic chemistry. In this mechanism, one bond is broken and one bond is formed synchronously, that is, in one step. S_N2 is a kind of nucleophilic substitution reaction mechanism. Since two reacting species are involved in the slow (rate-determining) step, this leads to the term **s**ubstitution **n**ucleophilic (bi-molecular) or S_N2; the other major kind is S_N1.] Because of the S_N2 substitution step, the reaction generally works well for primary and secondary alcohols, but fails for tertiary alcohols. If the reacting carbon center is chiral, the reaction generally occurs with inversion of configuration at the alcohol alpha carbon, as is usual with an S_N2 reaction. In a comparable reaction, PBr_3 also converts carboxylic acids to acyl bromides.

$$PBr_3 + 3RCOOH \rightarrow 3RCOBr + HP(O)(OH)_2$$

PBr_3 converts primary or secondary alcohols to alkyl bromides and catalyzes α-bromination of carboxylic acids, which are important intermediates in the synthesis of other important organic compounds.

Phosphorus pentabromide, PBr_5, is a reactive, yellow solid, which has the structure PBr_4^+ Br^- in the solid state but in the vapor phase is completely dissociated to PBr_3 and Br_2. Rapid cooling of this phase to -258°C (15 K) results in the formation of the ionic species phosphorus heptabromide ($[PBr_4]^+[Br_3]^-$). It can be utilized in organic chemistry to convert carboxylic acids to acyl bromides. It is highly corrosive. It decomposes above 100°C forming phosphorus tribromide and bromine:

$$PBr_5 \rightarrow PBr_3 + Br_2$$

Reversing this equilibrium to form PBr_5 by addition of Br_2 to PBr_3 is difficult in practice as the product is susceptible to further addition to produce phosphorus heptabromide (PBr_7). PBr_7 forms at normal conditions red prismatic crystals. PBr_7 can be formed via the sublimation of a mixture of phosphorus pentabromide and bromine.

$$PBr_5 + Br_2 \rightarrow PBr_7$$

The structure has of a PBr_4^+ cation paired with a tribromide (Br_3^-) anion, and the tribromide is non-symmetric.

Phosphorus triiodide, PI_3, is an unstable red solid which reacts violently with water. It is a common misconception that PI_3 is too unstable to be stored; it is, actually, commercially available. It is extensively utilized in organic chemistry to convert alcohols to alkyl iodides. In addition, it is a powerful reducing agent. Phosphorus also forms a lower iodide, P_2I_4, although the existence of PI_5 is doubtful at room temperature. The common method for the preparation of PI_5 is by the reaction of the elements, frequently through the addition of iodine to a solution of white phosphorus in carbon disulfide:

$$P_4 + 6I_2 \rightarrow 4PI_3$$

As another preparation method, PCl_3 may be converted to PI_3 by the action of hydrogen iodide or certain metal iodides. PI_3 reacts vigorously with water, forming phosphorous acid (H_3PO_3) and hydroiodic acid (HI), together with smaller amounts of phosphine (IUPAC name phosphane, PH_3) and P—P compounds. Alcohols similarly form alkyl iodides, this providing the most important application for PI_3. Phosphorus triiodide is a strong reducing agent, able to reduce sulfoxides to sulfides even at low temperatures.

4.5.3.3 Reactions with other elements

Phosphorus forms binary compounds with all elements except Sb, Bi, and the noble gases. Sulfur and the alkali metals also react vigorously with phosphorus on warming, and the element combines directly with all metals (except Bi, Hg, Pb) frequently with incandescence (e.g., Fe, Ni, Cu, Pt).

4.5.3.4 Reactions with aqueous acids and bases

Phosphorus reacts with aqueous alkali such as potassium hydroxide to produce phosphine and potassium hypophospite, $K_2H_2PO_2$. White phosphorus will react with concentrated nitric acid to produce phosphoric acid and nitrogen dioxide. The common oxides of phosphorus hydrolyze in water to give the weak acids, for example, H_3PO_3 from P_4O_6 and H_3PO_4 from P_4O_{10}.

4.5.3.5 Phosphate

In chemistry, a phosphate is an anion, salt, functional group or ester derived from a phosphoric acid. It generally denotes orthophosphate, a derivative of orthophosphoric acid H_3PO_4. The phosphate or orthophosphate ion $[PO_4]^{3-}$ is derived from phosphoric acid by the removal of three protons H^+. Removal of one or two protons results in the dihydrogen phosphate ion $[H_2PO_4]^-$ and the hydrogen phosphate ion $[HPO_4]^{2-}$ ion, respectively. These names are also applied to the salts of those anions, such as ammonium dihydrogen phosphate and trisodium phosphate. The phosphate ion contains a central phosphorus atom surrounded by four oxygen atoms in a tetrahedral coordination. It is the conjugate base of the hydrogen phosphate ion $H(PO_4)^{2-}$, which in its turn is the conjugate base of the dihydrogen phosphate ion $H_2(PO_4)^-$, which in its turn is the conjugate base of orthophosphoric acid, H_3PO_4. Numerous phosphates are not soluble in water at standard temperature and pressure. The sodium, potassium, rubidium, caesium, and ammonium phosphates are all water-soluble. Most other phosphates are only slightly soluble or are insoluble in water. As a rule, the hydrogen and dihydrogen phosphates are slightly more soluble than the corresponding phosphates. In water solution, orthophosphoric acid and its three derived anions coexist according to the dissociation and recombination equilibria below

$$H_3PO_4 \rightleftharpoons H_2PO_4^- + H^+ \; K_{a1} = [H^+][H_2PO_4^-]/[H_3PO_4] \approx 7.5 \times 10^{-3} \; pK_{a1} = 2.14$$
$$H_2PO_4^- \rightleftharpoons HPO_4^{2-} + H^+ \; K_{a2} = [H^+][HPO_4^{2-}]/[H_2PO_4^-] \approx 6.2 \times 10^{-8} \; pK_{a2} = 7.20$$
$$HPO_4^{2-} \rightleftharpoons PO_4^{3-} + H^+ \; K_{a3} = [H^+][PO_4^{3-}]/[HPO_4^{2-}] \approx 2.14 \times 10^{-13} \; pK_{a3} = 12.37$$

(values are at 25°C and 0 ionic strength). The pK_a values are the pH values where the concentration of each species is the same as that of its conjugate bases. At pH 1 or lower, the phosphoric acid is nearly undissociated. Around pH 4.7 (mid-way between the first two pK_a values) the dihydrogen phosphate ion, $[H_2PO_4]^-$, is basically the only species present. Around pH 9.8 (mid-way between the second and third pK_a values) the monohydrogen phosphate ion, $[HPO_4]^{2-}$, is the only species present. At or above pH 13, the acid is completely dissociated as the phosphate ion, $(PO_4)^{3-}$. This means that salts of the mono- and di-phosphate ions can be selectively crystallized from aqueous solution by setting the pH value to either 4.7 or 9.8. In effect, H_3PO_4, $H_2(PO_4)^-$ and $H(PO_4)^{2-}$ act as separate weak acids as the successive pK_a differ by more than 4. Phosphate can form many polymeric ions, for example, pyrophosphate), $(P_2O_7)^{4-}$, and triphosphate, $(P_3O_{10})^{5-}$. The various metaphosphate ions (which are typically long linear polymers) have an empirical formula of $(PO_3)^-$ and can be found in many compounds. In organic chemistry, phosphate or orthophosphate is an organophosphate, an ester of orthophosphoric acid of the form $PO_4RR'R''$ where one or more hydrogen atoms are substituted by organic groups, for example, trimethyl phosphate, $(CH_3)_3PO_4$. The term also refers to the trivalent functional group $OP(O\text{-})_3$ in such esters. Orthophosphates are especially important among the various phosphates due to their major roles in biochemistry, biogeochemistry, and ecology, and their economic importance for agriculture and industry. The addition and removal of phosphate groups (phosphorylation and dephosphorylation) form major steps in cell metabolism.

$$P_4 + 3KOH + 3H_2O \rightarrow PH_3 + 3KH_2PO_2$$
$$P_4 + 20HNO_3 \rightarrow 20NO_2 + 4H_3PO_4 + 4H_2O$$
$$P_4O_6 + 6H_2O \rightarrow 4H_3PO_3$$
$$P_4O_{10} + 6H_2O \rightarrow 4H_3PO_4$$

4.5.3.6 Organophosphorus compounds

Phosphorus shows a tendency to form innumerable organophosphorus compounds, which are defined as compounds containing P-C and P-O-C bonds. The numbers are many and are roughly categorized as derivatives of phosphorus(III) or (V) for the predominant classes or as derivatives of phosphorus(0), (I), or (II) for the less common classes. Examples of the derivatives of phosphorus(V) compounds include organophosphates, phosphonates and phosphinates, phosphine oxides, phosphonium salts, and phosphoranes. Phosphorus(III) derivatives include phosphites, phosphonites, phosphinites, phosphines, phosphaalkenes, and phosphaalkynes. The chemistry of these compounds is extensive and is beyond the scope of this book.

Phosphate esters have the general structure $P(=O)(OR)_3$ containing P(V). Missing a P − C bond, these compounds are in the technical sense not organophosphorus compounds but instead esters of phosphoric acid. Numerous derivatives are found naturally, for example, phosphatidylcholine. Phosphate esters are synthesized by alcoholysis of phosphorus oxychloride. A variety of mixed amido-alkoxo derivatives are known, one medically significant example being the anti-cancer drug cyclophosphamide ($C_{10}H_{19}O_6PS_2$). Also derivatives containing the thiophosphoryl group (P = S) include the pesticide malathion (diethyl 2-[(dimethoxyphosphorothioyl)sulfanyl]butanedioate, $C_{10}H_{19}O_6PS_2$). The organophosphates prepared on the largest scale are the zinc dithiophosphates. Phosphonates are esters of phosphonic acid and have the general formula $RP(=O)(OR')_2$. Phosphonates have many technical uses. A well-known example is glyphosate, better known as Roundup, $(HO)_2P(O)CH_2NHCH_2CO_2H$, a derivative of glycine and one of the most widely used herbicides. Bisphosphonates are a class of drugs to treat osteoporosis. The nerve gas agent sarin (IUPAC name (*RS*)-propan-2-yl methylphosphonofluoridate, $C_4H_{10}FO_2P$), containing both C−P and F−P bonds, is a phosphonate. Phosphinates have two P−C bonds, with the general formula $R_2P(=O)(OR')$. A commercially important example is the herbicide Glufosinate (IUPAC name (*RS*)-2-Amino-4-(hydroxy(methyl)phosphonoyl)butanoic acid, $C_5H_{12}NO_4P$). Like glyphosate, it has the structure $CH_3P(O)(OH)CH_2CH_2CH(NH_2)CO_2H$. The Michaelis−Arbuzov reaction is the main route for the synthesis of these compounds. For instance, dimethylmethylphosphonate is formed from the rearrangement of trimethylphosphite, which is catalyzed by methyl iodide. Phosphine oxides (designation $\sigma^4\lambda^5$) have the general structure $R_3P = O$ with formal oxidation state V. Phosphine oxides form hydrogen bonds and some are thus soluble in water. The P = O bond is very polar with a dipole moment of 4.51 D for triphenylphosphine oxide. Compounds related to phosphine oxides are for example, phosphine imides (R_3PNR') and related chalcogenides (R_3PE, E = S, Se, Te). These compounds are some of the most thermally stable organophosphorus compounds. Compounds with the general formula $[PR_4^+]X^-$ comprise the phosphonium salts. These species are tetrahedrally coordinated phosphorus(V) compounds. From a commercial perspective, the most important compound is tetrakis (hydroxymethyl)phosphonium chloride, $[P(CH_2OH)_4]Cl$, which is employed as a fire retardant in textiles. They are formed by the reaction of phosphine with formaldehyde in the presence of the mineral acid:

$$PH_3 + HX + 4CH_2O \rightarrow [P(CH_2OH)_4^+]X^-$$

Numerous phosphonium salts can be produced by alkylation and arylation of organophosphines:

$$PR_3 + R'X \rightarrow [PR_3R'^+]X^-$$

The methylation of triphenylphosphine is the first step in the preparation of the Wittig reagent. Phosphorus ylides are unsaturated phosphoranes, known as Wittig reagents, for example, $CH_2P(C_6H_5)_3$. The parent phosphorane ($\sigma^5\lambda^5$) is PH_5, which is unknown. Wittig reagents are typically produced from a phosphonium salt, which is in turn formed by the quaternization of triphenylphosphine with an alkyl halide. The alkylphosphonium salt is deprotonated with a strong base such as *n*-butyllithium:

$$[Ph_3P^+CH_2R]X^- + C_4H_9Li \rightarrow Ph_3P{=}CHR + LiX + C_4H_{10}$$

In addition to *n*-butyllithium (^{n}BuLi), other strong bases like sodium and potassium *t*-butoxide (^{t}BuONa, ^{t}BuOK), lithium, sodium and potassium hexamethyldisilazide (LiHMDS, NaHMDS, KHDMS, where HDMS = $N(SiMe_3)_2$), or sodium hydride (NaH) are often used. For stabilized Wittig reagents bearing conjugated electron-withdrawing groups, even relatively weak bases like aqueous sodium hydroxide or potassium carbonate can be used. Related compounds containing both halide and organic substituents on phosphorus are fairly common. Those with five organic substituents are rare, although $P(C_6H_5)_5$ is known, being derived from $P(C_6H_5)_4^+$ by reaction with phenyllithium. These compounds feature tetrahedral phosphorus(V) and are thought of as relatives of phosphine oxides. They are also derived from phosphonium salts, but by deprotonation not alkylation. Phosphites, occasionally labeled as phosphite esters, have the general structure $P(OR)_3$ with oxidation state +3. Such species are formed during the alcoholysis of phosphorus trichloride:

$$PCl_3 + 3ROH \rightarrow P(OR)_3 + 3HCl$$

The reaction is general, thus a large number of these species are known. Phosphites are used in the Perkow reaction (an organic reaction in which a trialkyl phosphite ester reacts with a haloketone to form a dialkyl vinyl phosphate and an alkyl halide) and the Michaelis–Arbuzov reaction (the chemical reaction of a trivalent phosphorus ester with an alkyl halide to form a pentavalent phosphorus species and another alkyl halide). They also act as ligands in organometallic chemistry. Intermediate between phosphites and phosphines are phosphonites ($P(OR)_2R'$) and phosphinite ($P(OR)R'_2$). These compounds are formed via alcoholysis reactions of the corresponding phosphinous and phosphonous chlorides (($PClR'_2$) and PCl_2R', respectively). The parent compound of the phosphines is PH_3, called phosphine in the US and British Commonwealth, but phosphane (IUPAC preferred name) elsewhere. Substitution by one or more hydrogen centers by an organic substituent (alkyl, aryl), results in $PH_{3-x}R_x$, an organophosphine, commonly known as phosphines. The phosphorus atom in phosphines has a formal oxidation state −3 ($\sigma^3\lambda^3$) and are the phosphorus analogs of amines. Similar to amines, phosphines have a trigonal pyramidal molecular geometry but often with smaller C-E-C angles (E=N, P), at least in the absence of steric effects. The C–P–C bond angle is 98.6° for trimethyl phosphine increasing to 109.7° when the methyl groups are replaced by *tert*-butyl groups. When utilized as ligands, the steric bulk of tertiary phosphines is evaluated by their cone angle. The barrier to pyramidal inversion is also much higher than nitrogen inversion to occur, and hence phosphines with three different substituents can be resolved into thermally stable optical isomers. Phosphines are frequently less basic than the corresponding amines, for example, the phosphonium ion itself has a pK_a of −14 compared to 9.21 for the ammonium ion; trimethyl phosphonium has a pK_a of 8.65 compared to 9.76 for trimethylammonium. Nevertheless, triphenylphosphine (pK_a 2.73) is more basic than triphenylamine (pK_a −5), mainly due to the fact that the lone pair of the nitrogen in NPh_3 is partially delocalized into the three phenyl rings. Whereas the lone pair on nitrogen is delocalized in pyrrole, the lone pair on phosphorus atom in the phosphorus equivalent of pyrrole (phosphole) is not. The reactivity of phosphines is similar to that of amines with regard to nucleophilicity in the formation of phosphonium salts with the general structure $PR_4^+X^-$. This property is utilized in the Appel reaction for converting alcohols to alkyl halides. Phosphines are readily oxidized to the corresponding phosphine oxides, while amine oxides are less easily formed. Partly for this reason, phosphines are very rarely found in nature. From a commercial perspective, the most important phosphine is triphenylphosphine (IUPAC name tiphenylphosphane, $C_{18}H_{15}P$). It is produced via the reaction of chlorobenzene, PCl_3, with sodium. Phosphines of a more specialized nature are typically produced by other routes.

Phosphorus halides undergo nucleophilic displacement by organometallic reagents such as Grignard reagents. Conversely, some syntheses involve nucleophilic displacement of phosphide anion equivalents ("R_2P^-") by aryl- and alkyl halides. Primary (RPH_2) and secondary phosphines (RR'PH and R_2PH) add to alkenes in the presence of a strong base (e.g., KOH in DMSO). Comparable reactions occur involving alkynes. Base is not needed for electron-deficient alkenes (e.g., derivatives of acrylonitrile) and alkynes. Under free-radical conditions the P-H bonds of primary and secondary phosphines add across alkenes. AIBN or organic peroxides are utilized as initiators. Tertiary phosphine oxides and sulfides can be reduced with chlorosilanes and other reagents. Compounds with carbon phosphorus(III) multiple bonds are known as phosphaalkenes ($R_2C = PR$) and phosphaalkynes ($RC{\equiv}P$). They are comparable in structure, but not in reactivity, to imines ($R_2C = NR$) and nitriles ($RC{\equiv}N$), respectively. In the compound phosphorine, one carbon atom in benzene is replaced by phosphorus. Species of this type are relatively rare but for that reason are of interest to researchers. A general route for the synthesis of phosphaalkenes is via the 1,2-elimination of suitable precursors, initiated thermally or by base such as DBU (2,3,4,6,7,8,9,10-octahydropyrimido[1,2-*a*]azepine, $C_9H_{16}N_2$), DABCO (1,4-diazabicyclo[2.2.2]octane, $C_6H_{12}N_2$), or triethylamine (IUPAC name *N,N*-dethylethanamine, $N(CH_2CH_3)_3$). Thermolysis of Me_2PH produces $CH_2 = PMe$, an unstable species in the condensed phase. Compounds where phosphorus exists in a formal oxidation state of less than III, are rare, but examples are known for each class. Organophosphorus(0) species are debatably illustrated by the carbene adducts, $[P(NHC)]_2$, where NHC is an N-heterocyclic carbene. With the formulae $(RP)_n$ and $(R_2P)_2$, respectively, compounds of phosphorus(I) and (II) are formed by reduction of the related organophosphorus(III) chlorides:

$$5PhPCl_2 + 5Mg \rightarrow (PhP)_5 + 5MgCl_2$$

$$2Ph_2PCl + Mg \rightarrow Ph_2P\text{-}PPh_2 + MgCl_2$$

Diphosphenes, with the formula R_2P_2, formally contain phosphorus–phosphorus double bonds. These phosphorus(I) species are uncommon but are stable provided that the organic substituents are large enough to prevent catenation. Numerous mixed-valence compounds are known, such as. the cage $P_7(CH_3)_3$.

4.5.4 Major uses

Phosphorus is an essential plant nutrient (often the limiting nutrient), and the bulk of all production is in concentrated phosphoric acids for agriculture fertilizers, containing as much as 70%–75% P_2O_5. Its yearly demand is increasing almost twice as fast as the growth of the human population that resulted in a large increase in phosphate (PO_4^{3-}) manufacture in the second half of the 20th century. Synthetic phosphate fertilization is essential since P is vital to all living organisms; natural P-bearing compounds are nearly insoluble and hence inaccessible to plants, and the natural P cycle is very slow. Fertilizer is frequently in the form of superphosphate of lime, a mixture of calcium dihydrogen phosphate ($Ca(H_2PO_4)_2$) with calcium sulfate dihydrate (gypsum, $CaSO_4{\cdot}2H_2O$) formed by reacting sulfuric acid and water with calcium phosphate. Processing phosphate minerals with sulfuric acid for producing fertilizer is so valuable to the global economy that this is the most important industrial market for sulfuric acid and the largest industrial application of elemental S. White phosphorus is extensively used to produce organophosphorus compounds via intermediate P-chlorides and P-sulfides, P-pentasulfide, and P-sesquisulfide. Organophosphorus compounds have numerous applications, such as in plasticizers, flame retardants, pesticides, extraction agents, nerve agents, and water treatment. In addition P is a significant component in steel production, in the making of phosphor bronze, and in many other related products. P is added to metallic Cu during its smelting process where it can react with any oxygen present as an impurity in Cu and to form P-containing Cu (CuOFP) alloys with a higher hydrogen embrittlement [Cu alloys which contain oxygen can be embrittled if exposed to hot hydrogen. The hydrogen diffuses through the copper and reacts with inclusions of Cu_2O, forming H_2O (water), which then forms pressurized bubbles at the grain boundaries. This process can cause the grains to literally be forced away from each other and is known as steam embrittlement (because steam is produced, not because exposure to steam causes the problem).] resistance compared to normal Cu.

The first striking match with a P head was invented by French chemist Charles Sauria (April 25, 1812 to August 22, 1895) in 1830. These matches (and subsequent modifications) were made with heads of white phosphorus, an oxygen-releasing compound (potassium chlorate, $KClO_3$, lead dioxide, PbO_2, or sometimes nitrate), and a binder. They were poisonous to the workers in the factories [phossy jaw, formally known as phosphorus necrosis of the jaw, was an occupational disease affecting those who worked with white phosphorus (also known as yellow phosphorus) without proper safeguards.], sensitive to storage conditions, toxic if ingested, and hazardous when accidentally ignited on a rough surface. The manufacturing was banned in several countries between 1872 and 1925. The international Berne Convention, ratified in 1906, prohibited the use of white phosphorus in matches. As a result, the "strike-anywhere" matches were progressively replaced by "safety matches," wherein the white phosphorus was replaced by phosphorus sesquisulfide (P_4S_3), sulfur, or antimony sulfide. These matches are hard to ignite on any surface other than a special strip. The strip contains red phosphorus which heats up upon striking and subsequently reacts with the oxygen-releasing compound in the head, thereby igniting the flammable material of the head.

$$16KCl_3 + 3P_4S_3 \rightarrow 16KCl + 9SO_2$$

Sodium tripolyphosphate ($Na_5P_3O_{10}$) made from phosphoric acid is used in laundry detergents in some countries but banned for this use in others. This compound softens the water to improve the performance of the detergents and to stop pipe/boiler tube corrosion. Phosphates are utilized to produce special glasses for sodium lamps. Bone ash, calcium phosphate, is used in the manufacture of fine China. Phosphoric acid made from elemental phosphorus is used in food applications such as soft drinks, and as a starting material for food-grade phosphates, for example, monocalcium phosphate for baking powder and sodium tripolyphosphate. Phosphates are used to improve the characteristics of processed meat and cheese and in toothpaste. White phosphorus is used in military applications as incendiary bombs, for smoke screening as smoke pots and smoke bombs and in tracer ammunition. These days, military uses of white phosphorus are limited by international law. In trace amounts, phosphorus is used as a dopant for n-type semiconductors. ^{32}P and ^{33}P are used as radioactive tracers in biochemical laboratories.

Phosphate is a strong complexing agent for the hexavalent uranyl (UO_2^{2+}) species; hence, apatite, $Ca_5(PO_4)_3(Cl,F,OH)$, and other natural phosphates can be very rich in U. Tributylphosphate, $(CH_3CH_2CH_2CH_2O)_3PO$, is an organophosphate soluble in kerosene used to extract uranium in the PUREX process for reprocessing spent nuclear fuel. PUREX is a chemical method used to purify fuel for nuclear reactors or nuclear weapons. It is an acronym standing for Plutonium Uranium Redox EXtraction. PUREX is the de facto standard aqueous nuclear reprocessing method for the recovery of U and Pu from used ("spent," or "irradiated") nuclear fuel. It is based on liquid–liquid extraction ion-exchange. First, the fuel is dissolved in nitric acid at a concentration of ca. 7 M. Solids are removed by filtration lest they form emulsions. The organic solvent consists of 30% tributylphosphate (TBP) in a hydrocarbon, such as kerosene. U ions are extracted as $UO_2(NO_3)_2 \cdot 2TBP$ complexes, and Pu as similar complexes. Other fission products remain in the aqueous phase, including americium (Am) and curium (Cm). Pu is separated from U by treating the kerosene solution with reducing agents to convert the Pu to the $+3$ oxidation state. Typical reducing agents include N,N-diethyl-hydroxylamine, ferrous sulfamate, and hydrazine. The Pu^{3+} passes into the aqueous phase. The U is stripped from the kerosene solution by back extraction into nitric acid at a concentration of *ca.* 0.2 M.

4.6 16 S — Sulfur

4.6.1 Discovery

Being widely available in elemental form, sulfur was known in ancient times and is even mentioned in the Torah (Genesis). English translations of the Bible usually referred to burning sulfur as "brimstone," resulting in the term "fire-and-brimstone" sermons, in which listeners are reminded of the fate of eternal damnation that waits for the unbelieving and unrepentant. It is from this part of the Bible that Hell is implied to "smell of sulfur" (likely because of its association with volcanic activity). As stated by the Ebers Papyrus (also known as Papyrus Ebers, an Egyptian medical papyrus of herbal knowledge dating to circa 1550 BCE), a sulfur balm was used in ancient Egypt to treat granular eyelids. Sulfur was employed for fumigation in preclassical Greece as is written in the Odyssey. Roman author, naturalist, and natural philosopher Pliny the Elder (CE 23–79) described sulfur in book 35 of his Natural History, stating that its best-known source was the island of Melos (Fig. 4.60). He indicated its use for fumigation, medicine, and bleaching cloth. A natural form of sulfur known as shiliuhuang (石硫黄) was recognized in China since the 6th century BCE and found in Hanzhong. By the 3rd century, the Chinese discovered that sulfur could be obtained from pyrite, FeS_2. Chinese Daoists (Taoism, or Daoism, is a religious or philosophical tradition of Chinese origin which emphasizes living in harmony with the Tao) were interested in sulfur's flammability and its reactivity with certain metals, however, its earliest practical applications originated in traditional Chinese medicine. A Song dynasty military treatize of 1044 CE defined different formulations for Chinese black powder, which is a mixture of potassium nitrate (KNO_3), charcoal, and sulfur. It is still an ingredient of black gunpowder. Indian alchemists, practitioners of "the science of mercury" (sanskrit rasaśāstra, रसशास्त्र), wrote widely about the use of sulfur in alchemical operations with mercury, from the 8th century CE onwards. In the rasaśāstra tradition, sulfur was known as "the smelly" (sanskrit gandhaka, गन्धक). Early European alchemists gave sulfur a unique alchemical symbol, a triangle at the top of a cross. In traditional skin treatment, elemental sulfur was applied (mainly in creams) to ease such conditions as scabies, ringworm, psoriasis, eczema, and acne. The mechanism of action is still not well understood—though elemental sulfur does oxidize

FIGURE 4.60 Pliny the Elder, by an unknown 19th-century artist: No contemporary depiction of Pliny is known to survive.

slowly to sulfurous acid, which is (through the action of sulfite) a mild reducing and antibacterial agent. In 1777 the French nobleman and chemist Antoine Lavoisier (August 26, 1743 to May 8, 1794) helped persuade scientists of that time that sulfur was an element, not a compound. Sulfur deposits in Sicily were the leading source for more than a century. By the late 18th century, about 2,000 tons per year of sulfur were imported into Marseilles, France, to produce sulfuric acid for use in the Leblanc process, an early industrial process for the production of soda ash (sodium carbonate) used throughout the 19th century, named after its inventor Nicolas Leblanc (French chemist and surgeon, December 6, 1742 to January 16, 1806). In industrializing Britain, with the rescinding of tariffs on salt in 1824, demand for sulfur from Sicily increased strongly. The growing British control and exploitation of the mining, refining, and transportation of the sulfur, together with the failure of this lucrative export to transform Sicily's backward and impoverished economy, resulted in the "Sulphur Crisis" of 1840, when King Ferdinand II (King of the Two Sicilies from 1830 till his death, January 12, 1810 to May 22, 1859) offered a monopoly of the sulfur industry to a French firm, violating a previous 1816 trade agreement with Britain. A peaceful solution was in the end negotiated by France. In 1867 elemental sulfur was found in underground deposits in Louisiana and Texas. The highly successful Frasch process was developed to extract this resource (see next section).

4.6.2 Mining, production, and major minerals

Sulfur (S) may be found by itself and historically was typically obtained in this form; pyrite, FeS_2, has also been a source of sulfur. In volcanic regions in Sicily, in historic times, it was found on the surface of the earth, and the "Sicilian process" was used: sulfur mounts were piled and stacked in brick kilns built on the sloping hillsides, with airspaces between them. Then, some sulfur was pulverized, spread over the stacked ore and ignited, causing the free sulfur to melt down the hills. In the end, the surface-borne deposits were exhausted, and miners quarried veins that in due course dotted the Sicilian landscape with labyrinthine mines. Elemental sulfur was later extracted from salt domes (in which it sometimes occurs in nearly pure form) until the late 20th century. Sulfur is now produced as a side product of other industrial processes such as in oil and gas refining, in which sulfur is undesired, from which it is obtained mainly as hydrogen sulfide. Organosulfur compounds, undesirable impurities in petroleum, may be upgraded by subjecting them to hydrodesulfurization, which cleaves the C–S bonds:

$$R - S - R + 2H_2 \rightarrow 2RH + H_2S$$

The resulting hydrogen sulfide from this process, and also as it occurs in natural gas, is converted into elemental sulfur by the Claus process invented by Carl Friedrich Claus (November 9, 1827 to August 29, 1900), a German chemist working in England. This process entails oxidation of some hydrogen sulfide to sulfur dioxide and then the comproportionation (a chemical reaction where two reactants, each containing the same element but with a different oxidation number, form a product in which the elements involved reach the same oxidation number) of the two:

$$3O_2 + 2H_2S \rightarrow 2SO_2 + 2H_2O$$
$$SO_2 + 2H_2S \rightarrow 3S + 2H_2O$$

As a mineral, native sulfur under salt domes is thought to be a fossil mineral resource, produced by the action of ancient bacteria on sulfate deposits. It is won from such salt dome mines mainly by the Frasch process developed by the German-born American chemist, Herman Frasch (December 25, 1851 to May 1, 1914) (Fig. 4.61). In this process, three concentric tubes are drilled into the sulfur deposit. Superheated water (i.e., steam) (165°C, 2.5–3 MPa) is injected

FIGURE 4.61 Herman Frasch, 1918.

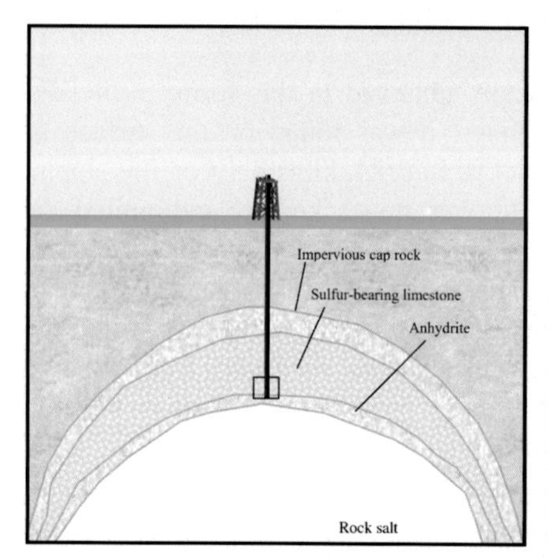

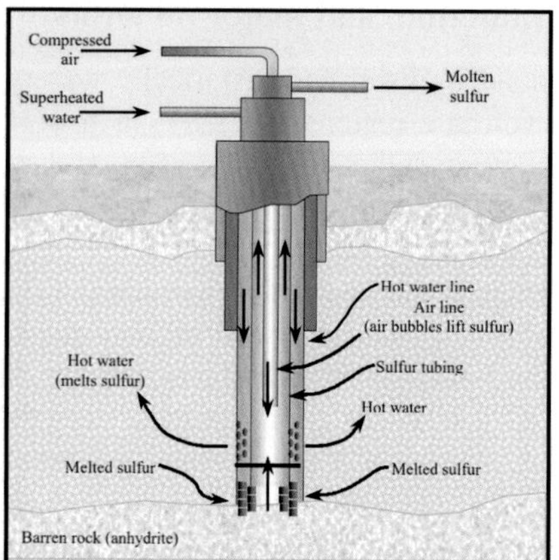

FIGURE 4.62 Illustration that shows the structure of a sulfur-containing salt dome and the details of the Frasch pump used to extract the sulfur from underground formations. Superheated water is pumped into the formation to melt the sulfur. The molten sulfur is lifted to the surface with compressed air. Materials Flow of Sulfur, USGS Open-File Report 02-298, Joyce A. Ober.

FIGURE 4.63 Well-formed crystals of lilac to colorless creedite, $Ca_3SO_4Al_2F_8(OH)_2{\cdot}2H_2O$, to 3 mm. Liberty (Hall) mine, Nye Co., Nevada, United States.

FIGURE 4.64 Chalcopyrite, $CuFeS_2$, crystal up to 8 mm with quartz, SiO_2, on galena, PbS, matrix. Kisbanya, Baia Mare, Romania.

FIGURE 4.65 Octahedral galena, PbS, crystals to 1.2 cm. Huanzala mine, Huallanca dist., Huánuco, Peru.

FIGURE 4.66 Marcasite, FeS_2, a nice example of cockscomb habit crystals to 8 mm. Joplin dist., Jasper Co., Missouri, United States.

FIGURE 4.67 Pyrite, FeS_2, specimen with octahedral and pentagono-dodecahedral forms, resulting in five faces creating a pyramidal termination. The smooth faces are octahedral, while the striated faces are the pentagono-dodecahedral faces. 4 x 3 x 2.5 cm. Butte dist., Silver Bow Co., Montana, United States.

into the deposit via the outermost tube. Sulfur (melting point 115°C) melts and flows into the middle tube. Water pressure alone is not enough to force the sulfur to the surface due to the molten sulfur's greater density; therefore, hot air is introduced via the innermost tube to froth the sulfur, making it less dense, and forcing it to the surface. The sulfur obtained can be very pure (99.7%–99.8%). In this form, it is light yellow in color. If contaminated by organic compounds, it can be dark(er)-colored; further purification is not economic and usually unnecessary. The Frasch process can be used for deposits from 50 to 800 m deep. A large volume of superheated water (3-38 m^3) is required to produce every ton of sulfur, and the associated energy cost is significant (Fig. 4.62).

In total there are slightly less than 1200 minerals, the majority forming sulfides (including sulfites) and sulfates. Sulfur occurs as a native element in two forms: sulfur (S_8) (Box 4.7) and rosickýite (S). Within the halides, 12 minerals contain

FIGURE 4.68 Pyrrhotite, Fe_7S_8, hexagonal, thin platy bronze-colored crystals to 2.5 cm associated with groups of discoidal siderite, $FeCO_3$, and quartz, SiO_2, crystals. Felsobanya, Baia Mare, Maramures Co., Romania.

FIGURE 4.69 Baryte, $BaSO_4$, crystal is 3.3 cm and nearly flawless on shale matrix. Elk Creek, Meade Co., South Dakota, United States.

S, 11 of which have SO_4 groups, for example, creedite ($Ca_3SO_4Al_2F_8(OH)_2{\cdot}2H_2O$) (Fig. 4.63), while radhakrishnaite has S substituting for Cl ($PbTe_3(Cl,S)_2$). In the oxide class, 13 minerals contain S in various forms (S, SO_4, SO_3), for example, hannebachite ($CaSO_3{\cdot}H_2O$), kuzelite ($Ca_4Al_2(OH)_{12}[SO_4]{\cdot}6H_2O$), and versiliaite ($Fe_2Fe_4Sb_6O_{16}S$). Fifteen carbonate minerals contain SO_4 groups, for example, leadhillite ($Pb_4(CO_3)_2(SO_4)(OH)_2$), schröckingerite ($NaCa_3(UO_2)(CO_3)_3(SO_4)F{\cdot}10H_2O$), and tychite ($Na_6Mg_2(CO_3)_4(SO_4)$). Under the borates, four minerals contain the sulfate group, for example, sulfoborite ($Mg_3[B(OH)_4]_2(SO_4)(OH,F)_2$). A larger group is found in the phosphate class with minerals such as beudantite ($PbFe_3(AsO_4)(SO_4)(OH)_6$), tsumebite ($Pb_2Cu(PO_4)(SO_4)(OH)$), and woodhouseite ($CaAl_3(PO_4)(SO_4)(OH)_6$). Within the large group of the silicate minerals, 41 contain S, for example, afghanite ($(Na,K)_{22}Ca_{10}(Si_{24}Al_{24}O_{96})(SO_4)_6Cl_6$), cancrinite ($(Na,Ca,\square)_8(Al_6Si_6O_{24})(CO_3,SO_4)_2{\cdot}2H_2O$), genthelvite ($Be_3Zn_4(SiO_4)_3S$), haüyne ($(Na,K)_3(Ca,Na)(Al_3Si_3O_{12})(SO_4,S,Cl)$), and nosean ($Na_8(Al_6Si_6O_{24})(SO_4){\cdot}H_2O$). Four organic compounds contain S, for example, julienite ($Na_2[Co(SCN)_4]{\cdot}8H_2O$)

FIGURE 4.70 Celestine, $SrSO_4$, light blue prismatic crystals to over 1 cm scattered on matrix. N'Chwaning I mine, Kalahari Mn fields, Northern Cape Province, South Africa.

FIGURE 4.71 Gypsum, $CaSO_4 \cdot 2H_2O$, water-clear crystal group 7 cm high. Naica, Mun. de Saucillo, Chihuahua, Mexico.

and levinsonite-(Y) ($(Y,Nd,La)Al(C_2O_4)(SO_4)_2 \cdot 12H_2O$). Some of the most important primary ore minerals containing metals such as Cu, Pb, and Zn are found as sulfides. Some of the most important sulfides of a total of nearly 500 sulfide minerals include acanthite (Ag_2S), bornite (Cu_5FeS_4), boulangerite ($Pb_5Sb_4S_{11}$), bournonite ($PbCuSbS_3$), chalcocite (Cu_2S), chalcopyrite ($CuFeS_2$) (Fig. 4.64), cinnabar (HgS), cobaltite (CoAsS), galena (PbS) (Fig. 4.65), marcasite (FeS_2) (Fig. 4.66), millerite (NiS), molybdenite (MoS_2), orpiment (As_2S_3), proustite (Ag_3AsS_3), pyrargyrite (Ag_3SbS_3), pyrite (FeS_2) (Fig. 4.67), pyrrhotite (Fe_7S_8) (Fig. 4.68), realgar (As_4S_4), sphalerite (ZnS), stibnite (Sb_2S_3), tennantite ($Cu_6[Cu_4(Fe,Zn)_2]As_4S_{13}$), tetrahedrite ($Cu_6[Cu_4(Fe,Zn)_2]Sb_4S_{13}$), and wurtzite ((Zn,Fe)S). The sulfate class contains just under 350 minerals. They include well-known minerals such as alunite ($KAl_3(SO_4)_2(OH)_6$), anglesite ($PbSO_4$), anhydrite ($CaSO_4$), baryte ($BaSO_4$) (Fig. 4.69), celestine ($SrSO_4$) (Fig. 4.70), cyanotrichite ($K_2Cu(SO_4)_2 \cdot 6H_2O$), gypsum ($CaSO_4 \cdot 2H_2O$) (Fig. 4.71), jarosite $\left(KFe_3^{3+}(SO_4)_2(OH)_6\right)$, linarite ($PbCu(SO_4)(OH)_2$), sturmanite ($Ca_6(Fe^{3+},Al,Mn^{3+})_2(SO_4)_2[B(OH)_4](OH)_{12} \cdot 25H_2O$) (Fig. 4.72), and thaumasite ($Ca_3(SO_4)[Si(OH)_6](CO_3) \cdot 12H_2O$) (Fig. 4.73).

FIGURE 4.72 Sturmanite, $Ca_6(Fe^{3+},Al,Mn^{3+})_2(SO_4)_2[B(OH)_4](OH)_{12}\cdot 25H_2O$, yellow, hexagonal prismatic crystals with dypyrimidal terminations to 7 mm. N'Chwaning mines, Kuruman, Northern Cape prov., South Africa.

FIGURE 4.73 Thaumasite, $Ca_3(SO_4)[Si(OH)_6](CO_3)\cdot 12H_2O$, pale-yellow hexagonal crystal to 1.2 cm. N'Chwaning II mine, Kalahari Mn fields, Northern Cape Province, South Africa.

BOX 4.7 Sulfur

Sulfur is a nonmetal that often occurs at or near the crater rims of active or extinct volcanoes, where it has been deposited from gases released in fumaroles, from volcanic vents and in associated hot springs. The sulfur may precipitate directly from the vapors or be produced as a result of bacterial action on sulfate minerals. Hydrothermal sulfide deposits may also contain native sulfur, usually in the near-surface oxidized zone, formed by oxidation of the sulfides. The largest concentrations of sulfur, however, are associated with salt domes formed from marine evaporite deposits. These evaporites consist predominantly of halite (NaCl) but typically also contain gypsum ($CaSO_4{\cdot}2H_2O$), anhydrite ($CaSO_4$), and calcite ($CaCO_3$). When the top of the salt dome comes in contact with fresh meteoric groundwater within roughly 1 km from the surface, halite is dissolved. Continuous upward movement of salt from its source allows a cap of less soluble calcite and gypsum to accumulate at the top of the salt dome. The cap commonly consists of an outer/upper zone of calcite, transitioning inward to gypsum and then anhydrite. Hydrogen sulfide is produced by anaerobic sulfur-reducing bacteria if hydrocarbons (oil/gas) are available according to the following general reaction:

$$CaSO_4 + CH_4(\text{hydrocarbons}) + \text{bacteria} \rightarrow H_2S + CaCO_3 + H_2O$$

The hydrogen sulfide is oxidized by either oxygen in the groundwater, hydrocarbon, or other chemical processes to produce elemental sulfur:

$$2H_2S + O_2 \rightarrow 2S + 2H_2O$$

The actual reaction paths are far more complex than these two overall reactions and may additionally involve aerobic bacteria. Because generation of sulfur involves the breakdown of sulfates and production of calcite, the sulfur is concentrated at the calcite-sulfate boundary in the caprock of the salt domes.

Sulfur has a distinctive yellow color. It burns in a candle flame. It is commonly found in irregular masses, imperfectly crystallized as well as massive, reniform, stalactites, and as encrustations. Crystals often show pyramidal or tabular habit, often with two dipyramids, prism {011} and pinacoids in combination. Crystallographically, sulfur is orthorhombic, *2/m2/m2/m* (space group *F2/d2/d2/d*). The hardness is 1½–2½. The luster is resinous, while the transparency ranges from transparent to translucent. It has a poor {101} and {110} cleavages and conchoidal fracture. The color, when pure, is bright yellow but can vary with impurities to yellow shades of green, gray, and red. The streak is white. Optically it is biaxial(+). Sulfur is a poor conductor of heat. When a crystal is held in the hand close to the ear, it can be heard to crack. This is caused by the expansion of the surface layers because of the heat from the hand, while the interior of the crystal is unaffected. Crystals of sulfur should therefore be handled with care (Fig. 4.74).

FIGURE 4.74 Sulfur, S, 1 cm euhedral crystal on aragonite, $CaCO_3$. Boling Dome, Newgulf, Wharton Co., Texas, United States.

4.6.3 Chemistry

Sulfur has 25 known isotopes, four of which are stable: ^{32}S (94.99% ± 0.26%), ^{33}S (0.75% ± 0.02%), ^{34}S (4.25% ± 0.24%), and ^{36}S (0.01% ± 0.01%). Besides ^{35}S, with a half-life of 87 days and formed in cosmic ray spallation of ^{40}Ar (also known as the x-process, is a set of naturally occurring nuclear reactions causing nucleosynthesis; it refers to the formation of chemical elements from the impact of cosmic rays on an object. Cosmic rays are highly energetic charged particles from beyond earth, ranging from protons, alpha particles, and nuclei of many heavier elements. About 1% of cosmic rays also consist of free electrons.), the other radioactive isotopes have half-lives shorter than 3 hours. When sulfide minerals are formed, isotopic equilibration among solids and liquid may result in small differences in the $\delta^{34}S$ values of cogenetic minerals. These differences between minerals can be utilized to approximate the equilibration temperature. Similarly, the $\delta^{13}C$ and $\delta^{34}S$ of coexisting carbonate and sulfide minerals can be utilized to obtain information about the pH and oxygen fugacity of the ore-bearing fluid during ore formation. In most forest ecosystems, sulfate results mostly from the atmosphere, while weathering of sulfidic ore minerals and evaporites (e.g., gypsum $CaSO_4{\cdot}2H_2O$ and anhydrite, $CaSO_4$) contribute an additional small amount of sulfur. Sulfur with a characteristic isotopic composition has been utilized to recognize pollution sources, and enriched sulfur has been used as a tracer in hydrologic studies.

A sulfur atom in the ground state has the electronic configuration $[Ne]3s^23p_x^23p_y^13p_z^1$ with two unpaired electrons just like oxygen above it in the periodic table. Despite the relatively low first and second ionization energies of sulfur with values of 999.6 and 2252 kJ/mol, respectively, the oxidation state of sulfur is commonly +4 or +6 instead of +2. The full range of oxidation states sulfur ions can assume is from −2 to +6 (Table 4.6).

Sulfur shows an extremely high tendency to form strong bonds with itself (catenation) and form linear and cyclic polyatomic sulfur molecules with varying degrees of crystallinity. The S-S bond character, distance and bond angle can lead to ring and chain molecules of two to more than 10^5 sulfur atoms. As such, sulfur is reported to have over 30 solid allotropes. Gaseous allotropes of sulfur, S_2 and S_3, analogous to O_2 and O_3 have been isolated and characterized. However, in contrast to oxygen, the most common is the bright yellow allotrope, S_8, which has a crown structure. S_8 can exist in various polymorphs, with the α-, β-, γ-S_8 as the three most fully identified polymorphs. The α-S_8 polymorph condenses into an orthorhombic crystal, is the most thermodynamically stable form, and is the most commonly found form in nature. It is insoluble in water, a good electrical insulator and poor thermal conductor. It converts to β-S_8 polymorph at 95.3°C. The β-S_8 polymorph is also a yellow solid with a monoclinic crystal. This form is stable only at a narrow range of temperatures. It reverts to the α-form below 95.3°C and decomposes to other forms at around 119°C. The γ-S_8 polymorph has a pale-yellow appearance and crystallizes in monoclinic needles. It is the densest of the S_8 polymorphs. This is the form of the rare mineral rosickýite. Other cyclic sulfur structures with 6, 7, 9–15, 18, and 20

TABLE 4.6 Sulfur properties.

Appearance	Lemon yellow
Allotropes	S_8, more than 30 known
Standard atomic weight $A_{r,std}$	32.06
Block	P-block
Element category	Reactive nonmetal
Electron configuration	$[Ne]\ 3s^2\ 3p^4$
Phase at STP	Solid
Melting point	115.21°C
Boiling point	444.6°C
Density (near r.t.)	Alpha: 2.07 g/cm^3 Beta: 1.96 g/cm^3 Gamma: 1.92 g/cm^3
Heat of fusion	Mono: 1.727 kJ/mol
Heat of vaporization	Mono: 45 kJ/mol
Molar heat capacity	22.75 J/(mol·K)
Oxidation states	**−2**, −1, +1, **+2**, +3, **+4**, +5, **+6**
Ionization energies	1st: 999.6 kJ/mol 2nd: 2252 kJ/mol 3rd: 3357 kJ/mol
Covalent radius	105 ± 3 pm
Van der Waals radius	180 pm

STP, Standard temperature and pressure.
Bold font indicates main oxidation state.

sulfur atoms have also been synthesized and characterized. Apart from the cyclic allotropes, sulfur also forms polymeric chains called collectively as polycatena sulfur. Solid polycatena sulfur comes in various forms including amorphous, laminar, fibrous, insoluble, etc.

The reactivity of sulfur depends on the form but very few studies have dealt with individual reactivity of the different allotropes. The cyclic forms are generally less reactive than the polycatena S forms. Atomic sulfur, which can readily be generated photolytically, is an extremely reactive species. Sulfur unites directly with all elements except the noble gases, nitrogen, tellurium, iodine, iridium, platinum, and gold. This does not preclude, however, the existence of compounds with S bonded directly to N, Te, I, Ir, Pt, and Au. Sulfur reacts in the cold with all the main group representatives of groups 1, 2, 13, Sn, Pb, and Bi, and also Cu, Ag, and Hg. The nonmetals B, C, Si, Ge, P, As, Sb, and Se react at elevated temperatures. The transition metals (except Ir, Pt, and Au) and the lanthanides and actinides also react on being heated with sulfur to form binary metal sulfides.

4.6.3.1 Oxides, nitrides, and hydrides

Reaction of sulfur in air is very slow at room temperature though traces of SO_2 can be found. It does not react with pure dry O_2 at room temperature but does with O_3. It ignites in air at a temperature of 250°C–260°C to give sulfur oxides.

$$S + O_2 \rightarrow SO_2 \text{ (sulfur dioxide)}$$
$$2SO_2 + O_2 \rightarrow 2SO_3 \text{ (sulfur trioxide)}$$

Multiple sulfur oxides are known; the sulfur-rich oxides include sulfur monoxide, disulfur monoxide, disulfur dioxides, and higher oxides containing peroxo groups. Sulfur forms sulfur oxoacids, some of which cannot be isolated and are only known through the salts. Sulfur dioxide and sulfites (SO_3^{2-}) are related to unstable sulfurous acid (H_2SO_3). Sulfur trioxide and sulfates (SO_4^{2-}) are related to sulfuric acid (H_2SO_4). Sulfuric acid and SO_3 combine to give oleum, a solution of pyrosulfuric acid ($H_2S_2O_7$) in sulfuric acid. Thiosulfate salts ($S_2O_3^{2-}$), sometimes referred as "hyposulfites", used in photographic fixing (hypo) and as reducing agents, feature sulfur in two oxidation states. Sodium dithionite ($Na_2S_2O_4$), contains the more highly reducing dithionite anion ($S_2O_4^{2-}$). The sulfate ion, SO_4^{2-} in salts, acid derivatives, and peroxides of sulfate are widely used in the industry. They occur widely in everyday life. Sulfates are salts of sulfuric acid and many are prepared from that acid. Preparation methods of metal sulfates include: (1) treating metal, metal oxide with sulfuric hydroxide or metal acid

$$Zn + H_2SO_4 \rightarrow ZnSO_4 + H_2$$
$$Cu(OH)_2 + H_2SO_4 \rightarrow CuSO_4 + 2H_2O$$
$$CdCO_3 + H_2SO_4 \rightarrow CdSO_4 + H_2O + CO_2$$

(2) oxidation of metal sulfides or sulfites. The sulfate anion consists of a central sulfur atom surrounded by four equivalent oxygen atoms in a tetrahedral coordination. The symmetry is the same as that of methane (CH_4). The sulfur atom is in the +6 oxidation state while the four oxygen atoms are each in the −2 state. The sulfate ion carries an overall charge of −2 and it is the conjugate base of the bisulfate (or hydrogen sulfate) ion, HSO_4^-, which is in turn the conjugate base of H_2SO_4, sulfuric acid. Organic sulfate esters, such as dimethyl sulfate, are covalent compounds and esters of sulfuric acid. The tetrahedral molecular geometry of the sulfate ion is as predicted by VSEPR theory. The first description of the bonding in modern terms was by Gilbert Lewis in his groundbreaking paper of 1916 where he discussed the bonding in terms of electron octets around each atom, that is no double bonds and a formal charge of +2 on the sulfur atom (Lewis, 1916). Later, Linus Pauling used valence bond theory to suggest that the most significant resonance canonicals had two π bonds involving d orbitals (Pauling, 1949). His logic was that the charge on sulfur had to be reduced, in accordance with his principle of electroneutrality. The S − O bond length of 149 pm is shorter than the bond lengths in sulfuric acid of 157 pm for S − OH. The double bonding was taken by Pauling to explain the shortness of the S − O bond. Pauling's use of d orbitals resulted in a discussion on the relative importance of π bonding and bond polarity (electrostatic attraction) in causing the shortening of the S − O bond. The result was a broad consensus that d orbitals play a role, but are not as important as Pauling had believed. A generally accepted description entailing pπ − dπ bonding was initially suggested by D. W. J. Cruickshank (Cotton and Wilkinson, 1966). In this model, fully occupied p orbitals on oxygen overlap with empty sulfur d orbitals (principally the d_{z^2} and $d^2_{x^2-y}$). Nevertheless, in this description, notwithstanding the fact that there was some π character to the S − O bonds, the bond has significant ionic

character. For sulfuric acid, computational analysis (with natural bond orbitals) confirms a clear positive charge on sulfur (theoretically +2.45) and a low 3d occupancy. Hence, the representation with four single bonds is the optimal Lewis structure instead of the one with two double bonds (thus the Lewis model, not the Pauling model). In this model, the structure obeys the octet rule and the charge distribution is in agreement with the electronegativity of the atoms. The difference between the S − O bond length in the sulfate ion and the S − OH bond length in sulfuric acid can be explained by the donation of p-orbital electrons from the terminal S = O bonds in sulfuric acid into the antibonding S − OH orbitals, weakening them causing the longer bond length of the latter. Nonetheless, the bonding representation of Pauling for sulfate and other main group compounds with oxygen is still a common way of representing the bonding in many chemistry textbooks. The apparent disagreement can be cleared by the realization that the covalent double bonds in the Lewis structure actually represent bonds that are strongly polarized by more than 90% towards the oxygen atom. On the other hand, in the structure with a dipolar bond, the charge is localized as a lone pair on the oxygen. Similar to the reaction with oxygen, it does not react directly with N_2 at room temperature. However, in a discharge tube, activated N reacts with sulfur. An important S–N compound is the cage structure tetrasulfur tetranitride (S_4N_4). Heating this compound results in polymeric sulfur nitride ($(SN)_x$), which has metallic properties even though it does not contain any metal atoms. Thiocyanates contain the SCN^- group. Oxidation of thiocyanate results in the formation of thiocyanogen, $(SCN)_2$ with the connectivity NCS-SCN. It reacts slowly with H_2 at 120°C and more rapidly above 200°C.

4.6.3.2 Halides

Sulfur ignites in F_2 and burns with a livid flame to give SF_6. The reaction with chlorine is less vigorous at room temperature but rapidly accelerates above this to give S_2Cl_2. Sulfur dissolves in liquid Br_2 to form S_2Br_2, which readily dissociates into its elements. Iodine does not form a binary compound directly even at elevated temperature. Iodine has been used as a cryoscopic solvent for sulfur.

$$S_8(s) + 24F_2(g) \rightarrow 8SF_6(l)$$
$$S_8(l) + 4Cl_2(g) \rightarrow 4S_2Cl_2(l)$$
$$S_8(l) + 4Br_2(g) \rightarrow 4S_2Br_2(l)$$

Several sulfur halides are important to modern industry. Sulfur hexafluoride is a dense gas used as an insulator gas in high voltage transformers; it is also a nonreactive and nontoxic propellant for pressurized containers. SF_6 can be produced from the elements by exposure of S_8 to F_2. Some other sulfur fluorides are cogenerated, but these are removed by heating the mixture to disproportionate any S_2F_{10} (which is highly toxic) and then scrubbing the product with NaOH to destroy remaining SF_4. Alternatively, using bromine, sulfur hexafluoride can be prepared from SF_4 and CoF_3 at lower temperatures (e.g. 100°C):

$$2CoF_3 + SF_4 + [Br_2] \rightarrow SF_6 + 2\,CoF_2 + [Br_2]$$

There is nearly no reaction chemistry for SF_6. A key contribution to the inertness of SF_6 is the steric hindrance of the sulfur atom, while its heavier group 16 counterparts, such as SeF_6 are more reactive than SF_6 as a result of less steric hindrance. It does not react with molten sodium below its boiling point, but reacts exothermically with lithium. SF_6 has an octahedral coordination geometry with six fluorine atoms attached to a central sulfur atom. It is a hypervalent molecule. Typical for a nonpolar gas, it is poorly soluble in water but quite soluble in nonpolar organic solvents. It is generally transported as a liquefied compressed gas.

Sulfur tetrafluoride is a rarely used organic reagent that is highly toxic. It is a corrosive species that releases dangerous HF upon exposure to water or moisture. SF_4 is formed by reacting SCl_2 with NaF in acetonitrile:

$$3SCl_2 + 4NaF \rightarrow SF_4 + S_2Cl_2 + 4NaCl.$$

In addition, SF_4 is formed in the absence of solvent at elevated temperatures. Alternatively, SF_4 at high yield can be prepared using sulfur (S), NaF and chlorine (Cl_2) in the absence of reaction medium, also at less-desirable elevated reaction temperatures (e.g., 225°C–450°C). A low temperature (e.g., 20°C–86°C) method of producing SF_4 at high yield, without the need for a reaction medium, has been shown using bromine (Br_2) instead of chlorine (Cl_2), S and KF:

$$S + (2 + x)Br_2 + 4KF \rightarrow SF_4\uparrow + xBr_2 + 4KBr$$

In organic chemistry, SF_4 is utilized to convert COH and C = O groups into CF and CF_2 groups, respectively. Certain alcohols easily form the corresponding fluorocarbon. Ketones and aldehydes give geminal difluorides. The presence of protons alpha to the carbonyl leads to side reactions and diminished (30%−40%) yield. Also diols can give cyclic sulfite esters, $(RO)_2SO$. Carboxylic acids convert to trifluoromethyl derivatives. For instance, the reaction of heptanoic acid with SF_4 at 100°C−130°C results in 1,1,1-trifluoroheptane. Hexafluoro-2-butyne can likewise be prepared from acetylenedicarboxylic acid. The coproducts from these fluorinations, including unreacted SF_4 together with SOF_2 and SO_2, are toxic but can be neutralized by their treatment with aqueous KOH. The use of SF_4 is being replaced in recent years by the more conveniently handled diethylaminosulfur trifluoride, Et_2NSF_3, "DAST", where Et = CH_3CH_2. This reagent is prepared from SF_4:

$$SF^4 + Me^3SiNEt^2 \rightarrow Et^2NSF^3 + Me^3SiF$$

Sulfur dichloride and disulfur dichloride are important industrial chemicals. Sulfur dichloride, SCl_2 is cherry-red liquid. It is the simplest sulfur chloride and one of the most common. It is utilized as a precursor to form organosulfur compounds. SCl_2 can be prepared by the chlorination of either elemental sulfur or disulfur dichloride. The reaction occurs in a series of steps, some of which are:

$$S_8 + 4\ Cl_2 \rightarrow 4\ S_2Cl_2;\ \Delta H = -58.2\ \text{kJ/mol}$$
$$S_2Cl_2 + Cl_2 \leftrightarrow 2\ SCl_2;\ \Delta H = -40.6\ \text{kJ/mol}$$

The addition of Cl_2 to S_2Cl_2 has been suggested to proceed via a mixed valence intermediate $Cl_3S{-}SCl$. SCl_2 undergoes even further chlorination to give SCl_4, but this species is unstable at near room temperature. It is likely that several S_xCl_2 exist where $x > 2$. Disulfur dichloride, S_2Cl_2, is the most common impurity in SCl_2. Removal of SCl_2 from S_2Cl_2 is possible via distillation with PCl_3 to form an azeotrope of 99% purity; however, sulfur dichloride loses chlorine slowly at room temperature and reverts to disulfur dichloride. Pure samples may be stored in sealed glass ampules which develop a slight positive pressure of chlorine, halting the decomposition. Disulfur dichloride, S_2Cl_2, has the structure implied by the formula Cl − S − S − Cl, wherein the angle between the $Cl^a - S - S$ and $S - S - Cl^b$ planes is 90°. This structure is referred to as *gauche*, and is similar to that for H_2O_2. A rare isomer of S_2Cl_2 is $S = SCl_2$; which forms transiently when S_2Cl_2 is exposed to UV-radiation. Pure disulfur dichloride is a yellow liquid that "smokes" in moist air due to reaction with water. An idealized (but complicated) equation is:

$$16S_2Cl_2 + 16H_2O \rightarrow 8SO_2 + 32HCl + 3S_8$$

It is prepared by partial chlorination of elemental sulfur. The reaction proceeds at usable rates at room temperature. In the laboratory, chlorine gas is led into a flask containing elemental sulfur. As disulfur dichloride is formed, the contents turn to a golden yellow liquid:

$$S_8 + 4Cl_2 \rightarrow 4S_2Cl_2;\ \Delta H = -58.2\ \text{kJ/mol}$$

Excess chlorine produces sulfur dichloride, which results in the liquid to become less yellow and more orange-red:

$$S_2Cl_2 + Cl_2 \leftrightarrow 2SCl_2;\ \Delta H = -40.6\ \text{kJ/mol.}$$

The reaction is reversible, and upon standing, SCl_2 releases chlorine to revert to the disulfur dichloride. Disulfur dichloride has the ability to dissolve large quantities of sulfur, which reflects in part the formation of polysulfanes:

$$S_2Cl_2 + nS \rightarrow S_{2+n}Cl_2$$

Disulfur dichloride can be purified from excess elemental sulfur through distillation. S_2Cl_2 also forms from the chlorination of CS_2 as in the synthesis of thiophosgene. S_2Cl_2 hydrolyzes to sulfur dioxide and elemental sulfur. When reacted with hydrogen sulfide, polysulfanes are formed as indicated in the following idealized formula:

$$2H_2S + S_2Cl_2 \rightarrow H_2S_4 + 2HCl$$

It reacts with ammonia to give heptasulfur imide (S_7NH) and related S − N rings $S_{8-x}(NH)_x$ ($x = 2, 3$). S_2Cl_2 has been employed to introduce C − S bonds. In the presence of aluminium chloride ($AlCl_3$), S_2Cl_2 reacts with benzene to give diphenyl sulfide:

$$8S_2Cl_2 + 16C_6H_6 \rightarrow 8(C_6H_5)_2S + 16HCl + S_8.$$

Anilines react with S_2Cl_2 in the presence of NaOH to give 1,2,3-benzodithiazolium salts (Herz reaction) which can be transformed into *ortho*-aminothiophenolates, these species are precursors to thioindigo dyes. It is also used to prepare sulfur mustard, commonly known as "mustard gas", by reaction with ethylene at 60°C (the Levinstein process):

$$8S_2Cl_2 + 16C_2H_4 \rightarrow 8(ClC_2H_4)_2S + S_8$$

Other applications of S_2Cl_2 include the manufacture of sulfur dyes, insecticides, and synthetic rubbers. It is also utilized in cold vulcanization of rubbers, as polymerization catalyst for vegetable oils and for hardening soft woods. Sulfur dichloride and disulfur dichloride are important industrial chemicals. Sulfuryl chloride and chlorosulfuric acid are derivatives of sulfuric acid; thionyl chloride ($SOCl_2$) is a common reagent in organic synthesis.

4.6.3.3 Reaction with acids and bases

Elemental sulfur does not react with acids. The sulfur trioxide, however, dissolves in water to form sulfuric acid (H_2SO_4). Sulfur reacts with hot aqueous potassium hydroxide, KOH, to form sulfide and thiosulfate species.

$$S_8 + 6KOH \rightarrow 2K_2S_3 + K_2S_2O_3 + 3H_2O$$

4.6.3.4 Organosulfur compounds

As explained above, sulfur compounds exhibit a rich variety owing to the numerous oxidation states and the range of bond types it can form. The chemistry of these compounds is even more extensive and will not be shown in detail in this book. Rather, a description of the different classes based on functional groups of organosulfur compounds will be given. Some of these groups of compounds are gaining resurgence in interests due to their roles in heterocyclic and radical chemistry especially in the stereo-controlled processes including asymmetric synthesis. Sulfides (formerly thioethers) are compounds with the C-S-C bond. An important member of this class is formed from the reaction called the Ferrario reaction where a phenyl ether is converted into the phenoxathiine compound. Phenoxatiin is important in the synthesis of polyamide and polyimide, which in turn are important in the production of materials such as nylons and aramids (e.g., Kevlar).

S_8, $AlCl_3$; $-H_2S$ heat

phenyl ether → phenoxathiine

Thioacetals and thioketals with the general structure, C-S-C-S-C, are a subclass of the sulfides. They are useful in the protection of carbonyl groups in organic synthesis. Thiols have the skeleton structure, R-SH. They are structurally similar to alcohols, R-OH, but exhibit better nucleophilicity, stronger acidity, and easier to oxidize than alcohols. Aliphatic thiols exhibit self-assembly behavior on gold surfaces. The disulfides, R-S-S-R, are useful in crosslinking processes. The thioesters, R-C(O)-S-R, are prominent in fatty acid synthesis. Sulfimides, $R_2S = NR'$, are of interest to pharmacology; sulfoximides, $R_2S(O) = NR'$, particularly methionine sulfoximide is of significance in inhibition of glutamine synthetase; and sulfonediimines, $R_2S(=NR')_2$, are biologically active and are used in heterocycle synthesis. S-nitrosothiols, $R - S - N{=}O$, can be a source of the nitrosonium ion, NO^+, and nitric oxide, NO, which may function as signaling molecules in living systems. The sulfoxides (R-S(O)-R), sulfone (R-S(O)$_2$-R), thiosulfinate (R-S(O)-S-R) and thiosulfonate (R-S(O)$_2$-S-R), sulfur halides (R-S-X, R-S(O)-X, R-SO$_2$X, RSCl$_3$, RSF$_3$ RSF$_5$), sulfines ($R_2C = S = O$), sulfenes ($R_2C = SO_2$), thiocarboxylic acids (R-C(O)-SH) and thioamides (R-C(S)-NR′R″), sulfonic acids (R-S(=O)$_2$ − OH), sulfinic acids (R-S(O)-H), sulfenic acids (R-S-OH), sulfonamides (R − SO$_2$NR′$_2$, (R − S(O) NR′$_2$), and sulfenamides, (R − SNR′$_2$) are well-known compounds with extensive chemistry. Other organosulfur compounds are less common. These include compounds with double and triple bonds between carbon and sulfur, although carbon disulfide, S=C=S, and carbon monosulfide, C≡S, are known.

Just like organic compounds of oxygen, which is above sulfur in the periodic table, organosulfur compounds are essential to life. The amino acids—cysteine, cystine, and methionine; the vitamins—biotin and thiamine and many cofactors (e.g., glutathione, thioredoxin) are organosulfur compounds. Core compounds needed for biochemical functioning have sulfur in them.

4.6.4 Major uses

Elemental sulfur is used mainly as a precursor to other chemicals. Most of it is converted to sulfuric acid (H_2SO_4):

$$2S + 3O_2 + 2H_2O \rightarrow 2H_2SO_4$$

The principal use for the acid is the extraction of phosphate ores for the production of fertilizer manufacturing. Other uses of sulfuric acid include oil refining, wastewater processing, and mineral extraction. S reacts directly with methane to give carbon disulfide (CS_2), used to manufacture cellophane (a thin, transparent sheet made of regenerated cellulose. Its low permeability to air, oils, greases, bacteria, and water makes it useful for food packaging.) and rayon (a manufactured fiber produced from regenerated cellulose fiber. The many types and grades of rayon can mimic the feel and texture of natural fibers such as silk, wool, cotton, and linen. The types that are similar to silk are frequently called artificial silk. Rayon is produced from naturally occurring polymers; therefore, it is not deemed to be synthetic. Technically, the term synthetic fiber is used for fully synthetic fibers. In production terms, rayon is categorized as "a fiber formed by regenerating natural materials into a usable form"). One of the uses of elemental S is in vulcanization of rubber, where polysulfide chains crosslink organic polymers. Large amounts of sulfites are used to bleach paper and to preserve dried fruit. Numerous surfactants and detergents [e.g., sodium lauryl sulfate, IUPAC name sodium dodecyl sulfate, $CH_3(CH_2)_{11}SO_4Na$] are sulfate derivatives. When silver-based photography was widespread, sodium and ammonium thiosulfate were widely used as "fixing agents." S is a component of gunpowder ("black powder").

Hydrated calcium sulfate, gypsum, ($CaSO_4 \cdot 2H_2O$) is mined in quantities of 100 million tons annually for use in Portland cement and fertilizers. S is more and more used as a component of fertilizers. The most important form of S for fertilizer is in the form of calcium sulfate. Elemental S is hydrophobic (not soluble in water) and cannot be taken up directly by plants. Over time, soil bacteria can convert it to soluble derivatives, which can subsequently be taken up by plants. Biologically formed sulfur particles are naturally hydrophilic because of a biopolymer coating and are easier to disperse over the land in a spray of diluted slurry, resulting in a faster uptake. The botanical requirement for S equals or exceeds the requirement for phosphorus. Sulfur deficiency has become widespread in many countries in Europe. Since atmospheric S inputs keep decreasing, the deficit in the S input/output is expected to increase unless S fertilizers are used.

Organosulfur compounds are utilized in, for example, pharmaceuticals, dyestuffs, and agrochemicals. Many drugs contain S, early examples being antibacterial sulfonamides, known as sulfa drugs. S is a part of many bacterial defense molecules. Most β-lactam antibiotics, including the penicillins, cephalosporins, and monolactams, contain S. Magnesium sulfate ($MgSO_4 \cdot 7H_2O$), known as Epsom salts when in hydrated crystal form, can be used as a laxative, a bath additive, an exfoliant, magnesium supplement for plants, or (when in dehydrated form) as a desiccant. S (specifically octasulfur, S_8) is used in pharmaceutical skin formulations for the treatment of acne and other skin conditions. It functions as a keratolytic agent and in addition kills bacteria, fungi, scabies mites, and other parasites. Precipitated S and colloidal S can be found in lotions, creams, powders, soaps, and bath additives, for the treatment of acne vulgaris, acne rosacea, and seborrhoeic dermatitis. Common adverse effects involve irritation of the skin at the application site, for example, dryness, stinging, itching, and peeling. S is converted to hydrogen sulfide (H_2S) via reduction, partly by bacteria. H_2S kills bacteria (perhaps including *propionibacterium acnes* which play a role in acne) fungi, and parasites such as scabies mites. Sulfur's keratolytic action is also mediated by H_2S; here the H_2S is formed by direct interaction with the target keratinocytes themselves. Mercaptans are a family of organosulfur compounds. Some are added to natural gas supplies because of their distinctive smell, so that gas leaks can be detected easily. Others are used in silver polish and in the production of pesticides and herbicides.

Elemental S is one of the oldest fungicides and pesticides. "Dusting S," elemental S in powdered form, is a widespread fungicide for grapes, strawberry, many vegetables, and several other crops. It is highly effective against a variety of powdery mildew diseases along with black spot. In organic production, S is the most important fungicide. It is the only fungicide used on organically farmed apples to protect against the main disease apple scab under colder conditions. Biosulfur (biologically produced elemental S with hydrophilic characteristics) can also be utilized for these applications. Standard-formulation dusting S is applied to crops with a S duster or from a dusting plane. Wettable sulfur is the commercial name for dusting sulfur formulated with additional components to make it water-miscible. It has similar applications and is used as a fungicide against mildew and other mold-related complications with plants and soil. Elemental S powder is also used as an "organic" (i.e., "green") insecticide (actually an acaricide) against ticks and mites. A common method of application is dusting the clothing or limbs with S powder. A diluted solution of lime sulfur [prepared by combining $Ca(OH)_2$ with elemental S in water] can be applied as a dip for pets to treat against ringworm (fungus), mange, and other dermatoses and parasites. In the past, sulfur candles of nearly pure S were burned to fumigate structures and wine barrels but are these days considered as being too toxic for use in residences. Small amounts of sulfur dioxide (SO_2) gas addition (or equivalent potassium metabisulfite, $K_2S_2O_5$, addition) to fermented wine to produce traces of sulfurous acid (formed when SO_2 reacts with water) and its sulfite salts in the mixture has become known as "the most powerful tool in winemaking." After the yeast fermentation stage in winemaking, sulfites absorb oxygen and prevent aerobic bacterial growth which could change ethanol to acetic acid, thereby souring the wine. Without this preservative step, indefinite refrigeration of the wine before consumption is typically needed. Comparable processes can be found since antiquity, but modern historical references of the practice can be found for the first time in the 15th century. The method is used by all wine producers including small organic wine producers. In addition, sulfur dioxide and various sulfites have been used for their antioxidant antibacterial preservative properties in various other parts of the food industry. The method has dropped in popularity due to reports of an allergy-like reaction of some persons to sulfites in foods.

4.7 17 Cl — Chlorine

4.7.1 Discovery

The most common compound of chlorine, sodium chloride (NaCl), has been known since historical times; archeologists have found proof that rock salt was used as early as 3000 BCE and brine as early as 6000 BCE. Its importance in food was very well known in classical antiquity and was from time to time used as payment for services for Roman generals and military tribunes. Elemental chlorine was possibly first isolated around 1200 with the discovery of aqua regia, a mixture of nitric acid and hydrochloric acid (optimally in a molar ratio of 1:3), and its ability to dissolve gold, since chlorine gas is one of the products of this reaction: though it was not identified as a new substance. Around 1630, chlorine was identified as a gas by the Flemish chemist, physiologist, and physician Jan Baptist van Helmont (January 12, 1580 to December 30, 1644) (Fig. 4.75). The element was first researched in detail in 1774 by Swedish chemist Carl Wilhelm Scheele, and the discovery is generally attributed to him (Scheele, 1774). Scheele produced chlorine by reacting MnO_2 (as the mineral pyrolusite) with HCl:

$$4HCl + MnO_2 \rightarrow MnCl_2 + 2H_2O + Cl_2$$

FIGURE 4.75 Jan Baptist van Helmont, painted c.1674 by Mary Beale.

FIGURE 4.76 Johann Salomo Christoph Schweigger, before 1857.

Scheele detected several of the properties of chlorine: the bleaching effect on litmus, the deadly effect on insects, the yellow-green color, and the smell similar to aqua regia. He called it "dephlogisticated muriatic acid air" since it was a gas (then called "airs") and it came from hydrochloric acid (then known as "muriatic acid"). He failed though to establish chlorine as an element. Common chemical theory during that period stated that an acid is a compound that contains oxygen (remnants of this survive in the German and Dutch names of oxygen: sauerstoff or zuurstof, both translating into English as acid substance), so a number of chemists, including Savoyard-French chemist Claude Berthollet (December 9, 1748 to November 6, 1822), proposed that Scheele's dephlogisticated muriatic acid air had to be a combination of oxygen and the yet undiscovered element, muriaticum. In 1809 French chemists Joseph Louis Gay-Lussac (December 6, 1778 to May 9, 1850) and Louis-Jacques Thénard (May 4, 1777 to June 21, 1857) attempted to decompose dephlogisticated muriatic acid air by reacting it with charcoal to release the free element muriaticum (and carbon dioxide) (Gay-Lussac and Thénard, 1809). They failed and published a report in which they discussed the option that dephlogisticated muriatic acid air is an element, but they were not convinced. In 1810 Cornish chemist and inventor Sir Humphry Davy (December 17, 1778 to May 29, 1829) attempted the same experiment again and determined that the substance was an element, and not a compound. He reported his results to the Royal Society on November 15 that same year (Davy, 1811). At that point, he called this new element "chlorine," from the Greek word χλωρος (chlōros), meaning green-yellow. The name "halogen," meaning "salt producer," was initially used for chlorine in 1811 by the German chemist, physicist, and professor of mathematics Johann Salomo Christoph Schweigger (April 8, 1779 to September 6, 1857) (Schweigger, 1811) (Fig. 4.76). This term was later used as a generic term to describe all the elements in the chlorine family (fluorine, bromine, iodine), after a proposition by Swedish chemist Jöns Jakob Berzelius (August 20, 1779 to August 7, 1848) in 1826. In 1823 English scientist Michael Faraday (September 22, 1791 −to August 25, 1867) liquefied chlorine for the first time and demonstrated that what was then known as "solid chlorine" had a structure of chlorine hydrate ($Cl_2{\cdot}H_2O$) (Faraday, 1823) (Fig. 4.77). Chlorine gas was first utilized in 1785 by Savoyard-French chemist Claude Berthollet (December 9, 1748 to November 6, 1822) to bleach textiles. Modern bleaches resulted from further work by Berthollet, who first made sodium hypochlorite (NaClO) in 1789 in his laboratory in the town of Javel (now part of Paris, France), by bubbling chlorine gas through a sodium carbonate solution. The resulting liquid, known as "Eau de Javel" ("Javel water"), was a weak solution of sodium hypochlorite. This method was not very effective, and other production methods were sought. Scottish chemist and industrialist Charles Tennant (May 3, 1768 to October 1, 1838) first produced a solution of calcium hypochlorite ("chlorinated lime"), then solid calcium hypochlorite (bleaching powder). These compounds produced low levels of elemental chlorine and could be more efficiently transported than sodium hypochlorite, which remained as dilute solutions due to the fact that when purified to eliminate water, it became a dangerously powerful and unstable oxidizer. Near the end of the 19th century, E. S. Smith patented a process to produce sodium hypochlorite using electrolysis of brine to produce sodium

FIGURE 4.77 Michael Faraday, c.1826, painted by H.W. Pickersgill (1782–1875), engraved by John Cochran (1821–65).

hydroxide and chlorine gas, which then mixed to form sodium hypochlorite. This is now known as the chloralkali process, first introduced on an industrial scale in 1892, and now the source of most elemental chlorine and sodium hydroxide. In 1884 Chemischen Fabrik Griesheim of Germany developed another chloralkali process that entered commercial production in 1888. Elemental chlorine solutions dissolved in chemically basic water (sodium and calcium hypochlorite) were first applied as antiputrefaction agents and disinfectants in the 1820s, in France, long before the formation of the germ theory of disease. This practice was pioneered by French chemist and pharmacist Antoine-Germain Labarraque (March 28, 1777 to December 9, 1850), who adapted Berthollet's "Javel water" bleach and other chlorine preparations. Elemental chlorine has since served a continuous function in topical antisepsis (wound irrigation solutions, etc.) and public sanitation, especially in swimming pool and drinking water. Chlorine gas was used for the first time as a chemical weapon on April 22, 1915, at Ypres by the German Army. The effect on the Allied forces was shocking as the existing gas masks were difficult to deploy and had not been broadly distributed.

4.7.2 Mining, production, and major minerals

Chlorine is too reactive to occur as the free element in nature but is very abundant in the form of its chloride salts. It is the 20th most abundant element in earth's crust and makes up 126 ppm of it, through the large deposits of chloride minerals, especially sodium chloride (halite, NaCl), that have been evaporated from water bodies. All of these pales in comparison with the reserves of chloride ions in seawater: smaller amounts at higher concentrations occur in some inland seas and underground brine wells, such as the Great Salt Lake in Utah and the Dead Sea in Israel.

Small volumes of chlorine gas can be produced in the laboratory by reacting hydrochloric acid (HCl) with manganese dioxide (MnO_2), but the need rarely arises as it is readily available. In industrial applications, elemental chlorine is usually produced by the electrolysis of sodium chloride dissolved in water. This method, the chloralkali process industrialized in 1892, now provides most industrial chlorine gas. Along with chlorine, the method yields hydrogen gas and sodium hydroxide, which is the most valuable product:

$$2NaCl + 2H_2O \rightarrow Cl_2 + H_2 + 2NaOH$$

The electrolysis of chloride solutions proceeds according to the following equations:

$$\text{Cathode: } 2H_2O + 2e^- \rightarrow H_2 + 2OH^-$$
$$\text{Anode: } 2Cl^- \rightarrow Cl_2 + 2e^-$$

In diaphragm cell electrolysis, an asbestos (or polymer-fiber) diaphragm separates a cathode and an anode, preventing the chlorine forming at the anode from remixing with the sodium hydroxide and the hydrogen formed at the cathode. The salt solution (brine) is uninterruptedly fed to the anode compartment and flows through the diaphragm to the cathode compartment, where the caustic alkali is formed, and the brine is partially depleted. Membrane cell electrolysis uses a permeable membrane as an ion exchanger. Saturated sodium (or potassium) chloride solution is passed through the anode compartment, leaving at a lower concentration. This method also produces very pure sodium (or potassium) hydroxide but has the disadvantage of requiring very pure brine at high concentrations. In the Deacon process, invented by English chemist and industrialist Henry Deacon (July 30, 1822 to July 23, 1876), hydrogen chloride recovered from the production of organochlorine compounds is recovered as chlorine (Fig. 4.78). The process relies on oxidation using oxygen:

$$4HCl + O_2 \rightarrow 2Cl_2 + 2H_2O$$

The reaction requires a catalyst. As introduced by Deacon, early catalysts were based on copper. Commercial processes, such as the Mitsui MT-Chlorine Process, have switched to chromium and ruthenium-based catalysts.

There are in total just under 450 minerals that contain chlorine in their crystal structure. Within the sulfide group 19 minerals contain Cl, such as ardaite ($Pb_{19}Sb_{13}S_{35}Cl_7$), corderoite ($Hg_3^{2+}S_2Cl_2$), and dadsonite ($Pb_{23}Sb_{25}S_{60}Cl$). Most common is Cl in the halide class with nearly 150 minerals, including well-known minerals such as atacamite ($Cu_2(OH)_3Cl$) (Fig. 4.79), boleite ($KPb_{26}Ag_9Cu_{24}(OH)_{48}Cl_{62}$)) (Fig. 4.80), chlorargyrite (AgCl) (Fig. 4.81), halite (NaCl) (Fig. 4.82), kleinite (($Hg_2N)(Cl,SO_4)\cdot nH_2O$) (Fig. 4.83), nadorite ($PbSbClO_2$) (Fig. 4.84), salammoniac (NH_4Cl) (Fig. 4.85), and sylvite (KCl) (Fig. 4.86). In the oxide class there are 27 minerals with Cl in their structure, for example, chloromenite ($Cu_9(SeO_3)_4O_2Cl_6$) and iowaite ($Mg_6Fe_2^{3+}(OH)_{16}Cl_2\cdot 4H_2O$). There are only 13 minerals among the carbonate minerals that contain Cl, such as northupite ($Na_3Mg(CO_3)_2Cl$) or phosgenite ($Pb_2CO_3Cl_2$). More minerals are found among the borates, a total of 28 different minerals, for example, boracite ($Mg_3(B_7O_{13})Cl$) and chambersite ($Mn_3^{2+}(B_7O_{13})Cl$). In the sulfate class 30 minerals are found that contain structural chlorine. These include minerals such as gordaite ($NaZn_4(SO_4)(OH)_6Cl\cdot 6H_2O$), hanksite ($Na_{22}K(SO_4)_9(CO_3)_2Cl$), and spangolite ($Cu_6Al(SO_4)(OH)_{12}Cl.3H_2O$). The phosphate class is represented by 28 different minerals containing Cl, including minerals such as: chlorapatite ($Ca_5(PO_4)_3Cl$), hedyphane ($Ca_2Pb_3(AsO_4)_3Cl$), lavendulan ($NaCaCu_5(AsO_4)_4Cl\cdot 5H_2O$), mimetite ($Pb_5(AsO_4)_3Cl$), pyromorphite ($Pb_5(PO_4)_3Cl$), and sampleite ($NaCaCu_5(PO_4)_4Cl\cdot 5H_2O$). More than 50 silicates contain structural chlorine, which includes minerals such as afghanite ($(Na,K)_{22}Ca_{10}(Si_{24}Al_{24}O_{96})(SO_4)_6Cl_6$), davyne ($(Na,K)_6Ca_2(Al_6Si_6O_{24})(Cl_2,SO_4)_2$), eudialyte ($Na_{15}Ca_6(Fe^{2+},Mn^{2+})_3Zr_3[Si_{25}O_{73}](O,OH,H_2O)_3(OH,Cl)_2$), haüyne ($(Na,K)_3(Ca,Na)(Al_3Si_3O_{12})(SO_4,S,Cl)$), marialite ("scapolite" endmember) ($Na_4Al_3Si_9O_{24}Cl$), sodalite

FIGURE 4.78 Henry Deacon.

FIGURE 4.79 Deep lustrous green lath-like atacamite, $Cu_2(OH)_3Cl$, crystals to 5 mm long in divergent flat-lying sprays on matrix. Associated with minor fibrous brochantite, $Cu_4(SO_4)(OH)_6$, and blue-green chrysocolla, $Cu_{2-x}Al_x(H_{2-x}Si_2O_5)(OH)_4{\cdot}nH_2O$ ($x < 1$). La Farola mine, Tierra Amarilla, Copiapó, Chile.

FIGURE 4.80 Boleite, $KPb_{26}Ag_9Cu_{24}(OH)_{48}Cl_{62}$, 1 cm on edge with several smaller crystals. Amelia mine, Boleo, Santa Rosalia, Baja California, Mexico.

FIGURE 4.81 Chlorargyrite, AgCl, yellow waxy masses, and well-formed crystals to 1 mm throughout spongy limonitic matrix. Broken Hill, New South Wales, Australia.

FIGURE 4.82 Plate of pure white cubic halite, NaCl, with the largest crystals 1−4 cm. San Jose, Santa Clara Co., California, United States.

FIGURE 4.83 Kleinite, $(Hg_2N)(Cl,SO_4).nH_2O$, orange well-formed crystals to 1 mm in 3−6 mm clusters McDermitt mine, Humboldt Co., Nevada, United States.

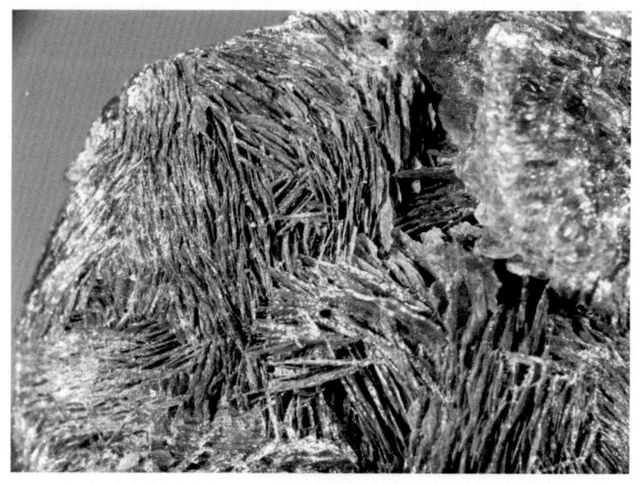

FIGURE 4.84 Deep amber-brown plates of nadorite, $PbSbClO_2$, about 1 cm. Djebel Nador, Constantine prov., Algeria.

FIGURE 4.85 A 6.5 cm group of white mostly formless crystals of salammoniac, NH_4Cl, on yellow rosickýite, S. Volcano Island, Lipari, Eolie Islands, Messina Prov., Sicily, Italy.

FIGURE 4.86 Cluster of nearly colorless, well-formed octahedral sylvite, KCl, crystals to 2.5 cm. Carlsbad Potash dist., Eddy Co., New Mexico, United States.

($Na_8(Al_6Si_6O_{24})Cl_2$) (Fig. 4.87), and tugtupite ($Na_4BeAlSi_4O_{12}Cl$). Finally, there are two organic compounds with chlorine in their structure, calclacite ($Ca(CH_3COO)Cl{\cdot}5H_2O$), and novgorodovaite ($Ca_2(C_2O_4)Cl_2{\cdot}2H_2O$).

4.7.3 Chemistry

Chlorine has two naturally occurring stable isotopes, ^{35}Cl making up 76% and ^{37}Cl making up the remaining 24%. Both are formed in stars in the oxygen-burning and silicon-burning processes. Both have nuclear spin 3/2 + and therefore can be used for NMR, though with the spin magnitude being larger than 1/2 resulting in a nonspherical nuclear charge distribution and hence resonance broadening as a result of a nonzero nuclear quadrupole moment and resultant quadrupolar relaxation. The other isotopes are all radioactive, with half-lives too short to occur in nature primordially (In geochemistry, geophysics, and geonuclear physics, primordial nuclides, also known as primordial isotopes, are nuclides found on earth that have existed in their current form since before earth was formed. Primordial nuclides were

FIGURE 4.87 Sodalite, $Na_8(Al_6Si_6O_{24})Cl_2$, neon blue, lustrous crystals to 2.5 mm. Lajuar Madan, Sar-e-Sang dist., Badakhshan, Afghanistan.

present in the interstellar medium from which the solar system was formed, and were formed in, or after, the Big Bang, by nucleosynthesis in stars and supernovae followed by mass ejection, by cosmic ray spallation, and potentially from other processes.). Of these, the most commonly used in the laboratory are ^{36}Cl (half-life 3.01×10^5 y) and ^{38}Cl (half-life 37.2 min), which may be formed through the neutron activation of natural chlorine. The most stable radioisotope is ^{36}Cl. The main decay mode of isotopes lighter than ^{35}Cl is electron capture to isotopes of sulfur; while that of isotopes heavier than ^{37}Cl is β decay to isotopes of argon; and ^{36}Cl may decay via either route to stable ^{36}S or ^{36}Ar. ^{36}Cl can be found in trace amounts in nature as a cosmogenic nuclide in a ratio of about $(7-10) \times 10^{-13}$ to 1 with stable chlorine isotopes: it is formed in the atmosphere by spallation of ^{36}Ar by interactions with cosmic ray protons. In the top meter of the lithosphere, ^{36}Cl is generated mostly by thermal neutron activation of ^{35}Cl and spallation of ^{39}K and ^{40}Ca. In the subsurface environment, muon capture by ^{40}Ca becomes more important to generate ^{36}Cl.

Chlorine is a member of the halogen group, positioned below fluorine and above bromine. Thus, its properties and reactivity are intermediate between these two elements. It is a greenish-yellow gas with an atom electron configuration of $[Ne]3s^23p^5$. Like all halogens, it has seven valence electrons and is thus one electron short to complete its octet. Elemental chlorine exists as a diatomic gas, forming a single bond with itself to complete its octet. It is a strong oxidizing agent, has the highest electron affinity, and the third-highest electronegativity on the Pauling scale (Table 4.7).

Chlorine is very reactive so that it is present in the earth's crust as ionic chloride compounds. It reacts, often vigorously, with all elements except the noble gases. It reacts with water to form hydrochloric acid and hypochlorous acid, which have been important and remain as such in the disinfectant industry. It reacts with CO, NO, and SO_2 to give $COCl_2$, NOCl, and SO_2Cl_2. It halogenates the metals, sometimes producing a higher metal oxidation state than the heavier halogens, Br_2 and I_2. For example, Re reacting with the halogens yields $ReCl_6$ $ReBr_5$ and ReI_4.

4.7.3.1 Reaction with air, oxygen, nitrogen, and hydrogen

Cl_2 does not react directly with oxygen, although chlorine oxides are known. Likewise, it is not reactive toward nitrogen and hydrogen at room temperature in the absence of light. Cl_2 reacts rapidly and violently with hydrogen at temperatures above 250°C or in the presence of ultraviolet light.

$$H_2 + Cl_2 \xrightarrow{UV\ light} 2HCl$$

Chlorine chemistry is important in the stratosphere and troposphere. The Cl_2 bond and the Cl bond with other elements are cleaved easily by UV light, which is available in the stratosphere, producing chlorine radicals responsible for ozone depletion. Cl radicals are highly reactive and can oxidize many constituents of the atmosphere including

TABLE 4.7 Chlorine properties.

Appearance	Pale yellow-green gas
Standard atomic weight $A_{r,std}$	35.45
Block	P-block
Element category	Reactive nonmetal
Electron configuration	[Ne] $3s^2$ $3p^5$
Phase at STP	Gas
Melting point	−101.5°C
Boiling point	−34.04°C
Density (at STP)	3.2 g/L
When liquid (at b.p.)	1.5625 g/cm^3
Heat of fusion	(Cl_2) 6.406 kJ/mol
Heat of vaporization	(Cl_2) 20.41 kJ/mol
Molar heat capacity	(Cl_2) 33.949 J/(mol·K)
Oxidation states	**−1**, **+1**, +2, **+3**, +4, **+5**, +6, **+7**
Ionization energies	1st: 1251.2 kJ/mol 2nd: 2298 kJ/mol 3rd: 3822 kJ/mol
Covalent radius	102 ± 4 pm
Van der Waals radius	175 pm

STP, Standard temperature and pressure.
Bold font indicates main oxidation state.

methane, NO_x, and elemental mercury. Cl radicals react with ozone to form ClO and O_2. If this was a simple one to one reaction, destroying one ozone molecule per chlorine radical present, the effect would have been less. However, the ClO can react with O atoms to generate O_2 and another Cl radical, which can then react with another ozone molecule. It is reported that one Cl radical can destroy around 100,000 ozone molecules that way, thus, thinning of the protective layer of ozone in the stratosphere is an unfortunate result. Cl_2 being reactive is not the main source of Cl radical in the stratosphere, but rather, the compounds called chlorofluorocarbons (CFCs), which were common in aerosols and refrigerants since the 1930s. They have become heavily regulated since the 1970s (By 1987, in response to a dramatic seasonal depletion of the ozone layer over Antarctica, diplomats in Montreal forged a treaty, the Montreal Protocol, which called for drastic reductions in the production of CFCs. On March 2, 1989, 12 European Community nations agreed to ban the production of all CFCs by the end of the century. In 1990, diplomats met in London and voted to significantly strengthen the Montreal Protocol by calling for a complete elimination of CFCs by the year 2000. By the year 2010, CFCs should have been completely eliminated from developing countries as well.).

$$Cl + O_3 \rightarrow ClO + O_2$$
$$ClO + O \rightarrow Cl + O_2$$

In the troposphere, harsh UV radiation is not readily available to produce chlorine atoms from stable species and only a few pathways for making chlorine atoms in the troposphere are known. Moreover, reactive chlorine atoms quickly react with moisture and end up as HCl and is removed by rainout. However, recent reports have shown that HCl reacts with NO_2 and N_2O_5 to produce, respectively, ClNO and $ClNO_2$. These are proposed to be the sources of chlorine atoms in the troposphere.

4.7.3.2 Reaction with other halogens

Chlorine, Cl_2, reacts with fluorine, F_2, at 225°C to form the interhalogen species, chlorine monofluoride (ClF), and chlorine trifluoride (ClF_3). Under higher temperatures and pressures (350°C and 225 atm), Cl_2 reacts with excess F_2 to form chlorine pentafluoride (ClF_5). Cl_2 reacts with bromine, Br_2, in the gas phase to form the unstable interhalogen species bromine monochloride (ClBr). Similarly, Cl_2 reacts with iodine, I_2, at room temperature to form the iodine monochloride (ICl).

$$Cl_2(g) + F_2(g) \rightarrow 2ClF(g)$$
$$Cl_2(g) + 3F_2(g) \rightarrow 2ClF_3(g)$$
$$Cl_2(g) + 5F_2(g) \rightarrow 2ClF_5(g)$$
$$Cl_2(g) + Br_2(g) \rightarrow 2BrCl(g)$$
$$Cl_2(g) + I_2(s) \rightarrow 2ICl(s)$$

4.7.3.3 *Reaction with acids and bases*

Cl_2 reacts with cold dilute sodium hydroxide to form sodium chloride (NaCl) and sodium chlorate(I) or sodium hypochlorite (NaClO). When Cl_2 is made to react with hot concentrated sodium hydroxide, the products are NaCl and sodium chlorate(V) or simply sodium chlorate ($NaClO_3$). Cl_2 reacting with water produces hydrochloric acid and hypochlorous acid.

$$Cl_2 + 2NaOH_{cold,dil.} \rightarrow NaCl + NaClO + H_2O$$
$$3Cl_2 + 6NaOH_{hot,\ conc.} \rightarrow 5NaCl + NaClO_3 + 3H_2O$$
$$Cl_2 + H_2O \rightarrow HClO + HCl$$

4.7.3.4 *Organic compounds of chlorine*

Organochlorines, organochlorides, chlorocarbon, or chlorinated hydrocarbon are organic compounds containing at least one covalently bonded atom of chlorine. The presence of the chlorine in the compound influences the chemical behavior of the molecule.

Organochlorines can be prepared from reactions with elemental chlorine, hydrogen chloride, or with other chlorinating agents such as thionyl chloride ($SOCl_2$), phosphorus pentachloride (PCl_5), sulfuryl chloride (SO_2Cl_2) and phosphorus trichloride (PCl_3). Using elemental chlorine, alkanes and aryl alkanes may be chlorinated under free radical conditions, with UV light. However, the extent of chlorination is difficult to control. Aryl chlorides may be prepared by the Friedel-Crafts reaction, using chlorine and a Lewis acid catalyst such as $FeCl_3$.

Chlorinated organic compounds are important building blocks in the chemical industry. They are found in alkaloids, terpenes, amino acids, flavonoids, steroids, and fatty acids. Many pharmaceutical agents and vitamins involve chlorine chemistry or require chlorine. Indeed, they are very useful compounds in many applications, but some are of profound environmental concern as having been seen in its chemistry in the stratosphere.

4.7.4 Major uses

Sodium chloride (NaCl) is by far the most common chlorine compound, and it is the principal source of chlorine and hydrochloric acid for the current massive chlorine-chemicals industry. Around 15000 Cl-containing compounds are commercially available, comprising a large variety of compounds, for example, chlorinated methanes and ethanes, vinyl chloride and its polymer polyvinyl chloride (PVC), aluminum trichloride ($AlCl_3$) for catalysis, the chlorides of Mg, Ti, Zr, and Hf that are precursors for manufacturing the pure elements, etc. Of all elemental Cl produced, around 63% is used in the production of organic compounds, and 18% in the production of inorganic chlorine compounds. It is used as an oxidizing agent and in substitution reactions. About 85% of pharmaceuticals use chlorine or its compounds at some point in their production. Historically, chlorine was commonly used to produce chloroform (an anesthetic, IUPAC name trichloromethane, $CHCl_3$) and carbon tetrachloride (CCl_4, a dry-cleaning solvent). However, both chemicals are now strictly controlled as they can cause liver damage. The remaining 19% of chlorine is used for bleaches and disinfection products. The most important of the Cl-containing organic compounds in terms of quantities are 1,2-dichloroethane ($C_2H_4Cl_2$) and vinyl chloride (IUPAC name chloroethene, C_2H_3Cl), intermediates in the production of PVC (IUPAC name poly(1-chloroethylene), $(C_2H_3Cl)_n$). This is a very versatile plastic used in window frames, car interiors, electrical wiring insulation, water pipes, blood bags, and vinyl flooring.

Chlorine was used for the first time to disinfect tap water at Maidstone, England in 1897 during a typhoid outbreak; the epidemic was brought under control. Ultimately chlorine became the method of purifying water throughout Britain and most of the developed world. Chlorine is generally used (in the form of hypochlorous acid, HClO) to kill bacteria and other microbes in drinking water supplies and public swimming pools. In most private swimming pools, chlorine itself is not used, but rather sodium hypochlorite (NaOCl), formed from chlorine and sodium hydroxide (NaOH), or solid tablets of chlorinated isocyanurates. The downside of using chlorine in swimming pools is that the chlorine reacts with the proteins in human hair and skin. The characteristic "chlorine aroma" linked with swimming pools is not due to the chlorine itself, but of chloramine (R_2NCl and $RNCl_2$ where R is an organic group), a chemical molecule formed through the reaction of free dissolved chlorine with amines in organic substances. As a disinfectant in water, chlorine is more than 3x as effective against *Escherichia coli* as bromine, and more than 6x as effective as iodine. More and more, chloramine is added directly to drinking water for disinfection purposes, a process called chloramination. Both chlorine and small amounts of ammonia are added to the water one at a time which react together to form chloramine (also called combined chlorine), a long-lasting disinfectant. Chloramine disinfection is used in both small and large water treatment plants. It is frequently unfeasible to store and utilize poisonous chlorine gas for water treatment; hence, other methods of adding chlorine are used, for example, hypochlorite (ClO^-) solutions, which gradually release chlorine into the water, and compounds like sodium dichloro-s-triazinetrione (dihydrate or anhydrous, IUPAC name sodium 3,5-dichloro-2,4,6-trioxo-1,3,5-triazinan-1-ide, $C_3Cl_2N_3NaO_3$), occasionally referred to as "dichlor," and trichloro-s-triazinetrione (IUPAC name 1,3,5-trichloro-1,3,5-triazinane-2,4,6-trione, $C_3Cl_3N_3O_3$), occasionally referred to as "trichlor." These materials are stable while solid and may be used in powdered, granular, or tablet form. When added in small quantities to pool water or industrial water systems, the chlorine atoms hydrolyze from the rest of the molecule forming hypochlorous acid (HOCl), which acts as a universal biocide, killing germs, microorganisms, algae, and so on.

Chlorine gas, also known as bertholite, was first used as a weapon in World War I by Germany on April 22, 1915 in the Second Battle of Ypres. Chlorine reacts with water in the mucosa of the lungs to form hydrochloric acid, destructive to living tissue and potentially lethal. During World War I it resulted in 5000 dead soldiers and some 15,000 soldiers were affected by it. Human respiratory systems can be protected from chlorine gas by gas masks with activated charcoal or other filters, which makes chlorine gas much less lethal than other chemical weapons. It was established by a German chemist later to be a (controversial because of his work on chlorine gas as an offensive weapon) Nobel laureate, Fritz Haber (December 9, 1868 to January 29, 1934) of the Kaiser Wilhelm Institute in Berlin, in collaboration with the German chemical industry IG Farben, which developed techniques for discharging chlorine gas against an entrenched enemy. After its first use, both sides in the war used chlorine gas as a chemical weapon, but it was quickly replaced by the more deadly phosgene (carbonyl dichloride, $COCl_2$) and mustard gas (1-chloro-2-[(2-chloroethyl)sulfanyl]ethane, $C_4H_8Cl_2S$). Chlorine gas was also used during the Iraq War in Anbar Province in 2007, with insurgents loading truck bombs with mortar shells and chlorine tanks. On October 24, 2014, it was reported that the Islamic State of Iraq and the Levant had used chlorine gas in the town of Duluiyah, Iraq. Laboratory analysis of clothing and soil samples corroborated the use of chlorine gas against Kurdish Peshmerga Forces in a vehicle-borne improvised explosive device attack on January 23, 2015 at the Highway 47 Kiske Junction near Mosul. Recently, the Syrian military has allegedly used chlorine as a chemical weapon delivered from barrel bombs and rockets.

References

Berthier, P. (1821). Analyse de l'alumine hydratée des Beaux, département des Bouches-du-Rhône. *Annales des mines,* 1st series, *6*, 531–534.

Berzelius, J. J. (1814). *Försök att, genom användandet af den electrokemiska theorien och de kemiska proportionerna: grundlägga ett rent vettensk. system för mineralogien.* Stockholm: A. Gadelius.

Berzelius, J. J. (1824a). II. Untersuchungen über Flussspathsäure und deren merkwürdigsten Verbindungen. *Annalen der Physik, 77*, 169–230.

Berzelius, J. J. (1824b). Undersökning af flusspatssyran och dess märkvärdigaste föreningar. *Kongliga Vetenskaps-Academiens Handlingar, 12*, 46–98.

Birch, A.J. (1944). Reduction by dissolving metals. Part 1. *Journal of the Chemical Society*, 430–436.

Cotton, F. A., & Wilkinson, G. (1966). *Advanced Inorganic Chemistry* (2nd ed.). New York, NY: Wiley.

Davy, H. (1808a). Electro-chemical researches on the decomposition of the earths; with observations on the metals obtained from the alkaline earths, and on the amalgam procured from ammonia. *Philosophical Transactions of the Royal Society of London, 98*, 333–370.

Davy, H. (1808b). On some new phenomena of chemical changes produced by electricity, particularly the decomposition of the fixed alkalies, and the exhibition of the new substances which constitute their bases; and on the general nature of alkaline bodies. *Philosophical Transactions of the Royal Society of London, 98*, 1–44.

Davy, H. (1809a). An account of some new analytical researches on the nature of certain bodies, particularly the alkalies, phosphorus, sulphur, carbonaceous matter, and the acids hitherto undecomposed: With some general observations on chemical theory. *Philosophical Transactions of the Royal Society of London, 99*, 39–104.

Davy, H. (1809b). Über einige neue Erscheinungen chemischer Veränderungen, welche durch die Electricität bewirkt werden; insbesondere über die Zersetzung der feuerbeständigen Alkalien, die Darstellung der neuen Körper, welche ihre Basen ausmachen, und die Natur der Alkalien überhaupt. *Annalen der Physik, 31*, 113–175.

Davy, H. (1811). The Bakerian Lecture. On some of the combinations of oxymuriatic gas and oxygene, and on the chemical relations of these principles, to inflammable bodies. *Philosophical Transactions of the Royal Society of London, 101*, 1–35.

Faraday, M. (1823). On fluid chlorine. *Philosophical Transactions of the Royal Society of London, 113*, 160–164.

Gay-Lussac, J. L., & Thenard, L. J. (1809). Extrait des mémoires lus à l'Institut national, depuis le 7 mars 1808 jusqu'au 27 février 1809. *Mémoires de Physique et de Chimie de la Société d'Arcueil, 2*, 295–358.

Gay-Lussac, J. L., & Thénard, L. J. (1811a). *Recherches physico-chimiques: faites sur la pile, sur la préparation chimique et les propriétés du potassium et du sodium, sur la décomposition de l'acide boracique, sur les acides fluorique, muriatique et muriatique oxigéné, sur l'action chimique de la lumière, sur l'analyse végétale et animale, etc* (vol 1). Paris: Deterville.

Gay-Lussac, J. L., & Thénard, L. J. (1811b). *Recherches physico-chimiques: faites sur la pile, sur la préparation chimique et les propriétés du potassium et du sodium, sur la décomposition de l'acide boracique, sur les acides fluorique, muriatique et muriatique oxigéné, sur l'action chimique de la lumière, sur l'analyse végétale et animale, etc* (vol 2). Paris: Deterville.

Lewis, G. N. (1916). The atom and the molecule. *Journal of the American Chemical Society, 38*, 762–785.

Ørsted, H.C. (1825). *Oversigt over det Kongelige Danske Videnskabernes Selskabs Forhanlingar og dets Medlemmerz Arbeider, fra 31 Mai 1824 til 31 Mai 1825*, 15–16, Tryt hos Kgl. Hofbogtrykker Bianco Luno, Kjövenhavn.
Pauling, L. (1949). The modern theory of valency. *Journal of the Chemical Society*, 1461–1467.
Sainte-Claire Deville, H. E. (1855). Du silicium et du titane. *Comptes rendus, 40*, 1034–1036.
Sainte-Claire Deville, H. E. (1859). *De l'aluminium, ses propriétés, sa fabrication.* Paris: Mallet-Bachelier.
Scheele, C. W. (1774). Om Brunsten, eller Magnesia, och dess Egenskaper. *Kongliga Vetenskaps Academiens Handlingar, 35*, 177–194.
Schweigger, J. S. C. (1811). Nachschreiben des Herausgebers, die neue Nomenclatur betreffend. *Journal für Chemie und Physik, 3*, 249–255.
Thomson, T. (1817). *A system of chemistry: In four volumes* (Vol 1). London: Baldwin, Craddock and Joy, Paternoster-Row.
Thomson, R. D. (1870). *Dictionary of chemistry with its applications to mineralogy, physiology and the arts.* London: Richard Griffin & Company.
Weeks, M. E. (1932). The discovery of the elements.II. Elements known to the alchemists. *Journal of chemical education, 9*, 11–21.
Wöhler, F. (1827). Ueber das Aluminium. *Annalen der Physik und Chemie, 2*, 146–161.
Wooster, C.B. & Godfrey, K.L. (1937). Mechanism of the reduction of unsaturated compounds with alkali metals and water. *Journal of the American Chemical Society*, 59, 596–597.

Further reading

Ainsworth, S. (2013). Epsom's deep bath. *Nurse Prescribing, 11*, 269.
Almqvist, E. (2003). *History of industrial gases.* New York: Springer US.
Amundsen, K., Aune, T. K., Bakke, P., Eklund, H. R., Haagensen, J. Ö., Nicolas, C., ... Wallevik, O. (2003). *Magnesium. Ullmann's encyclopedia of industrial chemistry.* Weinheim: Wiley-VCH.
Aspin, C. (1990). *The cotton industry.* Aylesbury: Shire.
Audi, G., Bersillon, O., Blachot, J., & Wapstra, A. H. (2003). The NUBASE evaluation of nuclear and decay properties. *Nuclear Physics A, 729*, 3–128.
Aufray, B., Kara, A., Vizzini, S. B., Oughaddou, H., LéAndri, C., Ealet, B., & Le Lay, G. (2010). Graphene-like silicon nanoribbons on Ag(110): A possible formation of silicene. *Applied Physics Letters, 96*, 183102.
Avedesian, M. M. (1999). *Magnesium and magnesium alloys.* Materials Park, OH: ASM International.
Banks, A. (1990). Sodium. *Journal of Chemical Education, 67*, 1046.
Bates, T. F. (1962). Halloysite and gibbsite formation in Hawaii. *Clays and Clay Minerals, 9*, 315–328. Berthier, P., Analyse de l'alumine hydratée des Beaux, département des Bouches-du-Rhóne, *Annales des mines*, 1st series, **6**, 1821, 531–534.
Berzelius, J. J. (1825). On the mode of obtaining silicium, and on the characters and properties of that substance. *Philosophical Magazine, 65*, 254–267.
Berzelius, J. N. J. (1832). *rsberättelse och framstegen i fysik og kemi, afgifven den 31 mars, 1832.* Stockholm: Tryckt hos P.A. Norstedt.
Besticker, A. C. (Ed.), (1963). *Proceedings of the First Symposium on Salt.* Cleveland, OH: Northern Ohio Geological Survey.
Birch, W. D., & Henry, D. A. (1993). *Phosphate minerals of Victoria.* Mineralogical Society of Victoria.
Botsch, W. (2001). Chemiker, Techniker, Unternehmer: Zum 150. Geburtstag von Hermann Frasch. *Chemie in Unserer Zeit, 35*, 324–331.
Bouvet, M. (1950). Les grands pharmaciens: Labarraque (1777–1850). *Revue d'Histoire de la Pharmacie, 38*, 97–107.
Cameron, A. G. W. (1973). Abundance of the elements in the solar system. *Space Science Reviews, 15*, 121–146.
Ceccotti, S. P. (1996). Plant nutrient sulphur-a review of nutrient balance, environmental impact and fertilizers. *Fertilizer Research, 43*, 117–125.
Clapham, J. H., & Power, E. E. (1977). *The Cambridge economic history of Europe from the decline of the Roman Empire.* Cambridge. Cotton, F.A., Wilkinson, G., Advanced Inorganic Chemistry (2nd ed.). 1966, Wiley, New York, NY.: Cambridge University Press.
Crass, M. F., Jr. (1941). A history of the match industry. Part 9. *Journal of Chemical Education, 18*, 428–431.
Cremlyn, R. J. W. (1996). *An introduction to organosulfur chemistry.* Chichester: Wiley.
Cullen, D. J. (1988). Mineralogy of nitrogenous guano on the Bounty Islands, SW Pacific Ocean. *Sedimentology, 93*, 421–428.
Dangic, A. (1995). Karst bauxite facies: A new conception and related systematic. *Geological Society (Greece), 4*, 694–698.
Dar, S. A., Khan, K. F., & Birch, W. D. (2017). *Sedimentary: Phosphates. Reference module in earth systems and environmental sciences* (pp. 1–16). Elsevier.
Dekker, R., Usechak, N., Först, M., & Driessen, A. (2008). Ultrafast nonlinear all-optical processes in silicon-on-insulator waveguides. *Journal of Physics D, 40*, R249–R271.
Denisenkov, P. A., & Ivanov, V. V. (1987). Sodium synthesis in hydrogen burning stars. *Soviet Astronomy Letters, 13*, 214–216.
Dickin, A. P. (2005). *In situ cosmogenic isotopes. Radiogenic isotope geology.* Cambridge: Cambridge University Press.
Diskowski, H., & Hofmann, T. (2000). *Phosphorus. Ullmann's encyclopedia of industrial chemistry.* Weinheim: Wiley-VCH.
Dodd, R. T. (1986). *Thunderstones and shooting stars: The meaning of meteorites.* Cambridge, MA; London: Harvard University Press.
Downs, A. J. (1993). *Chemistry of aluminium, gallium, indium and thallium.* London: Blackie Academic & Professional.
Dreizin, E. L., Berman, C. H., & Vicenzi, E. P. (2000). Condensed-phase modifications in magnesium particle combustion in air. *Scripta Materialia, 122*, 30–42.
Drozdov, A. (2007). *Aluminium: The thirteenth element* (B. Droitcour, D. Dynin, J. Melnikova et al., Trans). RUSAL Library.
Emsley, J. (2001). *Nature's building blocks.* Oxford: Oxford University Press.
Emsley, J. (2002). *The 13th element: The sordid tale of murder, fire, and phosphorus.* New York: John Wiley & Sons.

Eow, J. S. (2002). Recovery of sulfur from sour acid gas: A review of the technology. *Environmental Progress*, *21*, 143–162.

Ertl, G., Knözinger, H., & Weitkamp, J. (2010). *Preparation of solid catalysts*. Weinheim: Wiley-VCH.

Everts, S. (2017). When chemicals became weapons of war. <http://chemicalweapons.cenmag.org/when-chemicals-became-weapons-of-war/> Accessed 27.08.19.

Fontani, M., Costa, M., & Orna, M. V. (2014). *The lost elements: The periodic table's shadow side*. Oxford: Oxford University Press.

Gaines, R. V., Skinner, H. C., Foord, E. F., Mason, B., & Rosenzweig, A. (1997). *Dana's new mineralogy* (8th edn). New York: John Wiley & Sons, Inc.

Greenwood, N. N., & Earnshaw, A. (1996). *Chemistry of the elements*. Oxford: Elsevier Science & Technology Books.

Gregory, M. R., & Rodgers, K. A. (1989). Phosphate minerals from the Bounty Islands, South Pacific Ocean. *Neues Jahrbuch für Mineralogie, Abhandlungen*, *160*, 117–131.

Guilbert, J. M., & Park, C. F. (2007). *The geology of ore deposits*. Long Grove: Waveland Press.

Gupta, A. K., & Nicol, K. (2004). The use of sulfur in dermatology. *Journal of Drugs in Dermatology*, *3*, 427–431.

Harben, P. W., & Bates, R. L. (1990). *Industrial minerals, geology and world deposits*. London: Metal Bulletin.

Harder, E. C., & Greig, E. W. (1960). Bauxite. In J. L. Gillson (Ed.), *Industrial minerals and rocks* (pp. 65–85). New York: The American Institute of Mining, Metallurgical, and Petroleum Engineers.

Healy, J. F. (1999). *Pliny the elder on science and technology*. Oxford: Oxford University Press.

Helmboldt, O., Hudson, L. K., Misra, C., Wefers, K., Heck, W., Stark, H., … Rösch, N. (2007). Aluminum compounds, inorganic. In *Ullmann's Encyclopedia of Industrial Chemistry*.

Hill, C. A., Forti, P., & Shaw, T. R. (1997). *Cave minerals of the world* (Vol. 238). Huntsville: National Speleological Society.

Hill, V. G., & Ostojic, S. (1984). The characteristics and classification of bauxites. In L. Jacob (Ed.), *Bauxite: Proceedings of the 1984 Bauxite Symposium* (pp. 31–45). New York: SME, AIME.

Holmes, H. N. (1936). Fifty years of industrial aluminum. *The Scientific Monthly*, *42*, 236–239.

Hudson, L. K., Misra, C., Perrotta, A. J., Wefers, K., & Williams, F. S. (2000). *Aluminum oxide*. *Ullmann's Encyclopedia of Industrial Chemistry*. Weinheim: Wiley-VCH.

Hughes, J. P. W., Baron, R., Buckland, D. H., Cooke, M. A., Craig, J. D., Duffield, D. P., … Porter, A. (1962). Phosphorus necrosis of the jaw: A present-day study: With clinical and biochemical studies. *Occupational and Environmental Medicine*, *19*, 83–99.

Hyndman, A. W., Liu, J. K., & Denney, D. W. (1982). Sulfur recovery from oil sands. Sulfur: New sources and uses. *ACS Symposium Series*, *183*, 69–82.

Ihde, A. J. (1984). *The development of modern chemistry*. New York: Dover Publications.

Kainer, K. (2007). *Magnesium: Proceedings of the 7th International Conference on Magnesium Alloys and Their Applications*. Weinheim: Wiley-VCH.

Karkanas, P., & Goldberg, P. (2010). 23 - Phosphatic features. In G. Stoops, V. Marcelino, & F. Mees (Eds.), *Interpretation of Micromorphological Features of Soils and Regoliths* (pp. 521–541). Amsterdam: Elsevier.

Karkanas, P., Rigaud, J. P., Simek, J. F., Albert, R. A., & Weiner, S. (2002). Ash, bones and guano: A study of the minerals and phytoliths in the sediment of Grotte XVI, Dordogne, France. *Journal of Archaeological Science*, *29*, 721–732.

Kaufmann, D. W. (1960). Brine purification. In D. W. Kaufmann (Ed.), *The production and properties of salt and brine* (pp. 186–204). New York: Reinhold Publishing.

Kesler, S. E. (1994). *Mineral resources, economics and the environment*. New York: Macmillan College Publishing Company, Inc.

Kirchhoff, G., & Bunsen, R. (1860). Chemische Analyse durch Spectralbeobachtungen. *Annalen der Physik und Chemie*, *186*, 161–189.

Klein, C., & Dutrow, B. (2008). *The 23rd edition of the manual of mineral science (after James D. Dana)*. Hoboken, NJ: John Wiley & Sons, Inc.

Klemm, A., Hartmann, G., & Lange, L. (2000). *Sodium and sodium alloys*. *Ullmann's encyclopedia of industrial chemistry*. Weinheim: Wiley-VCH.

Kogel, J. E., Trivedi, N. C., Barker, J. M., & Krukowski, S. T. (2006). *Industrial minerals & rocks: Commodities, markets, and uses*. Littleton: Society for Mining, Metallurgy, and Exploration.

Korzhinsky, M. A., Tkachenko, S. I., Shmulovich, K. I., & Steinberg, G. S. (1995). Native AI and Si formation. *Nature*, *375*, 544.

Koski, T. A., Stuart, L. S., & Ortenzio, L. F. (1966). Comparison of chlorine, bromine, iodine as disinfectants for swimming pool water. *Applied Microbiology*, *14*, 276–279.

Krebs, R. E. (2006). *The history and use of our earth's chemical elements: A reference guide*. Westport: Greenwood Press.

Kutney, G. (2007). *Sulfur: History, technology, applications & industry*. Toronto: ChemTec Publishing.

Lalmi, B., Oughaddou, H., Enriquez, H., Kara, A., Vizzini, S., Ealet, B., & Aufray, B. (2010). Epitaxial growth of a silicene sheet. *Applied Physics Letters*, *97*, 223109.

Landis, C. A., & Craw, D. (2003). Phosphate minerals formed by reaction of bird guano with basalt at Cooks Head Rock and Green Island, Otago, New Zealand. *Journal of the Royal Society of New Zealand*, *33*, 487–495.

Lavoisier, A. (1799). *Elements of chemistry: In a new systematic order, containing all the modern discoveries* (4th ed.) (R. Kerr, Trans.). William Creech, Edinburgh.

Lefebure, V. (2014). *Riddle of the Rhine: Chemical strategy in peace and war*. Echo Library, [Place of publication not identified].

Lefond, S. J., & Jacoby, C. H. (1983). Salt. In S. J. Lefond (Ed.), *Industrial minerals and rocks*. New York: SME, AIME.

Lin, A. N., Reimer, R. J., & Carter, D. M. (1988). Sulfur revisited. *Journal of the American Academy of Dermatology*, *18*, 553–558.

Lumley, R. N. (2011). *Fundamentals of aluminium metallurgy: Production, processing and applications*. Oxford; Philadelphia: Woodhead Publication.

Lyle, J. P., Granger, D. A., & Sanders, R. E. (2000). *Aluminum alloys*. *Ullmann's encyclopedia of industrial chemistry*. Weinheim: Wiley-VCH.

Mackenzie, D. S., Totten, G. E., Marcel, D. (2003). *Handbook of aluminum*.

Maibach, H. I., Surber, C., & Orkin, M. (1990). Sulfur revisited. *Journal of the American Academy of Dermatology*, *23*, 154–156.

Nehb, W., & Vydra, K. (2006). *Sulfur, . Ullmann's encyclopedia of industrial chemistry* (Vol. 35, pp. 1–71). Weinheim: Wiley-VCH.

Nesse, W. D. (2000). *Introduction to mineralogy*. New York: Oxford University Press.

Newton, D. E., & Baker, L. W. (1999). *Chemical elements: From carbon to krypton*. Detroit: UXL.

Oesper, R. E., & Lemay, P. (1950). Henri Sainte-Claire Deville, 1818–1881. *Chymia, 3*, 205–221.

Paiva, A. P., & Malik, P. (2004). Recent advances on the chemistry of solvent extraction applied to the reprocessing of spent nuclear fuels and radioactive wastes. *Journal of Radioanalytical and Nuclear Chemistry, 261*, 485–496.

Pal, U. B., & Powell, A. C. (2007). The use of solid-oxide-membrane technology for electrometallurgy. *JOM, 59*, 44–49.

Patterson, S. H. (1984). Bauxite and nonbauxite aluminum resources of the world, an update. In L. Jacob (Ed.), *Bauxite: Proceedings of the 1984 Bauxite Symposium* (pp. 3–20). New York: SME, AIME.

Perkins, D. (2002). *Mineralogy* (2nd edn). Upper Saddle River, NJ: Prentice Hall.

Rapp, G. (2009). *Archaeomineralogy*. Berlin: Springer.

Riall, L. (1998). *Sicily and the unification of Italy: Liberal policy and local power, 1859-1866*. Oxford: Clarendon Press.

Richards, J. W. (1896). *Aluminium: Its history, occurrence, properties, metallurgy and applications, including its alloys* (3 edn). Philadelphia: Henry Carey Baird & Co.

Robb, L. (2005). *Introduction to ore-forming processes*. Malden: Blackwell Publishing.

Rossotti, H. (1998). *Diverse atoms. Profiles of the chemical elements*. Oxford: Oxford University Press.

Sainte-Claire Deville, H. E. (1854). Note sur deux procédés de préparation de l'aluminium et sur une nouvelle forme du silicium. *Comptes rendus, 39*, 321–326.

Sanders, F. W., & Auxier, J. A. (1962). Neutron activation of sodium in anthropomorphous phantoms. *Health Physics, 8*, 371–379.

Schrödter, K., Bettermann, G., Staffel, T., Wahl, F., Klein, T., & Hofmann, T. (2008). *Phosphoric acid and phosphates. Ullmann's encyclopedia of industrial chemistry*. Weinheim: Wiley-VCH.

Sciolist Online Etymology Dictionary. <https://www.etymonline.com/>. Accessed 21.06.19.

Segal, D. (2017). *Materials for the 21st century*. Oxford: Oxford University Press.

Shahack-Gross, R., Berna, F., Karkanas, P., & Weiner, S. (2004). Bat guano and preservation of archaeological remains in cave sites. *Journal of Archaeological Science, 31*, 1259–1272.

Sheller, M. (2014). *Aluminum dreams: The making of light modernity*. Cambridge, MA: MIT Press.

Sheppard, S. C., & Herod, M. (2012). Variation in background concentrations and specific activities of ^{36}Cl, ^{129}I and U/Th-series radionuclides in surface waters. *Journal of Environmental Radioactivity, 106*, 27–34.

Shortland, A., Schachner, L., Freestone, I., & Tite, M. (2006). Natron as a flux in the early vitreous materials industry: Sources, beginnings and reasons for decline. *Journal of Archaeological Science, 33*, 521–530.

Siegfried, R. (1963). The discovery of potassium and sodium, and the problem of the chemical elements. *Isis, 54*, 247–258.

Snelders, H. A. M. (1971). J. S. C. Schweigger: His romanticism and his crystal electrical theory of matter. *Isis, 62*, 328–338.

Stillman, J. M. (2008). *The story of alchemy and early chemistry*. Kila, MT: Kessinger Publication Co.

Thomson, D. W. (1995). Prelude to the sulphur war of 1840: The neapolitan perspective. *European History Quarterly, 25*, 163–180.

Threlfall, R. E. (1951). *The story of 100 years of phosphorus making, 1851-1951*. Oldbury, Worcs: Albright & Wilson.

Toy, A. D. F. (2013). *The chemistry of phosphorus: Vol 3. Pergamon texts in inorganic chemistry*. Oxford: Pergamon Press.

Venetski, S. (1969). "Silver" from clay. *Metallurgist, 13*, 451–453.

Voronkov, M. G. (2007). Silicon era. *Russian Journal of Applied Chemistry, 80*, 2190–2196.

Wade, K., & Banister, A. J. (2016). *The chemistry of aluminium, gallium, indium and thallium: Comprehensive inorganic chemistry, Vol 12. Pergamon texts in inorganic chemistry*. Oxford: Pergamon Press.

Wagner, J. R. (1900). *Manual of chemical technology* (W. Crookes, F, Fischer, Trans) (Vol xxiii), (968 p.) D. Appleton & Co., New York, NY.

Weeks, M. E., & Leichester, H. M. (1968). Discovery of the elements. *Journal of Chemical Education*, Easton, PA.

White, D. G. (1998). *The alchemical body: Siddha traditions in medieval India*. Chicago: University of Chicago Press.

Witt, M., & Roesky, H. W. (2000). Organoaluminum chemistry at the forefront of research and development. *Current Science, 78*, 410–430.

Zakharov, V. V. (2003). Effect of scandium on the structure and properties of aluminum alloys. *Metal Science and Heat Treatment, 45*, 246–253.

Zhang, Y. (1986). Ancient Chinese sulfur manufacturing processes. *Isis, 77*, 487–497.

Zreda, M. G., Phillips, F. M., Elmore, D., Kubik, P. W., Sharma, P., & Dorn, R. I. (1991). Cosmogenic chlorine-36 production rates in terrestrial rocks. *Earth and Planetary Science Letters, 105*, 94–109.

Zulehner, W., Neuer, B., & Rau, G. (2000). *Silicon. Ullmann's encyclopedia of industrial chemistry*. Weinheim: Wiley-VCH.

Chapter 5

Period 4

5.1 19 K — Potassium

5.1.1 Discovery

The English name for the element potassium comes from the word "potash," which comes from an early process of extracting various potassium salts: placing in a pot the ash of burnt wood or tree leaves, adding water, heating, and evaporating the solution. Potash is largely a combination of potassium salts since plants have little or no sodium content, and the rest of a plant's main mineral content contains calcium salts of relatively low solubility in water. Though potash has been utilized since ancient times, it was not understood for most of its history to be a fundamentally different substance from sodium mineral salts. German chemist, physician, and philosopher Georg Ernst Stahl (October 22, 1659 to May 24, 1734) (Fig. 5.1) got experimental proof that led him to propose a fundamental difference between sodium and potassium salts in 1702, and French physician, naval engineer, and botanist Henri Louis Duhamel du Monceau (July 20, 1700 to August 13, 1782) (Fig. 5.2) was able to prove this difference in 1736 (du Monceau, 1736). The exact chemical composition of potassium and sodium compounds, and the status as chemical element of potassium and sodium, was not known at that time, and consequently, French nobleman and chemist Antoine Lavoisier (August 26, 1743 to May 8, 1794) did not include the alkali in his list of chemical elements in 1789. For a long period, the only important uses for potash were in the production of glass, bleach, soap, and gunpowder as potassium nitrate (KNO_3). Potassium soaps from animal fats and vegetable oils were particularly valued since they tend to be more water soluble and of softer texture and are hence known as soft soaps. The discovery by German chemist Justus von Liebig (May 12, 1803 to April 18, 1873) in 1840 that potassium is an essential element for plants and that most soil types lack in potassium caused a sharp increase in demand for potassium salts (von Liebig, 1840). Wood ash from fir trees was at first utilized as a potassium salt source to produce fertilizer. However, with the discovery in 1868 of mineral deposits containing potassium chloride near Staßfurt, Germany, the production of potassium-containing fertilizers commenced at an industrial scale. Other potash deposits were discovered, and by the 1960s Canada became the leading producer. Potassium metal was first isolated in 1807 by Cornish chemist and inventor Sir Humphry Davy (December 17, 1778 to May 29, 1829), who derived it from caustic potash (KOH, potassium hydroxide) by electrolysis of molten KOH with the newly discovered voltaic pile (Davy, 1808a,b, 1809). Potassium was the first metal that was isolated by electrolysis. He named it potassium, which he derived from the word potash. The symbol "K" stems from kali, itself from the root word alkali, which in turn comes from Arabic: القَلْيَه al-qalyah "plant ashes." In 1797 the German chemist Martin Klaproth (December 1, 1743 to January 1, 1817) discovered "potash" in the minerals leucite ($KAlSi_2O_6$) and lepidolite (now a series name, compositionally between polylithionite and trilithionite, $KLi_2AlSi_4O_{10}F_2{-}KLi_{1.5}Al_{1.5}Si_3O_{10}F_2$) and understood that "potash" was not a product of plant growth but actually contained a new element, which he proposed to name kali (Klaproth, 1797). The German physicist and chemist Ludwig Wilhelm Gilbert (August 12, 1769 to March 7, 1824) suggested the name Kalium for Davy's "potassium." In 1814 the Swedish chemist Berzelius advocated the name Kalium for potassium, with the chemical symbol "K." The English and French speaking countries adopted the name potassium suggested by Sir Humphry Davy and French chemists Joseph Louis Gay-Lussac (December 6, 1778 to May 9, 1850) and Louis-Jacques Thénard (May 4, 1777 to June 21, 1857), while the Germanic countries adopted Gilbert/Klaproth's name Kalium. The "Gold Book" of the International Union of Physical and Applied Chemistry has designated the official chemical symbol as K.

5.1.2 Mining, production, and major minerals

Potassium salts, such as carnallite, $KMgCl_3{\cdot}6H_2O$, langbeinite, $K_2Mg_2(SO_4)_3$, polyhalite, $K_2Ca_2Mg(SO_4)_4{\cdot}2H_2O$, and sylvite, KCl, form widespread evaporite deposits in ancient lake bottoms and seabeds, making extraction of potassium

The Periodic Table: Nature's Building Blocks. DOI: https://doi.org/10.1016/B978-0-12-821279-0.00005-4

FIGURE 5.1 Georg Ernst Stahl, 1716, engraving by Johann Georg Mentzel (1677–1743).

FIGURE 5.2 Henri Louis Duhamel du Monceau.

salts in these environments commercially feasible. The main source of potassium—potash—is mined in Canada, Russia, Belarus, Kazakhstan, Germany, Israel, United States, Jordan, among other localities around the world. The first mined deposits were located near Staßfurt, Germany, but the deposits span from Great Britain over Germany into Poland. They are in the Zechstein and were deposited in the Middle to Late Permian. The largest deposits ever found lie 1000 m (3300 ft.) below the surface of the Canadian province of Saskatchewan. These deposits are in the Elk Point Group formed in the Middle Devonian. Saskatchewan, where several large mines have operated since the 1960s, pioneered the technique of freezing of wet sands (the Blairmore formation) to drive mine shafts through them. The water of the Dead Sea is used by Israel and Jordan as a source of potash, while the concentration in normal oceans is too low for commercial production at current prices.

Several methods are used to separate K salts from Na and Mg salts. The most used process is fractional precipitation based on the solubility differences of the salts at different temperatures. Electrostatic separation of the ground salt mixture is also utilized in some mines. The resulting Na and Mg waste is either stored in the mine or piled up in slag heaps. Most of the mined potassium mineral ends up as KCl after processing. The mineral industry refers to KCl either as potash, muriate of potash, or just MOP. Pure potassium metal can be obtained through electrolysis of KOH using a method that has changed little since it was first used by Cornish chemist and inventor Sir Humphry Davy (December 17, 1778 to May 29, 1829) in 1807. While the electrolysis process was developed and used on an industrial scale in the 1920s, the thermal process of reacting Na with KCl in a chemical equilibrium reaction became the leading process some 30 years later. The manufacture of sodium potassium alloys is accomplished by changing the reaction time and the amount of Na used in the reaction. The Griesheimer process employing the reaction of KF with CaC_2 was also used to produce potassium.

$$Na + KCl \rightarrow NaCl + K \text{ (Thermal method)}$$
$$2KF + CaC_2 \rightarrow 2K + CaF_2 + 2C \text{ (Griesheimer process)}$$

Reagent-grade potassium metal was about $22/kg in 2010 when purchased by the ton. Lower purity metal is significantly cheaper. The market is volatile as long-term storage of the metal is difficult. It has to be stored in a dry inert gas atmosphere or anhydrous mineral oil to prevent the formation of a surface layer of potassium superoxide (KO_2), a pressure-sensitive explosive that detonates when scratched. The resulting explosion often starts a fire difficult to extinguish.

Potassium is a common element in many minerals. So far, 553 minerals have been recognized as containing, or may contain through substitution, K in their structure. There are 11 sulfides that have K as part of their composition, for example, bartonite ($K_6Fe_{20}S_{26}S$), djerfisherite ($K_6(Fe,Cu,Ni)_{25}S_{26}Cl$), and ottensite ($(Na,K)_3Sb_6^{3+}(Sb^{3+}S_3)O_9{\cdot}3H_2O$). In the halide class one will find 20 minerals, such as carnallite ($KMgCl_3{\cdot}6H_2O$) and sylvite (KCl). Furthermore, 21 minerals are observed in the oxide class, including minerals such as carnotite ($K_2(UO_2)_2(VO_4)_2{\cdot}3H_2O$), cryptomelane ($K(Mn_7^{4+}Mn^{3+})O_{16}$), and todorokite ($(Na,Ca,K,Ba,Sr)_{1-x}(Mn,Mg,Al)_6O_{12}{\cdot}3\text{-}4H_2O$).

Only 11 carbonate minerals contain K, for example, fairchildite ($K_2Ca(CO_3)_2$) and niter (KNO_3). Within the borate class the number is limited to only seven minerals, including kaliborite ($KMg_2H[B_6O_8(OH)_5]_2(H_2O)_4$) and rhodizite ($(K,Cs)Al_4Be_4(B,Be)_{12}O_{28}$). A much larger group of more than 50 minerals is found under the sulfates. Here one finds well-known minerals like alunite ($KAl_3(SO_4)_2(OH)_6$), hanksite ($Na_{22}K(SO_4)_9(CO_3)_2Cl$), jarosite ($KFe_3^{3+}(SO_4)_2(OH)_6$), and polyhalite ($K_2Ca_2Mg(SO_4)_4{\cdot}2H_2O$). In the phosphate class a slightly smaller group of 40 minerals contain K, such as leucophosphite ($KFe_2^{3+}(PO_4)_2(OH){\cdot}2H_2O$), minyulite ($KAl_2(PO_4)_2(OH,F){\cdot}4H_2O$), and tiptopite ($K_2(Na,Ca)_2Li_3Be_6(PO_4)_6(OH)_2{\cdot}H_2O$). By far, the largest group of K-containing minerals, however, is found under

FIGURE 5.3 Translucent green apophyllite, $(Na,K)Ca_4(Si_8O_{20})(F,OH){\cdot}8H_2O$, crystal clusters to 1.5 cm with stilbite. Poonah dist., Maharashtra, India.

the silicates (more than 250 minerals). These include many very well-known minerals, for example, apophyllite ($(Na,K)Ca_4(Si_8O_{20})(F,OH)\cdot 8H_2O$) (Fig. 5.3), astrophyllite ($K_2NaFe_7^{2+}Ti_2Si_8O_{28}(OH)_4F$), fluorapophyllite-(K) ($KCa_4(Si_8O_{20})(F,OH)\cdot 8H_2O$), leucite ($K(AlSi_2O_6)$) (Fig. 5.4), microcline (Fig. 5.5) and orthoclase ($K(AlSi_3O_8)$), muscovite ($KAl_2(AlSi_3O_{10})(OH)_2$) (Fig. 5.6), phlogopite ($KMg_3(AlSi_3O_{10})(OH)_2$), and sanidine ($K(AlSi_3O_8)$) (Fig. 5.7). Finally, there are two organic compounds with K: antipinite ($KNa_3Cu_2(C_2O_4)_4$) and minguzzite ($K_3Fe^{3+}(C_2O_4)_3\cdot 3H_2O$).

5.1.3 Chemistry

There are 24 known isotopes of potassium, of which three are found naturally: ^{39}K (93.3%), ^{40}K (0.0117%), and ^{41}K (6.7%). ^{40}K with a half-life of 1.248×10^9 years decays to stable ^{40}Ar through electron capture or positron emission (11.2%) or to stable ^{40}Ca via β decay (88.8%). The decay of ^{40}K to ^{40}Ar forms the basis of common dating method of rocks. The conventional K—Ar dating method is based on the hypotheses that the rocks had no argon at the time of its formation and that all the subsequent radiogenic argon (^{40}Ar) was quantitatively retained within the rock. Minerals are dated through determination of the concentration of potassium and the amount of radiogenic ^{40}Ar that has accumulated.

FIGURE 5.4 Leucite, $K(AlSi_2O_6)$. The crystal in this specimen is 4 mm. Leucite crystals, like these, are trapezohedral, but mostly rounded. Vesuvius, Naples, Campania, Italy. This locality is often used for leucite photos in textbooks. Leucite is a relatively uncommon mineral found in potassium-rich igneous rocks, either lavas or shallow intrusives.

FIGURE 5.5 Microcline, $K(AlSi_3O_8)$. The crystals are 5–10 mm with white glossy albite. Moat Mountain, Conway, Carroll Co., New Hampshire, United States.

FIGURE 5.6 A 2.5-cm group of star-like (stellate), crystallized muscovite, $KAl_2(AlSi_3O_{10})(OH)_2$. Cornwall, England.

FIGURE 5.7 A single somewhat translucent, sharp sanidine crystal, $K(AlSi_3O_8)$, 3 × 2.5 × 1.5 cm. Sanidine is a high-temperature polymorph of the potassium feldspars. It is found in volcanic rocks as phenocrysts and easily weathered out. This is a common Carlsbad twin. Ophir dist., Oquirrh Mts, Tooele Co., Utah, United States.

Minerals best suited for dating are, for example, biotite, muscovite, metamorphic hornblende, and volcanic feldspar. Whole rock samples from volcanic flows and shallow intrusive bodies can also be dated provided they are unaltered. Besides dating, K isotopes have been used as tracers to study weathering and nutrient cycling as K is a macronutrient essential for life. ^{40}K is the radioisotope with the highest concentration in the body. In healthy animals and people, ^{40}K represents the largest source of radioactivity, larger even than ^{14}C. In an average human body weighing 70 kg, about 4400 nuclei of ^{40}K decay per second.

Potassium is a member of the alkali metal. It is a soft, silvery-white metal when freshly cut. The ease of involving the outermost ns^1 electron in bonding, coupled with the very high second stage ionization energy of the alkali metals, immediately explains both the great chemical reactivity of these elements and the fact that their oxidation state in compounds never exceeds +1 (Table 5.1). The reactivity of the alkali metals increases down the group. Thus potassium is far more reactive than the lighter lithium and sodium. It reacts with most nonmetals, such as sulfur, chlorine, fluorine, and phosphorus, and with proton donors such as alcohols, gaseous ammonia, and even alkynes. Its reaction with water

TABLE 5.1 Potassium properties.

Appearance	Silvery gray
Standard atomic weight $A_{r,std}$	39.0983
Block	s-Block
Element category	Alkali metal
Electron configuration	[Ar] $4s^1$
Phase at STP	Solid
Melting point	63.5°C
Boiling point	759°C
Density (near r.t.)	0.862 g/cm^3
When liquid (at m.p.)	0.828 g/cm^3
Heat of fusion	2.33 kJ/mol
Heat of vaporization	76.9 kJ/mol
Molar heat capacity	29.6 J/(mol·K)
Oxidation states	−1, **+1**
Ionization energies	1st: 418.8 kJ/mol 2nd: 3052 kJ/mol 3rd: 4420 kJ/mol
Atomic radius	Empirical: 227 pm
Covalent radius	203 ± 12 pm
Van der Waals radius	275 pm

STP, Standard temperature and pressure.
Bold font indicates main oxidation state.

is so violent and exothermic that the hydrogen gas produced catches fire and may explode. The other product produced is potassium hydroxide (KOH), which is basic.

$$2K(s) + 2H_2O(l) \rightarrow 2KOH(aq) + H_2(g)$$

5.1.3.1 Reaction with air, oxygen, hydrogen, and nitrogen

Potassium is very soft and easily cut resulting in a bright and shiny surface. However, this surface soon tarnishes because of reaction with oxygen and moisture from the air. If potassium is burned in air, the result is mainly formation of orange potassium superoxide, KO_2. Hydrogen combines directly with the metal to form potassium hydride (KH). This reaction was discovered by Cornish chemist and inventor Sir Humphry Davy (December 17, 1778 to May 29, 1829) when he noted that heating potassium just below its boiling point would cause its vaporization in a current of hydrogen. Despite the alkali metals being the most active group of metals, only lithium forms stable nitrides. The heavier alkali metals are chemically inactive to nitrogen at room temperature or even upon heating.

$$K + O_2 \rightarrow KO_2$$
$$2K + H_2 \rightarrow 2KH$$

5.1.3.2 Reaction of potassium with the halogens

Potassium metal can react directly with the halogens to form potassium halides. Potassium fluoride (KF), however, is often prepared by reacting potassium carbonate with excess hydrofluoric acid. The combination forms crystals of potassium bifluoride (KHF_2) upon evaporation of water. KHF_2 then yields KF upon heating.

$$2K + F_2 \rightarrow 2KF$$
$$2K + Cl_2 \rightarrow 2KCl$$
$$2K + Br_2 \rightarrow 2KBr$$
$$2K + I_2 \rightarrow 2KI$$
$$K_2CO_3 + 4HF \rightarrow 2KHF_2 + H_2O + CO_2\uparrow$$
$$KHF_2 \rightarrow KF + HF$$

5.1.3.3 Reaction of potassium with acids and bases

Potassium metal dissolves readily in dilute acids such as sulfuric acid and hydrochloric acid to form solutions containing the aquated K^+ ion together with hydrogen gas, H_2. With concentrated nitric acid, K reacts explosively to form potassium nitrate (KNO_3) and ammonium nitrate (NH_4NO_3). As discussed, potassium reacts very rapidly with water to form a colorless basic solution of potassium hydroxide (KOH) and hydrogen gas (H_2). The reaction is slower than that of rubidium (immediately below potassium in the periodic table) but faster than that of sodium (immediately above potassium in the periodic table).

$$2K(s) + H_2SO_4(aq) \rightarrow 2K^+(aq) + SO_4^{2-}(aq) + H_2(g)$$
$$8K + 10HNO_3 \rightarrow 8KNO_3 + NH_4NO_3 + 3H_2O$$
$$2K(s) + 2H_2O \rightarrow 2KOH(aq) + H_2(g)$$

5.1.3.4 Organopotassium compounds

The chemistry of organopotassium compounds is often discussed together with other organoalkali compounds especially with that of sodium. Organoalkali compounds have the most ionic bonds and are among the most reactive organometallic compounds known. They are very reactive that they burn spontaneously when exposed to air and react violently with water and carbon dioxide. The difficulty in preparation and handling of these compounds is probably one reason why they are overshadowed by organic compounds of the lighter lithium. Nevertheless, a number of preparation and reactions have been presented by M. Schlosser in 1964 and by Lochmann and Lim in 1971. Recent works are based on the original compounds synthesized by Schlosser and Lochmann and are collectively called Lochmann–Schlosser superbases. For example, B. Jennewein, S. Kimpel, D. Thalhei, and J. Klett reported in Chemistry, a European Journal, Vol. 24, May 2018 issue that they have isolated a new Lochmann–Schlosser superbase composed of an alkyl/alkoxy mixed aggregate of the formula, $K_4Np(OtAm)_3$, where Np is CH_2tBu and Am is OCEtMe$_2$. This organopotassium compound had been used for the tetrametalation of ferrocene. $K_4Np(OtAm)_3$ is thermally stable in solution allowing its use at elevated temperatures as shown in the case of tetrametalation of ferrocene at 60°C.

5.1.4 Major uses

Potassium is the eighth or ninth most common element by mass (0.2%) in the human body. K ions are present in a wide variety of proteins and enzymes. The adequate intakes (AIs) of K for males are as follows: 400 mg for 0–6-month-old males, 700 mg for 7–12-month-old males, 3000 mg for 1–3-year-old males, 3800 mg for 4–8-year-old males, 4500 mg for 9–13-year-old males, and 4700 mg for males older than 14 years. The AIs of K for females are as follows: 400 mg for 0–6-month-old females, 700 mg for 7–12-month-old females, 3000 mg for 1–3-year old females, 3800 mg for 4–8-year-old females, 4500 mg for 9–13-year-old females, and 4700 mg for females older than 14 years. The AIs of K are 4700 mg for 14–50-year-old pregnant females and 5100 mg for 14–50-year-old lactating females. Most people in Western countries consume insufficient amount of K per day. Potassium is present in all fruits, vegetables, meat, and fish. Foods with high K concentrations include yam, parsley, dried apricots, milk, chocolate, all nuts (especially almonds and pistachios), potatoes, bamboo shoots, bananas, avocados, coconut water, soybeans, and bran. The USDA lists tomato paste, orange juice, beet greens, white beans, potatoes, plantains, bananas, apricots, and many other dietary sources of K, ordered in decreasing K content. A day's worth of K is equivalent to 5 plantains or 11 bananas. Diets low in K can result in hypertension and hypokalemia. Hypokalemia is a low level of potassium (K^+) in the blood serum. Mild low K does not typically cause symptoms, for example, feeling tired, leg cramps, weakness, and constipation. Low potassium also increases the risk of an abnormal heart rhythm, which is often too slow and can cause cardiac arrest. Supplements of K are most commonly used in combination with diuretics, which block reabsorption of Na and water upstream from the distal tubule (thiazides and loop diuretics), as this stimulates increased distal tubular K secretion, with subsequent increased K excretion. A variety of prescription and over-the-counter supplements are available. Potassium chloride (KCl) may be dissolved in water, but the salty/bitter taste make liquid supplements unpalatable. Typical amounts range from 10 mmol (400 mg) to 20 mmol (800 mg). Potassium is also offered in tablets or capsules, which are formulated to let K leach slowly out of a matrix, since very high K^+ concentrations occurring next to a solid tablet can injure the gastric or intestinal mucosa. Potassium, in the form of KCl is used as a medication to treat and prevent low blood potassium. Low blood K may occur as a result of vomiting, diarrhea, or certain medications. It is given either by slow injection into a vein or by mouth. Potassium sodium tartrate ($KNaC_4H_4O_6$, Rochelle salt) is the main molecule in baking powder; it is also used in the silvering of mirrors. Potassium bromate ($KBrO_3$) is a strong oxidizer (E924), which is used to improve dough strength and rise height. Potassium bisulfite ($KHSO_3$) is used as a food preservative, for example, in wine and brewing (but not in meats). In addition, it is used to bleach textiles and straw and in the tanning of leathers.

K^+ ions are a critical component of plant nutrition and are found in most soil types. They are used as a fertilizer in agriculture, horticulture, and hydroponic culture in the form of chloride (KCl), sulfate (K_2SO_4), or nitrate (KNO_3). Agricultural fertilizers account for about 95% of global K production, and about 90% of this K is supplied as KCl (Box 5.1). The K content of most plants ranges between 0.5% and 2% of the harvested weight of crops, typically reported as the amount of K_2O. Modern high-yield agriculture depends upon fertilizers to replace the K lost at harvest. Though most agricultural fertilizers contain KCl, K_2SO_4 is used for Cl-sensitive crops or crops needing higher S content. The sulfate is produced mostly by decomposition of the minerals kainite ($KMg(SO_4)Cl{\cdot}3H_2O$) and langbeinite ($K_2Mg_2(SO_4)_3$). Only a very limited number of specialized fertilizers contain KNO_3. KOH is a strong base, which is used in neutralizing strong and weak acids, to control pH and to produce K salts. In addition, it is utilized to saponify fats and oils, in industrial cleaners, and in hydrolysis reactions, for example, of esters. KNO_3 or saltpeter is won from natural resources such as guano and evaporites or produced via the Haber process; it is the oxidant in gunpowder (black powder). Potassium cyanide (KCN) is used industrially to dissolve Cu and precious metals, especially Ag and Au, by

BOX 5.1 Potassium fertilizer

K, the third most important fertilizer element, is the eighth most abundant element in the rocks of the Earth's crust. K occurs in nearly all rocks and soils, although its quantity varies widely. It occurs in igneous and metamorphic rocks primarily as potassium feldspar (microcline and orthoclase, $KAlSi_3O_8$); weathering releases K, which is then incorporated in clay minerals such as illite ($KAl_2(Al,Si)_4O_{10}(OH)_2$). The K^+ ion of clay minerals is exchanged with plants by substitution of a H^+. Unlike N and P, K does not form an integral part of the plant components, but it is vital as a catalytic agent in numerous plant functions such as N metabolism, synthesis of proteins, activation of enzymes, and maintenance of water content. Although K occurs in most rocks, the only occurrences that can be economically extracted and processed into fertilizers are those in evaporite sequences. Nearly complete evaporation results in the deposition of large amounts of halite (NaCl) and smaller amounts of a variety of K and K-Mg salts, among which the most important are sylvite (KCl) (Fig. 5.8), langbeinite ($K_2Mg_2(SO_4)_3$), kainite ($KMg(SO_4)Cl{\cdot}3H_2O$), and carnallite ($KMgCl_3{\cdot}6H_2O$). Because the K salts are very soluble, they are only preserved in very arid regions, or in salt beds that are buried below zones of groundwater flow.

FIGURE 5.8 Cluster of nearly colorless, well-formed octahedral sylvite, KCl, crystals to 2.5 cm. Carlsbad Potash dist., Eddy Co., New Mexico, United States.

forming complexes. It is used in, for example, gold mining, electroplating, and electroforming of these metals. In addition, it is used in organic synthesis to produce nitriles (any organic compound that has a $-C\equiv N$ functional group). K_2CO_3 (also known as potash) is used in the making of glass, soap, color TV tubes, fluorescent lamps, textile dyes, and pigments. Potassium permanganate ($KMnO_4$) is an oxidizing, bleaching, and purification agent and is used for making saccharin (benzoic sulfimide, IUPAC name 1,1-dioxo-1,2-benzothiazol-3-one, $C_7H_5NO_3S$, an artificial sweetener). Potassium chlorate ($KClO_3$) is added to matches and explosives. Potassium bromide (KBr) was historically used as a sedative and in photography. Potassium chromate (K_2CrO_4) is used in a variety of applications, such as in inks, dyes, stains (bright yellowish-red color); in explosives and fireworks; in the tanning of leather; and in fly paper and safety matches, though all these applications are due to the chemistry of the $Cr_2O_4^{2-}$ ion, rather than the K^+ ion. There are many different uses of various K compounds, for example, potassium superoxide, KO_2, is an orange solid that functions as a portable O_2 source and a CO_2 absorber. It is extensively used in respiration systems in mines, submarines, and spacecraft as it takes less volume than the gaseous O_2.

$$4KO_2 + 2CO_2 \rightarrow 2K_2CO_3 + 3O_2$$

Another illustration is potassium cobaltinitrite, $K_3[Co(NO_2)_6]$, which is used as artist's pigment known as aureolin or cobalt yellow. An alloy of Na and K, NaK is a liquid used as a heat-transfer medium and a desiccant for producing dry and air-free solvents. In addition, it can be utilized in reactive distillation. The ternary alloy of 12% Na, 47% K, and 41% Cs has the lowest melting point of $-78°C$ of any metallic compound. Metallic K is used in several types of magnetometers (a device that measures magnetism—the direction, strength, or relative change of a magnetic field at a particular location).

5.2 20 Ca — Calcium

5.2.1 Discovery

Calcium compounds were known for thousands of years, even though their chemical composition was not understood until the 17th century. Lime as a building material and as plaster for statues was utilized as early as around 7000 BCE. The first dated lime kiln dates to 2500 BCE and was found in Khafajah, Mesopotamia. At about the same time, dehydrated gypsum ($CaSO_4 \cdot 2H_2O$) was being utilized in the Great Pyramid of Giza; this substance would later be used for the plaster in the tomb of Tutankhamun. The climate of present-day Italy being warmer than that of Egypt, the ancient Romans as an alternative used lime mortars formed by heating limestone (mainly the mineral calcite, $CaCO_3$); the name "calcium" itself originates from the Latin word calx "lime." Roman author, architect, civil engineer, and military engineer Vitruvius (c.80–70 BCE to after c.15 BCE) observed that the resulting lime was lighter than the original limestone, ascribing this to the boiling of the water (Fig. 5.9). In 1755 Scottish physicist and chemist Joseph Black (April

FIGURE 5.9 Presumed portrait of Vitruvius, engraving by Jacopo Bernardi (December 19, 1813 to October 9, 1897) based on a drawing by Vincenzo Ray.

FIGURE 5.10 Cubic green fluorite, CaF_2, crystals to 1.2 cm. Heights mine, Weardale, North Pennines, Durham Co., England.

16, 1728 to December 6, 1799) showed that this was due to the loss of carbon dioxide (CO_2), which as a gas had not been recognized by the ancient Romans. In 1787 French nobleman and chemist Antoine Lavoisier (August 26, 1743 to May 8, 1794) assumed that lime might be an oxide of a fundamental chemical element. In his table of the elements, Lavoisier recorded five "salifiable earths" [i.e., ores that could be made to react with acids to produce salts (salis = salt, in Latin): chaux (calcium oxide), magnésie (magnesia, magnesium oxide), baryte (barium sulfate), alumine (alumina, aluminum oxide), and silice (silica, silicon dioxide)]. Calcium, together with its congeners magnesium, strontium, and barium, was first isolated by Cornish chemist and inventor Sir Humphry Davy (December 17, 1778 to May 29, 1829) in 1808. Following the work of Swedish chemist Jöns Jakob Berzelius (August 20, 1779 to August 7, 1848) and Swedish doctor and writer Magnus Martin af Pontin (January 20, 1781 to January 30, 1858) on electrolysis, Davy isolated calcium and magnesium by putting a mixture of the respective metal oxides with mercury(II) oxide on a platinum plate that was used as the anode, the cathode being a platinum wire partially submerged into mercury. Electrolysis then gave calcium–mercury and magnesium–mercury amalgams, and distilling off the mercury produced the metal (Davy, 1808a,b). Nevertheless, pure calcium cannot be prepared in bulk by this method and a workable commercial process for its production was not found until over a century later.

5.2.2 Mining, production, and major minerals

At 3%, calcium is the fifth most abundant element in the Earth's crust, and the third most abundant metal after aluminum and iron. Sedimentary calcium carbonate ($CaCO_3$) deposits pervade the Earth's surface as fossilized remains of past marine life; they occur in two forms: the rhombohedral calcite (more common) and the orthorhombic aragonite (forming in more temperate seas) (Box 5.2). Rocks of the first type include limestone, dolomite, marble, and chalk; aragonite beds make up the Bahamas, the Florida Keys, and the Red Sea basins. Corals, seashells, and pearls are mostly made up of calcium carbonate. Among the other important minerals of calcium are gypsum ($CaSO_4{\cdot}2H_2O$), anhydrite ($CaSO_4$), fluorite (CaF_2), and apatite ($[Ca_5(PO_4)_3(OH,F]$). The major producers of calcium are China, Russia, and the United States. Canada and France are also among the minor producers. In Russia and China, Cornish chemist and inventor Sir Humphry Davy's (December 17, 1778 to May 29, 1829) process of electrolysis is still used but is, however, applied to molten calcium chloride. Because calcium is less reactive than strontium or barium, the oxide–nitride coating that results in air is stable, and lathe machining (a machine tool that rotates a workpiece about an axis of rotation to perform various operations such as cutting, sanding, knurling, drilling, deformation, facing, and turning, with tools that are applied to the workpiece to create an object with symmetry about that axis) and other standard metallurgical techniques are suitable for calcium. In the United States and Canada, calcium is instead produced by reducing lime with aluminum at high temperatures.

There are roughly 1400–1450 minerals that (can) contain calcium in their structure. In the sulfide class, six minerals contain Ca, for example, cronusite ($Ca_{0.2}CrS_2{\cdot}2H_2O$), oldhamite ($(Ca,Mg)S$), and sarabauite ($CaSb^{3+}_{10}S_6O_{10}$). Twenty-seven minerals are found in the halides class, such as creedite ($Ca_3SO_4Al_2F_8(OH)_2{\cdot}2H_2O$), fluorite ($CaF_2$) (Fig. 5.10), pachnolite ($NaCa[AlF_6]{\cdot}H_2O$), and thomsenolite ($NaCa[AlF_6]{\cdot}H_2O$). More than 100 oxide class minerals contain calcium, for example, cafarsite ($Ca_{5.9}Mn_{1.7}Fe_3Ti_3(AsO_3)_{12}{\cdot}4\text{-}5H_2O$), euxenite-(Y) ($(Y,Ca,Ce,U,Th)(Nb,Ta,Ti)_2O_6$),

BOX 5.2 $CaCO_3$ polymorphs

Calcium carbonate exists as three different polymorphs: calcite, aragonite, and vaterite.

The name calcite comes from the Latin word *calx*, meaning burnt lime. It is a common and widespread mineral (Fig. 5.11). It is an essential and major mineral in limestones and marbles and occurs in cave deposits and as vein mineral with other carbonates, sulfides, fluorite, baryte, and quartz. It also occurs in some rare carbonate-rich igneous rocks (carbonatite). Calcite is a common weathering product. Organic calcite is common in shells and skeletal material. Calcite is hexagonal, *2/m* (space group *R2/c*). It has a hardness of 3. The luster is vitreous to earthy, while its transparency varies from transparent to translucent. It has a perfect {101} cleavage (cleavage angle 74°55′) and parting along twin lamellae on {012}. Fracture is conchoidal. The color varies from colorless, white, to gray, but may be tinted brown, yellow, red, blue, and green. The streak is white. Optically it is uniaxial(−). Iceland spar refers to a clear calcite, usually in rhombohedral cleavage fragments. Calcite has many habits but the three most important are (1) prismatic, long or short prisms, in which the prism faces are prominent with base or rhombohedral terminations, (2) rhombohedral, and (3) scalenohedral, in which scalenohedrons predominate, often with prism faces and rhombohedral truncations. The most common scalenohedron is {211}. Twinning with the twin plane {011} is very common, often producing twin lamellae, {000} is also common.

Aragonite is named after the original locality Aragon, Spain. Aragonite is less stable than calcite under atmospheric conditions and much less common. It is precipitated in a narrow range of physicochemical conditions represented by low-temperature near-surface deposits (Fig. 5.12). It is found as disseminated carbonate in gypsum beds, as hot spring deposits, as

FIGURE 5.11 Calcite, $CaCO_3$, orange brown scalenohedral crystals. The largest crystal is 8 cm. Rosiclare, Hardin Co., Illinois, United States.

FIGURE 5.12 Aragonite, $CaCO_3$, a short 1.4-cm pseudohexagonal cyclic twinned crystal from the Type Locality. Molina de Aragón, Guadalajara, Castile-La Mancha, Spain.

(Continued)

BOX 5.2 (Continued)

precipitates from Ca-oversaturated waters, associated with sedimentary iron ore, in oxidized zones of ore deposits, in some cave formations, and in blueschist facies metamorphic rocks. It also occurs in shells and other organic carbonate materials. Floss ferri is a coral-like form of aragonite associated with iron deposits. Aragonite is orthorhombic *2/m2/m/2m* (space group $P2_1/m2_1/m2_1/m$). It has a hardness of 3½–4. Its luster is resinous, while the transparency varies from transparent to translucent. The color ranges from colorless, white, pale-yellow, to various other tints. The streak is white. Optically it is biaxial(−). Crystals are acicular, tabular, prismatic, or fibrous. Crystals may form radiating sprays, crusts, or masses of many different morphologies. Contact and cyclic twins are common, sometimes giving it a pseudohexagonal appearance.

Vaterite is a rare $CaCO_3$ modification that is metastable below approx. 400°C, which easily alters to calcite. It is named in honor of Heinrich August Vater (September 5, 1859 in Bremen, Germany to February 10, 1930 in Dresden, Germany), Professor of Mineralogy and Chemistry, Tharandt, Saxony (Germany). He was a pioneer in the areas of forest soil science, land evaluation, and forest fertilization. It has been found as a hydration product of larnite, also known from some carbonatites, with portlandite and bayerite, as fibrous masses with afwillite in manganese ore, and as tiny crystals in Alpine fissures. It is not uncommon as a biomineral. Vaterite is hexagonal *6/m2/m2/m* (space group $P6_3/mmc$). It has a hardness of ~3. The luster is subvitreous to waxy, transparent. The color is colorless to white with a white streak. Fracture is irregular/uneven/splintery. Optically it is uniaxial(+). It occurs as hexagonal crystals, and also as radial fibrous aggregates.

FIGURE 5.13 Scheelite, $Ca(WO_4)$, orange octahedral crystal to 6 mm. Ortiz mine, Santa Fe Co., New Mexico, United States.

lime (CaO), and perovskite ($CaTiO_3$). A total of 77 carbonates are found with Ca with well-known minerals such as alstonite ($BaCa(CO_3)_2$), ankerite ($Ca(Fe^{2+},Mg)(CO_3)_2$), aragonite ($CaCO_3$), calcite ($CaCO_3$), dolomite ($CaMg(CO_3)_2$), and kutnohorite ($CaMn^{2+}(CO_3)_2$). A slightly smaller group is found under the borates, including minerals such as colemanite ($Ca[B_3O_4(OH)_3]{\cdot}H_2O$), inderborite ($CaMg(H_3B_3O_7)_2{\cdot}8H_2O$), painite ($CaZrAl_9(BO_3)O_{15}$), and ulexite ($NaCa[B_5O_6(OH)_6]{\cdot}5H_2O$). The sulfate class contains 50 minerals, including well-known examples such as anhydrite ($CaSO_4$), ettringite ($Ca_6Al_2(SO_4)_3(OH)_{12}{\cdot}26H_2O$), glauberite ($Na_2Ca(SO_4)_2$), gypsum ($CaSO_4{\cdot}2H_2O$), powellite ($Ca(MoO_4)$), scheelite ($Ca(WO_4)$) (Fig. 5.13), and thaumasite ($Ca_3(SO_4)[Si(OH)_6](CO_3){\cdot}12H_2O$). A much larger group of more than 250 minerals can be found under the phosphate class, for example, anapaite ($Ca_2Fe^{2+}(PO_4)_2{\cdot}4H_2O$), autunite ($Ca(UO_2)_2(PO_4)_2{\cdot}11H_2O$), conichalcite ($CaCu(AsO_4)(OH)$), fluorapatite ($Ca_5(PO_4)_3F$), herderite ($CaBePO_4(F,OH)$), roselite ($Ca_2(Co^{2+},Mg)(AsO_4)_2{\cdot}2H_2O$), sampleite ($NaCaCu_5(PO_4)_4Cl{\cdot}5H_2O$), and scholzite ($CaZn_2(PO_4)_2{\cdot}2H_2O$) (Fig. 5.14). The largest group of minerals, 557, is found in the class comprising the silicates. These silicates include well-known minerals such as actinolite ($\{Ca_2\}\{Mg_{4.5-2.5}Fe_{0.5-2.5}\}(Si_8O_{22})(OH)_2$), andradite ($Ca_3Fe_2^{3+}(SiO_4)_3$), anorthite ($Ca(Al_2Si_2O_8)$), axinite-(Fe) ($Ca_2Fe^{2+}Al_2BSi_4O_{15}OH$), cavansite ($Ca(VO)Si_4O_{10}{\cdot}4H_2O$), chabazite-(Ca) ($((Ca,K_2,Na_2)_2[Al_2Si_4O_{12}]_2{\cdot}12H_2O$), danburite ($CaB_2Si_2O_8$), diopside ($CaMgSi_2O_6$), epidote ($\{Ca_2\}\{Al_2Fe^{3+}\}(Si_2O_7)(SiO_4)O(OH)$) (Fig. 5.15), grossular ($Ca_3Al_2(SiO_4)_3$) (Fig. 5.16), gyrolite

FIGURE 5.14 Scholzite, $CaZn_2(PO_4)_2{\cdot}2H_2O$, colorless bladed divergent crystals to 6 mm. Reaphook Hill, Flinders Ranges, South Australia, Australia.

FIGURE 5.15 A 5-cm pistachio green group of long prismatic and divergent epidote, $\{Ca_2\}\{Al_2Fe^{3+}\}(Si_2O_7)(SiO_4)O(OH)$, crystals. Toll Mtn. area, Jefferson Co., Montana, United States.

($NaCa_{16}Si_{23}AlO_{60}(OH)_8{\cdot}14H_2O$), hedenburgite ($CaFe^{2+}Si_2O_6$), heulandite-(Ca) ($(Ca,Na)_5(Si_{27}Al_9)O_{72}{\cdot}26H_2O$), ilvaite ($CaFe^{3+}Fe^{2+}_2(Si_2O_7)O(OH)$), kinoite ($Ca_2Cu_2(H_2O)_2[Si_3O_{10}]$), meionite ("scapolite" endmember) ($Ca_4Al_6Si_6O_{24}CO_3$), pectolite ($NaCa_2Si_3O_8(OH)$), prehnite ($Ca_2Al_2Si_3O_{10}(OH)_2$), (Fig. 5.17), pentagonite ($Ca(VO)Si_4O_{10}{\cdot}4H_2O$) (Fig. 5.18), stellerite ($Ca_4(Si_{28}Al_8)O_{72}{\cdot}28H_2O$) (Fig. 5.19), stilbite-(Ca) ($NaCa_4[Al_9Si_{27}O_{72}]{\cdot}nH_2O$), titanite ($CaTi(SiO_4)O$) (Fig. 5.20), tremolite ($\{Ca_2\}\{Mg_5\}(Si_8O_{22})(OH)_2$), wollastonite ($CaSiO_3$) (Fig. 5.21), and zoisite ($Ca_2Al_3[Si_2O_7][SiO_4]O(OH)$). Finally, there are eight organic compounds with Ca, for example, earlandite ($Ca_3(C_6H_5O_7)_2{\cdot}4H_2O$) and whewellite ($Ca(C_2O_4){\cdot}H_2O$).

5.2.3 Chemistry

Natural Ca is a mixture of five stable isotopes (^{40}Ca, ^{42}Ca, ^{43}Ca, ^{44}Ca, and ^{46}Ca) and one radioisotope with a half-life so long that it can be thought of as stable for all practical purposes (^{48}Ca, with a half-life of about 4.3×10^{19} years). Ca is the first (lightest) element to have six naturally occurring isotopes. By far the most common isotope is ^{40}Ca making

FIGURE 5.16 Grossular, $Ca_3Al_2(SiO_4)_3$, garnet to 8 mm from classic Canadian locality. Jeffrey mine, Asbestos, Quebec, Canada.

FIGURE 5.17 11 cm sea foam green botryoidal prehnite, $Ca_2Al_2Si_3O_{10}(OH)_2$. Paterson, Passiac Co., New Jersey, United States.

up 96.941%. It is formed in the silicon-burning process by fusion of alpha particles and is the heaviest stable nuclide with equal proton and neutron numbers; in addition, its occurrence is supplemented slowly by the decay of primordial ^{40}K. Adding another alpha particle would result in the formation of unstable ^{44}Ti, which rapidly decays via two successive electron captures to stable ^{44}Ca, making up 2.806% of all natural calcium and is the second most common isotope. The remaining four naturally occurring isotopes, ^{42}Ca, ^{43}Ca, ^{46}Ca, and ^{48}Ca, comprise less than 1% each. The four lighter isotopes are mostly formed by the oxygen-burning and silicon-burning processes, while the two heavier ones are formed by neutron-capturing processes. ^{46}Ca is mainly formed in a "hot" s-process, as it needs a rather high neutron flux so that short-lived ^{45}Ca can capture a neutron. ^{48}Ca is formed through electron capture in the r-process in type Ia supernovae, where high neutron excess and low enough entropy guarantee its survival. ^{46}Ca and ^{48}Ca are the first "classically stable" nuclides with a six-neutron or eight-neutron excess, respectively. While extremely neutron rich for such a light element, ^{48}Ca is very stable as it is a doubly magic nucleus with 20 protons and 28 neutrons arranged in closed shells. Its β decay to ^{48}Sc is strongly hindered due to the large mismatch in nuclear spin: ^{48}Ca has zero nuclear spin, being even–even, whereas ^{48}Sc has spin 6 +, so the decay is forbidden by the conservation of angular momentum. Although two excited states of ^{48}Sc are available for decay as well, they are also forbidden due to their high spins. Consequently, when ^{48}Ca does decay, it does so via double β decay to ^{48}Ti instead, making it the lightest nuclide known to undergo double β decay. Theoretically ^{46}Ca can also undergo double β decay to ^{46}Ti, but this has never been detected. The lightest and most common isotope ^{40}Ca is also doubly magic and could undergo double electron capture

FIGURE 5.18 Blue radial cluster of pentagonite, $Ca(VO)Si_4O_{10}{\cdot}4H_2O$, to 9 mm sitting in zeolite vug. Wagholi, near Pune, Maharashtra, India.

FIGURE 5.19 Light orange clusters of stellerite, $Ca_4(Si_{28}Al_8)O_{72}{\cdot}28H_2O$, to 1.3 cm. Sokolovskoe Iron mine, Kostanay prov., Kazakhstan.

FIGURE 5.20 Titanite, $CaTi(SiO_4)O$, 3.1-cm green twinned crystal. Ankarafa, Vohémar dist., Antsiranana, Madagascar.

to ^{40}Ar, but similarly this has never been detected. Ca is the only element to have two primordial doubly magic isotopes. The experimental lower limits for the half-lives of ^{40}Ca and ^{46}Ca are 5.9×10^{21} and 2.8×10^{15} years, respectively. Apart from the practically stable ^{48}Ca, the longest-lived radioisotope is ^{41}Ca, which decays through electron capture to stable ^{41}K with a half-life of 1.03×10^{5} years. Its presence in the early solar system as an extinct radionuclide has been deduced from excess of ^{41}K: trace amounts of ^{41}Ca also still currently exist, since it is a cosmogenic nuclide, continuously reformed through neutron activation of natural ^{40}Ca. A large number of other radioisotopes are known, ranging from ^{34}Ca to ^{57}Ca: they are all much shorter-lived than ^{41}Ca, the most stable among them are ^{45}Ca with a half-life of 163 days and ^{47}Ca with a half-life of 4.54 days. The isotopes lighter than ^{42}Ca typically undergo β^{+} decay to isotopes of K, while those heavier than ^{44}Ca typically undergo β^{-} decay to isotopes of Sc, although near the nuclear drip lines proton emission and neutron emission begin to be significant decay modes as well. Similar to other elements, a range of processes alter the relative abundance of Ca isotopes. Best known is the mass-dependent fractionation of Ca isotopes accompanying the precipitation of calcium-containing minerals such as calcite, aragonite (both $CaCO_3$), and apatite ($Ca_5(PO_4)_3(F,OH)$) from solution. The lighter isotopes are preferentially incorporated in these minerals' structures, leaving the surrounding solution enriched in the heavier isotopes at an extent of about 0.025% per atomic mass unit (amu) at room temperature. Mass-dependent Ca isotope differences in compositions are conventionally given as the ratio of two isotopes (usually $^{44}Ca/^{40}Ca$) in a sample in comparison to the same ratio in a standard reference material. $^{44}Ca/^{40}Ca$ varies by about 1% between common earth materials.

Calcium belongs to the alkaline earth metals with an atom electronic configuration of $[Ar]4s^2$. The two electrons in the outer s-orbitals are easily lost in a chemical reaction. Consequently, most calcium compounds are dipositive (Table 5.2). Its position in the fourth period means calcium has larger atomic size and ionic size than the lighter alkaline earth members. This large ionic size of Ca^{2+} (94 pm) lowers the charge/size ratio and enhances the ionic character of the bonding. Moreover, the lower (charge/size) ratio increases the kinetic lability of the ligands.

The heavier alkaline earth metals Ca, Sr, Ba (and Ra) are more reactive. They react more readily with nonmetals. Their reactivity follows a periodic trend with the heavier alkaline earths more reactive than the lighter members. Calcium, for instance, spontaneously reacts with water more quickly than magnesium but less quickly than strontium to produce calcium hydroxide and hydrogen gas.

5.2.3.1 Reaction in air, oxygen, nitrogen, and hydrogen

Bulk calcium is less reactive. It can be stored indefinitely at room temperature when the relative humidity is below 30%. At higher relative humidity values, it quickly forms a hydration coating in moist air. It reacts with oxygen and nitrogen in air to form a mixture of calcium oxide and calcium nitride. In pure oxygen, especially under high O_2

TABLE 5.2 Calcium properties.

Appearance	Dull gray, silver; with a pale-yellow tint
Standard atomic weight $A_{r,std}$	40.078
Block	s-Block
Element category	Alkaline earth metal
Electron configuration	[Ar] $4s^2$
Phase at STP	Solid
Melting point	842°C
Boiling point	1484°C
Density (near r.t.)	1.55 g/cm^3
When liquid (at m.p.)	1.378 g/cm^3
Heat of fusion	8.54 kJ/mol
Heat of vaporization	154.7 kJ/mol
Molar heat capacity	25.929 J/(mol·K)
Oxidation states	+1, **+2**
Ionization energies	1st: 589.8 kJ/mol
	2nd: 1145.4 kJ/mol
	3rd: 4912.4 kJ/mol
Atomic radius	Empirical: 197 pm
Covalent radius	176 ± 10 pm
Van der Waals radius	231 pm

STP, Standard temperature and pressure.
Bold font indicates main oxidation state.

FIGURE 5.21 Wollastonite, $CaSiO_3$, a compact fibrous but brittle, white mass 5 cm across. Commercial quarry, Crestmore, Riverside Co., California, United States.

pressure, the simple oxide, CaO, and peroxide, CaO_2, and probably the yellow superoxide, $Ca(O_2)_2$, can be formed by direct oxidation of calcium. Ca also reacts directly with H_2 to give calcium hydride, CaH_2.

5.2.3.2 Reaction with halogens

Calcium is very reactive toward the halogens fluorine, F_2, chlorine, Cl_2, bromine, Br_2, or iodine, I_2, and burns to form calcium(II) fluoride, CaF_2, calcium(II) chloride, $CaCl_2$, calcium(II) bromide, $CaBr_2$, and calcium(II) iodide, CaI_2, respectively. The reactions with bromine and iodine require heat to enable the formation of the products.

5.2.3.3 Reaction with acids and bases

Calcium metal dissolve in liquid NH_3 to give bronze-colored precipitates. With regard to its reactivity to acids, calcium metal dissolves readily in dilute acids to form solutions containing the aquated Ca(II) ion together with hydrogen gas, H_2.

5.2.3.4 Organocalcium compounds

Organometallic compounds of Ca are far more reactive than those of Mg and have been much less studied. Nevertheless, several organocalcium compounds have been prepared and characterized. Early works on reactive calcium dialkyls/diaryls CaR_2 (R = Me, Et, allyl, Ph, $PhCH_2$, etc.) have shown their preparation using HgR_2 even under mild conditions. Calcium dicyclopentadienyl was prepared by reacting the metal with $[Hg(C_5H_5)_2]$ or with cyclopentadiene, cyclo-C_5H_6. Calcium carbide in liquid ammonia and cyclo-C_5H_6 can also be used as precursors for $[Ca(C_5H_5)_2]$; the other product in this reaction is HC—CH. A highly reactive form of calcium was also prepared by the reduction of Ca(II) salts with preformed lithium biphenylide. This activated calcium was reported to undergo oxidative addition to organic bromides, chlorides, or even fluorides to form the organocalcium reagents under very mild conditions in high yields. The resulting RCaX compounds undergo Grignard-type reactions. Recently, reactions (addition, coupling, etc.) of element-H bonds (E-H, E=O, S, N, P, C, Si, B, H, etc.) with organic compounds catalyzed by organocalcium were reviewed by Yu et al. (2018). Harder (2010) gave a review of around 25 organocalcium compounds used so far in catalyzing reactions such as hydroamination, hydrophosphination, alkene hydrosilylation, and hydrogenation. Most of these reactions are important in the pharmaceutical industry and are often catalyzed by transition metals. It was rationalized that developing less expensive, more ecofriendly catalysts such as those based from calcium would have profound impact on the chemical industry.

5.2.4 Major uses

Calcium compounds are widely used. There are vast deposits of limestone ($CaCO_3$), which can be used directly as a building stone and indirectly for cement. When limestone is heated in kilns, it gives off carbon dioxide (CO_2) gas leaving behind quicklime (calcium oxide, CaO). This reacts vigorously with water to give slaked lime (calcium hydroxide, $Ca(OH)_2$). Slaked lime is used as a soil conditioner and in water treatment to reduce acidity, and in the chemical industry. Cement is made by heating a mixture of limestone and clay in a kiln at 1500°C, which drives of carbon dioxide, leaving a mixture of calcium (mainly) and aluminum silicates. These are then pulverized, and gypsum is added. When it is mixed with water, cement undergoes a complete series of reactions which have only recently been understood. When mixed with sand, slaked lime takes up carbon dioxide from the air and hardens as lime plaster. Gypsum ($CaSO_4{\cdot}2H_2O$) is used by builders as a plaster and in medical applications for setting bones, as "plaster of Paris."

The largest use of Ca is in the production of steel, because of its strong chemical affinity for oxygen and sulfur. Its oxides and sulfides, once formed, give liquid lime aluminate and sulfide inclusions in steel that float out. On treatment these inclusions disperse in the steel and become small and spherical, enhancing stability, cleanliness, and overall mechanical properties. In addition, Ca is utilized in maintenance-free automotive batteries, in which the use of 0.1% Ca-Pb alloys replacing the usual Sb-Pb alloys results in lower water loss and lower self-discharging. Because of the risk of expansion and cracking, Al is also added occasionally to these alloys. In addition, these Ca-Pb alloys are used in casting, substituting for Sb-Pb alloys. Calcium is also utilized to strengthen Al alloys used for bearings, for the control of graphitic C in cast iron, and to eliminate Bi impurities from Pb. Ca metal is present in some drain cleaners, where it serves to generate heat and $Ca(OH)_2$ that saponifies the fats and liquefies the proteins (e.g., those in hair) which can block drains. Besides metallurgy, the reactivity of Ca is exploited to eliminate nitrogen from high-purity argon gas and as a getter for oxygen and nitrogen. In addition, it is used as a reducing agent in the manufacturing of chromium, zirconium, thorium, and uranium. It can also be used to store H_2 gas, as it reacts with hydrogen to form solid calcium hydride, CaH_2, from which the hydrogen can easily be reextracted.

Many Ca compounds are used in food, as pharmaceuticals, and in medicine, for example, Ca and P are supplemented in food through the addition of calcium lactate (IUPAC name calcium 2-hydroxypropanoat, $C_6H_{10}CaO_6$), calcium diphosphate ($Ca_2P_2O_7$), and tricalcium phosphate (IUPAC name tricalcium bis(phosphate), $Ca_3(PO_4)_2$). The last can also be found as a polishing agent in toothpaste and in antacids. Calcium lactobionate (lactobionic acid = $C_{12}H_{22}O_{12}$, IUPAC name (2R,3R,4R)-2,3,5,6-tetrahydroxy-4-[[(2S,3R,4S,5R,6R)-3,4,5-trihydroxy-6-(hydroxymethyl)-2-tetrahydropyranyl]oxy]hexanoic acid) is a white powder that is used as a suspending agent for pharmaceuticals. In baking, monocalcium phosphate ($Ca(H_2PO_4)_2$) is used as a leavening agent (also known as a raising agent) since the 1930s. Two crystalline forms of calcium sulfite ($CaSO_3{\cdot}x(H_2O)$) are known: the hemihydrate and the tetrahydrate, ($CaSO_3{\cdot}½(H_2O)$ and $CaSO_3{\cdot}4(H_2O)$), and are used as a bleach in papermaking and as a disinfectant, respectively. As a food additive calcium sulfite is used as a preservative under the E number E226. Along with other antioxidant sulfites, it is commonly used in preserving wine, cider, fruit juice, canned fruit, and vegetables. Sulfites act as strong reducers in solution, and they serve as oxygen scavenger antioxidants to preserve food, but labeling is essential as some people might be hypersensitive. Calcium silicate (Ca_2SiO_4) is used as a reinforcing agent in rubber. It is also used as an anticaking agent in food preparation, including table salt (NaCl) and as an antacid. It is approved by the United Nation's FAO and WHO bodies as a safe food additive in a large variety of products with the E number E552. In addition, it is frequently used as a safe alternative to asbestos for high-temperature insulation materials. Industrial-grade piping and equipment insulation are often made from calcium silicate. It is used in passive fire protection and fireproofing as calcium silicate bricks or in roof tiles. Calcium acetate ($Ca(C_2H_3O_2)_2$) is a component of liming rosin (rosin reacted with lime; used as a binder in paints. Rosin is a brittle resin that can be crushed easily by hand. When rolled between the fingers, the body temperature is sufficient to make it sticky.) and is used to make metallic soaps and synthetic resins. Calcium acetate is used as a food additive, as a stabilizer, buffer, and sequestrant, mainly in candy products under the number E263. Because it is inexpensive, it was once a common starting material for the synthesis of acetone before the development of the cumene process:

$$Ca(CH_3COO)_2 \rightarrow CaCO_3(s) + (CH_3)_2CO(v)$$

A saturated solution of calcium acetate in alcohol forms a semisolid, flammable gel that is much like "canned heat" products such as Sterno. Its primary uses are in the food service industry for buffet heating and in the home for fondue and as a chafing fuel for heating chafing dishes. Other uses are for portable stoves and as an emergency heat source. Chemistry teachers often prepare "California Snowballs," a mixture of calcium acetate solution and ethanol. The resulting gel is whitish in color and can be formed to resemble a snowball. In kidney disease, blood levels of phosphate may rise (called hyperphosphatemia) leading to bone problems. Calcium acetate binds phosphate in the diet to lower blood phosphate levels.

Calcium is an essential element needed in large quantities in the human body. The ideal Ca intake is probably about 1000 mg per day, but, during childhood, pregnancy, and old age, when extra calcium is needed, it is probably best to increase this to 1500 mg as insurance. The Ca^{2+} ion functions as an electrolyte and is essential to the health of the muscular, circulatory, and digestive systems; is crucial to the building of bone; and supports synthesis and function of blood cells, for example, it regulates muscle contractions, nerve conduction, and blood clotting. Consequently, intra- and extracellular Ca levels are strictly regulated by the body. Ca can play this role since the Ca^{2+} ion easily reacts to form stable coordination complexes with many organic molecules, in particular proteins. In addition, it forms molecules with a wide range of solubilities, enabling the formation of the skeleton. About 75% of dietary Ca comes from dairy products and grains, the remaining 25% coming from vegetables, protein-rich foods, fruits, sugar, fats, and oil. Ca supplementation is controversial, since the bioavailability of Ca depends strongly on the solubility of the salt involved: calcium citrate (IUPAC name 2-hydroxy-1,2,3-propane-tricarboxylic acid calcium salt (2:3), $Ca_3(C_6H_5O_7)_2$), malate (IUPAC name calcium 2-hydroxybutanedioate, $Ca(C_2H_4O(COO)_2)$), and lactate (IUPAC name calcium 2-hydroxypropanoate, $C_6H_{10}CaO_6$) are highly bioavailable while the oxalate ($CaC_2O_4 \cdot x(H_2O)$, $0 < x < 3$) is much less so. The intestine absorbs around 33% of all Ca eaten as the free ion, and the plasma Ca level is subsequently regulated by the kidneys. For the body to be able utilize this Ca, the diet needs to contain plenty of vitamin D, which is provided by foods like fish liver oils, many types of fish, butter, margarine, and eggs.

5.3 21 Sc — Scandium

5.3.1 Discovery

Dmitri Mendeleev (February 8, 1834 to February 2, 1907), who is considered to be the father of the periodic table (See Chapter 1, section 1.1.8), predicted the existence of an element ekaboron, with an atomic mass between 40 and 48 in 1869. Swedish chemist Lars Fredrik Nilson (May 27, 1840 to May 14, 1899) and his team found this element in the minerals euxenite and gadolinite in 1879 (Fig. 5.22). Nilson separated 2 g of scandium oxide of high purity (Nilson 1879a,b). He called the element scandium, from the Latin Scandia meaning "Scandinavia." Nilson was seemingly unaware of Mendeleev's prediction, but Swedish chemist, biologist, mineralogist, and oceanographer Per Teodor Cleve (February 10, 1840 to June 18, 1905) (Fig. 5.23), best known for his discovery of the chemical elements holmium and thulium, recognized the correspondence between scandium and ekaboron and alerted Mendeleev (Cleve, 1879). Metallic scandium was synthesized for the first time in 1937 by electrolysis of a eutectic mixture of potassium, lithium, and scandium chlorides, at 700°C–800°C. The first pound of 99% pure scandium metal was produced in 1960. Production of aluminum alloys with scandium began in 1971, based on a US patent. Aluminum–scandium alloys were also developed in the former USSR. Laser crystals of gadolinium–scandium–gallium garnet were used in strategic defense applications developed for the Strategic Defense Initiative in the 1980s and 1990s.

5.3.2 Mining, production, and major minerals

In the Earth's crust, scandium is not rare. Estimates vary between 18 and 25 ppm, similar to the abundance of cobalt (20–30 ppm). Scandium is only the 50th most common element on Earth (35th most abundant in the crust), but it is the 23rd most common element in the Sun. Yet, scandium occurs only in trace amounts in many minerals. Rare minerals from Scandinavia and Madagascar such as thortveitite ($Sc_2Si_2O_7$), euxenite-(Y) ($(Y,Ca,Ce,U,Th)(Nb,Ta,Ti)_2O_6$), and

FIGURE 5.22 Lars Frederik Nilson, 1900.

FIGURE 5.23 Per Teodor Cleve, between 1880 and 1890.

BOX 5.3 Thortveitite, $(Sc,Y)_2Si_2O_7$, and kolbeckite, $ScPO_4{\cdot}2H_2O$

Thortveitite was named in 1911 by Jakob Grubbe Cock Schetelig in honor of Gunder Olaus Olsen Thortveit (October 1, 1872, Thortveit, Iveland, Aust-Agder, Norway to June 15, 1917 in Rossås, Iveland, Aust-Agder, Norway), farmer and miner from Iveland, Norway, who found the mineral. Thortveitite most often occurs in granitic pegmatites, but also in hydrothermal veins. It is monoclinic C2/m. Its hardness is 6–7. Its luster is vitreous to dull, transparency translucent to opaque. The color is grayish-green, black, gray, blue, yellow, or brown, and the streak is white. Optically it is biaxial(−). Cleavage on {110} is good, fracture is uneven to conchoidal. Crystals may occur singly or in radial aggregates, also flattened plates to elongate prismatic crystals. Simple rotational twinning on {110} is common, rare twins on {150}, polysynthetic twinning is sometimes observed.

FIGURE 5.24 Kolbeckite, $ScPO_4{\cdot}2H_2O$, a lime green crystal to 1 mm. Christy dep., Magnet Cove, Hot Spring Co., Arkansas, United States.

(Continued)

BOX 5.3 (Continued)

Kolbeckite was named in honor of Dr. Friedrich Ludwig Wilhelm Kolbeck (January 12, 1860, Dresden to February 6, 1943, Freiberg), German mineralogist, Mining Academy, Freiberg, Germany. It is a very rare secondary mineral in phosphate deposits and some hydrothermal veins. Kolbeckite is monoclinic 2/m (space group $P2_1/n$). Its hardness is 3½–4. Luster is vitreous to pearly, transparent. The cleavage is distinct/good on {010}, distinct on {100}, and poor on {001}, and the fracture is conchoidal. The color varies from colorless, light yellow; when impure: cyan-blue, blue-gray, to apple-green. It is optically biaxial(−). Crystals are commonly short prismatic [001], with prominent {110} and {011} and less prominent {010}, {101}, and {130}. Platy crystals in radiating aggregates are less common. Twinning on {100} is common (Fig. 5.24).

FIGURE 5.25 Kolbeckite, $ScPO_4{\cdot}2H_2O$, a spheroidal green crystal to 2 mm. Schlarbaum quarry, Klause, Bad Gleichenberg, Styria, Germany.

gadolinite ($(Ce,Y)_2FeBe_2Si_2O_{10}$) are the only known concentrated sources of this element. Thortveitite can contain up to 45% of scandium in the form of scandium oxide. The total world production of scandium is around 15 tons per annum, in the form of scandium oxide. The industrial demand is about 50% higher, and both the production and demand keep rising. In 2003 only three mines produced scandium: the uranium and iron mines in Zhovti Vody in Ukraine, the rare-earth mines in Bayan Obo, China, and the apatite mines in the Kola peninsula, Russia. Since then, numerous other countries have built scandium-producing facilities, where it is a by-product from the production of other elements and is sold as scandium oxide. To obtain metallic scandium, the oxide is converted to scandium fluoride and then reduced with metallic calcium. Madagascar and the Iveland-Evje region in Norway have the only deposits of minerals with high scandium content, thortveitite $(Sc,Y)_2Si_2O_7$ and kolbeckite $ScPO_4{\cdot}2H_2O$ (Box 5.3), but these are currently not exploited. The absence of reliable, secure, stable, long-term production has restricted the economic use of scandium.

There are only 19 minerals containing scandium in their structure. Two of them are oxides: allendeite ($Sc_4Zr_3O_{12}$) and heftetjernite ($ScTaO_4$). In the phosphate class, three minerals are present: juonniite ($CaMgSc(PO_4)_2(OH){\cdot}4H_2O$), kolbeckite ($ScPO_4{\cdot}2H_2O$) (Fig. 5.25), and pretulite ($Sc(PO_4)$). The others are all silicates, for example, bazzite ($Be_3Sc_2(Si_6O_{18})$), davisite ($CaScAlSiO_6$), and thortveitite ($Sc_2Si_2O_7$) (Fig. 5.26).

5.3.3 Chemistry

In nature, Sc exists only as the isotope ^{45}Sc, which has a nuclear spin of 7/2, and is the only stable isotope. Thirteen radioisotopes have been observed with the most stable being ^{46}Sc with a half-life of 83.79 days; ^{47}Sc with a half-life of 3.35 days; the positron emitter ^{44}Sc with a half-life of 4 hours; and ^{48}Sc with a half-life of 43.67 hours. All the other radioisotopes have half-lives less than 4 hours, and most have half-lives less than 2 minutes. In addition, Sc has five nuclear isomers, with the most stable being ^{44m}Sc (half-life of 58.6 hours). The isotopes of Sc range from ^{36}Sc to ^{60}Sc.

FIGURE 5.26 Slender greenish thortveitite, $Sc_2Si_2O_7$, crystals to 7–8 mm embedded throughout a piece of reddish Feldspar matrix. Thortveitite is a relatively rare REE silicate that was discovered in Norway. Flåt, Evje og Hornnes, Aust-Agder, Norway.

TABLE 5.3 Scandium properties.

Appearance	Silvery-white
Standard atomic weight $A_{r,std}$	44.956
Block	d-Block
Element category	Transition metal
Electron configuration	[Ar] $3d^1$ $4s^2$
Phase at STP	Solid
Melting point	1541°C
Boiling point	2836°C
Density (near r.t.)	2.985 g/cm^3
When liquid (at m.p.)	2.80 g/cm^3
Heat of fusion	14.1 kJ/mol
Heat of vaporization	332.7 kJ/mol
Molar heat capacity	25.52 J/(mol·K)
Oxidation states	+1, +2, **+3**
Ionization energies	1st: 633.1 kJ/mol
	2nd: 1235.0 kJ/mol
	3rd: 2388.6 kJ/mol
Atomic radius	Empirical: 162 pm
Covalent radius	170 ± 7 pm
Van der Waals radius	211 pm

STP, Standard temperature and pressure.
Bold font indicates main oxidation state.

The main decay mode at masses lower than ^{45}Sc is electron capture, and the principal mode at masses above it is β emission. The principal decay products at atomic weights below ^{45}Sc consist of Ca isotopes, while the principal products from higher atomic weights consist of Ti isotopes.

Scandium can usually be obtained as a by-product in the extraction of other materials. Scandium can also be obtained from thortveitite, which contains 35%–40% Sc_2O_3, but most is obtained as a by-product in the processing of uranium ores, which contain only about 0.02% Sc_2O_3, and in the production of tungsten. Its main applications in laser crystals and coatings are highly specialized, but the amount consumed is low. Scandium has an odd atomic number and has few stable isotopes. The metal is rather soft with a silvery-white color (Table 5.3).

5.3.3.1 Oxides and hydroxides

The hydroxides, $M(OH)_3$, or in the case of scandium possibly the hydrated oxide, are obtained as gelatinous precipitates from aqueous solutions of scandium salts by addition of an alkali hydroxide. The precipitate can be dissolved in an excess of concentrated NaOH to give an anionic species such as $[Sc(OH)_6]^{3-}$. Dissolution of the oxide or hydroxide in the appropriate acid provides the most convenient method for producing the salts of the colorless, diamagnetic M^{3+} ions. Such solutions, especially those of Sc^{3+}, are significantly hydrolyzed with the formation of polymeric hydroxy species. The oxide Sc_2O_3 and the hydroxide $Sc(OH)_3$ are amphoteric.

$$Sc(OH)_3 + 3OH^- \rightarrow [Sc(OH)_6]^{3-} \text{(scandate ion)}$$
$$Sc(OH)_3 + 3H^+ + 3H_2O \rightarrow [Sc(H_2O)_6]^{3+}$$

α- and γ-ScOOH are isostructural with their aluminum hydroxide oxide equivalents. Scandium(III) oxide has a cubic crystal structure (point group: tetrahedral, space group: $Ia\bar{3}$) containing six-coordinate metal centers. The Sc-O bond distances vary between 2.159 and 2.071 Å. Sc_2O_3 is an insulator with a band gap of 6.0 eV. Sc_2O_3 is the principal form of refined scandium produced by the mining industry, making it the starting compound for all scandium reactions. Sc_2O_3 reacts with most acids upon heating, to form the expected hydrated product. For instance, heating in excess aqueous HCl produces hydrated $ScCl_3{\cdot}nH_2O$. This can be converted to the anhydrous form by evaporation to dryness in the presence of NH_4Cl, with the mixture subsequently being purified to remove NH_4Cl by sublimation at 300°C–500°C. The presence of NH_4Cl is essential, as the hydrated $ScCl_3{\cdot}nH_2O$ would otherwise form a mixed oxychloride upon drying.

$$Sc_2O_3 + 6HCl + xH_2O \rightarrow 2ScCl_3{\cdot}nH_2O + 3H_2O$$
$$ScCl_3{\cdot}nH_2O + nNH_4Cl \rightarrow ScCl_3 + nH_2O + nNH_4Cl$$

Similarly, it is changed into hydrated scandium(III) triflate ($Sc(OTf)_3{\cdot}nH_2O$) by a reaction with triflic acid (trifluoromethanesulphonic acid, CF_3SO_3H). Metallic scandium is prepared commercially through the reduction of scandium oxide; this proceeds via transformation to scandium fluoride and subsequent reduction with metallic calcium. This method is in some ways comparable to the Kroll process for the production of metallic titanium. Scandium oxide forms scandate salts with alkalis, unlike its higher homologs yttrium oxide and lanthanum oxide, for example, forming $K_3Sc(OH)_6$ with KOH. In this compound, scandium oxide exhibits more similarity with aluminum oxide.

5.3.3.2 Halides

Except for the fluorides, the halides are very water soluble and deliquescent. The distinctive ability of Sc^{3+} to form complexes is illustrated by the fact that an excess of F causes the first-precipitated ScF_3 to redissolve as $[ScF_6]^{3-}$. The anhydrous halides are best prepared by direct reaction of the elements rather than by heating the hydrates, which causes hydrolysis. Heating the hydrated chlorides, for instance, gives Sc_2O_3. The anhydrous halides illustrate nicely the effects of ionic size on the coordination number of the metal. In all four of its halides, scandium is six coordinated.

Scandium(III) fluoride, ScF_3, is an ionic compound. ScF_3 can be prepared by reacting scandium and fluorine. It is also produced during the extraction from the ore thortveitite through the reaction of Sc_2O_3 with ammonium bifluoride at high temperature.

$$Sc_2O_3 + 6NH_4HF_2 \rightarrow 2ScF_3 + 6NH_4F + 3H_2O$$

The subsequent mixture contains several metal fluorides, which is reduced by reaction with calcium metal at high temperature. Additional purification steps are necessary to give usable metallic scandium. ScF_3 shows the rare property of negative thermal expansion, that is, it shrinks when heated. This phenomenon can be explained by the quartic oscillation of the fluoride ions. The energy stored in the bending strain of the fluoride ion is proportional to the fourth power of the displacement angle, in contrast to most other materials where it is proportional to the square of the displacement. A fluorine atom is bound to two scandium atoms, and as the temperature rises, the fluorine oscillates more perpendicularly to its bonds. This motion draws the scandium atoms closer together all through the bulk material, which then contracts. ScF_3 shows this behavior from at least −263°C to 827°C (10K–1100K), above which it exhibits the normal positive thermal expansion; moreover, the material has cubic symmetry over this entire temperature range, and up to at least 1327°C (1600K) at ambient pressure. The negative thermal expansion at very low temperatures is rather strong [coefficient of thermal expansion about −14 ppm/K from −213°C to −163°C (60K–110K)]. At ambient pressures, scandium trifluoride has the cubic crystal system, using the perovskite ($CaTiO_3$) structure with one metal position vacant. The unit cell dimension is 4.01 Å. Under pressure, it also forms different crystal structures with rhombohedral symmetry above 0.6 GPa and above 3 GPa tetrahedral symmetry.

Scandium(III) chloride is mainly of interest in the research laboratory. Both the anhydrous form and hexahydrate ($ScCl_3{\cdot}6H_2O$) are commercially available. $ScCl_3$ crystallizes with the layered BiI_3 structure, which consists of octahedral scandium centers. Monomeric $ScCl_3$ is the predominant species in the vapor phase at 627°C (900K), the dimer Sc_2Cl_6 accounts for around 8%. The electron diffraction pattern suggests that the monomer is planar and the dimer has two bridging Cl atoms with each Sc being four coordinated. $ScCl_3$ dissolves in water to give $[Sc(H_2O)_6]^{3+}$ ions. In fact, samples of $ScCl_3$ change to this hexahydrate upon exposure to air. The structure of the hexahydrate is *trans*-$[ScCl_2(H_2O)_4]Cl{\cdot}2H_2O$. With the less basic ligand tetrahydrofuran, $ScCl_3$ forms the adduct $ScCl_3(THF)_3$ as white crystals. This THF-soluble complex is employed in the synthesis of organoscandium compounds. $ScCl_3$ has been transformed to its dodecyl sulfate salt, which has been studied as a "Lewis acid-surfactant combined catalyst" in aldol-like reactions. $ScCl_3$ was used by Fischer et al. (1937) who first produced metallic scandium by electrolysis of a eutectic melt of $ScCl_3$ and other salts at 700°C–800°C. $ScCl_3$ reacts with scandium metal to give a number of chlorides where scandium has an oxidation state lower than +3, for example, ScCl, Sc_7Cl_{10}, Sc_2Cl_3, Sc_5Cl_8, and Sc_7Cl_{12}. For instance, the reduction of $ScCl_3$ with scandium metal in the presence of cesium chloride gives the compound $CsScCl_3$, which contains linear chains of composition $Sc(II)Cl_3^-$, consisting of $Sc(II)Cl_6$ octahedra sharing faces. Sc_7Cl_{12} consists of discrete $[Sc_6C1_{12}]^{3-}$ clusters, similar to the M_6C1_{12} clusters of Nb and Ta, along with separate Sc^{3+} ions. Sc_5Cl_8 is best regarded as $(ScC1_2^+)_n-(Sc_4CI_6^-)_n$ in which edge-sharing $ScC1_6$ octahedra and edge-sharing Sc_6 octahedra lie in parallel chains. Sc_2C1_3 and its Br analog are of unknown structure, but Sc_7C1_{10} is composed of a double chain of Sc_6 octahedra sharing edges and a parallel chain of $ScC1_6$ octahedra. ScCl, made up of close-packed layers of Sc and C1 atoms in the sequence C1-Sc-Sc-C1, has been shown to be stabilized by interstitial H impurity.

$ScBr_3$ is formed via the burning of scandium in bromine gas.

$$2Sc(s) + 3Br_2(g) \rightarrow 2ScBr_3(s)$$

$ScBr_3$ is utilized for solid-state synthesis of atypical clusters such as $Sc_{19}Br_{28}Z_4$ (Z = Mn, Fe, Os, or Ru). These clusters are of significance because of their structure and magnetic properties.

Scandium triiodide has a structure comparable to that of iron trichloride ($FeCl_3$), crystallizing in a rhombohedral lattice. Scandium is six coordinated, while iodine is three coordinated and forms a trigonal pyramidal. The purest scandium triiodide is prepared through direct reaction of the elements:

$$2Sc + 3I_2 \rightarrow 2ScI_3$$

Instead, but less effectively, it can be formed by dehydrating $ScI_3(H_2O)_6$.

5.3.3.3 *Hydrides and sulfides*

Scandium hydride, also known as scandium–hydrogen alloy, is an alloy formed by combining scandium and hydrogen. Hydrogen acts as a hardening agent, preventing dislocations in the scandium atom crystal lattice from sliding past one another. Varying the amount of hydrogen controls properties, for example, the hardness of the resulting scandium hydride. Scandium hydride with increased hydrogen content can be made harder than scandium. In the narrow range of concentrations, which make up scandium hydride, mixtures of hydrogen and scandium can form two different structures. At 25°C, the most stable form is the hexagonal close-packed (*hcp*) structure α-scandium. It is a rather soft metallic material that can dissolve a moderate percentage of hydrogen, no more than 0.89 wt.% at 22°C. If scandium hydride contains more than 0.89% hydrogen at room temperature, it converts into a face-centered cubic (*fcc*) structure, the δ-phase. It can dissolve substantially more hydrogen, as much as 4.29%, which indicates the upper hydrogen content of scandium hydride. Research suggests the presence of a third phase formed under extreme conditions called the η-phase. This phase can dissolve as much as 6.30% hydrogen.

Scandium trihydride is an unstable molecular compound with the chemical formula ScH_3. It has been produced as one of a number of other molecular scandium hydride products at low temperature employing laser ablation and characterized by infrared spectroscopy. Scandium trihydride has in recent times been the topic of Dirac–Hartree–Fock relativistic calculation studies, which examined the stabilities, geometries, and relative energies of hydrides of the formula MH_3, MH_2, or MH. Scandium trihydride is a quasitrigonal planar molecule with three equivalent Sc–H bonds. It has a (C_{3v}) structure with an equilibrium distance between Sc and H of 182.0 pm and the bond angle is 119.2°. By weight percent, the composition of scandium trihydride is 6.30% hydrogen and 93.70% scandium. In scandium trihydride, the formal oxidation states of H and Sc are −1 and +3, respectively, since the electronegativity of scandium is lower than that of hydrogen. The stability of metal hydrides with the formula MH_3 (M = Sc-Lu) increases as the atomic number of M increases. Early theoretical investigations of ScH_3 showed that the molecule is unstable, the bulk substance is probably a colorless gas with a low activation energy toward the transformation into trimeric clusters because of the electron-deficient nature of the monomer, not unlike the group 13 hydrides. One key difference is that the dimer is the most stable cluster for group 13 hydrides. This can be ascribed to the distortion produced by the d-orbitals. It cannot be prepared by methods employed to synthesize BH_3 or AlH_3.

Scandium monosulfide is the chemical compound ScS. Although its formula might indicate that it is a compound of scandium(II), that is, $[Sc^{2+}][S^{2-}]$, ScS is perhaps more accurately defined as a pseudo-ionic compound, containing $[Sc^{3+}][S^{2-}]$, with the residual electron occupying the conduction band of the solid. Scandium monosulfide can be formed by heating a mixture of scandium metal and powdered sulfur in the absence of air to 1150°C for 70 hours. Scandium monosulfide has the sodium chloride crystal structure type.

$$Sc + S \rightarrow ScS$$

Scandium(III) sulfide, Sc_2S_3, is a yellow solid. Metal sulfides are typically prepared by heating mixtures of the two elements, but in the case of scandium, this method results in the formation of scandium monosulfide, ScS. Sc_2S_3 can be produced by heating scandium(III) oxide under flowing hydrogen sulfide in a graphite crucible to 1550°C or higher for 2–3 hours. The crude product is subsequently purified by chemical vapor transport at 950°C utilizing iodine as the transport agent.

$$Sc_2O_3 + 3H_2S \rightarrow Sc_2S_3 + 3H_2O$$

Scandium(III) sulfide can also be produced through the reaction of scandium(III) chloride with dry hydrogen sulfide at elevated temperature.

$$2ScCl_3 + 3H_2S \rightarrow Sc_2S_3 + 6HCl$$

The crystal structure of Sc_2S_3 is similar to that of sodium chloride, in that it is based on a cubic close-packed array of anions. While NaCl has all the octahedral interstices in the anion lattice occupied by cations, Sc_2S_3 has one-third of them vacant. The vacancies are ordered, but in a very complicated pattern, resulting in a large, orthorhombic unit cell belonging to the space group *Fddd*. Above 1100°C, Sc_2S_3 loses sulfur, resulting in nonstoichiometric compounds such as $Sc_{1.37}S_2$.

5.3.3.4 Sulfates and nitrates

Sulfates and nitrates are known and in all cases they decompose to the oxides on heating. Scandium sulfate, $Sc_2(SO_4)_3$, is used in agriculture as a very dilute solution as a seed treatment to improve the germination of corn, peas, wheat, and other plants. Scandium(III) nitrate, $Sc(NO_3)_3$, is an ionic compound. It is an oxidizer, as all nitrates are. It is used in optical coatings, as catalysts, in electronic ceramics, and in the laser industry.

5.3.3.5 Complexes

Compared to later elements in their respective transition series, scandium, yttrium, and lanthanum have rather poorly developed coordination chemistries and form weaker coordinate bonds, lanthanum generally being even less inclined to form strong coordinate bonds than scandium. Sc, which shows more similarity with Al, is a class a acceptor, complexing most readily with 0-donor ligands particularly when chelating. The complex anion $[ScF_6]^{3-}$ and other halo complexes with a variety of stereochemistries must normally be prepared by dry methods to avoid hydrolysis, and iodo complexes are invariably unstable. Other complexes such as $[Sc(dmso)_6]^{3+}$, $[Sc(bipy)_3]^{3+}$, $[Sc(bipy)_2(NCS)_2]^+$, and $[Sc(bipy)_2Cl_2]^+$ exhibit scandium's usual coordination number of 6.

5.3.3.6 Organometallic compounds

In view of the electronic structures of the elements of this group, little interaction with n-acceptor ligands is to be expected although cocondensation of metal vapors with an excess of the bulky ligand, 1,3,5-tri-*tert*-butylbenzene at -196°C (77K), yields the unstable sandwich compounds $[M(\eta^6-Bu^t_3C_6H_3)_2]$, M = Sc, Y which are the first examples of these metals in oxidation state zero.

The organometallic chemistry of this group is dominated by compounds involving cyclopentadiene (Cp) and its methyl-substituted derivatives. The dominant species are $CpScX_2$, Cp_2ScX, and Cp_3Sc. The mononuclear compounds are of limited use due to poor steric shielding, which makes them prone to nucleophilic addition and solvent attack. Though many are thermally stable, they are invariably sensitive to moisture and oxygen. The first to be prepared were the ionic cyclopentadienides, $M(C_5H_5)_3$, formed by the reactions of anhydrous MCl_3 with NaC_5H_5 in tetrahydrofuran and purified by vacuum sublimation at 200°C–250°C. The solids are polymeric, $[Sc(C_5H_5)_3]$ being made up of zig-zag chains of $[Sc(\eta^5\text{-}C_5H_5)_2]$ groups joined by $\eta^1{:}\eta^1\text{-}C_5H_5$ bridges. Cp_2ScCl can be synthesized from sodium cyclopentadienide.

$$ScCl_3 + 2NaCp \rightarrow ScCp_2Cl + 2NaCl$$

Chlorine can be substituted by a multitude of other ligands, for example, by an allyl group in reaction with allylmagnesium bromide.

$$ScCp_2Cl + C_3H_5MgBr \rightarrow ScCp_2(\eta^3 - C_3H_5) + MgClBr$$

Cp_2ScX compounds are dimers with X forming a bridging ligand. Dimerization is prevented in the presence of coordinating solvents such as THF or MeCN or when the Cp group is replaced by a bulkier ligand such as the Cp* group (1,2,3,4,5-pentamethylcyclopentadiene, C_5Me_5H (Me = CH_3)). The compound Cp_2^* ScCl is a stable monomer. The trinuclear compound $ScCp_3$ can also be prepared from NaCp.

$$ScCl_3 + 3NaCp \rightarrow ScCp_3 + 3NaCl$$

This compound is sensitive to air and moisture and has a structure consisting of an infinite chain with one bridging Cp group and two pentahapto Cp units. Furthermore, the chain can be broken up with excess coordinating solvent. The formation of tris(allyl)scandium has been described using a reaction of $ScCl_3$ with allylpotassium in THF. In $Sc(C_3H_5)_3(THF)_2$ two allyl ligands are η^3 coordinated and one allyl ligand is η^1 coordinated. Lower oxidation states (+2, +1, 0) have also been detected in organoscandium compounds.

5.3.4 Major uses

Scandium is mainly used for research purposes. It has, however, great potential as it has almost the same low density as aluminum but a much higher melting point. Adding Sc to Al restricts the grain growth in the heat zone of welded Al components. This has two useful effects: (1) the precipitated Al_3Sc forms smaller crystals than in other Al alloys and (2) the volume of precipitate-free zones at the grain boundaries of age-hardening Al alloys is reduced. Both effects improve the alloy's usefulness. But, Ti alloys, which are comparable in lightness and strength, are cheaper and much more widely used. The alloy $Al_{20}Li_{20}Mg_{10}Sc_{20}Ti_{30}$ is as strong as Ti, light as Al, and hard as ceramic. The main use of Sc by weight is in Al-Sc alloys for small aerospace industry components. These alloys have 0.1%–0.5% Sc. They were used in Russian military aircraft, in particular in the MiG-21 and MiG-29. Some sport equipment, which depend on high-performance materials, have been made with Sc–Al alloys, including baseball bats and bicycle frames and components (apparently high-performance cricket bats incorporating Sc were deemed unsporting and were forbidden). Lacrosse sticks are also produced with Sc. The American firearm manufacturing company Smith & Wesson makes semiautomatic pistols and revolvers with frames of Sc alloy and cylinders of Ti or carbon steel. Some dentists use erbium–chromium-doped yttrium–scandium–gallium garnet (Er,Cr:YSGG) lasers for cavity preparation and in endodontics. The first Sc-based metal halide lamps were originally patented by General Electric and initially manufactured in North America, though they are now made in all major industrialized countries. About 20 kg of Sc (as Sc_2O_3) is utilized per year in the United States for high-intensity discharge lamps. One type of metal halide lamp, like the mercury vapor lamp, is made from scandium triiodide (ScI_3) and sodium iodide (NaI). This lamp is a white light source with high color rendering index that adequately resembles natural sunlight to permit good color reproduction with TV cameras. About 80 kg of Sc is used in metal halide lamps/light bulbs annually worldwide. The radioactive isotope ^{46}Sc, with a half-life of about 84 days, is used in oil refineries as a tracing agent to follow the movement of various fractions as the oil is refined. In a comparable manner it can detect leaks in underground pipes carrying liquids. Scandium triflate (IUPAC name scandium trifluoromethanesulphonate, $Sc(SO_3CF_3)_3$) is a catalytic Lewis acid used in organic chemistry. Compared with other Lewis acids, this reagent is stable toward water and can often be used in organic reactions as a true catalyst rather than one used in stoichiometric amounts. The compound is prepared by reaction of scandium oxide with trifluoromethanesulfonic acid. Another use of Sc is to induce germination of seeds. When it is applied as a dilute solution of scandium sulfate to corn, peas, and wheat, it increases the number of seeds successfully germinating.

5.4 22 Ti — Titanium

5.4.1 Discovery

Titanium was first found in 1791 by the clergyman and amateur geologist, William Gregor (December 25, 1761 to June 11, 1817), as an inclusion of a mineral in Cornwall, Great Britain (Fig. 5.27). He recognized the presence of a new element in the mineral ilmenite ($FeTiO_3$) after he found black sand by a stream and noticed the sand was attracted by a magnet. Analyzing the sand, he found two different metal oxides: iron oxide (explaining the attraction to the magnet)

FIGURE 5.27 William Gregor.

and 45.25% of a white metallic oxide he did not recognize (Gregor, 1791a,b). Understanding that the unidentified oxide contained a metal that did not match any known element, Gregor reported his observations to the Royal Geological Society of Cornwall and in the German science journal Crell's Annalen. Around the same time, Austrian mineralogist and mining engineer Franz-Joseph Müller von Reichenstein (July 1, 1740 or October 4, 1742 to October 12, 1825 or 1826) produced a similar substance but could not identify it. The oxide was independently rediscovered in 1795 by Prussian (German) chemist Martin Heinrich Klaproth (December 1, 1743 to January 1, 1817) in the mineral rutile (TiO_2) from Boinik (German name for Bajmócska), a village in Hungary (now Bojničky in Slovakia). Klaproth found that it contained a new element and named it for the Titans of Greek mythology (Klaproth, 1795). After hearing about Gregor's previous detection, he got a sample of menaccanite (variety of ilmenite) and confirmed that it was the same metal. The at present known procedures for extracting titanium from its various ores are difficult and expensive; it is not possible to reduce the ore by heating with carbon (as in iron smelting) as titanium combines with the carbon to produce titanium carbide. Pure metallic titanium (99.9%) was for the first time synthesized in 1910 by the New Zealand born metallurgist and inventor Matthew A. Hunter (1878–1961) at Rensselaer Polytechnic Institute by heating $TiCl_4$ with sodium at 700°C–800°C under high pressure in a batch process now known as the Hunter process. Titanium metal was not utilized outside the laboratory until 1932 when Luxembourgish metallurgist William Justin Kroll (November 24, 1889 to March 30, 1973) showed that it can be formed by reducing titanium tetrachloride ($TiCl_4$) with calcium. Eight years later, he refined this process with magnesium and even sodium in what is now known as the Kroll process. Although research continues into more efficient and cheaper processes, the Kroll process is still employed for commercial manufacturing. Titanium of very high purity was made in small quantities when Dutch chemist Anton Eduard van Arkel (November 19, 1893 to March 14, 1976) and Dutch physicist and chemist Jan Hendrik de Boer (March 19, 1899 to April 25, 1971) developed the iodide, or crystal bar, process in 1925, by reacting with iodine and decomposing the formed vapors over a hot filament to the pure titanium metal.

5.4.2 Mining, production, and major minerals

Titanium is the ninth most abundant element in the Earth's crust (0.63% by mass) and the seventh most abundant metal. It is present as oxides in most igneous rocks, sediments derived from them, living things, and natural bodies of water. Its proportion in soils is approximately 0.5%–1.5%. Common titanium-containing minerals are anatase (TiO_2), brookite (TiO_2), ilmenite ($FeTiO_3$), perovskite ($CaTiO_3$), rutile (TiO_2), and titanite (old discredited name sphene) ($CaTi(SiO_4)O$). Akaogiite is an extremely rare TiO_2 polymorph, resulting from high-pressure impact at a meteorite crater. Of these minerals, only rutile and ilmenite have economic importance, yet even they are difficult to find in high concentrations.

Significant titanium-bearing ilmenite deposits exist in western Australia, Canada, China, India, Mozambique, New Zealand, Norway, Sierra Leone, South Africa, and Ukraine. Total reserves of titanium are estimated to exceed 600 million tons.

The production of titanium metal involves four major steps: (1) reduction of titanium ore into "sponge," a porous form; (2) melting of the sponge, or sponge plus a master alloy to form an ingot (a piece of relatively pure material, usually metal, that is cast into a shape suitable for further processing. In steelmaking, it is the first step among semifinished casting products.); (3) primary fabrication, where an ingot is converted into general mill products such as billet, bar, plate, sheet, strip, and tube; and (4) secondary fabrication of finished shapes from mill products. Since it cannot be readily manufactured by reduction of its dioxide (TiO_2), titanium metal is produced by the reduction of $TiCl_4$ with magnesium metal in the Kroll process. The complexity of the batch production in the Kroll process explains the relatively high market value of titanium, despite the Kroll process being less expensive than the Hunter process (Box 5.4). To obtain $TiCl_4$ required by the Kroll process, TiO_2 is reduced in a carbothermic reaction (the reduction of substances, often metal oxides, using carbon as the reducing agent) in the presence of chlorine. In this process, the chlorine gas is

BOX 5.4 Hunter, Kroll and FCC processes

The Hunter process was the first industrial process to produce pure ductile metallic titanium. The method involves reducing titanium tetrachloride ($TiCl_4$) with sodium (Na) in a batch reactor with an inert atmosphere at a temperature of 1000°C. Dilute hydrochloric acid (HCl) is then used to leach the salt from the product.

$$TiCl_4 + 4Na \rightarrow 4NaCl + Ti$$

Prior to the Hunter process, all efforts to produce Ti metal resulted in highly impure material, often titanium nitride (which resembles a metal).

The Kroll process is a pyrometallurgical industrial process used to produce metallic titanium. The Kroll process substituted the Hunter process for almost all commercial manufacturing. In the Kroll process, $TiCl_4$ is reduced by liquid Mg at 800°C–850°C in a stainless-steel retort to ensure complete reduction.

$$TiCl_4 + 2Mg \rightarrow Ti + 2MgCl_2$$

Problems result from partial reduction of $TiCl_4$, giving lower chlorides $TiCl_2$ and $TiCl_3$. $MgCl_2$ can be further refined back to magnesium. The resulting porous metallic titanium sponge is purified by leaching or heated vacuum distillation. The sponge is jackhammered out, crushed, and pressed before it is melted in a consumable carbon electrode vacuum arc furnace. The melted ingot is allowed to solidify under vacuum. It is often remelted to remove inclusions and ensure uniformity. These melting steps add to the cost of the product. Titanium is about six times as expensive as stainless steel.

The FFC Cambridge Process is an electrochemical method in which solid metal compounds, particularly oxides, are cathodically reduced to the respective metals or alloys in molten salts. It is believed that this method will ultimately be capable of producing metals or alloys more efficiently than the current conventional processes, for example, the Kroll process. The process typically takes place between 900°C and 1100°C, with an anode (typically carbon) and a cathode (oxide being reduced) in a bath of molten $CaCl_2$. Depending on the nature of the oxide, it will exist at a specific potential relative to the anode, which is dependent on the quantity of CaO present in $CaCl_2$. The cathode is then polarized to a more negative voltage versus the anode. This is simply realized by imposing a voltage between the anode and cathode. When polarized to more negative voltages, the oxide releases oxygen ions into the $CaCl_2$ salt, which exists as CaO. To maintain charge neutrality, as oxygen ions are released from the cathode into the salt, oxygen ions must be released from the salt to the anode. This is observed as CO or CO_2 being evolved at the carbon anode. In theory, an inert anode could be used to produce oxygen. When negative voltages are reached, it is possible that the cathode would begin to produce Ca (which is soluble in $CaCl_2$). Ca is highly reductive and would further strip oxygen from the cathode, resulting in a calciothermic reduction. However, Ca dissolved into $CaCl_2$ produces a more conductive salt leading to reduced current efficiencies.

Cathode reaction mechanism

The electro-calciothermic reduction mechanism can be represented by the following sequence of reactions:

$$MO_x + xCa \rightarrow M + xCaO \quad (1)$$

When this reaction takes place on its own, it is referred to as the "calciothermic reduction" (or, more generally, an example of metallothermic reduction). For example, if the cathode is primarily made from TiO, then the calciothermic reduction would appear as follows:

$$TiO + Ca \rightarrow Ti + CaO$$

(Continued)

BOX 5.4 (Continued)

Though the cathode reaction can be written as above, it is in fact a gradual removal of oxygen from the oxide. For example, it has been proven that TiO_2 does not simply reduce to Ti. Actually, it reduces through the lower oxides (Ti_3O_5, Ti_2O_3, TiO, etc.) to Ti. The calcium oxide produced is then electrolyzed.

$$xCaO \rightarrow xCa^{2+} + xO^{2-} \quad (2a)$$

$$xCa^{2+} + 2xe^- \rightarrow xCa \quad (2b)$$

$$xO^{2-} \rightarrow x/2O_2 + 2xe^- \quad (2c)$$

Reaction (2b) defines the production of Ca metal from Ca^{2+} ions within the salt, at the cathode. The Ca would then react to reduce the cathode. The net result of reactions (1) and (2) is simply the reduction of the oxide into metal plus oxygen.

$$MOx \rightarrow M + x/2O_2 \quad (3)$$

Anode reaction mechanism

The use of molten $CaCl_2$ is important because this molten salt can dissolve and transport the O^{2-} ions to the anode to be discharged. The anode reaction hinges on the anode material. Depending on the system, it is possible to produce either CO or CO_2 or a mixture at the carbon anode.

$$C + 2O^{2-} \rightarrow CO_2 + 4e^-$$
$$C + O^{2-} \rightarrow CO + 2e^-$$

However, if an inert anode is used, such as that of high-density SnO_2, the discharge of the O^{2-} ions leads to the evolution of oxygen gas. However, the use of an inert anode has drawbacks. First, when the concentration of CaO is low, Cl_2 evolution at the anode becomes more favorable. Furthermore, when compared with a carbon anode, more energy is required to achieve the same reduced phase at the cathode. Inert anodes also tend to suffer from stability issues.

$$2O^{2-} \rightarrow O_2 + 4e^-$$

passed over red-hot rutile (TiO_2) or ilmenite ($FeTiO_3$) in the presence of carbon. After extensive purification by fractional distillation, the $TiCl_4$ is reduced with 800°C molten magnesium in an inert argon atmosphere. Titanium metal can be further purified using the van Arkel–de Boer method, which involves the thermal decomposition of titanium tetraiodide (TiI_4). A more recently developed batch production method, the FFC Cambridge process developed by Derek Fray, Tom Farthing, and George Chen between 1996 and 1997 at the University of Cambridge. (The name FFC comes from the first letters of their last names.), consumes titanium dioxide powder (a refined form of rutile) as feedstock and produces titanium metal, either powder or sponge. The process involves fewer steps than the Kroll process and takes less time. If mixed oxide powders are used, the product is an alloy. Common titanium alloys are produced by reduction. For example, cuprotitanium (rutile with copper added is reduced), ferrocarbon titanium (ilmenite reduced with coke in an electric furnace), and manganotitanium (rutile with manganese or manganese oxides) are reduced.

$$2FeTiO_3 + 7Cl_2 + 6C \rightarrow 2TiCl_4 + 2FeCl_3 + 6CO \ (900°C)$$
$$TiCl_4 + 2Mg \rightarrow 2MgCl_2 + Ti \ (1{,}100°C)$$

Titanium powder is produced utilizing a flow production method known as the Armstrong process similar to the batch production Hunter process. A stream of titanium tetrachloride ($TiCl_4$) gas is added to a stream of molten sodium metal, and the products (sodium chloride salt and titanium particles) are filtered from the extra sodium. Titanium is then separated from the salt by washing with water. Both sodium and chlorine are recycled and reintroduced into the process.

The number of minerals containing titanium comes up to just over 400. Titanium does occur as a natural native element (Ti) as well as nitride osbornite (TiN), phosphide florenskyite (FeTiP), carbide khamrabaevite ((Ti,V,Fe)C), and silicide zangboite ($TiFeSi_2$). The sulfide class contains only one mineral: heideite ($(Fe,Cr)_{1.15}(Ti,Fe)_2S_4$). Some of the best known minerals containing Ti are found in the oxide class (which contains more than 100 minerals), for example, anatase (TiO_2) (Fig. 5.28), brookite (TiO_2) (Fig. 5.29) (Box 5.5), ilmenite ($Fe^{2+}TiO_3$) (Fig. 5.30), perovskite ($CaTiO_3$) (Fig. 5.31), pseudobrookite (Fe_2TiO_5), rutile (TiO_2) (Fig. 5.32), and ulvöspinel ($TiFe_2O_4$). Within the carbonate class, only a single mineral contains Ti: sabinaite ($Na_4Zr_2TiO_4(CO_3)_4$). The borate class has two minerals with Ti in their structure: azoproite ($(Mg,Fe^{2+})_2(Fe^{3+},Ti,Mg)(BO_3)O_2$) and warwickite ($(Mg,Ti,Fe,Cr,Al)_2O(BO_3)$). The sulfate class is represented by a single mineral: alcaparrosaite

FIGURE 5.28 Dark blue dipyramidal anatase, TiO_2, crystals to 2 mm on water-clear quartz. Matskorhae, Odda, Hardangervidda, Hordaland, Norway.

FIGURE 5.29 Brookite, TiO_2, sharp, brownish red bladed crystal to 7 mm with adularia (K-feldspar). Intschi Tobel, Intschi, Reuss Valley, Uri, Switzerland.

($K_3Ti^{4+}Fe^{3+}(SO_4)_4O(H_2O)_2$). Five minerals are found in the phosphate class containing Ti, such as curetonite ($Ba(Al,Ti)(PO_4)(OH,O)F$) and paulkerrite ($K(Mg,Mn^{2+})_2(Fe^{3+},Al,Ti,Mg)_2\ Ti(PO_4)_4(OH)_3{\cdot}15H_2O$). By far, the largest group of minerals is found in the silicate class with nearly 200 different minerals. Some of the well-known minerals include astrophyllite ($K_2NaFe^{2+}_7Ti_2Si_8O_{28}(OH)_4F$), benitoite ($BaTi(Si_3O_9)$), kaersutite ($\{Na\}\{Ca_2\}\{Mg_3AlTi\}\ (Al_2Si_6O_{22})O_2$), neptunite ($Na_2KLiFe^{2+}_2Ti_2Si_8O_{24}$), titanite ($CaTi(SiO_4)O$) (Fig. 5.33), and yuksporite ($K_4(Ca,Na)_{14}(Sr,Ba)_2(\square,Mn,Fe)(Ti,Nb)_4(O,OH)_4(Si_6O_{17})_2(Si_2O_7)_3(H_2O,OH)_3$) (Fig. 5.34).

5.4.3 Chemistry

Naturally occurring titanium consists of five stable isotopes: ^{46}Ti, ^{47}Ti, ^{48}Ti, ^{49}Ti, and ^{50}Ti, with ^{48}Ti being the most common (73.72% natural abundance, with ^{46}Ti at 8.25%, ^{47}Ti at 7.44%, ^{49}Ti at 5.41%, and ^{50}Ti at 5.18%). More than 21 radioisotopes have been observed, the most stable of which are ^{44}Ti with a half-life of 63 years, ^{45}Ti with a half-life of 184.8 minutes, ^{51}Ti with a half-life of 5.76 minutes, and ^{52}Ti with a half-life of 1.7 minutes. The remaining radioisotopes have half-lives less than 33 seconds, with most of them less than half a second. The principal decay mode for isotopes below ^{48}Ti is electron capture and the principal mode above ^{48}Ti is β emission. The main decay products for the isotopes below ^{48}Ti are Sc isotopes and the primary products for the heavier Ti isotopes are V isotopes. Titanium becomes radioactive upon bombardment with deuterons, emitting mainly positrons and hard gamma rays.

As a metal, titanium is known for its high strength-to-weight ratio. It is a strong metal with low density that is rather ductile (in particular in the absence of oxygen), lustrous, and metallic-white in color (Table 5.4). The relatively high melting point ($>1650°C$) makes it useful as a refractory metal. It is paramagnetic and has rather low electrical and thermal conductivity than other metals. Titanium is superconducting when cooled below its critical temperature of

BOX 5.5 TiO_2 polymorphism

Titanium dioxide occurs in nature as the minerals rutile and anatase, and additionally as two high-pressure forms. One of these is a monoclinic baddeleyite-like form known as akaogiite, and the other is an orthorhombic α-PbO_2-like form known as brookite. Rutile's name comes from the Latin word *rutilus*, meaning red, an allusion to the color. Rutile, although not particularly abundant, is widespread. It is found typically as small grains in intermediate to mafic igneous rocks, some metamorphic rocks, veins, pegmatites, and some sediments. Rutile is tetragonal, *4/m2/m/2m* (space group $P4_2/m2_1/n2/m$). It has a hardness of 6–6½. The luster is adamantine to submetallic, the transparency is transparent to translucent. It has a distinct cleavage on {110}, less so on {100}, and the fracture is subconchoidal. The color varies from red, brown, yellow, to black in varieties high in Fe, Nb, or Ta. The streak is pale brown to yellow. Optically it is uniaxial(–). Crystals are commonly prismatic, often slender to acicular. Reticulated groups are known as the variety sagenite. Prism zone is vertically striated. Usually terminated by {101} or {111}. Rarely pyramidal or equant. Also occurs as granular, massive, coarse to fine masses. Twinning on {011} is common. Often geniculated, or contact twins of varied habit, sometimes sixlings or eightlings occasionally forming circular groupings. Epitaxial intergrowth known with hematite, magnetite, ilmenite, brookite, and anatase.

Anatase was named in 1801 by Rene Just Haüy from the Greek ανάτασις ("anatasis") for "extension," in allusion to the length of the pyramidal faces being longer in relation to their bases than in many tetragonal minerals. Anatase is typically found in veins or crevices of the Alpine type in gneiss or schist, where it is derived from leaching from the surrounding rock by hydrothermal solutions. Anatase is tetragonal, *4/m2/m2/m* (space group $I4_1/amd$). The hardness is 5½–6. The luster is adamantine, metallic-adamantine in dark varieties, transparent when light colored, to nearly opaque when deeply colored. Pyramidal crystals may appear opaque because of total reflection. Color varies from brown, pale-yellow or reddish-brown, indigo, black; pale green, pale lilac, gray, to rarely nearly colorless. The streak is white to pale-yellow. It has two perfect cleavages on {001} and {011} with a subconchoidal fracture. Optically it is uniaxial(–). Crystals are commonly acute pyramidal {011}, less commonly tabular or obtuse pyramidal, with {017} or {013}, often highly modified. Twining on {112} is rare.

Brookite was named in 1825 by Serve-Dieu Abailard "Armand" Lévy in honor of Henry James Brooke (May 25, 1771 in Exeter, England, UK to 1857 in London, England, UK), a wealthy textile manufacturer and actuary, amateur English crystallographer and mineralogist. Brooke discovered 12 minerals: annabergite, autunite, arfvedsonite, caledonite, childrenite, linarite, nitronatrite, susannite, thomsonite, and whewellite. He was the author of A Familiar Introduction to Crystallography and coauthor of Elementary Introduction to Mineralogy (1852). He wrote numerous scientific articles and was a fellow of the Royal Society, Geological Society of London, and the Linnaen Society. Brookite is found principally in metamorphic rocks in fissures and open cavities in veins of the Alpine cleft type, with anatase, rutile, titanite, albite, chlorite, etc. Also rarely found in ore (lead) on quartz with galena and chalcopyrite, in nepheline syenite pegmatites and volcanic tuff. It is orthorhombic, *2/m2/m2/m* (space group *Pbca*). It has a hardness of 5½–6. Its luster is adamantine, subadamantine, vitreous, subvitreous, to submetallic. Transparency ranges from transparent to opaque. The color shows various shades of brown to black; the streak is white to grayish to yellowish. Cleavage is indistinct on {120} with subconchoidal fracture. It is optically biaxial(+). Crystals are usually tabular {010} and elongated [001], also pseudohexagonal with {120} and {111} equally developed.

FIGURE 5.30 Ilmenite, $Fe^{2+}TiO_3$, octahedral looking submetallic crystal to 1.2 cm in yellow serpentine matrix. Snarum, Modum, Buskerud, Norway.

FIGURE 5.31 Single black lustrous, cubo-octahedral perovskite, $CaTiO_3$, crystal to 1.7 cm. Afrikanda Massif, Kola Peninsula, Russia.

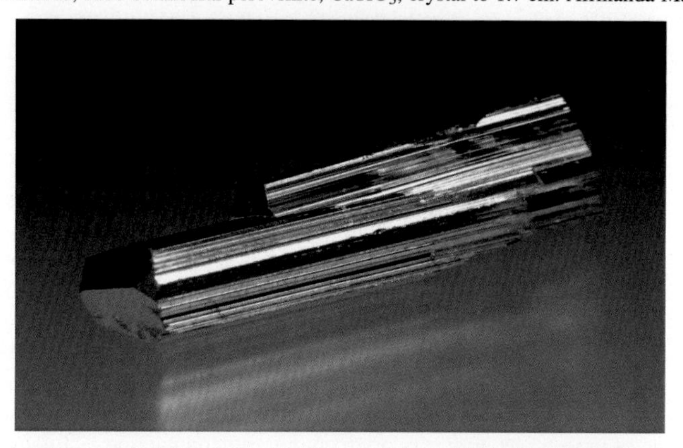

FIGURE 5.32 A terminated prismatic crystal of deep red rutile, TiO_2, to 2.5 cm. Hiddenite, Alexander Co., North Carolina, United States.

FIGURE 5.33 Titanite, $CaTi(SiO_4)O$, a 2.4 cm greenish yellow twinned crystal. Capelinha, Minas Gerais, Brazil.

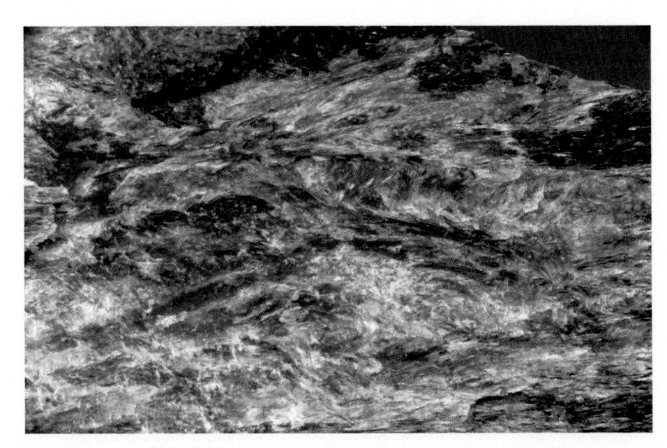

FIGURE 5.34 Yuksporite, $K_4(Ca,Na)_{14}(Sr,Ba)_2(\square,Mn,Fe)(Ti,Nb)_4(O,OH)_4(Si_6O_{17})_2(Si_2O_7)_3(H_2O,OH)_3$, red platy segregations in metamorphosed matrix. Overall 8 cm. Eveslogchorr Mt, Khibiny Massif, Kola Peninsula, Russia.

TABLE 5.4 Titanium properties.

Appearance	Silvery gray-white metallic
Standard atomic weight $A_{r,std}$	47.867
Block	d-Block
Element category	Transition metal
Electron configuration	[Ar] $3d^2\ 4s^2$
Phase at STP	Solid
Melting point	1668°C
Boiling point	3287°C
Density (near r.t.)	4.506 g/cm^3
When liquid (at m.p.)	4.11 g/cm^3
Heat of fusion	14.15 kJ/mol
Heat of vaporization	425 kJ/mol
Molar heat capacity	25.060 J/(mol·K)
Oxidation states	−2, −1, +1, +2, +3, **+4**
Ionization energies	1st: 658.8 kJ/mol
	2nd: 1309.8 kJ/mol
	3rd: 2652.5 kJ/mol
Atomic radius	Empirical: 147 pm
Covalent radius	160 ± 8 pm

STP, Standard temperature and pressure.
Bold font indicates main oxidation state.

0.49K. The metal is a dimorphic allotrope of a hexagonal α form that transforms into a body-centered cubic β form at 882°C. The specific heat of the α form increases dramatically as it is heated to this transition temperature but then drops and remains rather constant for the β form irrespective of temperature. The elements of this group are relatively electropositive, and if heated to high temperatures, they react directly with most nonmetals, particularly oxygen, hydrogen (reversibly), and, in the case of titanium, nitrogen (Ti actually burns in N_2). Ti is readily oxidized to the +4 state but in aqueous solution. Ti(III) can be prepared by the reduction of Ti(IV), either with Zn and dilute acid or electrolytically. It exists in dilute acids as violet, octahedral $[Ti(H_2O)_6]^{3+}$ ion. However, the M-C σ bonds are not strong, as might be expected for metals with so few d-electrons. Little help is available from synergic π bonding with simple carbonyls, only complexes like $Ti(CO)_6$ have been reported.

5.4.3.1 Oxides

TiO_2 is known primarily as a white pigment existing at room temperature in three forms—rutile, anatase, and brookite, each of which occurs naturally. Each contains six-coordinate titanium, but rutile is the most common form, both in nature and as produced commercially, and the others transform into it on heating. Titanium dioxide has eight modifications—in addition to the minerals rutile, anatase, akaogiite, and brookite, three metastable phases can be produced

TABLE 5.5 TiO_2 forms.

Form	Crystal system
Rutile	Tetragonal
Anatase	Tetragonal
Brookite	Orthorhombic
TiO_2(B)	Monoclinic
TiO_2(H), hollandite ($Ba(Mn_6^{4+}Mn_2^{3+})O_{16}$)-like form	Tetragonal
TiO_2(R), ramsdellite (MnO_2)-like form	Orthorhombic
TiO_2(II)-(α-PbO_2-like form)	Orthorhombic
Akaogiite (baddeleyite (ZrO_2)-like form, 7 coordinated Ti)	Monoclinic
TiO_2 -OI	Orthorhombic
Cubic form	Cubic
TiO_2 -OII, cotunnite ($PbCl_2$)-like	Orthorhombic

synthetically (monoclinic, tetragonal, and orthorhombic), and five high-pressure forms (α-PbO_2-like, baddeleyite-like, cotunnite-like, orthorhombic OI, and cubic phases) also exist (Table 5.5).

The production method depends on the ore. The most common ore mineral is ilmenite. The abundant rutile mineral sand can also be purified with the chloride process or other processes. Ilmenite is changed into pigment grade titanium dioxide via either the sulfate process or the chloride process. Both sulfate and chloride processes form the titanium dioxide pigment in the rutile crystal form, but the sulfate process can be modified to form the anatase form. Anatase, being softer, is employed in fiber and paper applications. The sulfate process is run as a batch process and the chloride process as a continuous process. Plants using the sulfate process need ilmenite concentrate (45%–60% TiO_2) or pretreated feedstocks as appropriate source of titanium. In the sulfate process, ilmenite is reacted with sulfuric acid to extract iron(II) sulfate pentahydrate. The resulting synthetic rutile is further treated based on the specifications of the end user, that is, pigment grade or otherwise. An additional method for the production of synthetic rutile from ilmenite is the Becher process, which first oxidizes ilmenite to remove the iron component. Another process, called the chloride process, changes ilmenite or other titanium sources to titanium tetrachloride via reaction with elemental chlorine, which is subsequently purified by distillation and reacted with oxygen to regenerate the chlorine and form titanium dioxide. Titanium dioxide pigment can also be prepared from higher titanium content feedstocks, for example, upgraded slag, rutile, and leucoxene (alteration product of mainly anatase or rutile, consisting of Fe-Ti oxides, including titanite, perovskite, titanian magnetite, but especially ilmenite.) via a chloride acid process. For specialty uses, TiO_2 films are produced by a variety of specialized reaction routes. Sol–gel routes comprise the hydrolysis of titanium alkoxides, for example, titanium ethoxide.

$$Ti(OEt)_4 + 2H_2O \rightarrow TiO_2 + 4EtOH$$

This method is appropriate for the preparation of films. A related method that likewise depends on molecular precursors entails chemical vapor deposition (CVD). In this process, the alkoxide is volatilized and subsequently decomposed on contact with a hot surface.

$$Ti(OEt)_4 \rightarrow TiO_2 + 2Et_2O$$

Of the lower oxides, Ti_3O_5 is a blue black material prepared by the reduction of TiO_2 with H_2 at 900°C. It shows a transition from semiconductor to metal at 175°C. Ti_2O_3 is a dark violet material with the corundum (Al_2O_3) structure. It is prepared by reacting TiO_2 and Ti metal at 1600°C and is generally inert, being resistant to most reagents except oxidizing acids.

TiO, a bronze-colored, readily oxidized material, is also prepared by the reaction of TiO_2 and Ti metal at 1500°C. It has a defect rock salt (NaCl) structure that tolerates a high proportion of vacancies (Schottky defects) in both Ti and O sites and so is highly nonstoichiometric. In pure TiO, 15% of both Ti and O sites are vacant. Careful annealing can result in ordering of the vacancies forming a monoclinic phase that has five TiO units in the primitive unit cell that shows lower resistivity. A high-temperature phase with Ti atoms in trigonal prismatic coordination has also been observed. Acid solutions of TiO are stable for a short period and subsequently decompose to produce hydrogen.

$$2Ti^{2+}(aq) + 2H^+(aq) \rightarrow 2Ti^{3+}(aq) + H_2(g)$$

Gas phase TiO exhibits strong bands in the optical spectra of cool (M-type) stars. Back in 2017, TiO was observed for the first time in the atmosphere of an exoplanet. Furthermore, it has been shown that the diatomic molecule TiO exists in the interstellar medium. Oxygen is soluble in metallic titanium up to a composition of $TiO_{0.5}$ with the oxygen atoms occupying octahedral sites in the *hcp* metal lattice: distinct phases that have been crystallographically characterized are Ti_6O, Ti_3O, and Ti_2O. It seems likely that in all these reduced oxide phases, there is extensive metal−metal bonding.

The dioxides are notable for their inertness particularly if they have been heated via fusion or firing at high temperatures up to 2500°C. With stoichiometric amounts of appropriate oxides produces a number of "titanates," "zirconates," and "hafnates." The titanates are of two main types: the orthotitanates $M(II)TiO_4$ and the metatitanates $M(II)TiO_3$. The names are misleading since the compounds almost never contain the discrete ions $[TiO_4]^{4-}$ and $[TiO_3]^{2-}$ analogous to phosphates or sulfites. Rather, the structures comprise three-dimensional networks of ions, which are of particular interest and importance because two of the metatitanates are the archetypes of common mixed metal oxide structures.

When M(II) is approximately the same size as Ti(IV) (i.e., M = Mg, Mn, Fe, Co, Ni), the structure is that of ilmenite, $FeTiO_3$, which consists of *hcp* oxygens with one-third of the octahedral interstices occupied by M(II) and another third by Ti(IV). If, however, M(II) is significantly larger than Ti(IV) (e.g., M = Ca, Sr, Ba), then the preferred structure is that of perovskite, $CaTiO_3$. $M(II)_2TiO_4$ (M = Mg, Zn, Mn, Fe, Co) has the spinel structure ($MgAl_2O_4$), which is the third important structure type adopted by many mixed metal oxides; in this, the cations occupy both octahedral and tetrahedral sites in a *ccp* array of oxide ions.

5.4.3.2 *Chalcogenides*

Titanium disulfide, TiS_2, is a golden yellow solid with high electrical conductivity and belongs to a group of compounds called transition metal dichalcogenides, ME_2. TiS_2 has been used as a cathode material in rechargeable batteries. Titanium(II) sulfide (TiS) was found in a meteorite, Yamato 691, as tiny flecks, making it a new mineral known as wassonite. TiS_2 is produced through the reaction of the elements at about 500°C.

$$Ti + 2S \rightarrow TiS_2$$

It can be more easily prepared from titanium tetrachloride, but this product is usually less pure than that formed from the elements.

$$TiCl_4 + 2H_2S \rightarrow TiS_2 + 4HCl$$

This method has been used for the preparation of TiS_2 films by CVD. Thiols and organic disulfides can be used instead of hydrogen sulfide. TiS_2 has a hexagonal close-packed (*hcp*) structure, similar to cadmium iodide (CdI_2). In this structure, half of the octahedral holes are filled with Ti^{4+}. Each Ti center is surrounded by six sulfur ligands in an octahedral coordination. Each sulfur is connected to three Ti centers, the geometry at S being pyramidal. A number of metal dichalcogenides have similar structures, but some, particularly MoS_2, do not. The layers of TiS_2 comprise covalent Ti-S bonds. The individual layers of TiS_2 are bonded together by van der Waals forces, which are rather weak intermolecular forces. It crystallizes in the space group $P\bar{3}m1$. The Ti-S bond length is 2.423 Å. TiS_2 is unstable in air. Upon heating, the solid oxidizes to titanium dioxide.

$$TiS_2 + O_2 \rightarrow TiO_2 + 2S$$

TiS_2 is similarly sensitive to water.

$$TiS_2 + 2H_2O \rightarrow TiO_2 + 2H_2S$$

Upon heating, TiS_2 releases sulfur, forming the titanium(III) derivative.

$$2TiS_2 \rightarrow Ti_2S_3 + S$$

Thin films of TiS_2 have been produced via sol-gel processes from titanium isopropoxide ($Ti(OPri)_4$) followed by spin coating. This technique forms amorphous material that crystallizes at high temperatures to hexagonal TiS_2, with crystallization orientations in the [001], [100], and [001] directions. Due to their high surface area, these films are attractive for battery applications. More specialized morphologies—nanotubes, nanoclusters, whiskers, nanodisks, thin films, and fullerenes—can be synthesized by combining the standard reagents, often $TiCl_4$ in unusual ways. For instance, flower-like morphologies were formed by reacting a solution of sulfur in 1-octadecene with titanium tetrachloride. A form of TiS_2 with a fullerene-like structure has been synthesized using the $TiCl_4/H_2S$ method. The spherical structures formed had diameters ranging from 30 to 80 nm. Due to their spherical shape, these fullerenes show a reduced friction coefficient and wear, which may be useful in some applications. Nanotubes of TiS_2 can be prepared utilizing a variation of the $TiCl_4/H_2S$ route. Based on transmission electron microscopy (TEM), these tubes have an outer diameter of 20 nm and an inner diameter of 10 nm. The average length of the nanotubes was 2–5 μm, and the nanotubes were found to be hollow. TiS_2 nanotubes with open-ended tips are described to store up to 2.5 wt.% hydrogen at 25°C and 4 MPa hydrogen gas pressure. Absorption and desorption rates are fast, which is desirable for hydrogen storage. The hydrogen atoms are suggested to bind to sulfur. Nanoclusters, or quantum dots of TiS_2, have unique electronic and chemical properties because of quantum confinement and very large surface to volume ratios. Nanoclusters can be formed using micelles. The nanoclusters are produced from a solution of $TiCl_4$ in tridodecylmethyl ammonium iodide (TDAI), which acted as the inverse micelle structure and seeded the growth of nanoclusters in the same general reaction as nanotubes. Nucleation takes place only inside the micelle cage because of the insolubility of the charged species in the continuous medium, which usually is a low dielectric constant inert oil. Similar to the bulk material, the nanocluster form of TiS_2 has a hexagonal layered structure. Quantum confinement produces well-separated electronic states and increases the band gap more than 1 eV compared to bulk TiS_2. A spectroscopic assessment indicates a large blueshift for the quantum dots of 0.85 eV. Nanodisks of TiS_2 are formed by reacting $TiCl_4$ with sulfur in oleylamine.

The single most useful and most investigated property of TiS_2 is its capacity to undergo intercalation upon treatment with electropositive elements. The process is a redox reaction, as shown in the following reaction with lithium.

$$TiS_2 + Li \rightarrow LiTiS_2$$

$LiTiS_2$ is usually defined as $Li^+\left[TiS_2^-\right]$. Throughout the intercalation and deintercalation, a range of stoichiometries are produced with the general formula Li_xTiS_2 ($x < 1$). During intercalation, the interlayer space expands (the lattice "swells") and the electrical conductivity of the material increases. Intercalation is aided by the weakness of the interlayer forces as well as the susceptibility of the Ti(IV) centers toward reduction. Intercalation can be performed by combining a suspension of the disulfide and a solution of the alkali metal in anhydrous ammonia. Instead, solid TiS_2 can also react with the alkali metal upon heating. The rigid-band model, which assumes that electronic band structure does not change with intercalation, defines changes in the electronic properties after intercalation. Deintercalation is the reverse of intercalation; the cations diffuse out from between the layers. This process is related to recharging a Li/TiS_2 battery. Intercalation and deintercalation can be examined by cyclic voltammetry. The microstructure of the titanium disulfide significantly influences the intercalation and deintercalation kinetics. Titanium disulfide nanotubes have been shown to have a higher uptake and discharge capacity than the corresponding polycrystalline structure. The higher surface area of the nanotubes has been suggested to offer more binding sites for the anode ions than the polycrystalline structure.

Titanium diselenide ($TiSe_2$) also known as titanium(IV) selenide is a material where selenium is viewed as selenide (Se^{2-}), which requires that titanium exists as Ti^{4+}. $TiSe_2$ is a member of metal dichalcogenides, which are compounds that consist of a metal and an element of the chalcogen column (S, Se, Te) within the periodic table. Many show properties of possible value in battery technology, for example, intercalation and electrical conductivity, but most applications concentrate on the less toxic and lighter disulfides, for example, TiS_2. A mixture of titanium and selenium is heated under argon atmosphere to form crude samples. The crude product is typically purified by chemical vapor transport employing iodine as the transport agent.

$$Ti + 2Se \rightarrow TiSe_2$$

Within the titanium–selenium system, numerous stoichiometries have been found. $TiSe_2$ crystallizes with the CdI_2-type structure similar to TiS_2, in which the octahedral holes between alternating hexagonal closely packed layer of Se^{2-} layers (i.e., half of the total number of octahedral holes) are occupied by Ti^{4+} centers. The CdI_2 structure is often described as a layer structure as the repeating layers of atoms perpendicular to the close-packed layer form the sequence Se-Ti-Se...Se-Ti-Se...Se-Ti-Se with weak van der Waals interactions between the selenium atoms in adjacent layers. The structure has (6,3)-coordination, being octahedral for the cation and trigonal pyramidal for the anions. This layered structure is known to undergo intercalation by alkali metals (M), resulting in the formation of M_xTiSe_2 ($x \leq 1$), thus expanding the weak van der Waals gaps between the two-dimensional layers.

5.4.3.3 Hydrides, carbides, borides, nitrides, and phosphides

Titanium hydride usually describes the inorganic compound TiH_2 and related nonstoichiometric compounds. It is industrially available as a stable gray/black powder, which is used as an additive in the manufacture of Alnico sintered magnets, sintering of powdered metals, manufacture of metal foam, manufacture of powdered titanium metal, and pyrotechnics. In the industrial process for making nonstoichiometric $TiH_{(2-x)}$, titanium metal sponge is reacted with hydrogen gas at atmospheric pressure at temperatures between 300°C and 500°C. Absorption of hydrogen is exothermic and rapid, changing the color of the sponge to gray/black. The brittle product is ground to a powder, which has a composition of approximately $TiH_{1.95}$. In the laboratory, titanium hydride is synthesized by heating titanium powder under flowing hydrogen at 700°C, the idealized reaction can be written as follows.

$$Ti + H_2 \rightarrow TiH_2$$

Other processes of preparing titanium hydride are, for example, electrochemical and ball milling methods. $TiH_{1.95}$ is not affected by water and air. It is slowly attacked by strong acids and degraded by hydrofluoric and hot sulfuric acids. It reacts quickly with oxidizing agents; this reactivity leads to the usage of titanium hydride in pyrotechnics. The compound has been employed to prepare high-purity hydrogen, which is released upon heating the solid compound starting at a temperature of 300°C. The dissociation completes only at the melting point of titanium. Titanium trihydride has been suggested for the long-term storage of tritium gas. As TiH_x approaches stoichiometry, it has a distorted body-centered tetragonal structure, called the ε-form with an axial ratio of less than 1. This composition is very unstable with regard to partial thermal decomposition, unless kept under a pure hydrogen atmosphere. If not, it will rapidly decompose at 25°C until an approximate composition of $TiH_{1.74}$ is reached. The compound with this composition has the fluorite (CaF_2) structure, and is called the δ-form, and only very slowly thermally decomposes at 25°C until an approximate composition of $TiH_{1.47}$ is reached, at which point, inclusions of the hexagonal close-packed α-form, which is the same form as pure titanium, start to form. The evolution of the dihydride from titanium metal and hydrogen has been studied in some detail. α-Titanium has a hexagonal close-packed (*hcp*) structure at 25°C. Hydrogen to begin with occupies tetrahedral interstitial sites in the titanium. As the H/Ti ratio approaches 2, the material changes from the β-form with a body-centered cubic (*bcc*) structure to the face-centered cubic (*fcc*) δ-form, the H atoms in the end filling all the tetrahedral sites to give the limiting stoichiometry of TiH_2. If titanium hydride contains 4.0% hydrogen at less than ~40°C it will transform into a body-centered tetragonal (*bct*) structure called ε-titanium. When titanium hydrides with less than 1.3% hydrogen, called hypoeutectoid titanium hydride, are cooled, the β-titanium phase of the mixture tries to restore the α-titanium phase, causing an excess of hydrogen. One option for hydrogen to leave the β-titanium phase is for the titanium to partially convert to δ-titanium, leaving behind titanium that is sufficiently low in hydrogen to take the form of α-titanium, producing an α-titanium matrix with δ-titanium inclusions. A metastable γ-titanium hydride phase has been described. When α-titanium hydride with a hydrogen content of 0.02%–0.06% is quenched quickly, it forms into γ-titanium hydride, as the atoms "freeze" in place when the cell structure changes from *hcp* to *fcc*. γ-Titanium has a body-centered tetragonal (*bct*) structure. Furthermore, there is no compositional change, so the atoms generally retain their same neighbors.

Titanium(IV) hydride (systematically named titanium tetrahydride), TiH_4, has not yet been obtained in bulk, so its bulk properties are still unknown. Nevertheless, molecular TiH_4 has been isolated in solid gas matrices. The molecular form is a colorless gas and very unstable toward thermal decomposition. As such the compound is not well characterized, but a number of its properties have been calculated via computational chemistry. TiH_4 was first formed in 1963 by the photodissociation of mixtures of $TiCl_4$ and H_2, followed by immediate mass spectrometry. Rapid analysis was necessary as TiH_4 is extremely unstable. Computational analysis of TiH_4 indicates a theoretical bond dissociation energy (BDE; relative to M + 4H) of 132 kcal/mole. Since the dissociation energy of H_2 is 104 kcal.mole^{-1}, the instability of TiH_4 can be thermodynamically anticipated to dissociate to metallic titanium and hydrogen.

$$TiH_4 \rightarrow Ti^0 + 2H_2 \left(76\ \text{kcal·mole}^{-1}\right)$$

TiH_4, along with other unstable molecular titanium hydrides, (TiH, TiH_2, TiH_3, and polymeric species), has been isolated at low temperature following laser ablation of titanium. It is believed that within solid TiH_4, the molecules form aggregations (polymers), connected through covalent bonds. Calculations indicate that TiH_4 is prone to dimerization. This is mainly ascribed to the electron deficiency of the monomer and the small size of the hydride ligands; which allows dimerization to occur with a very low-energy barrier since there is an insignificant increase in interligand repulsion. The dimer is calculated to be a fluxional molecule rapidly interconverting between a number of forms, all of which display bridging hydrogen atoms. This is an example of three-center two-electron bonding. Monomeric TiH_4 is the simplest transition metal molecule that displays sd^3 orbital hybridization.

Titanium carbide, TiC, is an extremely hard (Mohs 9–9.5) refractory ceramic material, comparable to tungsten carbide. It has the appearance of black powder with the sodium chloride (face-centered cubic, *fcc*) crystal structure. It is found naturally in very small crystals ranging in size from 0.1 to 0.3 mm as a very rare mineral khamrabaevite (Russian: Хамрабаевит)—(Ti,V,Fe)C. It was discovered in 1984 on Mount Arashan in the Chatkal District, USSR (modern-day Kyrgyzstan), near the Uzbek border. Tool bits without tungsten content can be manufactured of titanium carbide in nickel–cobalt matrix cermet, improving the cutting speed, precision, and smoothness of the workpiece. The resistance to wear, corrosion, and oxidation of a tungsten carbide–cobalt material can be enhanced by adding 6%–30% of titanium carbide to tungsten carbide. This results though in a solid solution that is more brittle and susceptible to breakage. Titanium carbide can be etched with reactive-ion etching. Titanium carbide is employed in the preparation of cermets, which are often used to machine steel materials at high cutting speeds. In addition, it is employed as an abrasion-resistant surface coating on metal parts, for example, tool bits and watch mechanisms. TiC is also employed as a heat shield coating for atmospheric reentry of spacecraft.

Titanium diboride (TiB_2) is an extremely hard ceramic, which has exceptional heat conductivity, oxidation stability, and resistance to mechanical erosion. TiB_2 is also a fairly good electrical conductor; thus it can be utilized as a cathode material in aluminum smelting and shaped by electrical discharge machining. TiB_2 does not occur naturally. TiB_2 powder can be produced by a number of high-temperature processes, for example, the direct reactions of titanium, or its oxides/hydrides, with elemental boron at temperatures above 1000°C, carbothermal reduction by thermite reaction of titanium oxide and boron oxide, or hydrogen reduction of boron halides in the presence of the metal or its halides. Among the different synthesis routes, electrochemical synthesis and solid-state reactions have been established to produce finer TiB_2 in large amounts. An illustration of a solid-state reaction is the borothermic reduction, which can be written as follows:

$$2TiO_2 + B_4C + 3C \rightarrow 2TiB_2 + 4CO \quad (5.1)$$

$$TiO_2 + 3NaBH_4 \rightarrow TiB_2 + 2Na(g, l) + NaBO_2 + 6H_2(g) \quad (5.2)$$

The first synthesis route (5.1), though, cannot form nanosized powders. Nanocrystalline (5–100 nm) TiB_2 was prepared using the reaction (5.2) or the following methods: solution phase reaction of $NaBH_4$ and $TiCl_4$, followed by annealing the amorphous precursor obtained at temperatures between 900°C and 1100°C; mechanical alloying of a mixture of elemental Ti and B powders; self-propagating high-temperature synthesis (SHS) comprising addition of varying amounts of NaCl; milling assisted SHS; and solvothermal reaction in benzene of metallic sodium with amorphous boron powder and $TiCl_4$ at a temperature of 400°C.

$$TiCl_4 + 2B + 4Na \rightarrow TiB_2 + 4NaCl$$

Many TiB_2 applications are hindered by economic factors, especially the costs of densifying a high melting point material—the melting point is approximately 2970°C, and, due to a layer of titanium dioxide that forms on the surface of the particles of a powder, it is extremely resistant to sintering. Admixture of around 10% silicon nitride helps the sintering, although sintering without silicon nitride has been described as well. Thin films of TiB_2 can be prepared by several methods. The electroplating of TiB_2 layers has two key advantages relative to physical vapor deposition (PVD) or CVD: the growing rate of the layer is 200 times higher (up to 5 μm/s), and the inconveniences of covering complex shaped products are significantly reduced. In regard to chemical stability, TiB_2 shows better stability in contact with pure iron than tungsten carbide or silicon nitride. TiB_2 is resistant to oxidation in air at temperatures up to 1100°C, and to hydrochloric and hydrofluoric acids, though it reacts with alkalis, nitric acid, and sulfuric acid.

Titanium nitride (TiN; sometimes called Tinite) is an extremely hard ceramic material, frequently employed as a coating on titanium alloys, steel, carbide, and aluminum components to enhance the substrate's surface properties. [In the form of a thin coating, TiN is utilized to harden and protect cutting and sliding surfaces, for decorative purposes (due to its golden appearance), and as a nontoxic exterior for medical implants.] In most uses, a coating of less than 5 μm (0.00020 in.) is applied. The most common techniques of TiN thin-film formation are PVD (typically sputter deposition, cathodic arc deposition or electron beam heating) and CVD. In both processes, pure titanium is sublimed and reacted with nitrogen in a high-energy, vacuum environment. TiN film can also be formed on Ti workpieces by reactive growth (e.g., annealing) in a nitrogen atmosphere. PVD is favored for steel parts since the deposition temperature surpasses the austenitizing temperature of steel. TiN layers are also sputtered on various higher melting point compounds such as stainless steels, titanium, and titanium alloys. Its high Young's modulus (values between 450 and 590 GPa have been observed) means that thick coatings have a tendency to flake away, rendering them much less resilient than thin layers. TiN coatings can likewise be deposited by thermal spraying, while TiN powders are formed by nitridation of titanium with nitrogen or ammonia at a temperature of 1200°C. Bulk ceramic objects can be produced by packing powdered metallic titanium into the required shape, compressing it to the proper density, and then igniting it in an atmosphere of pure nitrogen. The heat released by the chemical reaction between the metal and gas is enough to sinter the nitride reaction product into a hard, finished object. TiN will oxidize at 800°C in a normal atmosphere. It is chemically stable at 20°C but can be slowly attacked by concentrated acid solutions with rising temperatures. The typical TiN phase adopts a crystal structure of the NaCl type with a roughly 1:1 stoichiometry; TiN_x compounds with x ranging from 0.6 to 1.2 are, nevertheless, thermodynamically stable. TiN becomes superconducting at cryogenic temperatures, with a critical temperature up to 6.0K for single crystals. Superconductivity in thin-film TiN has been investigated comprehensively, with the superconducting properties strongly dependent on sample preparation, varying up to complete suppression of superconductivity at a superconductor–insulator transition. A thin film of TiN was chilled to near absolute zero, converting it into the first known superinsulator, with resistance suddenly increasing by a factor of 100,000. Osbornite (TiN) is a very rare naturally occurring mineral found nearly exclusively in meteorites.

Titanium(III) phosphide (TiP) is normally a gray powder. It is a metallic conductor with a high melting point. It is not attacked by common acids or water. Its physical properties are unlike those of the group 1 and group 2 phosphides that contain the P^{3-} anion (such as Na_3P), which are not metallic and are easily hydrolyzed. Titanium phosphide is categorized as a "metal-rich phosphide," where extra valence electrons from the metal are delocalized. Titanium phosphide can be produced via the reaction of $TiCl_4$ and PH_3. Other titanium phosphide phases exist, for example, Ti_3P, Ti_2P, Ti_7P_4, Ti_5P_3, and Ti_4P_3. Titanium phosphide should not be confused with titanium phosphate or titanium isopropoxide, both of which are also known by the acronym TIP.

5.4.3.4 Halides

The titanium tetrahalides show an interesting gradation in color as the charge transfer band moves steadily to lower energies (i.e., absorbing increasingly in the visible region of the spectrum) as the anion becomes more easily oxidized (F^- to I^-) by the small, highly polarizing titanium cation. Tetrahalides, but especially the chlorides and bromides, behave as Lewis acids dissolving in polar solvents to give rise to series of addition compounds; they also form complex anions with halides. They are all hygroscopic and hydrolysis follows the same pattern as complex formation, with the chlorides and bromides being more vulnerable than the fluorides and iodides.

Titanium(IV) fluoride, TiF_4, is a white hygroscopic solid. Unlike the other tetrahalides of titanium, it has a polymeric structure. The traditional process comprises the reaction of titanium tetrachloride with excess hydrogen fluoride.

$$TiCl_4 + 4HF \rightarrow TiF_4 + 4HCl$$

Purification is by sublimation, which includes reversible cracking of the polymeric structure. The Ti centers are octahedrally coordinated but conjoined in an unusual columnar structure. TiF_4 can form adducts with numerous ligands, for example, *cis*-$TiF_4(MeCN)_2$, which is produced through a reaction with acetonitrile. Titanium(III) fluoride (TiF_3) is a violet solid. It has a perovskite-like ($CaTiO_3$) structure in which each Ti center has octahedral coordination geometry and each F ligand is doubly bridging.

Titanium tetrachloride, $TiCl_4$, is an important intermediary in the manufacturing of titanium metal and the pigment titanium dioxide. $TiCl_4$ is a volatile liquid. On contact with humid air, it produces spectacular opaque clouds of titanium dioxide (TiO_2) and hydrated hydrogen chloride. It is at times referred to as "tickle" or "tickle 4" because of the phonetic similarity of its molecular formula ($TiCl_4$) to the word. $TiCl_4$ is prepared by the chloride method, which comprises the reduction of titanium oxide ores, usually ilmenite ($FeTiO_3$), with carbon under flowing chlorine at 900°C. Impurities are removed by distillation.

$$2FeTiO_3 + 7Cl_2 + 6C \rightarrow 2TiCl_4 + 2FeCl_3 + 6CO$$

The coproduction of $FeCl_3$ is unwanted, which has driven the development of alternative processes. As an alternative of directly using ilmenite, "rutile slag" can be used. This material, an impure form of TiO_2, is obtained from ilmenite by removal of iron, either employing carbon reduction or extraction with sulfuric acid. Crude $TiCl_4$ has a number of other volatile halides, for example, vanadyl chloride ($VOCl_3$), silicon tetrachloride ($SiCl_4$), and tin tetrachloride ($SnCl_4$), which must be removed. $TiCl_4$ is a dense, colorless distillable liquid, but crude samples can be yellow or even red-brown. It is one of the rare transition metal halides that forms a liquid at room temperature, VCl_4 being another example. This property reveals that molecules of $TiCl_4$ are weakly self-associated. Most metal chlorides are polymers, in which the chloride atoms bridge between the metals. Its melting and boiling points are comparable to those of CCl_4. Ti^{4+} has a "closed" electronic shell, with the same number of electrons as the inert gas argon. The tetrahedral structure for $TiCl_4$ is in agreement with its description as a d^0 metal center (Ti^{4+}) surrounded by four identical ligands. This configuration results in highly symmetrical structures and consequently the tetrahedral shape of the molecule. $TiCl_4$ has a structure similar to that of $TiBr_4$ and TiI_4. $TiCl_4$ and $TiBr_4$ react to give mixed halides $TiCl_{4-x}Br_x$, where $x = 0, 1, 2, 3, 4$. Magnetic resonance

(NMR) measurements also suggest that halide exchange is also fast between $TiCl_4$ and VCl_4. $TiCl_4$ is soluble in toluene and chlorocarbons. Some arenes can form complexes of the type $[(C_6R_6)TiCl_3]^+$. $TiCl_4$ reacts exothermically with donor solvents such as THF to form hexacoordinated adducts. Bulkier ligands (L) result in the formation of pentacoordinated adducts $TiCl_4L$. About 90% of the total $TiCl_4$ manufactured is utilized to produce the pigment titanium dioxide (TiO_2). The conversion comprises the hydrolysis of $TiCl_4$, a reaction that produces hydrogen chloride.

$$TiCl_4 + 2H_2O \rightarrow TiO_2 + 4HCl$$

In some instances, $TiCl_4$ is oxidized directly with oxygen.

$$TiCl_4 + O_2 \rightarrow TiO_2 + 2\,Cl_2$$

The world's supply of titanium metal is made from $TiCl_4$. The conversion entails the reduction of $TiCl_4$ with magnesium metal. This procedure is known as the Kroll process.

$$2Mg + TiCl_4 \rightarrow 2MgCl_2 + Ti$$

In the Hunter process, liquid sodium is the reducing agent instead of magnesium.

Titanium(III) chloride, $TiCl_3$, has at least four distinct species; additionally hydrated derivatives have also been described. $TiCl_3$ is one of the most common halides of titanium and is a vital catalyst to produce polyolefins. $TiCl_3$ is typically prepared by the reduction of titanium(IV) chloride. Older reduction methods use hydrogen.

$$2TiCl_4 + H_2 \rightarrow 2HCl + 2TiCl_3$$

It is conveniently reduced with aluminum and sold as a mixture with aluminum trichloride, $TiCl_3 \cdot AlCl_3$. This mixture can be separated to form $TiCl_3(THF)_3$. The complex has a meridional structure. Its hydrate can be prepared via dissolving titanium in aqueous hydrochloric acid.

$$2Ti + 6HCl + 6H_2O \rightarrow 2TiCl_3(H_2O)_3 + 3H_2$$

$TiCl_3$ forms numerous coordination complexes, the majority of which are octahedral. The light blue crystalline adduct $TiCl_3(THF)_3$ can be formed by reacting $TiCl_3$ with tetrahydrofuran.

$$TiCl_3 + 3C_4H_8O \rightarrow TiCl_3(OC_4H_8)_3$$

A similar dark green complex is formed from complexation with dimethylamine. As an example of a reaction where all ligands are exchanged, $TiCl_3$ is the precursor to form the tris(acetylacetonate) complex. The more reduced titanium(II) chloride is obtained by the thermal disproportionation of $TiCl_3$ at a temperature of 500°C. The reaction is driven by the loss of volatile $TiCl_4$.

$$2TiCl_3 \rightarrow TiCl_2 + TiCl_4$$

The ternary halides, such as A_3TiCl_6, have structures that are dependent on the nature of cation A^+. Cesium chloride treated with titanium(II) chloride and hexachlorobenzene forms crystalline $CsTi_2Cl_7$. In these structures, Ti^{3+} adopts an octahedral coordination geometry. In $TiCl_3$, each Ti atom has one d-electron, causing its derivatives to be paramagnetic, that is, the substance is attracted into a magnetic field. Solutions of titanium(III) chloride are violet, which is the result of excitations of its d-electron. The color is not very intense as the transition is forbidden by the Laporte selection rule. Four solid forms or polymorphs of $TiCl_3$ have been described. All have Ti in an octahedral coordination sphere. These forms can be characterized based on their crystallography together with their magnetic properties, which probes exchange interactions. β-$TiCl_3$ crystallizes as brown needles. Its structure comprises chains of $TiCl_6$ octahedra that share opposite faces so that the closest Ti-Ti distance is 2.91 Å. This short distance suggests strong metal–metal interactions. The three violet "layered" forms, so called for their color and their tendency to flake, are called α, γ, and δ. In α-$TiCl_3$, the chloride anions are hexagonal close-packed (*hcp*). In γ-$TiCl_3$, the chloride anions are cubic close-packed (*ccp*). Lastly, disorder in shift successions results in an intermediate between α and γ structures, known as the δ phase. $TiCl_6$ shares edges in each phase, with 3.60 Å being the shortest distance between the Ti cations. This large distance between Ti cations precludes direct metal–metal bonding. In contrast, the trihalides of the heavier metals hafnium and zirconium participate in metal–metal bonding. Direct Zr-Zr bonding is observed in zirconium(III) chloride. The difference between the Zr(III) and Ti(III) compounds is ascribed partly to the relative radii of these metal centers.

Titanium(II) chloride, $TiCl_2$, is a black solid hat has been investigated only moderately, perhaps due to its high reactivity. Ti(II) is a strong reducing agent: it has a high affinity for oxygen and reacts irreversibly with water to form H_2. The typical preparation is via the thermal disproportionation of $TiCl_3$ at a temperature of 500°C. Recall that the formation of $TiCl_2$ is driven by the loss of volatile $TiCl_4$. This reaction is comparable to that for the conversion of VCl_3 into VCl_2 and VCl_4. $TiCl_2$ crystallizes as the layered CdI_2 structure. Hence, the Ti(II) centers are octahedrally coordinated to six Cl ligands. Molecular complexes are known, for example, $TiCl_2(chel)_2$, where chel is DMPE $(CH_3)_2PCH_2CH_2P(CH_3)_2$ and TMEDA $((CH_3)_2NCH_2CH_2N(CH_3)_2)$. Such species are produced via the reduction of related Ti(III) and Ti(IV) complexes. Unusual electronic effects have been discovered in these species: $TiCl_2[(CH_3)_2PCH_2CH_2P(CH_3)_2]_2$ is paramagnetic with a triplet ground state, but in contrast $Ti(CH_3)_2[(CH_3)_2PCH_2CH_2P(CH_3)_2]_2$ is diamagnetic. A solid-state derivative of $TiCl_2$ is Na_2TiCl_4, which has been synthesized via the reaction of Ti metal with $TiCl_3$ in a NaCl flux. This species has a linear chain structure in which the Ti(II) centers once more are octahedrally coordinated with terminal, axial halides.

Titanium tetrabromide, $TiBr_4$, is the most volatile transition metal bromide. The properties of $TiBr_4$ are an average of $TiCl_4$ and TiI_4. $TiBr_4$ is diamagnetic, reflecting the d^0 configuration of the Ti center. This four-coordinated complex has a tetrahedral geometry. It can be produced via a number of different methods: from the elements, via the reaction of TiO_2 with carbon and bromine, and by reaction of $TiCl_4$ with HBr. $TiBr_4$ can form adducts, for example, $TiBr_4(THF)_2$ and $[TiBr_5]^-$. With bulky donor ligands, such as 2-methylpyridine (2-Mepy), five-coordinated adducts can be obtained. $TiBr_4$(2-MePy) is trigonal bipyramidal with the pyridine in the equatorial plane. $TiBr_4$ has been employed as a Lewis acid catalyst in organic synthesis. The tetrabromide and tetrachlorides of titanium react with each other to form a statistical mixture of the two tetrahalides, $TiBr_{4-x}Cl_x$ ($x = 0-4$). The mechanism of this redistribution reaction is not clear. One suggested pathway calls for the formation of dimer intermediaries.

Titanium(III) bromide, $TiBr_3$, is a blue black paramagnetic solid with a reddish reflection. It has a limited number of applications, even though it is a catalyst for the polymerization of alkenes. $TiBr_3$ can be prepared by heating the tetrabromide in a hydrogen atmosphere.

$$2TiBr_4 + H_2 \rightarrow 2TiBr_3 + 2HBr$$

It can likewise be formed through comproportionation of titanium metal and titanium tetrabromide.

$$Ti + 3TiBr_4 \rightarrow 4TiBr_3$$

There exist two different polymorphs, each with octahedral Ti centers. Heating the tribromide results in the formation of the dibromide together with the volatile tetrabromide.

$$TiBr_3 \rightarrow 2TiBr_4 + TiBr_2$$

The solid can dissolve in donor solvents (L) such as pyridine and nitriles to form 3:1 adducts.

$$TiBr_3 + 3L \rightarrow 2TiBr_3L_3$$

Titanium tetraiodide, TiI_4, is a black volatile solid. It is an intermediate in the Van Arkel process for the purification of titanium. Three processes are well known: (1) From the elements, typically using a tube furnace at 425°C:

$$Ti + 2I_2 \rightarrow TiI_4$$

This reaction can be reversed to form highly pure films of Ti metal. (2) Exchange reaction from titanium tetrachloride and HI.

$$TiCl_4 + 4HI \rightarrow TiI_4 + 4HCl$$

(3) Oxide-iodide exchange from aluminum iodide.

$$3TiO_2 + 4AlI_3 \rightarrow 3TiI_4 + 2Al_2O_3$$

TiI_4 is an unusual molecular binary metal iodide, comprising isolated molecules of tetrahedral Ti(IV) centers. The Ti-I distance is 261 pm. Reflecting its molecular character, TiI_4 can be distilled without decomposition at one atmosphere; this property is the basis of its usage in the Van Arkel process. The difference in melting point between $TiCl_4$ (m.p. −24°C) and TiI_4 (m.p. 150°C) is similar to the difference between the melting points of CCl_4 (m.p. −23°C) and CI_4 (m.p. 168°C), indicating the stronger intermolecular van der Waals bonding in the iodides. Two polymorphs of TiI_4 are known, one of which is highly soluble in organic solvents. In the less soluble cubic form, the Ti-I distance is 261 pm. Like $TiCl_4$ and $TiBr_4$, TiI_4 forms adducts with Lewis bases, and it can also be reduced. When the reduction is performed in the presence of Ti metal, polymeric Ti(III) and Ti(II) derivatives are formed, for example, $CsTi_2I_7$ and the chain structure $CsTiI_3$, respectively. TiI_4 shows extensive reactivity toward alkenes and alkynes producing the organoiodine derivatives. In addition, it affects pinacol couplings and other C—C bond-forming reactions.

Titanium(III) iodide, TiI_3, is a dark violet solid that is insoluble in solvents, except upon decomposition. TiI_3 can be produced through the reaction of titanium with iodine.

$$2Ti + 3I_2 \rightarrow 2TiI_3$$

In addition, it can be prepared by the reduction of TiI_4, for example, with aluminum. TiI_3 has a polymeric structure of face-sharing octahedra. Above 50°C, the Ti-Ti distances are equal, but below that temperature, it exhibits a phase transition. In the low-temperature phase, the Ti-Ti distances are alternating short and long. The low-temperature structure is comparable to that of molybdenum tribromide.

5.4.3.5 Oxyanions

Because of the high ratio of ionic charge to radius, normal salts of Ti(IV) cannot be prepared from aqueous solutions, which only yield basic, hydrolyzed species. Several oxometal(1V) compounds have been isolated but do not contain discrete MO^{2+} ions, being polymeric in the solid state. Thus $TiOSO_4{\cdot}H_2O$ contains chains of -Ti-O-Ti-O-with each Ti being approximately octahedrally coordinated to two bridging oxygen atoms, one water molecule, and an oxygen atom from each of three sulfates. Ion-exchange studies on aqueous solutions of Ti(IV) in 2M $HClO_4$ are consistent with the presence of monomeric doubly charged cationic species rather than polymers, though it is not clear whether the predominant species is $[TiO]^{2+}$ or $[Ti(OH)_2]^{2+}$.

5.4.3.6 Complexes

A very large number of titanium complexes have been prepared and, as expected for the d^0 configuration, they are diamagnetic. Hydrolysis, resulting in polymeric species with -OH- or -O- bridges, is common especially with titanium. A coordination number of six is the most usual for Ti(IV) but seven and even eight coordination is possible. TiF_4 forms six-coordinate adducts mainly with O- and N-donor ligands, and complexes of the type TiF_4L have all the appearances of fluorine-bridged polymers. $TiC1_4$ and $TiBr_4$ are especially prolific and are clearly "softer" acceptors than the fluoride. They form mainly yellow to red adducts of the types $[MXL_2]$ and [MX(L-L)] with ligands such as ethers, ketones, $OPC1_3$, amines, imines, nitriles, thiols, and thioethers.

The coordination chemistry of the tervalent oxidation state is virtually confined to that of titanium. Interpretation of the electronic spectrum of Ti(III) in aqueous solution was an early landmark in the development of Crystal Field Theory, the observed broad band being assigned to the $^2E_g \leftarrow {}^2T_{2g}$ transition (promotion of an electron from a t_2, to an e_g orbital). However, the absorption band actually observed for this and for other octahedral complexes of Ti(III) is never of the symmetrical shape expected for a single transition, but is rather an asymmetrical peak with a (usually) distinct shoulder on the low-energy side.

Due to inadequate or nonexistent back bonding, the only neutral, binary carbonyl so far reported is $Ti(CO)_6$, which has been produced by condensation of titanium metal vapor with CO in a matrix of inert gases at -263°C to -258°C (10K−15K), and identified spectroscopically. By contrast, if $MC1_4$ (M = Ti, Zr) in dimethoxyethane is reduced with potassium naphthalenide in the presence of a crown ether (to complex the K^+) under an atmosphere of CO, $[M(CO)_6]^{2-}$ salts are produced. These not only involve the metals in the exceptionally low formal oxidation state of −2 but are thermally stable up to 200°C and 130°C, respectively. However, the majority of their carbonyl compounds are stabilized by π-bonded ligands, usually cyclopentadienyl, as in $[M(\eta^5\text{-}C_5H_5)_2(CO)_2]$.

5.4.3.7 Organotitanium compounds

The development of the organometallics of titanium did not progress until after two important discoveries occurred in the 1950s. In 1951 ferrocene was discovered and provided the impetus for isolating the related cyclopentadienyl complexes of titanium. A true mononuclear titanocene itself was not isolated but the dichloride derivative, $[TiCl_2(\eta^5\text{-}C_5H_5)_2]$, turned out to be a convenient starting material for the preparation of countless other η-cyclopentadienyl derivatives of titanium and is widely used to the present day. Another important event in the early 1950s was the disclosure by Ziegler of catalysts made of a certain combinations of transition metal salts (particularly titanium halides) and aluminum alkyls or related main-group derivatives. The reaction mechanism was generally unclear, but these catalysts were shown to promote facile polymerization of ethylene at low pressures. Later work by Natta and others demonstrated that these catalysts and their closely related variations can stereoregularly polymerize α-olefins such as propene. The emergence of Ziegler−Natta catalysts had important commercial consequences which, in part, resulted in an enormous surge of interest in organotitanium chemistry. Since little was understood at how the catalysts work, a considerable amount of subsequent studies in industrial and academic laboratories were directed at clarifying the roles of titanium and aluminum in chain growth and termination reactions. Among the titanium complexes reported in the years following Ziegler's discovery included the σ-hydrocarbyl species $[Ti(CH_3)Cl_3]$ and $TiMe_4$.

Around the same time that interest in organotitanium exploded, Herman and Nelson (1953) isolated the first organotitanium compound, $TiPh(OPr^i)_3$, in 1952. This compound is a pale-yellow solid (melting point c. 88°C) that exhibits moderate thermal stability and could be stored without significant decomposition in the dark under nitrogen below 10°C. Several organotitanium compounds were since then prepared and characterized. Manfred Reetz summarized the properties of the Ti-C bond chemistry in a book chapter entitled, "Organotitanium Chemistry" in the book "Organometallics in Synthesis: A Manual" edited by Manfred Schlosser and first published in 2001. In this chapter, it was reiterated that the Ti-C bond is not particularly weak in a thermodynamic sense; however, kinetically low-energy decomposition pathways such as β-hydride elimination may limit the use of certain organotitanium compounds. It has been established that titanium is oxophilic and the TiO bond is fairly strong (~115 kcal. mol^{-1}). Thus any reaction in which a Ti-C bond converts to a Ti-O bond is expected to have a strong driving force. Further, the differences in bond lengths, Ti-C bond being typically 2.1 Å long while Ti-O is around 1.7 Å, mean that transition states in such reactions as Grignard or aldol additions are expected to be more compact in the case of titanium. Tetravalent titanium compounds have ligands that are tetrahedrally arranged around the metal. The behavior of organotitanium reagents varies according to the type of ligands attached to Ti. Reagents of the type $RTiCl_3$ and $RTi(NR_2)_3$ are monomeric in solution, while compounds such as $CH_3Ti(OR)_3$ show little aggregation or stay monomeric provided the alkoxy ligand is as bulky as isopropoxy (or bulkier). Analogs having smaller ligands (e.g., ethoxy) are dimeric. The type of ligands also affects the electronic nature of titanium. As electron-donating alkoxy ligands are introduced, Lewis acidity decreases in the series $TiCl_4$, Cl_3TiO^iPr, $Cl_2Ti(O^iPr)_2$, $ClTi(O^iPr)_3$, and $Ti(O^iPr)_4$. The same effect operates in reagents such as CH_3TiCl_3 versus $CH_3Ti(O^iPr)_3$. The proper choice of the ligands, thus, allows two parameters to be controlled in a predictable way—the electronic property of titanium, for example, Lewis acidity, and the steric environment around the metal. For example, the highly Lewis acidic compounds of the form $RTiCl_3$ can be useful in chelation-controlled additions to chiral alkoxy carbonyl compounds or in alkylation reactions of S_N1-active alkyl halides. The lesser Lewis acidity of the compounds $RTi(OR')$, $RTi(NR_2)_3$, or $RTiX_2Cp$ means that they can be ideal reagents for nonchelation-controlled additions. In all cases, basicity and reactivity are considerably lower than those of RMgX, RLi and resonance—stabilized "carbanions" and led to the hypothesis that titanium metalation of RMgX, RLi, or R_2Zn can promote chemo- and stereoselective C—C bond-forming reactions with carbonyl compounds and certain alkyl halides. Control of organic reactions by changing the metal in the organometallic reagent and by changing the ligands around the metal are now referred to as "metal tuning" and "ligand tuning," respectively. The possibilities are enormous and in the case of titanium has given rise to named reagents and reactions.

The Ziegler—Natta catalyst continues to be a powerful tool to polymerize α-olefins with high linearity and stereoselectivity and provides syndiotactic or isotactic polymers depending on the metal catalyst.

CH_3 — $TiCl_3$ / $Al(Et)_2Cl$ → isotactic

CH_3 — VCl_4, -78°C / $Al(iBu)_2Cl$ → syndiotactic

Lombardo's reagent is a carbenoid methylenation reagent developed to overcome a shortcoming of the Wittig reagent by methylenating enolizable carbonyl groups without loss of stereochemical integrity (Lombardo methylenation). It can be applied in a conversion of a ketene into an allene.

Et_3Si, C_7H_{15} C=C=O — CBr_2H_2, Zn, $TiCl_4$ / THF, −40°C to 5°C → Et_3Si, C_7H_{15} C=C=CH_2

Other important organotitanium reagents include the Tebbe reagent, Petasis reagent, and Nugent-RajanBabu reagent. The Tebbe reagent is synthesized from titanocene dichloride and trimethylaluminum in toluene solution. It is useful as a methylenation agent for carbonyl compounds (conversion of R_2C-O to R_2C-CH_2) especially when the carbonyl group is sterically challenged or easily forms the enol. The Petasis reagent or dimethyl titanocene is prepared from titanocene dichloride and methyllithium in diethyl ether. It is also a methylenation agent but is easier to prepare and handle than the Tebbe reagent. The Nugent-RajanBabu reagent is generated as a dimer $[(\eta^5\text{-Cp})2\text{Ti}(\mu\text{-Cl})]_2$ and used in situ from titanocene dichloride. It is a one-electron reductant used in synthetic organic chemistry for the generation of alcohols via anti-Markovnikov ring-opening of epoxides.

5.4.4 Major uses

Approximately 95% of all Ti ore is destined for refinement into titanium dioxide (TiO_2), a white permanent pigment used in paints, paper, toothpaste, and plastics. It is also used in a variety of other products, such as in cement, in gemstones, as an optical opacifier in paper, and a strengthening agent in graphite composite fishing rods and golf clubs. TiO_2 pigment is chemically inert, resists fading in sunlight, and is very opaque: it gives a pure and brilliant white color to most household plastics. Paint containing TiO_2 performs well in severe temperatures and marine environments. Pure TiO_2 has a very high refractive index and an optical dispersion higher than diamond. In addition, besides being a very important pigment, TiO_2 is also used in sunscreens because it prevents UV light from reaching the skin. Nanoparticles of TiO_2 appear invisible when applied to the skin. Titanium tetrachloride ($TiCl_4$), a colorless liquid, is important as an intermediate in the production of TiO_2 and is also used to make the Ziegler–Natta catalyst. In addition, $TiCl_4$ is used to iridize glass and, since it fumes strongly in moist air, make smoke screens.

Ti is used in steel as an alloying element (ferro-titanium) to decrease grain size and as a deoxidizer, while in stainless steel, it is used to reduce the carbon content. Ti is often alloyed with Al (to refine grain size), V, Cu (to harden), iron, Mn, Mo, and other metals. Ti mill products (sheet, plate, bar, wire, forgings, castings) are used in industrial, aerospace, recreational, and emerging markets. Powdered Ti is used in pyrotechnics as a source of bright-burning particles. Since Ti alloys have high tensile strength to density ratio, high corrosion resistance, fatigue resistance, high crack resistance, and ability to withstand moderately high temperatures without creeping, they find use in, for example, aircraft, armor plating, naval ships, spacecraft, and missiles. For these uses, Ti is alloyed with Al, Zr, Ni, V, and other elements to manufacture a variety of components. In fact, about 65% of all Ti metal produced is used in aircraft engines and frames. Since Ti is resistant to corrosion by sea water, it is used to produce propeller shafts, rigging, and heat exchangers in desalination plants; heater-chillers for saltwater aquariums, fishing line and leader, divers' knives, etc. Ti is used in the housings and components of ocean-deployed surveillance and monitoring devices for science and the military. The former Soviet Union manufactured submarine hulls of Ti alloys. Welded Ti pipes and process equipment (such as heat exchangers, tanks, process vessels, valves) are used in the chemical and petrochemical industries mainly for corrosion resistance. Particular alloys are used in oil and gas downhole applications and Ni hydrometallurgy for their high strength (e.g., Ti beta C alloy), corrosion resistance, or both. The pulp and paper industry uses Ti in process equipment in contact with corrosive media, such as sodium hypochlorite (NaOCl) or wet chlorine gas (in bleachery). Other uses comprise ultrasonic welding, wave soldering, and sputtering targets.

Ti metal can be found in automotive applications, particularly in automobile and motorcycle racing where low weight, high strength, and rigidity are critical. The metal is normally too expensive for the general consumer market. Ti is used in various sporting equipment, such as tennis rackets, golf clubs, lacrosse stick shafts; cricket, hockey, lacrosse and football helmet grills, and bicycle frames and components. While not a conventional metal for bicycle production, Ti bikes have been used by racing teams and adventure cyclists. Ti alloys are used in spectacle frames that are expensive though highly durable, long lasting, light weight, and cause no skin allergies. Many backpackers use Ti equipment, including cookware, eating utensils, lanterns, and tent stakes. Though marginally more expensive than traditional steel or Al alternatives, Ti products can be substantially lighter without compromising strength. Ti horseshoes are favored to steel by farriers as they are lighter and more durable. Due to its superior strength and light weight compared to other metals (steel, stainless steel, and aluminum), and because of recent advances in metalworking techniques, the use of Ti has become more common in the production of firearms, mainly pistol frames and revolver cylinders. For similar reasons, it is used in the body of laptop computers. Some high-end lightweight and corrosion-resistant tools, such as shovels and flashlights, are made of Ti or Ti alloys.

Due to its durability, Ti has become more popular for designer jewelery (especially, titanium rings). Its inertness makes it a good alternative for those with allergies or those who will be wearing the jewelery in environments such as swimming pools. Ti is also alloyed with gold to make an alloy that can be marketed as 24-karat gold as the 1% of alloyed Ti is not high enough to need a lower mark. The subsequent alloy is about the hardness of 14-karat gold and is more durable than pure 24-karat gold. Titanium's durability, light weight, and dent and corrosion resistance make it useful for watch cases. Some artists use Ti to make sculptures, decorative objects, and furniture. Ti may be anodized to vary the thickness of the surface oxide layer, causing optical interference fringes and a variety of bright colors. This coloration, combined with its chemical inertness, makes Ti a popular metal for body piercing.

Since Ti is biocompatible (nontoxic and not rejected by the body), it has many medical applications, for example, surgical implements and implants, such as hip balls and sockets (joint replacement) and dental implants that can stay in place for up to 20 years. The Ti is often alloyed with about 4% Al or 6% Al and 4% V. Ti has the intrinsic ability to osseointegrate, allowing use in dental implants that can last for more than 30 years. This property is also useful for orthopedic implant applications. These benefit from its lower modulus of elasticity (Young's modulus) to more closely match that of the bone that such devices are intended to repair. Consequently, skeletal loads are more evenly shared between bone and implant, resulting in a lower incidence of bone degradation due to stress shielding and periprosthetic bone fractures, which occur at the boundaries of orthopedic implants. Still, Ti alloy's stiffness is yet more than twice that of bone, so the adjacent bone bears a significantly reduced load and may deteriorate over time. Since Ti is nonferromagnetic, patients with Ti implants can be safely examined with magnetic resonance imaging (convenient for long-term implants). Preparing Ti for implantation in the body requires subjecting it to a high-temperature plasma arc removing the surface atoms, exposing fresh titanium that is immediately oxidized. Ti is also used in surgical instruments used in image-guided surgery, as well as wheelchairs, crutches, and any other products where high strength and low weight are advantageous.

5.5 23 V — Vanadium

5.5.1 Discovery

Vanadium was first found by Andrés Manuel del Río (November 10, 1764 to March 23, 1849), a Spanish-Mexican mineralogist, in 1801 (Fig. 5.35). Del Río separated the element from a sample of what he called "brown lead" ore, later named vanadinite ($Pb_5(VO_4)_3Cl$). He found that its salts showed a wide range of colors, and consequently he named the element panchromium (Greek: παγχρώμιο "all colors"). Later, Del Río gave the element the new name erythronium (Greek: ερυθρός "red") since most of the salts became red upon heating. In 1805 the French chemist Hippolyte Victor Collet-Descotils (November 21, 1773 to December 6, 1815), backed by del Río's friend Baron Alexander von Humboldt (Prussian polymath, geographer, naturalist, explorer, and influential proponent of Romantic philosophy and

FIGURE 5.35 Andrés Manuel del Río, 19th century, oil on canvas.

FIGURE 5.36 Nils Gabriel Sefström.

science, September 14, 1769 to May 6, 1859), incorrectly stated that del Río's new element was only a contaminated sample of chromium. Del Río believed Collet-Descotils' account and withdrew his claim. In 1831 the Swedish chemist Nils Gabriel Sefström (June 2, 1787 to November 30, 1845) (Fig. 5.36), a student of Swedish chemist Jöns Jacob Berzelius (August 20, 1779 to August 7, 1848), rediscovered the element in a new oxide he found while working with iron ores (Sefström, 1831). Later that same year, German chemist Friedrich Wöhler (July 31, 1800 to September 23, 1882) confirmed del Río's earlier work. Sefström selected a name beginning with V, which had not been assigned to any element yet. He named the element vanadium after Old Norse Vanadís (another name for the Norse Vanr goddess Freyja, whose attributes include beauty and fertility), for the numerous beautifully colored chemical compounds it produces. In 1831 the British-American geologist and geographer George William Featherstonhaugh (April 9, 1780 to September 28, 1866) proposed that vanadium should be renamed "rionium" after del Río, but this idea was not followed (Featherstonhaugh, 1831). The separation of V metal turned out to be difficult. In 1831 Berzelius reported the production of the metal, but British chemist Henry Enfield Roscoe (January 7, 1833 to December 18, 1915) proved that Berzelius had actually produced vanadium nitride (VN) (Roscoe, 1869-1870). Roscoe in the end formed the metal in 1867 by reduction of V(II) chloride, VCl_2, with hydrogen. In 1927 pure V was synthesized by reducing V_2O_5 with calcium. The initial large-scale industrial use of vanadium was in the steel alloy chassis of the Ford Model T, inspired by French race cars. Vanadium steel allowed for reduced weight while at the same time increasing tensile strength (ca.1905). For the first 10 years of the 20th century, most vanadium ore were mined by American Vanadium Company from the Minas Ragra in Peru. Later, the rising demand for uranium led to increased mining of that metal's ores. One major uranium ore was carnotite ($K_2(UO_2)_2(VO_4)_2 \cdot 3H_2O$), which also contains vanadium; thus vanadium became available as a by-product of uranium production. In the end, uranium mining began to source a large portion of the demand for vanadium (Box 5.6).

5.5.2 Mining, production, and major minerals

Vanadium is the 20th most abundant element in the Earth's crust; metallic vanadium is rare in nature. At the start of the 20th century, a large deposit of vanadium ore was discovered. For several years this patrónite (VS_4) deposit was an economically important source for vanadium ore. With the production of radium in the 1910s and 1920s from carnotite ($K_2(UO_2)_2(VO_4)_2 \cdot 3H_2O$) in which it is present in trace amounts, vanadium became available as a by-product of radium and uranium production. Vanadinite ($Pb_5(VO_4)_3Cl$) and other vanadium containing minerals are only mined in exceptional cases. With the increasing demand, now most of the world's vanadium metal is coming from vanadium-bearing magnetite found in ultramafic gabbro bodies. When this titanomagnetite is utilized to produce iron, most of the

BOX 5.6 Vanadium ore

Vanadium offers an example of the strong control that geochemistry exerts on the commercial use of an element. While it has an average abundance of 160 ppm, more than double that of Cu, it forms a limited number of minerals and is used much less widely. V-minerals are so rare because V^{3+}, the form in which it is found in much of the crust, is geochemically comparable to Fe^{3+}, an abundant ion that is part of many common minerals. As a result, V^{3+} usually substitutes for Fe^{3+} in minerals rather than forming separate V-minerals. The most common V-bearing Fe-mineral is magnetite (Fe_3O_4), which can have as much as 3% V_2O_5. The scarcity of V-minerals prohibited the metal from coming to the attention of early chemists and metallurgists until 1830, and even after that, it languished without important applications for another 60 years due to the difficulty in separating it from most ores. Hydrothermal deposits can form when V in rocks and minerals is oxidized during weathering to the pentavalent state (V^{5+}), which is highly soluble. Dissolved V^{5+} will be deposited if the solution is reduced or evaporated, for example, when groundwater encounters a reduced zone rich in organic matter. This geochemical process is very similar to that which forms uranium ores, and as a result V is a common constituent of many sandstone-type uranium deposits, often in the mineral carnotite ($K_2(UO_2)_2(VO_4)_2{\cdot}3H_2O$). V is also concentrated in many forms of organic material, including coal and oil. V as by-products from oil refining and ash from foil fuel combustion comprise >10% of world production. The largest resource of this type is in the tar sands of Alberta, Canada. Although the V content is only 0.0.2%–0.05%, fly ash from the processing plants contains up to 5% V_2O_5. However, attempts to recover this vanadium have not yet been commercially successful.

vanadium ends in the slag and is extracted from there. Vanadium is mined mostly in South Africa, north-western China, and eastern Russia. In addition, vanadium is present in bauxite and in deposits of crude oil, coal, oil shale, and tar sands. In crude oil, concentrations up to 1200 ppm have been recorded.

Vanadium metal is produced via a multistep process that starts with the roasting of crushed ore with NaCl or Na_2CO_3 at ~850°C to produce sodium metavanadate ($NaVO_3$). An aqueous extract of this solid is acidified to give "red cake," a polyvanadate salt, which is reduced with calcium metal. As an alternative for small-scale production, vanadium pentoxide is reduced with hydrogen or magnesium. Numerous other processes are also being used, in all of which vanadium is separated as a by-product of other processes. Purification of vanadium is possible utilizing the crystal bar process developed by Anton Eduard van Arkel and Jan Hendrik de Boer in 1925, which comprises the formation of the metal iodide, here vanadium(III) iodide, followed by decomposition to obtain the pure metal.

$$2V + 3I_2 \rightleftharpoons 2VI_3$$

Most vanadium is utilized in a steel alloy called ferrovanadium. Ferrovanadium is manufactured directly by reducing a combination of vanadium oxide, iron oxides, and iron in an electric furnace. The vanadium ends up in pig iron obtained from vanadium-bearing magnetite. Depending on the ore used, the slag can have up to 25% vanadium.

Vanadium is found in over 250 different minerals. It is found as a native element (V) and a carbide khamrabaevite ((Ti,V,Fe)C). The sulfide class is represented by eight different minerals, such as colusite ($Cu_{13}VAs_3S_{16}$) and sulvanite (Cu_3VS_4) (Fig. 5.37). Within the oxide class, 74 minerals contain V, for example, carnotite ($K_2(UO_2)_2(VO_4)_2{\cdot}3H_2O$), and pascoite ($Ca_3(V_{10}O_{28}){\cdot}17H_2O$). A total of eight sulfate minerals have V in their structure, for example, minasragite ($(V^{4+}O)(SO_4){\cdot}5H_2O$), rankachite ($Ca_{0.5}(V^{4+},V^{5+})(W^{6+},Fe^{3+})_2O_8(OH){\cdot}2H_2O$), and stanleyite ($(V^{4+}O)(SO_4){\cdot}6H_2O$). A large group of 71 different minerals can be found in the phosphate class (which includes the vanadates), such as descloizite ($PbZn(VO_4)(OH)$) (Fig. 5.38), francevillite ($Ba(UO_2)(VO_4)_2{\cdot}5H_2O$) (Fig. 5.39), mottramite ($PbCu(VO_4)(OH)$) (Fig. 5.40), namibite ($Cu(BiO)_2(VO_4)(OH)$), pucherite ($Bi(VO_4)$), sincosite ($Ca(VO)_2(PO_4)_2{\cdot}5H_2O$) (Fig. 5.41), tyuyamunite ($Ca(UO_2)(VO_4)_2{\cdot}5\text{-}8H_2O$) (Fig. 5.42), and vanadinite ($Pb_5(VO_4)_3Cl$) (Fig. 5.43) (Box 5.7). Twenty-nine minerals are known in the silicate class to contain V in their structures, for example, cavansite ($Ca(VO)Si_4O_{10}{\cdot}4H_2O$) (Fig. 5.44), pentagonite ($Ca(VO)Si_4O_{10}{\cdot}4H_2O$) (Fig. 5.45), and roscoelite ($K(V^{3+},Al)_2(AlSi_3O_{10})(OH)_2$).

5.5.3 Chemistry

Naturally occurring vanadium consists of one stable isotope, ^{51}V, and one radioactive isotope, ^{50}V. The latter has a half-life of 1.5×10^{17} years and a natural abundance of about 0.25%. ^{51}V has a nuclear spin of 7/2, making it a useful isotope for NMR spectroscopy. Twenty-four artificial radioisotopes have been observed, with mass number ranging from 40 to 65. The two most stable of these isotopes are ^{49}V with a half-life of 330 days and ^{48}V with a half-life of 16.0 days. The other radioisotopes have half-lives shorter than an hour, with most of them below 10 seconds. At least

FIGURE 5.37 Light brasy metallic crystal grains of sulvanite, Cu_3VS_4, to 8 mm in quartz, SiO_2, with alterations to green volborthite, $Cu_3(V_2O_7)(OH)_2{\cdot}2H_2O$. Thorpe Hills. Tooele Co., Utah, United States.

FIGURE 5.38 Descloizite, $PbZn(VO_4)(OH)$, brown to reddish-brown tabular crystals to over 1 cm. Berg Aukus, Grootfontein dist., Namibia.

FIGURE 5.39 Francevillite, $Ba(UO_2)_2(VO_4)_2{\cdot}5H_2O$, bright orange clusters and single crystals to 1.5 mm in diamond-shaped forms. Mounana mine, Franceville, Haut-Ogooué, Gabon.

FIGURE 5.40 6-cm group of soft, velvety, dark green growths of mottramite, $PbCu(VO_4)(OH)$. Mapimi dist., Durango, Mexico.

FIGURE 5.41 Green tabular sincosite, $Ca(VO)_2(PO_4)_2{\cdot}5H_2O$, crystals to 0.75 mm on matrix. Ross Hannibal mine, Lawrence Co., South Dakota, United States.

FIGURE 5.42 Tyuyamunite, $Ca(UO_2)_2(VO_4)_2{\cdot}5\text{-}8H_2O$, well-formed opaque bright yellow micro crystals with green malachite, $Cu_2(CO_3)(OH)_2$, about 2 mm. Mashamba West mine, Kolwezi dist., Katanga, Democratic Rep. of Congo.

BOX 5.7 Vanadinite $Pb_5(VO_4)_3Cl$

When one thinks of a typical vanadium mineral, the first one that springs to mind is vanadinite (Fig. 5.43). It is a rare secondary mineral found in the oxidized zone of lead ores associated with other secondary lead minerals such as cerussite, $PbCO_3$, and anglesite, $PbSO_4$. Vanadinite was named in reference to its vanadium content. Small amounts of PO_4 and AsO_4 may substitute for VO_4. In the variety *endlichite*, intermediate between vanadinite and mimetite ($Pb_5(AsO_4)_3Cl$), the proportion of V_2O_5 to As_2O_5 is nearly 1:1. Small amounts of Ca, Zn, and Cu can substitute for Pb. It is hexagonal, *6/m* (space group $P6_3/m$). The hardness is 3. It has a resinous to adamantine luster and its transparency is transparent to translucent. It has no cleavage and the fracture is subconchoidal. The color is usually ruby-red, orange-red, brown or yellow, and the streak is brownish yellow. Optically vanadinite is uniaxial(−). Vanadinite most commonly occurs in prismatic crystals with {100} and {0001}. It may have small pyramidal faces, rarely the hexagonal dipyramid. It also occurs in globular forms and as incrustations.

FIGURE 5.43 Red tabular, hexagonal vanadinite, $Pb_5(VO_4)_3Cl$, crystals from 5 to 15 mm. Acif mine, Mibladen, Khenifra Prov., Morocco.

FIGURE 5.44 Blue spherical crystal cluster of cavansite, $Ca(VO)Si_4O_{10}{\cdot}4H_2O$, to 1.5 cm set on matrix of pearly white stilbite, $(Na/Ca/K)_{6\text{-}7}[Al_{8\text{-}9}Si_{27\text{-}28}O_{72}]{\cdot}nH_2O$, crystals. Poonah dist., Maharashtra, India.

FIGURE 5.45 Blue radial cluster of pentagonite, $Ca(VO)Si_4O_{10}{\cdot}4H_2O$, to 9 mm sitting in zeolite vug. Wagholi, near Pune, Maharashtra, India.

TABLE 5.6 Vanadium properties.

Appearance	Blue-silver-gray metal
Standard atomic weight $A_{r,std}$	50.9415
Block	d-Block
Element category	Transition metal
Electron configuration	[Ar] $3d^3$ $4s^2$
Phase at STP	Solid
Melting point	1910°C
Boiling point	3407°C
Density (near r.t.)	6.0 g/cm^3
When liquid (at m.p.)	5.5 g/cm^3
Heat of fusion	21.5 kJ/mol
Heat of vaporization	444 kJ/mol
Molar heat capacity	24.89 J/(mol·K)
Oxidation states	−3, −1, +1, +2, +3, +4, **+5**
Ionization energies	1st: 650.9 kJ/mol
	2nd: 1414 kJ/mol
	3rd: 2830 kJ/mol
Atomic radius	Empirical: 134 pm
Covalent radius	153 ± 8 pm

STP, Standard temperature and pressure.
Bold font indicates main oxidation state.

four isotopes have metastable excited states. Electron capture is the principal decay mode for isotopes lighter than ^{51}V. For the heavier isotopes, the most common mode is β decay. The electron capture reactions result in the formation of Ti isotopes, while β decay results in Cr isotopes.

Vanadium is a medium-hard, ductile, steel-blue metal, which is electrically conductive and thermally insulating (Table 5.6). Some publications refer to vanadium as "soft," possibly since it is ductile, malleable, and not brittle. Vanadium is harder than most metals and steels. It has good corrosion resistance, and it is stable against attacks by alkalis and sulfuric and hydrochloric acids. It is oxidized in air at approximately 660°C (933K), but an oxide passivation layer forms even at room temperature. The chemistry of vanadium is remarkable for the accessibility of the four adjacent oxidation states 2–5. In aqueous solution, vanadium produces metal aquo complexes of which the colors are lilac $[V(H_2O)_6]^{2+}$, green $[V(H_2O)_6]^{3+}$, blue $[VO(H_2O)_5]^{2+}$, and yellow VO_3^-. Vanadium(II) compounds are reducing agents, while vanadium(V) compounds are oxidizing agents. Vanadium(IV) compounds frequently exist as vanadyl derivatives, which comprise the VO^{2+} center.

5.5.3.1 *Oxides*

Vanadium(II) oxide, VO, is one of the many oxides of vanadium. VO is a long-lived, electronically neutral reagent chemical. It has a distorted NaCl structure with weak V-V bonds. In accordance with band theory, VO is a conductor of electricity due to its partially filled conduction band and delocalization of electrons in the t_{2g} orbitals. VO is a nonstoichiometric compound; its composition can vary between $VO_{0.8}$ and $VO_{1.3}$.

Vanadium(III) oxide, V_2O_3, is a black solid produced via the reduction reaction of V_2O_5 with hydrogen or carbon monoxide. It is a basic oxide dissolving in acids to form solutions of vanadium (III) complexes. V_2O_3 has the corundum (Al_2O_3) structure. It is antiferromagnetic with a critical temperature of −113°C (160K). At this temperature, there is a sudden change in conductivity from metallic to insulating. Upon exposure to air, it slowly transforms to indigo blue V_2O_4. It occurs naturally as the rare mineral karelianite.

Vanadium(IV) oxide or vanadium dioxide, VO_2, is a dark blue solid. VO_2 is amphoteric, dissolving in nonoxidizing acids to form the blue vanadyl ion, $[VO]^{2+}$, in alkali to form the brown $[V_4O_9]^{2-}$ ion, or at high pH to form $[VO_4]^{4-}$. VO_2 has a phase transition at ~66°C. Electrical resistivity, opacity, etc., can change up to a number of orders. Due to these properties, it has been applied in surface coating, sensors, and imaging. Potential applications are, for example, use in memory devices, phase-change switches, aerospace communication systems, and neuromorphic computing. Following the method originally described by Swedish chemist Jöns Jacob Berzelius (August 20, 1779 to August 7, 1848), VO_2 is produced by comproportionation of vanadium(III) oxide and vanadium(V) oxide.

$$V_2O_5 + V_2O_3 \rightarrow 4VO_2$$

At 25°C, VO_2 has a distorted rutile (TiO_2) structure with shorter distances between pairs of V atoms pointing to metal–metal bonding. Above 68°C, the structure converts to an undistorted rutile structure and the metal–metal bonds are broken resulting in an increase in electrical conductivity and magnetic susceptibility as the bonding electrons are "released." The origin of this insulator to metal transition is still controversial and is of interest both for condensed matter physics and practical applications, such as electrical switches, tunable electrical filters, power limiters, nano-oscillators, memristors (a non-linear two-terminal electrical component relating electric charge and magnetic flux linkage), field-effect transistors, and metamaterials. At temperatures below Tc = 67°C (340K), VO_2 has a monoclinic (space group $P2_1/c$) crystal structure. Above Tc, the structure is tetragonal, like rutile (TiO_2). In the monoclinic phase, the V^{4+} ions form pairs along the *c* axis, resulting in alternate short and long V-V distances of 2.65 and 3.12 Å, respectively. In comparison, in the rutile phase the V^{4+} ions are separated by a fixed distance of 2.96 Å. Consequently, the number of V^{4+} ions in the crystallographic unit cell doubles from the rutile to the monoclinic phase. The equilibrium morphology of rutile VO_2 particles is acicular, laterally limited by (110) surfaces, which form the most stable termination planes. The surface tends to be oxidized relative to the stoichiometric composition, with the oxygen adsorbed on the (110) surface forming vanadyl species. The existence of V^{5+} ions at the surface of VO^2 films has been shown by X-ray photoelectron spectroscopy. At the rutile to monoclinic transition temperature (67°C), VO_2 also shows a metal to semiconductor transition in its electronic structure: the rutile phase is metallic while the monoclinic phase is semiconducting. The optical band gap of VO_2 in the low-temperature monoclinic phase is approximately 0.7 eV.

Vanadium(V) oxide (vanadia), V_2O_5, commonly known as vanadium pentoxide, is a brown/yellow solid, even though when freshly precipitated from aqueous solution it is deep orange in color. Due to its high oxidation state, it is both an amphoteric oxide and an oxidizing agent. From a commercial point of view, it is the most important compound of vanadium, as the principal precursor to alloys of vanadium, and is widely used as an industrial catalyst. The naturally occurring V_2O_5 mineral shcherbinaite is extremely rare, nearly always formed among fumaroles. A mineral trihydrate, $V_2O_5 \cdot 3H_2O$, also exists with the name of navajoite. Technical grade V_2O_5 is prepared as a black powder employed for the manufacture of vanadium metal and ferrovanadium. A vanadium ore or vanadium-rich residue is reacted with sodium carbonate and an ammonium salt to form sodium metavanadate, $NaVO_3$. This compound is subsequently acidified to pH 2–3 with H_2SO_4 to form a precipitate of "red cake." Next, the red cake is melted at 690°C to form the crude V_2O_5. Vanadium(V) oxide is formed when vanadium metal is heated with excess oxygen, but this material is contaminated with other lower oxides. A better suitable laboratory preparation route comprises the decomposition of ammonium metavanadate at 500°C–550°C.

$$2NH_4VO_3 \rightarrow V_2O_5 + 2NH_3 + H_2O$$

Upon heating a mixture of vanadium(V) oxide and vanadium(III) oxide, comproportionation results in the formation of vanadium(IV) oxide, as a deep blue solid.

$$V_2O_5 + V_2O_3 \rightarrow 4VO_2$$

The reduction can also be affected by oxalic acid, carbon monoxide, and sulfur dioxide. Further reduction utlizing hydrogen or excess CO can result in the formation of complex mixtures of oxides, for example, V_4O_7 and V_5O_9 before black V_2O_3 is reached. V_2O_5 is an amphoteric oxide. In contrast to most metal oxides, it dissolves slightly in water to produce a pale-yellow, acidic solution. Hence, V_2O_5 reacts with strong nonreducing acids to give solutions containing the pale-yellow salts with dioxovanadium(V) centers.

$$V_2O_5 + 2HNO_3 \rightarrow 2VO_2(NO_3) + H_2O$$

In addition, it reacts with strong alkali to produce polyoxovanadates, which have complex structures depending on the pH. If excess aqueous NaOH is used, the product is the colorless salt sodium orthovanadate, Na_3VO_4. If an acid is slowly added to a solution of Na_3VO_4, the color progressively deepens through orange to red before brown hydrated V_2O_5 precipitates at about pH 2. These solutions contain primarily the ions HVO_4^{2-} and $V_2O_7^{4-}$ in the pH range from 9 to 13, but below pH 9 more exotic species, for example, $V_4O_{12}^{4-}$ and $HV_{10}O_{28}^{5-}$ (decavanadate) are dominant. Upon treatment with thionyl chloride, V_2O_5 changes to the volatile liquid vanadium oxychloride, $VOCl_3$.

$$V_2O_5 + 3SOCl_2 \rightarrow 2VOCl_3 + 3SO_2$$

Hydrochloric acid and hydrobromic acid are oxidized to their corresponding halogen, for example,

$$V_2O_5 + 6HCl + 7H_2O \rightarrow 2\left[VO(H_2O)_5\right]^{2+} + 4Cl^- + Cl_2$$

Vanadates or vanadyl(V) compounds in acid solution are reduced by zinc amalgam through the colorful pathway:

$$VO_3^-(\text{yellow}) \rightarrow VO^{2+}(\text{blue}) \rightarrow V^{3+}(\text{green}) \rightarrow V^{2+}(\text{purple})$$

These ions are all hydrated to varying degrees. In terms of volume, the principal use for vanadium(V) oxide is in the manufacture of ferrovanadium. The oxide is heated with scrap iron and ferrosilicon, with lime added to form a calcium silicate slag. Aluminum can also be employed, forming the iron-vanadium alloy along with alumina as a by-product. One more significant application of vanadium(V) oxide is in the production of sulfuric acid, an important industrial chemical. Vanadium(V) oxide has a crucial role in catalyzing the mildly exothermic oxidation of sulfur dioxide to sulfur trioxide by air in the contact process.

$$2SO_2 + O_2 \rightleftharpoons 2SO_3$$

The discovery of this simple reaction, for which V_2O_5 is the most efficient catalyst, allowed sulfuric acid to grow into the inexpensive commodity chemical it currently is. The reaction is performed between 400°C and 620°C; below 400°C, V_2O_5 is catalytically inactive, while above 620°C, it starts to break down. Because it is known that V_2O_5 can be reduced to VO_2 by SO_2, one possible catalytic cycle is as follows:

$$SO_2 + V_2O_5 \rightarrow SO_3 + 2VO_2$$

followed by

$$4VO_2 + O_2 \rightarrow 2V_2O_5$$

In addition, it is employed as catalyst in the selective catalytic reduction (SCR) of NO_x emissions in some power plants. Because of its effectiveness in converting sulfur dioxide into sulfur trioxide, and thereby sulfuric acid, particular care must be taken with the operating temperatures and placement of a power plant's SCR unit when firing sulfur-containing fuels. Maleic anhydride is formed by the V_2O_5-catalyzed oxidation of butane with air.

$$C_4H_{10} + 4O_2 \rightarrow C_2H_2(CO)_2O + 8H_2O$$

Maleic anhydride is utilized to produce polyester resins and alkyd resins. Phthalic anhydride is formed in the same way by V_2O_5-catalyzed oxidation of ortho-xylene or naphthalene at 350°C–400°C. The equation for xylene oxidation is as follows:

$$C_6H_4(CH_3)_2 + 3O_2 \rightarrow C_6H_4(CO)_2O + 3H_2O$$

Phthalic anhydride is a precursor to plasticizers, employed to confer pliability to polymers. Various other commercial compounds are manufactured likewise, for example, adipic acid, acrylic acid, oxalic acid, and anthraquinone. Because of its high coefficient of thermal resistance, vanadium(V) oxide also finds use as a detector material in bolometers and microbolometer arrays for thermal imaging. It likewise can be applied as an ethanol sensor in ppm levels (up to 0.1 ppm).

5.5.3.2 *Carbide and nitride*

Vanadium carbide, VC, is an extremely hard refractory ceramic material. With a hardness of 9–9.5 Mohs, it is probably the hardest metal-carbide known. It is of significance as it is prevalent in vanadium metal and alloys. Being isomorphous with vanadium monoxide, it crystallizes in the rock salt (NaCl) structure. Since VC and VO are miscible, samples of VC usually have an impurity of the oxide. It is formed by heating vanadium oxides with carbon at approximately 1000°C. Vanadium carbide can be formed in the (111) orientation, when formed by radio frequency magnetron sputtering. While VC is thermodynamically stable, it transforms to V_2C at higher temperatures. Vanadium carbide is employed as an additive to tungsten carbide to refine the carbide crystals to enhance the properties of the cermet.

Vanadium nitride, VN, is formed during the nitriding of steel and improves wear resistance. Another phase, V_2N, also referred to as vanadium nitride, can be produced together with VN during nitriding. VN has a cubic, rock salt (NaCl) structure. A low-temperature form also exists, which has V_4 clusters. The low-temperature phase is the result of dynamic instability, when the energy of vibrational modes in the high-temperature NaCl-structure phase is reduced below zero. It is a strong-coupled superconductor. Nanocrystalline vanadium nitride has been asserted by Choi et al. (2006) to have potential for application in supercapacitors.

5.5.3.3 Halides

Vanadium(III) fluoride, VF_3, is a yellow-green, refractory solid formed in a two-step procedure from V_2O_3. Comparable to other transition metal fluorides (e.g., MnF_2), it shows magnetic ordering at low temperatures (e.g., $V_2F_6 \cdot 4H_2O$ orders below −261°C/12K). The first step involves the conversion to the hexafluorovanadate(III) salt using ammonium bifluoride.

$$V_2O_3 + 6(NH_4)HF_2 \rightarrow 2(NH_4)_3VF_6 + 3H_2O$$

In the second step, the hexafluorovanadate is thermally decomposed.

$$(NH_4)_3VF_6 \rightarrow 3NH_3 + 3HF + VF_3$$

The thermal decomposition of ammonium salts is a fairly common technique for the preparation of inorganic solids. VF_3 can also be produced by reacting V_2O_3 with HF. VF_3 is a crystalline solid with six-coordinated V atoms with bridging F atoms. The magnetic moment points to the presence of two unpaired electrons.

Vanadium(IV) fluoride (VF_4) is a paramagnetic yellow-brown solid that is extremely hygroscopic. In contrast to the corresponding vanadium tetrachloride, the tetrafluoride is not volatile as it has a polymeric structure. It decomposes before melting. VF_4 can be produced by reacting VCl_4 with HF.

$$VCl_4 + 4HF \rightarrow VF_4 + 4HCl$$

At 325°C, it decomposes, disproportionating to the tri- and pentafluorides.

$$2VF_4 \rightarrow VF_3 + VF_5$$

The structure of VF_4 is related to that of SnF_4. Each V center is octahedrally coordinated, surrounded by six F ligands. Four of the F centers bridge to adjacent V centers.

Vanadium(V) fluoride, VF_5, is a colorless volatile liquid. It is a highly reactive compound, as revealed by its ability to fluorinate organic substances. Vanadium pentafluoride can be produced via fluorination of vanadium metal.

$$2V + 5F_2 \rightarrow 2VF_5$$

As was previously noted, VF_5 can be prepared by the disproportionation of VF_4. But this conversion is performed at 650°C. Instead, it can be prepared by using elemental fluorine to fluorinate industrial concentrates and raw materials to form VF_5 on a commercial scale. VF_5 can be prepared from the reaction of raw materials such as metallic vanadium, ferrovanadium, vanadium(V) oxide, and vanadium tetrafluoride with elemental fluorine. VF_5 ionizes in the liquid state as manifested by the high values of Trouton's constant and electrical conductivities. VF_5 occurs entirely as a monomer in the gas phase. In the gas phase, it has a D_{3h} symmetric trigonal bipyramidal geometry. As a solid, VF_5 forms a polymeric structure with fluoride-bridged octahedrally coordinated V centers. It is the only known pentahalide of vanadium. It is a powerful fluorinating and oxidizing agent. It oxidizes elemental sulfur to sulfur tetrafluoride.

$$S + 4VF_5 \rightarrow 4VF_4 + SF_4$$

Similar to other electrophilic metal halides, it hydrolyzes, first to the oxyhalide.

$$VF_5 + H_2O \rightarrow VOF_3 + 2HF$$

Subsequently to the binary oxide:

$$2VOF_3 + 3H_2O \rightarrow V_2O_5 + 6HF$$

Hydrolysis is accelerated in the presence of base. Notwithstanding its tendency to hydrolyze, it can be dissolved in alcohols. It is a Lewis acid, as shown by its formation of the hexafluorovanadate.

$$VF_5 + KF \rightarrow KVF_6$$

VF_5 is a weaker acid and mostly undergoes oxidative and fluorinating reactions. It fluorinates unsaturated polyfluoroolefins forming polyfluoroalkanes. VF_5 dissolves without reaction in liquid Cl_2 and Br_2. It is moderately soluble in HF.

Vanadium(II) chloride, VCl_2, is the most reduced vanadium chloride. VCl_2 is an apple-green solid that dissolves in water to form purple solutions of the hexaaquo ion $[V(H_2O)_6]^{2+}$. Solid VCl_2 is produced by thermal decomposition of VCl_3, which leaves a residue of VCl_2.

$$2VCl_3 \rightarrow VCl_2 + VCl_4$$

Evaporation of $[V(H_2O)_6]^{2+}$ solutions forms crystals of $[V(H_2O)_6]Cl_2$. Solid VCl_2 has the cadmium iodide structure, with an octahedral coordination geometry. VBr_2 and VI_2 are structurally and chemically identical dihalides. Both have the d^3 configuration, with a quartet ground state, akin to Cr(III). VCl_2 is a powerful reducing agent, which can convert sulfoxides to sulfides, organic azides to amines, in addition to reductively coupling some alkyl halides.

Vanadium trichloride, VCl_3, is a purple salt and a common precursor to other vanadium(III) complexes. VCl_3 is formed by heating VCl_4 at 160°C–170°C under an inert gas flow, which removes Cl_2. The bright red liquid transforms to a purple solid. Heating of VCl_3 decomposes with volatilization of VCl_4, leaving VCl_2. Upon heating under H_2 at 675°C (but below 700°C), VCl_3 reduces to greenish VCl_2.

$$2VCl_3 + H_2 \rightarrow 2VCl_2 + 2HCl$$

Comproportionation of vanadium trichloride and vanadium(V) oxides results in the formation of vanadium oxydichloride.

$$V_2O_5 + VOCl_3 + 3VCl_3 \rightarrow 6VOCl_2$$

VCl_3 is a catalyst for the pinacol coupling reaction of benzaldehyde (PhCHO) to 1,2-diphenyl-1,2-ethanediol by a variety of reducing metals such as zinc.

$$Zn + 2H_2O + 2PhCHO \rightarrow (PhCH(OH))_2 + Zn(OH)_2$$

VCl_3 has the common BiI_3 structure, featuring hexagonally closest-packed Cl framework with V ions occupying the octahedral holes. VBr_3 and VI_3 have the same structure, though VF_3 has a structure more closely related to ReO_3. VCl_3 is paramagnetic and has 22 unpaired electrons. VCl_3 can produce colorful adducts and derivatives with a wide range of ligands. VCl_3 dissolves in water to form the hexahydrate, but the formula is misleading. The salt is described by the formula $[VCl_2(H_2O)_4]Cl.2H_2O$, that is, two of the water molecules are not bound to the V, whose structure resembles the equivalent Fe(III) derivative. Removal of the two bound Cl ligands from $[VCl_2(H_2O)_4]^+$ in aqueous solution produces the green $[V(H_2O)_6]^{3+}$ ion. With tetrahydrofuran, VCl_3 forms the red/pink adduct $VCl_3(THF)_3$. With acetonitrile, it forms the green adduct $VCl_3(MeCN)_3$. When treated with KCN, VCl_3 transforms to $[V(CN)_7]^{4-}$. It is common for early metals to have high coordination numbers (more than 6) with compact ligands. On the other hand, larger metals can produce complexes with relatively bulky ligands. This aspect is reflected by the isolation of $VCl_3(NMe_3)_2$, containing two bulky NMe_3 ligands. The reactive species V(mesityl)$_3$ forms from VCl_3. This compound can bind CO and, under suitable conditions, N_2.

$$VCl_3(THF)_3 + 3LiC_6H_2{-}2,4,6{-}Me_3 \rightarrow V(C_6H_2{-}2,4,6{-}Me_3)_3(THF) + 3LiCl$$

Vanadium tetrachloride, VCl_4, is a bright red liquid that acts as a useful reagent for the synthesis of other vanadium compounds. With one more valence electron than diamagnetic $TiCl_4$, VCl_4 is a paramagnetic liquid. It is one of only a few paramagnetic compounds that is liquid at room temperature. VCl_4 is produced by chlorination of vanadium metal. VCl_5 does not form in this reaction; Cl_2 lacks the oxidizing power to attack VCl_4. VCl_5 though can be formed indirectly from VF_5 at −78°C. Unlike the heavier analogs $NbCl_5$ and $TaCl_5$ that are stable and not particularly oxidizing. VF_5 can be formed directly by fluorination of vanadium metal, indicating the increased oxidizing power of F_2 compared to Cl_2. As an indication of its oxidizing power, VCl_4 releases Cl_2 at its boiling point (standard pressure) to form VCl_3. Consistent with its high oxidizing power, VCl_4 reacts with HBr at −50°C to form VBr_3. This reaction progresses via VBr_4, which releases Br_2 during warming to room temperature.

$$2VCl_4 + 8HBr \rightarrow 2VBr_3 + 8HCl + Br_2$$

VCl_4 can react to produce adducts with numerous donor ligands, for example, $VCl_4(THF)_2$. It is the precursor to vanadocene dichloride. In organic synthesis, VCl_4 is utilized for the oxidative coupling of phenols, for example, it converts phenol into a mixture of 4,4′-, 2,4′-, and 2,2′-biphenols.

$$2C_6H_5OH + 2VCl_4 \rightarrow HOC_6H_4 - C_6H_4OH + 2VCl_3 + 2HCl$$

VCl_4 acts as a catalyst for the polymerization of alkenes, in particular those useful in the rubber industry. The underlying technology is related to Ziegler–Natta catalysis, which comprises the intermediary of vanadium alkyls.

Vanadium(V) chloride, VCl_5, is a black diamagnetic solid. The molecules have a bioctahedral structure comparable to that of niobium(V) chloride. VCl_5 is produced from vanadium pentafluoride with excess boron trichloride.

$$2VF_5 + 10BCl_3 \rightarrow [VCl_5]_2 + 10BF_2Cl$$

It is unstable at room temperature with respect to vanadium(IV) chloride.

$$[VCl_5]_2 \rightarrow 2VCl_4 + Cl_2$$

Vanadium(II) bromide, also known as vanadium dibromide (VBr_2), is formed through the reduction of hydrogen from vanadium(III) bromide. Vanadium(III) bromide, also known as vanadium tribromide (VBr_3), is in the solid-state a polymeric species with octahedral vanadium(III) surrounded by six Br ligands.

As previously discussed, VBr_3 has been prepared by the reaction of VCl_4 with HBr. This reaction progresses via the unstable vanadium(IV) bromide, VBr_4, which releases Br_2 at room temperature. Like VCl_3, VBr_3 forms red-brown soluble complexes with dimethoxyethane and THF, for example, mer-$VBr_3(THF)_3$. Aqueous solutions prepared from VBr_3 contain *trans*-$[VBr_2(H_2O)_4]^+$. Evaporation of these solutions crystallizes the salt *trans*-$[VBr_2(H_2O)_4]Br$.

Vanadium(III) iodide, VI_3, is a paramagnetic solid formed by the reaction of vanadium powder with iodine at approximately 500°C. The black hygroscopic crystals dissolve in water to produce green solutions, distinctive for V(III) ions. The purification of vanadium metal by the chemical transport reaction comprises the reversible formation of vanadium(III) iodides in the presence of iodine and its subsequent decomposition to give pure metal.

$$2V + 3I_2 \rightleftharpoons 2VI_3$$

VI_3 crystallizes in a similar structure as bismuth(III) iodide. The iodides are hexagonal closest packed and the V centers occupy one-third of the octahedral holes. When solid samples are heated, the gas contains VI_4, which is possibly the volatile vanadium component in the vapor transport method. Thermal decomposition of the triiodide leaves a residue of vanadium(II) iodide.

$$2VI_3 \rightarrow VI_2 + VI_4$$

Vanadium oxydichloride, $VOCl_2$, is one of several oxychlorides of vanadium. It is a hygroscopic green solid. It is produced by comproportionation of vanadium trichloride and vanadium(V) oxides.

$$V_2O_5 + VOCl_3 + 3VCl_3 \rightarrow 6VOCl_2$$

Vanadium oxydichloride has a layered structure, with octahedrally coordinated V centers linked by doubly bridging oxide and chloride ligands. Vanadium oxytrichloride, $VOCl_3$, is a yellow distillable liquid, which hydrolyzes easily in air. It is an oxidizing agent. It is employed as a reagent in organic synthesis. Samples frequently appear red or orange due to an impurity of vanadium tetrachloride. $VOCl_3$ can be prepared through the chlorination of V_2O_5. The reaction takes place at around 600°C.

$$6Cl_2 + 2V_2O_5 \rightarrow 4VOCl_3 + 3O_2$$

Heating an intimate (well mixed with tiny particles) mixture of V_2O_5 and carbon with chlorine at 200°C–400°C results in $VOCl_3$. Here the carbon acts as a deoxygenation agent similar to its use in the Kroll process for the production of $TiCl_4$ from TiO_2. Vanadium(III) oxide can similarly be utilized as a precursor.

$$6Cl_2 + 2V_2O_3 \rightarrow 4VOCl_3 + O_2$$

A more typical laboratory synthesis comprises the chlorination of V_2O_5 with $SOCl_2$.

$$V_2O_5 + 3SOCl_2 \rightarrow 2VOCl_3 + 3SO_2$$

$VOCl_3$ is a compound with V in the +5 oxidation state and as such is diamagnetic. It is tetrahedral with O-V-Cl bond angles of 111° and Cl-V-Cl bond angles of 108°. The V-O and V-Cl bond lengths are 157 and 214 pm, respectively.

$VOCl_3$ is very reactive toward water and evolves HCl upon standing. It is soluble in nonpolar solvents, for example, benzene, CH_2Cl_2, and hexane. In some aspects, the chemical properties of $VOCl_3$ and $POCl_3$ are comparable. One difference is that $VOCl_3$ is a strong oxidizing agent, while $POCl_3$ is not. $VOCl_3$ rapidly hydrolyzes forming vanadium pentoxide and hydrochloric acid. An intermediate in this reaction is VO_2Cl.

$$2VOCl_3 + 3H_2O \rightarrow V_2O_5 + 6HCl$$

$VOCl_3$ reacts with alcohols, in particular, in the presence of a proton acceptor, to produce alkoxides, as shown by the synthesis of vanadyl isopropoxide.

$$VOCl_3 + 3HOCH(CH_3)_2 \rightarrow VO\left(OCH(CH_3)_2\right)_3 + 3HCl$$

$VOCl_3$ is similarly utilized in the synthesis of vanadium oxydichloride.

$$V_2O_5 + 3VCl_3 + VOCl_3 \rightarrow 6VOCl_2$$

VO_2Cl can be produced via an unusual reaction comprising Cl_2O.

$$VOCl_3 + Cl_2O \rightarrow VO_2Cl + 2Cl_2$$

Above 180°C, VO_2Cl decomposes to V_2O_5 and $VOCl_3$. Likewise, $VOCl_2$ also decomposes to form $VOCl_3$, together with VOCl. $VOCl_3$ is strongly Lewis acidic, as shown by its tendency to form adducts with a variety of bases, for example, acetonitrile and amines. In forming the adducts, vanadium converts from four-coordinated tetrahedral geometry to six-coordinated octahedral geometry, for example,

$$VOCl_3 + 2H_2NEt \rightarrow VOCl_3(H_2NEt)_2$$

$VOCl_3$ is employed as a catalyst or precatalytst in the manufacture of ethylene-propylene rubbers (EPDM).

Vanadyl perchlorate or vanadyl triperchlorate ($VO(ClO_4)_3$) is a golden yellow−colored liquid or crystalline compound. It consists of molecules covalently bonded and is rather volatile. Vanadyl perchlorate can be prepared by reacting vanadium pentoxide with dichlorine heptoxide at 5°C. It is purified via distillation under vacuum and recrystallization at 21°C. The reaction of barium perchlorate with vanadyl sulfate solution produces vanadyl perchlorate solution. A solution of vanadium(V) perchlorate can be prepared by dissolving vanadium pentoxide in perchloric acid. Pervanadyl perchlorate also called dioxovanadium perchlorate comprises VO_2^+ ions. Other perchlorates are, for example, vanadyl diperchlorate, oxovanadium perchlorate, and vanadium(IV) perchlorate $VO(ClO_4)_2$, which dissolve in water. Vanadic perchlorate also known as vanadium(III) perchlorate solution in water has a green-tinged blue color, substantially different from most other V(III) solutions, which are complexed.

5.5.3.4 Sulfate and phosphate

The term vanadium(II) sulfate covers a range of inorganic compounds with the general formula $VSO_4{\cdot}xH_2O$ where $0 \leq x \leq 7$. The hexahydrate is most frequently found. It is a violet solid that dissolves in water to form air-sensitive solutions of the aquo complex. The salt is isomorphous with $[Mg(H_2O)_6]SO_4$. Compared to the V-O bond length of 191 pm in $[V(H_2O)_6]^{3+}$, the V-O distance is 212 pm in the $[V(H_2O)_6]SO_4$. This close to 10% elongation signifies the effect of the lower charge and consequently weakened electrostatic attraction. The heptahydrate has also been crystallized. The compound is produced by electrolysis in sulfuric acid. The crystals also have $[V(H_2O)_6]^{2+}$ centers but with an extra water of crystallization. This salt is isomorphous with ferrous sulfate heptahydrate. A related salt is vanadous ammonium sulfate, $(NH_4)_2V(SO_4)_2 \cdot 6H_2O$, a Tutton's salt isomorphous with ferrous ammonium sulfate.

Vanadium(III) sulfate, $V_2(SO_4)_3$, is a pale-yellow solid that is stable in air, unlike most vanadium(III) compounds. It slowly dissolves in water to form the green aquo complex $[V(H_2O)_6]^{3+}$. It is formed by reacting V_2O_5 in sulfuric acid with elemental sulfur.

$$V_2O_5 + S + 3H_2SO_4 \rightarrow V_2(SO_4)_3 + SO_2 + 3H_2O$$

This conversion is a rare illustration of a reduction reaction by elemental sulfur. When heated in vacuum at or slightly below 410°C, it decomposes into vanadyl sulfate ($VOSO_4$) and SO_2. Vanadium(III) sulfate is stable in dry air but exposure to moist air for several weeks results in the formation of a green hydrate form. Vanadium(III) sulfate is a reducing agent.

Vanadyl(IV) sulfate covers a range of inorganic compounds with the formula $VOSO_4{\cdot}xH_2O$ where $0 \leq x \leq 6$. The pentahydrate is common. This hygroscopic blue solid is one of the most common sources of vanadium in the laboratory, indicating its high stability. It has the vanadyl ion, VO^{2+}, which has been labeled the "most stable diatomic ion." Vanadyl sulfate is most frequently prepared via the reduction of vanadium pentoxide with sulfur dioxide.

$$V_2O_5 + 7H_2O + SO_2 + H_2SO_4 \rightarrow 2[V(O)(H_2O)_4]SO_4$$

From aqueous solution, the salt crystallizes as the pentahydrate, the fifth water is not bonded to the V in the solid. Considered as a coordination complex, the ion is octahedral, with oxo, four equatorial water ligands, and a monodentate sulfate. The trihydrate has also been investigated. A hexahydrate exists below 13.6°C (286.8K). Two polymorphs of anhydrous $VOSO_4$ have been recognized. The V-O bond distance is 160 pm, about 50 pm shorter than the V-OH_2 bonds. In solution, the sulfate ion dissociates quickly. Being commonly available, vanadyl sulfate is a common precursor to form other vanadyl derivatives, for example, vanadyl acetylacetonate.

$$[V(O)(H_2O)_4]SO_4 + 2C_5H_8O_2 + Na_2CO_3 \rightarrow [V(O)(C_5H_7O_2)_2] + Na_2SO_4 + 5H_2O + CO_2$$

In acidic solution, oxidation of vanadyl sulfate results in the formation of yellow-colored vanadyl(V) derivatives. Reduction, for example, by zinc, forms vanadium(III) and vanadium(II) derivatives, which are typically green and violet, respectively. Similar to most water-soluble sulfates, vanadyl sulfate is only rarely found in nature. The anhydrous form is pauflerite, a mineral of fumarolic origin. Hydrated forms, also rare, include the hexahydrate (stanleyite), pentahydrates (minasragrite, orthominasragrite, and anorthominasragrite), and trihydrate (bobjonesite).

The term vanadium phosphates covers inorganic compounds with the formula VO_xPO_4 together with related hydrates with the formula $VO_xPO_4.nH_2O$. Some of these compounds are employed industrially as catalysts for oxidation reactions. Heating a suspension of vanadium pentoxide and phosphoric acid produces $VOPO_4{\cdot}2H_2O$, separated as a bright yellow solid. The V(V) centers are octahedral, with long, weak bonds to aquo ligands. Reduction of this compound with alcohols results in the formation of the vanadium(IV) phosphates. These are catalysts for the oxidation of butane to maleic anhydride. A crucial step in the activation of these catalysts is the transformation of $VO(HPO_4){\cdot}0.5H_2O$ to the pyrophosphate $(VO)_2(P_2O_7)$. This compound is known as vanadyl pyrophosphate along with vanadium oxide pyrophosphate. Seven polymorphs have been observed for anhydrous $VOPO_4$, denoted αI, αII, β, γ, δ, ω, and ε. These compounds contain the vanadyl group (VO) and phosphate (PO_4^{3-}). They are yellow, diamagnetic solids, but when contaminated with vanadium(IV) derivatives, samples show electron paramagnetic resonance (EPR) signals and have a bluish shine. For these compounds, vanadyl implies both vanadium(V) oxo and vanadium(IV) oxo centers, but conventionally vanadyl is only used for derivatives of VO^{2+}. Several vanadium(IV) phosphates have been recognized. These compounds are usually blue. In these materials, the phosphate anion is singly or doubly protonated. Examples are the hydrogenphosphates, $VOHPO_4.4H_2O$ and $VO(HPO_4){\cdot}0.5H_2O$, along with the dihydrogen phosphate $VO(H_2PO_4)_2$. Vanadium(III) phosphates missing the oxo ligand have the formula $VPO_4{\cdot}H_2O$ and $VPO_4{\cdot}2H_2O$. The monohydrate is isostructural with $MgSO_4{\cdot}H_2O$. It has the structure of the comparable hydrated aluminum phosphate. Oxidation of $VPO_4{\cdot}H_2O$ results in the formation of ε-$VOPO_4$.

Vanadyl nitrate, also known as vanadium oxytrinitrate or vanadium oxynitrate, is a compound of vanadium in the +5 oxidation state with nitrate groups and oxygen. The formula is $VO(NO_3)_3$. It is prepared from dinitrogen pentoxide and vanadium pentoxide. It is a nitrating agent, adding nitro groups to aromatic compounds, for example, benzene, phenol, chlorobenzene, anisole, acetanilide, benzoic acid, ethyl benzoate, and toluene. Vanadyl nitrate can be produced by soaking vanadium pentoxide in liquid dinitrogen pentoxide for around two days at 25°C. The yield for this reaction is around 85%.

$$V_2O_5 + 3N_2O_5 \rightarrow 2VO(NO_3)_3$$

Purification takes place via vacuum distillation. Vanadyl nitrate can also be prepared from vanadyl trichloride $VOCl_3$ and dinitrogen pentoxide. Vanadyl nitrate is a pale-yellow liquid, which is viscous though it can be poured. The $VO(NO_3)_3$ molecule has a distorted pentagonal bipyramid shape. The whole molecule has a C_s mirror symmetry. The vanadium oxygen double bond is in the same plane as a nitrate group roughly opposite. The other two nitrate groups are at an approximately 83° angle from the doubly bonded oxygen. The three nitrate groups are planar. Each nitrate is connected to the V atom through two oxygen atoms, but one is closer than the other. The nitrate on the other side of the oxygen is rather asymmetric, while the mirror image nitrate groups have more equivalent oxygen bond lengths. Vanadyl nitrate can dissolve in water, though it cannot be recrystallized, in its place a polymeric oxide is precipitated. Nitric acid is produced due to the reaction with water. Solvents are, for example, dichloromethane, nitromethane, carbon tetrachloride, cyclohexane, are trichlorofluoromethane. It is probably incompatible with amines, aromatic hydrocarbons, and ethers. Vanadyl nitrate is stable under nitrogen, ozone, or oxygen. The liquid is indefinitely stable at room temperature, in contrast to certain other transition metal nitrates that decompose to nitrogen oxides. At temperatures above 80°C it slowly decomposes. Vanadyl nitrate can react to form a solid pale-yellow adduct with boron trifluoride. An adduct is likewise formed with acetonitrile. When mixed with petrol or l-hexene, or other unsaturated hydrocarbons, vanadyl nitrate ignites. In contrast, it does not ignite with hexane. It reacts with paper, rubber, and wood and ignites many organic solvents. It can be employed to nitrate a variety of different organic compounds at high yield when diluted with dichloromethane. Nitrotoluene, methyl benzoate, and benzoic acid are nitrated by sustained exposure over a number of days. Benzonitrile does not react. Benzene, toluene, tert-butylbenzene, halo-benzenes, ortho-nitrotoluene, anisole, phenol, and acetanilide are all quickly nitrated in less than 30 minutes at 25°C. Hexammino-vanadium nitrate $V(NH_3)_6(NO_3)_3$ was stated to be produced by reacting hexamino vanadium trichloride with nitric acid. Yet, there is doubt about the existence of hexammino complexes of vanadium. Vanadyl(IV) nitrate $VO(NO_3)_2$ is prepared from vanadyl sulfate and barium nitrate, or vanadyl chloride and silver nitrate. It forms a blue solution, but cannot be crystallized, in its place forming vanadium pentoxide when evaporated. Another procedure to prepare it as green blue needles is via a reaction between vanadium pentoxide, oxalic acid, and dilute nitric acid at 90°C, but the solid form is not confirmed. Vanadium metal reacts with dinitrogen tetroxide with an acetonitrile catalyst at 0°C to form the brick red solid mononitratodioxovanadium(V), VO_2NO_3. Mononitratodioxovanadium readily dissolves in water to form an orange solution, but this is unstable converting to a deep brown gel after a day. VO_2NO_3 cannot be crystallized from the solution, in its place vanadium pentoxide is formed when the solution is evaporated. Additionally, if VO_2NO_3 is heated at 350°C, vanadium pentoxide solid is left behind.

5.5.3.5 Vanadates

In chemistry, a vanadate is a compound with an oxoanion of vanadium usually in its highest oxidation state of +5. The simplest vanadate ion is the tetrahedral, orthovanadate, VO_4^{3-} anion, which is present in, for example, sodium orthovanadate and in solutions of V_2O_5 in strong base (pH > 13). Conventionally this ion is characterized with a single double bond; nevertheless, this is a resonance form as the ion is a regular tetrahedron with four equivalent oxygen atoms.

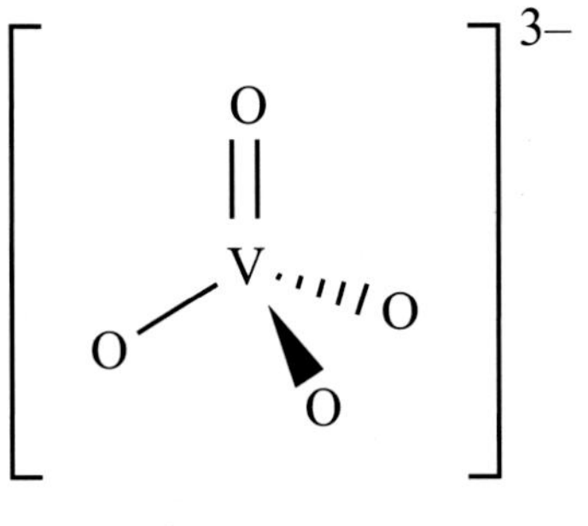

Furthermore, a variety of polyoxovanadate ions occur, which consist of discrete ions and "infinite" polymeric ions. There are also vanadates, such as rhodium vanadate, $RhVO_4$, which has a statistical rutile (TiO_2) structure where the Rh^{3+} and V^{5+} ions randomly occupy the Ti^{4+} positions in the rutile lattice, that do not contain a lattice of cations and balancing vanadate anions but are mixed oxides. In chemical classification when vanadate forms part of the name, it signifies that the compound has an anion in its structure with a central vanadium atom. Some examples of discrete ions are VO_4^{3-} ("orthovanadate," tetrahedral), $V_2O_7^{4-}$ ("pyrovanadate," corner-shared VO_4 tetrahedra, comparable to the dichromate ion), $V_3O_9^{3-}$ (cyclic with corner-shared VO_4 tetrahedra), $V_4O_{12}^{4-}$ (cyclic with corner-shared VO_4 tetrahedra), $V_5O_{14}^{3-}$ (corner-shared VO_4 tetrahedra), $V_6O_{18}^{6-}$ (ring), $V_{10}O_{28}^{6-}$ ("decavanadate," edge- and corner-shared VO_6 octahedra), $V_{12}O_{32}^{4-}$, $V_{13}O_{34}^{3-}$ (fused VO_6 octahedra), and $V_{18}O_{42}^{12-}$. Some examples of polymeric "infinite" ions are $[VO_3]_n^{n-}$ in, for example, $NaVO_3$, sodium metavanadate, and $[V_3O_8]_n^{n-}$ in CaV_6O_{16}. In these ions, vanadium has tetrahedral, square pyramidal, and octahedral coordination. In this respect, vanadium is parallel to tungstate and molybdate, whereas chromium though has a more limited range of ions. Dissolution of vanadium pentoxide in strongly basic aqueous solution results in the formation of the colorless VO_4^{3-} ion. On acidification, this solution's color slowly darkens through orange to red at around pH 7. Brown hydrated V_2O_5 precipitates around pH 2, redissolving to produce a light yellow solution containing the $[VO_2(H_2O)_4]^+$ ion. The number and character of the oxyanions that occur between pH 13 and 2 depend on pH along with concentration. For instance, the protonation of vanadate initiates a series of condensations to form polyoxovanadate ions.

$$\text{pH } 9-12: HVO_4^{2-},\ V_2O_7^{4-}$$
$$\text{pH } 4-9: H_2VO_4^-,\ V_4O_{12}^{4-},\ HV_{10}O_{28}^{5-}$$
$$\text{pH } 2-4: H_3VO_4,\ H_2V_{10}O_{28}^{4-}$$

5.5.3.6 *Organovanadium compounds*

Organovanadium compounds find only minor use as reagents in organic synthesis but are significant for polymer chemistry as catalysts. They are classed according to the type of ligand attached to vanadium such as carbonyls, cyclopentadienyls, alkyls and aryls, and arenes. Oxidation states for organovanadium compounds are mostly +2, +3, +4, and +5. Low valency vanadium is usually stabilized with carbonyl ligands.

The vanadium carbonyls, $V(CO)_6$, $V_2(CO)_{12}$, and $V(CO)_n$ (n = 1–5), can be directly synthesized by condensation of vanadium vapor with CO in a matrix of noble gases. They can also be prepared by reductive carbonylation of vanadium salts. For example, vanadium(III) chloride can be treated with carbon monoxide in the presence of a reducing agent such as Na. The resulting product can be oxidized to the 17e binary carbonyl $V(CO)_6$.

$$4Na + VCl_3 + 6CO \rightarrow Na\left[V(CO)_6\right] + 3NaCl$$

Vanadium forms the simple "sandwich" compound, vanadocene $[V(\eta^5\text{-}C_5H_5)_2]$, with the cyclopentadienyl ligand. Vanadocene is the lightest transition metal sandwich compound that is isolable at room temperature. It is a dark violet, paramagnetic (three unpaired electrons), and extremely air-sensitive compound. It reacts with dithioacetic acid to produce the dark brown tetramer $[V_4(\eta^5\text{-}C_5H_5)_4(\mu^3\text{-}S_4)$. It undergoes other oxidative addition reactions to provide compounds such as $[V(\eta^5\text{-}C_5H_5)_2Cl_n]$ (n = 1,2,3) and $[V(\eta^5\text{-}C_5H_5)_2R_2]$. The vanadocene dichloride can also be prepared from sodium cyclopentadienyl and vanadium tetrachloride.

$$2NaC_5H_5 + VCl_4 \rightarrow VCp_2Cl_2 + 2NaCl$$

Reduction of VCp_2Cl_2 gives the parent vanadocene $(Cp)_2V$.

$$VCp_2Cl_2 + LiAlH_4 \rightarrow V(Cp)_2$$

Monocyclopentadienyl vanadium compounds also exist such as tetravalent vanadium in $CpVCl_3$ and diamagnetic pentavalent vanadium in $CpV(O)Cl_2$.

Aside from vanadocene, vanadium also forms a variety of arene complexes including bis(benzene) vanadium complex, $V(\eta^6\text{-}C_6H_6)_2$. This can be prepared by the Friedel–Crafts method. Heating VCl_4 with $AlCl_3$ and aluminum powder in boiling benzene yields the golden yellow salt $[V(\eta^6\text{-}C_6H_6)_2][AlCl_4]$, which is then hydrolyzed to the neutral product.

$$VCl_4 + AlCl_3 + C_6H_6 \rightarrow \left[V\left(\eta^6-C_6H_6\right)_2\right][AlH_4]$$
$$\left[V\left(\eta^6-C_6H_6\right)_2\right][AlH_4] + H_2O \rightarrow V\left(\eta^6-C_6H_6\right)_2$$

Vanadium alkyls and aryls not supported by cyclopentadienyl ligands are rare. Only a handful exist and most of these involve bulky substituents such as the vanadate complexes $[VR_4]^-$ where R = 2,4,6-trimethylphenyl (mes), 2-methylphenyl, 2,6-dimethoxyphenyl), pentachlorophenyl, and 2,4,6-trichlorophenyl (tcp). The trialkyl vanadium compounds, $[VR_3]$, where R = mes or $CH(SiMe_3)_2$, are also known.

5.5.4 Major uses

Around 85% of total V produced is used as ferrovanadium containing 40%–50% vanadium or as a steel additive. The significant increase in strength of steel comprising small quantities of V was discovered at the beginning of the 20th century. V forms stable nitrides and carbides, causing a substantial increase in the strength of steel. Since then, V steel has been used for a variety of different applications, such as in axles, bicycle frames, crankshafts, gears, and other critical components. There exist two groups of V steel alloys. V high-carbon steel alloys contain between 0.15% and 0.25% V, while high-speed tool steels (HSS) have a much higher V concentration between 1% and 5%. HSS steel is used in surgical instruments and tools. Powder metallurgic alloys can contain up to 18% V. The high concentration of V carbides in these alloys improves wear resistance substantially. One particular use for these alloys is tools and knives. V stabilizes the beta form of Ti and increases the strength and temperature stability of Ti. Combined with Al in Ti alloys, it is used in jet engines, high-speed airframes, and dental implants. The most widespread alloy for seamless tubing is titanium 3/2.5 with 2.5% V. It is the main Ti alloy used in the aerospace, defense, and bicycle industries. A different common alloy, mainly produced in sheets, is titanium 6AL-4V, a Ti alloy containing 6% Al and 4% V. Several V alloys exhibit superconducting behavior. The first A15 phase superconductor was a V compound, V_3Si, which was discovered in 1952. V-Ga tape is used in superconducting magnets (17.5 teslas or 175,000 gauss). The structure of the superconducting A15 phase of V_3Ga is like that of the more common Nb_3Sn and Nb_3Ti. It has been suggested that a trace amount of 40 to 270 ppm V in Wootz steel and Damascus steel substantially improved the strength of the product, although the origin of V is uncertain.

V compounds are widely used as catalysts, for example, the most common oxide of vanadium, vanadium pentoxide V_2O_5, is used as a catalyst in producing H_2SO_4 by the contact process (Box 5.8) and as an oxidizer in maleic anhydride (preferred IUPAC name furan-2,5-dion, $C_2H_2(CO)_2O$) manufacture. V_2O_5 is utilized in ceramics to produce a golden color and is added to glass to produce a green or blue tint. V is a critical component of mixed metal oxide catalysts for the oxidation of propane (C_3H_8) and propylene (IUPAC name propene, C_3H_6) to acrolein (UPAC name prop-2-enal, C_3H_4O), acrylic acid (preferred IUPAC name prop-2-enoic acid, CH_2-CHCOOH) or the ammoxidation of propylene to acrylonitrile (preferred IUPAC name prop-2-enenitrile, CH_2CHCN). In use, the oxidation state of V changes dynamically and reversibly with the oxygen and the steam content of the reacting feed mixture.

A different V oxide, vanadium dioxide VO_2, is utilized in the manufacturing of glass coatings, which blocks infrared radiation (and not visible light) at a specific temperature. V oxide can be used to induce color centers in corundum (Al_2O_3) to create simulated alexandrite jewelery, although alexandrite in nature is a variety of the mineral chrysoberyl ($BeAl_2O_4$). The V redox battery, a type of flow battery, is an electrochemical cell comprising aqueous V ions in

BOX 5.8 Contact process for H_2SO_4 production

The contact process is the current method of making sulfuric acid in the high concentrations needed for industrial processes. Pt was the catalyst for this reaction; but, since it is prone to react with As impurities in the sulfur feedstock, V_2O_5 is now preferred. This process was patented in 1831 by British vinegar merchant, chemical manufacturer and inventor Peregrine Phillips (1800–1888). Besides being a much more efficient process for producing concentrated H_2SO_4 than the earlier lead chamber process, the contact process also produces sulfur trioxide and oleum. The process can be divided into five steps: (1) combining of sulfur and oxygen (O_2) to form SO_2, (2) purifying SO_2 in a purification unit, (3) adding an excess of oxygen to SO_2 in the presence of the catalyst V_2O_5 at 450°C and 1–2 atm., (4) adding the SO_3 formed to H_2SO_4, which gives rise to oleum (disulfuric acid, $H_2S_2O_7$), and (5) adding $H_2S_2O_7$ to water to form highly concentrated H_2SO_4. Purification of the air and SO_2 is essential to avoid catalyst poisoning (i.e., removing catalytic activities). The gas is then washed with water and dried with H_2SO_4. To conserve energy, the mixture is heated by exhaust gases from the catalytic converter by heat exchangers. SO_2 and O_2 then react as follows:

$$2SO_2(g) + O_2(g) \rightleftharpoons 2SO_3(g) \quad \Delta H = -197\ \text{kJ} \cdot \text{mol}^{-1}$$

Based on Le Chatelier's principle, a lower temperature should be used to shift the chemical equilibrium toward the right, thus increasing the yield. However, too low of a temperature will decrease the formation rate to an inefficient level. Therefore to increase the reaction rate, high temperatures (450°C), medium pressures (1–2 atm), and V_2O_5 are used to guarantee an adequate (>95%) conversion. The catalyst only serves to increase the rate of reaction as it does not alter the position of the thermodynamic equilibrium. The mechanism for the action of the catalyst consists of two steps:

(1) Oxidation of SO_2 into SO_3 by V^{5+}

$$2SO_2 + 4V^{5+} + 2O^{2-} \rightarrow 2SO_3 + 4V^{4+}$$

(2) Oxidation of V^{4+} back into V^{5+} by O_2 (catalyst regeneration)

$$4V^{4+} + O_2 \rightarrow 4V^{5+} + 2O^{2-}$$

Hot sulfur trioxide passes through the heat exchanger and is dissolved in concentrated H_2SO_4 in the absorption tower to form oleum.

$$H_2SO_4(l) + SO_3(g) \rightarrow H_2S_2O_7(l)$$

Note that directly dissolving SO_3 in water is impractical due to the highly exothermic nature of the reaction. Acidic vapor or mists are formed instead of a liquid. Oleum is reacted with water to form concentrated H_2SO_4.

$$H_2S_2O_7(l) + H_2O(l) \rightarrow 2H_2SO_4(l)$$

different oxidation states. This type of batteries was first suggested in the 1930s and developed commercially since the 1980s. Cells use 5 + and 2 + formal oxidization state ions. V redox batteries are applied commercially for grid energy storage. Vanadate can be used for protecting steel against rust and corrosion through conversion coating (Conversion coatings are coatings for metals where the part's surface is subjected to a chemical or electrochemical process by the coating material, which converts it into a decorative or protective substance. They are used for corrosion protection, to add decorative color and as paint primers). V foil is used in cladding Ti to steel because it is compatible with both Fe and Ti (bonding together of dissimilar metals. It is different from fusion welding or gluing as a method to fasten the metals together. Cladding is often achieved by extruding two metals through a die as well as pressing or rolling sheets together under high pressure. The moderate thermal neutron-capture cross-section and the short half-life of the V isotopes formed by neutron capture make V a suitable material for the inner structure of a fusion reactor.

5.6 24 Cr — Chromium

5.6.1 Discovery

Weapons found in burial pits from the late 3rd century BCE Qin Dynasty containing the Terracotta Warriors near Xi'an, China, have been chemically tested by archeologists. While these weapons were probably buried more than 2000 years ago, the ancient bronze tips of both the swords and crossbow bolts found in the pits exhibited surprisingly little corrosion, perhaps because the bronze was purposely coated with a thin layer of chromium oxide. Nevertheless, this oxide layer on the weapons did not consist of pure chromium metal or chrome plating as it is normally produced today, but a very thin 10−15 μm layer of chromium oxide molecules with up to 2% chromium was observed, which was sufficient to protect the bronze from corroding. Chromium minerals as pigments became known in the west in the 18th century. On July 26, 1761, German geologist and mineralogist Johann Gottlob Lehmann (August 4, 1719 to January 22, 1767) found an orange-red mineral in the Beryozovskoye mines in the Ural Mountains that he called Siberian red lead (Lehmanni, 1766) (Fig. 5.46). However, it was misidentified as a lead compound with selenium and iron components. The mineral in question was actually crocoite (or lead(II) chromate) with the chemical formula $PbCrO_4$. In 1770 Prussian zoologist and botanist Peter Simon Pallas (September 22, 1741 to September 8, 1811) (Fig. 5.47) visited the same locality as Lehmann and found a red lead mineral that was revealed to have useful properties as a pigment in paints. After Pallas, the use of Siberian red lead as a paint pigment rapidly advanced all over the region. In 1794 French pharmacist and chemist Louis Nicolas Vauquelin (May 16, 1763 to November 14, 1829) received some samples of this crocoite ore. He managed to create chromium trioxide (CrO_3) by reacting crocoite with hydrochloric acid. In 1797 he subsequently found that he could isolate metallic chromium by heating the oxide in a charcoal oven (Vauquelin, 1798). Because of this, he is generally credited as the person who truly discovered the element chromium. Vauquelin also managed to detect traces of chromium in precious gemstones, such as ruby (red variety of corundum) and emerald (green variety of beryl). Throughout the 19th century, chromium was mainly utilized not only as a pigment in paints but in tanning salts as well. For quite some time, the crocoite from Russia was the major source for such

FIGURE 5.46 Johann Gottlob Lehmann, January 2, 1761, unknown artist.

FIGURE 5.47 Peter Simon Pallas, a silhouette by A. Tardier, 18th century.

BOX 5.9 Chromite $FeCr_2O_4$

In every primary chrome ore district in the world, the chromite occurs in ultramafic rocks or rocks derived from them by alteration. Such rocks are peridotites, pyroxenites, dunites, serpentinites, and talc schists. Chromite ore deposits are classified into two major types based on the structure of the ore bodies and the regional pattern of ore distribution: stratiform and podiform deposits. Stratiform deposits occur in parallel layers with great lateral continuity. They are remarkably uniform and consist of high iron chrome ore. The Bushveld igneous complex (Transvaal, South Africa), the Great Dyke (Zimbabwe), and the Stillwater complex (Montana, United States) are examples of such stratiform deposits. Chromite, $FeCr_2O_4$, is one of the first minerals to crystallize and separate from a cooling magma, and large chromite deposits are thought to have been derived by such magmatic differentiation. Podiform deposits occur as pods, lenses, sack forms, slabs, and other irregular forms. The ore bodies are not continuous and follow no recognizable systematic distribution within the containing country rock. Examples are deposits in the Ural Mountains, Albania, Zimbabwe, and the Philippines. The name chromite refers to the mineral's composition. It is isometric (cubic) *4/m2/m* (space group *F4/d2/m*). Its hardness is 5½. The luster is metallic to submetallic, transparency is subtranslucent to opaque. It has no cleavage, while the fracture is subconchoidal. The color is brownish black to black, while the streak is brown to dark brown. Crystals are octahedral, but they are small and rare. They are more commonly massive, granular to compact.

tanning salts. In 1827 a larger chromite ore deposit (Box 5.9) was found near Baltimore, United States, which rapidly met the demand for tanning salts much more effectively than the crocoite that had been used before. This made the United States the principal manufacturer of chromium products till 1848, when larger chromite ore deposits were discovered near the city of Bursa, Turkey. Chromium metal is renowned for its reflective, metallic luster when polished. It is used as a protective and decorative coating on car parts, plumbing fixtures, furniture parts, and many other items, usually produced by electroplating. Chromium was utilized for electroplating as early as 1848, but this practice only became widespread with the development of an improved process in 1924.

5.6.2 Mining, production, and major minerals

Chromium is the 13th most abundant element in the Earth's crust with an average concentration of 100 ppm. Chromium compounds can be found in the environment from the erosion of chromium-containing rocks and can be redistributed by volcanic eruptions. About 40% of the chromite ores and concentrates in the world are coming from

South Africa, about 33% from Kazakhstan, while India, Russia, and Turkey are also considerable producers. Unexploited chromite ore deposits are abundant, but geographically concentrated in Kazakhstan and southern Africa.

The two main products of chromium ore refining are ferrochromium and metallic chromium. For these products, the ore smelter processes are quite different. To produce ferrochromium, the chromite ore ($FeCr_2O_4$) is reduced in large-scale electric arc furnaces or in smaller smelters with either aluminum or silicon in an aluminothermic reaction. For the manufacture of pure chromium, iron must be separated from chromium in a two-step roasting and leaching process. The chromite ore is heated with a mixture of calcium carbonate and sodium carbonate in the presence of air. Under these conditions, chromium is oxidized to the hexavalent form (Cr^{6+}), while the iron forms the stable Fe_2O_3. The following leaching step at higher elevated temperatures dissolves the chromates leaving the insoluble iron oxide behind. The chromate is then converted by sulfuric acid into the dichromate.

$$4FeCr_2O_4 + 8Na_2CO_3 + 7O_2 \rightarrow 8Na_2CrO_4 + 2Fe_2O_3 + 8CO_2$$
$$2Na_2CrO_4 + H_2SO_4 \rightarrow Na_2Cr_2O_7 + Na_2SO_4 + H_2O$$

The dichromate is subsequently converted to the Cr(III) oxide by reduction with carbon and then reduced in an aluminothermic reaction to chromium.

$$Na_2Cr_2O_7 + 2C \rightarrow Cr_2O_3 + Na_2CO_3 + CO$$
$$Cr_2O_3 + 2Al \rightarrow Al_2O_3 + 2Cr$$

Chromium is found in around 100 minerals. Within the elements class, it occurs as the native element (Cr), as a number of carbides, for example, tongbaite (Cr_3C_2), a nitride carlsbergite (CrN), a phosphide andreyivanovite (FeCrP), as well as two alloys with Fe, chromferide (Fe_3Cr_{1-x} ($x = 0.6$)) and ferchromide (Cr_3Fe_{1-x}). Nine sulfides are known to contain Cr, such as cronusite ($Ca_{0.2}CrS_2{\cdot}2H_2O$), daubréelite ($Fe^{2+}Cr_2^{3+}S_4$), and kalininite ($ZnCr_2S_4$). One halide is known to have Cr in its structure, yedlinite ($Pb_6Cr^{3+}Cl_6(O,OH,H_2O)_8$). The oxides are represented by 30 different minerals, such as chromite ($Fe^{2+}Cr_2^{3+}O_4$) (Box 5.9), magnesiochromite ($MgCr_2O_4$), and woodallite ($Mg_6Cr_2(OH)_{16}Cl_2{\cdot}4H_2O$). Only three carbonate minerals contain Cr, petterdite ($PbCr_2^{3+}(CO_3)_2(OH)_4{\cdot}H_2O$), putnisite ($SrCa_4Cr_8^{3+}(CO_3)_8SO_4(OH)_{16}{\cdot}23H_2O$) (Fig. 5.48), and stichtite ($Mg_6Cr_2^{3+}(OH)_{16}[CO_3]4H_2O$) (Fig. 5.49). Similarly, two borates are known with Cr, iquiqueite ($K_3Na_4Mg(CrO_4)B_{24}O_{39}(OH){\cdot}12H_2O$) and warwickite ($(Mg,Ti,Fe,Cr,Al)_2O(BO_3)$). The sulfate class includes the chromates and is represented by 20 minerals, including crocoite ($PbCrO_4$) (Fig. 5.50) (Box 5.10), edoylerite ($Hg_3^{2+}(CrO_4)S_2$) (Fig. 5.51), embreyite ($Pb_5(CrO_4)_2(PO_4)_2{\cdot}H_2O$), fornacite ($Pb_2Cu(CrO_4)(AsO_4)(OH)$), hemihedrite ($Pb_{10}Zn(CrO_4)_6(SiO_4)_2(OH)_2$), and vauquelinite ($Pb_2Cu(CrO_4)(PO_4)(OH)$). In the silicate class, 16 minerals are known to have Cr in their crystal structure, for example, kosmochlor ($NaCrSi_2O_6$) and uvarovite ($Ca_3Cr_2(SiO_4)_3$) (Fig 5.52).

FIGURE 5.48 Putnisite, $SrCa_4Cr_8^{3+}(CO_3)_8SO_4(OH)_{16}{\cdot}23H_2O$, an extremely rare strontium-chromium carbonate found only in Australia, tiny rectangular purple crystal to about 0.75 mm and many other crystal "grains" throughout matrix. Armstrong Mine, Widgiemooltha, Western Australia, Australia.

FIGURE 5.49 Stichtite, $Mg_6Cr_2^{3+}(OH)_{16}[CO_3]4H_2O$, compact purple mass 2.5 cm across. Tunnel Hill quarry, Serpentine Hill, Zeehan district, Tasmania, Australia.

FIGURE 5.50 Crocoite, $PbCrO_4$, long, needle-like crystals up to 8 mm. Adelaide mine, Dundas, Tasmania, Australia.

5.6.3 Chemistry

Naturally occurring chromium consists of three stable isotopes: ^{52}Cr, ^{53}Cr, and ^{54}Cr, with ^{52}Cr being the most common (83.789% natural abundance, followed by ^{53}Cr at 9.501%, ^{50}Cr at 4.345%, and ^{54}Cr at 2.365%). Nineteen radioisotopes have been observed, with the most stable ^{50}Cr having a half-life of (more than) 1.8×10^{17} years and ^{51}Cr having a half-life of 27.7 days. All the other radioisotopes have half-lives less than 24 hours and most less than 1 minute. In addition, Cr has two metastable nuclear isomers. ^{53}Cr is formed as the radiogenic decay product of ^{53}Mn (half-life of 3.74 million years). Cr isotopes are generally collocated (and compounded) with Mn isotopes. This condition is valuable in isotope geology. Mn–Cr isotope ratios support the evidence from ^{26}Al and ^{107}Pd about the solar system's early history. Variations in $^{53}Cr/^{52}Cr$ and Mn/Cr ratios from several meteorites point to an initial $^{53}Mn/^{55}Mn$ ratio that suggests the

BOX 5.10 Crocoite, $PbCrO_4$

The most colorful and to the imagination speaking chrome mineral is without a doubt crocoite, in particular the specimens coming from Dundas, Tasmania, Australia (Fig. 5.50). The name crocoite comes from the Greek word *krokos*, meaning "saffron," referring to its color. It is a rare mineral found in the oxidized zone of lead deposits in regions where lead veins have traversed rocks containing chromite. Crocoite is monoclinic, *2/m* (space group $P2_1/n$). It has a hardness of 2½–3. It has an imperfect cleavage on {110} and poor cleavage on {001}, and the fracture is subconchoidal. The luster is adamantine, and transparency is translucent. The color is bright hyacinth-red to orange-red, and the streak is orange-yellow. Optically it is biaxial(+). Crystals are commonly slender prismatic, vertically striated parallel to prism faces. Crocoite also occurs as columnar and granular aggregates. It is not abundant enough to be of commercial value as an ore, but of historic interest because the element chromium was first discovered in crocoite. It is also of interest to the mineral collector.

FIGURE 5.51 Edoylerite, $Hg_3^{2+}(CrO_4)S_2$, yellow, translucent, stubby crystals in aggregate in an area of 2 × 1 mm. Clear Creek mine, San Benito Co., California, United States.

FIGURE 5.52 Emerald green uvarovite, $Ca_3Cr_2(SiO_4)_3$, garnet. The single crystal measures 5 mm. Outokumpu, Finland.

Mn–Cr isotopic composition must be a consequence of the in situ decay of ^{53}Mn in differentiated planetary bodies. Therefore ^{53}Cr provides supplementary evidence for nucleosynthetic processes directly before coalescence of the solar system. The principal decay mode for isotopes lighter than ^{52}Cr is electron capture and the principal mode for the heavier ones is β decay.

Chromium is a steely gray, lustrous, hard and brittle transition metal highly valued for its corrosion resistance and polished esthetics. It has 24 electrons arranged in an electronic configuration of $[Ar]3d^5 4s^1$. It is the first element in the periodic table whose ground state electron configuration is an exception to the Aufbau principle (Table 5.7). The quantum mechanical explanation for this observed electron configuration of Cr is beyond the scope of this book. It is sufficient to know that many more exceptions to the Aufbau principle occur in *d* transition and *f* inner transition elements and are a consequence of the tendency of electrons to arrange in a manner that will minimize Coulombic repulsion energy and maximize exchange energy. All the six electrons in 3d and 4s subshells can be involved in reactions, and chromium can have oxidation states of up to +6. The stability of the oxidation state +6 can be noted in the fact that chromium(VI) tends to form poly oxoanions. However, the diversity of these is much smaller compared to that of the related polymolybdates and polytungstates. The oxidation states +5 and +4 are represented but mostly unstable while +3 is its most stable oxidation state. Chromium(II) is strongly reducing (Cr^{3+}/Cr^{2+}, E° −0.41 V) but features in an extensive cationic chemistry.

At ambient temperatures, Cr is resistant to atmospheric attack. Because of this reason Cr is widely used to protect other more reactive metals. Chromium becomes susceptible to attack only at high temperatures when it reacts with many nonmetals to often give interstitial and nonstoichiometric products. For example, chromium(III) nitride can be prepared by direct combination of chromium and nitrogen at 800°C, and chromium hydride (CrH_x; $x = 1-2$) has also been produced by exposing chromium films to hydrogen under high pressure and temperature. The chromium atom is too small to allow the ready insertion of carbon into its lattice, and its carbide is consequently more reactive than those of its predecessors.

$$2Cr + N_2 \rightarrow 2CrN$$

Chromium reacts more readily with acids though its reactivity depends on its purity and it can easily be rendered passive. It dissolves readily in dilute HCl. If it is very pure, it will often resist dilute H_2SO_4. Dilute or concentrated HNO_3 as well as aqua regia will render it passive, but the reasons for this are not well understood. In the presence of oxidizing agents such as KNO_3 or $KClO_3$, alkali melts rapidly attack chromium producing CrO_4^{2-}.

5.6.3.1 Oxides and oxoanions

Several oxides of chromium are known. Chromium(VI) oxide, CrO_3, is a strongly acidic and rather covalent oxide with a melting point of only 197°C. Its deep red crystals are made up of chains of corner-shared CrO_4 tetrahedra. It is

TABLE 5.7 Chromium properties.

Appearance	Silvery metallic
Standard atomic weight $A_{r,std}$	51.9962
Block	d-Block
Element category	Transition metal
Electron configuration	$[Ar]\ 3d^5\ 4s^1$
Phase at STP	Solid
Melting point	1907°C
Boiling point	2671°C
Density (near r.t.)	7.19 g/cm^3
When liquid (at m.p.)	6.3 g/cm^3
Heat of fusion	21.0 kJ/mol
Heat of vaporization	347 kJ/mol
Molar heat capacity	23.35 J/(mol·K)
Oxidation states	−4, −2, −1, +1, +2, **+3**, +4, +5, **+6**
Ionization energies	1st: 652.9 kJ/mol 2nd: 1590.6 kJ/mol 3rd: 2987 kJ/mol
Atomic radius	Empirical: 128 pm
Covalent radius	139 ± 5 pm

STP, Standard temperature and pressure.
Bold font indicates main oxidation state.

commonly called "chromic acid" and is generally prepared by the addition of concentrated H_2SO_4 to a saturated aqueous solution of a dichromate.

$$K_2Cr_2O_7(aq) + H_2SO_4(conc.) \rightarrow CrO_3 + K_2SO_4 + H_2O$$

CrO_3 is a strong oxidizing agent, a property widely used in organic chemistry. It melts with some decomposition and, if heated above 220°C–250°C, loses oxygen to give a succession of lower oxides, Cr_3O_8, Cr_2O_5, Cr_5O_{12} until the green chromium(III) oxide, Cr_2O_3, is formed. Cr_2O_3 is the main oxide of chromium and is found in nature in the form of a rare mineral, eskolaite. It is amphoteric, and while it is insoluble in water, it will dissolve in acid. It is used as a dark green pigment. It also has semiconductor properties and is antiferromagnetic below 35°C.

Cr_2O_3 is the most stable oxide of chromium and is the final product of combustion of the metal. However, it is more conveniently obtained by heating ammonium dichromate.

$$(NH_4)_2Cr_2O_7 \rightarrow Cr_2O_3 + N_2 + 4H_2O$$

An older process described the preparation of Cr_2O_3 from the mineral chromite, $(Fe,Mg)Cr_2O_4$. Chromite to Cr_2O_3 proceeds via $Na_2Cr_2O_7$, which is reduced with sulfur at high temperatures.

$$Na_2Cr_2O_7 + S \rightarrow Na_2SO_4 + Cr_2O_3$$

When produced by such dry methods, Cr_2O_3 is frequently unreactive but, if precipitated as hydrous oxide (or "hydroxide") from aqueous chromium(III) solutions, it is amphoteric. It dissolves readily in aqueous acids to give an extensive cationic chemistry based on the $[Cr(H_2O)_6]^{3+}$ ion and in alkalis to produce complicated, extensively hydrolyzed chromate(III) species ("chromites").

The third major oxide of chromium is the brown-black chromium(IV) oxide, CrO_2, which is an intermediate product in the decomposition of CrO_3 to Cr_2O_3. It has metallic conductivity, and its ferromagnetic properties lead to its commercial importance in the manufacture of magnetic recording tapes. Chromium (II) oxide, CrO, is a basic oxide and appears as an insoluble black powder. It can be prepared by the reduction of chromium(III) oxide by hypophosphites.

$$H_3PO_2 + 2Cr_2O_3 \rightarrow 4CrO + H_3PO_4$$

The mineral chromite can be heated with a mixture of calcium carbonate and sodium carbonate in the presence of air producing sodium chromate, Na_2CrO_4 and iron(III) oxide, Fe_2O_3. The sodium chromate formed in this process is the basis for the manufacture of all industrially important chromium chemicals.

$$4FeCr_2O_4 + 8Na_2CO_3 + 7O_2 \rightarrow 8Na_2CrO_4 + 2Fe_2O_3 + 8CO_2$$

Acidification of aqueous solutions of the yellow, tetrahedral chromate ion, CrO_4^{2-}, initiates a series of labile equilibria involving the formation of the orange-red dichromate ion, $Cr_2O_7^{2-}$.

$$\begin{gathered} HCrO_4^- \rightleftharpoons CrO_4^{2-} + H^+ \\ H_2CrO_4 \rightleftharpoons HCrO_4^- + H^+ \\ Cr_2O_7^{2-} + H_2O \rightleftharpoons 2\,HCrO_4^- \\ HCr_2O_7^- \rightleftharpoons Cr_2O_7^{2-} + H_2O \\ H_2CrO_4 \rightleftharpoons HCr_2O_7^- + H^+ \end{gathered}$$

The chromate ion is the predominant species in alkaline solutions and the dichromate can become the predominant ion in acidic solutions. Both ions are strong oxidizing agents. However, in basic solutions, the chromate is a weaker oxidizing agent.

$$\begin{aligned} Cr_2O_7^{2-} + 14H^+ + 6e^- &\rightarrow 2Cr^{3+} + 7H_2O \quad E^\circ = 1.33\text{ V} \\ CrO_4^{2-} + 4H_2O + 3e^- &\rightarrow Cr(OH)_3 + 5OH^- \quad E^\circ = -0.13\text{ V} \end{aligned}$$

5.6.3.2 *Hydrides, carbides, nitrides, and borides*

Chromium(II) hydride, systematic name chromium dihydride and poly(dihydridochromium), is a pale brown solid inorganic compound $(CrH_2)_n$ (sometimes given as $([CrH_2])_n$ or CrH_2). While it is thermodynamically unstable toward decomposition at ambient temperatures, it is kinetically metastable. Chromium(II) hydride is the second simplest polymeric chromium hydride (after chromium(I) hydride). In metallurgical chemistry, chromium(II) hydride is essential to specific forms of chromium-hydrogen alloys. The systematic name chromium dihydride, a valid IUPAC name, is constructed corresponding to the compositional nomenclature. Yet, as the name is compositional in nature, it does not differentiate between compounds of the same stoichiometry, for example, molecular species, which show distinct chemical properties. The systematic names poly(dihydridochromium) and poly[chromane(2)], also valid IUPAC names, are constructed corresponding to the additive and electron-deficient substitutive nomenclatures, respectively. They do differentiate the titular compound from the others. Dihydridochromium, also systematically known as chromane(2), is a related compound with the chemical formula CrH_2 (sometimes given as $[CrH_2]$). It is both thermodynamically and kinetically unstable at ambient temperature with the added tendency to autopolymerize and hence it cannot be concentrated. Dihydridochromium is the second simplest molecular chromium hydride (after hydridochromium) and is also the precursor for clusters with the same stoichiometry. Furthermore, it may be thought of as the chromium(II) hydride monomer. Molecular chromium(II) hydrides with the formulae CrH_2 and Cr_2H_4 have been isolated in solid gas matrices. The molecular hydrides are extremely unstable toward thermal decomposition. CrH_2 is the major principal product in the reaction of laser-ablated chromium with molecular hydrogen. Dihydridochromium is the most hydrogenated, ground state classical molecular hydride of chromium. In the presence of pure hydrogen, dihydridochromium easily transforms to bis(dihydrogen)dihydridochromium, $CrH_2(H_2)_2$, in an exothermic reaction. Dichromane(4) is prepared by hydrogenation. Here, chromium and hydrogen react according to the following reaction:

$$2Cr + 2H_2 \rightarrow HCr(\mu\text{-}H)_2CrH$$

This route comprises atomic chromium as an intermediate and occurs in two steps. The hydrogenation is a spontaneous reaction.

$$Cr(s) \rightarrow Cr(g)$$
$$2Cr(g) + 2H_2(g) \rightarrow HCr(\mu\text{-}H)_2CrH(g)$$

An electron pair of a Lewis base can join with the chromium center in dihydridochromium by adduction:

$$[CrH_2] + :L \rightarrow [CrH_2L]$$

Due to this capture of an adducted electron pair, dihydridochromium has Lewis acidic character. Dihydridochromium is a strong Lewis acid with the capability to capture at least five electron pairs from Lewis bases. It should be anticipated that chromium dihydride clusters and chromium(II) hydride have comparable acidic properties, even though reaction rates and equilibrium constants are different. In an inert gas matrix, atomic Cr reacts with H_2 to form the dihydride when it is irradiated with ultraviolet light between 320 and 380 nm. The reaction of chromium with molecular hydrogen is endothermic. 380 nm or greater wavelength radiation is necessary to form photochemically generated CrH_2. In diluted chromane(2), the molecules are known to oligomerize resulting in at least dichromane(4) (dimers), being linked by covalent bonds. CrH_2 is bent and is weakly repulsive to one hydrogen molecule, but attractive to two molecules of hydrogen. The bond angle is $118 \pm 5°$. The dimer has a distorted rhombus structure with C_{2h} symmetry.

Chromium(II) carbide is a ceramic compound that forms several different chemical compositions: Cr_3C_2, Cr_7C_3, and $Cr_{23}C_6$. At standard conditions it is as a gray solid. It is extremely hard and corrosion resistant. It is also a refractory compound, that is, it retains its strength at high temperatures as well. These characteristics make it useful as an additive to metal alloys. When chromium carbide crystals are integrated into the surface of a metal, they enhance the wear resistance and corrosion resistance of the metal and retain these properties at elevated temperatures. The hardest and most widely used carbide for this application is Cr_3C_2. Related naturally occurring minerals include tongbaite and isovite, $(Cr,Fe)_{23}C_6$, both extremely rare. Another chromium-rich carbide mineral is yarlongite, $Cr_4Fe_4NiC_4$. Chromium carbide can be prepared via mechanical alloying. In this type of method, metallic chromium and pure carbon in the form of graphite are loaded into a ball mill and ground into a fine powder. After the components have been ground, they are pressed into a pellet and exposed to hot isostatic pressing. Hot isostatic pressing uses an inert gas, mainly argon, in a sealed oven. This pressurized gas applies pressure to the sample from all directions while the oven is heated. The heat and pressure make the graphite and metallic chromium react with each other and form chromium carbide. Lowering the percentage of carbon content in the initial mixture causes an increase in the yield of Cr_7C_3 and $Cr_{23}C_6$. Another technique for the preparation of chromium carbide uses chromium oxide, pure aluminum, and graphite in a self-propagating exothermic reaction.

$$3Cr_2O_3 + 6Al + 4C \rightarrow 2Cr_3C_2 + 3Al_2O_3$$

In this method the reactants are ground and mixed in a ball mill. The mixed powder is subsequently pressed into a pellet and placed under an inert atmosphere of argon. The sample is then heated. A heated wire, a spark, a laser, or an oven may provide the heat. The exothermic reaction is started, and the resulting heat propagates the reaction throughout the rest of the material. There are three different crystal structures for chromium carbide corresponding to the three different chemical compositions. $Cr_{23}C_6$ adopts a cubic crystal structure, Cr_7C_3 crystallizes with a hexagonal structure, and Cr_3C_2, the most durable of the three compositions, adopts an orthorhombic crystal structure.

Chromium nitride, CrN, is very hard, and extremely resistant to corrosion. It is an interstitial compound, with N atoms occupying the octahedral holes in the Cr lattice. Therefore, it is not strictly a chromium(III) compound nor does it contain nitride ions (N^{3-}). Chromium exists as a second interstitial nitride, dichromium nitride, Cr_2N. Chromium(III) nitride can be formed by a direct reaction of chromium and nitrogen at 800°C:

$$2Cr + N_2 \rightarrow 2CrN$$

It can also be produced by PVD technique such as cathodic arc deposition. CrN is employed as a coating material for corrosion resistance and in metal forming and plastic molding applications. CrN is frequently utilized in medical implants and tools. In addition, CrN is an important component in advanced multicomponent coating systems, such as CrAlN, for hard, wear-resistant applications on cutting tools. The fundamental material physics of CrN, resulting in its favorable properties, has been discussed recently in high-profile scientific journals such as Nature Materials. Especially, the importance of magnetism in both the low-temperature and high-temperature phases has been shown through quantum mechanical calculations of the electronic structure of the compound. While rare, carlsbergite—the natural mineral form of chromium nitride—can be found in certain meteorites.

Chromium(III) boride, also known as chromium monoboride (CrB), is one of the six stable binary borides of chromium, which also comprises Cr_2B, Cr_5B_3, Cr_3B_4, CrB_2, and CrB_4. Similar to numerous other transition metal borides, it is extremely hard (21–23 GPa), has high strength (690 MPa bending strength), conducts heat and electricity comparable to many metallic alloys, and has a high melting point (~2100°C). In contrast to pure chromium, CrB is paramagnetic, with a magnetic susceptibility that is only weakly dependent on temperature. Because of these properties, together with others, CrB has been believed to be a potential candidate material for wear-resistant coatings and high-temperature diffusion barriers. It can be prepared as powders by numerous processes, for example, direct reaction of the constituent elemental powders, SHS, borothermic reduction, and molten salt growth. Slow cooling of molten, aluminum-saturated solutions from high temperatures has been shown to produce large single crystals, with a maximum size of $0.6 \times 0.6 \times 8.3$ mm. CrB crystallizes in the orthorhombic crystal structure (space group *Cmcm*). The crystal structure can be imagined as slabs face-sharing BCr_6 trigonal prisms, in the *ac*-plane, that are stacked parallel to the <010> crystallographic direction. Like Cr_3B_4 and Cr_2B_3, the B atoms in the structure form covalent bonds with each other and are characterized by unidirectional B–B–chains parallel to the <001> crystallographic direction. The transition metal monoborides VB, NbB, TaB, and NiB have a similar crystal structure.

5.6.3.3 Sulfides and chalcogenides

Chromium forms no trisulfide. Its compounds with sulfur are nonstoichiometric, most exhibit metallic character (or at least semiconducting properties), and also show a wide variety of magnetic behaviors encompassing diamagnetic, paramagnetic, antiferro-, ferri-, and ferromagnetic. Chromium(III) sulfide, Cr_2S_3, is formed by heating powdered Cr with sulfur, or by the action of $H_2S(g)$ on Cr_2O_3, $CrCl_3$, or Cr. It decomposes to CrS on being heated, via a number of intermediate phases, which approximate in composition to Cr_3S_4, Cr_5S_6, and Cr_7S_8. The structures are described as forming in this manner: if one-quarter of the Cr atoms are removed from alternate layers, Cr_7S_8 results; if one-third, Cr_5S_6 results; if two-thirds, Cr_4S_6 (i.e. Cr_2S_3) results; and if half, Cr_3S_4 results. These phases exhibit varying behavior; Cr_2S_3 and CrS are semiconductors, whereas Cr_7S_8, Cr_5S_6, and Cr_3S_4 are metallic, and all exhibit magnetic ordering. The corresponding selenides of chromium, CrSe, Cr_7Se_8, Cr_3Se_4, Cr_2Se_3, Cr_5Se_8, and Cr_7Se_{12} are broadly similar, as are the tellurides CrTe, Cr_7Te_8, Cr_5Te_6, Cr_3Te_4, Cr_2Te_3, Cr_5Te_8, and $CrTe_{\sim 2}$.

5.6.3.4 Halides

Chromium reacts with the halogens when heated to form chromium halides of varying oxidation states. The oxidation states of +6 and +5 are attained with fluorine only. The +4 oxidation state exists for all chromium halides, but chromium(IV) iodide has a doubtful or unstable existence. In the lower oxidation states, all the chromium halides are known. The yellow chromium(VI) fluoride, CrF_6, can be prepared by the direct action of the halogen on the metal but requires a temperature of 400°C and a pressure of 200–300 atm for its formation. Reduction of the pressure causes it to dissociate into the bright red, volatile chromium(V) fluoride, CrF_5, and F_2 even at temperatures as low as −100°C. The tetrahalides are better known than the hexa- and penta-halides, although chromium(IV) iodide, CrI_4, is of uncertain existence. The most stable representative is chromium(IV) fluoride, CrF_4. The tetrahalides, CrF_4, together with chromium(IV) bromide, $CrBr_4$, and chromium(IV) chloride, $CrCl_4$, are produced by treating the trihalides (CrX_3) with the corresponding halogen at elevated temperatures.

All four of the chromium trihalides are known and can be prepared by reacting the halogen and the metal, though CrF_3 is better obtained from HF and $CrCl_3$ at 500°C. The fluoride is green, the chloride red-violet, and the bromide and iodide dark green to black. Chromium(III) fluoride is the name for the inorganic compounds with the chemical formula CrF_3 together with a number of related hydrates. CrF_3 is a crystalline solid, which is insoluble in common solvents, but the colored hydrates $[Cr(H_2O)_6]F_3$ and $[Cr(H_2O)_6]F_3{\cdot}3H_2O$ are soluble in water. The trihydrate is green, and the hexahydrate is violet. The anhydrous form sublimes at 1100°C–1200°C. Chromium(III) fluoride is formed through the reaction of chromium(III) oxide with hydrofluoric acid.

$$Cr_2O_3 + 6HF + 9H_2O \rightarrow 2\left[Cr(H_2O)_6\right]F_3$$

As mentioned above, the anhydrous form can be prepared from hydrogen fluoride and chromic chloride.

$$CrCl_3 + 3HF \rightarrow CrF_3 + 3HCl$$

Another synthesis route for CrF_3 comprises the thermal decomposition of $(NH_3)CrF_6$.

$$(NH_4)_3CrF_6 \rightarrow CrF_3 + 3NH_3 + 3HF$$

A mixed valence Cr_2F_5 has also been found. Similar to nearly all compounds of chromium(III), these compounds have octahedral Cr centers. In the anhydrous form, the six coordination sites are occupied by F ligands that bridge to adjacent Cr centers. In the hydrates, a few or all the fluoride ligands are replaced by water. Chromium(III) fluoride has some applications as a mordant in textiles and as a corrosion inhibitor. Chromium(III) fluoride is known to catalyze the fluorination of chlorocarbons by HF.

The term chromium(III) chloride (also called chromic chloride) covers several compounds with the general formula $CrCl_3{\cdot}xH_2O$, where x can be 0, 5, and 6. The anhydrous $CrCl_3$ is a violet solid. The most common trichloride compound is the dark green hexahydrate, $CrCl_3{\cdot}6H_2O$. Chromium chlorides are used as catalysts and as precursors to dyes for wool. Anhydrous chromium(III) chloride can be produced via chlorination of chromium metal directly, or indirectly through carbothermic chlorination of chromium(III) oxide at 650°C–800°C.

$$Cr_2O_3 + 3C + 3Cl_2 \rightarrow 2CrCl_3 + 3CO$$

Dehydration with trimethylsilyl chloride in THF results in the formation of the solvate.

$$CrCl_3{\cdot}6H_2O + 12Me_3SiCl \rightarrow CrCl_3(THF)_3 + 6(Me_3Si)_2O + 12HCl$$

In addition, it can be formed by reacting the hexahydrate with thionyl chloride.

$$CrCl_3{\cdot}6H_2O + 6SOCl_2 \rightarrow CrCl_3 + 6SO_2 + 12HCl$$

The hydrated chlorides are produced by reaction of chromate with hydrochloric acid and methanol. In the laboratory, the hydrates are typically produced by dissolving the chromium metal or chromium(III) oxide in hydrochloric acid. Anhydrous chromium(III) chloride has the YCl_3 structure, with Cr^{3+} occupying one-third of the octahedral interstices in alternating layers of a pseudocubic close-packed lattice of Cl^- ions. The absence of cations in alternate layers results in weak bonding between adjacent layers. Consequently, crystals of $CrCl_3$ cleave easily along the planes between layers, which gives it a flaky (micaceous) appearance. Chromium(III) chlorides exhibit the slightly unusual property of existing in several distinct chemical forms (isomers), which differ with respect to the number of chloride anions that are coordinated to Cr(III) and the water of crystallization. The different forms exist both as solids and in aqueous solutions. A number of members are known of the series of $[CrCl_{3-n}(H_2O)_n]^{z+}$. The main hexahydrate can be more precisely written as $[CrCl_2(H_2O)_4]Cl{\cdot}2H_2O$. It comprises the cation *trans*-$[CrCl_2(H_2O)_4]^+$ and additional molecules of water and a chloride anion in the lattice. Two other hydrates have been observed, pale green $[CrCl(H_2O)_5]Cl_2{\cdot}H_2O$ and violet $[Cr(H_2O)_6]Cl_3$. Comparable behavior has been found for the other chromium(III) compounds. Slow reaction rates are common with chromium(III) complexes. The low reactivity of the d^3 Cr^{3+} ion can be described based on crystal field theory. One way of opening $CrCl_3$ up to substitution in solution is to reduce even a trace amount to $CrCl_2$, for example, using zinc in hydrochloric acid. This chromium(II) compound readily undergoes substitution and can exchange electrons with $CrCl_3$ via a chloride bridge, so all of the $CrCl_3$ can react rapidly. In the presence of some chromium(II) solid $CrCl_3$ dissolves quickly in water. Likewise, ligand substitution reactions of $[CrCl_2(H_2O)_4]^+$ solutions are accelerated by chromium(II) catalysts. With molten alkali metal chlorides, for example, potassium chloride, $CrCl_3$ forms salts of the type M_3CrCl_6 and $K_3Cr_2Cl_9$, which are also octahedral but the two Cr atoms are connected via three chloride bridges.

$CrCl_3$ is a Lewis acid, classified as "hard" according to the Hard-Soft Acid-Base theory. It can form a range of adducts of the type $[CrCl_3L_3]_z$, where L is a Lewis base. For instance, it reacts with pyridine (C_5H_5N) to form an adduct.

$$CrCl_3 + 3C_5H_5N \rightarrow CrCl_3(C_5H_5N)_3$$

The reaction with trimethylsilyl chloride in THF results in the anhydrous THF complex.

$$CrCl_3{\cdot}(H_2O)_6 + 12(CH_3)_3SiCl + 3THF \rightarrow CrCl_3(THF)_3 + 6\big((CH_3)_3Si\big)_2O + 12HCl$$

Chromium(III) chloride is employed as the precursor to numerous organochromium compounds, for example, bis(benzene)chromium, an analog of ferrocene.

2 [benzene] + $CrCl_3$ —($AlCl_3$, 140 °C, High pressure)→ Cr^+ [bis(benzene)] —($Na_2S_2O_4$, aq. NaOH)→ Cr [bis(benzene)]

Phosphine complexes derived from $CrCl_3$ catalyze the trimerization of ethylene to 1-hexene. One niche use of $CrCl_3$ in organic synthesis is for the in situ formation of chromium(II) chloride, a reagent for the reduction of alkyl halides and for the synthesis of (E)-alkenyl halides. The reaction is typically performed utilizing 2 moles of $CrCl_3$ per mole of lithium aluminum hydride, even though if aqueous acidic conditions are appropriate, zinc and hydrochloric acid may be enough.

(A) $Cl{-}CH_2{-}C(=O)OH$ —($CrCl_3$, Zn; HCl, H_2O)→ $CH_3{-}C(=O)OH$

(B) Ph–CHO —(CHI_3, THF; $CrCl_3$, $LiAlH_4$, 25 °C, 1 h.)→ Ph–CH=CH–I (67% yield, E/Z=94/6)

Chromium(III) chloride has also been employed as a Lewis acid in organic reactions, for example, as a catalyst for the nitroso Diels-Alder reaction.

Chromium(III) bromide, $CrBr_3$, is a dark colored solid that appears green in transmitted light but red with reflected light. It is employed as a precursor for catalysts for the oligomerization of ethylene. It is produced in a tube furnace by the reaction of bromine vapor and chromium powder at 1000°C. It is purified through extraction with absolute diethyl ether to remove any $CrBr_2$ and is then washed with absolute diethyl ether and absolute ethanol. Similar to the behavior of related chromium(III) halides, the tribromide dissolves in water to form $CrBr_3(H_2O)_3$ only upon the addition of catalytic quantities of a reducing agent, which forms $CrBr_2$. The reducing agent produces chromous bromide on the surface of the solid, which dissolves and re-oxidizes to Cr(III).

Chromium(III) iodide, also known as chromium triiodide (CrI_3), is a black solid that is employed to synthesize other chromium compounds. Since it is isomorphous with chromium(III) chloride ($CrCl_3$), it possesses a cubic-closest packing arrangement in a double-layer crystal lattice. In this structure, Cr is in an octahedral coordination geometry. Chromium triiodide is produced through the direct reaction of chromium metal with an excess of iodine. The reaction is performed at 500°C.

$$2Cr + 3I_2 \rightarrow 2CrI_3$$

To get high-purity samples, the product is thermally decomposed at 700°C to sublime out chromium(II) iodide. The diiodide is then reiodinated. At 25°C, CrI_3 is stable in contact with oxygen and moisture, but at temperatures approaching 200°C, it reacts with oxygen and releases iodine. Similar to $CrCl_3$, it shows slow solubility in water due to the kinetic inertness of Cr(III). Addition of small quantities of chromous iodide accelerates the dissolving dissolution.

Anhydrous chromium dihalides are conveniently prepared by reduction of trihalides with H_2 at 300°C–500°C, or by the action of HX (or I_2 for the diiodide) on the metal at temperatures of the order of 1000°C. They are all deliquescent, and the hydrates can be obtained by reduction of trihalides using pure chromium metal and aqueous HX. Chromium(II) fluoride, CrF_2, exists as a blue-green iridescent solid. CrF_2 is sparingly soluble in water, almost insoluble in alcohol, and soluble in boiling hydrochloric acid, though it is not attacked by hot distilled sulfuric acid or nitric acid. Similar to other chromous compounds, CrF_2 is oxidized to chromium(III) oxide in air. Similar to many difluorides, CrF_2 has a rutile (TiO_2) structure with octahedral molecular geometry about Cr(II) and trigonal geometry at F^-. Two of the six Cr-F bonds are long of about 2.43 Å, and four are short of about 2.00 Å. CrF_2 can be produced by passing anhydrous hydrogen fluoride over anhydrous chromium(II) chloride. The reaction will proceed at 25°C but is usually heated to 100°C–200°C to ensure completion.

$$CrCl_2 + 2HF \rightarrow CrF_2 + 2HCl$$

Chromium(II) chloride defines a group of inorganic compounds with the general formula $CrCl_2{\cdot}nH_2O$. The anhydrous solid is white when pure, but industrial samples are often gray or green; it is hygroscopic and easily dissolves in water to form bright blue air-sensitive solutions of the tetrahydrate $Cr(H_2O)_4Cl_2$. $CrCl_2$ has no industrial applications but is utilized on a laboratory scale for the synthesis of other chromium complexes. $CrCl_2$ is prepared either by reducing chromium(III) chloride with hydrogen at 500°C or by electrolysis.

$$2CrCl_3 + H_2 \rightarrow 2CrCl_2 + 2HCl$$

Small-scale preparations can also employ $LiAlH_4$, zinc, or related reagents, to reduce $CrCl_3$.

$$4CrCl_3 + LiAlH_4 \rightarrow 4CrCl_2 + LiCl + AlCl_3 + 2H_2$$
$$2CrCl_3 + Zn \rightarrow 2CrCl_2 + ZnCl_2$$

In addition, $CrCl_2$ can be produced by reacting a solution of chromium(II) acetate with hydrogen chloride.

$$Cr_2(OAc)_4 + 4HCl \rightarrow 2CrCl_2 + 4AcOH$$

$CrCl_2$ crystallizes in the *Pnnm* space group, which is an orthorhombically distorted variant of the rutile (TiO_2) structure; that is, it is isostructural with calcium chloride. The Cr centers are octahedrally coordinated, though distorted by the Jahn-Teller effect. The hydrated derivative, $CrCl_2(H_2O)_4$, forms monoclinic crystals with the $P2_1/c$ space group. The molecular geometry is approximately square planar with Cr-O distances of 2.078 Å and 2 Cr-Cl distances of 2.758 Å. The reduction potential for

$$Cr^{3+} + e^- \rightleftharpoons Cr^{2+}$$

is −0.41. Since the reduction potential of H^+ to H_2 in acidic conditions is +0.00, the chromous ion has enough potential to reduce acids to hydrogen, though this reaction does not occur without a catalyst. $CrCl_2$ is employed as precursor to prepare other inorganic and organometallic chromium complexes. Alkyl halides and nitroaromatics are reduced by $CrCl_2$. The moderate electronegativity of chromium and the range of substrates that $CrCl_2$ can accommodate result in organochromium reagents to be very synthetically versatile. It is a reagent in the Nozaki-Hiyama-Kishi reaction, a valuable method for preparing medium-size organic rings. It is also employed in the Takai olefination to produce vinyl iodides from aldehydes in the presence of iodoform.

5.6.3.5 Sulfate, nitrate, and phosphate

Chromium(III) sulfate normally refers to a series of compounds with the general formula $Cr_2(SO_4)_3{\cdot}x(H_2O)$, where x can range from 0 to 18. In addition, ill-defined but industrially important "basic chromium sulfates" are known. These salts are typically either violet or green solids that are soluble in water. They are widely used in tanning leather. Three chromium(III) sulfates have been well characterized. Anhydrous chromium(VI) sulfate, $Cr_2(SO_4)_3$, is a violet solid that dissolves in water upon addition of a reducing agent, forming chromium(II) sulfate. Hydrated chromium(III) sulfate, $Cr_2(SO_4)_3{\cdot}18H_2O$, is a violet solid that easily dissolves in water to form the metal aquo complex, $[Cr(H_2O)_6]^{3+}$. The formula of this compound can better be written as $[Cr(H_2O)_6]_2(SO_4)_3{\cdot}6H_2O$, sincc 6 of the 18 water molecules in this formula unit are water of crystallization. Hydrated chromium(III) sulfate, $Cr_2(SO_4)_3{\cdot}15H_2O$, is a green solid that also easily dissolves in water. It is formed by heating the $18H_2O$ compound above 70°C. Further heating results in anhydrous sulfate. A number of other chromium(VI) sulfates have been observed, but likewise contain hydroxide or oxide ligands. Most significant industrially is basic chromium sulfate, which is believed to be $[Cr_2(H_2O)_6(OH)_4]SO_4$. It is the product of the partial neutralization of the hexahydrates. Other chromium(III) hydroxides have been described. The most valuable sources of chromium(III) sulfate are the Cr(III) wastes from the chromate oxidation of numerous organic compounds. Anthroquinone and quinone are formed on a large scale via treatment of anthracene and phenol with chromic acid. Chromium(III) oxide coproduct is formed, which is easily extracted into sulfuric acid. Evaporation of these acidic solutions produces the hydrate salt described above. The hydrated salts of chromium sulfate can also be prepared, although impure, through extraction of numerous other chromium compounds, but these processes are not commercially viable. Extraction of chromite ore with sulfuric acid in the presence of some chromate results in the formation of chromium(III) sulfate solutions contaminated with other metal ions. Likewise, dissolution of chrome alloys results in the formation of chromium sulfate along with ferrous sulfate. Basic chromium sulfate is prepared from chromate salts via reduction with sulfur dioxide, but alternative processes exist. The reduction can formally be written as

$$Na_2Cr_2O_7 + 3SO_2 + H_2O \rightarrow Cr_2(SO_4)_3 + 2NaOH$$

Because 33% of the anion charges are caused by hydroxy ions, the basicity is 33% (though in tanning terminology, it is known as 33% reduced). Products with higher basicities, for example, 42% or 50% may be formed through the addition of sodium carbonate. They are frequently used together with sodium formate. The sodium sulfate is frequently left in the technical product as it is inert in connection with the tanning process. It is crucial to fully reduce the hexavalent chromium to trivalent chromium as the hexavalent form is more likely to result in health problems for tanners and leather consumers.

The term chromium(III) nitrate covers a series of inorganic compounds containing chromium, nitrate, and varying amounts of water. Most common is the dark violet hydrated solid, although an anhydrous green form also exists. Chromium(III) nitrate compounds are of a limited value industrially, having a few uses in the dyeing industry. It is common in research laboratories for the preparation of chromium coordination complexes. The fairly complicated formula—$[Cr(H_2O)_6](NO_3)_3{\cdot}3H_2O$—reflects the complex structure of this compound. The Cr centers are bonded to six water ligands, and the remaining volume of the solid is occupied by three nitrate anions and three molecules of water of crystallization. The anhydrous salt forms green crystals and is very soluble in water (unlike anhydrous chromium(III) chloride, which dissolves very slowly except under special conditions). At 100°C it decomposes. The red-violet hydrate is highly soluble in water. Chromium nitrate is employed in the preparation of alkali metal-free catalysts and in pickling. Chromium nitrate can be formed by dissolving chromium oxide in nitric acid.

Chromium(III) phosphates are a group of compounds with the general formula $CrPO_4{\cdot}nH_2O$, where $n = 0$, 4, or 6. All are deeply colored solids, where, for example, the anhydrous $CrPO_4$ is green and the hexahydrate $CrPO_4{\cdot}6H_2O$ is violet. Chromium phosphate is produced by reacting a phosphoric acid solution of chromium(VI) oxide with hydrazine. Hexahydrate chromium phosphate, $CrPO_4{\cdot}6H_2O$, is formed through the reduction of chromium trioxide, CrO_3, with ethanol in the presence of orthophosphoric acid, H_3PO_4, at temperatures between −24°C and +80°C. Gel-like chromium(III) phosphate is produced via the reduction of ammonium dichromate, $(NH_4)_2Cr_2O_7$, using ethanol, CH_3COOH, and nitric acid, HNO_3. This reaction is performed in the presence of ammonium dihydrogen phosphate and urea at an elevated temperature where tetradecyltrimethylammonium bromide (TTBr) is employed as structure directing agent. The synthesis of textured chromium phosphate is performed by mixing equimolar solutions of aqueous chromium nitrate and diammonium phosphate in a dish placed in a sealed chamber with the low-temperature ammonia vapor catalyst diffusing into the solution at a constant rate. After 24 hours, a purple film grows out from the liquid through the hydrolysis and polycondensation taking place in the reaction environment at the air/liquid and film/liquid interfaces. Surface tension makes the film compact making it easy to insert a microscope slide and lift the film from underneath the solution surface. Once formed the solution is washed with deionized water and ethanol and subsequently dried in a vacuum. The production of anhydrous chromium(III) phosphate starts with grinding a mixture of 75 mol% of chromium(III) oxide, Cr_2O_3, and 25 mol% of pure ammonium hydrogen phosphate, $(NH_4)_2HPO_4$. This mixture is pressed into pellets and then heated under air pressure at 400°C for 24 hours to remove the ammonia and water. Subsequently, a heating sequence of 450°C (24 hours), 700°C (3–24 hours), 800°C (24 hours), and 850°C (2–24 hours) takes place. The pellet mixture is slowly cooled after that.

5.6.3.6 *Organochromium compounds*

Organochromium chemistry had its beginnings from the works of Franz Hein when he isolated the first organochromium compounds in 1919 via the reaction of anhydrous $CrCl_3$ with PhMgBr in diethyl ether. A pyrophoric paramagnetic black intermediate was obtained, which on hydrolysis yielded little understood "phenylchromium compound" at that time. It was not until 1954 that the nature of these species was established and shown to be π-arene complexes. The Hein synthesis when performed in THF allowed the isolation of the red crystalline complex of chromium complexed to three THF molecule and three phenyl groups. Rearrangement of this σ-bonded arylchromium(III) species to the black paramagnetic intermediate and then subsequent hydrolysis yielded a mixture of arylchromium compounds. Accounts of this original reaction can be found in detail in a review by Peter Jolly in 1996 and in section 26.2 entitled,

"Chromium Compounds with η^2- η^8 Carbon Ligands," by R. Davis and L. A. P. Kane-Maguire of the 1982 reference book, Comprehensive Organometallic Chemistry Vol 3 edited by Wilkinson, Gordon, Stone and Abel.

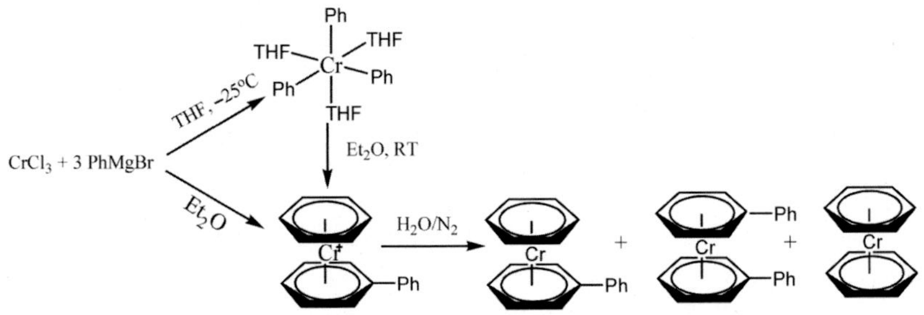

A wide variety of $[Cr(aryl)_3(THF)_3]$ complexes has since been synthesized via the reaction of a stoichiometric ratio of 3:1 ArMgBr to $CrCl_3$ in THF. Many of the resulting complexes rearrange to bis(arene) chromium sandwich compounds. However, ortho substituents and halogeno substituents, in general, inhibit this σ-to-π rearrangement. There is little or no π-complex obtained when aryl = 2-tolyl, mesitylene, 2-$MeOC_6H_4$, 2-$Me_2NC_6H_4$, 2-ClC_6H_5, 3-ClC_6H_5, or 4-(F,Cl,Br)C_6H_5. In addition to the bis(arene) complexes from the original Hein synthesis, chromocene and several other π-donating ligands with hapticity η^2 to η^8 have been synthesized and presented in the reference module, Comprehensive Organometallic Chemistry Volume 3.

Organochromium compounds also include chromium carbenes and carbynes as well as Cr-C σ-bonded compounds. The chemical bonding in a chromium carbene is based on σ-type electron donation of the filled lone pair orbital of the carbene atom to an empty chromium d-orbital and π back bonding of a filled chromium d-orbital to the empty p-orbital on carbon. An example is the complex $(CO)_5Cr$-$C(OCH_3)Ph$ shown below (left figure).

Chromium carbynes, on the other hand, have a triple bond consisting of a σ-bond and two π-bonds. The highest occupied molecular orbital (HOMO) of the carbyne ligand interacts with the lowest unoccupied molecular orbital (LUMO) of the metal to create the σ-bond. The two π-bonds are formed when the two HOMO orbitals of the metal back-donate to the LUMO of the carbyne. They are also called metal alkylidynes in which the carbon is a carbyne ligand (below, right figure).

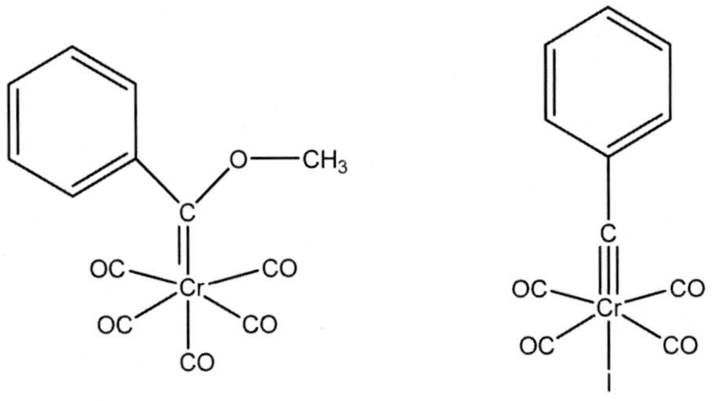

Organochromium compounds prepared from chromium(III) salts and organometallic reagents have been shown to possess carbanionic character and provide the first examples of organochromium additions to carbonyl substrates. For example, the reaction of triphenyltris(tetrahydrofuran)chromium(III) with carbon monoxide provided benzopinacol. Reactions of these materials with carbonyl substrates represent an elaborate array of chemoselective and stereoselective processes and make organochromium a useful reagent for organic synthesis.

5.6.4 Major uses

The production of metal alloys comprises around 85% of the total Cr usage. The rest is used in the chemical, refractory, and foundry industries. The strengthening effect of forming stable metal carbides at the grain boundaries and the strong increase in corrosion resistance made Cr an important alloying material for steel. The HSS (High Hardness Steel) have 3%–5% Cr. Stainless steel, the main corrosion-resistant metal alloy, is formed when Cr is added to Fe in high enough concentrations, typically the Cr content is above 11%. For stainless steel's formation, ferrochromium is added to the molten iron. In addition, Ni-based alloys increase in strength due to the formation of discrete, stable metal carbide particles at the grain boundaries. Due to the exceptional high-temperature properties of these Ni superalloys, they are used instead of common structural materials in jet engines and gas turbines. The comparatively high hardness and corrosion resistance of unalloyed Cr make chrome a reliable metal for surface coating. It is even now the most widespread metal for sheet coating due to its higher than average durability relative to other coating metals. A thin or thick Cr layer is deposited on pretreated metallic surfaces via electroplating techniques. Thin deposition entails a Cr layer below 1 μm thickness deposited by chrome plating and is used for decorative surfaces. Thicker Cr layers are deposited if wear-resistant surfaces are required. Both techniques use acidic chromate or dichromate solutions. To avoid the energy-consuming change in oxidation state, the use of Cr(III) sulfate is being developed. In the chromate conversion coating method, the strong oxidative properties of chromates are used to deposit a protective oxide layer on metals like Al, Zn, and Cd. This passivation and the self-healing properties by the chromate stored in the chromate conversion coating, which can migrate to local defects, are the advantages of this process. Due to environmental and health regulations on chromates, alternative coating methods are being developed. Chromic acid anodizing (or Type I anodizing) of Al is a different electrochemical technique, which does not result in the deposition of Cr, but uses chromic acid as electrolyte in the solution forming an oxide layer on Al. The use of chromic acid, instead of the normally used sulfuric acid, leads to a slight difference of these oxide layers. The high toxicity of Cr(VI) compounds, currently used in Cr electroplating process, and the strengthening of safety and environmental regulations require a search for alternatives for Cr or at least a change to the less toxic Cr(III) compounds. Chromium plating can be used to give a polished mirror finish to steel. Chromium-plated car and truck parts, such as bumpers, were once very common. It is also possible to chrome plate plastics, which were often used in bathroom fittings.

Crocoite (lead chromate, $PbCrO_4$) was utilized as a yellow pigment soon after its discovery. When a synthesis method became available starting from the more abundant chromite ($Fe^{2+}Cr_2^{3+}O_4$), chrome yellow became, together with cadmium yellow, one of the most used yellow pigments. The pigment does not photodegrade, but it has a tendency to become darker due to the formation of Cr(III) oxide. The use of chrome yellow has since declined because of environmental and safety concerns and was substituted by organic pigments or other alternatives that are free from Pb and Cr. Other pigments that are based around Cr are, for example, the deep shade of red pigment chrome red, which is simply Pb chromate with Pb(II) hydroxide ($PbCrO_4 \cdot Pb(OH)_2$). A very important chromate pigment, which was applied extensively in metal primer formulations, was Zn chromate, now substituted by Zn phosphate. A wash primer was formulated to replace the dangerous procedure of pretreating Al aircraft bodies with a phosphoric acid solution. This used zinc tetroxychromate ($4Zn(OH)_2 \cdot ZnCrO_4$) dispersed in a solution of polyvinyl butyral ($[C_8H_{14}O_2]_n$). An 8% solution of phosphoric acid in solvent was added just before application. It was discovered that an easily oxidized alcohol was a critical component. A thin layer of approximately 10–15 μm was applied, which changed from yellow to dark green when it was cured. It is still unclear what the correct mechanism is. Chrome green is a mixture of Prussian blue (IUPAC name iron(II,III) hexacyanoferrate(II,III), $Fe_4^{III}[Fe^{II}(CN)_6]_3$) and chrome yellow, while the chrome oxide green is Cr(III) oxide. In addition, Cr oxides are utilized as a green pigment in glassmaking and as a glaze for ceramics. Green Cr oxide is particularly lightfast and as such is applied in cladding coatings. It is also the key ingredient in infrared reflecting paints and used by the armed forces to paint vehicles to give them the same infrared reflectance as green plants.

Natural rubies are a variety of corundum (Al_2O_3) crystals that are red colored (the rarest type) as a result of Cr(III) ions (other colors of corundum gems are termed sapphires). A red-colored ruby can be synthesized by doping Cr(III) into synthetic corundum crystals. This type of synthetic ruby crystal was the basis for the first laser, made in 1960, which relied on stimulated emission of light from the Cr atoms in such a crystal. A ruby laser is lasing at 694.3 nm, in a deep red color. Due to their toxicity, Cr(VI) salts are utilized for wood preservation, for example, chromated copper arsenate (CCA) is used in timber treatment to protect wood from decay fungi, wood-attacking insects, including termites, and marine borers. The formulations contain Cr based on the oxide CrO_3 ranging from 35.3% to 65.5%. Cr(III) salts, in particular chrome alum and Cr(III) sulfate, are applied in leather tanning. Cr(III) stabilizes the leather through cross linking the collagen fibers. Cr tanned leather can have 4% to 5% of Cr, which is strongly bound to the proteins. While the Cr form used for tanning is not the toxic Cr(VI), there is still an interest in controlling Cr in the tanning industry. The high heat resistivity and high melting point make chromite and Cr(III) oxide materials suitable for high-temperature refractory uses, for example, in blast furnaces, cement kilns, molds for the firing of bricks, and as foundry sands for the casting of metals. In these applications, the refractory materials are made from mixtures of chromite and magnesite ($MgCO_3$). The use is decreasing due to the environmental regulations because of the possible risk of Cr(VI) formation. Various Cr compounds are used as catalysts for treating hydrocarbons, for example, the Phillips catalyst, prepared from Cr oxides, is used to make about 50% all polyethylene globally, Fe–Cr mixed oxides are used as high-temperature catalysts for the water gas shift reaction, and Cu chromite is a useful hydrogenation catalyst. Chromium is used in a variety of other applications. Cr(IV) oxide (CrO_2) is a magnetic compound. Its ideal shape anisotropy, which imparts high coercivity and remnant magnetization, made it much better than γ-Fe_2O_3. Cr(IV) oxide is used to make magnetic tape used in high-performance audio tape and standard audio cassettes. Chromates are added to drilling muds to avoid corrosion of steel under wet conditions. Cr(III) oxide (Cr_2O_3) is a metal polish ("green rouge"). Chromic acid is a powerful oxidizing agent and is a valuable compound for cleaning laboratory glassware of any trace of organic compounds. It is prepared by dissolving potassium dichromate ($K_2Cr_2O_7$) in concentrated sulfuric acid (H_2SO_4), which is then used to wash the apparatus. Sodium dichromate ($Na_2Cr_2O_7$) is occasionally used due to its four times higher solubility. The use of dichromate cleaning solutions is currently being phased out because of their high toxicity and environmental concerns. Modern alternative cleaning solutions are highly effective and Cr free. $K_2Cr_2O_7$ is a chemical reagent used as a titrating agent. Chrome alum is Cr(III) potassium sulfate ($KCr(SO_4)_2$, commonly found as $KCr(SO_4)_2{\cdot}12(H_2O)$), and is used as a mordant (i.e., a fixing agent) for dyes in fabric and in tanning.

5.7 25 Mn — Manganese

5.7.1 Discovery

The origin of the name manganese is multifaceted. In antiquity, two black minerals from Magnesia (located within modern Greece) were both called magnes from their place of origin but differed in gender. The masculine magnes attracted iron and was the iron ore now recognized as magnetite, Fe_3O_4, and which probably gave rise to the term magnet. The feminine magnes ore did not attract iron but was utilized to decolorize glass. This feminine magnes was later called magnesia, identified now as pyrolusite or manganese dioxide, MnO_2. Neither this mineral nor elemental manganese is magnetic. In the 16th century, manganese dioxide was known as manganesum by glassmakers, maybe as a corruption and concatenation of two words, as alchemists and glassmakers in due course had to distinguish between a magnesia negra (the black ore) from magnesia alba (a white ore, also from Magnesia, also useful in glassmaking). Superintendent of the Vatican Botanical Garden Michele Mercati (April 8, 1541 to June 25, 1593) named magnesia negra manganesa, and in the end the metal isolated from it became known as manganese (German: Mangan) (Fig. 5.53). The name magnesia ultimately was used to refer only to the white magnesia alba (magnesium oxide), which resulted in the name magnesium for the free element after it was isolated many years later. Several colorful oxides of manganese, for example, manganese dioxide, can be found in nature and have been used as pigments since the Stone Age. The cave paintings in Gargas, Haute-Garonne, southwestern France, that are between 30,000 and 24,000 years old

FIGURE 5.53 Michele Mercati by Petrus Nellus, before 1717.

FIGURE 5.54 Johan Gottlieb Gahn, 1842 by Johan Elias Cardon (1802–78).

comprise manganese pigments. Manganese compounds were utilized by Egyptian and Roman glassmakers, either to add to, or remove color from glass. "Glassmakers soap" was utilized uninterruptedly through the Middle Ages until modern times and is obvious in 14th-century glass from Venice. Since it was utilized in glassmaking, manganese dioxide was readily available for experiments by alchemists. Austrian alchemist Ignatius Gottfried Kaim (1746–1778) in 1770 and German-Dutch alchemist and chemist Johann Rudolf Glauber (March 10, 1604 to March 16, 1670) observed that manganese dioxide could be changed to permanganate, a useful laboratory reagent (Kaim, 1770). By the mid-18th century, the Swedish Pomeranian and German pharmaceutical chemist Carl Wilhelm Scheele (December 9, 1742 to May 21, 1786) used manganese dioxide to produce chlorine. First, hydrochloric acid, or a mixture of dilute sulfuric acid and sodium chloride, was made to react with manganese dioxide, later hydrochloric acid from the Leblanc process (an early industrial process to produce soda ash (sodium carbonate) used throughout the 19th century, named after its French inventor, chemist and surgeon Nicolas Leblanc (December 6, 1742 to January 16, 1806). It involved two stages: production of sodium sulfate from sodium chloride, followed by reaction of the sodium sulfate with coal and calcium carbonate to produce sodium carbonate.) was used and the manganese dioxide was recycled by the Weldon process developed by English industrial chemist and journalist Walter Weldon (October 31, 1832 to September 20, 1885). The manufacture of chlorine and hypochlorite bleaching agents was a large consumer of manganese ores. Scheele and other chemists were aware that manganese dioxide contained a new element, but they were not able to isolate it. Swedish chemist and metallurgist Johan Gottlieb Gahn (August 19, 1745 to December 8, 1818) was the first to produce an impure sample of manganese metal in 1774, by reducing the manganese dioxide with carbon (Fig. 5.54). The manganese concentration of some iron ores used in Greece resulted in speculations that steel produced from that ore has additional manganese, making the Spartan steel extremely hard. At the start of the 19th century, manganese was used in steelmaking and several patents were granted. In 1816 it was recognized that iron alloyed with manganese was harder but not more brittle.

5.7.2 Mining, production, and major minerals

Manganese comprises about 0.1% of the Earth's crust, the 12th most abundant of the crust's elements. Soil contains between 7 and 9000 ppm of manganese with an average of 440 ppm. Manganese occurs primarily as pyrolusite (MnO_2), braunite, $Mn^{2+}Mn_6^{3+}SiO_{12}$, psilomelane $(Ba,H_2O)_2Mn_5O_{10}$, and to a lesser extent as rhodochrosite ($MnCO_3$). The most important manganese ore is pyrolusite (MnO_2). Other economically significant manganese ores typically exhibit a close spatial relation to the iron ores. Land-based resources are large but irregularly distributed. Around 80% of the known world manganese ore deposits are in South Africa; other important manganese deposits are in Ukraine,

Australia, India, China, Gabon, and Brazil. According to a 1978 estimate, the ocean floor is covered with 500 billion tons of manganese nodules.

For the production of ferromanganese, the manganese ore is mixed with iron ore and carbon and subsequently reduced either in a blast furnace or in an electric arc furnace. The resulting ferromanganese has a manganese content of 30%–80%. Pure manganese used to produce iron-free alloys is obtained by leaching manganese ore with sulfuric acid followed by an electrowinning process. A more advanced extraction process comprises directly reducing manganese ore in a heap leach. This is done by percolating natural gas through the bottom of the heap; the natural gas provides the heat (needs to be at least 850°C) and the reducing agent (carbon monoxide). This reduces all the manganese ore to manganese oxide (MnO), which is a leachable form. The ore then moves through a grinding circuit to reduce the particle size of the ore to be between 150 and 250 μm, increasing the surface area to improve the leaching efficiency. The fine-grained ore is subsequently added to a leach tank of sulfuric acid and ferrous iron (Fe^{2+}) in a 1.6:1 ratio. The iron reacts with the manganese dioxide to form iron hydroxide and elemental manganese. This process results in about 92% recovery of the manganese. For further purification, the manganese can then be sent to an electrowinning facility.

Manganese is a relatively common element in minerals, with more than 650 minerals known that (can) contain this element. Within the elements class 2 silicide minerals are found, brownleeite (MnSi) and mavlyanovite (Mn_5Si_3). The sulfide class contains 11 different minerals with Mn, including minerals such as alabandite (MnS), hauerite (MnS_2), and samsonite ($Ag_4MnSb_2S_6$). The chlorides are represented by three minerals, chlormanganokalite ($K_4[MnCl_6]$), kempite ($Mn_2^{2+}(OH)_3Cl$), and scacchite ($MnCl_2$). More than 80 minerals are known in the oxide class, including some well-known minerals such as birnessite ($(Na,Ca)_{0.5}(Mn^{4+},Mn^{3+})_2O_4{\cdot}1.5H_2O$), bixbyite ($Mn_2^{3+}O_3$) (Fig. 5.55), coronadite ($Pb(Mn_6^{4+}Mn_2^{3+})O_{16}$), cryptomelane ($K(Mn_7^{4+}Mn^{3+})O_{16}$), hausmannite ($Mn^{2+}Mn_2^{3+}O_4$), hübnerite ($MnWO_4$) (Fig. 5.56), manganite ($Mn^{3+}O(OH)$) (Fig. 5.57), pyrolusite ($Mn^{4+}O_2$), ramsdellite ($Mn^{4+}O_2$), romanèchite ($(Ba,H_2O)_2(Mn^{4+},Mn^{3+})_5O_{10}$) (Box 5.11), and wodginite ($Mn^{2+}Sn^{4+}Ta_2O_8$) (Fig. 5.58). The carbonate class consists of 14 minerals with Mn in their structure, for example, kutnohorite ($CaMn^{2+}(CO_3)_2$) and rhodochrosite ($MnCO_3$) (Fig. 5.59). Within the borate class 15 minerals are known to contain Mn, such as gaudefroyite ($Ca_4Mn_{2-3}^{3+}(BO_3)_3(CO_3)(O,OH)_3$) and sussexite ($Mn^{2+}BO_2(OH)$). The sulfates are represented by 17 different minerals, including apjohnite ($Mn^{2+}Al_2(SO_4)_4{\cdot}22H_2O$), moorhouseite ($(Co,Ni,Mn)SO_4{\cdot}6H_2O$), and sturmanite ($Ca_6(Fe^{3+},Al,Mn^{3+})_2(SO_4)_2[B(OH)_4](OH)_{12}{\cdot}25H_2O$). Nearly 150 minerals are known in the phosphate class to contain Mn, examples are eosphorite ($Mn^{2+}Al(PO_4)(OH)_2{\cdot}H_2O$), fairfieldite ($Ca_2Mn^{2+}(PO_4)_2{\cdot}2H_2O$), hureaulite ($(Mn,Fe)_5(PO_4)_2(HPO_4)_2{\cdot}4H_2O$), lithiophilite ($LiMn^{2+}PO_4$), purpurite ($(Mn^{3+},Fe^{3+})PO_4$), and strunzite ($Mn^{2+}Fe_2^{3+}(PO_4)_2(OH)_2{\cdot}6H_2O$) (Fig. 5.60). Nearly 250 minerals are found in the silicate class with Mn in their crystal structure. These include minerals such as babingtonite ($Ca_2(Fe,Mn)FeSi_5O_{14}(OH)$), braunite ($Mn^{2+}Mn_6^{3+}(SiO_4)O_8$), bustamite ($CaMn^{2+}(Si_2O_6)$) (Fig. 5.61), eudialyite ($Na_{15}Ca_6(Fe^{2+},Mn^{2+})_3Zr_3[Si_{25}O_{73}](O,OH,H_2O)_3(OH,Cl)_2$), helvine ($Be_3Mn_4^{2+}(SiO_4)_3S$), inesite ($Ca_2(Mn,Fe)_7Si_{10}O_{28}(OH)_2{\cdot}5H_2O$) (Fig. 5.62), olmiite ($CaMn^{2+}[SiO_3(OH)](OH)$), piemontite ($\{Ca_2\}\{Al_2Mn^{3+}\}(Si_2O_7)(SiO_4)O(OH)$), rhodonite ($Mn^{2+}SiO_3$) (Fig. 5.63), serandite ($NaMn_2^{2+}Si_3O_8(OH)$),

FIGURE 5.55 Bixbyite, $Mn_2^{3+}O_3$, modified cube about 5 mm crystal, with a topaz, $Al_2(SiO_4)(F,OH)_2$, crystal in the background. Thomas Range, Juab Co., Utah, United States.

FIGURE 5.56 Hübnerite, $MnWO_4$, deep red to black bladed crystal to 3 cm. Black Pine mine, Philipsburg, Granite Co., Montana, United States.

FIGURE 5.57 Highly reflective and lustrous prismatic manganite, $Mn^{3+}O(OH)$, crystals to 4 cm. Ilfeld, Harz Mountains, Thuringia, Germany.

spessartine ($Mn_3^{2+}Al_2(SiO_4)_3$), and tephroite ($Mn_2^{2+}SiO_4$). Finally, two organic compounds are known with Mn, falottaite ($MnC_2O_4{\cdot}3H_2O$) and lindbergite ($Mn(C_2O_4){\cdot}2H_2O$).

5.7.3 Chemistry

Naturally occurring manganese is composed of one stable isotope, ^{55}Mn. Eighteen radioisotopes have been characterized, ranging from ^{46}Mn to ^{65}Mn. The most stable are ^{53}Mn with a half-life of 3.74 million years, ^{54}Mn with a half-life of 312.03 days, and ^{52}Mn with a half-life of 5.591 days. All the other radioisotopes have half-lives of less than 3 hours, and most of them less than 1 minute. The principal decay mode for isotopes lighter than ^{55}Mn is electron capture while the principal mode for the heavier isotopes is through β decay. Mn also has three meta states. Mn is part of the iron group of elements, which are believed to be formed in large stars just before a supernova explosion. ^{53}Mn decays to

FIGURE 5.58 Wodginite, $Mn^{2+}Sn^{4+}Ta_2O_8$, 1−2 mm black euhedral crystal in matrix. Tanco pegmatite, Bernic Lake, Manitoba, Canada.

FIGURE 5.59 Rhodochrosite, $MnCO_3$, a large pink rhombohedral crystal to 1.3 cm on edge perched perfectly at the center with several smaller crystals within the same vug lined with drusy quartz, SiO_2. American Tunnel, Silverton, San Juan Co., Colorado, United States.

^{53}Cr. Because of its relatively short half-life, ^{53}Mn is relatively rare, formed by cosmic rays impacting on iron. Mn isotopes are characteristically found together with Cr isotopes and are therefore useful in isotope geology and radiometric dating. Mn/Cr isotopic ratios strengthen the evidence from ^{26}Al and ^{107}Pd for the solar system's early history. Variations in $^{53}Cr/^{52}Cr$ and Mn/Cr ratios from several meteorites point to an initial $^{53}Mn/^{55}Mn$ ratio that suggest that the Mn/Cr isotopic composition must be due to the in situ decay of ^{53}Mn in differentiated planetary bodies. Consequently, ^{53}Mn provides further proof for nucleosynthetic processes just before coalescence of the solar system.

Manganese is a silvery-gray metal that resembles iron. It is hard and very brittle and difficult to fuse. It has 25 electrons arranged in an electronic configuration of $[Ar]3d^54s^2$. All the seven electrons in the 4s and 3d subshells can be involved in reactions; thus manganese can have oxidation states up to +7 in its compounds. The full range of oxidation states is −3 to +7 but +2 is the most stable (Table 5.8). Oxidation states of +2, +4 and +7 are common as exemplified by manganese ion (Mn^{2+}), manganese(IV) oxide (MnO_2), and permanaganate ion (MnO_4^-), respectively. Compounds in which manganese have the +5 and +6 oxidation states are prone to disproportionation reactions.

FIGURE 5.60 Strunzite, $Mn^{2+}Fe_2^{3+}(PO_4)_2(OH)_2{\cdot}6H_2O$, golden straw-like crystals scattered randomly to 4 mm in vuggy matrix with yellow laueite, $Mn^{2+}Fe_2^{3+}(PO_4)_2(OH)_2{\cdot}8H_2O$. Hagendorf-Süd, Oberpfälzer Wald, Bavaria, Germany.

FIGURE 5.61 Acicular or fibrous reddish-brown crystals of bustamite, $CaMn^{2+}(Si_2O_6)$, to 1.5 cm associated with green manganoan hedenbergite, $CaFe^{2+}Si_2O_6$. North Mine, Broken Hill, New South Wales, Australia.

5.7.3.1 *Oxides and hydroxides*

Manganese does not occur as a free element in nature but is found as a component of a number of minerals, for example, pyrolusite. The reactivity of manganese is similar to iron; it oxidizes superficially in air and forms rusts in moist air. It also burns in air or oxygen at elevated temperatures to give manganese(II,III) oxide (Mn_3O_4 or $MnO{\cdot}Mn_2O_3$). Other oxides of manganese are known. The mineral pyrolusite is manganese(IV) oxide (MnO_2). The mineral is normally purified by the action of dinitrogen tetroxide on pyrolusite to produce manganese(II) nitrate ($Mn(NO_3)_2$) which crystallizes upon evaporating the solvent. Heating the nitrate at 400°C releases N_2O_4 to give a purified MnO_2.

The main application of MnO_2 is in dry cell batteries, for example, the alkaline battery and the zinc-carbon battery. MnO_2 is also employed as a pigment and as a precursor to other manganese compounds, such as $KMnO_4$. It is utilized as a reagent in organic synthesis, for example, for the oxidation of allylic alcohols. Naturally occurring MnO_2 contains impurities and a substantial amount of manganese(III) oxide. Only a few deposits contain the γ modification in purity good enough for the battery industry. Production of batteries and ferrite (two of the principal applications of MnO_2)

FIGURE 5.62 Reddish thin bladed 3-mm crystal sprays of inesite, $Ca_2(Mn,Fe)_7Si_{10}O_{28}(OH)_2 5H_2O$. Wessels mine, Hotazel, Kalahari Mn fields, Northern Cape Province, South Africa.

FIGURE 5.63 Rhodonite, $Mn^{2+}SiO_3$, crystal about 1 cm in length, Broken Hill, New South Wales, Australia.

requires high-purity MnO_2. Batteries need what is known as "electrolytic manganese dioxide" (EMD) while ferrites need "chemical manganese dioxide" (CMD).

5.7.3.1.1 Chemical manganese dioxide

One route starts with natural MnO_2 and transforms it using dinitrogen tetroxide and water to a manganese(II) nitrate solution. Evaporation of the water results in the crystalline nitrate salt. At 400°C the salt decomposes, releasing N_2O_4 forming a residue of purified MnO_2. These two steps can be summarized as

$$MnO_2 + N_2O_4 \rightleftharpoons Mn(NO_3)_2$$

In an alternative method, MnO_2 is carbothermically reduced to manganese(II) oxide, which is dissolved in sulfuric acid. The filtered solution is reacted with ammonium carbonate to precipitate $MnCO_3$. The carbonate is calcined in air to produce a mixture of manganese(II) and manganese(IV) oxides. To finish the process, a suspension of this material in sulfuric acid is reacted with sodium chlorate. Chloric acid, which forms in situ, changes any Mn(III) and Mn(II) oxides to the dioxide, while releasing chlorine as a by-product. A third method comprises manganese heptoxide and manganese monoxide. The two reagents combine at a 1:3 ratio to form manganese dioxide.

$$Mn_2O_7 + 3MnO \rightarrow 5MnO_2$$

Finally, the reaction of potassium permanganate over manganese sulfate crystals forms the desired oxide.

$$2KMnO_4 + 3MnSO_4 + 2H_2O \rightarrow 5MnO_2 + K_2SO_4 + 2H_2SO_4$$

BOX 5.11 Manganese oxide dendrite pseudofossils

A dendrite is a crystal that develops with a typical multibranching tree-like form. Dendritic crystal growth is very common and illustrated by snowflake formation and frost patterns on a window. Dendritic crystallization forms a natural fractal pattern. These crystals can grow into a supercooled pure liquid or form from growth instabilities that occur when the growth rate is limited by the rate of diffusion of solute atoms to the interface. In the latter case, there must be a concentration gradient from the supersaturated value in the solution to the concentration in equilibrium with the crystal at the surface. Any protuberance that forms is then accompanied by an even steeper concentration gradient at its tip. This increases the diffusion rate to the tip. In opposition to this is the action of the surface tension tending to flatten the protuberance and setting up a flux of solute atoms from the protuberance out to the sides. However, overall, the protuberance becomes amplified. This process occurs again and again until a dendrite is produced. The term "dendrite" comes from the Greek word dendron (δενδρον), which means "tree." In paleontology, dendritic mineral crystal forms are often mistaken for fossils. These pseudofossils form as naturally occurring fissures in the rock and are filled by percolating mineral solutions. They form when water rich in manganese and iron flows along fractures and bedding planes between layers of limestone and other rock types, depositing dendritic crystals as the solution flows through. A variety of manganese oxides and hydroxides are involved, including birnessite ($Na_4Mn_{14}O_{27}{\cdot}9H_2O$), coronadite ($PbMn_8O_{16}$), cryptomelane ($KMn_8O_{16}$), hollandite ($BaMn_8O_{16}$), romanechite ($(Ba,H_2O)Mn_5O_{10}$), and todorokite ($(Ba,Mn,Mg,Ca,K,Na)_2Mn_3O_{12}.3H_2O$) (Fig. 5.64).

FIGURE 5.64 Manganese oxide dendrites on limestone from Solnhofen, Germany.

5.7.3.1.2 Electrolytic manganese dioxide

Electrolytic manganese dioxide (EMD) is employed in zinc-carbon batteries together with zinc chloride and ammonium chloride. EMD is also frequently utilized in zinc manganese dioxide rechargeable alkaline (Zn RAM) cells. For these uses, purity is particularly important. EMD is prepared in a comparable way as electrolytic tough pitch copper. The MnO_2 is dissolved in sulfuric acid (occasionally mixed with manganese sulfate) and subjected to a current between two electrodes. The MnO_2 dissolves, enters the solution as the sulfate, and is deposited on the anode.

TABLE 5.8 Manganese properties.

Appearance	silvery metallic
Standard atomic weight $A_{r,std}$	54.938
Block	d-Block
Element category	Transition metal
Electron configuration	[Ar] $3d^5$ $4s^2$
Phase at STP	Solid
Melting point	1246°C
Boiling point	2061°C
Density (near r.t.)	7.21 g/cm^3
When liquid (at m.p.)	5.95 g/cm^3
Heat of fusion	12.91 kJ/mol
Heat of vaporization	221 kJ/mol
Molar heat capacity	26.32 J/(mol·K)
Oxidation states	−3, −2, −1, +1, **+2**, +3, **+4**, +5, +6, **+7**
Ionization energies	1st: 717.3 kJ/mol
	2nd: 1509.0 kJ/mol
	3rd: 3248 kJ/mol
Atomic radius	Empirical: 127 pm
Covalent radius	Low spin: 139 ± 5 pm
	High spin: 161 ± 8 pm

STP, Standard temperature and pressure.
Bold font indicates main oxidation state.

A number of MnO_2 polymorphs are known, as well as a hydrated form. Similar to many other dioxides, MnO_2 crystallizes in the rutile (TiO_2) crystal structure (this polymorph is the mineral pyrolusite or β-MnO_2), with three-coordinate oxide and octahedral metal centers. MnO_2 is usually nonstoichiometric, being deficient in oxygen. The α-polymorph of MnO_2 has a very open structure with "channels" which can accommodate metal atoms such as silver or barium. α-MnO_2 is often called hollandite, after a closely related mineral.

The important reactions of MnO_2 are associated with its redox potential, both oxidation and reduction. MnO_2 is the main precursor to ferromanganese and related alloys, which are extensively used in the steel industry. Here, the conversions comprise carbothermal reduction using coke.

$$MnO_2 + 2C \rightarrow Mn + 2CO$$

The crucial reaction of MnO_2 in batteries is the one-electron reduction.

$$MnO_2 + e^- + H^+ \rightarrow MnO(OH)$$

MnO_2 catalyzes several reactions that produce O_2. In a classical laboratory demonstration, heating a mixture of potassium chlorate and manganese dioxide produces oxygen gas. Manganese dioxide likewise catalyzes the decomposition of hydrogen peroxide to oxygen and water.

$$2H_2O_2 \rightarrow 2H_2O + O_2$$

MnO_2 decomposes above approximately 530°C to manganese(III) oxide and oxygen. At temperatures close to 1000°C, the mixed valence compound Mn_3O_4 forms. Higher temperatures result in MnO. The reaction with hot concentrated sulfuric acid reduces the MnO_2 to manganese(II) sulfate:

$$2MnO_2 + 2H_2SO_4 \rightarrow 2MnSO_4 + O_2 + 2H_2O$$

The reaction of hydrogen chloride with MnO_2 was used by Carl Wilhelm Scheele in the original isolation of chlorine gas in 1774.

$$MnO_2 + 4HCl \rightarrow MnCl_2 + Cl_2 + 2H_2O$$

As a source of hydrogen chloride, Scheele reacted sodium chloride with concentrated sulfuric acid.

$$E^0\left(MnO_2(s) + 4H^+ + 2e^- \rightleftharpoons Mn^{2+} + 2H_2O\right) = +1.23\ V$$
$$E^0\left(Cl_2(g) + 2e^- \rightleftharpoons 2Cl^-\right) = +1.36\ V$$

The standard electrode potentials (SEPs) for the half reactions indicate that the reaction is endothermic at pH = 0 (1 M $[H^+]$), but it is favored by the lower pH as well as the evolution (and removal) of gaseous chlorine. This reaction is as well a useful way to remove the MnO_2 precipitate from the ground glass joints after running a reaction (i.e., an oxidation with potassium permanganate). Heating a mixture of KOH and MnO_2 in air gives green potassium manganate.

$$2MnO_2 + 4KOH + O_2 \rightarrow 2K_2MnO_4 + 2H_2O$$

Potassium manganate is the precursor to potassium permanganate, a common oxidant (with NaOH, the end product is not sodium manganate but an Mn(V) compound, which is one explanation why the potassium permanganate is more frequently used than sodium permanganate. Moreover, the potassium salt crystallizes better.) The potassium manganate is then transformed into permanganate via electrolytic oxidation in an alkaline media.

$$2K_2MnO_4 + 2H_2O \rightarrow 2KMnO_4 + 2KOH + H_2$$

While of no industrial importance, potassium manganate can be oxidized by chlorine or by disproportionation under acid conditions. The chlorine oxidation reaction can be written as

$$2K_2MnO_4 + Cl_2 \rightarrow 2KMnO_4 + 2KCl$$

And the acid-induced disproportionation reaction can be written as

$$3K_2MnO_4 + 4HCl \rightarrow 2KMnO_4 + MnO_2 + 2H_2O + 4KCl$$

A weak acid like carbonic acid is enough for this reaction.

$$3K_2MnO_4 + 2CO_2 \rightarrow 2KMnO_4 + MnO_2 + 2K_2CO_3$$

Permanganate salts can also be prepared by reacting a solution of Mn^{2+} ions with strong oxidants, for example, lead dioxide (PbO_2), sodium bismuthate ($NaBiO_3$), or peroxydisulfate. Tests for the presence of manganese use the vivid violet color of permanganate formed by these reagents. It dissolves in water to give intensely pink or purple solutions, the evaporation of which leaves prismatic purplish-black glistening crystals. In this compound, manganese is in the +7 oxidation state. Potassium hypomanganate, also known as potassium manganate(V), K_3MnO_4, is a bright blue solid and a rare illustration of a salt with the hypomanganate or manganate(V) anion where the manganese atom is in the +5 oxidation state. Preparative routes include (1) two-electron reduction of potassium permanganate with excess potassium sulfite

$$MnO_4^- + SO_3^{2-} + H_2O \rightarrow MnO_4^{3-} + SO_4^{2-} + 2H^+$$

(2) the single-electron reduction of potassium manganate with hydrogen peroxide in 10 M KOH solution

$$2MnO_4^{2-} + H_2O_2 + 2OH^- \rightarrow 2MnO_4^{3-} + O_2 + 2H_2O$$

(3) the single-electron reduction of potassium manganate with mandelate in 3–10 M KOH solution

$$2MnO_4^{2-} + C_8H_7O_3^- + 2OH^- \rightarrow 2MnO_4^{3-} + C_8H_5O_3^- + 2H_2O$$

and (4) disproportionation when MnO_2 is dissolved in a concentrated solution of KOH:

$$2MnO_2 + 3OH^- \rightarrow MnO_4^{3-} + MnO(OH) + H_2O$$

The compound is unstable because of the predisposition of the hypomanganate anion to disproportionate in all but the most alkaline solutions.

Manganese(IV) oxide can be used as starting material for the preparation of other oxides. Heating it at 800°C produces manganese sesquioxide (Mn_2O_3) and manganese(II,III) oxide when heated above 1000°C. Reacting MnO_2 with H_2, CO, or methane reduces it to MnO. The heptoxide of manganese (Mn_2O_7) also exists and has been described in early literature. However, it is not prepared intentionally as it is described to be able to explode even at 3°C. The explosive reaction of the heptoxide can be triggered by striking the sample or by contact with oxidizable organic compounds. Mn_2O_7 forms as a dark green oil through the addition of concentrated H_2SO_4 to $KMnO_4$. The reaction forms in the beginning permanganic acid, $HMnO_4$ (structurally, $HOMnO_3$), which is dehydrated by cold sulfuric acid to form its anhydride, Mn_2O_7.

$$2KMnO_4 + 2H_2SO_4 \rightarrow Mn_2O_7 + H_2O + 2KHSO_4$$

Mn_2O_7 can react further with sulfuric acid to form the remarkable manganyl(VII) cation MnO_3^+, which is isoelectronic with CrO_3

$$Mn_2O_7 + 2H_2SO_4 \rightarrow 2[MnO_3]^+[HSO_4]^- + H_2O$$

Mn_2O_7 decomposes near room temperature and explodes above 55°C. The products are MnO_2 and O_2. Ozone is also produced, giving a strong smell to the substance. The ozone can spontaneously ignite a piece of paper impregnated with an alcohol solution.

Manganese(II) oxide, MnO, forms green crystals. MnO occurs in nature as the rare mineral manganosite. Commercially it is produced via the reduction of MnO_2 with hydrogen, carbon monoxide or methane, for example,

$$MnO_2 + H_2 \rightarrow MnO + H_2O$$
$$MnO_2 + CO \rightarrow MnO + CO_2$$

MnO can also be produced through decarbonation of the carbonate.

$$MnCO_3 \rightarrow MnO + CO_2$$

This calcining process is performed anaerobically, otherwise Mn_2O_3 will form. Another method, mostly for demonstration purposes, is the oxalate method, which is also relevant to the synthesis of ferrous oxide and stannous oxide. Upon heating in an oxygen-free atmosphere (usually CO_2), manganese(II) oxalate decomposes into MnO.

$$MnC_2O_4{\cdot}2H_2O \rightarrow MnO + CO_2 + CO + 2H_2O$$

Similar to many monoxides, MnO has the rock salt (NaCl) structure, where cations and anions are both octahedrally coordinated. In addition, similar to many oxides, manganese(II) oxide is frequently nonstoichiometric: its composition can vary from MnO to $MnO_{1.045}$. Below −155°C (118K), MnO is antiferromagnetic. MnO has the honor of being one of the first compounds to have its magnetic structure revealed by neutron diffraction by Shull et al. in 1951. This study proved that the Mn^{2+} ions form a face-centered cubic magnetic sublattice where there are ferromagnetically coupled sheets that are antiparallel with adjacent sheets. Manganese(II) oxide undergoes the chemical reactions characteristic for an ionic oxide. Upon treatment with acids, it transforms to the corresponding manganese(II) salt and water. Oxidation of manganese(II) oxide results in the formation of manganese(III) oxide. Together with manganese sulfate, MnO is a component of fertilizers and food additives. Other applications are, for example, as a catalyst in the manufacture of allyl alcohol, ceramics, paints, colored glass, bleaching tallow, and textile printing.

5.7.3.2 Halides

With respect to reaction with other nonmetals, manganese is relatively unreactive at room temperature but combines with many at elevated temperatures. It burns in chlorine to give manganese(II) chloride ($MnCl_2$); reacts with fluorine to give manganese(II) fluoride (MnF_2) and manganese(III) fluoride (MnF_3). Manganese(II) chloride describes a series of compounds with the formula $MnCl_2.xH_2O$, where the value of x can be 0, 2, or 4. The tetrahydrate $MnCl_2{\cdot}4H_2O$ is the most common form of "manganese(II) chloride." The anhydrous form and a dihydrate $MnCl_2{\cdot}2H_2O$ also exist. Similar to many Mn(II) species, these salts are pink, with the paleness of the color characteristic of transition metal complexes with high-spin d^5 configurations. Manganese chloride is prepared by reacting manganese(IV) oxide with concentrated hydrochloric acid.

$$MnO_2 + 4HCl \rightarrow MnCl_2 + 2H_2O + Cl_2$$

This reaction was previously used for the production of chlorine. By carefully neutralizing the formed solution with $MnCO_3$, iron salts, which are common impurities in MnO_2, can be selectively precipitated. In the laboratory, manganese chloride can be formed by reacting manganese metal or manganese(II) carbonate with hydrochloric acid.

$$Mn + 2HCl + 4H_2O \rightarrow MnCl_2(H_2O)_4 + H_2$$
$$MnCO_3 + 2HCl + 3H_2O \rightarrow MnCl_2(H_2O)_4 + CO_2$$

Anhydrous $MnCl_2$ has a layered cadmium chloride-like structure. The tetrahydrate comprises octahedral *cis*-Mn$(H_2O)_4Cl_2$ molecules. The *trans* isomer, which is metastable, is also known. The dihydrate $MnCl_2(H_2O)_2$ forms a coordination polymer. Each Mn center is coordinated to four doubly bridging Cl ligands. The octahedron is completed by a pair of mutually *trans* aquo ligands. The hydrates dissolve in water to produce mildly acidic solutions with a pH of about 4. These solutions comprise the metal aquo complex $[Mn(H_2O)_6]^{2+}$. It is a weak Lewis acid, reacting with chloride ions to form a series of solids containing the following ions $[MnCl_3]^-$, $[MnCl_4]^{2-}$, and $[MnCl_6]^{4-}$. Both $[MnCl_3]^-$

and $[MnCl_4]^{2-}$ are polymeric. Upon treatment with typical organic ligands, manganese(II) undergoes oxidation by air to produce Mn(III) complexes, for example, $[Mn(EDTA)]^-$, $[Mn(CN)_6]^{3-}$, and $[Mn(acetylacetonate)_3]$. Triphenylphosphine forms a labile 2:1 adduct.

$$MnCl_2 + 2Ph_3P \rightarrow [MnCl_2(Ph_3P)_2]$$

Anhydrous manganese(II) chloride is a starting material for the synthesis of numerous manganese compounds. For instance, manganocene is formed through the reaction of $MnCl_2$ with a solution of sodium cyclopentadienide in THF.

$$MnCl_2 + 2NaC_5H_5 \rightarrow Mn(C_5H_5)_2 + 2NaCl$$

Manganese tetrafluoride, MnF_4, is the highest fluoride of manganese and is a powerful oxidizing agent that is used as a means of purifying elemental fluorine. MnF_4 was first unambiguously prepared in 1961 (reports of the preparation of MnF_4 date back to the 19th century, but are contradictory with the now-known chemistry of the actual compound.) through the reaction of MnF_2 (or other Mn(II) compounds) with a stream of fluorine gas at 550°C: the MnF_4 sublimed into the gas stream and was condensed onto a cold finger. This is still the most common preparation method, though the sublimation can be avoided by working at increased fluorine pressure (4.5–6 bar at 180°C–320°C) and mechanically agitating the powder to prevent sintering of the grains. The reaction can similarly be performed starting from manganese powder in a fluidized bed. Other preparation methods of MnF_4 are, for example, the fluorination of MnF_2 with krypton difluoride, or with F_2 in liquid hydrogen fluoride solution under ultraviolet light. MnF_4 has also been formed (but not isolated) in an acid-base reaction between antimony pentafluoride and K_2MnF_6 as part of a chemical synthesis of elemental fluorine.

$$K_2MnF_6 + 2SbF_5 \rightarrow MnF_4 + 2KSbF_6$$

Manganese tetrafluoride is in equilibrium with manganese(III) fluoride and elemental fluorine.

$$2MnF_4 \rightleftharpoons 2MnF_3 + F_2$$

Decomposition is favored by increasing the temperature and disfavored by the presence of fluorine gas, but the exact parameters of the equilibrium are uncertain, with some scientists saying that MnF_4 will decompose slowly at room temperature, while others place a practical lower temperature limit at 70°C, and another claiming that MnF_4 is basically stable up to 320°C. The yellow hexafluoromanganate(2 −) of alkali metal and alkaline earth metal cations have been recognized as early as 1899, and can be formed by the fluorination of MnF_2 in the presence of the fluoride of the appropriate cation. They are far more stable than manganese tetrafluoride. Potassium hexafluoromanganate(IV), K_2MnF_6, can also be produced through the controlled reduction of potassium permanganate in 50% aqueous hydrofluoric acid.

$$2KMnO_4 + 2KF + 10HF + 3H_2O_2 \rightarrow 2K_2MnF_6 + 8H_2O + 3O_2$$

The pentafluoromanganate(1 −) salts of potassium, rubidium, and cesium, $MMnF_5$, can be formed via fluorination of $MMnF_3$ or by the reaction of $[MnF_4(py)(H_2O)]$ with MF. The lemon-yellow heptafluoromanganate(3 −) salts of the same metals, M_3MnF_7, have also been synthesized.

Manganese(III) fluoride (also known as manganese trifluoride), MnF_3, is a red/purplish solid useful for converting hydrocarbons into fluorocarbons, that is, it is a fluorination agent. It forms a hydrate and numerous derivatives. MnF_3 can be produced by reacting a solution of MnF_2 in hydrogen fluoride with fluorine.

$$2MnF_2 + F_2 \rightarrow 2MnF_3$$

It can also be formed via the reaction of elemental fluorine with a manganese(II) halide at around 250°C. Like vanadium(III) fluoride, MnF_3 features octahedral metal centers with the same average M-F bond distances. In the solid Mn compound, though, it is distorted (and therefore a monoclinic unit cell instead of a higher-symmetry one) due to the Jahn-Teller effect, with pairs of Mn-F distances of 1.79, 1.91, and 2.09 Å. The hydrate $MnF_3{\cdot}3H_2O$ is formed through crystallization of MnF_3 from hydrofluoric acid. The hydrate has two known polymorphs, with space groups $P2_1/c$ and $P2_1/a$. Each comprises the salt $[Mn(H_2O)_4F_2]^+[Mn(H_2O)_2F_4]^-$. MnF_3 is Lewis acidic and forms various derivatives, for example, $K_2MnF_3(SO_4)$. MnF_3 reacts with sodium fluoride to give the octahedral hexafluoride.

$$3NaF + MnF_3 \rightarrow Na_3MnF_6$$

Related reactions produce salts of the anions MnF_5^{2-} or MnF_4^-. These anions have chain and layer structures, respectively, with bridging F. Manganese remains six coordinated, octahedral, and trivalent in all these compounds. MnF_3 fluorinates organic compounds such as aromatic hydrocarbons, cyclobutenes, and fullerenes. On heating, it decomposes to manganese(II) fluoride. Manganese(II) fluoride, MnF_2, is a light pink crystalline solid. It is prepared by reacting manganese and diverse compounds of manganese(II) with hydrofluoric acid. It is used in the production of special types of glass and lasers. It is a canonical example of uniaxial antiferromagnet (with Neel temperature of −205°C/68K) which has been experimentally researched since early on.

5.7.3.3 *Nitride and sulfide*

Manganese combines with nitrogen at temperatures around 1200°C to give manganese(II) nitride (Mn_3N_2). Manganese also combines directly with boron, carbon, sulfur, silicon, or phosphorus but not with hydrogen. Moreover, it decomposes water slowly when cold and rapidly on heating. Its reaction with dilute mineral acids produces hydrogen and the corresponding salts in the +2 oxidation state. Manganese(II) sulfide, MnS, occurs in nature as the mineral alabandite (isometric), rambergite (hexagonal), and the recently found browneite (isometric, with sphalerite-type, ZnS, structure, extremely rare and is only known from a meteorite). Manganese(II) sulfide can be formed through the reaction of a manganese(II) salt (such as manganese(II) chloride) with ammonium sulfide.

$$(NH_4)_2S + MnCl_2 \rightarrow 2NH_4Cl + MnS$$

The crystal structure of manganese(II) sulfide is comparable to that of sodium chloride. The pink color of MnS is probably caused by poor coupling between the lowest energy unoccupied Mn orbitals, resulting in discrete states rather than a delocalized band. Consequently, the lowest energy band-to-band electronic transition requires very high-energy (ultraviolet) photons.

5.7.3.4 *Sulfate and carbonate*

Manganese(II) sulfate, $MnSO_4{\cdot}H_2O$, is a pale pink deliquescent solid which is a commercially important manganese(II) salt. It is the precursor to manganese metal and many other chemical compounds. Similar to many metal sulfates, manganese sulfate forms a variety of hydrates: monohydrate, tetrahydrate, pentahydrate, and heptahydrate. The monohydrate is most common. Manganese(II) sulfate minerals are very rare in nature and always occur as hydrates. The monohydrate is called szmikite; tetrahydrate = ilesite; hexahydrate (the rarest) = chvaleticeite; pentahydrate = jôkokuite; heptahydrate = mallardite. Mn-deficient soil can be remediated with this salt. All these salts dissolve to form faintly pink solutions of the aquo complex $[Mn(H_2O)_6]^{2+}$. The pale pink color of Mn(II) salts is highly characteristic. Usually, manganese ores are purified through their conversion to manganese(II) sulfate. Reaction of aqueous solutions of the sulfate with sodium carbonate results in precipitation of manganese carbonate, which can be calcined to form the oxides MnO_x. In the laboratory, manganese sulfate can be prepared by reacting manganese dioxide with sulfur dioxide.

$$MnO_2 + SO_2 \rightarrow MnSO_4$$

It can also be produced by mixing potassium permanganate with sodium bisulfate and hydrogen peroxide. Manganese sulfate is a by-product of several industrially significant oxidation reactions that utilize MnO_2, including the production of hydroquinone and anisaldehyde. Electrolysis of manganese sulfate results in the formation of MnO_2, which is known as EMD. Instead, oxidation of manganese sulfate with potassium permanganate forms the so-called chemical manganese dioxide (CMD).

Manganese carbonate, $MnCO_3$, occurs naturally as the mineral rhodochrosite but it is normally manufactured industrially. It is a pale pink, water-insoluble solid. The carbonate is insoluble in water but, like most carbonates, hydrolyzes upon treatment with acids to give water-soluble salts. It decomposes with release of carbon dioxide, that is, calcining, at 200°C to give $MnO_{1.88}$.

$$MnCO_3 + 0.4O_2 \rightarrow MnO_{1.8} + CO_2$$

This method is occasionally used in the manufacture of manganese dioxide, which is used in dry cell batteries and for ferrites. $MnCO_3$ is extensively utilized as an additive to plant fertilizers to treat manganese deficient crops. It is also employed in health foods, in ceramics as a glaze colorant and flux, and in concrete stains. It is utilized in medicine as a hematinic (a nutrient required for the formation of blood cells in the process of haematopoiesis. The main hematinics are iron, B12, and folate). $MnCO_3$ has a similar structure as calcite ($CaCO_3$), comprising manganese(II) ions in an octahedral coordination geometry Treatment of aqueous solutions of manganese(II) nitrate with ammonia and carbon dioxide results in the precipitation of this faintly pink solid. The side product ammonium nitrate is used as fertilizer.

5.7.3.5 *Organomanganese compounds*

Despite the variety of oxidation states of manganese compounds, the most stable oxidation state is +2 and dominates the chemistry of manganese complexes. Organomanganese complexes are, therefore, manganese(II) or manganese(0) when combined with carbonyls and cyclopentadienyls as ligands. Two of the earliest organomanganese compounds are in the +2 oxidation state and was described in 1939 by Gilman and Bailie as cited by Cahiez et al. (2009). These are phenylmanganese iodide and diphenylmanganese, prepared from the reaction of phenyllithium and manganese iodide. Since then, reports on preparation and reactivity of organomanganese compound are sporadic and only picked up 40 years ago. The lagging development of organomanganese chemistry is probably due to the lower versatility of Mn complexes. They are considered not very different from complex to complex and are often viewed as less interesting than nearby first-row metals. Nevertheless, Mn is interesting since it is cheap and toxicologically benign. It is therefore expected that pursuit of knowledge on the fundamental coordination and reaction chemistry of organomanganese complexes picks up in the coming years.

The organomanganese(II) compounds synthesized thus far are prepared using transmetalation from the corresponding organolithium or organomagnesium reagents. Provided the starting organolithium or organomagnesium reagents are available or easily prepared, a vast array of alkyl, alkenyl, alkynyl, allyl, benzyl, and aryl or heteroarylmanganese compounds can be synthesized. The general scheme for preparing these compounds is shown below. The reactions are written in the order: (1) organomanganese halides, (2) symmetrical organomanganeses, and (3) organomaganates as the general class of compounds that can be prepared.

$$RM + MnX_2 \rightarrow RMnX \quad (M = Li, MgX')$$
$$RM + MnX_2 \rightarrow R_2Mn$$
$$RM + MnX_2 \rightarrow R_4MnM_2$$

The recent discovery that Mn(I) ions are useful in a variety of (de)hydrogenation reactions previously performed by ruthenium, iron, and cobalt pincer catalysts brought about a resurgence of organomanganese(I) compounds. Synthesis and reactivity of these type of compounds are presented by Kadassery et al. (2019). The stimulant to the group's interest in organomanganese(I) compounds had been the search for Mn complexes that can mediate photon-induced water splitting reactions. One such complex, the tetramer $[Mn(CO)_3(\mu_3\text{-}OH)]_4$, was shown to be the only complex that produced H_2 gas under rigorously dry conditions through a CO-photolysis-induced H-atom-transfer reaction. Wanting to understand the formation mechanism of the tetramer, the group unexpectedly created a novel tetranuclear butterfly complex with bridging carbonato ligands derived exclusively from $Mn_2(CO)_{10}$ and Me_3NO. The formation of the carbonate meant involvement of an intermediate Mn(I) compound. Indeed, the group established the formation of $Mn^{I}(CO)_3$ fragments charge-balanced by carbonato groups. The carbonato ligands in the butterfly complex are basic moieties. These were envisioned to play a role in direct activation of the Brønsted acidic bonds on various ligand platforms for generation of the corresponding manganese(I) tricarbonyl complexes. These manganese(I) tricarbonyl complexes are often precursors for generation of the manganese(I) dicarbonyl active catalyst. From their unexpected results, Kadassery et al. (2019) were able to develop better manganese(I) synthon (a synthon is a destructural unit within a molecule which is related to a possible synthetic operation such as carbocations and carbanions). Their works encountered challenges and they identified the three main difficulties. (1) There is a dearth of manganese precursors in forming active manganese(I) pincer catalysts. (2) The bifunctional ligands all rely on N atoms. (3) There is a lack of methods to prepare manganese(I) dicarbonyl complexes. Despite these challenges, the low cost and potential in biomedicine of manganese compounds can drive research in this area.

5.7.4 Major uses

Mn is too brittle to be of much use as a pure metal. Mn is vital to iron and steel manufacture due to its sulfur-fixing, deoxidizing, and alloying properties, as first documented by the British metallurgist and businessman Robert Forester Mushet (April 8, 1811 to January 29, 1891) who introduced the element in 1856, in the form of Spiegeleisen (literally "mirror-iron," German: Spiegel—mirror or specular; Eisen—iron, a ferromanganese alloy containing approximately 15% Mn and small quantities of C and Si), into steel to remove excess dissolved oxygen, sulfur, and phosphorus with the purpose to improve its malleability. Steel production, including its iron manufacture component, has accounted for most of the Mn usage, currently between 85% and 90% of total Mn usage. Mn is a major component of low-cost stainless steel. Ferromanganese (typically ~80% Mn) is the intermediary in many current processes. Small Mn quantities improve the workability of steel at high temperatures as they form a high-melting sulfide and inhibit the formation of a liquid Fe sulfide at the grain boundaries. When the Mn concentration reaches 4%, the embrittlement of the steel turns into a dominant feature. The embrittlement decreases at higher Mn contents and gets to an acceptable level at 8%. Steel with 8%–15% of Mn shows a high tensile strength as high as 863 MPa. Steel with 12% Mn was discovered by English metallurgist Sir Robert Abbott Hadfield (November 28, 1858 to September 30, 1940) in 1882 and is even now known as Hadfield steel (mangalloy). It was used for British military steel helmets and later by the US military. It is also used for railway tracks, safes, rifle barrels, and prison bars. The second major use for Mn is in Al alloys. Al with around 1.5% Mn exhibits increased resistance to corrosion through grains that absorb impurities which would lead to galvanic corrosion. The corrosion-resistant Al alloys 3004 and 3104 (0.8%–1.5% Mn) are used for most beverage cans. With Al, Sb, and Cu it forms highly magnetic alloys.

Methylcyclopentadienyl manganese tricarbonyl (IUPAC name tricarbonyl(methyl-η^5-cyclopentadienyl)manganese, $(C_5H_4CH_3)Mn(CO)_3$) is used as an additive in unleaded gasoline to improve octane rating and lessen engine knocking. Mn in this unusual organometallic molecule is in the 1 + oxidation state. Mn(IV) oxide (MnO_2) is used as a reagent in organic chemistry for the oxidation of benzylic alcohols (where the OH-group is adjacent to an aromatic ring). MnO_2 has been used throughout history to oxidize and neutralize the greenish tinge in glass due to trace amounts of Fe contamination. In addition, MnO_2 is used in the production of O_2 and Cl_2 as well as in drying black paints. In some formulations, it is a brown pigment for paint and is a constituent of natural umber (a natural brown or reddish-brown earth pigment that contains iron oxide and manganese oxide. Umber is darker than the other similar earth pigments, ocher and sienna). MnO_2 was also used in the original type of dry cell battery as an electron acceptor from Zn and is the blackish compound in C-Zn type flashlight cells. The MnO_2 is reduced to the Mn oxide hydroxide MnO(OH) during discharging, blocking the formation of hydrogen at the anode of the battery.

$$MnO_2 + H_2O + e^- \rightarrow MnO(OH) + OH^-$$

The same compound also functions in newer alkaline batteries (usually battery cells), which is based on the same basic reaction, but a different electrolyte mixture. Mn compounds have been applied as pigments and for the coloring of ceramics and glass. The brown color of ceramic is occasionally due to Mn compounds. In the glass industry, Mn compounds are used for two reasons. Mn(III) reacts with Fe(II) to produce a strong green color in glass by forming less-colored Fe(III) and slightly pink Mn(II), compensating for the residual color of the Fe(III). Larger amounts of Mn are used to produce pink-colored glass. Mn(IV) is used as an activator in red-emitting phosphors. While many materials are known to exhibit luminescence, most are not used in commercial applications because of their low efficiency or deep red emission. Nevertheless, a few Mn^{4+} activated fluorides have been described as possible red-emitting phosphors for warm-white light-emitting diodes (LEDs). Yet until now, $K_2SiF_6{:}Mn^{4+}$ is the only commercially available compound for use in warm-white LEDs. The metal is occasionally used in coins. Mn sulfate is used to make a fungicide. Mn(II) oxide is a powerful oxidizing agent and is used in quantitative analysis. It is also used to make fertilizer.

5.8 26 Fe — Iron

5.8.1 Discovery

As iron has been used for such a long time in history, it has numerous different names in different languages. The basis of its chemical symbol Fe is the Latin word ferrum, and its descendants are the names of the element in the Romance languages (for example, French fer, Spanish hierro, and Italian and Portuguese ferro). The word ferrum itself perhaps comes from the Semitic languages, via Etruscan, from a root that also gave rise to Old English bræs "brass." The English word iron originates ultimately from Proto-Germanic *isarnan, which is also the source of the German name Eisen and the Dutch ijzer. It was most likely borrowed from Celtic *isarnon, which originally comes from Proto-Indo-European *is-(e)ro- "powerful, holy" and lastly *eis "strong," referencing iron's strength as a metal. Iron is one of the elements undoubtedly known to antiquity. It has been worked, or wrought, for thousands of years. Nevertheless, iron objects of great age are much scarcer than objects made of gold or silver due to the ease with which iron corrodes. Beads made from meteoric iron in 3500 BCE or earlier were found in Gerzah, Egypt. The beads contain 7.5% nickel, which is a signature of meteoric origin as iron found in the Earth's crust generally has only infinitesimally low nickel content. Meteoric iron was highly regarded because of its origin from the heavens and was regularly used to forge weapons and tools. For example, a dagger made of meteoric iron was found in the tomb of Tutankhamun (c.1341–1323 BCE), containing comparable contents of iron, cobalt, and nickel as a meteorite found in the area, remnant of an ancient meteor shower. Items that were probably made of iron by the Egyptians date from 3000 to 2500 BCE. Meteoritic iron is comparably soft and ductile and easily forged by cold working but might get brittle when heated due to its nickel content. The first iron production began in the Middle Bronze Age but it took several hundreds of years before iron displaced bronze. Samples of smelted iron from Asmar, Mesopotamia and Tall Chagar Bazaar in northern Syria were produced sometime between 3000 and 2700 BCE. The Hittites established an empire in north-central Anatolia around 1600 BCE. They seem to be the first to understand the liberation of iron from its ores and regarded it highly in their culture. The Hittites started to smelt iron between 1500 and 1200 BCE and the practice spread to the rest of the Near East after their empire fell in 1180 BCE. The following period is known as the Iron Age. Smelted iron artifacts are found in India dating from 1800 to 1200 BCE, and in the Levant from around 1500 BCE (suggesting smelting in Anatolia or the Caucasus). Supposed references to iron in the Indian Vedas have been used for claims of a very early usage of iron in India respectively to date the texts as such. The rigveda term ayas (metal) likely refers to copper and bronze, while iron or śyāma ayas, literally "black metal," is first cited in the postrigvedic Atharvaveda. Some archeological evidence seems to indicate that iron was smelted in Zimbabwe and southeast Africa as far back as the 8th century BCE. Iron working was introduced to Greece in the late 11th century BCE, from where it spread quickly throughout the rest of Europe. This spread of ironworking in Central and Western Europe is related to the Celtic expansion. According to Roman author, naturalist and natural philosopher Pliny the Elder (CE 23–79), iron usage was common in the Roman era. The annual iron production of the Roman Empire is estimated at 84,750 tons, while the equally populous and contemporary Han China produced around 5000 tons. In China, iron only appeared between 700 and 500 BCE. Iron smelting may have been introduced into China through Central Asia. The oldest indication of the use of a blast furnace in China comes from the 1st century CE, and cupola furnaces were utilized as early as the Warring States period (403–221 BCE). Usage of the blast and cupola furnace continued to be widespread during the Song and Tang Dynasties. During the Industrial Revolution in Britain, English ironmaster Henry Cort (c.1740 to May 23, 1800) started refining iron from pig iron to wrought iron (or bar iron) applying innovative manufacturing systems (Fig. 5.65). In 1783 he patented the puddling process for refining iron ore. It was later improved by others, including metallurgist Joseph Hall (1789–1862), a former apprentice of Henry Cort. Cast iron was first made in China during the 5th century BCE but was scarcely found in Europe until the medieval period. The oldest cast iron artifacts were found by archeologists in what is now modern Luhe County, Jiangsu in China. Cast iron was used in ancient China for warfare, agriculture, and architecture. During the medieval period, methods were developed in Europe to produce wrought iron from cast iron (in this context known as pig iron) using finery forges. For all these processes, charcoal was essential as fuel. Medieval blast furnaces were about 10 ft. (3.0 m) tall and made of fireproof brick; forced air was generally produced by hand-operated bellows. Modern blast furnaces have become much larger, with hearths 14 m in diameter that permit the production of thousands of tons of iron per day, but basically work in much the same way as they did during medieval times. In 1709 English Quaker Abraham Darby I (April 14, 1678 to March 8, 1717) founded a coke-fired blast furnace to produce cast iron, replacing charcoal, while continuing to use blast furnaces. The resulting

FIGURE 5.65 Henry Cort, c.1780s.

FIGURE 5.66 Coalbrookdale by Night, 1801, painting by Philip James de Loutherbourg (1740–1812). Blast furnaces light the iron making town of Coalbrookdale.

availability of low-cost iron was one of the factors leading to the Industrial Revolution (Fig. 5.66). Near the end of the 18th century, cast iron began to substitute wrought iron for particular purposes, since it was cheaper. Carbon content in iron was not implicated as the cause for the differences in properties of wrought iron, cast iron, and steel until the 18th century. As iron was becoming cheaper and more abundant, it also became a key structural material after the building of the first ground-breaking iron bridge in 1778. This bridge still exists today as a monument to the part iron played in the Industrial Revolution. Subsequently, iron was used in rails, boats, ships, aqueducts, and buildings, as well as in iron cylinders in steam engines. Railways have been essential to the formation of modernity and ideas of progress and various languages (e.g., French chemin de fer, Spanish ferrocarril, Italian ferrovia, and German Eisenbahn) refer to railways as iron road.

5.8.2 Mining, production, and major minerals

Metallic or native iron is seldom found on the surface of the Earth as it tends to quickly oxidize when exposed to oxygen. Nevertheless, both the Earth's inner and outer cores, which account for 35% of the mass of the whole Earth, are believed to consist largely of an iron alloy, possibly with nickel. Electric currents in the liquid outer core are supposed to be the origin of the Earth's magnetic field. While iron is the most abundant element on Earth, it accounts for only 5% of the Earth's crust; consequently, being only the fourth most abundant element, after O, Si, and Al. Most of the iron in the crust is combined with various other elements to form many different minerals. An important class is the iron minerals such as hematite (Fe_2O_3), magnetite (Fe_3O_4), and siderite ($FeCO_3$), which are the major ores of iron. Many igneous rocks also contain the sulfide minerals pyrrhotite (Fe_7S_8), and pentlandite ($(Fe,Ni)_9S_8$). During weathering, iron tends to leach from sulfide deposits as the sulfate and from silicate deposits as the bicarbonate. Both of these are oxidized in aqueous solution and precipitate in even mildly elevated pH as iron(III) oxide. Large deposits of iron are the so-called banded iron formations (BIFs), a type of rock consisting of repeated thin layers of iron oxides alternating with bands of iron-poor shale and chert. These BIFs were formed between 3700 million years ago and 1800 million years ago (Box 5.12).

Currently, the industrial manufacture of iron or steel consists of two main steps. In the first step, iron ore is reduced with coke in a blast furnace, and the molten metal is separated from gross impurities such as silicate minerals. This step produces an alloy—pig iron—that has relatively large quantities of carbon. In the second step, the amount of carbon in the pig iron is decreased through oxidation to produce wrought iron, steel, or cast iron. Other metals can be added at this stage to produce alloy steels. The blast furnace is loaded with iron ores, generally hematite, Fe_2O_3, or magnetite, Fe_3O_4, plus coke (coal that has been separately baked to remove volatile components). Air preheated to 900°C is blown through the mixture in sufficient amount to allow the carbon to react to carbon monoxide.

$$2C + O_2 \rightarrow 2CO$$

This reaction increases the temperature to ~2000°C The carbon monoxide then reduces the iron ore to metallic iron.

$$Fe_2O_3 + 3CO \rightarrow 2Fe + 3CO_2$$

Some iron in the high-temperature lower region of the furnace can react directly with the coke.

$$2Fe_2O_3 + 3C \rightarrow 4Fe + 3CO_2$$

BOX 5.12 Banded iron formations (BIFs)

The largest concentrations of iron oxides are found in BIFs, which currently supply most of the world's iron ores and constitute the bulk of the world's iron ore reserves. The BIFs occur in Precambrian-aged rocks of all continents and are mined extensively in the United States, Canada, Brazil, Venezuela, Australia, South Africa, India, China, and Russia. They are generally between 30 and 700 m thick and often extend over hundreds to thousands of square kilometers. Many of these deposits have been exposed to some degree of metamorphosis, so that they now consist of fine-grained magnetite (Fe_3O_4) and/or hematite (Fe_2O_3) in a matrix of quartz, iron silicate, and iron carbonate minerals in a very compact, finely laminated rock. The BIFs generally display strong facies development, suggesting that at the same time differing conditions resulting in the formation of iron oxide-rich zones closest to the shore, and iron carbonate or iron silica zones farther out into the basin. The iron contents of the BIFs vary widely, but the presently mined deposits generally contain 20%–40% iron. The BIFs mode of origin is still not completely understood. BIFs have a number of major characteristics. The most important is that most (if not all) of them formed during the period of 3.6–1.8 billion years ago. They all show the typical banding and they are all very low in aluminum content and are nearly free of common detrital sedimentary debris. They resemble the Algoma-type deposits (after the locality in the Canadian shield, related to seafloor volcanism accompanied by submarine hot springs that issue forth solutions rich in iron and silica. The rapid cooling and oxidation of the solutions as they mix with sea water results in the precipitation of iron oxide and silica) in that they consist of fine layers of silica and iron oxide minerals, but the BIFs are much broader in extent and do not show any apparent relationship to submarine volcanism. Detailed research of the BIFs suggests that they formed in broad sedimentary basins following prolonged periods of continental weathering and erosion, and the inundation of the land surface by shallow seas. The erosion had earlier removed most of the detrital debris so that the deposition in the basins was largely that of a chemical precipitate. The origin of the iron is not clear, but it is possible that it originated from several different sources such as weathering of continental rocks, leaching of marine sediments, and submarine hydrothermal systems that discharged iron-rich fluids onto the seafloor. Currently it would be impossible to transport the huge quantities of iron from eroding land surfaces in rivers and streams or to disperse it widely from submarine vents, as it would be rapidly oxidized and precipitated in the insoluble ferric form. But, during the Precambrian when the BIFs formed, there was undoubtedly little free oxygen in Earth's atmosphere or dissolved in surface waters. The carbon dioxide concentration in the atmosphere was probably much higher, and gases such as methane, CH_4, might have been present. Under these conditions, rainwater, stream and lake waters, and ocean waters would have been slightly acidic and much less oxidizing, conditions that would have permitted the transport of iron in the soluble ferrous form. The iron accumulated in the broad shallow basins and slowly precipitated as iron oxides and hydroxides. The reason for the repetitive precipitation of layers of alternating iron oxides and silica has been a matter of intense discussions, with suggestions including annual climatic changes, cyclical periods of evaporation, the effects of microorganisms, changes in silica availability and releasing oxygen, and episodic volcanism. Whatever the final understanding of the origin of the BIFs, their formation was apparently directly controlled by the nature of the Precambrian atmosphere. As the atmosphere changed over time, so did the capacity of the ocean to act as a transporter of iron. When photosynthesis began to contribute large amounts of free oxygen into the atmosphere, banded iron deposits stopped forming; therefore there are no equivalent processes active in the world today.

A flux such as limestone ($CaCO_3$) or dolomite ($CaMg(CO_3)_2$) is also added to the furnace's load. Its purpose is to eliminate silica-containing minerals in the ore, which would otherwise clog the furnace. The furnace heat results in the decomposition of the carbonates to calcium oxide, which subsequently reacts with any excess silica present to produce a slag consisting of calcium silicate $CaSiO_3$ or other related products. At the furnace's temperature, the metal and the slag are both molten. They collect at the bottom as two immiscible liquid layers (with the slag on top), that can then without difficulty be separated. The slag can be used, for example, as a material in road construction or to improve mineral-poor soils for agriculture. Generally, the pig iron formed by the blast furnace process has up to 4%–5% carbon, with minor quantities of other impurities such as sulfur, magnesium, phosphorus, and manganese. This high level of carbon causes it to be relatively weak and brittle. Reducing the carbon amount to 0.002%–2.1% by mass produces steel, which can be as much as 1000 times harder than pure iron. Removing the other impurities, instead, produces cast iron, which is utilized to cast articles in foundries; for example, stoves, pipes, radiators, lampposts, and rails. Steel products are often exposed to various heat treatments after they are forged to shape. Annealing comprises heating the steel to 700°C–800°C for several hours followed by gradual cooling, making the steel softer and more workable. Due to environmental concerns, other methods of processing iron have been developed. "Direct iron reduction" reduces iron ore to a ferrous lump called "sponge" iron or "direct" iron that is appropriate for steelmaking. Two main reactions cover the direct reduction process: natural gas (methane) is partially oxidized (with heat and a catalyst).

$$2CH_4 + O_2 \rightarrow 2CO + 4H_2$$

The iron ore is subsequently reacted with these gases in a furnace, producing solid sponge iron.

$$Fe_2O_3 + CO + 2H_2 \rightarrow 2Fe + CO_2 + 2H_2O$$

Silica is removed by adding a limestone flux. Ignition of a mixture of aluminum powder and iron oxide produces metallic iron via the thermite reaction.

$$Fe_2O_3 + 2Al \rightarrow 2Fe + Al_2O_3$$

Instead, pig iron may be made into steel (with up to ~2% carbon) or wrought iron (commercially pure iron). Various processes have been utilized for this, such as finery forges (a forge used to produce wrought iron, from pig iron by decarburization. The process involved liquifying cast iron in a fining hearth and removing carbon from the molten cast iron through oxidation), puddling furnaces (the puddling furnace is a metalmaking technology used to create wrought iron or steel from the pig iron produced in a blast furnace. The furnace is constructed to pull the hot air over the iron without the fuel coming into direct contact with the iron, a system generally known as a reverberatory furnace or open-hearth furnace. The major advantage of this system is keeping the impurities of the fuel separated from the charge. The hearth is where the iron is charged, melted, and puddled.), Bessemer converters (the Bessemer process was the first inexpensive industrial process for the mass production of steel from molten pig iron before the development of the open-hearth furnace. The key principle is removal of impurities from the iron by oxidation with air being blown through the molten iron. The oxidation also raises the temperature of the iron mass and keeps it molten.), open-hearth furnaces, basic oxygen furnaces, and electric arc furnaces. In all cases, the objective is to oxidize some or all the carbon, together with other impurities.

Iron is a very common element in nature, which is reflected in the high number of minerals that (can) contain Fe in the crystal structure; between 1300 and 1350 minerals fall into this group. In the elements class 29 minerals are known, including the native element iron (Fe), as well as a variety of alloys, carbides, phosphides, etc., for example, awaruite (Ni_3Fe), barringerite ($(Fe,Ni)_2P$), cohenite (Fe_3C), hapkeite (Fe_2Si), and siderazot (Fe_5N_2). More than 100 sulfide class minerals are known to contain iron, including some very well-known minerals, for example, arsenopyrite (FeAsS), bornite (Cu_5FeS_4), chalcopyrite ($CuFeS_2$), löllingite ($FeAs_2$), marcasite (FeS_2) (Fig. 5.67), pyrite (FeS_2) (Fig. 5.68), pyrrhotite (Fe_7S_8) (Fig. 5.69), tetrahedrite ($Cu_6[Cu_4(Fe,Zn)_2]Sb_4S_{13}$), troilite (FeS), and wurtzite (Zn,Fe)S (Fig. 5.70). The halides are represented by 11 minerals with Fe, for example, hibbingite ($Fe^{2+}_2(OH)_3Cl$) and rinneite ($K_3Na[FeCl_6]$). In the oxide class one finds over 150 different minerals with structural iron, such as chromite ($Fe^{2+}Cr^{3+}_2O_4$), columbite-(Fe) ($Fe^{2+}Nb_2O_6$), ferrihydrite ($Fe^{3+}_{10}O_{14}(OH)_2$), hematite (Fe_2O_3) (Fig. 5.71), hercynite ($Fe^{2+}Al_2O_4$), ilmenite ($Fe^{2+}TiO_3$), magnetite ($Fe^{2+}Fe^{3+}_2O_4$) (Fig. 5.72), pseudobrookite (Fe_2TiO_5), tantalite-(Fe) ($Fe^{2+}Ta_2O_6$), trevorite ($Ni^{2+}Fe^{3+}_2O_4$), ulvöspinel ($TiFe_2O_4$), and wüstite (FeO). The carbonates contain 11 minerals with Fe, including well-known minerals such as ankerite ($Ca(Fe^{2+},Mg)(CO_3)_2$), pyroaurite ($Mg_6Fe^{3+}_2(OH)_{16}[CO_3]·4H_2O$), and siderite ($FeCO_3$) (Fig. 5.73). Fifteen different minerals have been found in the borate class to contain structural Fe, for example, ludwigite ($Mg_2Fe^{3+}(BO_3)O_2$) and vonsenite ($Fe^{2+}_2Fe^{3+}(BO_3)O_2$). The sulfate class consists of 100 different minerals with Fe in the crystal structure, among these are, for example, botryogen ($MgFe^{3+}(SO_4)_2(OH)·7H_2O$), copiapite ($Fe^{2+}Fe^{3+}_4(SO_4)_6(OH)_2·20H_2O$), halotrichite ($FeAl_2(SO_4)_4·22H_2O$), jarosite ($KFe^{3+}_3(SO_4)_2(OH)_6$), and melanterite

FIGURE 5.67 Marcasite, FeS_2, an excellent example of cockscomb habit with crystals to 1 cm on edge. Komořany, Most, Ústí, Bohemia, Czech Republic.

FIGURE 5.68 Pyrite, FeS_2, malformed flattened cubes. Crystals to 6 mm. near Belle Fourche, Butte Co., South Dakota, United States.

FIGURE 5.69 Pyrrhotite, Fe_7S_8, a group of three pseudo-octahedral crystals. The largest is 1.5 cm long, the two others are short prismatic and 1.5 cm. Santa Eulalia, Chihuahua, Mexico.

FIGURE 5.70 Wurtzite, (Zn,Fe)S, lustrous crystals to 7 mm showing orange internal reflections. 6.5 × 5 × 2.5 cm. Chiverton Great Consols, Zelah, Cornwall, England.

FIGURE 5.71 A thin 1-mm plate of hematite, Fe_2O_3, crystal. Uri, Switzerland.

FIGURE 5.72 Magnetite, $Fe^{2+}Fe_2^{3+}O_4$, lustrous octahedral crystals to 1 cm. Mina Huaquino, Cerro Huenquino, Potosi, Bolivia.

FIGURE 5.73 Orange brown rhombs of siderite, $FeCO_3$, to 1.5 cm on a bed of small quartz crystals. Neudorf, Harz, Germany.

FIGURE 5.74 Purple strengite, $FePO_4{\cdot}2H_2O$, grouping with crystal balls to 2 mm. Three Oaks Gap, Polk Co., Arkansas, United States.

FIGURE 5.75 Bladed crystal of deep green vivianite, $Fe_3^{2+}(PO_4)_2{\cdot}8H_2O$, terminated with a bit of siderite, $FeCO_3$, matrix 4.5 cm long. Siglo Viente mine, Llallagua, Bolivia.

($Fe^{2+}(H_2O)_6SO_4{\cdot}H_2O$). The group of phosphate minerals is significantly larger, with more than 250 minerals, including minerals such as anapaite ($Ca_2Fe^{2+}(PO_4)_2{\cdot}4H_2O$), arthurite ($CuFe_2^{3+}(AsO_4)_2(OH)_2{\cdot}4H_2O$), beraunite ($Fe^{2+}Fe_5^{3+}(PO_4)_4(OH)_5{\cdot}6H_2O$), beudantite ($PbFe_3(AsO_4)(SO_4)(OH)_6$), cacoxenite ($Fe_{24}^{3+}AlO_6(PO_4)_{17}(OH)_{12}{\cdot}75H_2O$), dufrénite ($Ca_{0.5}Fe^{2+}Fe_5^{3+}(PO_4)_4(OH)_6{\cdot}2H_2O$), ludlamite ($Fe_3^{2+}(PO_4)_2{\cdot}4H_2O$), pharmacosiderite ($KFe_4^{3+}(AsO_4)_3(OH)_4{\cdot}6-7H_2O$), rockbridgeite ($Fe^{2+}Fe_4^{3+}(PO_4)_3(OH)_5$), scorodite ($Fe^{3+}AsO_4{\cdot}2H_2O$), scorzalite ($Fe^{2+}Al_2(PO_4)_2(OH)_2$), strengite ($FePO_4{\cdot}2H_2O$) (Fig. 5.74), vauxite ($Fe^{2+}Al_2(PO_4)_2(OH)_2{\cdot}6H_2O$), and vivianite ($Fe_3^{2+}(PO_4)_2{\cdot}8H_2O$) (Fig. 5.75). Iron is a common element in many silicates with more than 300 minerals known to contain the element. Examples of these minerals include aegirine ($NaFe^{3+}Si_2O_6$), almandine ($Fe_3^{2+}Al_2(SiO_4)_3$), andradite ($Ca_3Fe_2^{3+}(SiO_4)_3$), cordierite ($(Mg,Fe)_2Al_3(AlSi_5O_{18})$), epidote ($\{Ca_2\}\{Al_2Fe^{3+}\}(Si_2O_7)(SiO_4)O(OH)$), fayalite ($Fe_2^{2+}SiO_4$), hedenbergite ($CaFe^{2+}Si_2O_6$), ilvaite ($CaFe^{3+}Fe_2^{2+}(Si_2O_7)O(OH)$), jadeite ($Na(Al,Fe^{3+})Si_2O_6$), schorl ($Na\left(Fe_3^{2+}\right)Al_6(Si_6O_{18})(BO_3)_3(OH)_3(OH)$), and staurolite ($Fe_2^{2+}Al_9Si_4O_{23}(OH)$). Finally, three oxalates are known within the class of organic compounds; humboldtine ($Fe^{2+}(C_2O_4){\cdot}2H_2O$), minguzzite ($K_3Fe^{3+}(C_2O_4)_3{\cdot}3H_2O$), and stepanovite ($NaMgFe^{3+}(C_2O_4)_3{\cdot}8\text{-}9H_2O$).

5.8.3 Chemistry

The Fe isotopes include four stable ones: ^{54}Fe (5.845% natural abundancy), ^{56}Fe (91.754%), ^{57}Fe (2.119%), and ^{58}Fe (0.282%). A couple dozen of isotopes have also been synthesized. Of these stable isotopes, only ^{57}Fe has a nuclear spin ($-1/2$). The nuclide ^{54}Fe hypothetically can undergo double electron capture to form ^{54}Cr, but the process has never been detected and only a lower limit on the half-life of 3.1×10^{22} years has been determined. ^{60}Fe is an extinct radionuclide with a long half-life of 2.6 million years). It is not present on Earth, but its final decay product is its granddaughter, the stable nuclide ^{60}Ni. Most of past research on the isotopic composition of Fe has focused on the nucleosynthesis of ^{60}Fe by studying meteorites and ore formation. In phases of meteorites such as Semarkona and Chervony Kut, a correlation between the concentration of ^{60}Ni, the granddaughter of ^{60}Fe, and the concentrations of the stable Fe isotopes provided proof for the existence of ^{60}Fe at the time of the solar system's formation. It is possible that the energy released through the decay of ^{60}Fe, together with that of ^{26}Al, contributed to the remelting and differentiation of asteroids after their formation 4.6 billion years ago. The most abundant isotope ^{56}Fe is of special interest to nuclear scientists as it represents the most common endpoint of nucleosynthesis. As ^{56}Ni (14 α particles) is easily formed from lighter nuclei in the α process in nuclear reactions in supernovae, it forms the endpoint of fusion chains inside extremely massive stars, since addition of another α particle, resulting in ^{60}Zn, involves a large amount of extra energy. The ^{56}Ni with a half-life of about 6 days is formed in significant amounts in these stars, but quickly decays by two successive positron emissions within supernova decay products in the supernova remnant gas cloud, first to radioactive ^{56}Co, and then to stable ^{56}Fe. Hence, Fe forms the most abundant element in the core of red giants and is the most abundant metal in iron meteorites and in the dense metal cores of planets such as Earth. In addition, it is rather common in the universe, compared to other stable metals of about the same atomic weight. Fe is present as the sixth most abundant element in the Universe and the most common refractory element. Though a further very small amount of energy gain could be extracted by forming ^{62}Ni, which has a marginally higher binding energy than ^{56}Fe, conditions in stars are not appropriate for this process. Element formation in supernovas and distribution on Earth significantly favor Fe over Ni, and besides, ^{56}Fe still has a lower mass per nucleon than ^{62}Ni due to its higher fraction of lighter protons. Consequently, formation of elements heavier than Fe necessitates supernova conditions, involving rapid neutron capture by starting ^{56}Fe nuclei.

At least four allotropes of iron are known, conventionally denoted α, γ, δ, and ε. The first three forms are observed at ordinary pressures. As molten iron cools below its freezing point of 1538°C, it crystallizes as its δ allotrope, which has a body-centered cubic (*bcc*) crystal structure. As it cools further to 1394°C, it converts to γ-iron, a face-centered cubic (*fcc*) crystal structure, or austenite. At 912°C and below, the crystal structure again changes to the *bcc* α-iron allotrope. Iron shows the typical chemical properties of the transition metals, that is, the ability to form variable oxidation states differing by steps of one and a very large coordination and organometallic chemistry. Its 26 electrons are arranged in the configuration [Ar]3d^{6}4s^2, of which the 3d and 4s electrons are relatively close in energy, and thus it can lose a variable number of electrons and there is no clear point where further ionization becomes unprofitable (Table 5.9).

TABLE 5.9 Iron properties.

Appearance	Lustrous metallic with a grayish tinge
Standard atomic weight $A_{r,std}$	55.845
Block	d-Block
Element category	Transition metal
Electron configuration	[Ar] 3d^6 4s^2
Phase at STP	Solid
Melting point	1538°C
Boiling point	2862°C
Density (near r.t.)	7.874 g/cm^3
When liquid (at m.p.)	6.98 g/cm^3
Heat of fusion	13.81 kJ/mol
Heat of vaporization	340 kJ/mol
Molar heat capacity	25.10 J/(mol·K)
Oxidation states	−4, −2, −1, +1, **+2**, **+3**, +4, +5, **+6**, +7
Ionization energies	1st: 762.5 kJ/mol
	2nd: 1561.9 kJ/mol
	3rd: 2957 kJ/mol
Atomic radius	Empirical: 126 pm
Covalent radius	Low spin: 132 ± 3 pm
	High spin: 152 ± 6 pm

STP, Standard temperature and pressure.
Bold font indicates main oxidation state.

Iron forms compounds mainly in the oxidation states +2 (iron(II), "ferrous") and +3 (iron(III), "ferric"). Iron also exists in higher oxidation states, for example, K_2FeO_4, which contains iron in its +6 oxidation state. Although iron(VIII) oxide (FeO_4) has been claimed, the result could not be reproduced and such a species (at least with iron in its +8 oxidation state) has been found to be questionable computationally. Yet, one form of anionic $[FeO_4]^-$ with iron in its +7 oxidation state, together with an iron(V)-peroxo isomer, has been observed by infrared spectroscopy at −269°C (4K) after cocondensation of laser-ablated Fe atoms with a mixture of O_2/Ar. Iron(IV) is a common intermediate in many biochemical oxidation reactions. Numerous organoiron compounds contain formal oxidation states of +1, 0, −1, or even −2. Many mixed valence compounds have both iron(II) and iron(III) centers, such as magnetite and Prussian blue ($Fe_4(Fe[CN]_6)_3$). Iron is the first of the transition metals that cannot reach its group oxidation state of +8, whereas its heavier congeners ruthenium and osmium can. In contrast to many other metals, iron does not form amalgams with mercury. Iron is by far the most reactive element in group 8; it is pyrophoric when finely divided and dissolves readily in dilute acids, forming Fe^{2+}. Yet it does not react with concentrated nitric acid and other oxidizing acids because of the formation of an impervious oxide layer, which can nevertheless react with hydrochloric acid.

5.8.3.1 *Oxides and hydroxides*

Iron forms several oxide and hydroxide compounds; the most common are iron(II,III) oxide (Fe_3O_4), and iron(III) oxide (Fe_2O_3). Iron(II) oxide, FeO, also exists. In total there are 16 known iron oxides and oxyhydroxides, which, besides the ones already mentioned include FeO_2, Fe_4O_5, Fe_5O_6, Fe_5O_7, $Fe_{25}O_{32}$, $Fe_{13}O_{19}$, $Fe(OH)_2$, $Fe(OH)_3$. FeOOH, $FeOOH{\cdot}0.4H_2O$, etc. Notwithstanding their names, they are actually all nonstoichiometric compounds whose compositions may vary. They are used in the production of ferrites, useful magnetic storage media in computers, and pigments. Iron(II) oxide or ferrous oxide is the inorganic compound with the formula FeO. It occurs naturally as the mineral wüstite. It is a black-colored powder that is occasionally mistaken for rust, the latter of which consists of hydrated iron(III) oxide (ferric oxide). Iron(II) oxide also describes a family of related nonstoichiometric compounds, which are typically iron deficient with compositions ranging from $Fe_{0.84}O$ to $Fe_{0.95}O$. FeO can be prepared by the thermal decomposition of iron(II) oxalate.

$$FeC_2O_4 \rightarrow FeO + CO_2 + CO$$

The procedure is conducted under an inert atmosphere to avoid the formation of iron(III) oxide (Fe_2O_3). Stoichiometric FeO can be prepared by heating $Fe_{0.95}O$ with metallic iron at 770°C and 36 kbar. FeO is thermodynamically unstable below 575°C, with a tendency to disproportionate to metal and Fe_3O_4.$4FeO \rightarrow Fe + Fe_3O_4$

As the mineral hematite, Fe_2O_3 is the main source of iron for the steel industry. Fe_2O_3 is easily attacked by acids. Iron(III) oxide is regularly called rust, and to some extent this label is useful, as rust shares several properties and has a comparable composition. To a chemist, though, rust is thought of as an ill-defined material, described as hydrated ferric oxide. Fe_2O_3 can be formed in various polymorphs. In the main ones, α and γ, iron has octahedral coordination geometry. That is, each Fe center is bound to 6 oxygen ligands. α-Fe_2O_3 (hematite) has the rhombohedral, corundum (α-Al_2O_3) structure and is the most common form. γ-Fe_2O_3 (mineral maghemite) has a cubic structure. It is metastable and converted from the α phase at high temperatures. Iron(III) oxide is a product of the oxidation of iron. It can be produced in the laboratory by electrolyzing a solution of sodium bicarbonate, an inert electrolyte, with an iron anode.

$$4Fe + 3O_2 + 2H_2O \rightarrow 4FeO(OH)$$

The resulting hydrated iron(III) oxide, written here as Fe(O)OH, dehydrates around 200°C.

$$2FeO(OH) \rightarrow Fe_2O_3 + H_2O$$

The most important reaction is its carbothermal reduction, which forms iron used in steelmaking.

$$Fe_2O_3 + 3CO \rightarrow 2Fe + 3CO_2$$

One more redox reaction is the extremely exothermic thermite reaction with aluminum.

$$2Al + Fe_2O_3 \rightarrow 2Fe + Al_2O_3$$

This process is employed to weld thick metals such as rails of train tracks by using a ceramic container to funnel the molten iron in between two sections of rail. Thermite is also utilized in weapons and making small-scale cast iron sculptures and tools. Partial reduction with hydrogen at approximately 400°C results in the formation of magnetite, a black magnetic material that contains both Fe(III) and Fe(II):

$$3Fe_2O_3 + H_2 \rightarrow 2Fe_3O_4 + H_2O$$

Iron(III) oxide is insoluble in water but dissolves readily in strong acid, for example, hydrochloric and sulfuric acids. It likewise dissolves well in solutions of chelating agents such as EDTA and oxalic acid. Heating iron(III) oxides with other metal oxides or carbonates yields materials known as ferrates (ferrate (III)).

$$ZnO + Fe_2O_3 \rightarrow Zn(FeO_2)_2$$

Under anaerobic conditions, ferrous hydroxide ($Fe(OH)_2$) can be oxidized by water to form magnetite and molecular hydrogen. This process is given by the Schikorr reaction (first studied by Gerhard Schikorr, a German specialist of iron corrosion, in his early works (~1928-1933) on iron(II) and iron(III) hydroxides).

$$3Fe(OH)_2 \rightarrow Fe_3O_4 + H_2 + 2H_2O$$

The well-crystallized Fe_3O_4 is thermodynamically more stable than the ferrous hydroxide ($Fe(OH)_2$). It can be prepared in the laboratory as a ferrofluid by the Massart method through mixing iron(II) chloride and iron(III) chloride in the presence of sodium hydroxide. It can in addition be prepared through chemical coprecipitation in the presence of ammonia, in a mixture of a solution 0.1 M of $FeCl_3 \cdot 6H_2O$ and $FeCl_2 \cdot 4H_2O$ with mechanic agitation of about 2000 rpm. The molar ratio of $FeCl_3$:$FeCl_2$ can be 2:1; this solution is heated at 70°C, and at once the agitation speed is increased to 7500 rpm and a solution of NH_4OH (10 volume %) is quickly added. Immediately a dark precipitate will be formed, which consists of nanoparticles of magnetite. In both cases, the precipitation reaction relies on a rapid transformation of acidic hydrolyzed iron ions into the spinel iron oxide structure, by hydrolysis at elevated pH values (above c. 10).

5.8.3.2 *Chalcogenide, hydride, boride, and nitride*

The best known sulfide is the mineral pyrite (FeS_2), also known as fool's gold because of its golden luster. It is not an iron(IV) compound, but is actually an iron(II) polysulfide containing Fe^{2+} and S_2^{2-} ions in a distorted sodium chloride structure. Iron sulfide can refer to range of chemical compounds composed of iron and sulfur. By increasing order of stability they are known as: iron(II) sulfide, FeS, the less stable amorphous form; greigite, a form of iron(II,III) sulfide (Fe_3S_4), analog to magnetite, Fe_3O_4; pyrrhotite, $Fe_{1-x}S$ (where $x = 0$ to 0.2), or Fe_7S_8; troilite, FeS, the endmember of pyrrhotite; mackinawite, $Fe_{1+x}S$ (where $x = 0$ to 0.1); marcasite, or iron(II) disulfide, FeS_2 (orthorhombic); and pyrite, or iron(II) disulfide, FeS_2 (cubic), the more stable endmember.

Iron(I) hydride, systematically named iron hydride and poly(hydridoiron), is a solid inorganic compound with the chemical formula $(FeH)_n$ (also written $([FeH])_n$ or FeH). It is both thermodynamically and kinetically unstable toward decomposition at ambient temperature, and as such, little is known about its bulk properties. Iron(I) hydride is the simplest polymeric iron hydride. Due to its instability, it has no practical industrial uses. However, in metallurgical chemistry, iron(I) hydride is fundamental to certain forms of iron-hydrogen alloys. FeH was first produced in the laboratory by heating iron to 2327°C (2600K) in a King-type furnace under a thin hydrogen atmosphere. Molecular FeH can also be formed (together with FeH_2 and other species) by vaporizing iron in an argon-hydrogen atmosphere and freezing the gas on a solid surface at −263°C (~10K). The compound can be observed by infrared spectroscopy, and about half of it disappears when the sample is momentarily warmed to −243°C (30K). A modified technique employs a pure hydrogen atmosphere condensed at −269°C (4K). This method also forms molecules that were believed to be FeH_3 (ferric hydride) but were later assigned to an association of FeH and molecular H_2. Molecular FeH has been formed by the decay of ^{57}Co embedded in solid hydrogen. FeH can also be prepared via the interaction of iron pentacarbonyl vapor and atomic hydrogen in a microwave discharge. Iron pentahydride, FeH_5, is a superhydride compound of iron and hydrogen, which is stable under high pressures. It is significant since it contains atomic hydrogen atoms that are not bonded into smaller molecular clusters and may be a superconductor. Pairs of hydrogen atoms are not bonded together into molecules. FeH_5 has been prepared by compressing a flake of iron with hydrogen in a diamond anvil cell to a pressure of 130 GPa and heating to below 1227°C (1500K). When decompressed to 66 GPa it decomposes to solid FeH_3. The unit cell is tetragonal with symmetry *I4/mmm*.

Iron boride describes several inorganic compounds with the formula Fe_xB_y. Two main iron borides are FeB and Fe_2B. Some iron borides have useful properties such as magnetism, electrical conductivity, corrosion resistance, and extreme hardness. Iron boride nanoparticles have been synthesized by reducing iron bromide salts in highly coordinating solvents using sodium borohydride. They have likewise been produced by reducing iron salts using sodium borohydride.

$$4FeSO_4 + 8NaBH_4 + 18H_2O \rightarrow 2Fe_2B + 6B(OH)_3 + 25H_2 + 4Na_2SO_4$$

FeB is orthorhombic and Fe_2B has a body-centered tetragonal structure. FeB has zig-zag chains of boron atoms that are coordinated by seven iron atoms. Boron atoms have a slightly distorted mono-capped trigonal prismatic iron atom coordination and two boron atom neighbors. Each trigonal prism shares two rectangular faces with the nearby prisms, forming infinite prism columns. Fe_2B contains single boron atoms in square antiprismatic iron atom coordination. Boron atoms are separated from each other. Iron tetraboride (FeB_4) is a superhard superconductor (Tc < 3K/-270°C).

Iron has five nitrides known at ambient conditions, Fe_2N, Fe_3N_4, Fe_4N, Fe_7N_3, and $Fe_{16}N_2$. They form crystalline, metallic solids. Group 8 transition metals form nitrides that decompose at relatively low temperatures—iron nitride, Fe_2N decomposes under loss of molecular nitrogen at ~400°C and formation of lower-nitrogen content iron nitrides. They are insoluble in water. At high pressure, stability and formation of new nitrogen-rich nitrides (N/Fe ratio ≥ 1) have been suggested and later discovered. These include the FeN, FeN_2, and FeN_4 solids which become thermodynamically stable from 17.7 GPa, 72 GPa, and 106 GPa, respectively.

5.8.3.3 Halides

The binary ferrous and ferric halides are well known. The ferrous halides typically are formed from reacting iron metal with the corresponding hydrohalic acid to produce the corresponding hydrated salts.

$$Fe + 2HX \rightarrow FeX_2 + H_2 \ (X = F, Cl, Br, I)$$

Iron reacts with fluorine, chlorine, and bromine to form the corresponding ferric halides, with ferric chloride as the most common.

$$2Fe + 3X_2 \rightarrow 2FeX_3 \ (X = F, Cl, Br)$$

Ferric iodide is an exception, being thermodynamically unstable because of the oxidizing power of Fe^{3+} and the high reducing power of I^-.

$$2I^- + 2Fe^{3+} \rightarrow I_2 + 2Fe^{2+} \left(E^0 = +0.23 \text{ V}\right)$$

Ferric iodide, a black solid, is not stable in ordinary conditions, but can be prepared via the reaction of iron pentacarbonyl with iodine and carbon monoxide in the presence of hexane and light at the temperature of −20°C, in the absence of oxygen and water.

5.8.3.4 Solution chemistry

The standard reduction potentials in acidic aqueous solution for some common iron ions can be given as

$$\begin{array}{ll} Fe^{2+} + 2e^- \rightleftharpoons Fe & E^0 = -0.447 \text{ V} \\ Fe^{3+} + 3e^- \rightleftharpoons Fe & E^0 = -0.037 \text{ V} \\ FeO_4^{2-} + 8H^+ + 3e^- \rightleftharpoons Fe^{3+} + 4H_2O & E^0 = +2.20 \text{ V} \end{array}$$

The red-purple tetrahedral ferrate(VI) anion is such a strong oxidizing agent that it oxidizes nitrogen and ammonia at room temperature, and even water itself in acidic or neutral solutions.

$$4FeO_4^{2-} + 10H_2O \rightarrow 4Fe^{3+} + 20OH^- + 3O_2$$

The Fe^{3+} ion has a large simple cationic chemistry, even though the pale-violet hexaquo ion $[Fe(H_2O)_6]^{3+}$ is very easily hydrolyzed when the pH increases above 0.

$$\begin{array}{ll} [Fe(H_2O)_6]^{3+} \rightleftharpoons [Fe(H_2O)_5(OH)]^{2+} + H^+ & K = 10^{-3.05} mol{\cdot}dm^{-3} \\ [Fe(H_2O)_5(OH)]^{2+} \rightleftharpoons [Fe(H_2O)_4(OH)_2]^+ + H^+ & K = 10^{-3.26} mol{\cdot}dm^{-3} \\ 2[Fe(H_2O)_6]^{3+} \rightleftharpoons [Fe(H_2O)_4(OH)]_4^{2+} + 2H^+ + 2H_2O & K = 10^{-2.91} mol{\cdot}dm^{-3} \end{array}$$

As the pH increases above 0, the yellow hydrolyzed species form and as it rises above 2–3, reddish-brown hydrous iron(III) oxide precipitates out of solution. While Fe^{3+} has an d^5 configuration, its absorption spectrum is not similar to that of Mn^{2+} with its weak, spin-forbidden d-d bands, as Fe^{3+} has higher positive charge and is more polarizing, lowering the energy of its ligand-to-metal charge transfer absorptions. Consequently, all the above complexes are relatively strongly colored, with the single exception of the hexaquo ion—and even that has a spectrum dominated by charge transfer in the near ultraviolet region. In contrast, the pale green iron(II) hexaquo ion $[Fe(H_2O)_6]^{2+}$ does not undergo significant hydrolysis. Carbon dioxide is not evolved when carbonate anions are added, but instead results in the precipitation of white iron(II) carbonate. In excess carbon dioxide, this forms the slightly soluble bicarbonate, which occurs frequently in groundwater, but it oxidizes rapidly in air to form iron(III) oxide that accounts for the brown deposits present in a sizable number of streams.

5.8.3.5 Organoiron compounds

Ferrocene, a stable compound of iron sandwiched between two cyclopentadienyl rings and discovered in 1951 by Kealy and Paulson, is considered to have started organometallic chemistry as a separate area of chemistry. Its structure and chemistry were deduced in 1952 independently by Wilkinson et al. and Fischer and Pfab. The impact of its discovery and subsequent chemistry studies was judged so significant that Wilkinson and Fischer were awarded the 1973 Nobel Prize for Chemistry for their independent pioneering work on the chemistry of sandwiched organometallic compounds.

Ferrocene is an orange solid with strong camphor-like odor. It is soluble in normal organic solvents but insoluble in water. It is stable in air to temperatures as high as 400°C. It exhibits a reversible oxidation at +0.4 V versus the saturated calomel electrode. It shows typical spectroscopic properties and reactivities of aromatic compounds such as benzene. Ferrocene, therefore, undergoes typical electrophilic aromatic substitutions on the cyclopentadienyl rings as shown in representative reaction below. The most classic and useful representative electrophilic reaction of ferrocene is its acylation in the presence of acyl chlorides and $AlCl_3$ (top right reaction). Ferrocene features Fe in the +2-oxidation state.

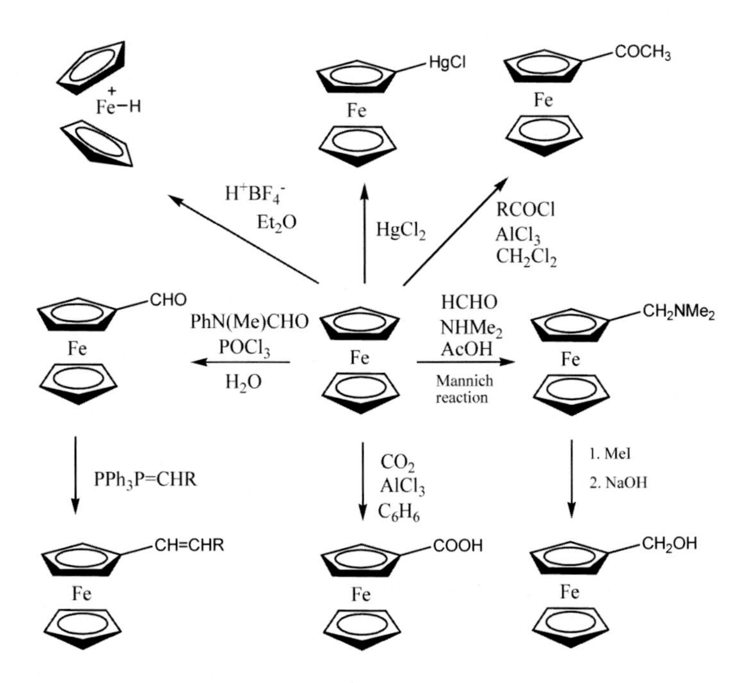

$$Na_2[Fe(CO)_4] + RCOCl \rightarrow Na[RC(O)Fe(CO)_4] + NaCl$$
$$Na[RC(O)Fe(CO)_4] + HCl \rightarrow RCHO + Fe(CO)_4 + NaCl$$

Other organoiron compounds feature Fe in −2 to +5 (even up to +7) oxidation states. As with other organometallics, a wide range of ligands are employed to support the Fe–C bond. These supporting ligands prominently include phosphines, carbon monoxide, and cyclopentadienyl, but hard ligands such as amines are employed as well. A typical organoiron(-II) is the disodium tetracarbonylferrate, $Na_2[Fe(CO)_4]$, now referred to as Collman's reagent. It can be used to convert acid chlorides to aldehydes via the intermediacy of iron acyl complex.

Organoiron(0) compounds are exemplified by alkene- and alkyne-Fe-CO derivatives. They are normally prepared by the reaction of iron(0) pentacarbonyl complex ($Fe(CO)_5$) with alkenes or alkynes. Organoiron(I) normally contains an Fe-Fe bond but exceptions have been found in complexes like $[Fe(anthracene)_2]^-$. Recent reactions involving tetramethylethylenediamine or TMEDA (for example, the reaction shown below) afford transformation of Fe(0) by the action of (a) Li, ethylene in THF at −78°C then letting it warm to room temperature and adding TMEDA to form the intermediate complex and adding (b) 1,2-dichloroethane and pentane results to the Fe(I) complex. Several other transformations involve different oxidation states of iron such as that of Fe(0)/Fe(II) system in the case of iron-catalyzed (using $[Fe(acac)_3$ with, TMEDA and hexamethylenetetramine or HMTA) cross-coupling reaction between alkenyl Grignard reagents and n- or s-alkyl bromides. This reaction and systems involving Fe(-II)/Fe(0), Fe(I)/Fe(III), Fe(II)/Fe (III), Fe(III)/Fe(IV), etc. are discussed in reviews by Fürstner et al. (2008), Neidig et al. (2019), and Parchomyk et al. (2018).

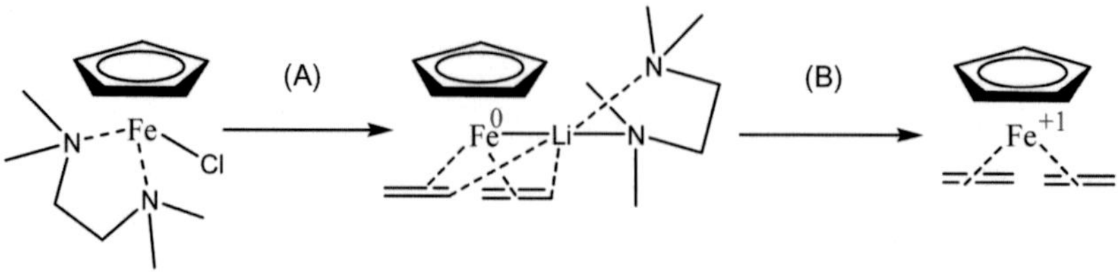

The organocompounds of iron are generally less active in many catalytic applications, however, they are abundant, less expensive and "greener" than other metals and are expected to continue to play a role in organic chemistry and other emerging applications

5.8.4 Major uses

Iron is the most commonly used of all the metals, about 90% of global metal production. Its low cost and high strength frequently make it the best material to withstand stress or transmit forces, such as the construction of machinery and machine tools, rails, automobiles, ship hulls, concrete reinforcing bars, and the load-carrying framework of buildings. As pure Fe is rather soft, it is generally combined with alloying elements to make steel. However, the mechanical properties of Fe are significantly affected by the sample's purity: pure, single crystals of Fe are actually softer than Al. An increase in the C concentration will result in a substantial increase in hardness and tensile strength of Fe. Maximum hardness is obtained with 0.6% C content, but the alloy has low tensile strength. Due to the softness of Fe, it is much easier to work with than its heavier congeners ruthenium and osmium. α-Fe is a relatively soft metal that can dissolve only a small amount of C (<0.021% by mass at 910°C). Austenite (γ-Fe) is equally soft and metallic but can dissolve substantially more C (up to 2.04% by mass at 1146°C). This form of Fe is used in the type of stainless steel used for cutlery, and hospital and food service equipment. Commercially available Fe is categorized based on purity and the abundance of additives. Pig iron has 3.5%−4.5% C and has variable quantities of contaminants, for example, S, Si, and P. Pig iron is not a marketable product, but rather an intermediate material in the production of cast iron and steel. The decrease in contaminants in pig iron which negatively affects material properties, such as S and P, yields cast iron containing 2%−4% C, 1%−6% Si, and trace amounts of Mn. Pig iron has a melting point in the range of 1145°C−1200°C, which is lower than either of its two main components and makes it the first product to be melted when C and Fe are heated together. Its mechanical properties differ enormously depending on the form of the C in the alloy. "White" cast irons has the C in the form of cementite, or iron carbide (Fe_3C). This hard, brittle compound dictates the mechanical properties of white cast irons, making them hard, but unresistant to shock. Cooling a mixture of Fe with 0.8% C slowly below 723°C to room temperature results in separate, alternating cementite and α-Fe layers, which is soft and malleable and is known as pearlite due to its appearance. Rapid cooling, in contrast, does not allow time for this separation and results in hard and brittle martensite. The steel can subsequently be tempered by reheating to a temperature in between, changing the proportions of pearlite and martensite. The final product below 0.8% C content is a pearlite-αFe mixture, and that above 0.8% C content is a pearlite-cementite mixture. In gray iron the C exists as separate, fine flakes of graphite, and makes the material brittle due to the sharp-edged graphite flakes that produce stress concentration sites within the material. A newer modification of gray iron, described as ductile iron, is specially treated with trace quantities of Mg to change the shape of graphite to spheroids, or nodules, lowering the stress concentrations and greatly increasing the toughness and strength of the material. Wrought iron contains <0.25% C but large amounts of slag that give it a fibrous characteristic. It is a tough, malleable material, but not as fusible as pig iron. Wrought iron contains fine fibers of slag entrapped within the metal. It is more corrosion resistant than steel. It has been almost completely superseded by mild steel for traditional "wrought iron" products and blacksmithing. Mild steel corrodes easier than wrought iron but is cheaper and more widely available. Carbon steel contains 2.0% C or less, with small quantities of Mn, S, P, and Si. Alloy steels contain variable quantities of C in addition to other metals, for example, Cr, V, Mo, Ni, W, etc. Their alloy content increases their cost, and hence they are typically only used for specialist applications. One common alloy steel, though, is stainless steel. Recent developments in ferrous metallurgy have produced a growing range of microalloyed steels, also called "HSLA" or high-strength, low alloy steels, having trace additions to produce high strengths and often spectacular toughness at negligible cost. Apart from conventional uses, Fe is also used for protection from ionizing radiation. Although it is lighter than another traditional protection material, Pb, it is mechanically much stronger. The major disadvantage of iron and steel is that pure iron, and most of its alloys, easily form rust if not protected in some form, a cost amounting to >1% of the global economy. Painting, galvanization, passivation, plastic coating, and bluing are all used to protect Fe from rust by excluding water and oxygen or by cathodic protection (CP). The mechanism of the rusting of iron can be envisaged as

$$\text{Cathode: } 3O_2 + 6H_2O + 12e^- \rightarrow 12OH^-$$
$$\text{Anode: } 4Fe \rightarrow 4Fe^{2+} + 8e^-;$$
$$4Fe^{2+} \rightarrow 4Fe^{3+} + 4e^-$$

Overall,

$$4Fe + 3O_2 + 6H_2O \rightarrow 4Fe^{3+} + 12OH^- \rightarrow 4Fe(OH)_3 \text{ or } 4FeO(OH) + 4H_2O$$

The electrolyte is typically Fe(II) sulfate in urban areas (formed when atmospheric SO_2 attacks Fe) and salt particles in the atmosphere in seaside areas.

Though the main use of Fe is in metallurgy, Fe compounds are also widespread in industry. Fe catalysts are traditionally used in the Haber-Bosch process to produce ammonia and in the Fischer–Tropsch process (first developed by Franz-Joseph Emil Fischer (March 19, 1877 to December 1, 1947) and Hans Tropsch (October 7, 1889 to October 8, 1935) at the Kaiser-Wilhelm-Institut für Kohlenforschung in Mülheim an der Ruhr, Germany, in 1925) for the conversion of carbon monoxide (CO) to hydrocarbons for fuels and lubricants. The general reaction can be given as

$$(2n + 1)H_2 + nCO \rightarrow C_nH_{2n+2} + nH_2O$$

Powdered Fe in an acidic solvent was used in the Bechamp reduction (the reaction was first used by French scientist Antoine Béchamp (October 16, 1816 to April 15, 1908) to reduce nitronaphthalene and nitrobenzene in 1854) of nitrobenzene ($C_6H_5NO_2$) to aniline ($C_6H_5NH_2$). Fe(III) oxide mixed with Al powder can be ignited to create a thermite reaction, used in welding large iron parts (like rails) and purifying ores. Fe(III) oxide and oxyhydroxide are used as reddish and ocher pigments. Fe(III) chloride is used in a number of different applications, for example, water purification and sewage treatment, in the dyeing of cloth, as a coloring agent in paints, as an additive in animal feed, and as an etchant for Cu in the production of printed circuit boards. It can also be dissolved in alcohol to form tincture of Fe, which is used as a medicine to stop bleeding in canaries. Fe(II) sulfate is utilized as a precursor for other Fe compounds. In addition, it is used for chromate reduction in cement. It is used to fortify foods and treat Fe deficiency (anaemia). Fe(III) sulfate is used to settle small sewage particles in tank water. Fe(II) chloride is utilized as a reducing flocculating agent, in the formation of Fe complexes and magnetic Fe oxides, and as a reducing agent in organic synthesis.

5.9 27 Co — Cobalt

5.9.1 Discovery

Cobalt compounds have been utilized for hundreds of years to produce a rich blue color in glass, glazes, and ceramics. Cobalt has been found in Egyptian sculpture, Persian jewelery as far back as the third millennium BCE, in the ruins of Pompeii, destroyed in CE 79, and in China, dating from the Tang dynasty (CE 618–907) and the Ming dynasty (CE 1368–1644). Cobalt has been applied to colorize glass since the Bronze Age. The excavation of the late Bronze Age Uluburun shipwreck produced an ingot of blue glass, cast during the 14th century BC. Blue glass from Egypt was either colored with copper, iron, or cobalt. The oldest cobalt-colored glass known originated from the 18th dynasty of Egypt (1550–1292 BCE). The origin of the cobalt the Egyptians used has not been identified so far. The word cobalt is originally derived from the German word Kobalt, from Kobold meaning "goblin," a superstitious term used for the ore of cobalt by miners. The first efforts to smelt those ores for copper or nickel were unsuccessful, producing simply a powder (cobalt(II) oxide) instead. Since the primary ores of cobalt always contain arsenic, smelting the cobalt ore oxidized the arsenic resulting in the highly toxic and volatile arsenic oxide, adding to the bad name of the ore. Swedish chemist and mineralogist Georg Brandt (June 26, 1694 to April 29, 1768) is credited with the discovery of cobalt around 1735, proving it to be a previously unidentified element, dissimilar from bismuth and other traditional metals (Brandt, 1735, 1746). Brandt called it a new "semimetal." He proved that compounds of cobalt metal were the origin of the blue color in glass, which earlier had been ascribed to the bismuth found with cobalt. Cobalt turns out to be the first metal to be discovered since the prehistorical era. All other known metals (Fe, Cu, Ag, Au, Zn, Hg, Sn, Pb, and Bi) have no known discoverers. Throughout the 19th century, a substantial amount of the world's production of cobalt blue (a dye made with cobalt compounds and alumina) and smalt (cobalt glass powdered for use for pigments in ceramics and painting) was produced at the Norwegian Blaafarveværket located at Åmot in Modum in Viken county (1776 to 1898). The first mines for the manufacture of smalt in the 16th century were situated in Norway, Sweden, Saxony and Hungary. With the discovery of cobalt ore in New Caledonia in 1864, the mining of cobalt in Europe declined. With the discovery of ore deposits in Ontario, Canada in 1904 and the discovery of even bigger deposits in the Katanga Province in the Congo in 1914, the mining operations moved again. When the Shaba conflict began in 1978, the copper mines of Katanga Province nearly ended mining. The influence on the global cobalt economy from this conflict was less than expected: cobalt is a rare metal, the pigment is very toxic, and the industry had by now found effective ways for recycling cobalt materials. In some instances, industry was able to change to cobalt-free replacements. In 1938 American nuclear physicist John Livingood (1903–1986) and chemist Glenn T. Seaborg (April 19, 1912 to February 25, 1999) discovered the radioisotope cobalt-60 (Livingood and Seaborg, 1938). This isotope was famously used at Columbia University in the 1950s to establish parity violation in radioactive beta decay.

5.9.2 Mining, production, and major minerals

Cobalt concentration in the Earth's crust is around 29 ppm. Cobalt as a native metal is not found on Earth because of its high reactivity toward oxygen in the atmosphere and chlorine in the ocean. The only exception is in recently fallen iron meteorites. The element has a medium abundance, but natural compounds of cobalt are numerous and small amounts of cobalt compounds are found in most rocks, soils, plants, and animals. In geological settings, cobalt is commonly accompanied by nickel. Both are characteristic components of iron meteorites, though cobalt is much less abundant than nickel. As with nickel, cobalt in meteoric iron alloys may have been well enough protected from oxygen and moisture to remain as the free (but alloyed) metal, though neither element is seen in that form in the ancient terrestrial crust. Cobalt in compound form occurs in copper and nickel minerals. It is the major metallic component that combines with sulfur and arsenic in the sulfidic cobaltite ($CoAsS$), safflorite ($CoAs_2$), glaucodot ($(Co,Fe)AsS$), and skutterudite ($CoAs_3$) minerals. The mineral cattierite (CoS_2) is similar to pyrite and occurs together with vaesite (NiS_2) in the copper deposits of the Katanga Province, Democratic Republic of Congo. When it is exposed to the atmosphere, weathering occurs and the sulfide minerals oxidize and form pink erythrite ("cobalt glance": $Co_3(AsO_4)_2{\cdot}8H_2O$) and sphaerocobaltite ($CoCO_3$).

The Democratic Republic of the Congo (DRC) today produces 63% of the world's cobalt. This market share may grow to 73% by 2025 if planned expansions take place as anticipated. In 2017 some exploration companies were planning to survey old silver and cobalt mines in the area of Cobalt, Ontario, Canada, where significant deposits are believed to be still present. The principal ores of cobalt are the minerals cobaltite ($CoAsS$), erythrite ($Co_3(AsO_4)_2{\cdot}8H_2O$), glaucodot ($(Co_{0.50}Fe_{0.50})AsS$), and skutterudite ($CoAs_3$), but most cobalt is obtained by reducing the cobalt by-products of nickel and copper mining and smelting. Since cobalt is typically formed as a by-product, the supply of cobalt depends largely on the economic viability of copper and nickel mining in a given market. Several processes exist to isolate cobalt from copper and nickel, contingent on the cobalt concentration and the exact composition of the used ore. One process is froth flotation, in which surfactants bind to different ore components, resulting in an enrichment of cobalt ores. Subsequent roasting converts the ores to cobalt sulfate, and the copper and the iron are oxidized to the oxide. Leaching with water extracts the sulfate and the arsenates. The residues are further leached with sulfuric acid, producing a solution of copper sulfate. Alternatively, cobalt can be leached from the slag of copper by smelting. The products of the above-mentioned processes are changed into the cobalt oxide (Co_3O_4). This oxide is reduced to metal by the aluminothermic reaction or reduction with carbon in a blast furnace (aluminothermic reactions are exothermic chemical reactions using aluminum as the reducing agent at high temperature. Aluminothermy started from the experiments of Ukranian and Russian physical chemist and metallurgist Nikolay Nikolayevich Beketov (January 13, 1827 to December 13, 1911) at the University of Kharkiv in Ukraine, who showed that Al restored metals from their oxides under high temperatures. The reaction was first used for the carbon-free reduction of metal oxides. The reaction is highly exothermic, but it has a high activation energy since strong interatomic bonds in the solids must be broken first. The oxide was heated with Al in a crucible in a furnace. The runaway reaction made it possible to produce only small quantities of material. German chemist Johannes Wilhelm "Hans" Goldschmidt (January 18, 1861 to May 21, 1923) improved the aluminothermic process between 1893 and 1898, by igniting the mixture of fine metal oxide and Al powder by a starter reaction without heating the mixture externally. The process was patented in 1895).

There are less than 70 minerals containing cobalt in their structure. Within the elements class one alloy exists, wairauite (CoFe). Twenty-nine sulfides contain Co in their crystal structure, for example, carrollite ($Cu(Co,Ni)_2S_4$), cobaltite (CoAsS) (Fig. 5.76), linnaeite ($Co^{2+}Co_2^{3+}S_4$) (Fig. 5.77), safflorite ($(Co,Ni,Fe)As_2$) (Fig. 5.78), and skutterudite ($CoAs_3$) (Fig. 5.79). The oxide class is represented by five minerals, for example, asbolane ($(Ni,Co)_{2-x}Mn^{4+}(O,OH)_4 \cdot nH_2O$) and heterogenite ($Co^{3+}O(OH)$) (Fig. 5.80). The carbonates contain a total of only three minerals, comblainite ($Ni_4Co_2(OH)_{12}[CO_3] \cdot 3H_2O$), kolwezite ($(Cu,Co)_2(CO_3)(OH)_2$), and spherocobaltite ($CoCO_3$). The sulfate class contains eight minerals with Co in the crystal structure, for example, bieberite ($CoSO_4 \cdot 7H_2O$) and cobaltzippeite ($Co(UO_2)_2(SO_4)O_2 \cdot 3.5H_2O$). One selenite is cobaltomenite ($CoSeO_3) \cdot 2H_2O$) (Fig. 5.81). A total of 19 minerals can be found in the phosphate class, such as cobaltaustinite ($CaCo(AsO_4)(OH)$), erythrite ($Co_3(AsO_4)_2 \cdot 8H_2O$) (Fig. 5.82), and roselite ($Ca_2(Co^{2+},Mg)(AsO_4)_2 \cdot 2H_2O$). Only a single silicate contains Co, oursinite ($Co(UO_2)_2(SiO_3OH)_2 \cdot 6H_2O$). Similarly, the organic compounds are represented by a single mineral, julienite ($Na_2[Co(SCN)_4] \cdot 8H_2O$).

FIGURE 5.76 Cobaltite, CoAsS, silvery 8 mm dodecahedral crystal in sulfide, probably pentlandite matrix. Håkansboda, Lindesberg, Västmanland, Sweden.

FIGURE 5.77 Linnaeite, $Co^{2+}Co_2^{3+}S_4$, an isometric cobalt sulfide in crystals less than 0.75 mm. Motycin, Bohemia, Czech Republic.

FIGURE 5.78 A 1 mm thick tangled network of veinlet safflorite, $(Co,Ni,Fe)As_2$. Nieder-Beerbach, Odenwald, Hesse, Germany.

FIGURE 5.79 Skutterudite, $CoAs_3$, shiny, silvery crystals to 5 mm in cluster. Bou Azzer dist., Anti-Atlas, Morocco.

FIGURE 5.80 Black botryoidal heterogenite, $Co^{3+}O(OH)$. Overall 4.5 cm. Kabolela mine, Katanga, Democratic Republic of Congo.

FIGURE 5.81 Cobaltomenite, $CoSeO_3{\cdot}2H_2O$, purple granular micro crystals across sandstone matrix, about 1 cm. Parco mine, Grand Co., Utah, United States.

FIGURE 5.82 Magenta, sharp crystals of erythrite, $Co_3(AsO_4)_2{\cdot}8H_2O$, from 1 to 1.5 cm. Bou Azzer dist., Anti-Atlas, Morocco.

5.9.3 Chemistry

The only stable cobalt isotope is ^{59}Co and the only one existing naturally on Earth. Twenty-two radioisotopes have been observed; ^{60}Co, the most stable isotope, has a half-life of 5.2714 years, while ^{57}Co has a half-life of 271.79 days, ^{56}Co a half-life of 77.27 days, and ^{58}Co a half-life of 70.86 days. The remaining radioisotopes of Co have half-lives shorter than 18 hours, and the majority of them shorter than 1 seconds. Cobalt also has four meta states, all of which have half-lives shorter than 15 minutes. The isotopes of cobalt range from ^{50}Co to ^{73}Co. The principal decay mode for isotopes lighter than ^{59}Co is electron capture and the main mode of decay for the heavier isotopes is via β decay. The main decay products for the lighter isotopes below ^{59}Co are Fe isotopes; for the heavier ones the decay products are Ni isotopes.

Cobalt is a ferromagnetic metal with a specific gravity of 8.9. It is a weakly reducing metal that is protected from oxidation by a passivating oxide film. It is attacked by halogens and sulfur. Heating in oxygen results in the formation of Co_3O_4 which loses oxygen at 900°C to form the monoxide CoO. The metal reacts with the halides (X_2) to give CoX_3. It does not react with hydrogen gas (H_2) or nitrogen gas (N_2) even when heated, but it does react with boron, carbon, phosphorus, arsenic and sulfur. At ordinary temperatures, it reacts slowly with mineral acids, and very slowly with moist, but not with dry, air. Common oxidation states of cobalt include +2 and +3, even though compounds with oxidation states ranging from −3 to +5 are also known (Table 5.10). A common oxidation state for simple compounds is +2 (cobalt(II)). These salts produce the pink-colored metal aquo complex $[Co(H_2O)_6]^{2+}$ in water. Addition of chloride produces the intensely blue $[CoCl_4]^{2-}$. In a borax bead flame test, cobalt exhibits a deep blue color in both oxidizing and reducing flames.

TABLE 5.10 Cobalt properties.

Standard atomic weight $A_{r,std}$	58.933
Block	d-Block
Element category	Transition metal
Electron configuration	[Ar] $3d^7$ $4s^2$
Phase at STP	Solid
Melting point	1495°C
Boiling point	2927°C
Density (near r.t.)	8.90 g/cm^3
When liquid (at m.p.)	8.86 g/cm^3
Heat of fusion	16.06 kJ/mol
Heat of vaporization	377 kJ/mol
Molar heat capacity	24.81 J/(mol·K)
Oxidation states	−3, −1, +1, **+2**, **+3**, +4, +5
Ionization energies	1st: 760.4 kJ/mol
	2nd: 1648 kJ/mol
	3rd: 3232 kJ/mol
Atomic radius	Empirical: 125 pm
Covalent radius	Low spin: 126 ± 3 pm
	High spin: 150 ± 7 pm

STP, Standard temperature and pressure.
Bold font indicates main oxidation state.

5.9.3.1 Oxides

Several oxides of cobalt are known. Green cobalt(II) oxide (CoO) has the rock salt (NaCl) structure. It is easily oxidized with water and oxygen to brown cobalt(III) hydroxide ($Co(OH)_3$). At temperatures of 600°C–700°C, CoO oxidizes to the blue cobalt (II,III) oxide (Co_3O_4), which has a spinel structure. Black cobalt(III) oxide (Co_2O_3) also exists. Cobalt oxides are antiferromagnetic at low temperature: CoO (Néel temperature 18°C/291K) and Co_3O_4 (Néel temperature: −233°C/40K), which is analogous to magnetite (Fe_3O_4), with a mixture of +2 and +3 oxidation states. Cobalt(II) oxide or cobalt monoxide is used widely in the ceramics industry as an additive to produce blue-colored glazes and enamels as well as in the chemical industry for producing cobalt(II) salts. Cobalt(II,III) oxide decomposes to cobalt(II) oxide at 950°C.

$$2Co_3O_4 \rightarrow 6CoO + O_2$$

While commercially available, cobalt(II) oxide can be produced in the laboratory by electrolyzing a solution of cobalt (II) chloride.

$$CoCl_2 + H_2O \rightarrow CoO + H_2 + Cl_2$$

In addition, it may be produced by precipitating the hydroxide, followed by thermal dehydration.

$$CoX + 2KOH \rightarrow Co(OH)_2 + K_2X$$
$$Co(OH)_2 \rightarrow CoO + H_2O$$

As can be expected, cobalt(II) oxide reacts with mineral acids to form the corresponding cobalt salts.

$$CoO + 2HX \rightarrow CoX_2 + H_2O$$

Cobalt(II) hydroxide or cobaltous hydroxide, $Co(OH)_2$, consists of divalent cobalt cations Co^{2+} and hydroxide anions OH^-. The pure compound, frequently called the "beta form" (β-$Co(OH)_2$) is a pink solid insoluble in water. The name is also used for a related compound, often named "alpha" or "blue" form (α-$Co(OH)_2$), which incorporates other anions in its molecular structure. This compound is blue and quite unstable. Cobalt(II) hydroxide is most employed as a drying agent for paints, varnishes, and inks, in the preparation of other cobalt compounds, as a catalyst and in the manufacture of battery electrodes. Cobalt(II) hydroxide precipitates as a solid when an alkali metal hydroxide is added to an aqueous solution of Co^{2+} salt, for example,

$$Co^{2+} + 2NaOH \rightarrow Co(OH)_2 + 2Na^+$$

The compound can be formed by reacting cobalt(II) nitrate in water with a solution of triethylamine $N(C_2H_5)_3$ as both the base and a complexing agent. In addition, it can be produced via electrolysis of a cobalt nitrate solution with a platinum cathode. Cobalt(II) hydroxide decomposes to cobalt(II) oxide at 168°C under vacuum and is oxidized by air. The thermal decomposition product in air above 300°C is Co_3O_4. Similar to iron(II) hydroxide, cobalt(II) hydroxide is a basic hydroxide and reacts with acids to form cobalt(II) salts. It likewise reacts with strong bases to form solutions with dark blue cobaltate(II) anions, $[Co(OH)_4]^{2-}$ and $[Co(OH)_6]^{4-}$. The pure (β) form of cobalt(II) hydroxide has the brucite ($Mg(OH)_2$) crystal structure. As such, the anion and cation packing are like those in cadmium iodide, in which the cobalt(II) cations have octahedral molecular geometry. The so-called "alpha form" (α-Co $(OH)_2$) is not a polymorph of the pure (β) form, but rather a more complex compound in which hydroxide-cobalt-hydroxide layers have a residual positive charge and alternate with layers of other anions such as nitrate, carbonate, chloride, etc. [the hydrotalcite or layered double hydroxide (LDH) structure]. It is usually formed as a blue precipitate when a base like sodium hydroxide is added to a solution of a cobalt(II) salt. The precipitate slowly converts to the β-form. Cobalt hydroxide can be obtained in the form of nanotubes, which may be of interest in nanotechnology and materials science. Cobalt(III) hydroxide or cobaltic hydroxide, $Co(OH)_3$ (sometimes written as H_3CoO_3), is an ionic compound, with trivalent cobalt cations Co^{3+} and hydroxyl anions OH^-. The compound is known in two structurally different forms, "brownish black" and "green." The brownish black form is a stable solid and can be produced through the reaction of water solutions of cobalt(II) chloride with sodium hydroxide, and subsequent oxidation with ozone. The green form, previously believed to be cobalt(II) peroxide, apparently needs carbon dioxide as a catalyst. It can be produced by adding hydrogen peroxide to a solution of cobalt(II) chloride in 96% ethanol at −30°C to −35°C, followed by adding a 15% solution of sodium carbonate in water with intense stirring. The subsequent dark green powder is relatively stable at liquid nitrogen temperature, but at room temperature it turns dark brown within a couple of days.

5.9.3.2 Chalcogenides

Cobalt sulfide is the name for chemical compounds with a formula Co_xS_y. Well-characterized compounds are CoS, CoS_2, Co_3S_4, and Co_9S_8. In general, the sulfides of cobalt are black, semiconducting, insoluble in water, and nonstoichiometric. The principal chalcogenides of cobalt include the black cobalt(II) sulfides, CoS_2, which has a pyrite-like (FeS_2) structure featuring disulfide groups, that is, $Co^{2+}S_2^{2-}$, and cobalt(III) sulfide (Co_2S_3). Cobalt sulfides occur widely as minerals, forming key sources of all cobalt compounds. Binary cobalt sulfide minerals include cattierite (CoS_2) and linnaeite (Co_3S_4). Linnaeite has the spinel structure. The Co_9S_8 compound is known as the very rare mineral cobaltpentlandite (the Co analog of pentlandite). In combination with molybdenum, the cobalt sulfides are employed as catalysts for the industrial process called hydrodesulfurization, which is applied on a large scale in refineries. Synthetic cobalt sulfides are widely investigated as electrocatalysts. Cobalt(II) selenide is an inorganic compound with the chemical formula CoSe.

5.9.3.3 Halides

Four dihalides of cobalt(II) are known: cobalt(II) fluoride (CoF_2, pink), cobalt(II) chloride ($CoCl_2$, blue), cobalt(II) bromide ($CoBr_2$, green), and cobalt(II) iodide (CoI_2, blue black). These halides exist in anhydrous and hydrated forms.

Cobalt(II) chloride is an inorganic compound of cobalt and chlorine, with the formula $CoCl_2$. It is a sky blue crystalline solid. The compound forms several hydrates $CoCl_2{\cdot}nH_2O$, for $n = 1, 2, 6$, and 9. Claims of the formation of tri- and tetrahydrates have not been confirmed. The dihydrate is purple and hexahydrate is pink. It is usually supplied as the hexahydrate $CoCl_2{\cdot}6H_2O$, which is one of the most commonly used cobalt compounds in the lab. At room temperature, anhydrous cobalt chloride has the $CdCl_2$ structure (*R3m*) in which the cobalt(II) ions are octahedrally coordinated. At about 706°C (20°C below the melting point), the coordination is thought to change to tetrahedral. Cobalt chloride is fairly soluble in water. Diluted aqueous solutions of $CoCl_2$ contain the species $[Co(H_2O)_6]^{2+}$, besides chloride ions. Concentrated solutions are red at room temperature but become blue at higher temperatures. The crystal unit of the solid hexahydrate $CoCl_26H_2O$ comprises the neutral molecule trans-$CoCl_2(H_2O)_4$ and two molecules of water of crystallization. This species dissolves easily in water and alcohol. The anhydrous salt is hygroscopic and the hexahydrate is deliquescent. Cobalt chloride can be produced in aqueous solution from cobalt(II) hydroxide or cobalt(II) carbonate and hydrochloric acid.

$$CoCO_3 + 2HCl(aq) \rightarrow CoCl_2(aq) + CO_2$$
$$Co(OH)_2 + 2HCl(aq) \rightarrow CoCl_2(aq) + 2H_2O$$

The solid dihydrate and hexahydrate can be prepared by evaporation. Cooling saturated aqueous solutions forms the dihydrate between 120.2°C and 51.25°C, and the hexahydrate below 51.25°C. Water ice, rather than cobalt chloride, will crystallize from solutions with concentration below 29%. The monohydrate and the anhydrous forms can be prepared by cooling solutions only under high pressure, above 206°C and 335°C, respectively. The anhydrous compound can be formed by heating the hydrates. On rapid heating or in a closed container, each of the 6-, 2-, and 1-hydrates partially melts into a mixture of the next lower hydrate and a saturated solution—at 51.25°C, 206°C, and 335°C, respectively. On slow heating in an open container, water evaporates out of each of the solid 6-, 2-, and 1-hydrates, leaving the next lower hydrate— at about 40°C, 89°C, and 126°C, respectively. Dehydration can also be affected with trimethylsilyl chloride.

$$CoCl_2{\cdot}6H_2O + 12(CH_3)_3SiCl \rightarrow CoCl_2 + 6\left[(CH_3)_3SiCl\right]_2O + 12HCl$$

The anhydrous compound can be purified by sublimation in vacuum. The hexahydrate and the anhydrous salt are weak Lewis acids. The adducts are typically either octahedral or tetrahedral. With pyridine (C_5H_5N), one obtains an octahedral complex.

$$CoCl_2{\cdot}6H_2O + 4C_5H_5N \rightarrow CoCl_2(C_5H_5N)_4 + 6H_2O$$

With triphenylphosphine ($P(C_6H_5)_3$), a tetrahedral complex is formed.

$$CoCl_2{\cdot}6H_2O + 2P(C_6H_5)_3 \rightarrow CoCl_2\left[P(C_6H_5)_3\right]_2 + 6H_2O$$

Salts of the anionic complex $CoCl_4^{2-}$ can be produced using tetraethylammonium chloride.

$$CoCl_2 + 2\left[(C_2H_5)_4N\right]Cl \rightarrow \left[(C_2H_5)_4N)\right]_2[CoCl_4]$$

The $CoCl_4^{2-}$ ion is the blue ion that forms upon addition of hydrochloric acid to aqueous solutions of hydrated cobalt chloride, which are pink. Reaction of the anhydrous compound with sodium cyclopentadienide gives cobaltocene $Co(C_5H_5)_2$. This 19-electron species is a good reducing agent, being readily oxidized to the yellow 18-electron cobaltacenium cation $[Co(C_5H_5)_2]^+$.

Cobalt(III) chloride or cobaltic chloride, $CoCl_3$, is an unstable and elusive compound. In this compound, the cobalt atoms have a formal charge of +3. The compound has been described to occur in the gas phase at high temperatures, in equilibrium with cobalt(II) chloride and chlorine gas. It has also been observed to be stable at very low temperatures, dispersed in a frozen argon matrix. Some publications from the 1920s and 1930s state the preparation of bulk quantities of this compound in pure form; but these results have never been reproduced, or have been attributed to other compounds like the hexachlorocobaltate(III) anion $CoCl_6^{3-}$. Those earlier papers state that it forms green solutions in anhydrous solvents such as ethanol and diethyl ether, and that it is stable only at very low temperatures (below −60°C). $CoCl_3$ was found in 1952 by Schäfer and Krehl in the gas phase when cobalt(II) chloride $CoCl_2$ was heated in an atmosphere of chlorine Cl_2. The trichloride is formed via the equilibrium reaction.

$$2CoCl_2 + Cl_2 \rightleftharpoons 2CoCl_3$$

At 645°C (918K, below the melting point of $CoCl_2$, 726°C/999K), the trichloride was the main cobalt species in the vapor, with partial pressure of 0.72 mm Hg versus 0.62 for the dichloride. However, equilibrium shifts to the left at higher temperatures. $CoCl_3$, in quantities large enough to study spectroscopically, was obtained by Green and others in 1983, by sputtering cobalt electrodes with chlorine atoms and trapping the formed molecules in frozen argon at −259°C (14K). A publication from 1969 states that treatment of solid cobalt(III) hydroxide $CoOOH{\cdot}H_2O$ with anhydrous ether saturated with HCl at −20°C resulted in a green solution (stable at −78°C) with the characteristic spectrum of $CoCl_3$. In a 1932 report, the compound was claimed to form during the electrolysis of cobalt(II) chloride in anhydrous ethanol. The infrared spectrum of the compound in frozen argon indicates that the isolated $CoCl_3$ molecule is planar with D_{3h} symmetry.

Cobalt(II) bromide ($CoBr_2$) is on its anhydrous form a green solid that is soluble in water, used primarily as a catalyst in some processes. Cobalt(II) bromide can be formed as a hydrate by the reaction of cobalt hydroxide with hydrobromic acid.

$$Co(OH)_2(s) + 2HBr(aq) \rightarrow CoBr_2{\cdot}6H_2O(aq)$$

Anhydrous cobalt(II) bromide may be produced through the direct reaction of elemental cobalt and liquid bromine. When anhydrous, cobalt(II) bromide appears as green crystals, while the hexahydrate appears as red-purple crystals. The hexahydrate loses four waters of crystallization molecules at 100°C forming the dihydrate.

$$CoBr_2{\cdot}6H_2O \rightarrow CoBr_2{\cdot}2H_2O + 4H_2O$$

Further heating to 130°C results in the anhydrous form.

$$CoBr_2{\cdot}2H_2O \rightarrow CoBr_2 + 2H_2O$$

The anhydrous form melts at 678°C. At higher temperatures, cobalt(II) bromide reacts with oxygen, forming cobalt(II, III) oxide and bromine vapor. The classical coordination compound bromopentaamminecobalt(III) bromide is synthesized by oxidation of a solution of cobalt(II) bromide in aqueous ammonia.

$$2CoBr_2 + 8NH_3 + 2NH_4Br + H_2O_2 \rightarrow 2\left[Co(NH_3)_5Br\right]Br_2 + 2H_2O$$

Triphenylphosphine complexes of cobalt(II) bromide have been used as a catalysts in organic synthesis.

Cobalt(II) iodide or cobaltous iodide are the anhydrous compound CoI_2 and the hexahydrate $CoI_2.6H_2O$. Cobalt(II) iodide is formed by reacting cobalt powder with gaseous hydrogen iodide. The hydrated form can be produced by the reaction of cobalt(II) oxide (or relate cobalt compounds) with hydroiodic acid. Cobalt(II) iodide crystallizes in two polymorphs, the α- and β-forms. The α-polymorph comprises black hexagonal crystals, which turn dark green when exposed to air. Under vacuum at 500°C, samples of α-CoI_2 sublime, forming the β-polymorph as yellow crystals. β-CoI_2 also easily absorbs moisture from the air, converting into green hydrate. Below 400°C, β-CoI_2 reverts back to the α-form. The anhydrous salt has the cadmium halide structure. The hexaaquo salt comprises separated $[Co(H_2O)_6]^{2+}$ and iodide ions.

Cobalt(II) fluoride is a chemical compound with the formula (CoF_2). It is a pink crystalline solid compound, which is antiferromagnetic at low temperatures (TN = 37.7K/235.5°C). The formula is given for both the red tetragonal crystal, (CoF_2), and the tetrahydrate red orthogonal crystal, ($CoF_2 4H_2O$). Cobalt(II) fluoride can be produced from anhydrous cobalt(II) chloride or cobalt(II) oxide in a stream of hydrogen fluoride.

$$CoCl_2 + 2HF \rightarrow CoF_2 + 2HCl$$
$$CoO + 2HF \rightarrow CoF_2 + H_2O$$

It is formed in the reaction of cobalt(III) fluoride with water. The tetrahydrate cobalt(II) fluoride is prepared by dissolving cobalt(II) in hydrofluoric acid. The anhydrous fluoride can be extracted by dehydration. Other reactions can occur at higher temperatures. It has been proven that at 500°C, fluorine will combine with cobalt producing a mixture of CoF_2 and $CoF_3{\cdot}CoF_2$ is a weak Lewis acid. Cobalt(II) complexes are typically octahedral or tetrahedral. As a 19-electron species it is a good reducing agent, reasonably oxidizable into an 18-electron compound. Cobalt(II) fluoride can be reduced by hydrogen at 300°C.
The reduction potential for the reaction

$$Co^{3+} + e^- \rightarrow Co^{2+}$$

is +1.92 V, beyond that for chlorine to chloride, +1.36 V. Consequently, cobalt(III) and chloride would result in the cobalt(III) being reduced to cobalt(II). Since the reduction potential for fluorine to fluoride is so high, +2.87 V, cobalt(III) fluoride is one of the few simple stable cobalt(III) compounds. Cobalt(III) fluoride, which is employed in some fluorination reactions, reacts vigorously with water. Cobalt(III) fluoride, CoF_3, exists as the anhydrate as well as several hydrates. The anhydrous compound is a hygroscopic brown solid. Cobalt trifluoride can be produced in the laboratory by reacting $CoCl_2$ with fluorine at 250°C.

$$2CoCl_2 + 3F_2 \rightarrow 2CoF_3 + 2Cl_2$$

In this redox reaction, Co^{2+} and Cl^- are oxidized to Co^{3+} and Cl_2, respectively, while F_2 is reduced to F^-. Cobalt(II) oxide (CoO) and cobalt(II) fluoride (CoF_2) can also be transformed to CoF_3 using fluorine. The compound can likewise be prepared by reacting $CoCl_2$ with ClF_3 or $BrF_3{\cdot}CoF_3$ decomposes upon contact with water to give oxygen.

$$4CoF_3 + 2H_2O \rightarrow HF + 4CoF_2 + O_2$$

It reacts with fluoride salts to form the anion $[CoF_6]^{3-}$, which contains a high-spin, octahedral cobalt(III) center. Anhydrous CoF_3 crystallizes in the rhombohedral system, in particular according to the aluminum trifluoride structure, with a = 527.9 pm, α = 56.97°. Each Co atom is bonded to six F atoms in octahedral geometry, with Co—F distance of 189 pm. Each F is a doubly bridging ligand. A hydrate $CoF_3{\cdot}3.5H_2O$ is known. It is thought to be better written as $[CoF_3(H_2O)_3]{\cdot}0.5H_2O$. It has been claimed that the hydrate $CoF_3{\cdot}3.5H_2O$ is isomorphic to $AlF_3{\cdot}3H_2O{\cdot}CoF_3$ is a powerful fluorinating agent. Utilized as a slurry, CoF_3 converts hydrocarbons to the perfluorocarbons.

$$2CoF_3 + R - H \rightarrow 2CoF_2 + R - F + HF$$

CoF_2 is the by-product of this reaction. Such reactions are occasionally accompanied by rearrangements or other reactions. The related reagent $KCoF_4$ is more selective. In the gas phase, CoF_3 is calculated to be planar in its ground state and possesses a threefold rotation axis (symmetry D_{3h}). The Co^{3+} ion has a ground state of $3d^6\ ^5D$. The F ligands split this state into, in energy order, $^5A'$, $^5E''$, and $^5E'$ states. The first energy difference is small and the $^5E''$ state is subject to the Jahn-Teller Effect, so this effect must be considered to be sure of the ground state. The energy lowering is small and does not change the energy order.

5.9.3.4 *Carbonate, sulfate, nitrate, and phosphate*

Cobalt(II) carbonate, $CoCO_3$, is a reddish paramagnetic solid and an intermediate in the hydrometallurgical purification of cobalt from its ores. It is an inorganic pigment, and a precursor to catalysts. Cobalt(II) carbonate also occurs as the rare red/pink mineral spherocobaltite. It is prepared by combining cobaltous sulfate and sodium bicarbonate solutions.

$$CoSO_4 + 2NaHCO_3 \rightarrow CoCO_3 + Na_2SO_4 + H_2O + CO_2$$

$CoCO_3$ has a structure similar to calcite ($CaCO_3$), comprising cobalt in an octahedral coordination geometry. Heating the carbonate takes place in a typical way for calcining, as the compound is partially oxidized.

$$6CoCO_3 + O_2 \rightarrow 2Co_3O_4 + 6CO_2$$

The resulting Co_3O_4 converts reversibly to CoO at high temperatures. Similar to most transition metal carbonates, cobalt carbonate is insoluble in water, but is easily attacked by mineral acids.

$$CoCO_3 + 2HCl + 5H_2O \rightarrow \left[Co(H_2O)_6\right]Cl_2 + CO_2$$

Cobalt carbonate is a precursor to cobalt carbonyl and various cobalt salts. It is a component of dietary supplements since cobalt is an essential element. It is a precursor to blue pottery glazes, famously in the case of Delftware.

Cobalt(II) sulfate is any of the inorganic compounds with the formula $CoSO_4{\cdot}xH_2O$. Normally cobalt sulfate refers to the hydrate $CoSO_4{\cdot}7H_2O$, which is one of the most commonly available salts of cobalt. Cobalt(II) sulfate heptahydrate appears as red monoclinic crystals that liquify around 100°C and become anhydrous at 250°C. It is soluble in water, slightly soluble in ethanol, and particularly soluble in methanol. The salts are paramagnetic. It forms by the reaction of metallic cobalt, its oxide, hydroxide, or carbonate with aqueous sulfuric acid. The hexahydrate is a metal aquo complex comprising octahedral $[Co(H_2O)_6]^{2+}$ ions together with sulfate anions. Hydrated cobalt(II) sulfate is employed in the making of pigments, together with the production of other cobalt salts. Cobalt pigment is utilized in porcelains and glass. Cobalt(II) sulfate is employed in storage batteries and electroplating baths, sympathetic inks, and as an additive to soils and animal feeds.

Cobalt nitrate is the cobalt(II) salt $Co(NO_3)_2{\cdot}xH_2O$. The most common form is the hexahydrate $Co(NO_3)_2{\cdot}6H_2O$, which is a red-brown deliquescent salt that is soluble in water and other polar solvents. The hexahydrate is formed by reacting metallic cobalt or one of its oxides, hydroxides, or carbonate with nitric acid.

$$Co + 4HNO_3 + 4H_2O \rightarrow Co(H_2O)_6(NO_3)_2 + 2NO_2$$
$$CoO + 2HNO_3 + 5H_2O \rightarrow Co(H_2O)_6(NO_3)_2$$
$$CoCO_3 + 2HNO_3 + 5H_2O \rightarrow Co(H_2O)_6(NO_3)_2 + CO_2$$

In addition to the anhydrous compound $Co(NO_3)_2$, several hydrates of cobalt(II) nitrate exist. These hydrates have the chemical formula $Co(NO_3)_2{\cdot}nH_2O$, where $n = 0, 2, 4, 6$. Anhydrous cobalt(II) nitrate has a three-dimensional polymeric network structure, with each cobalt(II) atom approximately octahedrally coordinated by six oxygen atoms, each from a different nitrate ion. Each nitrate ion coordinates to three cobalt atoms. The dihydrate is a two-dimensional polymer, where nitrate bridges between Co(II) centers and hydrogen bonding keep the layers together. The tetrahydrate comprises discrete, octahedral $[(H_2O)_4Co(NO_3)_2]$ molecules. The hexahydrate is better described as hexaaquacobalt(II) nitrate, $[Co(H_2O)_6][NO_3]_2$, as it comprises discrete $[Co(H_2O)_6]^{2+}$ and $[NO_3]^-$ ions. Above 55°C, the hexahydrate transforms to the trihydrate and at higher temperatures to the monohydrate. It is usually reduced to metallic high-purity cobalt. It can be absorbed on to a variety of catalyst supports for use in Fischer–Tropsch catalysis. It is also utilized in the manufacture of dyes and inks. Cobalt(III) nitrate, $Co(NO_3)_3$, is a stable green crystalline solid, that sublimates at ambient temperature. It is soluble in chloroform. Cobalt(III) nitrate can be produced through the reaction of dinitrogen pentoxide N_2O_5 with CoF_3. It can be purified by vacuum sublimation at 40°C. The molecular structure has three nitrate groups in bidentate coordination with the Co atom, which is therefore bonded to six oxygen atoms in a distorted octahedral arrangement. The nitrates are almost planar, and lie on three mutually perpendicular planes, giving a chiral molecule. The Co-O bond length is about 190 pm. The O-Co-O angle for the two oxygens in the same nitrate is about 68°. Cobalt(III) nitrate is soluble in water, forming a green solution that quickly turns pink, due to the formation of cobalt(II) ions and release of oxygen. Cobalt(III) nitrate can be intercalated in graphite, in the ratio of 1 molecule for each 12 carbon atoms, by heating the two substances at 40°C for 3 hours.

Cobalt phosphate, $Co_3(PO_4)_2$, is a commercial inorganic pigment known as cobalt violet. Thin films of this compound are water oxidation catalysts. The tetrahydrate $CoPO_4(H_2O)_4$ precipitates as a solid upon mixing aqueous solutions of cobalt(II) and phosphate salts. Upon heating, the tetrahydrate converts to the anhydrous compound. The anhydrous $CoPO_4$ comprises discrete phosphate (PO_4^{3-}) anions that link Co^{2+} centers. The cobalt ions occupy both octahedral (six-coordinate) and pentacoordinate sites in a 1:2 ratio.

5.9.3.5 Organocobalt compounds

Prior to 1950s organocobalt compounds were limited to ill-defined products of Grignard reagents (or other known organometallic reagents at that time) with cobalt halides. The simple alkyl derivatives produced were unstable, but species with bulky R = mesityl, biphenyl, or 2-methylnaphthyl could be characterized. Some of the alkyl complexes of cobalt(II) whose structures were determined with X-ray crystallography include the type $(Me_2NCH_2CH_2NMe_2)CoR_2$, $R = CH_2SiMe_3$, CH_2CMe_3, $[(Me_2NCH_2CH_2NMe_2)Li]_2[Co(CH_2SiMe_3)_4]$ and $[(Me_2NCH_2CH_2NMe_2)_2Li][CoCl\{CH(SiMe_3)_2\}_2]$. The discovery of ferrocene in this period initiated the great expansion of organometallic chemistry, including organocobalt chemistry. The first sandwich compounds of cobalt discovered around this time are cobaltocene itself, cobalticinium ion, alkylcobalt(I) carbonyls, and an η^3-allyl complex. The discovery that vitamin B_{12} coenzyme (see structure below) contains a cobalt(III)—carbon bond led to the preparation and characterization of many related organocobalt(III) complexes.

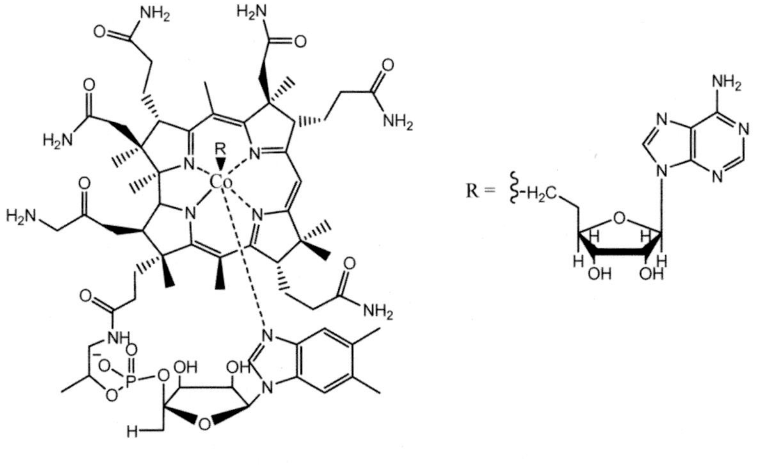

Much of the chemistry of vitamin β_{12} and related compounds is the chemistry of the Co(III)-C σ bond. This bond is easily broken homolytically as the BDE (Bond Dissociation Energy) of Co(III)-C in the coenzyme is estimated to be only 26 kcal/mol and those of related compounds are estimated to range from 18–30 kcal/mol as reported by Halperin et al. (2008). As such the Co(III)-C compounds serve as a radical precursors that pave way to cutting edge developments in the fields of biochemistry, medical research, and organic, polymer, and organometallic chemistries. The preparation and reactivity of Co(III)-C compounds such as vitamin B_{12}-like organocobalt(III); organocobalt(III) porphyrins; organocobaloxime(III); organocobalt(III) with Schiff base ligands; and organocobalt(III) complexes bearing bis(acetylacetonate)-type ligands are reviewed in Demarteau et al. (2019). In their review, they mentioned three different paths followed by the Co(III)-C bond breakage but focused only on the homolytic path and how this contribute to organic and polymer chemistries.

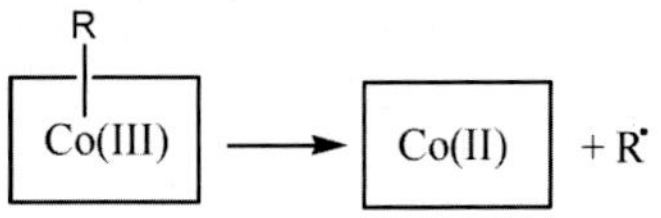

$$R\text{–Co(III)} \longrightarrow \text{Co(II)} + R^{\bullet}$$

The other two paths are heterolytic. In one of these paths, Co(III)-R leads to a supernucleophilic Co(I) with a lone electron pair. The other heterolytic path involves electrophilic attack such as with mercury salt (generally represented as MX_2) that can trap the alkyl group (R −) and produce an Co(III)-X complex. These types of reactions can be found in more detail in a review by Toscano and Marzilli (2007).

$$\text{R–Co(III)} \rightleftharpoons \text{Co(I)} + R^{+}$$

$$\text{R–Co(III)} \xrightarrow{MX_2} \text{X–Co(III)} + R^{-}$$

In addition to the cobalt(III) complexes, there is an extensive organometallic chemistry of cobalt(I). Most cobalt(I) complexes contain π-acid ligands such as Cp, CO, RNC, or R_3P. Cobalt possesses an odd number of valence electrons; thus it can only satisfy the 18-electron rule in their carbonyls if M—M bonds are present. In accord with this, mononuclear Co carbonyls are not formed. Instead $[Co_2(CO)_8]$ is the principal binary carbonyl of Co. But reduction of $[Co_2(CO)_8]$ with, for instance, sodium amalgam in benzene yields the monomeric and tetrahedral, 18-electron ion, $[Co(CO)_4]^-$. Acidification of this ion gives the pale-yellow hydride, $[HCo(CO)_4]$. Reductions employing Na metal in liquid NH_3 yield the "superreduced" $[Co(CO)_3]^{3-}$ containing Co in its lowest formal oxidation state.

5.9.4 Major uses

Co-based superalloys have traditionally used most of the Co produced. The temperature stability of these alloys makes them useful for turbine blades for gas turbines and aircraft jet engines, even though Ni-based single-crystal alloys exceed them in performance. In addition, Co-based alloys are corrosion- and wear-resistant, making them, like Ti, useful for manufacturing orthopedic implants that do not wear down over time. The development of wear-resistant Co alloys started in at the beginning of the 20th century with the stellite alloys, containing Cr with varying amounts of W and C. Alloys with Cr and W carbides are very hard and wear resistant. Special Co-Cr-Mo alloys such as Vitallium are used for making prosthetic parts (hip and knee replacements). Co alloys are also utilized for making dental prosthetics as a suitable alternative for Ni, which may be allergenic to some people. Some high-speed steels comprise Co for enhanced heat and wear resistance. The special alloys of Al, Ni, Co, and Fe, known as Alnico, and of Sm and Co (samarium-cobalt magnet) are being utilized in permanent magnets. It is also alloyed with 95% Pt for jewelery, producing an alloy appropriate for fine casting, which is also marginally magnetic. Co is used in electroplating for its attractive appearance, hardness, and resistance to oxidation. It is also used as a base primer coat for porcelain enamels. Lithium cobalt oxide ($LiCoO_2$) is extensively used in lithium-ion battery cathodes. The material is composed of Co oxide layers with intercalated Li. During discharge, the Li is released as Li ions. Nickel-cadmium (NiCd) and nickel metal hydride (NiMH) batteries likewise contain Co to enhance the oxidation of Ni in the battery. While in 2018 most Co in batteries was used in mobile devices, a more recent application for Co is in rechargeable batteries for electric cars.

Numerous Co compounds are used as oxidation catalysts. Co acetate ($Co(CH_3CO_2)_2{\cdot}4H_2O$) is used to convert xylene (dimethylbenzene, $(CH_3)_2C_6H_4$) to terephthalic acid (IUPAC name benzene-1,4-dicarboxylic acid, $C_6H_4(CO_2H)_2$), the precursor of the bulk polymer polyethylene terephthalate (PET, IUPAC name poly(ethyl benzene-1,4-dicarboxylate), $(C_{10}H_8O_4)_n$). Typical catalysts are the group of Co carboxylates (also known as cobalt soaps). They are also added to paints, varnishes, and inks as "drying agents" through the oxidation of drying oils. Co was once popular for making so-called "sympathetic ink," later better known as invisible ink. This stays invisible until it is warmed. Exactly who first discovered this phenomenon is unknown, but it was used for sending secret messages in the 17th century, and even after French chemist Jean Hellot (November 20, 1685 to February 15, 1766) disclosed its existence around 1700, it persisted in espionage. The ink was made by dissolving cobalt ore in aqua regia; cobalt chloride crystals were produced from this solution. These were dissolved in water with a little glycerin (glycerol, IUPAC name propane-1,2,3-triol, $C_3H_8O_3$) to give a nearly colorless (very pale pink) solution, which was then used to write with. Once this ink had dried another innocent letter with visible ink was written at right angles across the paper. When the letter reached its destination, the addressee just heated it to show the hidden writing. The heating drives off the water and glycerol molecules surrounding the cobalt, and chloride ions take their place instead, resulting in a dark blue molecule. A comparable trick was popular in the 19th century with artificial flowers that reacted to changes in the weather. White petals were dyed pink by coating them with cobalt chloride solution; they would stay this color while the weather was wet or the atmosphere humid. When the weather became sunny and dry, they turned violet, and when it was very dry they turned blue.

These same Co carboxylates are utilized to enhance the adhesion between steel and rubber in steel-belted radial tires. Furthermore, they are used as accelerators in polyester resin systems. Co-based catalysts are also used in reactions containing carbon monoxide (CO). Co is likewise a catalyst in the Fischer–Tropsch process for the hydrogenation of CO into liquid fuels. Hydroformylation of alkenes (an industrial process to produce aldehydes from alkenes. This chemical reaction entails the net addition of a formyl group (CHO) and a hydrogen atom to a carbon-carbon double bond.) frequently uses Co octacarbonyl ($Co_2(CO)_8$) as a catalyst, even though it is regularly substituted by more efficient Ir- and Rh-based catalysts. The hydrodesulfurization of petroleum uses a catalyst derived from Co and Mo, which facilitates the removal of sulfur impurities which inhibit the refining of liquid fuels.

Before the 19th century, Co was mostly used as a pigment. It has been used as early as the Middle Ages to manufacture smalt, a blue-colored glass. Smalt is manufactured by melting a mixture of roasted smaltite (a variety of the mineral skutterudite $(Co,Fe,Ni)As_2$), quartz (SiO_2), and potassium carbonate (K_2CO_3), which yields a dark blue silicate glass, which is finely ground after the production. Smalt was widely used to color glass and as pigment for paintings. In 1780 Swedish chemist and mineralogist Sven Rinman (June 23, 1720 to December 20, 1792) discovered cobalt green, and in 1802 French chemist Louis Jacques Thénard (May 4, 1777 to June 21, 1857) discovered cobalt blue. Co pigments such as cobalt blue (cobalt aluminate, $CoAl_2O_4$), cerulean blue (cobalt(II) stannate, $CoSnO_3$), various hues of cobalt green (doping cobalt(II) oxide in zinc oxide, the structure and color of compositions $Zn_{1-x}Co_xO$ depends on the value of x. For $x \leq 0.3$, the material adopts the wurzite structure (of ZnO) and is intensely green. For $x \geq 0.7$, the material has the sodium chloride structure (of CoO) and is pink. Intermediate values of x give a mixture of the two phases), and cobalt violet (cobalt phosphate, $Co_3(PO_4)_2$) are used as artist's pigments because of their superior chromatic stability. Aureolin (cobalt yellow, potassium hexanitritocobaltate(III), $K_3[Co(NO_2)_6]$) is now largely replaced by more lightfast yellow pigments.

Cobalt-60 (Co-60 or ^{60}Co) is useful as a γ-ray source since it can be produced in predictable amount and high activity by bombarding Co with neutrons. It produces γ rays with energies of 1.17 and 1.33 MeV. Co is utilized in external beam radiotherapy, sterilization of medical supplies and medical waste, radiation treatment of foods for sterilization (cold pasteurization), industrial radiography (e.g., weld integrity radiographs), density measurements (e.g., concrete density measurements), and tank fill height switches. ^{60}Co has a half-life of 5.27 years. Loss of potency necessitates the regular replacement of the source in radiotherapy and is one cause why Co machines have been mostly substituted by linear accelerators in modern radiation therapy. Cobalt-57 (Co-57 or ^{57}Co) is a radioisotope most frequently used in medical tests, as a radiolabel for vitamin B12 uptake, and for the Schilling test. ^{57}Co is used as a source in Mössbauer spectroscopy and is one of several potential sources in X-ray fluorescence devices. Nuclear weapon devices could deliberately incorporate ^{59}Co, some of which would be activated in a nuclear explosion to form ^{60}Co. The ^{60}Co, dispersed as nuclear fallout, is sometimes called a cobalt bomb.

5.10 28 Ni — Nickel

5.10.1 Discovery

Since nickel ores are easily mistaken for silver ores, understanding of this metal and its use dates to comparatively recent times. Nevertheless, the unintended use of nickel is ancient, and going back as far as 3500 BCE. Bronzes from what is known today as Syria have been discovered to have as much as 2% nickel. Some ancient Chinese manuscripts indicate that "white copper" (cupronickel, known as baitong) was used in China between 1700 and 1400 BCE. This Paktong white copper was exported to Britain as far back as the 17th century, but the nickel content of this alloy was not exposed until 1822. Coins of nickel-copper alloy were minted by the Bactrian kings Agathocles, Euthydemus II and Pantaleon in the 2nd century BCE, possibly out of the Chinese cupronickel. In medieval Germany, a red mineral was discovered in the Erzgebirge (Ore Mountains) that looked like copper ore. Yet, when miners were incapable to extract any copper from it, they blamed a naughty goblin of German mythology, Nickel (like Old Nick), for besetting the copper. They named this ore Kupfernickel from the German Kupfer for copper. This ore mineral is now known to be nickeline, NiAs. In 1751 Swedish mineralogist and chemist Baron Axel Fredrik Cronstedt (December 23, 1722 to August 19, 1765) attempted to extract copper from kupfernickel at a cobalt mine in the Swedish village of Los, and in its place formed a white metal that he named after the spirit that had given its name to the mineral, nickel (Cronstedt, 1751)

FIGURE 5.83 Axel Fredrik Cronstedt.

(Fig. 5.83). In modern German, Kupfernickel or Kupfer-Nickel designates the alloy cupronickel. In the beginning, the only ore for nickel was the rare Kupfernickel. Starting in 1824, nickel was obtained as a by-product of cobalt blue production. The initial large-scale smelting of nickel started in Norway in 1848 from nickel-rich pyrrhotite. The introduction of nickel in steel production in 1889 resulted in a larger demand for nickel, and the nickel ore deposits of New Caledonia, found in 1865, produced most of the world's supply between 1875 and 1915. The discovery of the large ore deposits in the Sudbury Basin, Canada in 1883, in Norilsk-Talnakh, Russia in 1920, and in the Merensky Reef, South Africa in 1924, made large-scale production of nickel possible.

5.10.2 Mining, production, and major minerals

On Earth, nickel occurs most commonly in combination with sulfur and iron in pentlandite ($(Fe,Ni)_9S_8$), with sulfur in millerite (NiS), with arsenic in the mineral nickeline (NiAs), and with arsenic and sulfur in nickel galena (PbS). Nickel is normally found in iron meteorites as the alloys kamacite, (Fe,Ni), and taenite, (Ni,Fe). The bulk of the nickel is mined from two types of ore deposits. The first is laterite, where the principal ore mineral mixtures are nickeliferous limonite, (Fe,Ni)OOH, and garnierite (a mixture of various hydrous nickel and nickel-rich silicates). The second type is magmatic sulfide deposits, where the primary ore mineral is pentlandite: $(Ni,Fe)_9S_8$. Australia and New Caledonia have the largest estimated nickel reserves, nearly half of the world's total. Identified land-based resources throughout the world averaging 1% nickel or more encompass at least 130 million tons of nickel. About 60% is present in laterites and 40% in sulfide deposits.

Nickel is produced through extractive metallurgy: it is extracted from the ore by conventional roasting and reduction methods that provide a metal of $>75\%$ purity. In many stainless-steel applications, 75% pure nickel can be utilized without the need for further purification, contingent on the impurities present. Historically, the majority of the sulfide ores have been processed using pyrometallurgical techniques to yield a matte (a term used in the field of pyrometallurgy given to the molten metal sulfide phases typically formed during smelting of copper, nickel, and other base metals) for subsequent refining. Recent developments in hydrometallurgical techniques have resulted in the production of considerably purer metallic nickel products. Most sulfide ores have traditionally been processed through concentration using froth flotation (a process for separating minerals from gangue by taking advantage of differences in their hydrophobicity. Hydrophobicity differences between valuable minerals and waste gangue are increased using surfactants and wetting agents) followed by pyrometallurgical extraction. In hydrometallurgical processes, nickel sulfide ores are concentrated with flotation (differential flotation if Ni/Fe ratio is too low. Bulk flotation is where a group of floatable minerals are targets together for maximum recovery whereas selective (or differential) flotation they are targeted sequentially with one of the minerals being depressed while the other floated.) and then smelted. The nickel matte is further processed with the Sherritt Gordon process, named for Sherritt Gordon Mines Ltd. (now Sherritt International) of Sherridon and Lynn Lake Manitoba Canada, based on the older Forward process developed by Canadian metallurgist and inventor Frank Forward (1902–1972). Hydrogen sulfide is added to the autoclave to remove nickel sulfide and copper sulfide which is fed back into the leaching process. Air is subsequently passed through the solution in the autoclave for oxyhydrolysis. In the following step the solution is reduced with hydrogen, again at high temperature and pressure, to precipitate nickel powder ($>99\%$). The remaining solution (containing about equal amounts of nickel and cobalt sulfides) is then adjusted to a lower temperature and pressure to precipitate the mixed sulfides. The fluid is concentrated and crystallized in ammonium sulfate ($(NH_4)_2SO_4$). The mixed sulfides are pressure leached with air and sulfuric acid. Ammonia is subsequently added to remove potassium and iron in the form of jarosite ($KFe_3^{3+}(OH)_6(SO_4)_2$). More ammonia and air are added for oxidation. The solution is removed from the autoclave and sulfuric acid is added to remove nickel as Ni(II) sulfate-ammonium sulfate hexahydrate, ($(NiSO_4)\cdot((NH_4)_2SO_4)\cdot 6H_2O$), which is then used to recover its nickel. The solution is then further reduced with more sulfuric acid and cobalt metal powder is added to assist in the nucleation of the precipitants (seeding). Finally, addition of hydrogen gas to saturation precipitates cobalt powder with a purity of approximately 99.6%. A second common refining method involves leaching of the metal matte into a nickel salt solution, followed by the electrowinning of the nickel from solution by plating it onto a cathode as electrolytic nickel. However, the purest metal is produced from nickel oxide by the Mond process, which achieves a purity of greater than 99.99%. The process was patented by German-born chemist and industrialist who took British nationality Ludwig Mond (March 7, 1839 to December 11, 1909) and has been in industrial use since before the beginning of the 20th century (Fig. 5.84). This process is based on the fact that carbon monoxide readily reacts reversibly with nickel to give nickel carbonyl. No other element forms a carbonyl compound under the mild conditions used in this process. This process has three steps.

1. Nickel oxide reacts with Syngas at 200°C to give nickel, together with impurities including iron and cobalt.

$$NiO(s) + H_2(g) \rightarrow Ni(s) + H_2O(g)$$

2. The impure nickel reacts with carbon monoxide at 50°C–60°C to form the gas nickel carbonyl in the presence of a sulfur catalyst, leaving the impurities as solids.

$$Ni(s) + 4CO(g) \rightarrow Ni(CO)_4(g)$$

FIGURE 5.84 Ludwig Mond, c.1909, oil on canvas painting by Solomon Joseph Solomon (1860–1927).

3. The mixture of nickel carbonyl and Syngas is heated to 220°C–250°C, resulting in decomposition back to nickel and carbon monoxide.

$$Ni(CO)_4(g) \rightarrow Ni(s) + 4CO(g)$$

Steps 2 and 3 demonstrate a chemical transport reaction, take advantage of the properties that (1) carbon monoxide and nickel readily combine to give a volatile complex and (2) this complex decomposes back to nickel and carbon monoxide at higher temperatures. The decomposition may be engineered to produce powder, but more commonly an existing substrate is coated with nickel. For example, nickel pellets are made by dropping small, hot pellets through the carbonyl gas resulting in the deposition of a layer of nickel onto the pellets. Iron gives iron pentacarbonyl, $Fe(CO)_5$, too, but this reaction is slow. If necessary, the nickel may be separated by distillation. Dicobalt octacarbonyl, $Co_2(CO)_8$, is also formed in nickel distillation as a by-product, but it decomposes to tetracobalt dodecacarbonyl at the reaction temperature to give a black nonvolatile solid.

$$2Co_2(CO)_8 \rightarrow Co_4(CO)_{12} + 4CO$$

Nickel is produced from nickel carbonyl using one of two different processes. It may be passed through a large chamber at high temperatures in which tens of thousands of nickel spheres, called pellets, are constantly stirred. The carbonyl decomposes and deposits pure nickel onto the nickel spheres. In the other process, nickel carbonyl is decomposed in a smaller chamber at 230°C to create a fine nickel powder. The by-product carbon monoxide is recirculated and reused. The highly pure nickel product is known under the name "carbonyl nickel" .

Just over 150 minerals are known to contain nickel as part of their crystal structure. The element class is represented by 18 minerals. Besides the native element nickel (Ni) it contains several alloys, for example, awaruite (Ni_3Fe), phosphides, for example, barringerite ($(Fe,Ni)_2P$), carbides, for example, haxonite ($(Fe,Ni)_{23}C_6$), one nitride, roaldite ($(Fe,Ni)_4N$), and one silicide, suessite ($(Fe,Ni)_3Si$). The sulfide class contains 63 minerals with Ni, such as gersdorffite (NiAsS), heazlewoodite (Ni_3S_2), millerite (NiS) (Fig. 5.85), nickeline (NiAs) (Fig. 5.86), pentlandite ($(Fe_xNi_y)_{\Sigma 9}S_8$), rammelsbergite ($NiAs_2$) (Fig. 5.87), siegenite ($CoNi_2S_4$), and violarite ($Fe^{2+}Ni_2^{3+}S_4$). Three halides with Ni are known, droninoite ($Ni_6Fe_3^{2+}(OH)_{16}Cl_2{\cdot}4H_2O$), gillardite ($Cu_3Ni(OH)_6Cl_2$), and nickelbischofite ($NiCl_2{\cdot}6H_2O$). The oxide class contains 12 minerals with Ni in their crystal structure, such as bunsenite (NiO) and trevorite ($Ni^{2+}Fe_2^{3+}O_4$). Fourteen Ni-containing minerals can be found in the carbonate class, for example, gaspéite ($Ni(CO_3)$), reevesite ($Ni_6Fe_2^{3+}(OH)_{16}(CO_3){\cdot}4H_2O$), and takovite ($Ni_6Al_2(OH)_{16}[CO_3]{\cdot}4H_2O$). Only a single borate has Ni in its structure, bonaccordite ($Ni_2Fe^{3+}(BO_3)O_2$). The sulfate class consists of 12 minerals with Ni, for example, carrboydite ($(Ni_{1-x}Al_x)(SO_4)_{x/2}(OH)_2{\cdot}nH_2O$) and morenosite ($NiSO_4{\cdot}7H_2O$). Nineteen Ni-containing minerals can be found in the phosphate class, for example, annabergite ($Ni_3(AsO_4)_2{\cdot}8H_2O$) (Box 5.13) and nickelaustinite ($CaNi(AsO_4)(OH)$). In the silicate class eight minerals contain Ni in the crystal structure, for example, népouite ($(Ni,Mg)_3(Si_2O_5)(OH)_4$) and pecoraite ($Ni_3(Si_2O_5)(OH)_4$). Finally, there is a single organic compound with Ni in its structure, abelsonite ($Ni(C_{31}H_{32}N_4)$).

FIGURE 5.85 Millerite, NiS, brassy acicular crystals to 3 mm. Halls Gap, Lincoln Co., Kentucky, United States.

FIGURE 5.86 Cut slab 4.5 cm across showing brassy nickeline, NiAs, with dull silver gersdorffite, NiAsS. Franklin, Sussex Co., New Jersey, United States.

FIGURE 5.87 Rammelsbergite, $NiAs_2$, silvery radiating crystals formed freely in seams of an open vug lined with micro quartz, SiO_2 crystals. Overall 5 cm. Ste Marie-aux-Mines, Haut-Rhin, Alsace, France.

5.10.3 Chemistry

Naturally occurring Ni is composed of five stable isotopes; ^{58}Ni, ^{60}Ni, ^{61}Ni, ^{62}Ni, and ^{64}Ni, with ^{58}Ni the most common (68.077% natural abundance). Isotopes heavier than ^{62}Ni cannot be formed by nuclear fusion without losing energy. ^{62}Ni has the highest mean nuclear binding energy per nucleon of any nuclide, at 8.7946 MeV/nucleon, even larger than both ^{56}Fe and ^{58}Fe (more abundant elements regularly wrongly cited as the most tightly bound nuclides). While this would appear to predict ^{62}Ni as the most abundant heavy element in the universe, the rather high rate of photodisintegration of Ni in stellar interiors results in Fe to be by far the most abundant. ^{60}Ni is the daughter product of the extinct radionuclide ^{60}Fe, which decays with a half-life of 2.6 million years. Since ^{60}Fe has such a long half-life, its persistence in materials in the solar system may create noticeable variations in the isotopic concentration of ^{60}Ni. Hence, the ^{60}Ni abundance in extra-terrestrial material may increase the understanding of the solar system's origin and early history. Some 18 nickel radioisotopes ranging from ^{48}Ni to ^{78}Ni have been observed, the most stable being ^{59}Ni with a half-life of 76,000 years, ^{63}Ni with 100 years, and ^{56}Ni with 6 days. All other radioisotopes have half-lives that are less than 60 hours, and most have half-lives less than 30 seconds. Ni has one meta state. Radioactive ^{56}Ni is formed in the silicon-burning process and later set free in large amounts during type Ia supernovae. The profile of the light curve of these supernovae at intermediate to late times is related to the decay via electron capture of ^{56}Ni to ^{56}Co and finally to ^{56}Fe. ^{59}Ni is a long-lived cosmogenic radionuclide with a half-life of 76,000 years. It is often used in isotope geology. ^{59}Ni has been utilized for dating the terrestrial age of meteorites and to determine abundances of extra-terrestrial dust in

BOX 5.13 Erythrite, $Co_3(AsO_4)_2{\cdot}8H_2O$, and annabergite, $Ni_3(AsO_4)_2{\cdot}8H_2O$

Erythrite and annabergite form the end-members of a complete solid solution. Both minerals are monoclinic, *2/m* (space group *I2/m*). Erythrite is a rare secondary mineral. In pink crusts it is known as *cobalt bloom* and occurs as an alteration product of cobalt arsenides. Erythrite was named in 1832 by François Sulpice Beaudant from the Greek ἐρυθρος, *"erythros"* for "red." It has a hardness of 1½–2½. It has a perfect cleavage on {010}, indistinct on {100} and {02}. Its luster adamantine to vitreous, pearly on cleavage, transparency is translucent. The color is crimson-red or peach-red, lighter colors indicate more Ni content. The streak is lighter than the color. Optically it is biaxial(+). Substitutions of Ca, Fe, Zn, and Mg have led to the description of several varieties. Crystals are prismatic to acicular [001], flattened {010}. Crystals are deeply striated, furrowed on [001] or {010} parallel to {*h0l*} or {0 *l*}. It also occurs as radial or stellate groups, globular or reniform masses with drusy surface, coarse-fibrous, earthy, or pulverulent.

Annabergite occurs in the oxide zone in nickel-bearing sulfide deposits. Annabergite was named by Henry J. Brooke and William Hallowes Miller in 1852 after one of the cotype localities, Annaberg, Saxony, Germany. It has a hardness similar to erythrite. The cleavage is perfect on {010}, indistinct on {100}, {02}. The luster is subadamantine, subvitreous, pearly, to earthy, transparency is transparent to translucent. The color is green, light gray to light apple-green, white; pale rose red when rich in cobalt. Its streak is pale green to white (paler than the mineral color). Optically it is biaxial(+). Mg-containing varieties were among the earliest described for this mineral. Annabergite is usually found as fine-crystalline coatings or earthy crusts. Crystals are usually rare and of poor quality, prismatic to acicular [001] and flattened {010}. Crystals are deeply striated or furrowed [001], also striated on {010} parallel to {*h0l*} or {0*l*} (Fig. 5.88).

FIGURE 5.88 Annabergite, $Ni_3(AsO_4)_2{\cdot}8H_2O$. Annabergite is isostructural with erythrite, $Co_3(AsO_4)_2{\cdot}8H_2O$, and forms a series with it. This specimen shows crystals to 4 mm. Laurium, Attika, Greece.

ice and sediments. ^{78}Ni half-life was recently determined at 110 ms and is thought to be an important isotope in supernova nucleosynthesis of elements heavier than Fe. The nuclide ^{48}Ni is the most proton-rich heavy element isotope known. With 28 protons and 20 neutrons ^{48}Ni is "double magic," as is ^{78}Ni with 28 protons and 50 neutrons and are hence remarkably stable for nuclides with such a large a proton-neutron imbalance.

Nickel is a silvery-white metal with a slight golden tinge that takes a high polish. It is one of only four elements that are magnetic at or near room temperature (the other three are iron, cobalt and gadolinium). The unit cell of nickel is a face-centered cube (*fcc*) with the lattice parameter of 0.352 nm, giving an atomic radius of 0.124 nm. This crystal structure is stable to pressures of at least 70 GPa. Nickel is a transition metal. It is hard, malleable, and ductile, and has for transition metals a rather high electrical and thermal conductivity. The nickel atom has two electron configurations, $[Ar]3d^84s^2$ and $[Ar]3d^94s^1$, which are very close in energy (Table 5.11). There is some discussion on which configuration has the lowest energy. Chemistry textbooks refer to the electron configuration of nickel as $[Ar]4s^23d^8$, which can also be given as $[Ar]3d^84s^2$. This configuration concurs with the Madelung energy ordering rule, which predicts that 4s

TABLE 5.11 Nickel properties.

Appearance	Lustrous, metallic, and silver with a gold tinge
Standard atomic weight $A_{r,std}$	58.6934
Block	d-Block
Element category	Transition metal
Electron configuration	[Ar] $3d^8$ $4s^2$ or [Ar] $3d^9$ $4s^1$
Phase at STP	Solid
Melting point	1455°C
Boiling point	2730°C
Density (near r.t.)	8.908 g/cm^3
When liquid (at m.p.)	7.81 g/cm^3
Heat of fusion	17.48 kJ/mol
Heat of vaporization	379 kJ/mol
Molar heat capacity	26.07 J/(mol·K)
Oxidation states	−2, −1, +1, **+2**, +3, +4
Ionization energies	1st: 737.1 kJ/mol 2nd: 1753.0 kJ/mol 3rd: 3395 kJ/mol
Atomic radius	Empirical: 124 pm
Covalent radius	124 ± 4 pm
Van der Waals radius	163 pm

STP, Standard temperature and pressure.
Bold font indicates main oxidation state.

is filled before 3d. It is supported by the experimental fact that the lowest energy state of the nickel atom is a $3d^84s^2$ energy level, in particular the $3d^8(^3F)4s^2$ ^3F, J = 4 level. Nevertheless, each of these two configurations splits into several energy levels due to fine structure, and the two sets of energy levels overlap. The average energy of states with configuration $[Ar]3d^94s^1$ is in reality lower than the average energy of states with configuration $[Ar]3d^84s^2$. Therefore the research literature on atomic calculations refer to the ground state configuration of nickel as $[Ar]3d^94s^1$. The most common oxidation state of nickel is +2, but compounds of Ni^0, Ni^+, and Ni^{3+} are well known, and the exotic oxidation states Ni^{2-}, Ni^{1-}, and Ni^{4+} have been observed and investigated.

5.10.3.1 Oxides and hydroxides

Nickel(II) oxide, NiO is the only well-characterized oxide of nickel (while nickel(III) oxide, Ni_2O_3 and NiO_2 have been claimed). Nickel(III) oxide, Ni_2O_3, is not well characterized and is sometimes referred to as black nickel oxide. Traces of Ni_2O_3 on nickel surfaces have been mentioned. The mineralogical form of NiO, bunsenite, is very rare. It is classified as a basic metal oxide. NiO can be produced by various methods. Upon heating above 400°C, nickel powder reacts with oxygen to form NiO. In a few industrial processes, green nickel oxide is prepared through heating a mixture of nickel powder and water at 1000°C, the rate for this reaction can be improved by adding NiO. The simplest and most successful route of preparation is by pyrolysis of a nickel(II) compounds, for example, the hydroxide, nitrate, and carbonate, which forms a light green powder. Preparation from the elements by heating the metal in oxygen can form gray to black powders which points to nonstoichiometry of the product. NiO has the NaCl structure, with octahedral Ni^{2+} and O^{2-} sites. Like numerous other binary metal oxides, NiO is frequently nonstoichiometric, that is, the Ni:O ratio deviates from 1:1. In nickel oxide this nonstoichiometry is accompanied by a color change, with the stoichiometrically correct NiO being green and the nonstoichiometric NiO being black. NiO has various specialized uses and normally applications differentiate between "chemical grade," which is relatively pure material for specialty applications, and "metallurgical grade," which is primarily used for the production of alloys. It is employed in the ceramic industry to make frits, ferrites, and porcelain glazes. The sintered oxide is utilized to manufacture nickel steel alloys. NiO was also a part in the nickel-iron battery, also called the Edison Battery, and is an element in fuel cells. It is the precursor to numerous nickel salts, for usage as specialty chemicals and catalysts. More recently, NiO was employed to produce the NiCd rechargeable batteries present in numerous electronic devices until the development of the environmentally superior NiMH battery. NiO, an anodic electrochromic material, has been extensively investigated as counter electrodes with tungsten oxide, cathodic electrochromic material, in complementary electrochromic devices. Black NiO is the precursor to nickel salts, which form through reactions with mineral acids. NiO is a versatile hydrogenation catalyst.

Heating nickel oxide with either hydrogen, carbon, or carbon monoxide reduces it to metallic nickel. It combines with the oxides of sodium and potassium at high temperatures ($>700°C$) to give the corresponding nickelate. Nickel oxide reacts with chromium(III) oxide in a basic moist environment to form nickel chromate.

$$2\,Cr_2O_3 + 4NiO + 3O_2 \rightarrow 4NiCrO_4$$

NiO is useful for demonstrating the failure of density functional theory and Hartree–Fock theory to explain the strong correlation. The term strong correlation refers to behavior of electrons in solids that is not well described (frequently not even in a qualitatively correct way) by simple one-electron theories such as the local-density approximation or Hartree–Fock theory. For example, the apparently simple material NiO has a partially filled 3d-band (the Ni atom has 8 of 10 possible 3d-electrons) and hence would be expected to be a good conductor. Nevertheless, strong Coulomb repulsion between d-electrons makes NiO instead a wide-bandgap Mott insulator. Consequently, strongly correlated materials have electronic structures that are neither simply free-electron-like nor completely ionic, but a combination of both.

Nickel oxide hydroxide, NiO(OH), is a black solid that is insoluble in all solvents but attacked by base and acid. It is a part of the nickel metal hydride battery. Its layered structure is similar to that of the brucite ($Mg(OH)_2$) polymorph of nickel(II) hydroxide, but with half as many hydrogens. The oxidation state of nickel is 3 +. It can be produced via the reaction of nickel(II) nitrate with aqueous potassium hydroxide and bromine as the oxidant.

$$Ni(OH)_2 + KOH + 0.5\,Br_2 \rightarrow KBr + H_2O + NiOOH$$

Nickel(II) hydroxide, $Ni(OH)_2$, is an apple-green solid which dissolves with decomposition in ammonia and amines and is attacked by acids. It is electroactive, being transformed to the Ni(III) oxy-hydroxide, leading to extensive usage in rechargeable batteries. The preparation involves reacting aqueous solutions of nickel(II) salts with potassium hydroxide. Nickel(II) hydroxide has two well-characterized polymorphs, α and β. The α structure comprises $Ni(OH)_2$ layers with intercalated anions or water. The β form has a hexagonal close-packed structure of Ni^{2+} and OH^- ions. In the presence of water, the α polymorph typically recrystallizes to the β form. Along with the α and β polymorphs, a number of γ nickel hydroxides have been observed, differentiated by crystal structures with much larger intersheet distances The naturally occurring $Ni(OH)_2$ mineral, theophrastite, was first identified in the Vermion region of northern Greece in 1980. It occurs as a translucent emerald green crystal formed in thin sheets near the boundaries of vesuvianite or chlorite crystals. A nickel-magnesium variant of the mineral, $(Ni,Mg)(OH)_2$ had been found earlier at Hagdale on the island of Unst in Scotland. Nickel(II) hydroxide is often employed in electrical car batteries. In particular, $Ni(OH)_2$ easily oxidizes to nickel oxyhydroxide, NiOOH, in combination with a reduction reaction, often of a metal hydride.

$$Ni(OH)_2 + OH^- \rightarrow NiO(OH) + H_2O + e^-$$
$$M + H_2O + e^- \rightarrow MH + OH^-$$

Net reaction (in H_2O).

$$Ni(OH)_2 + M \rightarrow NiOOH + MH$$

Of the two polymorphs, α-$Ni(OH)_2$ has a higher theoretical capacity and thus is commonly deemed to be superior in electrochemical applications. Nevertheless, it transforms to β-$Ni(OH)_2$ in alkaline solutions, leading to many studies into the probability of stabilized α-$Ni(OH)_2$ electrodes for industrial applications.

5.10.3.2 Chalcogenides

Nickel sulfide, NiS, is a black solid that is formed by reacting nickel(II) salts with hydrogen sulfide. Numerous nickel sulfides have been recognized. Aside from being useful ores, nickel sulfides are the products of desulfurization reactions, and are occasionally used as catalysts. Nonstoichiometric forms of nickel sulfide have been observed, for example, Ni_9S_8 and Ni_3S_2. The mineral millerite is also a nickel sulfide with the molecular formula NiS, even though its structure is different from synthetic stoichiometric NiS because of the conditions under which it forms. It occurs naturally in low-temperature hydrothermal systems, in cavities of carbonate rocks, and as a by-product of other nickel minerals. The precipitation of solid black nickel sulfide is a pillar of established qualitative inorganic analysis schemes, which starts with the separation of metals on the basis of the solubility of their sulfides. Such reactions can be written as

$$Ni^{2+}(aq) + H_2S(aq) \rightarrow NiS(s) + 2H^+(aq)$$

Numerous other better controlled methods have been established, for example, solid-state metathesis reactions (from $NiCl_2$ and Na_2S) and high temperature reactions of the elements. Similar to numerous related materials, nickel sulfide has the nickel arsenide structure. Here, Ni is octahedrally coordinated while the S centers are in trigonal prismatic sites. Nickel sulfide has two allotropes. The α form has a hexagonal unit cell, while the β form has a rhombohedral unit cell. The α phase is stable at temperatures over 379°C and transforms into the β phase at lower temperatures. That phase transition results in an increase in volume by 2%–4%. Float glass comprises a small quantity of nickel sulfide, formed from the sulfur in the fining agent Na_2SO_4 and the nickel contained in metallic alloy contaminants. Nickel sulfide inclusions are problematic for tempered glass applications. After the tempering process, nickel sulfide inclusions are in the metastable α phase. The inclusions eventually convert to the β phase (stable at low temperature), increasing in volume and initiating cracks in the glass. In the middle of tempered glass, the material is under tension, which results in the cracks to propagate and spontaneous glass fracture. Spontaneous fracturing can happen years or even decades after glass production.

Nickel selenide is the inorganic compound NiSe. As for numerous metal chalcogenides, the phase diagram for nickel(II) selenide is complex. Two other selenides of nickel are known, $NiSe_2$ with a pyrite (FeS_2) structure, and Ni_2Se_3. Additionally, NiSe is typically nonstoichiometric and is frequently written as $Ni_{1-x}Se$, with $0<x<0.15$. This material is a semiconducting solid and can be prepared in the form of a black fine powder, or silver if produced in the form of larger crystals. Nickel(II) selenide is insoluble in all solvents, although it can be degraded by strongly oxidizing acids. Usually, NiSe is produced by high-temperature reaction of the elements. Such reactions usually produce mixed phase products. In addition, milder methods have been described employing more specialized techniques, for example, reactions of the elements in liquid ammonia in a pressure vessel. Similar to many related compounds, nickel(II) selenide has the nickel arsenide structure. Here, Ni is octahedrally coordinated and the Se are in trigonal prismatic sites.

5.10.3.3 Hydrides, borides, and silicides

"The existence of definite hydrides of nickel and platinum is in doubt" according to Greenwood and Earnshaw (1996). However, this remark does not exclude the presence of nonstoichiometric hydrides. In fact, nickel is an extensively used hydrogenation catalyst. Experimental research on nickel hydrides is rare and mainly theoretical. Hydrogen hardens nickel, hindering dislocations in the nickel atom crystal lattice from sliding past one another. Varying the quantities of alloying hydrogen and the form of its presence in the nickel hydride (precipitated phase) controls properties like the hardness, ductility, and tensile strength of the nickel hydride. Nickel hydride with increased hydrogen content can be harder and stronger than nickel itself; however, such nickel hydride is also less ductile than nickel. Loss of ductility takes place because of cracks maintaining sharp points due to suppression of elastic deformation by the hydrogen, and voids forming under tension because of decomposition of the hydride. Hydrogen embrittlement can be problematic in nickel used in turbines at high temperatures. In the narrow range of stoichiometries adopted by nickel hydride, distinct structures are asserted. At room temperature, the most stable form of nickel is the face-centered cubic (*fcc*) structure α-nickel. It is a rather soft metallic material that can dissolve only a very small quantity of hydrogen, less than 0.002 wt.% at 1455°C, and only 0.00005% at 25°C. The solid solution phase with dissolved hydrogen that maintains the same structure as the original nickel is termed the α-phase. At 25°C, 6 kbar of hydrogen pressure is required to dissolve hydrogen in β-nickel, but the hydrogen desorbs at pressures below 3.4 kbar. Hydrogen dissociates on nickel surfaces. The dissociation takes place at different temperatures on each of these surfaces. Crystallographically distinct phases of nickel hydride are formed with high hydrogen gas pressure of 600 MPa. Instead, it can be formed electrolytically. The crystal form is face-centered cubic or β-nickel hydride. Hydrogen to nickel atomic ratios are up to one, with hydrogen occupying octahedral sites. The density of the β-hydride is 7.74 g/cm^3. It is colored gray. At a current density of 1 Amp/dm^2, in 0.5 mol/L of sulfuric acid and thiourea a surface layer of nickel will be converted to nickel hydride. This surface is full with cracks up to millimeters long. The direction of cracking is in the {001} plane of the original nickel crystals. The lattice constant of nickel hydride is 3.731 Å, which is 5.7% more than that of nickel. The near-stoichiometric NiH is unstable and loses hydrogen at pressures below 340 MPa. A large number of nickel hydride complexes are known, for example, the complex trans-$NiH(Cl)(P(C_6H_{11})_3)_2$.

Nickel borides are inorganic compounds with the chemical formula Ni_xB_y, where x and y vary. A common form is Ni_2B, which is available in two forms, known as P − 1 and P − 2. Other less common borides of nickel are NiB, Ni_3B, *o*-Ni_4B_3, and *m*-Ni_4B_3 (*o* for orthogonal, *m* for metastable). The synthesis of amorphous nickel boride is simple in comparison to other borides which need high temperatures, special techniques and equipment. The P − 1 form of Ni_2B can be prepared by mixing nickel(II) sulfate and sodium borohydride in alkaline aqueous solutions. The product is not nickel boride but nanoparticles of nickel dispersed in a boron compound matrix. The P − 2 form is formed likewise from nickel(II) acetate and sodium borohydride in ethanol. The product precipitates as a fine, black amorphous powder. These catalysts are generally produced in situ, which includes the usage of the $NiCl_2/NaBH_4$ mixture system. Ni_2B has been thought to be an amorphous compound, composed of Ni bonded to individual B centers. However, it consists of nanoparticles of Ni which on heating under inert conditions become more crystalline. The two forms P − 1 and P − 2 are different in terms of the amount of their contamination by $NaBO_2$ adsorbed on the surface. P − 1 Ni_2B has an oxide to boride ratio of 1:4, while that of P − 2 Ni_2B is 10:1. Their properties also are different with respect to their catalytic efficiency and substrate specificity. Nickel boride is in the form of black amorphous powder or black granules. It is insoluble in all solvents but reacts with concentrated mineral acids. The solid is air stable. As expected for a boride, it has a high melting point. Ni_2B is an efficient catalyst and reducing agent. It is employed as a heterogeneous hydrogenation catalyst.

Nickel silicides include a number of different intermetallic compounds of nickel and silicon. Nickel silicides include for example, Ni_3Si, $Ni_{31}Si_{12}$, Ni_2Si, Ni_3Si_2, NiSi, and $NiSi_2$. $Ni_{31}Si_{12}$, Ni_2Si, and NiSi have congruent melting points; the others form via a peritectic transformation (peritectic transformations are similar to eutectic reactions. Here, a liquid and solid phase of fixed proportions react at a fixed temperature to form a single solid phase. Since the solid product forms at the interface between the two reactants, it can form a diffusion barrier and generally causes such reactions to proceed much more slowly than eutectic or eutectoid transformations. Due to this, when a peritectic composition solidifies, it does not show the lamellar structure that is found with eutectic solidification.). The silicides can be produced via fusion or solid-state reaction between the elements, diffusion at a junction of the two elements, and other techniques such as ion beam mixing. Nickel silicides are usually chemically and thermally stable. They have low electrical resistivity; with NiSi 10.5−18 μΩ·cm, Ni_2Si 24−30 μΩ·cm, $NiSi_2$ 34−50 μΩ·cm; nickel-rich silicides have higher resistivity increasing to 90−150 μΩ·cm in $Ni_{31}Si_{12}$. Nickel silicides are essential in microelectronic devices—specific silicides are good conductors, with NiSi having a conductivity close to that of elemental nickel. With silicon carbide as the semiconductor, nickel reacts at elevated temperatures to produce nickel silicides and carbon. Nickel silicides have potential as coatings for nickel-based superalloys and stainless steel, because of their corrosion, oxidation, and wear resistance. In addition, NiSi has been studied as a hydrogenation catalyst for unsaturated hydrocarbons.

5.10.3.4 Halides

Nickel(II) fluoride, NiF_2, is an ionic compound and forms yellowish to green tetragonal crystals. In contrast to many fluorides, NiF_2 is stable in air. NiF_2 forms the passivating surface that forms on nickel alloys (e.g., monel) in the presence of hydrogen fluoride or elemental fluorine, which is the reason why nickel and its alloys are some of the few materials that can be utilized to store and/or transport these fluorine compounds. NiF_2 is also employed as a catalyst for the preparation of chlorine pentafluoride. NiF_2 is formed by the reaction of anhydrous nickel(II) chloride with fluorine at 350°C.

$$NiCl_2 + F_2 \rightarrow NiF_2 + Cl_2$$

The equivalent reaction of cobalt(II) chloride results in oxidation of the cobalt, while nickel remains in the +2 oxidation state after fluorination as its +3 oxidation state is less stable. Chloride is more easily oxidized than nickel(II). This is a characteristic halogen displacement reaction, where a halogen plus a less active halide makes the less active halogen and the more active halide. NiF_2 is also formed when fluorine is reacted with nickel metal. A melt of NiF_2 and KF reacts to form the green compound $K_2[NiF_4]$. The structure of this material is closely related to certain superconducting oxide materials. NiF_2 can react with strong bases to produce nickel(II) hydroxide, a green colored compound.

$$NiF_2 + 2NaOH \rightarrow Ni(OH)_2 + 2NaF$$

Nickel(II) chloride (or simply nickel chloride), $NiCl_2$, as the anhydrous salt it is yellow, but the more familiar hydrate $NiCl_2 \cdot 6H_2O$ is green. $NiCl_2$, in several forms, is the most important source of nickel for chemical synthesis. The nickel chlorides are deliquescent, absorbing moisture from the air to form a solution. The largest scale manufacture of nickel chloride comprises the extraction with hydrochloric acid of nickel matte and residues obtained from roasting refining nickel-containing ores. Nickel chloride is normally not prepared in the laboratory as it is inexpensive and has a long shelf-life. Heating the hexahydrate in the range 66°C−133°C produces the yellowish dihydrate, $NiCl_2 \cdot 2H_2O$. The hydrates transform to the anhydrous compound upon heating in thionyl chloride or by heating under a stream of HCl gas. Simply heating the hydrates does not result in the anhydrous dichloride.

$$NiCl_2.6H_2O + 6SOCl_2 \rightarrow NiCl_2 + 6SO_2 + 12HCl$$

The dehydration is accompanied by a color change from green to yellow. In case one needs a pure compound without presence of cobalt, nickel chloride can be obtained carefully heating hexaamminenickel chloride.

$$\left[Ni(NH_3)_6\right]Cl_2 \xrightarrow{175-200^\circ C} NiCl_2 + 6NH_3$$

$NiCl_2$ has the $CdCl_2$ structure. Here, each Ni^{2+} center is coordinated by six Cl^- centers, and each Cl is bonded to three Ni(II) centers. In $NiCl_2$ the Ni-Cl bonds have "ionic character." Yellow $NiBr_2$ and black NiI_2 have comparable structures, although with a different packing of the halides, having the CdI_2 structure instead. In contrast, $NiCl_2{\cdot}6H_2O$ comprises separate *trans*-$[NiCl_2(H_2O)_4]$ molecules linked more weakly to adjacent water molecules. Only four of the six water molecules in the formula are bonded to the Ni, and the remaining two are water of crystallization. Cobalt(II) chloride hexahydrate has a comparable structure. The hexahydrate occurs naturally as the very rare mineral nickelbischofite. The dihydrate $NiCl_2{\cdot}2H_2O$ has a structure intermediate between the hexahydrate and the anhydrous forms. It comprises infinite chains of $NiCl_2$, in which both Cl centers are bridging ligands. The *trans* sites on the octahedral centers are occupied by aquo ligands. A tetrahydrate $NiCl_2{\cdot}4H_2O$ has also been observed. Nickel(II) chloride solutions are acidic, with a pH of around 4 because of the hydrolysis of the Ni^{2+} ion. Most of the reactions attributed to "nickel chloride" actually consist of the hexahydrate, although some specialized reactions necessitate the use of the anhydrous form. Reactions starting from $NiCl_2{\cdot}6H_2O$ can be employed to produce various nickel coordination complexes since the H_2O ligands are quickly displaced by ammonia, amines, thioethers, thiolates, and organophosphines. In some derivatives, the chloride remains within the coordination sphere, whereas chloride is displaced with highly basic ligands. Some example of these complexes are shown in Table 5.12. Some nickel chloride complexes exist as an equilibrium mixture of two geometries; these are some of the most spectacular examples of structural isomerism for a given coordination number. For instance, $NiCl_2(PPh_3)_2$, containing four-coordinated Ni(II), exists in solution as a mixture of both the diamagnetic square planar and the paramagnetic tetrahedral isomers. Square planar complexes of nickel can frequently form five-coordinated adducts. $NiCl_2$ is the precursor to acetylacetonate complexes $Ni(acac)_2(H_2O)_2$ and the benzene-soluble (Ni $(acac)_2)_3$, which is a precursor to Ni(1,5-cyclooctadiene)$_2$, an essential reagent in organonickel chemistry. In the presence of water scavengers, hydrated nickel(II) chloride reacts with dimethoxyethane (dme) to produce the molecular complex $NiCl_2(dme)_2$. The dme ligands in this complex are labile. For instance, this complex can react with sodium cyclopentadienide to form the sandwich compound nickelocene. Hexammine nickel chloride complex is soluble while the corresponding cobalt complex is not, which allows for easy separating of these close-related metals under laboratory conditions. $NiCl_2$ and its hydrates are sometimes useful in organic synthesis. Nickel chloride solutions are employed for electroplating nickel onto other metal items.

Nickel(II) bromide is an inorganic compound $NiBr_2{\cdot}xH_2O$. The value of x can be 0 for the anhydrous material, in addition to 2, 3, or 6 for the three observed hydrated forms. The anhydrous material is a yellow-brown solid, which dissolves in water to form the blue-green hexahydrate. The structure of the nickel bromides depends on the degree of hydration. In all of these forms, the nickel(II) ion has an octahedral molecular geometry. Anhydrous $NiBr_2$ has the cadmium chloride structure with the interatomic distances for Ni-Br between 2.52 and 2.58 Å. The structure of the trihydrate has not been confirmed by X-ray crystallography. It is believed to have a chain structure. The di- and hexahydrates have structures similar to those of the corresponding chlorides. The dihydrate comprises a linear chain, while the hexahydrate contains isolated *trans*-$[NiBr_2(H_2O)_4]$ molecules along with two noncoordinated water molecules. $NiBr_2$ has Lewis acid character. It is used to produce catalysts for cross-coupling reactions and a variety of carbonylation reactions.

Nickel(II) iodide, NiI_2, is a paramagnetic black solid that dissolves easily in water to form blue-green solutions of the aquo complexes. This blue-green color is characteristic of hydrated nickel(II) compounds. Nickel iodides have some use in homogeneous catalysis. The anhydrous material crystallizes with the $CdCl_2$ structure, having octahedral coordination geometry at each Ni(II) center. NiI_2 is produced by dehydration of the pentahydrate. NiI_2 easily hydrates, and the hydrated form can be obtained by dissolution of nickel oxide, hydroxide, or carbonate in hydroiodic acid. The anhydrous form can be formed by reacting powdered nickel with iodine. NiI_2 has found some industrial uses as a catalyst in carbonylation reactions. It also has niche applications as a reagent in organic synthesis, in particular together with

TABLE 5.12 Properties of some Ni coordination complexes.

Complex	Color	Magnetism	Geometry
$[Ni(NH_3)_6]Cl_2$	Blue/violet	Paramagnetic	Octahedral
$[Ni(en)_3]^{2+}$	Violet	Paramagnetic	Octahedral
$NiCl_2(dppe)$	Orange	Diamagnetic	Square planar
$[Ni(CN)_4]^{2-}$	Colorless	Diamagnetic	Square planar
$[NiCl_4]^{2-}$	Yellowish-green	Paramagnetic	Tetrahedral

samarium(II) iodide. Similar to many nickel complexes, those derived from hydrated nickel iodide have been employed in cross-coupling reactions, for example, NiI_2 as the Ni(II) precatalyst in the aryl iodide and alkyl iodide cross-coupling.

5.10.3.5 Sulfate, carbonate, nitrate, nitrite, and chromate

Nickel(II) sulfate, or just nickel sulfate, generally refers to $NiSO_4{\cdot}6H_2O$. This highly soluble blue-colored salt is a frequent source of the Ni^{2+} ion for electroplating. Nickel sulfate occurs naturally as the rare mineral retgersite, which is a hexahydrate. The second hexahydrate is known as nickel hexahydrite $(Ni,Mg,Fe)SO_4{\cdot}6H_2O$. The heptahydrate, which is rather unstable in air, occurs as morenosite. The monohydrate can be found as the very rare mineral dwornikite $(Ni,Fe)SO_4{\cdot}H_2O$. The salt is generally prepared as a by-product of copper refining. It is also formed by dissolution of nickel metal or nickel oxides in sulfuric acid. Aqueous solutions of nickel sulfate react with sodium carbonate to precipitate nickel carbonate, which is a precursor for nickel-based catalysts and pigments. Addition of ammonium sulfate to concentrated aqueous solutions of nickel sulfate results in the precipitation of $Ni(NH_4)_2(SO_4)_2{\cdot}6H_2O$. This blue-colored solid is equivalent to Mohr's salt, $Fe(NH_4)_2(SO_4)_2{\cdot}6H_2O$. Nickel sulfate is used in the laboratory, for example, columns employed in polyhistidine-tagging, valuable in biochemistry and molecular biology, are regenerated with nickel sulfate. Aqueous solutions of $NiSO_4{\cdot}6H_2O$ and related hydrates react with ammonia to form $[Ni(NH_3)_6]SO_4$ and with ethylenediamine to form $[Ni(H_2NCH_2CH_2NH_2)_3]SO_4$. The latter is sometimes used as a calibrant for magnetic susceptibility measurements as it has no tendency to hydrate. At least seven sulfate salts of nickel(II) have been recognized. These salts differ in terms of their hydration and/or crystal habit. The common tetragonal hexahydrate crystallizes from aqueous solution between 30.7 and 53.8°C. Below these temperatures, a heptahydrate crystallizes, and above these temperatures an orthorhombic hexahydrate is formed. The yellow anhydrous form, $NiSO_4$, is a high-melting solid that is seldom encountered in the laboratory. This material is formed by heating the hydrates above 330°C. It decomposes at even higher temperatures to nickel oxide. $NiSO_4{\cdot}6H_2O$ comprises the octahedral $[Ni(H_2O)_6]^{2+}$ ions. These ions in turn are hydrogen bonded to sulfate ions. Dissolution of the salt in water produces solutions with the aquo complex $[Ni(H_2O)_6]^{2+}$. All nickel sulfates are paramagnetic.

Nickel(II) carbonate describes one or a mixture of inorganic compounds containing nickel and carbonate. From a commercial point of view, the most important nickel carbonate is basic nickel carbonate with the formula $Ni_4CO_3(OH)_6(H_2O)_4$. Simpler carbonates, ones more likely found in the laboratory, are $NiCO_3$ and its hexahydrate. All are paramagnetic green solids containing Ni^{2+} cations. The basic carbonate is an intermediary in the hydrometallurgical purification of nickel from its ores and is employed in electroplating of nickel. $NiCO_3$ has a structure similar to calcite $(CaCO_3)$, comprising Ni in an octahedral coordination geometry. Nickel carbonates are hydrolyzed upon contact with aqueous acids to form solutions with the $[Ni(H_2O)_6]^{2+}$ ion, liberating water and carbon dioxide in the process. Calcination (heating to drive off CO_2 and water) of these carbonates results in the formation of nickel oxide.

$$NiCO_3 \rightarrow NiO + CO_2$$

The nature of the resulting oxide depends on the nature of the precursor. The oxide formed from the basic carbonate is frequently most effective for catalysis. Basic nickel carbonate can be prepared by reacting solutions of nickel sulfate with sodium carbonate, shown below is the example of basic carbonate.

$$4Ni^{2+} + CO_3^{2-} + 6OH^- + 4H_2O \rightarrow Ni_4CO_3(OH)_6(H_2O)_4$$

The hydrated carbonate can be produced through electrolytic oxidation of nickel in the presence of carbon dioxide.

$$Ni + O + CO_2 + 6H_2O \rightarrow NiCO_3(H_2O)_4$$

Nickel carbonates are employed in some ceramic applications and as precursors for catalysts. The naturally occurring nickel carbonate is known as the rare mineral gaspéite. Basic Ni carbonates also have some natural representatives.

Nickel nitrate is the inorganic compound $Ni(NO_3)_2$ or any of its hydrates. The anhydrous form is not often encountered, so "nickel nitrate" generally refers to nickel(II) nitrate hexahydrate. The formula for this species is written in two ways: $Ni(NO_3)_2{\cdot}6H_2O$ and, more descriptively $[Ni(H_2O)_6](NO_3)_2$. The latter formula reflects that the nickel(II) center is coordinated by six water molecules in this hydrated salt. In the hexahydrate, the nitrate anions are not bonded to nickel. Also observed are three other hydrates: $Ni(NO_3)_2{\cdot}9H_2O$, $Ni(NO_3)_2{\cdot}4H_2O$, and $Ni(NO_3)_2{\cdot}2H_2O$. The hexahydrate is produced by the reaction of nickel oxide with nitric acid.

$$NiO + 2HNO_3 + 5H_2O \rightarrow Ni(NO_3)_2{\cdot}6H_2O$$

The anhydrous nickel nitrate is normally not produced by the heating the hydrates. Instead, it is formed by the reaction of the hydrates with dinitrogen pentoxide or of nickel carbonyl with dinitrogen tetroxide.

$$Ni(CO)_4 + 2N_2O_4 \rightarrow Ni(NO_3)_2 + 2NO + 4CO$$

The hydrated nitrate is frequently utilized as a precursor to prepare supported nickel catalysts. Nickel(II) nitrate is employed as the precursor for the explosive nickel hydrazine nitrate ($[Ni(N_2H_4)_3](NO_3)_2$), which is used as a lead-free and safer alternative to lead azide ($Pb(N_3)_2$) and lead styphnate (IUPAC name lead(II) 2,4,6-trinitrobenzene-1,3-bis (olate), $C_6HN_3O_8Pb$).

$$3N_2H_4 \cdot H_2O + Ni(NO_3)_2 \rightarrow \left[Ni(N_2H_4)_3\right](NO_3)_2 + 3H_2O$$

Nickel(II) nitrite is an inorganic compound with the chemical formula $Ni(NO_2)_2$. Anhydrous nickel nitrite was first found by Cyril Clifford Addison on 1961 when he reacted gaseous nickel tetracarbonyl with dinitrogen tetroxide, forming a green smoke. Nickel nitrite was the second transition element anhydrous nitrite discovered after silver nitrite. Nickel nitrite decomposes at a temperature of 220°C, but it can be heated up to 260°C in argon. The nitrite is covalently bonded to Ni, and the compound is slightly volatile. Liquid dinitrogentetroxide oxidizes nickel nitrite to nickel nitrate. The nitronickelates are related compounds where more nitro groups are attached to Ni to form an anion. They could be depicted as nickel double nitrites. In water when nickel nitrite dissolves, different mixed nitro-aqua complexes are produced, for example, $Ni(NO_2)_2(H_2O)_4$, $Ni(NO_2)_3(H_2O)_3^-$, and $Ni(NO_2)(H_2O)_5^+$. The aqueous complex $Ni(NO_2)_2(H_2O)_4$ is produced when an alkali metal nitrite is added to a nickel salt solution. This complex has a much more intense emerald green color compared to the $Ni(H_2O)_6^{2+}$ ion. The equilibrium constant for

$$Ni(H_2O)_6^{2+} + 2NO_2^- \rightleftharpoons Ni(NO_2)_2(H_2O)_4 + 2H_2O$$

is 0.16 at standard conditions. Brooker (1975) stated that intense light photocatalyzes the destruction of the ionic nitro complexes, leaving only $Ni(NO_2)_2(H_2O)_4$. Nickel nitrite slowly decomposes slightly in aqueous solution as the result of disproportionation.

$$3NO_2^- + 2H^+ \rightarrow 2NO(g) + NO_3^- + H_2O.$$

Nickel nitrite can react to produce complexes with other ligands. In a few of these the nitro groups are altered in their attachment to nickel, so that in place of linking via a nitrogen atom, they link via an oxygen atom. This is then called "nitrito." The change to nitrito occurs because the other ligands are bulky, resulting in steric hindrance. These complexes can be stable as solids. Some compounds are, for example, $Ni(pyridine)_4(ONO)_2$, $Ni(substituted ethylene diamines)_2(ONO)_2$, $Ni(N,N\text{-}diethylethylenediamine)_2(NO_2)_2$, $Ni(N,N'\text{-}diethylethylenediamine)_2(NO_2)_2$, $Ni(NO_2)_2(NH_3)_4$, $Ni(ethylenediamine)_2(NO_2)_2$, $Ni(N\text{-}monosubstituted\text{-}ethylenediamine)_2(NO_2)_2$, $Ni(1,2\text{-}diamino\text{-}2\text{-}methylpropane)_2(NO_2)_2$, $Ni(N\text{-}methylethylenediamine)_2(NO_2)_2$, $Ni(N\text{-}ethylethylenediamine)_2(NO_2)_2$, and $Ni(rac\text{-}diphenylethylenediamine)_2(NO_2)_2$. In a chloroform solution some of these nitro- complexes partially change into nitrito- complexes. This is known as linkage isomerism. Even more complexed nickel nitrites are, for example, $Ni(2\text{-}(aminomethyl)piperidine)_2(NO_2)_2$, $Ni(2\text{-}(aminomethyl)pyridine)_2(NO_2)_2$, $Ni(2\text{-}(methylaminomethyl)pyridine)_2(ONO)_2$, $Ni(2\text{-}(methylaminomethyl)piperidine)_2(ONO)_2$, $Ni(2\text{-}(aminomethyl)\text{-}6\text{-}methylpyridine)_2(ONO)_2$, $Ni(2\text{-}(aminomethyl)\text{-}6\text{-}methylpiperidine)_2(ONO)_2$, $Ni(N,N'\text{-}dimethylethylenediamine)_2(NO_2)_2 \cdot H_2O$, $Ni(N,N\text{-}dimethylethylenediamine)_2(NO_2)_2$, $Ni(\alpha\text{-}picoline)_2(NO_2)_2$, and $Ni(quinoline)_2(NO_2)_2$.

Nickel (II) chromate ($NiCrO_4$) is an acid-soluble compound, red-brown in color, with high tolerances for heat. Nickel (II) chromate can be prepared in the laboratory by heating a mixture of chromium(III) oxide and nickel oxide at between 700°C and 800°C under 1000 atmosphere's pressure of oxygen. It can also be formed at 535°C and 7.3 bar oxygen, but in that case the reaction takes days to complete. If the pressure is too low or temperature too high but above 660°C then the nickel chromium spinel $NiCr_2O_4$ is formed instead. Karin Brandt (1943) has stated to have produced nickel chromate using a hydrothermal technique. Precipitates of Ni^{2+} ions with chromate result in a brown substance that contains water. The structure of nickel chromate is similar to chromium vanadate, $CrVO_4$. Crystals have an orthorhombic structure with unit cell sizes a = 5.482 Å, b = 8.237 Å, c = 6.147 Å. The cell volume is 277.6 Å^3 with four formula per unit cell. Nickel chromate is dark in color, in contrast to most other chromates which are yellow. When heated at lower oxygen pressure around 600°C, nickel chromate decomposes to form the nickel chromite spinel, nickel oxide, and oxygen.

$$4NiCrO_4 \rightarrow 2NiCr_2O_4 + 2NiO + 3O_2(gas)$$

Nickel chromate can crystallize with ligands, for example, $[Ni(1,10\text{-phenanthroline})CrO_4 \cdot 3H_2O] \cdot H_2O$ which forms triclinic olive colored crystals, and orange crystals of $Ni(1,10\text{-phenanthroline})_3Cr_2O_7 \cdot 3H_2O$ and yellow powdered $Ni(1,10\text{-phenanthroline})_3Cr_2O_7 \cdot 8H_2O$.

5.10.3.6 *Organonickel compounds*

Organonickel compounds which include nickel carbonyls, allyls, cyclopentadienyls, carbenes, and alkyls or aryls are important as catalysts, as building blocks in organic chemistry and as precursors in CVD. They are also involved as short-lived intermediates in organic reactions. Organonickel complexes are prominent in numerous industrial processes including carbonylations, hydrocyanation, and the higher olefin process.

The first organonickel compound was nickel tetracarbonyl $Ni(CO)_4$, reported in 1890 when Ludwig Mond and coworkers observed that elemental nickel and CO reacted at room temperature to form $Ni(CO)_4$, which could be used to purify the metal (now referred as Mond process). Nickel with the d^8s^2 configuration and the 18-electron rule is expected to lead to carbonyls of formula $[M(CO)_4]$ and indeed $Ni(CO)_4$ exist. $Ni(CO)_4$ is an extremely toxic, colorless liquid (melting point $-19.3°C$, boiling point 42.2C°) which is tetrahedral in the vapor and in the solid (Ni-C 184 pm, C-O 115 pm). It is readily oxidized by air and can be reduced by alkali metals in liquid ammonia or THF to yield polynuclear carbonylate anion clusters consisting mainly of $[Ni_5(CO_{12})]^{2-}$ and $[Ni_6(CO)_{12}]^{2-}$.

Nickelocene, $[Ni^{II}(\eta^5\text{-}C_5H_5)_2]$, is a bright green, reactive solid first prepared by Fischer and Pfab in 1952, shortly after the discovery of the first sandwich metallorganic compound—ferrocene. It has been prepared in a one-pot reaction, by deprotonating cyclopentadiene with ethylmagnesium bromide, and adding anhydrous nickel(II) acetylacetonate. Its modern preparation entails treatment of anhydrous sources of $NiCl_2$ (such as hexaamminenickel chloride) in 1,2-dimethoxyethane with sodium or potassium cyclopentadienyl in dimethylsulfoxide (DMSO).

$$[Ni(NH_3)_6]Cl_2 + 2NaC_5H_5 \rightarrow Ni(C_5H_5)_2 + 2NaCl + 6NH_3$$

Nickelocene has the sandwich structure of ferrocene and show similar susceptibility to ring-addition reactions. However, the two extra electrons in nickelocene must be accommodated in an antibonding orbital; thus the orange-yellow, $[Ni(\eta^5\text{-}C_5H_5)_2]^+$, cation is easily obtained by oxidation. The "triple-decker sandwich" cation, $[Ni_2(\eta^5\text{-}C_5H_5)_3]^+$ is produced by reacting nickelocene with a Lewis acid such as BF_3. This triple-decker species is a useful synthon and has been used synthetically to cap monocyclopentadienide species with an NiCp group.

The alkene and alkyne complexes involve the Ni in the formally divalent state. The role of Ni complexes as intermediates in the oligomerization of conjugated dienes has been extensively studied, particularly by G. Wilke and his group. The history of its development is presented by G. Wilke in his 1988 review article entitled, "Contributions to Organonickel Chemistry" published in Angew. Chem. Int. Ed. Engl. Vol. 27, Issue number 1, pages 185–206. Highlights of this review include how the 1948 report of Reppe et al. on the unexpected discovery of the "cyclic polymerization of acetylene" to cyclooctatetraene (COT) using nickel catalysts ushered another milestone for organonickel chemistry. The original scheme for this reaction as presented by G. Wilke (1988) is shown below.

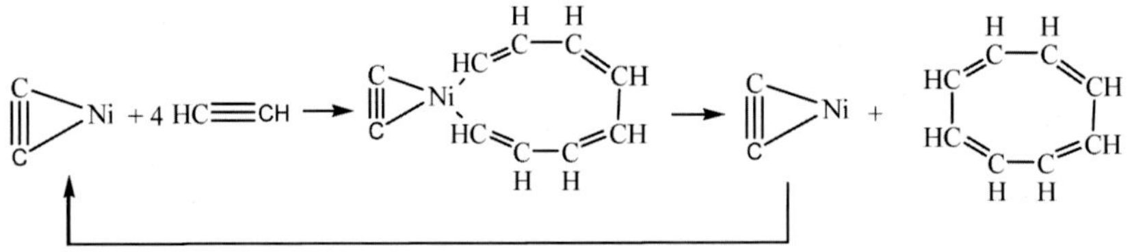

Another phase of organonickel chemistry development was ushered in by a failure of an otherwise routinely successful reaction between ethylaluminum and ethylene. Instead of forming higher aluminum alkyl compounds, 1-butene was formed. This was attributed to nickel impurities and was called the "nickel effect." The nickel effect necessitated the development of a stable nickel cocatalyst to favor the dimerization of ethylene to butene and this was achieved by treatment of nickel acetylacetonate in toluene with dietyhylaluminum ethoxide, Et_2AlOEt, in the presence of phenylacetylene. The discovery of cyclotrimerization of butadiene to 1,5,9-cyclododecatriene (CDT) using a Ziegler catalyst also led to a new research strategy—the controlled oligomerization of unsaturated compounds. Ni complexes have since been extensively studied as intermediates in the oligomerization of conjugated dienes. Other isomers of CDT have been prepared and led to observations that if a coordination site on the nickel is blocked by the addition of a ligand such as a tertiary phosphine, dimerization of the butadiene, rather than trimerization, occurs. Extensive mechanistic studies have been conducted for organonickel in the last decades which led to better understanding about the elementary steps and oxidation states of nickel in several varying reaction manifolds. This increased understanding is expected to direct future efforts toward the design of new nickel catalysts and the development of new transformations that can accomplish even more complex bond-forming reactions.

5.10.4 Major uses

The worldwide usage of Ni is currently: 68% in stainless steel; 10% in nonferrous alloys; 9% in electroplating; 7% in alloy steel; 3% in foundries; and 4% other uses (including batteries). About 27% of all Ni production is intended for engineering, 10% for building and construction, 14% for tubular products, 20% for metal goods, 14% for transport, 11% for electronic goods, and 5% for other applications. Ni is used in numerous specific and easily identifiable industrial and consumer products, such as stainless steel, alnico magnets, coinage, rechargeable batteries, electric guitar strings, microphone capsules, plating on plumbing fixtures, and special alloys. It is also used for plating and as a green tint in glass. Ni is predominantly an alloy metal, and its principal use is in Ni steels and Ni cast irons, in which it usually improves the tensile strength, toughness, and elastic limit. It is extensively used in numerous other alloys, for example, Ni brasses and bronzes, and alloys with Cu, Cr, Al, Pb, Co, Ag, and Au (Inconel, Incoloy, Monel, Nimonic). Due to its resistance to corrosion, Ni was sometimes used as an alternative for decorative silver. Ni was also sometimes used in some countries after 1859 as a low-cost coinage metal, but near the end of the 20th century it was superseded by cheaper stainless-steel (i.e., iron) alloys, except in the United States and Canada. Ni is an exceptional alloying agent for some precious metals and is used in the fire assay as a collector of platinum group elements (PGEs). In that role Ni is capable of completely collecting all six PGEs from ores, and of partly collecting Au. High-throughput Ni mines may also take part in PGE recovery (mainly Pt and Pd); for example, Norilsk in Russia and the Sudbury Basin in Canada. Ni foam or Ni mesh is used in gas diffusion electrodes for alkaline fuel cells. Rechargeable nickel-cadmium batteries are manufactured from sintered Ni powder and can be recharged and used more than a thousand times.

Ni and its alloys are commonly used as catalysts for hydrogenation reactions. Raney nickel, a finely divided Ni-Al alloy, is one familiar form, though related catalysts are similarly used, including Raney-type catalysts. Finely powdered Ni is used as a catalyst in the hydrogenation of oils to fats. In the 1890s two French chemists, Paul Sabatier (November 5, 1854 to August 14, 1941) and Jean-Baptiste Senderens (January 27, 1856 to September 26, 1937), showed that it was possible to convert edible oils into fats, in so doing making margarine, a low-cost alternative for butter that is just as nutritious (in terms of providing energy, but it contains the so-called *trans*-fats which some dietitians consider harmful), even though less tasty. Ni powder is also added to paints that coat the housing of sensitive electronic instruments as it shields them from stray radiation. Ni is a naturally magnetostrictive element that is, in the presence of a magnetic field, it undergoes a small change in length. (Internally, ferromagnetic materials have a structure that is divided into domains, each of which is a region of uniform magnetic polarization. When a magnetic field is applied, the boundaries between the domains shift and the domains rotate; both effects cause a change in the material's dimensions. The reason that a change in the magnetic domains of a material results in a change in the materials dimensions is a consequence of magnetocrystalline anisotropy, that it takes more energy to magnetize a crystalline material in one direction than another. If a magnetic field is applied to the material at an angle to an easy axis of magnetization, the material will tend to rearrange its structure so that an easy axis is aligned with the field to minimize the free energy of the system. Since different crystal directions are associated with different lengths this effect induces a strain in the material.) The magnetostriction of Ni is around -50 ppm, the negative sign indicating that it contracts. Ni is used as a binder in the cemented tungsten carbide (WC) or hard-metal industry and used in concentrations between 6% and 12% by weight. Ni makes the tungsten carbide magnetic and enhances the corrosion resistance of the cemented parts, though the hardness is less than those with a Co binder. ^{63}Ni, with its half-life of 100.1 years, is useful in a krytron device, a cold-cathode gas-filled tube intended for use as a very high-speed switch. It consists of a sealed glass tube with four electrodes. A small triggering pulse on the grid electrode switches the tube on, allowing a large current to flow between the cathode and anode electrodes. A krytron has four electrodes. Two are a conventional anode and cathode. One is a keep-alive electrode, arranged to be close to the cathode. The keep-alive has a low positive voltage applied, which causes a small area of gas to ionize near the cathode. High voltage is applied to the anode, but primary conduction does not occur until a positive pulse is applied to the trigger electrode. Once started, arc conduction carries a considerable current. The fourth is a control grid, usually wrapped around the anode, except for a small opening on its top. In place of or in addition to the keep-alive electrode some krytrons may contain a very tiny amount of radioactive material (usually less than 5 microcuries (180 kBq) of ^{63}Ni), which emits β particles (high-speed electrons) to make ionization easier. The radiation source serves to increase the reliability of ignition and formation of the keep-alive electrode discharge.

5.11 29 Cu — Copper

5.11.1 Discovery

Copper can be found naturally as native metallic copper and was known to some of the oldest civilizations on record. The history of copper use can be traced back to 9000 BCE in the Middle East; a copper pendant was discovered in northern Iraq that dates to 8700 BCE. Evidence indicates that gold and meteoric iron (but not smelted iron) were the only metals used by mankind before copper. The history of copper metallurgy has been suggested to follow this sequence: first, cold working of native copper (any metalworking process in which metal is shaped below its recrystallization temperature, usually at the ambient temperature), then annealing (a heat treatment that alters the physical and sometimes chemical properties of a material to increase its ductility and reduce its hardness, making it more workable), smelting (a process of applying heat to ore in order to extract a base metal), and, finally, lost-wax casting (the process by which a duplicate metal sculpture (often silver, gold, brass or bronze) is cast from an original sculpture). In south-eastern Anatolia, all four of these techniques appear around the same time at the beginning of the Neolithic era c.7500 BCE. Copper smelting was independently developed in different places. It was probably discovered in China before 2800 BCE, in Central America around CE 600, and in West Africa about the 9th or 10th century CE. Investment casting (lost-wax casting) was developed between 4500 and 4000 BCE in Southeast Asia and carbon dating has determined that mining took place at Alderley Edge in Cheshire, UK, around 2280–1890 BCE. Ötzi the Iceman, a male dated to have lived around 3300–3200 BCE, was discovered with an ax with a copper head 99.7% pure; high concentrations of arsenic in his hair seem to indicate an involvement in copper smelting. Knowledge of copper has helped the development of other metals; especially, copper smelting resulted in the discovery of iron smelting. Production in the Old Copper Complex in Michigan and Wisconsin is dated between 6000 and 3000 BCE. Natural bronze, a type of copper produced from ores rich in silicon, arsenic, and (rarely) tin, came into common use in the Balkans around 5500 BCE. Alloying copper with tin to produce bronze was first used about 4000 years after the development of copper smelting, and around 2000 years after "natural bronze" had come into general use. Bronze artifacts from the Vinča culture date to 4500 BCE. Sumerian and Egyptian artifacts of copper and bronze alloys date to 3000 BCE. The Bronze Age began in South-eastern Europe around 3700–3300 BCE, in North-western Europe about 2500 BCE. It ended with the beginning of the Iron Age, 2000–1000 BCE in the Near East, and 600 BCE in Northern Europe. The transition period between the Neolithic period and the Bronze Age was previously known as the Chalcolithic period (copper-stone), when copper tools were used together with stone tools. The term has slowly fallen out of favor since in some areas of the world, the Chalcolithic and Neolithic are coterminous at both ends. Brass, an alloy of copper and zinc, is of much more recent times. It was known to the Greeks but became an important addition to bronze during the Roman Empire. In Greece, copper was known as chalkos (χαλκός). It was a significant resource for the Romans, Greeks and other ancient peoples. In Roman times, it was called aes Cyprium, aes being the general Latin term for copper alloys and Cyprium from Cyprus, where much copper was mined. The phrase was shortened to cuprum, hence the English copper. Aphrodite (Venus in Rome) embodied copper in mythology and alchemy for its lustrous beauty and its ancient use in producing mirrors; Cyprus was sacred to the goddess. The seven heavenly bodies recognized by the ancients were related to the seven metals known in antiquity, and Venus was assigned to copper. Copper was first utilized in ancient Britain around the 3rd or 2nd century BCE. In North America, copper mining started with marginal workings by Native Americans. Native copper is known to have been mined from sites on Isle Royale with primitive stone tools between 800 and 1600. Copper metallurgy was thriving in South America, predominantly in Peru around CE 1000. Copper burial ornamentals from the 15th century have been discovered, but commercial manufacture of copper did not start until the early 20th century. The cultural role of copper has been significant, particularly in currency. Romans from the 6th till the 3rd century BCE used copper lumps as money. Initially, the copper itself was valued, but slowly the form and appearance of the copper became more important. Julius Caesar (July 12 or 13, 100 BCE to March 15, 44 BCE) had his own coins made from brass, while Octavianus Augustus (September 23, 63 BCE to August 19, CE 14) Caesar's coins were made from Cu-Pb-Sn alloys. With an estimated annual production of around 15,000 tons, Roman copper mining and smelting activities achieved a scale unrivaled until the period of the Industrial Revolution; the provinces most intensely mined were those of Hispania, Cyprus and in Central Europe. The gates of the Temple of Jerusalem were made of Corinthian bronze treated with depletion gilding. The process was most predominant in Alexandria, where alchemy is supposed to have started. In ancient India, copper was utilized in the holistic medical science Ayurveda to make surgical instruments and other medical equipment. Ancient Egyptians (~2400 BCE) applied copper for sterilizing wounds and drinking water, and later to treat headaches, burns, and itching. The Great Copper Mountain was a mine in Falun, Sweden, that was worked from the 10th century till its closure in 1992. It produced

FIGURE 5.89 Gottfried Osann.

two-thirds of Europe's copper consumption in the 17th century and assisted funding many of Sweden's wars throughout that period. It was referred to as the nation's treasury; Sweden had a copper backed currency. Copper is used in roofing, currency, and for photographic technology known as the daguerreotype. Copper was used in Renaissance sculptures and was used to build the Statue of Liberty; copper is still used in construction of various types. Copper plating and copper sheathing were extensively used to protect the underwater hulls of ships, a technique pioneered by the British Admiralty in the 18th century. The Norddeutsche Affinerie in Hamburg was the initial modern electroplating plant, beginning its production in 1876. The German chemist and physicist Gottfried Osann (October 26, 1796 to August 10, 1866) developed powder metallurgy in 1830 while measuring the metal's atomic mass (Fig. 5.89). Around that same time, it was found that the amount and type of alloying element (e.g., tin) to copper would affect bell tones. Through the increase in demand for copper for the Age of Electricity, from the 1880s until the Great Depression of the 1930s, the United States produced about one-third to half of the world's mined copper. Major districts comprised the Keweenaw district of northern Michigan, primarily native copper deposits, which was overtaken by the enormous sulfide deposits of Butte, Montana in the late 1880s, which itself was eclipsed by the porphyry deposits of the Southwest United States, especially at Bingham Canyon, Utah and Morenci, Arizona. The start of open pit steam shovel mining and innovations in smelting, refining, flotation concentration, and other processing steps resulted in mass production. Around the beginning of the 20th century, Arizona ranked first, followed by Montana, then Utah and Michigan. Flash smelting was established by Outokumpu in Finland and first used at Harjavalta in 1949; the energy-efficient process accounts for 50% of the world's primary copper production.

5.11.2 Mining, production, and major minerals

Copper is present in the Earth's crust in an average concentration of about 50 ppm. In nature, copper occurs in a large variety of minerals, including native copper (Cu), copper sulfides such as chalcopyrite ($CuFeS_2$), bornite (Cu_5FeS_4), digenite (Cu_9S_5), covellite (CuS), and chalcocite (Cu_2S), copper sulfosalts such as tetrahedite-tennantite ($Cu_6[Cu_4(Fe,Zn)_2]Sb_4S_{13}-(Cu,Fe)_{12}As_4S_{13}$), and enargite ($Cu_3AsS_4$), copper carbonates such as azurite ($Cu_3(CO_3)_2(OH)_2$) and malachite ($Cu_2(CO_3)(OH)_2$), and as copper(I) or copper(II) oxides such as cuprite (Cu_2O) and tenorite (CuO), respectively. The largest mass of elemental copper discovered weighed 420 tons and was found in 1857 on the Keweenaw Peninsula in Michigan, United States.

Most copper is mined or extracted as copper sulfides from large open pit mines in porphyry copper deposits that contain 0.4% to 1.0% copper. Sites include Chuquicamata, in Chile, Bingham Canyon Mine, in Utah, United States, and El Chino Mine, in New Mexico, United States. Copper can also be recovered through the in situ leach process.

Several locations in Arizona, United States, are thought to be prime candidates for this process. The amount of copper in use is increasing and the amount available is hardly enough to allow all countries to reach developed world levels of usage. The copper concentration in ores averages only ~0.6%, and the majority of commercial ores are in the form of sulfide minerals, in particular chalcopyrite ($CuFeS_2$), bornite (Cu_5FeS_4) and, to a lesser extent, covellite (CuS) and chalcocite (Cu_2S). These minerals are concentrated from crushed ores to the level of 10%–15% copper by froth flotation or bioleaching. Subsequent heating of this concentrated material with silica in a flash smelting furnace removes much of the iron as slag. The method makes use of the greater ease of transforming iron sulfides into oxides, which then react with the silica to form the silicate slag that floats on top of the heated mass. The resulting copper matte, consisting of Cu_2S, is roasted to change all sulfides into oxides (Fig. 5.90).

$$2Cu_2S + 3O_2 \rightarrow 2Cu_2O + 2SO_2$$

The cuprous oxide is converted to blister copper upon heating.

$$2Cu_2O \rightarrow 4Cu + O_2$$

The Sudbury matte process converts only half the sulfide to oxide and subsequently uses this oxide to remove the rest of the sulfur as oxide. It is then electrolytically refined and the anode mud is exploited for the platinum and gold it may contain. This step is based on the relatively easy reduction of copper oxides to copper metal. Natural gas is blown across the blister to eliminate most of the remaining oxygen and electrorefining produce pure copper from the resulting material.

$$Cu^{2+} + 2e^- \rightarrow Cu$$

Copper can be found in roughly 600 different minerals. The elements class is represented by native copper (Cu) (Fig. 5.91) and 18 alloys, for example, auricupride (Cu_3Au), danbaite ($CuZn_2$) and kolymite (Cu_7Hg_6). A large group of between 200 and 250 sulfide minerals, nearly one-third of all copper minerals, are known with Cu in their crystal structure. Among these are well-known minerals such as bornite (Cu_5FeS_4) (Fig. 5.92), bournonite ($PbCuSbS_3$) (Fig. 5.93), chalcocite (Cu_2S), chalcopyrite ($CuFeS_2$) (Fig. 5.94), covellite (CuS) (Fig. 5.95), digenite (Cu_9S_5), enargite (Cu_3AsS_4),

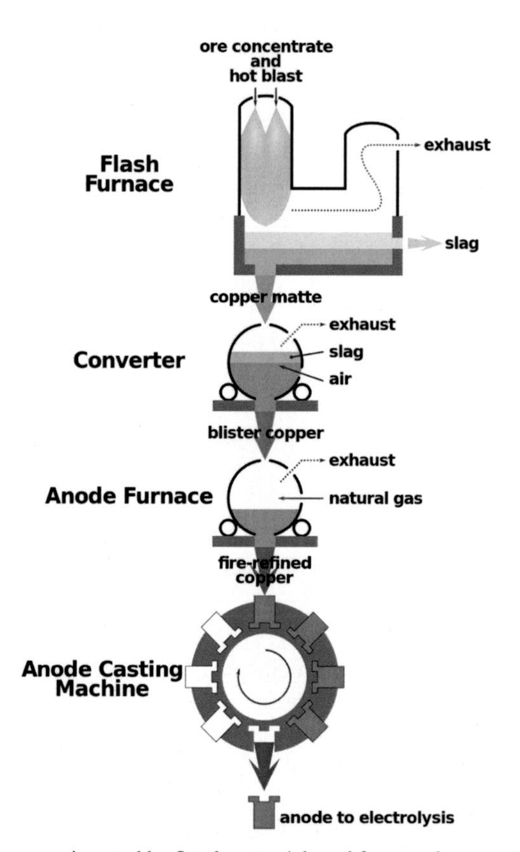

FIGURE 5.90 Copper flash smelting process, as invented by Outokumpu. Adapted from work prepared for US EPA (https://www3.epa.gov/ttnchie1/ap42/ch12/bgdocs/b12s03.pdf).

FIGURE 5.91 Arborescent copper, Cu, crystals to 2.5 cm. Ajo, Pima Co., Arizona, United States.

FIGURE 5.92 Iridescent blue purple bornite, Cu_5FeS_4, crystal to 1.2 cm. Dzhezkazgan mine, Karagandy, Kazakhstan.

tennantite ($Cu_6[Cu_4(Fe,Zn)_2]As_4S_{13}$), and tetrahedrite ($Cu_6[Cu_4(Fe,Zn)_2]Sb_4S_{13}$). A total of 35 copper-containing halides are known, for example, atacamite ($Cu_2(OH)_3Cl$), boleite ($KPb_{26}Ag_9Cu_{24}(OH)_{48}Cl_{62}$), connellite ($Cu_{19}(SO_4)(OH)_{32}Cl_4{\cdot}3H_2O$), and diaboleite ($Pb_2CuCl_2(OH)_4$) (Fig. 5.96). The oxide class contains 47 minerals with Cu, including minerals such as cuprite (Cu_2O) (Figs. 5.97 and 5.98), cuprospinel ($Cu^{2+}Fe^{3+}_2O_4$), and tenorite (CuO). Twenty-four carbonate minerals are known with Cu in their crystal structure, for example, aurichalcite ($(Zn,Cu)_5(CO_3)_2(OH)_6$) (Fig. 5.99), azurite ($Cu_3(CO_3)_2(OH)_2$), and malachite ($Cu_2(CO_3)(OH)_2$) (Fig. 5.100) (Box 5.14). The borates are represented by five minerals, for example, henmilite ($Ca_2Cu[B(OH)_4]_2(OH)_4$) and santarosaite (CuB_2O_4). More than 80 sulfate minerals contain structural copper, including minerals such as antlerite ($Cu_3(SO_4)(OH)_4$), brochantite ($Cu_4(SO_4)(OH)_6$), cyanotrichite ($Cu_4Al_2(SO_4)(OH)_{12}{\cdot}2H_2O$), linarite ($PbCu(SO_4)(OH)_2$).

FIGURE 5.93 A single 3 cm bournonite, $PbCuSbS_3$, crystal with deep striations and terminated on both ends. Yaogangxian mine, Chenzhou, Hunan prov., China.

FIGURE 5.94 Brassy gold colored chalcopyrite, $CuFeS_2$, crystal to 2 cm with black sphalerite, ZnS, and quartz, SiO_2. Casapalca, Lima Dept., Peru.

The phosphate class (including molybdates and vanadates) has 130 minerals with copper as an essential element, for example, arthurite ($CuFe^{3+}_2(AsO_4)_2(OH)_2{\cdot}4H_2O$), bayldonite ($PbCu_3(AsO_4)_2(OH)_2$), clinoclase ($Cu_3(AsO_4)(OH)_3$), conichalcite ($CaCu(AsO_4)(OH)$), duftite ($PbCu(AsO_4)(OH)$), mottramite ($PbCu(VO_4)(OH)$) (Fig. 5.101), olivenite ($Cu_2(AsO_4)(OH)$) (Fig. 5.102), pseudomalachite ($Cu_5(PO_4)_2(OH)_4$), szenicsite ($Cu_3(MoO_4)(OH)_4$) (Fig. 5.103), torbernite ($Cu(UO_2)_2(PO_4)_2{\cdot}12H_2O$), turquoise ($Cu(Al,Fe^{3+})_6(PO_4)_4(OH)_8{\cdot}4H_2O$) (Fig. 5.104), and zeunerite ($Cu(UO_2)_2(AsO_4)_2{\cdot}12H_2O$) (Fig. 5.105). Twenty-five minerals in the silicate class are known to contain copper in their crystal structure, such as chrysocolla ($Cu_{2-x}Al_x(H_{2-x}Si_2O_5)(OH)_4{\cdot}nH_2O$) (Fig. 5.106), dioptase ($CuSiO_3{\cdot}H_2O$) (Fig. 5.107), and kinoite ($Ca_2Cu_2(H_2O)_2[Si_3O_{10}]$). Finally, five minerals within the organic compounds class contain Cu, for example, hoganite ($Cu(CH_3COO)_2{\cdot}H_2O$) and moolooite ($Cu(C_2O_4){\cdot}0.4H_2O$).

5.11.3 Chemistry

There exist 29 known isotopes of copper. ^{63}Cu and ^{65}Cu are stable, with ^{63}Cu covering 69.15% and ^{65}Cu 30.85% of naturally occurring copper; both isotopes have a spin of 3/2. All the other isotopes are radioactive. ^{67}Cu is the

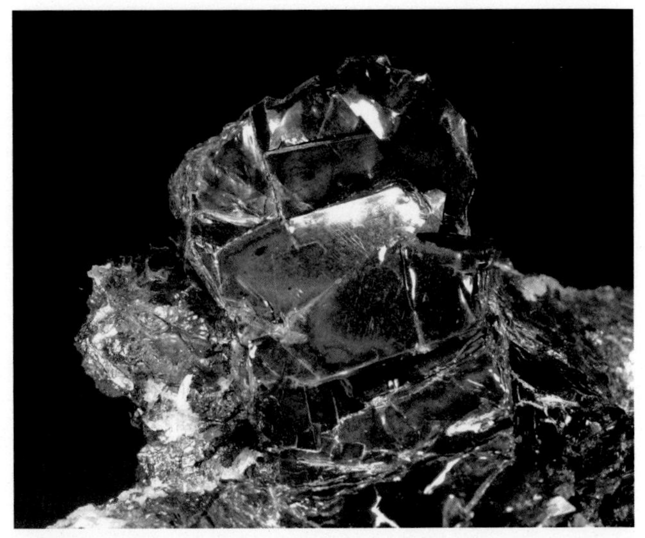

FIGURE 5.95 Covellite, CuS, platy iridescent purple crystals to 1.5 cm. Butte dist., Silver Bow Co., Montana, United States.

FIGURE 5.96 Finely crystalline diaboleite, $Pb_2CuCl_2(OH)_4$, specimen with baryte, $BaSO_4$, about 1 cm. Rowley mine, Maricopa Co., Arizona, United States.

FIGURE 5.97 Deep red cubo-octahedral crystals of cuprite, Cu_2O, to 2 mm. Gwennap, Cornwall, England.

FIGURE 5.98 Specimen with a pocket of filiform cuprite, Cu_2O, variety chalcotrichite. The bundle is 1.5 cm wide and the matrix is predominantly calcite, $CaCO_3$, and granular cuprite. Morenci mine, Greenlee Co., Arizona, United States.

FIGURE 5.99 Bladed crystals of aurichalcite, $(Zn,Cu)_5(CO_3)_2(OH)_6$, in divergent sprays to 3 mm. Mina Ojuela, Mapimi, Durango, Mexico.

FIGURE 5.100 Silky, shiny, deep green, fibrous malachite, $Cu_2(CO_3)(OH)_2$, 9 cm across. Luishia mine, Katanga, Democratic Republic of Congo.

BOX 5.14 Malachite, $Cu_2(CO_3)(OH)_2$, and azurite, $Cu_3(CO_3)_2(OH)_2$

Malachite was named in antiquity *molochitus* after the Greek μαλαχή, "mallows," in allusion to the green color of the leaves. Known in the new spelling, malachites, at least since 1661. Malachite is a very common secondary copper mineral with a widely variable habit. Malachite was extensively mined at the Great Orme mines in Britain 3800 years ago using stone and bone tools. Archeological evidence indicates that mining activity ended around 600 BCE with up to 1760 tons of copper being produced from the mined malachite. Archeological evidence indicates that the mineral has been mined and smelted to obtain copper at Timna Valley in Israel for over 3000 years. Since then, malachite has been used as both an ornamental stone and as a gemstone. In ancient Egypt the color green (wadj) was associated with death and the power of resurrection as well as new life and fertility. Ancient Egyptians believed that the afterlife contained an eternal paradise which resembled their lives but with no pain or suffering and referred to this place as the "Field of Malachite." Malachite was used as a mineral pigment in green paints from antiquity until about 1800. The pigment is moderately lightfast, very sensitive to acids, and varying in color. This natural form of green pigment has been replaced by its synthetic form, verditer, among other synthetic greens. Malachite is also used for decorative purposes, such as in the Malachite Room in the Hermitage Museum, which features a huge malachite vase, and the Malachite Room in Castillo de Chapultepec in Mexico City. "The Tazza," a large malachite vase, one of the largest pieces of malachite in North America and a gift from Tsar Nicholas II, stands as the focal point in the center of the room of Linda Hall Library. Malachite is monoclinic, *2/m* (space group $P2_1/a$). It has a hardness of 3½–4. Cleavage on {$\bar{2}01$} is perfect while on {010} is fair but rarely seen. The luster is adamantine to vitreous in crystals, often silky in fibrous varieties, dull in earth types. Transparency is transparent to translucent. Color is bright green, while the streak is pale green. Optically it is biaxial(–). Crystals are rare, slender prismatic but seldom distinct. Crystals may be pseudomorph after azurite. Malachite is usually massive, frequently colloform, botryoidal, stalactitic, or banded, commonly granular or earthy, and often intergrown with other secondary copper minerals.

Azurite is a soft, deep blue copper mineral produced by weathering of copper ore deposits. In the early 19th century, it was also known as *chessylite* after the type locality at Chessy-les-Mines near Lyon, France. It has been known since ancient times and was noted in Pliny the Elder's Natural History under the Greek name kuanos (κυανός: "deep blue," root of English cyan) and the Latin name caeruleum. The blue of azurite is extraordinarily deep and clear, and for that reason the mineral has been linked since antiquity with the deep blue color of low-humidity desert and winter skies. The modern English name of the mineral reflects this association, since both azurite and azure are derived via Arabic from the Persian lazhward (لاژورد), an area known for its deposits of another deep blue stone, lapis lazuli ("stone of azure"). Azurite is not a useful pigment since it is unstable in air. It was however used as a blue pigment in antiquity. Azurite is naturally occurring in Sinai and the Eastern Desert of Egypt. It was reported by Spurrell (1895) in, for example, a shell used as a pallet in a Fourth Dynasty (2613–2494 BCE) context in Meidum, a cloth over the face of a Fifth Dynasty (2494–2345 BCE) mummy also at Meidum and a number of Eighteenth Dynasty (1543–1292 BCE) wall paintings. Depending on the degree of fineness to which it was ground, and its basic content of copper carbonate, it gave a wide range of blues. It has been known as mountain blue or Armenian stone; in addition, it was formerly known as Azurro Della Magna (from Italian). When mixed with oil it turns slightly green. When mixed with egg yolk it turns green-gray. It is also known as blue bice and blue verditer, though verditer typically refers to a pigment made by a chemical process. Older examples of azurite pigment may show a more greenish tint due to changing into malachite. Much azurite was mislabeled lapis lazuli, a term applied to many blue pigments. As chemical analysis of paintings from the Middle Ages improves, azurite has been identified as a major source of the blues used by medieval painters. Lapis lazuli (the pigment ultramarine) was chiefly supplied from Afghanistan during the Middle Ages, whereas azurite was a common mineral in Europe at the time. Sizable deposits were discovered near Lyons, France. It was mined since the 12th century in the silver mines located in Saxony. Heating can be used to distinguish azurite from purified natural ultramarine blue, a more expensive but more stable blue pigment, as described by Cennino D'Andrea Cennini. Ultramarine withstands heat, whereas azurite converts to black copper oxide. However, gentle heating of azurite produces a deep blue pigment used in Japanese painting techniques. Azurite is used occasionally as beads and as jewelry, and as an ornamental stone. However, its softness and tendency to lose its deep blue color as it weathers limit such uses. Heating destroys azurite easily, so all mounting of azurite must be done at room temperature. Azurite is less common than malachite but has the same origin and associations. Pseudomorph alterations of malachite after azurite are commonly observed, less common after cuprite. Azurite is monoclinic, *2/m* (space group $P2_1/c$). It has a hardness of 3½–4. It has a perfect cleavage on {011} and fair on {100}. It is brittle with conchoidal fracture. The luster is vitreous, while the transparency is transparent to translucent. The color is intense azure-blue. The streak is light blue. Optically it is biaxial(+). The habit varies. Crystals are often complex and malformed, tabular or prismatic. Also found in radiating spherical groups. Typically, it is massive and earthy.

FIGURE 5.101 Dark green plumose mottramite, PbCu(VO4)(OH). Overall 8.5 mm across. Tsumeb mine, Tsumeb, Namibia.

FIGURE 5.102 Olivenite, $Cu_2(AsO_4)(OH)$, divergent bladed crystals to 2 mm with typical olive-green color. Majuba Hill mine, Pershing Co., Nevada, United States.

most stable with a half-life of 61.83 h. Seven metastable isotopes have been observed; ^{68m}Cu is the longest-lived with a half-life of 3.8 minutes. Isotopes with a mass number larger than 64 decay by β^-, whereas those with a mass number below 64 decay by β^+. ^{64}Cu, which has a half-life of 12.7 h, decays both ways. ^{62}Cu and ^{64}Cu have important applications. ^{62}Cu is, for example, used in ^{62}Cu-ATSM (^{62}Cu-diacetyl-bis(N4-methylthiosemicarbazone)) and ^{62}Cu-PTSM (^{62}Cu-pyruvaldehyde-bis(N4-methylthiosemicarbazone)) as a radioactive tracer for positron emission tomography (PET is a nuclear medicine functional imaging technique that is used to observe metabolic processes in the body as an aid to the diagnosis of disease. The system detects pairs of gamma rays emitted indirectly by a positron-emitting radioligand, which is introduced into the body on a biologically active molecule

FIGURE 5.103 Szenicsite, $Cu_3(MoO_4)(OH)_4$, lustrous green crystals to 2 cm arranged as a rosette. Jardinera No.1 mine, Inca de Oro, Atacama, Chile.

FIGURE 5.104 Bright blue turquoise, $Cu(Al,Fe^{3+})_6(PO_4)_4(OH)_8 4H_2O$, crystals in tiny clusters to 1.5 mm on quartz, SiO_2, matrix. Bishop mine, Lynch Station, Campbell Co., Virginia, United States.

called a radioactive tracer. Different ligands are used for different imaging purposes, depending on what the radiologist/researcher wants to detect.).

Copper is a soft, malleable, and ductile metal with very high thermal and electrical conductivity. It appears pinkish orange when its surface is freshly exposed. It is a member of the coinage metals and has 29 electrons arranged in an electronic configuration of $[Ar]3d^{10}4s^1$. The $d^{10}s^1$ electron configuration of Cu and its congeners is comparable to the p^6s^1 electron configurations of the alkali metals. The similarities, are, however confined to the stoichiometries of the compounds of the +1 oxidation state. The second and third ionization energies of copper and the other coinage metals are lower than those of the alkali metals. Thus they are able to adopt oxidation states higher than +1. In fact, copper forms a rich variety of compounds with +1 and +2 oxidation states, commonly called cuprous and cupric, respectively (Table 5.13).

Copper is stable in pure dry air at room temperature but slowly reacts with atmospheric oxygen to form a layer of brown-black copper oxide which protects the underlying metal from further corrosion (passivation). The green layer of

FIGURE 5.105 Emerald green plate of zeunerite, $Cu(UO_2)_2(AsO_4)_2{\cdot}12H_2O$, to 6 mm. Majuba Hill mine, Pershing Co., Nevada, United States.

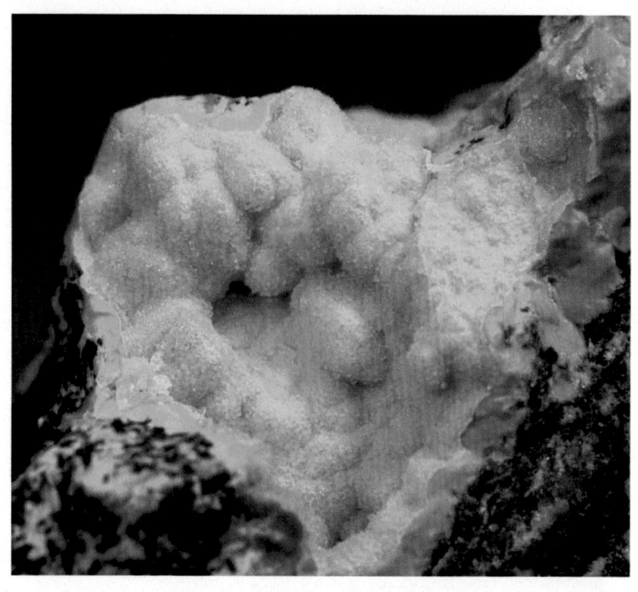

FIGURE 5.106 A deep turquoise blue–colored chrysocolla, $Cu_{2-x}Al_x(H_{2-x}Si_2O_5)(OH)_4{\cdot}nH_2O$ ($x < 1$), in the matrix with a large 2 cm drusy quartz, SiO_2, lined vug. Brown Monster mine, Inyo Co., California, United States.

verdigris seen on old copper structures, such as the roofing of many older buildings and the Statue of Liberty is a combination of copper chlorides, sulfides, sulfates and carbonates, depending upon environmental conditions such as sulfur-containing acid rain. In clean air, this green layer is brought about by the slow reaction of copper with carbon dioxide and water to produce the green copper carbonate ($CuCO_3$).

5.11.3.1 Oxides and hydroxides

The two stable oxides of copper, Cu_2O (yellow or red) and CuO (black) have narrow ranges of homogeneity and form when the metal is heated in air or O_2, Cu_2O being favored by high temperatures. Industrially or in the laboratory, both oxides are conveniently prepared by reducing copper salts.

$$4Cu + O_2 \rightarrow 2Cu_2O$$

Additives such as water and acids influence the rate of this process in addition to the further oxidation to copper(II) oxides. It is also prepared industrially through the reduction of copper(II) solutions with sulfur dioxide. Aqueous cuprous

FIGURE 5.107 A 1.5 cm cluster of lustrous emerald green dioptase, $CuSiO_3H_2O$, crystals perched on a piece of calcite, $CaCO_3$, matrix. Altyn-Tube, Kirghiz Steppes, Karagandy, Kazakhstan.

TABLE 5.13 Copper properties.

Appearance	Red orange metallic Luster
Standard atomic weight $A_{r,std}$	63.546
Block	d-Block
Element category	Transition metal
Electron configuration	[Ar] $3d^{10}$ $4s^1$
Phase at STP	Solid
Melting point	1084.62°C
Boiling point	2562°C
Density (near r.t.)	8.96 g/cm^3
When liquid (at m.p.)	8.02 g/cm^3
Heat of fusion	13.26 kJ/mol
Heat of vaporization	300.4 kJ/mol
Molar heat capacity	24.440 J/(mol·K)
Oxidation states	−2, +1, **+2**, +3, +4
Ionization energies	1st: 745.5 kJ/mol
	2nd: 1957.9 kJ/mol
	3rd: 3555 kJ/mol
Atomic radius	empirical: 128 pm
Covalent radius	132 ± 4 pm
Van der Waals	140 pm

STP, Standard temperature and pressure.
Bold font indicates main oxidation state.

chloride solutions react with base to form the same compound. In all cases, the color is highly sensitive to the procedural details. Formation of copper(I) oxide is the basis of the Fehling's test and Benedict's test for reducing sugars. These sugars reduce an alkaline solution of a copper(II) salt, forming a bright red precipitate of Cu_2O. It can be formed on silver-plated copper parts exposed to moisture when the silver layer is porous or damaged. This type of corrosion is known as red plague. Little evidence exists for cuprous hydroxide, which is anticipated to quickly undergo dehydration. Cu_2O crystallizes in a cubic structure with a lattice constant al = 4.2696 Å. The Cu atoms arrange in a *fcc* sublattice, the O atoms in a *bcc* sublattice. One sublattice is shifted by a quarter of the body diagonal. The space group is $Pn\bar{3}m$, which comprises the point group with full octahedral symmetry. Cu_2O is diamagnetic. In terms of their coordination spheres, Cu centers are 2-coordinated and the oxides are tetrahedral. The structure thus resembles in some sense the main polymorphs of SiO_2, and both structures exhibit interpenetrated lattices. Cu_2O dissolves in concentrated ammonia solution to form the colorless complex $[Cu(NH_3)_2]^+$, which is readily oxidized in air to the blue-colored $[Cu(NH_3)_4(H_2O)_2]^{2+}$. It dissolves in hydrochloric acid to form solutions of $CuCl_2^-$. Dilute sulfuric acid and nitric acid result in the formation of copper(II) sulfate and copper(II) nitrate, respectively. Cu_2O degrades to copper(II) oxide in moist air. Cu_2O is frequently used as a pigment, a fungicide, and an antifouling agent for marine paints. Rectifier diodes based on this compound have been employed industrially since 1924, long before silicon became the standard. It is also responsible for the pink color in a positive Benedict's test.

Copper(II) oxide, CuO, is a black solid that occur naturally as the mineral tenorite. It is a product of copper mining and the precursor to many other copper-containing products and chemical compounds. It is produced on a large scale by pyrometallurgy used to extract copper from ores. The ores are treated with an aqueous mixture of ammonium carbonate, ammonia, and oxygen to form copper(I) and copper(II) ammine complexes, which are extracted from the solids. These complexes are decomposed with steam to form CuO. It can be prepared by heating copper in air at around 300°C–800°C.

$$2Cu + O_2 \rightarrow 2CuO$$

For laboratory applications, pure copper(II) oxide is better produced through heating copper(II) nitrate, copper(II) hydroxide or basic copper(II) carbonate.

$$2Cu(NO_3)_2(s) \rightarrow 2CuO(s) + 4NO_2(g) + O_2(g)\ (180°C)$$
$$Cu(OH)_2(s) \rightarrow CuO(s) + H_2O(l)\ (80 - 100°C)$$
$$Cu_2CO_3(OH)_2(s) \rightarrow 2CuO(s) + CO_2(g) + H_2O(g)\ (290°C)$$

CuO has a monoclinic structure. The Cu atom is coordinated by 4 oxygen atoms in a nearly square planar configuration. The work function of bulk CuO is 5.3 eV CuO is a p-type semiconductor, with a narrow band gap of 1.2 eV. Cupric oxide can be used to manufacture dry cell batteries. Copper(II) oxide dissolves in mineral acids such as hydrochloric acid, sulfuric acid or nitric acid to form the corresponding copper(II) salts.

$$CuO + 2HNO_3 \rightarrow Cu(NO_3)_2 + H_2O$$
$$CuO + 2HCl \rightarrow CuCl_2 + H_2O$$
$$CuO + H_2SO_4 \rightarrow CuSO_4 + H_2O$$

It reacts with concentrated alkali to produce the corresponding cuprate salts.

$$2MOH + CuO + H_2O \rightarrow M_2[Cu(OH)_4]$$

In addition, it can be reduced to copper metal in reactions with hydrogen, carbon monoxide, or carbon.

$$CuO + H_2 \rightarrow Cu + H_2O$$
$$CuO + CO \rightarrow Cu + CO_2$$
$$2CuO + C \rightarrow 2Cu + CO_2$$

When cupric oxide replaces iron oxide in thermite the resulting mixture is a low explosive, not an incendiary. Cupric oxide can be employed to safely dispose of hazardous compounds such as cyanide, hydrocarbons, halogenated hydrocarbons and dioxins, through oxidation. For example, the decomposition reactions of phenol and pentachlorophenol follow these pathways.

$$C_6H_5OH + 14CuO \rightarrow 6CO_2 + 3H_2O + 14Cu$$
$$C_6Cl_5OH + 2H_2O + 9CuO \rightarrow 6CO_2 + 5HCl + 9Cu$$

Copper peroxide, CuO_2, appears as a dark olive-green solid or similarly colored suspension. It is unstable, decomposing to oxygen and other copper oxides. Copper peroxide is formed by the reaction of cold solutions of hydrogen peroxide and Schweizer's reagent (the chemical complex tetraamminediaquacopper dihydroxide, $[Cu(NH_3)_4(H_2O)_2](OH)_2)$, the latter prepared from copper hydroxide and dilute ammonia solution. The Schweizer's reagent utilized must not have excess ammonia. Copper peroxide may also be prepared via the reaction of an ice-cold solution of hydrogen peroxide with a suspension of copper hydroxide. It may also result from the very slow reaction of finely divided cupric oxide with cold hydrogen peroxide. When wet, copper peroxide decomposes at temperatures above 6°C; it "is far more stable when dry."

Copper(I) hydroxide, CuOH, is a mild, highly unstable alkali. The color of pure CuOH is yellow or orange-yellow, although it usually appears rather dark red due to impurities. It is extremely easily oxidized even at room temperature. Copper(I) hydroxide can be produced according to the following reaction.

$$3S_2O_3^{2-} + 32Cu^+ + 16H_2O \xrightarrow{ethanol} 32CuOH + 3S_8 + 8S_4O_6^{2-} + 8SO_4^{2-}$$

In this reaction, ethanol acts as a catalyst. It can also be a solvent for the by-product sulfur and thus remove it. Another route is via the double displacement of CuCl and NaOH:

$$CuCl + NaOH \rightleftharpoons NaCl + CuOH\downarrow$$

Remarkably, this method is rarely used as the CuOH produced will gradually dehydrate and eventually turn into Cu_2O. Comparable to iron(II) hydroxide, copper(I) hydroxide can readily oxidize into copper(II) hydroxide.

$$4CuOH + 2H_2O + O_2 \rightleftharpoons 4Cu(OH)_2$$

Copper(II) hydroxide, $Cu(OH)_2$, is a pale greenish blue or bluish green solid. Some forms of copper(II) hydroxide are sold as "stabilized" copper hydroxide, although they probably comprise a mixture of copper(II) carbonate and hydroxide. $Cu(OH)_2$ is a weak base. The mineral of the formula $Cu(OH)_2$ is called spertiniite. $Cu(OH)_2$ is rarely found as an uncombined mineral since it slowly reacts with carbon dioxide from the atmosphere to form a basic copper(II) carbonate. Therefore copper slowly forms a dull green coating in moist air by the reaction

$$2Cu(OH)_2 + CO_2 \rightarrow Cu_2CO_3(OH)_2 + H_2O$$

The green material is in theory a 1:1 mole mixture of $Cu(OH)_2$ and $CuCO_3$. This patina forms on bronze and other copper alloy statues, for example, the Statue of Liberty. $Cu(OH)_2$ can be prepared by adding sodium hydroxide to a dilute solution of copper(II) sulfate ($CuSO_4{\cdot}5H_2O$). The precipitate formed, though, often contains water and an substantial amount of sodium hydroxide impurity. A purer product can be obtained if in advance ammonium chloride is added to the solution. Instead, $Cu(OH)_2$ can easily be formed by electrolysis of water (containing a little electrolyte such as sodium sulfate, or magnesium sulfate) with a copper anode. The structure of $Cu(OH)_2$ has been determined by X-ray crystallography The Cu center is square pyramidal with 4 Cu-O distances in the plane of 1.96 Å, while the axial Cu-O distance is 2.36 Å. The hydroxide ligands in the plane are either doubly bridging or triply bridging. It is stable to approximately 100°C. $Cu(OH)_2$ reacts with a solution of ammonia to form a deep blue solution of tetramminecopper $[Cu(NH_3)_4]^{2+}$ complex ion. It catalyzes the oxidation of ammonia solutions in presence of dioxygen, forming copper ammine nitrites, such as $Cu(NO_2)_2(NH_3)_n$. $Cu(OH)_2$ is mildly amphoteric. It dissolves slightly in concentrated alkali, giving $[Cu(OH)_4]^{2-}$. Several copper(II)-containing minerals contain hydroxide. Prominent examples include azurite, malachite, antlerite, and brochantite. Azurite ($Cu_3(CO_3)(OH)_2$) and malachite ($Cu_2(CO_3)(OH)_2$) are hydroxy-carbonates, whereas antlerite ($Cu_3(SO_4)(OH)_4$) and brochantite ($Cu_4(SO_4)(OH)_6$) are hydroxy-sulfates. Many synthetic copper(II) hydroxide derivatives have been studied.

5.11.3.2 Chalcogenides

Copper also tarnishes when exposed to sulfur and some sulfur compounds, with which it reacts to form various copper sulfides. It is found in nature as the mineral chalcocite. It has a narrow range of stoichiometry ranging from $Cu_{1.997}S$ to $Cu_{2.000}S$. Cu_2S is formed when the metal is reacted with sulfur vapor while CuS is formed when Cu_2S is reacted with excess sulfur. The reaction of copper powder in molten sulfur quickly forms Cu_2S, whereas pellets of copper need much higher temperature. Cu_2S reacts with oxygen to form SO_2.

$$2Cu_2S + 3O_2 \rightarrow 2Cu_2O + 2SO_2$$

In the production of copper two-thirds of the molten copper sulfide is oxidized as above, and the Cu_2O reacts with unoxidized Cu_2S forming Cu metal.

$$Cu_2S + 2Cu_2O \rightarrow 6Cu + SO_2$$

There are two polymorphic forms of Cu_2S: a low-temperature monoclinic form ("low-chalcocite") which has a complex structure with 96 Cu atoms in the unit cell and a hexagonal form stable above 104°C. In the latter structure there are 24 crystallographically distinct Cu atoms and the structure has been described as resembling a hexagonal close-packed array of sulfur atoms with Cu atoms in planar three coordination. This structure was at first assigned an orthorhombic cell as a result of twinning in the sample crystal. There is also a crystallographically distinct phase (the mineral djurleite) with stoichiometry $Cu_{1.96}S$ which is nonstoichiometric (range $Cu_{1.934}S$-$Cu_{1.965}S$) and has a monoclinic structure with 248 Cu and 128 S atoms in the unit cell. Cu_2S and $Cu_{1.96}S$ have a similar appearance and are hard to distinguish from each other visually.

CuS is not a simple ionic salt of copper because it contains an S_2 unit and is better formulated as $Cu_2Cu(S_2)S$. It occurs in nature as the mineral covellite. Copper monosulfide can be produced by passing hydrogen sulfide gas through a solution of copper(II) salt. Instead, it can also be formed by melting an excess of sulfur with copper(I) sulfide or by precipitation with hydrogen sulfide from a solution of anhydrous copper(II) chloride in anhydrous ethanol. The reaction of copper with molten sulfur followed by boiling sodium hydroxide and the reaction of sodium sulfide with aqueous copper sulfate will also result in the formation of copper sulfide. Copper sulfide crystallizes in the hexagonal crystal system. There is also an amorphous high-pressure phase, which based on the Raman spectrum has been described as consisting of a distorted covellite structure. An amorphous room temperature semiconducting form can be obtained through the reaction of a Cu(II) ethylenediamine complex with thiourea, which converts to the crystalline covellite form at 30°C. The crystal structure of covellite has been reported a number of times, and though these studies generally agree that the space group is *P63/mmc*, there are small differences in bond lengths and angles between them. It is rather different from copper(II) oxide, but similar to CuSe (klockmannite). The covellite unit cell contains 6 formula units (12 atoms) in which 4 Cu atoms have tetrahedral coordination, 2 Cu atoms have trigonal planar coordination. Two pairs of S atoms are only 207.1 pm apart pointing to the existence of a S-S bond (a disulfide unit), while the 2 remaining S atoms form trigonal planar triangles around the Cu atoms and are surrounded by 5 Cu atoms in a pentagonal bipyramid. The S atoms at each end of a disulfide unit are tetrahedrally coordinated to 3 tetrahedrally coordinated Cu atoms and the other S atom in the disulfide unit. The formulation of copper sulfide as Cu(II)S (i.e., containing no S-S bond) is obviously contradictory to the crystal structure, and also in disagreement with the observed diamagnetism as a Cu(II) compound would have a d^9 configuration and be expected to be paramagnetic. XPS studies suggest that all of the Cu atoms have an oxidation state of +1. This is in disagreement with a formulation based on the crystal structure and obeying the octet rule that is found in many textbooks defining CuS as having both Cu(I) and Cu(II) that is, $(Cu^+)_2Cu^{2+}(S_2)^{2-}S^{2-}$. Another formulation as $(Cu^+)_3(S^{2-})(S_2)^-$ was suggested and supported by calculations. This formulation should not be construed as containing a radical anion, but rather that there is a delocalized valence "hole." EPR research on the precipitation of Cu(II) salts suggests that the reduction of Cu(II) to Cu(I) takes place in solution.

The other chalcogenides of Cu, the nonstoichiometric copper selenides and copper tellurides are metallic and show superconductivity at low temperatures. Copper(II) selenite is an inorganic salt often found as its dihydrate, $CuSeO_3{\cdot}2H_2O$, in the form of a blue powder. It can be produced from copper(II) acetate and selenous acid (H_2SeO_3 or more accurately $(HO)_2SeO$). Copper(II) selenite can be employed a catalyst for Kjeldahl digestion (a method for the quantitative determination of nitrogen contained in organic substances plus the nitrogen contained in the inorganic compounds ammonia and ammonium (NH_3/NH_4^+). Without modification, other forms of inorganic nitrogen, for instance nitrate, are not included in this measurement. This method was developed by Danish chemist Johan Gustav Christoffer Thorsager Kjeldahl (August 16, 1849 to July 18, 1900) in 1883.

5.11.3.3 Hydrides

Copper metal is not reactive to hydrogen even when heated and under pressure of up to 50 GPa. The hydride is known but is prepared by the reaction of cupric ions with hypophosphorous acid (H_3PO_2) to precipitate out the red CuH.

$$4Cu^{2+} + 6H_3PO_2 + 6H_2O \rightarrow 4CuH + 6H_3PO_3 + 8H^+$$

The reactions results in a red-colored precipitate of CuH, which is usually impure and slowly decomposes while liberating hydrogen, even at 0°C.

$$2CuH \rightarrow 2Cu + H_2$$

This slow decomposition also occurs underwater, although there are reports of the compound becoming pyrophoric if dried. A new synthesis method has been developed in 2017 by Lousada et al. Here, high-purity CuH nanoparticles were formed from basic copper carbonate, $CuCO_3{\cdot}Cu(OH)_2$. This reaction is faster and has a higher chemical yield than the copper sulfate-based synthesis and results in nanoparticles of CuH with higher purity and a smaller size distribution. The formed CuH can readily be transformed to conducting thin films of Cu. These films are prepared by spraying the CuH nanoparticles in their synthesis medium onto some insulating support. After drying, a conducting Cu film protected by a layer of mixed copper oxides is spontaneously formed. Copper hydride can also be prepared by reductive sonication. In this method, hexaaquacopper(II) and hydrogen(*) react to form copper hydride and oxonium according to the equation

$$\left[Cu(H_2O)_6\right]^{2+} + 3H^* \rightarrow 1/n(CuH)_n + 2[H_3O]^+ + 4H_2O$$

Hydrogen(*) is produced in situ from the homolytic sonication of water. Reductive sonication produces molecular copper hydride as an intermediate. In copper hydride, elements have the wurtzite (ZnS) crystal structure (polymeric), being connected by covalent bonds. Under certain conditions, a metastable amorphous solid forms. This solid decomposes above $-60°C$.

Hydridocopper (also systematically named cuprane(1)) is a related inorganic compound with the chemical formula CuH (also written as [CuH]). It is a gas that cannot be concentrated in pure form. Molecular copper hydride can be prepared through the reduction of copper iodide with lithium aluminum hydride in ether and pyridine.

$$4CuI + LiAlH_4 \rightarrow CuH + LiI + AlI_3$$

This was discovered in 1952 by E Wiberg and W Henle. The solution of this CuH in pyridine is characteristically dark red to dark orange. A precipitate is formed when ether is added to this solution. This will redissolve in pyridine. Impurities of the reaction products remain in the product. In this research, it was observed that the solidified diatomic substance is different from the wurtzite structure. The wurtzite substance was insoluble and was decomposed by lithium iodide, but not the solidified diatomic species. Furthermore, the wurtzite compound's decomposition is strongly base catalyzed, while the solidified diatomic species is not strongly affected at all. Some distinguishes between the two copper hydrides as the "insoluble-" and "soluble copper hydrides." The soluble hydride is susceptible to pyrolysis under vacuum and proceeds to completion under 100°C. Amorphous copper hydride is also formed by anhydrous reduction. In this process copper(I) and tetrahydroaluminate react to form molecular copper hydride and triiodoaluminum adducts. The molecular copper hydride is precipitated as amorphous copper hydride with the addition of diethyl ether. Amorphous copper hydride is converted into the wurtzite phase by annealing, accompanied by some decomposition. Hydridocopper is a hydrophilic (polar) solute, and thus dissolves in polar compounds. As hydridocopper is an electron-deficient compound, its dominant behavior is to polymerize, first to oligomers, then to copper hydride. A well-known oligomer is octahedro-hexacuprane(6), occurring in Stryker's reagent. Hydridocopper has acidic behavior for the same reason as normal copper hydride. Nevertheless, it does not form stable aqueous solutions, due partly to its autopolymerization, and its tendency to be oxidized by water. Copper hydride reversibly precipitates from pyridine solution, as an amorphous solid. However, repeated dissolution gives the regular crystalline form, which is insoluble. Under standard conditions, molecular copper hydride autopolymerizes to form the crystalline form, including under aqueous conditions, hence the aqueous production method devised by Wurtz.

Binary compounds of copper and nitrogen exists but these are not prepared from the elements at ordinary conditions or simple heating. Copper nitride (Cu_3N), is formed via pulsed laser deposition, radio frequency magnetron sputtering, or chemical synthesis routes using copper salts and various nitrogen sources.

5.11.3.4 Halides

The reaction of halogens with copper produce the dihalides, CuX_2 (X = F, Cl, Br). The existence of copper(I) fluoride or cuprous fluoride, CuF, is uncertain. It was reported in 1933 to have a sphalerite-type crystal structure. Modern textbooks, however, state that CuF is not known, since fluorine is so electronegative that it will always oxidize copper to its +2 oxidation state. Complexes of CuF such as $[(Ph_3P)_3CuF]$ are, though, known and well characterized. In contrast to other copper(I) halides such as copper(I) chloride, copper(I) fluoride tends to disproportionate into copper(II) fluoride and copper in a one-to-one ratio at ambient conditions, unless it is stabilized through complexation as in the example of $[Cu(N_2)F]$.

$$2CuF \rightarrow Cu + CuF_2$$

Copper(II) fluoride, CuF_2, is a white crystalline, hygroscopic solid with a rutile-type (TiO_2) crystal structure, comparable to other fluorides MF_2 (where M is a metal). CuF_2 can be prepared from copper and fluorine at temperatures of about 400°C. It forms as a direct reaction.

$$Cu + F_2 \rightarrow CuF_2$$

It loses fluorine in the molten stage at temperatures above 950°C.

$$2CuF_2 \rightarrow 2CuF + F_2$$
$$2CuF \rightarrow CuF_2 + Cu$$

The complex anions of CuF_3^-, CuF_4^{2-} and CuF_6^{4-} are produced when CuF_2 is exposed to compounds with fluoride ions F^-. CuF_2 has a monoclinic crystal structure and cannot attain a higher-symmetry structure. It forms rectangular prisms with a parallelogram base. CuF_2 is slightly soluble in water, but starts to decompose in hot water, forming basic F^- and Cu(OH) ions. CuF_2 can be utilized to synthesize fluorinated aromatic hydrocarbons through the reaction with aromatic hydrocarbons in an oxygen-containing atmosphere at temperatures over 450°C. This reaction is simpler than the Sandmeyer reaction but is only effective in forming compounds that can survive at the temperature used. A coupled reaction using oxygen and 2 HF regenerates the CuF_2, forming water. This method has been suggested as a "greener" method of preparing fluoroaromatics because it prevents the formation of toxic waste products, for example, NaF and NH_4F salts.

$$C_6H_6 + CuF_2 \xrightarrow{500°C} C_6H_5F + HF + Cu$$

$$H_2O + CuF_2 \xleftarrow{400°C} 2\,HF + Cu$$

Copper(I) chloride, commonly called cuprous chloride, is the lower chloride of copper, with the formula CuCl. The substance is a white solid sparingly soluble in water, but very soluble in concentrated hydrochloric acid. Impure samples appear green due to the presence of copper(II) chloride ($CuCl_2$). Copper(I) chloride is produced industrially by the direct combination of copper metal and chlorine at 450°C–900°C:

$$2Cu + Cl_2 \rightarrow 2CuCl$$

Copper(I) chloride can also be produced by reducing copper(II) chloride with sulfur dioxide, or with ascorbic acid (vitamin C) that acts as a reducing sugar.

$$2CuCl_2 + SO_2 + 2H_2O \rightarrow 2CuCl + H_2SO_4 + 2HCl$$
$$2CuCl_2 + C_6H_8O_6 \rightarrow 2CuCl + 2HCl + C_6H_6O_6$$

Many other reducing agents can be utilized. CuCl has the cubic sphalerite (ZnS) crystal structure at ambient conditions. Upon heating to 408°C the structure changes to hexagonal. Several other crystalline forms of CuCl appear at high pressures (several GPa). CuCl is a Lewis acid, which is classified as soft according to the Hard-Soft Acid-Base concept. Therefore it tends to produce stable complexes with soft Lewis bases, for example, triphenylphosphine

$$CuCl + P(C_6H_5)_3 \rightarrow \left[CuCl\left(P(C_6H_5)_3\right)\right]_4$$

While CuCl is insoluble in water, it dissolves in aqueous solutions containing suitable donor molecules. It forms complexes with halide ions, for example, forming H_3O^+ $CuCl_2^-$ with concentrated hydrochloric acid. It is attacked by CN^-, $S_2O_3^{2-}$, and NH_3 to form the corresponding complexes. Solutions of CuCl in HCl or NH_3 absorb carbon monoxide to form colorless complexes such as the chloride-bridged dimer $[CuCl(CO)]_2$. The same hydrochloric acid solutions likewise react with acetylene gas to produce $[CuCl(C_2H_2)]$. Ammoniacal solutions of CuCl react with acetylenes to give the explosive copper(I) acetylide, Cu_2C_2. Complexes of CuCl with alkenes can be produced by reduction of $CuCl_2$ by sulfur dioxide in the presence of the alkene in alcohol solution. Complexes with dienes, for example, 1,5-cyclooctadiene are especially stable: In the absence of other ligands, its aqueous solutions are unstable with respect to disproportionation into Cu and $CuCl_2$. Partly that is why samples in air assume a green coloration. The main application of CuCl is as a precursor to the fungicide copper oxychloride. For this applicatoin aqueous copper(I) chloride is formed by comproportionation and then air-oxidized.

$$Cu + CuCl_2 \rightarrow 2CuCl$$
$$4CuCl + O_2 + 2H_2O \rightarrow Cu_3Cl_2(OH)_4 + CuCl_2$$

CuCl is a catalyst for numerous organic reactions.

Copper(II) chloride, $CuCl_2$, is a light brown solid, which slowly absorbs moisture to convert to a blue-green dihydrate. Both the anhydrous and the dihydrate forms occur naturally as the very rare minerals tolbachite and eriochalcite, respectively. $CuCl_2$ is produced industrially through the chlorination of copper. Copper at red heat (300°C–400°C) reacts directly with chlorine gas, giving (molten) copper (II) chloride. The reaction is very exothermic.

$$Cu(s) + Cl_2(g) \rightarrow CuCl_2(l)$$

It is also industrially practical to react copper(II) oxide with an excess of ammonium chloride at similar temperatures.

$$CuO + 2NH_4Cl \rightarrow CuCl_2 + 2NH_3 + H_2O$$

While copper metal itself cannot be oxidized by hydrochloric acid, copper-containing bases, for example, the hydroxide, oxide, or copper(II) carbonate, can be. Once formed, a solution of $CuCl_2$ can be purified through crystallization. A standard technique takes the solution mixed in hot dilute hydrochloric acid and causes the crystals to form by cooling in a $CaCl_2$-ice bath. Anhydrous $CuCl_2$ has a distorted cadmium iodide structure. In this structure, the Cu centers are octahedrally coordinated. Most copper(II) compounds show distortions from idealized octahedral geometry because of the Jahn-Teller effect, which here describes the localization of one d-electron into a molecular orbital that is strongly antibonding with respect to a pair of Cl ligands. In $CuCl_2{\cdot}2H_2O$, the Cu again has a highly distorted octahedral geometry, the Cu(II) centers being coordinated by 2 water ligands and 4 Cl ligands, which bridge asymmetrically to other Cu centers. $CuCl_2$ is paramagnetic. Aqueous solution prepared from $CuCl_2$ comprise a range of copper(II) complexes depending on concentration, temperature, and the presence of additional chloride ions. These species include the blue-colored $[Cu(H_2O)_6]^{2+}$ and yellow- or red-colored halide complexes of the formula $[CuCl_{2+x}]^{x-}$. Copper(II) hydroxide precipitates after reacting copper(II) chloride solutions with a base such as NaOH.

$$CuCl_2 + 2NaOH \rightarrow Cu(OH)_2 + 2NaCl$$

Partial hydrolysis gives copper oxychloride, $Cu_2Cl(OH)_3$, a popular fungicide. $CuCl_2$ is a mild oxidant. It decomposes to CuCl and Cl_2 at 1000°C.

$$2CuCl_2 \rightarrow 2CuCl + Cl_2$$

$CuCl_2$ reacts with a number of other metals to form copper metal or copper(I) chloride with oxidation of the other metal. To transform copper(II) chloride to copper(I) derivatives, it can be useful to reduce an aqueous solution with sulfur dioxide as the reductant.

$$2CuCl_2 + SO_2 + 2H_2O \rightarrow 2CuCl + 2HCl + H_2SO_4$$

$CuCl_2$ reacts with HCl or other chloride sources to give complex ions: the red colored $CuCl_3^-$ (actually, it is a dimer, $Cu_2Cl_6^{2-}$, a couple of tetrahedrons that share an edge), and the green- or yellow-colored $CuCl_4^{2-}$.

$$CuCl_2 + Cl^- \rightleftharpoons CuCl_3^-$$
$$CuCl_2 + 2Cl^- \rightleftharpoons CuCl_4^{2-}$$

Several of these complexes can be crystallized from aqueous solution, and they have a wide range of different structures. $CuCl_2$ also forms various coordination complexes with ligands, for example, pyridine and triphenylphosphine oxide.

$$CuCl_2 + 2C_5H_5N \rightarrow [CuCl_2(C_5H_5N)_2](\text{tetragonal})$$
$$CuCl_2 + 2(C_6H_5)_3PO \rightarrow [CuCl_2((C_6H_5)_3PO)_2](\text{tetrahedral})$$

Nevertheless, “soft” ligands such as phosphines (e.g., triphenylphosphine), iodide, and cyanide together with some tertiary amines cause reduction to give copper(I) complexes. A major industrial application for $CuCl_2$ is as a cocatalyst with palladium(II) chloride in the Wacker process. Here, ethene (ethylene) is converted to ethanal (acetaldehyde) using water and air. During the reaction, $PdCl_2$ is reduced to Pd, and the $CuCl_2$ function is to re-oxidize this back to $PdCl_2$. Air can then oxidize the resultant CuCl back to $CuCl_2$, completing the cycle.

$$C_2H_4 + PdCl_2 + H_2O \rightarrow CH_3CHO + Pd + 2HCl$$
$$Pd + 2CuCl_2 \rightarrow 2CuCl + PdCl_2$$
$$4CuCl + 4HCl + O_2 \rightarrow 4CuCl_2 + 2H_2O$$

The overall process reaction can be written as

$$2C_2H_4 + O_2 \rightarrow 2CH_3CHO$$

Copper(I) bromide, CuBr, is a diamagnetic solid that has a polymeric structure akin to that for ZnS. It is widely used in the synthesis of organic compounds and as a lasing medium in copper bromide lasers. CuBr is white, although samples are often colored because of the presence of copper(II) impurities. The copper(I) ion also oxidizes readily in air. It is usually formed via the reduction of cupric salts with sulfite in the presence of bromide, for example, the reduction of copper(II) bromide with sulfite gives copper(I) bromide and hydrogen bromide.

$$2CuBr_2 + H_2O + SO_3^{2-} \rightarrow 2CuBr + SO_4^{2-} + 2HBr$$

CuBr is insoluble in most solvents because of its polymeric structure, which involves 4-coordinated, tetrahedral Cu centers interconnected by bromide ligands. Upon treatment with Lewis bases, CuBr transforms to molecular adducts, for example, with dimethyl sulfide, the colorless complex is produced.

$$CuBr + S(CH_3)_2 \rightarrow CuBr(S(CH_3)_2)$$

In this coordination complex, the Cu is 2-coordinated, with a linear geometry. Other soft ligands form related complexes. For instance, triphenylphosphine gives $CuBr(P(C_6H_5)_3)$, though this species has a more complex structure. Thermal excitation of copper(I) bromide vapor produces a blue violet emission which is of greater saturation than known copper(I) chloride emission. Copper(I) bromide is thus an advantageous emitter in pyrotechnic flames. In the Sandmeyer reaction, CuBr is used to convert diazonium salts into the corresponding aryl bromides.

$$ArN_2^+ + CuBr \rightarrow ArBr + N_2 + Cu^+$$

The earlier mentioned complex $CuBr(S(CH_3)_2)$ is extensively utilized to produce organocopper reagents. Related CuBr complexes are catalysts for Atom Transfer Radical Polymerization and Cu-catalyzed cross-dehydrogenative couplings.

Copper(II) bromide ($CuBr_2$) is used in photographic processing as an intensifier and as a brominating agent in organic synthesis. In addition, it is utilized in the copper vapor laser, a class of laser where the medium is copper bromide vapor formed in situ from hydrogen bromide reacting with the copper discharge tube. $CuBr_2$ lasers produce pulsed yellow and green light and have been investigated as a possible treatment for cutaneous lesions in dermatological applications. $CuBr_2$ can be prepared by combining copper oxide and hydrobromic acid.

$$CuO + 2HBr \rightarrow CuBr_2 + H_2O$$

$CuBr_2$ can be purified by crystallization twice from water, filtration to remove any CuBr and concentration under vacuum. The final product is then dehydrated using phosphorus pentoxide. In the solid-state $CuBr_2$ forms a polymeric structure, with $CuBr_4$ planar units connected on opposite sides to form chains. The crystal structure is monoclinic, space group *C2/m*, with lattice constants a = 714 pm, b = 346 pm, c = 718 pm, e ß = 121° 15′. $CuBr_2$ monomeric units are found in the gas phase at high temperature. $CuBr_2$ in chloroform-ethyl acetate reacts with ketones forming alpha-bromo ketones. The formed product can be directly utilized for the synthesis of its derivatives. This heterogeneous method is described as the most selective and direct method to form α-bromo ketones. Dibromination of NPGs, n-pentenyl glycosides, employing the $CuBr_2$/LiBr reagent combination was performed in order for an NPG to act as a glycosyl acceptor during halonium-promoted couplings. This type of reaction results in a high yield of the dibromides from alkenyl sugars which are resistant to a direct reaction with molecular bromine.

Copper(I) iodide, CuI, is also known as cuprous iodide. It is useful in various applications ranging from organic synthesis to cloud seeding. Pure CuI is white, but samples are often tan or even, when found in nature as rare mineral marshite, reddish-brown, but such color is caused by impurities. It is common for samples of iodide-containing compounds to become discolored because of the facile aerobic oxidation of the iodide anion to molecular iodine. CuI can be produced by heating iodine and copper in concentrated hydriodic acid, HI. In the laboratory though, copper(I) iodide is formed by simply mixing an aqueous solution of sodium or potassium iodide and a soluble copper(II) salt such copper sulfate.

$$Cu^{2+} + 2I^- \rightarrow CuI_2$$

The CuI_2 instantly decomposes to iodine and insoluble copper(I) iodide, while releasing I_2.

$$2CuI_2 \rightarrow 2CuI + I_2$$

This reaction has been used as a method for assaying copper(II) samples, since the evolved I_2 can be analyzed by redox titration. The reaction in itself may seem strange, as based on the rule of thumb for a proceeding redox reaction, $E^0_{oxidator} - E^0_{reductor} > 0$, this reaction fails. The quantity is below zero, so the reaction should not proceed. However, the equilibrium constant for the reaction is 1.38×10^{-13}. By using relatively moderate concentrations of 0.1 mol/L for both iodide and Cu^{2+}, the concentration of Cu^+ is calculated as 3×10^{-7}. Therefore the product of the concentrations is far in excess of the solubility product, so CuI precipitates. The process of precipitation lowers the copper(I) concentration, providing an entropic driving force based on Le Chatelier's principle, and allowing the redox reaction to proceed. CuI, like most binary metal halides, is an inorganic polymer. It has a rich phase diagram, that is, it exists in several crystalline forms. It has a sphalerite (ZnS) structure below 390°C (γ-CuI), a wurtzite (ZnS) structure between 390°C and 440°C (β-CuI), and a rock salt (NaCl) structure above 440°C (α-CuI). The ions are tetrahedrally coordinated when in the sphalerite or the wurtzite structure, with a Cu-I distance of 2.338 Å. CuBr and CuCl also transform from the sphalerite structure to the wurtzite structure at 405°C and 435°C, respectively. Hence, the longer the copper–halide bond length, the lower the temperature needs to be to change the structure from the sphalerite structure to the wurtzite structure. In comparison, the interatomic distances in CuBr and CuCl are 2.173 and 2.051 Å, respectively. CuI is poorly soluble in water (0.00042 g/L at 25°C), but it dissolves in the presence of NaI or KI to form the linear anion $[CuI_2]^-$. Dilution of such solutions with water reprecipitates CuI. This dissolution–precipitation process is used to purify CuI, forming colorless samples. CuI can be dissolved in acetonitrile, forming a solution of different complex compounds. Upon crystallization, molecular or polymeric compounds can be isolated. Dissolution also takes place when a solution of the appropriate complexing agent in acetone or chloroform is utilized, for example, thiourea and its derivatives. Solids that crystallize out of these solutions consist of hybrid inorganic chains. The diiodide of copper is unstable and forms the monoiodide easily.

5.11.3.5 Carbonate, sulfate, nitrate, and phosphate

Copper(II) carbonate or cupric carbonate, $CuCO_3$, is at ambient temperatures an ionic solid (a salt) consisting of copper (II) cations Cu^{2+} and carbonate anions CO_3^{2-}. It is rarely encountered as it is difficult to prepare and easily reacts with moisture in the air. The terms "copper carbonate," "copper(II) carbonate," and "cupric carbonate" nearly always refer (even in chemistry texts) to a basic copper carbonate (or copper(II) carbonate hydroxide), for example, $Cu_2(OH)_2CO_3$ (which occurs naturally as the mineral malachite) or $Cu_3(OH)_2(CO_3)_2$ (azurite). Thus the qualifier neutral may be used instead of "basic" to refer in particular to $CuCO_3$. Reactions that may be expected to form $CuCO_3$, for example, mixing solutions of copper(II) sulfate $CuSO_4$ and sodium carbonate Na_2CO_3 in ambient conditions, form instead a basic carbonate and CO_2, because of the great affinity of the Cu^{2+} ion for the hydroxide anion HO^-. Thermal decomposition of the basic carbonate at atmospheric pressure results in the formation of copper(II) oxide CuO rather than the carbonate. Back in 1960, Pistorius claimed successful formation of $CuCO_3$ by heating basic copper carbonate at 180°C in an atmosphere of carbon dioxide CO_2 (450 atm) and water (50 atm) for 36 hours. The bulk of the products was well-crystallized malachite $Cu_2CO_3(OH)_2$, but a small amount of a rhombohedral substance was also formed, thought to be $CuCO_3$. Yet, this synthesis was apparently not reproduced. Reliable formation of true copper(II) carbonate was published for the first time in 1973 by Hartmut Ehrhardt and coworkers. It was formed as a gray powder, by heating basic copper carbonate in an atmosphere of carbon dioxide (produced by the decomposition of silver oxalate $Ag_2C_2O_4$) at 500°C and 2 GPa (20,000 atm). This compound has a monoclinic structure. The stability of dry $CuCO_3$ depends critically on the partial carbon dioxide pressure (p_{co_2}). It is stable for months in dry air, but it decomposes slowly into CuO and CO_2 if p_{co_2} is less than 0.11 atm. In the presence of water or moist air at 25°C, $CuCO_3$ is stable only for p_{co_2} above 4.57 atmospheres and pH between approximately 4 and 8. Below that partial pressure, it reacts with water to form a basic carbonate (azurite, $Cu_3(CO_3)_2(OH)_2$).

$$3CuCO_3 + H_2O \rightarrow Cu_3(CO_3)_2(OH)_2 + CO_2$$

In highly basic solutions, the complex anion $Cu(CO_3)_2^{2-}$ is formed in its place.

Copper(II) sulfate, also known as copper sulfate, is the inorganic compound $CuSO_4{\cdot}xH_2O$, where x can range from 0 to 5. The pentahydrate (x = 5) is the most common form. Older names for this compound include blue vitriol, bluestone, vitriol of copper, and Roman vitriol. The pentahydrate ($CuSO_45H_2O$) is bright blue. It exothermically dissolves in water to form the aquo complex $[Cu(H_2O)_6]^{2+}$, which adopts an octahedral molecular geometry. The structure of the solid pentahydrate has a polymeric structure in which Cu is again octahedral but bonded to 4 water ligands. The Cu(II) $(H_2O)_4$ centers are interconnected by sulfate anions to form chains. Anhydrous $CuSO_4$ is a gray powder. Copper sulfate is manufactured commercially by treating copper metal with hot concentrated sulfuric acid or its oxides with dilute sulfuric acid. For laboratory usage, copper sulfate is commonly purchased. Commercial copper sulfate is typically about 98% pure copper sulfate and may have traces of water. Anhydrous $CuSO_4$ is 39.81 percent copper and 60.19 percent sulfate by mass, while in its blue, hydrous form, it is 25.47% copper, 38.47% sulfate (12.82% sulfur) and 36.06% water by mass. $CuSO_4{\cdot}5H_2O$ decomposes before melting. It loses two water molecules upon heating at 63°C, followed by two more at 109°C and the final water molecule at 200°C. Dehydration progresses by decomposition of the tetraaquacopper (2 +) moiety, two opposing aqua groups are lost to give a diaquacopper(2 +) moiety. The second dehydration step happens when the final two aqua groups are lost. Complete dehydration takes place when the final unbound water molecule is lost. At 650°C, $CuSO_4$ decomposes into copper(II) oxide (CuO) and sulfur trioxide (SO_3). Copper sulfate reacts with concentrated hydrochloric acid to give tetrachlorocuprate(II).

$$Cu^{2+} + 4Cl^- \rightarrow CuCl_4^{2-}$$

$CuSO_4$ is often included in children's chemistry sets. It is frequently utilized to grow crystals in schools and in copper plating experiments, despite its toxicity. $CuSO_4$ is regularly employed to demonstrate an exothermic reaction, in which steel wool or magnesium ribbon is placed in an aqueous solution of $CuSO_4$. It is utilized to show the principle of mineral hydration. The pentahydrate, which is blue, is heated, converting it into the anhydrous form, which is white, while the water that was present in the pentahydrate form evaporates. When water is subsequently added to the anhydrous compound, it converts back into the pentahydrate form, regaining its blue color. $CuSO_4{\cdot}5H_2O$, which can readily be formed by crystallization from solution as copper(II) sulfate, is quite hygroscopic. In an example of a "single metal replacement reaction," iron is submerged in a solution of $CuSO_4$. Iron reacts forming iron(II) sulfate and copper precipitates.

$$Fe + CuSO_4 \rightarrow FeSO_4 + Cu$$

In high school and general chemistry education, $CuSO_4$ is utilized as an electrolyte for galvanic cells, generally as a cathode solution. For instance, in a zinc/copper cell, copper ion in copper sulfate solution absorbs electrons from zinc forming metallic copper.

$$Cu^{2+} + 2e^- \rightarrow Cu\ (\text{cathode})\ E^0_{\text{cell}} = 0.34\text{V}$$

Copper(II) nitrate, $Cu(NO_3)_2$, forms a blue crystalline solid. Anhydrous $Cu(NO_3)_2$ forms deep blue-green crystals and sublimes in a vacuum at 150°C–200°C. $Cu(NO_3)_2$ also occurs as 5 different hydrates, the most common ones being the trihydrate and hexahydrate. These compounds are more generally encountered in industry than in the laboratory. Hydrated $Cu(NO_3)_2$ can be produced by hydration of the anhydrous compound or by treating copper metal with an aqueous solution of silver nitrate or concentrated nitric acid.

$$Cu + 4HNO_3 \rightarrow Cu(NO_3)_2 + 2H_2O + 2NO_2$$

Anhydrous $Cu(NO_3)_2$ is produced when copper metal is reacted with N_2O_4.

$$Cu + 2N_2O_4 \rightarrow Cu(NO_3)_2 + 2NO$$

Attempted dehydration of any of the hydrated copper(II) nitrates by heating instead forms the oxides and not $Cu(NO_3)_2$. At 80°C, the hydrates changes to "basic copper nitrate" ($Cu_2(NO_3)(OH)_3$), which converts to CuO at 180°C. Using this reactivity, copper nitrate can be employed to form nitric acid by heating it until decomposition and passing the fumes directly into water. This method is comparable to the last step in the Ostwald process.

$$2Cu(NO_3)_2 \rightarrow 2CuO + 4NO_2 + O_2$$
$$3NO_2 + H_2O \rightarrow 2HNO_3 + NO$$

Natural basic copper nitrates include the rare minerals gerhardtite and rouaite, both polymorphs of $Cu_2(NO_3)(OH)_3$. Anhydrous $Cu(NO_3)_2$ has been crystallized in two solvate-free polymorphs. α- and β-$Cu(NO_3)_2$ are fully 3-dimensional coordination polymer networks. The α form has only 1 Cu environment, with [4 + 1] coordination, but the β form has 2 different Cu centers, 1 with [4 + 1] and 1 that is square planar. The nitromethane solvate also features "[4 + 1] coordination," with 4 short Cu-O bonds of approximately 200 pm and 1 longer bond at 240 pm. They are coordination polymers, with infinite chains of copper(II) centers and nitrate groups. In the gas phase, copper(II) nitrate has 2 bidentate nitrate ligands. Therefore evaporation of the solid involves "cracking" to form the copper(II) nitrate molecule. Five different hydrates have been observed: the monohydrate ($Cu(NO_3)_2{\cdot}H_2O$), sesquihydrate ($Cu(NO_3)_2{\cdot}1.5H_2O$), hemipentahydrate ($Cu(NO_3)_2{\cdot}2.5H_2O$), trihydrate ($Cu(NO_3)_2{\cdot}3H_2O$), and hexahydrate ($[Cu(H_2O)_6](NO_3)_2$). The hexahydrate is noteworthy since the Cu-O distances are all equal, not exhibiting the usual effect of Jahn-Teller distortion that is otherwise typical of octahedral Cu(II) complexes. This noneffect is ascribed to the strong hydrogen bonding that limits the elasticity of the Cu-O bonds.

Copper(II) phosphate, $Cu_3(PO_4)_2$, may also be considered as the cupric salt of phosphoric acid. It is frequently encountered as the hydrated species $Cu_2(PO_4)OH$, which is green and occurs naturally as the mineral libethenite. Anhydrous copper(II) phosphate is blue with triclinic crystals and can be prepared by a high-temperature reaction between diammonium phosphate and copper(II) oxide.

$$2(NH_4)_2HPO_4 + 3CuO \rightarrow Cu_3(PO_4)_2 + 3H_2O + 4NH_3$$

5.11.3.6 Organocopper compounds

Copper is one of the oldest transition metals to be used in organic chemistry. The first organocopper compound was synthesized as early as 1859 by German chemist Rudolf Christian Böttger (April 28, 1806 to April 29, 1881). By passing acetylene gas through copper(I) chloride solution, he was able to form copper(I) acetylide Cu_2C_2 (Cu-C≡C-Cu), which is a very explosive product. Compounds of the form $LiCuR_2$ (R = alkyl and aryl) were prepared around 1936. These are now called Gilman reagents, which are important as they react with organic halides and replace the halide group with an R group. This allows for synthesis of complex products from simple building blocks. The generalized scheme for such reactions is shown below.

$$[R\text{-}Cu\text{-}R]^-Li^+ \xrightarrow{R'\text{-}X} \left[R\text{—}\overset{R'}{\underset{X}{Cu}}\text{—}R \right]^- Li^+ \longrightarrow R\text{-}Cu + R\text{-}R' + LiX$$

After Ukranian-American organic chemist Morris Selig Kharash (August 24, 1895 to October 9, 1957) and P.O. Tawney discovered in 1941 that the reaction of a Grignard reagent with cyclohexenone in the presence of Cu(I) resulted in 1,4-addition instead of 1,2-addition, the growth of catalytic asymmetric reactions involving organocopper has increased tremendously. Presently, the type of organocopper(I) reagents commonly described in literature are simple organocopper species, RCu and R_2Cu^-, or metal organocuprates such as R_2CuM (homocuprate) and RCu(X)M (heterocuprate), where R stands for a carbanion (e.g., alkyl, alkenyl, aryl); X is a nontransferable anion (e.g., halide, heteroatom anion, cyanide), and M is a main-group metal cation (e.g., Li^+, Mg^{2+}, Zn^{2+}). The importance of organocopper is seen in three reactions where they are more efficient than the precursor organolithium or organomagnesium reagents. These are conjugate addition to α,β-unsaturated carbonyl compounds, coupling with alkyl halides, epoxides or tosylates, and carbometalation of acetylenes.

5.11.4 Major uses

Historically, copper was the first metal to be worked by people. The discovery that it could be hardened with a little tin to form the alloy bronze gave name to the Bronze Age. Traditionally it has been one of the metals used to make coins, along with silver and gold. However, it is the most common of the three and therefore the least valued. All United States coins are now Cu alloys. The major uses of Cu are electrical wire (~60%), roofing and plumbing (~20%), and industrial machinery (~15%). Cu is used typically as the pure metal, but when greater hardness is necessary, it is combined with other metals in alloys such as brass and bronze (5% of total use). For more than 200 years, Cu paint has been used on boat hulls to curb the growth of plants and shellfish. A small portion of the Cu supply is used for nutritional supplements and fungicides in agriculture. Machining of Cu is possible, though alloys are preferred for good machinability in producing intricate parts.

Notwithstanding competition from other materials, Cu is still the preferred electrical conductor in almost all types of electrical wiring with the exception of overhead electric power transmission where Al is frequently preferred. Cu wire is used in a variety of applications, for example, in power generation, power transmission, power distribution, telecommunications, electronics circuitry, and countless types of electrical equipment. This comprises structural power wiring, power distribution cable, appliance wire, communications cable, automotive wire and cable, and magnet wire. About 50% of all Cu mined is used for electrical wire and cable conductors. Numerous electrical devices depend on Cu wiring due to its characteristic beneficial properties, for example, its high electrical conductivity, tensile strength, ductility, creep (deformation) resistance, corrosion resistance, low thermal expansion, high thermal conductivity, ease of soldering, malleability, and ease of installation.

Integrated circuits and printed circuit boards increasingly use Cu instead of Al due to its superior electrical conductivity. Heat sinks and heat exchangers use Cu due to its far better heat dissipation properties. Electromagnets, vacuum tubes, cathode ray tubes, and magnetrons in microwave ovens use Cu, as do waveguides for microwave radiation. Its superior conductivity improves the efficiency of electrical motors. This is crucial as motors and motor-driven systems make up 43%–46% of all global electricity consumption and 69% of all electricity used by industry. Increasing the mass and cross-section of Cu in a coil improves the efficiency of the motor.

Cu has been used since antiquity as a robust, corrosion resistant, and weatherproof architectural material. Roofs, flashings, rain gutters, downspouts, domes, spires, vaults, and doors have been made from Cu for centuries of not millennia. Recently, its architectural use has been expanded to comprise interior and exterior wall cladding, building expansion joints, radio frequency shielding, and antimicrobial and decorative indoor products, for example, handrails, bathroom fixtures, and counter tops. Some of its other key benefits as an architectural material consist of low thermal movement, light weight, lightning protection, and recyclability. Its unique natural green patina has long been popular with architects and designers. The final patina is a very durable layer that is extremely resistant to atmospheric corrosion, thus protecting the underlying metal against further weathering. It can consist of a mixture of carbonate and sulfate compounds in various amounts, dependent on the environmental conditions such as sulfur-containing acid rain. Architectural Cu and its alloys can also be "finished" to take on a specific look, feel, or color. Finishes can consist of, for example, mechanical surface treatments, chemical coloring, or coatings. Cu has very good brazing and soldering properties and can be welded; the best results are obtained with gas metal arc welding.

Cu is biostatic, meaning bacteria and many other forms of life will not grow on it. Consequently, it has been used extensively to line parts of ships to protect against barnacles and mussels. It was initially used pure but has since been replaced by Muntz metal (a form of alpha-beta brass with about 60% Cu, 40% Zn and a trace of Fe. It is named after George Fredrick Muntz (November 26, 1794 to July 30, 1857), an industrialist and Member of Parliament from Birmingham, England, who commercialized the alloy following his patent of 1832) and Cu-based paint. Cu alloys have become important netting materials in the aquaculture industry as they are antimicrobial and stop biofouling, even under extreme conditions and exhibit strong structural and corrosion-resistant properties in marine environments. Cu alloy touch surfaces have natural properties that kill a variety of microorganisms (e.g., *E. coli*, methicillin-resistant *Staphylococcus aureus*, *Staphylococcus, Clostridium difficile*, influenza A virus, adenovirus, and fungi). Hundreds of Cu alloys have been shown to kill more than 99.9% of disease-causing bacteria in less than 2 hours when cleaned frequently. The United States Environmental Protection Agency (EPA) has approved the registrations of these Cu alloys as "antimicrobial materials with public health benefits"; which permits manufacturers to legally claim the public health benefits of products made of these registered alloys. The EPA has also approved a long list of antimicrobial Cu products made from these alloys, such as bedrails, handrails, overbed tables, sinks, faucets, doorknobs, toilet hardware, computer keyboards, health club equipment, and shopping cart handles. For example, Cu doorknobs are used by some hospitals to reduce the transfer of disease, while Legionnaires' disease (also known as legionellosis, is a form of atypical pneumonia caused by any type of *Legionella* bacteria. Signs and symptoms include cough, shortness of breath, high fever, muscle pains, and headaches. Nausea, vomiting, and diarrhea may also occur. This often begins 2–10 days after exposure. The bacterium is found naturally in fresh water. It can contaminate hot water tanks, hot tubs, and cooling towers of large air conditioners. It is usually spread by breathing in mist that contains the bacteria. It can also occur when contaminated water is aspirated.) is suppressed by Cu tubing in plumbing systems. Antimicrobial Cu alloy products are now being installed in healthcare facilities in several countries around the world. Textile fibers can be blended with Cu to produce antimicrobial protective fabrics. Cu in the form of Bordeaux mixture (a blue gelatinous suspension of Cu sulfate and lime in water) was one of the first agrochemical pesticides, developed to control downy mildew on vines. This, and other variants using sodium carbonate or ammonium carbonate, continue to be used as fungicides.

Cu compounds, such as Fehling's solution, are used to differentiate between water-soluble carbohydrate and ketone functional groups, and as a test for reducing sugars and nonreducing sugars, supplementary to the Tollens' reagent test (a chemical reagent used to determine the presence of aldehyde, aromatic aldehyde, and alpha-hydroxy ketone functional groups. The reagent consists of a solution of silver nitrate, ammonia and some sodium hydroxide). The test was developed by German chemist Hermann von Fehling (June 9, 1812 to July 1, 1885) in 1849. But the solution has a drawback as it cannot differentiate between benzaldehyde and acetone. Fehling's solution is prepared by combining two separate solutions, known as Fehling's A and Fehling's B. Fehling's A is aqueous solution of Cu(II) sulfate, which is deep blue. Fehling's B is a colorless solution of aqueous potassium sodium tartrate (also known as Rochelle salt) made in a strong alkali, commonly with sodium hydroxide. Typically, the L-tartrate salt is used. The Cu(II) complex in Fehling's solution is an oxidizing agent and the active reagent in the test. Fehling's solution can be used to distinguish aldehyde vs ketone functional groups. The compound to be tested is added to the Fehling's solution and the mixture is heated. Aldehydes are oxidized, giving a positive result, but ketones do not react, unless they are α-hydroxy ketones. The bistartratocuprate(II) complex oxidizes the aldehyde to a carboxylate anion, and in the process the copper(II) ions of the complex are reduced to copper(I) ions. Red copper(I) oxide then precipitates out of the reaction mixture, which indicates a positive result that is, that redox has taken place The net reaction between an aldehyde and the copper(II) ions in Fehling's solution may be written as

$$RCHO + 2Cu^{2+} + 5OH^- \rightarrow RCOO^- + Cu_2O + 3H_2O$$

or with the tartrate included

$$RCHO + 2Cu(C_4H_4O_6)_2^{2-} + 5OH^- \rightarrow RCOO^- + Cu_2O + 4C_4H_4O_6^{2-} + 3H_2O$$

Fehling's test can be used as a generic test for monosaccharides and other reducing sugars (e.g., maltose). It will give a positive result for aldose monosaccharides (due to the oxidizable aldehyde group) but also for ketose monosaccharides, as they are converted to aldoses by the base in the reagent, and then give a positive result. Fehling's can be used to screen for glucose in urine, thus detecting diabetes. The green colored secondary copper mineral malachite, $Cu_2(CO_3)(OH)_2$, often found in the oxidized zone of Cu ore deposits, is used for polished slabs, tables, columns, etc.

5.12 30 Zn — Zinc

5.12.1 Discovery

Numerous isolated examples of the use of impure zinc in historical times have been found. Zinc ores were utilized to produce the zinc–copper alloy brass millennia before the discovery of zinc as a separate element. Judean brass from the 14th to 10th centuries BCE contains 23% zinc. Familiarity with the methods on how to make brass reached Ancient Greece by the 7th century BCE, but few varieties were made. Ornaments made of alloys containing 80%–90% zinc, with lead, iron, antimony, and other metals constituting the rest, have been discovered that are 2500 years old. A possibly prehistoric figurine containing 87.5% zinc was discovered in a Dacian archeological site. The oldest known pills were made of the zinc carbonates hydrozincite, $Zn_5(CO_3)_2(OH)_6$, and smithsonite, $ZnCO_3$. The pills were used for sore eyes and were discovered aboard the Roman ship Relitto del Pozzino, which sunk in 140 BCE. The production of brass was known to the Romans around 30 BCE. They manufactured brass by heating a mixture of powdered calamine (zinc silicate or carbonate), charcoal and copper in a crucible. The subsequent calamine brass was at that point either cast or hammered into shape to produce weapons. Some coins struck by Romans in the Christian era are manufactured of what is possibly calamine brass. The Greek geographer, philosopher, and historian Strabo (64 or 63 BCE to CE c.24) writing in the 1st century BCE (but quoting a now lost work of Greek historian and rhetorician Theopompus, c.380–315 BCE) discusses "drops of false silver" which when mixed with copper produce brass. This might refer to small amounts of zinc that were a by-product of smelting sulfide ores. Zinc in such remnants in smelting ovens was generally thrown away as it was regarded as useless. The Berne zinc tablet is a votive plaque dating to Roman Gaul (Gaul was a historical region of Western Europe during the Iron Age that was inhabited by Celtic tribes, encompassing present-day France, Luxembourg, Belgium, most of Switzerland, parts of Northern Italy, as well as the parts of the Netherlands and Germany on the west bank of the Rhine) consisting of an alloy that contains primarily zinc. The Charaka Samhita, supposed to have been written between CE 300 and 500, discusses a metal which, when oxidized, results in pushpanjan, supposed to be zinc oxide. Zinc mines at Zawar, near Udaipur in India, have been operating since the Mauryan period (c. 322 and 187 BCE). The smelting of metallic zinc here, though, seems to have started around the 12th century CE. It has been estimated that these mines produced around a million tons of metallic zinc and zinc oxide between the 12th and 16th centuries. Alternative estimates give an overall production of 60,000 tons of metallic zinc over this period. The Rasaratna Samuccaya, written around the 13th century CE, discusses two types of zinc-containing ores: one used for metal production and another used for medicinal applications. Zinc was clearly documented as a metal under the description of Yasada or Jasada in the medical Lexicon attributed to the Hindu king Madanapala (of Taka dynasty) and written around 1374. Smelting and production of impure zinc by reducing calamine with wool and other organic compounds was achieved in India in the 13th century. The Chinese did not manage this until the 17th century. Alchemists burned zinc metal in air and collected the resulting zinc oxide on a condenser. Some alchemists named this zinc oxide lana philosophica, Latin for "philosopher's wool," as it collected in wooly clusters, while others believed it resembled white snow and called it nix album. The name of the metal was perhaps first documented by Paracelsus (born Theophrastus von Hohenheim (full name Philippus Aureolus Theophrastus Bombastus von Hohenheim, 1493 or 1494 to September 24, 1541), a Swiss-born German alchemist, who mentioned the metal as "zincum" or "zinken" in his book Liber Mineralium II, in the 16th century. The word is likely a derivative from the German word zinke, which apparently meant "tooth-like, pointed or jagged" (metallic zinc crystals exhibit a needle-like habit). Zink could also suggest "tin-like" for its relation to the German word zinn denoting tin. An alternative possibility is that the word came from the Persian word سنگ seng meaning stone. The metal was also known as Indian tin, tutanego, calamine, and spinter. German physicist and chemist Andreas Libavius (c.1555 to July 25, 1616) got an amount of what he called "calay" of Malabar from a cargo ship seized from the Portuguese in 1596. Libavius described the properties of the sample, which may have been zinc. Zinc was frequently imported from the Orient to Europe in the 17th and early 18th centuries but was sometimes extremely expensive. Metallic zinc was produced in India by CE 1300, much earlier than in the West. Before the metal was isolated in Europe, it was shipped from India in about CE 1600. Malachy Postlewayt's Universal Dictionary of Trade and Commerce published in 1757, a contemporary source providing technical data in Europe, did not refer to zinc before 1751 but the element was researched before then. Flemish metallurgist and alchemist P. M. de Respour published that he had produced metallic zinc from zinc oxide in 1668. At the beginning of the 18th century, French physician and chemist Étienne François Geoffroy (February 13, 1672 to January 6, 1731) found that zinc oxide formed yellow crystals on iron bars positioned above the zinc ore that is being smelted. In Great Britain, doctor and metallurgist John Lane (c. October 1678 to 1741) is supposed to have carried out experiments to smelt zinc, probably at Landore (near Swansea), before his bankruptcy in 1726. In 1738 in Great Britain, metallurgist William Champion (1709–1789) patented a process to produce zinc from calamine in a vertical retort style smelter. His process was like that used at Zawar zinc mines in Rajasthan, but no evidence indicates he traveled to the Orient. Champion's method was applied through 1851. Generally, German chemist Andreas Marggraf (March 3, 1709 to August 7, 1782) is named as the discoverer of pure metallic zinc (Fig. 5.108), although Swedish chemist and mineralogist Anton von Swab (July 29, 1702 to January 28, 1768) had distilled zinc from calamine four years earlier (Fig. 5.109). In his 1746 experiment, Marggraf heated a combination of calamine and charcoal in a closed container without copper to produce a metal (Marggraf, 1746). This method became economically practical by 1752. William Champion's brother, John, patented a method in 1758 for calcining zinc sulfide (sphalerite, ZnS) into an oxide for use in the retort process. Earlier, only calamine could be used to isolate zinc. In 1798 German inventor and a pioneer in metallurgy Johann Christian Ruberg (date of birth unknown, baptized September 4, 1746 to September 5, 1807) enhanced the smelting technique by building the first horizontal retort smelter. Belgian inventor and industrialist Jean-Jacques Daniel Dony (February 24, 1759 to 1819) built a different type of horizontal zinc smelter that processed even more zinc. Italian physician, physicist, biologist and philosopher Luigi Galvani (September 9, 1737 to December 4, 1798) observed in 1780 that connecting the spinal cord of a freshly dissected frog to an iron rail attached by a brass hook made the frog's leg spasm. He erroneously believed he had found a capacity of nerves and muscles to produce electricity and named the effect "animal electricity" (Galvani, 1791). The galvanic cell and the process of galvanization were both named for him, and his findings paved the way for electrical batteries, galvanization, and CP. Galvani's friend, Italian physicist, chemist, and a pioneer of electricity and power Alessandro Volta (February 18, 1745 to March 5, 1827), continued studying the effect and created the Voltaic pile in 1800. The basic unit of Volta's pile was a simplified galvanic cell, made of plates of copper and zinc separated by an electrolyte and connected by a conductor externally. The units were stacked in series to produce the Voltaic cell, which produced electricity by movement of electrons from the zinc to the copper and allowing the zinc to corrode.

FIGURE 5.108 Andreas Marggraf, engraving c.1770.

FIGURE 5.109 Anton von Swab, 1849.

5.12.2 Mining, production, and major minerals

Zinc is the 4th most common metal in use, trailing only Fe, Al, and Cu. About 70% of the world's zinc originates from mining, while the remaining 30% comes from recycling secondary zinc. Worldwide, 95% of new zinc is mined from sulfide ore deposits, in which sphalerite (ZnS) is nearly always mixed with the sulfides of copper, lead and iron. Zinc mines are scattered throughout the world, with the main areas being China, Australia, and Peru. Zinc metal is produced through extractive metallurgy. The ore is finely ground, followed by froth flotation to separate ore minerals from gangue minerals (on the property of hydrophobicity), to get a zinc sulfide ore concentrate comprising around 50% zinc, 32% sulfur, 13% iron, and 5% SiO_2. Subsequent roasting converts the zinc sulfide concentrate to zinc oxide.

$$2ZnS + 3O_2 \rightarrow 2ZnO + 2SO_2$$

The sulfur dioxide is used to produce sulfuric acid, which is essential for the leaching process. If deposits of zinc carbonate (smithsonite, $ZnCO_3$), zinc silicate (willemite, Zn_2SiO_4, hemimorphite, $Zn_4Si_2O_7(OH)_2{\cdot}H_2O$), or zinc spinel (gahnite, $ZnAl_2O_4$) are used for zinc production, the roasting can be omitted. For further processing two basic methods can be employed: pyrometallurgy or electrowinning. Pyrometallurgy reduces the zinc oxide with carbon or carbon monoxide at 950°C into the metal, which is then distilled as zinc vapor to separate it from other metals, which are not volatile at those temperatures. The zinc vapor is collected in a condenser. The following equations describe the general reactions in this process.

$$2ZnO + C \rightarrow 2Zn + CO_2$$
$$ZnO + CO \rightarrow Zn + CO_2$$

In electrowinning, on the other hand, zinc is leached from the ore concentrate by sulfuric acid.

$$ZnO + H_2SO_4 \rightarrow ZnSO_4 + H_2O$$

Finally, the zinc is reduced by electrolysis.

$$2ZnSO_4 + 2H_2O \rightarrow 2Zn + 2H_2SO_4 + O_2$$

The sulfuric acid is regenerated and recycled into the leaching stage.

Zinc can be found in about 300 different minerals. Besides the native element (Zn), 5 alloys are known, such as bortnikovite (Pd_4Cu_3Zn) and danbaite ($CuZn_2$). The sulfide class contains 23 minerals with Zn in the crystal structure,

BOX 5.15 Sphalerite, ZnS

Sphalerite is the main ore of zinc and consists largely of ZnS but almost always contains some Fe (Fig. 5.110). When the Fe content is high it forms an opaque black variety known as marmatite. It is usually associated with galena (PbS), pyrite (FeS_2), and other sulfides along with calcite ($CaCO_3$), dolomite ($CaMg(CO_3)_2$), and fluorite (CaF_2). Around 95% of all primary zinc is extracted from sphaleritic ores. However, because of its variable trace element content, it is also a significant source of several other elements, for example, cadmium, gallium germanium, and indium. Crystals of suitable size and transparency have been cut into gemstones, usually featuring the brilliant cut to best display sphalerite's high dispersion which is over three times that of diamond. Due to their softness and fragility the gems are often left unset as collector's or museum pieces (although some have been set into pendants). Gem-quality material is usually a yellowish to honey brown, red to orange, or green. Sphalerite was named in 1847 by Ernst Friedrich Glocker from the Greek σφαλεροζ *"sphaleros"* = treacherous, in allusion to the ease with which dark varieties were mistaken for galena, but yielded no lead. Originally called blende in 1546 by Georgius Agricola (Georg Bauer). Sphalerite is isometric *3/m* (space group *F3m*). It has a hardness of 3½–4. Cleavage is perfect on {011}. The luster is nonmetallic and resinous to submetallic, also adamantine. The transparency varies from transparent to opaque. The color is colorless when pure, and green when nearly so, commonly yellow, brown to black, darkening with increasing iron content (called black jack). Also red (ruby zinc). The streak is white to yellow and brown. Optically it is isotropic. It has extremely high relief. Crystals may be distorted or rounded. Crystals show combinations of tetrahedra, dodecahedra, and cubes. Polysynthetic twinning is common. It commonly forms cleavable masses, coarse to fine granular, compact, botryoidal or cryptocrystalline.

FIGURE 5.110 Sphalerite, ZnS, dark green to yellow crystals to 1 cm with minor galena, PbS. Bachelor mine, Creede dist., Mineral Co., Colorado, United States.

for example, sphalerite (ZnS) (Fig. 5.110) (Box 5.15), tetrahedrite ($Cu_6[Cu_4(Fe,Zn)_2]Sb_4S_{13}$), and wurtzite ((Zn,Fe)S) (Fig. 5.111). Four halides contain Zn, for example, herbertsmithite ($Cu_3Zn(OH)_6Cl_2$) and simonkolleite ($Zn_5Cl_2(OH)_8 \cdot H_2O$). The oxide class consists of 42 minerals with Zn, such as franklinite ($Zn^{2+}Fe_2^{3+}O_4$), gahnite ($ZnAl_2O_4$), and zincite (ZnO) (Fig. 5.112). Twelve different carbonates are known to contain Zn in their structure, for example, aurichalcite ($(Zn,Cu)_5(CO_3)_2(OH)_6$) (Fig. 5.113), hydrozincite ($Zn_5(CO_3)_2(OH)_6$) and smithsonite ($ZnCO_3$). The sulfate class is represented by 40 different minerals with zinc in their crystal structure, including minerals such as goslarite ($ZnSO_4 \cdot 7H_2O$), hemihedrite ($Pb_{10}Zn(CrO_4)_6(SiO_4)_2(OH)_2$), and zinkosite ($ZnSO_4$). Seventy-eight minerals can be found in the phosphate class, for example, adamite ($Zn_2(AsO_4)(OH)$) (Fig. 5.114), austinite ($CaZn(AsO_4)(OH)$) (Fig. 5.115), descloizite ($PbZn(VO_4)(OH)$), köttigite ($Zn_3(AsO_4)_2 \cdot 8H_2O$), legrandite ($Zn_2(AsO_4)(OH) \cdot H_2O$) (Fig. 5.116), scholzite ($CaZn_2(PO_4)_2 \cdot 2H_2O$) (Fig. 5.117), and tarbuttite ($Zn_2(PO_4)(OH)$). The silicate class consists of 38 minerals with Zn in the crystal structure, for example, genthelvite ($Be_3Zn_4(SiO_4)_3S$), hemimorphite ($Zn_4Si_2O_7(OH)_2 \cdot H_2O$) (Fig. 5.118), and willemite (Zn_2SiO_4) (Fig. 5.119).

FIGURE 5.111 Wurtzite, (Zn,Fe)S, orange cone-shaped crystals to 1 mm on shale. Near Donohoe, Westmoreland Co., Pennsylvannia, United States.

FIGURE 5.112 Bright orange-red crystals of zincite, ZnO, to 1 mm scattered on fracture with creamy tan pyroaurite, $Mg_6Fe_2^{3+}(OH)_{16}[CO_3]{\cdot}4H_2O$. Sterling Hill mine, Ogdensburg, Sussex Co., New Jersey, United States.

5.12.3 Chemistry

Five stable isotopes of zinc are found in nature, with ^{64}Zn as the most common isotope (49.17% natural abundance). The others comprise ^{66}Zn (27.73%), ^{67}Zn (4.04%), ^{68}Zn (18.45%), and ^{70}Zn (0.61%). ^{64}Zn and the rare ^{70}Zn are theoretically unstable on energetic grounds, but their predicted half-lives are greater than 4.3×10^{18} years and 1.3×10^{16} years, hence their radioactivity could be disregarded for practical purposes. Several dozen radioisotopes have been observed. ^{65}Zn with a half-life of 243.66 days is the least active radioisotope, followed by ^{72}Zn with a half-life of 46.5 hours. Zn has 10 nuclear isomers. The metastable ^{69m}Zn has the longest half-life, 13.76 hours. The nucleus of a metastable isotope is in an excited state and will return to the ground state by emitting a photon in the form of a γ ray.

FIGURE 5.113 Bluish green sprays of aurichalcite, $(Zn,Cu)_5(CO_3)_2(OH)_6$, to 8 mm. Mina Ojuela, Mapimi, Durango, Mexico.

FIGURE 5.114 A yellow adamite, $Zn_2(AsO_4)(OH)$, group to 6 mm with manganoan terminations. Mina Ojuela, Mapimi, Durango, Mexico.

^{61}Zn has three excited metastable states, while ^{73}Zn has two and the isotopes ^{65}Zn, ^{71}Zn, ^{77}Zn and ^{78}Zn each have only one. The most common decay mode of a radioisotope of Zn lighter than ^{66}Zn is by electron capture. The decay product resulting from electron capture is an isotope of Cu.

$${}^{n}_{30}\mathrm{Zn} + \mathrm{e}^- \rightarrow {}^{n}_{29}\mathrm{Cu}$$

The most common decay mode of a radioisotope of zinc heavier than ^{66}Zn is β^- decay, which forms an isotope of Ga.

$${}^{n}_{30}\mathrm{Zn} \rightarrow {}^{n}_{31}\mathrm{Ga} + \mathrm{e}^- + \overline{\nu}_e$$

Zinc is a slightly brittle metal at room temperature and has a blue-silvery appearance when freshly cut. It is a member of group 12 of the periodic table and has 30 electrons arranged in an electronic configuration of $[Ar]3d^{10}4s^2$. As a result of the stability of filled d shell, compounds of zinc are characterized by the d^{10} configuration and show few characteristic properties of transition metals. With the exception of the derivatives of the Zn_2^{2+} ion which formally involve Zn(I), zinc chemistry almost exclusively involves Zn(II) (Table 5.14). In some respects, zinc is chemically similar to

FIGURE 5.115 Subadamantine, light green, glassy and elongated austinite, $CaZn(AsO_4)(OH)$, crystals in flat radial groups to 9 mm across. Gold Hill mine, Tooele Co., Utah, United States.

FIGURE 5.116 Legrandite, $Zn_2(AsO_4)(OH) \cdot H_2O$, lustrous golden yellow long prismatic crystals to over 1 cm. Mina Ojuela, Mapimi, Durango, Mexico.

magnesium as both elements exhibit the normal oxidation state (+2) and the Zn^{2+} and Mg^{2+} ions are of similar size. Many of their compounds are isomorphous and they both readily complex with O-donor ligands. However, zinc has a much greater tendency than magnesium to form covalent compounds. Zinc also resembles the transition elements in forming stable complexes not only with O-donor ligands but with N- and S-donor ligands, as well as with halides and CN^-.

Nonoxidizing acids dissolve Zn with the evolution of hydrogen. Its reaction with oxidizing acids such as nitric acid is more complex producing a variety of oxides of nitrogen depending on the concentration and temperature. Zinc is the only element in group 12 which dissolves in aqueous alkali to form ions such as aquated $[Zn(OH)_4]^{2-}$. Zinc is a strong reducing agent with a standard redox potential of −0.76 V. Pure zinc tarnishes quickly in air, forming a passive layer. The composition of this layer can be complex, but one constituent is probably basic zinc carbonate, $Zn_5(OH)_6CO_3$. The reaction of zinc with water is slowed by this passive layer. When this layer is corroded by acids such as hydrochloric acid and sulfuric acid, the reaction evolves forming hydrogen gas.

FIGURE 5.117 Scholzite, $CaZn_2(PO_4)_2{\cdot}2H_2O$, colorless prismatic crystals to 1 cm. Reaphook Hill, Flinders Ranges, South Australia, Australia.

FIGURE 5.118 Long water-clear blades of hemimorphite, $Zn_4Si_2O_7(OH)_2{\cdot}H_2O$, over 1 cm with yellow mimetite, $Pb_5(AsO_4)_3Cl$. Mina Ojuela, Mapimi, Durango, Mexico.

$$Zn + 2H^+ \rightarrow Zn^{2+} + H_2$$

Zinc reacts with alkalis as with acids. With oxidants such as chalcogens and halogens, Zn forms binary compounds, for example, ZnS and $ZnCl_2$.

5.12.3.1 *Oxide and hydroxide*

Zinc does combine easily with oxygen, tarnishing quickly in moist air and ignition in oxygen forms zinc oxide (ZnO). ZnO has been known longer than the metal itself as it is a by-product of the primitive production of brass. It is produced industrially by melting metallic zinc in a graphite crucible and vaporizing at temperatures typically around

FIGURE 5.119 Willemite, Zn_2SiO_4, colorless long, bladed radiating crystals to 1.5 mm. Berg Aukus, Grootfontein dist., Namibia.

TABLE 5.14 Zinc properties.

Appearance	Silver-gray
Standard atomic weight $A_{r,std}$	65.38
Block	d-Block
Element category	Posttransition metal, alternatively considered a transition metal
Electron configuration	[Ar] $3d^{10}$ $4s^2$
Phase at STP	Solid
Melting point	419.53°C
Boiling point	907°C
Density (near r.t.)	7.14 g/cm^3
When liquid (at m.p.)	6.57 g/cm^3
Heat of fusion	7.32 kJ/mol
Heat of vaporization	115 kJ/mol
Molar heat capacity	25.470 J/(mol·K)
Oxidation states	−2, 0, +1, **+2**
Ionization energies	1st: 906.4 kJ/mol 2nd: 1733.3 kJ/mol 3rd: 3833 kJ/mol
Atomic radius	Empirical: 134 pm
Covalent radius	122 ± 4 pm
Van der Waals radius	139 pm

STP, Standard temperature and pressure.
Bold font indicates main oxidation state.

1000°C. This is known as the indirect or French process. The direct or American process begins with diverse contaminated zinc composites, such as zinc ores or smelter by-products. The zinc precursors are reduced (carbothermal reduction) by heating with a source of carbon such as anthracite to form zinc vapor, which is subsequently oxidized as in the indirect process. Due to the lower purity of the source material, the final product is also of lower quality in the direct process as compared to the indirect one. A small amount of industrial production involves wet chemical processes, which begins with aqueous solutions of zinc salts, from which zinc carbonate or zinc hydroxide is precipitated. The solid precipitate is subsequently calcined at temperatures around 800°C. Numerous specialized methods exist for preparing ZnO for scientific research and niche applications. These methods can be categorized by the resulting ZnO form (bulk, thin film, nanowire), temperature ("low," that is close to room temperature or "high," that is T ~1000°C), process type (vapor deposition or growth from solution) and other parameters. Large single crystals (many cubic centimeters) can be grown by the gas transport (vapor-phase deposition), hydrothermal synthesis, or melt growth.

Nevertheless, due to the high vapor pressure of ZnO, growth from the melt is challenging. Growth by gas transport is hard to control, leaving the hydrothermal method as a preference. Thin films can be formed by, for example, CVD, metalorganic vapor-phase epitaxy, electrodeposition, pulsed laser deposition, sputtering, sol-gel synthesis, atomic layer deposition, spray pyrolysis, etc. White powdered zinc oxide can be prepared in the laboratory by electrolyzing a solution of sodium bicarbonate with a zinc anode. Zinc hydroxide and hydrogen gas are formed. The zinc hydroxide upon calcination decomposes to zinc oxide.

$$Zn + 2H_2O \rightarrow Zn(OH)_2 + H_2$$
$$Zn(OH)_2 \rightarrow ZnO + H_2O$$

Zinc vapor reacts with oxygen in the air to produce a normally white oxide. Zinc oxide crystallizes in two main forms, hexagonal wurtzite ([Zn,Fe]S) and cubic sphalerite (zincblende, ZnS). The wurtzite structure is most stable at ambient conditions and hence most common. The zincblende form can be stabilized by growing ZnO on substrates with a cubic lattice structure. In both instances, the zinc and oxide centers are tetrahedral, the most typical geometry for Zn(II). ZnO transforms to the rock salt (NaCl) structure at relatively high pressures about 10 GPa. Hexagonal and cubic polymorphs have no inversion symmetry (reflection of a crystal relative to any given point does not transform it into itself). This and other lattice symmetry properties result in piezoelectricity of the hexagonal and cubic ZnO, and pyroelectricity of hexagonal ZnO. The hexagonal structure has a point group *6mm* (Hermann-Mauguin notation) or C_{6v} (Schoenflies notation), and the space group is $P6_3mc$ or C_6v^4. The lattice constants are a = 3.25 Å and c = 5.2 Å; their ratio c/a ~1.60 is close to the ideal value for hexagonal cell c/a = 1.633. As in most group II–VI materials, the bonding in ZnO is largely ionic (Zn^{2+}–O^{2-}) with the corresponding radii of 0.074 nm for Zn^{2+} and 0.140 nm for O^{2-}. This property explains the preferential formation of wurtzite rather than the sphalerite structure, as well as the strong piezoelectricity of ZnO. Due to the polar Zn-O bonds, Zn and O planes are electrically charged. To preserve electrical neutrality, those planes reconstruct at the atomic level in most relative materials, but not in ZnO—its surfaces are atomically flat, stable and exhibit no reconstruction. This anomaly of ZnO is not yet fully understood. It is amphoteric, dissolving in acids to give the aqueous Zn^{2+} ion and in alkali to give the zincate (a.k.a. tetrahydroxozincate) ion, $[Zn(OH)_4]^{2-}$. Zinc hydroxide, Zn $(OH)_2$ is also amphoteric. Heating this oxide changes the color of the compound to yellow due to the evaporation of oxygen from the lattice to give a nonstoichiometric phase $Zn_{1+x}O$ ($x < 70$ ppm). The supernumerary Zn atoms in this nonstoichiometric phase produce lattice defects which trap electrons. The trapped electrons can then be excited by absorption of visible light. An excess of 0.02%–0.03% Zn metal can impart a whole range of colors—yellow, green, brown, red—to the oxide.

Zinc hydroxide, $Zn(OH)_2$, occurs naturally as 3 rare minerals: wülfingite (orthorhombic), ashoverite and sweetite (both tetragonal). It can be produced by adding sodium hydroxide solution, but not in excess, to a solution of any zinc salt. A white precipitate will be formed.

$$Zn^{2+} + 2OH^- \rightarrow Zn(OH)_2$$

Zn^{2+} is known to form hexa-aqua ions at high water concentrations and tetra-aqua ions at low concentrations of water and therefore the above reaction may be better described as the reaction of the aquated ion with hydroxide through donation of a proton.

$$Zn^{2+}(H_2O)_4(aq) + OH^-(aq) \rightarrow Zn^{2+}(H_2O)_3OH^-(aq) + H_2O(l)$$

Ensuing reactions discussed below can also, thus, be thought of as reactions with the aquated zinc ion and one can change them accordingly. But for simplicity, the water molecules are ignored from here-on. If excess sodium hydroxide is added, the precipitate of zinc hydroxide will dissolve, forming a colorless solution of zincate ion:

$$Zn(OH)_2 + 2OH^- \rightarrow Zn(OH)_4^{2-}$$

This property can be employed as a test for zinc ions in solution, but it is not exclusive, as aluminum and lead compounds react in a very similar way. In contrast to aluminum and lead hydroxides, zinc hydroxide also dissolves in excess aqueous ammonia to give a colorless, water-soluble ammine complex. Zinc hydroxide will dissolve since the ion is typically surrounded by water ligands; when excess sodium hydroxide is added to the solution the hydroxide ions will reduce the complex to a −2 charge and make it soluble. When excess ammonia is added, it forms an equilibrium providing hydroxide ions; the formation of hydroxide ions results in a comparable reaction as sodium hydroxide and forms a +2 charged complex with a coordination number of 4 with the ammonia ligands—this makes the complex soluble so that it dissolves. One principal application of zinc hydroxide is as an absorbent in surgical dressings. It is also employed to find zinc salts by mixing sodium hydroxide with the suspect salt.

5.12.3.2 Chalcogenides

Zinc is easily ignited with sulfur to produce a white luminescent pigment of zinc sulfide, ZnS. The natural form of the sulfide is called zinc blende or sphalerite. This is the most widespread form of ZnS, but it is also known in a second naturally occurring though much rarer form, wurtzite. The latter is the more stable at high temperatures. In both forms, Zn is tetrahedrally coordinated by 4 S and each S is tetrahedrally coordinated by 4 Zn; the structures differ significantly only in the type of close packing involved, being cubic in zincblende and hexagonal in wurtzite. Freshly precipitated ZnS dissolves readily in mineral acids with evolution of H_2S, but roasting renders it far less reactive. Zinc sulfide is typically prepared from waste materials from other applications. Typical sources are, for example, smelter, slag, and pickle liquors. It is also a by-product of the synthesis of ammonia from methane where zinc oxide is employed to scavenge hydrogen sulfide impurities in the natural gas.

$$ZnO + H_2S \rightarrow ZnS + H_2O$$

In the laboratory it is easily prepared by igniting a mixture of zinc and sulfur. Since zinc sulfide is insoluble in water, it can also be formed in a precipitation reaction. Solutions containing Zn^{2+} salts easily produce a precipitate of ZnS in the presence of sulfide ions (e.g., from H_2S).

$$Zn^{2+} + S^{2-} \rightarrow ZnS$$

This reaction is the basis of a gravimetric analysis for zinc. The conversion from the sphalerite form to the wurtzite form takes place at approximately 1020°C. A tetragonal form is also known as the very rare mineral polhemusite, with the formula (Zn,Hg)S. Zinc sulfide is used as an infrared optical material, transmitting from visible wavelengths to just over 12 μm. It can be employed planar as an optical window or shaped into a lens. Zinc sulfide is a common pigment, sometimes called sachtolith. When mixed with barium sulfate, zinc sulfide forms lithopone (C.I. Pigment White 5). Fine ZnS powder is an efficient photocatalyst, which generates H_2 gas from water upon irradiation. Sulfur vacancies can be introduced in the ZnS structure during its synthesis; this progressively turns the white-yellowish ZnS into a brown powder and enhances the photocatalytic activity through increased light absorption. Both sphalerite and wurtzite are intrinsic, wide-bandgap semiconductors. These are prototypical II–VI semiconductors, and they have structures similar to many of the other semiconductors, such as gallium arsenide. The cubic form of ZnS has a band gap of about 3.54 electron volts at 27°C (300K), while the hexagonal form has a band gap of about 3.91 electron volts. ZnS can be doped as either an n-type semiconductor or a p-type semiconductor

The other chalcogenides of zinc, yellow ZnSe and brown ZnTe are structurally akin to the sulfide and the former especially is used mainly in conjunction with ZnS as a phosphor. Zinc selenide (ZnSe) is a light-yellow solid. It is an intrinsic semiconductor with a band gap of about 2.70 eV at 25°C. ZnSe rarely occurs in nature and is found in the mineral that was named after German geologist Hans Stille (October 8, 1876 to December 26, 1966) called stilleite. ZnSe can be synthesized in both the hexagonal (wurtzite) and cubic (zincblende) crystal structure. ZnSe is insoluble in water but reacts with acids to form toxic hydrogen selenide gas. It can be deposited as a thin film via CVD techniques such as MOVPE and vacuum evaporation. It is a wide-bandgap semiconductor of the II–VI semiconductor group (since zinc and selenium belong to the 12th and 16th groups of the periodic table, respectively). The material can be doped n-type doping with, for example, halogen elements. P-type doping is more difficult although can be achieved by introducing gallium. ZnSe is utilized to produce II–VI LEDs and diode lasers. It emits blue light. ZnSe doped with chromium (ZnSe:Cr) has been employed as an infrared laser gain medium emitting at about 2.4 μm. It is applied as an infrared optical material with a unusually wide transmission wavelength range (0.45–21.5 μm). ZnSe activated with tellurium (ZnSe(Te)) is a scintillator with an emission peak at 640 nm, appropriate for matching with photodiodes. It is used in X-ray and γ ray detectors. ZnSe scintillators are significantly different from the ZnS ones.

Zinc telluride, ZnTe, is a semiconductor material with a direct band gap of 2.26 eV. It is generally a p-type semiconductor. Its crystal structure is cubic, like that for sphalerite (ZnS) and diamond (C). ZnTe forms a gray or brownish red powder, or ruby-red crystals when refined by sublimation. Zinc telluride usually had a cubic (sphalerite) crystal structure, though it can be also produced with the rock salt (NaCl) crystal structure or in hexagonal crystals (wurtzite structure). Its lattice constant is 0.6101 nm, allowing it to crystallize with or on aluminum antimonide, gallium antimonide, indium arsenide, and lead selenide. With some lattice mismatch, it can also be grown on other substrates, for example, GaAs, and it can be grown in thin-film polycrystalline (or nanocrystalline) form on substrates such as glass, for example, in the manufacture of thin-film solar cells. In the wurtzite (hexagonal) crystal structure, it has lattice parameters a = 0.427 and c = 0.699 nm. Zinc telluride can be easily doped and therefore it is one of the more common semiconducting materials used in optoelectronics. ZnTe is valuable for the development of numerous semiconductor devices, such as blue LEDs, laser diodes, solar cells, and components of microwave generators. It can be utilized for solar cells, for example, as a back-surface field layer and p-type semiconductor material for a CdTe/ZnTe structure or in PIN diode structures (a diode with a wide, undoped intrinsic semiconductor region between a p-type semiconductor and an n-type semiconductor region. The p-type and n-type regions are typically heavily doped because they are used for ohmic contacts.). The compound can also be applied as a component of ternary semiconductor compounds, such as $Cd_xZn_{(1-x)}Te$ (theoretically a mixture composed from the end-members ZnTe and CdTe), which can be produced with a varying composition x to allow the optical bandgap to be tuned as required. Zinc telluride together with lithium niobate is frequently employed for the generation of pulsed terahertz radiation in time-domain terahertz spectroscopy and terahertz imaging. When a crystal of such material is subjected to a high-intensity light pulse of subpicosecond duration, it emits a pulse of terahertz frequency through a nonlinear optical process known as optical rectification. On the other hand, subjecting a zinc telluride crystal to terahertz radiation causes it to exhibit optical birefringence and change the polarization of transmitting light, making it an electro-optic detector. Vanadium-doped zinc telluride, "ZnTe:V," is a nonlinear optical photorefractive material of potential use in the protection of sensors at visible wavelengths. ZnTe:V optical limiters are light and compact, without complex optics of conventional limiters. ZnTe:V can block a high-intensity jamming beam from a laser dazzler, while still passing the lower-intensity image of the observed scene. It can also be employed in holographic interferometry, in reconfigurable optical interconnections, and in laser optical phase conjugation devices. It provides a superior photorefractive performance at wavelengths between 600 and 1300 nm, in comparison with other III–V and II–VI compound semiconductors. By adding manganese as an additional dopant (ZnTe:V:Mn), its photorefractive yield can be substantially improved.

5.12.3.3 Nitride and hydride

Zinc does not combine with N_2 and H_2 in air but the hydride and nitride of zinc exist. Zinc nitride (Zn_3N_2) is typically obtained as (blue)gray crystals. It is a semiconductor. In pure form, it has the antibixbyite (bixbyite, $[Mn,Fe]_2O_3$) structure. Zinc nitride can be produced by thermally decomposing zincamide (zinc diamine) in an anaerobic environment, at temperatures over 200°C. The by-product of this reaction is ammonia.

$$3Zn(NH_2)_2 \rightarrow Zn_3N_2 + 4NH_3$$

It can also be prepared by heating zinc to 600°C in a current of ammonia; the by-product in this case is hydrogen gas.

$$3Zn + 2NH_3 \rightarrow Zn_3N_2 + 3H_2$$

The decomposition of zinc nitride into the elements at the same temperature is a competing reaction. At 700°C it decomposes. It has also been synthesized by producing an electric discharge between zinc electrodes in a nitrogen atmosphere. Thin films have been prepared via CVD of bis(bis(trimethylsilyl)amido]zinc with ammonia gas onto silica or ZnO coated alumina at temperatures from 275°C to 410°C. It is a semiconductor with a reported bandgap of c. 3.2 eV, though a thin zinc nitride film formed by electrolysis of molten salt mixture containing Li_3N with a zinc electrode exhibited a bandgap of 1.01 eV. Zinc nitride reacts violently with water to form ammonia and zinc oxide.

$$Zn_3N_2 + 3H_2O \rightarrow 3ZnO + 2NH_3$$

Zinc nitride reacts with lithium (formed in an electrochemical cell) by insertion. The initial reaction is the irreversible conversion into LiZn in a matrix of β-Li_3N. These products can subsequently be changed reversibly and electrochemically into LiZnN and metallic Zn.

Zinc hydride, ZnH_2, is a white, odorless solid which slowly decomposes into its elements at room temperature. A variety of coordination compounds containing Zn-H bonds are used as reducing agents, but ZnH_2 itself has no common applications. Zinc(II) hydride was first synthesized in 1947 by Hermann Schlesinger, through a reaction between dimethylzinc and lithium aluminum hydride; a reaction which was to some extent hazardous because of the pyrophoric nature of $Zn(CH_3)_2$.

$$Zn(CH_3)_2 + 2LiAlH_4 \rightarrow ZnH_2 + 2LiAlH_3CH_3$$

A significantly safer method is the precipitation of zinc hydride from the reaction of LiH with $ZnBr_2$ or NaH with ZnI_2.

$$2MH + ZnX_2 \xrightarrow{THF} ZnH_2 + 2MX$$

Recent research suggests that in zinc(II) hydride, Zn and H form a one-dimensional network (polymer) connected by covalent bonds. Solid zinc(II) hydride is the irreversible autopolymerization product of the molecular form, and the molecular form cannot be isolated in concentration. Solubilizing zinc(II) hydride in nonaqueous solvents, comprise adducts with molecular zinc(II) hydride, such as $ZnH_2(H_2)$ in liquid hydrogen. ZnH_2 slowly decomposes to metallic Zn and H_2 at 25°C, with decomposition becoming fast when it is heated above 90°C.

$$ZnH_2 \rightarrow H_2 + Zn^0$$

It is easily oxidized and is sensitive to both air and moisture; being hydrolyzed slowly by water but violently by aqueous acids, which points to possible passivation via the formation of a surface layer of ZnO. Notwithstanding this, older samples may be pyrophoric. Zinc hydride can thus be thought of as metastable at best, though it is still the most stable of all the binary first-row transition metal hydrides (c.f. titanium(IV) hydride). Molecular zinc(II) hydride, ZnH_2, was recently recognized as a volatile product of the acidified reduction of zinc ions with sodium borohydride. This reaction is comparable to the acidified reduction with lithium aluminum hydride, but a larger portion of the generated zinc(II) hydride is in the molecular form. This can be assigned to a slower reaction rate preventing a polymerizing concentration of building up over the progression of the reaction. This follows previous experiments in direct synthesis from the elements. Due to its relative thermal stability, molecular ZnH_2 is included in the short list of molecular metal hydrides, which have been successfully identified in the gas phase (i.e., not limited to matrix isolation). Molecular ZnH_2 in the gas phase is linear with a Zn-H bond length of 153.5 pm. The molecule can be found a singlet ground state of $^1\Sigma^{+}{}_{g}^{+}$. Quantum chemical calculations predict the molecular form to exist in a doubly hydrogen-bridged, dimeric ground state, with little or no formational energy barrier. The dimer can be depicted as di-μ-hydrido-bis(hydridozinc), according to IUPAC additive nomenclature.

5.12.3.4 Halides

Zinc react with halogens to form zinc halides, ZnX_2. Zinc difluoride can be prepared by the action of HF or F_2 on the metal. The other anhydrous halides are also best prepared by the dry methods of treating the heated metals with HCl or Cl_2, Br_2 or I_2 as appropriate. Aqueous methods of preparation produce the hydrated halides. Zinc fluoride (ZnF_2) exists as the anhydrous form and also as the tetrahydrate, $ZnF_2.4H_2O$ (rhombohedral crystal structure). It has a high melting point and has the rutile (TiO_2) structure containing 6 coordinated Zn, which indicates a substantial ionic nature in its chemical bonding. In contrast to the other zinc halides, $ZnCl_2$, $ZnBr_2$ and ZnI_2, it is not very soluble in water. ZnF_2 can be hydrolyzed by hot water to form zinc hydroxyfluoride, Zn(OH)F.

Zinc chlorides, $ZnCl_2$ and its hydrates of which nine crystalline forms are known, are colorless or white, and are highly soluble in water. $ZnCl_2$ itself is hygroscopic and even deliquescent. Samples should thus be protected from sources of moisture, including the water vapor present in ambient air. $ZnCl_2$ is extensively used in textile processing, metallurgical fluxes, and chemical synthesis. No mineral with this chemical composition is known aside from the very rare mineral simonkolleite, $Zn_5(OH)_8Cl_2{\cdot}H_2O$. Anhydrous $ZnCl_2$ can be produced from zinc and hydrogen chloride.

$$Zn(s) + 2HCl \rightarrow ZnCl_2 + H_2(g)$$

Hydrated forms and aqueous solutions can easily be produced in a comparable way by treating Zn metal with hydrochloric acid. Zinc oxide and zinc sulfide react with HCl.

$$ZnS(s) + 2HCl(aq) \rightarrow ZnCl_2(aq) + H_2S(g)$$

In contrast to many other elements, zinc basically exists in only one oxidation state, 2+, which simplifies purification of the chloride. Commercial zinc chloride usually contains water and products from hydrolysis as impurities. Such material can be purified by recrystallization from hot dioxane. Anhydrous samples can be purified by sublimation in a stream of hydrogen chloride gas, and subsequent heating the sublimate to 400°C in a stream of dry nitrogen gas. Lastly, the simplest method relies on treating the zinc chloride with thionyl chloride. Four crystalline forms (polymorphs) of $ZnCl_2$ are known: α, β, γ, and δ, and in each structure the Zn^{2+} ions are tetrahedrally coordinated to 4 chloride ions (Table 5.15). The pure anhydrous orthorhombic form (δ) quickly converts to one of the other forms on exposure to the atmosphere, and a possible reason is that the OH^- ions originating from absorbed water enable the rearrangement. Rapid cooling of molten $ZnCl_2$ gives a glass. The covalent nature of the anhydrous $ZnCl_2$ is reflected in its relatively low melting point of 275°C. Further proof for covalency is provided by the high solubility of the dichloride in ethereal solvents, where it forms adducts with the formula $ZnCl_2L_2$, where L = ligand, for example, $O(C_2H_5)_2$. In the gas phase, $ZnCl_2$ molecules are linear with a bond length of 205 pm. Molten $ZnCl_2$ has a high viscosity at its melting point and a

TABLE 5.15 Crystallographic properties of $ZnCl_2$ polymorphic forms.

Form	Symmetry	Group	*a* (nm)	*b* (nm)	c (nm)	ρ (g/cm³)
α	Tetragonal	*I42d*	0.5398	0.5398	0.64223	3.00
β	Tetragonal	*P4₂/nmc*	0.3696	0.3696	1.071	3.09
γ	Monoclinic	*P2₁/c*	0.654	1.131	1.23328	2.98
δ	Orthorhombic	*Pna2₁*	0.6125	0.6443	0.7693	2.98

With a, b, and c the lattice parameters and ρ the density calculated from the structure parameters.

comparatively low electrical conductivity, which increases significantly with temperature. A Raman scattering investigation of the melt pointed to the presence of polymeric structures, while a neutron scattering study revealed the existence of tetrahedral ($ZnCl_4$) complexes. Five hydrates of zinc chloride are known: $ZnCl_2(H_2O)_n$ with $n = 1$, 1.5, 2.5, 3 and 4. The tetrahydrate $ZnCl_2(H_2O)_4$ crystallizes from aqueous solutions of zinc chloride. Molten anhydrous $ZnCl_2$ at temperatures between 500°C and 700°C dissolves Zn metal, and, on rapid cooling of the melt, a yellow diamagnetic glass is formed, for which Raman spectroscopy shows it to contain the Zn_2^{2+} ion. A number of salts containing the tetrachlorozincate anion, $ZnCl_4^{2-}$, exist. "Caulton's reagent," $V_2Cl_3(thf)_6Zn_2Cl_6$ is an illustration of a salt with $Zn_2Cl_6^{2-}$. Cs_3ZnCl_5 contains tetrahedral $ZnCl_4^{2-}$ and Cl^- anions. No compounds containing the $ZnCl_6^{4-}$ ion have been characterized. Although zinc chloride is very soluble in water, solutions cannot be thought to contain simply solvated Zn^{2+} ions and Cl^- ions, $ZnCl_x(H_2O)_{(4-x)}$ species are also present. Aqueous solutions of $ZnCl_2$ are acidic: a 6 M aqueous solution has a pH of 1. The acidity of aqueous $ZnCl_2$ solutions relative to solutions of other Zn^{2+} salts is because of the formation of the tetrahedral chloro aqua complexes where the reduction in coordination number from 6 to 4 further decreases the strength of the O-H bonds in the solvated water molecules. In alkali solution in the presence of OH^- ions, a variety of zinc hydroxychloride anions are formed in solution, for example, $Zn(OH)_3Cl^{2-}$, $Zn(OH)_2Cl_2^{2-}$, $ZnOHCl_3^{2-}$, while $Zn_5(OH)_8Cl_2{\cdot}H_2O$ (mineral known as simonkolleite) precipitates. When ammonia is bubbled through a solution of zinc chloride, the hydroxide does not precipitate, in its place compounds containing complexed ammonia (ammines) are formed, $Zn(NH_3)_4Cl_2{\cdot}H_2O$ and on concentration $ZnCl_2(NH_3)_2$. The first contains the $Zn(NH_3)_6^{2+}$ ion, while the second is molecular with a distorted tetrahedral geometry. The main species in aqueous solution is $Zn(NH_3)_4^{2+}$ with $Zn(NH_3)_3Cl^+$ also present at lower NH_3:Zn ratio. Aqueous zinc chloride reacts with zinc oxide to produce an amorphous cement that was first investigated in the 1855 by French engineer Stanislas Sorel (1803 to March 18, 1871 in Paris). Sorel later went on to study the related magnesium oxychloride cement, which bears his name. When hydrated zinc chloride is heated, a residue of Zn(OH)Cl is formed, for example,

$$ZnCl_2{\cdot}2H_2O \rightarrow ZnCl(OH) + HCl + H_2O$$

The compound $ZnCl_2{\cdot}1/2HCl{\cdot}H_2O$ can be formed by careful precipitation from a solution of $ZnCl_2$ acidified with HCl. It comprises a polymeric anion $(Zn_2Cl_5^-)_n$ with balancing monohydrated hydronium ions, $H_5O_2^+$ ions. The formation of highly reactive anhydrous HCl gas produced when zinc chloride hydrates are heated is the basis of qualitative inorganic spot tests. The use of zinc chloride as a flux, occasionally mixed with ammonium chloride, involves the formation of HCl and its ensuing reaction with surface oxides. Zinc chloride forms two salts with ammonium chloride: $(NH_4)_2ZnCl_4$ and $(NH_4)_3ClZnCl_4$, which decompose on heating producing HCl, similar to zinc chloride hydrate. The action of zinc chloride/ammonium chloride fluxes, for example, in the hot-dip galvanizing process forms H_2 gas and ammonia fumes. Therefore while many zinc salts have different formulas and different crystal structures, these salts behave very comparably in aqueous solution, for example, solutions prepared from any of the polymorphs of $ZnCl_2$, as well as other halides (bromide, iodide), and the sulfate can frequently be employed interchangeably for the synthesis of other zinc compounds. A good example is the preparation of zinc carbonate.

$$ZnCl_2(aq) + Na_2CO_3(aq) \rightarrow ZnCO_3(s) + 2NaCl(aq)$$

Zinc bromide ($ZnBr_2$) is a colorless salt that shares many properties with zinc chloride ($ZnCl_2$), specifically a high solubility in water forming acidic solutions, and solubility in organic solvents. It is hygroscopic and forms a dihydrate $ZnBr_2{\cdot}2H_2O$. This dihydrate is produced through reacting zinc oxide or zinc metal with hydrobromic acid.

$$ZnO + 2HBr + H_2O \rightarrow ZnBr_2{\cdot}2H_2O$$

The anhydrous compound can be obtained by dehydration of the dihydrate with hot CO_2 or by reaction of zinc metal and bromine. Crystalline $ZnBr_2$ has the same structure as ZnI_2: 4 tetrahedral Zn centers share 3 vertices to form "super-tetrahedra" of nominal composition $(Zn_4Br_{10})^{2-}$, which are linked by their vertices to form a three-dimensional structure. $ZnBr_2{\cdot}2H_2O$ also has a usual structure and can be better written as $Zn(H_2O)_6\ Zn_2Br_6$, where the $Zn_2Br_6^{2-}$ ion has bromine bridges linking the 2 zinc atoms. A comparable structure is found in the dimeric form of aluminum bromide (Al_2Br_6). Gaseous $ZnBr_2$ is linear in accordance with VSEPR (Valence Shell Electron Pair Repulsion) theory with a Zn-Br bond length of 221 pm.

Zinc iodide, ZnI_2, in the anhydrous form is white and easily absorbs water from the atmosphere. It can be produced via the direct reaction of zinc and iodine in refluxing ether or by reacting zinc with iodine in aqueous solution.

$$Zn + I_2 \rightarrow ZnI_2$$

At 1150°C, zinc iodide vapor dissociates into zinc and iodine. In aqueous solution it forms various ions such as octahedral $Zn(H_2O)_6^{2+}$, $[ZnI(H_2O)_5]^+$ and tetrahedral $ZnI_2(H_2O)_2$, $ZnI_3(H_2O)^-$, and ZnI_4^{2-}. The structure of crystalline ZnI_2 is unusual, and although the Zn atoms are tetrahedrally coordinated, as in $ZnCl_2$, groups of 4 of these tetrahedra share 3 vertices to form "supertetrahedra" of composition (Zn_4I_{10}), which are linked by their vertices to form a three-dimensional structure similar to $ZnBr_2$ These "supertetrahedra" are comparable to the P_4O_{10} structure. Molecular ZnI_2 is linear as predicted by VSEPR theory with a Zn-I bond length of 238 pm.

5.12.3.5 Organozinc compounds

The stability of the filled d^{10} in Zn^{2+} characterize the organozinc compounds. Zn(II) ions do not form π complexes with CO, NO, olefins (alkenes) or other electron rich ligands. Moreover, the high ionization potential of the 3d electrons make it unlikely for use in dative π-bonding with electron-accepting ligands. Although, this cannot be completely excluded in complexes involving planar N-heterocyclic ligands. The case of an observed stability of zinc cyanide compounds is presumably due primarily to σ rather than π bonding. As mentioned, the filled d shell also prevents π acceptance, thus, complexes with cyclopentadienide ions are also σ- rather than π-bonded. Nevertheless, organozinc compounds are some of the earliest prepared. In fact, the discovery of zinc alkyls which was reported in 1849 by Sir Edward Frankland is considered by Greenwood and Earnshaw as the beginning of organometallic chemistry. Of course, just like other early organometallics, organozinc has taken a back seat to Grignard reagents after the latter were discovered in the 1900s. With around 50 years in between their discoveries, it is inevitable that many of the reactions in which Grignard reagents are eventually used had already been worked out on organozinc compounds.

The classes of organozinc compounds synthesized since the early work of Frankland can be classified into three based on the number of carbon atoms directly bound to zinc. These classes are: (1) ionic organozinc compounds divided into the organozincates (R_nZn^-) and organozinc cations ($RZnL_n^+$); (2) the diorganozinc compounds (R_2Zn) and their coordination complexes, which can be divided into subclasses depending on various types of coordinating ligands; and (3) heteroleptic (RZnX; X = H, halogen, OR′, NR'_2, PR'_2, SR′, etc.)) compounds in which X is an electronegative substituent.

The organometallic compounds of Zn are rather reactive and unstable to both air and water. Thus these compounds cannot be stored without special precautions and are generated immediately prior to use, which makes them inconvenient choice as reagent in synthesis. Nevertheless, cross-coupling with organozinc compounds or the Negishi reaction is indispensable when the combination of high reactivity and maximal tolerance to functionality in the coupling partners is required. In view of these important characteristics of organozinc, they remain relevant in organic reactions and continued development of new organozinc compounds and their chemistry are expected.

5.12.4 Major uses

Zinc is most frequently used as an anticorrosion agent, and galvanization (coating of iron or steel) is the most common application. Zn is more reactive than iron or steel and hence will attract nearly all local oxidation until it is entirely corroded away. A protective surface layer of Zn oxide and carbonate ($Zn_5(OH)_6(CO_3)_2$) is formed as the Zn corrodes. This protection persists even after the Zn layer is scratched but degrades over time as the Zn corrodes away. The Zn is applied electrochemically or as molten Zn by hot-dip galvanizing or spraying. Galvanization is used on a variety of materials, for example, chain-link fencing, guard rails, suspension bridges, lamp posts, metal roofs, heat exchangers, and car bodies. The relative reactivity of Zn and its capacity to attract oxidation to itself makes it an effective sacrificial anode in CP (Cathodic Protection), for example, the CP of a buried pipeline can be achieved by connecting anodes made from Zn to the pipe. Zn functions as the anode (negative terminus) by gradually corroding away while it passes electric current to the steel pipeline. Zn is also applied to cathodically protect metals that are exposed to sea water. A Zn disc affixed to a ship's Fe rudder will gradually corrode as the rudder remains undamaged. Likewise, a Zn plug connected to a propeller or the metal protective guard for the keel of the ship offers short-term protection. With an SEP (Standard Electrode Potential) of -0.76 volts, Zn is used as an anode material for batteries. (more reactive Li (SEP -3.04 V) is used for anodes in lithium batteries). Powdered Zn is used in alkaline batteries and the case (which at the same time serves as the anode) of zinc–carbon batteries is made from sheet Zn. Zn is also used as the anode or fuel of the zinc-air battery/fuel cell. The Zn–Ce redox flow battery also depend on a Zn-based negative half-cell.

A commonly used Zn alloy is brass, where Cu is alloyed with 3%–45% Zn, dependent on the type of brass. Brass is usually more ductile and stronger than Cu and has much better corrosion resistance. These properties make it valuable in applications, such as communication equipment, hardware, musical instruments, and water valves. Other commonly used Zn alloys comprise nickel silver, typewriter metal, soft and aluminum solder, and commercial bronze. Zn is also used in modern-day pipe organs as an alternative for the traditional Pb/Sn alloy in pipes. Alloys of 85%–88% Zn, 4%–10% Cu, and 2%–8% Al are used in a limited way in specific types of machine bearings. Zn is the major metal in American one cent coins (pennies) since 1982. The Zn core is coated with a thin layer of Cu to make it look like a copper coin. Alloys of Zn with minor quantities of Cu, Al, and Mg are useful in die casting as well as spin casting, in particular in the automotive, electrical, and hardware industries. These alloys are sold as Zamak. An illustration of this is zinc aluminum. The low melting point combined with the low viscosity of the alloy allows the creation of small and intricate shapes. The low working temperature results in rapid cooling of the cast products and fast manufacture for assembly. Another alloy, sold under the name Prestal, has 78% Zn and 22% Al, and is described as nearly as strong as steel but at the same time malleable as plastic. This superplasticity of the alloy lets it to be molded using die casts made of ceramics and cement. Comparable alloys containing a minor quantity of Pb can be cold-rolled into sheets. An alloy of 96% Zn and 4% Al is used to produce stamping dies for low production run uses for which ferrous metal dies would be too costly. For building facades, roofing, and other applications for sheet metal formed by deep drawing, roll forming, or bending, Zn alloys containing Ti and Cu are used. Unalloyed Zn is too brittle for these production methods. As a dense, inexpensive, easily worked material, Zn is used as a Pb substitute. Following Pb fears, Zn appears in weights for numerous applications, for example, fishing, tire balances, and flywheels. Cadmium zinc telluride (CZT) is a semiconductive alloy that can be split into a series of small sensing devices. These devices are like an integrated circuit and can detect the energy of incoming γ ray photons. When behind an absorbing mask, the CZT sensor array can ascertain the direction of the rays.

Zn oxide (ZnO) is commonly used as a white pigment in paints and as a catalyst in the production of rubber to distribute heat. ZnO is used to protect rubber polymers and plastics from ultraviolet (UV) radiation. The semiconductor properties of ZnO make it valuable in varistors and photocopying products. The zinc-oxide cycle is a 2-step thermochemical method based on Zn and ZnO for H_2 production. $ZnCl_2$ is frequently added to lumber as a fire retardant and occasionally as a wood preservative. It is also used in the production of other chemicals. Zinc methyl ($Zn(CH_3)_2$) is used in several organic syntheses. Zn sulfide (ZnS) is used in luminescent pigments, for example, on the hands of clocks, X-ray and television screens, and luminous paints. ZnS crystals are used in lasers operating in the mid-infrared region of the spectrum. Zn sulfate is a chemical used in dyes and pigments. Zn pyrithione (IUPAC name bis(2-pyridylthio)zinc 1,1′-dioxide, $C_{10}H_8N_2O_2S_2Zn$) is a fungistatic and bacteriostatic used in antifouling paints. Zn powder is occasionally used as a propellant in model rockets. When a compacted mixture of 70% Zn and 30% S powder is ignited there is a violent chemical reaction which produces Zn sulfide, along with a large volume of hot gas, heat, and light. Zn sheet metal is used to produce Zn bars. ^{64}Zn, the most abundant isotope of Zn, is very sensitive to neutron activation, being transmuted into the highly radioactive ^{65}Zn, which has a half-life of 244 days and produces intense γ radiation. Due to this, ZnO utilized in nuclear reactors as an anticorrosion agent is depleted of ^{64}Zn before use, this is known as depleted ZnO. For similar reasons, Zn has been suggested as a salting material for nuclear weapons (Co is another, better known salting material). A jacket of isotopically enriched ^{64}Zn would be irradiated by the intense high-energy neutron flux from an exploding thermonuclear weapon, creating a large quantity of ^{65}Zn substantially boosting the radioactivity of the weapon's fallout. This type of weapon is not known to have ever been built, tested, or used. ^{65}Zn has been applied as a tracer to research how alloys containing Zn wear out, or the path and the role of Zn in organisms. Zn dithiocarbamate ($C_2H_4N_2S_4Zn$) complexes are used as agricultural fungicides; these include Zineb, Metiram, Propineb and Ziram. It is also used to modify the crosslinking of certain polyolefins with sulfur, a process called vulcanization. Zn naphthenate ($C_{22}H_{14}O_4Zn$) is applied as wood preservative. Zn in the form of ZDDP (zinc dialkyldithiophosphates, $Zn[(S_2P(OR)_2]_2$ is a family of coordination compounds developed in the 1940s that feature Zn bound to the anion of a dialkyldithiophosphoric acid (e.g., ammonium diethyl dithiophosphate). Typically, R is a branched or linear alkyl between 1 and 14 carbons in length, for example, 2-butyl, pentyl, hexyl, 1,3-dimethylbutyl, heptyl, octyl, isooctyl (2-ethylhexyl), 6-methylheptyl, 1-methylpropyl, dodecylphenyl, and others. These uncharged compounds are not salts. They are soluble in nonpolar solvents, and the longer-chain derivatives easily dissolve in mineral and synthetic oils used as lubricants.), is used as an antiwear additive for metal parts in engine oil.

Organozinc chemistry is the science of compounds that contain carbon-zinc bonds, describing the physical properties, synthesis, and chemical reactions. Many organozinc compounds are important. Examples of some important applications include: the Frankland-Duppa Reaction, named after the British chemist Sir Edward Frankland (January 18, 1825 to August 9, 1899) and Baldwin Francis Duppa (February 18, 1828 to November 10, 1873), in which an oxalate ester (ROCOCOOR) reacts with an alkyl halide R'X, Zn and hydrochloric acid to form the α-hydroxycarboxylic esters RR'COHCOOR; the Reformatskii reaction, discovered by the Russian chemist Sergey Nikolaevich Reformatsky (April 1, 1860 to July 28, 1934), in which α-halo-esters and aldehydes are converted to β-hydroxy-esters; the Simmons–Smith reaction, named after the American chemists Howard Ensign Simmons, Jr. (June 17, 1929 to April 26, 1997) and coworker Ronald D. Smith, in which the carbenoid (iodomethyl)zinc iodide reacts with alkene(or alkyne) and converts them to cyclopropane; the addition reaction of organozinc compounds to form carbonyl compounds; the Barbier reaction (1899), named after Victor Grignard's teacher French organic chemist Philippe Barbier (March 2, 1848 to September 18, 1922), which is the Zn equivalent of the Mg Grignard reaction and is the better of the two. In presence of water, formation of the organomagnesium halide will fail, whereas the Barbier reaction can take place in water. On the downside, organozincs are much less nucleophilic than Grignards, and they are expensive and difficult to handle. Commercially available diorganozinc compounds are dimethylzinc, diethylzinc and diphenylzinc. In one study, the active organozinc compound is obtained from much cheaper organobromine precursors. The Negishi coupling (named after Manchurian-born Japanese chemist Ei-ichi Negishi (b. July 14, 1935) who was a corecipient of the 2010 Nobel Prize in Chemistry for the discovery and development of this reaction) is also an important reaction for the formation of new C-C bonds between unsaturated carbon atoms in alkenes, arenes and alkynes. The catalysts are Ni and Pd. A key step in the catalytic cycle is a transmetalation in which a Zn halide exchanges its organic substituent for another halogen with the Pd (Ni) metal center. The Fukuyama coupling, discovered by Japanese organic chemist Tohru Fukuyama (b. August 9, 1948)) in 1998, is another coupling reaction, but it uses a thioester as reactant and produces a ketone. Zn has found many uses as catalyst in organic synthesis including asymmetric synthesis, being inexpensive and readily available substitute for precious metal complexes. The results (yield and enantiomeric excess) achieved with chiral Zn catalysts are similar to those achieved with Pd, Ru, Ir and others.

In most single-tablet, over-the-counter, daily vitamin and mineral supplements, Zn is included in such forms as ZnO, Zn acetate, or Zn gluconate. Zn is generally thought to be an antioxidant. Yet, it is redox inert and thus can act as such only indirectly. Usually, Zn supplement is advised where there is high risk of Zn deficiency (e.g., in low- and middle-income countries) as a precautionary measure. Zn deficiency has been linked to major depressive disorder, and Zn supplements may be a useful treatment. Zn works as a simple, low-cost, and essential means to treat diarrheal episodes in children in the developing world. Zn can become depleted during diarrhea, but the latest research indicates that replacing Zn with a 10- to 14-day therapy can decrease the duration and severity of diarrheal episodes and may also inhibit future episodes up to 3 months. A Cochrane review indicated that people taking Zn supplement may be less prone to develop age-related macular degeneration. Supplementary Zn is also an effective therapy for acrodermatitis enteropathica, a genetic disorder affecting Zn absorption that used to be fatal to affected babies. Gastroenteritis is greatly reduced by ingestion of Zn, possibly through direct antimicrobial action of the ions in the gastrointestinal tract, or through the absorption of the Zn and rerelease from immune cells (all granulocytes secrete Zn), or both. Recently, a study described that adding large quantities of Zn to a urine sample masked the detection of drugs. The study did not test whether orally consuming a Zn dietary supplement could give the same result. Zn is a negative allosteric modulator of the $GABA_A$ receptor (an ionotropic receptor and ligand-gated ion channel. Its endogenous ligand is γ-aminobutyric acid (GABA), the major inhibitory neurotransmitter in the central nervous system.). Zn supplements (often Zn acetate or Zn gluconate lozenges) are a group of dietary supplements that are frequently used for the treatment of the common cold. Using Zn supplements at doses greater than 75 mg/day within 24 hours of the start of symptoms has been shown to decrease the duration of cold symptoms by around 1 day. Due to a lack of data, there is not enough evidence to establish whether the preventive use of Zn supplements lessens the likelihood of getting a cold. Unpleasant side-effects with Zn supplements by mouth consist of bad taste and nausea. The intranasal use of Zn-containing nasal sprays has been linked to the loss of the sense of smell; as a result, in June 2009, the United States Food and Drug Administration (USFDA) warned consumers to stop using intranasal Zn sprays. The human rhinovirus—the most common viral pathogen in humans—is the main cause of the common cold. The postulated mechanism through which Zn reduces the severity and/or duration of cold symptoms is the inhibition of nasal inflammation and the direct inhibition of rhinoviral receptor binding and rhinoviral replication in the nasal mucosa. Topical preparations of Zn include those used on the skin, often in the form of ZnO. Zinc preparations can protect against sunburn in the summer and windburn in the winter. Applied thinly to a baby's diaper area (perineum) with each diaper change, it can protect against diaper rash. Chelated Zn is used in toothpastes and mouthwashes to prevent bad breath. Zn pyrithione is widely included in shampoos to prevent dandruff.

5.13 31 Ga — Gallium

5.13.1 Discovery

In 1871 the presence of gallium was first foreseen by Russian chemist Dmitri Mendeleev (February 8, 1834 to February 2, 1907), who called it "eka-aluminum" based on its position in his periodic table. He also predicted some properties of eka-aluminum that agree closely with the real properties of gallium, for example, its density, melting point, oxide character and bonding in chloride. Mendeleev additionally predicted that eka-aluminum would be found by means of the spectroscope, and that metallic eka-aluminum would dissolve slowly in both acids and alkalis and would not react with air. He similarly foresaw that M_2O_3 would dissolve in acids to give MX_3 salts, that eka-aluminum salts would form basic salts, that eka-aluminum sulfate should form alums, and that anhydrous MCl_3 would have a higher volatility than $ZnCl_2$: all of these predictions proved to be correct. Gallium was found using spectroscopy by French chemist Paul-Emile Lecoq de Boisbaudran (April 18, 1838 to May 28, 1912) in 1875 from its characteristic spectrum (two violet lines) in a sample of sphalerite (ZnS) (Lecoq de Boisbaudran, 1875) (Fig. 5.120). Later that same year, he produced the free metal by electrolysis of the hydroxide in potassium hydroxide (KOH) solution. He called the element "gallia," from Latin Gallia meaning Gaul, after his native land of France. It was later claimed that, in one of those multilingual witticisms so cherished by men of science in the 19th century, he had also named gallium after himself: "Le coq" is French for "the rooster" and the Latin word for "rooster" is "gallus." In an 1877 article, Lecoq denied this inference. Initially, he measured the density of gallium as 4.7 g/cm^3, the only property that did not match Mendeleev's predictions; Mendeleev then wrote to him and recommended that he should remeasure the density, and subsequently Leqoc de Boisbaudran determined the correct value of 5.9 g/cm^3, that Mendeleev had predicted almost exactly. From its discovery in 1875 until the age of semiconductors, the main applications of gallium were high-temperature thermometrics and in metal alloys with uncommon properties of stability or ease of melting (some such being liquid at room temperature). The discovery of gallium arsenide as a direct band gap semiconductor in the 1960s started the most important stage in the applications of gallium.

FIGURE 5.120 Paul-Emile Lecoq de Boisbaudran.

5.13.2 Mining, production, and major minerals

Ga does not occur as a free element in the Earth's crust, and the few high-concentration minerals, such as gallite ($CuGaS_2$), are too rare to be mined as a primary source. The abundance in the Earth's crust is approximately 16.9 ppm. This is similar to the crustal abundances of Pb, Co, and Nb. However different to these elements, Ga does not form its own ore deposits with concentrations of >0.1 wt.% in ore. Instead it is found at trace concentrations comparable to the crustal value in zinc ores, and at slightly higher concentrations (around 50 ppm) in aluminum ores, from both of which it is produced as a by-product. This absence of independent ore deposits is due to gallium's geochemical behavior, exhibiting no strong enrichment in the processes pertinent to the formation of most ore deposits. It has been assessed that over 1 million tons of gallium can be obtained from known reserves of bauxite and zinc ores. Gallium is obtained entirely as a by-product during the processing of the ores of other metals. Its main source material is bauxite, the chief ore of aluminum, but small quantities are also extracted from zinc sulfide ores (sphalerite, ZnS, being the main host mineral). In the past, certain coals were an important source. During the processing of bauxite to alumina in the Bayer process, gallium accumulates in the sodium hydroxide liquor. From this liquor it can be isolated through several methods. The most current is the use of ion-exchange resins. Achievable extraction efficiencies critically hinge on the original gallium concentration in the feed bauxite. At a typical feed concentration of 50 ppm Ga, around 15% of the contained gallium is extractable. The remainder stays in the red mud and aluminum hydroxide streams. Gallium is removed from the ion-exchange resin in solution. Subsequent electrolysis then produces the gallium metal. For semiconductor use, it is necessary to further purify the gallium using zone melting or single-crystal extraction from a melt (Czochralski process). Purities of 99.9999% are routinely achieved and commercially available.

In total only seven minerals are known to contain gallium. Three sulfides exist with Ga, gallite ($CuGaS_2$), ishiharaite ((Cu,Ga,Fe,In,Zn)S), and zincobriartite ($Cu_2(Zn,Fe)(Ge,Ga)S_4$). Two oxide class minerals contain Ga, söhngeite ($Ga(OH)_3$) (Fig. 5.121) and tsumgallite (GaOOH) (Box 5.16). The phosphate class is represented by two minerals, gallobeudantite ($PbGa_3(AsO_4)(SO_4)(OH)_6$) and galloplumbogummite ($Pb(Ga,Al,Ge)_3(PO_4)_2(OH)_6$).

5.13.3 Chemistry

Ga has 31 known isotopes, ranging from ^{56}Ga to ^{86}Ga. Only ^{69}Ga and ^{71}Ga are stable isotopes and occur naturally. ^{69}Ga is more common with a natural abundancy of 60.1%, with ^{71}Ga the remaining 39.9%. All other isotopes are radioactive, with ^{67}Ga having the longest half-life of 3.261 days. Isotopes lighter than ^{69}Ga usually decay through β^+ decay (positron emission) or electron capture to isotopes of Zn, although the lightest few (with mass numbers 56 through 59)

FIGURE 5.121 Söhngeite, $Ga(OH)_3$, 2 mm crystals, extremely rare. Tsumeb mine, Tsumeb, Namibia.

BOX 5.16 Söhngeite, $Ga(OH)_3$ and tsumgallite, GaOOH

Söhngeite was named in 1965 by H. Strunz in honor of Adolf Paul Gerhard Söhnge (November 10, 1913 to July 14, 2006), chief geologist, Tsumeb corporation, Tsumeb, Namibia, who found the mineral. He was also a professor at the Stellenbosch University (1968–1978). It occurs as an alteration product of gallite-bearing germanite in an oxidation zone of a dolostone-hosted hydrothermal polymetallic ore deposit. It is orthorhombic, pseudocubic, *2/m2/m2/m* (space group $Pmn2_1$). It has a hydroxide perovskite structure with 5 independent OH groups and is based on $Ga(OH)_6$ octahedra. It is the Ga analog of bernalite, $Fe(OH)_3{\cdot}nH_2O$ ($n = 0.0 - 0.25$), and dzhalindite, $In(OH)_3$. Its hardness is 4–4½. It is transparent to semitransparent. The color is white, pale yellow, pale brown, to pale greenish yellow. It is optically isotropic. It occurs in crystalline twinned groups and aggregates, to 1 cm; very rarely as single crystals. Twinning as trillings of composite flattened and curved crystals, interpenetrant at right angles, according to unknown twinning law.

Tsumgallite is named for the type locality, the Tsumcor Mine, Tsumeb, Namibia, and the chemical composition, which contains gallium. It occurs in vugs of a tennantite-germanite ore. Tsumgallite is orthorhombic, *mmm* (space group *Pbnm*). It is a member of the diaspore (AlOOH) group. It has a hardness of 1–2. Cleavage is perfect on {010}. The luster is pearly, while the transparency is translucent. The color is pale greenish yellow to beige and the streak is white. Optically it is biaxial, sign unknown. It occurs as scaly mica-like aggregates less than 1 mm across. Individual thin, platy crystals not exceeding 40μm in size.

TABLE 5.16 Gallium properties.

Appearance	Silvery blue
Standard atomic weight $A_{r,std}$	69.723
Block	p-Block
Element category	Posttransition metal
Electron configuration	[Ar] $3d^{10}$ $4s^2$ $4p^1$
Phase at STP	solid
Melting point	29.7646°C
Boiling point	2400°C
Density (near r.t.)	5.91 g/cm^3
When liquid (at m.p.)	6.095 g/cm^3
Heat of fusion	5.59 kJ/mol
Heat of vaporization	256 kJ/mol
Molar heat capacity	25.86 J/(mol·K)
Oxidation states	−5, −4, −2, −1, +1, +2, **+3**
Ionization energies	1st: 578.8 kJ/mol
	2nd: 1979.3 kJ/mol
	3rd: 2963 kJ/mol
Atomic radius	Empirical: 135 pm
Covalent radius	122 ± 3 pm
Van der Waals radius	187 pm

STP, Standard temperature and pressure.
Bold font indicates main oxidation state.

decay through prompt proton emission. Isotopes heavier than ^{71}Ga decay through beta minus decay (electron emission), possibly with delayed neutron emission, to isotopes of germanium, while gallium-70 can decay through both β^- decay and electron capture. ^{67}Ga is unique among the light isotopes in having only electron capture as a decay mode, as its decay energy is not enough to allow positron emission. ^{67}Ga and ^{68}Ga (which has a half-life of 67.7 minutes) are both used in nuclear medicine.

Gallium has a silvery blue appearance. It has a low melting point at 29.7646°C and forms a liquid state that is denser than the solid. Gallium expands by around 3% when it solidifies and is normally not stored in glass or metal containers. This is to prevent rupture of the container when gallium changes state from liquid to solid. Gallium also diffuses into grain boundaries of materials like aluminum-zinc alloys and steel making them brittle. Nevertheless, this property of gallium to diffuse in the metal lattice of other metals is advantageous in some applications. For example, gallium is alloyed with plutonium cores of nuclear bombs to stabilize the plutonium crystal structure. Gallium is a member of Group 13 or Boron family with an atom electronic configuration of $[Ar]3d^{10}4s^24p^1$. Its compounds are found primarily in the +3 oxidation state, although the +1 oxidation state is also known (Table 5.16). Compounds considered as having the +2 oxidation state are shown to be mixed-oxidation state compounds containing both gallium(I) and gallium(III).

Gallium crystallizes in a unique orthorhombic (pseudotetragonal) structure with 8 atoms in the unit cell. Each gallium atom in the unit cell has one nearest neighbor at 244 pm while the other six atoms are farther away. The two nearest neighbors are covalently bonded. Hence, gallium clusters are considered as having Ga_2 dimers as their building block. This structure forms because of interactions between the single 4p electrons of gallium atoms which is farther away from the nucleus than the 4s electrons and the $[Ar]3d^{10}$ core. Gallium is fairly reactive, combining with most nonmetals at high temperatures. Interests in its chemistry are not only due to the important structure and properties of its compounds but also due to academic curiosity.

5.13.3.1 Reaction with air, hydrogen, oxygen, and nitrogen

The metal gallium is stable in dry air, oxidizing slowly until it forms an oxide film that prevents it from further oxidation. It burns in air or oxygen to form the white oxide, Ga_2O_3. This oxide exists in many polymorphs and can be reduced to the metal when heated at high temperatures in hydrogen. It gives the lower oxide Ga_2O when heated with gallium metal at 700°C. Gallium in its vapor form reacts with hydrogen under appropriate conditions to give metal hydrides with the formula MH or MH_3. These hydrides, however, are transient and convert back to its elements. GaH_3 cannot be prepared or isolated easily but Ga_2H_6 or $H_2Ga(H)_2GaH_2$ have been prepared and characterized to some extent. Like aluminum, gallium(III) hydride, GaH_3, known as gallane, may also be produced by reacting lithium gallanate ($LiGaH_4$) with gallium(III) chloride at −30°C.

$$3LiGaH_4 + GaCl_3 \rightarrow 3LiCl + 4GaH_3$$

In the presence of dimethyl ether as solvent, GaH_3 polymerizes to $(GaH_3)_n$. If no solvent is used, the dimer Ga_2H_6 (digallane) is formed as a gas. Its structure is similar to diborane, having two hydrogen atoms bridging the two gallium centers, unlike α-AlH_3 in which aluminum has a coordination number of 6. Gallane is unstable above −10°C, decomposing to elemental gallium and hydrogen. Gallium forms compounds with nitrogen but not through direct combination of the elements. GaN can be grown when gallium/sodium melt are subjected to 100 atmospheres of N_2 at 750°C. GaN thin films can be prepared by molecular-beam epitaxy on silicon carbide (SiC) or sapphire substrates. GaN crystals can also be formed by injecting ammonia gas into molten gallium at 900°C−980°C at normal atmospheric pressures or by reacting Ga_2O_3 with ammonia at elevated temperatures of the order of 1000°C.

5.13.3.2 Reactions with sulfur

Ga_2O is a very strong reducing agent, which can reduce H_2SO_4 to H_2S. It disproportionates at 800°C back to gallium and Ga_2O_3. Gallium(III) sulfide, Ga_2S_3, has 3 possible crystal modifications. It can be formed by the reaction of gallium with hydrogen sulfide (H_2S) at 950°C. Instead, $Ga(OH)_3$ can be used at 747°C.

$$2Ga(OH)_3 + 3H_2S \rightarrow Ga_2S_3 + 6H_2O$$

Reacting a mixture of alkali metal carbonates and Ga_2O_3 with H_2S results in the formation of thiogallates with the $[Ga_2S_4]^{2-}$ anion. Strong acids decompose these salts, releasing H_2S during the reaction. The mercury salt, $HgGa_2S_4$, can be utilized as a phosphor. Additionally, gallium can form sulfides in lower oxidation states, such as gallium(II) sulfide and the green gallium(I) sulfide, the latter of which is produced from the former by heating to 1000°C under a stream of nitrogen. The other binary chalcogenides, Ga_2Se_3 and Ga_2Te_3, have the zincblende structure (sphalerite, ZnS). They are all semiconductors but are easily hydrolyzed and have limited value.

5.13.3.3 Reaction with halogens

Gallium(III) halides of fluorine, chlorine, bromine and iodine are known. Gallium(III) fluoride can be prepared by the reaction of hydrogen fluoride and gallium at 440°C. Gallium(III) chloride can be prepared by the reaction of chlorine or hydrogen chloride gas on heated gallium. Unlike the trifluoride, gallium(III) chloride exists as dimeric molecules, Ga_2Cl_6, with a melting point of 78°C. Gallium(III) bromide can be prepared by reacting bromine/nitrogen gas mixture or hydrogen bromide with heated gallium. Gallium(III) iodide is prepared from stoichiometric quantities of the elements in a sealed tube. Similar to gallium(III) chloride, dimeric compounds are formed with bromine and iodine, Ga_2Br_6 and Ga_2I_6. Gallium(III) halides are important reagents in organic synthesis. They are considered effective Lewis acids which can activate several functional groups even under extremely mild conditions. Examples of organic reactions where gallium(III) halides have been successfully employed as Lewis acid catalysts include alkylation, allylation, radical reactions, cycloaddition reactions, Friedel−Craft's reactions and various coupling reactions. Like the other group 13 trihalides, gallium(III) halides are Lewis acids, reacting as halide acceptors with alkali metal halides to form salts containing GaX_4^- anions, where X is a halogen. They also react with alkyl halides to form carbocations and GaX_4^-. When heated to a high temperature, gallium(III) halides react with elemental gallium to form the respective gallium(I) halides. For example, $GaCl_3$ reacts with Ga to form GaCl.

$$2Ga + GaCl_3 \rightleftharpoons 3GaCl(g)$$

At lower temperatures, the equilibrium shifts toward the left and GaCl disproportionates back to elemental gallium and $GaCl_3$. GaCl can also be produced by reacting Ga with HCl at 950°C; the product can be condensed as a red solid. Gallium(I) compounds can be stabilized by forming adducts with Lewis acids. For example,

$$GaCl + AlCl_3 \rightarrow Ga^+[AlCl_4]^-$$

The so-called "gallium(II) halides," GaX_2, are actually adducts of gallium(I) halides with the respective gallium(III) halides, having the structure $Ga^+[GaX_4]^-$. For example,

$$GaCl + GaCl_3 \rightarrow Ga^+[GaCl_4]^-$$

5.13.3.4 Reactions with acids and bases

Metallic gallium dissolves slowly in common mineral acids to form gallium(III) salts such as $Ga_2(SO_4)_3$ and $Ga(NO_3)_3$. The reaction with hot, concentrated perchloric acid is faster and produce hydrated gallium perchlorate. Aqueous solutions of gallium(III) salts contain the hydrated gallium ion, $[Ga(H_2O)_6]^{3+}$. Adding NH_4OH to gallium solutions precipitates $Ga(OH)_3$. Heating $Ga(OH)_3$ at 100°C removes water and produces gallium oxide hydroxide, GaO(OH). Adding alkaline hydroxides such as NaOH solutions to gallium forms gallate salts containing the $Ga(OH)_4^-$ anion.

5.13.3.5 Organogallium compounds

Organogallium compounds contain a carbon to gallium chemical bond. These compounds are highly reactive. The lower trialkyl compounds of gallium are even spontaneously inflammable. The reactivity is, however, lower than that of the aluminum analogs. Despite their toxicity, they are important reagents or intermediates in several classes of organic reactions. They are especially useful in the preparation of gallium arsenide, for example, through the reaction of trimethyl gallium with arsine at 700°C.

$$Ga(CH_3)_3 + AsH_3 \rightarrow GaAs + 3CH_4$$

Alkylgalliums are monomeric. Lewis acidity decreases in the order Al > Ga > In and as a result organogallium compounds do not form bridged dimers similar to organoaluminum compounds. They do form stable peroxides. These alkylgallium compoounds are liquids at room temperature, with a low melting points, and are rather mobile and flammable. Triphenylgallium is monomeric in solution, but its crystals form chain structures due to weak intermolecular Ga···C interactions. Gallium trichloride is an often used starting reagent for the synthesis of organogallium compounds, such as in carbogallation reactions. Gallium trichloride reacts with lithium cyclopentadienide in diethyl ether to form the trigonal planar gallium cyclopentadienyl complex $GaCp_3$. Gallium(I) forms complexes with arene ligands, for example, hexamethylbenzene. Since this ligand is rather bulky, the structure of the $[Ga(\eta^6\text{-}C_6Me_6)]^+$ is that of a half-sandwich. Less bulky ligands such as mesitylene allow two ligands to be attached to the central gallium atom in a bent sandwich structure. Benzene is even less bulky and allows the formation of dimers, for example, $[Ga(\eta^6\text{-}C_6H_6)_2][GaCl_4]\cdot 3C_6H_6$.

5.13.4 Major uses

Extremely high-purity (>99.9999%) Ga is commercially available for the semiconductor industry. Gallium arsenide (GaAs) and gallium nitride (GaN) used in electronic components represented about 98% of the Ga consumption in the United States. Around 66% of semiconductor Ga is used in integrated circuits (mostly GaAs), for example, the manufacture of ultra-high-speed logic chips and MESFETs (metal–semiconductor field-effect transistor) for low-noise microwave preamplifiers in cell phones. Around 20% of this Ga is used in optoelectronics. Globally, GaAs makes up 95% of the annual Ga consumption with 53% originating from cell phones, 27% from wireless communications, and the remainder from other uses such as automotive, consumer, fiber-optic, and military applications. The current growth in GaAs consumption is mainly associated with the emergence of 3G and 4G smartphones, which use 10 times more GaAs than older models. In addition, GaAs and GaN can be found in various optoelectronic devices. Aluminum gallium arsenide (AlGaAs) is utilized in high-power infrared laser diodes. The semiconductors GaN and indium gallium nitride (InGaN) are used in blue and violet optoelectronic devices, mostly laser diodes and LEDs, for example, GaN 405 nm diode lasers are the violet light source for higher-density Blu-ray Disc compact data disc drives. Other key application of GaN are cable television transmission, commercial wireless infrastructure, power electronics, and satellites. Multijunction photovoltaic cells, developed for satellite power applications, are produced using molecular-beam epitaxy or metalorganic vapor-phase epitaxy of thin films of GaAs, indium gallium phosphide (InGaP), or indium gallium arsenide (InGaAs). The Mars Exploration Rovers and several satellites use triple-junction GaAs on Ge cells. Furthermore, Ga is a component in photovoltaic compounds (such as copper indium gallium selenium sulfide $Cu(In,Ga)(Se,S)_2$) used in solar panels as a cost-efficient substitute for crystalline silicon.

Ga easily alloys with most metals and is used for the production of low-melting alloys. The virtually eutectic alloy of Ga, In, and Sn is a room temperature liquid used in medical thermometers. This alloy, commercially known as Galinstan (with the "-stan" referring to the tin, stannum in Latin), has a low freezing point of $-19°C$. It has been proposed that this family of alloys could also be used to cool computer chips instead of water. Ga alloys have been assessed as substitutes for Hg dental amalgams, but these materials have not yet been widely accepted. Since Ga wets glass or porcelain, it can be used to produce brilliant mirrors. When the wetting action of Ga alloys is not needed (as in Galinstan glass thermometers), the glass must be covered with a transparent layer of Ga(III) oxide. The Pu used in nuclear weapon pits is stabilized in the δ phase and made machinable by alloying with Ga.

While Ga has no natural function in the human body, Ga ions interact with processes in the body in a way comparable to Fe(III). Since these processes include inflammation, a marker for many disease states, several Ga salts are used (or are in development) as pharmaceuticals and radiopharmaceuticals in medicine. Interest in the anticancer properties of Ga emerged when it was found that ^{67}Ga(III) citrate ($C_6H_5GaO_7$) injected in tumor-bearing animals localized to the tumor sites. Clinical trials have shown Ga nitrate ($Ga(NO_3)_3$) to have antineoplastic activity against non-Hodgkin's lymphoma and urothelial cancers. A new generation of Ga-ligand complexes such as tris(8-quinolinolato)Ga(III) (KP46, $C_{27}H_{18}GaN_3O_3$) and Ga maltolate (IUPAC name tris(3-hydroxy-2-methyl-4H-pyran-4-one)gallium, $Ga(C_6H_5O_3)_3$) has emerged. Ga nitrate (brand name Ganite) has been utilized as intravenous medication to treat hypercalcemia associated with tumor metastasis to bones. Ga is believed to interfere with osteoclast function, and the therapy may be effective after other treatments have failed. Ga maltolate, an oral, highly absorbable form of Ga(III) ion, is an antiproliferative to pathologically proliferating cells, particularly cancer cells and some bacteria that accept it instead of ferric iron (Fe^{3+}). Researchers are currently performing clinical and preclinical trials on this compound as a potential treatment for a number of cancers, infectious diseases, and inflammatory diseases. When Ga ions are erroneously taken up in place of Fe (III) by bacteria such as *Pseudomonas*, the ions interfere with respiration, and the bacteria die. This occurs since Fe is redox-active, allowing the transfer of electrons during respiration, whereas Ga is redox-inactive. A complex amine-phenol Ga(III) compound MR045 is selectively toxic for chloroquine (common malaria drug) resistant parasites. Both the Ga(III) complex and chloroquine work through preventing hemozoin crystallization, a waste product formed from the digestion of blood by the parasites. ^{67}Ga salts such as Ga citrate and Ga nitrate are applied as radiopharmaceutical agents in the nuclear medicine imaging known as gallium scan. The radioactive isotope ^{67}Ga is utilized, and the compound or salt of Ga is not important. The body handles Ga^{3+} in many ways as if it were Fe^{3+}, and the Ga^{3+} is bound (and concentrated) in areas of inflammation, such as infection, and in areas of rapid cell division, which allows these sites to be imaged with nuclear scan techniques. ^{68}Ga, a positron emitter with a half-life of 68 minutes, is these days used as a diagnostic radionuclide in PET-CT when linked to pharmacological preparations such as DOTATOC (IUPAC name 2-[4-[2-[[(2*R*)-1-[[(4*R*,7*S*,10*S*,13*R*,16*S*,19*R*)-10-(4-aminobutyl)-4-[[(2*R*,3*R*)-1,3-dihydroxybutan-2-yl]carbamoyl]-7-[(1*R*)-1-hydroxyethyl]-16-[(4-hydroxyphenyl)methyl]-13-(1*H*-indol-3-ylmethyl)-6,9,12,15,18-pentaoxo-1,2-dithia-5,8,11,14,17-pentazacycloicos-19-yl]amino]-1-oxo-3-phenylpropan-2-yl]amino]-2-oxoethyl]-7,10-bis(carboxymethyl)-1,4,7,10-tetrazacyclododec-1-yl]acetic acid, $C_{65}H_{92}N_{14}O_{18}S_2$), a somatostatin analog used for neuroendocrine tumors investigation, and DOTA-TATE (an amino acid peptide, with a covalently bonded DOTA bifunctional chelator, $C_{65}H_{90}N_{14}O_{19}S_2$), a newer one, used for neuroendocrine metastasis and lung neuroendocrine cancer, such as certain types of *microcytoma*. ^{68}Ga production as a pharmaceutical is chemically with the radionuclide being extracted by elution from ^{68}Ge, a synthetic radioisotope of Ge, in ^{68}Ga generators.

Ga is also used for neutrino detection in astronomy. Probably the largest volume of pure Ga yet collected in a single place is the Gallium-Germanium Neutrino Telescope used by the SAGE experiment at the Baksan Neutrino Observatory in Russia. This detector consists of 55–57 tons (equivalent to about 9 m^3) of liquid Ga. An earlier experiment was the GALLEX neutrino detector operated at the beginning of the 1990s in an Italian mountain tunnel. This detector consisted of 12.2 tons of watered ^{71}Ga. Solar neutrinos resulted in a few atoms of ^{71}Ga to react to radioactive ^{71}Ge, which were detected. This experiment showed that the solar neutrino flux is 40% less than anticipated based on theoretical calculations. This discrepancy was not explained until scientists came up with better solar neutrino detectors as well as better theories. Ga is also used as a liquid metal ion source for a focused ion beam, for example, a focused Ga-ion beam was used to create the world's smallest book, *Teeny Ted from Turnip Town* (2007), published by Robert Chaplin, and was certified by Guinness World Records as the world's smallest reproduction of a printed book. The book was produced in the Nano Imaging Laboratory at Simon Fraser University in Vancouver, British Columbia, Canada, with the assistance of SFU scientists Li Yang and Karen Kavanagh. The book's size is 0.07×0.10 mm. The letters are carved into 30 microtablets on a polished piece of single crystalline silicon, using a focused-gallium-ion beam with a minimum diameter of 7 nm. The book has its own ISBN, 978-1-894897-17-4. The story was written by Malcolm Douglas Chaplin and is "a fable about Teeny Ted's victory in the turnip contest at the annual county fair." The book has been published in a limited edition of 100 copies by the laboratory and requires a scanning electron microscope to read the text. An additional use of Ga is as an additive in glide wax for skis, and other low-friction surface materials. A well-known practical joke between chemists is to make Ga spoons and use them to serve tea to innocent visitors, as Ga has a similar look as its lighter homolog Al. The spoons then melt in the hot tea.

5.14 32 Ge — Germanium

5.14.1 Discovery

In his publication on The Periodic Law of the Chemical Elements in 1869, the Russian chemist Dmitri Mendeleev (February 8, 1834 to February 2, 1907) forecast the presence of several unknown chemical elements, including one that would fill a gap in the carbon family, located between silicon and tin. For its position in his periodic table, Mendeleev called it ekasilicon (Es), and he predicted its atomic weight to be 70 (later 72). In mid-1885, at a mine near Freiberg, Saxony, a new mineral was found and named argyrodite (Ag_8GeS_6) for its high silver content. The German chemist Clemens Winkler (December 26, 1838 to October 8, 1904) analyzed this new mineral, which showed to be a combination of silver, sulfur, and a new element (Winkler, 1887a,b) (Fig. 5.122). He managed to isolate this new element in 1886 and observed it to be similar to antimony. At first, he thought the new element to be eka-antimony but was quickly convinced that it was eka-silicon instead. Prior to Winkler publishing his findings on the new element, he decided that he would call his element neptunium, as the recent discovery of planet Neptune in 1846 had likewise been preceded by mathematical predictions of its existence. But, the name "neptunium" had already been suggested for another proposed chemical element (however not the element that today is now named neptunium, which was discovered in 1940). So, in its place, Winkler called the new element germanium (from the Latin word, Germania, for

FIGURE 5.122 Clemens Winkler, c.1895.

Germany) in honor of his homeland. Since this new element exhibited some parallels with the properties of the elements arsenic and antimony, its correct place in the periodic table was under consideration, but its similarities with Dmitri Mendeleev's predicted element "ekasilicon" established that place on the periodic table. With additional material from 500 kg of ore from the mines in Saxony, Winkler established the chemical properties of the new element in 1887. He also measured an atomic weight of 72.32 by analyzing pure germanium tetrachloride ($GeCl_4$), while French chemist Paul-Emile Lecoq de Boisbaudran (April 18, 1838 to May 28, 1912) inferred 72.3 by a comparison of the lines in the spark spectrum of the element (Lecoq de Boisbaudran, 1886). Winkler managed to prepare numerous new chemical compounds of germanium, including fluorides, chlorides, sulfides, dioxide, and tetraethyl germane ($Ge(C_2H_5)_4$), the first organogermane. The physical properties for those compounds—which matched well with Mendeleev's predictions—made the discovery a significant validation of Mendeleev's idea of element periodicity.

5.14.2 Mining, production, and major minerals

Ge's abundance in the Earth's crust is about 1.6 ppm. Only a few minerals like argyrodite (Ag_8GeS_6), briartite ($Cu_2(Zn,Fe)GeS_4$), germanite ($Cu_{13}Fe_2Ge_2S_{16}$) (Box 5.17), and renierite ($(Cu,Zn)_{11}(Ge,As)_2Fe_4S_{16}$) contain substantial concentrations of Ge. Only few of them (especially germanite) are, very infrequently, found in mineable amounts. Some copper-lead-zinc ore deposits have enough germanium to validate production from the final ore concentrate. An uncommon natural enrichment process results in a high concentration of germanium in some coal seams, discovered by Swiss-born Norwegian mineralogist Victor Moritz Goldschmidt (January 27, 1888 to March 20, 1947) during a comprehensive survey for germanium deposits. The maximum Ge concentration ever measured was in Hartley coal ash with as much as 1.6% germanium. The coal deposits near Xilinhaote, Inner Mongolia, are thought to comprise a projected 1600 tons of germanium.

About 118 tons of germanium was produced in 2011 worldwide, mostly in China, Russia, and United States. Germanium is produced mainly as a by-product from sphalerite (ZnS) ores where it can be present in concentrations as high as 0.3%, in particular from low-temperature sediment-hosted, massive Zn-Pb-Cu(-Ba) deposits and carbonate-hosted Zn-Pb deposits. It is estimated that at least 10,000 tons of extractable germanium is contained in known zinc reserves, particularly those hosted by Mississippi-Valley type deposits, while at least 112,000 tons will be found in coal reserves. As far back as 2007, 35% of the demand for germanium was met by recycling. Although it is obtained mainly from sphalerite, it is also found in silver, lead, and copper ores. Another source of germanium is fly ash of power plants fueled from coal deposits that contain germanium. Both Russia and China used this as a source for germanium. Russia's deposits are found in the far east of Sakhalin Island, and northeast of Vladivostok. The ore concentrates are mostly in the form of sulfides; they are roasted in by heating in air which converts it to the oxides.

$$GeS_2 + 3O_2 \rightarrow GeO_2 + 2SO_2$$

Some of the germanium remains in the dust produced, while the rest is converted to germanates, which are then leached (together with zinc) from the cinder with sulfuric acid. After neutralization, only the zinc stays in solution while germanium and other metals precipitate. After removing some of the zinc in the precipitate by the Waelz process (a method of recovering zinc and other relatively low boiling point metals from metallurgical waste (typically electric arc furnace flue dust) and other recycled materials using a rotary kiln), the residing Waelz oxide is leached for a second time. The dioxide (GeO_2) is obtained as precipitate and converted with chlorine gas or hydrochloric acid to germanium tetrachloride ($GeCl_4$), which has a low boiling point and can subsequently be obtained through distillation.

$$GeO_2 + 4HCl \rightarrow GeCl_4 + 2H_2O$$
$$GeO_2 + 2Cl_2 \rightarrow GeCl_4 + O_2$$

BOX 5.17 Germanite, $Cu_{13}Fe_2Ge_2S_{16}$

Germanite occurs in primary Cu-Pb-Zn ores in a dolostone-hosted hydrothermal polymetallic ore deposits (e.g., Tsumeb, Namibia). Germanite was named in allusions to the germanium content. It is isometric (cubic). *3m* (space group *F3n*). It has a hardness of 4. It has no cleavage, tenacity is brittle. It has a metallic to dull luster and is opaque. The color is reddish gray, tarnishes to dull brown. The streak is dark gray to black. Optically it is isotropic. It occurs mainly as minute cubic crystals, usually massive, intergrown with renierite. Small gray chalcocite crystals are often mistaken for germanite.

FIGURE 5.123 Argyrodite, Ag_8GeS_6, crude crystals to 4 mm. Himmelsfurst mine, Freiberg dist., Saxony, Germany.

FIGURE 5.124 Germanite, $Cu_{13}Fe_2Ge_2S_{16}$, 5 cm purplish, metallic, heavy mass. Tsumeb mine, Tsumeb, Namibia.

Germanium tetrachloride is either hydrolyzed to the oxide (GeO_2) or purified by fractional distillation and then hydrolyzed. The highly pure GeO_2 is now suitable to produce germanium glass. It is reduced to its element by reacting it with hydrogen, producing germanium suitable for infrared optics and semiconductor production.

$$GeO_2 + 2H_2 \rightarrow Ge + 2H_2O$$

The germanium used in steel production and other industrial processes is normally reduced using carbon.

$$GeO_2 + C \rightarrow Ge + CO_2$$

Thirty-two different minerals are known to contain germanium in their crystal structure. The sulfide class contains 16 different minerals, for example, argyrodite (Ag_8GeS_6) (Fig. 5.123), germanite ($Cu_{13}Fe_2Ge_2S_{16}$) (Fig. 5.124), and renierite ($(Cu^{1+},Zn)_{11}Fe_4(Ge^{4+},As^{5+})_2S_{16}$) (Fig. 5.125). Five different oxide class minerals are known, for example, argutite (GeO_2), eyselite ($Fe^{3+}Ge_3^{4+}O_7(OH)$), and stottite ($Fe^{2+}[Ge^{4+}(OH)_6]$) (Fig. 5.126). The sulfate class contains 4 different minerals with Ge, carraraite ($Ca_3(SO_4)[Ge(OH)_6](CO_3)\cdot 12H_2O$), fleischerite ($Pb_3Ge(SO_4)_2(OH)_6\cdot 3H_2O$), itoite ($Pb_3Ge^{4+}(SO_4)_2O_2(OH)_2$), and schaurteite ($Ca_3Ge(SO_4)_2(OH)_6\cdot 4H_2O$). A single phosphate is known to possibly contain Ge, galloplumbogummite ($Pb(Ga,Al,Ge)_3(PO_4)_2(OH)_6$). Six minerals are found in the silicate class, which also includes

FIGURE 5.125 A 7 cm specimen of purplish rose bronze renierite, $(Cu^{1+},Zn)_{11}Fe_4(Ge^{4+},As^{5+})_2S_{16}$. Prince Leopold, Kipushi mine, Katanga, Democratic Republic of Congo.

FIGURE 5.126 Stottite, $Fe^{2+}[Ge^{4+}(OH)_6]$, Glassy, perfect orange crystal to about 1 mm in a 1 cm vug with abundant dark orange-red needles of ludlockite, $PbFe_4^{3+}As_{10}^{3+}O_{22}$. Tsumeb mine, Tsumeb, Namibia.

the germanates, for example, brunogeierite ($Ge^{4+}Fe_2^{2+}O_4$), carboirite ($FeAl_2(GeO_4)O(OH)_2$), and krieselite ($(Al,Ga)_2(GeO_4)(OH)_2$).

5.14.3 Chemistry

Ge exists in 5 natural isotopes: ^{70}Ge, ^{72}Ge, ^{73}Ge, ^{74}Ge, and ^{76}Ge. Of these isotopes, ^{76}Ge is very slightly radioactive, decaying by double β decay with a half-life of 1.78×10^{21} years. ^{74}Ge is the most abundant isotope, with a natural abundance of 36.7%. ^{76}Ge is the rarest isotope with a natural abundance of 7.75%. Under bombardment with α particles, ^{72}Ge will produce stable ^{77}Se, releasing high-energy electrons in the process. Due to this, it is applied together with radon for nuclear batteries. In addition, at least 27 radioisotopes have been synthesized, ranging from ^{58}Ge to ^{89}Ge. The most stable of these radioisotopes is ^{68}Ge, which decays via electron capture with a half-life of 270.95 days. The least stable radioisotope is ^{60}Ge, which has only a half-life of 30 milliseconds. Although a majority of the Ge radioisotopes decay by β decay, ^{61}Ge and ^{64}Ge decay by β^+ delayed proton emission. ^{84}Ge through ^{87}Ge isotopes also show minor β^- delayed neutron emission decay paths.

Germanium is a lustrous, hard-brittle, grayish-white metalloid in the carbon group which is chemically similar to its group neighbors, silicon and tin. Its atom electron configuration is $[Ar]3d^{10}4s^24p^2$. It occurs mostly in the oxidation state +4 although many +2 compounds also exist. Other oxidation states are rare: +3 is found in compounds such as Ge_2Cl_6, and +3 and +1 are found on the surface of oxides, or negative oxidation states such as −4 in magnesium germanide, Mg_2Ge (Table 5.17).

5.14.3.1 Reaction with air, oxygen, nitrogen, and hydrogen

Germanium is a little more reactive than silicon, which is immediately above it in the periodic table. It starts to oxidize slowly in air at around 250°C, forming GeO_2. At red heat, germanium reacts with oxygen in air to from germanium dioxide.

$$Ge(s) + O_2(g) \rightarrow GeO_2(s)$$

Germanium metal can be made to react with hydrogen atoms generated in a high frequency plasma source to produce germane, GeH_4, and digermane, Ge_2H_6. However, for large-scale synthesis, germane is typically prepared by reduction of germanium compounds such as germanium dioxide, with hydride reagents. Commonly used hydride reagents for this purpose include sodium borohydride, potassium borohydride, lithium borohydride, lithium aluminum hydride, and sodium aluminum hydride. In the laboratory, germane can be generated by the reaction of Na_2GeO_3 or other Ge(IV) compounds with the aforementioned hydride reagents as shown below.

$$Na_2GeO_3 + NaBH_4 + H_2O \rightarrow GeH_4 + 2NaOH + NaBO_2$$

Germanium(IV) nitride, Ge_3N_4, is known but is typically produced through the reaction of germanium and ammonia rather than from the elements.

$$3Ge + 4NH_3 \rightarrow Ge_3N_4 + 6H_2$$

TABLE 5.17 Germanium properties.

Appearance	Grayish-white
Standard atomic weight $A_{r,std}$	72.631
Block	p-Block
Element category	Metalloid
Electron configuration	[Ar] $3d^{10}$ $4s^2$ $4p^2$
Phase at STP	Solid
Melting point	938.25°C
Boiling point	2833°C
Density (near r.t.)	5.323 g/cm^3
When liquid (at m.p.)	5.60 g/cm^3
Heat of fusion	36.94 kJ/mol
Heat of vaporization	334 kJ/mol
Molar heat capacity	23.222 J/(mol·K)
Oxidation states	**−4** −3, −2, −1, 0, +1, **+2**, +3, **+4** (amphoteric oxide)
Ionization energies	1st: 762 kJ/mol
	2nd: 1537.5 kJ/mol
	3rd: 3302.1 kJ/mol
Atomic radius	empirical: 122 pm
Covalent radius	122 pm
Van der Waals radius	211 pm

STP, Standard temperature and pressure.
Bold font indicates main oxidation state.

5.14.3.2 Reaction with halogens

Germanium tetrahalides, GeX_4, are readily prepared by direct action of the elements. They can also be prepared by the reaction of HX on GeO_2. Thus germanium tetrafluoride is formed by treating germanium with fluorine or by treating GeO_2 with hydrofluoric acid, HF.

$$Ge + 2F_2 \rightarrow GeF_4$$
$$GeO_2 + 4HF \rightarrow GeF_4 + 2H_2O$$

Similarly, $GeCl_4$, is prepared by heating germanium metal with chlorine or by the dissolution of GeO_2 with hydrochloric acid. $GeCl_4$ is especially important as it used in the production of organogermanium compounds.

$$Ge + 2Cl_2 \rightarrow GeCl_4$$
$$GeO_2 + 4HCl \rightarrow GeCl_4 + 2H_2O$$

Germanium tetrabromide, $GeBr_4$, can be obtained by reaction of germanium with bromine at 200°C or by the reaction of with hydrobromic acid at 180°C.

$$Ge + 2Br_2 \rightarrow GeBr_4$$
$$GeO_2 + 4HBr \rightarrow GeBr_4 + 2H_2O$$

Germanium tetraiodide is prepared from the reaction between germanium and iodine at temperatures above 220°C or by boiling GeO_2 with hydriodic acid.

$$Ge + 2I_2 \rightarrow GeI_4$$
$$GeO_2 + 4HI \rightarrow GeI_4 + 2H_2O$$

The germanium dihalides, GeX_2, are known and have been prepared by the reaction of elemental germanium with germanium tetrahalide. Germanium diiodide had been prepared by the action of hypophosphorus acid in the presence of hydriodic acid and water on germanium tetraiodide.

$$GeX_4 + Ge \rightarrow 2GeX_2 \ (X = F, Cl, Br)$$
$$GeI_4 + H_2O + H_3PO_2 \rightarrow GeI_2 + H_3PO_3 + 2HI$$

5.14.3.3 Reaction with acids and bases

Germanium does not dissolve in dilute acids and alkalis but dissolves slowly in hot concentrated sulfuric, nitric acids and aqua regia (mixture of nitric and hydrochloric acids). It reacts violently with molten alkalis such as sodium or potassium hydroxide to produce germanates ($[GeO_3]^{2-}$).

5.14.3.4 Organogermanium compounds

Germanium forms complexes with numerous elements such as carbon, oxygen, nitrogen, hydrogen, and phosphorous in several organic compounds. Interestingly, these reactions occur at ambient temperature usually in tetrahydrofuran under vacuum. The first organogermanium compound, tetraethylgermane ($Ge(C_2H_5)$, was synthesized by Winkler in 1887 through the reaction of germanium tetrachloride with diethylzinc. Organic germanium hydrides such as isobutylgermane ($(CH_3)_2CHCH_2GeH_3$) were found to be less hazardous and may be used as a liquid substitute for toxic germane gas in semiconductor applications. The synthesis of organogermanes have been reported in literature but their chemistry is less explored compared to their silicon and tin counterparts as they are more expensive. A review by Yorimitsu and Oshima in 2005 presents the chemistry of the readily prepared, easy to handle and storable organogermane—tri(2-furyl)germane (Fu_3GeH) along with others. This review posited that the 2-furyl group serves as a more electron withdrawing group than alkyl and phenyl groups. This is reflected by its ability to easily add via nucleophilic addition to carbonyl compounds. The authors believed that organogermanium opens ample possibilities for organic synthesis. The work of Keess and Oestrich in 2017 shows the emergence of organogermanium compounds, particularly the cyclohexa-2,5-dien-1-yl-substituted germanes, as state-of-the-art reagents for entry into the mild preparation of fully alkylated germanes.

5.14.4 Major uses

The significant properties of germania (GeO_2) are its high index of refraction and its low optical dispersion, which make it particularly valuable for wide-angle camera lenses, microscopy, and the core part of optical fibers. It has superseded titania as the dopant for silica fiber, removing the later heat treatment that used to make the fibers brittle. GeSbTe is a phase-change material utilized for its optic properties, for example, used in rewritable DVDs. Since Ge is transparent in the infrared wavelengths, it is a valuable infrared optical material that can be easily cut and polished into lenses and windows. It is specifically used as the front optic in thermal imaging cameras working in the 8 to 14-micron range for passive thermal imaging and for hot-spot detection in military, mobile night vision, and firefighting applications. It is used in infrared spectroscopes and other optical equipment that necessitate particularly sensitive infrared detectors. Since it has a very high refractive index (4.0) it has to be coated with antireflection agents. Specifically, a very hard special antireflection coating of diamond-like carbon, refractive index 2.0, forms a good match and produces a diamond-hard surface that is resistant against much environmental abuse.

SiGe alloys are fast developing into a valuable semiconductor material for high-speed integrated circuits. Circuits exploiting the properties of Si–SiGe junctions can be much faster than those using only Si. SiGe is starting to substitute for GaAs in wireless communications devices. The SiGe chips, with high-speed properties, can be produced with relatively cheap, well-known production methods of the Si chip industry. Solar panels form a main application of Ge. Ge forms the substrate of the wafers for high efficiency multijunction photovoltaic cells used for space applications. High-brightness LEDs, utilized for automobile headlights and to backlight LCD screens, are also an important application. Since Ge and GaAs have comparable lattice constants, Ge substrates can be utilized to produce GaAs solar cells, for example, the Mars Exploration Rovers and several satellites use triple-junction GaAs on Ge cells. Ge-on-insulator (GeOI) substrates are considered as a possible substitute for Si on miniaturized chips. Complementary metal-oxide semiconductor circuit based on GeOI substrates has recently been described. Other applications in electronics are, for example, phosphors in fluorescent lights and solid-state LEDs. Ge transistors are even now used in some effects pedals by musicians who want to replicate the unique tonal character of the "fuzz"-tone from the early rock and roll years.

GeO_2 is also used in catalysts for polymerization in the production of PET (IUPAC name poly(ethylbenzene-1,4-dicarboxylate), $(C_{10}H_8O_4)_n$). Due to the similarity between silica (SiO_2) and GeO_2, the SiO_2 stationary phase in some gas chromatography columns can be substituted by GeO_2. Recently, Ge has seen growing usage in precious metal alloys. In sterling silver alloys, for example, it reduces firescale (a layer of oxides that is visible on the surface of objects made of metal alloys containing Cu when the object is heated, as by a jeweler heating a ring to apply solder during a repair. On Cu-containing alloys of gold or of silver (such as sterling silver), it presents as a red or purple stain. This is because at high temperatures, oxygen mixes with the copper to form cuprous oxide and then cupric oxide, both of which disrupt the bright polished surface of the finished piece.), increases tarnish resistance, and improves precipitation hardening. A tarnish-proof silver alloy trademarked Argentium has 1.2% Ge. Semiconductor detectors produced from single-crystal high-purity Ge can accurately detect radiation sources, for example, in airport security. Ge is suitable for monochromators for beamlines utilized in single-crystal neutron scattering and synchrotron X-ray diffraction. Its reflectivity has advantages over Si in neutron and high-energy X-ray applications. High-purity Ge crystals are used in detectors for γ spectroscopy and the pursuit of dark matter. Furthermore, Ge crystals are utilized in X-ray spectrometers for the determination of P, Cl and S. Ge is developing into an important element for spintronics and spin-based quantum computing applications. Back in 2010, researchers showed room temperature spin transport and more recently donor electron spins in Ge has been proven to have very long coherence times.

5.15 33 As — Arsenic

5.15.1 Discovery

The word arsenic is derived from the Syriac word (al) zarniqa, from the Persian word زرنيخ zarnikh, meaning "yellow" (literally "gold-colored") and later "(yellow) orpiment." It was adopted into Greek as arsenikon (αρσενικόν), a form that is traditional etymology, being the neuter form of the Greek word arsenikos (αρσενικός), for "male," "virile." The Greek word was adopted in Latin as arsenicum, which in French turned into arsenic, from which the English word arsenic is taken. Arsenic sulfides (orpiment, As_2S_3, realgar, As_4S_4) and oxides have been known and used since historical times. Greek-Egyptian alchemist and Gnostic mystic Zosimos (CE c.300) gave a detailed account of roasting sandarach (realgar) to get cloud of arsenic (arsenic trioxide), which he subsequently reduced to gray arsenic. Since the symptoms of arsenic poisoning are not very specific, it was regularly used for murder until the arrival of the Marsh test (developed by the British chemist James Marsh, September 2, 1794 to June 21, 1846, and first published in 1836), a sensitive chemical test for its presence (Marsh, 1836). (An alternative less sensitive but more universal test is the Reinsch test, an initial indicator to detect the presence of antimony, arsenic, bismuth, selenium, thallium and mercury in a biological sample, discovered by Hugo Reinsch (1809–1884) in 1841). Due to its use by the ruling class to murder each other and its effectiveness and discreteness, arsenic has been known as the "poison of kings" and the "king of poisons." In the Bronze Age, arsenic was often added to bronze making the alloy harder (so-called "arsenical bronze"). German Catholic Dominican friar and bishop Albertus Magnus (also known as Saint Albert the Great and Albert of Cologne, c.1193 to November 15, 1280), known during his lifetime as *Doctor universalis* and *Doctor expertus*, is thought to have been the first to obtain the element from a compound in 1250, by heating soap in combination with arsenic trisulfide (Fig. 5.127). In 1649 German physician and pharmacologist Johann Schröder (1600–1664) published two ways of preparing arsenic (Fig. 5.128). He was the first person to recognize that arsenic was an element. Crystals of elemental (native) arsenic are found in nature, although rare. Cadet's fuming liquid (red-brown oily liquid consisting mostly of dicacodyl $(((CH_3)_2As)_2)$ and cacodyl oxide $(((CH_3)_2As)_2O)$.), often claimed as the first synthetic organometallic compound, was synthesized in 1760 by French chemist Louis Claude Cadet de Gassicourt (July 24, 1731 to October 17, 1799) by the reaction of potassium acetate with arsenic trioxide (Cadet de Gassicourt, 1760). In the Victorian era (1837–1901), "arsenic" ("white arsenic" or arsenic trioxide) was mixed with vinegar and chalk and eaten by women to

FIGURE 5.127 Albertus Magnus. 1352, Fresco by Tommaso da Modena (1326–1379) Sala del Capitolo (Seminario di Treviso).

FIGURE 5.128 Johann Schröder, 1641.

improve the skin tone of their faces, making their skin look paler to prove they did not work outside in the fields. Arsenic was also rubbed into the faces and arms of women to "improve their complexion." The unintentional use of arsenic in the adulteration of foodstuffs led to the Bradford sweet poisoning in 1858, which resulted in 200 cases of arsenic poisoning and around 20 deaths. Wallpaper manufacture also began to include dyes made from arsenic, which was supposed to increase the pigment's brightness. Two arsenic pigments have been widely used since their discovery—emerald green Paris Green ((copper(II) acetate triarsenite or copper(II) acetoarsenite), $Cu(C_2H_3O_2)_2{\cdot}3Cu(AsO_2)_2$) and yellowish-green Scheele's Green (cupric hydrogen arsenite, $CuHAsO_3$). Once the toxicity of arsenic became commonly known, these compounds were less frequently used as pigments and more often as insecticides. In the 1860s an arsenic by-product of dye production, London Purple was widely used. This was a solid mixture of arsenic trioxide, aniline, lime, and ferrous oxide, insoluble in water and very poisonous by inhalation or ingestion but it was later replaced with Paris Green, another arsenic-based dye. With better knowledge of the toxicology mechanism, two other compounds were used beginning in the 1890s. Arsenite of lime ($Ca_3(AsO_4)_2$) and arsenate of lead ($PbHAsO_4$) were used widely as insecticides until the discovery of DDT (dichlorodiphenyltrichloroethane, preferred IUPAC name 1,1′-(2,2,2-trichloroethane-1,1-diyl)bis(4-chlorobenzene), $C_{14}H_9Cl_5$) in 1942.

5.15.2 Mining, production, and major minerals

Arsenic comprises about 1.5 ppm of the Earth's crust and is the 53rd most abundant element. Minerals with the general formula MAsS and MAs_2 (M = Fe, Ni, Co) are the dominant commercial sources of arsenic, together with realgar (an arsenic sulfide mineral, As_4S_4) and native (elemental) arsenic (Box 5.18). An illustrative mineral is arsenopyrite (FeAsS), which is structurally related to pyrite (FeS_2). Many minor As-containing minerals are known. In addition, arsenic can be found in a variety of organic forms in the environment. China used to be the main producer of white arsenic with almost 70% world share, followed by Morocco, Russia, and Belgium. Most arsenic refinement operations in the US and Europe have been closed down due to environmental concerns. Arsenic is found in the smelter dust from copper, gold, and lead smelters, and is recovered primarily from copper refinement dust. Roasting arsenopyrite in air results in arsenic sublimation as As(III) oxide leaving iron oxides, whereas roasting without air produces gray arsenic. Additional purification from sulfur and other chalcogens is realized by sublimation in vacuum, in a hydrogen atmosphere, or by distillation from molten lead-arsenic mixture.

Arsenic is a common element in many minerals. Currently more than 500 minerals are known to contain As. As a native element it forms two different minerals arsenic (As) (Fig. 5.129) and arsenolamprite (As). In addition, three alloys are known, paradocrasite (Sb_3As), pararsenolamprite ((As,Sb)), and stibarsen (AsSb). A large group of just over 150 minerals can be found under the sulfides, for example, arsenopyrite (FeAsS) (Fig. 5.130), cobaltite (CoAsS), gersdorffite (NiAsS), getchellite ($AsSbS_3$) (Fig. 5.131), nickeline (NiAs), orpiment (As_2S_3) (Fig. 5.132), proustite (Ag_3AsS_3), realgar (As_4S_4), and safflorite ($(Co,Ni,Fe)As_2$) (Fig. 5.133). Two different halide polymorphs are known to contain As, ecdemite ($Pb_6Cl_4(As_2O_7)$) and heliophyllite ($Pb_6Cl_4(As_2O_7)$). In the oxide class, which includes the arsenites, 40 minerals contain As, for example, arsenolite (As_2O_3), claudetite (As_2O_3), and gebhardite

BOX 5.18 Arsenic

Arsenic deposits are rare, even though it forms a number of significant sulfide minerals, including arsenopyrite, FeAsS, realgar, As_4S_4, orpiment, As_2S_3, enargite, Cu_3AsS_4, and tennantite, $Cu_6Cu_4(Fe^{2+},Zn)_2As_4S_{12}S$. It is remarkably common in nearly all types of hydrothermal gold deposits, indicating that it is dissolved and precipitated according to the same processes that control gold movement. Arsenic is seldom produced from these deposits, however, since gold can be recovered without treating the arsenic-containing minerals. Enargite and tennantite are commonly found in the upper part of some porphyry copper deposits, where they can form large veins that are mined for copper, such as at Butte, Montana, US. It has been proposed that vapors escaping from underlying magmas carried the arsenic into these hydrothermal systems. In addition, arsenic occurs in veins with cobalt, nickel, and silver in the Cobalt-type silver ores, and in some tin deposits, even though it is seldom recovered from these deposits. Due to its extensive occurrence with other metals, arsenic is recovered mainly as a by-product. It vaporizes at a temperature of only 615°C and consequently is released while roasting gold or based metal ores. While modern smelters capture most of this vaporized arsenic in scrubbers, older smelters did not, and as a result they are surrounded by zones rich in arsenic. For example, around the Sudbury nickel smelters in Ontario, Canada, arsenic concentrations of lake sediments deposited during the last few decades are a number of times higher than those from earlier times before smelting began. Due to its toxic properties, there are now strict limits on arsenic emissions from smelters, and few smelters will take ore that contains arsenic.

FIGURE 5.129 Crystalline mass of dull gray arsenic, As, 10 × 7 × 5 cm. Pribram, Bohemia, Czech Republic.

FIGURE 5.130 Arsenopyrite, FeAsS, crystals to 4 mm. Hidalgo del Parral, Chihuahua, Mexico.

($Pb_8(As_2O_5)_2OCl_6$). A single carbonate is known, claraite ($(Cu,Zn)_{15}(CO_3)_4(AsO_4)_2(SO_4)(OH)_{14}{\cdot}7H_2O$). Two borates contain As, cahnite ($Ca_2[B(OH)_4](AsO_4)$) and teruggite ($Ca_4Mg[AsO_4|B_6O_7(OH)_6]_2{\cdot}12H_2O$). The sulfates are represented by five minerals, such as fornacite ($Pb_2Cu(CrO_4)(AsO_4)(OH)$) and paraniite-(Y) ($Ca_2Y(AsO_4)(WO_4)_2$). By far the largest group of minerals, more than 300, can be found in the phosphate class, which includes also the arsenates. This group includes minerals such as: adamite ($Zn_2(AsO_4)(OH)$) (Fig. 5.134), annabergite ($Ni_3(AsO_4)_2{\cdot}8H_2O$) (Fig. 5.135), bayldonite ($PbCu_3(AsO_4)_2(OH)_2$), beudantite ($PbFe_3(AsO_4)(SO_4)(OH)_6$), clinoclase ($Cu_3(AsO_4)(OH)_3$), conichalcite ($CaCu(AsO_4)(OH)$), duftite ($PbCu(AsO_4)(OH)$), erythrite ($Co_3(AsO_4)_2{\cdot}8H_2O$) (Fig. 5.136), legrandite ($Zn_2(AsO_4)(OH){\cdot}H_2O$), mimetite ($Pb_5(AsO_4)_3Cl$), olivenite ($Cu_2(AsO_4)(OH)$) (Fig. 5.137), scorodite ($Fe^{3+}AsO_4{\cdot}2H_2O$) (Fig. 5.138), and zeunerite ($Cu(UO_2)_2(AsO_4)_2{\cdot}12H_2O$) (Fig. 5.139). The silicate class contains eight minerals with As, for example, nelenite ($(Mn,Fe)_{16}(Si_{12}O_{30})(OH)_{14}\left[As_3^{3+}O_6(OH)_3\right]$) and schallerite ($Mn_{16}^{2+}As_3Si_{12}O_{36}(OH)_{17}$).

5.15.3 Chemistry

Arsenic exists naturally as a monoisotopic element, composed of one stable isotope, ^{75}As. As of 2003, at least 33 radioisotopes have also been synthesized, ranging from ^{60}As to ^{92}As. The most stable of these is ^{73}As with a half-life of 80.30 days. The remaining isotopes have much shorter half-lives of less than one day, except for ^{71}As with a half-life of 65.30 hours, ^{72}As with a half-life of 26.0 hours, ^{74}As with a half-life of 17.77 days, ^{76}As with a half-life of 1.0942 days, and ^{77}As with a half-life of 38.83 hours. Isotopes that are lighter than the stable ^{75}As generally decay by β^+

FIGURE 5.131 Getchellite, $AsSbS_3$, bright red, foliated crystals to 6 mm. Getchell mine, Humboldt Co., Nevada, United States.

FIGURE 5.132 Orpiment, As_2S_3, deep orange crystals to over 1 cm in a small group. Twin Creeks mine, Humboldt Co., Nevada, United States.

FIGURE 5.133 A 7 mm thick tangled network of veinlet safflorite, $(Co,Ni,Fe)As_2$. Nieder-Beerbach, Odenwald, Hesse, Germany.

FIGURE 5.134 Adamite, $Zn_2(AsO_4)(OH)$, divergent fan to 1 cm. Mina Ojuela, Mapimi, Durango, Mexico.

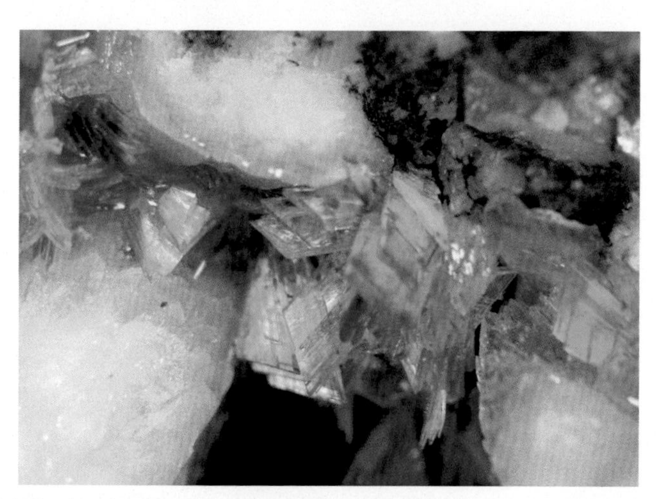

FIGURE 5.135 Annabergite, $Ni_3(AsO_4)_2{\cdot}8H_2O$, green crystals to 4 mm. Laurium, Attika, Greece.

decay, while those that are heavier than ^{75}As generally decay by β^- decay, though there are some exceptions. No less than 10 nuclear isomers have been characterized, ranging in atomic mass from 66 to 84. The most stable of these isomers is ^{68m}As with a half-life of 111 seconds.

Arsenic is a metalloid occurring in 3 allotropic forms and has a ground state electronic configuration of $[Ar]3d^{10}4s^24p^3$ with an unpaired electron in each of the three p orbitals. Much of the chemistry of arsenic can be interpreted directly on this basis. There are three crystalline forms of As, of which the ordinary, gray, semimetal, rhombohedral, α-form is the most stable at room temperature. Gray As becomes a semiconductor with a bandgap of 1.2–1.4 eV if made amorphous. Yellow arsenic forms by rapid cooling of As vapor and has a soft and waxy, somewhat like tetraphosphorus (P_4) appearance. This allotrope is unstable. It is molecular and is the most volatile, least dense, and most toxic. It is rapidly transformed into gray arsenic by light. The third allotrope is called black arsenic, which is similar in structure to black phosphorus. Black arsenic forms by cooling As vapor at around 100°C–220°C. It is glassy and brittle. It is also a poor electrical conductor (Table 5.18).

FIGURE 5.136 Magenta, sharp crystals of erythrite, $Co_3(AsO_4)_2 \cdot 8H_2O$, from 1 to 1.5 cm. Bou Azzer dist., Anti-Atlas, Morocco.

FIGURE 5.137 Olivenite, $Cu_2(AsO_4)(OH)$, divergent bladed crystals to 2 mm with typical olive-green color. Majuba Hill mine, Pershing Co., Nevada, United States.

FIGURE 5.138 A 7.5 cm scorodite, $Fe^{3+}AsO_4 \cdot 2H_2O$, cluster in matrix with a rich blue color. The crystals are to 1−2 mm throughout the large 4 cm vug. Trincheras, Sonora, Mexico.

FIGURE 5.139 Zeunerite, $Cu(UO_2)_2(AsO_4)_2 \cdot 12H_2O$, 4 mm crystal. Majuba Hill mine, Pershing Co., Nevada, United States.

TABLE 5.18 Arsenic properties.

Appearance	Metallic gray
Allotropes	Gray (most common), yellow, black
Standard atomic weight	74.922
Block	p-Block
Element category	Metalloid
Electron configuration	[Ar] $3d^{10}$ $4s^2$ $4p^3$
Phase at STP	Solid
Sublimation point	615°C
Density (near r.t.)	5.727 g/cm^3
When liquid (at m.p.)	5.22 g/cm^3
Heat of fusion	Gray: 24.44 kJ/mol
Heat of vaporization	34.76 kJ/mol (?)
Molar heat capacity	24.64 J/(mol·K)
Oxidation states	**−3**, −2, −1, +1, +2, **+3**, +4, **+5**
Ionization energies	1st: 947.0 kJ/mol
	2nd: 1798 kJ/mol
	3rd: 2735 kJ/mol
Atomic radius	Empirical: 119 pm
Covalent radius	119 ± 4 pm
Van der Waals radius	185 pm

STP, Standard temperature and pressure.
Bold font indicates main oxidation state.

Arsenic occurs in many minerals, usually in combination with sulfur and metals, but also as a pure elemental crystal. Despite the metalloidal character of the free element, the ionization energies and electronegativity of As are similar to those of phosphorus. Thus the element readily forms strong covalent bonds with most nonmetals. For instance, AsX_3 (X = H, hal, R, Ar etc.) are covalent molecules like PX_3. The structures of its compounds such as As_4O_6 and As_4O_{10} also resemble their P analogs in structure. The sulfides are covalent heterocyclic molecules though their stoichiometry and structure differ from those of P. Arsenic reacts with metals to form arsenides, though these are not ionic compounds containing the As^{3-} ion as the formation of such an anion would be highly endothermic. Even the group 1 arsenides such as trisodium arsenide (Na_3As) are not ionic but rather exhibit properties of intermetallic compounds. Arsenic is much less stable in the group oxidation state of +5 than its vertical neighbors, phosphorus and antimony. Hence, compounds like arsenic pentoxide and arsenic acid are potent oxidizers.

5.15.3.1 Reaction in air, oxygen, nitrogen, and hydrogen

Arsenic is stable in dry air but tarnishes upon exposure to humidity forming a golden-bronze surface layer that eventually becomes black. When heated in air, it burns to form arsenic trioxide (As_2O_3) and arsenic pentoxide (As_2O_5), releasing fumes with an odor resembling garlic. The simplest hydride of arsenic commonly known as arsine (AsH_3, IUPAC name: arsane) had been synthesized and characterized. However, its preparation is through reduction of As-containing compounds by nascent hydrogen. High yielding methods of preparation includes the low-temperature reduction of aluminum trichloride, $AsCl_3$, with lithium aluminum hydride, $LiAlH_4$, in diether solution. The dilute acid hydrolysis of arsenides of electropositive elements (e.g., Na_3As, Mg_3As_2, Zn_3As_2) also gives good yields of arsine gas.

$$4AsCl_3 + 3LiAlH_4 \rightarrow 4AsH_3 + 3LiCl + 3AlCl_3$$
$$Na_3As + 3H^+ \rightarrow AsH_3 + 3Na^+$$

5.15.3.2 Reaction with halogens

Arsenic reacts with fluorine, F_2, to form the arsenic pentafluoride gas. Under controlled conditions, it reacts with the halogens chlorine (Cl_2), bromine, Br_2, and iodine, I_2, to form the respective trihalides arsenic (III) chloride ($AsCl_3$), arsenic (III) bromide, $AsBr_3$, and arsenic (III) iodide, AsI_3.

$$2As(s) + 5F_2(g) \rightarrow 2AsF_5(g)$$
$$2As(s) + 3X_2(g) \rightarrow 2AsX_3(X = Cl, Br, I)$$

5.15.3.3 Reaction with acids and bases

Arsenic reacts with concentrated nitric acid to form arsenic acid, H_3AsO_4 and with dilute nitric acid to form arsenous acid, H_3AsO_3. It is oxidized by concentrated sulfuric acid to form arsenic trioxide, As_4O_6. However, it does not react with water, alkalis, or nonoxidizing acids. It does form the arsenite and liberate H_2 when reacted with molten sodium hydroxide.

$$2As + 6NaOH \rightarrow 2Na_3AsO_3 + 3H_2$$

5.15.3.4 Organoarsenic compounds

Organoarsenic compounds contain a chemical bond between arsenic and carbon. Arsenic in organometallic compounds are normally found in the +3 and the +5 state although the +1 state is also found in some compounds. The first organoarsenic compound was probably the foul-smelling cacodyl obtained by Louis Claude Cadet de Glassicourt (July 24, 1731 to October 17, 1799) in 1760. The compound was the product of the reaction between arsenic trioxide and potassium acetate and was only established as $Me_2AsAsMe_2$ in the mid-19th century. Organoarsenics are widely used in agriculture and plant protection agents. An organoarsenic compound called asphenamine (also known as Salvarsan) played an important role in medicine as the original chemotherapeutic treatment for syphilis. It is believed to have the structure containing rings with 3 or five arsenic atoms as shown below for the 3-membered ring.

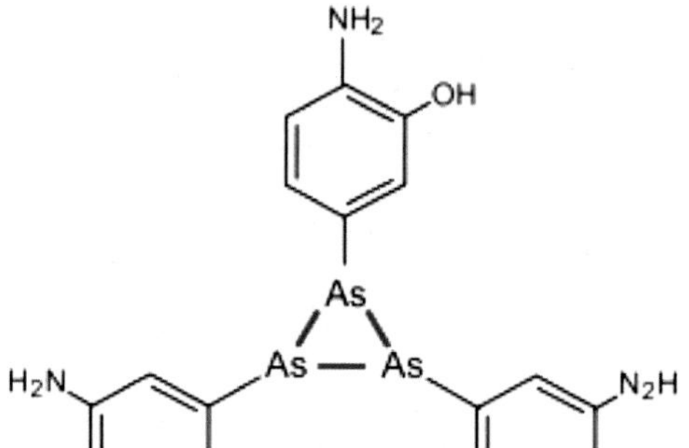

The use of organoarsenic compounds are now limited due to their toxicity. Handling them is always going to be potentially hazardous, requiring utmost care and stringent safety procedures. Nevertheless, several works are reinvigorating the field as presented in the review of Imoto and Naka in 2019. The key in these works is in the use of nonvolatile intermediate transformation technique to safely access functional organoarsenic compounds. Notable examples are cyclooligoarsines such as hexaphenylcyclohexaarsine. Cleavage of the As-As in these cyclooligoarsines generate reactive species (e.g., radicals, electrophiles and nucleophiles) that can then lead to the production of a variety of functional organoarsenic compounds. Imoto and Naka also reported the in situ generation of diiodoarsines. This in situ generation of the diiodoarsines is an effective method to perform subsequent substitution reactions without the need for isolation.

A series of arsole derivatives were prepared by this method and characterized for the first time. Interestingly, characterization of these arsole derivatives showed they have superior oxidative tolerance compared to that of the phosphorus analogs. This led to renewed interests in organoarsenic compounds and the expansion of the diversity of frameworks where arsenic atoms are incorporated.

5.15.4 Major uses

The As toxicity to insects, bacteria, and fungi led to its application as a wood preservative. In the 1930s a process of treating wood with CCA (also known as Tanalith. In the treated wood, As is believed to be in the form of Cr(III)arsenate $CrAsO_4$ and/or Cu(II)arsenate $Cu_3(AsO_4)_2$, or fairly stable Cr dimer-arsenic clusters) was invented, and for many years, this treatment was the most widespread industrial use of As. An increased awareness of As toxicity resulted in a CCA ban in consumer products in 2004, instigated by the European Union and United States. However, CCA is still heavily used elsewhere (e.g., on Malaysian rubber plantations). As was also used in several agricultural insecticides and poisons, for example, lead hydrogen arsenate ($PbHAsO_4$) was a widespread insecticide on fruit trees but exposure to it occasionally resulted in brain damage of those working the sprayers. In the second half of the 20th century, monosodium methyl arsenate (MSMA, IUPAC name sodium hydrogen methylarsonate, CH_4AsNaO_3) and disodium methyl arsenate (DSMA, IUPAC name disodium methyl-dioxido-oxoarsorane, $CH_3AsNa_2O_3$) —less toxic organic forms of As—replaced Pb arsenate in agriculture. These organic arsenicals were subsequently phased out by 2013 in all agricultural activities apart from cotton farming. As biogeochemistry is intricate and involves numerous adsorption and desorption processes. As toxicity is linked to its solubility and is affected by pH. Arsenite (AsO_3^{3-}) is more soluble than arsenate (AsO_4^{3-}) and is more toxic; but at a lower pH (more acidic), arsenate becomes more mobile and toxic. It was discovered that adding S, P, and Fe oxides to high-arsenite soils significantly lowers As phytotoxicity. As is used as a feed additive in poultry and swine production, especially in the United States to increase weight gain, improve feed efficiency, and to prevent disease. An example is roxarsone ((4-hydroxy-3-nitrophenyl)arsonic acid, $C_6AsNH_6O_6$), which had been used as a broiler starter by approximately 70% of US broiler growers. The Poison-Free Poultry Act of 2009 recommended to ban the use of roxarsone in industrial swine and poultry production. Alpharma, a subsidiary of Pfizer Inc., which produced roxarsone, voluntarily stopped sales of the drug in reaction to research showing elevated levels of inorganic As, a carcinogen, in treated chickens. A successor to Alpharma, Zoetis, still sells nitarsone ((4-nitrophenyl)arsonic acid, $C_6H_6AsNO_5$), primarily for use in turkeys. Organic As compounds show lower toxicity than pure As and promote the growth of chickens. Under some conditions, however, the As in chicken feed is transformed to the toxic inorganic form. A 2006 study of the remains of the well-known Australian racehorse, Phar Lap, showed that the 1932 death of the famous champion was the result of a huge overdose of As. Sydney veterinarian Percy Sykes indicated, “In those days, As was quite a common tonic, usually given in the form of a solution (Fowler’s Solution, a solution containing 1% potassium arsenite, $KAsO_2$) ... It was so common that I’d reckon 90% of the horses had arsenic in their system.”

During the 18th, 19th, and 20th centuries, several As compound were used as medicines, including arsphenamine, also known as Salvarsan (Salvarsan has long been assumed to have an As = As double bond, akin to the N=N linkage in azobenzene. However, in 2005, in an extensive mass spectrometric analysis Salvarsan was shown to have As-As single bonds, not As = As double bonds. The drug was also found to be a mixture consisting of cyclo-As_3 and cyclo-As_5 species) or compound 606, (by German-Jewish physician Paul Ehrlich, March 14, 1854 to August 20, 1915) and As trioxide (As_2O_3, by British pharmacist and physician Thomas Fowler, 1736–1801). Arsphenamine and neosalvarsan were prescribed for syphilis (because it is toxic to the bacterium *Treponema pallidum*, a spirochete that causes syphilis) and trypanosomiasis (name of several diseases in vertebrates caused by parasitic protozoan trypanosomes of the genus *Trypanosoma*. In humans this includes African trypanosomiasis and Chagas disease) but has since been replaced by modern antibiotics. As trioxide it has been used in a variety of ways over the past 500 years, most commonly in the treatment of cancer, but in medications as diverse as Fowler’s solution for psoriasis. The USFDA in 2000 approved this compound for the treatment of patients with acute promyelocytic leukemia that is resistant to all-trans retinoic acid. Lately, researchers have been locating tumors using ^{74}As (a positron emitter). This isotope provides clearer PET scan images than the preceding radioactive agent, ^{124}I, as the body tends to transport iodine to the thyroid gland producing signal noise.

The principal use of As is in alloying with Pb. Pb components in car batteries are strengthened by the presence of a very small amount of As. Dezincification of brass (a Cu–Zn alloy) is significantly reduced by adding some As. “Phosphorus Deoxidized Arsenical Copper” with an As content of 0.3% has an enhanced corrosion stability in specific environments. GaAs (see Section 5.13) is a valuable semiconductor material, used in integrated circuits. Circuits made from GaAs are much faster (but also far more costly) than those made from Si. Different to Si, GaAs has a direct bandgap, and can be used in laser diodes and LEDs to convert electrical energy straight into light. As compounds are used in making special glass types. Arsine gas (AsH_3) has become an essential dopant gas in the microchip industry, though this necessitates rigorous guidelines concerning its use as it is extremely toxic. The dopant is added in trace amounts so that a few As atoms become incorporated into the microchip and it is in those which define the degree of semiconductivity.

5.16 34 Se — Selenium

5.16.1 Discovery

Selenium (Greek σελήνη selene meaning "Moon") was found in 1817 by Swedish chemists Jöns Jakob Berzelius (August 20, 1779 to August 7, 1848) (Fig. 5.140) and Johan Gottlieb Gahn (August 19, 1745 to December 8, 1818) (Berzelius, 1818). Both chemists possessed a chemistry plant near Gripsholm, Sweden, producing sulfuric acid (H_2SO_4) by the lead chamber process. Sulfur dioxide (SO_2) is introduced with steam and nitrogen dioxide (NO_2) into large chambers lined with sheet lead where the gases are sprayed down with water and chamber acid (62%–70% sulfuric acid). The sulfur dioxide and nitrogen dioxide dissolve and over a period of approximately 30 minutes the sulfur dioxide is oxidized to sulfuric acid. The presence of nitrogen dioxide is essential for the reaction to progress at an acceptable rate. The process is highly exothermic (heat producing), and a key factor in the design of the chambers was to provide a way to dissipate the heat formed in the reactions. The pyrite (FeS_2) from the Falun mine produced a red precipitate in the lead chambers which was thought to be an arsenic compound, so the use of pyrite to make sulfuric acid was stopped. Berzelius and Gahn sought to use the pyrite and they detected that the red precipitate produced a smell like horseradish when burned. This smell was not characteristic of arsenic, but a comparable smell was known from tellurium compounds. Therefore, Berzelius's first letter to Alexander John Gaspard Marcet (1770–1822), a Genevan-born physician who became a British citizen in 1800, said that this was a tellurium compound. Yet, the absence of tellurium compounds in the Falun mine minerals ultimately led Berzelius to reanalyze the red precipitate, and in 1818 he sent a second letter to Marcet describing a newly found element similar to sulfur and tellurium. Due to its similarity to tellurium, named for the Earth, Berzelius named the new element after the Moon. In 1873 English electrical engineer Willoughby Smith (April 6, 1828 to July 17, 1891) observed that the electrical resistance of gray selenium was dependent on the ambient light (Smith 1873a,b) (Fig. 5.141). This resulted in its use as a cell for sensing light. The first commercial products containing selenium were developed by Werner Siemens in the mid-1870s. The selenium cell was used in the photophone (a telecommunications device that allows transmission of speech on a beam

FIGURE 5.140 Jöns Jakob Berzelius.

FIGURE 5.141 Willoughby Smith.

of light.) developed by Alexander Graham Bell and his assistant Charles Sumner Tainter on February 19, 1880, at Bell's laboratory at 1325L Street in Washington, DC (Bell, 1880). Selenium conducts an electric current proportional to the amount of light falling on its surface. This phenomenon was used in the development of light meters and comparable devices. Selenium's semiconductor properties resulted in many other applications in electronics. The development of selenium rectifiers started in the early 1930s, where they replaced copper oxide rectifiers as they were more efficient. These lasted in commercial applications until the 1970s, when they were substituted by cheaper and even more efficient silicon rectifiers. Selenium came to medical attention later due to its toxicity to industrial workers. Selenium was also documented as an important veterinary toxin, which can be observed in animals that have eaten high-selenium plants. In 1954 the first clues of specific biological functions of selenium were found in microorganisms by British-American microbiologist and biochemist Audrey Jane Gibson (née Pinsent; October 5, 1924 to June 10, 2008) (Pinsent, 1954). It was found to be critical for mammalian life in 1957. In the 1970s it was proven to be present in two independent sets of enzymes. This was followed by the detection of selenocysteine in proteins.

5.16.2 Mining, production, and major minerals

Native (i.e., elemental) selenium is a rare mineral, which normally does not occur in good crystals, but, when it does, they are steep rhombohedra or tiny acicular (hair-like) crystals. Separation of selenium is often difficult due to the presence of other compounds and elements. Selenium occurs naturally in a range of inorganic forms, including selenide, selenate, and selenite, but these minerals are rare. The common mineral variety selenite is not a selenium mineral at all, and contains no selenite ion, but is rather a variety of gypsum ($CaSO_4 2H_2O$) named like selenium for the moon well before the discovery of selenium. Selenium is most commonly found as an impurity, substituting a small part of the sulfur in sulfide ores of many metals. Selenium is most commonly produced from selenide in many sulfide ores, such as those of copper, nickel, or lead. Electrolytic metal refining is especially useful to produce selenium as a by-product, obtained from the anode mud of copper refineries. An alternative source has been the mud from the lead chambers of sulfuric acid plants, a method that is no longer in use. Selenium can be refined from these muds through several different methods. Nevertheless, most elemental selenium is produced as a by-product of refining copper or manufacturing sulfuric acid. Since its development, solvent extraction and electrowinning of copper provides a growing segment of the worldwide copper supply. This changes the availability of selenium as only a comparably small part of the selenium in the ore is leached with the copper. Industrial production of selenium generally comprises the extraction of selenium dioxide (SeO_2) from residues obtained during the purification of copper. Common production from the residue then starts with oxidation using sodium carbonate to produce selenium dioxide, which is subsequently mixed with water and

FIGURE 5.142 A 3.5 × 4 cm area of rich, gray, acicular crystals of native selenium, Se, on sandstone matrix. Ambrosia Lake, McKinley Co., New Mexico, United States.

FIGURE 5.143 Tarnished dark brownish gray berzelianite, Cu_2Se, about 2 cm. Habři mine, Rožná deposit, Vysočina, Moravia, Czech Republic.

acidified to form selenous acid, H_2SeO_3 (oxidation step). Finally, the selenous acid is bubbled with sulfur dioxide (reduction step) to give elemental selenium.

There are just over 100 minerals that contain selenium in their crystal structure. Though relatively rare, selenium (Se) does occur as a native element (Fig. 5.142). Se is a major element in more than 80 sulfides (which includes selenides), for example, antimonselite (Sb_2Se_3), berzelianite (Cu_2Se) (Fig. 5.143), clausthalite (PbSe) (Fig. 5.144), and penroseite ($(Ni,Co,Cu)Se_2$) (Fig. 5.145). Within the oxide class, which includes the selenites, there are 23 minerals with Se, such as ahlfeldite ($Ni(SeO_3)\cdot 2H_2O$) (Fig. 5.146), chalcomenite ($CuSeO_3\cdot 2H_2O$) (Fig. 5.147), cobaltomenite ($CoSeO_3\cdot 2H_2O$) (Fig. 5.148), and molybdomenite ($PbSeO_3$). The sulfates contain 6 different minerals, for example, carlosruizite ($K_6(Na,K)_4Na_6Mg_{10}(SeO_4)_{12}(IO_3)_{12}\cdot 12H_2O$), and olsacherite ($Pb_2(Se^{6+}O_4)(SO_4)$).

FIGURE 5.144 Silvery, metallic crystalline segregations of clausthalite, PbSe, in calcite, $CaCO_3$, about 2 cm. Possible pitchblende associated. Lašovice, Bohemia, Czech Republic.

FIGURE 5.145 Penroseite, $(Ni,Co,Cu)Se_2$, with clausthalite, PbSe, $4 \times 3 \times 3$ cm. Pacajake, Colquechaca, Potosi dept., Bolivia.

5.16.3 Chemistry

Selenium has 7 naturally occurring isotopes. Five of these, ^{74}Se, ^{76}Se, ^{77}Se, ^{78}Se, ^{80}Se, are stable, with ^{80}Se being the most abundant (49.8% natural abundance) followed by ^{78}Se at 23.69%, ^{76}Se at 9.23%, ^{77}Se at 7.60% and finally ^{74}Se at 0.86%. Also naturally occurring is the long-lived primordial radionuclide ^{82}Se (8.82% natural abundance), with a half-life of 9.2×10^{19} years (In geochemistry, geophysics and geonuclear physics, primordial nuclides, also known as primordial isotopes, are nuclides found on Earth that have existed in their current form since before Earth was formed. Primordial nuclides were present in the interstellar medium from which the solar system was formed, and were formed in, or after, the Big Bang, by nucleosynthesis in stars and supernovae followed by mass ejection, by cosmic ray spallation, and potentially from other processes. They are the stable nuclides plus the long-lived fraction of radionuclides surviving in the primordial solar nebula through planet accretion until the present. Only 286 such nuclides are known.). The nonprimordial radioisotope ^{79}Se also occurs in trace amounts in uranium ores as a product of nuclear fission. In addition, Se has a large number of unstable synthetic isotopes ranging from ^{64}Se to ^{95}Se; of which the most

FIGURE 5.146 Light pink, glassy crystals of cobaltoan ahlfeldite, $Ni(SeO_3)\cdot 2H_2O$, to 1 mm in vug of penroseite, $(Ni,Co,Cu)Se_2$, with bright blue chalcomenite, $CuSeO_3\cdot 2H_2O$. El Dragón mine, Quijarro, Potosí dept., Bolivia.

FIGURE 5.147 Chalcomenite, $CuSeO_3\cdot 2H_2O$, deep blue tabular micro crystals to 1 mm. Baccu Locci mine, Villaputzu, Cagliari, Sardinia, Italy.

stable isotopes observed are ^{75}Se with a half-life of 119.78 days and ^{72}Se with a half-life of 8.4 days. Isotopes lighter than the stable isotopes primarily undergo β^+ decay to isotopes of As, and isotopes heavier than the stable isotopes undergo β^- decay to isotopes of Br, with some minor neutron emission branches in the heaviest known isotopes.

Selenium is a reactive nonmetal, but it is also sometimes referred to as a metalloid. It belongs to group 16 or the chalcogens with properties that are intermediate between the elements above and below it in the periodic table, sulfur and tellurium, respectively. Its ground state electronic configuration is $[Ar]3d^{10}4s^24p^4$. Selenium compounds commonly exist in the oxidation states, -2, $+2$, $+4$, and $+6$. There are several allotropes of selenium but not as extensive as sulfur (Table 5.19). The red monoclinic crystals designated as α and β forms are composed of cyclic Se_8 molecules. The red amorphous form of selenium results from chemical reactions such as when selenious acid or one of its salts are treated with sulfur dioxide. A vitreous, black form is the result of rapid cooling of molten selenium. Gray Se results from mild heating then slow cooling of the other allotropes. It can also be formed by condensing Se vapor just below the

FIGURE 5.148 Cobaltomenite, $CoSeO_3{\cdot}2H_2O$, purple tiny crystal grains less than 0.5 mm. Parco mine, Grand Co., Utah, United States.

TABLE 5.19 Selenium properties.

Appearance	Black, red, and gray allotropes
Standard atomic weight $A_{r,std}$	78.972
Block	p-Block
Element category	Reactive nonmetal, sometimes considered a metalloid
Electron configuration	[Ar] $3d^{10}$ $4s^2$ $4p^4$
Phase at STP	Solid
Melting point	221°C
Boiling point	685°C
Density (near r.t.)	Gray: 4.81 g/cm^3 Alpha: 4.39 g/cm^3 Vitreous: 4.28 g/cm^3
When liquid (at m.p.)	3.99 g/cm^3
Heat of fusion	Gray: 6.69 kJ/mol
Heat of vaporization	95.48 kJ/mol
Molar heat capacity	25.363 J/(mol·K)
Oxidation states	**−2**, −1, +1, **+2**, +3, **+4**, +5, **+6**
Ionization energies	1st: 941.0 kJ/mol 2nd: 2045 kJ/mol 3rd: 2973.7 kJ/mol
Atomic radius	Empirical: 120 pm
Covalent radius	120 ± 4 pm
Van der Waals radius	190 pm

STP, Standard temperature and pressure.
Bold font indicates main oxidation state.

melting point. Gray Se is the most stable form of selenium. It has a hexagonal crystal lattice consisting of helical polymeric chains at a minimum distance of 343.6 pm between chain. The Se-Se distance is 237.3 pm and Se-Se-Se angle is 130.1° in the lattice. Gray Se is a semiconductor showing appreciable photoconductivity while the other Se forms are insulators. Unlike the other allotropes, it is insoluble in CS_2 and resists oxidation by air as well as attack by nonoxidizing acids. It does form polyselenides with strong reducing agents.

5.16.3.1 Reaction in air, oxygen, nitrogen, and hydrogen

The reaction of elemental selenium with oxygen forms selenium dioxide. Selenium trioxide is difficult to prepare from the elements as it is unstable with respect to the dioxide. It has been prepared through the dehydration of anhydrous selenic acid with phosphorus pentoxide at 150°C–160°C or through the reaction of liquid sulfur trioxide with potassium selenate. The simplest hydride of selenium, H_2Se, can be prepared industrially is produced by treating elemental selenium at T > 300°C with hydrogen gas.

$$Se_8 + 8O_2 \rightarrow 8SeO_2$$
$$P_4O_{10} + H_2SeO_4 \rightarrow P_4O_9(OH)_2 + SeO_3$$
$$SO_3 + K_2SeO_4 \rightarrow K_2SO_4 + SeO_3$$
$$Se_8 + 8H_2 \rightarrow 8SeH_2$$

5.16.3.2 *Reaction with halogens*

Selenium reacts with fluorine to form selenium hexafluoride. When this reaction is carefully controlled, say at 0°C and limited fluorine, the tetrafluoride can form. Fluorides where the oxidation state of Se is lower are difficult to isolate and characterize except for SeF_2, FSeSeF and Se = SeF_2 which can be trapped at lower temperatures.

$$Se_8 + 24F_2 \rightarrow 8SeF_6$$
$$Se_8 + 16F_2 \rightarrow 8SeF_4$$

Selenium tetrachloride, $SeCl_4$ or Se_4Cl_{16}, is prepared by heating selenium with chlorine forming a product that sublimes in the heated flask. This volatility of $SeCl_4$ can be exploited to purification of selenium. Selenium dichloride, $SeCl_2$, had been prepared by treating gray Se with sulfuryl chloride. However, solutions of $SeCl_2$ are unstable at room temperature, forming selenium(I) chloride, Se_2Cl_2, after several minutes at room temperature.

$$3SeCl_2 \rightarrow Se_2Cl_2 + SeCl_4$$

Apart from the disproportionation of the dichloride of Se to form Se(IV) and Se(I) chlorides, Se_2Cl_2 can be produced by reaction of the stoichiometric amounts of the elements or by adding chlorine to a suspension of powdered Se in CS_2. Selenium tetrabromide, $SeBr_4$ or its tetramer form, Se_4Br_{16}, as well as Se(I) bromide, Se_2Br_2 can be prepared similar to the chloride counterparts. While the lighter halides of selenium are known, iodides of selenium are not well known. The most recent study of the combination of iodine and selenium was conducted in 2018 by Voss et al. However, their results show that equal atomic fractions of the two elements form a eutectic mixture. While their study did not result to a binary iodide of selenium, the eutectic mixture has significant applications as it melts at 57°C while maintaining semiconducting behavior.

5.16.3.3 *Reaction with acids and bases*

Selenium is not attacked by nonoxidizing acids such as HCl. It is, however, oxidized by nitric acid to selenous acid. Se also reacts with boiling sulfuric acid to produce selenium(IV) oxide, sulfur dioxide and water. Boiling selenium and concentrated potassium or sodium hydroxide produces a mixture of selenide and selenite but the products immediately react with each other to form colloidal red selenium.

$$3Se + 4HNO_3 + H_2O \rightarrow 3H_2SeO_3 + 4NO$$
$$Se + 2H_2SO_4 \rightarrow SeO_2 + 2SO_2 + 2H_2O$$
$$6OH^- + 3Se \rightleftharpoons 2Se^{2-} + SeO_3^{2-} + 3H_2O$$

5.16.3.4 *Organoselenium compounds*

Organoselenium compounds are chemical compounds containing carbon-to-selenium chemical bonds. Selenium belongs to group 16 elements or chalcogens with sulfur and oxygen above it. Thus the structures of organoselenium compounds are similar to their sulfur counterparts but the properties can be different. This can be attributed to differences in bond lengths and strengths as well as to the differences in electronic behavior. The bond strength in selenium compounds are weaker compared to the lighter members of Group 16 (234 kJ/mol for the C-Se bond and 272 kJ/mol for the C-S bond); and the bond lengths are also longer (C-Se 198 pm, C-S 181 pm and C-O 141 pm). Organoselenium compounds are also more nucleophilic and more acidic than the corresponding sulfur compounds. Just like its sulfur and oxygen counterparts, organoselenium compounds are classified according to functional groups. The naming and general structure follows closely the scheme employed in oxygen and sulfur-containing organic compounds and will no longer be repeated. It is enough to know that if there is an alcohol (ROH) and thiol (RSH), there is a corresponding selenol (RSeH) and so on. Selenium compounds are known to exhibit toxicity. Nevertheless, interests in their chemistry continue due to their wide applications in organic synthesis and chemical biology. Recent works demonstrating that selenium-based catalysts can be used conveniently in a series of functional group transformations further adds to advances in organoselenium chemistry. The reader is referred to reviews by Freudendahl et al. (2009) and Singh and Wirth (2019) for a more detailed discussion of new developments in organic reactions and catalysis involving organoselenium compounds.

5.16.4 Major uses

The major commercial use of Se, approximately 50% of total consumption, is to produce glass. Se compounds give a red color to glass. This color compensates for the green or yellow tints that stems from Fe impurities typical for most glass. To achieve this various selenite and selenate salts are added. For other uses, a red color may be desired, produced by combinations of CdSe and CdS. Se can also be used to reduce the transmission of sunlight in architectural glass, giving it a bronze tint. In addition, Se is used to make pigments for ceramics, paint and plastics. In the electrowinning of Mn, adding SeO_2 reduces the power required to run the electrolysis cells. For every ton of Mn, an average 2 kg SeO_2 is used. Se is used with Bi in brasses to substitute for the more toxic Pb. The regulation of Pb in drinking water applications made a reduction of Pb in brass essential. The new brass is sold under the trade name EnviroBrass. Similar to Pb and S, Se enhances the machinability of steel at concentrations of about 0.15%. Se produces a similar machinability improvement in Cu alloys. The lithium–selenium (Li–Se) battery is seen as possibly one of the most promising energy storage systems within the family of lithium batteries. The Li–Se battery is an alternative to the Li–S battery, with an advantage of high electrical conductivity.

Copper indium gallium selenide (CIGS, a I–III–VI_2 compound semiconductor material composed of Cu, In, Ga, and Se. The material is a solid solution of Cu, In selenide (often abbreviated "CIS") and Cu Ga selenide, with a chemical formula of $CuIn_xGa_{(1-x)}Se_2$, where the value of x can vary from 1 (pure Cu In selenide) to 0 (pure Cu Ga selenide). It is a tetrahedrally bonded semiconductor, with the chalcopyrite crystal structure. The bandgap varies continuously with x from about 1.0 eV (for Cu In selenide) to about 1.7 eV (for Cu Ga selenide)) is a compound utilized in solar cells. Amorphous selenium (α-Se) thin films have been used as photoconductors in flat panel X-ray detectors (a class of solid-state X-ray digital radiography devices similar in principle to the image sensors used in digital photography and video. They are used in both projectional radiography and as an alternative to X-ray image intensifiers (IIs) in fluoroscopy equipment.). These detectors use amorphous Se to capture and convert incident X-ray photons directly into electric charge. Se rectifiers were used for the first time in 1933 and continued into the 1990s. Small quantities of organoselenium compounds have been used to modify the catalysts used for vulcanization in the production of rubber. The demand for Se by the electronics industry is falling. However, its photovoltaic and photoconductive properties are still valuable in photocopying, photocells, light meters and solar cells. Its application as a photoconductor in plain-paper copiers previously was a prominent application, but in the 1980s, the photoconductor application decreased (though it was still a significant end-use) as progressively more copiers shifted to organic photoconductors. While previously extensively used, Se rectifiers have mostly been superseded (or are being substituted) by Si-based devices. The most significant exception is in power DC surge protection, where the excellent energy capabilities of Se suppressors make them more suitable than metal oxide varistors. ZnSe was the first compound for blue LEDs, but GaN dominates that market. CdSe was an essential element in quantum dots. Sheets of amorphous Se convert X-ray images to patterns of charge in xeroradiography and in solid-state, flat panel X-ray cameras. Ionized Se (Se+24) is one of the active mediums applied in X-ray lasers. Se is only on a limited scale used as a catalyst in some chemical reactions, due to issues with its toxicity. In X-ray crystallography, incorporation of one or more Se atoms substituting for S helps with multiple-wavelength anomalous dispersion and single wavelength anomalous dispersion phasing. Se is utilized in the toning of photographic prints, and it is marketed as a toner by various photographic manufacturers. Se intensifies and extends the tonal range of black-and-white photographic images and enhances the permanence of prints. ^{75}Se is used as a γ source in industrial radiography. Se is toxic to the scalp fungus that causes dandruff, so it is used in some antidandruff shampoos.

5.17 35 Br — Bromine

5.17.1 Discovery

Bromine was independently discovered by German chemist Carl Jacob Löwig (March 17, 1803 to March 27, 1890) (Fig. 5.149) and French chemist Antoine Jérôme Balard (1802–1876) (Fig. 5.150), in 1825 and 1826, respectively

FIGURE 5.149 Carl Jacob Löwig, c.1850.

FIGURE 5.150 Antoine Jérôme Balard, 1870s.

(Löwig, 1829; Balard, 1826a,b). Löwig separated bromine from a mineral water spring in his hometown Bad Kreuznach (a well-known spa-town in Rhineland-Palatinate, Germany) in 1825. Löwig used a solution of the mineral salt saturated with chlorine and removed the bromine with diethyl ether. After evaporation of the diethyl ether a brown liquid was left. With this liquid as an example of his work he applied for a position in the laboratory of the German chemist Leopold Gmelin (August 2, 1788 to April 13, 1853) at the University of Heidelberg. The publication of the results was delayed and as a result Balard published his results first. Balard discovered bromine chemicals in the ash of seaweed from the salt marshes of Montpellier, France. The seaweed was used to obtain iodine, but in addition it contained bromine. Balard extracted the bromine from a solution of seaweed ash saturated with chlorine. The resulting substance exhibited properties intermediate between those of chlorine and iodine; consequently he attempted to prove that the substance was iodine monochloride (ICl), which was unsuccessful and he was convinced that instead he had discovered a new element that he called muride, derived from the Latin word muria for brine. After the French chemists

Louis Nicolas Vauquelin (May 16, 1763 to November 14, 1829), Louis Jacques Thénard (May 4, 1777 to June 21, 1857), and Joseph-Louis Gay-Lussac (December 6, 1778 to May 9, 1850) confirmed the experiments of the young pharmacist and chemist Balard, the results were reported at a lecture of the Académie des Sciences and published in Annales de Chimie et Physique. In his publication, Balard reported that he changed the name from muride to brôme on the proposal of M. Anglada. Brôme (bromine) derives from the Greek βρωμος (stench). Alternative sources state that the French chemist and physicist Joseph-Louis Gay-Lussac proposed the name brôme for the characteristic smell of the vapors. Bromine was not manufactured in large quantities until 1858, when the discovery of salt deposits in Staßfurt, Saxony-Anhalt, Germany, enabled its separation as a by-product of potash. Apart from some insignificant medical applications, the first commercial use was the daguerreotype. The daguerreotype process, or daguerreotypy, was the first publicly available photographic process, and for nearly 20 years it was the one most commonly used. It was invented by French artist and photographer Louis-Jacques-Mandé Daguerre (November 18, 1787 to July 10, 1851) and introduced worldwide in 1839. In 1840 bromine was found to have some advantages over the earlier used iodine vapor to create the light sensitive silver halide layer in daguerreotypy. Potassium bromide and sodium bromide were used as anticonvulsants and sedatives in the late 19th and early 20th centuries but were gradually superseded by chloral hydrate ($C_2H_3Cl_3O_2$) and then by the barbiturates (drugs that acts as a central nervous system depressant, based on the structure of barbituric acid, pyrimidine-2,4,6(1H,3H,5H)-trione, $C_4H_4N_2O_3$). In the early years of World War I, bromine substances such as xylyl bromide, $C_6H_4(CH_3)(CH_2Br)$, were used as poison gas.

5.17.2 Mining, production, and major minerals

Bromine is significantly less abundant in the crust than fluorine or chlorine, covering only 2.5 ppm of the Earth's crustal rocks, and then only as bromide salts. It is the 46th most abundant element in Earth's crust. It is far more abundant in the oceans, because of long-term leaching. There, it makes up 65 ppm, equivalent to a ratio of about one bromine atom for every 660 chlorine atoms. Salt lakes and brine wells can have higher bromine concentrations: for example, the Dead Sea contains 0.4% bromide ions. It is from these sources that bromine winning is typically economically achievable. The main sources of bromine are found in the US and Israel. The element is liberated by halogen exchange, using chlorine gas to oxidize Br^- to Br_2. This is subsequently removed with a blast of steam or air and is then condensed and purified.

In total there are only 16 minerals with bromine as part of the chemical composition. The sulfides are represented by four minerals, for example, perroudite ($Hg_5Ag_4S_5(I,Br)_2Cl_2$). The majority, 12 different minerals are found in the halides class, with minerals such as bromargyrite (AgBr) (Fig. 5.151) (Box 5.19), kadyrelite ($(Hg_2^{2+})_3OBr_3(OH)$), and vasilyevite ($(Hg^{2+})_{10}I_3Br_2Cl(CO_3)O_6$).

FIGURE 5.151 Bromargyrite, AgBr, bright yellow crystals to 4 mm in matrix. Kintore Opencut, Broken Hill, New South Wales, Australia.

BOX 5.19 Bromargyrite, AgBr

Bromargyrite is named in allusion to its composition, containing bromine and the Greek for silver, *argyros*. It is a rare secondary mineral in the oxidation zones of silver deposits, notably in arid regions, associated with silver (Ag), iodargyrite (AgI), smithsonite ($ZnCO_3$), and Fe-Mn oxides. It is isometric (cubic), *m3m (4/m2/m)*, space group *Fm3m*). It has a hardness of 2½. Its luster is resinous to adamantine, waxy, while its transparency is transparent to translucent. The color is pale yellow, greenish brown, to bright green. The streak is white to yellowish white. Its fracture is uneven to subconchoidal, while the tenacity is sectile, ductile, very plastic. It may give off a strong "medicinal" odor when exposed to air. Crystals are cubic, sometimes with {111} and {011}, rounded, to 1 cm; in parallel or subparallel groups; commonly as crusts and coatings, massive. Twinning on {111} is rare.

TABLE 5.20 Bromine properties.

Appearance	Reddish-brown
Standard atomic weight $A_{r,std}$	79.904
Block	p-Block
Element category	Reactive nonmetal
Electron configuration	[Ar] $3d^{10}$ $4s^2$ $4p^5$
Phase at STP	Liquid
Melting point	−7.2°C
Boiling point	58.8°C
Density (near r.t.)	Br_2, liquid: 3.1028 g/cm^3
Heat of fusion	(Br_2) 10.571 kJ/mol
Heat of vaporization	(Br_2) 29.96 kJ/mol
Molar heat capacity	(Br_2) 75.69 J/(mol·K)
Oxidation states	**−1**, **+1**, **+3**, +4, **+5**, **+7**
Ionization energies	1st: 1139.9 kJ/mol
	2nd: 2103 kJ/mol
	3rd: 3470 kJ/mol
Atomic radius	Empirical: 120 pm
Covalent radius	120 ± 3 pm
Van der Waals radius	185 pm

STP, Standard temperature and pressure.
Bold font indicates main oxidation state.

5.17.3 Chemistry

Bromine has two naturally occurring stable isotopes, ^{79}Br and ^{81}Br, with ^{79}Br having an abundancy of 51% with ^{81}Br making up the remaining 49%. Both have nuclear spin 3/2 − and thus may both be used for NMR, though ^{81}Br is more favorable. Other Br isotopes are all radioactive, exhibiting half-lives too short to occur in nature. Of these, the most important are ^{80}Br with a half-life of 17.7 minutes, ^{80m}Br with a half-life of 4.421 hours, and ^{82}Br with a half-life of 35.28 hours, which may be synthesized through neutron activation of natural Br. The most stable radioisotope is ^{77}Br which has a half-life of 57.04 hours. The primary decay mode of isotopes lighter than ^{79}Br is electron capture to isotopes of Se; while that of the heavier ones is by β decay to isotopes of Kr; while ^{80}Br may decay by either mode to stable ^{80}Se or ^{80}Kr.

Bromine is a red-brown liquid at room temperature that fumes readily to a similarly colored gas. It is a member of group 17 (halogens) and has the electron configuration [Ar]$3d^{10}4s^24p^5$, with the seven electrons in the fourth and outermost shell acting as its valence electrons (Table 5.20). Like all halogens, it is one-electron short of a full octet. Thus it is a strong oxidizing agent, abstracting an electron from many elements in order to complete its outer shell. It follows the corresponding periodic trend for the halogens. It is intermediate in electronegativity between chlorine, the element above it, and iodine, the element below it (F: 3.98, Cl: 3.16, Br: 2.96, I: 2.66). It is less reactive and a weaker oxidizing agent than chlorine and more reactive and stronger oxidizing agent than iodine. It reacts with compounds including M-M, M-H, or M-C bonds to form M-Br bonds. The high reactivity of bromine means it cannot be found as elemental bromine in nature but as colorless, soluble crystalline mineral bromide salts. Bond energies in bromine compounds tend to be lower than those in chlorine but higher than those in iodine.

5.17.3.1 Reaction with air, oxygen, nitrogen, and hydrogen

Bromine, Br_2, is not reactive toward oxygen, O_2, or nitrogen, N_2 at ordinary conditions. However, when an electric current is passed through a mixture of bromine and oxygen gases at low temperature and pressure, unstable yellow-brown crystals of bromine dioxide, BrO_2, are formed. This compound can also be formed when bromine reacts with ozone, O_3, in trichlorofluoromethane at −78°C. Ozonization of Br_2 in $CFCl_3$ at −90°C also yields an orange crystalline solid determined to have the formula $BrOBrO_2$. This compound decomposed above −40°C and detonates when warned rapidly to 0°C. Other oxides of bromine are known but these are not prepared from the elements.

$$Br_2(l) + 2O_3(g) \xrightarrow{CFCl_3,\ -78^\circ C} O_2(g) + 2BrO_2(s)$$
$$6Br_2(l) + 8O_3(g) \xrightarrow{CFCl_3,\ -90^\circ C} 3O_2(g) + 6BrOBrO_2(s)$$

The four oxoacids, hypobromous acid (HOBr), bromous acid (HOBrO), bromic acid ($HOBrO_2$), and perbromic acid ($HOBrO_3$), have been studied in greater details because of their greater stability, even though they are only so in aqueous solution. When bromine dissolves in aqueous solution, the following reactions occur.

$$Br_2 + H_2O \rightleftharpoons HOBr + H^+ + Br^- \quad K_{ac} = 7.2 \times 10^{-9} mol^2 l^{-2}$$
$$Br_2 + 2OH^- \rightleftharpoons OBr^- + H_2O + Br^- \quad K_{alk} = 2 \times 10^8 mol^{-1} l$$

Hypobromous acid is unstable and undergo slow disproportionation. The hypobromite ions hence formed disproportionate easily to produce bromide and bromate.

$$3BrO^- \rightleftharpoons 2Br^- + BrO_3^- \quad K = 10^{15}$$

Bromous acids and bromites are very unstable, though the strontium and barium bromites are known. More significant are the bromates, which are synthesized on a small scale through oxidation of bromide with aqueous hypochlorite and are strong oxidizing agents. Different to the chlorates, which very slowly disproportionate to chloride and perchlorate, the bromate anion is stable to disproportionation in both acidic and aqueous solutions. Bromic acid is a strong acid. Bromides and bromates may comproportionate to bromine.

$$BrO_3^- + 5Br^- + 6H^+ \rightarrow 3Br_2 + 3H_2O$$

The hydrogen compound of bromine, hydrogen bromide, is produced industrially by the reaction of hydrogen gas with bromine gas at 200°C–400°C in the presence of a platinum catalyst. Nevertheless, reduction of bromine with red phosphorus is a more practical way to form hydrogen bromide in the laboratory.

$$2P + 6H_2O + 3Br_2 \rightarrow 6HBr + 2H_3PO_3$$
$$H_3PO_3 + H_2O + Br_2 \rightarrow 2HBr + H_3PO_4$$

5.17.3.2 Reaction with halogens

Bromine, Br_2, reacts with other halogens to form interhalogen compounds. It reacts with F_2, Cl_2 and I_2 to from diatomic interhalogen compounds of the general formula BrX (X = F, Cl, I). The reaction with F_2 is carried out in the gas phase, the bromine fluoride product, however, is very unstable and disproportionates at room temperature to bromine molecule and bromine trifluoride, BrF_3 (and bromine pentafluoride, BrF_5). BrF_3 can also be prepared from the direct reaction of the elements but the conditions must be carefully chosen to avoid formation of other interhalogen compounds with varying stoichiometries. BrF_5 can be prepared by reacting excess fluorine with bromine, Br_2, at 150°C.

$$Br_2(g) + F_2(g) \rightarrow 2BrF(g)$$
$$3BrF(g) \rightarrow Br_2(l) + BrF_3(l)$$
$$5BrF(g) \rightarrow 2Br_2(l) + BrF_5(l)$$

Bromine reacts with chlorine in the gas phase to form the unstable interhalogen species bromine chloride, BrCl. However, BrCl cannot be isolated free of the starting elements. The reaction with iodine forms black crystals of iodine bromide, IBr.

5.17.3.3 *Reaction with acids and bases*

Bromine disproportionates to some extent in hot aqueous alkali to produce bromate, BrO_3^-, and bromide, Br^-, ions. It has been shown to oxidize sulfurous acid to sulfuric acid.

$$3Br_2(g) + 6OH^-(aq) \rightarrow BrO_3^-(aq) + 5Br^-(aq) + 3H_2O$$
$$Br_2 + H_2SO_3 + H_2O \rightarrow H_2SO_4 + 2HBr$$

5.17.3.4 *Organobromine compounds*

Organobromine or organobromide compounds contain carbon bonded to bromine. A variety of minor organobromine compounds are found in nature. The most pervasive is biogenic bromomethane (CH_3Br), which is produced in the marine environment along with other bromine-containing gases ($CHBr_3$, CH_2Br_2, CH_2IBr). This contributes to the tropospheric and stratospheric chemistry of bromine.

Organobromides are typically formed via additive or substitutive bromination of other organic precursors. Bromine itself can be used, but due to its toxicity and volatility safer brominating reagents are normally used, such as N-bromosuccinimide (IUPAC name 1-bromo-2,5-pyrrolidinedione, $C_4H_4BrNO_2$). The principal reactions for organobromides include dehydrobromination, Grignard reactions, reductive coupling, and nucleophilic substitution. The C-Br bond as with the other carbon-halogen bond is a common functional group that forms part of core organic chemistry. The differences in electronegativity between bromine (2.96) and carbon (2.55) makes the carbon in a C-Br bond electron-deficient and thus electrophilic. The reactivity of organobromine compounds is intermediate between the reactivity of organochlorine and organoiodine compounds. The choice of organobromides over the other organohalides takes into consideration their reactivity and cost. As much as reactions of organohalide compounds rank among the most important in organic chemistry, concerns on their effects on human health and the environment as well as their contributions to the destruction of ozone are mounting.

5.17.4 Major uses

Brominated flame retardants form a product of growing significance and cover the largest commercial use of Br. When the brominated material burns, the flame-retardant forms hydrobromic acid (HBr) which inhibits the radical chain reaction of the oxidation reaction of the fire. The mechanism involves the highly reactive hydrogen radicals, oxygen radicals, and hydroxy radicals reacting with hydrobromic acid to produce less reactive bromine radicals (i.e., free Br atoms). Br atoms may also react directly with other radicals to assist in terminating the free radical chain-reactions in combustion. To produce brominated polymers and plastics, Br-containing compounds can be integrated into the polymer during polymerization. One method comprises inclusion of a comparatively small quantity of brominated monomer during the polymerization process, for example, vinyl bromide (IUPAC name bromoethane, C_2H_3Br) can be used in the production of polyethylene (PE, IUPAC name polyethene or poly(methylene), $(C_2H_4)_n$), polyvinyl chloride (PVC, IUPAC name poly(1-chloroethylene), $(C_2H_3Cl)_n$) or polypropylene (PP, IUPAC name poly(1-methylethylene), $(C_3H_6)_n$). Exact highly brominated molecules can also be included that take part in the polymerization process, for example, tetrabromobisphenol A (IUPAC name 4,4′-(propane-2,2-diyl)bis(2,6-dibromophenol), $C_{15}H_{12}Br_4O_2$) can be added to polyesters or epoxy resins, where it becomes part of the polymer. Epoxys used in printed circuit boards are normally made from such flame-retardant resins, indicated by the FR in the abbreviation of the products. In some instances the Br-containing compound may be added after polymerization, for example, decabromodiphenyl ether (IUPAC name 2,3,4,5,6-Pentabromo-1-(2,3,4,5,6-pentabromophenoxy)benzene, $C_{12}Br_{10}O$) can be added to the final polymers. Several gaseous or highly volatile brominated halomethane compounds are nontoxic and make exceptional fire suppressant agents by this same mechanism and are especially efficient in enclosed spaces such as submarines, airplanes, and spacecraft. Yet, they are costly, and their manufacture and application has been greatly restricted because of their effect as ozone-depleting agents. As a consequence, they are no longer used in regular fire extinguishers, but still have niche uses in aerospace and military automatic fire-suppression applications. They comprise bromochloromethane (Halon 1011, CH_2BrCl), bromochlorodifluoromethane (Halon 1211, $CBrClF_2$), and bromotrifluoromethane (Halon 1301, $CBrF_3$).

AgBr is used, either isolated or in a mixture with AgCl and AgI, as the light sensitive component of photographic emulsions. Ethylene bromide (IUPAC name 1,2-dibromoethane, $C_2H_4Br_2$) was an additive in gasolines containing Pb antiengine knocking agents. It scavenges Pb by forming volatile $PbBr_2$, which is exhausted from the engine. This application accounted for 77% of the Br use in 1966 in the US. This usage has dropped since the 1970s due to environmental regulations. Toxic bromomethane (CH_3Br) was extensively used as pesticide to fumigate soil and to fumigate housing, by the tenting method. Ethylene bromide (IUPAC name 1,2-dibromoethane, $C_2H_4Br_2$) was likewise used. These volatile organobromine compounds are now all strictly controlled as ozone (O_3) depletion agents. The Montreal Protocol on Substances that Deplete the Ozone Layer programmed the phase out for these ozone-depleting chemicals by 2005, and organobromide pesticides are no longer applied (in housing fumigation they have been substituted by compounds such as sulfuryl fluoride, SO_2F_2, which contain neither the Cl or Br organics which damage ozone). In pharmacology, inorganic Br compounds, in particular KBr, were regularly used as general sedatives in the 19th and early 20th century. Bromides in the form of simple salts are even now used as anticonvulsants in both veterinary and human medicine, although the latter use differs from country to country, for example, the USFDA does not approve bromide for the treatment of any disease, and it was removed from over-the-counter sedative products like Bromo-Seltzer, in 1975. Commercially available organobromine drugs comprise the vasodilator nicergoline ($C_{24}H_{26}BrN_3O_3$), the sedative brotizolam ($C_{15}H_{10}BrClN_4S$), the anticancer agent pipobroman ($C_{10}H_{16}Br_2N_2O_2$), and the antiseptic merbromin ($C_{20}H_8Br_2HgNa_2O_6$). Otherwise, organobromine compounds are seldom pharmaceutically valuable, in contrast to organofluorine compounds. Some drugs are made as the bromide (or equivalents, hydrobromide) salts, but in these situations Br acts as an harmless counterion with no biological importance. Additional applications of organobromine compounds comprise high-density drilling fluids and dyes (such as Tyrian purple (main component 6,6′-dibromoindigo, IUPAC name 6-bromo-2-(6-bromo-3-hydroxy-1*H*-indol-2-yl)indol-3-one, $C_{16}H_8Br_2N_2O_2$) and the indicator bromothymol blue (IUPAC name 4,4′-(1,1-dioxido-3H-2,1-benzoxathiole-3,3-diyl)bis(2-bromo-6-isopropyl-3-methylphenol), $C_{27}H_{28}Br_2O_5S$)). Bromine itself, together with some of its compounds, are used in water treatment, and is the precursor of numerous inorganic compounds with a large number of applications. Zinc−bromine batteries are hybrid flow batteries utilized for stationary electrical power backup and storage; ranging from household scale to industrial scale.

References

Addison, C. C., Johnson, B. F. G., Logan, N., & Wojcicki, A. (1961). Transition-metal nitrites. *Proceedings of the Chemical Society*, 306−307.

Balard, A.J., (1826a). *Mémoire sur une substance particulière contenue dans l'eau de la mer, Annales de Chimie et de Physique, 2nd series, 32*, 337−381.

Balard, A.J., (1826b). Memoir on a peculiar Substance contained in Sea Water, *Annals of Philosophy, 28*, 381−387 and 411−426.

Bell, A.G., (1880). On the Production and Reproduction of Sound by Light, *American Journal of Science, Third Series, 20*, 305−324.

Brandt, G. (1735). Dissertatio de semimetallis. *Acta Literaria et Scientiarum Sveciae, 4*, 1−10.

Brandt, G. (1746). Rön och anmärkningar angäende en synnerlig färg—cobolt. *Kongliga Svenska vetenskapsakademiens handlingar, 7*, 119−130.

Brandt, K. (1943). X-ray analysis of $CrVO_4$ and isomorphous compounds. *Arkiv for Kemi, Mineralogi och Geologi, 17A*, 1−13.

Berzelius, J. J. (1818). Lettre de M. Berzelius à M. Berthollet sur deux métaux nouveaux. *Annales de Chimie et de Physique, 7*, 199−206, 2nd series.

Brooker, M. H. (1975). Infra-red and Raman spectral study of the aqueous nickel(II)−nitrite system. Evidence for photochemical alteration of the chemical equilibrium. *Journal of the Chemical Society, Faraday Transactions 1: Physical Chemistry in Condensed Phases, 71*, 647−656.

Cadet de Gassicourt, L. C., (1760). Suite d'Expériences Nouvelles sur l'Encre Sympathique de M. Hellot qui Peuvent Servir à l'Analyse du Cobolt; et Histoire d'une Liqueur Fumante, Tirée de l'Arsènic, *Memoires de Mathématique et de Physique. Presentés à l Académie Royale des Sciences par diverse Savans et lûs dans ses Assemblées, 13*, 623−637.

Cahiez, G., Duplais, C., & Buendia, J. (2009). Chemistry of organomanganese(II) compounds. *Chemical Reviews, 109*, 1434−1476.

Choi, D., Blomgren, G. E., & Kumta, P. N. (2006). Fast and reversible surface redox reaction in nanocrystalline vanadium nitride supercapacitors. *Advanced Materials, 18*, 1178−1182.

Cleve, P. T. (1879). Sur le scandium. *Comptes Rendus, 89*, 419−422.

Cronstedt, A.F., (1751). Rön och försök, Gjorde Med en Malm-art från Los Kobolt Grufvor i Farila Socken och Helsingeland, *Kongl. Svenska Veenskapas Academians Handlingar. 12*, 287−292.

Davis, R., & Kane-Maguire, L. A. P. (1982). 26.2 - Chromium compounds with $\eta^2-\eta^8$ carbon ligands. In G. Wilkinson, F. G. A. Stone, & E. W. Abel (Eds.), *Comprehensive organometallic chemistry* (pp. 953−1077). Oxford: Pergamon.

Davy, H. (1808a). Electro-chemical researches on the decomposition of the earths; with observations on the metals obtained from the alkaline earths, and on the amalgam procured from ammonia. *Philosophical Transactions of the Royal Society of London, 98*, 333−370.

Davy, H. (1808b). On some new phenomena of chemical changes produced by electricity, particularly the decomposition of the fixed alkalies, and the exhibition of the new substances which constitute their bases; and on the general nature of alkaline bodies. *Philosophical Transactions of the Royal Society of London, 98*, 1−44.

Davy, H. (1809). Ueber einige neue Erscheinungen chemischer Veränderungen, welche durch die Electricität bewirkt werden; insbesondere über die Zersetzung der feuerbeständigen Alkalien, die Darstellung der neuen Körper, welche ihre Basen ausmachen, und die Natur der Alkalien überhaupt. *Annalen der Physik, 31*, 113−175.

Demarteau, J., Debuigne, A., & Detrembleur, C. (2019). Organocobalt complexes as sources of carbon-centered radicals for organic and polymer chemistries. *Chemical Reviews, 119*, 6906−6955.

du Monceau, H. L. D. (1736). *Sur la Base de Sel Marin. Memoires de l'Academie Royale des Sciences* (pp. 65–68). Paris: J. Boudot.

Ehrhardt, H., Johannes, W., & Seidel, H. (1973). Notizen: Hochdrucksynthese von Kupfer(II)-Carbonat / High pressure synthesis of cupric carbonate. *Zeitschrift für Naturforschung B, 28*, 682.

Featherstonhaugh, G. W. (1831). New metal, provisionally called vanadium. *The Monthly American Journal of Geology and Natural Science, 1*, 67–69.

Fischer, W., Brünger, K., & Grieneisen, H. (1937). Über das metallische Scandium. *Zeitschrift für anorganische und allgemeine Chemie, 231*, 54–62.

Fischer, E. O., & Pfab, W. (1952). Cyclopentadien-Metallkomplexe, ein neuer Typ metallorganischer Verbindungen. *Zeitschrift für Naturforschung B, 7*. Available from https://doi.org/10.1515/znb-1952-0701.

Frankland, E., (1849). Notiz über eine neue Reihe organischer Körper, welche Metalle, Phosphor u. s. w. enthalten, *Liebig's Annalen der Chemie und Pharmacie 71*, 213–216.

Freudendahl, D. M., Santoro, S., Shahzad, S., Santi, C., & Wirth, T. (2009). Green chemistry with selenium reagents: Development of efficient catalytic reactions. *Angewandte Chemie International Edition, 48*, 8409–8411.

Fürstner, A., Martin, R., Krause, H., Seidel, G., Goddard, R., & Lehmann, C. W. (2008). Preparation, structure, and reactivity of nonstabilized organoiron compounds. Implications for iron-catalyzed cross coupling reactions. *Journal of the American Chemical Society, 130*, 8773–8787.

Galvani, L., (1791). De viribus electricitatis in motu musculari commentarius. In: De Bononiensi Scientiarum et Artium Instituto atque Academia Commentarii, vol. VII, pp. 363–418, Typographia Instituti Scientiarum, Bologna.

Green, D. W., McDermott, D. P., & Bergman, A. (1983). Infrared spectra of the matrix-isolated chlorides of iron, cobalt, and nickel. *Journal of Molecular Spectroscopy, 98*, 111–124.

Greenwood, N. N., & Earnshaw, A. (1996). *Chemistry of the elements.* Oxford: Elsevier Science & Technology Books.

Gregor, W. (1791a). Beobachtungen und Versuche über den Menakanit, einen in Cornwall gefundenen magnetischen Sand Chemische Annalen für die Freunde der Naturlehre. *Arzneygelahrtheit, Haushaltungskunst und Manufacturen, 1*, 40–54.

Gregor, W. (1791b). Sur le menakanite, espèce de sable attirable par l'aimant, trouvé dans la province de Cornouilles. *Observations et Mémoires sur la Physique, 39*, 72–78.

Halperin, E.C., Perez, C.A., Brady, L.W. (2008). *Perez and Brady's principles and practice of radiation oncology*; Wolters Kluwer/Lippincott Williams & Wilkins: Philadelphia, pp. 2106.

Harder, S. (2010). From limestone to catalysis: Application of calcium compounds as homogeneous catalysts. *Chemical Reviews, 110*, 3852–3876.

Hein, F. (1919). Notiz über Chromorganoverbindungen. *Berichte der deutschen chemischen Gesellschaft (A and B Series), 52*, 195–196.

Herman, D. F., & Nelson, W. K. (1953). Organotitanium Compounds. I. Isolation of a compound containing the titanium—carbon bond. *Journal of the American Chemical Society, 75*, 3877–3882.

Imoto, H., & Naka, K. (2019). The dawn of functional organoarsenic chemistry. *Chemistry – A European Journal, 25*, 1883–1894.

Jennewein, B., Kimpel, S., Thalheim, D., Klett, J., (2018). Towards the Next Generation of Lochmann–Schlosser Superbases: A Potassium Neopentyl/Alkoxy Aggregate used in the Tetra-Functionalization of Ferrocene. *Chemistry, A European Journal 24*, 7605–7609.

Jolly, P. W. (1996). From Hein to Hexene: Recent advances in the chemistry of organochromium π-complexes. *Accounts of Chemical Research, 29*, 544–551.

Kadassery, K. J., MacMillan, S. N., & Lacy, D. C. (2019). Resurgence of organomanganese(I) chemistry. Bidentate manganese(I) phosphine–phenol (ate) complexes. *Inorganic Chemistry, 58*, 10527–10535.

Kaim, I.G., (1770). De Metallis Dubiis. Schulz, Vienna.

Kealy, T. J., & Pauson, P. L. (1951). A new type of organo-iron compound. *Nature, 168*, 1039–1040.

Keess, S., & Oestrich, M. (2017). Access to fully alkylated germanes by $B(C_6F_5)_3$-catalyzed transfer hydrogermylation of alkenes. *Organic Letters, 19*, 1898–1901.

Klaproth, M. H. (1795). *Beiträge zur chemischen Kenntniss der Mineralkörper* (Vol 1). Berlin: Heinrich August Rottmann.

Klaproth, M. (1797). Nouvelles données relatives à l'histoire naturelle de l'alcali végétal. *Mémoires de l'Académie royale des sciences et belles-lettres (Berlin)*, 9–13.

Lecoq de Boisbaudran, P. É. (1875). Caractères chimiques et spectroscopiques d'un nouveau métal, le gallium, découvert dans une blende de la mine de Pierrefitte, vallée d'Argelès (Pyrénées). *Comptes Rendus, 81*, 493–495.

Lecoq de Boisbaudran, P. É. (1886). Sur le poids atomique du germanium. *Comptes Rendus, 103*, 452–453.

Lehmanni, I. G. (1766). *De Nova Minerae Plumbi Specie Crystallina Rubra, Epistola.* Petropoli: Academiae Scientiarum.

Livingood, J., & Seaborg, G. T. (1938). Long-lived radio cobalt isotopes. *Physical Review, 53*, 847–848.

Lochmann, L., Lím, D., (1971). Preparation of organopotassium compounds, *Journal of Organometallic Chemistry 28*, 153–158.

Lousada, C. M., Fernandes, R. M. F., Tarakina, N. V., & Soroka, I. L. (2017). Synthesis of copper hydride (CuH) from $CuCO_3 \cdot Cu(OH)_2$ – A path to electrically conductive thin films of Cu. *Dalton Transactions, 46*, 6533–6543.

Löwig, C.J., (1829). Das Brom und seine chemischen Verhältnisse, Carl Winter, Heidelberg.

Marggraf, A.S., (1746). Experiences sur la maniere de tirer le Zinc de sa veritable?miniere, c'est à dire, de la pierre calaminaire, *Histoire de l'Académie Royale des Sciences et Belles-Lettres de Berlin*, 49–57.

Marsh, J., (1836). Account of a method of separating small quantities of arsenic from substances with which it may be mixed, *Edinburgh New Philosophical Journal. 21*, 229–236.

Mond, L., Langer, K., & Quincke, F. (1890). Action of carbon monoxide on nickel. *Journal of the Chemical Society, 57*, 749–753.

Neidig, M. L., Carpenter, S. H., Curran, D. J., DeMuth, J. C., Fleischauer, V. E., Iannuzzi, T. E., … Wolford, N. J. (2019). Development and evolution of mechanistic understanding in iron-catalyzed cross-coupling. *Accounts of Chemical Research, 52*, 140–150.

Nilson, L. F. (1879a). Sur l'ytterbine, terre nouvelle de M. Marignac. *Comptes Rendus, 88*, 642–647.

Nilson, L. F. (1879b). Ueber Scandium, ein neues Erdmetall. *Berichte der deutschen chemischen Gesellschaft*, *12*, 554–557.

Parchomyk, T., Demeshko, S., Meyer, F., & Koszinowski, K. (2018). Oxidation states, stability, and reactivity of organoferrate complexes. *Journal of the American Chemical Society*, *140*, 9709–9720.

Pinsent, J. (1954). The need for selenite and molybdate in the formation of formic dehydrogenase by members of the Coliaerogenes group of bacteria. *Biochemical Journal*, *57*, 10–16.

Pistorius, C. W. F. T. (1960). Synthesis at high pressure and lattice constants of normal cupric carbonate. *Experientia*, *16*, 447–448.

Reetz, M. T. (2001). Organotitanium chemistry. In M. Schlosser (Ed.), *Organometallics in synthesis: A manual* (pp. 817–923). New York: John Wiley & Sons.

Roscoe, H. E. (1869–1870). Researches on vanadium. Part II. *Proceedings of the Royal Society of London*, *18*, 37–42.

Schäfer, H., & Krehl, K. (1952). Das gasförmige Kobalt(III)-chlorid und seine thermochemischen Eigenschaften. *Zeitschrift für anorganische und allgemeine Chemie*, *268*, 25–34.

Schlosser, M., (1964a). Natrium- und kalium-organische Verbindungen. Teil I: Eigenschaften und Reaktionsweisen. *Angewandte Chemie 76*, 124–143.

Schlosser, M., (1964b). Natrium- und kalium-organische Verbindungen. Teil II: Darstellung und präparative Anwendung. *Angewandte Chemie 76*, 258–269.

Sefström, N. G. (1831). Ueber das Vanadin, ein neues Metall, gefunden im Stangeneisen von Eckersholm, einer Eisenhütte, die ihr Erz von Taberg in Småland bezieht. *Annalen der Physik und Chemie*, *97*, 43–49.

Shull, C. G., Strauser, W. A., & Wollan, E. O. (1951). Neutron diffraction by paramagnetic and antiferromagnetic substances. *Physical Review*, *83*, 333–345.

Singh, F. V., & Wirth, T. (2019). Selenium reagents as catalysts. *Catalysis Science & Technology*, *9*, 1073–1091.

Smith, W. (1873a). The action of light on selenium. *Journal of the Society of Telegraph Engineers*, *2*, 31–33.

Smith, W. (1873b). Effect of light on selenium during the passage of an electric current. *Nature*, *7*, 303.

Toscano, P. J., & Marzilli, L. G. (2007). B12 and related organocobalt chemistry: Formation and cleavage of cobalt carbon bonds. *Progress in Inorganic Chemistry*, *31*, 105–204.

Vauquelin, L. N. (1798). Memoir on a new metallic acid which exists in the red lead of Siberia. *Journal of Natural Philosophy, Chemistry, and the Arts*, *3*, 145–146.

von Liebig, J. F. (1840). *Die organische Chemie in ihrer Anwendung auf Agricultur und Physiologie*. Braunschweig: Friedrich Bieweg und Sohn.

Voss, L. F., Murphy, J. W., Shao, Q., Henderson, R. A., Frye, C. D., Stoyer, M. A., & Nikolic, R. J. (2018). Selenium-iodide: A low melting point eutectic semiconductor. *Applied Physics Letters*, *113*, 242103.

Wiberg E., Henle, W. (1952). Über die Dämpfung der elektromagnetischen Eigenschwingungen des Systems Erde — Luft — Ionosphäre. *Zeitschrift für Naturforschung A. 7*, 250–252.

Wilke, G. (1988). Contributions to Organo-Nickel Chemistry. *Angewandte Chemie International Edition in English*, *27*, 185–206.

Wilkinson, G., Rosenblum, M., Whiting, M. C., & Woodward, R. B. (1952). The structure of iron bis-cyclopentadienyl. *Journal of the American Chemical Society*, *74*, 2125–2126.

Winkler, C. (1887a). Germanium, Ge, a new nonmetal element. *Berichte der Deutschen Chemischen Gesellschaft*, *19*, 210–211.

Winkler, C. (1887b). Mittheilungen über des Germanium. Zweite Abhandlung. *Praktische Chemie*, *36*, 177–209.

Yorimitsu, H., & Oshima, K. (2005). Recent advances in the use of tri(2-furyl)germane, triphenylgermane and their derivatives in organic synthesis. *Inorganc Chemistry Communications*, *8*, 131–142.

Yu, L., Qian, R., Deng, X., Wang, F., & Xu, Q. (2018). Calcium-catalyzed reactions of element-H bonds. *Science Bulletin*, *63*, 1010–1016.

Further reading

Abon, M., & Volta, J.-C. (1997). Vanadium phosphorus oxides for n-butane oxidation to maleic anhydride. *Applied Catalysis A: General*, *157*, 173–193.

Agricola, G., *De Natura Fossilium (Textbook of Mineralogy)*. Translated from the First Latin Ed. of 1546 by Mark Chance Bandy and Jean A. Bandy for the Mineralogical Society of America (M. C. Bandy, J. A. Bandy, Trans.). 1955 [1546] Boulder: Geological Society of America.

Ahmad, Z. (2003). The properties and application of scandium-reinforced aluminum. *JOM*, *55*, 35–39.

Ahmed, F. U., Yunus, S. M., Kamal, I., Begum, S., Khan, A. A., Ahsan, M. H., & Ahmad, A. A. Z. (1996). Optimization of germanium for neutron diffractometers. *International Journal of Modern Physics E*, *5*, 131–151.

Akimoto, J., Gotoh, Y., Oosawa, Y., Nonose, N., Kumagai, T., Aoki, K., & Takei, H. (1994). Topotactic oxidation of ramsdellite-Type $Li_{0.5}TiO_2$, a new polymorph of titanium dioxide: TiO_2(R). *Journal of Solid State Chemistry*, *113*, 27–36.

Aksoy, R., Selvi, E., Knudson, R., & Ma, Y. (2008). A high pressure x-ray diffraction study of titanium disulfide. *Journal of Physics: Condensed Matter*, *21*, 025403.

Alberts, V., Titus, J. T., & Birkmire, R. W. (2003). Material and device properties of single-phase $Cu(In,Ga)(Se,S)_2$ alloys prepared by selenization/sulfurization of metallic alloys. *Thin Solid Films*, *451–452*, 207–211.

Aleksandrov, K. S., Voronov, V. N., Vtyurin, A. N., Krylov, A. S., Molokeev, M. S., Pavlovskiĭ, M. S., ... Ancharov, A. I. (2009). Pressure-induced phase transition in the cubic ScF_3 crystal. *Physics of the Solid State*, *51*, 810–816.

Alessio, L., Campagna, M., & Lucchini, R. (2007). From lead to manganese through mercury: Mythology, science, and lessons for prevention. *American Journal of Industrial Medicine*, *50*, 779–787.

Al-Khatatbeh, Y., Lee, K. K. M., & Kiefer, B. (2009). High-pressure behavior of TiO_2 as determined by experiment and theory. *Physical Review B, 79*, 134114.

Alling, B., Marten, T., & Abrikosov, I. A. (2010a). Effect of magnetic disorder and strong electron correlations on the thermodynamics of CrN. *Physical Review B, 82*, 184430.

Alling, B., Marten, T., & Abrikosov, I. A. (2010b). Questionable collapse of the bulk modulus in CrN. *Nature Materials, 9*, 283–284.

Alonso, P. J., Forniés, J., García-Monforte, M. A., Martín, A., & Menjón, B. (2005). New homoleptic organometallic derivatives of vanadium(III) and vanadium(IV): Synthesis, characterization, and study of their electrochemical behaviour. *Chemistry – A European Journal, 11*, 4713–4724.

Anderson, F. J. (1981). *Riches of the earth: Ornamental, precious and semiprecious stones*. New York: Windward.

Angara, R., High frequency high amplitude magnetic field driving system for magnetostrictive actuators. 2009, Dissertation, University of Maryland, Baltimore.

Anger, G., Halstenberg, J., Hochgeschwender, K., Scherhag, C., Korallus, U., Knopf, H., ... Ohlinger, M. (2000). *Chromium compounds. Ullmann's encyclopedia of industrial chemistry*. Weinheim: Wiley-VCH.

Antman, K. H. (2001). The history of arsenic trioxide in cancer therapy. *The Oncologist, 6* (Suppl 2), 1–2.

Apelian, D., Paliwal, M., & Herrschaft, D. C. (1981). Casting with zinc alloys. *Journal of Metals, 33*, 12–19.

Archibald, S. J. (2003). 6.8 - Zinc. In J. A. McCleverty, & T. J. Meyer (Eds.), *Comprehensive Coordination Chemistry II*. Oxford: Pergamon.

Armstrong, R. D., Briggs, G. W. D., & Charles, E. A. (1988). Some effects of the addition of cobalt to the nickel hydroxide electrode. *Journal of Applied Electrochemistry, 18*, 215–219.

Arnold, P. L., Cloke, F. G. N., Hitchcock, P. B., & Nixon, J. F. (1996). The first example of a formal scandium(I) complex: Synthesis and molecular structure of a 22-electron scandium triple decker incorporating the novel 1,3,5-triphosphabenzene ring. *Journal of the American Chemical Society, 118*, 7630–7631.

Arnold, P. L., Geoffrey, N., Cloke, F., & Nixon, J. F. (1998). The first stable scandocene: Synthesis and characterisation of bis(η-2,4,5-tri-tert-butyl-1,3-diphosphacyclopentadienyl)scandium(II). *Chemical Communications*, 797–798.

Arnold, R., Augier, C., Bakalyarov, A. M., Baker, J. D., Barabash, A. S., Basharina-Freshville, A., ... Žukauskas, A. (2016). Measurement of the double-beta decay half-life and search for the neutrinoless double-beta decay of ^{48}Ca with the NEMO-3 detector. *Physical Review D, 93*, 112008.

Arny, H. V. (1917). *Principles of pharmacy*. Philadelphia, PA: W.B. Saunders, p. 483.

Astruc, D. (2017). Why is ferrocene so exceptional? *European Journal of Inorganic Chemistry, 2017*, 6–29.

Audi, G., Bersillon, O., Blachot, J., & Wapstra, A. H. (2003). The NUBASE evaluation of nuclear and decay properties. *Nuclear Physics A, 729*, 3–128.

Bagshaw, N. E. (1995). Lead alloys: Past, present and future. *Journal of Power Sources, 53*, 25–30.

Baldwin, W. H. (1931a). The story of nickel. I. How "Old Nick's" gnomes were outwitted. *Journal of Chemical Education, 8*, 1749–1760.

Baldwin, W. H. (1931b). The story of nickel. Part II. Nickel comes of age. *Journal of Chemical Education, 8*, 1954–1967.

Baldwin, W. H. (1931c). The story of nickel. Part III. Ore, matte, and metal. *Journal of Chemical Education, 8*, 2325–2339.

Ball, P. (2002). *The ingredients: A guided tour of the elements*. Oxford: Oxford University Press.

Banerjee, S. R., & Pomper, M. G. (2013). Clinical applications of gallium-68. *Applied Radiation and Isotopes, 76*, 2–13.

Baral, A., & Engelken, R. D. (2002). Chromium-based regulations and greening in metal finishing industries in the USA. *Environmental Science & Policy, 5*, 121–133.

Baran, E. J. (1998). Materials belonging to the $CrVO_4$ structure type: Preparation, crystal chemistry and physicochemical properties. *Journal of Materials Science, 33*, 2479–2497.

Barksdale, J. (1968). Titanium. In C. A. Hampel (Ed.), *The encyclopedia of the chemical elements* (pp. 732–738). New York: Reinhold Book Corporation.

Baturina, T. I., Mironov, A. Y., Vinokur, V. M., Baklanov, M. R., & Strunk, C. (2007). Localized superconductivity in the quantum-critical region of the disorder-driven superconductor-insulator transition in TiN thin films. *Physical Review Letters, 99*, 257003.

Bauer, G., Güther, V., Hess, H., Otto, A., Roidl, O., Roller, H., ... Beyer, T. (2017). *Vanadium and vanadium compounds. Ullmann's encyclopedia of industrial chemistry*. (pp. 1–22). Weinheim: Wiley-VCH.

Bauer, I., & Knölker, H.-J. (2015). Iron catalysis in organic synthesis. *Chemical Reviews, 115*, 3170–3387.

Baur, F., & Jüstel, T. (2016). Dependence of the optical properties of Mn^{4+} activated $A_2Ge_4O_9$ (A = K,Rb) on temperature and chemical environment. *Journal of Luminescence, 177*, 354–360.

Bautista, M. A., Pradhan, A. K., Iron and nickel abundances in H II regions and supernova remnants. Paper presented at the American Astronomical Society Meeting Abstracts #186, May 1, 1995.

Bekker, A., Slack, J. F., Planavsky, N., Krapez, B., Hofmann, A., Konhauser, K. O., & Rouxel, O. J. (2010). Iron formation: The sedimentary product of a complex interplay among mantle, tectonic, oceanic, and biospheric processes. *Economic Geology, 105*, 467–508.

Bentley, R., & Chasteen, T. G. (2002). Arsenic curiosa and humanity. *The Chemical Educator, 7*, 51–60.

Bernhardt, P. V., & Lawrance, G. A. (2003). 6.1 - Cobalt. In J. A. McCleverty, & T. J. Meyer (Eds.), *Comprehensive coordination chemistry II*. Oxford: Pergamon.

Bernstein, L. (1985). Germanium geochemistry and mineralogy. *Geochimica et Cosmochimica Acta, 49*, 2409–2422.

Bernstein, L. R., Tanner, T., Godfrey, C., & Noll, B. (2000). Chemistry and pharmacokinetics of gallium maltolate, a compound with high oral gallium bioavailability. *Metal-Based Drugs, 7*, 33–47.

Berzelius, J. J. (1814). *Försök att, genom användandet af den electrokemiska theorien och de kemiska proportionerna: grundlägga ett rent vettensk. system för mineralogien*. Stockholm: A. Gadelius.

Besenhard, J. O. (2010). *Handbook of battery materials*. Weinheim; New York, NY: Wiley-VCH.

Besmann, T. M. (2005). Thermochemical behavior of gallium in weapons-material-derived mixed-oxide light water reactor (LWR) fuel. *Journal of the American Ceramic Society, 81*, 3071–3076.

Bhutta, Z. A., Bird, S. M., Black, R. E., Brown, K. H., Gardner, J. M., Hidayat, A., … Shankar, A. (2000). Therapeutic effects of oral zinc in acute and persistent diarrhea in children in developing countries: Pooled analysis of randomized controlled trials. *The American Journal of Clinical Nutrition, 72*, 1516–1522.

Biggs, T., Taylor, S. S., & Van Der Lingen, E. (2005). The hardening of platinum alloys for potential jewellery application. *Platinum Metals Review, 49*, 2–15.

Biot, C., & Dive, D. (2010). Bioorganometallic chemistry and malaria. In G. Jaouen, & N. Metzler-Nolte (Eds.), *Medicinal organometallic chemistry* (pp. 155–193). Berlin, Heidelberg: Springer.

Birck, J. L., Rotaru, M., & Allegre, C. (1999). ^{53}Mn-^{53}Cr evolution of the early solar system. *Geochimica et Cosmochimica Acta, 63*, 4111–4117.

Birnbaum, K. (1869). *Die Kalidüngung in ihren Vortheilen und Gefahren*. Berlin: Wiegandt & Hempel.

Bishop, D. W., Thomas, P. S., & Ray, A. S. (1998). Raman spectra of nickel(II) sulfide. *Materials Research Bulletin, 33*, 1303–1306.

Biurrun, A., Caballero, L., Pelaz, C., León, E., & Gago, A. (1999). Treatment of a *Legionella pneumophila*-colonized water distribution system using copper-silver ionization and continuous chlorination. *Infection Control and Hospital Epidemiology, 20*, 426–428.

Bjerklie, S. (2006). A batty business: Anodized metal bats have revolutionized baseball. But are finishers losing the sweet spot? *Metal Finishing, 104*, 61–62.

Bjorkman, J. K. (1973). Meteors and meteorites in the ancient near east. *Meteoritics, 8*, 91–132.

Blachnik, R., & Müller, A. (2000). The formation of Cu_2S from the elements: I. Copper used in form of powders. *Thermochimica Acta, 361*, 31–52.

Boettger, R. (1859). Ueber die Einwirkung des Leuchtgases auf verschiedene Salzsolutionen, insbesondere auf eine ammoniakalische Kupferchlorürlösung. *Justus Liebigs Annalen der Chemie, 109*, 351–362.

Bonati, A., Pisano, G., & Royer Carfagni, G. (2019). A statistical model for the failure of glass plates due to nickel sulfide inclusions. *Journal of the American Ceramic Society, 102*, 2506–2521.

Borg, G., Kärner, K., Buxton, M., Armstrong, R., & van der Merwe, S. W. (2003). Geology of the Skorpion supergene zinc deposit, Southern Namibia. *Economic Geology, 98*, 749–771.

Bortz, M., Gutmann, M., & Yvon, K. (1999). Synthesis and structure determination of the first ternary cadmium hydride, Cs_3CdH_5. *Journal of Alloys and Compounds, 285*, L19–L21.

Bosacka, M., Jakubus, P., & Rychłowska-Himmel, I. (2007). Obtaining of chromium(III) phosphates(V) in the solid-state and their thermal stability. *Journal of Thermal Analysis and Calorimetry, 88*, 133–137.

Bottrill, M., Gavens, P. D., Kelland, J. W., & McMeeking, J. (1982). 22.4 - σ-Bonded hydrocarbyl complexes of titanium(IV). In G. Wilkinson, F. G. A. Stone, & E. W. Abel (Eds.), *Comprehensive Organometallic Chemistry* (pp. 433–474). Oxford: Pergamon.

Boudjouk, P., So, J.-H., Ackermann, M. N., Hawley, S. E., & Turk, B. E. (1992). Solvated and unsolvated anhydrous metal chlorides from metal chloride hydrates. In R. N. Grimes (Ed.), *Inorganic syntheses. Inorganic syntheses* (pp. 108–111). New York: Joh Wiley & Sons.

Bounoughaz, M., Salhi, E., Benzine, K., Ghali, E., & Dalard, F. (2003). A comparative study of the electrochemical behaviour of Algerian zinc and a zinc from a commercial sacrificial anode. *Journal of Materials Science, 38*, 1139–1145.

Bowen, R. (1994). *Isotopes in the earth sciences*. London: Chapman & Hall.

Bramfitt, B. L., & Benscoter, A. O. (2002). *The iron carbon phase diagram. Metallographer's guide practices and procedures for irons and steels* (pp. 24–28). Materials Park, OH: ASM International.

Brandán, S. A., Ben Altabef, A., & Varetti, E. L. (1995). Vibrational and electronic spectra of vanadyl nitrate, $VO(NO_3)_3$. *Spectrochimica Acta Part A: Molecular and Biomolecular Spectroscopy, 51*, 669–675.

Brandt, G. (1748). Cobalti nova species examinata et descripta. *Acta Regiae Societatis Scientiarum Upsaliensis, 3*, 33–41, 1st series.

Bray, F., Daniels, C., Golas, P. J., Harbsmeier, C., Huang, H. -T., Kerr, R., … Wagner, D. B. (2008). *Science and civilisation in China. Ferrous metallurgy Vol. 5 Chemistry and chemical technology*. Part 11 Chemistry and chemical technology. 2008, no publisher info.

Brencic, J. V., & Cotton, F. A. (1969). Octachlorodimolybdate(II) ion. Species with a quadruple metal-metal bond. *Inorganic Chemistry, 8*, 7–10.

Brignole, A. B., Cotton, F. A., Dori, Z., Dori, Z., Dori, Z., & Wilkinson, G. (1972). Rhenium and molybdenum compounds containing quadruple bonds. *Inorganic Syntheses*, 81–89.

Browne, C. A. (1926). Historical notes upon the domestic potash industry in early colonial and later times. *Journal of Chemical Education, 3*, 749–756.

Brownstein, S. (1980). The structure of VF_5 in solution. *Journal of Fluorine Chemistry, 15*, 539–540.

Brownstein, S., & Latremouille, G. (1974). Complex fluoroanions in solution. V. Vanadium pentafluoride. *Canadian Journal of Chemistry, 52*, 2236–2241.

Buckley, A. N. (1987). The surface oxidation of cobaltite. *Australian Journal of Chemistry, 40*, 231–239.

Burkhardt, E. R. (2006). *Potassium and potassium alloys. Ullmann's encyclopedia of industrial chemistry*. Weinheim: Wiley-VCH.

Burton, J. D., Culkin, F., & Riley, J. P. (2007). The abundances of gallium and germanium in terrestrial materials. *Geochimica et Cosmochimica Acta, 16*, 151–180.

Cahiez, G., Alami, M., Taylor, R. J. K., Reid, M., Foot, J. S., Fader, L., … Pabba, J. (2017). *Manganese dioxide. Encyclopedia of reagents for organic synthesis.* (pp. 1–16). New York: John Wiley & Sons.

Calderazzo, F., Maichle-Mossmer, C., Pampaloni, G., & Strähle, J. (1993). Low-temperature syntheses of vanadium(III) and molybdenum(IV) bromides by halide exchange. *Journal of the Chemical Society, Dalton Transactions*, 655–658.

Calvet, G., Dussaussois, M., Blanchard, N., & Kouklovsky, C. (2004). Lewis acid-promoted hetero Diels–Alder cycloaddition of α-acetoxynitroso dienophiles. *Organic Letters*, *6*, 2449–2451.

Cameron, A. G. W. (1973). Abundance of the elements in the solar system. *Space Science Reviews*, *15*, 121–146.

Campbell, F. C. (2008). *Elements of metallurgy and engineering alloys*. Materials Park, OH.: ASM International.

Canterford, J. H., & O'Donnell, T. A. (1967). Reactivity of transition metal fluorides. IV. Oxidation-reduction reactions of vanadium pentafluoride. *Inorganic Chemistry*, *6*, 541–544.

Cao, W., Wei, Y. N., Meng, X., Ji, Y., & Ran, S. (2017). A general method towards transition metal monoboride nanopowders. *International Journal of Materials Research*, *108*, 335–338.

Carlson, O. N., & Owen, C. V. (1961). Preparation of high-purity vanadium metals by the iodide refining process. *Journal of the Electrochemical Society*, *108*, 88–93.

Carney, M. J., & Smith, K. M. (2013). *Chromium compounds without CO or isocyanides. Reference Module in Chemistry, Molecular Sciences and Chemical Engineering*. Elsevier.

Chalmin, E., Menu, M., & Vignaud, C. (2003). Analysis of rock art painting and technology of Palaeolithic painters. *Measurement Science and Technology*, *14*, 1590–1597.

Chalmin, E., Vignaud, C., Salomon, H., Farges, F., Susini, J., & Menu, M. (2006). Minerals discovered in paleolithic black pigments by transmission electron microscopy and micro-X-ray absorption near-edge structure. *Applied Physics A*, *83*, 213–218.

Chandler, H. (2006). *Metallurgy for the non-metallurgist*. Materials Park, OH: ASM International.

Charles, J. A. (1967). Early arsenical bronzes-A metallurgical view. *American Journal of Archaeology*, *71*, 21–26.

Cheburaeva, R. F., Chaporova, I. N., & Krasina, T. I. (1992). Structure and properties of tungsten carbide hard alloys with an alloyed nickel binder. *Soviet Powder Metallurgy and Metal Ceramics*, *31*, 423–425.

Chen, D., Zhou, Y., & Zhong, J. (2016). A review on Mn^{4+} activators in solids for warm white light-emitting diodes. *RSC Advances*, *6*, 86285–86296.

Chen, G. Z., Fray, D. J., & Farthing, T. W. (2000). Direct electrochemical reduction of titanium dioxide to titanium in molten calcium chloride. *Nature*, *407*, 361–364.

Chertihin, G. V., & Andrews, L. (1994). Reactions of laser ablated Ti atoms with hydrogen during condensation in excess argon. Infrared spectra of the TiH, TiH_2, TiH_3, and TiH_4 molecules. *Journal of the American Chemical Society*, *116*, 8322–8327.

Chester, A. W., Heiba, E.-A., Dessau, R. M., & Koehl, W. J. (1969). The interaction of cobalt(III) with chloride ion in acetic acid. *Inorganic and Nuclear Chemistry Letters*, *5*, 277–283.

Chieh, C., & White, M. (1984). Crystal structure of anhydrous zinc bromide. *Zeitschrift für Kristallographie - Crystalline Materials*, *166*. Available from https://doi.org/10.1524/zkri.1984.166.3-4.189.

Chitambar, C. R. (2018). Chapter 10. Gallium complexes as anticancer drugs. In A. Sigel, H. Sigel, E. Freisinger, & R. K. O. Sigel (Eds.), *Metallodrugs: Development and action of anticancer agents. Metal ions in life sciences* (18, pp. 281–301). Berlin: de Gruyter GmbH.

Cintas, P. (2004). The road to chemical names and eponyms: Discovery, priority, and credit. *Angewandte Chemie International Edition*, *43*, 5888–5894.

Cintho, O. M., Favilla, E. A. P., & Capocchi, J. D. T. (2007). Mechanical–thermal synthesis of chromium carbides. *Journal of Alloys and Compounds*, *439*, 189–195.

Cipollina, A., Micale, G., & Rizzuti, L. (2010). *Seawater desalination: Conventional and renewable energy processes*. Milton Keynes: Lightning Source UK.

Clayton, D. D. (2003). *Handbook of isotopes in the cosmos*. Cambridge, New York: Cambridge University Press.

Cloke, F. G. N., Khan, K., & Perutz, R. N. (1991). η-Arene complexes of scandium(0) and scandium(II). *Journal of the Chemical Society, Chemical Communications*, 1372–1373.

Cloud, P. (1973). Paleoecological significance of the banded iron-formation. *Economic Geology*, *68*, 1135–1143.

Coe, P. L. (2001). *Cobalt(III) fluoride. Encyclopedia of reagents for organic synthesis*. New York: John Wiley & Sons.

Comelli, D., d'Orazio, M., Folco, L., El-Halwagy, M., Frizzi, T., Alberti, R., ... Valentini, G. (2016). The meteoritic origin of Tutankhamun's iron dagger blade. *Meteoritics & Planetary Science*, *51*, 1301–1309.

Comyns, A. E. (2007). *Encyclopedic dictionary of named processes in chemical technology*. Baton Rouge: CRC Press.

Corbett, J. D. (1981). Extended metal-metal bonding in halides of the early transition metals. *Accounts of Chemical Research*, *14*, 239–246.

Cordel, O. (1868). *Die Stassfurter Kalisalze in der Landwirthschalt: Eine Besprechung*. Aschersleben: L. Schnock.

Cotterell, M. (2004). *The Terracotta Warriors: The secret codes of the emperor's army*. Rochester: Bear & Company.

Cotton, F. A., Falvello, L. R., Llusar, R., Libby, E., Murillo, C. A., & Schwotzer, W. (1986). Synthesis and characterization of four vanadium(II) compounds, including vanadium(II) sulfate hexahydrate and vanadium(II) saccharinates. *Inorganic Chemistry*, *25*, 3423–3428.

Cotton, F. A., Falvello, L. R., Murillo, C. A., Pascual, I., Schultz, A. J., & Tomas, M. (1994). Neutron and X-ray structural characterization of the hexaaquavanadium(II) compound $VSO_4.7H_2O$. *Inorganic Chemistry*, *33*, 5391–5395.

Cotton, F. A., Koshevoy, I. O., Lahuerta, P., Murillo, C. A., Sanaú, M., Ubeda, M. A., & Zhao, Q. (2006). High yield syntheses of stable, singly bonded Pd_2^{6+} compounds. *Journal of the American Chemical Society*, *128*, 13674–13675.

Couper, J. (1837). On the effects of black oxide of manganese when inhaled into the lungs. *British Annals of Medicine, Pharmacy, Vital Statistics, and General Science*, *1*, 41–42.

Craddock, P. T. (1978). The composition of copper alloys used by the Greek, Etruscan and Roman civilizations. The origins and early use of brass. *Journal of Archaeological Science*, *5*, 1–16.

Craddock, P. T. (1998). *2000 years of zinc and brass*. London: British Museum.

Craddock, P. T., Gurjar, L. K., & Hegde, K. T. M. (1983). Zinc production in medieval India. *World Archaeology*, *15*, 211–217.

Crisp, D., Pathare, A., & Ewell, R. C. (2004a). The performance of gallium arsenide/germanium solar cells at the Martian surface. *Acta Astronautica*, *54*, 83–101.

Dahal, A., Gunasekera, J., Harringer, L., Singh, D. K., & Singh, D. J. (2016). Metallic nickel silicides: Experiments and theory for NiSi and first principles calculations for other phases. *Journal of Alloys and Compounds*, *672*, 110–116.

Dastur, Y. N., & Leslie, W. C. (1981). Mechanism of work hardening in Hadfield manganese steel. *Metallurgical Transactions A*, *12*, 749–759.

Davis, J. R. (1998). *Tool materials*. Materials Park, OH: ASM International.

Davis, J. R. (2007). *Nickel, cobalt, and their alloys*. Materials Park, OH: ASM International.

Davis, J. R. (2008). *Copper and copper alloys*. Materials Park, OH: ASM International.

Davis, J. R. (2010). *Metals handbook*. Materials Park, OH: ASM International.

de Callataÿ, F. (2005). The Graeco-Roman economy in the super long-run: Lead, copper, and shipwrecks. *Journal of Roman Archaeology*, *18*, 361–372.

Decanio, S. J., & Norman, C. S. (2008). Economics of the "critical use" of methyl bromide under the montreal protocol. *Contemporary Economic Policy*, *23*, 376–393.

de Fourcroy, A. F. (1804). *A general system of chemical knowledge: And its application to the phenomena of nature and art*. (W. Nicholson, Trans.), (11 Vols). London: Cadell and Davies.

Delahunt, J., & Lindeman, T. (2007). Review of the safety of potassium and potassium oxides, including deactivation by introduction into water. *Journal of Chemical Health and Safety*, *14*, 21–32.

Dell, R. M. (2000). Batteries: Fifty years of materials development. *Solid State Ionics*, *134*, 139–158.

DeMeo, S. (1995). Synthesis and decomposition of zinc iodide: Model reactions for investigating chemical change in the introductory laboratory. *Journal of Chemical Education*, *72*, 836.

Dennis, J. K., & Such, T. E. (1993). *Nickel and chromium plating*. Cambridge: Woodhead Publishing.

Dennis, W. H. (2010). *Metallurgy: 1863-1963*. New Brunswick: Aldine Transaction.

Dilman, A. D., & Levin, V. V. (2016). Advances in the chemistry of organozinc reagents. *Tetrahedron Letters*, *57*, 3986–3992.

Disegi, J. A., Kennedy, R. L., & Pilliar, R. (2011). *Cobalt-base alloys for biomedical applications*. West Conshohocken, PA: ASTM International.

Dixon, J. T., Green, M. J., Hess, F. M., & Morgan, D. H. (2004). Advances in selective ethylene trimerisation – A critical overview. *Journal of Organometallic Chemistry*, *689*, 3641–3668.

Donachie, M. J. (1988). *Titanium: A technical guide*. Metals Park, OH: ASM International.

Dove, M. F. A., Manz, B., Montgomery, J., Pattenden, G., & Wood, S. A. (1998). Vanadium(V) oxytrinitrate, $VO(NO_3)_3$. A powerful reagent for the nitration of aromatic compounds at room temperature under non-acidic conditions. *Journal of the Chemical Society, Perkin Transactions*, *1*, 1589–1590.

Downs, A. J. (1993). *Chemistry of aluminium, gallium, indium and thallium*. London: Blackie Academic & Professional.

Dubrovinskaia, N. A., Dubrovinsky, L. S., Ahuja, R., Prokopenko, V. B., Dmitriev, V., Weber, H. P., ... Johansson, B. (2001). Experimental and theoretical identification of a new high-pressure TiO_2 polymorph. *Physical Review Letters*, *87*, 275501.

Duhlev, R., Brown, I. D., & Faggiani, R. (1988). Zinc bromide dihydrate $ZnBr_2.2H_2O$: A double-salt structure. *Acta Crystallographica C*, *44*, 1696–1698.

Dyck, H. M., & Nordgren, T. E. (2002). The effect of TiO absorption on optical and infrared angular diameters of cool stars. *The Astronomical Journal*, *124*, 541–545.

Eberhardt, N. A., & Guan, H. (2016). Nickel hydride complexes. *Chemical Reviews*, *116*, 8373–8426.

Ebert, F., & Woitinek, H. (1933). Kristallstrukturen von Fluoriden. II. HgF, HgF_2, CuF und CuF_2. *Zeitschrift für anorganische und allgemeine Chemie*, *210*, 269–272.

Ebru, Ş. T., Hamide, K., & Ramazan, E. (2007). Structural and optical properties of zinc nitride films prepared by pulsed filtered cathodic vacuum arc deposition. *Chinese Physics Letters*, *24*, 3477–3480.

Eftekhari, A. (2017). The rise of lithium–selenium batteries. *Sustainable Energy & Fuels*, *1*, 14–29.

El Goresy, A., Chen, M., Gillet, P., Dubrovinsky, L., Graup, G., & Ahuja, R. (2001). A natural shock-induced dense polymorph of rutile with α-PbO_2 structure in the suevite from the Ries crater in Germany. *Earth and Planetary Science Letters*, *192*, 485–495.

Elliott, R., Coley, K., Mostaghel, S., & Barati, M. (2018). Review of manganese processing for production of TRIP/TWIP steels, Part 1: Current practice and processing fundamentals. *JOM*, *70*, 680–690.

Elschenbroich, C., Schmidt, E., Gondrum, R., Metz, B., Burghaus, O., Massa, W., & Wocadlo, S. (1997). Metal π complexes of benzene derivatives. Germanium in the periphery of bis(benzene)vanadium and bis(benzene)chromium. Synthesis and structure of new heterametallocyclophanes. *Organometallics*, *16*, 4589–4596.

Emerson, T. E., McElrath, D. L., & Fortier, A. C. (2012). *Archaic societies: Diversity and complexity across the midcontinent*. Albany: State University of New York Press.

Emsley, J. (2001). *Nature's building blocks*. Oxford: Oxford: Oxford University Press.

Enghag, P. (2004). *Encyclopedia of the elements: Technical data - History - Processing - Applications*. Weinheim: Wiley-VCH.

EPA. (2016). EPA announces updated draft efficacy protocol for copper surface sanitizer products. EPA. <https://archive.epa.gov/epa/pesticides/epa-announces-updated-draft-efficacy-protocol-copper-surface-sanitizer-products.html>.

Erickson, R. L. (1973). Crustal abundance of elements, and mineral reserves and resources. *US Geological Survey Professional Paper*, *820*, 21–25.

Evans, J. R., & Lawrenson, J. G. (2017). Antioxidant vitamin and mineral supplements for slowing the progression of age-related macular degeneration. *Cochrane Database of Systematic Reviews*, CD000254.

Fang, L., Kulkarni, S., Alhooshani, K., & Malik, A. (2007). Germania-based, Sol-Gel hybrid organic-inorganic coatings for capillary microextraction and gas chromatography. *Analytical Chemistry*, *79*, 9441–9451.

Farahani, S. K. V., Muñoz-Sanjosé, V., Zúñiga-Pérez, J., McConville, C. F., & Veal, T. D. (2013). Temperature dependence of the direct bandgap and transport properties of CdO. *Applied Physics Letters*, *102*, 022102.

Fewell, M. P. (1995). The atomic nuclide with the highest mean binding energy. *American Journal of Physics*, *63*, 653–658.

Figoni, P. (2011). *How baking works: Exploring the fundamentals of baking science*. Hoboken, NJ: John Wiley & Sons.

Filling, H. (1849). Die quantitative Bestimmung von Zucker und Stärkmehl mittelst Kupfervitriol. *Annalen der Chemie und Pharmacie*, *72*, 106–113.

Finney, A., Hitchman, M., & Raston, C. (1981). Structural and spectroscopic studies of transition metal nitrite complexes. I. Crystal structures and spectra of *trans*-Bis(ethane-1,2 diamine)dinitronickel(II), *trans*-Bis[N, N-dimethyl(ethane-1,2-diamine)]dinitritonickel(II) and *trans*-bis[N,N′-dimethyl(ethane-1,2 diamine)]dinitronickel(II) monohydrate. *Australian Journal of Chemistry*, *34*, 2047–2060.

Fitzgerald, K. P., Nairn, J., & Atrens, A. (1998). The chemistry of copper patination. *Corrosion Science*, *40*, 2029–2050.

Fjellvåg, H., & Karen, P. (1994). Crystal structure of $ScCl_3$ refined from powder neutron diffraction data. *Acta Chemica Scandinavica*, *48*, 294–297.

Flower, K. R. (2007). 5.07 - Molybdenum compounds without CO or isonitrile ligands. In D. M. P. Mingos, & R. H. Crabtree (Eds.), *Comprehensive organometallic chemistry III* (pp. 513–595). Oxford: Elsevier.

Folmer, J. C. W., & Jellinek, F. (1980). The valence of copper in sulphides and selenides: An X-ray photoelectron spectroscopy study. *Journal of the Less Common Metals*, *76*, 153–162.

Folmer, J. C. W., Jellinek, F., & Calis, G. H. M. (1988). The electronic structure of pyrites, particularly CuS_2 and $Fe_{1-x}CuxSe_2$: An XPS and Mössbauer study. *Journal of Solid State Chemistry*, *72*, 137–144.

Font, O., Querol, X., Juan, R., Casado, R., Ruiz, C. R., López-Soler, Á., ... Peña, F. G. (2007). Recovery of gallium and vanadium from gasification fly ash. *Journal of Hazardous Materials*, *139*, 413–423.

Fordyce, F. (2007). Selenium geochemistry and health. *AMBIO: A Journal of the Human Environment*, *36*, 94–97.

Frenzel, M. (2016). *The distribution of gallium, germanium and indium in conventional and non-conventional resources - Implications for global availability*. Freiberg: Technische Universität Bergakademie Freiberg.

Frenzel, M., Hirsch, T., & Gutzmer, J. (2016). Gallium, germanium, indium, and other trace and minor elements in sphalerite as a function of deposit type — A metaanalysis. *Ore Geology Reviews*, *76*, 52–78.

Gafner, G. (1989). The development of 990 gold-titanium: Its production, use and properties. *Gold Bulletin*, *22*, 112–122.

Gaines, R. V., Skinner, H. C., Foord, E. F., Mason, B., & Rosenzweig, A. (1997). *Dana's new mineralogy* (8th edn). New York: John Wiley & Sons, Inc.

Gardos, M. N., Soriano, B. L., Propst, S. H. (1990). Study on correlating rain erosion resistance with sliding abrasion resistance of DLC on germanium. Paper presented at the 34th Annual International Technical Symposium on Optical and Optoelectronic Applied Science and Engineering, San Diego, CA.

Garfinkel, Y. (1987). Burnt lime products and social implications in the pre-pottery neolithic B villages of the near east. *Paléorient*, *13*, 69–76.

Garrett, D. E. (1995). *Potash: Deposits, processing, properties and uses*. London: Chapman & Hall.

Geng, J., Jefferson, D. A., & Johnson, B. F. G. (2007). The unusual nanostructure of nickel–boron catalyst. *Chemical Communications*, 969–971.

Gettens, R. J., & Fitzhugh, E. W. (1993). Azurite and blue verditer. In A. Roy (Ed.), *Artists' pigments. A handbook of their history and characteristics* (Vol. 2, pp. 23–24). Oxford: Oxford University Press.

Gettens, R. J., Stout, G. L. (2015). *Painting materials: A short encyclopaedia*. pp. 109–110.

Giachi, G., Pallecchi, P., Romualdi, A., Ribechini, E., Lucejko, J. J., Colombini, M. P., & Mariotti Lippi, M. (2013). Ingredients of a 2,000-y-old medicine revealed by chemical, mineralogical, and botanical investigations. *Proceedings of the National Academy of Sciences*, *110*, 1193–1196.

Giannichedda, E. (2007). Metal production in late antiquity. In L. Lavan, E. Zanini, & A. Sarantis (Eds.), *Technology in transition AD 300–650* (p. 200). Leiden: Brill.

Gibaud, S., & Jaouen, G. (2010). Arsenic-based drugs: From Fowler's solution to modern anticancer chemotherapy. *Topics in Organometallic Chemistry*, *32*, 1–20.

Girgsdies, F., Schneider, M., Brückner, A., Ressler, T., & Schlögl, R. (2009). The crystal structure of δ-$VOPO_4$ and its relationship to ω-$VOPO_4$. *Solid State Sciences*, *11*, 1258–1264.

Glenn, W. (1895). *Chrome in the southern Appalachian region, . Transactions of the American Institute of Mining, Metallurgical and Petroleum Engineers* (25, pp. 481–499).

Goh, S. W., Buckley, A. N., & Lamb, R. N. (2006). Copper(II) sulfide? *Minerals Engineering*, *19*, 204–208.

Goldberg, D. E., Sharma, V., Oksman, A., Gluzman, I. Y., Wellems, T. E., & Piwnica-Worms, D. (1997). Probing the chloroquine resistance locus of *Plasmodium falciparum* with a novel class of multidentate metal(III) coordination complexes. *Journal of Biological Chemistry*, *272*, 6567–6572.

Goldschmidt, H. (1898a). Über ein neues Verfahren zur Darstellung von Metallen und Legirungen mittelst Aluminiums. *Justus Liebigs Annalen der Chemie*, *301*, 19–28.

Goldschmidt, H. (1898b). Über ein neues Verfahren zur Erzeugung von hohen Temperaturen und zur Darstellung von schwer schmelzbaren kohlefreien Metallen. *Zeitschrift für Elektrochemie*, *4*, 494–499.

Goldschmidt, H., & Vautin, C. (1898). Aluminium as a heating and reducing agent. *Journal of the Society of Chemical Industry*, *6*, 543–545.

Goldschmidt, V. M. (1930). Ueber das Vorkommen des Germaniums in Steinkohlen und Steinkohlenprodukten. Nachrichten von der Gesellschaft der Wissenschaften zu Göttingen. *Mathematisch-Physikalische Klasse*, 398–402.

Goldschmidt, V. M., & Peters, C. (1933). Zur Geochemie des Germaniums Nachrichten von der Gesellschaft der Wissenschaften zu Göttingen. *Mathematisch-Physikalische Klasse*, 141–166.

Gol'Dshtein, Y. E., Mushtakova, T. L., & Komissarova, T. A. (1979). Effect of selenium on the structure and properties of structural steel. *Metal Science and Heat Treatment*, *21*, 741–746.

Goodgame, D. M. L., & Hitchman, M. A. (1964). Studies of nitro and nitrito complexes. I. Some nitrito complexes of nickel(II). *Inorganic Chemistry*, *3*, 1389–1394.

Goodgame, D. M. L., & Hitchman, M. A. (1965). Studies of nitro and nitrito complexes. II. Complexes containing chelating NO_2 groups. *Inorganic Chemistry*, *4*, 721–725.

Goodgame, D. M. L., & Hitchman, M. A. (1966). Studies of nitro and nitrito complexes. III. Some nitro complexes of nickel(II) and a nitro-nitrito equilibrium. *Inorganic Chemistry*, *5*, 1303–1307.

Gordon, R. B., Bertram, M., & Graedel, T. E. (2006). Metal stocks and sustainability. *Proceedings of the National Academy of Sciences*, *103*, 1209–1214.

Gray, L. (2006). *Zinc*. New York: Marshall Cavendish Benchmark.

Gray, T. W., & Mann, N. (2012). *The elements: A visual exploration of every known atom in the universe*. New York: Black Dog & Leventhal Publishers: Distributed by Workman Publication.

Greber, J. F. (2000). *Gallium and gallium compounds. Ullmann's encyclopedia of industrial chemistry*. Weinheim: Wiley-VCH.

Green, J. (1996). Mechanisms for flame retardancy and smoke suppression – A review. *Journal of Fire Sciences*, *14*, 426–442.

Greve, B. K., Martin, K. L., Lee, P. L., Chupas, P. J., Chapman, K. W., & Wilkinson, A. P. (2010). Pronounced negative thermal expansion from a simple structure: Cubic ScF_3. *Journal of the American Chemical Society*, *132*, 15496–15498.

Grochala, W., & Edwards, P. P. (2004). Thermal decomposition of the non-interstitial hydrides for the storage and production of hydrogen. *Chemical Reviews*, *104*, 1283–1316.

Grund, S. C., Hanusch, K., & Wolf, H. U. (2008). *Arsenic and arsenic compounds. Ullmann's encyclopedia of industrial chemistry*. Weinheim: Wiley-VCH.

Gu, Y., Qian, Y., Chen, L., & Zhou, F. (2003). A mild solvothermal route to nanocrystalline titanium diboride. *Journal of Alloys and Compounds*, *352*, 325–327.

Guan, H., & Buchheit, R. G. (2004). Corrosion protection of aluminum alloy 2024-T3 by vanadate conversion coatings. *Corrosion*, *60*, 284–296.

Gulliver, D. J., Levason, W., & Webster, M. (1981). Coordination stabilised copper(I) flouride. Crystal and molecular structure of fluorotris(triphenylphosphine)copper(I)-ethanol (1/2), $Cu(PPh_3)_3$, P·2EtOH. *Inorganica Chimica Acta*, *52*, 153–159.

Gupta, P., Fang, F., Rubanov, S., Loho, T., Koo, A., Swift, N., ... Kennedy, J. (2019). Decorative black coatings on titanium surfaces based on hard bi-layered carbon coatings synthesized by carbon implantation. *Surface and Coatings Technology*, *358*, 386–393.

Habashi, F. Discovering the 8th metal. International Zinc Association (IZA). <https://web.archive.org/web/20090304154217/> <http://www.iza.com/Documents/Communications/Publications/History.pdf> Accessed 22.07.19.

Habashi, F. (1997). *Handbook of extractive metallurgy*. Weinheim: Wiley-VCH.

Haidemenopoulos, G. N. (2018). *Physical metallurgy: Principles and design*. Boca Raton: CRC Press.

Hampel, C. A. (1968). *The encyclopedia of the chemical elements*. New York: Reinhold Book Corp.

Hasin, P., & Wu, Y. (2012). Sonochemical synthesis of copper hydride (CuH). *Chemical Communications*, *48*, 1302–1304.

Hawksley, L., & National, A. (2016). *Bitten by witch fever: Wallpaper & arsenic in the Victorian home*. New York: Thames & Hudson.

Headlee, A. J. W., & Hunter, R. G. (1953). Elements in coal ash and their industrial significance. *Industrial and Engineering Chemistry*, *45*, 548–551.

Hebrard, F., & Kalck, P. (2009). Cobalt-catalyzed hydroformylation of alkenes: Generation and recycling of the carbonyl species, and catalytic cycle. *Chemical Reviews*, *109*, 4272–4282.

Heiserman, D. L. (1992). *Exploring chemical elements and their compounds*. Blue Ridge Summit, PA: Tab Books.

Hemmingsen, L., Bauer, R., Bjerrum, M. J., Schwarz, K., Blaha, P., & Andersen, P. (1999). Structure, chemical bonding, and nuclear quadrupole interactions of β-$Cd(OH)_2$: Experiment and first principles calculations. *Inorganic Chemistry*, *38*, 2860–2867.

Henderson, J. (2007). *The science and archaeology of materials: An investigation of inorganic materials*. London, New York: Routledge.

Hesse, R. W. (2007). *Jewelry making through history: An encyclopedia*. Westport: Greenwood Press.

Hingston, J., Collins, C. D., Murphy, R. J., & Lester, J. N. (2001). Leaching of chromated copper arsenate wood preservatives: A review. *Environmental Pollution*, *111*, 53–66.

Hluchan, S. E., & Pomerantz, K. (2006). *Calcium and calcium alloys. Ullmann's encyclopedia of industrial chemistry*. Weinheim: Wiley-VCH.

Hoffmann, J. E. (1989). Recovering selenium and tellurium from copper refinery slimes. *JOM*, *41*, 33–38.

Höll, R., Kling, M., & Schroll, E. (2007). Metallogenesis of germanium – A review. *Ore Geology Reviews*, *30*, 145–180.

Holleman, A. F., Wiberg, E., & Wiberg, N. (2007). Cobalt. In A. F. Holleman, E. Wiberg, & N. Wiberg (Eds.), *Lehrbuch der anorganischen Chemie* (pp. 1146–1152). Berlin: Walter de Gruyter.

Holmyard, E. J. (1962). *Makers of chemistry*. Oxford: Clarendon Press.

Holton, E. C. (1926). Insecticides and fungicides. *Industrial & Engineering Chemistry*, *18*, 931–933.

Hong, S., Candelone, J.-P., Patterson, C. C., & Boutron, C. F. (1996). History of ancient copper smelting pollution during Roman and medieval times recorded in Greenland ice. *Science*, *272*, 246–249.

Hörner, T. G., & Klüfers, P. (2016). The species of Fehling's solution. *European Journal of Inorganic Chemistry*, *2016*, 1798–1807.

Hüfner, S. (1994). Electronic structure of NiO and related 3D-transition-metal compounds. *Advances in Physics, 43*, 183–356.

Hukkanen, E., & Walden, H. (1985). The production of vanadium and steel from titanomagnetites. *International Journal of Mineral Processing, 15*, 89–102.

Hurd, L. C., Kemmerer, G. I., & Meloche, V. W. (1930). The ammonates of copper selenite. *Journal of the American Chemical Society, 52*, 3881–3886.

Hyde, C. K. (1998). *Copper for America: The United States copper industry from colonial times to the 1990s*. Tucson: University of Arizona Press.

Hyvärinen, O., Lindroos, L., & Yllö, E. (1989). Recovering selenium from copper refinery slimes. *JOM, 41*, 42–43.

Ioffe, D., & Frim, R. (2011). *Bromine, organic compounds. Kirk-Othmer encyclopedia of chemical technology* (pp. 1–26). New York: John Wiley & Sons.

Itahara, H., Simanullang, W. F., Takahashi, N., Kosaka, S., & Furukawa, S. (2019). Na-melt synthesis of fine Ni_3Si powders as a hydrogenation catalyst. *Inorganic Chemistry, 58*, 5406–5409.

Jackson, C. B., & Werner, R. C. (1957). *The manufacture of potassium and NaK, . Handling and uses of the alkali metals, Vol 19. Advances in chemistry* (Vol 19, pp. 169–173). American Chemical Society.

Jansen, T., Gorobez, J., Kirm, M., Brik, M. G., Vielhauer, S. O. M., Khaidukov, N. M., … Jüstel, T. (2018). Narrow band deep red photoluminescence of $Y_2Mg_3Ge_3O_{12}$:Mn^{4+},Li^+ inverse garnet for high power phosphor converted LEDs. *ECS Journal of Solid State Science and Technology, 7*, R3086–R3092.

Jeevanandam, P., Koltypin, Y., & Gedanken, A. (2001). Synthesis of nanosized α-nickel hydroxide by a sonochemical method. *Nano Letters, 1*, 263–266.

Jenkins, R. (1945). The zinc industry in England: The early years up to 1850. *Transactions of the Newcomen Society, 25*, 41–52.

Jennewein, M., Lewis, M. A., Zhao, D., Tsyganov, E., Slavine, N., He, J., … Thorpe, P. (2008). Vascular imaging of solid tumors in rats with a radioactive arsenic-labeled antibody that binds exposed phosphatidylserine. *Journal of Clinical Cancer, 14*, 1377–1385.

Joerissen, L., Garche, J., Fabjan, C., & Tomazic, G. (2004). Possible use of vanadium redox-flow batteries for energy storage in small grids and stand-alone photovoltaic systems. *Journal of Power Sources, 127*, 98–104.

Johnson, B., The Great Orme Mines. 2014, Historic UK Ltd. <https://www.historic-uk.com/HistoryUK/HistoryofWales/The-Great-Orme-Mines/> Accessed 22.07.19.

Jones, F. T. (2007). A broad view of arsenic. *Poultry Science, 86*, 2–14.

Jordan, A. J., Lalic, G., & Sadighi, J. P. (2016). Coinage metal hydrides: Synthesis, characterization, and reactivity. *Chemical Reviews, 116*, 8318–8372.

Jorgensen, U. G. (1994). Effects of TiO in stellar atmospheres. *Astronomy and Astrophysics, 284*, 179–186.

Joseph, G. (1999). *Copper: Its trade, manufacture, use, and environmental status*. Materials Park, OH: ASM International.

Kabata-Pendias, A. (1998). Geochemistry of selenium. *Journal of Environmental Pathology, Toxicology and Oncology, 17*, 173–177.

Kaji, M. (2002). D. I. Mendeleev's concept of chemical elements and the principles of chemistry. *Bulletin for the History of Chemistry, 27*, 4–16.

Kalnicky, D., & Singhvi, R. (2001). Field portable XRF analysis of environmental samples. *Journal of Hazardous Materials, 83*, 93–122.

Kamath, P. V., Dixit, M., Indira, L., Shukla, A. K., Kumar, V. G., & Munichandraiah, N. (1994). Stabilized α-$Ni(OH)_2$ as electrode material for alkaline secondary cells. *Journal of the Electrochemical Society, 141*, 2956–2959.

Kanazawa, K., Yoshida, S., Shigekawa, H., & Kuroda, S. (2015). Dynamic probe of ZnTe(110) surface by scanning tunneling microscopy. *Science and Technology of Advanced Materials, 16*, 015002.

Kappler, A., Pasquero, C., Konhauser, K. O., & Newman, D. K. (2005). Deposition of banded iron formations by anoxygenic phototrophic Fe (II)-oxidizing bacteria. *Geology, 33*, 865–868.

Karlsson, S. (2017). Spontaneous fracture in thermally strengthened glass - A review & outlook. *Ceramics - Silikaty, 61*, 188–201.

Kasap, S., Frey, J. B., Belev, G., Tousignant, O., Mani, H., Laperriere, L., … Rowlands, J. A. (2009). Amorphous selenium and its alloys from early xeroradiography to high resolution X-ray image detectors and ultrasensitive imaging tubes. *Physica Status Solidi B, 246*, 1794–1805.

Kaspersma, J., Doumena, C., Munrob, S., & Prinsa, A.-M. (2002). Fire retardant mechanism of aliphatic bromine compounds in polystyrene and polypropylene. *Polymer Degradation and Stability, 77*, 325–331.

Katsuta, N., Shimizu, I., Helmstaedt, H., Takano, M., Kawakami, S., & Kumazawa, M. (2012). Major element distribution in Archean banded iron formation (BIF): Influence of metamorphic differentiation. *Journal of Metamorphic Geology, 30*, 457–472.

Kauffman, G. B., Fang, L. Y., Viswanathan, N., & Townsend, G. (1984). Purification of copper (i) iodide. In S. L. Holt (Ed.), *Inorganic syntheses. Inorganic syntheses* (pp. 101–103). New York: John Wiley & Sons.

Kearey, P., Brooks, M., & Hill, I. (2011). *Optical pumped magnetometer. An introduction to geophysical exploration* (p. 164) Malden, MA: Blackwell Publishing.

Kerr, P. F. (1945). Cattierite and vaesite: New Co-Ni minerals from the Belgian Kongo. *American Mineralogist, 30*, 483–492.

Khanra, A. K., Pathak, L. C., Mishra, S. K., & Godkhindi, M. M. (2004). Effect of NaCl on the synthesis of TiB_2 powder by a self-propagating high-temperature synthesis technique. *Materials Letters, 58*, 733–738.

Kharasch, M. S., & Fields, E. K. (1941). Factors determining the course and mechanisms of grignard reactions. IV. The effect of metallic halides on the reaction of aryl grignard reagents and organic halides. *Journal of the American Chemical Society, 63*, 2316–2320.

Kharasch, M. S., & Tawney, P. O. (1941). Factors determining the course and mechanisms of grignard reactions. II. The effect of metallic compounds on the reaction between isophorone and methylmagnesium bromide. *Journal of the American Chemical Society, 63*, 2308–2316.

Kharton, V. V. (2011). *Solid state electrochemistry II: Electrodes, interfaces and ceramic membranes*. Weinheim: Wiley-VCH.

Khodakov, A. Y., Chu, W., & Fongarland, P. (2007). Advances in the development of novel cobalt Fischer-Tropsch catalysts for synthesis of long-chain hydrocarbons and clean fuels. *Chemical Reviews, 107*, 1692–1744.

Khusnutdinova, J. R., Rath, N. P., & Mirica, L. M. (2010). Stable mononuclear organometallic Pd(III) complexes and their C − C bond formation reactivity. *Journal of the American Chemical Society*, *132*, 7303−7305.

Kim, J. G., & Walsh, P. J. (2006). From aryl bromides to enantioenriched benzylic alcohols in a single flask: Catalytic asymmetric arylation of aldehydes. *Angewandte Chemie International Edition*, *45*, 4175−4178.

Kirtley, S. W. (1982). 26.1 - Chromium compounds with η^1-carbon ligands. In G. Wilkinson, F. G. A. Stone, & E. W. Abel (Eds.), *Comprehensive organometallic chemistry* (pp. 783−951). Oxford: Pergamon.

Kleefisch, E. W. (1981). *Industrial applications of titanium and zirconium*. Philadelphia, PA: American Society for Testing and Materials.

Klein, C. (2005). Some Precambrian banded iron-formations (BIFs) from around the world: Their age, geologic setting, mineralogy, metamorphism, geochemistry, and origins. *American Mineralogist*, *90*, 1473−1499.

Klein, C., & Dutrow, B. (2008). *The 23rd edition of the manual of mineral science (after James D. Dana)*. Hoboken, NJ: John Wiley & Sons, Inc.

Knop, O., & Hartley, J. M. (1968). Refinement of the crystal structure of scandium oxide. *Canadian Journal of Chemistry*, *46*, 1446−1450.

Kobayashi, S., & Manabe, K. (2000). Green Lewis acid catalysis in organic synthesis. *Pure and Applied Chemistry*, *72*, 1373−1380.

Koch, E.-C. (2015). Spectral investigation and color properties of copper(I) halides CuX (X = F, Cl, Br, I) in pyrotechnic combustion flames. *Propellants, Explosives, Pyrotechnics*, *40*, 799−802.

Kogel, J. E., Trivedi, N. C., Barker, J. M., & Krukowski, S. T. (2006). *Industrial minerals & rocks: Commodities, markets, and uses*. Littleton: Society for Mining, Metallurgy, and Exploration.

Kolasinski, K. W. (2004). *Surface science: Foundations of catalysis and nanoscience*. Chichester: Wiley.

Kondinski, A., & Monakhov, K. Y. (2017). Breaking the gordian knot in the structural chemistry of polyoxometalates: Copper(II)− oxo/hydroxo clusters. *Chemistry − A European Journal*, *23*, 7841−7852.

Konstantinou, I. K., & Albanis, T. A. (2004). Worldwide occurrence and effects of antifouling paint booster biocides in the aquatic environment: A review. *Environment International*, *30*, 235−248.

Krebs, R. E. (2006). *The history and use of our earth's chemical elements: A reference guide*. Westport: Greenwood Press.

Kristiansen, R. (2003). Scandium − Mineraler I Norge. *Stein*, 14−23.

Kuzovnikov, M. A., & Tkacz, M. (2016). Synthesis of ruthenium hydride. *Physical Review B*, *93*, 064103.

Lach, D. F. (2010). *Asia in the making of Europe, Volume II: A century of wonder. Book 3: The scholarly disciplines* (Vol. 2). Chicago: University of Chicago Press.

Langeslay, R. R., Kaphan, D. M., Marshall, C. L., Stair, P. C., Sattelberger, A. P., & Delferro, M. (2019). Catalytic applications of vanadium: A mechanistic perspective. *Chemical Reviews*, *119*, 2128−2191.

Langner, B. E. (2000). *Selenium and selenium compounds. Ullmann's encyclopedia of industrial chemistry*. Weinheim: Wiley-VCH. Available from https://doi.org/10.1002/14356007.a23_525.

Lanman, S. W. (2000). Colour in the garden: Malignant magenta. *Garden History*, *28*, 209−221.

Larson, M. L., Nannelli, P., Block, B. P., Edwards, D. A., & Mallock, A. K. (1970). II. Preparation of some metal halides anhydrous molybdenum halides and oxide halides—A summary: Molybdenum(II) halides. *Inorganic Syntheses*, 165−178.

Lascelles, K., Morgan, L. G., Nicholls, D., Beyersmann, D., & Institute N. (2019). *Nickel compounds. Ullmann's encyclopedia of industrial chemistry.* (pp. 1−17). Weinheim: Wiley-VCH.

Latroche, M., Brohan, L., Marchand, R., & Tournoux, M. (1989). *New hollandite oxides: $TiO_2(H)$ and $K_{0.06}TiO_2$. Journal of Solid State Chemistry* (81, pp. 78−82).

Lavine, M. S. (2018). Make no bones about titanium. *Science*, *359*, 173−174.

Lavoisier, A. (1799). *Elements of chemistry: In a new systematic order; containing all the modern discoveries* (4th ed.). Edinburgh: William Creech. (R. Kerr, Trans).

Lechtman, H. (1996). Arsenic bronze: Dirty copper or chosen alloy? A view from the Americas. *Journal of Field Archaeology*, *23*, 477−514.

Lee, S., Hippalgaonkar, K., Yang, F., Hong, J., Ko, C., Suh, J., ... Wu, J. (2017). Anomalously low electronic thermal conductivity in metallic vanadium dioxide. *Science*, *355*, 371−374.

Liu, L., & Xi, Z. (2018). Organocopper(III) compounds with well-defined structures undergo reductive elimination to form C-C or C−heteroatom bonds. *Chinese Journal of Chemistry*, *36*, 1213−1221.

Livingstone, A., & Bish, D. L. (2018). On the new mineral theophrastite, a nickel hydroxide, from Unst, Shetland, Scotland. *Mineralogical Magazine*, *46*, 1−5.

Livingstone, E. S., & Maitland, A. (1991). A high power, segmented metal, copper bromide laser. *Measurement Science and Technology*, *2*, 1119−1120.

Łowicki, D., Baś, S., & Mlynarski, J. (2015). Chiral zinc catalysts for asymmetric synthesis. *Tetrahedron*, *71*, 1339−1394.

Lucas, A. (2012). *Ancient Egyptian materials and industries*. Whitefish, MT: Kessinger Legacy Reprints.

Lynch, M. (2003). *Mining in World History*. London: Reaktion Books.

Lyons, T. W., & Reinhard, C. T. (2009). Early Earth: Oxygen for heavy-metal fans. *Nature*, *461*, 179−181.

Maile, E., & Fischer, R. A. (2005). MOCVD of the cubic zinc nitride phase, Zn_3N_2, using $Zn[N(SiMe_3)_2]_2$ and ammonia as precursors. *Chemical Vapor Deposition*, *11*, 409−414.

Mallinson, J. C. (1993). *Chromium dioxide. The foundations of magnetic recording* (p. 32) Boston, MA: Academic Press.

Mandal, B. K., & Suzuki, K. T. (2002). Arsenic round the world: A review. *Talanta*, *58*, 201−235.

Mandeville, C., & Fulbright, H. (1943). The energies of the γ-rays from Sb^{122}, Cd^{115}, Ir^{192}, Mn^{54}, Zn^{65}, and Co^{60}. *Physical Review*, *64*, 265−267.

Manning, T. D., Parkin, I. P., Clark, R. J. H., Sheel, D., Pemble, M. E., & Vernadou, D. (2002). Intelligent window coatings: Atmospheric pressure chemical vapour deposition of vanadium oxides. *Journal of Materials Chemistry, 12*, 2936–2939.

Marchand, R., Brohan, L., & Tournoux, M. (1980). TiO_2(B) a new form of titanium dioxide and the potassium octatitanate $K_2Ti_8O_{17}$. *Materials Research Bulletin, 15*, 1129–1133.

Marcopoulos, T., & Economou, M. (1980). Theophrastite, $Ni(OH)_2$, a new mineral from northern Greece. *American Mineralogist, 66*, 1020–1021.

Marden, J. W., & Rich, M. N. (1927). Vanadium. *Industrial and Engineering Chemistry, 19*, 786–788.

Marggraf, A. S. (1761). *Chymische Schriften* (Vol 1). Berlin: Arnold Weber.

Margolin, A., Popovitz-Biro, R., Albu-Yaron, A., Rapoport, L., & Tenne, R. (2005). Inorganic fullerene-like nanoparticles of TiS_2. *Chemical Physics Letters, 411*, 162–166.

Markiewicz, W., Mains, E., Vankeuren, R., Wilcox, R., Rosner, C., Inoue, H., ... Tachikawa, K. (1977). A 17.5 Tesla superconducting concentric Nb_3Sn and V_3Ga magnet system. *IEEE Transactions on Magnetics, 13*, 35–37.

Marks, R., Pearse, A. D., & Walker, A. P. (1985). The effects of a shampoo containing zinc pyrithione on the control of dandruff. *British Journal of Dermatology, 112*, 415–422.

Marrion, A. R. (2004). *The chemistry and physics of coatings*. Cambridge: The Royal Society of Chemistry.

Martin, S. R. (1995). The state of our knowledge about ancient copper mining in Michigan. *The Michigan Archaeologist, 41*, 119–138.

Matschullat, J. (2000). Arsenic in the geosphere—A review. *The Science of the Total Environment, 249*, 297–312.

Matsui, H., Fukumoto, K., Smith, D. L., Chung, H. M., van Witzenburg, W., & Votinov, S. N. (1996). Status of vanadium alloys for fusion reactors. *Journal of Nuclear Materials, 233–237*, 92–99.

Mattesini, M., de Almeida, J. S., Dubrovinsky, L., Dubrovinskaia, N., Johansson, B., & Ahuja, R. (2004). High-pressure and high-temperature synthesis of the cubic TiO_2 polymorph. *Physical Review B, 70*, 212101.

Mayo-Wilson, E., Junior, J. A., Imdad, A., Dean, S., Chan, X. H. S., Chan, E. S., ... Bhutta, Z. A. (2014). Zinc supplementation for preventing mortality, morbidity, and growth failure in children aged 6 months to 12 years of age. *Cochrane Database of Systematic Reviews*, CD009384.

Mazej, Z. (2002). Room temperature syntheses of MnF_3, MnF_4 and hexafluoromanganete(IV) salts of alkali cations. *Journal of Fluorine Chemistry, 114*, 75–80.

Mccray, W. P. (1998). Glassmaking in renaissance Italy: The innovation of venetian cristallo. *The Journal of the Minerals, 50*, 14–19.

McDonald, I., Sloan, G. C., Zijlstra, A. A., Matsunaga, N., Matsuura, M., Kraemer, K. E., ... Markwick, A. J. (2010). Rusty old stars: A source of the missing interstellar iron? *The Astrophysical Journal Letters, 717*, L92–L97.

McNeil, I. (1990). The emergence of nickel. In I. McNeil (Ed.), *An encyclopaedia of the history of technology* (pp. 96–100). New York: Routledge.

McNeil, I. (2002). *An encyclopedia of the history of technology*. New York: Taylor & Francis.

McNutt, P. M. (1990). *The forging of Israel: Iron technology, symbolism and tradition in ancient society*. Sheffield: Almond Press.

Mei, A. B., Hellman, O., Wireklint, N., Schlepütz, C. M., Sangiovanni, D. G., Alling, B., ... Greene, J. E. (2015). Dynamic and structural stability of cubic vanadium nitride. *Physical Review B, 91*, 054101.

Meyer, R. J. (1962). *Chrom: Teil A - Lieferung 1. Geschichtliches. Vorkommen. Technologie. Element bis Physikalische Eigenschaften*. Berlin: Springer.

Meylan, G. (2016). The anthropogenic cycle of zinc: Status quo and perspectives. *Resources, Conservation and Recycling, 123*, 1–10.

Miller, H. (1980). Potash from wood ashes: Frontier technology in Canada and the United States. *Technology and Culture, 21*, 187–208.

Mioduski, T., Gumiński, C., & Zeng, D. (2012). IUPAC-NIST solubility data series. 94. Rare Earth metal iodides and bromides in water and aqueous systems. Part 1. Iodides. *Journal of Physical and Chemical Reference Data, 41*, 013104.

Moiseyev, V. N. (2006). *Titanium alloys Russian aircraft and aerospace applications*. Boca Raton: Taylor & Francis.

Morgan, J. W., & Anders, E. (1980). Chemical composition of Earth, Venus, and Mercury. *Proceedings of the National Academy of Sciences, 77*, 6973–6977.

Morris, P. R. (2008). *A history of the world semiconductor industry*. London: Institution of Engineering and Technology in association with the Science Museum.

Morrison, R. D., & Murphy, B. (2010). *Environmental forensics: Contaminant specific guide*. Academic Press.

Moskalyk, R. R. (2003). Gallium: The backbone of the electronics industry. *Minerals Engineering, 16*, 921–929.

Moskalyk, R. R. (2004). Review of germanium processing worldwide. *Minerals Engineering, 17*, 393–402.

Moskalyk, R. R., & Alfantazi, A. M. (2003). Processing of vanadium: A review. *Minerals Engineering, 16*, 793–805.

Moss, S. C., & Newnham, R. E. (1964). The chromium position in ruby. *Zeitschrift für Kristallographie, 120*, 359–363.

Mostefaoui, S., Lugmair, G. W., Hoppe, P., & El Goresy, A. (2004). Evidence for live ^{60}Fe in meteorites. *New Astronomy Reviews, 48*, 155–159.

Mott, R. A. (2014). Dry and wet puddling. *Transactions of the Newcomen Society, 49*, 156–157.

Muhlethaler, B., Thissen, J., & Muhlethaler, B. (1969). Smalt. *Studies in Conservation, 14*, 47–61.

Muller, O., & Roy, R. (1968). Formation and stability of the platinum and rhodium oxides at high oxygen pressures and the structures of Pt_3O_4, β-PtO_2 and RhO_2. *Journal of the Less-Common Metals, 16*, 129–146.

Muller, O., White, W. B., & Roy, R. (1969). Infrared spectra of the chromates of magnesium, nickel and cadmium. *Spectrochimica Acta Part A: Molecular Spectroscopy, 25*, 1491–1499.

Muller, O., White, W. B., & Roy, R. (1969). X-ray diffraction study of the chromates of nickel, magnesium and cadmium. *Zeitschrift für Kristallographie, 130*, 112–120.

Murphy, E. A., & Aucott, M. (1998). An assessment of the amounts of arsenical pesticides used historically in a geographical area. *Science of the Total Environment, 218*, 89–101.

Murray, C. B., Norris, D. J., & Bawendi, M. G. (1993). Synthesis and characterization of nearly monodisperse CdE (E = sulfur, selenium, tellurium) semiconductor nanocrystallites. *Journal of the American Chemical Society, 115*, 8706–8715.

Nachman, K. E., Graham, J. P., Price, L. B., & Silbergeld, E. K. (2005). Arsenic: A roadblock to potential animal waste management solutions. *Environmental Health Perspectives, 113*, 1123–1124.

Nakamura-Messenger, K., Clemett, S. J., Rubin, A. E., Choi, B.-G., Zhang, S., Rahman, Z., ... Keller, L. P. (2012). Wassonite: A new titanium monosulfide mineral in the Yamato 691 enstatite chondrite. *American Mineralogist, 97*, 807–815.

Nakhal, S., Weber, D., Irran, E., Lerch, M., Schwarz, B., & Ehrenberg, H. (2013). Synthesis, crystal structure, and magnetic properties of a new vanadium fluoride hydrate $V_2F_64\ H_2O$. *Zeitschrift für Kristallographie - Crystalline Materials, vol 228*. Available from https://doi.org/10.1524/zkri.2013.1664.

Naumann d'Alnoncourt, R., Csepei, L.-I., Hävecker, M., Girgsdies, F., Schuster, M. E., Schlögl, R., & Trunschke, A. (2014). The reaction network in propane oxidation over phase-pure MoVTeNb M1 oxide catalysts. *Journal of Catalysis, 311*, 369–385.

Naumov, A. V. (2007). World market of germanium and its prospects. *Russian Journal of Non-Ferrous Metals, 48*, 265–272.

Naumov, A. V. (2010). Selenium and tellurium: State of the markets, the crisis, and its consequences. *Metallurgist, 54*, 197–200.

Needham, J., Wang, L., Robinson, K. G., Lu, G.-D., Tsien, T.-H., Ho, P.-Y., ... Wagner, D. B. (1999). *Science and civilisation in China* (Vol. 5). Chemistry and Chemical Technology, Part 13 Mining.

Neikov, O. D., Naboychenko, S., & Mourachova, I. (2009). *Handbook of non-ferrous metal powders: Technologies and applications*. Burlington: Elsevier.

Nesse, W. D. (2000). *Introduction to mineralogy*. New York: Oxford University Press.

Nikiforov, G. B., Roesky, H. W., & Koley, D. (2014). A survey of titanium fluoride complexes, their preparation, reactivity, and applications. *Coordination Chemistry Reviews, 258-259*, 16–57.

Nikitin, M. I., & Zbezhneva, S. G. (2014). Thermochemistry of vanadium fluorides: The formation enthalpies of vanadium fluorides. *High Temperature, 52*, 809–813.

Nishio-Hamane, D., Shimizu, A., Nakahira, R., Niwa, K., Sano-Furukawa, A., Okada, T., ... Kikegawa, T. (2010). The stability and equation of state for the cotunnite phase of TiO_2 up to 70 GPa. *Physics and Chemistry of Minerals, 37*, 129–136.

Nouri, K. (2012). *Lasers in dermatology and medicine*. London: Springer-Verlag.

Nozari, A., Ataie, A., & Heshmati-Manesh, S. (2012). Synthesis and characterization of nano-structured TiB_2 processed by milling assisted SHS route. *Materials Characterization, 73*, 96–103.

Nuccio, P. M., & Valenza, M. (1979). Determination of metallic iron, nickel and cobalt in meteorites. *Rendiconti Societa Italiana di Mineralogia e Petrografia, 35*, 355–360.

Oates, J. A. H. (2008). *Lime and limestone: Chemistry and technology, production and uses*. Weinheim: Wiley-VCH.

Ohishi, Y., Sugizaki, M., Sun, Y., Muta, H., & Kurosaki, K. (2019). Thermophysical and mechanical properties of CrB and FeB. *Journal of Nuclear Science and Technology, 56*, 859–865.

Ojima, I., Athan, A. A., Chaterpaul, S. J., Kaloko, J. J., & Teng, Y.-H. G. (2013). Organorhodium chemistry. In B. H. Lipshutz (Ed.), *Organometallics in synthesis* (pp. 135–318). New York: Wiley Online Books. John Wiley & Sons.

Okada, S., Kudou, K., Iizumi, K., Kudaka, K., Higashi, I., & Lundström, T. (1996). Single-crystal growth and properties of CrB, Cr_3B_4, Cr_2B_3 and CrB_2 from high-temperature aluminum solutions. *Journal of Crystal Growth, 166*, 429–435.

Okazawad, H., Yonekura, Y., Fujibayashi, Y., Nishizawa, S., Magata, Y., Ishizu, K., ... Konishi, J. (1994). Clinical application and quantitative evaluation of generator-produced copper-62-PTSM as a brain perfusion tracer for PET. *Journal of Nuclear Medicine, 35*, 1910–1915.

Olsen, S. E., Tangstad, M., & Lindstad, T. (2007). *Production of manganese ferroalloys*. Trondheim: Tapir Akademisk Forlag.

Orchin, G. J., Fazio, D. D., Bernardo, A. D., Hamer, M., Yoon, D., Cadore, A. R., ... Hadfield, R. H. (2019). Niobium diselenide superconducting photodetectors. *Applied Physics Letters, 114*, 251103.

Ostrooumov, M., & Taran, Y. (2015). Discovery of native vanadium, a new mineral from the Colima Volcano, State of Colima (Mexico). *Revista de la Sociedad Española de Mineralogía, 20*, 109–110.

Oswald, H. R., Reller, A., Schmalle, H. W., & Dubler, E. (1990). Structure of copper(II) hydroxide, $Cu(OH)_2$. *Acta Crystallographica C, 46*, 2279–2284.

Ou, G.-C., Jiang, L., Feng, X.-L., & Lu, T.-B. (2009). Vanadium polyoxoanion-bridged macrocyclic metal complexes: From one-dimensional to three-dimensional structures. *Dalton Transactions*, 71–76.

Pagel, B. E. J. (1997). *Nucleosynthesis evolution galaxies*. Cambridge: Cambridge University Press.

Pankratov, D. A., Veligzhanin, A. A., & Zubavichus, Y. V. (2013). Structural features of green cobalt(III) hydroxide. *Russian Journal of Inorganic Chemistry, 58*, 67–73.

Parr, P. J. (1974). *Review of "timma: Valley of the biblical copper mines" by Beno Rothenberg, . Bulletin of the School of Oriental and African Studies, University of London* (37, pp. 223–224).

Partin, D. E., Williams, D. J., & O'Keeffe, M. (1997). The crystal structures of Mg_3N_2 and Zn_3N_2. *Journal of Solid State Chemistry, 132*, 56–59.

Pearson, C. D., & Green, J. B. (1993). Vanadium and nickel complexes in petroleum resin acid, base, and neutral fractions. *Energy Fuels, 7*, 338–346.

Penichon, S. (1999). Differences in image tonality produced by different toning protocols for matte collodion photographs. *Journal of the American Institute for Conservation, 38*, 124–143.

Perkins, D. (2002). *Mineralogy* (2nd edn). Upper Saddle River, NJ: Prentice Hall.

Pfab, W., & Fischer, E. O. (1953). Zur Kristallstruktur der Di-cyclopentadienyl-verbindungen des zweiwertigen Eisens, Kobalts und Nickels. *Zeitschrift für anorganische und allgemeine Chemie*, *274*, 316–322.

Photos, E. (1989). The question of meteoritic versus smelted nickel-rich iron: Archaeological evidence and experimental results. *World Archaeology*, *20*, 403–421.

Pigott, V. C. (1999). *The archaeometallurgy of the Asian old world*. Ephrata: Museum University of Pennsylvania.

Pleger, T. C. (2003). A brief introduction to the old copper complex of the Western Great Lakes: 4000–1000 BC. Paper presented at the The Historic Oconto Area: Proceedings of Twenty-Seventh Annual Meeting of Forest History Association of Wisconsin, Inc., 4–5 October, 2002, Oconto, WI.

Pomerantz, M., Combs, G. L., Jr, Dassanayake, N. L., Olah, G., & Surya Prakash, G. K. (1982). 42. Vanadium dichloride solution. In J. P. Fackler (Ed.), *Inorganic syntheses. Inorganic syntheses* (pp. 185–187). New York: John Wiley & Sons.

Pops, H. (2008). Processing of wire from antiquity to the future. *Wire Journal International*, 58–66.

Porter, F. (1991). *Zinc handbook: Properties, processing, and use in design*. New York: M. Dekker.

Porter, F. C. (1994). *Corrosion resistance of zinc and zinc alloys*. New York: Marcel Dekker.

Potter, R. M., & Rossman, G. R. (1979). The mineralogy of manganese dendrites and coatings. *American Mineralogist*, *64*, 1219–1226.

Powers, D. C., Benitez, D., Tkatchouk, E., Goddard, W. A., & Ritter, T. (2010). Bimetallic reductive elimination from dinuclear Pd(III) complexes. *Journal of the American Chemical Society*, *132*, 14092–14103.

Powers, D. C., & Ritter, T. (2009). Bimetallic Pd(III) complexes in palladium-catalysed carbon–heteroatom bond formation. *Nature Chemistry*, *1*, 302–309.

Pray, A. R., Heitmiller, R. F., Strycker, S., Aftandilian, V. D., Muniyappan, T., Choudhury, D., & Tamres, M. (1990). Anhydrous metal chlorides. *Inorganic Syntheses*, 321–323.

Pulak, C. (1998). The Uluburun shipwreck: An overview. *International Journal of Nautical Archaeology*, *27*, 188–224.

Pye, C. C., Corbeil, C. R., & Rudolph, W. W. (2006). An ab initio investigation of zinc chloro complexes. *Physical Chemistry Chemical Physics*, *8*, 5428–5436.

Radivojević, M., Rehren, T., Kuzmanović-Cvetković, J., Jovanović, M., & Northover, J. P. (2013). Tainted ores and the rise of tin bronzes in Eurasia, c. 6500 years ago. *Antiquity*, *87*, 1030–1045.

Rahman, F. A., Allan, D. L., Rosen, C. J., & Sadowsky, M. J. (2004). Arsenic availability from chromated copper arsenate (CCA)-treated wood. *Journal of Environmental Quality*, *33*, 173–180.

Ramberg, I. B., Bryhny, I., Nottvedt, A., & Rangnes, K. (2008). *The making of a land: Geology of Norway*. Trondheim: Norks Geologisk Forening.

Rancke-Madsen, E. (1975). The discovery of an element. *Centaurus*, *19*, 299–313.

Rappoport, Z., & Marek, I. (2007). *The chemistry of organozinc compounds: R-Zn*. Chichester: John Wiley & Sons.

Rau, J. V., Latini, A., Generosi, A., Rossi Albertini, V., Ferro, D., Teghil, R., & Barinov, S. M. (2009). Deposition and characterization of superhard biphasic ruthenium boride films. *Acta Materialia*, *57*, 673–681.

Ray, P. C. (1903). *A history of Hindu chemistry from the earliest times to the middle of the Sixteenth Century, AD: With Sanskrit texts, variants, translation and illustrations* (Vol 1). Bengal Chemical & Pharmaceutical Works, Limited.

Rehren, T. (1996). A Roman zinc tablet from Bern, Switzerland: Reconstruction of the manufacture. Paper presented at the Archaeometry 94. The Proceedings of the 29th International Symposium on Archaeometry, Ankara, May 9–14, 1994.

Rehren, T. (2003). Aspects of the production of cobalt-blue glass in Egypt. *Archaeometry*, *43*, 483–489.

Reidies, A. H. (2000). *Manganese compounds. Ullmann's encyclopedia of industrial chemistry*. Weinheim: Wiley-VCH.

Renfrew, C. (2011). *Before civilization: The radiocarbon revolution and prehistoric europe*. New York, NY: Random House.

Reppe, W., Schlichting, O., Klager, K., & Toepel, T. (1948). Cyclisierende polymerisation von acetylen I Über cyclooctatetraen. *Justus Liebigs Annalen der Chemie*, *560*, 1–92.

Rickard, T. A. (1932). The nomenclature of copper and its alloys. *Journal of the Royal Anthropological Institute*, *62*, 281–290.

Rieke, G. H. (2007). Infrared detector arrays for astronomy. *Annual Review of Astronomy and Astrophysics*, *45*, 77–115.

Rittenhouse, P. A. (1979). Potash and politics. *Economic Geology*, *74*, 353–357.

Rivadulla, F., Bañobre-López, M., Quintela, C. X., Piñeiro, A., Pardo, V., Baldomir, D., … Goodenough, J. B. (2009). Reduction of the bulk modulus at high pressure in CrN. *Nature Materials*, *8*, 947–951.

Roberts, R. O. (1951). *Dr John Lane and the foundation of the non-ferrous metal industry in the Swansea valley. Gower (Gower Society)* (pp. 19–24).

Roberts-Austen, W. C. (1898). The extraction of nickel from its ores by the Mond process. *Nature*, *59*, 63–64.

Rohe, D. M. M., Battersby, R. V., & Wolf, H. U. (2014). *Zinc compounds. Ullmann's encyclopedia of industrial chemistry* (pp. 1–8). Weinheim: Wiley-VCH.

Roldán, S., Winkel, E. G., Herrera, D., Sanz, M., & Van Winkelhoff, A. J. (2003). The effects of a new mouthrinse containing chlorhexidine, cetylpyridinium chloride and zinc lactate on the microflora of oral halitosis patients: A dual-centre, double-blind placebo controlled study. *Journal of Clinical Periodontology*, *30*, 427–434.

Rosenqvist, T. (2004). *Principles of extractive metallurgy*. Trondheim: Tapir Academic Press.

Rosiak, D., Okuniewski, A., & Chojnacki, J. (2018). Copper(I) iodide ribbons coordinated with thiourea derivatives. *Acta Crystallographica Section C, Structural Chemistry*, *74*, 1650–1655.

Rossotti, H. (1998). *Diverse atoms. Profiles of the chemical elements*. Oxford: Oxford University Press.

Roza, G. (2008). *Titanium*. New York: Rosen Central.

Rugel, G., Faestermann, T., Knie, K., Korschinek, G., Poutivtsev, M., Schumann, D., ... Wohlmuther, M. (2009). New measurement of the ^{60}Fe half-life. *Physical Review Letters, 103*, 072502–072505.

Russell, W. A., Papanastassiou, D. A., & Tombrello, T. A. (1978). Ca isotope fractionation on the earth and other solar system materials. *Geochimica et Cosmochimica Acta, 42*, 1075–1090.

Sato, F., Urabe, H., & Okamoto, S. (2000). Synthesis of organotitanium complexes from alkenes and alkynes and their synthetic applications. *Chemical Reviews, 100*, 2835–2886.

Sato, H., Endo, S., Sugiyama, M., Kikegawa, T., Shimomura, O., & Kusaba, K. (1991). Baddeleyite-type high-pressure phase of TiO_2. *Science, 251*, 786–788.

Savenije, T. J., Warman, J. M., Barentsen, H. M., van Dijk, M., Zuilhof, H., & Sudhölter, E. J. R. (2000). Corinthian bronze and the gold of the alchemists. *Macromolecules, 33*, 60–66.

Sawyer, R. D., & Sawyer, M. (2007). *The seven military classics of Ancient China*. Boulder, CO: Basic Books.

Sayre, E. V., & Smith, R. W. (1961). Compositional categories of ancient glass. *Science, 133*, 1824–1826.

Schaefer, J., Faestermann, T., Herzog, G. F., Knie, K., Korschinek, G., Masarik, J., ... Winckler, G. (2006). Terrestrial manganese-53—A new monitor of Earth surface processes. *Earth and Planetary Science Letters, 251*, 334–345.

Schall, C. (1932). Zur anodischen Oxydation von Co und Ni-Dichlorid (Nachtrag). *Zeitschrift für Elektrochemie und angewandte physikalische Chemie, 38*, 27–31.

Schivelbusch, W. (2014). *The railway journey: The Industrialization of time and space in the nineteenth century*. Berkeley: University of California Press.

Schmidbaur, H., & Bayler, A. (2009). Synthesis and uses of organosilver compounds. In Z. Rappoport (Ed.), *PATAI'S chemistry of functional groups*. New York: John Wiley & Sons.

Schultz, H., Bauer, G., Schachl, E., Hagedorn, F., & Schmittinger, P. (2000). *Potassium compounds. Ullmann's encyclopedia of industrial chemistry*. Weinheim: Wiley-VCH.

Schwarz, K., & Foltz, C. M. (1957). Selenium as an integral part of factor 3 against dietary necrotic liver degeneration. *Journal of the American Chemical Society, 79*, 3292–3293.

Sciolist Online Etymology Dictionary. <https://www.etymonline.com/> Accessed 21.06.19.

Sedaghati, E., Boffin, H. M. J., MacDonald, R. J., Gandhi, S., Madhusudhan, N., Gibson, N. P., ... Rauer, H. (2017). Detection of titanium oxide in the atmosphere of a hot Jupiter. *Nature, 549*, 238.

Seyferth, D. (2001). Cadet's fuming arsenical liquid and the cacodyl compounds of bunsen. *Organometallics, 20*, 1488–1498.

Shannon, R. D., & Prewitt, C. T. (1970). Synthesis and structure of a new high-pressure form of Rh_2O_3. *Journal of Solid State Chemistry, 2*, 134–136.

Shaposhnik, V. A. (2007). History of the discovery of potassium and sodium (on the 200th anniversary of the discovery of potassium and sodium. *Journal of Analytical Chemistry, 62*, 1100–1102.

Shaw, I., & Nicholson, P. T. (2009). *Ancient Egyptian materials and technology*. Cambridge: Cambridge University Press.

Shayesteh, A., Appadoo, D. R. T., Gordon, I. E., & Bernath, P. F. (2004). Vibration – rotation emission spectra of gaseous ZnH_2 and ZnD_2. *Journal of the American Chemical Society, 126*, 14356–14357.

Shayesteh, A., Yu, S., & Bernath, P. F. (2005). Gaseous HgH_2, CdH_2, and ZnH_2. *Chemistry – A European Journal, 11*, 4709–4712.

Shen, C., Trypiniotis, T., Lee, K. Y., Holmes, S. N., Mansell, R., Husain, M., ... Barnes, C. H. W. (2010). Spin transport in germanium at room temperature. *Applied Physics Letters, 97*, 162104.

Siegfried, R. (1963). The discovery of potassium and sodium, and the problem of the chemical elements. *Isis, 54*, 247–258.

Sigel, A., Sigel, H., & Sigel, R. K. O. (2016). *The alkali metal ions: Their role for life*. Cham: Springer.

Sigillito, A. J., Jock, R. M., Tyryshkin, A. M., Beeman, J. W., Haller, E. E., Itoh, K. M., & Lyon, S. A. (2015). Electron spin coherence of shallow donors in natural and isotopically enriched germanium. *Physical Review Letters, 115*, 247601.

Simon, A., Dronskowski, R., Krebs, B., & Hettich, B. (1987). The crystal structure of Mn_2O_7. *Angewandte Chemie International Edition in English, 26*, 139–140.

Simons, P. Y., & Dachille, F. (1967). The structure of TiO_2II, a high-pressure phase of TiO_2. Acta Crystallographica Section C. *Structural Chemistry, 23*, 334–336.

Simpson, R. S. (2003). *Lighting control: Technology and applications*. Oxford: Focal Press.

Singh, M., & Das, R. R. (2013). Zinc for the common cold. *Cochrane Database of Systematic Reviews*, CD001364.

Skulan, J., & DePaolo, D. J. (1999). Calcium isotope fractionation between soft and mineralized tissues as a monitor of calcium use in vertebrates. *Proceedings of the National Academy of Sciences, 96*, 13709–13713.

Smith, W. F., & Hashemi, J. (2006). *Foundations of materials science and engineering*. Boston: McGraw-Hill.

Smook, G. A. (2012). *Handbook for pulp & paper technologists*. Vancouver: Angus Wilde Publications.

Springett, B. E. (1988). Application of selenium-tellurium photoconductors to the xerographic copying and printing processes. *Phosphorus and Sulfur and the Related Elements, 38*, 341–350.

Sreeram, K., & Ramasami, T. (2003). Sustaining tanning process through conservation, recovery and better utilization of chromium. *Resources, Conservation and Recycling, 38*, 185–212.

Stadtman, T. C. (2002). Some functions of the essential trace element, selenium. In A. M. Roussel, R. A. Anderson, & A. E. Favier (Eds.), *Trace elements in man and animals* (10, pp. 831–836). Boston, MA: Springer.

Standfuss, S., Abinet, E., Spaniol, T. P., & Okuda, J. (2011). Allyl complexes of scandium: Synthesis and structure of neutral, cationic and anionic derivatives. *Chemical Communications*, *47*, 11441–11443.

Stansbie, J. H. (2012). *Iron and steel*. Hardpress Publishing.

Stelter, M., & Bombach, H. (2004). Process optimization in copper electrorefining. *Advanced Engineering Materials*, *6*, 558–562.

Stwertka, A. (1999). *A guide to the elements*. Oxford: Oxford University Press.

Subramanian, M. A., & Manzer, L. E. (2002). A "greener" synthetic route for fluoroaromatics via copper (II) fluoride. *Science*, *297*, 1665-1665.

Sun, Y., Tian, X., He, B., Yang, C., Pi, Z., Wang, Y., & Zhang, S. (2011). Studies of the reduction mechanism of selenium dioxide and its impact on the microstructure of manganese electrodeposit. *Electrochimica Acta*, *56*, 8305–8310.

Swardfager, W., Herrmann, N., McIntyre, R. S., Mazereeuw, G., Goldberger, K., Cha, D. S., ... Lanctôt, K. L. (2013). Potential roles of zinc in the pathophysiology and treatment of major depressive disorder. *Neuroscience & Biobehavioral Reviews*, *37*, 911–929.

Szczęsny, R., Szłyk, E., Wiśniewski, M. A., Hoang, T. K. A., & Gregory, D. H. (2016). Facile preparation of copper nitride powders and nanostructured films. *Journal of Materials Chemistry C*, *4*, 5031–5037.

Tamadon, F., & Seppelt, K. (2013). The elusive halides VCl_5, $MoCl_6$, and $ReCl_6$. *Angewandte Chemie International Edition*, *52*, 767–769.

Tamm, M., & Baker, R. J. (2007). Molybdenum compounds with CO or isocyanides. In D. M. P. Mingos, & R. H. Crabtree (Eds.), *Comprehensive Organometallic Chemistry III* (pp. 391–512). Oxford: Elsevier.

Tang, H., Li, C., Yang, X., Mo, C., Cao, K., & Yan, F. (2011). Synthesis and tribological properties of $NbSe_3$ nanofibers and $NbSe_2$ microsheets. *Crystal Research and Technology*, *46*, 400–404.

Tao, Z.-L., Xu, L.-N., Gou, X.-L., Chen, J., & Yuan, H.-T. (2004). TiS_2 nanotubes as the cathode materials of Mg-ion batteries. *Chemical Communications*, 2080–2081.

Tate, B. K., Jordan, A. J., Bacsa, J., & Sadighi, J. P. (2017). Stable mono- and dinuclear organosilver complexes. *Organometallics*, *36*, 964–974.

Thiele, U. K. (2001). The current status of catalysis and catalyst development for the industrial process of poly(ethylene terephthalate) polycondensation. *International Journal of Polymeric Materials*, *50*, 387–394.

Tietze, H. (1981). The crystal and molecular structure of oxovanadium(V) orthophosphate dihydrate, $VOPO_4.2H_2O$. *Australian Journal of Chemistry*, *34*, 2035–2038.

Timberlake, S., & Prag, J. (2005). *The archaeology of Alderley Edge: Survey, excavation and experiment in an ancient mining landscape*. Oxford, England: J. and E. Hedges.

Toedt, J., Koza, D., & Cleef-Toedt, K. V. (2005). *Chemical composition of everyday products*. Westport, CT: Greenwood Press.

Tompsett, D. A., & Islam, M. S. (2013). Electrochemistry of hollandite α-MnO_2: Li-ion and Na-ion insertion and Li_2O incorporation. *Chemistry of Materials*, *25*, 2515–2526.

Toyoura, K., Tsujimura, H., Goto, T., Hachiya, K., Hagiwara, R., & Ito, Y. (2005). Optical properties of zinc nitride formed by molten salt electrochemical process. *Thin Solid Films*, *492*, 88–92.

Trofast, J. (2011). Berzelius' discovery of selenium. *Chemistry International*, *33*, 16–19.

Troyanov, S. I., Morozov, I. V., Znamenkov, K. O., & Korenev, Y. M. (1995). Synthesis and X-ray structure of new copper(II) nitrates: $Cu(NO_3)_2H_2O$ and β-modification of $Cu(NO_3)_2$. *Zeitschrift für anorganische und allgemeine Chemie*, *621*, 1261–1265.

Tsuji, J. (2004). *The basic chemistry of organopalladium compounds. Palladium reagents and catalysts* (pp. 1–26). New York: John Wiley & Sons.

Tudela, D. (2008). Silver(II) oxide or silver(I,III) oxide? *Journal of Chemical Education*, *85*, 863.

Tunes, M. A., da Silva, F. C., Camara, O., Schön, C. G., Sagás, J. C., Fontana, L. C., ... Edmondson, P. D. (2018). Energetic particle irradiation study of TiN coatings: Are these films appropriate for accident tolerant fuels? *Journal of Nuclear Materials*, *512*, 239–245.

Turner, A. (1999). Viewpoint: The story so far: An overview of developments in UK food regulation and associated advisory committees. *British Food Journal*, *101*, 274–283.

Tweedale, G. (1985). Sir Robert Abbott Hadfield F.R.S. (1858–1940), and the discovery of manganese steel. *Notes and Records of the Royal Society of London*, *40*, 63–74.

Urbain, G. (1911). Sur un nouvel élément qui accompagne le lutécium et le scandium dans les terres de la gadolinite: le celtium. *Comptes Rendus*, *152*, 141–143.

Ure, A., & Karmarsch, K. (1843). *Technisches wörterbuch oder Handbuch der Gewerbskunde ...: Bearb. nach Dr. Andrew Ure's Dictionary of arts, manufactures and mines* (Vol 1). Prague: G. Haase.

Vahidnia, A., Van Der Voet, G. B., & De Wolff, F. A. (2007). Arsenic neurotoxicity – A review. *Human & Experimental Toxicology*, *26*, 823–832.

van Arkel, A. E., & de Boer, J. H. (1925). Darstellung von reinem Titanium-, Zirkonium-, Hafnium- und Thoriummetall. *Zeitschrift für anorganische und allgemeine Chemie*, *148*, 345–350.

Venkatratnam, A., & Lents, N. (2011). Zinc reduces the detection of cocaine, methamphetamine, and THC by ELISA urine testing. *Journal of Analytical Toxicology*, *35*, 333–340.

Verhoeven, J. D. (1975). *Fundamentals of physical metallurgy*. New York: John Wiley & Sons.

Verhoeven, J. D., Pendray, A. H., & Dauksch, W. E. (1998). The key role of impurities in ancient damascus steel blades. *Journal of the Minerals, Metals and Materials Society*, *50*, 58–64.

Wade, K., & Banister, A. J. (2016). *The chemistry of aluminium, gallium, indium and thallium: Comprehensive inorganic chemistry* (vol 12). Oxford: Pergamon Press, Pergamon Texts in Inorganic Chemistry.

Wagner, D. B. (1993). *Iron and steel in ancient China*. Leiden: Brill.

Wagner, D. B. (2003). Chinese blast furnaces from the 10th to the 14th century. *Historical Metallurgy*, *37*, 25–37.

Waitkins, G. R., Bearse, A. E., & Shutt, R. (1942). Industrial utilization of selenium and tellurium. *Industrial & Engineering Chemistry, 34*, 899–910.

Wan, C., Wang, Y., Wang, N., Norimatsu, W., Kusunoki, M., & Koumoto, K. (2011). Intercalation: Building a natural superlattice for better thermoelectric performance in layered chalcogenides. *Journal of Electronic Materials, 40*, 1271–1280.

Wang, G., Huang, B., Li, Z., Lou, Z., Wang, Z., Dai, Y., & Whangbo, M.-H. (2015). Synthesis and characterization of ZnS with controlled amount of S vacancies for photocatalytic H_2 production under visible light. *Scientific Reports, 5*, 8544.

Wang, H., Huang, X., Lin, J., Cui, J., Chen, Y., Zhu, C., ... Liu, Z. (2017). High-quality monolayer superconductor $NbSe_2$ grown by chemical vapour deposition. *Nature Communications, 8*, 394.

Wang, S. (2006). Cobalt—Its recovery, recycling, and application. *JOM, 58*, 47–50.

Wang, X., & Andrews, L. (2003). Chromium hydrides and dihydrogen complexes in solid neon, argon, and hydrogen: Matrix infrared spectra and quantum chemical calculations. *The Journal of Physical Chemistry A, 107*, 570–578.

Wang, X., & Andrews, L. (2004). Infrared spectra of Zn and Cd hydride molecules and solids. *The Journal of Physical Chemistry A, 108*, 11006–11013.

Wang, X., Chertihin, G. V., & Andrews, L. (2002). Matrix infrared spectra and DFT calculations of the reactive MH_x (x = 1, 2, and 3), $(H_2)MH_2$, MH_2^+, and MH_4^- (M = Sc, Y, and La) species. *The Journal of Physical Chemistry A, 106*, 9213–9225.

Washio, K. (2003). SiGe HBT and BiCMOS technologies for optical transmission and wireless communication systems. *IEEE Transactions on Electron Devices, 50*, 656–668.

Watling, H. R. (2006). The bioleaching of sulphide minerals with emphasis on copper sulphides — A review. *Hydrometallurgy, 84*, 81–108.

Weckhuysen, B. M., & Schoonheydt, R. A. (1999). Olefin polymerization over supported chromium oxide catalysts. *Catalysis Today, 51*, 215–221.

Weeks, M. E., & Leichester, H. M. (1968). Discovery of the elements. *Journal of Chemical Education*, Easton, PA.

Weil, E. D., & Levchik, S. (2004). A review of current flame retardant systems for epoxy resins. *Journal of Fire Sciences, 22*, 25–40.

Westing, A. H. (1986). *Global resources and international conflict: Environmental factors in strategic policy and action.* Oxford: Oxford University Press.

Whelan, J. M., Struthers, J. D., & Ditzenberger, J. A. (1960). Separation of sulfur, selenium, and tellurium from arsenic. *Journal of the Electrochemical Society, 107*, 982–985.

White, R. E., & Hanusa, T. P. (2005). Ruthenium: Organometallic chemistry based in part on the article ruthenium: Organometallic chemistry by Ulrich Koelle which appeared in the encyclopedia of inorganic chemistry. In R. B. King, R. H. Crabtree, C. M. Lukehart, D. A. Atwood, & R. A. Scott (Eds.), *Encyclopedia of Inorganic Chemistry* (1st Edition). New York: John Wiley & Sons.

Wietelmann, U., Felderhoff, M., & Rittmeyer, P. (2016). *Hydrides. Ullmann's encyclopedia of industrial chemistry.* Weinheim: Wiley-VCH.

Wildermuth, E., Stark, H., Friedrich, G., Ebenhöch, F. L., Kühborth, B., Silver, J., & Rituper, R. (2000). *Iron Compounds. Ullmann's Encyclopedia of Industrial Chemistry.* Weinheim: Wiley-VCH.

Wilkinson, V. M., & Gould, G. W. (1998). *Food irradiation a reference guide.* Cambridge, England: Woodhead Publication.

Williams, R. (2008). *Limekilns and limeburning.* London: Bloomsbury Publishing.

Willies, L., Craddock, P. T., Gurjar, L. J., & Hegde, K. T. M. (1984). Ancient lead and zinc mining in Rajasthan, India. *World Archaeology, 16*, 222–233.

Wishart, D. J. (2005). *Encyclopedia of the great plains* (p. 433) University of Nebraska.

Witteveen, H. J., & Farnau, E. F. (1921). Colors developed by cobalt oxides. *Industrial & Engineering Chemistry, 13*, 1061–1066.

Witzel, M. (2001). Autochthonous Aryans? The evidence from old Indian and Iranian texts. *Electronic Journal of Vedic Studies (EJVS), 7-3*, 1–93.

Woosley, S., & Janka, T. (2006). The physics of core collapse supernovae. *Nature Physics, 1*, 147–154.

Wróblewski, A. K. (2008). The downfall of parity—The revolution that happened fifty years ago. *Acta Physica Polonica B, 39*, 251–264.

Wu, C. S. (1957). Experimental test of parity conservation in beta decay. *Physical Review, 105*, 1413–1415.

Wu, H., & Ye, P. D. (2016). Fully depleted Ge CMOS devices and logic circuits on Si. *IEEE Transactions on Electron Devices, 63*, 3028–3035.

Wurtz, A. (1844). Sur l'hydrure de cuivre. *Comptes rendus, 18*, 702–704.

Xiao-quan, S., Wen, W., & Bei, W. (1992). Determination of gallium in coal and coal fly ash by electrothermal atomic absorption spectrometry using slurry sampling and nickel chemical modification. *Journal of Analytical Atomic Spectrometry, 7*, 761–764.

Xie, Z., Liu, Q., Chang, Z., & Zhang, X. (2013). The developments and challenges of cerium half-cell in zinc–cerium redox flow battery for energy storage. *Electrochimica Acta, 90*, 695–704.

Yamaguchi, T., Hayashi, S., & Ohtaki, H. (1989). X-ray diffraction and Raman studies of zinc(II) chloride hydrate melts, $ZnCl_2.rH_2O$ (r = 1.8, 2.5, 3.0, 4.0, and 6.2). *The Journal of Physical Chemistry, 93*, 2620–2625.

Yan, R., Khalsa, G., Schaefer, B. T., Jarjour, A., Rouvimov, S., Nowack, K. C., ... Jena, D. (2019). Thickness dependence of superconductivity in ultrathin NbS_2. *Applied Physics Express, 12*, 023008.

Yoshikai, N., & Nakamura, E. (2012). Mechanisms of nucleophilic organocopper(I) reactions. *Chemical Reviews, 112*, 2339–2372.

Young, R. (1957). The geochemistry of cobalt. *Geochimica et Cosmochimica Acta, 13*, 28–41.

Youssef, K. M., Zaddach, A. J., Niu, C., Irving, D. L., & Koch, C. C. (2015). A novel low-density, high-hardness, high-entropy alloy with close-packed single-phase nanocrystalline structures. *Materials Research Letters, 3*, 95–99.

Zakharov, V. V. (2003). Effect of scandium on the structure and properties of aluminum alloys. *Metal Science and Heat Treatment, 45*, 246–253.

Zee, R. J. V., DeVore, T. C., Jr., & Weltner, W. (1979). CrH and CrH_2 molecules: ESR and optical spectroscopy at 4°K. *The Journal of Chemical Physics, 71*, 2051–2056.

Zhang, J., & Richardson, H. W. (2016). *Copper compounds. Ullmann's encyclopedia of industrial chemistry* (pp. 1–31). Weinheim: Wiley-VCH.

Zhang, T., Das, S. K., Fels, D. R., Hansen, K. S., Wong, T. Z., Dewhirst, M. W., & Vlahovic, G. (2013). PET With ^{62}Cu-ATSM and ^{62}Cu-PTSM is a useful imaging tool for hypoxia and perfusion in pulmonary lesions. *American Journal of Roentgenology*, *201*, W698–W706.

Zhang, W., & Cheng, C. Y. (2007). Manganese metallurgy review. Part I: Leaching of ores/secondary materials and recovery of electrolytic/chemical manganese dioxide. *Hydrometallurgy*, *89*, 137–159.

Zhang, X. G. (1996). *Corrosion and electrochemistry of zinc*. New York: Plenum Press.

Zhao, J., Xia, L., Sehgal, A., Lu, D., McCreery, R. L., & Frankel, G. S. (2001). Effects of chromate and chromate conversion coatings on corrosion of aluminum alloy 2024-T3. *Surface and Coatings Technology*, *140*, 51–57.

Zhou, Z., Zhou, N., Xia, M., Yokoyama, M., & Hintzen, H. T. (2016). Research progress and application prospects of transition metal Mn4 + -activated luminescent materials. *Journal of Materials Chemistry C*, *4*, 9143–9161.

Zhu, X. Q., Tang, H. S., & Sun, X. H. (2014). Genesis of banded iron formations: A series of experimental simulations. *Ore Geology Reviews*, *63*, 465–469.

Zhuang, Z., Peng, Q., Zhuang, J., Wang, X., & Li, Y. (2006). Controlled hydrothermal synthesis and structural characterization of a nickel selenide series. *Chemistry – A European Journal*, *12*, 211–217.

Zibaseresht, R., & Hartshorn, R. M. (2006). Hexaaquacopper(II) dinitrate: Absence of Jahn-Teller distortion. *Acta Crystallographica E*, *62*, i19–i22.

Chapter 6

Period 5

6.1 37 Rb — Rubidium

6.1.1 Discovery

Rubidium was discovered in 1861 by German chemist Robert Wilhelm Eberhard Bunsen (March 30, 1811 to August 16, 1899) (Fig. 6.1) and German physicist Gustav Robert Kirchhoff (March 12, 1824 to October 17, 1887) (Fig. 6.2), in Heidelberg, Germany, in the mineral lepidolite ($KLi_2AlSi_4O_{10}F_2$—$KLi_{1.5}Al_{1.5}Si_3O_{10}F_2$) by flame spectroscopy. For the bright red lines in its emission spectrum, they selected a name derived from the Latin word rubidus, meaning "deep red." Rubidium is an insignificant component in lepidolite (Kirchhoff and Bunsen, 1861). Kirchhoff and Bunsen processed 150 kg of lepidolite containing only 0.24% rubidium oxide (Rb_2O). Both potassium and rubidium formed insoluble salts with chloroplatinic acid, $[H_3O]_2[PtCl_6](H_2O)x$ ($0 \leq x \leq 6$), but those salts exhibited a slight difference in solubility in hot water. Consequently, the less-soluble rubidium hexachloroplatinate (Rb_2PtCl_6) could be separated through fractional crystallization. After reduction of the hexachloroplatinate with hydrogen, the process produced 0.51 g of rubidium chloride for further studies. Bunsen and Kirchhoff started their first large-scale separation of cesium and rubidium compounds with 44,000 L of mineral water, which resulted in 7.3 g of cesium chloride and 9.2 g of rubidium chloride. Rubidium was the second element, soon after cesium, to be discovered by spectroscopy, just 1 year after the invention of the spectroscope by Bunsen and Kirchhoff. The two scientists used the rubidium chloride to approximate that the atomic weight of the new element was 85.36 (at this time the recognized value is 85.47). They attempted to produce elemental rubidium by electrolysis of molten rubidium chloride, but instead of a metal, they found a blue homogeneous substance which "neither under the naked eye nor under the microscope showed the slightest trace of metallic substance." They assumed it was a subchloride (Rb_2Cl), but the product was possibly a colloidal mixture of the metal and rubidium chloride. In a second experiment to obtain metallic rubidium, Bunsen managed to reduce rubidium by heating charred rubidium tartrate ($C_4H_4O_6Rb_2$) (Bunsen, 1863). Although the distilled rubidium was pyrophoric (ignites spontaneously in air at or below 55°C), they managed to measure the density and the melting point. The high quality of this research in the 1860s can be judged by the fact that their obtained density differs by less than 0.1 g/cm^3 and the melting point by less than 1°C from the currently accepted values. The low radioactivity of rubidium was observed for the first time in 1908, but that was before the theory of isotopes was established in 1910, and the low level of activity (half-life greater than 10^{10} years) made understanding the process difficult (Campbell and Wood, 1908). The now proven beta decay of ^{87}Rb to stable ^{87}Sr was still a matter of debate in the late 1940s (Lewis, 1952). Rubidium had negligible industrial value before the 1920s. Since then, the most significant use of rubidium is in research and development, mainly in chemical and electronic applications. In 1995, ^{87}Rb was used to produce a Bose–Einstein condensate (A state of matter of a dilute gas of bosons cooled to temperatures very close to absolute zero (−273.15°C). Under such conditions, a large fraction of bosons occupy the lowest quantum state, at which point microscopic quantum phenomena, particularly wavefunction interference, become apparent macroscopically.), for which the discoverers, American physicists Eric Allin Cornell (born December 19, 1961) and Carl Edwin Wieman (born March 26, 1951) and German physicist Wolfgang Ketterle (born October 21, 1957, won the 2001 Nobel Prize in Physics (Levi, 2001).

6.1.2 Mining, production, and major minerals

Rubidium is the 23rd most abundant element in the Earth's crust, nearly as abundant as Zn and rather more common than Cu. It is found naturally in the minerals leucite, $KAlSi_2O_6$, pollucite, $(Cs,Na)AlSi_2O_6{\cdot}nH_2O$, carnallite, $KMgCl_3{\cdot}6H_2O$, and zinnwaldite (series name for polylithionite-siderophyllite), $KLi_2AlSi_4O_{10}F_2$—$KFe_2AlAl_2Si_2O_{10}(OH)_2$, which can have as much as 1% rubidium oxide. In these minerals the Rb substitutes for K in the crystals structure. Lepidolite (series name, compositionally between polylithionite and trilithionite,

The Periodic Table: Nature's Building Blocks. DOI: https://doi.org/10.1016/B978-0-12-821279-0.00006-6

FIGURE 6.1 Robert Wilhelm Eberhart Bunsen (c.1860).

FIGURE 6.2 Gustav Robert Kirchhoff.

$KLi_2AlSi_4O_{10}F_2$—$KLi_{1.5}Al_{1.5}Si_3O_{10}F_2$) contains between 0.3% and 3.5% Rb and is the commercial source of the element. Some potassium minerals and potassium chlorides also contain the element in commercially important quantities. Although rubidium is more abundant in Earth's crust than cesium, the limited applications and the lack of a mineral rich in rubidium restricts the production of rubidium compounds to about 2–4 tons per year. A number of different methods are available for separating potassium, rubidium, and cesium. The fractional crystallization of a rubidium and cesium alum $(Cs,Rb)Al(SO_4)_2 \cdot 12H_2O$ yields after 30 (!) subsequent steps pure rubidium alum. Two other methods have been reported, the chlorostannate process yielding pure rubidium chloride and the ferrocyanide process yielding pure rubidium carbonate. Rb_2CO_3 with potassium and cesium impurities is obtained from roasted rubidium ferrocyanide. For

FIGURE 6.3 Rubicline, $Rb(AlSi_3O_8)$, is the rubidium analog of microcline. On this specimen, the rubicline is a small 2 mm crystal grain. Vasin-Myl'k Mt, Voron'i Tundry, Kola Peninsula, Russia.

several years in the 1950s and 1960s, a by-product of potassium production called Alkarb was a main source for rubidium. Alkarb contained 21% rubidium, with the rest being potassium and a small amount of cesium. Today the largest producers of cesium, such as the Tanco Mine, Manitoba, Canada, produce rubidium as a by-product from pollucite, $(Cs,Na)_2(Al_2Si_4O_{12})\cdot 2H_2O$.

Rubidium is extremely rare in minerals with only three minerals known, the borate ramanite-(Rb) ($Rb[B_5O_6(OH)_4]\cdot 2H_2O$), and the two silicates rubicline ($Rb(AlSi_3O_8)$) (Fig. 6.3) and voloshinite ($Rb(LiAl_{1.5}\square_{1.5})(Al_{0.5}Si_{3.5})O_{10}F_2$).

6.1.3 Chemistry

Although Rb is monoisotopic, Rb in the Earth's crust is composed of two isotopes: the stable ^{85}Rb (natural abundancy 72.2%) and the radioactive ^{87}Rb (27.8%). Twenty-four further Rb isotopes have been synthesized with half-lives of less than 3 months; the majority of them are highly radioactive and have few applications. ^{87}Rb has a half-life of 48.8×10^9 years, which is more than three times the age of the universe of $(13.799 \pm 0.021) \times 10^9$ years, making it a primordial nuclide. It easily substitutes for K in minerals and is hence widespread. Rb has been utilized extensively in dating rocks; ^{87}Rb β decays to stable ^{87}Sr. In the process of fractional crystallization, Sr will be concentrated in plagioclase, leaving Rb in the liquid phase. Consequently, the Rb/Sr ratio in residual magma may increase over time, and the progressing differentiation produces rocks with increasing Rb/Sr ratios (Box 6.1). The highest ratios (10 or more) are found in pegmatites. If the initial Sr concentration is known or can be extrapolated, then the age can be determined by measurement of the Rb and Sr concentrations and of the $^{87}Sr/^{86}Sr$ ratio. The obtained dates indicate the true age of the minerals only, provided that the rocks have not been subsequently altered. ^{82}Rb, one of the element's nonnatural isotopes, is produced by electron capture decay of ^{82}Sr with a half-life of 25.36 days. ^{82}Rb with a half-life of 76 seconds decays via positron emission to stable ^{82}Kr.

Rubidium is a very soft, silvery-white metal belonging to the alkali metal group or Group 1. Its electronic configuration is $[Kr]5s^1$. Following the periodic trend for Group 1, its atomic and ionic radii are bigger than potassium but smaller than cesium, and the first ionization energy is lower than the lighter alkali metals. As a result of this low ionization energy, its oxidation state, as with other members of the group, is always +1 (Table 6.1).

6.1.3.1 Reactions with air and water

Rubidium is more reactive than potassium. Its reaction with water to form rubidium hydroxide (RbOH) and hydrogen gas (H_2) is more violent than the lighter alkali metals, often igniting the H_2 gas that is produced. Rubidium ignites spontaneously in air making, handling of the element difficult. The partial oxidation of rubidium in air gives the colored suboxides, Rb_6O and Rb_9O_2. The final oxygenation product is the superoxide (RbO_2) and this can be reduced to rubidium oxide (Rb_2O) by adding more Rb.

BOX 6.1 Rubidium-strontium dating.

The Rb-Sr dating method is a radiometric dating technique utilized by geoscientists to determine the age of rocks and minerals from the quantities they contain of ^{87}Rb, ^{87}Sr, and ^{86}Sr. The usefulness of the Rb-Sr isotope system is based on the fact that ^{87}Rb decays to ^{87}Sr with a half-life of 49.23 billion years. Furthermore, Rb is a highly incompatible element that, during partial melting of the mantle, preferentially joins the magmatic melt rather than remain in mantle minerals. Therefore Rb is enriched in crustal rocks. The radiogenic daughter, ^{87}Sr, is formed in the decay process and was formed in rounds of stellar nucleosynthesis predating the solar system's creation. Different minerals in a given geologic setting can take up distinctly different ratios of radiogenic ^{87}Sr to naturally occurring ^{86}Sr ($^{87}Sr/^{86}Sr$) through time; and their age can be determined by measuring the $^{87}Sr/^{86}Sr$ using a mass spectrometer, provided that the amount of ^{87}Sr present when the rock or mineral formed is known, and calculating the amount of ^{87}Rb from a measurement of the Rb present and knowledge of the $^{85}Rb/^{87}Rb$ weight ratio. If these minerals formed from the same silicic melt, then each mineral had identical initial $^{87}Sr/^{86}Sr$ ratios as the parent melt. However, since Rb substitutes for K in minerals and these minerals have different K/Ca ratios, these minerals will have had different Rb/Sr ratios. In the process of fractional crystallization, Sr has a tendency to become concentrated in plagioclase, leaving Rb in the liquid phase. Consequently, the Rb/Sr ratio in residual magma may increase over time, producing rocks with increasing Rb/Sr ratios with increasing differentiation. Typically, Rb/Sr increases in the order of plagioclase, hornblende, K-feldspar, biotite, and muscovite. Therefore given enough time for significant production of radiogenic ^{87}Sr, measured $^{87}Sr/^{86}Sr$ values will be different in the minerals, increasing in the same order. The age of a sample is obtained by analyzing several minerals within multiple subsamples from different parts of the original sample. The $^{87}Sr/^{86}Sr$ ratio for each subsample is plotted against its $^{87}Rb/^{86}Sr$ ratio on a graph called an isochron. If these form a straight line, then the subsamples are consistent, and the age is probably reliable. The slope of the line dictates the age of the sample. Rb-Sr dating depends on correctly measuring the Rb-Sr ratio of a mineral or whole rock sample, plus deriving an accurate $^{87}Sr/^{86}Sr$ ratio for the mineral or whole rock sample. Several prerequisites must be fulfilled before a Rb-Sr date can be interpreted as representing the time of emplacement or formation of a rock: (1) the system must have remained closed to Rb and Sr diffusion from the time at which the rock formed or fell below the closure temperature (commonly considered as 650°C); (2) the minerals which are taken from a rock to obtain an isochron must have formed in chemical equilibrium with each other or in the case of sediments, be deposited at the same time; and (3) the rock must not have undergone any metasomatism (the chemical alteration of a rock by hydrothermal and other fluids. It is the replacement of one rock by another of different mineralogical and chemical composition. The minerals which compose the rocks are dissolved and new mineral formations are deposited in their place. Dissolution and deposition occur simultaneously, and the rock remains solid.), which could have disturbed the Rb-Sr system either thermally or chemically. One of the major disadvantages (and, conversely, the most important use) of using Rb and Sr to obtain a radiometric date is their relative mobility, especially in hydrothermal fluids. Rb and Sr are relatively mobile alkaline elements and as such are rather easily moved around by the hot, often carbonated hydrothermal fluids present during metamorphism or magmatism. On the other hand, these fluids may metasomatically alter a rock, introducing new Rb and Sr into the rock (mostly during potassic alteration or calcic (albitization) alteration. Rb-Sr can then be used on the altered mineralogy to date the time of this alteration, but not the date at which the rock initially formed. Therefore, in order to assign age significance to a result it is necessary to carefully study the metasomatic and thermal history of the rock, any metamorphic events, and any evidence of fluid movement. A Rb-Sr date which is at variance with other geochronometers may not be useless, it may be providing data on another event and not the formation age of the rock.

Rb_2O is a yellow colored solid. The alkali metal oxides M_2O (M = Li, Na, K, Rb) crystallize in the antifluorite structure. In the antifluorite structure the positions of the anions and cations are reversed relative to their positions in CaF_2, with rubidium ions in 8 coordination (cubic) and oxide ions in 4 coordination (tetrahedral). For laboratory use, RbOH is normally utilized instead of the oxide. The hydroxide is more useful, less reactive toward atmospheric moisture, and less expensive than the oxide. As for most alkali metal oxides, the best synthesis of Rb_2O does not involve oxidation of the metal but reduction of the anhydrous nitrate:

$$10Rb + 2RbNO_3 \rightarrow 6Rb_2O + N_2$$

Typical for alkali metal hydroxides, RbOH cannot be dehydrated to the oxide. Instead, the hydroxide can be decomposed to the oxide (by reduction of the hydrogen ion) using Rb metal:

$$2Rb + 2RbOH \rightarrow 2Rb_2O + H_2$$

Metallic Rb reacts with O_2, as shown by its tendency to quickly tarnish in air. The tarnishing process is relatively colorful as it proceeds via bronze-colored Rb_6O and copper-colored Rb_9O_2. The suboxides of rubidium that have been

TABLE 6.1 Rubidium properties.

Appearance	Gray white
Standard atomic weight ($A_{r,std}$)	85.4678
Block	s-Block
Element category	Alkali metal
Electron configuration	[Kr] $5s^1$
Phase at STP	Solid
Melting point	39.30°C
Boiling point	688°C
Density (near r.t.)	1.532 g/cm^3
When liquid (at m.p.)	1.46 g/cm^3
Heat of fusion	2.19 kJ/mol
Heat of vaporization	69 kJ/mol
Molar heat capacity	31.060 J/(mol·K)
Oxidation states	−1, **+1**
Ionization energies	First: 403 kJ/mol Second: 2632.1 kJ/mol Third: 3859.4 kJ/mol
Atomic radius	Empirical: 248 pm
Covalent radius	220 ± 9 pm
Van der Waals radius	303 pm

STP, Standard temperature and pressure.
Bold font indicates main oxidation state.

characterized by X-ray crystallography are, for example, Rb_9O_2 and Rb_6O. The final product of oxygenation of Rb is mainly RbO_2, rubidium superoxide:

$$Rb + O_2 \rightarrow RbO_2$$

This superoxide can then be reduced to Rb_2O by reacting it with excess rubidium metal:

$$3Rb + RbO_2 \rightarrow 2Rb_2O$$

Similar to other alkali metal oxides, Rb_2O is a strong base. Hence, Rb_2O reacts exothermically with water to form rubidium hydroxide.

$$Rb_2O + H_2O \rightarrow 2RbOH\ Rb_2O$$

is so reactive toward water that it is thought of as hygroscopic. Upon heating, Rb_2O reacts with hydrogen to rubidium hydroxide and rubidium hydride:

$$Rb_2O + H_2 \rightarrow RbOH + RbH$$

6.1.3.2 Halides

Rubidium reacts with halogens to produce rubidium halides (RbX, X=F, Cl, Br, I). Rubidium fluoride (RbF) forms cubic crystals with the rock-salt (NaCl) structure. There are a number of different routes for synthesizing rubidium fluoride. One entails reacting rubidium hydroxide with hydrofluoric acid:

$$RbOH + HF \rightarrow RbF + H_2O$$

A second method is to neutralize rubidium carbonate with hydrofluoric acid:

$$Rb_2CO_3 + 2HF \rightarrow 2RbF + H_2O + CO_2$$

A third possible route is through the reaction of rubidium hydroxide with ammonium fluoride:

$$RbOH + NH_4F \rightarrow RbF + H_2O + NH_3$$

The least used route because of the expense of rubidium metal is to react it directly with fluorine gas, as rubidium reacts violently with halogens:

$$2Rb + F_2 \rightarrow 2RbF$$

The most common method to produce pure rubidium chloride is the reaction of its hydroxide with hydrochloric acid, followed by recrystallization:

$$RbOH(aq) + HCl(aq) \rightarrow RbCl(aq) + H_2O(l)$$

Since RbCl is hygroscopic, it must be protected from atmospheric moisture, e.g. using a desiccator. RbCl is mainly used in laboratories. Rubidium chloride reacts with sulfuric acid to rubidium hydrogen sulfate, $RbHSO_4$. There are a number of different routes for synthesizing rubidium bromide. Similar to the fluoride and the chloride RbOH can react with hydrobromic acid:

$$RbOH + HBr \rightarrow RbBr + H_2O$$

A different method is to neutralize rubidium carbonate with hydrobromic acid:

$$Rb_2CO_3 + 2HBr \rightarrow 2RbBr + H_2O + CO_2$$

6.1.3.3 *Other compounds*

It dissolves in mercury forming amalgams in stoichiometries such as Rb_3Hg_{20}, Rb_5Hg_{19}, and Rb_7Hg_{31}. It also forms alloys with gold, iron, cesium, sodium, and potassium. It does not form alloys with lithium. The hydride of rubidium exists and has been synthesized from the elements using a mechanical alloying method.

Rubidium is the second most reactive metal and can form many compounds. However, high cost coupled with an uncertain and limited supply limits its exploration for commercial purposes. Nevertheless, several potentially important compounds of rubidium have been prepared and characterized. These include rubidium hydroxide (RbOH), rubidium carbonate (Rb_2CO_3), rubidium copper sulfate ($Rb_2SO_4 \cdot CuSO_4 \cdot 6H_2O$), and rubidium silver iodide ($RbAg_4I_5$).

Rubidium carbonate, Rb_2CO_3, is a convenient compound of rubidium; it is stable, not particularly reactive, and easily soluble in water, and is the form in which rubidium is usually sold. It can be produced by reacting ammonium carbonate with rubidium hydroxide. Rubidium nitrate, $RbNO_3$, is a white crystalline powder that is highly soluble in water and very slightly soluble in acetone. In a flame test, $RbNO_3$ gives a mauve/light purple color. $RbNO_3$ can be produced either by dissolving rubidium metal, its hydroxide or carbonate in nitric acid.

$$RbOH + HNO_3 \rightarrow RbNO_3 + H_2O$$
$$2Rb + 2HNO_3 \rightarrow 2RbNO_3 + H_2$$

Rubidium hydride, RbH, can be synthesized by reacting rubidium metal with hydrogen gas. As a hydride of an alkali metal, it is reactive towards even weak oxidizing agents. A redox reaction will occur with chlorine or fluorine and produce a lot of heat. Rubidium hydride will react violently with water or air and careful storage is necessary. Rubidium cyanide (RbCN) is a white solid, easily soluble in water, with a smell reminiscent of bitter almonds, and somewhat similar in appearance to sugar. Rubidium cyanide has chemical properties similar to potassium cyanide and hence it is a very toxic substance. Rubidium cyanide can be synthesized through the reaction of hydrogen cyanide and rubidium hydroxide in alcohol or ether:

$$HCN + RbOH \rightarrow RbCN + H_2O$$

6.1.3.4 *Organorubidium compounds*

Several organometallic compounds of rubidium have also been synthesized. For example, rubidium dihexadecylphosphates ($RbO_2P(OC_{16}H_{33})_2$), was synthesized and shown to exhibit thermotropic behavior. Other compounds of rubidium have been prepared for academic interests. Some of the methods of preparing organorubidium compounds are: (1) by the reaction of the corresponding organolithium reagent and an alkoxide, (2) by alkyl group exchange, and (3) by using alkali metal hydrides as metallating agents.

1) $LiR + MOR' \rightarrow MR + LiOR'$ (M = Na, K, Rb, Cs)
2) $MR + R''H \rightarrow MR'' + RH$
3) $MH + RH \rightarrow MR + H_2$

An example is the preparation of the bis(trimethylsilyl)methyl derivative of Rb from the organolithium compound and the alkoxide, $RbOC_6H_2^tBu_2$-2, 6. Methyl, butyl, or trimethylsilylmethyl compounds of the heavy alkali metals can be used to generate other organometallic compounds. The metalation by metal hydride was used in the synthesis of base-free cyclopentadienyl metal derivatives. Organic compounds of the heavier alkali metals are rarely reported in organic synthesis. However, in view of their greater reactivity compared with organolithium reagents, it is possible for reactions to proceed at lower temperatures without interference from unwanted rearrangement of reactive intermediates.

6.1.4 Major uses

Rb compounds (e.g., nitrate) are occasionally used in fireworks to produce a purple color. Rb has also been studied for application in a thermoelectric generator utilizing the magnetohydrodynamic principle, where hot Rb ions are passed through a magnetic field. These conduct electricity and function like a framework of a generator, thus generating electric current. Rb, especially vaporized ^{87}Rb, is one of the most frequently used atomic species utilized for laser cooling and Bose–Einstein condensation. Its advantageous characteristics for this use comprise the ready availability of low-cost diode laser light at the appropriate wavelength, and the reasonable temperatures needed to achieve substantial vapor pressures. For cold atom applications necessitating tunable interactions, ^{85}Rb is better due to its rich Feshbach spectrum (In physics, a Feshbach resonance can occur upon collision of two slow atoms, when they temporarily stick together forming an unstable compound with short lifetime (so-called resonance). It is a feature of many-body systems in which a bound state is achieved if the coupling(s) between at least one internal degree of freedom and the reaction coordinates, which lead to dissociation, vanish. The opposite situation, when a bound state is not formed, is a shape resonance. It is named after American physicist Herman Feshbach (February 2, 1917 to December 22, 2000), a physicist at MIT.). Rb has been utilized for polarizing ^{3}He, creating volumes of magnetized ^{3}He gas, with the nuclear spins aligned instead of random. Rb vapor is optically pumped using a laser and the polarized Rb polarizes ^{3}He via the hyperfine interaction. These spin-polarized ^{3}He cells are valuable for neutron polarization measurements and for creating polarized neutron beams for other applications.

The resonant element in atomic clocks uses the hyperfine structure of Rb's energy levels, and Rb is valuable for high-precision timing. It is used as the principal component of secondary frequency references (Rb oscillators) in cell site transmitters and other electronic transmitting, networking, and test equipment. Such Rb standards are frequently used with GPS to create a "primary frequency standard" that has superior accuracy and is cheaper than Cs standards. These Rb standards are frequently mass-produced for the telecommunication industry. Other possible or existing uses of Rb are for example, as a working fluid in vapor turbines, as a getter in vacuum tubes, and as a photocell component. In addition, Rb is utilized as a compound in special types of glass, in the manufacture of superoxide by burning in oxygen, in the research of K ion channels in biology, and as the vapor in atomic magnetometers.

^{87}Rb is utilized with other alkali metals in the advancement of spin-exchange relaxation-free magnetometers. ^{82}Rb is utilized in positron emission tomography (PET). Rb is in many aspects comparable to K and tissue with high K content will also gather the radioactive Rb. One of the principal applications is for myocardial perfusion imaging. Due to changes in the blood–brain barrier in brain tumors, Rb accumulates more in brain tumors than normal brain tissue, permitting the use of radioisotope ^{82}Rb in nuclear medicine to detect and image brain tumors. ^{82}Rb has a very short half-life of only 76 seconds, and the creation from ^{82}Sr decay has to be done near the patient. Rb was assessed for the effect on manic depression and depression. Dialysis patients experiencing depression exhibit Rb depletion and consequently a supplementation may help during depression. In some experiments the Rb was dispensed as RbCl with up to 720 mg per day for 60 days. The ions are not very toxic; a 70 kg person has on average 0.36 g of Rb, and an increase of this amount by 50–100 times did not show any negative effects in test persons. The biological half-life of Rb in humans is 31–46 days. Rb is easily ionized; hence it was considered for use in ion engines, but was found to be less effective than cesium. It has also been proposed for use as a working fluid in vapor turbines.

6.2 38 Sr — Strontium

6.2.1 Discovery

Strontium is named after the Scottish village of Strontian (Gaelic Sròn an t-Sìthein), where it was first found in the ores of the lead mines. In 1790, Adair Crawford (1748 to July 29, 1795), a Northern Irish physician and chemist engaged in the preparation of barium, and his colleague Scottish military surgeon and chemist, and professor of chemistry at the Royal Military Academy, Woolwich, William Cruickshank (died 1810 or 1811), recognized that the Strontian ores showed properties that were different from those in other "heavy spars" sources. This permitted Crawford to conclude on page 355 of his 1790 publication "... it is probable indeed, that the Scotch mineral is a new species of earth which has not hitherto been sufficiently examined" (Crawford, 1790). The German physician and mineral collector Friedrich Gabriel Sulzer (October 10, 1749 to December 14, 1830) analyzed together with German physician, naturalist, physiologist, and anthropologist Johann Friedrich Blumenbach (May 11, 1752 to January 22, 1840) the mineral from Strontian and named it strontianite (Fig. 6.4). He also concluded that it was different from witherite ($BaCO_3$) and contained a new earth (neue Grunderde). In 1793 the Scottish physician Thomas Charles Hope (July 21, 1766 to June 13, 1844), a professor of chemistry at the University of Glasgow, suggested the name strontites (Fig. 6.5). He confirmed prior work of

FIGURE 6.4 Johann Friedrich Blumenbach, Zweihundert Bildnisse und Lebensabrisse berühmter deutscher Männer, first ed., Leipzig 1854, editor Ludwig Bechstein.

FIGURE 6.5 Thomas Charles Hope. Mezzotint by T. Hodgetts after H. Raeburn (CC-BY 4.0).

Crawford and reported: "... Considering it a peculiar earth I thought it necessary to give it a name. I have called it Strontites, from the place it was found; a mode of derivation in my opinion, fully as proper as any quality it may possess, which is the present fashion" (Hope, 1798). The element was ultimately separated by Cornish chemist and inventor Sir Humphry Davy (December 17, 1778 to May 29, 1829) in 1808 through electrolysis of a mixture containing strontium chloride and mercuric oxide and reported by him in a lecture to the Royal Society on June 30, 1808. In line with the nomenclature of the other alkaline earths, he changed the name to strontium. Many other early investigators examined strontium ore, including Martin Heinrich Klaproth (1793) and Richard Kirwan (1794) among others. The first large-scale application of strontium was in the production of sugar from sugar beet. Although a crystallization process using strontium hydroxide was patented by French chemist Augustin-Pierre Dubrunfaut (September 1, 1797, Lille to October 7, 1881) in 1849, the large-scale application started with the improvement of the production process in the early 1870s. The German sugar industry utilized the process well into the 20th century. Before the First World War the sugar beet industry used 100,000–150,000 tons of strontium hydroxide for this process per annum. The strontium hydroxide was recycled in the sugar production, but the demand to replace losses during production was sufficiently high to generate a substantial demand starting the mining of strontianite in the Münsterland region, North Rhine-Westphalia, Germany. The mining of strontianite in Germany came to an end when mining of the celestine ($SrSO_4$) deposits in Gloucestershire, South West England, started. These mines provided most of the world strontium supply from 1884 to 1941. Though the celestine deposits in the Granada Basin in Andalusia, Spain, were identified some time ago, the large-scale mining did not start before the 1950s.

6.2.2 Mining, production, and major minerals

Strontium commonly occurs in nature, being the 15th most abundant element on Earth (its heavier congener barium being the 14th), at an average of approximately 360 ppm in the Earth's crust and is found chiefly as the sulfate mineral celestine ($SrSO_4$) (Box 6.2) and the carbonate mineral strontianite ($SrCO_3$). Of these two minerals, celestine occurs much more regularly in deposits of large enough size to be economical for mining. Because strontium is used most frequently in the carbonate form, strontianite would be the more suitable of the two common minerals, but few deposits have been found that are suitable for mining.

The three main producers of strontium as celestine as of 2015 are China, Spain, and Mexico, while Argentina and Morocco are smaller producers. While strontium deposits occur widely in the United States, they have not been in production since 1959. A large percentage of mined celestine ($SrSO_4$) is converted into the carbonate using two different processes. Either the celestine is directly leached with sodium carbonate solution or the celestine is roasted with coal to form the sulfide. The second step results in a dark-colored material consisting mostly of strontium sulfide. This so-called "black ash" is subsequently dissolved in water and filtered. The sulfate is reduced to the sulfide by the carbothermic reduction:

$$SrSO_4 + 2C \rightarrow SrS + 2CO_2$$

BOX 6.2 Celestine ($SrSO_4$).

Celestine is the sole mineral of industrial importance. It occurs mainly in sedimentary rocks, particularly dolomite, limestone, and marl, either as primary precipitate from aqueous solution or, more usually, by the interaction of gypsum or anhydrite with Sr-rich waters. It may be a primary mineral in hydrothermal veins and is also found filling veins and cavities in basic eruptive rocks. It was originally named fasriger schwerspath by Andreas Gotthelf Schütz in 1791, which was renamed schwefelsaurer strontianite aus Pennsylvania by Martin Klaproth in 1797. It was renamed by Abraham Gottlieb Werner in 1798 in German zoelestin from the Greek *cœlestis* for celestial, in allusion to the faint blue color of the original specimen, and finally renamed Schützit by Dietrich Ludwig Gustav Karsten in 1800. Although far superior celestine crystals were previously known from Sicily, they were thought to be barium-rich as the element strontium was not discovered until the late 1780s and not formally described until 1792. Celestine is orthorhombic, *2/m2/m2/m* (space group *Pnma*). It has a hardness of 3–3.5. It has perfect cleavage on {001} and good on {210}, fracture is uneven, and tenacity is brittle. Parting on twin gliding and translation gliding planes. The luster is adamantine to vitreous to pearly, while the transparency is transparent to translucent. The color is colorless, white, gray, often faintly blue, or rarely light red. The streak is white. Optically it is biaxial (+). Crystals are usually blockier than those of barite ($BaSO_4$). It is commonly tabular parallel to {001} or prismatic parallel to *a* or *b* with prominent development of {0*kl*} and {*h*0*l*} prisms. Crystals elongated parallel to *a* are frequently terminated by nearly equally developed faces of d{101} and m{210}. It also occurs as radiating fibrous, reniform, or granular masses. Twinning has been reported on {210}, {101}, and other planes (doubtful).

Strontium carbonate is then precipitated from the strontium sulfide solution through the introduction of carbon dioxide. About 300,000 tons per year are processed using this method. The metal is commercially produced by reducing strontium oxide with aluminum. The strontium is distilled from the mixture. Strontium metal can also be produced on a small scale through electrolysis of a strontium chloride solution in molten potassium chloride:

$$Sr^{2+} + 2e^- \rightarrow Sr$$
$$2Cl^- \rightarrow Cl_2 + 2e^-$$

The element strontium can be found in just over 100 different minerals. Ten different halides are known with Sr in their structure, for example, bøggildite ($Na_2Sr_2Al_2PO_4F_9$), jarlite ($Na(Sr,Na)_7MgAl_6F_{32}(OH,H_2O)_2$), and strontiofluorite (SrF_2). The oxide class contains 11 minerals with Sr, for example, crichtonite ($Sr(Mn,Y,U)Fe_2(Ti,Fe,Cr,V)_{18}(O,OH)_{38}$) and tausonite ($SrTiO_3$). A larger group of 17 minerals is found under the carbonates, such as putnisite ($SrCa_4Cr_8^{3+}(CO_3)_8)SO_4((OH)_{16}\cdot 23H_2O$) (Fig. 6.6), strontianite ($SrCO_3$) (Fig. 6.7) and welagonite ($Na_2Sr_3Zr(CO_3)_6\cdot 3H_2O$) (Fig. 6.8). Five borate minerals are known with Sr, for example, tunellite ($SrB_6O_9(OH)_2\cdot 3H_2O$) and veatchite

FIGURE 6.6 Putnisite, $SrCa_4Cr_8^{3+}(CO_3)_8SO_4(OH)_{16}\cdot 23H_2O$, an extremely rare strontium-chromium carbonate found only in Australia. This specimen is very rich with three tiny rectangular purple crystals to about 0.75 mm and many other crystal "grains" throughout matrix. Armstrong Mine, Widgiemooltha, Western Australia, Australia.

FIGURE 6.7 One-centimeter groups of white strontianite, $SrCO_3$, in radial bursts on white barite. Minerva #1 Mine, Hardin Co., Illinois, United States.

FIGURE 6.8 Weloganite, $Na_2Sr_3Zr(CO_3)_6{\cdot}3H_2O$, stacked discoidal crystals in funnel-shaped clusters to 1.3 cm. Francon Quarry, Montreal, Quebec, Canada.

FIGURE 6.9 Light blue crystals of celestine, $SrSO_4$, to over 1 cm scattered on matrix. N'Chwaning I Mine, Kalahari Mn fields, Northern Cape Province, South Africa.

($Sr_2B_{11}O_{16}(OH)_5{\cdot}H_2O$). Only two sulfates contain Sr, celestine ($SrSO_4$) (Fig. 6.9) and kalistrontite ($K_2Sr(SO_4)_2$). Twenty-five minerals can be found in the phosphate class, for example, goyazite ($SrAl_3(PO_4)(PO_3OH)(OH)_6$) (Fig. 6.10), palermoite ($(Li,Na)_2(Sr,Ca)Al_4(PO_4)_4(OH)_4$), and svanbergite ($SrAl_3(PO_4)(SO_4)(OH)_6$). The largest group of Sr minerals is found, however, under the silicates with 54 minerals. These include minerals such as: brewsterite-(Sr) ($(Sr,Ba,Ca)[Al_2Si_6O_{16}]{\cdot}5H_2O$), charoite ($(K,Sr)_{15-16}(Ca,Na)_{32}[Si_6O_{11}(O,OH)_6]_2[Si_{12}O_{18}(O,OH)_{12}]_2[Si_{17}O_{25}(O,OH)_{18}]_2(OH,F)_4{\cdot}{\sim}3H_2O$), lamprophyllite ($(Na,Mn^{2+})_3(Sr,Na)_2(Ti,Fe^{3+})_3(Si_2O_7)_2O_2(OH,O,F)_2$), stronalsite ($Na_2SrAl_4Si_4O_{16}$), and yuksporite ($K_4(Ca,Na)_{14}(Sr,Ba)_2\ (\square,Mn,Fe)(Ti,Nb)_4(O,OH)_4(Si_6O_{17})_2\ (Si_2O_7)_3(H_2O,OH)_3$).

6.2.3 Chemistry

Naturally occurring Sr is a mixture of four stable isotopes: ^{84}Sr, ^{86}Sr, ^{87}Sr, and ^{88}Sr. Their abundance increases with increasing mass number and the heaviest, ^{88}Sr, makes up about 82.6% of all naturally occurring Sr; nevertheless, the abundance varies due to the formation of radiogenic ^{87}Sr as the daughter of long-lived β-decaying ^{87}Rb. Of the radioisotopes, the principal decay mode of the isotopes lighter than ^{85}Sr is electron capture or positron emission to isotopes of

FIGURE 6.10 Crystals of white goyazite, $SrAl_3(PO_4)(PO_3OH)(OH)_6$, to 0.5 mm with yellowish herderite, $CaBePO_4(F,OH)$. Alto Bernardino, Picui, Paraiba, Brazil.

TABLE 6.2 Strontium properties.

Appearance	Silvery-white metallic, with a pale-yellow tint
Standard atomic weight ($A_{r,std}$)	87.62
Block	s-Block
Element category	Alkaline earth metal
Electron configuration	[Kr] $5s^2$
Phase at STP	Solid
Melting point	777°C
Boiling point	1377°C
Density (near r.t.)	2.64 g/cm^3
when liquid (at m.p.)	2.375 g/cm^3
Heat of fusion	7.43 kJ/mol
Heat of vaporization	141 kJ/mol
Molar heat capacity	26.4 J/(mol·K)
Oxidation states	+1, **+2**
Ionization energies	First: 549.5 kJ/mol
	Second: 1064.2 kJ/mol
	Third: 4138 kJ/mol
Atomic radius	Empirical: 215 pm
Covalent radius	195 ± 10 pm
Van der Waals radius	249 pm

STP, Standard temperature and pressure.
Bold font indicates main oxidation state.

Rb, while that of the isotopes heavier than ^{88}Sr is electron emission to isotopes of Y. Of special interest are ^{89}Sr and ^{90}Sr. The first has a half-life of 50.6 days and is used to treat bone cancer due to strontium's chemical similarity and hence ability to replace calcium. Although ^{90}Sr with a half-life of 28.90 years has been utilized in a similar manner, it is also an isotope of concern in the fallout from nuclear weapons and nuclear accidents because of its formation as a fission product. Its presence in bones can result in bone cancer, cancer of nearby tissues, and leukemia. The 1986 Chernobyl nuclear accident contaminated about 30,000 km^2 with higher than 10 kBq per square meter with ^{90}Sr, accounting for 5% of the core inventory of ^{90}Sr.

Strontium (Sr) is a soft silver-white yellowish metallic element. It is a member of alkaline earth elements (Group 2) and occupies an intermediate position between Ca and Ba. Its electron configuration is $[Kr]5s^2$ and it easily loses two electrons from the 5s shell to form the Sr^{2+} ion, which has a filled electron shell that is isoelectronic with the noble gas, krypton (Table 6.2). Along with other members of the heavier alkaline earth group

[Ca, Sr, Ba, (Ra)], the chemical properties of Sr are dominated by its strong reducing power. It reacts readily with nonmetals even with nitrogen.

6.2.3.1 Reactions with air, nitrogen, and hydrogen

The surface of strontium metal is silver-white when freshly cut but rapidly turn yellowish as it gets covered with a thin layer of oxide that helps protect the metal from attack by air. This oxide protection, however, is less effective than the corresponding layer in magnesium. Finely powdered strontium ignites spontaneously in air to give white strontium oxide, SrO. Heating strontium in air at temperatures above 380°C will also form strontium nitride, Sr_3N_2. As Sr is very reactive and does not occur as the metal in nature, strontium oxide is more normally made by heating strontium carbonate, $SrCO_3$. Strontium hydroxide, $Sr(OH)_2$, is a caustic alkali. Since $Sr(OH)_2$ is slightly soluble in cold water, its preparation can easily be performed through the addition of a strong base such as NaOH or KOH, drop by drop to a solution of any soluble strontium salt, most commonly $Sr(NO_3)_2$. The $Sr(OH)_2$ will precipitate as a fine white powder. From here, the solution is filtered, and the $Sr(OH)_2$ is washed with cold water and dried. Strontium reacts with hydrogen at high temperatures to form strontium hydride, SrH_2.

$$2Sr(s) + O_2(g) \rightarrow 2SrO(s)$$
$$3Sr(s) + N_2(g) \rightarrow Sr_3N_2(s)$$
$$2Sr(s) + H_2(g) \rightarrow 2SrH_2(s)$$

6.2.3.2 Reactions with the halogens

Strontium burns with the halogens fluorine (F_2), chlorine (Cl_2), bromine (Br_2), and iodine (I_2) to form the halide salts, SrX_2 (X=Cl, Br, I). Bromine reacts at about 400°C and iodine at a dull red heat. Since strontium is very reactive and does not occur as the free metal, the halides of strontium are usually prepared from strontium salts. The common starting salt for the preparation of the halides is strontium carbonate ($SrCO_3$). When $SrCO_3$ is treated with hydrohalic acids (HF, HCl, HBr, or HI), it forms the corresponding halide with release of CO_2.

$$Sr + X_2 \rightarrow SrX_2 \quad (X = F,\ Cl,\ Br,\ I)$$
$$SrCO_3 + HX \rightarrow SrX_2 + H_2O + CO_2$$

Strontium fluoride, SrF_2, also known as strontium difluoride and strontium(II) fluoride, is a stable brittle white crystalline solid with melting point of 1477°C and boiling point of 2460°C. It occurs naturally as the mineral strontiofluorite. Strontium chloride ($SrCl_2$) is a typical salt, forming neutral aqueous solutions. Similar to all compounds of Sr, it emits a bright red color in a flame; in fact it is used as a source of redness in fireworks. Its chemical properties are intermediate between those for barium chloride, which is more toxic, and calcium chloride. Strontium chloride can be produced by reacting strontium hydroxide or strontium carbonate with hydrochloric acid:

$$Sr(OH)_2 + 2HCl \rightarrow SrCl_2 + 2H_2O$$

Crystallization from cold aqueous solution results in the hexahydrate, $SrCl_2{\cdot}6H_2O$. Dehydration of this salt occurs in stages, commencing above 61°C. Full dehydration occurs at 320°C. Strontium chloride is the precursor to other strontium compounds, for example, yellow strontium chromate, strontium carbonate, and strontium sulfate. Reaction of strontium chloride with the sodium salt of the desired anion (or alternately carbon dioxide gas to form the carbonate) results in the precipitation of the salt:

$$SrCl_2 + Na_2CrO_4 \rightarrow SrCrO_4 + 2NaCl$$
$$SrCl_2 + Na_2CO_3 \rightarrow SrCO_3 + 2NaCl$$
$$SrCl_2 + H_2O + CO_2 \rightarrow SrCO_3 + 2HCl$$
$$SrCl_2 + Na_2SO_4 \rightarrow SrSO_4 + 2NaCl$$

The solid has a deformed rutile (TiO_2) structure. In the vapor phase the $SrCl_2$ molecule is non-linear with a Cl—Sr—Cl angle of about 130°. This is an exception to VSEPR theory which would predict a linear structure. Ab initio calculations have been used to propose that contributions from d orbitals in the shell below the valence shell are responsible. Another suggestion has been that polarization of the electron core of the strontium atom results in a distortion of the core electron density that interacts with the Sr—Cl bonds. At room temperature, strontium bromide, $SrBr_2$, has a crystal structure with a tetragonal unit cell and space group *P4/n*. This structure is known as α-$SrBr_2$ and is isostructural with $EuBr_2$ and USe_2. Around 650°C, α-$SrBr_2$ undergoes a first-order solid-solid phase transition to a much less ordered phase, β-$SrBr_2$, which has the cubic fluorite (CaF_2) structure. The β-phase of $SrBr_2$ has a much higher ionic conductivity of about 1 S cm^{-1}, comparable to that of molten $SrBr_2$, due to extensive disorder in the bromide sublattice. Strontium bromide melts at 657°C. Strontium iodide (SrI_2) is an ionic, water-soluble, and deliquescent compound that can be utilized in medicine as an alternative for potassium iodide. It is also employed as a scintillation gamma radiation detector, typically doped with europium, because of its optical clarity, relatively high density, high effective atomic number ($Z = 48$), and high scintillation light yield. Recently, europium-doped strontium iodide ($SrI_2:Eu^{2+}$) has emerged as a promising scintillation material for γ-ray spectroscopy with extremely high light yield and proportional response, exceeding that of the widely used high performance commercial scintillator $LaBr_3:Ce^{3+}$.

6.2.3.3 *Reactions with acids*

Strontium reacts slowly with water to form strontium hydroxide ($Sr(OH)_2$) and hydrogen gas (H_2). The reaction is quicker than that of calcium but slower than that of barium in accordance with the periodic trend of reactivity for Group 2.

$$Sr(s) + 2H_2O(g) \rightarrow Sr(OH)_2(aq) + H_2(g)$$

Strontium metal dissolves readily in dilute or concentrated hydrochloric acid to form solutions containing the aquated Sr^{2+} ion together with hydrogen gas (H_2).

$$Sr(s) + 2H^+ \rightarrow Sr^{2+}(aq) + H_2(g)$$

Strontium metal also dissolves directly in liquid ammonia to give a dark blue solution. The Sr^{2+} ion is also capable of reacting with Lewis bases, especially those considered as hard donor atoms such as oxygen, nitrogen, or fluorine.

$$Sr + NH_3 \rightarrow Sr(NH_3)_6$$

6.2.3.4 *Organostrontium compounds and related complexes*

Organostrontium compounds contain one or more strontium–carbon bonds. One such compound is the strontium dicyclopentadienyl, $Sr(C_5H_5)_2$, prepared from the direct reaction of strontium metal with mercurocene or cyclopentadiene itself. Related complexes of strontium include the seven (7) new salen compounds recently reported by Sanchez et al. (2002). Salen is a tetradentate C_2-symmetric ligand synthesized from salicylaldehyde (sal) and ethylenediamine (en), which coordinates with Sr via the lone pairs in the oxygen atom. The complexes were prepared by the reaction of various salen(t-Bu)H_2 ligands with the metals in ethanol. Maudez et al. (2007) also explored the coordination chemistry of strontium by preparing mixed-metal alkoxide aggregates of the form [IM(OtBu)$_4${Li(thf)}$_4$(OH)] (M=Ca, Sr, Ba, Eu). The authors investigated the various clusters' behavior in nonaqueous solvents, their analogy to transition metals, and possible applications in oxide materials, investigating cluster compounds and coordination networks. There are only a few publications on organostrontium compounds and their related complexes because they are more difficult to make and are more reactive. Moreover, the corresponding calcium and barium compounds generally serve the same purpose yet are cheaper and easier to make.

6.2.4 Major uses

Using about 75% of total production, the main use for Sr historically was in glass for color television cathode ray tubes (CRTs), where it prevents X-ray emission. This use of Sr is strongly decreasing as CRTs are being replaced by other display methods. This decrease has a considerable impact on the mining and refining of Sr. All components of the CRT had to absorb X-rays. In the neck and the funnel of the tube, Pb glass was utilized for this reason, but this type of glass exhibited a browning effect as a result of the interaction of the X-rays with the glass. Consequently, the front panel was produced from a different glass mixture with Sr and Ba to absorb the X-rays. The typical values for the glass mixture determined for a recycling study in 2005 were 8.5% Sr oxide and 10% Ba oxide. Since Sr is so like Ca, it is also assimilated in the bone. All four stable isotopes are absorbed, in nearly the same ratios they exist in nature. However, the actual isotope distribution depends largely on the geographical location. Hence, examining the bone of an individual can help to establish the region it came from. This method allows the identification of ancient migration patterns and the sources of commingled human remains in battlefield burial sites. Sr aluminate ($SrAl_2O_4$) is regularly used in glow-in-the-dark toys, as it is chemically and biologically inert. When activated with a suitable dopant (e.g., europium, written as $Eu:SrAl_2O_4$), it acts as a photoluminescent phosphor with long persistence of phosphorescence [phosphorescence is a type of photoluminescence related to fluorescence. Unlike fluorescence, a phosphorescent material does not immediately re-emit the radiation it absorbs. Most photoluminescent events, in which a chemical substrate absorbs and then re-emits a photon of light, are fast, in the order of 10 nanoseconds. Light is absorbed and emitted at these fast time scales in cases where the energy of the photons involved matches the available energy states and allowed transitions of the substrate. In the special case of phosphorescence, the electron which absorbed the photon (energy) undergoes an unusual intersystem crossing into an energy state of higher spin multiplicity, usually a triplet state. As a result, the excited electron can become trapped in the triplet state with only "forbidden" transitions available to return to the lower energy singlet state. These transitions, although "forbidden," will still occur in quantum mechanics but are kinetically unfavored and thus progress at significantly slower time scales. Most phosphorescent compounds are still relatively fast emitters, with triplet lifetimes on the order of milliseconds. However, some compounds have triplet lifetimes up to minutes or even hours, allowing these substances to effectively store light energy in the form of very slowly degrading excited electron states. If the phosphorescent quantum yield is high, these substances will release significant amounts of light over long time scales, creating so-called "glow-in-the-dark" materials.]. $SrCO_3$ and other Sr salts are added to fireworks to produce a deep red color. Fireworks use around 5% of the global production.

$SrCO_3$ is used in the production of hard ferrite magnets [a ceramic material made by mixing and firing large proportions of Fe(III) oxide (Fe_2O_3) blended with small proportions of one or more additional metallic elements such as Ba, Mn, Ni, and Zn. They are both electrically nonconductive, signifying that they are insulators, and ferrimagnetic, meaning they can easily be magnetized or attracted to a magnet. Hard ferrites have high coercivity, so they are difficult to demagnetize. They are used to make permanent magnets for applications such as refrigerator magnets, loudspeakers, and small electric motors.]. $SrCl_2$ is occasionally used in toothpaste for sensitive teeth. Small quantities are used in Zn refining to remove small quantities of Pb impurities. The metal itself has a limited application as a getter, to remove unwanted gases in vacuums by reacting with them, though Ba can likewise be used for this purpose. The ultra-narrow optical transition between the $[Kr]5s_2^1S_0$ electronic ground state and the metastable $[Kr]5s5p^3P_0$ excited state of ^{87}Sr is one of the foremost contenders for the future re-definition of the second in terms of an optical transition set against the present definition resulting from a microwave transition between different hyperfine ground states of ^{133}Cs. Present optical atomic clocks based on this transition even now exceed the precision and accuracy of the existing definition of the second. ^{89}Sr is the active ingredient in Metastron (a preparation of ^{89}Sr chloride made by GE Healthcare), a radiopharmaceutical used for bone pain secondary to metastatic bone cancer. The Sr is treated by the body like Ca, preferentially absorbing it into bone at sites of increased osteogenesis. This localization concentrates the radiation exposure on the cancerous lesion. ^{90}Sr has been utilized as a power source for radioisotope thermoelectric generators (RTGs). ^{90}Sr generates about 0.93 watts of heat per gram (it is lower for the form of ^{90}Sr used in RTGs, which is SrF_2). Still, ^{90}Sr has one-third the lifetime and a lower density than ^{238}Pu, another RTG fuel. The major benefit of ^{90}Sr is that it is more cost-effective than ^{238}Pu which is found in nuclear waste. The former Soviet Union used almost 1000 of these RTGs on its northern coast as a power source for lighthouses and meteorology stations.

6.3 39 Y — Yttrium

6.3.1 Discovery

In 1787 Swedish army lieutenant and part-time chemist Carl Axel Arrhenius (March 29, 1757 to November 20, 1824) collected a heavy black rock in an old quarry near the Swedish village of Ytterby (now part of the Stockholm Archipelago). Thinking that it was an unidentified mineral with the newly discovered element tungsten, he called it ytterbite and sent samples to various chemists for analysis. Finnish chemist, physicist, and mineralogist Johan Gadolin (June 5, 1760 to August 15, 1852) (Fig. 6.11) at the University of Åbo recognized a new oxide (or "earth") in Arrhenius' sample in 1789 and published his completed analysis in 1794 (Gadolin, 1794). Swedish chemist Anders Gustaf Ekeberg (January 16, 1767 to February 11, 1813) in 1797 was able to support the identification and called the new oxide yttria. In the years after French nobleman and chemist Antoine Lavoisier (August 26, 1743 to May 8, 1794) established the first modern description of chemical elements, it was thought that earths could be reduced to their elements, meaning that finding a new earth was equal to finding the element within, which here would have been yttrium. In 1843, Swedish chemist Carl Gustaf Mosander (September 10, 1797 to October 15, 1858) discovered that samples of yttria actually consisted of three different oxides: white yttrium oxide (yttria), yellow terbium oxide (confusingly, this was called "erbia" at the time), and rose-colored erbium oxide (called "terbia" at the time) (Mosander, 1843). A fourth oxide, ytterbium oxide, was separated in 1878 by Swiss chemist Jean Charles Galissard de Marignac (April 24, 1817 to April 15, 1894). New elements were later separated from each of those oxides, and each element was named, in some way, after the village of Ytterby, near the quarry where they were initially discovered (see ytterbium, terbium, and

FIGURE 6.11 Johan Gadolin at age 19, around 1779, "it is likely that this portrait was painted to ease his mother's heartache when her oldest surviving child first left home."

erbium). In the following years, a total of seven other new metals were found in "Gadolin's yttria." As yttria was found to be a mineral and not an oxide, German chemist Martin Heinrich Klaproth (December 1, 1743 to January 1, 1817) give it the new name gadolinite in honor of Gadolin. German chemist Friedrich Wöhler (July 31, 1800 to September 23, 1882) erroneously believed he had obtained the metal in 1828 from a volatile chloride he thought to be yttrium chloride, but German mineralogist and analytical chemist Heinrich Rose (August 6, 1795 to January 27, 1864) demonstrated otherwise in 1843 and properly separated the element himself in that same year (Wöhler, 1828; Rose, 1843). Up until the early 1920s, the chemical symbol Yt was used for the element; later Y came into general use. In 1987 yttrium barium copper oxide was found to achieve high-temperature superconductivity. It was only the second material identified to show this property, and it was the first known compound to achieve superconductivity above the (economically significant) boiling point of nitrogen.

6.3.2 Mining, production, and major minerals

Yttrium is found in most rare-earth minerals, it is found in some uranium ores, but is never found in the Earth's crust as a free element. The Earth's crust contains about 31 ppm yttrium, making it the 28th most abundant element, about 400 times more common than silver. As of April 2018, there are reports of the discovery of very large reserves of rare-earth elements (REEs) on a tiny Japanese island. Minami-Torishima Island, also known as Marcus Island, has been labeled as having "enormous potential" for REEs and yttrium. Since Y is chemically so similar to the lanthanides, it occurs in the same ores (rare-earth minerals) and is extracted by the same refinement processes. Carbonate- and fluoride-containing ores such as bastnäsite ([(Ce, La, etc.)(CO_3)F]) contain an average of 0.1% of Y compared to the 99.9% for the 16 other REEs. The main producer for bastnäsite from the 1960s to the 1990s was the Mountain Pass rare-earth mine in California, making the United States the largest producer of REEs during that period. Monazite ([(Ce, La, etc.)PO_4]) is mainly found in placer deposits of sand formed by the transportation and gravitational separation of eroded granite. Monazite can contain up to 3% Y. The biggest ore deposits were discovered in India and Brazil at the beginning of the 20th century, resulting in these countries being the largest producers of Y in the first half of the 20th century. Xenotime, a REE phosphate, is the main REE ore containing as much as 60% Y as yttrium phosphate (YPO_4). This applies to xenotime-(Y) (Box 6.3). The largest mine is the Bayan Obo deposit in China, making China the largest exporter for Y since the shutting down of the Mountain Pass mine in the 1990s. Ion absorption clays or Lognan clays are the weathering products of granite and contain only 1% of REEs. The final ore concentrate can have as much as 8% Y. Ion absorption clays are mostly found in southern China. One process to produce pure yttrium from the mixed oxide ores is to dissolve the oxide in sulfuric acid and fractionate it by ion-exchange chromatography. After the addition of oxalic acid, yttrium oxalate is precipitated. The oxalate is subsequently converted into the oxide by heating under

BOX 6.3 Xenotime-(Y) (YPO_4).

Xenotime-(Y) was found by the Norwegian mineralogist Nils Otto Tank (1800–1864) and first described by the Swedish chemist Jøns Jacob Berzelius in 1824 as Phosphorsyrad Ytterjord from a granite pegmatite at Hidra, Flekkefjord, Norway. Later, Glocker (1831) introduced the name Ytterspath. The name xenotime was finally coined by the French mineralogist François Sulpice Beudant (1832) in his second edition of *Traité élémentaire de Minéralogie*. The name is from the Greek κευός = "vain" and τιμή = "honor," in allusion to the fact that the yttrium in it was first mistaken by Berzelius for a new element (Yttrium was discovered already in 1794). It was later renamed as xenotime-(Y), with a suffix, -(Y), according to the "Levinson modifiers." Xenotime-(Y) occurs widely as a minor accessory mineral in both acidic and alkalic igneous rocks and in large crystals in the associated pegmatites, also in mica and quartz-rich gneisses and in the quartz veinlets that form the pegmatitic portions of these rocks. It is commonly associated with monazite, zircon, ilmenite, rutile, anatase, magnetite, hematite, sillimanite, feldspars, and occasionally with fergusonite and columbite-tantalite. It is a common detrital mineral in placers. Xenotime-(Y) is tetragonal, *4/m2/m2/m* (space group $I4_1/amd$). It is isostructural with zircon ($ZrSiO_4$) and may contain minor heavy rare earth elements, Ca, U, Th, Si, F, and other elements. Its hardness is 4–5. The luster is vitreous to resinous and its transparency is translucent to opaque. The cleavage on {100} is good; the fracture is uneven to splintery, while the tenacity is brittle. The color varies from yellowish brown, reddish-brown, flesh-red, grayish-white, wine-yellow, pale-yellow, to greenish. The streak is pale brown, yellowish, or reddish, white. It is paramagnetic. It may exhibit yellow cathodoluminescence. Optically it is uniaxial(+). It occurs as short to long [001] prismatic crystals, with {010}, {110}, may be pyramidal {011}, equant, several other minor forms noted, to 5 cm. In radial or rosette-like aggregates of coarse crystals. Twinning on {111} is rare.

FIGURE 6.12 Euxenite-(Y), $(Y,Ca,Ce,U,Th)(Nb,Ta,Ti)_2O_6$, a small group of crystals with outer yellow crust, 2 × 1.5 × 0.8 cm. Ankazobe Pegmatite Field, Analamanga, Antananarivo, Madagascar.

oxygen. The reaction of the resulting yttrium oxide with hydrogen fluoride produces yttrium fluoride. When quaternary ammonium salts are utilized as extractants, most of the yttrium will stay in the aqueous phase. When the counter-ion is nitrate, the light lanthanides are removed, and when the counter-ion is thiocyanate, the heavy lanthanides are removed. In this manner, yttrium salts of 99.999% purity are produced. Only a few tons of yttrium metal are produced annually through reduction of yttrium fluoride to a metal sponge with a calcium-magnesium alloy. The temperature of an arc furnace at >1600°C is enough to melt the yttrium. Yearly world production of yttrium oxide had reached about 600 tons by 2001, which increased to around 7000 tons in 2014. Worldwide reserves of yttrium oxide were thought to be more than 500,000 tons in 2014. The main countries for these reserves are Australia, Brazil, China, India, and the United States.

About a hundred different minerals are known to contain Y in their crystal structure. Three halides are known to contain Y, chukhrovite-(Y) ($Ca_3(Y,Ce)[F|SO_4|(AlF_6)_2]\cdot 10H_2O$), gagarinite-(Y) ($NaCaYF_6$), and tveitite-(Y) ($(Y, Na)_6Ca_6Ca_6F_{42}$). In the oxide class more than 25 minerals are known to have Y in their crystal structure, for example, euxenite-(Y) ($(Y,Ca,Ce,U,Th)(Nb,Ta,Ti)_2O_6$) (Fig. 6.12), fergusonite-(Y) ($YNbO_4$), polycrase-(Y) ($Y(Ti,Nb)_2(O,OH)_6$), samarskite-(Y) ($YFe^{3+}Nb_2O_8$) (Fig. 6.13), and yttriaite-(Y) (Y_2O_3). About 20 minerals are found in the ate class containing Y, such as adamsite-(Y) ($NaY[CO_3]_2{\cdot}6H_2O$) (Fig. 6.14), bastnäsite-(Y) ($Y(CO_3)F$) (Fig. 6.15), lokkaite-(Y) ($Ca(Y,Gd,Nd,Dy)_4(CO_3)_7{\cdot}9H_2O$) (Fig. 6.16), and synchysite-(Y) ($CaY(CO_3)_2F$) (Fig. 6.17). One borate is known, moydite-(Y) ($Y[B(OH)_4](CO_3)$). The sulfate class includes two minerals with Y, paraniite-(Y) ($Ca_2Y(AsO_4)(WO_4)_2$) and sejkoraite-(Y) ($Y_2(UO_2)_8(SO_4)_4O_6(OH)_2{\cdot}26H_2O$). Nine different phosphate class minerals are known to have Y in their crystal structure, for example, agardite-(Y) ($YCu_6(AsO_4)_3(OH)_6{\cdot}3H_2O$) (Fig. 6.18), churchite-(Y) ($Y(PO_4){\cdot}2H_2O$), and xenotime-(Y) ($Y(PO_4)$) (Fig. 6.19). The largest group of about 40 minerals containing Y is found in the silicate class, such as allanite-(Y) ($\{CaY\}\{Al_2Fe^{2+}\}(Si_2O_7)(SiO_4)O(OH)$), gadolinite-(Y) ($Y_2Fe^{2+}Be_2Si_2O_{10}$) (Fig. 6.20), and yttrialite-(Y) ($(Y,Th)_2Si_2O_7$). A single organic compound exists, levinsonite-(Y) ($(Y,Nd,La)Al(C_2O_4)(SO_4)_2{\cdot}12H_2O$), with Y in its structure.

FIGURE 6.13 Group of parallel stacked samarskite-(Y), $YFe^{3+}Nb_2O_8$, crystals with terminations, 2.5 × 1.5 × 1 cm. Burroughs Mine, Jefferson Co., Colorado, United States.

FIGURE 6.14 Adamsite-(Y), $NaY(CO_3)_2{\cdot}6H_2O$, white prismatic crystals in a 4 mm spray. Mont Saint-Hilaire, Rouville Co., Quebec, Canada.

FIGURE 6.15 Bastnäsite-(Y), $Y(CO_3)F$, a thick tabular deep orange-brown crystal, 1.5 × 1.5 × 1 cm. Zegi Mountain, Khyber Agency, FATA, Pakistan.

FIGURE 6.16 Lokkaite-(Y), $Ca(Y,Gd,Nd,Dy)_4(CO_3)_7 \cdot 9H_2O$, pearly, snow-white, radial groups of flat-lying crystals with pearly, platy tengerite-(Y), $Y_2(CO_3)_3 \cdot 2-3H_2O$, crystals. Each is not more than 1 mm across. Pyörönmaa pegmatite, Kangasala, Finland.

FIGURE 6.17 Red masses of synchysite-(Y), $CaY(CO_3)_2F$, in magnetite, Fe_3O_4, about 5 mm. Scrub Oaks Mine, Morris Co., New Jersey, United States.

FIGURE 6.18 Agardite-(Y), $YCu_6(AsO_4)_3(OH)_6{\cdot}3H_2O$, bluish-green acicular sprays to 2 mm. Imperial Mine, Esmeralda Co., Nevada, United States.

FIGURE 6.19 Xenotime-(Y), $Y(PO_4)$, highly lustrous deep root-beer brown crystal group 4.5 cm long. Bahia, Brazil.

FIGURE 6.20 Gadolinite-(Y), $Y_2Fe^{2+}Be_2Si_2O_{10}$, a dark brown to black single crystal to 3 cm showing many faces. Hitterø, Vest-Agder, Norway.

6.3.3 Chemistry

Yttrium isotopes are among the most common products of uranium nuclear fission in nuclear explosions and nuclear reactors. In the framework of nuclear waste management, the most significant isotopes of yttrium are ^{91}Y and ^{90}Y, with half-lives of 58.51 days and 64 hours, respectively. While ^{90}Y has a short half-life, it exists in secular equilibrium with its long-lived parent isotope ^{90}Sr with a half-life of 29 years [in nuclear physics, secular equilibrium is a situation in which the quantity of a radioactive isotope remains constant because its production rate (e.g., due to decay of a parent isotope) is equal to its decay rate]. All group 3 elements have an odd atomic number and therefore few stable isotopes. Yttrium has only a single stable isotope, ^{89}Y, which is also the only naturally occurring isotope. ^{89}Y is believed to be more abundant than it otherwise would be, due in part to the s-process, which allows sufficient time for isotopes formed by other processes to decay by electron emission (essentially, a neutron becomes a proton while an electron and anti-neutrino are emitted). Such a slow process has a tendency to favor isotopes with atomic mass numbers around 90, 138, and 208, which have remarkably stable atomic nuclei with 50, 82, and 126 neutrons, respectively. ^{89}Y has a mass number close to 90 and has 50 neutrons in its nucleus. No less than 32 synthetic isotopes of Y have been found, and these range from ^{76}Y to ^{108}Y. The least stable of these isotopes is ^{106}Y with a half-life of >150 ns (^{76}Y has a half-life of >200 ns), while the most stable isotope is ^{88}Y with a half-life of 106.626 days. Apart from the isotopes ^{91}Y, ^{87}Y, and ^{90}Y, with half-lives of 58.51 days, 79.8 hours, and 64 hours, respectively, the remaining isotopes have half-lives of less than a day and the majority of less than an hour. Y isotopes lighter than ^{88}Y decay primarily by positron emission (proton → neutron) to form Sr isotopes. Yttrium isotopes with mass numbers at or above 90 decay mainly by electron emission (neutron → proton) to form Zr isotopes. In addition, isotopes with mass numbers at or above 97 have been shown to exhibit minor decay paths of β^- delayed neutron emission. Yttrium has as a minimum 20 metastable ("excited") isomers ranging in mass number from 78 to 102. Multiple excitation states have been found for ^{80}Y and ^{97}Y. Though the majority of yttrium's isomers are anticipated to be less stable than their ground state, ^{78m}Y, ^{84m}Y, ^{85m}Y, ^{96m}Y, ^{98m1}Y, ^{100m}Y, and ^{102m}Y have longer half-lives compared to their corresponding ground states, as these isomers decay via β decay instead of isomeric transition.

Yttrium is a silvery metallic transition metal, chemically similar to the lanthanides, and has often been included in discussions about "rare-earth elements." It has 39 electrons arranged in an electronic configuration of $[Kr]4d^15s^2$. Its chemistry concerns the formation of a predominantly ionic $+3$ oxidation state arising from the loss of the two electrons from the 5s subshell and one electron from the 4d subshell (Table 6.3).

6.3.3.1 *Oxides*

Yttrium reacts with oxygen but slowly because of the formation of a protective oxide coating. It does burn in air or oxygen above 450°C to give the oxide, Y_2O_3. Yttriaite-(Y), approved as a new mineral species in 2010, is the natural form of yttria. It is extremely rare, occurring as inclusions in native tungsten particles in a placer deposit of the Bol'shaja

TABLE 6.3 Yttrium properties.

Appearance	Silvery-white
Standard atomic weight ($A_{r,std}$)	88.906
Block	d-Block
Element category	Transition metal
Electron configuration	[Kr] $4d^1$ $5s^2$
Phase at STP	Solid
Melting point	1526°C
Boiling point	2930°C
Density (near r.t.)	4.472 g/cm^3
When liquid (at m.p.)	4.24 g/cm^3
Heat of fusion	11.42 kJ/mol
Heat of vaporization	363 kJ/mol
Molar heat capacity	26.53 J/(mol·K)
Oxidation states	+1, +2, **+3**
Ionization energies	First: 600 kJ/mol
	Second: 1180 kJ/mol
	Third: 1980 kJ/mol
Atomic radius	Empirical: 180 pm
Covalent radius	190 ± 7 pm

STP, Standard temperature and pressure.
Bold font indicates main oxidation state.

Pol'ja (Russian: Вольшая Полья) river, Prepolar Ural, Siberia. Yttrium oxide is an essential starting point for inorganic compounds. For organometallic chemistry it is reacted with concentrated hydrochloric acid and ammonium chloride to produce YCl_3. Yttria is widely used to manufacture Eu:YVO_4 and Eu:Y_2O_3 phosphors that give the red color in color TV picture tubes. Yttrium oxide is used to stabilize the zirconia in late-generation porcelain-free metal-free dental ceramics. This is a very hard ceramic used as a strong base material in some full ceramic restorations. The full name of zirconia employed in dentistry is "yttria-stabilized zirconia" or YSZ. Y_2O_3 is used to produce the high-temperature superconductor $YBa_2Cu_3O_7$, known as "1−2−3" to indicate the ratio of the metal constituents:

$$2Y_2O_3 + 8BaO + 12CuO + O_2 \rightarrow 4YBa_2Cu_3O_7$$

This synthesis is typically performed at 800°C.

6.3.3.2 *Halides*

It reacts with halogens to form the trihalides, YX_3 (F, Cl, Br, I). The halides, except the fluorides, are all very water-soluble and deliquescent. In general, the anhydrous halides are therefore best prepared by direct reaction of the elements rather than by heating the hydrates, which produces the oxyhalides, YOX. The reaction of heated yttrium with hydrogen produces highly conducting materials with the composition.

Yttrium(III) fluoride, YF_3, is not known naturally in "pure" form. YF_3 can be prepared via the reaction of fluorine with yttria or yttrium hydroxide with hydrofluoric acid.

$$Y(OH)_3 + 3HF \rightarrow YF_3 + 3H_2O$$

Yttrium(III) fluoride can be utilized for the manufacture of metallic yttrium, thin films, glasses, and ceramics. It occurs naturally as the mineral waimirite-(Y).

Yttrium(III) chloride exists in two forms, the hydrate (YCl_36H_2O) and an anhydrous form (YCl_3). Both are colorless solids that are highly soluble in water and deliquescent. YCl_3 is often synthesized via the "ammonium chloride route," starting from either Y_2O_3 or hydrated chloride or oxychloride, or $YCl_3 \cdot 6H_2O$. These reactions form $(NH_4)_2[YCl_5]$:

$$10NH_4Cl + Y_2O_3 \rightarrow 2(NH_4)_2[YCl_5] + 6NH_3 + 3H_2O$$

$$YCl_3 \cdot 6H_2O + 2NH_4Cl \rightarrow (NH_4)_2[YCl_5] + 6H_2O$$

The pentachloride is then thermally decomposed:

$$(NH_4)_2[YCl_5] \rightarrow 2NH_4Cl + YCl_3$$

The thermolysis reaction proceeds via the intermediary $(NH_4)[Y_2Cl_7]$. Reacting Y_2O_3 with aqueous HCl forms the hydrated chloride ($YCl_3 \cdot 6H_2O$). When heated, this salt converts to yttrium oxychloride rather than reverting to the anhydrous form. Solid YCl_3 has a cubic structure with close-packed chloride ions and yttrium ions filling one-third of the octahedral holes and the resulting YCl_6 octahedra sharing three edges with adjacent octahedra, forming a layered structure. This structure is shared by a range of compounds, particularly $AlCl_3$.

Yttrium oxyfluoride, YOF, is under normal conditions a colorless solid. The decomposition of crystalline hydrate of yttrium fluoride upon heating (900°C) in vacuum results in the formation of YOF:

$$4YF_3 \cdot H_2O \rightarrow 2YOF + 2YF_3 + 4HF$$

Alternatively, hydrolysis of yttrium fluoride with superheated steam (800°C) also gives YOF:

$$2YF_3 + H_2O \rightarrow YOF + 2HF$$

Yttrium oxyfluoride forms colorless tetragonal crystals with cell parameters $a = 0.3910$ nm and $c = 0.5431$ nm. According to the hexagonal crystal family, the cell parameters are: $a = 0.38727$ nm, $c = 1.897$ nm, and $Z = 6$.

6.3.3.3 Hydrides, nitrides, and phosphides

YH_2 transitions into an insulator with an addition of one hydrogen to form YH_3. Heated yttrium also reacts with nitrogen to form yttrium nitride, YN, but the yield is low. To increase the yield, YN had been prepared by converting yttrium to the hydride, first, then converting YH_2 to YN by heating the hydride at 900°C in the presence of nitrogen. YH_2 adopts a face-centered cubic (*fcc*) structure and is a metallic compound. Under high pressure, extra hydrogen can combine to form an insulator with a hexagonal structure, with a composition close to YH_3. Hexagonal YH_3 has a bandgap of 2.6 eV. Under pressure of 12 GPa YH_3 converts to an intermediate state, and when the pressure increases further to 22 GPa another metallic face-centered cubic phase is formed. Meanwhile, yttrium's reaction with water produces aqueous Y^{3+} ions with evolution of hydrogen gas. Its reaction with weak acids produces the corresponding salts. For example, hydrofluoric acid, phosphoric acid, and oxalic acid produce the insoluble salts, YF_3, YPO_4, and $Y_2(C_2O_4)$, respectively. Its reaction with strong acids produces the soluble salts. Yttrium nitride is a hard ceramic material comparable to titanium nitride and zirconium nitride. The nitrides of lanthanum, scandium, and yttrium exhibit semiconducting properties and moreover the lattice structure of YN differs only by 8% from that of gallium nitride. This makes YN a potential buffer layer between a substrate and the GaN layer during GaN crystal growth. Yttrium phosphide is an inorganic compound of yttrium and phosphorus with the chemical formula YP. The compound may be also classified as yttrium(III) phosphide. Heating (500°C–1000°C) of pure elements in a vacuum form YP:

$$Y + P \rightarrow YP$$

Yttrium phosphide forms cubic crystals. Yttrium phosphide is a semiconductor employed in laser diodes and in high power and frequency applications.

6.3.3.4 Borides

Yttrium boride refers to a crystalline material composed of different ratios of yttrium to boron, for example, YB_2, YB_4, YB_6, YB_{12}, YB_{25}, YB_{50}, and YB_{66}. They are all gray-colored, hard solids with high-melting temperatures. The most common form is the yttrium hexaboride YB_6. It shows superconductivity at relatively high temperature of −264.8°C (8.4K) and, similar to LaB_6, is an electron cathode. Another notable yttrium boride is YB_{66}. It has a large lattice constant (2.344 nm), high thermal and mechanical stability, and consequently is employed as a diffraction grating for low-energy synchrotron radiation (1–2 keV). Yttrium diboride has the same hexagonal crystal structure as aluminum diboride and magnesium diboride—an important superconducting material. Its space group is *P6/mmm* (No. 191), $a = 0.33041$ nm, $c = 0.38465$ nm, and the calculated density is 5.05 g/cm^3. In this structure, the boron atoms form graphite-like sheets with yttrium atoms between them. YB_2 crystals are unstable to moderate heating in air—they start oxidizing at 400°C and completely oxidize at 800°C. It melts at ~2100°C. YB_4 has tetragonal crystal structure with space group *P4/mbm* (No. 127), $a = 0.711$ nm, $c = 0.4019$ nm, and a calculated density of 4.32 g/cm^3. YB_6 is a black odorless powder with a density of 3.67 g/cm^3; it has the same cubic crystalline structure as other hexaborides (CaB_6, LaB_6, etc.). YB_6 is a superconductor with the relatively high transition temperature (onset) of 8.4K. YB_{12} crystals have a cubic structure with a density of 3.44 g/cm^3, space group *Fm3m* (No. 225), and $a = 0.7468$ nm. Its structural unit is 12 cuboctahedron. The Debye temperature of YB_{12} is ~767°C (1040K), and it is not superconducting at temperatures above −270.6°C (2.5K). The structure of yttrium borides with B/Y ratio of 25 and above consists of a network of B_{12} icosahedra. The boron framework of YB_{25} is one of the simplest among icosahedron-based borides—it consists of only one kind of icosahedra and one bridging boron site. The yttrium sites have partial occupancies of c. 60%–70%, and the YB_{25} formula purely indicates the average atomic ratio $[B]/[Y] = 25$. The bonding distance between the bridging boron and the equatorial boron atoms is 0.1755 nm, which is typical for the strong covalent B–B bond (bond length 0.17–0.18 nm); thus the bridging boron atoms strengthen the individual network planes. On the other hand, the large distance between the boron atoms within the bridge (0.2041 nm) reveals a weaker interaction, and thus the bridging sites contribute little to the bonding between the network planes. YB_{25} crystals can be grown by heating a compressed pellet of yttria (Y_2O_3) and boron powder to $\sim$1700°C. The YB_{25} phase is stable up to 1850°C. Above this temperature it decomposes into YB_{12} and YB_{66} without melting. This makes it difficult to grow a single crystal of YB_{25} by the melt growth method. YB_{50} crystals have orthorhombic structure with space group $P2_12_12$ (No. 18), $a = 1.66251$ nm, $b = 1.76198$ nm, and $c = 0.94797$ nm. Similar to YB_{25}, they can also be grown by heating a compressed pellet of yttria (Y_2O_3) and boron powder to $\sim$1700°C. Above this temperature YB_{50} decomposes into YB_{12} and YB_{66} without melting. This also makes it difficult to grow a single crystal of YB_{50} by the melt growth method. The YB_{66} structure is face-centered cubic, with space group $Fm3c$ (No. 226) and lattice constant $a = 2.3440(6)$ nm. There are 13 boron sites B1–B13 and one yttrium site. The B1 site forms one B_{12} icosahedron and the B2–B9 sites make up another icosahedron. These icosahedra arrange in a 13-icosahedron unit $(B_{12})_{12}B_{12}$, which is known as a supericosahedron. The icosahedron formed by the B1 site atoms is situated at the center of the supericosahedron. The supericosahedron is one of the basic units of the boron framework of YB_{66}. There are two types of supericosahedra: one occupies the cubic face centers and another, which is rotated by 90 degrees, is situated at the center of the cell and at the cell edges. Hence, there are eight supericosahedra (1248 boron atoms) in the unit cell.

6.3.3.5 *Sulfate and nitrate*

Yttrium(III) sulfate is a compound with the composition $Y_2(SO_4)_3$. The most common forms are the anhydrate and octahydrate. Yttrium(III) sulfate can be formed using either the corresponding oxide, hydroxide, or carbonate.

$$2Y(OH)_3 + 3H_2SO_4 \rightarrow Y_2(SO_4)_3 + 6H_2O$$

Yttrium sulfate can form double salts such as $MY(SO_4)_2$ and $M_3Y(SO_4)_3$:

$$Y_2(SO_4)_3 + M_2SO_4 \rightarrow 2MY(SO_4)_2$$
$$Y_2(SO_4)_3 + 3M_2SO_4 \rightarrow 2M_3Y(SO_4)_3 \quad (M = \text{alkali metals})$$

Yttrium(III) nitrate is a compound with the composition $Y(NO_3)_3$. The hexahydrate is the most common form commercially available. Yttrium(III) nitrate can be produced by dissolving the corresponding metal oxide in 6 mol/L nitric acid:

$$Y_2O_3 + 6HNO_3 \rightarrow 2Y(NO_3)_3 + 3H_2O$$

Yttrium(III) nitrate hexahydrate dehydrates at relatively low temperature. Upon further heating, the basic salt $YONO_3$ is formed. At 600°C, the thermal decomposition is complete. Y_2O_3 is the final reaction product. $Y(NO_3)_3 \cdot 3TBP$ is produced when tributyl phosphate is used as the extracting solvent. Yttrium(III) nitrate is primarily employed as a source of Y^{3+} cations. It is a precursor for some yttrium-containing compounds, for example, $Y_4Al_2O_9$, $YBa_2Cu_3O_{6.5+x}$, and yttrium-based metal-organic framework compounds. It can also be used as a catalyst in organic synthesis.

6.3.3.6 *Organoyttrium compounds*

The organometallic chemistry of yttrium, as of the lanthanides, is dominated by compounds involving cyclopentadiene and its methyl-substituted derivative. Nevertheless, organoyttrium compounds with alkyl, aryl, alkenyl, allyl, alkynyl, hydrido, carbene, indenyl, fluorenyl, and cyclooctatetraenyl have also been prepared. These are presented in an article entitled, "Scandium, Yttrium & The Lanthanides: Organometallic Chemistry" by Schumann and Fedushkin (2005), which is based in part on an earlier version of the article of the same title by R.D. Köhn et al. in the *Encyclopedia of Inorganic Chemistry*, First Edition.

The alkoxides and aryloxides of yttrium have garnered interest when observations of the high-temperature superconductivity of yttrium barium copper oxide (YBCO) were published in an article by Wu et al. (1987). In that article, YBCO was deposited from the solid-state reaction of Y_2O_3, $BaCO_3$, and CuO in appropriate ratios. The solid-state reaction process for the preparation of mixed oxides is well established owing to its ease of use. However, it has some disadvantages including high reaction temperatures. An alternative method of deposition that shows promise for scaling up of high-temperature superconductors is metallo-organic chemical vapor deposition (MOCVD). For MOCVD to work, the precursor must have the requisite volatility, stability, reactivity, and decomposition characteristics. Yttrium compounds are polymeric and not easily volatilized but with the use of bulky ligands such as in the compounds of the form,

$[Y(OR)_3(thf)_3]$, where R = $(CF_3)_2MeC$-, volatile octahedral compounds were produced. A few more studies have been focused on the synthesis of new yttrium organocompounds that are volatile and thermally stable. Recently, Milanov et al. (2012) synthesized a series of volatile guanidinate rare-earth precursors including tris(*N*,*N'*-diisopropyl-2-dimethylamidoguanidinato)-yttrium(III), $[Y(DPDMG)_3]$. These series of compounds were reported to have high reactivity and thermal stability and are envisioned to have potential applications in atomic layer deposition of gate dielectric materials.

6.3.4 Major uses

The red color of old color television cathode ray tubes was typically emitted from a yttria (Y_2O_3) or yttrium oxide sulfide (Y_2O_2S) host lattice doped with Eu^{3+} phosphors. The red color itself is produced by the Eu as the Y collects energy from the electron gun and passes it to the phosphor. Y compounds can act as host lattices for doping with various lanthanide cations, for example, Tb^{3+} can be used as a doping agent to emit green luminescence. Yttria is used as a sintering additive in the manufacturing of porous silicon nitride. It is often used as a general starting compound for material science and for creating other Y compounds. Y compounds are, for example, used as a catalyst for ethylene polymerization. In its metal form, Y is utilized on the electrodes of some high-performance spark plugs. Y is applied in gas mantles for propane lanterns as a substitute for Th, which is radioactive. Presently being developed is Y-stabilized zirconia as a solid electrolyte and as an oxygen sensor in automobile exhaust systems. Y is also utilized in the synthesis of a large variety of synthetic garnets, for example, Y-Fe-garnets ($Y_3Fe_5O_{12}$, "YIG"), which are very effective microwave filters; Y, Fe, Al, and Gd garnets (e.g., $Y_3(Fe,Al)_5O_{12}$ and $Y_3(Fe,Gd)_5O_{12}$), which have significant magnetic properties. In addition, YIG is very effective as an acoustic energy transmitter and transducer. Y-Al-garnet ($Y_3Al_5O_{12}$, YAG) has a hardness of 8.5 and is finding use as a gemstone in jewelry (simulated diamond). Ce-doped Y-Al-garnet (YAG:Ce) crystals are utilized as phosphors to produce white LEDs. YAG, yttria, yttrium lithium fluoride ($LiYF_4$), and yttrium orthovanadate (YVO_4) are used together with dopants such as Nd, Er, and Yb in near-infrared (IR) lasers. YAG lasers can run at high power and are used for drilling and cutting metal. The single crystals of doped YAG are usually manufactured via the Czochralski process. Small quantities of Y (0.1%–0.2%) have been utilized to decrease the grain sizes of Cr, Mo, Ti, and Zr. Y is also used to improve the strength of Al and Mg alloys. The addition of Y to alloys commonly enhances workability, adds resistance to high-temperature recrystallization, and substantially improves resistance to high-temperature oxidation. In addition, Y can be utilized to deoxidize V and other nonferrous metals. Yttria stabilizes cubic zirconia used in jewelry. Y has been researched as a nodulizer in ductile cast iron, changing the morphology of the graphite into compact nodules instead of flakes, which improves ductility and fatigue resistance. Since it has a high-melting point, Y oxide is applied in some ceramics and glass to impart shock resistance and low thermal expansion properties. These same properties make such glass useful in camera lenses.

Y is a major ingredient in the yttrium barium copper oxide ($YBa_2Cu_3O_7$, aka "YBCO" or "1–2–3") superconductor. This superconductor is noteworthy since the operating superconductivity temperature is above liquid nitrogen's boiling point (−196.05°C). As liquid nitrogen is cheaper than the liquid helium (−270°C) required for metallic superconductors, the operating costs for applications would be far lower. The actual superconducting material is often reported as $YBa_2Cu_3O_{7-d}$, with $d < 0.7$ for superconductivity. The explanation for this is still unclear, though it is understood that the vacancies appear only in certain positions in the crystal, the Cu-oxide planes, and chains, resulting in a particular oxidation state of the Cu atoms, which in some way leads to the superconducting behavior. The theory of low-temperature superconductivity has been well known since the BCS theory in 1957 {or Bardeen–Cooper–Schrieffer theory [named after American physicists John Bardeen (May 23, 1908 to January 30, 1991), Leon Cooper (born February 28, 1930), and John Robert Schrieffer (May 31, 1931 to July 27, 2019)], the first microscopic theory of superconductivity since Dutch physicist Heike Kamerlingh Onnes's 1911 discovery (September 21, 1853 to February 21, 1926). The theory describes superconductivity as a microscopic effect caused by a condensation of Cooper pairs into a boson-like state.}. It is centered on a peculiarity of the interaction between two electrons in a crystal lattice. Still, the BCS theory does not clarify high-temperature superconductivity, and its exact mechanism is still unknown. What has been recognized is that the composition of the Cu-oxide materials must be accurately controlled for superconductivity to arise. This superconductor is a black and green, multicrystal, multiphase material. Scientists are working on a class of materials known as perovskites that are alternative combinations of these elements, hoping to create a practical high-temperature superconductor.

The radioactive isotope ^{90}Y is used in drugs such as Yttrium Y 90-DOTA-tyr3-octreotide and Yttrium Y 90 ibritumomab tiuxetan for the treatment of a variety of cancers, such as lymphoma, leukemia, liver, ovarian, colorectal, pancreatic, and bone cancers. It works by binding to monoclonal antibodies, which subsequently bind to the cancer cells and destroy them via intense β-radiation from the ^{90}Y. A procedure called radioembolization is used to treat hepatocellular carcinoma and liver metastasis. Radioembolization is a low-toxicity, targeted liver cancer therapy that makes use of millions of tiny beads made of glass or resin containing radioactive ^{90}Y. These radioactive microspheres are distributed directly into the blood vessels feeding specific liver tumors/segments or lobes. It is minimally invasive, and patients can typically go home after just a few hours. This procedure may not eradicate all tumors all over the entire liver but works on one segment or one lobe at a time and may need multiple sessions. Needles made of ^{90}Y, which can cut more precisely than scalpels, have been utilized to sever pain-transmitting nerves in the spinal cord, while ^{90}Y is similarly used to carry out radionuclide synovectomy in the treatment of inflamed joints, especially knees, in patients suffering from conditions such as rheumatoid arthritis. A Nd-doped Y-Al-garnet laser has been used in an experimental, robot-assisted radical prostatectomy in canines trying to reduce collateral nerve and tissue damage, while Er-doped lasers are starting to be used for cosmetic skin resurfacing.

6.4 40 Zr — Zirconium

6.4.1 Discovery

The name zirconium comes from the name of the mineral zircon, $Zr(SiO_4)$ [the word is related to Persian zargun (zircon;zar-gun, "gold-like" or "as gold")]. The zirconium-containing mineral zircon and its color varieties (jargoon, hyacinth, jacinth, ligure) were already mentioned in biblical writings. The mineral was not known to have a new element in its composition until 1789, when German chemist Martin Heinrich Klaproth (December 1, 1743 to January 1, 1817) analyzed a jargoon from the island of Ceylon (now Sri Lanka) (Fig. 6.21). He called the new element Zirkonerde (zirconia) (Klaproth, 1789). Cornish chemist and inventor Sir Humphry Davy (December 17, 1778 to May 29, 1829) tried to separate this new element in 1808 using electrolysis but was unsuccessful. Zirconium metal was first isolated in an impure form in 1824 by Swedish chemist Jöns Jacob Berzelius (August 20, 1779 to August 7, 1848) by heating a mixture of potassium and potassium zirconium fluoride in an iron tube (Berzelius, 1824). The crystal bar process (also known as the Iodide Process), developed by Dutch chemists Anton Eduard van Arkel (November 19, 1893 to March 14, 1976) and Jan Hendrik de Boer (March 19, 1899 to April 25, 1971) in 1925, was the first industrial procedure for the commercial manufacture of metallic zirconium (van Arkel and de Boer, 1924a,b, 1925). It comprises the formation followed by the thermal decomposition of zirconium tetraiodide (ZrI_4), and was replaced in 1945 by the much cheaper Kroll process established by Luxembourgish metallurgist William Justin Kroll (born Guillaume Justin Kroll; November 24, 1889 to March 30, 1973), in which zirconium tetrachloride ($ZrCl_4$) is reduced by magnesium:

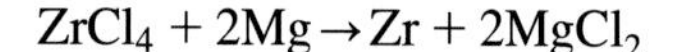

$$ZrCl_4 + 2Mg \rightarrow Zr + 2MgCl_2$$

FIGURE 6.21 Martin Heinrich Klaproth, engraving Ambroise Tardieu (1788–1841) after original portrait by Eberhard-Siegfried Henne.

6.4.2 Mining, production, and major minerals

Zirconium has a concentration of about 130 ppm within the Earth's crust. It is not found in nature as a native metal, due to its inherent instability with respect to water. The main commercial source of Zr is zircon ($ZrSiO_4$) (Box 6.4), which is found largely in Australia, Brazil, India, Russia, South Africa, and the United States, in addition to smaller deposits around the world. In recent years, approximately two-thirds of zircon is mined in Australia and South Africa. Zircon resources are estimated to be more than 60 million tons worldwide and current annual worldwide zirconium production is around 900,000 tons. Zr can be found in numerous other minerals, including the commercially useful ores baddeleyite (ZrO_2) and kosnarite ($KZr_2(PO_4)_3$).

Zirconium is a by-product of the mining and processing of the titanium minerals ilmenite ($FeTiO_3$) and rutile (TiO_2), along with tin mining. Zirconium metal is much more expensive than zircon since the reduction processes are expensive. Obtained from coastal waters, zircon-bearing sand is purified by spiral concentrators to remove the lighter materials, which are subsequently returned to the water as they are natural components of beach sand. Using magnetic separation, the titanium ores ilmenite and rutile are separated. Most zircon is utilized without further treatment in commercial applications (mainly in refractory products, where it is valued for its high-melting temperature of 2550°C), but a small percentage is converted to the metal.

Recall that Zr metal is prepared from the reaction of zirconium(IV) chloride with magnesium in the Kroll process (vide supra). The resulting Zr is sintered until sufficiently ductile for metalworking. Commercial zirconium metal typically comprises 1%–3% of hafnium, which is usually not a problem since the chemical properties of hafnium and zirconium are very similar. However, their neutron-absorbing properties are very different, making it necessary to separate the hafnium from zirconium for usage in nuclear reactors. A number of different separation processes are being used to achieve this. The liquid–liquid extraction of the thiocyanate (SCN^-)-oxide derivatives exploits the fact that the hafnium derivative is slightly more soluble in methyl isobutyl ketone [preferred International Union of Pure and Applied Chemistry (IUPAC) name 4-methylpentan-2-one, $(CH_3)_2CHCH_2C(O)CH_3$] than in water [liquid-liquid extraction, also known as solvent extraction and partitioning, is a method to separate compounds or metal complexes, based on their relative solubilities in two different immiscible liquids, usually water (polar) and an organic solvent (nonpolar). There is a net transfer of one or more species from one liquid into another liquid phase, generally from aqueous to organic. The transfer is driven by chemical potential, that is, once the transfer is complete, the overall system of protons and electrons that make up the solutes and the solvents are in a more stable configuration (lower free energy)]. This method is used mainly in the USA. Zr and Hf can also be separated using fractional crystallization of potassium hexafluorozirconate (K_2ZrF_6), which is less soluble in water than the analogous hafnium derivative. Fractional distillation of the tetrachlorides, also known as extractive distillation, is used mainly in Europe. The zirconium product of a quadruple vacuum arc melting process in combination with hot extruding and different rolling applications, followed by curing utilizing high-pressure, high-temperature gas

BOX 6.4 Zircon ($ZrSiO_4$).

Zircon is a common accessory mineral of igneous rocks, particularly in siliceous and alkaline plutonic rocks, including granite, granodiorite, syenite, monzonite, and pegmatite. Economic deposits of zircon, however, are of the placer type, which formed as a result of long-term weathering of the igneous rocks and is distributed in riverbeds and ocean beaches. As a member of the heavy mineral sands, zircon is always found in association with rutile and ilmenite. Zircon was renamed in 1783 by Abraham Gottlob Werner from the Arabic (and, in turn, from the Persian "azargun") "zar," gold, plus "gun," colored, referring to one of the many colors that the mineral may display. It was originally named λυγκύριον "lyncurion" in ~300 BCE by Theophrastus. A mineral that may have been today's zircon was called chrysolithos by Pliny in 37, which was called jacinth by Georgius Agricola in 1555, mentioned as jargon by Axel Cronstedt in 1758, and called hyacinte by Barthélemy Faujas de Saint-Fond in 1772. Numerous later synonyms have been advanced. Zircon is tetragonal, *4/m2/m2/m* (space group $I4_1/amd$). It has a hardness of 7.5. Cleavage on {010} and {111} is poor, the fracture is conchoidal, and the tenacity is brittle. The luster is vitreous to adamantine, while the transparency is transparent to translucent. The color can be reddish-brown, yellow, green, blue, gray, to colorless. The streak is white. When transparent, it can be cut as a gemstone. Blue is not a natural color for cut zircon but is due to heat treatment. The colorless, yellowish, or smoky stones are known as jargon, because, although resembling diamond, they have little value. Besides Hf, zircon can contain some thorium and uranium, making the mineral radioactive. As a result, the zircon undergoes metamictization due to the radioactive decay destroying the crystal structure converting it in amorphous zircon. Metamict zircon has a greasy luster. Optically zircon is uniaxial (+), but will turn isotropic upon metamictization. It shows thermoluminescence, cathodoluminescence, and may fluoresce under UV light. Crystals are most commonly tabular to prismatic, with square cross sections, terminated by {111}, to 30 cm; as irregular grains, massive. Twinning can occur on {101}, geniculated.

autoclaving, produces reactor-grade zirconium, which is around 10 times more expensive than the hafnium-contaminated commercial grade. Hafnium has to be removed from zirconium for nuclear applications as hafnium has a neutron absorption cross section 600 times greater than zirconium. The separated hafnium can then be utilized for reactor control rods.

More than 100 minerals are known with Zr in their chemical composition. Fifteen of those minerals are found in the oxide class, for example, baddeleyite (ZrO_2) and zirconolite ($CaZrTi_2O_7$). Only two carbonates are known to contain Zr, sabinaite ($Na_4Zr_2TiO_4(CO_3)_4$) and welagonite ($Na_2Sr_3Zr(CO_3)_6 \cdot 3H_2O$). Painite ($CaZrAl_9(BO_3)O_{15}$) is the only borate with Zr in its structure. Within the sulfate class zircosulfate ($(Zr,Ti)(SO_4)_2 \cdot 4H_2O$) is the only mineral with Zr. Eight phosphate class minerals are known to contain Zr, for example, gainesite ($Na(Na,K)(Be,Li)Zr_2(PO_4)_4 \cdot 1.5{-}2H_2O$), and kosnarite ($KZr_2(PO_4)_3$). The largest group of more than about 80 minerals can be found under the silicates, for example, catapleiite ($Na_2Zr(Si_3O_9) \cdot 2H_2O$) (Fig. 6.22), elpidite ($Na_2ZrSi_6O_{15} \cdot 3H_2O$) (Fig. 6.23), eudialyte ($Na_{15}Ca_6(Fe^{2+},Mn^{2+})_3Zr_3[Si_{25}O_{73}](O,OH,H_2O)_3(OH,Cl)_2$) (Fig. 6.24), peterasite ($Na_5Zr_2(Si_6O_{18})(Cl,OH) \cdot 2H_2O$), and zircon ($Zr(SiO_4)$) (Fig. 6.25).

FIGURE 6.22 Catapleiite, $Na_2Zr(Si_3O_9) \cdot 2H_2O$, two creamy light brown crystals from 5 to 7 mm across. Mont Saint-Hilaire, Rouville Co., Quebec, Canada.

FIGURE 6.23 Elpidite, $Na_2ZrSi_6O_{15} \cdot 3H_2O$, white splintery crystals to 1.5 cm. Mont Saint-Hilaire, Rouville Co., Quebec, Canada.

FIGURE 6.24 Eudialyte, $Na_{15}Ca_6(Fe^{2+},Mn^{2+})_3Zr_3[Si_{25}O_{73}](O,OH,H_2O)_3(OH,Cl)_2$, well-formed deep red, euhedral crystal to 1 cm across. Ilimaussaq complex, Narsaq, Kitaa prov., Greenland.

FIGURE 6.25 Terminated creamy brown zircon, $Zr(SiO_4)$, crystal to 2.5 cm. Betroka dist., Anosy, Tuléar prov., Madagascar.

6.4.3 Chemistry

Naturally occurring Zr is composed of five isotopes. ^{90}Zr (51.45% natural abundance), ^{91}Zr (11.22%), ^{92}Zr (17.15%), and ^{94}Zr (17.38%) are stable, although ^{94}Zr is predicted to undergo double β decay (not observed experimentally) with a half-life of more than 1.10×10^{17} years. ^{96}Zr (2.80%) with a half-life of 2.4×10^{19} years is the longest-lived radioisotope of Zr. Twenty-eight isotopes of Zr have been synthesized, ranging from ^{78}Zr to ^{110}Zr. ^{93}Zr is the longest-lived of

TABLE 6.4 Zirconium properties.

Appearance	Silvery-white
Standard atomic weight ($A_{r,std}$)	91.224
Block	d-Block
Element category	Transition metal
Electron configuration	[Kr] $4d^2$ $5s^2$
Phase at STP	Solid
Melting point	1855°C
Boiling point	4377°C
Density (near r.t.)	6.52 g/cm^3
When liquid (at m.p.)	5.8 g/cm^3
Heat of fusion	14 kJ/mol
Heat of vaporization	591 kJ/mol
Molar heat capacity	25.36 J/(mol·K)
Oxidation states	−2, +1, +2, +3, **+4**
Ionization energies	First: 640.1 kJ/mol
	Second: 1270 kJ/mol
	Third: 2218 kJ/mol
Atomic radius	Empirical: 160 pm
Covalent radius	175 ± 7 pm

STP, Standard temperature and pressure.
Bold font indicates main oxidation state.

these isotopes with a half-life of 1.53×10^6 years. ^{110}Zr, the heaviest isotope of Zr, is also the most radioactive, and has an estimated half-life of 30 milliseconds. Radioisotopes at or above mass number 93 decay via electron emission, while those at or below 89 decay via positron emission. The single exception is ^{88}Zr, which decays via electron capture. In addition, five isotopes of Zr are known to exist as metastable isomers: ^{83m}Zr, ^{85m}Zr, ^{89m}Zr, ^{90m1}Zr, ^{90m2}Zr, and ^{91m}Zr. Of these metastable isotopes, ^{90m2}Zr has the shortest half-life at 131 nanoseconds, while ^{89m}Zr has the longest half-life of 4.161 minutes.

Zirconium is a lustrous, grayish-white transition metal that also appears in the form of a bluish-black powder. It has 40 electrons arranged in an electronic configuration of [Kr]$4d^2 5s^2$. Zirconium forms a wide range of inorganic compounds and coordination complexes wherein zirconium has the oxidation state of +4. Oxidation states of +3 such as in zirconium(III) chloride, $ZrCl_3$, and +2 such as in zirconium(II) hydride, ZrH_2, can also be found but are rare (Table 6.4).

6.4.3.1 *Oxides, nitrides, carbides, and hydrides*

Zirconium in bulk form is fairly inactive, resisting corrosion due to the formation of an adherent layer of zirconium(IV) oxide, ZrO_2. Zirconium is also resistant to attack by most cold acids, alkalis, or with water. It does react with hot acids such as sulfuric and nitric acids, although the formation of the oxide in these oxidizing acids reduces their reactivity. When finely divided, Zr is pyrophoric. When heated to high temperatures, it reacts with nonmetals such as oxygen, hydrogen, carbon, nitrogen, and the halogens to form ZrO_2 and ZrH_2, ZrC, ZrN, and ZrX_4 (X = F, Cl, Br, I), respectively. Zirconia can also form mixed oxides with lead oxide and titania to form lead zirconate titanate (PZT), a valuable piezoelectric material.

Most of the compounds of zirconium are mostly valued for their inert chemistries. ZrO_2, commonly known as zirconia and the most important oxide of Zr, is exceptionally tough with respect to fracture and has high chemical resistance. Zirconium carbide and zirconium nitride are also valued for their refractory properties. Zirconia is prepared by calcining zirconium compounds, exploiting its high thermal stability. Three different crystalline phases are known: monoclinic below 1170°C, tetragonal between 1170°C and 2370°C, and cubic above 2370°C. The trend is for higher symmetry at higher temperatures. A small quantity of the oxides of calcium or yttrium stabilizes the cubic phase. The very rare mineral tazheranite $(Zr,Ti,Ca)O_2$ is cubic. In contrast to TiO_2, which contains six-coordinate Ti in all phases, monoclinic zirconia consists of seven-coordinate zirconium centers. This difference is due to the larger size of Zr atom relative to the Ti atom. Zirconium tungstate ($Zr(WO_4)_2$) has the rare property of shrinking in all dimensions when heated, whereas most other substances expand when heated. Zirconyl chloride is a rare water-soluble zirconium complex with the rather complex formula $[Zr_4(OH)_{12}(H_2O)_{16}]Cl_8$.

ZrN grown by physical vapor deposition (PVD) has a light gold color comparable to elemental gold. ZrN possesses a room-temperature electrical resistivity of 12.0 μΩ·cm, a temperature coefficient of resistivity of $5.6 \cdot 10^{-8}$ Ω·cm/K, a superconducting transition temperature of 10.4K (− 262.8°C), and a relaxed lattice parameter of 0.4575 nm. The hardness of single-crystal ZrN is 22.7 ± 1.7 GPa and elastic modulus is 450 GPa. Zirconium nitride is a hard ceramic material comparable to titanium nitride and is a cement-like refractory material. Therefore it is employed in refractories, cermets, and laboratory crucibles. When applied using the PVD coating process it is commonly utilized for coating medical devices, industrial parts (notably drill bits), car and aerospace components, and other parts subject to high-wear and corrosive environments. Zirconium nitride was proposed as a hydrogen peroxide fuel tank liner for rockets and aircraft.

Zirconium carbide (ZrC) is an extremely hard refractory ceramic material, commercially used in tool bits for cutting tools. It is typically processed by sintering. Zirconium carbide is produced by carbothermal reduction of zirconia by graphite. Densified ZrC is made by sintering powder of ZrC at temperatures upwards of 2000°C. Hot pressing of ZrC can bring down the sintering temperature and thus helps in producing fine-grained fully densified ZrC. Spark plasma sintering has also been utilized to form fully densified ZrC. Poor oxidation resistance over 800°C limits the applications of ZrC. One method to enhance the oxidation resistance of ZrC is to produce composites. Important composites proposed are ZrC-ZrB_2 and ZrC-ZrB_2-SiC composites. These composites can work up to 1800°C. ZrC has the appearance of a gray metallic powder with cubic crystal structure. It is highly corrosion-resistant. This group IV interstitial transition-metal carbide is also a member of ultrahigh temperature ceramics or UHTC. Because of the presence of metallic bonding, ZrC has a thermal conductivity of 20.5 W/m·K and an electrical conductivity (resistivity ~43 μΩ·cm), both of which are comparable to that of zirconium metal. The strong covalent Zr−C bond gives this material a very high-melting point (~3530°C), high modulus (~440 GPa) and hardness (25 GPa). ZrC has a lower density (6.73 g/cm^3) compared to other carbides for example, WC (15.8 g/cm^3), TaC (14.5 g/cm^3), or HfC (12.67 g/cm^3). ZrC appears suitable for application in re-entry vehicles, rocket/scramjet engines, or supersonic vehicles in which low-densities and high-temperature load-bearing capabilities are vital constraints. Similar to most carbides of refractory metals, zirconium carbide is substoichiometric, that is, it contains carbon vacancies. At carbon contents higher than about $ZrC_{0.98}$ the material contains free carbon. ZrC is stable for a carbon-to-metal ratio ranging from 0.65 to 0.98. The group IVA metal carbides, TiC, ZrC, and SiC, are virtually inert toward attack by strong aqueous acids (HCl) and strong aqueous bases (NaOH) even at 100°C; nevertheless, ZrC does react with HF. The mixture of zirconium carbide and tantalum carbide is an important cermet material.

Zirconium(II) hydride, ZrH_2, is a molecular chemical compound, which has been produced by laser ablation and isolated at low temperature. Zirconium(II) hydride has frequently been the subject of Dirac−Hartree−Fock relativistic calculation studies, which explore the stabilities, geometries, and relative energies of hydrides of the formula MH_4, MH_3, MH_2, or MH. Zirconium(II) hydride has a dihedral (C_{2v}) structure. In zirconium(II) hydride, the formal oxidation states of hydrogen and zirconium are −1 and +2, respectively, since the electronegativity of zirconium is lower than that of hydrogen. The stability of metal hydrides with the formula MH_2 (M = Ti-Hf) decreases as the atomic number increases.

6.4.3.2 Borides

Zirconium diboride (ZrB_2) is a highly covalent refractory ceramic material with a hexagonal crystal structure. ZrB_2 is known as a UHTC with a melting point of 3246°C. This together with its relatively low density of ~6.09 g/cm^3 (measured density may be higher due to hafnium impurities) and good high-temperature strength makes it a contender for high-temperature aerospace applications, for example, hypersonic flight or rocket propulsion systems. It is an unusual ceramic with relatively high thermal and electrical conductivities. ZrB_2 parts are usually hot pressed and then machined into shape. Sintering of ZrB_2 is impeded by the compound's covalent nature and presence of surface oxides, which increase grain coarsening before densification during sintering. Pressureless sintering of ZrB_2 is feasible with sintering additives such as boron carbide and carbon, which react with the surface oxides to increase the driving force for sintering, but mechanical properties are reduced compared to hot-pressed ZrB_2. Addition of ~30 vol.% SiC to ZrB_2 improves oxidation resistance through SiC forming a protective oxide layer. ZrB_2 can be prepared by stoichiometric reaction between constituent elements Zr and B. This reaction allows precise stoichiometric control of the compound. At 2000 K, the formation of ZrB_2 via stoichiometric reaction is thermodynamically favorable ($\Delta G = -279.6$ kJ/mol) and therefore this method can be applied to prepare ZrB_2 by self-propagating high-temperature synthesis (SHS). This method takes advantage of the high exothermic energy of the reaction to produce high-temperature, fast combustion reactions. Benefits of SHS comprise higher purity of ceramic products, increased sinterability, and shorter processing times. Nevertheless, the very rapid heating rates can result in incomplete reactions between Zr and B, the formation of stable oxides of Zr, and the retention of porosity. Stoichiometric reactions have also been carried out by reaction of attrition-milled (wearing materials by grinding) Zr and B powder (and then hot pressing at 600°C for 6 hours), and nanoscale particles have been prepared by reaction of attrition-milled Zr and B precursor crystallites (10 nm in size). Reduction of ZrO_2 to its diboride can also be accomplished via metallothermic reduction. Low-cost precursor materials are used and reacted:

$$ZrO_2 + B_2O_3 + 5Mg \rightarrow ZrB_2 + 5MgO$$

Mg is used as a reactant to allow for acid leaching of undesirable oxide products. Stoichiometric excesses of Mg and B_2O_3 are frequently necessary during metallothermic reductions in order to react all available ZrO_2. These reactions are exothermic and can be utilized to prepare the diborides by SHS. Synthesis of ZrB_2 from ZrO_2 via SHS frequently results in incomplete conversion of reactants, and hence double SHS has been used. A second SHS reaction with Mg and H_3BO_3 as reactants along with the ZrB_2/ZrO_2 mixture results in increased conversion to the diboride, and particle sizes of 25–40 nm at 800°C. Synthesis of UHTCs by boron carbide reduction is one of the most popular methods for UHTC synthesis. The precursor materials for this reaction (ZrO_2 and B_4C) are more cost-effective than those necessary for the stoichiometric and borothermic reactions. ZrB_2 is prepared at temperatures higher than 1600°C for at least 1 hour by the following reaction:

$$2ZrO_2 + B_4C + 3C \rightarrow 2ZrB_2 + 4CO$$

This method involves a slight excess of boron, as a small amount of boron is oxidized during boron carbide reduction. ZrC has also been found as a product from this reaction, but if the reaction is performed with 20%–25% excess B_4C, the ZrC phase disappears, and only ZrB_2 remains. Lower synthesis temperatures ($\sim$1600°C) produce UHTCs that have finer grain sizes and better sinterability. Boron carbide has to be ground prior to the boron carbide reduction in order to promote oxide reduction and diffusion processes. Boron carbide reductions can also be performed via reactive plasma spraying if a UHTC coating is required. Precursor or powder particles react with the plasma at high temperatures (6000°C–15,000°C), which considerably lowers the reaction time. ZrB_2 and ZrO_2 phases have been formed using a plasma voltage and current of 50 V and 500 A, respectively. These coating materials exhibit uniform distribution of fine particles and porous microstructures, which increased hydrogen flow rates. Another route for the synthesis of UHTCs is the borothermic reduction of ZrO_2 with B. At temperatures over 1600°C, pure diboride can be obtained from this technique. Because of the loss of some boron as boron oxide, excess boron is required during borothermic reduction. Mechanical milling can lower the reaction temperature needed during borothermic reduction. This is because of the increased particle mixing and lattice defects that result from decreased particle sizes of ZnO_2 and B after milling. This method is also not very useful for industrial applications because of the loss of expensive boron as boron oxide during the reaction. Nanocrystals of ZrB_2 were successfully formed by Zoli's reaction, a reduction of ZrO_2 with $NaBH_4$ using a molar ratio M:B of 1:4 at 700°C for 30 minutes under argon flow.

$$ZrO_2 + 3NaBH_4 \rightarrow ZrB_2 + 2Na(g, l) + NaBO_2 + 6H_2(g)$$

ZrB_2 can be formed by solution-based synthesis methods as well, even though few significant studies have been performed. Solution-based routes allow for low-temperature synthesis of ultrafine UHTC powders. For example, ZrB_2 powders were synthesized using the inorganic-organic precursors $ZrOCl_2{\cdot}8H_2O$, boric acid and phenolic resin at 1500°C. The synthesized powders showed 200 nm crystallite size and low oxygen content ($\sim$1.0 wt.%). ZrB_2 preparation from polymeric precursors has also been studied recently. ZrO_2 can be dispersed in boron carbide polymeric precursors before the reaction. Heating the reaction mixture to 1500°C causes the in situ generation of boron carbide and carbon, and the reduction of ZrO_2 to ZrB_2 quickly follows. The polymer has to be stable, processable, and contain boron and carbon in order to be useful for the reaction. Dinitrile polymers formed from the condensation of dinitrile with decaborane meet these conditions. CVD can be employed to synthesize ZrB_2. Hydrogen gas is used to reduce vapors of zirconium tetrachloride and boron trichloride at substrate temperatures over 800°C. Recently, high-quality thin films of ZrB_2 were likewise prepared by PVD. ZrB_2 gets its high-temperature mechanical stability from the high atomic defect energies (i.e., the atoms do not deviate easily from their lattice sites). This means that the defects concentration will remain low, even at high temperatures, preventing failure of the material. The layered bonding between each layer is also very strong but means that the ceramic is highly anisotropic, having different thermal expansions in the "z" <001> direction. While the material has excellent high-temperature properties, the ceramic has to be prepared extremely carefully as any excess of either zirconium or boron will not be accommodated in the ZrB_2 lattice (i.e., the material does not deviate from stoichiometry). Instead it will form extra lower melting point phases that may initiate failure under extreme conditions. Zirconium diboride is also a topic of research as a potential material for nuclear reactor control rods because of the presence of boron and hafnium.

$$^{10}B + n_{th} \rightarrow \left[^{11}B\right] \rightarrow \alpha + {}^{7}Li + 2.31\ MeV.$$

The layered structure provides a plane for helium diffusion to occur. Helium is formed as a transmutation product of ^{10}B—it is the α particle in the above reaction—and will rapidly migrate through the lattice between the layers of zirconium and boron, however, not in the "z" direction. Of interest, the other transmutation product, lithium, is expected to be trapped in the boron vacancies that are produced by the ^{10}B transmutation and not to be released from the lattice.

6.4.3.3 Halides and pseudohalides

All four common halides are known, ZrF_4, $ZrCl_4$, $ZrBr_4$, and ZrI_4. All have polymeric structures and are far less volatile than the corresponding monomeric titanium tetrahalides. All tend to hydrolyze to give the so-called oxyhalides and dioxides. Three crystalline phases of ZrF_4 have been observed, α (monoclinic), β (tetragonal, space group *P42/m*, No. 84), and γ (unknown structure). β and γ phases are unstable and irreversibly convert to the α-phase at 400°C. Zirconium fluoride is used as a zirconium source in oxygen-sensitive applications, for example, metal production. Zirconium fluoride can be purified through distillation or sublimation. Conditions/substances to avoid include moisture, active metals, acids, and oxidizing agents. Zirconium fluoride in a mixture with other fluorides is used as a coolant for molten salt reactors. In the mixture with sodium fluoride it is a candidate coolant for the advanced high-temperature reactor. Together with uranium salt, zirconium fluoride can be a component of fuel coolant in molten salt reactors.

Zirconium(IV) bromide, $ZrBr_4$, is a colorless solid and the principal precursor to other Zr-Br compounds. $ZrBr_4$ is produced by the carbotherm reaction of bromine on zirconium oxide:

$$ZrO_2 + 2C + 2Br_2 \rightarrow ZrBr_4 + 2CO$$

Like many related tetrahalides, it is purified by sublimation. Like related tetrabromides of Ti and Hf, $ZrBr_4$ hydrolyzes easily to give oxybromide, with release of hydrogen bromide.

Zirconium(IV) iodide, ZrI_4, is the most readily available iodide of zirconium. It is an orange-colored solid that degrades in the presence of water. The compound was once renowned as an intermediate in the purification of zirconium metal. This compound is volatile, subliming as intact tetrahedral ZrI_4 molecules. It is prepared by the direct reaction of powdered zirconium metal and iodine. Pyrolysis of zirconium(IV) iodide gas by contact of hot wire was the first industrial process for the commercial production of pure ductile metallic zirconium. Like most binary metal halides, it has a polymeric structure. The compound exists as a polymer consisting of octahedral Zr(IV) centers, each with a pair of terminal iodide ligands and four doubly bridging iodide ligands. The Zr-I distances are 2.692 (terminal) and 3.030 Å.

Zirconium(IV) chloride, also known as zirconium tetrachloride, ($ZrCl_4$) is an inorganic compound regularly used as a precursor to other compounds of zirconium. This white high-melting solid hydrolyzes quickly in humid air. The formation of $ZrCl_4$ involves a reaction of the oxide with carbon as the oxide "getter" and chlorine.

$$ZrO_2 + 2C + 2Cl_2 \rightarrow ZrCl_4 + 2CO$$

A laboratory-scale process uses carbon tetrachloride instead of carbon and chlorine:

$$ZrO_2 + 2CCl_4 \rightarrow ZrCl_4 + 2COCl_2$$

In contrast to molecular $TiCl_4$, solid $ZrCl_4$ has a polymeric structure in which each Zr is octahedrally coordinated. This difference in structures is the reason for the difference in their properties: $TiCl_4$ is distillable, but $ZrCl_4$ is a solid. In the solid state, $ZrCl_4$ has a tape-like linear polymeric structure—the same structure adopted by $HfCl_4$. This polymer degrades easily upon treatment with Lewis bases, which cleave the Zr-Cl-Zr linkages. $ZrCl_4$ is an intermediate in the conversion of zirconium minerals to metallic zirconium by the Kroll process. In nature, zirconium minerals always exist as oxides. For their conversion to bulk metal, these refractory oxides are first converted to the tetrachloride, which can then be distilled at high temperatures. The purified $ZrCl_4$ is subsequently reduced with Zr metal to produce zirconium (III) chloride. $ZrCl_4$ is the most common precursor for CVD of zirconium dioxide and zirconium diboride. In organic synthesis zirconium tetrachloride is used as a weak Lewis acid for the Friedel–Crafts reaction, the Diels–Alder reaction, and intramolecular cyclization reactions. Zirconium(III) chloride, $ZrCl_3$, is a blue-black solid that is highly sensitive to air. The material was first claimed by Ruff and Wallstein who reduced zirconium tetrachloride with aluminum to form impure samples. Later, the problem with aluminum contamination was solved when it was produced by reduction utilizing zirconium metal:

$$Zr + 3ZrCl_4 \rightarrow 4ZrCl_3$$

When aluminum is utilized as the reducing agent with zirconium tetrachloride, a series of choloroaluminates are the result, for example, $[Zr(AlCl_4)_2(AlCl_4)_2]$ and $Zr(AlCl_4)_3$. Because the trihalides, such as zirconium trichloride, are relatively nonvolatile, contamination can be prevented by using a gaseous reductant. For instance, zirconium trichloride can be formed by reduction of zirconium tetrachloride with hydrogen.

$$2ZrCl_4 + H_2 \rightarrow 2ZrCl_3 + 2HCl$$

Some zirconium halides ($ZrCl_3$, $ZrBr_3$, and ZrI_3) have structures comparable to HfI_3. They also have the same space group (*P63/mcm*) and hexagonal structure with two molecules in the cell. The magnetic susceptibility of zirconium trichloride implies metal–metal interactions of the unpaired electron on each Zr(III) center. The magnetic moment of $ZrCl_3$ (0.4 BM) suggests considerable overlap of metal orbitals.

The corresponding tetraalkoxides are also known. Unlike the halides, the alkoxides dissolve in nonpolar solvents. Dihydrogen hexafluorozirconate is used in the metal finishing industry as an etching agent to promote paint adhesion.

6.4.3.4 Organozirconium compounds

Organozirconium chemistry is the study of compounds containing a carbon–zirconium bond. The vast majority of work on organozirconium relates to zirconocene hydrochloride, $[\{ZrHCl(\eta\text{-}C_5H_5)_2\}_n]$, or Schwartz's reagent and its role in hydrozirconation. Hydrozirconation is a term used to describe a sequence of reactions whereby an unsaturated organic substrate is reacted with $[\{ZrHCl(\eta\text{-}C_5H_5)_2\}_n]$ and the resultant organozirconium product (which is generally not isolated) is converted into a variety of organic products. Hydrozirconation of alkynes or alkenes in organic synthesis offers the advantage of allowing reactions to proceed at room temperature and no stringent requirements such as special glassware or glove-box techniques. As such, the technique has enjoyed applications in the synthesis of natural products and several pharmaceuticals.

The steps in hydrozirconation involve Zr–C bond formation by insertion of the unsaturated substrate into the Zr–H bond followed by Zr–C bond scission. Zr–C scission can proceed directly to provide the organic product or indirectly, either by prior insertion of CO and final acyl-Zr cleavage or by transfer of the hydrocarbyl group to another metal (i.e., the organozirconium compound becomes a carbo-metalating reagent), followed by a transformation of the newly formed hydrocarbylmetal complex. An example of the general scheme for the formation of alkenyl zirconocene intermediates and the subsequent transformation to various organic compounds is shown below. A more detailed discussion of the reactions of alkenyl zirconocene can be found in reviews by Pinheiro et al. (2018) and Wipf and Jahn (1996).

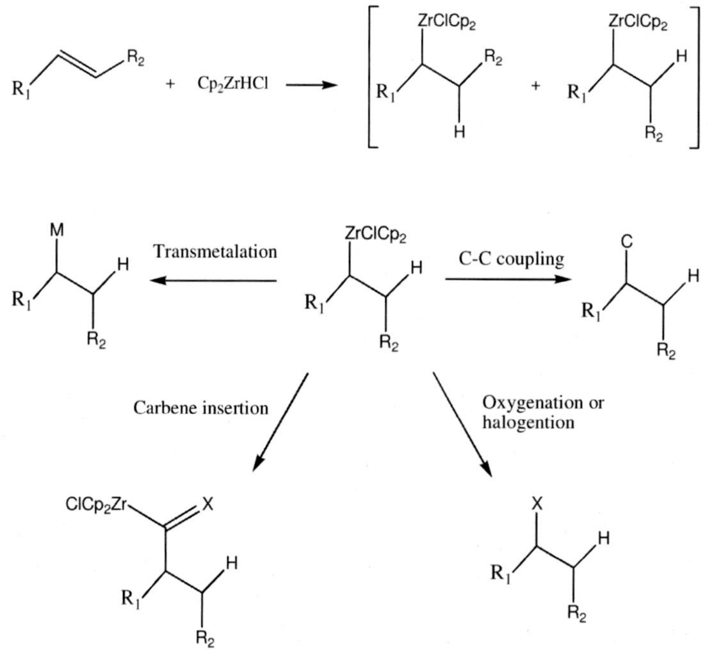

+ Cp_2ZrHCl →

Transmetalation

C-C coupling

Carbene insertion

Oxygenation or halogention

6.4.4 Major uses

Nearly all of the mined zircon ($ZrSiO_4$) is used as is in high-temperature applications. This mineral is refractory, hard, and resistant to chemical attacks. Due to these properties, zircon has many applications, only some of which are highly publicized. Its principal application is as an opacifier, giving a white, opaque appearance to ceramic materials. Zircon mixed with vanadium or praseodymium makes blue and yellow pigments for glazing pottery. Due to its chemical resistance, zircon is utilized in aggressive environments, such as molds for molten metals. Zirconium dioxide (ZrO_2) is utilized for laboratory crucibles, in metallurgical furnaces, and as a refractory material. Since it is mechanically strong and flexible, it can be sintered to make ceramic knives and other blades. Zircon and cubic zirconia (ZrO_2) are cut for use in jewelry. Zircon is a natural semiprecious gemstone found in a variety of colors. The most desirable have a golden hue. Colorless cubic zirconia, when cut, resemble diamonds. Zirconia is a component in certain abrasives, for example, grinding wheels and sandpaper. A minor portion of zircon is converted to Zr, which is used in a variety of highly specialized applications. Due to its excellent corrosion resistance, it is frequently utilized as an alloying compound in materials that are exposed to aggressive environments, for example, surgical appliances, light filaments, and watch cases. The high reactivity of Zr with O_2 at high temperatures is used in some specific applications such as explosive primers and as getters in vacuum tubes. The same property is (in all probability) the objective for adding Zr nanoparticles as pyrophoric material in explosive weapons such as the BLU-97/B Combined Effects Bomb. Burning Zr was used as a light source in some photographic flashbulbs. Zr powder with a mesh size from 10 to 80 is sometimes used in pyrotechnic compositions to produce sparks. The high reactivity of Zr results in bright white sparks. Materials made from Zr metal and ZrO_2 are utilized in space vehicles where heat resistance is required. High-temperature parts such as combustors, blades, and vanes in jet engines and stationary gas turbines are increasingly being protected by thin ceramic layers, usually composed of a mixture of zirconia and yttria. With niobium, zirconium is superconductive at low temperatures and is used to make superconducting magnets.

Cladding for nuclear reactor fuels uses around 1% of the Zr supply, mostly in the form of zircaloys. The required properties of these alloys are a low neutron-capture cross section and corrosion resistance under standard operating conditions. Effective methods for removing the Hf impurities were established for this reason. One drawback of Zr alloys is that Zr reacts with water at high temperatures, producing H_2 gas and accelerated degradation of the fuel rod cladding:

$$Zr + 2H_2O \rightarrow ZrO_2 + 2H_2$$

This exothermic reaction is very slow below 100°C, but at temperature above 900°C the reaction is rapid. The redox reaction is important to the instability of fuel assemblies at high temperatures. This reaction was responsible for a small H_2 explosion inside the reactor building of Three Mile Island nuclear power plant in 1979, but at that time, the containment building was not damaged. The same reaction happened in the reactors 1, 2, and 3 of the Fukushima I Nuclear Power Plant (Japan) after the reactor cooling was interrupted by the earthquake and tsunami disaster of March 11, 2011, leading to the Fukushima I nuclear accidents. After venting the H_2 in the maintenance hall of those reactors, the mixture of H_2 with atmospheric O_2 exploded, severely damaging the installations and at least one of the containment buildings. To prevent explosion, the direct venting of H_2 to the open atmosphere would have been a better design solution. These days, to avoid the danger of explosion in many pressurized water reactor (PWR) containment buildings, a catalyst-based recombiner is installed that converts H_2 and O_2 into H_2O at room temperature before the hazard arises.

The isotope ^{89}Zr has been used for the tracking and quantification of molecular antibodies with PET cameras (a procedure called "immuno-PET"). Immuno-PET is currently entering the phase of wide-scale clinical applications. Until a short time ago, radiolabelling with ^{89}Zr was a difficult process involving multiple steps. In 2001–03 an enhanced multistep method was established using a succinylated derivative of desferrioxamine B ($C_{25}H_{48}N_6O_8$)(N-sucDf) as a bifunctional chelate, and an improved way of binding ^{89}Zr to mAbs (monoclonal antibodies) was published in 2009. The new procedure is fast, comprises only two steps, and uses two commonly available ingredients: ^{89}Zr and the appropriate chelate. Ongoing developments also include the use of siderophore derivatives (small, high-affinity iron-chelating compounds) to bind $^{89}Zr(IV)$. Zr-containing compounds are utilized in numerous biomedical applications such as dental implants and crowns, knee and hip replacements, middle-ear ossicular chain reconstruction, and other restorative and prosthetic devices. Zr binds urea (IUPAC name carbonyl diamide, $CO(NH_2)_2$), a property that has been widely used to the benefit of patients with chronic kidney disease. For instance, Zr is a key element of the sorbent column dependent dialysate regeneration and recirculation system known as the REDY system, which was originally established in 1973. Over 2,000,000 dialysis treatments have been completed using the sorbent column in the REDY system. While the REDY system was replaced in the 1990s by cheaper substitutes, new sorbent-based dialysis systems are being assessed and approved by the US Food and Drug Administration (FDA). Renal solutions established the DIALISORB technology, a portable, low-water dialysis system. Also, developmental versions of a wearable artificial kidney have integrated sorbent-based technologies. Na-Zr-cyclosilicate is being studied for oral therapy in the treatment of hyperkalemia. It is a highly selective oral sorbent intended to trap in particular K ions in preference to other ions in the gastrointestinal tract. A combination of monomeric and polymeric Zr^{4+} and Al^{3+} complexes with hydroxide, chloride, and glycine, called Al-Zr-tetrachlorohydrex gly or AZG, is used in a formulation as an antiperspirant in many deodorant products. It is chosen for its capability to block pores in the skin preventing sweat from leaving the body. Zr carbonate ($3ZrO_2 \cdot CO_2 \cdot H_2O$) was once used in lotions to treat poison ivy but was terminated as it sometimes caused skin reactions.

6.5 41 Nb — Niobium

6.5.1 Discovery

Niobium was recognized by British chemist Charles Hatchett (January 2, 1765 to March 10, 1847) in 1801 (Hatchett, 1802a–c) (Fig. 6.26). He discovered a new element in a mineral sample that had been sent to England from Connecticut, United States in 1734 by John Winthrop F.R.S. [(1681–1747, grandson of early governor of the Connecticut Colony John Winthrop the Younger, February 12, 1606 to April 6, 1676) and called the mineral columbite and the new element columbium after Columbia, the poetical name for the United States. The columbium discovered by Hatchett was most likely a mixture of the new element with tantalum. As a result, there was significant misunderstanding over the difference between columbium (niobium) and the closely related tantalum. In 1809, English chemist and physicist William Hyde Wollaston (August 6, 1766 to December 22, 1828) compared the oxides derived from both columbium—columbite, with a density of 5.918 g/cm^3, and tantalum—tantalite, with a density over 8 g/cm^3, and decided that the two oxides, notwithstanding the substantial difference in density, were the same; therefore he kept the name tantalum (Wollaston, 1809). This was disputed in 1846 by German mineralogist and analytical chemist Heinrich Rose (August 6, 1795 to January 27, 1864), who reasoned that there were two different elements in the tantalite sample and named them after children of Tantalus: niobium (from Niobe) and pelopium (from Pelops) (Rose, 1847) (Fig. 6.27). This misunderstanding was a result of the observed small differences between tantalum and niobium. The claimed new elements pelopium, ilmenium, and dianium were similar to niobium or mixtures of niobium and tantalum. The differences between tantalum and niobium were unambiguously confirmed in 1864 by Swedish chemist and mineralogist Christian Wilhelm Blomstrand (October 20, 1826 to November 5, 1897) and French chemist Henri Etienne Sainte-Claire Deville (March 11, 1818 to July 1, 1881), along with French chemist Louis J. Troost (October 17, 1825 to September 30, 1911), who determined the formulas of some of the compounds in 1865 and lastly by Swiss chemist Jean Charles Galissard de Marignac (April 24, 1817 to April 15, 1894) in 1866, who all demonstrated that there were

FIGURE 6.26 Charles Hatchett, engraving by F.C. Lewis after the painting by T. Phillips. *From Faulkner's History of Chelsea (1829).*

FIGURE 6.27 Heinrich Rose, c.1825, engraved by W.C. Sharpe (either 1749–1824 or 1803–75), drawn by l'Alemand.

only two elements. Papers on ilmenium continued to be published until 1871. de Marignac was the first to isolate the metal in 1864, when he reduced niobium chloride by heating it in an atmosphere of hydrogen. Even though de Marignac managed to isolate tantalum-free niobium on a larger scale by 1866 (de Marignac, 1866; de Marignac et al., 1866), it was not until the early 20th century that niobium was utilized in incandescent lamp filaments, the first commercial use. This use was rapidly superseded through the replacement of niobium with tungsten, which has a higher melting point. That niobium increases the strength of steel was first found in the 1920s, and this application is still its main use. In 1961, the American physicist Eugene Kunzler and coworkers at Bell Labs found that niobium-tin continues to exhibit superconductivity in the presence of strong electric currents and magnetic fields, making it the first material to support the high currents and fields needed for suitable high-power magnets and electrical power machinery. This discovery allowed—two decades later—the manufacture of long multistrand cables wound into coils to generate large, powerful electromagnets for rotating machinery, particle accelerators, and particle detectors. Columbium (symbol "Cb") was the name initially given by Hatchett after his discovery of the metal in 1801. The name reflected that the type specimen of the ore came from America (Columbia). This name persisted in American journals—the last paper published by the American Chemical Society with columbium in its title dates from 1953—while niobium was used in Europe. To end this confusion, the name niobium was selected for element 41 at the 15th Conference of the Union of Chemistry in Amsterdam in 1949. A year later this name was formally approved by IUPAC after more than a century of controversy, notwithstanding the chronological precedence of the name columbium. This was a sort of concession; IUPAC recognized tungsten instead of wolfram in regard to North American usage; and niobium instead of columbium in regard to European usage. Although most US chemical societies and government organizations generally use the official IUPAC name, some metallurgists and metal societies still use the original American name, "columbium".

6.5.2 Mining, production, and major minerals

Niobium is estimated to be the 34th most common element in the Earth's crust, with 20 ppm. Some think that the abundance on Earth is much higher, and that the element's high density has concentrated it in the Earth's core. Free Nb is not found in nature. Minerals that contain niobium frequently also contain tantalum. A good example is the columbite-tantalite series ($(Fe,Mg,Mn)(Ta,Nb)_2O_6$) (Box 6.5). Columbite-tantalite minerals are most usually found as accessory minerals in pegmatite intrusions, and in alkaline intrusive rocks. Less common are the niobates of calcium, uranium, thorium, and the REEs. Examples of such niobates are pyrochlore ($(Na,Ca)_2Nb_2O_6(OH,F)$) (now a group name, e.g., fluorcalciopyrochlore) and euxenite-(Y) ($(Y,Ca,Ce,U,Th)(Nb,Ta,Ti)_2O_6$). Large deposits of niobium have been discovered accompanying carbonatites (carbonate-silicate igneous rocks). The three largest currently mined deposits of

BOX 6.5 Columbite solid solution series.

Columbite-(Fe) was named in 1801 by Charles Hatchett (1765–1847) from a specimen collected in the mid-17th century in a place called Nautneague (now regarded as Haddam, Connecticut, United States). The specimen was from the mineral collection of John Winthrop (1606–76), governor of Connecticut and was given to Hans Sloane by John Winthrop (1681–1747) who was the grandson of the governor with the same name. From the type specimen, Hatchett named the new element columbium (now called niobium). The columbite group name refers to columbium, the old name for Nb. The species, columbite-(Fe) (Fig. 6.28) retains the old root name and has the suffix -(Fe) to indicate the dominance of iron in the species, idem for columbite-(Mn) and columbite-(Mg). It has an end-member composition $Fe^{2+}Nb_2O_6$ but a complete solid solution exists to columbite-(Mn) (originally referred to as manganocolumbite, it was changed to the current form in 2008 under IMA rules) with end-member composition $Mn^{2+}Nb_2O_6$, as well as columbite-(Mg) (originally referred to as magnocolumbite) with dominant Mg instead of Fe or Mg. In the other structural position the Nb can be substituted by Ta forming a solid solution with the tantalite series. Columbite-(Fe) and columbite-(Mn) are typical of granitic pegmatites and other environments where Nb > Ta. They may also rarely be found in alkalic pegmatites as alteration product of pyrochlore and other Nb minerals and as disseminations in albitites and apogranites, associated with zircon, xenotime-(Y), cassiterite, and ilmenorutile. Columbite-(Mg) forms under very unusual conditions in pegmatite genesis of Mg saturation and virtual absence of Fe. All three are orthorhombic, *2/m2/m2/m* (space group *Pbcn*). Columbite-(Fe) and columbite-(Mn) have a hardness of 6, while that of columbite-(Mg) is 6.5. Cleavage on {010} is distinct, on {100} less so, the fracture is subconchoidal to uneven, while the tenacity is brittle. The luster is vitreous to submetallic. It commonly tarnishes to iridescent for columbite-(Fe). Most are opaque. The color for all three is black to brownish black, while the streak is black to dark brown. Optically they are biaxial (–). Columbite-(Fe) and columbite-(Mn) crystals are prismatic, often flattened {010} or thick tabular {100}, sometimes nearly equant, or massive. Twinning on twin plane {201} is common, usually contact twins, flattened, and heart-shaped. Columbite-(Mg) generally occurs as small prismatic crystals.

FIGURE 6.28 Columbite-(Fe), $Fe^{2+}Nb_2O_6$, 1.5 cm black, tabular crystal. King Lithia Mine, Pennington Co., South Dakota, United States.

pyrochlore, two in Brazil and one in Canada, were found in the 1950s and are still the main suppliers of niobium mineral concentrates. The largest deposit is found within a carbonatite intrusion in Araxá, Minas Gerais, Brazil; the other active Brazilian deposit is located near Catalão, Goiás, also found within a carbonatite intrusion. Collectively, these two mines yield about 88% of the world's supply. In addition, Brazil has a large but still unexploited deposit near São Gabriel da Cachoeira, Amazonas, as well as several smaller deposits, notably in the state of Roraima. The third largest supplier of niobium is a carbonatite-hosted mine in Saint-Honoré, near Chicoutimi, Quebec, Canada, which produces between 7% and 10% of the world's supply.

After the separation from other minerals, the mixed oxides of tantalum Ta_2O_5 and niobium Nb_2O_5 are obtained. The first stage in the processing of these oxides is the reaction with hydrofluoric acid (HF):

$$Ta_2O_5 + 14HF \rightarrow 2H_2[TaF_7] + 5H_2O$$
$$Nb_2O_5 + 10HF \rightarrow 2H_2[NbOF_5] + 3H_2O$$

The first industrial-scale separation, developed by Swiss chemist Jean Charles Galissard de Marignac (April 24, 1817 to April 15, 1894), made use of the different solubilities of the complex niobium and tantalum fluorides, dipotassium oxypentafluoroniobate monohydrate ($K_2[NbOF_5]\ H_2O$), and dipotassium heptafluorotantalate ($K_2[TaF_7]$) in water. Newer methods utilize the liquid extraction of the fluorides from aqueous solution by organic solvents such as cyclohexanone ($(CH_2)_5CO$). The complex niobium and tantalum fluorides are extracted separately from the organic solvent with water and are subsequently either precipitated by the addition of potassium fluoride (KF) to produce a potassium fluoride complex, or precipitated with ammonia (NH_4OH) as the pentoxide:

$$H_2[NbOF_5] + 2KF \rightarrow K_2[NbOF_5]\downarrow + 2HF$$

Or:

$$2H_2[NbOF_5] + 10NH_4OH \rightarrow Nb_2O_5\downarrow + 10NH_4F + 7H_2O$$

A number of different processes are used for the reduction to metallic niobium. One example is the electrolysis of a molten mixture of $K_2[NbOF_5]$ and sodium chloride; another is the reduction of the fluoride with sodium. With this process, a relatively high-purity niobium can be produced. In large-scale production, Nb_2O_5 is reduced with hydrogen or carbon. In the aluminothermic reaction, a mixture of iron oxide and niobium oxide is reacted with aluminum:

$$3Nb_2O_5 + Fe_2O_3 + 12Al \rightarrow (6Nb + 2Fe)_{alloy} + 6Al_2O_3$$

Small quantities of oxidizers such as sodium nitrate ($NaNO_3$) are added to enhance the reaction. The products of the reaction are aluminum oxide and ferroniobium, an alloy of iron and niobium used in steel production. Ferroniobium contains between 60% and 70% niobium. Without iron oxide, the aluminothermic process is used to produce niobium metal:

$$3Nb_2O_5 + 10Al \rightarrow 6Nb + 5Al_2O_3$$

Additional purification is required to reach the grade for superconductive alloys. To achieve this electron beam melting under vacuum is the method used by the two major distributors of niobium.

More than 100 different minerals are known to contain niobium in their crystal structure. Within the elements class a single carbide, niobocarbide ($(Nb,Ta)C$), is known. Similarly, only a single sulfide, edgarite ($FeNb_3S_6$), is known to contain Nb. In the oxide class more than 50 minerals are known with Nb in their structure, for example, aeschynite-(Nd) ($(Nd,Ln,Ca)(Ti,Nb)_2(O,OH)_6$), columbite-(Fe) ($Fe^{2+}Nb_2O_6$), fergusonite-(Y) ($YNbO_4$), hydroxycalciopyrochlore ($(Ca,Na,U,\square)_2(Nb,Ti)_2O_6(OH)$) (Fig. 6.29), Ixiolite ($(Ta,Nb,Sn,Fe,Mn)_4O_8$) (Fig. 6.30), lueshite ($NaNbO_3$) (Fig. 6.31), polycrase-(Y) ($Y(Ti,Nb)_2(O,OH)_6$), and pyrochlore ($A_2Nb_2(O,OH)_6Z$ with A=Na, Ca, Sn^{2+}, Sr, Pb^{2+}, Sb^{3+} Y,

FIGURE 6.29 Several sharp, dull, brown, hydroxycalciopyrochlore, $(Ca,Na,U,\square)_2(Nb,Ti)_2O_6(OH)$, crystals to 4 mm. Silver Crater Mine, Bancroft, Ontario, Canada.

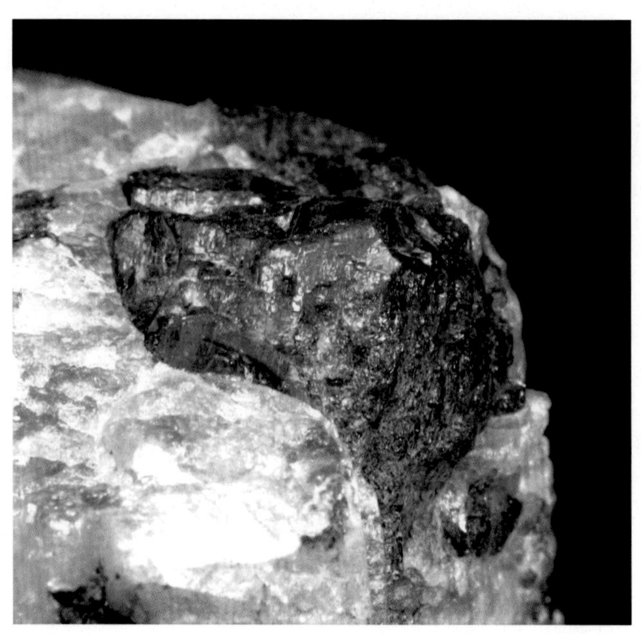

FIGURE 6.30 Ixiolite, $(Ta,Nb,Sn,Fe,Mn)_4O_8$, black crude crystals from 5 to 10 mm embedded in quartz. Zinnwald, Erzgebirge, Bohemia, Czech Republic.

FIGURE 6.31 Lueshite, $NaNbO_3$, a 0.7 mm pseudocubic crystal. Basically, sodium niobate found in pegmatites, rather rare. Lueshe Mine, Kivu, Democratic Republic of Congo.

U^{4+}, H_2O or □ and Z=OH, F, O, H_2O or □) (Fig. 6.32). A single borate mineral, schiavinatoite ($(Nb,Ta)(BO_4)$), is known to contain Nb. The phosphates are represented by two different minerals, johnwalkite ($K(Mn^{2+},Fe^{2+},Fe^{3+})_2(Nb^{5+},Ta^{5+})(PO_4)_2O_2\cdot 2(H_2O,OH)$), and olmsteadite ($KFe_2^{2+}(Nb^{5+},Ta^{5+})(PO_4)_2O_2\cdot 2H_2O$) (Fig. 6.33). However, the majority of Nb-containing minerals are found in the group of the silicates with 67 minerals, for example, epistolite ($Na_2(Nb,Ti)_2(Si_2O_7)O_2\cdot nH_2O$), fersmanite ($Ca_4(Na,Ca)_4(Ti,Nb)_4(Si_2O_7)_2O_8F_3$), and yuksporite ($K_4(Ca,Na)_{14}(Sr,Ba)_2(\square,Mn,Fe)(Ti,Nb)_4(O,OH)_4(Si_6O_{17})_2(Si_2O_7)_3(H_2O,OH)_3$).

FIGURE 6.32 Pyrochlore, $A_2Nb_2(O,OH)_6Z$ (A = Na, Ca, Sn^{2+}, Sr, Pb^{2+}, Sb^{3+}, Y, U^{4+}, H_2O or □. Z = OH, F, O, H_2O or □), large deep red octahedral crystal measuring to 1.3 cm on edge. Vishnovogorsk, Chelyabinsk Oblast, Urals, Russia.

FIGURE 6.33 Olmsteadite, $KFe_2^{2+}(Nb^{5+},Ta^{5+})(PO_4)_2O_2{\cdot}2H_2O$, single deep reddish-brown, terminated crystal to 5.5 mm. White Cap Mine, Pennington Co., South Dakota, United States.

6.5.3 Chemistry

Niobium in the Earth's crust exists only as a single stable isotope, ^{93}Nb. By 2003, no less than 32 radioisotopes had been synthesized, ranging from ^{81}Nb to ^{113}Nb. The most stable of these radioisotopes is ^{92}Nb with a half-life of 34.7 million years, while one of the least stable is ^{113}Nb with an estimated half-life of 30 milliseconds. Isotopes that are lighter than the stable ^{93}Nb have a tendency to decay via β^+ decay, while those isotopes that are heavier have a tendency to decay via β^- decay, though there are some exceptions. ^{81}Nb, ^{82}Nb, and ^{84}Nb have minor β^+ delayed proton emission decay paths; ^{91}Nb decays by electron capture and positron emission, whereas ^{92}Nb decays by both β^+ and β^- decay. At least 25 nuclear isomers have been observed, ranging from ^{84}Nb to ^{104}Nb. Within this range, only ^{96}Nb, ^{101}Nb, and ^{103}Nb do not have any isomers. The most stable of these isomers is ^{93m}Nb with a half-life of 16.13 years, while the least

TABLE 6.5 Niobium properties.

Appearance	Gray metallic, bluish when oxidized
Standard atomic weight ($A_{r,std}$)	92.906
Block	d-Block
Element category	Transition metal
Electron configuration	[Kr] $4d^4$ $5s^1$
Phase at STP	Solid
Melting point	2477°C
Boiling point	4744°C
Density (near r.t.)	8.57 g/cm^3
Heat of fusion	30 kJ/mol
Heat of vaporization	689.9 kJ/mol
Molar heat capacity	24.60 J/(mol·K)
Oxidation states	−3, −1, +1, +2, +3, +4, **+5**
Ionization energies	First: 652.1 kJ/mol Second: 1380 kJ/mol Third: 2416 kJ/mol
Atomic radius	Empirical: 146 pm
Covalent radius	164 ± 6 pm

STP, Standard temperature and pressure.
Bold font indicates main oxidation state.

stable isomer is ^{84m}Nb with a half-life of 103 ns. All these isomers decay via isomeric transition or β decay except for ^{92m1}Nb, which has a minor electron capture branch.

Niobium is a light gray, crystalline, and ductile transition metal. It has 41 electrons arranged in an electronic configuration of [Kr]$4d^4 5s^1$. This electron configuration that does not follow Aufbau principle is not unique to niobium. It can be observed in other transition metals such as chromium, copper, and molybdenum. The underlying quantum mechanical calculations behind this will not be discussed here but it is sufficient to say that this is a result of electron–electron interactions. All five electrons from the 5s and 4d subshells participate in reactions resulting in compounds with niobium having an oxidation state of +5. Nevertheless, other oxidation states ranging from +4 to −1 can also be found (Table 6.5). For example, niobium carbide, where Nb takes a +4 oxidation state, is known. Niobium resists attack by fused alkalis and by acids, including aqua regia, hydrochloric, sulfuric, nitric, and phosphoric acids. However, it is attacked by hydrofluoric acid and hydrofluoric/nitric acid mixtures to form the fluoride complexes.

6.5.3.1 Oxides

Niobium is corrosion-resistant, beginning to oxidize in air only when temperature rises above 400°C. Niobium forms oxides in the oxidation state +5 (Nb_2O_5). The pentoxide can be used to prepare the other oxides of Nb such as niobium (IV) oxide (NbO_2), niobium sesquioxide (Nb_2O_3), and the rarer niobium(II) oxide (NbO). Nb_2O_5 is produced through hydrolysis of alkali metal niobates and alkoxides and the fluorides using base. Such seemingly simple procedures allow hydrated oxides that are calcined. Given that Nb_2O_5 is the most common and robust compound of niobium, many reactions, both practical and esoteric, exist for its preparation. The oxide for instance forms when niobium metal is oxidized in air. The oxidation of niobium dioxide, NbO_2, in air forms the polymorph, L-Nb_2O_5. Pure Nb_2O_5 can be obtained through hydrolysis of $NbCl_5$:

$$2NbCl_5 + 5H_2O \rightarrow Nb_2O_5 + 10HCl$$

A production route using sol–gel techniques has been published in which niobium alkoxides are hydrolyzed in the presence of acetic acid, followed by calcination of the gels to form the polymorphic phase, T-Nb_2O_5. Nanosized niobium pentoxide particles have been prepared by LiH reduction of $NbCl_5$, followed by aerial oxidation as part of a synthesis of nanostructured niobates. Nb_2O_5 has numerous polymorphic forms all based largely on octahedrally coordinated niobium atoms. The polymorphs are identified with a number of prefixes. The form most commonly encountered is monoclinic H-Nb_2O_5, which has a complex structure, with a unit cell containing 28 niobium atoms and 70 oxygen, where 27 of the niobium atoms are octahedrally coordinated and one tetrahedrally. There is an uncharacterized solid hydrate, $Nb_2O_5{\cdot}nH_2O$, the so-called niobic acid (previously called columbic acid), which can be produced through hydrolysis of a basic solution of niobium pentachloride or Nb_2O_5 dissolved in HF. Molten niobium pentoxide has a lower mean coordination number than the crystalline forms, with a structure consisting mostly of NbO_5 and NbO_6 polyhedra. Nb_2O_5 is attacked by HF and dissolves in fused alkali. The conversion of Nb_2O_5 is the main route for the industrial production of niobium metal. The main technique is reduction of this oxide with aluminum:

$$3Nb_2O_5 + 10Al \rightarrow 6Nb + 5Al_2O_3$$

Another but less used reaction comprises carbothermal reduction, which proceeds via reduction with carbon and forms the basis of the two-stage Balke process:

$$Nb_2O_5 + 7C \rightarrow 2NbC + 5CO \text{ (heated under vacuum at 1800°C)}$$
$$5NbC + Nb_2O_5 \rightarrow 7Nb + 5CO$$

Various routes are known for the transformation of Nb_2O_5 to its halides. The main obstacle is incomplete reaction to give the oxyhalides. In the laboratory, the conversion can be effected with thionyl chloride:

$$Nb_2O_5 + 5SOCl_2 \rightarrow 2NbCl_5 + 5SO_2$$

Nb_2O_5 reacts with CCl_4 to give niobium oxychloride $NbOCl_3$. High-temperature reduction with H_2 results in the formation of NbO_2:

$$Nb_2O_5 + H_2 \rightarrow 2NbO_2 + H_2O$$

Niobium monoxide forms as a result of a comproportionation using an arc furnace:

$$Nb_2O_5 + 3Nb \rightarrow 5NbO$$

The burgundy-colored niobium(III) oxide, one of the first superconducting oxides, can also be synthesized by comproportionation:

$$Li_3NbO_4 + 2NbO \rightarrow 3LiNbO_2$$

Treating Nb_2O_5 with aqueous NaOH at 200°C can produce crystalline sodium niobate, $NaNbO_3$, while the reaction with KOH may form soluble Lindqvist-type hexaniobates, $Nb_6O_{19}^{8-}$. Lithium niobates such as $LiNbO_3$ and Li_3NbO_4 can be formed by reacting lithium carbonate with Nb_2O_5. Niobium also reacts with hydrogen at 200°C and with nitrogen at 400°C to form interstitial and nonstoichiometric hydrides and nitrides, respectively.

Niobium dioxide, NbO_2, is a bluish-black nonstoichiometric solid within a compositional range of $NbO_{1.94}$-$NbO_{2.09}$ It can be formed by reacting Nb_2O_5 with H_2 at 800°C–1350°C. Another route is the reaction of Nb_2O_5 with Nb powder at 1100°C. The room temperature form NbO_2 has a tetragonal, rutile-like (TiO_2) structure with short Nb–Nb distances pointing to Nb–Nb bonding. The high-temperature phase also has a rutile-like structure with short Nb–Nb distances. Two high-pressure phases have been published; one with a rutile-like structure, again with short Nb–Nb distances, and a higher pressure with baddeleyite-related (ZrO_2) structure. NbO_2 is insoluble in water and is a strong reducing agent, reducing carbon dioxide to carbon and sulfur dioxide to sulfur. In an industrial process to produce niobium metal, NbO_2 is produced as an intermediary, by the hydrogen reduction of Nb_2O_5. The NbO_2 is then reacted with magnesium vapor to produce the niobium metal.

Niobium monoxide, NbO, is a gray solid with metallic conductivity. NbO can be formed via reduction of Nb_2O_5 by H_2. More commonly, it is produced through comproportionation:

$$Nb_2O_5 + 3Nb \rightarrow 5NbO$$

NbO has an unusual structure, being cubic as is rock-salt (NaCl) structure, but where both niobium and oxygen atoms are four-coordinate square planar. The niobium centers are arranged in octahedra, and there is a structural similarity to the octahedral niobium clusters in lower halides of niobium. In NbO the Nb–Nb bond length is 298 pm, which is close to the 285 pm in the metal. One analysis of the bonding concludes that strong and nearly covalent bonds exist between the metal centers. It is a superconductor with a transition temperature of 1.38K (−271.77°C). It is used in capacitors where a layer of Nb_2O_5 is created around NbO grains as the dielectric.

6.5.3.2 Halides

Niobium reacts with halogens to form halides in the oxidation states of +5 and +4 as well as diverse substoichiometric compounds. It reacts with fluorine at room temperature to form niobium pentafluoride (NbF_5), a white solid with a melting point of 79.0°C. The solid consists of tetramers $[NbF_5]_4$. It is a solid that is rarely used. Niobium pentafluoride is prepared by treatment of any niobium compound with fluorine:

$$2Nb + 5F_2 \rightarrow 2NbF_5$$
$$2NbCl_5 + 5F_2 \rightarrow 2NbF_5 + 5Cl_2$$

It reacts with hydrogen fluoride to form H_2NbF_7, a superacid. In hydrofluoric acid, NbF_5 converts to $[[NbF_7]^{2-}$ and $[[NbF_5O]^{2-}$. The relative solubility of these potassium salts and related tantalum fluorides is the basis of the Marignac method for the separation of Nb and Ta. Niobium(IV) fluoride, NbF_4, is a nonvolatile black solid. It absorbs vapor strongly. It reacts to NbO_2F in moist air. NbF_4 reacts with water to form a brown solution and a brown precipitate, whose components are not known. It is stable between 275°C and 325°C when heated in a vacuum. However, it disproportionates at 350°C rapidly to form niobium(V) fluoride and niobium(III) fluoride:

$$2NbF_4 \rightarrow NbF_5 + NbF_3 \text{ (at 350°C)}$$

Niobium(IV) fluoride has a crystal structure similar to that of tin(IV) fluoride, in which each niobium atom is surrounded by six fluorine atoms forming an octahedron. Of the six fluorine atoms surrounding a single niobium atom, four are bridging to nearby octahedra, leading to a structure of octahedra connected in layers.

$NbCl_5$ forms a dimeric structure (edge-shared bioctahedron) in contrast to the corner-shared tetrameric structure of the fluoride. It is a yellow crystalline solid. It hydrolyzes in air, and samples are often contaminated with small amounts of $NbOCl_3$. It is often employed as a precursor to prepare other compounds of niobium. $NbCl_5$ can be purified by sublimation. Industrially, niobium pentachloride is prepared by direct chlorination of niobium metal at 300°C–350°C:

$$2Nb + 5Cl_2 \rightarrow 2NbCl_5$$

In the laboratory, it is frequently prepared from Nb_2O_5, the main obstacle being incomplete reaction to give the oxyhalides. The conversion can be effected with thionyl chloride. It can also be formed by chlorination of niobium pentoxide in the presence of carbon at 300°C. The products, though, contain small amounts of $NbOCl_3$. Niobium(V) chloride forms chloro-bridged dimers in the solid state. Each niobium center is six-coordinated, but the octahedral coordination is substantially distorted. The equatorial niobium–chlorine bond lengths are 225 pm (terminal) and 256 pm (bridging), whereas the axial niobium–chlorine bonds are 229.2 pm and are deflected inwards to form an angle of 83.7° with the equatorial plane of the molecule. The Nb-Cl-Nb angle at the bridge is 101.3°. The Nb–Nb distance is 398.8 pm, which is too long for any metal–metal interaction. $NbBr_5$, $TaCl_5$, and $TaBr_5$ are isostructural with $NbCl_5$, but NbI_5 and TaI_5 have different structures. It reacts with chlorine around 200°C to form niobium pentachloride ($NbCl_5$), a yellow solid with a melting point of 203.4°C. Niobium(V) chloride is the principal precursor for the synthesis of the alkoxides of niobium, which find niche uses in sol–gel processing. It is also the precursor to many other laboratory reagents. In organic synthesis, $NbCl_5$ is a specialized Lewis acid in activating alkenes for the carbonyl-ene reaction and the Diels–Alder reaction. Niobium chloride can also produce N-acyliminium compounds from certain pyrrolidines, which are substrates for nucleophiles such as allyltrimethylsilane, indole, or the silyl enol ether of benzophenone. Niobium(IV) chloride, also known as niobium tetrachloride ($NbCl_4$), is a solid that forms dark violet crystals, is highly sensitive to air and moisture, and disproportionates into niobium(III) chloride and niobium(V) chloride when heated. Niobium(IV) chloride is usually prepared by allowing elemental niobium and niobium(V) chloride crystals to react over several days in a temperature gradient, with the metal at about 400°C and the salt at about 250°C.

$$4NbCl_5 + Nb \rightarrow 5NbCl_4$$

Niobium(IV) chloride can also be formed by a similar reduction of niobium pentachloride with powdered aluminum.

$$3NbCl_5 + Al \rightarrow 3NbCl_4 + AlCl_3$$

A similar route is also applied in the synthesis of niobium(IV) bromide and tantalum(IV) chloride. Niobium(IV) iodide exists as well and may be synthesized by thermal decomposition of niobium(V) iodide. At 400°C $NbCl_4$ disproportionates:

$$2NbCl_4 \rightarrow NbCl_3 + NbCl_5$$

The disproportionation of niobium(IV) chloride can be employed to prepare tetrachlorobis(tetrahydrofuran) niobium, a useful synthon in Nb(IV) chemistry because of the lability of the attached tetrahydrofuran ligands. This compound can be prepared by first reacting $NbCl_5$ with aluminum in acetonitrile followed by the addition of tetrahydrofuran (thf) to the formed solid.

$$3NbCl_5 + Al + 3CH_3CN \rightarrow 3NbCl_4(NCCH_3)_3 + AlCl_3$$
$$3NbCl_4(NCCH_3)_3 + AlCl_3 + 3C_4H_8O \rightarrow 3NbCl_4(thf)_2 + 9MeCN + AlCl_3(thf)$$

Niobium chloride has been employed in the syntheses of indazoles, which can be applied as LXRs or liver X receptor inhibitors.

Niobium(V) bromide is the inorganic compound with the formula Nb_2Br_{10}. Its name comes from the compound's empirical formula, $NbBr_5$. It is a diamagnetic, orange solid that hydrolyzes easily. The compound has an edge-shared bioctahedral structure, which means that two $NbBr_5$ units are joined by two bromide bridges. The pentachloride and pentaiodides of Nb and Ta have a similar structural motif. There is no bond between the Nb centers. It is produced by the reaction of bromine with niobium metal at high temperature in a tube furnace.

Both pentahalides (F, Cl) are hydrolyzed to give oxides and oxyhalides, such as $NbOCl_3$. It is a white, crystalline, diamagnetic solid. It is regularly present as an impurity in samples of niobium pentachloride, a common reagent in niobium chemistry. Niobium oxychloride is formed by reacting the pentachloride with oxygen:

$$2NbCl_5 + O_2 \rightarrow 2NbOCl_3 + 2Cl_2$$

This reaction is conducted at about 200°C. $NbOCl_3$ additionally forms as a key side-product in the reaction of niobium pentoxide with various chlorinating agents, for example, carbon tetrachloride and thionyl chloride.

$$2Nb_2O_5 + 6CCl_4 \rightarrow 4NbOCl_3 + 6COCl_2$$

In the solid state the coordination sphere for niobium is a distorted octahedron. The Nb–O bonds and Nb–Cl bonds are unequal. This structure can be depicted as a planar Nb_2Cl_6 core connected by O-Nb-O bridges. Thus the compound is best described as a polymer, comprising a double-stranded chain. In the gas phase above 320°C the Raman spectrum indicates the presence of a pyramidal monomer containing a niobium–oxygen double bond.

6.5.3.3 *Chalcogenides*

Niobium disulfide, NbS_2, is a black layered solid that can be exfoliated into ultrathin grayish sheets comparable to other transition-metal dichalcogenides. These layers show superconductivity, where the transition temperature increases from approximately 2K to 6K (−271.2°C to −267.2°C) with the layer thickness increasing from 6 to 12 nm and then saturates with thickness. Niobium diselenide or niobium(IV) selenide, $NbSe_2$, is a lubricant, and a superconductor at temperatures below 7.2K (−266.0°C) that show a charge density wave (CDW). $NbSe_2$ crystallizes in several related forms and can be mechanically exfoliated into monatomic layers, like other transition-metal dichalcogenide monolayers. Monolayer $NbSe_2$ shows very different properties from the bulk material, for example, of Ising superconductivity, quantum metallic state, and strong enhancement of the CDW. $NbSe_2$ crystals and thin films can be formed by CVD. Nb oxide, selenium, and NaCl powders are heated to different temperatures in the range between 300°C and 800°C at ambient pressure in a furnace that allows maintaining a temperature gradient along its axis. Powders are placed in different locations in the furnace, and a mixture of argon and hydrogen is used as the carrier gas. The $NbSe_2$ thickness can be accurately controlled by adjusting the temperature of Se powder. In addition, $NbSe_2$ monolayers can be exfoliated from the bulk or deposited by molecular beam epitaxy. $NbSe_2$ occurs in a number of forms, such as 1H, 2H, 4H, and 3R, where H stands for hexagonal and R for rhombohedral, and the number 1, 2, etc., refers to the number of Se-Nb-Se layers in a unit cell. The Se-Nb-Se layers are bonded together with relatively weak van der Waals forces and can be exfoliated into 1H monolayers. They can be offset in a number of ways to create different crystal structures, the most stable being 2H. Since the layers in $NbSe_2$ are only weakly bonded together, different substances can penetrate between the layers to form well-defined intercalation compounds. Compounds with helium, rubidium, transition metals, and posttransition metals have been synthesized. Extra niobium atoms, up to one-third extra, can also be added between the layers. Extra metal atoms from first transition-metal series can intercalate up to 1:3 ratio. Niobium triselenide, $NbSe_3$, belongs to the class of transition-metal trichalcogenides. It was the first published example of one-dimensional (1D) compound to exhibit the phenomenon of sliding CDWs. Because of its many studies and exhibited phenomena in quantum mechanics, $NbSe_3$ has become the model system for quasi-1D CDWs. The compound is produced via the solid-state reaction by heating niobium and selenium at 600°C–700°C:

$$Nb + 3Se \rightarrow NbSe_3$$

The formed black crystals can comprise $NbSe_2$ impurities. It can be purified by chemical vapor transport (CVT) between 650°C and 700°C. The lower limit of CVT was determined by the temperature at which $NbSe_2$ is no longer stable. $NbSe_3$ has a highly anisotropic structure. The Nb centers are bonded within trigonal prisms characterized by six Se ligands. The $NbSe_6$ prisms form infinite chains lined up in parallel. While the prisms share the same coordination, the cell consists of three chain types repeated twice, where each chain is defined by its Se−Se bond length. The Se−Se bond lengths are 2.37, 2.48, and 2.91 Å.

6.5.3.4 *Carbides and nitrides*

Niobium carbide (NbC and Nb_2C) is an extremely hard refractory ceramic material, commercially used in tool bits for cutting tools. It is typically processed by sintering and is a regular additive as grain growth inhibitor in cemented carbides. It is a brown-gray metallic powder with purple luster. It is extremely corrosion-resistant. Niobium carbide can be prepared by heating of niobium oxide in a vacuum at 1800°C together with coke. Niobium carbide has a Young's modulus of about 452 GPa and a shear modulus of 182 GPa. It has a Poisson's ratio of 0.227. Niobocarbide—the naturally occurring mineral of niobium carbide—is extremely rare.

Niobium nitride, NbN, becomes a superconductor at low temperatures (about 16K/ − 257.2°C) and is utilized in detectors for IR light. NbN's key application is as a superconductor. Detectors based on NbN can detect a single photon in the 1−10-μm section of the IR spectrum, which is important for astronomy and telecommunications. It can detect changes up to 25 GHz. NbN is also employed in absorbing antireflective coatings. In 2015 Panasonic Corp. reported that it had created a photocatalyst based on NbN that could absorb 57% of sunlight to support the decomposition of water to produce hydrogen gas as fuel for electrochemical fuel cells.

6.5.3.5 *Organoniobium compounds*

Several organometallic compounds of niobium have been prepared. These include salts of alkyl niobium ($[Nb(CH_3)_6]^-$), alkylidenes (Nb=CHR), cyclopentadienyl derivatives, and alkyne derivatives. Organoniobium compounds find limited use in organic synthesis. It does appear that substantial amount of work has been done on the alkylidenes of Nb. These are presented in the chapter authored by Zhang et al. (2017) entitled, "Synthesis and Reaction Chemistry of Alkylidene Complexes with Titanium, Zirconium, Vanadium, and Niobium: Effective Catalysts for Olefin Metathesis Polymerization and Other Organic Transformations" in the book series *Advances in Organometallic Chemistry* (Vol. 68).

6.5.4 Major uses

Nb is a useful microalloying element for steel, inside which it forms Nb carbide and Nb nitride. These compounds enhance the grain refining and retard recrystallization and precipitation hardening. These effects subsequently improve the toughness, strength, formability, and weldability. Within microalloyed stainless steels, the Nb concentration is a low (<0.1%) but crucial addition to high-strength low-alloy steels that are extensively used structurally in modern vehicles. Nb is occasionally used in substantially higher amounts for highly wear-resistant machine components and knives. These same Nb alloys are frequently utilized in pipeline construction. Amounts of Nb in Ni-, Co-, and Fe-based superalloys can be as high as 6.5% for specific applications such as jet engine components, gas turbines, rocket subassemblies, turbocharger systems, heat resisting, and combustion equipment. Nb precipitates a hardening γ''-phase within the grain structure of Ni-, Co-, and Fe-based superalloys. One example superalloy is Inconel 718, consisting of roughly 50% Ni, 18.6% Cr, 18.5% Fe, 5% Nb, 3.1% Mo, 0.9% Ti, and 0.4% Al. These superalloys were used, for example, in advanced air frame systems for the Gemini program. Another Nb alloy was utilized for the nozzle of the Apollo Service Module. Since Nb is oxidized at temperatures above 400°C, a protective coating is necessary for these applications to prevent the alloy from becoming brittle. C-103 alloy was created at the start of the 1960s jointly by the Wah Chang Corporation and Boeing Co., DuPont, Union Carbide Corp., and General Electric Co., while at the same time some other companies were also developing Nb-base alloys, largely driven by the Cold War and Space Race. It consists of 89% Nb, 10% Hf, and 1% Ti and is used for liquid rocket thruster nozzles, for example, the main engine of the Apollo Lunar Modules. The nozzle of the Merlin Vacuum series of engines built by SpaceX for the upper stage of its Falcon 9 rocket consists of a Nb alloy. The reactivity of Nb with oxygen necessitates it to be worked in a vacuum or inert atmosphere, which substantially increases the cost and difficulty of manufacture. Vacuum arc remelting and electron beam melting, novel methods at the time, allowed the development of Nb and other reactive metals. The project that produced C-103 started in 1959 with no less than 256 experimental Nb alloys in the "C-series" (possibly from columbium) that could be melted as buttons and rolled into sheet. Wah Chang had a supply of Hf, refined from nuclear-grade Zr alloys, that it would like to put to a profitable use. The 103rd experimental composition of the C-series alloys, Nb-10Hf-1Ti, had the best mixture of formability and high-temperature properties.

Niobium-germanium (Nb_3Ge), niobium-tin (Nb_3Sn), and the Nb-Ti alloys are utilized as a type II superconductor wire for superconducting magnets, which are used in magnetic resonance imaging and nuclear magnetic resonance (NMR) instruments, as well as in particle accelerators. For instance, the Large Hadron Collider employs 600 tons of superconducting strands, whilst the International Thermonuclear Experimental Reactor utilizes approximately 600 tons of Nb_3Sn strands and 250 tons of Nb-Ti strands. The superconducting radio-frequency (SRF) cavities used in the free-electron lasers FLASH (Free electron LASer in Hamburg, result of the canceled TESLA linear accelerator project) and XFEL (X-Ray Free-Electron Laser) are produced from pure Nb. A cryomodule team at Fermilab used the same SRF technology from the FLASH project to build 1.3 GHz nine-cell SRF cavities made from pure Nb. The cavities will be used in the 30-km linear particle accelerator of the International Linear Collider. Identical technology will be used in LCLS-II (Linac Coherent Light Source II) at SLAC (Stanford Linear Accelerator Center) National Accelerator Laboratory and PIP-II (Proton Improvement Plan-II) at Fermilab. The high sensitivity of superconducting Nb nitride bolometers makes them a perfect detector for electromagnetic radiation in the THz frequency band. These detectors were tested at the Submillimeter Telescope, the South Pole Telescope, the Receiver Lab Telescope, and at APEX (Atacama Pathfinder EXperiment), and are now employed in the HIFI (Heterodyne Instrument for the Far Infrared) instrument on board the Herschel Space Observatory.

Lithium niobate ($LiNbO_3$), which is a ferroelectric (ferroelectricity is a characteristic of certain materials that have a spontaneous electric polarization that can be reversed by the application of an external electric field), is widely used in mobile telephones and optical modulators, and for the manufacture of surface acoustic wave devices. It fits in the group of the ABO_3 structure ferroelectrics like lithium tantalate ($LiTaO_3$) and barium titanate ($BaTiO_3$). Nb capacitors exist as alternative to Ta capacitors, but Ta capacitors still dominate. Nb is added to glass to achieve a higher refractive index, making it possible to create thinner and lighter corrective glasses. Nb and some Nb alloys are physiologically inert and hypoallergenic. Hence, Nb finds use in prosthetics and implant devices, such as pacemakers. Nb treated with NaOH forms a porous layer that facilitates osseointegration. Similar to Ti, Ta, and Al, Nb can be heated and anodized ("reactive metal anodization") to create a wide variety of iridescent colors for jewelry, where its hypoallergenic property is highly desired. Nb is also used as a precious metal in some commemorative coins, often with silver or gold. The arc-tube seals of high-pressure Na-vapor lamps are produced from Nb, occasionally alloyed with 1% Zr; Nb has a comparable thermal expansion coefficient, matching sintered alumina arc-tube ceramic, a translucent material that is resistant to chemical attack or reduction by the hot liquid Na and Na vapor contained inside the operating lamp. Nb is utilized in arc welding rods for certain stabilized grades of stainless steel and in anodes for cathodic protection systems on several types of water tanks, which are then typically plated with Pt. Nb is a critical component of high-performance heterogeneous catalysts for the production of acrylic acid (IUPAC name prop-2-enoic acid, $C_3H_4O_2$) by selective oxidation of propane (C_3H_8). Nb has been used to produce the high voltage wire of the Solar Corona particles receptor module of the Parker Solar Probe (a NASA robotic spacecraft launched in 2018, with the mission of repeatedly probing and making observations of the outer corona of the Sun. It will approach to within 9.86 solar radii (6.9 million km) from the center of the Sun and by 2025 will travel, at closest approach, as fast as 690,000 km/hour, or 0.064% the speed of light).

6.6 42 Mo — Molybdenum

6.6.1 Discovery

Molybdenite—the principal ore from which molybdenum is now separated—was formerly known as molybdena. Molybdena was confused with and often used as if it were graphite (a form of carbon, C). Similar to graphite, molybdenite can be applied to blacken a surface or as a solid lubricant. Even when molybdena was distinguishable from graphite, it was still confused with the common lead ore now known as galena (PbS); the name comes from Ancient Greek Μόλυβδος molybdos, meaning lead. (The Greek word itself has been suggested as a loanword from Anatolian Luvian and Lydian languages.) Even though (supposedly) molybdenum was purposely alloyed with steel in one 14th-century Japanese sword (manufactured around 1330), that art was never widely used and was later lost. In the West in 1754, Swedish chemist and mineralogist Bengt Andersson Qvist (October 21, 1726 to October 14, 1799) analyzed a sample of molybdenite and showed that it did not contain lead and therefore was not galena (Quist, 1754). By 1778, Swedish Pomeranian and German pharmaceutical chemist Carl Wilhelm Scheele (December 9, 1742 to May 21, 1786) stated

FIGURE 6.34 Carl Wilhelm Scheele (1887 Popular Science Monthly Volume 31).

firmly that molybdena was neither galena nor graphite (Fig. 6.34). In its place, Scheele correctly put forward that molybdena was an ore of a separate new element, which he called molybdenum for the mineral in which it was found, and from which it might be separated (Scheele, 1779). Swedish chemist Peter Jacob Hjelm (October 2, 1746 to October 7, 1813) successfully separated molybdenum using carbon and linseed oil in 1781 (Hjelm, 1789). For the next 100 years, molybdenum had no commercial use. It was rather uncommon, the pure metal was difficult to isolate, and the required techniques of metallurgy were not developed yet. Early molybdenum steel alloys exhibited great potential of increased hardness but attempts to produce the alloys on a large scale were hindered by variable results, a tendency toward brittleness, and recrystallization. In 1906, American physicist and engineer William D. Coolidge (October 23, 1873 to February 3, 1975) filed a patent for rendering molybdenum ductile, resulting in applications such as a heating element for high-temperature furnaces and as a support for tungsten-filament light bulbs; oxide formation and degradation necessitate that molybdenum is physically sealed or held in an inert gas. In 1913 Frank E. Elmore established a froth flotation process to recover molybdenite from ores; flotation is still the primary isolation process. During the First World War, demand for molybdenum spiked; it was applied both in armor plating and as a substitute for tungsten in high-speed steels. Some British tanks were protected by 75 mm manganese steel plating, but this was shown to be ineffective. The manganese steel plates were substituted with much lighter 25 mm molybdenum steel plates permitting higher speed, greater maneuverability, and better protection. The Germans likewise used molybdenum-doped steel for heavy artillery, for example, in the superheavy howitzer nicknamed Big Bertha (42 cm kurze Marinekanone 14 L/12 (short naval cannon), or Minenwerfer-Gerät), as traditional steel melts at the temperatures produced by the propellant of the one-ton shell. After the war, demand fell until metallurgical developments permitted wide-ranging development of peacetime applications. In the Second World War molybdenum once more became strategically important as a replacement for tungsten in steel alloys.

6.6.2 Mining, production, and major minerals

Molybdenum is the 54th most abundant element in the Earth's crust. The relative scarcity of molybdenum in the Earth's crust is counterbalanced by its concentration in several water-insoluble ores, frequently in combination with sulfur similar to copper, with which it is often found. Although molybdenum can be found in molybdate minerals such as wulfenite ($PbMoO_4$) and powellite ($CaMoO_4$), the major commercial source is molybdenite (MoS_2). Molybdenum is mined as a principal ore, while also being recovered as a by-product of copper and tungsten mining. The largest producers are China, the United States, Chile, Peru, and Mexico. The total reserves are estimated at 10 million tons and are mostly concentrated in China, the United States, and Chile. By continent, 93% of world molybdenum production is about evenly shared between North America, South America (mainly in Chile), and China. Europe and the rest of Asia (mostly Armenia, Russia, Iran, and Mongolia) produce the remainder. Historically, the Knaben Mine in southern Norway, started in 1885, was the first dedicated molybdenum mine. It was shut down in 1973 but was reopened in 2007, and today produces 100,000 kg of molybdenum disulfide annually. Large mines in Colorado, United States (such as the Henderson Mine and the Climax Mine) and in British Columbia, Canada, produce molybdenite as their primary product, whereas numerous porphyry copper ore deposits such as the Bingham Canyon Mine in Utah, United States, and the Chuquicamata Mine in northern Chile yield molybdenum as a by-product of copper mining.

In molybdenite processing, the first step is for the ore to be roasted in air at a temperature of 700°C. The process produces gaseous sulfur dioxide (SO_2) and the Mo(VI) oxide:

$$2MoS_2 + 7O_2 \rightarrow 2MoO_3 + 4SO_2$$

The oxidized ore is then typically extracted with aqueous ammonia to give ammonium molybdate, $(NH_4)_2(MoO_4)$:

$$MoO_3 + 2NH_3 + H_2O \rightarrow (NH_4)_2(MoO_4)$$

Copper, an impurity often present in molybdenite, is less soluble in ammonia. To completely remove copper from the solution, it is precipitated with hydrogen sulfide (H_2S). Ammonium molybdate converts to ammonium dimolybdate, $(NH_4)_2Mo_2O_7$, which is isolated as a solid. This solid transforms into molybdenum trioxide upon heating:

$$(NH_4)_2Mo_2O_7 \rightarrow 2MoO_3 + 2NH_3 + H_2O$$

The crude molybdenum trioxide can then be further purified through sublimation at 1100°C. Metallic molybdenum is produced by reduction of the oxide with hydrogen:

$$MoO_3 + 3H_2 \rightarrow Mo + 3H_2O$$

The molybdenum for steel production is reduced by the aluminothermic reaction with addition of iron to produce ferromolybdenum. A common form of ferromolybdenum contains 60% molybdenum.

Molybdenum is found in about 50 different minerals. A single phosphide, monipite (MoNiP), is found in the elements class. Six different sulfides are known, including minerals such as hemusite (Cu_6SnMoS_8) and the polymorphs jordisite and molybdenite (MoS_2) (Fig. 6.35) (Box 6.6). Within the halides class two minerals are known with Mo, mosesite ($(Hg_2N)(Cl,SO_4,MoO_4)\ H_2O$) and parkinsonite ($Pb_7MoO_9Cl_2$). The oxide class contains 15 different minerals

FIGURE 6.35 A large, silvery and soft crystal group of molybdenite, MoS_2, 6 cm across, showing the classic nature of lamellar molybdenite in shiny sheets developed as hexagonal crystal fans. Wolfram Camp, Tablelands, Queensland, Australia.

BOX 6.6 Molybdenite (MoS_2).

Perhaps over 95% of the world's supply of molybdenite has been produced from porphyry molybdenum or porphyry copper-molybdenum deposits, even though the molybdenite also occurs in pegmatite and aplite dikes, contact metamorphic zones of limestone, and bedded deposits in sedimentary rocks. In these porphyry deposits chalcopyrite and molybdenite are found as disseminated grains, in stockworks of quartz veins in fractured or brecciated hydrothermally altered granitic intrusive rocks, and in intruded igneous or sedimentary country rocks. One or more major faults passing through or close to the ore bodies are common features and continued fracturing of host intrusives and of enclosing country rocks is a common characteristic. Variations of the name molybdaena and molybdenite were used for lead ores by Dioscorides (CE 50–70), Pliny the Elder (CE 79), and Agricola (1556), but the modern use of molybdenite did not begin until Johan Gottschalk Wallerius wrote about molybdenite in Mineralogia, eller Mineralriket published in 1747. There was still a multiplicity of minerals receiving the same name, but modern molybdenite and graphite were the most common minerals given this name. The element molybdenum was discovered by Carl Wilhelm Scheele in 1778 and he provided molybdenite to Peter Jacob Hjelm who was able to isolate molybdenum in 1781. Scheele showed that molybdenite, in the modern sense, was soluble in acid, while graphite was not. From the Greek μόλυβδος meaning "lead," but a name having a new usage unlike that of former times. Molybdenite has three polymorphic forms: (1) molybdenite-2H (hexagonal), (2) molybdenite-3R (trigonal), and (3) jordisite (amorphous). Generally, when molybdenite is used without suffix, it refers to molybdenite-2H. Molybdenite is hexagonal, *6/m2/m2/m* (space group $P6_3/m2/m2/m$). Its hardness is 3.5–4. Molybdenite has a perfect cleavage on {0001}, laminae flexible but not elastic. The luster is metallic and the transparency is opaque. The color is lead-gray to silver, while the streak is green-gray to grayish black with a bluish tint. Crystals occur in hexagonal plates, thin to thick tabular on {0001}, or short, slightly tapering prisms, commonly foliated, massive, or in scales. Molybdenite-3R is trigonal, *3m* (space group *R3m*). Jordisite was named after Eduard Friedrich Alexander Jordis (1868–1917), colloid chemist. Although generally thought to be amorphous, jordisite gives electron diffraction powder patterns (SAED) indicating at least short range atomic ordering. It has layered structure, and the layer spacing is c. 6 Å as in molybdenite. It has a hardness of 1–2. The luster is dull, earthy to submetallic. It is sectile. The color is blue-gray to black, while the streak is gray-black. It has a massive habit.

FIGURE 6.36 Powellite, $Ca(MoO_4)$, light green long-pointed crystals to 2 mm. Jardinera No. 1 Mine, Inca de Oro, Atacama, Chile.

with Mo in their crystal structure, such as bamfordite ($Fe^{3+}Mo_2O_6(OH)_3 \cdot H_2O$), molybdite ($MoO_3$), and umohoite ($(UO_2)MoO_4 \cdot 2H_2O$). The sulfates, which include the molybdates, are represented by 17 minerals, for example, ferrimolybdite ($Fe_2(MoO_4)_3 \cdot nH_2O$), powellite ($Ca(MoO_4)$) (Fig. 6.36), szenicsite ($Cu_3(MoO_4)(OH)_4$) (Fig. 6.37), and wulfenite ($Pb(MoO_4)$) (Fig. 6.38). Only a few phosphates are known to contain Mo, such as the betpakdalite series (e.g., betpakdalite-CaCa: $[Ca_2(H_2O)_{17}Ca(H_2O)_6][Mo^{6+}_8As^{5+}_2Fe^{3+}_3O_{36}(OH)]$) and schlegelite ($Bi_7(AsO_4)_3(MoO_4)_2O_4$).

FIGURE 6.37 Szenicsite, $Cu_3(MoO_4)(OH)_4$, lustrous green crystals to 2 cm arranged as a rosette. Jardinera No. 1 Mine, Inca de Oro, Atacama, Chile.

FIGURE 6.38 Deep orange crystal of wulfenite, $Pb(MoO_4)$, to 5 mm. Tiger, Pinal Co., Arizona, United States.

6.6.3 Chemistry

Thirty-five isotopes of molybdenum are known, ranging from ^{83}Mo to ^{117}Mo, as well as four metastable nuclear isomers. Seven isotopes occur naturally, ^{92}Mo, ^{94}Mo, ^{95}Mo, ^{96}Mo, ^{97}Mo, ^{98}Mo, and ^{100}Mo. Of these naturally occurring isotopes, only ^{100}Mo is unstable. ^{98}Mo is the most common isotope (24.14% natural abundancy). ^{100}Mo with a half-life of about 10^{19} years undergoes double β decay into ^{100}Ru. Mo isotopes with mass numbers from 111 to 117 all have half-lives of around 150 nanoseconds. All radioisotopes of Mo decay into niobium (Nb), technetium (Tc), and ruthenium (Ru) isotopes. The most common isotopic Mo use involves ^{99}Mo, a fission product that is a parent radioisotope to the short-lived gamma-emitting daughter radioisotope ^{99m}Tc, a nuclear isomer used in various imaging applications in medicine. In 2008 the Delft University of Technology, The Netherlands, applied for a patent on the ^{98}Mo-based production of ^{99}Mo.

Molybdenum is a silvery-white metal but commonly occurs as a dark gray or black powder with a metallic luster. It has 42 electrons arranged in an electronic configuration of $[Kr]4d^5 5s^1$. The oxidation states of Mo in its compounds range from -4 to $+6$ showing its extraordinary chemical versatility. The most stable of these oxidation states are $+4$ and $+6$ (Table 6.6). The rich chemistry of Mo is shown in its ability to: (1) combine with an element (e.g., halogen or

TABLE 6.6 Molybdenum properties.

Appearance	Gray metallic
Standard atomic weight ($A_{r,std}$)	95.95
Block	d-Block
Element category	Transition metal
Electron configuration	[Kr] $4d^5$ $5s^1$
Phase at STP	Solid
Melting point	2623°C
Boiling point	4639°C
Density (near r.t.)	10.28 g/cm^3
When liquid (at m.p.)	9.33 g/cm^3
Heat of fusion	37.48 kJ/mol
Heat of vaporization	598 kJ/mol
Molar heat capacity	24.06 J/(mol·K)
Oxidation states	−4, −2, −1, +1, +2, +3, **+4**, +5, **+6**
Ionization energies	First: 684.3 kJ/mol
	Second: 1560 kJ/mol
	Third: 2618 kJ/mol
Atomic radius	Empirical: 139 pm
Covalent radius	154 ± 5 pm

STP, Standard temperature and pressure.
Bold font indicates main oxidation state.

oxygen) to form a series of compounds with readily interconvertible oxidation states, (2) form complexes with many inorganic and organic ligands including physiologically important compounds, and (3) form direct metal–metal bonds between molybdenum atoms.

6.6.3.1 Oxide, carbide, nitride, and silicide

Notwithstanding the versatility of Mo, it does not react with air or oxygen at room temperature. Mo starts to slowly oxidize at 300°C and oxidizes rapidly at temperatures above 600°C, resulting in molybdenum(VI) oxide, MoO_3. This Mo(VI) oxide, which is volatile at high temperatures, is the precursor to virtually all other Mo compounds as well as alloys. MoO_3 dissolves in strong alkaline water to form molybdates (MoO_4^{2-}). These molybdates then tend to form structurally complex oxyanions by condensation at lower pH values, such as $[Mo_7O_{24}]^{6-}$ and $[Mo_8O_{26}]^{4-}$. Molybdenum is also unreactive toward most acids and alkali such as hydrochloric acid, hydrofluoric acid, ammonia, sodium hydroxide, or dilute sulfuric acid. It does dissolve in hot strong sulfuric or nitric acids releasing SO_2 and NO, respectively. From an industrial perspective, the most important compounds are molybdenum disulfide (MoS_2) and molybdenum trioxide (MoO_3). The black disulfide molybdenite is the main mineral. It is roasted in air to give the trioxide:

$$2MoS_2 + 7O_2 \rightarrow 2MoO_3 + 4SO_2$$

At high temperatures, its reactions with certain gases such as carbon monoxide and nitrogen are slow. It produces a surface film of carbide with CO and nitride film with nitrogen. Molybdenum carbide (MoC and Mo_2C) is an extremely hard refractory ceramic material, commercially used in tool bits for cutting tools. Molybdenum disilicide ($MoSi_2$, or molybdenum silicide), an intermetallic compound, a silicide of molybdenum, is a refractory ceramic with main application in heating elements. It has a moderate density, melting point of 2030°C, and is electrically conductive. At high temperatures it forms a passivation layer of silicon dioxide, shielding it from further oxidation. The thermal stability of $MoSi_2$ together with its high emissivity makes this material, besides WSi_2, attractive for use as a high emissivity coating in heat shields for atmospheric entry. $MoSi_2$ is a gray metallic-looking material with tetragonal crystal structure (α-modification); its β-modification is hexagonal and unstable. It is insoluble in most acids but soluble in nitric acid and hydrofluoric acid. $MoSi_2$ and $MoSi_2$-based materials are typically produced by sintering. Plasma spraying can be utilized for making its dense monolithic and composite forms; material formed by this method may have a percentage of β-$MoSi_2$ as a result of its rapid cooling.

6.6.3.2 Halides

Molybdenum reacts with fluorine to form primarily the hexafluoride (MoF_6). In addition, MoF_4, a green ionic fluid, and MoF_5, a hygroscopic yellow solid, can also be synthesized. Molybdenum(V) fluoride is prepared through the reaction of molybdenum and molybdenum hexafluoride:

$$Mo + 5MoF_6 \rightarrow 6MoF_5$$

At about 120°C, it disproportionates to the tetra- and hexafluoride:

$$2MoF_5 \rightarrow MoF_4 + MoF_6$$

Molybdenum hexafluoride, also molybdenum(VI) fluoride, MoF_6, is a colorless solid, which melts just below room temperature and boils in 34°C. It is one of the 17 known binary hexafluorides.

Molybdenum(II) bromide, $MoBr_2$, is formed by the reaction of elemental molybdenum(II) chloride with lithium bromide. Instead, it can be produced by the disproportionation of molybdenum(III) bromide in vacuum at 600°C. Molybdenum(III) bromide, $MoBr_3$, is a black solid that is insoluble in most solvents but dissolves in donor solvents such as pyridine. $MoBr_3$ is formed by the reaction of elemental molybdenum and bromine at 350°C.

$$2Mo + 3Br_2 \rightarrow 2MoBr_3$$

It adopts a structure comprising infinite chains of face-sharing octahedra with alternatingly short and long Mo–Mo contacts. The same structure is present in the tribromides of ruthenium and technetium. In contrast, in the high-temperature phase of titanium(III) iodide, the Ti–Ti separation is invariant. Molybdenum(IV) bromide, also known as molybdenum tetrabromide, $MoBr_4$, is a black solid. $MoBr_4$ can be produced by reacting molybdenum(V) chloride with hydrogen bromide:

$$2MoCl_5 + 10HBr \rightarrow 2MoBr_4 + 10HCl + Br_2$$

The reaction proceeds via the unstable molybdenum(V) bromide, which releases bromine at room temperature. Molybdenum(IV) bromide can also be synthesized by oxidation of molybdenum(III) bromide with bromine. Molybdenum(III) iodide, MoI_3, is formed by the reaction of molybdenum hexacarbonyl with iodine gas at 105°C.

The reaction with chlorine yields primarily the pentahalide ($MoCl_5$). Molybdenum chlorides reflect the broad range of oxidation states of molybdenum. Chlorides with Mo having oxidation states of +2 to +6 exists. Molybdenum(VI) chloride is known, but it cannot be prepared by addition of chlorine to $MoCl_5$. $MoCl_5$ is a polymeric black solid, which can be prepared by reacting the pentahalide with tetrachloroethene. $MoCl_5$ is produced by chlorination of Mo metal but likewise by chlorination of MoO_3. $MoCl_5$ is a good Lewis acid toward non-oxidizable ligands. It produces an adduct with chloride to form $[MoCl_6]^-$. In organic synthesis, the compound is occasionally used in chlorinations, deoxygenation, and oxidative coupling reactions. $MoCl_5$ is a strong oxidant. It is reduced by MeCN to form an orange complex, $MoCl_4(MeCN)_2$, which in turn reacts with THF to form $MoCl_4(THF)_2$, which is a precursor to other molybdenum-containing complexes. $MoCl_5$ is reduced by HBr to give $MoBr_4$. The bromination reaction is conducted in ethylbromide at −50°C, resulting in $MoBr_5$ as an intermediary. Upon warming to 20°C, Br_2 is produced and the formal oxidation state of molybdenum changes from +5 to +4. The net transformation is shown in the equation:

$$2MoCl_5 + 10HBr \rightarrow 2MoBr_4 + 10HCl + Br_2$$

$MoBr_4$ reacts with THF to give the Mo(III) species mer-$MoBr_3(THF)_3$. $MoCl_5$ is a good Lewis acid toward nonoxidizable ligands. It forms an adduct with chloride to give $[MoCl_6]^-$. In organic synthesis, the compound finds occasional use in chlorinations, deoxygenation, and oxidative coupling reactions.

Molybdenum tetrachloride, $MoCl_4$, exists as two polymorphs, a polymeric ("α") and a hexameric ("β") structures, though neither form is soluble in any solvent without degradation. In each polymorph, the Mo center is octahedral with two terminal chloride ligands and four doubly bridging ligands. It can be produced through dechlorination of molybdenum pentachloride using tetrachloroethene:

$$2MoCl_5 + C_2Cl_4 \rightarrow 2MoCl_4 + C_2Cl_6$$

The acetonitrile adduct, which is a useful intermediary, can be formed directly from the pentachloride:

$$2MoCl_5 + 5CH_3CN \rightarrow 2MoCl_4(CH_3CN)_2 + ClCH_2CN + HCl$$

The MeCN ligands can be exchanged with other ligands, for example:

$$MoCl_4(CH_3CN)_2 + 2THF \rightarrow MoCl_4(THF)_2 + 2CH_3CN$$

Molybdenum(III) chloride, $MoCl_3$, is a dark red solid, which can be prepared by the reduction of molybdenum pentachloride with hydrogen. The greatest yield is, however, produced via the reduction of pure molybdenum(V) chloride, utilizing anhydrous tin(II) chloride as a reducing agent. This reaction has to be performed under a moisture-free and oxygen-free atmosphere of prepurified nitrogen. There are two forms of $MoCl_3$—alpha (α) and beta (β). The α structure is comparable to that of aluminum chloride ($AlCl_3$). In this structure, molybdenum has octahedral coordination geometry and shows cubic close packing in its crystalline structure. The β structure, in contrast, shows hexagonal close packing. The THF complex of molybdenum trichloride is trichlorotris(tetrahydrofuran) molybdenum (III), with the chemical formula $C_{12}H_{24}O_3Cl_3Mo$. It is prepared by stirring $MoCl_4(THF)_2$, THF, and coarse tin powder. It is a light orange solid that is very moisture-sensitive; therefore it must be stored under dry argon in a freezer and in the dark. The complex adopts an octahedral geometry.

Molybdenum(II) chloride has at least two forms, and both have attracted much attention from academic researchers because of the unexpected structures seen for these compounds and the fact that they give rise to hundreds of derivatives. The other molybdenum(II) chloride is potassium octachlorodimolybdate. Mo_6Cl_{12} is prepared by the reaction of molybdenum(V) chloride with molybdenum metal:

$$12MoCl_5 + 18Mo \rightarrow 5Mo_6Cl_{12}$$

This reaction proceeds via the intermediaries $MoCl_3$ and $MoCl_4$, which are also reduced by the presence of excess Mo metal. The reaction is performed in a tube furnace at 600°C−650°C. Once isolated, Mo_6Cl_{12} can undergo numerous reactions while retaining the Mo_6^{12+} core. Heating in concentrated HCl results in $(H_3O)_2[Mo_6Cl_{14}]$. The terminal chloride ligands, labeled "ausser" (German for on the outside), are readily exchanged:

$$(H_3O)_2[Mo_6Cl_{14}] + 6HI \rightarrow (H_3O)_2[Mo_6Cl_8I_6] + 6HCl$$

Under more forcing conditions, all 14 ligands can be exchanged, to form salts of $[Mo_6Br_{14}]^{2-}$ and $[Mo_6I_{14}]^{2-}$. Rather than having a close-packed structure typical of metal dihalides, for example, cadmium chloride, molybdenum(II) chloride has a structure based on clusters. Molybdenum(II), which is a rather large ion, prefers to form compounds with metal−metal bonds, that is, metal clusters. The species Mo_6Cl_{12} is polymeric, comprising cubic $Mo_6Cl_8^{4+}$ clusters interconnected by chloride ligands that bridge from cluster to cluster. This compound transforms easily into salts of the dianion $[Mo_6Cl_{14}]^{2-}$. In this anion, each Mo has one terminal chloride but is otherwise part of an Mo_6 octahedron embedded inside a cube defined by eight chloride centers. Therefore the coordination environment of each Mo is four triply bridging chloride ligands, four Mo neighbors, and one terminal Cl. The cluster has $24e^-$, four being supplied by each Mo^{2+}.

Potassium octachlorodimolybdate (systematic name potassium bis(tetrachloridomolybdate)(Mo−Mo)(4 −)), $K_4Mo_2Cl_8$ (also written as $K_4[Cl_4MoMoCl_4]$), is known as a red-colored, microcrystalline solid. The anion is of historic interest as one of the earliest examples of a quadruple bonding. The salt is typically formed as the pink-colored dihydrate. The compound is produced in two steps from molybdenum hexacarbonyl:

$$2Mo(CO)_6 + 4HO_2CCH_3 \rightarrow Mo_2(O_2CCH_3)_4 + 2H_2 + 12CO$$
$$Mo_2(O_2CCH_3)_4 + 4HCl + 4KCl \rightarrow K_4Mo_2Cl_8 + 4HO_2CCH_3$$

The reaction of the acetate with HCl was first defined as providing trimolybdenum compounds, but later crystallographic analysis proved that the product comprises the $MoCl_8^{4-}$ ion with D_{4h} symmetry, in which the Mo−Mo distance is 2.14 Å.

The elusive hexachloride was made in 2012 by Tamadon and Seppelt and published in 2013. They reacted MoF_6 with boron trichloride at room temperature to get the shiny black hexagonal plates of $MoCl_6$. However, $MoCl_6$ is unstable.

$$2MoCl_5 + C_2Cl_4 \rightarrow 2MoCl_4 + C_2Cl_6$$
$$MoCl_5 + H_2 \rightarrow MoCl_3 + 2HCl$$
$$12MoCl_5 + 18Mo \rightarrow 5Mo_6Cl_{12}$$
$$2MoF_6 + 6BCl_3 \rightarrow MoCl_6 + 6BF_2Cl$$

6.6.3.3 *Organomolybdenum compounds*

The organometallic compounds of Mo are dominated by bonding the ligands lacking β-hydrogen atoms; by dimerizing to form compounds with Mo−Mo bonds; or by complexing with π-bonding ligands such as CO, cyclopentadienyl, and η^6-arenes. M-C σ-bonded compounds are also known but not as dominant as the aforementioned types of compounds. At the time Greenwood and Earnshaw wrote their book on *Chemistry of the Elements* in 1996, little was known about homoleptic Mo-C σ-bonded compounds. In fact, their account only mentioned the existence of the dimer type compound with trimethylsilymethyl (tms) as ligand, $[(tms)_3Mo=Mo(tms)_3]$. The other compounds reported were with π-bonding ligands. Examples of compounds reported in their book were the hexacarbonyl compound $[Mo(CO)_6]$, which reacts with halogens at low temperature to yield carbonyl halides $[Mo(CO)_4X_2]$. From the carbonyl halide, many adducts of the type $[Mo(CO)_3L_2X_2]$ can be produced. The metallocene of Mo is not produced when the hexacarbonyl is reacted with sodium cyclopentadienide but rather form the mixed ligand type $[(\eta^5\text{-}C_5H_5)Mo^I(CO)_3]_2$. This dimerizes to achieve the 18-valence electron configuration by means of an M−M bond. The dibenzene sandwich compound, $[Mo(\eta^6\text{-}C_6H_6)_2]$, is more stable than the monomeric metallocene compounds of Mo. Mo does form complexes with a C_7-ring compound. When the halide of Mo in Na/Hg in THF is reacted with cycloheptatriene, $[Mo(\eta^7\text{-}C_7H_7)(\eta^7\text{-}C_7H_9)]$ is produced and a variety of derivatives have already been obtained.

Since the discussion of Greenwood and Earnshaw (1996), several publications have reported the synthesis and characterization of σ-bonded Mo-C compounds. These are discussed for instance in a book chapter authored by Flower in the book *Comprehensive Organometallic Chemistry III* published in 2007. Some of the compounds featured in the book include compounds $[Mo(CH_3)_5]$, $[Mo(CH_3)_6]$, and $Li[Mo(CH_3)_7]$. The blue, paramagnetic compound $[Mo(CH_3)_5]$ is prepared by reacting $MoCl_5$ with $ZnMe_2$. The orange-brown, oxygen-sensitive $[Mo(CH_3)_6]$ is prepared by treating MoF_6 with $Zn(CH_3)_2$. The hexamethyl compound is only stable up to 10°C. The other σ-bonded Mo-C compounds reported are those containing cyclopentadienyl-supporting ligands, nitrogen-supporting ligands, oxygen-supporting ligands, and alkynyl-containing complexes. Other complexes not containing the carbonyl ligand are also reported and the reader is directed to the book chapter by Flower (2007) for more details.

6.6.4 Major uses

Approximately 86% of all Mo produced is used in metallurgy, with the remainder used in chemical applications. The estimated worldwide use is around 35% structural steel, 25% stainless steel, 14% chemicals, 9% tool and high-speed steels, 6% cast iron, 6% Mo elemental metal, and 5% superalloys. Mo can tolerate extreme temperatures without substantially expanding or softening, making it valuable in environments of intense heat, for example, military armor, aircraft parts, electrical contacts, industrial motors, and filaments. The majority of high-strength steel alloys contain between 0.25% and 8% Mo. Even in these low concentrations, over 43,000 tons of Mo are used annually in stainless steels, tool steels, cast irons, and high-temperature superalloys. Mo is also useful in steel alloys due to its high corrosion resistance and weldability. Mo provides corrosion resistance to type-300 stainless steels (specifically type-316) and in particular in the so-called superaustenitic stainless steels (such as alloy AL-6XN, 254SMO, and 1925hMo) (Austenite, also known as gamma-phase Fe (γ-Fe), is a metallic, nonmagnetic allotrope of Fe or a solid solution of Fe, with an alloying element. In plain-carbon steel, austenite exists above the critical eutectoid temperature of 727°C; other alloys of steel have different eutectoid temperatures. The austenite allotrope is named after English metallurgist Sir William Chandler Roberts-Austen (March 3, 1843 to November 22, 1902); it exists at room temperature in stainless steel. From 912°C to 1394°C α-Fe undergoes a phase transition from body-centered cubic (*bcc*) to the *fcc* configuration of γ-Fe, also called austenite. This is similarly soft and ductile but can dissolve considerably more carbon (as much as 2.03% by mass at 1146°C).). Mo increases lattice strain, consequently increasing the energy necessary to dissolve Fe atoms from the surface. Mo is likewise used to improve the corrosion resistance of ferritic (e.g., grade 444) and martensitic (e.g., 1.4122 and 1.4418) stainless steels (ferritic stainless-steel forms one of the four stainless steel families, the other three being austenitic, martensitic, and duplex stainless steels). Due to its lower density and more stable price, Mo is occasionally used in place of W. An illustration is the "M" series of high-speed steels, for example, M2, M4, and M42, as substitution for the "T" steel series, which have W. Mo can as well be utilized as a flame-resistant coating for other metals. Even though its melting point is 2623°C, Mo quickly oxidizes at temperatures above 760°C, making it more applicable for use in vacuum environments. TZM (Mo (~99%), Ti (~0.5%), Zr (~0.08%), and some C) is a corrosion-resisting Mo superalloy that is resistant to molten fluoride salts at temperatures over 1300°C. It has approximately double the strength of pure Mo and is more ductile and more weldable, yet in tests it was resistant against corrosion of a standard eutectic salt (FLiBe) and salt vapors used in molten salt reactors for 1100 hours with so little corrosion that it was challenging to quantify. Other Mo-based alloys that do not have Fe have only limited uses. For instance, due to its resistance to molten Zn, both pure Mo and Mo-W alloys (70%/30%) are used for the manufacture of piping, stirrers, and pump impellers that come into contact with molten Zn. Mo powder is used as a fertilizer for specific plants, such as cauliflower. Elemental Mo is utilized in NO, NO_2, and NO_x analyzers in power plants for pollution controls. At 350°C, Mo acts as a catalyst for NO_2/NO_x to form NO molecules that can be detected by IR light. Mo anodes substitute for W in some low-voltage X-ray sources for specialized uses such as mammography. The radioactive isotope ^{99}Mo is used to create ^{99m}Tc, used for medical imaging. The isotope is handled and stored as the molybdate.

Molybdenum disulfide (MoS_2) is utilized as a solid lubricant and a high-pressure high-temperature (HPHT) anti-wear agent. It forms strong films on metallic surfaces and is a frequent additive to HPHT greases—when a catastrophic grease failure occurs, a thin layer of Mo inhibits contact of the lubricated parts. In addition, it has semiconducting properties with definite advantages compared to traditional silicon or graphene in electronics applications. MoS_2 is also utilized as a catalyst in hydrocracking of petroleum fractions containing nitrogen, sulfur, and oxygen. Molybdenum disilicide ($MoSi_2$) is an electrically conducting ceramic with primary use in heating elements operating at temperatures above 1500°C in air. Molybdenum trioxide (MoO_3) is applied as an adhesive between enamels and metals. Lead molybdate (wulfenite, $Pb(MoO_4)$) coprecipitated with lead chromate ($PbCrO_4$) and lead sulfate ($PbSO_4$) is a bright-orange pigment utilized with ceramics and plastics. The Mo-based mixed oxides are useful catalysts in the chemical industry, for example, catalysts for the selective oxidation of propylene (IUPAC name propene, C_3H_6) to acrolein (IUPAC name prop-2-enal, $CH_2{=}CHCHO$ or C_3H_4O) and acrylic acid (IUPAC name prop-2-enoic acid, $CH_2{=}CHCOOH$ or $C_3H_4O_2$) and the ammoxidation of propylene to acrylonitrile (IUPAC name prop-2-enenitrile, CH_2CHCN or C_3H_3N). Appropriate catalysts and methods for the direct selective oxidation of propane (C_3H_8) to acrylic acid are being studied. Ammonium heptamolybdate (IUPAC name ammonium docosaoxoheptamolybdate(6−) $(NH_4)_6Mo_7O_{24}$) is utilized for biological staining. Mo-coated soda-lime glass is used in copper indium gallium selenide (CIGS) solar cells, called CIGS solar cells. Phosphomolybdic acid ($H_3PMo_{12}O_{40}$) functions as a stain used in thin-layer chromatography.

6.7 44 Ru — Ruthenium

6.7.1 Discovery

Although naturally found platinum alloys comprising all six platinum-group metals (PGMs) were utilized for an extended period by pre-Columbian Americans and familiar as a material to European chemists starting at the mid-16th century, it was not until the mid-18th century that platinum was recognized as a pure element. That naturally occurring platinum also contained palladium, rhodium, osmium, and iridium was found at the beginning of the 19th century. Platinum discovered in alluvial sands of Russian rivers provided raw material for use in plates and medals and for the minting of rouble coins, beginning in 1828. Leftovers from the platinum production for coinage were obtainable in the Russian Empire, and consequently most of the research on platinum and the metals associated with it was performed in Eastern Europe. It is conceivable that the Polish writer, physician, chemist, and biologist Jędrzej Śniadecki (November 30, 1768 to May 12, 1838) separated element 44 (which he called "vestium" after the asteroid Vesta discovered shortly before) from South American platinum ores in 1807 (Fig. 6.39). He published an account of his find in 1808 (Śniadecki, 1808). Three years later, Académie de France published a note stating that the discovery could not be reproduced.

FIGURE 6.39 Jędrzej Śniadecki (1843).

FIGURE 6.40 Karl Ernst Claus.

Dejected by this, Śniadecki dropped all his claims and did not discuss vestium any longer. Swedish chemist Jöns Jacob Berzelius (August 20, 1779 to August 7, 1848) and German chemist and physicist Gottfried Osann (October 26, 1796 to August 10, 1866) almost discovered ruthenium in 1827. They studied residues that were left after dissolving crude platinum from the Ural Mountains in aqua regia (a mixture of nitric acid and hydrochloric acid, optimally in a molar ratio of 1:3). Berzelius did not observe any unusual metals, but Osann believed he had discovered three new metals, which he named pluranium, ruthenium, and polinium. This inconsistency resulted in a long-lasting argument between Berzelius and Osann about the composition of the residues. Since Osann could not reproduce his separation of ruthenium, he ultimately abandoned his claims (Osann, 1829). The name "ruthenium" was selected by Osann since the analyzed samples originated from the Ural Mountains in Russia (Osann, 1828). The name itself comes from Ruthenia, the Latin word for Rus', a historical area that included present-day Ukraine, Belarus, western Russia, and parts of Slovakia and Poland. In 1844, Karl Ernst Claus (January 23, 1796 to March 24, 1864), a Russian chemist and naturalist of Baltic German descent, proved that the materials isolated by Gottfried Osann had small amounts of ruthenium, which Claus had discovered the same year (Claus, 1845) (Fig. 6.40). Claus separated ruthenium from the platinum residues of rouble production while he was working in Kazan University, Kazan, similar to how its heavier congener osmium had been discovered 40 years earlier. Claus proved that ruthenium oxide contained a new metal and obtained 6 g of ruthenium from the part of crude platinum that was insoluble in aqua regia. Selecting the name for the new element, Claus stated: "I named the new body, in honor of my Motherland, ruthenium. I had every right to call it by this name because Mr. Osann abandoned his ruthenium and the word does not yet exist in chemistry."

6.7.2 Mining, production, and major minerals

As the 74th most abundant element in Earth's crust, ruthenium is comparatively rare, found in about 100 parts per trillion. This element is mostly found in ores with the other PGMs in the Ural Mountains and in North and South America. Small but economically significant amounts are also found in pentlandite extracted from Sudbury, Ontario, Canada, and in pyroxenite deposits in South Africa. Native metallic ruthenium is a very rare mineral (Ir commonly replaces part of Ru in its structure). Around 12 tons of ruthenium are mined annually with world reserves estimated at about 5,000 tons. The mined PGM mixtures show a wide variation in composition, depending on their geochemical formation conditions. For example, the PGMs mined in South Africa have on average 11% ruthenium, whereas the PGMs mined in the former USSR have only 2%. Ruthenium, osmium, and iridium are thought of as the minor PGMs. Ruthenium, like the other PGMs, is commercially produced as a by-product from nickel, and copper, and platinum metals ore processing. In the process of electrorefining copper and nickel, noble metals such as silver, gold, and the PGMs form a precipitate as

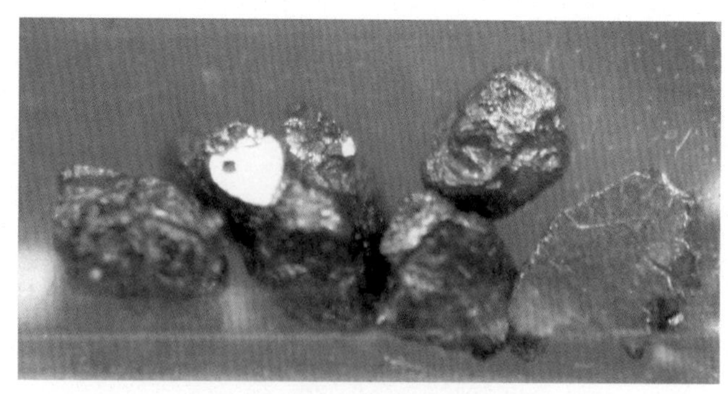

FIGURE 6.41 Rutheniridosmine, (Ir,Os,Ru), five 1 mm crude platy crystals. Nizhne-Tagilskii massif, Urals, Russia.

FIGURE 6.42 Laurite, RuS_2, dozens of tiny placer nuggets less than 1 mm. Tanah Laut, Kalimantan Selatan, Borneo Island, Indonesia.

anode mud, which serves as the feedstock for the extraction. The metals are converted to ionized solutes by any of a number of different methods, depending on the feedstock composition. One typical process starts with fusion with sodium peroxide (Na_2O_2) followed by dissolution in aqua regia (a mixture of nitric acid and hydrochloric acid, optimally in a molar ratio of 1:3), and solution in a mixture of chlorine with hydrochloric acid (HCl). Osmium, ruthenium, rhodium, and iridium are insoluble in aqua regia and readily precipitate, leaving the other metals in solution. Rhodium is removed from the residue by treatment with molten sodium bisulfate ($NaHSO_4$). The remaining insoluble residue, now containing Ru, Os, and Ir is treated with sodium oxide (Na_2O), in which Ir is insoluble, producing a solution containing only the dissolved Ru and Os salts. After oxidation to the volatile oxides, RuO_4 is separated from OsO_4 by precipitation of $(NH_4)_3RuCl_6$ with ammonium chloride (NH_4Cl) or by distillation or extraction with organic solvents of the volatile osmium tetroxide (OsO_4). Hydrogen is subsequently used to reduce the ammonium ruthenium chloride producing a powder. The product is reduced using hydrogen, providing the metal as a powder or sponge metal that can then be further treated with powder metallurgy techniques or argon-arc welding.

Ruthenium is only found in a total of 12 minerals. Four of these minerals are alloys, hexaferrum ((Fe,Os,Ru,Ir)), hexamolybdenum ((Mo,Ru,Fe,Ir,Os)), rutheniridosmine ((Ir,Os,Ru)) (Fig. 6.41), and ruthenium ((Ru,Ir)). Within the sulfides class one finds eight minerals, anduoite ($(Ru,Os)As_2$), irarsite ((Ir,Ru,Rh,Pt)AsS), iridarsenite ($(Ir,Ru)As_2$), laurite (RuS_2) (Fig. 6.42), omeiite ($(Os,Ru)As_2$), osarsite ((Os,Ru)AsS), ruarsite ((Ru,Os)AsS), and ruthenarsenite ((Ru,Ni)As).

6.7.3 Chemistry

Naturally occurring Ru consists of seven stable isotopes, ^{96}Ru (5.54% natural abundance), ^{98}Ru (1.87%), ^{99}Ru (12.76%), ^{100}Ru (12.60%), ^{101}Ru (17.06%), ^{102}Ru (31.55%), and ^{104}Ru (18.62%). Furthermore, 34 radioisotopes have been observed. Of these radioisotopes, the most stable are ^{106}Ru with a half-life of 373.59 days, ^{103}Ru with a half-life of 39.26 days, and ^{97}Ru with a half-life of 2.9 days. Fifteen other radioisotopes have been observed ranging from ^{90}Ru to ^{115}Ru. The majority of these radioisotopes have half-lives that are less than 5 minutes, except ^{95}Ru which has a half-

TABLE 6.7 Ruthenium properties.

Appearance	Silvery-white metallic
Standard atomic weight ($A_{r,std}$)	101.07
Block	d-Block
Element category	Transition metal
Electron configuration	[Kr] $4d^7$ $5s^1$
Phase at STP	Solid
Melting point	2334°C
Boiling point	4150°C
Density (near r.t.)	12.45 g/cm^3
When liquid (at m.p.)	10.65 g/cm^3
Heat of fusion	38.59 kJ/mol
Heat of vaporization	619 kJ/mol
Molar heat capacity	24.06 J/(mol·K)
Oxidation states	−4, −2, +1, +2, **+3**, **+4**, +5, +6, +7, +8
Ionization energies	First: 710.2 kJ/mol
	Second: 1620 kJ/mol
	Third: 2747 kJ/mol
Atomic radius	Empirical: 134 pm
Covalent radius	146 ± 7 pm

STP, Standard temperature and pressure.
Bold font indicates main oxidation state.

life of 1.643 hours and ^{105}Ru which has a half-life of 4.44 hours. The principal decay mode for isotopes lighter than the most abundant isotope, ^{102}Ru, is electron capture and the principal mode for the heavier isotopes is by β emission. The main decay product for the isotopes lighter than ^{102}Ru is technetium (Tc) and the main decay product for the heavier isotopes is rhodium (Rh).

Ruthenium is a silvery-white lustrous metal belonging to the platinum group of the periodic table. It has 44 electrons arranged in an electronic configuration of $[Kr]4d^75s^1$. All eight electrons in the 5s subshell and 4d subshell can participate in chemical reactions and thus ruthenium can have an oxidation state of +8, the highest for any element. The full range of oxidation states of ruthenium is from 0 to +8 with +2, +3, and +4 states being the most common. Compounds in which Ru has a −2 oxidation state also exist (Table 6.7). Ruthenium is unreactive to acids but will be oxidized by potassium hydroxide in the presence of potassium nitrate to form potassium ruthenate, K_2RuO_4. It is also readily attacked by solutions of alkali metal hypochlorites, alkali melts, and peroxides.

6.7.3.1 Oxides, chalcogenide, hydride, nitride, and boride

Ruthenium does not tarnish as it tends to form a protective coating of the dioxide. It reacts with oxygen only when subjected to temperatures above 600°C. The oxidation of ruthenium results in ruthenium(IV) oxide (RuO_2). This in turn can be oxidized by sodium metaperiodate to the volatile yellow tetrahedral ruthenium tetroxide, RuO_4, which is an aggressive and strong oxidizing agent. RuO_4 can oxidize dilute hydrochloric acid and organic solvents such as ethanol at room temperature and is easily reduced to ruthenate (RuO_4^{2-}) in aqueous alkaline solutions. Ruthenium(IV) oxide, RuO_2, is a black solid and the most common oxide of ruthenium. It is extensively employed as an electrocatalyst for producing chlorine, chlorine oxides, and O_2. Similar to various other dioxides, RuO_2 has the rutile (TiO_2) structure. It is typically prepared by oxidation of ruthenium trichloride. Approximately stoichiometric single crystals of RuO_2 can be formed by CVT, using O_2 as the transport agent:

$$RuO_2 + O_2 \rightleftharpoons RuO_4$$

RuO_2 films can be produced through CVD from volatile ruthenium compounds. RuO_2 can also be formed by electroplating from a solution of ruthenium trichloride. Electrostatically stabilized hydrosols of pure ruthenium dioxide hydrate have been synthesized by using the autocatalytic reduction of ruthenium tetroxide in aqueous solution. The formed particle populations may be controlled to consist of considerably monodisperse, uniform spheres with diameters in the range between 40 and 160 nm. Ruthenium tetroxide, RuO_4, is a yellow volatile solid that melts near room temperature. Samples are usually black because of impurities. One of the few solvents in which RuO_4 forms stable solutions is CCl_4. RuO_4 is produced through oxidation of ruthenium(III) chloride with $NaIO_4$.

$$8Ru^{3+}(aq) + 5IO_4^-(aq) + 12H_2O(l) \rightarrow 8RuO_4(s) + 5I^-(aq) + 24H^+(aq)$$

Because of the cost, toxicity, and high reactivity of RuO_4, it is frequently produced in situ and employed in catalytic amounts in organic reactions, by using a substoichiometric amount of ruthenium(III) or -(IV) precursor and a stoichiometric amount of sodium metaperiodate as the terminal oxidant to uninterruptedly regenerate small quantities of RuO_4. In typical reactions using RuO_4 as the oxidant, numerous forms of ruthenium conveniently act as precursors to RuO_4, most generally used are $RuCl_3{\cdot}xH_2O$ or $RuO_2{\cdot}xH_2O$. The molecule has a tetrahedral geometry, with the Ru–O distances ranging between 169 and 170 pm.

Ruthenium also forms dichalcogenides, in which the most common is ruthenium sulfide (RuS_2), the form of the mineral laurite. Like the other metals of the platinum group, ruthenium is unreactive, inert to most other chemicals, and insoluble in all acids, even aqua regia. Ruthenium does not react with hydrogen. The hydride RuH has been produced but only at hydrogen pressures of 14 GPa in a diamond anvil cell. The nitride, on the other hand, had been deposited as films on silicon substrate using a radio frequency–sputtering technique. Ruthenium borides' most remarkable property is potentially high hardness. A Vickers hardness HV = 50 GPa has been published for thin films consisting of RuB_2 and Ru_2B_3 phases. This value is substantially higher than those of bulk RuB_2 or Ru_2B_3, but it has yet to be corroborated independently, as measurements on superhard materials are inherently challenging. Ruthenium diboride was first believed to consist of a hexagonal structure, similar to rhenium diboride, but it was later tentatively determined to have an orthorhombic structure instead.

6.7.3.2 Halides

Ruthenium is attacked by halogens at elevated temperatures to form the whole range of trihalides. The fluoride and iodide are, however, often prepared using other routes. The dark brown powder of ruthenium(III) fluoride, RuF_3, can be made by iodine reduction of RuF_5. The black solid of ruthenium(III) iodide, meanwhile, can be prepared by reacting RuO_4 with hydriodic acid or by thermal decomposition of $Ru(NH_3)_5I_3$.

$$2Ru + 3X_2 \rightarrow 2RuX_3 \quad (X = Cl, Br)$$

Fluorine reacts to form the fluoride with Ru in +5 oxidation state, RuF_5. An extremely moisture-sensitive brown solid of ruthenium(VI) fluoride, RuF_6, can be made by fluorination of RuF_5 under forcing conditions. It can also be prepared by a direct reaction of ruthenium metal in a gas stream of fluorine and argon at 400°C–450°C. The yield of this reaction is low though and usually less than 10%.

$$Ru + 3F_2 \rightarrow RuF_6$$

Deep pink polycrystalline solid of ruthenium (IV) fluoride can be prepared by the reaction of arsenic pentafluoride, AsF_5 and $RuFe_6^{2-}$. Ruthenium(II) halides, except the difluoride, are also known.

$$2RuF_5 + F_2 \xrightarrow{50\ \text{atm},\ 230^\circ\text{C}} 2RuF_6$$

RuF_6 is a dark brown crystalline solid that melts at 54°C. The solid structure measured at −140°C is orthorhombic space group *Pnma*; lattice parameters are $a = 9.313$ Å, $b = 8.484$ Å, and $c = 4.910$ Å. There are four formula units (here, discrete molecules) per unit cell, resulting in a density of 3.68 g/cm^3. The RuF_6 molecule itself (the form important for the liquid or gas phase) adopts an octahedral molecular geometry, which has point group O_h. The Ru–F bond length is 1.818 Å. Ruthenium pentafluoride, RuF_5, is a green volatile solid that has rarely been investigated but is of interest as a binary fluoride of ruthenium. It is sensitive toward hydrolysis. Its structure comprises Ru_4F_{20} tetramers, similar to the isostructural platinum pentafluoride. Within the tetramers, each Ru has an octahedral molecular geometry, with two bridging fluoride ligands.

Ruthenium(III) chloride, $RuCl_3$, more commonly refers to the hydrate $RuCl_3{\cdot}xH_2O$. Both the anhydrous and hydrated compounds are dark brown or black solids. The hydrate, with a variable amount of water of crystallization, regularly approximating to a trihydrate, is frequently used as a starting compound in ruthenium chemistry. Anhydrous ruthenium(III) chloride is typically produced by heating powdered ruthenium metal with chlorine. In the initial synthesis, the chlorination was performed in the presence of carbon monoxide, the product being carried by the gas stream and crystallizing upon cooling. Two allotropes of $RuCl_3$ are known. The black α-form has the $CrCl_3$-type structure with long Ru–Ru contacts of 346 pm. This allotrope has honeycomb layers of Ru^{3+}, which are enveloped with an octahedral cage of Cl^- anions. The Ru cations are magnetic residing in a low-spin $J \sim 1/2$ ground state with net angular momentum L = 1. Layers of α-$RuCl_3$ are bonded to each other with weak van der Waals forces. These can be cleaved to form monolayers using scotch tape. The dark brown metastable β-form crystallizes in a hexagonal cell; this form comprises infinite chains of face-sharing octahedra with Ru–Ru contacts of 283 pm, comparable to the structure of zirconium trichloride. The β-form is irreversibly converted to the α-form at 450°C–600°C. The β-form is diamagnetic, while α-$RuCl_3$ is paramagnetic at room temperature. $RuCl_3$ vapor decomposes into the elements at high temperatures. As the most commonly available ruthenium compound, $RuCl_3 \cdot xH_2O$ is the precursor to many hundreds of chemical compounds. The important property of ruthenium complexes, chlorides and others, is the existence of more than one oxidation state, some of which are kinetically inert. All second and third-row transition metals form exclusively low-spin complexes, while ruthenium is special in the stability of adjacent oxidation states, in particular Ru(II), Ru(III) (as in the parent $RuCl_3 \cdot xH_2O$), and Ru(IV).

6.7.3.3 *Organoruthenium compounds*

The chemistry of organoruthenium compounds did not develop until the discovery of ferrocene in 1951. Ferrocene consists of an iron ion sandwiched in between two cyclopentadienyl rings. After its discovery, similar compounds were developed and ruthenocene was reported in 1952. Ferrocene and ruthenocene exhibit similar chemical properties but differ in their redox properties. Ruthenocene is more stable toward oxidation compared with ferrocene. Sandwich structures with arenes are also common. The bis(arene) Ru complex show interesting chemistry through its ability to undergo two-electron redox change that converts one arene from a η^6- into a bent η^4-ligand. In addition to the many variations in ruthenocene, bis(arene), and the (arene)(cyclopentadienyl) complexes, Ru also forms complexes with carbonyls (CO) and phosphines especially the triphenylphosphines (PPh_3). A phosphine complex of ruthenium, $RuCl_2(PPh_3)_3$, has been used as starting material for the synthesis of another important class of organoruthenium complexes now called Grubbs first-generation catalysts. These catalysts are useful in olefin metathesis and prepared by one-pot synthesis using $RuCl_2(PPh_3)_3$, phenyldiazomethane, and tricyclohexylphosphine. This first-generation catalyst is an important precursor for the succeeding catalysts that were developed.

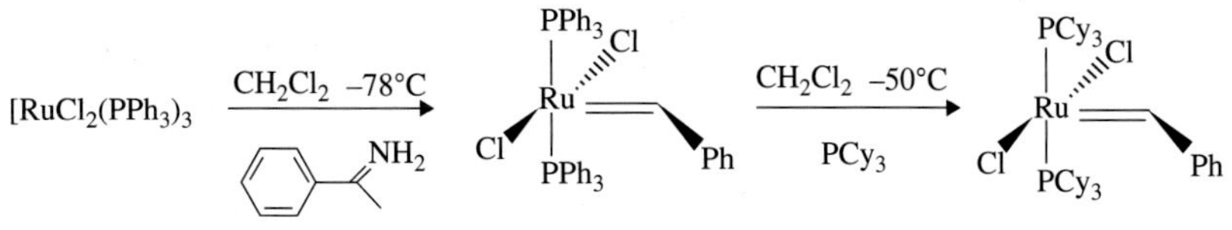

6.7.4 Major uses

Since it hardens Pt and Pd alloys, Ru is utilized in electrical contacts, where a thin film is adequate to realize the required durability. With comparable properties and cheaper than Rh, electric contacts form a key use of Ru. The plate is added to the base via electroplating or sputtering. RuO_2 together with Pb and Bi ruthenates are utilized in thick-film chip resistors. These two electronic uses cover about 50% of the total Ru consumption. Ru is rarely alloyed with metals other than from the Pt group, where small amounts enhance some properties. The increased corrosion resistance in Ti alloys resulted in the development of a special alloy containing 0.1% Ru. In addition, Ru is used in several advanced high-temperature single-crystal superalloys, with uses in, for example, the turbines in jet engines. Some Ni-based superalloy compositions have been reported, such as EPM-102 (with 3% Ru), TMS-162 (with 6% Ru), TMS-138, and TMS-174, the latter two having 6% Re. Fountain pen nibs are often tipped with Ru alloy. Since 1944 the famous Parker 51 fountain pen has been fitted with the "RU" nib, a 14K gold nib tipped with 96.2% Ru and 3.8% Ir. Ru is an element of mixed-metal oxide anodes utilized for cathodic protection of underground and submerged structures, and for electrolytic cells for methods such as generating chlorine from saltwater. The fluorescence of some Ru complexes is quenched by oxygen, finding use in optode sensors for oxygen. Ruthenium red, $[(NH_3)_5Ru\text{-}O\text{-}Ru(NH_3)_4\text{-}O\text{-}Ru(NH_3)_5]^{6+}$, is a biological stain used to stain polyanionic molecules such as pectin and nucleic acids for light microscopy and electron microscopy. The β-decaying isotope ^{106}Ru is used in radiotherapy of eye tumors, mainly malignant melanomas of the uvea (the middle layer of the eye. It lies beneath the white part of the eye). Ru-centered complexes are being researched for possible anticancer properties. Compared with Pt complexes, those of Ru show greater resistance to hydrolysis and more selective action on tumors. Ru tetroxide exposes latent fingerprints by reacting on contact with fatty oils or fats with sebaceous contaminants and producing brown/black Ru dioxide pigment. Some Ru complexes absorb light throughout the visible spectrum and are being actively researched for solar energy technologies. For example, Ru-based compounds have been used for light absorption in dye-sensitized solar cells, a promising new low-cost solar cell system. Many Ru-based oxides show very unusual properties, such as a quantum critical point behavior, exotic superconductivity (in its Sr ruthenate form), and high-temperature ferromagnetism.

Many Ru-containing compounds exhibit useful catalytic properties. The catalysts are conveniently divided into those that are soluble in the reaction medium, homogeneous catalysts, and those that are not, which are called heterogeneous catalysts. Ru nanoparticles can be formed inside halloysite. This abundant naturally occurring clay mineral has a structure of rolled nanosheets (nanotubes), which can support both the Ru nanocluster synthesis and its products for later use in industrial catalysis. Ru-promoted Co catalysts are utilized in Fischer–Tropsch synthesis. Solutions containing Ru trichloride are highly active for olefin metathesis [an organic reaction that entails the redistribution of fragments of alkenes (olefins) by the scission and regeneration of C=C double bonds]. Such catalysts are used commercially to produce, for example, polynorbornene (norbornene IUPAC name bicyclo[2.2.1]hept-2-ene, C_7H_{10}). Well-defined Ru carbene and alkylidene complexes exhibit similar reactivity and offer a mechanistic understanding of these industrial processes. For instance, the Grubbs' catalysts have been used in the manufacture of drugs and advanced materials. Grubbs catalysts are a series of transition-metal carbene complexes used as catalysts for olefin metathesis. They are named after American chemist Robert H. Grubbs (born February 27, 1942) who supervised their synthesis. Several generations of the catalyst have been developed. Grubbs catalysts tolerate many functional groups in the alkene substrates, are air-tolerant, and are compatible with a wide range of solvents. Consequently, Grubbs catalysts have become widespread in synthetic organic chemistry. Grubbs, together with American chemist Richard R. Schrock (born January 4, 1945) and French chemist Yves Chauvin (October 10, 1930 to January 27, 2015), won the 2005 Nobel Prize in Chemistry in recognition of their contributions to the development of olefin metathesis. Ru complexes are very active catalysts for transfer hydrogenations (at times described as "borrowing hydrogen" reactions). This process is employed for the enantioselective hydrogenation of ketones, aldehydes, and imines [Asymmetric hydrogenation is a chemical reaction that adds two atoms of hydrogen preferentially to one of two faces of an unsaturated substrate molecule, such as an alkene or ketone. The selectivity derives from the manner that the substrate binds to the chiral catalysts. In jargon, this binding transmits spatial information (what chemists refer to as chirality) from the catalyst to the target, favoring the product as a single enantiomer. This enzyme-like selectivity is particularly applied to bioactive products such as pharmaceutical agents and agrochemicals.]. This reaction exploits using chiral Ru complexes introduced by Japanese chemist Ryoji Noyori (born September 3, 1938). For example, (cymene)Ru(S,S-TsDPEN) catalyzes the hydrogenation of benzil (IUPAC name 1,2-diphenylethane-1,2-dione, $(C_6H_5CO)_2$) into (*R,R*)-hydrobenzoin (IUPAC name (1*R*,2*R*)-1,2-diphenylethane-1,2-diol, $C_{14}H_{14}O_2$). In this reaction, formate and water/alcohol serve as the source of H_2: Ryōji Noyori received the 2001 Nobel Prize in Chemistry for contributions to the field of asymmetric hydrogenation. In 2012, Japanese chemist Masaaki Kitano and colleagues, using an organic Ru catalyst, showed the possibility of ammonia synthesis using a stable electride as an electron donor and reversible hydrogen store. Small-scale, intermittent production of ammonia, for local agricultural use, may be a viable substitute for electrical grid attachment as a sink for power generated by wind turbines in isolated rural installations.

6.8 45 Rh — Rhodium

6.8.1 Discovery

Rhodium [Greek rhodon (ῥόδον) meaning "rose"] was found in 1803 by English chemist and physicist William Hyde Wollaston (August 6, 1766 to December 22, 1828), shortly after his discovery of palladium (Wollaston, 1804) (Fig. 6.43). He used crude platinum ore probably from South America. His procedure involved dissolving the ore in aqua regia (a mixture of nitric acid and hydrochloric acid, optimally in a molar ratio of 1:3) and neutralizing the acid with sodium hydroxide (NaOH). He subsequently precipitated the platinum as ammonium chloroplatinate through reaction with ammonium chloride (NH_4Cl). Most other metals such as Cu, Pb, Pd, and Rh were precipitated with Zn. Diluted nitric acid dissolved all metals except palladium and rhodium. Of these two metals, palladium dissolved in aqua regia but rhodium did not, and the rhodium was precipitated through the reaction with sodium chloride as $Na_3[RhCl_6]\cdot nH_2O$. After washing with ethanol, the rose-red precipitate was reacted with zinc, which substituted for the rhodium in the ionic compound and in that way released the rhodium as free metal. After the discovery, the rare element had only insignificant applications; for example, by the turn of the century, Rh-containing thermocouples were used to measure temperatures up to 1800°C. The first major application was electroplating for decorative uses and as corrosion-resistant coating. The commercial introduction of the three-way catalytic converter by Volvo in 1976 resulted in a strong increase in the demand for rhodium metal.

FIGURE 6.43 William Hyde Wollaston, painting by John Jackson (1778–1831).

The earlier catalytic converters utilized platinum or palladium, while the three-way catalytic converter used rhodium to reduce the amount of NO_x in the car exhaust gases.

6.8.2 Mining, production, and major minerals

Rhodium is one of the rarest elements in the Earth's crust, comprising an estimated 0.0002 ppm. Its rarity affects its price and its use in commercial applications. The industrial production of rhodium is complex because the ores are mixed with other metals such as Pd, Ag, Pt, and Au, and only a few rhodium-bearing minerals exist (see Section 6.7.2 on ruthenium production for more details). It is found in platinum ores and isolated as a white inert metal that is difficult to fuse. Principal deposits are found in South Africa; in river sands of the Ural Mountains; and in North America, including the copper-nickel sulfide mining area of the Sudbury, Ontario, region. Even though the amount at Sudbury is very small, the large volume of processed nickel ore makes rhodium recovery cost-effective. The main exporter of rhodium is South Africa followed by Russia. The world production is about 30 tons per year.

Rhodium is found in only 17 different minerals. Within the elements class only a single mineral, rhodium ((Rh,Pt)), is found. The remaining 16 minerals are all sulfides and arsenides. These include bowieite ($(Rh,Ir,Pt)_2S_3$) (Fig. 6.44), cherepanovite (RhAs), cuprorhodsite ($(Cu,Fe)Rh_2S_4$), ferhodsite ($(Fe,Rh,Ni,Ir,Cu,Pt)_9S_8$), hollingworthite ((Rh,Pt,Pd)AsS), irarsite ((Ir,Ru,Rh,Pt)AsS), kashinite ($(Ir,Rh)_2S_3$), kingstonite ($(Rh,Ir,Pt)_3S_4$), konderite ($Cu_3Pb(Rh,Pt,Ir)_8S_{16}$), miassite ($Rh_{17}S_{15}$), oberthürite ($Rh_3Ni_{32}S_{32}$), palladodymite ($(Pd,Rh)_2As$), polkanovite ($Rh_{12}As_7$), rhodarsenide ($(Rh,Pd)_2As$), rhodplumsite ($Rh_3Pb_2S_2$), and zaccariniite (RhNiAs).

6.8.3 Chemistry

Naturally occurring Rh consist of only a single isotope, ^{103}Rh. The most stable synthetic radioisotopes are ^{101}Rh with a half-life of 3.3 years, ^{102}Rh with a half-life of 207 days, ^{102m}Rh with a half-life of 2.9 years, and ^{99}Rh with a half-life of 16.1 days. Twenty other radioisotopes have been synthesized ranging from ^{93}Rh to ^{117}Rh. The majority of these radioisotopes have half-lives shorter than an hour, except for ^{100}Rh which has a half-life of 20.8 hours and ^{105}Rh which has a half-life of 35.36 hours. Rh has a significant number of metastates, the most stable being ^{102m}Rh with a half-life of about 2.9 years and ^{101m}Rh with a half-life of 4.34 days. In isotopes lighter than the stable isotope ^{103}Rh, the principal decay mode is electron capture and the main decay product is Ru. In isotopes heavier than ^{103}Rh, the principal decay mode is beta emission and the primary product is palladium.

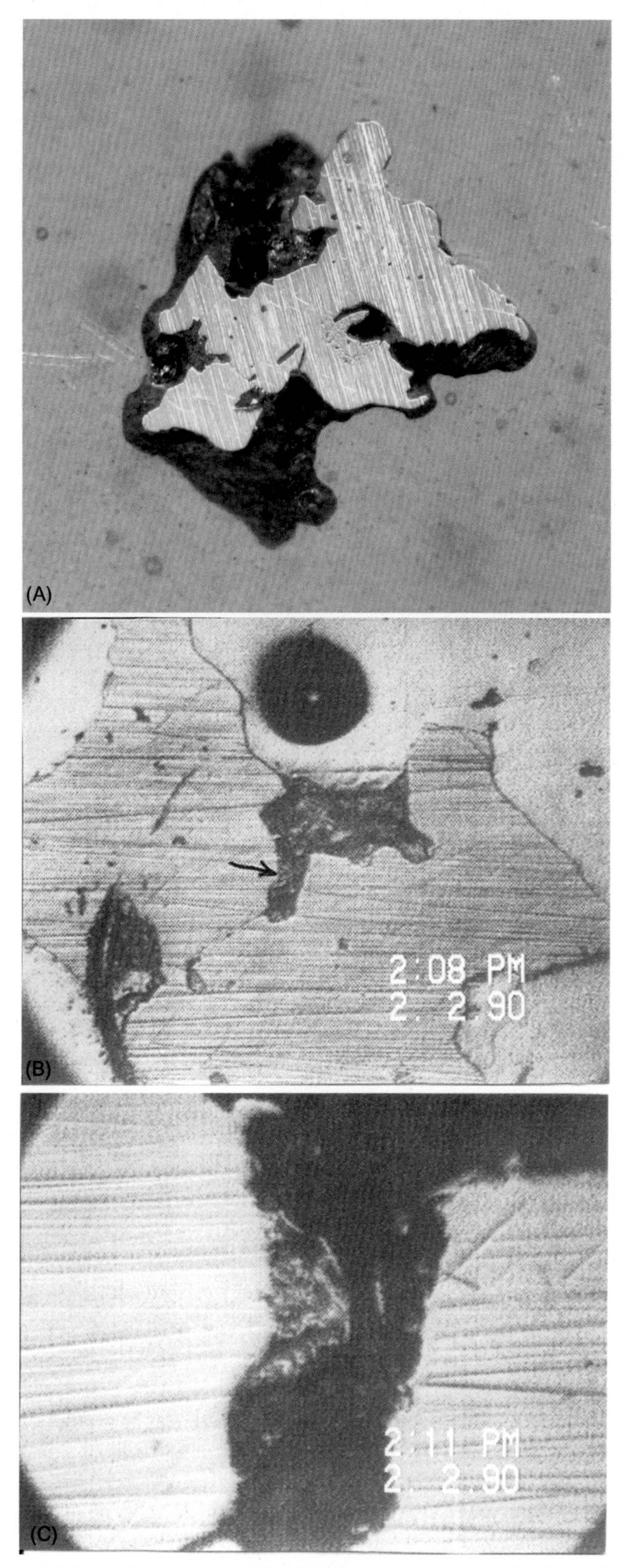

FIGURE 6.44 (A) Bowieite, $(Rh,Ir,Pt)_2S_3$, in platinum, Pt, polished microprobe section (top left). (B and C) bowieite in platinum, microprobe images (middle and bottom). Goodnews Bay, Alaska, United States.

TABLE 6.8 Rhodium properties.

Appearance	Silvery-white metallic
Standard atomic weight ($A_{r,std}$)	102.905
Block	d-Block
Element category	Transition metal
Electron configuration	[Kr] $4d^8$ $5s^1$
Phase at STP	Solid
Melting point	1964°C
Boiling point	3695°C
Density (near r.t.)	12.41 g/cm^3
When liquid (at m.p.)	10.7 g/cm^3
Heat of fusion	26.59 kJ/mol
Heat of vaporization	493 kJ/mol
Molar heat capacity	24.98 J/(mol·K)
Oxidation states	−3, −1, +1, +2, **+3**, +4, +5, +6
Ionization energies	First: 719.7 kJ/mol
	Second: 1740 kJ/mol
	Third: 2997 kJ/mol
Atomic radius	Empirical: 134 pm
Covalent radius	142 ± 7 pm

STP, Standard temperature and pressure.
Bold font indicates main oxidation state.

Rhodium is a rare, silvery-white, hard transition metal. It has 45 electrons arranged in an electronic configuration of $[Kr]4d^85s^1$. Anomalies in electronic configuration like this are also observed in the neighboring elements, niobium, molybdenum, ruthenium, and several others in the transition elements. Rh can take oxidation states from 0 to +6 but the most common is +3 (Table 6.8).

6.8.3.1 Oxides

Rhodium is corrosion-resistant and chemically inert, unreactive to air or oxygen, water, and acids at room temperature. It does form nonvolatile oxides. Rhodium(III) oxide, Rh_2O_3, is normally prepared from the reaction of ruthenium(III) chloride, $RuCl_3$, at high temperatures. Alternative methods include: Rh metal powder is fused with potassium hydrogen sulfate. The addition of sodium hydroxide forms the hydrated rhodium oxide, which upon heating converts to Rh_2O_3; rhodium oxide thin films can be prepared by exposing a Rh layer to oxygen plasma; nanoparticles can be prepared through hydrothermal synthesis. Rhodium oxide films behave as a fast two-color electrochromic system: Reversible yellow ↔ dark green or yellow ↔ brown-purple color changes are obtained in KOH solutions by applying a voltage of ~1 V. Rhodium oxide films are transparent and conductive, similar to indium tin oxide (ITO)—the common transparent electrode, but Rh_2O_3 has 0.2 eV lower work function than ITO. Therefore deposition of rhodium oxide on ITO enhances the carrier injection from ITO, thus enhancing the electrical properties of organic light-emitting diodes (LEDs). It converts to rhodium(IV) oxide, RhO_2 when heated above 750°C. RhO_2 is highly insoluble even in hot aqua regia. RhO_2 adopts the tetragonal rutile (TiO_2) structure. RhO_2 has metallic resistivity with values $<10^{-4}$ Ω·cm. It converts in air to Rh_2O_3 at 850°C and subsequently to the metal and oxygen at 1050°C. Rhodium(V) oxides stabilized by alkali and alkaline earth metals such as Sr_3LiRhO_6 and Sr_3NaRhO_6 are also known. These are prepared by reacting RhO_2 with the hydroxides of Sr, Li, or Na at 600°C.

6.8.3.2 Halides

Under anhydrous dry conditions, Rh can react with the halogens to form ruthenium(III) halides, RuX_3 (X = F, Cl, Br, I). In excess of elemental fluorine and at elevated temperatures, a highly corrosive rhodium hexafluoride, RhF_6, is formed.

$$Rh + 3F_2 \rightarrow RhF_6$$

The RhF_6 molecule adopts an octahedral molecular geometry. Consistent with its d^3 configuration, the six Rh−F bond lengths are equivalent, at 1.824 Å. It crystallizes in an orthorhombic space group *Pnma* with lattice parameters of $a = 9.323$ Å, $b = 8.474$ Å, and $c = 4.910$ Å. Similar to some other metal fluorides, RhF_6 is highly oxidizing. It attacks glass even in the absence of water. It can even react with elemental oxygen. Rhodium pentafluoride, RhF_5, is prepared by fluorination of rhodium trifluoride at 400°C. Rhodium(III) chloride, $RhCl_3(H_2O)_n$ (where n varies from 0 to 3), is diamagnetic solid featuring octahedral Rh(III) centers. Depending on the value of n, the material is either a dense brown solid or a soluble reddish salt. Anhydrous rhodium(III) chloride, $RhCl_3$, is prepared by reaction of chlorine with rhodium sponge metal at 200°C–300°C. The hydrated form of the trichloride, such as $RhCl_3(H_2O)_3$, is produced from salts such as Na_3RhCl_6. Aqueous solutions of $RhCl_3(H_2O)_3$ have been studied by ^{103}Rh NMR spectroscopy indicating the presence of several species. The percentages of these species vary with time and depend on the concentration of chloride. The relative distribution of these species affects the color of the solutions, which can range from yellow (the hexaaquo ion) to "raspberry-red." Some of these species include $[Rh(H_2O)_6]^{3+}$, $[RhCl(H_2O)_5]^{2+}$, *cis*- and *trans*-$[RhCl_2(H_2O)_4]^+$, and $[RhCl_3(H_2O)_3]$. Individual ions have been isolated using ion-exchange chromatography. Anhydrous rhodium chloride crystallizes with the YCl_3 and $AlCl_3$ structure. The metal centers are octahedral, and the halides are doubly bridging. It is a dense brown solid that is insoluble in common solvents and of little value in the laboratory. This hydrate is often used as precursor for the preparation of complexes and organorhodium compounds. For example, it is used as the precursor for the synthesis of di-μ-chloro-bis(η^4-1,5-cyclooctadiene)-dirhodium(I), $[Rh(C_8H_{12})CI]_2$.

$$2RhCI_3 \cdot 3H_2O \xrightarrow{2C_8H_{12},\ 2CH_3CH_2OH,\ 2Na_2CO_3} [Rh(C_8H_{12})CI]_2 + 2CH_3CHO + 4NaCI + 2CO_2 + 8H_2O$$

The best-known rhodium-halogen compound is the Wilkinson's catalyst chlorotris(triphenylphosphine)rhodium(I), $RhCl(PPh_3)_3$. It is a catalyst used in the hydroformylation or hydrogenation of alkenes and is prepared by reacting the trichloride hydrate with triphenylphosphine.

$$RhCI_3 \cdot 3H_2O + 4PPh_3 \rightarrow RhCI(PPh_3)_3 + 2H_2O + OPPh_3 + 2HCI$$

6.8.3.3 Organorhodium compounds

Most organorhodium compounds are found in oxidation states of +1 and +3 although oxidation states from +6 to −3 can also be found. Rhodium exhibits reversible oxidation states, making it useful in a broad range of organic-catalytic transformations. Below are examples of organorhodium compounds with Rh(0), Rh(I), and Rh(III).

The most common homoleptic Rh(0) complexes are $Rh_4(CO)_{12}$ and $Rh_6(CO)_{16}$. These complexes, especially, $Rh_4(CO)_{12}$ can be used in catalysis directly or can serve as a catalyst precursor. Bis(triphenylphosphine)rhodium carbonyl chloride, $RhCl(CO)[PPh_3]_2$, represents the Rh(I) group and is usually prepared by the carbonylation of Wilkinson's catalyst.

$$RhCl\left[P(C_6H_5)_3\right]_3 + CO \rightarrow RhCl(CO)\left[P(C_6H_5)_3\right]_2 + P(C_6H_5)_3$$

Pentamethylcyclopentadienyl rhodium dichloride is an Rh(III) organometallic compound with the formula $[(C_5(CH_3)_5RhCl_2)]_2$, commonly abbreviated $[Cp^*RhCl_2]_2$. It is normally prepared by the reaction of rhodium trichloride trihydrate and pentamethylcyclopentadiene in hot methanol.

$$2Cp^* - H + 2RhCl_3(H_2O)_3 \xrightarrow{CH_3OH,\Delta} [Cp^*RhCl_2]_2 + 2HCl + 6H_2O$$

6.8.4 Major uses

The principal use of Rh is in cars as a catalytic converter, replacing harmful unburned hydrocarbons, carbon monoxide, and nitrogen oxide exhaust emissions with less noxious gases. Of the roughly 30,000 kg of Rh used globally in 2012, 81% (24,300 kg) was used for this application, while 8060 kg was recovered from old converters. Around 964 kg of Rh was utilized in the glass industry, generally for the manufacture of fiberglass and flat-panel glass, and 2520 kg was used in the chemical industry. Rh is preferable to the other PGMs in the reduction of nitrogen oxides to nitrogen and oxygen:

$$2NO_x \rightarrow xO_2 + N_2$$

Rh catalysts are utilized in several industrial processes, especially in catalytic carbonylation of methanol (CH_3OH) to produce acetic acid (CH_3COOH, also written as CH_3CO_2H or $C_2H_4O_2$) by the Monsanto process. The Monsanto process has mostly been replaced by the Cativa process, a similar Ir-based process developed by BP Chemicals Ltd., which is more economical and environmentally friendly. It is as well utilized to catalyze addition of hydrosilanes to molecular double bonds, a process crucial in the production of certain silicone rubbers. Rh catalysts are additionally used to reduce benzene

(C_6H_6) to cyclohexane (C_6H_{12}). The complex of an Rh ion with BINAP (IUPAC name 2,2′-bis(diphenylphosphino)-1,1′-binaphthyl, $C_{44}H_{32}P_2$) is an extensively used chiral catalyst for chiral synthesis, as in the synthesis of menthol.

Rh is used in jewelry and for decorations. It is electroplated on white gold and platinum to produce a reflective white surface at time of selling, after which the thin layer wears away with use. This is called Rh flashing in the jewelry trade. It may similarly be utilized as coating on sterling silver to protect it from tarnishing (formation of silver sulfide, Ag_2S, produced from atmospheric hydrogen sulfide, H_2S). Solid (pure) Rh jewelry is extremely uncommon, more due to the difficulty of creation (high-melting point and poor malleability) than because of its high price. The high cost ensures that Rh is used only as an electroplate. Rh has as well been used for honors or to indicate elite status, when more frequently used metals such as Ag, Au, or Pt were judged insufficient. For example, in 1979 the Guinness Book of World Records presented Paul McCartney with an Rh-plated disk for being history's all-time bestselling songwriter and recording artist.

Rh is utilized as an alloying agent for hardening and increasing the corrosion resistance of Pt and Pd. These alloys can be found in, for example, furnace windings, bushings for glass fiber production, thermocouple elements, electrodes for aircraft spark plugs, and laboratory crucibles. Additional uses are, for example, electrical contacts, where it is used because of its small electrical resistance, small and stable contact resistance, and great corrosion resistance; Rh plated through either electroplating or evaporation is very hard and suitable for optical instruments; filters in mammography systems for the characteristic X-rays it produces; Rh neutron detectors used in combustion engineering nuclear reactors to measure neutron flux levels—this method necessitates a digital filter to regulate the current neutron flux level, producing three distinct signals: immediate, a few seconds delay, and a minute delay, each with its own signal level; all three are combined in the Rh detector signal. The three Palo Verde nuclear reactors each have 305 Rh neutron detectors, 61 detectors on each of five vertical levels, giving an accurate three-dimensional (3D) "image" of reactivity and permitting fine-tuning to use the nuclear fuel most efficiently.

6.9 46 Pd — Palladium

6.9.1 Discovery

English chemist and physicist William Hyde Wollaston (August 6, 1766 to December 22, 1828) described the discovery of a new noble metal in July 1802 in his lab-book and named it palladium in August of the same year. Wollaston purified an amount of the material and presented it, without revealing the discoverer, in a small shop in Soho in April 1803. After severe criticism from Irish chemist, mineralogist and playwright Richard Chenevix (c.1774 to April 5, 1830) (Fig. 6.45) that palladium was an alloy of platinum and mercury, Wollaston anonymously offered a reward of £20 for 20 grains of synthetic palladium alloy. Chenevix was awarded the Copley Medal (award given by the Royal Society,

FIGURE 6.45 Richard Chenevix.

for "outstanding achievements in research in any branch of science." It alternates between the physical and the biological sciences) in 1803 after he published his research on palladium. Wollaston published the discovery of rhodium in 1804 and referenced some of his work on palladium (Wollaston, 1804). He revealed that he was the discoverer of palladium in a paper in 1805 (Wollaston, 1805). Wollaston named it in 1802 after the asteroid 2 Pallas, which had been discovered 2 months earlier. Wollaston discovered palladium in crude platinum ore from South America by dissolving the ore in aqua regia (a mixture of nitric acid and hydrochloric acid, optimally in a molar ratio of 1:3), neutralizing the solution with sodium hydroxide and precipitating platinum as ammonium chloroplatinate after reaction with ammonium chloride. He added mercuric cyanide to form the compound palladium(II) cyanide, which was then heated to isolate the palladium metal. Palladium chloride was once upon a time prescribed as a tuberculosis treatment at a daily rate of 0.065 g (approximately 1 mg/kg of body weight). This treatment had numerous negative side-effects and was later substituted by more effective drugs.

6.9.2 Mining, production, and major minerals

Palladium can be found as a metal alloyed with Au and other PGMs in placer deposits of the Ural Mountains, Australia, Ethiopia, North and South America. To produce palladium, these deposits play only a minor role. The most significant commercial sources are nickel-copper deposits in the Sudbury Basin, Ontario, Canada, and the

FIGURE 6.46 Vial filled with tiny grains of placer palladium, Pd. Minas Gerais, Brazil.

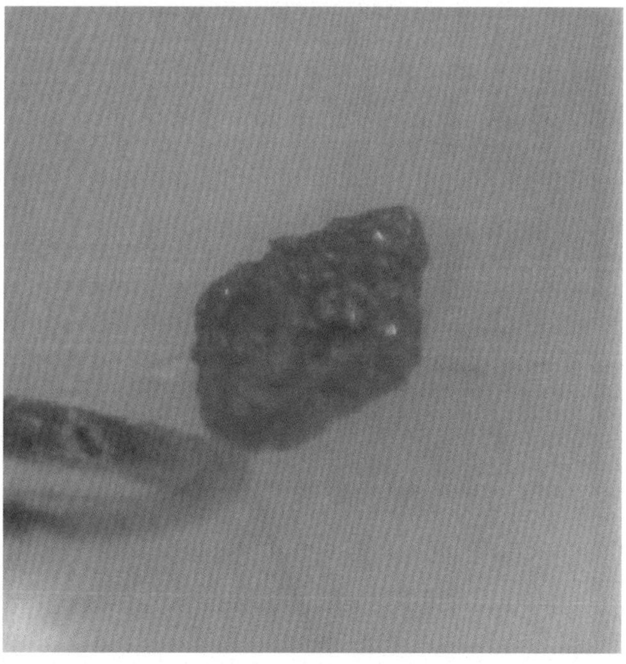

FIGURE 6.47 Arsenopalladinite, $Pd_8(As,Sb)_3$, a tiny, less than 0.75 mm grain. Itabira, Iron Quadrangle, Minas Gerais, Brazil.

Norilsk–Talnakh deposits in Siberia. The other large deposit is the Merensky Reef PGMs deposit within the Bushveld Igneous Complex, South Africa. The Stillwater igneous complex of Montana and the Roby zone ore body of the Lac des Îles igneous complex of Ontario are the two other sources of palladium in Canada and the United States. Palladium is found in the rare minerals cooperate, (Pt,Pd,Ni)S, and polarite, Pd(Bi,Pb). Russia is the top producer, followed by South Africa, Canada, and the United States. Palladium is also produced in nuclear fission reactors and can be isolated from the spent nuclear fuel; however, this palladium source is not utilized. None of the currently working nuclear reprocessing facilities are capable to extract palladium from the high-level radioactive waste.

More than 50 minerals are known to contain palladium in their structure. Within the element class 13 alloys can be found, such as atokite ($(Pd,Pt)_3Sn$), palladium ((Pd,Pt)) (Fig. 6.46), and potarite (PdHg). In the sulfide class 48 minerals are known to contain Pd, for example, arsenopalladinite ($Pd_8(As,Sb)_3$) (Fig. 6.47), coldwellite (Pd_3Ag_2S), isomertieite ($Pd_{11}Sb_2As_2$), palladoarsenide (Pd_2As), and telluropalladinite (Pd_9Te_4). A single oxide, palladinite (PdO), is known.

6.9.3 Chemistry

Naturally occurring Pd consists of seven isotopes, six of which are stable. The most stable radioisotopes are ^{107}Pd with a half-life of 6.5 million years (found in nature), ^{103}Pd with 16.991 days (synthetic), and ^{100}Pd with 3.63 days (synthetic). Eighteen other radioisotopes have been observed ranging from ^{91}Pd to ^{123}Pd. These radioisotopes have half-lives of less than 30 minutes, except for ^{101}Pd which has a half-life of 8.47 hours, ^{109}Pd which has a half-life of 13.7 hours, and ^{112}Pd which has a half-life of 21 hours. For isotopes lighter than the most abundant stable isotope, ^{106}Pd, the principal decay mode is electron capture with the main decay product being Rh. The primary mode of decay for the isotopes heavier than ^{106}Pd is via β decay with the main product being Ag. Radiogenic ^{107}Ag as a decay product of ^{107}Pd was first discovered in 1978 in the Santa Clara meteorite of 1976. The discoverers proposed that the coalescence and differentiation of iron-cored small planets may have happened 10 million years after a nucleosynthetic event. ^{107}Pd versus Ag correlations found in bodies, which have been melted since accretion of the solar system, must be a reflection of the existence of short-lived nuclides in the early solar system.

Palladium is a rare and lustrous silvery-white metal. It has 46 electrons arranged in an electronic configuration of $[Kr]4d^{10}$. Electrons in the s-shell fill the d orbitals because they have less energy. Palladium compounds primarily exist in the 0 and +2 oxidation states, but other less common oxidation states are also known (Table 6.9). Palladium dissolves slowly in concentrated oxidizing acids, nitric acid and sulfuric acid. When it is finely ground, it can be dissolved in hydrochloric acid. It dissolves readily at room temperature in aqua regia. It also dissolves in fused alkali metal oxides and peroxides.

TABLE 6.9 Palladium properties.

Appearance	Silvery-white
Standard atomic weight ($A_{r,std}$)	106.42
Block	d-Block
Element category	Transition metal
Electron configuration	$[Kr]\ 4d^{10}$
Phase at STP	Solid
Melting point	1554.9°C
Boiling point	2963°C
Density (near r.t.)	12.023 g/cm^3
When liquid (at m.p.)	10.38 g/cm^3
Heat of fusion	16.74 kJ/mol
Heat of vaporization	358 kJ/mol
Molar heat capacity	25.98 J/(mol·K)
Oxidation states	0, +1, **+2**, +3, **+4**
Ionization energies	First: 804.4 kJ/mol Second: 1870 kJ/mol Third: 3177 kJ/mol
Atomic radius	Empirical: 137 pm
Covalent radius	139 ± 6 pm
Van der Waals radius	163 pm

STP, Standard temperature and pressure.
Bold font indicates main oxidation state.

6.9.3.1 Oxide

Palladium does not react with oxygen at standard temperature. It does form a thin layer of palladium(II) oxide (PdO) when heated to 800°C. It is the only well-characterized oxide of palladium. Above approximately 900°C, the oxide reacts to palladium metal and oxygen gas. It is not attacked by acids. PdO is frequently formed as a poorly defined compound for applications as a catalyst. Palladium oxide is produced by heating palladium sponge metal in oxygen at 350°C.

$$2Pd + O_2 \rightarrow 2PdO$$

The oxide is formed as a black powder. It may also be produced specifically for catalytic applications by heating variously a mixture of palladium(II) chloride and potassium nitrate,

$$2PdCl_2 + 4KNO_3 \rightarrow 2PdO + 4KCl + 4NO_2 + O_2 \text{ (possible reaction)},$$

or the product of dissolving palladium in aqua regia, followed by the addition of sodium nitrate at 600°C. A hydrated form of the oxide (which dissolves in acid) can be produced through precipitation from solution, for example, by hydrolysis of palladium nitrate or reaction of a soluble palladium compound with a strong base. The brown hydrated oxide transforms into black anhydrous oxide on heating. Its susceptibility to attack by acids diminishes at lower water content. The hydrated oxide (hydroxide), $PdO{\cdot}nH_2O$ can be formed as a dark-yellow precipitate by adding alkali to a solution of palladium nitrate, $Pd(NO_3)_2$. The structure of PdO is tetragonal (space group *P42/mmc*) with unit cell parameters $a = 3.044$ and $c = 5.328$ Å. The Pd atoms are square planar as anticipated for a d^8 metal ion and the oxygen atoms are approximately tetrahedral. The closest Pd–Pd distance is 3.044 Å and is almost within the range, which can be considered a bonding distance.

6.9.3.2 Hydrides

Palladium hydride is metallic palladium that has a substantial amount of hydrogen within its crystal lattice. Notwithstanding its name, it is not an ionic hydride but rather an alloy of palladium with metallic hydrogen that can be written as PdH_x. Its interaction with hydrogen had been a subject of intense interest both theoretically and experimentally. Pd can reversibly absorb a large amount of hydrogen. Upon hydrogen absorption, the conductivity of Pd decreases until it becomes a semiconductor at a composition of $PdH_{0.5}$. The hydrogen is first chemisorbed at the surface of Pd but is incorporated in the metal lattice at increased hydrogen pressures to form the α- and β-phase hydrides. The absorption of hydrogen produces two different phases, both of which contain palladium metal atoms in a face-centered cubic (*fcc*, rock salt, or NaCl) lattice, which is the same structure as pure palladium metal. At low concentrations up to $PdH_{0.02}$ the palladium lattice expands slightly, from 388.9 to 389.5 pm. Above this concentration the second phase appears with a lattice constant of 402.5 pm. Both phases coexist until a composition of $PdH_{0.58}$ when the α-phase disappears. The β-phase causes distortion and hardening of Pd membranes and its formation is avoided by maintaining the temperature above 300°C. Neutron diffraction studies have shown that hydrogen atoms randomly occupy the octahedral interstices in the metal lattice (in an *fcc* lattice there is one octahedral hole per metal atom). The limit of absorption at normal pressures is $PdH_{0.7}$, indicating that approximately 70% of the octahedral holes are occupied. The absorption of hydrogen is reversible, and hydrogen rapidly diffuses through the metal lattice. Metallic conductivity reduces as hydrogen is absorbed; until at around $PdH_{0.5}$ the solid becomes a semiconductor.

6.9.3.3 Halides

Palladium is oxidized by F_2 and Cl_2 and Br_2 at red heat. The most stable product of the action of fluorine on metallic palladium is actually palladium(II,IV) fluoride, also known as palladium trifluoride. It has the empirical formula PdF_3, but is better described as the mixed-valence compound palladium(II) hexafluoropalladate(IV), $Pd(II)[Pd(IV)F_6]$, and is often written as $Pd[PdF_6]$ or Pd_2F_6. $Pd[PdF_6]$ is the most stable product of the reaction of fluorine and metallic palladium.

$$2Pd + 3F_2 \rightarrow Pd[PdF_6]$$

$Pd[PdF_6]$ is paramagnetic and both Pd(II) and Pd(IV) fill octahedral sites in the crystal structure. The Pd^{II}-F distance is 2.17 Å, while the Pd^{IV}-F distance is slightly shorter at 1.90 Å. PdF_4 has been produced by reacting palladium(II,IV) fluoride with fluorine gas at pressures around 7 atm and at 300°C for several days. It is a strong oxidizing agent and undergoes fast hydrolysis in moist air. PdF_2, a violet and easily hydrolyzed compound, is produced when $Pd^{II}[Pd^{IV}F_6]$ is refluxed with SeF_4 and is notable as one of the very few paramagnetic compounds of Pd^{II}.

$$Pd[PdF_6] + SeF_4 \rightarrow 2PdF_2 + SeF_6$$

Similar to its lighter congener nickel(II) fluoride, PdF_2 gas has a rutile-type (TiO_2) crystal structure, containing octahedrally coordinated Pd, which has the electronic configuration $t_{2g}^6\, e_g^2$. This configuration causes PdF_2 to be paramagnetic because of two unpaired electrons, one in each e_g-symmetry orbital of palladium.

The other known halides are the dihalides, palladium(II) chloride, $PdCl_2$, palladium(II) bromide, $PdBr_2$, and palladium(II) iodide, PdI_2. $PdCl_2$ is a common starting material in palladium chemistry—palladium-based catalysts are of value in organic synthesis. Palladium(II) chloride is produced by dissolving palladium metal in aqua regia or hydrochloric acid in the presence of chlorine. It can also be formed by heating palladium sponge metal with chlorine gas at 500°C. Two forms of $PdCl_2$ exist, denoted α and β. In both forms, the palladium centers have a square-planar coordination geometry that is typical of Pd(II). Moreover, in both forms, the Pd(II) centers are linked by μ_2-chloride bridges. The α-form is a polymer, consisting of "infinite" slabs or chains. In contrast, the β-form is molecular, comprising an octahedral cluster of six Pd atoms. Each of the 12 edges of this octahedron is spanned by Cl^-. $PtCl_2$ has similar structures, while $NiCl_2$ has the $CdCl_2$ structure, with hexacoordinated Ni(II). $PdCl_2$ is a common starting material in the synthesis of other palladium compounds. It is not very soluble in water or noncoordinating solvents, so the first step in its use is frequently the synthesis of labile but soluble Lewis base adducts, such as bis(benzonitrile)palladium dichloride and bis(acetonitrile)palladium dichloride. These complexes are produced by reacting $PdCl_2$ with hot solutions of the nitriles:

$$PdCl_2 + 2RCN \rightarrow PdCl_2(RCN)_2$$

Although sometimes suggested, inert-gas methods are not required if the complex is to be used in situ. For instance, bis(triphenylphosphine)palladium(II) dichloride may be formed from palladium(II) chloride by reacting it with triphenylphosphine in benzonitrile:

$$PdCl_2 + 2PPh_3 \rightarrow PdCl_2(PPh_3)_2$$

Further reduction in the presence of more triphenylphosphine forms tetrakis(triphenylphosphine)palladium(0); the second reaction may be performed without purifying the intermediary dichloride:

$$PdCl_2(PPh_3)_2 + 2PPh_3 + 2.5N_2H_4 \rightarrow Pd(PPh_3)_4 + 0.5N_2 + 2N_2H_5^+Cl^-$$

Otherwise, palladium(II) chloride can be solubilized in the form of the tetrachloropalladate anion, for example, sodium tetrachloropalladate, through the reaction with the appropriate alkali metal chloride in water: Palladium(II) chloride is insoluble in water, while the product dissolves:

$$PdCl_2 + 2MCl \rightarrow M_2PdCl_4$$

Additionally, this compound may further react with phosphines to produce phosphine complexes of palladium. Palladium chloride may also be employed to form heterogeneous palladium catalysts: palladium on barium sulfate, palladium on carbon, and palladium chloride on carbon. Even when dry, $PdCl_2$ is able to quickly stain stainless steel. Consequently, palladium(II) chloride solutions are sometimes employed to test for the corrosion resistance of stainless steel. Palladium(II) chloride is occasionally utilized in carbon monoxide detectors. Carbon monoxide reduces palladium (II) chloride to palladium:

$$PdCl_2 + CO + H_2O \rightarrow Pd + CO_2 + 2HCl$$

Residual $PdCl_2$ is converted to red PdI_2, the concentration of which may be determined colorimetrically:

$$PdCl_2 + 2KI \rightarrow PdI_2 + 2KCl$$

Palladium(II) chloride is employed in the Wacker process for the manufacture of aldehydes and ketones from alkenes. Palladium(II) chloride can also be utilized for the cosmetic tattooing of leukomas in the cornea. $PdBr_2$ is commercially available, though less common than $PdCl_2$, the usual starting material for palladium chemistry. Unlike the chloride, $PdBr_2$ is insoluble in water, but dissolves when heated in acetonitrile to form monomeric acetonitrile adducts:

$$PdBr_2 + 2MeCN \rightarrow PdBr_2(MeCN)_2$$

Palladium(II) iodide, PdI_2, is commercially available, though less common than palladium(II) chloride. Historically, the quantity of palladium in a sample was determined gravimetrically by precipitation as palladium (II) iodide. In contrast to the Pd(II) chloride and bromide, palladium(II) iodide is not quite as soluble in excess iodide.

6.9.3.4 Pd(III) compounds

In chemistry, palladium(III) compounds include palladium in the unusual +3 oxidation state (in most of its compounds, it has the oxidation state II). Pd(III) compounds occur in mononuclear and dinuclear forms. Palladium(III) is most often invoked, not observed in mechanistic organometallic chemistry. Pd(III) has a d^7 electronic configuration, which results in a Jahn–Teller distorted octahedral geometry. This geometry could also be considered as being intermediate between square planar and octahedral. These complexes are low-spin and paramagnetic. The first Pd(III) mononuclear complex determined by X-ray crystallography was published in 1987 by Blake et al. It was formed by oxidation of the 1,4,7-trithiacyclononane (ttcn) complex $[Pd(ttcn)_2]^{3+}$. X-ray crystallography showed the expected Jahn–Teller distorted octahedral geometry, despite the highly symmetric structure of the ligand. The first organometallic Pd(III) complex determined by X-Ray crystallography was published in 2010 by Khusnutdinova et al. Organopalladium complexes supported with a macrocyclic tetradentate ligand undergo single-electron oxidation to produce a Pd(III) species that is stabilized by the axially positioned amine. The authors suggested that although the axial nitrogen stabilizes a distorted octahedral geometry, the t-Bu group and the rigidity of the macrocyclic structure prevents the oxidation to a more conventional octahedral Pd(IV).

Pairs of Pd(III) centers can couple, resulting in a Pd–Pd bond order of 1, for example, a two-electron oxidation of two Pd(II) species can form a diamagnetic, Pd(III)–Pd(III) dimer with a bond order of 1. The first illustration of a dipalladium(III) complex was produced through oxidation of the dinuclear Pd(II) complex of triazabicyclodecene. The first organometallic dinuclear Pd(III) complexes were published by Cotton and coworkers in 2006. These complexes catalyze the diborylation of terminal olefins. Because of the facile reduction of these complexes to Pd(II) species by diborane, the authors indicated that the dinuclear Pd(III) complexes act as precatalysts for active Pd(II) catalysts. The reactivity of dinuclear Pd(III) species as an active catalytic intermediary is generally discussed in the context of C–H activation. Though it was suggested that Pd-catalyzed oxidative C–H functionalization reactions involve a Pd(IV) intermediate, Ritter and coworkers first suggested that these oxidative reactions could include a dinuclear Pd(III) intermediary instead of Pd(IV). Dinuclear Pd species are involved in Pd-catalyzed C–H chlorination. Through X-ray crystallography, Ritter unequivocally proved that a dinuclear Pd(III) complex is produced when the palladacycle is reacted with a two-electron oxidant, and such a dinuclear complex undergoes C–Cl reductive elimination at ambient temperature. Both experimental and computational data were consistent with a concerted 1,1-reductive elimination mechanism for the C–Cl forming step. The authors demonstrated that such a bimetallic participation of redox event lowers the activation barrier for the reductive elimination step by around 30 kcal/mol in comparison to a monometallic pathway. Acetoxylation of 2-phenylpyridine has also been shown to include a dinuclear Pd(III) intermediate.

6.9.3.5 Organopalladium compounds

In organic synthesis, two kinds of Pd compounds, namely Pd(II) salts and Pd(0) complexes, are used. Pd(II) compounds are mainly used as oxidizing reagents, or catalysts for a few reactions. Pd(0) complexes are always used as catalysts. The prime starting material for making palladium catalysts or precursors is palladium(II) chloride. It is prepared from the reaction of palladium with chlorine. Heterogeneous palladium catalysts such as palladium on barium sulfate, palladium on carbon, and palladium chloride on carbon are prepared from $PdCl_2$ as precursor. Solutions of $PdCl_2$ in nitric acid react with acetic acid to give palladium(II) acetate, also a versatile reagent. Complexation in the second and third row of transition elements such as Pd has a pronounced effect on the splitting of d orbitals. As a result, complexes of Pd are mostly diamagnetic and the vast majority are planar. For example, $PdCl_2$ reacts with ligands (L) to give square-planar complexes of the type $PdCl_2L_2$ such as the benzonitrile derivative $PdX_2(PhCN)_2$.

$$PdCl_2 + 2L \rightarrow PdCl_2L_2 \quad (L = PhCN,\ PPh_3,\ NH_3,\ etc)$$

Palladium forms a range of zerovalent complexes with the formula PdL_4, PdL_3, and PdL_2. Palladium(0) and palladium(II) are catalysts in coupling reactions. These reactions are widely practiced for the synthesis of fine chemicals and their importance has been recognized by the 2010 Nobel Prize in Chemistry to Richard F. Heck, Ei-ichi Negishi, and Akira Suzuki. Reactions involving Pd(0) include the reduction of a mixture of $PdCl_2(PPh_3)_2$ and PPh_3 gives tetrakis(triphenylphosphine)palladium(0).

$$2PdCl_2(PPh_3)_2 + 4PPh_3 + 5N_2H_4 \rightarrow 2Pd(PPh_3)_4 + N_2 + 4N_2H_5^+Cl^-$$

Another major palladium(0) complex, tris(dibenzylideneacetone)dipalladium(0) ($Pd_2(dba)_3$), is prepared by reducing sodium tetrachloropalladate in the presence of dibenzylideneacetone. ($Pd(PPh_3)_4$, and ($Pd_2(dba)_3$) serve either as catalysts or precatalysts in reactions.

Palladium(II) compounds are equally important as catalysts in organic synthesis. One prime example is dichloro (1,5−cyclooctadiene)palladium, a favored precursor to catalysts owing to the ease of displacing the diene. Another iconic palladium(II) complex is allylpalladium chloride dimer (APC). Allyl compounds with suitable leaving groups react with palladium(II) salts to π-allyl complexes having hapticity 3. These intermediates react with nucleophiles such as carbanions derived from malonates or amines in allylic amination.

6.9.4 Major uses

The leading application of Pd currently is in catalytic converters. Pd is also utilized in, for example, jewelry, dentistry, watch-making, blood sugar test strips, aircraft spark plugs, surgical instruments, and electrical contacts. In addition, Pd is used to manufacture professional transverse (concert or classical) flutes. As a commodity, Pd bullion has ISO currency codes of XPD and 964. Pd is one of only four metals to have such codes, together with Au, Ag, and Pt. Since it absorbs hydrogen, Pd is a key component of the controversial cold fusion experiments that began in 1989.

When it is finely distributed, as with Pd on carbon, Pd forms a useful catalyst in, for example, heterogeneous catalytic processes such as hydrogenation, dehydrogenation, and petroleum cracking. Pd is also essential to the Lindlar catalyst, also called Lindlar's Palladium [A Lindlar catalyst is a heterogeneous catalyst that consists of Pd deposited on $CaCO_3$, which is then poisoned with various forms of Pb or S. It is used for the hydrogenation of alkynes to alkenes (i.e., without further reduction into alkanes) and is named after its inventor British-Swiss chemist Herbert Lindlar (March 15, 1909 to June 27, 2009).]. Many C−C bonding reactions in organic chemistry are aided by Pd compound catalysts. For instance, Mizoroki−Heck reaction [the chemical reaction of an unsaturated halide (or triflate) with an alkene in the presence of a base and a Pd catalyst (or Pd nanomaterial-based catalyst) to form a substituted alkene. It is named after Japanese chemist Tsutomu Mizoroki and American chemist Richard F. Heck (August 15, 1931 to October 10, 2015).]; Suzuki coupling [an organic reaction, classified as a cross-coupling reaction, where the coupling partners are a boronic acid and an organohalide catalyzed by a Pd(0) complex. It was first published in 1979 by Japanese chemist Akira Suzuki (born September 12, 1930) and he shared the 2010 Nobel Prize in Chemistry with American chemist Richard F. Heck (August 15, 1931 to October 10, 2015) and Manchurian-born Japanese chemist Ei-ichi Negishi (born July 14, 1935) for their effort for discovery and development of Pd-catalyzed cross-couplings in organic synthesis]; Tsuji−Trost reactions [also known as the Trost allylic alkylation or allylic alkylation, is a Pd-catalyzed substitution reaction involving a substrate that contains a leaving group in an allylic position. The Pd catalyst first coordinates with the allyl group and then undergoes oxidative addition, forming the π-allyl complex. This allyl complex can then be attacked by a nucleophile, resulting in the substituted product. This research was first pioneered by Japanese chemist Jiro Tsuji (born May 11, 1927) in 1965 and, later, adapted by American chemist Barry Trost (born June 13, 1941) in 1973 with the introduction of phosphine ligands. The scope of this reaction has been expanded to many different carbon, nitrogen, and oxygen-based nucleophiles; many different leaving groups; many different phosphorus, nitrogen, and sulfur-based ligands; and many different metals, although palladium is still preferred.]; Wacker process [or the Hoechst-Wacker process (named after the chemical companies of the same name) refers to the oxidation of ethylene to acetaldehyde in the presence of Pd(II) chloride as the catalyst. This chemical reaction was one of the first homogeneous catalysts with organopalladium chemistry applied on an industrial scale.]; Negishi reaction [a widely employed transition metal−catalyzed cross-coupling reaction. The reaction couples organic halides or triflates with organozinc compounds, forming C−C bonds in the process. A Pd(0) species is generally utilized as the metal catalyst, though nickel is sometimes used. The reaction is named after Nobel Prize winner Manchurian-born Japanese chemist Ei-ichi Negishi (born July 14, 1935)]; Stille coupling [or the Migita−Kosugi−Stille coupling, is a chemical reaction widely used in organic synthesis which involves the coupling of an organotin compound (also known as organostannanes) with a variety of organic electrophiles via Pd-catalyzed coupling reaction. The groundwork for the Stille reaction was laid in 1976 and 1977 by British chemist Colin Eaborn (March 15, 1923 to February 22, 2004), and Japanese chemists Toshihiko Migita, and Masanori Kosugi, who explored numerous Pd-catalyzed couplings involving organotin reagents. American chemist John Stille (May 8, 1930 to July 19, 1989) and Israeli chemist David Milstein (born June 4, 1947) developed a much milder and more broadly applicable procedure in 1978. Stille's work on this area might have earned him a share of the 2010 Nobel Prize, which was awarded to Richard Heck, Ei-ichi Negishi, and Akira Suzuki for their work on the Heck, Negishi, and Suzuki coupling reactions. However, Stille died in the plane crash of United Airlines Flight 232 in 1989.]; and Sonogashira coupling [a cross-coupling reaction used in organic synthesis to form C−C bonds. It employs a Pd catalyst as well as copper cocatalyst to form a C−C bond between a terminal alkyne and an aryl or vinyl halide. The Sonogashira cross-coupling reaction has been employed in a wide variety of areas, due to its usefulness in the formation of C−C bonds. The reaction can be carried out under mild conditions, such as at room temperature, in aqueous media, and with a mild base, which has allowed for the use of the Sonogashira cross-coupling reaction in the synthesis of complex molecules. Its applications include pharmaceuticals, natural products, organic materials, and nanomaterials. The alkynylation reaction of aryl halides using aromatic acetylenes was described in 1975 in three independent contributions by Cassar, Dieck, and Heck, as well as Japanese chemist Kenkichi Sonogashira (born October 25, 1931), Tohda and Hagihara. All the reactions use Pd catalysts to afford the same reaction products. However, the protocols of Cassar and Heck are performed solely using Pd and involve harsh reaction conditions (i.e., high reaction temperatures). The use of Cu cocatalyst in addition to Pd complexes in Sonogashira's procedure allowed the reactions to be carried under mild reaction conditions in excellent yields. A rapid development of the Pd/Cu systems followed and enabled numerous synthetic applications, while Cassar–Heck conditions were left, maybe unfairly, all but overlooked.]. When dispersed on conductive materials, Pd is an exceptional electrocatalyst for oxidation of primary alcohols in alkaline media. Pd is likewise a versatile metal for homogeneous catalysis, utilized together with a broad range of ligands for highly selective chemical transformations. A 2008 study proved that Pd is an efficient catalyst for C–F bonds. Pd catalysis is mainly employed in organic chemistry and industrial applications, while its application is increasing as a tool for synthetic biology; for example, in 2017, effective in vivo catalytic activity of Pd nanoparticles was shown in mammals to treat disease.

The second largest use of Pd in electronics is in multilayer ceramic capacitors where Pd (and Pd-Au alloy) is utilized for electrodes. Pd (occasionally alloyed with Ni) is employed for component and connector plating in consumer electronics and in soldering materials. Hydrogen easily diffuses through heated Pd, and membrane reactors containing Pd membranes are used to produce high-purity H_2. Pd is utilized in palladium-hydrogen electrodes in electrochemical research. Pd(II) chloride easily catalyzes carbon monoxide (CO) gas to carbon dioxide (CO_2) and is therefore valuable in carbon monoxide detectors. Pd easily absorbs H_2 at room temperature, forming Pd hydride PdH_x with $x < 1$. Although this property is common to various transition metals, Pd has an exceptionally high absorption capacity while not losing its ductility up until x approaches 1. This property has been studied in creating an efficient, low-cost, and safe H_2 fuel storage medium, despite the fact that Pd itself is at present too expensive for this application. The hydrogen content in Pd can be related to its magnetic susceptibility, which decreases with the increase of hydrogen and becomes zero for $PdH_{0.62}$. At higher ratios, the solid solution becomes diamagnetic. Pd is utilized in small quantities (around 0.5%) in some alloys of dental amalgam to reduce corrosion and enhance the metallic luster of the final restoration.

Pd has been utilized as a precious metal in jewelry starting in 1939 as an alternative to Pt in the alloys called "white gold," where the naturally white color of Pd does not necessitate Rh plating. Pd is much less dense than Pt. Like Au, Pd can be beaten into leaf form as thin as 100 nm. Unlike Pt, Pd may discolor at temperatures above 400°C and it is relatively brittle. Before 2004, the primary use of Pd in jewelry was the production of white gold. Pd is one of the three most common alloying metals in white gold (Ni and Ag can also be used). Pd-gold is more costly than Ni-gold, but rarely triggers allergic reactions (although some cross-allergies with Ni may occur). When Pt became a strategic resource during World War II, many jewelry bands were made from Pd instead. Up till then Pd was hardly used in jewelry due to the technical difficulty of casting. With the casting problem solved the use of Pd in jewelry increased, initially as the Pt price increased while the price of Pd decreased. At the start of 2004, when Au and Pt prices increased steeply, China began making large volumes of Pd jewelry, using some 37 tons in 2005. Later changes in the relative price of Pt decreased the demand for Pd to 17.4 tons in 2009. Demand for Pd as a catalyst has increased the price of Pd to about 50% higher than that of Pt at the start of 2019. In January 2010, hallmarks for Pd were created by assay offices in the United Kingdom, and hallmarking became compulsory for all jewelry advertising pure or alloyed Pd. Articles can be marked as 500, 950, or 999 parts of Pd per thousand of the alloy. Fountain pen nibs made from Au are occasionally plated with Pd when an Ag (rather than Au) look is preferred. Sheaffer has used Pd plating for years, either as an accent on otherwise Au nibs or covering the Au entirely. In the platinotype printing process, photographers make fine-art black-and-white prints using Pt or Pd salts. Often used with Pt, Pd provides an alternative to Ag.

6.10 47 Ag — Silver

6.10.1 Discovery

The word "silver" can be found in Anglo-Saxon in a variety of spellings, such as seolfor and siolfor. A comparable form exists in the Germanic languages (compare Old High German silabar and silbir). The currently used chemical symbol Ag comes from the Latin word for "silver," argentum (compare Ancient Greek ἄργυρος, árgyros), from the Proto-Indo-European root *h₂erǵ- (formerly reconstructed as *arǵ-), meaning "white" or "shining": this was the usual Proto-Indo-European word for the metal, whose reflexes are missing in Germanic and Balto-Slavic. The Balto-Slavic words for silver are rather alike to the Germanic ones (e.g., Russian серебро [serebro], Polish srebro, Lithuanian sidabras) and they may have a shared origin, though this is ambiguous: some scholars have proposed the

Akkadian sarpu "refined silver" as its origin, connected to the word sarapu "to refine or smelt." Silver was one of the seven metals of ancient times that were known to prehistoric humans and whose discovery is therefore lost to history. Especially, the three metals of group 11, copper, silver, and gold, are found as native elements in nature and were possibly used as the first primitive forms of money in contrast to simple bartering. Yet, unlike copper, silver did not result in the growth of metallurgy due to its low structural strength and was more often used ornamentally or as money. As silver is more reactive than gold, supplies of native silver were much more restricted than those of gold. For example, silver was more expensive than gold in Egypt until around the 15th century BCE: the Egyptians are believed to have isolated gold from silver by heating the metals with salt, and subsequently reducing the silver chloride to obtain the metal. The circumstances changed with the discovery of cupellation, a process that permitted silver metal to be separated from its ores (cupellation is a refining process in metallurgy, where ores or alloyed metals are treated under very high temperatures and have controlled operations to separate noble metals, such as gold and silver, from base metals such as lead, copper, zinc, arsenic, antimony, or bismuth, present in the ore. The process is based on the principle that precious metals do not oxidize or react chemically, unlike the base metals; so, when they are heated at high temperatures, the precious metals remain apart, and the others react forming slags or other compounds). Despite the fact that slag heaps discovered in Asia Minor and on the islands of the Aegean Sea suggest that silver was being separated from lead as early as the fourth millennium BCE, and one of the oldest silver extraction centers in Europe was Sardinia in the early Chalcolithic period [The Chalcolithic, a name derived from the Greek: χαλκός khalkós, "copper" and from λίθος líthos, "stone" or Copper Age, also known as the Eneolithic or Aeneolithic (from Latin aeneus "of copper") is an archeological period which researchers usually regard as part of the broader Neolithic (although scholars originally defined it as a transition between the Neolithic and the Bronze Age).], these processes did not spread extensively until later, when it expanded all over the region and beyond. The roots of silver manufacture in India, China, and Japan were almost certainly correspondingly ancient, but are not well-documented because of their great age. After the Phoenicians originally landed in what is now Spain, they got so much silver that they could not load it all on their ships, and consequently used silver to weight their anchors instead of lead (Phoenicia from the Ancient Greek: Φοινίκη, Phoiníkē) was a thalassocratic, ancient Semitic-speaking Mediterranean civilization that originated in the Levant, specifically Lebanon, in the west of the Fertile Crescent. Scholars generally agree that it was centered on the coastal areas of Lebanon and included northern Israel, and southern Syria reaching as far north as Arwad, but there is some dispute as to how far south it went, the furthest suggested area being Ashkelon. Its colonies later reached the Western Mediterranean, such as Cádiz in Spain and most notably Carthage in North Africa, and even the Atlantic Ocean. The civilization spread across the Mediterranean between 1500 and 300 BCE). During the period of the Greek and Roman civilizations, silver coins were an essential part of the economy: the Greeks were already separating silver from galena (PbS) by the 7th century BCE, and the rise of Athens was partially made possible because of the close by silver mines at Laurium, from which they mined around 30 tons per annum from 600 to 300 BCE. The stability of the Roman currency depended to a high degree on the sourcing of silver bullion, mainly from Spain, where Roman miners produced on a scale unmatched before the discovery of the New World. Attaining a maximum production of 200 tons per annum, an estimated silver stock of 10,000 tons was dispersed in the Roman economy around the middle of the 2nd century CE, 5 to 10 times more than the total amount of silver available to medieval Europe and the Abbasid Caliphate around CE 800. The Romans also documented the mining of silver in central and northern Europe in that same period. This production stopped almost completely with the fall of the Roman Empire and did not start again until the time of Charlemagne, or Charles the Great (April 2, 742 to January 28, 814), king of the Franks from 768, king of the Lombards from 774, and emperor of the Romans from 800: at that time, tens of thousands of tons of silver had already been mined. Central Europe turned into the center of silver production during the Middle Ages, as the Mediterranean silver ore deposits exploited by the ancient civilizations had been exhausted. Silver mining started in Bohemia, Saxony, Erzgebirge, Alsace, the Lahn region, Siegerland, Silesia, Hungary (Fig. 6.48), Norway, Steiermark, Salzburg, and the southern Black Forest. The majority of these ores were relatively rich in silver and could easily be separated by hand from the remaining rock and then smelted; some deposits of native silver were also discovered. Numerous of these mines were quickly exhausted, but some of them remained active until the Industrial Revolution (the transition to new manufacturing processes in Europe and the United States, in the period from about 1760 to some time between 1820 and 1840.), before which the world production of silver was around a paltry 50 tons per annum. In the Americas, high-temperature silver-lead cupellation processing was established by pre-Inca civilizations as early as CE 60–120; while silver deposits in India, China, Japan, and pre-Columbian America continued to be mined during this period. With the discovery of America and the plundering of silver by the Spanish conquistadors, Central and South America turned into the dominant producers of silver until around the start of the 18th century, predominantly Peru, Bolivia, Chile, and Argentina: the last of these countries later took its

FIGURE 6.48 Silver mining in Kutná Hora (size 645 × 886 mm; 1490s).

name from that of the metal that composed so much of its mineral wealth. The silver trade became part of a global network of exchange. As one historian noted, silver "went round the world and made the world go round." Much of this silver ended up in the hands of the Chinese. A Portuguese merchant in 1621 observed that silver "wanders throughout all the world. before flocking to China, where it remains as if at its natural center." Nevertheless, much of it went to Spain, permitting Spanish rulers to follow military and political ambitions in both Europe and the Americas. "New World mines," concluded several historians, "supported the Spanish empire." In the 19th century, the main production of silver moved to North America, particularly Canada, Mexico, and Nevada in the United States: some secondary production from lead and zinc ores took place in Europe, while some silver ore deposits in Siberia and the Russian Far East as well as in Australia were also mined. Poland appeared as a significant silver producer during the 1970s after the discovery of copper deposits rich in silver, before the center of production returned to the Americas in the 1980s. Currently, Peru and Mexico are still amongst the principal silver producers, but the spread of silver production around the world is relatively balanced and around 20% of the silver supply comes from recycling as an alternative to new production.

6.10.2 Mining, production, and major minerals

The abundance of silver in the Earth's crust is 0.08 ppm, about the same as that of mercury. It is mostly found in sulfide ores, in particular acanthite, Ag_2S. Acanthite deposits occasionally also contain native silver when they occur in reducing environments, and when in contact with saltwater they react to form chlorargyrite, AgCl, which is ubiquitous in Chile and New South Wales, Australia. Most other silver minerals are silver pnictides [A pnictogen is one of the chemical elements in group 15 of the periodic table. This group is also known as the nitrogen family. It consists of the elements nitrogen (N), phosphorus (P), arsenic (As), antimony (Sb), bismuth (Bi)] or chalcogenides (A chalcogenide is a chemical compound consisting of at least one chalcogen anion and at least one more electropositive element. Although all group 16 elements of the periodic table are defined as chalcogens, the term chalcogenide is more commonly reserved for sulfides, selenides, tellurides, and polonides); they are generally lustrous semiconductors. Most true silver deposits, in contrast to argentiferous deposits of other metals, came from Tertiary period vulcanism. The main sources of silver are copper, copper-nickel, lead, and lead-zinc ores from Peru, Bolivia, Mexico, China, Australia, Chile, Poland, and Serbia. Countries such as Peru, Bolivia, and Mexico have been mining silver since 1546 and are still among the most important world producers. Major silver-producing mines are Cannington (Australia), Fresnillo (Mexico), San Cristóbal (Bolivia), Antamina (Peru), Rudna (Poland), and Penasquito (Mexico). Tajikistan is recognized as having some of the largest silver deposits in the world.

Silver is typically found in nature in combination with other metals, or in minerals that contain Ag compounds, generally in the form of sulfides such as galena (PbS) or cerussite ($PbCO_3$) (Box 6.7). Hence, the principal production of Ag involves the smelting and subsequent cupellation of argentiferous lead ores, a historically important process. Cupellation is a refining process in metallurgy, where ores or alloyed metals are treated under very high temperatures and have controlled operations to separate noble metals, such as gold and silver, from base metals such as lead, copper, zinc, arsenic, antimony, or bismuth, present in the ore. The process is based on the principle that precious metals do not oxidize or react chemically, unlike the base metals; therefore, when they are heated at high temperatures, the precious metals remain separate and the others react forming slags or other compounds. Lead

BOX 6.7 Geology of silver deposits.

Silver occurs, besides the native element (Fig. 6.49), in electrum, (Au,Ag), in silver sulfide, acanthite, Ag_2S, and several complex sulfide minerals containing lead, copper, antimony, and arsenic, such as tennantite-tetrahedrite, $Cu_6Cu_4(Fe^{2+},Zn)_2As_4S_{12}S$-$Cu_6Cu_4(Fe^{2+},Zn)_2Sb_4S_{12}S$, and the rather rare argentotennantite-argentotetrahedrite series, $Ag_6Cu_4(Fe^{2+},Zn)_2As_4S_{12}S$-$Ag_6Cu_4(Fe^{2+},Zn)_2Sb_4S_{12}S$. It has been discovered in a wide variety of hydrothermal deposits as well as in a few placer deposits containing electrum. It is the most important metal in three deposit types but is more commonly a by-product. The most familiar Ag-dominant deposits are epithermal vein deposits (major sources of gold). Many of these types of veins in the United States and Mexico have Ag-to-Au ratios >100, resulting in their economics to be dominated by silver. The other two types of Ag-dominant deposits are less well known. The Cerro Rico type, which includes the largest silver deposit in the world, is best developed in the Bolivian tin-silver belt. Cerro Rico, at Potosi, Bolivia, comprises quartz-Ag-Sn veins that cut a dome of silica-rich volcanic rocks. The veins started to deposit ore minerals at temperatures around 500°C from highly saline fluids and continued to precipitate ore down to temperatures below 100°C as the solutions became less saline. This pattern indicates that the hydrothermal vein system started with hot magmatic waters and that cooler meteoric water invaded later and finally dominated the hydrothermal system. Cobalt-nickel-arsenide veins are also significant sources of silver. The largest deposit, at Cobalt, Ontario, Canada, consists of veins with native silver and Co, Ni, and Fe arsenides in calcite and quartz. The veins occur in sediments above and below a large tabular gabbro intrusion and they appear to have formed from basinal brines that were heated by the intrusion. By-product Ag comes largely from gold and base-metal deposits. In most of these deposits Ag minerals form small inclusions in base-metal sulfide minerals and they are difficult to separate.

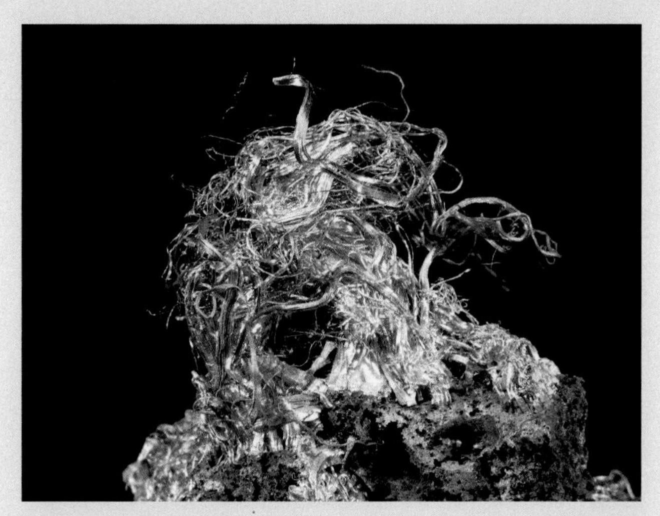

FIGURE 6.49 Curled silver, Ag, wires, 4 cm, Linqiu, Shanxi Province, China.

melts at 327°C, lead oxide at 888°C, and silver melts at 960°C. To separate the silver, the alloy is melted again at the high temperature of 960°C–1000°C in an oxidizing environment. The lead oxidizes to lead monoxide, known as litharge [which is also known as a naturally occurring mineral with the same composition PbO, the name comes from the Greek λιθάργυρος, the name given by Greek physician, pharmacologist, botanist, and author of *De Materia Medica*, Pedanius Dioscorides (c.CE 40 to CE 90) to a material obtained in the process of separating lead from silver by fire metallurgy], which captures the oxygen from the other metals present. The liquid lead oxide is separated or absorbed by capillary action (the ability of a liquid to flow in narrow spaces without the assistance of, or even in opposition to, external forces like gravity. It occurs because of intermolecular forces between the liquid and surrounding solid surfaces. If the diameter of the tube is sufficiently small, then the combination of surface tension (which is caused by cohesion within the liquid) and adhesive forces between the liquid and container wall act to propel the liquid] into the hearth linings.

$$Ag(s) + 2Pb(s) + O_2(g) \rightarrow 2PbO(absorbed) + Ag(l)$$

Currently, silver metal is instead chiefly obtained as a secondary by-product of electrolytic refining [a technique that uses a direct electric current to drive an otherwise nonspontaneous chemical reaction. Electrolysis is commercially important as a stage in the separation of elements from naturally occurring sources such as ores using an electrolytic cell] of copper, lead, and zinc, and by application of the Parkes process, patented by metallurgist and inventor from Birmingham, England, Alexander Parkes (December 29, 1813 to June 2/9, 1890) in 1850 (he received two additional patents in 1852), on lead bullion from ore that also contains silver. In these processes, silver follows the nonferrous metal such as lead or copper through its concentration and smelting and is later purified out. For example, in the production of copper, purified copper is electrolytically deposited on the cathode, whereas the less reactive precious metals such as silver and gold are collected under the anode as what is known as the "anode slime." This is subsequently isolated and purified of remaining base metals through treatment with hot aerated dilute sulfuric acid (H_2SO_4) and heating with lime or silica flux, before the silver is purified to over 99.9% purity via electrolysis in nitrate solution. Commercial-grade fine silver is at least 99.9% pure, and purities greater than 99.999% are available. In 2014, Mexico was the number one producer of silver, followed by China and Peru.

Silver is found in nearly different 200 minerals. Besides the native element silver (Ag) (Fig. 6.49), it is known in six alloys with mercury, for example, eugenite ($Ag_{11}Hg_2$) and moschellandsbergite (Ag_2Hg_3). More than 100 minerals can be found in the sulfide class, including well-known minerals such as acanthite (Ag_2S), allargentum ($Ag_{1-x}Sb_x$), argentopyrite ($AgFe_2S_3$), argyrodite (Ag_8GeS_6) (Fig. 6.50), dyscrasite (Ag_3Sb) (Fig. 6.51), hessite (Ag_2Te) (Fig. 6.52), imiterite (Ag_2HgS_2) (Fig. 6.53), miargyrite ($AgSbS_2$), petzite (Ag_3AuTe_2) (Fig. 6.54), polybasite ($[(Ag,Cu)_6(Sb,As)_2S_7][Ag_9CuS_4]$) (Fig. 6.55), proustite (Ag_3AsS_3), pyrargyrite (Ag_3SbS_3), samsonite ($Ag_4MnSb_2S_6$) (Fig. 6.56), stephanite (Ag_5SbS_4), and sylvanite ($(Au,Ag)_2Te_4$) (Fig. 6.57). Seven halide minerals contain Ag, for example, boleite ($KPb_{26}Ag_9Cu_{24}(OH)_{48}Cl_{62}$), bromargyrite (AgBr), chlorargyrite (AgCl), miersite ($(Ag,Cu)I$), and perroudite ($Hg_5Ag_4S_5(I,Br)_2Cl_2$) (Fig. 6.58). Three different oxide minerals, aurorite ($(Mn^{2+},Ag,Ca)Mn^{4+}_3O_7\cdot 3H_2O$), quetzalcoatlite ($Zn_6Cu_3(TeO_6)_2(OH)_6\cdot Ag_xPb_yCl_{x+2y}$), and stetefeldite ($Ag_2Sb_2(O,OH)_7$), are known with Ag in their structure. The only sulfate-containing Ag is argentojarosite ($AgFe^{3+}_3(SO_4)_2(OH)_6$), while the only phosphate class mineral is tillmannsite ($(Ag_3Hg)(VO_4,AsO_4)$).

FIGURE 6.50 A rich mass of silvery argyrodite, Ag_8GeS_6, 3 × 2 × 1.5 cm. Porco, Potosi, Bolivia.

FIGURE 6.51 Dyscrasite, Ag_3Sb, dull gray, deeply striated and typical, dyscrasite crystals to 1.5 cm in etched calcite matrix. St. Andreasberg dist., Harz, Lower Saxony, Germany.

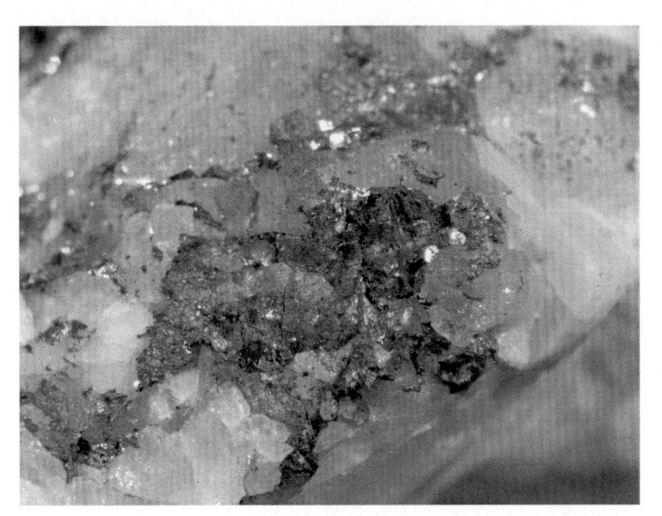

FIGURE 6.52 Hessite, Ag_2Te, silvery platy crystals to 2 mm with galena, PbS, on quartz, SiO_2, matrix. Oliver, British Columbia, Canada.

FIGURE 6.53 Imiterite, Ag_2HgS_2, silvery needles with rounded terminations to about 1 mm in calcite, $CaCO_3$, vugs. Imiter Mine, Anti-Atlas Mts., Dades Region, Morocco.

FIGURE 6.54 Two-centimeter specimen rich in gold, Au, with silvery petzite, Ag_3AuTe_2, crystals throughout the matrix, about 1 cm. Sunnyside Mine, Silverton, San Juan Co., Colorado, United States.

FIGURE 6.55 Polybasite, $[(Ag,Cu)_6(Sb,As)_2S_7][Ag_9CuS_4]$, 5 mm pseudohexagonal crystal. Double Rainbow Mine, Galena, Lawrence Co., South Dakota, United States.

FIGURE 6.56 Samsonite, $Ag_4MnSb_2S_6$, short prismatic crystals from 1 to 8 mm. Some crystals show deep internal red reflections. Samson Mine, St Andreasberg dist., Harz, Lower Saxony, Germany.

FIGURE 6.57 Sylvanite, $(Au,Ag)_2Te_4$ (Au:Ag ratio usually is close to 1:1), sharply terminated, goldish prismatic crystals, the largest of which is 4 mm. Emperor Mine, Tavua Gold Field, Viti Levu, Fiji.

FIGURE 6.58 Perroudite, $Hg_5Ag_4S_5(I,Br)_2Cl_2$, red needles less than 0.5 mm throughout the vug. Coppin Pool, Ashburton Shire, Western Australia, Australia.

6.10.3 Chemistry

Naturally occurring Ag is composed of two stable isotopes, ^{107}Ag and ^{109}Ag, with ^{107}Ag being slightly more common (51.839% natural abundancy). This almost equal abundance is rare in the periodic table. Both Ag isotopes are formed in stars via the s-process (slow neutron capture), as well as in supernovas via the r-process (rapid neutron capture). Twenty-eight radioisotopes have been observed, the most stable being ^{105}Ag which has a half-life of 41.29 days, ^{111}Ag which has a half-life of 7.45 days, and ^{112}Ag which has a half-life of 3.13 hours. Ag has a

TABLE 6.10 Silver properties.

Appearance	Lustrous white metal
Standard atomic weight ($A_{r,std}$)	107.8682
Block	d-Block
Element category	Transition metal
Electron configuration	[Kr] $4d^{10}$ $5s^1$
Phase at STP	Solid
Melting point	961.78°C
Boiling point	2162°C
Density (near r.t.)	10.49 g/cm^3
When liquid (at m.p.)	9.320 g/cm^3
Heat of fusion	11.28 kJ/mol
Heat of vaporization	254 kJ/mol
Molar heat capacity	25.350 J/(mol·K)
Oxidation states	−2, −1, **+1**, +2, +3 (amphoteric oxide)
Ionization energies	First: 731.0 kJ/mol
	Second: 2070 kJ/mol
	Third: 3361 kJ/mol
Atomic radius	Empirical: 144 pm
Covalent radius	145 ± 5 pm
Van der Waals radius	172 pm

STP, Standard temperature and pressure.
Bold font indicates main oxidation state.

number of nuclear isomers, the most stable being ^{108m}Ag with a half-life of 418 years, ^{110m}Ag with a half-life of 249.95 days, and ^{106m}Ag with a half-life of 8.28 days. All other radioisotopes have half-lives of less than an hour, and most of these radioisotopes have half-lives of less than 3 minutes. Ag isotopes range from ^{93}Ag to ^{130}Ag; the principal decay mode for the isotopes lighter than the stable isotope ^{107}Ag is electron capture and the primary mode for the heavier radioisotopes is β decay. The main decay products for the radioisotopes lighter than ^{107}Ag are Pd isotopes, and the main products for the heavier radioisotopes are Cd isotopes. The ^{107}Pd isotope decays via β emission to ^{107}Ag with a half-life of 6.5 million years. Iron meteorites are the only objects with a high-enough Pd/Ag ratio to produce quantifiable variations in ^{107}Ag abundance. Radiogenic ^{107}Ag was first discovered in the Santa Clara meteorite in 1978. The discoverers proposed that the coalescence and differentiation of iron-cored small planets may have happened 10 million years after a nucleosynthetic event. ^{107}Pd-^{107}Ag correlations found in bodies that have clearly been melted since the accretion of the solar system must be a reflection of the existence of unstable nuclides in the early solar system.

Silver is a soft, white, lustrous transition metal, it exhibits the highest electrical conductivity, thermal conductivity, and reflectivity of any metal. It has 47 electrons arranged in an electronic configuration of [Kr]$4d^{10}5s^1$. The chemistry of silver is predominantly that of the +1 oxidation state, reflecting the increasingly limited range of oxidation states along the transition series as the d orbitals fill and stabilize (Table 6.10). Silver is not attacked by nonoxidizing acids but the metal dissolves readily in hot concentrated sulfuric acid, as well as dilute or concentrated nitric acid. Silver dissolves easily in aqueous solutions of cyanide, in the presence of air, and especially in the presence of hydrogen peroxide. Silver forms cyanide complexes (silver cyanide) that are soluble in water in the presence of an excess of cyanide ions. These silver cyanide solutions are useful in electroplating applications. Silver metal is also attacked by strong oxidizers such as potassium permanganate ($KMnO_4$) and potassium dichromate ($K_2Cr_2O_7$), and in the presence of potassium bromide (KBr). The reaction of silver with these compounds is used in photography to bleach silver images, converting them to silver bromide that can either be fixed with thiosulfate or redeveloped to intensify the original image.

6.10.3.1 Oxides and chalcogenides

Silver is a rather unreactive metal. Silver and gold have rather low chemical affinities for oxygen, lower than copper, and it is thus expected that silver oxides are thermally quite unstable. Soluble silver(I) salts precipitate dark-brown silver(I) oxide, Ag_2O, after the addition of alkali. (The hydroxide AgOH exists only in solution; otherwise, it spontaneously decomposes to the oxide.) Silver oxide can be produced by reacting aqueous solutions of silver nitrate and an alkali hydroxide. This reaction does not provide significant quantities of silver hydroxide because of the favorable energetics for the following reaction:

$$2AgOH \rightarrow Ag_2O + H_2O \ (pK = 2.875)$$

Ag_2O contains linear, two-coordinate Ag centers linked by tetrahedral oxides. It is isostructural with Cu_2O. It "dissolves" in solvents that degrade it. It is slightly soluble in water because of the formation of the ion $Ag(OH)_2^-$ and perhaps related hydrolysis products. It dissolves in ammonia solution to form soluble derivatives. A slurry of Ag_2O is easily attacked by acids:

$$Ag_2O + 2HX \rightarrow 2AgX + H_2O \quad (HX = HF,\ HCl,\ HBr,\ \text{or}\ HI,\ HO_2CCF_3)$$

It will also react with solutions of alkali chlorides to precipitate silver chloride, leaving a solution of the corresponding alkali hydroxide. Similar to many silver compounds, Ag_2O is photosensitive. Ag_2O is very easily reduced to metallic silver and decomposes to silver and oxygen above 280°C. This and other silver(I) compounds may be oxidized by the strong oxidizing agent peroxodisulfate to black AgO, a mixed silver(I,III) oxide of formula $Ag(I)Ag(III)O_2$. Some other mixed oxides with silver in nonintegral oxidation states, such as Ag_2O_3 and Ag_3O_4, are also known, as is Ag_3O which behaves as a metallic conductor. Silver(I,III) oxide, Ag_4O_4, is a component of silver-zinc batteries. It can be formed by the slow addition of a silver(I) salt to a persulfate solution, for example, $AgNO_3$ to a $Na_2S_2O_8$ solution. It has an unusual structure, being a mixed-valence compound. It is a dark brown solid that decomposes with evolution of O_2 in water. It dissolves in concentrated nitric acid to give brown solutions containing the Ag^{2+} ion. Although its empirical formula, AgO, implies that silver is in the +2 oxidation state in this compound, AgO is actually diamagnetic. X-ray diffraction studies show that the silver atoms are present in two different coordination environments, one having two collinear oxide neighbors and the other four coplanar oxide neighbors. AgO is therefore formulated as $Ag(I)Ag(III)O_2$ or $Ag_2O \cdot Ag_2O_3$. It is a 1:1 molar mixture of silver(I) oxide, Ag_2O, and silver(III) oxide, Ag_2O_3. It has previously been called silver peroxide, which is incorrect since it does not contain the peroxide ion, O_2^{2-}.

Silver does react with sulfur and its compounds to form the black silver sulfide Ag_2S. It is the cause of the black tarnish on some old silver objects. It is useful as a photosensitizer in photography. It may also be formed from the reaction of hydrogen sulfide with silver metal or aqueous Ag^+ ions. Three forms have been observed: monoclinic acanthite (β-form), stable below 179°C, body-centered cubic so-called argentite (α-form) stable above 180°C, and a high-temperature face-centered cubic (γ-form) stable above 586°C. The higher temperature forms are electrical conductors. It is found in nature as the relatively low-temperature mineral acanthite, an important silver ore. The acanthite, monoclinic, form has two kinds of silver centers, one with two and the other with three near neighbor sulfur atoms. Argentite refers to a cubic form, which, due to instability at "normal" temperatures, is found in the form of the pseudomorph of acanthite after argentite (i.e., it still shows the crystal habit of argentite but has the crystal structure of acanthite, see Kloprogge and Lavinsky, 2017). Many nonstoichiometric selenides and tellurides are also known; especially, AgTe ~ 3 is a low-temperature superconductor.

6.10.3.2 Halides

Silver(II) does not react with the halogens, except for fluorine gas, with which it forms the difluoride, AgF_2. A strong yet thermally stable and therefore safe fluorinating agent, silver(II) fluoride is often used to synthesize hydrofluorocarbons. In stark contrast to this, all four silver(I) halides exist. The fluoride, chloride, and bromide have the sodium chloride structure, but the iodide has three known stable forms at different temperatures; that at room temperature is the cubic zinc blende (old name for the mineral sphalerite, ZnS) structure. They can all be prepared through the direct reaction of their respective elements. Silver nitrate can be utilized to precipitate halides; this application is useful in quantitative analysis of halides. As the halogen group is descended, the silver halide gains a more and more covalent character, solubility decreases, and the color changes from the white chloride to the yellow iodide as the energy required for ligand-metal charge transfer ($X^-Ag^+ \rightarrow XAg$) decreases. However, close attention is required for other compounds in the test solution. Some compounds can significantly increase or decrease the solubility of AgX. Examples of compounds that increase the solubility are, for example, cyanide, thiocyanate, thiosulfate, thiourea, amines, ammonia, sulfite, thioether, and crown ether. Examples of compounds that reduces the solubility are, for example, many organic thiols and nitrogen compounds that do not have a solubilizing group other than mercapto group or the nitrogen site, such as mercaptooxazoles, mercaptotetrazoles (in particular 1-phenyl-5-mercaptotetrazole), benzimidazoles (in particular 2-mercaptobenzimidazole), benzotriazole, and these compounds further substituted by hydrophobic groups. Compounds such as thiocyanate and thiosulfate increase solubility when they are present in a sufficiently large amount, because of the formation of highly soluble complex ions, but they also significantly depress solubility when present in a very small amount, because of the formation of sparingly soluble complex ions. The fluoride is anomalous, as the fluoride ion is so small that it has substantial solvation energy and hence is highly water-soluble and creates di- and tetrahydrates. The other three silver halides are highly insoluble in aqueous solutions and are very commonly used in gravimetric analytical methods. All four are photosensitive (even though the monofluoride is so only to ultraviolet (UV) light), in particular the bromide and iodide which photodecompose to silver metal, and hence were employed in traditional photography. The reaction involved is:

$$X^- + h\nu \rightarrow X + e^- \text{(excitation of the halide ion, which gives up its extra electron into the conduction band)}$$
$$Ag^+ + e^- \rightarrow Ag \text{ (liberation of a silver ion, which gains an electron to become a silver atom)}$$

The process is not reversible since the silver atom liberated is usually found at a crystal defect or an impurity site, so that the electron's energy is lowered enough that it is "trapped."

6.10.3.3 Nitrate, carbonate, and fulminate

White silver nitrate, $AgNO_3$, is a versatile precursor to numerous other silver compounds, in particular the halides, and is much less sensitive to light. It was once known as lunar caustic as silver was called luna by the ancient alchemists, who believed that silver was associated with the moon. It is frequently used for gravimetric analysis, utilizing the insolubility of the heavier silver halides for which it is a common precursor. Silver nitrate is used in many ways in organic synthesis, for example, for deprotection and oxidations. Ag^+ binds alkenes reversibly, and silver nitrate has been employed to separate mixtures of alkenes by selective absorption. The resulting adduct can be decomposed with ammonia to release the free alkene.

Yellow silver carbonate, Ag_2CO_3 can simply be produced through the reaction of aqueous solutions of sodium carbonate with a deficiency of silver nitrate. Its principal application is for the manufacture of silver powder for use in microelectronics. It is reduced with formaldehyde, forming silver free of alkali metals:

$$Ag_2CO_3 + CH_2O \rightarrow 2Ag + 2CO_2 + H_2$$

Silver carbonate is also utilized as a reagent in organic synthesis such as the Koenigs–Knorr reaction. In the Fétizon oxidation, silver carbonate on celite (diatomaceous earth) serves as an oxidizing agent to produce lactones from diols. It is also used to convert alkyl bromides into alcohols.

Silver fulminate, AgCNO, a powerful, touch-sensitive explosive used in percussion caps, is prepared through the reaction of silver metal with nitric acid in the presence of ethanol. Additional dangerously explosive silver compounds include silver azide, AgN_3, prepared by reaction of silver nitrate with sodium azide, and silver acetylide, Ag_2C_2, produced when silver reacts with acetylene gas in ammonia solution. In its most characteristic reaction, silver azide decomposes explosively, releasing nitrogen gas; given the photosensitivity of silver salts, this behavior may be produced by shining a light on its crystals.

$$2AgN_3(s) \rightarrow 3N_2(g) + 2Ag(s)$$

6.10.3.4 Organosilver compounds

Under standard conditions, silver does not form simple alkyl-silver compounds, due to the weakness of the Ag–C bond. Ag–C σ bonds with alkyls and aryls are less stable than those of copper(I) and tend to explode under ambient conditions. The poor thermal stability of simply alkyl- and aryl-silver compounds is reflected in the relative decomposition temperatures of AgMe (−50°C) and CuMe (−15°C) as well as those of PhAg (74°C) and PhCu (100°C). Perfluoroalkyl ligands provide stability to C–Ag bond and this is seen in the compound $AgCF(CF_3)_2$. Alkenyl-silver compounds are also more stable than their alkyl-silver counterparts. Interestingly, alkyl-silver compounds have also been reported in which the organic group is a di- or tri-substituted anionic $(sp^3)C^-$. For example, the transmetalation reaction between $AgBF_4$ and a tertiary bis(silyl)pyridylmethyllithium compound afforded a dinuclear alkyl-silver species, which is stable at room temperature for several days. Another organosilver compound obtained by the reaction of 1,3-dithiane 1,1,3,3-tetraoxide with silver nitrate is stable in alkaline aqueous solutions.

Silver alkynyls such as silver acetylides have been prepared and used in Pd-catalyzed coupling reactions. The silver acetylides in these reactions are seen as playing roles of halide scavenger and an external oxidant. At the same time, it also acts as nucleophile to promote innovative transformations by reacting with Pd(II) intermediates through a transmetalation step. Interest in the role of organosilver in promoting carbon–carbon bond formation is growing but there remain key challenges in developing robust and reliable organosilver reagents. Organosilver compounds are susceptible to thermal and photochemical decomposition and these reactions can compete with the desired coupling reaction. Recent works focus on the stabilization of organosilver compounds against thermal decomposition. For example, Tate et al. (2017) synthesized a series of mononuclear and dinuclear complexes of silver(I), supported by an *N*-heterocyclic carbene and bound to sp^3-, sp^2-, and sp-hybridized carbanions. The compounds showed good thermal stability compared to typical organosilver reagent. Given that organosilver compounds, under the right conditions, have been effectively used as sources of carbon-based radicals or carbanions, works on more thermally and photochemically stable organosilver compounds are expected to increase.

6.10.4 Major uses

The most important use of Ag besides in coins during most of history was in the production of jewelry and other general-use items, and this remains a major use these days. Some examples are table silver for cutlery, for which Ag is well suited because of its antibacterial properties. Western concert flutes are typically plated with or made from sterling silver; actually, most silverware is merely Ag-plated instead of made from pure Ag; the Ag is generally deposited by electroplating. Ag-plated glass is used for mirrors, vacuum flasks, and Christmas tree decorations. Since pure Ag is very soft, most Ag utilized for these uses is alloyed with Cu, with finenesses of 925/1000, 835/1000, and 800/1000 being common. One disadvantage is the rapid tarnishing of Ag when exposed to hydrogen sulfide (H_2S) and its derivatives. Incorporating precious metals such as Pd, Pt, and Au increases resistance to tarnishing but is rather expensive; base metals such as Zn, Cd, Si, and Ge do not completely prevent corrosion and have a tendency to affect the luster and color of the alloy. Electrolytically refined pure Ag plating is effective at enhancing resistance to tarnishing. The common solutions for re-establishing the luster of tarnished silver are dipping baths that reduce the silver sulfide surface to metallic Ag and cleaning off the layer of tarnish with a paste; the latter method also has the additional side effect of polishing the Ag at the same time. An easy chemical approach to remove the sulfide tarnish is to bring Ag items into contact with Al foil while immersed in water with a dissolved conducting salt, for example, NaCl.

In medicine, Ag is integrated into wound dressings and utilized as an antibiotic coating in medical devices. Wound dressings containing silver sulfadiazine ((4-amino-*N*-2-pyrimidinylbenzenesulfonamidato-*NN*,01)-silver, $C_{10}H_9AgN_4O_2S$) or Ag nanomaterials are applied to treat external infections. Ag is also utilized in several medical applications, for example, urinary catheters (where preliminary evidence suggests it lessens catheter-related urinary tract infections) and in endotracheal breathing tubes (where evidence indicates it lessens ventilator-associated pneumonia). The Ag^+ ion is bioactive and in high-enough concentration easily kills bacteria in vitro. Ag^+ ions impede enzymes in the bacteria which transport nutrients, form structures, and synthesize cell walls; these ions additionally bond with the bacteria's genetic material. Ag and Ag nanoparticles are utilized as an antimicrobial in a range of industrial, health care, and domestic uses, for example, infusing clothing with nanosilver particles which lets them remain odorless for longer. Bacteria can over time, though, develop resistance to the antimicrobial action of Ag. Ag compounds are taken up by the body like Hg compounds but do not have the toxicity of the latter. Ag and its alloys are employed in cranial surgery to replace bone, and Ag-Sn-Hg amalgams are utilized in dentistry. Ag diamine fluoride, the fluoride salt of a coordination complex with the formula $[Ag(NH_3)_2]F$, is a topical pharmaceutical used to treat and prevent dental caries (cavities) and relieve dentinal hypersensitivity.

Ag is especially valuable in electronics for conductors and electrodes due to its high electrical conductivity even when tarnished. Bulk Ag and Ag foils were used to produce vacuum tubes and are still used now in the production of semiconductor devices, circuits, and their components. For instance, Ag is utilized in high-quality connectors for RF, VHF (very high frequency), and higher frequencies, especially in tuned circuits such as cavity filters where conductors cannot be scaled by more than 6%. Printed circuits and RFID (radio-frequency identification) antennas are manufactured with Ag paints. Powdered Ag and its alloys are employed in, for example, paste preparations for conductor layers and electrodes, ceramic capacitors, and other ceramic components. Ag-containing brazing alloys are employed for brazing metallic compounds, mainly Co-, Ni-, and Cu-based alloys, tool steels, and precious metals. The basic elements are Ag and Cu, with other elements selected based on the application desired: for example, Zn, Sn, Cd, Pd, Mn, and P. Ag increases the workability and corrosion resistance during usage. Ag is useful in making chemical equipment due to its low chemical reactivity, high thermal conductivity, and workability. Ag crucibles (alloyed with 0.15% Ni to prevent recrystallization of the metal at red heat) are used for carrying out alkaline fusions. In addition, Cu and Ag are utilized when performing F chemistry. Equipment for work at high temperatures is frequently Ag-plated. Ag and its alloys with Au are employed as wire or ring seals for oxygen compressors and vacuum equipment.

Ag metal forms a good catalyst for oxidation reactions; actually it is to some extent too good for most applications, as finely divided Ag has a tendency to result in complete oxidation of organic substances to CO_2 and H_2O, and therefore coarser-grained Ag tends to be used in its place. For example, 15% Ag supported on α-Al_2O_3 or silicates is a catalyst for the oxidation of ethene (C_2H_4) to ethylene oxide (IUPAC name oxirane, C_2H_4O) at 230°C–270°C. Dehydrogenation of methanol (CH_3OH) to formaldehyde (CH_2O) is conducted at 600°C−720°C over an Ag gauze or crystals as the catalyst, as is dehydrogenation of isopropanol (IUPAC name propan-2-ol, C_3H_8O) to acetone (IUPAC name propan-2-one, $(CH_3)_2CO$). In the gas phase, glycol (ethylene glycol, IUPAC name: ethane-1,2-diol, $(CH_2OH)_2$) yields glyoxal (preferred IUPAC name oxaldehyde, $C_2H_2O_2$) and ethanol (C_2H_5OH) yields acetaldehyde (CH_3CHO), while organic amines are dehydrated to nitriles. The photosensitivity of the Ag halides permitted their use in traditional photography, but digital photography, which does not use Ag, is now dominant. The photosensitive emulsion employed in black-and-white photography is a suspension of Ag halide crystals in gelatin, possibly mixed in with some noble metal compounds for better photosensitivity, developing, and tuning. Color photography involves adding of special dye components and sensitizers, so that the original black-and-white Ag image couples with a different dye component. The initial Ag images are bleached off and the Ag is subsequently recovered and recycled. Ag nitrate is the starting compound in all cases. Nanosilver particles, between 10 and 100 nm in size, are employed in many applications. They are found in conductive inks for printed electronics and have a much lower melting point than larger Ag particles of micrometer size. In addition, they are used pharmaceutically in antibacterials and antifungals very similar to the larger Ag particles. As stated by the European Union Observatory for Nanomaterials (EUON) (https://euon.echa.europa.eu/), Ag nanoparticles are also utilized in pigments, as well as in cosmetics.

Pure Ag metal can be employed as a food coloring. It has the designation E174 and is approved in the European Union. Traditional Pakistani and Indian dishes occasionally involve decorative silver foil called vark, as well as in several other cultures, silver dragée are utilized to decorate cakes, cookies, and other dessert items. Photochromic lenses include Ag halides, as a result UV light in natural daylight liberates metallic Ag, darkening the lenses. The Ag halides are reformed in lower light intensities. Colorless AgCl films are employed in radiation detectors. Zeolite sieves containing Ag^+ ions are used to desalinate seawater during rescues, using Ag ions to precipitate chloride as AgCl. Ag is likewise utilized for its antibacterial properties for water sanitization, but the use of this is limited by restrictions on Ag consumption. Colloidal Ag is likewise employed for closed swimming pools disinfection; while it has the advantage of not giving off a smell like hypochlorite treatments do, colloidal Ag is not sufficiently active for more contaminated open swimming pools. Small AgI crystals are utilized in cloud seeding to trigger rainfall.

6.11 48 Cd — Cadmium

6.11.1 Discovery

Cadmium (Latin cadmia, Greek καδμεία meaning "calamine," a cadmium-bearing mixture of minerals that was named after the Greek mythological character Κάδμος, Cadmus, the founder of Thebes) was discovered at the same time in 1817 by German chemists Friedrich Stromeyer (August 2, 1776 to August 18, 1835) (Fig. 6.59) and Karl Samuel Leberecht Hermann

FIGURE 6.59 Friedrich Stromeyer (c.1820).

FIGURE 6.60 Karl Samuel Leberecht Hermann.

(January 20, 1765 to September 1, 1846) (Fig. 6.60), as an impurity in zinc carbonate (calamine, now known as smithsonite, $ZnCO_3$), and, for a century, Germany was the only significant producer of the metal (Hermann, 1818). The metal was named after the Latin word for calamine, as it was originally discovered in this zinc ore. Stromeyer observed that some impure samples of calamine changed color when heated but pure calamine did not. He was determined in researching these results and ultimately separated cadmium metal by roasting and reducing the sulfide. The possibility to use cadmium yellow as a pigment was recognized in the 1840s, but low availability of cadmium restricted this application (Waterston and Burton, 1844; Rowbotham and Rowbotham, 1850). While cadmium and its compounds are toxic in certain forms and concentrations, the British Pharmaceutical Codex from 1907 shows that cadmium iodide was applied as a medication to treat "enlarged joints, scrofulous glands, and chilblains." In 1907, the International Astronomical Union defined the international length unit ångström (10^{-10} m) in terms of a red cadmium spectral line (1 wavelength = 6438.46963 Å). This was approved by the seventh General Conference on Weights and Measures in 1927. In 1960, the definitions of both the meter and ångström were changed to use krypton instead of cadmium. After the industrial-scale manufacture of cadmium started in the 1930s and 1940s, the most important utilization of cadmium was the coating of iron and steel to stop corrosion; in 1944, 62% and in 1956, 59% of the cadmium in the United States was used for plating (Plating is a surface covering in which a metal is deposited on a conductive surface. Plating has been done for hundreds of years; it is also critical for modern technology. Plating is used to decorate objects, for corrosion inhibition, to improve solderability, to harden, to improve wearability, to reduce friction, to improve paint adhesion, to alter conductivity, to improve IR reflectivity, for radiation shielding, and for other purposes. Jewelry typically uses plating to give a silver or gold finish.). In 1956, 24% of the cadmium in the United States was used for a second application in red, orange, and yellow pigments from sulfides and selenides of cadmium. The heat stabilizing effect of cadmium chemicals like the carboxylates cadmium laurate ($C_{24}H_{46}CdO_4$) and cadmium stearate ($C_{36}H_{70}CdO_4$) on PVC (polyvinyl chloride, $(C_2H_3Cl)_n$) resulted in growing demand for those compounds in the 1970s and 1980s. The use of cadmium in pigments, coatings, stabilizers, and alloys sharply dropped due to environmental and health regulations in the 1980s and 1990s; in 2006, only 7% of the total cadmium use was for plating, and only 10% for pigments.

6.11.2 Mining, production, and major minerals

Cadmium makes up about 0.1 ppm of Earth's crust. It is far rarer than zinc, with which it is commonly found and makes up about 65 ppm. No significant deposits of cadmium-containing ores are known. The only cadmium mineral of significance, greenockite (CdS), is almost always found together with sphalerite (ZnS). This association is due to the geochemical similarity between zinc and cadmium, with no probable geological process to separate them. Consequently, cadmium is produced primarily as a by-product of mining, smelting, and refining sulfidic ores of zinc, and, to a lesser degree, lead and copper. Small volumes of cadmium, about 10% of usage, are produced from secondary sources, mostly from dust created by recycling iron and steel scrap. Production in the United States started back in 1907, but widespread use started after the First World War. Metallic cadmium can be found in the Vilyuy River basin in Siberia. Rocks mined for phosphate fertilizers comprise fluctuating amounts of cadmium, resulting in a cadmium concentration of as high as 300 ppm in the fertilizers and a high cadmium concentration in agricultural soils. Coal can have substantial concentrations of cadmium, which ends up generally in flue dust. Cadmium is a commonly occurring impurity in many zinc ores, and it is most often separated during the production of zinc. Some zinc ore concentrates from sulfidic zinc ores can have up to 1.4% of cadmium. In the 1970s, the average production of cadmium was 6.5 pounds per ton of zinc. Zinc sulfide ores are roasted in the presence of oxygen, converting the zinc sulfide to the oxide. Subsequently the zinc metal is obtained either by smelting the oxide with carbon or by electrolysis in sulfuric acid. Cadmium is isolated from the zinc metal by vacuum distillation if the zinc is smelted, or cadmium sulfate is precipitated from the electrolysis solution.

Only 27 minerals are known to contain cadmium. The native metal cadmium (Cd) is the only mineral in the element class. Eleven different minerals are found in the sulfide class, for example, greenockite (CdS) (Fig. 6.61) and quadratite ($Ag(Cd,Pb)AsS_3$). One oxide, monteponite (CdO), is known to contain Cd. A single carbonate, otavite ($CdCO_3$) (Fig. 6.62), is known with Cd. Within the sulfate class there are seven minerals with Cd, for example, burnsite

FIGURE 6.61 Greenockite, CdS, bright yellow pulverulent coating, about 1 cm. Sterling Hill Mine, Ogdensburg, Sussex Co., New Jersey, United States.

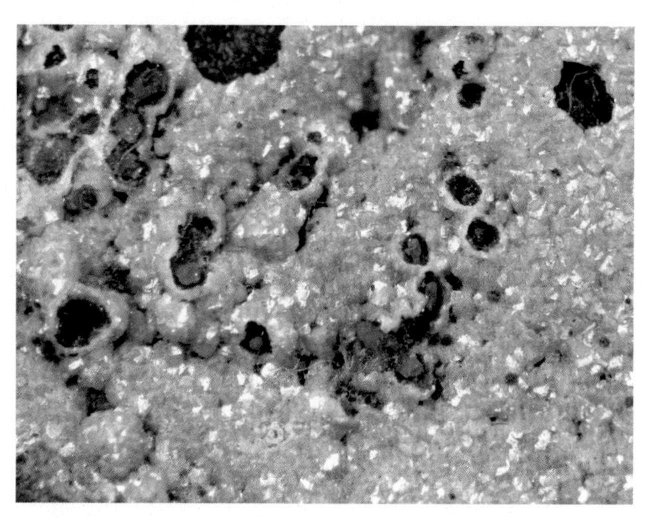

FIGURE 6.62 Otavite, $CdCO_3$, pearly white crystals to 0.5 mm on cerussite, $PbCO_3$. Tsumeb Mine, Tsumeb, Namibia.

FIGURE 6.63 Keyite, $Cu_3^{2+}Zn_4Cd_2(AsO_4)_6{\cdot}2H_2O$, single cobalt-blue crystals to 0.4 mm. Associated with colorless schultenite, $Pb(HAsO_4)$, and green cuprian adamite, $(Zn,Cu)_2AsO_4OH$. Tsumeb Mine, Tsumeb, Namibia.

($KCdCu_7^{2+}(SeO_3)_2O_2Cl_9$) and edwardsite ($Cu_3Cd_2(SO_4)_2(OH)_6{\cdot}4H_2O$). The phosphate class is represented by six different minerals, such as birchite ($Cd_2Cu_2(PO_4)_2(SO_4){\cdot}5H_2O$) and keyite ($Cu_3^{2+}Zn_4Cd_2(AsO_4)_6{\cdot}2H_2O$) (Fig 6.63).

6.11.3 Chemistry

Naturally occurring Cd consists of eight isotopes. Two of them are radioactive, ^{113}Cd (12.13% abundancy, β decay, half-life of 7.7×10^{15} years) and ^{116}Cd (7.51% abundancy two-neutrino double β decay, half-life of 2.9×10^{19} years), and three are expected to decay but have not been observed to do so under laboratory conditions. The remaining three isotopes are ^{106}Cd (1.25% abundancy), ^{108}Cd (0.89% abundancy, both double electron capture), and ^{114}Cd (28.75% abundancy, double β decay); only the lower limits on these half-lives have been obtained. As a minimum three isotopes—^{110}Cd, ^{111}Cd, and ^{112}Cd—are stable. Among the synthetic isotopes, the most long-lived are ^{109}Cd with a half-life of 462.6 days, and ^{115}Cd with a half-life of 53.46 hours. All other radioisotopes have half-lives of less than 2.5 hours, and most have half-lives of less than 5 minutes. Cd has eight known metastates, with the most stable being ^{113m}Cd with a half-life of 14.1 years, ^{115m}Cd with a half-life of 44.6 days, and ^{117m}Cd with a half-life of 3.36 hours. The known Cd isotopes range from ^{95}Cd to ^{132}Cd. For isotopes lighter than ^{112}Cd, the principal decay mode is electron capture and the dominant decay product is Ag. Heavier isotopes decay mostly via β emission producing In. One isotope, ^{113}Cd, absorbs neutrons with high selectivity: with very high probability, neutrons with energy below the cadmium cut-off will be absorbed, while those higher than the cut-off will be transmitted. The Cd cut-off is about 0.5 eV, and neutrons below that level are considered to be slow neutrons, distinct from intermediate and fast neutrons. Cd is created via the s-process in low- to medium-mass stars with masses of 0.6–10 solar masses, over thousands of years, in which an Ag atom captures a neutron and then undergoes β decay.

Cadmium is a soft, silvery-white metal. It has 48 electrons arranged in an electronic configuration of $[Kr]4d^{10}5s^2$. The two outermost electrons participate in reactions and cadmium usually has an oxidation state of +2 but also exists in the +1 state (Table 6.11). Cadmium, along with its congeners, is characterized by the stability of filled d shell. Thus these elements show few of the characteristic properties of transition metals despite their position in the d-block of the periodic table. As such, they are often not treated as transition metals. Cadmium dissolves in hydrochloric acid, sulfuric acid, and nitric acid to form cadmium chloride ($CdCl_2$), cadmium sulfate ($CdSO_4$), or cadmium nitrate ($Cd(NO_3)_2$). The oxidation state +1 can be produced by dissolving cadmium in a mixture of cadmium chloride and aluminum chloride, forming the Cd_2^{2+} cation, which is similar to the Hg_2^{2+} cation in mercury(I) chloride.

$$Cd + CdCl_2 + 2AlCl_3 \rightarrow Cd_2(AlCl_4)_2$$

TABLE 6.11 Cadmium properties.

Appearance	Silvery bluish-gray metallic
Standard atomic weight ($A_{r,std}$)	112.414
Block	d-Block
Element category	Posttransition metal, alternatively considered a transition metal
Electron configuration	[Kr] $4d^{10}$ $5s^2$
Phase at STP	Solid
Melting point	321.07°C
Boiling point	767°C
Density (near r.t.)	8.65 g/cm^3
When liquid (at m.p.)	7.996 g/cm^3
Heat of fusion	6.21 kJ/mol
Heat of vaporization	99.87 kJ/mol
Molar heat capacity	26.020 J/(mol·K)
Oxidation states	−2, +1, **+2**
Ionization energies	First: 867.8 kJ/mol Second: 1631.4 kJ/mol Third: 3616 kJ/mol
Atomic radius	Empirical: 151 pm
Covalent radius	144 ± 9 pm
Van der Waals radius	158 pm

STP, Standard temperature and pressure.
Bold font indicates main oxidation state.

6.11.3.1 Oxide

Cadmium tarnishes quickly in moist air and combines with oxygen to form brown amorphous cadmium oxide (CdO). This oxide has a crystalline form that appears dark red and changes color when heated. It is one of the key precursors to other cadmium compounds. It crystallizes in a cubic rock-salt (NaCl) lattice, with octahedral cation and anion centers. It occurs naturally as the rare mineral monteponite. Cadmium oxide can be formed as a colorless amorphous powder or as brown or red crystals. Cadmium oxide is an n-type semiconductor with a bandgap of 2.18 eV (2.31 eV) at room temperature. It is prepared by burning elemental cadmium in air. Pyrolysis of other cadmium compounds, such as the nitrate or the carbonate, also gives this oxide. When pure, it is red, but CdO is unusual in being available in many differing colors because of its tendency to form defect structures resulting from anion vacancies. Cadmium oxide is produced industrially by oxidizing cadmium vapor in air. Cadmium oxide is employed in cadmium plating baths, electrodes for storage batteries, cadmium salts, catalyst, ceramic glazes, phosphors, and nematocide. CdO is utilized as a transparent conductive material, which was produced as a transparent conducting film as early as 1907 by German physicist Karl Wilhelm Sali Baedeker (February, 3 1877 to August 6, 1914, grandson of Karl Baedeker, the founder of the eponymous travel guide publishing house). Cadmium oxide in the form of thin films has been employed in a variety of applications, for example, photodiodes, phototransistors, photovoltaic cells, transparent electrodes, liquid-crystal displays (LCD), IR detectors, and antireflection coatings. CdO microparticles undergo bandgap excitation when exposed to UV-A light and are also selective in phenol photodegradation. CdO is a basic oxide and is hence attacked by aqueous acids to produce solutions of $[Cd(H_2O)_6]^{2+}$. After treatment with strong alkaline solutions, $[Cd(OH)_4]^{2-}$ forms. A thin coat of cadmium oxide forms on the surface of cadmium in moist air at room temperature. Cadmium will oxidize at 25°C to form CdO. Cadmium vapor and steam will form CdO and hydrogen in a reversible reaction.

Cadmium hydroxide, $Cd(OH)_2$, is a white crystalline ionic compound that is a key component of Ni-Cd batteries. It is formed in storage battery anodes, in nickel-cadmium and silver-cadmium storage batteries in its discharge:

$$2NiO(OH) + 2H_2O + Cd \rightarrow Cd(OH)_2 + Ni(OH)_2$$

Cadmium hydroxide has the same structure as $Mg(OH)_2$, comprising slabs of octahedral metal centers surrounded by octahedra of hydroxide ligands. It is prepared by treating cadmium nitrate with sodium hydroxide:

$$Cd(NO_3)_2 + 2NaOH \rightarrow Cd(OH)_2 + 2NaNO_3$$

Preparation from other cadmium salts is more complicated. Cadmium hydroxide is more basic than zinc hydroxide. It forms the anionic complex $Cd(OH)_4^{2-}$ when reacted with concentrated caustic soda solution. It produces complexes with cyanide, thiocyanate, and ammonium ions when added to the solutions of these ions. Cadmium hydroxide loses water on heating, giving cadmium oxide. Decomposition starts at 130°C and is complete at 300°C. Reactions with mineral acids (HX) form the corresponding cadmium salts (CdX_2). With hydrochloric acid, sulfuric acid, and nitric acid, the products are cadmium chloride, cadmium sulfate, and cadmium nitrate, respectively.

6.11.3.2 Chalcogenides and hydride

Cd also burns in sulfur, phosphorus and the halogens on being heated. The metal does not react with hydrogen, carbon, or nitrogen. It is known to form alloys with a variety of other metals. Cadmium sulfide, CdS, is a yellow solid. It occurs in nature with two different crystal structures as the rare minerals greenockite and hawleyite but is more prevalent as an impurity substituent in the similarly structured zinc ores sphalerite and wurtzite, which are the major economic sources of cadmium. As a compound that is simple to separate and purify, it is the main source of cadmium for all industrial applications. Its vivid yellow color led to its use as a pigment for the yellow paint "cadmium yellow" in the 18th century. CdS can be produced through the precipitation from soluble cadmium(II) salts with sulfide ions. This reaction has been employed for gravimetric analysis and qualitative inorganic analysis. The preparative route and the following treatment of the product affect the polymorphic phase that is formed (i.e., cubic vs hexagonal). It has been claimed that chemical precipitation methods produce the cubic zinc blende form. Pigment production typically involves the precipitation of CdS, followed by washing of the solid precipitate to remove soluble cadmium salts and finally calcination (roasting) to transform it to the hexagonal form and subsequent milling to form a powder. When cadmium sulfide selenides are needed the CdSe is coprecipitated with CdS and the cadmium sulfoselenide is formed during the calcination step. Specialized techniques are employed to prepare CdS films as components in some photoresistors and solar cells. In the chemical bath deposition method, thin films of CdS have been produced utilizing thiourea as the sulfide anion source and an ammonium buffer solution to control pH:

$$Cd^{2+} + H_2O + (NH_2)_2CS + 2NH_3 \rightarrow CdS + (NH_2)_2CO + 2NH_4^+$$

Cadmium sulfide can also be formed using metal-organic vapor-phase epitaxy (MOVPE) and MOCVD methods via the reaction of dimethylcadmium with diethyl sulfide:

$$Cd(CH_3)_2 + Et_2S \rightarrow CdS + CH_3CH_3 + C_4H_{10}$$

Other methods to form films of CdS are, for example, sol–gel techniques, sputtering, electrochemical deposition, spraying with precursor cadmium salt, sulfur compound and dopant, and screen printing using a slurry containing dispersed CdS. CdS can be dissolved in acids.

$$CdS + 2HCl \rightarrow CdCl_2 + H_2S$$

When solutions of sulfide containing dispersed CdS particles are irradiated with light hydrogen gas is formed:

$$H_2S \rightarrow H_2 + S \ \Delta H_f = +9.4 \text{ kcal/mol}$$

The suggested mechanism entails the creation of electron–hole pairs when incident light is absorbed by the cadmium sulfide followed by these reacting with water and sulfide. Production of an electron–hole pair is as follows:

$$CdS + h\nu \rightarrow e^- + hole^+$$

Reaction of electron

$$2e^- + 2H_2O \rightarrow H_2 + 2OH^-$$

Reaction of hole

$$2hole^+ + S^{2-} \rightarrow S$$

CdS has, similar to ZnS, two crystal forms, the more stable hexagonal wurtzite structure (found in the mineral greenockite) and the cubic sphalerite structure (found in the mineral hawleyite). In both of these forms the cadmium and sulfur atoms are four-coordinated. There exists also a high-pressure phase with the NaCl rock-salt structure. CdS is a direct bandgap semiconductor (gap 2.42 eV). The proximity of its bandgap to visible light wavelengths gives it a colored appearance. CdS is utilized as pigment in plastics, showing good thermal stability, light and weather fastness, chemical resistance, and high opacity. As a pigment, CdS is called cadmium yellow (CI pigment yellow 37). The general commercial availability of cadmium sulfide from the 1840s led to it being embraced by artists, particularly Van

Gogh, Monet (in his London series and other works), and Matisse (Bathers by a river 1916–19). The presence of cadmium in paints has been used to find forgeries in paintings purported to have been made prior to the 19th century.

Cadmium selenide, CdSe, is a black to red-black solid that is classified as a II–VI semiconductor of the n-type. Most of recent research on CdSe is focused on its nanoparticles. The production of cadmium selenide has been performed in two different ways. Bulk crystalline CdSe is prepared by the high-pressure vertical Bridgman method or high-pressure vertical zone melting. CdSe can also be prepared in the form of nanoparticles. A number of different methods for the preparation of CdSe nanoparticles have been developed, for example, arrested precipitation in solution, synthesis in structured media, high-temperature pyrolysis, sonochemical and radiolytic methods. Preparation of CdSe by arrested precipitation in solution is done by introducing alkylcadmium and trioctylphosphine selenide (TOPSe) precursors into a heated solvent under controlled conditions.

$$Me_2Cd + TOPSe \rightarrow CdSe + (byproducts)$$

CdSe nanoparticles can be modified by the formation of two-phase materials with ZnS coatings. The surfaces can be further modified, for example, with mercaptoacetic acid, to confer solubility. Synthesis in structured environments describes the formation of CdSe in liquid crystal or surfactant solutions. The addition of surfactants to solutions frequently leads to a phase change in the solution resulting in liquid crystallinity. A liquid crystal is comparable to a solid crystal in that the solution has long-range translational order. Examples of this ordering are layered alternating sheets of solution and surfactant, micelles, or even a hexagonal arrangement of rods. High-temperature pyrolysis synthesis is generally performed utilizing an aerosol comprising a mixture of volatile cadmium and selenium precursors. The precursor aerosol is then carried through a furnace with an inert gas, such as hydrogen, nitrogen, or argon. In the furnace the precursors react to give CdSe as well as a number of by-products. Three crystalline forms of CdSe exist, which follow the structures of: wurtzite (hexagonal), sphalerite (cubic), and rock salt (cubic). The sphalerite CdSe structure is unstable and converts to the wurtzite structure upon moderate heating. The transition starts at about 130°C, and at 700°C it completes in less than a day. The rock-salt (NaCl) structure is only observed under high pressure.

Cadmium telluride, CdTe, is mainly used as the semiconducting material in cadmium telluride photovoltaics and an IR optical window. It is usually sandwiched with cadmium sulfide to form a p-n junction solar PV cell. Typically, CdTe PV cells use an n-i-p structure. CdTe is insoluble in water. CdTe has a high-melting point of 1041°C with evaporation starting at 1050°C. CdTe is more stable than its parent compounds cadmium and tellurium and most other Cd compounds, because of its high-melting point and insolubility. CdTe is commercially available as a powder, or as crystals. It can be produced into nanocrystals. Bulk CdTe is transparent in the IR, from close to its bandgap energy (1.5 eV at 300K/27°C, which corresponds to IR wavelength of about 830 nm) out to wavelengths greater than 20 μm; correspondingly, CdTe is fluorescent at 790 nm. As the size of CdTe crystals is reduced to a few nanometers or less, hence making them CdTe quantum dots, the fluorescence peak shifts through the visible range into the UV. CdTe can be alloyed with mercury to form a versatile IR detector compound (HgCdTe). CdTe alloyed with a small amount of zinc produces an exceptional solid-state X-ray and γ-ray detector (CdZnTe). CdTe is also employed as an IR optical compound for optical windows and lenses and is shown to provide a good performance across a wide range of temperatures. In addition, CdTe is used for electro-optic modulators. It has the greatest electro-optic coefficient of the linear electro-optic effect among II–VI compound crystals. CdTe doped with chlorine is used as a radiation detector for X-rays, γ-rays, β particles, and α particles. CdTe can operate at room temperature permitting the construction of compact detectors for a wide variety of applications in nuclear spectroscopy.

Cadmium hydride (systematically named cadmium dihydride) is an inorganic compound with the chemical formula $(CdH_2)_n$ (also written as $([CdH_2])_n$ or CdH_2). It is a solid, known only as a thermally unstable, insoluble white powder. The systematic name cadmium dihydride, a valid IUPAC name, is constructed according to the compositional nomenclature. Cadmium dihydride is also used to refer to the related molecular compound dihydridocadmium and its oligomers. Care should be taken to prevent confusing the two compounds. Cadmium hydride is also used as a compositional IUPAC name for the compound with the chemical formula CdH. Solid cadmium hydride, based on its IR spectrum, is thought to contain hydrogen-bridge bonds. Other lower metal hydrides polymerize in a similar way. Unless cooled below −20°C, cadmium hydride rapidly decomposes to form cadmium and hydrogen:

$$(CdH_2)_n \rightarrow nCd + nH_2$$

Dihydridocadmium is the monomeric, molecular form with the chemical formula CdH_2 (also written $[CdH_2]$). It is a colorless gas that does not persist undiluted. It was formed by the gas-phase reaction of excited cadmium atoms with dihydrogen, H_2, and the structure was obtained based on high-resolution IR emission spectra. The molecule is linear, with a bond length of 168.3 pm. The two-coordinate hydridocadmium group (–CdH) in hydridocadmiums such as dihydridocadmium can accept an electron-pair donating ligand into the molecule through adduction:

$$[CdH_2] + L \rightarrow [CdH_2L]$$

Due to this acceptance of the electron-pair donating ligand (L), dihydridocadmium exhibits a Lewis-acidic character. Dihydridocadmium can accept two electron pairs from ligands, as in the case of the tetrahydridocadmate(2 –) anion (CdH_4^{2-}). The compound, Cs_3CdH_5, produced via the reaction of cesium hydride, CsH, and cadmium metal powder at high temperature contains the CdH_4^{2-} ion, along with cesium cations, Cs^+, and hydride anions, H^-. The tetrahedral anion is an illustration of an ionic complex of CdH_2. The average Cd–H bond length in CdH_4^{2-} is 182 pm. In gaseous dihydridocadmium, the molecules form groups (trimers), being connected by van der Waals forces.

6.11.3.3 Halides

Cadmium fluoride (CdF_2) is a mostly water-insoluble source of cadmium used in oxygen-sensitive applications, such as the production of metallic alloys. In extremely low concentrations (ppm), this and other fluoride compounds are employed in limited medical treatment protocols. Fluoride compounds have substantial uses in synthetic organic chemistry. Cadmium fluoride is produced through the reaction of gaseous fluorine or hydrogen fluoride with cadmium metal or its salts, such as the chloride, oxide, or sulfate. It can also be formed by dissolving cadmium carbonate in 40% hydrofluoric acid solution, evaporating the solution followed by drying in a vacuum at 150°C. An alternative method of producing it is to mix cadmium chloride and ammonium fluoride solutions, followed by crystallization. The insoluble cadmium fluoride is filtered from solution. Additionally, cadmium fluoride has also been produced by reacting fluorine with cadmium sulfide. This reaction happens very rapidly and forms nearly pure fluoride at much lower temperatures than other reactions.

Cadmium chloride is the white crystalline compound, $CdCl_2$. It is a hygroscopic solid that is highly soluble in water and slightly soluble in alcohol. Although it is thought to be ionic, it has a substantial covalent character in its bonding. The crystal structure of $CdCl_2$, composed of two-dimensional layers of ions, is a reference commonly used for describing other crystal structures. Also known are the hydrated forms $CdCl_2{\cdot}H_2O$ and $CdCl_2{\cdot}5H_2O$. Anhydrous cadmium chloride can be produced through the reaction of anhydrous chlorine or hydrogen chloride gas with heated cadmium metal.

$$Cd + 2HCl \rightarrow CdCl_2 + H_2$$

Hydrochloric acid may be employed to prepare hydrated $CdCl_2$ from the metal, or from cadmium oxide or cadmium carbonate. $CdCl_2$ forms crystals with rhombohedral symmetry. Cadmium iodide, CdI_2, has a very similar crystal structure to $CdCl_2$. The individual layers in the two structures are identical, but in $CdCl_2$ the chloride ions are arranged in a *ccp* lattice, while in CdI_2 the iodide ions are arranged in an *hcp* lattice. $CdCl_2$ dissolves well in water and other polar solvents. In water, its high solubility is caused in part by the formation of complex ions such as $[CdCl_4]^{2-}$. Due to this behavior, $CdCl_2$ is a mild Lewis acid.

$$CdCl_2 + 2Cl^- \rightarrow [CdCl_4]^{2-}$$

With large cations, the trigonal bipyramidal $[CdCl_5]^{3-}$ ion can be isolated. $CdCl_2$ is utilized for the manufacture of cadmium sulfide, used as "Cadmium Yellow," a brilliant-yellow stable inorganic pigment.

$$CdCl_2 + H_2S \rightarrow CdS + 2HCl$$

In the laboratory, anhydrous $CdCl_2$ can be employed for the synthesis of organocadmium compounds of the type R_2Cd, where R is an aryl or a primary alkyl. These were previously used in the synthesis of ketones from acyl chlorides:

$$CdCl_2 + 2RMgX \rightarrow R_2Cd + MgCl_2 + MgX_2$$
$$R_2Cd + 2R'COCl \rightarrow 2R'COR + CdCl_2$$

Such reagents have largely been replaced by much less toxic organocopper compounds. $CdCl_2$ is also utilized for photocopying, dyeing, and electroplating.

Cadmium bromide is a cream-colored crystalline ionic cadmium salt of hydrobromic acid that is soluble in water. It is very toxic, along with other cadmium compounds. Cadmium bromide can be produced by heating cadmium with bromine vapor. Likewise, the compound can be formed by the reaction of dry cadmium acetate with glacial acetic acid and acetyl bromide. Alternatively, it can be prepared by dissolving cadmium or cadmium oxide in hydrobromic acid and evaporating the solution to dryness under helium in an inert atmosphere. It is employed in the production of photographic film, engraving, and lithography.

Cadmium iodide, CdI_2, is notable for its crystal structure, which is typical for compounds of the form MX_2 with strong polarization effects. Cadmium iodide is produced through the addition of cadmium metal, or its oxide, hydroxide, or carbonate to hydroiodic acid. In addition, the compound can be formed by heating cadmium with iodine. In CdI_2 the iodide anions form a hexagonal close-packed arrangement, while the cadmium cations fill all the octahedral sites in alternate layers. The resulting structure comprises a layered lattice. This same basic structure can be found in many other salts and minerals. CdI_2 is mostly ionically bonded though with a partial covalent character. CdI_2 is utilized in lithography, photography, electroplating, and the manufacturing of phosphors.

6.11.3.4 Organocadmium compounds

Organocadmium compounds have found far less synthetic application than zinc organometallics. This may be due to their somewhat reduced reactivity and their thermal and photochemical instability, as well as their toxicity. The reduced reactivity of organocadmium is demonstrated in the conversion of acyl chlorides to ketones. Other organometallics would have converted the acyl chlorides to the alcohol. This reduced activity is attributed to the less nucleophilic character of the alkyl groups bonded to Cd. This mild reactivity is exploited in selective reactions such as in the conversion of diisoamyl-cadmium with β-carbomethoxypropionyl chloride to methyl 4-keto-7-methyloctanoate without reacting further with the ketone group or the ester group. These selective reactions often provide higher and cleaner yields than the corresponding organozinc reagents.

Organocadmium compounds, and especially dimethyl-cadmium, CH_3-Cd-CH_3, have also become important reagents for the synthesis of cadmium chalcogenide films and nanoparticles. $CdMe_2$ together with diethyl-cadmium, CH_3CH_2-Cd-CH_2CH_3, were the first synthesized organocadmium compounds and were prepared in 1917 by Erich Krause. Nowadays, they are generally prepared by transmetalation or by an exchange reaction between an organometallic reagent and a cadmium salt. A host of other organocadmium compounds are prepared using this method. For example, diphenyl-cadmium was prepared by the reaction of phenyl-lithium with cadmium bromide and bis(trifluoromethyl)cadmium was prepared by transmetalation of $(CF_3)_2Hg$ with Me_2Cd in dimethoxyethane.

6.11.4 Major uses

Back in 2009, around 85% of total Cd was used in batteries, mainly in rechargeable Ni-Cd batteries. Ni-Cd cells have a nominal cell potential of 1.2 V. The cell comprises a positive Ni hydroxide electrode and a negative Cd electrode plate separated by an alkaline electrolyte (KOH). The European Union limited Cd in electronics in 2004 to 0.01%, with some exemptions, and decreased the general limit on Cd content to 0.002%. An additional type of battery is the Ag-Cd battery. Cd electroplating, using around 6% of worldwide Cd production, is employed in the aircraft industry to decrease corrosion of steel components. This coating is passivated by chromate salts A drawback of Cd plating is hydrogen embrittlement of high-strength steels from the electroplating process. Consequently, steel parts heat-treated to tensile strength above 1300 MPa should be coated by an alternate process (e.g., special low-embrittlement Cd electroplating processes or PVD). Ti embrittlement from Cd-plated tool residues caused the exclusion of those tools (and the application of routine tool assessment to detect Cd contamination) in the A-12/SR-71, U-2, and subsequent aircraft programs that use Ti. Cd is employed in the control rods of nuclear reactors, serving as a very efficient "neutron poison" to regulate the neutron flux in nuclear fission. As Cd rods are inserted in the core of a nuclear reactor, the Cd absorbs neutrons, stopping them from producing additional fission events, and hence regulating the amount of reactivity. The PWR (Pressurized Water Reactor) devised by Westinghouse Electric Company utilizes an alloy comprising 80% Ag, 15% In, and 5% Cd.

Cd oxide was historically utilized in black-and-white television phosphors and in the blue and green phosphors of color television cathode ray tubes. Cd sulfide (CdS) is employed as a photoconductive surface coating for photocopier drums. Several Cd salts are utilized in paint pigments, with CdS as a yellow pigment being the most frequently used. CdSe is a red pigment, generally known as cadmium red. To painters who work with the pigment, Cd delivers the most brilliant and durable yellows, oranges, and reds—so much so that during manufacture, these colors are extensively toned down prior to being ground with oils and binders or blended into watercolors, gouaches, acrylics, and other paint and pigment formulations. Since these pigments are potentially toxic, users should employ a barrier cream on the hands to avoid absorption through the skin although the amount of Cd absorbed into the body through the skin is reported to be $<1\%$. In PVC (polyvinyl chloride, IUPAC name poly(1-chloroethylene),

$(C_2H_3Cl)_n$), Cd compounds were added as heat, light, and weathering stabilizers. At present, Cd stabilizers have been entirely substituted with Ba-Zn, Ca-Zn, and organo-tin stabilizers. Cd is utilized in numerous types of solder and bearing alloys, as it has a low coefficient of friction and fatigue resistance. It is also present in several of the lowest-melting alloys, such as Wood's metal [also known as Lipowitz's alloy or by the commercial names Cerrobend, Bendalloy, Pewtalloy and MCP 158, is a eutectic, fusible alloy with a melting point of ~70°C. It is a eutectic alloy of 50% Bi, 26.7% Pb, 13.3% Sn, and 10% Cd by weight. The alloy is named for American dentist and inventor Barnabas Wood (May 17, 1819 to May 30, 1875).].

He-Cd lasers are a frequent source of blue-UV laser light. They work at either 325 or 422 nm in fluorescence microscopes and various laboratory experiments. CdSe quantum dots emit bright luminescence under UV excitation (e.g., He-Cd laser). The color of this luminescence is green, yellow, or red, changing with the particle size. Colloidal solutions of those particles are employed for imaging of biological tissues and solutions under a fluorescence microscope. Cd is an element of some compound semiconductors, such as CdS, CdSe, and CdTe, utilized for light detection and solar cells. HgCdTe is sensitive to IR light and can be employed as an IR detector, motion detector, or switch in remote control devices. In molecular biology, Cd can be utilized to block voltage-dependent Ca channels from fluxing Ca ions, and in hypoxia studies to stimulate proteasome-dependent degradation of HIF-1α [Hypoxia-inducible factor 1-alpha is a subunit of a heterodimeric transcription factor hypoxia-inducible factor 1 (HIF-1) that is encoded by the HIF1A gene. It is a basic helix-loop-helix PAS (Per-Arnt-Sim) domain containing protein and is considered as the master transcriptional regulator of cellular and developmental response to hypoxia. The dysregulation and overexpression of HIF1A by either hypoxia or genetic alternations have been strongly associated with cancer biology, as well as several other pathophysiologies, in particular in areas of vascularization and angiogenesis, energy metabolism, cell survival, and tumor invasion.]. Cd-selective sensors based on the fluorophore BODIPY (boron-dipyrromethene, IUPAC name 4,4-difluoro-4-bora-3a,4a-diaza-s-indacene, $C_9H_7BF_2N_2$) have been developed for imaging and sensing of Cd in cells. One of the most popular methods to monitor Cd in aqueous environments is the usage of electrochemistry; one illustration is by attaching a self-assembled monolayer that can help achieve a Cd-selective electrode with a ppt-level sensitivity.

6.12 49 In — Indium

6.12.1 Discovery

In 1863, the German chemist Ferdinand Reich (February 19, 1799 to April 27, 1882) (Fig. 6.64) and chemist and mineralogist Hieronymous Theodor Richter (November 21, 1824 to September 25, 1898) (Fig. 6.65) at the Freiberg

FIGURE 6.64 Ferdinand Reich (1932). *Weeks, M. E., Journal of Chemical Education (page 1429).*

FIGURE 6.65 Hieronymous Theodor Richter (c.1873).

University of Mining and Technology in Germany were testing ores from the mines around Freiberg, Saxony. For this they reacted the minerals pyrite (FeS_2), arsenopyrite (FeAsS), galena (PbS), and sphalerite (ZnS) with hydrochloric acid to distill raw zinc chloride. Reich, who was color-blind, had hired Richter as an assistant for distinguishing the colored spectral lines. Since they knew that ores from that region occasionally contain thallium, they investigated the emission spectra for the green thallium emission lines. In its place, they observed a bright blue line. Since that blue line did not match any known element, they theorized that a new element had to be present in the minerals. They called the element indium, from the indigo color seen in its spectrum, after the Latin indicum, meaning "of India." Richter went on to separate the metal in 1864 (Reich and Richer, 1863, 1864). An ingot of 0.5 kg (1.1 lb) was presented at the World Fair 1867. Reich and Richter later on had a major fall out after Richter claimed to be the sole discoverer.

6.12.2 Mining, production, and major minerals

Indium is the 68th most abundant element in Earth's crust at about 50 ppb. This is comparable to the crustal abundance of Ag, Bi, and Hg. It hardly ever forms its own minerals or occurs in elemental form. Only a few indium minerals such as roquesite ($CuInS_2$) are known, and none occur at high-enough concentrations for economic production. Instead, indium is generally a trace component of more common ore minerals, such as sphalerite (ZnS) and chalcopyrite ($CuFeS_2$). From these minerals, it can be isolated as a by-product during smelting. Even though the enrichment of indium in these ore deposits is high relative to its crustal abundance, it is not enough, at current prices, to support winning of indium as the main product. Indium is obtained solely as a by-product during the processing of the ores of other metals. Its main source material consists of sulfidic zinc ores, where it is mostly hosted by sphalerite (ZnS). Minor amounts are possibly also produced from sulfidic copper ores. During the roast-leach-electrowinning process steps of zinc smelting, indium is accumulated in the iron-rich residues. From these residues, it can be isolated via different processes. Additionally, it may be recovered straight from the process solutions. Further purification is performed through electrolysis. The exact method varies with the mode of operation of the smelter. China is the top producer of indium, followed by South Korea, Japan, and Canada.

In total 14 minerals are known to contain indium. Besides the native element indium (In) there are two alloys, damiaoite ($PtIn_2$) and yixunite (Pt_3In). The largest group, nine different minerals, is found in the sulfide class, for example, indite ($FeIn_2S_4$) and roquesite ($CuInS_2$) (Fig. 6.66). The only oxide mineral is dzhalindite ($In(OH)_3$) (Fig. 6.67). In the phosphate class there is a single arsenate mineral, yanomamite ($InAsO_4{\cdot}2H_2O$) (Fig. 6.68).

FIGURE 6.66 A slice of microscopic gray roquesite, $CuInS_2$, in bornite, Cu_5FeS_4, 2 mm. Akenobe Mine, Oya-cho, Hyogo, Kinki, Honshu Island, Japan.

FIGURE 6.67 Dzhalindite, $In(OH)_3$, yellow to brownish-yellow veinlets (1 mm) and patches. Dzhalinda Sn deposit, Malyi Khingan Range, Siberia, Russia.

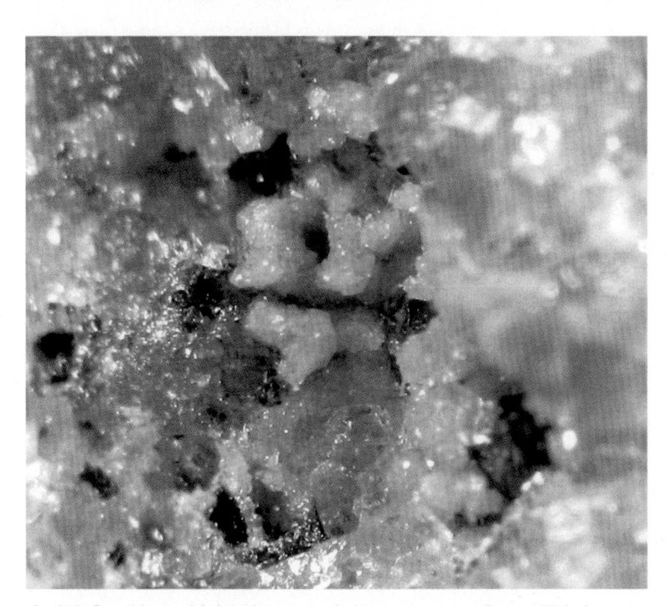

FIGURE 6.68 Yanomamite, $InAsO_4{\cdot}2H_2O$, white to bluish tiny crystals in aggregate to 2 mm. This is an extremely rare indium arsenate found at only this mine. Mangabeira Tin deposit, Monte Alegre de Goiás, Goiás, Brazil.

6.12.3 Chemistry

Thirty-nine In isotopes are known, ranging from ^{97}In to ^{135}In. Only two isotopes occur naturally as primordial nuclides: ^{113}In, the only stable isotope, and ^{115}In, which has a half-life of 4.41×10^{14} years, four orders of magnitude greater than the age of the universe and nearly 30,000 times greater than that of natural Th. The half-life of ^{115}In is extremely long since the β decay to ^{115}Sn is spin-forbidden. ^{115}In forms 95.72% of all naturally occurring indium, with only 4.28% as the stable ^{113}In. In is one of only three known elements (the others being Te and Re) of which the stable isotope is less abundant in nature than the long-lived primordial radioisotopes. The most stable synthetic isotope is ^{111}In, with a half-life of about 2.8 days. All remaining radioisotopes have half-lives shorter than 5 hours. Indium also has 47 metastates, with $^{114m1}In$ as the most stable isotope with a half-life of approximately 49.51 days, more stable than the ground state of any In isotope other than the primordial isotope. All decay is via isomeric transition. The In radioisotopes lighter than ^{115}In principally decay through electron capture or positron emission to form Cd isotopes, while the other In isotopes from ^{115}In and heavier principally decay via β^- decay to form Sn isotopes.

Indium is a member of group 13 on the periodic table and its properties are mostly intermediate between its vertical neighbors gallium and thallium. Its 49 electrons are arranged in an electronic configuration of $[Kr]4d^{10}5s^25p^1$. In most compounds, the three outermost electrons are donated to produce indium(III), In^{3+} ions. In some cases, indium takes an oxidation state lower than +3 (Table 6.12). Indium(I) is more common owing to the special stabilization of the monovalent state due to the inert pair effect, in which relativistic effects stabilize the 5s orbital, observed in heavier elements of groups 13–16. Other oxidation states such as indium(0) and indium(II) are also known and the chemistry of these In(0) and In(II) compounds are treated in more detail in a review by Pardoe and Downs (2007).

6.12.3.1 *Oxides and hydroxides*

Indium metal does not react with water, but it is oxidized by stronger oxidizing agents such as halogens to give indium (III) compounds. It does not form a boride, silicide, or carbide, and the hydride (InH_3) is too unstable unless coordinated and exists only in ethereal solutions at low temperatures. Indium is rather basic in aqueous solution, showing only slight amphoteric characteristics, and unlike the lighter aluminum and gallium, it is insoluble in aqueous alkaline solutions. Indium(III) oxide, In_2O_3, forms when indium metal is burned in air or when the hydroxide or nitrate is heated.

$$4In + 3O_2 \xrightarrow{800^\circ C} 2In_2O_3$$

TABLE 6.12 Indium properties.

Property	Value
Appearance	Silvery lustrous gray
Standard atomic weight ($A_{r,std}$)	114.818
Block	p-Block
Element category	Posttransition metal
Electron configuration	$[Kr]\ 4d^{10}\ 5s^2\ 5p^1$
Phase at STP	Solid
Melting point	156.60°C
Boiling point	2072°C
Density (near r.t.)	7.31 g/cm^3
When liquid (at m.p.)	7.02 g/cm^3
Heat of fusion	3.281 kJ/mol
Heat of vaporization	231.8 kJ/mol
Molar heat capacity	26.74 J/(mol·K)
Oxidation states	−5, −2, −1, +1, +2, **+3**
Ionization energies	First: 558.3 kJ/mol Second: 1820.7 kJ/mol Third: 2704 kJ/mol
Atomic radius	Empirical: 167 pm
Covalent radius	142 ± 5 pm
Van der Waals radius	193 pm

STP, Standard temperature and pressure.
Bold font indicates main oxidation state.

Bulk amounts can be produced by heating indium(III) hydroxide or the nitrate, carbonate or sulfate. Thin films of indium oxide can be synthesized by sputtering of indium target in argon/oxygen atmosphere. They can be employed as diffusion barriers ("barrier metals") in semiconductors, for example, to inhibit diffusion between aluminium and silicon. Monocrystalline nanowires were synthesized from indium oxide by laser ablation, allowing precise diameter control down to 10 nm. Amorphous indium oxide is insoluble in water but soluble in acids, while crystalline indium oxide is insoluble in both water and acids. The crystalline form exist in two phases, the cubic (bixbyite type, $(Mn,Fe)_2O_3$) and rhombohedral (corundum type, Al_2O_3). Both phases have a band gap of ~3 eV. The rhombohedral phase is formed at high temperatures and pressures or when using non-equilibrium growth methods. It has a space group $R\bar{3}c$ When heated to 700°C, indium(III) oxide is produced In_2O, (called indium(I) oxide or indium suboxide), which decomposes at 2000°C. It is soluble in acids but not in alkali. With ammonia at high temperature indium nitride is formed

$$In_2O_3 + 2NH_3 \rightarrow 2InN + 3H_2O$$

With K_2O and indium metal the compound K_5InO_4 containing tetrahedral InO_4^{5-} ions has been synthesized. Reacting with a range of metal trioxides produced perovskites, for example,

$$In_2O_3 + Cr_2O_3 \rightarrow 2InCrO_3$$

Indium(III) hydroxide, $In(OH)_3$, is primarily used as a precursor to indium(III) oxide, In_2O_3. It is occasionally found as the rare mineral dzhalindite. Neutralizing a solution of an In^{3+} salt such as $In(NO_3)_3$ or a solution of $InCl_3$ results in a white precipitate that on ageing forms $In(OH)_3$. A thermal decomposition of freshly prepared $In(OH)_3$ shows the first step is the conversion of $In(OH)_3{\cdot}xH_2O$ to cubic $In(OH)_3$ The precipitation of indium hydroxide was a step in the separation of indium from sphalerite ore by Reich and Richter, the discoverers of indium (Reich and Richter, 1863, 1864). $In(OH)_3$ is amphoteric, similar to $Ga(OH)_3$ and $Al(OH)_3$ but is much less acidic than $Ga(OH)_3$ having a lower solubility in alkali than in acid and is for all intents and purposes a basic hydroxide. Dissolving $In(OH)_3$ in strong alkali results in solutions that probably consists of either four coordinate $In(OH)_4^-$ or $In(OH)_4(H_2O)^-$. Reaction with acetic acid or carboxylic acids is likely to produce the basic acetate or carboxylate salt, such as $In(OH)(OOCCH_3)_2$. At 10 MPa pressure and 250-400°C $In(OH)_3$ converts to indium oxide hydroxide, InO(OH) (which has a distorted rutile, TiO_2, structure). Laser ablation of $In(OH)_3$ gives InOH, indium(I) hydroxide, a bent molecule with an In—O—H angle of around 132° and an In—O bond length of 201.7 pm.

6.12.3.2 Hydrides, nitrides, and phosphides

Indium trihydride is an inorganic compound with the chemical formula $(InH_3)_n$ (also written as $([InH_3])_n$ or InH_3). It is a covalent network solid. Moreover, it is unstable at standard temperature and pressure. The systematic name indium trihydride, a valid IUPAC name, is constructed according to the compositional nomenclature. Indium trihydride is also used to refer to the related molecular compound indigane and its oligomers. Care should be taken to avoid confusing the two compounds. For solid InH_3 a three-dimensional network polymeric structure, where In atoms are connected by In—H—In bridging bonds, has been proposed to account for the growth of broad infrared bands when samples of InH_3 and InD_3 produced on a solid hydrogen matrix are warmed. Such a structure is known for solid AlH_3. When heated above −90°C, indium trihydride decomposes to form indium—hydrogen alloy and elemental hydrogen. As of 2013, the only known method of synthesizing indium trihydride is the autopolymerization of indigane below −90°C. Indigane (also called trihydridoindium or indane, not to be confused with the hydrocarbon indane) is the monomer with the chemical formula InH_3 (also written as $[InH_3]$). It is a colorless gas that cannot persist undiluted. Indigane is the simplest of the indiganes. Unsolvated indigane will spontaneously autopolymerize, first to oligomers, and finally indium trihydride. It has been observed in matrix isolation. Gas phase stability has been predicted. The infrared spectrum was obtained in the gas phase by laser ablation of indium in the presence of hydrogen gas. Several indium hydride complexes have been reported, for example, complexes with two hydride ligands replaced by other ligands are $K_3[K(Me_2SiO)_7][HIn(Me_3CCH_2)_3]_4$ and $HIn(2\text{-}Me_2NCH_2\text{-}C_6H_4)_2$. Indium nitride, InN, is much more difficult to prepare than other group III nitrides for its low thermal stability. But, its potential importance in the optoelectronic and high-speed electronic devices is driving interests in new methods of preparation. Typical precursor materials for InN are the halides and oxides. For example, Purdy (1994) presented a method of preparing InN via decomposition of indium amide, $In(NH_2)_3$, which in turn, was synthesized from indium(III) iodide, InI_3 and potassium amide, KNH_2. Hertrampf et al. (2018), on the other hand, prepared the material using $InCl_3$ and KNH_2 in supercritical ammonia at pressures up to 3000 bar.

$$3KNH_2 + InI_3 \rightarrow 3KI + In(NH_2)_3$$
$$In(NH_2)_3 \rightarrow InN + 2NH_3$$

Schwenzer et al. (2004) reported the high-temperature (600°C–730°C) ammonolysis of In_2O_3. While Wang et al. (2016) presented a low-temperature (190°C) synthesis using indium oxide and sodium amide with the presence of sulfur.

$$In_2O_3 + NH_3 \rightarrow InN + 3H_2O$$
$$In_2O_3 + 3S + 6NaNH_2 \rightarrow 2InN + 2NH_3 + 3NaHS + 3NaOH$$

InN, when alloyed with GaN to form the ternary system InGaN, has a direct bandgap spanning from the IR (0.69 eV) to the UV (3.4 eV) and optically matches the solar spectrum. There are challenges in applying this property to actual devices as p-type doping of InN and InGaN and the heteroepitaxial growth of InN with GaN are very challenging. It is therefore expected that interest in the materials chemistry and engineering of InN will continue in the future.

Indium phosphide (InP) is a binary semiconductor with a face-centered cubic (sphalerite, ZnS) crystal structure, identical to that of GaAs and most of the III–V semiconductors. Indium phosphide can be prepared from the reaction of white phosphorus and indium iodide at 400°C, but also by direct reaction of the purified elements at high temperature and pressure, or by thermal decomposition of a mixture of a trialkyl indium compound and phosphine.

6.12.3.3 Halides

Indium(III) fluoride or indium trifluoride, InF_3, is a white solid with a rhombohedral crystal structure very similar to that of rhodium(III) fluoride. Each In center is octahedrally coordinated. It is prepared by reacting indium(III) oxide with hydrogen fluoride or hydrofluoric acid. Indium(III) fluoride is utilized in the synthesis of non-oxide glasses. It catalyzes the addition of trimethylsilyl cyanide (TMSCN) to aldehydes to form cyanohydrins. Indium(III) chloride, $InCl_3$, is a white, flaky solid with applications in organic synthesis as a Lewis acid. It is also the most available soluble derivative of indium. Being a relatively electropositive metal, indium reacts rapidly with chlorine to form the trichloride. Indium trichloride is very soluble and deliquescent. A synthesis method has been published using an electrochemical cell in a mixed methanol-benzene solution. Like $AlCl_3$ and $TlCl_3$, $InCl_3$ crystallizes as a layered structure consisting of a close-packed chloride arrangement containing layers of octahedrally coordinated In(III) centers, a structure similar to that observed for YCl_3. In contrast, $GaCl_3$ crystallizes as dimers containing Ga_2Cl_6. Molten $InCl_3$ conducts electricity, while $AlCl_3$ does not as it converts to the molecular dimer, Al_2Cl_6. Indium(III) bromide, (indium tribromide), $InBr_3$, is a Lewis acid and has been used in organic synthesis. It is prepared by reacting indium with bromine. $InBr_3$ forms complexes with ligands, L, $InBr_3L$, $InBr_3L_2$, $InBr_3L_3$. Reaction with indium metal forms lower valent indium bromides, $InBr_2$, In_4Br_7, In_2Br_3, In_5Br_7, In_7Br_9, indium(I) bromide. In refluxing xylene solution $InBr_3$ and In metal react to produce $InBr_2$. It has the same crystal structure as aluminium trichloride, with 6 coordinated indium atoms. When molten it is dimeric, In_2Br_6, and it is mainly dimeric in the gas phase. The dimer has bridging bromine atoms with a structure similar to dimeric aluminium trichloride Al_2Cl_6.

A surprising number of intermediate chlorides and bromides are known, but only one iodide, and no difluoride. Rather than the apparent oxidation state of +2, these compounds contain indium in the +1 and +3 oxidation states. The solid monohalides InCl, InBr and InI are all unstable with respect to water, decomposing to the metal and indium (III) species. They fall between gallium(I) compounds, which are more reactive and thallium(I) that are stable with respect to water. InI is the most stable. Up until relatively recently the monohalides have been scientific curiosities, however with the discovery that they can be used to prepare indium cluster and chain compounds they are now attracting much more interest. Salts of $InCl^-$, $InBr^-$ and InI^- are known. The salt $LiInF_4$ has been synthesized, although it does not contain tetrahedral anions but has an unusual layer structure with octahedrally coordinated Indium atoms. Salts of InF_6^{3-}, $InCl^{3-}$ and $InBr^{3-}$ have all been synthesized. The $InCl^{2-}$ ion has been found to be square pyramidal in the salt $(NEt_4)_2InCl_5$, with the same structure as $(NEt_4)_2TlCl_5$, but is trigonal bipyramidal in tetraphenylphosphonium pentachloroindate acetonitrile solvate. The $InBr^{2-}$ ion has similarly been found square pyramidal, although distorted, in the bis(4-chloropyridinium) salt and trigonal bipyramidal in $Bi_{37}InBr_{48}$. The In_2X^- ions contain a single bridging halogen atom. Whether the bridge is bent or linear has not been determined yet. The chloride and bromide have been detected using electrospray mass spectrometry. The In_2I^- ion has been found in the salt $CsIn_2I_7$. The caesium salts of In_2Cl^{3-} and In_2Br^{3-} both consist of binuclear anions with octahedrally coordinated Indium atoms. $In^IX_2^-$ is produced when the $In_2X_6^{2-}$ ion disproportionates. Salts containing the In^IX^{2-} ions have been produced and their vibrational spectra interpreted as exhibiting C_{3v} symmetry, due to a trigonal pyramidal geometry, with structures similar to the isoelectronic SnX^- ions. Salts of the chloride, bromide and iodide ions $(Bu_4N)_2In_2X_6$ have been synthesized. In non aqueous solvents this ion disproportionates to produce In^IX^- and $In^{III}X^-$. Following the discovery of the $In_2Br_6^{2-}$ a number of related neutral compounds containing the $In_2^{II}X_4$ unit have been produced from the reaction of indium dihalides with neutral ligands. Some chemists refer to these adducts, when used as the starting point for the synthesis of cluster compounds as 'In_2X_4' e.g. the TMEDA adduct.

6.12.3.4 Chalcogenides

Indium sulfide is typically produced by direct reaction of the elements. Production from volatile complexes of indium and sulfur, for example, dithiocarbamates (e.g. $Et_2In^{III}S_2CNEt_2$), has been studied for vapor deposition techniques. Thin films of the β complex can be produced by chemical spray pyrolysis. Solutions of In(III) salts and organic sulfur compounds (often thiourea) are sprayed onto preheated glass plates, where the chemicals react to form thin films of indium sulfide. Changing the temperature at which the chemicals are deposited and the In:S ratio can affect the optical band gap of the film. Single-walled indium sulfide nanotubes can be prepared by the use of two solvents (one in which the compound dissolves poorly and one in which it dissolves well). There is partial replacement of the sulfido ligands with O^{2-}, and the compound forms thin nanocoils, which self-assemble into arrays of nanotubes with diameters of ~10 nm, and walls approximately 0.6 nm thick. The process mimics protein crystallization. Indium(III) selenide has potential for use in photovoltaic devices and it has been studied extensively. The two most common phases, α and β, have a layered structure, while γ is a "defect wurtzite (ZnS) structure." In total, five forms (α, β, γ, δ, κ) have been observed. The α- β phase transition is accompanied by a change in electrical conductivity. The band-gap of γ-In_2Se_3 is ~1.9 eV. The crystalline form of a sample can depend on the preparation method, for example thin films of pure γ-In_2Se_3 have been synthesized from trimethylindium, $InMe_3$, and hydrogen selenide, H_2Se, applying MOCVD methods.

6.12.3.5 Organoindium compounds

Prior to the 1980s the studies on organoindium compounds are sporadic. It is only very recently that interests in the chemistry of organoindium compounds exploded. Araki et al. (1988) showed that powdered indium was superior in mediating allylation of carbonyl compounds than the existing ones using other metals (Li, Mg, Zn, Mg, Sn, Sb, Ce, Pb, and Bi) with regard to its generality, high yields, and mildness of reaction conditions. The scheme of this reaction is shown below:

3 RI + 2 In → $R_2In(\mu\text{-}I)_2InI_2$ (R, R, In, I, I, In, R, I) —(1. 2 R'COR"; 2. H^+)→ 2 R'C(OH)R"

R = allyl

Since then, various researchers have contributed to knowledge in this field and presented additional advantages of the use of indium and organoindium reagents in organic synthesis. This includes the ability of indium and organoindium mediated reactions to be carried out in water, which is cheap and environmentally friendly. Many of the reactions are also highly stereo- and regioselective with few by-products, making it easy to purify the desired product. It is therefore expected that contributions to the chemistry of organoindium reagents and intermediates will continue to grow.

6.12.4 Major uses

In 1924, In was found to stabilize nonferrous metals, and that developed into the first major application. The first large-scale use for In was coating bearings in high-performance aircraft engines during the Second World War, to shield against damage and corrosion; currently this is no longer a key use of In. New applications were found in fusible alloys, solders, and electronics. Back in the 1950s, tiny In beads were employed for the emitters and collectors of PNP alloy-junction transistors. In the mid- and late 1980s, the advancement of InP semiconductors and In-Sn oxide thin films for LCD resulted in growing interest. By 1992, the applications of thin films had grown into the leading end-use. In_2O_3 and In-Sn oxide (ITO) are utilized as transparent conductive coatings on glass substrates in electroluminescent panels. In-Sn oxide is also employed as a light filter in low-pressure Na-vapor lamps. The IR radiation is reflected back into the lamp, which raises the temperature within the tube and enhances the lamp's performance. In has numerous semiconductor-related uses. Several In compounds, for example, InSb and InP, are semiconductors with valuable properties: one precursor is usually trimethylindium (TMI, $In(CH_3)_3$), which is likewise used as the semiconductor dopant in II–VI compound semiconductors. InAs and InSb are employed for low-temperature transistors while InP is used for high-temperature transistors. The compound semiconductors InGaN and InGaP are utilized in LEDs and laser diodes. In is used in photovoltaics as the semiconductor CIGS (copper indium gallium selenide), also known as CIGS solar cells, a type of second-generation thin-film solar cell. In is used in PNP bipolar junction transistors with Ge: when soldered at low temperature, In does not stress the Ge. In wire is utilized as a vacuum seal as well as a thermal conductor in cryogenics and ultrahigh-vacuum uses, in production applications like gaskets that deform to fill gaps. In is a component in the Ga-In-Sn alloy galinstan, which is liquid at room temperature and substitutes for Hg in some thermometers. Other alloys of In with Bi, Cd, Pb, and Sn, which have higher but still low melting points (between 50°C and 100°C), are employed in fire sprinkler systems and heat regulators. In is one of the numerous substitutes for Hg in alkaline batteries to stop the Zn from corroding and releasing H_2 gas. In is added to a few dental amalgam alloys to reduce the surface tension of the Hg and permit the use of less Hg and easier amalgamation. In's high neutron-capture cross section for thermal neutrons makes it appropriate for use in control rods for nuclear reactors, usually in an alloy of 80% Ag, 15% In, and 5% Cd. In nuclear engineering, the (n,n′) reactions of ^{113}In and ^{115}In are utilized to ascertain magnitudes of neutron fluxes.

6.13 50 Sn — Tin

6.13.1 Discovery

The word tin is common among Germanic languages and can be traced back to reconstructed Proto-Germanic *tin-om; cognates comprise German Zinn, Swedish tenn, and Dutch tin. It is not found in other branches of Indo-European, except by borrowing from Germanic (e.g., Irish tinne from English). The Latin name stannum initially meant an alloy of silver and lead and came to mean "tin" in the 4th century—the prior Latin word for it was plumbum candidum, or "white lead." Stannum apparently came from an earlier stāgnum (meaning the same substance), the origin of the Romance (modern languages that evolved from Vulgar Latin between the third and eighth centuries) and Celtic terms for tin. The source of stannum/stāgnum is not known; it could be pre-Indo-European. The Meyers Konversations-Lexikon (Meyers Konversations-Lexikon or Meyers Lexikon was a major encyclopedia in the German language that existed in various editions, and by several titles, from 1839 to 1984, when it merged with the Brockhaus Enzyklopädie.) speculates quite the opposite that stannum is a derivative from (the ancestor of) the Cornish word stean, and is proof that Cornwall in the first centuries CE was the main source of tin. Tin mining and use can be dated to the beginnings of the Bronze Age around 3000 BCE, when it was found that copper objects formed of polymetallic ores with varying metal contents had different physical properties. The oldest bronze objects had a tin or arsenic content of less than 2% and are hence thought to be the result of accidental alloying due to trace amounts of other metals present in the original copper ore. The addition of a second metal to copper increases its hardness, lowers the melting temperature, and improves the casting process by producing a more fluid melt that cools to a denser, less spongy metal. This was a significant invention that permitted for the much more intricate shapes cast in closed molds of the Bronze Age. Arsenical bronze objects appeared first in the Near East where arsenic is frequently found in association with copper ore, but the health risks were quickly understood and the search for sources of the much less harmful tin ores started early in the Bronze Age. This created the demand for rare tin metal and resulted in a trade network that connected the distant sources of tin to the markets of Bronze Age cultures. Cassiterite (SnO_2), the tin oxide form of tin, was almost certainly the original source of tin in ancient times. The mineral name is derived from the term "Cassiterides" which was applied to "islands off the western coast of Europe" in pre-Roman times (the exact location of these "islands" has been hotly debated over the years, current thought is that the source was probably mainland Spain and that even 2000 years ago, traders had a habit of providing misleading locality information to protect their sources) (Fig. 6.69). Other forms of tin ores

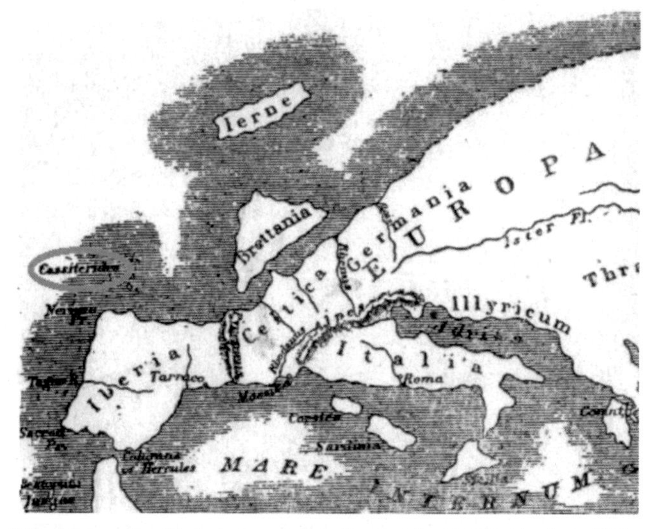

FIGURE 6.69 Map of Europe based on Strabo's geography, showing the Cassiterides just off the northwest tip of Iberia. Strabo (64 or 63 BCE to c. CE 24) was a Greek geographer, philosopher, and historian who lived in Asia Minor during the transitional period of the Roman Republic into the Roman Empire.

are less abundant; sulfide minerals such as stannite (Cu_2FeSnS_4) necessitate a more complicated smelting process. Cassiterite frequently accumulates in alluvial channels as placer deposits as it is harder, heavier, and more chemically resistant than the accompanying granite source rock. Cassiterite is generally black or commonly dark in color, and these placer deposits can be without difficulty seen in riverbanks. Alluvial (placer) deposits could be easily collected and isolated by methods similar to gold panning.

6.13.2 Mining, production, and major minerals

Tin is the 49th most abundant element in Earth's crust, representing 2 ppm compared with 75 ppm for Zn, 50 ppm for Cu, and 14 ppm for Pb. Tin does not occur as the native element. Cassiterite (SnO_2) is the only economically important source of tin, though small amounts of tin are recovered from complex sulfides such as stannite (Cu_2FeSnS_4), cylindrite ($Pb_3Sn_4FeSb_2S_{14}$), franckeite ($(Pb,Sn)_6FeSn_2Sb_2S_{14}$), canfieldite ($Ag_8SnS_6$), and teallite ($PbSnS_2$). Minerals containing tin are nearly always associated with granite rock (a common type of felsic intrusive igneous rock that is granular and phaneritic in texture), typically at a level of around 1% tin oxide content. Due to the higher specific gravity of tin dioxide, about 80% of mined tin is from secondary deposits found downstream from the primary lodes (Box 6.8). Tin is frequently recovered from granules washed downstream in the past and deposited in valleys or the sea. The most economical ways of mining tin are dredging, hydraulicking, or open pits. The majority of the world's tin is produced from placer deposits, which can contain as little as 0.015% tin. Most of the tin has been mined in China, Indonesia, Peru, Bolivia, and Brazil. Approximations of tin production have historically varied with the dynamics of economic viability and the progress of mining technologies, but it is estimated that, at recent consumption rates and technologies, the Earth will run out of mineable tin in 40 years. Secondary, or scrap, tin is now increasingly an important source of the metal. Recovery of tin through secondary production, or recycling of scrap tin, is growing fast. While the United States has neither mined since 1993 nor smelted tin since 1989, it was the largest secondary producer, recycling nearly 14,000 tons in 2006. New deposits have been reported in southern Mongolia, and in 2009, new deposits of tin were discovered in Colombia. Tin is obtained through carbothermic reduction of the oxide ore with carbon or coke. Both reverberatory (a metallurgical or process furnace that isolates the material being processed from contact with the fuel, but not from contact with combustion gases. The term reverberation is used here in a generic sense of rebounding or reflecting, not in the acoustic sense of echoing.) and electric furnace can be used.

About 100 minerals are known with tin in their crystal structure. Besides the native metal tin (Sn) there are 10 different alloys known to occur in nature, for example, niggliite (PtSn) and yuanjiangite (AuSn). In the sulfide class about

BOX 6.8 Geology of tin deposits.

The most important tin mineral is cassiterite (SnO_2), which forms ore deposits in only several parts of the world. Around 75% of the global production originates from only six countries and three of these, Malaysia, Thailand, and Indonesia, essentially represent a single tin-bearing region. The reason that tin is formed in so few places has been discussed for years with two major competing theories. One states that deposits are found only in Sn-rich parts of the crust, possibly inherited from the Earth's early history. The other theory states that tin can be found anywhere where distinct Sn-concentrating processes are active. Irrespective of which is correct, Sn deposits are nearly always found in veins associated with granite intrusions that probably formed by partial melting of older sedimentary rocks. The most common deposits, lode deposits, are found in pegmatites, quartz veins, stockworks, and disseminations clustered around protrusions known as cupolas at the top of these intrusions. In Cornwall, United Kingdom, Sn ore consists of coarse-grained muscovite, quartz, and other minerals known as greisen, which formed by high-temperature ($\sim 500°C$) alteration of granite. Where wall rocks consist of limestone, such as at Renison, Tasmania, Australia, and Dachang, China, sulfide-rich, Sn-bearing replacement deposits are formed. Tin can also be found in veinlets and stockworks related to volcanic domes in Mexico and Bolivia, including the huge Cerro Rico Sn-Ag deposit (see Box 6.7). All these types of deposits have a very close association with igneous rocks in common, suggesting that the hydrothermal fluids from which they are formed are largely of magmatic origin. Tin is an important by-product from tungsten vein and stockwork deposits and porphyry molybdenum deposits, where it probably has a similar magmatic origin. It is also found as a trace constituent in some volcanogenic massive sulfide (VMS) and sedimentary exhalative deposits. Cassiterite is highly resistant to weathering, forming placer deposits during the erosion of tin deposits. Placer tin deposits are the leading source of world tin. Those that have not moved far from their source and consist largely of cassiterite in regolith are known as residual placers, those that have moved a short distance downslope as elluvial placers, and those that are in the drainage system as alluvial, stream, or beach placers.

40 different minerals contain Sn, for example, canfieldite (Ag_8SnS_6), cylindrite ($Pb_3Sn_4FeSb_2S_{14}$), herzenbergite (SnS), kësterite (Cu_2ZnSnS_4), and stannite (Cu_2FeSnS_4) (Fig. 6.70). Two halides, abhurite ($Sn_{21}Cl_{16}(OH)_{14}O_6$) and panichiite ($(NH_4)_2SnCl_6$), have Sn in their crystal structure. The oxides are represented by around 25 different minerals, for example, cassiterite (SnO_2) (Fig. 6.71) (Box 6.9), varlamoffite (Sn_2FeO_5OH) (Fig. 6.72), wickmanite ($Mn^{2+}[Sn(OH)_6]$) (Fig. 6.73), and wodginite ($Mn^{2+}Sn^{4+}Ta_2O_8$) (Fig. 6.74). Two borates are known with Sn, nordenskiöldine ($CaSn^{4+}[BO_3]_2$) and tusionite ($Mn^{2+}Sn^{4+}[BO_3]_2$). In the silicate class 12 minerals are found with Sn, for example, eakerite ($Ca_2SnAl_2Si_6O_{18}(OH)_2{\cdot}2H_2O$) and stokesite ($CaSn[Si_3O_9]{\cdot}2H_2O$).

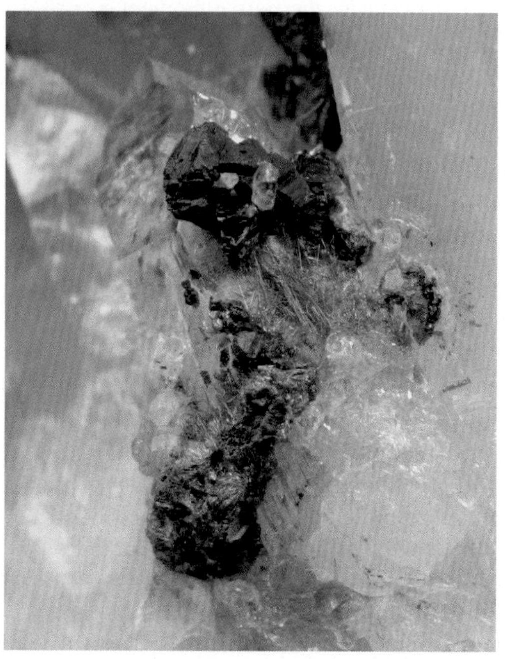

FIGURE 6.70 Combination specimen with brassy 3–5 mm stannite, Cu_2FeSnS_4, crystal groups, black sphalerite, ZnS, and arsenopyrite, FeAsS, with clear quartz, SiO_2. Yaogangxian Mine, Chenzhou, Hunan prov., China.

FIGURE 6.71 Translucent brown twinned cassiterite, SnO_2, crystals from 1.3 to 2 cm. Viloco, Araca dist., La Paz dept., Bolivia.

BOX 6.9 Cassiterite (SnO_2).

Cassiterite is the major ore mineral for tin production (see Box 6.8). The name is derived from the term "Cassiterides" which was applied to "islands off the western coast of Europe" in pre-Roman times (the exact location of these "islands" has been hotly debated over the years, current thought is that the source was probably mainland Spain and that even 2000 years ago, traders had a habit of providing misleading locality information to protect their sources). Cassiterite is tetragonal, *4/m2/m2/m* (space group $P4_2/mmm$). It has a hardness of 6–7. The cleavage on {100} is imperfect, while {110} indistinct; partings on {111} or {011}. The fracture is subconchoidal to uneven, while the tenacity is brittle. The luster is adamantine to submetallic, transparency is transparent to translucent. The color is usually brown or black, rarely yellow or white. The streak is white. Optically it is uniaxial (+). Crystals are short to long prismatic k [001], with {110} and {100} well developed, terminated by steep pyramidal forms, to 10 cm; less commonly pyramidal. In radially fibrous botryoidal crusts and concretionary masses; coarse to fine granular, massive. Twinning is very common on {011}, as contact and penetration twins, geniculated; lamellar. Frequently in elbow-shaped twins with a characteristic notch, giving rise to the miner's term *visor tin*. Crystals are occasionally cut for collectors.

FIGURE 6.72 Varlamoffite, $(Sn,Fe)(O,OH)_2$, yellow crusty alterations on matrix, overall 3.5 cm. Hingston Down Quarry, Gunnislake, Cornwall, England.

FIGURE 6.73 Tiny yellow wickmanite, $Mn^{2+}[Sn(OH)_6]$, crystals on quartz, SiO_2, crystals. San Martín Mine, Sombrerete, Zacatecas, Mexico.

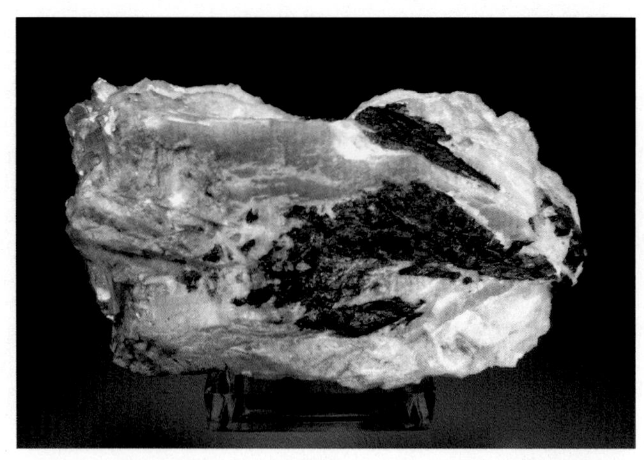

FIGURE 6.74 Six-centimeter specimen with dark reddish-brown to black wodginite, $Mn^{2+}Sn^{4+}Ta_2O_8$, crystals to 4 cm in albite, $Na(AlSi_3O_8)$, matrix. Peerless pegmatite, Pennington Co., South Dakota, United States.

6.13.3 Chemistry

Tin has 10 stable isotopes, ranging from ^{112}Sn to ^{124}Sn, the highest number for any element. Of these, the most abundant are ^{120}Sn (32.58% natural abundancy), ^{118}Sn (24.22%), and ^{116}Sn (14.54%), while the least abundant is ^{115}Sn (0.34%). The isotopes with even mass numbers have no nuclear spin, while those with odd have a spin of +1/2. Tin, with its three common isotopes ^{116}Sn, ^{118}Sn, and ^{120}Sn, is among the easiest elements to measure by NMR spectroscopy, and its chemical shifts are referenced against $SnMe_4$ (only H, F, P, Tl, and Xe have a higher receptivity for NMR spectroscopy for samples containing isotopes at their natural abundances). The large number of stable Sn isotopes is believed to be a direct consequence of the atomic number 50, a "magic number" in nuclear physics. It also can be found as 29 unstable isotopes, including all the remaining atomic masses between 99 and 137. Except for ^{126}Sn which has a half-life of 230,000 years, all the other radioisotopes have half-lives of less than a year. The radioactive ^{100}Sn and ^{132}Sn are two of the few nuclides with a "doubly magic" nucleus: notwithstanding being unstable, having very lopsided proton-neutron ratios, they represent endpoints beyond which stability drops off very quickly. Additionally, 30 metastable isomers have been observed for isotopes between ^{111}Sn and ^{131}Sn, the most stable being ^{121m}Sn which has a half-life of 43.9 years. The relative differences in the Sn stable isotope abundances can be described by their different ways of production in stellar nucleosynthesis. ^{116}Sn through ^{120}Sn inclusive are formed in the s-process (slow neutron capture) in most stars and hence they are the most common isotopes, whereas ^{122}Sn and ^{124}Sn are only formed in the r-process (rapid neutron capture) in supernovae and are less common. In addition, the isotopes ^{117}Sn through ^{120}Sn receive contributions from the r-process. Finally, the rarest proton-rich isotopes, ^{112}Sn, ^{114}Sn, and ^{115}Sn, cannot be formed in substantial quantities in either the s- or r-process and are considered among the p-nuclei, whose origins are not well understood yet. Some speculated mechanisms for their formation comprise proton capture as well as photodisintegration, while ^{115}Sn might also be partially formed in the s-process, both directly, and as the daughter of long-lived ^{115}In.

Tin (Sn) is a silvery metal with a faint yellow hue. It has two allotropes. The white, tetragonal β-Sn is stable at room temperature. This transforms into gray, cubic diamond structure α-Sn at low temperatures (13.2°C). It has 50 electrons arranged in electronic configuration of $[Kr]4d^{10}5s^25p^2$. The four electrons in the outer 5s shell are normally involved in bonding. As such, tin compounds exist normally in oxidation states of +4. The "inert pair effect" is also observed in tin and +2 oxidation states are common. In coordination compounds and transitory species of tin, other oxidation states can be found varying from −4, −3, −2, −1, +1, and +3 (Table 6.13.).

6.13.3.1 *Oxides, hydrides and nitrates*

Tin is notably more reactive and electropositive than Ge, the element directly above it, though it is still markedly amphoteric in its aqueous chemistry. It is stable toward water at ordinary temperatures but reacts with steam to give SnO_2 plus H_2. It is unreactive in air, oxygen, nitrogen, and hydrogen at ordinary conditions. But it does burn in air or oxygen to give SnO_2. In an electric discharge, it combines with nitrogen to form tin nitrides but not with hydrogen. It does form tin

TABLE 6.13 Tin properties.

Appearance	Silvery-white (β) or gray (α)
Allotropes	Alpha, α (gray); beta, β (white)
Standard atomic weight ($A_{r,std}$)	118.711
Block	p-Block
Element category	Posttransition metal
Electron configuration	[Kr] $4d^{10}$ $5s^2$ $5p^2$
Phase at STP	Solid
Melting point	231.93°C
Boiling point	2602°C
Density (near r.t.)	α: 5.769 g/cm^3, β: 7.265 g/cm^3
When liquid (at m.p.)	6.99 g/cm^3
Heat of fusion	β: 7.03 kJ/mol
Heat of vaporization	β: 296.1 kJ/mol
Molar heat capacity	β: 27.112 J/(mol·K)
Oxidation states	**−4**, −3, −2, −1, +1, **+2**, +3, **+4**
Ionization energies	First: 708.6 kJ/mol
	Second: 1411.8 kJ/mol
	Third: 2943.0 kJ/mol
Atomic radius	Empirical: 140 pm
Covalent radius	139 ± 4 pm
Van der Waals radius	217 pm

STP, Standard temperature and pressure.
Bold font indicates main oxidation state.

hydrides, SnH_n (n = 1, 2, 3 4) when tin atoms are generated by laser ablation and reacted with hydrogen in excess neon or argon. Tin shows little or no reaction with cold dilute HCl and H_2SO_4 but it reacts with dilute HNO_3 to produce $Sn(NO_3)_2$ and NH_4NO_3. Its reaction with hot concentrated hydrohalic acids yields tin(II) halides SnX_2 and H_2.

$$Sn + 2HX \rightarrow SnX_2 + H_2$$

Whereas, its reaction with hot concentrated H_2SO_4 forms $SnSO_4$ and SO_2. By contrast, the action of hot aqueous alkali yields hydroxostannate(IV) compounds.

$$Sn + 2OH^- + 4H_2O \rightarrow \left[Sn(OH)_6\right]^{2-} + 2H_2$$

6.13.3.2 Halides

Tin(II) fluoride, SnF_2, frequently referred to commercially as stannous fluoride (from Latin stannum, "tin"), is a colorless solid used as an ingredient in toothpastes. It can be produced by evaporating a SnO solution in 40% HF. Readily soluble in water, SnF_2 is hydrolyzed. At low concentration, it forms species such as $SnOH^+$, $Sn(OH)_2$, and $Sn(OH)^{3-}$. At higher concentrations, mostly polynuclear species are formed, for example, $Sn_2(OH)_2^{2+}$ and $Sn_3(OH)_4^{2+}$. Aqueous solutions easily oxidize forming insoluble Sn(IV) precipitates, which are ineffective as a dental prophylactic. Research on the oxidation with Mössbauer spectroscopy on frozen samples indicates that O_2 is the oxidizing species. SnF_2 acts as a Lewis acid. For instance, it reacts to form a 1:1 complex $(CH_3)_3NSnF_2$ and 2:1 complex $[(CH_3)_3N]_2SnF_2$ with trimethylamine, and a 1:1 complex with dimethylsulfoxide, $(CH_3)_2SO{\cdot}SnF_2$. In solutions with the fluoride ion, F^-, it produces fluoride complexes SnF_3^-, $Sn_2F_5^-$, and $SnF_2(OH_2)$. Crystallization from an aqueous NaF solution results in compounds with polynuclear anions, for example, $NaSn_2F_5$ or $Na_4Sn_3F_{10}$ depending on the reaction conditions, rather than $NaSnF_3$. $NaSnF_3$, containing the pyramidal SnF_3^- anion, can be prepared from a pyridine−water solution. Other compounds with the pyramidal SnF_3^- anion are also known, for example, $Ca(SnF_3)_2$. SnF_2 is a reducing agent, with a standard reduction potential of $E°(Sn(IV)/Sn(II)) = +0.15$ V. Solutions in HF are easily oxidized by a number of different oxidizing agents (O_2, SO_2, or F_2) to form the mixed-valence compound Sn_3F_8 (containing Sn(II) and Sn(IV) and no Sn−Sn bonds). The monoclinic form consists of tetramers, Sn_4F_8, where there are two discrete coordination environments for the Sn atoms. In each case, there are three nearest neighbors, with Sn at the apex of a trigonal pyramid, and the lone pair of electrons sterically active. Other described forms have the GeF_2 and paratellurite structures. In the vapor phase, SnF_2 forms monomers, dimers, and trimers. Monomeric SnF_2 is a nonlinear molecule with an Sn − F bond length of 206 pm. Complexes of SnF_2, sometimes called difluorostannylene, with an alkyne and aromatic compounds deposited in an argon matrix at −261°C (12K) have been described.

Tin(IV) fluoride, SnF_4, is a white solid with a melting point above 700°C. It can be produced by the reaction of tin metal with fluorine gas:

$$Sn + 2F_2 \rightarrow SnF_4$$

However, a passivating metal fluoride layer will be formed, and the surface will ultimately become unreactive. Another synthesis is the reaction of $SnCl_4$ with anhydrous hydrogen fluoride:

$$SnCl_4 + 4HF \rightarrow SnF_4 + 4HCl$$

With alkali metal fluorides (e.g., KF) hexafluorostannates are formed (e.g., K_2SnF_6), which contain the octahedral SnF_6^{2-} anion. SnF_4 behaves as a Lewis acid and adducts $L_2{\cdot}SnF_4$ and $L{\cdot}SnF_4$ have been prepared. In contrast to the other Sn tetrahalides, tin(IV) chloride, tin(IV) bromide, and tin(IV) iodide, which all contain tetrahedrally coordinated Sn, Sn(IV) fluoride consists of planar layers of octahedrally coordinated Sn, where the octahedra share four corners and there are two terminal, unshared, F atoms trans to one another. The melting point of SnF_4 is much higher (700°C) than the other tin(IV) halides which are relatively low melting, ($SnCl_4$, −33.3°C; $SnBr_4$, 31°C; SnI_4, 144°C). The structure can also be contrasted with the tetrafluorides of the lighter members of group 14, (CF_4, SiF_4, and GeF_4) which in the solid-state form molecular crystals.

Tin(II) chloride, $SnCl_2$, also known as stannous chloride, is a white crystalline solid. It forms a stable dihydrate, but aqueous solutions tend to hydrolyze, especially at higher temperatures. $SnCl_2$ is extensively used as a reducing agent (in acid solution), and in electrolytic baths for tin-plating. $SnCl_2$ possesses a lone pair of electrons, such that the molecule in the gas phase is bent. In the solid state, crystalline $SnCl_2$ forms chains linked via chloride bridges. The dihydrate is also three-coordinate, with one H_2O coordinated on to the Sn, and a second H_2O coordinated to the first. The main part of the molecule is stacked into double layers in the crystal lattice, with the "second" H_2O sandwiched between the layers. Anhydrous $SnCl_2$ is formed by the reaction of dry hydrogen chloride gas with tin metal. The dihydrate is produced by a comparable reaction, using hydrochloric acid:

$$Sn(s) + 2HCl(aq) \rightarrow SnCl_2(aq) + H_2(g)$$

The water is then carefully evaporated from the acidic solution to form crystals of $SnCl_2{\cdot}2H_2O$. This dihydrate can be dehydrated to the anhydrous form using acetic anhydride. Sn(II) chloride can dissolve in less than its own mass of water without apparent decomposition, but as the solution is diluted, hydrolysis occurs to form an insoluble basic salt:

$$SnCl_2(aq) + H_2O(l) \rightleftharpoons Sn(OH)Cl(s) + HCl(aq)$$

Consequently, if clear solutions of Sn(II) chloride are to be used, it must be dissolved in hydrochloric acid (characteristically of the same or higher molarity as the stannous chloride) to maintain the equilibrium of the reaction toward the left-hand side (using Le Chatelier's principle). $SnCl_2$ solutions are likewise unstable toward oxidation by oxygen in the air:

$$6SnCl_2(aq) + O_2(g) + 2H_2O(l) \rightarrow 2SnCl_4(aq) + 4Sn(OH)Cl(s)$$

This can be prevented by storing the solution over lumps of tin metal. There are many instances where Sn(II) chloride acts as a reducing agent, reducing silver, and gold salts to the metal, and Fe(III) salts to Fe(II), for example,

$$SnCl_2(aq) + 2FeCl_3(aq) \rightarrow SnCl_4(aq) + 2FeCl_2(aq)$$

It also reduces Cu(II) to Cu(I). Solutions of Sn(II) chloride can similarly serve just as a source of Sn^{2+} ions, which can produce other tin(II) compounds via precipitation reactions. For instance, the reaction with sodium sulfide results in the brown/black Sn(II) sulfide:

$$SnCl_2(aq) + Na_2S(aq) \rightarrow SnS(s) + 2NaCl(aq)$$

If alkali is added to a solution of $SnCl_2$, a white precipitate of hydrated tin(II) oxide is initially formed; this subsequently dissolves in excess base to produce a stannite salt such as sodium stannite:

$$SnCl_2(aq) + 2NaOH(aq) \rightarrow SnO{\cdot}H_2O(s) + 2NaCl(aq)$$
$$SnO{\cdot}H_2O(s) + NaOH(aq) \rightarrow NaSn(OH)_3(aq)$$

Anhydrous $SnCl_2$ can be used to prepare various Sn(II) compounds in nonaqueous solvents. For instance, the lithium salt of 4-methyl-2,6-di-tert-butylphenol reacts with $SnCl_2$ in THF to form the yellow linear two-coordinate compound $Sn(OAr)_2$ (Ar=aryl). Sn(II) chloride also acts as a Lewis acid, forming complexes with ligands such as chloride ion, for example,

$$SnCl_2(aq) + CsCl(aq) \rightarrow CsSnCl_3(aq)$$

The majority of these complexes are pyramidal, and as complexes such as $SnCl_3$ have a full octet, there is a little tendency to add more than one ligand. The lone pair of electrons in these complexes is available for bonding, though, and consequently the complex itself can act as a Lewis base or ligand. This can be observed in the ferrocene-related product of the following reaction:

$$SnCl_2 + Fe(\eta^5 - C_5H_5)(CO)_2HgCl \rightarrow Fe(\eta^5 - C_5H_5)(CO)_2SnCl_3 + Hg$$

$SnCl_2$ can be utilized to produce a variety of such compounds containing metal–metal bonds. For instance, the reaction with dicobalt octacarbonyl:

$$SnCl_2 + Co_2(CO)_8 \rightarrow (CO)_4Co - (SnCl_2) - Co(CO)_4$$

Tin(IV) chloride, $SnCl_4$, also known as tin tetrachloride or stannic chloride, is a colorless hygroscopic liquid that fumes on contact with air. It is utilized as a precursor to other tin compounds. It was first discovered by German historian, poet, and physician (and alchemist) Andreas Libavius (c. 1550 to July 1616) and was known as *spiritus fumans libavii*. It is produced through the reaction of chlorine gas with tin at 115°C:

$$Sn + 2Cl_2 \rightarrow SnCl_4$$

Besides water, other Lewis bases form adducts, such as ammonia and organophosphines. With hydrochloric acid the complex $[SnCl_6]^{2-}$ is formed producing the so-called hexachlorostannic acid. Anhydrous Sn(IV) chloride is a key precursor in organotin chemistry. Upon treatment with Grignard reagents, Sn(IV) chloride results in tetraalkyltin compounds:

$$SnCl_4 + 4RMgCl \rightarrow SnR_4 + 4MgCl_2$$

Anhydrous tin(IV) chloride reacts with tetraorganotin compounds in redistribution reactions:

$$SnCl_4 + SnR_4 \rightarrow 2SnCl_2R_2$$

These organotin halides are more useful than the tetraorganotin derivatives. While a specialized application, $SnCl_4$ is used in Friedel–Crafts reactions as a Lewis acid catalyst for alkylation and cyclisation. Stannic chloride is utilized in reactions with fuming (90%) nitric acid for the selective nitration of activated aromatic rings in the presence of inactivated ones. Anhydrous Sn(IV) chloride solidifies at -33°C to give monoclinic crystals with the $P2_1/c$ space group. It is isostructural with $SnBr_4$. The molecules have a near-perfect tetrahedral symmetry with average Sn–Cl distances of 227.9(3) pm. A number of hydrates of $SnCl_4$ are known. The pentahydrate, $SnCl_4 \cdot 5H_2O$ was previously known as butter of tin. They all contain $[SnCl_4(H_2O)_2]$ molecules together with varying amounts of H_2O of crystallization. The additional H_2O molecules link together the molecules of $[SnCl_4(H_2O)_2]$ through hydrogen bonds. Although the pentahydrate is the most common hydrate, lower hydrates have also been described.

Tin(II) bromide, $SnBr_2$, can be prepared via the reaction of metallic tin with HBr distilling off the H_2O/HBr and cooling:

$$Sn + 2HBr \rightarrow SnBr_2 + H_2$$

But the reaction will form tin (IV) bromide in the presence of oxygen. In the gas phase $SnBr_2$ is nonlinear with a bent configuration like $SnCl_2$ in the gas phase. The Br–Sn–Br angle is 95° and the Sn–Br bond length is 255 pm. There is evidence of dimerization in the gaseous phase. The solid-state structure is related to that of $SnCl_2$ and $PbCl_2$ and the tin atoms have five near bromine atom neighbors in an approximately trigonal bipyramidal configuration. $SnBr_2$ is soluble in donor solvents such as acetone, pyridine, and dimethylsulfoxide to give pyramidal adducts. Several hydrates are known, $2SnBr_2 \cdot H_2O$, $3SnBr_2 \cdot H_2O$, and $6SnBr_2 \cdot 5H_2O$, which in the solid phase have tin coordinated by a distorted trigonal prism of six bromine atoms with Br or H_2O capping one or two faces. When dissolved in HBr the pyramidal $SnBr_3^-$ ion is formed. Similar to $SnCl_2$ it is a reducing agent. With various alkyl bromides oxidative addition can occur to form the alkyltin tribromide, from example,

$$SnBr_2 + RBr \rightarrow RSnBr_3$$

Sn(II) bromide can act as a Lewis acid forming adducts with donor molecules, for example, trimethylamine where it produces $NMe_3{\cdot}SnBr_2$ and $2NMe_3{\cdot}SnBr_2$. It can also act as both donor and acceptor in, for example, the complex $F_3B{\cdot}SnBr_2{\cdot}NMe_3$ where it is a donor to boron trifluoride and an acceptor to trimethylamine.

Tin(IV) bromide, $SnBr_4$, is a colorless low melting solid. It can be produced by reaction of the elements at normal temperatures:

$$Sn + 2Br_2 \rightarrow SnBr_4$$

In aqueous solution $Sn(H_2O)_6^{4+}$ is the main ionic species among a range of six coordinate ions with from 0 to 6 bromide ligands (e.g., $Sn(H_2O)_6^{4+}$, $SnBr(H_2O)_5^{3+}$). In basic solution the $Sn(OH)_6^{2-}$ ion is present. $SnBr_4$ forms 1:1 and 1:2 complexes with ligands, for example, with trimethylphosphine $SnBr_4{\cdot}P(CH_3)_3$ and $SnBr_4{\cdot}2P(CH_3)_3$ can be formed. $SnBr_4$ crystallizes in a monoclinic form with molecular $SnBr_4$ units that have a distorted tetrahedral geometry.

Tin(II) iodide, SnI_2, also known as stannous iodide, is an ionic tin salt. It has a formula weight of 372.519 g/mol. It is a red to red-orange solid. Its melting point is 320°C, and its boiling point is 714°C. Tin(IV) iodide, also known as stannic iodide, is the chemical compound with the formula SnI_4. This tetrahedral molecule crystallizes as a bright orange solid that readily dissolves in nonpolar solvents such as benzene. The compound is typically produced by the reaction of iodine and tin:

$$Sn + 2I_2 \rightarrow SnI_4$$

The compound hydrolyzes in water. In aqueous hydroiodic acid, it reacts to produce a rare example of a hexaiodometallate:

$$SnI_4 + 2I^- \rightarrow [SnI_6]^{2-}$$

6.13.3.3 Chalcogenides

Tin(II) sulfide, SnS, occurs naturally as the rare mineral herzenbergite (α-SnS). At elevated temperatures above 632°C (905K), SnS undergoes a second order phase transition to β-SnS (space group: *Cmcm*). Recently, it has become apparent that a new polymorph of SnS exist based upon the cubic crystal system, π-SnS (space group: *P213*). Sn(II) sulfide can be produced through the reaction of tin with sulfur, or tin(II) chloride with hydrogen sulfide.

$$Sn + S \rightarrow SnS$$
$$SnCl_2 + H_2S \rightarrow SnS + 2HCl$$

Sn(II) sulfide is a dark brown or black solid, insoluble in water, but soluble in concentrated hydrochloric acid. Sn(II) sulfide is soluble in $(NH_4)_2S$. It has a layer structure like that of black phosphorus. Similar to black phosphorus, tin(II) sulfide can be ultrasonically exfoliated in liquids to produce atomically thin semiconducting SnS sheets that have a wider optical band gap (>1.5 eV) than the bulk crystal. Sn(II) sulfide is an interesting potential candidate for the next generation thin film solar cells. At present, both Cadmium Telluride and Copper Indium Gallium Sulfide (CIGS) are used as p-type absorber layers, but they are formed from toxic, scarce constituents. Sn(II) sulfide, however, is produced from cheap, abundant elements, and is nontoxic. It also has a high optical absorption coefficient, p-type conductivity, and a mid-range direct band gap of 1.3–1.4 eV, essential electronic properties for this type of absorber layer. Based on the detailed balance calculation using the material bandgap, the power conversion efficiency of a solar cell using a Sn(II) sulfide absorber layer could be as high as 32%, comparable to crystalline silicon. Lastly, Sn(II) sulfide is stable in both alkaline and acidic conditions. All these characteristics suggest tin(II) sulfide as an interesting material to be used as a solar cell absorber layer. Currently, Sn(II) sulfide thin films for application in photovoltaic cells are still in the research phase of development with power conversion efficiencies less than 5%. Barriers for application comprise a low open circuit voltage and an inability to realize many of the above properties due to challenges in fabrication, but S(II) sulfide remains a promising material if these technical challenges are overcome.

Tin(IV) sulfide, SnS_2, crystallizes in the cadmium iodide motif, with the Sn(IV) situated in "octahedral holes" defined by six sulfide centers. It occurs naturally as the rare mineral berndtite. It is useful as a semiconductor material with a band gap of 2.2 eV. It precipitates as a brown solid upon the addition of H_2S to solutions of Sn(IV) species. This reaction is reversed at low pH. Crystalline SnS_2 has a bronze color and is used in decorative coating where it is known as mosaic gold. The material also reacts with sulfide salts to produce a series of thiostannates with the general formula $[SnS_2]_m[S]_n^{2n-}$. A simplified equation for this depolymerization reaction is:

$$SnS_2 + S^{2-} \rightarrow 1/x\left[SnS_3^{2-}\right]_x$$

Tin selenide, SnSe, is also known as stannous selenide. Sn(II) selenide is a narrow band-gap (IV–VI) semiconductor and has received significant attention for uses such as low-cost photovoltaics and memory-switching devices. Sn (II) selenide is a typical layered metal chalcogenide; that is, it includes a Group 16 anion (Se^{2-}) and an electropositive element (Sn^{2+}), and it is arranged in a layered structure. Sn(II) selenide shows low thermal conductivity as well as reasonable electrical conductivity, creating the prospect of it being utilized in thermoelectric materials. Sn(II) selenide can be produced by reacting the elements tin and selenium above 350°C. Problems with the composition are often encountered during synthesis. Two phases exist—the hexagonal $SnSe_2$ phase and the orthorhombic SnSe phase. Specific nanostructures can be produced, but few 2D nanostructures have been formed. Both square SnSe nanostructures and single-layer SnSe nanostructures have been synthesized. Historically, phase-controlled synthesis of 2D tin selenide nanostructures has been rather difficult. Sheet-like nanocrystalline SnSe with an orthorhombic phase has been produced with good purity and crystallization via a reaction between a selenium alkaline aqueous solution and Sn(II) complex at room temperature under atmospheric pressure. SnSe nanocrystals have likewise been formed by a gas-phase laser photolysis reaction using $Sn(CH_3)_4$ and $Se(CH_3)_2$ as precursors. A few-atom-thick SnSe nanowires can be grown inside narrow (~1 nm diameter) single-wall carbon nanotubes by heating the nanotubes with SnSe powder in vacuum at 960°C. In contrast to the bulk SnSe, they have the cubic crystal structure. Sn(II) selenide has a layered orthorhombic crystal structure at room temperature, which can be derived from a three-dimensional distortion of the NaCl structure. There are two-atom-thick SnSe slabs (along the b–c plane) with strong Sn–Se bonding within the plane of the slabs, which are then linked with weaker Sn–Se bonding along the a-axis direction. The structure consists of highly distorted $SnSe_7$ coordination polyhedra, which have three short and four very long Sn–Se bonds, and a lone pair of the Sn^{2+} sterically accommodated between the four long Sn–Se bonds. The two-atom-thick SnSe slabs are corrugated, generating a zig-zag accordion-like projection along the b-axis. The easy cleavage in this system is along the (100) planes. While cooling from its high-temperature, the higher symmetry phase (space group *Cmcm*), SnSe undergoes a displacive (shear) phase transition at ~477°C–527°C (~750K–800K), resulting in a lower symmetry *Pnma* space group. Due to this layered, zig-zag accordion-like structure, SnSe demonstrates low anharmonicity and an intrinsically ultralow lattice thermal conductivity, making SnSe one of the world's least thermally conductive crystalline materials. Heat cannot travel well through this material because of its very "soft," accordion-like layered structure, which does not transmit vibrations well either.

Tin telluride, SnTe, is a IV–VI narrow band gap semiconductor and has direct band gap of 0.18 eV. It is frequently alloyed with Pb to make lead tin telluride, which is used as an infrared detector material. Tin telluride normally forms a p-type semiconductor (extrinsic semiconductor) due to Sn vacancies and is a low temperature superconductor. SnTe exists in three crystal phases. At Low temperatures, where the concentration of hole carriers is less than $1.5 \times 10^{20}\ cm^{-3}$, it exists as a rhombohedral phase also known as α-SnTe. At room temperature and atmospheric pressure, SnTe exists in NaCl-like cubic crystal phase, known as β-SnTe. While at 18 kbar pressure, β-SnTe transforms to γ-SnTe, orthorhombic phase, space group *Pnma*. This phase change is characterized by 11% increase in density and 360% increase in resistance for γ-SnTe. SnTe is a thermoelectric material. Theoretical studies infer that the n-type performance may be particularly good. Commonly, Pb is alloyed with SnTe in order to access interesting optical and electronic properties. In addition, as a result of quantum confinement, the band gap of the SnTe increases beyond the bulk band gap, covering the mid-IR wavelength range.

6.13.3.4 Organotin compounds

As early as 1860, a diversity of organotin compounds and their chemistry was already established. It started in 1849 with Edward Frankland trying to isolate the ethyl radical and ended up accidentally discovering an organostannane compound. This was later analyzed in 1852 to be diethyltin diiodide, $(C_2H_5)_2SnI_2$. Inspired by this work, many researchers including Frankland and his group made variations in the halides and alkyl groups and a number of organotin compounds with general formulas, R_2SnX_2 and R_3SnX ($R = C_2H_5$, CH_3; $X = I$, Cl, Br), were produced. Moreover, applying the known chemistry of trialkylstibine and trialkyl bismuth compounds at that time to tin yielded the polystannanes, $[(C_2H_5)_2Sn]_n$. This was found to be converted to $[(C_2H_5)_2SnO]_n$ when a solution of polystannanes in diethylether was left in air and then the solvent evaporated. The above compounds were then used as starting materials for preparation of other organotin compounds. The discovery of Grignard reagents in the 1900s further increased the number of organotin compounds that can be synthesized. The Grignard synthesis has become a classic method of preparing organotin such as in the preparation of tetraalkyl-tin, R_4Sn, as shown below.

$$4RMgBr + SnCl_4 \rightarrow R_4Sn + 4MgClBr$$

6.13.4 Major uses

Tin has long been used in alloys with Pb as solder, in quantities between 5 and 70% w/w. Sn with Pb forms a eutectic mixture of 61.9% Sn and 38.1% Pb by weight (or atomic proportions: 73.9% Sn and 26.1% Pb), with a melting temperature of 183°C. These solders are mainly utilized for joining pipes or electric circuits. After the European Union Waste Electrical and Electronic Equipment Directive (WEEE Directive) and Restriction of Hazardous Substances Directive came into effect on July 1, 2006, the amount of Pb in these types of alloys has been reduced. Substituting Pb results in numerous problems, not in the least a higher melting point, and the development of Sn whiskers triggering electrical problems. Sn pest can appear in Pb-free solders, resulting in loss of the soldered joint. Alternative alloys are fast being discovered, though difficulties related to joint integrity persist.

Sn bonds easily to Fe and is employed for coating Pb, Zn, and steel to inhibit corrosion. Tinplated steel containers are extensively used for food preservation, and this forms a large segment of the market for metallic Sn. A tinplate canister for preserving food was first produced in 1812 in London. The British call them "tins," while Americans call them "cans" or "tin cans." One derivation of the word is the slang term "tinnie" or "tinny," meaning "can of beer" in Australia. The tin whistle is so called as it was originally mass-produced in Sn-plated steel. Cu cooking vessels such as saucepans and frying pans are often coated with a thin Sn plating, as interaction of acid foods with Cu can form toxic compounds.

Sn together with other elements forms a wide range of valuable alloys. Sn is most frequently alloyed with Cu. Pewter contains 85%–99% Sn; bearing metal has a high Sn percentage too. Bronze is mainly Cu (12% tin), while adding P results in phosphor bronze. Bell metal is likewise a Cu-Sn alloy, comprising 22% Sn. Sn has occasionally been employed in coins; for example, it once formed a single-digit percentage (typically 5% or less) of American and Canadian pennies. As Cu is frequently the major metal in these coins, occasionally including Zn, these could be termed bronze and/or brass alloys. The Nb-Sn compound Nb_3Sn is commercially utilized in coils of superconducting magnets for its high critical temperature (−255.15°C/18K) and critical magnetic field (25 T). A superconducting magnet like this weighing as little as 2 kg is capable to produce a magnetic field like that of a conventional electromagnet weighing tons. A small amount of Sn is included in Zr alloys for the cladding of nuclear fuel. The majority of metal pipes in a pipe organ are made of an Sn-Pb alloy, with 50/50 as the most frequent composition. The percentage of Sn in the pipe characterizes the pipe's tone, as Sn has a desirable tonal resonance. When an Sn-Pb alloy cools, the Pb cools marginally faster and creates a mottled or spotted effect. This metal alloy is known as spotted metal. Key benefits of using Sn for pipes comprise its appearance, its workability, and resistance to corrosion.

In and Sn oxides are electrically conductive and transparent and are utilized to produce transparent electrically conducting films with uses in optoelectronics devices for instance LCD. Punched Sn-plated steel, also known as pierced tin, is an artisan technique originating in central Europe for creating housewares that are both functional and decorative. In America, pie safes and food safes were used in history before refrigeration. These safes consisted of wooden cupboards of various styles and sizes—either floor-standing or hanging cupboards intended to deter vermin and insects and to keep dust from perishable products. These cupboards had tinplate inserts in the doors and occasionally in the sides, punched out by the homeowner, cabinetmaker, or a tinsmith in various shapes to let air circulate while at the same time excluding flies. Modern reproductions are still popular in North America. Window glass is most often produced by floating molten glass on molten Sn (float glass), resulting in a flat and flawless surface. This is also called the "Pilkington process," named after the British glass manufacturer Pilkington, which pioneered the technique (invented by British engineer and businessman Sir Lionel Alexander Bethune Pilkington, January 7, 1920 to May 5, 1995) in the 1950s at their production site in St Helens, Merseyside. Sn is also applied as the negative electrode in advanced Li-ion batteries. Its use is to some extent restricted since some tin surfaces catalyze decomposition of carbonate-based electrolytes used in Li-ion batteries. Sn(II) fluoride is added to certain dental care products as stannous fluoride (SnF_2). SnF_2 can be mixed with Ca abrasives, while the more common NaF progressively becomes biologically inactive in the presence of Ca compounds. In addition, it has been proven to be more efficient than NaF in controlling gingivitis (a nondestructive disease that causes inflammation of the gums). Sn forms a number of different intermetallic phases with Li metal, rendering it a potentially attractive material for battery applications. Large volumetric expansion of Sn after alloying with Li and instability of the Sn-organic electrolyte interface at low electrochemical potentials are the major challenges to usage in commercial cells. The issues were partly resolved by Sony. An Sn intermetallic compound with Co and C has been applied by Sony in its Nexelion cells released into the market in the late 2000s. The composition of the active material is roughly $Sn_{0.3}Co_{0.4}C_{0.3}$. Current research revealed that only some crystalline aspects of tetragonal (beta) Sn are to blame for the adverse electrochemical activity.

Of all the Sn compounds, the organotin compounds are most heavily used. The most important industrial use of organotin compounds is in the stabilization of PVC (polyvinyl chloride, $(C_2H_3Cl)_n$) plastics. In the absence of stabilizers, PVC would quickly degrade under heat, light, and atmospheric oxygen, resulting in discolored, brittle products. Sn scavenges labile chloride ions (Cl^-), which would otherwise initiate loss of HCl from the plastic material. Once dehydrochlorination starts, it is autocatalytic (a single chemical reaction is said to be autocatalytic if one of the reaction products is also a catalyst for the same or a coupled reaction). Characteristic Sn compounds comprise carboxylic acid derivatives of dibutyltin dichloride (IUPAC name dibutyl(dichloro)stannane, $(C_4H_9)_2Cl_2Sn$), such as the dilaurate (lauric acid, IUPAC name dodecanoic acid, $C_{12}H_{24}O_2$). Some organotin compounds are fairly toxic, with both advantages and disadvantages, and are utilized for their biocidal properties, for example, as fungicides, pesticides, algaecides, wood preservatives, antifouling agents, etc. Tributyltin oxide is used as a wood preservative. Tributyltin ($(C_4H_9)_3Sn$) was employed as additive in ship paint to stop growth of marine organisms on ship hulls, with use decreasing after organotin compounds were identified as persistent organic pollutants with an exceptionally high toxicity for some marine organisms. The European Union banned the use of organotin compounds in 2003, while concerns over the toxicity of these compounds to marine life and damage to the reproduction and growth of some marine species (a few publications report biological effects to marine life at a concentration as low as 1 ng/L) have resulted in a worldwide ban by the International Maritime Organization. Many countries now limit the use of organotin compounds to vessels larger than 25 m. Some Sn reagents are valuable in organic chemistry. In the most important application, $SnCl_2$ is a frequently used reducing agent for the conversion of nitro and oxime groups to amines. The Stille reaction couples organotin compounds with organic halides or pseudohalides [or the Migita–Kosugi–Stille coupling, is a chemical reaction widely used in organic synthesis which involves the coupling of an organotin compound (also known as organostannanes) with a variety of organic electrophiles via Pd-catalyzed coupling reaction. The groundwork for the Stille reaction was laid in 1976 and 1977 by British chemist Colin Eaborn (March 15, 1923 to February 22, 2004) and Japanese chemists Toshihiko Migita and Masanori Kosugi, who explored numerous Pd-catalyzed couplings involving organotin reagents. American chemist John Stille (May 8, 1930 to July 19, 1989) and Israeli chemist David Milstein (born June 4, 1947) developed a much milder and more broadly applicable procedure in 1978].

6.14 51 Sb — Antimony

6.14.1 Discovery

Antimony(III) sulfide, Sb_2S_3, was known in predynastic Egypt as an eye cosmetic (kohl) as early as about 3100 BCE, when the cosmetic palette was invented (The cosmetic palettes are archeological artifacts, initially used in predynastic Egypt to grind and apply ingredients for facial or body cosmetics. The decorative palettes of the late fourth millennium BCE seem to have lost this function and became commemorative, ornamental, and possibly ceremonial. They were made almost solely out of siltstone with a few exceptions. The siltstone originated from quarries in the Wadi Hammamat.). An artifact, supposed to be part of a vase, made of antimony dating to about 3000 BCE was found at Telloh, Chaldea (part of present-day Iraq), and a copper object plated with antimony dating between 2500 and 2200 BCE was discovered in Egypt. The Roman author, naturalist, and natural philosopher Pliny the Elder (born Gaius Plinius Secundus, CE 23–79) wrote about several ways of formulating antimony sulfide for medical purposes in his treatise Naturalis Historia (Natural History, he began it in 77, and had not made a final revision at the time of his death during the CE 79 eruption of Vesuvius). In addition, Pliny the Elder distinguished between "male" and "female" forms of antimony; the male form is possibly the sulfide, while the female form, which is superior, heavier, and less friable, has been alleged to be native metallic antimony. The Greek physician, pharmacologist, and botanist Pedanius Dioscorides (c. CE 40–90) stated that antimony sulfide could be roasted by heating with a current of air (Fig. 6.75). It is believed that this produced metallic antimony. The first account of a technique for separating antimony can be found in the 1540 book *De la pirotechnia* by Italian metallurgist Vannoccio Biringuccio (c.1480–1539) (Biringuccio, 1540), preceding the more famous 1556 book by German Humanist scholar and one of the leading experts on mineralogy and metallurgy Georgius Agricola (March 24, 1494 to November 21, 1555), *De re metallica*. In this context, Agricola has been repeatedly wrongly credited with the discovery of metallic antimony. The book *Currus Triumphalis Antimonii* (*The Triumphal Chariot of Antimony*), relating the preparation of metallic antimony, was published in Germany in 1604. It was supposed to be written by a Benedictine monk under the name Basilius Valentinus in the 15th century; if it were authentic, which it is not, it would predate Biringuccio. There is no evidence of such a name on the rolls in Germany or Rome and no mention of this name before 1600. His putative history, like his imaginary portrait, appears to be of later creation than the writings themselves. During the 18th century it was suggested that the author of the works attributed to Basil Valentine was Johann Thölde, a salt manufacturer in Germany who lived roughly 1565–1624.

FIGURE 6.75 Pedanius Dioscorides from "Les vrais pourtraits et vies des hommes illustres" by André Thévet (1516 to November 23, 1590) published in 1594.

FIGURE 6.76 Andreas Libavius.

Modern scholarship now suggests that one author was Thölde, but that others were involved. Thölde published the first five books under Valentine's name. The metal antimony was known to German physician and chemist Andreas Libavius (c.1555 to July 25, 1616) in 1615 who got it by adding iron to a molten mixture of antimony sulfide, salt, and potassium tartrate (Fig. 6.76). This method produced antimony with a crystalline or starred surface. With the start of challenges to the phlogiston theory [a superseded scientific theory that postulated that a fire-like element called phlogiston is contained within combustible bodies and released during combustion. The name comes from the Ancient Greek φλογιστόν phlogistón (burning up), from φλόξ phlóx (flame). It was first stated in 1667 by German physician and alchemist Johann Joachim Becher (May 6, 1635 to October 1682) and then put together more formally by German chemist, physician, and philosopher Georg Ernst Stahl (22 October 1659 to May 24, 1734). The theory attempted to explain processes such as combustion and rusting, which are now collectively known as oxidation.], it was realized that antimony is an element forming sulfides, oxides, and other compounds, as do other metals. The first discovery of naturally occurring pure antimony in the Earth's crust was described by the Swedish scientist and local mine district engineer Anton von Swab in 1783; the type-sample was collected from the Sala Silver Mine in the Bergslagen mining district of Sala, Västmanland, Sweden (Klaproth, 1803). The medieval Latin form, from which the modern languages and late Byzantine Greek got their names for antimony, is antimonium. The source of this is ambiguous; all propositions have some difficulty either of form or interpretation. The prevalent etymology, from ἀντίμοναχός antimonachos or French antimoine, still has adherents; this would mean "monk-killer," and is based on many early alchemists being monks, and antimony being poisonous. An alternative popular etymology is the hypothetical Greek word ἀντίμόνος antimonos, "against aloneness," explained as "not found as metal," or "not found unalloyed." Another conjecture is a hypothetical Greek word ανθήμόνιον anthemonion, which would mean "floret," based on several examples of related Greek words (but not that one) which describe chemical or biological efflorescence. The early uses of antimonium include the translations, in 1050–1100, by physician Constantine the African (died before 1098/1099) of Arabic medical treatises. Several authorities believe antimonium is a scribal corruption of some Arabic form; others think it derives from ithmid; other possibilities include athimar, the Arabic name of the metalloid, and a hypothetical as-stimmi, derived from or parallel to the Greek. The standard chemical symbol for antimony (Sb) is attributed to Swedish chemist Jöns Jakob Berzelius (August 20, 1779 to August 7, 1848), who derived the abbreviation from stibium (though he first used the abbreviation St) (Berzelius, 1813, 1814). The Egyptians called antimony mśdmt; in hieroglyphs, the vowels are uncertain, but the Coptic form of the word is ⲥⲧⲏⲙ (stēm). The Greek word, στίμμι stimmi, is maybe a loan word from Arabic or from Egyptian stm and is used by Attic tragic poets of the 5th century BCE. Later Greeks also used στἰβι stibi, as did Roman encyclopedist Aulus Cornelius Celsus (c.25 BCE to c. CE 50) and Roman author, a naturalist, and natural philosopher Pliny the Elder (CE 23–79, writing in Latin, in the 1st century CE. Pliny also gives the names stimi [sic], larbaris, alabaster, and the "very common" platyophthalmos, "wide-eye" (from the effect of the cosmetic). Later Latin authors modified the word to Latin as stibium. The Arabic word for the substance, in contrast to the cosmetic, can appear as إثمد ithmid, athmoud, othmod, or uthmod. Littré (The Dictionnaire de la langue française by Émile Littré, commonly called simply the "Littré," is a four-volume dictionary of the French language published in Paris by Hachette. The dictionary was originally issued in 30 parts, 1863–72; a second edition is dated 1872–77. A further edition is reported in 1877, published by Hachette.) proposes the first form, which is the earliest, derives from stimmida, an accusative for stimmi.

6.14.2 Mining, production, and major minerals

The abundance of antimony in the Earth's crust is estimated to be between 0.2 and 0.5 ppm, comparable to thallium at 0.5 ppm and silver at 0.08 ppm. Although this element is not abundant, it is found in more than 200 mineral species. Antimony is occasionally found as the native metal (e.g., on Antimony Peak, Kern Co., California, USA), but more commonly it is found in the sulfide mineral stibnite (Sb_2S_3), which is the predominant ore mineral. China is the top producer of antimony followed at a distance by Russia, South Africa, Bolivia, and Tajikistan. Xikuangshan Mine in Hunan province has the largest deposits in China with an estimated deposit of 2.1 million metric tons. The extraction of antimony from ores depends on the quality and composition of the ore. Most of the antimony is mined in the form of its sulfides (e.g., stibnite, Sb_2S_3); lower-grade ores are concentrated by froth flotation, while higher-grade ores are heated to 500°C–600°C, at which temperature the stibnite melts and separates from the gangue minerals. Antimony can be separated from the crude antimony sulfide through reduction with scrap iron according to the reaction:

$$Sb_2S_3 + 3Fe \rightarrow 2Sb + 3FeS$$

The sulfide is subsequently converted to an oxide, which is then roasted, occasionally for the purpose of vaporizing the volatile Sb(III) oxide, which is recovered. This material is often used directly for its main applications, impurities being arsenic and sulfide. Antimony is obtained from the oxide by a carbothermal reduction:

$$2Sb_2O_3 + 3C \rightarrow 4Sb + 3CO_2$$

The lower-grade ores are reduced in blast furnaces, whereas the higher-grade ores are reduced in reverberatory furnaces.

FIGURE 6.77 Crystalline and granular, silvery native antimony, Sb, $4 \times 2.5 \times 2$ cm. Nuevo Tepache, Sonora, Mexico.

FIGURE 6.78 Nest of boulangerite, $Pb_5Sb_4S_{11}$, crystals covering the face in a 3.5×3 cm area associated with chalcopyrite, $CuFeS_2$, crystals to 6 mm and dolomite, $CaMg(CO_3)_2$. Felsobanya, Baia Mare, Maramures Co., Romania.

Antimony is found in over 200 different minerals. Four alloys, paradocrasite (Sb_3As), pararsenolamprite (As,Sb), sorosite ($Cu_{1+x}(Sn,Sb)$, where $0.1 \leq x \leq 0.2$, and stibarsen (AsSb), are known in addition to the native metal antimony (Sb) (Fig. 6.77). Over 150 minerals are found in the sulfide class containing Sb in their crystal structures, for example, allargentum ($Ag_{1-x}Sb_x$), aurostibite ($AuSb_2$), berthierite ($FeSb_2S_4$), boulangerite ($Pb_5Sb_4S_{11}$) (Fig. 6.78), bournonite ($PbCuSbS_3$) (Fig. 6.79), cylindrite ($Pb_3Sn_4FeSb_2S_{14}$), getchellite ($AsSbS_3$) (Fig. 6.80), jamesonite ($Pb_4FeSb_6S_{14}$) (Fig. 6.81), kermesite (Sb_2S_2O), miargyrite ($AgSbS_2$), pyrargyrite (Ag_3SbS_3), stephanite (Ag_5SbS_4), stibnite (Sb_2S_3) (Box 6.10), and tetrahedrite ($Cu_6[Cu_4(Fe,Zn)_2]Sb_4S_{13}$) (Fig. 6.82). Four halides are known with Sn, such as kelyanite ($(Hg^{2+})_6Sb^{3+}BrCl_2O_6$) and nadorite ($PbSbClO_2$) (Fig. 6.83). Within the oxide class just under 50 minerals are found with Sb in their structure, for example, cervantite ($Sb^{3+}Sb^{5+}O_4$), senarmonite (Sb_2O_3) (Fig. 6.84), and stibiconite ($Sb^{3+}Sb_2^{5+}O_6(OH)$). Two borates are known that (can) contain Sb, blatterite ($Sb^{5+}(Mn^{3+},Fe^{3+})_9(Mn^{2+},Mg)_{35}(BO_3)_{16}O_{32}$) and chestermanite ($Mg_2(Fe^{3+},Mn^{3+},Al,Sb^{3+})(BO_3)O_2$). Six sulfate class minerals are known, such as coquandite $\left(Sb_x^{6+}O_{8+x}(SO_4)(OH)_x(H_2O)_{1-x}(x \approx 0.3)\right)$ and klebelsbergite ($Sb_4O_4(SO_4)(OH)_2$). In the phosphate class five minerals are found with Sb in their structure, for example, parwelite ($Mn_{10}Sb_2As_2Si_2O_{24}$) and richelsdorfite ($Ca_2Cu_5Sb(AsO_4)_4(OH)_6Cl{\cdot}6H_2O$). The silicates are represented by six minerals such as chapmanite $\left(Fe_2^{3+}Sb^{3+}(Si_2O_5)O_3(OH)\right)$ and welshite ($Ca_4Mg_9Sb_3O_4[Si_6Be_3AlFe_2O_{36}]$).

FIGURE 6.79 Bournonite, $PbCuSbS_3$, silvery twinned crystals to 5 mm with quartz, SiO_2. Pachapaqui dist, Bolognesi prov., Ancash dept., Peru.

FIGURE 6.80 Lamellar red getchellite, $AsSbS_3$, to 4 mm. Getchell Mine, Humboldt Co., Nevada, United States.

FIGURE 6.81 Steely gray acicular or filiform crystals of jamesonite, $Pb_4FeSb_6S_{14}$, to over 1 cm. Santa Eulalia, Chihuahua, Mexico.

FIGURE 6.82 Tetrahedrite, $Cu_6Cu_4(Fe^{2+},Zn)_2Sb_4S_{12}S$, crystal faces from 1.5 to 3.0 cm. Black Pine Mine, Philipsburg, Granite Co., Montana, United States.

FIGURE 6.83 Deep amber-brown plates of nadorite, $PbSbClO_2$, about 1 cm. Djebel Nador, Constantine prov., Algeria.

FIGURE 6.84 Senarmontite, Sb_2O_3, turbid white octahedral crystals from about 0.75 to 1 mm. Manhattan dist., Nye Co., Nevada, United States.

BOX 6.10 Stibnite, Sb_2S_3.

Antimony forms the sulfide mineral stibnite, as well as the copper-lead-antimony sulfides, tetrahedrite, $Cu_6[Cu_4(Fe,Zn)_2]Sb_4S_{13}$, and jamesonite, $Pb_4FeSb_6S_{14}$, all of which are formed by precipitation from hydrothermal solutions. Even though stibnite is the main mineral in a few deposits, typically with arsenic or mercury, it is more commonly a by-product of lead-zinc-silver mining. Tetrahedrite is abundant in, for example, the Coeur d'Alene veins in Idaho, United States, while several different antimony minerals are found in smaller amounts in Kuroko-type VMS deposits, chimney-manto lead-zinc-silver deposits, tin-tungsten veins, and tungsten skarn deposits. Stibnite is orthorhombic, *2/m2/m2/m* (space group $P2_1/b2_1/n2_1/m$). It has a hardness of 2. The cleavage on {010} is perfect, while the fracture is uneven. It is highly flexible but not elastic, slightly sectile. The luster is metallic, splendent on cleavage surfaces, while the transparency is opaque. The color is silver or lead-gray to black, while the streak is lead-gray. Crystals are slender to stout, complexly terminated, elongated along [001], to 0.65 m; bent crystals not uncommon, rarely twisted. In radiating and confused groups of acicular crystals, or bladed forms with prominent cleavage; also occurring as columnar, granular, or very fine masses. Twinning is rare on twin planes {130}, {120}, and perhaps {310} (Fig. 6.85).

FIGURE 6.85 Stibnite, Sb_2S_3, several terminated crystals, the longest to 7 cm, all diverging from a common point in the matrix. Wuning Mine, Qingjiang, Wuning Co., Jiangxi prov., China.

6.14.3 Chemistry

Antimony consists of two naturally occurring stable isotopes: ^{121}Sb with an abundance of 57.36% and ^{123}Sb with an abundance of 42.64%. In addition, it has 35 radioisotopes., The longest-lived radioisotope is ^{125}Sb with a half-life of 2.7582 years (and decays by β^- to ^{125}Te). In addition, 29 metastable states have been observed. The most stable of these metastable states is $^{120m1}Sb$ with a half-life of 5.76 days. Isotopes that are lighter than the stable isotope ^{123}Sb generally decay via β^+ decay, while those that are heavier than ^{123}Sb generally decay via β^- decay, though there are some exceptions.

Antimony is a silvery, lustrous gray metalloid with five known forms. The α-form crystallizes in a trigonal cell, isomorphic with the gray allotrope of arsenic (Table 6.14). The other forms are metastable. Yellow Sb is unstable above −90°C; black Sb is formed by cooling gaseous Sb; a rare explosive form can be made by electrolysis of antimony(III) chloride. The remaining two forms are formed by high-pressure techniques. Antimony is a member of group 15 or pnictogens. It has 51 electrons arranged in an electronic configuration of $[Kr]4d^{10}5s^25p^3$. As with other posttransition elements in period 5, the "inert pair effect" is observed and the common oxidation states are +3 and +5. Antimonide compounds where Sb takes an oxidation state of −3 are common. These compounds usually occur when antimony is combined with more electropositive metals. Some antimonides are notable for semiconducting properties (e.g., GaSb and InSb). In some of its compounds, oxidation states of −2, −1, +1, +2 +4 can also be found.

TABLE 6.14 Antimony properties.

Appearance	Silvery lustrous gray
Standard atomic weight ($A_{r,std}$)	121.760
Block	p-Block
Element category	Metalloid
Electron configuration	[Kr] '$4d^{10}$ $5s^2$ $5p^3$
Phase at STP	Solid
Melting point	630.63°C
Boiling point	1635°C
Density (near r.t.)	6.697 g/cm^3
When liquid (at m.p.)	6.53 g/cm^3
Heat of fusion	19.79 kJ/mol
Heat of vaporization	193.43 kJ/mol
Molar heat capacity	25.23 J/(mol·K)
Oxidation states	**−3**, −2, −1, +1, +2, **+3**, +4, **+5**
Ionization energies	First: 834 kJ/mol Second: 1594.9 kJ/mol Third: 2440 kJ/mol
Atomic radius	Empirical: 140 pm
Covalent radius	139 ± 5 pm
Van der Waals radius	206 pm

STP, Standard temperature and pressure.
Bold font indicates main oxidation state.

6.14.3.1 *Reactions with oxygen and hydrogen*

Antimony is less reactive than arsenic. It is stable to air and moisture at room temperature but oxidizes when heated under controlled conditions to give antimony(III) oxide (Sb_2O_3), antimony(II) oxide (Sb_2O_4), or antimony(V) oxide (Sb_2O_5). It combines with sulfur when heated at temperatures of 500°C–900°C to form antimony trisulfide, Sb_2S_3. This is the form of antimony mostly found in minerals. Antimony(III) oxide, Sb_2O_3, is the most important commercial compound of antimony. It occurs naturally as the minerals valentinite and senarmontite. Similar to most polymeric oxides, Sb_2O_3 dissolves in aqueous solutions. It is prepared via two methods, re-volatilizing of crude antimony(III) oxide and by oxidation of antimony metal. In the first method step 1 entails the oxidation of crude stibnite to crude antimony (III) oxide using furnaces operating at approximately 500 to 1,000°C.

$$2Sb_2S_3 + 9O_2 \rightarrow 2Sb_2O_3 + 6SO_2$$

In step 2 the crude antimony(III) oxide is purified by sublimation. In the second method antimony metal is oxidized to antimony(III) oxide in furnaces. This reaction is exothermic. Antimony(III) oxide is formed through sublimation and recovered in bag filters. The size of the formed particles is controlled by process conditions in furnace and gas flow. The reaction can be schematically given as:

$$4Sb + 3O_2 \rightarrow 2Sb_2O_3$$

with hydrolysis. Sb_2O_3 is an amphoteric oxide, it dissolves in aqueous NaOH solution to give the meta-antimonite $NaSbO_2$, which can be isolated as the trihydrate. Sb_2O_3 likewise dissolves in concentrated mineral acids to form the corresponding salts, which hydrolyze upon dilution with water. With nitric acid, the trioxide is oxidized to antimony(V) oxide. When heated with carbon, the oxide is reduced to antimony metal. With other reducing agents, e.g. sodium borohydride or lithium aluminium hydride, the unstable and very toxic gas stibine is formed. When heated with potassium bitartrate, a complex salt potassium antimony tartrate, $KSb(OH)_2 \cdot C_4H_2O_6$ is produced. The structure of Sb_2O_3 depends on the temperature of the sample. Dimeric Sb_4O_6 is the high temperature (1560°C) gas. Sb_4O_6 molecules are bicyclic cages, similar to the related oxide of phosphorus(III), phosphorus trioxide. The cage structure is retained in a solid that crystallizes with a cubic structure. The Sb-O distance is 197.7 pm and the O–Sb–O angle of 95.6°. This form occurs naturally as the mineral senarmontite. Above 606°C, the more stable form has an orthorhombic structure, consisting of pairs of -Sb-O-Sb-O- chains that are linked by oxide bridges between the Sb centers. This form occurs naturally as the mineral valentinite.

Antimony tetroxide, Sb_2O_4, occuring naturally as the mineral cervantite, is white but reversibly yellows upon heating. The material, with empirical formula SbO_2, is known as antimony tetroxide to indicate the presence of two kinds of Sb centers. The compound is produced when Sb_2O_3 is heated in air:

$$2Sb_2O_3 + O_2 \rightarrow 2Sb_2O_4 \ \Delta H = -187 \ kJ/mol$$

At 800°C, antimony(V) oxide loses oxygen to form the same compound:

$$2Sb_2O_5 \rightarrow 2Sb_2O_4 + O_2 \ \Delta H = -64 \ kJ/mol$$

The material is mixed valence, containing both Sb(V) and Sb(III) centers. Two different polymorphs have been observed, one orthorhombic and one monoclinic. Both forms feature octahedral Sb(V) centers arranged in sheets with distorted Sb(III) centers bound to four oxides.

Antimony pentoxide (Sb_2O_5) always occurs in hydrated form, $Sb_2O_5 \cdot nH_2O$. It contains antimony in the +5 oxidation state. The hydrated oxide is produced via hydrolysis of antimony pentachloride; or by acidification of potassium hexahydroxoantimonate(V). Alternatively, it may be prepared by oxidation of antimony trioxide with nitric acid. The hydrated oxide is insoluble in nitric acid, but dissolves in a concentrated KOH solution to form potassium hexahydroxoantimonate(V), or $KSb(OH)_6$. When heated to 700°C, the yellow hydrated pentoxide converts to an anhydrous white solid with the formula Sb_6O_{13}, containing both antimony(III) and antimony(V). Heating to 900°C results in a white, insoluble powder of Sb_2O_4 in both α and β forms. The β form consists of antimony(V) in octahedral interstices and pyramidal $Sb^{III}O_4$ units. In these compounds, the antimony(V) atom is octahedrally coordinated to six hydroxy groups. The pentoxide can be reduced to antimony metal by heating with hydrogen or potassium cyanide. Antimony does not react directly with H_2. The hydride of antimony, called stibine (SbH_3) has been synthesized and characterized but the compound is unstable with respect to the elements at room temperature. SbH_3 is an important source of Sb dopants and is often prepared by the reaction of Sb(III) compounds with hydride sources as shown below.

$$4Sb_2O_3 + 6LiAlH_4 \rightarrow 8SbH_3 + 3Li_2O + 3Al_2O_3$$
$$4SbCl_3 + 3NaBH_4 \rightarrow 4SbH_3 + 3NaCl + 3BCl_3$$

6.14.3.2 Reactions with halides

Antimony forms two series of halides: SbX_3 and SbX_5. The trihalides SbF_3, $SbCl_3$, $SbBr_3$, and SbI_3 are all molecular compounds with a trigonal pyramidal molecular geometry. Antimony reactions with the lighter halogens are vigorous and more sedate with the heavier Br_2 and I_2. The trifluoride SbF_3 is formed through the reaction of Sb_2O_3 with HF:

$$Sb_2O_3 + 6HF \rightarrow 2SbF_3 + 3H_2O$$

It is Lewis acidic and readily accepts fluoride ions to form the complex anions SbF_4^- and SbF_5^{2-}. With fluorine, it is oxidized to give antimony pentafluoride.

$$SbF_3 + F_2 \rightarrow SbF_5$$

It is used as a fluorination reagent in organic chemistry. This application was reported by the Belgian chemist Frédéric Jean Edmond Swarts (September 2, 1866 to September 6, 1940) in 1892, who showed its usefulness for converting chloride compounds to fluorides. The method entails reaction of antimony trifluoride with chlorine or with antimony pentachloride to form the active species antimony trifluorodichloride ($SbCl_2F_3$). This compound can also be produced in bulk. The Swarts reaction is typically applied to the synthesis of organofluorine compounds, but experiments have been performed using silanes. It was once used for the industrial production of freon (IUPAC name dichloro(difluoro) methane, CCl_2F_2). Other fluorine-containing Lewis acids serve as fluorinating agents in conjunction with hydrogen fluoride. Molten SbF_3 is a weak electrical conductor. Antimony pentafluoride, SbF_5, is a colorless, viscous liquid and a valuable Lewis acid. It is a component of the superacid fluoroantimonic acid, the strongest known acid, produced when mixing liquid HF with liquid SbF_5 in a 2:1 ratio. It is notable for its Lewis acidity and its ability to react with almost all known compounds. Antimony pentafluoride is prepared by reacting antimony pentachloride with anhydrous hydrogen fluoride:

$$SbCl_5 + 5HF \rightarrow SbF_5 + 5HCl$$

It can also be prepared from antimony trifluoride and fluorine. Similar to SbF_5 enhancing the Brønsted acidity of HF, it augments the oxidizing power of F_2. This effect is shown by the oxidation of oxygen:

$$2SbF_5 + F_2 + 2O_2 \rightarrow 2[O_2]^+[SbF_6]^-$$

Antimony pentafluoride has also been employed in the first discovered chemical reaction that produces fluorine gas from fluoride compounds:

$$4SbF_5 + 2K_2MnF_6 \rightarrow 4KSbF_6 + 2MnF_3 + F_2$$

The driving force for this reaction is the high affinity of SbF_5 for F^-, which is the same property that recommends the use of SbF_5 to produce superacids. SbF_5 is a strong Lewis acid, exceptionally so toward sources of F^- to form the very stable anion $[SbF_6]^-$, known as hexafluoroantimonate. $[SbF_6]^-$ is a weakly coordinating anion comparable to PF_6^-. Although it is only weakly basic, $[SbF_6]^-$ does react with additional SbF_5 to give a centrosymmetric adduct:

$$SbF_5 + [SbF_6]^- \rightarrow [Sb_2F_{11}]^-$$

Antimony trichloride, $SbCl_3$, a soft colorless solid with a pungent odor was known to the alchemists as butter of antimony. The trichloride $SbCl_3$ is formed by dissolving Sb_2S_3 in hydrochloric acid:

$$Sb_2S_3 + 6HCl \rightarrow 2SbCl_3 + 3H_2S$$

$SbCl_3$ is easily hydrolyzed and samples of $SbCl_3$ must be protected from moisture. With a limited amount of water it forms antimony oxychloride releasing hydrogen chloride:

$$SbCl_3 + H_2O \rightarrow SbOCl + 2HCl$$

With more water it forms $Sb_4O_5Cl_2$ which on heating to 460°C under argon changes to $Sb_8O_{11}Cl_{12}$. $SbCl_3$ easily forms the $(C_5H_5NH)SbCl_4$ containing a chain anion with distorted Sb^{III} octahedra. Likewise, the salt $(C_4H_9NH_3)_2SbCl_5$ contains a polymeric anion of composition $[SbCl_5^{2-}]_n$ with distorted octahedral Sb^{III}. With nitrogen donor ligands, L, complexes with a stereochemically active lone-pair are produced, e.g. Ψ-trigonal bipyramidal $LSbCl_3$ and Ψ-octahedral L_2SbCl_3. $SbCl_3$ is only a feeble Lewis base, although some complexes have been observed, e.g. the carbonyl complexes, $Fe(CO)_3(SbCl_3)_2$ and $Ni(CO)_3SbCl_3$. Antimony pentachloride, $SbCl_5$, is a colorless oil, but typical samples are yellowish due to impurities. Due to its tendency to hydrolyze to hydrochloric acid, $SbCl_5$ is a highly corrosive substance and carbonizes non-fluorinated plastics. Antimony pentachloride is produced by passing chlorine gas into molten antimony trichloride:

$$SbCl_3 + Cl_2 \rightarrow SbCl_5$$

Gaseous $SbCl_5$ has a trigonal bipyramidal structure. $SbCl_5$ hydrolyzes to produce hydrochloric acid and antimony oxychlorides. This reaction is suppressed in the presence of a large excess of chloride, due to the formation of the hexachloroantimonate complex ion:

$$SbCl_5 + Cl^- \rightarrow [SbCl_6]^-$$

The mono- and tetrahydrates are known, $SbCl_5 \cdot H_2O$ and $SbCl_5 \cdot 4H_2O$. This compound can produce adducts with many Lewis bases. It is utilized as the standard Lewis acid in the Gutmann scale of Lewis basicity (In chemistry a donor number (DN) is a quantitative measure of Lewis basicity. A donor number is defined as the negative enthalpy value for the 1:1 adduct formation between a Lewis base and the standard Lewis acid $SbCl_5$ (antimony pentachloride), in dilute solution in the noncoordinating solvent 1,2-dichloroethane with a zero DN. The units are kilocalories per mole for historical reasons. The donor number is a measure of the ability of a solvent to solvate cations and Lewis acids. The method was developed by V. Gutmann in 1976. Likewise Lewis acids are characterized by acceptor numbers.). In addition it is a strong oxidizing agent. The pentahalides SbF_5 and $SbCl_5$ both exhibit a trigonal bipyramidal molecular geometry in the gas phase; however, in the liquid phase, SbF_5 is polymeric, whereas $SbCl_5$ is monomeric. SbF_5 is a powerful Lewis acid utilized to synthesize the superacid fluoroantimonic acid ("H_2SbF_7"). Oxyhalides are more common for antimony than for arsenic and phosphorus. Antimony trioxide can be dissolved in concentrated acid to form oxoantimonyl compounds, for example, SbOCl and $(SbO)_2SO_4$.

Antimony tribromide ($SbBr_3$) may be prepared via the reaction of antimony with elemental bromine, or by the reaction of antimony trioxide with hydrobromic acid. It can also be produced by the reaction of bromine with a mixture of antimony sulfide and antimony trioxide at 250°C. Antimony tribromide has two crystalline forms, both having orthorhombic symmetries. When a warm carbon disulfide solution of $SbBr_3$ is rapidly cooled, it crystallizes into the needle-like α-$SbBr_3$, which then slowly converts to the more stable β form. Antimony tribromide hydrolyzes in water to form hydrobromic acid and antimony trioxide:

$$2SbBr_3 + 3H_2O \rightarrow Sb_2O_3 + 6HBr$$

Antimony triiodide, SbI_3 is a ruby-red solid and is the only characterized "binary" iodide of antimony, that is, the sole compound isolated with the formula Sb_xI_y. It may be produced by the reaction of antimony with elemental iodine, or the reaction of antimony trioxide with hydroiodic acid. In addition, it may be formed by the reaction of antimony and iodine in boiling benzene or tetrachloroethane. Similar to many iodides of the heavier main group elements, its structure depends on the phase. Gaseous SbI_3 is a molecular, pyramidal species as anticipated by VSEPR theory. In the solid state, however, the Sb center is octahedrally coordinated by six iodide ligands, three of which are closer and three more distant. For the related compound BiI_3, all six Bi-I distances are equal.

6.14.3.3 Chalcogenides

Antimony pentasulfide is an inorganic compound of antimony and sulfur, also known as antimony red. It is a nonstoichiometric compound with a variable composition. Its exact structure is unknown. Commercial samples are usually contaminated with sulfur, which may be removed by washing with carbon disulfide in a Soxhlet extractor. Antimony pentasulfide can be prepared via the reaction of antimony with sulfur at a temperature between 250 and 400°C in an inert atmosphere. It may be used as a red pigment and is one possible precursor to Schlippe's Salt, Na_3SbS_4, which can be synthesized according to the equation:

$$3Na_2S + Sb_2S_5 + 9H_2O \rightarrow 2Na_3SbS_4 \cdot 9H_2O$$

Similar to numerous sulfides, this compound liberates hydrogen sulfide upon reaction with strong acids like hydrochloric acid.

$$6HCl + Sb_2S_5 \rightarrow 2SbCl_3 + 3H_2S + 2S$$

Mössbauer spectroscopy suggests that this compound is a derivative antimony(III), explaining the production of antimony(III) chloride, instead of antimony(V) chloride, upon acidification. It is thus not similar to the phosphorus(V) compound phosphorus pentasulfide.

Antimony trisulfide (Sb_2S_3) occurs naturally as the mineral stibnite and the amorphous red mineral metastibnite. It is produced for use in safety matches, military ammunition, explosives and fireworks. It also is employed in the production of ruby-colored glass and in plastics as a flame retardant. Historically the stibnite form was utilized as a grey pigment in paintings from in the 16th century. It is a semiconductor with a direct band gap of 1.8–2.5 eV. With suitable doping, p and n type materials can be prepared. Sb_2S_3 can be formed from the elements at a temperature between 500 and 900°C:

$$2Sb + 3S \rightarrow Sb_2S_3$$

Sb_2S_3 is precipitated when H_2S is passed through an acidified solution of Sb(III). This reaction has been employed as a gravimetric method for determining antimony, bubbling H_2S through a solution of Sb(III) compound in hot HCl deposits an orange form of Sb_2S_3 which turns black under the reaction conditions. Sb_2S_3 is easily oxidized, reacting vigorously with oxidizing agents. It burns in air with a blue flame. It reacts with incandescence with cadmium, magnesium and zinc chlorates. Mixtures of Sb_2S_3 and chlorates may explode. In the extraction of antimony from antimony ores the alkaline sulfide process is used where Sb_2S_3 reacts to form thioantimonate(III) salts (also called thioantimonite):

$$3Na_2S + Sb_2S_3 \rightarrow 2Na_3SbS_3$$

A number of salts containing different thioantimonate(III) ions can be formed from Sb_2S_3, such as: $[SbS_3]^{3-}$, $[SbS_2]^-$, $[Sb_2S_5]^{4-}$, $[Sb_4S_9]^{6-}$, $[Sb_4S_7]^{2-}$ and $[Sb_8S_{17}]^{10-}$.

"Schlippe's salt", $Na_3SbS_4 \cdot 9H_2O$, a thioantimonate(V) salt is formed when Sb_2S_3 is boiled with sulfur and sodium hydroxide. The reaction can be visualized as:

$$Sb_2S_3 + 3S^{2-} + 2S \rightarrow 2[SbS_4]^{3-}$$

The structure of the needle-like form of Sb_2S_3, stibnite, contains linked ribbons in which antimony atoms are in two different coordination environments, trigonal pyramidal and square pyramidal. Similar ribbons are found in Bi_2S_3 and

Sb_2Se_3. The red form, metastibnite, is amorphous. Recent research seems to indicate that there are a number of closely related temperature-dependent structures of stibnite which have been termed stibnite (I) the high-temperature form, identified previously, stibnite(II), and stibnite(III). Other research suggests that the actual coordination polyhedra of antimony are in fact SbS_7, with (3 + 4) coordination at the M1 site and (5 + 2) at the M2 site. These coordination polyhedra suggest the presence of secondary bonds. Some of the secondary bonds impart cohesion and are connected with packing.

Antimony triselenide, Sb_2Se_3, occurs naturally as the sulfosalt mineral antimonselite, which crystallizes in an orthorhombic space group. In this compound, antimony has a formal oxidation state +3 and selenium −2. The bonding in this compound has covalent character as evidenced by the black color and semiconducting properties of this and related materials. The low-frequency dielectric constant (ε_0) has been measured to be 133 along the *c* axis of the crystal at room temperature, which is unusually large. Its band gap is 1.18 eV at room temperature. The compound can be prepared by the reaction of antimony with selenium and has a melting point of 612°C (885 K). Sb_2Se_3 is currently actively researched for application in thin-film solar cells. A record light-to-electricity conversion efficiency of 9.2% has been published. Antimony telluride, Sb_2Te_3, as is true of other pnictogen chalcogenide layered materials, is a grey crystalline solid with layered structure. The layers contain two atomic sheets of antimony and three atomic sheets of tellurium and are held together by weak van der Waals forces. Sb_2Te_3 is a narrow-gap semiconductor with a band gap 0.21 eV; it is also a topological insulator, and thus exhibits thickness-dependent physical properties. Select stoichiometric compounds may be produced through the reaction of antimony with tellurium at 500−900°C.

$$2Sb(l) + 3Te(l) \rightarrow Sb_2Te_3(l)$$

Like other binary chalcogenides of antimony and bismuth, Sb_2Te_3 has been studied for its semiconductor properties. It can be transformed into both *n*-type and *p*-type semiconductors by doping with an appropriate dopant. Doping Sb_2Te_3 with iron introduces multiple Fermi pockets, in contrast to the single frequency detected for pure Sb_2Te_3, and results in reduced carrier density and mobility.

6.14.3.4 Reactions with acids

Dilute acids do not attack Sb but concentrated oxidizing acids react readily, especially at elevated temperatures. Concentrated HNO_3 gives hydrated Sb_2O_5; aqua regia gives a solution of $[SbCl_6]^-$ hot concentrated H_2SO_4 gives the salt $Sb_2(SO_4)_3$.

$$2Sb + 10HNO_3 \rightarrow Sb_2O_5 + 5H_2O + 10NO_2$$
$$3Sb + 18HCl + 5HNO_3 \rightarrow 3H[SbCl_6] + 5NO_2 + 10H_2O$$
$$2Sb + 6H_2SO_4 \rightarrow 3SO_2 + Sb_2(SO_4)_3 + 6H_2O$$

6.14.3.5 Organoantimony compounds

At the same time that the organotin compounds were developed in 1850s, the laboratories of Frankland and Löwig also prepared the first trialkyl antimony compounds, trimethyl- and triethylstibine. Since then, a wide range of organoantimony compounds especially the Sb(III) and Sb(V) organocompounds have been studied. The stibines or compounds of the type R_3Sb can be formed from antimony trichloride with organolithium compounds or Grignard reagents. The organoantimony halides can also be synthesized from $SbCl_3$ with organolithium reagents.

$$SbCl_3 + 3RLi(\text{or } RMgCl) \rightarrow R_3Sb + LiCl$$
$$SbCl_3 + RLi \rightarrow RSbCl_2 + LiCl$$
$$SbCl_3 + 2RLi \rightarrow R_2SbCl + 2LiCl$$

Trialkylstibines are usually used as precursors for the synthesis of various alkyl-antimony compounds. For example, the stiboranes or compounds of the type R_5Sb can be synthesized using the trivalent Sb precursors. Despite the ease at which the trialkylstibine and its halide derivatives can be used as precursors to prepare a host of other organoantimony compounds, progress in this area is low.

$$R_3Sb + Cl_2 \rightarrow R_3SbCl_2$$
$$R_3SbCl_2 + 2RLi \rightarrow R_5Sb$$

6.14.4 Major uses

Antimony is primarily employed in the trioxide form (Sb_2O_3) in flame-proofing compounds, always together with halogenated flame retardants apart from halogen-containing polymers. The flame retarding effect of Sb_2O_3 is the result of the creation of halogenated Sb compounds, which react with hydrogen atoms, and probably also with oxygen atoms and OH radicals, consequently inhibiting fire. These flame retardants can be found in, for example, children's clothing, toys, aircraft, and automobile seat covers. They are also included in polyester resins in fiberglass composites for items such as light aircraft engine covers. The resin will burn in the presence of an externally generated flame; however, it will extinguish once the external flame is removed. Sb forms a very beneficial alloy with Pb, improving its hardness and mechanical strength. For most uses involving Pb, variable quantities of Sb are added as alloying metal. In lead-acid batteries, this addition enhances plate strength and charging properties. It is employed in, for example, antifriction alloys [such as Babbitt metal. The original Babbitt alloy was invented in 1839 by American inventor Isaac Babbitt (July 26, 1799 to May 26, 1862) in Taunton, Massachusetts, United States. He disclosed one of his alloy recipes but kept others as trade secrets. Other formulations were developed later], in bullets and lead shot, electrical cable sheathing, type metal (e.g., for linotype printing machines), solder (certain "lead-free" solders comprise 5% Sb), in pewter, and in hardening alloys with low Sn content in the making of organ pipes.

Three other uses consume almost all the rest of global Sb supply. One such application is as a stabilizer and catalyst to produce polyethylene terephthalate (PET, IUPAC name poly(ethyl benzene-1,4-dicarboxylate), $(C_{10}H_8O_4)_n$). A second application is as a fining agent to eliminate microscopic bubbles in glass, mostly for TV screens; Sb ions interact with oxygen, inhibiting the tendency of the latter to form bubbles. The third use is pigments. Sb is more and more being utilized in semiconductors as a dopant in n-type Si wafers for diodes, IR detectors, and Hall-effect devices. Back in the 1950s, the emitters and collectors of n-p-n alloy-junction transistors were doped with tiny beads of a Pb-Sb alloy. Indium antimonide (InSb) is utilized as a compound for mid-IR detectors. Biology and medicine have few uses for Sb. Treatments comprising Sb, called antimonials, are used as emetics [a substance that induces vomiting when administered orally or by injection. An emetic is used medically when a substance has been ingested and must be expelled from the body immediately (for this reason, many toxic and easily digestible products such as rat poison contain an emetic). Inducing vomiting can remove the substance before it is absorbed into the body.]. Sb compounds are employed as antiprotozoan drugs (a class of pharmaceuticals used in treatment of protozoan infection. Protozoans have little in common with each other (e.g., *Entamoeba histolytica*, a unikont eukaryotic organism, is less closely related to *Naegleria fowleri*, a bikont eukaryotic organism, than it is to *Homo sapiens*, which belongs to the unikont phylogenetic group) and so agents effective against one pathogen may not be effective against another.]. Potassium antimonyl tartrate ($K_2Sb_2(C_4H_2O_6)_2$), or tartar emetic, was historically used as an antischistosomal drug from 1919 on (Schistosomiasis, also known as snail fever and bilharzia, is a disease caused by parasitic flatworms called schistosomes. The urinary tract or the intestines may be infected). It was later superseded by praziquantel ($C_{19}H_{24}N_2O_2$). Sb and its compounds are utilized in a number of veterinary formulations, for example, anthiomaline (lithium antimony thiomalate, $C_{12}H_9Li_6O_{12}S_3Sb$) as a skin conditioner in ruminants [mammals that are able to acquire nutrients from plant-based food by fermenting it in a specialized stomach prior to digestion, principally through microbial actions. The process, which takes place in the front part of the digestive system and therefore is called foregut fermentation, typically requires the fermented ingesta (known as cud) to be regurgitated and chewed again. The process of rechewing the cud to further break down plant matter and stimulate digestion is called rumination. The word "ruminant" comes from the Latin *ruminare*, which means "to chew over again."]. Sb has a nourishing or conditioning effect on keratinized tissues in animals (keratin is one of a family of fibrous structural proteins. It is the key structural material making up hair, nails, feathers, horns, claws, hooves, and the outer layer of skin.). Sb-based drugs, such as meglumine antimoniate ($C_7H_{18}NO_8Sb$), are also considered the drugs of choice for treatment of leishmaniasis in domestic animals (a disease caused by parasites of the *Leishmania* type. It is spread by the bite of certain types of sandflies. The disease can present in three main ways: cutaneous, mucocutaneous, or visceral. The cutaneous form presents with skin ulcers, while the mucocutaneous form presents with ulcers of the skin, mouth, and nose, and the visceral form starts with skin ulcers and then later presents with fever, low red blood cells, and enlarged spleen and liver). Unfortunately, in addition to having low therapeutic indices, the drugs have insignificant penetration of the bone marrow, where some of the *Leishmania* amastigotes (a protist cell that does not have visible external flagella or cilia) reside and curing the disease—especially the visceral form—is extremely difficult. Elemental Sb as an antimony pill was once used as a drug. It could be used again by others after ingestion and elimination(!). Sb(III) sulfide is employed in the heads of certain safety matches. In addition, Sb sulfides assist in stabilizing the friction coefficient in car brake pad materials. Sb is utilized in, for example, bullets, bullet tracers, paint, glass art, and as an opacifier in enamel. ^{124}Sb is used in combination with Be in neutron sources; the γ-rays emitted by ^{124}Sb start the photodisintegration of Be. The emitted neutrons have an average energy of 24 keV. Natural Sb is employed in start-up neutron sources.

6.15 52 Te — Tellurium

6.15.1 Discovery

Tellurium (Latin tellus meaning "earth") was first found in the 18th century in a gold ore from the mines in Kleinschlatten (today Zlatna), near today's city of Alba Iulia, Romania. This ore was known as "Faczebajer weißes blättriges Golderz" (white leafy gold ore from Faczebaja, German name of Facebánya, now Faţa Băii in Alba County) or antimonalischer Goldkies (antimonic gold pyrite), and according to Hungarian chemist Anton Leopold von Rupprecht (also spelled as Ruprecht, November 14, 1748 to October 6, 1814), was Spießglaskönig (argent molybdique), containing native antimony (von Rupprecht, 1783). In 1782 Austrian mineralogist and mining engineer Franz-Joseph Müller von Reichenstein (July 1, 1740 or October 4, 1742 to October 12, 1825 or 1826), who was then serving as the Austrian chief inspector of mines in Transylvania, determined that the ore did not have antimony but was instead bismuth sulfide. The following year, he stated that this was wrong, and that the ore instead consisted mostly of gold together with an unknown metal very similar to antimony (Müller von Reichenstein, 1783a,b). After a thorough study that took 3 years and involved more than 50 tests, he determined the specific gravity of the mineral and observed that when heated, the new metal gave off a white smoke with a radish-like odor; that it colored sulfuric acid red; and that when this solution was diluted with water, it gave a black precipitate. Yet, he was not able to identify this metal and called it aurum paradoxium (paradoxical gold) and metallum problematicum (problem metal), because it did not show the properties predicted for antimony. In 1789 a Hungarian botanist and chemist, Pál Kitaibel (February 3, 1757 to December 13, 1817) (Fig. 6.86), found the element independently in an ore from Deutsch-Pilsen that had been thought to be argentiferous molybdenite, but later he attributed the new metal to Müller (Döbling, 1928). In 1798 it was named by German chemist Martin Heinrich Klaproth (December 1, 1743 to January 1, 1817), who had earlier isolated it from the mineral calaverite ($AuTe_2$) (Klaproth, 1798).

FIGURE 6.86 Pál Kitaibel (1790).

BOX 6.11 Calaverite ($AuTe_2$).

Tellurium forms a large family of minerals known as the tellurides. The best-known of these minerals are the precious metal tellurides, including calaverite, named in 1868 by Fredrick August Ludwig Karl Wilhelm Genth for the type locality at the Stanislaus Mine, Calaveras County, California, United States, the site of Mark Twain's story of the famous jumping frog and an area of gold-telluride-bearing quartz veins. Although calaverite and other telluride minerals are found in many precious metal deposits, particularly the famous Emperor Mine in Fiji, they are rarely a source of tellurium because the ores are not processed in a way that permits recovery. Calaverite is monoclinic, *2/m* or 2 (space group *C2/m* or *C2*). It is dimorphous with krennerite (orthorhombic, *mm2*, space group *Pma2*). It has a hardness of 2.5–3. The luster is metallic, and transparency is opaque. It has no cleavage; fracture is subconchoidal to uneven, while the tenacity is very brittle. The color is grass-yellow to silver-white, while the streak is greenish to yellowish gray. Crystals are bladed and short to slender prisms elongated k [010], striated k [010], to 1 cm, also occurring as massive or granular. Twinning is common on {110}, and less common on {031} and {111} (Fig. 6.87).

FIGURE 6.87 Large plates of calaverite, $AuTe_2$, crystals to 4 cm showing skeletal growth patterns. Independence Mine, Cripple Creek dist., Teller Co., Colorado, United States.

6.15.2 Mining, production, and major minerals

With an abundance in the Earth's crust similar to that of Pt (about 1 μg/kg), tellurium is one of the rarest stable solid elements. In comparison, even the rarest of the stable lanthanides have crustal abundances of 500 μg/kg. Tellurium is sometimes found in its native (i.e., elemental) form, but is more often found as the tellurides of gold such as calaverite (Box 6.11) and krennerite (two different polymorphs of $AuTe_2$), petzite, Ag_3AuTe_2, and sylvanite, $AgAuTe_4$. Gold itself is usually found on its own, but when found as a chemical compound, it is most frequently together with tellurium. Although tellurium is found with gold more often than in its native form, it is observed to occur even more often as tellurides in combination with more common metals (e.g., melonite, $NiTe_2$). Natural tellurite (TeO_2) and tellurate (containing an oxyanion of tellurium where tellurium has an oxidation number of +6) minerals also occur, formed by oxidation of tellurides near the Earth's surface. In contrast to selenium, tellurium does not typically replace sulfur in minerals because of the large difference in ion radii. Thus many common sulfide minerals can have considerable quantities of selenium and only traces of tellurium. In the gold rush of 1893, miners in Kalgoorlie rejected a pyritic material as they searched for pure gold, and it was used to fill in potholes and build sidewalks. In 1896 that tailing was found to be the mineral calaverite ($AuTe_2$) and not pyrite (FeS_2) which started a second gold rush that included mining the streets.

The primary tellurium source comes from anode sludges from the electrolytic refining of blister copper. It is also a component of dust from blast furnace refining of lead. Processing of 1000 tons of copper ore typically produces 1 kg of tellurium. The anode sludges have the selenides and tellurides of the noble metals present in compounds with the general formula M_2Se or M_2Te (M=Cu, Ag, Au). At temperatures of ~500°C the anode sludges are roasted with sodium carbonate (Na_2CO_3) under air. In this reaction the metal ions are reduced to the metals, while at the same time the telluride is converted to sodium tellurite.

$$M_2Te + O_2 + Na_2CO_3 \rightarrow Na_2TeO_3 + 2M + CO_2$$

Tellurites can be leached from the mixture with water and are usually present as hydrotellurites $HTeO_3^-$ in solution. Selenites are also formed in this reaction, but they can be isolated by the addition of sulfuric acid. This addition causes the hydrotellurites to be converted into the insoluble tellurium dioxide (TeO_2) while the selenites stay in solution.

$$HTeO_3^- + OH^- + H_2SO_4 \rightarrow TeO_2 + SO_4^{2-} + 2H_2O$$

Finally, the metal is obtained from the oxide through reduction either by electrolysis or by reacting the tellurium dioxide with sulfur dioxide in sulfuric acid.

$$TeO_2 + 2SO_2 + 2H_2O \rightarrow Te + 2SO_4^{2-} + 4H^+$$

Commercial-grade tellurium is usually marketed as 200-mesh powder (Mesh is a measurement of particle size often used in determining the particle-size distribution of a granular material. Standard US 200 mesh is equivalent to 75 μm) but is also available as slabs, ingots, sticks, or lumps (Box 6.11).

About 150 minerals are known to contain tellurium in their crystal structures. Tellurium (Te) occurs as a native element (Fig. 6.88). The largest group is found in the sulfide class with 76 different minerals. These include minerals such as altaite (PbTe) (Fig. 6.89), calaverite ($AuTe_2$), coloradoite (HgTe) (Fig. 6.90), empressite (AgTe), hessite (Ag_2Te) (Fig. 6.91), joséite (Bi_4TeS_2) (Fig. 6.92), krennerite (Au_3AgTe_8), petzite (Ag_3AuTe_2) (Fig. 6.93), sylvanite ($(Au,Ag)_2Te_4$) (Fig. 6.94), and tetradymite (Bi_2Te_2S) (Fig. 6.95). Two halides, kolarite ($PbTeCl_2$) and radhakrishnaite ($PbTe_3(Cl,S)_2$), have Te in their structure. Nearly 50 minerals in the oxide class are known to contain Te, for example, emmonsite ($Fe_2^{3+}(TeO_3)_3 \cdot 2H_2O$) (Fig. 6.96), fairbankite ($Pb(TeO_3)$), spiroffite ($Mn_2^{2+},Te_3^{4+}O_8$) (Fig. 6.97), and tellurite (TeO_2). Twelve

FIGURE 6.88 A 6 mm skeletal silvery tellurium, Te, crystal with several smaller and some associated goldish sylvanite, $(Au,Ag)_2Te_4$. Emperor Mine, Tavua Gold Field, Viti Levu, Fiji.

FIGURE 6.89 Altaite, PbTe, silvery crystal grains to 5 mm. Organ dist., Dona Ana Co., New Mexico, United States.

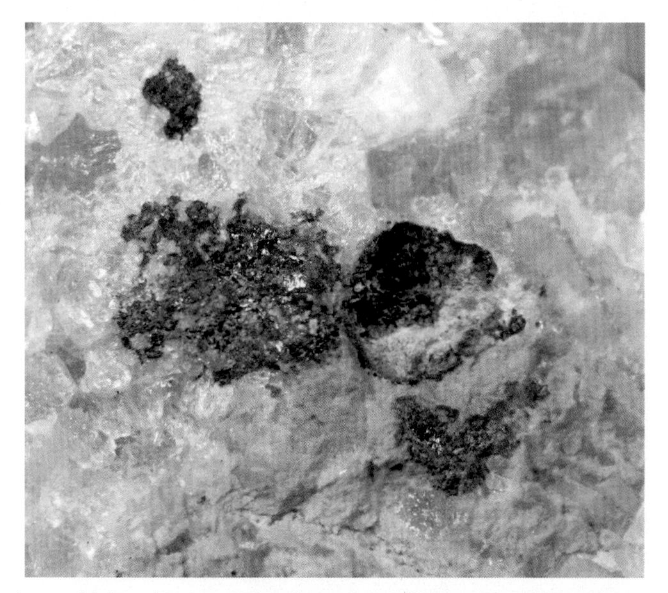

FIGURE 6.90 Specimen with large black crystal grains to 5 mm of coloradoite, HgTe, with green crusts of tlapallite, $(Ca,Pb)_3CaCu_6[Te^{4+}_3Te^{6+}O_{12}]_2\,(Te^{4+}O_3)_2(SO_4)_2{\cdot}3H_2O$. Bambollita Mine, Moctezuma, Sonora, Mexico.

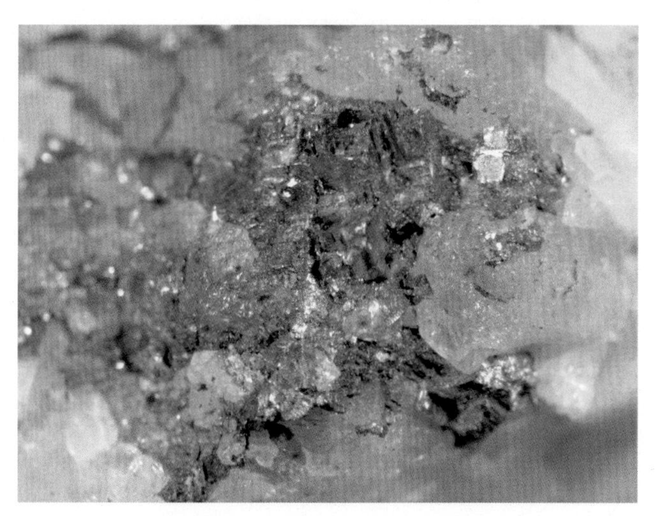

FIGURE 6.91 Silvery platy hessite, Ag_2Te, crystals to 2 mm with galena, PbS, on quartz, SiO_2, matrix. Oliver, British Columbia, Canada.

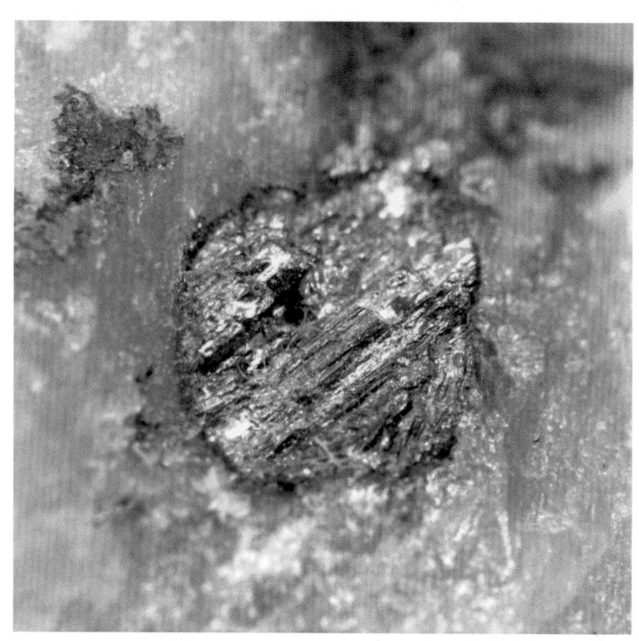

FIGURE 6.92 Joséite, Bi_4TeS_2, platy silvery crystal, elongated to 2 mm in quartz, SiO_2. Carrock Mine, Caldbeck Fells, Cumbria, England.

FIGURE 6.93 Specimen with silvery petzite, Ag_3AuTe_2, and galena, PbS, occurring with gold, Au, in quartz, SiO_2, matrix. American Tunnel, Silverton, San Juan Co., Colorado, United States.

minerals are found in the sulfate class, which includes the tellurates, for example, montanite ($Bi_2(TeO_6)\cdot 2H_2O$) and utahite ($Cu_5Zn_3(TeO_4)_4(OH)_8\cdot 7H_2O$). Three phosphate class minerals, cheremnykhite ($Pb_3Zn_3(VO_4)_2(TeO_6)$), dugganite ($Pb_3Zn_3(AsO_4)_2(TeO_6)$), and kuksite ($Pb_3Zn_3(PO_4)_2(TeO_6)$), are known to contain Te. The only known silicate is burckhardite ($Pb_2(Fe^{3+}Te^{6+})[AlSi_3O_8]O_6$).

6.15.3 Chemistry

Naturally occurring tellurium consists of eight isotopes. Six of these isotopes, ^{120}Te (0.09% abundance), ^{122}Te (2.55% abundance), ^{123}Te (0.89% abundance), ^{124}Te (4.74% abundance), ^{125}Te (7.07% abundance), and ^{126}Te (18.84% abundance), are stable, while the other two, ^{128}Te (31.74% abundance) and ^{130}Te (34.08% abundance), have been observed to be somewhat radioactive, with extremely long half-lives, such as 2.2×10^{24} years for ^{128}Te. This is the longest identified half-life among all radionuclides and is about 160 trillion (10^{12}) times the age of the known universe. All the

FIGURE 6.94 Sylvanite, $(Au,Ag)_2Te_4$ (Au:Ag ratio usually is close to 1:1), brassy gold, striated and tabular crystal to 4 mm. Cresson Mine, Cripple Creek, Teller Co., Colorado, United States.

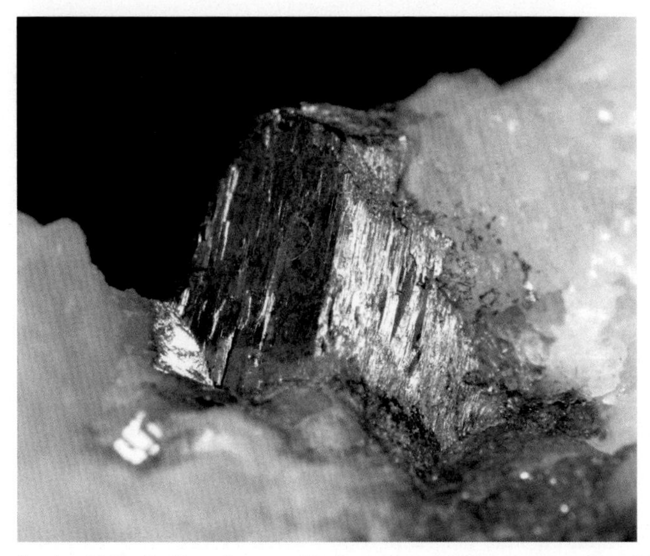

FIGURE 6.95 Tetradymite, Bi_2Te_2S, silvery lustrous thin laminae or foliated crystal to about 1.2 cm. Carrock Mine, Caldbeck Fells, Cumbria, England.

stable isotopes together cover only 33.2% of all naturally occurring Te. A further 30 artificial radioisotopes are known, ranging from ^{105}Te to ^{142}Te and with half-lives of 19 days or less. In addition, 17 nuclear isomers are known, with half-lives up to 154 days. Tellurium (^{106}Te to ^{110}Te) is among the lightest elements known to undergo α decay. The atomic mass of Te (127.60 g/mol) exceeds that of I (126.90 g/mol), the following element in the periodic table.

Tellurium is a member of the chalcogens and is chemically related to other members of the group selenium and sulfur. Tellurium has two forms, crystalline and amorphous. Crystalline Te is the common silvery-white, lustrous form. It is a brittle and easily pulverized metalloid. Amorphous Te appears as a black-brown powder and forms when precipitated out of tellurous acid or telluric acid ($Te(OH)_6$). Its atomic number is 52 and its electrons are arranged in an electronic configuration of $[Kr]4d^{10}5s^25p^4$. There are six electrons in the outer shell, but as with its posttransition neighbors, the inert pair effect makes an oxidation state that is less than the valence especially stable. Compounds of tellurium are thus seen in the +4 oxidation states. Nevertheless, compounds with oxidation states of −2, +2, and +6

FIGURE 6.96 Worm-like crystals of rare emmonsite, $Fe_2^{3+}(TeO_3)_3 \cdot 2H_2O$, to 3 mm. Moctezuma Mine, Moctezuma, Sonora, Mexico.

FIGURE 6.97 Pink spiroffite, $Mn_2^{2+}Te_3^{4+}O_8$, crystal broken out of a large vug with yellow tellurite, TeO_2, overall 4.5 cm. Moctezuma Mine, Moctezuma, Sonora, Mexico.

are common (Table 6.15). The oxidation state of −2 is found when Te is combined with strongly electropositive metals such as group 1, group 2, and lanthanides. The +2, +4, and +6 oxidation states are found when Te is combined with electronegative elements, Cl, O, and F. It combines with most elements, although less readily than O and S. For instance, it can corrode iron, copper, and steel but only in its molten state.

6.15.3.1 Oxides and hydrides

Tellurium adopts a polymeric structure that resists oxidation by air and is not volatile. When burned in oxygen, it forms tellurium dioxide. It does not form the hydride efficiently from the elements in contrast to the lighter chalcogens. Tellurium monoxide was first described in 1883 as a black amorphous solid produced by the heat decomposition of $TeSO_3$ in vacuum, disproportionating into tellurium dioxide, TeO_2 and elemental tellurium upon heating. Since then, though, the existence in the solid phase has been doubted and disputed, even though it is known as a vapor fragment. The black solid may be merely an equimolar mixture of elemental tellurium and tellurium dioxide. Tellurium dioxide is produced by heating tellurium in air, where it burns with a blue flame.

$$Te + O_2 \rightarrow TeO_2$$

TABLE 6.15 Tellurium properties.

Property	Value
Appearance	Silvery lustrous gray (crystalline), brown-black powder (amorphous)
Standard atomic weight ($A_{r,std}$)	127.60
Block	p-Block
Element category	Metalloid
Electron configuration	[Kr] $4d^{10}$ $5s^2$ $5p^4$
Phase at STP	Solid
Melting point	449.51°C
Boiling point	988°C
Density (near r.t.)	6.24 g/cm^3
When liquid (at m.p.)	5.70 g/cm^3
Heat of fusion	17.49 kJ/mol
Heat of vaporization	114.1 kJ/mol
Molar heat capacity	25.73 J/(mol·K)
Oxidation states	**−2**, −1, +1, **+2**, +3, **+4**, +5, **+6**
Ionization energies	First: 869.3 kJ/mol Second: 1790 kJ/mol Third: 2698 kJ/mol
Atomic radius	Empirical: 140 pm
Covalent radius	138 ± 4 pm
Van der Waals radius	206 pm

STP, Standard temperature and pressure.
Bold font indicates main oxidation state.

Tellurium trioxide, β-TeO_3, is formed via thermal decomposition of $Te(OH)_6$. The other two forms of trioxide described in the literature, the α- and γ- forms, were determined not to be true oxides of tellurium in the +6 oxidation state, but instead a mixture of Te^{4+}, OH^- and O_2^-. Tellurium also forms mixed-valence oxides, Te_2O_5 and Te_4O_9. The tellurium oxides and hydrated oxides form a series of acids, such as tellurous acid (H_2TeO_3), orthotelluric acid ($Te(OH)_6$) and meta-telluric acid ($(H_2TeO_4)_n$). The two forms of telluric acid form tellurate salts containing the TeO_4^{2-} and TeO_6^{6-} anions, respectively. Tellurous acid forms tellurite salts containing the anion TeO_3^{2-}. It does not form the hydride efficiently from the elements in contrast to the lighter chalcogens. It does not form the hydride efficiently from the elements in contrast to the lighter chalcogens. Thus hydrogen telluride is normally prepared by hydrolysis of a 15%−50% H_2SO_4 by a Te electrode. It is also prepared by hydrolysis of the tellurides of electropositive metals such as aluminum telluride, Al_2Te_3.

6.15.3.2 *Halides*

Tellurium burns in fluorine, F_2, to form tellurium hexafluoride, TeF_6. In the +6 oxidation state, the $-OTeF_5$ structural group can be found in several compounds e.g. $HOTeF_5$, $B(OTeF_5)_3$, $Xe(OTeF_5)_2$, $Te(OTeF_5)_4$ and $Te(OTeF_5)_6$. The existence of the square antiprismatic anion TeF_8^{2-} has also been confirmed. However, when conditions are controlled such as using F_2 mixed with nitrogen gas at 0°C, the tetrafluoride, TeF_4 is formed. In carefully controlled conditions, it can also combine with the other halogens to form the respective tetrahalides. In the +4 oxidation state, halotellurate anions are known, e.g. $TeCl_6^{2-}$ and $Te_2Cl_{10}^{2-}$. Halotellurium cations have also been shown to exist, e.g. TeI_3^+, found in TeI_3AsF_6. Tellurium can also form Te_3Cl_2 when stoichiometric quantities of Te and Cl_2 are heated in a sealed tube.

$$Te + 3F_2 \rightarrow TeF_6$$
$$Te + 2X_2 \rightarrow TeX_4 \ \ (X = F, \ Cl, \ Br, \ I)$$
$$3Te + Cl_2 \rightarrow Te_3Cl_2$$

6.15.3.3 *Tellurides*

Reduction of Te metal results in the formation the tellurides and polytellurides, Te_n^{2-}. The −2 oxidation state is found in binary compounds with many metals, such as zinc telluride, ZnTe, produced by heating tellurium with zinc. Decomposition of ZnTe with hydrochloric acid produces hydrogen telluride (H_2Te), a highly unstable analogue of the other chalcogen hydrides, H_2O, H_2S and H_2Se:

$$ZnTe + 2HCl \rightarrow ZnCl_2 + H_2Te$$

H_2Te is unstable, while salts of its conjugate base $[TeH]^-$ are stable.

6.15.3.4 *Reactions with acids*

Tellurium is resistant to attack by dilute acids and bases but is attacked by oxidizing acids. It is oxidized to tellurous acid (H_2TeO_3) by nitric acid. When tellurium is reacted with concentrated sulfuric acid, a red solution of the Zintl ion, Te_4^{2+} is formed (Zintl compounds feature naked anionic clusters that are generated by reduction of heavy main group *p* elements, mostly metals or semimetals, with alkali metals, often as a solution in anhydrous liquid ammonia or ethylenediamine. Examples of Zintl anions are $[Bi_3]^{3-}$, $[Sn_9]^{4-}$, $[Pb_9]^{4-}$, and $[Sb_7]^{3-}$. Although these species are called "naked clusters," they are usually strongly associated with alkali metal cations.). The oxidation of tellurium by AsF_5 in liquid SO_2 forms the same square planar cation, in addition to the trigonal prismatic, yellow-orange Te_6^{4+}:

$$4Te + 3AsF_5 \rightarrow Te_4^{2+}(AsF_6^-)_2 + AsF_3$$
$$6Te + 6AsF_5 \rightarrow Te_6^{4+}(AsF_6^-)_4 + 2AsF_3$$

Other tellurium Zintl cations include the polymeric Te_7^{2+} and the blue-black Te_8^{2+}, containing two fused 5-membered tellurium rings. The latter cation is produced in the reaction of tellurium with tungsten hexachloride:

$$8Te + 2WCl_6 \rightarrow Te_8^{2+}(WCl_6^-)_2$$

Interchalcogen cations also exist, e.g. $Te_2Se_6^{2+}$ (distorted cubic geometry) and $Te_2Se_8^{2+}$. These are produced by oxidizing mixtures of tellurium and selenium with AsF_5 or SbF_5.

6.15.3.5 *Organotellurium compounds*

The tellurium analogs of the simple organosulfur compounds are known. However, the rarity of the metalloid and the limited applications hampered growth in this field. Tellurols are unstable. Diorganomono- and ditellurides are the most frequently found organotellurium compounds. Telluroxides (R_2TeO) are also known. Frequently used tellurium based reagents are hydrogen telluride, NaHTe, sodium telluride, and PhTeLi. Since Te is insoluble and polymeric, it is often not a useful precursor to organotellurium compounds, however, it is attacked by hydride reducing agents:

$$Te + 2LiBHEt_3 \rightarrow Li_2Te + H_2 + 2\ BEt_3$$

and organolithium compounds:

$$Te + RLi \rightarrow RTeLi$$

One difference with sulfur and selenium chemistry is the availability of the tetrachloride $TeCl_4$. It reacts with alkenes and alkynes to form the chloro tellurium trichloride addition product:

$$RCH = CH_2 + TeCl_4 \rightarrow RCH(Cl){-}CH_2TeCl_3$$

These organotellurium derivatives are susceptible to further reactions. Diphenyl ditelluride is used as a source of $PhTe^-$ in organic synthesis. Some of its reactions are: organic reduction of aldehydes, alkenes, alkynes, nitro compounds, oxiranes to alkenes; debromination of vicinal dibromides with E2 elimination (An elimination reaction is a type of organic reaction in which two substituents are removed from a molecule in either a one or two-step mechanism. The one-step mechanism is known as the E2 reaction, and the two-step mechanism is known as the E1 reaction. The numbers refer not to the number of steps in the mechanism, but rather to the kinetics of the reaction: E2 is bimolecular (second-order) while E1 is unimolecular (first-order).). Other methods in organotellurium chemistry include: tellurium in vinylic tellurium trichlorides can be replaced by halides with a variety of reagents (iodine, NBS [N-bromosuccinimide]); detellurative cross-coupling reaction; compounds of the type Ar_2TeCl_2 engage in a coupling reaction to the corresponding biaryls with Raney nickel or palladium; Stille reaction; hydrotelluration: Compounds of the type RTeH react with alkynes R'CCH to R'HCCTeR with anti addition to a Z-alkene. In contrast hydrostannylation, hydrozirconation and hydroalumination in similar reactions react with syn addition; Te/Li exchange in transmetallation is used in the synthesis of lithium reagents with demanding functional groups; allylic oxidation: like the selenium counterpart selenoxide oxidation, allylic telluroxides undergo [2,3]-sigmatropic rearrangements forming allylic alcohols after hydrolysis; olefin synthesis: like the selenium counterpart selenoxide elimination, certain telluroxides (RTeOR) can form alkenes on heating. The best-known organotelluride is dimethyl telluride, $(CH_3)_2Te$, as it is used in MOVPE as source of Te. It is prepared by the methylation of telluride salts as shown below.

$$2CH_3I + Na_2Te \rightarrow (CH_3)_2Te + 2NaI$$

6.15.4 Major uses

The largest use of Te is in the metallurgy in Fe, stainless steel, Cu, and Pb alloys. The addition to steel and Cu creates an alloy that is easier machinable. It is alloyed into cast iron for promoting chill for spectroscopy, where the presence of electrically conductive free graphite has a tendency to interfere with spark emission testing outcomes. In Pb, Te enhances strength and durability and reduces the corrosive action of sulfuric acid. Te is employed in CdTe solar panels. National Renewable Energy Laboratory lab tests of Te showed some of the best efficiencies for solar cell electric power generators. Substituting some of the Cd in CdTe by Zn, creating (Cd,Zn)Te, produces a solid-state X-ray detector, offering an alternative to single-use film badges. IR-sensitive semiconductor material is created by alloying Te with Cd and Hg to form CdHgTe. Organotellurium compounds such as dimethyl telluride ($(CH_3)_2Te$), diethyl telluride ($(C_2H_5)_2Te$), diisopropyl telluride ($(C_3H_7)_2Te$), diallyl telluride ($(C_3H_5)_2Te$), and methyl allyl telluride ($CH_3(C_3H_5)Te$) are starting compounds for synthesizing MOVPE growth of II–VI compound semiconductors. Diisopropyl telluride (DIPTe) is the preferred starting material for low-temperature growth of CdHgTe by MOVPE. The highest purity metalorganics of both Se and Te are employed in these processes. The compounds for semiconductor industry are prepared by adduct purification (a technique for preparing extremely pure simple organometallic compounds, which are generally unstable and hard to handle, by purifying a stable adduct with a Lewis acid and then obtaining the desired product from the pure adduct by thermal decomposition). Te, as Te suboxide, is employed in the media layer of rewritable optical disks, comprising ReWritable Compact Disks, ReWritable Digital Video Disks, and ReWritable Blu-ray Disks. TeO_2 is utilized to make acousto-optic modulators (AOTFs, Acousto-optic tunable filters, and AOBSs, acousto-optical beam splitters) for confocal microscopy (most often confocal laser scanning microscopy or laser confocal scanning microscopy, an optical imaging technique for increasing optical resolution and contrast of a micrograph by means of using a spatial pinhole to block out-of-focus light in image formation. Capturing multiple two-dimensional images at different depths in a sample enables the reconstruction of 3D structures (a process known as optical sectioning) within an object. This technique is used extensively in the scientific and industrial communities and typical applications are in life sciences, semiconductor inspection, and materials science.).

Te is utilized in the latest phase-change memory chips created by Intel. Bismuth telluride (Bi_2Te_3) and lead telluride (PbTe) are working elements of thermoelectric devices. PbTe is also utilized in far-IR detectors. Te compounds are applied as pigments for ceramics. Selenides and tellurides significantly increase the optical refraction of glass extensively used in optical glass fibers for telecommunications. Mixtures of Se and Te are employed together with Ba peroxide (BaO_2) as an oxidizer in the delay powder of electric blasting caps. Organic tellurides have been used as initiators for living radical polymerization and electron-rich mono- and di-tellurides possess antioxidant activity. Rubber can be vulcanized with Te instead of S or Se. The rubber manufactured in this way exhibits better heat resistance. Tellurite agar (a selective medium that uses tellurite. The medium appears cream to yellow-colored and takes the form of a free-floating powder. It is a modification of Neill's medium) is utilized to identify members of the Corynebacterium genus, most typically *Corynebacterium diphtheriae*, the pathogen responsible for diphtheria. Te is a principal component of high-performing mixed oxide catalysts for the heterogeneous catalytic selective oxidation of propane (C_3H_8) to acrylic acid (IUPAC name prop-2-enoic acid, $CH_2 = CHCOOH$). The surface elemental composition changes dynamically and reversibly with the reaction conditions. In the presence of steam, the surface of the catalyst is enriched in Te and V, which results in the improvement of the acrylic acid production. Neutron bombardment of Te is the most popular method to produce ^{131}I, which in turn is employed to treat some thyroid conditions, and as a tracer compound in hydraulic fracturing.

6.16 53 I — Iodine

6.16.1 Discovery

Iodine was discovered in 1811 by French chemist Bernard Courtois (February 8, 1777 to September 27, 1838), the son of a saltpeter (KNO_3, an essential component of gunpowder) producer (Courtois, 1813). During the Napoleonic Wars [1803–15, a series of major conflicts pitting the French Empire and its allies, led by Napoleon I (or Napoleon Bonaparte, August 15, 1769 to May 5, 1821), against a fluctuating array of European powers formed into various coalitions, financed and usually led by the United Kingdom.], saltpeter was in great demand in France. Saltpeter produced from French niter beds needed sodium carbonate (Na_2CO_3), which could be obtained from seaweed collected on the coasts of Normandy and Brittany. To isolate the sodium carbonate, seaweed was burned, and the ash washed with water. The remaining waste was destroyed by adding sulfuric acid (H_2SO_4). Courtois at one time added too much sulfuric acid and a cloud of purple vapor formed. He observed that the vapor crystallized on cold surfaces, making dark crystals. He assumed that these crystals were a new element but did not have the money to study it further. Courtois gave samples to his friends, French physicists and chemists Charles Bernard Desormes (June 3, 1777 to August 30, 1862) and Nicolas Clément (January 12, 1779 to November 21, 1841), to continue the study of this material. He also presented some of the material to French chemist and physicist Joseph Louis Gay-Lussac (December 6, 1778 to May 9, 1850) (Fig. 6.98), and to French physicist and mathematician André-Marie Ampère (January 20, 1775 to June 10, 1836) (Fig. 6.99). On November 29, 1813, Desormes and Clément made Courtois' discovery public. They reported the substance to a meeting of the Imperial Institute of France [a

FIGURE 6.98 Joseph Louis Gay-Lussac, engraving by Ambroise Tardieu (1788–1841).

FIGURE 6.99 André-Marie Ampère, after steel engraving by Tardieu (1825).

FIGURE 6.100 Sir Humphry Davy, painting by Sir Thomas Lawrence (1769–1830).

French learned society, grouping five académies, including Académie française (French Academy, concerning the French language), Académie des inscriptions et belles-lettres (Academy of Humanities), Académie des sciences (Academy of Sciences), Académie des beaux-arts (Academy of Fine Arts), Académie des sciences morales et politiques (Academy of Moral and Political Sciences)], a summary of their announcement appeared in the *Gazette nationale ou Le Moniteur Universel* of 2 December 1813. On December 6 of that same year, Gay-Lussac declared that the new compound was either an element or a compound of oxygen (Gay-Lussac, 1813a,b). It was Gay-Lussac who proposed the name "iode," from the Greek word ἰοειδής (ioeidēs) for violet (for the color of the iodine vapor). Ampère had given some of his material to Cornish chemist and inventor Sir Humphry Davy (December 17, 1778 to May 29, 1829), who investigated the substance and observed its resemblance to chlorine (Davy, 1813) (Fig. 6.100). Davy sent a letter dated 10 December 1813 to the Royal Society of London reporting that he had discovered a new element. Arguments broke out between Davy and Gay-Lussac over who found iodine first; nonetheless, both chemists accredited Courtois as the first to separate the elements. Although Davy is typically given credit for this discovery, most of his work was hurried and incomplete. Gay-Lussac presented a much more complete study of iodine in a long memoir presented to the National Institute on August 1, 1814, and subsequently published in the Annales de Chimie (Gay-Lussac, 1814). In early periodic tables, iodine is often given the symbol J, for jod, its name in German.

6.16.2 Mining, production, and major minerals

Iodine is the least abundant of the stable halogens, encompassing only 0.46 ppm of Earth's crustal rocks (in comparison: F 544 ppm, Cl 126 ppm, Br 2.5 ppm). Of the 84 elements that occur in significant quantities in nature, it ranks 61st in abundance. Iodide minerals are rare, and most deposits that are sufficiently concentrated for cost-effective winning are iodate minerals instead, for example, lautarite, $Ca(IO_3)_2$, and dietzeite, $H_2Ca_2(IO_3)_2CrO_4O$. These are the minerals that can be found as trace impurities in the caliche (a sedimentary rock, a hardened natural cement of calcium carbonate that binds other materials—such as gravel, sand, clay, and silt. It occurs worldwide, in aridisol and mollisol soil orders—generally in arid or semiarid regions), found in Chile, where the main product is sodium nitrate ($NaNO_3$). Overall, they can have at least 0.02% and at most 1% iodine by weight. Sodium iodate ($NaIO_3$) is isolated from the caliche and reduced to iodide by sodium bisulfite ($NaHSO_3$). This solution is subsequently reacted with freshly extracted iodate, resulting in comproportionation (or synproportionation, a chemical reaction where two reactants, each containing the same element but with a different oxidation number, form a product in which the elements involved reach the same oxidation number) to iodine, which can be filtered off. The caliche was the major source of iodine in the 19th century and is still important today, replacing kelp (which is no longer a commercially practical source), but in the late 20th century brines developed as a comparable source. The Minami Kanto gas field east of Tokyo, Japan, and the Anadarko Basin gas field in northwest Oklahoma, United States, are the two principal sources. The brine is hotter than 60°C at the depth of the source. The brine is first purified and acidified using sulfuric acid (H_2SO_4), then the iodide present is oxidized to iodine with chlorine. An iodine solution is produced but is dilute and must be concentrated. Air is blown into the solution to evaporate the iodine, which is passed into an absorbing tower where sulfur dioxide reduces the iodine. The hydrogen iodide (HI) is reacted with chlorine to precipitate the iodine. After filtering and purification, the iodine is packed.

$$2HI + Cl_2 \rightarrow I_2\uparrow + 2HCl$$
$$I_2 + 2H_2O + SO_2 \rightarrow 2HI + H_2SO_4$$
$$2HI + Cl_2 \rightarrow I_2\downarrow + 2HCl$$

Instead, the brine can be reacted with silver nitrate ($AgNO_3$) to precipitate out iodine as silver iodide (AgI), which is subsequently decomposed by reaction with iron to form metallic silver and a solution of Fe(II) iodide (FeI_2). The iodine may then be liberated by displacement with chlorine. These sources ensure that Chile and Japan are currently the largest producers of iodine.

Only 29 minerals are known to contain iodine. Of these five minerals are sulfides, for example, demicheleite-(I) (BiSI) and radtkeite ($Hg_3^{2+}S_2ICl$). The halides are represented by 11 minerals, such as coccinite (HgI_2), ioargyrite (AgI) (Fig. 6.101), marshite (CuI) (Fig. 6.102), and miersite ((Ag,Cu)I). The oxide class, which includes the iodates,

FIGURE 6.101 Iodargyrite, AgI, light honey yellow crude crystals associated with dark blue azurite, $Cu_3(CO_3)_2(OH)_2$, to 2 mm. Broken Hill, New South Wales, Australia.

FIGURE 6.102 Marshite, CuI, octahedral crystals to 4 mm. Rubtsovskiy Mine, Altaiskiy Krai, W. Siberia, Russia.

FIGURE 6.103 Salesite, $Cu(IO_3)(OH)$, euhedral single crystals from 1 to 2 mm. Chuquicamata Mine, El Loa Prov., Antofagasta, Chile.

contains 10 minerals, for example, bellingerite ($Cu_3(IO_3)_6{\cdot}2H_2O$) and salesite ($Cu(IO_3)(OH)$) (Fig. 6.103). Three sulfates are known with I in their structure, carlosruizite ($K_6(Na,K)_4Na_6Mg_{10}(SeO_4)_{12}(IO_3)_{12}{\cdot}12H_2O$), fuenzalidaite ($K_6(Na,K)_4Na_6Mg_{10}(SO_4)_{12}(IO_3)_{12}{\cdot}12H_2O$), and hectorfloresite ($Na_9(SO_4)_4(IO_3)$).

6.16.3 Chemistry

Of the 37 known isotopes of iodine, only one exists in nature, ^{127}I. The other isotopes are radioactive and have half-lives too short to be primordial. As such, iodine is monoisotopic and its atomic weight is known to great precision. ^{129}I is the longest-lived radioisotope with a half-life of 15.7 million years, decaying via β^- decay to stable ^{129}Xe. Some ^{129}I was produced together with ^{127}I before the formation of the solar system, but it has by this time completely decayed away, making it an extinct radionuclide that is however still valuable in dating the history of the early solar system or very old groundwaters, because of its mobility in the environment. Its previous existence may be determined from an excess of its daughter ^{129}Xe. ^{129}I traces still currently exist, since it is also a cosmogenic nuclide, produced from cosmic ray spallation of atmospheric Xe: these trace amounts form 10^{-14} to 10^{-10} of all terrestrial iodine. It was also formed as a result of open-air nuclear testing and is not hazardous due to its incredibly long half-life, the longest of all fission products. At the height of thermonuclear testing in the 1960s and 1970s, ^{129}I still made up only about 10^{-7} of all terrestrial iodine. Excited states of ^{127}I and ^{129}I are frequently used in Mössbauer spectroscopy. The radioisotopes lighter than the stable ^{127}I principally decay via ε decay to Te isotopes, while the heavier isotopes principally decay via β^- to Xe isotopes (both ^{123}I and ^{131}I can also decay via γ decay).

Iodine is a member of group 17 in the periodic table, below fluorine, chlorine, and bromine. It is the heaviest stable member of its group. it exists as a lustrous, purple-black nonmetallic solid at standard conditions that melts to form a deep violet liquid at 114°C and boils to a violet gas at 184°C. Iodine has 53 electrons arranged in an electron configuration of $[Kr]4d^{10}5s^25p^5$. Like its lighter counterparts, it has seven valence electrons, one short of an octet. However, its electronegativity is lower, so it takes many oxidation states with -1, $+1$, $+3$, $+5$, and $+7$ as the most common ones (Table 6.16). Iodide compounds in which iodine is paired with less electronegative elements are more common. When paired with the more electronegative elements, iodine takes positive oxidation states.

Iodine is the least reactive of the halogens but is still one of the more reactive elements. It is not able to halogenate carbon monoxide, nitric oxide, and sulfur dioxide in contrast with chlorine which is able to react with said compounds to form phosgene, nitrosyl chloride, and sulfuryl chloride, respectively. Iodination of metals, which can take on more than one oxidation state, also results in the lower oxidation state for that metal. Following a periodic trend, iodine has the lowest ionization energy among the halogens and thus is more easily oxidized by them. For example, reaction with fluorine at elevated temperatures oxidizes iodine to the $+7$ state to form iodine heptafluoride. At room temperature, it is oxidized to the $+5$ state. At lower temperatures (-45°C) and in $CFCl_3$ suspensions, the IF compound can be formed fleetingly and the compound IF_3 can be isolated. IF_3 however is very unstable and decomposes above -28°C. Interhalogen compounds can also be formed with Br and Cl. Like the other halogens, iodine is diatomic, but the I–I bond is much weaker. Molecules of I_2 dissociate into two iodine atoms when irradiated with a wavelength of light that matches the dissociation energy of I_2, about 578 nm. The iodine atom is highly reactive and can attach to a side of a

TABLE 6.16 Iodine properties.

Appearance	Lustrous metallic gray, violet as a gas
Standard atomic weight ($A_{r,std}$)	126.904
Block	p-Block
Element category	Reactive nonmetal
Electron configuration	[Kr] $4d^{10}$ $5s^2$ $5p^5$
Phase at STP	Solid
Melting point	113.7°C
Boiling point	184.3°C
Density	4.933 g/cm^3
Heat of fusion	(I_2) 15.52 kJ/mol
Heat of vaporization	(I_2) 41.57 kJ/mol
Molar heat capacity	(I_2) 54.44 J/(mol·K)
Oxidation states	**−1**, **+1**, **+3**, +4, **+5**, +6, +7
Ionization energies	First: 1008.4 kJ/mol
	Second: 1845.9 kJ/mol
	Third: 3180 kJ/mol
Atomic radius	Empirical: 140 pm
Covalent radius	139 ± 3 pm
Van der Waals radius	198 pm

STP, Standard temperature and pressure.
Bold font indicates main oxidation state.

hydrogen molecule and break the H−H bond. This is usually the reaction employed when high-purity hydrogen iodide, HI, is desired. Otherwise, HI can be prepared industrially by the reaction of I_2 with hydrazine, N_2H_4.

$$I_2 \xrightarrow{hv,\ 578\ \text{nm}} 2I \xrightarrow{+2H_2} 2H_2I \xrightarrow{+2I} 4HI$$
$$2I_2 + N_2H_4 \rightarrow 4HI + N_2$$

6.16.3.1 Iodides

Iodine also forms iodides with most elements in the periodic table. Known exceptions are inertness as with the noble gases; nuclear instability as with most elements heavier than bismuth and higher electronegativity than iodine as with O, N, and lighter halogens in which the binary compounds are not called iodides but oxides, nitrides, bromides, chlorides, or fluorides of iodine. As an element in period 5, when iodine takes an electron to complete an octet, it forms iodide characterized by a large radius. This coupled with the weak oxidizing power of iodine results in binary iodides with lower oxidation states for the other element compared with those formed with the lighter halogens. The maximum oxidation states in iodide are reported for niobium, tantalum, and protactinium.

$$2\ Ta + 5I_2(\text{excess}) \rightarrow Ta_2I_{10}$$

Binary iodides can be prepared in several ways. These include high-temperature oxidative iodination of the element with iodine or hydrogen iodide, high-temperature iodination of a metal oxide or other halide by iodine, a volatile metal iodide, carbon tetraiodide, or an organic iodide. When the iodide product is known to be stable to hydrolysis, it can be prepared by the reaction of the element or its oxide, hydroxide, or carbonate with hydroiodic acid, and then dehydrated by mildly high temperatures combined with either low-pressure or anhydrous hydrogen iodide gas.

6.16.3.2 Nitrides

Iodine does not react with nitrogen. Its binary nitride, NI_3, is known and can be produced, albeit in low yield, by the reaction of boron nitride with iodine monofluoride in trichlorofluoromethane at −30°C. Perhaps, it is fortunate that the amount produced by this reaction is low since NI_3 is extremely shock-sensitive detonating even with the slightest air current to produce N_2 and I_2 with release of heat.

$$BN + 3IF \rightarrow NI_3 + BF_3$$
$$2NI_3 \rightarrow N_2 + 3I_2 \quad (-290\ \text{kJ/mol})$$

6.16.3.3 *Reactions with oxygen*

Iodine forms the most stable oxides of the halogens. Diiodine pentoxide, I_2O_s, can be synthesized through the direct oxidation of I_2 with oxygen in a glow discharge but is most conveniently prepared by dehydrating iodic acid at 200°C in a stream of dry air. Other oxides are known but of less importance than I_2O_5. For example, tetraiodine nonoxide, I_4O_9, has been prepared by the gas-phase reaction between I_2 and ozone but its properties are not extensively characterized. It is known to decompose to I_2O_5 at 75°C. The other isolable oxide, diiodine tetroxide is prepared by the action of concentrated sulfuric acid on HIO_3 or a solid iodate. It also decomposes to I_2O_5 at 135°C.

$$2I_2 + 5O_2 \rightarrow 2I_2O_5$$
$$4I_4O_9 \rightarrow 6I_2O_5 + 3O_2 + 2I_2$$

The chemistry of diiodine pentoxide is extensive. It can oxidize nitrogen oxide, ethylene, and hydrogen sulfide. It adds fluorine by reacting with the element itself or compounds of fluorine such as bromine trifluoride, sulfur tetrafluoride, or chloryl fluoride, resulting in iodine pentafluoride. It reacts with sulfur trioxide and peroxydisulfuryl difluoride ($S_2O_6F_2$) to form salts of the iodyl cation, $[IO_2]$, and is reduced by concentrated sulfuric acids to iodosyl salts involving $[IO]^+$.

6.16.3.4 *Reactions with acids*

The formation of oxoacids and its salts from the reaction of iodine with water, acids, and bases are well known. Detailed discussion on the oxidation-reduction potentials, thermodynamics, and kinetics affecting these reactions will not be discussed. Only the simple reactions will be described here. The reaction of iodine with water is reactant favored with $K_{ac} = 2.0 \times 10^{-13}$ and the hypoiodous acid is only stable in aqueous solutions. The disproportionation of IO^- to the iodide and iodate ions is, however, very favorable with $K \sim 10^{20}$ and IO^- rapidly disproportionates to I^- and IO_3^-. When the reactant is not water but a hot oxidizing acid such as nitric acid, I_2 is oxidized to iodic acid, which crystallizes out on cooling. Iodate is also produced when I_2 reacts with hot aqueous alkali.

$$I_2 + H_2O \rightleftharpoons HIO + H^+ + I^-$$
$$3IO^- \rightarrow 2I^- + IO_3^-$$
$$3I_2 + 10HNO_3 \rightarrow 6HIO_3 + 10NO + 2H_2O$$
$$3I_2 + 6OH^- \rightarrow IO_3^- + 5I^- + 3H_2O$$

When iodine is dissolved in strong acids, for example, fuming sulfuric acid, a bright blue paramagnetic solution containing I_2^+ cations is formed. A solid salt of the diiodine cation may be formed by oxidizing iodine with antimony pentafluoride:

$$2I_2 + 5SbF_5 \xrightarrow{SO_2\ 20^\circ C} 2I_2Sb_2F_{11} + SbF_3$$

The salt $I_2Sb_2F_{11}$ is dark blue, and the blue tantalum analog $I_2Ta_2F_{11}$ is also known. While the I—I bond length in I_2 is 267 pm, that in I_2^+ is only 256 pm because the missing electron in the latter has been removed from an antibonding orbital, resulting in a stronger and consequently shorter bond. In fluorosulfuric acid solution, deep-blue I_2^+ reversibly forms dimers at a temperature below −60°C, forming red rectangular diamagnetic I_4^{2+}. Other polyiodine cations are not as well-characterized, including bent dark-brown or black I_3^+ and centrosymmetric C_{2h} green or black I_5^+, known in the AsF_6^- and $AlCl_4^-$ salts among others.

The periodates are not made from elemental iodine but rather from the oxidation of I^-, I_2, or IO_3^- in aqueous solution. Their industrial preparation involves oxidation of alkaline $NaIO_3$ either electrochemically (using a PbO_2 anode) or with Cl_2. They are strong oxidizing agents and important in reactions involving breaking a carbon—carbon bond.

$$IO_3^- + 6OH^- + Cl_2 \rightarrow IO_6^{5-} + 3H_2O + 2Cl^-$$

6.16.3.5 *Organoiodine compounds*

Very few organoiodine compounds are by themselves industrially important. However, they are fundamentally important in the development of organic synthesis. The simplest organoiodine compound, methyl iodide (CH_3I), remains relevant in the industrial preparation of acetic acid and acetic anhydride. CH_3I and most alkyl iodides are synthesized by the reaction of alcohols with phosphorus triiodide. These alkyl iodides then serve as reagents in nucleophilic substitution reactions, or for preparing Grignard reagents. Hypervalent organoiodine compounds (compounds with iodine having an oxidation state higher than −1) are also gaining interest in organic synthesis. Examples of these are 2-iodoxybenzoic acid (IBX) and the Dess–Martin periodinane (DMP), first reported in 1983 by Dess and Martin and are used as a common reagent for the oxidation of alcohols to aldehydes.

$KBrO_3$, H_2SO_4 → HOAc, Ac_2O, 100°C →

2-Iodobenzoic acid IBX DMP

IBX and DMP were found to have drawbacks: IBX is potentially explosive and shows limited solubility in organic solvents and DMP is moisture-sensitive. Nevertheless, this paper spurred the growth of the preparation and chemistry of hypervalent organoiodine compounds. Current advances in the applications of these compounds are given in a review by Yoshimura and Zhdankin (2016). In this review, the authors presented a general description of the structure and reactivity of polyvalent iodine compounds and gave several examples of the use of trivalent and pentavalent organoiodine compounds in organic syntheses. They summarized that the hypervalent iodine compounds show similar reactivities to that of the transition metal derivatives. These compounds participate in ligand exchange, oxidative addition, reductive elimination, and ligand coupling, typical of the chemistry of transition metal. Iodine, however, is considered more environmentally benign and relatively inexpensive compared with the transition metals. In addition to these, new discoveries of hypervalent catalytic systems, recyclable reagents, and development of new enantioselective reactions using chiral hypervalent iodine compounds were also presented. All these are regarded as particularly important achievements in the field of hypervalent iodine chemistry.

6.16.4 Major uses

In contrast to chlorine and bromine, which have one substantial major use dwarfing all others, iodine is employed in many applications of variable significance. Around 50% of all produced I_2 is used in a variety of organoiodine compounds; about 15% remains as the pure element, about 15% is employed to make KI, while another 15% is used to produce other inorganic iodine compounds. The last 5% is for minor other applications. The major uses of iodine compounds include catalysts, animal feed supplements, stabilizers, dyes, colorants and pigments, pharmaceuticals, sanitation (from tincture of iodine), and photography, while minor uses comprise smog inhibition, cloud seeding, and a variety of uses in analytical chemistry.

Potassium tetraiodomercurate(II), K_2HgI_4, also known as Nessler's reagent (named after German chemist Julius Nessler, June 6, 1827 to March 19, 1905), is frequently utilized as a sensitive spot test for ammonia. This pale solution becomes deeper yellow in the presence of ammonia. At higher concentrations, a brown precipitate may form. The sensitivity as a spot test is about 0.3 μg NH_3 in 2 μL.

$$NH_4^+ + 2\left[HgI_4\right]^{2-} + 4OH^- \rightarrow \underset{\text{(iodide of Millon's base)}}{HgO\cdot Hg(NH_2)I} \downarrow + 7I^- + 3H_2O$$

The formula for the brown precipitate is given as $3HgO{\cdot}Hg(NH_3)_2I_2$ or as $NH_2{\cdot}Hg_2I_3$. Likewise, Cu_2HgI_4 is employed as a precipitating reagent to test for alkaloids. The iodide and iodate anions are frequently utilized for quantitative volumetric analysis, for example, in iodometry and the iodine clock reaction. The iodine clock reaction is a classical chemical clock demonstration experiment to display chemical kinetics in action; it was discovered by Swiss chemist Hans Heinrich Landolt (December 5, 1831 to March 15, 1910) in 1886. The iodine clock reaction exists in a few variations, which each involve iodine species (iodide ion, free iodine, or iodate ion) and redox reagents in the presence of starch. Two colorless solutions are mixed and at first there is no visible reaction. After a short time-delay, the liquid suddenly turns to a shade of dark blue due to the formation of a triiodide-starch complex. In some variations, the solution will repeatedly cycle from colorless to blue and back to colorless, until the reagents are depleted. For example, the hydrogen peroxide variation starts from a solution of hydrogen peroxide with sulfuric acid. To this is added a solution containing potassium iodide, sodium thiosulfate, and starch. There are two reactions occurring in the solution. In the first, slow reaction, triiodide is produced:

$$H_2O_2 + 3I^- + 2H^+ \rightarrow I_3^- + 2H_2O$$

In the second, fast reaction, triiodide is reconverted to three iodide ions by the thiosulfate:

$$2S_2O_3^{2-} + I_3^- \rightarrow S_4O_6^{2-} + 3I^-$$

After some time, the solution always changes color to a very dark blue, almost black. When the solutions are mixed, the second reaction causes the triiodide ion to be consumed much faster than it is generated, and only a small amount of triiodide is present in the dynamic equilibrium. Once the thiosulfate ion has been exhausted, this reaction stops, and the blue color caused by the triiodide—starch complex appears. Anything that accelerates the first reaction will shorten the time until the solution changes color. Decreasing the pH (increasing H^+ concentration) or increasing the concentration of iodide or hydrogen peroxide will shorten the time. Adding more thiosulfate will have the opposite effect; it will take longer for the blue color to appear. An aqueous alkaline iodine solution is employed in the iodoform test for methyl ketones. The iodine test for starch is even today utilized as a method to detect counterfeit banknotes printed on starch-containing paper. The spectra of the I_2 molecule comprises (not exclusively) tens of thousands of sharp spectral lines in the wavelength range between 500 and 700 nm. It is hence a frequently used wavelength reference (secondary standard). By measuring with a spectroscopic Doppler-free technique while centering on one of these lines, the hyperfine structure of the I_2 molecule is revealed. A line is then resolved in such a manner that either 15 components (from even rotational quantum numbers, J_{even}) or 21 components (from odd rotational quantum numbers, J_{odd}) are measurable.

Elemental I_2 is utilized as a disinfectant either as the element or as the water-soluble triiodide anion I_3^- formed in situ by adding iodide to poorly water-soluble elemental iodine (the opposite chemical reaction makes some free elemental I_2 available for antisepsis). Elemental I_2 may also be employed to treat iodine deficiency. In the alternative, iodine may be produced from iodophors, such as a surfactant or povidone [polyvinylpyrrolidone (PVP), IUPAC name 1-ethenylpyrrolidin-2-one, $(C_6H_9NO)_n$, forming povidone-iodine, $(C_6H_9NO)_n \cdot xI$], which contain I_2 complexed with a solubilizing agent (iodide ion may be thought of loosely as the iodophor in triiodide water solutions). Examples of these preparations comprise tincture of iodine: I_2 in ethanol, or iodine and NaI in a mixture of ethanol and water; and Lugol's iodine: iodine and iodide in water alone, forming mostly triiodide. In contrast to tincture of iodine, Lugol's iodine has a minimized amount of the free iodine (I_2) component. The antimicrobial action of iodine is fast and works at low concentrations, and consequently it is employed in operating theaters. Its specific mode of action is not known. It penetrates microorganisms and attacks specific amino acids (such as cysteine and methionine), nucleotides, and fatty acids, eventually resulting in cell death. In addition, it has an antiviral action, but nonlipid viruses and parvoviruses are less susceptible than lipid enveloped viruses. I_2 most likely attacks surface proteins of the enveloped viruses, and in addition it might destabilize membrane fatty acids through reaction with unsaturated carbon bonds. In medicine, a saturated solution of KI is used to treat acute thyrotoxicosis (Thyrotoxicosis is the condition that occurs due to excessive thyroid hormone of any cause. Signs and symptoms vary between people and may include irritability, muscle weakness, sleeping problems, a fast heartbeat, heat intolerance, diarrhea, enlargement of the thyroid, hand tremor, and weight loss.). It is also employed to block uptake of ^{131}I in the thyroid gland, when this isotope is used as part of radiopharmaceuticals (such as iobenguane, meta-iodobenzylguanidine, $C_8H_{10}IN_3$) that are not targeted to the thyroid or thyroid-type tissues. ^{131}I (usually as iodide) is a component of nuclear fallout and is especially dangerous due to the thyroid gland's tendency to concentrate ingested iodine and hold it for periods longer than this isotope's radiological half-life of 8.02 days. Therefore people at risk of exposure to environmental radioactive iodine (^{131}I) in fallout may be prescribed nonradioactive KI tablets. The typical adult dose is one 130 mg tablet per 24 hours, supplying 100 mg of ionic iodine (the typical daily dose of iodine for normal health is of order 100 μg). Ingestion of this high dose of nonradioactive iodine reduces the uptake of radioactive iodine by the thyroid gland. As an element with high electron density and atomic number, iodine absorbs X-rays weaker than 33.3 keV due to the photoelectric effect of the innermost electrons. Organoiodine compounds are utilized with intravenous injection as X-ray radiocontrast agents. This application is often used in combination with advanced X-ray techniques such as angiography and computed tomography scanning. Currently, all water-soluble radiocontrast agents rely on iodine.

The manufacture of ethylenediamine dihydroiodide (IUPAC name ethane-1,2-diamine dihydroiodide, $C_2H_{10}I_2N_2$), provided as a nutritional supplement for livestock, uses a large quantity of available iodine. Another significant application is a catalyst to produce acetic acid by the Monsanto process (The catalytically active species is the anion *cis*-$[Rh(CO)_2I_2]^-$. The first organometallic step is the oxidative addition of methyl iodide to *cis*-$[Rh(CO)_2I_2]^-$ to form the hexacoordinate species $[(CH_3)Rh(CO)_2I_3]^-$. This anion rapidly transforms, via the migration of a methyl group to an adjacent carbonyl ligand, affording the pentacoordinate acetyl complex $[(CH_3CO)Rh(CO)I_3]^-$. This five-coordinate complex then reacts with carbon monoxide to form the six-coordinate dicarbonyl complex, which undergoes reductive elimination to release acetyl iodide ($CH_3C(O)I$). The catalytic cycle involves two nonorganometallic steps: conversion of methanol to methyl iodide and the hydrolysis of the acetyl iodide to acetic acid and hydrogen iodide. The reaction has been shown to be first-order with respect to methyl iodide and $[Rh(CO)_2I_2]^-$. Hence, the oxidative addition of methyl iodide is proposed as the rate-determining step.) and Cativa process [The catalytic cycle begins with the reaction of methyl iodide with the square-planar active catalyst species to form the octahedral Ir(III) species, the *fac*-isomer of $[Ir(CO)_2(CH_3)I_3]^-$. This oxidative addition reaction involves the formal insertion of the Ir(I) center into the C–I bond of methyl iodide. After ligand exchange (iodide for carbon monoxide), the migratory insertion of carbon monoxide into the Ir–C bond, the final two steps result in the formation of a square pyramidal species with a bound acetyl ligand. The active catalyst species is regenerated by the reductive elimination of acetyl iodide from the final step, a de-insertion reaction. The acetyl iodide is hydrolyzed to produce the acetic acid product, in the process generating hydroiodic acid which is in turn used to convert the starting material (methanol) to the methyl iodide used in the first step.]. In these technologies, which support the world's demand for acetic acid, hydroiodic acid converts the methanol feedstock into methyl iodide, which undergoes carbonylation. Hydrolysis of the resulting acetyl iodide regenerates hydroiodic acid and gives acetic acid. Inorganic iodides find specialized uses. Ti, Zr, Hf, and Th are purified by the van Arkel process, which involves the reversible formation of the tetraiodides of these elements. Silver iodide (AgI) is a major ingredient to traditional photographic film. Thousands of kilograms of AgI are used annually for cloud seeding to induce rain. The organoiodine compound erythrosine (IUPAC name 2-(6-hydroxy-2,4,5,7-tetraiodo-3-oxo-xanthen-9-yl)benzoic acid, $C_{20}H_6I_4Na_2O_5$) is an important food coloring agent. Perfluoroalkyl iodides are precursors to important surfactants, such as perfluorooctanesulfonic acid (PFOS, IUPAC name 1,1,2,2,3,3,4,4,5,5,6,6,7,7,8,8,8-heptadecafluoro-1-octanesulfonic acid, $C_8HF_{17}O_3S$).

References

Araki, S., Ito, H., & Butsugan, Y. (1988). Indium in organic synthesis: Indium-mediated allylation of carbonyl compounds. *Journal of Organic Chemistry*, *53*, 1831–1833.

Berzelius, J. J. (1813). Essay on the cause of chemical proportions, and on some circumstances relating to them: Together with a short and easy method of expressing them. *Annals of Philosophy*, *2*, 443–454.

Berzelius, J. J. (1814). Essay on the cause of chemical proportions, and on some circumstances relating to them: Together with a short and easy method of expressing them. *The Annals of Philosophy*, *3*, 51–62, 93–106, 244–255, 353–364.

Berzelius, J. J. (1824). Extrait d'une Lettre de M. Berzelius a M. Duloug. *Annales de chimie et de physique*, *26*, 39–43.

Biringuccio, V. (1540). Del antimonio & sua miniera, Capitolo terzo. In *De la Pirotechnia* (pp. 27–28), Vol. 2. Curtio Navo e fratelli, Venice, Italy (Note: Only every second page of this book is numbered, so the relevant passage is to be found on the 74th and 75th pages of the text).

Blake, Al. J., Holder, A. J., Hyde, T. I., & Schröder, M. (1987). Stabilisation of mononuclear palladium(III). The single crystal X-ray structure of the $[Pd(L)_2]^{3+}$ cation (L = 1,4,7-trithiacyclononane). *Journal of Chemical Society, Chemical Communications*, 987–988.

Bunsen, R. (1863). Ueber die darstellung und die eigenschaften des rubidiums. *Annalen der Chemie und Pharmacie*, *125*, 367–368.

Campbell, N. R., & Wood, A. (1908). The radioactivity of rubidium. *Proceedings of the Cambridge Philosophical Society*, *14*, 15–21.

Claus, K. (1845). О способе добывания чистой платины из руд. Горный журнал. *Mining Journal*, *7*, 157–163.

Cotton, F. A., Koshevoy, I. O., Lahuerta, P., Murillo, C. A., Sanaú, M., Ubeda, M. A., & Zhao, Q. (2006). High yield syntheses of stable, singly bonded Pd_2^{6+} compounds. *Journal of the American Chemical Society*, *128*, 13674–13675.

Courtois, B. (1813). Découverte d'une substance nouvelle dans le Vareck. *Annales de Chimie*, *88*, 304–310.

Crawford, A. (1790). On the medicinal properties of the muriated barytes. *Medical Communications*, *2*, 301–359.

Davy, H. (1808). Electro-chemical researches on the decomposition of the earths; with observations on the metals obtained from the alkaline earths, and on the amalgam procured from ammonia. *Philosophical Transactions of the Royal Society of London*, *98*, 333–370.

Davy, H. (1813). Sur la nouvelle substance découverte par M. Courtois, dans le sel de Vareck. *Annales de Chimie*, *88*, 322–329.

Davy, H. (1814). Some experiments and observations on a new substance which becomes a violet coloured gas by heat. *Philosophical Transactions of the Royal Society of London*, *104*, 74–93.

de Marignac, J. C. G. (1866). Recherches sur les combinaisons du niobium. *Annales de Chimie et de Physique*, *4*, 7–75.

de Marignac, J. C. G., Blomstrand, C. W., Deville, H., Troost, L., & Hermann, R. (1866). Tantalsäure, niobsäure (ilmensäure) und titansäure. *Fresenius' Journal of Analytical Chemistry*, *5*, 384–389.

Dess, D. B., & Martin, J. C. (1983). Readily accessible 12-I-51 oxidant for the conversion of primary and secondary alcohols to aldehydes and ketones. *Journal of Organic Chemistry*, *48*, 4155–4156.

Döbling, H. (1928). *Die Chemie in Jena zur Goethezeit*. Jena: Gustav Fischer.

Flower, K. R. (2007). 5.07 – Molybdenum compounds without CO or isonitrile ligands. In D. M. P. Mingos, & R. H. Crabtree (Eds.), *Comprehensive organometallic chemistry III* (pp. 513–595). Oxford: Elsevier.

Gadolin, J. (1794). Undersökning af en svart tung stenart ifrån ytterby stenbrott i roslagen. *Kongliga Vetenskaps Academiens Nya Handlingar*, *15*, 137–155.

Gay-Lussac, J. L. (1813a). Sur la combination de l'iode avec d'oxigène. *Annales de Chimie*, *88*, 319–321.

Gay-Lussac, J. L. (1813b). Sur un nouvel acide formé avec la substance décourverte par M. Courtois. *Annales de Chimie*, *88*, 311–318.

Gay-Lussac, J. L. (1814). Mémoire sur l'iode. *Annales de Chimie, 91*, 5–160.

Greenwood, N. N., & Earnshaw, A. (1996). *Chemistry of the elements.* Oxford: Elsevier Science & Technology Books.

Hatchett, C. (1802a). An analysis of a mineral substance from North America, containing a metal hitherto unknown. *Philosophical Transactions of the Royal Society of London, 92*, 49–66.

Hatchett, C. (1802b). Eigenschaften und chemisches verhalten des von Charles Hatchett entdeckten neuen metalls, columbium. *Annalen der Physik, 11*, 120–122.

Hatchett, C. (1802c). Outline of the properties and habitudes of the metallic substance, lately discovered by Charles Hatchett, Esq. and by him denominated columbium. *Journal of Natural Philosophy, Chemistry, and the Arts*, 32–34.

Hermann, C. S. (1818). Noch ein schreiben über das neue Metall. *Annalen der Physik, 59*, 113–116.

Hertrampf, J., Becker, P., Widenmeyer, M., Weidenkaff, A., Schlücker, E., & Niewa, R. (2018). Ammonothermal crystal growth of indium nitride. *Crystal Growth & Design, 18*, 2365–2369.

Hjelm, P. J. (1789). Versuche mit molybdäna, und reduction der selben erde der Königl Schwedischen Akademie der Wissenschaften neue abhandlungen aus der naturlehre. *Haushaltungskunst und Mechanik, 49*, 268–278.

Hope, T. C. (1798). Account of a mineral from Strontian and of a particular species of earth which it contains. *Transactions of the Royal Society of Edinburgh, 4*, 3–39.

Kirchhoff, G., & Bunsen, R. (1861). Chemische analyse durch spectralbeobachtungen. *Annalen der Physik und Chemie, 189*, 337–381.

Kirwan, R. (1794). Experiments on a new earth found near Stronthian in Scotland. *The Transactions of the Royal Irish Academy, 5*, 243–256.

Kitano, M., Inoue, Y., Yamazaki, Y., Hayashi, F., Kanbara, S., Matsuishi, S., ... Hosono, H. (2012). Ammonia synthesis using a stable electride as an electron donor and reversible hydrogen store. *Nature Chemistry, 4*, 934–940.

Klaproth, M. H. (1789). Kleine mineralogische Beitrage. *Crell's Annalen (Chemische Annalen für die Freunde der Naturlehre, Arzneygelahrtheit, Haushaltungskunst und Manufakturen)*, 7–12.

Klaproth, M. H. (1793). Chemische versuche über die strontianerde. *Crell's Annalen, 2*, 189–202.

Klaproth, M. (1798). Ueber die siebenbürgischen golderze, und das in selbigen enthaltene neue metall. *Chemische Annalen für die Freunde der Naturlehre, Arzneygelahrtheit, Haushaltungskunst und Manufacturen, 1*, 91–104.

Klaproth, M. (1803). XL. Extracts from the third volume of the analyses. *The Philosophical Magazine, 17*, 230–243.

Kloprogge, J. T., & Lavinsky, R. (2017). *Photo atlas of mineral pseudomorphism.* Amsterdam: Elsevier.

Khusnutdinova, J. R., Rath, N. P., & Mirica, L. M. (2010). Stable mononuclear organometallic Pd(III) complexes and their C – C bond formation reactivity. *Journal of the American Chemical Society, 132*, 7303–7305.

Levi, B. G. (2001). Cornell, Ketterle, and Wieman share Nobel prize for Bose-Einstein condensates. *Physics Today, 54*, 14–16.

Lewis, G. M. (1952). The natural radioactivity of rubidium. *Philosophical Magazine Series 7, 43*, 1070–1074.

Maudez, W., Meuwly, M., & Fromm, K. M. (2007). Analogy of the coordination chemistry of alkaline earth metal and lanthanide Ln^{2+} ions: The isostructural zoo of mixed metal cages $[IM(OtBu)_4\{Li(thf)\}_4(OH)]$ (M = Ca, Sr, Ba, Eu), $[MM'_6(OPh)_8(thf)_6]$ (M = Ca, Sr, Ba, Sm, Eu, M' = Li, Na), and their derivatives with 1,2-dimethoxyethane. *Chemistry – A European Journal, 13*, 8302–8316.

Milanov, A. P., Xu, K., Cwik, S., Parala, H., de los Arcos, T., Becker, H.-W., ... Devi, A. (2012). Sc_2O_3, Er_2O_3, and Y_2O_3 thin films by MOCVD from volatile guanidinate class of rare-earth precursors. *Dalton Transactions, 41*, 13936–13947.

Mosander, C. G. (1843). Ueber die das cerium begleitenden neuen metalle lathanium und didymium, so wie über die mit der yttererde vorkommenden neuen metalle erbium und terbium. *Annalen der Physik und Chemie, 60*, 297–315.

Müller von Reichenstein, F. J. (1783a). Über den vermeintlichen natürlichen Spiessglaskönig. *Physikalische arbeiten der einträchtigen freunde in Wien, 1*, 57–59.

Müller von Reichenstein, F. J. (1783b). *Versuche mit dem in der Grube Mariahilf in dem Gebirge Fazebay bey Zalathna vorkommenden vermeinten gediegenen Spiesglaskönig. Physikalische arbeiten der einträchtigen freunde in Wien* (pp. 63–69).

Osann, G. (1828). Fortsetzung der untersuchung des Platins vom Ural. *Annalen der Physik, 89*, 283–297.

Osann, G. (1829). Berichtigung, meine analyse des uralschen Platins betreffend. *Annalen der Physik, 91*, 158.

Pardoe, J. A. J., & Downs, A. J. (2007). Development of the chemistry of indium in formal oxidation states lower than +3. *Chemical Reviews, 107*, 2–45.

Pinheiro, D. L. J., de Castro, P. P., & Amarante, G. W. (2018). Recent developments and synthetic applications of nucleophilic zirconocene complexes from Schwartz's reagent. *European Journal of Organic Chemistry, 2018*, 4828–4844.

Powers, D. C., & Ritter, T. (2009). Bimetallic Pd(III) complexes in palladium-catalysed carbon–heteroatom bond formation. *Nature Chemistry, 1*, 302–309.

Powers, D. C., Benitez, D., Tkatchouk, E., Goddard, W. A., & Ritter, T. (2010). Bimetallic reductive elimination from dinuclear Pd(III) complexes. *Journal of the American Chemical Society, 132*, 14092–14103.

Purdy, A. P. (1994). Indium(III) amides and nitrides. *Inorganic Chemistry, 33*, 282–286.

Quist, B. A. (1754). Rön om bly-erts. *Kongliga Vetenskaps Academiens Handlingar, 15*, 189–210.

Reich, F., & Richter, T. (1863). Ueber das Indium. *Journal für Praktische Chemie, 90*, 172–176.

Reich, F., & Richter, T. (1864). Ueber das Indium. *Journal für Praktische Chemie, 92*, 480–485.

Rose, H. (1843). Einige Bemerkungen über die Yttererde. *Annalen der Physik, 135*, 101–111.

Rose, H. (1847). Ueber die Säure im Columbit von Nordamérika. *Annalen der Physik, 146*, 572–577.

Rowbotham, T., & Rowbotham, T. L. (1850). *The Art of Landscape Painting in Water Colours* (p. 10) London: Windsor and Newton.

Sánchez, M., Harvey, M. J., Nordstrom, F., Parkin, S., & Atwood, D. A. (2002). Salen-type compounds of calcium and strontium. *Inorganic Chemistry, 41*, 5397–5402.

Scheele, C. W. K. (1779). Versuche mit Wasserbley; Molybdaena. *Der Königl Schwedischen Akademie der Wissenschaften neue Abhandlungen aus der Naturlehre, Haushaltungskunst und Mechanik, 40*, 238–248.
Schumann, H., & Fedushkin, I. L. (2005). Scandium, yttrium & the lanthanides: Organometallic chemistry based in part on the article scandium, yttrium & the lanthanides: Organometallic chemistry by R.D. Köhn, G. Kociok-Köhn, & H. Schumann which appeared in the encyclopedia of inorganic chemistry. In R. B. King, R. H. Crabtree, C. M. Lukehart, D. A. Atwood, & R. A. Scott (Eds.), *Encyclopedia of inorganic chemistry* (1st ed.). New York: John Wiley & Sons.
Schwenzer, B., Loeffler, L., Seshadri, R., Keller, S., Lange, F. F., Den Baars, S. P., & Mishra, U. K. (2004). Preparation of indium nitride micro- and nanostructures by ammonolysis of indium oxide. *Journal of Materials Chemistry, 14*, 637–641.
Śniadecki, J. (1808). Rosprawa o nowym metallu w surowey platynie odkrytym przez Jędrzeia Śniadeckiego. Wilno: Nakł. i Drukiem J. Zawadzkiego.
Tamadon, F., & Seppelt, K. (2013). The elusive halides VCl_5, $MoCl_6$, and $ReCl_6$. *Angewandte Chemie International Edition, 52*, 767–769.
Tate, B. K., Jordan, A. J., Bacsa, J., & Sadighi, J. P. (2017). Stable mono- and dinuclear organosilver complexes. *Organometallics, 36*, 964–974.
van Arkel, A. E., & de Boer, J. H. (1924a). Die trennung des Zirkoniums von anderen Metallen, einschließlich Hafnium, durch fraktionierte Distillation. *Zeitschrift für Anorganische und Allgemeine Chemie, 141*, 289–296.
van Arkel, A. E., & de Boer, J. H. (1924b). Die trennung von Zirkonium und Hafnium durch Kristallisation ihrer Ammoniumdoppelfluoride. *Zeitschrift für Anorganische und Allgemeine Chemie, 141*, 284–288.
van Arkel, A. E., & de Boer, J. H. (1925). Darstellung von reinem Titanium-, Zirkonium-, Hafnium- und Thoriummetall. *Zeitschrift für Anorganische und Allgemeine Chemie, 148*, 345–350.
von Rupprecht, A. (1783). Über den vermeintlichen siebenbürgischen natürlichen Spiessglaskönig. *Physikalische Arbeiten der einträchtigen Freunde in Wien, 1*, 70–74.
Wang, L., Pan, Y., Shen, Q., Zhang, J., Bao, K., Lou, Z., ... Zhou, Q. (2016). Sulfur-assisted synthesis of indium nitride nanoplates from indium oxide. *RSC Advances, 6*, 98153–98156.
Waterston, W., & Burton, J. H. (1844). *Cyclopædia of commerce, mercantile law, finance, commercial geography and navigation* (p. 122) London: H. G. Bohn.
Wipf, P., & Jahn, H. (1996). Synthetic applications of organochlorozirconocene complexes. *Tetrahedron, 52*, 12853–12910.
Wöhler, F. (1828). Ueber das Beryllium und Yttrium. *Annalen der Physik und Chemie, 89*, 577–582.
Wollaston, W. H. (1804). On a new metal, found in crude platina. *Philosophical Transactions of the Royal Society of London, 94*, 419–430.
Wollaston, W. H. (1805). On the discovery of palladium; with observations on other substances found with platina. *Philosophical Transactions of the Royal Society of London, 95*, 316–330.
Wollaston, W. H. (1809). On the identity of columbium and tantalum. *Philosophical Transactions of the Royal Society, 99*, 246–252.
Wu, M. K., Ashburn, J. R., Torng, C. J., Hor, P. H., Meng, R. L., Gao, L., ... Chu, C. W. (1987). Superconductivity at 93 K in a new mixed-phase Y-Ba-Cu-O compound system at ambient pressure. *Physical Review Letters, 58*, 908–910.
Yoshimura, A., & Zhdankin, V. V. (2016). Advances in synthetic applications of hypervalent iodine compounds. *Chemical Reviews, 116*, 3328–3435.
Zhang, S., Zhang, W., & Nomura, K. (2017). Chapter two – Synthesis and reaction chemistry of alkylidene complexes with titanium, zirconium, vanadium, and niobium: Effective catalysts for olefin metathesis polymerization and other organic transformations. In P. J. Pérez (Ed.), *Advances in organometallic chemistry* (Vol. 68, pp. 93–136). Academic Press.

Further reading

Adams, G. P., Shaller, C. C., Dadachova, E., Simmons, H. H., Horak, E. M., Tesfaye, A., ... Weiner, L. M. (2004). A single treatment of yttrium-90-labeled CHX-A″–C6.5 diabody inhibits the growth of established human tumor xenografts in immunodeficient mice. *Cancer Research, 64*, 6200–6206.
Ade, P. A. R., Aghanim, N., Arnaud, M., Ashdown, M., Aumont, J., Baccigalupi, C., ... Zonca, A. (2016). Planck 2015 results. *Astronomy & Astrophysics, 594*, A13–A75.
Agricola, G. (1556, 1912). *De re metallica; tr. from the 1st Latin ed. of 1556, with biographical introduction, annotations and appendices upon the development of mining methods, metallurgical processes, geology, mineralogy & mining law, from the earliest times to the 16th century* (H. Hoover, & L. H. Hoover, Trans.). London: Mining Magazine.
Agulyansky, A. (2004). *Chemistry of tantalum and niobium fluoride compounds*. Amsterdam: Elsevier Science.
Akutagawa, S. (1995). Asymmetric synthesis by metal BINAP catalysts. *Applied Catalysis A: General, 128*, 171–207.
Albrecht, S., Cymorek, C., & Eckert, J. (2011). *Niobium and niobium compounds. Ullmann's encyclopedia of industrial chemistry*. Weinheim: Wiley-VCH.
Albright, W. F. (1918). Notes on egypto-semitic etymology. II. *The American Journal of Semitic Languages and Literatures, 34*, 215–255.
Alessandrello, A., Arnaboldi, C., Brofferio, C., Capelli, S., Cremonesi, O., Fiorini, E., ... Pobes, C. (2003). New limits on naturally occurring electron capture of ^{123}Te. *Physical Review C, 67*, 014323.
Alfantazi, A. M., & Moskalyk, R. R. (2003). Processing of indium: A review. *Minerals Engineering, 16*, 687–694.
Amatayakul, W., & Ramnäs, O. (2001). Life cycle assessment of a catalytic converter for passenger cars. *Journal of Cleaner Production, 9*, 395–403.
Andrews, A. J., Owsiacki, L., Kerrich, R., & Strong, D. F. (1986). The silver deposits at Cobalt and Gowganda, Ontario. I: Geology, petrography and whole rock geochemistry. *Canadian Journal of Earth Science, 1*, 519–525.

Audi, G., Bersillon, O., Blachot, J., & Wapstra, A. H. (2003). The NUBASE evaluation of nuclear and decay properties. *Nuclear Physics A*, *729*, 3–128.

Ayres, R. U., & Ayres, L. (2002). *A handbook of industrial ecology*. Cheltenham: Edward Elgar Publishing.

Bachmann, K., Frenzel, M., Krause, J., & Gutzmer, J. (2017). Advanced identification and quantification of In-bearing minerals by scanning electron microscope-based image analysis. *Microscopy and Microanalysis*, *23*, 527–537.

Bachmann, K. J. (1981). Properties, preparation, and device applications of indium phosphide. *Annual Review of Materials Science*, *11*, 441–484.

Bauman, G., Charette, M., Reid, R., & Sathya, J. (2005). Radiopharmaceuticals for the palliation of painful bone metastases – A systematic review. *Radiotherapy and Oncology*, *75*, 258–270.

Bayley, J., & Crossley, D. (2008). *Metals and metalworking: A research framework for archaeometallurgy*. London: The Historical Metallurgy Society.

Beattie, M., & Taylor, J. (2011). Silver alloy vs. uncoated urinary catheters: A systematic review of the literature. *Journal of Clinical Nursing*, *20*, 2098–2108.

Bettinelli, M., Baroni, U., & Pastorelli, N. (1988). Determination of arsenic, cadmium, lead, antimony, selenium and thallium in coal fly ash using the stabilised temperature platform furnace and Zeeman-effect background correction. *Journal of Analytical Atomic Spectrometry*, *3*, 1005–1011.

Biason Gomes, M. A., Onofre, S., Juanto, S., de, S., & Bulhões, L. O. (1991). Anodization of niobium in sulphuric acid media. *Journal of Applied Electrochemistry*, *21*, 1023–1026.

Black, H. (2005). Getting the lead out of electronics. *Environmental Health Perspectives*, *113*, A682–A685.

Bortz, M., Gutmann, M., & Yvon, K. (1999). Synthesis and structure determination of the first ternary cadmium hydride, Cs_3CdH_5. *Journal of Alloys and Compounds*, *285*, L19–L21.

Bowen, R. (1994). *Isotopes in the earth sciences*. London: Chapman & Hall.

Boyle, R. W., *The geochemistry of silver and its deposits*, Vol. 160. 1968, Geological Survey of Canada Bulletin. Geological Survey of Canada.

Brannt, W. T., *The metallic alloys: A practical guide for the manufacture of all kinds of alloys, amalgams, and solders, used by metal workers. With an appendix on the coloring of alloys* (A. Krupp, & A. Wildberger, Trans.). 1888, Philadelphia: H.C. Baird & Company.

Brencic, J. V., & Cotton, F. A. (1969). Octachlorodimolybdate(II) ion. Species with a quadruple metal-metal bond. *Inorganic Chemistry*, *8*, 7–10.

Brignole, A. B., Cotton, F. A., Dori, Z., Dori, Z., Dori, Z., & Wilkinson, G. (1972). Rhenium and molybdenum compounds containing quadruple bonds. *Inorganic Syntheses*, 81–89.

Brumby, A., Braumann, P., Zimmermann, K., Broeck, F., Vandevelde, T., Goia, D., ... Schiele, R. (2008). *Silver, silver compounds, and silver alloys. Ullmann's encyclopedia of industrial chemistry*. Weinheim: Wiley-VCH.

Burdun, G. D. (1958). On the new determination of the meter. *Measurement Techniques*, *1*, 259–264.

Busana, M. G., Prudenziati, M., & Hormadaly, J. (2006). Microstructure development and electrical properties of RuO_2-based lead-free thick film resistors. *Journal of Materials Science: Materials in Electronics*, *17*, 951–962.

Buxbaum, G., & Pfaff, G. (2005). *Cadmium pigments. Industrial inorganic pigments* (pp. 121–123). Weinheim: Wiley-VCH.

Cameron, A. G. W. (1973). Abundance of the elements in the solar system. *Space Science Reviews*, *15*, 121–146.

Canavese, C., Decostanzi, E., Branciforte, L., Caropreso, A., Nonnato, A., & Sabbioni, E. (2001). Depression in dialysis patients: Rubidium supplementation before other drugs and encouragement? *Kidney International*, *60*, 1201–1202.

Captain, I., Deblonde, G. J.-P., Rupert, P. B., An, D. D., Illy, M.-C., Rostan, E., ... Abergel, R. J. (2016). Engineered recognition of tetravalent zirconium and thorium by chelator–protein systems: Toward flexible radiotherapy and imaging platforms. *Inorganic Chemistry*, *55*, 11930–11936.

Cartlidge, E. (2018). *With better atomic clocks, scientists prepare to redefine the second. American Association for the Advancement of Science*. <https://www.sciencemag.org/news/2018/03/better-atomic-clocks-scientists-prepare-redefine-second> Accessed 04.09.19.

Caseri, W. (2014). Initial organotin chemistry. *Journal of Organometallic Chemistry*, *751*, 20–24.

Caswell, L. R., & Stone Daley, R. W. (1999). The Delhuyar brothers, tungsten, and Spanish silver. *Bulletin for the History of Chemistry*, *23*, 11–19.

Chamberlain, A. L., Fahrenholtz, W. G., & Hilmas, G. E. (2009). Reactive hot pressing of zirconium diboride. *Journal of the European Ceramic Society*, *29*, 3401–3408.

Charles, J. A. (1979). The development of the usage of tin and tin-bronze: Some problems. In *Paper presented at The Search for Ancient Tin, Washington D.C.: A seminar organized by Theodore A. Wertime and held at the Smithsonian Institution and the National Bureau of Standards, Washington DC*. March 14–15, 1977. Washington, DC.

Chattaway, F. D. (1909). The discovery of iodine. *The Chemical News and Journal of Industrial Science; with which is Incorporated the "Chemical Gazette"*, *99*, 193–195.

Chen, J. H., & Wasserburg, G. J. (1990). The isotopic composition of Ag in meteorites and the presence of ^{107}Pd in protoplanets. *Geochimica et Cosmochimica Acta*, *54*, 1729–1743.

Cherednichenko, S., Drakinskiy, V., Berg, T., Khosropanah, P., & Kollberg, E. (2008). Hot-electron bolometer terahertz mixers for the Herschel Space Observatory. *Review of Scientific Instruments*, *79*, 034501.

Chertihin, G. V., & Andrews, L. (1995). Reactions of laser-ablated Zr and Hf atoms with hydrogen. Matrix infrared spectra of the MH, MH_2, MH_3, and MH_4 molecules. *The Journal of Physical Chemistry*, *99*, 15004–15010.

Chichester Vajargah, S. H., Madaahhosseini, H., & Nemati, Z. (2007). Preparation and characterization of yttrium iron garnet (YIG) nanocrystalline powders by auto-combustion of nitrate-citrate gel. *Journal of Alloys and Compounds*, *430*, 339–343.

Chin, C., Grimm, R., Julienne, P., & Tiesinga, E. (2010). Feshbach resonances in ultracold gases. *Reviews of Modern Physics*, *82*, 1225–1286.

Choudhury, A., & Hengsberger, E. (1992). Electron beam melting and refining of metals and alloys. *The Iron and Steel Institute of Japan International, 32*, 673–681.

Cierny, J., & Weisberger, G. (2003). Bronze Age tin mines in Central Asia. *BAR International Series (Supplementary), 1199*, 23–34.

Clarke, F. W. (1914). Columbium versus niobium. *Science, 39*, 139–140.

Clayton, D. D. (2003). *Handbook of isotopes in the cosmos*. Cambridge, New York: Cambridge University Press.

Clayton, D. D. (2007). *Principles of stellar evolution and nucleosynthesis*. Chicago: University of Chicago Press.

Colon, P., Pradelle-Plasse, N., & Galland, J. (2003). Evaluation of the long-term corrosion behavior of dental amalgams: Influence of palladium addition and particle morphology. *Dental Materials, 19*, 232–239.

Considine, G. D. (2005). *Van Nostrand's encyclopedia of chemistry*. Wiley-Interscience.

Coplen Tyler, B., & Peiser, H. S. (1998). History of the recommended atomic-weight values from 1882 to 1997: A comparison of differences from current values to the estimated uncertainties of earlier values (technical report). *Pure and Applied Chemistry, 70*, 237–257.

Cotton, S. A. (1997). *Chemistry of precious metals*. London; New York: Blackie Academic & Professional.

Cotton, S. A. (2006). *Scandium, yttrium & the lanthanides: Inorganic & coordination chemistry. Encyclopedia of inorganic chemistry*. Weinheim: Wiley-VCH.

Craddock, P. T. (2010). *Early metal mining and production*. London: Archetype Publications Ltd.

de Callataÿ, F. (2005). The Graeco-Roman Economy in the super long-run: Lead, copper, and shipwrecks. *Journal of Roman Archaeology, 18*, 361–372.

Deri, M. A., Ponnala, S., Zeglis, B. M., Pohl, G., Dannenberg, J. J., Lewis, J. S., & Francesconi, L. C. (2014). Alternative chelator for ^{89}Zr radiopharmaceuticals: Radiolabeling and evaluation of 3,4,3-(LI-1,2-HOPO). *Journal of Medicinal Chemistry, 57*, 4849–4860.

Diemann, E., Müller, A., & Barbu, H. (2002). Die spannende entdeckungsgeschichte des tellurs (1782–1798) bedeutung und komplexität von elemententdeckungen. *Chemie in unserer Zeit, 36*, 334–337.

Downs, A. J. (1993). *Chemistry of aluminium, gallium, indium and thallium*. London: Blackie Academic & Professional.

Drahl, C. (2008). Palladium's hidden talent. *Chemical & Engineering News, 86*, 53–56.

Dürr, S. (2010). *Biofouling*. Oxford: Wiley-Blackwell.

Dvornický, R., & Šimkovic, F. (2011). Second unique forbidden β decay of ^{115}In and neutrino mass. *AIP Conference Proceedings, 1417*, 33–36.

Ebdon, L. (2010). *Organotin in industrial and domestic products. Trace element speciation for environment, food, and health* (p. 144) Milton Keynes: Lightning Source UK Ltd.

Eggert, P., Priem, J., & Wettig, E. (1982). Niobium: A steel additive with a future. *Economic Bulletin, 19*, 8–11.

Eichelbrönner, G. (1998). Refractory metals: Crucial components for light sources. *International Journal of Refractory Metals and Hard Materials, 16*, 5–11.

El-Mallawany, R. A. H. (2002). *Tellurite glasses handbook: Physical properties and data*. Boca Raton: CRC Press.

Emsley, J. (2001). *Nature's building blocks*. Oxford: Oxford University Press.

Endlich, F. M. (1888). On some interesting derivations of mineral names. *The American Naturalist, 22*, 21–32.

Farahani, S. K. V., Muñoz-Sanjosé, V., Zúñiga-Pérez, J., McConville, C. F., & Veal, T. D. (2013). Temperature dependence of the direct bandgap and transport properties of CdO. *Applied Physics Letters, 102*, 022102.

Farina, V., Krishnamurthy, V., & Scott, W. J. (1998). *The Stille reaction*. New York: Wiley.

Fernando, D. (1998). *Alchemy: An illustrated A to Z*. Blandford.

Fischer, R. C. (2016). Antimony: Organometallic chemistry. In R. A. Scott (Ed.), *Encyclopedia of inorganic and bioinorganic chemistry*. (pp. 1–13). New York: John Wiley & Sons Ltd.

Flynn, D. O., & Giraldez, A. (1995). Born with a "Silver Spoon: The origin of world trade in 1571. *Journal of World History, 2*, 201–221.

Fortey, R. A. (2004). *The earth: An intimate history (501 pp)*.

Frank, A. G. (1998). *ReORIENT: Global economy in the Asian Age*. Berkeley: University of California Press.

French, S. J. (1934). A story of indium. *Journal of Chemical Education, 11*, 270–272.

Frenzel, M. (2016). *The distribution of gallium, germanium and indium in conventional and non-conventional resources – Implications for global availability*. Freiberg: Technische Universität Bergakademie Freiberg.

Frenzel, M., Hirsch, T., & Gutzmer, J. (2016). Gallium, germanium, indium, and other trace and minor elements in sphalerite as a function of deposit type—A metaanalysis. *Ore Geology Reviews, 76*, 52–78.

Fthenakis, V. M. (2004). Life cycle impact analysis of cadmium in CdTe PV production. *Renewable and Sustainable Energy Reviews, 8*, 303–334.

Fürstner, A. (2000). Olefin metathesis and beyond. *Angewandte Chemie International Edition, 39*, 3012–3043.

Gaines, R. V., Skinner, H. C., Foord, E. F., Mason, B., & Rosenzweig, A. (1997). *Dana's new mineralogy* (8th ed.). New York: John Wiley & Sons, Inc.

Garrett, C. E., & Prasad, K. (2004). The art of meeting palladium specifications in active pharmaceutical ingredients produced by Pd-catalyzed reactions. *Advanced Synthesis & Catalysis, 346*, 889–900.

Geballe, T. H. (1993). Superconductivity: From physics to technology. *Physics Today, 46*, 52–56.

Genkin, A. D., & Evstigneeva, T. L. (1986). Associations of platinum-group minerals of the Norilsk copper-nickel sulfide ores. *Economic Geology, 81*, 1203–1212.

Gentile, T. R., Chen, W. C., Jones, G. L., Babcock, E., & Walker, T. G. (2005). Polarized ^{3}He spin filters for slow neutron physics. *Journal of Research of the National Institute of Standards and Technology, 110*, 299–304.

Gianduzzo, T., Colombo, J. R., Jr, Haber, G.-P., Hafron, J., Magi-Galluzzi, C., Aron, M., ... Kaouk, J. H. (2008). Laser robotically assisted nerve-sparing radical prostatectomy: A pilot study of technical feasibility in the canine model. *BJU International*, *102*, 598–602.

Gilchrist, R. (1943). The platinum metals. *Chemical Reviews*, *32*, 277–372.

Godley, R., Starosvetsky, D., & Gotman, I. (2004). Bonelike apatite formation on niobium metal treated in aqueous NaOH. *Journal of Materials Science: Materials in Medicine*, *1*, 1073–1077.

Gottschalk, A. (1969). Technetium-99m in clinical nuclear medicine. *Annual Review of Medicine*, *20*, 131–140.

Grant, C. A., & Sheppard, S. C. (2008). Fertilizer impacts on cadmium availability in agricultural soils and crops. *Human and Ecological Risk Assessment*, *14*, 210–228.

Gray, T. W., & Mann, N. (2012). *The elements: A visual exploration of every known atom in the universe*. New York: Black Dog & Leventhal Publishers: Distributed by Workman Publishing.

Greenwood, N. N. (2003). Vanadium to dubnium: From confusion through clarity to complexity. *Catalysis Today*, *78*, 5–11.

Griffith, W. P. (2003a). Bicentenary of four platinum group metals: Osmium and iridium – Events surrounding their discoveries. *Platinum Metals Review*, *47*, 175–183.

Griffith, W. P. (2003b). Rhodium and palladium – Events surrounding its discovery. *Platinum Metals Review*, *47*, 175–183.

Griffith, W. P., & Morris, P. J. T. (2003). Charles Hatchett FRS (1765–1847), chemist and discoverer of niobium. *Notes and Records of the Royal Society of London*, *57*, 299–316.

Grochala, W., & Edwards, P. P. (2004). Thermal decomposition of the non-interstitial hydrides for the storage and production of hydrogen. *Chemical Reviews*, *104*, 1283–1316.

Grund, S. C., Hanusch, K., Breunig, H. J., & Wolf, H. U. (2006). *Antimony and antimony compounds. Ullmann's encyclopedia of industrial chemistry*. Weinheim: Wiley-VCH.

Guloy, A. M., & Corbett, J. D. (1996). Synthesis, structure, and bonding of two lanthanum indium germanides with novel structures and properties. *Inorganic Chemistry*, *35*, 2616–2622.

Guo, W. X., Shu, D., Chen, H. Y., Li, A. J., Wang, H., Xiao, G. M., ... Chen, S. (2009). Study on the structure and property of lead tellurium alloy as the positive grid of lead-acid batteries. *Journal of Alloys and Compounds*, *475*, 102–109.

Gupta, C. K. (1992). *Extractive metallurgy of molybdenum*. Boca Raton: CRC Press.

Gupta, C. K., & Suri, A. K. (1993). *Extractive metallurgy of niobium*. Boca Raton: Taylor & Francis.

Halligudi, S. B., Bajaj, H. C., Bhatt, K. N., & Krishnaratnam, M. (1992). Hydrogenation of benzene to cyclohexane catalyzed by rhodium(I) complex supported on montmorillonite clay. *Reaction Kinetics and Catalysis Letters*, *48*, 547–552.

Halperin, E. C., Perez, C. A., & Brady, L. W. (2008). *Perez and Brady's principles and practice of radiation oncology*. Philadelphia: Wolters Kluwer/Lippincott Williams & Wilkins.

Hampel, C. A. (1968). *The encyclopedia of the chemical elements*. New York: Reinhold Book Corp.

Harder, A. (2002). Chemotherapeutic approaches to schistosomes: Current knowledge and outlook. *Parasitology Research*, *88*, 395–397.

Harris, D. C., & Cabri, L. J. (1973). The nomenclature of the natural alloys of osmium, iridium and ruthenium based on new compositional data of alloys from world-wide occurrences. *Canadian Mineralogist*, *12*, 104–112.

Hartman, H. L. (1992). *SME mining engineering handbook* (Vol. 1, pp. 3–42). Littleton: Society for Mining, Metallurgy & Exploration.

Hastie, J. W. (1973). Mass spectrometric studies of flame inhibition: Analysis of antimony trihalides in flames. *Combustion and Flame*, *21*, 49–54.

Heck, R., & Farrauto, R. J. (2001). Automobile exhaust catalysts. *Applied Catalysis A: General*, *221*, 443–457.

Heck, R., Gulati, S., & Farrauto, R. J. (2001). The application of monoliths for gas phase catalytic reactions. *Chemical Engineering Journal*, *82*, 149–156.

Heidingsfeldová, M., & Čapka, M. (1985). Rhodium complexes as catalysts for hydrosilylation crosslinking of silicone rubber. *Journal of Applied Polymer Science*, *30*, 1837–1846.

Heiserman, D. L. (1992). *Exploring chemical elements and their compounds*. Blue Ridge Summit: Tab Books.

Hemmingsen, L., Bauer, R., Bjerrum, M. J., Schwarz, K., Blaha, P., & Andersen, P. (1999). Structure, chemical bonding, and nuclear quadrupole interactions of β-$Cd(OH)_2$: Experiment and first principles calculations. *Inorganic Chemistry*, *38*, 2860–2867.

Heriot, T. H. P. (1920). *Manufacture of sugar from the cane and beet*. London: Longmans, Green & Co.

Hermann, R. (1871). Fortgesetzte untersuchungen über die verbindungen von ilmenium und niobium, sowie über die zusammensetzung der niobmineralien. *Journal für Praktische Chemie*, *3*, 373–427.

Hesse, R. W. (2007). *Jewelrymaking through history: An encyclopedia*. Westport: Greenwood Press.

Heuveling, D. A., Visser, G. W. M., Baclayon, M., Roos, W. H., Wuite, G. J. L., Hoekstra, O. S., ... van Dongen, G. A. M. S. (2011). ^{89}Zr-nanocolloidal albumin–based PET/CT lymphoscintigraphy for sentinel node detection in head and neck cancer: preclinical results. *The Journal of Nuclear Medicine*, *52*, 1580–1584.

Hudgens, S., & Johnson, B. (2004). Overview of phase-change chalcogenide nonvolatile memory technology. *Material Research Society Bulletin*, *29*, 1–4.

Hulett, G. A., & Berger, H. W. (1904). Volatilization of platinum. *Journal of the American Chemical Society*, *26*, 1512–1515.

Hull, C. (2005). *Pewter*. Buckinghamshire: Shire.

Hunt, L. B., & Lever, F. M. (1969). Platinum metals: A survey of productive resources to industrial uses. *Platinum Metals Review*, *13*, 126–138.

Hutchison, C. S. (1988). Geology of tin deposits in Asia and the Pacific selected papers from the international symposium on the geology of tin deposits held in Nanning, China, October 26–30, 1984, jointly sponsored by ESCAP/RMRDC and the Ministry of Geology, People's Republic of China. Berlin: Springer.

Ipser, H., Flandorfer, H., Luef, C., Schmetterer, C., & Saeed, U. (2007). Thermodynamics and phase diagrams of lead-free solder materials. *Journal of Materials Science: Materials in Electronics, 18*, 3–17.

Jang, H., & Kim, S. J. (2000). The effects of antimony trisulfide (Sb_2S_3) and zirconium silicate ($ZrSiO_4$) in the automotive brake friction material on friction characteristics. *Wear, 239*, 229–236.

Jennings, T. C. (2005). Cadmium environmental concerns. In C. E. Wilkes, J. W. Summers, C. A. Daniels, & M. T. Berard (Eds.), *PVC handbook* (p. 149). Munich; Cincinnati: Hanser.

Jiao, Y., Grant, C. A., & Bailey, L. D. (2004). Effects of phosphorus and zinc fertilizer on cadmium uptake and distribution in flax and durum wheat. *Journal of the Science of Food and Agriculture, 84*, 777–785.

Kanazawa, Y., & Kamitani, M. (2006). Rare earth minerals and resources in the world. *Journal of Alloys and Compounds, 408–412*, 1339–1343.

Kao, L., Peacor, D. R., Coveney, J. R. M., Zhao, G., Dungey, K. E., Curtis, M. D., & Penner-Hahn, J. E. (2001). A C/MoS_2 mixed-layer phase (MoSC) occurring in metalliferous black shales from southern China, and new data on jordisite. *American Mineralogist, 86*, 852–861.

Karuna Purnapu Rupa, P., Sharma, P., Mohanty, R. M., & Balasubramanian, K. (2010). Microstructure and phase composition of composite coatings formed by plasma spraying of ZrO_2 and B_4C powders. *Journal of Thermal Spray Technology, 19*, 816–823.

Kassianidou, V. (2003). Early extraction of silver from complex polymetallic ores. In P. T. Craddock, & J. Lang (Eds.), *Mining and metal production through the ages* (pp. 198–206). London: British Museum Press.

Kelly, W. R., Gounelle, G. J., & Hutchison, R. (1978). Evidence for the existence of ^{107}Pd in the early solar system. *Geophysical Research Letters, 5*, 1079–1082.

Kemal, M., Arslan, V., Akar, A., & Canbazoglu, M. (1996). Production of $SrCO_3$ by black ash process: Determination of reductive roasting parameters. In *Paper presented at the Symposium: Changing scopes in mineral processing: Proceedings of the 6th international mineral processing symposium*, Kusadasi, Turkey, 24–26 September, 1996. Rotterdam.

Khusnutdinova, J. R., Rath, N. P., & Mirica, L. M. (2010). Stable mononuclear organometallic Pd(III) complexes and their C – C bond formation reactivity. *Journal of the American Chemical Society, 132*, 7303–7305.

Kielhorn, J., Melber, C., Keller, D., & Mangelsdorf, I. (2002). Palladium – A review of exposure and effects to human health. *International Journal of Hygiene and Environmental Health, 205*, 417–432.

Kim, H., Gilmore, C. M., Piqué, A., Horwitz, J. S., Mattoussi, H., Murata, H., ... Chrisey, D. B. (1999). Electrical, optical, and structural properties of indium–tin–oxide thin films for organic light-emitting devices. *Journal of Applied Physics, 86*, 6451–6461.

Kinch, M. A. (2007). *Fundamentals of infrared detector materials*. Bellingham: SPIE.

Klein, C., & Dutrow, B. (2008). *The 23rd edition of the manual of mineral science (after James D. Dana)*. Hoboken: John Wiley & Sons, Inc.

Knoll, G. F. (2000). *Radiation detection and measurement*. New York: John Wiley.

Knorr, K. E. (1973). *Tin under control*. Stanford.: Stanford University Library.

Kobell, V. (1860). Ueber eine eigenthümliche säure, diansäure, in der gruppe der tantal- und niob-verbindungen. *Journal für Praktische Chemie, 79*, 291–303.

Koch, E.-C. (2002). Special materials in pyrotechnics, part II: Application of caesium and rubidium compounds in pyrotechnics. *Journal Pyrotechnics, 15*, 9–24.

Kogel, J. E., Trivedi, N. C., Barker, J. M., & Krukowski, S. T. (2006). *Industrial minerals & rocks: Commodities, markets, and uses*. Littleton: Society for Mining, Metallurgy, and Exploration.

Kolarik, Z., & Renard, E. V. (2003). Recovery of value fission platinoids from spent nuclear fuel. Part I: general considerations and basic chemistry. *Platinum Metals Review, 47*, 74–87.

Kong, J., Tang, D. Y., Zhao, B., Lu, J., Ueda, K., Yagi, H., & Yanagitani, T. (2005). 9.2-W diode-end-pumped Yb:Y_2O_3 ceramic laser. *Applied Physics Letters, 86*, 161116.

Kòrösy, F. (1939). Reaction of tantalum, columbium and vanadium with iodine. *Journal of the American Chemical Society, 61*, 838–843.

Kovalenko, V. I., & Yarmolyuk, V. V. (1995). Endogenous rare metal ore formations and rare metal metallogeny of Mongolia. *Economic Geology, 90*, 520–529.

Krebs, R. E. (2006). *The history and use of our earth's chemical elements: A reference guide*. Westport: Greenwood Press.

Kuang, D., Ito, S., Wenger, B., Klein, C., Moser, J.-E., Humphry-Baker, R., ... Grätzel, M. (2006). High molar extinction coefficient heteroleptic ruthenium complexes for thin film dye-sensitized solar cells. *Journal of the American Chemical Society, 128*, 4146–4154.

Kushner, J. B. (1940). Modern rhodium plating. *Metals and Alloys, 11*, 137–140.

Kuzovnikov, M. A., & Tkacz, M. (2016). Synthesis of ruthenium hydride. *Physical Review B, 93*, 064103.

Lalovic, M., & Werle, H. (1970). The energy distribution of antimonyberyllium photoneutrons. *Journal of Nuclear Energy, 24*, 123–132.

Langedal, M. (1997). Dispersion of tailings in the Knabena—Kvina drainage basin, Norway, 1: Evaluation of overbank sediments as sampling medium for regional geochemical mapping. *Journal of Geochemical Exploration, 58*, 157–172.

Lansdown, A. R. (2006). *Molybdenum disulphide lubrication*. Amsterdam: Elsevier.

Lehmann, B. (1990). *Metallogeny of tin*. Berlin: Springer.

Li, Z., Wakai, R. T., & Walker, T. G. (2006). Parametric modulation of an atomic magnetometer. *Applied Physics Letters, 89*, 2357553-1–2357553-3.

Lilje, L., Kako, E., Kostin, D., Matheisen, A., Möller, W. D., Proch, D., ... Twarowski, K. (2004). Achievement of 35MV/m in the superconducting nine-cell cavities for TESLA. *Nuclear Instruments and Methods in Physics Research Section A: Accelerators, Spectrometers, Detectors and Associated Equipment, 524*, 1–12.

Lindenhovius, J. L. H., Hornsveld, E. M., den Ouden, A., Wessel, W. A. J., & ten Kate, H. H. J. (2000). Powder-in-tube (PIT) Nb_3Sn conductors for high-field magnets. *IEEE Transactions on Applied Superconductivity, 10*, 975–978.

Louis, H. (1911). *Metallurgy of tin*. New York: McGraw-Hill.

Lucas, I. T., Syzdek, J., & Kostecki, R. (2011). Interfacial processes at single-crystal β-Sn electrodes in organic carbonate electrolytes. *Electrochemistry Communications, 13*, 1271–1275.

Lyday, P. A., & Kaiho, T. (2015). *Iodine and iodine compounds. Ullmann's encyclopedia of industrial chemistry* (pp. 1–13). Weinheim: Wiley-VCH.

Mackenzie, D. H., & Davies, R. H. (1990). Broken Hill lead-silver-zinc deposit at Z.C. mines. In F. E. Hughes (Ed.), *Geology of the Mineral Deposits of Australia and Papua New Guinea* (pp. 1079–1084). Melbourne: The Australasian Institute of Mining and Metallurgy.

MacMillan, J. P., Park, J. W., Gerstenberg, R., Wagner, H., Köhler, K., & Wallbrecht, P. (2002). *Strontium and strontium compounds. Ullmann's encyclopedia of industrial chemistry*. Weinheim: Wiley-VCH.

Maekawa, T., Igari, S.-I., & Kaneko, N. (2006). Chemical and isotopic compositions of brines from dissolved-in-water type natural gas fields in Chiba, Japan. *Geochemical Journal, 40*, 475–484.

Maeno, Y., Rice, T. M., & Sigrist, M. (2001). The intriguing superconductivity of strontium ruthenate. *Physics Today, 54*, 42–47.

Mahrova, T. V., Fukin, G. K., Cherkasov, A. V., Trifonov, A. A., Ajellal, N., & Carpentier, J.-F. (2009). Yttrium complexes supported by linked bis(amide) ligand: Synthesis, structure, and catalytic activity in the ring-opening polymerization of cyclic esters. *Inorganic Chemistry, 48*, 4258–4266.

Maillard, J.-Y., & Hartemann, P. (2012). Silver as an antimicrobial: Facts and gaps in knowledge. *Critical Reviews in Microbiology, 39*, 373–383.

Manchester, F. D., San-Martin, A., & Pitre, J. M. (1994). The H-Pd (hydrogen-palladium) system. *Journal of Phase Equilibria, 15*, 62–83.

Martin, J., Ortega-Huertas, M., & Torres-Ruiz, J. (1984). Genesis and evolution of strontium deposits of the Granada basin (Southeastern Spain): Evidence of diagenetic replacement of a stromatolite belt. *Sedimentary Geology, 39*, 281–298.

Martin, J. L., McKenzie, C. R., Thomas, N. R., Sharpe, J. C., Warrington, D. M., Manson, P. J., ... Wilson, A. C. (1999). Output coupling of a Bose-Einstein condensate formed in a TOP trap. *Journal of Physics B: Atomic, Molecular and Optical Physics, 32*, 3065–3075.

Martínez-Abad, A., Ocio, M. J., Lagarón, J. M., & Sánchez, G. (2013). Evaluation of silver-infused polylactide films for inactivation of Salmonella and feline calicivirus in vitro and on fresh-cut vegetables. *International Journal of Food Microbiology, 162*, 89–94.

McCallum, R. I. (1999). *Antimony in medical history: An account of the medical uses of antimony and its compounds since early times to the present.* Edinburgh; Durham: Pentland Press.

McDonnell, G., & Russell, A. D. (1999). Antiseptics and disinfectants: Activity, action, and resistance. *Clinical Microbiology Reviews, 12*, 147–179.

Méar, F., Yot, P., Cambon, M., & Ribes, M. (2006). The characterization of waste cathode-ray tube glass. *Waste Management, 26*, 1468–1476.

Meier, S. M., & Gupta, D. K. (1994). The evolution of thermal barrier coatings in gas turbine engine applications. *Journal of Engineering for Gas Turbines and Power, 116*, 250–257.

Mellor, J. W. (1964). *Antimony, . A comprehensive treatise on inorganic and theoretical chemistry* (Vol. 9, p. 339). New York: John Wiley & Sons.

Miller, M. A., Askevold, B., Mikula, H., Kohler, R. H., Pirovich, D., & Weissleder, R. (2017). Nano-palladium is a cellular catalyst for in vivo chemistry. *Nature Communications, 8*, 15906.

Moorey, P. R. S. (1999). *Ancient Mesopotamian materials and industries: The archaeological evidence.* Winona Lake: Eisenbrauns.

Morrow, H. (2010). *Cadmium and cadmium alloys. Kirk-Othmer encyclopedia of chemical technology* (pp. 1–36). John Wiley & Sons.

Morteani, G. (1991). The rare earths: their minerals, production and technical use. *European Journal of Mineralogy, 3*, 641–650.

Moulson, A. J., & Herbert, J. M. (2008). *Electroceramics: Materials, properties, applications.* Chichester; New York: Wiley.

Muller, O., & Roy, R. (1968). Formation and stability of the platinum and rhodium oxides at high oxygen pressures and the structures of Pt_3O_4, β-PtO_2 and RhO_2. *Journal of the Less-Common Metals, 16*, 129–146.

Mullin, J. B., & Miller, L. S. (1991). *Crystalline cadmium sulfide. Electronic materials: From silicon to organics* (p. 273) New York: Plenum.

Murray, C. B., Norris, D. J., & Bawendi, M. G. (1993). Synthesis and characterization of nearly monodisperse CdE (E = sulfur, selenium, tellurium) semiconductor nanocrystallites. *Journal of the American Chemical Society, 115*, 8706–8715.

Murray, T. (1993). Elementary scots: The discovery of strontium. *Scottish Medical Journal, 38*, 188–189.

Naumann d'Alnoncourt, R., Csepei, L.-I., Hävecker, M., Girgsdies, F., Schuster, M. E., Schlögl, R., & Trunschke, A. (2014). The reaction network in propane oxidation over phase-pure MoVTeNb M1 oxide catalysts. *Journal of Catalysis, 311*, 369–385.

Naumov, A. V. (2008). Review of the world market of rare-earth metals. *Russian Journal of Non-Ferrous Metals, 49*, 14–22.

Nekrasov, I. Y. (1996). *Geochemistry, mineralogy and genesis of gold deposits* (P. M. Rao, Trans.). Rotterdam: A.A. Balkema.

Nesse, W. D. (2000). *Introduction to mineralogy.* New York: Oxford University Press.

Nielsen, R. H., & Wilfing, G. (2010). *Zirconium and zirconium compounds. Ullmann's encyclopedia of industrial chemistry.* Weinheim: Wiley-VCH.

Minerals, Metals and Materials Society (Ed.). (2002). Niobium, science & technology: Proceedings of the international symposium niobium 2001: held in Orlando, Florida, USA, December 2–5, 2001. Bridgeville: Niobium 2001 Ltd.

Nishii, J., Morimoto, S., Inagawa, I., Iizuka, R., Yamashita, T., & Yamagishi, T. (1992). Recent advances and trends in chalcogenide glass fiber technology: A review. *Journal of Non-Crystalline Solids, 140*, 199–208.

Nishiuchi, K., Kitaura, H., Yamada, N., & Akahira, N. (1998). Dual-layer optical disk with Te–O–Pd phase-change film. *Japanese Journal of Applied Physics, 37*.

Nishiyama, K., Nakamur, T., Utsumi, S., Sakai, H., & Abe, M. (2009). Preparation of ultrafine boride powders by metallothermic reduction method. *Journal of Physics: Conference Series, 176*, 012043.

Norman, N. C. (1997). *Chemistry of arsenic, antimony and bismuth.* London: Blackie Academic & Professional.

Norton, J. J. (1973). Lithium, cesium, and rubidium—The rare alkali metals. In D. A. Brobst, & W. P. Pratt (Eds.), *United States mineral resources* (pp. 365–378). Washington, D.C.: U.S. Geological Survey Professional, Paper no. 820.

Noyhouzer, T., & Mandler, D. (2011). Determination of low levels of cadmium ions by the under potential deposition on a self-assembled monolayer on gold electrode. *Analytica Chimica Acta, 684*, 1–7.

O'Brien, R. C., Ambrosi, R. M., Bannister, N. P., Howe, S. D., & Atkinson, H. V. (2008). Safe radioisotope thermoelectric generators and heat sources for space applications. *Journal of Nuclear Materials, 377*, 506–521.

Ojima, I., Athan, A. A., Chaterpaul, S. J., Kaloko, J. J., & Teng, Y.-H. G. (2013). Organorhodium chemistry. In B. H. Lipshutz (Ed.), *Organometallics in synthesis* (pp. 135–318). New York: John Wiley & Sons.

Orchin, G. J., Fazio, D. D., Bernardo, A. D., Hamer, M., Yoon, D., Cadore, A. R., ... Hadfield, R. H. (2019). Niobium diselenide superconducting photodetectors. *Applied Physics Letters, 114*, 251103.

Padmanabhan, T. (2001). *Theoretical astrophysics*. Cambridge: Cambridge University Press.

Panáček, A., Kvítek, L., Smékalová, M., Večeřová, R., Kolář, M., Röderová, M., ... Zbořil, R. (2018). Bacterial resistance to silver nanoparticles and how to overcome it. *Nature Nanotechnology, 13*, 65–71.

Park, J. W., Chun, Y. S., Choi, E., Kim, G. T., Choi, H., Kim, C. H., ... Park, J. W. (2000). Cadmium blocks hypoxia-inducible factor (HIF)-1-mediated response to hypoxia by stimulating the proteasome-dependent degradation of HIF-1alpha. *European Journal of Biochemistry, 267*, 4198–4204.

Partington, J. R. (1942). The early history of strontium. *Annals of Science, 5*, 157–166.

Paschalis, C., Jenner, F. A., & Lee, C. R. (1978). Effects of rubidium chloride on the course of manic-depressive illness. *Journal of the Royal Society of Medicine, 71*, 343–352.

Patel, Z., & Khul'ka, K. (2001). Niobium for steelmaking. *Metallurgist, 45*, 477–480.

Patterson, C. C. (1972). Silver stocks and losses in ancient and medieval times. *The Economic History Review, 25*, 205–235.

Penhallurick, R. D. (1986). *Tin in antiquity: Its mining and trade throughout the ancient world with particular reference to Cornwall*. Leeds: Maney Publishing.

Perkins, D. (2002). *Mineralogy* (2nd ed.). Upper Saddle River: Prentice Hall.

Pernicka, E., Rehren, T., & Schmitt-Strecker, S. (1998). Late Uruk silver production by cupellation at Habuba Kabira, Syria. In *paper presented at the metallurgica antiqua: In honour of Hans-Gert Bachmann and Robert Maddin*. Bochum, 1998.

Perry, R. S., Kitagawa, K., Grigera, S. A., Borzi, R. A., Mackenzie, A. P., Ishida, K., & Maeno, Y. (2004). Multiple first-order metamagnetic transitions and quantum oscillations in ultrapure $Sr_3Ru_2O_7$. *Physical Review Letters, 92*, 166602.

Pilkington, L. A. B. (1969). Review lecture. The float glass process. *Proceedings of the Royal Society of London Series A: Mathematical and Physical Sciences, 314*, 1–25.

Pitchkov, V. N. (1996). The discovery of ruthenium. *Platinum Metals Review*, 40, 181–188.

Pottlacher, G., Hosaeus, H., Wilthan, B., Kaschnitz, E., & Seifter, A. (2002). Thermophysikalische eigenschaften von festem und flüssigem Inconel 718. *Thermochimica Acta, 382*, 55–267.

Powalla, M., & Dimmler, B. (2000). Scaling up issues of CIGS solar cells. *Thin Solid Films, 361–362*, 540–546.

Powell, L. V., Johnson, G. H., & Bales, D. J. (1989). Effect of admixed indium on mercury vapor release from dental amalgam. *Journal of Dental Research, 68*, 1231–1233.

Price, T. D., Schoeninger, M. J., & Armelagos, G. J. (1985). Bone chemistry and past behavior: An overview. *Journal of Human Evolution, 14*, 419–447.

Rane, S., Prudenziati, M., & Morten, B. (2007). Environment friendly perovskite ruthenate based thick film resistors. *Materials Letters, 61*, 595–599.

Rao, C. R. K., & Trivedi, D. C. (2005). Chemical and electrochemical depositions of platinum group metals and their applications. *Coordination Chemistry Reviews, 249*, 613–631.

Rau, J. V., Latini, A., Generosi, A., Rossi Albertini, V., Ferro, D., Teghil, R., & Barinov, S. M. (2009). Deposition and characterization of superhard biphasic ruthenium boride films. *Acta Materialia, 57*, 673–681.

Raub, C. J. (2004). The minting of platinum roubles. Part I: History and current investigations. *Platinum Metals Review, 48*, 66–69.

Rayner-Canham, G., & Zheng, Z. (2008). Naming elements after scientists: An account of a controversy. *Foundations of Chemistry, 10*, 13–18.

Reardon, A. C. (2011). *Metallurgy for the non-metallurgist* (2nd ed.). Materials Park: ASM International.

Renner, H., Schlamp, G., Kleinwächter, I., Drost, E., Lüschow, H. M., Tews, P., ... Drieselman, R. (2002). *Platinum group metals and compounds. Ullmann's encyclopedia of industrial chemistry*. Weinheim: Wiley-VCH.

Rieuwerts, J. (2015). *The elements of environmental pollution*. London: Routledge, Taylor & Francis Group, Earthscan.

Rose, H. (1844). Ueber die zusammensetzung der Tantalite und ein im Tantalite von Baiern enthaltenes neues Metall. *Annalen der Physik, 139*, 317–341.

Rosenblatt, A., Stamford, T. C. M., & Niederman, R. (2009). Silver diamine fluoride: A caries "silver-fluoride bullet". *Journal of Dental Research, 88*, 116–125.

Rossotti, H. (1998). *Diverse atoms. Profiles of the chemical elements*. Oxford: Oxford University Press.

Roth, J. F. (1975). Rhodium catalysed carbonylation of methanol. *Platinum Metals Review, 19*, 12–14.

Ruggiero, A., Holland, J. P., Hudolin, T., Shenker, L., Koulova, A., Bander, N. H., ... Grimm, J. (2011). Targeting the internal epitope of prostate-specific membrane antigen with 89Zr-7E11 immuno-PET. *The Journal of Nuclear Medicine, 52*, 1608–1615.

Rushforth, R. (2004). Palladium in restorative dentistry: Superior physical properties make palladium an ideal dental metal. *Platinum Metals Review, 48*, 30–31.

Russell, C. A. (2000). Antimony's curious history. *Notes and Records of the Royal Society of London, 54*, 115–116.

Russell, S. S., Gounelle, M., & Hutchison, R. (2001). Origin of short-lived radionuclides. *Philosophical Transactions of the Royal Society of London Series A: Mathematical, Physical and Engineering Sciences, 359*, 1991–2004.

Salem, R., & Lewandowski, R. J. (2013). Chemoembolization and radioembolization for hepatocellular carcinoma. *Clinical Gastroenterology and Hepatology, 11*, 604–611.

Schmidbaur, H., & Bayler, A. (2009). Synthesis and uses of organosilver compounds. In Z. Rappoport (Ed.), *PATAI'S chemistry of functional groups*. New York: John Wiley & Sons.

Schmitt, H. W. (1960). Determination of the energy of antimony-beryllium photoneutrons. *Nuclear Physics, 20*, 220−226.

Schrader, G. F., Elshennawy, A. K., & Doyle, L. E. (2000). *Manufacturing processes & materials*. Dearborn: Society of Manufacturing Engineers.

Schultze, C. A., Stanish, C., Scott, D. A., Rehren, T., Kuehner, S., & Feathers, J. K. (2009). Direct evidence of 1900 years of indigenous silver production in the Lake Titicaca Basin of Southern Peru. *Proceedings of the National Academy of Sciences of the United States of America, 106*, 17280−17283.

Schutz, R. W. (1996). Ruthenium enhanced titanium alloys. *Platinum Metals Review, 40*, 54−61.

Schwarz-Schampera, U., & Herzig, P. M. (2002). *Indium: Geology, mineralogy, and economics*. Berlin: Springer Verlag.

Schweissing, M. M., & Grupe, G. (2003). Stable strontium isotopes in human teeth and bone: A key to migration events of the late Roman period in Bavaria. *Journal of Archaeological Science, 30*, 1373−1383.

Sciolist Online Etymology Dictionary. <https://www.etymonline.com/> Accessed 21.06.19.

Scoullos, M. J. (2001). *Mercury, cadmium, lead: Handbook for sustainable heavy metals policy and regulation*. Dordrecht; London: Kluwer Academic.

Sebenik, R. F., Burkin, A. R., Dorfler, R. R., Laferty, J. M., Leichtfried, G., Meyer-Grünow, H., ... Thielke, S. A. (2000). *Molybdenum and molybdenum compounds. Ullmann's encyclopedia of industrial chemistry*. Weinheim: Wiley-VCH.

Setoyama, D., Ito, M., Matsunaga, J., Muta, H., Uno, M., & Yamanaka, S. (2005). Mechanical properties of yttrium hydride. *Journal of Alloys and Compounds, 394*, 207−210.

Setterberg, C. (1882). Ueber die darstellung von rubidium- und cäsiumverbindungen und über die gewinnung der metalle selbst. *Justus Liebig's Annalen der Chemie, 211*, 100−116.

Shannon, R. D., & Prewitt, C. T. (1970). Synthesis and structure of a new high-pressure form of Rh_2O_3. *Journal of Solid State Chemistry, 2*, 134−136.

Shayesteh, A., Yu, S., & Bernath, P. F. (2005). Gaseous HgH_2, CdH_2, and ZnH_2. *Chemistry − A European Journal, 11*, 4709−4712.

Shelef, M., & Graham, G. W. (1994). Why rhodium in automotive three-way catalysts? *Catalysis Reviews, 36*, 433−457.

Shenai, D. V., Timmons, M. L., DiCarlo, R. L., Jr., & Marsman, C. J. (2004). Correlation of film properties and reduced impurity concentrations in sources for III/V-MOVPE using high-purity trimethylindium and tertiarybutylphosphine. *Journal of Crystal Growth, 272*, 603−608.

Shenai-Khatkhate, D. V., Webb, P., Cole-Hamilton, D. J., Blackmore, G. W., & Mullin, B. J. (1988). Ultra-pure organotellurium precursors for the low-temperature MOVPE growth of II/VI compound semiconductors. *Journal of Crystal Growth, 93*, 744−749.

Shlyk, L., Kryukov, S., Schüpp-Niewa, B., Niewa, R., & De Long, L. E. (2008). High-temperature ferromagnetism and tunable semiconductivity of (Ba, Sr)$M_{2\pm x}Ru_{4\mp x}O_{11}$ (M = Fe, Co): A new paradigm for spintronics. *Advanced Materials, 20*, 1315−1320.

Shortland, A. J. (2006). Application of lead isotope analysis to a wide range of late bronze age Egyptian materials. *Archaeometry, 48*, 657−669.

Shu, J., Grandjean, B. P. A., Neste, A. V., & Kaliaguine, S. (1991). Catalytic palladium-based membrane reactors: A review. *The Canadian Journal of Chemical Engineering, 69*, 1036−1060.

Sillitoe, R. H., Halls, C., & Grant, J. N. (1975). Porphyry tin deposits in Bolivia. *Economic Geology, 70*, 913−927.

Simon, W. (1991). Mexican silver mining. *Engineering Mining Journal, 192*, 18−22.

Sinha, P., Kriegner, C. J., Schew, W. A., Kaczmar, S. W., Traister, M., & Wilson, D. J. (2008). Regulatory policy governing cadmium-telluride photovoltaics: A case study contrasting life cycle management with the precautionary principle. *Energy Policy, 36*.

Snyder, G., & Fabryka-Martin, J. (2007). I-129 and Cl-36 in dilute hydrocarbon waters: Marine-cosmogenic, in situ, and anthropogenic sources. *Applied Geochemistry, 22*, 692−714.

Sokolov, A. P., Pochivalin, G. P., Shipovskikh, Y. M., Garusov, Y. V., Chernikov, O. G., & Shevchenko, V. G. (1993). Rhodium self-powered detector for monitoring neutron fluence, energy production, and isotopic composition of fuel. *Atomic Energy, 74*, 365−367.

Soisson, D. J., McLafferty, J. J., & Pierret, J. A. (1961). Staff-industry collaborative report: Tantalum and niobium. *Industrial and Engineering Chemistry, 53*, 861−868.

Stanford, E. C. C. (1862). On the economic applications of seaweed. *Journal of the Society of Arts*, 185−189.

Staples, L. W. (1951). Ilsemannite and jordisite. *American Mineralogist, 36*, 609−614.

Stwertka, A. (1999). *A guide to the elements*. Oxford: Oxford University Press.

Su, K., & Sneddon, L. G. (1993). A polymer precursor route to metal borides. *Chemistry of Materials, 5*, 1659−1668.

Suess, H., & Urey, H. (1956). Abundances of the elements. *Reviews of Modern Physics, 28*, 53−74.

Sulzer, F. G., & Blumenbach, J. F. (1791). Über den Strontianit, ein Schottisches Foßil, das ebenfalls eine neue Grunderde zu enthalten scheint. *Bergmännisches Journal*, 433−436.

Surmann, P., & Zeyat, H. (2005). Voltammetric analysis using a self-renewable nonmercury electrode. *Analytical and Bioanalytical Chemistry, 383*, 1009−1013.

Sutphin, D. M., Sabin, A. E., & Reed, B. L. (1992). *International strategic minerals inventory summary report: Tin*. Diane Publishing Company.

Suttill, K. R. (1988). A fabulous silver porphyry: Cerro Rico de Potosi. *Engineering Mining Journal, 189*, 50−53.

Swain, P. A. (2005). Bernard Courtois (1777−1838) famed for discovering iodine (1811), and his life in Paris from 1798. *Bulletin for the History of Chemistry, 30*, 103−111.

Taki, M. (2013). Imaging and sensing of cadmium in cells. In A. Sigel, H. Sigel, & R. K. O. Sigel (Eds.), *Cadmium: From toxicity to essentiality* (pp. 99−115). Dordrecht: Springer.

Tamm, M., & Baker, R. J. (2007). Molybdenum compounds with CO or isocyanides. In D. M. P. Mingos, & R. H. Crabtree (Eds.), *Comprehensive organometallic chemistry III* (pp. 391–512). Oxford: Elsevier.

Tang, H., Li, C., Yang, X., Mo, C., Cao, K., & Yan, F. (2011). Synthesis and tribological properties of $NbSe_3$ nanofibers and $NbSe_2$ microsheets. *Crystal Research and Technology*, *46*, 400–404.

Tokurakawa, M., Takaichi, K., Shirakawa, A., Ueda, K.-i, Yagi, H., Yanagitani, T., & Kaminskii, A. A. (2007). Diode-pumped 188fs mode-locked Yb^{3+}:Y_2O_3 ceramic laser. *Applied Physics Letters*, *90*, 071101.

Tornieporth-Oetting, I., & Klapötke, T. (1990). Nitrogen triiodide. *Angewandte Chemie International Edition*, *29*, 677–679.

Tsuji, J. (2004). *Palladium reagents and catalysts: New perspectives for the 21st century*. Chichester: Wiley.

Tsuji, J. (2004). *The basic chemistry of organopalladium compounds. Palladium reagents and catalysts* (pp. 1–26). New York: John Wiley & Sons.

Tsunoura, T., Yoshida, K., Yano, T., & Kishi, Y. (2017). Fabrication, characterization, and fluorine-plasma exposure behavior of dense yttrium oxyfluoride ceramics. *Japanese Journal of Applied Physics*, *56*, 06HC02.

Tudela, D. (2008). Silver(II) oxide or silver(I,III) oxide? *Journal of Chemical Education*, *85*, 863.

Turekian, K. K., & Wedepohl, K. H. (1961). Distribution of the elements in some major units of the earth's crust. *Geological Society of America Bulletin*, *72*, 175–192.

Turneaure, F. S. (1971). The Bolivian tin-silver province. *Economic Geology*, *66*, 215–225.

Usselman, M. (1978). The Wollaston/Chenevix controversy over the elemental nature of palladium: A curious episode in the history of chemistry. *Annals of Science*, *35*, 551–579.

van der Heyden, A., & Edgecombe, D. R. (1990). Silver-lead-zinc deposit at South Mine, Broken Hill. In F. E. Hughes (Ed.), *Geology of the mineral deposits of Australia and Papua New Guinea* (pp. 1073–1077). Melbourne: The Australasian Institute of Mining and Metallurgy.

van Rij, C. M., Sharkey, R. M., Goldenberg, D. M., Frielink, C., Molkenboer, J. D. M., Franssen, G. M., … Boerman, O. C. (2011). Imaging of prostate cancer with immuno-PET and immuno-SPECT using a radiolabeled anti-EGP-1 monoclonal antibody. *The Journal of Nuclear Medicine*, *52*, 1601–1607.

Venetskii, S. (1971). Indium. *Metallurgist*, *15*, 148–150.

Verryn, S. M. C., & Merkle, R. K. W. (1994). Compositional variation of cooperite, braggite, and vysotskite from the Bushveld Complex. *Mineralogical Magazine*, *58*, 223–234.

Vinokurov, V. A., Stavitskaya, A. V., Chudakov, Y. A., Ivanov, E. V., Shrestha, L. K., Ariga, K., … Lvov, Y. M. (2017). Formation of metal clusters in halloysite clay nanotubes. *Science and Technology of Advanced Materials*, *18*, 147–151.

Volk, T., & Wöhlecke, M. (2009). *Lithium niobate: Defects, photorefraction and ferroelectric switching*. Berlin: Springer.

von Glahn, R. (1996). Myth and reality of China's seventeenth century monetary crisis. *Journal of Economic History*, *56*, 429–454.

von Lippmann, E. O. (1919). *Entstehung und ausbreitung der Alchemie; mit einem Anhange: Zur älteren Geschichte der Metalle; ein beitrag zur kulturgeschichte*. Berlin: Julius Springer.

Wade, K., & Banister, A. J. (2016). *The chemistry of aluminium, gallium, indium and thallium: Comprehensive inorganic chemistry, . Pergamon texts in inorganic chemistry* (Vol. 12). Oxford: Pergamon Press.

Waight, P. (2002). *Chernobyl: Assessment of radiological and health impact, 2002 update; Chapter I – The site and accident sequence*. Paris: OECD-NEA.

Walther, J. V. (2009). *Essentials of geochemistry*. Sudbury: Jones and Bartlett Publishers.

Wang, C. W. (1919). *Antimony: Its history, chemistry, mineralogy, geology, metallurgy, uses, preparation, analysis, production and valuation with complete bibliographies*. London: Charles Griffin and Co. Ltd.

Wang, H., Huang, X., Lin, J., Cui, J., Chen, Y., Zhu, C., … Liu, Z. (2017). High-quality monolayer superconductor $NbSe_2$ grown by chemical vapour deposition. *Nature Communications*, *8*, 394.

Wang, X., Andrews, L., Chertihin, G. V., & Souter, P. F. (2002). Infrared spectra of the novel Sn_2H_2 species and the reactive $SnH_{1,2,3}$ and $PbH_{1,2,3}$ intermediates in solid neon, deuterium, and argon. *The Journal of Physical Chemistry A*, *106*, 6302–6308.

Ware, M. (2005). Book review: Photography in platinum and palladium. *Platinum Metals Review*, *49*, 190–195.

Watson, J. T., Roe, D. K., & Selenkow, H. A. (1965). Iodine-129 as a nonradioactive tracer. *Radiation Research*, *26*, 159–163.

Wedepohl, H. K. (1995). The composition of the continental crust. *Geochimica et Cosmochimica Acta*, *59*, 1217–1232.

Weeks, M. E., & Leichester, H. M. (1968). Discovery of the Elements. *Journal of Chemical Education*, Easton, PA.

Wei, X. B. C., Izhvanov, O., Haines, C. D., & Olevsky, E. A. (2016). Zirconium carbide produced by spark plasma sintering and hot pressing: Densification kinetics, grain growth, and thermal properties. *Materials Characterization*, *9*, 577.

Weil, E. D., & Levchik, S. V. (2009). *Flame retardants for plastics and textiles: Practical applications*. Munich: Hanser Verlag.

Weisberg, A. M. (1999a). Rhodium plating. *Metal Finishing*, *97*, 296.

Weisberg, A. M. (1999b). Ruthenium plating. *Metal Finishing*, *97*, 297.

White, R. E., & Hanusa, T. P. (2005). Ruthenium: Organometallic chemistry based in part on the article ruthenium: Organometallic chemistry by Ulrich Koelle which appeared in the encyclopedia of inorganic chemistry. In R. B. King, R. H. Crabtree, C. M. Lukehart, D. A. Atwood, & R. A. Scott (Eds.), *Encyclopedia of inorganic chemistry* (1st ed.). New York: John Wiley & Sons.

Willardson, R. K., & Beer, A. C. (1981). *Mercury cadmium telluride*. New York: Academic Press.

Wilson, N. J., Craw, D., & Hunter, K. (2004). Antimony distribution and environmental mobility at an historic antimony smelter site, New Zealand. *Environmental Pollution*, *129*, 257–266.

Winer, W. (1967). Molybdenum disulfide as a lubricant: A review of the fundamental knowledge. *Wear*, *10*, 422–452.

Wolterbeek, H. T., & Bode, P. (2011). A process for the production of no-carrier added 99Mo. European Patent EP2301041 (A1) — 2011-03-30.

Yamashita, H., Yamaguchi, S., Nishimura, R., & Maekawa, T. (2001). Voltammetric studies of antimony ions in soda-lime-silica glass melts up to 1873K. *Analytical Sciences*, *17*, 45–50.

Yan, R., Khalsa, G., Schaefer, B. T., Jarjour, A., Rouvimov, S., Nowack, K. C., ... Jena, D. (2019). Thickness dependence of superconductivity in ultrathin NbS_2. *Applied Physics Express*, *12*, 023008.

Yen, C. K., Yano, Y., Budinger, T. F., Friedland, R. P., Derenzo, S. E., Huesman, R. H., & O'Brien, H. A. (1982). Brain tumor evaluation using Rb-82 and positron emission tomography. *Journal of Nuclear Medicine*, *23*, 532–537.

Young, O. E. (1965). The Spanish tradition in gold and silver mining. *Arizona and the West*, *7*, 299–314.

Zhdankin, V. V. (2011). Organoiodine(V) reagents in organic synthesis. *The Journal of Organic Chemistry*, *76*, 1185–1197.

Zheng, Z., & Lin, C. (1996). The behaviour of rare-earth elements (REE) during weathering of granites in southern Guangxi, China. *Chinese Journal of Geochemistry*, *15*, 344–352.

Zweibel, K. (2010). The impact of tellurium supply on cadmium telluride photovoltaics. *Science*, *328*, 699–701.

Chapter 7

Periods 6 and 7 (including lanthanides and actinides)

7.1 55 Cs — Cesium

7.1.1 Discovery

In 1860 German chemist Robert Wilhelm Eberhard Bunsen (March 30, 1811 to August 16, 1899) and physicist Gustav Robert Kirchhoff (March 12, 1824 to October 17, 1887) discovered cesium in the mineral water from the spa-town of Bad Dürkheim, Rhineland-Palatinate, Germany (Kirchhoff and Bunsen, 1861) (Fig. 7.1). Due to the bright blue emission lines in the emission spectrum, they derived the element's name from the Latin word caesius, meaning sky blue. Cesium was the first element to be revealed with a spectroscope, an instrument invented by Bunsen and Kirchhoff only a year earlier. To get a pure sample of cesium, 44,000 L of mineral water had to be evaporated to produce 240 kg of concentrated salt solution. The alkaline earth metals (group 2: Be, Mg, Ca, Sr, Ba, and Ra) were precipitated either as sulfates (SO_4^{2-}) or oxalates ($C_2O_4^{2-}$), leaving the alkali metal in the solution. After reaction to the nitrates (NO_3^-) and extraction with ethanol, a Na-free mixture was the result. From this mixture, the lithium was precipitated by ammonium carbonate ($[NH_4]_2CO_3$). Potassium, rubidium, and cesium form insoluble salts with chloroplatinic acid ($[H_3O]_2[PtCl_6](H_2O)x$ ($0 \leq x \leq 6$)), but these salts show a small difference in solubility in hot water, and the less-soluble cesium and rubidium hexachloroplatinate ($(Cs,Rb)_2PtCl_6$) were obtained by fractional crystallization. After reduction of the hexachloroplatinate with hydrogen, cesium and rubidium were isolated based on their differences in solubility of their carbonates in alcohol. Finally, the process resulted in 9.2 g of rubidium chloride (RbCl) and 7.3 g of cesium chloride (CsCl) from the initial 44,000 L of mineral water. From the cesium chloride, Bunsen and Kirchhoff estimated the atomic weight of the newly obtained element at 123.35 (compared to the now accepted value of 132.9). They attempted to produce elemental cesium by electrolysis of molten cesium chloride, however instead of a metal, they produced a blue homogeneous substance, which "neither under the naked eye nor under the microscope showed the slightest trace of metallic substance"; consequently, they thought it was a subchloride (Cs_2Cl). In reality, the product was perhaps a colloidal mixture of the metal and cesium chloride. The electrolysis of the aqueous solution of chloride with a mercury cathode resulted in a cesium amalgam that readily decomposed under the aqueous conditions. The pure metal was ultimately obtained by the German chemist Carl Theodor Setterberg (April 30, 1853 in Järpås, Skaraborg, Sweden to April 7, 1941) while working on his doctorate with German organic and theoretical chemist Friedrich August Kekulé (September 7, 1829 to July 13, 1896) and Bunsen. In 1882 he managed to isolate cesium metal through electrolysis of cesium cyanide, circumventing the difficulties with cesium chloride (Setterberg, 1882). Setterberg went on, in 1882, to describe some of the properties of the metal: melting point 26°C–27°C, density 1.88 g cm^{-3} (today's values: 28.4°C and 1.93 g cm^{-3}, respectively). Historically, the most significant use of cesium has been in research and development, mainly in chemical and electrical fields. Very limited applications were in existence for cesium until the 1920s, when it came into use in radio vacuum tubes, where it had two purposes: as a getter (a deposit of reactive material that is placed inside a vacuum system, for the purpose of completing and maintaining the vacuum), it removed excess oxygen after manufacture, and as a coating on the heated cathode, it increased the electrical conductivity. Since 1967 the International System of Measurements has based the primary unit of time, the second, on the properties of cesium. The International System of Units (SI) defines the second as the duration of 9,192,631,770 cycles at the microwave frequency of the spectral line corresponding to the transition between two hyperfine energy levels of the ground state of cesium-133. The 13th General Conference on Weights and Measures of 1967 defined a second as "the duration of 9,192,631,770 cycles of microwave light absorbed or emitted by the hyperfine transition of cesium-133 atoms in their ground state undisturbed by external fields".

The Periodic Table: Nature's Building Blocks. DOI: https://doi.org/10.1016/B978-0-12-821279-0.00007-8

FIGURE 7.1 Gustav Robert Kirchhoff (left) and Robert Bunsen (right), c.1850.

7.1.2 Mining, production, and major minerals

Cesium is a relatively rare element, about 3 ppm in the Earth's crust. It is the 45th most abundant element and the 36th among the metals. It is 3.3% as abundant as Rb, with which it is closely associated, chemically. Due to its large ionic radius, cesium is one of the "incompatible elements." During magma cooling and crystallization, Cs is concentrated in the remaining liquid phase and crystallizes last. Therefore the largest deposits of cesium are found in pegmatite ore bodies formed by this enrichment process. Because cesium does not substitute for potassium as readily as rubidium does, the alkali evaporite minerals sylvite (KCl) and carnallite ($KMgCl_3\cdot6H_2O$) may contain only 0.002% cesium. Therefore Cs is found in only a limited number of minerals. Small amounts of cesium may be found in beryl ($Be_3Al_2(SiO_3)_6$) and avogadrite ($(K,Cs)BF_4$), up to 15 wt.% Cs_2O in the closely to beryl related mineral pezzottaite ($Cs(Be_2Li)Al_2Si_6O_{18}$), up to 8.4 wt.% Cs_2O in the rare mineral londonite ($(Cs,K)Al_4Be_4(B,Be)_{12}O_{28}$), and less in the more widespread rhodizite ($(K,Cs)Al_4Be_4(B,Be)_{12}O_{28}$). The only commercially significant ore for cesium is pollucite $Cs(AlSi_2O_6)$ (Box 7.1), which is found in a few places around the world in zoned pegmatites, together with the more economically important lithium minerals, lepidolite (polylithionite–trilithionite series, $KLi_2AlSi_4O_{10}F_2$–$KLi_{1.5}Al_{1.5}Si_3O_{10}F_2$) and petalite ($LiAlSi_4O_{10}$). Within the pegmatites, the large grain size and the strong separation of the minerals result in high-grade ore for mining. One of the world's most important and richest sources of cesium is the Tanco Mine at Bernic Lake in Manitoba, Canada, estimated to have 350,000 metric tons of pollucite ore, or more than two-thirds of the world's reserve base. Although the stoichiometric content of cesium in pollucite is 42.6%, pure pollucite samples from this deposit have only about 34% cesium, while the average concentration is 24 wt.%. Commercial pollucite has more than 19% cesium. The Bikita pegmatite deposit in Zimbabwe is mined for its petalite, but in addition it contains a substantial amount of pollucite. Another prominent source of pollucite is found in the Karibib Desert, Namibia. At the current rate of world production between 5 and 10 metric tons per year, Cs reserves will last for thousands of years.

Mining and refining pollucite ore is a selective process at a smaller scale than for most other metals. The ore is first crushed, hand sorted, but generally not concentrated, and subsequently ground. Cesium is then extracted from pollucite mostly by three

BOX 7.1 Pollucite, $(Cs,Na)_2(Al_2Si_4O_{12})·2H_2O$.

Pollucite is a zeolite mineral found in lithium-rich granite pegmatites; it may occur in kiloton amounts. It was named by Breithaupt in 1846 and named for Pollux, one of the twin brothers (Gemini) of Greek and Roman mythology, when two new minerals were found together that resembled one another, for its common association with "castorite" (petalite, $LiAlSi_4O_{10}$). Pollucite forms a complete solid solution series with analcime ($Na(AlSi_2O_6)·H_2O$). It is isometric (cubic), *m*3*m* (4/*m*2/*m*) (space group *Ia*3*d*). Pollucite has a hardness of 6½–7. Cleavage occurs in traces, the fracture is uneven to conchoidal, and the tenacity is brittle. The luster is vitreous; greasy to dull on the exterior, transparency is transparent to translucent. The color is colorless, white, gray, or tinted pink, blue, to violet, while the streak is white. Optically it is isotropic; it may show weak sectional birefringence. Crystals are rare, as cubes, dodecahedra, or trapezohedra, forms {100} and {211} common, {110} and {210} rare; commonly striated, up to 12 cm; as rounded grains, fine grained or massive.

different processes: acid digestion, alkaline decomposition, and direct reduction. In the acid digestion, the pollucite is dissolved with concentrated acids, for example, hydrochloric (HCl), sulfuric (H_2SO_4), hydrobromic (HBr), or hydrofluoric (HF) acids. With hydrochloric acid, a mixture of soluble chlorides is produced, and the insoluble chloride double salts of cesium are precipitated as cesium antimony chloride (Cs_4SbCl_7), cesium iodine chloride (Cs_2ICl), or cesium hexachlorocerate ($Cs_2(CeCl_6)$). After isolation, the pure precipitated double salt is decomposed, and pure CsCl is precipitated by evaporating the water. The sulfuric acid method yields the insoluble double salt directly as cesium alum ($CsAl(SO_4)_2·10H_2O$). The aluminum sulfate component is reacted to insoluble aluminum oxide by roasting the alum with carbon at increased temperatures, and the subsequent product is leached with water to yield a cesium sulfate (Cs_2SO_4) solution. Roasting pollucite with calcium carbonate ($CaCO_3$) and calcium chloride ($CaCl_2$) yields insoluble calcium silicates and soluble cesium chloride. Leaching with water or dilute ammonia (NH_4OH) yields a dilute chloride (CsCl) solution. This solution can be evaporated to produce cesium chloride or transformed into cesium alum or cesium carbonate. Though not economically viable, the ore can be directly reduced with potassium, sodium, or calcium in vacuum to obtain the cesium metal directly. Most of the mined cesium (as salts) is directly converted into cesium formate ($HCOO^-Cs^+$) for uses such as in oil drilling. The principal smaller-scale marketable compounds of cesium are cesium chloride and nitrate. Alternatively, cesium metal may be produced from the purified compounds derived from the ore. Cesium chloride and the other cesium halides can be reduced at 700°C–800°C with calcium or barium, and cesium metal distilled. Similarly, the aluminate, carbonate, or hydroxide can be reduced by magnesium. The metal can also be isolated by electrolysis of fused cesium cyanide (CsCN). Extremely pure and gas-free cesium can be obtained through 390°C thermal decomposition of cesium azide CsN_3, which can be produced from a reaction between aqueous cesium sulfate and barium azide. In vacuum applications, cesium dichromate can be reacted with zirconium to manufacture pure cesium metal without other gaseous products.

$$Cs_2Cr_2O_7 + 2Zr \rightarrow 2Cs + 2ZrO_2 + Cr_2O_3$$

Only 27 minerals are known in nature to have structural Cs. Two sulfides, galkhaite ($(Cs,Tl)(Hg,Cu,Zn)_6(As,Sb)_4S_{12}$) (Fig. 7.2) and pautovite ($CsFe_2S_3$), contain Cs. In the oxide class three minerals are known, cesiokenopyrochlore ($(\square Nb_2(O,OH)_6Cs_{1-x}$ ($x \sim 0.20$)), cesplumtantite ($(Cs,Na)_2(Pb,Sb^{3+})_3Ta_8O_{24}$), and margaritasite ($(Cs,K,H_3O)_2(UO_2)_2(VO_4)_2·H_2O$). Two borates, londonite ($(Cs,K,Rb)Al_4Be_4(B,Be)_{12}O_{28}$) (Fig. 7.3) and ramanite-(Cs) ($Cs[B_5O_6(OH)_4]·2H_2O$), have structural Cs. The only two Cs-containing sulfates are cesiodymite ($CsKCu_5O(SO_4)_5$) and phyllotungstite ($(H_2O,M)_x(W,Fe)(O,OH)_3·yH_2O$ (M = Ca, Cs, Pb, or K)). Within the phosphate class two minerals are found with Cs, cesiumpharmacosiderite ($CsFe_4[(AsO_4)_3(OH)_4]·4H_2O$) and mccrillisite ($NaCs(Be,Li)Zr_2(PO_4)_4·1-2H_2O$). The silicate class contains 16 minerals such as pezzottaite ($Cs(Be_2Li)Al_2(Si_6O_{18})$) (Fig. 7.4) and pollucite ($(Cs,Na)_2(Al_2Si_4O_{12})·2H_2O$).

7.1.3 Chemistry

Cs has 39 known isotopes, ranging from ^{112}Cs to ^{151}Cs. Some are formed from lighter elements by slow neutron capture (s-process) inside old stars and by the r-process in supernova explosions. The only stable isotope is ^{133}Cs. While it has a large nuclear spin (7/2 +), nuclear magnetic resonance spectroscopy (NMR) studies can use this isotope at a resonating frequency of 11.7 MHz. Radioactive ^{135}Cs has a very long half-life of about 2.3 million years, the longest of all Cs radioisotopes. ^{137}Cs and ^{134}Cs have half-lives of 30.17 and 2.0648 years, respectively. ^{137}Cs decays to a short-lived ^{137m}Ba by β^- decay, and then to nonradioactive Ba, whereas ^{134}Cs decays by ε decay into ^{134}Ba directly. ^{129}Cs, ^{131}Cs, ^{132}Cs, and ^{136}Cs have half-lives between 1 day and 2 weeks, while the majority of the other isotopes have much shorter

FIGURE 7.2 Galkhaite, $(Cs,Tl)(Hg,Cu,Zn)_6(As,Sb)_4S_{12}$, red cubic crystals, the largest to 1 mm. Getchell mine, Humboldt Co., Nevada, United States.

FIGURE 7.3 Londonite, $(Cs,K,Rb)Al_4Be_4(B,Be)_{12}O_{28}$, 8-mm crystal in rhodizite, $(K,Cs)Al_4Be_4(B,Be)_{12}O_{28}$. Antsongomvato mine, Betroka dist., Antananarivo, Madagascar.

half-lives ranging from a few seconds to fractions of a second. No less than 21 metastable nuclear isomers exist. Other than ^{134m}Cs (with a half-life of just under 3 h), all are extremely unstable and decay with half-lives of a few minutes or less. ^{135}Cs is one of the long-lived fission products of U formed in nuclear reactors. But the quantity of this fission product is reduced in most reactors since its predecessor, ^{135}Xe, is a strong neutron poison and regularly transmutes to stable ^{136}Xe before it can decay to ^{135}Cs. The β decay from ^{137}Cs to ^{137m}Ba is by strong emission of γ radiation. ^{137}Cs and ^{90}Sr are the main medium-lived products of nuclear fission and the major cause of radioactivity from spent nuclear fuel after several years of cooling, lasting several centuries. These two isotopes form the main source of residual radioactivity in the area of the Chernobyl disaster. Due to the low capture rate, disposing of ^{137}Cs through neutron capture is not practicable and the single current solution is to let it decay over time. Nearly all Cs formed from nuclear fission comes from the β decay of originally more neutron-rich fission products, passing through various isotopes of I and Xe. Since I and Xe are volatile and can diffuse through nuclear fuel or air, radioactive Cs is frequently formed far from the original site of fission. During nuclear weapons testing between the 1950s and 1980s, ^{137}Cs was released into the atmosphere and returned to the Earth's surface as part of radioactive fallout.

FIGURE 7.4 Pezzottaite, $Cs(Be_2Li)Al_2(Si_6O_{18})$, pink to red hexagonal crystals to 5 mm showing good terminations. Very rare. Sakavalana mine, Ambatovita, Fianarantsoa Prov., Madagascar.

TABLE 7.1 Cesium properties.

Appearance	Pale gold
Standard atomic weight $A_{r,std}$	132.905
Block	s-Block
Element category	Alkali metal
Electron configuration	[Xe] $6s^1$
Phase at STP	Solid
Melting point	28.5°C
Boiling point	671°C
Density	1.93 g/cm^3
When liquid (at m.p.)	1.843 g/cm^3
Heat of fusion	2.09 kJ/mol
Heat of vaporization	63.9 kJ/mol
Molar heat capacity	32.210 J/(mol·K)
Oxidation states	−1, **+1**
Ionization energies	1st: 375.7 kJ/mol 2nd: 2234.3 kJ/mol 3rd: 3400 kJ/mol
Atomic radius	Empirical: 265 pm
Covalent radius	244 ± 11 pm
Van der Waals radius	343 pm

STP, Standard temperature and pressure.
Bold font indicates main oxidation state.

Cesium is a soft, golden-yellow metal with a melting point of 28.5°C, which makes it a liquid at temperatures close to room temperature. It has 55 electrons arranged in an electronic configuration of [Xe] $6s^1$. It is a member of the alkali metals and exhibits physical and chemical properties similar to other members of the group especially of rubidium and potassium above it. The ease of abstracting the lone electron from the outer shell dominates most of the chemistry of the alkali metal group and as such their oxidation states in isolable compounds are always +1 (Table 7.1). Following the periodic trend of chemical behavior, Cs is more reactive than the lighter alkali metals and is in fact regarded as the most reactive of all metals. It is so reactive that it reacts with solid water even at −116°C. It explodes upon contact with liquid water.

7.1.3.1 Alloys

Just like Rb and K above it, Cs dissolves in mercury forming amalgams in stoichiometries such as Cs_3Hg_{20} and Cs_5Hg_{19}. A few amalgams have been subject of research: $CsHg_2$ is black with a purple metallic luster, while CsHg is golden colored, also with a metallic luster. Cesium forms alloys with the other alkali metals and gold. At temperatures below 650°C, it does not alloy with cobalt, iron, molybdenum, nickel, platinum, tantalum, or tungsten. It can form well-defined intermetallic compounds with antimony, gallium, indium, and thorium, which are all photosensitive. It forms mixtures with all the other alkali metals (except lithium). The alloy with a molar distribution of 41% Cs, 47% K, and 12% Na has the lowest melting point of any known metal alloy, at −78°C.

7.1.3.2 Reactions with oxygen

Like other alkali metals, Cs exhibits a high luster when freshly cut but rapidly tarnishes in air due to reaction with O_2 and moisture. The highly reactive nature of Cs makes it ignite spontaneously in air producing primarily the superoxide, CsO_2. The combination of Cs with oxygen results in extensive Cs-O compounds. The "normal" cesium oxide (Cs_2O) forms yellow-orange hexagonal crystals and is the only oxide of the anti-$CdCl_2$ type. It vaporizes at 250°C and decomposes to cesium metal. The peroxide, Cs_2O_2, can be made by decomposing CsO_2 at 400°C. The ozonide, CsO_3, can be prepared by reacting anhydrous cesium hydroxide with ozonide and extracting the product with liquid ammonia. Cesium oxides and suboxides were also isolated and studied, among them the bronze-colored Cs_7O and violet-colored Cs_4O. In addition to the superoxide and the ozonide CsO_3, several other brightly colored suboxides have also been studied. These include $Cs_{11}O_3$, Cs_3O (dark green), CsO, Cs_3O_2, and Cs_7O_2. The latter may be heated in a vacuum to generate Cs_2O.

Cesium hydroxide (CsOH) is hygroscopic and strongly basic. It rapidly etches the surface of semiconductors such as silicon. CsOH has been previously regarded by chemists as the "strongest base," reflecting the relatively weak attraction between the large Cs^+ ion and OH^-; it is indeed the strongest Arrhenius base, but a number of compounds that do not dissolve in water, such as n-butyllithium and sodium amide, are more basic.

7.1.3.3 Reactions with halogens and chalcogenides

Cs forms compounds with hydrogen, chalcogens (S, Se, and Te), and halogens to form the hydride, CsH, calchogenides, Cs_2E (E = S, Se, Te), and halides, CsX (X = F, Cl, Br, I), respectively. Cesium fluoride (CsF) is a hygroscopic white solid that is extensively used in organofluorine chemistry as a source of fluoride anions. Cesium fluoride has the halite (NaCl) structure, which means that the Cs^+ and F^- pack in a cubic closest packed array as do Na^+ and Cl^- in sodium chloride. It should be noted that cesium and fluorine have the lowest and highest electronegativities, respectively, among all the known elements. Cesium chloride (CsCl) crystallizes in the simple cubic crystal system. Also called the "cesium chloride structure," this structural motif is composed of a primitive cubic lattice with a two-atom basis, each with an eightfold coordination; the chloride atoms lie upon the lattice points at the edges of the cube, while the cesium atoms lie in the holes in the center of the cubes. This structure is shared with CsBr and CsI and many other compounds that do not contain Cs. In contrast, the majority of other alkaline halides have the sodium chloride (NaCl) structure. The CsCl structure is preferred because Cs^+ has an ionic radius of 174 pm and Cl^- 181 pm. Anecdotally, CsH was reported to form when resonant laser radiation impinges on a cell with Cs vapors and minute amounts of H_2 present. The alkali metals share the ability to dissolve in liquid ammonia to form blue-colored solutions. These solutions contain solvated cations and quasi free electrons and had been valuable as powerful and selective reducing agents. These had been used in reducing various metal oxides and halides to prepare the corresponding metallic elements (or their alloys).

7.1.3.4 Salts

Most cesium compounds contain the cation Cs^+, which forms ionic bonds to a wide variety of anions. One notable exception is the ceside anion (Cs^-), and others are the several suboxides (See section 7.1.3.2). Salts of Cs^+ are typically colorless except if the anion itself is colored. Numerous simple salts are hygroscopic, but less so than the corresponding salts of lighter alkali metals. The phosphate, acetate, carbonate, halides, oxide, nitrate, and sulfate salts are all water soluble. Double salts are frequently less soluble, and the low solubility of cesium aluminum sulfate is used in refining Cs from ores. In addition, the double salts with antimony (such as $CsSbCl_4$), bismuth, cadmium, copper, iron, and lead are also poorly soluble.

7.1.3.5 Organocesium compounds

Organocesium compounds are less common than the lighter alkali metals. This is not surprising given the fact that cesium is the most reactive metal, extremely air- and moisture sensitive and highly pyrophoric. Using the metal in synthesis requires utmost care and is carried out in low temperature and air- and moisture-free inert environments. Nevertheless, several organocesium compounds have been synthesized and studied. Some of these were synthesized along with the other two heavy alkali metals K and Rb to create closely related series of compounds with the aim of understanding the effects of incremental changes in properties of the starting materials. Recent organocesium compounds prepared are noteworthy for their stability to air, moisture, and silica gel flash chromatography. The work of Martinez-Ariza et al. published in Andgewandte Chemie (2015) described the preparation of such compounds as mixing in the same pot four Ugi reagents, *p*-methyloxyaniline, ethyl glyoxylate, 2,6-dimethylphenylisonitrile, and cyanoacetic acid, yielding an intermediate product, which was then dried by evaporating the solvent. This crude product was then added with cesium carbonate in dimethylformamide (DMF) at room temperature resulting to product 1 below. In a similar fashion, ethyl glyoxylate was replaced with methylacetoacetate as the carbonyl carbon to give product 2. Several analogs were prepared for both products by varying the R_1 and R_2 groups. The possible variations to the structure that can be prepared are many as the amine and isonitrile inputs provide two more points of diversification.

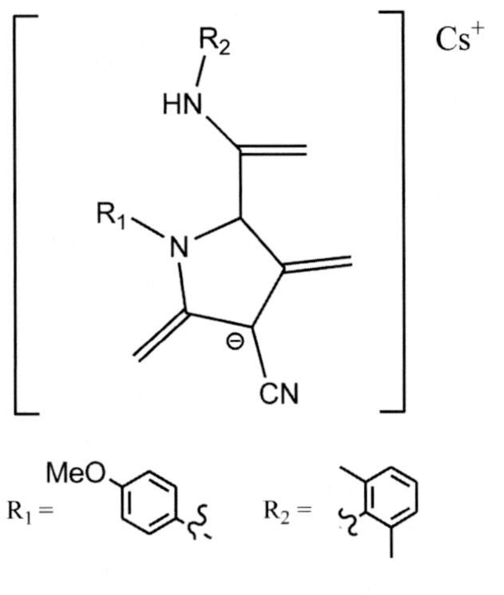

7.1.4 Major uses

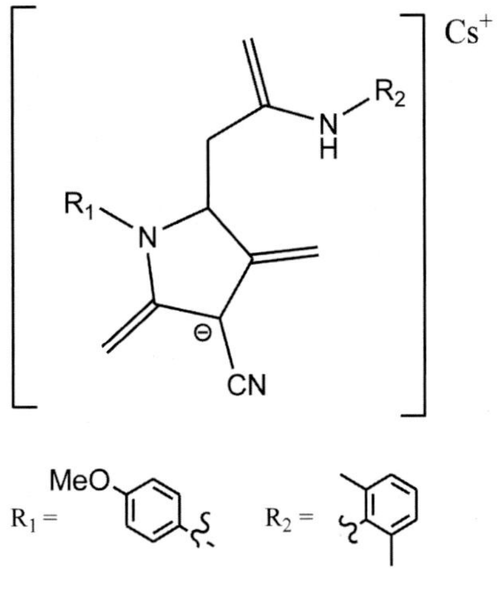

The principal current application of Cs is in Cs formate drilling fluids for the extractive oil industry. Aqueous solutions of Cs formate ($HCOO^-Cs^+$)—formed through the reaction of CsOH with formic acid—were established in the mid-1990s for usage as oil well drilling and completion fluids. The role of a drilling fluid is to lubricate drill bits, bring rock cuttings to the surface, and maintain pressure on the rock formation during drilling of the well. Completion fluids support the emplacement of control hardware after drilling is finished but before production by maintaining the pressure. The high density of the Cs formate brine (up to 2.3 g/cm^3), in combination with the rather benign nature of most Cs compounds, decreases the need for toxic high-density suspended solids in the drilling fluid—an important technological, engineering, and environmental advantage. In contrast to the components of numerous other heavy liquids, Cs formate is comparatively environment friendly. Cs formate brine can be mixed with K and Na formates to adjust the density of the fluids to as low as that of water (1.0 g/cm^3). Moreover, it is biodegradable and may be recycled, which is significant considering its high cost. Alkali formates are safe to handle and do not damage the producing formation or downhole metals as other corrosive, high-density brines (e.g., $ZnBr_2$ solutions) occasionally do; they also need less clean-up and lower disposal costs.

Cs-based atomic clocks employ the electromagnetic transitions in the hyperfine structure of ^{133}Cs atoms as a reference point. The initial accurate Cs clock was created by English physicist Louis Essen (September 6, 1908 to August 24, 1997) in 1955 at the National Physical Laboratory (United Kingdom). Cs clocks have improved over the past 65 years and are looked upon as "the most accurate realization of a unit that mankind has yet achieved." These clocks measure frequency with an error of 2−3 parts in 10^{14}, which is equivalent to an accuracy of 2 ns per day, or 1 s in 1.4 million years. The newest versions are more accurate than 1 part in 10^{15}, equivalent to an accuracy of approximately 1 s in 20 million years. The Cs standard is the primary standard for standards-compliant time and frequency measurements. Cs clocks control the timing of cell phone networks and the Internet. It was recently (2018) proposed by the International Committee for Weights and Measures that the second, symbol s, the SI unit of time, be defined utilizing the fixed numerical value of the Cs frequency $\Delta\nu_{Cs}$, the unperturbed ground-state hyperfine transition frequency of the ^{133}Cs atom.

Cs vapor thermionic generators are low-power devices that convert heat energy to electrical energy. In the two-electrode vacuum tube converter, Cs neutralizes the space charge close to the cathode and increases the current flow. In addition, Cs is important for its photo-emissive properties, converting light to electricity. It is employed in photoelectric cells since Cs-based cathodes, for example, the intermetallic compound K_2CsSb, have a low threshold voltage for emission of electrons. Different photo-emissive devices using Cs include optical character recognition devices, photomultiplier tubes, and video camera tubes. However, Ge, Rb, Se, Si, Te, and some other elements can be substituted for Cs in photosensitive materials. Cesium iodide (CsI), bromide (CsBr), and fluoride (CsF) crystals are used for scintillators in scintillation counters extensively used in mineral exploration and particle physics research to detect γ- and X-ray radiation. Being a heavy element, Cs provides good stopping power with better detection. Cs compounds may provide a faster response (CsF) and be less hygroscopic (CsI). Cs vapor is employed in numerous common magnetometers. Cs is also utilized as an internal standard in spectrophotometry. Similar to other alkali metals, Cs has a great affinity for oxygen and is utilized as a "getter" in vacuum tubes. Other applications of Cs metal are, for example, high-energy lasers, vapor glow lamps, and vapor rectifiers.

The high density of the Cs ion makes solutions of CsCl, cesium sulfate (Cs_2SO_4), and cesium trifluoroacetate (Cs (O_2CCF_3)) advantageous in molecular biology for density gradient ultracentrifugation. This technology is used chiefly to isolate viral particles, subcellular organelles and fractions, and nucleic acids from biological samples. Comparatively few chemical applications employ Cs. Doping with Cs compounds is known to improve the effectiveness of some metal-ion catalysts for chemical synthesis, such as acrylic acid (IUPAC name prop-2-enoic acid, CH_2-CHCOOH), anthraquinone ($C_{14}H_8O_2$), ethylene oxide (IUPAC name oxirane, C_2H_4O), methanol (CH_3OH), phthalic anhydride (IUPAC name 2-benzofuran-1,3-dione, $C_6H_4(CO)_2O$), styrene (IUPAC name ethenylbenzene, C_6H_5CH-CH_2), methyl methacrylate (IUPAC name methyl 2-methylprop-2-enoate, CH_2-$C(CH_3)COOCH_3$) monomers, and numerous olefins. It is likewise utilized in the catalytic conversion of SO_2 into SO_3 in the manufacture of sulfuric acid (H_2SO_4). CsF finds a niche application in organic chemistry as a base and as an anhydrous source of F ions. Cs salts occasionally substitute for K or Na salts in organic synthesis, for example, cyclization, esterification, and polymerization. Cs has been employed in thermoluminescent radiation dosimetry. When exposed to radiation, it acquires crystal defects that, upon heating, heal with emission of light directly proportional to the received dose. Therefore measuring the light pulse with a photomultiplier tube allows the accumulated radiation dose to be calculated. $CsNO_3$ is utilized as an oxidizer and pyrotechnic colorant to burn Si in infrared flares, since it emits a considerable amount of its light in the near-infrared spectrum. Cs is employed to decrease the radar signature of exhaust plumes in the Lockheed SR-71 Blackbird military aircraft (a long-range, high-altitude, Mach 3 + strategic reconnaissance aircraft that was operated by the US Air Force. The shape of the SR-71 was based on the A-12, which was one of the first aircraft to be designed with a reduced radar cross-section. The SR-71 served with the US Air Force from 1964 to 1998.) Cs and Rb carbonates have been added to glass as they decrease electrical conductivity and enhance stability and durability of fiber optics and night vision devices. CsF or CsAl fluoride ($CsAlF_4$) are utilized in fluxes formulated for brazing Al alloys that comprise Mg. Cs salts have been assessed as antishock reagents after the administration of arsenical drugs. Due to their effect on heart rhythms, though, they are less likely to be used than K and Rb salts. They have also been employed to treat epilepsy.

^{137}Cs is a radioisotope frequently used as a γ-emitter in industrial applications. Its advantages comprise a half-life of roughly 30 years, its availability from the nuclear fuel cycle, and producing ^{137}Ba as a stable end product. Its high water solubility is a drawback that makes it unsuited for large pool irradiators for food and medical supplies. It has been employed in agriculture, cancer treatment, and for the sterilization of, for example, food, sewage sludge, and surgical equipment. Radioactive Cs isotope in radiation devices were applied in the medical field to treat certain types of cancer, but the development of better alternatives and the use of water-soluble CsCl in the sources, which could generate wide-ranging contamination, increasingly put some of these Cs sources out of commission. ^{137}Cs has been used in a number of industrial measurement gages, such as moisture, density, leveling, and thickness gages. In addition, it has been utilized in well logging devices for determining the electron density of the rock formations, which is analogous to the bulk density of the formations. ^{137}Cs has been employed in hydrologic studies analogous to those with tritium (T or ^{3}H). As a daughter product of fission bomb testing from the 1950s through the mid-1980s, ^{137}Cs was released into the atmosphere, where it was easily absorbed into solution. Identified year-to-year variation within that period permits correlation with soil and sediment layers. ^{134}Cs and to a lesser extent ^{135}Cs have likewise been used in hydrology to quantify the Cs produced by the nuclear power industry. While they are less prevalent than either ^{133}Cs or ^{137}Cs, these tracer isotopes are created exclusively from anthropogenic sources. ^{133}Cs can be laser cooled and employed to probe fundamental and technological problems in quantum physics. It has an especially suitable Feshbach spectrum to allow studies of ultracold atoms requiring tunable interactions (In physics, a Feshbach resonance can occur upon collision of two slow atoms, when they temporarily stick together forming an unstable compound with short lifetime (so-called resonance). It is a feature of many-body systems in which a bound state is achieved if the coupling(s) between at least one internal degree of freedom and the reaction coordinates, which lead to dissociation, vanish. The opposite situation, when a bound state is not formed, is a shape resonance. It is named after American physicist Herman Feshbach (February 2, 1917 to December, 22, 2000), a physicist at MIT (Massachusetts Institute of Technology).

7.2 56 Ba — Barium

7.2.1 Discovery

Alchemists in the early Middle Ages were familiar with some barium minerals. Smooth pebble-like stones of the mineral baryte ($BaSO_4$) were discovered in volcanic rock near Bologna, Italy, and subsequently were known as "Bologna stones." Alchemists were fascinated by them since after exposure to light they would glow for years. The phosphorescent properties of baryte heated with organics were described by V. Casciorolus, a shoemaker from Bologna, in 1602. Swedish Pomeranian and German pharmaceutical chemist Carl Scheele (December 9, 1742 to May 21, 1786) found that baryte contained a new element in 1774, but was unable to isolate barium, only barium oxide (BaO). Swedish chemist and metallurgist Johan Gottlieb Gahn (August 19, 1745 to December 8, 1818) also isolated barium oxide 2 years later in similar studies. Oxidized barium was at first named "barote" by French chemist and politician Louis Bernard Guyton de Morveau (January 4, 1737 to January 2, 1816) (Fig. 7.5), a name that was changed by French nobleman and chemist Antoine Lavoisier (August 26, 1743 to May 8, 1794) to baryta. Also, in the 18th century, English botanist, geologist, chemist, physician, and first systematic investigator of the bioactivity of digitalis William Withering (March 17, 1741 to October 6, 1799) (Fig. 7.6) observed a heavy mineral in the lead mines of Cumberland, now known to be witherite ($BaCO_3$). Barium was first isolated by electrolysis of molten barium salts in 1808 by Cornish chemist and inventor Sir Humphry Davy (December 17, 1778 to May 29, 1829) in England (Davy, 1808). Davy, by analogy with calcium, named "barium" after baryta, with the "-ium" ending signifying a metallic element. German chemist Robert Wilhelm Eberhard Bunsen (March 30, 1811 to August 16, 1899) and British

FIGURE 7.5 Louis Bernard Guyton de Morveau, stipple engraving by Edme Quenedey (1756–1830).

FIGURE 7.6 William Withering, painting by Carl Frederik von Breda (1759–1818).

chemist and physicist Augustus Matthiessen (January 2, 1831 to October 6, 1870) isolated pure barium through electrolysis of a molten mixture of barium chloride ($BaCl_2$) and ammonium chloride (NH_4Cl). The production of pure oxygen in the Brin process was a large-scale use of barium peroxide in the 1880s, before it was replaced by electrolysis and fractional distillation of liquefied air in the early 1900s. In this Brin process barium oxide reacts at 500°C–600°C with air to form barium peroxide, which decomposes above 700°C by releasing oxygen.

$$2BaO + O_2 \rightleftharpoons 2BaO_2$$

Barium sulfate ($BaSO_4$) was first used as a radiocontrast agent in X-ray imaging of the digestive system in 1908.

7.2.2 Mining, production, and major minerals

The abundance of barium is 0.0425% in the Earth's crust. The main commercial source of barium is baryte (also called barite, barytes or heavy spar, $BaSO_4$) with deposits in numerous parts of the world. An alternative commercial source, far less significant than baryte, is witherite ($BaCO_3$). The main deposits can be found in England, Romania, and the former USSR. The baryte reserves are estimated between 0.7 and 2 billion tons. Baryte production has risen since the second half of the 1990s. China accounts for more than 50% of this output, followed by India, Morocco, the United States, Turkey, Iran, and Kazakhstan. The barium mineral, benitoite ($BaTi(Si_3O_9)$), occurs as a very rare blue fluorescent gemstone and is the official state gem of California.

The mined ore is washed, crushed, classified, and separated from quartz. If the quartz penetrates too deeply into the ore, or the iron, zinc, or lead content is unusually high, then froth flotation is utilized instead. The product is a 98% pure baryte (by mass); the purity should be no less than 95%, with a minimal content of iron and silicon dioxide. It is subsequently reduced by carbon to form barium sulfide.

$$BaSO_4 + 2C \rightarrow BaS + 2CO_2\uparrow$$

The water-soluble barium sulfide forms the starting material for other compounds: reacting BaS with oxygen produces the sulfate, with nitric acid the nitrate, with carbon dioxide the carbonate, etc. The nitrate can be thermally decomposed to produce the oxide. Barium metal is produced by reduction with aluminum at 1100°C. The first step is the production of the intermetallic compound $BaAl_4$.

$$3BaO + 14Al \rightarrow 3BaAl_4 + Al_2O_3$$

$BaAl_4$ as an intermediate, which is then reacted with barium oxide to produce the metal. In this reaction not all barium is reduced though.

$$8BaO + BaAl_4 \rightarrow Ba\downarrow + 7BaAl_2O_4$$

The remaining barium oxide reacts with the formed aluminum oxide.

$$BaO + Al_2O_3 \rightarrow BaAl_2O_4$$

The overall reaction is

$$4\,BaO + 2Al \rightarrow 3Ba\downarrow + BaAl_2O_4.$$

Barium vapor is condensed and packed into molds in an atmosphere of argon. This process is used commercially, producing ultrapure barium. Normally sold barium has a purity of about 99%, with the major impurities being strontium and calcium (up to 0.8% and 0.25%, respectively) while other impurities are less than 0.1%. A comparable reaction with silicon at 1200°C produces barium and barium metasilicate. Electrolysis is not utilized as barium readily dissolves in molten halides and the product is rather impure.

There are roughly 250 minerals with barium in their crystal structure. Only a single sulfide has been found in nature with Ba: owensite ($(Ba,Pb)_6(Cu,Fe,Ni)_{25}S_{27}$). Four halides are known with Ba, bøgvadite ($Na_2SrBa_2Al_4F_{20}$), frankdicksonite (BaF_2), usovite ($Ba_2CaMgAl_2F_{14}$), and zhangpeishanite (BaFCl). Just over 20 minerals can be found among the oxides, for example, hollandite ($Ba(Mn^{4+}_6Mn^{3+}_2)O_{16}$), and romanèchite ($(Ba,H_2O)_2(Mn^{4+},Mn^{3+})_5O_{10}$). A total of 25 carbonate minerals are known to contain Ba, for example, alstonite ($BaCa(CO_3)_2$) (Fig. 7.7), benstonite ($Ba_6Ca_6Mg(CO_3)_{13}$), and witherite ($BaCO_3$) (Fig. 7.8). The sulfates are represented by only three different minerals, baryte ($BaSO_4$) (Box 7.2), hashemite ($Ba(CrO_4)$), and walthierite ($Ba_{0.5}Al_3(SO_4)_2(OH)_6$). About 40 different minerals are found in the phosphate class, for example, bariopharmacosiderite ($Ba_{0.5}Fe^{3+}_4(AsO_4)_3(OH)_4{\cdot}5H_2O$), francevillite ($Ba(UO_2)_2(VO_4)_2{\cdot}5H_2O$) (Fig. 7.9), gorceixite ($BaAl_3(PO_4)(PO_3OH)(OH)_6$), and uranocircite ($Ba(UO_2)_2(PO_4)_2{\cdot}10H_2O$). The largest group of more than 150 different minerals is found under the silicates. These include minerals such as benitoite ($BaTi(Si_3O_9)$), celsian ($Ba(Al_2Si_2O_8)$), harmotome ($(Ba_{0.5},Ca_{0.5},K,Na)_5[Al_5Si_{11}O_{32}]{\cdot}10H_2O$), hyalophane ($(K,Ba)[Al(Si,Al)Si_2O_8]$), and yuksporite ($K_4(Ca,Na)_{14}(Sr,Ba)_2(\square,Mn,Fe)(Ti,Nb)_4(O,OH)_4(Si_6O_{17})_2(Si_2O_7)_3(H_2O,OH)_3$).

FIGURE 7.7 Alstonite, $BaCa(CO_3)_2$, colorless to white, 1–2 mm, pseudohexagonal dipyramids showing horizontal striations. New Brancepeth Colliery, Deerness Valley, Durham, England.

FIGURE 7.8 A 4.5-cm crystal of witherite, $BaCO_3$, showing classic cyclic twinning on {110} producing a pseudohexagonal prism. Minerva #1 mine, Hardin Co., Illinois, United States.

FIGURE 7.9 Francevillite, $Ba(UO_2)_2(VO_4)_2 \cdot 5H_2O$, bright orange clusters and single crystals to 1.5 mm in diamond-shaped forms. Mounana mine, Franceville, Haut-Ogooué, Gabon.

7.2.3 Chemistry

Barium found in the Earth's crust is a mixture of seven primordial nuclides, ^{130}Ba, ^{132}Ba, and ^{134}Ba through ^{138}Ba. ^{130}Ba decays very slowly to ^{130}Xe by double β^+ decay, while ^{132}Ba theoretically decays similarly to ^{132}Xe, with half-lives a thousand times greater than the age of the universe. The abundance of both radioisotopes is about 0.1% of all naturally occurring Ba. The radioactivity of these isotopes is so weak that they form no danger to life. Of the stable isotopes, ^{138}Ba has a natural abundance of 71.70%; the remaining isotopes have decreasing abundance with decreasing mass number (^{137}Ba 11.23%, ^{136}Ba 7.85%, ^{135}Ba 6.59%, and ^{134}Ba 2.42%). In total, Ba has about 40 synthetic isotopes, ranging from ^{114}Ba to ^{153}Ba. The most stable synthetic radioisotope is ^{133}Ba with a half-life of approximately 10.51 years. Five other isotopes have half-lives of more than 1 day. In addition, Ba has 10 meta states, of which $^{133m1}Ba$ with a half-life of around 39 h is the most stable.

BOX 7.2 Baryte, $BaSO_4$.

Baryte deposits are generally classified by their mode of occurrence in three major types: (1) vein and cavity filling, (2) residual, and (3) bedded. The vein- and cavity-filling deposits occur mostly in fault and breccia zones at low temperatures, associated with dolomite, fluorite, and metalliferous ores. The deposits of the north Pennine ore fields of the United Kingdom involves veins and replacements in limestone originating from hydrothermal solutions derived from magma. Vein- and cavity-filling deposits are widely present in the United States. Baryte deposits in collapse and sink structures are common in the Central district of Missouri and in the Appalachian states. Many of these deposits in the western states are associated with igneous rocks of the Tertiary age. The baryte of the residual deposits commonly occurs in clays derived from weathering of impure limestone. In SE Missouri, extensive low-grade baryte deposits occur in residual clay derived from underlying dolomite. In central Missouri, high-grade baryte deposits of both residual and vein fillings are present in carbonate host rocks associated with galena, sphalerite, and chalcopyrite. The bedded deposits occur generally in shale and limestone, in which the baryte is found together with chert, pyrite, iron oxides, carbonates, and clay minerals. The baryte of this type, in contrast to the generally white baryte from vein fillings and residual deposits, is typically dark gray to black and fine grained. Large bedded baryte deposits occur in the Ouachita Mountains of Arkansas. The deposits occur at the base of the Stanley shale of Mississippian age, overlying a thick sequence of bedded cherts. Pb and Zn mineralization is absent in the deposits. Bedded deposits of Middle Devonian age can be found in Germany at Meggen in the Rhenish Schiefergebirge. Unlike the Arkansas occurrence, there is major accumulation of Pb-Zn sulfides here. Host sediments are black shales with some sandstone, like the shale of the Arkansas deposits, but chert is lacking in this deposit. Substantial deposits of baryte in association with shale hosted Pb and Zn mineralization occur in the lower Paleozoic strata of the Selwyn basin in the Canadian Cordillera. Finally, large bedded baryte deposits occur in the lower Cambrian black shale series of south China in the Yangtze platform area and the adjacent Qinling geosyncline to the north. The baryte ore can be divided in four textural types: (1) massive, (2) laminated, (3) banded, and (4) nodular. There are carbonaceous mudstone interlayers in the baryte ore beds, and organic carbon content can be as high as 11 wt.% (Fig. 7.10).

FIGURE 7.10 Tabular golden baryte, $BaSO_4$, crystals to 2 cm. Iglesias, Sardinia, Italy.

Barium is a soft, silvery metal with 56 electrons arranged in an electronic configuration of $[Xe]6s^2$. It is the fifth element in group 2 or alkaline earths, which are characterized by two valence electrons in their outer shell. As such, it resembles the physical and chemical properties of group 2. Barium is, however, more reactive than magnesium, calcium, and strontium and always exhibits the oxidation state of +2 (Table 7.2). Its reactivity, in fact, resembles that of the neighboring alkali metals. Reports of oxidation states less than +2 are reported but these are for compounds that are rare and unstable molecular species, characterized only in the gas phase such as BaF. As with other very reactive metals (alkali and alkaline earths), barium is never found as the free element but as compound with other elements. For example, it is found as the sulfate form in the mineral baryte.

TABLE 7.2 Barium properties.

Appearance	Silvery gray, with a pale-yellow tint
Standard atomic weight	137.328
Block	s-Block
Element category	Alkaline earth metals
Electron configuration	[Xe] $6s^2$
Phase at STP	Solid
Melting point	727°C
Boiling point	1845°C
Density (near r.t.)	3.51 g/cm^3
When liquid (at m.p.)	3.338 g/cm^3
Heat of fusion	7.12 kJ/mol
Heat of vaporization	142 kJ/mol
Molar heat capacity	28.07 J/(mol·K)
Oxidation states	+1, **+2**
Ionization energies	1st: 502.9 kJ/mol 2nd: 965.2 kJ/mol 3rd: 3600 kJ/mol
Atomic radius	Empirical: 222 pm
Covalent radius	215 ± 11 pm
Van der Waals radius	268 pm

STP, Standard temperature and pressure.
Bold font indicates main oxidation state.

7.2.3.1 *Oxides, nitrides and hydrides*

Barium exhibits typical reactions of the alkaline earths but with more vigor. Reactions with water and its derivatives (e.g., alcohols) are very exothermic and release hydrogen gas.

$$Ba + 2ROH \rightarrow Ba(OR)_2 + H_2 \text{ (R is an alkyl group or a hydrogen atom)}$$

Its silver luster is easily lost as its reaction with oxygen or air occurs at room temperature. Thus storage and handling of the metal is similar to the reactive group 1 metals, that is, it must be stored in oil or in an inert atmosphere. Burning barium in air typically forms a mixture of the oxide and nitride. When temperature reaches 500°C, the peroxide also forms. Barium ozonide, $Ba(O_3)_2$, was also reported to form when ozone in freon 12 was reacted with barium peroxide in the same. The ozonide decomposes to the superoxide, $Ba(O_2)_2$, at −70°C.

$$2Ba + O_2 \rightarrow 2BaO$$
$$Ba + O_2 \rightarrow BaO_2$$
$$3Ba + N_2 \rightarrow Ba_3N_2$$

Barium reacts with hydrogen at elevated temperatures to form the hydride, BaH_2, which is very reactive and potentially explosive when mixed with a solid oxidant such as chromates. The reaction with other nonmetals such as the halogens, chalcogens, carbon, phosphorus, and silicon to form the respective halides, chalcogenides, carbides, phosphides, and silicides generally proceed with heating. Barium also combines with several metals, including aluminum, zinc, lead, and tin, forming intermetallic phases and alloys.

Differences in the properties of compounds of group 2 metals as one goes down the group continue to attract attention in the chemistry of these compounds. In a recent paper, barium is described to form subnitride, Ba_2N, when it is dissolved in sodium melt and heated with nitrogen gas in a sealed stainless-steel crucible. This barium subnitride was then reacted with barium halides as shown in the reaction scheme below to form ternary and quaternary barium nitride halides. Handling of the nitride and nitride halide products was difficult owing to their extreme air and moisture sensitivity.

$$2Ba_2N + BaX_2 \rightarrow 2Ba_2NX + Ba$$
$$2Ba_2N + (1-y)BaX_2 + yBaX'_2 \rightarrow 2BaNX_{1-y}X'_y + Ba \; (X, X' = Cl, Br, I;\; 0<y<1)$$

7.2.3.2 *Ammonia*

As with the reactive alkali metals, barium reacts with ammonia to form a deep blue complex of the form $Ba(NH_3)_6$. This complex slowly decomposes to the amide, $Ba(NH_2)_2$, with release of NH_3 and H_2 gases. With respect to its reactivity with acids, it is readily attacked by most acids except by sulfuric acid, which forms a passivation layer of barium sulfate, that prevents the reaction from proceeding.

7.2.3.3 Halides

Barium chloride, $BaCl_2$, is one of the most common water-soluble salts of barium. Similar to most other barium salts, it is white, toxic, and gives a yellow-green coloration to a flame. It is also hygroscopic, converting first to the dihydrate $BaCl_2(H_2O)_2$. It has limited use in both the laboratory and industry. On an industrial scale, it is prepared via a two step process from baryte (barium sulfate):

$$BaSO_4 + 4C \rightarrow BaS + 4CO$$

This first step needs to be performed at high temperatures.

$$BaS + 2HCl \rightarrow BaCl_2 + H_2S.$$

Instead of HCl, chlorine can be used. Barium chloride can in principle be produced from barium hydroxide or barium carbonate. These basic salts react with hydrochloric acid to form hydrated barium chloride. $BaCl_2$ crystallizes in two forms (polymorphs). One form has the cubic fluorite (CaF_2) structure and the other the orthorhombic cotunnite ($PbCl_2$) structure. Both polymorphs allow for the large Ba^{2+} ion to have coordination numbers greater than six. The coordination of Ba^{2+} is 8 in the fluorite (CaF_2) structure and 9 in the cotunnite structure. When cotunnite-structure $BaCl_2$ is subjected to pressures of 7–10 GPa, it transforms to a third structure, a monoclinic post-cotunnite phase. The coordination number of Ba^{2+} increases from 9 to 10. In aqueous solution $BaCl_2$ behaves as a simple salt; in water it is a 1:2 electrolyte and the solution exhibits a neutral pH. Its solutions react with sulfate ion to form a white precipitate of barium sulfate.

$$Ba^{2+} + SO_4^{2-} \rightarrow BaSO_4.$$

Oxalate reacts in a similar way. Reacted with sodium hydroxide (NaOH), it forms the dihydroxide [$Ba(OH)_2$], which is moderately soluble in water.

Barium fluoride, BaF_2, is a colorless solid that occurs in nature as the rare mineral frankdicksonite. Similar to $BaCl_2$, it has the fluorite (CaF_2) structure and at high pressure the cotunnite ($PbCl_2$) structure. Like CaF_2, it is resilient to and insoluble in water. Above c. 500°C, BaF_2 is corroded by moisture, but in dry environments it can be utilized up to 800°C. Prolonged exposure to moisture degrades transmission in the vacuum UV range. It is less resistant to water than CaF_2, but it is the most resistant of all the optical fluorides to high-energy radiation, although its far ultraviolet transmittance is lower. It is relatively hard, very sensitive to thermal shock and fractures rather easily. In the vapor phase the BaF_2 molecule is non-linear with an F-Ba-F angle of approximately 108°. Its nonlinearity is an exception to VSEPR theory (Valence Shell Electron Pair Repulsion theory). Ab initio calculations suggest that contributions from d orbitals in the shell below the valence shell are responsible. Another suggestion is that polarization of the electron core of the barium atom creates an approximately tetrahedral distribution of charge that interacts with the Ba-F bonds.

Barium bromide ($BaBr_2$), like barium chloride, dissolves well in water and is toxic. Barium bromide can be poduced by reacting barium sulfide or barium carbonate with hydrobromic acid:

$$BaS + 2HBr \rightarrow BaBr_2 + H_2S$$
$$BaCO_3 + 2HBr \rightarrow BaBr_2 + CO_2 + H_2O$$

Barium bromide crystallizes from concentrated aqueous solution in its dihydrate, $BaBr_2 \cdot 2H_2O$. Heating this dihydrate to 120°C produces the anhydrous form. $BaBr_2$ crystallizes in a lead chloride (cotunnite, $PbCl_2$) motif, giving white orthorhombic crystals that are deliquescent. In aqueous solution $BaBr_2$ reacts as a simple salt. Solutions of barium bromide reacts with the sulfate salts to form a solid precipitate of barium sulfate.

$$BaBr_2 + SO_4^{2-} \rightarrow BaSO_4 + 2Br^-$$

Similar reactions take place with oxalic acid, hydrofluoric acid, and phosphoric acid, resulting in solid precipitates of barium oxalate, fluoride, and phosphate, respectively.

Barium iodide, BaI_2, exists as an anhydrous compound and a hydrated one ($BaI_2(H_2O)_2$), both of these are white solids. When heated, hydrated barium iodide converts to the anhydrous salt. The hydrated form is freely soluble in water, ethanol, and acetone. Anhydrous BaI_2 can be produced by reacting Ba metal with 1,2-diiodoethane in ether. BaI_2 reacts with alkyl potassium compounds to form organobarium compounds. BaI_2 can be reduced with lithium biphenyl, to form a highly active form of barium metal. The structure of the anhydrous form resembles that of $PbCl_2$ with each Ba center bound to 9 iodide ligands and has a crystalline packing structure that is ratjer similar to $BaCl_2$.

7.2.3.4 *Organobarium compounds*

The organometallic chemistry of group 2 metals is dominated by Be and Mg. Organocalcium compounds are gaining interest for their potential in cost-effective, eco-friendly applications in organic synthesis. Less is known for the heavier Sr and Ba, although reactive barium had been reported in 1994 by Yanagisawa et al. This reactive barium was prepared from the reduction of barium iodide with two equivalents of lithium biphenylide in dry THF. This was then directly used to prepare allylic barium that showed unexpected selective allylation reactions with carbonyl compounds as shown in the reaction scheme below.

$$BaI_2 + 2Li^+[Ph\text{-}Ph]^- \xrightarrow[25^{\circ}C,\ 30\ min]{THF} Ba^*$$

Ba^* + R_1, R_2, Cl $\xrightarrow[-78^{\circ}C]{THF}$ [R_1, R_2, BaCl] $\xrightarrow{R_3COR_4}$ R_1, R_2, HO, R_4, R_3, α

[R_1, R_2, BaCl] $\xrightarrow{CO_2}$ R_1, R_2, COOH, α

Following this, a variety of selective carbon–carbon bond-forming reactions, including cross-coupling reactions with allylic halides, additions to aldehydes and imines, and ring opening reaction of epoxides, have been developed. The effectivity of organobarium in forming carbon–carbon bonds with regio- and stereoselectivity is expected to drive further studies on the chemistry of organobarium compounds.

7.2.4 Major uses

Barium, as a metal or when alloyed with Al, is employed to eliminate undesirable gases (gettering) from vacuum tubes, such as old TV picture tubes. Ba is appropriate for this application due to its low vapor pressure and reactivity toward oxygen, nitrogen, carbon dioxide, and water; it can even partially eliminate noble gases by dissolving them in the crystal lattice. This use is progressively disappearing due to the replacement with new tubeless LCD and plasma sets. Other applications of elemental Ba are insignificant and include as additive to silumin (Al–Si alloys) which improves their structure, as well as bearing alloys; Pb–Sn soldering alloys—to improve the creep resistance; alloy with Ni for spark plugs; additive to steel and cast iron as an inoculant; and alloys with Ca, Mn, Si, and Al as high-grade steel deoxidizers.

Barium sulfate (the mineral baryte, $BaSO_4$) is vital to the petroleum industry as a drilling fluid in oil and gas wells. The precipitate of $BaSO_4$ (known as "blanc fixe," from the French for "permanent white") is employed in paints and varnishes; as a filler in ringing ink, plastics, and rubbers; as a paper coating pigment; and in nanoparticles, to enhance the physical properties of some polymers, such as epoxies. $BaSO_4$ has a low toxicity and relatively high density of around 4.5 g/cm^3 (and consequently opacity to X-rays). Therefore it is utilized as a radiocontrast agent in X-ray imaging of the digestive system ("barium meals" and "barium enemas"). Lithopone, a pigment that contains $BaSO_4$ and ZnS blended with organic compounds, is a permanent white with good covering power that does not darken when exposed to sulfides. Lithopone was discovered in the 1870s by DuPont. It was manufactured by Krebs Pigments and Chemical Company and other companies. The material came in different "seals," which varied in the content of ZnS. Gold and bronze seals contain 40%–50% ZnS, offering more hiding power and strength. Although the popularity of ZnS peaked around 1920, approximately 223,352 tons were produced in 1990. It is mainly used in paints, putty, and plastics.

Other Ba compounds are limited to certain niche applications, due to the toxicity of Ba^{2+} ions ($BaCO_3$ is a rat poison), which is not a problem for the insoluble $BaSO_4$. BaO coating on the electrodes of fluorescent lamps enables the release of electrons. Due to its high atomic density, $BaCO_3$ increases the refractive index and luster of glass and was used to reduce leaks of X-rays from cathode ray tubes (CRT) TV sets. Ba, typically as $Ba(NO_3)_2$, imparts a yellow or apple green color to fireworks; for brilliant green, barium monochloride (BaCl) is used. Barium peroxide (BaO_2) is a catalyst in the aluminothermic reaction (thermite) for welding rail tracks. It is also a green flare in tracer ammunition and a bleaching agent. Barium titanate ($BaTiO_3$) forms a promising electroceramic. BaF_2 is utilized for optics in infrared applications due to its wide transparency range between 0.15 and 12 μm. Yttrium barium copper oxide (YBCO) was the first high-temperature superconductor cooled by liquid nitrogen, with a transition temperature of $-180.2^{\circ}C$

(93K) that exceeded the boiling point of nitrogen (77K or −196.2°C). Many YBCO compounds have the general formula $YBa_2Cu_3O_{7-x}$ (also known as Y123), although materials with other Y:Ba:Cu ratios exist, such as $YBa_2Cu_4O_y$ (Y124) or $Y_2Ba_4Cu_7O_y$ (Y247). Ferrite, a type of sintered ceramic comprising Fe_2O_3 and BaO, is both electrically nonconductive and ferrimagnetic and can be temporarily or permanently magnetized. $Ba(OH)_2$ is employed to produce oil and grease additives, in the refining of beet sugar, as dehairing agent, and in the production of special types of glass.

7.3 57 La — Lanthanum

7.3.1 Discovery

In 1751 the Swedish mineralogist and chemist Axel Fredrik Cronstedt (December 23, 1722 to August 19, 1765) found a heavy mineral at the mine at Bastnäs, an ore field near Riddarhyttan, Västmanland, Sweden, which was later called cerite. Thirty years later, the 15-year-old Vilhelm (Wilhelm) Hisinger (who later became a physicist and chemist, December 23, 1766 to June 28, 1852) (Fig. 7.11) from the family owning the mine, sent a sample of it to Swedish Pomeranian and German pharmaceutical chemist Carl Scheele (December 9, 1742 to May 21, 1786), who did not discover any new elements in it. In 1803 after Vilhelm Hisinger had become an ironmaster, he returned to study the mineral with Swedish chemist Jöns Jacob Berzelius (August 20, 1779 to August 7, 1848) and obtained a new oxide which they called ceria after the dwarf planet Ceres, which had been discovered 2 years earlier. Ceria was at the same time independently isolated in Germany by chemist Martin Heinrich Klaproth (December 1, 1743 to January 1, 1817). Between 1839 and 1843, ceria was proven to be a mixture of different oxides by the Swedish surgeon and chemist Carl Gustaf Mosander (September 10, 1797 to October 15, 1858) (Fig. 7.12), who resided in the same house as Berzelius. Mosander separated out two other oxides, which he named lanthana and didymia (Mosander, 1843). He partially decomposed a sample of cerium nitrate by roasting it in air and then treating the resulting oxide with dilute nitric acid (HNO_3) (Berzelius, 1839a,b). Because the properties of lanthanum were only slightly different from those of cerium

FIGURE 7.11 Vilhelm Hisinger. This image was first published in the first (1876−1899), second (1904−1926), and third (1923−1937) edition of Nordisk familjebok.

FIGURE 7.12 Carl Gustaf Mosander. 1859. Painting by Karl Gustaf Plagemann (1805–1868).

and occurred along with it in its salts, he named it from the Ancient Greek λανθάνειν [lanthanein] (lit. to lie hidden). Relatively pure lanthanum metal was first isolated in 1923.

7.3.2 Mining, production, and major minerals

Lanthanum is the third most abundant of all the lanthanides, making up 39 ppm of the Earth's crust, after neodymium at 41.5 ppm and cerium at 66.5 ppm. Notwithstanding being among the so-called "rare earth metals," lanthanum is hence not that rare at all, but it is historically so called since it is rarer than the "common earths" such as lime (Ca) and magnesia (Mg), and historically only a limited number of deposits were known. Lanthanum is considered a rare earth metal because the method to mine it is difficult, time consuming, and expensive. It is rarely the main lanthanide found in the rare earth minerals, and in their chemical formulae it is usually preceded by cerium. Rare examples of La-dominant minerals are monazite-(La) ($La(PO_4)$) and lanthanite-(La) ($(La,Ce)_2(CO_3)_3 \cdot 8H_2O$). The La^{3+} ionic radius is similar to those of the early lanthanides of the cerium group (those up to samarium and europium) that immediately follow in the periodic table, and consequently it tends to occur together with them in phosphate, silicate, and carbonate minerals, such as monazite ($M^{III}PO_4$) and bastnäsite ($M^{III}CO_3F$), where M refers to all the rare earth metals except scandium and the radioactive promethium (mostly Ce, La, and Y). Bastnäsite is generally deficient in thorium and the heavy lanthanides, and the purification of the light lanthanides from it is less complicated.

The first step involves, after the ore has been crushed and ground, treatment with hot concentrated sulfuric acid, evolving carbon dioxide, hydrogen fluoride, and silicon tetrafluoride. The resulting product is subsequently dried and leached with water, leaving the early lanthanide ions, including lanthanum, in solution. The process for monazite, which typically contains all the rare earths as well as thorium, is more complicated. Monazite, due to its magnetic properties, can be isolated by repeated electromagnetic separation. After this separation step, it is then reacted with hot concentrated sulfuric acid to form water-soluble sulfates of the rare earths. The acidic filtrates are partially neutralized with sodium hydroxide to pH 3–4. As a result, thorium precipitates out of solution as its hydroxide, which can then be removed. In the next step the solution is reacted with ammonium oxalate to convert the rare earths to their insoluble oxalates. The oxalates are converted to oxides by heating. The oxides are subsequently dissolved in nitric acid that excludes one of the main components, cerium, whose oxide is insoluble in HNO_3. Lanthanum is isolated as a double salt with ammonium nitrate through crystallization. This salt is relatively less soluble than the other rare earth double

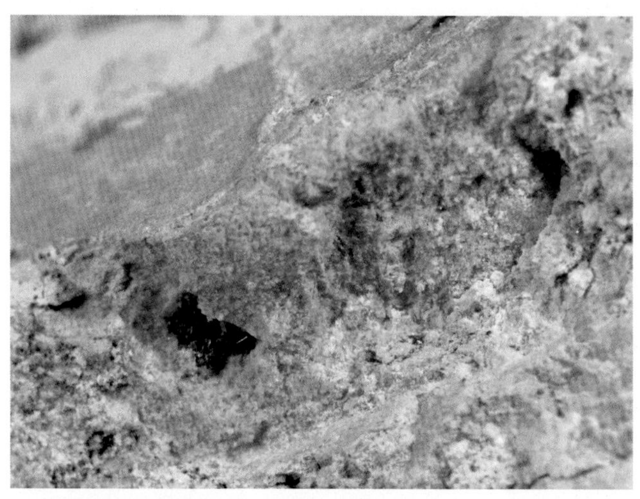

FIGURE 7.13 Sky blue patches of Bastnäsite-(La), $La(CO_3)F$, throughout matrix, about 1 cm. Paratoo Cu mine, Mt Lofty Ranges, South Australia, Australia.

FIGURE 7.14 Florencite-(La), $LaAl_3(PO_4)_2(OH)_6$, grayish-blue spindly and formless crystals growing outward from a matrix of botyroidal goethite, FeOOH. Overall 1.5 cm. Igarape Bahia mine, Serra dos Carajas, Para, Brazil.

salts and therefore stays in the residue. One has to be careful when handling some of these residues as they may contain ^{228}Ra, the daughter product of ^{232}Th, which is a strong gamma emitter. Lanthanum is comparatively easy to isolate as it has only one neighboring lanthanide, cerium, which can be removed using its ability to be oxidized to the +4 state; subsequently, lanthanum may be separated out by the classical process of fractional crystallization of $La(NO_3)_3 \cdot 2NH_4NO_3 \cdot 4H_2O$, or by ion exchange techniques when higher purity is needed. Lanthanum metal is obtained from its oxide by heating it with ammonium chloride or fluoride and hydrofluoric acid at 300°C–400°C to produce the chloride or fluoride.

$$La_2O_3 + 6NH_4Cl \rightarrow 2LaCl_3 + 6NH_3 + 3H_2O$$

This is followed by reduction with alkali or alkaline earth metals in vacuum or argon atmosphere.

$$LaCl_3 + 3Li \rightarrow La + 3LiCl$$

FIGURE 7.15 Rhabdophane-(La), $La(PO_4)\cdot H_2O$, dark brown crystalline grain to 4 mm. Pegmatite #60, Karnasurt Mt., Lovozero Massif, Kola Peninsula, Russia.

FIGURE 7.16 Cerite-(La), $(La,Ce,Ca)_9(Fe,Ca,Mg)(SiO_4)_3(HSiO_4)_4(OH)_3$, small pinkish lamellar crystals associated with earthy brown ancylite-(Ce), $CeSr(CO_3)_2(OH)\cdot H_2O$, about 8 mm. Hackmann Valley, Yukspor Mt, Khibiny Massif, Kola Peninsula, Russia.

Alternatively, pure lanthanum can be obtained through electrolysis of a molten mixture of anhydrous $LaCl_3$ and NaCl or KCl at elevated temperatures.

There are about 70 minerals with lanthanum in the crystal structure. Two halides, fluocerite-(La) ($(La,Ce)F_3$) and håleniusite-(La) ($(La,Ce)OF$), are known to contain La. The oxides are represented by four minerals, including davidite-(La) ($(La,Ce)(Y,U,Fe)(Ti,Fe)_{20}(O,OH)_{38}$) and fergusonite-(Ce) ($(Ce,La,Nd)NbO_4$). The carbonate class contains around 20 minerals, such as ancylite-(La) ($LaSr(CO_3)_2(OH)\cdot H_2O$), bastnäsite-(La) ($La(CO_3)F$) (Fig. 7.13), and parasite-(La) ($CaLa_2(CO_3)_3F_2$). Two borate minerals are known to be able to accommodate La in their crystal structure, braitschite-(Ce) ($(Ca,Na_2)_7(Ce,La)_2B_{22}O_{43}\cdot 7H_2O$) and peprossiite-(Ce) ($(Ce,La)(Al_3O)_{2/3}B_4O_{10}$). The phosphate class contains 11 minerals with La, for example, agardite-(La) ($LaCu_6(AsO_4)_3(OH)_6\cdot 3H_2O$), florencite-(La) ($LaAl_3(PO_4)_2(OH)_6$) (Fig. 7.14), monazite-(La) ($La(PO_4)$) and rhabdophane-(La) ($La(PO_4)\cdot H_2O$) (Fig. 7.15). About 30 minerals are known in the silicate class with La in their crystal structure, such as allanite-(La) ($\{CaLa\}\{Al_2Fe^{2+}\}(Si_2O_7)(SiO_4)O(OH)$), cerite-(La) ($(La,Ce,Ca)_9(Fe,Ca,Mg)(SiO_4)_3(HSiO_4)_4(OH)_3$) (Fig. 7.16), and gadolinite-(Ce) ($(Ce,La,Nd,Y)_2Fe^{2+}Be_2Si_2O_{10}$).

7.3.3 Chemistry

Naturally occurring La exists as two isotopes, the stable ^{139}La and the primordial long-lived radioisotope ^{138}La. ^{139}La is by far the most abundant, with a natural abundance of 99.911%: it is produced in the s-process (slow neutron capture, which occurs in low- to medium-mass stars) and the r-process (rapid neutron capture, which occurs in core-collapse supernovae). The very rare isotope ^{138}La is one of the few primordial odd-odd nuclei, with a long half-life of 1.05×10^{11} years: it is one of the proton-rich p-nuclei that cannot be formed in either the s-process or r-process. ^{138}La, together with the even scarcer ^{180m}Ta, is formed in the ν-process, where neutrinos interact with stable nuclei. All other La isotopes are only observed under laboratory conditions: except for ^{137}La with a half-life of about 60,000 years, all exhibit half-lives less than a day, and the majority of them have half-lives of less than a minute. The isotopes ^{139}La and ^{140}La are found as fission products of uranium.

Lanthanum is a soft, ductile, silvery-white metal that is soft enough to be cut with a knife. It is the eponym of the lanthanide series, a group of 15 similar elements between lanthanum and lutetium in the periodic table, of which lanthanum is the first and the prototype. It is on occasion considered as the first element of the sixth period transition metals, which would put it in group 3, though lutetium is occasionally put in this position instead. Lanthanum is traditionally counted among the REEs. The usual oxidation state is +3 (Table 7.3).

Similar to scandium, yttrium, and actinium atoms, the 57 electrons of a lanthanum atom are arranged in the configuration $[Xe]5d^1 6s^2$, with three valence electrons outside the noble gas core. In chemical reactions, lanthanum nearly always gives up these three valence electrons from the 5d and 6s subshells to form the +3 oxidation state, achieving the stable configuration of the preceding noble gas xenon. A few lanthanum(II) compounds are also known, but they are much less stable. Within the lanthanide series, lanthanum is unique as it does not have any 4f electrons; indeed, the abrupt contraction and lowering of energy of the 4f orbital that are important for the chemistry of the lanthanides only begin to happen at cerium. Consequently, lanthanum is only very weakly paramagnetic, in contrast to the strongly paramagnetic later lanthanides (with the exceptions of the last two, ytterbium and lutetium, where the 4f shell is completely full). Moreover, as the melting points of the trivalent lanthanides are related to the extent of hybridization of the 6s, 5d, and 4f electrons, lanthanum has the second lowest (after cerium) melting point of all the lanthanides: 920°C. Lanthanum is the least volatile of the lanthanides. Similar to most of the lanthanides, lanthanum has a hexagonal crystal structure at room temperature. At 310°C, lanthanum changes to a face-centered cubic (fcc) structure, and at 865°C, it changes to a body-centered cubic (bcc) structure.

As anticipated from periodic trends, lanthanum has the largest atomic radius among the lanthanides and the stable group 3 elements. Therefore it is the most reactive among these elements. It tarnishes slowly in air and burns readily to form lanthanum(III) oxide, La_2O_3, which is nearly as basic as calcium oxide. A centimeter-sized piece of lanthanum will completely corrode in a year as its oxide spalls off similar to iron rust, rather than forming a protective oxide coating like aluminum and lanthanum's lighter congeners scandium and yttrium. Lanthanum reacts with the halogens at 25°C to form the trihalides, and upon warming, it will form binary compounds with the nonmetals nitrogen,

TABLE 7.3 Lanthanum properties.

Appearance	Silvery white
Standard atomic weight $A_{r,std}$	138.905
Block	d-Block
Element category	Lanthanide, sometimes considered a transition metal
Electron configuration	$[Xe]\ 5d^1\ 6s^2$
Phase at STP	Solid
Melting point	920°C
Boiling point	3464°C
Density (near r.t.)	6.162 g/cm^3
When liquid (at m.p.)	5.94 g/cm^3
Heat of fusion	6.20 kJ/mol
Heat of vaporization	400 kJ/mol
Molar heat capacity	27.11 J/(mol·K)
Oxidation states	+1, +2, +3
Ionization energies	1st: 538.1 kJ/mol 2nd: 1067 kJ/mol 3rd: 1850.3 kJ/mol
Atomic radius	Empirical: 187 pm
Covalent radius	207 ± 8 pm

STP, Standard temperature and pressure.
Bold font indicates main oxidation state.

carbon, sulfur, phosphorus, boron, selenium, silicon, and arsenic. Lanthanum reacts slowly with water to form lanthanum(III) hydroxide, $La(OH)_3$. In dilute sulfuric acid, lanthanum readily forms the aquated +3 ion $[La(H_2O)_9]^{3+}$: this is colorless in aqueous solution since La^{3+} has no f electrons. Lanthanum is the strongest and hardest base among the lanthanides and group 3 elements, which is once more expected from its being the largest of them.

7.3.3.1 Oxide

Lanthanum oxide is a white solid that can be synthesized by direct reaction of its constituent elements. Lanthanum oxide can be crystallized in several polymorphs. Because of the large size of the La^{3+} ion, La_2O_3 adopts a hexagonal seven-coordinate structure that transforms to the six-coordinate structure of scandium oxide (Sc_2O_3) and yttrium oxide (Y_2O_3) at higher temperature. To prepare hexagonal La_2O_3, a 0.1 M solution of $LaCl_3$ is sprayed onto a preheated substrate, commonly made of metal chalcogenides. The process can be thought of as taking place in two steps – hydrolysis followed by dehydration:

$$2LaCl_3 + 3H_2O \rightarrow La(OH)_3 + 3HCl$$
$$2La(OH)_3 \rightarrow La_2O_3 + 3H_2O$$

Another method to obtain hexagonal La_2O_3 involves precipitation of nominal $La(OH)_3$ from aqueous solution using a combination of 2.5% NH_3 and the surfactant sodium dodecyl sulfate followed by heating and stirring for 24 hours at 80°C:

$$2LaCl_3 + 3H_2O + 3NH_3 \rightarrow La(OH)_3 + 3NH_4Cl$$

Another route is e.g.

$$2La_2S_3 + 3CO_2 \rightarrow 2La_2O_3 + 3CS_2$$

When it reacts with water, lanthanum hydroxide, $La(OH)_3$, is formed while at the same time producing a lot of heat in the reaction, and a hissing sound can be heard. Lanthanum hydroxide will react with atmospheric carbon dioxide to form the basic carbonate.

7.3.3.2 Halides

Lanthanum fluoride, LaF_3, is insoluble in water, and its formation can be utilized as a qualitative test for the presence of La^{3+}. The heavier halides are all very soluble deliquescent compounds. The anhydrous halides are produced by direct reaction of their elements, since heating the hydrates results in hydrolysis; for example, heating hydrated $LaCl_3$ produces LaOCl. Lanthanum chloride, $LaCl_3$, is a common salt of lanthanum primarily used in research. It is a white solid that is highly soluble in water and alcohols. It forms upon reaction of the elements, but a more often utilized route involves heating a mixture of lanthanum(III) oxide and ammonium chloride at 200–250°C:

$$La_2O_3 + 6NH_4Cl \rightarrow 2LaCl_3 + 6NH_3 + 3H_2O$$

From the trichloride the other trihalides can be produced by exchange. Reduction with potassium produces metallic lanthanum. Lanthanum metal the chloride or fluoride by reduction with alkali or alkaline earth metals in vacuum or argon atmosphere:

$$LaCl_3 + 3Li \rightarrow La + 3LiCl$$

In addition, pure lanthanum can be formed by electrolysis of molten mixture of anhydrous $LaCl_3$ and NaCl or KCl at elevated temperatures. Lanthanum(III) bromide ($LaBr_3$), when pure, is a colorless white powder. Single crystals of $LaBr_3$ are hexagonal with a melting point of 783°C. It is highly hygroscopic and water-soluble. There are several known hydrates, $La_3Br{\cdot}xH_2O$, of the salt is also known. It is frquently employed as a source of lanthanum in chemical synthesis and as a scintillation material in certain applications.

7.3.3.3 Hydride

Lanthanum reacts exothermically with hydrogen to form the dihydride LaH_2, a black, pyrophoric, brittle, conducting compound with the calcium fluoride (CaF_2) structure. This is a nonstoichiometric compound, and further absorption of hydrogen is possible, with an associated loss of electrical conductivity, until the more salt-like LaH_3 is reached. Similar to LaI_2 and LaI, LaH_2 is probably an electride compound (An electride is an ionic compound in which an electron is the anion. Solutions of alkali metals in ammonia are electride salts.)

7.3.3.4 Coordination compounds

Because of the large ionic radius and great electropositivity of La^{3+}, there is not much covalent contribution to its bonding and consequently it has a limited coordination chemistry, like yttrium and the other lanthanides. Lanthanum oxalate does not dissolve very much in alkali–metal oxalate solutions and $[La(acac)_3(H_2O)_2]$ decomposes around 500°C. Oxygen is the most common donor atom in lanthanum complexes, which are mostly ionic and often have high coordination numbers over 6: 8 is the most characteristic, forming square antiprismatic (In geometry, the square antiprism is the second in an infinite set of antiprisms formed by an even-numbered sequence of triangle sides closed by two polygon caps. It is also known as an anticube. A square antiprism is the favored geometry when eight pairs of electrons surround a central atom. One molecule with this geometry is the octafluoroxenate(VI) ion (XeF_8^{2-}) in the salt nitrosonium octafluoroxenate(VI); however, the molecule is distorted away from the idealized square antiprism. Very few ions are cubical because such a shape would cause large repulsion between ligands; protactinium octafluoride PaF_8^{3-} is one of the few examples.) and dodecadeltahedral structures (In geometry, the snub disphenoid, Siamese dodecahedron, triangular dodecahedron, trigonal dodecahedron, or dodecadeltahedron is a three-dimensional (3D) convex polyhedron with 12 equilateral triangles as its faces. It is not a regular polyhedron because some vertices have four faces and others have five. It is a dodecahedron, one of the eight deltahedra (convex polyhedra with equilateral triangle faces) and one of the 92 Johnson solids (nonuniform convex polyhedra with regular faces).) These high-coordinate species, reaching up to coordination number 12 with the use of chelating ligands such as in $La_2(SO_4)_3 \cdot 9H_2O$, often have a low degree of symmetry because of stereo-chemical factors.

7.3.3.5 Organolanthanum compounds

Lanthanum chemistry tends not to involve π bonding due to the electron configuration of the element; therefore its organometallic chemistry is rather limited. The best characterized organolanthanum compounds are the cyclopentadienyl complex $La(C_5H_5)_3$, which is produced by reacting anhydrous $LaCl_3$ with NaC_5H_5 in tetrahydrofuran (THF; IUPAC name oxolane, $(CH_2)_4O$) and its methyl-substituted derivatives.

7.3.4 Major uses

Modern uses of La are, for example, as material used for anodic material of nickel-metal hydride batteries is La $(Ni_{3.6}Mn_{0.4}Al_{0.3}Co_{0.7})$. Because of the high cost to remove the other lanthanides, a mischmetal with more than 50% of La is employed instead of pure La. The compound is an intermetallic component of the AB_5 type. Since the majority of hybrid cars utilize nickel-metal hydride batteries, huge amounts of La are needed for the manufacture of hybrid cars. A typical hybrid car battery for a Toyota Prius needs 10–15 kg La. As engineers push the technology to improve fuel efficiency, twice that amount of La possibly will be necessary per car. Hydrogen sponge alloys can contain La. These alloys can store up to 400 times their own volume of H_2 gas in a reversible adsorption process. Heat energy is released every time they do so; consequently these alloys have potential in energy conservation systems. Mischmetal, a pyrophoric alloy used in lighter flints, contains 25%–45% La. A pyrophoric substance (from Greek πυροφόρος, pyrophoros, "fire-bearing") is a substance that ignites spontaneously in air at or below 54°C (for gases) or within 5 min after coming into contact with air (for liquids and solids).

La oxide and the boride are used in electronic vacuum tubes as hot cathode materials with strong emissivity of electrons. Crystals of LaB_6 are used in high-brightness, extended-life, thermionic electron emission sources for electron microscopes and Hall effect thrusters (HETs). In spacecraft propulsion, a HET is a type of ion thruster in which the propellant is accelerated by an electric field. HETs use a magnetic field to limit the electrons' axial motion and then use them to ionize propellant, efficiently accelerate the ions to produce thrust, and neutralize the ions in the plume. HETs (based on the discovery by American physicist Edwin Hall, November 7, 1855 to November 20, 1938) are sometimes referred to as Hall thrusters or Hall-current thrusters. The HET is classed as a moderate specific impulse (1600 s) space propulsion technology and has benefited from considerable theoretical and experimental research since the 1960s.

LaF_3 is an important constituent of a heavy fluoride glass named ZBLAN (Heavy metal fluoride glasses were accidentally discovered in 1975 by M. Poulain and J. Lucas at the University of Rennes in France, including a family of glasses ZBLAN with a composition ZrF_4-BaF_2-LaF_3-AlF_3-NaF.). This glass has excellent transmittance in the infrared range and is consequently used in fiber-optical communication systems. Ce-doped $LaBr_3$ and $LaCl_3$ are new inorganic scintillators, which combine a high light yield, best energy resolution, and fast response. Their high yield translates into superior energy resolution; furthermore the light output is very stable and rather high over a very wide temperature range, making it especially interesting for high-temperature applications. These scintillators are even now extensively used commercially in neutron or γ ray detectors. Carbon arc lamps employ a mixture of REEs to enhance the light quality. This use, in particular by the motion picture industry for studio lighting and projection, used up around 25% of the rare-earth compounds produced until the carbon arc lamps were phased out. La (III) oxide (La_2O_3) enhances the alkali resistance of glass and is utilized in producing special optical glasses, for example, infrared-absorbing glass, in addition to camera and telescope lenses, due to the high refractive index and low dispersion of rare-earth glasses. La oxide is also employed as a grain-growth additive during the liquid-phase sintering of SiN and ZrB_2. Small quantities of La added to steel increase its malleability, resistance to impact, and ductility, while addition of La to Mo reduces its hardness and sensitivity to temperature variations. Minor quantities of La are present in many pool products to eliminate the phosphates that feed algae. La oxide additive to W is utilized in gas tungsten arc welding (GTAW) electrodes, as a replacement for radioactive Th. Several compounds of La and other rare-earth elements [(REEs) oxides, chlorides, etc.] are components for various catalysts, for example, petroleum cracking catalysts. La–Ba radiometric dating is used to approximate ages of rocks and ores; however the technique is rarely used. La carbonate ($La_2(CO_3)_3$) was approved as a medication (Fosrenol, Shire Pharmaceuticals) to absorb excess phosphate in cases of end-stage renal failure. LaF_3 is used in phosphor lamp coatings. Mixed with Eu fluoride, it is also used in the crystal membrane of F ion-selective electrodes. Similar to horseradish peroxidase, La is employed as an electron-dense tracer in molecular biology. La-modified bentonite (or phoslock) is utilized to eliminate phosphates from water in lake treatments.

7.4 58 Ce — Cerium

7.4.1 Discovery

Cerium was discovered in 1803 in Bastnäs, Sweden, by Swedish chemist Jöns Jacob Berzelius (August 20, 1779 to August 7, 1848) (Fig. 7.17) and mine owner Wilhelm Hisinger (December 23, 1766 to June 28, 1852), and at the same time independently in Germany by chemist Martin Heinrich Klaproth (December 1, 1743 to January 1, 1817). Cerium was named by Berzelius after the dwarf planet Ceres, discovered 2 years earlier. The dwarf planet itself is named after the Roman goddess of agriculture, grain crops, fertility, and motherly relationships, Ceres. Cerium was initially separated in the form of its oxide, which was called ceria, a term that is still used today. The metal itself was too electropositive to be separated using, at that time, the current smelting technology, a property of all rare-earth metals. After the development of electrochemistry by Cornish chemist and inventor Sir Humphry Davy (December 17, 1778 to May 29, 1829) 5 years later, the earths soon produced the metals they contained. Ceria, as isolated in 1803, contained all the lanthanides present in the cerite ore from Bastnäs, Sweden, and therefore only had about 45% of what is now known to

FIGURE 7.17 Jöns Jacob Berzelius, daguerreotype.

be pure ceria. It was not until Swedish surgeon and chemist Carl Gustaf Mosander (September 10, 1797 to October 15, 1858) was successful in removing lanthana and "didymia" in the late 1830s that ceria was obtained in its pure form (Mosander, 1843). Wilhelm Hisinger was a wealthy mine owner and amateur scientist, and sponsor of Berzelius. He owned and controlled the mine at Bastnäs and had been attempting for years to discover the chemical composition of the abundant heavy gangue rock (the "Tungsten of Bastnäs," which notwithstanding its name contained no tungsten), now known as cerite, that he had in his mine. Mosander and his family lived for many years in the same house as Berzelius, and Mosander was certainly convinced by Berzelius to study ceria further.

7.4.2 Mining, production, and major minerals

Cerium is the most abundant of all the lanthanides, making up 66 ppm of the Earth's crust; this value is just behind that of Cu (68 ppm), and Ce is even more abundant than common metals such as Pb (13 ppm) and Sn (2.1 ppm). Hence, notwithstanding being one of the so-called rare-earth metals, cerium is not rare at all. Cerium can be found in various minerals, but the most significant commercial sources are the minerals of the monazite and bastnäsite groups, where it makes up about 50% of the total lanthanide content. Monazite-(Ce) ($Ce(PO_4)$) is the most common representative of the monazite group, with "-Ce" being the Levinson suffix indicating the dominance of this particular REE representative. Also the cerium-dominant bastnäsite-(Ce) ($Ce(CO_3)F$) is the most important of the bastnasite group. Cerium is the easiest lanthanide to isolate from its minerals as it is the only lanthanide that can attain a stable +4 oxidation state in aqueous solution. Due to the decreased solubility of cerium in the +4 oxidation state, cerium is sometimes depleted from rocks relative to other REEs and is incorporated into zircon, because Ce^{4+} and Zr^{4+} have the same charge and comparable ionic radii. In extreme cases, cerium(IV) can form its own minerals separated from other REEs, such as cerianite-(Ce), $(Ce,Th)O_2$.

Bastnäsite, $Ln^{III}CO_3F$ (Ln = lanthanide), is generally lacking in thorium and the heavy lanthanides beyond samarium and europium, and therefore the production of cerium from it is relatively direct. In a first step, the bastnäsite is purified, using dilute hydrochloric acid to dissolve calcium carbonate impurities. The purified ore is subsequently roasted in air to oxidize it to the lanthanide oxides: while most of the lanthanides will be oxidized to the sesquioxides Ln_2O_3, cerium will instead be oxidized to the dioxide CeO_2. This is insoluble in water and can be leached out with 0.5 M

FIGURE 7.18 Ancylite-(Ce), $CeSr(CO_3)_2(OH)\cdot H_2O$, 5- to 6-mm pink crystals. Mont Saint-Hilaire, Rouville Co., Quebec, Canada.

FIGURE 7.19 Bastnäsite-(Ce), $Ce(CO_3)F$, a well-crystallized rosette 1 cm across. Trimouns Talc mine, Luzenac, Ariège, Midi-Pyrénées, France.

FIGURE 7.20 Parisite-(Ce), $CaCe_2(CO_3)_3F_2$, translucent crystal to 9 mm. Trimouns Talc mine, Luzenac, Ariège, Midi-Pyrénées, France.

hydrochloric acid, leaving the other lanthanides behind. The process for monazite, $(Ln,Th)PO_4$, which typically contains all the rare earths, as well as thorium, is more complex. Monazite, due to its magnetic properties, can be isolated through repeated electromagnetic separation. After the separation step, the monazite is reacted with hot concentrated sulfuric acid to obtain water-soluble sulfates of the rare earths. The acidic filtrates are partially neutralized with sodium hydroxide to pH 3–4. Thorium is precipitated out of solution as its hydroxide and is removed. Subsequently, the remaining solution is treated with ammonium oxalate to precipitate the rare earths as their insoluble oxalates. The oxalates are then converted to oxides by heating. The oxides are dissolved in nitric acid, except for cerium oxide, which is insoluble in HNO_3 and therefore precipitates out. One has to be careful when handling some of the residues as they may contain ^{228}Ra, the daughter product of ^{232}Th, which is a strong gamma emitter.

There are around 150 minerals containing cerium as part of their chemical composition. Five halides are known to contain Ce, for example, chukhrovite-(Ce) ($Ca_3(Ce,Y)[F|SO_4|(AlF_6)_2]\cdot 10H_2O$) and fluocerite-(Ce) ($(Ce,La)F_3$). The oxide class contains 16 minerals with Ce, such as aeschynite-(Ce) ($(Ce,Ca,Fe,Th)(Ti,Nb)_2(O,OH)_6$), cerianite-(Ce) ($(Ce^{4+},Th)O_2$),

FIGURE 7.21 Synchysite-(Ce), $CaCe(CO_3)_2F$, an orange prismatic, pseudohexagonal (monoclinic) crystal to 2 mm. Monte Cervandone, Ossola Valley, Piedmont, Italy.

FIGURE 7.22 Florencite-(Ce), $CeAl_3(PO_4)_2(OH)_6$, deep orange crystal to 5 mm on colorless magnesite. Brumado, Bahia, Brazil.

FIGURE 7.23 Monazite-(Ce), $Ce(PO_4)$ (always contains major to minor amounts of other REEs (Nd, La, Sm, etc.) replacing Ce. Also often contains trace amounts of U and Th (coupled with Ca).), good twinned crystals to 8 mm in aggregate. São João da Chapada, Diamantina, Minas Gerais, Brazil.

FIGURE 7.24 Allanite-(Ce,Nd), {Ca(Ce,Nd)}(Al_2Fe^{2+})(Si_2O_7)(SiO_4)O(OH), reddish brown crystal group with terminations, 1 × 0.6 × 0.1 cm. Trimouns Talc mine, Luzenac, Ariège, Midi-Pyrénées, France.

FIGURE 7.25 Undifferentiated black minerals of allanite, {Ca(Ce,La,Nd)}(Al_2Fe^{2+})(Si_2O_7)(SiO_4)O(OH), and thorianite, ThO_2, with massive, fine-grained purplish britholite-(Ce), $(Ce,Ca)_5(SiO_4)_3OH$. Jamestown District, Boulder Co., Colorado, United States.

davidite-(Ce) ($(Ce,La)(Y,U,Fe)(Ti,Fe)_{20}(O,OH)_{38}$), and fergusonite-(Ce) ($(Ce,La,Y)NbO_4$). Some 30 minerals are present in the carbonate class, for example, ancylite-(Ce) ($CeSr(CO_3)_2(OH)\cdot H_2O$) (Fig. 7.18), bastnäsite-(Ce) ($Ce(CO_3)F$) (Fig. 7.19), hydroxylbastnäsite-(Ce) ($Ce(CO_3)(OH)$), parasite-(Ce) ($CaCe_2(CO_3)_3F_2$) (Fig. 7.20), and synchysite-(Ce) ($CaCe(CO_3)_2F$) (Fig. 7.21). Two borates, braitschite-(Ce) ($(Ca,Na_2)_7(Ce,La)_2B_{22}O_{43}\cdot 7H_2O$) and peprossiite-(Ce) ($(Ce,La)(Al_3O)_{2/3}B_4O_{10}$), are known with Ce in their crystal structure. Tancaite-(Ce) ($FeCe(MoO_4)_3\cdot 3H_2O$) is the only molybdate in the sulfate class with Ce. About 20 different minerals are known in the phosphate class to contain Ce, for example, agardite-(Ce) ($CeCu_6(AsO_4)_3(OH)_6\cdot 3H_2O$), florencite-(Ce) ($CeAl_3(PO_4)_2(OH)_6$) (Fig. 7.22), gasparite-(Ce) ($Ce(AsO_4)$), monazite-(Ce) ($Ce(PO_4)$) (Fig. 7.23), and rhabdophane-(Ce) ($Ce(PO_4)\cdot H_2O$). The largest group of around 70 minerals is found in the silicates class, for example, allanite-(Ce) ({CaCe}{Al_2Fe^{2+}}(Si_2O_7)(SiO_4)O(OH))

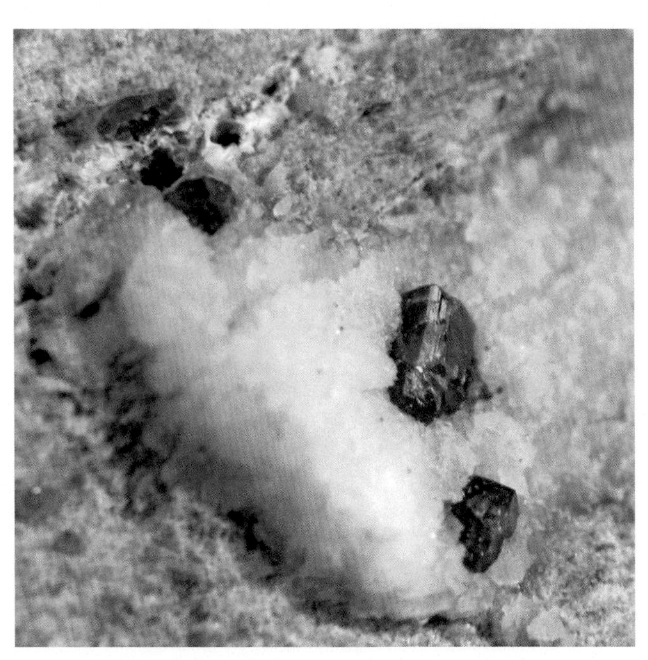

FIGURE 7.26 Joaquinite-(Ce), $NaBa_2Ce_2FeTi_2[Si_4O_{12}]_2O_2(OH,F)\cdot H_2O$, 1 mm or less orange crystals on white natrolite, $Na_2Al_2Si_3O_{10}\cdot 2H_2O$. Gem mine, San Benito Co., California, United States.

(Fig. 7.24), britholite-(Ce) ($(Ce,Ca)_5(SiO_4)_3OH$) (Fig. 7.25), cerite-(Ce) ($(Ce,Ca)_9(Mg,Fe)(SiO_4)_3(HSiO_4)_4(OH)_3$), gadolinite-(Ce) ($(Ce,La,Nd,Y)_2Fe^{2+}Be_2Si_2O_{10}$), joaquinite-(Ce) ($NaBa_2Ce_2FeTi_2[Si_4O_{12}]_2O_2(OH,F)\cdot H_2O$) (Fig. 7.26), and steenstrupine-(Ce) ($Na_{14}Mn_2^{2+}(Fe^{3+},Mn^{3+})_2Ce_6(Zr,Th)(Si_6O_{18})_2(PO_4)_6(HPO_4)(OH)_2\cdot 2H_2O$).

7.4.3 Chemistry

Naturally occurring Ce exists as four isotopes: ^{136}Ce (0.19%), ^{138}Ce (0.25%), ^{140}Ce (88.4%), and ^{142}Ce (11.1%). All four are observationally stable, nevertheless the light isotopes ^{136}Ce and ^{138}Ce are theoretically predicted to undergo inverse double β decay to Ba isotopes, and the heaviest isotope ^{142}Ce is predicted to undergo double β decay to ^{142}Nd or α decay to ^{138}Ba. Furthermore, ^{140}Ce would release energy upon spontaneous fission. None of these decay modes have up till now been detected, even though the double β decay of ^{136}Ce, ^{138}Ce, and ^{142}Ce has been experimentally searched for. The present experimentally known limits for their half-lives are $^{136}Ce > 3.8 \times 10^{16}$ years, $^{138}Ce > 1.5 \times 10^{14}$ years, and $^{142}Ce > 5 \times 10^{16}$ years. The remaining isotopes are synthetic and radioactive. The most stable of them are ^{144}Ce with a half-life of 284.9 days, ^{139}Ce with a half-life of 137.6 days, ^{143}Ce with a half-life of 33.04 days, and ^{141}Ce with a half-life of 32.5 days. The remaining radioisotopes have half-lives less than 4 days, and the majority of them have half-lives less than 10 min. The isotopes from ^{140}Ce to ^{144}Ce are formed as fission products of uranium. The main decay mode of the isotopes lighter than ^{140}Ce is inverse β decay or electron capture to La isotopes, whereas that of the heavier isotopes is β decay to Pr isotopes. The extreme rarity of the proton-rich ^{136}Ce and ^{138}Ce has been ascribed to the fact that they cannot be formed in the most common stellar nucleosynthesis processes for elements beyond iron, the s-process (slow neutron capture) and the r-process (rapid neutron capture). This is so since they are bypassed by the reaction flow of the s-process, and the r-process nuclides are blocked from decaying to them by more neutron-rich stable nuclides. Such nuclei are known as p-nuclei, and their origin is up till now not well understood: some speculated processes for their formation comprise proton capture as well as photodisintegration. ^{140}Ce is the most common isotope, since it can be formed in both the s- and r-processes, whereas ^{142}Ce can only be formed in the r-process. Another reason for the abundance of ^{140}Ce is that it is a magic nucleus, having a closed neutron shell (it has 82 neutrons), and therefore it has a very low cross-section toward further neutron capture. While its proton number of 58 is not magic, it is granted additional stability, as its eight additional protons past the magic number 50 enter and complete the 1 $g_{7/2}$ proton orbital. The abundances of the Ce isotopes can vary very slightly in natural sources, as ^{138}Ce and ^{140}Ce are the daughter isotopes of the long-lived primordial radionuclides ^{138}La and ^{144}Nd, respectively.

Cerium is a soft, ductile, and silvery-white metal that it is soft enough to be cut with a knife. Cerium is the second element in the lanthanide series, and though it frequently shows the +3 oxidation state typical for the series, it also

TABLE 7.4 Cerium properties.

Appearance	Silvery white
Standard atomic weight $A_{r,std}$	140.116
Block	f-Block
Element category	Lanthanide
Electron configuration	[Xe] $4f^1$ $5d^1$ $6s^2$
Phase at STP	Solid
Melting point	795°C
Boiling point	3443°C
Density (near r.t.)	6.770 g/cm^3
When liquid (at m.p.)	6.55 g/cm^3
Heat of fusion	5.46 kJ/mol
Heat of vaporization	398 kJ/mol
Molar heat capacity	26.94 J/(mol·K)
Oxidation states	+1, +2, **+3**, **+4**
Ionization energies	1st: 534.4 kJ/mol
	2nd: 1050 kJ/mol
	3rd: 1949 kJ/mol
Atomic radius	Empirical: 181.8 pm
Covalent radius	204 ± 9 pm

STP, Standard temperature and pressure.
Bold font indicates main oxidation state.

exceptionally has a stable +4 state that does not oxidize water (Table 7.4). It is also considered one of the REEs. In the periodic table, it appears between the lanthanides lanthanum to its left and praseodymium to its right, and above the actinide thorium. It is a ductile metal with a hardness similar to that of silver. Its 58 electrons are arranged in the configuration $[Xe]4f^1 5d^1 6s^2$, of which the four outer electrons are valence electrons. Directly after lanthanum, the 4f orbitals abruptly contract and are lowered in energy to the point that they participate readily in chemical reactions; yet, this effect is not yet strong enough at cerium and thus the 5d subshell is still occupied. Most lanthanides can utilize just three electrons as valence electrons, as afterward the remaining 4f electrons are too strongly bound. Cerium is an exception due to the stability of the empty f shell in Ce^{4+} and the fact that it comes very early in the lanthanide series, where the nuclear charge is still low enough until neodymium to allow the removal of the fourth valence electron by chemical means.

Four allotropic forms of cerium are known to exist at standard pressure and are given the common labels of α to δ: the high-temperature form, δ-cerium, has a *bcc* crystal structure and exists above 726°C, the stable form below 726°C to approximately room temperature is γ-cerium, with an *fcc* crystal structure, the double-hexagonally close-packed (*dhcp*) form β-cerium is the equilibrium structure approximately from room temperature to −150°C, and the *fcc* form α-cerium is stable below about −150°C; it has a density of 8.16 g/cm^3. Additional solid phases occur only at high pressures. Both γ and β forms are quite stable at room temperature, although the equilibrium transformation temperature is estimated at approximately 75°C.

Cerium has a variable electronic structure. The energy of the 4f electron is almost the same as that of the outer 5d and 6s electrons, which are delocalized in the metallic state, and only a small amount of energy is needed to change the relative occupancy of these electronic levels. This results in dual valence states. For instance, a volume change of around 10% is observed when cerium is subjected to high pressures or low temperatures. It seems that the valence changes from about 3–4 when it is cooled or compressed. At lower temperatures the behavior of cerium is complicated by the slow rates of transformation. Transformation temperatures are subject to substantial hysteresis and values given here are approximate. Upon cooling below −15°C, γ-cerium starts to change to β-cerium, but the transformation involves a volume increase and, as more β forms, the internal stresses build up and suppress further transformation. Cooling below approximately −160°C will start formation of α-cerium but this is only from remaining γ-cerium. β-cerium does not significantly transform to α-cerium except in the presence of stress or deformation. At atmospheric pressure, liquid cerium is denser than its solid form at the melting point.

7.4.3.1 Reactions with oxygen and chalcogens

Cerium tarnishes in air, forming a spalling oxide layer like iron rust; a centimeter-sized piece of cerium metal corrodes completely in about a year. It burns readily at 150°C to form the pale-yellow cerium(IV) oxide, also known as ceria.

$$Ce + O_2 \rightarrow CeO_2$$

This may be reduced to cerium(III) oxide with hydrogen gas. Cerium metal is highly pyrophoric, that is, when it is ground or scratched, the resulting shavings catch fire. This reactivity conforms to periodic trends, as cerium is one of the first and therefore one of the largest lanthanides. Cerium(IV) oxide has the fluorite (CaF_2) structure, comparable to the dioxides of praseodymium and terbium. Cerium oxide is utilized as a catalytic converter for the minimization of CO emissions in the exhaust gases from cars. When there is a shortage of oxygen, cerium(IV) oxide is reduced by carbon monoxide to cerium(III) oxide:

$$2CeO_2 + CO \rightarrow Ce_2O_3 + CO_2$$

When there is an oxygen surplus, the process is reversed and cerium(III) oxide is oxidized to cerium(IV) oxide:

$$2Ce_2O_3 + O_2 \rightarrow 4CeO_2$$

Major automotive applications for cerium(III) oxide are, as a catalytic converter for the oxidation of CO and NOx emissions in the exhaust gases from cars, and in addition, cerium oxide is applied as a fuel additive to diesel fuels, resulting in increased fuel efficiency and decreased hydrocarbon derived particulate matter emissions, although the health effects of the cerium oxide bearing engine exhaust is a point of research and debate. Many nonstoichiometric chalcogenides are also known, along with the trivalent Ce_2Z_3 (Z = S, Se, Te). The monochalcogenides CeZ conduct electricity and would better be formulated as $Ce^{3+}Z^{2-}e^{-}$. While CeZ_2 compounds are known, they are polychalcogenides with cerium (III): cerium(IV) chalcogenides are still unknown. Cerium is a highly electropositive metal and reacts with water. The reaction is slow with cold water but speeds up with increasing temperature, producing cerium(III) hydroxide and hydrogen gas.

$$2Ce\,(s) + 6H_2O\,(l) \rightarrow 2Ce\,(OH)_3(aq) + 3H_2(g)$$

7.4.3.2 Reactions with halogens

Cerium metal reacts with all the halogens to produce trihalides.

$$2Ce(s) + 3F_2(g) \rightarrow 2CeF_3(s)\ \text{(white)}$$
$$2Ce(s) + 3Cl_2(g) \rightarrow 2CeCl_3(s)\ \text{(white)}$$
$$2Ce(s) + 3Br_2(g) \rightarrow 2CeBr_3(s)\ \text{(white)}$$
$$2Ce(s) + 3I_2(g) \rightarrow 2CeI_3(s)\ \text{(yellow)}$$

Reaction with excess fluorine forms the stable white tetrafluoride CeF_4; the other tetrahalides are not known. Cerium(III) chloride ($CeCl_3$), also known as cerous chloride or cerium trichloride, is a white hygroscopic solid; it rapidly absorbs water on exposure to moist air to form a hydrate, which appears to be of variable composition, though the heptahydrate $CeCl_3 \cdot 7H_2O$ is known. It is highly soluble in water, and (when anhydrous) it is soluble in ethanol and acetone. A useful form of anhydrous $CeCl_3$ can be produced if care is taken to heat the heptahydrate gradually to 140°C over many hours under vacuum. This may or may not contain a small amount of CeOCl from hydrolysis, but it is suitable for use with organolithium and Grignard reagents. Pure waterless $CeCl_3$ can be prepared by dehydration of the hydrate either by slowly heating to 400°C with 4–6 equivalents of ammonium chloride under high vacuum, or by heating with an excess of thionyl chloride for three hours. Alternatively, the waterless halide may be formed from cerium metal and hydrogen chloride. It is commonly purified by high temperature sublimation under high vacuum. Cerium(III) bromide, $CeBr_3$, is a white hygroscopic solid that is of interest as a component of scintillation counters. It has been known since at least 1899, when Muthman and Stützel published its preparation from cerium sulfide and gaseous HBr (Muthman and Stützel, 1899). Aqueous solutions of $CeBr_3$ can be produced through the reaction of $Ce_2(CO_3)_3 \cdot H_2O$ with HBr. The product, $CeBr_3 \cdot H_2O$ can be dehydrated by heating with NH_4Br followed by sublimation of residual NH_4Br. $CeBr_3$ can be distilled at reduced pressure (~0.1 Pa) in a quartz ampoule at 875–880°C. Like the related salt $CeCl_3$, the bromide absorbs water on exposure to moist air. The compound melts congruently at 722°C. Solid $CeBr_3$ hass the hexagonal, UCl_3-type crystal structure with $P6_3/m$ space group. Of the dihalides, only the bronze diiodide CeI_2 is known; like the diiodides of lanthanum, praseodymium, and gadolinium, this is a cerium(III) electride compound. True cerium(II) compounds are restricted to a few unusual organocerium complexes.

7.4.3.3 Reactions with acids and oxidizing agents

Cerium dissolves readily in dilute sulfuric acid to form solutions containing the colorless Ce^{3+} ions, which exist as $[Ce(H_2O)_9]^{3+}$ complexes.

$$2Ce(s) + 3H_2SO_4(aq) \rightarrow 2Ce^{3+}(aq) + 3SO_4^{2-}(aq) + 3H_2(g)$$

The solubility of cerium is much higher in methanesulfonic acid (CH_3SO_3H). Cerium(III) and terbium(III) have ultraviolet absorption bands of relatively high intensity compared with the other lanthanides, as their configurations (one electron more than an empty or half-filled f-subshell, respectively) make it easier for the extra f electron to undergo f→d transitions instead of the forbidden f→f transitions of the other lanthanides. Cerium(III) sulfate is one of the few salts whose solubility in water decreases with increasing temperature.

Cerium(IV) aqueous solutions may be formed by reacting cerium(III) solutions with the strong oxidizing agents peroxodisulfate ($S_2O_8^{2-}$) or bismuthate (BiO_3^-). The value of $E^o(Ce^{4+}/Ce^{3+})$ varies extensively depending on conditions because of the relative ease of complexation and hydrolysis with various anions, though +1.72 V is a typically representative value, that for $E^o(Ce^{3+}/Ce)$ is −2.34 V. Cerium is the only lanthanide that has significant aqueous and coordination chemistry in the +4 oxidation state. Because of the ligand-to-metal charge transfer, aqueous cerium(IV) ions are orange yellow. Aqueous cerium(IV) is metastable in water and is a strong oxidizing agent that oxidizes hydrochloric acid to produce chlorine gas. For instance, ceric ammonium nitrate (IUPAC name diammonium cerium(IV) nitrate, $(NH_4)_2Ce(NO_3)_6$) is a common oxidizing agent in organic chemistry, releasing organic ligands from metal carbonyls. In the Belousov–Zhabotinsky reaction, cerium oscillates between the +4 and +3 oxidation states to catalyze the reaction. The discovery of the phenomenon is credited to Soviet chemist and biophysicist Boris Pavlovich Belousov (February 19, 1893 to June 12, 1970). In 1951 while trying to find the nonorganic analogs to the Krebs cycle, he noted that in a mix of potassium bromate, cerium(IV) sulfate, malonic acid, and citric acid in dilute sulfuric acid, the ratio of concentration of the cerium(IV) and cerium(III) ions oscillated, causing the color of the solution to oscillate between a yellow solution and a colorless solution. This is due to the cerium(IV) ions being reduced by malonic acid to cerium(III) ions, which are then oxidized back to cerium(IV) ions by bromate(V) ions. Belousov made two attempts to publish his finding but was rejected because he could not explain his results to the satisfaction of the editors of the journals to which he submitted his results. Soviet biochemist Simon El'evich Shnoll (born March 21, 1930 in Moscow) encouraged Belousov to continue his efforts to publish his results. In 1959 his work was finally published in a less respectable, non-reviewed journal. After Belousov's publication, Shnoll gave the project in 1961 to a graduate student, Soviet biophysicist Anatol Markovich Zhabotinsky (January 17, 1938 to September 16, 2008), who investigated the reaction sequence in detail; however the results of these men's work were still not widely disseminated and were not known in the West until a conference in Prague in 1968. Cerium(IV) salts, in particular cerium(IV) sulfate, are often used as standard reagents for volumetric analysis in cerimetric titrations.

7.4.3.4 Nitrates

The nitrate complex $[Ce(NO_3)_6]^{2-}$ is the most common cerium complex encountered when utilizing cerium(IV) as an oxidizing agent. $[Ce(NO_3)_6]^{2-}$ and its cerium(III) analog $[Ce(NO_3)_6]^{3-}$ have 12-coordinate icosahedral molecular geometry, whereas $[Ce(NO_3)_5]^{2-}$ has 10-coordinate bicapped dodecadeltahedral molecular geometry. Cerium nitrates can in addition form 4:3 and 1:1 complexes with 18-crown-6 (the ratio referring to that between cerium and the crown ether, IUPAC name of 1,4,7,10,13,16-hexaoxacyclooctadecane, $[C_2H_4O]_6$). Halogen-containing complex ions, for example, CeF_8^{4-}, CeF_6^{2-}, and the orange $CeCl_6^{2-}$ are also known.

7.4.3.5 Organocerium compounds

Organocerium chemistry is comparable to that of other lanthanides, being primarily that of the cyclopentadienyl and cyclooctatetraenyl compounds. The cerium(III) cyclooctatetraenyl compound exhibits the uranocene structure. In general, organocerium compounds cannot be isolated, and are therefore studied in solution through their reactions with other species. There are some notable exceptions though, e.g. the $Cp_3^*Ce(III)$ complex, but they are rather rare. Complexes involving Ce of various oxidation states have been observed: though lanthanides are most stable in the +3 state, complexes of cerium(IV) have been described. These latter compounds have found less widespread application because of their oxidizing nature, and the majority of literature regarding organometallic cerium complexes involves the +3 oxidation state. Especially, organocerium compounds have been developed widely as non-basic carbon nucleophiles in organic synthesis. Since cerium is relatively non-toxic, they serve as an "environmentally friendly" alternative to other organometallic reagents. Organocerium compounds are typically produced through transmetallation of the respective organolithium or Grignard reagent. The most common cerium source for this use is cerium(III) chloride, which can be formed in anhydrous form via dehydration of the commercially available heptahydrate. Precomplexation with tetrahydrofuran is critical for the success of the transmetallation, with most reactions requiring "vigorous stirring for a period of no less than 2 hours". Reagents derived from alkyl, alkynyl, and alkenyl organometallic reagents as well as cerium enolates have been described. The stability of each is approximately the same regardless of origin (i.e. lithiate or Grignard), with the exception of alkenyl reagents, which tend to be more stable when derived from the corresponding lithiate. The reasons for this are still poorly understood. Functional groups compatible with the parent organometallic compound are generally also stable upon transmetallation to cerium. The solution structure of organocerium reagents is still unclear, even though there is agreement that it depends heavily on the conditions under which it is produced. Especially, those derived from organolithium reagents likely are thought to form something comparable to a 'true' organocerium structure, "$R\text{-}CeCl_2$," whereas those derived from Grignard reagents are more appropriately characterized as -ate complexes of the form "$R\text{-}MgX \cdot CeCl_3$". In addition, the solvent appears to alter the solution structure of the complex, with differences observed between reagents prepared in diethyl ether and tetrahydrofuran. There is evidence that the parent chloride forms a polymeric species in THF solution, of the form $[Ce(\mu\text{-}Cl)_2(H_2O)(THF)_2]_n$, but whether this type of polymer exists once the organometallic reagent is formed is not yet known.

7.4.3.6 Ce(IV)

Notwithstanding the common name of cerium(IV) compounds, the Japanese spectroscopist Akio Kotani stated "there is no genuine example of cerium(IV)." The explanation for this can be found in the structure of ceria itself, which always has some octahedral vacancies where oxygen atoms would be expected to be positioned and could be better considered a non-stoichiometric compound with chemical formula CeO_{2-x}. Moreover, each cerium atom in ceria does not lose all four of its valence electrons, but retains a partial hold on the last one, resulting in an oxidation state intermediate between $+3$ and $+4$. Even allegedly purely tetravalent compounds such as $CeRh_3$, $CeCo_5$, or ceria itself have X-ray photoemission and X-ray absorption spectra more characteristic of intermediate-valence compounds. The 4f electron in cerocene, $Ce(C_8H_8)_2$, is poised ambiguously between being localized and delocalized, and this compound is also considered intermediate valent.

7.4.4 Major uses

The first application of Ce was in gas mantles, developed by the Austrian chemist Carl Auer von Welsbach (September 1, 1858 to August 4, 1929). In 1885 he had earlier investigated mixtures of Mg, La, and Y oxides, but these produced green-tinted light and were unsuccessful. Six years later, he found that pure ThO_2 gave a much better, yet blue, light, and that mixing it with CeO_2 produced a bright white light. Furthermore, CeO_2 serves as a catalyst for the combustion of ThO_2. This produced significant commercial success for von Welsbach and his discovery and created substantial demand for Th; its mining resulted in a significant quantity of lanthanides being extracted at the same time as by-products. Various uses were quickly found for them, in particular in the pyrophoric alloy called "mischmetall" composed of 50% Ce, 25% La, with the remainder made up by other lanthanides, which is used extensively for lighter flints. Typically, Fe is also added to create an alloy called ferrocerium, also developed by von Welsbach. Because of the chemical similarities of the lanthanides, chemical separation is generally not essential for their applications, for example, the mixing of mischmetall into steel to enhance its strength and workability, or as catalysts for the cracking of petroleum. This property of Ce saved the life of Italian Jewish chemist, writer, and Holocaust survivor Primo Levi (July 31, 1919 to April 11, 1987) at the Auschwitz concentration camp during World War II, when he found a supply of ferrocerium alloy and bartered it for food (Levi's professional qualifications were useful: in mid-November 1944, he secured a position as an assistant in IG Farben's Buna Werke laboratory that was intended to produce synthetic rubber. By avoiding hard labor in freezing outdoor temperatures, he was able to survive; also, by stealing materials from the laboratory and trading them for extra food.)

Ceria (CeO_2) is the most extensively used compound of Ce. The principal use of ceria is as a polishing compound, for example, in chemical-mechanical planarization. Here, ceria has substituted other metal oxides for the manufacture of high-quality optical surfaces. Chief applications for the lower sesquioxide are as a catalytic converter for the oxidation of CO and NO_x emissions in the exhaust gases from cars. In addition, ceria has been employed as a replacement for its radioactive congener thoria (ThO_2), for example, in the production of electrodes utilized in GTAW, where ceria as an alloying element enhances arc stability and ease of starting while reducing burn-off. Ce(IV) sulfate (an inorganic compound. It exists as the anhydrous salt $Ce(SO_4)_2$ as well as a few hydrated forms: $Ce(SO_4)_2 \cdot xH_2O$, with x equal to 4, 8, or 12) is employed as an oxidizing agent in quantitative analysis. Ce(IV) (See comment in section 7.4.3.6 on Ce(IV)) in methanesulfonic acid (CH_3SO_3H) solutions is used in industrial-scale electrosynthesis as a recyclable oxidant. Ceric ammonium nitrate (IUPAC name diammonium Ce(IV) nitrate, $(NH_4)_2Ce(NO_3)_6$), is employed as an oxidant in organic chemistry and in etching electronic components, and as a primary standard for quantitative chemical analysis. The photostability of pigments can be improved by adding Ce. It provides pigments with light fastness and stops clear polymers from darkening in sunlight. Television glass plates are subject to electron bombardment, which tends to darken them through the formation of F-centers as color centers. This phenomenon is suppressed through the addition of CeO_2. In addition, Ce is a crucial constituent of phosphors employed in TV screens and fluorescent lamps. Ce sulfide (Ce_xS_y, commonly Ce_2S_3) is a red pigment that is stable up to 350°C. The pigment forms a nontoxic substitute for CdS pigments. Ce is employed as alloying element in Al to form castable eutectic alloys (Al–Ce alloys with 6–16 wt.% Ce) to which Mg and/or Si can be further added; these alloys show outstanding high temperature strength.

7.5 60 Nd — Neodymium

7.5.1 Discovery

Didymium was discovered by Swedish surgeon and chemist Carl Gustaf Mosander (September 10, 1797 to October 15, 1858) in 1841 and was so called since it was very similar to lanthanum, with which it was found (Mosander, 1843). Mosander incorrectly thought didymium to be an element, under the impression that "ceria" (sometimes called cerite) separated by Swedish chemist Jöns Jacob Berzelius (August 20, 1779 to August 7, 1848) in 1803 was actually a mixture of cerium, lanthanum, and didymium. He was right about lanthanum as an element, but not about didymium. Mosander did as well as could be anticipated for that time, since spectroscopy had not yet been invented. His three "elements" accounted for no less than 95% of the rare earths in the original cerite from Bastnäs, Sweden. Didymium had not been hard to find, because it was giving the pinkish tinge to the salts of ceria when in trivalent (3 +) form. During the period when didymium was thought to be an element, the symbol Di was used for it. In the illustration of Mendeleev's first attempt at a periodic table, it will be noted that the atomic weights assigned to the various lanthanides, including didymium, reflect the original belief that they were divalent (2 +) (Mendeleev, 1869) (Fig. 7.27). Their real trivalency (3 +) meant that his atomic weights

ОПЫТЪ СИСТЕМЫ ЭЛЕМЕНТОВЪ

ОСНОВАННОЙ НА ИХЪ АТОМНОМЪ ВѢСѢ И ХИМИЧЕСКОМЪ СХОДСТВѢ

			Ti = 50	Zr = 90	? = 180.
			V = 51	Nb = 94	Ta = 182
			Cr = 52	Mo = 96	W = 186.
			Mn = 55	Rh = 104,4	Pt = 197,4.
			Fe = 56	Ru = 104,4	Ir = 198
			Ni = Co = 59	Pl = 106,6	Os = 199.
H = 1			Cu = 63,4	Ag = 108	Hg = 200
	Be = 9,4	Mg = 24	Zn = 65,2	Cd = 112	
	B = 11	Al = 27,4	? = 68	Ur = 116	Au = 197?
	C = 12	Si = 28	? = 70	Sn = 118	
	N = 14	P = 31	As = 75	Sb = 122	Bi = 210?
	O = 16	S = 32	Se = 79,4	Te = 128?	
	F = 19	Cl = 35	Br = 80	I = 127	
Li = 7	Na = 23	K = 39	Rb = 85,4	Cs = 133	Tl = 204
		Ca = 40	Sr = 87,6	Ba = 137	Pb = 207
		? = 45	Ce = 92		
		?Er = 56	La = 94		
		?Yt = 60	Di = 95		
		?In = 75,6	Th = 118?		

Д. Менделѣевъ

FIGURE 7.27 Didymium ("Di = 95") in the first edition periodic table of Mendeleev (1869).

FIGURE 7.28 Carl Auer von Welsbach.

for them were only about 67% of their true values. In 1874 Swedish chemist, biologist, mineralogist, and oceanographer Per Teodor Cleve (February 10, 1840 to June 18, 1905, best known for his discovery of the chemical elements holmium, Ho, and thulium, Tm) reasoned that didymium was made up of at least two elements. In 1879 French chemist Paul-Émile Lecoq de Boisbaudran (April 18, 1838 to May 28, 1912, known for his discoveries of the chemical elements gallium, Ga, samarium, Sm, and dysprosium, Dy) was successful in separating samarium, from didymium contained in North Carolinian samarskite (Lecoq de Boisbaudran, 1879). Neodymium was discovered by Baron Carl Auer von Welsbach (September 1, 1858 to 4 August 1929), an Austrian chemist and inventor, in Vienna in 1885 (von Welsbach, 1885) (Fig. 7.28). He isolated neodymium, as well as the element praseodymium (Pr), from didymium through fractional crystallization of the double ammonium nitrate tetrahydrates from nitric acid (HNO_3), while following the separation by spectroscopic analysis; nevertheless, it was not obtained in relatively pure form until 1925. The name neodymium is derived from the Greek words neos (νέος), new, and didymos (διδύμος), twin. Double nitrate crystallization was the method of commercial neodymium purification until the 1950s. Lindsay Chemical Division was the first to commercialize large-scale ion exchange purification of neodymium. Beginning in the 1950s, high-purity (above 99%) neodymium was mainly obtained through an ion exchange process from monazite, a mineral rich in REEs (monazite-(Nd), $Nd(PO_4)$). The metal itself is attained by means of electrolysis of its halide salts. These days, most neodymium is obtained from bastnäsite, $(Ce,La,Nd,Pr)CO_3F$, and purified by solvent extraction. Ion exchange purification is reserved for preparing the highest purities (typically >99.99%). The developing technology, and enhanced purity of commercially available neodymium oxide, was shown in the presence of neodymium glasses that can be found in collections today. Early neodymium glasses made in the 1930s have a more reddish or orange tinge than modern versions, which are more cleanly purple, because of the difficulties in removing the last traces of praseodymium in the era when production depended on the fractional crystallization process.

7.5.2 Mining, production, and major minerals

Although it belongs to the rare-earth metals, neodymium is not rare at all. Its abundance in the Earth's crust is about 38 ppm, which is the second highest among REEs, following cerium. Neodymium is not found in nature as a free

FIGURE 7.29 Kozoite-(Nd), $(Nd,La,Sm,Pr)(CO_3)(OH)$, pink platy crystal group to 2 mm. Mitsukoshi, Hizen-cho, Karatsu City, Saga, Kyushu, Japan.

FIGURE 7.30 Lanthanite-(Nd), $(Nd,La)_2(CO_3)_3{\cdot}8H_2O$, a rare mineral that occurs as a secondary mineral, typically formed by the alteration of weathering of REE-bearing minerals. Overall 2.5 cm. Curitiba, Paraná, Brazil.

element, but instead it occurs in ore minerals such as monazite and bastnäsite (these are actually mineral group names rather than single mineral names) that contain small amounts of all rare-earth metals. In these minerals neodymium is hardly ever dominant (as in the case of La), with Ce as the most abundant lanthanide; some exceptions comprise monazite-(Nd) ($Nd(PO_4)$) and kozoite-(Nd) ($(Nd,La,Sm,Pr)(CO_3)(OH)$). The main mining areas are found in China, the United States, Brazil, India, Sri Lanka, and Australia. The reserves of neodymium are estimated at about 8 million tons. Most of the today's production comes from China. Neodymium is typically between 10% and 18% of the rare-earth content of commercial ore deposits of the light REE minerals bastnäsite and monazite. With neodymium compounds being the most strongly colored of the trivalent lanthanides, it can sporadically dominate the coloration of rare-earth minerals in the case that competing chromophores are absent. It usually gives a pink coloration. Great examples of this include monazite crystals from the tin deposits in Llallagua, Bolivia; ancylite ($Sr(Ce,La)(CO_3)_2(OH){\cdot}H_2O$) from Mont Saint-Hilaire, Quebec, Canada; or lanthanite from the Saucon Valley, Pennsylvania, United States. As with neodymium glasses, such minerals change their colors under different lighting conditions. The absorption bands of neodymium interact with the visible emission spectrum of mercury vapor, with the unfiltered shortwave UV light causing neodymium-containing minerals to reflect a characteristic green color. This can be seen with monazite-containing sands or bastnäsite-containing ore.

FIGURE 7.31 Blue schuilingite-(Nd), $PbCu(Nd,Gd,Sm,Y)(CO_3)_3(OH)\cdot 1.5H_2O$. Field of view about 1 cm. Kasompi mine, Swambo, Katanga, Democratic Republic of Congo.

FIGURE 7.32 Sprays of light blue agardite-(Nd), $NdCu_6(AsO_4)_3(OH)_6\cdot 3H_2O$, to 1 mm across with great coverage on limonitic matrix. Imperial mine, Esmeralda Co., Nevada, United States.

Only 37 minerals are known with neodymium in their chemical composition. Chukhrovite-(Nd) ($Ca_3(Nd,Y)Al_2(SO_4)F_{13}\cdot 12H_2O$) is the only known halide. Four oxides are known with Nd, including aeschynite-(Nd) ($(Nd,Ln,Ca)(Ti,Nb)_2(O,OH)_6$) and fergusite-(Ce)/fergusonite-(Nd) ($(Nd,Ce)NbO_4$). The carbonates are represented by 12 different minerals, such as bastnäsite-(Nd) ($Nd(CO_3)F$) kozoite-(Nd) ($(Nd,La,Sm,Pr)(CO_3)(OH)$) (Fig. 7.29, lanthanide-(Nd) ($(Nd,La)_2(CO_3)_3\cdot 8H_2O$) (Fig. 7.30), schuilingite-(Nd) ($PbCu(Nd,Gd,Sm,Y)(CO_3)_3(OH)\cdot 1.5H_2O$) (Fig. 7.31), and synchysite-(Nd) ($CaNd(CO_3)_2F$). The same number of minerals is found in the phosphate class, for example, agardite-(Nd) ($NdCu_6(AsO_4)_3(OH)_6\cdot 3H_2O$) (Fig. 7.32), florencite-(Nd) ($NdAl_3(PO_4)_2(OH)_6$), monazite-(Nd) ($Nd(PO_4)$), rhabdophane-(Nd) ($Nd(PO_4)\cdot H_2O$), and wakefieldite ($Nd(VO_4)$). The silicate class contains eight minerals with Nd, for example, allanite-(Nd) ($\{CaNd\}\{Al_2Fe^{2+}\}(Si_2O_7)(SiO_4)O(OH)$) (Fig. 7.33), åskagenite-(Nd) ($\{Mn^{2+}Nd\}\{Al_2Fe^{3+}\}(Si_2O_7)(SiO_4)O_2$), and swamboite-(Nd) ($Nd_{0.333}[(UO_2)(SiO_3OH)](H_2O)_{\sim 2.5}$).

7.5.3 Chemistry

Naturally occurring Nd consists of a mixture of five stable isotopes, ^{142}Nd (27.2%), ^{143}Nd (12.2%), ^{145}Nd (8.5%), ^{146}Nd (17.2%), and ^{148}Nd (5.8%), and two radioisotopes, ^{144}Nd (23.8%) and ^{150}Nd (5.6%). In total 31 Nd radioisotopes have

FIGURE 7.33 Allanite-(Nd), $(CaNd)(Al_2Fe^{2+})(Si_2O_7)(SiO_4)O(OH)$, sharp single crystal. 1 × 0.4 × 0.1 cm. Luzenac, Ariège, Midi-Pyrénées, France.

been observed as of 2010, with the most stable radioisotopes being the naturally occurring ones: ^{144}Nd (α decay with a half-life of 2.29×10^{15} years to ^{140}Ce) and ^{150}Nd (double β decay with a half-life of 6.8×10^{18} years to ^{150}Sm). The other radioisotopes have half-lives less than 11 days, and most of them have half-lives less than 70 s. In addition, Nd has 13 known meta states, with the most stable ones being ^{139m}Nd with a half-life of 5.5 h, ^{135m}Nd with a half-life of 5.5 min, and $^{133m1}Nd$ with a half-life of ~70 s. The principal decay modes for isotopes lighter than ^{142}Nd are electron capture and positron decay to Pr isotopes, and the principal mode for the heavier isotopes is β^- decay to Pm (promethium) isotopes.

Neodymium belongs to the lanthanide series and is an REE. It is a hard, slightly malleable silvery metal. When oxidized, neodymium reacts rapidly to form pink, purple/blue, and yellow compounds in the +2, +3, and +4 oxidation states, respectively (Table 7.5). Metallic neodymium has a bright, silvery metallic luster, but as one of the more reactive lanthanide rare-earth metals, it rapidly oxidizes in air. The oxide layer that forms then peels off, exposing the metal to further oxidation. As a consequence, a centimeter-sized piece of neodymium completely oxidizes in less than a year. Neodymium commonly exists in two allotropic forms, with a transformation from a dhcp to a bcc structure occurring at approximately 863°C.

7.5.3.1 Reactions with oxygen and water

Neodymium metal tarnishes slowly in air and it burns readily at about 150°C to form neodymium(III) oxide. It forms very light grayish-blue hexagonal crystals.

$$4Nd + 3O_2 \rightarrow 2Nd_2O_3$$

Neodymium(III) oxide is formed when neodymium(III) nitride or neodymium(III) hydroxide is burned in air. It is utilized to dope glass, including sunglasses, to manufacture solid-state lasers, and to color glasses and enamels. Neodymium-doped glass turns purple because of the absorbance of yellow and green light, and is employed in welding

TABLE 7.5 Neodymium properties.

Appearance	Silvery white
Standard atomic weight $A_{r,std}$	144.242
Block	f-Block
Element category	Lanthanide
Electron configuration	[Xe] $4f^4$ $6s^2$
Phase at STP	Solid
Melting point	1024°C
Boiling point	3074°C
Density (near r.t.)	7.01 g/cm^3
When liquid (at m.p.)	6.89 g/cm^3
Heat of fusion	7.14 kJ/mol
Heat of vaporization	289 kJ/mol
Molar heat capacity	27.45 J/(mol·K)
Oxidation states	+2, **+3**, +4
Ionization energies	1st: 533.1 kJ/mol 2nd: 1040 kJ/mol 3rd: 2130 kJ/mol
Atomic radius	Empirical: 181 pm
Covalent radius	201 ± 6 pm

STP, Standard temperature and pressure.
Bold font indicates main oxidation state.

goggles. Some neodymium-doped glass is dichroic; meaning that it changes color depending on the lighting. One kind of glass named for the mineral alexandrite appears blue in sunlight and red in artificial light. In addition, neodymium(III) oxide is used as a polymerization catalyst. Neodymium is a fairly electropositive element, and it reacts slowly with cold water, but rather rapidly with hot water to form neodymium(III) hydroxide.

$$2\,Nd(s) + 6\,H_2O(l) \rightarrow 2Nd(OH)_3(aq) + 3H_2(g)$$

7.5.3.2 Reactions with halogens

Neodymium metal reacts vigorously with all the halogens.

$$2Nd(s) + 3F_2(g) \rightarrow 2NdF_3\,(s)\ \text{(violet)}$$
$$2Nd(s) + 3Cl_2(g) \rightarrow 2NdCl_3(s)\ \text{(mauve)}$$
$$2Nd(s) + 3Br_2(g) \rightarrow 2NdBr_3(s)\ \text{(violet)}$$
$$2Nd(s) + 3I_2(g) \rightarrow 2NdI_3(s)\ \text{(green)}$$

The addition of hydrofluoric acid to $NdCl_3$ produces neodymium fluoride.

Neodymium(III) fluoride, NdF_3, is a purplish pink colored solid with a high melting point. Like other lanthanide fluorides it is highly insoluble in water, which allows it to be formed from aqueous neodymium nitrate through a reaction with hydrofluoric acid, from which it precipitates as a hydrate.

$$Nd(NO_3)_3(aq) + 3HF \rightarrow NdF_3{\cdot}0.5H_2O + 3HNO_3$$

Anhydrous NdF_3 can be obtained by the simple drying of the hydrate, contrary to the hydrates of other neodymium halides, which form mixed oxyhalides if heated.

$NdCl_3$ is a mauve colored hygroscopic solid whose color changes to purple after absorption of atmospheric water. The resulting hydrate, like many other neodymium salts, has the interesting property that it shows different colors under fluorescent light, and in the chloride's case, light yellow. The structure of neodymium(III) chloride in solution critically depends on the solvent. In water, the major species is $Nd(H_2O)_8^{3+}$, and this situation is common for most rare earth chlorides and bromides. In methanol, the species is $NdCl_2(CH_3OH)_6^+$ and in hydrochloric acid $NdCl(H_2O)_7^{2+}$. The coordination of neodymium is octahedral (eightfold) in all cases, but the ligand structure is different. $NdCl_3$ is a soft paramagnetic solid, which turns ferromagnetic at very low temperature of -272.7°C (0.5K). Its electrical conductivity is about 240 S/m and heat capacity is ~100 J/(mol·K). $NdCl_3$ is readily soluble in water and ethanol, but not in chloroform or ether. The reduction reaction of $NdCl_3$ with Nd metal at temperatures above 650°C results in the formation of $NdCl_2$.

$$2NdCl_3 + Nd \rightarrow 3NdCl_2$$

Heating of $NdCl_3$ with water vapors or silica results in the formation of neodymium oxochloride.

$$NdCl_3 + H_2O \rightarrow NdOCl + 2HCl$$
$$2NdCl_3 + SiO_2 \rightarrow 2NdOCl + SiCl_4$$

The reaction of $NdCl_3$ with hydrogen sulfide at around 1100°C results in the formation of neodymium sulfide.

$$2\,NdCl_3 + 3H_2S \rightarrow 2Nd_2S_3 + 6HCl$$

7.5.3.3 Nitrides, phophides, carbides, and sulfides

The reactions with ammonia and phosphine at high temperatures produce neodymium nitride and phosphide, respectively.

$$NdCl_3 + NH_3 \rightarrow NdN + 3HCl$$
$$NdCl_3 + PH_3 \rightarrow NdP + 3HCl$$

Other related compounds include the carbide, NdC, and neodium(II) sulfide, NdS, as well as the neodium(III) sulfide, Nd_2S_3.

7.5.3.4 Reactions with acids

Neodymium dissolves readily in dilute sulfuric acid to form solutions that contain the lilac Nd(III) ion. These exist as $[Nd(H_2O)_9]^{3+}$ complexes.

$$2Nd(s) + 3H_2SO_4(aq) \rightarrow 2Nd^{3+}(aq) + 3SO_4^{2-}(aq) + 3H_2(g)$$

Lanthanides including neodymium are famous for their bright luminescence and therefore are widely used as fluorescent labels. In particular, $NdCl_3$ has been incorporated into organic molecules, such as DNA, which could be then easily traced using a fluorescence microscope during various physical and chemical reactions.

7.5.3.5 Organoneodymium compounds

Some organoneodymium compounds have found use in applications such as for near-infrared electroluminescence applications, for example, neodymium-quinolate complex, and neodymium-containing polymethylmethacrylate (PMMA) films as organic light emitting diode (OLED) color filters. O'Riordan et al. (2008) for example synthesized structurally pure, near-infrared emissive Nd-(5,7-dichloro-8-hydroxyquinoline)$_4$ tetrakis complex. Cho et al. (2006) fabricated by a solvent casting method neodymium(Nd)-containing PMMA films in order to improve the primary color purity of the OLED, and their optical properties were investigated as functions of Nd content in films. This study showed that absorption in Nd-containing PMMA films was due to intratransition within the 4f shell of the Nd^{3+} ion, and that it becomes larger with an increase in the Nd concentration in the films. In OLED devices with color filters, luminance from unnecessary emissive light is adequately reduced, producing a wider color gamut and higher color purity.

7.5.4 Major uses

Nd magnets (in fact an alloy, $Nd_2Fe_{14}B$) are the strongest permanent magnets in existence. A Nd magnet can lift a thousand times its own weight. These magnets are less expensive, lighter, and stronger than Sm–Co magnets. Still, they are not better in all aspects, as Nd-based magnets lose their magnetism at lower temperatures and have a tendency to rust, whereas Sm–Co magnets do not. Nd magnets can be found in products such as microphones, professional loudspeakers, in-ear headphones, guitar and bass guitar pickups, and computer hard disks where low mass, small volume, or strong magnetic fields are essential. Nd magnet electric motors have also been responsible for the creation of completely electrical model aircraft at the beginning of the 21st century, so that these are substituting internal combustion–powered models globally. Similarly, because of this high magnetic capacity per weight, Nd is utilized in the electric motors of hybrid and electric cars, and in the electricity generators of certain designs of commercial wind turbines (only wind turbines with "permanent magnet" generators employ Nd). Nd has a remarkably large specific heat capacity at liquid-helium temperatures, so it is convenient in cryocoolers. Probably due to its similarities to Ca^{2+}, Nd^{3+} has been described to promote plant growth. REE compounds are regularly used in China as fertilizer. Sm–Nd dating is valuable for determining the age relationships of rocks and meteorites.

Some transparent materials with a small concentration of Nd ions can be employed in lasers as gain media for infrared wavelengths (1054–1064 nm), such as Nd:YAG (yttrium aluminum garnet), Nd:YLF (yttrium lithium fluoride), Nd:YVO_4 (yttrium orthovanadate), and Nd:glass. Nd-doped crystals (typically Nd:YVO_4) produce high-powered infrared laser beams that are transformed to green laser light in commercial diode-pumped solid-state (DPSS) hand-held lasers and laser pointers (DPSS lasers are solid-state lasers made by pumping a solid gain medium, for example, a ruby or a Nd-doped YAG crystal, with a laser diode. The most common DPSS laser in use is the 532 nm wavelength green laser pointer. A powerful (>200 mW) 808 nm wavelength infrared GaAlAs laser diode pumps a Nd-doped yttrium aluminum garnet (Nd:YAG) or a Nd-doped yttrium orthovanadate (Nd:YVO_4) crystal, which produces 1064 nm wavelength light from the main spectral transition of the neodymium ion. This light is then frequency doubled using a nonlinear optical process in a KTP (potassium titanyl phosphate, $KTiOPO_4$) crystal, producing 532 nm light. Green diode-pumped solid-state lasers are usually around 20% efficient, although some lasers can reach up to 35% efficiency. In other words, a green DPSS laser using a 2.5 W pump diode would be expected to output around 500–900 mW of 532 nm light.) The current laser at the UK Atomic Weapons Establishment, the High Energy Laser Embodying Neodymium (HELEN) 1-TW Nd-glass laser, is used to acquire data for modeling on how density, temperature, and pressure interact inside warheads. HELEN can form plasmas at about −167°C (106K), from which the opacity and transmission of radiation are determined. Nd glass solid-state lasers are employed in very high-power (terawatt scale), high-energy (megajoules) multiple beam systems for inertial confinement fusion. Nd:glass lasers are typically frequency tripled to the third harmonic at 351 nm in laser fusion devices.

Nd glass is manufactured via the inclusion of Nd oxide (Nd_2O_3) in the glass melt. Typically, in daylight or incandescent light Nd glass looks lavender, but it looks pale blue under fluorescent lighting. Nd may be utilized to color glass in subtle shades ranging from pure violet through wine-red and warm gray. The first marketable application of purified Nd was in glass coloration, beginning with research by Leo Moser in November 1927. The resulting "Alexandrite" glass remains a signature color of the Moser glassworks to this day (Moser a.s. is a luxury glass manufacturer based in Karlovy Vary, Czech Republic, previously Karlsbad in Bohemia, Austria-Hungary. The original company of Ludwig Moser & Söhne, founded in 1857 by Ludwig Moser in Karlovy Vary, was a glass workshop initially devoted to polishing and engraving glass blanks; only later did the company begin designing and making its own art glass products. Following the death of his father in 1916, Leo Moser took over the direction and the company expanded significantly resulting in their recognition by a Grand Prize award at the Paris International Exhibition of Decorative Art in 1925.) Nd glass was extensively emulated in the early 1930s by American glasshouses, in particular Heisey, Fostoria ("wisteria"), Cambridge ("heatherbloom"), and Steuben ("wisteria"), and elsewhere (e.g., Lalique, in France, or Murano). Tiffin's "twilight" was produced from around 1950–1980. Present sources comprise glass-makers in the Czech Republic, the United States, and China. The sharp absorption bands of Nd results in the glass color to change under different lighting conditions, being reddish purple under daylight or yellow incandescent light, but blue under white fluorescent lighting, or greenish under trichromatic lighting. Mixed with Au or Se, beautiful red colors result. Because Nd coloration hinges on "forbidden" f–f transitions deep within the atom, there is comparatively little influence on the color from the chemical environment, so the color is not affected by the thermal history of the glass. Nevertheless, for the best color, Fe-containing impurities must be minimized in the silica used to make the glass. The identical forbidden nature of the f–f transitions makes rare-earth colorants less intense than those of most d-transition elements, thus more colorant must be added to a glass to accomplish the desired color intensity. The original Moser formula used around 5% of Nd oxide in the glass melt, enough that Moser denoted these as "rare-earth doped" glasses. Since it is a strong base, that level of Nd would have influenced the melting properties of the glass, and the lime content of the glass might have had to be corrected accordingly. Light transmitted through Nd glasses exhibits remarkably sharp absorption bands; the glass is employed in astronomical research to create sharp bands by which spectral lines may be calibrated. One more use is the production of selective astronomical filters to reduce the effect of light pollution from Na and fluorescent lighting while passing other colors, in particular dark red hydrogen-α emission from nebulae. In addition, Nd is employed to eliminate the green color produced by Fe contaminants from the glass. Nd is a component of "didymium" (denoting a mixture of salts of Nd and Pr) used for coloring glass to make welder's and glass-blower's goggles; the sharp absorption bands eliminate the strong Na emission at 589 nm. The comparable absorption of the yellow Hg emission line at 578 nm is the main cause of the blue color observed for Nd glass under traditional white fluorescent lighting. Nd and didymium glass are utilized in color-enhancing filters in indoor photography, especially in filtering out the yellow hues from incandescent lighting. Likewise, Nd glass is starting to be extensively employed more directly in incandescent light bulbs. These lamps have Nd in the glass to filter out yellow light, producing a whiter light which more resembles sunlight. The use of Nd in car rear-view mirrors, to lessen the glare at night, has been patented. Like its use in glasses, Nd salts are also employed as a colorant for enamels.

7.6 62 Sm — Samarium

7.6.1 Discovery

Discovery of samarium and related elements was reported by several chemists in the second half of the 19th century; nevertheless, most sources give the priority to the French chemist Paul Émile Lecoq de Boisbaudran (April 18, 1838 to May 28, 1912) (Fig. 7.34). Boisbaudran separated samarium oxide and/or hydroxide in Paris in 1879 from the mineral samarskite ($(Y,Ce,U,Fe)_3(Nb,Ta,Ti)_5O_{16}$) and recognized a new element in it through its sharp optical absorption lines (Lecoq de Boisbaudran, 1879). The Swiss chemist Marc Abraham Delafontaine (March 31,1837 to 1911) (Fig. 7.35) reported a new element decipium (from Latin: decipiens meaning "deceptive, misleading") in 1878, but in 1880–1881 showed that it was a mixture of several elements, one being the same as Boisbaudran's samarium (Delafontaine, 1878a,b, 1881). While samarskite was first discovered in the distant Russian region of Urals, by the late 1870s its deposits had been found in other places making the mineral accessible to many researchers. Especially, it was discovered that the samarium obtained by Boisbaudran was also impure and had a similar amount of europium (Eu). The pure element was isolated only in 1901 by French chemist Eugène-Anatole Demarçay (January 1, 1852 to March 5, 1903) (Fig. 7.36). Boisbaudran called his element samaria after the mineral samarskite, which in turn honored Vassili Samarsky-Bykhovets (November 7, 1803 to May 31, 1870). Samarsky-Bykhovets, as the Chief of Staff of the Russian Corps of Mining Engineers, had granted access for two German mineralogists, the brothers Gustav Rose (March 18, 1798 to July 15, 1873) and Heinrich Rose (who was also an analytical chemist, August 6, 1795 to January 27, 1864), to study the mineral samples from the Urals. In this sense samarium was the first chemical element to be named after a person. Later the name samaria used by Boisbaudran was changed into samarium, to conform with other element names, and samaria currently is occasionally used to refer to samarium oxide, similar to yttria, zirconia, alumina, ceria, holmia, etc. The symbol Sm was proposed for samarium; though an alternative Sa was regularly used instead until the 1920s. Before the advent of ion exchange separation technology in the 1950s, samarium had no commercial applications in its pure form. However, a by-product of the fractional crystallization purification of neodymium was a mixture of samarium and gadolinium that was named "Lindsay Mix" after the company that produced it. This material is believed to have been used for nuclear control rods in some early nuclear reactors. These days, a comparable

FIGURE 7.34 Paul Émile Lecoq de Boisbaudran.

FIGURE 7.35 Marc Abraham Delafontaine. Chicago Tribune (May 30, 1897).

FIGURE 7.36 Eugène-Anatole Demarçay.

commodity product has the name "samarium–europium–gadolinium" (SEG) concentrate. It is prepared by solvent extraction from the mixed lanthanides separated from bastnäsite (or monazite). Since the heavier lanthanides have a greater affinity for the solvent used, they are with no trouble extracted from the bulk using comparatively small proportions of solvent. Not all rare-earth manufacturers who process bastnäsite do so on a sufficiently large scale to continue onward with the isolation of the components of SEG, which typically account for only 1% or 2% of the original ore. Such manufacturers will consequently be making SEG with the intention to selling it to specialized processors. This way, the valuable europium (Eu) content of the ore is saved for use in phosphor production. Samarium purification follows the removal of the europium.

7.6.2 Mining, production, and major minerals

With the average concentration of about 8 ppm, samarium is the 40th most abundant element in the Earth's crust. It is the fifth most abundant lanthanide. Samarium is not found free in nature, but, like other REEs, is contained in many minerals, including monazite, bastnäsite, cerite, gadolinite, and samarskite; monazite (in which samarium occurs at concentrations of up to 2.8%) and bastnäsite are mostly used as commercial sources. World resources of samarium are estimated at 2 million tons; they are mostly located in China, the United States, Brazil, India, Sri Lanka, and Australia. By far, China has the largest production; it is followed by the United States and India. Domination of samarium in minerals is unique. Minerals with essential (dominant) samarium include monazite-(Sm) and florencite-(Sm). They are very rare. Samarium-151 is produced in nuclear fission of uranium with the yield of about 0.4% of the total number of fission events. It is also synthesized upon neutron capture by samarium-149, which is added to the control rods of nuclear reactors. Consequently, samarium-151 is present in spent nuclear fuel and radioactive waste. Samarium is typically sold in its oxide form, which is one of the cheapest lanthanide oxides. While mischmetal—a mixture of rare earth metals containing about 1% of samarium—has long been utilized, comparatively pure samarium has been obtained only in recent times, using ion exchange methods, solvent extraction processes, and electrochemical deposition. The metal is frequently obtained through electrolysis of a molten mixture of Sm(III) chloride with sodium chloride (NaCl) or calcium chloride ($CaCl_2$). Alternatively, samarium can be prepared through reduction of its oxide with lanthanum. The product is subsequently distilled to isolate samarium (boiling point 1794°C) from lanthanum (boiling point 3464°C).

Samarium is found in only five minerals. These include three carbonate minerals, kozoite-(Nd) ($(Nd,La,Sm,Pr)(CO_3)(OH)$), schuilingite-(Nd) ($PbCu(Nd,Gd,Sm,Y)(CO_3)_3(OH)\cdot 1.5H_2O$) (Fig. 7.37), and shabaite-(Nd) ($Ca(Nd,Sm,Y)_2(UO_2)(CO_3)_4(OH)_2\cdot 6H_2O$), and two phosphate minerals, florencite-(Sm) ($SmAl_3(PO_4)_2(OH)_6$) and monazite-(Sm) ($Sm(PO_4)$).

FIGURE 7.37 Blue schuilingite-(Nd), $PbCu(Nd,Gd,Sm,Y)(CO_3)_3(OH)\cdot 1.5H_2O$. FOV about 1 cm. Kasompi mine, Swambo, Katanga, Democratic Republic of Congo.

7.6.3 Chemistry

Naturally occurring Sm consists of four stable isotopes, ^{144}Sm, ^{150}Sm, ^{152}Sm, and ^{154}Sm, and three extremely long-lived radioisotopes, ^{147}Sm (half-life 1.06×10^{11} years), ^{148}Sm (7×10^{15} years), and ^{149}Sm ($> 2 \times 10^{15}$ years), with ^{152}Sm the most common isotope (natural abundance 26.75%). ^{149}Sm is listed by different literature sources either as a stable isotope or radioisotope. The long-lived radioisotopes, ^{146}Sm (synthetic), ^{147}Sm, and ^{148}Sm, mainly decay by α particle emission to Nd isotopes. Lighter unstable Sm isotopes mainly decay by electron capture to isotopes of Pm, whereas the heavier isotopes convert via β^- decay to Eu isotopes. The α decay of ^{147}Sm to ^{143}Nd with a half-life of 1.06×10^{11} years is used for Sm–Nd dating. The half-lives of ^{151}Sm and ^{145}Sm are 90 years and 340 days, respectively. The other radioisotopes have half-lives less than 2 days, and most have half-lives less than 48 s. In addition, Sm has five nuclear isomers with the most stable ^{141m}Sm having a half-life of 22.6 min), while ^{143m1}Sm has a half-life of 66 s and ^{139m}Sm a half-life of 10.7 s.

Sm is a relatively hard silvery metal that slowly oxidizes in air. As a typical member of the lanthanide series, samarium typically has the oxidation state +3 (Table 7.6). Compounds of samarium(II) are also known though, most notably the monoxide SmO, monochalcogenides SmS, SmSe, and SmTe, as well as samarium(II) iodide. The last compound is a common reducing agent in chemical synthesis. Samarium is a rare-earth metal with hardness and density comparable to those of zinc. With a boiling point of 1794°C, samarium is the third most volatile lanthanide after ytterbium and europium. At ambient conditions, samarium generally has a rhombohedral structure (α form). Upon heating to 731°C, its crystal symmetry changes into hexagonally close-packed (*hcp*), though the transition temperature depends on the metal purity. Further heating to 922°C changes the crystal symmetry into *bcc*. Heating to 300°C combined with compression to 40 kbar produces a *dhcp*. Applying higher pressure of the order of hundreds or thousands of kilobars induces a series of phase transformations, especially with a tetragonal phase appearing at about 900 kbar. In one study, the *dhcp* phase could be formed without compression, employing a nonequilibrium annealing regime with a rapid temperature change between about 400°C and 700°C, confirming the transient character of this samarium phase. Also, thin films of samarium formed by vapor deposition may contain the *hcp* or *dhcp* phases at ambient conditions.

Freshly prepared samarium has a silvery luster. In air, it slowly oxidizes at room temperature and spontaneously ignites at 150°C. Even when stored under mineral oil, samarium slowly oxidizes and forms a grayish-yellow powder of the oxide–hydroxide mixture at the surface. The metallic exterior of a sample can be protected by sealing it under an inert gas such as argon. Samarium is relatively electropositive and reacts slowly with cold water and rather rapidly with hot water to form samarium hydroxide.

$$2Sm(s) + 6H_2O(l) \rightarrow 2Sm(OH)_3(aq) + 3H_2(g)$$

Samarium dissolves easily in dilute sulfuric acid to form solutions containing the yellow to pale green Sm(III) ions, which exist as $[Sm(H_2O)_9]^{3+}$ complexes.

$$2Sm(s) + 3H_2SO_4(aq) \rightarrow 2Sm^{3+}(aq) + 3SO_4^{2-}(aq) + H_2(g)$$

TABLE 7.6 Samarium properties.

Appearance	Silvery white
Standard atomic weight $A_{r,std}$	150.36
Block	f-Block
Element category	Lanthanide
Electron configuration	[Xe] $4f^6$ $6s^2$
Phase at STP	Solid
Melting point	1072°C
Boiling point	1900°C
Density (near r.t.)	7.52 g/cm^3
When liquid (at m.p.)	7.16 g/cm^3
Heat of fusion	8.62 kJ/mol
Heat of vaporization	192 kJ/mol
Molar heat capacity	29.54 J/(mol·K)
Oxidation states	+1, +2, +3, +4
Ionization energies	1st: 544.5 kJ/mol
	2nd: 1070 kJ/mol
	3rd: 2260 kJ/mol
Atomic radius	Empirical: 180 pm
Covalent radius	198 ± 8 pm

STP, Standard temperature and pressure.
Bold font indicates main oxidation state.

Samarium is one of the few lanthanides that exhibit the oxidation state +2. The Sm^{2+} ions are blood-red in aqueous solution.

7.6.3.1 Oxides

The most stable oxide of samarium is the sesquioxide Sm_2O_3. Samarium(III) oxide may be prepared by two reactions: (a) thermal decomposition of samarium(III) carbonate, hydroxide, nitrate, oxalate, or sulfate.

$$Sm_2(CO_3)_3 \rightarrow Sm_2O_3 + 3CO_2$$

(b) By burning the metal in air or oxygen at a temperature above 150°C.

$$4Sm + 3O_2 \rightarrow 2Sm_2O_3$$

Samarium(III) oxide will dissolve in mineral acids, such as HCl, producing salts upon evaporation and crystallization.

$$Sm_2O_3 + 6HCl \rightarrow 2SmCl_3 + 3H_2O$$

The oxide can be reduced to metallic samarium by heating with a reducing agent, such as hydrogen or carbon monoxide, at elevated temperatures. Similar to many other samarium compounds, it exists in several crystalline phases. The trigonal form is formed by slow cooling from the melt. The melting point of Sm_2O_3 is quite high (2345°C) and therefore melting is typically accomplished not by direct heating, but with induction heating, through a radio frequency coil. The Sm_2O_3 crystals of monoclinic symmetry can be grown by the flame fusion method (Verneuil process) from the Sm_2O_3 powder, which yields cylindrical boules up to several centimeters long and about 1 cm in diameter. The boules are transparent when pure and defect free and orange otherwise. Heating the metastable trigonal Sm_2O_3 to 1900°C transforms it to the more stable monoclinic phase. Cubic Sm_2O_3 has also been characterized. Samarium is one of the few lanthanides that form a monoxide, SmO. This lustrous golden-yellow compound was synthesized by reducing Sm_2O_3 with samarium metal at high temperature (1000°C) and increased pressure above 50 kbar; lowering the pressure produced an incomplete reaction. SmO has the cubic rock salt (NaCl) lattice structure.

7.6.3.2 Chalcogenides

Samarium forms the trivalent sulfide, selenide, and telluride (Sm_2X_3, X = S, Se, Te). Divalent chalcogenides SmS, SmSe, and SmTe with cubic rock salt (NaCl) crystal structure are also known. Single crystals or polycrystals of samarium monochalcogenides can be produced by reacting the metal with sulfur, selenium, or tellurium vapors at high temperature. Thin films can be formed by magnetron sputtering or electron beam physical vapor deposition (PVD), that is, bombardment of samarium metal target with electrons in and appropriate gas atmosphere (e.g., hydrogen disulfide for SmS). They are noteworthy by converting from semiconducting to metallic state at room temperature upon application of pressure. Whereas the transition is continuous and occurs at about 20–30 kbar in SmSe and SmTe, it is sudden in SmS and needs only 6.5 kbar. This effect produces a dramatic color change in SmS from black to golden yellow when its crystals of films are scratched or polished. The transition does not alter the lattice symmetry, but there is a sharp decrease (~15%) in the crystal volume. It shows hysteresis, that is, when the pressure is released, SmS returns to the semiconducting state at a much lower pressure of about 0.4 kbar.

7.6.3.3 Halides

Samarium metal reacts with all the halogens, producing trihalides.

$$2Sm(s) + 3X_2(g) \rightarrow 2SmX_3(s) \ (X = F,\ Cl,\ Br \text{ or } I)$$

$SmCl_3$ is usually synthesized by the "ammonium chloride" method, which entails the initial formation of $(NH_4)_2[SmCl_5]$. This compound can be prepared from common starting materials at a moderate reaction temperature of 230°C from samarium oxide.

$$10NH_4Cl + Sm_2O_3 \rightarrow 2(NH_4)_2[SmCl_5] + 6NH_3 + 3H_2O$$

The pentachloride is subsequently heated to 350°C–400°C resulting in the evolution of ammonium chloride and leaving a residue of the anhydrous trichloride.

$$(NH_4)_2[SmCl_5] \rightarrow 2NH_4Cl + SmCl_3$$

Aqueous solutions of samarium(III) chloride can be obtained by dissolving metallic samarium or samarium carbonate in hydrochloric acid. Samarium(III) chloride is a moderately strong Lewis acid, which ranks as "hard" according to the HSAB concept (HSAB concept is an initialism for "hard and soft (Lewis) acids and bases." Also known as the Pearson acid-base concept, it is widely used in chemistry for explaining stability of compounds, reaction mechanisms, and pathways. It assigns the terms "hard" or "soft," and "acid" or "base" to chemical species. "Hard" applies to species that are small, have high charge states (the charge criterion applies mainly to acids, to a lesser extent to bases), and are weakly polarizable. "Soft" applies to species that are big, have low charge states, and are strongly polarizable.) Aqueous solutions of samarium chloride can be utilized to synthesize samarium trifluoride.

$$SmCl_3 + 3KF \rightarrow SmF_3 + 3KCl$$

Further reduction of the trihalides with samarium, lithium, or sodium metals at elevated temperatures (around 700°C–900°C) yields dihalides. The diiodide can also be synthesized by heating SmI_3, or by reacting the metal with 1,2-diiodoethane or -methane in anhydrous THF (IUPAC name oxolane, $(CH_2)_4O$) at room temperature.

$$Sm(s) + ICH_2{-}CH_2I \xrightarrow{THF} SmI_2 + CH_2 = CH_2.$$
$$Sm + I(CH_2)_2I \xrightarrow{THF} SmI_2 + H_2C = CH_2$$

Samarium(II) iodide, SmI_2, when employed as a solution for organic synthesis, is known as "Kagan's reagent." SmI_2 is a green solid and its solutions are green as well. It is a strong one-electron reducing agent that is used in organic synthesis. Solid, solvent-free SmI_2 forms by high-temperature decomposition of samarium(III) iodide (SmI_3).

In addition to dihalides, the reduction also forms numerous nonstoichiometric samarium halides with a well-defined crystal structure, such as Sm_3F_7, Sm_5Br_{11}, Sm_6Br_{13}, $Sm_{11}Br_{24}$, $Sm_{14}F_{33}$, and $Sm_{27}F_{64}$. Samarium halides change their crystal structures when one type of halide atoms is substituted for another, which is an uncommon behavior for most elements (e.g., actinides). Numerous halides have two major crystal phases for one composition, one being considerably more stable and the other being metastable. The latter is formed upon compression or heating, followed by quenching to ambient conditions. For example, compressing the normally monoclinic samarium diiodide and releasing the pressure results in a $PbCl_2$-type orthorhombic structure (density 5.90 g/cm^3), and similar treatment produces a new phase of samarium triiodide (density 5.97 g/cm^3).

7.6.3.4 Borides

Sintering powders of samarium oxide and boron, in vacuum, yields a powder containing several samarium boride phases, and their volume ratio can be controlled through the mixing proportion. The powder can be converted into larger crystals of a certain samarium boride using arc melting or zone melting methods, relying on the differences in melting/crystallization temperatures of SmB_6 (2580°C), SmB_4 (about 2300°C), and SmB_{66} (2150°C). All these materials are hard, brittle, dark-gray solids with the hardness increasing with the boron concentration. Samarium diboride is too volatile to be synthesized with these methods and involves high pressure (about 65 kbar) and relatively low temperatures between 1140°C and 1240°C to stabilize its crystallization. Increasing the temperature causes the preferential formation of SmB_6. Samarium hexaboride is a typical intermediate-valence compound where samarium is present both as Sm^{2+} and Sm^{3+} ions at the ratio 3:7. It belongs to a class of Kondo insulators, that is, at high temperatures (above −223°C/50K), its properties are typical of a Kondo metal, with metallic electrical conductivity characterized by strong electron scattering, but at low temperatures, it behaves as a nonmagnetic insulator with a narrow band gap of about 4–14 meV. The cooling-induced metal–insulator transition in SmB_6 is accompanied by a sharp increase in the thermal conductivity, with a maximum at approximately −258°C (15K). The explanation for this increase is that electrons themselves do not contribute to the thermal conductivity at low temperatures, which is dominated by phonons, but the decrease in electron concentration reduced the rate of electron–phonon scattering. New research seems to indicate that it may be a topological insulator.

7.6.3.5 Other inorganic compounds

Samarium carbides are formed by melting a graphite–metal mixture in an inert atmosphere. After the synthesis, they are unstable in air and have to be researched also under inert atmosphere. Samarium monophosphide, SmP, is a semiconductor with a bandgap of 1.10 eV, the same as for silicon, and high electrical conductivity of n-type. It can be formed by annealing at 1100°C mixed powders of phosphorus and samarium in an evacuated quartz ampule. Phosphorus is extremely volatile at high temperatures and may explode, therefore the heating rate must be kept well below 1°C/min. A similar method is used for the monarsenide SmAs, but the synthesis temperature is higher at 1800°C.

Many crystalline binary compounds are known for samarium and one of the group 14, 15, or 16 element X, where X is Si, Ge, Sn, Pb, Sb, or Te, while metallic alloys of samarium form another large group. They are all produced by annealing mixed powders of the corresponding elements. Many of the resulting compounds are nonstoichiometric and have nominal compositions Sm_aX_b, where the b/a ratio varies between 0.5 and 3.

7.6.3.6 Organosamarium compounds

Samarium forms a cyclopentadienide $Sm(C_5H_5)_3$ and its chloro derivatives $Sm(C_5H_5)_2Cl$ and $Sm(C_5H_5)Cl_2$. They are synthesized by reacting samarium trichloride with NaC_5H_5 in THF. Contrary to cyclopentadienides of most other lanthanides, in $Sm(C_5H_5)_3$ some C_5H_5 rings bridge each other by forming ring vertexes η^1 or edges η^2 toward another neighboring samarium atom, thus forming polymeric chains. The chloro derivative $Sm(C_5H_5)_2Cl$ has a dimer structure, which is more precisely expressed as $(\eta^5\text{-}C_5H_5)_2Sm(\mu\text{-}Cl)_2(\eta^5\text{-}C_5H_5)_2$. Here, the chlorine bridges can be substituted, for example, by iodine, hydrogen, or nitrogen atoms or by CN groups. The $(C_5H_5)^-$ ion in samarium cyclopentadienides can be substituted by the indenide $(C_9H_7)^-$ or cyclooctatetraenide $(C_8H_8)^{2-}$ ring, resulting in $Sm(C_9H_7)_3$ or $KSm(\eta^8\text{-}C_8H_8)_2$. The latter compound has a structure similar to that of uranocene (IUPAC name bis(η^8-cyclooctatetraenyl)uranium(IV), $U(C_8H_8)_2$). There is also a cyclopentadienide of divalent samarium, $Sm(C_5H_5)_2$—a solid that sublimates at about 85°C. Contrary to ferrocene (bis(η^5-cyclopentadienyl)iron, $Fe(C_5H_5)_2$), the C_5H_5 rings in $Sm(C_5H_5)_2$ are not parallel but are tilted by 40°. Alkyls and aryls of samarium are formed using a metathesis reaction in THF or ether, for example,

$$SmCl_3 + 3LiR \rightarrow SmR_3 + 3LiCl$$
$$Sm(OR)_3 + 3LiCH(SiMe_3)_2 \rightarrow Sm\{CH(SiMe_3)_2\}_3 + 3LiOR$$

Here R is a hydrocarbon group and Me stands for methyl.

7.6.4 Major uses

One of the chief uses of Sm is in Sm–Co magnets with a nominal composition of $SmCo_5$ or Sm_2Co_{17}. They possess high permanent magnetization, which is around 10,000 × that of Fe and is second only to that of Nd magnets. Yet, Sm-based magnets show better resistance to demagnetization, since they are stable to temperatures over 700°C (cf. 300°C–400°C for Nd magnets). These magnets are found in, for example, small motors, headphones, and high-end magnetic pickups for guitars and related musical instruments. One more important use of Sm and its compounds is as catalyst and chemical reagent. Sm catalysts assist decomposition of plastics, dechlorination of pollutants such as polychlorinated biphenyls (PCBs), as well as the dehydration and dehydrogenation of ethanol. Sm(III) triflate ($Sm(OTf)_3$, that is, $Sm(CF_3SO_3)_3$), is one of the most efficient Lewis acid catalysts for a halogen-promoted Friedel–Crafts reaction with alkenes (a set of reactions developed by French chemist and mineralogist Charles Friedel (March 12, 1832 to April 20, 1899) (Fig. 7.38) and American chemist James Crafts (March 8, 1839 to June 20, 1917) (Fig. 7.39) in 1877 to attach substituents to an aromatic ring. The Friedel–Crafts reactions are of two main types: alkylation reactions and acylation reactions. Both proceed by electrophilic aromatic substitution.). SmI_2 is a very often used reducing and coupling agent in organic synthesis, for example, in the desulfonylation reactions (chemical reactions leading to the removal of a sulfonyl group from organic compounds. As the sulfonyl functional group is electron-withdrawing, methods for cleaving the S-C bonds of sulfones are typically reductive in nature.); annulation (In organic chemistry annulation (from the Latin *anellus* for "little ring"; occasionally annelation) is a chemical reaction in which a new ring is constructed on a molecule); Danishefsky (published by the group of American chemist Samuel Danishefsky (born March 10, 1936) in 1996), Kuwajima (by the group of Japanese chemist Isao Kuwajima (born November 11, 1937) of the Tokyo Institute of Technology published in the 1990s), Mukaiyama (published by the group of Japanese chemist Teruaki Mukaiyama (January 5, 1927 to November 17, 2018) of the Tokyo University of Science between 1997 and 1999) and Holton (published by American chemist Robert A. Holton (born 1944) and his group at Florida State University) in 1994 was the first total synthesis of Taxol). Taxol total syntheses (Paclitaxel (PTX), $C_{47}H_{51}NO_{14}$, sold under the brand name Taxol among others, is a chemotherapy medication used to treat a number of types of cancer. This includes ovarian cancer, breast cancer, lung cancer, Kaposi sarcoma, cervical cancer, and pancreatic cancer. It is given by injection into a vein. There is also an albumin-bound formulation.); strychnine total synthesis; Barbier reaction; and other reductions with SmI_2. In its typical oxidized form, Sm is added to ceramics and glasses where it improves the infrared light absorption. As a (minor) part of mischmetal, Sm is present in "flint" ignition device of numerous lighters and torches.

FIGURE 7.38 Charles Friedel, c.1890s.

Radioactive ^{153}Sm is a β emitter with a half-life of 46.3 h. It is used to destroy cancer cells in the treatment of lung cancer, prostate cancer, breast cancer, and osteosarcoma. To achieve this, ^{153}Sm is chelated with ethylene diamine tetramethylene phosphonate and injected intravenously. The chelation stops buildup of radioactive Sm in the body that would cause excessive irradiation and possible formation of new cancer cells. ^{149}Sm exhibits a high cross-section for neutron capture (41,000 barns) and is thus used in the control rods of nuclear reactors. Its advantage in comparison to competing elements, for example, B and Cd, is stability of absorption—the majority of the fusion and decay products of ^{149}Sm are other isotopes of Sm that are also good neutron absorbers. For instance, the cross-section of ^{151}Sm is 15,000 barns, it is on the order of hundreds of barns for ^{150}Sm, ^{152}Sm, and ^{153}Sm, and is 6800 barns for naturally occurring (mixed-isotope) Sm. Among the decay products in a nuclear reactor, ^{149}Sm is seen as the second most important for reactor design and operation after ^{135}Xe.

Sm hexaboride (SmB_6), an intermediate-valence compound where Sm is present both as Sm^{2+} and Sm^{3+} ions at the ratio 3:7, has recently been shown to be a topological insulator (Kondo insulator. In solid-state physics, Kondo insulators (also referred as heavy fermion semiconductors) are understood as materials with strongly correlated electrons, which open up a narrow band gap (about 10 meV) at low temperatures with the chemical potential lying in the gap, whereas in heavy fermions the chemical potential is located in the conduction band. The band gap opens up at low temperatures due to hybridization of localized electrons (mostly f electrons) with conduction electrons, a correlation effect known as the Kondo effect. As a consequence, a transition from metallic behavior to insulating behavior is seen in resistivity measurements. The band gap could be either direct or indirect.) with possible uses in quantum computing. Sm-doped CaF_2 crystals were employed as an active medium in one of the first solid-state lasers designed and built by American physicist Peter Pitirimovich Sorokin (July 10, 1931 to September 24, 2015) (coinventor of the dye laser) and Mirek Stevenson at IBM research labs in early 1961. This Sm laser emitted pulses of red light at 708.5 nm. It had to be cooled with liquid He and therefore did not find practical use. Another Sm-based laser turned into the first saturated X-ray laser operating at wavelengths below 10 nm. It provided 50-ps pulses at 7.3 and 6.8 nm appropriate for use in holography, high-resolution microscopy of biological specimens, deflectometry, interferometry, and radiography of dense plasmas related to confinement fusion and astrophysics. Saturated operation meant that the maximum possible

FIGURE 7.39 James Mason Crafts.

power was extracted from the lasing medium, producing the high peak energy of 0.3 mJ. The active medium was Sm plasma created by irradiating Sm-coated glass with a pulsed infrared Nd-glass laser (wavelength $\sim 1.05\ \mu m$). The change in electrical resistivity in Sm monochalcogenides can be employed in a pressure sensor or in a memory device triggered between a low-resistance and high-resistance state by external pressure, and these types of devices are currently being developed commercially. In addition, Sm monosulfide creates an electric voltage upon moderate heating to $\sim 150°C$, which can be used in thermoelectric power converters.

The measurement of relative concentrations of Sm and Nd isotopes ^{147}Sm, ^{144}Nd, and ^{143}Nd allows the determination of the age and origin of rocks and meteorites in Sm–Nd dating. Both elements are lanthanides and have very similar physical and chemical properties. Consequently, Sm–Nd dating is either insensitive to partitioning of the marker elements during various geological processes, or such partitioning can well be understood and modeled from the ionic radii of the involved elements. The Sm^{3+} ion is a possible activator for application in warm-white light emitting diodes. It provides high luminous efficacy due to the narrow emission bands, though the generally low quantum efficiency and inadequate absorption in the UV-A to blue spectral region hamper commercial application. In the last few years it has been shown that nanocrystalline $BaFCl:Sm^{3+}$ as produced by coprecipitation can act as a very efficient X-ray storage phosphor. The coprecipitation results in nanocrystallites of about 100–200 nm in size and their sensitivity as X-ray storage phosphors is amplified an astonishing ~500,000 times due to the specific arrangements and density of defect centers compared to microcrystalline samples produced by sintering at high temperature. The mechanism is based on the reduction of Sm^{3+} to Sm^{2+} by trapping electrons that are formed during exposure to ionizing radiation in the BaFCl host. The 5 DJ-7 FJ f-f luminescence lines can be very efficiently excited via the parity allowed $4f^6 \rightarrow 4f^5$ 5d transition at ~417 nm. This wavelength is perfect for efficient excitation by blue-violet laser diodes as the transition is electric dipole allowed and therefore relatively intense (400 l/(mol·cm)). The phosphor has possible uses in personal dosimetry, dosimetry and imaging in radiotherapy, and medical imaging.

7.7 64 Gd — Gadolinium

7.7.1 Discovery

Gadolinium is named after the mineral gadolinite ($(Ce,Y)_2FeBe_2Si_2O_{10}$), which was named after Finnish chemist, physicist, and geologist Johan Gadolin (June 5, 1760 to August 15, 1852). In 1880 the Swiss chemist Jean Charles Galissard de Marignac (April 24, 1817 to April 15, 1894) (Fig. 7.40) detected the spectroscopic lines from gadolinium in samples of gadolinite (which in reality contains rather little gadolinium, but enough to exhibit a spectrum) and in the separate silicate mineral cerite ($(La,Ce,Ca)_9(Fe,Ca,Mg)(SiO_4)_3[SiO_3(OH)]_4(OH)_3$). The latter mineral turned out to contain a much higher concentration of the element with the new spectral line. De Marignac ultimately isolated a mineral oxide from cerite, which he understood was the oxide of this new element. He called the oxide "gadolinia." Since he understood that "gadolinia" was the oxide of a new element, the discovery of gadolinium is generally attributed to him. The French chemist Paul Émile Lecoq de Boisbaudran (April 18, 1838 to May 28, 1912) managed to obtain the gadolinium metal from gadolinia in 1886.

7.7.2 Mining, production, and major minerals

Gadolinium is a constituent in many minerals such as monazite and bastnäsite. The metal is too reactive to exist naturally as a native element. Paradoxically, as indicated above, the mineral gadolinite ($(Ce,Y)_2FeBe_2Si_2O_{10}$) essentially has only traces of this element. The abundance in the Earth's crust is around 6.2 ppm. The major mining areas are found in China, the United States, Brazil, Sri Lanka, India, and Australia with reserves predicted to surpass 1 million tons. World production of pure gadolinium currently is around 400 tons per annum.

FIGURE 7.40 Jean Charles Galissard de Marignac, c.1850.

FIGURE 7.41 Blue schuilingite-(Nd), $PbCu(Nd,Gd,Sm,Y)(CO_3)_3(OH)\cdot 1.5H_2O$. FOV about 2 mm. Kasompi mine, Swambo, Katanga, Democratic Republic of Congo.

Gadolinium is produced both from monazite and bastnäsite after crushing and transformation to their oxides. The oxides are then reacted with hydrochloric acid or sulfuric acid, which converts the insoluble oxides into soluble chlorides or sulfates. Subsequently, the acidic filtrates are partially neutralized using sodium hydroxide to pH 3–4. As a result, thorium will precipitate as its hydroxide and can then be separated. The remaining solution is reacted with ammonium oxalate ($(NH_4)_2C_2O_4$) to change the rare earths into their insoluble oxalates. The oxalates are then decomposed to their oxides by heating. The oxides are dissolved in nitric acid that eliminates one of the main components, cerium, as its oxide is insoluble in HNO_3. The remaining solution is subsequently reacted with magnesium nitrate ($Mg(NO_3)_2$) to obtain a crystallized mixture of double salts of gadolinium, samarium, and europium. These salts are separated by ion exchange chromatography. The rare earth ions are then selectively washed out by an appropriate complexing agent. Finally, gadolinium metal is produced from its oxide or salts by heating in the presence of calcium at 1450°C in an argon atmosphere. Sponge gadolinium can be obtained through reduction of molten $GdCl_3$ with a suitable metal at temperatures below 1312°C (the melting point of Gd) at reduced pressure.

Only three carbonate minerals are known with gadolinium in their chemical composition, lepersonnite-(Gd) ($Ca(Gd,Dy)_2(UO_2)_{24}(SiO_4)_4(CO_3)_8(OH)_{24}\cdot 48H_2O$), lokkaite-(Y) ($Ca(Y,Gd,Nd,Dy)_4(CO_3)_7\cdot 9H_2O$), and schuilingite-(Nd) ($PbCu(Nd,Gd,Sm,Y)(CO_3)_3(OH)\cdot 1.5H_2O$) (Fig. 7.41).

7.7.3 Chemistry

Naturally occurring Gd consists of six stable isotopes, ^{154}Gd (2.18% natural abundance), ^{155}Gd (14.80%), ^{156}Gd (2047%), ^{157}Gd (15.65%), ^{158}Gd (24.84%), and ^{160}Gd (21.86%), and one radioisotope, ^{152}Gd (0.20%). The predicted double β decay of ^{160}Gd has never been detected (only the lower limit on its half-life of more than 1.3×10^{21} years has been set experimentally). A total of 29 radioisotopes have been detected, with the most stable being the naturally occurring ^{152}Gd, with a half-life of about 1.08×10^{14} years, and synthetic ^{150}Gd, with a half-life of 1.79×10^6 years. The other radioisotopes have half-lives of less than 75 years, with most having half-lives of less than 25 s. Gd has four metastable isomers, with the most stable being ^{143m}Gd with a half-life of 110 s, ^{145m}Gd with a half-life of 85 s, and ^{141m}Gd with a half-life of 24.5 s). The isotopes lighter than ^{158}Gd mainly decay by electron capture to Eu isotopes, while the heavier ones decay mainly by β decay to Tb (terbium) isotopes.

Gadolinium is a silvery-white metal when oxidation is eliminated (Table 7.7). It is only slightly malleable and is a ductile REE. Gadolinium reacts with atmospheric oxygen or moisture slowly to produce a black surface coating. Gadolinium below its Curie point of 20°C is ferromagnetic, with an attraction to a magnetic field higher than that of nickel. Above this temperature it is the most paramagnetic element. It crystallizes in the *hcp* α-form at room temperature, but when heated to temperatures above 1235°C, it changes into its β-form, which has a *bcc* structure. Gadolinium can be reacted with most elements to form Gd(III) derivatives. It also combines with nitrogen, carbon, sulfur, phosphorus, boron, selenium, silicon, and arsenic at elevated temperatures, producing binary compounds.

TABLE 7.7 Gadolinium properties.

Appearance	Silvery white
Standard atomic weight $A_{r,std}$	157.25
Block	f-Block
Element category	Lanthanide
Electron configuration	[Xe] $4f^7$ $5d^1$ $6s^2$
Phase at STP	Solid
Melting point	1312°C
Boiling point	3000°C
Density (near r.t.)	7.90 g/cm^3
When liquid (at m.p.)	7.4 g/cm^3
Heat of fusion	10.05 kJ/mol
Heat of vaporization	301.3 kJ/mol
Molar heat capacity	37.03 J/(mol·K)
Oxidation states	+1, +2, **+3**
Ionization energies	1st: 593.4 kJ/mol
	2nd: 1170 kJ/mol
	3rd: 1990 kJ/mol
Atomic radius	Empirical: 180 pm
Covalent radius	196 ± 6 pm

STP, Standard temperature and pressure.
Bold font indicates main oxidation state.

7.7.3.1 Reactions with oxygen and water

In contrast to other REEs, metallic gadolinium is quite stable in dry air. Nevertheless, it tarnishes rapidly in moist air, forming a loosely adhering gadolinium(III) oxide (Gd_2O_3), which spalls off, exposing more surface to oxidation.

$$4Gd + 3O_2 \rightarrow 2Gd_2O_3$$

Gadolinium oxide can be synthesized through thermal decomposition of the hydroxide, nitrate, carbonate, or oxalates. Several techniques are known to produce gadolinium oxide nanoparticles, mostly based on precipitation of the hydroxide by the reaction of gadolinium ions with hydroxide and subsequent thermal dehydration to the oxide. The nanoparticles are always coated with a protective material to avoid the formation of larger polycrystalline aggregates. Nanoparticles of gadolinium oxide are used as a potential contrast agent for magnetic resonance imaging (MRI). Gadolinium is a strong reducing agent, which reduces oxides of several metals into their elements. Gadolinium is fairly electropositive and reacts slowly with cold water and rather quickly with hot water to form gadolinium hydroxide.

$$2Gd + 6H_2O \rightarrow 2Gd(OH)_3 + 3H_2.$$

7.7.3.2 Reactions with acids

Gadolinium metal is attacked quickly by dilute sulfuric acid to produce solutions containing the colorless Gd(III) ions, which exist as $[Gd(H_2O)_9]^{3+}$ complexes.

$$2Gd + 3H_2SO_4 + 18H_2O \rightarrow 2\left[Gd(H_2O)_9\right]^{3+} + 3SO_4^{2-} + 3H_2$$

Gadolinium oxide, Gd_2O_3, dissolves in acids to give the salts, such as gadolinium(III) nitrate.

7.7.3.3 Reactions with halogens

Gadolinium metal reacts with the halogens (X_2, X = F, Cl, Br, I) at a temperature of about 200°C.

$$2Gd + 3X_2 \rightarrow 2GdX_3$$

In almost all of its compounds, gadolinium adopts the oxidation state +3. All four trihalides are known. All are white, except for the iodide, which is yellow. The most commonly used halide form is gadolinium(III) chloride ($GdCl_3$). It is a colorless, hygroscopic, water-soluble solid. $GdCl_3$ is usually prepared by the "ammonium chloride" route, which involves the initial synthesis of $(NH_4)_2[GdCl_5]$. This material can be synthesized at a reaction temperature of 230°C from gadolinium oxide,

$$10NH_4Cl + Gd_2O_3 \rightarrow 2(NH_4)_2[GdCl_5] + 6NH_3 + 3H_2O$$

from hydrated gadolinium chloride,

$$4NH_4Cl + 2GdCl_3 \cdot 6H_2O \rightarrow 2(NH_4)_2[GdCl_5] + 12H_2O$$

or from gadolinium metal.

$$10NH_4Cl + 2Gd \rightarrow 2(NH_4)_2[GdCl_5] + 6NH_3 + 3H_2$$

In the next step the pentachloride is decomposed at 300°C.

$$(NH_4)_2[GdCl_5] \rightarrow GdCl_3 + 2NH_4Cl$$

This pyrolysis reaction proceeds via the intermediary $NH_4[Gd_2Cl_7]$. The ammonium chloride route is more popular and less expensive than other methods. $GdCl_3$ can, however, also be synthesized by the reaction of solid Gd at 600°C in a flowing stream of HCl.

$$2Gd + 6HCl \rightarrow 2GdCl_3 + 3H_2$$

Gadolinium(III) chloride also forms a hexahydrate, $GdCl_3{\cdot}6H_2O$. The hexahydrate is prepared by gadolinium(III) oxide (or chloride) in concentrated HCl followed by evaporation.

Reduced gadolinium compounds are known, especially in the solid state. Gadolinium(II) halides are formed by heating Gd(III) halides in the presence of metallic Gd in tantalum containers. Gadolinium also forms sesquichloride Gd_2Cl_3, which can be further reduced to GdCl by annealing at 800°C. This gadolinium(I) chloride forms platelets with layered graphite-like structure.

7.7.3.4 Other inorganic compounds

Gadolinium oxysulfide (Gd_2O_2S), also known as gadolinium sulfoxylate, GOS, or Gadox, is an inorganic compound, a mixed oxide–sulfide of gadolinium. The crystal structure of gadolinium oxysulfide has trigonal symmetry (space group $P\bar{3}m1$). Each gadolinium ion is coordinated by four oxygen atoms and three sulfur atoms in a noninversion symmetric arrangement. The Gd_2O_2S structure is a sulfur layer with double layers of gadolinium and oxygen in between. The principal application of gadolinium oxysulfide is in ceramic scintillators. Scintillators are employed in radiation detectors for medical diagnostics. The scintillator is the main radiation sensor that emits light when struck by high-energy photons. There are two main drawbacks to this scintillator: (1) the hexagonal crystal structure, which emits only optical translucency and (2) low external light collection at the photodiode. The other disadvantage is the high X-ray damage to the sample. Gadolinium gallium garnet (GGG, $Gd_3Ga_5O_{12}$) is a synthetic crystalline material of the garnet group, with good mechanical, thermal, and optical properties. It is typically colorless. It has a cubic lattice, a density of 7.08 g/cm^3, and Mohs hardness between 6.5 and 7.5. Its crystals are formed using the Czochralski method. During the synthesis, various dopants can be added for color modification. The material is also employed in the manufacture of various optical components and as a substrate material for magneto-optical films (magnetic bubble memory). It is also used in jewelry as a diamond simulant. GGG can also be used as a seed substrate for the growth of other garnets such as YIG (yttrium iron garnet, $Y_3Fe_5O_{12}$). Gadolinium oxyorthosilicate (GSO) is a type of scintillating inorganic crystal used for imaging in nuclear medicine and for calorimetry in particle physics. The formula is Gd_2SiO_5.

7.7.3.5 Organogadolinium compounds

Gadolinium(III), like most lanthanide ions, forms complexes with high coordination numbers. This tendency is illustrated by the use of the chelating agent DOTA (dodecane tetraacetic acid, IUPAC name 1,4,7,10-tetraazacyclododecane-1,4,7,10-tetraacetic acid, $C_{16}H_{28}N_4O_8$), an octadentate ligand. Salts of $[Gd(DOTA)]^-$ are useful in MRI. A variety of related chelate complexes have been developed, including gadodiamide. Gadobenic acid (INN, trade name MultiHance) is a complex of gadolinium with the ligand BOPTA (4-carboxy-5,8,ll-tris(carboxymethyl)-l-phenyl-2-oxa-5,8,ll-triazatridecan-13-oic acid. See structure below). In the form of the methylglucamine salt meglumine gadobenate (INNm) or gadobenate dimeglumine (USAN), it is used as a gadolinium-based MRI contrast medium. BOPTA is a derivative of DTPA (pentetic acid or diethylenetriaminepentaacetic acid, IUPAC name 2-[bis[2-[bis(carboxymethyl)amino]ethyl]amino]acetic acid, $C_{14}H_{23}N_3O_{10}$) in which the one-terminal carboxyl group, -C(O)OH, is replaced by $-C-O-CH_2C_6H_5$.

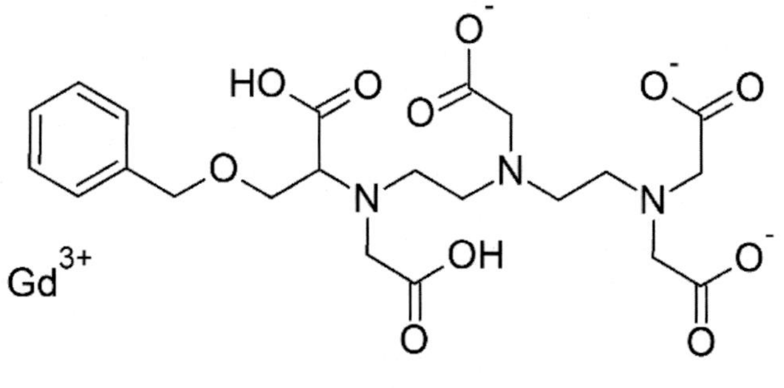

Thus gadobenic acid is closely related to gadopentetic acid. Gadopentetic acid is one of the trade names for a gadolinium-based MRI contrast agent (GBCA), usually administered as a salt of a complex of gadolinium with DTPA (diethylenetriaminepentacetate) with the chemical formula $A_2[Gd(DTPA)(H_2O)]$; where cation A is the protonated form of the amino sugar meglumine; the salt goes under the name "gadopentetate dimeglumine." In the "gadobenate" ion gadolinium ion is nine-coordinated with BOPTA acting as an eight-coordinating ligand. The ninth position is occupied by a water molecule, which exchanges rapidly with water molecules in the immediate vicinity of the strongly paramagnetic complex, providing a mechanism for MRI contrast enhancement. Gadoteric acid (gadoterate meglumine, trade names Artirem, Dotarem and Clariscan) is a macrocycle-structured GBCA. It consists of the organic acid DOTA as a chelating agent, and gadolinium (Gd^{3+}), and is used in form of the meglumine salt (Gadoterate meglumine).

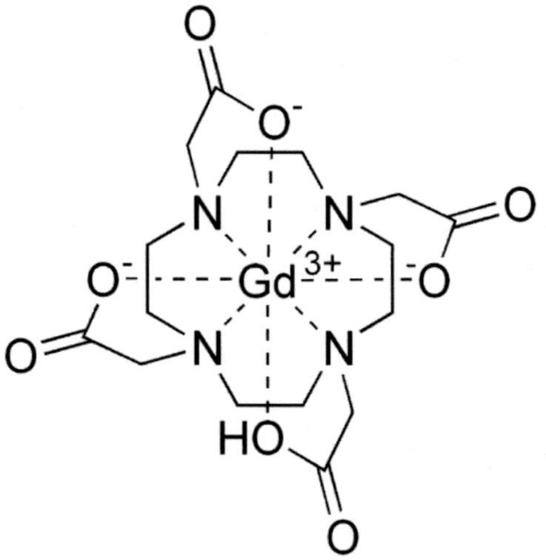

There exist a number of other gadolinium-based organic compounds that are used as MRI contrast agents, for example, gadobutrol, gadodiamide, gadofosveset, and gadoteridol. Gadolinium acetylacetonate is a compound with formula $Gd(C_5H_7O_2)_3$. It is the gadolinium complex of acetylacetone. Gadolinium acetylacetonate, along with cerium acetylacetonate, can be used as precursors to synthesize gadolinia-doped ceria (GDC) gel powders using the sol-gel method.

7.7.4 Major uses

Gd has no large-scale uses, but it has a number of specialized applications. Since ^{157}Gd possesses a high neutron cross-section, it is utilized to target tumors in neutron therapy. Gd is effective for use with neutron radiography and in shielding of nuclear reactors. It is uniquely suited for this purpose with two isotopes, ^{155}Gd and ^{157}Gd, that have excellent neutron-absorbing capabilities. The latter is four times more absorbing than the former and more than 300 times more absorbing than B, the other element that is employed for this purpose. Therefore about 5% of Gd oxide is intimately blended with the ^{235}U oxide pellets used in the reactors; this makes much more effective use of the nuclear fuel. It is employed as a secondary, emergency shut down measure in certain nuclear reactors, in particular of the CANDU reactor type (The CANDU (Canada Deuterium Uranium) is a Canadian pressurized heavy-water reactor design used to generate electric power. The acronym refers to its deuterium oxide (heavy water, D_2O or ^2H_2O) moderator and its use of (originally, natural) uranium fuel.). Gd is also utilized in nuclear marine propulsion systems as a burnable poison. To control large quantities of excess fuel reactivity without control rods, burnable poisons are loaded into the core. Burnable poisons are compounds that have a high neutron absorption cross-section that are converted into compounds of relatively low absorption cross-section due to neutron absorption. Due to the burn-up of the poison material, the negative reactivity of the burnable poison decreases over core life. If possible, these poisons should decrease their negative reactivity at the same rate that the fuel's excess positive reactivity is depleted.

Gd has unusual metallurgic properties, with as little as 1% of Gd enhancing the workability and resistance of Fe, Cr, and related alloys to high temperatures and oxidation. Metallic Gd is seldom used as the metal itself, but its alloys are required to make magnets and electronic components, for example, the recording heads for video recorders. In addition, the alloys are employed for magnetic data storage disks, which can be either permanent or erasable. These consist of ultrathin layers of several metals deposited on glass or plastic of which Gd is an important component, along with other rare-earth metals, Fe and Co. Gd is paramagnetic at room temperature, with a ferromagnetic Curie point of 20°C (In physics and materials science, the Curie temperature (TC), or Curie point, is the temperature above which certain materials lose their permanent magnetic properties, which can (in most cases) be replaced by induced magnetism. The Curie temperature is named after French physicist Pierre Curie (May 15, 1859 to April 19, 1906), who showed that magnetism was lost at a critical temperature.). Paramagnetic ions, such as Gd, improve nuclear relaxation rates, making Gd valuable for MRI. Solutions of organic Gd complexes and Gd compounds are employed as intravenous MRI contrast agent to improve images in medical MRI and magnetic resonance angiography (MRA) procedures. Magnevist (brand of gadopentetate dimeglumine) is the most common example. Nanotubes packed with Gd, known as "gadonanotubes," are 40 times more effective than the normal Gd contrast agents. After injection, Gd-based contrast agents accumulate in abnormal tissues of the brain and body providing a better image contrast between normal and abnormal tissues, helping the location of abnormal cell growths and tumors. Gd as a phosphor is also employed in other imaging methods. In X-ray systems Gd is present in the phosphor layer, suspended in a polymer matrix at the detector. Tb-doped Gd oxysulfide (Gd_2O_2S:Tb) at the phosphor layer changes the X-rays released from the source into light. This material emits green light at 540 nm because of the presence of Tb^{3+}, which is very beneficial for improving the imaging quality. The energy conversion of Gd is up to 20%, that is, one-fifth of the X-ray energy hitting the phosphor layer can be changed into visible photons. Gd oxyorthosilicate (Gd_2SiO_5, GSO; typically doped with 0.1%–1% of Ce) is a single crystal that is utilized as a scintillator in medical imaging such as positron emission tomography (PET) or for detecting neutrons.

Gd compounds were also used for making green phosphors for color TV tubes. ^{153}Gd is formed in a nuclear reactor from elemental Eu or enriched Gd targets. It has a half-life of 240 ± 10 days and emits γ radiation with strong peaks at 41 and 102 keV. It is used in a large number of quality-assurance applications, for example, line sources and calibration phantoms, to ensure that nuclear medicine imaging systems function correctly and produce useful images of radioisotope distribution inside the patient. It is also employed as a γ-ray source in X-ray absorption measurements or in bone density gages for osteoporosis screening (Dual-energy X-ray absorptiometry (DXA, previously DEXA) is a means of measuring bone mineral density (BMD). Two X-ray beams, with different energy levels, are aimed at the patient's bones. When soft tissue absorption is subtracted out, BMD can be determined from the absorption of each beam by bone. DXA is the most widely used and most thoroughly studied bone density measurement technology.), in addition to its use in the Lixiscope (Low Intensity X-ray Imaging Scope) portable X-ray imaging system developed by NASA. Gd is used for making gadolinium yttrium garnet (Gd:$Y_3Al_5O_{12}$); it has microwave uses and is utilized in making several optical components and as substrate material for magneto-optical films. GGG ($Gd_3Ga_5O_{12}$) was used for imitation diamonds and for computer bubble memory (a type of nonvolatile computer memory, which uses a thin film of a magnetic material to hold small magnetized areas, called bubbles or domains, each storing one bit of data. The material is arranged to form a series of parallel tracks that the bubbles can move along under the action of an external magnetic field. The bubbles are read by moving them to the edge of the material where they can be read by a conventional magnetic pickup and then rewritten on the far edge to keep the memory cycling through the material. In operation, bubble memories are similar to delay line memory systems. Bubble memory started out as a promising technology in the 1980s, offering memory density of an order similar to hard drives but performance more comparable to core memory while lacking any moving parts. This led many to contemplate it a candidate for a "universal memory" that could be used for all storage needs. The introduction of dramatically faster semiconductor memory chips pushed bubble into the slow end of the scale, and similarly dramatic increases in hard drive capacity made it uncompetitive in price terms.). Gd can also act as an electrolyte in solid oxide fuel cells (SOFCs, an electrochemical conversion device that produces electricity directly from oxidizing a fuel. Fuel cells are characterized by their electrolyte material; the SOFC has a solid oxide or ceramic electrolyte. SOFCs use a solid oxide electrolyte to conduct negative oxygen ions from the cathode to the anode. The electrochemical oxidation of the oxygen ions with hydrogen or carbon monoxide occurs on the anode side. More recently, proton-conducting SOFCs (PC-SOFC) are being developed which transport protons instead of oxygen ions through the electrolyte with the advantage of being able to be run at lower temperatures than traditional SOFCs.) Using Gd as a dopant for materials like Ce oxide (as Gd-doped ceria) produces an electrolyte with both high ionic conductivity and low operating temperatures, which are optimal for cost-effective production of fuel cells. Currently, magnetic refrigeration near room temperature is being studied, which could provide important efficiency and environmental advantages compared to conventional refrigeration methods (Magnetic refrigeration, or adiabatic demagnetization, is a cooling technology based on the magnetocaloric effect, an intrinsic property of magnetic solids. The refrigerant is often a paramagnetic salt, such as cerium magnesium nitrate. The active magnetic dipoles in this case are those of the electron shells of the paramagnetic atoms. A strong magnetic field is applied to the refrigerant, forcing its various magnetic dipoles to align and putting these degrees of freedom of the refrigerant into a state of lowered entropy. A heat sink then absorbs the heat released by the refrigerant due to its loss of entropy. Thermal contact with the heat sink is then broken so that the system is insulated, and the magnetic field is switched off. This increases the heat capacity of the refrigerant, thus decreasing its temperature below the temperature of the heat sink.). Gd-based compounds, such as $Gd_5(Si_xGe_{1-x})_4$, are at present the most promising materials, due to their high Curie temperature and huge magnetocaloric effect. Pure Gd itself exhibits a large magnetocaloric effect near its Curie temperature of 20°C, and this has started a major search for Gd alloys with a larger effect and tunable Curie temperature. In $Gd_5(Si_xGe_{1-x})_4$, Si and Ge ratios can be varied to change the Curie temperature. This technology is still very early in development, and substantial material developments still are necessary before it is commercially viable.

7.8 66 Dy — Dysprosium

7.8.1 Discovery

In 1878 erbium ores were discovered to comprise the oxides of holmium and thulium. French chemist Paul-Émile Lecoq de Boisbaudran (April 18, 1838 to May 28, 1912), while studying holmium oxide, separated dysprosium oxide from it in Paris in 1886 (Lecoq de Boisbaudran, 1886) (Fig. 7.42). His procedure for separating the dysprosium involved dissolving dysprosium oxide in acid and then adding ammonia to precipitate the hydroxide. He was only successful in separating dysprosium from its oxide after more than 30 attempts at his procedure. Subsequently, he called the element dysprosium from the Greek dysprositos (δυσπρόσιτος), meaning "hard to get." The element was not obtained in relatively pure form until after the development of ion exchange techniques using ion exchange resins (a resin or polymer that acts as a medium for ion exchange. It is an insoluble matrix (or support structure) normally in the form of small (0.25–0.5 mm radius) microbeads fabricated from an organic polymer substrate. The beads are typically porous, providing a large surface area on and inside them. The trapping of ions occurs along with the accompanying release of other ions, and thus the process is called ion exchange.) by Canadian American chemist Frank Spedding (October 22, 1902 to December 15, 1984) at Iowa State University in the early 1950s (Spedding et al., 1958). He was a renowned expert on REEs and on extraction of metals from minerals. The uranium extraction process helped make it possible for the Manhattan Project to build the first atomic bombs.

7.8.2 Mining, production, and major minerals

The concentration of Dy in the Earth's crust is about 5.2 ppm. Although dysprosium has never been found as the element, it is found in small amounts in many minerals, for example, xenotime ($(Y,Yb)PO_4$), fergusonite ($(Ce,La,Y)NbO_4$), gadolinite ($(Ce,Y)_2FeBe_2Si_2O_{10}$), euxenite-(Y) ($(Y,Ca,Ce,U,Th)(Nb,Ta,Ti)_2O_6$), polycrase-(Y) ($(Y,Ca,Ce,U,Th)(Ti,Nb,Ta)_2O_6$), aeschynite ($(Ce,Nd,Y,Ca,Fe,Th)(Ti,Nb)_2(O,OH)_6$), monazite ($(Ce,La,Nd,Th,Sm,Gd,Ca)PO_4$), and bastnäsite ($(Ce,La,Y)CO_3F$), frequently together with erbium and holmium or other REEs. No dysprosium-dominant mineral (i.e., with dysprosium prevailing over other rare earths in the composition) has yet been found. In the high-yttrium varieties of these minerals, dysprosium tends to occur as the most abundant of the heavy lanthanides, comprising up to 7%–8% of the concentrate (as compared to about 65% for yttrium). Dysprosium is produced mainly from monazite sand, a mixture of various phosphates. The metal is produced as a by-product in the commercial extraction of yttrium. In separating dysprosium, most of the unwanted metals can be removed magnetically or by a flotation method. Dysprosium can subsequently be isolated from other rare earth metals present by an ion exchange displacement method. The ensuing dysprosium ions can then react with either fluorine or chlorine to form dysprosium fluoride, DyF_3, or dysprosium chloride, $DyCl_3$. Finally, these fluoride or chloride compounds are reduced by either calcium or lithium metals.

$$3Ca + 2DyF_3 \rightarrow 2Dy + 3CaF_2$$
$$3Li + DyCl_3 \rightarrow Dy + 3LiCl$$

FIGURE 7.42 Paul-Émile Lecoq de Boisbaudran.

The components are placed in a tantalum crucible and heated in a helium atmosphere. As the reaction progresses, the subsequent halide compounds and molten dysprosium separate because of their differences in density. After the mixture has cooled, the dysprosium can be cut away from the impurities.

Three minerals are known with significant amounts of dysprosium in the crystal structure. Among these are two carbonates, lepersonnite-(Gd) ($Ca(Gd,Dy)_2(UO_2)_{24}(SiO_4)_4(CO_3)_8(OH)_{24}\cdot 48H_2O$) and lokkaite-(Y) ($Ca(Y,Gd,Nd,Dy)_4(CO_3)_7\cdot 9H_2O$), and one silicate, yftisite-(Y) ($(Y,Dy,Er)_4(Ti,Sn)(SiO_4)_2O(F,OH)_6$).

7.8.3 Chemistry

Naturally occurring Dy consists of seven isotopes: ^{156}Dy (0.056% natural abundance), ^{158}Dy (0.095%), ^{160}Dy (2.329%), ^{161}Dy (18.889%), ^{162}Dy (25.475%), ^{163}Dy (24.896%), and ^{164}Dy (28.260%). They are all thought to be stable, though ^{156}Dy may theoretically decay via α decay with a half-life of more than 1×10^{18} years. Twenty-nine synthetic radioisotopes have also been characterized, ranging from ^{138}Dy to ^{173}Dy. The most stable of these radioisotopes is ^{154}Dy, which has a half-life of about 3×10^6 years, and ^{159}Dy, which has a half-life of 144.4 days. The least stable is ^{138}Dy, with a half-life of 200 ms. As a rule, radioisotopes lighter than the stable isotopes decay mainly via β^+ decay, while those that are heavier decay predominantly by β^- decay. Nevertheless, ^{154}Dy decays principally by α decay, and ^{152}Dy and ^{159}Dy decay principally by electron capture. In addition, Dy is known to have at least 11 metastable isomers, ranging in atomic mass from 140 to 165. The most stable of these is ^{165m}Dy with a half-life of 1.257 min. ^{149}Dy has two metastable isomers, the second of which, $^{149m2}Dy$, has a half-life of 28 ns.

Dysprosium is a REE that has a metallic, bright silver luster (Table 7.8). It is quite soft. Its physical characteristics can be significantly affected by even trace amounts of impurities. Dysprosium together with holmium has the highest magnetic strengths of the elements, in particular at low temperatures. Dysprosium has a simple

TABLE 7.8 Dysprosium properties.

Appearance	Silvery white
Standard atomic weight $A_{r,std}$	162.500
Block	f-Block
Element category	Lanthanide
Electron configuration	[Xe] $4f^{10}$ $6s^2$
Phase at STP	Solid
Melting point	1407°C
Boiling point	2562°C
Density (near r.t.)	8.540 g/cm^3
When liquid (at m.p.)	8.37 g/cm^3
Heat of fusion	11.06 kJ/mol
Heat of vaporization	280 kJ/mol
Molar heat capacity	27.7 J/(mol·K)
Oxidation states	+1, +2, **+3**, +4
Ionization energies	1st: 573.0 kJ/mol
	2nd: 1130 kJ/mol
	3rd: 2200 kJ/mol
Atomic radius	Empirical: 178 pm
Covalent radius	192 ± 7 pm

STP, Standard temperature and pressure.
Bold font indicates main oxidation state.

ferromagnetic ordering at temperatures below −188.2°C (85K). Above that temperature, it turns into a helical antiferromagnetic state in which all of the atomic moments in a particular basal plane layer are parallel and oriented at a fixed angle to the moments of adjacent layers. This unusual antiferromagnetism changes into a disordered (paramagnetic) state at −94°C (179K).

7.8.3.1 *Reactions with oxygen and water*

Dysprosium metal tarnishes slowly in air and burns readily to produce dysprosium(III) oxide.

$$4Dy + 3O_2 \rightarrow 2Dy_2O_3$$

Dysprosium oxide (Dy_2O_3) is a sesquioxide compound of the rare earth metal dysprosium. It is a pastel yellowish-greenish, slightly hygroscopic powder having specialized uses in ceramics, glass, phosphors, lasers, and dysprosium metal halide lamps. Dysprosium oxide, also known as dysprosia, is a white powder that is highly magnetic, more so than iron oxide. It can react with acids to produce the corresponding dysprosium(III) salts, e.g.

$$Dy_2O_3 + 6HCl \rightarrow 2DyCl_3 + 3H_2O$$

Dysprosium is fairly electropositive and reacts slowly with cold water (and rather quickly with hot water) to form dysprosium hydroxide.

$$2Dy(s) + 6H_2O(l) \rightarrow 2Dy(OH)_3(aq) + 3H_2(g)$$

7.8.3.2 *Reactions with halogens*

Dysprosium metal vigorously reacts with all the halogens at temperatures above 200°C.

$$2Dy\ (s) + 3F_2(g) \rightarrow 2DyF_3(s)\ (green)$$
$$2Dy\ (s) + 3Cl_2(g) \rightarrow 2DyCl_3(s)\ (white)$$
$$2Dy\ (s) + 3Br_2(g) \rightarrow 2DyBr_3(s)\ (white)$$
$$2Dy\ (s) + 3\ I_2(g) \rightarrow 2DyI_3(s)\ (green)$$

Dysprosium(III) chloride ($DyCl_3$) is a white to yellow solid, which rapidly absorbs water on exposure to moist air to form a hexahydrate, $DyCl_3 \cdot 6H_2O$. Simple rapid heating of the hydrate causes partial hydrolysis to an oxychloride, DyOCl. Dysprosium(III) chloride can be used as a starting point for the preparation of other dysprosium salts. Dysprosium metal is formed when a molten mixture of $DyCl_3$ in eutectic LiCl–KCl is electrolyzed. The reduction happens via Dy^{2+}, at a tungsten cathode. $DyCl_3$ is frequently synthesized using the "ammonium chloride route," starting from either Dy_2O_3 or hydrated chloride or oxychloride or $DyCl_3 \cdot 6H_2O$. These methods form $(NH_4)_2[DyCl_5]$ in the first step.

$$10NH_4Cl + Dy_2O_3 \rightarrow 2(NH_4)_2[DyCl_5] + 6NH_3 + 3H_2O$$
$$DyCl_3 \cdot 6H_2O + 2NH_4Cl \rightarrow (NH_4)_2[DyCl_5] + 6H_2O$$

In the next step the pentachloride is thermally decomposed.

$$(NH_4)_2[DyCl_5] \rightarrow 2NH_4Cl + DyCl_3$$

The thermolysis reaction proceeds via the intermediary $(NH_4)[Dy_2Cl_7]$. Reacting Dy_2O_3 with aqueous HCl results in the formation of hydrated chloride ($DyCl_3 \cdot 6H_2O$). This salt cannot be rendered anhydrous by heating. Instead it forms an oxychloride. Dysprosium(III) chloride is a moderately strong Lewis acid, which ranks as "hard" according to the HSAB concept. Aqueous solutions of dysprosium chloride can be used to prepare other dysprosium(III) compounds, for example, dysprosium(III) fluoride.

$$DyCl_3 + 3NaF \rightarrow DyF_3 + 3NaCl$$

7.8.3.3 Reactions with acids

Dysprosium dissolves readily in dilute sulfuric acid forming solutions with the yellow Dy(III) ions, which exist as a $[Dy(H_2O)_9]^{3+}$ complex.

$$2Dy(s) + 3H_2SO_4(aq) \rightarrow 2Dy^{3+}(aq) + 3SO_4^{2-}(aq) + 3H_2(g)$$

The formed compound, dysprosium(III) sulfate, is noticeably paramagnetic.

7.8.3.4 Other compounds

Dysprosium reacts with a variety of nonmetals at high temperatures to form binary compounds with varying composition and oxidation states +3 and sometimes +2, such as DyN, DyP, DyH_2, and DyH_3; DyS, DyS_2, Dy_2S_3, and Dy_5S_7; DyB_2, DyB_4, DyB_6, and DyB_{12}, as well as Dy_3C and Dy_2C_3. Dysprosium carbonate, $Dy_2(CO_3)_3$, and dysprosium sulfate, $Dy_2(SO_4)_3$, result from similar reactions. The majority of dysprosium compounds are soluble in water; however dysprosium carbonate tetrahydrate ($Dy_2(CO_3)_3 \cdot 4H_2O$) and dysprosium oxalate decahydrate ($Dy_2(C_2O_4)_3 \cdot 10H_2O$) are both insoluble in water. Two dysprosium carbonate compounds ($Dy_2(CO_3)_3 \cdot 2{-}3H_2O$) and ($DyCO_3(OH)$), are known to form via a poorly ordered (amorphous) precursor phase with a formula of $Dy_2(CO_3)_3 \cdot 4H_2O$. This amorphous precursor contains highly hydrated spherical nanoparticles of 10–20 nm diameter that are extremely stable under dry treatment at ambient and high temperatures. Dysprosium titanate ($Dy_2Ti_2O_7$) is an inorganic compound, a ceramic of the titanate family, with pyrochlore structure (Pyrochlore $(Na,Ca)_2Nb_2O_6(OH,F)$ is a mineral group of the niobium end member of the pyrochlore supergroup. The general formula, $A_2B_2O_7$ (A and B are metals), represents a family of phases isostructural to the mineral pyrochlore.). Dysprosium titanate, like holmium titanate and holmium stannate, is a spin ice material (A spin ice is a magnetic substance that does not have a single minimal-energy state. It has magnetic moments (i.e., "spin") as elementary degrees of freedom, which are subject to frustrated interactions. By their nature, these interactions prevent the moments from exhibiting a periodic pattern in their orientation down to a temperature much below the energy scale set by the said interactions. Spin ices show low-temperature properties, residual entropy in particular, closely related to those of common crystalline water ice. The most prominent compounds with such properties are dysprosium titanate ($Dy_2Ti_2O_7$) and holmium titanate ($Ho_2Ti_2O_7$). The orientation of the magnetic moments in spin ice resembles the positional organization of hydrogen atoms (more accurately, ionized hydrogen, or protons) in conventional water ice.) In 2009 quasiparticles resembling magnetic monopoles were observed at low temperature and high magnetic field (In particle physics, a magnetic monopole is a hypothetical elementary particle that is an isolated magnet with only one magnetic pole (a north pole without a south pole or vice versa). A magnetic monopole would have a net "magnetic charge." Modern interest in the concept stems from particle theories, notably the grand unified and superstring theories, which predict their existence.) Dysprosium titanate (Dy_2TiO_5) has been employed since 1995 as material for control rods of commercial nuclear reactors. Dysprosium acetylacetonate, also known as $Dy(acac)_3$, is a compound with formula $Dy(C_5H_7O_2)_3$. It is the dysprosium complex of acetylacetone.

7.8.4 Major uses

Dy is used, in combination with V and other elements, in producing laser materials and commercial lighting. Due to its high thermal-neutron absorption cross-section, Dy-oxide–Ni cermets (a composite material composed of ceramic (cer) and metal (met) materials) are employed in neutron-absorbing control rods in nuclear reactors. Dy–Cd chalcogenides form sources of infrared radiation, which is suitable for the research of chemical reactions. Since Dy and its compounds are highly susceptible to magnetization, they are used in many data storage applications, such as in hard disks. Dy is more and more sought-after for the permanent magnets used in electric car motors and wind turbine generators. Nd–Fe–B magnets can have up to 6% of the Nd replaced by Dy to increase the coercivity for demanding applications (in electrical engineering and materials science, the coercivity, also called the magnetic coercivity, coercive field, or coercive force, is a measure of the ability of a ferromagnetic material to withstand an external magnetic field without becoming demagnetized), such as drive motors for electric vehicles and generators for wind turbines. This substitution would require up to 100 g of Dy per electric car produced. This may rapidly exhaust its available supply in the near future. The Dy substitution may similarly be valuable in other applications as it enhances the corrosion resistance of the magnets. Dy is one of the elements of Terfenol-D, along with Fe and Tb (an alloy of the formula $Tb_xDy_{1-x}Fe_2$ ($x \sim 0.3$), is a magnetostrictive material. It was initially developed in the 1970s by the Naval Ordnance Laboratory (NOL) in the United States. The technology for manufacturing the material efficiently was developed in the 1980s at Ames Laboratory under a US Navy funded program. It is named after terbium (Ter), iron (Fe), NOL, and the D comes from dysprosium.) Terfenol-D has the highest room-temperature magnetostriction of any known material, which is employed in transducers, wide-band mechanical resonators, and high-precision liquid-fuel injectors. Magnetostriction is a property of ferromagnetic materials that causes them to change their shape or dimensions during the process of magnetization. The variation of materials' magnetization due to the applied magnetic field changes the magnetostrictive strain until reaching its saturation value, λ. The effect was first identified in 1842 by English physicist, mathematician, and brewer James Prescott Joule (December 24, 1818 to October 11, 1889) when observing a sample of iron. Joule studied the nature of heat and discovered its relationship to mechanical work. This led to the law of conservation of energy, which in turn led to the development of the first law of thermodynamics. The SI derived unit of energy, the joule, is named after him.

Dy is utilized in dosimeters for determining ionizing radiation. $CaSO_4$ or CaF_2 crystals are doped with Dy. When these crystals are exposed to radiation, the Dy atoms become excited and luminescent. The luminescence can be measured to calculate the degree of exposure to which the dosimeter was subjected. Nanofibers of Dy compounds exhibit high strength and a large surface area. Hence, they can be employed to reinforce other materials and serve as a catalyst. Fibers of Dy oxide fluoride can be created by heating an aqueous solution of $DyBr_3$ and NaF to 450°C at 450 bars for 17 h. This material is unusually strong, surviving for more than 100 h in a variety of aqueous solutions at temperatures over 400°C without redissolving or aggregating. DyI_3 and $DyBr_3$ is utilized in high-intensity metal halide lamps. These compounds dissociate close to the hot center of the lamp, discharging isolated Dy atoms. The latter re-emit light in the green and red part of the spectrum, thus effectively producing bright light. Some paramagnetic crystal salts of Dy (dysprosium gallium garnet, dysprosium aluminum garnet, and dysprosium iron garnet) are employed in adiabatic demagnetization refrigerators. Magnetic refrigeration is a cooling technology based on the magnetocaloric effect. This technique can be used to attain extremely low temperatures, as well as the ranges used in common refrigerators. The effect was first observed by a German physicist Emil Gabriel Warburg (March 9, 1846 to July 28, 1931) in 1881 and subsequently by French physicist Pierre-Ernest Weiss (March 25, 1865 to October 24, 1940) and Swiss physicist, inventor, and explorer Auguste Piccard (January 28, 1884 to March 24, 1962) in 1917. The fundamental principle was suggested by Dutch-American physicist and physical chemist, and Nobel laureate in Chemistry Peter Joseph William Debye (March 24, 1884 to November 2, 1966) in 1926 and American chemist and Nobel laureate William Francis Giauque (May 12, 1895 to March 28, 1982) in 1927. The first working magnetic refrigerators were constructed by several groups beginning in 1933. Magnetic refrigeration was the first method developed for cooling below about -272.9°C (0.3K) (a temperature attainable by ^3He refrigeration, i.e., pumping on the ^3He vapors). The trivalent Dy ion (Dy^{3+}) has been subject of research because of its downshifting luminescence properties. Dy-doped yttrium aluminum garnet (YAG:Dy) excited in the UV region of the electromagnetic spectrum results in the emission of photons of longer wavelength in the visible region. This idea forms the foundation for a new generation of UV-pumped white light emitting diodes. Dy oxide is employed as a dopant in special ceramics, for example, $BaTiO_3$, which is used to make small capacitors with high capacitance, for electronic applications. Other applications of Dy are in erasable optical laser-read disks and in temperature-compensating capacitors.

7.9 70 Yb — Ytterbium

7.9.1 Discovery

Ytterbium was first found by the Swiss chemist Jean Charles Galissard de Marignac (April 24, 1817 to April 15, 1894) in 1878. While studying gadolinite ($(Ce,Y)_2FeBe_2Si_2O_{10}$), Marignac discovered a new component in the earth at that time known as erbia, and he called it ytterbia, for Ytterby, the Swedish village near the mine where he found the new component of erbium. Marignac thought that ytterbia was a compound of a new element that he named "ytterbium." In 1907 the French chemist Georges Urbain (April 12, 1872 to November 5, 1938) separated Marignac's ytterbia into two different components, which he named neoytterbia and lutecia (Urbain, 1908). Neoytterbia later was renamed to ytterbium, while lutecia became known as the element lutetium. Simultaneously, the Austrian chemist and inventor Carl Auer von Welsbach (September 1, 1858 to August 4, 1929) (Fig. 7.43) independently separated these elements from ytterbia, but he named them aldebaranium and cassiopeium; the American (of British origin) chemist Charles James (April 27, 1880 to December 10, 1928) likewise separated these elements around the same time. Urbain and Welsbach pointed the finger at each other for publishing results based on the work of the other party (Urbain, 1909, von Welsbach, 1908). The Commission on Atomic Mass, consisting of American chemist Frank Wigglesworth Clarke (March 19, 1847 to May 23, 1931), German chemist Friedrich Wilhelm Ostwald (September 2, 1853 to April 4, 1932), and Georges Urbain, which was at that time in charge of the attribution of new element names, settled the disagreement in 1909 by granting priority to Urbain and approving his names as the official element names, since the separation of lutetium from Marignac's ytterbium was first described by Urbain. After Urbain's names were officially accepted,

FIGURE 7.43 Carl Auer von Welsbach, 1910.

neoytterbium was reverted to ytterbium. The chemical and physical properties of ytterbium could not be obtained with any precision until 1953, when the first almost pure ytterbium metal was obtained by using ion exchange processes.

7.9.2 Mining, production, and major minerals

The abundance of ytterbium in the Earth's crust is around 3 ppm. Ytterbium is found with other REEs in several rare minerals. It is most often recovered commercially from monazite ($(Ce,La,Nd,Th,Sm,Gd,Ca)PO_4$) sand (0.03% ytterbium). The element is also found in euxenite-(Y) ($(Y,Ca,Ce,U,Th)(Nb,Ta,Ti)_2O_6$) and xenotime ($(Y,Yb)PO_4$). The major mining areas are found in China, the United States, Brazil, India, Sri Lanka, and Australia. The world production of ytterbium is only about 50 tons per year, reflecting that it has limited commercial applications. Reserves of ytterbium are estimated as 1 million tons. As an even-numbered lanthanide, in accordance with the Oddo–Harkins rule (The Oddo–Harkins rule holds that an element with an even atomic number (such as C: element 6) is more abundant than both elements with the adjacently larger and smaller odd atomic numbers (such as B: element 5 and N: element 7, respectively for C). This tendency of the abundance of the chemical elements was first reported by Italian chemist Giuseppe Oddo (June 9, 1865 to November 5, 1954) in 1914 and American chemist William Draper Harkins (December 28, 1873 to March 7, 1951) in 1917.), ytterbium is significantly more abundant than its immediate neighbors, thulium (Tm) and lutetium (Lu), which occur in the same concentration at levels of about 0.5% each. Ytterbium is often the most common substitute in yttrium minerals. In very few known cases/occurrences ytterbium prevails over yttrium, as, for example, in xenotime-(Yb). A report of native ytterbium from the Moon's regolith is known. It is quite difficult to isolate ytterbium from other lanthanides because of its similar properties. Consequently, the process is rather long. In a first step, minerals such as monazite or xenotime are dissolved into various acids, such as sulfuric acid. Ytterbium can subsequently be isolated from other lanthanides by ion exchange, as can other lanthanides. The solution is applied to a resin, which binds the various lanthanides in different ways. This is subsequently exposed to various complexing agents, and because of the different bonding types exhibited by the different lanthanides, the compounds can be separated. Ytterbium is isolated from other REEs either by ion exchange or through reduction with sodium amalgam. In the latter process, a buffered acidic solution of trivalent (+3) REEs is reacted with a molten sodium–mercury alloy, which results in the reduction and dissolution of Yb^{3+}. The resulting alloy is then reacted with hydrochloric acid (HCl). The metal is separated from the solution as its oxalate and subsequently decomposed to its oxide upon heating. Finally, the oxide is reduced to metal by heating in the presence of lanthanum, aluminum, cerium, or zirconium in high vacuum. The metal is purified through sublimation and then collected over a condensed plate.

Five minerals are known with ytterbium. These include one oxide, samarskite-(Yb) ($(Yb,Y,Fe^{3+},Fe^{2+},U,Th,Ca)_2(Nb,Ta)_2O_8$), one phosphate, xenotime-(Yb) ($Yb(PO_4)$), and three silicates, hingganite-(Yb) ($(Yb,Y,REE)_2\square Be_2[SiO_4]_2(OH)_2$) (Fig. 7.44), keiviite-(Yb) ($Yb_2Si_2O_7$) (Fig. 7.45), and vyuntspakhkite-(Y) ($(Y,Yb)_4Al_{2.5\text{-}1.5}(Si,Al)_{1.5\text{-}2.5}(SiO_4)_4O(OH)_7$).

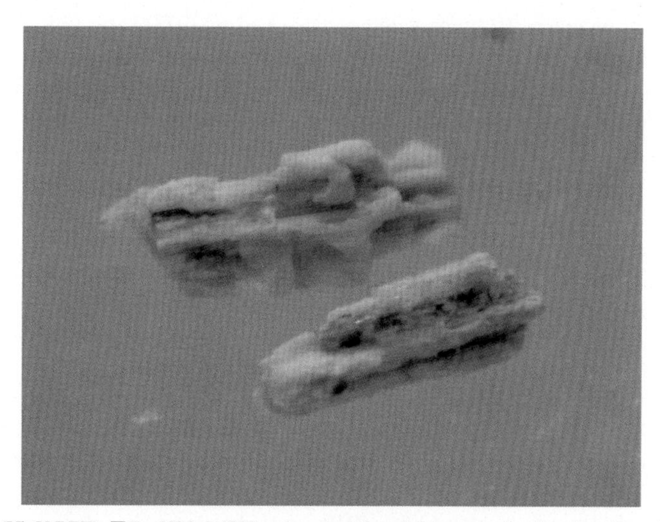

FIGURE 7.44 Hingganite-(Yb), $(Yb,Y,REE)_2\square Be_2[SiO_4]_2(OH)_2$, tiny, less than 0.75 mm grains. Two locality mineral—very rare. Tangen Feldspar quarry, Kragerø, Telemark, Norway.

FIGURE 7.45 Keiviite-(Yb), $Yb_2Si_2O_7$, tiny white crystal grains less than 0.4 mm in fluorite, CaF_2. Ploskaya Mt, Western Keivy Massif, Kola Peninsula, Russia.

7.9.3 Chemistry

Natural occurring Yb consists of seven stable isotopes: ^{168}Yb (0.126% abundance), ^{170}Yb (3.023%), ^{171}Yb (14.216%), ^{172}Yb (21.754%), ^{173}Yb (16.098%), ^{174}Yb (31.896%), and ^{176}Yb (12.887%). Twenty-seven synthetic radioisotopes have been characterized, with the most stable ones being ^{169}Yb which has a half-life of 32.026 days, ^{175}Yb which has a half-life of 4.185 days, and ^{166}Yb which has a half-life of 56.7 h. All other radioisotopes have half-lives less than 2 h, and the majority of them have half-lives less than 20 min. In addition, Yb has 12 meta states, with the most stable being ^{169m}Yb with a half-life of 46 s. The Yb isotopes range from ^{148}Yb to ^{181}Yb. The principal decay mode of radioisotopes lighter than ^{174}Yb is electron capture to produce Tm (thulium) isotopes, while the principal decay mode for the radioisotopes heavier than ^{174}Yb is β^- decay to produce Lu (lutetium) isotopes.

Ytterbium is a soft, malleable, and ductile chemical element that displays a bright silvery luster when pure (Table 7.9). It is a REE, and it is readily dissolved by strong mineral acids. Ytterbium has three allotropes labeled by the Greek letters α, β, and γ; their transformation temperatures are $-13°C$ and 795°C, though the exact transformation temperature depends on the pressure and stress. The β allotrope (6.966 g/cm^3) exists at room temperature, and it has a *fcc* crystal structure. The high-temperature γ allotrope (6.57 g/cm^3) has a *bcc* crystalline structure. The α allotrope (6.903 g/cm^3) has a hexagonal crystalline structure and is stable at low temperatures. The β allotrope possesses a metallic electrical conductivity at normal atmospheric pressure, but it becomes a semiconductor when exposed to a pressure of about 16,000 atm. (1.6 GPa). Its electrical resistivity increases 10 times when compressed to 39,000 atm (3.9 GPa), but subsequently drops to about 10% of its room-temperature resistivity at about 40,000 atm. (4.0 GPa). Unlike other rare-earth metals, which typically have antiferromagnetic and/or ferromagnetic properties at low temperatures, ytterbium is paramagnetic at temperatures above $-272°C$ (1.0K). Nevertheless, the α allotrope is diamagnetic. With a melting point of 824°C and a boiling point of 1196°C, ytterbium has the smallest liquid range of all the metals. Unlike most other lanthanides, which have a *hcp* lattice, ytterbium crystallizes in the *fcc* system. Ytterbium has a density of 6.973 g/cm^3, which is substantially lower than those of the neighboring lanthanides, thulium (9.32 g/cm^3) and lutetium (9.841 g/cm^3). Its melting and boiling points are also considerably lower than those of thulium and lutetium. This is caused by the closed-shell electron configuration of ytterbium ([Xe]$4f^{14}6s^2$), which causes only the two 6s electrons to be available for metallic bonding (unlike the other lanthanides where three electrons are available) and increases ytterbium's metallic radius.

7.9.3.1 Reactions with oxygen and water

Ytterbium metal tarnishes slowly in air. Finely dispersed ytterbium readily oxidizes in air and under oxygen to form ytterbium(III) oxide (Yb_2O_3), which crystallizes in the "rare-earth C-type sesquioxide" structure, which is related to the fluorite (CaF_2) structure with one-quarter of the anions removed, leading to ytterbium atoms in two different six-coordinate (nonoctahedral) environments. Ytterbium(III) oxide can be reduced to ytterbium(II) oxide (YbO) with elemental ytterbium, which crystallizes in the same structure as sodium chloride (NaCl). Mixtures of powdered ytterbium with polytetrafluoroethylene or hexachloroethane burn with a luminous emerald-green flame. Ytterbium reacts

TABLE 7.9 Ytterbium properties.

Appearance	Silvery white, with a pale-yellow tint
Standard atomic weight $A_{r,std}$	173.045
Block	f-Block
Element category	Lanthanide
Electron configuration	[Xe] $4f^{14}$ $6s^2$
Phase at STP	Solid
Melting point	824°C
Boiling point	1196°C
Density (near r.t.)	6.90 g/cm^3
When liquid (at m.p.)	6.21 g/cm^3
Heat of fusion	7.66 kJ/mol
Heat of vaporization	129 kJ/mol
Molar heat capacity	26.74 J/(mol·K)
Oxidation states	+1, +2, **+3**
Ionization energies	1st: 603.4 kJ/mol 2nd: 1174.8 kJ/mol 3rd: 2417 kJ/mol
Atomic radius	Empirical: 176 pm
Covalent radius	187 ± 8 pm

STP, Standard temperature and pressure.
Bold font indicates main oxidation state.

with hydrogen to form various nonstoichiometric hydrides. Ytterbium dissolves slowly in water, but rapidly in acids, liberating hydrogen gas. Ytterbium is fairly electropositive, and it reacts slowly with cold water and rather rapidly with hot water to form ytterbium(III) hydroxide.

$$2Yb(s) + 6H_2O(l) \rightarrow 2Yb(OH)_3(aq) + 3H_2(g)$$

The ytterbium(III) ion absorbs light in the near-infrared range of wavelengths, but not in visible light, so ytterbia, Yb_2O_3, is white in color and the salts of ytterbium are also colorless.

7.9.3.2 Reactions with halogens

Ytterbium reacts with all the halogens.

$$2Yb\ (s) + 3F_2(g) \rightarrow 2YbF_3(s)\ (white)$$
$$2Yb\ (s) + 3Cl_2(g) \rightarrow 2YbCl_3(s)\ (white)$$
$$2Yb\ (s) + 3Br_2(g) \rightarrow 2YbBr_3(s)\ (white)$$
$$2Yb\ (s) + 3I_2(g) \rightarrow 2YbI_3(s)\ (white)$$

$YbCl_3$ can be prepared from Yb_2O_3 with either high-temperature carbon tetrachloride gas, or hot hydrochloric acid followed by drying at high temperature.

$$2Yb_2O_3 + 3CCl_4(g) \rightarrow 4YbCl_3(s) + 3CO_2(g)$$
$$Yb_2O_3 + 6HCl(g) \rightarrow 2YbCl_3(s) + 3H_2O$$

In practice there are better ways to prepare $YbCl_3$ for lab use. The aqueous HCl/ammonium chloride route is very effective.

$$10NH_4Cl + Yb_2O_3 \rightarrow 2(NH_4)_2[YbCl_5] + 6NH_3 + 3H_2O$$

In the next step the pentachloride is thermally decomposed.

$$(NH_4)_2[YbCl_5] \rightarrow 2NH_4Cl + YbCl_3$$

Alternatively, hydrated $YbCl_3$ may be dehydrated using a variety of reagents, particularly trimethylsilyl chloride. Other methods have been reported including reacting the finely powdered metal with mercuric chloride at high temperature in a sealed tube. A number of routes to solvated $YbCl_3$ have been published, including reaction of the metal with various halocarbons in the presence of a donor solvent such as THF or dehydration of the hydrated chloride using trimethylsilyl or thionyl chloride, again in a solvent such as THF.

The valence electron configuration of Yb^{3+} (from $YbCl_3$) is $4f^{13}5s^{2}5p^{6}$, which has critical implications for the chemical behavior of Yb^{3+}. In addition, the size of Yb^{3+} dictates its catalytic behavior and biological applications. For instance, while both Ce^{3+} and Yb^{3+} have a single unpaired f electron, Ce^{3+} is much larger than Yb^{3+} as lanthanides become much smaller with increasing effective nuclear charge because of the f electrons not being as well shielded as the d electrons. This behavior is known as the lanthanide contraction. The small size of Yb^{3+} results in fast catalytic behavior and an atomic radius (0.99 Å) similar to many biologically important ions. The gas-phase thermodynamic properties of this compound are difficult to ascertain since the compound can disproportionate to form $[YbCl_6]^{3-}$ or dimerize. The Yb_2Cl_6 species was detected by electron impact (EI) mass spectrometry as ($Yb_2Cl_5^+$). Other difficulties in obtaining experimental data arise from the numerous low-lying f–d and f–f electronic transitions. Notwithstanding these problems, the thermodynamic properties of $YbCl_3$ have been determined and the C_{3V} symmetry group has been assigned based on the four active infrared vibrations.

Ytterbium forms both dihalides and trihalides with the halogens fluorine, chlorine, bromine, and iodine. The dihalides are susceptible to oxidation to the trihalides at room temperature and disproportionate to the trihalides and metallic ytterbium at high temperature.

$$3YbX_2 \rightarrow 2YbX_3 + Yb\ (X = F,\ Cl,\ Br,\ I)$$

Ytterbium(II) chloride ($YbCl_2$) is an inorganic chemical compound. It was first prepared in 1929 through reduction of ytterbium(III) chloride, $YbCl_3$, using hydrogen.

$$2YbCl_3 + H_2 \rightarrow 2YbCl_2 + 2HCl$$

Like other Yb(II) compounds and other low-valence rare earth compounds, $YbCl_2$ is a strong reducing agent. It is unstable in aqueous solution, reducing water to hydrogen gas. Some ytterbium halides are used as reagents in organic synthesis. For instance, ytterbium(III) chloride ($YbCl_3$) is a Lewis acid and can be used as a catalyst in the Aldol and Diels–Alder reactions. Ytterbium(II) iodide (YbI_2) may be employed, like samarium(II) iodide, as a reducing agent for coupling reactions. Ytterbium(III) fluoride (YbF_3) is utilized as an inert and nontoxic tooth filling because it continuously releases fluoride ions, which are thought to be good for dental health, and is also a good X-ray contrast agent.

7.9.3.3 Reactions with acids

Ytterbium dissolves readily in dilute sulfuric acid forming solutions with the colorless Yb(III) ions, which exist as nonahydrate complexes.

$$2Yb(s) + 3H_2SO_4(aq) + 18H_2O(l) \rightarrow 2\left[Yb(H_2O)_9\right]^{3+}(aq) + 3SO_4^{2-}(aq) + 3H_2(g)$$

7.9.3.4 Yb(II) versus Yb(III)

While generally trivalent, ytterbium readily forms divalent compounds. This behavior is uncommon for lanthanides, which almost completely form compounds with an oxidation state of +3. The +2 state has a valence electron configuration of $4f^{14}$ since the fully filled f shell gives more stability. The yellow-green ytterbium(II) ion is a very strong reducing agent and decomposes water, releasing hydrogen gas, and thus only the colorless ytterbium(III) ion occurs in aqueous solution. Samarium and thulium behave in a similar way in the +2 state, but europium(II) is stable in aqueous solution. Ytterbium metal behaves comparable to europium metal and the alkaline earth metals, dissolving in ammonia to form blue electride salts (an electride is an ionic compound in which an electron is the anion; solutions of alkali metals in ammonia are electride salts).

7.9.4 Major uses

The ^{169}Yb isotope (with a half-life of 32 days), which is formed together with the short-lived ^{175}Yb isotope (half-life 4.2 days) through neutron activation during the irradiation of Yb in nuclear reactors, has been employed as a radiation source in portable X-ray machines. Similar to X-rays, the γ rays emitted by the source pass through soft tissues of the body but are blocked by bones and other dense materials. Hence, small ^{169}Yb samples emitting γ rays serve like miniature X-ray machines valuable for radiography of small objects. Experiments have proven that radiographs obtained with a ^{169}Yb source are about equal to those obtained with X-rays with energies between 250 and 350 keV. In addition, ^{169}Yb is utilized in nuclear medicine.

Yb clocks are the record holders for stability with ticks stable to within less than two parts in 1 quintillion (2×10^{-18}). The clocks developed at the National Institute of Standards and Technology (NIST) depend on about

10,000 rare-earth atoms cooled to 10 μK (10 millionths of a degree above absolute zero) and trapped in an optical lattice—a series of pancake-shaped wells made of laser light. Another laser that "ticks" 518 trillion times per second provokes a transition between two energy levels in the atoms. The high number of atoms is crucial to the clocks' high stability. Visible light waves oscillate faster than microwaves, and consequently optical clocks can be more accurate than Cs atomic clocks. The model with one single Yb ion caught in an ion trap is extremely accurate. The optical clock based on it is accurate to 17 digits after the decimal point. A set of two experimental atomic clocks based on Yb atoms at the NIST has set a record for stability. NIST physicists published in the August 22, 2013 issue of Science Express that the Yb clocks' ticks are stable to within less than two parts in 1 quintillion (1 followed by 18 zeros), about 10 times better than the earlier best results published for other atomic clocks. For comparison, these clocks would be accurate to less than a second for a period similar to the age of the universe.

Yb can also be employed as a dopant to help enhance the grain refinement, strength, and other mechanical properties of stainless steel. Certain Yb alloys have infrequently been utilized in dentistry. The Yb^{3+} ion is employed as a doping ion in active laser media, in particular in solid-state lasers (a laser that uses a gain medium that is a solid, rather than a liquid such as in dye lasers or a gas as in gas lasers. Semiconductor-based lasers are also in the solid state but are generally considered as a separate class from solid-state lasers) and double-clad fiber (DCF) lasers (DCF is a class of optical fiber with a structure comprising three layers of optical material rather than the usual 2. The inner-most layer is called the core. It is surrounded by the inner cladding, which is surrounded by the outer cladding. The three layers are made of materials with different refractive indices. There are two different types of DCFs. The first was established early in optical fiber history with the purpose of engineering the dispersion of optical fibers. In these fibers, the core carries the majority of the light, and the inner and outer cladding change the waveguide dispersion of the core-guided signal. The second kind of fiber was established in the late 1980s for application with high-power fiber amplifiers and fiber lasers. In these fibers, the core is doped with active dopant material; it both guides and amplifies the signal light. The inner cladding and core together guide the pump light, which delivers the energy necessary to allow amplification in the core. In these fibers, the core has the highest refractive index and the outer cladding has the lowest. In most cases the outer cladding is made of a polymer material instead of glass.) Yb lasers are extremely efficient, have long lifetimes and can produce short pulses; in addition, Yb can with no trouble be incorporated into the material used to make the laser. Yb lasers usually radiate in the 1.06−1.12 μm band being optically pumped at wavelength 900 nm−1 μm, depending on the host and the application. The small quantum defect makes Yb a potential dopant for efficient lasers and power scaling [In laser science, the term "quantum defect" refers to the fact that the energy of a pump photon is generally higher than that of a signal photon (photon of the output radiation). The difference of energies goes to the heat; this heat may carry away the excess of entropy delivered with the multimode incoherent pump. The quantum defect of a laser can be defined as part of the energy of the pumping photon, which is lost (not turned into photons at the lasing wavelength) in the gain medium at the lasing. At a given frequency ω_p of pump and given frequency ω_s of lasing, the quantum defect $q = \hbar\omega_p - \hbar\omega_s$. Such a quantum defect has the dimension of energy; for the efficient operation, the temperature of the gain medium (measured in units of energy) should be small compared to the quantum defect. At a fixed pump frequency, the higher the quantum defect, the lower is the upper bound for the power efficiency.] The kinetic of excitations in Yb-doped compounds is simple and can be explained within the concept of effective cross-sections; for most Yb-doped laser compounds (as for numerous other optically pumped gain media), the McCumber relation holds [The McCumber relation or the McCumber theory is a relationship between the effective cross-sections of absorption and emission of light in the physics of solid-state lasers. It is named after American physicist Dean Everett McCumber (born November 25, 1930), who proposed the relationship in 1964. It is typical that the lasing properties of a medium are determined by the temperature and the population at the excited laser level and are not sensitive to the method of excitation used to achieve it. In this case, the absorption cross-section $\sigma_a(\omega)$ and the emission cross-section $\sigma_e(\omega)$ at frequency ω can be related to the lasers gain $G(\omega)$ in such a way that the gain at this frequency can be determined as follows:

$$G(\omega) = N_2\sigma_e(\omega) - N_1\sigma_a(\omega)$$

D.E. McCumber had postulated these properties and found that the emission and absorption cross-sections are not independent], even though the application to the Yb-doped composite materials has been a matter of debate. Typically, low Yb concentrations are used. At high concentrations, the Yb-doped compounds show photodarkening (glass fibers) or even a switch to broadband emission (crystals and ceramics) rather than efficient laser action [Photodarkening is an optical effect detected in the interaction of laser radiation with amorphous media (glasses) in optical fibers. Up till now, such formation of color centers was reported only in glass fibers. Photodarkening limits the density of excitations in fiber lasers and amplifiers. Experimental results indicate that operating at a saturated regime helps to reduce photodarkening.] This effect may be connected to not only overheating but also conditions of charge compensation at high concentrations of Yb ions. Significant progress has been achieved in the power scaling lasers and amplifiers produced with Yb-doped optical fibers. Power levels have increased from the 1 kW regimes because of the improvements in components along with the Yb-doped fibers. Production of low numerical apperture, large mode area (LMA) fibers allows to achieve near-perfect beam qualities ($M2 < 1.1$) at power levels of 1.5 kW to higher than 2 kW at ~ 1064 nm in a broadband configuration. In addition, Yb-doped LMA fibers have the advantages of a larger mode field diameter, which compensates for the impacts of nonlinear effects such as stimulated Brillouin scattering [named after French physicist Léon Brillouin (August 7, 1889 to October 4, 1969), refers to the interaction of light with the material waves in a medium. It is mediated by the refractive index dependence on the material properties of the medium; as described in optics, the index of refraction of a transparent material changes under deformation (compression-distension or shear-skewing). The result of the interaction between the light-wave and the carrier-deformation wave is that a fraction of the transmitted light-wave changes its momentum (and thus its frequency and energy) in preferential directions, as if by diffraction caused by an oscillating 3D diffraction grating.] and stimulated Raman scattering (Raman scattering or the Raman effect is the inelastic scattering of photons by matter, meaning that there is an exchange of energy and a change in the light's direction. Typically, this involves vibrational energy being gained by a molecule as incident photons from a visible laser are shifted to lower energy. This is called normal Stokes Raman scattering. The effect is exploited by chemists and physicists to gain information about materials for a variety of purposes by performing various forms of Raman spectroscopy. Rayleigh scattering was discovered and explained in the 19th century. The weaker Raman effect is named after Indian physicist Sir Chandrashekhara Venkata Raman (November 7, 1888 to November 21, 1970), who discovered it in 1928 with assistance from his student Sir Kariamanickam Srinivasa Krishnan (December 4, 1898 to June 14, 1961). Raman was awarded the Nobel prize in Physics in 1930 for his discovery although Soviet physicist Grigory Samuilovich Landsberg (January 22, 1890 to February 2, 1957) and Soviet physicist of Belarusian-Jewish background Leonid Isaakovich Mandelstam (May 4, 1879 to November 27, 1944) observed the effect in crystals the same year as Raman. The effect had been predicted theoretically by Austrian theoretical physicist Adolf Gustav Stephan Smekal (September 12, 1895 to March 7, 1959) in 1923.), which limit the achievement of higher power levels and offer a discrete advantage over single mode Yb-doped fibers. So as to attain even higher power levels in Yb-based fiber systems, all aspects of the fiber must be taken into account. These can only be realized via optimization of all the Yb fiber parameters, ranging from the core background losses to the geometrical properties, with the purpose of reducing the splice losses within the cavity. Power scaling also requires optimization of matching passive fibers within the optical cavity. The optimization of the Yb-doped glass itself via host glass alteration with various dopants also plays a large role in decreasing the background loss of the glass, enhancements in slope efficiency of the fiber, and increased photodarkening performance, all of which contribute to increased power levels in 1-μm systems.

At present, Yb is being studied as a potential substitute for Mg in high-density pyrotechnic payloads for kinematic infrared decoy flares [A flare or decoy flare is an aerial infrared countermeasure used by a plane or helicopter to counter an infrared homing ("heat-seeking") surface-to-air missile or air-to-air missile. Flares are commonly composed of a pyrotechnic composition based on magnesium or other hot-burning metal, with burning temperature equal to or hotter than engine exhaust. The aim is to make the infrared-guided missile seek out the heat signature from the flare rather than the aircraft's engines.]. As Yb(III) oxide (Yb_2O_3) has a much higher emissivity in the infrared range than MgO, a higher radiant intensity is achieved with Yb-based payloads than those normally based on Mg/Teflon/Viton [MTV, a pyrolant, that is, an energetic material that generates hot flames upon combustion. Teflon and Viton are trademarks of DuPont for polytetrafluoroethylene, $(C_2F_4)_n$, and fluoroelastomer, $(CH_2CF_2)_n(CF(CF_3)CF_2)_n$.]. Similar to other REEs, Yb can be employed to dope phosphors, or for ceramic capacitors and other electronic devices, and it can even serve as an industrial catalyst. Yb has a single absorption band at 985 nm in the infrared spectral region and this has been utilized to convert radiant energy into electrical energy in devices that couple it to silicon photocells. The electrical conductivity of Yb varies rather unusual as pressure on the metal increases. At 16,000 atm it becomes a semiconductor and subsequently its electrical resistance increases as the pressure increases until it reaches 40,000 atm when the trend abruptly reverses and Yb starts to become more conducting. This property of Yb was used in stress gages designed to monitor the intense shock waves of nuclear explosions, as well as to monitor ground deformations from earthquakes.

7.10 72 Hf — Hafnium

7.10.1 Discovery

In his report on The Periodic Law of the Chemical Elements, in 1869, Russian chemist Dmitri Mendeleev (February 8, 1834 to February 2, 1907) had indirectly projected the presence of a heavier analog of titanium and zirconium. At the time of his formulation in 1871, Mendeleev thought that the elements were ordered by their atomic masses and positioned lanthanum (element 57) in the place below zirconium. The exact placement of the elements in the periodic table and the location of missing elements was achieved by measuring the specific weight of the elements and comparing the chemical and physical properties. The X-ray spectroscopic work performed by English physicist Henry Moseley (November 23, 1887 to August 10, 1915) in 1914 provided a direct relationship between spectral line and effective nuclear charge (Moseley, 1913,1914). This resulted in the nuclear charge, or atomic number of an element, being used to determine its place within the periodic table. Using this method, Moseley determined the number of lanthanides and showed the gaps in the atomic number sequence at numbers 43, 61, 72, and 75. The discovery of the gaps resulted in a widespread search for these missing elements. In 1914 a number of scientists claimed the discovery after Henry Moseley predicted the gap in the periodic table for the then-undiscovered element 72. French chemist Georges Urbain (April 12, 1872 to November 5, 1938) proclaimed that he discovered element 72 in REEs in 1907 and published his results on celtium in 1911 (Urbain, 1911). Neither the spectra nor the chemical behavior he reported agreed with those of the element found later, and hence his claim was denied after a long-lasting argument. The disagreement was partly due to the fact that the chemists preferred the chemical techniques that led to the discovery of celtium, whereas the physicists count on the use of the new X-ray spectroscopy method that showed that the compounds found by Urbain did not contain element 72. By early 1923 a number of scientists such as Danish physicist Niels Bohr (October 7, 1885 to November 18, 1962) and English chemist Charles R. Bury (1890–1968) suggested that element 72 should be similar to zirconium and thus was not part of the group of REEs. These ideas were based on Bohr's theories of the atom, the X-ray spectroscopy of Moseley, and the chemical arguments of Austrian-born British chemist Friedrich Paneth (August 31, 1887 to September 17, 1958). Stimulated by these ideas and by the re-emergence in 1922 of Urbain's claims that element 72 was a REE discovered in 1911, Dutch physicist Dirk Coster (October 5, 1889 to February 12, 1950) (Fig. 7.46) and Hungarian radiochemist Georg von Hevesy (György Károly Hevesy, August 1, 1885 to July 5, 1966) (Fig. 7.47) were inspired to look for the new element in zirconium ores. Hafnium was found by the two scientists in 1923 in Copenhagen, Denmark, confirming the original 1869 prediction of Mendeleev (Coster and von Hevesy, 1923; von Hevesy, 1923, 1925). It was in the end found in zircon from Norway through X-ray spectroscopy analysis. The city

FIGURE 7.46 Dirk Coster, c.1930s.

FIGURE 7.47 Georg von Hevesy, c.1913.

where the discovery took place led to the element being named for the Latin name for "Copenhagen," Hafnia, the hometown of Niels Bohr. Currently, the Faculty of Science of the University of Copenhagen uses in its seal a stylized image of the hafnium atom. Hafnium was isolated from zirconium through repeated recrystallization of the double ammonium or potassium fluorides by Valdemar Thal Jantzen and von Hevesey. Dutch chemists Anton Eduard van Arkel (November 19, 1893 to March 14, 1976, Leiden) and Dutch chemist and physicist Jan Hendrik de Boer (March 19, 1899 to April 25, 1971) were the first to obtain metallic hafnium by passing hafnium tetraiodide vapor over a heated tungsten filament in 1924 (van Arkel and de Boer, 1924a,b, 1925). This method for differential purification of zirconium and hafnium is still in use today.

7.10.2 Mining, production, and major minerals

Hafnium is estimated to make up around 5.8 ppm of the Earth's upper crust by mass. It does not exist as a native element on Earth but is found as a solid solution with zirconium in natural zirconium compounds such as zircon, $ZrSiO_4$, which usually has about 1%–4% of the Zr replaced by Hf. Infrequently, the Hf/Zr ratio increases during crystallization wto give the isostructural mineral hafnon $(Hf,Zr)SiO_4$, with atomic Hf > Zr (Box 7.3). Major sources of zircon (and hence hafnium) ores are heavy mineral sands ore deposits, pegmatites, especially in Brazil and Malawi, and carbonatite intrusions, predominantly the Crown Polymetallic Deposit at Mount Weld, Western Australia. A potential source of hafnium is trachyte tuffs containing the rare zircon-hafnium silicates eudialyte, $Na_{15}Ca_6(Fe,Mn)_3Zr_3SiSi_{25}O_{73}(OH,O,H_2O)_3$, or armstrongite, $CaZrSi_6O_{15}\cdot 2.5H_2O$, near Dubbo in New South Wales, Australia.

The heavy mineral sand ore deposits of the titanium ore minerals ilmenite ($FeTiO_3$) and rutile (TiO_2) also produce most of the mined zirconium (zircon, $ZrSiO_4$) and consequently also most of the hafnium. Zirconium is a good nuclear fuel-rod cladding metal, with the required properties of a very low neutron capture cross-section and good chemical

BOX 7.3 Hafnon $HfSiO_4$.

Only a single hafnium mineral is known, the silicate hafnon ($HfSiO_4$). It was named in allusion to its composition, being a silicate of HAFNium, and ending in "ON" in allusion to, and to be consistent with, its relationship to ZircON in 1974 by Correia-Neves et al. It forms a solid solution series with zircon. It occurs in tantalum-bearing granite pegmatites (Zambézia district, Mozambique), in a weathered pegmatite (Mt. Holland, Western Australia). Hafnon is tetragonal, *4/m2/m2/m* (space group *I4₁/amd*). Its hardness has not been determined but is probably similar to zircon, that is, 7½. It is transparent to translucent, the color is red-orange, brown-yellow, or colorless (rare). Optically it is uniaxial. It occurs as euhedral to irregular crystals and fragments, to 1 cm, heavily zoned with zircon, the outermost portions of which represent this species.

stability at high temperatures. Nevertheless, due to hafnium's neutron-absorbing properties, the presence of hafnium impurities in zirconium would result in it to be far less suitable for nuclear reactor applications. Therefore an almost complete separation of zirconium and hafnium is essential for the use of zirconium in nuclear power applications. The manufacture of hafnium-free zirconium is the principal source for hafnium. The chemical properties of hafnium and zirconium are virtually identical, which makes the separation of the two elements extremely difficult. The processes previously utilized—fractional crystallization of ammonium fluoride salts or the fractional distillation of the chloride—have been shown to be unsuitable for large-scale production. Since zirconium was selected as material for nuclear reactor programs in the 1940s, an isolation process had to be developed. Liquid–liquid extraction methods with a wide variety of solvents were established and are currently still applied for the manufacture of hafnium. About 50% of all hafnium metal produced is obtained as a by-product of zirconium refinement. The final product of the isolation is Hf(IV) chloride. The purified Hf(IV) chloride is converted to the metal by reduction with magnesium or sodium, as in the Kroll process.

$$HfCl_4 + 2Mg \xrightarrow{1100^\circ C} 2MgCl_2 + Hf$$

Further purification is achieved through a chemical transport reaction developed by van Arkel and de Boer: Hafnium reacts with iodine in a closed vessel at temperatures of 500°C, forming Hf(IV) iodide; at a tungsten filament of 1700°C the reverse reaction occurs, and the iodine and hafnium are released. The hafnium forms a solid coating at the tungsten filament, and the iodine can react with additional hafnium, resulting in a steady turn over.

$$Hf + 2I_2 \xrightarrow{500^\circ C} HfI_4$$
$$HfI_4 \xrightarrow{1700^\circ C} Hf + 2I_2$$

7.10.3 Chemistry

No less than 34 isotopes of Hf have been characterized, ranging from ^{153}Hf to ^{186}Hf. The five stable isotopes are found in the range of ^{176}Hf to ^{180}Hf, with ^{180}Hf having the highest natural abundance at just over 35%. The radioactive isotopes have half-lives ranging from only 400 ms for ^{153}Hf, to 2.0×10^{15} years for the most stable radioisotope ^{174}Hf.

Hafnium is a shiny, silvery, ductile metal that is corrosion resistant and chemically comparable to zirconium (since it has the same number of valence electrons, being in the same group, but also to relativistic effects; the expected expansion of atomic radii from period 5–6 is practically canceled out by the lanthanide contraction) (Table 7.10). The physical properties of hafnium metal are substantially affected by zirconium impurities, particularly the nuclear properties, as these two elements are among the most difficult to separate due their chemical similarity. A noticeable physical difference between these metals is their density, with zirconium having about one-half the density of hafnium. The chemistry of hafnium and zirconium is similar that the two cannot be separated based on differing chemical reactions. The melting points and boiling points of the compounds and the solubility in solvents are the major differences in the chemistry of these twin elements.

7.10.3.1 Reaction with oxygen

Hafnium reacts in air forming a protective film, which prevents further corrosion. Hafnium(IV) oxide is the inorganic compound with the formula HfO_2. Also known as hafnia, this colorless solid is one of the most common and stable compounds of hafnium. It is an electrical insulator with a band gap of 5.3–5.7 eV. Hafnium(IV) oxide is fairly inert. It reacts with strong acids such as concentrated sulfuric acid and with strong bases. It dissolves slowly in

TABLE 7.10 Hafnon properties.

Appearance	Steel gray
Standard atomic weight $A_{r,std}$	178.49
Block	d-Block
Element category	Transition metal
Electron configuration	[Xe] $4f^{14}$ $5d^2$ $6s^2$
Phase at STP	Solid
Melting point	2233°C
Boiling point	4603°C
Density (near r.t.)	13.31 g/cm^3
When liquid (at m.p.)	12 g/cm^3
Heat of fusion	27.2 kJ/mol
Heat of vaporization	648 kJ/mol
Molar heat capacity	25.73 J/(mol·K)
Oxidation states	−2, +1, +2, +3, **+4**
Ionization energies	1st: 658.5 kJ/mol
	2nd: 1440 kJ/mol
	3rd: 2250 kJ/mol
Atomic radius	Empirical: 159 pm
Covalent radius	175 ± 10 pm

STP, Standard temperature and pressure.
Bold font indicates main oxidation state.

hydrofluoric acid to form fluorohafnate anions. At elevated temperatures, it reacts with chlorine in the presence of graphite or carbon tetrachloride to form hafnium tetrachloride. Hafnia has the same structure as zirconia (ZrO_2). Unlike TiO_2, which features six-coordinate Ti in all phases, zirconia and hafnia have seven-coordinate metal centers. Various crystalline phases have been experimentally observed, for example, cubic ($Fm\bar{3}m$), tetragonal ($P4_2/nmc$), monoclinic ($P2_1/c$), and orthorhombic (*Pbca* and *Pnma*). In addition, it is known that hafnia may adopt two additional orthorhombic metastable phases (space group $Pca2_1$ and $Pmn2_1$) over a wide range of pressures and temperatures, probably being the sources of the ferroelectricity recently observed in thin films of hafnia. Thin films of hafnium oxides deposited by atomic layer deposition are usually crystalline. Since semiconductor devices benefit from having amorphous films present, hafnium oxide has been alloyed with aluminum or silicon (forming hafnium silicates), which possesses a higher crystallization temperature than hafnium oxide.

7.10.3.2 Halides

The metal is not readily attacked by acids, but it can be oxidized with halogens or burnt in air. Like zirconium, finely divided hafnium powder can ignite spontaneously in air. The metal is resistant to concentrated alkalis. Hafnium tetrafluoride, HfF_4, is a white solid. It has the same structure as zirconium tetrafluoride, with eight-coordinate Hf(IV) centers. Hafnium tetrabromide, $HfBr_4$, is the most common bromide of hafnium. It is a colorless, diamagnetic moisture-sensitive solid that sublimes in vacuum. It has a structure similar to that of zirconium tetrabromide, featuring tetrahedral Hf centers, contrary to the polymeric nature of hafnium tetrachloride. Hafnium tetraiodide, HfI_4, is a red-orange, moisture-sensitive, sublimable solid that is formed by heating a mixture of hafnium with excess iodine. It is an intermediate in the crystal bar process for producing hafnium metal. In HfI_4, the hafnium centers have an octahedral coordination geometry. Like most binary metal halides, the compound is polymeric. It is a one-dimensional (1D) polymer consisting of chains of edge-shared bioctahedral Hf_2I_8 subunits, like the motif found in $HfCl_4$. The nonbridging iodide ligands have shorter bond lengths to Hf than the bridging iodide ligands.

Hafnium(IV) chloride, $HfCl_4$, is a colorless solid and is the precursor to most hafnium organometallic compounds. It has various highly specialized uses, primarily in materials science and as a catalyst. $HfCl_4$ can be prepared by several related methods: (1) the reaction of carbon tetrachloride and hafnium oxide at above 450°C;

$$HfO_2 + 2CCl_4 \rightarrow HfCl_4 + 2COCl_2$$

(2) Chlorination of a mixture of HfO_2 and carbon above 600°C using chlorine gas or sulfur monochloride:

$$HfO_2 + 2Cl_2 + C \rightarrow HfCl_4 + CO_2$$

(3) Chlorination of hafnium carbide above 250°C. The compound is highly reactive toward water, evolving hydrogen chloride:

$$HfCl_4 + H_2O \rightarrow HfOCl_2 + 2HCl$$

Aged samples are therefore often contaminated with oxychlorides, which are also colorless. THF forms a monomeric 2:1 complex.

$$HfCl_4 + 2OC_4H_8 \rightarrow HfCl_4(OC_4H_8)_2$$

Since this complex is soluble in organic solvents, it is a valuable reagent for preparing other complexes of hafnium. In the gas phase, both $ZrCl_4$ and $HfCl_4$ have the monomeric tetrahedral structure seen for $TiCl_4$. Electronographic investigations of $HfCl_4$ in gas phase showed that the Hf-Cl internuclear distance is 2.33 Å and the Cl...Cl internuclear distance is 3.80 Å. The ratio of internuclear distances r(Me-Cl)/r(Cl...Cl) is 1.630 and this value is in good agreement with the value for the regular tetrahedron model (1.633). Hafnium tetrachloride is the precursor to highly active catalysts for the Ziegler-Natta polymerization of alkenes, in particular propylene. Typical catalysts are derived from tetrabenzylhafnium. $HfCl_4$ is an effective Lewis acid for various applications in organic synthesis. For instance, ferrocene is alkylated with allyldimethylchlorosilane more efficiently using hafnium chloride relative to aluminum trichloride. The larger size of Hf may decrease $HfCl_4$'s tendency to complex to ferrocene. $HfCl_4$ increases the rate and control of 1,3-dipolar cycloadditions. It was observed to produce better results than other Lewis acids when used with aryl and aliphatic aldoximes, allowing specific exo-isomer formation.

7.10.3.3 Borides, sulfides, and carbides

Hafnium diboride belongs to the class of ultrahigh-temperature ceramics, a type of ceramic composed of hafnium and boron. It has a melting temperature of about 3250°C. It is an unusual ceramic, having relatively high thermal and electrical conductivities, properties it has in common with isostructural titanium diboride and zirconium diboride. It is a gray, metallic looking material. Hafnium diboride has a hexagonal crystal structure, a molar mass of 200.11 g/mole, and a density of 10.5 g/cm^3. Hafnium diboride is often combined with carbon, boron, silicon, silicon carbide, and/or nickel to improve the consolidation of the hafnium diboride powder (sintering). It is frequently formed into a solid by a process called hot pressing, where the powders are pressed together using both heat and pressure. The material has potential for use in hypervelocity re-entry vehicles such as intercontinental ballistic missile heat shields or aerodynamic leading-edges, due to its strength and thermal properties. Unlike polymer and composite material, HfB_2 can be formed into aerodynamic shapes that will not ablate during re-entry. Hafnium diboride is also being studied as a potential new material for nuclear reactor control rods. In addition, it is being investigated as a microchip diffusion barrier. If prepared correctly, the barrier can be less than 7 nm in thickness. Nanocrystals of HfB_2 with rose-like morphology were obtained reacting HfO_2 and $NaBH_4$ at 700°C−900°C under argon flow.

$$HfO_2 + 3NaBH_4 \rightarrow HfB_2 + 2Na(g,l) + NaBO_2 + 6H_2(g)$$

Hafnium disulfide is an inorganic compound of hafnium and sulfur. It is a layered dichalcogenide with the chemical formula HfS_2. A few atomic layers of this material can be exfoliated using the standard Scotch Tape technique similar to graphene and utilized for the fabrication of a field-effect transistor. High-yield production of HfS_2 has also been proven using liquid-phase exfoliation, resulting in the production of stable few-layer HfS_2 flakes. Hafnium disulfide powder can be prepared by reacting hydrogen sulfide and hafnium oxides at 500°C−1300°C. Hafnium carbide (HfC) is a chemical compound of hafnium and carbon. With a melting point of around 3900°C, it is one of the most refractory binary compounds known. But it has a low oxidation resistance, with oxidation starting at temperatures as low as 430°C. Hafnium carbide is usually carbon deficient, and hence its composition is frequently expressed as HfC_x ($x = 0.5$ to 1.0). It has a cubic (rock salt, NaCl) crystal structure at any value of x. Hafnium carbide powder is prepared by the reduction of hafnium(IV) oxide with carbon at a temperature between 1800°C and 2000°C. A long processing time is necessary to remove all oxygen. Instead, high-purity HfC coatings can be formed by chemical vapor deposition (CVD) from a gas mixture of methane, hydrogen, and vaporized hafnium(IV) chloride. Because of the technical complexity and high cost of the synthesis, HfC has a very limited application, notwithstanding its favorable properties such as high hardness (Mohs hardness >9, compare diamond = 10) and melting point. The magnetic properties of HfC_x change from paramagnetic for $x \leq 0.8$ to diamagnetic at larger x. An inverse behavior (dia- to paramagnetic transition with increasing x) is observed for TaC_x, even though having the same crystal structure as HfC_x. Tantalum hafnium carbide is a refractory chemical compound with the general formula $Ta_xHf_{y-x}C_y$, which can be considered as a solid solution of tantalum carbide and hafnium carbide. Separately, these two carbides have the highest melting points among the binary compounds, 3880°C (415 K) and 3928°C

(4201K), respectively, and their "alloy" with a composition Ta_4HfC_5 is thought to have a melting point of 3990°C (4263K). A very limited number of measurements of the melting point of tantalum hafnium carbide have been reported, due to the obvious experimental difficulties at these extreme temperatures. A 1965 study of the TaC–HfC solid solutions found a minimum in the vaporization rate at temperatures 2225°C–2275°C and thus maximum in the thermal stability for Ta_4HfC_5. This rate was comparable to that of tungsten and was slightly dependent on the initial density of the samples, which were sintered from TaC–HfC powder mixtures, also at 2225°C–2275°C. In a different study, Ta_4HfC_5 was found to have the minimum oxidation rate among the TaC–HfC solid solutions. Individual tantalum and hafnium carbides have a rock salt (NaCl) cubic lattice structure. They are usually carbon deficient and have nominal formulas TaC_x and HfC_x, with x = 0.7–1.0 for Ta and x = 0.56–1.0 for Hf. The same structure has also been found for at least some of their solid solutions. The density calculated from X-ray diffraction data is 13.6 g/cm^3 for $Ta_{0.5}Hf_{0.5}C$. A hexagonal NiAs-type structure (space group *P6₃/mmc*) with a density of 14.76 g/cm^3 has been published for $Ta_{0.9}Hf_{0.1}C_{0.5}$.

7.10.3.4 Organohafnium compounds

Organohafnium compounds behave almost identically to organozirconium compounds, as hafnium, which is just below zirconium on the periodic table, has chemical properties comparable to those of zirconium. A large group of Hf analogs of zirconium compounds have been reported, such as bis(cyclopentadienyl)hafnium(IV) dichloride ($(C_5H_5)_2Cl_2Hf$), bis(cyclopentadienyl)hafnium(IV) dihydride ($(C_5H_5)_2HfH_2$), and dimethylbis(cyclopentadienyl)hafnium(IV) ($(C_5H_5)_2Hf(CH_3)_2$).

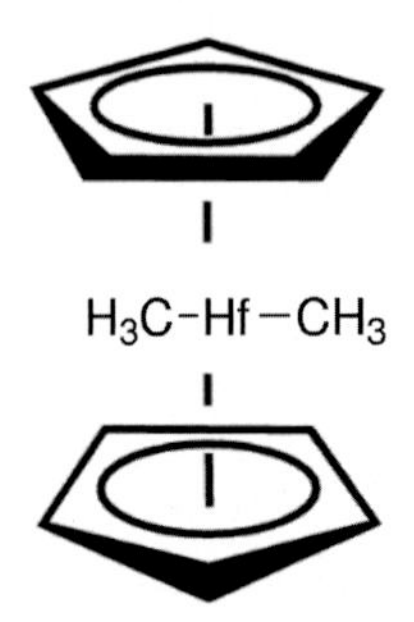

Dimethylbis(cyclopentadienyl)hafnium(IV)

Hafnium acetylacetonate, also known as $Hf(acac)_4$, is a coordination compound with formula $Hf(C_5H_7O_2)_4$. This white solid is the most important hafnium complex of acetylacetonate. The complex has a square antiprismatic geometry with eight almost equivalent Hf–O bonds. The molecular symmetry is D_2, that is, the complex is chiral. It is formed from hafnium tetrachloride and acetylacetone, and base. $Zr(acac)_4$ is very similar in structure and properties. Along with titanium tetrabutoxide (TBT), hafnium acetylacetonate serves as a catalyst for the production of poly(butylene terephthalate). $HfCl_4$ undergoes salt metathesis with Grignard reagents. Tetrabenzylhafnium ($(C_6H_5CH_2)_4Hf$) can be produced via this route. With alcohols, alkoxides are formed.

$$HfCl_4 + 4ROH \rightarrow Hf(OR)_4 + 4HCl$$

These compounds have complicated structures. Reduction of $HfCl_4$ is particularly difficult. In the presence of phosphine ligands, reduction can be achieved with potassium–sodium alloy.

$$2HfCl_4 + 2K + 4P(C_2H_5)_3 \rightarrow Hf_2Cl_6\left[P(C_2H_5)_3\right]_4 + 2KCl$$

The deep green dihafnium product is diamagnetic. X-ray crystallography proves that the complex has an edge-shared bioctahedral structure, similar to the Zr analog.

7.10.4 Major uses

The nuclei of some Hf isotopes can each absorb multiple neutrons, which makes it a good element for usage in the control rods for nuclear reactors. Its neutron-capture cross-section is approximately 600 times that of Zr (other elements that are good neutron absorbers are Cd and B.) Exceptional mechanical properties and excellent corrosion-resistance properties allow it to be used in the unforgiving environment of pressurized water reactors. The German research reactor FRM II (Forschungs-Neutronenquelle Heinz Maier-Leibnitz, research reactor Munich II) employs Hf as a neutron absorber. It is likewise common in military reactors, especially in the US naval reactors, though rarely found in civilian reactors, the first core of the Shippingport Atomic Power Station (a conversion of a naval reactor) being a prominent exception. The Shippingport Atomic Power Station was (according to the US Nuclear Regulatory Commission) the world's first full-scale atomic electric power plant devoted exclusively to peacetime uses. It was located near the present-day Beaver Valley Nuclear Generating Station on the Ohio River in Beaver County, Pennsylvania, United States, about 25 miles (40 km) from Pittsburgh. The reactor reached criticality on December 2, 1957, and aside from stoppages for three core changes, it remained in operation until October 1982. The first electrical power was produced on December 18, 1957 as engineers synchronized the plant with the distribution grid of Duquesne Light Company. The first core used at Shippingport originated from a canceled nuclear-powered aircraft carrier and used highly enriched uranium (93% ^{235}U) as "seed" fuel surrounded by a "blanket" of natural ^{238}U, in a so-called seed-and-blanket design; in the first reactor about half the power came from the seed.

Isotopes of Hf and Lu (along with Yb) are also utilized in isotope geochemistry and geochronological applications, for example, in Lu–Hf dating. It is frequently employed as a tracer of isotopic evolution of Earth's mantle through time. This is because ^{176}Lu decays to ^{176}Hf with a half-life of about 37 billion years. In most geologic materials, the mineral zircon ($ZrSiO_4$) is the principal host of Hf ($>$ 10,000 ppm) and is frequently the focus of Hf research in geology. Hf is easily substituted in the zircon crystal structure and is consequently very resistant to Hf mobility and contamination. In addition, zircon has an extremely low Lu/Hf ratio, making any correction for initial Lu minimal. While the Lu/Hf system can be utilized to determine a "model age," that is, the time at which it was derived from a given isotopic reservoir, for example, the depleted mantle, these "ages" do not carry the same geologic import as do other geochronological dating techniques as the results often produce isotopic mixtures and thus produce an average age of the material from which it was derived. Garnets are another mineral group that contains considerable quantities of Hf to serve as a geochronometer. The high and variable Lu/Hf ratios observed in garnet make it suitable for dating metamorphic events.

Hf is employed in alloys with Fe, Ti, Nb, Ta, and other metals. An alloy used for liquid rocket thruster nozzles, for example, used in the main engine of the Apollo Lunar Modules, is C103 that comprises 89% Nb, 10% Hf, and 1% Ti. Small additions of Hf improve the adherence of protective oxide scales on Ni-based alloys. It thus enhances the corrosion resistance in particular under cyclic temperature conditions that tend to break oxide scales by creating thermal stresses between the bulk material and the oxide layer. Hf-based compounds are used in gate insulators in the 45 nm generation of integrated circuits from Intel, IBM, etc. Hf oxide-based compounds are practical high-k dielectrics, permitting a decrease of the gate leakage current, which enhances performance at these IC scales. Because of its heat resistance and its affinity to oxygen and nitrogen, Hf is a good scavenger for oxygen and nitrogen in gas-filled and incandescent lamps. In addition, Hf is employed as the electrode in plasma cutting due to its ability to shed electrons into air. Hf is used in high-temperature ceramics. Hf carbide and nitride are some of the most refractory materials identified, that is, they will not melt except under the most extreme temperatures, for example, for Hf nitride the melting point is 3310°C.

7.11 73 Ta — Tantalum

7.11.1 Discovery

Tantalum was first found in Sweden in 1802 by Swedish chemist Anders Gustaf Ekeberg (January 16, 1767 to February 11, 1813), in two mineral samples—one from Sweden and the other from Finland (Ekeberg, 1802a,b) (Fig. 7.48). In 1801 British chemist Charles Hatchett (January 2, 1765 to March 10, 1847) had found columbium (now niobium), and in 1809 the English chemist and physicist William Hyde Wollaston (August 6, 1766 to December 22, 1828) compared its oxide, columbite with a density of 5.918 g/cm^3, to that of tantalum, tantalite with a density of 7.935 g/cm^3 (Wollaston, 1809). He decided that the two oxides, notwithstanding their difference in measured density, were the same and kept the name tantalum. After German chemist Friedrich Wöhler (July 31, 1800 to September 23, 1882) obtained the same results, it was believed that columbium and tantalum were the same element. This conclusion was questioned in 1846 by the German mineralogist and analytical chemist Heinrich Rose (August 6, 1795 to January 27, 1864), who claimed that there were two extra elements in the tantalite sample, and he named them after the children of Tantalus: niobium (from Niobe, the goddess of tears) and pelopium (from Pelops) (Rose, 1847). The alleged element "pelopium" was later shown to be a mixture of tantalum and niobium, and it was observed that the niobium was indistinguishable from the columbium already discovered in 1801 by Hatchett. The differences between tantalum and niobium were verified without any doubt in 1864 by Swedish mineralogist and chemist Christian Wilhelm Blomstrand (October 20, 1826 to November 5, 1897), and French chemist Henri Etienne Sainte-Claire Deville (March 11, 1818 to July 1, 1881), as well as by French chemist Louis Joseph Troost (October 17, 1825, Paris to September 30, 1911) (Fig. 7.49), who in 1865 obtained the experimental formulas of some of their compounds (de Marignac et al., 1866). Additional confirmation came from the Swiss chemist Jean Charles Galissard de Marignac (April 24, 1817 to April 15, 1894) in 1866, who showed that there were only two elements

FIGURE 7.48 Anders Gustaf Ekeberg, 1847.

FIGURE 7.49 Louis Joseph Troost, 1912, Popular Science Monthly Volume 81.

FIGURE 7.50 Werner von Bolton.

(de Marignac, 1866). These findings did not prohibit scientists from publishing papers about the so-called ilmenium up till 1871. De Marignac was the first to isolate the metallic form of tantalum in 1864, when he reduced tantalum chloride by heating it in an atmosphere of hydrogen. Early chemists had only been able to obtain impure tantalum, and the first relatively pure ductile metal was isolated by German chemist and materials scientist Werner von Bolton (April 8, 1868 to October 28, 1912) in Charlottenburg in 1903 (Fig. 7.50). Wires made with metallic tantalum were used to make filaments for light bulb until tungsten replaced it in widespread use. The name tantalum came from the name of the mythological Tantalus, the father of Niobe in Greek mythology. In the fable, he had been punished after death by being convicted to stand knee-deep in water with perfect fruit growing above his head, both of which eternally tantalized him. (If he bent to drink the water, it drained below the level he could reach, and if he reached for the fruit, the branches moved out of his grasp.) Anders Ekeberg wrote "This metal I call tantalum ... partly in allusion to its incapacity, when immersed in acid, to absorb any and be saturated." For tens of years, the commercial method for isolating tantalum from niobium used the fractional crystallization of potassium heptafluorotantalate (K_2TaF_7) away from potassium oxypentafluoroniobate monohydrate, a method that was developed by Jean Charles Galissard de Marignac in 1866. This process has been replaced by solvent extraction from fluoride-containing solutions of tantalum.

7.11.2 Mining, production, and major minerals

Tantalum is estimated to make up between 1 and 2 ppm of the Earth's crust by weight. There are many different Ta-containing minerals, only a few are being used as ore minerals: tantalite (a series consisting of tantalite-(Fe), tantalite-(Mn), and tantalite-(Mg)), microlite (now a group name), $(Ca,Na)_2Ta_2O_6(O,OH,F)$, wodginite, $Mn(Sn,Ta)Ta_2O_8$, euxenite-(Y), $(Y,Ca,Ce,U,Th)(Nb,Ta,Ti)_2O_6$, and polycrase-(Y), $(Y,Ca,Ce,U,Th)(Ti,Nb,Ta)_2O_6$. Tantalite $(Fe,Mn)Ta_2O_6$ is the most important mineral for tantalum production (Box 7.4). Tantalite forms a series with columbite $(Fe, Mn)(Ta,Nb)_2O_6$; when

BOX 7.4 Tantalum deposits.

Tantalum is found in the microlite group, which is the tantalum-rich equivalent of the columbium pyrochlore group, and tantalite, the tantalum-rich equivalent of the columbite series. Tantalum deposits occur mainly in pegmatites and veins associated with granitic intrusive rocks. Tantalite resists weathering and can accumulate in placer deposits. Tantalum production comes mostly from pegmatites and veins in Brazil, Canada, and central Africa, although important quantities also come from the slags of tin-producing operations in southeast Asia. Similar to the columbite series, tantalite as such is an obsolete group name for minerals with the general formula $(Mn,Fe)(Ta,Nb)_2O_6$, equivalent to the tantalite-(Fe)-tantalite-(Mn) solid solution series. Generally, manganese-dominant members are more abundant in nature. Tantalite-(Mn) is orthorhombic, *2/m2/m2/m* (space group *Pbcn*). It has a hardness of 6. The cleavage is distinct on {100}, less distinct {010}, the fracture is irregular/uneven, subconchoidal, while the tenacity is brittle. It is paramagnetic. The luster is vitreous to submetallic, while the transparency is opaque. The color varies from pink to nearly colorless, reddish brown to black, while the streak is red, scarlet to black. Optically it is biaxial(+). Crystals are short prismatic {001} to tabular/flattened {010} or {100}, also equant. Twinning can be observed on {021} and {023}, pseudohexagonal trillings.

there is more tantalum than niobium it is called tantalite and when there is more niobium than tantalum it is called columbite. Other minerals include samarskite -(Y), $(Y,Ce,U,Fe)_3(Nb,Ta,Ti)_5O_{16}$, and fergusonite, $(Ce,La,Y)NbO_4$. The main mining of tantalum was in Australia, in Western Australia, Greenbushes in the Southwest and Wodgina in the Pilbara region. The Wodgina mine produced a primary Ta concentrate, which was further upgraded at the Greenbushes operation. While the large-scale producers of Nb can be found in Brazil and Canada, the ore there also produces a small amount of Ta as a by-product. Some other countries, for example, China, Ethiopia, and Mozambique, produce ores with a higher percentage of tantalum, and they produce a substantial quantity of the world's total output of Ta. Ta is also mined in Thailand and Malaysia as a by-product of tin mining. In the process of gravitational separation of the ores from these placer deposits, not only is cassiterite (SnO_2) obtained but also a small amount of tantalite. The slag from the tin smelters then contains commercially valuable amounts of Ta, which is leached from the slag. World Ta mining has shown a significant geographical shift since the beginning of the 21st century when mining was predominantly in Australia and Brazil. Starting in 2007 and through 2014, the most important sources of Ta production shifted to mainly the Democratic Republic of the Congo, Rwanda, and some other African countries. Future resources of Ta, in order of estimated volume, are being explored in Saudi Arabia, Egypt, Greenland, China, Mozambique, Canada, Australia, the United States, Finland, and Brazil. It is estimated that there are less than 50 years left of Ta resources, based on present-day extraction rates, indicating the necessity for increased recycling of Ta.

A number of different steps are involved in the isolation of tantalum from tantalite. In the first step, the tantalite is crushed and concentrated by gravity separation. This is normally carried out near the mine site. The refining of tantalum from its ores is one of the more challenging separation processes in industrial metallurgy. The main problem is that tantalum ores contain significant amounts of niobium, which has chemical properties nearly identical to those of tantalum. Numerous methods have been developed to tackle this problem. These days, the isolation is achieved through hydrometallurgy. The isolation starts with leaching the ore with hydrofluoric acid (HF) together with sulfuric acid (H_2SO_4) or hydrochloric acid (HCl). This leaching reaction separates the tantalum and niobium from the various nonmetallic impurities in the crushed ore. Even though Ta occurs in various minerals, it is conveniently represented as the pentoxide, as most oxides of Ta(V) behave in the same way under these conditions. A simplified equation for its extraction with HF can therefore be given as

$$Ta_2O_5 + 14HF \rightarrow 2H_2[TaF_7] + 5H_2O$$

The same reactions occur for the Nb oxides, though the hexafluoride is typically predominant under the extraction conditions.

$$Nb_2O_5 + 12HF \rightarrow 2H[NbF_6] + 5H_2O$$

The above reaction equations are simplified: it is thought that bisulfate (HSO_4^-) and chloride compete as ligands for the Nb(V) and Ta(V) ions, when sulfuric and hydrochloric acids are used, respectively. The tantalum and niobium fluoride complexes are subsequently separated from the aqueous solution through liquid–liquid extraction into organic solvents, such as cyclohexanone ($(CH_2)_5CO$), octanol ($C_8H_{17}OH$), and methyl isobutyl ketone (preferred IUPAC name 4-methylpentan-2-one, $(CH_3)_2CHCH_2C(O)CH_3$). This simple procedure allows the removal of most metal-containing impurities (such as Fe, Mn, Ti, and Zr), which stay in the aqueous phase in the form of their fluorides and other complexes. Separation of the tantalum from niobium is subsequently realized through lowering the ionic strength of the

FIGURE 7.51 Fluornatromicrolite, $(Na_{1.5}Bi_{0.5})Ta_2O_6F$, euhedral deep green crystal to 3 mm in albite, $Na(AlSi_3O_8)$, matrix. Quixaba pegmatite, Frei Martinho, Paraiba, Brazil.

FIGURE 7.52 Several sharp, dull, brown, very good hydroxycalciopyrochlore, $(Ca,Na,U,\square)_2(Nb,Ti)_2O_6(OH)$, crystals to 4 mm. Silver Crater mine, Bancroft, Ontario, Canada.

acid mixture, which causes the niobium to dissolve in the aqueous phase. It has been suggested that the oxyfluoride $H_2[NbOF_5]$ is formed under these conditions. Following the removal of the niobium, the solution of purified $H_2[TaF_7]$ is neutralized with aqueous ammonia to precipitate the solid hydrated tantalum oxide, which can then be calcined to form tantalum pentoxide (Ta_2O_5). As an alternative for the hydrolysis, the $H_2[TaF_7]$ can be reacted with potassium fluoride (KF) to produce potassium heptafluorotantalate according to the reaction.

$$H_2[TaF_7] + 2KF \rightarrow K_2[TaF_7] + 2HF$$

FIGURE 7.53 Ixiolite, $(Ta,Nb,Sn,Fe,Mn)_4O_8$, black crude crystals from 5 to 10 mm embedded in quartz, SiO_2. Zinnwald, Erzgebirge, Bohemia, Czech Republic.

FIGURE 7.54 Tantalite-(Mn), $Mn^{2+}Ta_2O_6$, a thick tabular, lustrous, crystal. The crystal tapers downward and has many slightly striated faces. $5 \times 4.5 \times 3$ cm. Alto do Giz pegmatite, Rio Grande do Norte, Brazil.

In contrast to $H_2[TaF_7]$, the potassium salt is easily crystallized and handled as a solid. $K_2[TaF_7]$ can be transformed to metallic tantalum through reduction with sodium, at around 800°C in molten salt.

$$K_2[TaF_7] + 5Na \rightarrow Ta + 5NaF + 2KF$$

In an older process, called the Marignac process, the mixture of $H_2[TaF_7]$ and $H_2[NbOF_5]$ was reacted to form a mixture of $K_2[TaF_7]$ and $K_2[NbOF_5]$, which was subsequently separated through fractional crystallization, exploiting their differences in solubility in water. Alternatively, tantalum can be refined using electrolysis, applying a modified form of the Hall–Héroult method. Instead of necessitating the input oxide and output metal to be in liquid form, tantalum electrolysis works on nonliquid powdered oxides. The original discovery happened in 1997 when scientists of

FIGURE 7.55 Wodginite, $Mn^{2+}Sn^{4+}Ta_2O_8$, 1–2-mm black euhedral crystals in matrix. Tanco pegmatite, Bernic Lake, Manitoba, Canada.

FIGURE 7.56 Olmsteadite, $KFe_2^{2+}(Nb^{5+},Ta^{5+})(PO_4)_2O_2{\cdot}2H_2O$, single deep reddish brown, terminated crystal to 5.5 mm. White Cap mine, Pennington Co., South Dakota, United States.

Cambridge University immersed small samples of certain oxides in baths of molten salt and reduced the oxide by applying an electric current. The cathode uses powdered metal oxide, while the anode is made of carbon. The molten salt at 1000°C acts as the electrolyte.

Tantalum is found in more than 50 different minerals. Within the elements class, one alloy jedwabite ($Fe_7(Ta,Nb)_3$), and one carbide tantalcarbide (TaC) are known. About 50 minerals are found in the oxide class, which includes tantalates, for example, alumotantite ($AlTaO_4$), bismutotantalite ($Bi(Ta,Nb)O_4$), calciotantite ($CaTa_4O_{11}$), ferrowodginite ($Fe^{2+}Sn^{4+}Ta_2O_8$), fluornatromicrolite ($(Na_{1.5}Bi_{0.5})Ta_2O_6F$) (Fig. 7.51), heftetjernite ($ScTaO_4$), hydroxycalciopyrochlore

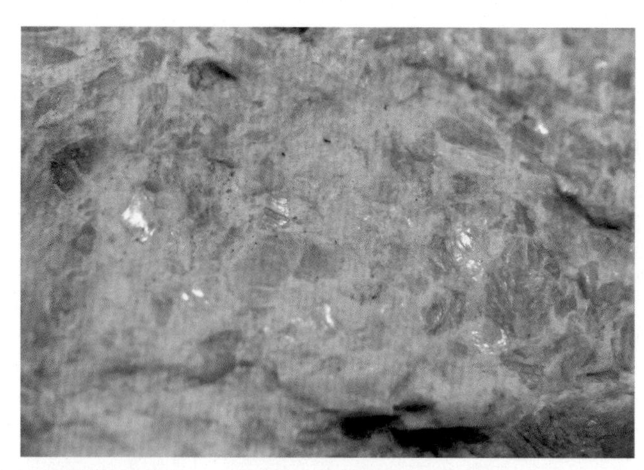

FIGURE 7.57 Holtite, $(Ta_{0.6}\square_{0.4})Al_6BSi_3O_{18}(O,OH)_{2.25}$, colorless to white pearly crystal plates to 3 cm. Vasin-Myl'k Mt, Voron'i Tundry, Kola Peninsula, Russia.

$((Ca,Na, U,\square)_2(Nb,Ti)_2O_6(OH)$ (Fig. 7.52), ixiolite $((Ta,Nb,Sn,Fe,Mn)_4O_8)$ (Fig. 7.53), tantalite-(Mn) $(Mn^{2+}Ta_2O_6)$ (Fig. 7.54), tantite (Ta_2O_5), and wodginite $(Mn^{2+}Sn^{4+}Ta_2O_8)$ (Fig. 7.55). Béhierite $((Ta^{5+},Nb^{5+})(BO_4))$ and schiavinatoite $((Nb,Ta)(BO_4))$ are the only two borates with Ta. The same number of minerals is known in the phosphate class, johnwalkite $(K(Mn^{2+},Fe^{2+},Fe^{3+})_2(Nb^{5+},Ta^{5+})(PO_4)_2O_2\cdot 2(H_2O,OH))$ and olmsteadite $(KFe_2^{2+}(Nb^{5+},Ta^{5+})(PO_4)_2O_2\cdot 2H_2O)$ (Fig. 7.56). Only two silicates have Ta in their crystal structure, holtite $((Ta_{0.6}\square_{0.4})Al_6BSi_3O_{18}(O,OH)_{2.25})$ (Fig. 7.57) and żabińskiite $(Ca[Al_{0.5}(Ta,Nb)_{0.5}](SiO_4)O)$.

7.11.3 Chemistry

Natural Ta occurs as two isotopes: metastable ^{180m}Ta (0.012%) and stable ^{181}Ta (99.988%). ^{180m}Ta is predicted to decay in three different processes: isomeric transition to the ground state of ^{180}Ta, β decay to ^{180}W, or electron capture to ^{180}Hf. Yet, decay of this nuclear isomer has not been detected, and hence only a lower limit on its half-life of 2.0×10^{16} years has been set. The ground state of ^{180}Ta has a half-life of only 8.125 h. ^{180m}Ta is the only naturally existing nuclear isomer (not including radiogenic and cosmogenic short-lived nuclides). It is also the rarest isotope in the universe, with regard to the elemental abundance of Ta and isotopic abundance of ^{180m}Ta in the natural mixture of isotopes (not including radiogenic and cosmogenic short-lived nuclides). Ta has been studied theoretically as a potential "salting" material for nuclear weapons (Co is the better-known theoretical salting material). An external shell of ^{181}Ta would be irradiated by the intensive high-energy neutron flux from a proposed exploding nuclear weapon. This would transmute Ta into the radioactive isotope ^{182}Ta with a half-life of 114.4 days and produce γ rays with about 1.12 MeV of energy apiece, which would strongly increase the radioactivity of the nuclear fallout from the explosion for a number of months. This type of "salted" weapons has never been produced or tested, insofar as is publicly known, and certainly never used as weapons. Ta can be applied as a target material for accelerated proton beams to produce various short-lived isotopes including ^8Li, ^{80}Rb, and ^{160}Yb.

Tantalum is dark (blue-gray), dense, ductile, very hard, and highly conductive of heat and electricity. The metal is well known for its resistance to corrosion by acids; in fact, at temperatures below 150°C tantalum is nearly completely resistant to attack by the normally aggressive aqua regia (a mixture of nitric acid and hydrochloric acid, optimally in a molar ratio of 1:3). It can be dissolved with hydrofluoric acid or acidic solutions with the fluoride ion and sulfur trioxide, as well as with a solution of potassium hydroxide. Its high melting point of 3017°C (and boiling point 5458°C) is surpassed among the elements only by tungsten, rhenium, and osmium for metals, and carbon (Table 7.11). Tantalum occurs in two crystalline phases, α and β. The α phase is fairly ductile and soft; it has *bcc* structure (space group *Im3m*, lattice constant a = 0.33058 nm). The β phase is hard and brittle; its crystal symmetry is tetragonal (space group $P4_2/mnm$, a = 1.0194 nm, c = 0.5313 nm). The β phase is metastable and transforms to the α phase upon heating to 750°C–775°C. Bulk tantalum is nearly completely α phase, and the β phase typically exists as thin films formed by magnetron sputtering, CVD, or electrochemical deposition from a eutectic molten salt solution.

Tantalum forms compounds in oxidation states −3 to +5. Most commonly encountered are oxides of Ta(V), which includes all minerals. The chemical properties of Ta and Nb are very similar. In aqueous media, Ta exhibits only the

TABLE 7.11 Tantalum properties.

Appearance	Gray blue
Standard atomic weight $A_{r,std}$	180.948
Block	d-Block
Element category	Transition metal
Electron configuration	[Xe] $4f^{14}$ $5d^{3}$ $6s^{2}$
Phase at STP	Solid
Melting point	3017°C
Boiling point	5458°C
Density (near r.t.)	16.69 g/cm^3
When liquid (at m.p.)	15 g/cm^3
Heat of fusion	36.57 kJ/mol
Heat of vaporization	753 kJ/mol
Molar heat capacity	25.36 J/(mol·K)
Oxidation states	−3, −1, +1, +2, +3, +4, **+5**
Ionization energies	1st: 761 kJ/mol
	2nd: 1500 kJ/mol
Atomic radius	Empirical: 146 pm
Covalent radius	170 ± 8 pm

STP, Standard temperature and pressure.
Bold font indicates main oxidation state.

+5 oxidation state. Similar to niobium, tantalum is hardly soluble in dilute solutions of hydrochloric, sulfuric, nitric, and phosphoric acids because of the precipitation of hydrous Ta(V) oxide. In basic media, Ta can be solubilized because of the formation of polyoxotantalate species.

7.11.3.1 Oxides, nitrides, carbides, and sulfides

Tantalum is noticeably less prone than niobium to form oxides in lower oxidation states. The rutile (TiO_2) phase TaO_2 is known but has not been studied, and a cubic rock salt (NaCl) type phase TaO with a narrow homogeneity range has also been reported but not fully characterized. Ta_2O_5 has two well-established polymorphs, which have a reversible transition temperature at 1355°C. Tantalum pentoxide (Ta_2O_5) is the most valuable compound from the viewpoint of applications. There are many oxides of tantalum in lower oxidation states, including many defect structures, lightly studied, or poorly characterized. Tantalates, compounds having $[TaO_4]^{3-}$ or $[TaO_3]^{-}$, are numerous. Lithium tantalate ($LiTaO_3$) has a perovskite ($CaTiO_3$) structure. Lanthanum tantalate ($LaTaO_4$) has isolated TaO_4^{3-} tetrahedra in its structure. Tantalum oxide is regularly employed in electronics, frequently in the form of thin films. For these applications it can be prepared by MOCVD (or related techniques), which involves the hydrolysis of its volatile halides or alkoxides:

$$Ta_2(OEt)_{10} + 5H_2O \rightarrow Ta_2O_5 + 10EtOH$$
$$2TaCl_5 + 5H_2O \rightarrow Ta_2O_5 + 10HCl$$

Ta_2O_5 does not react appreciably with either HCl or HBr, although it will dissolve in hydrofluoric acid, and reacts with potassium bifluoride and HF:

$$Ta_2O_5 + 4KHF_2 + 6HF \rightarrow 2K_2[TaF_7] + 5H_2O$$

Ta_2O_5 can be reduced to metallic Ta using metallic reductants such as calcium and aluminium.

$$Ta_2O_5 + 5Ca \rightarrow 2Ta + 5CaO$$

As in the cases of other refractory metals, the hardest known compounds of tantalum are nitrides and carbides. Tantalum carbide, TaC, like the more commonly used tungsten carbide, is a hard ceramic that is used in cutting tools. The melting points of tantalum carbides has a maximum at ~3880°C depending on the purity and measurement conditions; this temperature is among the highest for binary compounds. Only tantalum hafnium carbide may have a slightly higher melting point of ~3942°C, while the melting point of hafnium carbide is similar to that of TaC. TaC_x compounds have a cubic (NaCl) crystal structure for x = 0.7–1.0; the lattice parameter increases with x. $TaC_{0.5}$ has two major crystalline forms. The more stable one has an anti-cadmium iodide-type trigonal structure, which transforms upon heating to about 2000°C into a hexagonal structure with no long-range order for the carbon atoms. There are multiple phases of compounds, stoichimetrically from Ta_2N to Ta_3N_5 including TaN. The tantalum–nitrogen system includes several states among which a nitrogen solid solution in tantalum, as well as several nitride phases, which can vary from the expected stoichiometry due to lattice vacancies. Annealing of nitrogen rich "TaN" can result in a conversion to a two phase mixture of TaN and Ta_5N_6. Ta_5N_6 is believed to be the more thermally stable compound - though it decomposes in vacuum at 2500°C to Ta_2N. It has been reported that the decomposition in vacuum results in a change from Ta_3N_5 via Ta_4N_5, Ta_5N_6, ε-TaN, to Ta_2N. Tantalum(III) nitride is used as a thin film insulator in some microelectronic fabrication processes.

Tantalum forms a wide variety of binary chalcogenides (sulfides, selenides, and tellurides), which frequently differ both in stoichiometry and in structure from the oxides, for example, $Ta_{1+x}S_2$, TaS_2, TaS, Ta_2Se_3, $Ta_{1+x}Se_2$, $TaSe_2$, $TaSe_3$, TaTe, $Ta_{1-x}Te_2$, $TaTe_2$, and $TaTe_4$. Phases approximating to the stoichiometry MS have the NiAs-type structure, whereas MS_2 have layer lattices related to MoS_2, CdI_2, or $CdCl_2$. Sometimes complex layer-sequences occur in which the six-coordinate metal atom is alternatively octahedral and trigonal prismatic. The best studied chalcogenide is TaS_2, a layered semiconductor, as seen for other transition metal dichalcogenides. A tantalum–tellurium alloy forms quasicrystals.

7.11.3.2 Halides

Tantalum halides span the oxidation states of +5, +4, and +3. All pentahalides of Ta can be prepared easily by direct reaction of the appropriate halogen with the heated tantalum metal. They are all comparatively volatile, hydrolysable solids (indicative of the covalency to be expected in such a high oxidation state) in which the metals attain octahedral coordination through halide bridges. TaF_5 is a tetramer, while the chlorides and bromides are dimers. The colors vary from white fluorides, yellow chlorides, and orange bromides, to brown iodides. The decreasing energy of the charge-transfer bands responsible for these colors reflects the increasing polarizability of the anions from F^- to I^-. Tantalum pentafluoride (TaF_5) is a white solid with a melting point of 97.0°C. The anion $[TaF_7]^{2-}$ is used for its separation from niobium. The chloride $TaCl_5$, which exists as a dimer, is the most important reagent in the synthesis of new Ta compounds. It hydrolyzes easily to an oxychloride. The lower halides TaX_4 and TaX_3 include Ta-Ta bonds. The tetrahalides can be prepared by direct action of the elements. The Ta tetrahalides (except TaF_4, which is unknown) are usually prepared by reduction of the corresponding pentahalide and are all readily hydrolyzed. Notwithstanding claims for the existence of TaF_3, it is possible that this blue material is actually an oxide fluoride but, since O^{2-} and F^- are isoelectronic and similar in size, they are difficult to distinguish by X-ray methods. The remaining known trihalides, $TaCl_3$ and $TaBr_3$, are dark colored, rather unreactive materials. High-temperature reductions of TaX_5 with the metals (or Na or Al) yield a series of phases based on $[Ta_6X_{12}]^{n+}$ units consisting of octahedral clusters of metal atoms with the halogen atoms situated above each edge of the octahedra. The known oxohalides are normally prepared from the oxides but are not very well known and are restricted almost completely to the oxidation states of +4 and +5. This group includes $TaOF_3$, TaO_2F, $TaOCl_3$, $TaOCl_2$, TaO_2Cl, $TaOBr_3$, $TaOBr_2$, TaO_2Br, $TaOI_3$, $TaOI_2$, and TaO_2I. The Ta(V) compounds are quite volatile, though less so than the pentahalides.

7.11.3.3 Organotantalum compounds

Organotantalum compounds include pentamethyltantalum ($Ta(CH_3)_5$), mixed alkyltantalum chlorides, alkyltantalum hydrides, alkylidene complexes, and cyclopentadienyl derivatives of the same. Pentamethyltantalum can be prepared from the reaction of methyllithium with $Ta(CH_3)_3Cl_2$. $Ta(CH_3)_3Cl_2$ is in turn made from tantalum pentachloride and dimethylzinc. The pentamethyltantalum has a square pyramid shape. Ignoring the C-H bonds, the molecule has C_{4v} symmetry. The four carbon atoms at the base of the pyramid are called basal, while the carbon atom at the top is called apical or apex. The distance from tantalum to the apical carbon atom is 2.110 Å and that to the basal carbon atoms is 2.180 Å. The distance from hydrogen to carbon in the methyl groups is 1.106 Å. The angle subtended by the two basal carbon bonds is 82.2°, and the angle between the bonds to the apex and a carbon on the base is about 111.7°. With many carbon–hydrogen bonds near Ta, equivalents of pentamethyltantalum are susceptible to alpha elimination. Excess methyllithium reacts to form higher coordinated methyl tantalum ions $[Ta(CH_3)_6]^-$ and $[Ta(CH_3)_7]^{2-}$. Pentamethyltantalum in solution forms a stable insoluble complex material when mixed with dmpe (1,2-bis(dimethylphosphino)ethane, IUPAC name ethane-1,2-diylbis(dimethylphosphane) $(CH_3)_2PCH_2CH_2P(CH_3)_2$. With nitric oxide it forms a white colored dimer with formula $\{TaMe_3[ON(Me)NO]_2\}_2$ (Me = CH_3). Complexes such as $[TaMe_5(dmpe)]$ decompose spontaneously above room temperature and, although free $TaMe_5$ has been isolated, it can explode spontaneously at room temperature even in the absence of air. Notwithstanding this instability, the Ta–Me bond itself is quite strong:

thermochemical studies have shown that the mean bond dissociation energy D(Ta-Me) in $TaMe_5$ is $261 \pm 6\,kJ \cdot mol^{-1}$. Reduction of $TaCl_5$ or $TaCl_3$ under an atmosphere of CO yields salts of the $[Ta(CO)_6]^-$ ions, which have the noble gas electron configuration. Diverse salts and substituted derivatives are known for the hexacarbonyl $[Ta(CO)_6]^-$ and related isocyanides. An important compound is the mixed methylmethylene derivative of bis(cylopentadienyl)tantalum(V) synthesized according to the following series of high-yield reactions.

$$[Ta(\eta^5-C_5H_5)_2Me_3)] \xrightarrow{Ph_3C^+F_4^-} [Ta(\eta^5-C_5H_5)_2Me_2)]^+ \xrightarrow{+Me_3P=CH_3} [Ta(\eta^5-C_5H_5)_2(CH_3)(=CH_2)]$$

The three Ta−C distances for this complex are Ta = CH_2 203 pm, Ta-CH_3 225 pm, and Ta-C(C_5H_5) 216 pm. The two cyclopentadienyl rings are eclipsed and the CH_2 group orients perpendicular to the C-Ta-C plane. Mono(cyclopentadienyl) or "half-sandwich" poly-oxo complexes are of significance as hydrocarbon-soluble models for oxide catalysts. The reaction of water with $[Ta(\eta^5\text{-}C_5Me_5)(PMe_3)_2]$ produces the colorless compound $[Ta_4(\eta^5\text{-}C_5Me_5)_4O_7(OH)_2]$, which has a tetranuclear "butterfly" core. The chemistry of these metals with ring systems other than cyclopentadienyl has not been developed much, but as larger rings afford more bonding electrons, it could be expected that the relatively electron-poor, early transition elements should provide an interesting future research area.

7.11.4 Major uses

The most important use for Ta, as the metal powder, is in the manufacture of electronic components, primarily capacitors and certain high-power resistors. Ta electrolytic capacitors make use of the tendency of Ta to form a protective oxide surface layer, using Ta powder, pressed into a pellet shape, as one "plate" of the capacitor, the oxide as the dielectric, and an electrolytic solution or conductive solid as the other "plate." Since the dielectric layer can be very thin (thinner than the comparable layer in, e.g., an Al electrolytic capacitor), a high capacitance can be reached in a small volume. Due to the size and weight advantages, Ta capacitors are of interest for use in portable telephones, personal computers, car electronics, and cameras.

In addition, Ta is employed to make various alloys with high melting points, strength, and ductility. Alloyed with other metals, it is utilized in producing carbide tools for metalworking equipment and in the manufacture of superalloys, for example, jet engine components, chemical process equipment, nuclear reactors, missile parts, heat exchangers, tanks, and vessels. Due to its ductility, Ta can be drawn into fine wires or filaments, which are used for evaporating metals such as Al. As it resists attack by body fluids and is bioinert and nonirritating, Ta is extensively employed in producing surgical instruments and implants. For instance, porous Ta coating is utilized in the construction of orthopedic implants because of its capacity to form a direct bond to hard tissue. The high stiffness of Ta makes it necessary to employ it as highly porous foam or scaffold with lower stiffness for hip replacement implants to avoid stress shielding. Since Ta is a nonferrous, nonmagnetic metal, these implants are thought to be acceptable for patients undergoing MRI procedures.

Ta is inert against most acids, except HF and hot H_2SO_4, and hot alkaline solutions likewise result in Ta to corrode. This property makes it a suitable metal for chemical reaction vessels and pipes for corrosive liquids. Heat exchanging coils for the steam heating of HCl are produced from Ta. Ta was widely employed in the manufacture of ultrahigh-frequency electron tubes for radio transmitters. Ta is able to capture oxygen and nitrogen by forming nitrides and oxides and consequently assists to withstand the high vacuum needed for the tubes when used for internal parts such as grids and plates. The high melting point and oxidation resistance resulted in the application of Ta in the manufacture of vacuum furnace parts. Ta is highly inert and hence is formed into various corrosion-resistant parts, for example, thermowells, valve bodies, and Ta fasteners. Because of its high density, shaped charge and explosively formed penetrator liners have been made from Ta. Ta significantly increases the armor penetration capabilities of a shaped charge because of its high density and high melting point. In addition, it is sporadically used in precious watches. Ta oxide is used to manufacture special high refractive index glass for camera lenses. Other ways in which Ta has been used are as the electrodes in neon lights, in a.c./d.c. (alternating current/direct current) rectifiers, and for the spinneret dies through which cellulose solutions are forced in the process of making rayon fibers (The many types and grades of rayon can imitate the feel and texture of natural fibers such as silk, wool, cotton, and linen. The types that resemble silk are often called artificial silk. Rayon is made from purified cellulose, harvested primarily from wood pulp, which is chemically converted into a soluble compound. It is then dissolved and forced through a spinneret to produce filaments that are chemically solidified, resulting in fibers of nearly pure cellulose.)

7.12 74 W — Tungsten

7.12.1 Discovery

In 1781 Swedish Pomeranian and German pharmaceutical chemist Carl Wilhelm Scheele (December 9, 1742 to May 21, 1786) found that a new acid, tungstic acid, could be obtained from scheelite (at the time known as tungsten, $CaWO_4$) (Scheele, 1781). Scheele and Swedish mineralogist and chemist Torbern Olof Bergman (March 20, 1735 to July 8, 1784) proposed that it could be possible to isolate a new metal by reducing this acid (Fig. 7.58). In 1783 Spanish chemist and mineralogist Juan José (June 15, 1754 to September 20, 1796) and his brother chemist Fausto de Luyar (Elhuyar) (October 11, 1755 to February 6, 1833) (Fig. 7.59) discovered an acid made from wolframite (now known as the ferberite–hübnerite series, $FeWO_4$–$MnWO_4$) that was the same as tungstic acid. Later in that same year, at the Royal Basque Society (in Basque Euskalerriaren Adiskideen Elkartea and in Spanish Real Sociedad Bascongada de Amigos del País, also known as La Bascongada or Bascongada Society, was founded in the mid-18th century to encourage the scientific, cultural, and economic development of the Basque Country) in the town of Bergara (located in the province of Gipuzkoa, in the autonomous community of Basque Country, in the north of, Spain) the brothers managed to separate tungsten by reduction of this acid with charcoal, and the discovery of the element is attributed to them (they called it "wolfram" or "volfram") (de Luyar and de Luyar, 1783, 1785). The strategic value of tungsten became clear in the early 20th century. The British government decided in 1912 to free the Carrock mine, Mungrisdale, Eden, Cumbria, from the German owned Cumbrian Mining Company and, during the First World War, restrict German access elsewhere. In the Second World War, tungsten played a more substantial role in background political dealings. Portugal, as the foremost European source of tungsten, was pressured by both German and Allied governments, because of its wolframite ore deposits at Panasqueira. Tungsten's desirable properties, for example, resistance to high temperatures, its hardness and density, and its strengthening of alloys, made it a significant metal for the arms industry, both as a component of weapons and equipment and used in production itself, for example, in tungsten carbide cutting tools for machining steel. The name "tungsten" (from the Swedish tung sten, "heavy stone") is used in English, French, and numerous other languages as the name of the element, but not in the Nordic countries. "Tungsten" was the old Swedish name for the mineral scheelite ($CaWO_4$). "Wolfram" (or "volfram") is used in most European (especially Germanic,

FIGURE 7.58 Torbern Olof Bergman, painting on canvas by Ulrika Pasch (1735–1796).

FIGURE 7.59 Fausto de Luyar, Vienna 1788.

Spanish, and Slavic) languages and is derived from the mineral wolframite (now discredited and replaced by the series ferberite–hübnerite), which is the origin of the chemical symbol W. The name "wolframite" originates from the German word "wolf rahm" ("wolf soot" or "wolf cream"), the name given to tungsten by Swedish chemist and mineralogist Johannes Gottschalk Wallerius (July 11, 1709 to November 16, 1785) in 1747 (Fig. 7.60). This, in turn, originated from Latin "lupi spuma," the name German Humanist scholar and one of the leading experts on mineralogy and metallurgy Georg Agricola (March 24, 1494 to November 21, 1555) used for the element in 1546, which translates into English as "wolf's froth" and is a reference to the large quantities of tin consumed by the mineral during its extraction.

7.12.2 Mining, production, and major minerals

Tungsten is found mainly in the minerals constituting the solid solution series between ferberite, $FeWO_4$, and hübnerite, $MnWO_4$, and scheelite, $CaWO_4$ (Box 7.5). Other tungsten minerals range in their level of abundance from moderate to very rare and have almost no economic value. The world's tungsten reserves are estimated at 3,200,000 tons; they are mostly found in China, Canada, Russia, Vietnam, and Bolivia. Currently, China, Vietnam, and Russia are the foremost producers. Canada stopped production in late 2015 as a result of the closure of its only tungsten mine. In the meantime, Vietnam had significantly increased production in the 2010s, due to the major optimization of its local refining operations, and surpassed Russia and Bolivia. China is still the world's leader not only in mining but also in export and

FIGURE 7.60 Johannes Gottschalk Wallerius, 1849 lithograph by Otto Henrik Wallgren (1795–1857).

BOX 7.5 Tungsten deposits.

The main sources of tungsten are scheelite ($CaWO_4$) and the ferberite–hübnerite series (formerly known as wolframite, $FeWO_4$–$MnWO_4$), which are deposited by hydrothermal solutions. Most tungsten deposits are closely associated with granitic intrusions from which the metal-bearing hydrothermal fluids were probably originated. There are two major types of hydrothermal tungsten deposits: (1) scheelite-bearing skarns and (2) ferberite–hübnerite–bearing quartz veins. Scheelite-bearing skarns are formed where granitic intrusions come into contact with limestones. These deposits generally have a complex metal content, including Mo, Cu, Pb, Fe, and even Bi. Ferberite–hübnerite–bearing veins tend to occur in swarms of veins that cut granite and nearby, noncarbonate sedimentary rocks. Many tungsten veins are closely associated with Sn and Mo. Tungsten can also form deposits in low-temperature environments, though these have not been important metal sources. It has been found in hot-spring deposits and the Searles Lake evaporite brines that contain 70 mg/L dissolved tungsten. Tungsten production and reserves are about equally divided between scheelite skarns and ferberite–hübnerite veins.

consumption of tungsten products. The tungsten production slowly increases in other countries besides China due to the rising demand for tungsten. At the same time its supply by China is severely controlled by the Chinese Government, fighting illegal mining and extreme pollution due to mining and refining methods. Tungsten is considered to be a conflict mineral due to the unethical mining methods in the Democratic Republic of the Congo. There exists a large tungsten ore deposit on the edge of Dartmoor in the United Kingdom, which was mined during World Wars I and II as the Hemerdon Mine. Due to tungsten price increases, this mine was reopened in 2014, but ceased production in 2018.

Tungsten is produced from its ores in several steps. The ore is sooner or later converted to W(VI) oxide (WO_3), which is heated with hydrogen or carbon to produce powdered tungsten. Since tungsten has a high melting point, it is not commercially feasible to cast tungsten ingots. In its place, powdered tungsten is mixed with small quantities of powdered nickel or other metals and sintered. During the sintering process, the nickel or other metal diffuses into the tungsten, resulting in the formation of an alloy. Alternatively, tungsten can be obtained through hydrogen reduction of WF_6.

$$WF_6 + 3H_2 \rightarrow W + 6HF$$

or through pyrolytic decomposition

$$WF_6 \rightarrow W + 3F_2 \ (\Delta H_r = +)$$

FIGURE 7.61 Group of black, shiny ferberite, $FeWO_4$, crystals, the largest is 3 cm. Tazna mine, Cerro Tazna, Potosí, Bolivia.

FIGURE 7.62 Hübnerite, $MnWO_4$, dark red to brownish red elongated, wafer thin plates reaching 2.5 cm. Adams mine, Silverton, San Juan Co., Colorado, United States.

Around 50 minerals containing tungsten are known to occur in nature. Besides the native metal tungsten (W), a single carbide is known, qusongite (WC) in the elements class. Four sulfides are known: catamarcaite (Cu_6GeWS_8), kiddcreekite (Cu_6SnWS_8), ovamboite ($Cu_{20}(Fe,Cu,Zn)_6W_2Ge_6S_{32}$), and tungstite ($WS_2$). A single halide, pinalite ($Pb_3WO_5Cl_2$), contains W. Most minerals are found in the sulfate class, which includes the tungstates (wolframates) with around 35 minerals, for example, ferberite ($FeWO_4$) (Fig. 7.61)–hübnerite ($MnWO_4$) (Fig. 7.62) series (formerly known as wolframite), raspite ($PbWO_4$), scheelite ($CaWO_4$) (Fig. 7.63), stolzite ($PbWO_4$) (Fig. 7.64), and tungstibite ($Sb^{3+}_2WO_6$). The silicates are represented by four minerals: johnsenite-(Ce) ($Na_{12}Ce_3Ca_6Mn_3Zr_3WSi_{25}O_{73}(CO_3)(OH)_2$), khomyakovite ($Na_{12}Sr_3Ca_6Fe_3Zr_3W[Si_{25}O_{73}](O,OH,H_2O)_3(OH,Cl)_2$), manganokhomyakovite ($Na_{12}Sr_3Ca_6Mn_3Zr_3W[Si_{25}O_{73}](O,OH,H_2O)_3(OH,Cl)_2$), and welinite ($Mn^{2+}_6(W^{6+},Mg)_2(SiO_4)_2(O,OH)_6$).

FIGURE 7.63 Large orange octahedral scheelite, $Ca(WO_4)$, crystal to 4.5 cm on muscovite, $KAl_2(AlSi_3O_{10})(OH)_2$. Pingwu Co., Mianyang pref., Sichuan prov., China.

FIGURE 7.64 Stolzite, $Pb(WO_4)$, honey yellow tetragonal dipyramidal crystals about 0.75 mm. Cordillera mine, Georgiana Co., New South Wales, Australia.

7.12.3 Chemistry

Naturally occurring tungsten exists as four stable isotopes, ^{182}W (26.50% abundance), ^{183}W (14.31%), ^{184}W (30.64%), and ^{186}W (28.43%), and one very long-lived radioisotope, ^{180}W (0.12%). Theoretically, all five can decay into isotopes of Hf via α emission, but only ^{180}W decay has been detected, with a half-life of 1.8 ($\pm$ 0.2) $\times 10^{18}$ years; this means that on average this produces around 2 α decays of ^{180}W per gram of natural W annually. The other naturally occurring isotopes have not been detected to decay, constraining their half-lives to be no less than 4×10^{21} years. An additional 30 synthetic W radioisotopes have been observed, the most stable are ^{181}W which has a half-life of 121.2 days (ε decay to ^{181}Ta), ^{185}W which has a half-life of 75.1 days (β^- decay to ^{185}Re), ^{188}W which has a half-life of 69.4 days, ^{178}W which has a half-life of 21.6 days, and ^{187}W which has a half-life of 23.72 h. All other radioisotopes have half-lives of less than 3 h, and the majority of them have half-lives less than 8 min. In addition, W also has 11 meta states, with the most stable being ^{179m}W with a half-life of 6.4 min.

In its raw form, tungsten is a hard steel-gray metal that is often brittle and hard to work (Table 7.12). If produced very pure, tungsten retains its hardness and becomes malleable enough that it can be worked easily. Of all metals in pure form, tungsten has the highest melting point (3422°C), lowest vapor pressure (at temperatures >1650°C), and the highest tensile

TABLE 7.12 Tungsten properties.

Appearance	Grayish white, lustrous
Standard atomic weight $A_{r,std}$	183.84
Block	d-Block
Element category	Transition metal
Electron configuration	[Xe] $4f^{14}$ $5d^4$ $6s^2$
Phase at STP	Solid
Melting point	3422°C
Boiling point	5930°C
Density (near r.t.)	19.3 g/cm^3
When liquid (at m.p.)	17.6 g/cm^3
Heat of fusion	52.31 kJ/mol
Heat of vaporization	774 kJ/mol
Molar heat capacity	24.27 J/(mol·K)
Oxidation states	−4, −2, −1, 0, +1, +2, +3, **+4**, +5, **+6**
Ionization energies	1st: 770 kJ/mol 2nd: 1700 kJ/mol
Atomic radius	Empirical: 139 pm
Covalent radius	162 ± 7 pm

STP, Standard temperature and pressure.
Bold font indicates main oxidation state.

strength. While carbon remains solid at higher temperatures than tungsten, carbon sublimes at atmospheric pressure instead of melting, so it has no melting point. Tungsten has the lowest coefficient of thermal expansion of any pure metal. The low thermal expansion and high melting point and tensile strength of tungsten are due to strong covalent bonds formed between tungsten atoms by the 5d electrons. Tungsten forms two major crystalline forms: α and β. The former has a *bcc* structure and is the more stable form. The structure of the β phase is called A15 cubic [also known as β-W or Cr_3Si structure types, a series of intermetallic compounds with the chemical formula A_3B (where A is a transition metal and B can be any element) and a specific structure]; it is metastable but can coexist with the α phase at ambient conditions because of nonequilibrium synthesis or stabilization by impurities. Unlike the α phase, which crystallizes in isometric grains, the β form shows a columnar habit. The α phase has one-third of the electrical resistivity and a much lower superconducting transition temperature TC relative to the β phase: c. 0.015K (−272.135°C) versus 1–4K (−272.15– −269.15°C); mixing the two phases results in intermediate TC values. The TC value can also be increased by alloying tungsten with another metal (e.g., 7.9K/−265.3°C for W-Tc). Elemental tungsten resists attack by oxygen, acids, and alkalis. The most common formal oxidation state of tungsten is +6, but it displays all oxidation states from −4 to +6.

7.12.3.1 Reactions with oxygen and water

Tungsten typically combines with oxygen to form the yellow tungstic oxide, WO_3, which dissolves in aqueous alkaline solutions as tungstate ions, WO_4^{2-}. WO_3 can form intercalation compounds with alkali metals, which are known as bronzes; for example, sodium tungsten bronze. It is formed as an intermediate in the recovery of tungsten from its minerals. Tungsten ores are reacted with alkalis to produce WO_3. Further reaction with carbon or hydrogen gas reduces WO_3 to the pure metal. WO_3 is a strong oxidative agent, and it reacts with REEs, iron, copper, aluminum, manganese, zinc, chromium, molybdenum, carbon, hydrogen, and silver to form the pure tungsten metal, and gold and platinum to form tungsten dioxide.

$$2WO_3 + 3C \rightarrow 2W + 3CO_2 \text{ (high temperature)}$$
$$WO_3 + 3H_2 \rightarrow W + 3H_2O \text{ (550 − 850°C)}$$
$$WO_3 + 2Fe \rightarrow W + Fe_2O_3$$
$$2WO_3 + Pt \rightarrow 2WO_2 + PtO_2$$

Tungsten(VI) oxide occurs naturally in the form of hydrates, which include minerals tungstite $WO_3 \cdot H_2O$, meymacite $WO_3 \cdot 2H_2O$, and hydrotungstite (of the same composition as meymacite). These minerals are rare to very rare secondary tungsten minerals. Tungsten trioxide can be prepared in several different ways. $CaWO_4$, or scheelite, can react with HCl to form tungstic acid, which decomposes to WO_3 and water at high temperatures.

$$CaWO_4 + 2HCl \rightarrow CaCl_2 + H_2WO_4$$
$$H_2WO_4 \rightarrow H_2O + WO_3$$

Another typical route to obtain WO_3 is by calcination of ammonium paratungstate (APT) under oxidizing conditions.

$$(NH_4)_{10}[H_2W_{12}O_{42}]\cdot 4H_2O \rightarrow 12WO_3 + 10NH_3 + 10H_2O$$

The WO_3 crystal structure is temperature dependent: tetragonal at temperatures above 740°C, orthorhombic from 330°C to 740°C, monoclinic from 17°C to 330°C, triclinic from −50°C to 17°C, and monoclinic again at temperatures below −50°C. The most common structure of WO_3 is monoclinic with space group *P2₁/n*.

Tungsten(IV) oxide, WO_2, is a bronze-colored solid that crystallizes in a monoclinic cell. The rutile-like structure features distorted octahedral WO_6 centers with alternate short W-W bonds (248 pm). Each tungsten center has the d^2 configuration, which gives the material a high electrical conductivity. WO_2 is prepared by reduction of WO_3 with tungsten powder in about 40 h at 900°C. An intermediate in this reaction is the partially reduced, mixed valence species $W_{18}O_{49}$.

$$2WO_3 + W \rightarrow 3WO_2$$

Single crystals are obtained by chemical transport technique using iodine. Iodine transports WO_2 in the form of the volatile species WO_2I_2.

Tungsten(III) oxide (W_2O_3) has been reported by Dezelah et al. (2006) as being grown as a thin film by atomic layer deposition at temperatures between 140°C and 240°C using $W_2(N(CH_3)_2)_6$ as a precursor. It is not referred to in major textbooks. Some older literature refers to the compound W_2O_3, but as the atomic weight of tungsten was believed at the time to be 92, that is, approximately half the modern accepted value of 183.84, the compound being referred to was WO_3. Tungsten pentoxide (W_2O_5) was reported in early literature but was shown to be $W_{18}O_{49}$. Sometimes called mineral blue, it is a blue solid obtained by the reaction of WO_3 and tungsten metal at 700°C. There are a number of these unusual intermediate oxides known that can form from the reaction of the metal and WO_3; for example, $W_{20}O_{58}$, $W_{24}O_{70}$, and $W_{18}O_{49}$ contain both octahedral and pentagonal bipyramidal coordinations of the metal atoms by oxygen.

In aqueous solution, tungstate forms the heteropoly acids and polyoxometalate anions under neutral and acidic conditions. As tungstate is increasingly treated with acid, it first forms the soluble, metastable "paratungstate A" anionic group, $W_7O_{24}^{6-}$, which over time transforms to the less-soluble "paratungstate B" anion, $H_2W_{12}O_{42}^{10-}$. Further acidification forms the very soluble metatungstate anion, $H_2W_{12}O_{40}^{6-}$, after which equilibrium is reached. The metatungstate ion exists as a symmetric cluster of 12 tungsten−oxygen octahedra known as the Keggin anion. Many other polyoxometalate anions are known to exist as metastable species. The substitution of a different atom such as phosphorus in place of the two central hydrogens in metatungstate produces a wide variety of heteropoly acids, such as phosphotungstic acid $H_3PW_{12}O_{40}$.

7.12.3.2 Halides

The broad range of oxidation states of tungsten is shown in its variety of chlorides: tungsten(II) chloride, which exists as the hexamer W_6Cl_{12}, is a polymeric cluster compound. The material dissolves in concentrated HCl, forming $(H_3O)_2[W_6Cl_{14}](H_2O)_x$. Heating this salt results in yellow-brown W_6Cl_{12}. Tungsten(II) chloride is prepared by reduction of the hexachloride with bismuth as a typical reductant.

$$6WCl_6 + 8Bi \rightarrow W_6Cl_{12} + 8BiCl_3$$

Tungsten(III) chloride, which exists as the hexamer W_6Cl_{18}, is a cluster compound. It is a brown solid, formed by chlorination of tungsten(II) chloride. Containing 12 doubly bridging chloride ligands, the cluster takes on a structure related to the corresponding chlorides of niobium and tantalum. In contrast, W_6Cl_{12} contains eight triply bridging chlorides. A related mixed valence W(III)−W(IV) chloride is produced by reduction of the hexachloride with bismuth.

$$9WCl_6 + 8Bi \rightarrow 3W_3Cl_{10} + 8BiCl_3$$

Tungsten(IV) chloride, WCl_4, is a black diamagnetic solid with a polymeric structure. It contains linear chains of tungsten atoms each in octahedral geometry. Of the six chloride ligands linked to each W center, four are bridging ligands. The W−W separations are alternatingly bonding (2.688 Å) and nonbonding (3.787 Å). WCl_4 is typically produced by reducing WCl_6. Many reductants have been reported, including red phosphorus, tungsten hexacarbonyl, gallium, tin, and antimony. The latter is reported to be optimal.

$$3WCl_6 + 2Sb \rightarrow 3WCl_4 + 2SbCl_3$$

Tungsten(V) chloride, WCl_5, is a black solid with a dimeric structure, with a pair of octahedral tungsten(V) centers bridged by two chloride ligands. The W−W separation is 3.814 Å, which is nonbonding. The compound is isostructural with Nb_2Cl_{10} and Mo_2Cl_{10}. The compound evaporates to give trigonal bipyramidal WCl_5 monomers. This compound is analogous in many ways to the more familiar molybdenum pentachloride. It is prepared by reducing WCl_6. One method involves the use of tetrachloroethylene as the reductant.

$$2WCl_6 + C_2Cl_4 \rightarrow W_2Cl_{10} + C_2Cl_6$$

The blue-green solid is volatile under vacuum and slightly soluble in nonpolar solvents. The compound is oxophilic and is very reactive toward Lewis bases.

Tungsten(VI) chloride, WCl_6, is a dark violet blue species that exists as a volatile solid under standard conditions. It is an important starting reagent in the preparation of tungsten compounds. As a d^0 ion, W(VI) forms diamagnetic derivatives. WCl_6 is octahedral with equivalent W–Cl distances of 2.24–2.26 Å. WCl_6 can be prepared by chlorinating tungsten metal in a sealed tube at 600°C.

$$W + 3Cl_2 \rightarrow WCl_6$$

Tungsten(VI) chloride is a blue-black solid at room temperature. At lower temperatures, it turns into wine red. A red form of the compound can be formed by quickly condensing its vapor, which transforms to the blue-black form on gentle heating. It is easily hydrolyzed, even by moist air, giving the orange oxychlorides $WOCl_4$ and WO_2Cl_2, and then, tungsten trioxide. WCl_6 is soluble in carbon disulfide, carbon tetrachloride, and phosphorus oxychloride. Methylation with trimethylaluminum forms hexamethyl tungsten.

$$WCl_6 + 3Al_2(CH_3)_6 \rightarrow W(CH_3)_6 + 3Al_2(CH_3)_4Cl_2$$

Treatment with butyl lithium produces a reagent that is useful for deoxygenation of epoxides. The chloride ligands in WCl_6 can be substituted by many anionic ligands including Br^-, NCS^-, and RO^- (R = alkyl, aryl). Reduction of WCl_6 gives, successively, tungsten(V) chloride and tungsten(IV) chloride. WCl_6 is an aggressively corrosive oxidant and hydrolyzes to release hydrogen chloride.

Tungsten(VI) fluoride, also known as tungsten hexafluoride, WF_6, is a toxic, corrosive, colorless gas, with a density of about 13 g/L (roughly 11 times heavier than air). It is one of the densest gases known under standard conditions. WF_6 is frequently used by the semiconductor industry to form tungsten films, through CVD. The WF_6 molecule is octahedral with the symmetry point group of O_h. The W-F bond distances are 183.2 pm. Between 2.3°C and 17°C, tungsten hexafluoride condenses into a pale-yellow liquid with a density of 3.44 g/cm^3 at 15°C. At 2.3°C it freezes into a white solid with a cubic crystalline structure, with lattice constant of 628 pm and a calculated density of 3.99 g/cm^3. At −9°C this structure converts to an orthorhombic solid with lattice constants of a = 960.3 pm, b = 871.3 pm, and c = 504.4 pm and a density of 4.56 g/cm^3. In this phase, the W-F distance is 181 pm, and the mean closest intermolecular contacts are 312 pm. WF_6 is usually produced by the exothermic reaction of F_2 gas with tungsten powder between 350°C and 400°C.

$$W + 3F_2 \rightarrow WF_6$$

The gaseous product is isolated from WOF_4, a common impurity, by distillation. As a variant of the direct fluorination, the metal can be placed in a heated reactor, slightly pressurized to 1.2 to 2.0 psi (8.3 to 13.8 kPa), with a constant flow of WF_6 infused with a small amount of fluorine gas. The fluorine gas in the above method can be replaced by ClF, ClF_3, or BrF_3. Another procedure for producing WF_6 is to react WO_3 with HF, BrF_3, or SF_4. WF_6 can also be obtained by conversion of WCl_6.

$$WCl_6 + 6HF \rightarrow WF_6 + 6HCl \text{ or}$$
$$WCl_6 + 2AsF_3 \rightarrow WF_6 + 2AsCl_3 \text{ or}$$
$$WCl_6 + 3SbF_5 \rightarrow WF_6 + 3SbF_3Cl_2$$

On contact with water, WF_6 produces hydrogen fluoride (HF) and tungsten oxyfluorides, eventually forming tungsten trioxide.

$$WF_6 + 3H_2O \rightarrow WO_3 + 6HF$$

Unlike some other metal fluorides, WF_6 is neither a useful fluorinating agent nor a powerful oxidant. It can be reduced to the yellow WF_4. WF_6 reacts upon contact with a silicon substrate. The WF_6 decomposition on silicon is temperature dependent.

$$2WF_6 + 3Si \rightarrow 2W + 3SiF_4 \quad \text{below 400°C and}$$
$$WF_6 + 3Si \rightarrow W + 3SiF_2 \quad \text{above 400°C}$$

This dependence is crucial, as twice as much silicon is being consumed above 400°C. The deposition occurs selectively on pure Si only, but not on silicon oxide or nitride, therefore the reaction is highly susceptible to contamination or substrate pretreatment. The decomposition reaction is fast but saturates when the tungsten layer reaches a thickness of 10–15 μm. The saturation occurs since the tungsten layer stops diffusion of WF_6 molecules to the Si substrate, which is the only catalyst of molecular decomposition in this process. If the deposition occurs not in an inert but in an oxygen-containing atmosphere (air), then instead of tungsten, a tungsten oxide layer is produced. WF_6 is an extremely corrosive compound that attacks any tissue. Because of the formation of HF upon reaction of WF_6 with humidity, WF_6 storage vessels have Teflon gaskets.

Tungsten(V) fluoride, WF_5, is a hygroscopic yellow solid. Like most pentafluorides, it adopts a tetrameric structure, consisting of $[WF_5]_4$ molecules, where each W center achieves octahedral coordination. WF_5 is produced by the reaction of tungsten and tungsten hexafluoride.

$$W + 5WF_6 \rightarrow 6WF_5$$

At 25°C, it disproportionates to the tetra- and hexafluorides.

$$2WF_5 \rightarrow WF_4 + WF_6$$

Tungsten tetrafluoride is an inorganic compound with the formula WF_4. This little-known solid has been invoked, along with WF_5, as an intermediary in the CVD of tungsten films using WF_6. WF_4 was shown to have a polymeric structure based on Mössbauer spectroscopy. It has been produced by the reaction of the coordination complex $WCl_4(MeCN)_2$ with AsF_3. It has also been formed by the reaction of WF_6 and a W filament at 600°C–800°C. The compound can be re-oxidized to W(VI) compounds by treatment with fluorine and chlorine.

$$WF_4 + X_2 \rightarrow WF_4X_2$$

Upon heating, it disproportionates to WF_6 and tungsten metal.

Tungsten(V) bromide, WBr_5, consists of an edge-shared bioctahedral structure, with two bridging bromide ligands, so its molecular formula is actually W_2Br_{10}. Tungsten(V) bromide is formed by reacting tungsten powder with bromine in between 650°C and 1000°C. The product is frequently contaminated with tungsten hexabromide. Tungsten(V) bromide is the precursor to other tungsten compounds by reduction reactions, for example, tungsten(IV) bromide can be prepared by reduction with aluminum or tungsten. WBr_4 can be purified by chemical vapor transport.

$$3WBr_5 + Al \rightarrow 3WBr_4 + AlBr_3$$

Excess WBr_5 and $AlBr_3$ are then removed through sublimation at 240°C.

Tungsten(II) bromide can then be prepared by heating the tetrabromide. At 450°C–500°C, gaseous pentabromide is evolved leaving yellow-green residue of WBr_2. A similar method can also be used for the synthesis of tungsten(II) chloride. Since it is relatively easy to reduce tungsten pentahalides, they can be employed as alternative synthesis routes to tungsten (IV) halide adducts, for example, the reaction of WBr_5 with pyridine gives $WBr_4(py)_2$.

$$2WBr_5 + 7C_5H_5N \rightarrow 2WBr_4(C_5H_5N)_2 + \text{bipyridine} + C_5H_5NHBr$$

Several tungsten oxyhalides are known. Tungsten(VI) oxytetrachloride, $WOCl_4$, is a diamagnetic solid used to prepare other tungsten complexes. The orange-colored compound is soluble in nonpolar solvents, while it reacts with alcohols and water and forms adducts with Lewis bases. It consists of weakly associated square pyramidal monomers. $WOCl_4$ is formed from WO_3 or WCl_6.

$$WO_3 + 2SOCl_2 \rightarrow WOCl_4 + 2SO_2$$
$$WCl_6 + (Me_3Si)_2O \rightarrow WOCl_4 + 2Me_3SiCl$$

$WOCl_4$ is Lewis acidic. It is a precursor to catalysts used for polymerization of alkynes. Tungsten dichloride dioxide, WO_2Cl_2, is a yellow-colored solid that is used as a precursor for other tungsten compounds. Like other tungsten halides, WO_2Cl_2 is sensitive to moisture, undergoing hydrolysis. WO_2Cl_2 is formed by ligand redistribution reaction from WO_3 and WCl_6.

$$2WO_3 + WCl_6 \rightarrow 3WO_2Cl_2$$

Using a two-zone tube furnace, a vacuum-sealed tube containing these solids is heated to 350°C. The yellow product sublimes to the cooler end of the reaction tube. No redox reaction occurs in this process. An alternative route emphasizes the oxophilicity of tungsten.

$$WCl_6 + 2O(Si(CH_3)_3)_2 \rightarrow 3WO_2Cl_2 + 4ClSi(CH_3)_3$$

This reaction, like the preceding one, proceeds via the intermediary $WOCl_4$. The compound is a polymer consisting of distorted octahedral W centers. The monomer is characterized by two short W-O distances, typical for a multiple W-O bond, and two long W-O distances more typical of a single or dative W-O bond. Tungsten forms several other oxyhalides including $WOCl_3$ and $WOCl_2$. The corresponding bromides ($WOBr_4$, $WOBr_3$, $WOBr_2$, and WO_2Br_2) are also known as are WOF_4, WO_2F_2, WO_2I_2, and WO_2I.

7.12.3.3 Carbides, borides, nitrides, chalcogenides, and silicides

Tungsten carbides (W_2C and WC) are formed by heating powdered tungsten with carbon at 1400°C–2000°C. Other routes include a lower temperature fluid bed process that reacts either tungsten metal or blue WO_3 with CO/CO_2 mixture and H_2 between 900°C and 1200°C. WC can also be formed by heating WO_3 with graphite: directly at 900°C or in hydrogen at 670°C following by carburization in argon at 1000°C. Chemical vapor deposition methods that have been investigated are, for example, reacting WCl_6 with H_2 (as a reducing agent) and methane (as the source of carbon) at 670°C

$$WCl_6 + H_2 + CH_4 \rightarrow WC + 6HCl$$

or reacting WF_6 with H_2 (as reducing agent) and methanol (as source of carbon) at 350°C

$$WF_6 + 2H_2 + CH_3OH \rightarrow WC + 6HF + H_2O$$

W_2C is resistant to chemical attack, although it reacts strongly with chlorine to form tungsten hexachloride (WCl_6).

Tungsten borides are compounds of tungsten and boron. Their most remarkable property is high hardness. The Vickers hardness of WB or WB_2 crystals is ~20 GPa and that of WB_4 is ~30 GPa for loads exceeding 3 N. Single crystals of WB_{2-x}, $x = 0.07-0.17$ (about 1 cm diameter, 6 cm length) can be formed by the floating zone method, while WB_4 crystals can be grown by arc melting a mixture of elemental tungsten and boron. WB_2 has the same hexagonal structure as most diborides (AlB_2, MgB_2, etc.) WB has several forms, α (tetragonal), β (orthorhombic), and δ (tetragonal). The oxidation of W_2B, WB, and WB_2 is substantial at temperatures above 600°C. The final oxidation products contain WO_3 and possibly amorphous B_2O_3 or H_3BO_3. The melting temperatures of W_2B, WB, and WB_2 are 2670°C, 2655°C, and 2365°C, respectively.

Tungsten nitride (W_2N, WN, WN_2) is an inorganic compound, a nitride of tungsten. It is a hard, solid, brown-colored ceramic material that is electrically conductive and decomposes in water. It is employed in microelectronics as a contact material, for conductive layers, and barrier layers between silicon and other metals, for example, tungsten or copper. It is less frequently used than titanium nitride or tungsten films. Tungsten nitride forms together with tungsten dioxide, tungsten trioxide, and tungsten pentoxide when a light bulb breaks while the filament is heated.

Tungsten disulfide is an inorganic chemical compound composed of tungsten and sulfur with the chemical formula WS_2. It occurs naturally as the rare mineral tungstenite. This compound is part of certain catalysts used for hydrodesulfurization and hydrodenitrification. WS_2 adopts a layered structure similar, or isotypic with MoS_2, instead with W atoms situated in trigonal prismatic coordination sphere. Owing to this layered structure, WS_2 forms inorganic nanotubes, which were discovered after heating a thin sample of WS_2. WS_2 is formed via a number of routes. Many of these routes comprise treating oxides with sources of sulfide or hydrosulfide, supplied as hydrogen sulfide or generated in situ. Widely used methods for the growth of monolayer WS_2 are, for example, CVD, PVD, or metal organic CVD (MOCVD), though most current methods result in sulfur vacancy defects in excess of 1×10^{13} cm^{-2}. Other methods involve thermolysis of tungsten(VI) sulfides (e.g., $(R_4N)_2WS_4$) or the equivalent (e.g., WS_3).

Tungsten diselenide, WSe_2, has a hexagonal crystalline structure comparable to MoS_2. Every tungsten atom is covalently bonded to six selenium ligands in a trigonal prismatic coordination sphere, and each selenium is bonded to three tungsten atoms in a pyramidal geometry. Heating thin films of tungsten under pressure from gaseous selenium and high temperatures (>527°C/800K) using the sputter deposition technique leads to the films crystallizing in hexagonal structures with the correct stoichiometric ratio.

$$W + 2Se \rightarrow WSe_2$$

Tungsten(IV) telluride (WTe_2) is an inorganic semimetallic chemical compound. In October 2014, WTe_2 was found to exhibit an extremely large magnetoresistance: 13 million percent with no known saturation point. The resistance is proportional to the square of the magnetic field. This may be because of the compound being the first example of a compensated semimetal, in which the number of mobile holes is the same as the number of electrons. The structure of WTe_2 is layered, and it can be exfoliated into thin sheets down to monolayers. Nevertheless, electrons can easily move between the layers, in contrast to other two-dimensional (2D) semiconductors. The fraction of charge carriers is 0.005 per formula unit (WTe_2). When subjected to pressure, the magnetoresistance effect in WTe_2 is reduced. At a pressure of 10.5 GPa magnetoresistance disappears. Above this same pressure of 10.5 GPa, WTe_2 can become a superconductor. At 13.0 GPa the transition to superconductivity occurs below −266,7°C (6.5K). It has also been reported by Sie et al.

(2019) that terahertz-frequency light pulses can switch the crystal structure of WTe_2 between orthorhombic and monoclinic by altering the material's atomic lattice.

Tungsten silicide (WSi_2) is an inorganic compound, a silicide of tungsten. It is an electrically conductive ceramic material. Tungsten silicide can react violently with substances such as strong acids, fluorine, oxidizers, and interhalogens. It is employed in microelectronics as a contact material, with a resistivity of 60–80 μΩ cm; it forms at 1000°C. It is frequently used as a shunt over polysilicon lines to increase their conductivity and increase signal speed. Tungsten silicide layers can be formed by CVD, for example, using monosilane or dichlorosilane with WF_6 as source gases. The deposited film is nonstoichiometric and needs annealing to convert to more conductive stoichiometric form. Tungsten silicide is a replacement for earlier tungsten films. Tungsten silicide is also applied as a barrier layer between silicon and other metals. Tungsten silicide is also useful in microelectromechanical systems, where it is mostly used as thin films for the manufacture of microscale circuits. For this, films of tungsten silicide can be plasma-etched using, for example, nitrogen trifluoride gas. WSi_2 works well in applications as oxidation-resistant coatings. Similar to $MoSi_2$ the high emissivity of WSi_2 makes this material appealing for high-temperature radiative cooling, with implications in heat shields.

7.12.3.4 Organotungsten compounds

Organotungsten compounds are numerous and span a range of oxidation states. Noteworthy examples are the trigonal prismatic $W(CH_3)_6$ and octahedral $W(CO)_6$. Hexamethyltungsten is the chemical compound $W(CH_3)_6$ also written WMe_6. Categorized as a transition metal alkyl complex, hexamethyltungsten is an air-sensitive, red, crystalline solid at room temperature; but it is extremely volatile and sublimes at −30°C. Due to its six methyl groups it is very soluble in petroleum, aromatic hydrocarbons, ethers, carbon disulfide, and carbon tetrachloride. Hexamethyltungsten was first described in 1973 by Shortland and Wilkinson, who published its formation by the reaction of methyllithium with tungsten hexachloride in diethyl ether. The synthesis was inspired in part by earlier work that suggested that tetrahedral methyl transition metal compounds are thermally unstable, in the hopes that an octahedral methyl compound would prove to be more robust. In 1976 Galyer and Wilkinson revealed an improved synthesis route using trimethylaluminum in conjunction with trimethylamine, rather than methyllithium. The stoichiometry of the improved synthesis can be represented as:

$$WCl_6 + 6Al(CH_3)_3 \rightarrow W(CH_3)_6 + 6Al(CH_3)_2Cl$$

Alternatively, the alkylation can employ dimethylzinc.

$$WX_6 + 3Zn(CH_3)_2 \rightarrow W(CH_3)_6 + 3ZnX_2 \ (X = F, Cl)$$

$W(CH_3)_6$ has a distorted trigonal prismatic geometry with C_{3v} symmetry for the WC_6 framework and C_3 symmetry including the hydrogen atoms. The structure (excluding the hydrogen atoms) can be envisaged as consisting of a central atom, capped on either side by two eclipsing sets of three carbon atoms, with one triangular set slightly larger but also closer to the central atom than the other. The trigonal prismatic geometry is unusual in that most six-coordinate organometallic compounds have an octahedral molecular geometry. At room temperature, hexamethyltungsten decomposes, releasing methane and trace amounts of ethane. The black residue is purported to contain polymethylene and tungsten, but the decomposition of $W(CH_3)_6$ to form tungsten metal is highly unlikely. The following equation is the approximate stoichiometry as originally proposed by Shortland and Wilkinson (1973).

$$W(CH_3)_6 \rightarrow 3CH_4 + (CH_2)_3 + W$$

Like many organometallic complexes, WMe_6 is destroyed by oxygen. Likewise, acids give methane and unknown tungsten derivatives, while halogens give the methyl halide and leave the tungsten halide. Reaction of $W(CH_3)_6$ with F_2 diluted with Ne at −90°C results in $W(CF_3)_6$ in 50% yield as an extremely volatile white solid. Hexamethyltungsten (VI) reacts with trimethylphosphine in light petroleum to form $WMe_6(PMe_3)$, which in neat PMe_3, with UV irradiation produces the carbyne complex trans-WMe(:::CMe)$(PMe_3)_4$ in high yield. Serious explosions have been reported as a result of working with $W(CH_3)_6$, even in the absence of air.

Tungsten hexacarbonyl is the chemical compound with the formula $W(CO)_6$. This colorless compound is noteworthy as a volatile, air-stable derivative of tungsten in its zero oxidation state. $W(CO)_6$ is formed by the reduction of WCl_6 under pressure of carbon monoxide. The compound is relatively air stable. It is slightly soluble in nonpolar organic solvents. $W(CO)_6$ is widely used in electron beam–induced deposition technique. It is easily vaporized and decomposed by the electron beam providing a convenient source of tungsten atoms. $W(CO)_6$ has an octahedral geometry consisting of six rod-like CO ligands radiating from the central W atom with dipole moment 0 D. All reactions of $W(CO)_6$ start with displacement of some CO ligands in $W(CO)_6$. $W(CO)_6$ behaves comparable to $Mo(CO)_6$ but tends to form compounds that are kinetically more robust. One derivative is the dihydrogen complex $W(CO)_3[P(C_6H_{11})_3]_2(H_2)$. Three of these CO ligands can be substituted by acetonitrile. $W(CO)_6$ has been utilized to desulfurize organosulfur compounds and as a precursor to catalysts for alkene metathesis. Replacement of the carbonyl groups by either π-donor or σ-donor ligands is possible, giving a host of materials of the form $[W(CO)_{6-x}M_x]$ or $[W(CO)_{6-2x}(M-M)_x]$ (e.g., M = NO, NH_3, CN, PF_3; M-M = bipy, butadiene). $[W(CO)_5X^-]$ ions (X = halogen, CN, or SCN) are formed in this way. The low-temperature reaction ($-78°$) of the halogens with $[W(CO)_6]$ results in the W(II) carbonyl halides, $[W(CO)_4X_2]$ from which many adducts, $[W(CO)_3L_2X_2]$, are obtained. While not all of these have been fully characterized, those that have are seven coordinated and mostly capped octahedral. Reduction of the hexacarbonyls with a borohydride in liquid ammonia results in the formation of dimeric $[W_2(CO)_{10}]^{2-}$, which is isostructural with the isoelectronic $[Mn_2(CO)_{10}]$. Hydrolysis of these dimers produces the yellow hydrides $[(CO)_5W\text{-}H\text{-}W(CO)_5]$, which maintain the 18 valence electron configuration through a three-center, two-electron W-H-W bond. A related compound $[NEt_4]_2^+[H_2W_2(CO)_8]^{2-}$ is of interest as it has two hydrogen bridges and a W-W distance indicative of a W— W double bond (301.6 pm compared to about 320 pm for a W-W single bond).

Tetrakis(hexahydropyrimidinopyrimidine)ditungsten(II), known as ditungsten tetra(hpp), is the name of the coordination compound with the formula $W_2(hpp)_4$. This compound consists of a pair of tungsten centers linked by the conjugate base of four hexahydropyrimidopyrimidine (hpp) ligands. It adopts a structure sometimes called a Chinese lantern structure or paddlewheel compound, the prototype being copper(II) acetate. The molecule is of research interest because it has the lowest ionization energy (3.51 eV) of all stable chemical elements or chemical compounds since 2005. This value is even lower than of cesium with 3.89 eV (or 375 kJ/mol) located at the extreme left lower corner of the periodic table (although francium is at a lower position in the periodic table compared to cesium, it has a higher ionization energy and is radioactive) or known metallocene reducing agents such as decamethylcobaltocene with 4.71 eV. This coordination compound is formed by the reaction of tungsten hexacarbonyl with 1,3,4,6,7,8-hexahydro-2H-pyrimido [1,2-a]pyrimidine (Hhpp) in o-dichlorobenzene at 200°C.

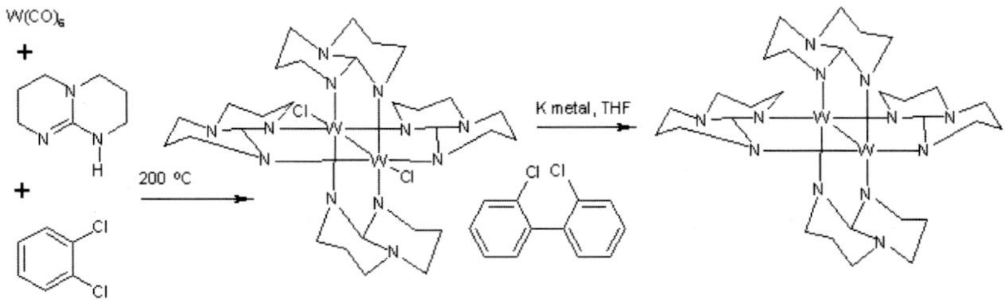

The reaction produces $W_2(hpp)_4Cl_2$. Dichlorobenzene provides the chlorine atoms and is itself reductively coupled to 2,2′-dichlorobiphenyl. The bond order between the tungsten centers in $W_2(hpp)_4Cl_2$ is three. This dichloride is further reduced by potassium metal in THF to $W_2(hpp)_4$. This species has a quadruple bond between the two tungsten centers. Related quadruple bonded complexes include $[W_2Cl_8]^{4-}$ and $[Mo_2Cl_8]^{4-}$. Due to its low ionization energy, $W_2(hpp)_4$ is easily oxidized back to the dichloride by dichloromethane. It is easily oxidized to the corresponding cation with the oxidants fullerene or with tetracyanoquinodimethane.

7.12.4 Major uses

Tungsten is primarily utilized in the manufacture of hard materials based on W carbide, one of the hardest carbides, with a melting point of 2770°C. WC is an effective electrical conductor, though W_2C is less so. WC is employed to produce wear-resistant abrasives, and "carbide" cutting tools, for example, knives, drills, circular saws, milling and turning tools utilized by the metalworking, woodworking, and mining, petroleum, and construction industries. Carbide tooling is essentially a ceramic/metal composite, in which metallic Co acts as a binding (matrix) material to hold the WC particles in place. This type of industrial usage covers around 60% of present-day W consumption. The jewelry business creates rings of sintered W carbide, W carbide/metal composites, and also metallic W. WC/metal composite rings utilize Ni as the metal matrix instead of Co as it has a higher luster after polishing. Occasionally producers or retailers refer to W carbide as a metal, however it is a ceramic. Due to its hardness, rings created of WC are very abrasion resistant and will keep a polished finish longer than rings made of metallic W. WC rings are brittle, though, and can crack under a sharp blow.

The hardness and density of W are used in making heavy metal alloys, for example, high-speed steel, which can have as much as 18% W. Its high melting point makes W a good element for uses such as rocket nozzles, for example, in the UGM-27 Polaris submarine-launched ballistic missile. W alloys are employed in a variety of different applications, including the aerospace and car industries and radiation shielding. Superalloys with W such as Hastelloy (Haynes International, Inc., headquartered in Kokomo, Indiana, is one of the largest producers of corrosion-resistant and high-temperature alloys. In addition to Kokomo, Haynes has manufacturing facilities in Arcadia, Louisiana, and Mountain Home, North Carolina.) and Stellite (a range of cobalt–chromium alloys designed for wear resistance. The alloys may also contain tungsten or molybdenum and a small but important amount of carbon. Stellite is a trademarked name of Kennametal Inc. Prior to that it was owned by Union Carbide, Stellite Division. Invented by American inventor, metallurgist, automotive pioneer, entrepreneur, and industrialist Elwood Haynes (October 14, 1857 to April 13, 1925) in the early 1900s as a substitute for cutlery that stained (or that had to be constantly cleaned).) are used in turbine blades and wear-resistant parts and coatings.

Quenched (martensitic) W steel (around 5.5% to 7.0% W and 0.5% to 0.7% C) was employed for making hard permanent magnets, because of its high remanence (or remanent magnetization or residual magnetism is the magnetization left behind in a ferromagnetic material (such as Fe) after an external magnetic field is removed) and coercivity (in electrical engineering and materials science, the coercivity, also called the magnetic coercivity, coercive field or coercive force, is a measure of the ability of a ferromagnetic material to withstand an external magnetic field without becoming demagnetized), as observed by British physicist and electrical engineer John Hopkinson (July 27, 1849 to August 27, 1898) as early as 1886. The magnetic properties of a metal or an alloy are extremely sensitive to its microstructure. For instance, though the element W is not ferromagnetic (but Fe is), when added to steel in these proportions, it stabilizes the martensite phase (a very hard form of steel crystalline structure. It is named after the German metallurgist Adolf Martens (March 6, 1850 to July 24, 1914). By analogy the term can also refer to any crystal structure that is formed by diffusionless transformation), which has an increased ferromagnetism, in comparison to the ferrite (iron) phase, because of its larger resistance to magnetic domain wall motion (A domain wall is a term used in physics that can have similar meanings in magnetism, optics, or string theory. These phenomena can all be generically described as topological solitons, which occur whenever a discrete symmetry is spontaneously broken. Nonmagnetic inclusions in the volume of a ferromagnetic material, or dislocations in crystallographic structure, can cause "pinning" of the domain walls. Such pinning sites cause the domain wall to sit in a local energy minimum and an external field is required to "unpin" the domain wall from its pinned position. The act of unpinning will cause sudden movement of the domain wall and sudden change of the volume of both neighboring domains; this causes Barkhausen noise. The Barkhausen effect is a name given to the noise in the magnetic output of a ferromagnet when the magnetizing force applied to it is changed. It was discovered by German physicist Heinrich Georg Barkhausen (December 2, 1881 to February 20, 1956) in 1919).

Tungsten's heat resistance makes it valuable in arc welding applications when used in combination with another highly conductive metal such as Ag or Cu. The Ag or Cu provides the required conductivity and the W permits the welding rod to endure the high temperatures of the arc welding environment. W, typically alloyed with Ni and Fe or Co to form heavy alloys, is employed in kinetic energy penetrators as a substitute for depleted U, in uses where its radioactivity is problematic even in depleted form, or where its additional pyrophoric properties are not wanted (e.g., in ordinary small arms bullets intended to penetrate body armor). Likewise, W alloys have also been employed in cannon shells, grenades, and missiles, to produce supersonic shrapnel. Germany utilized W during World War II to make shells for antitank gun designs employing the Gerlich squeeze bore principle to attain very high muzzle velocity and improved armor penetration from relatively small caliber and light weight field artillery. The weapons were highly effective, but a scarcity of W used in the shell core restricted that effectiveness. W has also been employed in Dense Inert Metal Explosives (DIMEs), which use it as dense powder to decrease collateral damage while increasing the lethality of explosives within a small radius (DIME is an experimental type of explosive that has a relatively small but effective blast radius. It is manufactured by producing a homogeneous mixture of an explosive material (such as phlegmatized HMX or RDX) and small particles of a chemically inert material such as tungsten.)

W(IV) sulfide is used as a high-temperature lubricant and forms a component of catalysts for hydrodesulfurization (a catalytic chemical process widely used to remove S from natural gas and from refined petroleum products, such as gasoline or petrol, jet fuel, kerosene, diesel fuel, and fuel oils). However, MoS_2 is more frequently employed for such applications. W oxides are employed in ceramic glazes and Ca/Mg-tungstates are extensively utilized in fluorescent lighting. Crystal tungstates are utilized as scintillation detectors in nuclear physics and nuclear medicine. Other salts that comprise W are employed in the chemical and tanning industries. WO_3 is included in selective catalytic reduction catalysts used in coal-fired power plants. These catalysts convert nitrogen oxides (NO_x) to nitrogen (N_2) and water (H_2O) using ammonia (NH_3). The WO_3 supports the physical strength of the catalyst and increases the catalyst life. Na_2WO_4 is used in the

Folin–Ciocalteu reagent, a mixture of different chemicals used in the "Lowry Assay" for protein content analysis. The Folin–Ciocalteu reagent or Folin's phenol reagent or Folin–Denis reagent, also called the gallic acid equivalence method, is a mixture of phosphomolybdate and phosphotungstate used for the colorimetric in vitro assay of phenolic and polyphenolic antioxidants. It is named after Swedish-born American chemist Otto Knut Olof Folin (April 4, 1867 to October 25, 1934), Romanian physician, researcher, professor, and author Vintilă Ciocâlteu (April 12, 1890 to February 3, 1947) and American biochemist and physiologist Willey Glover Denis (February 26, 1879 to January 9, 1929). The reagent not only measures phenols but will react with any reducing substance. It thus measures the total reducing capacity of a sample, not just phenolic compounds. This reagent is part of the Lowry protein assay and will also react with some N-containing compounds such as hydroxylamine and guanidine. The reagent has also been proven to be reactive toward thiols, many vitamins, the nucleotide base guanine, the trioses glyceraldehyde and dihydroxyacetone, and some inorganic ions. Cu complexation increases the reactivity of phenols toward this reagent.

Uses necessitating W's high density comprise, for example, weights, counterweights, ballast keels for yachts, tail ballast for commercial aircraft, and as ballast in race cars for NASCAR and Formula One. Though currently in Formula One a much more advanced material is used: A W alloy trademarked Densamet. Depleted U is also employed for these applications, because of its similarly high density. Seventy-five-kg W blocks were utilized as "cruise balance mass devices" in the entry vehicle part of the 2012 Mars Science Laboratory spacecraft. It is an ideal element to employ as a dolly for riveting, where the mass required for good results can be realized in a compact bar. High-density alloys of W with Ni, Cu, or Fe are employed in high-quality darts (to allow for a smaller diameter and thus tighter groupings) or for fishing lures (W beads allow the fly to sink quickly). W has been applied lately in nozzles for 3D printing; the high-wear resistance and thermal conductivity of WC enhance the printing of abrasive filaments. Some cello C strings are wound with W. The extra density gives this string more projection and frequently cellists will buy just this string and use it with three strings from a different set. W is utilized as an absorber on the electron telescope on the Cosmic Ray System of the two Voyager spacecraft. Its density, similar to that of gold, allows W to be employed in jewelry as a substitute for Au or Pt. Metallic W is hypoallergenic and is harder than Au alloys (yet not as hard as WC), making it valuable for rings that will resist scratching, in particular in designs with a polished finish. Since the density is similar to that of Au (W is only 0.36% less dense) and its price of the order of one-thousandth, W can also be utilized in forging of gold bars, for example, by plating a W bar with Au, which has been found since the 1980s, or by taking an existing gold bar, drilling holes, and exchanging the removed Au with W rods. The densities are not precisely identical, and other properties of Au and W differ, but Au-plated W will pass superficial tests. Au-plated W is offered commercially in China (the main source of W), both in jewelry and as bars.

Since it keeps its strength at high temperatures and has a high melting point, elemental W is employed in a variety of high-temperature applications, for example, light bulb, cathode ray tube, vacuum tube filaments, heating elements, and rocket engine nozzles. Its high melting point also makes W useful for aerospace and high-temperature applications, for example, electrical, heating, and welding applications, particularly in the GTAW process (gas tungsten arc welding, also called tungsten inert gas welding). Due to its conductive properties and comparatively chemical inertness, W is likewise utilized in electrodes, and in the emitter tips in electron beam instruments that use field emission guns, such as electron microscopes. In electronics, W is employed as an interconnect material in integrated circuits, between the SiO_2 dielectric material and the transistors. It is employed in metallic films, which substitute the wiring utilized in conventional electronics with a coat of W (or Mo) on Si. The electronic structure of W makes it one of the chief sources for X-ray targets, as well as for shielding from high-energy radiations (e.g., in the radiopharmaceutical industry for shielding radioactive samples of FDG (fluorodeoxyglucose). Fludeoxyglucose (18F) (INN), or fludeoxyglucose F 18, also commonly called fluorodeoxyglucose and abbreviated [^{18}F]FDG, ^{18}F-FDG or FDG, is a radiopharmaceutical used in the medical imaging modality PET, IUPAC name 2-deoxy-2-[^{18}F]fluoroglucose, $C_6H_{11}{}^{18}FO_5$). It is also employed in γ imaging as a material from which coded apertures are produced, because of its excellent shielding properties. W powder is applied as a filler material in plastic composites, which are employed as a nontoxic alternative for Pb in bullets, shot, and radiation shields. Because W's thermal expansion is comparable to borosilicate glass, it is employed for making glass-to-metal seals. As well as a high melting point, when W is doped with K, it has an improved shape stability (compared to nondoped W). This guarantees that the filament does not sag, and no undesired changes occur.

Via top-down nanofabrication methods, W nanowires have been created and researched as far back as 2002. Because of an extremely high surface to volume ratio, the development of a surface oxide layer, and the single crystal nature of this material, the mechanical properties are fundamentally different from those of bulk W. These W nanowires have possible uses in nanoelectronics and importantly as pH probes and gas sensors. Like Si nanowires, W nanowires are often created from a bulk W precursor followed by a thermal oxidation step to control morphology with respect to length and aspect ratio. Using the Deal–Grove model it is possible to predict the oxidation kinetics of nanowires made through such thermal oxidation processing (The Deal–Grove model mathematically describes the growth of an oxide layer on the surface of a material. In particular, it is used to predict and interpret thermal oxidation of silicon in semiconductor device fabrication. The model was first published in 1965 by American chemist Bruce E. Deal (September 20, 1927 to April 17, 2007) and Hungarian-born American businessman, engineer, author Andrew Grove (born András István Gróf; September 2, 1936 to March 21, 2016), of Fairchild Semiconductor, and served as a step in the development of CMOS devices and the fabrication of integrated circuits.)

7.13 75 Re — Rhenium

7.13.1 Discovery

Rhenium (Latin: Rhenus meaning: "Rhine," one of the major rivers in Europe) was the second last-found of the elements that have a stable isotope (other new elements later found in nature, such as francium (Fr), are radioactive). The presence of a yet-undiscovered element at this place in the periodic table had originally been projected by Russian chemist Dmitri Mendeleev (February 8, 1834 to February 2, 1907). Further calculated data were determined by English physicist Henry Moseley (November 23, 1887 to August 10, 1915) in 1914 (Moseley, 1913,1914). In 1908 Japanese chemist Masataka Ogawa (February 21, 1865 to July 11, 1930) reported that he had found the 43rd element and called it nipponium (Np) after Japan (Nippon in Japanese) (Ogawa, 1908a,b) (Fig. 7.65). However, recent examination showed the occurrence of rhenium (element 75), not element 43, although this reinterpretation has been doubted by American (of Maltese origin) chemist, writer, and philosopher of science Eric Scerri (born 1953) (Scerri, 2013). The symbol Np was later used for the element

FIGURE 7.65 Masataka Ogawa, early 1900s.

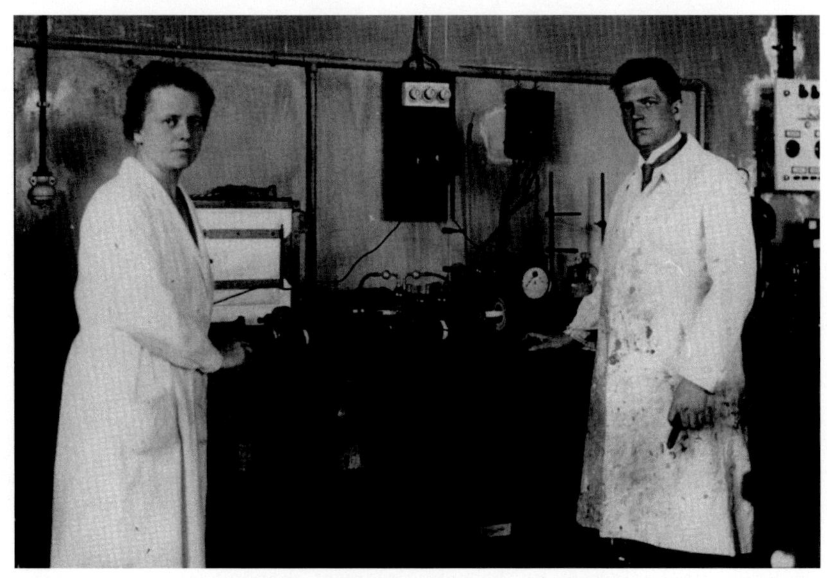

FIGURE 7.66 Ida and Walter Noddack in their laboratory c.1927.

neptunium, and the name "nihonium," also named after Japan, along with symbol Nh, was later used for the extremely radioactive synthetic element 113 (the most stable isotope nihonium-286 has a half-life of about 10 s). Element 113 was also first found by a group of Japanese scientists and was named in respectful deference to Ogawa's research. Rhenium is usually thought to have been found by German chemists Walter Noddack (August 17, 1893 to December 7, 1960), his wife, chemist and physicist Ida Eva Noddack (born Ida Eva Tacke, February 25, 1896 to September 24, 1978) (Fig. 7.66), and Otto Berg (November 23, 1873 to 1939) in Germany. In 1925 they announced that they had found the element in platinum ore and in the mineral columbite, $(Fe,Mg,Mn)Nb_2O_6$. They also found rhenium in gadolinite, $(Ce,Y)_2FeBe_2Si_2O_{10}$, and molybdenite, MoS_2. In 1928 they managed to isolate 1 g of the element by processing 660 kg of molybdenite (Noddack and Noddack, 1929). It was assessed in 1968 that 75% of rhenium metal in the United States was utilized for research and for development of refractory metal alloys. Several years later the superalloys became extensively used.

7.13.2 Mining, production, and major minerals

Rhenium is one of the rarest elements in the Earth's crust with an average concentration of 0.5 to 1 ppb making it the 77th most abundant element in the Earth's crust. Rhenium occurs in concentrations up to 0.2% in molybdenite (MoS_2), the major commercial source, even though single molybdenite samples with up to 1.88% have been discovered. Chile has the world's largest rhenium reserves, part of their copper ore deposits, and was the leading producer as of 2005. It was only recently that the first rhenium mineral was found and described (in 1994, IMA approved 2004), a rhenium sulfide mineral rheniite (ReS_2) condensing from a fumarole on Kudriavy volcano, Iturup island, in the Kuril Islands. Kudriavy releases up to 20–60 kg rhenium per year primarily as the mineral rheniite.

Industrial rhenium is separated from molybdenum roaster-flue gas obtained from copper sulfide ores. Some molybdenum ores contain 0.001% to 0.2% rhenium. Re(VII) oxide and perrhenic acid (this is the chemical compound with the formula $Re_2O_7(OH_2)_2$. It is obtained by evaporating aqueous solutions of Re_2O_7. Conventionally, perrhenic acid is considered to have the formula $HReO_4$, and a species of this formula forms when Re(VII) oxide sublimes in the presence of water or steam) easily dissolve in water; they are leached from flue dusts and gases and obtained through precipitation with potassium or ammonium chloride as the perrhenate salts, followed by purification through recrystallization. Total world production is around 40 to 50 tons annually; the major producers are Chile, the United States, Peru, and Poland. Recycling of used Pt-Re catalyst and special alloys permit the recovery of an additional 10 tons annually. The metal form is obtained through a reduction reaction of ammonium perrhenate with hydrogen gas at high temperatures.

$$2NH_4ReO_4 + 7H_2 \rightarrow 2Re + 8H_2O + 2NH_3$$

FIGURE 7.67 Highly reflective silvery flaky rheniite, ReS_2, crystals covering matrix. 1 cm area. Kudriavy volcano, Iturup Island, Far-Eastern Region, Russia.

Rhenium is only known in three minerals, as the native element rhenium (Re) and two sulfides, rheniite (ReS_2) (Fig. 7.67) and tarkianite ($(Cu,Fe)(Re,Mo)_4S_8$).

7.13.3 Chemistry

Rhenium exists as one stable isotope, ^{185}Re, occurring in minority abundance, a condition found only in two other elements (In and Te). Naturally occurring rhenium consists only of 37.4% ^{185}Re, with 62.6% ^{187}Re, which is unstable but has a very long half-life of 4.12×10^{10} years. This lifetime can be significantly affected by the charge state of the Re atom. The β^- decay of ^{187}Re to ^{187}Os is utilized for Re-Os dating of ore deposits. The available energy for this β decay (2.6 keV) is among the lowest identified for all radionuclides. The isotope ^{186m}Re is noteworthy as it is one of the longest-lived metastable isotopes with a half-life of about 200,000 years. There exist 25 other documented Re radioisotopes.

Rhenium is a silvery-white metal with one of the highest melting points of all elements, exceeded by only tungsten and carbon. It also has one of the highest boiling points of all elements. In addition, it is one of the densest, exceeded only by platinum, iridium, and osmium. Rhenium has a *hcp* crystal structure, with lattice parameters a = 276.1 pm and c = 445.6 pm. In bulk form and at ambient conditions, it resists alkalis, sulfuric acid, hydrochloric acid, dilute (but not concentrated) nitric acid, and aqua regia. Rhenium compounds are known for all the oxidation states between −3 and +7 except −2. The oxidation states +7, +6, +4, and +2 are the most common (Table 7.13).

7.13.3.1 Oxides and chalcogenides

The most common oxide is the volatile colorless Re_2O_7. Rhenium trioxide ReO_3 adopts a perovskite-like structure. Other oxides include Re_2O_5, ReO_2, and Re_2O_3. Rhenium(VII) oxide, Re_2O_7, is a yellowish solid that is the anhydride of $HOReO_3$. Re_2O_7 is the raw material for all rhenium compounds, being the volatile fraction formed during roasting the host ore. Crystalline Re_2O_7 is an inorganic polymer, which consists of alternating octahedral and tetrahedral Re centers. Upon heating, the polymer cracks to give molecular (nonpolymeric) Re_2O_7. This molecular species is very similar to manganese heptoxide, consisting of a pair of ReO_4 tetrahedra sharing a vertex, that is, $O_3Re\text{-}O\text{-}ReO_3$. Rhenium(VII) oxide is produced when metallic rhenium or its oxides or sulfides are oxidized at 500°C–700°C in air. Re_2O_7 is very reactive toward water. It dissolves in water to form perrhenic acid. It is a precursor to methylrhenium trioxide (MTO), a catalyst for oxidation reactions. Rhenium(VII) oxide finds some use in organic synthesis as a catalyst for ethenolysis, carbonyl reduction and amide reduction.

Rhenium trioxide or rhenium(VI) oxide, ReO_3, is a red solid with a metallic luster, which resembles copper in appearance. It is the only stable trioxide of the group 7 elements (Mn, Tc, Re). ReO_3 can be prepared by reducing rhenium(VII) oxide with carbon monoxide.

$$Re_2O_7 + CO \rightarrow 2ReO_3 + CO_2$$

Re_2O_7 can also be reduced with dioxane (IUPAC name 1,4-dioxacyclohexane, $C_4H_8O_2$). Rhenium oxide crystallizes with a primitive cubic unit cell, with a lattice parameter of 3.742 Å (374.2 pm). The structure of ReO_3 is similar to that of perovskite (ABO_3), without the large A cation at the center of the unit cell. Each rhenium center is surrounded by an

TABLE 7.13 Rhenium properties.

Appearance	Silvery grayish
Standard atomic weight $A_{r,std}$	186.207
Block	d-Block
Element category	Transition metal
Electron configuration	[Xe] $4f^{14}$ $5d^5$ $6s^2$
Phase at STP	Solid
Melting point	3186°C
Boiling point	5630°C
Density (near r.t.)	21.02 g/cm^3
When liquid (at m.p.)	18.9 g/cm^3
Heat of fusion	60.43 kJ/mol
Heat of vaporization	704 kJ/mol
Molar heat capacity	25.48 J/(mol·K)
Oxidation states	−3, −1, 0, +1, +2, +3, **+4**, +5, +6, +7
Ionization energies	1st: 760 kJ/mol 2nd: 1260 kJ/mol 3rd: 2510 kJ/mol
Atomic radius	Empirical: 137 pm
Covalent radius	151 ± 7 pm

STP, Standard temperature and pressure.
Bold font indicates main oxidation state.

octahedron defined by six oxygen centers. These octahedra share corners to form the 3D structure. The coordination number of O is 2 since each oxygen atom has two neighboring Re atoms. Upon heating to 400°C under vacuum, it undergoes disproportionation.

$$3ReO_3 \rightarrow Re_2O_7 + ReO_2$$

ReO_3 is unusual for an oxide since it shows very low resistivity. It behaves like a metal in that its resistivity drops as its temperature decreases. At 27°C (300K), its resistivity is 100.0 nΩ·m, whereas at − 173°C (100K), this decreases to 6.0 nΩ·m, about 17 times less than at 27°C (300K). ReO_3 finds limited use in organic synthesis as a catalyst for amide reduction.

Rhenium(IV) oxide or rhenium dioxide, ReO_2, is a gray to black crystalline solid, a laboratory reagent that can be utilized as a catalyst. It has the rutile (TiO_2) structure. It can be prepared via comproportionation.

$$2Re_2O_7 + 3Re \rightarrow 7ReO_2$$

At high temperatures it undergoes disproportionation.

$$7ReO_2 \rightarrow 2Re_2O_7 + 3Re$$

It produces perrhenates with alkaline hydrogen peroxide and oxidizing acids. In molten sodium hydroxide it produces sodium rhenate.

$$2NaOH + ReO_2 \rightarrow Na_2ReO_3 + H_2O$$

The perrhenate ion is the anion with the formula ReO_4^- or a compound containing this ion. The perrhenate anion is tetrahedral, being comparable in both size and shape to perchlorate and the valence isoelectronic permanganate. The perrhenate anion is stable over a broad pH range and can be precipitated from solutions using organic cations. At normal pH, perrhenate exists as metaperrhenate (ReO_4^-), but at high pH mesoperrhenate (ReO_5^{3-}) forms. Perrhenate, like its conjugate acid perrhenic acid, contains rhenium in the oxidation state of +7 with a d^0 configuration. Solid perrhenate salts take on the color of the cation. Typical perrhenate salts are the alkali metal derivatives and ammonium perrhenate, NH_4ReO_4. These salts are formed by oxidation of rhenium compounds with nitric acid followed by neutralization of the resulting perrhenic acid. In inorganic chemistry, the perrhenate anion is also utilized as a weakly coordinating anion. It is a weaker base than Cl^- or Br^- but stronger than ClO_4^- or BF_4^-. In contrast to the related permanganate, perrhenate is nonoxidizing. Silver perrhenate reacts with trimethylsilyl chloride to form the silyl "ester" $(CH_3)_3SiOReO_3$. The perrhenate ion reacts with cyanide to form *trans*-$[ReO_2(CN)_4]^{3-}$. With sulfide compound such as hydrogen sulfide, it forms black ReS_2 and Re_2S_7. These species form via the intermediary ReO_3S^-. Heating ammonium perrhenate (APR) gives Re_2O_7 and then ReO_2. Pure rhenium powder can be formed from APR by heating it in the presence of H_2.

$$2NH_4ReO_4 + 7H_2 \rightarrow 2Re + 8H_2O + 2NH_3$$

Heating must be done slowly as APR decomposes to volatile Re_2O_7 starting at 250°C. When heated in a sealed tube at 500°C, APR decomposes to ReO_2.

$$2NH_4ReO_4 \rightarrow 2ReO_2 + N_2 + 4H_2O$$

The chemistry of the perrhenate ion is comparable to that of the pertechnetate ion Tc_4O^-. Hence, perrhenate is occasionally used as a carrier for trace levels of pertechnetate, for example, in nuclear medicine scanning procedures. Perrhenate is also utilized as a safer option than pertechnetate for nuclear waste vitrification research, like volatility or encapsulation in solids. Perrhenic acid, $Re_2O_7(OH_2)_2$, is formed by evaporating aqueous solutions of Re_2O_7. Conventionally, perrhenic acid is thought to have the formula $HReO_4$, and a species of this formula is formed when rhenium(VII) oxide is sublimated in the presence of water or steam. When a solution of Re_2O_7 is stored for a period of months, it breaks down and crystals of $HReO_4 \cdot H_2O$ are formed, which have tetrahedral ReO_4^- units. For most applications, perrhenic acid and rhenium(VII) oxide are used interchangeably. Rhenium can be dissolved in nitric or concentrated sulfuric acid to form perrhenic acid. The structure of solid perrhenic acid is $[O_3Re\text{-}O\text{-}ReO_3(H_2O)_2]$. This species is an unusual example of a metal oxide coordinated to water—most often metal-oxo-aquo species are unstable relative to the corresponding hydroxides.

$$M(O)(H_2O) \rightarrow M(OH)_2$$

The two rhenium atoms show different bonding geometries, with one being tetrahedral and the other octahedral, and with the water ligands coordinated to the latter. Gaseous perrhenic acid is tetrahedral, as suggested by its formula $HReO_4$. Perrhenic acid or the related anhydrous oxide Re_2O_7 reacts to dirhenium heptasulfide upon treatment with hydrogen sulfide.

$$Re_2O_7 + 7H_2S \rightarrow Re_2S_7 + 7H_2O$$

The heptasulfide, which has a complex structure, catalyzes the hydrogenation of double bonds and is useful as it tolerates sulfur compounds, which poison noble metal catalysts. In addition, Re_2S_7 catalyzes the reduction of nitric oxide to N_2O. Perrhenic acid in the presence of HCl is reduced in the presence of thioethers and tertiary phosphines to form Re (V) complexes with the formula $ReOCl_3L_2$. Perrhenic acid together with platinum on a support produces a useful hydrogenation and hydrocracking catalyst for the petroleum industry. For instance, silica impregnated with a solution of perrhenic acid is reduced with hydrogen at 500°C. This catalyst is used in the dehydrogenation of alcohols and in addition promotes the decomposition of ammonia.

The known sulfides consist of ReS_2 and Re_2S_7. Rhenium disulfide, ReS_2, has a layered structure where atoms are strongly bonded within each layer. The layers are held together by weak Van der Waals bonds and can easily be peeled off from the bulk material. It is a 2D group VII transition metal dichalcogenide (TMD). In contrast to other TMDs, ReS_2 has shown layer-independent electrical, optical, and vibrational properties. Nanostructured ReS_2 can typically be obtained via mechanical exfoliation, CVD, and chemical and liquid exfoliations. Larger crystals can be grown with the assistance of liquid carbonate flux at high pressure. It is extensively utilized in electronic and optoelectronic device, energy storage, photocatalytic and electrocatalytic reactions.

Rhenium(VII) sulfide, Re_2S_7, can be produced through the reaction of ReO_4^- and H_2S in 4N HCl. It can also be prepared by direct combination of rhenium and sulfur:

$$2Re + 7S \xrightarrow{\Delta} Re_2S_7$$

Or by treating rhenium(VII) oxide with hydrogen sulfide:

$$Re_2O_7 + 7H_2S \xrightarrow{\Delta} Re_2S_7 + 7H_2O$$

Rhenium(VII) sulfide decomposes when heated in vacuum:

$$Re_2S_7 \xrightarrow{600^\circ C} 2ReS_2 + 3S$$

It is converted to oxide when heated in air:

$$2Re_2S_7 + 21O_2 \xrightarrow{\Delta} 2Re_2O_7 + 14SO_2$$

Perrhenate salts can be transformed to tetrathioperrhenate by the action of ammonium hydrosulfide.

Rhenium diselenide, $ReSe_2$, has a layered structure where atoms are strongly bonded within each layer. The layers are held together by weak van der Waals bonds and can be easily peeled off from the bulk material similar to ReS_2.

However, whereas most other layered dichalcogenides have a high (hexagonal) symmetry, $ReSe_2$ has a very low triclinic symmetry, and this symmetry does not change from the bulk to monolayers. $ReSe_2$ with a thickness as small as a triple-atomic layer can be formed by CVD at ambient pressure. A mixture of Ar and hydrogen gases is blown through a tube whose ends are kept at different temperatures. The substrate and ReO_3 powder are placed at the hot end which is heated to 750°C, and selenium powder is located at the cold end which is kept at 250°C.

$$2ReO_3 + 7Se \rightarrow 2ReSe_2 + 3SeO_2$$

Rhenium ditelluride is an inorganic compound of rhenium and tellurium with the formula $ReTe_2$. Unlike rhenium disulfide and diselenide, it does not have a layered structure.

7.13.3.2 Borides and hydrides

Rhenium diboride (ReB_2) is a hard compound having the hardness comparable to that of tungsten carbide, silicon carbide, titanium diboride or zirconium diboride. Rhenium diboride (ReB_2) is a synthetic superhard material. It was first synthesized in 1962 and re-emerged recently hoping to achieve a hardness similar to that of diamond (Mohs hardness 10, hardest known mineral). The described ultrahigh hardness has been doubted, though this is a matter of definition as in the original test rhenium diboride was able to scratch diamond. The preparation method of this material does not require high pressures as with other hard synthetic materials, for example, cubic boron nitride, which makes production relatively inexpensive. Nevertheless, rhenium itself is an expensive metal. The compound is prepared from a mixture of rhenium, noted for its resistance to high pressure, and boron, which forms short, strong covalent bonds with rhenium. ReB_2 can be prepared via at least three different routes at standard atmospheric pressure: solid-state metathesis, melting in an electric arc, and direct heating of the elements. In the metathesis reaction, rhenium trichloride and magnesium diboride are mixed and heated in an inert atmosphere and the magnesium chloride by-product is washed away. Excess boron is required to avoid formation of other phases such as Re_7B_3 and Re_3B. In the arc melting method, rhenium and boron powders are mixed and a large electric current is passed through the mixture, also in an inert atmosphere. In the direct reaction method, the rhenium-boron mixture is sealed in a vacuum and held at a high temperature over a longer period of time (1000°C for 5 days). At least the last two routes can produce pure ReB_2 without any other phases. The hardness of ReB_2 exhibits considerable anisotropy because of its hexagonal layered structure, being greatest along the *c* axis. In contrast to the scratch hardness test, its indentation hardness (HV $\sim$ 22 GPa) is much lower than that of diamond and is similar to that of tungsten carbide, silicon carbide, titanium diboride or zirconium diboride. ReB_2 slowly reacts with water, transforming into a hydroxide. Two features contribute to the high hardness of ReB_2: (1) a high density of valence electrons, and (2) an abundance of short covalent bonds. Rhenium has one of the highest valence electron densities of any transition metal (476 electrons/nm^3, compared to 572 electrons/nm^3 for osmium and 705 electrons/nm^3 for diamond). The addition of boron involves only a 5% expansion of the rhenium lattice, since the small boron atoms fill the existing spaces between the rhenium atoms. Moreover, the electronegativities of rhenium and boron are sufficiently close (1.9 and 2.04 on the Pauling scale) that they form covalent bonds in which the electrons are shared almost equally.

A distinctive derivative of rhenium is nonahydridorhenate, originally thought to be the rhenide anion, Re^-, but in fact comprising the ReH_9^{2-} anion in which the oxidation state of rhenium is +7. Potassium nonahydridorhenate(VII), K_2ReH_9, is a colorless salt that is soluble in water but only poorly soluble in most alcohols. The ReH_9^{2-} anion is a rare illustration of a coordination complex bearing only hydride ligands. ReH_9^{2-} is an uncommon example of a nonacoordinated complex, its high coordination number being credited to the small size of the hydride ligand and the high positive charge on the Re(VII) center. Its structure comprises a tricapped trigonal prism. The diamagnetic sodium salt, like the analogous technetium compound, is produced by treating an ethanol solution of sodium perrhenate, $NaReO_4$, with sodium metal. Via cation exchange, it can be transformed to the corresponding tetraethylammonium salt, $(N(C_2H_5)_4)_2ReH_9$. Isostructural with TcH_9^{2-}, ReH_9^{2-} comprises a trigonal prism with Re atom in the center and six hydrogen atoms at the corners. Three more hydrogen ligands define a triangle lying parallel to the base and crossing the prism in its center. While those hydride ligands are not the same, their electronic structure is nearly the same. The coordination number of 9 in this complex is the highest known for any rhenium complex.

7.13.3.3 Halides and oxyhalides

The most common rhenium chlorides are $ReCl_6$, $ReCl_5$, $ReCl_4$, and $ReCl_3$. The structures of these compounds frequently include extensive Re–Re bonding, which is typical of Re in oxidation states lower than VII. Salts of $[Re_2Cl_8]^{2-}$ include a quadruple metal–metal bond. Although the highest rhenium chloride contains Re(VI), fluorine gives the d^0 Re(VII) derivative rhenium heptafluoride. Rhenium(VI) chloride, $ReCl_6$, is a black paramagnetic solid. The molecules adopt an octahedral structure similar to tungsten(VI) chloride. Rhenium(VI) chloride is formed from the rhenium hexafluoride with excess boron trichloride.

$$2ReF_6 + 6BCl_3 \rightarrow ReCl_6 + 6BF_2Cl$$

It is unstable at room temperature with respect to rhenium(V) chloride.

$$2ReCl_6 \rightarrow [ReCl_5]_2 + Cl_2$$

Rhenium pentachloride has the formula Re_2Cl_{10} but it is usually referred to as rhenium pentachloride. It is a red-brown solid. Rhenium pentachloride has a bioctahedral structure and can better be written as $Cl_4Re(\mu\text{-}Cl)_2ReCl_4$. The Re−Re distance is 3.74 Å. The motif is like that observed for tantalum pentachloride. This compound was first synthesized in 1933 by Geilmann et al., just a couple of years after the discovery of rhenium. The preparation comprises chlorination of rhenium at temperatures up to 900°C. The compound can be purified by sublimation. $ReCl_5$ is one of the most oxidized binary chlorides of Re. It does not undergo additional chlorination. $ReCl_6$ has been formed from rhenium hexafluoride. Rhenium heptafluoride is known but not the corresponding heptachloride. It degrades in air to a brown liquid. Although rhenium pentachloride has no commercial uses, it is of historical importance as one of the initial catalysts for olefin metathesis. Reduction produces trirhenium nonachloride. Oxygenation forms the Re(VII) oxychloride.

$$ReCl_5 + 3Cl_2O \rightarrow ReO_3Cl + 5Cl_2$$

Comproportionation of the penta- and trichlorides produces rhenium tetrachloride.

Rhenium(IV) chloride, $ReCl_4$, is a black solid, which is of interest as a binary phase although else of little practical importance. A second polymorph of $ReCl_4$ is also known. $ReCl_4$ can be formed by comproportionation of rhenium(V) chloride and rhenium(III) chloride. It can also be prepared by reduction of rhenium(V) chloride with antimony trichloride.

$$2ReCl_5 + SbCl_3 \rightarrow 2ReCl_4 + SbCl_5$$

Tetrachloroethylene at 120°C is also effective as a reductant.

$$2ReCl_5 + C_2Cl_4 \rightarrow 2ReCl_4 + C_2Cl_6$$

X-ray crystallography has shown that it has a polymeric structure. The Re−Re bonding distance is 2.728 Å. Re centers are octahedral, being surrounded by six chloride ligands. Pairs of octahedra share faces. The Re_2Cl_9 subunits are linked by bridging chloride ligands. The structural motif—corner-shared bioctahedra—is uncommon in the binary metal halides.

Trirhenium nonachloride is a compound with the formula $ReCl_3$, occasionally also written Re_3Cl_9. It is a dark red hygroscopic solid that is insoluble in ordinary solvents. It is significant in the history of inorganic chemistry as an early instance of a cluster compound with metal−metal bonds. It is employed as a starting molecule for the preparation of other rhenium complexes. Trirhenium nonachloride comprises Re_3Cl_{12} subunits that share three chloride bridges with adjacent clusters. The interconnected network of clusters forms sheets. Surrounding each Re center are seven ligands, four bridging chlorides, one terminal chloride, and two Re−Re bonds. Trirhenium nonachloride is efficiently formed by thermal decomposition of rhenium pentachloride or hexachlororhenic(IV) acid.

$$3ReCl_5 \rightarrow Re_3Cl_9 + 3Cl_2$$

If the product is vacuum sublimed at 500°C, the resulting compound is relatively unreactive, although the partially hydrated compound can be more useful synthetically. Other production methods comprise reacting rhenium with sulfuryl chloride. This reaction is occasionally performed with the addition of aluminum chloride. It is also formed by heating $Re_2(O_2CCH_3)_4Cl_2$ under HCl.

$$3Re_2(O_2CCH_3)_4Cl_2 + 12HCl \rightarrow 2Re_3Cl_9 + 12HO_2CCH_3$$

Reaction of the tri- and pentachlorides produces rhenium tetrachloride.

$$3ReCl_5 + Re_3Cl_9 \rightarrow 6ReCl_4$$

Rhenium hexafluoride, also rhenium(VI) fluoride, (ReF_6) is a compound of rhenium and fluorine and one of the 17 known binary hexafluorides. ReF_6 is prepared by reacting ReF_7 with Re metal at 300°C in a pressure vessel.

$$6ReF_7 + Re \rightarrow 7ReF_6$$

Rhenium hexafluoride (ReF_6) is a liquid at room temperature. At 18.5°C, it freezes into a yellow solid. The boiling point is 33.7°C. The solid structure determined at −140°C is orthorhombic space group *Pnma*. Lattice parameters are a = 9.417 Å, b = 8.570 Å, and c = 4.965 Å. There are four formula units (in this case, discrete molecules) per unit cell, resulting in a density of 4.94 $g \cdot cm^{-3}$. The ReF_6 molecule itself (the form important for the liquid or gas phase) has octahedral molecular geometry, which has point group O_h. The Re-F bond length is 1.823 Å. Rhenium heptafluoride, ReF_7, is a yellow low-melting solid and is the only thermally stable metal heptafluoride. It has a distorted pentagonal bipyramidal structure comparable to IF_7. The structure is nonrigid as shown by electron diffraction studies. It can be formed from the elements at 400°C.

$$2Re + 7F_2 \rightarrow 2ReF_7$$

With fluoride donors like CsF, the ReF_8^- anion is formed, which has a square antiprismatic structure. With antimony pentafluoride, SbF_5, a fluoride acceptor, the $ReF_6{}^+$ cation is formed.

Bromides and iodides of rhenium are also well known ($ReBr_5$, $ReBr_4$, $[ReBr_3]_3$, ReI_4, and $[ReI_3]_3$). Like tungsten and molybdenum, with which it shares chemical similarities, rhenium forms a variety of oxyhalides. The oxychlorides are most common and include $ReOCl_4$ and $ReOCl_3$. Also known are $ReOF_5$, $ReOF_4$, $ReOF_3$, ReO_2F_2, ReO_3F, ReO_3Cl, $ReOBr_4$, and ReO_3Br.

7.13.3.4 Organorhenium compounds

Re exists in ten known oxidation states from −3 to +7 except −2, and all but Re(−3) are represented by organorhenium compounds. Most are formed from salts of perrhenate and related binary oxides. The halides, for example $ReCl_5$, are also useful precursors as are some oxychlorides. A notable characteristic of organorhenium chemistry is the coexistence of oxide and organic ligands in the same coordination sphere.

Dirhenium decacarbonyl is the most common starting compound to organorhenium chemistry. Its reduction with sodium amalgam forms $Na[Re(CO)_5]$ with rhenium in the formal oxidation state −1. Dirhenium decacarbonyl can be oxidized with bromine to bromopentacarbonylrhenium(I).

$$Re_2(CO)_{10} + Br_2 \rightarrow 2Re(CO)_5Br$$

Reduction of this pentacarbonyl with zinc and acetic acid produces pentacarbonylhydridorhenium.

$$Re(CO)_5Br + Zn + HOAc \rightarrow Re(CO)_5H + ZnBr(OAc)$$

Bromopentacarbonylrhenium(I) is easily decarbonylated. In refluxing water, it forms the triaquo cation.

$$Re(CO)_5Br + 3H_2O \rightarrow \left[Re(CO)_3(H_2O)_3\right]Br + 2CO$$

With tetraethylammonium bromide $Re(CO)_5Br$ reacts to form the anionic tribromide.

$$Re(CO)_5Br + 2NEt_4Br \rightarrow [NEt_4]_2\left[Re(CO)_3Br_3\right] + 2CO$$

One of the first transition metal hydride complexes to be described was $(C_5H_5)_2ReH$. A variety of half-sandwich compounds have been prepared from $(C_5H_5)Re(CO)_3$ and $(C_5Me_5)Re(CO)_3$. Significant derivatives include the electron-precise oxide $(C_5Me_5)ReO_3$ and $(C_5H_5)_2Re_2(CO)_4$.

Methylrhenium trioxide (MTO), CH_3ReO_3, a volatile, colorless solid has been employed as a catalyst in some laboratory experiments. It can be formed by many routes; a typical reaction involves Re_2O_7 and tetramethyltin.

$$Re_2O_7 + (CH_3)_4Sn \rightarrow CH_3ReO_3 + (CH_3)_3SnOReO_3$$

Analogous alkyl and aryl derivatives are known. While $PhReO_3$ is unstable and decomposes at −30°C, the corresponding sterically hindered mesityl and 2,6-xylyl derivatives ($MesReO_3$ and $2,6\text{-}(CH_3)_2C_6H_3ReO_3$) are stable at room temperature. The electron-poor 4-trifluoromethylphenylrhenium trioxide ($4\text{-}CF_3C_6H_4ReO_3$) is similarly relatively stable. MTO catalyzes for the oxidations with hydrogen peroxide. Terminal alkynes produce the corresponding acid or ester, internal alkynes yield diketones, and alkenes give epoxides. In addition, MTO catalyzes the conversion of aldehydes and diazoalkanes into an alkene. Rhenium is also able to form complexes with fullerene ligands such as $Re_2(PMe_3)_4H_8(\eta^2{:}\eta^2C_{60})$.

Oxotrichlorobis(triphenylphosphine)rhenium(V), $ReOCl_3(PPh_3)_2$, is a yellow, air-stable solid and a precursor to various other rhenium complexes. In this diamagnetic compound, Re has an octahedral coordination environment with one oxo, three chloro, and two mutually *trans*-triphenylphosphine ligands. The oxidation state of rhenium is +5 and its configuration is d^2. $ReOCl_3(PPh_3)_2$ is commercially available, but it is easily prepared by reacting perrhenic acid with triphenylphosphine in a mixture of hydrochloric acid and acetic acid. In this reaction, Re(VII) is reduced to Re(V), and triphenylphosphine is oxidized to its oxide.

$$HReO_4 + 3HCl + 3PPh_3 \rightarrow ReOCl_3(PPh_3)_2 + Ph_3PO + 2H_2O$$

The required perrhenic acid solution can be produced in situ from rhenium(VII) oxide. $ReOCl_3(PPh_3)_2$ is used as a precursor to various other oxo-, nitridio, and hydrido complexes. It transforms to $ReH_7(PPh_3)_2$ by a treatment with $LiAlH_4$. $ReOCl_3(PPh_3)_2$ catalyzes the selective oxidation of secondary alcohols by dimethyl sulfoxide (DMSO), forming the corresponding ketals.

7.13.4 Major uses

The Ni-based superalloys show enhanced creep strength after addition of Re. The alloy usually has 3% or 6% of Re. Second-generation alloys contain 3%; these alloys were used in the engines for the F-15 and F-16 fighter aircraft, while the newer single-crystal third-generation alloys contain 6% of Re; they are used in the F-22 and F-35 engines. Re is also used in the superalloys, such as CMSX-4 (second generation) and CMSX-10 (third generation) that are employed in industrial gas turbine engines (A superalloy, or high-performance alloy, is an alloy that exhibits several key characteristics: excellent mechanical strength, resistance to thermal creep deformation, good surface stability, and resistance to corrosion or oxidation. The crystal structure is typically face-centered cubic austenitic). Re can result in superalloys to develop microstructural instability, forming unwanted topologically close-packed phases. In the fourth and fifth generation superalloys, Ru is utilized to avoid this effect. Back in 2006, the global consumption was 68% for superalloys, while the use for catalysts only accounted for 14% and the remaining applications for 18%. The growing demand for military jet engines and the constant supply necessitated the development of superalloys with a lower Re concentrations. For instance, the newer CFM International CFM56 high-pressure turbine blades will use Rene N515 with a Re content of 1.5% replacing Rene N5 with 3%. Re enhances the properties of W. W-Re alloys are more ductile at low temperature, letting them to be more readily machined. The high-temperature stability is also increased. The effect increases with the Re content, and hence W alloys are manufactured with up to 27% of Re, which is the solubility limit. W-Re wire was initially made in attempts to create a wire that was more ductile after recrystallization. This allows the wire to meet precise performance objectives, such as superior vibration resistance, better ductility, and higher resistivity. One use for the W-Re alloys is X-ray sources. The high melting point of both elements, combined with their high atomic mass, makes them stable against the prolonged electron impact. In addition, Re-W alloys are used as thermocouples to measure temperatures up to 2200°C. The high temperature stability, low vapor pressure, good wear resistance, and the ability to withstand arc corrosion of Re are advantageous in self-cleaning electrical contacts. Especially the discharge occurring during the switching oxidizes the contacts. But Re_2O_7 has poor stability (sublimes at $\sim$360°C) and consequently is removed during the discharge. Re possesses a high melting point and a low vapor pressure like Ta and W. Hence, Re filaments show a higher stability if the filament is operated not in vacuum, but in oxygen-containing atmosphere. Those filaments are extensively employed in mass spectrometers, in ion gages and in photoflash lamps in photography.

Re in the form of Re-Pt alloy is used as catalyst for catalytic reforming, that is, a chemical process to convert petroleum refinery naphthas (Petroleum naphtha is an intermediate hydrocarbon liquid stream derived from the refining of crude oil) with low octane ratings into high-octane liquid products. The process converts low-octane linear hydrocarbons (paraffins) into branched alkanes (isoparaffins) and cyclic naphthenes, which are then partially dehydrogenated to produce high-octane aromatic hydrocarbons. The dehydrogenation also produces significant amounts of by-product H_2 gas, which is fed into other refinery processes such as hydrocracking. A side reaction is hydrogenolysis, which produces light hydrocarbons of lower value, such as methane (CH_4), ethane (C_2H_6), propane (C_3H_8) and butane (C_4H_{10}). Globally, around 30% of catalysts used for this process have Re. The olefin metathesis is the other reaction for which Re is used as catalyst. Normally Re_2O_7 on alumina (Al_2O_3) is used for this process. Olefin metathesis is an organic reaction that entails the redistribution of fragments of alkenes (olefins) by the scission and regeneration of C-C double bonds. Because of the relative simplicity of olefin metathesis, it often creates fewer undesired by-products and hazardous wastes than alternative organic reactions. For their elucidation of the reaction mechanism and their discovery of a variety of highly active catalysts, French chemist Yves Chauvin (10 October 27, 1930 to January 27, 2015), together with American chemists Robert Howard Grubbs (February 1942), and Richard Royce Schrock (born January 4, 1945) were collectively awarded the 2005 Nobel Prize in Chemistry. Re catalysts are very resistant to chemical poisoning from N, S, and P and therefore are employed in certain types of hydrogenation reactions. Catalyst poisoning refers to the partial or total deactivation of a catalyst by a chemical compound. Poisoning refers explicitly to chemical deactivation, rather than other mechanisms of catalyst degradation such as thermal decomposition or physical damage.

^{188}Re and ^{186}Re are radioactive isotopes, which are utilized for treatment of liver cancer. They both have comparable penetration depth in tissues (5 mm for ^{186}Re and 11 mm for ^{188}Re), but ^{186}Re has the advantage of a longer half-life (90 vs 17 h). ^{188}Re is experimentally being trialed in a new treatment of pancreatic cancer where it is delivered through the bacterium *Listeria monocytogenes*. Correlated by periodic trends, Re has a similar chemistry to that of Tc; work done to label Re onto target compounds can frequently be translated to Tc. This is valuable for radiopharmacy, where it is problematic to work with Tc—in particular ^{99m}Tc employed in medicine—because of its cost and short half-life.

7.14 76 Os — Osmium

7.14.1 Discovery

Osmium was first found in 1803 by English chemists Smithson Tennant (November 30, 1761 to February 22, 1815) (Fig. 7.68) and William Hyde Wollaston (August 6, 1766 to December 22, 1828) in London, England (Tennant, 1804). The discovery of osmium is intimately connected with that of platinum and the other metals of the platinum group. Platinum arrived in Europe as platina ("small silver"), first found in the late 17th century in silver mines around the Chocó Department, Colombia. The finding that this metal was not an alloy, but a distinct new element, was published in 1748 (de Ulloa, 1748, page 606). Chemists who researched platinum dissolved it in aqua regia (a mixture of hydrochloric and nitric acids) to produce soluble salts. They constantly detected a small amount of a dark, insoluble residue. French chemist Joseph Louis Proust (September 26, 1754 to July 5, 1826) was of the opinion that the residue was graphite (a form of carbon). French chemists Hippolyte-Victor Collet-Descotils (November 21, 1773 to December 6, 1815), Antoine François, comte de Fourcroy (June 15, 1755 to December 16, 1809) (Fig. 7.69), and pharmacist and chemist Louis Nicolas Vauquelin (May 16, 1763 to November 14, 1829) (Fig. 7.70) also detected iridium in the black platinum residue in 1803, but did not separate enough material for further study (Collet-Descotils, 1803, Fourcroy and Vauquelin, 1803). Later Antoine François, comte de Fourcroy and Nicolas-Louis Vauquelin recognized a metal in the platinum residue they named 'ptène'. In 1803 Smithson Tennant tested the insoluble residue and determined that it must contain a new metal (Tennant, 1804). Vauquelin reacted the powder consecutively with alkali and acids and got a volatile new oxide, which he thought was of this new metal—which he called ptene, from the Greek word πτηνος (ptènos) for winged. However, Tennant, who

FIGURE 7.68 Smithson Tennant.

FIGURE 7.69 Antoine François, comte de Fourcroy, portrait by François Séraphin Delpech (1778–1825).

FIGURE 7.70 Nicolas-Louis Vauquelin, 1866, after the engraving by François-Jacques Dequevauviller (c.1783–c.1848).

had the advantage of a much larger volume of residue, continued his studies and recognized two before undiscovered elements in the black residue, iridium and osmium. He produced a yellow solution (probably of cis$-[Os(OH)_2O_4]^{2-}$) by reactions with sodium hydroxide (NaOH) at red heat. After acidification he was able to distill the formed OsO_4. He christened it osmium after Greek osme meaning "a smell," because of the ashy and smoky smell of the volatile osmium tetroxide. The findings on these new elements were reported in a letter to the Royal Society on June 21, 1804. Uranium and osmium were initial effective catalysts in the Haber process, the nitrogen fixation reaction of nitrogen and hydrogen to produce ammonia, giving enough product to make the process commercially successful. During that period, a group at BASF led by German chemist and engineer Carl Bosch (August 27, 1874 to April 26, 1940) bought most of the world's supply of osmium to use as a catalyst. A short time later, in 1908, cheaper catalysts based on iron and iron oxides were developed by the same group for the first pilot plants, eliminating the need for the expensive and rare osmium. BASF SE is a German chemical company and the largest chemical producer in the world. The BASF Group comprises subsidiaries and joint ventures in more than 80 countries and operates six integrated production sites and 390 other production sites in Europe, Asia, Australia, the Americas, and Africa. Currently, osmium is obtained mainly as a by-product from the processing of platinum and nickel ores.

7.14.2 Mining, production, and major minerals

Osmium is the least abundant stable element in the Earth's crust, with an average of 50 parts per trillion in the continental crust. Osmium is found in nature as a native element or in natural alloys; especially the iridium–osmium alloys, osmiridium (Os rich), and iridosmium (Ir rich). In nickel and copper deposits, the platinum group metals occur as sulfides (i.e., (Pt,Pd)S)), tellurides (e.g., PtBiTe), antimonides (e.g., PdSb), and arsenides (e.g., $PtAs_2$); in all these minerals Pt is substituted by a small amount of Ir and Os. As with all the Pt-group elements, Os can be found naturally in alloys with Ni or Cu. Within the Earth's crust, Os, like Ir, is found at highest concentrations in three types of geologic structure: igneous deposits (crustal intrusions from below), impact craters, and deposits reworked from one of the former structures. The largest identified primary reserves are in the Bushveld Igneous Complex in South Africa, nevertheless the large Cu–Ni deposits near Norilsk in Russia, and the Sudbury Basin in Canada are also important sources of Os. Smaller reserves can be found in the United States. The alluvial deposits used by pre-Columbian people in the Chocó Department, Colombia are even today a resource for Pt-group elements. The second large alluvial deposit was found in the Ural Mountains, Russia, which is currently being mined as well.

Osmium is commercially produced as a by-product from nickel and copper mining and processing. In the process of electrorefining of copper and nickel, noble metals such as silver, gold and the platinum group metals, as well as with nonmetallic elements such as selenium and tellurium settle at the bottom of the cell as anode mud, which then forms the starting material for their production. Isolating the metals involves that they first be dissolved. A number of different methods can be applied to accomplish this, depending on the separation process and the composition of the mixture. Two typical processes are fusion with sodium peroxide (Na_2O_2) followed by dissolution in aqua regia (a mixture of nitric acid and hydrochloric acid, optimally in a molar ratio of 1:3), and dissolution in a mixture of chlorine with hydrochloric acid. Osmium, ruthenium, rhodium and iridium can be separated from platinum, gold and base metals by their insolubility in aqua regia, leaving a solid residue. Rhodium can be isolated from the residue by reaction with molten sodium bisulfate. The insoluble residue, containing Ru, Os, and Ir, is then reacted with sodium oxide, in which Ir is insoluble, producing water-soluble Ru and Os salts. After the oxidation reaction to their volatile oxides, RuO_4 is isolated from OsO_4 by precipitation of $(NH_4)_3RuCl_6$ after reacting with ammonium chloride. Finally, after it is dissolved, osmium is isolated from the other platinum group metals through distillation or extraction with organic solvents of the volatile osmium tetroxide. The first process is comparable to the procedure used by Tennant and Wollaston. Both processes are appropriate for industrial scale production. In either case, the product is reduced using hydrogen, producing the metal as a powder or sponge that can be treated utilizing known powder metallurgical methods.

Osmium can be found in nine different minerals. Five of these minerals are in the elements class and are alloys: hexaferrum (Fe,Os,Ru,Ir), hexamolybdenum (Mo,Ru,Fe,Ir,Os), iridium (varieties iridosmine and rutheniridosminium) (Ir,Os,Ru), osmium (Os,Ir,Ru), and rutheniridosmine (Ir,Os,Ru). The remaining four minerals are in the sulfide class, erlichmanite (OsS_2) (Fig. 7.71), omeiite ($(Os,Ru)As_2$), osarsite ((Os,Ru)AsS), and ruarsite ((Ru,Os)AsS).

7.14.3 Chemistry

Osmium consists of seven naturally occurring isotopes, six of which are stable: ^{184}Os, ^{187}Os, ^{188}Os, ^{189}Os, ^{190}Os, and (most abundant at 40.78%) ^{192}Os. ^{186}Os decays via α decay with such a long half-life of 2.0 (± 1.1) $\times 10^{15}$ years, about 140,000 times the age of the universe, that for practical purposes it can be considered stable. α decay is predicted for all seven naturally occurring isotopes, but it has been detected only for ^{186}Os, probably because of the extremely long half-lives. It has been predicted that ^{184}Os and ^{192}Os can decay via double β decay, but this has not been detected up till now. ^{187}Os is the daughter of ^{187}Re (half-life 4.56×10^{10} years) and is widely applied in dating terrestrial as well as meteoric rocks. In addition, it has been applied to determine the intensity of continental weathering over geologic time and to fix minimum ages for stabilization of the mantle roots of continental cratons. This decay causes Re-rich

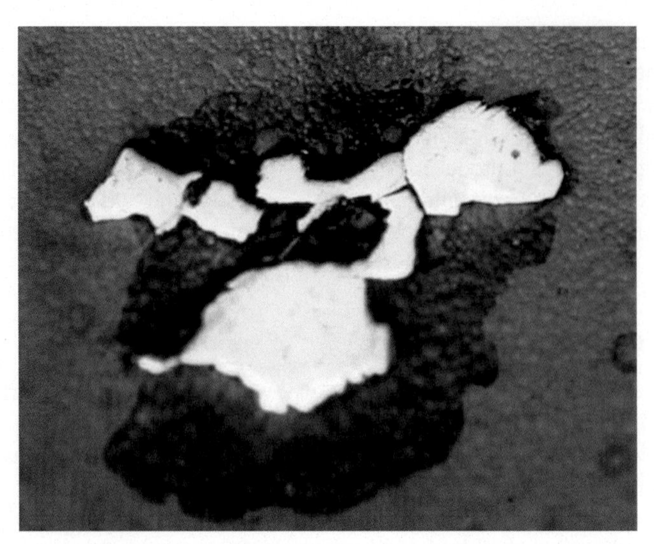

FIGURE 7.71 Erlichmanite, OsS_2, polished section embedded in epoxy (for microprobe study). Goodnews Bay, Alaska, United States.

TABLE 7.14 Osmium properties.

Appearance	Silvery, blue cast
Standard atomic weight $A_{r,std}$	190.23
Block	d-Block
Element category	Transition metal
Electron configuration	[Xe] $4f^{14}$ $5d^6$ $6s^2$
Phase at STP	Solid
Melting point	3033°C
Boiling point	5012°C
Density (near r.t.)	22.59 g/cm^3
When liquid (at m.p.)	20 g/cm^3
Heat of fusion	31 kJ/mol
Heat of vaporization	378 kJ/mol
Molar heat capacity	24.7 J/(mol·K)
Oxidation states	−4, −2, −1, 0, +1, +2, +3, **+4**, +5, +6, +7, +8
Ionization energies	1st: 840 kJ/mol
	2nd: 1600 kJ/mol
Atomic radius	Empirical: 135 pm
Covalent radius	144 ± 4 pm

STP, Standard temperature and pressure.
Bold font indicates main oxidation state.

minerals to be unusually rich in ^{187}Os. Though, the most noteworthy application of Os isotopes in geology has been in combination with the abundance of Ir, to characterize the layer of shocked quartz along the Cretaceous–Paleogene boundary that marks the extinction (among others) of the nonavian dinosaurs 66 million years ago.

Osmium has a blue-gray tint and is the densest stable element; it is approximately twice as dense as lead and slightly denser than iridium. Calculations of density from X-ray diffraction data have produced the most reliable data for these elements, resulting in a value of 22.587 ± 0.009 g/cm^3 for osmium, somewhat denser than the 22.562 ± 0.009 g/cm^3 of iridium; both metals are nearly 23 times as dense as water (Table 7.14). Osmium is a hard but brittle metal that remains lustrous even at high temperatures. It has an extremely low compressibility. Correspondingly, its bulk modulus is very high, between 395 and 462 GPa, which is in the same range as that of diamond (443 GPa). The hardness of osmium is moderately high at 4 GPa. Due to its hardness, brittleness, low vapor pressure (the lowest of the platinum group metals), and very high melting point (the fourth highest of all elements, after only carbon, tungsten, and rhenium), solid osmium is hard to machine, form, or work.

Osmium forms compounds with oxidation states varying between −4 to +8. The most common oxidation states are +2, +3, +4, and +8. The +8 oxidation state is significant as it is the highest attained by any chemical element aside from iridium's +9 and is found only in xenon, ruthenium, hassium, and iridium. The oxidation states −1 and −2 exemplified by the two reactive compounds $Na_2[Os_4(CO)_{13}]$ and $Na_2[Os(CO)_4]$ are employed in the synthesis of osmium cluster compounds. Notwithstanding its broad range of compounds in numerous oxidation states, osmium in bulk form at normal temperatures and pressures resists attack by all acids, including aqua regia but is attacked by fused alkalis.

7.14.3.1 Oxides

The most common compound with the +8 oxidation state is osmium tetroxide. This toxic compound is formed when powdered osmium is exposed to air. The compound is significant for its many applications, notwithstanding its toxicity and the rarity of osmium. It also possesses several unusual properties, one being that the solid is water soluble, and it is very volatile with a strong smell. The compound is colorless, although most samples appear yellow. This is most probably because of the presence of the impurity OsO_2, which has a yellow-brown color. Osmium(VIII) oxide crystalizes in the monoclinic system. It has a distinctive acrid chlorine-like odor. The element name osmium stems from *osme*, Greek for odor. OsO_4 is volatile: it sublimes at room temperature. It is soluble in a wide range of organic solvents. In addition, it is moderately soluble in water, with which it reacts reversibly to form osmic acid. The osmium tetroxide molecule is tetrahedral and hence nonpolar. This nonpolarity helps OsO_4 penetrate charged cell membranes. OsO_4 is 518 times more soluble in carbon tetrachloride than in water. The osmium of OsO_4 has an oxidation number of VIII; yet, the metal does not have a corresponding 8+ charge as the bonding in the compound is mostly covalent in character (the ionization energy needed to produce a formal 8+ charge also far surpasses the energies available in normal chemical reactions). The osmium atom has eight valence electrons ($6s^2$, $5d^6$) with double bonds to the four oxide ligands resulting in a 16-electron complex. This is isoelectronic with permanganate and chromate ions. OsO_4 is produced gradually when osmium powder reacts with O_2 at ambient temperature. Reaction of bulk solid involves heating to 400°C.

$$Os + 2O_2 \xrightarrow{\Delta T} OsO_4$$

OsO_4 is a Lewis acid and a mild oxidant. It reacts with alkaline aqueous solution to form the perosmate anion $OsO_4(OH)_2^{2-}$. This anion is readily reduced to osmate anion, $OsO_2(OH)_4^{4-}$. When the Lewis base is an amine, adducts are likewise formed. Consequently, OsO_4 can be stored in the form of osmeth, in which OsO_4 is complexed with hexamine. Osmeth can be dissolved in THF and diluted in an aqueous buffer solution to produce a dilute (0.25%) working solution of OsO_4. With tert-BuNH_2 ($(CH_3)_3CNH_2$), the imido derivative is formed.

$$OsO_4 + Me_3CNH_2 \rightarrow OsO_3(NCMe_3) + H_2O$$

Likewise, with NH_3 the nitrido complex is produced.

$$OsO_4 + NH_3 + KOH \rightarrow K[Os(N)O_3] + 2H_2O$$

The $[Os(N)O_3]^-$ anion is isoelectronic and isostructural with OsO_4. OsO_4 is very soluble in tert-butyl alcohol. In solution, it is easily reduced by hydrogen to osmium metal. The suspended osmium metal can be employed as a catalyst for hydrogenation of a wide range of organic chemicals with double or triple bonds.

$$OsO_4 + 4H_2 \rightarrow Os + 4H_2O$$

OsO_4 undergoes "reductive carbonylation" with carbon monoxide in methanol at 127°C (400K) and 200 bar to form the triangular cluster $Os_3(CO)_{12}$.

$$3OsO_4 + 24CO \rightarrow Os_3(CO)_{12} + 12CO_2$$

Osmium tetroxide reacts to form red osmates $OsO_4(OH)_2^{2-}$ with a base. With ammonia, it produces the nitrido-osmates OsO_3N^-. Osmium tetroxide boils at 130°C and is a strongly oxidizing agent. Potassium osmate, $K_2[OsO_2(OH)_4]$, is a diamagnetic purple salt with osmium in the 6+ oxidation state. When dissolved in water a pink solution is produced, while when dissolved in methanol, the salt forms a blue solution. The salt gained interest as a catalyst for the asymmetric dihydroxylation of olefins. The complex anion is octahedral. Like related d^2 dioxo complexes, the oxo ligands are *trans*. The Os-O and Os-OH distances are 1.75(2) and 1.99(2) Å, respectively. It is a comparatively rare illustration of a metal oxo complex that follows the 18e rule. The compound was first described by French chemist Edmond Frémy (February 28, 1814 to February 3, 1894) in 1844. Potassium osmate is formed by reducing perosmates using alcohol.

$$2OsO_4 + C_2H_5OH + 3KOH \rightarrow CH_3CO_2K + 2K_2[OsO_2(OH)_4]$$

Alkaline oxidative fusion of osmium metal also produces this salt.

Unlike OsO_4, osmium dioxide (OsO_2) is nonvolatile, and much less reactive and toxic. It occurs as a brown to black crystalline powder, but single crystals are golden and have metallic conductivity. The compound crystallizes in the rutile (TiO_2) structural motif, that is, the connectivity is very comparable to that in the mineral rutile. OsO_2 can be prepared by the reaction of osmium with various oxidizing agents, for example, sodium chlorate, osmium tetroxide, and nitric oxide at approximately 600°C. Using chemical transport, one can produce large crystals of OsO_2, sizes up to $7 \times 5 \times 3$ mm have been reported. Single crystals exhibit a metallic resistivity of ~15 μΩ cm. A typical transport agent is O_2 via the reversible formation of volatile OsO_4.

$$OsO_2 + O_2 \rightleftharpoons OsO_4$$

OsO_2 does not dissolve in water but is attacked by dilute hydrochloric acid.

7.14.3.2 Halides

Osmium pentafluoride (OsF_5) as well as osmium heptafluoride (OsF_7), are known, but osmium trifluoride (OsF_3) has not yet been synthesized. Osmium hexafluoride can be prepared by a direct reaction of osmium metal exposed to an excess of elemental fluorine gas at 300°C.

$$Os + 3F_2 \rightarrow OsF_6$$

Osmium hexafluoride (OsF_6) is a yellow crystalline solid that melts at 33.4°C and boils at 47.5°C. The solid structure measured at −140°C is orthorhombic with space group *Pnma*. Lattice parameters are a = 9.387 Å, b = 8.543 Å, and c = 4.944 Å. There are four formula units (here, discrete molecules) per unit cell, resulting in a density of 5.09 $g \cdot cm^{-3}$. The OsF_6 molecule itself (the form important for the liquid or gas phase) has octahedral molecular geometry, which has point group O_h. The Os-F bond length is 1.827 Å. Partial hydrolysis of OsF_6 results in the formation of $OsOF_4$. Osmium pentafluoride, OsF_5, is a blue-green solid. Like the pentafluorides of Ru, Rh, and Ir, OsF_5 occurs as a tetramer in the solid state. OsF_5 can be produced by reduction of OsF_6 with iodine as a solution in iodine pentafluoride.

$$10OsF_6 + I_2 \rightarrow 10OsF_5 + 2IF_5$$

Osmium(IV) chloride or osmium tetrachloride, $OsCl_4$, exists in two polymorphs (crystalline forms). The compound is employed to synthesize other osmium complexes. It was first described in 1910 by German chemists Otto Ruff (December 12, 1871 to September 17, 1939) and Ferdinand Bornemann as the product of chlorination of osmium metal. This route forms the high-temperature polymorph.

$$Os + 2Cl_2 \rightarrow OsCl_4$$

This reddish-black polymorph is orthorhombic and has a structure in which osmium centers are octahedrally coordinated, sharing opposite edges of the $OsCl_6$ octahedra to form a chain. A brown, apparently cubic polymorph is produced upon reduction of osmium tetroxide with thionyl chloride.

$$OsO_4 + 4SOCl_2 \rightarrow OsCl_4 + 2Cl_2 + 4SO_2$$

Osmium tetraoxide dissolves in hydrochloric acid forming the hexachloroosmate anion.

$$OsO_4 + 10HCl \rightarrow H_2OsCl_6 + 2Cl_2 + 4H_2O$$

The lower oxidation states are stabilized by the larger halogens, so that the trichloride, tribromide, triiodide, and even diiodide are known. The oxidation state +1 is known only for osmium iodide (OsI), whereas several carbonyl complexes of osmium, such as triosmium dodecacarbonyl ($Os_3(CO)_{12}$), represent oxidation state 0.

Osmium forms a number of oxofluorides. All of them are very sensitive to moisture. Purple *cis*-OsO_2F_4 forms at −196°C (77K) in an anhydrous HF solution.

$$OsO_4 + 2KrF_2 \rightarrow cis\text{-}OsO_2F_4 + 2Kr + O_2$$

OsO_4 similarly reacts with F_2 to form yellow OsO_3F_2.

$$2OsO_4 + 2F_2 \rightarrow 2OsO_3F_2 + O_2$$

OsO_4 reacts with one equivalent of $[Me_4N]F$ at 25°C and two equivalents at −20°C.

$$OsO_4 + [Me_4N]F \rightarrow [Me_4N][OsO_4F]$$
$$OsO_4 + 2[Me_4N]F \rightarrow [Me_4N]_2[cis\text{-}OsO_4F_2]$$

7.14.3.3 Borides

Osmium borides are compounds of osmium and boron. Their most notable property is their potentially high hardness. It is believed that a combination of high electron density of osmium with the strength of boron-osmium covalent bonds will make osmium borides superhard compounds, however this has not been shown yet. For instance, OsB_2 is hard (hardness comparable to that of corundum), but not superhard. Osmium borides are formed in vacuum or inert atmosphere to inhibit the formation of osmium tetroxide, which is a toxic compound. The reaction takes place at high temperatures ($\sim 1000°C$) in a mixture of MgB_2 and $OsCl_3$. Three osmium borides are known: OsB, Os_2B_3, and OsB_2. The first two have a hexagonal structure, comparable to that of rhenium diboride. Osmium diboride was first also sought as hexagonal, but one of its phases was later reassigned as orthorhombic. In recent synthesis methods, it has also been determined that a hexagonal phase of OsB_2 exists with a similar structure to ReB_2.

7.14.3.4 Organo-osmium compounds

In the group 8 elements osmium resembles ruthenium in its complexes. Since Os is more expensive than Ru, the chemistry is less developed and has fewer applications. Of course, the expense of the catalyst is offset if turnover numbers are high. Consequently, osmium tetroxide is a valuable oxidizing agent in organic chemistry in particular in the conversion of alkenes to 1,2-diols. The 5d-orbitals in Os are higher in energy than the 4d-orbitals in Ru. Thus, π back-bonding to alkenes and CO is stronger for Os compounds, which results in more stable organic derivatives. This effect is demonstrated by the stability of the alkene derivatives such as $[Os(NH_3)_5(alkene)]^{2+}$ or $[Os(NH_3)_5(arene)]^{2+}$. Generally, the lower oxidation states of osmium are stabilized by ligands that are good σ-donors (such as amines) and π-acceptors (heterocycles containing nitrogen). The higher oxidation states are stabilized by strong σ- and π-donors, such as O^{2-} and N^{3-}. Important compounds, at least for research purposes, are the carbonyls such as triosmium dodecacarbonyl and decacarbonyldihydridotriosmium. The phosphine complexes are similar to those of ruthenium, but hydride derivatives, for example, $OsHCl(CO)(PPh_3)_3$, have a tendency to be more stable.

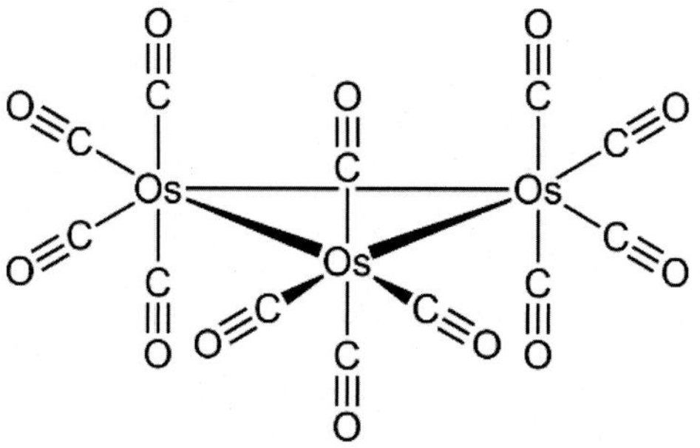

triosmium dodecacarbonyl

Triosmium dodecacarbonyl, $Os_3(CO)_{12}$, is a yellow-colored metal carbonyl cluster that forms an important precursor to organo-osmium compounds. Many of the advances in cluster chemistry are the result of research on derivatives of $Os_3(CO)_{12}$ and its lighter analog $Ru_3(CO)_{12}$. The cluster has D_{3h} symmetry, comprising an equilateral triangle of Os atoms, each of which bears two axial and two equatorial CO ligands. The Os-Os bond distance is 2.88 Å. $Ru_3(CO)_{12}$ has the same structure, while $Fe_3(CO)_{12}$ is different, with two bridging CO ligands resulting in C_{2v} symmetry. $Os_3(CO)_{12}$ is prepared by the direct reaction of OsO_4 with carbon monoxide at 175°C.

$$3OsO_4 + 24CO \rightarrow Os_3(CO)_{12} + 12CO_2$$

The yield is almost quantitative. The chemical properties of $Os_3(CO)_{12}$ have been extensively examined. Direct reactions of ligands with the cluster frequently result in complex product distributions as the inert Os-CO bonds involve high temperatures to break, and at such high temperatures, the initially formed adducts react further. More effectively, $Os_3(CO)_{12}$ is transformed to more labile derivatives such as $Os_3(CO)_{11}(MeCN)$ and $Os_3(CO)_{10}(MeCN)_2$ using Me_3NO as a decarbonylating agent. $Os_3(CO)_{11}(MeCN)$ reacts with various even weakly basic ligands to produce adducts such as $Os_3(CO)_{11}(ethylene)$ and $Os_3(CO)_{11}(pyridine)$. The direct reaction of $Os_3(CO)_{12}$ with ethylene and pyridine causes degradation of these organic ligands to give the vinyl hydride $HOs_3(CO)_{10}(\eta^1,\eta^2\text{-}C_2H_3)$ and the pyridyl-hydride $HOs_3(CO)_{10}(NC_5H_4)$. These products demonstrate the stability of Os-H and Os-C bonds. $Os_3(CO)_{12}$ is a platform to study the manners that hydrocarbons can interact with ensembles of metals. For instance, $CH_3(H)Os_3(CO)_{10}$ offered one of the first unambiguous illustrations of agostic bonding (a term in organometallic chemistry for the interaction of a coordinatively unsaturated transition metal with a

C-H bond, when the two electrons involved in the C-H bond enter the empty d-orbital of a transition metal, resulting in a 3-center 2-electron bond). From the perspective of bonding, $H_2Os_3(CO)_{10}$ is significant. In this molecule, the 2 hydride ligands bridge 1 Os-Os edge. The molecule exhibits reactivity similar to diborane.

Decacarbonyldihydridotriosmium, $H_2Os_3(CO)_{10}$, a purple-violet crystalline air-stable cluster is notable as it is electronically unsaturated and therefore adds various substrates. The trinuclear cluster contains an isosceles triangular array of metals with one short edge (rOs-Os = 2.68 Å), which is spanned by the two hydride ligands, and two longer edges (rOs-Os = 2.81 Å). It can be written as $Os(CO)_4[Os(CO)_3(\mu\text{-}H)]_2$. The bonding in the Os_2H_2 subunit has been compared to the three-center, 2e bonding in diborane. It is formed by purging a solution of $Os_3(CO)_{12}$ in octane (or other inert solvent of similar boiling point) with H_2.

$$Os_3(CO)_{12} + H_2 \rightarrow Os_3H_2(CO)_{10} + 2CO$$

The cluster reacts with a wide variety of reagents under mild conditions. A good example is its reaction with diazomethane to give $Os_3(CO)_{10}(\mu\text{-}H)(\mu\text{-}CH_3)$, showing an agostic interaction, the first discovered in a metal cluster. Osmocene, $Os(\eta\text{-}C_5H_5)_2$, is an organo-osmium compound found as a white solid. It is a metallocene. Osmocene is commercially available. It may be formed by the reaction of osmium tetroxide with hydrobromic acid, followed by zinc and cyclopentadiene.

7.14.4 Major uses

Due to the volatility and extreme toxicity of its oxide, Os is seldom utilized in its pure state, but is instead frequently alloyed with other metals for high-wear applications. Os alloys such as osmiridium are very hard and, together with other Pt-group metals, are employed in, for example, fountain pen tips, instrument pivots, and electrical contacts, since they can resist wear from everyday operation. Historically they were utilized for the tips of phonograph styli during the late 78 rpm and early "LP" and "45" record era, between roughly 1945 and 1955. Os alloy tips were much more durable than steel and Cr needle points, but wore out much faster than competing, and more expensive, sapphire and diamond tips, consequently they were discontinued. OsO_4 has been employed in fingerprint detection and in staining fatty tissue for optical and electron microscopy. Since it is a strong oxidant, it cross-links lipids principally by reacting with unsaturated C−C bonds and thus both fixes biological membranes in place in tissue samples and at the same time stains them. Since Os atoms are very electron dense, Os staining significantly improves image contrast in transmission electron microscopy (TEM) studies of biological materials. Those carbon materials otherwise have extremely weak TEM contrast. Another Os compound, Os ferricyanide (OsFeCN), shows comparable fixing and staining action.

OsO_4 and its derivative K-osmate are significant oxidants in organic synthesis. For the Sharpless asymmetric dihydroxylation (also called the Sharpless bishydroxylation), which uses osmate for the conversion of a double bond into a vicinal diol. It is common practice to perform this reaction using a catalytic amount of OsO_4, which after reaction is regenerated with reoxidants such as potassium ferricyanide (IUPAC name potassium hexacyanoferrate(III), $K_3[Fe(CN)_6]$) or N-methylmorpholine N-oxide ((more correctly 4-methylmorpholine 4-oxide, $C_5H_{11}NO_2$). This significantly decreases the amount of the highly toxic and very expensive OsO_4 required. Such chiral diols are important in organic synthesis. The introduction of chirality into nonchiral reactants using chiral catalysts is an important concept in organic synthesis. This reaction was developed principally by American chemist Karl Barry Sharpless (born 28 April 1941) building on the already known racemic Upjohn dihydroxylation (an organic reaction which converts an alkene to a cis vicinal diol. It was developed by the Upjohn Company in 1976), for which he was awarded a share of the 2001 Nobel Prize in Chemistry. OsO_4 is very costly for this application, therefore $KMnO_4$ is often used in its place, although the yields are lower for this less expensive chemical reagent.

In 1898 an Austrian chemist Auer von Welsbach (September 1, 1858 to August 4, 1929) created the Oslamp with a filament made of Os, which he commercialized in 1902. After only a couple of years, Os was replaced by the more stable metal W. W has the highest melting point of all metals, and its application in light bulbs increases both the luminous efficacy as well as the life of incandescent lamps. The light bulb producer Osram (founded in 1906, when three German companies, Auer-Gesellschaft, AEG and Siemens & Halske, united their lamp production facilities) derived its name from the elements of osmium and Wolfram (the latter is German for tungsten). Similar to Pd, powdered Os efficiently absorbs hydrogen atoms. This could make Os a possible material for a metal hydride battery electrode. But Os is costly and would react with KOH, the most common battery electrolyte. Os posseses high reflectivity in the ultraviolet range of the electromagnetic spectrum; for example, at 600 Å Os has a reflectivity twice that of Au. This high reflectivity is very useful in space-based UV spectrometers, which have minimized mirror sizes because of space limitations. Os-coated mirrors were used in a couple of space missions aboard the Space Shuttle, but it quickly became clear that the oxygen radicals in the low Earth orbit are sufficiently abundant to significantly deteriorate the Os layer.

The only known medical application of Os is synovectomy in arthritic patients in Scandinavia. Synovectomy is a procedure where the synovial tissue surrounding a joint is removed. This procedure is typically recommended to provide relief from a condition in which the synovial membrane or the joint lining becomes inflamed and irritated and is not controlled by medication alone. If arthritis (inflammation of the joint) is not controlled, it can lead to irreversible joint damage. The synovial membrane or "synovium" encloses each joint and also secretes a lubricating fluid that allows different joint motions such as rolling, folding and stretching. When the synovium becomes inflamed or irritated, it increases fluid production, resulting in warmth, tenderness, and swelling in and around the joint. It comprises the local administration of OsO_4, which is a highly toxic compound. The absence of reports of long-term side effects seems to indicate that Os itself can be biocompatible, though this is dependent on the Os compound administered. In recent years, Os(VI) and Os(II) compounds were reported to exhibit anticancer activity *in vivo*, showing promise for utilizing Os compounds as anticancer drugs.

7.15 77 Ir — Iridium

7.15.1 Discovery

The discovery of iridium is closely connected with that of platinum and the other elements of the platinum group. Native platinum used by ancient Ethiopians and by South American cultures always had a minor quantity of the other platinum group elements, including iridium. Platinum arrived in Europe as platina ("silverette"), discovered in the 17th century by the Spanish conquerors in a district currently known as the department of Chocó in Colombia. The discovery that this metal was not an alloy of already identified elements, but instead a separate new element, did not happen until 1748 (de Ulloa, 1748). Chemists who examined platinum dissolved it in aqua regia (a mixture of hydrochloric and nitric acids) to create soluble salts. They continuously found a small amount of a dark, insoluble residue. French chemist Joseph Louis Proust (September 26, 1754 to July 5, 1826) believed that the residue was graphite. The French chemists Hippolyte-Victor Collet-Descotils (November 21, 1773 to December 6, 1815), Antoine François, comte de Fourcroy (June 15, 1755 to December 16, 1809), and pharmacist and chemist Louis Nicolas Vauquelin (May 16, 1763 to November 14, 1829) also detected the black residue in 1803 but did not have enough material for any further study (Collet-Descotils, 1803, Fourcroy and Vauquelin, 1803). In 1803 English chemists Smithson Tennant (November 30, 1761 to February 22, 1815) studied the insoluble residue and decided that it must comprise a new metal. Vauquelin reacted the powder alternately with alkali and acids and obtained a volatile new oxide, which he believed to be of this new metal—which he named ptene, from the Greek word πτηνός ptēnós, "winged." (Tennant, 1804), who had the benefit of a much larger amount of residue, continued his study and identified the two beforehand undiscovered elements in the black residue, iridium and osmium. He managed to obtain dark red crystals (probably of $Na_2[IrCl_6]{\cdot}nH_2O$) through a sequence of reactions with sodium hydroxide (NaOH) and hydrochloric acid (HCl). He called it iridium after Iris (Ἶρις), the Greek winged goddess of the rainbow and the messenger of the Olympian gods, as many of the salts he got were strongly colored. Detection of the new elements was reported in a letter to the Royal Society on June 21, 1804. British chemist, mineralogist and zoologist John George Children (May 18, 1777 to January 1, 1852) was the first to melt a sample of iridium in 1813 using "the greatest galvanic battery that has ever been constructed" (at that time) (Children, 1815) (Fig. 7.72). The first to separate high-purity iridium was American chemist Robert Hare (January 17, 1781 to May 15, 1858) in 1842 (Fig. 7.73). He determined that it had a density of about 21.8 g/cm^3 and observed that the metal is nearly unmalleable and very hard. The first melting in a significant amount was done by French chemists Henri Étienne Sainte-Claire Deville (March 11, 1818 to July 1, 1881) (Fig. 7.74) and Jules Henri Debray (July 26, 1827, in Amiens to July 19, 1888) in 1860 (Sainte-Claire Deville and Debray, 1861) (Fig. 7.75). They had to burn more than 300 L of pure O_2 and H_2 gas for each kilogram of iridium. First a pupil then an assistant and later a collaborator of Deville's, Debray eventually succeeded him as Professor at the École Normale. Their association was extremely close, and their joint work on the melting of platinum and its alloys extended over many years. These extreme complications in melting the metal restricted the options for handling iridium. English inventor John Isaac Hawkins (1772–1855) was trying to obtain a fine and hard point for fountain pen nibs, and in 1834 succeeded to produce an iridium-pointed gold pen. In 1880 American businessman and industrialist John Holland (August 25, 1838–1917) ("John Holland Gold Pen Company" was a large maker of pens and related products during the late 19th century) and American chemist William Lofland Dudley (April 16, 1859 to September 8, 1914) managed to melt iridium by adding phosphorus and patented the process in the United States; British company Johnson Matthey later maintained they had been applying a comparable process since 1837 and had already presented fused iridium at a number of World Fairs. The first application of an alloy of iridium with ruthenium in thermocouples was made by German physicist Otto Feussner (1890–1934) in 1933. These thermocouples permitted the measurement of high temperatures in air up to 2000°C. In Munich, Germany in 1957 physicist Rudolf Mössbauer (January 31, 1929 to September 14, 2011), in what has been recognized as one of the "landmark experiments in 20th-century physics," discovered the resonant and recoil-free

FIGURE 7.72 John George Children, 1826 portrait by Benjamin Rawlinson Faulkner (1787–1849).

FIGURE 7.73 Robert Hare, between 1892 and 1893 Popular Science Monthly Volume 42.

FIGURE 7.74 Étienne Sainte-Claire Deville.

emission and absorption of gamma rays by atoms in a solid metal sample containing only ^{191}Ir (Mössbauer, 1958). This phenomenon, now named the Mössbauer effect (which has subsequently been discovered for other nuclei, such as ^{57}Fe), and developed as Mössbauer spectroscopy, has made significant contributions to research in physics, chemistry, biochemistry, metallurgy, and mineralogy. Mössbauer was awarded the Nobel Prize in Physics in 1961, at the age of 32, just 3 years after he published his discovery. In 1986 Rudolf Mössbauer was awarded the Albert Einstein Medal and the Elliot Cresson Medal for his achievements.

7.15.2 Mining, production, and major minerals

Iridium is one of the nine least abundant stable elements in the Earth's crust, having an average concentration of 0.001 ppm in crustal rock. In contrast to its low abundance in crustal rock, iridium is comparatively common in meteorites, with concentrations of 0.5 ppm or higher (Box 7.6). The overall concentration of iridium on Earth is believed to be much higher than what is found in crustal rocks, but because of the density and siderophilic ("iron-loving") character of iridium, it has accumulated below the crust in the Earth's core when the planet was still molten. Ir is found in nature as a native element and in natural alloys; in particular the Ir—Os alloys, osmiridium (Os rich), and iridosmium (Ir rich). In the Ni and Cu deposits, the platinum group metals can be found as sulfides (i.e., (Pt,Pd)S), tellurides (i.e., PtBiTe), antimonides (PdSb), and arsenides (i.e. $PtAs_2$). In all these compounds, platinum is substituted by a minor amount of Ir and Os. As with all the platinum group elements, Ir occurs naturally in alloys with raw Ni or Cu. Several Ir-dominant minerals, with Ir as the species-forming element, are known. They are extremely rare and frequently embody the Ir equivalents of the above-given minerals. Some examples are irarsite ((Ir,Ru,Rh,Pt)AsS) and cuproiridsite ($(Cu,Fe)Ir_2S_4$). In the Earth's crust, iridium is observed at highest concentrations in three different types of geological structures: igneous deposits (crustal intrusions from below), impact

FIGURE 7.75 Jules Henri Debray.

BOX 7.6 Mass extinction at the Cretaceous–Paleogene boundary.

The boundary between the Cretaceous and Paleogene 66 million years ago has been identified by a thin stratum of Ir-rich clay. A research team led by Luis Alvarez (June 13, 1911 to September 1, 1988) and his son Walter (born October 3, 1940) proposed in 1980 an extra-terrestrial origin for this iridium, ascribing it to an asteroid or comet impact. Their theory, known as the Alvarez hypothesis, is now generally recognized to explain the extinction of the nonavian dinosaurs. A large buried impact crater structure with an estimated age of about 66 million years was later discovered under what is currently the Yucatán Peninsula, Mexico (the Chicxulub crater). Others though claim that the Ir could have been of volcanic origin instead, as the Earth's core is rich in Ir, and active volcanoes such as Piton de la Fournaise, on the island of Réunion, are still releasing Ir.

craters, and deposits reworked from one of the former structures. The largest known primary reserves are found in the Bushveld igneous complex in South Africa, (near the largest known impact crater, the Vredefort crater) although the large Cu–Ni ore deposits near Norilsk in Russia, and the Sudbury Basin (also an impact crater) in Canada contain also substantial reserves of Ir. Smaller reserves are found in the United States. In addition, Ir can be found in secondary deposits, combined with Pt and other Pt-group elements in alluvial deposits. The alluvial deposits used by pre-Columbian people in the Chocó Department of Colombia even today produce Pt-group metals.

In addition, iridium is commercially produced as a by-product from nickel and copper mining and processing. In the process of copper and nickel electrorefining, noble metals such as gold, silver, and the platinum group elements, as well as selenium and tellurium, settle to the bottom of the cell in what is known as the anode mud, which forms the starting material for their extraction. To isolate the different metals, they must first be dissolved. A number of different separation processes can be applied depending on the nature of the mixture; two illustrative methods are fusion with

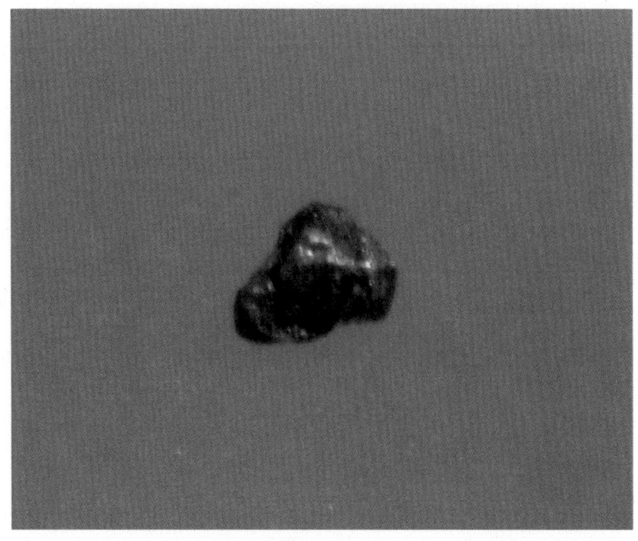

FIGURE 7.76 0.5 mm platinum, Pt, nugget with included irarsite, (Ir,Ru,Rh,Pt)AsS. Goodnews Bay, Alaska, United States.

sodium peroxide (Na_2O_2) followed by dissolution in aqua regia (a mixture of nitric acid and hydrochloric acid, optimally in a molar ratio of 1:3), and dissolution in a mixture of chlorine with hydrochloric acid. After the mixture is dissolved, iridium is isolated from the other platinum group elements through precipitation of the ammonium hexachloroiridate ($(NH_4)_2IrCl_6$) or through extraction of $IrCl_6^{2-}$ using organic amines. The first method is similar to the method Tennant and Wollaston applied for their isolation. The second method can be used as a continuous liquid–liquid extraction process and is hence more appropriate for industrial scale production of iridium. In both cases, the last step involves the reduction of the product with hydrogen, producing the metal as a powder or sponge that can subsequently be treated using common powder metallurgical methods.

Iridium can be found in 25 different minerals. Ten of these minerals fall within the elements class: for example, chengdeite (Ir_3Fe), garutiite (Ni,Fe,Ir), hexaferrum (Fe,Os,Ru,Ir), iridium (Ir,Os,Ru), and rhuteniridosmine (Ir,Os,Ru). The sulfide class contains 15 different minerals, for example, changchengite (IrBiS), cuproiridsite ($(Cu,Fe)Ir_2S_4$), irarsite ((Ir,Ru,Rh,Pt)AsS) (Fig. 7.76), shuangfengite ($IrTe_2$), tolovkite (IrSbS) and xingzhongite ($Pb^{2+}Ir_2^{3+}S_4$) (Fig. 7.77).

7.15.3 Chemistry

Iridium consists of two naturally occurring, stable isotopes, ^{191}Ir and ^{193}Ir, with natural abundances of 37.3% and 62.7%, respectively. In addition, no less than 37 artificial radioisotopes have been characterized, ranging from ^{164}Ir to ^{202}Ir. ^{192}Ir is the most stable radioisotope with a half-life of 73.827 days and is used in brachytherapy (a form of radiotherapy where a sealed radiation source is placed inside or next to the area requiring treatment. Brachytherapy is commonly used as an effective treatment for cervical, prostate, breast, esophageal, and skin cancer and can also be used to treat tumors in many other body sites) and in industrial radiography, mainly for nondestructive testing of welds in steel in the oil and gas industries; ^{192}Ir sources have played a part in several radiological accidents. Another three radioisotopes have half-lives of more than 1 day: ^{188}Ir (1.73 days), ^{189}Ir (13.2 days), and ^{190}Ir (11.8 days). Radioisotopes lighter than ^{191}Ir decay by some combination of β^+ decay, α decay, and (rare) proton emission, except for ^{189}Ir, which decays by electron capture. Synthetic radioisotopes heavier than ^{191}Ir decay by β^- decay, while ^{192}Ir in addition has a minor electron capture decay path. All currently known Ir isotopes were found between 1934 and 2008, with the most recent discoveries being $^{200-202}Ir$. No less than 32 metastable isomers have been detected, ranging in mass number from 164 to 197. The most stable of these is $^{192m2}Ir$ with a half-life of 241 years that decays by isomeric transition, making it more stable than any of the artificial Ir isotopes in their ground states. The least stable isomer is $^{190m3}Ir$ that has a half-life of only 2 μs. ^{191}Ir was the first one of any elements to be revealed to present the Mössbauer effect. This makes it useful for Mössbauer spectroscopy in research areas such as physics, chemistry, biochemistry, metallurgy, and mineralogy.

A member of the platinum group metals, iridium is white, resembling platinum, but with a slight yellowish cast (Table 7.15). Due to its hardness, brittleness, and very high melting point, solid iridium is hard to machine, form, or work; so powder metallurgy is generally used instead. It is the only metal to retain good mechanical properties in air at

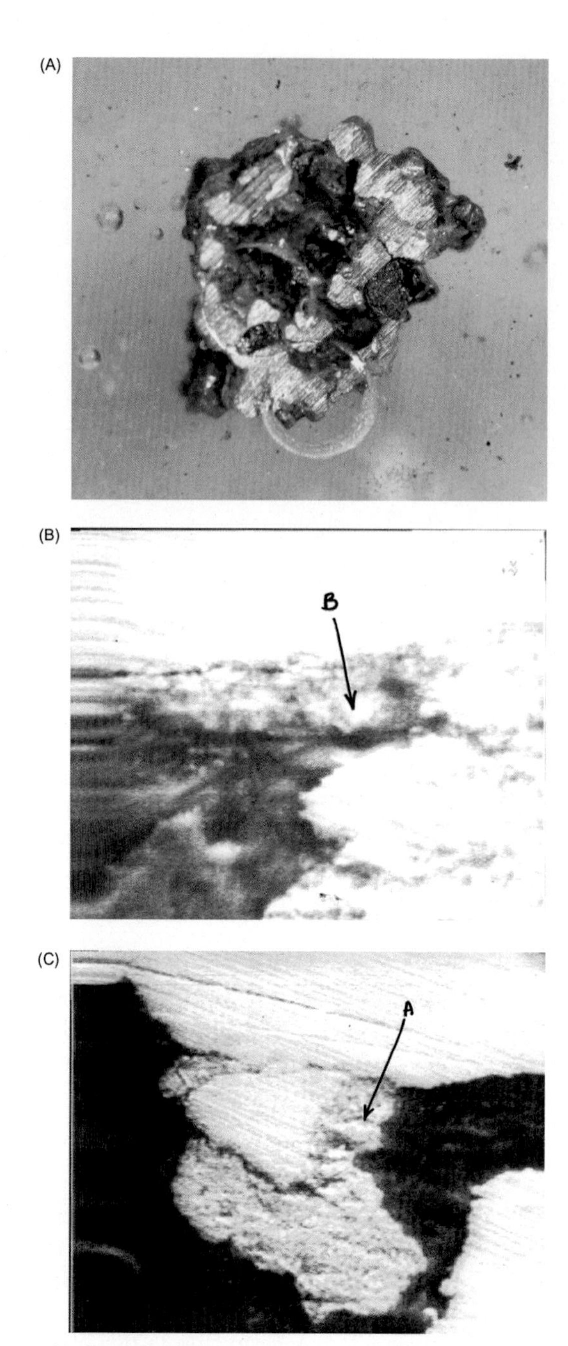

FIGURE 7.77 (A) (top) Xingzhongite, $Pb^{2+}Ir_2^{3+}S_4$. Microprobe mount with a 2-mm polished grain containing submicroscopic rims of this rare iridium–platinum mineral found in three localities worldwide. Typically occurs as rims around irisdosmine, (Os,Ir), (B) and (C) Microprobe image of xingzhongite (bottom), Goodnews Bay, Alaska, United States.

temperatures above, 600°C. It has the 10th highest boiling point of all elements and becomes a superconductor at temperatures below −273.01°C (0.14K). Iridium is the most corrosion-resistant metal known: it is not attacked by nearly any acid, aqua regia, molten metals, or silicates at high temperatures. It can, though, be attacked by some molten salts, such as sodium cyanide and potassium cyanide, as well as oxygen and the halogens (especially fluorine) at higher temperatures. Iridium can also react directly with sulfur at atmospheric pressure to yield iridium disulfide.

Iridium forms compounds in oxidation states between −3 and +9; the most common oxidation states are +3 and +4. Well-described examples of the high +6 oxidation state are uncommon but include IrF_6 and two mixed oxides Sr_2MgIrO_6 and Sr_2CaIrO_6. In addition, it was described in 2009 that iridium(VIII) oxide (IrO_4) was synthesized under matrix isolation conditions (−276.2°C/6K in Ar) by UV irradiation of an iridium-peroxo complex. This species, though, is not anticipated to be stable as a bulk solid at higher temperatures. The highest oxidation state (+9), which

TABLE 7.15 Iridium properties.

Appearance	Silvery white
Standard atomic weight $A_{r,std}$	192.217
Block	d-Block
Element category	Transition metal
Electron configuration	[Xe] $4f^{14}$ $5d^7$ $6s^2$
Phase at STP	Solid
Melting point	2446°C
Boiling point	4130°C
Density (near r.t.)	22.56 g/cm^3
When liquid (at m.p.)	19 g/cm^3
Heat of fusion	41.12 kJ/mol
Heat of vaporization	564 kJ/mol
Molar heat capacity	25.10 J/(mol·K)
Oxidation states	−3, −1, 0, +1, +2, **+3**, **+4**, +5, +6, +7, +8, +9
Ionization energies	1st: 880 kJ/mol
	2nd: 1600 kJ/mol
Atomic radius	Empirical: 136 pm
Covalent radius	141 ± 6 pm

STP, Standard temperature and pressure.
Bold font indicates main oxidation state.

is also the highest documented for any element, is only identified in one cation, IrO_4^+; it is only identified as gas-phase species and is not known to produce any salts.

7.15.3.1 Oxides, chalcogenides, and hydrides

Iridium dioxide, IrO_2, a blue-black solid, is the only well-characterized oxide of iridium. The compound has the rutile (TiO_2) structure, containing six coordinate iridium and three coordinate oxygen. It is employed with other rare oxides in the coating of anode electrodes for industrial electrolysis and in microelectrodes for electrophysiology research. It can be prepared by treating the green form of iridium trichloride with oxygen at high temperatures.

$$2IrCl_3 + 2O_2 \rightarrow 2IrO_2 + 3Cl_2$$

A hydrated form is also known. A sesquioxide, Ir_2O_3, has been described as a blue-black powder which is oxidized to IrO_2 by HNO_3. Iridium tetroxide (IrO_4, Iridium(VIII) oxide) is a binary compound of oxygen and iridium in oxidation state +8. This compound was prepared by photochemical rearrangement of $[(\eta^1\text{-}O_2)IrO_2]$ in solid argon at a temperature of 6K (−267°C). At higher temperatures the oxide is unstable. The recognition of the iridium tetroxide cation by infrared photodissociation spectroscopy with formal oxidation state IX has been described. Iridium also forms iridates with oxidation states +4 and +5, such as K_2IrO_3 and $KIrO_3$, which can be prepared from the reaction of potassium oxide or potassium superoxide with iridium at high temperatures. Lithium iridate, Li_2IrO_3, forms black crystals with three slightly different layered atomic structures, α, β, and sometimes γ. Lithium iridate shows metal-like, temperature-independent electrical conductivity, and changes its magnetic ordering from paramagnetic to antiferromagnetic upon cooling to − 258°C (15K). Li_2IrO_3 typically crystallizes in the α or β phase, and a uncommon γ phase has been described. The α-Li_2IrO_3 crystal structure comprises an alternate stacking of hexagonal Li layers and honeycombs of edge-sharing IrO_6 octahedra with Li in the center. The offset in adjacent layers results in a relatively low (monoclinic) crystal symmetry. Li_2IrO_3 crystals have abundant twinning defects where the *ab* crystal planes are rotated by 120° around the *c* axis. Li_2IrO_3 crystals can be formed by direct sintering of Ir and Li metals, which both oxidize during heating in ambient atmosphere. The α phase is formed at 750°C−1050°C, while heating to higher temperatures results in the β phase. The use of Li metal rather than more traditional lithium carbonate, which is easier to handle and store, produces larger crystals. The γ phase can be formed by the calcination of lithium carbonate and iridium(IV) oxide, followed by annealing in molten lithium hydroxide at 700°C−800°C.

The corresponding disulfides, diselenides, sesquisulfides, and sesquiselenides are known, and IrS_3 has also been reported. Iridium disulfide is the binary inorganic compound with the formula IrS_2. Synthesized by the direct reaction of the elements, the compound has the pyrite (FeS_2) crystal structure at high pressure. At normal atmospheric pressures, an orthorhombic polymorph is formed. The high- and low-pressure forms both have octahedral Ir centers, but the S−S distances are pressure dependent. While not practical, IrS_2 is a highly active catalyst for hydrodesulfurization. Iridium(III) sulfide, Ir_2S_3, is an insoluble black solid, formed by heating a mixture of elemental iridium and sulfur. Crystals can be grown by chemical vapor transport using bromine as the transporting agent. The structure comprises octahedral and tetrahedral Ir and S centers, respectively. No close Ir-Ir contacts are present. Rh_2S_3 and Rh_2Se_3 have the same structure. While no binary hydrides of iridium, Ir_xH_y, are known, complexes have been described that contain IrH_5^{4-} and IrH_6^{3-}, where iridium has the +1 and +3 oxidation states, respectively. The ternary hydride $Mg_6Ir_2H_{11}$ is believed to have both the IrH_5^{4-} and the 18-electron IrH_4^{5-} anions.

7.15.3.2 Halides

No monohalides or dihalides are known, whereas trihalides, IrX_3, are known for all of the halogens. For oxidation states +4 and above, only the tetrafluoride, pentafluoride, and hexafluoride are known. Iridium hexafluoride, IrF_6, is a volatile and highly reactive yellow solid composed of octahedral molecules. Iridium hexafluoride is formed by a direct reaction of iridium metal in an excess of elemental fluorine gas at 300°C. But it is thermally unstable and has to be frozen out of the gaseous reaction mixture to prevent dissociation.

$$Ir + 3F_2 \rightarrow IrF_6$$

IrF_6 melts at 44°C and boils at 53.6°C. The solid structure determined at −140°C is orthorhombic with space group *Pnma*. Lattice parameters are a = 9.411 Å, b = 8.547 Å, and c = 4.952 Å. There are four formula units (here, discrete molecules) per unit cell, resulting in a density of 5.11 $g \cdot cm^{-3}$. The IrF_6 molecule itself (the form important for the liquid or gas phase) has octahedral molecular geometry, which has point group O_h. The Ir-F bond length is 1.833 Å. It decomposes in water and is reduced to IrF_4, a crystalline solid, by iridium black. Iridium pentafluoride (IrF_5), a highly reactive yellow low-melting solid, has similar properties but it is in reality a tetramer, Ir_4F_{20}, formed by four corner-sharing octahedra. This structure is similar to RuF_5 and OsF_5. It can be produced by the controlled decomposition of IrF_6 or the reduction of IrF_6 with silicon powder or H_2 in anhydrous HF. Iridium(IV) fluoride, IrF_4, is a dark brown solid. Early reports of IrF_4 before 1965 are questionable and seem to describe the compound IrF_5 instead. The solid can be formed by reduction of IrF_5 with iridium black or reduction with H_2 in aqueous HF. The crystal structure of the solid is noteworthy as it was the first instance of a 3D lattice structure found for a metal tetrafluoride and since then RhF_4, PdF_4, and PtF_4 have been observed to have the same structure. The structure has six-coordinated, octahedral, iridium where two edges of the octahedra are shared and the two unshared fluorine atoms are *cis* to one another.

Hexachloroiridic(IV) acid, H_2IrCl_6, and its ammonium salt are the most important iridium compounds from an industrial standpoint. They are involved in the purification of iridium and employed as precursors for most other iridium compounds, as well as in the preparation of anode coatings. The $IrCl_6^{2-}$ ion has an intense dark brown color and can be readily reduced to the lighter-colored $IrCl_6^{3-}$ and vice versa. Iridium tetrachloride, approximate formula $IrCl_4(H_2O)_n$, is a water-soluble dark brown amorphous solid. Iridium trichloride, $IrCl_3$, which can be prepared in anhydrous form from direct oxidation of iridium powder by chlorine at 650°C, or in hydrated form by dissolving Ir_2O_3 in hydrochloric acid, is frequently utilized as a starting compound for the preparation of other Ir(III) compounds. The anhydrous compound is quite rare, but the related hydrate is useful for synthesizing other iridium compounds. The anhydrous salt is a dark green crystalline solid. More often found is the trihydrate $IrCl_3(H_2O)_3$. Iridium is isolated from the other platinum group metals in the form of crystalline ammonium hexachloroiridate, $(NH_4)_2[IrCl_6]$, which can be reduced to iridium metal in a stream of H_2. The spongy Ir so formed reacts with chlorine at 300°C–400°C to produce iridium(III) chloride. Similar to the related rhodium compound, $IrCl_3$ has the structure of $AlCl_3$. Hydrated iridium trichloride is formed by heating hydrated iridium(III) oxide with hydrochloric acid. On an industrial scale, most iridium complexes are prepared from ammonium hexachloroiridate or the related chloroiridic acid (H_2IrCl_6) since these salts are the most common commercial forms of iridium chlorides available. Hydrated iridium(III) chloride is employed in the laboratory for the synthesis of other iridium compounds such as Vaska's complex, *trans*-$[IrCl(CO)(PPh_3)_2]$. The complex was first reported by American chemist John W. DiLuzio (August 22, 1935 to July 19, 2019) and Estonian-American chemist Lauri Vaska (May 7, 1925 to November 15, 2015) in 1961. Alkene complexes, for example, cyclooctadiene iridium chloride dimer and chlorobis(cyclooctene)iridium dimer, can also be synthesisd by heating the trichloride with the appropriate alkene in water/alcohol mixtures. A well-defined $IrCl_4(H_2O)_n$ derivative is ammonium hexachloroiridate(III) ($(NH_4)_2IrCl_6$). It is used to prepare catalysts, such as the Henbest Catalyst (named after English chemist H. Bernard Henbest, 1924 to 2004) for transfer hydrogenation of cyclohexanones. Iridium(III) complexes are diamagnetic (low-spin) and generally have an octahedral molecular geometry. The salt crystallizes in a cubic motif like that of ammonium hexachloroplatinate. The $[IrCl_6]^{2-}$ centers have octahedral molecular geometry. It is a major intermediate in the separation of iridium from ores. Most other metals form insoluble sulfides when aqueous solutions of their chlorides are treated with hydrogen sulfide, but $[IrCl_6]^{2-}$ resists ligand substitution. Upon heating under H_2, the solid salt transforms to the metal.

$$(NH_4)_2[IrCl_6] + 2H_2 \rightarrow Ir + 6HCl + 2NH_3$$

7.15.3.3 Organoiridium compounds

Organoiridium compounds share many characteristics with those of rhodium, but less so with cobalt. Iridium can exist in oxidation states of − III to + V, but iridium(I) and iridium(III) are the more common. iridium(I) compounds (d^8 configuration) usually occur with square planar or trigonal bipyramidal geometries, whereas iridium(III) compounds (d^6 configuration) typically have an octahedral geometry. Organoiridium compounds contain Ir–C bonds where the metal is typically in lower oxidation states.

7.15.3.3.1 Iridium(0)

Iridium(0) complexes are binary carbonyls, the principal member being tetrairidium dodecacarbonyl, $Ir_4(CO)_{12}$. In contrast to the related $Rh_4(CO)_{12}$, all CO ligands are terminal in $Ir_4(CO)_{12}$, analogous to the difference between $Fe_3(CO)_{12}$ and $Ru_3(CO)_{12}$. This air-stable species is only poorly soluble in organic solvents. It has been employed to synthesize bimetallic clusters and catalysts, for example, for the water gas shift reaction, and reforming, but these studies are of purely basic research interest This is the most common and stable binary carbonyl of iridium. Each Ir center is in octahedral coordination, being bonded to three other iridium atoms and three terminal CO ligands. $Ir_4(CO)_{12}$ has T_d symmetry with an average Ir–Ir distances of 2.693 Å. The related clusters $Rh_4(CO)_{12}$ and $Co_4(CO)_{12}$ have C_{3v} symmetry due to the presence of three bridging CO ligands in each. It is synthesized in two steps by reductive carbonylation of hydrated iridium trichloride. The first step forms $[Ir(CO)_2Cl_2]^-$.

$$IrCl_3 + 3CO + H_2O \rightarrow \left[Ir(CO)_2Cl_2\right]^- + CO_2 + 2H^+ + Cl^-$$
$$4\left[Ir(CO)_2Cl_2\right]^- + 6CO + 2H_2O \rightarrow Ir_4(CO)_{12} + 2CO_2 + 4H^+ + 8Cl^-$$

7.15.3.3.2 Iridium(I)

A well-known example in this group is Vaska's complex, *trans*-carbonylchlorobis(triphenylphosphine)iridium(I), which has the formula $IrCl(CO)[P(C_6H_5)_3]_2$. This square planar diamagnetic organometallic complex comprises a central iridium atom bound to two mutually *trans*-triphenylphosphine ligands, carbon monoxide, and a chloride ion. Vaska's complex can undergo oxidative addition and is noteworthy for its ability to bind to O_2 reversibly. It is a bright yellow crystalline solid. The synthesis comprises heating virtually any iridium chloride salt with triphenylphosphine and a carbon monoxide source. The most widespread method utilizes DMF (dimethylformamide) as a solvent, and sometimes aniline is added to accelerate the reaction. Another common solvent is 2-methoxyethanol. The reaction is typically performed under nitrogen. In the synthesis, triphenylphosphine serves as both a ligand and a reductant, and the carbonyl ligand is derived by decomposition of DMF, possibly by a de-insertion of an intermediate Ir-C(O)H species. The following is a potential balanced equation for this complex reaction.

$$IrCl_3(H_2O)_3 + 3P(C_6H_5)_3 + HCON(CH_3)_2 + C_6H_5NH_2 \rightarrow IrCl(CO)\left[P(C_6H_5)_3\right]_2$$
$$+ \left[(CH_3)_2NH_2\right]Cl + OP(C_6H_5)_3 + [C_6H_5NH_3]Cl + 2H_2O$$

Typical sources of iridium used in this preparation are $IrCl_3 \cdot xH_2O$ and H_2IrCl_6. While iridium(I) complexes are often useful homogeneous catalysts, Vaska' complex is not. Instead it is iconic in the diversity of its reactions. Research on Vaska's complex helped provide the conceptual context for homogeneous catalysis. Vaska's complex, with 16 valence electrons, is thought to be "coordinatively unsaturated" and can therefore bind to one two-electron or two one-electron ligands to become electronically saturated with 18 valence electrons. The addition of two one-electron ligands is known as oxidative addition. Upon oxidative addition, the oxidation state of the iridium changes from Ir(I) to Ir(III). The four-coordinated square planar arrangement in the starting complex transforms to an octahedral, six-coordinate product. Vaska's complex undergoes oxidative addition with conventional oxidants such as halogens, strong acids, for example, HCl, and other molecules known to react as electrophiles, such as iodomethane (CH_3I). Vaska's complex binds O_2 reversibly.

$$IrCl(CO)\left[P(C_6H_5)_3\right]_2 + O_2 \rightleftharpoons IrCl(CO)\left[P(C_6H_5)_3\right]_2O_2$$

The dioxygen ligand is bonded to Ir by both oxygen atoms, known as side-on bonding. In myoglobin and hemoglobin, by contrast, O_2 binds end-on, attaching to the metal via only one of the two oxygen atoms. The resulting dioxygen adduct reverts to the parent complex upon heating or purging the solution with an inert gas, shown by a color change from orange back to yellow.

Other common complexes include Ir_2Cl_2(cyclooctadiene)$_2$ ($Ir_2Cl_2(C_8H_{12})_2$) and chlorobis(cyclooctene)iridium ($Ir_2Cl_2(C_8H_{14})_4$) dimer. Cyclooctadiene iridium chloride dimer is an organoiridium compound with the formula $Ir_2Cl_2(C_8H_{12})_2$, where C_8H_{12} is the diene 1,5-cyclooctadiene. It is an orange solid that is soluble in organic solvents. The complex is employed as a precursor for other iridium complexes, some of which are utilized as homogeneous catalysts. The solid is air-stable though its solutions degrades in air. The compound is formed by heating hydrated iridium trichloride and cyclooctadiene in alcohol solvent. In the reaction Ir(III) is reduced to Ir(I). The iridium centers are square planar as is typical for a d^8 complex. The Ir_2Cl_2 core is folded with a dihedral angle of 86°. The molecule crystallizes in yellow-orange and red-orange polymorphs; the second one is more common. A prominent derivative is Crabtree's catalyst (developed by British-American chemist Robert H. Crabtree (b. 17 April 1948)), (SP-4)tris(cyclohexyl)phosphane[(1-2-η:5-6-η)-cycloocta-1,5-diene]pyridineiridium(1 +) hexafluoridophosphate(1 −), [$C_8H_{12}IrP(C_6H_{11})_3C_5H_5N]PF_6$. The chloride ligands can also be substituted to give diiridium complexes showing modified reactivity, for example, $Ir_2(OCH_3)_2(C_8H_{12})_2$. A closely associated though even more reactive complex is chlorobis(cyclooctene)iridium dimer. The analog of Wilkinson's catalyst (named after English chemist and Nobel laureate Sir Geoffrey Wilkinson (July, 14 1921 to September, 26 1996)), $IrCl(PPh_3)_3$, undergoes orthometalation.

$$IrCl(PPh_3)_3 \rightarrow HIrCl(PPh_3)_2(PPh_2C_6H_4)$$

This difference between $RhCl(PPh_3)_3$ and $IrCl(PPh_3)_3$ signifies the generally greater tendency of iridium to undergo oxidative addition. A comparable trend is exhibited by $RhCl(CO)(PPh_3)_2$ and $IrCl(CO)(PPh_3)_2$, only the second oxidatively adds O_2 and H_2. The olefin complexes chlorobis(cyclooctene)iridium dimer and cyclooctadiene iridium chloride dimer are frequently employed as sources of "IrCl," using the lability of the alkene ligands or their susceptibility to removal by hydrogenation. Crabtree's catalyst ([$Ir(P(C_6H_{11})_3)$(pyridine)(cyclooctadiene)]PF_6) is a useful homogeneous catalyst for hydrogenation of alkenes. (η^5-Cp)$Ir(CO)_2$ oxidatively adds C-H bonds upon photolytic dissociation of one CO ligand.

7.15.3.3.3 Iridium(II), (III), and (V)

As is the case for rhodium(II), iridium(II) is rarely encountered. One example is iridocene, $IrCp_2$ ($Ir(C_5H_5)_2$). As with rhodocene, iridocene dimerizes at room temperature. Iridium is usually supplied commercially in the Ir(III) and Ir(IV) oxidation states. Important starting reagents are hydrated iridium trichloride and ammonium hexachloroiridate. These salts are reduced upon reaction with CO, hydrogen, and alkenes. An example is the carbonylation of the trichloride.

$$IrCl_3(H_2O)_x + 3CO \rightarrow \left[Ir(CO)_2Cl_2\right]^- + CO_2 + 2H^+ + Cl^- + (x-1)H_2O$$

Numerous organoiridium(III) compounds are synthesized from the pentamethylcyclopentadienyl iridium dichloride dimer. Many of these derivatives have kinetically inert cyclometalated ligands. Pentamethylcyclopentadienyl iridium dichloride, $[(C_5(CH_3)_5IrCl_2)]_2$, or abbreviated $[Cp^*IrCl_2]_2$, is a bright orange air-stable diamagnetic solid. The compound has C_{2h} symmetry. Each Ir is pseudo-octahedral. It was first synthesized by the reaction of hydrated iridium trichloride with hexamethyl Dewar benzene. More conveniently, the compound is formed by the reaction of hydrated iridium trichloride and pentamethylcyclopentadiene in hot methanol, from which the product precipitates.

$$2Cp^*H + 2IrCl_3(H_2O)_3 \rightarrow [Cp^*IrCl_2]_2 + 2HCl + 6H_2O$$

The two Ir-μ-Cl bonds are labile and can be cleaved to form various adducts of the general formula Cp^*IrCl_2L. Such adducts undergo further substitution to form cations $[Cp^*IrClL_2]^+$ and $[Cp^*IrL_3]^{2+}$. The chloride ligands can also be substituted by other anions such as carboxylates, nitrite, and azide. Reduction of $[Cp^*IrCl_2]_2$ under a CO atmosphere produces the dicarbonyl $[Cp^*Ir(CO)_2]$, which can be decarbonylated to form the unsaturated derivative $[Cp^*Ir(CO)]_2$. Reaction of $[Cp^*IrCl_2]_2$ with borohydride under a H_2 atmosphere produces the iridium(V) derivative Cp^*IrH_4. $[Cp^*IrCl_2]_2$ is a precursor to catalysts for the asymmetric transfer hydrogenation of ketones. The chemistry of this compound is like that of its rhodium analog, the pentamethylcyclopentadienyl rhodium dichloride dimer. Related half-sandwich complexes have played a central role in the development of C-H activation. Oxidation states greater than III are more common for iridium than rhodium. They typically have strong-field ligands. One often cited example is oxotrimesityliridium(V).

7.15.4 Major uses

The high melting point, hardness, and corrosion resistance of Ir and its alloys govern most of its uses. Ir (or occasionally Pt alloys or Os) and typically Ir alloys possess a low wear and are employed, for example, for multipored spinnerets, through which a plastic polymer melt is extruded to form fibers, such as rayon. Os–Ir is utilized for compass bearings and for balances. Their resistance to arc erosion makes Ir alloys perfect for electrical contacts for spark plugs, and Ir-based spark plugs are predominantly employed in aviation. Pure Ir is very brittle, to the point of being hard to weld since the heat-affected zone cracks, but it can be made more ductile by adding small amounts of Ti and Zr (0.2% of each seemingly works well). Corrosion and heat resistance turn Ir into a significant alloying agent. Some long-life aircraft engine parts are manufactured from an Ir alloy, while an Ir–Ti alloy is employed for deep-water pipes due to its corrosion resistance. Ir is as well utilized as a hardening agent in Pt alloys. The Vickers hardness of 100% Pt is 56 HV, but Pt with 50% of Ir can reach more than 500 HV. Devices that have to resist very high temperatures are frequently made from Ir. For instance, high-temperature Ir crucibles are employed in the Czochralski process to produce oxide single crystals (such as sapphires) for use in computer memory devices and in solid-state lasers. The crystals, for example, GGG and yttrium gallium garnet, are grown by melting presintered charges of mixed oxides under oxidizing conditions at temperatures up to 2100°C.

An alloy of 90% Pt and 10% Ir was used in 1889 to obtain the International Prototype Metre and kilogram mass, stored by the International Bureau of Weights and Measures near Paris. The meter bar was substituted as the definition of the fundamental unit of length in 1960 by a line in the atomic spectrum of Kr, but the kilogram prototype is still the international standard of mass (until 20 May 2019). Ir has been employed in the radioisotope thermoelectric generators of unmanned spacecraft such as the Voyager, Viking, Pioneer, Cassini, Galileo, and New Horizons. Ir was selected to encapsulate the ^{238}Pu fuel in the generator as it can withstand the operating temperatures of up to 2000°C and for its great strength. One more application concerns X-ray optics, in particular X-ray telescopes. For example, the mirrors of the Chandra X-ray Observatory are coated with a 60 nm thick Ir layer. Ir was shown to be the best element for reflecting X-rays after Ni, Au, and Pt were also tested. The Ir layer, which had to be flat to within a few atoms, was formed by depositing Ir vapor under high vacuum on a substrate of Cr. Ir is also employed in particle physics to create antiprotons, a form of antimatter. Antiprotons are formed by shooting a high-intensity proton beam at a conversion target, which must consist of an extremely high-density material. While W may be utilized instead, Ir has the advantage of greater stability under the shock waves formed by the temperature rise because of the incident beam.

Ir compounds are employed as catalysts in the Cativa process for carbonylation (refers to reactions that introduce carbon monoxide into organic and inorganic substrates. CO is abundantly available and conveniently reactive, so it is widely used as a reactant in industrial chemistry. The term carbonylation also refers to oxidation of protein side chains.) of methanol (CH_3OH) to produce acetic acid (ethanoic acid, CH_3COOH). C-H bond activation is a research area on reactions that cleave C-H bonds, which were historically thought of as being unreactive. The first published successes (1982) at activating C-H bonds in saturated hydrocarbons employed organometallic Ir complexes that undergo an oxidative addition with the hydrocarbon. Ir complexes are being studied as catalysts for asymmetric hydrogenation. Asymmetric hydrogenation is a chemical reaction that adds two atoms of hydrogen preferentially to one of two faces of an unsaturated substrate molecule, such as an alkene or ketone. The selectivity derives from the manner that the substrate binds to the chiral catalysts. In jargon, this binding transmits spatial information (what chemists refer to as chirality) from the catalyst to the target, favoring the product as a single enantiomer. This enzyme-like selectivity is particularly applied to bioactive products such as pharmaceutical agents and agrochemicals. These catalysts have been employed in the synthesis of natural products and able to hydrogenate some difficult substrates, such as unfunctionalized alkenes, enantioselectively (generating only one of the two possible enantiomers). Ir forms a variety of complexes of fundamental interest in triplet harvesting in OLEDs. Ir is a good catalyst for the decomposition of hydrazine (N_2H_4) (into hot nitrogen and ammonia), which is used in practice in low-thrust rocket engines.

The radioisotope ^{192}Ir is one of the two most significant energy sources applied in industrial γ-radiography for nondestructive testing of metals. Furthermore, ^{192}Ir is employed as a source of γ radiation to treat cancer using brachytherapy, a type of radiotherapy where a sealed radioactive source is placed inside or next to the area needing treatment. Specific treatments comprise high-dose-rate prostate brachytherapy, bilary duct brachytherapy, and intracavitary cervix brachytherapy. Very recently, medical researchers reported that Ir linked to albumin, forming a photosensitized molecule, can penetrate cancer cells and, after being irradiated with light (a process called photodynamic therapy), kill the cancer cells.

7.16 78 Pt — Platinum

7.16.1 Discovery

Archeologists have found traces of platinum in the gold used in ancient Egyptian burials as far back as 1200 BCE. Nevertheless, the level of early Egyptians' knowledge of the metal is uncertain. It is quite possible they did not know there was platinum in their gold. The metal was used by pre-Columbian Americans near the current coastal city Esmeraldas, Ecuador to make artifacts of a white gold-platinum alloy. Archeologists typically associate the tradition of platinum-working in South America with the La Tolita Culture (c.600 BCE to 200 CE), but determining exact dates and location is problematic, as most platinum artifacts from the area were obtained second-hand through the antiquities trade instead of through direct archeological excavation. To work the platinum, the pre-Columbian Americans used a relatively sophisticated system of powder metallurgy. The platinum used in their objects was not the pure metal, but instead a naturally occurring alloy of the platinum group metals, with small amounts of palladium, rhodium, and iridium. The first European mention of platinum appears in 1557 in the writings of the Italian humanist, scholar and physician Julius Caesar Scaliger (April 23, 1484 to October 21, 1558) as an account of an unknown noble metal found between Darién and Mexico, "which no fire nor any Spanish artifice has yet been able to liquefy." From their first experiences with platinum, the Spanish mostly saw the metal as a type of impurity in gold, and it was treated as such. It was frequently just thrown out, and there was an official decree forbidding the adulteration of gold with platinum impurities. In 1735 Spanish general of the navy, explorer, scientist, author, astronomer, colonial administrator, and the first Spanish governor of Louisiana Antonio de Ulloa y de la Torre-Giral, (January 12, 1716 to July 3, 1795) (Fig. 7.78) and mathematician, scientist, naval officer, and mariner Don Jorge Juan y Santacilia (January 5, 1713 to June 21, 1773 Madrid) observed Native Americans mining platinum while the Spaniards were traveling through Colombia and Peru for eight years. Ulloa and Juan discovered mines with the whitish metal nuggets and brought them home to Spain. Antonio de Ulloa returned to Spain and started the first mineralogy laboratory in Spain and was the first to scientifically study platinum in 1748 (de Ulloa, 1748). His historical description of the expedition encompassed a description of platinum as being neither separable nor calcinable (The

FIGURE 7.78 Antonio de Ulloa, c.1856, painting by Andrés Cortés y Aguilar (1812–1879).

IUPAC defines calcination as "heating to high temperatures in air or oxygen." However, calcination is also used to mean a thermal treatment process in the absence or limited supply of air or oxygen applied to ores and other solid materials to bring about a thermal decomposition.) Ulloa also predicted the discovery of platinum mines. After publishing the report in 1748, Ulloa did not continue to study the new metal. In 1758 he was sent to supervise mercury mining operations in Huancavelica. In 1741 Charles Wood (1702 to October 1774), a British metallurgist (ironmaster), found numerous samples of Colombian platinum in Jamaica, which he sent to British doctor and scientist William Brownrigg (March 24, 1712 to January 6, 1800) for further research. In 1750 after examining the platinum sent to him by Wood, Brownrigg presented a comprehensive report of the metal to the Royal Society, asserting that he had seen no mention of it in any preceding records of known minerals (Watson and Brownrigg, 1749). Brownrigg also pointed to platinum's extremely high melting point and refractoriness toward borax. Other chemists all over Europe soon started to study platinum, for example, German chemist Andreas Sigismund Marggraf (March 3, 1709 to August 7, 1782) (Marggraf, 1760), Swedish chemist and mineralogist Torbern Bergman (March 20, 1735 to July 8, 1784), Swedish chemist Jöns Jakob Berzelius (August 20, 1779 to August 7, 1848), British chemist and physician William Lewis (c.1708–1781), and French chemist Pierre Macquer (October 9, 1718 to February 15, 1784). In 1752 Swedish chemist Henrik Scheffer (December 28, 1710 to August 10, 1759) published a comprehensive scientific account of the metal, which he referred to as "white gold," together with an explanation of how he managed to fuse platinum ore with the aid of arsenic. Scheffer described platinum as being less pliable than gold, but with comparable resistance to corrosion (Scheffer, 1752). German diplomat and chemist Karl Heinrich Joseph von Sickingen (1737–1791) researched platinum extensively in 1772. He managed to obtain malleable platinum by alloying it with gold, dissolving the alloy in hot aqua regia (a mixture of nitric acid and hydrochloric acid, optimally in a molar ratio of 1:3), precipitating the platinum with ammonium chloride (NH_4Cl), burning the ammonium chloroplatinate, and hammering the subsequent finely divided platinum to make it cohere (von Sicklingen, 1782). German (Prussian) chemist, physicist and biologist Franz Karl Achard (April 28, 1753 to April 20, 1821) produced the first platinum crucible in 1784 (Fig. 7.79). He worked with the platinum by fusing it with arsenic, followed by volatilizing the arsenic. Since the other platinum group elements were not discovered yet (platinum was the first in the list), Scheffer and Sickingen wrongly hypothesized that because of its hardness—which is somewhat more than for pure iron—platinum would be a fairly nonpliable material, even brittle on occasion, when actually its ductility and malleability are close to that of gold. Their

FIGURE 7.79 Franz Karl Achard.

FIGURE 7.80 Pierre-François Chabaneau, drawing by L. Lascuráin.

assumptions were to be expected since the platinum they worked with was highly contaminated with trace amounts of platinum group elements such as osmium and iridium, among others, which embrittled the platinum alloy. Alloying this impure platinum residue called "plyoxen" with gold was the only way back then to obtain a pliable compound. These days though, very pure platinum can be obtained, and extremely long wires can be very easily drawn from pure platinum, because of its crystalline structure comparable to that of many other soft metals. In 1786 King Charles III of Spain (January 20, 1716 to December 14, 1788) provided a library and laboratory to French chemist Pierre-François Chabaneau (June 27, 1754 to February 18, 1842) (Fig. 7.80) to support his study of platinum. Chabaneau managed to remove several impurities from the ore, including gold, mercury, lead, copper, and iron. This led him to think he was studying a single metal, while actually the ore still contained the yet-undiscovered platinum group elements. This resulted in inconsistent outcomes in his experiments. Sometimes, the platinum seemed malleable, but when it was alloyed with iridium, it would become much more brittle. Other times the metal was completely incombustible, but when alloyed with osmium, it would volatilize. After numerous months, Chabaneau managed to obtain 23 kg of pure, malleable platinum by hammering and compressing the sponge form while white-hot. Chabeneau understood the infusibility of platinum would make objects made of it valuable, and so started a business with Joaquín Cabezas making platinum ingots and utensils. This became the beginning of what is now recognized as the "platinum age" in Spain.

7.16.2 Mining, production, and major minerals

Platinum is a very rare metal, occurring at a concentration of only around 0.005 ppm in the Earth's crust. Pt is often found as the native element and as alloys with the other Pt-group elements and Fe mostly. Most often the native Pt is found in secondary deposits in alluvial deposits. The alluvial deposits used by pre-Columbian people in the Chocó Department, Colombia is even today mined for Pt-group metals. Another large alluvial deposit exists in the Ural Mountains, Russia, and it is still producing. In Ni and Cu deposits, Pt-group metals occur as sulfides (e.g., (Pt,Pd)S), tellurides (e.g., PtBiTe), antimonides (e.g. PtSb), and arsenides (e.g., $PtAs_2$), and as alloys with Ni or Cu. Platinum arsenide, sperrylite ($PtAs_2$), is a major source of platinum associated with Ni ores in the Sudbury Basin deposit in Ontario, Canada. At Platinum, Alaska, about 17,000 kg was mined between 1927 and 1975. It ceased operations in 1990. The rare sulfide mineral cooperite, (Pt,Pd,Ni)S, occurs in the Merensky Reef within the Bushveld complex, Gauteng, South Africa. In 1865, chromites ($FeCr_2O_4$) were identified in the Bushveld region of South Africa, followed by the discovery of Pt in 1906. In 1924 the South African geologist, prospector, scientist, conservationist, and philanthropist Hans

FIGURE 7.81 Hans Merensky.

Merensky (March 16, 1871 in Botshabelo to October 21, 1952) found a large reserve of Pt in the Bushveld Igneous Complex in South Africa (Fig. 7.81). The specific layer he found, named the Merensky Reef, covers about 75% of the world's known platinum. The large Cu–Ni deposits near Norilsk in Russia, and the Sudbury Basin, Canada, form the two other large deposits. In the Sudbury Basin, the enormous amount of nickel ore processed compensate for the fact platinum is only present at a concentration of about 0.5 ppm in the ore. Smaller reserves can be found in the United States, for example in the Absaroka Range in Montana. In 2010 South Africa was the top producer of Pt at almost 77%, followed by Russia at 13%. Large Pt deposits exist in the state of Tamil Nadu, India.

When pure platinum is obtained from placer deposits or other ores, it can be isolated from them by different methods to remove impurities. Since platinum is significantly denser than many of its impurities, the lighter impurities can be taken out by simply floating them in a liquid. In addition, platinum is paramagnetic, while nickel and iron are both ferromagnetic and can therefore be removed by running an electromagnet over the mixture. As platinum has a higher melting point than most other impurities, many of these can be burned or melted away short of melting the platinum. Lastly, platinum is resistant to hydrochloric and sulfuric acids, while most other compounds are readily attacked by them. Metal impurities can be separated by adding the mixture to either of the two acids and recovering the remaining platinum. One appropriate method for purification of the platinum ore, which consists of platinum, gold, and the other platinum group elements, is to react it with aqua regia (a mixture of nitric acid and hydrochloric acid, optimally in a molar ratio of 1:3), in which palladium, gold, and platinum are dissolved, whereas osmium, iridium, ruthenium, and rhodium stay unreacted. The gold is subsequently precipitated through the addition of Fe(II) chloride and after filtering off the gold, the platinum is then precipitated as ammonium chloroplatinate through the reaction with ammonium chloride (NH_4Cl). Finally, the ammonium chloroplatinate can be converted to platinum through heat treatment. Unprecipitated hexachloroplatinate(IV) may be reduced trough a reaction with elemental zinc, and a similar method is suitable for small-scale recovery of platinum from laboratory residues.

A total of 39 minerals are known to contain platinum in their crystal structure. The element class is represented by 12 minerals, including the native metal platinum (Pt) (Fig. 7.82) and 16 alloys such as isoferroplatinum (Pt_3Fe) (Fig. 7.83), niggliite (PtSn), and tetraferroplatinum (PtFe). The sulfides class contains 22 different minerals with Pt, for example, braggite ((Pt,Pd,Ni)S) (Fig. 7.84), cooperite (PtS) (Fig. 7.85), and sperrylite ($PtAs_2$) (Fig. 7.86).

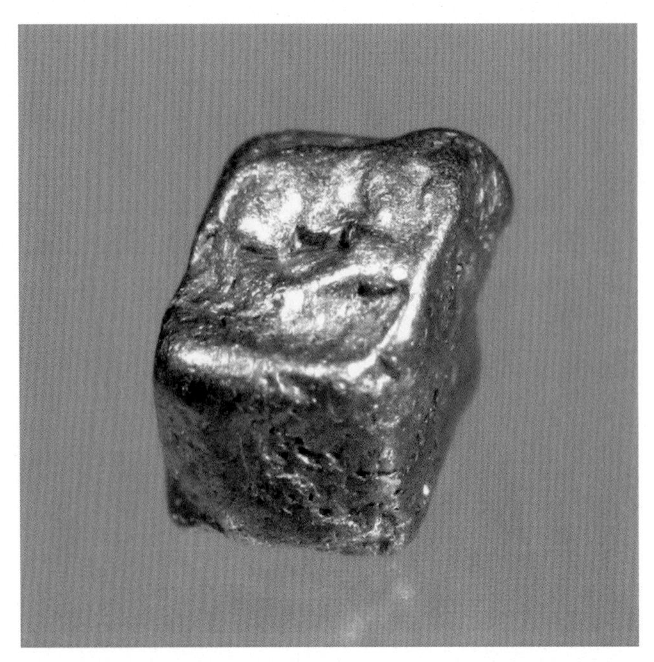

FIGURE 7.82 A 3-mm cubic crystal of platinum, Pt. Konder massif, Aldan shield, Ayan-Maya, Khabarovskiy Kray, Russia.

FIGURE 7.83 Isoferroplatinum, Pt_3Fe, grayish gold cubic crystal to 4 mm. Konder massif, Aldan shield, Ayan-Maya, Khabarovskiy Kray, Russia.

7.16.3 Chemistry

Platinum consists of six naturally occurring isotopes: ^{190}Pt, ^{192}Pt, ^{194}Pt, ^{195}Pt, ^{196}Pt, and ^{198}Pt. The most abundant of these is ^{195}Pt at 33.775% of all Pt. It is the only stable isotope with a nonzero spin; with a spin of 1/2, ^{195}Pt satellite peaks are commonly detected in ^{1}H and ^{31}P NMR spectroscopy of organo-Pt compounds (i.e., Pt-phosphine and Pt-alkyl complexes). ^{190}Pt is the least abundant isotope with an abundance of only 0.012%. Of the naturally occurring Pt isotopes, only ^{190}Pt is unstable, though it decays via α decay to ^{186}Os with a very long half-life of 6.5×10^{11} years, causing an activity of 15 Bq/kg of natural Pt. ^{198}Pt can undergo α decay, but its decay has never been detected (the half-life is known to be at least 3.2×10^{14} years); consequently, it can be for practical purposes be considered as stable. In addition, platinum has 31 artificial isotopes ranging in atomic mass from 166 to 204, resulting in a total number of

FIGURE 7.84 A polished slab showing the large crystal grains of ultramafic rock, Numerous metallic crystals measuring from 1 to 5 mm are braggite, (Pt,Pd,Ni)S, crystals. Merensky Reef, Bushveld Complex, Limpopo prov., South Africa.

FIGURE 7.85 Microscopic cooperite, PtS, interfaced with coarsely crystalline chromite, $Fe^{2+}Cr_2^{3+}O_4$, in a matrix of pentlandite, $(Fe_xNi_y)_{\Sigma 9}S_8$ (x + y = 9), and pyrrhotite, Fe_7S_8, about 1 cm. Associated with these minerals, though microscopic, are the platinum minerals of braggite, (Pt,Pd,Ni)S, and probable moncheite, $(Pt,Pd)(Te,Bi)_2$. Rustenburg, Bushveld Complex, Limpopo prov., South Africa.

known isotopes of 39. The least stable artificial radioisotope is ^{166}Pt which has a half-life of 300 μs, whereas the most stable artificial radioisotope is ^{193}Pt which has a half-life of 50 years. Most Pt isotopes decay by some combination of β decay and α decay. ^{188}Pt, ^{191}Pt, and ^{193}Pt decay principally by electron capture. ^{190}Pt and ^{198}Pt have been predicted to have energetically favorable double β decay paths.

Pure platinum is a lustrous, ductile, and malleable, silver-white metal (Table 7.16). Platinum is more ductile than gold, silver, or copper, hence being the most ductile of pure metals, but it is less malleable than gold. The metal has exceptional resistance to corrosion, is stable at high temperatures and has stable electrical properties. Platinum does oxidize, forming PtO_2, at 500°C; this oxide can be readily removed thermally. It reacts vigorously with fluorine at 500°C to form platinum tetrafluoride. It is attacked by chlorine, bromine, iodine, and sulfur. Platinum is insoluble in hydrochloric and nitric acid, but dissolves in hot aqua regia (a mixture of nitric and hydrochloric acids), to form chloroplatinic acid, H_2PtCl_6.

The most common oxidation states of platinum are +2 and +4. The +1 and +3 oxidation states are less common and are often stabilized by metal bonding in bimetallic (or polymetallic) species. Tetracoordinate platinum(II)

FIGURE 7.86 Well-formed, complex 3 mm lustrous silvery crystal of sperrylite, $PtAs_2$. Broken Hammer dep., Sudbury, Ontario, Canada.

TABLE 7.16 Platinum properties.

Appearance	Silvery white
Standard atomic weight $A_{r,std}$	195.085
Block	d-Block
Element category	Transition metal
Electron configuration	[Xe] $4f^{14}$ $5d^9$ $6s^1$
Phase at STP	Solid
Melting point	1768.3°C
Boiling point	3825°C
Density (near r.t.)	21.45 g/cm^3
When liquid (at m.p.)	19.77 g/cm^3
Heat of fusion	22.17 kJ/mol
Heat of vaporization	510 kJ/mol
Molar heat capacity	25.86 J/(mol·K)
Oxidation states	−3, −2, −1, +1, **+2**, +3, **+4**, +5, +6
Ionization energies	1st: 870 kJ/mol
	2nd: 1791 kJ/mol
Atomic radius	Empirical: 139 pm
Covalent radius	136 ± 5 pm
Van der Waals radius	175 pm

STP, Standard temperature and pressure.
Bold font indicates main oxidation state.

compounds tend to have 16-electron square planar geometries. While elemental platinum is generally unreactive, it dissolves in hot aqua regia to form aqueous chloroplatinic acid (H_2PtCl_6).

$$Pt + 4HNO_3 + 6HCl \rightarrow H_2PtCl_6 + 4NO_2 + 4H_2O$$

As a soft acid, platinum has a great affinity for sulfur, such as on DMSO (dimethyl sulfoxide); many DMSO complexes have been described and care should be taken in the choice of reaction solvent.

7.16.3.1 Oxides and sulfides

Addition of alkali to an aqueous solution of $[Pt(H_2O)_4](ClO_4)_2$ under an Ar atmosphere produces a white amphoteric hydroxide of Pt(II) at pH 4 which redissolves at pH 10. The precipitate slowly turns black at room temperature (more quickly when dried at 100°C) and has been formulated as $PtO_x \cdot H_2O$, but it is too unstable to be fully characterized. The stable oxide of platinum is found, instead, in the higher oxidation state. Addition of alkali to aqueous solutions of $PtCl_4$ produces a yellow amphoteric precipitate of the hydrated dioxide which redissolves on boiling with an excess of strong alkali to form solutions of $[Pt(OH)_6]^{2-}$; it also dissolves in acids. Dehydration by heating results in nearly black

PtO_2 but this and the less common PtO both decompose to the elements above 650°C and cannot be completely dehydrated without some loss of oxygen. Platinum(II,IV) oxide, Pt_3O_4, is formed in the following reaction:

$$2Pt^{2+} + Pt^{4+} + 4O^{2-} \rightarrow Pt_3O_4$$

Adams' catalyst, also known as platinum dioxide, is commolly represented as platinum(IV) oxide hydrate, $PtO_2 \cdot H_2O$. It is used as a catalyst for hydrogenation and hydrogenolysis in organic synthesis. This dark brown powder is commercially available. The oxide itself is not an active catalyst, but it becomes active after exposure to hydrogen whereupon it converts to platinum black (a fine powder of platinum with good catalytic properties. The name of platinum black is due to its black color.), which is responsible for reactions. Adams' catalyst is produced from chloroplatinic acid H_2PtCl_6 or ammonium chloroplatinate, $(NH_4)_2PtCl_6$, by fusion with sodium nitrate. The method involves first preparing a platinum nitrate which is subsequently heated to expel nitrogen oxides.

$$H_2PtCl_6 + 6NaNO_3 \rightarrow Pt(NO_3)_4 + 6NaCl_{(aq)} + 2HNO_3$$
$$Pt(NO_3)_4 \rightarrow PtO_2 + 4NO_2 + O_2$$

The resulting brown cake is washed with water to remove the nitrates. The catalyst can either be used as is or dried and stored in a desiccator for later use. Platinum can be recovered from spent catalyst by conversion to ammonium chloroplatinate using aqua regia followed by ammonia. Platinum forms a mono- and a disulfide. Black PtS_2 is formed when H_2S is passed through an aqueous solution of Pt(IV). Green PtS is best prepared by heating $PtCl_2$, Na_2CO_3 and S. The crystal chemistry and electrical (and magnetic) properties of this phase and the many selenides and tellurides of Pt are complex.

7.16.3.2 Halides

The hexa- and penta-halides are the dark-red PtF_6 and $(PtF_5)_4$, which are both formed by controlled heating of Pt and F_2. The former is a volatile solid and, after RhF_6, is the least stable platinum-group metal hexafluoride. It is one of the strongest oxidizing agents identified, oxidizing both O_2 (to $O_2^+[PtF_6^-]$) and Xe (to $XePtF_6$). The pentafluoride is also very reactive and has a similar tetrameric structure as the pentafluorides of Ru, Os, Rh and Ir. It easily disproportionates into the hexa- and tetrafluorides. Platinum alone forms all four tetrahalides and these vary in color from the light brown PtF_4 to the very dark brown PtI_4. PtF_4 is formed by the reaction of BrF_3 on $PtCl_2$ at 200°C and is violently hydrolyzed by water. The others can be formed directly from the elements, the chloride being recrystallizable from water while the bromide and iodide are more soluble in alcohol and ether. The diamagnetic "trichloride" and "tribromide" of Pt contain both Pt(II) and Pt(IV) and the triiodide probably does also. All the dihalides, with the exception of PtF_2, have been identified; fluorine possibly being too strongly oxidizing to be easily compatible with the metal in the lower of its two major oxidation states. The dichlorides of Pt are prepared from the elements and exist in two isomeric forms, and which form is produced depends on the exact experimental conditions applied. The high temperature phase, α-$PtCl_2$ is insoluble in water but dissolves in hydrochloric acid forming $[PtCl_4]^{2-}$ ions. It has been described as both olive-green and black, the latter consisting of edge- and corner-sharing $PtCl_4$ units. The dark red β-$PtCl_2$ is isomorphous with β-$PdCl_2$ and the Pt_6Cl_{12} unit is retained on dissolution in benzene. Hexachloroplatinic acid is probably the most important platinum halide, as it serves as the precursor for many other platinum compounds. By itself, it has various applications in photography, zinc etchings, indelible ink, plating, mirrors, porcelain coloring, and as a catalyst. The reaction of hexachloroplatinic acid with an ammonium salt, such as ammonium chloride, produces ammonium hexachloroplatinate, which is relatively insoluble in ammonium solutions. Heating this ammonium salt in the presence of hydrogen reduces it to elemental platinum. Potassium hexachloroplatinate is likewise insoluble, and hexachloroplatinic acid has been utilized in the determination of potassium ions by gravimetry. When hexachloroplatinic acid is heated, it decomposes through platinum (IV) chloride and platinum(II) chloride to elemental platinum, although the reactions do not occur stepwise:

$$(H_3O)_2PtCl_6 \cdot nH_2O \rightleftharpoons PtCl_4 + 2HCl + (n+2)H_2O$$
$$PtCl_4 \rightleftharpoons PtCl_2 + Cl_2 \; PtCl_2 \rightleftharpoons Pt + Cl_2$$

All three reactions are reversible. Oxohalides in this group are seemingly limited to the strongly oxidizing $PtOF_3$. The compound reported to be $PtOF_4$ is in reality $O_2^+[PtF_6]$.

7.16.3.3 Complexes

Pt(IV) complexes rival those of Pt(II) in number, and are both thermodynamically stable and kinetically inert. Those with halides, pseudo-halides, and N-donor ligands are especially numerous. Of the large number of conceivable compounds ranging from $[PtX_6]^{2-}$ through $[Pt_4L_2]$ to $[PtL_6]^{4+}$, (X = F, Cl, Br, I, CN, SCN, SeCN; L = NH_3, amines) a large variety have been synthesized and characterized. K_2PtCl_6 is commercially the most common compound of platinum and the brownish red, "chloroplatinic acid," $H_2[PtCl_6](aq)$, is the typical starting molecule in Pt (IV) chemistry. It is formed by dissolving platinum metal sponge in aqua regia, followed by one or more evaporation steps with hydrochloric acid. An alternative route to Pt(II) chemistry is provided by precipitation of the slightly soluble K_2PtCl_6 followed by its reduction with hydrazine to K_2PtCl_4. O-donor ligands such as OH^- and acetylacetonate also coordinate to Pt(IV), but S- and Se-, and more in particular P- and As-donor ligands, tend to reduce it to Pt(II).

Of the very few monomeric trivalent compounds of Pt, the blue-colored $(NBu_4)(Pt(C_6Cl_5)_4]$ can be formed by oxidizing the Pt(II) salt. The most abundant examples of this oxidation state, however, are the dinuclear Pt compounds of the type, $[Pt_2(L\text{-}L)_4L_2]^{n-}$ having a single Pt-Pt bonds and the same tetra-bridged structure as Mo(II) and Cr(II). The first of these described was $K_2[Pt_2(SO_4)_4(H_2O)_2]$, formed from $[Pt(NO_2)_2(NH_3)_2]$ and sulfuric acid, but those with phosphate or P-donor, pyrophosphito, $(P_2O_5H_2)^{2-}$, bridges are more abundant. Pt- Pt distances vary between 278.2 pm, found with pyrophosphito bridges, and 245.1 pm in $Cs_3[Pt_2(\mu.\text{-}O_2CMe)_2(\mu.\text{-}O_2CCH_2)_2]$. This yellow complex is formed by a intricate procedure from K_2PtCl_4 and MeCOOAg and, in addition to a pair of O,O-donor acetate bridges, contains a pair of unique C,O-donor, -O, CO,CH_2-, bridges. Stable tetra-acetato bridged dimers are not found. Several compounds which have in the past been reported to contain the trivalent metals have later been shown to contain them in more than one oxidation state. One such is H. Wolfram's red salt, $Pt(EtNH_2)_4Cl_3 \cdot 2H_2O$, which has a structure comprising alternate octahedral Pt(IV) and square planar Pt(II) linked by Cl bridges, that is, $[Pt(II)(EtNH_2)_4]^{2+}[$*trans*-$(\mu.\text{-}Cl)^{2-}Pt(IV)(EtNH_2)_4]^{2+}Cl_4{}^{-} \cdot 4H_2O$. Other examples are the 1D conductors of platinum, of which the cyano complexes are the best understood. $K_2[Pt(CN)_4] \cdot 3H_2O$ is a very stable colorless solid, though with appropriate partial oxidation it is possible to form a bronze-colored, "cation deficient," $K_{1.75}Pt(CN)_4 \cdot 1.5H_2O$, and other partially oxidized compounds such as $K_2Pt(CN)_4Cl_{0.3} \cdot 3H_2O$. In these compounds, square planar $[Pt(CN)_4]^{n-}$ ions are stacked resulting in a linear chain of Pt atoms in which the Pt-Pt distances of 280–300 pm (compared to 348 pm in the original $K_2[Pt(CN)_4] \cdot 3H_2O$ and 278 pm in the metal) allow strong overlap of the d_{z^2} orbitals. This explains the metallic conductance of these compounds along the crystal axis. Oxalato complexes [e.g., $K_{1.6}Pt(C_2O_4)_2 \cdot 1.2H_2O$] originally prepared as early as 1888 by the Swedish chemist and Nobel laureate (1912) H. G. Söderbaum (March, 12 1862 to September 22, 1933), are also 1D conductors with similar structures.

The effect of complexation on the splitting of d orbitals is much greater in the case of second and third than for first row transition elements, and the associated effects are marked for Pt(II). As a result, its complexes are, with rare exceptions, diamagnetic and the vast majority are planar also. Not many complexes are formed with O-donor ligands but, of the few that are, $[Pt(H_2O)_4]^{2+}$ ions, and the polymeric anhydrous acetate $[Pt(O_2CMe)_2]_4$. are the most important. Almost square planar $[M(NO_3)_4]^{2-}$ anions containing the unusual unidentate nitrato ion have also been described. Fluoro complexes are even less widespread, the preference of these cations being for the other halides, cyanide, nitride, and heavy atom-donor ligands. The complexes $[PtX_4]^{2-}$ (X = Cl, Br, I, SCN, CN) are all readily formed and can be crystallized as salts of $[NH_4]^+$ and the alkali metals. By employing $[NR_4]^+$ cations it is possible to separate binuclear halogen-bridged anions $[Pt_2X_6]^{2-}$ (X = Br, I) which keep the square planar coordination of Pt. The aqueous solution of red $[PtCl_4]^{2-}$ is a common starting material for the synthesis of other Pt(II) complexes by consecutive substitutions of the chloride ligands. In the $[Pt(SCN)_4]^{2-}$ complex the ligands bond through the π-acceptor (S) ends, while in the presence of stronger π-acceptor ligands such as PR_3 and AsR_3 they tend to bond through their N ends. Not surprisingly, therefore a number of examples of linkage isomerism have been observed in compounds of the type *trans*-$[M(PR_3)_2(SCN)_2]$. Complexes with ammonia and amines, especially those of the types $[PtL_4]^{2+}$ and $[PtL_2X_2]$, are numerous for Pt(II). For instance, the colorless $[Pt(NH_3)_4]Cl_2 \cdot H_2O$ can be formed by adding NH_3 to an aqueous solution of $PtCl_2$ and, in 1828, was the first of the platinum ammines to be discovered. Other salts of the $[Pt(NH_3)_4]^{2+}$ ion are readily derived, the most familiar being Magnus's green salt $[Pt(NH_3)_4][PtCl_4]$ (named after German experimental scientist Heinrich Gustav Magnus (May, 2 1802 to April, 4 1870). His training was mostly in chemistry but his later research was mostly in physics.). That a green salt formed from the combination of a colorless cation and a red anion was surprising and is a result of the crystal structure, which comprises the square planar anions and cations stacked alternately to form a linear chain of Pt atoms only 325 pm apart. Interaction between these metal atoms shifts the d-d absorption of the $[PtCl_4]^{2-}$ ion from the green region (thus the normal red color) toward the red, so giving the green color. Magnus's salt is an electrolyte and nonionized polymerization isomers of the stoichiometry $PtCl_2(NH_3)_2$ are also known which can be formed as monomeric *cis* and *trans*-isomers:

$$[PtCl_4]^{2-} \xrightarrow{NH_3} cis\text{-}\left[PtCl_2(NH_3)_2\right]$$
$$\left[Pt(NH_3)_4\right]^{2+} \xrightarrow{HCl} trans\text{-}\left[PtCl_2(NH_3)_2\right]$$

Many substitution reactions are possible starting with these ammines. A resurgence of interest in these apparently simple complexes of platinum started in 1969 when the anti-tumor activity of *cis*-$[PtCl_2(NH_3)_2]$ ("cisplatin") was discovered. Binding of cisplatin to DNA appeared to be the key property of the action and, since the *trans*-isomer is inactive, it was evident that chelation (or at least coordination to donor atoms in close proximity) is a critical part of the activity. Extensive research, using in particular proton NMR, indicated that Pt loses the Cl^- ligands and binds to N-7 atoms of a pair of guanine bases on adjacent strands of DNA. In order to circumvent serious side effects of cisplatin (kidney- and neurotoxicity) replacement Pt compounds have been developed. The most significant of these is "carboplatin" in which the *cis*-chlorides have been substituted by the O-chelate, cyclobutanedicarboxylate but all of these compounds have ligands with NH groups that enable the hydrogen bonding believed to stabilize the distortions of the DNA structure.

Compounds of the type $[Pt(PR_3)_4]$, of which $[Pt(PPh_3)_4]$ has been most comprehensively researched, are in general yellow, air-stable solids or liquids formed by reducing Pt(II) complexes in H_2O or H_2O/EtOH solutions with hydrazine or sodium borohydride. They are tetrahedral molecules whose most important property is their inclination to dissociate in solution to form 3-coordinate, planar $[Pt(PR_3)_3]$ and, in traces, perhaps also $[Pt(PR_3)_2]$ species. The latter are intermediates in a wide range of addition reactions (several of which may correctly be considered as oxidative additions) producing compounds such as $[Pt(II)(PPh_3)_2L_2]$, (L = O, CN, N_3) and $[Pt(II)(PPh_3)_2LL']$, (L,L′ = H,Cl; R,I) as well as $[Pt(0)(C_2H_4)(PPh_3)_2]$ and $[Pt(0)(CO)_2(PPh_3)_2)$. The mechanism by which this low oxidation state is stabilized has been an ongoing matter of discussion. That it is not simple is obvious from the fact that, unlike nickel, platinum needs the presence of phosphines for the formation of stable carbonyls. For most transition metals the π-acceptor properties of the ligand are thought to be of considerable importance and there is no reason to doubt that this is true for Ni(0). For Pd(0) and Pt(0), though, it seems that σ-bonding ability is also significant, and the smaller importance of π back-bonding which this implies is in accord with the higher ionization energy of Pt (865 $kJ \cdot mol^{-1}$) compared tot that for Ni (737 $kJ \cdot mol^{-1}$).

7.16.3.4 Organoplatinum compounds

Organoplatinum compounds exist in oxidation state 0 to IV, with oxidation state II most abundant. The general order in bond strength is Pt-C (sp) > Pt-O > Pt-N > Pt-C (sp^3). Organoplatinum and organopalladium chemistry are comparable, but organoplatinum compounds are more stable and thus less useful as catalysts.

7.16.3.5 Organoplatinum(0)

Most organoplatinum(0) compounds contain alkene and alkyne ligands. Carbonyl complexes are rare, and the analog of $Ni(CO)_4$ is elusive. The alkene and alkyne ligands act as two-electron donors, e.g in the complexes $(PPh_3)_2Pt(C_2H_4)$ and $(PPh_3)_2Pt(C_2Ph_2)$. The ethylene ligand in $(PPh_3)_2Pt(C_2H_4)$ is labile and exchanges with alkynes and electrophilic alkenes, even C_{60} a fullerene. A general synthesis method to form $(PPh_3)_2Pt(un)$ (un = alkene, alkyne) is reduction of potassium tetrachloroplatinate with ethanolic potassium hydroxide or hydrazine in the presence of a phosphine ligand such as triphenylphosphine and the alkene or alkyne. Such reactions proceed via the intermediary *cis*-dichlorobis(triphenylphosphine)platinum(II). Nitrogen-based ligands rarely support the formation of platinum complexes of alkenes and alkynes. Zerovalent organoplatinum complexes without phosphine ligands are frequently synthesized via $PtCl_2(COD)$.

$$Li_2C_8H_8 + PtCl_2(COD) + 3C_7H_{10} \rightarrow \left[Pt(C_7H_{10})_3\right] + 2LiCl + C_8H_8 + C_8H_{12}$$
$$Pt(C_7H_{10})_3 + 2COD \rightarrow Pt(COD)_2 + 3C_7H_{10}$$

where C_7H_{10} is norbornene.

7.16.3.6 Organoplatinum(I)

Platinum(I) compounds are rare but usually are diamagnetic as they have Pt-Pt bonds. An illustration is the dication $[Pt_2(CO)_6]^{2+}$. A historically noteworthy organoplatinum(II) compound is Zeise's salt (named for its discoverer Danish organic chemist William Christopher Zeise (October 15, 1789 to November 12, 1847)), which is formed from ethylene and potassium tetrachloroplatinate.

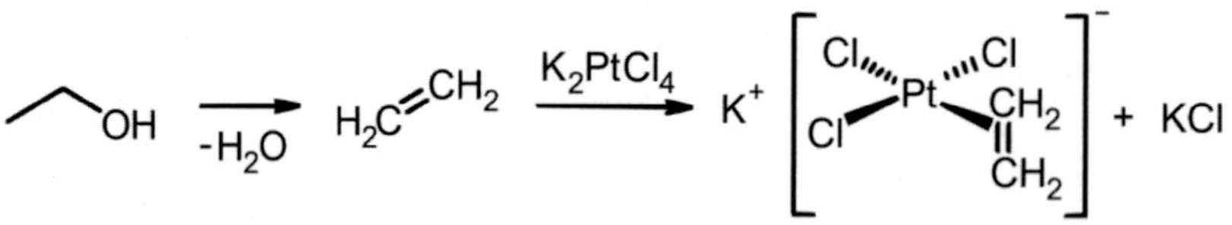

$$\text{CH}_3\text{CH}_2\text{OH} \xrightarrow{-H_2O} H_2C{=}CH_2 \xrightarrow{K_2PtCl_4} K^+[PtCl_3(C_2H_4)]^- + KCl$$

The colorless diolefin complex dichloro(cycloocta-1,5-diene)platinum(II) is a more modern related compound, and is more extensively utilized.

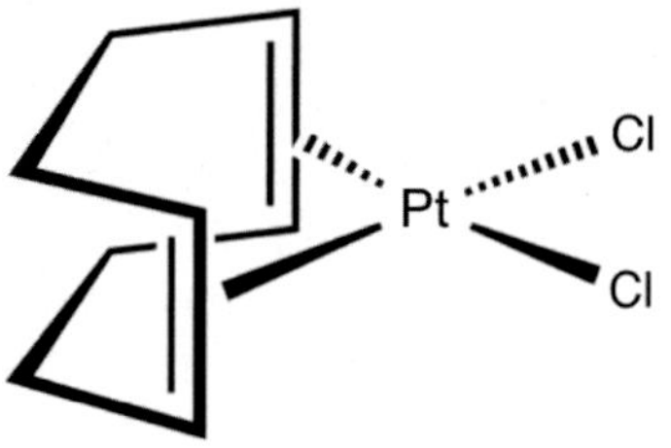

dichloro(cycloocta-1,5-diene)platinum(II)

The stability and variation of platinum(II) alkene complexes differs from the rarity of alkene complexes of nickel (II). Platinum allyl complexes are also common. Unlike nickel chemistry, where compounds such as CpNi(L)X are common, cyclopentadienyl derivatives of Pt(II) are uncommon, consistent with the reluctance of Pt(II) to change to pentacoordinate. Alkyl and aryl platinum(II) complexes are frequently produced by oxidative addition of an alkyl halide or aryl halide to a Pt(0) precursor such as tetrakis(triphenylphosphine)platinum(0) or $Pt(C_2H_4)(PPh_3)_2$. Instead, platinum (II) chlorides are prone to alkylation.

$$PtCl_2(SMe_2)_2 + 2MeLi \rightarrow PtMe_2(SMe_2)_2 + 2LiCl$$

The dimethylsulfide ligands in $PtMe_2(SMe_2)_2$ can be substituted by other ligands. Many organoplatinum(II) complexes form via ortho-metalation and related intramolecular C-H activation processes.

7.16.3.7 Organoplatinum(IV)

The first organoplatinum compound ever synthesized was trimethylplatinum iodide from platinum(IV) chloride and methylmagnesium iodide, reported by English chemists Sir William Jackson Pope (October, 31 1870 to October, 17 1939) and Stanley John Peachy in 1907 (Pope and Peachy, 1907). The compound has a cubane-like structure with four triply bridging iodide ligands. "Tetramethylplatinum" was claimed in 1952 by American organic chemist Henry Gilman (May 9, 1893 to November 7, 1986) as a derivative of this tetramer, but this claim was later shown to be incorrect ("tetramethylplatinum" was shown to be $[PtMe_3OH]_4$). Salts of $[PtMe_6]^{2-}$ and $[PtMe_4]^{2-}$ have been characterized. Organoplatinum(IV) hydrides are uncommon. The first isolated examples were prepared from organotin halides or acids with orthometalated arylplatinum(II) compounds. $Me(PEt_3)_2PtOTf$ reacts reversibly with triflic acid between $-60°C$ and $-80°C$, producing methane and $(PEt_3)_2Pt(OTf)_2$ at $-20°C$.

7.16.4 Major uses

The most common application of Pt is as a catalyst in chemical reactions, frequently as platinum black (a fine powder of platinum with good catalytic properties, the name of platinum black is due to its black color.) It has been used as a catalyst since the beginning of the 19th century, when Pt powder was employed to catalyze the ignition of hydrogen. Its most significant use is in cars as a catalytic converter, which allows the complete combustion of low concentrations of unburned hydrocarbons from the exhaust into carbon dioxide (CO_2) and water vapor. In addition, Pt is employed in the petroleum industry as a catalyst in a range of separate processes, but especially in catalytic reforming of naphthas into higher-octane gasoline that becomes rich in aromatic compounds. PtO_2, also known as Adams' catalyst, is used as a hydrogenation catalyst, specifically for vegetable oils. Adams' catalyst, also known as platinum dioxide, is usually represented as Pt(IV) oxide hydrate,

$PtO_2 \cdot H_2O$. It is a catalyst for hydrogenation and hydrogenolysis in organic synthesis. This dark brown powder is commercially available. The oxide itself is not an active catalyst, but it becomes active after exposure to hydrogen whereupon it converts to platinum black, which is responsible for reactions. Adams' catalyst is prepared from chloroplatinic acid H_2PtCl_6 or ammonium chloroplatinate, $(NH_4)_2PtCl_6$, by fusion with sodium nitrate. The first published preparation was reported by V. Voorhees and American organic chemist Roger Adams (January 2, 1889 to July 6, 1971) in 1922. The procedure involves first preparing a Pt nitrate which is then heated to expel nitrogen oxides.

$$H_2PtCl_6 + 6NaNO_3 \rightarrow Pt(NO_3)_4 + 6NaCl(aq) + 2HNO_3$$
$$Pt(NO_3)_4 \rightarrow PtO_2 + 4NO_2 + O_2$$

The resulting brown cake is washed with water to free it from nitrates. The catalyst can either be used as is or dried and stored in a desiccator for later use. Pt can be recovered from spent catalyst by conversion to ammonium chloroplatinate using aqua regia followed by ammonia. Pt also strongly catalyzes the decomposition of hydrogen peroxide (H_2O_2) into water and oxygen and it is used in fuel cells as a catalyst for the reduction of oxygen. The standard hydrogen electrode also employs a platinized Pt electrode because of its corrosion resistance, and other qualities. In the laboratory, Pt pans and supports are employed in thermogravimetric analysis since the strict necessity of chemical inertness upon heating to high temperatures ($\sim 1000°C$). Pt is utilized as an alloying agent for a variety of metal products, such as fine wires, noncorrosive laboratory containers, medical instruments, dental prostheses, electrical contacts, and thermocouples. Pt–Co, an alloy of about three parts Pt and one part Co, is employed to produce relatively strong permanent magnets. Pt-based anodes are used in ships, pipelines, and steel piers. From 1889 to 1960, the meter was defined as the length of a Pt–Ir (90:10) alloy bar, called the International Prototype Meter bar. The preceding bar was made of Pt in 1799. Until May 2019, the kilogram was defined by the International Prototype Kilogram; a cylinder of the same Pt–Ir alloy made in 1879. The International Committee for Weights and Measures (CIPM) accepted a redefinition of the SI base units in November 2018 that defines the kilogram by specifying the Planck constant as exactly $6.62607015 \times 10^{-34}$ $kg \cdot m^2 \cdot s^{-1}$. This approach effectively determines the kilogram in terms of the second and the meter.

Pt is a precious metal commodity (similar to Au); its bullion has the ISO currency code of XPT. Coins, bars, and ingots are traded or collected. Pt is used in jewelry, typically as a 90%–95% alloy, because of its inertness. It is utilized, besides this property, for its prestige and inherent bullion value. Jewelry trade journals recommend jewelers to present minute surface scratches (which they term patina) as a desirable feature in an effort to increase the value of Pt products. In watchmaking, high-end companies, for example, Rolex, Breitling, etc., use Pt for making their limited-edition watch series. Watchmakers appreciate the exceptional properties of Pt, as it neither tarnishes nor wears out (the latter quality relative to Au). The price of Pt, similar to other industrial commodities, is more volatile than that of Au. During periods of continued economic stability and growth, the price of Pt has a tendency to to be up to twice that of Au, while during periods of economic uncertainty, the price of Pt tends to drop because of decreased industrial demand, falling below that of Au. Gold prices are more stable in slow economic times, as gold is considered a safe haven. While Au is utilized in industrial applications, its demand is not so determined by industrial applications. Back in the 18th century, Pt's rarity made King Louis XV of France (February 15, 1710 to May 10, 1774) declare it the only metal fit for a king.

The electrical industry employs Pt coatings for computer hard disks, thermocouples and fuel cells. The first of these is the most important, with around 80% of disks now having Pt. Modern PCs have a hard disk drive with several disks, produced from either Al or high-quality glass coated with several layers, one of which is a Co-Pt alloy with magnetic properties in which information is stored. Proton exchange membrane (PEM) fuel cells are expected to become important for electric vehicles this century, since they operate at low temperatures, start up cold, and are compact. PEM fuel cells comprise Pt-coated carbon electrodes separated by a polymer membrane. At the anode, the fuel (H_2 gas) releases its electrons and the stripped H nuclei (the protons, H^+) then move through the polymer to the cathode where they react with oxygen from the air to form water and pick up electrons. The net outcome is a flow of current. The glass industry employs Pt for optical fibers and liquid crystal display glass, in particular for laptop and hand-held computers. Pt is used in balloon catheters and pacemakers, where a 90% Pt-10% Os alloy is used.

7.17 79 Au — Gold

7.17.1 Discovery

Since 1990, gold artifacts found at the Nahal Qana cave cemetery of the 4th millennium BCE were the oldest from the Levant (an approximate historical geographical term referring to a large area in the Eastern Mediterranean, primarily in Western Asia. In its narrowest sense, it is equivalent to the historical region of Syria. In its widest historical sense, the

Levant included all of the eastern Mediterranean with its islands; that is, it included all of the countries along the Eastern Mediterranean shores, extending from Greece to Cyrenaica.). Gold artifacts in the Balkans have their origin also from the 4th millennium BCE, for example those discovered in the Varna Necropolis (a burial site in the western industrial zone of Varna (approximately half a kilometer from Lake Varna and 4 km from the city center)) in Bulgaria. The oldest gold treasure in the world, dating from 4600 BCE to 4200 BCE, was discovered at this site. Gold artifacts such as the golden hats and the Nebra disk appeared in Central Europe from the 2nd millennium BCE Bronze Age. The Nebra sky disk is a bronze disk of about 30 cm diameter and a weight of 2.2 kg inlaid with gold symbols. These are taken to represent the sun or full moon, a lunar crescent, and stars (including a cluster thought to be the Pleiades). Two golden arcs along the sides, marking the angle between the solstices, were added later. The last addition was one more arc at the bottom surrounded with multiple strokes (of uncertain meaning, variously supposed to represent a Solar Barge with numerous oars, as the Milky Way, or as a rainbow). The disk is attributed to a site near Nebra, Saxony-Anhalt, in Germany, and associatively dated to c.1600 BCE. It has been associated with the Bronze Age Unetice culture, an archeological culture at the start of the Central European Bronze Age, dated roughly to about 2300–1600 BCE. The eponymous site for this culture, the village of Únětice, is in the central Czech Republic, northwest of Prague. The style in which the disk is made differs from any artistic style then known from the period, hence the disk was at first assumed to be a forgery, but it is now generally believed to be authentic. The Nebra sky disk shows the oldest representation of the cosmos up till now known from anywhere in the world. In June 2013 it was included in the UNESCO Memory of the World Register and called "one of the most important archeological finds of the twentieth century." The earliest identified map of a gold mine was drawn in the 19th Dynasty of Ancient Egypt (1320–1200 BCE), while the first written reference to gold was documented in the 12th Dynasty around 1900 BCE. Egyptian hieroglyphs dating back to as early as 2600 BCE describe gold, which King Tushratta of the Mitanni stated was "more plentiful than dirt" in Egypt. Egypt and in particular Nubia (a region along the Nile river encompassing the area between Aswan in southern Egypt and Khartoum in central Sudan. It was the seat of one of the earliest civilizations of ancient Africa, with a history that can be traced from at least 2500 BCE onward with the Kerma culture.) had the resources to make them major gold-producing areas for most of their history. One of the oldest known maps, known as the Turin Papyrus Map, shows the plan of a gold mine in Nubia together with some of the local geology (reportedly discovered at Deir el-Medina in Thebes, collected by Italian antiquities collector, diplomat, and politician Bernardino Drovetti (known as Napoleon's Proconsul, January 7, 1776 to March 5, 1852) in Egypt sometime before 1824 and now preserved in Turin's Museo Egizio. The map was drawn about 1150 BCE by the well-known Scribe-of-the-Tomb Amennakhte, son of Ipuy. It was prepared for Ramesses (also written Ramses or Rameses) IV's quarrying expedition to the Wadi Hammamat in the Eastern Desert, which exposes Precambrian rocks of the Arabian-Nubian Shield (he reigned from 1155 to 1149 BCE as the third pharaoh of the Twentieth Dynasty of the New Kingdom of Ancient Egypt). The purpose of the expedition was to obtain blocks of bekhen-stone (metagraywacke sandstone) to be used for statues of the king.). The primitive working techniques are reported by both Greek geographer, philosopher, and historian Strabo (64 or 63 BCE to c. 24 CE) (Fig. 7.87) and Greek historian Diodorus Siculus (fl. 1st century BCE) (Fig. 7.88) and included fire-setting (a method of traditional mining used most commonly from prehistoric times up to the Middle Ages. Fires were set against a rock face to heat the stone, which was then doused with liquid, causing the stone to fracture by thermal shock. Some experiments have suggested that the water (or any other liquid) did not have a noticeable effect on the rock, but rather helped the miners' progress by quickly cooling down the area after the fire. This technique was best performed in opencast mines where the smoke and fumes could dissipate safely. The technique was very dangerous in underground workings without adequate ventilation. The method became largely redundant with the growth in use of explosives.). Large mines could also be found across the Red Sea in what is these days known as Saudi Arabia. Gold is mentioned in the Amarna letters (an archive, written on clay tablets, mainly consisting of diplomatic correspondence between the Egyptian administration and its representatives in Canaan and Amurru during the New Kingdom, between c. 1360–1332 BCE. The letters were found in Upper Egypt at el-Amarna, the current name for the ancient Egyptian capital of Akhetaten, founded by pharaoh Akhenaten (1350s–1330s BCE) during the 18th dynasty of Egypt. The Amarna letters are unusual, since they are mostly written in a script known as Akkadian cuneiform, the writing system of ancient Mesopotamia, instead of that of ancient Egypt, and the language used has occasionally been characterized as a mixed language, Canaanite-Akkadian. The letters span a period of no more than 30 years. The number of known tablets total 382 with gold mentioned in the tablets numbered 19 and 26 from around the 14th century BCE. Gold is noted regularly in the Bible's Old Testament, beginning with Genesis 2:11 (at Havilah), the story of The Golden Calf and many parts of the temple including the Menorah and the golden altar. In the New Testament, it is included with the gifts of the magi in the first chapters of Matthew. The Book of Revelation 21:21 describes the city of New Jerusalem as having streets "made of pure gold, clear as crystal." Exploitation of gold in the south-east corner of the Black Sea is thought to date from the time of Midas (Midas is the name of at least three members of the royal house of Phrygia. The most famous King Midas is

FIGURE 7.87 Strabo.

FIGURE 7.88 Diodorus Siculus, c.800s.

popularly remembered in Greek mythology for his ability to turn everything he touched into gold. This came to be called the golden touch, or the Midas touch), and this gold was important in the establishment of what is probably the world's earliest coinage in Lydia (an Iron Age kingdom of western Asia Minor located generally east of ancient Ionia in the

FIGURE 7.89 Detail from the Catalan Atlas Sheet 6 showing Mansa Musa sitting on a throne and holding a gold coin. Pen with colored inks on parchment, 1375.

modern western Turkish provinces of Uşak, Manisa and inland İzmir) around 610 BCE. The tale of the golden fleece dating from 8th century BCE may refer to the use of fleeces to trap gold dust from placer deposits in the ancient world. From the 6th or 5th century BCE, the Chu (Chinese state, c. 1030 BCE to 223 BCE, a state during the Zhou dynasty) used the Ying Yuan, a type of square gold coin. In Roman metallurgy, new processes for extracting gold on a major scale were established by introducing hydraulic mining methods (Hydraulic mining originated out of ancient Roman techniques that used water to excavate soft underground deposits. Its modern form, using pressurized water jets produced by a nozzle called a "monitor," came about in the 1850s during the California Gold Rush in the United States.), especially in Hispania (the Roman name for the Iberian Peninsula and its provinces) from 25 BCE onward and in Dacia (Dacia was bounded in the south approximately by the Danubius river (Danube), in the east it was bounded by the Pontus Euxinus (Black Sea) and the river Danastris (Dniester). At times Dacia included areas between the Tisa and the Middle Danube. The Carpathian Mountains are located in the middle of Dacia. It thus corresponds to the present-day countries of Romania and Moldova, as well as smaller parts of Bulgaria, Serbia, Hungary, Poland, Slovakia and Ukraine.) from 106 CE onward. One of the biggest mines was located at Las Medulas in León, north-western Spain, where seven long aqueducts allowed them to sluice most of a large alluvial deposit. The mines at Roşia Montană in Alba County, Transylvania, Romania, were also very large, and until very recently, still mined by opencast techniques. The Romans also mined smaller deposits in Britain, such as placer and hard-rock deposits at Dolaucothi, in the valley of the River Cothi, near Pumsaint, Carmarthenshire, Wales. The different techniques they used are well defined by Roman author, naturalist and natural philosopher Pliny the Elder (23–79 CE) in his encyclopedia Naturalis Historia. During Mansa Musa's (c.1280–1337, ruler of the Mali Empire from 1312 to 1337) (Fig. 7.89) hajj to Mecca in 1324, he came through Cairo in July 1324, and was supposedly escorted by a camel train that comprised thousands of people and nearly a hundred camels, where he gave away so much gold that it lowered the price of gold in Egypt for more than 10 years, resulting in high inflation. A contemporary of Mansa Musa, an Arab historian remarked:

"Gold was at a high price in Egypt until they came in that year. The mithqal did not go below 25 dirhams and was generally above, but from that time its value fell, and it cheapened in price and has remained cheap till now. The mithqal does not exceed 22 dirhams. This has been the state of affairs for about 12 years until this day by reason of the large amount of gold which they brought into Egypt and spent there [...]." —Chihab Al-Umari (Arab historian 1300–1349), Kingdom of Mali

The European exploration of the Americas was strongly influenced by rumors about the gold ornaments displayed in great abundance by Native American peoples, particularly in Mesoamerica, Peru, Ecuador and Colombia. The Aztecs looked upon gold as the product of the gods, calling it literally "god excrement" (teocuitlatl in Nahuatl), and after

Moctezuma II (c.1466 to June 29, 1520, the ninth tlatoani or ruler of Tenochtitlán, reigning from 1502 to 1520) was killed, most of this gold was sent to Spain. Nevertheless, for the native peoples of North America gold was considered useless and they attributed much more value to other minerals that were directly related to their usefulness, such as obsidian, flint, and slate. Rumors of cities full of gold fueled legends of El Dorado. One major goal of the alchemists was to make gold from other substances, such as lead—apparently by the interaction with a mythical substance called the philosopher's stone. Although they were never successful in this effort, the alchemists did encourage an interest in methodically finding out what can be done with substances, thereby laying the foundation for today's chemistry. Their symbol for gold was the circle with a point at its center (⊙), which was also the astrological symbol and the ancient Chinese character for the Sun. "Gold" is related to similar words in many Germanic languages, deriving via Proto-Germanic *gulþą from Proto-Indo-European *$\acute{g}^h$elh$_3$- ("to shine, to gleam; to be yellow or green"). The symbol Au is from the Latin: aurum, the Latin word for "gold." The Proto-Indo-European ancestor of aurum was *h$_2$é-h$_2$us-o-, meaning "glow." This word is derived from the same root (Proto-Indo-European *h$_2$u̯es- "to dawn") as *h$_2$éu̯;sōs, the ancestor of the Latin word Aurora, "dawn." This etymological relationship is presumably behind the frequent claim in scientific publications that aurum meant "shining dawn."

7.17.2 Mining, production, and major minerals

On Earth, gold is found in ores in rock formed from the Precambrian time onward. It most often occurs as a native metal, typically in a metal solid solution with silver, that is, as an Au−Ag alloy with usually a Ag content of 8%–10%. Electrum is alloyed Au with more than 20% Ag. Native gold occurs as very small to microscopic particles embedded in rock, often together with quartz or sulfide minerals such as "Fool's Gold," which is a pyrite (FeS_2) in so-called lode deposits. The metal in a native state is also found in the form of free flakes, grains, or larger nuggets that have been eroded from rocks and transported to alluvial deposits called placer deposits. Such free gold is always richer at the surface of gold-bearing veins owing to the oxidation of accompanying minerals followed by weathering, and washing of the dust into streams and rivers, where it collects and can be welded by water action to form nuggets. Gold can occasionally be found together with tellurium as the minerals calaverite ($AuTe_2$), krennerite (Au_3AgTe_8), nagyagite ($[Pb_3(Pb,Sb)_3S_6](Au,Te)_3$), petzite ($Ag_3AuTe_2$), and sylvanite ($(Au,Ag)_2Te_4$), and as the rare bismuthide maldonite (Au_2Bi) and antimonide aurostibite ($AuSb_2$). Gold also occurs in rare alloys with Cu, Pb, and Hg as the minerals auricupride (Cu_3Au), novodneprite ($AuPb_3$) and weishanite ($(Au, Ag)_3Hg_2$). Recent study results indicate that microbes can sometimes play a significant role in forming gold deposits, transporting and precipitating gold to form grains and nuggets that collect in alluvial deposits. Since the 1880s, South Africa has been the producer of a large percentage of the global gold supply, and about 50% of the gold presently mined is from South Africa. During the 19th century, gold rushes occurred whenever large gold deposits were found. The first documented discovery of gold in the United States was in 1803 at the Reed Gold Mine near Georgeville, North Carolina. The first major gold strike in the US found place in a small town in north Georgia called Dahlonega. Other gold rushes occurred in California, Colorado, the Black Hills, Otago in New Zealand, Australia, Witwatersrand in South Africa, and the Klondike in Canada. In 2007 China surpassed South Africa as the largest gold producer globally, the first time since 1905 that South Africa was the largest producer. Since 2017, China has been the world's number one gold-mining country, followed in order by Australia, Russia, the United States, Canada, and Peru. South Africa, which dominated world gold production for most of the 20th century, has dropped to sixth place. Other major producers are Ghana, Burkina Faso, Mali, Indonesia, and Uzbekistan.

Gold extraction is most economical in large, easily mined deposits. Ore grades as low as 0.5 parts per million (ppm) can be economical. Characteristic ore grades in open-pit mines are in the order of 1 to 5 ppm; ore grades in underground or hard-rock mines are typically no less than 3 ppm. Since ore grades of 30 ppm or more are generally required before gold becomes visible to the naked eye, in most current gold mines the gold is invisible. After initial production, gold is often subsequently refined industrially by the Wohlwill process which is based on electrolysis or by the Miller process, that is chlorination in the melt. The Wohlwill process was invented in 1874 by German-Jewish engineer of electrochemistry Emil Wohlwill (November 24, 1835 to February 2, 1912). This electrochemical method comprises using a cast gold ingot, often called a Doré bar, of 95% + gold to act as an anode. Lower percentages of gold in the anode will interfere with the reaction, especially when the contaminating metal is silver or one of the platinum group metals. The cathode(s) for this reaction consist(s) of small sheets of pure (24k) gold sheeting or stainless steel. A current is applied to the system, and electricity travels through the electrolyte of chloroauric acid ($HAuCl_4$). Gold and other metals are dissolved at the anode, and pure gold (coming through the chloroauric acid by ion transfer) is plated onto the gold cathode. When the anode is dissolved, the cathode is removed and melted or otherwise

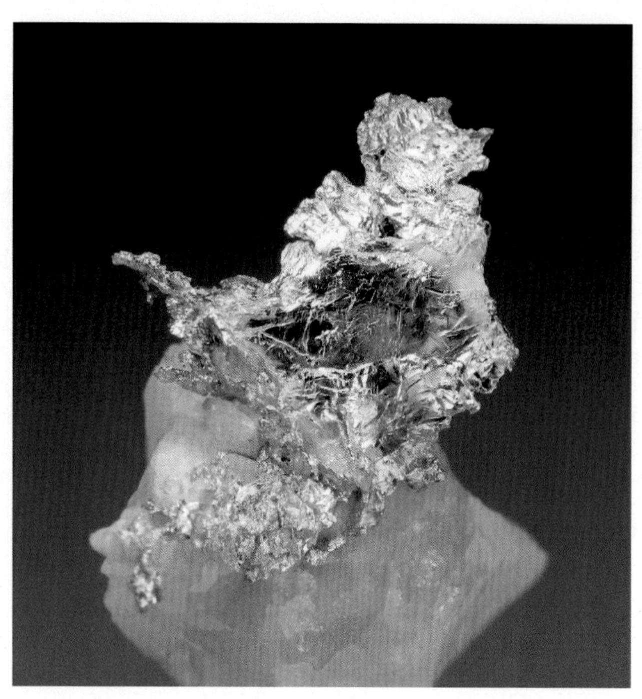

FIGURE 7.90 A large gold, Au, leaf protruding from its white quartz, SiO_2, matrix. The gold leaf measures 2.1 × 1.4 cm. Harvard mine, Jamestown, Tuolumne Co., California, United States.

FIGURE 7.91 Leaf gold, Au, to 1.2 × 0.9 cm on quartz, SiO_2. Harvard mine, Jamestown, Tuolumne Co., California, United States.

processed in the manner essential for sale or use. The resulting gold is 99.999% pure, and of higher purity than gold produced by the other common refining method, the Miller process, which produces gold of 99.95% purity. The Miller process is an industrial scale chemical procedure used to refine gold to a high degree of purity. It was invented by English-Australian assayer and inventor Francis Bowyer Miller (December 18, 1828 to September 17, 1887) who

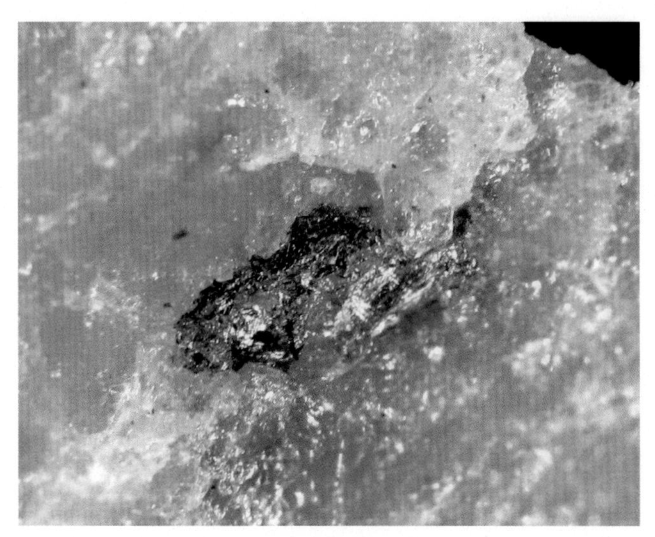

FIGURE 7.92 Aurostibite, $AuSb_2$, metallic gray crystal grain to 2 mm. Krásná Hora nad Vltavou, Bohemia, Czech Republic.

FIGURE 7.93 Calaverite, $AuTe_2$, elongated, striated bladed crystal to 6 mm. Cripple Creek dist., Teller Co., Colorado, United States.

patented it in 1867). This chemical process involves blowing a stream of pure chlorine gas over and through a crucible filled with molten, but impure, gold. This process purifies the gold since almost all other elements will form chlorides before gold does, and they can then be removed as salts that are insoluble in the molten metal. When all impurities have been removed from the gold (observable by a change in flame color) the gold is removed and processed in the manner essential for sale or use. The resulting gold is 99.95% pure. The Wohlwill process results in higher purity but is more complex and is only applied in small-scale installations. Other methods of assaying and purifying smaller amounts of gold include parting and inquartation as well as cupellation (a refining process in metallurgy, where ores or alloyed metals are treated under very high temperatures and have controlled operations to separate noble metals, like gold and silver, from base metals like lead, copper, zinc, arsenic, antimony or bismuth, present in the ore. The process is based on the principle that precious metals do not oxidize or react chemically, unlike the base metals; so when they are heated at high temperatures, the precious metals remain apart and the others react

FIGURE 7.94 Nagyágite, $[Pb_3(Pb,Sb)_3S_6](Au,Te)_3$, silvery lead-gray somewhat foliated metallic plates to 7 mm. Sacarîmb (Nagyag in Hungary), Romania.

FIGURE 7.95 Petzite, Ag_3AuTe_2, formless silvery gray masses on quartz, SiO_2, 8 mm. Cornucopia mine, Baker Co., Oregon, United States.

forming slags or other compounds.), or refining methods based on the dissolution of gold in aqua regia (a mixture of nitric acid and hydrochloric acid, optimally in a molar ratio of 1:3). The global consumption of gold produced is approximately 50% in jewelry, 40% in investments, and 10% in industry.

The number of naturally occurring minerals containing gold is 35. Of these 35, 13 are in the elements class, besides the native metal gold (Au) (Figs. 7.90 and 7.91), the others are all alloys, for example, auricupride (Cu_3Au), tetra-auricupride (AuCu), and yuanjiangite (AuSn). The sulfide class contains 22 different minerals, such as aurostibite ($AuSb_2$) (Fig. 7.92), calaverite ($AuTe_2$) (Fig. 7.93), krennerite (Au_3AgTe_8), nagyágite ($[Pb_3(Pb,Sb)_3S_6](Au,Te)_3$) (Fig. 7.94), petzite ($Ag_3AuTe_2$) (Fig. 7.95), and sylvanite ($(Au,Ag)_2Te_4$) (Fig. 7.96).

7.17.3 Chemistry

Gold has only one naturally occurring stable isotope, ^{197}Au, so gold is both a mononuclidic and monoisotopic element. Thirty-six radioisotopes have been synthesized, ranging from ^{169}Au to ^{205}Au. The most stable of these radioisotopes is ^{195}Au which has a half-life of 186.1 days. The least stable radioisotope is ^{171}Au, which has a half-life of 30 μs and decays by proton emission. Most of radioisotopes lighter than ^{197}Au decay by some combination of proton emission, α decay, and β^+ decay. The exceptions are ^{195}Au, which decays via electron capture, and ^{196}Au, which decays frequently via electron capture (93%) with a minor β^- decay path (7%). All radioisotopes heavier than ^{197}Au decay via β^- decay. In addition, no

FIGURE 7.96 Sylvanite, $(Au,Ag)_2Te_4$ (Au:Ag ratio usually is close to 1:1), streaks and thin sheets to 1cm. Portland mine, Lawrence Co., South Dakota, United States.

TABLE 7.17 Gold properties.

Appearance	Metallic yellow
Standard atomic weight $A_{r,std}$	196.967
Block	d-Block
Element category	Transition metal
Electron configuration	[Xe] $4f^{14}$ $5d^{10}$ $6s^1$
Phase at STP	Solid
Melting point	1064.18°C
Boiling point	2970°C
Density (near r.t.)	19.30 g/cm^3
When liquid (at m.p.)	17.31 g/cm^3
Heat of fusion	12.55 kJ/mol
Heat of vaporization	342 kJ/mol
Molar heat capacity	25.418 J/(mol·K)
Oxidation states	−3, −2, −1, **+1**, +2, **+3**, +5
Ionization energies	1st: 890.1 kJ/mol
	2nd: 1980 kJ/mol
Atomic radius	Empirical: 144 pm
Covalent radius	136 ± 6 pm
Van der Waals radius	166 pm

STP, Standard temperature and pressure.
Bold font indicates main oxidation state.

less than 32 nuclear isomers have been observed, ranging in atomic mass from 170 to 200. Within that range, ^{178}Au, ^{180}Au, ^{181}Au, ^{182}Au, and ^{188}Au do not have isomers. The most stable isomer is ^{198m2}Au which has a half-life of 2.27 days, while the least stable isomer is ^{177m2}Au which has a half-life of only 7 ns. ^{184m1}Au has three different decay paths: β^+ decay, isomeric transition, and α decay. No other isomer or isotope of gold exists that has three different decay paths.

Gold is the most malleable of all metals. It can be drawn into a monoatomic wire and subsequently stretched to about twice its length before it breaks. These nanowires are distorted through the formation, reorientation and migration of dislocations and crystal twins without noticeable hardening. Gold leaf can be beaten thin enough to become semi-transparent. The transmitted light appears greenish blue, as gold strongly reflects yellow and red. These semitransparent sheets likewise strongly reflect infrared light, rendering them useful as infrared (radiant heat) shields in visors of heat-resistant suits, and in sun-visors for spacesuits. Gold is a good conductor of heat and electricity. Gold has a density of 19.3 g/cm^3, nearly identical to that of tungsten at 19.25 g/cm^3. While most metals are gray or silvery white, gold is slightly reddish-yellow (Table 7.17). This color is determined by the frequency of plasma oscillations among the metal's valence electrons, in the ultraviolet range for most metals but in the visible range for gold due to relativistic effects affecting the orbitals around gold atoms.

Though gold is the most noble of the noble metals, it still forms a variety of compounds. The oxidation state of gold in its compounds varies between -1 to $+5$, but Au(I) and Au(III) dominate its chemistry. Au(I), described as the aurous ion, is the most common oxidation state with soft ligands such as thioethers, thiolates, and tertiary phosphines. Au(I) compounds are usually linear. A good illustration is $Au(CN)_2^-$. Gold ions in solution are easily reduced and precipitated as metal by adding any other metal as the reducing agent. The added metal is oxidized and dissolves, allowing the gold to be displaced from solution and be recovered as a solid precipitate.

7.17.3.1 Oxides and chalcogenides

Gold does not react with oxygen at any temperature and is resistant to attack from ozone up to 100°C. The action of alkali on aqueous Au(III) solutions produces a precipitate, probably of $Au_2O_3 \cdot xH_2O$, which on dehydration yields brown Au_2O_3. This is the only verified oxide of gold. It decomposes when heated above approximately 160°C and, when hydrous, is weakly acidic, dissolving in concentrated alkali and probably forming salts of the $[Au(OH)_4]^-$ ion.

Gold does not react with sulfur directly. The reaction of H_2S with aqueous Au(I) precipitates Au_2S, while passing H_2S through cold solutions of $AuCl_3$ in dry ether produces Au_2S_3, which is rapidly reduced to Au(I) or the metal upon addition of water. Gold chalcogenides, such as gold sulfide, contain equal amounts of Au(I) and Au(III). Gold(III) sulfide is not described in textbooks or reviews. The selenides and tellurides are all metallic and some, for example, Au_3Te_5, are superconductors at low temperature. Other phases are Au_2Te_3 and $AuTe_2$. Most of these are nonstoichiometric.

7.17.3.2 Halides

Some free halogens react with gold. Gold is strongly attacked by fluorine at dull-red heat to form gold(III) fluoride. AuF_5 is an unstable, polymeric, diamagnetic, dark-red powder, produced by heating $[O_2][AuF_6]$ under reduced pressure and condensing the product on to a "cold finger."

$$2Au + 2O_2 + 6F_2 \xrightarrow{370^\circ C, 8\ atm} 2O_2AuF_6 \xrightarrow{180^\circ C/20^\circ C\ (hot/cold)} 2AuF_5 + 2O_2 + F_2$$

The compound has a tendency to dissociate into AuF_3 and, when treated with XeF_2 in anhydrous HF solution below 25°C, forms yellow-orange crystals of the $[Xe_2F_3][AuF_6]$ complex.

$$AuF_5 + 2XeF_2 \xrightarrow{HF/0^\circ C} [Xe_2F_3][AuF_6] \xrightarrow{>60^\circ C} AuF_3 + XeF_2 + XeF_4$$

In the $+3$ oxidation state, gold is known to form binary halides, though AuI_3 has not been isolated. The chloride and the bromide are red-brown solids produced directly from the elements and have a planar dimeric structure in both the solid and vapor phases. Powdered gold reacts with chlorine at 180°C to form $AuCl_3$. Gold(III) chloride is most frequently produced by passing chlorine gas over gold powder at 180°C:

$$2Au + 3Cl_2 \rightarrow Au_2Cl_6$$

A different preparation route is by reacting Au^{3+} species with chloride to produce tetrachloroaurate. Its acid, chloroauric acid, is then heated to eliminate hydrogen chloride gas. Reaction with aqua regia produces gold(III) chloride:

$$Au(s) + 3NO_3^-(aq) + 6H^+(aq) \rightleftharpoons Au^{3+}(aq) + 3O_2(g) + 3H_2O(l)$$
$$Au^{3+}(aq) + 3NOCl(g) + 3NO_3^-(aq) \rightarrow AuCl_3(aq) + 6NO_2(g)$$
$$AuCl_3(aq) + Cl^-(aq) \rightleftharpoons AuCl_4^-(aq)$$
$$2HAuCl_4(s) \rightarrow Au_2Cl_6(s) + 2HCl(g)$$

$AuCl_3$ exists as a chloride-bridged dimer both as a solid and as a vapor, at least at low temperatures. Gold(III) bromide behaves similarly. The structure is comparable to that of iodine(III) chloride. In gold(III) chloride, each gold center is square planar, which is typical of a metal complex with a d^8 electron count. The bonding in $AuCl_3$ is thought to be somewhat covalent. Gold reacts with bromine at 140°C to form gold(III) bromide, but reacts only very slowly with iodine to form the monoiodide.

$$2\ Au + 3Br_2 \rightarrow Au_2Br_6$$

Likewise, the halide-exchange reaction of gold(III) chloride with hydrobromic acid has also been shown to be successful in synthesizing gold(III) bromide:

$$Au_2Cl_6 + 6HBr \rightarrow 6HCl + Au_2Br_6$$

This reaction is driven by the formation of the relatively more stable hydrochloric acid compared with hydrobromic acid. The neutral monomer $AuBr_3$, as well as the other neutral gold trihalide species, has not been isolated in the gas phase which suggests that the coordination number three is not favored. Predominantly, gold(III) displays square planar coordination corresponding to a preferred coordination number of four. Specifically, in solution gold(III) trihalides have the tendency to add a fourth ligand to form the more preferred four-coordinate complex. With respect to gold tribromide, it is common to purchase gold(III) bromide hydrate, $AuBr_3 \cdot H_2O$, where the central gold atom exhibits a coordination number of four, rather than the anhydrous form of the compound, which exhibits a coordination number of three. Alternatively, if there is no addition of a fourth ligand, gold tribromide will oligomerize to form the halogen-bridged dimer complex mentioned previously for gold(III) chloride.

$$2AuBr_3 \rightarrow Au_2Br_6$$

In addition, like gold(III) chloride, gold tribromide is a Lewis acid and can form several complexes. For instance, in the presence of hydrobromic acid, the dimer dissolves and bromoauric acid is formed.

$$HBr(aq) + AuBr_3(aq) \rightarrow H^+AuBr_4^-(aq)$$

The dimer also undergoes hydrolysis rapidly in moist air.

All four monohalides of gold have been synthesized but the fluoride only by mass spectrometric methods. AuCl and AuBr can be prepared by heating the trihalides to no more than 150°C and AuI by heating the metal and iodine. At higher temperatures they dissociate into the elements. AuI is a chain polymer which features linear two-coordinate Au with Au–I 262 pm and the angle Au-I-Au 72°. The binary gold halides, such as AuCl, form zigzag polymeric chains, again featuring linear coordination at Au. Au(III) (auric) is a common oxidation state, and is illustrated by gold(III) chloride, Au_2Cl_6. The gold atom centers in the Au(III) complexes, like other d^8 compounds, are usually square planar, with chemical bonds that have both a covalent and a ionic character. It forms a useful starting compound for a lot of coordination chemistry. Dissolving it in hydrochloric acid forms the stable $[AuCl_4]^-$ ion. Treatment of Au_2Cl_6 with F_2 or BrF_3 also provides a route to AuF_3, a strong ftuorinating agent. This orange solid compound comprises square planar AuF_4 units which share *cis*-F atoms with two adjacent AuF_4 units so as to form a helical chain. Gold pentafluoride, along with its derivative anion, AuF_6^-, and its difluorine complex, gold heptafluoride, is the sole example of gold(V), the highest proven oxidation state. No halides are known for gold in the +2 oxidation state.

7.17.3.3 Amalgam

Gold readily dissolves in mercury at room temperature to form an amalgam, and forms alloys with many other metals at higher temperatures. These alloys can be produced to modify the hardness and other metallurgical properties, to control melting point or to create exotic colors. Gold amalgam has been shown to be effective where gold fines ("flour gold") would not be extractable from ore using hydro-mechanical methods. Large quantities of mercury were used in placer mining, where deposits composed largely of decomposed granite slurry were separated in long runs of "riffle boxes," with mercury dumped in at the head of the run. The amalgam formed is a heavy solid mass of dull gray color. (The use of mercury in 19th century placer mining in California, now banned, has resulted in extensive pollution problems in riverine and estuarine environments, ongoing to this day.) Occasionally sizable slugs of amalgam are collected from downstream river and creek bottoms by amateur wet-suited miners seeking gold nuggets with the aid of an engine-powered water vacuum mounted on a float. Currently, mercury amalgamation has been replaced by other methods to recuperate gold and silver from ore in developed nations. Dangers of mercurial toxic waste have played a key role in the phasing out of the mercury amalgamation processes. Nevertheless, mercury amalgamation is still often used by small-scale gold placer miners (often illegally), especially in developing countries.

7.17.3.4 Reactions with acids and bases

Gold is nonreactive to most acids. It does not react with hydrofluoric, hydrochloric, hydrobromic, hydriodic, sulfuric, or nitric acid. It does react with selenic acid and is dissolved by aqua regia (a 1:3 mixture of nitric acid and hydrochloric acid). Nitric acid oxidizes the metal to +3 ions, but only in trace amounts, typically undetectable in the pure acid due to the chemical equilibrium of the reaction. Nevertheless, the ions are removed from the equilibrium by hydrochloric acid, forming $AuCl_4^-$ ions, or chloroauric acid, thus facilitating further oxidation. Gold is likewise unaffected by most bases. It does not react with aqueous, solid, or molten sodium or potassium hydroxide. It does, however, react with sodium or potassium cyanide under alkaline conditions in the presence of oxygen to form soluble complexes.

7.17.3.5 Gold(III) complexes

For gold +3 is the best known oxidation state. The typical route to gold(III) chemistry is by dissolving the metal in aqua regia, or the compound Au_2Cl_6 in concentrated HCl, after which evaporation produces yellow chloroauric acid, $HAuCl_4 \cdot 4H_2O$, from which various salts of the square planar ion $[AuCl_4]^-$ can be obtained. Other square planar ions of the type $[AuX_4]^-$ can subsequently be derived in which X = F, Br, I, CN, SCN and NO_3, the last of these being of interest as one of the few verified examples of the unidentate nitrate ion. $[Au(SCN)_4]^-$ contains S-bonded SCN but this ligand also gives rise to linkage isomers, here in the K^+ and $(NEt_4)^+$ salts of $[Au(CN_2(SCN_2]^-$ and $[Au(CN_2(NCS_2]^-$. Various cationic complexes have been synthesized with amines, both unidentate (e.g., py, quinoline, as well as NH_3) and chelating (e.g., en, bipy, phen). $[Au(C_6H_4\text{-}CH_2NMe_2\text{-}2)(phen)(PPh_3]^{2+}$ is a good illustration with the further interest that its distorted square pyramidal structure provides a uncommon example of Au(III) with a coordination number larger than of 4. Octahedral $[AuI_2(diars_2]^+$ also has a "high" coordination number, although phosphine and arsine complexes are usually easily reduced to Au(I) species. Reduction of Au(III) to Au(I) in aqueous solution by nucleophiles such as I^-, SCN^- and other S-donor ligands have been a matter of extensive research. Most occur by rapid ligand substitution followed by the rate determining electron transfer, although some reductions by I^- occur without substitution. With SCN^- the rates of substitution and electron transfer are finely balanced. In forming the ftuoro complex $[AuF_4]^-$, and in fact in forming the simple fluoride AuF_3, Au(III) differs from the isoelectronic Pt(II) since the corresponding $[PtF_4]^{2-}$ and PtF_2 are unknown.

7.17.3.6 Gold(I) complexes

Au(I) easily forms linear 2-coordinate complexes such as $[AuX_2]^-$ (X = Cl, Br, I) and also the technologically significant $[Au(CN)_2]^-$. However, it is much more susceptible to oxidation and to disproportionation into Au(III) and Au(0) which makes all its binary compounds, except AuCN, unstable to water. It is also more clearly a class b or "soft" metal with a preference for the heavier donor atoms P, As and S. Stable, linear complexes are formed when tertiary phosphines reduce Au (III) in ethanol.

$$[AuCl_4]^- \xrightarrow{PR_3/EtOH} [AuCl(PR_3)]$$

The Cl ligand can be substituted by other halides and pseudo-halides through metathetical reactions. Trigonal planar coordination is observed in phosphine complexes of the stoichiometry $[AuL_2X]$ but 4-coordination, though possible, is less widespread. Diarsine produce the nearly tetrahedral complex $[Au(diars)_2]^+$ but, for some unknown reason, the colorless complexes $[AuL_4]^+[BPh_4]^-$ with monodentate phosphines fail to achieve a regular tetrahedral geometry. Complexes with dithiocarbamates have linear S-Au-S coordination but are dimeric and the Au—Au distance of 276 pm compared with 288 pm in the metal and 250 pm in gaseous Au_2 is an indication of metal—metal bonding.

7.17.3.7 Rare oxidation states

Less common oxidation states of gold include −1, +2, and +5. Some gold compounds exhibit aurophilic bonding, which describes the tendency of gold ions to interact at distances that are too long to be a typical Au—Au bond but shorter than van der Waals bonding. The interaction is estimated to be similar in strength to that of a hydrogen bond. The −1 oxidation state occurs in aurides, compounds containing the Au^- anion. Cesium auride (CsAu), for instance, crystallizes in the cesium chloride motif; rubidium, potassium, and tetramethylammonium aurides are also known. Gold has the highest electron affinity of any metal, at 222.8 kJ/mol, making Au^- a stable species.

Most of the compounds supposedly of Au(II) are in fact mixed valence Au(I)/Au(III) compounds. Examples include the sulfate $Au(I)Au(III)(SO_4)_2$ (evaporation of a solution of $Au(OH)_3$ in concentrated H_2SO_4 produces red crystals of this sulfate) and the chlorocomplex, $Cs_2[Au(I)Cl_2][Au(III)Cl_4]$, the anions of the latter being arranged forming linearly coordinated Au(I) and tetragonally distorted, octahedral Au(III). One of the few confirmed examples of Au(II) is the maleonitriledithiolato complex which has a magnetic moment at room temperature of 1.85 BM. Even here, however, ESR evidence suggests substantial delocalization of the unpaired electron on to the ligands and, in solution, the complex is easily oxidized to Au(III).

7.17.3.8 Cluster compounds

Polymeric complexes of the types formed by copper and silver are not observed for gold but in its place a range of variously colored cluster compounds, with gold in an average oxidation state < 1 and including M-M bonds, can be formed by the general process of reducing a gold phosphine halide, typically with sodium borohydride. Yellow $[Au_6P(C_6H_4\text{-}4\text{-}Me)_3\}_6]^{2+}$ consists of an octahedron of six gold atoms with a phosphine attached to each Au. Red $[Au_8(PPh_3)_8]^{2+}$ can be seen as a chair-like, centered hexagon of gold atoms with an eighth gold atom located above the chair with each gold atom having a phosphine attached to it. Clusters are known in which more gold atoms are added to the chair in a roughly spherical manner (e.g., $[Au_{11}\{P(C_6H_4\text{-}4\text{-}F)_3\}_7I_3]$ in which the central gold has no attached ligands and giving ultimately a centered icosahedron as observed in the dark-red $[Au_{13}Cl_2(PMe_2Ph)_{10}]^{3+}$). Another group of clusters can be distinguished with flatter, ring or torus shapes, for example, in the red-brown $[Au_8(PPh_3)_7]^{2+}$ and green $[Au_9P(C_6H_4\text{-}4\text{-}Me)_3\}_8]^{3+}$. This last group is characterized by lower electron counts than the first, indicating a lower involvement of p-orbitals in M-M bonding and consequently less tangential skeletal bonding. This is in agreement with the observation that only clusters with an icosahedral structure (stabilized by both tangential and radial skeletal bonding) are stereochemically rigid on the NMR time scale at room temperature. Heteronuclear clusters containing a range of other transition metals can be formed by the general reaction of $AuPR_3$ with a carbonyl anion of the appropriate metal. "Clusters of clusters" of Au -Ag have been prepared with metal frameworks based on vertex sharing icosahedra, the basic unit being an Au-centered $\{Au_7Ag_6)$ icosahedron. The largest of these is $[Au_{22}Ag_{24}(PPh_3)_{12}Cl_{10}]$ comprising 4 $\{Au_7Ag_6\}$ icosahedra in a tetrahedral arrangement with six shared vertices. The red-brown $[Au_{55}(PPh_3)_{12}Cl_6]$ is formed by reducing $Au(PPh_3)Cl$ with B_2H_6 and is perhaps best regarded as a cubo-octahedral fragment of close-packed Au atoms. From it, water-soluble $[Au_{55}(Ph_2PC_6H_4SO_3Na\cdot 2H_2O)_{12}Cl_6]$ can be formed through ligand exchange.

7.17.3.9 Organogold compounds

Au(I) alkyls can be obtained with an appropriate ligand present, for example,

$$[Au(PEt_3)X] + LiR \rightarrow [Au(PEt_3)R] + LiX$$

The colorless solids are linear monomers. Some anionic Au(I) alkyls are known, for example $[N(PPh_3)_2]^+[Au(acac)_2]$. In this compound it is the central C of the ligand, $HC(COMe)_2$ which is attached to the gold. The alkyl derivatives of Au(III) were discovered as early as 1907 by W. J. Pope and C. S. Gibson and they comprise some of the most common and stable organo compounds of the group and are significant for not needing the stabilizing presence of π-bonding ligands (Pope and Gibson, 1907). They are three types: (1) AuR_3 (stable, when they occur at all, only in ether below $-35°C$); (2) AuR_2X (largely the most stable); X = anionic ligand, in particular Br; and (3) $AuRX_2$ (unstable, only dibromides have been characterized). Corresponding aryl derivatives are uncommon and unstable. Thus, though $AuMe_3$ decomposes above $-35°C$, it is stabilized in $[AuMe_3(PPh_3)]$, while $AuPh_3$ is not known. The dialkylgold(III) halides are generally synthesized from the tribromide and a Grignard reagent.

$$AuBr_3 + 2RMgBr \rightarrow AuR_2Br + 2\ MgBr_2$$

Numerous other anions can subsequently be introduced by metathetical reactions with the appropriate silver salt.

$$AuR_2Br + AgX \rightarrow AuR_2X + AgBr$$

In all instances where the structure has been established, the Au(III) has planar four-fold coordination and polymerizes as appropriate to achieve this. The halides for example are dimeric but with the cyanide, which forms linear rather than bent bridges, tetramers are formed.

7.17.4 Major uses

Gold has been extensively used globally as money, for effective indirect exchange, and to store wealth in hoards. For exchange uses, mints provide standardized gold bullion coins, bars and other units of fixed weight and purity. Bills (that mature into gold coin) and gold certificates (convertible into gold coin at the issuing bank) supplemented the circulating stock of gold standard money in nearly all 19th century industrial economies. In the lead-up to World War I the warring nations changed to fractional gold standards, inflating their currencies to pay for the war effort. After the war, the victorious countries, most particularly Britain, progressively restored gold-convertibility, but international flows of gold via bills of exchange continued to be embargoed; international shipments were made solely for bilateral trades or to pay war reparations. After World War II gold was substituted with a system of nominally convertible currencies related by fixed exchange rates following the Bretton Woods system. The Bretton Woods system of monetary management established the rules for commercial and financial relations among the United States, Canada, Western European countries, Australia, and Japan after the 1944 Bretton Woods Agreement. The Bretton Woods system was the first example of a fully negotiated monetary order intended to govern monetary relations among independent states. The chief features of the Bretton Woods system were an obligation for each country to adopt a monetary policy that maintained its external exchange rates within 1% by tying its currency to gold and the ability of the International Monetary Fund (IMF) to bridge temporary imbalances of payments. Gold standards and the direct convertibility of currencies to gold have been abandoned by national governments, starting in 1971 by the United States' refusal to redeem its dollars in gold. Fiat currency now fills most monetary roles (Fiat money is a currency without intrinsic value that has been established as money, often by government regulation. Fiat money does not have use value and has value only because a government maintains its value, or because parties engaging in exchange agree on its value. It was introduced as an alternative to commodity money and representative money. Commodity money is created from a good, often a precious metal such as gold or silver, which has uses other than as a medium of exchange (such a good is called a commodity). Representative money is like fiat money, but it represents a claim on a commodity (which can be redeemed to a greater or lesser extent)). Switzerland was the last country to tie its currency to gold; it supported 40% of its value up until the Swiss became an IMF member in 1999. Central banks still keep a part of their liquid reserves as gold in some form, and metals exchanges, for example, the London Bullion Market Association, continues to clear transactions denominated in gold, including future delivery contracts. Even though the gold stock increases by only 1%–2% annually, very little gold is irreversibly consumed. Inventory above ground is enough to satisfy many years of industrial and even artisan applications at current prices. While the prices of some Pt-group metals can be much higher, gold has long been thought of as the most desirable of the precious metals, and therefore its value has been used as the standard for many currencies. The ISO 4217 currency code of gold is XAU. Modern bullion coins for investment or collector purposes do not require good mechanical wear properties; they are typically fine gold at 24k, although the American Gold Eagle and the British gold sovereign are still minted in 22k (0.916) metal in historical tradition, and the South African Krugerrand, first released in 1967, is also 22k (0.916). The special issue Canadian Gold Maple Leaf coin comprises the highest purity gold of any bullion coin, at 99.999% or 0.99999, while the popular issue Canadian Gold Maple Leaf coin has a purity of 99.99%. In 2006, the United States Mint started minting the American Buffalo gold bullion coin with a purity of 99.99%.

Due to the softness of pure (24k) gold, it is typically alloyed with base metals for use in jewelry, improving its hardness and ductility, melting point, color and other properties. Alloys with lower karat rating, characteristically 22k (0.916), 18k (0.750), 14k (0.585), or 1 k (0.417), have higher percentages of Cu or other base metals or Ag or Pd in the alloy. Ni is toxic, and its release from nickel white gold is governed by law in Europe. Pd–Au alloys cost more than those using Ni. High-karat white gold alloys are better resistant to corrosion than are either pure silver or sterling silver. The Japanese craft of Mokume-gane is based on the color contrasts between laminated colored gold alloys to produce decorative wood-grain effects. Gold solder is employed for joining the components of gold jewelry via high-temperature hard soldering or brazing. If the work is to be of hallmarking quality, the gold solder alloy must be of the same fineness (purity) as the rest of the jewelry, and alloy formulas are produced to color-match yellow and white gold. Gold solder is typically produced in at least three melting point ranges known as Easy, Medium and Hard. By applying the hard, high-melting point solder first, followed by solders with increasingly lower melting points, goldsmiths can create complex items with several separate soldered joints. In addition, gold can be produced in thread form and used in embroidery.

Only 10% of the global use of new gold produced goes to industry, but by far the most significant industrial use for new gold is in the manufacture of corrosion-free electrical connectors in computers and other electrical devices. For instance, as stated by the World Gold Council, a representative cell phone may have 50 mg of gold, worth about 50 cents. But since almost one billion cell phones are produced each year, this adds up to $500 million in gold from just this application alone. Though Au is attacked by free Cl_2, its good conductivity and general resistance to oxidation and corrosion in other environments (including resistance to nonchlorinated acids) has resulted in its widespread industrial use in the electronic era as a thin-layer coating on electrical connectors, thus guaranteeing good connection. For instance, Au is employed in the connectors of the more high-end electronics cables, such as audio, video and USB cables. The advantage of utilizing Au over other connector metals such as Sn in these applications has been a matter of discussion; Au connectors are often criticized by audio-visual experts as pointless for most people and seen as just a marketing ploy. But, the use of Au in other uses in electronic sliding contacts in highly humid or corrosive atmospheres, and in use for contacts with a very high failure cost (certain computers, communications equipment, spacecraft, jet aircraft engines) is still very common. In addition to sliding electrical contacts, Au is also employed in electrical contacts due to its resistance to corrosion, electrical conductivity, ductility and lack of toxicity. Switch contacts are usually exposed to more intense corrosion stress than are sliding contacts. Fine Au wires are employed to connect semiconductor devices to their packages via a technique called wire bonding. The concentration of free electrons in Au metal is 5.91×10^{22} cm^{-3}. Gold is highly conductive to electricity and has been utilized for electrical wiring in certain high-energy applications (only Ag and Cu are more conductive per volume, but Au has the advantage of corrosion resistance). For instance, Au electrical wires were employed during certain of the Manhattan Project's atomic experiments, but large high-current Ag wires were used in the calutron isotope separator magnets in the project. It has been projected that about 16% of global Au and 22% of global Ag is contained in electronics technology in Japan.

Metallic and Au compounds have long been employed for therapeutic applications. Gold, usually as the metal, is perhaps the oldest administered medicine (apparently by shamans) and known to Greek physician, pharmacologist, botanist, and author Pedanius Dioscorides (c.40−90 CE). In medieval times, Au was often believed to be advantageous for the health, in the belief that something so rare and beautiful could not be anything but healthy. Even some current esotericists and forms of alternative medicine ascribe metallic gold a healing power. During the 19th century Au had a reputation as a "nervine," a therapy for nervous disorders. Depression, epilepsy, migraine, and glandular problems such as amenorrhea and impotence were treated, and most notably alcoholism (Keeley, 1897). The apparent inconsistency of the actual toxicology of Au points to the possibility of serious gaps in the understanding of the action of Au in physiology. Only salts and radioisotopes of Au are of pharmacological value, since elemental (metallic) Au is inert to all chemicals it encounters inside the body (i.e., ingested Au cannot be attacked by stomach acid). Some Au salts do have anti-inflammatory properties and at present two are still used as pharmaceuticals in the treatment of arthritis and other related conditions in the United States (sodium aurothiomalate, $C_4H_4AuNaO_4S$, and auranofin, $C_{20}H_{34}AuO_9PS^0$). These drugs have been explored to help to reduce the pain and swelling of rheumatoid arthritis, and in addition (historically) against tuberculosis and some parasites. Gold alloys are employed in restorative dentistry, in particular in tooth restorations, for example, crowns and permanent bridges. The gold alloys' slight malleability helps in the formation of a much better molar mating surface with other teeth and produces results that are usually more satisfactory than those produced by the creation of porcelain crowns. The use of gold crowns in more prominent teeth such as incisors is favored in some cultures and discouraged in others.

Colloidal Au preparations (suspensions of Au nanoparticles) in water are intensely red-colored and can be produced with closely controlled particle sizes up to a few tens of nanometers across via reduction of gold chloride (chloroauric acid, $HAuCl_4$) with citrate $\left(C_3H_5O(COO)_3^{3-}\right)$ or ascorbate (ascorbic acid, $C_6H_8O_6$) ions. Colloidal Au is employed in research applications in medicine, biology and materials science. The technique of immunogold labeling uses the capability of the gold nanoparticles to adsorb protein molecules onto their surfaces. Colloidal Au nanoparticles coated with specific antibodies can be employed as probes for the existence and position of antigens on the surfaces of cells. In ultrathin sections of tissues viewed by electron microscopy, the immunogold labels are visible as extremely dense round spots at the position of the antigen. Gold, or alloys of Au and Pd, are used as conductive coating to biological specimens and other nonconducting materials such as plastics, minerals, polished rock sections and glass to be viewed in a scanning electron microscope. The coating, which is usually applied by sputtering with an Ar plasma, has a triple role in this application. Its very high electrical conductivity drains electrical charge to earth, and its very high density provides stopping power for electrons in the electron beam, helping to limit the depth to which the electron beam penetrates the specimen. This increases definition of the position and topography of the specimen surface and improves the spatial resolution of the image produced. Gold also creates a high output of secondary electrons when irradiated by an electron beam, and these low-energy electrons are the most frequently used signal source used in the scanning electron microscope. The isotope ^{198}Au (half-life 2.7 days) is employed, in nuclear medicine, in some cancer treatments and for treating other diseases.

Gold can be utilized in food and has the E number 175. Back in 2016, the European Food Safety Authority printed an opinion on the re-evaluation of gold used as a food additive. Concerns comprised the probable presence of trace quantities of gold nanoparticles in the food additive, and that these nanoparticles have been proven to be genotoxic in mammalian cells in vitro. Gold leaf, flake, or dust is used on and in some gourmet foods, particularly sweets and drinks as decorative ingredient. Gold flake was employed by the nobility in medieval Europe as an embellishment in food and drinks, in the form of leaf, flakes or dust, either to establish the host's wealth or in the belief that something that expensive and rare must be good for one's health. Danziger Goldwasser (German for Gold water of Danzig) or Goldwasser (German for Goldwater) is a traditional German herbal liqueur made in what is currently Gdańsk, Poland, and Schwabach, Germany, and has flakes of gold leaf. There are also some exclusive (c.$1000) cocktails which have flakes of gold leaf. Nevertheless, because metallic gold is inert to all body chemistry, it has no taste, it provides no nutrition, and it leaves the body intact. Vark is a foil consisting of a pure metal that is occasionally gold and is employed for garnishing sweets in South Asian cuisine.

Gold creates a deep, powerful red color when employed as a coloring agent in cranberry glass. In photography, Au toners are utilized to shift the color of AgBr black-and-white prints toward brown or blue tones, or to improve their stability. Applied on sepia-toned prints, Au toners create red tones. Kodak reported formulas for various kinds of Au toners, which use gold as the chloride. Gold forms a good reflector of electromagnetic radiation such as infrared and visible light, as well as radio waves. It is employed, for example, for the protective coatings on numerous artificial satellites, in infrared protective faceplates in thermal-protection suits and astronauts' helmets, and in electronic warfare planes such as the EA-6B Prowler. Gold is applied as the reflective layer on some high-end CDs. Cars may utilize Au for heat shielding, for example, McLaren employs Au foil in the engine compartment of its F1 model. Gold can be produced so thin that it looks semitransparent. It is employed in certain aircraft cockpit windows for de-icing or anti-icing by passing electricity through it. The heat formed because of the resistance of the Au is enough to inhibit ice from forming. Gold is attacked by and dissolves in alkaline solutions of potassium or sodium cyanide (KCN and NaCN), to form the salt gold cyanide—a process that has been applied to extract metallic Au from ores in the cyanide process. The chemical reaction for the dissolution of gold, the "Elsner Equation," follows

$$4Au(s) + 8NaCN(aq) + O_2(g) + 2H_2O(l) \rightarrow 4Na\left[Au(CN)_2\right](aq) + 4NaOH(aq)$$

(A paper published by L. Elsner in 1846 first correctly identified the chemical reaction) In this redox process, oxygen removes, via a two-step reaction, one electron from each gold atom to form the complex $Au(CN)_2^-$ ion. Although cyanide is cheap, effective, and biodegradable, its high toxicity has led to new methods for extracting gold using less toxic reagents. Other extractants have been examined including thiosulfate ($S_2O_3^{2-}$), thiourea ($SC(NH_2)_2$), iodine/iodide (I_2), ammonia (NH_4OH), liquid mercury (Hg), and alpha-cyclodextrin (with six glucose units, IUPAC name cyclomaltohexaose, $C_{36}H_{60}O_{30}$). Challenges include reagent cost and the efficiency of gold recovery. Thiourea has been implemented commercially for ores containing stibnite (Sb_2S_3). In addition, gold cyanide is the electrolyte employed in commercial electroplating of Au onto base metals and electroforming. Gold, when dispersed in nanoparticles, can serve as a heterogeneous catalyst of chemical reactions.

7.18 80 Hg — Mercury

7.18.1 Discovery

Mercury was discovered in Egyptian tombs that date back to 1500 BCE. In China and Tibet, mercury use was believed to prolong life, heal fractures, and maintain generally good health, while it is now known that exposure to mercury vapor leads to serious adverse health effects. The first emperor of China, Qín ShǐHuáng Dì (February 18, 259 BCE to September 10, 210 BCE)—supposedly buried in a tomb that contained rivers of flowing mercury on a model of the land he ruled, representative of the rivers of China—was killed by drinking a mercury and powdered jade mixture made by Qin alchemists (causing liver failure, mercury poisoning, and brain death) whose plan was to give him eternal life. The tomb was built at the foot of Mount Li, 30 km away from Xi'an. Modern archeologists have located the tomb and have inserted probes deep into it. The probes revealed abnormally high quantities of mercury, some 100 times the naturally occurring rate, suggesting that some parts of the legend are credible. Secrets were maintained, as most of the workmen who built the tomb were killed. Khumarawayh ibn Ahmad ibn Tulun (864 to January 18, 896 CE), the second Tulunid ruler of Egypt (ruled 884–896), notorious for his overindulgence and wastefulness, apparently built a basin filled with mercury, on which he would lie on top of air-filled cushions and be rocked to sleep. In November 2014 "large quantities" of mercury were discovered in a chamber 60 feet below the 1800-year-old pyramid known as the "Temple of the Feathered Serpent," "the third largest pyramid of Teotihuacan,", a pre-Columbian site in central Mexico along with "jade statues, jaguar remains, a box filled with carved shells and rubber balls." This structure is noteworthy partially due to the finding in the 1980s of more than a hundred possibly sacrificial victims found buried beneath the pyramid. The burials, like the pyramid, are dated to between 150 and 200 BCE. The ancient Greeks used cinnabar (HgS) in ointments; the ancient Egyptians and the Romans used it in cosmetics. In Lamanai (in the north of Belize), once a major city of the Maya civilization, a pool of mercury was found under a marker in a Mesoamerican ballcourt. By 500 BCE, mercury was used to make amalgams (Medieval Latin amalgama, "alloy of mercury") with other metals. Alchemists believed mercury to be the "First Matter" from which all metals were formed. They thought that different metals could be made by changing the quality and quantity of sulfur contained within the mercury. The purest of these was gold, and mercury was necessary in experiments to transmutate base (or impure) metals into gold, which was the ultimate goal of many alchemists. The deposits in

FIGURE 7.97 Cinnabar from Almaden, Spain. Hand-colored copper-plate engraving (1811) depicting actual-size a 3.7-inch specimen in the collection of the Royal Institute, donated by I. G. Children. Published as Plate 18, Volume 1, of Exotic Mineralogy by James Sowerby (1757–1822).

Almadén (Spain), Monte Amiata (Italy) (Fig. 7.97), and Idrija (now Slovenia) dominated mercury mining from the opening of the mine in Almadén 2500 years ago (closed in 2002), until new deposits were found at the end of the 19th century. Hg is the modern chemical symbol for mercury. It is derived from the word hydrargyrum, a Latinized form of the Greek word ύδράργυρος (hydrargyros), which is a compound word meaning "water-silver" (from ύδρ- hydr-, the root of ὕδωρ, "water," and ἄργυρος argyros "silver") —because it is liquid like water and shiny like silver. The element was named after the Roman god Mercury, famous for his speed and mobility. It is associated with the planet Mercury; the astrological symbol for the planet is also one of the alchemical symbols for the metal; the Sanskrit word for alchemy is Rasavātam which means "the way of mercury." Mercury is the only metal for which the alchemical planetary name became the common name.

7.18.2 Mining, production, and major minerals

Mercury is an extremely rare element in Earth's crust, having an average crustal abundance of only 0.08 ppm by mass. Since it does not blend geochemically with those elements that constitute the majority of the crustal mass, mercury ores can be extremely concentrated considering the element's average crustal abundance. The richest mercury ores contain up to 2.5% mercury by mass, and even the leanest concentrated deposits are at least 0.1% mercury (12,000 × average crustal abundance). It is found either as a native metal (rare) or in cinnabar (HgS), metacinnabar (HgS), corderoite ($Hg_3S_2Cl_2$), livingstonite ($HgSb_4S_8$) and other minerals, with cinnabar (HgS) being the most common ore. Mercury ores typically occur in very young orogenic belts where rocks of high density are forced to the crust of the Earth, often in hot springs or other volcanic regions (Box 7.7). Beginning in 1558, with the invention of the patio process to extract Ag from ore using Hg, Hg became an essential resource in the economy of Spain and its American colonies. The process was invented by Spanish merchant and mining specialist Bartolomé de Medina (born around 1504 in Seville) in Pachuca, Mexico, in 1554. Silver ores were crushed (typically either in "arrastras" or stamp mills) to a fine slime which was mixed with salt, water, magistral (basically an impure form of copper sulfate), and Hg, and spread in a 1-to-2-foot-thick (0.30–0.61 m) layer in a patio, (a shallow-walled, open enclosure). Horses were driven around on the patio to further mix the ingredients. After weeks of mixing and soaking in the sun, a complex reaction converted the silver to native metal, which formed an amalgam with the mercury and was recovered. The amount of salt and copper sulfate varied from one-quarter to ten pounds of one or the other, or both, per ton of ore treated. The decision of how much of each ingredient to add, how much mixing was needed, and when to halt the process depended on the skill of an azoguero (English: quicksilver man). The loss of Hg in amalgamation processes was usually one to two times the weight of silver recovered. Mercury was used to extract silver from the lucrative mines in New Spain and Peru. Initially, the

BOX 7.7 Mercury deposits.

Mercury frequently forms deposits on the margins of larger hydrothermal systems, related to its relatively high solubility in low-temperature fluids, especially alkaline ones. It also easily vaporizes and therefore can be carried in vapors that rise above boiling hydrothermal fluids to condense in overlying cooler groundwater. Most Hg deposits formed at temperatures below 200°C from dilute, meteoric waters and some deposits even contain petroleum residues. On a worldwide scale, they are most commonly found in areas of young volcanism at convergent tectonic margins. For example, in north-western Nevada mercury deposits are present in the McDermitt caldera formed by explosive rhyolitic volcanism only several million years ago, while the Monte Amiata Hg district in Italy is on the slopes of a young volcano. Interestedly, the world's largest Hg deposits at Almadén, Spain, do not follow these generalizations. Mercury at Almadén, which hosts more than one-third of global reserves, is present principally in pores in an extensive sandstone layer. There is no visible presence of a large hydrothermal system in the area and igneous rocks are also absent. It has been proposed that this Hg deposit formed when Hg-bearing hot springs flowed out on the sea floor, they could be associated with a buried hydrothermal system. Mercury is also recovered as a by-product of gold and some base metal ores. The large McLaughlin gold mine in the Napa Valley, California, was discovered beneath a Hg mine. Often, Hg minerals are not observed in the ores, and much of the Hg may substitute for other metals in sulfide minerals, in particular sphalerite (ZnS).

FIGURE 7.98 Cinnabar, HgS, cherry red crystal to 5 mm among snow white rhombic dolomite, $CaMg(CO_3)_2$. Wanshan mine, Tongren Pref., Guizhou prov., China.

Spanish Crown's mines in Almadén in Southern Spain provided all the mercury for the colonies. New mercury deposits were found in the New World, and more than 100,000 tons of Hg were produced from the region of Huancavelica, Peru, during a period of three centuries following the discovery of deposits there in 1563. The patio process and later the pan amalgamation process continued to create great demand for Hg to treat Ag ores until the late 19th century. Former mines in Italy, the United States and Mexico, which once produced a large percentage of the world supply, have now been entirely exhausted or, in the case of Slovenia (Idrija) and Spain (Almadén), shut down due to the fall of the price of mercury. Nevada's McDermitt Mine, the last operating Hg mine in the United States, stopped production in 1992. Mercury is extracted by heating cinnabar (HgS) in a current of air and condensing the vapor.

$$HgS + O_2 \rightarrow Hg + SO_2$$

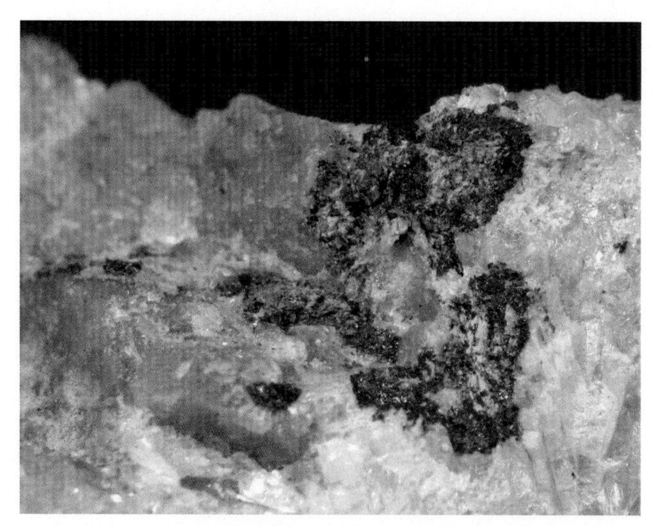

FIGURE 7.99 Specimen with large black crystal grains to 5 mm of coloradoite, HgTe, with green crusts of tlapallite, $(Ca,Pb)_3CaCu_6[Te^{4+}{}_3Te^{6+}O_{12}]_2(Te^{4+}O_3)_2(SO_4)_2\cdot 3H_2O$. Bambollita mine, Moctezuma, Sonora, Mexico.

FIGURE 7.100 Galkhaite, $(Cs,Tl)(Hg,Cu,Zn)_6(As,Sb)_4S_{12}$, several cherry red glassy crystals to less than 0.3 mm. Getchell mine, Humboldt Co., Nevada, United States.

There are about 100 minerals that contain mercury in their chemical composition. Mercury as a native metal is a liquid at room temperature but is considered a mineral. There are 12 alloys known, for example, eugenite ($Ag_{11}Hg_2$), moschellandsbergite (Ag_2Hg_3), and potarite (PdHg). The sulfides are represented by around 45 minerals, such as cinnabar (HgS) (Fig. 7.98), coloradoite (HgTe) (Fig. 7.99), galkhaite ($(Cs,Tl)(Hg,Cu,Zn)_6(As,Sb)_4S_{12}$) (Fig. 7.100), imiterite (Ag_2HgS_2) (Fig. 7.101), metacinnabar (HgS) and perroudite ($Hg_5Ag_4S_5(I,Br)_2Cl_2$) (Fig. 7.102). The halides contain around 25 different minerals with Hg, for example, calomel ($(Hg_2)^{2+}Cl_2$) (Fig. 7.103), coccinite (HgI_2), eglestonite ($(Hg_2^{2+})_3OCl_3(OH)$), kleinite ($(Hg_2N)(Cl,SO_4)\cdot nH_2O$) (Fig. 7.104) and terlinguaite ($(Hg_2^{2+})Hg_2^{2+}Cl_2O_2$). Only three oxides, magnolite ($(Hg^{2+})(Te^{4+}O_3)$), montroydite (HgO), and shakhovite ($(Hg^{2+})Hg_2^{2+}(Sb^{3+}O_3)(OH)_3$), are known to contain Hg. The same number of carbonates with Hg are known: clearcreekite ($(Hg^{2+})_{1.5}(CO_3)(OH)\cdot 2H_2O$), peterbaylissite ($(Hg^{2+})_{1.5}(CO_3)(OH)\cdot 2H_2O$), and szymańskiite ($Hg_{16}(Ni,Mg)_6(CO_3)_{12}(OH)_{12}(H_3O)_8\cdot 3H_2O$). The sulfate class contains six different minerals, for example, deanesmithite ($(Hg^{2+})Hg_3^{2+}(CrO_4)S_2O$), edoylerite ($Hg_3^{2+}(CrO_4)S_2$) (Fig. 7.105), schuettite ($Hg_3^{2+}(SO_4)O_2$), and wattersite ($(Hg^{2+})_2Hg^{2+}(CrO_4)O_2$). The same number of phosphate class minerals is known to contain Hg, for example, artsmithite ($(Hg^{2+})_2Al(PO_4)_{1.74}(OH)_{1.78}$) (Fig. 7.106), chursinite

FIGURE 7.101 Imiterite, Ag_2HgS_2, silvery needles with rounded terminations to about 1mm in calcite, $CaCO_3$, vugs. Imiter mine, Anti-Atlas Mountains, Dades Region, Morocco.

FIGURE 7.102 Perroudite, $Hg_5Ag_4S_5(I,Br)_2Cl_2$, crystal sprays of well-defined acicular crystals to 0.75 mm radiating out from the center. Cap Garonne mine, Le Pradet, Var, France.

$((Hg^{2+})_{0.5}Hg^{2+}(AsO_4))$, kuznetsovite $((Hg^{2+})Hg^{2+}(AsO_4)Cl)$, and tillmannsite $((Ag_3Hg)(VO_4,AsO_4))$. The only silicate is edgarbaileyite $((Hg\text{-}Hg)_3Si_2O_7)$.

7.18.3 Chemistry

There are seven naturally occurring stable isotopes of Hg, with ^{202}Hg being the most abundant (29.74%), followed by ^{200}Hg (23.14%), ^{199}Hg (16.94%), ^{198}Hg (10.04%), ^{204}Hg (6.82%), and ^{196}Hg (0.15%). The longest-lived synthetic radioisotopes are ^{194}Hg with a half-life of 444 years and ε decay to ^{194}Au, and ^{203}Hg with a half-life of 46.612 days and β^- to ^{203}Tl decay. Most of the other radioisotopes have half-lives less than a day (except for ^{197}Hg, which has a half-life of 64.14 h). ^{199}Hg and ^{201}Hg are the most frequently used NMR-active nuclei, with spins of 1/2 and 3/2, respectively.

Mercury is a heavy, silvery-white liquid metal. Relative to other metals, it is a poor conductor of heat, but a reasonable conductor of electricity. It has a freezing point of −38.83°C and a boiling point of 356.73°C (Table 7.18),

FIGURE 7.103 Grayish, waxy masses of calomel, $(Hg_2)^{2+}Cl_2$, altered from red cinnabar, HgS, about 5 mm. Palatinate, Germany.

FIGURE 7.104 Kleinite, $(Hg_2N)(Cl,SO_4)\cdot nH_2O$, short prismatic orange-yellow crystals to 2 mm in nice clusters. McDermitt mine, Humboldt Co., Nevada, United States.

both the lowest of any stable metal, even though initial experiments on copernicium and flerovium have suggested that they have even lower boiling points (copernicium being the element below mercury in the periodic table, following the trend of decreasing boiling points down group 12). Upon freezing, the volume of mercury decreases by 3.59% and its density changes from 13.69 g/cm^3 when liquid to 14.184 g/cm^3 when solid. Solid mercury is malleable and ductile and can be cut with a knife.

A complete description of mercury's extreme volatility can only be obtained from a deep understanding of its quantum physics, but it can be summarized as follows: mercury has a unique electron configuration where electrons fill up all the available 1s, 2s, 2p, 3s, 3p, 3d, 4s, 4p, 4d, 4f, 5s, 5p, 5d, and 6s subshells. Since this configuration firmly resists removal of an electron, mercury behaves comparable to the noble gases, which form weak bonds and therefore melt at low temperatures. The stability of the 6s shell is caused by the presence of a filled 4f shell. An f shell inadequately screens the nuclear charge that increases the attractive Coulomb interaction of the 6s shell and the nucleus. The absence of a filled inner f shell is the explanation for the somewhat higher melting temperature of cadmium and zinc in group 12, even though both cadmium and zinc still melt easily and also have unusually low boiling points. Mercury exists in two oxidation states, I and II.

FIGURE 7.105 Yellow dendritic edoylerite, $Hg_3^{2+}(CrO_4)S_2$, in a 3 × 5 mm area on cinnabar, HgS. Clear Creek mine, San Benito Co., California, United States.

FIGURE 7.106 The 2-mm cluster of creamy white artsmithite, $(Hg^{2+})_2Al(PO_4)_{1.74}(OH)_{1.78}$, is pseudomorph after former fluorapatite, $Ca_5(PO_4)_3F$. Funderburk prospect, Cowhide Cove, Pike Co., Arkansas, United States.

Mercury does not react with most acids, for example, dilute sulfuric acid, while oxidizing acids, for example, concentrated sulfuric acid and nitric acid or aqua regia, dissolve it to form sulfate, nitrate, and chloride. Similar to silver, mercury reacts with atmospheric hydrogen sulfide. Mercury reacts with solid sulfur flakes, which are employed in mercury spill kits to absorb mercury (spill kits also utilize activated carbon and powdered zinc).

Mercury dissolves numerous metals such as gold and silver to form amalgams. Iron is an exception, and iron flasks have historically been utilized to trade mercury. A number of other first row transition metals with the exception of manganese, copper and zinc are also resistant to forming amalgams. Another element that does not easily form amalgams with mercury is platinum. Sodium amalgam is an often used reducing agent in organic synthesis and is also employed in high-pressure sodium lamps. Mercury easily reacts with aluminum to produce a mercury-aluminum amalgam when the two pure metals come into contact. Because the amalgam destroys the aluminum oxide layer which shields metallic aluminum from oxidizing in-depth (as in iron rusting), even trace amounts of mercury can badly corrode aluminum. Therefore mercury is not permitted on an aircraft under most circumstances due to the risk of it

TABLE 7.18 Mercury properties.

Appearance	Silvery
Standard atomic weight $A_{r,std}$	200.592
Block	d-Block
Element category	Posttransition metal, alternatively considered a transition metal
Electron configuration	[Xe] $4f^{14}$ $5d^{10}$ $6s^2$
Phase at STP	Liquid
Melting point	−38.829°C
Boiling point	356.73°C
Density (near r.t.)	13.534 g/cm^3
Heat of fusion	2.29 kJ/mol
Heat of vaporization	59.11 kJ/mol
Molar heat capacity	27.983 J/(mol·K)
Oxidation states	−2, +1 (mercurous), **+2** (mercuric)
Ionization energies	1st: 1007.1 kJ/mol
	2nd: 1810 kJ/mol
	3rd: 3300 kJ/mol
Atomic radius	Empirical: 151 pm
Covalent radius	132 ± 5 pm

STP, Standard temperature and pressure.
Bold font indicates main oxidation state.

forming an amalgam with exposed aluminum parts in the aircraft. Mercury embrittlement is the most common type of liquid metal embrittlement.

7.18.3.1 Compounds of mercury(I)

In contrast to its lighter neighbors in group 12, cadmium and zinc, mercury typically forms simple stable compounds with metal−metal bonds. Most mercury(I) compounds are diamagnetic and have the dimeric cation, Hg_2^{2+}. Stable derivatives consist of, for example, the chloride and nitrate. Reaction of Hg(I) compounds with strong ligands, sulfide, cyanide, etc., causes disproportionation to Hg^{2+} and elemental mercury. Mercury(I) oxide, also known as mercurous oxide, Hg_2O, is a brown/black powder, insoluble in water, toxic but without taste or smell. It is chemically unstable and reacts to mercury(II) oxide and mercury metal.

7.18.3.1.1 Halides

Mercury(I) chloride, a colorless solid also known as the mineral calomel, is the compound with the formula Hg_2Cl_2, with the connectivity Cl-Hg-Hg-Cl. It reacts with chlorine to form mercuric chloride, which resists further oxidation. Hg_2Cl_2 is a linear molecule. Mercurous chloride forms through the reaction of elemental mercury and mercuric chloride.

$$Hg + HgCl_2 \rightarrow Hg_2Cl_2$$

It can also be produced via a metathesis reaction of aqueous mercury(I) nitrate with various chloride sources such as NaCl or HCl.

$$2HCl + Hg_2(NO_3)_2 \rightarrow Hg_2Cl_2 + 2HNO_3$$

The reaction with ammonia results in Hg_2Cl_2 disproportionation.

$$Hg_2Cl_2 + 2NH_3 \rightarrow Hg + Hg(NH_2)Cl + NH_4Cl$$

Mercurous chloride is used widely in electrochemistry, taking advantage of the ease of its oxidation and reduction reactions. The calomel electrode is a reference electrode, in particular in older publications. Over the past 50 years, it has been replaced by the silver/silver chloride (Ag/AgCl) electrode. While the mercury electrodes have been generally abandoned because of the dangerous nature of mercury, numerous chemists still consider them to be more accurate and not dangerous as long as they are handled properly. The differences in experimental potentials differ little from literature values. Other electrodes can vary by 70 to 100 mV. Mercurous chloride decomposes into mercury(II) chloride and elemental mercury upon exposure to UV light.

$$Hg_2Cl_2 \rightarrow HgCl_2 + Hg$$

The formation of Hg can be employed to determine the number of photons in a light beam, by the method of actinometry. By using a light reaction in the presence of mercury(II) chloride and ammonium oxalate, mercury(I) chloride, ammonium chloride and carbon dioxide are formed.

$$2HgCl_2 + (NH_4)_2C_2O_4 + Light \rightarrow Hg_2Cl_2(s) + 2[NH_4^+][Cl^-] + 2CO_2$$

This reaction was discovered by Austrian chemist Joseph Maria Eder (March 16, 1855 to October 18, 1944) (thus the name Eder reaction) in 1880 and reinvestigated by American chemist William Earl Roseveare in 1929 (Roseveare, 1930).

Mercury(I) fluoride or mercurous fluoride, Hg_2F_2, consists of small yellow cubic crystals, which turn black when exposed to light. Mercury(I) fluoride is formed by the reaction of mercury(I) carbonate with hydrofluoric acid.

$$Hg_2CO_3 + 2HF \rightarrow Hg_2F_2 + CO_2 + H_2O$$

When added to water, mercury(I) fluoride hydrolyzes to elemental liquid mercury, mercury(II) oxide, and hydrofluoric acid.

$$Hg_2F_2 + H_2O \rightarrow Hg + HgO + 2HF$$

It can be used in the Swarts reaction to react alkyl halides into alkyl fluorides.

$$2R{-}X + Hg_2F_2 \rightarrow 2R{-}F + Hg_2X_2 \ (X = Cl,\ Br,\ I)$$

Mercury(I) bromide or mercurous bromide, Hg_2Br_2, is a compound that changes color from white to yellow when heated and fluoresces a salmon color under UV light. It is used in acousto-optical devices. A very rare mineral is kuzminite, $Hg_2(Br,Cl)_2$. Mercury(I) bromide is formed through the oxidation of elemental mercury with elemental bromine or via the addition of sodium bromide to a solution of mercury(I) nitrate. Mercury(I) iodide, Hg_2I_2, is photosensitive and decomposes readily to mercury and HgI_2. Mercury(I) iodide can be formed through directly reacting mercury and iodine.

7.18.3.1.2 *Hydrides and chalcogenides*

Mercury(I) hydride, a colorless gas, has the formula HgH, containing no Hg-Hg bond. It has not yet been obtained in bulk; hence its bulk properties are still unknown. Nevertheless, molecular mercury(I) hydrides with the formulae HgH and Hg_2H_2 have been separated in solid gas matrices. These molecular hydrides are extremely unstable toward thermal decomposition. Mercury(I) hydride is the heaviest group 12 monohydride. The composition of mercury(I) hydride is 0.50% hydrogen and 99.50% mercury. In mercury(I) hydride, the formal oxidation states of hydrogen and mercury are -1 and $+1$, respectively, as the electronegativity of mercury is lower than that of hydrogen. The stability of metal hydrides with the formula MH (M = Zn-Hg) increases as the atomic number of M increases. The Hg-H bond is very weak and consequently the compound has only been matrix isolated at temperatures below $-267°C$ (6K). The dihydride, HgH_2, has also been observed this way. A related compound is bis(hydridomercury)(Hg—Hg) with the formula Hg_2H_2, which can be thought of as dimeric mercury(I) hydride. It spontaneously decomposes into the monomeric form. The mercury center in mercury complexes such as hydridomercury can accept or donate a single electron by association.

$$HgH + R \rightarrow HHgR$$

Due to this acceptance or donation of the electron, hydridomercury has a radical character. It is a moderately reactive monoradical.

Mercury(I) sulfide or mercurous sulfide is a hypothetical chemical compound of mercury and sulfur, with elemental formula Hg_2S. Its existence has been questioned; it may be stable below 0°C or in suitable environments, but is unstable at room temperature, decomposing into metallic mercury and mercury(II) sulfide. This compound was described in the 19th century by Jöns Jacob Berzelius (August 20, 1779 to August 7, 1848) as a black precipitate formed by passing H_2S through solutions of mercury(I) salts. However, its existence was disputed in 1816 by French pharmacist Nicolas-Jean-Baptiste-Gaston Guibourt (July 2, 1790 to August 22, 1867). In his thesis he stated that the precipitate formed in such a manner was nothing more than an intimate mixture of mercury(II) sulfide HgS and metallic mercury Hg_2, which could be separated by heating or grinding. (He also denied the reality of mercurous oxide Hg_2O, for the same reason.) Reviewing Guibourt's article in 1825, British chemist William Thomas Brande (January 11, 1788 to February 11, 1866) doubted his conclusions. He observed that the proportions of mercury and sulfur in the precipitate were stoichiometric for the formula Hg_2S; and that nitrogen triiodide, silver fulminate, and mercury fulminate were accepted compounds, even though they were decomposed by slight friction. He stated that the black precipitate did not exhibit any sign of metallic mercury or cinnabar (though it was easily decomposed into them). He also stated that hot nitric acid did not attack cinnabar, whereas it quickly turns precipitated "mercurous sulfide" to mercuric nitrate without leaving any residue. In 1894 Italian chemists Ubaldo Antony and Quirino Sestini claimed to have found that mercurous sulfide was stable at $-10°C$, but disproportionated into Hg_2 and HgS when heated to 0°C. New insights that might result in the successful synthesis of Hg_2S has been coming to light since 1958 through the research of Klaus Brodersen and others (e.g. Brodersen et al., 1975). The reaction between dimercury(I) salts and Lewis bases in polar solvents normally breaks the Hg–Hg bond. The successful formation of S–Hg–Hg–S compounds can be accomplished with nonpolar solvents, weak Lewis bases, and NH acidic nitrogen compounds.

7.18.3.1.3 Sulfate

Mercury(I) sulfate, commonly called mercurous sulfate, Hg_2SO_4, is a metallic compound that forms a white, pale yellow or beige powder. It is a metallic salt of sulfuric acid produced by replacing both hydrogen atoms with mercury(I). It is very toxic; it can be fatal if inhaled, ingested, or absorbed by skin. The crystal structure of mercurous sulfate is made up of Hg_2^{2+} dumbbells and SO_4^{2-} anions as main building units. The Hg_2^{2+} dumbbell is surrounded by 4 oxygen atoms with Hg-O distance between 2.23 and 2.93 Å, while the Hg-Hg distance is about 2.50 Å. Mercury(I) sulfate has the mercury atoms arranged in doublets with a bond distance of 2.50 Å. The metal atom doublets are oriented parallel to the *a*-axis in a unit cell. Mercury doublets form part of infinite chain SO_4-Hg-Hg-SO_4-Hg-Hg-... The Hg-Hg-O bond angle is 165° ± 1°. The chain crosses the unit cell diagonally. The mercury sulfate structure is held together by weak Hg-O interactions. The SO_4 does not act as a single anion, but rather coordinated to the mercury metal. One method to produce mercury(I) sulfate is to react the acidic solution of mercury(I) nitrate with a concentrated sulfuric acid solution.

$$Hg_2(NO_3)_2 + H_2SO_4 \rightarrow Hg_2SO_4 + 2HNO_3$$

It can also be formed by reacting an excess of mercury with concentrated sulfuric acid.

$$2Hg + 2H_2SO_4 \rightarrow Hg_2SO_4 + 2H_2O + SO_2$$

7.18.3.2 Compounds of mercury(II)

Mercury(II) is the most common oxidation state and is the main one found in nature as well. Mercury(II) oxide, the main oxide of mercury, forms when the metal is exposed to air for long periods at elevated temperatures. Mercury(II) oxide, also known as mercuric oxide or simply mercury oxide, has a formula of HgO. It has a red or orange color. Mercury(II) oxide is a solid at room temperature and pressure. The mineral montroydite is very rarely found. It transforms to the elements upon heating near 400°C, as was shown by English chemist Joseph Priestley (March 24, 1733 to February 6, 1804) in an early synthesis of pure oxygen. The red form of HgO can be produced by heating Hg in oxygen at about 350°C, or by pyrolysis of $Hg(NO_3)_2$. The yellow form can be prepared through precipitation of aqueous Hg^{2+} with alkali. The difference in color is caused by the particle size, both forms have the same structure consisting of near linear O-Hg-O units linked in zigzag chains with an Hg-O-Hg angle of 108°. Under atmospheric pressure mercuric oxide has two crystalline phases: one is called montroydite (orthorhombic, *2/m2/m2/m*, *Pnma*), and the second is analogous to the sulfide mineral cinnabar (hexagonal, *hP6*, *P3221*); both are characterized by Hg-O chains. At pressures over 10 GPa both structures transform to a tetragonal phase. Hydroxides of mercury are poorly characterized. Mercury(II) hydroxide or mercury dihydroxide is an inorganic metal hydroxide with the chemical formula HgH_2O_2 or $Hg(OH)_2$. It is impossible to make the compound by adding OH^- to Hg^{2+} in aqueous solution, since this simply precipitates yellow solid HgO, even though it is possible that $Hg(OH)_2$ is momentarily formed during the reaction. The first experimental proof for the existence of the molecule was reported by Wang and Andrews in 2005. They formed it through irradiating a frozen mixture of mercury, oxygen and hydrogen with light from a mercury arc lamp. The mixture had been prepared by evaporating mercury atoms at 50°C into a gas consisting of neon, argon or deuterium (in separate experiments) plus 2%–8% hydrogen and 0.2%–2.0% oxygen. This mixture was subsequently condensed at 5K onto a cesium iodide window, through which it could be irradiated.

7.18.3.2.1 Halides

All four mercuric halides are known. They form tetrahedral complexes with other ligands but the halides have a linear coordination geometry. Best known is mercury(II) chloride or mercuric chloride (historically "corrosive sublimate"), an easily sublimating white solid. $HgCl_2$ forms coordination complexes that are typically tetrahedral, such as. $HgCl_4^{2-}$. It is a laboratory reagent and a molecular compound that is extremely toxic to humans. Once used as a treatment for syphilis, it is no longer used for medicinal purposes due to mercury toxicity and the availability of superior treatments. Mercuric chloride occurs not as a salt consisting of discrete ions, but rather is composed of linear triatomic molecules, hence its tendency to sublime. In the crystal form, each mercury atom is bonded to two close chloride ligands with Hg-Cl distance of 2.38 Å; six more chlorides are more distant at 3.38 Å. Mercuric chloride is formed by the reaction of chlorine with mercury or mercury(I) chloride, through the addition of hydrochloric acid to a hot, concentrated solution of mercury(I) compounds such as the nitrate.

$$HgNO_3 + 2HCl \rightarrow HgCl_2 + H_2O + NO_2$$

Heating a mixture of solid mercury(II) sulfate and sodium chloride also results in volatile $HgCl_2$, which sublimes and condenses in the form of small rhombic crystals. Its solubility increases from 6% at 20°C to 36% at 100°C. In the presence of chloride ions, it dissolves to give the tetrahedral coordination complex $[HgCl_4]^{2-}$.

Mercury(II) fluoride, HgF_2, is most commonly prepared through the reaction of mercury(II) oxide with hydrogen fluoride.

$$HgO + 2HF \rightarrow HgF_2 + H_2O$$

It can also be produced through the fluorination of mercury(II) chloride

$$HgCl_2 + F_2 \rightarrow HgF_2 + Cl_2$$

or of mercury(II) oxide

$$2HgO + 2F_2 \rightarrow 2HgF_2 + O_2$$

with oxygen as by-product. Mercury(II) fluoride is known as a selective fluorination agent.

Mercury(II) bromide or mercuric bromide is the chemical compound $HgBr_2$. This white crystalline solid is a laboratory reagent. Similar to mercury(II) chloride, it is highly toxic. Mercury(II) bromide can be prepared by: adding potassium bromide to a solution of mercuric salt and crystallizing; by precipitation using a mercury(II) nitrate and sodium bromide solution; or by dissolving mercury(II) oxide in hydrobromic acid. Also, mercury(II) bromide can be formed by reacting mercury with bromine. Mercury(II) bromide is used as a reagent in the Koenigs–Knorr reaction (named after German chemist Wilhelm Koenigs (April 22, 1851 to December 16,1906) and Edward Knorr, a student of Koenigs), which forms glycoside linkages on carbohydrates. It is also employed to test for the presence of arsenic. The arsenic in the sample is first converted to arsine gas by treatment with hydrogen. Arsine reacts with mercury(II) bromide.

$$AsH_3 + 3HgBr_2 \rightarrow As(HgBr)_3 + 3HBr$$

The white mercury(II) bromide will turn yellow, brown, or black if arsenic is present in the sample. Mercury(II) bromide reacts violently with elemental indium at high temperatures and, when exposed to potassium, can form shock-sensitive explosive mixtures.

Mercury(II) iodide, HgI_2, is typically formed synthetically but can also be found in nature as the extremely rare mineral coccinite. In contrast to the related mercury(II) chloride it is hardly soluble in water (<100 ppm). Mercury(II) iodide is prepared by adding an aqueous solution of potassium iodide to an aqueous solution of mercury(II) chloride with stirring; the precipitate is filtered off, washed, and dried at 70°C.

$$HgCl_2 + 2KI \rightarrow HgI_2 + KCl$$

Mercury(II) iodide displays thermochromism; when heated above 126°C it undergoes a phase transition, from the red α crystalline form to a pale yellow β form. As the sample cools, it slowly reacquires its original color. A third orange form is also known; this can be prepared by recrystallization and is also metastable, eventually converting back to the red α form. The various forms can exist in a diverse range of crystal structures and as a result mercury(II) iodide shows a remarkably complex phase diagram. Known as Nessler's reagent (named after German chemist Julius Neßler (Nessler), June 6, 1827 to March 19, 1905), potassium tetraiodomercurate(II) (HgI_4^{2-}) is still occasionally used to test for ammonia due to its tendency to form the deeply colored iodide salt of Millon's base.

7.18.3.2.2 Hydrides and chalcogenides

Mercury(II) hydride (systematically named mercurane(2) and dihydridomercury), HgH_2 (also written as $[HgH_2]$), is both thermodynamically and kinetically unstable at ambient temperature, and as such, little is known about its bulk properties. Nonetheless, it forms a white, crystalline solid, which is kinetically stable at temperatures below −125°C. Mercury(II) hydride is the second simplest mercury hydride (after mercury(I) hydride). Because of its instability, it has no practical industrial applications. Yet, in analytical chemistry, mercury(II) hydride is essential to certain forms of spectrometric techniques employed to determine mercury content. In addition, it is studied for its effect on high sensitivity isotope-ratio mass spectrometry methods which comprise mercury, such as MC-ICP-MS (Multicollector-Inductively Coupled Plasma-Mass Spectrometry), when used to compare thallium to mercury. In solid mercury(II) hydride, the HgH_2 molecules are linked through mercurophilic bonds. Trimers and a lower amount of dimers are detected in the vapor phase. In contrast to solid zinc(II), and cadmium(II) hydride, which are network solids, solid mercury(II) hydride is a covalently bound molecular solid. This is caused by relativistic effects, which also explain the relatively low decomposition temperature of −125°C. The HgH_2 molecule is linear and symmetric in the form H-Hg-H. The bond length is 1.646543 Å. The energy required to break the Hg-H bond in HgH_2 is 70 kcal/mol. The second bond in the resulting HgH is much weaker only requiring 8.6 kcal/mol to break. Reacting two hydrogen atoms releases 103.3 kcal/mol, and therefore HgH_2 formation from hydrogen molecules and Hg gas is endothermic at 24.2 kcal/mol (Shayesteh et al., 2005). Mercury(II) hydride may be formed by the reduction of mercury(II) chloride. To achieve this, mercury(II) chloride and a hydride salt equivalent react to form mercury(II) hydride according to the following equations, depending on the stoichiometry of the reaction.

$$2HgCl_2 + 2H^- \rightarrow HgCl_4^{2-} + HgH_2$$
$$HgCl_2 + 2H^- \rightarrow HgH_2 + 2Cl^-$$

Adaptations of this method exist where mercury(II) chloride is replaced by its heavier halide homologs. Mercury(II) hydride can also be formed by direct synthesis from the elements in the gas phase or in cryogenic inert gas matrices.

$$Hg \rightarrow Hg^*$$
$$Hg^* + H_2 \rightarrow [HgH_2]^*$$
$$[HgH_2]^* \rightarrow HgH_2$$

This necessitates excitation of the mercury atom to the ^1P or ^3P state, since atomic mercury in its ground state does not insert into the dihydrogen bond. Excitation is achieved using an ultraviolet-laser, or electric discharge. The initial yield is high; but, due to the product being in an excited state, a significant quantity dissociates quickly into mercury(I) hydride, then back into the initial reagents.

$$[HgH_2]^* \rightarrow HgH + H$$
$$HgH \rightarrow Hg + H$$
$$2H \rightarrow H_2$$

This is the preferred method for matrix isolation research. In addition to mercury(II) hydride, it also forms other mercury hydrides in lesser amounts, such as the mercury(I) hydrides (HgH and Hg_2H_2). Upon reaction with a Lewis base, mercury(II) hydride transforms to an adduct. Upon reaction with a standard acid, mercury(II) hydride and its adducts change either to a mercury salt or a mercuran(2)yl derivative and elemental hydrogen. Oxidation of mercury(II) hydride gives elemental mercury. Unless cooled below −125°C, mercury(II) hydride decomposes to elemental mercury and hydrogen.

$$HgH_2 \rightarrow Hg + H_2$$

Being a soft metal, mercury forms very stable derivatives with the heavier chalcogens. Well known is mercury (II) sulfide, HgS, which occurs in nature as the mineral cinnabar and forms the brilliant pigment vermilion. Similar to ZnS, HgS is dimorphic with two crystal forms: red cinnabar (α-HgS, trigonal, *hP6*, *P3221*), is the form in which mercury is most commonly found in nature. Black, metacinnabar (β-HgS), is less common in nature and adopts the zinc blende (T_d^2-$F\bar{4}3m$) crystal structure. Crystals of red, α-HgS, are optically active. This is due to the Hg-S helices in the structure. β-HgS is precipitated as a black powder when H_2S is bubbled through solutions of Hg(II) salts. β-HgS does not react with all but concentrated acids. Mercury metal is formed from the cinnabar ore by roasting in air and condensing the vapor. When α-HgS is used as a red pigment, it is called vermilion. The tendency of vermilion to darken has traditionally been ascribed to conversion from red α-HgS to black β-HgS. Yet, β-HgS was not observed at excavations in Pompeii, where originally red walls darkened, and was instead attributed to the formation of Hg-Cl compounds (e.g., corderoite, $Hg_3S_2Cl_2$, calomel, $HgCl_2$, and terlinguaite, $Hg_4Cl_2O_2$) and gypsum ($CaSO_4 \cdot 2H_2O$).

Mercury(II) selenide (HgSe) and mercury(II) telluride (HgTe) are also known, in addition to a number of derivatives, for example, mercury cadmium telluride and mercury zinc telluride which are semiconductors useful as infrared detector materials. Mercury selenide (HgSe) is a gray-black crystalline solid semimetal with a sphalerite (ZnS) structure. The lattice constant is 0.608 nm. Mercury selenide can also refer to the following chemical compounds: $HgSe_2$ and $HgSe_8$. HgSe occurs naturally as the mineral tiemannite. Together with other II–VI compounds, colloidal nanocrystals of HgSe can be produced. Mercury telluride (HgTe) is a semimetal related to the II–VI group of semiconductor materials. Alternative names are mercuric telluride and mercury(II) telluride. HgTe occurs in nature as the mineral coloradoite. HgTe bonds are weak. Their enthalpy of formation, around −32 kJ/mol, is less than a third of the value for the related compound cadmium telluride. HgTe is readily etched by acids, such as hydrobromic acid. Bulk growth can be accomplished from a mercury and tellurium melt at a high mercury vapor pressure. In addition, HgTe can be grown epitaxially, for example, by sputtering or by metalorganic vapor phase epitaxy. Recently it was proven both theoretically and experimentally, that mercury telluride quantum well shows a unique new state of matter—the "topological insulator." In this phase, though the bulk is an insulator, current can be carried by electronic states confined close to the sample edges. In contrast to the quantum hall effect, here no magnetic field is necessary to create this unique behavior. Furthermore, oppositely directed edge states carry opposite spin projections.

7.18.3.2.3 Sulfate and nitrate

Mercury(II) sulfate, commonly called mercuric sulfate, $HgSO_4$, is an odorless solid that forms white granules or crystalline powder. In water, it separates into an insoluble sulfate with a yellow color and sulfuric acid. It can be formed by heating concentrated H_2SO_4 with elemental mercury.

$$Hg + 2H_2SO_4 \rightarrow HgSO_4 + SO_2 + 2H_2O$$

It can also be formed by dissolving solid yellow mercuric oxide in concentrated sulfuric acid and water. An acidic solution of mercury sulfate is known as Denigés' reagent. It was frequently used during the 20th century as a qualitative analysis reagent. If Denigés' reagent is added to a solution containing compounds that have tertiary alcohols, a yellow or red precipitate will form.

Mercury(II) nitrate is a toxic colorless or white soluble crystalline mercury(II) salt of nitric acid. It was historically employed to treat fur to make felt in a process known as "carroting." The phrase "mad as a hatter" is linked to psychological illness due to excessive exposure to mercury(II) nitrate. The use continued in the United States until it was banned in December 1941 by the United States Public Health Service. While this sounds positive for one's health, the ban in fact freed mercury(II) nitrate to be utilized in the production of detonators in World War II. Mercury(II) nitrate is formed through the reaction of hot concentrated nitric acid with mercury metal; under these conditions the nitric acid acts as an oxidizing agent. Dilute nitric acid would form mercury(I) nitrate instead. Mercuric nitrate is utilized in mercuration reactions. It is used especially in reactions involving ketones. One of the chemicals with which it is most effective is acetone. This reaction utilizes mercuric nitrate, mercuric oxide, and calcium sulfate to convert acetone, $CH_3C(O)CH_3$, into $CH_3C(O)CH_2Hg$. Acetone is a compound for which most other mercuration routes prove unsuccessful. The mercuric nitrate compound works since it is a strong oxidizing agent. Furthermore, when mercury is dissolved in nitric acid the acid form of mercuric nitrate is formed. The acidic form is capable of inverting molecules of sucrose.

7.18.3.2.4 Cyanide and thiocyanate

Mercury(II) cyanide, also called mercuric cyanide, is a coordination compound of nitrogen, carbon and mercury. It is a colorless, odorless, toxic white powder with a bitter metallic taste. It has a melting point of 320°C, at which it decomposes and releases toxic mercury fumes. It is very soluble in polar solvents, for example, water, alcohol, and ammonia; slightly soluble in ether; and insoluble in benzene and other hydrophobic solvents. It quickly decomposes in acid to form hydrogen cyanide. In addition, it decomposes when exposed to light, becoming darker in color. It reacts vigorously with oxidizing agents; fusion with metal chlorates, perchlorates, nitrates, or nitrites can result in a violent explosion. Mercuric cyanide can be formed by reacting yellow mercury oxide with hydrocyanic acid, which is generally performed by passing HCN gas into HgO in water. When soluble $Hg(CN)_2$ is formed, the solution is evaporated to crystallize the product.

$$HgO + 2HCN \rightarrow Hg(CN)_2 + H_2O$$

$Hg(CN)_2$ can also be produced by mixing HgO with finely powdered Prussian blue ($Fe(III)_4[Fe(II)(CN)_6]_3$). Furthermore, it can be formed by reacting mercuric sulfate with potassium ferrocyanide in water.

$$K_4Fe(CN)_6 + 3HgSO_4 \rightarrow 3Hg(CN)_2 + 2K_2SO_4 + FeSO_4$$

Another route to produce mercuric cyanide is through the disproportionation of mercury(I) derivatives. In these reactions, metallic mercury precipitates, while $Hg(CN)_2$ stays in solution.

$$Hg_2(NO_3)_2 + 2KCN \rightarrow Hg + Hg(CN)_2 + 2KNO_3$$

Mercury cyanide can be employed as a promoter in the Koenigs–Knorr reaction for the synthesis of glycosides. Cyanogen, $(CN)_2$, forms upon heating dry mercury cyanide but the method is substandard compared to other routes.

$$Hg(CN)_2 \rightarrow (CN)_2 + Hg$$

Coordination polymers can be prepared from $Hg(CN)_2$ building blocks. Large single crystals of [(TMEDA) Cu-$[Hg(CN)_2]_2$][$HgCl_4$] form upon reacting the labile transition metal halide $CuCl_2$, the soft Lewis acid Hg $(CN)_2$, and N,N,N',N'-tetramethylethylenediamine (TMEDA). The movement of two labile chloride ligands from the harder Cu(II) to the softer Hg(II) drives the formation of the crystal.

Mercury(II) thiocyanate ($Hg(SCN)_2$) is an inorganic chemical compound, the coordination complex of Hg^{2+} and the thiocyanate anion. It is a white powder. The first synthesis of mercury thiocyanate was probably achieved in 1821 by Jöns Jacob Berzelius.

$$HgO + 2HSCN \rightarrow Hg(SCN)_2 + H_2O$$

It is formed by reacting solutions containing mercury(II) and thiocyanate ions. The low solubility product of mercury thiocyanate results in its precipitation from solution.

$$Hg(NO_3)_2 + 2KSCN \rightarrow Hg(SCN)_2 + 2KNO_3$$

The compound has a polymeric structure with Hg^{2+} centers linearly coordinated to two S atoms with a distance of 2.381 Å. Four weak Hg^{2+}-N interactions are indicated with distances of 2.81 Å. Mercury thiocyanate was in the past used in pyrotechnics producing an effect known as the Pharaoh's serpent or Pharaoh's snake. When the compound is in the presence of a strong enough heat source, a fast, exothermic reaction that forms a large mass of coiling, serpent-like solid is started. An inconspicuous flame, which is frequently blue though it can also be yellow/orange, accompanies the combustion. The resulting solid can range from dark graphite gray to light tan in color with the inside usually much darker than the outside. The reaction has several stages: igniting mercury thiocyanate results in the formation of an insoluble brown mass that is primarily carbon nitride, C_3N_4 plus mercury sulfide and carbon disulfide.

$$2Hg(SCN)_2 \rightarrow 2HgS + CS_2 + C_3N_4$$

Carbon disulfide combusts to carbon dioxide and sulfur dioxide.

$$CS_2 + 3O_2 \rightarrow CO_2 + 2SO_2$$

The heated C_3N_4 partially breaks down to form nitrogen gas and cyanogen.

$$2C_3N_4 \rightarrow 3(CN)_2 + N_2$$

Mercury sulfide reacts with oxygen to produce mercury vapor and sulfur dioxide. If the reaction is carried out inside a container, a gray film of mercury coating on its inner surface can be observed.

$$HgS + O_2 \rightarrow Hg + SO_2$$

This reaction was discovered by German chemist Friedrich Wöhler (July 31, 1800 to September 23, 1882) in 1821, soon after the first synthesis of mercury thiocyanate: "winding out from itself at the same time worm-like processes, to many times its former bulk, a very light material the color of graphite...." For a period, a firework product known as "Pharaoschlangen" was available to the public in Germany but was eventually banned when the toxic properties of the material were discovered through the death of several children mistakenly eating the resulting solid. A comparable, though less extreme, effect to the Pharaoh's serpent can be obtained using a firework known as a black snake. These are usually benign products, typically consisting of sodium bicarbonate or a mixture of linseed oil and naphthalenes. Mercury thiocyanate has a number of applications in chemical synthesis. It is the precursor to potassium tris(thiocyanato)mercurate(II) (K[$Hg(SCN)_3$]) and cesium tris(thiocyanato)mercurate(II) (Cs[$Hg(SCN)_3$]). The $Hg(SCN)_3^-$ ion can also exist independently and is readily formed from the compounds above, as well as from others. Its reactions with organic halides result in two products, one with the sulfur bound to the organic compound and one with the nitrogen bound to the organic compound.

Mercury(II) salts form a variety of complex derivatives with ammonia. These include Millon's base (Hg_2N^+), the 1D polymer (salts of $HgNH_2^+)_n$), and "fusible white precipitate" or $[Hg(NH_3)_2]Cl_2$. Mercury fulminate is a detonator widely used in explosives.

7.18.3.3 Other valence states

Mercury(IV) fluoride, HgF_4, is the first mercury compound to be reported with mercury in the oxidation state IV. Mercury, like the other group 12 elements (cadmium and zinc), has an s^2d^{10} electron configuration and usually only forms bonds involving its 6s orbital. This indicates that the highest oxidation state mercury normally has is II, and therefore it is typically deemed a posttransition metal instead of a transition metal. HgF_4 was first reported from experiments in 2007, but its presence is still a matter of discussion; experiments performed in 2008 could not replicate the result. Speculation about higher oxidation states for mercury has been around since the 1970s, and theoretical calculations in the 1990s predicted that it should be stable in the gas phase, with a square planar geometry in agreement with a formal d^8 configuration. Nevertheless, experimental evidence remained elusive until 2007, when HgF_4 was first produced utilizing solid neon and argon for matrix isolation at a temperature of $-269°C$ (4K). HgF_4 was observed using infrared spectroscopy. Analysis of density functional theory and coupled cluster calculations has shown that the d orbitals are involved in bonding, suggesting that mercury should be considered a transition metal after all. Yet, that conclusion has been disputed by W. B. Jensen (2008) with the argument that HgF_4 only exists under highly atypical nonequilibrium conditions and should best be considered as an exception. HgF_4 is formed by the reaction of elemental mercury with fluorine.

$$Hg + 2F_2 \rightarrow HgF_4$$

HgF_4 is only stable in matrix isolation at 4K ($-269°C$); upon heating, or if the HgF_4 molecules touch each other, it decomposes to mercury(II) fluoride and fluorine.

$$HgF_4 \rightarrow HgF_2 + F_2$$

HgF_4 is a diamagnetic, square planar molecule. The mercury atom has a formal $6s^25d^86p^6$ electron configuration, and as such it obeys the octet rule but not the 18-electron rule. HgF_4 is isoelectronic with the tetrafluoroaurate anion, AuF_4^-, and is valence isoelectronic with the tetrachloroaurate ($AuCl_4^-$), tetrabromoaurate ($AuBr_4^-$), and tetrachloroplatinate ($PtCl_4^{2-}$) anions.

Linear trimercury and tetramercury cations: compounds containing the linear Hg_3^{2+} (mercury(2/3)) and Hg_4^{2+} (mercury(1/2)) cations have been prepared. These ions are only found in the solid state in compounds such as $Hg_3(AlCl_4)_2$ and $Hg_4(AsF_6)_2$. The Hg–Hg bond length is 255 pm in Hg_3^{2+}, and 255–262 pm in Hg_4^{2+}. The bonding comprises two-center two-electron bonds formed by 6s orbitals.

Cyclic mercury cations: the existence of the triangular Hg_3^{4+} cation was established in a reinvestigation of the mineral terlinguaite in 1989 and has since been prepared in several compounds. The bonding has been described in terms of a three-center two-electron bond where overlap of the 6s orbitals on the mercury atoms gives (in D_{3h} symmetry) a bonding "a_1" orbital.

Chain and layer polycations: the golden-yellow compound $Hg_{2.86}(AsF_6)$, called "alchemists' gold" by its discoverers, consists of perpendicular chains of Hg atoms. The "metallic" compounds Hg_3NbF_6 and Hg_3TaF_6 have hexagonal layers of mercury atoms separated by layers of MF_6^- anions. Both are superconductors below 7K.

7.18.3.4 Organomercury compounds

Organic mercury compounds are historically significant but are of limited industrial use in the western world. Typically, the Hg–C bond is stable toward air and moisture but sensitive to light. Mercury(II) salts are an unusual example of simple metal complexes that can react directly with aromatic rings. Organomercury compounds are always divalent and typically two-coordinate with a linear geometry. In contrast to organocadmium and organozinc compounds, organomercury compounds do not react with water. They typically have the formula HgR_2, which are regularly volatile, or HgRX, which are frequently solids, where R is aryl or alkyl and X is usually halide or acetate. Methylmercury, a general term for compounds with the formula CH_3HgX, is a hazardous family of compounds that are often found in polluted water. They form through a process known as biomethylation.

Organomercury compounds are formed by a variety of methods, for example, the direct reaction of hydrocarbons and mercury(II) salts. In this regard, organomercury chemistry is rather similar to organopalladium chemistry and differs from organocadmium compounds. Electron-rich arenes undergo direct mercuration upon reaction with mercury(II) acetate, $Hg(O_2CCH_3)_2$. The one acetate group that remains on mercury can be substituted by chloride.

$$C_6H_5OH + Hg(O_2CCH_3)_2 \rightarrow C_6H_4(OH)\text{-}2\text{-}HgO_2CCH_3 + CH_3CO_2H$$

$$C_6H_4(OH)\text{-}2\text{-}HgO_2CCH_3 + NaCl \rightarrow C_6H_4(OH)\text{-}2\text{-}HgCl + NaO_2CCH_3$$

The first such reaction, including a mercuration of benzene itself was described by German chemist Otto Dimroth (March 28, 1872 to May 16, 1940) between 1898 and 1902. Mercury(II) has a high affinity for sulfur ligands. Hg $(OAc)_2$ can be utilized as a reagent to remove the acetamidomethyl protecting group, which is used to "protect" thiol groups in organic synthesis. Likewise, $Hg(OAc)_2$ is a standard reagent to convert thiocarbonate esters into dithiocarbonates.

$$(RS)_2C = S + H_2O + Hg(OAc)_2 \rightarrow (RS)_2C = O + HgS + 2HOAc$$

The Hg^{2+} center can bind to alkenes, causing the addition of hydroxide and alkoxide. For instance, the reaction of methyl acrylate with mercuric acetate in methanol forms an α-mercuri ester.

$$Hg(O_2CCH_3)_2 + CH_2 = CHCO_2CH_3 \rightarrow CH_3OCH_2CH(HgO_2CCH_3)CO_2CH_3$$

The resulting Hg–C bond can be cleaved with bromine to give the matching alkyl bromide.

$$CH_3OCH_2CH(HgO_2CCH_3)CO_2CH_3 + Br_2 \rightarrow CH_3OCH_2CHBrCO_2CH_3 + BrHgO_2CCH_3$$

This reaction is known as the Hofmann–Sand Reaction (Hofmann and Sand, 1900). A general synthetic route to organomercury compounds comprises alkylation with Grignard reagents and organolithium compounds. Diethylmercury can be formed from the reaction of mercury chloride with two equivalents of ethylmagnesium bromide, a conversion that would typically be conducted in diethyl ether solution. The ensuing $(CH_3CH_2)_2Hg$ is a dense liquid (2.466 g/cm^3) that boils at 57°C at 16 torr. The compound is slightly soluble in ethanol and soluble in ether. Likewise, diphenylmercury (m.p. 121°C–123°C) can be synthesized through the reaction of mercury chloride and phenylmagnesium bromide. A related synthesis involves formation of phenylsodium in the presence of mercury(II) salts. Hg(II) can be alkylated through the treatment with diazonium salts in the presence of copper metal. In this manner 2-chloromercuri-naphthalene has been synthesized. Phenyl(trichloromethyl)mercury can be produced by generating dichlorocarbene in the presence of phenylmercuric chloride. A useful carbene source is sodium trichloroacetate. This compound releases dichlorocarbene on heating.

$$C_6H_5HgCCl_3 \rightarrow C_6H_5HgCl + CCl_2$$

Organomercury compounds are useful synthetic intermediates because of the well-controlled conditions under which they undergo cleavage of the Hg-C bonds. Diphenylmercury is a source of the phenyl radical in specific syntheses. Treatment with aluminum results in triphenyl aluminum.

$$3Ph_2Hg + 2Al \rightarrow (AlPh_3)_2 + 3Hg$$

As pointed out above, organomercury compounds react with halogens to form the corresponding organic halide. Organomercurials are frequently employed in transmetalation reactions with lanthanides and alkaline earth metals. Cross-coupling of organomercurials with organic halides is catalyzed by palladium, which offers a method for C-C bond formation. Typically of low selectivity, but if performed in the presence of halides, selectivity increases. Carbonylation of lactones has been proven to employ Hg(II) reagents under palladium catalyzed conditions. (C-C bond formation and *cis* ester formation). Due to their toxicity and low nucleophilicity, organomercury compounds are only used on a limited scale. The oxymercuration reaction of alkenes to alcohols using mercuric acetate progresses via organomercury intermediates. A related reaction producing phenols is the Wolffenstein–Böters reaction (named after German chemists Richard Wolffenstein (21 August 1864 – 5 June 1926) and Oskar Böters (November 19, 1848 to February 5, 1912)). The toxicity is useful in antiseptics such as thiomersal and merbromin, and fungicides such as ethylmercury chloride and phenylmercury acetate. Mercurial diuretics such as mersalyl acid were previously in regular use but have been replaced by the thiazides and loop diuretics, which are safer and longer acting, as well as being orally active.

7.18.4 Major uses

Hg and its compounds have been utilized in medical treatment; however, they are currently much less widespread than before, now that their toxic effects are much better understood. Hg forms an element in dental amalgams. Thiomersal (known as Thimerosal in the United States, IUPAC name ethyl(2-mercaptobenzoato-(2-)-O,S) mercurate(1-) sodium, $C_9H_9HgNaO_2S$) is an organic compound employed as a preservative in vaccines, but this use is decreasing. Thiomersal is metabolized to ethyl mercury ($C_2H_5Hg^+$). While it was extensively hypothesized that this Hg-based preservative could cause or trigger autism in children, scientific research showed no evidence to support this. Still, thiomersal has been eliminated from, or decreased to trace quantities in all US vaccines advised for children up to 6 years of age, except for inactivated influenza vaccine. Another Hg compound, merbromin (Mercurochrome and several other brand names, IUPAC name dibromohydroxymercurifluorescein, $C_{20}H_8Br_2HgNa_2O_6$), is a topical antiseptic applied for minor cuts and scrapes that is still used in some countries. Hg in the form of one of its common ore minerals, cinnabar (HgS), is used in a number of traditional medicines, in particular in traditional Chinese medicine. Analysis of its safety has concluded that it can result in substantial Hg intoxication when heated, consumed in overdose, or taken long term, and can have adverse effects at therapeutic doses, yet effects from therapeutic doses are usually reversible. While this form of Hg seems to be less toxic than other forms, its application in traditional Chinese medicine has not yet been justified, since the therapeutic basis for the use of cinnabar is not understood. Currently the usage of Hg in medicine has significantly decreased in all respects, in particular in developed countries. Thermometers and sphygmomanometers (blood pressure meters) containing Hg were developed in the early 18th and late 19th centuries, respectively. In the early 21st century, their use is decreasing and has been banned in certain countries, states and medical institutions. Hg compounds are present in some over-the-counter drugs, for example, topical antiseptics, stimulant laxatives, diaper-rash ointment, eye drops, and nasal sprays. The Food and Drug Administration has "inadequate data to establish general recognition of the safety and effectiveness" of the Hg compounds in these products. Hg is still employed in certain diuretics even though alternatives now exist for most therapeutic uses.

Certainly the largest use of Hg near the end of the 20th century was in the Hg-cell process (also called the Castner-Kellner process, invented by American industrial chemist Hamilton Young Castner (September 11, 1858 to October 11, 1899) and Austrian chemist, inventor and industrialist Karl Kellner (September 1, 1851 to June 7, 1905) in the 1890s) where metallic Na is produced as an amalgam at a cathode made from Hg; this Na is afterwards reacted with water to produce NaOH. Much of the industrial Hg releases of the 20th century resulted from this process, though modern plants are asserted to be safe in this respect. After approximately 1985, all new chloralkali manufacturing facilities built in the United States employed membrane cell or diaphragm cell technologies to make Cl_2.

Hg is also found in liquid mirror telescopes. Some transit telescopes use a basin of Hg to form a flat and perfectly horizontal mirror, valuable in determining an absolute vertical or perpendicular reference. Concave horizontal parabolic mirrors may be created by rotating liquid Hg on a disk, the parabolic form of the liquid formed in this manner reflecting and focusing incident light. These telescopes are less expensive than traditional large mirror telescopes by up to a factor of 100, but the mirror cannot be tilted and must always point straight up. Liquid Hg is a part of widespread secondary reference electrode (known as the calomel electrode) in electrochemistry as a substitute for the standard hydrogen electrode. The calomel electrode is employed to calculate the electrode potential of half cells. Last, but not least, the triple point of Hg, $-38.8344°C$, is a fixed point employed as a temperature standard for the International Temperature Scale (ITS-90). In polarography, a type of voltammetry, both the dropping Hg electrode and the hanging Hg drop electrode use elemental Hg. This use permits a new uncontaminated electrode to be available for each analysis or each new experiment. In addition, Hg-containing compounds are useful in the area of structural biology. Mercuric compounds such as $Hg(II)Cl_2$ or potassium tetraiodomercurate (II) ($K_2[HgI_4]$) can be added to protein crystals in an attempt to form heavy atom derivatives that can be utilized to solve the phase problem in X-ray crystallography through isomorphous replacement or anomalous scattering methods.

Gaseous Hg is employed in Hg-vapor lamps and some "neon sign" type advertising signs and fluorescent lamps. Those low-pressure lamps emit very spectrally narrow lines, which are by tradition utilized in optical spectroscopy for calibration of spectral positions. Gaseous Hg is also found in some electron tubes, including ignitrons (a type of gas-filled tube used as a controlled rectifier and dating from the 1930s. Invented by American electrical engineer Joseph Slepian (February 11, 1891 to December 19, 1969) while employed by Westinghouse, Westinghouse was the original manufacturer and owned trademark rights to the name "Ignitron." Ignitrons are closely related to Hg-arc valves but differ in the way the arc is ignited. They function similarly to thyratrons; a triggering pulse to the igniter electrode turns the device "on," allowing a high current to flow between the cathode and anode electrodes. After it is turned on, the current through the anode must be reduced to zero to restore the device to its nonconducting state. They are used to switch high currents in heavy industrial applications.), thyratrons (a type of gas-filled tube used as a high-power electrical switch and controlled rectifier. Thyratrons can handle much greater currents than similar hard-vacuum tubes. Electron multiplication occurs when the gas becomes ionized, producing a phenomenon known as Townsend discharge. Gases used include Hg vapor, Xe, Ne, and (in special high-voltage applications or applications requiring very short switching times) H_2. Unlike a vacuum tube (valve), a thyratron cannot be used to amplify signals linearly.), and Hg arc rectifiers. In addition, it is utilized in specialist medical care lamps for skin tanning and disinfection. Gaseous Hg is added to cold cathode Ar-filled lamps to improve the ionization and electrical conductivity. An Ar-filled lamp with no Hg will have dull spots and will fail to light correctly. Lighting with Hg can be bombarded/oven pumped only one time. When added to Ne-filled tubes the light emitted will be inconsistent red/blue spots until the initial burning-in process is finished; in the end it will emit a stable dull off-blue color. The Deep Space Atomic Clock being developed by the Jet Propulsion Laboratory uses Hg in a linear ion-trap-based clock. This new application of Hg permits to make very compact atomic clocks, with low energy requirements, and is thus ideal for space probes and Mars missions.

Hg, as thiomersal, is widely used in the making of mascara. Back in 2008, Minnesota became the first state in the United States to ban deliberately added Hg in cosmetics, giving it a tougher standard than the federal government. Research in geometric mean urine Hg concentrations found an earlier unrecognized source of exposure (skin care products) to inorganic Hg among New York City residents. Population-based biomonitoring also indicated that Hg concentration levels are higher in people who ate seafood and fish meals. Hg(II) fulminate ($Hg(CNO)_2$) is a primary explosive which is mainly used as a primer of a cartridge in firearms. Hg(II) fulminate is prepared by dissolving Hg in nitric acid and adding ethanol to the solution. It was first prepared by Edward Charles Howard in 1800. The crystal structure of this compound was only determined in 2007. Silver fulminate can be prepared in a similar way, but this salt is even more unstable than mercury fulminate; it can even explode under water and is impossible to accumulate in large amounts because it detonates under its own weight. The thermal decomposition of Hg(II) fulminate can begin at temperatures as low as 100°C, though it proceeds at a much higher rate with increasing temperature. A possible reaction for the decomposition of Hg(II) fulminate yields carbon dioxide gas, nitrogen gas, and a combination of relatively stable mercury salts.

$$4Hg(CNO)_2 \rightarrow 2CO_2 + N_2 + HgO + 3Hg(OCN)CN$$
$$Hg(CNO)_2 \rightarrow 2CO + N_2 + Hg$$
$$Hg(CNO)_2 \rightarrow :Hg(OCN)_2 \left(\text{cyanate or/and isocyanate}\right)$$
$$2Hg(CNO)_2 \rightarrow 2CO_2 + N_2 + Hg + Hg(CN)_2 \text{ (mercury(II) cyanide)}$$

Historically, Hg was widely employed in hydraulic gold mining in order to help the gold to sink through the flowing water-gravel mixture. Thin gold particles may form Hg–Au amalgam and hence increase the gold recovery rates. Large-scale use of Hg stopped in the 1960s. However, Hg is still used in small scale, often clandestine, gold prospecting and mining (e.g., in South America). It has been estimated that some 45,000 metric tons of Hg used in California for placer mining have never been recovered. Hg was similarly employed in historic silver mining.

7.19 81 Tl — Thallium

7.19.1 Discovery

Thallium (Greek θαλλός, thallos, meaning "a green shoot or twig") was first found using flame spectroscopy in March 1861. The name is based on thallium's bright green spectral emission lines. After the publication of the enhanced method of flame spectroscopy by German chemist Robert Wilhelm Eberhard Bunsen (March 30, 1811 to August 16, 1899) and German physicist Gustav Robert Kirchhoff (March 12, 1824 to October 17, 1887) and the discovery of cesium and rubidium in 1859–1860, flame spectroscopy became an appropriate method to analyze the chemical composition of minerals and other chemical products. British chemist and physicist Sir William Crookes (June 17, 1832 to April 4, 1919) (Fig. 7.107) and French chemist Claude-Auguste Lamy (June 15, 1820 to March 20, 1878) (Fig. 7.108) both began to use this new analytical technique. William Crookes used it to obtain spectroscopic analyses of tellurium in selenium compounds deposited in the lead chamber of a sulfuric acid production plant near Tilkerode in the Harz mountains, Germany. He had gotten the samples for his study on selenium cyanide from German chemist August Wilhelm von Hofmann (April 8, 1818 to May 5, 1892) years earlier. By 1862 Crookes managed to separate small amounts of the new element and measure the properties of a few compounds (Crookes, 1861a,b, 1862). Claude-Auguste Lamy used a spectrometer that was like the one used by Crookes' to obtain the chemical composition of a selenium-containing compound that was deposited during the production of sulfuric acid from pyrite (FeS_2). He also observed the new green line in the spectra and was convinced that a new element was present (Lamy, 1862). Lamy had gotten his material from the sulfuric acid plant of his friend French chemist Charles Fréd Kuhlmann (May 22, 1803 to January 27, 1881) and this by-product was available in large amounts. Lamy started to separate the new element from that by-product. The fact that Lamy could work with sufficient amounts of thallium allowed him to obtain the properties of several compounds and furthermore he was successful

FIGURE 7.107 Sir William Crookes, 1906 photo by George Charles Beresford (1864–1938).

in preparing a small ingot of metallic thallium through remelting thallium he had obtained by electrolysis of thallium salts. As both chemists found thallium independently and a large part of the work, in particular the separation of the metallic thallium was done by Lamy, Crookes tried to protect his own priority on the research results. Lamy was presented a medal at the International Exhibition in London 1862: *For the discovery of a new and abundant source of thallium*, and after strong objections Crookes also received a medal: *Thallium, for the discovery of the new element*. The argument between the two chemists continued through 1862 and 1863. Most of the argument ended after Crookes was elected Fellow of the Royal Society in June 1863. The main use of thallium was as poison for rodents. After several accidents the use as poison was banned in the United States by Presidential Executive Order 11643 in February 1972 by Richard Nixon. In the following years a number of other countries also banned its application.

7.19.2 Mining, production, and major minerals

While thallium is a modestly abundant element in the Earth's crust, with an estimated concentration of around 0.7 ppm, generally in association with K-based minerals in clays, soils, and granites, Tl is not commonly commercially recoverable from these sources. The major source of Tl for practical purposes is the trace amount present in Cu, Pb, Zn, and other heavy metal-sulfide ores. Tl is found in minerals such as crookesite $TlCu_7Se_4$, hutchinsonite $TlPbAs_5S_9$, and lorándite $TlAsS_2$. It can also be found as a trace element in pyrite (FeS_2), and Tl is obtained as a by-product of roasting this mineral for the production of sulfuric acid. Tl can also be obtained from the smelting of Pb and Zn ores. Manganese nodules present on the ocean floor contain some Tl, but the collection of these nodules has so far been too expensive. In addition, there is the possibility of long-term damage to the oceanic environment. Furthermore, a number of other Tl minerals, containing between 16% and 60% Tl, can be found in nature as complexes of sulfides or selenides that primarily contain Sb, As, Cu, Pb, and/or Ag. These minerals are rare and have had no economic importance as sources of Tl. The Allchar deposit in southern Macedonia was the only area where Tl was actively mined. This deposit still has an estimated reserve of around 500 tons of Tl, and it is a source for several rare thallium minerals, for example

FIGURE 7.108 Claude-Auguste Lamy.

FIGURE 7.109 Slender prismatic crystals to 6 mm of dark metallic-black hutchsonite, $TlPbAs_5S_9$, with a dark yellow druse of crystalline orpiment, As_2S_3, with minor realgar, As_4S_4. Quiruvilca mine, Santiago de Chuco prov., La Libertad dept., Peru.

FIGURE 7.110 Lorándite, $TlAsS_2$, deep carmine red crystals to 5 mm in vugs of orpiment, As_2S_3. Allchar, Roszdan, Macedonia.

FIGURE 7.111 Lafossaite, Tl(Cl,Br), is a sublimation product of fumarolic gases at volcanoes such as this locality. The crystals occur as extremely small yellowish-white cubes, about 5 mm vug. La Fossa crater, Vulcano Island, Messina prov., Sicily, Italy.

lorándite. It is estimated that the worldwide production of Tl is around 10 metric tons per year, the large majority of it as a by-product from the smelting of Cu, Zn, and Pb ores.

Thallium is either produced from the dusts from the smelter flues or from residues such as slag that are collected at the end of the smelting process. The raw materials used for thallium production contain large amounts of other compounds and hence purification forms the first step. The thallium is leached either using a base or sulfuric acid from the starting material. Subsequently, the thallium is precipitated a number of times from the solution to remove impurities. Finally, it is reacted to form thallium sulfate and the thallium is extracted through electrolysis on platinum or stainless-steel plates. The thallium production dropped by about one-third between 1995 and 2009—from about 15 to 10 metric tons. Because there exist a number of small ore deposits with relatively high thallium concentration, it is technically possible to increase the production if a new application becomes practical for widespread use.

There are about 70 different minerals that contain thallium in their crystal structure. About 60 of those minerals can be found in the class of the sulfides and include minerals such as carlinite (Tl_2S), hutchinsonite ($TlPbAs_5S_9$) (Fig. 7.109), lorándite ($TlAsS_2$) (Fig. 7.110), pierrotite ($Tl_2(Sb,As)_{10}S_{16}$), and rathite ($Ag_2Pb_{12-x}Tl_{x/2}As_{18+x/2}S_{40}$). Five halides, for example, hephaistosite ($TlPb_2Cl_5$), lafossaite (Tl(Cl,Br)) (Fig. 7.111), and steropesite (Tl_3BiCl_6), are known to contain Tl. Avicennite (Tl_2O_3) (Fig. 7.112) is the only known oxide with Tl. Four sulfates, for example, dorallcharite

FIGURE 7.112 The avicennite, Tl_2O_3, is present in minute voids of jasperoidal pods weathered out of limestone. It forms as earthy pink coatings from the oxidation of thallium sulfide crystals. Lookout Pass, Tooele Co., Utah, United States.

($TlFe_3^{3+}(SO_4)_2(OH)_6$) and lanmuchangite ($Tl^+Al(SO_4)_2 \cdot 10H_2O$) have structure Tl. Thalliumpharmacosiderite ($TlFe_4[(AsO_4)_3(OH)_4] \cdot 4H_2O$) is the only arsenate mineral in the phosphate class.

7.19.3 Chemistry

Thallium consists of 25 isotopes with atomic masses ranging from 184 to 210. ^{203}Tl (29.5% natural abundance) and ^{205}Tl (70.5% natural abundance) are the only stable isotopes and form nearly all of the naturally occurring thallium. ^{204}Tl is the most stable radioisotope, which has a half-life of 3.78 years. It is synthesized through the neutron activation of stable Tl in a nuclear reactor. The most useful radioisotope, ^{201}Tl with a half-life of 73 h, decays by electron capture, emitting X-rays ($\sim$70–80 keV), and photons of 135 and 167 keV in 10% total abundance. Consequently, this radioisotope has good imaging properties without disproportionate patient radiation dose. It is the most popular isotope used for Tl nuclear cardiac stress tests (The best known example of a nuclear stress test is myocardial perfusion imaging. Typically, a radiotracer (Tc-99 sestamibi, Myoview or thallous chloride 201) is injected during the test. After a suitable waiting period to ensure proper distribution of the radiotracer, scans are acquired with a gamma camera to capture images of the blood flow. Scans acquired before and after exercise are examined to assess the state of the coronary arteries of the patient. Showing the relative amounts of radioisotope within the heart muscle, the nuclear stress tests more accurately identify regional areas of reduced blood flow.)

Thallium is a gray posttransition metal with 81 electrons arranged in an electronic configuration of [Xe] $4f^{14}5d^{10}6s^26p^1$. The three valence electrons in the outermost shell participate in chemical reactions and so, thallium tends to oxidize to the +3 oxidation states as ionic salts. The +3 state resembles that of the other elements in group 13 (boron, aluminum, gallium, indium). However, the inert pair effect is pronounced for thallium and the +1 state is more prominent in its compounds. Thallium(I) resembles the chemistry of alkali metals (Table 7.19).

7.19.3.1 Oxides and hydroxides

Thallium resembles tin when pristine and stored in inert environments, but quickly discolors when exposed to air to form a gray oxide. It reacts with steam or moist air to give yellow TlOH. TlOH is very soluble and is a strong base resembling those of the alkali hydroxides. Tl(I) forms colorless, well-crystallized salts of many oxoacids, and these tend to be anhydrous like those of the similarly sized Rb and Cs. Thallium can form many oxides using various preparation methods. Tl_2O can be made by heating Tl_2CO_3 in N_2 at 700°C; Tl_2O_3 can be made by oxidation of aqueous $TlNO_3$ with Cl_2 or Br_2 then precipitating the hydrated oxide. A mixed valence oxide Tl_4O_3 ($3Tl_2O \cdot Tl_2O_3$) can be made by heating 3:1 mixture of Tl_2CO_3 and Tl_2O_3 at 450°C in an inert atmosphere and TlO_2 can be made by electrolysis of an aqueous solution of Tl_2SO_4 and oxalic acid between Pt electrodes. Thallium(III) oxide, also called thallic oxide, occurs in nature as the rare mineral avicennite. Its structure is related to that of Mn_2O_3. Tl_2O_3 is metallic with high conductivity and is a degenerate n-type semiconductor which may have potential application in solar cells. A method of

TABLE 7.19 Thallium properties.

Appearance	Silvery white
Standard atomic weight $A_{r,std}$	204.38
Block	p-Block
Element category	Posttransition metal
Electron configuration	[Xe] $4f^{14}$ $5d^{10}$ $6s^2$ $6p^1$
Phase at STP	Solid
Melting point	304°C
Boiling point	1473°C
Density (near r.t.)	11.85 g/cm^3
When liquid (at m.p.)	11.22 g/cm^3
Heat of fusion	4.14 kJ/mol
Heat of vaporization	165 kJ/mol
Molar heat capacity	26.32 J/(mol·K)
Oxidation states	−5, −2, −1, **+1**, +2, **+3**
Ionization energies	1st: 589.4 kJ/mol
	2nd: 1971 kJ/mol
	3rd: 2878 kJ/mol
Atomic radius	Empirical: 170 pm
Covalent radius	145 ± 7 pm
Van der Waals radius	196 pm

STP, Standard temperature and pressure.
Bold font indicates main oxidation state.

producing Tl_2O_3 by MOCVD is known. Any practical use of thallium(III) oxide will always have to take into account the poisonous nature of thallium. Contact with moisture and acids may produce poisonous thallium compounds. It is prepared by the reaction of thallium with oxygen or hydrogen peroxide in an alkaline thallium(I) solution. Instead, it can be formed by the oxidation of thallium(I) nitrate by chlorine in an aqueous potassium hydroxide solution. Thallium (III) hydroxide, $Tl(OH)_3$, also known as thallic hydroxide, is a white solid. Thallium(III) hydroxide is a very weak base; it is changed to thallium(III) ion, Tl^{3+}, only in strongly acid conditions.

7.19.3.2 Chalcogenides

Thallium reacts with the chalcogenides sulfur, selenium, and tellurium to form the respective sulfides, selenides and tellurides. These compounds are well studied because many of them are semiconductors, semimetals, photoconductors, or light emitters, and Tl_2Te_3 has been found to be a superconductor at low temperatures. Thallium(I) sulfide, Tl_2S, was used in some of the earliest photoelectric detectors by American chemist, physicist, and inventor Theodore Willard Case (December 12, 1888 to May 13, 1944) who created the so-called thalofide (sometimes spelt thallofide) cell, used in early film projectors (Case, 1920). Case defined the detector material as consisting of thallium, oxygen and sulfur, and this was mistakenly described by others as being thallium oxysulfide, which is a compound that is not known. Case's work was then built upon by American physicist Robert Joseph Cashman (September 27, 1906 to September 27, 1988) who recognized that the controlled oxidation of the Tl_2S film was key to the operation of the cell. Cashman's work ended in the development of long wave infrared detectors used during the Second World War. Reliable Tl_2S detectors were also developed in Germany at the same time. Tl_2S is found in nature as the mineral carlinite which is the only sulfide mineral of thallium that does not contain at least two metals. Tl_2S has a distorted anti-CdI_2 structure. Tl_2S can be formed from the elements or by precipitating the sulfide from a solution of thallium(I), for example, the sulfate or nitrate. Thin films have been deposited, produced from a mixture of citratothallium complex and thiourea. Heating the film in nitrogen at 300°C converts all the product into Tl_2S.

7.19.3.3 Hydrides and nitrides

The reaction of Tl with hydrogen was studied solely for academic purposes. The reaction sequence involved atomization of thallium, followed by cryogenic codeposition with hydrogen, and concluded with shortwave ultraviolet irradiation. TlH_3 appeared too unstable and never isolated in bulk. The minute information known for the compound was derived from matrix isolation studies and IR spectroscopies of laser ablated thallium in the presence of hydrogen gas. Thallium hydride (systematically named thallium trihydride) is an inorganic compound with the empirical chemical formula TlH_3. It has not yet been obtained in bulk; consequently, its bulk properties are still unknown. Nevertheless, molecular thallium hydride has been isolated in solid gas matrices. In 2004 American chemist, American chemist William Lester Self Andrews (b. January 31, 1942) synthesized thallium hydride for the first time (Wang and Andrews, 2004). Thallium hydride is mainly prepared for research purposes. Thallium hydride is the simplest thallane. Thallium is the heaviest member of the Group 13 metals; the stability of group 13 hydrides decreases with increasing periodic number. This is commonly attributed to poor overlap of the metal valence orbitals with that of the 1s orbital of hydrogen. Despite encouraging early reports, it is doubtful that a thallium hydride species has been isolated. Thallium hydrides have been observed only in matrix isolation studies; the infrared spectrum was determined in the gas phase by laser ablation of thallium in the presence of hydrogen gas. This study confirmed aspects of ab initio calculations conducted by Hunt and Schwerdtfeger (1996) which indicated the comparable stabilities of thallium and indium hydrides. There has not been a confirmed isolation of a thallium hydride complex to date.

Thallium does not form simple nitride compounds. Black thallium(I) nitride is known but the compound is explosive and not much is known for thallium(III) nitride. Perhaps, the toxicity of thallium, raising health concerns in its use in industry and research, accounts for this scarcity of information. Nevertheless, interest in completing the preparation of the nitrides of group 13 motivated Shaginyan (2019) to conduct a study on the synthesis and characterization of TlN in film form. The synthesis involved direct current (DC) diode reactive sputtering of metal thallium target in a glow discharge in nitrogen environment. The TlN was determined to crystallize in a wurtzite structure. The TlN film, however, is easily oxidized in open air. Thallium azide, TlN_3, is a yellow-brown crystalline solid poorly soluble in water. While it is not nearly as sensitive to shock or friction as lead azide, it can easily be detonated by a flame or spark. It can be kept safely dry in a closed nonmetallic container. Thallium azide can be formed reacting an aqueous solution of thallium(I) sulfate with sodium azide. Thallium azide will precipitate; the yield can be optimized by cooling. TlN_3 has the same structure as KN_3, RbN_3, and CsN_3. The azide is bound to eight cations in an eclipsed orientation. The cations form bonds to eight terminal N centers.

7.19.3.4 Halides

The trihalides, TlX_3 (X = F, Cl, Br), of thallium are known but are not the stable forms of the thallium halides. Often, they are prepared by other means. TlF_3 is best prepared by the direct fluorination of Tl_2O_3 with F_2, BrF_3 or SF_4 at 300°C. TlF forms hard white orthorhombic crystals which are slightly deliquescent in humid air but revert to the anhydrous form in dry air. It has a distorted sodium chloride (rock salt) crystal structure, due to the $6s^2$ inert pair on Tl^+. Thallium(I) fluoride is unusual among the thallium(I) halides in that it is very soluble in water, while the others are not. Thallium(I) fluoride can be prepared by the reaction of thallium(I) carbonate with hydrofluoric acid. At 655°C, TlF decomposes to its elements.

$$2TlF \rightarrow 2Tl + F_2$$

$TlCl_3$ and $TlBr_3$ are obtained from aqueous solution as the stable tetrahydrates and $TlCl_3 \cdot 4H_2O$ can be dehydrated with $SOCl_2$ to give anhydrous $TlCl_3$. Thallium(I) bromide, TlBr, is used in room-temperature detectors of X-rays, γ-rays and blue light, as well as in near-infrared optics. It is a semiconductor with a band gap of 2.68 eV. The crystalline structure is of cubic CsCl type at 25°C, but it lowers to the orthorhombic thallium iodide type upon cooling, the transition temperature probably being affected by the presence of impurities. Nanometer-thin TlBr films grown on LiF, NaCl or KBr substrates exhibit a rock salt structure. Thallium is very toxic and a cumulative poison which can be absorbed through the skin. Acute and chronic effects of ingesting thallium compounds include fatigue, limb pain, peripheral neuritis, joint pain, loss of hair, diarrhea, vomiting and damage to central nervous system, liver and kidneys. Thallium(I) chloride, also known as thallous chloride, TlCl, is a colorless solid and an intermediate in the isolation of thallium from its ores. Typically, an acidic solution of thallium(I) sulfate is reacted with hydrochloric acid to precipitate insoluble thallium(I) chloride. The low solubility of TlCl is used in chemical synthesis: treatment of metal chloride complexes with $TlPF_6$, forms the corresponding metal hexafluorophosphate derivative. The resulting TlCl precipitate is isolated by filtration of the reaction mixture. The overall methodology is comparable to the use of $AgPF_6$, except that Tl^+ is much less oxidizing. The crystalline structure is of cubic cesium chloride type at 25°C, but it lowers to the orthorhombic thallium iodide type upon cooling, the transition temperature probably being affected by the impurities. Nanometer-thin TlCl films grown on KBr substrates exhibit a rock salt structure, while the films deposited on mica or NaCl are of the regular CsCl type. A very rare mineral lafossaite, Tl(Cl,Br), is a natural form of thallium(I) chloride. Thallium(I) chloride, like all thallium compounds, is extremely toxic, although its low solubility limits its toxicity. TlI_3 is an intriguing compound which is obtained as black crystals by evaporating an equimolar solution of TlI and I_2 in concentrated aqueous HI. Thallium(I) is the stable oxidation state for the halides. TlF is readily obtained by the action of aqueous HF on Tl_2CO_3. TlCl, TlBr, and TlI are all prepared by acidification of Tl(I) salts. It is also formed as a by-product in thallium-promoted iodination of phenols with thallium

(I) acetate. Attempts to oxidize TlI to thallium(III) iodide failed, since oxidation produces the thallium(I) triiodide, $Tl^+I_3^-$. The room temperature form of TlI is yellow and has an orthorhombic structure, which can be seen as a distorted NaCl structure. The distorted structure is thought to be caused by favorable thallium-thallium interactions, the closest Tl-Tl distance is 383 pm. At 175°C the yellow form transforms to a red TlCl form. This phase transition is accompanied by about two orders of magnitude increase in electrical conductivity. The TlI structure can be stabilized down to room temperature by doping TlI with other iodides, for example, RbI, CsI, KI, AgI, TlBr, and TlCl. Therefore the presence of impurities might be responsible for the coexistence of the cubic and orthorhombic TlI phases at ambient temperature and pressure. Under high pressure, 160 kbar, TlI becomes a metallic conductor. Nanometer-thin TlI films grown on LiF, NaCl, or KBr substrates exhibit the cubic rock salt structure.

7.19.3.5 Organothallium compounds

Thallium(I) and thallium(III) salts are widely used in many organic transformations despite their toxic nature. Thallium (III) nitrate, also known as thallic nitrate, is a thallium compound with chemical formula $Tl(NO_3)_3$. It is normally found as the trihydrate. It is a colorless and highly toxic solid. It is a strong oxidizing agent useful in organic synthesis. Among its many transformations, it oxidizes methoxyl phenols to quinone acetals, alkenes to acetals, and cyclic alkenes to ring-contracted aldehydes. The few studies published are presented in a review by Casas et al. in 1999. Some simple compounds include: thallous malonate (dithallium(1+) propanedioate, $C_3H_2O_4Tl_2$), thallous acetate ($TlCH_3COO$), and cyclopentadienylthallium (C_5H_5Tl). The last is a light yellow solid and is insoluble in most organic solvents, but sublimes easily. It is employed as a precursor to transition metal and main group cyclopentadienyl complexes, together with organic cyclopentadiene derivatives. Cyclopentadienylthallium is produced by the reaction of thallium(I) sulfate, sodium hydroxide, and cyclopentadiene.

$$Tl_2SO_4 + 2NaOH \rightarrow 2TlOH + Na_2SO_4$$
$$TlOH + C_5H_6 \rightarrow TlC_5H_5 + H_2O$$

The compound has a polymeric structure, comprising infinite chains of bent metallocenes. The Tl—T—Tl angles are 130°. Upon sublimation, the polymer breaks into monomers of C_{5v} symmetry. In comparison to other cyclopentadienyl (Cp) transfer reagents, such as cyclopentadienyl sodium, CpMgBr and Cp_2Mg, cyclopentadienylthallium is less air sensitive. It is also much less of a reducing agent.

7.19.4 Major uses

Tl(I)Br and Tl(I)I crystals have been employed as infrared optical materials, since they are harder than other often used infrared optics, and since they have transmission at considerably longer wavelengths. Tl(I) oxide has been utilized to produce glasses with a high index of refraction. In combination with S or Se and As, Tl has been employed in the manufacture of high-density glasses that have low melting points in the range between 125°C and 150°C. These glasses have room temperature properties that are like ordinary glasses and are durable, insoluble in water and have unique refractive indices. Tl(I) sulfide's electrical conductivity is a function of exposure to infrared light thus making this material useful in photoresistors. Tl selenide has been employed in a bolometer for infrared detection (A bolometer is a device for measuring the power of incident electromagnetic radiation via the heating of a material with a temperature-dependent electrical resistance. It was invented in 1878 by the American astronomer, physicist, inventor of the bolometer and aviation pioneer Samuel Pierpont Langley (August 22, 1834 to February 27, 1906)). Doping Se semiconductors with Tl enhances their performance, hence it is employed in trace quantities in Se rectifiers. Another use of Tl doping is the NaI crystals in γ radiation detection instruments. Here, the NaI crystals are doped with a trace quantity of Tl to enhance their efficiency as scintillation generators. Certain electrodes in dissolved oxygen analyzers have Tl. The study of Tl is ongoing to create high-temperature superconducting compounds for such uses as MRI, storage of magnetic energy, magnetic propulsion, and electric power generation and transmission. This work began after the discovery of the first Tl-Ba-Ca-Cu oxide superconductor in 1988. Tl-cuprate superconductors have been found that have transition temperatures above − 153.15°C (120K). Certain Hg-doped Tl-cuprate superconductors even show transition temperatures above − 143.15°C (130K) at ambient pressure, almost as high as the world-record-holding Hg-cuprates.

Before the extensive use of ^{99m}Tc in nuclear medicine, the radioactive isotope ^{201}Tl, with a half-life of 73 h, was the principal isotope for nuclear cardiography. The nuclide is even now utilized for stress tests for risk stratification in patients with coronary artery disease (CAD). This radioisotope of Tl can be formed using a transportable generator comparable to the ^{99m}Tc generator. The generator comprises ^{201}Pb (half-life 9.33 h), which decays by electron capture to ^{201}Tl. The ^{201}Pb can be created in a cyclotron by the bombardment of Tl with protons or deuterons by the (p,3n) and (d,4n) reactions. A Tl-stress test is a type of scintigraphy in which the amount of Tl in tissues correlates with tissue blood supply (from Latin *scintilla*, "spark," also known as a Gamma scan, is a diagnostic test in nuclear medicine, where radioisotopes attached to drugs that travel to a specific organ or tissue (radiopharmaceuticals) are taken internally and the emitted gamma radiation is captured by external detectors (gamma cameras) to form 2D images in a similar process to the capture of X-ray images.). Viable cardiac cells have normal Na^+/K^+ ion exchange pumps. The Tl^+ cation binds the K^+ pumps and is transported into the cells. Exercise or dipyridamole (trademarked as Persantine and others, is a medication that inhibits blood clot formation when given chronically and causes blood vessel dilation when given at high doses over a short time, $C_{24}H_{40}N_8O_4$) induces widening (vasodilation) of arteries in the body. This creates coronary steal (a phenomenon where an alteration of circulation patterns leads to a reduction in the blood directed to the coronary circulation) by areas where arteries are maximally dilated. Areas of infarct or ischemic tissue will remain "cold." Pre- and poststress Tl may point to areas that will benefit from myocardial revascularization. Redistribution points to the presence of coronary steal and the existence of ischemic CAD.

A Hg–Tl alloy, which forms a eutectic at 8.5% Tl, has been published to freeze at −60°C, about 20°C below the freezing point of Hg. This alloy is employed in thermometers and low-temperature switches. In organic synthesis, Tl (III) salts, as Tl trinitrate or triacetate, are valuable reagents for performing different transformations in aromatics, ketones, olefins, etc. Tl is a component of the alloy in the anode plates of Mg-seawater batteries. Soluble Tl-salts are added to Au plating baths to increase the speed of plating and to reduce grain size within the Au layer. A saturated solution of equal parts of Tl(I) formate ($Tl(CHO_2)$) and Tl(I) malonate ($Tl(C_3H_3O_4)$) in water is known as Clerici solution. It is a mobile, odorless liquid which changes from yellowish to colorless upon reducing the concentration of the Tl salts. The solution was invented in 1907 by the Italian geologist and mineralogist Enrico Clerici (October 15, 1862 to August 26, 1938). With a density of 4.25 g/cm^3 at 20°C, Clerici solution is one of the heaviest aqueous solutions known. It was used in the 20th century for measuring the density of minerals by the flotation method, but its use has discontinued due to the high toxicity and corrosiveness of the solution. Saturated Clerici solution is denser than spinel, garnet, diamond, and corundum, as well as many other minerals. A saturated Clerici solution at 20°C can separate densities up to 4.2 g/cm^3, while a saturated solution at 90°C can separate densities up to 5.0 g/cm^3. The change in density is due to the increased solubility of the heavy Tl salts at the higher temperature. A range of solution densities between 1.0 and 5.0 g/cm^3 can be achieved by diluting with water. The refractive index shows significant, linear and well reproducible variation with the density; it changes from 1.44 for 2 g/cm^3 to 1.70 for 4.28 g/cm^3. Therefore the density can be easily measured by optical techniques. Tl iodide is often employed as an additive in metal halide lamps, often in combination with one or two halides of other metals. It allows optimization of the lamp temperature and color rendering, and shifts the spectral output to the green region, which is beneficial for underwater lighting.

7.20 82 Pb — Lead

7.20.1 Discovery

Metallic lead beads as old as 7000–6500 BCE have been discovered in Asia Minor (Anatolia, also known as Asia Minor, Asian Turkey, the Anatolian peninsula or the Anatolian plateau, is the westernmost protrusion of Asia, which makes up the majority of modern-day Turkey) and possibly represents the first instance of metal smelting. In that period lead had few (if any) applications due to its softness and dull appearance. The main cause for the spread of lead production was its association with silver, which may be obtained by burning galena (a common lead mineral, PbS) (Fig. 7.113). The Ancient Egyptians were the first to apply lead minerals in cosmetics, a use that spread to Ancient Greece and beyond; the Egyptians may well have used lead also to make sinkers in fishing nets, glazes, glasses, enamels, and ornaments. Several civilizations of the Fertile Crescent (The Fertile Crescent is a crescent-shaped region in the Middle East, spanning modern-day Iraq, Israel, Palestinian Territories, Syria, Lebanon, Egypt, and Jordan as well as the south-eastern fringe of Turkey and the western fringes of Iran. Some authors also include Cyprus.) used lead as a writing material, as currency, and as a construction material. Lead was used in the Ancient Chinese royal court as a stimulant, as currency, and as a contraceptive; the Indus Valley civilization (The Indus Valley Civilization was a Bronze Age civilization in the north-western regions of South Asia, lasting from 3300 to 1300 BCE, and in its mature form from 2600 to 1900 BCE. Along with ancient Egypt and Mesopotamia it was one of three early civilizations of the region comprising North Africa, West Asia and South Asia, and of the three, the most widespread, its sites spanning an area stretching from northeast Afghanistan, through much of Pakistan, and into western and north-western India.) and the Mesoamericans (Mesoamerica is a historical region and cultural area in North America. It extends from approximately central Mexico through Belize, Guatemala, El Salvador,

FIGURE 7.113 PLUMBUM Galaena. Sulphure of Lead; Galena. Plate no. 24. *From: Sowerby, James. 1802–1817. British Mineralogy: Or Colored figures intended to elucidate the mineralogy of Great Britain. Plate from vol: 1. page no.55.*

Honduras, Nicaragua, and northern Costa Rica, and within this region pre-Columbian societies flourished before the Spanish colonization of the Americas.) used it for making amulets; and the eastern and southern African peoples used lead in wire drawing. Since silver was broadly used as a decorative material and an exchange medium, lead ore deposits were mined in Asia Minor from 3000 BCE onward; later, lead deposits were mined in the Aegean (The Aegean Islands are the group of islands in the Aegean Sea, with mainland Greece to the west and north and Turkey to the east; the island of Crete delimits the sea to the south, those of Rhodes, Karpathos and Kasos to the southeast.) and Laurion (Laurium or Lavrio or Lavrion, is a town in south-eastern part of Attica, Greece. Laurium was famous in Classical antiquity for its silver mines, which was one of the chief sources of revenue of the Athenian state.) These three regions together dominated production of mined lead until c.1200 BCE. Since 2000 BCE, the Phoenicians (Phoenicia was a thalassocratic (a state with primarily maritime realms, an empire at sea), ancient Semitic-speaking Mediterranean civilization that originated in the Levant, specifically Lebanon, in the west of the Fertile Crescent.) mined deposits in the Iberian Peninsula; by 1600 BCE, lead mining could also be found in Cyprus, Greece, and Sardinia. Rome's territorial conquests in Europe and across the Mediterranean, and its development of mining, resulted in it becoming the largest producer of lead during the classical era, with an estimated production topping at 80,000 tons per year. Similar to their predecessors, the Romans obtained lead generally as a by-product of silver smelting. Lead mining could be found in Central Europe, Britain, the Balkans, Greece, Anatolia, and Hispania, the last at that time covering about 40% of total production. Lead tablets were frequently used as a material for letters. Lead coffins, cast in flat sand forms, with interchangeable motifs to suit the faith of the deceased were used in ancient Judea (The Roman province of Judea incorporated the regions of Judea, Samaria and Idumea, and extended over parts of the former regions of the Hasmonean and Herodian kingdoms of Judea. It was named after Herod Archelaus's Tetrarchy of Judea, but the Roman province encompassed a much larger territory. The name "Judea" was derived from the Kingdom of Judah of the 6th century BCE). Lead was used for manufacturing water pipes in the Roman Empire; the Latin word for the metal, plumbum, is the origin of the English word "plumbing." Its ease of working and resistance to corrosion guaranteed its widespread use in other applications including pharmaceuticals, roofing, currency, and warfare. Writers of that time, such as senator and historian Cato the Elder (Marcus Porcius Cato, 234 to 149 BCE), agricultural writer Columella (Lucius Junius Moderatus Columella, 4 to c.70 CE), and naturalist and natural philosopher Pliny the Elder (Gaius Plinius Secundus, 23 to 79 CE), recommended lead (or lead-coated) vessels for the preparation of sweeteners and preservatives added to wine and food. The lead conferred an agreeable taste due to the formation of "sugar of lead" (lead (II) acetate), whereas copper or bronze vessels could impart a bitter flavor through verdigris formation (Verdigris is the common name for a green pigment obtained through the application of acetic acid to copper plates or the natural patina formed when copper, brass or bronze is weathered and exposed to air or seawater over time. It is usually a basic copper carbonate (known as the malachite, $Cu_2CO_3(OH)_2$), but near the sea will be a basic copper chloride (known as the mineral atacamite, $Cu_2(OH)_3Cl$). If acetic acid is present at the time of weathering, it may consist of copper(II) acetate.). The Roman author Vitruvius (Marcus Vitruvius Pollio, c.80–70 BCE to after c.15 BCE)) noted the health dangers of lead and modern researchers have proposed that lead poisoning may have played a crucial role in the decline of the Roman Empire. Others have criticized such strong statements, pointing out, for example, that not all abdominal pain is the result of lead poisoning. Based on archeological studies, Roman lead pipes resulted in higher lead levels in tap water, but such an effect was "unlikely to have been truly harmful." When lead poisoning did happen, victims were called "saturnine," dark and cynical, after the ghoulish father of the gods, Saturn. By association, lead was considered the father of all metals. Its status in Roman society was low as it was easily accessible and inexpensive. During the classical era (and even up to the 17th century), tin was often not distinguished from lead: Romans called lead plumbum nigrum ("black lead"), and tin plumbum candidum ("bright lead"). The association of lead and tin can also be observed in other languages: the word olovo in Czech translates to "lead," but in Russian, its cognate олово (olovo) means "tin." To add to the confusion, lead exhibit a close relationship with antimony: both elements generally occur as sulfides (galena, PbS, and stibnite, Sb_2S_3), frequently together. Pliny erroneously noted that stibnite would give lead on heating, instead of antimony. In countries such as Turkey and India, the originally Persian name surma came to refer to either antimony sulfide or lead sulfide, and in some languages, such as Russian, gave its name to antimony (сурьма). Lead mining in Western Europe dropped after the fall of the Western Roman Empire, with Arabian Iberia as the only region still producing a significant amount. The largest production of lead occurred in South and East Asia, particularly China and India, where lead mining grew fast. In Europe, lead mining started to increase in the 11th and 12th centuries, when it was again used for roofing and piping. Beginning in the 13th century, lead was utilized to make stained glass. In the European and Arabian traditions of alchemy, lead was thought of as an impure base metal which, by the isolation, purification and balancing of its constituent essences, could be transformed to pure and incorruptible gold. During that period, lead was used more and more for adulterating wine (An adulterant is a substance found within other substances such as food, cosmetics, pharmaceuticals, fuel or other chemicals that compromises the safety or effectiveness of said substance.). The use of such wine was prohibited for use in Christian rites by a papal bull in 1498, but it continued to be drank and caused mass poisonings up to the late 18th century. Lead was an important material in parts of the printing press, which was invented around 1440; lead dust was commonly breathed in by print workers, resulting in lead poisoning. Firearms were developed around that same period, and lead, despite being more expensive than iron, became the principal material for making bullets. It was less damaging to iron gun barrels, had a higher density (which allowed for better preservation of the bullet's velocity), and its lower melting point made the manufacture of bullets easier as they could be made using a wood fire. Lead, in the form of Venetian ceruse (also known as blanc de ceruse de Venise and Spirits of Saturn), was broadly used in cosmetics by Western European aristocracy since whitened faces were seen as a sign of modesty (Fig. 7.114). This practice later extended to white wigs and eyeliners, and only slowly disappeared with the French Revolution in the late 18th century. A similar style appeared in Japan in the 18th century with the emergence of the geisha (Japanese women who entertain through performing the ancient traditions of art, dance, and singing, and are distinctively characterized by their wearing of kimono and oshiroi makeup. Contrary to popular belief, geisha are not the Eastern equivalent of a prostitute; a misconception originating in the West due to interactions with Japanese oiran courtesans, whose traditional attire is similar to that of geisha.), a tradition that existed long into the 20th century. The white faces of women "came to represent their feminine virtue as Japanese women," with lead generally used in the whitener. In the New World, lead production was reported soon after the arrival of European settlers. The oldest written record dates to 1621 in the English Colony of Virginia, 14 years after its foundation. In Australia the first mine opened by colonists was a lead mine, in 1841. In Africa, lead mining and smelting were known in the Benue Trough (a major geological structure underlying a large part of Nigeria and extending about 1000 km northeast from the Bight of Benin to Lake Chad. It is part of the broader Central African Rift System.) and the lower Congo Basin (the sedimentary basin of the Congo River. The Congo Basin is located in Central Africa, in a region known as west equatorial Africa.), where lead was traded with Europeans, and as a currency by the 17th century, well before the occupation, division, and colonization of African territory by Western European powers during the period of the New Imperialism, between 1881 and 1914. The second half of the 18th century, in Britain, and later continental Europe and the United States, was the period now known as the period of the Industrial Revolution. This was the first time in which the amounts of lead mined surpassed those of Rome. Britain was the leading producer, losing this position by the mid-19th century with the exhaustion of its mines and the development of lead mining in Germany, Spain, and the United States. By 1900, the United States was the leader in worldwide lead production, and other non-European countries, such as Canada, Mexico, and Australia, had started significant mining; lead production outside Europe by then topped that from within Europe. A large portion of the demand for lead originated from the plumbing and painting industry—lead paints were in regular use. At this point in time, more (working class) people were exposed to lead and lead poisoning cases spiraled. This resulted in extensive study of the

FIGURE 7.114 Portrait of Queen Elizabeth I, she was commonly depicted with a whitened face. Lead in face whiteners is thought to have contributed to her death, about 1573–1575, Nicholas Hilliard (1547–1619).

effects of lead exposure. Lead was shown to be more hazardous in its fume form than as a solid metal. Lead poisoning and gout (a form of inflammatory arthritis characterized by recurrent attacks of a red, tender, hot, and swollen joint) were connected; British physician Alfred Baring Garrod (May 3, 1819 to December 28, 1907) observed a third of his gout patients were plumbers and painters. The effects of chronic ingestion of lead, including mental disorders, were also subject of research in the 19th century. The first laws intended to decrease lead poisoning in factories were ratified during the 1870s and 1880s in the United Kingdom. Additional evidence of the risk that lead posed to humans was found in the late 19th and early 20th centuries.

7.20.2 Mining, production, and major minerals

Lead is classified as a chalcophile under the Goldschmidt classification (developed by Victor Goldschmidt (1888–1947), a geochemical classification which groups the chemical elements within the Earth according to their preferred host phases into lithophile (rock-loving), siderophile (iron-loving), chalcophile (ore-loving or chalcogen-loving), and atmophile (gas-loving) or volatile (the element, or a compound in which it occurs, is liquid or gaseous at ambient surface conditions)), meaning it is normally found in combination with sulfur. It rarely occurs as its native metal. Many lead minerals are relatively light and, in Earth's history, have stayed in the crust instead of sinking deeper into the Earth's interior. This explains the relatively high crustal abundance of 14 ppm; it is the 38th most abundant element in the crust. The most common lead-bearing mineral is galena (PbS), which is typically found together with zinc ores (mainly as sphalerite ZnS). Most other lead minerals are related to galena in some way; anglesite, $PbSO_4$, is a product of galena oxidation; and cerussite or white lead ore, $PbCO_3$, is a decomposition product of galena. Sb, Sn, As, Ag, Au, Cu, and Bi are common impurities in Pb minerals. Global Pb resources surpass 2 billion tons. Significant deposits are found in Australia, China, Ireland, Mexico, Peru, Portugal, Russia, and the United States. Worldwide reserves are estimated to total 88 million tons.

The primary and secondary lead production methods are comparable. Some primary production plants currently supplement their processes with recycled scrap lead, and this tendency will probably increase in the future. Given adequate techniques, lead produced using secondary methods is indistinguishable from lead produced using primary methods.

Scrap lead from the building trade is generally fairly clean and can be re-melted without the necessity for smelting, yet refining is sometimes required. Secondary lead production is hence cheaper, especially regarding energy requirements, than is primary production, frequently by >50%. Most lead ores have a low proportion of Pb (rich ores have a typical content of 3%–8%) that have to be concentrated for extraction. The initial processing of the ores typically involves a range of steps such as crushing, dense-medium separation, grinding, froth flotation, and drying. The subsequent concentrate, which has a Pb content between 30% and 80% by mass (regularly 50%–60%), is then turned into (impure) Pb metal. There are two principal methods to achieve this: a two-stage process involving roasting followed by blast furnace extraction, carried out in separate vessels; or a direct process in which the extraction of the concentrate occurs in a single vessel. The last is the most common route these days, though the former is still important. First, the sulfide concentrate is roasted in air to oxidize the lead sulfide.

$$2PbS(s) + 3O_2(g) \rightarrow 2PbO(s) + 2SO_2(g)\uparrow$$

Since the original concentrate did not consist of pure lead sulfide, roasting produces not only the desired Pb(II) oxide but also a mixture of oxides, sulfates, and silicates of lead and of the other metals contained in the ore. This impure lead oxide is then reduced in a coke-fired blast furnace to the (again, impure) metal.

$$2PbO(s) + C(s) \rightarrow 2Pb(s) + CO_2(g)\uparrow$$

FIGURE 7.115 Bournonite, $PbCuSbS_3$, silvery cyclic twinned crystals to 5 mm with quartz, SiO_2. Pachapaqui dist, Bolognesi prov., Ancash dept., Peru.

FIGURE 7.116 Cubic galena, PbS, crystal to 2 cm showing corners which grew faster than the centers during crystal growth. In association are brassy gold chalcopyrite, $CuFeS_2$, and quartz, SiO_2. Septemvri mine, Madan ore field, Rhodope Mts, Smolyan Oblast, Bulgaria.

FIGURE 7.117 Hutchinsonite, $TlPbAs_5S_9$, deep red crystals to 1.1 cm. Quiruvilca mine, Santiago de Chuco prov., La Libertad dept., Peru.

FIGURE 7.118 Deep amber brown plates of nadorite, $PbSbClO_2$, about 1 cm. Djebel Nador, Constantine prov., Algeria.

Impurities present are mostly metals such as arsenic, antimony, bismuth, zinc, copper, silver, and gold. Typically, they are removed in a sequence of pyrometallurgical processes. The melt is reacted in a reverberatory furnace with air, steam, and sulfur, which oxidizes the impurities except for silver, gold, and bismuth. Oxidized metal contaminants float to the top of the melt and are skimmed off. Metallic silver and gold are removed and recovered applying the Parkes process (an example of liquid–liquid extraction, patented by English metallurgist and inventor Alexander Parkes (December 29, 1813 to June 29, 1890) in 1850. He received two additional patents in 1852), in which zinc is added to the lead. Zinc, which is immiscible in lead, dissolves the silver and gold. The subsequent zinc solution containing the silver and gold can then be separated from the lead, and the silver and gold retrieved. Bismuth is then removed from the lead using the Betterton–Kroll process (developed by Luxembourgish metallurgist William Justin Kroll (November 24, 1889 to March 30, 1973)) and patented in 1922. Further improvements were developed by American metallurgist Jesse Oatman Betterton (1884–1960) in the 1930s.), through treatment with metallic calcium and magnesium. The resulting bismuth compounds have higher melting points and lower densities than the lead and can be removed as dross. The compounds are treated with chlorine to free up the bismuth. Temperature used in this process is around 380°C to 500°C. As an alternative to the pyrometallurgical processes, very pure lead can be produced by electrolytical processing the smelted lead via the Betts process (named for its inventor American chemist Anson Gardner Betts (14 April 1876 - 3 February 1976) who filed several patents for this method starting in 1901). Anodes of impure lead and cathodes of pure lead are placed in an electrolyte of lead

FIGURE 7.119 Cerussite, $PbCO_3$, crystal group to 2 cm. Silver Queen mine, near Galena, Lawrence Co., South Dakota, United States.

FIGURE 7.120 Reddish brown stubby phosgenite, $Pb_2CO_3Cl_2$, crystal to 2.5 cm with a flat termination and some beveled edges. Monteponi mine, Iglesias, Sardinia, Italy.

fluorosilicate ($PbSiF_6$). At the anode, lead dissolves into solution, as do metal impurities that are less noble than lead. Impurities that are more noble than lead, such as silver, gold, and bismuth, flake from the anode as it dissolves and settle to the bottom of the vessel as "anode mud." Pure metallic lead plates onto the cathode, with the less noble metals remaining in solution. Due to the high cost of electrolysis, this process is used only when very pure lead is needed. In the direct method, lead bullion and slag are produced straight from the lead concentrates. The lead sulfide concentrate is melted in a furnace and oxidized, forming lead monoxide (PbO). Carbon (as coke or coal gas) is added to the molten charge along with fluxing agents (a chemical cleaning agent, flowing agent, or purifying agent. Fluxes may have more than one function at a time). The lead monoxide (PbO) is in that way reduced to metallic lead (Pb), together with a slag rich in lead monoxide. If the input lead sulfide concentrate is rich in lead, as much as 80% of the original lead content can be produced as bullion; the residual 20% forms a slag rich in lead monoxide. For a low-grade feed, all the lead can be oxidized to a high-lead slag. Metallic lead is then produced from the high-lead (25%–40%) slags via submerged fuel combustion or injection, reduction assisted by an electric furnace, or a combination of both. Smelting, which is a critical part of the primary production, is frequently skipped in the secondary production process. It is only applied when metallic lead has experienced significant

FIGURE 7.121 Anglesite, $PbSO_4$, several crystals to 1.5 cm with an adamantine luster. Touissit-Bou Beker dist., Jerada prov., L'oriental, Morocco.

FIGURE 7.122 Linarite, $PbCu(SO_4)(OH)_2$, slender crystal to 2.5 mm. Grand Reef mine, Graham Co., Arizona, United States.

FIGURE 7.123 Wulfenite, $Pb(MoO_4)$, a small group of orange, tabular crystals, the largest to 5 mm in a cluster. Red Cloud mine, La Paz Co., Arizona, United States.

FIGURE 7.124 Beudantite, $PbFe_3(AsO_4)(SO_4)(OH)_6$, amber brown to brownish red tabular, pseudohexagonal crystals to 2 mm in clusters. Tsumeb mine, Tsumeb, Namibia.

oxidation. The method is comparable to that of primary production in either a blast furnace or a rotary furnace, with the critical difference being the larger variation in yields: blast furnaces produce hard lead (10% antimony) whereas reverberatory and rotary kiln furnaces produce semisoft lead (3%–4% antimony). The Isasmelt process (an energy-efficient smelting process that was jointly developed from the 1970s to the 1990s by Mount Isa Mines Limited (a subsidiary of MIM Holdings Limited and now part of Glencore plc) and the Australian government's Commonwealth Scientific and Industrial Research Organization ("CSIRO"). It has relatively low capital and operating costs for a smelting process.) is a more recent smelting method that can be used as an extension to primary production; battery paste from spent lead–acid batteries (having lead sulfate and lead oxides) has its sulfate removed through treatment with alkali, and subsequent treatment in a coal-fueled furnace in the presence of oxygen, which produces impure lead, with antimony as the most commonly occurring impurity. Refining of secondary lead is comparable to that of primary lead; some of the refining processes may be skipped dependent on the material recycled and its potential contaminations present. Of the sources of lead for recycling, lead–acid batteries are the most significant; in addition, lead pipe, sheet, and cable sheathing are also substantial.

FIGURE 7.125 Carminite, $PbFe_2^{3+}(AsO_4)_2(OH)_2$, 1 mm acicular crystals. Kintore Opencut, Broken Hill, New South Wales, Australia.

FIGURE 7.126 Descloizite, $PbZn(VO_4)(OH)$, brown to reddish brown tabular crystals to over 1 cm. Berg Aukus, Grootfontein dist., Namibia.

Over 450 minerals are known with lead in their crystal structure. In the element class, besides the native metal lead (Pb) there are six known alloys, for example, leadamalgam ($Pb_{0.7}Hg_{0.3}$) and plumbopalladinite (Pd_3Pb_2). Around 175 of the Pb-containing minerals are found in the sulfide class, for example, boulangerite ($Pb_5Sb_4S_{11}$), bournonite ($PbCuSbS_3$) (Fig. 7.115), cylindrite ($Pb_3Sn_4FeSb_2S_{14}$), galena (PbS) (Fig. 7.116), hutchinsonite ($TlPbAs_5S_9$) (Fig. 7.117), jamesonite

FIGURE 7.127 Large crystalline 8 cm mass of light purple or lilac hedyphane, $Ca_2Pb_3(AsO_4)_3Cl$. Långban, Filipstad, Värmland, Sweden.

FIGURE 7.128 Mimetite, Pb5$(AsO_4)_3Cl$, A 5-cm group of bright yellow spheroidal shaped clusters to 7−9 mm. San Pedro Corralitos, Chihuahua, Mexico.

($Pb_4FeSb_6S_{14}$), and zinkenite ($Pb_9Sb_{22}S_{42}$). The halides contain about 40 minerals with Pb in their chemical composition, for example, boleite ($KPb_{26}Ag_9Cu_{24}(OH)_{48}Cl_{62}$), cotunnite ($PbCl_2$), diaboleite ($Pb_2CuCl_2(OH)_4$), mendipite ($Pb_3Cl_2O_2$), and nadorite ($PbSbClO_2$) (Fig. 7.118). About 70 minerals are found in the oxide class, including minerals such as coronadite ($Pb(Mn_6^{4+}Mn_2^{3+})O_{16}$), curite ($Pb_3(UO_2)_8O_8(OH)_6 \cdot 3H_2O$), litharge ($PbO$), minium ($Pb_3O_4$), and plattnerite ($PbO_2$). The carbonate class is represented by 17 minerals, for example, cerussite ($PbCO_3$) (Fig. 7.119), dudasite ($PbAl_2(CO_3)_2(OH)_4 \cdot H_2O$), leadhillite ($Pb_4(CO_3)_2(SO_4)(OH)_2$), and phosgenite ($Pb_2CO_3Cl_2$) (Fig. 7.120). Leucostaurite ($Pb_2[B_5O_9]Cl \cdot 0.5H_2O$) is the only known borate. Nearly 50 minerals can be found in the sulfate class, including well-known minerals such as anglesite ($PbSO_4$) (Fig. 7.121), crocoite ($PbCrO_4$), linarite ($PbCu(SO_4)(OH)_2$) (Fig. 7.122), plumbojarosite ($Pb_{0.5}Fe_3^{3+}(SO_4)_2(OH)_6$), stolzite ($PbWO_4$), and wulfenite ($PbMoO_4$) (Fig. 7.123). The phosphate class is represented by about 75 different minerals, for example, bayldonite ($PbCu_3(AsO_4)_2(OH)_2$), beudantite ($PbFe_3(AsO_4)(SO_4)(OH)_6$) (Fig. 7.124), carminite ($PbFe_2^{3+}(AsO_4)_2(OH)_2$) (Fig. 7.125), descloizite ($PbZn(VO_4)(OH)$) (Fig. 7.126), duftite ($PbCu(AsO_4)(OH)$), hedyphane ($Ca_2Pb_3(AsO_4)_3Cl$) (Fig. 7.127), mimetite ($Pb_5(AsO_4)_3Cl$) (Fig. 7.128), mottramite ($PbCu(VO_4)(OH)$) (Fig. 7.129), and pyromorphite ($Pb_5(PO_4)_3Cl$) (Fig. 7.130). The silicates are represented by around 35 minerals, such as alamosite ($PbSiO_3$), esperite ($PbCa_2Zn_3(SiO_4)_3$), kasolite ($Pb(UO_2)[SiO_4] \cdot H_2O$), melanotekite ($Pb_2Fe_2^{3+}(Si_2O_7)O_2$), and wickenburgite ($CaPb_3Al_2Si_{10}O_{24}(OH)_6$).

FIGURE 7.129 Soft velvety dark green growths of mottramite, $PbCu(VO_4)(OH)$, 5 mm across. Mapimi dist., Durango, Mexico.

FIGURE 7.130 Pyromorphite, $Pb_5(PO_4)_3Cl$, dark grass green, lustrous, stubby 1 mm crystals, Chaillac, Indre, Centre, France.

7.20.3 Chemistry

Natural Pb has four stable isotopes: ^{204}Pb, ^{206}Pb, ^{207}Pb, and ^{208}Pb, and traces of five short-lived radioisotopes. The high number of isotopes is consistent with lead's atomic number being even (An even number of either protons or neutrons generally increases the nuclear stability of isotopes, compared to isotopes with odd numbers. No elements with odd atomic numbers have more than two stable isotopes; even-numbered elements have multiple stable isotopes, with tin (element 50) having the highest number of isotopes of all elements, ten). Pb has a magic number of protons (82), for which the nuclear shell model precisely predicts a particularly stable nucleus. ^{208}Pb has 126 neutrons, again a magic number, which may explain the reason why ^{208}Pb is extremely stable. With its high atomic number, Pb forms the heaviest element with natural isotopes that are regarded as stable; ^{208}Pb is the heaviest stable nucleus (this distinction previously fell to Bi, with an atomic number of 83, until its single primordial isotope, ^{209}Bi, was observed in 2003 to decay very slowly, with a half-life of 1.9×10^{19} years.). The four stable Pb isotopes could theoretically decay via α decay to Hg isotopes with a release of energy, but this has not been detected for any of these stable isotopes; their predicted half-lives range from 10^{35} to 10^{189} years (at least 10^{25} times the currently known age of the universe). Three of the stable isotopes are found in three of the four major decay chains: ^{206}Pb, ^{207}Pb, and ^{208}Pb form the last decay products of ^{238}U, ^{235}U, and ^{232}Th, respectively. These decay chains are generally known as the uranium chain, the actinium chain, and the thorium chain. Their isotopic concentrations in natural rock samples strongly depend on the occurrence of these three parent U and Th isotopes. For example, the relative abundance of ^{208}Pb can vary between 52% in normal samples and 90% in thorium ores; for this reason, the standard atomic weight of lead is generally reported with only one decimal place. With increasing age of, for example, rocks or minerals, the ^{206}Pb and ^{207}Pb to ^{204}Pb ratio increase, as the former two are supplemented by radioactive decay of heavier elements while the latter is not; this allows for what is known as lead–lead dating. As U decays into Pb, their relative concentrations change; this is the basis for uranium–lead dating. ^{207}Pb shows nuclear magnetic resonance, a property that has been utilized for research on its compounds in both solution and solid state, including in the human body. Next to the stable isotopes, which forms nearly all naturally existing lead, there exist trace amounts of some radioactive isotopes. One of them is ^{210}Pb. Even though it has a relatively short half-life of only 22.3 years, small amounts are found in nature as ^{210}Pb is formed by a long decay series that starts with ^{238}U (which has existed for billions of years on Earth). ^{211}Pb, ^{212}Pb, and ^{214}Pb form part of the decay chains of ^{235}U, ^{232}Th, and ^{238}U, respectively, hence traces of all three of these Pb isotopes occur naturally. Minute traces of ^{209}Pb are formed by the very rare cluster decay of ^{223}Ra, one of the daughter products of naturally occurring ^{235}U, and the decay chain of ^{237}Np, trace amounts of which are formed via neutron capture in uranium ores. ^{210}Pb is especially valuable for helping to identify the ages of rock and mineral samples by determining its ratio to ^{206}Pb (both these isotopes exist in a single decay chain). In total, 43 artificial Pb isotopes have been observed, with mass numbers ranging from 178 to 220. ^{205}Pb is the most stable radioisotope, with a half-life of approximately 1.5×10^7 years. ^{205}Pb decays exclusively via electron capture, which means when there are no electrons available and Pb is fully ionized with all 82 electrons removed it cannot decay. Fully ionized ^{205}Tl, the isotope ^{205}Pb would decay to, becomes unstable and can decay into a bound state of ^{205}Pb. The second most stable is ^{202}Pb, which has a half-life of approximately 53,000 years, longer than any of the naturally occurring trace Pb radioisotopes.

Lead is a soft, malleable, and low-melting point metal. It is a heavy metal that is denser than most common materials. It has 82 electrons arranged in an electronic configuration of $[Xe]4f^{14}5d^{10}6s^26p^2$. There are four valence electrons in the outermost shell but the inert pair effect, common to posttransition elements, is important and thus, lead takes the +2 oxidation state in most of its compounds. Nevertheless, the +4 oxidation state is also common in lead compounds (Table 7.20).

Lead appears silvery with a hint of blue when freshly cut but tarnishes to a dull gray color when exposed to moist air. Finely divided Pb powder is pyrophoric but the bulk metal is relatively inert. The reduction in reactivity of the metal is due to the formation of a thin, protective layer of insoluble product such as oxide, oxocarbonate, sulfate or chloride. The latter two passivating layers, $PbSO_4$ and $PbCl_2$, normally form in urban and marine environments. The inertness of lead metal has been exploited as one of the main assets of the metal since early times. The bulk metal does

TABLE 7.20 Lead properties.

Appearance	Metallic gray
Standard atomic weight $A_{r,std}$	207.2
Block	p-Block
Element category	Posttransition metal
Electron configuration	[Xe] $4f^{14}$ $5d^{10}$ $6s^2$ $6p^2$
Phase at STP	Solid
Melting point	327.46°C
Boiling point	1749°C
Density (near r.t.)	11.34 g/cm^3
When liquid (at m.p.)	10.66 g/cm^3
Heat of fusion	4.77 kJ/mol
Heat of vaporization	179.5 kJ/mol
Molar heat capacity	26.650 J/(mol·K)
Oxidation states	−4, −2, −1, +1, **+2**, +3, **+4**
Ionization energies	1st: 715.6 kJ/mol 2nd: 1450.5 kJ/mol 3rd: 3081.5 kJ/mol
Atomic radius	Empirical: 175 pm
Covalent radius	146 ± 5 pm
Van der Waals radius	202 pm

STP, Standard temperature and pressure.
Bold font indicates main oxidation state.

not burn easily, requiring a temperature of 600°C–800°C to form PbO in air. It is also resistant to sulfuric acid attack and is widely used for handling hot concentrated H_2SO_4. It does react with other acids. Aqueous HCl reacts slowly to give the sparingly soluble $PbCl_2$ (< 1% at room temperature) and nitric acid reacts quite rapidly to liberate oxides of nitrogen and form the very soluble $Pb(NO_3)_2$. Lead is also attacked by organic acids such as acetic acid in the presence of air to give $Pb(OAc)_2$. This apparent reaction with organic acids is fortunate as this precludes contact with the metal when processing or storing wine, fruit juices, and other drinks in those times when the toxicity of the metal was not very well known. It did not prevent its use in ancient times, however. Its use and the corresponding toxicity effects are mentioned in historical accounts of Vitruvius' observation on lead. He stated that, "water is much more wholesome from earthenware pipes than from lead pipes. For it seems to be made injurious by lead, because white lead is produced by it, and this is said to be harmful to the human body."

7.20.3.1 Oxides

PbO produced by reacting molten Pb with air or O_2 above 600°C exists as a red tetragonal β form (known as the mineral litharge) stable at room temperature and a yellow orthorhombic α form (known as the mineral massicot) stable above 488°C. Litharge is the most important oxide and most widely used inorganic compound of Pb. A mixed valence oxide Pb_3O_4 (red lead) is made by heating PbO in air in a reverberatory furnace at 450°C–500°C and used to be an important commercial pigment and primer. Other oxides of lead are well known. Lead(IV) oxide, PbO_2, can be found in nature as the minerals plattnerite and scrutinyite. When heated in air it decomposes as follows:

$$PbO_2 \xrightarrow{293^\circ C} Pb_{12}O_{19} \xrightarrow{351^\circ C} Pb_{12}O_{19} \xrightarrow{374^\circ C} Pb_3O_4 \xrightarrow{605^\circ C} PbO$$

In addition, thermal decomposition of lead(II) nitrate or lead(II) carbonate leads to the formation of PbO:

$$2Pb(NO_3)_2 \rightarrow 2PbO + 4NO_2 + O_2$$
$$PbCO_3 \rightarrow PbO + CO_2$$

PbO is manufactured on a large scale as an intermediate product in refining raw lead ores, usually galena (PbS), into metallic lead. At a temperature of ~1,000°C the sulfide is converted to the oxide:

$$2PbS + 3O_2 \rightarrow 2PbO + 2SO_2$$

Metallic lead is formed by reducing PbO with carbon monoxide at ~1,200°C:

$$PbO + CO \rightarrow Pb + CO_2$$

PbO is amphoteric, which means that it reacts with both acids and with bases. With acids, it forms salts of Pb^{2+} via the intermediacy of oxo clusters such as $[Pb_6O(OH)_6]^{4+}$. With strong bases, PbO dissolves to form plumbite (also called plumbate(II)) salts:

$$PbO + H_2O + OH^- \rightarrow [Pb(OH)_3]^-$$

or

$$PbO_2 + 2OH^- + 2H_2O \rightarrow Pb(OH)_6^{2-}$$

The sesquioxide Pb_2O_3 can also be obtained by controlled decomposition of PbO_2 at 580°C–620°C under 1.4 kbar oxygen pressure, along with several nonstoichiometric phases. Many of them show defective fluorite (CaF_2) structures in which some oxygen atoms are substituted by vacancies: PbO can be considered as having such a structure, with every alternate layer of oxygen atoms missing.

7.20.3.2 Chalcogenides

Pb also reacts with the other members of group 16. The heavier members of group 16 (S, Se, and Te referred to as chalcogens) react with molten Pb give the chalcogenides PbS, PbSe and PbTe. Pure PbS can be made by direct reaction of the elements or by reaction of $Pb(OAc)_2$ with thiourea. Reaction of lead with sulfur or hydrogen sulfide produces lead sulfide. The solid has the NaCl-like structure (simple cubic) up to its melting point at 1114°C. If the heating occurs in presence of air, it decomposes to form the monoxide and the sulfate. The compound is nearly insoluble in water, weak acids, and $(NH_4)_2S/(NH_4)_2S_2$ solution is the key for separation of lead from analytical groups I to III elements, tin, arsenic, and antimony. The compound dissolves in nitric and hydrochloric acids, to form elemental sulfur and hydrogen sulfide, respectively. Heating mixtures of the monoxide and the sulfide produces the metal.

$$2PbO + PbS \rightarrow 3Pb + SO_2\uparrow$$

PbSe (m.p. 1075°C) can be obtained by volatilizing $PbCl_2$ with H_2Se, by reacting $PbEt_4$ with H_2Se in organic solvents, or by reducing $PbSeO_4$ with H_2 or C in an electric furnace; PbTe is best made by heating Pb with the stoichiometric amount of Te.

7.20.3.3 Halides

Lead reacts with fluorine even at room temperatures to give PbF_2, while it requires heating to react with Cl_2 to give $PbCl_2$. Heating lead carbonate with hydrogen fluoride yields the hydrofluoride, which decomposes to the difluoride when it melts. This white crystalline powder is more soluble than the diiodide, but less than the dibromide and the dichloride. No coordinated lead fluorides exist (except the unstable PbF^+ cation). The tetrafluoride, a yellow crystalline powder, is unstable. Lead(II) bromide is typically formed by heating lead salts with bromide salts but can also be formed as a result of the use of leaded gasoline. Tetraethyl lead, $PbEt_4$, was once widely used to improve the combustion of gasoline. Leaded gasoline is mixed with organobromine compounds so that lead oxides can be converted to volatile lead(IV) bromide. Lead(IV) iodide, on the other hand, is often prepared from the precipitation reactions of soluble lead and iodide salts. Pb can form two series of halides: MX_2 and MX_4. PbX_2 are more stable than PbX_4. The only stable tetrahalide is the yellow PbF_4. $PbCl_4$ is stable below 0°C but decomposes to $PbCl_2$ and Cl_2 above 50°C; $PbBr_4$ is even less stable and PbI_4 is of doubtful existence. The tetrachloride is formed upon dissolving the dioxide in hydrochloric acid; to prevent the exothermic decomposition, it must be kept under concentrated sulfuric acid. Their solubility increases with temperature; adding more halides first decreases the solubility, but subsequently increases because of complexation, with the maximum coordination number being 6. The complexation depends on halide ion numbers, atomic number of the alkali metal, the halide of which is added, temperature and solution ionic strength. Lead(II) forms a series of complexes with chloride, the formation of which alters the corrosion chemistry of the lead. This will tend to limit the solubility of lead in saline media. The equilibrium constants for aqueous lead chloride complexes at 25°C are as follows:

$$Pb^{2+} + Cl^- \rightarrow PbCl^+ \qquad K = 12.59$$
$$PbCl^+ + Cl^- \rightarrow PbCl_2 \qquad K = 14.45$$
$$PbCl_2 + Cl^- \rightarrow PbCl_3^- \qquad K = 3.98 \times 10^{-1}$$
$$PbCl_3^- + Cl^- \rightarrow PbCl_4^{2-} \qquad K = 8.92 \times 10^2$$

Many lead(II) pseudohalides are known, e.g. the cyanide, cyanate, and thiocyanate (pseudohalides are polyatomic analogues of halogens, whose chemistry, resembling that of the true halogens, allows them to substitute for halogens in several classes of chemical compounds). Lead(II) forms an extensive variety of halide coordination complexes, such as $[PbCl_4]^{2-}$, $[PbCl_6]^{4-}$, and the $[Pb_2Cl_9]_n^{5n-}$ chain anion.

7.20.3.4 Hydrides

The simple hydride, plumbane, PbH_4 is the least well-characterized group 14 hydride. It is likely that it has never been prepared except perhaps in trace amounts at high dilution. The methods that yielded MH_4 for other group 14 elements all fail even at low temperatures. The alkyl derivatives of lead hydride have been successfully prepared, though. The alkyl derivatives R_2PbH_2 and R_3PbH were prepared from the corresponding halides and $LiAlH_4$ at −78°C.

7.20.3.5 Negative oxidation states

Negative oxidation states can occur as Zintl phases, as either free lead anions, as in Ba_2Pb, with lead formally being lead(−IV), or in oxygen-sensitive ring-shaped or polyhedral cluster ions such as the trigonal bipyramidal Pb_5^{2-} ion, where two lead atoms are lead(−I) and three are lead(0) (In chemistry, a Zintl phase is the product of a reaction between a group 1 (alkali metal) or group 2 (alkaline earth) and any post-transition metal or metalloid (i.e. from group 13, 14, 15 or 16). It is named after the German chemist Eduard Zintl (January 21, 1898 to January 17, 1941) who investigated them in the 1930s). In such anions, each atom is at a polyhedral vertex and contributes two electrons to each covalent bond along an edge from their sp^3 hybrid orbitals, the other two being an external lone pair. They may be made in liquid ammonia via the reduction of lead by sodium.

7.20.3.6 *Organolead compounds*

The organic chemistry of Pb is much less extensive than that of Sn, though over 2000 organolead compounds were known even in the late 1990s. Lead can form multiply-bonded chains, a property it shares with its lighter homologs in the carbon group. Its capacity to do so is much less because the Pb–Pb bond energy is over three and a half times lower than that of the C–C bond. With itself, lead can build metal–metal bonds of an order up to three. With carbon, lead forms organolead compounds comparable to, but usually less stable than, typical organic compounds (because the Pb–C bond is rather weak). Lead mainly forms organolead(IV) compounds, even when starting with inorganic lead(II) reactants; very few organolead(II) compounds are known. The most well-characterized exceptions known today are Pb $[CH(SiMe_3)_2]_2$ and $Pb(\eta^5\text{-}C_5H_5)_2$.

The lead analog of the simplest organic compound, methane, is plumbane. Two simple derivatives, tetramethyllead and tetraethyllead, are the best known organolead compounds. These compounds are relatively stable: tetraethyllead only starts to decompose if heated or if exposed to sunlight or ultraviolet light (tetraphenyllead is even more thermally stable, decomposing at 270°C). With sodium metal, lead easily forms an equimolar alloy that reacts with alkyl halides to produce organometallic compounds such as tetraethyllead.

$$4NaPb + 4CH_3CH_2Cl \rightarrow (CH_3CH_2)_4Pb + 4NaCl + 3Pb$$

A significant aspect of $PbEt_4$ is the weakness of its 4 C–Pb bonds. At the working temperatures of internal combustion engines, $(CH_3CH_2)_4Pb$ fully decomposes into lead and lead oxides along with combustible, short-lived ethyl radicals. Lead and lead oxide scavenge radical intermediates in the combustion reactions. Engine knock is the result of a cool flame, an oscillating low-temperature combustion reaction that happens before the proper, hot ignition. Lead quenches the pyrolyzed radicals and thus stops the radical chain reaction that would maintain a cool flame, inhibiting it from disturbing the smooth ignition of the hot flame front. Lead itself is the reactive antiknock agent, and $PbEt_4$ serves as a gasoline-soluble lead carrier. When $PbEt_4$ burns, it forms not only carbon dioxide and water but also lead.

$$(CH_3CH_2)_4Pb + 13O_2 \rightarrow 8CO_2 + 10H_2O + Pb$$

This lead can oxidize further to form compounds such as lead(II) oxide.

$$2Pb + O_2 \rightarrow 2PbO$$

Pb and PbO would rapidly over-accumulate and destroy an engine. Therefore the lead scavengers 1,2-dibromoethane and 1,2-dichloroethane are employed together with $PbEt_4$—these agents form volatile lead(II) bromide and lead(II) chloride, respectively, which are flushed from the engine and into the air. $PbEt_4$ used to be produced on a larger tonnage than any other single organometallic compound for its use in leaded gasoline. Its production has been reduced in the 1970s and has since slowly been phased out.

The oxidizing nature of many organolead compounds is usefully exploited: lead tetraacetate is an important laboratory reagent for oxidation in organic synthesis. Other organolead compounds are less chemically stable. For many organic compounds, a lead analog does not exist. The most useful laboratory-scale routes to organoleads involve the use of Grignard reagents, RMgX, and lead(II) chloride. Organolithium, LiR, or organoaluminum, AlR_3, on lead(II) compounds such as $PbCl_2$, or lead(IV) compounds such as R_2PbX_2, R_3PbX, or K_2PbCl_6 have also been used. On the industrial scale the reaction of an alkyl halide, RX, on a Pb/Na alloy is much used. For example, methylmagnesium chloride reacts with lead chloride to tetramethyllead. Reaction of a lead(II) source with sodium cyclopentadienide produces the lead metallocene, plumbocene ($C_{10}H_{10}Pb$). Some arene compounds react directly with lead tetraacetate to aryl lead compounds in an electrophilic aromatic substitution. For example, anisole with lead tetraacetate results in 'p-methoxyphenyllead triacetate in chloroform and dichloroacetic acid. Other compounds of lead are organolead halides of the type $R_nPbX_{(4-n)}$, organolead sulphinates ($R_nPb(OSOR)_{(4-n)}$) and organolead hydroxides ($R_nPb(OH)_{(4-n)}$). Some characteristic examples are

$$R_4Pb + HCl \rightarrow R_3PbCl + RH$$
$$R_4Pb + SO_2 \rightarrow R_3PbO(SO)R$$
$$R_3PbCl + 1/2Ag_2O(aq) \rightarrow R_3PbOH + AgCl$$
$$R_2PbCl_2 + 2OH^- \rightarrow R_2Pb(OH)_2 + 2Cl^-$$

$R_2Pb(OH)_2$ compounds are amphoteric. At pH lower than 8 they form R_2Pb^{2+} cations and with pH higher than 10, $R_2Pb(OH)_3^-$ anions. Derived from the hydroxides are the plumboxanes.

$$4R_3PbOH + 2Na \rightarrow 2(R_3Pb)_2O + 2NaOH + H_2$$

These give access to polymeric alkoxides.

$$(R_3Pb)_2O + R'OH \rightarrow 1/n(R_3PbOR')_n - nH_2O$$

The C−Pb bond is weak and therefore homolytic cleavage of organolead compounds to free radicals is easy. General reaction types of aryl and vinyl organoleads are transmetalation, for example, with boronic acids and acid-catalyzed heterocyclic cleavage. Organolead compounds find use in coupling reactions between arene compounds. They are more reactive than the comparable organotin compounds and can thus be used to synthesize sterically crowded biaryls. In oxyplumbation, organolead alkoxides are added to polar alkenes.

$$H_2C = CH{-}CN + (Et_3PbOMe)_n \rightarrow$$
$$MeO{-}CH_2{-}HC\,(PbEt_3) - CN \rightarrow MeO{-}CH_2{-}CH_2{-}CN$$

The alkoxide is regenerated in the subsequent methanolysis and therefore acts as a catalyst. Organolead compounds form various reactive intermediates such as lead-free radicals

$$Me_3PbCl + Na\ (77\ K) \rightarrow Me_3Pb.$$

and plumbylenes, the lead carbene equivalents.

$$Me_3Pb - Pb - Me_3 \rightarrow [Me2Pb]$$
$$[Me_2Pb] + (Me_3Pb)_2 \rightarrow Me_3Pb - Pb(Me)_2{-}PbMe_3$$
$$Me_3Pb - Pb(Me)_2{-}PbMe_3 \rightarrow Pb(0) + 2Me_4Pb$$

These intermediates break up by disproportionation.

7.20.4 Major uses

Different to popular belief, pencil leads in wooden pencils have never been manufactured from Pb. When the pencil originated as a wrapped graphite (carbon, C) writing tool, the graphite used was called *plumbago* (literally, act for lead or lead mock-up). Pb metal has a number of useful mechanical properties, for example, high density, low melting point, ductility, and relative inertness. Various metals are superior to Pb in some of these properties but are usually less common and more challenging to extract from parent ores. Its toxicity has resulted in its phasing out for some applications. Pb has been used for bullets since their creation in the Middle Ages. It is cheap; its low melting point means that small arms ammunition and shotgun pellets can be molded with minimal technical equipment; and it is denser than other popular metals, which allows for better preservation of velocity. It is still the key material for bullets, alloyed with other metals as hardeners. Concerns have been raised that Pb bullets employed for hunting can cause environmental damage. Its high density and corrosion resistance have been used in several related applications. It is employed as ballast in, for example, sailboat keels where its density permits it to occupy a small volume and minimize water resistance, consequently counterbalancing the heeling effect of wind on the sails. It is also utilized in scuba diving weight belts to offset the diver's buoyancy. Back in 1993, the base of the Leaning Tower of Pisa was stabilized with 600 tons of Pb. Due to its corrosion resistance, Pb is employed as a protective sheath around underwater cables. Pb has numerous applications in the construction industry; Pb sheets are employed as architectural metals in, for example, roofing material, cladding, flashing, gutters and gutter joints, and on roof parapets. Detailed Pb moldings are utilized as decorative motifs to fix Pb sheet. Pb is even today utilized in statues and sculptures, including for armatures. Historically it was regularly used to balance the wheels of cars; due to environmental concerns this application is now being phased out in favor of alternative materials.

Pb is added to Cu alloys, such as brass and bronze, to enhance machinability and for its lubricating properties. Being almost insoluble in Cu the Pb forms solid globules in imperfections all through the alloy, such as grain boundaries. In low concentrations, as well as acting as a lubricant, the globules impede the formation of swarf (also known as chips or by other process-specific names (such as turnings, filings, or shavings), are pieces of metal, wood, or plastic that are the debris or waste resulting from machining, woodworking, or similar subtractive (material-removing) manufacturing processes.) as the alloy is worked, thus enhancing machinability. Cu alloys with higher amounts of Pb are used in bearings. The Pb delivers lubrication, and the Cu offers the load-bearing support. Its high density, atomic number, and formability are the basis for the application of Pb as a barrier that absorbs sound, vibrations, and radiation. Pb possesses no natural resonance frequencies; therefore Pb sheet is employed as a sound deadening layer in the walls, floors, and ceilings of sound studios. Organ pipes are frequently made from a Pb alloy, mixed with different amounts of Sn to influence the tone of each pipe. Pb is a well-recognized shielding material for radiation in nuclear science and in X-ray rooms due to its denseness and high attenuation coefficient (characterizes how easily a volume of material can be penetrated by a beam of light, sound, particles, or other energy or matter. A large attenuation coefficient means that the beam is quickly "attenuated" (weakened) as it passes through the medium, and a small attenuation coefficient means that the medium is relatively transparent to the beam.). Molten Pb has been utilized as a coolant in Pb-cooled fast reactors.

The principal use of Pb at the beginning of the 21st century is in Pb−acid batteries. The Pb in these batteries does not come in direct contact with humans, hence there are fewer toxicity problems. People who work in battery manufacturing plants though may be exposed to Pb dust and inhale it. The reactions in the battery between Pb, PbO_2, and sulfuric acid deliver a dependable source of voltage. Supercapacitors incorporating Pb−acid batteries, for instance, have been installed in kilowatt and megawatt scale applications in various countries such as Australia, Japan, and the United States in frequency regulation, solar smoothing and shifting, wind smoothing, and other uses. These batteries possess lower energy density and charge-discharge efficiency than Li-ion batteries but are considerably less expensive. Pb is utilized in high-voltage power cables as sheathing metal to stop water diffusion into the insulation; this application is declining as Pb is being phased out. Similarly, its application in solder for electronics is also being phased out by certain countries in order to decrease the amount of environmentally hazardous waste. Pb is one of three metals used in the Oddy test for museum materials, helping detect organic acids, aldehydes, and acidic gases (a procedure created at the British Museum by conservation scientist William Andrew Oddy (born 6 January 1942) in 1973). This test calls for a sample of the material in question to be placed in an airtight container with three coupons of different metals—silver, lead, and copper—that are not touching each other or the sample of the material. The container is sealed with a small amount of de-ionized water to maintain a high humidity, then heated at 60°C for 28 days. An identical container with three metal coupons acts as a control. If the metal coupons show no signs of corrosion, then the material is believed suitable to be placed in and around art objects. The Oddy test is not a contact test but is for testing off-gassing. Each metal detects a different set of corrosive agents. The silver is for detecting reduced sulfur compounds and carbonyl sulfides. The lead is for detecting organic acids, aldehyde, and acidic gases. The copper is for detecting chloride, oxide, and sulfur compounds. Besides being the main use for Pb metal, Pb−acid batteries are also the main consumer of Pb compounds. The energy storage/release reaction used in these batteries involves $PbSO_4$ and PbO_2.

$$Pb(s) + PbO_2(s) + 2H_2SO_4(aq) \rightarrow 2PbSO_4(s) + 2H_2O(l)$$

Additional uses of Pb compounds are very specialized and often fading. Pb-based coloring agents are employed in ceramic glazes and glass, especially for red and yellow shades. While Pb paints are phased out in Europe and North America, they are still in use in less developed countries such as China, India, and Indonesia. Pb(IV)-tetraacetate ($Pb(C_2H_3O_2)_4$) and PbO_2 are used as oxidizing agents in organic chemistry. Pb is regularly utilized in the polyvinyl chloride (PVC) coating of electrical cords. It can be employed to treat candle wicks to ensure a longer, more even burn. Due to its toxicity, European and North American producers employ substitutes such as Zn. Lead glass contains between 12% and 28% Pb oxide, altering its optical characteristics and decreasing the transmission of ionizing radiation. Pb-based semiconductors such as PbTe and PbSe can be found in photovoltaic cells and infrared detectors.

7.21 83 Bi — Bismuth

7.21.1 Discovery

The name bismuth originates from around the 1660s and is of ambiguous etymology. It is one of the first 10 metals to have been found. Bismuth appears in the 1660s, from obsolete German Bismuth, Wismut, Wissmuth (early 16th century); possibly associated with Old High German hwiz ("white"). The New Latin bisemutum (because of German Humanist scholar and one of the leading experts on mineralogy and metallurgy Georgius Agricola (March 24, 1494 to November 21, 1555) (Fig. 7.131), who Latinized many German mining and technical terms) is from the German Wismuth, perhaps from weiße Masse, "white mass." The element was often mistaken in early times for tin or lead due to its similarity to these elements. Bismuth has been recognized since ancient times, so its discovery cannot be attributed to a single person. Agricola, in his De Natura Fossilium (Agricola, 1546) wrote that bismuth is a distinct metal in a family of metals including tin and lead. This was based on study of the metals and their physical properties. Miners in the period of alchemy also named bismuth tectum argenti, or "silver being made," in the sense of silver still in the process of being formed within the Earth. Starting with German chemist Johann Heinrich Pott (1692 to March 29, 1777) in 1738 (Pott, 1738), Swedish Pomeranian and German pharmaceutical chemist Carl Wilhelm Scheele (December 9, 1742 to May 21, 1786) and Swedish chemist and mineralogist Torbern Olof Bergman (March 20, 1735 to July 8, 1784), the difference between lead and bismuth became clear, and French chemist Claude François Geoffroy (1729 to June 18,

FIGURE 7.131 Georg Agricola, De re metallica 1556.

FIGURE 7.132 Deeply "striated" large bismuth, Bi, crystal to 3.5 cm across. Sorata, Larecaja prov., La Paz Dept., Bolivia.

FIGURE 7.133 Steely gray, long bladed crystals of bismuthinite, Bi_2S_3, to 1.5 cm associated with chalcopyrite, $CuFeS_2$. Creole mine, Beaver Co., Utah, United States.

FIGURE 7.134 Specimen with tiny acicular crystals of emplectite, $CuBiS_2$, to less than 1 mm. Tannenbaum mine, Breitenbrunn dist., Erzgebirge, Saxony, Germany.

FIGURE 7.135 Joséite, Bi_4TeS_2, platy silvery crystals, elongated to 2 mm in quartz, SiO_2. Carrock mine, Caldbeck Fells, Cumbria, England.

1753) proved in 1753 that this metal is different from both lead and tin (Geoffroy, 1753). Bismuth was also familiar to the Incas and utilized (together with the usual copper and tin) in a special bronze alloy for knives.

7.21.2 Mining, production, and major minerals

In the Earth's crust, bismuth is about twice as abundant as gold. The most important ores of bismuth are bismuthinite (Bi_2S_3) and bismite (Bi_2O_3). Native bismuth is known from Australia, Bolivia, and China. The world mining production of bismuth in 2016 was about 10,000 tons, with the major contributions from China, Vietnam and Mexico. Bismuth is mainly a by-product of extraction of other metals such as lead, copper, tin, molybdenum and tungsten. Bismuth stays in crude lead bullion (which can have up to 10% Bi) through a number of different refining steps, until it is removed by the Kroll-Betterton process (developed by Luxembourgish metallurgist William Justin Kroll (November 24, 1889 to March 30, 1973)) and patented in 1922. Further improvements were developed by American metallurgist Jesse Oatman

FIGURE 7.136 Tetradymite, Bi_2Te_2S, silvery thin laminae or foliated crystal to about 1.2 cm. Carrock mine, Caldbeck Fells, Cumbria, England.

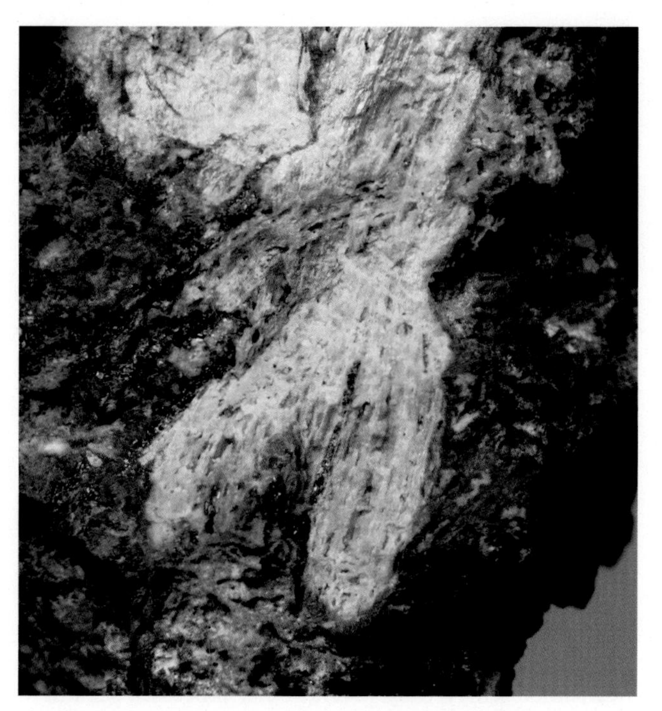

FIGURE 7.137 Total replacement of bismuthinite, Bi_2S_3, by yellow bismite, Bi_2O_3, about 3 cm. Llallagua, Bolivia.

Betterton (1884–1960) in the 1930s.), through treatment with metallic calcium and magnesium. The resulting bismuth compounds have higher melting points and lower densities than the lead and can be removed as dross (a mass of solid impurities floating on a molten metal or dispersed in the metal. The term dross comes from the Old English word dros, meaning the scum produced when smelting metals.). The compounds are treated with chlorine to free up the bismuth. Temperature used in this process is around 380°C to 500°C. Alternatively it is removed using the electrolytic Betts process (named for its inventor American chemist Anson Gardner Betts (April 14, 1876 to February 3, 1976) who filled several patents for this method starting in 1901). Bismuth will behave in the same way as another of its major metals, copper. The raw bismuth metal from both methods still has substantial quantities of other metals, principally lead. By

FIGURE 7.138 A specimen composed mostly of bismuth, Bi, internally but with alterations of yellow beyerite, $Ca(BiO)_2(CO_3)_2$, and green bismutite, $(BiO)_2CO_3$, about 2 cm. Outpost mine, Yavapai Co., Arizona, United States.

FIGURE 7.139 Mixite, $BiCu_6(AsO_4)_3(OH)_6{\cdot}3H_2O$, bluish green acicular sprays to 3 mm in baryte, $BaSO_4$. Mammoth mine, Tintic dist., Juab Co., Utah, United States.

reacting the molten mixture with chlorine gas, the metals are reacted to form their chlorides while bismuth remains unchanged. In addition, impurities can be removed by several other processes such as with fluxes and treatments producing high-purity bismuth metal (over 99% Bi).

Over 150 minerals are known in nature to contain bismuth. Bismuth (Bi) (Fig. 7.132) occurs as a native element. More than 100 minerals containing Bi can be found in the sulfide class, for example, aikinite ($PbCuBiS_3$), bismuthinite (Bi_2S_3) (Fig. 7.133), cosalite ($Pb_2Bi_2S_5$), emplectite ($CuBiS_2$) (Fig. 7.134), joséite-B (Bi_4Te_2S) (Fig. 7.135), and tetradymite (Bi_2Te_2S) (Fig. 7.136). Six halides with Bi are known, for example, bismoclite (BiOCl) and perite ($PbBiClO_2$). About 15 minerals are known in the oxide class, such as bismite (Bi_2O_3) (Fig. 7.137), bismutocolumbite ($Bi(Nb,Ta)O_4$), and uranosphaerite ($Bi(UO_2)O_2(OH)$). Three carbonates, beyerite ($Ca(BiO)_2(CO_3)_2$), bismutite ($(BiO)_2CO_3$) (Fig. 7.138), and kettnerite ($CaBiCO_3OF$), contain Bi in their structure. The sulfate class is represented by about 10 minerals, for example, cannonite ($Bi_2(SO_4)O(OH)_2$), chiluite ($Bi_3Te^{6+}Mo^{6+}O_{10.5}$), and montanite ($Bi_2(TeO_6){\cdot}2H_2O$). In the phosphate class around 35 different minerals containing Bi can be found, such as clinobisvanite ($Bi(VO_4)$), mixite ($BiCu_6(AsO_4)_3(OH)_6{\cdot}3H_2O$) (Fig. 7.139), namibite ($Cu(BiO)_2(VO_4)(OH)$), pucherite ($Bi(VO_4)$) and clinobisvanite

FIGURE 7.140 Flaming reddish orange crusts of tightly packed crystals of clinobisvanite, $Bi(VO_4)$, on yellow pottsite, $(Pb_{3x}Bi_{4-2x})(VO_4)_4 \cdot H_2O$ ($0.8 < x < 1.0$) (ideal form: $(Pb_3Bi)Bi(VO_4)_4 \cdot H_2O$), about 2 cm. Linka mine, Lander Co., Nevada, United States.

FIGURE 7.141 Eulytine, $Bi_4(SiO_4)_3$, deep reddish crystals to 1mm lining vug. Schneeberg dist., Erzgebirge, Saxony, Germany.

($Bi(VO_4)$) (Fig. 7.140). Three silicates, bismutoferrite ($Fe_2^{3+}Bi(SiO_4)_2(OH)$), eulytine ($Bi_4(SiO_4)_3$) (Fig. 7.141), and minasgeraisite-(Y) ($BiCa(Y,Ln)_2(\square,Mn)_2(Be,B,Si)_4Si_4O_{16}[(OH),O]_4$), are known to contain structural bismuth.

7.21.3 Chemistry

The only primordial isotope of bismuth, ^{209}Bi, was historically thought to be the heaviest stable isotope, but it had long been suspected to be unstable on theoretical grounds. This was finally established in 2003, when scientists at the Institut d'Astrophysique Spatiale in Orsay, France, determined the α emission half-life of ^{209}Bi to be 1.9×10^{19} years, over a billion times longer than the estimated age of the universe (current estimate about 13.8×10^9 years). Due to its extremely long half-life, for all currently known medical and industrial applications, Bi can be dealt with as if it is stable and nonradioactive. It is only of academic interest since Bi is one of a few elements whose radioactivity was suspected and theoretically predicted before being observed in the laboratory. Bi has the longest known α decay half-life, even though ^{128}Te has a double β decay half-life of more than 2.2×10^{24} years. The extremely long half-life means that

TABLE 7.21 Bismuth properties.

Appearance	Lustrous brownish silver
Standard atomic weight $A_{r,std}$	208.980
Block	p-Block
Element category	Posttransition metal
Electron configuration	[Xe] $4f^{14}$ $5d^{10}$ $6s^2$ $6p^3$
Phase at STP	solid
Melting point	271.5°C
Boiling point	1564°C
Density (near r.t.)	9.78 g/cm^3
When liquid (at m.p.)	10.05 g/cm^3
Heat of fusion	11.30 kJ/mol
Heat of vaporization	179 kJ/mol
Molar heat capacity	25.52 J/(mol·K)
Oxidation states	−3, −2, −1, +1, +2, **+3**, +4, +5
Ionization energies	1st: 703 kJ/mol
	2nd: 1610 kJ/mol
	3rd: 2466 kJ/mol
Atomic radius	Empirical: 156 pm
Covalent radius	148 ± 4 pm
Van der Waals radius	207 pm

STP, Standard temperature and pressure.
Bold font indicates main oxidation state.

less than one billionth of all the Bi present at the formation of the planet Earth would have decayed into ^{205}Tl since then. A number of Bi radioisotopes with short half-lives occur within the radioactive disintegration chains of actinium, radium, and thorium, and more have been synthesized in the laboratory. ^{213}Bi is also found on the decay chain of ^{233}U. Commercially, the radioisotope ^{213}Bi can be formed by bombarding radium with bremsstrahlung photons from a linear particle accelerator. In 1997 an antibody conjugate containing ^{213}Bi, which has a half-life of 45 min and decays with the emission of an α particle, was used to treat leukemia patients. The use of this isotope has also been attempted in cancer treatment, for example, in the targeted alpha therapy (TAT) program.

Bismuth is a brittle metal with a white, silver-pink hue, often with an iridescent oxide layer showing many colors from yellow to blue. It has 83 electrons arranged in an electronic configuration of [Xe] $4f^{14}5d^{10}6s^26p^3$. It has five valence electrons, but the inert pair effect is more pronounced for bismuth and the +5 oxidation state is less stable. Most compounds of Bi exist in the +3 oxidation state (Table 7.21).

7.21.3.1 Oxides

Bismuth is stable to both dry and moist air at ordinary temperatures. Direct reaction of Bi with O_2 at elevated temperatures yields Bi_2O_3. It also reacts with water when red-hot to make bismuth(III) oxide.

$$2Bi + 3H_2O \rightarrow Bi_2O_3 + 3H_2$$

When molten, at temperatures over 710°C, this oxide corrodes any metal oxide, and even platinum. On reaction with base, it forms two series of oxyanions: BiO_2^-, which is polymeric and forms linear chains, and BiO_3^{3-}. The anion in Li_3BiO_3 is in reality a cubic octameric anion, $Bi_8O_{24}^{24-}$, while the anion in Na_3BiO_3 is tetrameric. The structures of Bi_2O_3 differ significantly from those of arsenic(III) oxide, As_2O_3, and antimony(III) oxide, Sb_2O_3. Bi_2O_3 has five crystallographic polymorphs. The room temperature phase, α-Bi_2O_3, has a monoclinic crystal structure. There are three high temperature phases, a tetragonal β-phase, a *bcc* γ-phase, a cubic δ-phase, and an ε-phase. The room temperature α-phase possesses a complex structure with layers of oxygen atoms with layers of bismuth atoms in between. The bismuth atoms are in two different environments which can be described as distorted six and five coordinated, respectively. β-Bi_2O_3 possesses a structure related to fluorite (CaF_2). γ-Bi_2O_3 has a structure related to that of $Bi_{12}SiO_{20}$ (a sillenite), where a fraction of the Bi atoms occupy the position occupied by Si(IV), and may be written as $Bi_{12}Bi_{0.8}O_{19.2}$. δ-Bi_2O_3 possesses a defective fluorite-type crystal structure in which two of the eight oxygen sites in the unit cell are vacant. Finally, ε-Bi_2O_3 possesses a structure related to the α- and β-phases, but as the structure is fully ordered, it is an ionic insulator. It can be produced by hydrothermal means and converts to the α-phase at 400°C. The monoclinic α-phase changes to the cubic δ-Bi_2O_3 when heated above 729°C, which remains the structure up to the melting point at 824°C. The behavior of Bi_2O_3 on cooling from the δ-phase is more intricate, with the possible formation of two intermediate metastable phases: the tetragonal β-phase or the *bcc* γ-phase. The γ-phase can exist at 25°C with very slow cooling rates, but α-Bi_2O_3 is always produced on cooling the β-phase. While formed by heating, it reverts to α-Bi_2O_3 when the system is cooled below 727°C, δ-Bi_2O_3 can be produced directly through electrodeposition and remains relatively stable at room temperature, in an electrolyte of bismuth compounds that is also rich in sodium or potassium hydroxide so as to have a pH near 14. Bismuth trioxide is commercially produced from bismuth subnitrate. The latter is formed by dissolving bismuth in hot nitric acid. Addition of excess sodium hydroxide followed by continuous heating of the mixture precipitates bismuth(III) oxide as a heavy yellow powder. In addition, the trioxide can be prepared by ignition of bismuth hydroxide. Bismuth trioxide can also be prepared by heating bismuth subcarbonate at approximately 400°C. Oxidation with ammonium persulfate and dilute caustic soda produces bismuth tetroxide. The same product can be formed by utilizing other oxidizing agents, for example, potassium ferricyanide or concentrated caustic potash solution. Electrolysis of bismuth(III) oxide in hot concentrated alkali solution results in a scarlet red precipitate of bismuth(V) oxide. Bismuth(III) oxide reacts with mineral acids to produce the corresponding bismuth(III) salts. Reaction with acetic anhydride and oleic acid forms bismuth trioleate. Atmospheric carbon dioxide (CO_2) dissolved in water easily reacts with Bi_2O_3 to form bismuth subcarbonate, $(BiO)_2CO_3$. Bismuth oxide is considered a basic oxide, which explains the high reactivity with CO_2. But when acidic cations, for example, Si(IV), are introduced within the structure of the bismuth oxide, the reaction with CO_2 do not occur. The dark red bismuth(V) oxide, Bi_2O_5, is unstable, liberating O_2 gas upon heating. The compound $NaBiO_3$ is a strong oxidizing agent. In aqueous solution, the Bi^{3+} ion is solvated to form the aqua ion $Bi(H_2O)_8^{3+}$ in strongly acidic conditions. At pH > 0 polynuclear species exist, the most significant of which is thought to be the octahedral complex $[Bi_6O_4(OH)_4]^{6+}$. Bismuth continues the trend to electropositive behavior and Bi_2O_3 is definitely basic, in contrast to the acidic oxides of the lighter group 15 members, nitrogen and phosphorus and the ampotheric oxides of arsenic and antimony. There is also a growing tendency to form salts of oxoacids by reaction of either the metal or its oxide with the acid.

$$2Bi + 6H_2SO_4 \rightarrow Bi_2(SO_4)_3 + 3SO_2 + 6H_2O$$
$$Bi + 4HNO_3 \rightarrow Bi(NO_3)_3 + NO + 2H_2O$$

7.21.3.2 *Halides*

The halides of bismuth in low oxidation states have been shown to have unusual structures. What was originally thought to be bismuth(I) chloride, BiCl, turns out to be a complex compound consisting of Bi_9^{5+} cations and $BiCl_5^{2-}$ and $Bi_2Cl_8^{2-}$ anions. The Bi_9^{5+} cation adopts a distorted tricapped trigonal prismatic molecular geometry, and is also observed in $Bi_{10}Hf_3Cl_{18}$, which is formed by reducing a mixture of hafnium(IV) chloride and bismuth chloride with elemental bismuth, having the structure $[Bi^+][Bi_9^{5+}][HfCl_6^{2-}]_3$. Other polyatomic bismuth cations have also been found, for example, Bi_8^{2+}, present in $Bi_8(AlCl_4)_2$. Bismuth in addition forms a low-valence bromide with the same structure as "BiCl." Bismuth(III) chloride can be prepared directly by passing chlorine over bismuth.

$$2Bi + 3Cl_2 \rightarrow 2BiCl_3$$

It can also be prepared by dissolving bismuth metal in aqua regia, evaporating the mixture to give $BiCl_3 \cdot 2H_2O$, which can then be distilled to produce the anhydrous trichloride. Instead, it may also be formed by reacting hydrochloric acid with bismuth oxide and evaporating the solution.

$$Bi_2O_3 + 6HCl \rightarrow 2BiCl_3 + 3H_2O$$

Additionally, the compound can be formed by dissolving bismuth in concentrated nitric acid followed by adding solid sodium chloride into this solution.

$$Bi + 6HNO_3 \rightarrow Bi(NO_3)_3 + 3H_2O + 3NO_2$$
$$Bi(NO_3)_3 + 3NaCl \rightarrow BiCl_3 + 3NaNO_3$$

In the gas-phase $BiCl_3$ is pyramidal with a Cl-Bi-Cl angle of 97.5° and a bond length of 242 pm. In the solid state, each Bi atom has three near neighbors at 250 pm, two at 324 pm, and three at a mean of 336 pm. This structure is the same as that of $AsCl_3$, $AsBr_3$, $SbCl_3$, and $SbBr_3$. Bismuth chloride is easily hydrolyzed to bismuth oxychloride, BiOCl.

$$Bi^{3+}(aq) + Cl^-(aq) + H_2O(l) \rightleftharpoons BiOCl(s) + 2H^+(aq)$$

This reaction can be reversed by adding an acid, for example, hydrochloric acid. Reaction of solid $BiCl_3$ with water vapor below 50°C forms the intermediate monohydrate, $BiCl_3 \cdot H_2O$. Bismuth chloride is an oxidizing agent, easily reducing to metallic bismuth by reducing agents. In contrast to the usual expectation by consistency with periodic trends, $BiCl_3$ is a Lewis acid, forming various chloro complexes for example, $[BiCl_6]^{3-}$ that strongly violates the octet rule. Also, the octahedral structure of this coordination complex does not follow the predictions of VSEPR theory, since the lone pair on bismuth is unexpectedly stereochemically inactive. The dianionic complex $[BiCl_5]^{2-}$ does however have the expected square pyramidal structure.

There is a true monoiodide, BiI, which has chains of Bi_4I_4 units. BiI disproportionates upon heating to the triiodide, BiI_3, and elemental bismuth. A monobromide of identical structure also exists. In oxidation state +3, bismuth forms trihalides with all of the halogens: BiF_3, $BiCl_3$, $BiBr_3$, and BiI_3. All of these with the exception of BiF_3 are hydrolyzed by water. Bismuth(III) chloride reacts with hydrogen chloride in ether solution to form the acid $HBiCl_4$. The oxidation state +5 is less often encountered. One such compound is BiF_5, a strong oxidizing and fluorinating agent. It is also a strong fluoride acceptor, reacting with xenon tetrafluoride to form the XeF_3^+ cation.

$$BiF_5 + XeF_4 \rightarrow XeF_3^+BiF_6^-$$

Bismuth(III) iodide has a distinctive crystal structure, with iodide centers occupying a hexagonally closest packed lattice, and bismuth centers occupying either none or two-thirds of the octahedral holes (alternating by layer), therefore it is said to occupy one-third of the total octahedral holes. Bismuth(III) iodide forms upon heating a mixture of iodine and bismuth powder.

$$2Bi + 3I_2 \rightarrow 2BiI_3$$

It can also be prepared by the reaction of bismuth oxide with aqueous hydroiodic acid.

$$Bi_2O_3(s) + 6HI(aq) \rightarrow 2BiI_3(s) + 3H_2O(l)$$

$BiBr_3$ may be prepared through the reaction of bismuth oxide and hydrobromic acid.

$$Bi_2O_3 + 6HBr \rightleftharpoons 2BiBr_3 + 3H_2O$$

Bismuth tribromide can also be formed through the direct oxidation of bismuth in bromine. Bismuth tribromide is a highly water-soluble crystalline bismuth source for applications compatible with bromides and lower (acidic) pH. Most metal bromide compounds are water soluble for applications in water treatment, chemical analysis and in ultrahigh purity for certain crystal growth applications.

Bismuth trifluoride can be formed by reacting bismuth(III) oxide with hydrofluoric acid.

$$Bi_2O_3 + 6HF \rightarrow 2BiF_3 + 3H_2O$$

It occurs in nature as the rare mineral gananite. α-BiF_3 has a cubic crystalline structure (space group $Fm\bar{3}$). BiF_3 is the prototype for the DO_3 structure, which is adopted by some intermetallics, for example, Mg_3Pr, Cu_3Sb, Fe_3Si, and $AlFe_3$, as well as by the hydride $LaH_{3.0}$. The unit cell is face centered cubic with Bi at the face centers and vertices, and F at the octahedral site (mid-edges, center), and tetrahedral sites (centers of the eight subcubes)—thus the primitive cell contains 4 Bi and 12 F. The edge length of the BiF_3 cell is 0.5853 nm. β-BiF_3 has the YF_3 structure where the bismuth atom has distorted nine coordination, tricapped trigonal prism. This structure is generally considered to be ionic, and contrasts with fluorides of the lighter members of group 15, PF_3, AsF_3, and SbF_3, where MX_3 molecular units are present in the solid.

Bi reacts with fluorine to form bismuth(V) fluoride at 500°C or bismuth(III) fluoride at lower temperatures. Reaction with other halogens results in only bismuth(III) halides.

$$BiF_3 + F_2 \rightarrow BiF_5$$

In an alternative reaction, ClF_3 is used as the fluorinating agent at 350°C.

$$BiF_3 + ClF_3 \rightarrow BiF_5 + ClF$$

BiF_5 is the most reactive of the pnictogen pentafluorides and is an extremely strong fluorinating agent. It reacts vigorously with water to form ozone and oxygen difluoride, and with iodine or sulfur at 25°C. BiF_5 fluorinates paraffin oil (hydrocarbons) to fluorocarbons above 50°C and oxidizes UF_4 to UF_6 at 150°C. At 180°C, BiF_5 fluorinates Br_2 to BrF_3 and Cl_2 to ClF. In addition, BiF_5 reacts with alkali metal fluorides, MF, to form hexafluorobismuthates, $M[BiF_6]$, containing the hexafluorobismuthate anion, $[BiF_6]^-$.

The trihalides are corrosive and easily react with moisture, forming oxyhalides with the formula BiOX. The structure of bismuth oxychloride can be thought of as consisting of layers of Cl^-, Bi^{3+} and O^{2-} ions. These ions are ordered as Cl-Bi-O-Bi-Cl-Cl-Bi-O-Bi-Cl, that is, with alternating anions (Cl^-, O^{2-}) and cations (Bi^{3+}). The layered structure gives rise to the pearlescent properties of this material. The bismuth centers have a distorted square antiprismatic coordination geometry. The Bi^{3+} ion is coordinated to four chloride atoms, forming one of the square faces, each at a distance of 3.06 Å from Bi, and four oxygen atoms forming the other square face, each at a distance of 2.32 Å from Bi. The oxygen atoms are tetrahedrally coordinated by four bismuth atoms. BiOCl is formed during the reaction of bismuth chloride with water, that is, the hydrolysis reaction.

$$BiCl_3 + H_2O \rightarrow BiOCl + 2HCl$$

When heated above 600°C, BiOCl converts to $Bi_{24}O_{31}Cl_{10}$, known as the "Arppe compound" (named after Finnish chemist (organic chemistry, mineralogy) Adolf Edvard Arppe, June 9, 1818 to April 14, 1894), which has a complex layer structure. BiOCl exists in nature as the rare mineral bismoclite. It has been used in cosmetics since the days of ancient Egypt. It is part of the "pearly pigment present in eye shadow, hair sprays, powders, nail polishes, and other cosmetic products." Due to the plate-like structure of the BiOCl, its suspensions exhibit optical properties like nacre. An analogous compound, bismuth oxynitrate, is used as a white pigment.

7.21.3.3 Hydrides and chalcogenides

The hydride of Bi, BiH_3, first detected in minute traces by Austrian born English chemist Friedrich Adolf Paneth (August 31,1887 to September 17, 1958) in 1918 using a radiochemical technique involving $^{212}Bi_2Mg_3$, is unstable and cannot be kept above −45°C. This is a result of the decrease in the strength of covalent linkages to the large Bi. Many organobismuth compounds are also unstable for the same reason.

Bismuth(III) sulfide occurs in nature as the mineral bismuthinite. Bismuth(III) sulfide can be formed by reacting a bismuth(III) salt with hydrogen sulfide.

$$2Bi^{3+} + 3H_2S \rightarrow Bi_2S_3 + 6H^+$$

Bismuth (III) sulfide can also be synthesized by reacting elemental bismuth with elemental sulfur in an evacuated silica tube at 500°C for 96 h.

$$2Bi + 3S \rightarrow Bi_2S_3$$

Bismuth(III) sulfide is isostructural with Sb_2S_3 (mineral stibnite). Bismuth atoms are in two different environments, both of which have seven coordinate bismuth atoms, four in a near planar rectangle, and three more distant forming an irregular seven-coordination group. It can react with acids to form the odoriferous hydrogen sulfide gas. Bismuth(III) sulfide may be formed in the body by the reaction of the common gastrointestinal drug bismuth subsalicylate with naturally occurring sulfides; this causes temporary black tongue when the sulfides are in the mouth and black feces when the sulfides are in the colon.

Bismuth selenide (Bi_2Se_3), a gray compound, is a semiconductor and a thermoelectric material. When perfect stoichiometric bismuth selenide should be a semiconductor (with a gap of 0.3 eV), but naturally occurring selenium vacancies act as electron donors and it often acts as a semimetal. Bismuth telluride (Bi_2Te_3), a gray powder, is a semiconductor. When it is alloyed with antimony or selenium, it forms an efficient thermoelectric material for refrigeration or portable power generation. Bi_2Te_3 is a topological insulator, and thus exhibits thickness-dependent physical properties. The mineral form of Bi_2Te_3 is tellurobismuthite, which is moderately rare. There are many natural bismuth tellurides of different stoichiometry, as well as compounds of the Bi-Te-S-(Se) system, like Bi_2Te_2S (tetradymite). Bismuth telluride can be produced simply by sealing mixed powders of bismuth and tellurium metal in a quartz tube under vacuum (critical, as an unsealed or leaking sample may explode in a furnace) and heating it to 800°C in a muffle furnace.

7.21.3.4 Organobismuth compounds

Organobismuth compounds are reagents that contain a C-Bi bond. These reagents have recently gained popularity due to their ease of synthesis and the surprisingly low toxicity and have been useful for the preparation of pharmaceuticals. Similar to As and Sb above it in group 15, Bi forms a wide range of organometallic compounds in both the +3 and the +5 state. Their preparation follows typical routes discussed in the preparation of other organometallics such as using Grignard reagents. There are several books and review papers (e.g., Gagnon et al., 2017; Suzuki & Matano, 2001) discussing the chemistry of organometallic compounds of bismuth and the reader is referred to these for a detailed treatment of the subject.

7.21.4 Major uses

Bismuth is an element in some drugs, though the usage of some of these substances is decreasing. Bi-subsalicylate ($C_7H_5BiO_4$) is employed as an antidiarrheal. It is also prescribed to treat some other gastrointestinal illnesses and Cd poisoning. The mechanism of action of this compound is still not well understood, though an oligodynamic effect (the toxic effect of small quantities of heavy metal ions on microbes) may be involved in at least some instances. Salicylic acid (IUPAC name 2-hydroxybenzoic acid, $C_7H_6O_3$) from hydrolysis of the compound is antimicrobial for toxogenic *Escherichia coli*, a significant pathogen responsible for traveler's diarrhea. A combination of Bi-subsalicylate and Bi-subcitrate ($C_{12}H_8BiK_5O_{14}$) is employed to treat the bacteria triggering peptic ulcers. Bibrocathol (4,5,6,7-tetrabrom-1,3,2-benzodioxabismol-2-ol, $C_6HBiBr_4O_3$) is an organic Bi-containing chemical employed to treat eye infections. Bi-subgallate ($C_7H_5BiO_6$) is utilized as an internal deodorant to treat the bad odor from flatulence and feces. Bi-compounds (including sodium bismuth tartrate, $C_8H_8BiNaO_{12}$) were in the past utilized to treat syphilis (a sexually transmitted infection caused by the bacterium *Treponema pallidum* subspecies *pallidum*). "Milk of bismuth" (an aqueous suspension of Bi-hydroxide, $Bi(OH)_3$, and Bi-subcarbonate, $Bi_2O_2(CO_3)$) was sold as an alimentary remedy at the beginning of the 20th century. Bi-subnitrate ($Bi_5O(OH)_9(NO_3)_4$) and Bi-subcarbonate ($Bi_2O_2(CO_3)$) are likewise employed in pharmaceuticals. Bi-oxychloride (BiOCl) is occasionally added to cosmetics, for example, as a pigment in paint for eye shadows, hair sprays and nail polishes. This cheimcal is found as the mineral bismoclite (BiOCl) and in crystal form consists of layers of atoms that refract light chromatically, producing an iridescent look comparable to nacre of pearl. It was employed as a cosmetic in ancient Egypt and in many other places later. Bi white (also "Spanish white") can indicate either Bi-oxychloride (BiOCl) or Bi-oxynitrate ($BiONO_3$), when utilized as a white pigment. Bi-vanadate ($BiVO_4$) is used as a light-stable nonreactive paint pigment (especially in artists' paints), frequently as a substitute for the more toxic CdS yellow and orange-yellow pigments. The most widespread variety in artists' paints is a lemon yellow, visually identical to its Cd-containing alternative. It performs practically identically to the Cd pigments, for example, in terms of resistance to degradation from UV exposure, opacity, tinting strength, and lack of reactivity when mixed with other pigments. Besides being a substitute for several Cd yellows, it also acts as a nontoxic visual substitute for the older chromate pigments made with Zn, Pb, and Sr. If a green pigment and $BaSO_4$ (for increased transparency) are added it can also serve as a replacement for Ba-chromate, which possesses a more greenish cast than the others. Compared to Pb-chromates, it does not blacken upon exposure to hydrogen sulfide in the air (a process accelerated by UV exposure) and has a particularly brighter color than them, in particular the lemon yellow (which is the most translucent) dull, and fastest to blacken due to the higher percentage of $PbSO_4$ needed to create that shade. It is also employed, on a limited basis due to its cost, as a car paint pigment.

Bi is employed in metal alloys with other metals such as Fe, to form alloys to be used in automatic sprinkler systems for fires. It was also utilized to form Bi-bronze which was employed as early as the Bronze Age. The difference in density between Pb (11.32 g/cm^3) and Bi (9.78 g/cm^3) is sufficiently small that for numerous ballistics and weighting uses, Bi can replace Pb. For instance, it can substitute for Pb as a dense material in fishing sinkers. It has been employed as a substitute for Pb in shot, bullets and less-lethal riot gun ammunition. Many countries now ban the application of Pb-shot for the hunting of wetland birds, as many birds are susceptible to Pb poisoning due to mistaken ingestion of Pb (instead of small stones and grit) to help digestion, or even ban the application of Pb for all hunting, for example, in the Netherlands. Bi-Sn alloy shot is one substitute that offers comparable ballistic performance as Pb. (A further cheaper but also more poorly performing option is "steel" shot, which is in fact soft Fe.) Bi's lack of malleability does, though, make it unsuitable for application in expanding hunting bullets. Bi, as a dense element of high atomic weight, is utilized in Bi-impregnated latex shields to protect from X-rays in medical examinations, such as X-ray Computed Tomography (CT) scans, primarily as it is believed to be nontoxic. The European Union's Restriction of Hazardous Substances Directive for reduction of Pb has expanded Bi's application in electronics as an element of low-melting point solders, as a substitute for traditional Sn-Pb solders. Its low toxicity will be particularly significant for solders to be employed in food processing machines and Cu water pipes, though it can also be employed in other uses, for example, in the car industry, in the EU. Bi has been assessed as a substitute for Pb in free-machining brasses for plumbing uses, even though it does not completely approximate the performance of leaded steels. Numerous Bi alloys possess low melting points and can be found in specialty applications such as solders. Various automatic sprinklers, electric fuses, and safety devices in fire detection and suppression systems consist of the eutectic In19.1-Cd5.3-Pb22.6-Sn8.3-Bi44.7 alloy, which melts at a low temperature of 47°C. This is a suitable temperature as it is doubtful to be surpassed under everyday living conditions. Low-melting alloys, such as Bi-Cd-Pb-Sn alloy which melts at 70°C, are likewise employed in car and aviation industries. Prior to deforming a thin-walled metal part, it is filled with a melt or covered with a thin layer of the alloy to minimize the chance of breaking. Subsequently the alloy is eliminated by submerging the part in boiling water. Bi is utilized to produce free-machining steels and free-machining Al alloys for precision machining properties. It has a comparable effect on Pb and enhances the chip breaking during machining. The shrinking on solidification in Pb and the expansion of Bi offset each other and hence Pb and Bi are frequently used in comparable amounts. Likewise, alloys consisting of similar parts of Bi and Pb show a very small change (on the order 0.01%) upon melting, solidification or aging. These alloys are employed in high-precision casting, for example, in dentistry, to create models and molds. Bi is also employed as an alloying element in the manufacture of malleable irons and as a thermocouple material. Bi is similarly employed in Al−Si cast alloys to refine the Si morphology. On the other hand, it showed a poisoning effect on the modification of Sr. Certain Bi alloys, for example, Bi35-Pb37-Sn25, are mixed with nonsticking compounds such as mica, glass and enamels since they easily wet them permitting to create joints to other parts. Adding Bi to Cs improves the quantum yield of Cs cathodes. Sintering of Bi and Mn powders at 300°C creates a permanent magnet and magnetostrictive material, which is employed in ultrasonic generators and receivers operating in the range between 10 and 100 kHz and in magnetic memory devices.

Bi is included in BSCCO (Bi-Sr-Ca-Cu-oxide) which is a group of comparable superconducting compounds discovered in 1988 which show the highest superconducting transition temperatures found so far. Bi subnitrate (bismuth oxynitrate is commercially available as $Bi_5O(OH)_9(NO_3)_4$) is a constituent of glazes that creates an iridescence and is employed as a pigment in paint. Bi telluride (Bi_2Te_3) is a semiconductor and an outstanding thermoelectric material. Bi_2Te_3 diodes are employed in, for example, mobile refrigerators, CPU coolers, and as detectors in infrared spectrophotometers. Bi oxide, in its delta form (one of five crystallographic polymorphs, cubic δ-Bi_2O_3 phase, it has a defective fluorite-type (CaF_2) crystal structure in which two of the eight oxygen sites in the unit cell are vacant), is a solid electrolyte for oxygen. This form usually breaks down below a high temperature limit, but can be electrodeposited well below this temperature in a highly alkaline solution. Interest has focused on δ-Bi_2O_3 as it is chiefly an ionic conductor. Along with its electrical properties, thermal expansion properties are very important when considering possible uses for solid electrolytes. High thermal expansion coefficients signify large dimensional variations under heating and cooling, which would restrict the performance of an electrolyte. The transition from the high-temperature δ-Bi_2O_3 to the intermediate tetragonal β-Bi_2O_3 is associated with a large volume change and therefore a deterioration of the mechanical properties of the material. This, combined with the very narrow temperature stability range of the δ-phase (727°C−824°C) has resulted in research on its stabilization to room temperature. Bi germanate (Most commonly the term refers to the compound with chemical formula $Bi_4Ge_3O_{12}$ (BGO), with the cubic evlitine crystal structure. The term may also refer to a different compound with formula $Bi_{12}GeO_{20}$, an electro-optical material with sillenite ($Bi_{12}SiO_{20}$) structure, and $Bi_2Ge_3O_9$) is a scintillator, extensively used in X-ray and γ ray detectors. Additional uses are, for example, as a catalyst for making acrylic fibers, as an electrocatalyst in the conversion of CO_2 to CO, as ingredient in lubricating greases, and in crackling microstars (dragon's eggs) in pyrotechnics, as the oxide, subcarbonate or subnitrate.

7.22 90 Th — Thorium

7.22.1 Discovery

In 1815 the Swedish chemist Jöns Jacob Berzelius (August 20, 1779 to August 7, 1848) analyzed a rare sample of gadolinite from a copper mine in Falun, central Sweden. He observed impregnated traces of an unidentified white mineral, which he cautiously thought to be an earth (oxide in modern chemical nomenclature) of an unknown element. Berzelius had previously found two elements, cerium and selenium, but he had made a public error once before, proclaiming the finding of a new element, gahnium, that was later shown to be zinc oxide. Berzelius privately called the hypothetical element "thorium" in 1817 and its supposed oxide "thorina" after Thor, the Norse god of thunder. In 1824 after more deposits of the same mineral in Vest-Agder, Norway, were found, he withdrew his discovery, as the mineral (later named xenotime, (Y,Yb)PO_4) proved to be mostly yttrium orthophosphate. In 1828 Norwegian priest and mineralogist Morten Thrane Esmark (August 21, 1801 to April 24, 1882) found a black mineral on Løvøya island, Telemark county, Norway. He studied the minerals in Telemark, where he served as vicar. He commonly sent the most interesting specimens, such as this one, to his father, Jens Esmark (January 31, 1763 to January 26, 1839) (Fig. 7.142), a noted mineralogist and professor of mineralogy and geology at the Royal Frederick University in Christiania (today called Oslo). The elder Esmark determined that it was an unknown mineral and sent a sample to Berzelius for study. Berzelius showed that it contained a new element. He published his results in 1829, having separated an impure sample by reducing $KThF_5$ with potassium metal (Berzelius, 1829a,b). Berzelius reused the name of the earlier hypothetical element discovery and called the source mineral thorite ($ThSiO_4$). In the periodic table published by Russian chemist Dmitri Mendeleev (February 8, 1834 to February 2, 1907) in 1869, thorium and

FIGURE 7.142 Jens Esmark, in Carl Schnitler: Slegten fra 1814: studier over Norsk embedsmandskultur I klassicismens tidsalder 1814–1840: kulturformene, p. 380. Kristiania: Aschehoug, 1911.

the rare-earth elements were located outside the main body of the table, at the end of each vertical period after the alkaline earth metals. This was based on the assumption at that time that thorium and the rare-earth metals were divalent (2 +). With the later discovery that the REEs were typically trivalent (3 +) and thorium was tetravalent (4 +), Mendeleev shifted cerium and thorium to group IV in 1871, which also contained the modern carbon group (group 14) and titanium group (group 4), because their maximum oxidation state was 4 + . Cerium was shortly after removed from the main body of the table and shifted to a separate lanthanide series; thorium was left with group 4 as it had comparable properties to its supposed lighter congeners in that group, such as titanium and zirconium. Berzelius determined some initial properties of the new metal and its chemical compounds: he correctly determined that the thorium–oxygen mass ratio of thorium oxide was 7.5 (its actual value is close to that, ~7.3), but he thought the new element was divalent rather than tetravalent, and so calculated that the atomic mass was 7.5 times that of oxygen (120 amu); it is in fact 15 times as large. He noted that thorium was a very electropositive metal, ahead of cerium and behind zirconium. Metallic thorium was separated for the first time in 1914 by Dutch entrepreneurs Dirk Lely Jr. and Lodewijk Hamburger. Lely and Hamburger obtained 99% pure thorium metal by reducing thorium chloride with sodium metal. While thorium was discovered in 1828 its first application did not appear till 1885, when Austrian chemist and inventor Carl Auer von Welsbach (September 1, 1858 to August 4, 1929) invented the gas mantle, a portable source of light which produces light from the incandescence of thorium oxide when heated by burning gaseous fuels. Many more uses were next found for thorium and its compounds, including ceramics, carbon arc lamps, heat-resistant crucibles, and as catalysts for industrial chemical reactions such as the oxidation of ammonia to nitric acid. Thorium was first observed to be radioactive in 1898, by the German chemist Gerhard Carl Schmidt (July 5, 1865 to October 16, 1949 and later that year, independently, by the Polish and naturalized French physicist Marie Skłodowska Curie (November 7, 1867 to July 4, 1934) (Fig. 7.143) (Curie, 1898; Schmidt 1898a,b). It was the second element that was discovered to be radioactive, after the 1896 observation of radioactivity in uranium by French engineer and physicist Henri Becquerel (December 15, 1852 to August 25, 1908). Beginning in 1899, the New Zealand physicist Ernest Rutherford (August 30, 1871 to October 19, 1937) and the American electrical engineer Robert Bowie Owens (October 29, 1870 to November 3, 1940) studied the radiation from thorium; preliminary observations

FIGURE 7.143 Marie Skłodowska Curie, c.1920.

exposed that it varied significantly (Rutherford and Owens, 1899). It was noted that these variations were the result of a short-lived gaseous daughter of thorium, which they found to be a new element. This element is now called radon, the only one of the rare radioelements to be found in nature as a daughter of thorium rather than uranium. After accounting for the contribution of radon, Rutherford, now working with the British physicist/radiochemist Frederick Soddy (September 2, 1877 to September 22, 1956), determined how thorium decayed at a fixed rate over time into a series of other elements between 1900 and 1903. This observation resulted in the recognition of the half-life as one of the outcomes of the alpha particle experiments that led to the disintegration theory of radioactivity. Up to late in the 19th century, chemists universally thought that thorium and uranium were similar to hafnium and tungsten; the presence of the lanthanides in the sixth row was considered to be a one-off coincidence. In 1892 British chemist Henry Bassett (1838–1920) suggested a second extra-long periodic table row to fit in known and undiscovered elements, considering thorium and uranium to be analogous to the lanthanides. In 1913 Danish physicist Niels Bohr (October 7, 1885 to November 18, 1962) developed a theoretical model of the atom and its electron orbitals, which was quickly widely accepted. The model showed that the seventh row of the periodic table should also have f shells filling before the d-shells that were filled in the transition elements, like the sixth row with the lanthanides preceding the 5d transition metals. The presence of a second inner transition series, in the form of the actinides, was not believed to exist until parallels with the electron structures of the lanthanides had been recognized; Bohr proposed that the filling of the 5f orbitals may be delayed to after uranium (U). It was only with the finding of the first transuranic elements, which from plutonium (Pu) onward have foremost 3 + and 4 + oxidation states like the lanthanides, that it was accepted that the actinides were undeniably filling f-orbitals rather than d orbitals, with the transition-metal-like chemistry of the early actinides being the exception and not the rule. In 1945, when American chemist Glenn T. Seaborg (April 19, 1912 to February 25, 1999) and his research group had found the transuranic elements americium (Am) and curium (Cm), he realized that thorium was the second member of the actinide series and was filling an f-block row, rather than being the heavier congener of hafnium (Hf) filling a fourth d-block row.

7.22.2 Mining, production, and major minerals

Natural thorium is typically nearly pure ^{232}Th, which is the longest-lived and most stable isotope of thorium, having a half-life comparable to the age of the universe. Its decay accounts for a gradual decrease of Th content of the Earth: the planet these days has about 85% of the amount of Th present at the formation of the Earth. The other naturally occurring Th isotopes are much shorter-lived; of those, only ^{230}Th is typically measurable, occurring in secular equilibrium (a situation in which the quantity of a radioactive isotope remains constant because its production rate (e.g., due to decay of a parent isotope) is equal to its decay rate.) with its parent ^{238}U, and accounts for at most 0.04% of natural Th. Thorium is only found as a minor constituent of most minerals, and was for this reason earlier supposed to be rare. In nature, Th occurs in the 4 + oxidation state, together with U(IV), Zr(IV), Hf(IV), and Ce(IV), and also with Sc, Y, and the trivalent lanthanides which have similar ionic radii. Due to Th's radioactivity, minerals containing it are often metamict (amorphous), their crystal structures having been destroyed by the alpha radiation produced during thorium decay. An extreme example is ekanite, $(Ca,Fe,Pb)_2(Th,U)Si_8O_{20}$, which nearly never is found in crystalline form because of the Th it contains. Monazite, $(Ce,La,Nd,Th,Sm,Gd,Ca)PO_4$, is the most important economic source of Th as it is found in large deposits worldwide, mainly in India, South Africa, Brazil, Australia, and Malaysia. It has an average of around 2.5% Th, though some deposits may hold up to 20%. Allanite, $Ca(Ce,Y,La)FeAl_2(Si_2O_7)(SiO_4)O(OH)$, can contain 0.1%–2% Th and zircon, $ZrSiO_4$, up to 0.4% Th. ThO_2 occurs as the rare mineral thorianite. Due to its being isotypic with the uranium dioxide (UO_2) mineral uraninite, these two common actinide dioxides can form solid-state solutions and the name of the mineral changes according to the ThO_2 content. Thorite ($ThSiO_4$) also has a high Th concentration and is the mineral in which Th was first found. In Th-silicate minerals, the Th^{4+} and SiO_4^{4-} ions are frequently replaced with M^{3+} (where M = Sc, Y, or Ln) and phosphate (PO_4^{3-}) ions, respectively.

The limited demand makes mining purely for the extraction of thorium not profitable, and it is nearly always extracted together with the REEs, which themselves may be by-products of the mining of other minerals. The current dependence on monazite ($(Ce,La,Nd,Th)PO_4$) for production is because thorium is largely produced as a by-product; other minerals such as thorite ($ThSiO_4$) have more thorium and can easily be used for production if demand increases. Current understanding of the distribution of thorium resources is poor, since the low demand has led to very limited exploration efforts. In 2014 world production of the monazite concentrate, from which thorium could be won, was 2700 tons. The general production process of thorium involves concentration of thorium minerals; separation of thorium from the concentrate; purification of thorium; and (optionally) conversion to compounds, such as thorium dioxide (ThO_2). There are two groups of Th minerals for Th production: primary and secondary deposits. Primary deposits are found in acidic granitic magmas and pegmatites. They are concentrated, but of small size. Secondary deposits occur at the mouths of rivers in granitic mountain regions. In these secondary deposits, thorium-containing minerals are enriched along with other heavy minerals. Initial concentration steps depend on the type of deposit. For the primary deposits, the source pegmatites, which are usually obtained by open cut mining, are crushed and then undergo flotation. Alkaline earth metal carbonates may be dissolved through reaction with hydrogen chloride (HCl); subsequently followed by thickening, filtration, and calcination. The result is a concentrate with rare-earth content of up to 90%. Secondary materials (e.g., coastal sands) undergo gravity separation, followed by magnetic separation using a set of magnets of increasing strength. Monazite obtained via this process can be as pure as 98%. Industrial production in the 20th century was based on treatment with hot, concentrated sulfuric acid (H_2SO_4) in cast iron vessels, and subsequent selective precipitation by dilution with water. Many alternative methods have been suggested, but only one so far has shown to be effective commercially: alkaline digestion with hot sodium hydroxide (NaOH) solution. This is more expensive than the original process but produces a higher purity of thorium; especially, it eliminates phosphates from the concentrate. Acid digestion is a two-step method, involving usage of up to 93% sulfuric acid at a temperature of 210°C–230°C. First, 60% sulfuric acid is added, thickening the mixture as reaction products are formed. Then, fuming sulfuric acid (sulfuric acid reacts with its anhydride, SO_3, to form $H_2S_2O_7$, called pyrosulfuric acid, fuming sulfuric acid, disulfuric acid or oleum or, less commonly, Nordhausen acid.) is added and the reaction mixture is maintained at the same temperature for another five hours to reduce the volume of solution remaining after dilution. The concentration of the sulfuric acid is determined based on reaction rate and viscosity, which both increase with concentration, although with viscosity retarding the reaction. Increasing the temperature similarly speeds up the reaction, but temperatures over 300°C must be avoided, since they can result in insoluble thorium pyrophosphate to form. Because the dissolution is very exothermic (an exothermic reaction is a chemical reaction that releases energy through light or heat), the monazite sand cannot be fed into the sulfuric acid too fast. On the other hand, at temperatures $<200°C$ the reaction does not go sufficiently fast for the method to be practical. To guarantee that no precipitates form to block the reactive monazite surface, the mass of sulfuric acid used has to be double that of the sand, rather than the 60% that would be anticipated from the reaction stoichiometry. The mixture is subsequently cooled to 70°C and diluted with 10 times its volume of cold water, so that any remaining unreacted monazite settles on the bottom whereas the REEs and thorium remain in solution. Thorium can then be isolated through precipitation as the phosphate at pH 1.3, as the REEs do not precipitate until the pH is dropped to 2. On the other hand, the alkaline digestion process is carried out in 30%–45% sodium hydroxide (NaOH) solution at approximately 140°C for about three hours. Too high a temperature results in the formation of poorly soluble thorium oxide and an excess of uranium in the filtrate, while too low a concentration of NaOH produces a very slow reaction. These reaction conditions are relatively mild and necessitate monazite sand with a particle size under 45 μm. After filtration, the filter cake comprises thorium and the REEs as their hydroxides, uranium as sodium diuranate ($Na_2U_2O_7 \cdot 6H_2O$), and phosphate as trisodium phosphate (Na_3PO_4). This crystallizes as trisodium phosphate decahydrate when cooled below 60°C. Uranium impurities in the filter product tend to increase with the amount of silicon dioxide (SiO_2) in the reaction mixture, requiring recrystallization before commercial applications. The hydroxides are dissolved at 80°C in 37% hydrochloric acid (HCl). Subsequent filtration of the remaining precipitates followed by adding 47% sodium hydroxide (NaOH) produces the precipitation of thorium and uranium at approximately pH 5.8. Complete drying of the precipitate has to be avoided, as air may oxidize cerium from the +3 to the +4 oxidation state, and the resulting Ce(IV) can liberate free chlorine from the hydrochloric acid. The REEs once more precipitate out at higher pH. The precipitates are then neutralized by the original sodium hydroxide solution, though most of the phosphate has to be removed first to avoid precipitation of the rare-earth phosphates. Solvent extraction may also be utilized to separate out the thorium and uranium, through dissolution of the resultant filter cake in nitric acid (HNO_3). The presence of titanium hydroxide ($Ti(OH)_4$) is deleterious as it can bind the thorium and prevent it from dissolving completely. High thorium concentrations are needed in nuclear applications. Concentrations of atoms with high neutron capture cross-sections must be very low (e.g., gadolinium concentration must be <1 ppm by weight). In earlier days, repeated dissolution and recrystallization was applied to achieve high purity. Currently, liquid solvent extraction methods using selective complexation of Th^{4+} are employed. For example, after alkaline digestion and the removal of phosphate, the subsequent nitrato complexes of thorium, uranium, and the REEs can be isolated through extraction with tributyl phosphate ($(CH_3CH_2CH_2CH_2O)_3PO$) in kerosene (a low viscosity, clear liquid formed from hydrocarbons obtained from the fractional distillation of petroleum between 150°C and 275°C, mixture composed of carbon chains that typically contain between 10 and 16 carbon atoms per molecule).

About 45 minerals are found in nature to contain structural thorium. Cabvinite ($Th_2F_7(OH) \cdot 3H_2O$) is the only known halide. The oxide class is represented by 14 different minerals, such as cerianite-(Ce) ((Ce^{4+},Th)O_2) and thorianite (ThO_2) (Fig. 7.144). Two carbonates are known, thorbastnäsite ($ThCa(CO_3)_2F_2 \cdot 3H_2O$) and tuliokite ($Na_6BaTh(CO_3)_6 \cdot 6H_2O$). Two recently discovered molybdates, ichnusaite ($Th(MoO_4)_2 \cdot H_2O$) and nuragheite ($Th(MoO_4)_2 \cdot H_2O$) can be found in the sulfate class. A total of 10 phosphate class minerals have Th, for example, althupite ($AlTh(UO_2)_7(PO_4)_4(OH)_5O_2 \cdot 15H_2O$) and cheralite ($CaTh(PO_4)_2$). The silicate class contains 16 minerals with Th, such as ciprianiite ($Ca_4[(Th,U),Ca]_{\Sigma 2}Al(Be_{0.5}\square_{1.5})_{\Sigma 2}[B_4Si_4O_{22}](OH)_2$), coutinhoite

FIGURE 7.144 Thorianite, ThO_2, cubic crystal to 8 mm. Andranondambo, Tulear prov., Madagascar.

FIGURE 7.145 Thorite, $Th(SiO_4)$, 6mm crystal. Commercial quarry, Crestmore, Riverside Co., California, United States.

$(Th_xBa_{(1-2x)}(UO_2)_2Si_5O_{13}\cdot(H_2O)_{1+y}$ [$0 < x < 0.5$ and $0 < y < (2+x)$]), ekanite ($Ca_2ThSi_8O_{20}$), huttonite ($ThSiO_4$), thorite ($ThSiO_4$) (Fig. 7.145), and thornasite ($(Na,K)_{12}Th_3[Si_8O_{19}]_4\cdot 18H_2O$).

7.22.3 Chemistry

All except for two elements up to Bi have an isotope that is practically stable for all applications ("classically stable"), with the exceptions being Tc (technetium) and Pm (promethium). All elements from Po (polonium) onward are measurably radioactive. ^{232}Th is one of the three nuclides beyond Bi (the remaining two being ^{235}U and ^{238}U) that have half-lives measured in billions of years; its half-life is 14.05 billion years, about three times the age of our planet Earth, and slightly longer than the age of the universe (about 13.8 billion years). About 80% of the Th present at Earth's formation has survived to the present. ^{232}Th is the only Th isotope occurring in quantity (99.98%) in nature, with a trace of ^{230}Th (0.02%). The stability of ^{232}Th is ascribed to its closed nuclear shell with 142 neutrons. Th has a typical terrestrial isotopic composition, with atomic weight 232.0377(4). It is one of only three radioactive elements (along with protactinium and uranium) that exist in large enough quantities on Earth for a standard atomic weight to be measured. Th nuclei are prone to α decay since the strong nuclear force cannot overcome the electromagnetic repulsion between their protons. The α decay of ^{232}Th starts the $4n$ decay chain which includes isotopes with a mass number divisible by 4 (hence the name; it is also known as the thorium series after its progenitor). This chain of consecutive α and β decays starts with the decay of ^{232}Th to ^{228}Ra and stops at ^{208}Pb. Any sample of Th or its compounds contains traces of these daughter elements, which are isotopes of thallium (Tl), lead (Pb), bismuth (Bi), polonium (Po), radon (Rn), radium (Ra), and actinium (Ac). Natural Th samples can be chemically purified to isolate useful daughter nuclides, such as ^{212}Pb, which is utilized in nuclear medicine for cancer therapy. ^{227}Th (an α emitter with a half-life of 18.68 days) can also be used in cancer treatments such as TATs (Targeted Alpha-particle Therapy). In addition, ^{232}Th very infrequently undergoes spontaneous fission instead of α decay, and leaves evidence of this in its minerals (as trapped xenon gas formed as a fission product), but the partial half-life of this process is very long at more than 1021 years and α decay is dominant. Thirty radioisotopes have been observed, which range from ^{209}Th to ^{238}Th. After ^{232}Th, the most stable of them are ^{230}Th with a half-life of 75,380 years, ^{229}Th (half-life 7340 years), ^{228}Th (half-life 1.92 years), ^{234}Th (half-life 24.10 days), and ^{227}Th (half-life 18.68 days). All these isotopes can be observed in nature as trace radioisotopes because of their existence in the decay chains of ^{232}Th, ^{235}U, ^{238}U, and ^{237}Np: the last of these is long extinct in nature because of its short half-life (2.14 million years), but is constantly formed in trace amounts from neutron capture in uranium

BOX 7.8 Radiometric dating

Two radiometric dating methods involve Th isotopes: U-Th dating, based on the decay of ^{234}U to ^{230}Th, and ionium–thorium dating, which measures the ^{232}Th to ^{230}Th ratio (the name ionium for ^{230}Th is a remnant from a period when different isotopes were not recognized to be the same element and were given different names). These are based on the fact that ^{232}Th is a primordial radioisotope, but ^{230}Th only forms as intermediate decay product in the decay chain of ^{238}U. U-Th dating is a rather short-range process due to the short half-lives of ^{234}U and ^{230}Th relative to the age of the Earth: it is also accompanied by a sister process concerning the α decay of ^{235}U into ^{231}Th, which very rapidly becomes the longer-lived ^{231}Pa (this process is often used to check the results of U-Th dating). U-Th dating is frequently used to obtain the age of calcium carbonate materials such as speleothem or coral, since U is more soluble in water than Th and Pa, which are selectively precipitated into ocean-floor sediments, where their ratios are determined. The scheme has a range of several hundred thousand years. Ionium–thorium dating is a related process, which uses the insolubility of Th (both ^{232}Th and ^{230}Th) and hence its presence in ocean sediments to date these sediments by determining the ^{232}Th to ^{230}Th ratio. Both these dating methods assume that the proportion of ^{230}Th to ^{232}Th does not change during the period when the sediment layer was formed, that the sediment did not already contain Th before contributions from the decay of U, and that the Th cannot migrate within the sediment layer.

ores (Box 7.8). All of the remaining Th isotopes have half-lives less than 30 days and most have half-lives less than 10 min. In deep seawaters the isotope ^{230}Th forms to 0.04% of natural Th. This is due to its parent ^{238}U being soluble in water, however ^{230}Th is insoluble and as a result precipitates into the sediment. Uranium ores with low Th concentrations can be purified to produce gram-sized Th samples of which over 25% is the ^{230}Th isotope, as ^{230}Th is one of the daughters of ^{238}U. The International Union of Pure and Applied Chemistry (IUPAC) reclassified Th as a binuclidic element in 2013; it had previously been classified as a mononuclidic element. Three Th nuclear isomers (or metastable states) have been observed, $^{216m1}Th$, $^{216m2}Th$, and ^{229m}Th. ^{229m}Th has the lowest known excitation energy of any isomer, measured at 7.6 ± 0.5 eV. This is so low that when it undergoes isomeric transition, the emitted γ radiation is in the ultraviolet range (γ rays are distinguished by their origin in the nucleus, not their wavelength; hence there is no lower limit to γ energy derived from radioactive decay). Different Th isotopes are chemically identical, but have slightly differing physical properties: for example, the densities of pure ^{228}Th, ^{229}Th, ^{230}Th, and ^{232}Th are expected to be 11.5, 11.6, 11.6, and 11.7 g/cm^3, respectively. The isotope ^{229}Th is expected to be fissionable with a bare critical mass of 2839 kg, yet with steel reflectors this value could drop to 994 kg (A fissionable nuclide is capable of undergoing fission (even with a low probability) after capturing a high-energy neutron. Some of these nuclides can be induced to fission with low-energy thermal neutrons with a high probability; they are referred to as fissile. A fertile nuclide is one that could be bombarded with neutrons to produce a fissile nuclide. Critical mass is the mass of a ball of a material which could undergo a sustained nuclear chain reaction.) ^{232}Th is not fissionable, however it is fertile as it can be converted to fissile ^{233}U by neutron capture and ensuing β decay.

Thorium is a moderately soft, paramagnetic, bright silvery radioactive actinide metal (Table 7.22). In the periodic table, it lies to the right of actinium, to the left of protactinium, and below cerium. Pure thorium is very ductile and, as normal for metals, can be cold-rolled, swaged, and drawn. At room temperature, thorium metal has a face-centered cubic (*fcc*) crystal structure; it has two other forms, one at high temperature (above 1360°C; body-centered cubic (*bcc*)) and one at high pressure (around 100 GPa; body-centerd tetragonal). Thorium is almost half as dense as uranium and plutonium and is harder than either of them. It becomes superconductive below −271.8°C (1.4K). Its melting point of 1750°C is above both those of actinium (1227°C) and protactinium (1568°C). At the start of period 7, from francium to thorium, the melting points of the elements increase (as in other periods), because the number of delocalized electrons each atom contributes increases from one in francium to four in thorium, resulting greater attraction between these electrons and the metal ions as their charge increases from one to four. After thorium, there is a new downward trend in melting points from thorium to plutonium, where the number of f electrons increases from about 0.4 to about 6: this is caused by the increasing hybridization of the 5f and 6d orbitals and the formation of directional bonds resulting in more complex crystal structures and weakened metallic bonding (The f electron count for thorium is a noninteger due to a 5f–6d overlap.). Thorium has the highest melting and boiling points and second-lowest density; only actinium is lighter. Thorium's boiling point of 4788°C is the fifth highest of all the elements with known boiling points.

A thorium atom has 90 electrons, of which four are valence electrons. Three atomic orbitals are theoretically available for the valence electrons to occupy: 5f, 6d, and 7s. Notwithstanding its position in the f-block of the periodic table, it has an anomalous [Rn]6d^{2}7s^2 electron configuration in the ground state, as the 5f and 6d subshells in the early actinides are very close in energy, even more so than the 4f and 5d subshells of the lanthanides: thorium's 6d subshells are lower in energy

TABLE 7.22 Thorium properties.

Appearance	Silvery, often with black tarnish
Standard atomic weight $A_{r,std}$	232.0377
Block	f-Block
Element category	Actinide
Electron configuration	[Rn] $6d^2$ $7s^2$
Phase at STP	Solid
Melting point	1750°C
Boiling point	4788°C
Density (near r.t.)	11.7 g/cm^3
Heat of fusion	13.81 kJ/mol
Heat of vaporization	514 kJ/mol
Molar heat capacity	26.230 J/(mol·K)
Oxidation states	+1, +2, +3, **+4**
Ionization energies	1st: 587 kJ/mol
	2nd: 1110 kJ/mol
	3rd: 1930 kJ/mol
Atomic radius	Empirical: 179.8 pm
Covalent radius	206 ± 6 pm

STP, Standard temperature and pressure.
Bold font indicates main oxidation state.

than its 5f subshells, as its 5f subshells are not well shielded by the filled 6s and 6p subshells and are destabilized. This is caused by relativistic effects, which become stronger near the bottom of the periodic table, in particular the relativistic spin–orbit interaction. The closeness in energy levels of the 5f, 6d, and 7s energy levels of thorium results in thorium nearly always losing all four valence electrons and occurring in its highest possible oxidation state of +4. This is different from its lanthanide congener cerium, in which +4 is also the highest possible state, but +3 plays a significant role and is more stable. Thorium is much more like the transition metals zirconium and hafnium than like cerium in its ionization energies and redox potentials, and hence also in its chemistry: this transition-metal-like behavior is the norm in the first half of the actinide series. Notwithstanding the anomalous electron configuration for gaseous thorium atoms, metallic thorium shows substantial 5f involvement. This was first understood in 1995, when it was pointed out that a hypothetical metallic state of thorium that had the [Rn] $6d^27s^2$ configuration with the 5f orbitals above the Fermi level should be hexagonal close (*hcp*) packed like the group 4 elements titanium, zirconium, and hafnium, and not face-centered cubic (*fcc*) as it actually is. The actual crystal structure can only be explained when the 5f states are invoked, confirming that thorium, and not protactinium, acts as the first actinide metallurgically. The 5f character of thorium is also clear in the uncommon and very unstable + 3 oxidation state, in which thorium shows the electron configuration [Rn] $5f^1$. Tetravalent thorium compounds are typically colorless or yellow, like those of silver or lead, since the Th^{4+} ion has no 5f or 6d electrons. Thorium chemistry is consequently essentially that of an electropositive metal forming a single diamagnetic ion with a stable noble gas configuration, pointing to a similarity between thorium and the main group elements of the s-block. Thorium and uranium are the most studied of the radioactive elements since their radioactivity is low enough not to require special handling in the laboratory.

7.22.3.1 Reactions with oxygen and water

Thorium is a highly reactive and electropositive metal. With a standard reduction potential of −1.90 V for the Th^{4+}/Th couple, it is slightly more electropositive than zirconium or aluminum. Finely divided thorium metal can show pyrophoricity, spontaneously igniting in air. When heated in air, thorium turnings ignite and burn with a brilliant white light to produce the dioxide ThO_2, also called thoria or thorina, which has the fluorite (CaF_2) structure. In bulk, the reaction of pure thorium with air is slow, though corrosion may take place after several months; the majority of thorium samples are contaminated with varying degrees of the dioxide, which significantly accelerates corrosion. Such samples slowly tarnish, turning gray and finally black at the surface. Thoria, a refractory material, has the highest melting point (3390°C) of all known oxides. It is slightly hygroscopic and reacts readily with water and many gases, while it dissolves easily in concentrated nitric acid in the presence of fluoride. Reports of thorium peroxide, initially thought to be Th_2O_7 and formed from reacting thorium salts with hydrogen peroxide, were later revealed to have both peroxide anions and the anions of the reacting thorium salt. Thorium monoxide has been formed through laser ablation of thorium in the presence of oxygen. This highly polar molecule is calculated to have one of the largest known internal electric fields.

At standard temperature and pressure, thorium is slowly attacked by water, but does not readily dissolve in most common acids, with the exception of hydrochloric acid, in which it dissolves leaving a black insoluble residue of ThO(OH,Cl)H. It dissolves in concentrated nitric acid with a small amount of catalytic fluoride or fluorosilicate ions; if these are absent, passivation by the nitrate can take place, as with uranium and plutonium. Thorium hydroxide, $Th(OH)_4$, can be formed by adding a hydroxide of ammonium or an alkali metal to a thorium salt solution, where it emerges as a gelatinous precipitate that will dissolve in dilute acids, among other substances. It can also be produced by electrolysis of thorium nitrates. It is stable from 260–450°C; at 470°C and above it continuously decomposes to convert to thoria. It easily absorbs atmospheric carbon dioxide to form the hydrated carbonate $ThOCO_3 \cdot xH_2O$ and, under high-pressure conditions in a carbon dioxide atmosphere, $Th(CO_3)_2 \cdot 0.5H_2O$ or $Th(OH)_2CO_3 \cdot 2H_2O$. A number of mixed oxides are known, for example, $BaThO_3$, which has the perovskite ($CaTiO_3$) structure.

When heated in air, thorium dioxide emits intense blue light; the light turns to white when ThO_2 is mixed with its lighter homolog cerium dioxide (CeO_2, ceria): this is the basis for its previously common application in gas mantles. The light emitted by thorium dioxide is higher in wavelength than the blackbody emission expected from incandescence at the same temperature, an effect called candoluminescence. It occurs as ThO_2:Ce acts as a catalyst for the recombination of free radicals that appear in high concentration in a flame, whose deexcitation releases large amounts of energy. Adding 1% cerium dioxide heightens the effect by increasing emissivity in the visible region of the spectrum; and because cerium, unlike thorium, can occur in multiple oxidation states, its charge and thus visible emissivity will be contingent on the region in the flame it is found in (as such regions vary in their chemical composition and therefore how oxidizing or reducing they are).

7.22.3.2 Reactions with chalcogenides and pnictides

A number of binary thorium chalcogenides and oxychalcogenides are known with sulfur, selenium, and tellurium. As well as several binary compounds, the oxychalcogenides ThOS (yellow), ThOSe, and ThOTe also exist. The five binary thorium sulfides—ThS (lustrous metallic), Th_2S_3 (brown metallic), Th_7S_{12} (black), ThS_2 (purple-brown), and Th_2S_5 (orange-brown)—can be formed by reacting hydrogen sulfide with thorium, its halides, or thoria (the last if carbon is present): they all hydrolyze in acidic solutions. The six selenides are equivalent to the sulfides, with the addition of $ThSe_3$. The five tellurides are also analogous to the sulfides and selenides (though Th_2Te_5 is not known), but have slightly different crystal structures: for example, ThS has the sodium chloride structure, but ThTe has the cesium chloride structure, as the Th^{4+} and Te^{2-} ions are similar in size whereas the S^{2-} ions are much smaller.

All five chemically characterized pnictogens (nitrogen, phosphorus, arsenic, antimony, and bismuth) also produce compounds with thorium. Three thorium nitrides are known: ThN, Th_3N_4, and Th_2N_3. The brass colored Th_3N_4 is most easily formed by heating thorium metal in a nitrogen atmosphere. Th_3N_4 and Th_2N_3 decompose to the golden-yellow ThN, and indeed ThN can frequently be observed covering the surface of Th_3N_4 samples as Th_3N_4 is hygroscopic and water vapor in the air can decompose it. Thin films of ThN are metallic in nature and, similar to all other actinide mononitrides, have the sodium chloride structure. In addition, ThN is a low-temperature superconductor. All three nitrides can react with thorium halides to produce halide nitrides ThNX (X = F, Cl, Br, I). The heavier pnictogens also form similar monopnictides, except ThBi which has not yet been structurally characterized. The other well-characterized thorium pnictides are Th_3P_4, Th_2P_{11}, ThP_7, Th_3As_4, $ThAs_2$, Th_3Sb_4, $ThSb_2$, and $ThBi_2$. Thorium germanides are also known.

7.22.3.3 Hydrides, borides, and carbides

Thorium reacts with hydrogen to produce the thorium hydrides ThH_2 and Th_4H_{15}, the latter of which is superconducting below −265.7 to −265.2°C (7.5K–8K); at standard temperature and pressure, it conducts electricity like a metal. The hydrides are thermally unstable and easily decompose upon exposure to air or moisture. Thorium is the only element that forms a hydride higher than MH_3. Finely divided thorium metal reacts very readily with hydrogen at standard conditions, but large pieces may need to be heated to 300–400°C for a reaction to take place. At approximately 850°C, the reaction forming first ThH_2 and then Th_4H_{15} occurs without breaking up the structure of the thorium metal. Thorium hydrides react easily with oxygen or steam to convert to thoria, and at 250–350°C rapidly react with hydrogen halides, sulfides, phosphides, and nitrides to form the corresponding thorium binary compounds.

Three binary thorium borides are known: ThB_6, ThB_4, and ThB_{12}. The last is isotypic with UB_{12}. While reports of ThB_{66} and ThB_{76} exist, they may merely be thorium-stabilized boron allotropes. ThB_6 and ThB_{12} may be formed by heating thorium with boron. The three known binary thorium carbides are ThC_2, Th_2C_3, and ThC: all are formed by reacting thorium or thoria with carbon. ThC and ThC_2 are refractory solids and have melting points above 2600°C. Thorium borides, carbides, silicides, and nitrates are all refractory materials and are therefore of interest as possible nuclear fuels.

7.22.3.4 Reactions with halogens

All four thorium tetrahalides are known, as are some low-valent bromides and iodides: the tetrahalides are all 8-coordinated hygroscopic compounds that dissolve easily in polar solvents such as water. Many associated polyhalide ions are also known. Thorium tetrafluoride (ThF_4, white, m.p. 1068°C) is most easily produced by reacting various thorium salts, thoria, or thorium hydroxide with hydrogen fluoride: procedures that involve steps in the aqueous phase are more difficult as they result in hydroxide and oxide fluorides that must be reduced with hydrogen fluoride or fluorine gas. Thorium tetrafluoride has a monoclinic crystal structure similar to zirconium tetrafluoride and hafnium tetrafluoride, where the Th^{4+} ions are coordinated with F^- ions in slightly distorted square antiprisms.

Thorium tetrachloride ($ThCl_4$, white, m.p. 770°C) is formed by heating thoria in an organochloride compound such as carbon tetrachloride. The typical method of purification is crystallization from an aqueous solution and subsequently heating the product above 100°C to dehydrate it. Additional purification can be achieved by subliming it. Its melting and boiling points are respectively 770°C and 921°C. It undergoes a phase transition at 405°C, with a low-temperature α phase and high-temperature β phase. However, the β phase generally persists below the transition temperature. Both phases crystallize in the tetragonal crystal system and the structural differences are minor. Below -203°C, a low-temperature form exists with a complex structure.

Thorium tetrabromide ($ThBr_4$, white, m.p. 679°C) can be formed either by reacting thorium(IV) hydroxide with hydrobromic acid (which has the drawback of frequently resulting in products contaminated with oxybromides) or by directly reacting bromine or hydrogen bromide with thorium metal or compounds. The product can subsequently be purified by sublimation at 600°C in a vacuum. The melting and boiling points are 679°C and 857°C. Similar to the tetrachloride, both an α and a β form exist and both are isotypic to the tetrachloride forms, though the phase transition here happens at 426°C. There is also a low-temperature form. Incomplete reports of the lower bromides $ThBr_3$, $ThBr_2$, and ThBr are known (the last only known as a gas-phase molecular species): $ThBr_3$ and $ThBr_2$ are known to be very reactive and at high temperatures disproportionate.

Thorium tetraiodide (ThI_4, yellow, m.p. 556°C) is produced by direct reaction of the elements in a sealed silica ampule. Water and oxygen must be absent, or else $ThOI_2$ and ThO_2 can contaminate the product. It has a different crystal structure compared to the other tetrahalides, as it is monoclinic. The lower iodides ThI_3 (black) and ThI_2 (gold) can be produced by reducing the tetraiodide with thorium metal. (ThI is also anticipated to form as an intermediate in the dissociation of ThI_4 to thorium metal.) These do not contain Th(III) and Th(II), but instead contain Th^{4+} and could be more clearly written as $Th^{4+}(I^-)_3(e^-)$ and $Th^{4+}(I^-)_2(e^-)_2$, respectively. Subject to the amount of time allowed for the reaction between ThI_4 and thorium, two modifications of ThI_3 can be formed: shorter times give thin lustrous rods of α-ThI_3, while longer times give small β-ThI_3 crystals with green to brass colored luster. Both forms are rapidly oxidized by air and reduce water, swiftly forming large quantities of hydrogen gas. ThI_2 also has two modifications, which can be formed by adjusting the reaction temperature: at 600°C, α-ThI_2 is formed, while a reaction temperature of 700–850°C produces β-ThI_2, which has a golden luster.

Numerous polynary halides with the alkali metals, barium, thallium, and ammonium are known for thorium fluorides, chlorides, and bromides. For instance, when treated with potassium fluoride and hydrofluoric acid, Th^{4+} forms the complex anion ThF_6^{2-}, which precipitates as an insoluble salt, K_2ThF_6.

7.22.3.5 Coordination compounds

In an acidic aqueous solution, thorium exists as the tetrapositive aqua ion $[Th(H_2O)_9]^{4+}$, which has tricapped trigonal prismatic molecular geometry: at pH < 3, the solutions of thorium salts are dominated by this cation. The Th^{4+} ion is the largest of the tetrapositive actinide ions, and contingent on the coordination number can have a radius between 0.95 and 1.14 Å. It is quite acidic due to its high charge, slightly stronger than sulfurous acid: consequently it tends to undergo hydrolysis and polymerization (but to a lesser extent than Fe^{3+}), mainly to $[Th_2(OH)_2]^{6+}$ in solutions with pH 3 or below, but in more alkaline solution polymerization continues until the gelatinous hydroxide $Th(OH)_4$ forms and precipitates out (however equilibrium may take weeks to be reached, as the polymerization usually slows down before the precipitation). As a hard Lewis acid, Th^{4+} favors hard ligands with oxygen atoms as donors: complexes with sulfur atoms as donors are less stable and are more susceptible to hydrolysis. High coordination numbers are the norm for thorium because of its large size. Thorium nitrate pentahydrate was the first known compound with coordination number 11, the oxalate tetrahydrate has coordination number 10, and the borohydride (first prepared in the Manhattan Project) has coordination number 14. These thorium salts are known for their high solubility in water and polar organic solvents. Numerous other inorganic thorium compounds with polyatomic anions are known, such as the perchlorates, sulfates, sulphites, nitrates, carbonates, phosphates, vanadates, molybdates, and chromates, and their hydrated forms. They are essential in thorium purification and the disposal of nuclear waste, but the majority of them have not yet been fully characterized, in particular with respect to their structural properties. For instance, thorium nitrate is formed by reacting thorium hydroxide with nitric acid: it is soluble in water and alcohols and is an important intermediate in the purification of thorium and its compounds. Thorium perchlorate is very water-soluble and crystallizes from acidic solutions as the tetrahydrate $Th(ClO_4)_4 \cdot 4H_2O$, whereas thorium nitrate forms tetra- and pentahydrates, is soluble in water and alcohols, and is an essential intermediate in the purification of thorium and its compounds. Due to its great tendency toward hydrolysis, thorium does not form simple carbonates, but instead carbonato complexes such as $[Th(CO_3)_5]^{6-}$, comparable to uranium(IV) and plutonium(IV). Thorium forms a stable tetranitrate, $Th(NO_3)_4 \cdot 5H_2O$, a property shared only by plutonium(IV) among the actinides: it is the most common thorium salt. Another illustration of the high coordination characteristic of thorium is $[Th(C_5H_5NO)_6(NO_3)_2]^{2+}$, a 10-coordinated complex with distorted bicapped antiprismatic molecular geometry. The anionic $[Th(NO_3)_6]^{2-}$ is isotypic to its cerium, uranium, neptunium, and plutonium equivalents and has a distorted icosahedral structure. Especially important is the borohydride, $Th(BH_4)_4$, first prepared in the Manhattan Project along with its uranium(IV) analog. It is prepared as follows:

$$ThF_4 + 2Al(BH_4)_3 \rightarrow Th(BH_4)_4 + 2AlF_2BH_4$$

following which thorium borohydride can be simply isolated, as it sublimes out of the reaction mixture. Like its protactinium(IV) and uranium(IV) equivalents, it is a thermally and chemically stable compound where thorium has a coordination number of 14 with a bicapped hexagonal antiprismatic molecular geometry. Thorium complexes with organic ligands, such as oxalate, citrate, and EDTA, are much more stable. In natural thorium-containing waters, organic thorium complexes usually occur in concentrations orders of magnitude higher than the inorganic complexes, even when the concentrations of inorganic ligands are much higher than those of organic ligands.

7.22.3.6 Organothorium compounds

Most of the research on organothorium compounds has concentrated on the cyclopentadienyl complexes and cyclooctatetraenyls. Like many of the early and middle actinides (up to americium, and also expected for curium), thorium forms a cyclooctatetraenide complex: the yellow $Th(C_8H_8)_2$, thorocene. It is isotypic with the better-known equivalent uranium compound uranocene. It can be formed by reacting $K_2C_8H_8$ with thorium tetrachloride in THF at the temperature of dry ice, or by reacting thorium tetrafluoride with MgC_8H_8. It is unstable in air and decomposes in water or at 190°C. Half-sandwich compounds are also known, such as $(\eta^8\text{-}C_8H_8)ThCl_2(THF)_2$, which has a piano-stool structure and is produced by reacting thorocene with thorium tetrachloride in THF.

The simplest of the cyclopentadienyls are $Th(C_5H_5)_3$ and $Th(C_5H_5)_4$: many derivatives are known. The former (which has two forms, one purple and one green) is an uncommon example of thorium in the formal +3 oxidation state; a formal +2 oxidation state occurs in a derivative. The chloride derivative $[Th(C_5H_5)_3Cl]$ is produced by heating thorium tetrachloride with limiting $K(C_5H_5)$ used (other univalent metal cyclopentadienyls can also be utilized). The alkyl and aryl derivatives are formed from the chloride derivative and have been employed to research the nature of the Th–C σ bond. Of special interest is the dimer $[Th(\eta^5\text{-}C_5H_5)_2\text{-}\mu\text{-}(\eta^5,\eta^1\text{-}C_5H_5)]_2$, where the two thorium atoms are bridged by two cyclopentadienyl rings, similarly to the structure of niobocene. Other organothorium compounds are not well characterized. Tetrabenzylthorium, $Th(CH_2C_6H_5)$, and tetraallylthorium, $Th(C_3H_5)_4$, are known, but their structures have not been determined. They decompose gradually at room temperature. Thorium forms the monocapped trigonal prismatic anion $[Th(CH_3)_7]^{3-}$, while heptamethylthorate forms the salt $[Li(tmeda)]_3[ThMe_7]$ (tmeda = $Me_2NCH_2CH_2NMe_2$). Although one methyl group is only attached to the thorium atom (Th–C distance 257.1 pm) and the other six connect the lithium and thorium atoms (Th–C distances 265.5–276.5 pm), they behave equivalently in solution. Tetramethylthorium, $Th(CH_3)_4$, is not known, but its adducts are stabilized by phosphine ligands.

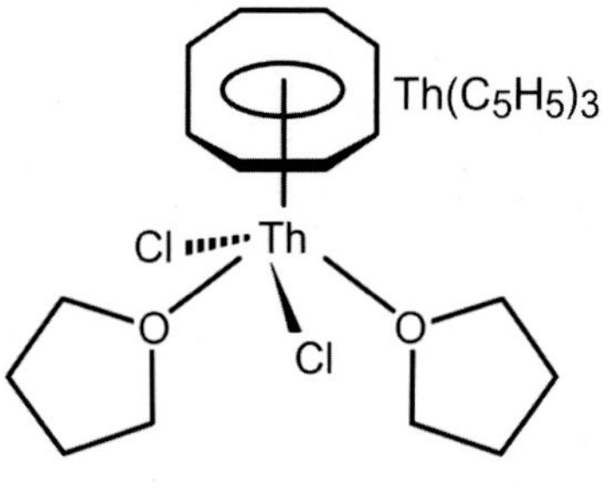

Th(C5H5)3
Cl
Th
O
Cl
O

7.22.4 Major uses

Nonradioactivity-associated applications of Th have been decreasing since the 1950s because of environmental concerns mainly resulting from the radioactivity of Th and its decay products. Most Th uses employ its dioxide (ThO_2, occasionally called "thoria" in the industry), instead of the metal. This compound has a melting point of 3300°C, the highest of all known oxides; only some compounds have higher melting points. This improves the compound to stay solid in a flame, and it substantially improves the brightness of the flame; this is the most important reason why Th is used in gas mantles. All materials emit energy (glow) at high temperatures, but the light emitted by Th is almost completely in the visible spectrum, hence the brightness of Th mantles. Energy, part of it in the way of visible light, is emitted when Th is exposed to a source of energy itself, such as a cathode ray, heat or ultraviolet light. This effect it has in common with CeO_2, which converts ultraviolet light into visible light more effectively, however ThO_2 gives a higher flame temperature, emitting less infrared light. Th in mantles, although still widespread, has been gradually substituted by Y since the end of the 1990s. While manufacturing incandescent filaments, recrystallization of W is substantially reduced by adding small quantitiess of ThO_2 to the W sintering powder prior to drawing the filaments. Adding a small amount of Th to W thermocathodes significantly decreases the work function of electrons; hence, electrons are emitted at considerably lower temperatures. Th forms a one-atom-thick layer on the surface of W. The work function from a Th surface is decreased probably due to the electric field on the interface between Th and W formed because of Th's higher electropositivity. As far back as the 1920s, thoriated W wires have been employed in electronic tubes and in the cathodes and anticathodes of X-ray tubes and rectifiers. Due to the reactivity of Th with atmospheric oxygen and nitrogen, Th also marks impurities in the evacuated tubes. The development of transistors in the 1950s substantially decreased this application, but not completely. ThO_2 is employed in GTAW (gas tungsten arc welding) to improve the high-temperature strength of W electrodes and increase arc stability. ThO_2 is being substituted in this application with other oxides, for example, those of Zr, Ce, and La. ThO_2 can be found in heat-resistant ceramics, for example, high-temperature laboratory crucibles, either as the principal compound or as an addition to ZrO_2. An alloy of 90% Pt and 10% Th is an efficient catalyst for oxidizing ammonia to nitrogen oxides, but this has been substituted by an alloy of 95% Pt and 5% Rh due to its better mechanical properties and durability. When added to glass, ThO_2 increases its refractive index and decreases dispersion. This type of glass is used in high-quality lenses for cameras and scientific instruments. The radiation from these lenses can darken them over time and turn them yellow over a period of years and degrade film, though the health risks are negligible. Yellowed lenses may be returned to their original colorless state through prolonged exposure to intense ultraviolet radiation. ThO_2 has since been substituted by rare-earth oxides as they offer comparable effects and on top of that are not radioactive. ThF_4 is employed as an antireflection material in multilayered optical coatings. It is transparent to electromagnetic waves in the wavelength range between 0.35 and 12 μm, a range that comprises near ultraviolet, visible and mid-infrared light. Its radiation is mainly due to α particles, which can simply be stopped by a thin cover layer of a different material. Alternatives for ThF_4 are being researched since the 2010s.

The most important nuclear power source in a reactor is the neutron-induced fission of a nuclide; the synthetic fissile nuclei ^{233}U and ^{239}Pu can be bred from neutron capture by the naturally occurring nuclides ^{232}Th and ^{238}U. ^{235}U can be found naturally and is also fissile. In the Th fuel cycle, the fertile isotope ^{232}Th is bombarded with slow neutrons (n), undergoing neutron capture to form ^{233}Th, which then undergoes two successive β^- decays to become first ^{233}Pa and subsequently the fissile ^{233}U:

$$^{232}_{90}\mathrm{Th} + \mathrm{n} \longrightarrow ^{233}_{90}\mathrm{Th} + \gamma \xrightarrow{\beta^-,\ 21.8\ \mathrm{min}} ^{233}_{91}\mathrm{Pa} \xrightarrow{\beta^-,\ 27.0\ \mathrm{days}} ^{233}_{92}\mathrm{U}$$

^{233}U is fissile and can be employed as a nuclear fuel similar to ^{235}U or ^{239}Pu. When ^{233}U undergoes nuclear fission, the neutrons emitted can hit other ^{232}Th nuclei, maintaining the cycle. This is analog to the U fuel cycle in fast breeder reactors where ^{238}U undergoes neutron capture to become ^{239}U, β decaying to first ^{239}Np and then fissile ^{239}Pu. Th is more abundant than U and can fulfill the global energy requirements for much longer. ^{232}Th absorbs neutrons more easily than ^{238}U, and ^{233}U has a higher probability of fission upon neutron capture (92.0%) than ^{235}U (85.5%) or ^{239}Pu (73.5%). In addition, it releases on average more neutrons upon fission. A single neutron capture by ^{238}U forms transuranic waste along with the fissile ^{239}Pu, but ^{232}Th only forms this waste after five captures, creating ^{237}Np. This number of captures does not happen for 98%–99% of the ^{232}Th nuclei as the intermediate products ^{233}U or ^{235}U undergo fission, and less long-lived transuranics are formed. Hence, Th is a potentially interesting alternative to U in mixed oxide fuels to reduce the formation of transuranics and optimize the destruction of Pu. Th fuels additionally result in a safer and better-performing reactor core since ThO_2 has a higher melting point, better thermal conductivity, and a lower coefficient of thermal expansion and is chemically more stable than the current generally used fuel UO_2, which can further oxidize to U_3O_8. The used fuel is hard and hazardous to reprocess due to many of the daughter products of ^{232}Th

and ^{233}U being strong γ emitters. All ^{233}U production processes give rise to impurities of ^{232}U, either from parasitic knock-out (n,2n) reactions on ^{232}Th, ^{233}Pa, or ^{233}U that result in the loss of a neutron, or from double neutron capture of ^{230}Th, an impurity in natural ^{232}Th:

$$^{230}_{90}\mathrm{Th} + \mathrm{n} \longrightarrow {}^{231}_{90}\mathrm{Th} + \gamma \xrightarrow{\beta^-, 25.5\ \mathrm{h}} {}^{231}_{91}\mathrm{Pa} \xrightarrow{\alpha, 3.28 \times 10^4\ \mathrm{y}} {}^{231}_{91}\mathrm{Pa} + \mathrm{n} \longrightarrow {}^{232}_{91}\mathrm{Pa} + \gamma \xrightarrow{\beta^-, 1.3\ \mathrm{d}} {}^{232}_{92}\mathrm{U}$$

^{232}U by itself is not especially dangerous, but rapidly decays to form the strong γ emitter ^{208}Tl. (^{232}Th follows the same decay chain, but its much longer half-life means that the amounts of ^{208}Tl formed are insignificant.) These impurities of ^{232}U make ^{233}U simple to detect and hazardous to work on, and the impracticality of their separation restricts the options of nuclear proliferation using ^{233}U as the fissile material. ^{233}Pa has a fairly long half-life of 27 days and a high cross-section for neutron capture. Consequently, it is a neutron poison: rather than quickly decaying to the useful ^{233}U, a substantial quantity of ^{233}Pa converts to ^{234}U and consumes neutrons, reducing the reactor efficiency. To prevent this, ^{233}Pa is removed from the active zone of Th molten salt reactors throughout their operation, so that it just decays to ^{233}U. The irradiation of ^{232}Th with neutrons, and subsequently its processing, need to be understood before these benefits can be achieved, and this necessitates more advanced technology than the currently used U and Pu fuel cycle; research in this area continues. The isotopes formed in the Th fuel cycle are primarily not transuranic, though some of them are still very hazardous, for example, ^{231}Pa, which has a half-life of 32,760 years and is a key contributor to the long-term radiotoxicity of spent nuclear fuel.

7.23 92 U — Uranium

7.23.1 Discovery

The application of uranium in its natural oxide form goes back to at least the year 79 BCE, when it was utilized to impart a yellow color to ceramic glazes. Yellow glass with 1% uranium oxide was found in a Roman villa on Cape Posillipo in the Bay of Naples, Italy in 1912. From the start of the late Middle Ages, pitchblende (mainly UO_2, the mineral uraninite) was produced in the Habsburg silver mines in Joachimsthal, Bohemia (now Jáchymov in the Czech Republic), and was utilized as a coloring agent in the local glass manufacturing. In the early 19th century, the world's only known resource of uranium ore were these mines. The discovery of the element is attributed to the German chemist Martin Heinrich Klaproth (December 1, 1743 to January 1, 1817) (Fig. 7.146). While he was experimenting in his laboratory in Berlin in 1789, Klaproth managed to precipitate a yellow compound (probably sodium diuranate, $Na_2U_2O_7{\cdot}6H_2O$) by dissolving pitchblende in nitric acid (HNO_3) and neutralizing the solution with sodium hydroxide (NaOH). Klaproth expected the yellow substance to be the oxide of a new element and heated it with charcoal resulting in a black powder, which he believed was the newly discovered metal itself (actually, that powder was an oxide of uranium) (Klaproth, 1879). He named the newly found element after the planet Uranus (named after the primordial Greek god of the sky), which had been discovered eight years before by German-born British astronomer and composer William Herschel (November 15, 1738 to August 25, 1822). In 1841 Eugène-Melchior Péligot (March 24, 1811 to April 15, 1890), Professor of Analytical Chemistry at the Conservatoire National des Arts et Métiers (Central School of Arts and Manufactures) in Paris, obtained the first sample of uranium metal by heating uranium tetrachloride (UCl_4) with potassium (Péligot, 1842) (Fig. 7.147). French engineer and physicist Antoine Henri Becquerel (December 15, 1852 to August 25, 1908) discovered radioactivity by using uranium in 1896 (Fig. 7.148). Becquerel made the discovery in Paris by leaving an amount of potassium uranyl sulfate, $K_2UO_2(SO_4)_2$, on top of an unexposed photographic plate in a drawer and observed that the plate had become "fogged." He deduced that a form of invisible light or rays emitted by uranium had exposed the plate. A research group led by Italian and naturalized-American physicist Enrico Fermi (September 29, 1901 to November 28, 1954) in 1934 found that bombarding uranium with neutrons produced the emission of beta rays (electrons or positrons from the elements produced) (Fermi, 1938). The fission products were in the beginning mistaken for new elements with atomic numbers 93 and 94, which the Dean of the Faculty of Rome, physicist Orso Mario Corbino (April 30, 1876 to January 23, 1937), named ausonium (atomic symbol Ao, now known as neptunium. It was named after a Greek name of Italy, Ausonia) and hesperium (atomic symbol Es, now known as plutonium. It was named in Italian Esperio after a Greek name of Italy, Hesperia, "the land of the West"), respectively. The research resulting in the discovery of uranium's ability to fission (break apart) into lighter elements and release binding energy was performed by German chemists Otto Hahn (March 8, 1879 to July 28, 1968) and Fritz Strassmann (February 22, 1902 to

FIGURE 7.146 Martin Heinrich Klaproth, Mezzotint. Wellcome Trust, UK, https://wellcomeimages.org/indexplus/image/V0003234.html. CC-BY4.0.

FIGURE 7.147 Eugène-Melchior Péligot, c.1860, photo by Gaspard Félix Tournachon (1820–1910).

FIGURE 7.148 Antoine Henri Becquerel, c.1905, photo by Paul Nadar (1856–1939).

April 22, 1980) in Hahn's laboratory in Berlin. Austrian-Swedish physicist Lise Meitner (November 7, 1878 to October 27, 1968) and her nephew, the Austrian physicist Otto Robert Frisch (October 1, 1904 to September 22, 1979), came up with the physical explanation published in February 1939 and called the process "nuclear fission" (Meitner and Frisch, 1939). Later on, Fermi theorized that the fission of uranium could produce enough neutrons to sustain a fission reaction. Validation of this hypothesis came in 1939, and later study determined that on average around 2.5 neutrons are released by each fission of the rare uranium isotope uranium-235. Fermi urged American physicist Alfred O. C. Nier (May 28, 1911 to May 16, 1994) to isolate uranium isotopes for determination of the fissile component, and on February 29, 1940, Nier utilized an instrument he had constructed at the University of Minnesota to isolate the world's first uranium-235 sample in the Tate Laboratory. After it was sent to Columbia University's cyclotron (a type of particle accelerator invented by American nuclear physicist Ernest O. Lawrence (August 8, 1901 to August 27, 1958) in 1929–1930 at the University of California, Berkeley, and patented in 1932. A cyclotron accelerates charged particles outwards from the center along a spiral path. The particles are held to a spiral trajectory by a static magnetic field and accelerated by a rapidly varying (radio frequency) electric field.), American physicist John Ray Dunning (September 24, 1907 to August 25, 1975) established the sample to be the isolated fissile material on March 1. Further research found that the far more common uranium-238 isotope can be transmuted into plutonium (Nuclear transmutation is the conversion of one chemical element or an isotope into another chemical element. Because any element (or isotope of one) is defined by its number of protons (and neutrons) in its atoms, that is, in the atomic nucleus, nuclear transmutation occurs in any process where the number of protons or neutrons in the nucleus is changed.), which, like uranium-235, is also fissile by thermal neutrons. These findings led several countries to start research with the aim to develop nuclear weapons and nuclear power. On December 2, 1942, as part of the Manhattan Project, another research group led by Enrico Fermi managed to initiate the first artificial self-sustained nuclear chain reaction, Chicago Pile-1. An early plan using enriched uranium-235 was abandoned as it was not yet obtainable in adequate amounts. Working in a laboratory under the stands of Stagg Field at the University of Chicago, the team created the conditions needed for such a reaction by piling together 400 short tons (360 metric tons) of graphite, 58 short tons (53 metric tons) of uranium oxide, and six short tons (5.5 metric tons) of uranium metal, most of which was supplied by the Westinghouse Lamp Plant in an improvised production process.

7.23.2 Mining, production, and major minerals

Uranium is the 51st element in order of abundance in the Earth's crust. Uranium is also the highest-numbered element to be found naturally in significant quantities on Earth and is almost always found combined with other elements. The decay of uranium, thorium, and potassium-40 in the Earth's mantle is believed to be the principal heat source that keeps the outer core liquid and drives mantle convection, which subsequently drives plate tectonics of the Earth's crust. Uranium's average concentration in the Earth's crust is between 2 and 4 ppm. Uranium is found in hundreds of minerals, including uraninite (UO_2, the most common uranium ore), carnotite ($K_2(UO_2)_2(VO_4)_2 \cdot 3H_2O$), autunite ($Ca(UO_2)_2(PO_4)_2 \cdot 11H_2O$), uranophane ($Ca(UO_2)_2(SiO_3OH)_2 \cdot 5H_2O$), torbernite ($Cu(UO_2)_2(PO_4)_2 \cdot 10H_2O$), and coffinite ($U(SiO_4) \cdot nH_2O$). Significant concentrations of uranium occur in some deposits such as phosphate rock deposits, and lignite (often referred to as brown coal, is a soft, brown, combustible, sedimentary rock formed from naturally compressed peat. It is considered the lowest rank of coal due to its relatively low heat content.), and monazite ($(Ce,La,Nd,Th)PO_4$) sands in uranium-rich ores (it is recovered commercially from deposits with as low as 0.1% uranium) (Box 7.9).

BOX 7.9 The Oklo Fossil Reactor

In 1972 the French physicist Francis Perrin (August 17, 1901 to July 4, 1992) found 15 ancient and no longer active natural nuclear fission reactors in three separate ore deposits at the Oklo mine in Gabon, West Africa, collectively known as the Oklo Fossil Reactors. In May 1972 at the Pierrelatte uranium enrichment facility in France, routine mass spectrometry comparing UF_6 samples from the Oklo Mine, located in Gabon, showed a discrepancy in the amount of the ^{235}U isotope. Normally the concentration is 0.72% while these samples had only 0.60%, a significant difference. This discrepancy required explanation, as all civilian uranium handling facilities must meticulously account for all fissionable isotopes to ensure that none are diverted for weapons purposes. Thus the French Commissariat à l'énergie atomique (CEA) began an investigation. A series of measurements of the relative abundances of the two most significant isotopes of the uranium mined at Oklo showed anomalous results compared to those obtained for uranium from other mines. Further investigations into this uranium deposit discovered uranium ore with a ^{235}U concentration as low as 0.44%. Subsequent examination of isotopes of fission products such as neodymium and ruthenium also showed anomalies. This loss in ^{235}U is exactly what happens in a nuclear reactor. A possible explanation, therefore, was that the uranium ore had operated as a natural fission reactor. Other observations led to the same conclusion, and on September 25, 1972, the CEA announced their finding that self-sustaining nuclear chain reactions had occurred on Earth about 2 billion years ago. The natural nuclear reactor formed when a uranium-rich mineral deposit became inundated with groundwater that acted as a neutron moderator, and a nuclear chain reaction took place. The heat generated from the nuclear fission caused the groundwater to boil away, which slowed or stopped the reaction. After cooling of the mineral deposit, the water returned, and the reaction restarted, completing a full cycle every 3 hours. The fission reaction cycles continued for hundreds of thousands of years and ended when the ever-decreasing fissile materials no longer could sustain a chain reaction. Fission of uranium normally produces five known isotopes of the fission-product gas xenon; all five have been found trapped in the remnants of the natural reactor, in varying concentrations. The concentrations of xenon isotopes, found trapped in mineral formations 2 billion years later, make it possible to calculate the specific time intervals of reactor operation: approximately 30 minutes of criticality followed by 2 hours and 30 minutes of cooling down to complete a 3-hour cycle. A key factor that made the reaction possible was that, at the time the reactor went critical 1.7 billion years ago, the fissile isotope ^{235}U made up about 3.1% of the natural uranium, which is comparable to the amount used in some of today's reactors. (The remaining 96.9% was non-fissile ^{238}U). Because ^{235}U has a shorter half-life than ^{238}U, and thus decays more rapidly, the current abundance of ^{235}U in natural uranium is about 0.70–0.72%. A natural nuclear reactor is therefore no longer possible on Earth without heavy water or graphite. The Oklo uranium ore deposits are the only known sites in which natural nuclear reactors existed. Other rich uranium ore bodies would also have had sufficient uranium to support nuclear reactions at that time, but the combination of uranium, water and physical conditions needed to support the chain reaction was unique, as far as is currently known, to the Oklo ore bodies. Another factor which probably contributed to the start of the Oklo natural nuclear reactor at 2 billion years, rather than earlier, was the increasing oxygen content in the Earth's atmosphere. Uranium is naturally present in the rocks of the earth, and the abundance of fissile ^{235}U was at least 3% or higher at all times prior to reactor startup. Uranium is soluble in water only in the presence of oxygen. Therefore, the rising oxygen levels during the aging of the Earth may have allowed uranium to be dissolved and transported with groundwater to places where a high enough concentration could accumulate to form rich uranium ore bodies. Without the new aerobic environment available on Earth at the time, these concentrations probably could not have taken place. It is estimated that nuclear reactions in the uranium in centimeter- to meter-sized veins consumed about five tons of ^{235}U and elevated temperatures to a few hundred degrees Celsius. Most of the non-volatile fission products and actinides have only moved centimeters in the veins during the last 2 billion years. The capacity of the surrounding sediment to contain the nuclear waste products has been mentioned by the US federal government as supporting evidence for the viability to store spent nuclear fuel at the Yucca Mountain nuclear waste repository.

FIGURE 7.149 Uraninite, UO_2, a 7 mm cubic crystal in calcite, $CaCO_3$, with fluorite, CaF_2. Renfrew Co., Ontario, Canada.

FIGURE 7.150 Autunite, $Ca(UO_2)_2(PO_4)_2 \cdot 11H_2O$, bright yellow platy crystals to 3 mm. Apex mine, Lander Co., Nevada, United States.

Worldwide production of U_3O_8 (yellowcake) in 2013 amounted to 70,015 tons, of which 32% was mined in Kazakhstan. Other important U producing countries are Canada, Australia, Niger, Namibia, and Russia. Low-grade U ore produced typically has between 0.01% and 0.25% U oxides. Wide-ranging procedures have to be used to extract the metal from its ore. High-grade ores found in Athabasca Basin deposits in Saskatchewan, Canada can have up to 23% U oxides on average. It is estimated that 5.5 million tons of U exist in ore reserves that are economically viable at US$59 per lb of uranium, while 35 million tons are classed as mineral resources (reasonable prospects for eventual commercial mining). Australia has 31% of the world's known U ore reserves and the world's largest single U deposit, located at the Olympic Dam Mine in South Australia. There exists a substantial U reserve in Bakouma a sub-prefecture in the prefecture of Mbomou in Central African Republic. In nature, U(VI) forms highly soluble carbonate complexes at alkaline pH. This results in an increase in mobility and availability of uranium to groundwater and soil from nuclear wastes which results in health hazards. Nevertheless, it is difficult to precipitate uranium as phosphate in the presence of excess carbonate at

FIGURE 7.151 Francevillite, $Ba(UO_2)_2(VO_4)_2{\cdot}5H_2O$, bright orange clusters and single crystals to 1.5 mm in diamond-shaped forms. Mounana mine, Franceville, Haut-Ogooué, Gabon.

FIGURE 7.152 Torbernite, $Cu(UO_2)_2(PO_4)_2{\cdot}10H_2O$, pastel green square tabular crystals from 1–2 mm. Contencas mine, Portugal.

alkaline pH. Some bacteria strains, such as *Shewanella putrefaciens*, *Geobacter metallireducens* and some strains of *Burkholderia fungorum*, use uranium for their growth while converting U(VI) to U(IV).

Uranium ore is mined in a number of ways: by open-pit, underground, in situ leaching, and borehole mining. Low-grade uranium ore typically has between 0.01% and 0.25% in uranium oxides. Wide-ranging methods must be used to obtain the uranium from its ore. In contrast, high-grade ores such as those found in for example, the Athabasca Basin deposits in Saskatchewan, Canada, can have up to 23% uranium oxides on average. In a first step the uranium ore is crushed and rendered into a fine powder and then leached with either an acid or alkali. The leachate is subjected to one of several series of precipitation, solvent extraction, and ion exchange. The subsequent mixture, generally known as yellowcake, contains at least 75% uranium oxides U_3O_8. The yellowcake is then calcined at increased temperatures to remove impurities from the milling process before refining and conversion. Commercial-grade uranium can be manufactured via the reduction reaction of uranium halides with alkali or alkaline earth metals. Uranium metal can also be obtained using electrolysis of KUF_5 or UF_4, dissolved in molten calcium chloride ($CaCl_2$) and sodium chloride (NaCl). Very pure uranium can be obtained using the thermal decomposition reactions of uranium halides on a hot filament.

FIGURE 7.153 Tyuyamunite, $Ca(UO_2)_2(VO_4)_2{\cdot}5{-}8H_2O$, well-formed opaque bright yellow micro crystals with green malachite, $Cu_2(CO_3)(OH)_2$, about 4 mm. Mashamba West mine, Kolwezi dist., Katanga, Democratic Rep. of Congo.

FIGURE 7.154 Spray of bright green ulrichite, $CaCu(UO_2)(PO_4)_2{\cdot}4H_2O$, in a 1 mm vug. Lake Boga granite quarry, Lake Boga, Victoria, Australia.

Uranium can be found in over 200, often very colorful minerals. About 85 of these minerals are found in the oxide class, for example, carnotite ($K_2(UO_2)_2(VO_4)_2{\cdot}3H_2O$), curite ($Pb_3(UO_2)_8O_8(OH)_6{\cdot}3H_2O$), francevillite ($Ba(UO_2)_2(VO_4)_2{\cdot}5H_2O$), schoepite ($(UO_2)_8O_2(OH)_{12}{\cdot}10H_2O$), tyuyamunite ($Ca(UO_2)_2(VO_4)_2{\cdot}5{-}8H_2O$), and uraninite ($UO_2$) (Fig. 7.149). The carbonates are represented by about 35 different minerals, such as andersonite ($Na_2Ca(UO_2)(CO_3)_3{\cdot}6H_2O$), liebigite ($Ca_2(UO_2)(CO_3)_3{\cdot}11H_2O$), rutherfordine ($(UO_2)CO_3$), and zellerite ($Ca(UO_2)(CO_3)_2{\cdot}5H_2O$). Within the sulfate class 22 different minerals can be found that contain U, for example, cousinite ($MgU_2Mo_2O_{11}{\cdot}6H_2O$), uranopilite ($(UO_2)_6(SO_4)O_2(OH)_6{\cdot}14H_2O$), and zippeite ($K_3(UO_2)_4(SO_4)_2O_3(OH){\cdot}3H_2O$). About 75 different minerals are found in the phosphate class, for example, althupite ($AlTh(UO_2)_7(PO_4)_4(OH)_5O_2{\cdot}15H_2O$),

FIGURE 7.155 Zeunerite, $Cu(UO_2)_2(AsO_4)_2\cdot 10H_2O$, emerald-green, translucent, tabular crystals to 5 mm. Majuba Hill mine, Pershing Co., Nevada, United States.

FIGURE 7.156 Boltwoodite, $(K,Na)(UO_2)(SiO_3OH)\cdot 1.5H_2O$, 2–3-mm golden-yellow sprays. Rossing mine, Rossing, Swakopmund, Namibia.

autunite ($Ca(UO_2)_2(PO_4)_2\cdot 11H_2O$) (Fig. 7.150), francevillite ($Ba(UO_2)_2(VO_4)_2\cdot 5H_2O$) (Fig. 7.151), phurcalite ($Ca_2(UO_2)_3(PO_4)_2O_2\cdot 7H_2O$), torbernite ($Cu(UO_2)_2(PO_4)_2\cdot 10H_2O$) (Fig. 7.152), tyuyamunite ($Ca(UO_2)_2(VO_4)_2\cdot 5\text{-}8H_2O$) (Fig. 7.153), ulrichite ($CaCu(UO_2)(PO_4)_2\cdot 4H_2O$) (Fig. 7.154), and zeunerite ($Cu(UO_2)_2(AsO_4)_2\cdot 10H_2O$) (Fig. 7.155). The silicate class contains about 20 different minerals with structural U, such as boltwoodite ($(K,Na)(UO_2)(SiO_3OH)\cdot 1.5H_2O$) (Fig. 7.156), cuprosklodowskite ($Cu(UO_2)_2(SiO_3OH]_2\cdot 6H_2O$) (Fig. 7.157), kasolite ($Pb(UO_2)[SiO_4]\cdot H_2O$) (Fig. 7.158), sklodowskite ($Mg(UO_2)_2(SiO_3OH)_2\cdot 6H_2O$) (Fig. 7.159), uranophane ($Ca(UO_2)_2(SiO_3OH)_2\cdot 5H_2O$), and weeksite ($K_2(UO_2)_2(Si_5O_{13})\cdot 4H_2O$) (Fig. 7.160).

FIGURE 7.157 Cuprosklodowskite, $Cu(UO_2)_2(SiO_3OH]_2{\cdot}6H_2O$, bright green acicular crystal sprays in 1 cm vug. Shinkolobwe, Katanga, Democratic Republic of Congo.

FIGURE 7.158 Kasolite, $Pb(UO_2)[SiO_4]{\cdot}H_2O$, thick orange bladed spray to 1 cm across. Shinkolobwe, Katanga, Democratic Republic of Congo.

7.23.3 Chemistry

Natural uranium contains three major isotopes: ^{238}U (99.28% natural abundance), ^{235}U (0.71%), and ^{234}U (0.0054%). All three are radioactive, emitting α particles, with the exception that all three radioisotopes have small probabilities of undergoing spontaneous fission, instead of α emission. In addition, there exist five other trace isotopes: ^{239}U, which is formed when ^{238}U undergoes spontaneous fission, releasing neutrons that are captured by another ^{238}U atom; ^{237}U, which is formed when ^{238}U captures a neutron but emits two more, which subsequently decays to ^{237}Np; and finally, ^{233}U, which is formed in the decay chain of ^{237}Np. It is also expected that ^{232}Th can undergo double β decay, which would produce ^{232}U, but this has not yet been detected. ^{238}U is the most stable U radioisotope with a half-life of around 4.468×10^9 years (about the age of the Earth). ^{235}U has a half-life of approximately 7.13×10^8 years, while ^{234}U has a half-life of around 2.48×10^5 years. For natural U, about 49% of its α rays are emitted by each of ^{238}U atom and the same by ^{234}U (as the ^{234}U is formed from ^{238}U) and about 2.0% of them by the ^{235}U. When the Earth was young,

FIGURE 7.159 Sklodowskite, $Mg(UO_2)_2(SiO_3OH)_2 \cdot 6H_2O$, canary yellow acicular crystals to 4 mm, free-standing on gypsum, $CaSO_4 \cdot 2H_2O$. Animas mine, Francisco Portillo, Santa Eulalia dist., Chihuahua, Mexico.

FIGURE 7.160 Weeksite, $K_2(UO_2)_2(Si_5O_{13}) \cdot 4H_2O$, canary yellow spherical crystals to 2 mm with radial cross-sections. Topaz Mountain, Thomas Range, Juab Co., Utah, United States.

probably about 20% of its uranium was in the form of ^{235}U, but the amount of ^{234}U was probably much lower than this. ^{238}U is generally an α emitter (sporadically, it undergoes spontaneous fission), decaying through the uranium series, which has 18 members, into ^{206}Pb, via a number of different decay paths. The decay chain of ^{235}U, which is called the actinium series, has 15 members and eventually ends with ^{207}Pb. The constant rates of decay in these decay series makes the comparison of the ratios of parent to daughter elements useful in radiometric dating. ^{234}U, a member of the uranium series (i.e., the decay chain of ^{238}U), decays to ^{206}Pb via a series of comparatively short-lived isotopes. ^{233}U is formed from ^{232}Th by neutron bombardment, generally in a nuclear reactor. In addition, ^{233}U is fissile. Its decay chain forms part of the neptunium series and ends with ^{209}Bi and ^{205}Tl. ^{235}U is important for both nuclear reactors and nuclear weapons, as it is the only U isotope existing in nature on Earth in any substantial quantity that is fissile, that is, it can be split into two or three fragments (fission products) by thermal neutrons. ^{238}U is not fissile, though is a fertile isotope, as after neutron activation it can form ^{239}Pu, another fissile isotope. The ^{238}U nucleus can absorb one neutron to form the radioactive isotope ^{239}U, which decays by β emission to ^{239}Np, also a β-emitter, that then decays within a few days into ^{239}Pu.

TABLE 7.23 Uranium properties.

Appearance	Silvery gray metallic; corrodes to a spalling black oxide coat in air
Standard atomic weight $A_{r,std}$	238.029
Block	f-Block
Element category	Actinide
Electron configuration	[Rn] $5f^3$ $6d^1$ $7s^2$
Phase at STP	Solid
Melting point	1132.2°C
Boiling point	4131°C
Density (near r.t.)	19.1 g/cm^3
When liquid (at m.p.)	17.3 g/cm^3
Heat of fusion	9.14 kJ/mol
Heat of vaporization	417.1 kJ/mol
Molar heat capacity	27.665 J/(mol·K)
Oxidation states	+1, +2, +3, +4, +5, **+6**
Ionization energies	1st: 597.6 kJ/mol
	2nd: 1420 kJ/mol
Atomic radius	Empirical: 156 pm
Covalent radius	196 ± 7 pm
Van der Waals radius	186 pm

STP, Standard temperature and pressure.
Bold font indicates main oxidation state.

After refining, uranium is a silvery-white, weakly radioactive metal. It has a Mohs hardness of 6, about the same as that of titanium, rhodium, manganese and niobium. It is malleable, ductile, slightly paramagnetic, strongly electropositive and a poor electrical conductor. Uranium metal has an extremely high density of 19.1 g/cm^3, denser than lead (11.3 g/cm^3), but slightly less dense than tungsten and gold (19.3 g/cm^3) (Table 7.23). Uranium metal has three allotropic forms: α stable up to 668°C (orthorhombic, space group *Cmcm*, lattice parameters a = 285.4 pm, b = 587 pm, c = 495.5 pm), β stable from 668°C to 775°C (tetragonal, space group $P4_2/mnm$, $P4_2nm$, or $P4n2$, lattice parameters a = 565.6 pm, b = c = 1075.9 pm), and γ from 775°C to melting point—this is the most malleable and ductile state (body-centered cubic, lattice parameter a = 352.4 pm).

7.23.3.1 Oxidation states and oxides

Calcined uranium yellowcake, as manufactured in many large mills, consists of a range of uranium oxidation species in various forms from most oxidized to least oxidized. Particles with short residence times in a calciner will usually be less oxidized than those with long retention times or particles recovered in the stack scrubber. Uranium content is generally referenced to U_3O_8, which goes back to the days of the Manhattan Project when U_3O_8 was employed as an analytical chemistry reporting standard. Phase relationships in the uranium−oxygen system are complex. The most important oxidation states of uranium are uranium(IV) and uranium(VI), and their two corresponding oxides are, respectively, uranium dioxide (UO_2) and uranium trioxide (UO_3). In addition, other uranium oxides such as uranium monoxide (UO), diuranium pentoxide (U_2O_5), and uranium peroxide ($UO_4 \cdot 2H_2O$) do occur. The most common types of uranium oxide are triuranium octoxide (U_3O_8) and UO_2. Both compounds are solids that have low solubility in water and are relatively stable over a wide range of environmental conditions. Triuranium octoxide is (depending on conditions) the most stable compound of uranium and is the one most commonly formed in nature. Uranium dioxide is the form in which uranium is most frequently used as a nuclear reactor fuel. At ambient temperatures, UO_2 will slowly convert to U_3O_8. Due to their stability, uranium oxides are commonly considered the preferred chemical form for storage or disposal. Uranium dioxide or uranium(IV) oxide (UO_2), also known as urania or uranous oxide, is a black, radioactive, crystalline powder that naturally occurs in the mineral uraninite. It is used in nuclear fuel rods in nuclear reactors. A mixture of uranium and plutonium dioxides is used as MOX fuel (Mixed oxide fuel is nuclear fuel that contains more than one oxide of fissile material, usually consisting of plutonium blended with natural uranium, reprocessed uranium, or depleted uranium.). Prior to 1960, it was used as yellow and black color in ceramic glazes and glass.Uranium dioxide is formed by reducing uranium trioxide with hydrogen.

$$UO_3 + H_2 \xrightarrow{700°C} UO_2 + H_2O$$

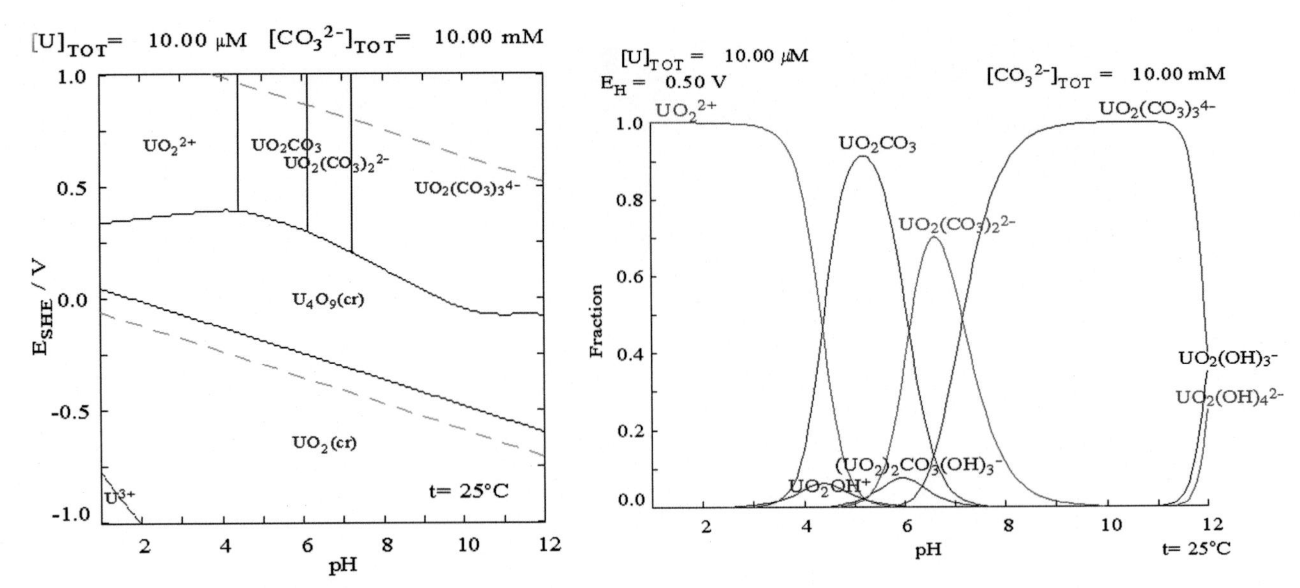

FIGURE 7.161 (Left) Uranium pourbaix diagram in carbonate media. (Right) Uranium fraction diagram with carbonate present.

This reaction plays an important role in the production of nuclear fuel through nuclear reprocessing and uranium enrichment. Triuranium octoxide (U_3O_8) is an olive green to black, odorless solid. It is one of the more popular forms of yellowcake and is shipped between mills and refineries in this form. U_3O_8 has potential long-term stability in a geologic environment. In the presence of oxygen (O_2), uranium dioxide (UO_2) is oxidized to U_3O_8, whereas uranium trioxide (UO_3) loses oxygen at temperatures over 500°C and is reduced to U_3O_8. The compound can be formed by any one of three primary chemical conversion methods, involving either uranium tetrafluoride (UF_4) or uranyl fluoride (UO_2F_2) as intermediates. It is commonly considered to be the more attractive form for disposal purposes since, under normal environmental conditions, U_3O_8 is one of the most kinetically and thermodynamically stable forms of uranium. Triuranium octoxide is converted to uranium hexafluoride for the purpose of uranium enrichment.

Salts of many oxidation states of uranium are water soluble and may be examined in aqueous solutions. The most common ions are U^{3+} (brown-red), U^{4+} (green), UO_2^+ (unstable), and UO_2^{2+} (yellow), for U(III), U(IV), U(V), and U (VI), respectively. Some solid and semimetallic compounds such as UO and US exist for the formal oxidation state uranium(II), but no simple ions are known to exist in solution for that state. U^{3+} ions liberate hydrogen from water and are thus deemed to be highly unstable. The UO_2^{2+} ion represents the uranium(VI) state and is known to form compounds such as uranyl carbonate, uranyl chloride and uranyl sulfate. UO_2^{2+} also forms complexes with a number of organic chelating agents, the most commonly encountered of which is uranyl acetate, ($UO_2(CH_3COO)_2{\cdot}2H_2O$). In contrast to the uranyl salts of uranium and polyatomic ion uranium oxide cationic forms, the uranates, salts containing a polyatomic uranium oxide anion, are usually not water soluble.

The interactions of carbonate anions with uranium(VI) cause the Pourbaix diagram to change dramatically when the medium is changed from water to a carbonate containing solution (Fig. 7.161). Though most carbonates are insoluble in water (students are often taught that all carbonates other than those of alkali metals are insoluble in water), uranium carbonates are often soluble in water. This is because a U(VI) cation can bind two terminal oxides and three or more carbonates to form anionic complexes. When the pH of a uranium(VI) solution increases, the uranium is transformed to a hydrated uranium oxide hydroxide and at high pH values it becomes an anionic hydroxide complex. When carbonate is added, uranium is changed to a series of carbonate complexes when the pH is increased. One consequence of these reactions is increased solubility of uranium in the pH range 6 to 8, something that has a direct impact on the long-term stability of spent uranium dioxide nuclear fuels.

7.23.3.2 Hydrides, carbides, and nitrides

Uranium metal heated to 250°C–300°C reacts with hydrogen to form uranium hydride

$$2U + 3H_2 \rightleftharpoons 2UH_3$$

Uranium hydride is not an interstitial compound, causing the metal to expand upon hydride formation. In its lattice, each uranium atom is surrounded by 6 other uranium atoms and 12 atoms of hydrogen; each hydrogen atom occupies a large tetrahedral hole in the lattice. The density of hydrogen in uranium hydride is approximately the same as in liquid water or in liquid hydrogen. The U-H-U linkage through a hydrogen atom is present in the structure. Uranium hydride forms when uranium metal becomes exposed to water:

$$7U + 6H_2O \rightarrow 3UO_2 + 4UH_3$$

The resulting uranium hydride is pyrophoric; if the metal (e.g. a damaged fuel rod) is exposed to air afterwards, excessive heat may be generated and the bulk uranium metal itself can ignite. Hydride-contaminated uranium can be passivated by exposure to a gaseous mixture of 98% helium with 2% oxygen. Condensed moisture on uranium metal promotes formation of hydrogen and uranium hydride; a pyrophoric surface may be formed in absence of oxygen. This is a possible problem with underwater storage of spent nuclear fuel in spent fuel ponds. Depending on the size and distribution on the hydride particles, self-ignition can occur after an indeterminate length of exposure to air. Such exposure poses risk of self-ignition of fuel debris in radioactive waste storage vaults. Uranium metal exposed to steam produces a mixture of uranium hydride and uranium dioxide. Uranium hydride exposed to water evolves hydrogen. In contact with strong oxidizers this may cause fire and explosions. Contact with halocarbons may cause a violent reaction. Even higher temperatures will reversibly remove the hydrogen. This behavior makes uranium hydrides convenient starting compounds to produce reactive uranium powder together with a variety of uranium carbide, nitride, and halide compounds. Two crystal phases of uranium hydride exist: an α form that is formed at low temperatures and a β form that is formed when the formation temperature is above 250°C. Uranium carbides and uranium nitrides are both relatively inert semimetallic compounds that are marginally soluble in acids, react with water, and can ignite in air to form U_3O_8. Carbides of uranium include uranium monocarbide (UC), uranium dicarbide (UC_2), and diuranium tricarbide (U_2C_3). Both UC and UC_2 are produced by adding carbon to molten uranium or by exposing the metal to carbon monoxide at high temperatures. Stable below 1800°C, U_2C_3 is prepared by subjecting a heated mixture of UC and UC_2 to mechanical stress. Uranium nitrides formed by direct exposure of the metal to nitrogen include uranium mononitride (UN), uranium dinitride (UN_2), and diuranium trinitride (U_2N_3). The common route for producing UN is carbothermic reduction of uranium oxide (UO_2) in a 2 step method:

$$3UO_2 + 6C \rightarrow 2UC + UO_2 + 4CO$$

(in argon, > 1450°C for 10 to 20 hours)

$$4UC + 2UO_2 + 3N_2 \rightarrow 6UN + 4CO$$

Sol-gel methods and arc melting of pure uranium under nitrogen atmosphere can also be used. A different common method for producing UN_2 is the ammonolysis of uranium tetrafluoride. Uranium tetrafluoride is exposed to ammonia gas under high pressure and temperature, which replaces the fluorine with nitrogen and generates hydrogen fluoride. Hydrogen fluoride is a colorless gas at this temperature and mixes with the ammonia gas. Another method of UN synthesis uses direct formation of UH_3 from metallic uranium by exposing it to hydrogen gas at temperatures above 280°C. Since UH_3 has a higher specific volume than the metallic phase, hybridation can be utilized to physically decompose otherwise solid uranium. Following hybridation, UH_3 can be exposed to a nitrogen atmosphere at a temperature of ~500°C forming U_2N_3. By additional heating to temperatures over 1150°C, the sesquinitride can then be decomposed to UN.

$$2U + 3H_2 \rightarrow 2UH_3$$
$$4UH_3 + 3N_2 \rightarrow 2U_2N_3$$
$$2U_2N_3 \rightarrow 4UN + N_2$$

Each uranium dinitride complex is thought to have three distinct compounds present simultaneously since decomposing of uranium dinitride (UN_2) results in uranium sesquinitride (U_2N_3), and the uranium mononitride (UN). Uranium dinitrides decompose to uranium mononitride by the following sequence:

$$4UN_2 \rightarrow 2U_2N_3 + N_2$$
$$2U_2N_3 \rightarrow 4UN + N_2$$

Decomposition of UN_2 is the most common method for isolating uranium sesquinitride (U_2N_3).

7.23.3.3 Halides

Uranium hexafluoride is the feedstock used to separate ^{235}U from natural uranium. All uranium fluorides are produced using uranium tetrafluoride (UF_4); UF_4 itself is formed by hydrofluorination of uranium dioxide. Reduction of UF_4 with hydrogen at 1000°C forms uranium trifluoride (UF_3). Under the right conditions of temperature and pressure, the reaction of solid UF_4 with gaseous uranium hexafluoride (UF_6) can produce the intermediate fluorides of U_2F_9, U_4F_{17}, and UF_5. At room temperature, UF_6 has a high vapor pressure, making it useful in the gaseous diffusion process to separate the rare ^{235}U from the common ^{238}U isotope. This compound can be produced from uranium dioxide or uranium hydride, for example,

$$UO_2 + 4HF \rightarrow UF_4 + 2H_2O \text{ (500°C, endothermic)}$$
$$UF_4 + F_2 \rightarrow UF_6 \text{ (350°C, endothermic)}$$

Milled uranium ore—U_3O_8 or "yellowcake"—is dissolved in nitric acid, resulting in a solution of uranyl nitrate $UO_2(NO_3)_2$. Pure uranyl nitrate is obtained by solvent extraction, which is subsequently treated with ammonia to produce ammonium diuranate ("ADU", $(NH_4)_2U_2O_7$). Reduction with hydrogen produces UO_2, which is converted with hydrofluoric acid (HF) to uranium tetrafluoride, UF_4. Oxidation with fluorine yields UF_6. During nuclear reprocessing, uranium is reacted with chlorine trifluoride to give UF_6:

$$U + 2ClF_3 \rightarrow UF_6 + Cl_2$$

The UF_6, a white solid, is extremely reactive (by fluorination), easily sublimes (producing a vapor that behaves as a nearly ideal gas) and is the most volatile compound of uranium known to exist.

One route of producing uranium tetrachloride (UCl_4) is to directly combine chlorine with either uranium metal or uranium hydride. Uranium tetrachloride is commonly synthesized by the reaction of uranium trioxide (UO_3) and hexachloropropene. Solvent UCl_4 adducts can be produced by a simpler reaction of UI_4 with hydrogen chloride in organic solvents. Uranium tetrachloride is a hygroscopic, dark green solid, which sublimes in a high vacuum at ~500°C. In the crystal structure the uranium is surrounded by eight chlorine atoms, four at 264 pm and the other four at 287pm. The molecule UCl_4 is a Lewis acid and dissolves in solvents that can act as non-protic Lewis bases. Dissolution in protic solvents is more complicated. When UCl_4 is added to water the uranium aqua ion is formed.

$$UCl_4 + xH_2O \rightarrow [U(H_2O)_x]^{4+} + 4Cl^-$$

The aqua ion $[U(H_2O)_x]^{4+}$, (x is 8 or 9) is strongly hydrolyzed.

$$[U(H_2O)_x]^{4+} \rightleftharpoons [U(H_2O)_{x-1}(OH)]^{3+} + H^+$$

The pK_a for this reaction is ca. 1.6, so hydrolysis is absent only in solutions of acid strength 1 $mol \cdot dm^{-3}$ or stronger ($pH < 0$). Further hydrolysis occurs at $pH > 3$. Weak chloro complexes of the aqua ion may be formed. Published estimates of the log K value for the formation of $[UCl]^{3+}(aq)$ vary from -0.5 to $+3$ because of difficulty in dealing with simultaneous hydrolysis. With alcohols, partial solvolysis may occur.

$$UCl_4 + xROH \rightleftharpoons UCl_{4-x}(OR)_x + xHCl$$

Uranium tetrachloride dissolves in non-protic solvents such as tetrahydrofuran, acetonitrile, dimethyl formamide etc. that can act as Lewis bases. Solvates of formula UCl_4L_x are formed which may be isolated. The solvent has to be completely free of dissolved water, or hydrolysis will occur, with the solvent, S, picking up the released proton.

$$UCl_4 + H_2O + S \rightleftharpoons UCl_3(OH) + SH^+ + Cl^-$$

The solvent molecules may be substituted by another ligand in a reaction such as

$$UCl_4 + 2Cl^- \rightarrow [UCl_6]^{2-}.$$

The solvent is not shown, just as when complexes of other metal ions are formed in aqueous solution. Solutions of UCl_4 are susceptible to oxidation by air, causing the formation of complexes of the uranyl ion. The reduction of UCl_4 by hydrogen forms uranium trichloride (UCl_3) while the higher chlorides of uranium are produced by reaction with additional chlorine. All uranium chlorides react with water and air. Bromides and iodides of uranium are produced by direct reaction of, respectively, bromine and iodine with uranium or by adding UH_3 to those element's acids. Known examples are, for example, UBr_3, UBr_4, UI_3, and UI_4. Uranium oxyhalides are water soluble and consist of, for example, UO_2F_2, $UOCl_2$, UO_2Cl_2, and UO_2Br_2. The stability of the oxyhalides decreases as the atomic weight of the component halide increases.

7.23.3.4 Organouranium compounds

Efforts in the 1940s to produce volatile carbonyls and alkyls of uranium for isotopic separations failed, though, as with the lanthanides, simple carbonyls of uranium have since been obtained in argon matrices quenched to $-269°C$ (4K). Subsequent research has mainly focused on cyclopentadienyls and, to a lesser extent, cyclooctatetraenyls; σ-bonded alkyl and aryl derivatives of the cyclopentadienyls have also been produced. The cyclopentadienyls are of the several main types, for example, $[U^{III}(C_5H_5)_3]$, this uranium compound is formed directly from UCl_3 and $K(C_5H_5)$, and $[U^{IV}(C_5H_5)_4]$ for which the general method of preparation can be given as:

$$UCl_4 + 4K(C_5H_5) \xrightarrow{\text{Reflux in } C_6H_6} [U(C_5H_5)_4] + 4\,KCl$$

It contains four identical η^5 rings arranged tetrahedrally around the uranium atom. Halide derivatives, the most plentiful are of the type $[U^{IV}(C_5H_5)_3X]$, can be prepared by the general reaction

$$UX_4 + 3M(C_5H_5) \rightarrow [U(C_5H_5)_3X] + 3MX$$

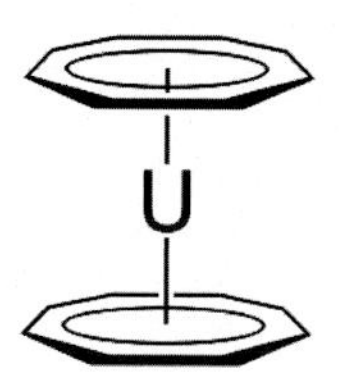

The first publication of an organoactinide was that of the pale brown $[U(C_5H_5)_3Cl]$ by L. T. Reynolds and English chemist and Nobel laureate Geoffrey Wilkinson (July 14, 1921 to September 26, 1996) in 1956. They showed that, in contrast to $Ln(C_6H_5)_3$, this compound does not form ferrocene when reacted with $FeCl_2$, implying greater covalency in the bonding of $C_5H_5^-$ to U^{IV} than to Ln^{III}. Substitution of Cl in $[U(C_5H_5)_3Cl]$ and $[Th(C_5H_5)_3Cl]$, by other halogens or by alkoxy, alkyl, aryl or BH_4 groups, offers the most extensive synthetic route in this area. A basically tetrahedral disposition of three $(\eta^5\text{-}C_5H_5)$ rings and the fourth group around the metal appears to be common. The alkyl and aryl derivatives $[U(\eta^5\text{-}C_5H_5)_3R]$ are of interest because they provide a way to study the U-C σ bond, and mechanistic research of their thermal decomposition (thermolysis) has been prominent. The exact mechanism is not yet clear but it is evidently not β-elimination as the eliminated molecule is RH, the H of which comes from a cyclopentadienyl ring. Until now it has not been possible to produce either U^{III} or U^{IV} compounds with three C_5Me_5 rings around a single uranium atom. However, $[U(\eta^5\text{-}C_5Me_4H)_3Cl]$ and $[U(\eta^5\text{-}C_4Me_4PhCl]$ have been synthesized. The complex $[U(\eta^8\text{-}C_8H_8)_2]$ of cyclooctatetraene has been prepared by the reaction

$$UCl_4 + 2K_2C_8H_8 \xrightarrow{THF} [U(C_8H_8)_2] + 4KCl$$

followed by sublimation under vacuum. They are "sandwich" molecules with parallel and eclipsed rings. This structure is strikingly like that of ferrocene, and extensive debate on the nature of the metal-ring bonding indicates that this too is very similar. In order to highlight these similarities with the cyclopentadienyls, the names "uranocene," etc., have been coined. While thermally stable, these compounds are particularly sensitive to air and, except for uranocene, are also decomposed by water. Nevertheless, uranocene can be made more air-stable through the use of sufficiently bulky substituents, for example, 1,3,5,7-tetraphenylcyclo-octatetraene forms the completely air-stable $[U(\eta^8\text{-}C_8H_4Ph_4)_2]$, in which the parallel ligands are virtually eclipsed but the phenyl substituents staggered and rotated on average 42° out of the C_8 ring plane.

7.23.4 Major uses

The principal use of U in the military sector is in high-density penetrators. This ammunition comprises depleted U (DU) alloyed with 1%–2% of other metals, for example, Ti or Mo. At high impact speed, the density, hardness, and pyrophoricity of the projectile facilitate to destroy heavily armored targets. Tank armor and other removable vehicle armor can likewise be hardened using DU plates. The use of DU became politically and environmentally controversial after the usage of this type of munitions by the US, UK and other countries during wars in the Persian Gulf and the Balkans raised issues with respect to U compounds left in the soil. DU is also employed as a shielding material in certain containers utilized to store and transport radioactive materials. Though the U itself is radioactive, its high density makes it more efficient than Pb in stopping radiation from strong sources such as Ra. Other applications of DU are, for example, as counterweights for aircraft control surfaces and as ballast for missile re-entry vehicles. Because of its high density, DU can be found in inertial guidance systems and in gyroscopic compasses. DU is selected over comparably dense metals because of its capacity to be readily machined and cast as well as being relatively cheap. The most important danger of exposure to DU is chemical poisoning by U oxide rather than radioactivity (U being only a weak α emitter). During the final stages of World War II, the complete Cold War period, and to some extent afterward, ^{235}U has been employed as the fissile explosive isotope to make nuclear weapons. At first, two key types of fission bombs were built: a fairly simple device that utilizes ^{235}U and a more complex mechanism that utilizes ^{239}Pu obtained from ^{238}U. Later, a much more complex and far more potent kind of fission/fusion bomb (thermonuclear weapon) was constructed, that employs a Pu-based device to trigger a mixture of tritium (T or ^{3}H) and deuterium (D or ^{2}H) to undergo nuclear fusion. Such bombs are jacketed in a nonfissile (unenriched) U case, and they get more than half their power from the fission of this material by fast neutrons from the nuclear fusion process.

The most important use of U in the civilian sector is to fuel nuclear power plants. One kilogram of ^{235}U can in theory produce about 20 TJ of energy (2×10^{13} J), if complete fission takes place; the equivalent amount of energy as 1500 tons of coal. Commercial nuclear power plants employ fuel that is usually enriched to approximately 3% ^{235}U. The CANDU (CANada Deuterium Uranium, a Canadian pressurized heavy-water reactor) and Magnox (a type of nuclear power/production reactor that was designed to run on natural U with graphite as the moderator and CO_2 gas as the heat exchange coolant. It belongs to the wider class of gas cooled reactors. The name comes from the Mg—Al alloy used to clad the fuel rods inside the reactor. Like most other "Generation I nuclear reactors," the Magnox was designed with the dual purpose of producing electrical power and ^{239}Pu for the nascent nuclear weapons program in Britain. The name refers specifically to the United Kingdom design but is sometimes used generically to refer to any similar reactor.) designs are the only commercial reactors able to utilize unenriched U fuel. Fuel used for US Navy reactors is normally highly enriched in ^{235}U (the exact amounts are classified though). In a breeder reactor, ^{238}U can also be converted into Pu.

$$^{238}_{92}U + n \rightarrow ^{239}_{92}U + \gamma \xrightarrow{\beta^-} ^{239}_{93}Np \xrightarrow{\beta^-} ^{239}_{93}Pu$$

Before (and, sometimes, after) the discovery of radioactivity, U was mainly used in small quantities for yellow glass and pottery glazes, such as U-glass and in Fiestaware. The discovery and isolation of Ra in U ore (uraninite, formerly pitchblende, is a radioactive, U-rich mineral and ore with a chemical composition that is principally UO_2, but due to oxidation the mineral usually contains variable amounts of U_3O_8. Also, due to radioactive decay, the ore also contains oxides of Pb and trace quantities of He. It may also comprise Th and REEs.) by Polish and naturalized-French physicist and chemist Marie Skłodowska Curie (November 7, 1867 to July 4, 1934) started large-scale U mining in order to extract the Ra, which was utilized to produce glow-in-the-dark paints for clock and aircraft dials. This left a massive amount of U as a waste product, as it takes 3 tons of U to isolate a single gram of Ra. This waste product was redirected to the glazing industry, making U-glazes very cheap and available in large amounts. In addition to the pottery glazes, U tile glazes accounted for most of the usage, for example, common bathroom and kitchen tiles which could be manufactured in green, yellow, mauve, black, blue, red and other colors. U was also employed in photographic chemicals (especially U nitrate (uranyl nitrate [$UO_2(NO_3)_2$]) as a toner. During the first half of the 19th century, many photosensitive metal salts had been identified as candidates for photographic processes, among them uranyl nitrate. The prints thus produced were known as uranium prints, urbanities, or more commonly, uranotypes. The first uranium printing processes were invented by Scotsman Charles J. Burnett between 1855 and 1857 and used this compound as the sensitive salt. Burnett authored a 1858 article comparing "Printing by the Salts of the Uranic and Ferric Oxides". The process employs the ability of the uranyl ion to pick up two electrons and reduce to the lower oxidation state of U(IV) under ultraviolet light. Uranotypes can vary from print to print from a more neutral, brown russet to strong Bartolozzi red, with a very long tone grade. Surviving prints are slightly radioactive, a property which can be used as a means to nondestructively identify them. Several other more elaborate photographic processes employing the compound appeared and vanished during the second half of the 19th century with names like Wothlytype, Mercuro-Uranotype and the Auro-Uranium process. U-papers were produced commercially at least until the end of the 19th century, vanishing due to the superior sensitivity and practical advantages of silver halides. From the 1930s through the 1950s Kodak Books described a U-toner (Kodak T-9) using uranium nitrate hexahydrate. Some alternative process photographers including American Blake Ferris (February 19, 1968 to April 26, 2011) and American Robert Wolfgang Schramm (born November 13, 1934) continue(d) to make uranotype prints.). In addition, it was used in lamp filaments for stage lighting bulbs, to enhance the look of dentures, and in the leather and wood industries for stains and dyes. U-salts are mordants of silk or wool (a mordant or dye fixative is a substance used to set (i.e., bind) dyes on fabrics by forming a coordination complex with the dye, which then attaches to the fabric). Uranyl acetate ($UO_2(CH_3COO)_2 \cdot 2H_2O$) and uranyl formate ($UO_2(CHO_2)_2 \cdot H_2O$) are employed as electron-dense "stains" in TEM, to improve the contrast of biological specimens in ultrathin sections and in negative staining of viruses, isolated cell organelles and macromolecules. The discovery of the radioactivity of U led to further scientific and practical uses of U. The long half-life of the isotope ^{238}U (4.51×10^9 years) makes it suitable for use in assessing the age of the earliest igneous rocks and for other forms of radiometric dating, such as U–Th dating, U–Pb dating, and U–U dating. U metal is utilized for X-ray targets in the creation of high-energy X-rays.

References

Agricola, G. (1546, 1955). De Natura Fossilium (Textbook of Mineralogy): Translated from the First Latin Ed. of 1546 by Mark Chance Bandy and Jean A. Bandy for the Mineralogical Society of America (M.C. Bandy, J.A. Bandy, Trans.). Boulder: Geological Society of America.

Berzelius, J. J. (1829a). Undersökning af ett nytt mineral (Thorit), som innehåller en förut obekant jord. *Kungliga Svenska Vetenskaps Akademiens Handlingar*, 1–30.

Berzelius, J. J. (1829b). Untersuchung eines neues Minerals und einer darin erhalten zuvor unbekannten Erde. *Annalen der Physik und Chemie, 16*, 385–415.

Berzelius, J. J. (1839a). Latanium — A new metal. *Philosophical Magazine, 14*, 390–391.

Berzelius, J. J. (1839b). Nouveau métal. *Comptes rendus, 8*, 356–357.

Bohr, N. (1922). *The theory of spectra and atomic constitution: Three essays*. London: Cambridge University press.

Brodersen, K., Liehr, G., & Rölz, W. (1975). Stabile Quecksilber(I)-Schwefel-Verbindungen, 1. *Chemische Berichte, 108*, 3243–3246.

Burnett, C. J. (1858). On the production of direct positives-on printing by the salts of the uranic and ferric oxides, with observations climetic and chemical. *The Liverpool Manchester Photographic J*, 99–101, 111–112, 127–129, 155–157.

Casas, J. S., García-Tasende, M. S., & Sordo, J. (1999). Structural aspects of the coordination chemistry of organothallium(III) and organomercury(II) derivatives. *Coordination Chemistry Reviews, 193-195*, 283–359.

Case, T. W. (1920). "Thalofide Cell"—A new photo-electric substance. *Physical Review, 15*, 289–292.

Children, J. G. (1815). An Account of Some Experiments with a Large Voltaic Battery. *Philosophical Transactions of the Royal Society of London, 105*, 363–374.

Cho, Y., Choi, Y. K., & Sohn, S. H. (2006). Optical properties of neodymium-containing polymethylmethacrylate films for the organic light emitting diode color filter. *Applied Physics Letters, 89*, 051102.

Collet-Descotils, H. V. (1803). Sur la cause des couleurs différentes qu'affectent certains sels de platine. *Annales de chimie, 48*, 153–176.

Correia-Neves, J. M., Lopes Nunes, J. E., & Sahama, T. G. (1974). High hafnium members of the zircon-hafnon series from the granite pegmatites of Zambézia, Mozambique. *Contributions to Mineralagy and Petrology, 48*, 73–80.

Coster, D., & von Hevesy, G. (1923). On the missing element of atomic number 72. *Nature, 111*, 79.

Crookes, W. (1861a). Further remarks on the supposed new metalloid. *Chemical News, 3*, 303.

Crookes, W. (1861b). On the existence of a new element, probably of the sulphur group. *Chemical News, 3*, 193–194.

Crookes, W. (1862). Preliminary researches on thallium. *Proceedings of the Royal Society of London, 12*, 150–159.

Curie, M. (1898). Rayons émis par les composés de l'uranium et du thorium. *Comptes Rendus, 126*, 1101–1103.

Davy, H. (1808). Electro-chemical researches on the decomposition of the earths; with observations on the metals obtained from the alkaline earths, and on the amalgam procured from ammonia. *Philosophical Transactions of the Royal Society of London, 98*, 333–370.

Dezelah, E.-K. O. M., Szilágyi, I. M., Campbell, J. M., Arstila, K., Niinistö, L., & Winter, C. H. (2006). Atomic layer deposition of tungsten(III) oxide thin films from $W_2(NMe_2)_6$ and water: Precursor-based control of oxidation state in the thin film material. *Journal of the American Chemical Society, 128*, 9638–9639.

de Luyart, J. J., & de Luyart, F. (1783). Análisis químico del volfram, y examen de un nuevo metal, que entra en su composición. *Generales celebradas por la Real Sociedad Bascongada de los Amigos del País en la ciudad de Vitoria por setiembre de, 1783*, 46–88.

de Luyart, J. J., & de Luyart, F. (1785). *A chemical analysis of Wolfram and examination of a new metal, which enters its composition*. London, England: G. Nicol. (C. Cullen, Trans.).

de Marignac, J. C. G. (1866). Recherches sur les combinaisons du niobium. *Annales de chimie et de physique, 4*, 7–75.

de Marignac, J. C. G., Blomstrand, C. W., Deville, H., Troost, L., & Hermann, R. (1866). Tantalsäure, Niobsäure, (Ilmensäure) und Titansäure. *Fresenius' Journal of Analytical Chemistry, 5*, 384–389.

de Ulloa, A. J. (1748). *Relación histórica del viaje a la América meridional* (vol 1). Fundación Universitaria Española.

Delafontaine, M. (1878a). Sur le décepium, métal nouveau de la samarskite. *Journal de pharmacie et de chimie, 28*, 540–542.

Delafontaine, M. (1878b). Sur le décepium, métal nouveau de la samarskite. *Comptes rendus hebdomadaires des séances de l'Académie des Sciences, 87*, 632–634.

Delafontaine, M. (1881). Sur le décipium et le samarium. *Comptes rendus hebdomadaires des séances de l'Académie des Sciences, 93*, 63–64.

Ekeberg, A. (1802a). Of the Properties of the Earth Yttria, compared with those of Glucine; of Fossils, in which the first of these Earths in contained; and of the Discovery of a metallic Nature (Tantalium). *Journal of Natural Philosophy, Chemistry, and the Arts, 3*, 251–255.

Ekeberg, A. (1802b). Uplysning om Ytterjorden egenskaper, i synnerhet i aemforelse med Berylljorden:om de Fossilier, havari förstnemnde jord innehales, samt om en ny uptäckt kropp af metallik natur. *Kungliga Svenska Vetenskapsakademiens Handlingar, 23*, 68–83.

Elsner, L. (1846). Beobachtungen über das Verhalten regulinischer Metalle in einer wässrigen Lösung van Cyankalium. *Journal für Praktische Chemie, 37*, 441–446.

Fermi, E. (1938). *Artificial radioactivity produced by neutron bombardment: Nobel Lecture. Royal Swedish Academy of Sciences.* <https://www.nobelprize.org/uploads/2018/06/fermi-lecture.pdf> Accessed 26.06.19.

Fourcroy, A. F., & Vauquelin, N. L. (1803). Mémoire sur le platin. *Annales de Chimie, 1803*(48), 177–185.

Gagnon, A., Dansereau, J., & Le Roch, A. (2017). Organobismuth reagents: synthesis, properties and applications in organic synthesis. *Synthesis, 49*, 1707–1745.

Galyer, A. L., & Wilkinson, G. (1976). New synthesis of hexamethyltungsten(VI). The octamethyltungstate-(VI) lon. *Journal of the Chemical Society, Dalton Transactions*, 2235–2238.

Geilmann, W., Wrigge, F. W., & Biltz, W. (1933). Rheniumpentachlorid. *Zeitschrift für anorganische und allgemeine Chemie, 214*, 244–247.

Geoffroy, C.F. (1753). Sur Bismuth. Histoire de l'Académie Royale des Sciences Avec les Mémoires de Mathématique & de Physique Tirez des Registres de Cette Académie, 190–194.

Hofmann, K. A., & Sand, J. (1900). Ueber das Verhalten von Mercurisalzen gegen Olefine. *Berichte der deutschen chemischen Gesellschaft, 33*, 1340–1353.

Hunt, P., & Schwerdtfeger, P. (1996). Are the compounds InH_3 and TlH_3 stable gas phase or solid state species? *Inorganic Chemistry, 35*, 2085–2088.

Jensen, W. B. (2008). Is mercury now a transition element? *Journal of Chemical Education, 85*, 1182.

Keeley, L. E. (1897). *Opium: Its use abuse and cure; or, from bondage to freedom.* Chicago, IL: The Banner of Gold, Co.

Kirchhoff, G., & Bunsen, R. (1861). Chemische Analyse durch Spectralbeobachtungen. *Annalen der Physik und Chemie, 189*, 337–381.

Klaproth, M. H. (1789). Chemische Untersuchung des Uranits, einer neuentdeckten metallischen Substanz. *Chemische Annalen, 2*, 387–403.

Lamy, A. (1862). De l'existencè d'un nouveau métal, le thallium. *Comptes Rendus, 54*, 1255–1262.

Lecoq de Boisbaudran, P. É. (1886). L'holmine (ou terre X de M Soret) contient au moins deux radicaux métallique. *Comptes Rendus, 143*, 1003–1006.

Lecoq de Boisbaudran, P. E. (1879). Recherches sur le samarium, radical d'une terre nouvelle extraite de la samarskite. *Comptes rendus hebdomadaires des séances de l'Académie des sciences, 89*, 212–214.

Marggraf, A. S. (1760). *Versuche mit dem neuen mineralischen Körper Platina del pinto genannt.* Graz: Widmannstätter.

Martinez-Ariza, G., Ayaz, M., Roberts, S. A., Rabanal-León, W. A., Arratia-Pérez, R., & Hulme, C. (2015). The synthesis of stable, complex organocesium tetramic acids through the Ugi reaction and cesium-carbonate-promoted cascades. *Angewandte Chemie International Edition, 54*, 11672–11676.

Meitner, L., & Frisch, O. R. (1939). Disintegration of uranium by neutrons: A new type of nuclear reaction. *Nature, 143*, 239–240.

Mendeleev, D. (1869). Ueber die beziehungen der eigenschaften zu den atomgewichten der elemente. *Zeitschrift für Chemie, 12*, 405–406.

Mosander, C.G., 1843. On the new metals lanthanum and didymium, which are associated with cerium; and on erbium and terbium, new [m]etals associated with yttria, Report of the Thirteens meeting of the British Association for the Advancement of Science held at Cork in August 1843, John Murray, 1844, 25–32.

Moseley, H. (1914). The high-frequency spectra of the elements, Part II. *Philosophical Magazine Series 7, 27*, 703–713.

Moseley, H. G. J. (1913). The high-frequency spectra of the elements. *Philosophical Magazine 6th series, 26*, 1024–1034.

Mössbauer, R.L., (1958). Gammastrahlung in Ir^{191}. Zeitschrift für Physik A, 151, 124–143.

Muthmann, W., & Stützel, L. (1899). Eine einfache Methode zur Darstellung der Schwefel-, Chlor- und Brom-Verbindungen der Ceritmetalle". *Berichte der deutschen chemischen Gesellschaft, 32*, 3413–3419.

Noddack, W., & Noddack, I. (1929). Die Herstellung von einem Gram Rhenium. *Zeitschrift für Anorganische und Allgemeine Chemie, 183*, 353–375.

Ogawa, M. (1908a). Preliminary Note on a New Element in Thorianite. *Chemical News, 98*, 249–251.

Ogawa, M. (1908b). Preliminary Note on a New Element allied to Molybdenum. *Chemical News, 98*, 261–264.

O'Riordan, A., Van Deun, R., Mairiaux, E., Moynihan, S., Fias, P., Nockemann, P., ... Redmond, G. (2008). Synthesis of a neodymium-quinolate complex for near-infrared electroluminescence applications. *Thin Solid Films, 516*, 5098–5102.

Péligot, E.-M. (1842). Recherches Sur L'Uranium. *Annales de chimie et de physique*, 5–47.

Pope, W. J., & Peachy, S. J. (1907). A new class of organo-metallic compounds. Prelimary notice. Trimethylplatinimethyl hydroxide and its salts. *Proceedings Chem. Soc. London, 23*, 86–87.

Pope, W. J., & Gibson, C. S. (1907). CCII.-The alkyl compounds of gold. *J. Chem. Society, Trans, 91*, 2061–2066.

Pott, J. H. (1738). *De Wismutho. Exercitationes chymicae, vol Part 2* (pp. 134–197). Berolini: Rüdigerus.

Reynolds, L. T., & Wilkinson, G. (1956). π-cyclopentadienyl compounds of uranium-IV and thorium-IV. *Journal of Inorganic and Nuclear Chemistry, 2*, 246–253.

Rose, H. (1847). Ueber die Säure im Columbit von Nordamérika. *Annalen der Physik*, *146*, 572–577.

Roseveare, W. E. (1930). The X-ray photochemical reaction between potassium oxalate and mercuric chloride. *Journal of the American Chemical Society*, *52*, 2612–2619.

Rutherford, E., & Owens, R. B. (1899). Thorium and uranium radiation. *Transactions of the Royal Society of Canada*, *2*, 9–12.

Sainte-Claire Deville, H., & Debray, H. (1861). De la métallurgie du platine et des métaux qui l'accompagnent, Annales de Chemie et de Physique. *Series*, *3*(61), 5–160.

Scheffer, H.T., 1752. Det hvita gullet, eller sjunde metallen, kalladt i Spanien, Platina del Pinto, pintos små silfver, beskrifvit til fin sin natur, Kungliga Vetenskaps Academiens Handlingar, 13, 1752, 269 – 275.

Scerri, E. (2013). *A tale of seven elements*. New York: OUP USA.

Scheele, C. W. (1781). Tungstens bestånds-delar. *Kungliga Vetenskaps Academiens Nya Handlingar*, *2*, 89–95.

Schmidt, G. C. (1898a). Über die vom Thorium und den Thoriumverbindungen ausgehende Strahlung. *Verhandlungen der Physikalischen Gesellschaft zu Berlin*, *17*, 14–16.

Schmidt, G. C. (1898b). Über die von den Thorverbindungen und einigen anderen Substanzen ausgehende Strahlung. *Annalen der Physik und Chemie*, *65*, 141–151.

Setterberg, C. (1882). Ueber die Darstellung von Rubidium- und Cäsiumverbindungen und über die Gewinnung der Metalle selbst. *Justus Liebig's Annalen der Chemie*, *211*, 100–116.

Shaginyan, L. R. (2019). Synthesis and properties of thallium nitride films. *Materials Chemistry and Physics*, *227*, 157–162.

Shayesteh, A., Yu, S., & Bernath, P. F. (2005). Infrared emission spectra and equilibrium structures of gaseous HgH_2 and HgD_2. *The Journal of Physical Chemistry A*, *109*, 10280–10286.

Shortland, A. J., & Wilkinson, G. (1973). Preparation and properties of hexamethyltungsten. *Journal of the Chemical Society, Dalton Transactions*, 872–876.

Sie, E. J., Nyby, C. M., Pemmaraju, C. D., Park, S. J., Shen, X., Yang, J., ... Lindenberg, A. M. (2019). An ultrafast symmetry switch in a Weyl semi-metal. *Nature*, *565*, 61–66.

Spedding, F. H., Gschneidner, K., & Daane, A. H. (1958). The crystal structures of some of the rare earth carbides. *Journal of the American Chemical Society*, *80*, 4499–4503.

Suzuki, H., & Matano, Y. (2001). *Organobismuth chemistry*. Amsterdam: Elsevier.

Tennant, S. (1804). On two metals, found in the black powder remaining after the solution of platina. *Philosophical Transactions of the Royal Society*, *94*, 411–418.

Urbain, G. (1908). Un nouvel élément, le lutécium, résultant du dédoublement de l'ytterbium de Marignac. *Comptes Rendus*, *145*, 759–762.

Urbain, G. (1909). Lutetium und Neoytterbium oder Cassiopeium und Aldebaranium – Erwiderung auf den Artikel des Herrn Auer v. Welsbach. *Monatshefte für Chemie*, *31*, I–VI.

Urbain, G. (1911). Sur un nouvel élément qui accompagne le lutécium et le scandium dans les terres de la gadolinite: le celtium. *Comptes Rendus*, *152*, 141–143.

van Arkel, A. E., & de Boer, J. H. (1924a). Die Trennung des Zirkoniums von anderen Metallen, einschließlich Hafnium, durch fraktionierte Distillation. *Zeitschrift für Anorganische und Allgemeine Chemie*, *141*, 289–296.

van Arkel, A. E., & de Boer, J. H. (1924b). Die Trennung von Zirkonium und Hafnium durch Kristallisation ihrer Ammoniumdoppelfluoride. *Zeitschrift für Anorganische und Allgemeine Chemie*, *141*, 284–288.

van Arkel, A. E., & de Boer, J. H. (1925). Darstellung von reinem Titanium-, Zirkonium-, Hafnium- und Thoriummetall. *Zeitschrift für anorganische und allgemeine Chemie*, *148*, 345–350.

von Hevesy, G. (1923). Über die Auffindung des Hafniums und den gegenwärtigen Stand unserer Kenntnisse von diesem Element. *Berichte der Deutschen Chemischen Gesellschaft (A and B Series)*, *56*, 1503–1516.

von Hevesy, G. (1925). The discovery and properties of hafnium. *Chemical Reviews*, *2*, 1–41.

von Welsbach, C. A. (1885). Die Zerlegung des Didyms in seine Elemente. *Monatshefte für Chemie und verwandte Teile anderer Wissenschaften*, *6*, 477–491.

von Welsbach, C. A. (1908). Die Zerlegung des Ytterbiums in seine Elemente. *Monatshefte für Chemie*, *29*, 181–225.

von Sickingen, C., 1782. Versuche über Platina". Mannheim.

Wang, X., & Andrews, L. (2004). Infrared spectra of thallium hydrides in solid neon, hydrogen, and argon. *The Journal of Physical Chemistry A*, *108*, 3396–3402.

Wang, X., & Andrews, L. (2005a). Infrared spectrum of $Hg(OH)_2$ in solid neon and argon. *Inorganic Chemistry*, *44*, 108–113.

Wang, X., & Andrews, L. (2005b). Mercury dihydride forms a covalent molecular solid. *Physical Chemistry Chemical Physics*, *7*, 750–759.

Watson, W., & Brownrigg, W. (1749). Several papers concerning a new semi-metal, called platina; Communicated to the Royal Society by Mr. Wm. Watson F. R. S. *Philosophical Transactions*, *46*, 584–596.

Wollaston, W. H. (1809). On the identity of columbium and tantalum. *Philosophical Transactions of the Royal Society*, *99*, 246–252.

Yanagisawa, A., Habaue, S., Yasue, K., & Yamamoto, H. (1994). Allylbarium reagents: Unprecedented regio- and stereoselective allylation reactions of carbonyl compounds. *Journal of the American Chemical Society*, *116*, 6130–6141.

Further reading

Abdur-Rashid, K., Lough, A. J., & Morris, R. H. (2001). Intra- and inter-ion-pair protonic-hydridic bonding in polyhydridobis(phosphine)rhenates. *Canadian Journal of Chemistry*, *79*, 964–976.

Abrahams, S. C., Ginsberg, A. P., & Knox, K. (1964). Transition metal-hydrogen compounds. II. The crystal and molecular structure of potassium rhenium hydride, K_2ReH_9. *Inorganic Chemistry*, *3*, 558–567.

Action, Q. A. (2012). *Issues in cancer epidemiology and research* (Scholarly Editions). Georgia: Atlanta.

Agricola, G. (1556, 1912). De re metallica; tr. from the 1st Latin ed. of 1556, with biographical introduction, annotations and appendices upon the development of mining methods, metallurgical processes, geology, mineralogy & mining law, from the earliest times to the 16th century (H. Hoover L.H. Hoover Trans.). London: Mining magazine.

Agulyansky, A. (2004). *Chemistry of tantalum and niobium fluoride compounds*. Amsterdam: Elsevier Science.

Alberto, R., Egli, A., Abram, U., Hegetschweiler, K., Gramlich, V., & Schubiger, P. A. (1994). Synthesis and reactivity of $[NEt_4]_2[ReBr_3(CO)_3]$. Formation and structural characterization of the clusters $[NEt_4][Re_3(\mu_3\text{-}OH)(\mu\text{-}OH)_3(CO)_9]$ and $[NEt_4][Re_2(\mu\text{-}OH)_3(CO)_6]$ by alkaline titration. *Journal of the Chemical Society, Dalton Transactions*, 2815–2820.

Albright, T. A., & Tang, H. (1992). The structure of pentamethyltantalum. *Angewandte Chemie International Edition in English*, *31*, 1462–1464.

Aldridge, S., & Downs, A. J. (2001). Hydrides of the main-group metals: New variations on an old theme. *Chemical Reviews*, *101*, 3305–3366.

Alpers, C. N., Hunerlach, M. P., May, J. Y., & Hothem, R. L. (2005). *Mercury contamination from historical gold mining in California*. U.S. Geological Survey. https://pubs.usgs.gov/fs/2005/3014/ Accessed 22.07.19.

Alvarez, L. W., Alvarez, W., Asaro, F., & Michel, H. V. (1980). Extraterrestrial cause for the Cretaceous–Tertiary extinction. *Science*, *208*, 1095–1108.

Amit, S., & Sharma, B. P. (2005). Development of dysprosium titanate based ceramics. *Journal of the American Ceramic Society*, *88*, 1064–1066.

Arenas, L. F., Ponce de León, C., & Walsh, F. C. (2016). Electrochemical redox processes involving soluble cerium species. *Electrochimica Acta*, *205*, 226–247.

Arnett, D. (1996). *Supernovae and nucleosynthesis: An investigation of the history of matter, from the Big Bang to the present*. Princeton, NJ: Princeton University Press.

Asamoto, R. R., & Novak, P. E. (1968). Tungsten-Rhenium thermocouples for use at high temperatures. *Review of Scientific Instruments*, *39*, 1233.

Ashikari, M. (2003). The memory of the women's white faces: Japaneseness and the ideal image of women. *Japan Forum*, *15*, 55–79.

Audi, G., Bersillon, O., Blachot, J., & Wapstra, A. H. (2003). The NUBASE evaluation of nuclear and decay properties. *Nuclear Physics A*, *729*, 3–128.

Aycan, S. (2005). Chemistry education and mythology. *Journal of Social Sciences*, *1*, 238–239.

Bailey, A. S., Hughes, R. W., Hubberstey, P., Ritter, C., Smith, R. I., & Gregory, D. H. (2011). New ternary and quaternary barium nitride halides: Synthesis and crystal chemistry. *Inorganic Chemistry*, *50*, 9545–9553.

Baker, M. (2013). *Radioactive bacteria attack cancer*. Springer Nature. https://www.nature.com/news/radioactive-bacteria-attack-cancer-1.12841> Accessed 19.07.2019.

Balke, C. W. (1935). Columbium and tantalum. *Industrial & Engineering Chemistry*, *27*, 1166–1169.

Band, A., Albu-Yaron, A., Livneh, T., Cohen, H., Feldman, Y., Shimon, L., ... Tenne, R. (2004). Characterization of oxides of cesium. *The Journal of Physical Chemistry B*, *108*, 12360–12367.

Barnard, C., & Bennett, S. (2004). Oxidation states of ruthenium and osmium. *Platinum Metals Review*, *48*, 157–158.

Bäuerlein, P., Antonius, C., Löffler, J., & Kümpers, J. (2008). Progress in high-power nickel–metal hydride batteries. *Journal of Power Sources*, *176*, 547–554.

Baur, F., Katelnikovas, A., Sazirnakovas, S., & Jüstel, T. (2014). Synthesis and optical properties of $Li_3Ba_2La_3(MoO_4)_8{:}Sm^{3+}$. *Zeitschrift für Naturforschung*, *69b*, 183–192.

Beck, B. R., Becker, J. A., Beiersdorfer, P., Brown, G. V., Moody, K. J., Wilhelmy, J. B., ... Kelley, R. L. (2007). Energy splitting of the ground-state doublet in the nucleus ^{229}Th. *Physical Review Letters*, *98*, 142501.

Beck, H. P., & Gladrow, E. (1979). Zur Hochdruckpolymorphie der Seltenerd-Trihalogenide. *Zeitschrift für anorganische und allgemeine Chemie*, *453*, 79–92.

Beeken, R. B., & Schweitzer, J. W. (1981). Intermediate valence in alloys of SmSe with SmAs. *Physical Review B*, *23*, 3620–3626.

Beeman, J. W., Bellini, F., Cardani, L., Casali, N., Di Domizio, S., Fiorini, E., ... Vignati, M. (2013). New experimental limits on the α decays of lead isotopes. *The European Physical Journal A*, *49*, 50.

Beglov, V. M., Pisarev, B. K., & Reznikova, G. G. (1992). Effect of boron and hafnium on the corrosion resistance of high-temperature nickel alloys. *Metal Science and Heat Treatment*, *34*, 251–254.

Beiner, G. G., Lavi, M., Seri, H., Rossin, A., Lev, O., Gun, J., & Rabinovich, R. (2015). Oddy tests: Adding the analytical dimension. *Collection Forum*, *29*, 22–36.

Benjamin, S. L., Levason, W., Pugh, D., Reid, G., & Zhang, W. (2012). Preparation and structures of coordination complexes of the very hard Lewis acids ZrF_4 and HfF_4. *Dalton Transactions*, *41*, 12548–12557.

Bentel, G. C. (1996). *Caesium-137 machines. Radiation therapy planning* (pp. 22–23). New York, NY; London: McGraw-Hill, Health Professions Divison.

Berdan, F. F., & Anawalt, P. R. (1992). *The essential codex mendoza*. Berkeley, CA: University of California Press.

Berzelius, J. J. (1824). Undersökning af några Mineralier. 1. Phosphorsyrad Ytterjord. *Kungliga Svenska Vetenskapsakademiens Handlingar*, *2*, 334–338.

Bick, M., & Prinz, H. (2000). *Cesium and cesium compounds. Ullmann's encyclopedia of industrial chemistry*. Weinheim: Wiley-VCH.

Bisel, S. C., & Bisel, J. F. (2002). Health and nutrition at Herculaneum. In W. F. Jashemski, & F. G. Meyer (Eds.), *The natural history of Pompeii* (pp. 451–475). Cambridge: Cambridge University Press.

Bisson, J.-F., Kouznetsov, D., Ueda, K.-I., Fredrich-Thornton, S. T., Petermann, K., & Huber, G. (2007). Switching of emissivity and photoconductivity in highly doped Yb^{3+}:Y_2O_3 and Lu_2O_3 ceramics. *Applied Physics Letters*, *90*, 201901.

Bisson, M. S., Childs, S. T., Vogel, J. O., De Barros, P., & Holl, A. F. C. (2000). *Ancient African metallurgy: The sociocultural context*. Walnut Creek: AltaMira Press.

Black, J. (1994). Biological performance of tantalum. *Clinical Materials*, *16*, 167–173.

Blackburn, P. E. (1966). The vapor pressure of rhenium. *The Journal of Physical Chemistry*, *70*, 311–312.

Bobyens, J. C. A., Levendis, D. C., Bruce, M. I., & Williams, M. L. (1986). Crystal structure of osmocene, $Os(\eta\text{-}C_5H_5)_2$. *Journal of Crystallographic and Spectroscopic Research*, *16*, 519–524.

Bochkarev, M. N. (2004). Molecular compounds of "new" divalent lanthanides. *Coordination Chemistry Reviews*, *248*, 835–851.

Bondarenko, Y. A., Kablov, E. N., Surova, V. A., & Echin, A. B. (2006). Effect of high-gradient directed crystallization on the structure and properties of rhenium-bearing single-crystal alloy. *Metal Science and Heat Treatment*, *48*, 360–363.

Bonetti, R., Chiesa, C., Guglielmetti, A., Matheoud, R., Poli, G., Mikheev, V. L., & Tretyakova, S. P. (1995). First observation of spontaneous fission and search for cluster decay of ^{232}Th. *Physical Review C*, *51*, 2530–2533.

Bosch, F., Faestermann, T., Friese, J., Heine, F., Kienle, P., Wefers, E., … Takahashi, K. (1996). Observation of bound-state $\beta-$ decay of fully ionized ^{187}Re: ^{187}Re-^{187}Os cosmochronometry. *Physical Review Letters*, *77*, 5190–5193.

Bowers, B. (2001). Scanning our past from London: The filament lamp and new materials. *Proceedings of the IEEE*, *89*, 413–415.

Bowmaker, G. A., Churakov, A. V., Harris, R. K., Howard, J. A. K., & Apperley, D. C. (1998). Solid-state ^{199}Hg MAS NMR studies of mercury(II) thiocyanate complexes and related compounds. Crystal structure of $Hg(SeCN)_2$. *Inorganic Chemistry*, *37*, 1734–1743.

Bray, C. (2001). *Dictionary of glass: Materials and techniques*. London; Philadelphia, PA: A & C Black; University of Pennsylvania Press.

Brent, G. F., & Harding, M. D. (1995). Surfactant coatings for the stabilization of barium peroxide and lead dioxide in pyrotechnic compositions. *Propellants, Explosives, Pyrotechnics*, *20*, 300–303.

Brobst, D. A. (1984). The geological framework of barite resources. *Transactions of the Institute of Mining and Metallurgy (London): Section A, Mining Industry*, *93*, 123–130.

Brodersen, K. (1981). Dimercury(I)-nitrogen compounds and other addition complexes of the $^{+}Hg\text{-}Hg^{+}$ ion. *Comments on Inorganic Chemistry*, *1*, 207–225.

Brodersen, K., Göbel, G., & Liehr, G. (1989). Terlinguait $Hg_4O_2Cl_2$ – ein Mineral mit ungewöhnlichen Hg_3-Baueinheiten. *Zeitschrift für anorganische und allgemeine Chemie*, *575*, 145–153.

Bronger, W., Brassard, L. à, Müller, P., Lebech, B., & Schultz, T. (1999). K_2ReH_9, eine Neubestimmung der Struktur. *Zeitschrift für anorganische und allgemeine Chemie*, *625*, 1143–1146.

Brown, F., Hall, G. R., & Walter, A. J. (1955). The half-life of Cs^{137}. *Journal of Inorganic and Nuclear Chemistry*, *1*, 241–247.

Brown, I. D., Gillespie, R. J., Morgan, K. R., Tun, Z., & Ummat, P. K. (1984). Preparation and crystal structure of mercury hexafluoroniobate (Hg_3NbF_6) and mercury hexafluorotantalate (Hg_3TaF_6): Mercury layer compounds. *Inorganic Chemistry*, *23*, 4506–4508.

Brown, I., Cutforth, B., Davies, C., Gillespie, R., Ireland, P., & Vekris, J. (2011). Alchemists' Gold, $Hg_{2.86}AsF_6$: An X-ray crystallographic study of a novel disordered mercury compound containing metallically bonded infinite cations. *Canadian Journal of Chemistry*, *52*, 791–793.

Bucher, E., Schmidt, P. H., Jayaraman, A., Andres, K., Maita, J. P., Nassau, K., & Dernier, P. D. (1970). New first-order phase transition in high-purity ytterbium metal. *Physical Review B*, *2*, 3911–3917.

Buckman, R. W. (2000). New applications for tantalum and tantalum alloys. *JOM*, *52*, 40–41.

Bünzli, J. G., & Mcgill, I. (2019). *Rare earth elements. Ullmann's encyclopedia of industrial chemistry*. Weinheim: Wiley-VCH.

Burkholder, M. A., & Johnson, L. L. (1990). *Colonial Latin America* (pp. 157–159). Oxford: Oxford University Press.

Burt, R. O. (1993). *Caesium and cesium compounds*, . (4 edn, pp. 749–764). *Kirk-Othmer encyclopedia of chemical technology*, (vol 5, pp. 749–764). New York, NY: John Wiley & Sons, Inc.

Bury, C. R. (1921). Langmuir's theory of the arrangement of electrons in atoms and molecules. *Journal of the American Chemical Society*, *43*, 1602–1609.

Busso, M., Gallino, R., & Wasserburg, G. J. (1999). Nucleosynthesis in asymptotic giant branch stars: Relevance for galactic enrichment and solar system formation. *Annual Review of Astronomy and Astrophysics*, *37*, 239–309.

Cameron, A. G. W. (1973). Abundance of the elements in the solar system. *Space Science Reviews*, *15*, 121–146.

Carreira, J. F. C. (2017). YAG:Dy – Based single white light emitting phosphor produced by solution combustion synthesis. *Journal of Luminescence*, *183*, 251–258.

Caswell, L. R., & Stone Daley, R. W. (1999). The Delhuyar brothers, tungsten, and Spanish silver. *Bulletin for the History of Chemistry*, *23*, 11–19.

Cawthorn, R. G. (1999). Seventy-fifth anniversary of the discovery of the Platiniferous Merensky Reef. *Platinum Metals Review*, *43*, 146–148.

Ceccarelli, P. (2013). *Ancient Greek letter writing: A cultural history (600 BC–150 BC)*. Oxford: Oxford University Press.

Černý, P., & Simpson, F. M. (1978). The Tanco Pegmatite at Bernic Lake, Manitoba: X. Pollucite. *Canadian Mineralogist*, *16*, 325–333.

Charrier, E., Charsley, E. L., Laye, P. G., Markham, H. M., Berger, B., & Griffiths, T. T. (2006). Determination of the temperature and enthalpy of the solid–solid phase transition of caesium nitrate by differential scanning calorimetry. *Thermochimica Acta*, *445*, 36–39.

Chaston, J. C. (1980). The powder metallurgy of platinum. *Platinum Metals Review*, *24*, 70–79.

Chau, Y. P., & Lu, K. S. (1995). Investigation of the blood-ganglion barrier properties in rat sympathetic ganglia by using lanthanum ion and horse-radish peroxidase as tracers. *Acta Anatomica*, *153*, 135–144.

Chen, J.-T. (2005). Platinum: Organometallic chemistry. Based in part on the article platinum: Organometallic chemistry by Charles M. Lukehart which appeared in the encyclopedia of inorganic chemistry. In R. B. King, R. H. Crabtree, C. M. Lukehart, D. A. Atwood, & R. A. Scott (Eds.), *Encyclopedia of inorganic chemistry* (First Edition). New York, NY: John Wiley & Sons.

Chin, C., Grimm, R., Julienne, P., & Tiesinga, E. (2010). Feshbach resonances in ultracold gases. *Reviews of Modern Physics*, *82*, 1225–1286.

Choi, J. H., Mao, Y., & Chang, J. P. (2011). Development of hafnium based high-k materials—A review. *Materials Science and Engineering: R: Reports*, *72*, 97–136.

Chu, Y. (1989). Genetic types of barite deposits in China. *Mineral Deposits*, *8*, 91–96.

Chung, H.-Y., Weinberger, M. B., Levine, J. B., Kavner, A., Yang, J.-M., Tolbert, S. H., & Kaner, R. B. (2007). Synthesis of ultra-incompressible superhard rhenium diboride at ambient pressure. *Science*, *316*, 436–439.

Cimalla, V. (2008). Nanomechanics of single crystalline tungsten nanowires. *Journal of Nanomaterials*, *2008*, 1–9.

Cohen, R., Della Valle, C. J., & Jacobs, J. J. (2006). Applications of porous tantalum in total hip arthroplasty. *Journal of the American Academy of Orthopaedic Surgeons*, *14*, 646–655.

Conroy, L. E., Ben-Dor, L., Kershaw, R., & Wold, A. (1995). Molybdenum(IV) oxide and tungsten(IV) oxide single crystals. *Inorganic Syntheses*, 105–107.

Coplen Tyler, B., & Peiser, H. S. (1998). History of the recommended atomic-weight values from 1882 to 1997: A comparison of differences from current values to the estimated uncertainties of earlier values (Technical Report). *Pure and Applied Chemistry*, *70*, 237–257.

Costa, E., Pringle, P. G., Ravetz, M., & Puddephatt, R. J. (1996). [(1,2,5,6-N)-1,5-cyclooctadiene]dimethyl-platinum(II). In A. H. Cowley (Ed.), *Inorganic syntheses*. New York, NY: John Wiley & Sons.

Cotnoir, B. (2006). *The Weiser concise guide to alchemy*. San Francisco, CA: Red Wheel Weiser.

Cotton, F. A., Donahue, J. P., Lichtenberger, D. L., Murillo, C. A., & Villagrán, D. (2005). Expeditious access to the most easily ionized closed-shell molecule, $W_2(hpp)_4$. *Journal of the American Chemical Society*, *127*, 10808–10809.

Cotton, F. A., & Rice, C. E. (1978). Tungsten pentachloride. *Acta Crystallographica B*, *34*, 2833–2834.

Cox, R. (1997). *The pillar of celestial fire: And the lost Science of the ancient seers*. Fairfield: 1st World Publishing, Incorporated.

Cozzini, C., Angloher, G., Bucci, C., von Feilitzsch, F., Hauff, D., Henry, S., ... Wulandari, H. (2004). Detection of the natural α decay of tungsten. *Physical Review C*, *70*, 064606.

Crabtree, R. (1979). Iridium compounds in catalysis. *Accounts of Chemical Research*, *12*, 331–337.

Cristina Cassani, M., Gun'ko, Y. K., Hitchcock, P. B., Hulkes, A. G., Khvostov, A. V., Lappert, M. F., & Protchenko, A. V. (2002). Aspects of non-classical organolanthanide chemistry. *Journal of Organometallic Chemistry*, *647*, 71–83.

Crookes, W. (1863). On thallium. *Philosophical Transactions of the Royal Society of London*, *153*, 173–192.

Crookes, W. (1908). On the use of iridium crucibles in chemical operations. *Proceedings of the Royal Society of London Series A, Containing Papers of a Mathematical and Physical Character*, *80*, 535–536.

Croteau, G., Dills, R., Beaudreau, M., & Davis, M. (2010). Emission factors and exposures from ground-level pyrotechnics. *Atmospheric Environment*, *44*, 3295–3303.

Cumberland, R. W., Weinberger, M. B., Gilman, J. J., Clark, S. M., Tolbert, S. H., & Kaner, R. B. (2005). Osmium diboride, an ultra-incompressible, hard material. *Journal of the American Chemical Society*, *127*, 7264–7265.

Dąbek, J., & Halas, S. (2007). Physical foundations of rhenium-osmium method – A review. *Geochronometria*, *27*, 23–26.

Danevich, F. A., Georgadze, A. S., Kobychev, V. V., Nagorny, S. S., Nikolaiko, A. S., Ponkratenko, O. A., ... Maurenzig, P. R. (2003). α activity of natural tungsten isotopes. *Physical Review C*, *67*, 014310.

Danevich, F. A., Kobychev, V. V., Ponkratenko, O. A., Tretyak, V. I., & Zdesenko, Y. G. (2001). Quest for double beta decay of ^{160}Gd and Ce isotopes. *Nuclear Physics A*, *694*, 375–391.

Darling, A. S. (1960). Iridium platinum alloys. *Platinum Metals Review*, *4*, 18–26.

Davidson, A., Ryman, J., Sutherland, C. A., Milner, E. F., Kerby, R. C., Teindl, H., ... Bolt, H. M. (2014). *Lead. Ullmann's encyclopedia of industrial chemistry*. Weinheim: Wiley-VCH.

Davis, T. L. (1940). Pyrotechnic snakes. *Journal of Chemical Education*, *17*, 268.

de Callataÿ, F. (2005). The Graeco-Roman economy in the super long-run: Lead, copper, and shipwrecks. *Journal of Roman Archaeology*, *18*, 361–372.

de Carvalho, H. G., & de Araújo Penna, M. (1972). Alpha-activity of ^{209}Bi. *Lettere al Nuovo Cimento (1971-1985)*, *3*, 720–722.

De Gregorio, A. (2003). A historical note about how the property was discovered that hydrogenated substances increase the radioactivity induced by neutrons. *Nuovo Saggiatore*, *19*, 41–47.

de Laeter John, R., Böhlke John, K., De Bièvre, P., Hidaka, H., Peiser, H. S., Rosman, K. J. R., & Taylor, P. D. P. (2003). Atomic weights of the elements. Review 2000 (IUPAC Technical Report). *Pure and Applied Chemistry*, *75*, 683.

de Marcillac, P., Coron, N., Dambier, G., Leblanc, J., & Moalic, J. P. (2003). Experimental detection of α-particles from the radioactive decay of natural bismuth. *Nature*, *422*, 876–878.

Deadmore, D. L. (1965). Vaporization of tantalum carbide-hafnium carbide solid solutions. *Journal of the American Ceramic Society*, *48*, 357–359.

Deblonde, G. J. P., Chagnes, A., Bélair, S., & Cote, G. (2015). Solubility of niobium(V) and tantalum(V) under mild alkaline conditions. *Hydrometallurgy*, *156*, 99–106.

Deblonde, G. J.-P., Lohrey, T. D., Booth, C. H., Carter, K. P., Parker, B. F., Larsen, Å., ... Cuthbertson, A. S. (2018). Solution thermodynamics and kinetics of metal complexation with a hydroxypyridinone chelator designed for thorium-227 targeted alpha therapy. *Inorganic Chemistry, 57*, 14337–14346.

Deiseroth, H. J. (1997). Alkali metal amalgams, a group of unusual alloys. *Progress in Solid State Chemistry, 25*, 73–123.

DeKosky, R. K. (1973). Spectroscopy and the elements in the late nineteenth century: The work of Sir William Crookes. *The British Journal for the History of Science, 6*, 400–423.

Deliens, M., & Piret, P. (1982). Bijvoetite et lepersonnite, carbonates hydrates d'uranyle et des terres rares de Shinkolobwe, Zaïre. *Canadian Mineralogist, 20*, 231–238.

Delile, H., Blichert-Toft, J., Goiran, J.-P., Keay, S., & Albarède, F. (2014). Lead in ancient Rome's city waters. *Proceedings of the National Academy of Sciences, 111*, 6594–6599.

Dewberry, C. T., Etchison, K. C., & Cooke, S. A. (2007). The pure rotational spectrum of the actinide-containing compound thorium monoxide. *Physical Chemistry Chemical Physics, 9*, 4895–4897.

Dilworth, J. R., & Parrott, S. J. (1998). The biomedical chemistry of technetium and rhenium. *Chemical Society Reviews, 27*, 43–55.

DiMeglio, J. L., & Rosenthal, J. (2013). Selective conversion of CO_2 to CO with high efficiency using an bismuth-based electrocatalyst. *Journal of the American Chemical Society, 135*, 8798–8801.

Downs, A. J. (1993). *Chemistry of aluminium, gallium, indium and thallium*. London: Blackie Academic & Professional.

Downs, J. D., Blaszczynski, M., Turner, J., Harris, M. (2006). Drilling and Completing Difficult HP/HT Wells With the Aid of Cesium Formate Brines-A Performance Review. Paper presented at the IADC/SPE Drilling Conference, Miami, Florida, USA.

Drews, T., Supeł, J., Hagenbach, A., & Seppelt, K. (2006). Solid State Molecular Structures of Transition Metal Hexafluorides. *Inorganic Chemistry, 45*, 3782–3788.

Druding, L. F., & Corbett, J. D. (1961). Lower Oxidation States of the Lanthanides. Neodymium(II) Chloride and Iodide. *Journal of the American Chemical Society, 83*, 2462–2467.

Dunn, P. J., Graham, A. B., Grigg, R., & Higginson, P. (2000). Tandem 1,3-azaprotiocyclotransfer–cycloaddition reactions between aldoximes and divinyl ketone. Remarkable rate enhancement and control of cycloaddition regiochemistry by hafnium(IV) chloride. *Chemical Communications*, 2035–2036.

Dyckhoff, F., Li, S., Reich, R. M., Hofmann, B. J., Herdtweck, E., & Kühn, F. E. (2018). Synthesis, characterization and application of organorhenium(VII) trioxides in metathesis reactions and epoxidation catalysis. *Dalton Transactions, 47*, 9755–9764.

Earle, G. D., Medikonduri, R., Rajagopal, N., Narayanan, V., & Roddy, P. A. (2005). Tungsten-Rhenium Filament Lifetime Variability in Low Pressure Oxygen Environments. *IEEE Transactions on Plasma Science, 33*, 1736–1737.

Ebert, K. H., Massa, W., Donath, H., Lorberth, J., Seo, B. S., & Herdtweck, E. (1998). Organoplatinum compounds: VI. Trimethylplatinum thiomethylate and trimethylplatinum iodide. The crystal structures of $[(CH_3)_3PtS(CH_3)]_4$ and $[(CH_3)_3PtI]_4 \cdot 0.5CH_3I$. *Journal of Organometallic Chemistry, 559*, 203–207.

Ede, A. (2006). *The chemical element: a historical perspective*. Westport, Conn: Greenwood Press.

Edwards, A. J. (1969). Crystal structure of tungsten pentafluoride. *Journal of the Chemical Society A: Inorganic, Physical, Theoretical*, 909–909.

EFSA Panel on Food Additives and Nutrient Sources added to Food (ANS). (2016). Scientific opinion on the re-evaluation of gold (E 175) as a food additive. *EFSA Journal, 14*, 4362.

Eisler, R. (2006). *Mercury hazards to living organisms*. Boca Raton, FL: CRC/Taylor & Francis.

Emsley, J. (2001). *Nature's building blocks*. Oxford: Oxford University Press, Oxford.

Enghag, P. (2004). *Encyclopedia of the elements: Technical data - History - Processing - Applications*. Weinheim: Wiley-VCH.

Engström, M., Klasson, A., Pedersen, H., Vahlberg, C., Käll, P.-O., & Uvdal, K. (2006). High proton relaxivity for gadolinium oxide nanoparticles. *Magnetic Resonance Materials in Physics, Biology and Medicine, 19*, 180–186.

Essen, L., & Parry, J. V. L. (1955). An atomic standard of frequency and time interval: A caesium resonator. *Nature, 176*, 280–282.

Evans, W. J., Gummersheimer, T. S., & Ziller, J. W. (1995). Coordination chemistry of samarium diiodide with ethers including the crystal structure of tetrahydrofuran-solvated samarium diiodide, $SmI_2(THF)_5$. *Journal of the American Chemical Society, 117*, 8999–9002.

Farahany, S., Ourdjini, A., Idris, M. H., & Thai, L. T. (2011a). Effect of bismuth on the microstructure of unmodified and Sr-modified Al-7%Si-0.4Mg alloy. *Journal of Transactions of Nonferrous Metals Society of China, 21*, 1455–1464.

Farahany, S., Ourdjini, A., Idris, M. H., & Thai, L. T. (2011b). Poisoning effect of bismuth on modification behavior of strontium in LM25 alloy. *Journal of Bulletin of Materials Science, 34*, 1223–1231.

Feltham, A. M., & Spiro, M. (1971). Platinized platinum electrodes. *Chemical Reviews, 71*, 177–193.

Fernelius, W. C. (1982). Hafnium. *Journal of Chemical Education, 59*, 242.

Finger, S. (2006). *Doctor Franklin's Medicine*. Philadelphia, PA: University of Pennsylvania Press, Incorporated.

Fink, P. J., Miller, J. L., & Konitzer, D. G. (2010). Rhenium reduction—Alloy design using an economically strategic element. *JOM, 62*, 55–57.

Finlay, I. G., Mason, M. D., & Shelley, M. (2005). Radioisotopes for the palliation of metastatic bone cancer: A systematic review. *The Lancet Oncology, 6*, 392–400.

Fontani, M., Costa, M., & Orna, M. V. (2014). *The lost elements: The periodic table's shadow side*. Oxford: Oxford University Press.

Frankel, C. (1999). *The end of the dinosaurs: Chicxulub crater and mass extinctions*. Cambridge: Cambridge University Press.

Frémy, E. (1844). Ueber das Osmium. *Journal für Praktische Chemie, 33*, 407–416.

Friestad, G. K., Branchaud, B. P., Navarrini, W., & Sansotera, M. (2007). *Cesium fluoride. Encyclopedia of reagents for organic synthesis*. New York, NY: John Wiley & Sons.

Fu, Y., Romero, M. J., Habtemariam, A., Snowden, M. E., Song, L., Clarkson, G. J., ... Sadler, P. J. (2012). The contrasting chemical reactivity of potent isoelectronic iminopyridine and azopyridine osmium(ii) arene anticancer complexes. *Chemical Science*, *3*, 2485–2494.

Furuta, E., Yoshizawa, Y., & Aburai, T. (2000). Comparisons between radioactive and non-radioactive gas lantern mantles. *Journal of Radiological Protection*, *20*, 423–431.

Gaines, R. V., Skinner, H. C., Foord, E. F., Mason, B., & Rosenzweig, A. (1997). *Dana's new mineralogy* (8th edn). New York, NY: John Wiley & Sons, Inc.

Gallo, C. F. (1967). The effect of thallium iodide on the arc temperature of Hg discharges. *Applied Optics*, *6*, 1563–1565.

Gando, A., Gando, Y., Ichimura, K., Ikeda, H., Inoue, K., Kibe, Y., ... Decowski, M. P. (2011). Partial radiogenic heat model for Earth revealed by geoneutrino measurements. *Nature Geoscience*, *4*, 647–651.

Gelis, V. M., Chuveleva, E. A., Firsova, L. A., Kozlitin, E. A., & Barabanov, I. R. (2005). Optimization of separation of ytterbium and lutetium by displacement complexing chromatography. *Russian Journal of Applied Chemistry*, *78*, 1420–1426.

Gibson, B. K. (1991). Liquid mirror telescopes: history. *Journal of the Royal Astronomical Society of Canada*, *85*, 158–171.

Gilbert, H. L., & Barr, M. M. (1955). Preliminary investigation of hafnium metal by the Kroll process. *Journal of The Electrochemical Society*, *102*, 243–245.

Gilchrist, R. (1943). The platinum metals. *Chemical Reviews*, *32*, 277–372.

Gilfillan, S. C. (1965). Lead poisoning and the fall of Rome. *Journal of Occupational Medicine*, *7*, 53–60.

Girard, P., Namy, J. L., & Kagan, H. B. (1980). Divalent lanthanide derivatives in organic synthesis. 1. Mild preparation of samarium iodide and ytterbium iodide and their use as reducing or coupling agents. *Journal of the American Chemical Society*, *102*, 2693–2698.

Gladyshevskii, E. I., & Kripyakevich, P. I. (1965). Monosilicides of rare earth metals and their crystal structures. *Journal of Structural Chemistry*, *5*, 789–794.

Gong, Y., Zhou, M., Kaupp, M., & Riedel, S. (2009). Formation and characterization of the iridium tetroxide molecule with iridium in the oxidation state + VIII. *Angewandte Chemie International Edition*, *48*, 7879–7883.

Gopher, A., Tsuk, T., Shalev, S., & Gophna, R. (1990). Earliest gold artifacts in the levant. *Current Anthropology*, *31*, 436–443.

Gordon, R. B., & Rutledge, J. W. (1984). Bismuth bronze from Machu Picchu, Peru. *Science*, *223*, 585–586.

Graham, A. R. (1955). Cerianite CeO_2: A new rare-earth oxide mineral. *American Mineralogist*, *40*, 560–564.

Gray, T. W., & Mann, N. (2012). *The elements: A visual exploration of every known atom in the universe*. New York, NY: Black Dog & Leventhal Publishers: Distributed by Workman Pub.

Greenwood, N. N., & Earnshaw, A. (1996). *Chemistry of the elements*. Oxford: Elsevier Science & Technology Books.

Griffin, H. C. (2010). Natural radioactive decay chains. In A. Vértes, S. Nagy, Z. Klencsár, R. G. Lovas, & F. Rösch (Eds.), *Handbook of nuclear chemistry* (p. 668). Dordrecht: Springer Science + Business Media B.V.

Griffith, W. P. (2004). Bicentenary of four platinum group metals. Part II: Osmium and iridium – Events surrounding their discoveries. *Platinum Metals Review*, *48*, 182–189.

Griffith, W. P., & Morris, P. J. T. (2003). Charles Hatchett FRS (1765–1847), chemist and discoverer of niobium. *Notes and Records of the Royal Society of London*, *57*, 299–316.

Groeger, S., Pazgalev, A. S., & Weis, A. (2005). Comparison of discharge lamp and laser pumped cesium magnetometers. *Applied Physics B*, *80*, 645–654.

Grout, J. (2017). *Lead poisoning and Rome*. University of Chicago. <http://penelope.uchicago.edu/~grout/encyclopaedia_romana/wine/leadpoisoning.html> Accessed 26.06.19.

Grukh, D. A., Bogatyrev, V. A., Sysolyatin, A. A., Paramonov, V. M., Kurkov, A. S., & Dianov, E. M. (2004). Broadband radiation source based on an ytterbium-doped fibre with fibre-length-distributed pumping. *Quantum Electronics*, *34*, 247–248.

Gschneidner, K., Pecharsky, V., & Tsokol, A. (2005). Recent developments in magnetocaloric materials. *Reports on Progress in Physics*, *68*, 1479–1539.

Gull, T. R., Herzig, H., Osantowski, J. F., & Toft, A. R. (1985). Low earth orbit environmental effects on osmium and related optical thin-film coatings. *Applied Optics*, *24*, 2660–2665.

Gulliver, D. J., & Levason, W. (1982). The chemistry of ruthenium, osmium, rhodium, iridium, palladium and platinum in the higher oxidation states. *Coordination Chemistry Reviews*, *46*, 1–127.

Gupta, C. K., & Krishnamurthy, N. (2010). *Extractive metallurgy of rare earths*. Milton Keynes UK; Boca Raton, Fla: Lightning Source UK Ltd.; CRC Press.

Gupta, C. K., & Mukherjee, T. K. (1990). *Hydrometallurgy in extraction processes*. Boca Raton, Florida: CRC Press.

Gupta, C. K., & Suri, A. K. (1993). *Extractive metallurgy of niobium*. Boca Raton: Taylor & Francis.

Haaland, A., Hammel, A., Rypdal, K., Verne, H. P., Volden, H. V., & Pulham, C. (1992). The structures of pentamethyltantalum and antimony: One square pyramid and one trigonal bipyramid. *Angewandte Chemie International Edition in English*, *31*, 1464–1467.

Haghseresht, F., Wang, S., & Do, D. D. (2009). A novel lanthanum-modified bentonite, Phoslock, for phosphate removal from wastewaters. *Applied Clay Science*, *46*, 369–375.

Hainfeld, J. F., Dilmanian, F. A., Slatkin, D. N., & Smilowitz, H. M. (2008). Radiotherapy enhancement with gold nanoparticles. *Journal of Pharmacy and Pharmacology*, *60*, 977–985.

Hajra, S., Maji, B., & Bar, S. (2007). Samarium triflate-catalyzed halogen-Promoted Friedel-Crafts alkylation with alkenes. *Organic Letters*, *9*, 2783–2786.

Hála, J., Johnson, S. A., Clever, H. L., Oshiba, T., & Salomon, M. (1989). *Halides, oxyhalides and salts of halogen complexes of titanium, zirconium, hafnium, vanadium, niobium and tantalum*. Oxford: Pergamon Press.

Halmshaw, R. (1954). The use and scope of Iridium 192 for the radiography of steel. *British Journal of Applied Physics, 5*, 238–243.

Han, Y., Huynh, H. V., & Tan, G. K. (2007). Mono- vs Bis(carbene) complexes: A detailed study on platinum(II) – benzimidazolin-2-ylidenes. *Organometallics, 26*, 4612–4617.

Handley, J. R. (1986). Increasing applications for iridium. *Platinum Metals Review, 30*, 12–13.

Haubrichs, R., & Zaffalon, P.-L. (2017). Osmium vs. 'Ptène'; The naming of the densest metal. *Johnson Matthey Technology Review, 61*, 190–196.

Heilbron, J. L. (1966). The work of H. G. J. Moseley. *Isis, 57*, 336–364.

Heimann, P. M. (1967). Moseley and celtium: The search for a missing element. *Annals of Science, 23*, 249–260.

Heiserman, D. L. (1992). *Exploring chemical elements and their compounds*. Blue Ridge Summit, PA: Tab Books.

Hermann, R. (1871). Fortgesetzte Untersuchungen über die Verbindungen von Ilmenium und Niobium, sowie über die Zusammensetzung der Niobmineralien. *Journal für Praktische Chemie, 3*, 373–427.

Hernberg, S. (2000). Lead poisoning in a historical perspective. *American Journal of Industrial Medicine, 38*, 244–254.

Hesse, R. W. (2007). *Jewelrymaking through history: An encyclopedia*. Westport: Greenwood Press.

Hildebrand, A. R., Penfield, G. T., Kring, D. A., Pilkington, M., Zanoguera, A. C., Jacobsen, S. B., & Boynton, V. W. (1991). Chicxulub crater; A possible cretaceous/tertiary boundary impact crater on the Yucatan Peninsula, Mexico. *Geology, 19*, 867–871.

Hodge, T. A. (1981). Vitruvius, lead pipes and lead poisoning. *American Journal of Archaeology, 85*, 486–491.

Hoffmann, M. M., Young, J. S., & Fulton, J. L. (2000). Unusual dysprosium ceramic nano-fiber growth in a supercritical aqueous solution. *Journal of Materials Science, 35*, 4177–4183.

Hofstadter, R. (1949). The detection of gamma-rays with thallium-activated sodium iodide crystals. *Physical Review, 75*, 796–810.

Holder, E., Langefeld, B. M. W., & Schubert, U. S. (2005). New trends in the use of transition metal-ligand complexes for applications in electroluminescent devices. *Advanced Materials, 17*, 1109–1121.

Hopper, K. D., King, S. H., Lobell, M. E., TenHave, T. R., & Weaver, J. S. (1997). The breast: Inplane x-ray protection during diagnostic thoracic CT—Shielding with bismuth radioprotective garments. *Radiology, 205*, 853–858.

Horsley, G. W. (1957). The preparation of bismuth for use in a liquid-metal fuelled reactor. *Journal of Nuclear Energy (1954), 6*, 41–52.

Hostettler, M., & Schwarzenbach, D. (2005). Phase diagrams and structures of HgX_2 (X = I, Br, Cl, F). *Comptes Rendus Chimie, 8*, 147–156.

Hoyano, J. K., & Graham, W. A. G. (1982). Oxidative addition of the carbon-hydrogen bonds of neopentane and cyclohexane to a photochemically generated iridium(I) complex. *Journal of the American Chemical Society, 104*, 3723–3725.

Huan, T. D., Sharma, V., Rossetti, G. A., & Ramprasad, R. (2014). Pathways towards ferroelectricity in hafnia. *Physical Review B, 90*, 064111.

Hubicka, H., & Drobek, D. (1997). Anion-exchange method for separation of ytterbium from holmium and erbium. *Hydrometallurgy, 47*, 127–136.

Hult, M., Wieslander, J. S. E., Marissens, G., Gasparro, J., Wätjen, U., & Misiaszek, M. (2009). Search for the radioactivity of ^{180m}Ta using an underground HPGe sandwich spectrometer. *Applied Radiation and Isotopes, 67*, 918–921.

Hunt, L. B. (1987). A history of iridium. *Platinum Metals Review, 31*, 32–41.

Hunt, L. B., & Lever, F. M. (1969). Platinum metals: A survey of productive resources to industrial uses. *Platinum Metals Review, 13*, 126–138.

Hwang, I.-C., & Seppelt, K. (2000). The structures of ReF_8^- and UF_8^{2-}. *Journal of Fluorine Chemistry, 102*, 69–72.

Ihde, A. J. (1984). *The development of modern chemistry*. New York: Dover Publications.

Ikezoe, H., Ikuta, T., Hamada, S., Nagame, Y., Nishinaka, I., Tsukada, K., ... Ohtsuki, T. (1996). A decay of a new isotope of ^{209}Th. *Physical Review C, 54*, 2043–2046.

Imam, S. K. (2001). Advancements in cancer therapy with alpha-emitters: A review. *International Journal of Radiation Oncology · Biology · Physics, 51*, 271–278.

Ishiwatari, N., & Nagai, H. (1981). Release of xenon-137 and iodine-137 from UO_2 pellet by pulse neutron irradiation at NSRR. *Nippon Genshiryoku Gakkaishi, 23*, 843–850.

Ivey, H. F. (1974). Candoluminescence and radical-excited luminescence. *Journal of Luminescence, 8*, 271–307.

Jahns, R. H. (1939). Clerici solution for the specific gravity determination of small mineral grains. *American Mineralogist, 24*, 116–122.

James, F. A. J. L. (1984). Of 'medals and muddles' the context of the discovery of thallium: William Crookes's early. *Notes and Records of the Royal Society of London, 39*, 65–90.

Jankovic, S. (1988). The Allchar Tl–As–Sb deposit, Yugoslavia and its specific metallogenic features. *Nuclear Instruments and Methods in Physics Research Section A: Accelerators, Spectrometers, Detectors and Associated Equipment, 271*, 286.

Jansen, M. (2008). The chemistry of gold as an anion. *Chemical Society Reviews, 37*, 1826–1835.

Jensen, W. B. (2003). The place of zinc, cadmium, and mercury in the periodic table. *Journal of Chemical Education, 80*, 952–961.

Jensen, W. B. (2009). The origin of the Brin process for the manufacture of oxygen. *Journal of Chemical Education, 86*, 1266–1268.

Jia, Y. X., Lee, C. S., & Zettl, A. (1994). Stabilization of the $Tl_2Ba_2Ca_2Cu_3O_{10}$ superconductor by Hg doping. *Physica C, 234*, 24–28.

Johansson, B., Ahuja, R., Eriksson, O., & Wills, J. M. (1995). Anomalous fcc crystal structure of thorium metal. *Physical Review Letters, 75*, 280–283.

Jones, C. J., & Thornback, J. (2009). *Medicinal applications of coordination chemistry*. Cambridge: Royal Society of Chemistry.

Kaesz, H. D., Glavee, G. N., & Angelici, R. J. (1990). Decacarbonyldi-μ-hydridotriosmium: $Os_3(\mu\text{-}H)_2(Co)_{10}$. *Inorganic Syntheses*, 238–239.

Kaiser, N., & Pulker, H. K. (2011). *Optical interference coatings*. Berlin; London: Springer.

Kaji, M. (2002). D. I. Mendeleev's concept of chemical elements and The principles of chemistry. *Bulletin for the History of Chemistry, 27*, 4–16.

Källström, K., Munslow, I., & Andersson, P. G. (2006). Ir-catalysed asymmetric hydrogenation: Ligands, substrates and mechanism. *Chemistry: A European Journal, 12*, 3194–3200.

Kaminskii, V. V., Solov'ev, S. M., & Golubkov, A. V. (2002). Electromotive force generation in homogeneously heated semiconducting samarium monosulfide. *Technical Physics Letters*, *28*, 229–231.

Kanazawa, T., Amemiya, T., Ishikawa, A., Upadhyaya, V., Tsuruta, K., Tanaka, T., & Miyamoto, Y. (2016). Few-layer HfS_2 transistors. *Scientific Reports*, *6*, 22277.

Kase, T., Konashi, K., Takahashi, H., & Hirao, Y. (1993). Transmutation of Cesium-137 using proton accelerator. *Journal of Nuclear Science and Technology*, *30*, 911–918.

Kauffman, G. B., Teter, L. A., & Rhoda, R. N. (1963). Inorganic syntheses. 61. Recovery of platinum from laboratory residues. In J. Kleinberg (Ed.), *Inorganic Syntheses* (7, pp. 232–236). McGraw-Hill Book Company, Inc.

Kauffman, G. B., Thurner, J. J., & Zatko, D. A. (1967). Ammonium hexachloroplatinate(IV). In S. Y. Tyree (Ed.), *Inorganic Syntheses* (9). Inorganic Syntheses. McGraw-Hill Book Co., Inc.

Kaur, H., Yadav, S., Srivastava, A. K., Singh, N., Rath, S., Schneider, J. J., ... Srivastava, R. (2018). High-yield synthesis and liquid-exfoliation of two-dimensional belt-like hafnium disulphide. *Nano Research*, *11*, 343–353.

Kean, W. F., & Kean, I. R. L. (2008). Clinical pharmacology of gold. *Inflammopharmacology*, *16*, 112–125.

Keister, J. B., & Shapley, J. R. (1982). Solution structures and dynamics of complexes of decacarbonyldihydrotriosmium with Lewis bases. *Inorganic Chemistry*, *21*, 3304–3310.

Kellogg, R., & Flatau, A. (2004). Wide band tunable mechanical resonator employing the ΔE effect of terfenol-D. *Journal of Intelligent Material Systems & Structures*, *15*, 355–368.

Kim, D. J., Thomas, S., Grant, T., Botimer, J., Fisk, Z., & Xia, J. (2013). Surface hall effect and nonlocal transport in SmB_6: Evidence for surface conduction. *Scientific Reports*, *3*, 3150.

Kim, K., & Shim, K. B. (2003). The effect of lanthanum on the fabrication of ZrB_2–ZrC composites by spark plasma sintering. *Materials Characterization*, *50*, 31–37.

Klein, C., & Dutrow, B. (2008). *The 23rd edition of the manual of mineral science (after James D. Dana)*. Hoboken, NJ: John Wiley & Sons, Inc.

Knief, R. A. (1992). *Nuclear engineering: Theory and technology of commercial nuclear power*. New York, NY; London: Taylor & Francis.

Koch, E.-C., & Hahma, A. (2012). Metal-fluorocarbon pyrolants. XIV: High density-high performance decoy flare compositions based on ytterbium/polytetrafluoroethylene/Viton®. *Zeitschrift für anorganische und allgemeine Chemie*, *638*, 721–724.

Koch, E.-C., Weiser, V., Roth, E., Knapp, S., & Kelzenberg, S. (2012). Combustion of ytterbium metal. *Propellants, Explosives, Pyrotechnics*, *37*, 9–11.

Kogel, J. E., Trivedi, N. C., Barker, J. M., & Krukowski, S. T. (2006). *Industrial minerals & rocks: Commodities, markets, and uses*. Littleton: Society for Mining, Metallurgy, and Exploration.

Kolb, H. C., Van Nieuwenhze, M. S., & Sharpless, K. B. (1994). Catalytic asymmetric dihydroxylation. *Chemical Reviews*, *94*, 2483–2547.

König, M., Wiedmann, S., Brüne, C., Roth, A., Buhmann, H., Molenkamp, L. W., ... Zhang, S.-C. (2007). Quantum spin hall insulator state in HgTe quantum wells. *Science*, *318*, 766–770.

Koponen, J. J., Söderlund, M. J., Hoffman, H. J., & Tammela, S. K. T. (2006). Measuring photodarkening from singlemode ytterbium doped silica fibers. *Optics Express*, *14*, 11539–11544.

Korzhinsky, M. A., Tkachenko, S. I., Shmulovich, K. I., Taran, Y. A., & Steinberg, G. S. (2004). Discovery of a pure rhenium mineral at Kudriavy volcano. *Nature*, *369*, 51–52.

Kouznetsov, D. (2007). Comment on "Efficient diode-pumped Yb:Gd_2SiO_5 laser" [Appl. Phys. Lett. 88, 221117 (2006)]. *Applied Physics Letters*, *90*, 066101.

Krause, J. A., Siriwardane, U., Salupo, T. A., Wermer, J. R., Knoeppel, D. W., & Shore, S. G. (1993). Preparation of $[Os_3(CO)_{11}]^{2-}$ and its reactions with $OS_3(CO)_{12}$; structures of $[Et_4N][HOs_3(CO)_{11}]$ and $H_2OS_4(CO)$. *Journal of Organometallic Chemistry*, *454*, 263–271.

Krebs, B., & Sinram, D. (1980). Hafniumtetrajodid HfI_4: Struktur und eigenschaften. Ein neuer AB_4-strukturtyp. *Journal of the Less Common Metals*, *76*, 7–16.

Krebs, R. E. (2006). *The history and use of our Earth's chemical elements: A reference guide*. Westport: Greenwood Press.

Krebs, W. (1981). The geology of the Meggen ore deposit. In K. H. Wolf (Ed.), *Handbook of Stratabound and Stratiform Deposits* (pp. 509–549). Amsterdam: Elsevier.

Kremenetsky, A. A., & Chaplygin, I. V. (2010). Concentration of rhenium and other rare metals in gases of the Kudryavy Volcano (Iturup Island, Kurile Islands). *Doklady Earth Sciences*, *430*, 114–119.

Kresse, R., Baudis, U., Jäger, P., Riechers, H. H., Wagner, H., Winkler, J., & Wolf, H. U. (2007). *Barium and barium compounds. Ullmann's encyclopedia of industrial chemistry*. Weinheim: Wiley-VCH.

Krishnamurthy, N. (2004). *Extractive metallurgy of rare earths*. Boca Raton: CRC Press.

Krüger, J., Winkler, P., Lüderitz, E., Lück, M., & Wolf, H. U. (2003). *Bismuth, bismuth alloys, and bismuth compounds. Ullmann's encyclopedia of industrial chemistry. Major reference works*. Weinheim: Wiley-VCH.

Kwak, T. A. P. (1987). *W-Sn skarn deposits and related metamorphic skarns and granitoids*. Amsterdam: Elsevier.

Lagunas-Solar, M. C., Little, F. E., & Goodart, C. D. (1982). An integrally shielded transportable generator system for thallium-201 production. *International Journal of Applied Radiation and Isotopes*, *33*, 1439–1443.

Lah, N. A. C., & Trigueros, S. (2019). Synthesis and modelling of the mechanical properties of Ag, Au and Cu nanowires. *Science and Technology of Advanced Materials*, *20*, 225–261.

Langeslay, R. R., Fieser, M. E., Ziller, J. W., Furche, F., & Evans, W. J. (2015). Synthesis, structure, and reactivity of crystalline molecular complexes of the $\{[C_5H_3(SiMe_3)_2]_3Th\}^{1-}$ anion containing thorium in the formal +2 oxidation state. *Chemical Science*, *6*, 517–521.

Langmuir, C. H., & Broecker, W. (2012). *How to build a habitable planet: The story of earth from the big bang to humankind.* Princeton: Princeton University Press.

Larminie, J., & Dicks, A. (2011). *Fuel cell systems explained.* Chichester, West Sussex: J. Wiley.

Larsen, E., Fernelius, W. C., & Quill, L. (1943). Concentration of Hafnium. Preparation of Hafnium-Free Zirconia. *Industrial & Engineering Chemistry Analytical Edition*, *15*, 512–515.

Lassner, E., & Schubert, W.-D. (1999). *Tungsten: properties, chemistry, technology of the elements, alloys and chemical compounds.* New York, NY; London: Kluwer Academic.

Lavrentyev, A. A., Gabrelian, B. V., Vorzhev, V. B., Nikiforov, I. Y., Khyzhun, O. Y., & Rehr, J. J. (2008). Electronic structure of cubic $Hf_xTa_{1-x}C_y$ carbides from X-ray spectroscopy studies and cluster self-consistent calculations. *Journal of Alloys and Compounds*, *462*, 4–10.

Lazarova, N., James, S., Babich, J., & Zubieta, J. (2004). A convenient synthesis, chemical characterization and reactivity of $[Re(CO)_3(H_2O)_3]Br$: The crystal and molecular structure of $[Re(CO)_3(CH_3CN)_2Br]$. *Inorganic Chemistry Communications*, *7*, 1023–1026.

Leach, D. L. (1980). Nature of mineralizing fluids in the barite deposits of central and southeast Missouri. *Economic Geology*, *75*, 1168–1180.

Lebowitz, E., Greene, M. W., Fairchild, R., Bradley-Moore, P. R., Atkins, H. L., Ansari, A. N., ... Belgrave, E. (1975). Thallium-201 for medical use. *The Journal of Nuclear Medicine*, *16*, 151–155.

Lee, O. I. (1928). The mineralogy of hafnium. *Chemical Reviews*, *5*, 17–37.

Lee, S. L., Doxbeck, M., Mueller, J., Cipollo, M., & Cote, P. (2004). Texture, structure and phase transformation in sputter beta tantalum coating. *Surface and Coatings Technology*, *177-178*, 44–51.

Leger, J. M., Yacoubi, N., & Loriers, J. (1981). Synthesis of rare earth monoxides. *Journal of Solid State Chemistry*, *36*, 261–270.

Levi, P. (2012). *The periodic table (trans: Rosenthal R).* London: Penguin.

Levin, H. L. (2010). *The Earth through time.* Hoboken (N.J.): J. Wiley.

Lewis, J. (1985). Lead poisoning: A historical perspective. *EPA Journal*, *11*, 15–18.

Li, W., Pan, H., Ding, L. e, Zeng, H., Lu, W., Zhao, G., ... Xu, J. (2006). Efficient diode-pumped $Yb:Gd_2SiO_5$ laser. *Applied Physics Letters*, *88*, 221117.

Lietzke, M. H., Holt, M. L., Weyden, A. J. V., & Callison, J. H. (1950). Tungsten(VI) chloride (tungsten hexachloride). In L. F. Audrieth (Ed.), *Inorganic Syntheses.* Wiley: Inorganic Syntheses.

Liu, J., Shi, J.-Z., Yu, L.-M., Goyer, R. A., & Waalkes, M. P. (2008). Mercury in traditional medicines: Is cinnabar toxicologically similar to common mercurials? *Experimental Biology and Medicine*, *233*, 810–817.

Liu, M., Jin, P., Xu, Z., Hanaor, D. A. H., Gan, Y., & Chen, C. (2016). Two-dimensional modeling of the self-limiting oxidation in silicon and tungsten nanowires. *Theoretical and Applied Mechanics Letters*, *6*, 195–199.

Liu, Z., Stevens-Kalceff, M., & Riesen, H. (2012). Photoluminescence and cathodoluminescence properties of nanocrystalline $BaFCl:Sm^{3+}$ X-ray storage phosphor. *Journal of Physical Chemistry C*, *116*, 8322–8331.

Louis, C., Pluchery, O. (2017). *Gold nanoparticles for physics, chemistry, and biology*, London: World Scientific Publishing Europe Ltd.

Loveland, W. D., Morrissey, D. J., & Seaborg, G. T. (2006). *Modern nuclear chemistry.* Hoboken, NJ: Wiley-Interscience.

Maddahi, J., & Berman, D. (2001). Detection, evaluation, and risk stratification of coronary artery disease by thallium-201 myocardial perfusion scintigraphy 155. In E. G. DePuey, D. S. Berman, & E. V. Garcia (Eds.), *Cardiac SPECT imaging* (pp. 155–178). Philadelphia, PA: Lippincott Williams & Wilkins.

Maile, F. J., Pfaff, G., & Reynders, P. (2005). Effect pigments—past, present and future. *Progress in Organic Coatings*, *54*, 150–163.

Marczenko, Z., Balcerzak, M., & Kloczko, E. (2000). *Separation, preconcentration and spectrophotometry in inorganic analysis.* Amsterdam: Elsevier.

Markowitz, W., Hall, R., Essen, L., & Parry, J. (1958). Frequency of cesium in terms of ephemeris time. *Physical Review Letters*, *1*, 105–107.

Marsac, R., Réal, F., Banik, N. L., Pédrot, M., Pourret, O., & Vallet, V. (2017). Aqueous chemistry of Ce(iv): Estimations using actinide analogues. *Dalton Transactions*, *46*, 13553–13561.

Masau, M., Černý, P., Cooper, M. A., & Chapman, R. (2002). Monazite-(Sm), a new member of the monazite group from the Annie claim #3 granitic pegmatite, Southeastern Manitoba. *Canadian Mineralogist*, *40*, 1649.

Maslenkov, S. B., Burova, N. N., & Khangulov, V. V. (1980). Effect of hafnium on the structure and properties of nickel alloys. *Metal Science and Heat Treatment*, *22*, 283–285.

McCann, E. L., III, Brown, T. M., Djordjevic, C., & Morris, R. E. (1972). Tungsten(V) Chloride (Tungsten Pentachloride). *Inorganic Syntheses*, *13*, 150–154.

McCumber, D. E. (1964). Einstein relations connecting broadband emission and absorption spectra. *Physical Review B*, *136*, 954–957.

McDonald, D., & Hunt, L. B. (1982). *A history of platinum and its allied metals.* London: Johnson Matthey.

McDonald, M. (1959). The platinum of new Granada: Mining and metallurgy in the Spanish colonial empire. *Platinum Metals Review*, *3*, 140–145.

McKelvey, W., Jeffery, N., Clark, N., Kass, D., & Parsons, P. J. (2011). Population-based inorganic mercury biomonitoring and the identification of skin care products as a source of exposure in New York City. *Environmental Health Perspectives*, *119*, 203–209.

Meinert, H., & Dimitrov, A. (1976). On the chemistry of tungsten tetrafluoride. *Zeitschrift für Chemie*, *16*, 29–30.

Mel'nikov, V. P. (1982). Some details in the prehistory of the discovery of element 72. *Centaurus*, *26*, 317–322.

Meng, J., & Ren, Y. (1991). Studies on the electrical properties of rare earth monophosphides. *Journal of Solid State Chemistry*, *95*, 346–351.

Meyer, G., Garcia, E., & Corbett, J. D. (1989). The ammonium chloride route to anhydrous rare earth chlorides—The example of YCl_3. *Inorganic Syntheses, 25*, 146–150.

Meyer, G., & Morss, L. R. (1991). *Synthesis of lanthanide and actinide compounds*. Dordrecht: Kluwer.

Meyer, G., & Schleid, T. (1986). The metallothermic reduction of several rare-earth trichlorides with lithium and sodium. *Journal of the Less Common Metals, 116*, 187–197.

Min, M., Xu, H., Chen, J., & Fayek, M. (2005). Evidence of uranium biomineralization in sandstone-hosted roll-front uranium deposits, northwestern China. *Ore Geology Reviews, 26*, 198–206.

Mohammadi, R., Lech, A. T., Xie, M., Weaver, B. E., Yeung, M. T., Tolbert, S. H., & Kaner, R. B. (2011). Tungsten tetraboride, an inexpensive superhard material. *Proceedings of the National Academy of Sciences, 108*, 10958–10962.

Möhl, D. (1997). Production of low-energy antiprotons. *Zeitschrift Hyperfine Interactions, 109*, 33–41.

Mol, J. C. (1999). Olefin metathesis over supported rhenium oxide catalysts. *Catalysis Today, 51*, 289–299.

Möller, P., Cerny, P., & Saupe, F. (2014). *Lanthanides, tantalum and niobium mineralogy, geochemistry, characteristics of primary ore deposits, prospecting, processing and applications proceedings of a workshop in Berlin, November 1986*. Berlin: Springer Berlin.

Montserrat, D. (2002). *Akhenaten: History, fantasy and ancient Egypt*. London: Routledge.

Morris, D. J. P., Tennant, D. A., Grigera, S. A., Klemke, B., Castelnovo, C., Moessner, R., ... Perry, R. S. (2009). Dirac strings and magnetic monopoles in the spin ice $Dy_2Ti_2O_7$. *Science, 326*, 411–414.

Morse, P. M., Shelby, Q. D., Kim, D. Y., & Girolami, G. S. (2008). Ethylene complexes of the early transition metals: Crystal structures of $[HfEt_4(C_2H_4)^{2-}]$ and the negative-oxidation-state species $[TaHEt(C_2H_4)_3{}^{3-}]$ and $[WH(C_2H_4)_4{}^{3-}]$. *Organometallics, 27*, 984–993.

Mosseri, S., Henglein, A., & Janata, E. (1990). Trivalent lead as an intermediate in the oxidation of lead(II) and the reduction of lead(IV) species. *The Journal of Physical Chemistry, 94*, 2722–2726.

Munson, R. A. (1968). The synthesis of iridium disulfide and nickel diarsenide having the pyrite structure. *Inorganic Chemistry, 7*, 389–390.

Murmann, R. K., & Barnes, C. L. (2002). Redetermination of the crystal structure of potassium trans-(dioxo) tetra(hydroxo)osmate(VI), $K_2[Os(OH)_4(O)_2]$. *Zeitschrift für Kristallographie - New Crystal Structures, 217*.

Nadler, H. G. (2000). *Rhenium and rhenium compounds. Ullmann's encyclopedia of industrial chemistry*. Wiley-VCH, Weinheim.

Nagy, S. (2009). *Radiochemistry and nuclear chemistry*. Oxford, UK: Eolss Publishers.

Nakai, S., Masuda, A., & Lehmann, B. (1988). La-Ba dating of bastnaesite. *American Mineralogist, 73*, 1111–1113.

Nakashima, T., Hayashi, H., Tashiro, H., & Matsushita, T. (1998). Gender and hierarchical differences in lead-contaminated Japanese bone from the Edo period. *Journal of Occupational Health, 40*, 55–60.

Naumov, A. V. (2007). Rhythms of rhenium. *Russian Journal of Non-Ferrous Metals, 48*, 418–423.

Naumov, A. V. (2008). Review of the world market of rare-earth metals. *Russian Journal of Non-Ferrous Metals, 49*, 14–22.

Nayar, P. S., & Hamilton, W. O. (1977). Thallium selenide infrared detector. *Applied Optics, 16*, 2942–2944.

Nemat-Nasser, S., Isaacs, J. B., & Liu, M. (1998). Microstructure of high-strain, high-strain-rate deformed tantalum. *Acta Materialia, 46*, 1307–1325.

Nesse, W. D. (2000). *Introduction to Mineralogy*. New York: Oxford University Press.

Ni, W.-X., Man, W.-L., Cheung, M. T.-W., Sun, R. W.-Y., Shu, Y.-L., Lam, Y.-W., ... Lau, T.-C. (2011). Osmium(VI) complexes as a new class of potential anti-cancer agents. *Chemical Communications, 47*, 2140–2142.

Nilson, L. F. (1882). Über metallisches thorium. *Berichte der Deutschen Chemischen Gesellschaft, 15*, 2537–2547.

NIST. (2013). *Ytterbium atomic clocks set record for stability*. NIST. <https://www.nist.gov/news-events/news/2013/08/nist-ytterbium-atomic-clocks-set-record-stability> Accessed 18.09.19.

Noddack, W., Tacke, I., & Berg, O. (1925). Die Ekamangane. *Naturwissenschaften, 13*, 567–574.

Nolting, D. D., Messerle, L., Yuan, M., Disalvo, F. J. (2014) Octahedral hexatungsten halide clusters. *Inorganic Syntheses 36*, 19–23.

Norman, M. J., Andrew, J. E., Bett, T. H., Clifford, R. K., England, J. E., Hopps, N. W., ... Stevenson, M. (2002). Multipass reconfiguration of the HELEN Nd: Glass laser at the atomic weapons establishment. *Applied Optics, 41*, 3497–3505.

Norman, N. C. (1997). *Chemistry of arsenic, antimony and bismuth*. London: Blackie Academic & Professional.

Nriagu, J. O. (1983). Saturnine gout among Roman aristocrats — Did lead poisoning contribute to the fall of the Empire? *The New England Journal of Medicine, 308*, 660–663.

Ohriner, E. K. (2008). Processing of iridium and iridium alloys. *Platinum Metals Review, 52*, 186–197.

Öhrström, L., & Reedijk, J. (2016). Names and symbols of the elements with atomic numbers 113, 115, 117 and 118 (IUPAC Recommendations 2016). *Pure and Applied Chemistry, 88*, 1225–1229.

Ojebuoboh, F. K. (1992). Bismuth—Production, properties, and applications. *JOM, 44*, 46–49.

Okabe, T. H., & Sadoway, D. R. (1998). Metallothermic reduction as an electronically mediated reaction. *Journal of Materials Research, 13*, 3372–3377.

Okada, S., Kudou, K., & Lundström, T. (1995). Preparations and some properties of W_2B, δ -WB and WB_2 crystals from high-temperature metal solutions. *Japanese Journal of Applied Physics, 34*, 226–231.

Okamoto, H. (2010). Cs-O (cesium-oxygen). *Journal of Phase Equilibria and Diffusion, 31*, 86–87.

Paganias, C. G., Tsakotos, G. A., Koutsostathis, S. D., & Macheras, G. A. (2012). Osseous integration in porous tantalum implants. *Indian Journal of Orthopaedics, 46*, 505–513.

Paneth, F. (1922). *Das periodische System der chemischen Elemente, 1. Ergebnisse der Exakten Naturwissenschaften*. Berlin: Springer-Verlag.

Parker, S. K., Schwartz, B., Todd, J., & Pickering, L. K. (2004). Thimerosal-containing vaccines and autistic spectrum disorder: A critical review of published original data. *Pediatrics, 114*, 793–804.

Parnell, R. J. G. (1924). Bismuth in the treatment of syphilis. *Journal of the Royal Society of Medicine, 17*, 19–26 (War section).

Patchett, P. J. (1983). Importance of the Lu-Hf isotopic system in studies of planetary chronology and chemical evolution. *Geochimica et Cosmochimica Acta, 47*, 81–91.

Patchett, P. J., & Tatsumoto, M. (1980). Lu–Hf total-rock isochron for the eucrite meteorites. *Nature, 288*, 571–574.

Pattison, J. E. (1999). Finger doses received during 153Sm injections. *Health Physics, 77*, 530–535.

Pendergast, D. M. (1982). Ancient maya mercury. *Science, 217*, 533–535.

Perkins, D. (2002). *Mineralogy* (2nd edn). Upper Saddle River, NJ: Prentice Hall.

Peter, A., & Viraraghavan, T. (2005). Thallium: A review of public health and environmental concerns. *Environment International, 31*, 493–501.

Pinhey, J. T. (1996). Organolead(IV) triacetates in organic synthesis. *Pure and Applied Chemistry, 68*, 819–824.

Pletcher, D., & Walsh, F. C. (1990). *Industrial electrochemistry*. London: Chapman and Hall.

Pletikosić, I., Ali, M. N., Fedorov, A. V., Cava, R. J., & Valla, T. (2014). Electronic structure basis for the extraordinary magnetoresistance in WTe_2. *Physical Review Letters, 113*, 216601.

Pohl, W. L. (2011). *Economic geology of metals. Economic geology principles and practice* (pp. 149–284). Chichester: John Wiley & Sons.

Pombeiro, A. J. L., Guedes da Silva, M. F. C., & Crabtree, R. H. (2005). Technetium & rhenium: Inorganic & coordination chemistry. In R. B. King, R. H. Crabtree, C. M. Lukehart, D. A. Atwood, & R. A. Scott (Eds.), *Encyclopedia of inorganic chemistry*. New York, NY: Wiley.

Procter, D. J., Flowers, R. A., & Skrydstrup, T. (2010). *Organic synthesis using samarium diiodide: A practical guide*. Cambridge: Royal Society of Chemistry.

Puddephatt, R. J. (2001). Platinum(IV) hydride chemistry. *Coordination Chemistry Reviews, 219–221*, 157–185.

Qin, J., He, D., Wang, J., Fang, L., Lei, L., Li, Y., ... Bi, Y. (2008). Is rhenium diboride a superhard material? *Advanced Materials, 20*, 4780–4783.

Rabinowitz, M. B. (1995). Imputing lead sources from blood lead isotope ratios. In M. E. Beard, & S. D. Allen Iske (Eds.), *Lead in paint, soil, and dust: Health risks, exposure studies, control measures, measurement methods, and quality assurance* (pp. 63–75). Philadelphia, PA: American Society for Testing and Materials.

Rafferty, J. P. (2011). *Geochronology, dating, and Precambrian time: the beginning of the world as we know it*. New York, NY: Britannica Educational Publishing.

Ramage, C. K. (1980). *Lyman cast bullet handbook* (3rd edn). Middletown, CT: Lyman Products Corporation.

Ramdahl, T., Bonge-Hansen, H. T., Ryan, O. B., Larsen, Å., Herstad, G., Sandberg, M., ... Brevik, E. M. (2016). An efficient chelator for complexation of thorium-227. *Bioorganic & Medicinal Chemistry Letters, 26*, 4318–4321.

Raymond, K. N., & Pierre, V. C. (2005). Next generation, high relaxivity gadolinium MRI agents. *Bioconjugate Chemistry, 16*, 3–8.

Reddy, A., & Braun, C. L. (2010). Lead and the Romans. *Journal of Chemical Education, 87*, 1052–1055.

Reinhardt, K., & Winkler, H. (2000). *Cerium mischmetal, cerium alloys, and cerium compounds. Ullmann's encyclopedia of industrial chemistry*. Weinheim: Wiley-VCH.

Renner, H., Schlamp, G., Kleinwächter, I., Drost, E., Lüschow, H. M., Tews, P., ... Drieselman, R. (2002). *Platinum group metals and compounds. Ullmann's encyclopedia of industrial chemistry*. Weinheim: Wiley-VCH.

Repina, S. A., Popova, V. I., Churin, E. I., Belogub, E. V., & Khiller, V. V. (2014). Florencite-(Sm)—(Sm,Nd)$Al_3(PO_4)_2(OH)_6$: A new mineral species of the alunite-jarosite group from the Subpolar Urals. *Geology of Ore Deposits, 5*, 564–574.

Retief, F., & Cilliers, L. P. (2006). Lead poisoning in ancient Rome. *Acta Theologica, 26*, 147–164.

Rich, V. (2014). *The international lead trade*. Cambridge: Elsevier Science.

Richards, D. G., McMillin, D. L., Mein, E. A., & Nelson, C. D. (2002). Gold and its relationship to neurological/glandular conditions. *The International Journal of Neuroscience, 112*, 31–53.

Riehl, M. E., Wilson, S. R., & Girolami, G. S. (1993). Synthesis, x-ray crystal structure, and phosphine-exchange reactions of the hafnium(III)-hafnium(III) dimer $Hf_2Cl_6(PEt_3)_4$. *Inorganic Chemistry, 32*, 218–222.

Rieuwerts, J. (2015). *The elements of environmental pollution*. London: Routledge, Taylor & Francis Group, Earthscan.

Risovany, V. D., Varlashova, E. E., & Suslov, D. N. (2000). Dysprosium titanate as an absorber material for control rods. *Journal of Nuclear Materials, 281*, 84–89.

Riva, M. A., Lafranconi, A., & d'Orso, M. I. G. C. (2012). Lead poisoning: Historical aspects of a paradigmatic "occupational and environmental disease". *Safety and Health at Work, 3*, 11–16.

Robinson, R., & Thoennessen, M. (2012). Discovery of tantalum, rhenium, osmium, and iridium isotopes. *Atomic Data and Nuclear Data Tables, 98*, 911–932.

Rodney, W. S., & Malitson, I. H. (1956). Refraction and Dispersion of Thallium Bromide Iodide. *Journal of the Optical Society of America, 46*, 338–346.

Ronen, Y. (2006). A rule for determining fissile isotopes. *Nuclear Science and Engineering, 152*, 334–335.

Ronen, Y. (2010). Some remarks on the fissile isotopes. *Annals of Nuclear Energy, 37*, 1783–1784.

Rose, H. (1844). Ueber die Zusammensetzung der Tantalite und ein im Tantalite von Baiern enthaltenes neues Metall. *Annalen der Physik, 139*, 317–341.

Roseblade, S. J., & Pfaltz, A. (2007). Iridium-catalyzed asymmetric hydrogenation of olefins. *Accounts of Chemical Research, 40*, 1402–1411.

Rossotti, H. (1998). *Diverse Atoms. Profiles of the chemical elements*. Oxford: Oxford University Press.

Roth, J., Bendayan, M., & Orci, L. (1980). FITC-protein A-gold complex for light and electron microscopic immunocytochemistry. *Journal of Histochemistry and Cytochemistry, 28*, 55–57.

Rouschias, G. (1974). Recent advances in the chemistry of rhenium. *Chemical Reviews, 74*, 531–566.

Ruchowska, E., Płóciennik, W. A., Żylicz, J., Mach, H., Kvasil, J., Algora, A., ... Weissman, L. (2006). Nuclear structure of ^{229}Th. *Physical Review C, 73*, 044326.

Ruff, O., & Bornemann, F. (1910). über das Osmium, seine analytische Bestimmung, seine Oxyde und seine Chloride. *Zeitschrift für anorganische Chemie, 65*, 429–456.

Russell, M. S. (2009). *The chemistry of fireworks*. Cambridge, UK: RSC Publishing.

Ryabchikov, D. I., & Gol'Braikh, E. K. (2013). *The analytical chemistry of thorium: International series of monographs on analytical chemistry*. Oxford: Pergamon Press.

Ryashentseva, M. A. (1998). Rhenium-containing catalysts in reactions of organic compounds. *Russian Chemical Reviews, 67*, 157–177.

Rytuba, J. J. (2003). Mercury from mineral deposits and potential environmental impact. *Environmental Geology, 43*, 326–338.

Ryzhikov, V. D., Grinev, B. V., Pirogov, E. N., Onyshchenko, G. M., Ivanov, A. I., Bondar, V. G., ... Kostyukevych, S. A. (2005). Use of gadolinium oxyorthosilicate scintillators in x-ray radiometers. *Optical Engineering, 44*, 1–6.

Samson, B., Carter, A., & Tankala, K. (2011). Rare-earth fibres power up. *Nature Photonics, 5*, 466–467.

Samson, G. W. (1885). *The divine law as to wines: established by the testimony of sages, physicians, and legislators against the use of fermented and intoxicating wines: confirmed by Egyptian, Greek, and Roman methods of preparing unfermented wines for festal, medicinal, and sacramental uses*. Philadelphia, PA: J.B. Lippincott.

Saunders, N. (2003). *Tungsten and the elements of Groups 3 to 7*. Chicago, Ill: Heinemann Library.

Scarborough, J. (1984). The myth of lead poisoning among the Romans: An essay review. *Journal of the History of Medicine and Allied Sciences, 39*, 469–475.

Scerri, E. R. (1994). Prediction of the nature of hafnium from chemistry, Bohr's theory and quantum theory. *Annals of Science, 51*, 137–150.

Schemel, J. H. (1977). *ASTM manual on zirconium and hafnium*. Philadelphia, PA: American Society for Testing and Materials.

Schilling, J. (1902). Die eigentlichen Thorit-Mineralien (Thorit und Orangit). *Zeitschrift für Angewandte Chemie, 15*, 921–929.

Schmidbaur, H., Cronje, S., Djordjevic, B., & Schuster, O. (2005). Understanding gold chemistry through relativity. *Chemical Physics, 311*, 151–161.

Schott, G. D. (2012). Some observations on the history of the use of barium salts in medicine. *Medical History, 18*, 9–21.

Sciolist Online Etymology Dictionary. <https://www.etymonline.com/> Accessed 21.06.19.

Scott, D. A., & Bray, W. (1980). Ancient platinum technology in South America: Its use by the Indians in pre-Hispanic times. *Platinum Metals Review, 24*, 147–157.

Seaborg, G. T. (1968). Uranium. In C. A. Hampel (Ed.), *The encyclopedia of the chemical elements* (pp. 773–786). New York, NY: Reinhold Book Corporation.

Selbekk, R. S. (2007). *Morten Thrane Esmark*. Kunnskapsforlaget. <http://www.snl.no/Morten_Thrane_Esmark> Accessed 26.06.19.

Seppelt, K. (1996). Response: Structure of $W(CH_3)_6$. *Science, 272*, 182–183.

Sera, M., Kobayashi, S., Hiroi, M., Kobayashi, N., & Kunii, S. (1996). Thermal conductivity of RB_6 (R = Ce, Pr, Nd, Sm, Gd) single crystals. *Physical Review B, 54*, R5207–R5210.

Seymour, R. J., & O'Farrelly, J. (2012). *Platinum-group metals. Kirk-Othmer Encyclopedia of Chemical Technology* (pp. 1–37). John Wiley & Sons.

Shaw, D. (1952). The geochemistry of thallium. *Geochimica et Cosmochimica Acta, 2*, 118–154.

Sheng, Z. Z., & Hermann, A. M. (1988). Bulk superconductivity at 120 K in the Tl–Ca/Ba–Cu–O system. *Nature, 332*, 138–139.

Sheppeard, H., & Ward, D. J. (1980). Intra-articular osmic acid in rheumatoid arthritis: Five years' experience. *Rheumatology, 19*, 25–29.

Shevtsov, Y. V., & Beizel', N. F. (2011). Pb distribution in multistep bismuth refining products. *Inorganic Materials, 47*, 139–142.

Shi, N. L., & Fort, D. (1985). Preparation of samarium in the double hexagonal close packed form. *Journal of the Less Common Metals, 113*, 21–23.

Shimada, S., Inagaki, M., & Matsui, K. (1992). Oxidation kinetics of hafnium carbide in the temperature range of 480° to 600°C. *Journal of the American Ceramic Society, 75*, 2671–2678.

Shnyder, S. D., Fu, Y., Habtemariam, A., van Rijt, S. H., Cooper, P. A., Loadman, P. M., & Sadler, P. J. (2011). Anti-colorectal cancer activity of an organometallic osmium arene azopyridine complex. *MedChemComm, 2*, 666–668.

Sibi, M. P., Silva, L. F., Jr, & Carneiro, V. M. T. (2008). *Thallium(III) nitrate trihydrate. Encyclopedia of Reagents for Organic Synthesis. Major Reference Works*. New York: John Wiley & Sons.

Sigel, A., & Sigel, H. (2004). *Metal ions and their complexes in medication, 41. Metal ions in biological systems*. New York; London: Marcel Dekker; Taylor & Francis.

Silverman, M. S. (1966). High-pressure (70-kbar) synthesis of new crystalline lead dichalcogenides. *Inorganic Chemistry, 5*, 2067–2069.

Simmons, J. G. (2000). *The scientific 100: A ranking of the most influential scientists, past and present*. New York, NY: Kensington Publishing Corporation.

Sims, Z. (2016). Cerium-based, intermetallic-strengthened aluminum casting alloy: High-volume co-product development. *JOM, 68*, 1940–1947.

Sinha, S. P., Shelly., Sharma, V., Meenakshi., Srivastava, S., & Srivastava, M. M. (1993). Neurotoxic effects of lead exposure among printing press workers. *Bulletin of Environmental Contamination and Toxicology, 51*, 490–493.

Smedley, J., Rao, T., Wang, E. (2009) K_2CsSb cathode development. *AIP Conference Proceedings 1149*, 1062–1066.

Smil, V. (2004). *Enriching the Earth: Fritz Haber, Carl Bosch, and the transformation of world food production*. Cambridge, MA: MIT Press.

Smirnov, A. Y., Borisevich, V. D., & Sulaberidze, A. (2012). Evaluation of specific cost of obtainment of lead-208 isotope by gas centrifuges using various raw materials. *Theoretical Foundations of Chemical Engineering, 46*, 373–378.

Sobernheim, M. (1993). Khumārawaih. In M. T. Houtsma, A. J. Wensinck, T. W. Arnold, W. Heffening, & E. Lévi-Provençal (Eds.), *E. J. Brill's first encyclopaedia of islam* (4, pp. 1913–1936). Leiden: E.J. Brill, 973.

Sochinskii, N. V., Abellán, M., Rodríguez-Fernández, J., Saucedo, E., Ruiz, C. M., & Bermúdez, V. (2007). Effect of Yb concentration on the resistivity and lifetime of CdTe:Ge:Yb codoped crystals. *Applied Physics Letters, 91*, 202112.

Söderlund, U., Patchett, P. J., Vervoort, J. D., & Isachsen, C. E. (2004). The ^{176}Lu decay constant determined by Lu–Hf and U–Pb isotope systematics of Precambrian mafic intrusions. *Earth and Planetary Science Letters, 219*, 311–324.

Solovyev, G. I., & Spear, K. E. (1972). Phase behavior in the Sm-B system. *Journal of the American Ceramic Society, 55*, 475–479.

Sorokin, P. P. (1979). Contributions of IBM to laser science—1960 to the present. *IBM Journal of Research and Development, 23*, 476–489.

Sox, T. E., & Olson, C. A. (1989). Binding and killing of bacteria by bismuth subsalicylate. *Antimicrob Agents and Chemotherapy, 33*, 2075–2082.

Spencer, J. H., Nesbitt, J. M., Trewhitt, H., Kashtiban, R. J., Bell, G., Ivanov, V. G., ... Smith, D. C. (2014). Raman spectroscopy of optical transitions and vibrational energies of ~1 nm HgTe extreme nanowires within single walled carbon nanotubes. *ACS Nano, 8*, 9044–9052.

Stabenow, F., Saak, W., & Weidenbruch, M. (2003). Tris(triphenylplumbyl)plumbate: An anion with three stretched lead–lead bonds. *Chemical Communications*, 2342–2343.

Steele, M. L., & Wertz, D. L. (1977). Solvent effects on the coordination of neodymium(3 +) ions in concentrated neodymium trichloride solutions. *Inorganic Chemistry, 16*, 1225–1228.

Stellman, J. M., & International Labour Office. (1998). *Encyclopaedia of occupational health and safety*. Geneva: International Labor Office.

Stevens, D. G. (1999). World War II economic warfare: The United States, Britain, and Portuguese Wolfram. *Historian, 61*, 539–556.

Stillman, J. M. (2008). *The story of alchemy and early chemistry*. Kila, MT: Kessinger Pub. Co.

Stoll, W. (2000). *Thorium and thorium compounds. Ullmann's encyclopedia of industrial chemistry*. Weinheim: Wiley-VCH.

Stone, R. (1997). An element of stability. *Science, 278*, 571–572.

Strod, A. J. (1957). Cesium—A new industrial metal. *American Ceramic Bulletin, 36*, 212–213.

Stwertka, A. (1999). *A guide to the elements*. Oxford: Oxford University Press.

Szymański, J. T. (1982). A mineralogical study and crystal-structure determination of nonmetamict ekanite, $ThCa_2Si_8O_{20}$. *Canadian Mineralogist, 20*, 65–75.

Takahashi, K., Boyd, R. N., Mathews, G. J., & Yokoi, K. (1987). Bound-state beta decay of highly ionized atoms. *Physical Review C, 36*, 1522–1528.

Tamadon, F., & Seppelt, K. (2013). The elusive halides VCl_5, $MoCl_6$, and $ReCl_6$. *Angewandte Chemie International Edition, 52*, 767–769.

Taylor, E. C., & McKillop, A. (1970). Thallium in organic synthesis. *Accounts of Chemical Research, 3*, 956–960.

Taylor, J. C., & Wilson, P. W. (1974). The structure of β-tungsten hexachloride by powder neutron and X-ray diffraction. *Acta Crystallographica B, 30*, 1216–1220.

Taylor, M. D., & Carter, C. P. (1962). Preparation of anhydrous lanthanide halides, especially iodides. *Journal of Inorganic and Nuclear Chemistry, 24*, 387–391.

Taylor, V. F., Evans, R. D., & Cornett, R. J. (2008). Preliminary evaluation of $^{135}Cs/^{137}Cs$ as a forensic tool for identifying source of radioactive contamination. *Journal of Environmental Radioactivity, 99*, 109–118.

Tenne, R., Margulis, L., Genut, M., & Hodes, G. (1992). Polyhedral and cylindrical structures of tungsten disulphide. *Nature, 360*, 444–446.

Tessalina, S. G., Yudovskaya, M. A., Chaplygin, I. V., Birck, J.-L., & Capmas, F. (2008). Sources of unique rhenium enrichment in fumaroles and sulphides at Kudryavy volcano. *Geochimica et Cosmochimica Acta, 72*, 889–909.

Thomas, W., & Elias, H. (1976). Darstellung von $HfCl_4$ und $HfBr_4$ durch umsetzung von hafnium mit geschmolzenen metallhalogeniden. *Journal of Inorganic and Nuclear Chemistry, 38*, 2227–2229.

Thomson, T. (1830). *The history of chemistry, 1-2*. London: H. Colburn and R. Bentley.

Thomson, T. (1831). *A system of chemistry of inorganic bodies. 1*. London: Baldwin & Cradock and William Blackwood.

Thong, J. T. L. (2010). Thermal oxidation of polycrystalline tungsten nanowire. *Journal of Applied Physics, 108*, 094312-1–094312-6.

Thornton, I., Rautiu, R., & Brush, S. M. (2001). *Lead: The Facts*. London: International Lead Association.

Thurston, J. H., Kolesnichenko, V., Messerle, L., Latturner, S. E., & Ainsworth, W. (2014). Trinuclear Tungsten Halide Clusters. *Inorganic Syntheses, 36*, 24–30.

Tilgner, H. G. (2000). *Forschen Suche und Sucht*. Hamburg: Libri Books on Demand.

Timur, A., & Toksoz, M. N. (1985). Downhole geophysical logging. *Annual Review of Earth and Planetary Sciences, 13*, 315–344.

Tjoelker, R. L., Prestage, J. D., Burt, E. A., Chen, P., Chong, Y. J., Chung, S. K., ... Wang, R. (2016). Mercury ion clock for a NASA technology demonstration mission. *IEEE Transactions on Ultrasonics, Ferroelectrics, and Frequency Control, 63*, 1034–1043.

Tliha, M., Mathlouthi, H., Lamloumi, J., & Percheronguegan, A. (2007). AB_5-type hydrogen storage alloy used as anodic materials in Ni-MH batteries. *Journal of Alloys and Compounds, 436*, 221–225.

Tokareva, S. A. (1971). Alkali and alkaline earth metal ozonides. *Russian Chemical Reviews, 40*, 165–174.

Torr, M. R. (1985). Osmium coated diffraction grating in the space shuttle environment: Performance. *Applied Optics, 24*, 2959–2961.

Toutain, J.-P., & Meyer, G. (1989). Iridium-bearing sublimates at a hot-spot volcano (Piton De La Fournaise, Indian Ocean). *Geophysical Research Letters, 16*, 1391–1394.

Trigg, G. L. (1995). *Landmark experiments in twentieth century physics*. New York, NY: Dover Publications.

Trovarelli, A. (2005). *Catalysis by ceria and related materials*. London: Imperial College Press.

Tsai, K.-R., Harris, P. M., & Lassettre, E. N. (1956). The crystal structure of cesium monoxide. *The Journal of Physical Chemistry, 60*, 338–344.

Tuček, K., Carlsson, J., & Wider, H. (2006). Comparison of sodium and lead-cooled fast reactors regarding reactor physics aspects, severe safety and economical issues. *Nuclear Engineering and Design*, *236*, 1589−1598.

Tücks, A., & Beck, H. P. (2007). The photochromic effect of bismuth vanadate pigments: Investigations on the photochromic mechanism. *Dyes and Pigments*, *72*, 163−177.

Tungate, M. (2011). *Branded beauty: How marketing changed the way we look*. London: Kogan Page.

Turekian, K. K., & Wedepohl, K. H. (1961). Distribution of the elements in some major units of the Earth's crust. *Geological Society of America Bulletin*, *72*, 175−192.

Uchida, H. (1999). Hydrogen solubility in rare earth based hydrogen storage alloys. *International Journal of Hydrogen Energy*, *24*, 871−877.

Urbain, G. (1922). Sur les séries L du lutécium et de l'ytterbium et sur l'identification d'un celtium avec l'élément de nombre atomique 72. *Comptes Rendus*, *174*, 1347−1349.

Urbancic, M. A., Shapley, J. R., Sauer, N. N., & Angelici, R. J. (1990). Pentacarbonylhydridorhenium. *Inorganic Syntheses*, 165−168.

Vallina, B., Rodriguez-Blanco, J. D., Brown, A. P., Blanco, J. A., & Benning, L. G. (2013). Amorphous dysprosium carbonate: Characterization, stability and crystallization pathways. *Journal of Nanoparticle Research*, *15*, 1438−1450.

Vaska, L., & DiLuzio, J. W. (1961). Carbonyl and hydrido-carbonyl complexes of iridium by reaction with alcohols. hydrido complexes by reaction with acid. *Journal of the American Chemical Society*, *83*, 2784−2785.

Venetskii, S. I. (1974). Osmium. *Metallurgist*, *18*, 155−157.

Voitovich, R. F., & Golovko, É. I. (1975). Oxidation of hafnium alloys with nickel. *Metal Science and Heat Treatment*, *17*, 207−209.

Vol'nov, I. I., & Matveev, V. V. (1963). Synthesis of cesium ozonide through cesium superoxide. *Bulletin of the Academy of Sciences of the USSR, Division of chemical science*, *12*, 1040−1043.

von der Wense, L., Seiferle, B., Laatiaoui, M., Neumayr, J. B., Maier, H.-J., Wirth, H.-F., ... Thirolf, P. G. (2016). Direct detection of the ^{229}Th nuclear clock transition. *Nature*, *533*, 47.

Voorhees, V., & Adams, R. (1922). The use of the oxides of platinum for the catalytic reduction of organic compounds. I. *Journal of the American Chemical Society*, *44*, 1397−1405.

Wagner, R., Neubauer, C., Deville, H. S.-C., Sorel., Wagenmann, L., & Girard, A. (1856). Notizen. Ueber die vermeintliche Identität der Oxyphensäure mit dem farblosen Hydrochinon. *Journal für Praktische Chemie*, *67*, 490−508.

Waldron, H. A. (1985). Lead and lead poisoning in antiquity. *Medical History*, *29*, 107−108.

Walsh, J. (1981). A Manhattan project postscript. *Science*, *212*, 1369−1371.

Walters, W., Gooch, W., & Burkins, M. (2001). The penetration resistance of a titanium alloy against jets from tantalum shaped charge liners. *International Journal of Impact Engineering*, *26*, 823−830.

Wang, C., Daimon, H., Onodera, T., Koda, T., & Sun, S. (2008). A general approach to the size- and shape-controlled synthesis of platinum nanoparticles and their catalytic reduction of oxygen. *Angewandte Chemie International Edition*, *47*, 3588−3591.

Wang, G., Zhou, M., Goettel, J. T., Schrobilgen, G. J., Su, J., Li, J., ... Riedel, S. (2014). Identification of an iridium-containing compound with a formal oxidation state of IX. *Nature*, *514*, 475.

Wang, X., Andersson, M. R., Thompson, M. E., & Inganäsa, O. (2004). Electrophosphorescence from substituted poly(thiophene) doped with iridium or platinum complex. *Thin Solid Films*, *468*, 226−233.

Wang, X., Andrews, L., Riedel, S., & Kaupp, M. (2007). Mercury is a transition metal: The first experimental evidence for HgF_4. *Angewandte Chemie International Edition*, *46*, 8371−8375.

Wang, X., Liu, Z., Stevens-Kalceff, M., & Riesen, H. (2014). Mechanochemical preparation of nanocrystalline BaFCl doped with samarium in the 2+ oxidation state. *Inorganic Chemistry*, *53*, 8839−8841.

Wang, Z. C., & Li, G. (1991). Barite and witherite deposits in Lower Cambrian shales of South China: Stratigraphic distribution and geochemical characterization. *Economic Geology*, *86*, 354−363.

Wedepohl, H. K. (1995). The composition of the continental crust. *Geochimica et Cosmochimica Acta*, *59*, 1217−1232.

Weeks, M. E., & Leichester, H. M. (1968). *Discovery of the elements*. Easton, PA: Journal of Chemical Education.

Wheeler, L. D. (1986). The price of neutrality: Portugal, the Wolfram Question, and World War II. *Luso-Brazilian Review*, *23*, 107−127.

White, C., Yates, A., Maitlis, P. M., & Heinekey, D. M. (1992). (η5-pentamethylcyclopentadienyl) rhodium and -iridium compounds. *Inorganic Syntheses*, 228−234.

Wickleder, M. S., Fourest, B., & Dorhout, P. K. (2006). Thorium. In L. R. Morss, N. M. Edelstein, & J. Fuger (Eds.), *The chemistry of the actinide and transactinide elements* (3rd edn, pp. 52−160). Berlin: Springer-Verlag.

Wildervanck, J. C., & Jellinek, F. (1971). The dichalcogenides of technetium and rhenium. *Journal of the Less Common Metals*, *24*, 73−81.

Winder, C. (1993). *The history of lead — Part 1. LEAD Action News* **2**(1), https://lead.org.au/lanv2n1/lanv2n1-11.html.

Winder, C. (1994a). *The history of lead — Part 2. LEAD Action News* **2**(2), https://lead.org.au/lanv2n2/lanv2n2-17.html.

Winder, C. (1994b). *The history of lead — Part 3. LEAD Action News* **2**(3) https://lead.org.au/lanv2n3/lanv2n3-22.html.

Winder, C. (1994c). *The history of lead — Part 4. LEAD Action News* **2**(4) https://lead.org.au/lanv2n4/lanv2n4-14.html.

Wolgast, S., Kurdak, Ç., Sun, K., Allen, J. W., Kim, D.-J., & Fisk, Z. (2013). Low-temperature surface conduction in the Kondo insulator SmB_6. *Physical Review B*, *88*, 180405.

Woodhead, J. A. (1991). The metamictization of zircon: Radiation dose dependent structural characteristics. *American Mineralogist*, *76*, 74−82.

Woosley, S. E., Hartmann, D. H., Hoffman, R. D., & Haxton, W. C. (1990). The ν-process. *The Astrophysical Journal*, *356*, 272−301.

Wrackmeyer, B., & Horchler, K. (1990). ^{207}Pb-NMR Parameters. In G. A. Webb (Ed.), *Annual Reports on NMR Spectroscopy* (22, pp. 249–306). Academic Press.

Wriedt, H. A. (1989). The O-W (oxygen-tungsten) system. *Bulletin of Alloy Phase Diagrams, 10*, 368–384.

Wright, D. C. (2001). *The History of China*. Westport: Greenwood Press.

Wu, M. K., Ashburn, J. R., Torng, C. J., Hor, P. H., Meng, R. L., Gao, L., ... Chu, C. W. (1987). Superconductivity at 93 K in a new mixed-phase Y-Ba-Cu-O compound system at ambient pressure. *Physical Review Letters, 58*, 908–910.

Xiao, Z., & Laplante, A. R. (2004). Characterizing and recovering the platinum group minerals—A review. *Minerals Engineering, 17*, 961–979.

Xie, Z., Blair, R. G., Orlovskaya, N., Cullen, D. A., & Andrew Payzant, E. (2014). Thermal stability of hexagonal OsB_2. *Journal of Solid State Chemistry, 219*, 210–219.

Yen, P. C., Chen, R. S., Chen, C. C., Huang, Y. S., & Tiong, K. K. (2004). Growth and characterization of OsO_2 single crystals. *Journal of Crystal Growth, 262*, 271–276.

Yoshihara, H. K. (2004). Discovery of a new element 'nipponium': Re-evaluation of pioneering works of Masataka Ogawa and his son Eijiro Ogawa. *Spectrochimica Acta Part B Atomic Spectroscopy, 59*, 1305–1310.

Yu, L., & Yu, H. (2004). *Chinese coins: Money in history and society*. San Francisco, CA: Long River Press.

Yu, L. Q., Wen, Y., & Yan, M. (2004). Effects of Dy and Nb on the magnetic properties and corrosion resistance of sintered NdFeB. *Journal of Magnetism and Magnetic Materials, 283*, 353–356.

Yuhas, A. (2015). *Liquid mercury found under Mexican pyramid could lead to king's tomb*. The Guardian, 24 April.

Zhang, J., MacPhee, A. G., Lin, J., Wolfrum, E., Smith, R., Danson, C., ... Wark, J. S. (1997). A saturated X-ray laser beam at 7 nanometers. *Science, 276*, 1097–1100.

Zhang, P., Huang, H., Banerjee, S., Clarkson, G. J., Ge, C., Imberti, C., & Sadler, P. J. (2019). Nucleus-targeted organoiridium–albumin conjugate for photodynamic cancer therapy. *Angewandte Chemie International Edition, 58*, 2350–2354.

Zhang, X., Butch, N. P., Syers, P., Ziemak, S., Greene, R. L., & Paglione, J. (2013). Hybridization, inter-ion correlation, and surface states in the kondo insulator SmB_6. *Physical Review X, 3*, 011011.

Zhao, G., Su, L., Xu, J., & Zeng, H. (2007). Response to "Comment on 'Efficient diode-pumped Yb:Gd_2SiO_5 laser' [Appl. Phys. Lett. 90, 066101 (2007)]". *Applied Physics Letters, 90*, 066103.

Zheng, Y.-Q., Jonas, E., Nuss, J., & Schnering, H. G. V. (1998). The DMSO solvated octahedro-[W_6Cl]Cl cluster molecule. *Zeitschrift für anorganische und allgemeine Chemie, 624*, 1400–1404.

Zheng, Y. Q., Peters, K., & Schnering, H. G. V. (1998). Crystal structure of tungsten pentabromide, WBr_5. *Zeitschrift für Kristallographie - New Crystal Structures, 213*. Available from https://doi.org/10.1524/ncrs.1998.213.14.499.

Zhou, Y., Kolesnichenko, V., Messerle, L., Alayoglu, S., & Eichhorn, B. (2014). Crystalline and amorphous forms of tungsten tetrachloride. *Inorganic Syntheses, 36*, 30–34.

Zhu, Z., & Cheng, C. Y. (2011). Solvent extraction technology for the separation and purification of niobium and tantalum: A review. *Hydrometallurgy, 107*, 1–12.

Ziegler, E., Hignette, O., Morawe, C., & Tucoulou, R. (2001). High-efficiency tunable X-ray focusing optics using mirrors and laterally-graded multilayers. *Nuclear Instruments and Methods in Physics Research Section A: Accelerators, Spectrometers, Detectors and Associated Equipment, 467–468*, 954–957.

Zitko, V., Carson, W. V., & Carson, W. G. (1975). Thallium: Occurrence in the environment and toxicity to fish. *Bulletin of Environmental Contamination and Toxicology, 13*, 23–30.

Zoli, L., Galizia, P., Silvestroni, L., & Sciti, D. (2018). Synthesis of group IV and V metal diboride nanocrystals via borothermal reduction with sodium borohydride. *Journal of the American Ceramic Society, 101*, 2627–2637.

Name Index

Note: Page numbers followed by "*f*" refer to figures.

A

B

C

D

Mineral Index

Note: Page numbers followed by "*f*," "*t*," and "*b*" refer to figures, tables, and boxes, respectively.

D

E

F

G

H

M

N

O

P

T

U

V

W

X

Y

Z

Subject Index

Note: Page numbers followed by "*f*" and "*b*" refer to figures and boxes, respectively.

C

D

E

F

G

J

K

L

P

Q

R

T

U

V

W

X

Y

Z

Printed in the United States
By Bookmasters